Get Connected to

ConnectED

connectED.mcgraw-hill.com

D1308509

Your online portal to everything you need!

For Students
Leave your books at school. Now you can go online and interact with your eStudentEdition from any place, any time!

For Teachers
ConnectED is your one-stop online center for everything you need to teach, including: digital eTeacherEdition, lesson planning and scheduling tools, pacing, and assessment.

For Parents
Get homework help, help your student prepare for testing, and review math topics.

(t)Richard Hutchings, (c)STOCK4B-RF/Getty Images. (b)Jupiterimages/Thinkstock/Alamy

Today's Program for Tomorrow's Leaders

TRUSTED CONTENT

The Glencoe High School Series was built for the Common Core State Standards. Throughout the program, teachers will find support for implementing the Standards for Mathematical Content and the Standards for Mathematical Practice.

 Education

McGraw-Hill Education is built on more than a century of innovation, trust and leadership. Because of our extraordinary success and unparalleled experience, we have developed relationships that ensure future growth, the prominence to shape our industry, and the size to make a difference.

RESOURCES AT YOUR FINGERTIPS

A variety of supporting resources are supplied both in print and online.

Teacher resources such as leveled worksheets, assessment resources, and answer keys

Student workbooks such as the Study Notebook, the Homework Practice Workbook, and others

Supplemental resources such as Real-World Problem Solving Graphic Novels and Foldables™ Study Organizers

ONLINE ANYTIME, ANYWHERE

ConnectED contains all the digital resources to support your Glencoe program. Students and teachers can navigate a variety of tools to support, reinforce, or extend the content from class.

Ready to take the next digital step? Using tablets? Setting up 1-to-1 computing?

Innovation offers a great opportunity in education: the promise of a fully engaged class, improved student achievement, and a greater connection between teachers and students. We are combining McGraw-Hill's proven experience in educational content with innovative technology to help fulfill this promise of education for students around the world.

McGraw-Hill is helping teachers and districts navigate new digital innovation. Learn more about resources for your digital transition at **http://www.mheonline.com/experiencematters/**

Glencoe

ALGEBRA 2

Dodecahedral Knot. By Paul Nylander (http://bugman123.com)

Education

Bothell, WA • Chicago, IL • Columbus, OH • New York, NY

---

 connectED.mcgraw-hill.com **Your Digital Math Portal**

Animation	Vocabulary	eGlossary	Personal Tutor	Virtual Manipulatives	Graphing Calculator	Audio	Foldables	Self-Check Practice	Worksheets

In this Common Core State Standards edition of *Glencoe Algebra 2,* students are challenged to develop 21st century skills such as critical thinking and creative problem solving while engaging with exciting careers within Science, Technology, Engineering, and Mathematics (STEM) related fields.

Photo Credit:

Cover Dodecahedral Knot. By Paul Nylander (http://bugman123.com)

Common Core State Standards ©2010. National Governors Association Center for Best Practices and Council of Chief State School Officers. All rights reserved.

Understanding by Design® is a registered trademark of the Association for Supervision and Curriculum Development ("ASCD").

TI-Nspire is a trademark of Texas Instruments Incorporated.
Texas Instruments images used by permission.

The Math Forum®, Ask Dr. Math®, Dr. Math®, and T2T® are registered trademarks of

Drexel University

connectED.mcgraw-hill.com

The McGraw·Hill Companies

Education

Copyright © 2014 by The McGraw-Hill Companies, Inc.

All rights reserved. No part of this publication may be reproduced or distributed in any form or by any means, or stored in a database or retrieval system, without the prior written consent of The McGraw-Hill Companies, Inc., including, but not limited to, network storage or transmission, or broadcast for distance learning.

Send all inquiries to:
McGraw-Hill Education
STEM Learning Solutions Center
8787 Orion Place
Columbus, OH 43240

ISBN: 978-0-07-663991-5 *(Teacher Edition)*
MHID: 0-07-663991-6 *(Teacher Edition)*
ISBN: 978-0-07-663990-8 *(Student Edition)*
MHID: 0-07-663990-8 *(Student Edition)*

Printed in the United States of America.

2 3 4 5 6 7 8 9 QVR 17 16 15 14 13

McGraw-Hill is committed to providing instructional materials in Science, Technology, Engineering, and Mathematics (STEM) that give students a solid foundation, one that prepares them for college and careers in the 21st Century.

Contents in Brief

The correlation of the Common Core State Standards, Traditional Algebra II Pathway to *Glencoe Algebra 2* begins on page T12.

Chapter 0 **Preparing for Advanced Algebra**

Unit 1 Linear Relations and Functions

    1 **Equations and Inequalities**

    2 **Linear Relations and Functions**

    3 **Systems of Equations and Inequalities**

Unit 2 Quadratic, Polynomial, and Radical Functions and Relations

    4 **Quadratic Functions and Relations**

    5 **Polynomials and Polynomial Functions**

    6 **Inverses and Radical Functions and Relations**

Unit 3 Advanced Functions and Relations

    7 **Exponential and Logarithmic Functions and Relations**

    8 **Rational Functions and Relations**

    9 **Conic Sections**

Unit 4 Discrete Mathematics

    10 **Sequences and Series**

    11 **Probability and Statistics**

Unit 5 Trigonometry

    12 **Trigonometric Functions**

    13 **Trigonometric Identities and Equations**

Student Handbook

Suggested Pacing Guide	
Each chapter includes multiple days for review and assessment.	
Chapter 0	Optional
Chapter 1	9 days
Chapter 2	13 days
Chapter 3	15 days
Chapter 4	14 days
Chapter 5	14 days
Chapter 6	11 days
Chapter 7	14 days
Chapter 8	11 days
Chapter 9	12 days
Chapter 10	13 days
Chapter 11	11 days
Chapter 12	15 days
Chapter 13	8 days
Total	**160 days**

Our lead authors ensure that the Macmillan/McGraw-Hill and Glencoe/McGraw-Hill mathematics programs are truly vertically aligned by beginning with the end in mind—success in Algebra 1 and beyond. By "backmapping" the content from the high school programs, all of our mathematics programs are well articulated in their scope and sequence.

Lead Authors

John A. Carter, Ph.D.
Principal
Adlai E. Stevenson High School
Lincolnshire, Illinois

Areas of Expertise: Using technology and manipulatives to visualize concepts; mathematics achievement of English-language learners

Gilbert J. Cuevas, Ph.D.
Professor of Mathematics Education
Texas State University—San Marcos
San Marcos, Texas

Areas of Expertise: Applying concepts and skills in mathematically rich contexts; mathematical representations

Roger Day, Ph.D., NBCT
Mathematics Department Chairperson
Pontiac Township High School
Pontiac, Illinois

Areas of Expertise: Understanding and applying probability and statistics; mathematics teacher education

Carol Malloy, Ph.D.
Associate Professor Emerita
University of North Carolina at Chapel Hill
Chapel Hill, North Carolina

Areas of Expertise: Representations and critical thinking; student success in Algebra 1

The McGraw-Hill Companies

Program Authors

Berchie Holliday, Ed.D.
National Mathematics Consultant
Silver Spring, Maryland

Areas of Expertise: Using mathematics to model and understand real-world data; the effect of graphics on mathematical understanding

Ruth Casey
Mathematics Consultant
Regional Teacher Partner
University of Kentucky
Lexington, Kentucky

Areas of Expertise: Graphing technology and mathematics

Contributing Authors

Dinah Zike **FOLDABLES**
Educational Consultant
Dinah-Might Activities, Inc.
San Antonio, Texas

Jay McTighe
Educational Author and Consultant
Columbia, MD

(bl) Courtesy of Dinah Zike, (others) The McGraw-Hill Companies

These professionals were instrumental in providing valuable input and suggestions for improving the effectiveness of the mathematics instruction.

Lead Consultant

Viken Hovsepian
Professor of Mathematics
Rio Hondo College
Whittier, California

Consultants

Mathematical Content

Grant A. Fraser, Ph.D.
Professor of Mathematics
California State University, Los Angeles
Los Angeles, California

Arthur K. Wayman, Ph.D.
Professor of Mathematics Emeritus
California State University, Long Beach
Long Beach, California

Gifted and Talented

Shelbi K. Cole
Research Assistant
University of Connecticut
Storrs, Connecticut

College Readiness

Robert Lee Kimball, Jr.
Department Head, Math and Physics
Wake Technical Community College
Raleigh, North Carolina

Differentiation for English-Language Learners

Susana Davidenko
State University of New York
Cortland, New York
Alfredo Gómez
Mathematics/ESL Teacher
George W. Fowler High School
Syracuse, New York

Graphing Calculator

Jerry Cummins
Former President
National Council of Supervisors of
   Mathematics
Western Springs, Illinois

Mathematical Fluency

Robert M. Capraro
Associate Professor
Texas A&M University
College Station, Texas

Pre-AP

Dixie Ross
Lead Teacher for Advanced Placement
   Mathematics
Pflugerville High School
Pflugerville, Texas

Reading and Writing

ReLeah Cossett Lent
Author and Educational Consultant
Morganton, Georgia

Lynn T. Havens
Director of Project CRISS
Kalispell, Montana

The McGraw-Hill Companies

Reviewers

Corey Andreasen
Mathematics Teacher
North High School
Sheboygan, Wisconsin

Mark B. Baetz
Mathematics Coordinating Teacher
Salem City Schools
Salem, Virginia

Kathryn Ballin
Mathematics Supervisor
Newark Public Schools
Newark, New Jersey

Kevin C. Barhorst
Mathematics Department Chair
Independence High School
Columbus, Ohio

Brenda S. Berg
Mathematics Teacher
Carbondale Community High School
Carbondale, Illinois

Dawn Brown
Mathematics Department Chair
Kenmore West High School
Buffalo, New York

Sheryl Pernell Clayton
Mathematics Teacher
Hume Fogg Magnet School
Nashville, Tennessee

Bob Coleman
Mathematics Teacher
Cobb Middle School
Tallahassee, Florida

Jane E. Cotts
Mathematics Teacher
O'Fallon Township High School
O'Fallon, Illinois

Michael D. Cuddy
Mathematics Instructor
Zypherhills High School
Zypherhills, Florida

Melissa M. Dalton, NBCT
Mathematics Instructor
Rural Retreat High School
Rural Retreat, Virginia

Tina S. Dohm
Mathematics Teacher
Naperville Central High School
Naperville, Illinois

Laurie L.E. Ferrari
Mathematics Teacher
L'Anse Creuse High School–North
Macomb, Michigan

Steve Freshour
Mathematics Teacher
Parkersburg South High School
Parkersburg, West Virginia

Shirley D. Glover
Mathematics Teacher
TC Roberson High School
Asheville, North Carolina

Caroline W. Greenough
Mathematics Teacher
Cape Fear Academy
Wilmington, North Carolina

Susan Hack, NBCT
Mathematics Teacher
Oldham County High School
Buckner, Kentucky

Michelle Hanneman
Mathematics Teacher
Moore High School
Moore, Oklahoma

Theresalynn Haynes
Mathematics Teacher
Glenbard East High School
Lombard, Illinois

Sandra Hester
Mathematics Teacher/AIG Specialist
North Henderson High School
Hendersonville, North Carolina

Jacob K. Holloway
Mathematics Teacher
Capitol Heights Junior High School
Montgomery, Alabama

Robert Hopp
Mathematics Teacher
Harrison High School
Harrison, Michigan

Eileen Howanitz
Mathematics Teacher/Department
 Chairperson
Valley View High School
Archbald, Pennsylvania

Charles R. Howard, NBCT
Mathematics Teacher
Tuscola High School
Waynesville, North Carolina

Sue Hvizdos
Mathematics Department
 Chairperson
Wheeling Park High School
Wheeling, West Virginia

Elaine Keller
Mathematics Teacher
Mathematics Curriculum Director
 K-12
Northwest Local Schools
Canal Fulton, Ohio

Sheila A. Kotter
Mathematics Educator
River Ridge High School
New Port Richey, Florida

Frank Lear
Mathematics Department Chair
Cleveland High School
Cleveland, Tennessee

Jennifer Lewis
Mathematics Teacher
Triad High School
Troy, Illinois

Catherine McCarthy
Mathematics Teacher
Glen Ridge High School
Glen Ridge, New Jersey

Jacqueline Palmquist
Mathematics Department Chair
Waubonsie Valley High School
Aurora, Illinois

Thom Schacher
Mathematics Teacher
Otsego High School
Otsego, Michigan

Laurie Shappee
Teacher/Mathematics Coordinator
Larson Middle School
Troy, Michigan

Jennifer J. Southers
Mathematics Teacher
Hillcrest High School
Simpsonville, South Carolina

Sue Steinbeck
Mathematics Department Chair
Parkersburg High School
Parkersburg, West Virginia

Kathleen D. Van Sise
Mathematics Teacher
Mandarin High School
Jacksonville, Florida

Karen Wiedman
Mathematics Teacher
Taylorville High School
Taylorville, Illinois

Willard Clay/Getty Images

What do teachers want in a mathematics program? *Glencoe Algebra 2* answers that question with a program that meets your needs.

SECTION 1 | Trusted Content

- **Built for the Common Core State Standards**
- **Multiple Paths to Learning**
- **Understanding by Design**

SECTION 2 | Resources at Your Fingertips

- **Use ConnectED as Your Portal**
- **Teach from Your eTeacherEdition**
- **Assessment that Validates Learning**
- **Built-In Differentiated Instruction**

SECTION 3 | 21st Century Technology

- **Connect with Digital Natives**
- **College and Work Readiness**
- **STEM Careers**
- **21st Century Skills**

Willard Clay/Getty Images

Common Core State Standards

*W*ith American students fully prepared for the future, our communities will be best positioned to compete successfully in the global economy. — *Common Core State Standards Initiative*

What is the goal of the Common Core State Standards?

The mission of the *Common Core State Standards* is to provide a consistent, clear understanding of of what students are expected to learn, so teachers and parents know what they need to do to help them. The standards are designed to be robust and relevant to the real world, reflecting the knowledge and skills that students need for success in college and careers.

Who wrote the standards?

The National Governors Association Center for Best Practices and the Council of Chief State School Officers worked with representatives from participating states, a wide range of educators, content experts, researchers, national organizations, and community groups.

What are the major points of the standards?

The standards seek to develop both students' mathematical understanding and their procedural skill. The *Standards for Mathematical Practice* describe varieties of expertise that mathematics teachers at all levels should seek to develop in their students. The *Standards for Mathematical Content* define what students should understand and be able to do at each level in their study of mathematics.

How do I implement the standards?

The *Common Core State Standards* are shared goals and expectations for what knowledge and skills your students need to succeed. You as a teacher, in partnership with your colleagues, principals, superintendents, decide how the standards are to be met. *Glencoe Algebra 2* is designed to help you devise lesson plans and tailor instruction to the individual needs of the students in your classroom as you meet the *Common Core State Standards*.

At the high school level the *Common Core State Standards* are organized by conceptual category. To ease implementation four model course pathways were created: traditional, integrated, accelerated traditional and accelerated integrated. *Glencoe Algebra 1, Glencoe Geometry,* and *Glencoe Algebra 2* follow the traditional pathway.

How do I decode the standards?

This diagram provides clarity for decoding the standard identifiers.

A.REI.2

Conceptual Category
N = Number and Quantity
A = Algebra
F = Functions
S = Statistics and Probability

Domain

Standard

Domain Names	Abbreviations
The **C**omplex **N**umber System	**CN**
Seeing **S**tructure in **E**xpressions	**SSE**
Arithmetic with **P**olynomials and **R**ational Expressions	**APR**
Creating **E**quations	**CED**
Reasoning with **E**quations and **I**nequalities	**REI**
Interpreting **F**unctions	**IF**
Building **F**unctions	**BF**
Linear, Quadratic, and **E**xponential Models	**LE**
Trigonometric **F**unctions	**TF**
Interpreting Categorical and Quantitative **D**ata	**ID**
Making **I**nferences and Justifying **C**onclusions	**IC**
Using Probability to **M**ake **D**ecisions	**MD**

There are numerous tools for implementing the Common Core State Standards available throughout the program, including:

- **Standards at point-of-use** in the Chapter Planner and in each lesson of the Teacher Edition,

- **Complete standards coverage** in *Glencoe Algebra 2* ensures that you have all the content you need to teach the standards,

- **Correlations** that show at a glance where each standard is addressed in *Glencoe Algebra 2.*

You can also visit connectED.mcgraw-hill.com to learn more about the Common Core State Standards. There you can choose from an extensive collection of resources to use when planning instruction.

Common Core State Standards, Traditional Algebra II Pathway, Correlated to *Glencoe Algebra 2*, Common Core Edition

Lessons in which the standard is the primary focus are indicated in **bold**.

Standards	Student Edition Lesson(s)	Student Edition Page(s)
Number and Quantity		
The Complex Number System N-CN		
Perform arithmetic operations with complex numbers. 1. Know there is a complex number i such that $i^2 = -1$, and every complex number has the form $a + bi$ with a and b real.	**4-4**	**246–252**
2. Use the relation $i^2 = -1$ and the commutative, associative, and distributive properties to add, subtract, and multiply complex numbers.	**4-4**	**246–252**
Use complex numbers in polynomial identities and equations. 7. Solve quadratic equations with real coefficients that have complex solutions.	**4-5,** Extend 4-5, **4-6**	**256–262,** 263, **264–272**
8. (+) Extend polynomial identities to the complex numbers.	4-4, **4-6**	246–252, **264–272**
9. (+) Know the Fundamental Theorem of Algebra; show that it is true for quadratic polynomials.	**5-7**	**358–365**
Algebra		
Seeing Structure in Expressions A-SSE		
Interpret the structure of expressions. 1. Interpret expressions that represent a quantity in terms of its context. ★ a. Interpret parts of an expression, such as terms, factors, and coefficients.	1-1, **4-1**	5–10, **219–227**
b. Interpret complicated expressions by viewing one or more of their parts as a single entity.	1-1, 1-4, 2-2, 2-4, **2-6,** 2-7, 4-1, **4-6,** 4-7, 5-4, 9-2, 9-3, 9-4, 9-5, **9-6,** 10-7	5–10, 27–32, 69–74, 83–89, **101–107,** 109–116, 219–227, **264–272,** 275–280, 330–337, 599–605, 607–613, 615–622, 624–631, **632–636,** 705–709

(+) Advanced Mathematics Standards   ★ Mathematical Modeling Standards

Standards	Student Edition Lesson(s)	Student Edition Page(s)
2. Use the structure of an expression to identify ways to rewrite it.	1-2, 4-3, **4-5**, 6-4, **6-5**, 7-2, 7-3, 7-4, 7-7, 7-8	11–17, 238–245, **256–262**, 407–411, **415–421**, 461–467, 468–475, 478–483, 501–507, 509–515
Write expressions in equivalent forms to solve problems. 4. Derive the formula for the sum of a finite geometric series (when the common ratio is not 1), and use the formula to solve problems. ★	**10-3**	**674–680**
Arithmetic with Polynomials and Rational Expressions A-APR		
Perform arithmetic operations on polynomials. 1. Understand that polynomials form a system analogous to the integers, namely, they are closed under the operations of addition, subtraction, and multiplication; add, subtract, and multiply polynomials.	4-3, **5-1**	238–245, **303–309**
Understand the relationship between zeros and factors of polynomials. 2. Know and apply the Remainder Theorem: For a polynomial $p(x)$ and a number a, the remainder on division by $x - a$ is $p(a)$, so $p(a) = 0$ if and only if $(x - a)$ is a factor of $p(x)$.	**5-6**	**352–357**
3. Identify zeros of polynomials when suitable factorizations are available, and use the zeros to construct a rough graph of the function defined by the polynomial.	**5-7**	**358–365**
Use polynomial identities to solve problems. 4. Prove polynomial identities and use them to describe numerical relationships.	**4-3**, 4-5, **4-6**, Extend 5-5	**238–245**, 256–262, **264–272**, 366
5. (+) Know and apply the Binomial Theorem for the expansion of $(x + y)^n$ in powers of x and y for a positive integer n, where x and y are any numbers, with coefficients determined for example by Pascal's Triangle.	**10-6**, Extend 10-6, 10-7, **11-4**	**699–703**, 704, 705–709, **752–759**
Rewrite rational expressions. 6. Rewrite simple rational expressions in different forms; write $\frac{a(x)}{b(x)}$ in the form $q(x) + \frac{r(x)}{b(x)}$, where $a(x)$, $b(x)$, $q(x)$, and $r(x)$ are polynomials with the degree of $r(x)$ less than the degree of $b(x)$, using inspection, long division, or, for the more complicated examples, a computer algebra system.	**5-2**, Extend 5-2	**311–317**, 318–319
7. (+) Understand that rational expressions form a system analogous to the rational numbers, closed under addition, subtraction, multiplication, and division by a nonzero rational expression; add, subtract, multiply, and divide rational expressions.	**8-1, 8-2**, 8-4, 8-6	**529–537, 538–544**, 553–560, 570–578

Standards	Student Edition Lesson(s)	Student Edition Page(s)
Creating Equations★ A-CED		
Create equations that describe numbers or relationships. 1. Create equations and inequalities in one variable and use them to solve problems.	1-3, 1-4, 1-5, 1-6, 4-3, 4-5, 4-6, 4-8, 5-5, 5-6, 5-7, 7-2, 7-4, 7-5, 7-6, 7-8, 8-6	18–25, 27–32, 33–39, 41–48, 238–245, 256–262, 264–272, 282–288, 342–349, 352–357, 358–365, 461–467, 478–483, 485–491, 492–498, 508–515, 570–578
2. Create equations in two or more variables to represent relationships between quantities; graph equations on coordinate axes with labels and scales.	Throughout the text; for example, 2-4, 3-1, 4-2, 6-3, 9-3, 12-7	Throughout the text; for example, 83–89, 136–145, 229–236, 400–406, 607–613, 837–843
3. Represent constraints by equations or inequalities, and by systems of equations and/or inequalities, and interpret solutions as viable or non-viable options in a modeling context.	1-4, 1-5, 1-6, 2-6, 2-7, 2-8, **3-1**, 3-2, **3-3**, 3-4, 3-7, 3-8, 4-8, 7-8, 8-6	27–32, 33–39, 41–48, 101–107, 109–116, **117–121**, 136–145, **146–152**, 154–160, 189–197, 198–204, 282–288, 509–515, 570–578
4. Rearrange formulas to highlight a quantity of interest, using the same reasoning as in solving equations.	**4-6**, 9-1, 9-3, 10-2	**264–272**, 593–598, 607–613, 666–673
Reasoning with Equations and Inequalities A-REI		
Understand solving equations as a process of reasoning and explain the reasoning. 2. Solve simple rational and radical equations in one variable, and give examples showing how extraneous solutions may arise.	**6-7, Extend 6-7, 8-6, Extend 8-6**	**429–435, 436–437, 570–578, 579–580**
Represent and solve equations and inequalities graphically. 11. Explain why the x-coordinates of the points where the graphs of the equations $y = f(x)$ and $y = g(x)$ intersect are the solutions of the equation $f(x) = g(x)$; find the solutions approximately, e.g., using technology to graph the functions, make tables of values, or find successive approximations. Include cases where $f(x)$ and/or $g(x)$ are linear, polynomial, rational, absolute value, exponential, and logarithmic functions. ★	**3-1**, Extend 4-2, Extend 5-7, Extend 6-7, Explore 7-2, Extend 7-6, Extend 8-6, **9-7**	**136–145**, 237, 350–351, 436–437, 459–460, 499–500, 579–580, **640–645**
Functions		
Interpreting Functions F-IF		
Interpret functions that arise in applications in terms of the context. 4. For a function that models a relationship between two quantities, interpret key features of graphs and tables in terms of the quantities, and sketch graphs showing key features given a verbal description of the relationship. ★	Throughout the text; for example, 2-1, Extend 2-1, Extend 2-2, 2-6, 5-3, 8-4, 10-1, 12-6	Throughout the text; for example, 61–67, 68, 75, 101–107, 322–329, 553–560, 659–665, 830–836

(+) Advanced Mathematics Standards   ★ Mathematical Modeling Standards

Standards	Student Edition Lesson(s)	Student Edition Page(s)
5. Relate the domain of a function to its graph and, where applicable, to the quantitative relationship it describes.	**2-1,** 2-6, 4-1, 5-3, 6-2, 6-3, 7-1, 7-3, **8-3,** 8-4, 12-7	**61–67,** 101–107, 219–227, 322–329, 393–398, 400–406, 451–458, 468–475, **545–551,** 553–560, 837–843
6. Calculate and interpret the average rate of change of a function (presented symbolically or as a table) over a specified interval. Estimate the rate of change from a graph. ★	**2-3,** Extend 4-7	**76–82,** 281
Analyze functions using different representations. 7. Graph functions expressed symbolically and show key features of the graph, by hand in simple cases and using technology for more complicated cases. ★ b. Graph square root, cube root, and piecewise-defined functions, including step functions and absolute value functions.	**2-6, 6-3, Extend 6-4**	**101–107, 400–406, 413**
c. Graph polynomial functions, identifying zeros when suitable factorizations are available, and showing end behavior.	5-3, **5-4,** Extend 5-4, 5-6, **5-7**	322–329, **330–337,** 338–339, 352–357, **358–365**
e. Graph exponential and logarithmic functions, showing intercepts and end behavior, and trigonometric functions, showing period, midline, and amplitude.	**7-1, 7-3, 12-7, 12-8**	**451–458, 468–475, 837–843, 845–852**
8. Write a function defined by an expression in different but equivalent forms to reveal and explain different properties of the function. a. Use the process of factoring and completing the square in a quadratic function to show zeros, extreme values, and symmetry of the graph, and interpret these in terms of a context.	**4-3, 4-5,** 4-7	**238–245, 256–262,** 275–280
b. Use the properties of exponents to interpret expressions for exponential functions.	**7-1, 7-8**	**451–458, 509–515**
9. Compare properties of two functions each represented in a different way (algebraically, graphically, numerically in tables, or by verbal descriptions).	**2-2, 2-7,** 4-1, 5-3, 6-1, 6-3, 7-1, 8-4, 9-6	**69–74, 109–116,** 219–227, 322–329, 385–392, 400–406, 451–458, 553–560, 632–636
Building Functions F-BF		
1. Build a function that models a relationship between two quantities. b. Combine standard function types using arithmetic operations.	**6-1,** Extend 7-8, 12-8	**385–392,** 516, 845–852
Build new functions from existing functions. 3. Identify the effect on the graph of replacing $f(x)$ by $f(x) + k$, $k\, f(x)$, $f(kx)$, and $f(x + k)$ for specific values of k (both positive and negative); find the value of k given the graphs. Experiment with cases and illustrate an explanation of the effects on the graph using technology.	Explore 2-7, **2-7,** Explore 4-7, **4-7, 6-3,** Extend 6-4, **7-1,** **7-3, 8-3,** Explore 12-8, **12-8**	108, **109–116,** 273–274, **275–280, 400–406,** 416, **451–458, 468–475, 545–551,** 844, **845–852**

Common Core State Standards *Continued*

Standards	Student Edition Lesson(s)	Student Edition Page(s)
4. Find inverse functions. a. Solve an equation of the form $f(x) = c$ for a simple function f that has an inverse and write an expression for the inverse.	6-2	393–398
Linear, Quadratic, and Exponential Models F-LE		
Construct and compare linear and exponential models and solve problems. 4. For exponential models, express as a logarithm the solution to $ab^{ct} = d$ where a, c, and d are numbers and the base b is 2, 10, or e; evaluate the logarithm using technology.	7-2, 7-8	461–467, 509–515
Trigonometric Functions F-TF		
Extend the domain of trigonometric functions using the unit circle. 1. Understand radian measure of an angle as the length of the arc on the unit circle subtended by the angle.	12-2, 12-6	799–805, 830–836
2. Explain how the unit circle in the coordinate plane enables the extension of trigonometric functions to all real numbers, interpreted as radian measures of angles traversed counterclockwise around the unit circle.	12-6	830–836
Model periodic phenomena with trigonometric functions. 5. Choose trigonometric functions to model periodic phenomena with specified amplitude, frequency, and midline. ★	12-7, 12-8	837–843, 845–852
Prove and apply trigonometric identities. 8. Prove the Pythagorean identity $\sin^2(\theta) + \cos^2(\theta) = 1$ and use it to calculate trigonometric ratios.	**13-1**, 13-2, 13-3, 13-4, 13-5	**873–879**, 880–885, 886–891, 893–899, 901–907
Statistics and Probability		
Interpreting Categorical and Quantitative Data S-ID		
Summarize, represent, and interpret data on a single count or measurement variable. 4. Use the mean and standard deviation of a data set to fit it to a normal distribution and to estimate population percentages. Recognize that there are data sets for which such a procedure is not appropriate. Use calculators, spreadsheets, and tables to estimate areas under the normal curve.	11-5, Extend 11-5	760–766, 767–768

(+) Advanced Mathematics Standards   ★ Mathematical Modeling Standards

Standards	Student Edition Lesson(s)	Student Edition Page(s)
Making Inferences and Justifying Conclusions S-IC		
Understand and evaluate random processes underlying statistical experiments 1. Understand statistics as a process for making inferences about population parameters based on a random sample from that population.	**11-2, 11-6**	**733–741, 769–776**
2. Decide if a specified model is consistent with results from a given data-generating process, e.g., using simulation.	**11-1, Extend 11-1**	**723–730, 731–732**
Make inferences and justify conclusions from sample surveys, experiments, and observational studies 3. Recognize the purposes of and differences among sample surveys, experiments, and observational studies; explain how randomization relates to each.	**11-1**	**723–730**
4. Use data from a sample survey to estimate a population mean or proportion; develop a margin of error through the use of simulation models for random sampling.	**Extend 11-1, 11-6**	**731–732, 769–775**
5. Use data from a randomized experiment to compare two treatments; use simulations to decide if differences between parameters are significant.	**11-1**	**723–730**
6. Evaluate reports based on data.	**Extend 11-1**	**731–732**
Using Probability to Make Decisions S-MD		
6. (+) Use probabilities to make fair decisions (e.g., drawing by lots, using a random number generator).	11-4	752–759
7. (+) Analyze decisions and strategies using probability concepts (e.g., product testing, medical testing, pulling a hockey goalie at the end of a game).	11-3, 11-4, 11-6	742–750, 752–759, 769–776

Common Core State Standards for Mathematical Practice, Correlated to *Glencoe Algebra 2*, Common Core Edition

1. **Make sense of problems and persevere in solving them.**
   Glencoe Algebra 2 exhibits these practices throughout the entire program. Some specific lessons for review are: Lessons 1-1, 2-8, 3-2, 4-1, 5-3, 6-5, 7-8, 8-5, 9-2, 10-7, 11-2, 12-7, and 13-2.

2. **Reason abstractly and quantitatively.**
   Glencoe Algebra 2 exhibits these practices throughout the entire program. Some specific lessons for review are: Lessons 1-2, 2-4, 3-1, 4-3, 5-1, 6-1, 7-2, 8-3, 9-1, 10-1, 11-3, 12-2, and 13-1.

3. **Construct viable arguments and critique the reasoning of others.**
   Glencoe Algebra 2 exhibits these practices throughout the entire program. Some specific lessons for review are: Lessons 1-3, 2-2, 3-4, 4-2, 5-4, 6-3, 7-1, 8-2, 9-6, 10-7, 11-1, 12-4, and 13-3.

4. **Model with mathematics.**
   Glencoe Algebra 2 exhibits these practices throughout the entire program. Some specific lessons for review are: Lessons 1-5, 2-5, 3-3, Extend 4-8, 5-5, 6-7, Extend 7-3, 8-5, 9-3, 10-6, 11-4, 12-8, and 13-5.

5. **Use appropriate tools strategically.**
   Glencoe Algebra 2 exhibits these practices throughout the entire program. Some specific lessons for review are: Lessons 1-6, 2-5, 3-8, Extend 4-1, Extend 5-4, Extend 6-4, Explore 7-8, Extend 8-6, Explore 9-4, Extend 10-5, Extend 11-1, Explore 12-8, and Explore 13-5.

6. **Attend to precision.**
   Glencoe Algebra 2 exhibits these practices throughout the entire program. Some specific lessons for review are: Lessons 1-4, 2-7, 3-1, 4-4, 5-2, 6-4, 7-3, 8-6, 9-5, 10-4, 11-5, 12-3, and 13-5.

7. **Look for and make use of structure.**
   Glencoe Algebra 2 exhibits these practices throughout the entire program. Some specific lessons for review are: Lessons 1-2, 2-1, 3-7, 4-7, 5-6, 6-2, 7-7, 8-4, 9-4, 10-1, 11-6, 12-6, and 13-1.

8. **Look for and express regularity in repeated reasoning.**
   Glencoe Algebra 2 exhibits these practices throughout the entire program. Some specific lessons for review are: Lessons 1-3, 2-3, 3-3, 4-6, 5-8, 6-2, 7-5, 8-1, 9-6, 10-2, 11-5, Explore 12-1, and 13-2.

Students will engage with tasks they find interesting, challenging, and important. In the classroom, key factors are teacher-student relationships, pedagogy, and classroom climate.

Glencoe Algebra 2 is designed to provide your students with a balanced approach to mathematics. Your students become successful through a variety of teaching modalities.

Keeping It Interesting

Students who are engaged are more likely to pay attention in class. **Glencoe Algebra 2** discusses topics that relate to today's students.

Content that Connects

Students use a Then, Now, Why instructional plan that connects what they know with what they are learning. New Vocabulary and Review Vocabulary help students learn to "talk math."

SE Lesson Opener, p. 615

Mike Kemp/Getty Images

Problem-Based Learning

A wealth of problem-solving opportunities include:

- **Multiple Representations** in every chapter
- **H**igher **O**rder **T**hinking **problems** in every lesson
- Worked-out **Examples** that follow a four-step plan
- **Problem-Solving Strategy Tips** throughout
- **Test-Taking Strategies** in every chapter
- **Word-Problem Practice** masters for every lesson

Learning by Doing

Labs help to maintain learner motivation. Algebra Labs introduce and reinforce concepts using manipulatives like algebra tiles. Graphing Technology Labs allow students to explore concepts using graphing calculators, including TI-Nspire Technology™.

SE Multiple Representations, p. 356

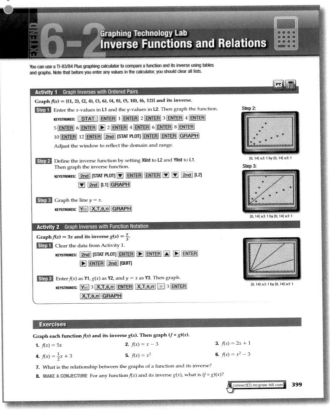

SE Graphing Technology Lab, p. 399

Understanding by Design

*W*hat should students know and be able to do? Understanding by Design can be used to help teachers identify learning goals, develop revealing assessments of student understanding, and plan effective and engaging learning activities.

Backward Design

Understanding by Design (UbD) is a framework that uses backward design to create a coherent curriculum by considering the desired results first and then planning instruction.

The backward design process guided the development of **Glencoe Algebra 1, Glencoe Geometry,** and **Glencoe Algebra 2.**

Identifying Desired Results

The first step in developing an effective curriculum using the UbD framework is to consider the goals. What should students know and be able to do?

Glencoe Algebra 2 addresses the big ideas of algebra and focuses student attention on **Essential Questions**, which are located within each chapter of the Student and Teacher Editions.

An Essential Question is provided at the beginning of each chapter. These thought provoking questions can be used as:

- a discussion starter for your class; throughout the discussion identify what the students already know and what they would like to know about the topic. Revisit these notes throughout the chapter.
- a benchmark of understanding; post these questions in a prominent place and have students expand upon their initial response as their understanding of the subject material grows.

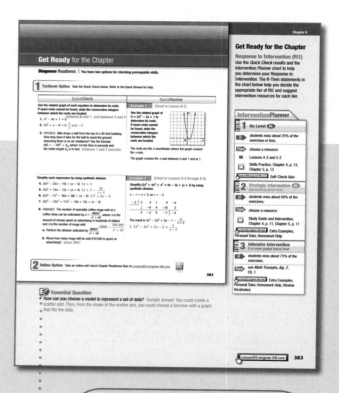

Essential Question
- How can you choose a mode
  scatter plot. Then, from the s

> *In mathematics, essential questions are used to develop students' understanding of key concepts as well as core processes.*
>
> — Jay McTighe,
> co-author of *Understanding by Design*

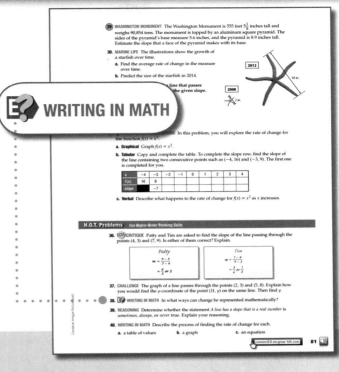

Follow-up Essential Questions can be found throughout each chapter. These questions challenge students to apply specific knowledge to a broader context, thus deepening their understanding.

Determine Acceptable Evidence

A variety of assessment opportunities are available that enable students to show evidence of their understanding.

- **Practice and Problem Solving** and **H.O.T. Problems** allow students to explain, interpret, and apply mathematical concepts.

- **Mid-Chapter Quizzes** and **Chapter Tests** offer more traditional methods of assessment.

- **McGraw-Hill eAssessment** can also be used to customize and create assessments.

Plan Learning Experiences and Instruction

There are numerous performance activities available throughout the program, including:

- **Algebra Labs** that offer students hands-on learning experiences, and

- **Graphing Technology Labs** that use graphing calculators to aid student understanding.

You can also visit <u>connectED.mcgraw-hill.com</u> to choose from an extensive collection of resources to use when planning instruction, such as presentation tools, chapter projects, editable worksheets, digital animations, and eTools.

The McGraw-Hill Companies

Use ConnectED as Your Portal

connectED.mcgraw-hill.com

Nearly 84% of teachers say that they spend only about half of their time teaching, as opposed to disciplining or doing administrative work.

ConnectED is a time-saving online portal that has all of your digital program resources in one place. Through this portal, you can access the complete Student Edition and Teacher Edition, digital animations and tutorials, editable worksheets, presentation tools, and assessment resources.

With ConnectED Mobile you can browse through your course content on the go. The app includes a powerful eBook engine where you can download, view, and interact with your books.

allows you to:

- build lesson plans with easy-to-find print and digital resources
- search for activities to meet a variety of learning modalities
- teach with technology by providing virtual manipulatives, lesson animations, whole-class presentations, and more
- personalize instruction with print and digital resources

- provide students with anywhere, anytime access to student resources and tools, including eBooks, tutorials, animations, and the eGlossary
- access AdvanceTracker, which allows you to assign online assessments, track student progress, generate reports, and differentiate instruction

***N**early 90% of teachers agree that technology enhances their ability to teach.*

Imagine being able to prepare and plan your instruction without having to carry your Teacher Edition home each night. Your eTeacherEdition has all of the content of your Teacher Edition, plus links to thousands of assets right at point of use.

⋯ eTeacherEdition

Sharing your eBook

The eTeacher Edition allows for the teacher wrap and answers to be shown or hidden allowing you to use your book on a projector and interactive whiteboard with your class.

Resources

All Resources on the Page

The resource dock at the bottom of the screen allows you to quickly navigate to resources for planning or presenting a lesson. Teachers can jump to personal tutors, virtual manipulatives, and animations more quickly and easily.

Virtual Manipulatives

Tools provide your students with a digital way to explore concepts in and out of class. While you have a way to help students visualize concepts and to create problem based learning opportunities.

Digital Tools

From animations to assessment, all the tools you need are now housed in one location making planning simpler and presenting easier.

Assessment that Validates Learning

*O*nly 48% of teachers feel that standardized tests are effective in helping them track student performance.

Glencoe Algebra 2 offers a variety of frequent and meaningful assessments built right into the curriculum structure and teacher support materials. The program includes both traditional and nontraditional methods of assessment, including quizzes and tests, performance tasks, and open-ended assessments.

Digital assessment solutions offer additional options for creating, customizing, administering, and instantly grading assessments.

TE Assessment and Intervention, p. 58E

1 Diagnostic

Initial Assessment

Assess students' knowledge at the beginning of the year with *Diagnostic and Placement Tests.* The results and scoring guides identify students who may need additional resources to meet grade-level standards.

Entry-Level Assessment

Assess students' prior knowledge at the beginning of a chapter or lesson.

Student Edition
- Get Ready for the Chapter

Teacher Edition
- Intervention Planner

Online Resources
connectED.mcgraw-hill.com
- Chapter Readiness Quiz

2 Formative

Progress Monitoring
Determine if students are progressing adequately as you teach each lesson. Use the assessments to differentiate lesson instruction and practice.

Student Edition
- Guided Practice
- Check Your Understanding
- H.O.T. Problems
- Mid-Chapter Quiz
- Study Guide and Review

Additional Resources
Chapter Resource Masters
- Mid-Chapter Test
- 4 Quizzes
- Standardized Test Practice

Digital Resources
- McGraw-Hill eAssessment

Online Resources
connectED.mcgraw-hill.com
- Self-Check Quizzes
- Mid-Chapter Test
- Study Guide and Review

McGraw-Hill eAssessment:
- Offers both online and print capabilities
- Provides full assessment reporting
- Generates additional homework assignments and study guides (including practice worksheets and quizzes)
- Includes a complete set of pre-made tests and additional editable question banks
- Allows custom assessment questions to be easily added
- Provides questions in English and Spanish

3 Summative

Summative Assessment
Assess student success in learning the concepts in each chapter. Use remediation suggestions to address problem areas.

Student Edition
- Practice Test
- Standardized Test Practice

Additional Resources
Chapter Resource Masters
- Vocabulary Test
- 6 Leveled Chapter Tests
- Extended-Response Test

Digital Resources
- McGraw-Hill eAssessment
- Chapter Tests
- Cumulative Standardized Test Practice

Online Resources
connectED.mcgraw-hill.com
- Chapter Tests
- Cumulative Standardized Test Practice

Built-In Differentiated Instruction

Approximately 43% of teachers feel their classes are so mixed in terms of students' learning abilities that they can't teach them effectively.

AL	Approaching Grade Level
OL	On Grade Level
BL	Beyond Grade Level
ELL	English Language Learners

Glencoe Algebra 2 fully supports the 3-tier RtI model with print and digital resources to diagnose students, identify areas of need, and conduct short, frequent assessments for accurate data-driven decision making. Every lesson provides easy-to-use resources that consider the needs of all students.

RtI: Response to Intervention

TIER 1 DailyIntervention

OL On grade level

Core instruction targets on-level students. Comprehensive instructional materials help you personalize instruction for every student.

- Diagnostic Teaching
- Options for Differentiated Instruction
- Leveled Exercise Sets, Resources, and Technology
- Data-Driven Decision Making

BL Beyond grade level

At every step, resources and assignments are available for advanced learners.

- Higher-Order Thinking Questions
- Differentiated Homework Options
- Enrichment Masters
- Differentiated Instruction: Extension

TE Get Ready for the Chapter, p. 133

TE Differentiated Instruction, p. 337

SE H.O.T. Problems, p. 371

ELL English Language Learners

Comprehensive resources are found throughout the program.

- Teacher Edition with strategies to modify activities and lesson content
- Multilingual eGlossary with definitions for each vocabulary word in 13 languages

TIER 2 StrategicIntervention

AL Approaching grade level

You can choose from a myriad of intervention tips and ancillary materials to support struggling learners.

- Using Manipulatives
- Alternate Teaching Strategies
- Online resources, including animations, examples, and Personal Tutors

TIER 3 IntensiveIntervention

AL Significantly below grade level

For students who are far below grade level, *Math Triumphs* provides step-by-step instruction, vocabulary support, and meaningful practice.

Mike Kemp/Getty Images

The Math Forum @ Drexel University

The unique partnership between McGraw-Hill and the Math Forum provides additional support beyond your curriculum. The Math Forum at Drexel University is an online resource that offers support for you and your students. Access their rich database of resources directly from ConnectED.

mathforum.org
The Math Forum
@ Drexel University

Student Resources

The Math Forum is the leading online resource for improving math learning, teaching, and communications using the collaborative power of the Internet.

The Math Forum improves math education through:

- online mentoring
- integration of technology into the math classroom
- development of mathematical practices
- action research
- team problem solving
- professional development

Ask Dr. Math®

Ask Dr. Math is a question and answer service for math students and their teachers. Questions have been submitted from all over the world. Answers are provided via email and a selection of them are archived online.

The Ask Dr. Math archive is fully searchable by keyword, or you can browse by topic or grade band. Over 10,000 questions and answers are available. The questions asked most frequently are posted on the Ask Dr. Math FAQ page.

Ask Dr. Math is a good resource for students who need clarification on a difficult concept, or a different perspective on a math lesson.

Students can link to Ask Dr. Math directly from ConnectED.

(c) Goodshoot/Punchstock

Teacher Resources

The Teacher2Teacher (T2T®) service is an outgrowth of the Ask Dr. Math service at the Math Forum. T2T is a peer-mentored question-and-answer service. Teachers and parents ask questions about mathematics teaching and learning. As questions are answered, an archive is built, so that everyone can benefit from past work.

The archive includes questions on diverse topics such as: classroom management techniques, effective grading practices, incorporating math history into lessons, methods of factoring trinomials, and assessment ideas. The questions asked most often are stored on a Frequently Asked Questions page.

Links to T2T are available in ConnectED.

Teacher2Teacher
- Browse the archives
- Search by keyword, grade level, math topic, or education topic
- Ask a question

The Teacher's Lounge
- Discuss current math education topics with other teachers
- Contribute your own ideas
- Ask related questions

Get educator support from the Math Forum's T2T. You may want to search, or ask, for graphing activities. Start your search at connectED.mcgraw-hill.com.

Most equations can be solved by adding the same number to each side of the equation, by subtracting the same number from each side, or by multiplying or dividing each side by the same nonzero number.

1-4 Solving Absolute Value Equations

The absolute value of a number is its distance from zero on a number line. The absolute value symbols are grouping symbols like parentheses. For example, to evaluate the expression $2 \cdot |15 - 31|$, first calculate inside the absolute value symbols.

$$2 \cdot |15 - 31| = 2 \cdot |-16|$$
$$= 2 \cdot (16)$$
$$= 32$$

The equation $|a - 6| = 4$ can be interpreted as *the distance between some number a and 6 is 4 units*. The value $a - 6$ can be 4 or -4. So, if $a - 6 = 4$, then $a = 10$. If $a - 6 = -4$, then $a = 2$. The solutions are 10 and 2. An equation like $|x| = -2$ is *never* true, so there is no solution.

1-5 Solving Inequalities

An open sentence that contains the symbol $<$, $\leq$, $>$, or $\geq$ is called an *inequality*. For any two real numbers, a and b, exactly one of the following statements is true: $a < b$, $a = b$, or $a > b$.

Solving an inequality involves using the Properties of Inequality to write a series of equivalent inequalities, ending with one that isolates the variable. These properties include:

- the Addition Property of Inequality,
- the Subtraction Property of Inequality,
- the Multiplication Property of Inequality, and
- the Division Property of Inequality.

In general, adding the same number to (or subtracting the same number from) each side of an inequality, or multiplying or dividing each side by a positive number, does not change the truth of an inequality. Multiplying or dividing an inequality by a negative number *reverses* the order of the inequality.

There are infinitely many solutions to an inequality. These solutions can be expressed in more than one way:

- **Graphs** When the solution is graphed on a number line, an open circle indicates a value that is not included in the solution, and a closed circle indicates a value that is included. Circles are used with $<$ and $>$, and dots are used with $\leq$ and $\geq$.
- **Set Builder Notation** Solutions can be written using set-builder notation. A solution such as $x \geq 4$ would be written $\{x \mid x \geq 4\}$, and read *the set of values x such that x is greater than or equal to 4*.

1-6 Solving Compound and Absolute Value Inequalities

An inequality that consists of two inequalities joined by the word *and* or the word *or* is called a *compound inequality*. To solve a compound inequality, each part of the inequality must be solved.

- The graph of a compound inequality that contains *and* is the *intersection* of the solution sets of the two inequalities.
- The graph of a compound inequality that contains *or* is the *union* of the solution sets of the two inequalities.
- An absolute value inequality using $<$ or $\leq$ is related to a compound inequality using the word *and*. For example, thinking of $|a| < 7$ as $|a - 0| < 7$, then the value of a is any number whose distance from 0 is less than 7 units.

Possible values for a

- An absolute value inequality using $>$ or $\geq$ is related to a compound inequality using the word *or*. For example, thinking of $|b| > 5$ as $|b - 0| > 5$, then the value of b is any number whose distance from 0 is greater than 5.

Possible values for b

Mc Graw Hill | Professional Development

Targeted professional development has been articulated throughout Algebra 2. More quality, customized professional development is available from McGraw-Hill Professional Development. Visit connectED.mcgraw-hill.com for details on each product.

- Online Lessons emphasize the strategies and techniques used to teach Algebra 2. Includes streaming video, interactive pages, and online tools.
- Video Workshops allow mentors, coaches, or leadership personnel to facilitate on-site workshops on educational strategies in mathematics and mathematical concepts.
- MHPD Online (www.mhpdonline.com) offers online professional development with video clips of instructional strategies, links, student activities, and news and issues in education.
- Teaching Today (teachingtoday.glencoe.com) gives secondary teachers practical strategies and materials that inspire excellence and innovation in teaching.

connectED.mcgraw-hill.com   2F

Connect with Digital Natives

connectED.mcgraw-hill.com

Today's students have an unprecedented access to and appetite for technology and new media. They perceive technology as their friend and rely on it to study, work, play, relax, and communicate.

Your students are accustomed to the role that computers play in today's world. They may well be the first generation whose primary educational tool is a computer or a cell phone. The eStudentEdition allows your students to access their math curriculum anytime, anywhere. Along with your student edition, *Glencoe Algebra 2* provides a blended instructional solution for your next-generation students.

Investigate

Animations
illustrate key concepts through step-by-step tutorials and videos.

Vocabulary
tools include fun Vocabulary Review Games.

Multilingual eGlossary
presents key vocabulary in 13 languages.

Learn

Personal Tutor
presents an experienced educator explaining step-by-step solutions to problems.

Virtual Manipulatives
are outstanding tools for enhancing understanding.

Graphing Calculator
provides other calculator keystrokes for each Graphing Technology Lab.

Audio
is provided to enhance accessibility for all students.

Foldables
provide a unique way to enhance student study skills.

Practice

Self-Check Practice
allows students to check their understanding and send results to their teacher.

Worksheets
provide additional practice and reteaching opportunities.

College and Work Readiness

*T*he U.S. Department of Labor estimates that 90% of 21st-century skilled workforce jobs will require post-secondary education. To be competitive in tomorrow's global workforce, American students must be prepared to succeed in college.

A strong high school curriculum is a good predictor of college readiness (Adelman, 2006). Students who take at least three years of college-preparatory mathematics using programs like **Glencoe Algebra 1**, **Glencoe Geometry**, and **Glencoe Algebra 2** are less likely to need remedial courses in college than students who do not (Abraham & Creech, 2002).

College Readiness

David Conley at the University of Oregon developed the following criteria for college readiness.

Key Content Knowledge

Glencoe Algebra 1, Glencoe Geometry, and **Glencoe Algebra 2** have been aligned to rigorous state and national standards, including the *NCTM Principles & Standards for School Mathematics*, the College Board Standards for College Success, and the American Diploma Project's Benchmarks. Correlations to these standards can be found at connectED.mcgraw-hill.com.

Habits of Mind

These include critical thinking skills such as analysis, interpretation, problem solving, and reasoning. Students can hone critical higher-order thinking skills through the use of **H.O.T. (Higher Order Thinking) Problems.**

Contextual Skills

These are practical skills like understanding the admissions process and financial aid, placement testing, and communicating with professors. Throughout each Glencoe mathematics program, students are required to write, explain, justify, prove, and analyze.

Academic Behaviors

These include general skills such as reading comprehension, time management, note-taking, and metacognition. Reading Math tips and Vocabulary Links help students with reading comprehension. Study Notebooks and Anticipation Guides help students build note-taking skills and aid with metacognition.

Ariel Skelley/Photographer's Choice/Getty Images

Sources:
Abraham, A. & Creech, J. (2002). *Reducing Remedial Education.* Atlanta, GA: Southern Regional Education Board.
ACT, Inc. (2006). *Readiness for College and Readiness for Work: Same or Different?* Iowa City, IA
The Conference Board, Corporate Voices for Working Families, the Partnership for 21st Century Skills, and the Society for Human Resource Management (2006).
   Are They Really Ready to Work?
Conley, D. (2007). *Toward a More Comprehensive Conception of College Readiness.* Eugene, OR: Educational Policy Improvement Center.

Work Readiness

Work Readiness is the ability of entry-level employees to add value in front-line jobs.

Does College Readiness Lead to Success in the Workplace?

The U.S. Chamber of Commerce has created a National Work Readiness Credential that involves nine sub-skills, one of which is using math to solve problems. In a recent study, 53.5% of employers who responded reported that high school graduate entrants into the workforce are "deficient" in mathematics, while 30.4% felt that knowledge of mathematics is "very important." (Are They Really Ready to Work?, 2006)

What About Students Who Don't Plan to Go to College?

In today's technological world, math is no longer just for students who go to college. ACT compared the skills needed to succeed as a freshman in college and those needed for job-training programs. They found that students need to be educated to a comparable level in algebra, geometry, data analysis, and statistics for success in either situation (ACT, 2006).

College & Career READINESS   http://ready.ccr.mcgraw-hill.com/

McGraw-Hill is committed to helping educators ensure that all students graduate from high school equipped to succeed in college or in their chosen careers. McGraw-Hill programs bridge the gap between secondary and postsecondary curricula by including pre-college content as well as the study skills and transitional life skills that are necessary for both secondary and college academic success.

Getty Images

STEM Careers

*S*tudent intentions to go to college have increased over the past 20 years. In 1988, 80% said they were likely to go to college, compared to 90% today.

Developing STEM Careers

With **Glencoe Algebra 2,** you can unleash your students' curiosity about the world around them and prepare them for exciting **STEM** (**S**cience, **T**echnology, **E**ngineering, and **M**ath) careers.

Real-World Careers
Real-World Careers are engaging, providing information on exciting careers.

Examples
Examples are relevant, connecting in-class experiences to the world beyond the classroom.

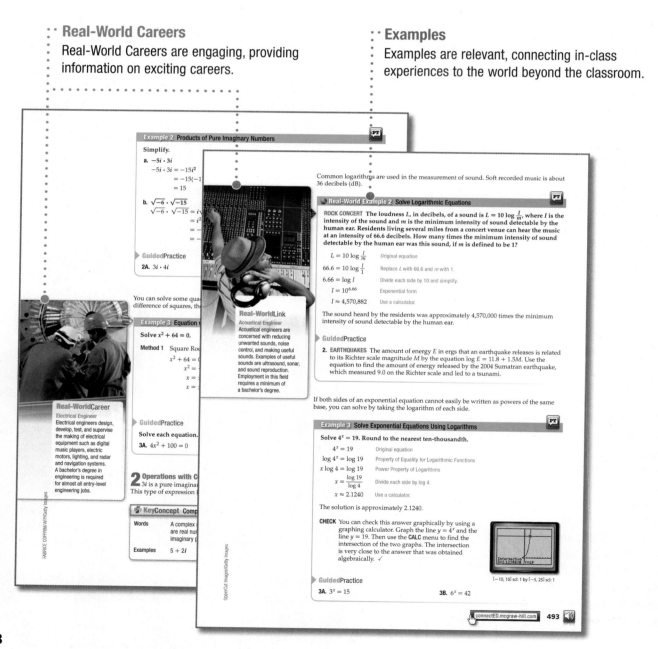

21st Century Skills

The current and future health of America's 21st Century Economy depends directly on how broadly and deeply Americans reach a new level of literacy—'21st Century Literacy'—that includes strong academic skills, thinking, reasoning, teamwork skills, and proficiency in using technology.

— 21st Century Workforce Commission National Alliance of Business

Developing 21st Century Skills

The Partnership for 21st Century Skills identifies the following key student elements of a 21st century education.

Core Subjects and 21st Century Themes

Glencoe Algebra 1, Glencoe Geometry, and *Glencoe Algebra 2* have been aligned to rigorous state and national standards, including the Common Core State Standards, NCTM Principles & Standards for School Mathematics, and the American Diploma Project's Benchmarks. Correlations to these standards can be found at underlinedconnectED.mcgraw-hill.com.

Throughout each Glencoe mathematics program, students solve problems that incorporate 21st century themes, such as financial literacy. In addition, there is also a project within each chapter that incorporates 21st Century Skills.

21st Century Student Outcomes and Support Systems

Learning and Innovation Skills

Students who are prepared for increasingly complex life and work environments are creative and innovative critical thinkers, problem solvers, effective communicators, and know how to work collaboratively. Throughout each Glencoe mathematics program, students are required to write, explain, justify, prove, and analyze. Students can hone critical thinking skills through the use of **H.O.T. (Higher Order Thinking) Problems** and are encouraged to work collaboratively in labs.

Life and Career Skills

McGraw-Hill is committed to helping educators ensure that all students graduate from high school equipped to succeed in college or in their chosen careers. McGraw-Hill programs bridge the gap between secondary and postsecondary curricula by including pre-college content as well as the study skills and transitional life skills that are necessary for both secondary and college academic success.

Information, Media, and Technology Skills

Throughout each Glencoe mathematics program, students use technology, including graphing calculators and the Internet, to develop 21st century mathematics knowledge and skills.

21st Century Assessments

Glencoe Algebra 1, Glencoe Geometry, and *Glencoe Algebra 2* offer a variety of frequent and meaningful assessments built right into the curriculum structure and teacher support materials. These programs include both traditional and nontraditional methods of assessment, including quizzes and tests, performance tasks, and open-ended assessments. Digital assessment solutions offer additional options for creating, customizing, administering, and instantly grading assessments.

CHAPTER 0

Preparing for Advanced Algebra

Get Started On Chapter 0 . . . . **P2**

■ **Pretest** **P3**

0-1   **Representing Functions** **P4**

0-2   **FOIL** **P6**

0-3   **Factoring Polynomials** **P7**

0-4   **Counting Techniques** **P9**

0-5   **Adding Probabilities** **P13**

0-6   **Multiplying Probabilities** **P16**

0-7   **Congruent and Similar Figures** **P20**

0-8   **The Pythagorean Theorem** **P22**

0-9   **Measures of Center, Spread, and Position** **P24**

■ **Posttest** **P29**

Table of Contents

 connectED.mcgraw-hill.com   **Your Digital Math Portal**

 Vocabulary
p. P4

 Multilingual eGlossary
p. P2

 Personal Tutor
p. P6

 Foldables
p. P2

Ed Freeman/Stone/Getty Images

i

CHAPTER 1
Equations and Inequalities

Get Ready for Chapter 1 3

CCSS Mathematical Content

1-1 Expressions and Formulas 5 A.SSE.1.a, A.SSE.1.b

1-2 Properties of Real Numbers 11 A.SSE.2

1-3 Solving Equations 18 A.CED.1

■ **Mid-Chapter Quiz** 26

1-4 Solving Absolute Value Equations 27 A.SSE.1.b, A.CED.1

1-5 Solving Inequalities 33 A.CED.1, A.CED.3

📖 **Explore: Algebra Lab** Interval Notation 40

1-6 Solving Compound and Absolute Value Inequalities 41 A.CED.1, A.CED.3

Assessment
■ **Study Guide and Review** 49
■ **Practice Test** 53
■ **Preparing for Standardized Tests** 54
■ **Standardized Test Practice, Chapter 1** 56

Jon Boyes/Photographer's Choice/Getty Images

 Animation pp. 12, 84

 Vocabulary pp. 5, 69

 Multilingual eGlossary pp. 4, 60

 Personal Tutor pp. 6, 63

Linear Relations and Functions

Get Ready for Chapter 2 59

CCSS **Mathematical Content**

2-1 **Relations and Functions** 61   F.IF.4, F.IF.5
   📖 **Extend: Algebra Lab** Discrete and Continuous Functions 68   F.IF.4

2-2 **Linear Relations and Functions** 69   F.IF.4, F.IF.9
   📖 **Extend: Algebra Lab** Roots of Equations and Zeros of Functions 75   F.IF.4

2-3 **Rate of Change and Slope** 76   F.IF.4, F.IF.6

2-4 **Writing Linear Equations** 83   A.SSE.1.b, A.CED.2
   🖩 **Extend: Graphing Technology Lab** Direct Variation 90   F.IF.4

   ■ **Mid-Chapter Quiz** 91

2-5 **Scatter Plots and Lines of Regression** 92   F.IF.4
   📖 **Extend: Algebra Lab** Correlation and Causation 99   F.IF.4

2-6 **Special Functions** 101   F.IF.4, F.IF.7.b

   🖩 **Explore: Graphing Technology Lab** Families of Lines 108   F.BF.3
2-7 **Parent Functions and Transformations** 109   F.IF.4, F.BF.3
2-8 **Graphing Linear and Absolute Value Inequalities** 117   A.CED.3

Assessment
■ **Study Guide and Review** 122
■ **Practice Test** 127
■ **Preparing for Standardized Tests** 128
■ **Standardized Test Practice, Chapters 1–2** 130

Roine Magnusson/Photodisc/Getty Images

 Virtual Manipulatives pp. 28, 61

 Graphing Calculator p. 90

 Foldables pp. 4, 60

 Self-Check Practice pp. 7, 127

Get Ready for Chapter 3 .... **133**

   ▦ **Explore: Graphing Technology Lab** Intersections of Graphs **135**

3-1 Solving Systems of Equations **136**   A.CED.3, A.REI.11

3-2 Solving Systems of Inequalities by Graphing **146**   A.CED.3

   ▦ **Extend: Graphing Technology Lab** Systems of Linear Inequalities **153**

3-3 Optimization with Linear Programming **154**   A.CED.3

3-4 Systems of Equations in Three Variables **161**   A.CED.3

   ■ **Mid-Chapter Quiz** **168**

   ▭ **Explore: Spreadsheet Lab** Organizing Data with Matrices **169**

3-5 Operations with Matrices **171**

3-6 Multiplying Matrices **179**

   ▭ **Extend: Graphing Technology Lab** Operations with Matrices **187**

3-7 Solving Systems of Equations Using Cramer's Rule **189**   A.CED.3

3-8 Solving Systems of Equations Using Inverse Matrices **198**   A.CED.3

   ▦ **Extend: Graphing Technology Lab** Augmented Matrices **205**

   Assessment

   ■ **Study Guide and Review** **206**

   ■ **Practice Test** **211**

   ■ **Preparing for Standardized Tests** **212**

   ■ **Standardized Test Practice, Chapters 1–3** **214**

(CCSS) **Mathematical Content**

DreamPictures/Blend Images/Getty Images

 connectED.mcgraw-hill.com   **Your Digital Math Portal**

 Animation pp. 146, 247

Vocabulary pp. 206, 290

 Multilingual eGlossary pp. 134, 218

 Personal Tutor pp. 136, 222

Quadratic Functions and Relations

Get Ready for Chapter 4 217

4-1 Graphing Quadratic Functions ..... 219 — A.SSE.1.a, F.IF.9
  ▦ **Extend: Graphing Technology Lab** Modeling Real-World Data ..... 228 — F.IF.4

4-2 Solving Quadratic Equations by Graphing ..... 229 — A.CED.2, F.IF.4
  ▦ **Extend: Graphing Technology Lab** Solving Quadratic Equations by Graphing ..... 237 — A.REI.11

4-3 Solving Quadratic Equations by Factoring ..... 238 — A.SSE.2, F.IF.8.a

4-4 Complex Numbers ..... 246 — N.CN.1, N.CN.2
  ✏ **Extend: Algebra Lab** The Complex Plane 253

  ■ **Mid-Chapter Quiz** 255

4-5 Completing the Square ..... 256 — N.CN.7, F.IF.8.a
  ▦ **Extend: Graphing Technology Lab** Solving Quadratic Equations ..... 263 — N.CN.7

4-6 The Quadratic Formula and the Discriminant ..... 264 — N.CN.7, A.SSE.1.b

  ▦ **Explore: Graphing Technology Lab** Families of Parabolas ..... 273 — F.IF.4, F.BF.3

4-7 Transformations of Quadratic Graphs ..... 275 — F.IF.8.a, F.BF.3
  📖 **Extend: Algebra Lab** Quadratics and Rate of Change ..... 281 — F.IF.4, F.IF.6

4-8 Quadratic Inequalities ..... 282 — A.CED.1, A.CED.3
  ▦ **Extend: Graphing Technology Lab** Modeling Motion 289

Assessment
  ■ **Study Guide and Review** 290
  ■ **Practice Test** 295
  ■ **Preparing for Standardized Tests** 296
  ■ **Standardized Test Practice, Chapters 1–4** 298

CCSS Mathematical Content

Virtual Manipulatives pp. 137, 219

Graphing Calculator pp. 205, 237

Foldables pp. 134, 218

Self-Check Practice pp. 183, 295

Black 100/Allsport Concepts/Getty Images

v

CHAPTER 5 Polynomials and Polynomial Functions

Get Ready for Chapter 5 301

(CCSS) **Mathematical Content**

5-1 Operations with Polynomials 303   A.APR.1
 📖 **Extend: Algebra Lab** Dimensional Analysis 310

5-2 Dividing Polynomials 311   A.APR.6
 🖩 **Extend: Graphing Technology Lab** Dividing Polynomials 318   A.APR.6

 🖩 **Explore: Graphing Technology Lab** Power Functions 320   F.IF.4
5-3 Polynomial Functions 322   F.IF.4, F.IF.7.c

5-4 Analyzing Graphs of Polynomial Functions 330   F.IF.4, F.IF.7.c
 🖩 **Extend: Graphing Technology Lab** Modeling Data Using Polynomial Functions 338   F.IF.4, F.IF.7.c

 ■ **Mid-Chapter Quiz** 340

 🖩 **Explore: Graphing Technology Lab** Solving Polynomial Equations by Graphing 341
5-5 Solving Polynomial Equations 342   A.CED.1
 🖩 **Extend: Graphing Technology Lab** Polynomial Identities 350   A.REI.11

5-6 The Remainder and Factor Theorems 352   A.APR.2, F.IF.7.c

5-7 Roots and Zeros 358   N.CN.9, A.APR.3
 🖩 **Extend: Graphing Technology Lab** Analyzing Polynomial Functions 366   A.APR.4

5-8 Rational Zero Theorem 367

 ■ **Study Guide and Review** 373
 ■ **Practice Test** 377
 ■ **Preparing for Standardized Tests** 378
 ■ **Standardized Test Practice, Chapters 1–5** 380

connectED.mcgraw-hill.com **Your Digital Math Portal**

 Animation
pp. 306, 387

 Vocabulary
pp. 303, 400

 Multilingual eGlossary
pp. 302, 384

PT **Personal Tutor**
pp. 304, 386

Photodisc/Getty Images

CHAPTER 6

Inverses and Radical Functions and Relations

Get Ready for Chapter 6 . . . . **383**

6-1 Operations on Functions **385**

6-2 Inverse Functions and Relations **393**
  ▦ Extend: **Graphing Technology Lab** Inverse Functions and Relations **399**

6-3 Square Root Functions and Inequalities **400**

6-4 *n*th Roots **407**
  ▦ Extend: **Graphing Technology Lab** Graphing *n*th Root Functions **413**

  ■ **Mid-Chapter Quiz** **414**

6-5 Operations with Radical Expressions **415**

6-6 Rational Exponents **422**

6-7 Solving Radical Equations and Inequalities **429**
  ▦ Extend: **Graphing Technology Lab** Solving Radical Equations and Inequalities **436**

  Assessment
  ■ **Study Guide and Review** **438**
  ■ **Practice Test** **443**
  ■ **Preparing for Standardized Tests** **444**
  ■ **Standardized Test Practice, Chapters 1–6** **446**

CCSS **Mathematical Content**

	F.IF.9, F.BF.1.b
	F.IF.4, F.BF.4.a
	F.IF.7.b, F.BF.3
	A.SSE.2
	F.IF.7.b, F.BF.3
	A.SSE.2
	A.REI.2
	A.REI.2, A.REI.11

Getty Images

CHAPTER 7
Exponential and Logarithmic Functions and Relations

Get Ready for Chapter 7 449

Mathematical Content

7-1 Graphing Exponential Functions 451   F.IF.7.e, F.IF.8.b

    🖩 **Explore: Graphing Technology Lab** Solving Exponential Equations and Inequalities ... 459   A.REI.11

7-2 Solving Exponential Equations and Inequalities 461   A.CED.1, F.LE.4

7-3 Logarithms and Logarithmic Functions 468   F.IF.7.e, F.BF.3

    🖩 **Extend: Graphing Technology Lab** Choosing the Best Model 476

7-4 Solving Logarithmic Equations and Inequalities 478   A.SSE.2, A.CED.1

    ■ **Mid-Chapter Quiz** 484

7-5 Properties of Logarithms 485   A.CED.1

7-6 Common Logarithms 492   A.CED.1

    🖩 **Extend: Graphing Technology Lab** Solving Logarithmic Equations and Inequalities ... 499   A.REI.11

7-7 Base e and Natural Logarithms 501   A.SSE.2

    💻 **Explore: Spreadsheet Lab** Compound Interest 508

7-8 Using Exponential and Logarithmic Functions 509   F.IF.8.b, F.LE.4

    🖩 **Extend: Graphing Technology Lab** Cooling 516   F.BF.1.b

Assessment

■ **Study Guide and Review** 517

■ **Practice Test** 521

■ **Preparing for Standardized Tests** 522

■ **Standardized Test Practice, Chapters 1–7** 524

connectED.mcgraw-hill.com   Your Digital Math Portal

 Animation pp. 453, 532

 Vocabulary pp. 517, 581

 Multilingual eGlossary pp. 450, 528

 Personal Tutor pp. 455, 531

Purestock/Getty Images

CHAPTER 8

Rational Functions and Relations

Get Ready for Chapter 8 527

 Mathematical Content

8-1 Multiplying and Dividing Rational Expressions 529   A.APR.7

8-2 Adding and Subtracting Rational Expressions 538   A.APR.7

8-3 Graphing Reciprocal Functions 545   A.CED.2, F.BF.3

    ■ **Mid-Chapter Quiz** 552

8-4 Graphing Rational Functions 553   A.CED.2, F.IF.9

    ▦ **Extend: Graphing Technology Lab** Graphing Rational Functions 561

8-5 Variation Functions 562   A.CED.2

8-6 Solving Rational Equations and Inequalities 570   A.CED.1, A.REI.2

    ▦ **Extend: Graphing Technology Lab** Solving Rational Equations and Inequalities 579   A.REI.2, A.REI.11

Assessment
■ **Study Guide and Review** 581
■ **Practice Test** 585
■ **Preparing for Standardized Tests** 586
■ **Standardized Test Practice, Chapters 1–8** 588

Image Source/Masterfile

 Virtual Manipulatives pp. 451, 557

 Graphing Calculator pp. 516, 579

 Foldables pp. 450, 528

 Self-Check Practice pp. 464, 585

CHAPTER 9

Conic Sections

Get Ready for Chapter 9 . . . . **591**

 Mathematical Content

9-1 Midpoint and Distance Formulas **593**   A.CED.4

9-2 Parabolas **599**   A.SSE.1.b, A.CED.2

☐ **Explore: Graphing Technology Lab** Equations of Circles **606**

9-3 Circles **607**   A.SSE.1.b, A.CED.4

✋ **Explore: Algebra Lab** Investigating Ellipses **614**

9-4 Ellipses **615**   A.SSE.1.b, A.CED.2

■ **Mid-Chapter Quiz** **623**

9-5 Hyperbolas **624**   A.SSE.1.b, A.CED.2

9-6 Identifying Conic Sections **632**   A.SSE.1.b, F.IF.9

☐ **Extend: Graphing Technology Lab** Analyzing Quadratic Relations **637**

☐ **Explore: Graphing Technology Lab** Linear-Nonlinear Systems **639**

9-7 Solving Linear-Nonlinear Systems **640**   A.REI.11

Assessment
■ **Study Guide and Review** **646**
■ **Practice Test** **651**
■ **Preparing for Standardized Tests** **652**
■ **Standardized Test Practice, Chapters 1–9** **654**

 connectED.mcgraw-hill.com   **Your Digital Math Portal**

 Animation
pp. 641, 659

 Vocabulary
pp. 599, 710

 Multilingual eGlossary
pp. 592, 658

 Personal Tutor
pp. 600, 660

Brand X Pictures/PunchStock

CHAPTER 10 Sequences and Series

Get Ready for Chapter 10 657

10-1 Sequences as Functions 659   F.IF.4

10-2 Arithmetic Sequences and Series 666   A.CED.4

10-3 Geometric Sequences and Series 674   A.SSE.4

📖 **Explore: Algebra Lab** Area Under a Curve 681

10-4 Infinite Geometric Series 683

🖩 **Extend: Graphing Technology Lab** Limits 690

■ **Mid-Chapter Quiz** 691

10-5 Recursion and Iteration 692

🖥 **Extend: Spreadsheet Lab** Amortizing Loans 698

10-6 The Binomial Theorem 699   A.APR.5

📖 **Extend: Algebra Lab** Combinations and Pascal's Triangle 704   A.APR.5

10-7 Proof by Mathematical Induction 705   A.SSE.1.b, A.APR.5

Assessment
■ **Study Guide and Review** 710
■ **Practice Test** 715
■ **Preparing for Standardized Tests** 716
■ **Standardized Test Practice, Chapters 1–10** 718

CCSS Mathematical Content

Martin Child/Photodisc/Getty Images

 Virtual Manipulatives pp. 594, 681

 Graphing Calculator pp. 639, 690

 Foldables pp. 592, 658

 Self-Check Practice pp. 603, 715

CHAPTER 11

Statistics and Probability

Get Ready for Chapter 11 .... **721**

			CCSS **Mathematical Content**
11-1	**Designing a Study**	**723**	S.IC.3, S.IC.5
	▦ Extend: **Graphing Technology Lab** Simulations and Margin of Error	**731**	S.IC.4, S.IC.6
11-2	**Distributions of Data**	**733**	S.IC.1
11-3	**Probability Distributions**	**742**	S.MD.7
	■ **Mid-Chapter Quiz**	**751**	
11-4	**The Binomial Distribution**	**752**	S.MD.6, S.MD.7
11-5	**The Normal Distribution**	**760**	S.ID.4
	▭ Extend: **Spreadsheet Lab** Normal Approximation of Binomial Distributions	**767**	S.ID.4
11-6	**Confidence Intervals and Hypothesis Testing**	**769**	S.IC.1, S.IC.4

Assessment
■ **Study Guide and Review** **777**
■ **Practice Test** **781**
■ **Preparing for Standardized Tests** **782**
■ **Standardized Test Practice, Chapters 1–11** **784**

🖑 connectED.mcgraw-hill.com **Your Digital Math Portal**

 Animation
pp. 720, 799

 Vocabulary
pp. 742, 859

 Multilingual eGlossary
pp. 722, 788

 Personal Tutor
pp. 724, 789

MIXA/Getty Images

xii

CHAPTER 12 Trigonometric Functions

Get Ready for Chapter 12 787

    🖥 **Explore: Spreadsheet Lab** Investigating Special Right Triangles 789

12-1   **Trigonometric Functions in Right Triangles** 790

12-2   **Angles and Angle Measure** 799   F.TF.1
    📖 **Extend: Geometry Lab** Areas of Parallelograms 806

12-3   **Trigonometric Functions of General Angles** 807

12-4   **Law of Sines** 814
    ✋ **Extend: Geometry Lab** Regular Polygons 822

12-5   **Law of Cosines** 823

    ■ **Mid-Chapter Quiz** 829

12-6   **Circular and Periodic Functions** 830   F.TF.1, F.TF.2

12-7   **Graphing Trigonometric Functions** 837   F.IF.7.e, F.TF.5

    🖩 **Explore: Graphing Technology Lab** Trigonometric Graphs 844   F.BF.3
12-8   **Translations of Trigonometric Graphs** 845   F.IF.7.e, F.BF.3

12-9   **Inverse Trigonometric Functions** 853   A.CED.2

Assessment
■ **Study Guide and Review** 859
■ **Practice Test** 865
■ **Preparing for Standardized Tests** 866
■ **Standardized Test Practice, Chapters 1–12** 868

CCSS **Mathematical Content**

Dennis Welsh/UpperCut Images/Getty Images

  Virtual Manipulatives pp. 737, 822

 Graphing Calculator pp. 731, 844

 Foldables pp. 722, 788

 Self-Check Practice pp. 738, 865

CHAPTER 13

Trigonometric Identities and Equations

Get Ready for Chapter 13 871

13-1 Trigonometric Identities 873 F.TF.8

13-2 Verifying Trigonometric Identities 880 F.TF.8

13-3 Sum and Difference of Angles Identities 886 F.TF.8

■ **Mid-Chapter Quiz** 892

13-4 Double-Angle and Half-Angle Identities 893 F.TF.8

▦ **Explore: Graphing Technology Lab** Solving Trigonometric Equations 900

13-5 Solving Trigonometric Equations 901 F.TF.8

Assessment
■ **Study Guide and Review** 908
■ **Practice Test** 911
■ **Preparing for Standardized Tests** 912
■ **Standardized Test Practice, Chapters 1–13** 914

CCSS Mathematical Content

Maria Teijeiro/Photodisc/Getty Images

🖑 **connectED.mcgraw-hill.com**   **Your Digital Math Portal**

Animation p. 870	Vocabulary p. 873	Multilingual eGlossary p. 872	Personal Tutor p. 875	Graphing Calculator p. 900	Foldables p. 872	Self-Check Practice p. 897

Student Handbook

Contents

How to Use the Student Handbook

Extra Practice R1

Selected Answers and Solutions R14

Glossary/Glosario R124

Index R149

Formulas . . . . **Inside Back Cover**

Symbols and Parent Functions . . . . **Inside Back Cover**

Rubberball/Getty Images

CHAPTER 0

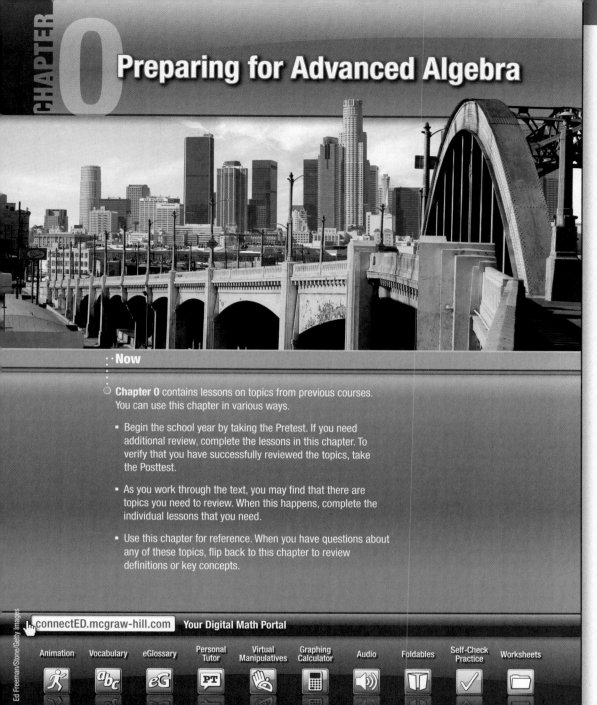

Preparing for Advanced Algebra

Ed Freeman/Stone/Getty Images

Now

○ **Chapter 0** contains lessons on topics from previous courses. You can use this chapter in various ways.

- Begin the school year by taking the Pretest. If you need additional review, complete the lessons in this chapter. To verify that you have successfully reviewed the topics, take the Posttest.

- As you work through the text, you may find that there are topics you need to review. When this happens, complete the individual lessons that you need.

- Use this chapter for reference. When you have questions about any of these topics, flip back to this chapter to review definitions or key concepts.

connectED.mcgraw-hill.com **Your Digital Math Portal**

| Animation | Vocabulary | eGlossary | Personal Tutor | Virtual Manipulatives | Graphing Calculator | Audio | Foldables | Self-Check Practice | Worksheets |

Using Chapter 0

The concepts presented in Chapter 0 are review from previous courses. You may wish to use all or some of the chapter at the beginning of the school year to refresh students' skills. Or you may wish to begin with Chapter 1 and use the Chapter 0 lessons as needed to reinforce prerequisite skills as you progress through the program.

FOLDABLES StudyOrganizer

Each chapter of *Glencoe Algebra 2* features a Foldables Study Organizer students can make to organize their notes. Encourage students to use these tools to make their study time more productive.

Get Started on the Chapter

You will review several concepts, skills, and vocabulary terms as you study Chapter 0. To get ready, identify important terms and organize your resources.

FOLDABLES StudyOrganizer

Throughout this text, you will be invited to use Foldables to organize your notes.

Why should you use them?

- They help you organize, display, and arrange information.
- They make great study guides, specifically designed for you.
- You can use them as your math journal for recording main ideas, problem-solving strategies, examples, or questions you may have.
- They give you a chance to improve your math vocabulary.

How should you use them?

- Write general information—titles, vocabulary terms, concepts, questions, and main ideas—on the front tabs of your Foldable.
- Write specific information—ideas, your thoughts, answers to questions, steps, notes, and definitions—under the tabs.
- Use the tabs for:
  - math concepts in parts, like types of triangles,
  - steps to follow, or
  - parts of a problem, like *compare and contrast* (2 parts) or *what, where, when, why*, and *how* (5 parts).
- You may want to store your Foldables in a plastic zipper bag that you have three-hole punched to fit in your notebook.

When should you use them?

- Set up your Foldable as you begin a chapter, or when you start learning a new concept.
- Write in your Foldable every day.
- Use your Foldable to review for homework, quizzes, and tests.

ReviewVocabulary

English		Español
domain	p. P4	dominio
range	p. P4	rango
quadrants	p. P4	cuadrantes
mapping	p. P4	transformaciones
function	p. P4	función
outcome	p. P9	resultados
sample space	p. P9	espacio muestral
permutation	p. P9	permutación
factorial	p. P10	factorial
combination	p. P11	combinación
theoretical probability	p. P13	probabilidad teórica
experimental probability	p. P13	probabilidad experimental
simple event	p. P14	evento simple
compound event	p. P14	suceso compuesto
mutually exclusive	p. P14	mutuamente exclusiva
independent events	p. P16	eventos independientes
dependent events	p. P16	eventos dependientes
conditional probability	p. P16	probabilidad condicional
two-way frequency table	p. P18	tabla de doble entrada o de contingencia
population	p. P24	población
sample	p. P24	muestra
mean	p. P24	media
median	p. P24	mediana
mode	p. P24	moda
range	p. P25	rango
variance	p. P25	varianza
standard deviation	p. P25	desviación estándar
five-number summary	p. P26	resumen de cinco números
outlier	p. P27	valor atípico

CHAPTER 0 Pretest

State the domain and range of each relation. Then determine whether each relation is a function. Write *yes* or *no*. 1. D = {−3, 8, 14}, R = {1, 4, 6}; yes

1. {(14, 1), (−3, 6), (8, 4)}

2. D = {−2, 3}, R = {−4, −2, 0, 2}; no

Domain   Range

Name the quadrant in which each point is located.

3. (−6, −2) III   4. (4, −3) IV   5. (−5, 7) II

6. $x^2 + 5x + 4$

Find each product. 7. $a^2 + 3a − 18$

6. $(x + 1)(x + 4)$ 7. $(a − 3)(a + 6)$

8. $(m − 2)(m − 5)$ 9. $(c + 8)(c − 8)$
   $m^2 − 7m + 10$ $c^2 − 64$

10. **NUMBER THEORY** There are two integers. One is 5 more than a number, and the other is 1 less than the same number. **a.** $n + 5, n − 1$

   a. Write expressions for the two numbers.

   b. Write a polynomial expression for the product of the numbers. $n^2 + 4n − 5$

Factor each polynomial.

11. $10ab^2 + 5b$ $5b(2ab + 1)$ 12. $15d − 12cd^2$ $3d(5 − 4cd)$

13. $y^2 + 6y − 7$ 14. $a^2 − 13a + 36$
   $(y − 1)(y + 7)$ $(a − 4)(a − 9)$

15. **ICE CREAM** How many different 1-scoop ice cream cones can you order if you have a choice of two cones types and 15 flavors of ice cream? **30 cones**

Determine whether each situation involves *permutations* or *combinations*. Then solve.

16. How many ways are there to place an algebra book, a geometry book, a chemistry book, an English book, and a health book on a shelf? **permutations; 120**

17. How many ways are there to select 3 of 15 flavors of juice at the grocery store? **combinations; 455**

A card is drawn at random from a standard deck. Determine whether the events are *mutually exclusive* or *not mutually exclusive*. Then find each probability.

18. P(2 or black card)   19. P(jack or 4)
   not mutually exclusive; $\frac{7}{13}$ mutually exclusive; $\frac{2}{13}$

Determine whether the events are *independent* or *dependent*. Then find the probability.

20. Two cards are selected at random one after the other from a standard deck without replacement. What is the probability that both cards are aces? **dependent;** $\frac{1}{221}$

21. A coin is tossed and a die is rolled. What is the probability that the coin lands on heads and the number rolled is greater than 4? **independent;** $\frac{1}{6}$

22. **SHOPPING** During one month, 18% of *INET Clothing* customers purchased item #345A. If 5% of customers order this item and item #345B, find the probability that someone who ordered item #345A also ordered item #345B. **about 28%**

23. Determine whether the triangles are *similar*, *congruent*, or *neither*. **similar**

24. **PHOTOGRAPHS** A photo that is 3 inches wide by 5 inches long is being enlarged so that it is 12 inches long. How wide will the enlarged photo be? **7.2 in.**

Find each missing measure. Round to the nearest tenth, if necessary.

25. 29 cm

   c cm
   20 cm
   21 cm

26. $a = 6$ yd,
   $b = 9$ yd,
   $c = ?$ **10.8 yd**

The lengths of three sides of a triangle are given. Determine whether each triangle is a right triangle.

27. 12 yd, 14 yd, 16 yd **no**   28. 15 km, 20 km, 25 km **yes**

Find the mean, median, mode, range, and standard deviation of each data set. Then identify any outliers.

29. scores for 16 students on a math test: 85, 100, 92, 36, 74, 88, 92, 86, 88, 82, 98, 70, 78, 92, 84, 100
   ≈ 84.1, 87, 92, 64, ≈ 15.4; 36

30. weight to the nearest pound of 17 cats: 10, 15, 9, 8, 11, 10, 9, 8, 13, 9, 12, 10, 13, 11, 9, 12, 10
   ≈ 10.5 lb, 10 lb, 9 lb and 10 lb, 7 lb, ≈ 1.9 lb; no outliers

Using the Pretest
The Chapter 0 Pretest assesses students' understanding of the concepts presented in Chapter 0. You may use the pretest to determine whether students need to complete each lesson in Chapter 0 before beginning the content in Chapter 1.

1 Focus

VerticalAlignment

Lesson 0-1 Identify the domain and range of functions.

After Lesson 0-1 Identify the domain and range of linear, quadratic, radical, exponential, logarithmic, and rational functions.

2 Teach

Example 1 shows how to identify the domain and range of a relation.
Example 2 shows how to identify the quadrant in which a point is located.
Example 3 shows how to use the domain and range to determine whether a relation is a function.

Additional Examples

1 State the domain and range of the relation.
{(−2, −1), (−1, 0), (1, −5), (2, 7)}
D = {−2, −1, 1, 2}
R = {−5, −1, 0, 7}

2 Name the quadrant in which the point $P(2, -6)$ is located.
Quadrant IV

WatchOut!

Student Misconceptions Remind students that a function can pair more than one element of the domain with a single element of the range. However, the reverse is not true.

LESSON 0-1 Representing Functions

::·**Objective**

● Identify the domain and range of functions.

 New Vocabulary
domain
range
quadrants
mapping
function

Recall that a *relation* is a set of ordered pairs. The **domain** of a relation is the set of all first coordinates (*x*-coordinates) from the ordered pairs, and the **range** is the set of all second coordinates (*y*-coordinates) from the ordered pairs.

Example 1 Domain and Range

State the domain and the range of the relation.
{(−3, 3), (0, −7), (1, −5), (2, 4)}

The domain is the set of *x*-coordinates.
D = {−3, 0, 1, 2}

The range is the set of *y*-coordinates.
R = {−7, −5, 3, 4}

A relation can be graphed on a coordinate plane. A coordinate plane is formed by the intersection of the horizontal axis, or *x*-axis, and the vertical axis, or *y*-axis. The axes meet at the origin (0, 0) and divide the plane into four **quadrants**. Any ordered pair in the coordinate plane can be written in the form (*x*, *y*).

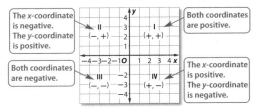

Example 2 Locate Coordinates

Name the quadrant in which $T(-8, 5)$ is located.

Point *T* has a negative *x*-coordinate and a positive *y*-coordinate. The point is located in Quadrant II.

A relation can also be represented by a table or a mapping. A **mapping** illustrates how each element of the domain is paired with an element in the range.

Ordered Pairs	Table	Graph	Mapping

Ordered Pairs
(1, 2)
(−2, 3)
(0, −3)

Table

x	y
1	2
−2	3
0	−3

A **function** is a relation in which each element of the domain is paired with *exactly one* element of the range.

 P4 | Lesson 0-1

Teach with Tech

Document Camera Select a student and give him or her a set of ordered pairs. Have the student work on the document camera to show and explain how to create a table, a graph, and a mapping from the ordered pairs.

Example 3 Identify Domain and Range

WatchOut!

Functions Remember that in a function, an element of the range can be paired with more than one element of the domain. But an element of the domain cannot be paired with more than one element of the range.

State the domain and range of each relation. Then determine whether each relation is a function.

a. {(10, 3), (6, −2), (7, 4), (−8, −9)}

D = {−8, 6, 7, 10}
R = {−9, −2, 3, 4}
For each element of the domain, there is only one corresponding element in the range. So, this relation is a function.

b.

x	y
1	3, 4
2	7
3	4

D = {1, 2, 3}
R = {3, 4, 7}

Because 1 is paired with 3 and 4, this is not a function.

c.

Domain   Range

D = {−3, 0, 1, 5}
R = {−4, −2, 0, 2}
Because 0 is paired with −2 and 0, this is not a function.

d.

D = {−1, 0, 1, 4}
R = {−3, −1, 0, 2}
This is a function.

Exercises

State the domain and range of each relation. Then determine whether each relation is a function. Write *yes* or *no*.

1. {(2, 7), (3, 10), (1, 6)}   D = {1, 2, 3}, R = {6, 7, 10}; yes

2. {(−6, 0), (5, 5), (9, −2), (−2, −9)}   **2.** D = {−6, −2, 5, 9}, R = {−9, −2, 0, 5}; yes

3.

x	y
1	5
2	7
1	9

D = {1, 2}, R = {5, 7, 9}; no

4.

x	y
−12	0
−10	1
−8	2
−6	4

D = {−12, −10, −8, −6}, R = {0, 1, 2, 4}; yes

5.

Domain   Range

D = {−2, −1, 0, 3}, R = {−3, −2, 2}; yes

6.

Domain   Range

D = {−8, −7, −4}, R = {2, 3, 5, 6}; no

7.

D = {−1, 0, 1, 2, 3}, R = {−3, −2, −1, 2, 3, 4}; no

8.

D = {−3, −2, 0, 1, 2, 3, 4}, R = {−1, 0, 1, 2}; yes

Name the quadrant in which each point is located.

9. (5, 3) I   **10.** (8, −6) IV   **11.** (2, 0) none   **12.** (−7, −1) III

Additional Example

3 State the domain and range of each relation. Then determine whether each relation is a function.

a. {(3, 2), (4, 1), (6, 2), (8, 0)}
D = {3, 4, 6, 8}
R = {0, 1, 2}, yes

b.

x	y
0	−2
3	2
6	2

D = {0, 3, 6}
R = {−2, 2}, yes

c.

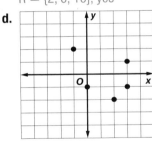

Domain   Range
x   y

D = {−4, −1, 5}
R = {2, 6, 10}, yes

d.

D = {−1, 0, 2, 3}
R = {−2, −1, 1, 2}, no

3 Assess

Formative Assessment
Use Exercises 1–12 to assess whether students can identify the domain and range of a relation and determine whether the relation is a function.

Ticket Out the Door Ask students to list four ordered pairs that could be elements of a single function.

1 Focus

VerticalAlignment

▼

Lesson 0-2 Use the FOIL method to multiply binomials.

▼

After Lesson 0-2 Factor quadratic trinomials.

2 Teach

Example 1 shows how to use the FOIL method to find the product of two binomials.

Additional Example

1 Find each product.

a. $(d - 1)(d + 6)$ $d^2 + 5d - 6$

b. $(2p - 3)(6p - 1)$ $12p^2 - 20p + 3$

WatchOut!

Preventing Errors It may help some students to write a binomial such as $(x - 5)$ in the form $(x + (-5))$ to emphasize that the sign is included as part of the constant term.

3 Assess

Formative Assessment

Use Exercises 1–18 to assess whether students understand how to use FOIL to multiply binomials.

Crystal Ball Ask students how they think today's lesson on multiplying binomials will help with tomorrow's lesson on factoring polynomials.

LESSON 0-2 FOIL

··Objective

● Use the FOIL method to multiply binomials.

The product of two binomials is the sum of the products of the *first* terms, the *outer* terms, the *inner* terms, and the *last* terms.

F   O   I   L

Example 1 Use the FOIL Method

Find each product.

a. $(x + 3)(x - 5)$

$$\begin{array}{cc} \text{F} \quad \text{L} & \text{First} \;\; \text{Outer} \quad\; \text{Inner} \;\; \text{Last} \end{array}$$
$$(x + 3)(x - 5) = x \cdot x + x \cdot (-5) + 3 \cdot x + 3 \cdot (-5)$$
$$\text{I}$$
$$\text{O}$$
$$= x^2 - 5x + 3x - 15$$
$$= x^2 - 2x - 15$$

b. $(3y + 2)(5y + 4)$

$$(3y + 2)(5y + 4) = 3y \cdot 5y + 3y \cdot 4 + 2 \cdot 5y + 2 \cdot 4$$
$$= 15y^2 + 12y + 10y + 8$$
$$= 15y^2 + 22y + 8$$

Exercises

Find each product.

1. $(a + 2)(a + 4)$ $a^2 + 6a + 8$

2. $(v - 7)(v - 1)$ $v^2 - 8v + 7$

3. $(h + 4)(h - 4)$ $h^2 - 16$

4. $(d - 1)(d + 1)$ $d^2 - 1$

5. $(b + 4)(b - 3)$ $b^2 + b - 12$

6. $(t - 9)(t + 11)$ $t^2 + 2t - 99$

7. $(r + 3)(r - 8)$ $r^2 - 5r - 24$

8. $(k - 2)(k + 5)$ $k^2 + 3k - 10$

9. $(p + 8)(p + 8)$ $p^2 + 16p + 64$

10. $(x - 15)(x - 15)$ $x^2 - 30x + 225$

11. $(2c + 1)(c - 5)$ $2c^2 - 9c - 5$

12. $(7n - 2)(n + 3)$ $7n^2 + 19n - 6$

13. $(3m + 4)(2m - 5)$ $6m^2 - 7m - 20$

14. $(5g + 1)(6g + 9)$ $30g^2 + 51g + 9$

15. $(2q - 17)(q + 2)$ $2q^2 - 13q - 34$

16. $(4t - 7)(3t - 12)$ $12t^2 - 69t + 84$

17. NUMBERS I am thinking of two integers. One is 7 less than a number, and the other is 2 greater than the same number.

a. Write expressions for the two numbers. $n - 7, \; n + 2$

b. Write a polynomial expression for the product of the numbers. $n^2 - 5n - 14$

18. OFFICE SPACE Monica's current office is square. Her office in the company's new building will be 3 feet wider and 5 feet longer.

a. Write expressions for the dimensions of Monica's new office. $x + 3, \; x + 5$

b. Write a polynomial expression for the area of Monica's new office. $x^2 + 8x + 15$

c. Suppose Monica's current office is 7 feet by 7 feet. How much larger will her new office be? 71 ft^2

 P6 | Lesson 0-2

Teach with Tech

Interactive Whiteboard Work through an example of multiplying two binomials. Write an example on the board and use a different color for each of the four terms. Create a table with four columns for the four parts of FOIL. In each column, write the expression for the two terms to be multiplied using the same colors as before.

LESSON 0-3 Factoring Polynomials

Objective

- Use various techniques to factor polynomials.

Some polynomials can be factored using the Distributive Property.

Example 1 Use the Distributive Property

Factor $4a^2 + 8a$.

Find the GCF of $4a^2$ and $8a$.

$4a^2 = 2 \cdot 2 \cdot a \cdot a$ $\quad 8a = 2 \cdot 2 \cdot 2 \cdot a$ $\quad \rightarrow \quad$ GCF: $2 \cdot 2 \cdot a$ or $4a$

$4a^2 + 8a = 4a(a) + 4a(2)$ Rewrite each term using the GCF.

$\qquad\quad = 4a(a + 2)$ Distributive Property

Thus, the completely factored form of $4a^2 + 8a$ is $4a(a + 2)$.

To factor trinomials of the form $x^2 + bx + c$, find two integers m and p with a product equal to c and a sum equal to b. Then write $x^2 + bx + c$ using the pattern $(x + m)(x + p)$.

Example 2 Use Factors and Sums

Factor each polynomial.

a. $x^2 + 5x + 6$ [Both b and c are positive.]

b is 5 and c is 6. Find two numbers with a product of 6 and a sum of 5.

Factors of 6	Sum of factors	
1, 6	7	
2, 3	5	The correct factors are 2 and 3.

$x^2 + 5x + 6 = (x + m)(x + p)$ Write the pattern.

$\qquad\qquad\quad = (x + 2)(x + 3)$ $m = 2$ and $p = 3$

b. $x^2 - 8x + 12$ [b is negative, and c is positive.]

$b = -8$ and $c = 12$. This means that $m + p$ is negative and mp is positive. So m and p must both be negative.

Factors of 12	Sum of factors	
-1, -12	-13	
-2, -6	-8	The correct factors are -2 and -6.

$x^2 - 8x + 12 = (x + m)(x + p)$ Write the pattern.

$\qquad\qquad\quad = [x + (-2)][x + (-6)]$ $m = -2$ and $p = -6$

$\qquad\qquad\quad = (x - 2)(x - 6)$ Simplify.

c. $x^2 + 14x - 15$ [b is positive, and c is negative.]

$b = 14$ and $c = -15$. This means that $m + p$ is positive and mp is negative. So either m or p must be negative, but not both.

Factors of -15	Sum of factors	
1, -15	-14	
-1, 15	14	The correct factors are -1 and 15.

$x^2 + 14x - 15 = (x + m)(x + p)$ Write the pattern.

$\qquad\qquad\quad = [x + (-1)](x + 15)$ $m = -1$ and $p = 15$

$\qquad\qquad\quad = (x - 1)(x + 15)$ Simplify.

---

1 Focus

VerticalAlignment

Lesson 0-3 Use various techniques to factor polynomials.

After Lesson 0-3 Simplify rational expressions. Solve polynomial equations.

2 Teach

Example 1 shows how to use the Distributive Property to factor a polynomial. **Example 2** shows how to factor quadratic trinomials of the form $x^2 + bx + c$. **Example 3** shows how to factor quadratic trinomials of the form $ax^2 + bx + c$. **Example 4** shows how to factor special products such as perfect square trinomials and the difference of squares.

Additional Examples

1 Factor $9cd^2 - 6cd^3$.
$3cd^2(3 - 2d)$

2 Factor each polynomial.
   a. $q^2 + 7q + 10$ $(q + 5)(q + 2)$
   b. $x^2 - 7x + 12$ $(x - 3)(x - 4)$
   c. $x^2 + 4x - 21$ $(x + 7)(x - 3)$

Tips for New Teachers

Sense-Making Remind students that in Algebra 1 they learned that a monomial is a number, a variable, or the product of a number and one or more variables; a binomial is the sum of two monomials; and a trinomial is the sum of three monomials. Ask students to give an example of each.

3 Factor $20y^2 - 11y - 3$.
$(4y - 3)(5y + 1)$

4 Factor each polynomial.
   a. $9h^2 - 42h + 49$ $(3h - 7)^2$
   b. $9d^2 - 64$ $(3d + 8)(3d - 8)$

WatchOut!

Preventing Errors Remind students that it is easy to confuse signs when factoring trinomials. To help prevent these mistakes, suggest that students use FOIL to check the factors.

Teach with Tech

RSS Create a class Web page about how to factor different types of polynomials. After each class period, update the page with information from the most recent lesson, such as notes from class, video clips, and additional resources. Have students subscribe to an RSS feed so they can easily receive these updates.

3 Assess

Formative Assessment
Use Exercises 1–24 to assess whether students understand how to factor polynomials.

Yesterday's News Ask students to explain how yesterday's lesson on multiplying binomials helped them with today's lesson on factoring polynomials.

To factor quadratic trinomials of the form $ax^2 + bx + c$, find two integers m and p with a product equal to ac and a sum equal to b. Write $ax^2 + bx + c$ using the pattern $ax^2 + mx + px + c$. Then factor by grouping.

Example 3 Use Factors and Sums

Factor $6x^2 + 7x - 3$.

$a = 6$, $b = 7$, and $c = -3$. This means that $m + p$ is positive and mp is negative. So either m or p must be negative, but not both.

Factors of -18	Sum of factors
1, -18	-17
-1, 18	17
2, -9	-7
-2, 9	7

The correct factors are -2 and 9.

$$
\begin{aligned}
6x^2 + 7x - 3 &= 6x^2 + mx + px - 3 && \text{Write the pattern.} \\
&= 6x^2 + (-2)x + 9x - 3 && m = -2 \text{ and } p = 9 \\
&= (6x^2 - 2x) + (9x - 3) && \text{Group terms with common factors.} \\
&= 2x(3x - 1) + 3(3x - 1) && \text{Factor the GCF from each group.} \\
&= (2x + 3)(3x - 1) && \text{Distributive Property}
\end{aligned}
$$

StudyTip

Checking Solutions You can check to see that you have factored correctly by multiplying the factors and comparing the product to the original polynomial.

Here are some special products.

Perfect Square Trinomials		**Difference of Squares**
$(a + b)^2 = (a + b)(a + b)$	$(a - b)^2 = (a - b)(a - b)$	$a^2 - b^2 = (a + b)(a - b)$
$\quad\quad = a^2 + 2ab + b^2$	$\quad\quad = a^2 - 2ab + b^2$	

Example 4 Use Special Products

Factor each polynomial.

a. $4x^2 + 20x + 25$ ← The first and last terms are perfect squares. The middle term is equal to $2(2x)(5)$. This is a perfect square trinomial of the form $(a + b)^2$.

$$
\begin{aligned}
4x^2 + 20x + 25 &= (2x)^2 + 2(2x)(5) + 5^2 && \text{Write as } a^2 + 2ab + b^2. \\
&= (2x + 5)^2 && \text{Factor using the pattern.}
\end{aligned}
$$

b. $x^2 - 4$ ← This is a difference of squares.

$$
\begin{aligned}
x^2 - 4 &= x^2 - (2)^2 && \text{Write in the form } a^2 - b^2. \\
&= (x + 2)(x - 2) && \text{Factor the difference of squares.}
\end{aligned}
$$

5. $(y + 3)(y + 9)$
7. $(3y + 1)(y + 4)$
8. $(7x + 2)(x + 7)$
9. $(3x + 4)(x + 8)$
10. $(x - 3)(x - 2)$
11. $(y - 4)(y - 1)$
12. $(3x - 5)(2x - 1)$
13. $2(3a - b)(a - 8b)$
14. $(11x - 1)(x - 7)$
15. $(2x - 3y)(9x - 2y)$
18. $(2a + 3b)^2$
19. $(x + 12)(x - 12)$
20. $(2c + 3)(2c - 3)$

Exercises

Factor each polynomial.

1. $12x^2 + 4x$ $4x(3x + 1)$
2. $6x^2y + 2x$ $2x(3xy + 1)$
3. $8ab^2 - 12ab$ $4ab(2b - 3)$
4. $x^2 + 5x + 4$ $(x + 1)(x + 4)$
5. $y^2 + 12y + 27$
6. $x^2 + 6x + 8$ $(x + 2)(x + 4)$
7. $3y^2 + 13y + 4$
8. $7x^2 + 51x + 14$
9. $3x^2 + 28x + 32$
10. $x^2 - 5x + 6$
11. $y^2 - 5y + 4$
12. $6x^2 - 13x + 5$
13. $6a^2 - 50ab + 16b^2$
14. $11x^2 - 78x + 7$
15. $18x^2 - 31xy + 6y^2$
16. $x^2 + 4xy + 4y^2$ $(x + 2y)^2$
17. $9x^2 - 24x + 16$ $(3x - 4)^2$
18. $4a^2 + 12ab + 9b^2$
19. $x^2 - 144$
20. $4c^2 - 9$
21. $16y^2 - 1$ $(4y + 1)(4y - 1)$
22. $25x^2 - 4y^2$ $(5x + 2y)(5x - 2y)$
23. $36y^2 - 16$ $4(3y + 2)(3y - 2)$
24. $9a^2 - 49b^2$ $(3a + 7b)(3a - 7b)$

LESSON 0-4 Counting Techniques

Objective

- Find the total number of outcomes using a variety of methods.

 NewVocabulary
outcome
probability experiment
sample space
tree diagram
permutation
factorial
combination

Recall that an **outcome** is the result of a single trial of a process involving chance, called a **probability experiment**. The set of all possible outcomes of such an experiment is called a **sample space**. A **tree diagram**, like the one shown, can be used to systematically list the outcomes in a sample space. Notice that there are 8 total outcomes in this sample space.

When you only need to know the total number of outcomes, you can use the **Fundamental Counting Principle**.

Blood Type	Rh Factor	Sample Space
A	+	A+
	−	A−
B	+	B+
	−	B−
AB	+	AB+
	−	AB−
O	+	O+
	−	O−

KeyConcept Fundamental Counting Principle

Words	The number of possible outcomes in a sample space can be found by multiplying the number of possible outcomes for each event.
Symbols	If event M can occur in m ways and is followed by event N that can occur in n ways, then the event M followed by N can occur in $m \cdot n$ ways.

This rule can be extended to three or more events.

If there are 4 possible blood types and 2 possible Rh-factors, there are $4 \cdot 2$ or 8 possible ways to label blood donations.

Example 1 Fundamental Counting Principle

BICYCLES A bicycle manufacturer makes five- and ten-speed bikes in seven different colors and four different frame sizes. How many different bicycles does the manufacturer make?

There are 2 gear choices, 7 color choices, and 4 frame choices. Apply the Fundamental Counting Principle.

gear		color		frame		number of bicycles
2	·	7	·	4	=	56

56 different bicycles can be made.

Counting problems can also involve determining the number of different arrangements of objects. An arrangement of a group of distinct objects in a certain order is called a **permutation**. For example, there are 6 permutations of the letters A, B, and C.

ABC   ACB   BAC   BCA   CAB   CBA

To determine this number mathematically, apply the Fundamental Counting Principle. There are 3 choices for the first letter. Once this letter is chosen, 2 choices remain for the second letter. After the first two letters are chosen, only one choice remains.

first letter		second letter		third letter		permutations
3	·	2	·	1	=	6

The product $3 \cdot 2 \cdot 1$ can also be written as 3!, which is read *3 factorial*.

 connectED.mcgraw-hill.com   **P9**

Focus on Mathematical Content

Counting Outcomes Tree diagrams and the Fundamental Counting Principle are two methods used for counting outcomes. Tree diagrams are helpful when you need to know each individual outcome and the number of choices is small. The Fundamental Counting Principle is useful when the number of choices is large, a list of specific outcomes is not needed, or the number of possible outcomes is needed quickly.

1 Focus

VerticalAlignment

Lesson 0-4 Find the total number of outcomes using a variety of methods.

After Lesson 0-4 Compute theoretical and experimental probabilities and compute probabilities of compound events.

2 Teach

Example 1 shows how to use the Fundamental Counting Principle to determine the number of possible outcomes for an event. **Example 2** shows how to use the Fundamental Counting Principle to find the number of permutations. **Example 3** shows how to use the Permutation Formula to find the number of permutations. **Example 4** shows how to use the Combination Formula to find the number of combinations. **Example 5** shows how to determine whether a counting situation involves a permutation or a combination.

Additional Example

1 **SHOES** A shoe manufacturer makes brown and black shoes in nine different sizes. How many different shoes does the manufacturer make? 18

2 **BASEBALL** There are 9 players in a baseball lineup. No two players can share the same spot in the lineup. How many lineups are possible? 362,880

3 **PHOTOGRAPHY** A photographer has matted and framed 15 photographs and needs to select 10 for a gallery show. How many ways can the photographer arrange the photographs for the show? 10,897,286,400

WatchOut!

Student Misconceptions Inform students that words like *arrangement*, *sequence*, and *order*, can be clues that a situation involves a permutation.

Tips for New Teachers

Pacing Permutations and combinations are frequently confused. Consider discussing multiple examples of each. For each example, discuss how to find the number of combinations or permutations. Then change the problem slightly to solve for the other. For instance, in Example 2 if we just wanted to know the top 8 finalists who were moving to the next round, it would be a combination, not a permutation.

 KeyConcept Factorial

Words	*n* **factorial** is the product of all counting numbers beginning with *n* and counting backward to 1. Zero factorial is defined to equal 1.
Symbols	$n! = n \cdot (n-1) \cdot (n-2) \cdot (n-3) \cdot \cdots \cdot 3 \cdot 2 \cdot 1$ and $0! = 1$

The number of permutations of *n* distinct objects is *n*!.

Example 2 Permutations of *n* Objects

BAND There are 8 finalists in a band competition. In how many different ways can the bands be ranked if they cannot receive the same ranking?

$8! = 8 \cdot 7 \cdot 6 \cdot 5 \cdot 4 \cdot 3 \cdot 2 \cdot 1$ or 40,320   Permutations of *n* distinct objects

The bands can be ranked in 40,320 different ways.

Suppose in Example 2 you wanted to know how many different ways the first-, second-, and third-place rankings in the competition could be awarded. This also involves permutations. It is the number of permutations of 8 finalists taken 3 at a time.

first place		second place		third place		permutations
8	·	7	·	6	=	336

The number of permutations can also be found using factorials by finding the number of permutations of all 8 bands and dividing out the number of arrangements of bands that finished with rankings other than first, second, or third.

$$\frac{8!}{(8-3)!} = \frac{8!}{5!} = \frac{8 \cdot 7 \cdot 6 \cdot 5 \cdot 4 \cdot 3 \cdot 2 \cdot 1}{5 \cdot 4 \cdot 3 \cdot 2 \cdot 1} = 8 \cdot 7 \cdot 6$$

This result is generalized below.

ReadingMath

Permutations The number of permutations of *n* objects taken *r* at a time can also be denoted as $P(n, r)$.

KeyConcept **Permutations of *n* Objects Taken *r* at a Time**

The number of permutations of *r* objects taken from a group of *n* distinct objects is given by

$$_nP_r = \frac{n!}{(n-r)!}.$$

Example 3 Permutation of *n* Objects Taken *r* at a Time

TUTORING How many different ways can two students be assigned to five tutors if only one student is assigned to each tutor?

Since the order in which the students are arranged determines the tutor to which a student is assigned, this situation involves permutations.

$_5P_2 = \dfrac{5!}{(5-2)!}$ Permutations of 5 things taken 2 at a time

$= \dfrac{5!}{3!}$ Subtract.

$= \dfrac{5 \cdot 4 \cdot 3 \cdot 2 \cdot 1}{3 \cdot 2 \cdot 1}$ or 20   Divide out common factors and simplify.

The students can be assigned in 20 different ways.

StudyTip

With Repetition The number of permutations of *n* objects in which one object is repeated *p* times and another object is repeated *q* times is $\frac{n!}{p!q!}$. This rule can be extended to permutations involving three or more repeated objects.

 P10 | Lesson 0-4 | Counting Techniques

Teach with Tech

Portable Media Player Have students look at their media players to find the number of songs they have. Ask them to determine how many 5-song playlists they could make if order *does* matter. Have students discuss their results with the class.

A selection of distinct objects in which the order of the objects selected is *not* important is called a **combination**. Suppose you select 3 out of 4 books, A, B, C, and D, for a book report. Selecting books A, B, and C is not different from selecting books C, B, A, or any other arrangements of these three books.

To find the number of combinations of *n* objects taken *r* at a time, divide the number of permutations by the number of ways to arrange each selection of *r* objects.

KeyConcept Combinations of *n* Objects Taken *r* at a Time

The number of combinations of *r* objects taken from a group of *n* distinct objects is given by

$$_nC_r = \frac{_nP_r}{r!} = \frac{n!}{(n-r)!\,r!}.$$

Example 4 Combinations of *n* Objects Taken *r* at a Time

CARDS How many ways are there to choose 5 cards from a standard deck of 52 playing cards?

Since the order in which the cards are chosen is not important, this situation involves combinations.

$$_{52}C_5 = \frac{52!}{(52-5)!\,5!}$$ Combinations of 52 things taken 5 at a time

$$= \frac{52!}{47!\,5!}$$ Subtract.

$$= 2{,}598{,}960$$ Use a calculator.

There are 2,598,960 ways to choose 5 cards from a standard deck of playing cards.

> **StudyTip**
> Evaluating Permutations and Combinations You can also use the nPr and nCr features on your graphing calculator to evaluate permutations and combinations, respectively.

A critical first step in solving a counting problem is deciding whether order is important. If the order of objects selected is important, then the problem involves permutations. If objects are selected without regard to order, the problem involves combinations.

Example 5 Permutations or Combinations?

Twenty-five students write their names on slips of paper. Then three different names are chosen at random to receive prizes. Determine whether each situation involves *permutations* or *combinations*.

a. choosing 3 people to each receive a "no homework" coupon

Since the prizes are identical, the order in which the students' names are drawn *is not* important. This situation involves combinations.

b. choosing 3 people to each receive one of the following prizes: 1st prize, a new graphing calculator; 2nd prize, a "no homework" coupon; 3rd prize, a new pencil

Because the prizes are different, the order in which the students' names are drawn *is* important. This situation involves permutations.

  P11

Additional Examples

4 **DINING** A restaurant offers a total of 8 side dishes. How many different ways can a customer choose 3 side dishes? 56

5 Three hundred people buy tickets for a raffle. Then five different raffle tickets are drawn at random to claim prizes. Determine whether each situation involves *permutations* or *combinations*.

a. choosing 5 people to each receive one of the following prizes: 1st ticket drawn, a keychain; 2nd ticket drawn, a mug; 3rd ticket drawn, a T-shirt; 4th ticket drawn, a television; and 5th ticket drawn, a new car. permutation

b. choosing 5 people to each receive a T-shirt. combination

Teach with Tech
Interactive Whiteboard When computing permutations and combinations, use the board to show the cancellation of the factorials.

Formative Assessment

Use Exercises 1–19 to assess whether students can find the total number of possible outcomes for an event using a variety of methods.

Crystal Ball Have students explain how today's lesson on permutations and combinations might help them with computing probabilities of mutually exclusive and not mutually exclusive events in the next lesson.

Additional Answer

19b. 308,915,776; 712,882,560; The password with one digit is more secure, because the chance of someone guessing this password at random is $\dfrac{1}{712,882,560}$, which is less than the chance of someone guessing a 6-character password that contains only letters, $\dfrac{1}{308,915,776}$.

Exercises

Use the Fundamental Counting Principle to determine the number of outcomes.

1. FOOD How many different combinations of sandwich, side, and beverage are possible? **60**

Sandwiches	Sides	Beverages
hot dog	chips	bottled water
hamburger	apple	soda
veggie burger	pasta salad	juice
bratwurst		milk
grilled chicken		

2. QUIZZES Each question on a five-question multiple-choice quiz has answer choices labeled A, B, C, and D. How many different ways can a student answer the five questions? **1024**

3. DANCES Dane is renting a tuxedo for prom. Once he has chosen his jacket, he must choose from three types of pants and six colors of vests. How many different ways can he select his attire for prom? **18**

4. MANUFACTURING A baseball glove manufacturer makes a glove with the different options shown in the table. How many different gloves are possible? **48**

Option	Number of Choices
sizes	4
types by position	3
materials	2
levels of quality	2

Evaluate each permutation or combination.

5. $_6P_3$ **120**   **6.** $_7P_5$ **2520**   **7.** $_4C_2$ **6**

8. $_{12}C_7$ **792**   **9.** $_6C_1$ **6**   **10.** $_9P_5$ **15,120**

Determine whether each situation involves *permutations* or *combinations*. Then solve the problem.

11. SCHOOL Charlita wants to take 6 different classes next year. Assuming that each class is offered each period, how many different schedules could she have? **permutations, 720**

12. BALLOONS How many 4-colored groups can be selected from 13 different colored balloons? **combinations, 715**

13. CONTEST How many ways are there to choose the winner and first, second, and third runners-up in a contest with 10 finalists? **permutations, 5040**

14. BANDS A band is choosing 3 new backup singers from a group of 18 who try out. How many ways can they choose the new singers? **combinations, 816**

15. PIZZA How many different two-topping pizzas can be made if there are 6 options for toppings? **combinations, 15**

16. SOFTBALL How many ways can the manager of a softball team choose players for the top 4 spots in the lineup if she has 7 possible players in mind? **permutations, 840**

17. NEWSPAPERS A newspaper has 9 reporters available to cover 4 different stories. How many ways can the reporters be assigned? **permutations, 3024**

18. READING Jack has a reading list of 12 books. How many ways can he select 9 books from the list to check out of the library? **combinations, 220**

19. CHALLENGE Abby is registering at a Web site and must select a six-character password. The password can contain either letters or digits.

   a. How many passwords are possible if characters can be repeated? if no characters can be repeated? **2,176,782,336; 1,402,410,240**

   b. How many passwords are possible if all characters are letters that can be repeated? if the password must contain exactly one digit? Which type of password is more secure? Explain. **See margin.**

 P12 | Lesson 0-4 | Counting Techniques

Focus on Mathematical Content

Permutations and Combinations Ask students which would have more possible outcomes, the number of combinations of *n* items taken *r* at a time or the number of permutations of *n* items taken *r* at a time. Explain your reasoning.

LESSON 0-5 — Adding Probabilities

Objectives

- Compute theoretical and experimental probabilities.
- Compute probabilities of compound events.

NewVocabulary
probability
probability model
uniform or simple
  probability model
theoretical probability
experimental probability
simple event
compound event
mutually exclusive events
odds

Probability is a measure of the chance that a given event E will occur. The probability $P(E)$ that an event will occur is always given as a ratio between 0 and 1, inclusive.

E will not occur.	E is equally likely to occur or not occur.	E is certain to occur.
$P(E) = 0$	$P(E) = \frac{1}{2}$	$P(E) = 1$

A **probability model** is a mathematical model used to represent the outcomes of an experiment. A **uniform** or **simple probability model** is used to describe an experiment for which all outcomes are equally likely or have the same probability of occurring.

If we assume that all outcomes of an experiment are equally likely, then we can calculate the **theoretical probability** that an event will occur using the sample space of possible outcomes. We can also calculate the empirical or **experimental probability** using outcomes obtained by actually performing trials of the experiment.

KeyConcept Theoretical and Experimental Probability

If each outcome is assumed to be equally likely, the theoretical probability P of an event E is given by

$$P(E) = \frac{\text{number of favorable outcomes}}{\text{number of possible outcomes}}.$$

Given the frequency of outcomes from a certain number of trials of an experiment, the experimental probability P of an event E is given by

$$P(E) = \frac{\text{number of favorable trials}}{\text{number of trials}}.$$

Example 1 Theoretical and Experimental Probability

The graph shows the results of several trials of an experiment in which a single die is rolled.

a. What is the experimental probability of rolling a 6?

From the graph we can see that there were 10 trials that resulted in 6. The total number of trials was $9 + 6 + 12 + 6 + 7 + 10$ or 50.

$$P(6) = \frac{\text{number of favorable trials}}{\text{number of trials}}$$
$$= \frac{10}{50} \text{ or } \frac{1}{5}, \text{ which is } 20\%$$

b. What is the theoretical probability of rolling a 6?

List the sample space. The possible outcomes for this experiment are 1, 2, 3, 4, 5, or 6. Of these 6 possible outcomes, only one is a 6.

$$P(6) = \frac{\text{number of favorable outcomes}}{\text{number of possible outcomes}} = \frac{1}{6} \text{ or about } 17\%$$

 connectED.mcgraw-hill.com **P13**

Focus on Mathematical Content

Probability Theoretical probability is what *should* happen, and experimental probability is what *does* happen. Experimental events are actually performed and are therefore subject to outside influences such as human error or environmental conditions. The more an experiment is performed, the closer the experimental probability will get to the theoretical probability.

---

1 Focus

VerticalAlignment

Lesson 0-5 Compute theoretical and experimental probabilities and compute probabilities of compound events.

After Lesson 0-5 Compute probabilities of independent and dependent events.

2 Teach

Example 1 shows how to find the theoretical and experimental probability of a simple event. **Example 2** shows how to find the probability of a compound event.

Additional Example

1 The graph shows the results of several trials of an experiment in which a spinner with eight equal sections of different colors is spun.

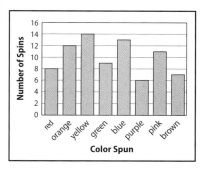

a. What is the experimental probability of spinning yellow? $\frac{14}{80}$ or 17.5%

b. What is the theoretical probability of spinning yellow? $\frac{1}{8}$ or 12.5%

Additional Example

2 Determine whether the events are *mutually exclusive* or *not mutually exclusive.* Then find the probability.

a. Nancy has a box of dog toys that contains 8 squeaky toys, 5 plush toys, and 2 bones. If she selects a toy at random from the box to give to her dog, what is the probability that it is a plush toy or a bone? mutually exclusive; $\frac{7}{15}$

b. Suppose that of 900 pints of ice cream available for sale, 200 are sugarless, 500 are vanilla, and 50 are sugarless vanilla. What is the probability that a pint selected at random is either sugarless or vanilla? not mutually exclusive; $\frac{13}{18}$

Tips for New Teachers

Kinesthetic Learners When students are having difficulty understanding what a question is asking, encourage them to act out the event. In Example 2, have students use real, or homemade, sports cards to represent the situation described in the problem.

An event that has a single outcome, such as rolling a 6 on a die, is called a **simple event**. Many problems involve finding the probability of a composite or **compound event**, which consists of two or more simple events. To find the probability of a compound event, you must consider whether the events can occur at the same time.

Events that cannot occur at the same time are said to be **mutually exclusive**. Mutually exclusive events have no outcomes in common. For example, because it is not possible to draw a card from a standard deck that is both a king and a queen, these two events are mutually exclusive. It is possible to draw a card that is both a king and a spade, so these events are *not* mutually exclusive.

StudyTip

Mutually and Not Mutually Exclusive Events that are mutually exclusive are sometimes said to be *disjoint*. Events that are not mutually exclusive are sometimes said to be *inclusive* events.

Mutually Exclusive

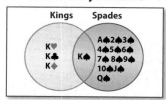

Not Mutually Exclusive

StudyTip

Why Subtract? Notice that when two events are not mutually exclusive, you add the probabilities and then *subtract* the probabilities of the outcomes that are common to both. If these common probabilities are not subtracted, you have actually counted them twice.

KeyConcept Addition Rules for Probability

If two events A and B are **mutually exclusive**, the probability that A or B will occur is	If two events A and B are **not mutually exclusive**, the probability that A or B will occur is
$P(A \text{ or } B) = P(A) + P(B).$	$P(A \text{ or } B) = P(A) + P(B) - P(A \text{ and } B).$

Example 2 Add Probabilities

Determine whether the events are *mutually exclusive* or *not mutually exclusive.* Then find the probability.

a. Keisha has a stack of 8 baseball cards, 5 basketball cards, and 6 hockey cards. If she selects a card at random from the stack, what is the probability that it is a baseball or a hockey card?

These events are mutually exclusive, because the card cannot be both a baseball card *and* a hockey card. There is a total of $8 + 5 + 6$ or 19 cards.

$P(\text{baseball or hockey}) = P(\text{baseball}) + P(\text{hockey})$ Mutually exclusive events

$\qquad\qquad\qquad\qquad = \frac{8}{19} + \frac{6}{19} \text{ or } \frac{14}{19}$ Substitute and add.

b. Suppose that of 1400 students, 550 take Spanish, 700 take biology, and 400 take both Spanish and biology. What is the probability that a student selected at random takes Spanish or biology?

Because some students take both Spanish S and biology B, the events are not mutually exclusive.

$P(S) = \frac{550}{1400} \text{ or } \frac{11}{28}$ $P(B) = \frac{700}{1400} \text{ or } \frac{14}{28}$ $P(S \text{ and } B) = \frac{400}{1400} \text{ or } \frac{8}{28}$

$P(S \text{ or } B) = P(S) + P(B) - P(S \text{ and } B)$ Not mutually exclusive events

$\qquad\qquad\quad = \frac{11}{28} + \frac{14}{28} - \frac{8}{28} \text{ or } \frac{17}{28}$ Substitute and simplify.

WatchOut!

Reading Math The expression *A or B* does allow for the possibility of both *A* and *B* occurring. This may confuse some students because in nonmathematical language this is not the case.

1. **CARNIVAL GAMES** A spinner has sections of equal size. The table shows the results of several spins. **See Chapter 0 Answer Appendix.**

   a. Copy the table and add a column to show the experimental probability of the spinner landing on each of the colors with the next spin.

   b. Create a bar graph that shows these experimental probabilities.

   c. Add a column to your table that shows the theoretical probability of the spinner landing on each of the colors with the next spin.

   d. Create a bar graph that shows these theoretical probabilities.

   e. Interpret and compare the graphs you created in parts **b** and **d**.

Color	Frequency
red	6
blue	7
yellow	9
orange	12
purple	5
green	11

Determine whether the events are *mutually exclusive* or *not mutually exclusive*. Then find the probability. 2–5. See margin.

2. Two dice are rolled.

   a. P(sum of 10 or doubles) b. P(sum of 6 or 7) c. P(sum < 3 or sum > 10)

3. A card is drawn at random from a standard deck of cards.

   a. P(club or diamond) b. P(ace or spade) c. P(jack or red card)

4. In a French class, there are 10 freshmen, 8 sophomores, and 2 juniors. Of these students, 9 freshmen, 2 sophomores, and 1 junior are female. A student is selected at random.

   a. P(freshman or female) b. P(sophomore or male) c. P(freshman or sophomore)

5. There are 40 vehicles on a rental car lot. All are either sedans or SUVs. There are 18 red vehicles, and 3 of them are sedans. There are 15 blue vehicles, and 9 of them are SUVs. Of the remaining vehicles, all are black and 2 are SUVs. A vehicle is selected at random.

   a. P(blue or black) b. P(red or SUV) c. P(black or sedan)

6. **DRIVING** A survey of Longview High School students found that the probability of a student driving while texting was 0.16, the probability of a student getting into an accident while driving was 0.07, and the probability of a student getting into an accident while driving and texting was 0.05. What is the probability of a student driving while texting or getting into an accident while driving? **0.18 or 18%**

7. **REASONING** Explain why the rule $P(A \text{ or } B) = P(A) + P(B) - P(A \text{ and } B)$ can be used for both mutually exclusive and not mutually exclusive events. **When events are mutually exclusive, $P(A$ and $B)$ will always equal 0, so the probability will simplify to $P(A) + P(B)$.**

   ODDS Another measure of the chance that an event will occur is called **odds**. The odds of an event occurring is a ratio that compares the number of ways an event can occur s (successes) to the number of ways it cannot occur f (failure), or s to f. The sum of the number of success and failures equals the number of possible outcomes. 8–10. See margin.

8. A card is drawn from a standard deck of 52 cards. Find the odds in favor of drawing a heart. Then find the odds against drawing an ace.

9. Two fair coins are tossed. Find the odds in favor of both landing on heads. Then find the odds in favor of exactly one landing on tails.

10. The results of rolling a die 120 times are shown.

Roll	·	··	···	····	·····	······
Frequency	16	24	17	25	30	8

Find the experimental odds against rolling a 1 or a 6. Then find the experimental odds in favor of rolling a number less than 3.

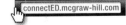

3 Assess

Formative Assessment
Use Exercises 1–10 to assess whether students can compute theoretical probabilities, experimental probabilities, and probabilities of compound events.

Ticket Out the Door Write examples of events involving probability on index cards. Give each student one index card. Before students can leave the room, have them state whether their event is simple, compound, mutually exclusive, or inclusive, and have them give the probability of the event.

Additional Answers

2a. not mutually exclusive, $\frac{2}{9}$

2b. mutually exclusive, $\frac{11}{36}$

2c. mutually exclusive, $\frac{1}{9}$

3a. mutually exclusive, $\frac{1}{2}$

3b. not mutually exclusive, $\frac{4}{13}$

3c. not mutually exclusive, $\frac{7}{13}$

4a. not mutually exclusive, $\frac{13}{20}$

4b. not mutually exclusive, $\frac{1}{2}$

4c. mutually exclusive, $\frac{9}{10}$

5a. mutually exclusive, $\frac{11}{20}$

5b. not mutually exclusive, $\frac{29}{40}$

5c. not mutually exclusive, $\frac{2}{5}$

8. 13 to 39 or 1 to 3; 48 to 4 or 12 to 1

9. 1 to 3; 1 to 1

10. 96 to 24 or 4 to 1; 40 to 80 or 1 to 2

Teach with Tech
Interactive Whiteboard Help students visualize an example involving events that are not mutually exclusive with a Venn diagram. Label one circle with the event for A and the other circle with the event for B. Have students help you list all the possible outcomes on the board next to the diagram. Have students take turns coming to the board and dragging the outcomes to the appropriate spot on the diagram. Show students that you must subtract $P(A$ and $B)$ because it is the intersection of the two sets.

1 Focus

VerticalAlignment

Lesson 0-6 Find probabilities of independent events, dependent events, and use two-way frequency tables to find conditional probabilities.

After Lesson 0-6 Model data using distribution curves.

2 Teach

Example 1 shows how to find the probability of an independent event. **Example 2** shows how to find the probability of a dependent event. **Example 3** shows how to find conditional probability. **Example 4** shows how to use two-way frequency tables to find conditional probabilities.

Additional Example

1 A card is drawn from a standard deck of 52 cards. A spinner with four equal sections of different colors is spun. What is the probability of drawing a jack and the spinner landing on red? $\frac{1}{52}$

 LESSON 0-6 Multiplying Probabilities

Objectives

- Find probabilities of independent and dependent events.
- Use two-way frequency tables to find conditional probabilities.

NewVocabulary
independent events
dependent events
conditional probability
two-way frequency table

If the occurrence of one event does not affect the probability of a second event occurring, then the two events are **independent events**.

KeyConcept Probability of Independent Events

If two events A and B are independent, then the probability that A and B will occur is

$$P(A \text{ and } B) = P(A) \cdot P(B).$$

This rule can be extended to three or more independent events.

Example 1 Probability of Independent Events

A coin is tossed and a die is rolled. What is the probability of the coin landing on tails and rolling a 3?

Since the outcome of tossing the coin does not affect the outcome of rolling the die, these events are independent.

$P(\text{tails and 3}) = P(\text{tails}) \cdot P(3)$ Probability of independent events

$\qquad = \frac{1}{2} \cdot \frac{1}{6}$ or $\frac{1}{12}$ $P(\text{tails}) = \frac{1}{2}$ and $P(3) = \frac{1}{6}$

CHECK Use a listing of the possible outcomes from this compound event.
H1, H2, H3, H4, H5, H6, T1, T2, T3, T4, T5, T6

There is only 1 outcome out of the 12 possible outcomes that indicates landing on tails and rolling a 3, so $P(\text{tails and 3})$ is $\frac{1}{12}$. ✓

If the occurrence of the first event *does* affect the probability of the second event occurring, then the events are **dependent events**. An example of dependent events is drawing a card from a standard deck of cards, not putting it back, and then drawing a second card. To find the probability of drawing a jack and then a queen in this situation, we modify the multiplication rule.

Probability of drawing a jack on the first draw

Probability of drawing a queen on second draw given that a jack was drawn first

$P(\text{jack and queen}) = \dfrac{4}{52} \cdot \dfrac{4}{51}$

Out of 52 cards, 4 are jacks.   Out of the 51 cards left, 4 are queens.

The probability of an event A occurring given that event B has already occurred is called a **conditional probability** and is represented by $P(B \mid A)$, read *the probability of B given A*. This notation is used in the rule for the probability of two dependent events.

KeyConcept Probability of Dependent Events

If two events A and B are dependent, then the probability that A and B will occur is

$$P(A \text{ and } B) = P(A) \cdot P(B \mid A).$$

This rule can be extended to three or more dependent events.

 P16 | Lesson 0-6

Teach with Tech

Student Response System Present several real-world examples that represent independent or dependent events to the class. Before discussing each example, check for students' understanding by asking them whether the example models an independent or a dependent event. You can have them respond with a 1 for independent and a 2 for dependent.

Example 2 Probability of Dependent Events

A bag contains 12 red, 9 blue, 11 yellow, and 8 green marbles. If two marbles are drawn at random and not replaced, what is the probability that a red and then a blue marble are drawn?

The event of drawing the first marble affects the probability of drawing the second marble, because there is one fewer marble from which to choose. So, the events are dependent.

$$P(\text{red}) = \frac{12}{40} \quad \begin{array}{l}\leftarrow \text{ number of red marbles} \\ \leftarrow \text{ total number of marbles}\end{array} \qquad P(\text{blue} \mid \text{red}) = \frac{9}{39} \quad \begin{array}{l}\leftarrow \text{ number of blue marbles} \\ \leftarrow \text{ number of marbles remaining}\end{array}$$

$$= \frac{3}{10} \qquad\qquad\qquad\qquad\qquad\qquad = \frac{3}{13}$$

$$P(\text{red and blue}) = P(\text{red}) \cdot P(\text{blue} \mid \text{red}) \qquad \text{Probability of dependent events}$$

$$= \frac{3}{10} \cdot \frac{3}{13} \qquad\qquad \text{Substitute.}$$

$$= \frac{9}{130} \qquad\qquad \text{Multiply.}$$

The probability that a red and then a blue marble are drawn is $\frac{9}{130}$.

Solving the equation for the probability of two dependent events for $P(A \mid B)$ gives a rule for computing the conditional probability of an event.

$$P(A \text{ and } B) = P(A) \cdot P(B \mid A) \qquad \text{Probability of dependent events}$$

$$\frac{P(A \text{ and } B)}{P(A)} = P(B \mid A) \qquad \text{Divide each side by } P(A).$$

> **StudyTip**
> **Independent Events** If A and B are independent events, then $P(B \mid A) = P(B)$.

KeyConcept Conditional Probability

If A and B are dependent events, then the conditional probability of event B occurring, given that event A has already occurred, is

$$P(B \mid A) = \frac{P(A \text{ and } B)}{P(A)}, \text{ where } P(A) \neq 0.$$

Example 3 Conditional Probability

FOOD At a restaurant, 25% of customers order chili. If 4% of customers order chili and a baked potato, find the probability that someone who orders chili also orders a baked potato.

$$P(\text{baked potato} \mid \text{chili}) = \frac{P(\text{chili and baked potato})}{P(\text{chili})} \qquad \text{Conditional probability}$$

$$= \frac{0.04}{0.25} \qquad\qquad \begin{array}{l}P(\text{chili and baked potato}) = 0.04 \\ \text{and } P(\text{chili}) = 0.25\end{array}$$

$$= 0.16 \qquad\qquad \text{Simplify.}$$

The probability that someone who orders chili also orders a baked potato is 16%.

Additional Examples

2 On the last day of the grading period, three students with the highest average get to pick a prize from a bag that contains 4 pairs of sunglasses, 5 key chains, and 6 car air fresheners. What is the probability that the first student draws a pair of sunglasses, the second student draws a car air freshener, and the third student draws a keychain? $\frac{4}{91}$

3 A bag contains 8 red pencils, 6 yellow pencils, and 5 purple pencils. Two pencils are drawn one at a time and not replaced. What is the probability that the second pencil is purple, given that the first pencil is red? $\frac{5}{18}$

Tips for New Teachers

Use a Diagram or Model It is sometimes helpful to sketch objects or use models, such as counters, to represent independent or dependent events. In Example 2, different colored counters could represent the marbles.

Additional Example

4 **ATHLETES** The table below shows the number of students who are varsity and junior varsity athletes.

Class	Varsity	Junior Varsity
freshman	7	269
sophomore	22	262
junior	36	276
senior	51	257

a. Find the probability that a student is a varsity athlete given that he or she is a junior. 0.115 or 11.5%

b. Find the probability that a student is a junior varsity athlete given that he or she is a senior. 0.834 or 83.4%

Focus on Mathematical Content

Conditional Probability When event *A* influences the outcome of event *B*, then conditional probability must be used to find the probability of event *B*. The occurrence of *A* can be thought of as reducing the sample space for event *B*.

ReviewVocabulary

relative frequency the ratio of the number of observations in a statistical category to the total number of observations

StudyTip

Joint and Marginal Frequencies The totals row and totals column in a two-way table report the *marginal frequencies*, while the cells in the interior of the table report the *joint frequencies*. The total number of observations is reported in the bottom right corner of the table.

A *contingency* or **two-way frequency table** is often used to show the observed or relative frequencies of data from an experiment classified according to two variables, with the rows indicating one variable and the columns indicating the other. These tables can be used to find conditional probabilities.

● Real-World Example 4 Two-Way Frequency Table

MEDICINE A drug company conducted an experiment to determine the effectiveness of a certain new drug. Test subjects were randomly assigned to one of two groups: a treatment group, which received the drug, or a control group, which received a placebo instead of the drug. The contingency table below shows the results.

Group	Condition Improves (Y)	Condition Does Not Improve (N)
Treatment (T)	1600	400
Control (C)	1200	800

a. Find the probability that a test subject's condition improved given that he or she was in the treatment group.

Step 1 Add a row and a column to the table, and calculate the totals.

Group	Condition Improves (Y)	Condition Does Not Improve (N)	Totals
Treatment (T)	1600	400	2000
Control (C)	1200	800	2000
Totals	1800	1200	4000

Step 2 Divide each observed frequency by the total number of people in the study, 4000, to obtain a table of relative frequencies.

Group	Condition Improves (Y)	Condition Does Not Improve (N)	Totals
Treatment (T)	0.4	0.1	0.5
Control (C)	0.3	0.2	0.5
Totals	0.45	0.3	1

Step 3 Find the conditional probability $P(Y \mid T)$.

$$P(Y \mid T) = \frac{P(T \text{ and } Y)}{P(T)} \qquad \text{Conditional probability}$$

$$= \frac{0.4}{0.5} \text{ or } 0.8 \qquad P(T \text{ and } Y) = 0.4 \text{ and } P(T) = 0.5$$

The probability that a subject's condition improved given that he or she was in the treatment group is 0.8 or 80%.

b. Find the probability that a test subject was in the control group given that his or her condition did not improve.

$$P(C \mid N) = \frac{P(N \text{ and } C)}{P(N)} \qquad \text{Conditional probability}$$

$$= \frac{0.2}{0.3} \text{ or about } 0.667 \qquad P(N \text{ and } C) = 0.2 \text{ and } P(N) = 0.3$$

The probability that a test subject was in the control group given that his or her condition did not improve is about 0.667 or about 66.7%.

 P18 | Lesson 0-6 | Multiplying Probabilities

Teach with Tech

Interactive Whiteboard Draw a tree diagram on the board to show students how to find the conditional probabilities. Save your work and send the diagram to students for reference.

Determine whether the events are *independent* or *dependent*. Then find the probability.

1. A red die and a blue die are rolled. What is the probability of getting the result shown? **independent; $\frac{1}{36}$**

2. Yana has 4 black socks, 6 blue socks, and 8 white socks in his drawer. If he selects three socks at random with no replacement, what is the probability that he will first select a blue sock, then a black sock, and then another blue sock? **dependent; $\frac{5}{204}$ or about 0.025**

A die is rolled twice. Find each probability.

3. $P(2$ and $3)$ $\frac{1}{36}$
4. $P($two 4s$)$ $\frac{1}{36}$
5. $P($no 6s$)$ $\frac{25}{36}$
6. $P($two of the same number$)$ $\frac{1}{6}$

A bag contains 8 blue marbles, 6 red marbles, and 5 green marbles. Three marbles are drawn one at a time. Find each probability.

7. The second marble is green, given that the first marble is blue and not replaced. $\frac{5}{18}$

8. The second marble is red, given that the first marble is green and is replaced. $\frac{6}{19}$

9. The third marble is red, given that the first two are red and blue and not replaced. $\frac{5}{17}$

10. The third marble is green, given that the first two are red and are replaced. $\frac{5}{19}$

DVDS There are 8 action, 3 comedy, and 5 drama DVDs on a shelf. Suppose three DVDs are selected at random from the shelf. Find each probability.

11. $P(3$ action$)$, with replacement $\frac{1}{8}$

12. $P(2$ action, then a comedy$)$, without replacement $\frac{1}{20}$

13. **CARDS** You draw a card from a standard deck of cards and show it to a friend. The friend tells you that the card is red. What is the probability that you correctly guess that the card is the ace of diamonds? $\frac{1}{26}$

14. **HONOR ROLL** Suppose the probability that a student takes AP Calculus and is on the honor roll is 0.0035, and the probability that a student is on the honor roll is 0.23. Find the probability that a student takes AP Calculus given that he or she is on the honor roll. **about 0.015**

15. **DRIVING TESTS** The table shows how students in Mr. Diaz's class fared on their first driving test. Some took a class to prepare, while others did not. Find each probability.

Status	Class	No Class
passed	64	48
failed	18	32

   a. Paige passed, given that she took the class. $\frac{32}{41}$

   b. Madison failed, given that she did not take the class. $\frac{2}{5}$

   c. Jamal did not take the class, given that he passed. $\frac{3}{7}$

16a. $\frac{78}{199}$ or about 39.2%

16. **SCHOOL CLUBS** King High School tallied the number of males and females that were members of at least one after school club. Find each probability.

Gender	Clubs	No Clubs
male	156	242
female	312	108

   a. A student is a member of a club given that he is male.

   b. A student is not a member of a club given that she is female. $\frac{9}{35}$ or 25.7%

   c. A student is a male given that he is not a member of a club. $\frac{121}{175}$ or 69.1%

17. **FOOTBALL ATTENDANCE** The number of students who have attended a football game at North Coast High School is shown. Find each probability.

Class	Freshman	Sophomore	Junior	Senior
attended	48	90	224	254
not attended	182	141	36	8

   a. Given that a student is a freshman, the student has not attended a game. $\frac{91}{115}$ or about 79.1%

   b. Given that a student has attended a game, the student is an upperclassman (a junior or senior). $\frac{239}{308}$ or about 77.6%

3 Assess

Formative Assessment

Use Exercises 1–17 to assess whether students can find probabilities of independent events, dependent events, and use two-way frequency tables to find conditional probabilities.

Ticket Out the Door Put a contingency table on the board that compares the numbers of girls and boys in the class with respect to a specific trait, like participation in school sports, playing in the band, or working after school. Then ask students to write a conditional probability related to the data in the table.

LESSON 0-7 Congruent and Similar Figures

1 Focus

VerticalAlignment

Lesson 0-7 Identify and use congruent and similar figures.

After Lesson 0-7 Use the Pythagorean Theorem to solve problems.

2 Teach

Example 1 shows how to write congruence statements. **Example 2** shows how to determine whether two polygons are similar. **Example 3** shows how to use similar triangles to solve a real-world problem.

Additional Examples

1 The corresponding parts of two congruent triangles are marked on the figure. Write a congruence statement for the two triangles.
$\triangle ABD \cong \triangle CBD$

2 Determine whether the polygons are similar. Justify your answer.

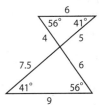

Yes; the corresponding angles are equal and the corresponding sides are proportional.

∵Objective

- Identify and use congruent and similar figures.

 NewVocabulary
congruent
similar

Congruent figures have the same size and the same shape. Two polygons are congruent if their corresponding sides and their corresponding angles are congruent.

$\triangle ABC \cong \triangle EFD$

Congruent Angles	Congruent Sides
$\angle A \cong \angle E$	$\overline{AB} \cong \overline{EF}$
$\angle B \cong \angle F$	$\overline{BC} \cong \overline{FD}$
$\angle C \cong \angle D$	$\overline{AC} \cong \overline{ED}$

Read the symbol $\cong$ as *is congruent to.*

The order of the vertices indicates the corresponding parts.

Example 1 Congruence Statements

The corresponding parts of two congruent triangles are marked on the figure. Write a congruence statement for the two triangles.

List the congruent angles and sides.

$\angle A \cong \angle D$ $\angle B \cong \angle E$ $\angle ACB \cong \angle DCE$

$\overline{AB} \cong \overline{DE}$ $\overline{AC} \cong \overline{DC}$ $\overline{BC} \cong \overline{EC}$

Match the vertices of the congruent angles. Therefore, $\triangle ABC \cong \triangle DEC$.

Similar figures have the same shape, but not necessarily the same size.

In similar figures, corresponding angles are congruent, and the measures of corresponding sides are proportional. (They have equivalent ratios.)

Congruent Angles
$\angle A \cong \angle D, \angle B \cong \angle E, \angle C \cong \angle F$

Proportional Sides
$\dfrac{AB}{DE} = \dfrac{BC}{EF} = \dfrac{AC}{DF}$

$\triangle ABC \sim \triangle DEF$ ← Read the symbol $\sim$ as *is similar to.*

Example 2 Determine Similarity

Determine whether the polygons are similar. Justify your answer.

Because $\frac{4}{3} = \frac{8}{6} = \frac{4}{3} = \frac{8}{6}$, the measures of the sides are proportional. However, the corresponding angles are not congruent. The polygons are not similar.

Tips for New Teachers

Reasoning Show students that they can separate a figure with two polygons and draw a new diagram to clarify which side and angle measures of the polygon are similar.

Example 3 Solve a Problem Involving Similarity

CIVIL ENGINEERING The city of Mansfield plans to build a bridge across Pine Lake. Use the information in the diagram at the right to find the distance across Pine Lake.

$\triangle ABC \sim \triangle ADE$

$\dfrac{AB}{AD} = \dfrac{BC}{DE}$ Definition of similar polygons

$\dfrac{100}{220} = \dfrac{55}{DE}$ $AB = 100$, $AD = 100 + 120$ or 220, $BC = 55$

$100DE = 220(55)$ Cross products

$100DE = 12{,}100$ Simplify.

$DE = 121$ Divide each side by 100.

The distance across the lake is 121 meters.

StudyTip

Reasonableness When solving a problem using a proportion, examine your solution for reasonableness. In Example 3, *DA* is more than twice *BA*, so *DE* should be more than twice *BC*. The solution is reasonable.

Exercises

Determine whether each pair of figures is *similar*, *congruent*, or *neither*.

1. **similar**

2. **congruent**

3. **neither**

4. **neither**

5. **similar**

6. **congruent**

Each pair of polygons is similar. Find the values of x and y.

7. **8; 21**

8. **7.2; 20.8**

9. **10.2; 13.6**

10. **SHADOWS** On a sunny day, Jason measures the length of his shadow and the length of a tree's shadow. Use the figures at the right to find the height of the tree. **4.5 m**

11. **PHOTOGRAPHY** A photo that is 4 inches wide by 6 inches long must be reduced to fit in a space 3 inches wide. How long will the reduced photo be? $4\frac{1}{2}$ **in.**

12. **SURVEYING** Surveyors use instruments to measure objects that are too large or too far away to measure by hand. They can use the shadows that objects cast to find the height of the objects without measuring them. A surveyor finds that a telephone pole that is 25 feet tall is casting a shadow 20 feet long. A nearby building is casting a shadow 52 feet long. What is the height of the building? **65 ft**

Additional Example

3 **GEOGRAPHY** A person is standing at the edge of the water, looking at an island that is 2 miles off shore. When a 5-inch long card is held 20 inches from the person's eye, the length of the island is covered exactly. How long is the island? 0.5 mile

Teach with Tech

Wiki Have students work in pairs to create a wiki page about the different ways to prove two figures are congruent or similar. Have students review and revise each others' work as they create the page.

WatchOut!

Preventing Errors When using similarity to solve problems, students may set up the proportion incorrectly. Remind them that the order in which the vertices are named tells you which parts of the figure correspond.

3 Assess

Formative Assessment

Use Exercises 1–12 to assess whether students understand the concepts of congruence and similarity.

Crystal Ball Ask students how they think the concept of similarity in today's lesson will help them with tomorrow's lesson on the Pythagorean Theorem.

1 Focus

VerticalAlignment

Lesson 0-8 Use the Pythagorean Theorem and its converse.

After Lesson 0-8 Find the distance between two points on the coordinate plane. Find the equation of a circle.

2 Teach

Example 1 shows how to use the Pythagorean Theorem to find the length of the hypotenuse of a right triangle. **Example 2** shows how to use the Pythagorean Theorem to find the length of one of the legs of a right triangle. **Example 3** shows how to use the Pythagorean Theorem to determine whether a triangle is a right triangle.

Additional Example

1 Find the length of the hypotenuse of each right triangle.

a.

17 cm

b.

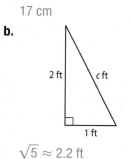

$\sqrt{5} \approx 2.2$ ft

LESSON 0-8 The Pythagorean Theorem

·:·Objective

● Use the Pythagorean Theorem and its converse.

The **Pythagorean Theorem** states that in a right triangle, the square of the length of the hypotenuse c is equal to the sum of the squares of the lengths of the legs a and b.

That is, in any right triangle, $c^2 = a^2 + b^2$.

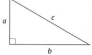

Example 1 Find Hypotenuse Measures

Find the length of the hypotenuse of each right triangle.

a.

$c^2 = a^2 + b^2$	Pythagorean Theorem
$c^2 = 5^2 + 12^2$	$a = 5$ and $b = 12$
$c^2 = 25 + 144$	Simplify.
$c^2 = 169$	Add.
$c = \sqrt{169}$	Take the positive square root of each side.
$c = 13$	The length of the hypotenuse is 13 inches.

b.

$c^2 = a^2 + b^2$	Pythagorean Theorem
$c^2 = 6^2 + 10^2$	$a = 6$ and $b = 10$
$c^2 = 36 + 100$	Simplify.
$c^2 = 136$	Add.
$c = \sqrt{136}$	Take the positive square root of each side.
$c \approx 11.7$	Use a calculator.

To the nearest tenth, the length of the hypotenuse is 11.7 centimeters.

You can also find the length of a leg of a right triangle given the lengths of the hypotenuse and the other leg.

Example 2 Find Leg Measures

Find the length of the missing leg in each right triangle.

a.

$c^2 = a^2 + b^2$	Pythagorean Theorem
$25^2 = a^2 + 7^2$	$c = 25$ and $b = 7$
$625 = a^2 + 49$	Simplify.
$625 - 49 = a^2 + 49 - 49$	Subtract 49 from each side.
$576 = a^2$	Simplify.
$\sqrt{576} = a$	Take the positive square root of each side.
$24 = a$	The length of the leg is 24 feet.

 P22 | Lesson 0-8

WatchOut!

Positive Square Root When finding the length of a side using the Pythagorean Theorem, use only the positive and not the negative square root, since length cannot be negative.

b.

$$c^2 = a^2 + b^2 \qquad \text{Pythagorean Theorem}$$
$$4^2 = 2^2 + b^2 \qquad c = 4 \text{ and } a = 2$$
$$16 = 4 + b^2 \qquad \text{Simplify.}$$
$$16 - 4 = 4 - 4 + b^2 \qquad \text{Subtract 4 from each side.}$$
$$12 = b^2 \qquad \text{Simplify.}$$
$$\sqrt{12} = b \qquad \text{Take the positive square root of each side.}$$
$$3.5 \approx b \qquad \text{Use a calculator.}$$

To the nearest tenth, the length of the leg is 3.5 meters.

The **converse of the Pythagorean Theorem** states that if the sides of a triangle have lengths a, b, and c, and $c^2 = a^2 + b^2$, then the triangle is a right triangle.

Example 3 Identify a Right Triangle

The lengths of the three sides of a triangle are 5, 7, and 9 inches. Determine whether this triangle is a right triangle.

Because the longest side is 9 inches, use 9 as c, the measure of the hypotenuse.

$$c^2 = a^2 + b^2 \qquad \text{Pythagorean Theorem}$$
$$9^2 \stackrel{?}{=} 5^2 + 7^2 \qquad c = 9, a = 5, \text{ and } b = 7$$
$$81 \stackrel{?}{=} 25 + 49 \qquad \text{Evaluate } 9^2, 5^2, \text{ and } 7^2.$$
$$81 \neq 74 \qquad \text{Simplify.}$$

Because $c^2 \neq a^2 + b^2$, the triangle is *not* a right triangle.

Exercises

Find each missing measure. Round to the nearest tenth, if necessary.

1. 39 ft

2. 24 km

3. 8.3 cm

4. $a = 3, b = 4, c = ?$ **5**
5. $a = ?, b = 12, c = 13$ **5**
6. $a = 14, b = ?, c = 50$ **48**

7. $a = 2, b = 9, c = ?$ **9.2**
8. $a = 6, b = ?, c = 13$ **11.5**
9. $a = ?, b = 7, c = 11$ **8.5**

The lengths of three sides of a triangle are given. Determine whether each triangle is a right triangle.

10. 5 in., 7 in., 8 in. **no**
11. 9 m, 12 m, 15 m **yes**
12. 6 cm, 7 cm, 12 cm **no**

13. 11 ft, 12 ft, 16 ft **no**
14. 10 yd, 24 yd, 26 yd **yes**
15. 11 km, 60 km, 61 km **yes**

16. **FLAGPOLES** If a flagpole is 30 feet tall and Mai-Lin is standing a distance of 15 feet from the flagpole, what is the distance from her feet to the top of the flagpole? **about 33.5 ft**

? ft   30 ft
15 ft

17. **CONSTRUCTION** The walls of a recreation center are being covered with paneling. A doorway is 0.9 meter wide and 2.5 meters high. What is the length of the widest rectangular panel that can be taken through this doorway? **about 2.66 m**

18. **OPEN ENDED** Create an application problem involving right triangles and the Pythagorean Theorem. Then solve your problem, drawing diagrams if necessary. **See students' work.**

Additional Examples

2 Find the length of the missing leg in each right triangle.

a.
8 m   10 m
b m
6 m

b.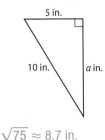
5 in.
10 in. a in.
$\sqrt{75} \approx 8.7$ in.

3 The lengths of the three sides of a right triangle are 10, 24, and 26 centimeters. Determine whether this triangle is a right triangle. **yes**

WatchOut!

Preventing Errors Point out to students that when given three numbers that represent the lengths of the sides of a right triangle, the greatest value is always the measure of the hypotenuse, and the two lesser values are the measures of the two legs.

3 Assess

Formative Assessment

Use Exercises 1–18 to assess whether students understand how to use the Pythagorean Theorem.

Name the Math Have students describe the Pythagorean Theorem, including examples of when to use it or its converse.

Teach with Tech

Student Response System Show students a set of three numbers, and ask them if these could be the side lengths of a right triangle. Have students respond with A for yes and B for no.

LESSON 0-9 Measures of Center, Spread, and Position

1 Focus

VerticalAlignment

▼

Lesson 0-9 Find measures of center, spread, and position.

▼

After Lesson 0-9 Model data using distribution curves.

2 Teach

Example 1 shows how to determine measures of center for a data set.
Example 2 shows how to determine measures of variation for a data set.
Example 3 shows how to find and interpret a five-number summary.
Example 4 shows how to identify and examine the effect of an outlier on a data set.

Additional Example

1 **EXERCISE** The numbers of minutes per day that a student spends exercising for a week are shown below. Find the mean, median, and mode.

50, 75, 45, 60, 45, 75, 45

about 56 min, 50 min, 45 min

WatchOut!

Common Misconceptions Remind students that measures of central tendency must fall within the range of the data, between the minimum and maximum.

·· Objective

● Find measures of center, spread, and position.

 NewVocabulary
statistics
descriptive statistics
population
variable
data
univariate data
sample
measures of center or central tendency
mean
median
mode
measures of spread or variation
range
variance
standard deviation
quartile
lower quartile
upper quartile
five-number summary
interquartile range
outlier

 P24 | Lesson 0-9

Statistics is the science of collecting, organizing, displaying, and analyzing data in order to draw conclusions and make predictions. The branch of statistics that focuses on collecting, summarizing, and displaying data is called **descriptive statistics**.

The entire group of interest to a statistician is called a **population**. A **variable** is a characteristic of a population that can assume different values called **data**. Data that include only one variable are called **univariate data**. When it is not possible to obtain data about every member of a population, a representative **sample** or subset of the population is selected.

Population

Sample

Univariate data are often summarized using a single number to represent what is average or typical. Measures of average are also called **measures of center** or **central tendency**. The most common measures of center are mean, median, and mode.

KeyConcept Measures of Center

- The **mean** is the sum of the values in a set of data $x_1, x_2, \ldots, x_n$ divided by the total number of values n in the set. The formula for population mean μ is $\mu = \frac{x_1 + x_2 + \ldots + x_n}{n}$.

- The **median** is the middle value or the mean of the two middle values in a set of data when the data are arranged in numerical order.

- The **mode** is the value or values that appear most often in a set of data. A set of data can have no mode, one mode, or more than one mode.

For sample mean, replace μ with $\bar{x}$.

Example 1 Measures of Center

SOFT DRINKS The number of milligrams of sodium in a 12-ounce can of ten different brands of regular cola are shown below. Find the mean, median, and mode.

$$50, 30, 25, 20, 40, 35, 35, 10, 15, 35$$

Mean $\bar{x} = \dfrac{50 + 30 + 25 + 20 + 40 + 35 + 35 + 10 + 15 + 35}{10}$ Find the sum of the data values and divide by the number of values, 10.

$= \dfrac{295}{10}$ or 29.5 milligrams

Median 10, 15, 20, 25, 30, 35, 35, 35, 40, 50   Arrange the data in order.

$\dfrac{30 + 35}{2}$ or 32.5 milligrams   Find the mean of the middle two values.

Mode The value that occurs most often in the set is 35, so the mode of the data set is 35 milligrams.

Focus on Mathematical Content

Mode In a set of data, there can be more than one mode. A set with two modes is bimodal. A set with three modes is trimodal, and so on. There can also be data sets with no mode.

Because two very different data sets can have the same mean, statisticians also use **measures of spread** or **variation** to describe how widely the data values vary and how much the values differ from what is typical. The most common of these measures are listed below.

KeyConcept Measures of Spread

- The **range** is the difference between the greatest and least values in a set of data.

- The **variance** in a set of data $x_1, x_2, \ldots, x_n$ is the mean of the squares of the deviations or differences from the mean. The formula for population variance σ^2 is

$$\sigma^2 = \frac{(x_1 - \mu)^2 + (x_2 - \mu)^2 + \ldots + (x_n - \mu)^2}{n}.$$

- The **standard deviation** in a set of data $x_1, x_2, \ldots, x_n$ is the average amount by which each individual value deviates or differs from the mean. It is the square root of the variance. The formula for population standard deviation σ is

$$\sigma = \sqrt{\sigma^2} \text{ or } \sqrt{\frac{(x_1 - \mu)^2 + (x_2 - \mu)^2 + \ldots + (x_n - \mu)^2}{n}}.$$

For sample variance and standard deviation, replace σ^2 with s^2, μ with $\bar{x}$, and n with $n - 1$.

StudyTip
Unbiased Estimator Dividing by $n - 1$ instead of n when finding a sample variance and standard deviation gives a slightly larger value, and therefore an unbiased estimate, of the population standard deviation or variance.

Example 2 Measures of Spread

MIDTERM EXAMS Two classes took the same midterm exam. The scores of five students from each class are shown. Both sets of scores have a mean of 84.2.

Class A
85, 76, 92, 88, 80

a. Find the range, variance, and standard deviation for the sample scores from Class A.

Class B
75, 85, 95, 98, 68

Range $92 - 76$ or 16   range = greatest value − least value

Variance   Subtract the mean of the data, 84.2, from each value. Then square each result and find the sum of these squares. Finally, since these data were obtained from a sample of all scores in Class A, divide the sum by $n - 1$.

X	$X - \bar{X}$	$(X - \bar{X})^2$
85	$85 - 84.2 = 0.8$	$0.8^2 = 0.64$
76	$76 - 84.2 = -8.2$	$(-8.2)^2 = 67.24$
92	$92 - 84.2 = 7.8$	$7.8^2 = 60.84$
88	$88 - 84.2 = 3.8$	$3.8^2 = 14.44$
80	$80 - 84.2 = -4.2$	$(-4.2)^2 = 17.64$
		Sum = 160.8

StudyTip
Squared Differences The sum of the differences of the data values from the mean of the set will always be zero. For this reason, variance is calculated using the sum of the *squares* of the differences.

$$s^2 = \frac{(x_1 - \bar{x})^2 + \ldots + (x_n - \bar{x})^2}{n - 1} \qquad \text{Sample variance}$$

$$= \frac{160.8}{5 - 1} \text{ or about } 40.2 \qquad (x_1 - \bar{x})^2 + \ldots + (x_5 - \bar{x})^2 = 160.8 \text{ and } n = 5$$

Standard Deviation $s = \sqrt{s^2} = \sqrt{\dfrac{160.8}{5 - 1}}$ or about 6.3   Sample standard deviation

(continued on the next page)

Additional Example

2 WATER PARK Carla took a poll in her morning and afternoon study halls to see how many times her classmates visited a water park during summer vacation. Both sets of data have a mean of 1.5.

Water Park Visits					
Study Hall A			**Study Hall B**		
1	2	1	3	0	2
1	1	1	1	3	2
3	2	3	4	0	5
3	2	1	3	2	1

a. Find the range, variance, and standard deviation for the sample scores from Study Hall A. 2, 0.87, 0.76

b. Use a calculator to find the range, variance, and standard deviation for the sample scores from Study Hall B. 5, 1.5, 2.25

c. Compare the sample standard deviations of Study Hall A and Study Hall B. Since the sample standard deviation of Study Hall A is 0.87 and the sample standard deviation of Study Hall B is 1.5, there is more variability in the sample scores from Study Hall B than from Study Hall A.

Tips for New Teachers

Common Misconceptions Explain to students that the standard deviation is a number representing the typical or representative variation for the data items in that set. It tells how far a data value will typically be from the mean of the entire data set.

3 DISCUS The distances in feet of Josey's last 10 discus throws were 90, 110, 103, 99, 95, 101, 107, 105, 92, and 109.

a. Find the minimum, lower quartile, median, upper quartile, and maximum of the data set. 90, 95, 102, 107, 110

b. Interpret this five-number summary. During her last 10 throws, Josey threw a minimum of 90 feet and a maximum of 120 feet. She threw less than 95 feet 25% of the time, less than 102 feet 50% of the time, and less than 107 feet 75% of the time.

WatchOut!

Preventing Errors Students may be confused and mistakenly try to find the quartiles by dividing the range into four equal sections. Stress that the upper and lower quartiles are medians of a subset of the original data set.

Focus on Mathematical Content

Quartiles You may wish to mention to students that the lower quartile is also referred to as the first quartile and the upper quartile as the third quartile.

b. Use a calculator to find the range, variance, and standard deviation for the sample scores from Class B.

To find measures of center and spread using a graphing calculator, first clear all lists. Press $\boxed{\text{STAT}}$ $\boxed{\text{ENTER}}$ to enter each data value. Then press $\boxed{\text{STAT}}$ and select **1-Var Stats** from the **CALC** menu. Scroll to see all one-variable statistics.

The range **maxX − minX** for Class B is 98 − 68 or 30. The sample standard deviation **Sx** is about 12.8, so the variance is about 12.8^2 or 163.8.

c. Compare the sample standard deviations of Class A and Class B.

Since the sample standard deviation of Class A is 6.3 and the sample standard deviation of Class B is 12.8, there is more variability in the sample scores from Class B than from Class A.

> **StudyTip**
>
> **Comparing Standard Deviations** Standard deviation can only be compared in this way when the units of both data sets are the same.

Statisticians use measures of position to describe where specific values fall within a data set. **Quartiles** are three position measures that divide a data set arranged in ascending order into four groups, each containing about 25% of the data. The median marks the second quartile Q_2 and separates the data into upper and lower halves. The first or **lower quartile** Q_1 is the median of the lower half, while the third or **upper quartile** Q_3 is the median of the upper half.

> **StudyTip**
>
> **Calculating Quartiles** When the number of values in a set of data is odd, the median is not included in either half of the data when calculating Q_1 or Q_3.

The three quartiles, along with the minimum and maximum values, are called a **five-number summary** of a data set.

Example 3 Five-Number Summary

PART-TIME JOB The number of hours Liana worked each week for the last 12 weeks were 21, 10, 18, 12, 15, 13, 20, 20, 19, 16, 18, and 14.

a. Find the minimum, lower quartile, median, upper quartile, and maximum of the data set.

Use a calculator to find the one-variable statistics for the data set. The minimum **minX** of the data set is 10, the lower quartile Q_1 is 13.5, the median **Med** is 17, the upper quartile Q_3 is 19.5, and the maximum **maxX** is 21.

b. Interpret this five-number summary.

Over the last 12 weeks, Liana worked a minimum of 10 hours and a maximum of 21 hours. She worked less than 13.5 hours 25% of the time, less than 17 hours 50% of the time, and less than 19.5 hours 75% of the time.

Teach with Tech

Interactive Whiteboard Write each data value in a box on the board. Drag the boxes to rearrange them in order from least to greatest. Group and drag the lower half of the data to separate it from the upper half of the data. If there are an odd number of data values, do not move the middle value and identify it as the median. Repeat this process similarly to find the lower and upper quartiles.

StudyTip

Interquartile Range A large interquartile range means that the data are spread out.

The difference between Q_3 and Q_1, is called the **interquartile range** IQR. The interquartile range contains about 50% of the values. Before deciding on which measures best describe a set of data, check the data set for outliers. An **outlier** is an extremely high or extremely low value when compared with the rest of the values in the set. To check for outliers, look for data values that are beyond the upper or lower quartiles by more than 1.5 times the interquartile range.

Example 4 Effect of an Outlier

HOMEWORK The number of minutes each of the 22 students in a class spent working on the same algebra assignment is shown below.

15, 12, 25, 15, 27, 10, 16, 18, 30, 35, 22, 25, 65, 20, 18, 25, 15, 13, 25, 22, 15, 28

a. Identify any outliers in the data.

Calculate the interquartile range IQR by using a graphing calculator to find Q_1 and Q_3.

$$IQR = Q_3 - Q_1 = 25 - 15 \text{ or } 10$$

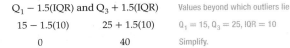

Use the interquartile range to find the values beyond which any outliers would lie.

$Q_1 - 1.5(IQR)$ and	$Q_3 + 1.5(IQR)$	Values beyond which outliers lie
$15 - 1.5(10)$	$25 + 1.5(10)$	$Q_1 = 15, Q_3 = 25, IQR = 10$
0	40	Simplify.

There are no values less than 0, but there is one value greater than 40. The value 65 can be considered an outlier for this data set.

b. Find the mean, median, mode, range, and standard deviation of the data set with and without the outlier. Describe the effect on each measure.

You can use a graphing calculator. The data set represents the entire population of the class, so use the population standard deviation σ.

Data Set	Mean	Median	Mode	Range	Standard Deviation
with outlier	≈22.5	21	15	55	≈11.2
without outlier	≈20.5	20	15	25	≈6.5

Removal of the outlier causes the mean, median, range, and standard deviation to decrease. Notice that the mean is affected more by the outlier than the median.

Exercises

Find the mean, median, and mode for each set of data.

1. number of pages in each novel assigned for summer reading:
   224, 272, 374, 478, 960, 394, 404, 308, 480, 624 **451.8 min, 399 min, no mode**

2. height in centimeters of bean plants at the end of an experiment:
   14.5, 12, 16, 11, 14, 11, 10.5, 14, 11.5, 15, 13.5 **13 cm, 13.5 cm, 11 cm and 14 cm**

3. number of text messages sent each day during the last two weeks:
   18, 35, 53, 44, 26, 57, 23, 27, 47, 33, 4, 35, 39, 41 **≈34.4 text messages, 35 text messages, 35 text messages**

Additional Example

4 GAMES The number of minutes each of the 26 students in a class spent to complete an obstacle course is shown below.

5, 2, 5, 5, 8, 12, 6, 7, 5, 5, 6, 5, 5, 5, 6, 10, 7, 5, 5, 7, 5, 7, 5, 7, 6, 6

a. Identify any outliers in the data. 12

b. Find the mean, median, mode, range, and standard deviation of the data set with and without the outlier. Describe the effect on each measure.

Data Set	With Outlier	Without Outlier
mean	≈6.04	5.8
median	5.5	5
mode	5	5
range	10	8
standard deviation	≈1.89	≈1.47

The mean, median, range, and standard deviation are affected by the removal of the outlier from the data set. Removal of the outlier causes all four measures to decrease.

Focus on Mathematical Content

Extreme Values Extreme values in a set of data are called outliers because they lie outside the overall pattern of distribution. Outliers affect the mean because the mean is sensitive to each value in the set. The median and mode are only slightly affected by an individual number. Therefore, the median and mode will not be greatly affected by outliers.

Formative Assessment

Use Exercises 1–11 to assess whether students can find measures of center, spread, and position.

Ticket Out the Door Give students a data set to analyze. Ask students to write on the back of an index card the mean, median, mode, variance, standard deviation, range, lower quartile, upper quartile, and identify any outliers. As they leave the class, they should turn in their index cards.

Additional Answers

4. Sample; Ride A: 49 min, ≈210.4 min, ≈14.5 min; Ride B: 19 min, ≈42.3 min, ≈6.5 min; since the sample standard deviation of Ride A is greater than that of Ride B, there is more variability in the sample wait times for Ride A than Ride B.

5. Sample; Walk A: 47, ≈242.0, ≈15.6; Walk B: 92, ≈1115.4, ≈33.4; since the sample standard deviation of Walk B is greater than that of Walk A, there is more variability in the number of sponsors obtained by participants in Walk B than in Walk A.

6. Population; Class A: 15, ≈13.7, 3.7; Class B: 12, ≈10.5, ≈3.2; since the sample standard deviation of Class A is greater than that of Class B, there is more variability in the number of days that students missed during the school year for Class A than for Class B.

7. 18, 23, 25, 27, 29; Sample answer: There are 18 students in the smallest math class at Central High and 29 students in the largest class. 25% of the classes have less than 23 students, 50% of the classes have less than 25 students, and 75% of the classes have less than 27 students.

State whether the data in sets A and B represent *sample* or *population* data. Then find the range, variance, and standard deviation of each set. Use the standard deviations to compare the variability between the data sets. **4–6. See margin.**

4.

Wait Times (min)					
Ride A			Ride B		
45	22	40	35	50	32
48	11	51	31	35	45
36	55	60	45	49	40
32	24	37	43	37	45

5.

Number of Sponsors Obtained by Participants					
Charity Walk A			Charity Walk B		
44	14	61	8	28	15
22	27	25	100	42	19
38	50	49	25	75	82

6.

Number of Days Each Student Missed This Year													
Class A													
10	8	5	9	7	3	6	8	14	11	8	4	7	8	2
5	13	0	15	9	7	9	10	9	11	14	8	12	10	1
Class B													
5	8	13	7	9	4	10	2	12	9	6	11	3	8	5
12	6	7	8	11	12	8	9	3	10	5	13	9	1	8

Find the minimum, lower quartile, median, upper quartile, and maximum of each data set. Then interpret this five-number summary. **7–8. See margin.**

7.

Number of Students in Each Math Class at Central High														
25	27	26	26	19	27	24	23	19	28	25	24	20	22	22
24	26	18	28	29	29	26	24	24	23	23	25	25	29	28

8.

State Mean ACT Scores									
20.2	21.3	21.5	20.4	21.6	20.3	22.5	21.5	17.8	20.5
20.0	21.7	21.3	20.2	21.6	22.0	21.6	20.3	19.8	22.6
20.8	22.4	21.4	22.2	18.8	21.5	21.7	21.2	22.5	21.2
20.1	22.3	20.3	21.2	21.4	20.6	22.5	21.8	21.9	19.3
21.5	20.5	20.3	21.5	22.7	20.9	22.5	22.2	21.4	20.7

Identify any outliers in each data set, and explain your reasoning. Then find the mean, median, mode, range, and standard deviation of the data set with and without the outlier. Describe the effect on each measure. **9–10. See Chapter 0 Answer Appendix.**

9. fuel efficiency in miles per gallon of 15 randomly selected automobiles:
40, 36, 29, 45, 51, 36, 48, 34, 36, 22, 13, 42, 31, 44, 32, 34

10. number of posts to a certain blog each month during a particular year:
25, 23, 21, 27, 29, 19, 10, 21, 20, 18, 26, 23

11. **CEREAL** The weights, in ounces, of 20 randomly selected boxes of a certain brand of cereal are shown. **a–c. See margin.**
16.7, 16.8, 15.9, 16.1, 16.5, 16.6, 16.5, 15.9, 16.7, 16.5,
16.6, 14.9, 16.5, 16.1, 15.8, 16.7, 16.2, 16.5, 16.4, 16.6

a. Identify any outliers in the data set, and explain your reasoning.

b. If the outlier was removed and an additional cereal box that was 17.35 ounces was added, would this value be an outlier of the new data set? Explain.

c. What are some possible causes of outliers in this situation?

 P28 | Lesson 0-9 | Measures of Center, Spread, and Position

8. 17.8, 20.4, 21.4, 21.8, 22.7; Sample answer: The lowest mean score for a state is 17.8 and the highest mean score is 22.7. 25% of the states have a mean score that is less than 20.4, 50% of the states have a mean score that is less than 21.4, and 75% of the states have a mean score that is less than 22.7.

11a. Yes; sample answer: Any outliers would be less than 15.35 or greater than 17.35. Since 14.9 < 15.35, it is an outlier.

11b. No; sample answer: Any outliers would be less than 15.4 or greater than 17.4. Since 17.4 > 17.35, it would not be an outlier.

11c. Sample answers: data recording errors, manufacturing errors

Posttest

State the domain and range of each relation. Then determine whether each relation is a function. Write *yes* **or** *no.* **1.** $D = \{0, 4, 5, 7\}$, $R = \{-2, -1, 5, 9, 12\}$; no

1. $\{(4, 5), (5, -1), (0, 12), (0, -2), (7, 9)\}$

2.

2. $D = \{-4, -2, 1, 2\}$, $R = \{-3, -1, 1, 2, 3\}$; no

Name the quadrant in which each point is located.

3. $(-3, 7)$ II **4.** $(10, -11)$ IV **5.** $(-15, -3)$ III

6. $8n^2 + 2n - 6$
Find each product. 7. $30p^2 - 56p + 10$

6. $(4n - 3)(2n + 2)$ **7.** $(5p - 1)(6p - 10)$

8. $(7x + 4)(7x + 4)$ **9.** $(3k - 2)(6k + 9)$
$49x^2 + 56x + 16$ $18k^2 + 15k - 18$

10. GEOMETRY The height of a rectangle is 3 millimeters less than twice the width.

  a. Write an expression for each measure. w; $2w - 3$

  b. Write a polynomial expression for the area of the rectangle. $2w^2 - 3w$

11. $(2x + y)^2$ **12.** $(5a - 2)^2$

Factor each polynomial.

11. $4x^2 + 4xy + y^2$ **12.** $25a^2 - 20a + 4$

13. $4a^2 + 16ab + 16b^2$ **14.** $81t^2 - 36$ $9(3t - 2)(3t + 2)$
$4(a + 2b)^2$

15. STUDENT COUNCIL A student council has 6 seniors, 5 juniors, and 1 sophomore as members. How many ways can a 3-member committee be formed that includes one member from each class?
30 ways

Determine whether each situation involves *permutations* **or** *combinations.* **Then solve.**

16. How many ways are there to select one competitor and one alternate out of 8 students? **permutations; 56**

17. How many ways are there to form a team of 7 athletes from a group of 15 who try out?
combinations; 6435

RESTAURANT There are 24 male and 36 female patrons in a restaurant. Of the 11 patrons under 10 years old, 6 are male. Of the 14 patrons over 55 years old, 9 are female. A patron is selected at random. Determine whether the events are *mutually exclusive* or *not mutually exclusive.* Then find each probability.

18. P(female or under 10) **19.** P(under 10 or over 55)
not mutually exclusive; $\frac{7}{10}$ **mutually exclusive;** $\frac{5}{12}$

Determine whether the events are *independent* **or** *dependent.* **Then find the probability.**

20. Slips of paper numbered 1 through 10 are placed into a bag. What is the probability of drawing the number 10 three times in a row if a slip is drawn at random and then replaced? **independent;** $\frac{1}{1000}$

21. Two students are selected at random from a class that consists of 13 males and 7 females. What is the probability that both students are female? **dependent;** $\frac{21}{190}$

22. TESTING Of the students who took both the Mid-Chapter 4 Quiz and the Chapter 4 Test, 56% passed the quiz and 48% passed both the quiz and the test. If a student passed the quiz, find the probability that he or she also passed the test. **about 86%**

23. Determine whether the rectangles are *similar*, *congruent*, or *neither.* **neither**

24. COMPUTERS A computer image of a painting 320 pixels wide by 240 pixels high. If the actual painting is 42 inches wide, how high is it? **31.5 in.**

Find each missing measure. Round to the nearest tenth, if necessary.

25.

a in.   **24 in.**
30 in.   18 in.

26. $a = 33$ cm, $b = ?$ cm, $c = 45$ cm **30.6 cm**

The lengths of three sides of a triangle are given. Determine whether each triangle is a right triangle.

27. 6 in., 8 in., 12 in. **no** **28.** 30 m, 34 m, 16 m **yes**
29. ≈ 36.1 students, 33.5 students, 35 students, 59 students, ≈ 16.8 students; 75 students
Find the mean, median, mode, range, and standard deviation of each data set. Then identify any outliers.

29. number of students present at 8 student council meetings: 23, 45, 16, 75, 32, 35, 28, 35

30. running time in minutes for 17 movies: 95, 102, 148, 140, 110, 103, 107, 104, 99, 111, 109, 124, 109, 90, 92, 110, 129 ≈ **110.7 min, 109 min, 109 and 111 min, 58 min, ≈ 16.1 min; 148 min**

Using the Posttest
Use the Chapter 0 Posttest to assess students' understanding of the concepts after you have presented the lessons in Chapter 0. If students are still having difficulty with one or more concepts, refer to *Math Triumphs: Foundations for Algebra 2* for strategies for reteaching.

Lesson 0-5

1a.

Color	Frequency	Experimental Probability
red	6	0.12
blue	7	0.14
yellow	9	0.18
orange	12	0.24
purple	6	0.10
green	11	0.22

1b.

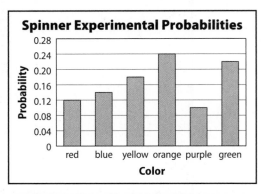

1c.

Color	Frequency	Experimental Probability	Theoretical Probability
red	6	0.12	$0.1\overline{6}$
blue	7	0.14	$0.1\overline{6}$
yellow	9	0.18	$0.1\overline{6}$
orange	12	0.24	$0.1\overline{6}$
purple	6	0.10	$0.1\overline{6}$
green	11	0.22	$0.1\overline{6}$

1d.

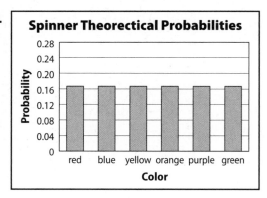

1e. Sample answer: Since all bars in the graph of the theoretical probabilities are the same height, the graph represents a uniform distribution. This means that in theory, the chance of landing on any one of the colors is equally likely. The graph of the experimental probabilities indicates that in practice, it is more likely that the spinner will land on orange or green than on any of the other colors, since the heights of those bars are taller than any others in the graph.

Lesson 0-9

9. 13; Sample answer: Any outliers would be less than 14.25 or greater than 60.25. Since 13 < 14.25, it is an outlier.

Data Set	Mean	Median	Mode	Range	Standard Deviation
with outlier	≈35.8	36	36	38	≈9.3
without outlier	≈37.3	36	36	29	≈7.5

Removing the outlier did not affect the median or mode. However, the removal did affect the mean, standard deviation, and range. The mean and standard deviation increased, and the range decreased.

10. 10; Sample answer: Any outliers would be less than 10.5 or greater than 34.5. Since 10 < 10.5, it is an outlier.

Data Set	Mean	Median	Mode	Range	Standard Deviation
with outlier	≈21.8	22	21	19	≈4.8
without outlier	≈22.9	23	21	11	≈3.3

Removing the outlier did not affect the mode. However, the removal did affect the mean, median, standard deviation, and range. The mean and median increased, and the standard deviation and range decreased.

	Diagnostic Assessment Quick Check			
	LESSON 1-1 45 min: 1 day / 90 min: 0.5 day	**LESSON 1-2** 45 min: 1 day / 90 min: 0.5 day	**LESSON 1-3** 45 min: 1 day / 90 min: 0.5 day	
Title	Expressions and Formulas	Properties of Real Numbers	Solving Equations	
Objectives	▪ Use the order of operations to evaluate expressions. ▪ Use formulas.	▪ Classify real numbers. ▪ Use the properties of real numbers to evaluate expressions.	▪ Translate verbal expressions into algebraic expressions and equations, and vice versa. ▪ Solve equations using the Properties of equality.	
Key Vocabulary	variables algebraic expressions order of operations formula	real numbers rational numbers irrational numbers integers whole numbers natural numbers	open sentence equation solution	
CCSS	A.SSE.1a, A.SSE.1b	A.SSE.2	A.CED.1	
Multiple Representations				
Lesson Resources	connectED.mcgraw-hill.com ▫ Leveled Worksheets ▫ Vocabulary ▫ Personal Tutor ▫ Virtual Manipulatives ▫ Self-Check Quiz ▪ *5-Minute Check* ▪ *Study Notebook*	connectED.mcgraw-hill.com ▫ Leveled Worksheets ▫ Quiz 1 ▫ Vocabulary ▫ Animations ▫ Personal Tutor ▫ Self-Check Quiz ▪ *5-Minute Check* ▪ *Study Notebook*	connectED.mcgraw-hill.com ▫ Leveled Worksheets ▫ Quiz 2 ▫ Vocabulary ▫ Personal Tutor ▫ Self-Check Quiz ▪ *5-Minute Check* ▪ *Study Notebook* ▪ *Teaching Algebra with Manipulatives*	
Resources for Every Lesson	IWB eStudent Edition IWB Interactive Classroom	▪ eTeacher Edition ▪ eSolutions Manual ▪ *McGraw-Hill* eAssessment		
Differentiated Instruction	pp. 6, 8, 10	pp. 12, 13, 17	pp. 20, 21	
			Formative Assessment Mid-Chapter Quiz	

IWB All digital assets are Interactive Whiteboard ready.

Suggested Pacing			
Time Periods	Instruction	Review & Assess	Total
45-minute	7 days	2 days	9 days
90-minute	4 days	1 day	5 days

LESSON 1-4 45 min: 1 day / 90 min: 0.75 day	**LESSON 1-5** 45 min: 1 day / 90 min: 0.75 day	**EXPLORE 1-6** 45 min: 0.5 day / 90 min: 0.25 day	**LESSON 1-6** 45 min: 1.5 days / 90 min: 0.75 day
Solving Absolute Value Equations	**Solving Inequalities**	**Algebra Lab: Interval Notation**	**Solving Compound and Absolute Value Inequalities**
■ Evaluate expressions involving absolute values. ■ Solve absolute value equations.	■ Solve one-step inequalities. ■ Solve multi-step inequalities.	■ Use interval notation to describe sets of numbers.	■ Solve compound inequalities. ■ Solve absolute value inequalities.
absolute value empty set extraneous solution	set-builder notation	interval notation infinity	compound inequality intersection union
A.SSE.1b, A.CED.1	A.CED.1, A.CED.3		A.CED.1, A.CED.3
connectED.mcgraw-hill.com	connectED.mcgraw-hill.com	connectED.mcgraw-hill.com	connectED.mcgraw-hill.com
📁 Leveled Worksheets 🔤 Vocabulary PT Personal Tutor ✋ Virtual Manipulatives ✓ Self-Check Quiz ■ *5-Minute Check* ■ *Study Notebook*	📁 Leveled Worksheets 📁 Quiz 3 🔤 Vocabulary PT Personal Tutor ✋ Virtual Manipulatives ✓ Self-Check Quiz ■ *5-Minute Check* ■ *Study Notebook*	🔤 Vocabulary ■ *Teaching Algebra with Manipulatives*	📁 Leveled Worksheets 📁 Quiz 4 🔤 Vocabulary 🏃 Animations PT Personal Tutor ✋ Virtual Manipulatives ✓ Self-Check Quiz ■ *5-Minute Check* ■ *Study Notebook* ■ *Teaching Algebra with Manipulatives*
IWB eStudent Edition IWB Interactive Classroom	■ eTeacher Edition ■ eSolutions Manual ■ McGraw-Hill eAssessment		
pp. 29, 32	pp. 35, 39		pp. 42, 44
			Summative Assessment Study Guide and Review Practice Test

Assessment and Intervention

SE = Student Edition, TE = Teacher Edition, CRM = Chapter Resource Masters

Diagnosis	Prescription
Beginning Chapter 1	
Get Ready for Chapter 1 **SE**	Response to Intervention **TE**
Beginning Every Lesson	
Then, Now, Why? **SE** 5-Minute Checks	Chapter 0 **SE**

DIAGNOSTIC ASSESSMENT

During/After Every Lesson	
Guided Practice **SE**, every example Check Your Understanding **SE** H.O.T. Problems **SE** Spiral Review **SE** Additional Examples **TE** Watch Out! **TE** Step 4, Assess **TE** Chapter 1 Quizzes **CRM**, pp. 45–46 Self-Check Quizzes connectED.mcgraw-hill.com	TIER 1 Intervention Skills Practice **CRM**, Ch. 1 connectED.mcgraw-hill.com TIER 2 Intervention Differentiated Instruction **TE**; Study Guide and Intervention **CRM**, Ch. 1 TIER 3 Intervention *Math Triumphs, Alg. 2,* Ch. 1 and 2
Mid-Chapter	
Mid-Chapter Quiz **SE** Mid-Chapter Test **CRM**, p. 47 McGraw-Hill eAssessment	TIER 1 Intervention Skills Practice **CRM**, Ch. 1 connectED.mcgraw-hill.com TIER 2 Intervention Study Guide and Intervention **CRM**, Ch. 1 TIER 3 Intervention *Math Triumphs, Alg. 2,* Ch. 1 and 2
Before Chapter Test	
Chapter Study Guide and Review **SE** Practice Test **SE** Standardized Test Practice **SE** Chapter Test connectED.mcgraw-hill.com Standardized Test Practice connectED.mcgraw-hill.com Vocabulary Review connectED.mcgraw-hill.com McGraw-Hill eAssessment	TIER 1 Intervention Skills Practice **CRM**, Ch. 1 connectED.mcgraw-hill.com TIER 2 Intervention Study Guide and Intervention **CRM**, Ch. 1 TIER 3 Intervention *Math Triumphs, Alg. 2,* Chs. 1 and 2

FORMATIVE ASSESSMENT

After Chapter 1	
Multiple-Choice Tests, Forms 1, 2A, 2B **CRM**, pp. 49–54 Free-Response Tests, Forms 2C, 2D, 3 **CRM**, pp. 55–60 Vocabulary Test **CRM**, p. 48 Extended Response Test **CRM**, p. 61 Standardized Test Practice **CRM**, pp. 62–64 McGraw-Hill eAssessment	Study Guide and Intervention **CRM**, Ch. 1 connectED.mcgraw-hill.com

SUMMATIVE ASSESSMENT

Option 1 Reaching All Learners

Kinesthetic Have pairs of students make up a set of real number cards. For each type of real number (rational, irrational, integer, natural, or whole), pairs should make up four cards that represent that type of real number. Then pairs will play a game of "Concentration." The first partner will turn over two cards. The two cards are removed if they represent the same type of real number. If not, both are turned back over. The other partner then takes a turn.

Interpersonal Place students in pairs or small groups and assign inequalities such as those in Exercises 10–21 on page 37 for them to solve. Tell students to use the Key Concept Summaries on pages 33 and 34 and discuss which property is best to use to solve the inequalities they have been assigned. Make sure all group members participate in the discussion.

Option 2 Approaching Level

Write $x - 6 = 15$ on the board. Review the Addition and Subtraction Properties of Equality as you solve for x. Then erase all the equals signs and replace them with $>$.

Discuss the similarities. Repeat, using $<$.

Option 3 English Learners

Have students work in pairs to read, discuss, and plan a solution strategy for real-world problems. This interaction can help students to identify individual difficulties with word problems and also to discover new strategies used by other students.

Option 4 Beyond Level

Ask groups of students to write their own set of instructions for solving one- and two-step inequalities, including when to reverse the inequality sign and how to tell when the graph begins with a circle or a dot.

Focus on Mathematical Content

VerticalAlignment

Before Chapter 1

Related Topics from Algebra 1

- use symbols to represent unknowns and variables
- use the Commutative, Associative, and Distributive Properties to simplify algebraic expressions
- formulate linear equations and inequalities to solve problems, and solve the equations and inequalities

Chapter 1

- use the properties of real numbers to evaluate expressions and formulas
- classify real numbers
- use the properties of equality to solve equations
- solve absolute value equations
- solve inequalities, compound inequalities, and absolute value inequalities

After Chapter 1

Preparation for Precalculus

- define functions, describe characteristics of functions, and translate among verbal, numerical, graphical, and symbolic representations of functions

Lesson-by-LessonPreview

1-1 Expressions and Formulas

An algebraic expression is an expression that contains at least one variable. It may also contain numbers and operations. When evaluating an algebraic expression, each variable must be replaced with a given value, and the order of operations must be followed. The order of operations is

- evaluate expressions inside grouping symbols,
- evaluate all powers,
- multiply and/or divide from left to right, and then
- add and/or subtract from left to right.

1-2 Properties of Real Numbers

Every real number corresponds to exactly one point on the number line, and every point on the number line represents exactly one real number. Real numbers are classified as either rational or irrational.

- A rational number can be expressed as a ratio $\frac{m}{n}$, where m and n are integers and $n \neq 0$.
- Any real number that is not rational is irrational.

Properties of real numbers are used to justify the steps used in solving equations and to describe mathematical relationships.

The properties of real numbers include the following:

- the Commutative Properties of Addition and Multiplication,
- the Associative Properties of Addition and Multiplication, and
- the Distributive Property.

The sets of real numbers include identity elements for the operations of addition and multiplication, an additive inverse for every real number, and a multiplicative inverse for every real number except 0.

1-3 Solving Equations

An equation is a mathematical sentence stating that two mathematical expressions are equal. Solving an equation requires a series of equations, equivalent to the given equation, that result in a final equation that isolates the variable on one side of the equals sign. The Properties of Equality can be used to solve equations. The Properties of Equality include

- the Reflexive Property of Equality,
- the Symmetric Property of Equality,
- the Transitive Property of Equality, and
- Substitution.

The Math Forum
@ Drexel University

Get educator support from the Math Forum's T2T. You may want to search, or ask, for graphing activities. Start your search at connectED.mcgraw-hill.com.

Most equations can be solved by adding the same number to each side of the equation, by subtracting the same number from each side, or by multiplying or dividing each side by the same nonzero number.

1-4 Solving Absolute Value Equations

The absolute value of a number is its distance from zero on a number line. The absolute value symbols are grouping symbols like parentheses. For example, to evaluate the expression $2 \cdot |15 - 31|$, first calculate inside the absolute value symbols.

$$2 \cdot |15 - 31| = 2 \cdot |-16|$$
$$= 2 \cdot (16)$$
$$= 32$$

The equation $|a - 6| = 4$ can be interpreted as *the distance between some number* a *and 6 is 4 units.* The value $a - 6$ can be 4 or -4. So, if $a - 6 = 4$, then $a = 10$. If $a - 6 = -4$, then $a = 2$. The solutions are 10 and 2. An equation like $|x| = -2$ is *never* true, so there is no solution.

1-5 Solving Inequalities

An open sentence that contains the symbol $<$, $\leq$, $>$, or $\geq$ is called an *inequality.* For any two real numbers, a and b, exactly one of the following statements is true: $a < b$, $a = b$, or $a > b$.

Solving an inequality involves using the Properties of Inequality to write a series of equivalent inequalities, ending with one that isolates the variable. These properties include:

- the Addition Property of Inequality,
- the Subtraction Property of Inequality,
- the Multiplication Property of Inequality, and
- the Division Property of Inequality.

In general, adding the same number to (or subtracting the same number from) each side of an inequality, or multiplying or dividing each side by a positive number, does not change the truth of an inequality. Multiplying or dividing an inequality by a negative number *reverses* the order of the inequality.

There are infinitely many solutions to an inequality. These solutions can be expressed in more than one way:

- **Graphs** When the solution is graphed on a number line, an open circle indicates a value that is not included in the solution, and a closed circle indicates a value that is included. Circles are used with $<$ and $>$, and dots are used with $\leq$ and $\geq$.

- **Set Builder Notation** Solutions can be written using set-builder notation. A solution such as $x \geq 4$ would be written $\{x \mid x \geq 4\}$, and read *the set of values* x *such that* x *is greater than or equal to 4.*

1-6 Solving Compound and Absolute Value Inequalities

An inequality that consists of two inequalities joined by the word *and* or the word *or* is called a *compound inequality.* To solve a compound inequality, each part of the inequality must be solved.

- The graph of a compound inequality that contains *and* is the *intersection* of the solution sets of the two inequalities.

- The graph of a compound inequality that contains *or* is the *union* of the solution sets of the two inequalities.

- An absolute value inequality using $<$ or $\leq$ is related to a compound inequality using the word *and.* For example, thinking of $|a| < 7$ as $|a - 0| < 7$, then the value of a is any number whose distance from 0 is less than 7 units.

Possible values for a

- An absolute value inequality using $>$ or $\geq$ is related to a compound inequality using the word *or.* For example, thinking of $|b| > 5$ as $|b - 0| > 5$, then the value of b is any number whose distance from 0 is greater than 5.

Possible values for b

Professional Development

Targeted professional development has been articulated throughout Algebra 2. More quality, customized professional development is available from McGraw-Hill Professional Development. Visit connectED.mcgraw-hill.com for details on each product.

- Online Lessons emphasize the strategies and techniques used to teach Algebra 2. Includes streaming video, interactive pages, and online tools.

- Video Workshops allow mentors, coaches, or leadership personnel to facilitate on-site workshops on educational strategies in mathematics and mathematical concepts.

- MHPD Online (www.mhpdonline.com) offers online professional development with video clips of instructional strategies, links, student activities, and news and issues in education.

- Teaching Today (teachingtoday.glencoe.com) gives secondary teachers practical strategies and materials that inspire excellence and innovation in teaching.

Chapter Project

Take Me Out to the Ball Game

Students use what they have learned about equations and inequalities to complete a project.

This chapter project addresses business literacy, as well as several specific skills identified as being essential to student success by the Framework for 21st Century Learning.

Visit connectED.mcgraw-hill.com for student and teacher handouts.

KeyVocabulary Introduce the key vocabulary in the chapter using the routine below.

Define: The absolute value of a number is its distance from zero on the number line, represented by $|x|$.

Example: $|-7| = 7$

Ask: Can the absolute value of a number ever be a negative number? Explain. No, because distance is nonnegative.

CHAPTER 1

Equations and Inequalities

Then

You wrote expressions with variables.

Now

You will:

- Simplify and evaluate algebraic expressions.
- Solve linear and absolute value equations.
- Solve and graph inequalities.

Why? ▲

MONEY Connecting money to mathematics is one of the most practical skills you can learn. As long as you use money, you will be using mathematics. In this chapter, you will explore money topics such as sales tax, income, and budgeting for your first apartment.

connectED.mcgraw-hill.com **Your Digital Math Portal**

| Animation | Vocabulary | eGlossary | Personal Tutor | Virtual Manipulatives | Graphing Calculator | Audio | Foldables | Self-Check Practice | Worksheets |

Jon Boyes/Photographer's Choice/Getty Images

Get Ready for the Chapter

Diagnose Readiness | You have two options for checking prerequisite skills.

1 **Textbook Option** Take the Quick Check below. Refer to the Quick Review for help.

QuickCheck	QuickReview

Simplify.

1. $15.7 + (-3.45)$ **12.25**
2. $-18.54 - (-32.05)$ **13.51**
3. $-9.8 \cdot 6.75$ **−66.15**
4. $4 \div (-0.5)$ **−8**
5. $3\frac{2}{3} + \left(-1\frac{4}{5}\right)$ **$1\frac{13}{15}$**
6. $\frac{54}{7} - \frac{26}{6}$ **$3\frac{8}{21}$**
7. $\left(\frac{6}{5}\right)\left(-\frac{10}{9}\right)$ **$-1\frac{1}{3}$**
8. $-3 \div \frac{7}{8}$ **$-3\frac{3}{7}$**
9. **CRAFTS** Felisa needs $\frac{7}{8}$ yard of one type of material to make a quilt. How much of this material will she need to make 12 quilts? **$10\frac{1}{2}$ yd**

Example 1 (Used in Lessons 1-1 through 1-3)

Simplify $\left(\frac{3}{16}\right)\left(-\frac{4}{5}\right)$.

$\left(\frac{3}{16}\right)\left(-\frac{4}{5}\right) = -\frac{3(4)}{16(5)}$ Multiply the numerators and the denominators.

$= -\frac{12}{80}$ Simplify.

$= -\frac{12 \div 4}{80 \div 4}$ Divide the numerator and denominator by the GCF, 4.

$= -\frac{3}{20}$ Simplify.

Evaluate each power.

10. 6^3 **216**
11. $(-4)^3$ **−64**
12. $-(0.6)^2$ **−0.36**
13. $-(-2.5)^3$ **15.625**
14. $\left(\frac{4}{5}\right)^2$ **$\frac{16}{25}$**
15. $\left(\frac{7}{3}\right)^4$ **$\frac{2401}{81}$**
16. $\left(-\frac{7}{10}\right)^2$ **$\frac{49}{100}$**
17. $-\left(\frac{15}{2}\right)^3$ **$-\frac{3375}{8}$**
18. **FOOD** Nate's Deli offers 3 types of bread, 3 types of meat, and 3 types of cheese. How many different sandwiches can be made with 1 type each of bread, meat, and cheese? **3^3 or 27**

Example 2 (Used in Lesson 1-1)

Evaluate $(-1.5)^3$.

$(-1.5)^3 = (-1.5)(-1.5)(-1.5)$ $(-1.5)^3$ means 1.5 is a factor 3 times.

$= -3.375$ Simplify.

Identify each statement as *true* or *false*.

19. $-6 \geq -7$ **true**
20. $8 > -5$ **true**
21. $\frac{1}{7} \leq \frac{1}{9}$ **false**
22. $\frac{5}{6} \leq \frac{25}{30}$ **true**
23. **MEASUREMENT** Christy has a board that is 0.6 yard long. Marissa has a board that is $\frac{2}{3}$ yard long. Marissa states that $\frac{2}{3} > 0.6$. Is she correct? **yes**

Example 3 (Used in Lesson 1-5)

Identify $\frac{3}{8} > \frac{12}{24}$ as *true* or *false*.

$\frac{3}{8} \overset{?}{>} \frac{12 \div 3}{24 \div 3}$ Divide 12 and 24 by 3 to get a denominator of 8.

$\frac{3}{8} \not> \frac{4}{8}$ Simplify.

False; $\frac{3}{8} \not> \frac{4}{8}$ because $\frac{3}{8} < \frac{4}{8}$.

2 **Online Option** Take an online self-check Chapter Readiness Quiz at connectED.mcgraw-hill.com.

3

Get Ready for the Chapter

Response to Intervention (RtI)
Use the *Quick Check* results and the Intervention Planner chart to help you determine your Response to Intervention. The If-Then statements in the chart below help you decide the appropriate tier of RtI and suggest intervention resources for each tier.

InterventionPlanner

TIER 1 **On Level** **OL**

If students miss about 25% of the exercises or less,

Then choose a resource:

connectED.mcgraw-hill.com   Self-Check Quiz

TIER 2 **Strategic Intervention** **AL**
approaching grade level

If students miss about 50% of the exercises,

Then choose a resource:

connectED.mcgraw-hill.com   Extra Examples, Personal Tutor, Homework Help

TIER 3 **Intensive Intervention**
2 or more grades below level

If students miss about 75% of the exercises,

Then use *Math Triumphs, Alg. 2,* Ch. 1 and 2

connectED.mcgraw-hill.com   Extra Examples, Personal Tutor, Homework Help, Review Vocabulary

Diagnose students' readiness for each chapter by using either the in-text **Quick Check** or the online **Chapter Readiness Quiz**. Then use the **Intervention Planner** to choose the correct program resource to reinforce each student's prerequisite skills.

E? **Essential Question**

- How are symbols useful in mathematics?
  Sample answer: Symbols allow you to express mathematical concepts in a condensed form.

Essential Questions frame each chapter. These open-ended questions guide student inquiry into the "big ideas" central to each chapter. Throughout the Student Edition, Exercises that refer to the chapter or follow-up Essential Questions are indicated by the icon **E?**.

- What mathematical symbols do you know?
  Sample answers: addition, subtraction, multiplication, division, equals sign, greater than, less than, grouping symbols, pi

FOLDABLES StudyOrganizer

Dinah Zike's Foldables®

Focus As students work through this chapter, they write notes and show examples about equations and inequalities on cards to be placed in the pockets of the Foldables.

Teach Have students make and label their Foldables as illustrated. At the end of each lesson, ask students to write about their experiences with real numbers, computational skills, equations, inequalities, or other concepts presented in the lessons. Have them create study cards with notes and examples and place them in the appropriate pocket.

When to Use It Encourage students to add to their Foldables as they work through the chapter and to use them to review for the chapter test.

Differentiated Instruction

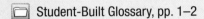 Student-Built Glossary, pp. 1–2

Students should complete the chart by providing a definition of each term and an example as they progress through Chapter 1. This study tool can also be used to review for the chapter test.

Get Started on the Chapter

You will learn several new concepts, skills, and vocabulary terms as you study Chapter 1. To get ready, identify important terms and organize your resources. You may wish to refer to Chapter 0 to review prerequisite skills.

FOLDABLES StudyOrganizer

Equations and Inequalities Make this Foldable to help you organize your Chapter 1 notes about equations and inequalities. Begin with one sheet of 11" × 17" paper.

1 **Fold** 2" tabs on each of the short sides.

2 **Then** fold in half in both directions. Open and cut as shown.

3 **Refold** along the width. Staple each pocket. Label pockets as *Algebraic Expressions, Properties of Real Numbers, Solving Equations,* and *Solve and Graph Inequalities.* Place index cards for notes in each pocket.

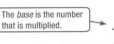

Foldables® are a unique way to enhance students' study skills. Encourage students to add to their Foldable as they work through the chapter and to use it for review for the chapter test.

NewVocabulary

English		Español
variable	p. 5	variable
algebraic expression	p. 5	expressión algebraica
order of operations	p. 5	orden de las operaciones
formula	p. 6	fórmula
real numbers	p. 11	números reales
rational numbers	p. 11	números racional
irrational numbers	p. 11	números irracional
integers	p. 11	enteros
whole numbers	p. 11	números enteros
natural numbers	p. 11	números naturales
open sentence	p. 18	enunciado abierto
equation	p. 18	ecuación
solution	p. 18	solución
absolute value	p. 27	valor absoluto
empty set	p. 28	conjunto vacío
set-builder notation	p. 35	notación de construcción de conjuntos
compound inequality	p. 41	desigualdad compuesta
intersection	p. 41	intersección
union	p. 42	unión

ReviewVocabulary

evaluate evaluar to find the value of an expression

inequality desigualdad an open sentence that contains the symbol $<$, $\leq$, $>$, or $\geq$

power potencia an expression of the form x^n, read *x to the nth power*

The *base* is the number that is multiplied.

The *exponent* tells how many times the base is used as a factor.

x^n

The number that can be expressed using an exponent is called a *power.*

LESSON 1-1

Expressions and Formulas

:: Then	:: Now	:: Why?
● You used the rules of exponents.	**1** Use the order of operations to evaluate expressions. **2** Use formulas.	● The following formula can be used to calculate a baseball player's on-base percentage x.

$$x = \frac{h + w + p}{b + w + p + s}$$

- h is the number of hits.
- w is the number of walks.
- p is the number of times the player has been hit by a pitch.
- b is the number of times at bat.
- s is the number of sacrifice flies and sacrifice bunts.

During the first twenty games of a season, Ian has 9 hits, 2 walks, 38 at bats, 5 sacrifice flies, and he is hit by 1 pitch. The expression $\frac{9 + 2 + 1}{38 + 2 + 1 + 5}$ gives Ian's on-base percentage.

> **New Vocabulary** is listed at the beginning of every lesson.

 NewVocabulary
variables
algebraic expressions
order of operations
formula

 Common Core State Standards

Content Standards
A.SSE.1.a Interpret parts of an expression, such as terms, factors, and coefficients.

A.SSE.1.b Interpret complicated expressions by viewing one or more of their parts as a single entity.

Mathematical Practices
1 Make sense of problems and persevere in solving them.

1 Order of Operations **Variables** are letters used to represent unknown quantities. Expressions that contain at least one variable are called **algebraic expressions**. You can evaluate an algebraic expression by replacing each variable with a number and then applying the **order of operations**.

KeyConcept Order of Operations

Step 1 Evaluate the expressions inside grouping symbols.

Step 2 Evaluate all powers.

Step 3 Multiply and/or divide from left to right.

Step 4 Add and/or subtract from left to right.

Example 1 Evaluate Algebraic Expressions

Evaluate $m + (p - 1)^2$ if $m = 3$ and $p = -4$.

$m + (p - 1)^2 = 3 + (-4 - 1)^2$ Replace m with 3 and p with −4.

$\quad\quad\quad\quad = 3 + (-5)^2$ Add −4 and −1.

$\quad\quad\quad\quad = 3 + 25$ Evaluate $(-5)^2$.

$\quad\quad\quad\quad = 28$ Add 3 and 25.

▶ **Guided**Practice

Evaluate each expression if $m = 12$ and $q = -1$.

1A. $m + (3 - q)^2$ 28   **1B.** $m \div 2q + 4$ −2

 connectED.mcgraw-hill.com   **5**

Taxi/Getty Images

1 Focus

VerticalAlignment

▼ **Before Lesson 1-1** Use the rules of exponents.

▼ **Lesson 1-1** Use the order of operations to evaluate expressions. Use formulas.

▼ **After Lesson 1-1** Solve linear equations.

2 Teach

Scaffolding Questions
Have students read the **Why?** section of the lesson.

Ask:
- What does the variable b represent? number of times at bat

- What is Ian's on-base percentage? 0.261

- How would Ian's on-base percentage change if he had not been hit by any pitches? It would decrease from 0.261 to 0.244.

- Why does the formula represent on-base percentage? The numerator represents all the ways that a batter can reach first base. The denominator represents all the batter's plate appearances. (Note that at-bats does not include walks, hits by pitches, sacrifice flies, or sacrifice bunts.)

Lesson 1-1 Resources

Resource	Approaching Level **AL**	On Level **OL**	Beyond Level **BL**	English Learners **ELL**
Teacher Edition	■ Differentiated Instruction, p. 6	■ Differentiated Instruction, pp. 6, 8, 10	■ Differentiated Instruction, pp. 8, 10	■ Differentiated Instruction, p. 6
Chapter Resource Masters	■ Study Guide and Intervention, pp. 5–6 ■ Skills Practice, p. 7 ■ Practice, p. 8 ■ Word Problem Practice, p. 9	■ Study Guide and Intervention, pp. 5–6 ■ Skills Practice, p. 7 ■ Practice, p. 8 ■ Word Problem Practice, p. 9 ■ Enrichment, p. 10	■ Practice, p. 8 ■ Word Problem Practice, p. 9 ■ Enrichment, p. 10	■ Study Guide and Intervention, pp. 5–6 ■ Skills Practice, p. 7 ■ Practice, p. 8 ■ Word Problem Practice, p. 9
Other	■ 5-Minute Check 1-1 ■ Study Notebook	■ 5-Minute Check 1-1 ■ Study Notebook	■ 5-Minute Check 1-1 ■ Study Notebook	■ 5-Minute Check 1-1 ■ Study Notebook

1 Order of Operations

Examples 1 and 2 show how to evaluate three different algebraic expressions by replacing the variables with numbers and then applying the order of operations.

Teach with Tech

Student Response System Write expressions on the board that use several different operations (such as $6 \times [5 - 4] \div 2 + 1$). Ask students which operation should be performed first. Give them a numerical key to use for their responses (for example, 1 for $+$, 2 for $-$, 3 for $\times$, and 4 for $\div$).

Formative Assessment

Use the Guided Practice exercises after each example to determine students' understanding of concepts.

Additional Examples

1 Evaluate $(x - y)^3 + 3$ if $x = 1$ and $y = 4$. -24

2 a. Evaluate $n - t(n^2 - t)$ if $n = 2$ and $t = 3.4$. -0.04

   b. Evaluate $\dfrac{8xy + z^3}{y^2 + 5}$ if $x = 5$, $y = -2$, and $z = -1$. -9

Additional Examples also in Interactive Classroom PowerPoint® Presentations

 IWB Interactive White Board READY

 Teaching the Mathematical Practices

Structure Mathematically proficient students look closely to discern structure. They can see complicated things as being composed of several objects. Remind students that the value of an expression can be determined only if the order of operations is followed.

 Example 2 Evaluate Algebraic Expressions

a. Evaluate $a + b^2(b - a)$ if $a = 5$ and $b = -3.2$.

$$
\begin{aligned}
a + b^2(b - a) &= 5 + (-3.2)^2(-3.2 - 5) &&\quad a = 5 \text{ and } b = -3.2 \\
&= 5 + (-3.2)^2(-8.2) &&\quad \text{Subtract 5 from } -3.2. \\
&= 5 + 10.24(-8.2) &&\quad \text{Evaluate } (-3.2)^2. \\
&= 5 + (-83.968) &&\quad \text{Multiply 10.24 and } -8.2. \\
&= -78.968 &&\quad \text{Add 5 and } -83.968.
\end{aligned}
$$

> **StudyTip**
>
> **CCSS** Structure
>
> Remember that a fraction bar is a type of grouping symbol. Evaluate the expressions in the numerator and denominator separately before dividing.

b. Evaluate $\dfrac{x^4 - 3wy}{y^3 + 2w}$ if $w = 4$, $x = -3$, and $y = -5$.

$$
\begin{aligned}
\frac{x^4 - 3wy}{y^3 + 2w} &= \frac{(-3)^4 - 3(4)(-5)}{(-5)^3 + 2(4)} &&\quad w = 4, x = -3, \text{ and } y = -5 \\
&= \frac{81 - 3(4)(-5)}{-125 + 2(4)} &&\quad \text{Evaluate the numerator and denominator separately.} \\
&= \frac{81 - (-60)}{-125 + 8} &&\quad \text{Multiply in the numerator and denominator.} \\
&= \frac{141}{-117} \text{ or } -\frac{47}{39} &&\quad \text{Simplify the numerator and denominator. Then simplify the fraction.}
\end{aligned}
$$

> **Guided**Practice

Evaluate each expression if $h = 4$, $j = -1$, and $k = 0.5$.

2A. $h^2k + h(h - k)$ 22   **2B.** $j + (3 - h)^2$ 0   **2C.** $\dfrac{j^2 - 3h^2k}{j^3 + 2}$ -23

---

2 Formulas A **formula** is a mathematical sentence that expresses the relationship between certain quantities. If you know the value of every variable in the formula except one, you can find the value of the remaining variable.

> **Reading**Math
>
> Exponents x^2 may be read as *x squared* or *x to the second power*. x^3 may be read as *x cubed* or *x to the third power*.

 Real-World Example 3 Use a Formula

TORNADOES The formula for the volume of a cone, $V = \frac{1}{3}\pi r^2 h$, can be used to approximate the volume of a tornado. Find the approximate volume of the tornado at the right.

$$
\begin{aligned}
V &= \tfrac{1}{3}\pi r^2 h &&\quad \text{Volume of a cone} \\
&= \tfrac{1}{3}\pi(75)^2(225) &&\quad r = 75 \text{ and } h = 225 \\
&= \tfrac{1}{3}\pi(5625)(225) &&\quad \text{Evaluate } 75^2. \\
&\approx 1{,}325{,}359 &&\quad \text{Multiply.}
\end{aligned}
$$

75 m

225 m

The approximate volume of the tornado is about 1,325,359 cubic meters.

> **Guided**Practice

3. GEOMETRY The formula for the volume V of a rectangular prism is $V = \ell wh$, where ℓ represents the length, w represents the width, and h represents the height. Find the volume of a rectangular prism with a length of 4 feet, a width of 2 feet, and a height of 3.5 feet. 28 ft³

Differentiated Instruction AL OL ELL

If students have trouble memorizing the order of operations,

Then ask them to create a song, rhyme, or mnemonic device that will help them to remember the correct order of operations. **P**lease **E**xcuse **M**y **D**ear **A**unt **S**ally is an example of a mnemonic device.

Example 1 Evaluate each expression if $a = -2$, $b = 3$, and $c = 4.2$.

1. $a - 2b + 3c$ 4.6   **2.** $2a + (b + 3)^2$ 32   **3.** $a + 3[b^2 - (a + c)]$ 18.4

Example 2 **4.** $5c - 2[(b - a) + c]$ 2.6   **5.** $4(2a + 3b) - 2c$ 11.6   **6.** $\dfrac{a^2 + 4c}{3b + 2a}$ 4.16

7. $\dfrac{b^3 + ac}{ab + 2bc}$ 0.96875   **8.** $\dfrac{3b + 2a}{5 - c}$ 6.25   **9.** $\dfrac{3a - 2c}{4ab}$ 0.6

Example 3 **10. VOLLEYBALL** A player's attack percentage A is calculated using the formula $A = \dfrac{k - e}{t}$, where k represents the number of kills, e represents the number of attack errors including blocks, and t represents the total attacks attempted. Find the attack percentage given each set of values.

    a. $k = 22$, $e = 11$, $t = 35$ **b.** $k = 33$, $e = 9$, $t = 50$ 0.48 or 48%
      about 0.314 or 31.4%

Practice and Problem Solving   Extra Practice is on page R1.

Example 1 Evaluate each expression if $w = -3$, $x = 4$, $y = 2.6$, and $z = \frac{1}{3}$.

11. $y + x - z$ $6\frac{4}{15}$ **12.** $w - 2x + y \div 2$ -9.7 **13.** $4(x - w)$ 28

14. $6(y + x)$ 39.6   **15.** $9z - 4y + 2w$ -13.4 **16.** $3y - 4z + x$ $10\frac{7}{15}$

17. GAS MILEAGE The gasoline used by a car is measured in miles per gallon and is related to the distance traveled by the following formula.

         miles per gallon × number of gallons = distance traveled

    a. During a trip your car used a total of 46.2 gallons of gasoline. If your car gets 33 miles to the gallon, how far did you travel? 1524.6 mi

    b. Your friend has decided to buy a hybrid car that gets 60 miles per gallon. The gasoline tank holds 12 gallons. How far can the car go on one tank of gasoline? 720 mi

Example 2 Evaluate each expression if $a = -4$, $b = -0.8$, $c = 5$, and $d = \frac{1}{5}$.

18. $\dfrac{a + b}{c - d}$ -1 **19.** $\dfrac{a - b}{bd}$ 20   **20.** $\dfrac{ac}{d + b}$ $33\frac{1}{3}$

21 $\dfrac{b^2 c^2}{ad}$ -20 **22.** $\dfrac{b + 6}{4(d + c)}$ 0.25   **23.** $\dfrac{5(d + a)}{2ab^2}$ ≈ 3.71

24b. Yes; Sample answer: 98.6°F = 37°C, so a temperature above 37°C indicates a fever.

24. **SENSE-MAKING** The formula $C = \dfrac{5(F - 32)}{9}$ can be used to convert temperatures in degrees Fahrenheit to degrees Celsius.

    a. Room temperature commonly ranges from 64°F to 73°F. Determine the room temperature range in degrees Celsius. 17.8°C to 22.8°C

    b. The normal average human body temperature is 98.6°F. A temperature above this indicates a fever. If your temperature is 42°C, do you have a fever? Explain your reasoning.

Example 3 **25. GEOMETRY** The formula for the area A of a triangle with height h and base b is $A = \frac{1}{2}bh$. Write an expression to represent the area of the triangle. $\frac{1}{2}(x + 7)(2x)$

B **26. FINANCIAL LITERACY** The profit that a business made during a year is $536,897,000. If the business divides the profit evenly for each share, estimate how much each share made if there are 10,995,000 shares. $48.83

---

2 Formulas

Example 3 shows how to find the value of a variable in a formula when the values of all the other variables are known.

Additional Example

3 **GEOMETRY** The formula for the area A of a trapezoid is $A = \frac{1}{2}h(b_1 + b_2)$, where h represents the height, and b_1 and b_2 represent the measures of the bases. Find the area of a trapezoid with base lengths of 13 meters and 25 meters and a height of 8 meters. 152 m²

---

3 Practice

Formative Assessment
Use Exercises 1–10 to check for understanding.

Use the chart at the bottom of this page to customize assignments for your students.

 Teaching the Mathematical Practices

Sense-Making Mathematically proficient students start by explaining to themselves the meaning of a problem and looking for entry points to its solution. They analyze givens, constraints, relationships, and goals, and make conjectures about the form and meaning of the solution and plan a solution pathway rather than simply jumping into a solution attempt.

> The **Differentiated Homework Options** provide leveled assignments. Many of the homework exercises are paired, so that students can do the odds one day and the evens the next day.

Differentiated Homework Options

Level	Assignment	Two-Day Option	
AL Basic	11–25, 43, 45, 47–67	11–25 odd, 50–53	12–24 even, 43, 45, 47–49, 54–67
OL Core	11–25 odd, 26–28, 29–33 odd, 35–43, 45, 47–67	11–25, 50–53	26–43, 45, 47–49, 54–67
BL Advanced	26–59, (optional: 60–67)		

Multiple Representations

In Exercise 42, students use geometric drawings, a ruler, a table of values, and an algebraic expression to compare the volumes of two cylinders.

CCSS Teaching the Mathematical Practices

Reasoning Mathematically proficient students make sense of quantities and their relationships in problem situations. Quantitative reasoning entails habits of creating a coherent representation of the problem at hand; considering the units involved; attending to the meaning of quantities, not just how to compute them; and knowing and flexibly using different properties of operations and objects.

Additional Answers

42a. Sample answer:

42b. Sample answer:

cylinder	radius	height	volume
1	2 in.	5 in.	$20\pi \approx 62.8$ in^3
2	4 in.	1 in.	$16\pi \approx 50.3$ in^3

45. Sample answer: b; Since t is time, it must be nonnegative. So $-2t(2t + 1)$ will be negative for all values of t other than 0. The maximum value of $-2t(2t + 1)$ is 0, which occurs when $t = 0$. Thus, the maximum value of $-2t(2t + 1) + 6$ is 6.

 Follow-up

Students have explored absolute value.

Ask:
- Why is it important to understand what the symbols in a formula represent? Sample answer: If you are unsure of what the symbols represent, you may not use the right formula or you may use the right formula, but not in the right way. You also may not know if the answer that you find is reasonable after using the formula.

27. CCSS REASONING The radius of Earth's orbit is 93,000,000 miles.

  a. Find the circumference of Earth's orbit assuming that the orbit is a circle. The formula for the circumference of a circle is $2\pi r$. **584,336,233.6 mi**

  b. Earth travels at a speed of 66,698 miles per hour around the Sun. Use the formula $T = \frac{C}{V}$, where T is time in hours, C is circumference, and V is velocity to find the number of hours it takes Earth to revolve around the Sun. **8761 h**

27c. Yes; $\frac{8761}{24}$ = 365 days or 1 year.

  c. Did you prove that it takes 1 year for Earth to go around the Sun? Explain.

28. ANCIENT PYRAMID The Great Pyramid in Cairo, Egypt, is approximately 146.7 meters high, and each side of its base is approximately 230 meters.

  a. Find the area of the base of the pyramid. Remember $A = \ell w$. **52,900 m^2**

  b. The volume of a pyramid is $\frac{1}{3}Bh$, where B is the area of the base and h is the height. What is the volume of the Great Pyramid? **2,586,810 m^3**

Evaluate each expression if $w = \frac{3}{4}$, $x = 8$, $y = -2$, and $z = 0.4$.

29. $x^3 + 2y^4$ **544**   **30.** $(x - 6z)^2$ **31.36**   **31.** $2(6w - 2y) - 8z$ **13.8**

32. $\frac{(y + z)^2}{xw}$ **≈0.427**   **33.** $\frac{12w - 6y}{z^2}$ **131.25**   **34.** $\frac{wx + yz}{wx - yz}$ **0.765**

35. GEOMETRY The formula for the volume V of a cone with radius r and height h is $V = \frac{1}{3}\pi r^2 h$. Write an expression for the volume of the cone at the right. **$6\pi x^3$**

36. SEARCH ENGINES Page rank is a numerical value that represents how important a page is on the Web. One formula used to calculate the page rank for a page is $PR = 0.15 + 0.85L$, where L is the page rank of the linking page divided by the number of outbound links on the page. Determine the page rank of a page in which $L = 10$. **8.65**

37 WEATHER In 1898, A.E. Dolbear studied various species of crickets to determine their "chirp rate" based on temperatures. He determined that the formula $t = 50 + \frac{n - 40}{4}$, where n is the number of chirps per minute, could be used to find the temperature t in degrees Fahrenheit. What is the temperature if the number of chirps is 120? **70°F**

 38. FOOTBALL The following formula can be used to calculate a quarterback efficiency rating.

$$\left(\frac{\frac{C}{A} - 0.3}{0.2} + \frac{\frac{Y}{A} - 3}{4} + \frac{\frac{T}{A}}{0.05} + \frac{0.095 - \frac{I}{A}}{0.04}\right) \cdot \frac{100}{6}$$

  - C is the number of passes completed.
  - A is the number of passes attempted.
  - Y is passing yardage.
  - T is the number of touchdown passes.
  - I is the number of interceptions.

  Find Peyton Manning's efficiency rating to the nearest tenth for the season statistics shown. **95.0**

Peyton Manning
2008 Quarterback
Indianapolis Colts

STATISTICS

Passes Completed	371
Passes Attempted	555
Passing Yardage	4002
Touchdown Passes	27
Interceptions	12

39b. $4.42; $6.62; $11.62; Sample answer: The average prices found in part a become increasingly higher with time.

39. MOVIES The average price for a movie ticket can be represented by $P = \frac{y^2}{400} + \frac{7y}{100} + 2.96$ where y is the number of years since 1980.

  a. Find the average price of a ticket in 1990, 2000, and 2010. **$3.91; $5.36; $7.31**

  b. Another equation that can be used to represent ticket prices is $P = \frac{y^3}{2500} - \frac{y^2}{100} + \frac{6y}{25} + 2.62$. Find the price of a ticket in 1990, 2000, and 2010. How do these values compare to those you found in part **a**?

DifferentiatedInstruction OL BL

Extension Challenge students by having them convert square and cubic units. After completing Exercise 28a, convert the area of the base to square centimeters and square kilometers. **529,000,000 cm^2; 0.0529 km^2** After completing Exercise 28b, convert the volume to cubic centimeters and cubic kilometers. **2,586,810,000,000 cm^3; 0.00258681 km^3**

Differentiated Instruction: Extension activities help you cultivate skills that students will need to have success in higher mathematics.

40. GEOMETRY The area of a triangle can be found using Heron's Formula, $A = \sqrt{s(s-a)(s-b)(s-c)}$, where a, b, and c are the lengths of the three sides of the triangle, and $s = \dfrac{a+b+c}{2}$. Find the area of the triangle at the right. **31.5 in²**

6 in.
11 in.
14 in.

41 Evaluate $y = \sqrt{b^2\left(1 - \dfrac{x^2}{a^2}\right)}$ if $a = 6$, $b = 8$, and $x = 3$. Round to the nearest tenth. **6.9**

42. MULTIPLE REPRESENTATIONS You will write expressions using the formula for the volume of a cylinder. Recall that the volume of a cylinder can be found using the formula $v = \pi r^2 h$, in which v = volume, r = radius, and h = height.

42c. Sample answer: π times 2 squared times 5 minus π times 4 squared times 1

a. **Geometric** Draw two cylinders of different sizes. **See margin.**

b. **Tabular** Use a ruler to measure the radius and height of each cylinder. Organize the measures for each cylinder into a table. Include a column in your table to calculate the volume of each cylinder. **See margin.**

c. **Verbal** Write a verbal expression for the difference in volume of the two cylinders.

d. **Algebraic** Write and solve an algebraic expression for the difference in volume of the two cylinders. **Sample answer:** $\pi(2)^2(5) - \pi(4)^2(1) = 4\pi \approx 12.5 \text{ in}^3$

H.O.T. Problems   Use Higher-Order Thinking Skills

46. Halfway between $\dfrac{m}{n}$ and $\dfrac{p}{q}$ is the average of the two: $\dfrac{1}{2}\left(\dfrac{m}{n} + \dfrac{p}{q}\right) = \dfrac{qm+pn}{2nq}$.

48. Sample answer: A formula is used to calculate the price of filling a gasoline tank in which the price = number of gallons × price per gallon. If calculated incorrectly, you may underestimate or overestimate how much you will need to pay.

43. CCSS CRITIQUE Lauren and Rico are evaluating $\dfrac{-3d - 4c}{2ab}$ for $a = -2$, $b = -3$, $c = 5$, and $d = 4$. Is either of them correct? Explain your reasoning. **Lauren; $-12 - 20 = -32$**

Lauren
$\dfrac{-3d - 4c}{2ab} = \dfrac{-3(4) - 4(5)}{2(-2)(-3)}$
$= \dfrac{-12 - 20}{12} = \dfrac{-32}{12} = -\dfrac{8}{3}$

Rico
$\dfrac{-3d - 4c}{2ab} = \dfrac{-3(4) - 4(5)}{2(-2)(-3)}$
$= \dfrac{-12 - 20}{12} = \dfrac{8}{12} = \dfrac{2}{3}$

44. CHALLENGE For any three distinct numbers a, b, and c, $a\$b\$c$ is defined as $a\$b\$c = \dfrac{-a - b - c}{c - b - a}$. Find $-2\$(-4)\$5$. $\dfrac{1}{11}$

45. REASONING The following equivalent expressions represent the height in feet of a stone thrown downward off a bridge where t is the time in seconds after release. Which do you find most useful for finding the maximum height of the stone? Explain. **See margin.**

a. $-4t^2 - 2t + 6$ b. $-2t(2t + 1) + 6$ c. $-2(t - 1)(2t + 3)$

46. CHALLENGE Let m, n, p, and q represent nonzero positive integers. Find a number in terms of m, n, p, and q that is halfway between $\dfrac{m}{n}$ and $\dfrac{p}{q}$.

47. OPEN ENDED Write an algebraic expression using $x = -2$, $y = -3$, and $z = 4$ and all four operations for which the value of the expression is 10. **Sample answer:** $y\left(\dfrac{-4z}{x^2} - x\right) + z$

48. WRITING IN MATH Provide an example of a formula used in everyday situations. Explain its usefulness and what happens if the formula is not used correctly.

49. WRITING IN MATH Use the information for on-base percentage given at the beginning of the lesson to explain why a formula for on-base percentage is more useful than a table of specific percentages. **A table of on-base percentages is limited to those situations listed, while a formula can be used to find any on-base percentage.**

A **Study Guide and Intervention, Practice, Word Problem Practice,** and **Enrichment Master** are shown for every lesson. These masters can be found in the Chapter Resource Masters.

CCSS Teaching the Mathematical Practices

Critique Mathematically proficient students are also able to compare the effectiveness of two plausible arguments, distinguish correct logic or reasoning from that which is flawed, and—if there is a flaw in an argument—explain what it is.

Study Guide and Intervention

Practice

Enrichment

Word Problem Practice

Standardized Test Practice exercises help students solidify their knowledge of the standards using exercises in a multiple-choice format.

WatchOut!

Preventing Errors Students may be reluctant to take time to show all the steps they use when evaluating an expression, such as showing the substituted values before doing the computations. Help them to see that these steps enable them to diagnose errors and prevent calculation errors that may keep them from getting correct values.

4 Assess

Name the Math Have each student tell a partner or write the order of operations.

The **Four-Step Teaching Plan** shows you how to **Focus, Teach, Practice,** and **Assess** each lesson. Each lesson ends with a creative strategy for closing the lesson.

Standardized Test Practice

50. SAT/ACT If the area of a square with side x is 9, what is the area of a square of side $4x$? **B**

   A 36   **D** 324

   B 144   **E** 1296

   C 212

51. SHORT RESPONSE A coffee shop owner wants to open a second shop when his daily customer average reaches 800 people. He has calculated the daily customer average in the table below for each month since he has opened.

Month	Daily Customer Average
1	225
2	298
3	371
4	444

If the trend continues, during what month can he open a second shop? **month 9**

52. GEOMETRY In $\triangle DFG$, $\overline{FH}$ and $\overline{HG}$ are angle bisectors and $m\angle D = 84°$. How many degrees are in $\angle FHG$? **G**

   F 96

   G 132

   H 145

   J 192

53. A skydiver in a computer game free-falls from a height of 3000 meters at a rate of 55 meters per second. Which equation can be used to find h, the height of the skydiver after t seconds of free fall? **B**

   A $h = -55t - 3000$

   B $h = -55t + 3000$

   C $h = 3000t - 55$

   D $h = 3000t + 55$

Spiral Review

54. The lengths of the three sides of a triangle are 10, 14, and 18 inches. Determine whether this triangle is a right triangle. (Lesson 0-8) **no**

55. The legs of a right triangle measure 6 centimeters and 8 centimeters. Find the length of the hypotenuse. (Lesson 0-8) **10 cm**

56. MAPS On a map of the U.S., the cities of Milwaukee, Wisconsin, and Charlotte, North Carolina, are $6\frac{1}{2}$ inches apart. The actual distance between Milwaukee and Charlotte is 670 miles. If Birmingham, Alabama, and St. Petersburg, Florida, are 465 miles apart, how far apart are they on the map? (Lesson 0-7) **about $4\frac{1}{2}$ in.**

57. Factor $6x^2 + 12x$. (Lesson 0-3) **$6x(x + 2)$**

58. Find the product of $(a + 2)(a - 4)$. (Lesson 0-2) **$a^2 - 2a - 8$**

59. NUMBER An integer is 2 less than a number, and another integer is 1 greater than double that same number. What are the two integers if their sum is 14? (Lesson 0-2) **3 and 11**

Skills Review

Evaluate each expression.

60. $\sqrt{4}$ **2**

61. $\sqrt{25}$ **5**

62. $\sqrt{81}$ **9**

63. $\sqrt{121}$ **11**

64. $-\sqrt{9}$ **−3**

65. $-\sqrt{16}$ **−4**

66. $\sqrt{\dfrac{49}{100}}$ **$\dfrac{7}{10}$**

67. $\sqrt{\dfrac{25}{64}}$ **$\dfrac{5}{8}$**

 10 | Lesson 1-1 | Expressions and Formulas

DifferentiatedInstruction **OL BL**

Extension Write five numbers, all of which are between 1 and 10, on the board. Give students a "power number" between 1 and 10. Tell them to use each of the five numbers exactly once along with any operation and grouping symbols to write an expression that will make the "power number." For example, suppose the five numbers are 2, 3, 4, 7, and 8 and the "power number" is 1. Students can produce the "power number" by using the expression $4 + 7 - 2(8 - 3)$.

1-2 Properties of Real Numbers

:: Then	:: Now	:: Why?
• You identified and used the arithmetic properties of real numbers.	**1** Classify real numbers. **2** Use the properties of real numbers to evaluate expressions.	• The Central High School Boosters sell snacks and beverages at school functions. The items are priced the same to make determining the total cost easy. You can use the Distributive Property to calculate the total cost when multiple items are purchased.

 NewVocabulary
real numbers
rational numbers
irrational numbers
integers
whole numbers
natural numbers

Common Core State Standards

Content Standards
A.SSE.2 Use the structure of an expression to identify ways to rewrite it.

Mathematical Practices
2 Reason abstractly and quantitatively.
7 Look for and make use of structure.

1 Real Numbers **Real numbers** consist of several different kinds of numbers.

- **Rational numbers** can be expressed as a ratio $\frac{a}{b}$, where a and b are integers and b is not zero. The decimal form of a rational number is either a terminating or repeating decimal.

- The decimal form of an **irrational number** neither terminates nor repeats. Square roots of numbers that are not perfect squares are irrational numbers.

- The sets of **integers**, $\{\ldots, -3, -2, -1, 0, 1, 2, 3, \ldots\}$, **whole numbers**, $\{0, 1, 2, 3, 4, \ldots\}$, and **natural numbers**, $\{1, 2, 3, 4, 5, \ldots\}$, are subsets of the rational numbers. These numbers are subsets of the rational numbers because every integer n is equal to $\frac{n}{1}$.

KeyConcept Real Numbers (R)

Letter	Set	Examples
Q	rationals	$0.125, -\frac{7}{8}, \frac{2}{3} = 0.66\ldots$
I	irrationals	$\pi = 3.14159 \ldots$ $\sqrt{3} = 1.73205 \ldots$
Z	integers	$-5, 17, -23, 8$
W	wholes	$2, 96, 0, \sqrt{36}$
N	naturals	$3, 17, 6, 86$

Example 1 Classify Numbers

Name the sets of numbers to which each number belongs.

a. -23 integers (Z), rationals (Q), reals (R)

b. $\sqrt{50}$ irrationals (I), reals (R)

c. $-\frac{4}{9}$ rationals (Q), reals (R)

▸ **GuidedPractice**

1A. -185 Z, Q, R   **1B.** $-\sqrt{49}$ Z, Q, R   **1C.** $\sqrt{95}$ I, R   **1D.** $-\frac{7}{8}$ Q, R

connectED.mcgraw-hill.com   **11**

---

1 Focus

VerticalAlignment

Before Lesson 1-2 Identify and use the arithmetic properties of real numbers.

Lesson 1-2 Classify real numbers. Use the properties of real numbers to evaluate expressions.

After Lesson 1-2 Use real number properties to solve equations and inequalities.

2 Teach

Scaffolding Questions
Have students read the **Why?** section of the lesson.

Ask:
- If each item costs $1.50, how much would four items cost? $6.00

- Write an expression using the Distributive Property that could be used to calculate the cost of two protein bars and one popcorn if each item costs $1.50. $2(1.50) + 1.50$ or $3(1.50)$

Scaffolding Questions give direction and momentum to the lesson, clarify its purpose, and keep students on task.

Lesson 1-2 Resources

Resource	Approaching Level **AL**	On Level **OL**	Beyond Level **BL**	English Learners **ELL**
Teacher Edition	• Differentiated Instruction, p. 13	• Differentiated Instruction, pp. 12, 13, 17	• Differentiated Instruction, pp. 12, 17	• Differentiated Instruction, p. 13
Chapter Resource Masters	• Study Guide and Intervention, pp. 11–12 • Skills Practice, p. 13 • Practice, p. 14 • Word Problem Practice, p. 15	• Study Guide and Intervention, pp. 11–12 • Skills Practice, p. 13 • Practice, p. 14 • Word Problem Practice, p. 15 • Enrichment, p. 16	• Practice, p. 14 • Word Problem Practice, p. 15 • Enrichment, p. 16	• Study Guide and Intervention, pp. 11–12 • Skills Practice, p. 13 • Practice, p. 14 • Word Problem Practice, p. 15
Other	• 5-Minute Check 1-2 • Study Notebook	• 5-Minute Check 1-2 • Study Notebook	• 5-Minute Check 1-2 • Study Notebook	• 5-Minute Check 1-2 • Study Notebook

Eisenhut & Mayer Wien/FoodPix/Getty Images

1 Real Numbers

Example 1 shows how to classify numbers.

Formative Assessment

Use the Guided Practice exercises after each example to determine students' understanding of concepts.

Additional Example

1 Name the sets of numbers to which each number belongs.

a. $\sqrt{6}$ I, R

b. 5 N, W, Z, Q, R

c. $-\dfrac{2}{3}$ Q, R

Additional Examples also in Interactive Classroom PowerPoint® Presentations

IWB Interactive White Board READY

Additional Examples parallel the examples in the text exactly. Step-by-step solutions for these examples are included in **Interactive Classroom.**

Teach with Tech

Interactive Whiteboard Draw a set diagram on the board showing how the set of real numbers is separated into rational and irrational numbers, integers, whole numbers, etc. Write a list of 12 real numbers, and choose students to come to the board to drag them into the correct set in the diagram.

WatchOut!

Student Misconceptions Remind students that $\sqrt{9}$, for example, means only the positive root, if one exists. So, $\sqrt{9} = 3$. To indicate both roots of the equation $x^2 = 9$, the mathematical notation is $x = \pm\sqrt{9}$ or $x = \pm 3$.

StudyTip

Real Numbers A number can belong to more than one set of numbers. For example, if a number is natural, it is also whole, an integer, rational, and real.

2 Properties of Real Numbers

Some of the properties of real numbers are summarized below.

ConceptSummary Real Number Properties

For any real numbers a, b, and c:

Property	Addition	Multiplication
Commutative	$a + b = b + a$	$a \cdot b = b \cdot a$
Associative	$(a + b) + c = a + (b + c)$	$(a \cdot b) \cdot c = a \cdot (b \cdot c)$
Identity	$a + 0 = a = 0 + a$	$a \cdot 1 = a = 1 \cdot a$
Inverse	$a + (-a) = 0 = (-a) + a$	$a \cdot \dfrac{1}{a} = 1 = \dfrac{1}{a} \cdot a, a \neq 0$
Closure	$a + b$ is a real number.	$a \cdot b$ is a real number.
Distributive	$a(b + c) = ab + ac$ and $(b + c)a = ba + ca$	

Example 2 Name Properties of Real Numbers

Name the property illustrated by $5 \cdot (4 \cdot 13) = (5 \cdot 4) \cdot 13$.

Associative Property of Multiplication

The Associative Property of Multiplication states that the way in which you group factors does not affect the product.

> **Guided**Practice

2. Name the property illustrated by $2(x + 3) = 2x + 6$. **Distributive Property**

You can use the properties of real numbers to identify related values.

StudyTip

Additive and Multiplicative Inverses The additive inverse of a number has the opposite sign as the number. The multiplicative inverse of a number has the same sign as the number.

Example 3 Additive and Multiplicative Inverses

Find the additive inverse and multiplicative inverse for $-\dfrac{5}{8}$.

Since $-\dfrac{5}{8} + \dfrac{5}{8} = 0$, the additive inverse of $-\dfrac{5}{8}$ is $\dfrac{5}{8}$.

Since $\left(-\dfrac{5}{8}\right)\left(-\dfrac{8}{5}\right) = 1$, the multiplicative inverse of $-\dfrac{5}{8}$ is $-\dfrac{8}{5}$.

> **Guided**Practice

Find the additive and multiplicative inverse for each number.

3A. 1.25 $-1.25; 0.8$

3B. $2\dfrac{1}{2}$ $-2\dfrac{1}{2}, \dfrac{2}{5}$

Many real-world applications involve working with real numbers.

 12 | Lesson 1-2 | Properties of Real Numbers

DifferentiatedInstruction OL BL

Extension Discuss with students why π is irrational. Have students investigate the history of π.

Real-World Example 4 Distributive Property

MONEY The prices of the components of a computer package offered by Computer Depot are shown in the table. If a 6% sales tax is added to the purchase price, how much sales tax is charged for this computer package?

Component	Price ($)
Computer	359.95
Monitor	219.99
Printer	79.00
Digital Camera	149.50
Software Bundle	99.00

There are two ways to determine the total sales tax.

Method 1 Multiply, then add.

Multiply each dollar amount by 6% or 0.06 and then add.

$T = 0.06(359.95) + 0.06(219.99) + 0.06(79.00) + 0.06(149.50) + 0.06(99.00)$

$= 21.60 + 13.20 + 4.74 + 8.97 + 5.94$

$= 54.45$

Method 2 Add, then multiply.

Find the total cost of the computer package, and then multiply the total by 0.06.

$T = 0.06(359.95 + 219.99 + 79.00 + 149.50 + 99.00)$

$= 0.06(907.44)$

$= 54.45$

The sales tax charged is $54.45. Notice that both methods result in the same answer.

▶ **Guided**Practice

4. **JOBS** Kayla makes $8 per hour working at a grocery store. The number of hours Kayla worked each day in one week are 3, 2.5, 2, 1, and 4. How much money did Kayla earn this week? **$100**

The properties of real numbers can be used to simplify algebraic expressions.

Example 5 Simplify an Expression

Simplify $3(2q + r) + 5(4q - 7r)$.

$3(2q + r) + 5(4q - 7r)$

$= 3(2q) + 3(r) + 5(4q) - 5(7r)$ Distributive Property

$= 6q + 3r + 20q - 35r$ Multiply.

$= 6q + 20q + 3r - 35r$ Commutative Property (+)

$= (6 + 20)q + (3 - 35)r$ Distributive Property

$= 26q - 32r$ Simplify.

▶ **Guided**Practice

5. Simplify $3(4x - 2y) - 2(3x + y)$. **$6x - 8y$**

 connectED.mcgraw-hill.com   **13**

Real-WorldCareer

Retail Store Manager Store managers are responsible for the day-to-day operations of a retail store. A store manager may range from being a high school graduate to having a 4-year degree, depending on the business.

2 Properties of Real Numbers

Example 2 shows how to identify properties of real numbers. **Example 3** shows how to find additive and multiplicative inverses of real numbers. **Example 4** shows how to use the Distributive Property to solve a real-world problem. **Example 5** shows how properties of real numbers can be used to simplify algebraic expressions.

Additional Examples

2 Name the property illustrated by $(-8 + 8) + 15 = 0 + 15$.
Additive Inverse Property

3 Find the additive inverse and multiplicative inverse for -7.
additive: 7; multiplicative: $-\frac{1}{7}$

4 **STAMPS** Audrey went to a post office and bought eight 45¢ stamps and eight 32¢ postcard stamps. What was the total amount of money Audrey spent on stamps? $6.16

5 Simplify $4(3a - b) + 2(b + 3a)$.
$18a - 2b$

> **Examples** illustrate lesson concepts and closely mirror the exercises. **Guided Practice** exercises give students an opportunity to try a similar problem on their own.

Tips for New Teachers

Reading Help students to recall the Distributive Property by connecting the name to *distributing,* "handing out papers, for example, one to each person." Point out that the factor outside the parentheses acts as a multiplier for each term within the parentheses.

> New teachers, or teachers new to teaching mathematics, may especially appreciate the **Tips for New Teachers.**

DifferentiatedInstruction   AL   OL   ELL

If ▶ students have trouble remembering the names of properties,

Then ▶ have students connect the term *commutative* with *commuting,* or moving from one place to another. Have them connect the term *associative* with *associate,* "the people you associate with, or your group."

connectED.mcgraw-hill.com   **13**

3 Practice

Formative Assessment

Use Exercises 1–17 to check for understanding.

Use the chart on the next page to customize assignments for your students.

 Teaching the Mathematical Practices

Reasoning Mathematically proficient students use quantitative reasoning to create a coherent representation of the problem at hand; consider the units involved; attend to the meaning of quantities, and flexibly use different properties of operations and objects.

Additional Answer

13c. If she continues to mow the same number of lawns, at the end of next week she will have the money. This may not be reasonable because not all the lawns she mowed this week may need to be mowed again next week.

> Every effort is made to show **answers** on the reduced Student Edition page, or in the margin of the Teacher Edition.

Example 1 Name the sets of numbers to which each number belongs.

1. 62 N, W, Z, Q, R   **2.** $\frac{5}{4}$ Q, R   **3.** $\sqrt{11}$ I, R   **4.** −12 Z, Q, R

Example 2 Name the property illustrated by each equation.

5. $(6 \cdot 8) \cdot 5 = 6 \cdot (8 \cdot 5)$ Assoc. (×)   **6.** $7(9 - 5) = 7 \cdot 9 - 7 \cdot 5$ Dist.

7. $84 + 16 = 16 + 84$ Comm. (+)   **8.** $(12 + 5)6 = 12 \cdot 6 + 5 \cdot 6$ Dist.

Example 3 Find the additive inverse and multiplicative inverse for each number.

9. -7 $7; -\frac{1}{7}$ **10.** $\frac{4}{9}$ $-\frac{4}{9}, \frac{9}{4}$ **11.** 3.8 $-3.8; \frac{1}{3.8}$ **12.** $\sqrt{5}$ $-\sqrt{5}; \frac{1}{\sqrt{5}}$

Example 4

13. CCSS REASONING Melba is mowing lawns for $22 each to earn money for a video game console that costs $550.

13a. $22(2 + 4 + 3 + 1 + 5 + 6 + 7)$ or $22(2) + 22(4) + 22(3) + 22(1) + 22(5) + 22(6) + 22(7)$

a. Write an expression to represent the total amount of money Melba earned during this week.

b. Evaluate the expression from part **a** by using the Distributive Property. $616

c. When do you think Melba will earn enough for the video game console? Is this reasonable? Explain. **See margin.**

Lawns Mowed in One Week

Day	Lawns Mowed
Monday	2
Tuesday	4
Wednesday	3
Thursday	1
Friday	5
Saturday	6
Sunday	7

Example 5 Simplify each expression.

14. $5(3x + 6y) + 4(2x - 9y)$ $23x - 6y$ **15.** $6(6a + 5b) - 3(4a + 7b)$ $24a + 9b$

16. $-4(6c - 3d) - 5(-2c - 4d)$ $-14c + 32d$ **17.** $-5(8x - 2y) - 4(-6x - 3y)$ $-16x + 22y$

Practice and Problem Solving
Extra Practice is on page R1.

Example 1 Name the sets of numbers to which each number belongs.

18. $-\frac{4}{3}$ Q, R   **19.** -8.13 Q, R   **20.** $\sqrt{25}$ N, W, Z, Q, R   **21.** $0.\overline{61}$ Q, R

22. $\frac{9}{3}$ N, W, Z, Q, R   **23** $-\sqrt{144}$ Z, Q, R   **24.** $\frac{21}{7}$ N, W, Z, Q, R   **25.** $\sqrt{17}$ I, R

Example 2 Name the property illustrated by each equation.

26. $-7y + 7y = 0$ Inverse (+)   **27.** $8\sqrt{11} + 5\sqrt{11} = (8 + 5)\sqrt{11}$ Dist.

28. $(16 + 7) + 23 = 16 + (7 + 23)$ Assoc. (+)   **29.** $\left(\frac{22}{7}\right)\left(\frac{7}{22}\right) = 1$ Inverse (×)

Example 3 Find the additive inverse and multiplicative inverse for each number.

30. -8 $8; -\frac{1}{8}$ **31.** 12.1 $-12.1; \frac{1}{12.1}$ **32.** -0.25 $0.25; -4$

33. $\frac{6}{13}$ $-\frac{6}{13}, \frac{13}{6}$ **34.** $-\frac{3}{8}$ $\frac{3}{8}; -\frac{8}{3}$ **35.** $\sqrt{15}$ $-\sqrt{15}; \frac{1}{\sqrt{15}}$

Example 4

36. CONSTRUCTION Jorge needs two different kinds of concrete: quick drying and slow drying. The quick-drying concrete mix calls for $2\frac{1}{2}$ pounds of dry cement, and the slow-drying concrete mix calls for $1\frac{1}{4}$ pounds of dry cement. He needs 5 times more quick-drying concrete and 3 times more slow-drying concrete than the mixes make.

a. How many pounds of dry cement mix will he need? $16\frac{1}{4}$ lb

b. Use the properties of real numbers to show how Jorge could compute this amount mentally. Justify each step. **See Chapter 1 Answer Appendix.**

 14 │ Lesson 1-2 │ Properties of Real Numbers

Differentiated Homework Options

Level	Assignment	Two-Day Option	
AL Basic	18–42, 53, 55–89	19–41 odd, 62–65	18–42 even, 53, 55–66–89
OL Core	19–41 odd, 43–45, 47, 49–53, 55–89	18–42, 62–65	43–53, 55–61, 66–8
BL Advanced	43–81, (optional: 82–89)		

Example 5

Simplify each expression.

37. $8b - 3c + 4b + 9c$ $12b + 6c$

38. $-2a + 9d - 5a - 6d$ $-7a + 3d$

39. $4(4x - 9y) + 8(3x + 2y)$ $40x - 20y$

40. $6(9a - 3b) - 8(2a + 4b)$ $38a - 50b$

41. $-2(-5g + 6k) - 9(-2g + 4k)$ $28g - 48k$

42. $-5(10x + 8z) - 6(4x - 7z)$ $-74x + 2z$

 43. **FOOTBALL** Illustrate the Distributive Property by writing two expressions for the area of a college football field. Then find the area of the football field. $53(60 + 60)$; $53(60) + 53(60)$; 6360 yd^2

60 yds ← → 60 yds

44a. $870{,}192(0.142 + 0.056 + 0.05 + 0.049)$; $870{,}192(0.142) + 870{,}192(0.056) + 870{,}192(0.05) + 870{,}192(0.049)$

44. **PETS** The chart shows the percent of dogs registered with the American Kennel Club that are of the eight most popular breeds.

   a. Illustrate the Distributive Property by writing two expressions to represent the number of registered dogs of the top four breeds.

   b. Evaluate the expressions you wrote to find the number of registered dogs of the top four breeds. **258,447**

 45 **FINANCIAL LITERACY** Billie is given $20 in lunch money by her parents once every two weeks. On some days, she packs her lunch, and on other days, she buys her lunch. A hot lunch from the cafeteria costs $4.50, and a cold sandwich from the lunch line costs $2.

   a. Billie decides that she wants to buy a hot lunch on Thursday and Friday of the first week and on Wednesday of the second week. Use the Distributive Property to determine how much that will cost. **$13.50**

   b. How many cold sandwiches can Billie buy with the amount left over? **3**

   c. Assuming that both weeks are Monday through Friday, how many times will Billie have to pack her lunch? **4 times**

Top Dogs	
Breed	**Percent of Registered Dogs**
Labrador Retrievers	14.2
Yorkshire Terriers	5.6
German Shepherds	5.0
Golden Retrievers	4.9
Beagles	4.5
Dachshunds	4.1
Boxers	4.1
Poodles	3.4
Total Registered Dogs	**870,192**

Source: American Kennel Club

Simplify each expression. **48.** $-36a - 30b + 24c$ **49.** $-42x - 72y - 30z$

46. $\frac{1}{3}(5x + 8y) + \frac{1}{4}(6x - 2y)$ $\frac{19}{6}x + \frac{13}{6}y$

47. $\frac{2}{5}(6c - 8d) + \frac{3}{4}(4c - 9d)$ $\frac{27}{5}c - \frac{199}{20}d$

48. $-6(3a + 5b) - 3(6a - 8c)$

49. $-9(3x + 8y) - 3(5x + 10z)$

50. **MODELING** Mary is making curtains out of the same fabric for 5 windows. The two larger windows are the same size, and the three smaller windows are the same size.

One larger window requires $3\frac{3}{4}$ yards of fabric, and one smaller window needs $2\frac{1}{3}$ yards of fabric.

   a. How many yards of material will Mary need? $14\frac{1}{2}$ yd

   b. Use the properties of real numbers to show how Mary could compute this amount mentally. **See margin.**

connectED.mcgraw-hill.com 15

WatchOut!

Common Errors If students are having difficulty identifying properties, suggest that they look first at the change from one expression to the other and ask themselves if it is a change in grouping (associativity) or in position (commutativity).

CCSS **Teaching the Mathematical Practices**

Modeling Mathematically proficient students can apply the mathematics they know to solve problems arising in everyday life, analyze relationships mathematically to draw conclusions, and interpret their mathematical results in the context of a situation.

Additional Answer

50b. $2\left(3\frac{3}{4}\right) + 3\left(2\frac{1}{3}\right)$

$= 2\left(3 + \frac{3}{4}\right) + 3\left(2 + \frac{1}{3}\right)$
(Definition of a mixed number)

$= 2(3) + 2\left(\frac{3}{4}\right) + 3(2) + 3\left(\frac{1}{3}\right)$
(Distributive Property)

$= 6 + \frac{3}{2} + 6 + 1$ (Multiply.)

$= 6 + 6 + 1 + \frac{3}{2}$
(Commutative Property (+))

$= 13 + \frac{3}{2}$ (Add.)

$= 14\frac{1}{2}$ (Add.)

Students can access online resources such as **Personal Tutor, Extra Examples,** and **Self-Check Quizzes** at connectED.mcgraw-hill.com.

Study Guide and Intervention
AL OL ELL

Practice
AL OL BL ELL

Word Problem Practice
AL OL BL ELL

51. MULTIPLE REPRESENTATIONS Consider the following real numbers.

$$-\sqrt{6}, 3, \frac{-15}{3}, 4.1, \pi, 0, \frac{3}{8}, \sqrt{36}$$ **51a–c. See Chapter 1 Answer Appendix.**

a. **Tabular** Organize the numbers into a table according to the sets of numbers to which each belongs.

b. **Algebraic** Convert each number to decimal form. Then list the numbers from least to greatest.

c. **Graphical** Graph the numbers on a number line.

d. **Verbal** Make a conjecture about using decimal form to list real numbers in order.

Sample answer: By converting the real numbers into decimal form, they can be easily lined up and compared.

52. CLOTHING A department store sells shirts for $12.50 each. Dalila buys 2, Latisha buys 3, and Pilar buys 1.

a. Illustrate the Distributive Property by writing two expressions to represent the cost of these shirts. $12.50(2 + 3 + 1); 12.50 \cdot 2 + 12.50 \cdot 3 + 12.50 \cdot 1$

b. Use the Distributive Property to find how much money the store received from selling these shirts. **$75**

H.O.T. Problems   Use Higher-Order Thinking Skills

53. WHICH ONE DOESN'T BELONG? Identify the number that does not belong with the other three. Explain your reasoning.

| $\sqrt{21}$ | $\sqrt{35}$ | $\sqrt{67}$ | $\sqrt{81}$ |

$\sqrt{81}$; It is a rational number, while the other three are irrational numbers.

54. CHALLENGE If $12(5r + 6t) = w$, then in terms of w, what is $48(30r + 36t)$? **24w**

55. CCSS CRITIQUE Luna and Sophia are simplifying $4(14a - 10b) - 6(b + 4a)$. Is either of them correct? Explain your reasoning.

55. No; Luna did not distribute the negative sign to the second term and Sophia switched the a and b terms because usually a comes first. The correct answer is $32a - 46b$.

Luna	Sophia
$4(14a - 10b) - 6(b + 4a)$	$4(14a - 10b) - 6(b + 4a)$
$56a - 40b - 6b + 24a$	$56a - 40b - 6a - 24b$
$80a - 46b$	$50a - 64b$

56. REASONING Determine whether the following statement is *sometimes*, *always*, or *never* true. Explain your reasoning.

An irrational number is a real number within a radical sign.

Sometimes; π and e are two examples of irrational numbers that do not involve the radical symbol.

57. OPEN ENDED Determine whether the Closure Property of Multiplication applies to irrational numbers. If not, provide a counterexample. **Sample answer: $\sqrt{5} \cdot \sqrt{5} = \sqrt{25}$ or 5, which is not irrational.**

OPEN ENDED The set of all real numbers is *dense*, meaning between any two distinct members of the set there lies infinitely many other members of the set. Find an example of (a) a rational number, and (b) an irrational number between the given numbers. **58–60. See margin.**

58. 2.45 and 2.5   **59.** π and $\frac{10}{3}$ **60.** $1.\overline{9}$ and 2.01

61. WRITING IN MATH Explain and provide examples to show why the Commutative Property does not hold true for subtraction or division. **See margin.**

Enrichment
OL BL

H.O.T. Problems require students to use **H**igher **O**rder **T**hinking skills to solve problems.

Multiple Representations

In Exercise 51, students use information organized in a table, in a number line, and algebra to compare and order real numbers.

62. EXTENDED RESPONSE Lenora bought several pounds of cashews and several pounds of almonds for a party. The cashews cost $8 per pound, and the almonds cost $6 per pound. Lenora bought a total of 7 pounds and paid a total of $48. Write and solve a system of equations to determine the pounds of cashews and the pounds of almonds that Lenora purchased.

62. $8c + 6a = 48$ and $c + a = 7$; $c = 3$ lb; $a = 4$ lb

63. SAT/ACT Find the 10th term in the series 2, 4, 7, 11, 16, ... **C**

  A 41
  B 46
  C 56
  D 67
  E 72

64. GEOMETRY What are the coordinates of point A in the parallelogram? **G**

  F $(b - a, c)$ H (b, c)
  G $(a - b, c)$ J (c, c)

65. What is the domain of the function that contains the points $(-3, 0)$, $(0, 4)$, $(-2, 5)$, and $(6, 4)$? **B**

  A $\{-3, 6\}$ C $\{0, 4, 5, 6\}$
  B $\{-3, -2, 0, 6\}$ D $\{-3, -2, 0, 4, 5, 6\}$

66. Evaluate $8(4 - 2)^3$. (Lesson 1-1) **64**

67. Evaluate $a + 3(b + c) - d$, if $a = 5$, $b = 4$, $c = 3$, and $d = 2$. (Lesson 1-1) **24**

68. GEOMETRY The formula for the area A of a circle with diameter d is $A = \pi\left(\dfrac{d}{2}\right)^2$.

Write an expression to represent the area of the circle. (Lesson 1-1) $\pi\left(\dfrac{x+3}{2}\right)^2$

69. CONSTRUCTION A 10-meter ladder leans against a building so that the top is 9.64 meters above the ground. How far from the base of the wall is the bottom of the ladder? (Lesson 0-8) **about 2.66 m**

Factor each polynomial. (Lesson 0-3)

70. $14x^2 + 10x - 8$ $2(7x^2 + 5x - 4)$ **71.** $9x^2 - 3x + 18$ $3(3x^2 - x + 6)$ **72.** $8x^2 + 16x + 12$ $4(2x^2 + 4x + 3)$

73. $10x^2 - 20x$ $10x(x - 2)$ **74.** $7x^2 - 14x - 21$ $7(x - 3)(x + 1)$ **75.** $12x^2 - 18x - 24$ $6(2x^2 - 3x - 4)$

Find each product. (Lesson 0-2)

76. $(x + 2)(x - 3)$ $x^2 - x - 6$ **77.** $(y + 2)(y - 1)$ $y^2 + y - 2$ **78.** $(a - 5)(a + 4)$ $a^2 - a - 20$

79. $(b - 7)(b - 3)$ $b^2 - 10b + 21$ **80.** $(n + 6)(n + 8)$ $n^2 + 14n + 48$ **81.** $(p - 9)(p + 1)$ $p^2 - 8p - 9$

Evaluate each expression if $a = 3$, $b = \dfrac{2}{3}$, and $c = -1.7$.

82. $6b - 5$ -1 **83.** $\dfrac{1}{6}b + 1$ $\dfrac{10}{9}$ **84.** $2.3c - 7$ -10.91 **85.** $-8(a - 4)$ 8

86. $a + b + c$ ≈ 1.967 **87.** $\dfrac{a \cdot b}{c}$ ≈ -1.176 **88.** $a^2 - c$ 10.7 **89.** $\dfrac{a \cdot c}{a}$ -1.7

DifferentiatedInstruction OL BL

Extension Since the product of any two whole numbers is always a whole number, the set of whole numbers is said to be *closed* under multiplication. This is an example of the Closure Property. State whether each statement is true or false. If false, give a counterexample.

a. The set of integers is closed under multiplication. true

b. The set of whole numbers is closed under subtraction. false; $1 - 6 = -5$

c. The set of whole numbers is closed under division. false; $1 \div 5 = 0.2$

Critique Mathematically proficient students are also able to compare the effectiveness of two plausible arguments, distinguish correct logic or reasoning from that which is flawed, and—if there is a flaw in an argument—explain what it is.

WatchOut!

Error Analysis For Exercise 55, point out to students that they should pay close attention to the order of operations and like terms as they simplify expressions.

4 Assess

Ticket Out the Door Have each student write the name and an example of one of the properties in this lesson.

Formative Assessment
Check for student understanding of concepts in Lessons 1-1 and 1-2.

📁 Quiz 1, p. 45

Additional Answers

58. Sample answer: (a) 2.46 and (b) 2.48448444844448 . . .

59. Sample answer: (a) 3.2 and (b) $\sqrt{10}$

60. Sample answer: (a) 2.001 and (b) 2.001000100001 . . .

61. Sample answer: The Commutative Property does not hold for subtraction or division because order matters with these two operations. In addition or multiplication, the order does not matter.

For example, $2 + 4 = 4 + 2$ and $2 \cdot 4 = 4 \cdot 2$. However, with subtraction, $2 - 4 \neq 4 - 2$, and with division, $\dfrac{2}{4} \neq \dfrac{4}{2}$.

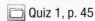

1 Focus

VerticalAlignment

Before Lesson 1-3 Use properties of real numbers to evaluate expressions.

Lesson 1-3 Translate verbal expressions into algebraic expressions and equations, and vice versa.
Solve equations using the properties of equality.

After Lesson 1-3 Solve systems of equations.

2 Teach

Scaffolding Questions

Have students read the **Why?** section of the lesson.

Ask:

- What does the variable *m* represent? mile

- Is the length of a kilometer greater than or less than the length of a mile? less than

- About how many kilometers are in 12 miles? about 19.3 kilometers

All of the **Lesson Resources** are leveled for students who are **below grade level, on grade level,** and **above grade level,** and for students who are **English language learners.**

LESSON 1-3 Solving Equations

::Then	::Now	::Why?
● You used properties of real numbers to evaluate expressions.	**1** Translate verbal expressions into algebraic expressions and equations, and vice versa. **2** Solve equations using the properties of equality.	● The United States is one of the few countries in the world that measures distances in miles. When traveling by car in different countries, it is often useful to convert miles to kilometers. To find the approximate number of kilometers *k* in miles *m*, divide the number of miles by 0.62137.

$$m \text{ miles} \times \frac{1 \text{ kilometer}}{0.62137 \text{ mile}} \approx k \text{ kilometers}$$
$$\frac{m}{0.62137} \approx k \text{ kilometers}$$

NewVocabulary
open sentence
equation
solution

Common Core State Standards

Content Standards
A.CED.1 Create equations and inequalities in one variable and use them to solve problems.

Mathematical Practices
3 Construct viable arguments and critique the reasoning of others.

8 Look for and express regularity in repeated reasoning.

1 Verbal Expressions and Algebraic Expressions Verbal expressions can be translated into algebraic expressions by using the language of algebra.

Example 1 Verbal to Algebraic Expression

Write an algebraic expression to represent each verbal expression.

a. 2 more than 4 times the cube of a number $\quad 4x^3 + 2$

b. the quotient of 5 less than a number and 12 $\quad \dfrac{n-5}{12}$

GuidedPractice

1A. the cube of a number increased by 4 times the same number $\quad p^3 + 4p$

1B. three times the difference of a number and 8 $\quad 3(x-8)$

A mathematical sentence containing one or more variables is called an **open sentence**. A mathematical sentence stating that two mathematical expressions are equal is called an **equation**.

Example 2 Algebraic to Verbal Sentence

Write a verbal sentence to represent each equation.

a. $6x = 72$ The product of 6 and a number is 72.

b. $n + 15 = 91$ The sum of a number and 15 is ninety-one.

GuidedPractice   **2A, 2B.** See margin.

2A. $g - 5 = -2$ **2B.** $2c = c^2 - 4$

Open sentences are neither true nor false until the variables have been replaced by numbers. Each replacement that results in a true sentence is called a **solution** of the open sentence.

 18 | Lesson 1-3

Jeremy Woodhouse/Photodisc/Getty Images

Lesson 1-3 Resources

Resource	Approaching Level **AL**	On Level **OL**	Beyond Level **BL**	English Learners **ELL**
Teacher Edition	▪ Differentiated Instruction, p. 20	▪ Differentiated Instruction, p. 21	▪ Differentiated Instruction, p. 21	▪ Differentiated Instruction, p. 20
Chapter Resource Masters	▪ Study Guide and Intervention, pp. 17–18 ▪ Skills Practice, p. 19 ▪ Practice, p. 20 ▪ Word Problem Practice, p. 21	▪ Study Guide and Intervention, pp. 17–18 ▪ Skills Practice, p. 19 ▪ Practice, p. 20 ▪ Word Problem Practice, p. 21 ▪ Enrichment, p. 22 ▪ Graphing Calculator Activity, p. 23	▪ Practice, p. 20 ▪ Word Problem Practice, p. 21 ▪ Enrichment, p. 22	▪ Study Guide and Intervention, pp. 17–18 ▪ Skills Practice, p. 19 ▪ Practice, p. 20 ▪ Word Problem Practice, p. 21
Other	▪ 5-Minute Check 1-3 ▪ Study Notebook ▪ Teaching Algebra with Manipulatives	▪ 5-Minute Check 1-3 ▪ Study Notebook ▪ Teaching Algebra with Manipulatives	▪ 5-Minute Check 1-3 ▪ Study Notebook	▪ 5-Minute Check 1-3 ▪ Study Notebook ▪ Teaching Algebra with Manipulatives

2 Properties of Equality To solve equations, we can use properties of equality. Some of these properties are listed below.

Math HistoryLink

Diophantus of Alexandria (c. 200–284)
Diophantus was famous for his work in algebra. His main work was titled *Arithmetica* and introduced symbolism to Greek algebra as well as propositions in number theory and polygonal numbers.

KeyConcept Properties of Equality

Property	Symbols	Examples
Reflexive	For any real number a, $a = a$.	$b + 12 = b + 12$
Symmetric	For all real numbers a and b, if $a = b$, then $b = a$.	If $18 = -2n + 4$, then $-2n + 4 = 18$.
Transitive	For all real numbers a, b, and c, if $a = b$ and $b = c$, then $a = c$.	If $5p + 3 = 48$ and $48 = 7p - 15$, then $5p + 3 = 7p - 15$.
Substitution	If $a = b$, then a may be replaced by b and b may be replaced by a.	If $(6 + 1)x = 21$, then $7x = 21$.

Example 3 Identify Properties of Equality

Name the property illustrated by each statement.

a. If $3a - 4 = b$, and $b = a + 17$, then $3a - 4 = a + 17$.

Transitive Property of Equality

b. If $2g - h = 62$, and $h = 24$, then $2g - 24 = 62$.

Substitution Property of Equality

▶ **Guided Practice**

3. If $-11a + 2 = -3a$, then $-3a = -11a + 2$. **Symmetric**

Solving most equations requires assuming that the original equation has a solution, and performing the same operations on each side of the equals sign. The properties of equality allow for the equation to be solved in this way.

KeyConcept

	Addition and Subtraction Properties of Equality
Symbols	For any real numbers, a, b, and c, if $a = b$, then $a + c = b + c$ and $a - c = b - c$.
Examples	If $x - 6 = 14$, then $x - 6 + 6 = 14 + 6$. If $n + 5 = -32$, then $n + 5 - 5 = -32 - 5$.

	Multiplication and Division Properties of Equality
Symbols	For any real numbers, a, b, and c, $c \neq 0$, if $a = b$, then $a \cdot c = b \cdot c$ and $\frac{a}{c} = \frac{b}{c}$.
Examples	If $\frac{m}{8} = -7$, then $8 \cdot \frac{m}{8} = 8 \cdot (-7)$. If $-2y = 12$, then $\frac{-2y}{-2} = \frac{12}{-2}$.

Additional Answers (Guided Practice)

2A. The difference of a number and 5 is -2.

2B. Two times a number is equal to the difference of that number squared and 4.

akg-images

---

1 Verbal Expressions and Algebraic Expressions
Example 1 shows how to translate verbal expressions into algebraic expressions. **Example 2** shows how to translate algebraic expressions into verbal expressions.

Formative Assessment
Use the Guided Practice exercises after each example to determine students' understanding of concepts.

Additional Examples

1 Write an algebraic expression to represent each verbal expression.

a. 7 less than a number $n - 7$

b. the square of a number decreased by the product of 5 and the number $x^2 - 5x$

2 Write a verbal sentence to represent each equation.

a. $6 = -5 + x$ Six is equal to -5 plus a number.

b. $7y - 2 = 19$ Seven times a number minus 2 is 19.

▶ **Additional Examples** also in Interactive Classroom PowerPoint® Presentations

IWB Interactive White Board READY

2 Properties of Equality
Example 3 shows how to identify properties of equality.

Additional Example

3 Name the property illustrated by each statement.

a. $a - 2.03 = a - 2.03$
Reflexive Property of Equality

b. If $9 = x$, then $x = 9$.
Symmetric Property of Equality

4 Solve each equation. Check your solution.
  a. $m - 5.48 = 0.02$ 5.5
  b. $18 = \frac{1}{2}t$ 36

5 Solve
$53 = 3(y - 2) - 2(3y - 1)$.
  -19

![CCSS] **Teaching the Mathematical Practices**

Regularity Mathematically proficient students notice if calculations are repeated, and look both for general methods and for shortcuts. As they work to solve a problem, mathematically proficient students maintain oversight of the process, while attending to the details. Encourage students to recognize the relationship between the Multiplication and Division Properties of Equality when they are applied using fractions.

Focus on Mathematical Content

Rules for Solving Equations The rules used to solve equations are based on the Properties of Equality. When a number is added to or subtracted from each side of an equation, the result is an equivalent equation. This equivalent equation will have the same solution as the original.

Tips for New Teachers

Checking Solutions Explain that checking solutions to discover possible errors is a vital procedure when you use math on the job.

StudyTip

CCSS Regularity
In Example 4b, notice that multiplying both sides of the equation by $-\frac{8}{5}$ is the same as dividing both sides by $-\frac{5}{8}$.

Study Tips offer students helpful information about the topics they are studying.

Example 4 Solve One-Step Equations

Solve each equation. Check your solution.

a. $n - 3.24 = 42.1$

$n - 3.24 = 42.1$	Original equation
$n - 3.24 + 3.24 = 42.1 + 3.24$	Add 3.24 to each side.
$n = 45.34$	Simplify.

The solution is 45.34.

CHECK	$n - 3.24 = 42.1$	Original equation
	$45.34 - 3.24 \stackrel{?}{=} 42.1$	Substitute 45.34 for n.
	$42.1 = 42.1$ ✓	Simplify.

b. $-\frac{5}{8}x = 20$

$-\frac{5}{8}x = 20$	Original equation
$-\frac{8}{5}\left(-\frac{5}{8}\right)x = -\frac{8}{5}(20)$	Multiply each side by $-\frac{8}{5}$.
$x = -32$	Simplify.

The solution is -32.

CHECK	$-\frac{5}{8}x = 20$	Original equation
	$-\frac{5}{8}(-32) \stackrel{?}{=} 20$	Replace x with -32.
	$20 = 20$ ✓	Simplify.

▶ **Guided**Practice

4A. $x - 14.29 = 25$ 39.29 **4B.** $\frac{2}{3}y = -18$ -27

To solve an equation with more than one operation, undo operations by working backward.

Example 5 Solve a Multi-Step Equation

Solve $5(x + 3) + 2(1 - x) = 14$.

$5(x + 3) + 2(1 - x) = 14$	Original equation
$5x + 15 + 2 - 2x = 14$	Apply the Distributive Property.
$3x + 17 = 14$	Simplify the left side.
$3x = -3$	Subtract 17 from each side.
$x = -1$	Divide each side by 3.

StudyTip

Checking Answers
When solving for a variable, you can use substitution to check your answer by replacing the variable in the original equation with your answer.

▶ **Guided**Practice

Solve each equation.

5A. $-10x + 3(4x - 2) = 6$ 6 **5B.** $2(2x - 1) - 4(3x + 1) = 2$ -1

 20 | Lesson 1-3 | Solving Equations

DifferentiatedInstruction **AL** **ELL**

If ▶ students have difficulty transitioning from verbal expressions to algebraic expressions and vice versa,

Then ▶ pair these students with students who are not having trouble. Let them act as a mentor to help the students having difficulties.

You can use properties to solve an equation for a variable.

Example 6 Solve for a Variable

GEOMETRY The formula for the area A of a trapezoid is $A = \frac{1}{2}h(b_1 + b_2)$, where h represents the height, and b_1 and b_2 represent the measures of the bases. Solve the formula for b_2.

$A = \frac{1}{2}h(b_1 + b_2)$	Area formula
$2A = 2\left[\frac{1}{2}h(b_1 + b_2)\right]$	Multiply each side by 2.
$2A = h(b_1 + b_2)$	Simplify.
$\frac{2A}{h} = \frac{h(b_1 + b_2)}{h}$	Divide each side by h.
$\frac{2A}{h} = b_1 + b_2$	Simplify.
$\frac{2A}{h} - b_1 = b_1 + b_2 - b_1$	Subtract b_1 from each side.
$\frac{2A}{h} - b_1 = b_2$	Simplify.

6. $h = \frac{S - 2\pi r^2}{2\pi r}$

▶ **Guided**Practice

6. The formula for the surface area S of a cylinder is $S = 2\pi r^2 + 2\pi rh$, where r is the radius of the base and h is the height of the cylinder. Solve the formula for h.

There are often many ways to solve a problem. Using the properties of equality can help you find a simpler way.

Standardized Test Example 7 Use Properties of Equality

If $6x - 12 = 18$, what is the value of $6x + 5$?

A 5   **B** 11   **C** 35   **D** 41

Read the Test Item

You are asked to find the value of $6x + 5$. Note that you do not have to find the value of x. Instead, you can use the Addition Property of Equality to make the left side of the equation $6x + 5$.

Solve the Test Item

$6x - 12 = 18$	Original equation
$6x - 12 + 17 = 18 + 17$	Add 17 to each side because $-12 + 17 = 5$.
$6x + 5 = 35$	Simplify.

The answer is C.

▶ **Guided**Practice

7. If $5y + 2 = \frac{8}{3}$, what is the value of $5y - 6$? **G**

F $\frac{-20}{3}$ **G** $\frac{-16}{3}$ **H** $\frac{16}{3}$ **J** $\frac{32}{3}$

connectED.mcgraw-hill.com **21**

Test-Taking Tip

Read the Question Read the question carefully before solving the equation. In Example 7, you are to find the value of $6x + 5$, not the value of x.

Example 6 shows how to use properties to solve a formula for a specified variable. **Example 7** shows how to solve a standardized test question using the Addition Property of Equality.

Additional Examples

6 **GEOMETRY** The formula for the surface area S of a cone is $S = \pi r\ell + \pi r^2$, where ℓ is the slant height of the cone and r is the radius of the base. Solve the formula for ℓ. $\ell = \frac{S - \pi r^2}{\pi r}$

7 **STANDARDIZED TEST PRACTICE**

If $4g + 5 = \frac{4}{9}$, what is the value of $4g - 2$? C

A $-\frac{41}{36}$ **C** $-\frac{59}{9}$

B $-\frac{41}{9}$ **D** $-\frac{67}{7}$

Every chapter includes one or more worked-out **Standardized Test Examples** that are similar to problems found on state assessments.

Teach with Tech

Blog Have students write a blog entry to summarize how to solve one-step equations. Make sure that students use the concept of inverse operations in their explanations.

Teach with Tech features throughout the Teacher Edition offer tips on using various types of technology such as interactive whiteboards, document cameras, blogs, and more, to enhance your teaching.

DifferentiatedInstruction **OL** **BL**

Extension The formula for the perimeter of a rectangle is $P = 2\ell + 2w$. Find the area of a rectangle that has a perimeter P of 22 inches and a width w of 3 inches. (*Hint:* Begin by solving the perimeter formula for ℓ.) 24 in^2

Check Your Understanding exercises are intended to be completed in class. **Example references** show students where to look back for review.

3 Practice

Formative Assessment

Use Exercises 1–21 to check for understanding.

Use the chart at the bottom of this page to customize assignments for your students.

 Teaching the Mathematical Practices

Precision Mathematically proficient students try to use clear definitions in their reasoning, calculate accurately and efficiently, and make explicit use of definitions.

Additional Answers

3. The sum of five times a number and 7 equals 18.

4. The difference between the square of a number and 9 is 27.

5. The difference between five times a number and the cube of that number is 12.

6. Eight more than the quotient of a number and four is −16.

Example 1   Write an algebraic expression to represent each verbal expression.

1. the product of 12 and the sum of a number and negative 3 $12\,[x + (-3)]$

2. the difference between the product of 4 and a number and the square of the number $4x - x^2$

Example 2   Write a verbal sentence to represent each equation. **3–6. See margin.**

3. $5x + 7 = 18$ 4. $x^2 - 9 = 27$

5. $5y - y^3 = 12$ 6. $\frac{x}{4} + 8 = -16$

Example 3   Name the property illustrated by each statement.

7. $(8x - 3) + 12 = (8x - 3) + 12$ **Reflexive Property**   8. If $a = -3$ and $-3 = d$, then $a = d$.   **Transitive Property**

Examples 4–5   **CCSS PRECISION**   Solve each equation. Check your solution.

9. $z - 19 = 34$ **53**   10. $x + 13 = 7$ **−6**   11. $-y = 8$ **−8**

12. $-6x = 42$ **−7**   13. $5x - 3 = -33$ **−6**   14. $-6y - 8 = 16$ **−4**

15. $3(2a + 3) - 4(3a - 6) = 15$ **3**   16. $5(c - 8) - 3(2c + 12) = -84$ **8**

17. $-3(-2x + 20) + 8(x + 12) = 92$ **4**   18. $-4(3m - 10) - 6(-7m - 6) = -74$ **−5**

Example 6   Solve each equation or formula for the specified variable.

19. $8r - 5q = 3$, for q $q = \dfrac{8r - 3}{5}$ 20. $Pv = nrt$, for n $\dfrac{Pv}{rt} = n$

Example 7   21. **MULTIPLE CHOICE** If $\frac{y}{5} + 8 = 7$, what is the value of $\frac{y}{5} - 2$?   **B**

    A −10   B −3   C 1   D 5

Practice and Problem Solving

Extra Practice is on page R1.

Example 1

27. The quotient of the sum of 3 and a number and 4 is 5.

28. Three less than four times the square of a number is 13.

Write an algebraic expression to represent each verbal expression.

22. the difference between the product of four and a number and 6 $4n - 6$

23. the product of the square of a number and 8 $8x^2$

24. fifteen less than the cube of a number $x^3 - 15$

25. five more than the quotient of a number and 4 $\frac{x}{4} + 5$

29. $n =$ number of home runs Jacobs hit; $n + 6 =$ number of home runs Cabrera hit; $2n + 6 = 46$; Jacobs: 20 home runs, Cabrera: 26 home runs.

Example 2   Write a verbal sentence to represent each equation.

26. $8x - 4 = 16$ 27. $\frac{x + 3}{4} = 5$ 28. $4y^2 - 3 = 13$

   26. Four less than 8 times a number is 16.

 29. BASEBALL During a recent season, Miguel Cabrera and Mike Jacobs hit a combined total of 46 home runs. Cabrera hit 6 more home runs than Jacobs. How many home runs did each player hit? Define a variable, write an equation, and solve the problem.

Example 3   Name the property illustrated by each statement. **30. Subtr. (=)**

30. If $x + 9 = 2$, then $x + 9 - 9 = 2 - 9$ 31. If $y = -3$, then $7y = 7(-3)$ **Subst.**

32. If $g = 3h$ and $3h = 16$, then $g = 16$ 33. If $-y = 13$, then $-(-y) = -13$ **Mult. (=)**
Transitive Property

22 | Lesson 1-3 | Solving Equations

Differentiated Homework Options

Level	Assignment	Two-Day Option	
AL Basic	22–50, 62, 64–82	23–49 odd, 67–70	22–50 even, 62, 64–66, 71–82
OL Core	23–51 odd, 52, 53–57 odd, 59–62, 64–82	22–50, 67–70	51–62, 64–66, 71–82
BL Advanced	51–74, (optional: 75–82)		

34. MONEY Aiko and Kendra arrive at the state fair with $32.50. What is the total number of rides they can go on if they each pay the entrance fee?
n = number of rides; $2(7.50) + n(2.50) = 32.50$; 7

Entrance Fee: $7.50
Rides: $2.50 each

Examples 4–5 **CCSS PRECISION** Solve each equation. Check your solution.

35. $3y + 4 = 19$ 5

36. $-9x - 8 = 55$ -7

37. $7y - 2y + 4 + 3y = -20$ -3

38. $5g + 18 - 7g + 4g = 8$ -5

39 $5(-2x - 4) - 3(4x + 5) = 97$ -6

40. $-2(3y - 6) + 4(5y - 8) = 92$ 8

41. $\frac{2}{3}(6c - 18) + \frac{3}{4}(8c + 32) = -18$ -3

42. $\frac{3}{5}(15d + 20) - \frac{1}{6}(18d - 12) = 38$ 4

43. GEOMETRY The perimeter of a regular pentagon is 100 inches. Find the length of each side. s = length of a side; $5s = 100$; 20 in.

44. MEDICINE For Nina's illness her doctor gives her a prescription for 28 pills. The doctor says that she should take 4 pills the first day and then 2 pills each day until her prescription runs out. For how many days does she take 2 pills?
x = the number of days she takes 2 pills; $4 + 2x = 28$; 12 days

Example 6 **Solve each equation or formula for the specified variable.**

45. $E = mc^2$, for m $m = \dfrac{E}{c^2}$

46. $c(a + b) - d = f$, for a $a = \dfrac{f + d}{c} - b$

47. $z = \pi q^3 h$, for h $h = \dfrac{z}{\pi q^3}$

48. $\dfrac{x + y}{z} - a = b$, for y $y = z(a + b) - x$

49. $y = ax^2 + bx + c$, for a $a = \dfrac{y - bx - c}{x^2}$

50. $wx + yz = bc$, for z $z = \dfrac{bc - wx}{y}$

51. GEOMETRY The formula for the volume of a cylinder with radius r and height h is π times the radius times the radius times the height.

   a. Write this as an algebraic expression. $V = \pi \times r \times r \times h$

   b. Solve the expression in part **a** for h. $h = \dfrac{V}{\pi r^2}$

52. AWARDS BANQUET A banquet room can seat a maximum of 69 people. The coach, principal, and vice principal have invited the award-winning girls' tennis team to the banquet. If the tennis team consists of 22 girls, how many guests can each student bring? n = number of guests that each student can bring; $22n + 25 = 69$; 2 guests

CCSS PRECISION Solve each equation. Check your solution.

53. $5x - 9 = 11x + 3$ -2

54. $\dfrac{1}{x} + \dfrac{1}{4} = \dfrac{7}{12}$ 3

55. $5.4(3k - 12) + 3.2(2k + 6) = -136$ -4

56. $8.2p - 33.4 = 1.7 - 3.5p$ -4

57. $\dfrac{4}{9}y + 5 = -\dfrac{7}{9}y - 8$ $-\dfrac{117}{11}$

58. $\dfrac{3}{4}z - \dfrac{1}{3} = \dfrac{2}{3}z + \dfrac{1}{5}$ $\dfrac{32}{5}$

x = the cost of rent each month; $622 + 428 + 240 + 144 + 12x = 10,734$; $775 per month

59. FINANCIAL LITERACY Benjamin spent $10,734 on his living expenses last year. Most of these expenses are listed at the right. Benjamin's only other expense last year was rent. If he paid rent 12 times last year, how much was Benjamin's rent each month?

Expense	Annual Cost
Electric	$622
Gas	$428
Water	$240
Renter's Insurance	$144

Study Guide and Intervention
AL OL ELL

NAME _____ DATE _____ PERIOD _____

1-3 Study Guide and Intervention
Solving Equations

Verbal Expressions and Algebraic Expressions The chart suggests some ways to help you translate word expressions into algebraic expressions. Any letter can be used to represent a number that is not known.

Word Expression	Operation
and, plus, sum, increased by, more than	addition
minus, difference, decreased by, less than	subtraction
times, product, of (as in $\frac{1}{2}$ of a number)	multiplication
divided by, quotient	division

Example 1 Write an algebraic expression to represent 18 less than the quotient of a number and 3.
$\frac{n}{3} - 18$

Example 2 Write a verbal sentence to represent $6(n - 2) = 14$.
Six times the difference of a number and two is equal to 14.

Exercises

Write an algebraic expression to represent each verbal expression.

1. the sum of six times a number and 25
2. four times the sum of a number and 3
3. 7 less than fifteen times a number
4. the difference of nine times a number and the quotient of 6 and the same number
5. the sum of 100 and four times a number
6. the product of 3 and the sum of 11 and a number
7. four times the square of a number increased by five times the same number
8. 23 more than the product of 7 and a number

Write a verbal sentence to represent each equation.

9. $3n - 35 = 79$
10. $2(n^2 + 3n^2) = 4n$
11. $\frac{5n}{n+3} = n - 8$

Lesson 1-3

Practice
AL OL BL ELL

NAME _____ DATE _____ PERIOD _____

1-3 Practice
Solving Equations

Write an algebraic expression to represent each verbal expression.

1. 5 more than the quotient of a number and 5
2. the sum of two consecutive integers
3. 5 times the sum of a number and 1
4. 1 less than twice the square of a number

Write a verbal sentence to represent each equation.

5. $5 - 2x = 4$
6. $3y = 4y^2$
7. $3c = 2(c - 1)$
8. $\frac{m}{5} = 3(2m + 1)$

Name the property illustrated by each statement.

9. If $t - 13 = 52$, then $52 = t - 13$.
10. If $8(2q + 1) = 4$, then $2(2q + 1) = 1$.
11. If $h + 12 = 22$, then $h = 10$.
12. If $4m = -15$, then $-12m = 45$.

Solve each equation. Check your solution.

13. $14 - 8 = 6r$
14. $9 + 4n = -59$
15. $\frac{3}{4} - \frac{1}{2}n = \frac{5}{8}$
16. $\frac{5}{6}c + \frac{3}{4} = \frac{11}{12}$
17. $-1.6r + 5 = -7.8$
18. $6z - 5 = 7 - 9z$
19. $5(6 - 4v) = v + 21$
20. $6v - 5 = -3(2y + 1)$

Solve each equation or formula for the specified variable.

21. $E = mc^2$, for m
22. $c = \frac{3d + 1}{3}$, for d
23. $h = vt - gt^2$, for v
24. $E = \frac{1}{2}Iw^2 + U$, for I

25. **GEOMETRY** The length of a rectangle is twice the width. Find the width if the perimeter is 60 centimeters. Define a variable, write an equation, and solve the problem.

26. **GOLF** Luis and three friends went golfing. Two of the friends rented clubs for $6 each. The total cost of the rented clubs and the green fees for each person was $76. What was the cost of the green fees for each person? Define a variable, write an equation, and solve the problem.

Word Problem Practice
AL OL BL ELL

NAME _____ DATE _____ PERIOD _____

1-3 Word Problem Practice
Solving Equations

1. **AGES** Robert's father is 5 years older than 3 times Robert's age. Let Robert's age be denoted by R and let Robert's father's age be denoted by F. Write an equation that relates Robert's age and his father's age.

2. **AIRPLANES** The Citation Sovereign is a small jet that can carry up to 2650 pounds. The number of passengers p and the number of suitcases c that the airplane can carry are estimated by the equation $180p + 60c = 2650$. If 10 people board the aircraft, how many suitcases can the airplane carry?

3. **GEOMETRY** The length of a rectangle is 10 units longer than its width. If the total perimeter of the rectangle is 44 units, what is the width?

4. **SAVINGS** Jason started with d dollars in his piggy bank. One week later, Jason doubled the amount in his piggy bank. Another week later, Jason was able to add $20 to his piggy bank. At this point, the piggy bank had $50 in it. What is d?

5. **DOMINOES** Nancy is setting up a train of dominoes from the front entrance straight down the hall to the kitchen entrance. The thickness of each domino is t. Nancy places the dominoes so that the space separating consecutive dominoes is $3t$. The total distance that N dominoes takes up is given by $d = t(4N + 1)$.

   a. Nancy measures her dominoes and finds that $t = 1$ centimeter. She measures the distance of her hallway and finds that $d = 321$ centimeters. Rewrite the equation that relates d, t, and N with the given values substituted for t and d.

   b. How many dominoes did Nancy have in her hallway?

Lesson 1-3

Enrichment
OL BL

NAME _____ DATE _____ PERIOD _____

1-3 Enrichment
United States' Gross National Product

The Gross National Product, GNP, is an important indicator of the U.S. economy. The GNP contains information about the inflation rate, the Bond market, and the Stock market. It is composed of consumer goods, investments, government expenditures, exports, and imports.

Calculated from $GNP = C + I + G + X - M$, where

C is consumer goods (e.g. TVs, cars, food, furniture, clothes, doctors' fees, and dining)
I is investments (e.g. factories, computers, airlines, and housing)
G is government spending and investments (e.g. ships, roads, schools, NASA, and bombs)
X is exports (e.g. corn, wheat, cars, and computers)
M is for imports, (e.g. cars, computer chips, clothes, and oil)
$X - M$ is exports minus imports and equals trade surplus or deficit.

1. The most important sector of the U.S. economy is consumption. It makes up about 60% of the entire GNP. In 2000, the U.S.'s GNP was 10.5 trillion dollars. In the same year, there were 1 trillion dollars in investments, but a 1 trillion dollar trade deficit. Assuming that consumption made up 60% of the GNP, how much did the government budget for spending?

Multiple Representations

In Exercise 61, students use a number line and a table of values to illustrate the workings of absolute values.

WatchOut!

Error Analysis In Exercise 62, encourage students to give a complete explanation of the error. For example, "Steven should have subtracted b from the entire left side, $\frac{2A}{h}$, in the last step of the equation."

Error Analysis exercises help students identify and address common errors before they occur. Prompts in the margin of the Teacher Edition help you guide students toward understanding.

Additional Answers

60c. Yes; it seems reasonable that two crews working 4 miles apart would be able to complete the same amount of miles in the same amount of time.

61b.

Integer	Distance from Zero
−5	5
−4	4
−3	3
−2	2
−1	1
0	0
1	1
2	2
3	3
4	4
5	5

61c.

---

60 CCSS **SENSE-MAKING** The Sunshine Skyway Bridge spans Tampa Bay, Florida. Suppose one crew began building south from St. Petersburg, and another crew began building north from Bradenton. The two crews met 10,560 feet south of St. Petersburg approximately 5 years after construction began.

a. Suppose the St. Petersburg crew built an average of 176 feet per month. Together the two crews built 21,120 feet of bridge. Determine the average number of feet built per month by the Bradenton crew. **176 ft**

b. About how many miles of bridge did each crew build? **2 mi**

c. Is this answer reasonable? Explain. **See margin.**

61 **MULTIPLE REPRESENTATIONS** The absolute value of a number describes the distance of the number from zero. **a.**

a. Geometric Draw a number line. Label the integers from −5 to 5.

b. Tabular Create a table of the integers on the number line and their distance from zero. **b–d. See margin.**

c. Graphical Make a graph of each integer x and its distance from zero y using the data points in the table.

d. Verbal Make a conjecture about the integer and its distance from zero. Explain the reason for any changes in sign.

H.O.T. Problems Use Higher-Order Thinking Skills

62. ERROR ANALYSIS Steven and Jade are solving $A = \frac{1}{2}h(b_1 + b_2)$ for b_2. Is either of them correct? Explain your reasoning.

66. Sample answer: The Transitive Property utilizes the Substitution Property. While the Substitution Property is done with two values, that is, one being substituted for another, the Transitive Property deals with three values, determining that since two values are equal to a third value, then they must be equal.

Steven	Jade
$A = \frac{1}{2}h(b_1 + b_2)$	$A = \frac{1}{2}h(b_1 + b_2)$
$\frac{2A}{h} = (b_1 + b_2)$	$\frac{2A}{h} = (b_1 + b_2)$
$\frac{2A - b_1}{h} = b_2$	$\frac{2A}{h} - b_1 = b_2$

63. CHALLENGE Solve $d = \sqrt{(x_2 - x_1)^2 + (y_2 - y_1)^2}$ for y_1. $y_1 = y_2 - \sqrt{d^2 - (x_2 - x_1)^2}$

64. REASONING Use what you have learned in this lesson to explain why the following number trick works. **Translating this number trick into an expression yields:**

- Take any number.
- Multiply it by ten.
- Subtract 30 from the result.
- Divide the new result by 5.
- Add 6 to the result.
- Your new number is twice your original.

$\frac{(10x - 30)}{5} + 6 = 2x$

$\frac{(10x - 30)}{5} = 2x - 6$

$(2x - 6) + 6 = 2x$

65. OPEN ENDED Provide one example of an equation involving the Distributive Property that has no solution and another example that has infinitely many solutions.
Sample answer: $3(x - 4) = 3x + 5$; $2(3x - 1) = 6x - 2$

66. WRITING IN MATH Compare and contrast the Substitution Property of Equality and the Transitive Property of Equality.

24 | Lesson 1-3 | Solving Equations

---

62. Jade; Sample answer: in the last step, when Steven subtracted b_1 from each side, he mistakenly put the $- b_1$ in the numerator instead of after the entire fraction.

61d. For positive integers, the distance from zero is the same as the integer. For negative integers, the distance is the integer with the opposite sign because distance is always positive.

CCSS **Teaching the Mathematical Practices**

Sense-Making Mathematically proficient students start by explaining to themselves the meaning of a problem and looking for entry points to its solution. They analyze givens, constraints, relationships, and goals. They check their answers to problems using a different method, and they continually ask themselves, "Does this make sense?"

67. The graph shows the solution of which inequality? **D**

A $y < \frac{2}{3}x + 4$ **C** $y < \frac{3}{2}x + 4$

B $y > \frac{2}{3}x + 4$ **D** $y > \frac{3}{2}x + 4$

68. SAT/ACT What is $1\frac{1}{3}$ subtracted from its reciprocal? **G**

F $-2\frac{2}{3}$ **J** $\frac{1}{4}$

G $-\frac{7}{12}$ **K** $\frac{3}{4}$

H $-\frac{1}{12}$

69. GEOMETRY Which of the following describes the transformation of $\triangle ABC$ to $\triangle A'B'C'$? **A**

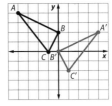

A a reflection across the y-axis and a translation down 2 units

B a reflection across the x-axis and a translation down 2 units

C a rotation 90° to the right and a translation down 2 units

D a rotation 90° to the right and a translation right 2 units

70. SHORT RESPONSE A local theater sold 1200 tickets during the opening weekend of a movie. On the following weekend, 840 tickets were sold. What was the percent decrease of tickets sold? **30%**

Spiral Review

71. Simplify $3x + 8y + 5z - 2y - 6x + z$. (Lesson 1-2) **$-3x + 6y + 6z$**

72. BAKING Tamera is making two types of bread. The first type of bread needs $2\frac{1}{2}$ cups of flour, and the second needs $1\frac{3}{4}$ cups of flour. Tamera wants to make 2 loaves of the first recipe and 3 loaves of the second recipe. How many cups of flour does she need? (Lesson 1-2) **$10\frac{1}{4}$ c**

73. LANDMARKS Suppose the Space Needle in Seattle, Washington, casts a 220-foot shadow at the same time a nearby tourist casts a 2-foot shadow. If the tourist is $5\frac{1}{2}$ feet tall, how tall is the Space Needle? (Lesson 0-7) **605 ft**

h ft

5.5 ft

2 ft   220 ft

74. Evaluate $a - [c(b - a)]$, if $a = 5$, $b = 7$, and $c = 2$. (Lesson 1-1) **1**

Skills Review

Identify the additive inverse for each number or expression.

75. $-4\frac{1}{5}$ **$4\frac{1}{5}$** **76.** 3.5 **-3.5** **77.** $-2x$ **$2x$** **78.** $6 - 7y$ **$-6 + 7y$**

79. $3\frac{2}{3}$ **$-3\frac{2}{3}$** **80.** -1.25 **1.25** **81.** $5x$ **$-5x$** **82.** $4 - 9x$ **$-4 + 9x$**

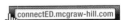

4 Assess

Yesterday's News Have each student write how yesterday's concepts helped with today's new material.

Formative Assessment
Check for student understanding of concepts in Lesson 1-3.

📁 Quiz 2, p. 45

Formative Assessment

Use the Mid-Chapter Quiz to assess students' progress in the first half of the chapter.

For problems answered incorrectly, have students review the lessons indicated in parentheses.

Summative Assessment
📁 Mid-Chapter Test, p. 47

McGraw-Hill eAssessment Customize and create multiple versions of your Mid-Chapter Quiz and their answer keys.

FOLDABLES StudyOrganizer

Dinah Zike's Foldables®
Before students complete the Mid-Chapter Quiz, encourage them to review the information for Lessons 1-1 through 1-3 in their Foldables.

The **Mid-Chapter Quiz** reviews skills and concepts presented in previous lessons.

Mid-Chapter Quiz
Lessons 1-1 through 1-3

1. Evaluate $3c - 4(a + b)$ if $a = -1$, $b = 2$ and $c = \frac{1}{3}$. (Lesson 1-1) **−3**

2. **TRAVEL** The distance that Maurice traveled in 2.5 hours riding his bicycle can be found by using the formula $d = rt$, where d is the distance traveled, r is the rate, and t is the time. How far did Maurice travel if he traveled at a rate of 16 miles per hour? (Lesson 1-1) **40 m**

3. Evaluate $(5 - m)^3 + n(m - n)$ if $m = 6$ and $n = -3$. (Lesson 1-1) **−28**

4. **GEOMETRY** The formula for the surface area of the rectangular prism below is given by the formula $S = 2xy + 2yz + 2xz$. What is the surface area of the prism if $x = 2.2$, $y = 3.5$, and $z = 5.1$? (Lesson 1-1) **73.54 units²**

5. **MULTIPLE CHOICE** What is the value of $\frac{q^2 + rt}{qr - 2t}$ if $q = -4$, $r = 3$, and $t = 8$? (Lesson 1-1) **B**

   A $-\frac{17}{6}$

   B $-\frac{10}{7}$

   C $-\frac{2}{7}$

   D $-\frac{1}{6}$

Name the sets of numbers to which each number belongs. (Lesson 1-2)

6. $\frac{25}{11}$ **Q, R**

7. $-\frac{128}{32}$ **Z, Q, R**

8. $\sqrt{50}$ **I, R**

9. -32.4 **Q, R**

10. What is the property illustrated by the equation $(4 + 15)7 = 4 \cdot 7 + 15 \cdot 7$? (Lesson 1-2) **Dist.**

11. Simplify $-3(7a - 4b) + 2(-3a + b)$. (Lesson 1-2) **−27a + 14b**

12. **CLOTHES** Brittany is buying T-shirts and jeans for her new job. T-shirts cost $10.50, and jeans cost $26.50. She buys 3 T-shirts and 3 pairs of jeans. Illustrate the Distributive Property by writing two expressions representing how much Brittany spent. (Lesson 1-2) **3(10.50 + 26.50) or 3(10.50) + 3(26.50)**

13. **MULTIPLE CHOICE** Which expression is equivalent to $\frac{2}{3}(4m - 5n) + \frac{1}{5}(2m + n)$? (Lesson 1-2) **F**

    F $\frac{46}{15}m - \frac{47}{15}n$

    G $46m - 47n$

    H $-\frac{mn}{15}$

    J $\frac{5}{4}m - \frac{9}{8}n$

14. Identify the additive inverse and the multiplicative inverse for $\frac{7}{6}$. (Lesson 1-2) **additive: $-\frac{7}{6}$; mult.: $\frac{6}{7}$**

15. Write a verbal sentence to represent the equation $\frac{a}{a-3} = 1$. (Lesson 1-3) **The quotient of a number a and the difference of a number a and 3 is equal to 1.**

16. Solve $6x + 4y = -1$ for x. (Lesson 1-3) $x = -\frac{2}{3}y - \frac{1}{6}$

17. **MULTIPLE CHOICE** Which algebraic expression represents the verbal expression, the product of 4 and the difference of a number and 13? (Lesson 1-3) **B**

    A $4n - 13$

    B $4(n - 13)$

    C $\frac{4}{n - 13}$

    D $\frac{4n}{13}$

18. Solve $-3(6x + 5) + 2(4x) = 20$. (Lesson 1-3) $-\frac{7}{2}$

19. What is the height of the trapezoid below? (Lesson 1-3) **7.5 units**

20. **GEOMETRY** The formula for the surface area of a sphere is $SA = 4\pi r^2$, and the formula for the volume of a sphere is $V = \frac{4}{3}\pi r^3$. (Lesson 1-3) **a. $\frac{32}{3}\pi$ in³; 16π in²**

    a. Find the volume and surface area of a sphere with radius 2 inches. Write your answers in terms of π.

    b. Is it possible for a sphere to have the same numerical value for the surface area and volume? If so, find the radius of such a sphere. **yes; 3 units**

InterventionPlanner

TIER 1 On Level **OL**	**TIER 2** Strategic Intervention **AL** approaching grade level	**TIER 3** Intensive Intervention 2 or more grades below level
If students miss about 25% of the exercises or less,	**If** students miss about 50% of the exercises,	**If** students miss about 75% of the exercises,
Then choose a resource:	**Then** choose a resource:	**Then** use *Math Triumphs, Alg. 2*, Ch. 1 and 2
SE Lessons 1-1, 1-2, and 1-3	📁 Study Guide and Intervention, pp. 5, 11, and 17	🖱 connectED.mcgraw-hill.com Extra Examples, Personal Tutor, Homework Help, Review Vocabulary
📁 Skills Practice, pp. 7, 13, and 19	🖱 connectED.mcgraw-hill.com Extra Examples, Personal Tutor, Homework Help	
TE Chapter Project		
🖱 connectED.mcgraw-hill.com Self-Check Quiz		

LESSON 1-4 Solving Absolute Value Equations

::Then	::Now	::Why?
• You solved equations using properties of equality.	**1** Evaluate expressions involving absolute values. **2** Solve absolute value equations.	• Sailors sometimes use a laser range finder to determine distances. Suppose one such range finder is accurate to within ±0.5 yard. This means that if a sailor estimating the distance to shore reads 323.1 yards on the laser range finder, the distance to shore might actually be as close as 322.6 or as far away as 323.6 yards. These extremes can be described by the equation $|E - 323.1| = 0.5$.

New Vocabulary
absolute value
empty set
constraint
extraneous solution

CCSS Common Core State Standards

Content Standards
A.SSE.1.b Interpret complicated expressions by viewing one or more of their parts as a single entity.
A.CED.1 Create equations and inequalities in one variable and use them to solve problems.

Mathematical Practices
6 Attend to precision.

1 Absolute Value Expressions The **absolute value** of a number is its distance from 0 on a number line. Since distance is nonnegative, the absolute value of a number is always nonnegative. The symbol $|x|$ is used to represent the absolute value of a number x.

KeyConcept Absolute Value

Words	For any real number a, if a is positive or zero, the absolute value of a is a. If a is negative, the absolute value of a is the opposite of a.
Symbols	For any real number a, $|a| = a$ if $a \geq 0$, and $|a| = -a$ if $a < 0$.
Model	$|-4| = 4$ and $|4| = 4$

When evaluating expressions, absolute value bars act as a grouping symbol. Perform any operations inside the absolute value bars first.

Example 1 Evaluate an Expression with Absolute Value

Evaluate $8.4 - |2n + 5|$ if $n = -7.5$.

$$8.4 - |2n + 5| = 8.4 - |2(-7.5) + 5| \quad \text{Replace } n \text{ with } -7.5.$$
$$= 8.4 - |-15 + 5| \quad \text{Multiply 2 and } -7.5.$$
$$= 8.4 - |-10| \quad \text{Add } -15 \text{ and } 5.$$
$$= 8.4 - 10 \quad |-10| = 10$$
$$= -1.6 \quad \text{Subtract 10 from 8.4.}$$

GuidedPractice

1A. Evaluate $|4x + 3| - 3\frac{1}{2}$ if $x = -2$. $1\frac{1}{2}$ **1B.** Evaluate $1\frac{1}{3} - |2y + 1|$ if $y = -\frac{2}{3}$. 1

---

1 Focus

VerticalAlignment

Before Lesson 1-4 Solve equations using properties of equality.

Lesson 1-4 Evaluate expressions involving absolute values. Solve absolute value equations.

After Lesson 1-4 Solve absolute value inequalities.

2 Teach

Scaffolding Questions

Have students read the **Why?** section of the lesson.

Ask:

- In $|E - 323.1| = 0.5$, what does E represent? the actual distance to shore

- What is the meaning of the number 0.5 in the equation? the degree of accuracy of the range finder

- What would be the equation for the distance to shore estimated at 962.3 yards? $|E - 962.3| = 0.5$

- How can this be shown on a number line?

961.8 962.3 962.8

Lesson 1-4 Resources

Resource	Approaching Level **AL**	On Level **OL**	Beyond Level **BL**	English Learners **ELL**
Teacher Edition	• Differentiated Instruction, p. 29	• Differentiated Instruction, pp. 29, 32	• Differentiated Instruction, p. 32	• Differentiated Instruction, p. 29
Chapter Resource Masters	• Study Guide and Intervention, pp. 24–25 • Skills Practice, p. 26 • Practice, p. 27 • Word Problem Practice, p. 28	• Study Guide and Intervention, pp. 24–25 • Skills Practice, p. 26 • Practice, p. 27 • Word Problem Practice, p. 28 • Enrichment, p. 29 • Spreadsheet Activity, p. 30	• Practice, p. 27 • Word Problem Practice, p. 28 • Enrichment, p. 29	• Study Guide and Intervention, pp. 24–25 • Skills Practice, p. 26 • Practice, p. 27 • Word Problem Practice, p. 28
Other	• 5-Minute Check 1-4 • Study Notebook	• 5-Minute Check 1-4 • Study Notebook	• 5-Minute Check 1-4 • Study Notebook	• 5-Minute Check 1-4 • Study Notebook

OLIVIER MORIN/AFP/Getty Images

1 Absolute Value Expressions

Example 1 shows how to evaluate expressions that contain absolute values.

Formative Assessment

Use the Guided Practice exercises after each example to determine students' understanding of concepts.

Additional Example

1 Evaluate $2.7 + |6 - 2x|$ if $x = 4$. 4.7

Additional Examples also in Interactive Classroom PowerPoint® Presentations

 Interactive White Board READY

Tips for New Teachers

Reading Students may find it helpful to read the first absolute value bar as "the distance of" and the last absolute value bar as "from zero, without regard to direction." So, the expression $|6 - 2x|$ would be read as "the distance of the value of $6 - 2x$ from zero, without regard to direction."

2 Absolute Value Equations

Examples 2–4 show how to solve absolute value equations with two solutions, no solution, and one solution.

Additional Example

2 Solve $|y + 3| = 8$. Check your solutions. $\{-11, 5\}$

Real-WorldLink

Originally, players used leather gloves to hit tennis balls. Soon after, the glove was placed at the end of a stick to extend the reach of the "hand."

Source: The Cliff Richard Tennis Foundation

Problem-SolvingTip

Write an Equation
Frequently, the best way to solve a problem is to use the given information to write and solve an equation.

---

2 Absolute Value Equations

Some equations contain absolute value expressions. The definition of absolute value is used in solving these equations. For any real numbers a and b, where $b \geq 0$, if $|a| = b$, then $a = b$ or $-a = b$. This second case is often written as $a = -b$.

 Real-World Example 2 Solve an Absolute Value Equation

TENNIS A standard adult tennis racket has a 100-square-inch head, plus or minus 20 square inches. Write and solve an absolute value equation to determine the least and greatest possible sizes for the head of an adult tennis racket.

Understand We need to determine the greatest and least possible sizes for the head of a tennis racket given the middle size and the range in sizes.

Plan When writing an absolute value equation, the middle or *central* value is always placed inside the absolute value symbols. The *range* is always placed on the other side of the equality symbol.

$$\underset{\text{central value}}{\underbrace{\qquad}} \qquad \underset{\text{range}}{\underbrace{\qquad}}$$
$$|x - c| = r$$

Solve
$|x - c| = r$ *Absolute value equation*

$|x - 100| = 20$ *$c = 100$, and $r = 20$*

Case 1 $\quad a = b$	**Case 2** $\quad a = -b$
$x - 100 = 20$	$x - 100 = -20$
$x - 100 + 100 = 20 + 100$	$x - 100 + 100 = -20 + 100$
$x = 120$	$x = 80$

Check

$|x - 100| = 20$	$|x - 100| = 20$
$|120 - 100| \overset{?}{=} 20$	$|80 - 100| \overset{?}{=} 20$
$|20| \overset{?}{=} 20$	$|-20| \overset{?}{=} 20$
$20 = 20$ ✓	$20 = 20$ ✓

On a number line, you can see that both solutions are 20 units away from 100.

The solutions are 120 and 80. The greatest size is 120 square inches and the least is 80 square inches.

▶ **Guided**Practice

Solve each equation. Check your solutions.

2A. $9 = |x + 12|$ $\{-21, -3\}$

2B. $8 = |y + 5|$ $\{-13, 3\}$

Because the absolute value of a number is always positive or zero, an equation like $|x| = -4$ is never true. Thus, it has no solution. The solution set for this type of equation is the **empty set**, symbolized by $\{ \}$ or $\varnothing$.

Rischgitz/Hulton Archive/Getty Images

Example 3 No Solution

Solve $|3x - 2| + 8 = 1$.

$\|3x - 2\| + 8 = 1$	Original equation
$\|3x - 2\| + 8 - 8 = 1 - 8$	Subtract 8 from each side.
$\|3x - 2\| = -7$	Simplify.

This sentence is *never* true. The solution set is ∅.

▶ **Guided**Practice Solve each equation. Check your solutions.

3A. $-2|3a| = 6$ ∅   **3B.** $|4b + 1| + 8 = 0$ ∅

In mathematics, a **constraint** is a condition that a solution must satisfy. Equations can be viewed as constraints in a problem situation. The solutions of the equation meet the constraints of the problem.

Even if the correct procedure for solving the equation is used, the answers may not be actual solutions to the original equation. Such a number is called an **extraneous solution**.

StudyTip

CCSS Precision
It is possible for an absolute value equation to have only one solution. Remember to set up two cases. Then check your solutions.

Example 4 One Solution

Solve $|x + 10| = 4x - 8$. Check your solutions.

Case 1 $a = b$

$x + 10 = 4x - 8$
$10 = 3x - 8$
$18 = 3x$
$6 = x$

Case 2 $a = -b$

$x + 10 = -(4x - 8)$
$x + 10 = -4x + 8$
$5x + 10 = 8$
$5x = -2$
$x = -\frac{2}{5}$

There appear to be two solutions, 6 and $-\frac{2}{5}$.

CHECK Substitute each value in the original equation.

$|x + 10| = 4x - 8$
$|6 + 10| \stackrel{?}{=} 4(6) - 8$
$|16| \stackrel{?}{=} 24 - 8$
$16 = 16$ ✔

$|x + 10| = 4x - 8$
$\left|-\frac{2}{5} + 10\right| \stackrel{?}{=} 4\left(-\frac{2}{5}\right) - 8$
$\left|9\frac{3}{5}\right| \stackrel{?}{=} -1\frac{3}{5} - 8$
$9\frac{3}{5} \neq -9\frac{3}{5}$ ✗

Because $9\frac{3}{5} \neq -9\frac{3}{5}$, the only solution is 6. The solution set is {6}.

▶ **Guided**Practice Solve each equation. Check your solutions.

4A. $2|x + 1| - x = 3x - 4$ {3}   **4B.** $3|2x + 2| - 2x = x + 3$ {−1}

Additional Examples

3 Solve $|6 - 4t| + 5 = 0$. ∅

4 Solve $|8 + y| = 2y - 3$. Check your solutions. {11}

CCSS Teaching the Mathematical Practices

Precision Mathematically proficient students calculate accurately and efficiently, express numerical answers with a degree of precision appropriate for the problem context. Encourage students to verify their answers by substituting them into the original equations.

WatchOut!

Preventing Errors Remind students to think about the meaning of the mathematical sentence before they begin their calculations and again when they evaluate the reasonableness of their solution. Explain to students that they can solve verbal problems when they ask questions about words they do not understand, take time to read, understand, and plan, using a sketch to help.

Teach with Tech

Interactive Whiteboard On the board, work through several examples solving absolute value equations. Save your work to a file and send it to your students so they can use it as an additional reference.

DifferentiatedInstruction

▶ **If** students found anything from the lesson confusing,

▶ **Then** ask them to record two or three of the confusing items separately on an index card. Have them write an explanation or example for each item in their own words. This will help them review in the future.

Formative Assessment
Use Exercises 1–13 to check for understanding.

Use the chart at the bottom of this page to customize assignments for your students.

 Teaching the Mathematical Practices

Modeling Mathematically proficient students can apply the mathematics they know to solve problems arising in everyday life, analyze relationships mathematically to draw conclusions, and interpret their mathematical results in the context of a situation.

Multiple Representations
In Exercise 44, students use a number line, information organized in a table, and algebra to analyze inequalities.

Additional Answers

43. $|x - 100| = 245$; $x = 345$ or -145; maximum: 345 ft above sea level; minimum: 145 ft below sea level. No, the maximum is reasonable but the minimum is not. Florida's lowest point should be at sea level where Florida meets the Atlantic Ocean and the Gulf of Mexico.

44a. Sample answer:

44c. Sample answer: If A is less than B, then any number added to or subtracted from A will be less than the same number added to or subtracted from B. If B is greater than A, then any number added to or subtracted from B is greater than the same number added to or subtracted from A.

Check Your Understanding

  = Step-by-Step Solutions begin on page R14.

Example 1 Evaluate each expression if $x = -4$ and $y = -9$. 4.

1. $|x - 8|$ 12   **2.** $|7y|$ 63   **3.** $-3|xy|$ -108 **4.** $-2|3x + 8| - 4$ -12

5. CCSS **MODELING** Most freshwater tropical fish thrive if the water is within 2°F of 78°F.

   a. Write an equation to determine the least and greatest optimal temperatures. $|x - 78| = 2$

5b. least: 76°F, greatest: 80°F

   b. Solve the equation you wrote in part **a.**

   c. If your aquarium's thermometer is accurate to within plus or minus 1°F, what should the temperature of the water be to ensure that it reaches the minimum temperature? Explain. 77°F; This would ensure a minimum temperature of 76°F.

Examples 2–4 Solve each equation. Check your solutions.

6. $|x + 8| = 12$ $\{4, -20\}$ **7.** $|y - 4| = 11$ $\{15, -7\}$

8. $|a - 5| + 4 = 9$ $\{10, 0\}$ **9.** $|b - 3| + 8 = 3$ $\varnothing$

10. $3|2x - 3| - 5 = 4$ $\{3, 0\}$ **11.** $-2|5y - 1| = -10$ $\left\{\frac{6}{5}, -\frac{4}{5}\right\}$

12. $|a - 4| = 3a - 6$ $\{2.5\}$ **13.** $|b + 5| = 2b + 3$ $\{2\}$

Practice and Problem Solving

Extra Practice is on page R1.

Example 1 Evaluate each expression if $a = -3$, $b = -5$, and $c = 4.2$.

14. $|-3c|$ 12.6   **15.** $|5b|$ 25   **16.** $|a - b|$ 2   **17.** $|b - c|$ 9.2

18. $|3b - 4a|$ 3   **19.** $2|4a - 3c|$ 49.2   **20.** $-|3c - a|$ -15.6 **21.** $-|abc|$ -63

22. **FOOD** To make cocoa powder, cocoa beans are roasted. The ideal temperature for roasting is 300°F, plus or minus 25°. Write and solve an equation describing the maximum and minimum roasting temperatures for cocoa beans. $|x - 300| = 25$; maximum: 325°F; minimum: 275°F

Examples 2–4 Solve each equation. Check your solutions.

23. $|z - 13| = 21$ $\{34, -8\}$ **24.** $|w + 9| = 17$ $\{8, -26\}$

25. $9 = |d + 5|$ $\{4, -14\}$ **26.** $35 = |x - 6|$ $\{-29, 41\}$

27. $5|q + 6| = 20$ $\{-2, -10\}$ **28.** $-3|r + 4| = -21$ $\{3, -11\}$

29. $3|2a - 4| = 0$ $\{2\}$ **30.** $8|5w - 1| = 0$ $\left\{\frac{1}{5}\right\}$

31 $2|3x - 4| + 8 = 6$ $\varnothing$ **32.** $4|7y + 2| - 8 = -7$ $\left\{-\frac{1}{4}, -\frac{9}{28}\right\}$

33. $-3|3t - 2| - 12 = -6$ $\varnothing$ **34.** $-5|3z + 8| - 5 = -20$ $\left\{-\frac{5}{3}, -\frac{11}{3}\right\}$

B **35.** **MONEY** The U.S. Mint produces quarters that weigh about 5.67 grams each. After the quarters are produced, a machine weighs them. If the quarter weighs 0.02 gram more or less than the desired weight, the quarter is rejected. Write and solve an equation to find the heaviest and lightest quarters the machine will approve. $|x - 5.67| = 0.02$; heaviest: 5.69 g; lightest: 5.65 g

Evaluate each expression if $q = -8$, $r = -6$, and $t = 3$.

36. $12 - t|3r + 2|$ -36 **37.** $2q + |2rt + q|$ 28   **38.** $-5t - q|8r - t|$ 393

Differentiated Homework Options

Level	Assignment	Two-Day Option	
AL Basic	14–34, 45, 47–74	15–33 odd, 52–55	14–34 even, 45, 47–51, 56–74
OL Core	15–43 odd, 44, 45, 47–74	14–34, 52–55	35–45, 47–51, 56–74
BL Advanced	35–68, (optional: 69–74)		

Solve each equation. Check your solutions.

39. $8x = 2|6x - 2|$ $\left\{1, \dfrac{1}{5}\right\}$

40. $-6y + 4 = |4y + 12|$ $\left\{-\dfrac{4}{5}\right\}$

41. $8z + 20 = -|2z + 4|$ $\left\{-\dfrac{8}{3}\right\}$

42. $-3y - 2 = |6y + 25|$ $\left\{-3, -\dfrac{23}{3}\right\}$

43. **SEA LEVEL** Florida is on average 100 feet above sea level. This level varies by as much as 245 feet depending on precipitation and your location. Write and solve an equation describing the maximum and minimum sea levels for Florida. Is this solution reasonable? Explain. **See margin.**

44. **MULTIPLE REPRESENTATIONS** Draw a number line. **a. See margin.**

   a. Geometric Label any 5 integers on the number line points A, B, C, D, and F.

   b. Tabular Fill in each blank in the table with either $>$ or $<$ using the points from the number line.

$A \underline{<} B$	$A + C \underline{<} B + C$ $A + D \underline{<} B + D$ $A + F \underline{<} B + F$
$B \underline{>} A$	$B + C \underline{>} A + C$ $B + D \underline{>} A + D$ $B + F \underline{>} A + F$

$A \underline{<} B$	$A - C \underline{<} B - C$ $A - D \underline{<} B - D$ $A - F \underline{<} B - F$
$B \underline{>} A$	$B - C \underline{>} A - C$ $B - D \underline{>} A - D$ $B - F \underline{>} A - F$

   c. Verbal Describe the patterns in the table. **See margin.**

   d. Algebraic Describe the patterns algebraically, using the variable x to replace C, D, and F. If $A < B$, then $A + x < B + x$. If $A < B$, then $A - x < B - x$. If $B > A$, then $B + x > A + x$. If $B > A$, then $B - x > A - x$.

H.O.T. Problems — *Use Higher-Order Thinking Skills*

45. **CCSS CRITIQUE** Ana and Ling are solving $|3x + 14| = -6x$. Is either of them correct? Explain your reasoning.

Ana
$|3x + 14| = -6x$
$3x + 14 = -6x$ or $3x + 14 = 6x$
$9x = -14$ $14 = 3x$
$x = -\dfrac{14}{9}$ ✓ $x = \dfrac{14}{3}$ ✓

Ling
$|3x + 14| = -6x$
$3x + 14 = -6x$ or $3x + 14 = 6x$
$9x = -14$ $14 = 3x$
$x = -\dfrac{14}{9}$ ✓ $x = \dfrac{14}{3}$ ✗

45. Ling; Ana included an extraneous solution. She would have caught this error if she had checked to see if her answers were correct by substituting the values into the original equation.

46. **CHALLENGE** Solve $|2x - 1| + 3 = |5 - x|$. List all cases and resulting equations. (*Hint*: There are four possible cases to examine as potential solutions.) **See Chapter 1 Answer Appendix.**

REASONING If a, x, and y are real numbers, determine whether each statement is *sometimes*, *always*, or *never* true. Explain your reasoning.

47. If $|a| > 7$, then $|a + 3| > 10$. **47–49. See Chapter 1 Answer Appendix.**

48. If $|x| < 3$, then $|x| + 3 > 0$.

49. If y is between 1 and 5, then $|y - 3| \le 2$.

50. **OPEN ENDED** Write an absolute value equation of the form $|ax + b| = cx + d$ that has no solution. Assume that a, b, c, and $d \ne 0$.

50. Sample answer: $|2x + 1| = x - 3$, or $|3x + 10| = x - 5$, or $|x - 1| = \dfrac{1}{2}x - 4$

51. **WRITING IN MATH** How are symbols used to represent mathematical ideas? Use an example to justify your reasoning. **See Chapter 1 Answer Appendix.**

CCSS Teaching the Mathematical Practices

Critique Mathematically proficient students are also able to compare the effectiveness of two plausible arguments, distinguish correct logic or reasoning from that which is flawed, and—if there is a flaw in an argument—explain what it is.

Crystal Ball Have students write how they think what they have learned in Lessons 1-3 and 1-4 will connect with Lesson 1-5, *Solving Inequalities.*

Tips for New Teachers

Looking Ahead Lesson 1-5 presents solving inequalities using steps similar to those for solving equations. Exercises 69–74 should be used to determine students' familiarity with solving equations.

Standardized Test Practice

52. If $4x - y = 3$ and $2x + 3y = 19$, what is the value of y? **D**

A 2
B 3
C 4
D 5

53. **GRIDDED RESPONSE** Two male and 2 female students from each of the 9th, 10th, 11th, and 12th grades comprise the Student Council. If a Student Council representative is chosen at random to attend a board meeting, what is the probability that the student will be either an 11th grader or male? **5/8**

54. Which equation is equivalent to $4(9 - 3x) = 7 - 2(6 - 5x)$? **G**

F $8x = 41$
G $22x = 41$
H $8x = 24$
J $22x = 24$

55. **SAT/ACT** A square with side length 4 units has one vertex at the point $(1, 2)$. Which of the following points *cannot* be diagonally opposite that vertex? **E**

A $(-3, -2)$ D $(5, 6)$
B $(-3, 6)$ E $(1, 6)$
C $(5, -2)$

Spiral Review

Solve each equation. Check your solution. (Lesson 1-3)

56. $4x + 6 = 30$ **6**

57. $5p - 10 = 4(7 + 6p)$ **−2**

58. $\frac{3}{5}y - 7 = \frac{2}{5}y + 3$ **50**

59. **MONEY** Nhu is saving to buy a car. In the first 6 months, his savings were $80 less than $\frac{3}{4}$ the price of the car. In the second six months, Nhu saved $50 more than $\frac{1}{5}$ the price of the car. He still needs $370. (Lesson 1-3)

a. What is the price of the car? **$6800**

b. What is the average amount of money Nhu saved each month? **$535.83**

c. If Nhu continues to save the average amount each month, in how many months will he be able to afford the car? **1 mo**

Name the property illustrated by each equation. (Lesson 1-2)

60. $(1 + 8) + 11 = 11 + (1 + 8)$ **Comm. (+)**

61. $z(9 - 4) = z \cdot 9 - z \cdot 4$ **Distributive**

Simplify each expression. (Lesson 1-2)

62. $7a + 3b - 4a - 5b$ **$3a - 2b$**

63. $3x + 5y + 7x - 3y$ **$10x + 2y$**

64. $3(15x - 9y) + 5(4y - x)$ **$40x - 7y$**

65. $2(10m - 7a) + 3(8a - 3m)$ **$11m + 10a$**

66. $8(r + 7t) - 4(13t + 5r)$ **$-12r + 4t$**

67. $4(14c - 10d) - 6(d + 4c)$ **$32c - 46d$**

68. **GEOMETRY** The formula for the surface area of a rectangular prism is $SA = 2\ell w + 2\ell h + 2wh$, where ℓ represents the length, w represents the width, and h represents the height. Find the surface area of the rectangular prism at the right. (Lesson 1-1) **$358\ \text{in}^2$**

7 in.
5 in.
12 in.

Skills Review

Solve each equation.

69. $15x + 5 = 35$ **2**

70. $2.4y + 4.6 = 20$ **≈6.417**

71. $8a + 9 = 6a - 7$ **−8**

72. $3(w - 1) = 2w - 6$ **−3**

73. $\frac{1}{2}(2b - 4) = 2 + 8b$ **$-\frac{4}{7}$**

74. $\frac{1}{3}(6p - 24) = 18 + 3p$ **−26**

DifferentiatedInstruction OL BL

Extension For equations with one set of absolute value symbols, two cases must be considered. For an equation with two sets of absolute value symbols, four cases must be considered. How many cases must be considered for an equation containing three sets of absolute value symbols? **8**

LESSON 1-5

Solving Inequalities

:: Then	:: Now	:: Why?
• You solved equations involving absolute values.	**1** Solve one-step inequalities. **2** Solve multi-step inequalities.	• Josh is trying to decide between two text messaging plans offered by a wireless telephone company. To compare these two rate plans, we can use inequalities. The monthly access fee for Plan 1 is less than the fee for Plan 2, $55 < $60. However, the additional text messaging fee for Plan 1 is greater than that of Plan 2, $0.25 > $0.20.

	Plan 1	Plan 2
Monthly Fee	$55	$60
Text Messages Included	400	650
Additional Text Messages	$0.25	$0.20

NewVocabulary
set-builder notation

Common Core State Standards

Content Standards
A.CED.1 Create equations and inequalities in one variable and use them to solve problems.

A.CED.3 Represent constraints by equations or inequalities, and by systems of equations and/or inequalities, and interpret solutions as viable or nonviable options in a modeling context.

Mathematical Practices
4 Model with mathematics.

1 One-Step Inequalities For any two real numbers, a and b, exactly one of the following statements is true.

$$a < b \qquad a = b \qquad a > b$$

Adding the same number to, or subtracting the same number from, each side of an inequality does not change the truth of the inequality.

KeyConcept

Addition Property of Inequality

Words

For any real numbers, a, b, and c:

If $a > b$, then $a + c > b + c$.

If $a < b$, then $a + c < b + c$.

Models

Subtraction Property of Inequality

Words

For any real numbers, a, b, and c:

If $a > b$, then $a - c > b - c$.

If $a < b$, then $a - c < b - c$.

Models

These properties are also true for $\leq$, $\geq$, and $\neq$.

These properties can be used to solve inequalities. The solution sets of inequalities in one variable can then be graphed on number lines.

connectED.mcgraw-hill.com   33

McGraw-Hill Companies, Inc.

---

1 Focus

VerticalAlignment

Before Lesson 1-5 Solve equations involving absolute values.

Lesson 1-5 Solve one-step inequalities.
Solve multi-step inequalities.

After Lesson 1-5 Solve systems of inequalities.

2 Teach

Scaffolding Questions

Have students read the **Why?** section of the lesson.

Ask:
- If Josh knows that he will text no more than 400 messages per month, which plan is better for him? Plan 1
- How much would he pay if he texts 800 messages under Plan 1? Plan 2? $155; $90
- If he texts 500 messages per month, which plan is better for him? Plan 2

Lesson 1-5 Resources

Resource	Approaching Level **AL**	On Level **OL**	Beyond Level **BL**	English Learners **ELL**
Teacher Edition	• Differentiated Instruction, p. 35	• Differentiated Instruction, pp. 35, 39	• Differentiated Instruction, pp. 35, 39	• Differentiated Instruction, p. 35
Chapter Resource Masters	• Study Guide and Intervention, pp. 31–32 • Skills Practice, p. 33 • Practice, p. 34 • Word Problem Practice, p. 35	• Study Guide and Intervention, pp. 31–32 • Skills Practice, p. 33 • Practice, p. 34 • Word Problem Practice, p. 35 • Enrichment, p. 36	• Practice, p. 34 • Word Problem Practice, p. 35 • Enrichment, p. 36	• Study Guide and Intervention, pp. 31–32 • Skills Practice, p. 33 • Practice, p. 34 • Word Problem Practice, p. 35
Other	• 5-Minute Check 1-5 • Study Notebook	• 5-Minute Check 1-5 • Study Notebook	• 5-Minute Check 1-5 • Study Notebook	• 5-Minute Check 1-5 • Study Notebook

1 One-Step Inequalities

Example 1 shows how to solve an inequality using addition or subtraction.
Example 2 shows how to solve an inequality using multiplication or division.

Formative Assessment

Use the Guided Practice exercises after each example to determine students' understanding of concepts.

Additional Example

1 Solve $4y - 3 < 5y + 2$. Graph the solution set on a number line. $y > -5$

Additional Examples also in Interactive Classroom PowerPoint® Presentations

IWB Interactive White Board READY

WatchOut!

Preventing Errors Ask students if it makes a difference whether the inequality symbol is $<$, $>$, $\leq$, or $\geq$ when they use the Addition and Subtraction Properties of Inequality. There is no difference in calculations, but there is a difference in the direction and beginning of the graph of the solution set.

Additional Answers (Guided Practice)

1A. $w > 6$

1B. $x > 5$

2A. $x \leq 6$

2B. $y > -2.5$

3A. $x \geq -2$

3B. $y \leq -\dfrac{1}{3}$

3C. $v \leq 6$

3D. $d < -\dfrac{1}{3}$

ReviewVocabulary

Inequality Symbols

> greater than; is more than

< less than; is fewer than

≥ greater than or equal to; is at least; is no less than

≤ less than or equal to; is at most; is no more than

Example 1 Solve an Inequality Using Addition or Subtraction

Solve $y - 6 < 3$. Graph the solution set on a number line.

$y - 6 < 3$	Original inequality
$y - 6 + 6 < 3 + 6$	Add 6 to each side.
$y < 9$	Simplify.

Any real number less than 9 is a solution of this inequality. The graph of the solution set is shown at the right.

> A circle means that this point is *not* included in the solution set.

CHECK Substitute 8 and then 10 for y in $y - 6 < 3$. The inequality should be true for $y = 8$ and false for $y = 10$. ✓

▶ **Guided**Practice 1A, 1B. See margin.

Solve each inequality. Graph the solution set on a number line.

1A. $5w + 3 > 4w + 9$
1B. $5x - 3 > 4x + 2$

StudyTip

Graphing Inequalities A circle is used for $<$ and $>$. A dot is used for $\leq$ and $\geq$.

Multiplying or dividing each side of an inequality by a positive number does not change the truth of the inequality. However, multiplying or dividing each side of an inequality by a *negative* number requires that the order of the inequality be *reversed*. For example, to reverse $\leq$, replace it with $\geq$.

KeyConcept

Multiplication Property of Inequality	
Words	**Examples**
For any real numbers, a, b, and c,	$-5 < -3$
where c is positive:	$-5(6) < -3(6)$
If $a > b$, then $ac > bc$.	$-30 < -18$
If $a < b$, then $ac < bc$.	
where c is negative:	$12 > -7$
If $a > b$, then $ac < bc$.	$12(-4) < -7(-4)$
If $a < b$, then $ac > bc$.	$-48 < 28$

Division Property of Inequality	
Words	**Examples**
For any real numbers, a, b, and c,	$-12 < -8$
where c is positive:	$\dfrac{-12}{4} < \dfrac{-8}{4}$
If $a > b$, then $\dfrac{a}{c} > \dfrac{b}{c}$.	$-3 < -2$
If $a < b$, then $\dfrac{a}{c} < \dfrac{b}{c}$.	
where c is negative:	$-21 > -14$
If $a > b$, then $\dfrac{a}{c} < \dfrac{b}{c}$.	$\dfrac{-21}{-7} < \dfrac{-14}{-7}$
If $a < b$, then $\dfrac{a}{c} > \dfrac{b}{c}$.	$3 > 2$

These properties are also true for $\leq$, $\geq$, and $\neq$.

Key Concept boxes highlight definitions, formulas, and other important ideas. Multiple representations—words, symbols, examples, models—aid students' understanding.

 34 | Lesson 1-5 | Solving Inequalities

ReadingMath

Set-Builder Notation
$\{y \mid y < 9\}$ is read *the set of all numbers y such that y is less than 9.*

The solution set of an inequality can be expressed by using **set-builder notation**. For example, the solution set in Example 1 can be expressed as $\{y \mid y < 9\}$.

Example 2 Solve an Inequality Using Multiplication or Division

Solve $-4.2x \leq -29.4$. Graph the solution set on a number line.

$-4.2x \leq -29.4$	Original inequality
$\dfrac{-4.2x}{-4.2} \geq \dfrac{-29.4}{-4.2}$	Divide each side by -4.2, reversing the inequality symbol.
$x \geq 7$	Simplify.

The solution set is $\{x \mid x \geq 7\}$. The graph of the solution is shown below.

A dot means that this point is included in the solution set.

CHECK Substitute 6 and then 8 for x in $-4.2x \leq -29.4$. The inequality should be true for $x = 8$ and false for $x = 6$. ✓

GuidedPractice **2A, 2B.** See margin.

Solve each inequality. Graph the solution set on a number line.

2A. $-4x \geq -24$ **2B.** $-9.2y < 23$

---

2 Multi-Step Inequalities

Solving multi-step inequalities is similar to solving multi-step equations.

Example 3 Solve Multi-Step Inequalities

Solve $-4c \leq \dfrac{5c + 58}{6}$. Graph the solution set on a number line.

$-4c \leq \dfrac{5c + 58}{6}$	Original inequality
$-24c \leq 5c + 58$	Multiply each side by 6.
$-29c \leq 58$	Add $-5c$ to each side.
$c \geq -2$	Divide each side by -29, reversing the inequality symbol.

The solution set is $\{c \mid c \geq -2\}$ and is graphed below.

CHECK Substitute -3 and then -1 for x in $-4c \leq \dfrac{5c + 58}{6}$. The inequality should be true for $x = -1$ and false for $x = -3$. ✓

GuidedPractice **3A–3D.** See margin.

Solve each inequality. Graph the solution set on a number line.

3A. $-3x \leq \dfrac{-4x + 22}{5}$ **3B.** $8y \geq \dfrac{-5y + 9}{-4}$

3C. $-6(-4v + 3) \leq 2(10v + 3)$ **3D.** $-5(3d - 7) > 3(2d + 14)$

StudyTip

When multiplying or dividing by a negative number, remember to reverse the inequality symbol.

connectED.mcgraw-hill.com   **35**

Additional Example

2 Solve $12 \geq -0.3p$. Graph the solution set on a number line.
$\{p \mid p \geq -40\}$

2 Multi-Step Inequalities

Example 3 shows how to solve a multi-step inequality. **Example 4** shows how to write and use an inequality to solve a real-world problem.

Additional Example

3 Solve $-x > \dfrac{x - 7}{2}$. Graph the solution set on a number line.
$\left\{x \mid x < \dfrac{7}{3}\right\}$

WatchOut!

Preventing Errors Remind students that in order to keep each intermediate inequality equivalent to the original, they must show the division by a negative **and** the reversal of the inequality sign in the same step.

Focus on Mathematical Content

Properties of Inequality Adding the same number to, or subtracting the same number from, each side of an inequality does not change the truth of the inequality. Multiplying or dividing each side of an inequality by a positive number also does not change the truth of the inequality. However, multiplying or dividing each side of an inequality by a negative number does change the truth of the inequality.

DifferentiatedInstruction   (AL) (OL) (BL) (ELL)

Social Learners Have students first discuss the differences between solving an equation and solving an inequality and then discuss how the solution processes are the same.

Additional Example

4 **CONSUMER COSTS** Javier has at most $15.00 to spend today. He buys a bag of pretzels and a bottle of juice for $1.59. If gasoline at this store costs $2.89 per gallon, how many gallons of gasoline, to the nearest tenth of a gallon, can Javier buy for his car? no more than 4.6 gal

3 Practice

Formative Assessment

Use Exercises 1–9 to check for understanding.

Use the chart on the bottom of the next page to customize assignments for your students.

Additional Answers

1. $b < 8$

$$-10\ -8\ -6\ -4\ -2\ \ 0\ \ 2\ \ 4\ \ 6\ \ 8\ \ 10$$

2. $d < 20$

$$-20\ -16\ -12\ -8\ -4\ \ 0\ \ 4\ \ 8\ \ 12\ \ 16\ \ 20$$

3. $x \le -6$

$$-10\ -8\ -6\ -4\ -2\ \ 0\ \ 2\ \ 4\ \ 6\ \ 8\ \ 10$$

4. $y \le 7$

$$-10\ -8\ -6\ -4\ -2\ \ 0\ \ 2\ \ 4\ \ 6\ \ 8\ \ 10$$

5. $w < 2$

$$-5\ -4\ -3\ -2\ -1\ \ 0\ \ 1\ \ 2\ \ 3\ \ 4\ \ 5$$

6. $z \ge -\dfrac{3}{4}$

$$-5\ -4\ -3\ -2\ -1\ \ 0\ \ 1\ \ 2\ \ 3\ \ 4\ \ 5$$

7. $s \ge \dfrac{3}{2}$

$$-5\ -4\ -3\ -2\ -1\ \ 0\ \ 1\ \ 2\ \ 3\ \ 4\ \ 5$$

8. $x \ge -8.5$

$$-10\ -8\ -6\ -4\ -2\ \ 0\ \ 2\ \ 4\ \ 6\ \ 8\ \ 10$$

Real-World Link
In 2007, the Netcraft Web Server Survey found over 108,000,000 distinct Web sites.
Source: Netcraft

Real-World Example 4 Write and Solve an Inequality

WEB SITES Enrique's company pays Salim to advertise on Salim's Web site. Salim's Web site earns $15 per month plus $0.05 every time a visitor clicks on the advertisement. What is the least number of clicks per month that Salim needs in order to earn $50 per month or more?

Understand Let c = the number of clicks on the advertisement. Salim earns $15 per month and $0.05 per click, and he wants to earn a minimum of $50 for the advertisement.

Plan Write an inequality.

Words	The monthly income is $15 plus $0.05 per click, and the total should be at least $50.

▼

Variable	Let c represent the number of clicks per month.

▼

Inequality	Flat fee	plus	fee per click	is at least	$50.
	15	+	$0.05c$	≥	50

Solve

$15 + 0.05c \ge 50$	Original inequality
$0.05c \ge 35$	Subtract 15 from each side.
$c \ge 700$	Divide each side by 0.05.

Check

$15 + 0.05c \ge 50$	Original inequality
$5 + 0.05(700) \overset{?}{\ge} 50$	Replace c with 700.
$15 + 35 \overset{?}{\ge} 50$	Multiply.
$50 \ge 50$ ✓	Add.

Visitors to Salim's Web site need to click on Enrique's advertisement at least 700 times per month in order for Salim to earn $50 or more from Enrique's company.

▶ **Guided**Practice
Rosa can go less than 80 minutes past her free minutes and still pay less than $70.

4. Rosa's cell phone plan costs her $50 per month plus $0.25 for each minute she goes beyond her free minutes. How many minutes can she go beyond her free minutes and still pay less than a total of $70?

Check Your Understanding

⬤ = Step-by-Step Solutions begin on page R14. ☑

Examples 1–3 Solve each inequality. Then graph the solution set on a number line.

1. $b + 6 < 14$ **2.** $12 - d > -8$

1–8.
See margin.

3. $18 \le -3x$ **4.** $-5y \ge -35$

5. $-4w - 13 > -21$ **6.** $8z - 9 \ge -15$

(7) $s \ge \dfrac{s + 6}{5}$ **8.** $\dfrac{2x - 9}{4} \le x + 2$

Example 4

9. **CCSS** **MODELING** Tara is delivering bags of mulch. Each bag weighs 48 pounds, and the push cart weighs 65 pounds. If her flat-bed truck is capable of hauling 2000 pounds, how many bags of mulch can Tara safely take on each trip? **40 bags**

🔊 **36** | Lesson 1-5 | Solving Inequalities

---

10. $m > -4$

$$-5\ -4\ -3\ -2\ -1\ \ 0\ \ 1\ \ 2\ \ 3\ \ 4\ \ 5$$

11. $n \le -3$

$$-5\ -4\ -3\ -2\ -1\ \ 0\ \ 1\ \ 2\ \ 3\ \ 4\ \ 5$$

12. $r < -6$

$$-10\ -8\ -6\ -4\ -2\ \ 0\ \ 2\ \ 4\ \ 6\ \ 8\ \ 10$$

13. $t \le \dfrac{1}{2}$

$$-5\ -4\ -3\ -2\ -1\ \ 0\ \ 1\ \ 2\ \ 3\ \ 4\ \ 5$$

14. $w \ge 28$

$$25\ \ 26\ \ 27\ \ 28\ \ 29\ \ 30\ \ 31\ \ 32\ \ 33\ \ 34\ \ 35$$

15. $k < 27$

$$25\ \ 26\ \ 27\ \ 28\ \ 29\ \ 30\ \ 31\ \ 32\ \ 33\ \ 34\ \ 35$$

Ed Kashi/Terra/Corbis

Examples 1–3 Solve each inequality. Then graph the solution set on a number line.

10–21.
See margin.

10. $m - 8 > -12$

11. $n + 6 \leq 3$

12. $6r < -36$

13. $-12t \geq -6$

14. $-\dfrac{w}{4} \leq -7$

15. $\dfrac{k}{3} - 14 < -5$

16. $4x - 15 \leq 21$

17. $-6z - 14 > -32$

18. $-16 \geq 5(2z - 11)$

19. $12 < -4(3c - 6)$

20. $\dfrac{3y - 4}{0.2} - 8 > 12$

21. $\dfrac{9z + 5}{4} + 18 < 26$

Example 4

22. **GYMNASTICS** In a gymnastics competition, an athlete's final score is calculated by taking 75% of the average technical score and adding 25% of the artistic score. All scores are out of 10, and one gymnast has a 7.6 average technical score. What artistic score does the gymnast need to have a final score of at least 8.0? **9.2**

B Define a variable and write an inequality for each problem. Then solve.

23. Twelve less than the product of three and a number is less than 21. $3x - 12 < 21$; $x < 11$

24. The quotient of three times a number and 4 is at least -16. $\dfrac{3x}{4} \geq -16$; $x \geq -\dfrac{64}{3}$

25. The difference of 5 times a number and 6 is greater than the number. $5x - 6 > x$; $x > 1.5$

26. The quotient of the sum of 3 and a number and 6 is less than -2. $\dfrac{x + 3}{6} < -2$; $x < -15$

27. **HIKING** Danielle can hike 3 miles in an hour, but she has to take a one-hour break for lunch and a one-hour break for dinner. If Danielle wants to hike at least 18 miles, solve $3(x - 2) \geq 18$ to determine how many hours the hike should take. **at least 8 hours**

Solve each inequality. Then graph the solution set on a number line. 28–36. See margin.

28. $18 - 3x < 12$

29. $-8(4x + 6) < -24$

30. $\dfrac{1}{4}n + 12 \geq \dfrac{3}{4}n - 4$

31. $0.24y - 0.64 > 3.86$

32. $10x - 6 \leq 4x + 42$

33. $-6v + 8 > -14v - 28$

34. $n > \dfrac{-3n - 15}{8}$

35. $-2r < \dfrac{6 - 2r}{5}$

36. $\dfrac{9z - 4}{5} \leq \dfrac{7z + 2}{4}$

37. **MONEY** Jin is selling advertising space in *Central City Magazine* to local businesses. Jin earns 3% commission for every advertisement he sells plus a salary of $250 a week. If the average amount of money that a business spends on an advertisement is $500, how many advertisements must he sell each week to make a salary of at least $700 that week? **37b.** $a \geq 30$; He must sell at least 30 advertisements.

a. Write an inequality to describe this situation. $250 + 0.03(500a) \geq 700$

b. Solve the inequality and interpret the solution.

C Define a variable and write an inequality for each problem. Then solve.

38. One third of the sum of 5 times a number and 3 is less than one fourth the sum of six times that number and 5. $\dfrac{5n + 3}{3} < \dfrac{6n + 5}{4}$; $n < 1.5$

39. The sum of one third a number and 4 is at most the sum of twice that number and 12. $\dfrac{x}{3} + 4 \leq 2x + 12$; $x \geq -4.8$

40. **CCSS SENSE-MAKING** The sides of square $ABCD$ are extended to form rectangle $DEFG$. If the perimeter of the rectangle is at least twice the perimeter of the square, what is the maximum length of a side of square $ABCD$? **9 in.**

E   F

8 in.

A   B

D   C   10 in.   G

Differentiated Homework Options

Level	Assignment	Two-Day Option	
AL Basic	10–22, 45, 46, 48–66	11–21 odd, 50–53	10–22 even, 45, 46, 48, 49, 54–66
OL Core	11–21 odd, 22–27, 29–35 odd, 37–43, 45, 46, 48–66	10–22, 50–53	23–43, 45, 46, 48, 49, 54–66
BL Advanced	23–60, (optional: 61–66)		

16. $x \leq 9$

17. $z < 3$

18. $z \leq 3.9$

19. $c < 1$

20. $y > \dfrac{8}{3}$

21. $z < 3$

28. $x > 2$

29. $x > -\dfrac{3}{4}$

30. $n \leq 32$

31. $y > 18.75$

32. $x \leq 8$

33. $v > -4.5$

34. $n > -\dfrac{15}{11}$

35. $r > -\dfrac{3}{4}$

36. $z \leq 26$

1 Focus

Objective Describe the solution set of an inequality using interval notation.

Teaching Tip

Before starting this activity, ask students to think of symbols they have used in their math studies so far.

Some examples might be a radical sign, a line segment symbol, a division sign, and so on.

2 Teach

Working in Cooperative Groups

Put students in pairs, mixing abilities. Have groups complete Exercises 1–5.

Because symbols are used in math to represent numerical language concepts, language skills are important. One way to provide assistance with math symbols is to connect new symbols with symbols students already know. Point out, for example, that $+\infty$ has a familiar symbol attached to it, the $+$ sign. Have students read the symbols in their exercises aloud. Be sure students use language for the symbols. For example, students should say "positive infinity" when they see $+\infty$ in an exercise.

Practice Have students complete Exercises 6–18.

3 Assess

Formative Assessment

Use Exercise 18 to assess whether students understand interval notation and can write a compound inequality when given interval notation.

From Concrete to Abstract

Ask students to summarize what they have learned about the symbols used in interval notation. Ask them to cite an example of each.

The solution set of an inequality can be described by using **interval notation**. The **infinity** symbols below are used to indicate that a set is unbounded in the positive or negative direction, respectively.

CCSS **Common Core State Standards**
Content Standards
A.CED.1 Create equations and inequalities in one variable and use them to solve problems.
A.CED.3 Represent constraints by equations or inequalities, and by systems of equations and/or inequalities, and interpret solutions as viable or nonviable options in a modeling context.

To indicate that an endpoint is *not* included in the set, a parenthesis, (or), is used. Parentheses are always used with the symbols $+\infty$ and $-\infty$, because they do not include endpoints.

$x < 2$

interval notation
$(-\infty, 2)$

A bracket is used to indicate that the endpoint, -2, *is* included in the solution set below.

$x \geq -2$

interval notation
$[-2, +\infty)$

In interval notation, the symbol for the union of the two sets is $\cup$. The compound inequality $y \leq -7$ or $y > -1$ is written as $(-\infty, -7] \cup (-1, +\infty)$.

Exercises

Write each inequality using interval notation.

1. $\{a \mid a \leq -3\}$ $(-\infty, -3]$

2. $\{n \mid n > -8\}$ $(-8, +\infty)$

3. $\{y \mid y < 2 \text{ or } y \geq 14\}$ $(-\infty, 2) \cup [14, +\infty)$

4. $\{b \mid b \leq -9 \text{ or } b > 1\}$ $(-\infty, -9] \cup (1, +\infty)$

5. $\{t \mid 1 < t < 3\}$ $(1, 3)$

6. $\{m \mid m \geq 4 \text{ or } m \leq -7\}$ $(-\infty, -7] \cup [4, +\infty)$

7. $\{x \mid x \geq 0\}$ $[0, +\infty)$

8. $\{r \mid -3 < r < 4\}$ $(-3, 4)$

9. $(-\infty, -1) \cup (1, +\infty)$

10. $[-5, 5]$

11. $(-8, 8)$

12. $(-\infty, -1.5] \cup [1.5, +\infty)$

13. $(-\infty, 11]$

14. $(-\infty, -5) \cup (25, +\infty)$

Graph each solution set on a number line. 15–17. See margin.

15. $(-1, -\infty)$

16. $(-\infty, 4]$

17. $(-\infty, 5] \cup (7, +\infty)$

18. **WRITING IN MATH** Write in words the meaning of $(-\infty, 3) \cup [10, +\infty)$. Then write the compound inequality that this notation represents.

19. **WRITING IN MATH** How are symbols used to write solution sets for inequalities? Explain. See margin.

18. All values less than 3, not including 3, are part of the solution set. All values greater than and including 10 are part of the solution set; $x < 3$ or $x \geq 10$.

Additional Answers

15.

16.

17.

19. Sample answer: Infinity symbols are used to indicate whether a set is unbounded in the positive or negative direction; a parenthesis is used to indicate that an endpoint is not included in the set; a bracket is used to indicate that an endpoint is included in the set; a union symbol is used to indicate the union of two sets.

LESSON 1-6

Solving Compound and Absolute Value Inequalities

∶∶ Then	∶∶ Now	∶∶ Why?
● You solved one-step and multi-step inequalities.	**1** Solve compound inequalities. **2** Solve absolute value inequalities.	● Marine biologists often have to transplant a dolphin from its natural habitat to a pool. Dolphins prefer the temperature of water to be at least 22°C but no more than 29°C. The acceptable temperature of water t for dolphins can be described by the following compound inequality. $t \geq 22$ and $t \leq 29$

NewVocabulary
compound inequality
intersection
union

Common Core State Standards

Mathematical Practices
5 Use appropriate tools strategically.

1 Compound Inequalities A **compound inequality** consists of two inequalities joined by the word *and* or the word *or*. To solve a compound inequality, you must solve each part of the inequality. The graph of a compound inequality containing *and* is the **intersection** of the solution sets of the two inequalities.

KeyConcept "And" Compound Inequalities

Words	A compound inequality containing the word *and* is true if and only if *both* inequalities are true.
Example	$x \geq -4$

$x < 3$

$x \geq -4$ and $x < 3$

Another way of writing $x \geq -4$ and $x < 3$ is $-4 \leq x < 3$.
Both forms are read *x is greater than or equal to −4 and less than 3.*

Example 1 Solve an "And" Compound Inequality

Solve $8 < 3y - 7 \leq 23$. Graph the solution set on a number line.

Method 1 Solve separately.	**Method 2** Solve both together.
Write the compound inequality using the word *and*. Then solve each inequality.	Solve both parts at the same time by adding 7 to each part. Then divide each part by 3.
$8 < 3y - 7$ and $3y - 7 \leq 23$ $15 < 3y$ $\qquad\qquad 3y \leq 30$ $5 < y$ $\qquad\qquad\quad y \leq 10$ $\qquad\qquad 5 < y \leq 10$	$8 < 3y - 7 \leq 23$ $15 < \quad 3y \quad \leq 30$ $5 < \quad\;\; y \quad \leq 10$ $\qquad\quad 5 < y \leq 10$

(continued on the next page)

connectED.mcgraw-hill.com   **41**

Paul Katz/Photodisc/Getty Images

1 Focus

VerticalAlignment

▼

Before Lesson 1-6 Solve one-step and multi-step inequalities.

▼

Lesson 1-6 Solve compound inequalities.
Solve absolute value inequalities.

▼

After Lesson 1-6 Solve systems of inequalities.

2 Teach

Scaffolding Questions
Have students read the **Why?** section of the lesson.

Ask:

■ Do dolphins prefer the water temperature to be greater than or less than 22°C? greater than

■ Do they prefer the water temperature to be greater than or less than 29°C? less than

■ Write a verbal statement describing how dolphins prefer the water that represents the compound inequality $t \geq 22$ and $t \leq 29$. Dolphins prefer the water to be between 22°C and 29°C inclusive.

Lesson 1-6 Resources

Resource	Approaching Level **AL**	On Level **OL**	Beyond Level **BL**	English Learners **ELL**
Teacher Edition	■ Differentiated Instruction, p. 42	■ Differentiated Instruction, pp. 42, 44	■ Differentiated Instruction, p. 44	■ Differentiated Instruction, p. 42
Chapter Resource Masters	■ Study Guide and Intervention, pp. 37–38 ■ Skills Practice, p. 39 ■ Practice, p. 40 ■ Word Problem Practice, p. 41	■ Study Guide and Intervention, pp. 37–38 ■ Skills Practice, p. 39 ■ Practice, p. 40 ■ Word Problem Practice, p. 41 ■ Enrichment, p. 42	■ Practice, p. 40 ■ Word Problem Practice, p. 41 ■ Enrichment, p. 42	■ Study Guide and Intervention, pp. 37–38 ■ Skills Practice, p. 39 ■ Practice, p. 40 ■ Word Problem Practice, p. 41
Other	■ 5-Minute Check 1-6 ■ Study Notebook ■ Teaching Algebra with Manipulatives	■ 5-Minute Check 1-6 ■ Study Notebook ■ Teaching Algebra with Manipulatives	■ 5-Minute Check 1-6 ■ Study Notebook	■ 5-Minute Check 1-6 ■ Study Notebook ■ Teaching Algebra with Manipulatives

Example 1 shows how to solve an "and" compound inequality. **Example 2** shows how to solve an "or" compound inequality.

Formative Assessment

Use the Guided Practice exercises after each example to determine students' understanding of concepts.

Additional Examples

1 Solve $10 \leq 3y - 2 < 19$. Graph the solution set on a number line. $\{y \mid 4 \leq y < 7\}$

2 Solve $x + 3 < 2$ or $-x \leq -4$. Graph the solution set on a number line. $\{x \mid x < -1 \text{ or } x \geq 4\}$

▸ **Additional Examples** also in Interactive Classroom PowerPoint® Presentations

IWB **Interactive White Board** READY

WatchOut!

Preventing Errors Remind students that the word *and* used in Method 1 means the value for $3y - 7$ must meet *both* conditions. That is, the value must be greater than 8 *and* less than or equal to 23. An *or* compound inequality must be written as two or more separate inequalities separated by the word *or*. They cannot be shortened into one "double" inequality. Explain why it does not make sense to rewrite $x < 2$ *or* $x > 7$ as $2 > x > 7$.

Graph the solution set for each inequality and find their intersection.

The solution set is $\{y \mid 5 < y \leq 10\}$ or $(5, 10]$.

▸ **Guided Practice** **1A, 1B.** See Chapter 1 Answer Appendix.

Solve each inequality. Graph the solution set on a number line.

1A. $-12 \leq 4x + 8 \leq 32$

1B. $-5 \geq 3z - 2 > -14$

The graph of a compound inequality containing *or* is the **union** of the solution sets of the two inequalities.

Get Animated activities are online illustrations of key concepts through animations, Interactive Labs, and BrainPOPs®.

KeyConcept "Or" Compound Inequalities

Words	A compound inequality containing the word *or* is true if one or more of the inequalities is true.

Example:
$x \geq 5$

$x < -3$

$x \geq 5$ or $x < -3$

Example 2 Solve an "Or" Compound Inequality

Solve $k + 6 < -4$ or $3k \geq 14$. Graph the solution set.

Solve each inequality separately.

$k + 6 < -4$ or $3k \geq 14$

$k < -10$ $k \geq \dfrac{14}{3}$

$k < -10$

$k \geq \dfrac{14}{3}$

$k < -10$ or $k \geq \dfrac{14}{3}$

▸ **Guided Practice** **2A, 2B.** See Chapter 1 Answer Appendix.

Solve each inequality. Graph the solution set on a number line.

2A. $5j \geq 15$ or $-3j \geq 21$

2B. $g - 6 > -11$ or $2g + 4 < -15$

DifferentiatedInstruction **AL** **OL** **ELL**

If ▸ students make the mistake of wanting to associate *union* with the word *and* because union often indicates the joining of two or more things,

Then ▸ as a memory device, point out that the word *or* begins with the letter *o*, which is found in *union*, but *and* begins with the letter *a*, which is not found in *union*.

2 **Absolute Value Inequalities** In Lesson 1-4, you learned that the absolute value of a number is its distance from 0 on the number line. You can use this definition to solve inequalities involving absolute value.

Example 3 Solve Absolute Value Inequalities

Solve each inequality. Graph the solution set on a number line.

a. $|x| < 3$

$|x| < 3$ means that the distance between x and 0 on a number line is less than 3 units. To make $|x| < 3$ true, substitute numbers for x that are fewer than 3 units from 0.

> Notice that the graph of $|x| < 3$ is the same as the graph of $x > -3$ and $x < 3$.

All of the numbers between -3 and 3 are less than 3 units from 0. The solution set is $\{x | -3 < x < 3\}$ or $(-3, 3)$.

b. $|x| > 5$

$|x| > 5$ means that the distance between x and 0 on a number line is more than 5 units. To make $|x| > 5$ true, substitute numbers for x that are more than 5 units from 0.

> Notice that the graph of $|x| > 5$ is the same as the graph of $x < -5$ or $x > 5$.

All of the numbers between and including -5 and 5 are no more than 5 units from 0, so they are not part of the solution set. The solution set is $\{x | -5 > x \text{ or } x > 5\}$ or $(-\infty, -5) \cup (5, \infty)$.

GuidedPractice **3A–3D.** See margin.

Solve each inequality. Graph the solution set on a number line.

3A. $|t| < 6$ **3B.** $|u| < -3$

3C. $|t| > 3$ **3D.** $|u| > -2$

An absolute value inequality can be solved by rewriting it as a compound inequality.

ReadingMath

within and *between* When solving problems involving inequalities, *within* is meant to be inclusive. Use ≤ or ≥.

Between is meant to be exclusive. Use < or >.

StudyTip

Absolute Value Inequalities Because the absolute value of a number is never negative, solutions involving negative numbers are as follows.

$|x| < -5$ is the empty set.

$|x| > -5$ is infinite solutions.

KeyConcept Absolute Value Inequalities

For all real numbers a, b, c, and x, $c > 0$, the following statements are true.

Absolute Value Inequality	Compound Inequality	Example
$|ax + b| > c$	$ax + b > c$ or $ax + b < -c$	If $|4x + 5| > 7$, then $4x + 5 > 7$ or $4x + 5 < -7$.
$|ax + b| < c$	$-c < ax + b < c$	If $|4x + 5| < 7$, then $-7 < 4x + 5 < 7$.

These statements are also true for ≤ and ≥, respectively.

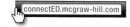

Examples 3–4 show how to solve absolute value inequalities.

Additional Example

3 **a.** Solve $2 > |d|$. Graph the solution set on a number line. $\{d | -2 < d < 2\}$

b. Solve $3 < |d|$. Graph the solution set on a number line. $\{d | d < -3 \text{ or } d > 3\}$

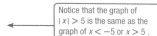

Tips for New Teachers

Reading Make sure students understand the meaning of Examples 3 and 4 before they go on. Have them say the problem in words (Example 3: "The distance of x from 0 without regard to direction is less than 3.") and demonstrate where x can be located on a number line.

Focus on Mathematical Content

Compound Inequalities A number is a solution to a compound inequality with *and* if the number is a solution to both inequalities. A number is a solution to a compound inequality with *or* if the number is a solution to either inequality.

WatchOut!

Preventing Errors Absolute value signs need to be interpreted so that the statement containing the absolute value signs can be rewritten as an equivalent statement without absolute value signs.

Additional Answers (Guided Practice)

3A. $\{t | -6 < t < 6\}$

3D. $\{$all real numbers$\}$

3B. ∅

3C. $\{t | t > 3 \text{ or } t < -3\}$

Example 5 shows how to write and solve an absolute value inequality in a real-world situation.

Additional Examples

4 Solve $|2x - 2| \geq 4$. Graph the solution set on a number line.
$\{x \mid x \leq -1 \text{ or } x \geq 3\}$

$$\begin{array}{c}\overleftarrow{} \quad \bullet \quad \quad \bullet \quad \overrightarrow{} \\ -3 \;-2\; -1 \;\; 0 \;\; 1 \;\; 2 \;\; 3 \;\; 4 \;\; 5\end{array}$$

5 **JOB HUNTING** To prepare for a job interview, Hinda researches the position's requirements and pay. She discovers that the average starting salary for the position is $38,500, but her actual starting salary could differ from the average by as much as $2450.

a. Write an absolute value inequality to describe this situation.
$|38{,}500 - x| \leq 2450$

b. Solve the inequality to find the range of Hinda's starting salary.
$\{x \mid 36{,}050 \leq x \leq 40{,}950\}$; Hinda's starting salary will fall within $36,050 and $40,950.

Teach with Tech

Document Camera Choose a student to work through an example using the document camera. Be sure the student uses a graph to find the solution to the inequality.

> **Teach with Tech** features throughout the Teacher Edition offer tips on using various types of technology such as interactive whiteboards, document cameras, blogs, and more, to enhance your teaching.

Example 4 Solve a Multi-Step Absolute Value Inequality

Solve $|6y - 5| \geq 13$. Graph the solution set on a number line.

$|6y - 5| \geq 13$ is equivalent to $6y - 5 \geq 13$ or $6y - 5 \leq -13$. Solve the inequality.

$6y - 5 \geq 13$	or $\quad 6y - 5 \leq -13$	Rewrite the inequality.
$6y \geq 18$	$6y \leq -8$	Add 5 to each side.
$y \geq 3$	$y \leq -\dfrac{8}{6}$ or $-\dfrac{4}{3}$	Divide each side by 6.

The solution set is $\left\{ y \mid y \leq -\dfrac{4}{3} \text{ or } y \geq 3 \right\}$ or $\left(-\infty, -\dfrac{4}{3}\right] \cup [3, \infty)$.

$$\begin{array}{c}\overleftarrow{} \quad \bullet \bullet \quad \overrightarrow{} \\ -5\,-4\,-3\,-2\,-1\;\;0\;\;1\;\;2\;\;3\;\;4\;\;5\end{array}$$

GuidedPractice

Solve each inequality. Graph the solution set on a number line.

4A. $|4x - 7| > 13$

4B. $|5z + 2| \leq 17$

4A. $\left\{ x \mid x < -\dfrac{3}{2} \text{ or } x > 5 \right\}$

$$\begin{array}{c}\overleftarrow{} \quad \circ \circ \quad \overrightarrow{} \\ -5\,-4\,-3\,-2\,-1\;\;0\;\;1\;\;2\;\;3\;\;4\;\;5\end{array}$$

4B. $\{z \mid -3.8 \leq z \leq 3\}$

$$\begin{array}{c}\overleftarrow{} \quad \bullet \bullet \quad \overrightarrow{} \\ -5\,-4\,-3\,-2\,-1\;\;0\;\;1\;\;2\;\;3\;\;4\;\;5\end{array}$$

An inequality can be viewed as a constraint in a problem situation. Each solution of the inequality represents a combination that meets the constraint.

In real-world problems, the domain and range are often restricted to nonnegative or whole numbers.

Real-World Example 5 Write and Solve an Absolute Value Inequality

MONEY Amanda is apartment hunting in a specific area. She discovers that the average monthly rent for a 2-bedroom apartment is $600 a month, but the actual price could differ from the average as much as $225 a month.

a. Write an absolute value inequality to describe this situation.

Let r = average monthly rent. $|600 - r| \leq 225$

b. Solve the inequality to find the range of monthly rent.

Rewrite the absolute value inequality as a compound inequality. Then solve for r.

$$-225 \leq 600 - r \qquad \leq 225$$
$$-225 - 600 \leq 600 - r - 600 \leq 225 - 600$$
$$-825 \leq \qquad -r \qquad \leq -375$$
$$825 \geq \qquad r \qquad \geq 375$$

The solution set is $\{r \mid 375 \leq r \leq 825\}$ or $[375, 825]$. Thus, monthly rent could fall between $375 and $825, inclusive.

GuidedPractice

5. **TUITION** Rachel is considering colleges to attend and determines that the average tuition among her choices is $3725 per year, but the tuition at a school could differ by as much as $1650 from the average. Write and solve an absolute value inequality to find the range of tuition. $|t - 3725| \leq 1650$; $\{t \mid \$2075 \leq t \leq \$5375\}$

 44 | Lesson 1-6 | Solving Compound and Absolute Value Inequalities

DifferentiatedInstruction OL BL

Extension Solve $|x - 2| > x + 4$. $\{x \mid x < -1\}$

Real-WorldLink

Apartment costs vary greatly depending on location. Of the major U.S. cities, New York has the highest average monthly rent of $2400, while Oklahoma City is lowest at $543.

Source: MSN

Tom McGhee/Workbook Stock/Getty Images

Examples 1–4 Solve each inequality. Graph the solution set on a number line. **1–10. See margin.**

1. $-4 < g + 8 < 6$

2. $-9 \le 4y - 3 \le 13$

3. $z + 6 > 3$ or $2z < -12$

4. $m - 7 \ge -3$ or $-2m + 1 \ge 11$

5. $|c| \ge 8$

6. $|q| \ge -1$

7. $|z| < 6$

8. $|x| \le -4$

9. $|3v + 5| > 14$

10. $|4t - 3| \le 7$

Example 5 **11. MONEY** Khalid is considering several types of paint for his bedroom. He estimates that he will need between 2 and 3 gallons. The table at the right shows the price per gallon for each type of paint Khalid is considering. Write a compound inequality and determine how much he could be spending.
$43.96 \le c \le 77.94$; between $43.96 and $77.94

Paint Type	Price per Gallon
Flat	$21.98
Satin	$23.98
Semi-Gloss	$24.98
Gloss	$25.98

Practice and Problem Solving Extra Practice is on page R1.

12–21. See Chapter 1 Answer Appendix.

Examples 1–4 Solve each inequality. Graph the solution set on a number line.

12. $8 < 2v - 4 < 16$

13. $-7 \le 4d - 3 \le -1$

14. $4r + 3 < -6$ or $3r - 7 > 2$

15. $6y - 3 < -27$ or $-4y + 2 < -26$

16. $|6h| < 12$

17. $|-4k| > 16$

18. $|3x - 4| > 10$

19 $|8t + 3| \le 4$

20. $|-9n - 3| < 6$

21. $|-5j - 4| \ge 12$

Example 5 **22. CCSS MODELING** Forensic scientists use the equation $h = 2.6f + 47.2$ to estimate the height h of a woman given the length in centimeters f of her femur bone.

a. Suppose the equation has a margin of error of ±3 centimeters. Write an inequality to represent the height of a woman given the length of her femur bone. $|2.6f + 47.2| < 3$

b. If the length of a female skeleton's femur is 50 centimeters, write and solve an absolute value inequality that describes the woman's height in centimeters.
$|h - 177.2| < 3$; 174.2 cm < h < 180.2 cm

B Write an absolute value inequality for each graph.

23.
$|x - 1| \le 5$

24.
$|x - 1| \ge 5$

25.
$|x + 9| \le 3$

26.
$|x - 1| > 2$

27.
$|x - 2| \ge 10$

28.
$|x - 4| < 6$

29.
$|x + 3| > 1$

30.
$|x - 5| \le 3$

 connectED.mcgraw-hill.com **45**

Differentiated Homework Options

Level	Assignment	Two-Day Option	
AL Basic	12–22, 53, 55–74	13–21 odd, 63–66	12–22 even, 53, 55–62, 67–74
OL Core	13–21 odd, 22, 23–31 odd, 32, 33–43 odd, 44, 45–51 odd, 52, 53, 55–74	12–22, 63–66	23–53, 55–62, 67–74
BL Advanced	23–71, (optional: 72–74)		

---

3 Practice

Formative Assessment

Use Exercises 1–11 to check for understanding.

Use the chart on the bottom of the page to customize assignments for your students.

Additional Answers

1. $\{g \mid -12 < g < -2\}$

2. $\{y \mid -1.5 \le y \le 4\}$

3. $\{z \mid z > -3$ or $z < -6\}$

4. $\{m \mid m \ge 4$ or $m \le -5\}$

5. $\{c \mid c \ge 8$ or $c \le -8\}$

6. {all real numbers}

7. $\{z \mid -6 < z < 6\}$

8. $\varnothing$

9. $\left\{v \mid v > 3 \text{ or } v < -\dfrac{19}{3}\right\}$

10. $\{t \mid -1 \le t \le 2.5\}$

Study Guide and Intervention

AL OL ELL

NAME _____ DATE _____ PERIOD _____

1-6 Study Guide and Intervention

Solving Compound and Absolute Value Inequalities

Compound Inequalities A compound inequality consists of two inequalities joined by the word *and* or the word *or*. To solve a compound inequality, you must solve each part separately.

| And Compound Inequalities | The graph is the intersection of solution sets of two inequalities. |
| Or Compound Inequalities | The graph is the union of solution sets of two inequalities. |

Example 1 Solve $-3 \le 2x + 5 \le 19$. Graph the solution set on a number line.

Example 2 Solve $3y - 2 \ge 7$ or $2y - 1 \le -9$. Graph the solution set on a number line.

Exercises

Solve each inequality. Graph the solution set on a number line.

1. $-10 < 3x + 2 \le 14$
2. $3n + 8 < 23$ or $\frac{n}{4} - 6 > 7$
3. $18 < 4x - 10 < 50$
4. $5k + 2 < -13$ or $8k - 1 > 19$
5. $100 \ge 5y - 45 \ge 225$
6. $\frac{2}{3}b - 2 > 10$ or $\frac{2}{3}b + 5 < -4$
7. $22 < 6w - 2 < 82$
8. $4d - 1 > -9$ or $2d + 5 < 11$

Practice

AL OL BL ELL

NAME _____ DATE _____ PERIOD _____

1-6 Practice

Solving Compound and Absolute Value Inequalities

Write an absolute value inequality for each graph.

Solve each inequality. Graph the solution set on a number line.

5. $-8 \le 3y - 20 < 52$
6. $3(5x - 2) < 24$ or $6x - 4 > 4 + 5x$
7. $2x - 3 > 15$ or $3 - 7x < 17$
8. $15 - 5x \le 0$ and $5x + 6 \ge -14$
9. $|2x| \ge 5$
10. $|y + 5| < 2$
11. $|x - 8| \ge 3$
12. $|2x - 2| \le 3$
13. $|2x + 2| - 7 \le -5$
14. $|x| > x - 1$
15. $|3b + 5| \le -2$
16. $|3x - 2| - 2 < 1$

17. **RAINFALL** In 90% of the last 30 years, the rainfall at Shell Beach has varied no more than 6.5 inches from its mean value of 24 inches. Write and solve an absolute value inequality to describe the rainfall in the other 10% of the last 30 years.

18. **MANUFACTURING** A company's guidelines call for each can of soup produced not to vary from its stated volume of 14.5 fluid ounces by more than 0.08 ounces. Write and solve an absolute value inequality to describe acceptable can volumes.

Word Problem Practice

AL OL BL ELL

NAME _____ DATE _____ PERIOD _____

1-6 Word Problem Practice

Solving Compound and Absolute Value Inequalities

1. **AQUARIUM** The depth d of an aquarium tank for dolphins satisfies $|d - 50| < 5$. Rewrite this as a compound inequality that does not involve the absolute value function.

2. **HIKING** For a hiking trip, everybody must bring at least one backpack. However, because of space limitations, nobody is allowed to bring more than two backpacks. Let n be the number of people going on the hiking trip and b be the number of backpacks allowed. Write a compound inequality that describes how b and n are related.

3. **CONCERT** Jacinta is organizing a large fund-raiser concert in a space with a maximum capacity of 10,000 people. Her goal is to raise at least $100,000. Tickets cost $20 per person. Jacinta spends $50,000 to put the event together. Write and solve a compound inequality that describes N, the number of attendees needed to achieve Jacinta's goal.

4. **NUMBERS** Amy is thinking of two numbers a and b. The sum of the two numbers must be within 10 units of zero. If a is between −100 and 100, write a compound inequality that describes the possible values of b.

5. **AIRLINE BAGGAGE** Many airlines have a size limitation for carry-on luggage. The limitation states that the sum of the length, width, and height of the suitcase must not exceed 45 inches.

a. Write an inequality that describes the airlines' carry-on size limitation.

b. A passenger needs to bring a soil sample on the plane that is at least 1 cubic foot. The passenger is bringing it in a suitcase that is in the shape of a cube with side length n inches. Write an inequality that gives the minimum length for n.

c. Write a compound inequality for n using parts a and b. Find the maximum and minimum values for n.

Enrichment

OL BL

NAME _____ DATE _____ PERIOD _____

1-6 Enrichment

Conjunctions and Disjunctions

The compound sentence that solves an absolute value inequality is called either a *conjunction* or a *disjunction*.

Example 1 Solve $|2x| < 10$.

$|2x| < 10$ means $2x < 10$ and $2x > -10$.

Solve each inequality. $x < 5$ and $x > -5$.

Every solution for $|2x| < 10$ is a replacement for x that makes both $x < 5$ and $x > -5$ true.

A compound sentence that combines two statements by the word *and* is a *conjunction*.

Example 2 Solve $|3x - 7| \ge 11$.

$|3x - 7| \ge 11$ means $3x - 7 \ge 11$ or $3x - 7 \le -11$.

Solve each inequality. $3x \ge 18$ or $3x \le -4$

31. **DOGS** The Labrador retriever is one of the most recognized and popular dogs kept as a pet. Using the information given, write a compound inequality to describe the range of healthy weights for a fully grown female Labrador retriever. $55 \le w \le 70$

Healthy Heights and Weights for Labrador Retrievers		
Gender	Height (in.)	Weight (lb)
Male	22.5–24.5	65–80
Female	21.5–23.5	55–70

32. **GEOMETRY** The *Exterior Angle Inequality Theorem* states that an exterior angle measure is greater than the measure of either of its corresponding remote interior angles. Write two inequalities to express the relationships among the measures of the angles of $\triangle ABC$. $m\angle 4 > m\angle 1$, $m\angle 4 > m\angle 2$

33–40. See Chapter 1 Answer Appendix.

Solve each inequality. Graph the solution set on a number line.

33. $28 > 6k + 4 > 16$

34. $m - 7 > -12$ or $-3m + 2 > 38$

35. $|-6h| > 90$

36. $-|-5k| > 15$

37. $3|2z - 4| - 6 > 12$

38. $6|4p + 2| - 8 < 34$

39. $\dfrac{|5f - 2|}{6} > 4$

40. $\dfrac{|2w + 8|}{5} \ge 3$

Write an algebraic expression to represent each verbal expression.

41. numbers that are at least 4 units from −5 $|x + 5| \ge 4$

42. numbers that are no more than $\frac{3}{8}$ unit from 1 $|x - 1| \le \frac{3}{8}$

43. numbers that are at least 6 units but no more than 10 units from 2 $6 \le |x - 2| \le 10$

44a. red: $|x - 24.42| \le 0.07$; blue: $|x - 24.42| \le 0.25$; green: $|x - 24.42| \le 0.5$

44. **CCSS REASONING** NASCAR rules stipulate that a car must conform to a set of 32 templates, each shaped to fit a different contour of the car. When a template is placed on a car, the gap between it and the car cannot exceed the specified tolerance. Each template is marked on its edge with a colored line that indicates the tolerance for the template.

a. Suppose a certain template is 24.42 inches long. Use the information in the table at the right to write an absolute value inequality for templates with each line color.

Line Color	Tolerance (in.)
Red	0.07
Blue	0.25
Green	0.5

44b. red: $24.35 \le x \le 24.49$; blue: $24.17 \le x \le 24.67$; green: $23.92 \le x \le 24.92$

b. Find the acceptable lengths for that part of a car if the template has each line color.

c. Graph the solution set for each line color on a number line. **See margin.**

d. The tolerance of which line color includes the tolerances of the other line colors? Explain your reasoning. **Red; the red line color has the smallest tolerance, $0.07 < 0.25 < 0.5$, so the other line colors would be well within their tolerances.**

Solve each inequality. Graph the solution set on a number line. **45–50. See margin.**

45. $n + 6 > 2n + 5 > n - 2$

46. $y + 7 < 2y + 2 < 0$

47. $2x + 6 < 3(x - 1) \le 2(x + 3)$

48. $a - 16 \le 2(a - 4) < a + 2$

49. $4g + 8 \ge g + 6$ or $7g - 14 \ge 2g - 4$

50. $5t + 7 > 2t + 4$ and $3t + 3 < 24 - 4t$

51. $|s - 88| > 38$; $\{s \mid s > 126 \text{ or } s < 50\}$

51. **HEALTH** Hypoglycemia (low blood sugar) and hyperglycemia (high blood sugar) are potentially dangerous and occur when a person's blood sugar fluctuates by more than 38 mg from the normal blood sugar level of 88 mg. Write and solve an absolute value inequality to describe blood sugar levels that are considered potentially dangerous.

Additional Answer

44c.

Red

Blue

Green

52. AIR TRAVEL The airline on which Drew is flying has weight restrictions for checked baggage. Drew is checking one bag.

  a. Describe the ranges of weights that would classify Drew's bag as free, $25, $50, and unacceptable. $x \le 50; 50 < x \le 70; 70 < x \le 100; x > 100$

  b. If Drew's bag weighs 68 pounds, how much will he pay to take it on the plane? **$25**

Cost for Checked Baggage	
Weight	**Cost**
Up to 50 lb limit	free
20 lb over limit	$25
More than 20, but less than 50 lb over limit	$50
More than 50 lb over limit	not accepted

WatchOut!

Error Analysis For Exercise 53, have students pay special attention to the direction of each inequality symbol.

H.O.T. Problems · Use Higher-Order Thinking Skills

53. **ARGUMENTS** David and Sarah are solving $4|-5x - 3| - 6 \ge 34$. Is either of them correct? Explain your reasoning.

53. Sample answer: David; when Sarah converted the absolute value into two inequalities, she mistakenly switched the inequality symbols.

David	Sarah
$4\|-5x - 3\| - 6 \ge 34$	$4\|-5x - 3\| - 6 \ge 34$
$\|-5x - 3\| \ge 10$	$\|-5x - 3\| \ge 10$
$-5x - 3 \ge 10 \text{ or } -5x - 3 \le -10$	$-5x - 3 \le 10 \text{ or } -5x - 3 \ge -10$
$-5x \ge 13 \qquad -5x \le -7$	$-5x \le 13 \qquad -5x \ge -7$
$x \le -\dfrac{13}{5} \qquad x \ge \dfrac{7}{5}$	$x \ge -\dfrac{13}{5} \qquad x \le \dfrac{7}{5}$

59. Sample answer: The graph on the left indicates a solution set from −3 to 5. The graph on the right indicates a solution set of all numbers less than or equal to −3 or greater than or equal to 5.

61. Each of these has a non-empty solution set except for $x > 5$ and $x < 1$. There are no values of x that are simultaneously greater than 5 and less than 1.

54. CHALLENGE Solve $|x - 2| - |x + 2| > x.$ $x < 0$

60. Sample answer: $\left| x - \dfrac{a+b}{2} \right| \le b - \dfrac{a+b}{2}$

REASONING Determine whether each statement is *true* or *false*. If false, provide a counterexample. **55–57. See margin.**

55. The graph of a compound inequality involving an *and* statement is bounded on the left and right by two values of x.

56. The graph of a compound inequality involving an *or* statement contains a region of values that are not solutions.

57. The graph of a compound inequality involving an *and* statement includes values that make all parts of the given statement true.

58. WRITING IN MATH An alternate definition of absolute value is to define $|a - b|$ as the distance between a and b on the number line. Explain how this definition can be used to solve inequalities of the form $|x - c| < r$. **See margin.**

59. REASONING The graphs of the solutions of two different absolute value inequalities are shown. Compare and contrast the absolute value inequalities.

60. OPEN ENDED Write an absolute value inequality with a solution of $a \le x \le b$.

61. WHICH ONE DOESN'T BELONG? Identify the compound inequality that is not the same as the other three. Explain your reasoning.

$-3 < x < 5$	$x > 2$ and $x < 3$	$x > 5$ and $x < 1$	$x > -4$ and $x > -2$

62. WRITING IN MATH Summarize the difference between *and* compound inequalities and *or* compound inequalities. **See margin.**

CCSS **Teaching the Mathematical Practices**

Arguments Mathematically proficient students understand and use stated assumptions, definitions, and previously established results in constructing arguments. They make conjectures and build a logical progression of statements to explore the truth of their conjectures. And they are able to analyze situations by breaking them into cases, and can recognize and use counterexamples.

Additional Answers

55. False; sample answer: the graph of $x > 2$ and $x > 5$ is a ray bounded only on one end.

56. False; sample answer: the graph of $x > 2$ or $x < 3$ includes the entire number line.

57. true

58. Sample answer: $|x - c|$ represents the distance between some unknown value of the variable x and a point c on the number line. The solution set of the inequality is the set of all numbers such that the distance from the numbers to c is less than r units. Use a number line to find the numbers that are r units from c in either direction.

62. Sample answer: A compound inequality that contains *and* is true if and only if both individual inequalities are true, while an inequality containing *or* only needs one of the individual inequalities to be true.

---

45. $\{n \mid -7 < n < 1\}$

46. $\varnothing$

47. $\varnothing$

48. $\{a \mid -8 \le a < 10\}$

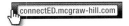

49. $\left\{ g \mid g \ge -\dfrac{2}{3} \right\}$

50. $\{t \mid -1 < t < 3\}$

Ticket Out the Door Have each student tell a partner or write the steps used to solve $|x - 4| < 3$.

Formative Assessment

Check for student understanding of concepts in Lesson 1-6.

📁 Quiz 4, p. 46

Standardized Test Practice

63. Which of the following best describes the graph of the equations below? **C**

$$24y = 8x + 11$$
$$36y = 12x + 11$$

A The lines have the same x-intercept.

B The lines have the same y-intercept.

C The lines are parallel.

D The lines are perpendicular.

64. SAT/ACT Find an expression equivalent to $\left(\dfrac{3x^3}{y^3}\right)^3$. **K**

F $\dfrac{9x^6}{3y}$ **J** $\dfrac{27x^6}{3y}$

G $\dfrac{9x^9}{y^3}$ **K** $\dfrac{27x^9}{y^3}$

H $\dfrac{9x^6}{y^3}$

65. GRIDDED RESPONSE How many cubes that measure 4 centimeters on each side can be placed completely inside the box below? **60**

66. Which graph represents the solution set for $|3x - 6| + 8 \geq 17$? **A**

Spiral Review

67. HEALTH The National Heart Association recommends that less than 30% of a person's total daily caloric intake come from fat. One gram of fat yields nine Calories. Consider a healthy 21-year-old whose average caloric intake is between 2500 and 3300 Calories. (Lesson 1-5)

    a. Write an inequality that represents the suggested fat intake for the person. $750 \leq x \leq 990$

    b. What is the greatest suggested fat intake for the person? **110 g**

68. TRAVEL Maggie is planning a 5-day, 5-night, trip to a family reunion. She wants to spend no more than $1000. Her plane ticket is $375, and the hotel is $85 per night. (Lesson 1-5)

    a. Let f represent the cost of food for one day. Write an inequality to represent this situation. $800 + 5f \leq 1000$

    b. Solve the inequality and interpret the solution. **She can spend no more than $40 per day on food.**

Solve each equation. Check your solutions. (Lesson 1-4)

69. $4|x - 5| = 20$ $\{0, 10\}$ **70.** $|3y + 10| = 25$ $\left\{-\dfrac{35}{3}, 5\right\}$ **71.** $|7z + 8| = -9$ $\varnothing$

Skills Review

Name the property illustrated by each statement.

72. If $5x = 7$, then $5x + 3 = 7 + 3$. **Addition (=)**

73. If $-3x + 9 = 11$ and $6x + 2 = 11$, then $-3x + 9 = 6x + 2$. **Transitive (=)**

74. If $[x + (-2)] + (-4) = 5$, then $x + [-2 + (-4)] = 5$. **Assoc. (+)**

🄴❓ Follow-up

Students have explored expressions, equations, and inequalities.

Ask:

- How are symbols used to write expressions, equations, and inequalities? Sample answer: If only operational symbols are used without a symbol that denotes equality or inequality, the statement is an expression. If an equals sign is used to show that two or more expressions are equal, the statement is an equation. If an inequality symbol is used to show that two or more expressions are unequal, the statement is an inequality.

Study Guide and Review

Study Guide

KeyConcepts

Expressions and Formulas (Lesson 1-1)

- Use the order of operations to solve equations.

Properties of Real Numbers (Lesson 1-2)

- Real numbers can be classified as rational (Q) or irrational (I). Rational numbers can be classified as integers (Z), whole numbers (W), natural numbers (N), and/or quotients of these.

Solving Equations (Lessons 1-3 and 1-4)

- Verbal expressions can be translated into algebraic expressions.
- The absolute value of a number is the number of units it is from 0 on a number line.
- For any real numbers a and b, where $b \geq 0$, if $|a| = b$, then $a = b$ or $a = -b$.

Solving Inequalities (Lessons 1-5 and 1-6)

- Adding or subtracting the same number from each side of an inequality does not change the truth of the inequality.
- When you multiply or divide each side of an inequality by a negative number, the direction of the inequality symbol must be reversed.
- The graph of an *and* compound inequality is the intersection of the solution sets of the two inequalities. The graph of an *or* compound inequality is the union of the solution sets of the two inequalities.
- An *and* compound inequality can be expressed in two different ways. For example, $-2 \leq x \leq 3$ is equivalent to $x \geq -2$ and $x \leq 3$.
- For all real numbers a and b, where $b > 0$, the following statements are true.
  1. If $|a| < b$ then $-b < a < b$.
  2. If $|a| > b$ then $a > b$ or $a < -b$.

FOLDABLES StudyOrganizer

Be sure the Key Concepts are noted in your Foldable.

KeyVocabulary

absolute value (p. 27)	irrational numbers (p. 11)
algebraic expressions (p. 5)	natural numbers (p. 11)
compound inequality (p. 41)	open sentence (p. 18)
constraint (p. 29)	order of operations (p. 5)
empty set (p. 28)	rational numbers (p. 11)
equation (p. 18)	real numbers (p. 11)
extraneous solution (p. 29)	set-builder notation (p. 35)
formula (p. 6)	solution (p. 18)
infinity (p. 40)	union (p. 42)
integers (p. 11)	variables (p. 5)
intersection (p. 41)	whole numbers (p. 11)
interval notation (p. 40)	

VocabularyCheck

State whether each sentence is *true* or *false*. If *false*, replace the underlined term to make a true sentence.

1. The absolute value of a number is always <u>negative</u>. **false; nonnegative**

2. $\sqrt{12}$ belongs to the set of <u>rational</u> numbers. **false; irrational**

3. An <u>equation</u> is a statement that two expressions have the same value. **true**

4. A solution of an equation is a value that makes the equation <u>false</u>. **false; true**

5. The empty set contains <u>no</u> elements. **true**

6. A mathematical sentence containing one or more variables is called an <u>open sentence</u>. **true**

7. The graph of a compound inequality containing <u>*and*</u> is the union of the solution sets of the two inequalities. **false; or**

8. Variables are used to represent <u>unknown</u> quantities. **true**

9. The set of <u>rational</u> numbers includes terminating and repeating decimals. **true**

10. Expressions that contain at least one variable are called <u>algebraic expressions</u>. **true**

Formative Assessment

KeyVocabulary The page references after each word denote where that term was first introduced. If students have difficulty answering questions 1–10, remind them that they can use these page references to refresh their memories about the vocabulary.

Summative Assessment

☐ Vocabulary Test, p. 48

Vocabulary Review provides students the opportunity to check their understanding of important concepts and terminology in an online game format.

> Students can use the **Vocabulary Check** in the Study Guide and Review and the **Vocabulary Review Puzzle** to review the vocabulary of the chapter.

FOLDABLES StudyOrganizer

Dinah Zike's Foldables®
Have students look through the chapter to make sure they have included examples in their Foldables. Suggest that students keep their Foldables handy while completing the Study Guide and Review pages. Point out that their Foldables can serve as a quick review tool when studying for the chapter test.

Lesson-by-Lesson Review

Intervention If the given examples are not sufficient to review the topics covered by the questions, remind students that the lesson references tell them where to review that topic in their textbook.

Two-Day Option Have students complete the Lesson-by-Lesson Review. Then you can use McGraw-Hill eAsessment to customize another review worksheet that practices all the objectives of this chapter or only the objectives on which your students need more help.

Lesson-by-Lesson Review

1-1 Expressions and Formulas

Evaluate each expression.

11. $[28 - (16 + 3)] \div 3$ **3**

12. $\frac{2}{3}(3^3 + 12)$ **26**

13. $\frac{15(9 - 7)}{3}$ **10**

Evaluate each expression if $w = 0.2$, $x = 10$, $y = \frac{1}{2}$, and $z = -4$.

14. $4w - 8y$ **−3.2**

15. $z^2 + xy$ **21**

16. $\frac{5w - xy}{z}$ **1**

17. **GEOMETRY** The formula for the volume of a cylinder is $V = \pi r^2 h$, where V is the volume, r is the radius, and h is the height. What is the volume of a cylinder that is 6 inches high and has a radius of 3 inches? **≈169.65 in³**

Example 1

Evaluate $(12 - 15) \div 3^2$.

$(12 - 15) \div 3^2 = -3 \div 3^2$ Subtract.

$\qquad\qquad\qquad = -3 \div 9$ $3^2 = 9$

$\qquad\qquad\qquad = -\frac{1}{3}$ Divide.

Example 2

Evaluate $\frac{a^2}{2ac - b}$ if $a = -6$, $b = 5$, and $c = 0.25$.

$\frac{a^2}{2ac - b} = \frac{(-6)^2}{2(-6)(0.25) - 5}$ $a = -6$, $b = 5$, and $c = 0.25$

$\qquad = \frac{36}{2(-1.5) - 5}$ Evaluate the numerator and denominator separately.

$\qquad = \frac{36}{-8}$ or $-\frac{9}{2}$ Simplify.

1-2 Properties of Real Numbers

Name the sets of numbers to which each value belongs.

18. $1.\overline{3}$ **Q, R**   **19.** $\sqrt{4}$ **20.** $-\frac{3}{4}$ **Q, R**
N, W, Z, Q, R

Simplify each expression.

21. $4x - 3y + 7x + 5y$ **$11x + 2y$**

22. $2(a + 3) - 4a + 8b$ **$-2a + 8b + 6$**

23. $4(2m + 5n) - 3(m - 7n)$ **$5m + 41n$**

24. **MONEY** At Fun City Amusement Park, hot dogs sell for $3.50 and sodas sell for $2.50. Dion bought 3 hot dogs and 3 sodas during one day at the park.

a. Illustrate the Distributive Property by writing two expressions to represent the cost of the hot dogs and the sodas. **$3(3.50 + 2.50)$ or $3(3.50) + 3(2.50)$**

b. Use the Distributive Property to find how much money Dion spent on food and drinks. **$18**

Example 3

Name the sets of numbers to which $\sqrt{50}$ belongs.

$\sqrt{50} = 5\sqrt{2}$ Irrationals (I), and reals (R)

Example 4

Simplify $-4(a + 3b) + 5b$.

$-4(a + 3b) + 5b$ Original expression

$= -4(a) + -4(3b) + 5b$ Distributive Property

$= -4a - 12b + 5b$ Multiply.

$= -4a - 7b$ Simplify.

Students can complete the exercises in the **Lesson-by-Lesson Review** as they prepare for the chapter test. If they need extra help, examples are provided.

1-3 Solving Equations

Solve each equation. Check your solution.

25. $8 + 5r = -27$ -7

26. $4w + 10 = 6w - 13$ $\frac{23}{2}$

27. $\frac{x}{6} + \frac{x}{3} = \frac{3}{4}$ $\frac{3}{2}$

28. $6b - 5 = 3(b + 2)$ $\frac{11}{3}$

29. **MONEY** It cost Lori $14 to go to the movies. She bought popcorn for $3.50 and a soda for $2.50. How much was her ticket? $8

Solve each equation or formula for the specified variable.

30. $2k - 3m = 16$ for k $k = \frac{16 + 3m}{2}$

31. $\frac{r + 5}{mn} = p$ for m $m = \frac{r + 5}{pn}$

32. $A = \frac{1}{2}h(a + b)$ for h $h = \frac{2A}{a + b}$

33. **GEOMETRY** Yu-Jun wants to fill the water container at the right. He knows that the radius is 2 inches and the volume is 100.48 cubic inches. What is the height of the water bottle? Use the formula for the volume of a cylinder, $V = \pi r^2 h$, to find the height of the bottle. 8 in.

2 in.

h

Example 5

Solve $-3(a - 3) + 2(3a - 2) = 14$.

$-3(a - 3) + 2(3a - 2) = 14$	Original equation
$-3a + 9 + 6a - 4 = 14$	Distributive Property
$-3a + 6a + 9 - 4 = 14$	Commutative Property
$3a + 5 = 14$	Substitution Property
$3a = 9$	Subtraction Property
$a = 3$	Division Property

Example 6

Solve each equation or formula for the specified variable.

a. $y = 2x + 3z$ for x

$y = 2x + 3z$	Original equation
$y - 3z = 2x$	Subtract $3z$ from each side.
$\frac{y - 3z}{2} = x$	Divide each side by 2.

b. $V = \frac{\pi r^2 h}{3}$ for h

$V = \frac{\pi r^2 h}{3}$	Original equation
$3V = \pi r^2 h$	Multiply each side by 3.
$\frac{3V}{\pi r^2} = h$	Divide each side by πr^2.

1-4 Solving Absolute Value Equations

Solve each equation. Check your solution.

34. $|r + 5| = 12$ $\{-17, 7\}$

35. $4|a - 6| = 16$ $\{2, 10\}$

36. $|3x + 7| = -15$ $\varnothing$

37. $|b + 5| = 2b - 9$ $\{14\}$

38. **MEASUREMENT** Marcos is cutting ribbons for a craft project. Each ribbon needs to be $\frac{3}{4}$ yard long. If each piece is always within plus or minus $\frac{1}{16}$ yard, how long are the shortest and longest pieces of ribbon? $\frac{11}{16}$ yd; $\frac{13}{16}$ yd

Example 7

Solve $|3m + 7| = 13$.

Case 1	Case 2
$a = b$	$a = -b$
$3m + 7 = 13$	$3m + 7 = -13$
$3m = 6$	$3m = -20$
$m = 2$	$m = -\frac{20}{3}$

The solutions are 2 and $-\frac{20}{3}$.

Anticipation Guide

Have students complete the Chapter 1 Anticipation Guide and discuss how their responses have changed now that they have completed Chapter 1.

Additional Answers

39. $a \geq -6$

40. $r > 55$

41. $x \leq -\dfrac{2}{9}$

42. $p > -2$

44. $\left\{ m \mid m < \dfrac{3}{2} \text{ or } m > 3 \right\}$

45. $\{ x \mid -2 < x < 4 \}$

46. $\left\{ y \mid y \leq -\dfrac{1}{2} \text{ or } y > 3 \right\}$

47. $\left\{ m \mid \dfrac{13}{5} \leq m < \dfrac{24}{5} \right\}$

48. $\{ a \mid -13 < a < 13 \}$

49. $\{ p \mid -5 \leq p \leq 33 \}$

50. $\left\{ k \mid -\dfrac{7}{3} < k < \dfrac{8}{3} \right\}$

1-5 Solving Inequalities

Solve each inequality. Then graph the solution set on a number line.

39. $-4a \leq 24$ **39–42. See margin.**

40. $\dfrac{r}{5} - 8 > 3$

41. $4 - 7x \geq 2(x + 3)$

42. $-p - 13 < 3(5 + 4p) - 2$

43. **MONEY** Ms. Hawkins is taking her science class on a field trip to a museum. She has \$572 to spend on the trip. There are 52 students that will go to the museum. The museum charges \$5 per student, and Ms. Hawkins gets in for free. If the students will have slices of pizza for lunch that cost \$2 each, how many slices can each student have?
3 or fewer slices each

Example 8

Solve $2m - 7 < -11$. Graph the solution set on a number line.

$2m - 7 < -11$ Original inequality

$2m < -4$ Add 7 to each side.

$m < -2$ Divide each side by 2.

The solution set is $\{ m \mid m < -2 \}$.

The graph of the solution set is shown below.

1-6 Solving Compound and Absolute Value Inequalities

Solve each inequality. Graph the solution set on a number line. **44–52. See margin.**

44. $2m + 4 < 7$ or $3m + 5 > 14$

45. $-5 < 4x + 3 < 19$

46. $6y - 1 > 17$ or $8y - 6 \leq -10$

47. $-2 \leq 5(m - 3) < 9$

48. $|a| + 2 < 15$

49. $|p - 14| \leq 19$

50. $|6k - 1| < 15$

51. $|2r + 7| < -1$

52. $\dfrac{1}{3}|8q + 5| \geq 7$

53. **MONEY** Cara is making a beaded necklace for a gift. She wants to spend between \$20 and \$30 on the necklace. The bead store charges \$2.50 for large beads and \$1.25 for small beads. If she buys 3 large beads, how many small beads can she buy to stay within her budget? Write and solve a compound inequality to describe the range of possible beads.
$20 \leq 2.50(3) + 1.25b \leq 30; \ 10 \leq b \leq 18$

Example 9

Solve each inequality. Graph the solution set on a number line.

a. $-14 \leq 3x - 8 < 16$

$-14 \leq 3x - 8 < 16$ Original inequality

$-6 \leq 3x < 24$ Add 8 to each part.

$-2 \leq x < 8$ Divide each part by 3.

The solution set is $\{ x \mid -2 \leq x < 8 \}$.

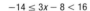

b. $|3a - 5| > 13$

$|3a - 5| > 13$ is equivalent to $3a - 5 > 13$ or $3a - 5 < -13$.

$3a - 5 > 13$ or $3a - 5 < -13$

$3a > 18$ $3a < -8$ Subtract.

$a > 6$ $a < -\dfrac{8}{3}$ Divide.

The solution set is $\left\{ a \mid a > 6 \text{ or } a < -\dfrac{8}{3} \right\}$.

---

51. $\varnothing$

52. $\left\{ q \mid q \leq -\dfrac{13}{4} \text{ or } q \geq 2 \right\}$

Additional Answers (Practice Test)

8. $b < -2$

10. $\left\{ r \mid r < -2 \text{ or } r > \dfrac{7}{2} \right\}$

11. $\{ p \mid -7 \leq p \leq 15 \}$

Practice Test

CHAPTER 1

1. Evaluate $x + y^2(2 + x)$ if $x = 3$ and $y = -1$. **8**

2. Simplify $-4(3a + b) - 2(a - 5b)$. **$-14a + 6b$**

3. **MULTIPLE CHOICE** If $3m + 5 = 23$, what is the value of $2m - 3$? **B**

   A 105

   B 9

   C $\frac{47}{3}$

   D 6

4. Solve $r = \frac{1}{2}m^2p$ for p. **$p = \frac{2r}{m^2}$**

Write an algebraic expression to represent each verbal expression.

5. twice the difference of a number and 11 **$2(n - 11)$**

6. the product of the square of a number and 5 **$5n^2$**

7. Evaluate $2|3y - 8| + y$ if $y = 2.5$. **3.5**

8. Solve $-2b > \frac{18 - b}{5}$. Graph the solution set on a number line. **See margin.**

9. **MONEY** Carson has $35 to spend at the water park. The admission price is $25 and each soda is $2.50. Write an inequality to show how many sodas he can buy. **$35 \geq 25 + 2.50s$**

10. Solve $r - 3 < -5$ or $4r + 1 > 15$. Graph the solution set. **See margin.**

11. Solve $|p - 4| \leq 11$. Graph the solution set on a number line. **See margin.**

12. **MULTIPLE CHOICE** Which graph represents the solution set for $4 < 6t + 1 \leq 43$? **F**

F, G, H, J number lines from −2 to 8

13. **MONEY** Sofia is buying new skis. She finds that the average price of skis is $500 but the actual price could differ from the average by as much as $250. Write and solve an absolute value inequality to describe this situation.
   $|p - 500| \leq 250$; $250 \leq p \leq 750$

14. **GARDENING** Andy is making 3 trapezoidal garden boxes for his backyard. Each trapezoid will be the size of the trapezoid below. He will place stone blocks around the borders of the boxes. How many feet of stones will Andy need? **102 ft**

8 ft
7 ft 7 ft
12 ft

Solve each equation.

15. $|x + 4| = 3$ **$\{-7, -1\}$**

16. $|3m + 2| = 1$ **$\{-1, -\frac{1}{3}\}$**

17. $|3a + 2| = -4$ **$\varnothing$**

18. $|2t + 5| - 7 = 4$ **$\{-8, 3\}$**

19. $|5n - 2| - 6 = -3$ **$\{-\frac{1}{5}, 1\}$**

20. $|p + 6| + 9 = 8$ **$\varnothing$**

21. **GEOMETRY** The volume of a cylinder is given by the formula $V = \pi r^2 h$. What is the volume of the cylinder below? **about 1017.88 cm³**

12 cm
9 cm

22. Solve $-3b - 5 \geq -6b - 13$. Graph the solution set on a number line. **See margin.**

23. Evaluate $\frac{3(x + y)}{4xy^2}$ if $x = \frac{2}{3}$ and $y = -2$. **$-\frac{3}{8}$**

24. Name the set(s) of numbers to which $-\frac{1}{3}$ belongs. **Q, R**

25. **MONEY** The costs for making necklaces at two craft stores are shown in the table. For what quantity of beads does The Accessory Store have a better deal? Use the inequality $15 + 3.25b < 20 + 2.50b$.

Shop	Cost per Chain	Cost per Bead
The Accessory Store	$15	$3.25
Finishing Touch	$20	$2.50

25. When you buy 6 or fewer beads, The Accessory Store is a better deal.

 connectED.mcgraw-hill.com   **53** ✓

Summative Assessment

Use these alternate leveled chapter tests to differentiate assessment for your students.

Leveled Chapter 1 Tests

Form	Type	Level	📁 Page(s)
1	MC	AL	49–50
2A	MC	OL	51–52
2B	MC	OL	53–54
2C	FR	OL	55–56
2D	FR	OL	57–58
3	FR	BL	59–60
Vocabulary Test			48
Extended-Response Test			61

MC = multiple-choice questions
FR = free-response questions

Six forms of **chapter tests**, a **Vocabulary Test**, and an **Extended-Response Test** are provided for each chapter. These are available in the Chapter Resource Masters.

McGraw-Hill eAssessment Customize and create multiple versions of your chapter tests and their answer keys. All of the questions from the leveled chapter tests in the *Chapter 1 Resource Masters* are also available on McGraw-Hill eAssessment.

Additional Answer

22. $b \geq -\frac{8}{3}$

−4 −3 −2 −1 0 1 2

InterventionPlanner

TIER 1 — On Level OL

If ▶ students miss about 25% of the exercises or less,

Then ▶ choose a resource:

   SE Lessons 1-1, 1-2, 1-3, 1-4, 1-5, and 1-6

   📁 Skills Practice, pp. 7, 13, 19, 26, 33, and 39

   TE Chapter Project

   connectED.mcgraw-hill.com Self-Check Quiz

TIER 2 — Strategic Intervention AL
approaching grade level

If ▶ students miss about 50% of the exercises,

Then ▶ choose a resource:

   📁 Study Guide and Intervention, pp. 5, 11, 17, 24, 31, and 37

   connectED.mcgraw-hill.com Extra Examples, Personal Tutor, Homework Help

TIER 3 — Intensive Intervention
2 or more grades below level

If ▶ students miss about 75% of the exercises,

Then ▶ use *Math Triumphs, Alg. 2,* Ch. 1 and 2

   connectED.mcgraw-hill.com Extra Examples, Personal Tutor, Homework Help, Review Vocabulary

1 Focus

Objective Use the strategy of eliminating unreasonable answers to solve standardized test problems.

2 Teach

Scaffolding Questions

Ask:

- Have you ever noticed that some answer choices are clearly incorrect? Answers will vary.

- What are some indications that an answer choice is not the correct answer? Sample answer: the wrong power is used, the wrong equality or inequality symbol is used.

Eliminate Unreasonable Answers

You can eliminate unreasonable answers to help you find the correct answer when solving multiple-choice test items.

Strategies for Eliminating Unreasonable Answers

Step 1

Read the problem statement carefully to determine exactly what you are being asked to find.

Ask yourself:

- What am I being asked to solve?

- In what format (that is, fraction, number, decimal, percent, type of graph) will the correct answer be?

- What units (if any) will the correct answer have?

Step 2

Carefully look over each possible answer choice and evaluate for reasonableness.

- Identify any answer choices that are clearly incorrect and eliminate them.

- Eliminate any answer choices that are not in the proper format.

- Eliminate any answer choices that do not have the correct units.

Step 3

Solve the problem and choose the correct answer from those remaining.
Check your answer.

Standardized Test Example

Read the problem. Identify what you need to know. Then use the information in the problem to solve.

The formula for the area A of a trapezoid with height h and bases b_1 and b_2 is $A = \frac{h}{2}(b_1 + b_2)$. Write an expression to represent the area of the trapezoid at the right.

A $26x^2 + 2x$ **C** $13x + 1$

B $52x^2 + 4x$ **D** $28x + 10$

BananaStock/PictureQuest

To compute the area of the trapezoid, you need to multiply half the height, $2x$, by another linear factor in x. So, the correct answer will contain an x^2 term. Since choices C and D are both linear, they can be eliminated. The correct answer is either A or B. Multiply to find the expression for the area.

$A = \frac{h}{2}(b_1 + b_2)$

$\quad = \frac{4x}{2}(8x + 3 + 5x - 2)$

$\quad = 2x(13x + 1)$

$\quad = 26x^2 + 2x$

The correct answer is A.

The formula for the volume V of a rectangular prism with length ℓ, width w, and height h is $V = \ell wh$. Write an expression to represent the volume of the rectangular prism below. B

x + 1

2x

x − 3

A $4x - 2$

B $2x^3 - 4x^2 - 6x$

C $2x^2 - 4x - 6$

D $2x^3 - 6x^2 - 4x$

3 Assess

Use Exercises 1–6 to assess students' understanding.

Exercises

Read each problem. Eliminate any unreasonable answers. Then use the information in the problem to solve.

1. The graph below shows the solution to which inequality? **C**

$-6\ -5\ -4\ -3\ -2\ -1\ \ 0\ \ 1\ \ 2\ \ 3\ \ 4\ \ 5\ \ 6$

A $8x - 9 \leq 5x - 3$

B $8x - 9 < 5x - 3$

C $8x - 9 \geq 5x - 3$

D $8x - 9 > 5x - 3$

2. Einstein's theory of relativity relates the energy E of an object to its mass m and the speed of light c. This relationship can be represented by the formula $E = mc^2$. Solve the formula for m. **G**

F $m = \frac{c^2}{E}$ **H** $m = \frac{c}{E^2}$

G $m = \frac{E}{c^2}$ **J** $m = \frac{E^2}{c}$

3. A rectangle has a width of 8 inches and a perimeter of 30 inches. What is the perimeter, in inches, of a similar rectangle with a width of 12 inches? **B**

A 40   **C** 48

B 45   **D** 360

4. The rectangular prism below has a volume of 82 cubic inches. What will the volume be if the length, width, and height of the prism are all doubled? **J**

$V = 82 \text{ in}^3$

F 41 in^3

G 164 in^3

H 482 in^3

J 656 in^3

5. Evaluate $a + (b + 1)^2$ if $a = 3$ and $b = 2$. **C**

A -6

B -1

C 12

D 15

6. At a veterinarian's office, 2 cats and 4 dogs are seen in a random order. What is the probability that the 2 cats are seen in a row? **F**

F $\frac{1}{3}$ **H** $\frac{1}{2}$

G $\frac{2}{3}$ **J** $\frac{3}{5}$

Diagnose Student Errors

Survey student responses for each item. Class trends may indicate common errors and misconceptions.

1. A did not include the denominator
   B correct
   C did not multiply before adding
   D guess

2. F correct
   G multiplied by 3 instead of $\frac{1}{3}$
   H did not square r
   J squared h instead of r

3. A this property says $x = x$
   B guess
   C correct
   D this property says if $x = y$ and $y = z$, then $x = z$.

4. F used the wrong inequality
   G correct
   H guess
   J reversed the placement of 0.2 and 81.5

5. A identified a set in which −25 does belong
   B identified a set in which −25 does belong
   C identified a set in which −25 does belong
   D correct

6. F correct
   G simplified the inequality incorrectly
   H reversed the inequality sign
   J simplified the inequality incorrectly

7. A wrote 2 more than a quotient of a number and 5
   B wrote 5 more than the product of a number and 2
   C correct
   D wrote 5 more than a quotient of a number and 2

Multiple Choice

Read each question. Then fill in the correct answer on the answer document provided by your teacher or on a sheet of paper.

1. Evaluate $\frac{m^2 + 2mn}{n^2 - 1}$ if $m = -3$ and $n = 2$. **B**

   A −3

   B −1

   C 2

   D 4

2. The volume of a cone with height h and radius r can be found by multiplying one-third π by the product of the height and the square of the radius. Which equation represents the volume of a cone? **F**

   F $V = \frac{1}{3}\pi r^2 h$

   G $V = 3\pi r^2 h$

   H $V = \frac{1}{3}\pi rh$

   J $V = \frac{1}{3}\pi rh^2$

3. Which property of equality is illustrated by the equation below? **C**

$$a + 2 = 4 \quad \rightarrow \quad 4 = a + 2$$

   A Reflexive

   B Substitution

   C Symmetric

   D Transitive

Test-Taking Tip

Question 1 Substitute −3 for m and 2 for n in the expression. Then use the order of operations to evaluate the expression.

---

4. Suppose a thermometer is accurate to within plus or minus 0.2°F. If the thermometer reads 81.5°F, which absolute value inequality represents the actual temperature T? **G**

   F $|T - 81.5| < 0.2$

   G $|T - 81.5| \leq 0.2$

   H $|T - 0.2| < 81.5$

   J $|T - 0.2| \leq 81.5$

5. To which set of numbers does −25 *not* belong? **D**

   A integers

   B rationals

   C reals

   D wholes

6. Which number line shows the solution of the inequality $2n - 3 \geq 5n - 6$? **F**

   F

   G

   H

   J

7. Write an algebraic expression to represent the verbal expression below. **C**

   | two more than the product of a number and 5 |

   A $\frac{n}{5} + 2$

   B $2n + 5$

   C $5n + 2$

   D $\frac{n}{2} + 5$

Short Response/Gridded Response

Record your answers on the answer sheet provided by your teacher or on a sheet of paper.

8. Use the absolute value equation below to answer each question.

$$|x - 3| - 2 = 0$$

a. How many solutions are there of the absolute value equation? **2**

b. Solve the equation. **$x = 1, 5$**

9. GRIDDED RESPONSE The table below shows the fill amounts and tolerances of different size soft drinks at a fountain drink vending machine. What is the maximum acceptable fill amount, in fluid ounces, for a medium drink? **21.35**

Size	Amount (fl. oz)	Tolerance (fl. oz)
small	16	0.25
medium	21	0.35
large	32	0.4

10. Simplify the expression below. Show your work. **$-18a + 19b$**

$$-4(3a - b) + 3(-2a + 5b)$$

11. While grilling steaks, Washington likes to keep the grill temperature at 425°, plus or minus 15°.

a. Write an absolute value inequality to model this situation. Let t represent the temperature of the grill. **$|t - 425| \leq 15$**

b. Within what range of temperatures does Washington like the grill to be when he cooks his steaks? **$410 \leq t \leq 440$**

12. GRIDDED RESPONSE Cameron uses a laser range finder to determine distances on the golf course. Her range finder is accurate to within 0.5 yard. If Cameron measures the distance from the tee to the flag on a par 3 to be 136 yards, what is the minimum number of yards that the distance could actually be? **135.5**

Extended Response

Record your answers on a sheet of paper. Show your work.

13. Cindy is evaluating the expression $\dfrac{-5m - 3n}{-2p + r}$ for $m = 1, n = -4, p = -3$, and $r = -2$. Her work is shown below.

$$\frac{-5m - 3n}{-2p + r} = \frac{-5(1) - 3(-4)}{-2(-3) + (-2)}$$

$$= \frac{-5 - 12}{6 - 2} = -\frac{17}{4} = -4\frac{1}{4}$$

a. What error did Cindy make in her computation? **See margin.**

b. What is the correct answer? **$1\frac{3}{4}$**

14. The table at the right shows Ricardo's scores on the first 5 math quizzes this quarter. Each quiz is worth 100 points. There will be 1 more quiz this quarter.

Quiz	Score
1	86
2	79
3	80
4	85
5	77

a. In order to receive a B, Ricardo must have a quiz average of 82 or better. Write an inequality that can be solved to find the minimum score he must earn on Quiz 6. **See margin.**

b. Solve the inequality you wrote in part **a**. **$n \geq 85$**

c. What does the solution mean? **Ricardo must score at least an 85 on his last quiz to earn a B.**

Need Extra Help?

If you missed Question...	1	2	3	4	5	6	7	8	9	10	11	12	13	14
Go to Lesson...	1-1	1-3	1-3	1-6	1-2	1-5	1-3	1-4	1-6	1-2	1-6	1-4	1-1	1-5

Additional Answers

13a. Sample answer: she did not simplify the numerator of the expression correctly. She has a sign error.

14a. $\dfrac{86 + 79 + 80 + 85 + 77 + n}{6} \geq 82$

The items in the **Standardized Test Practice** help students practice the kinds of questions found on state assessments. The **Need Extra Help?** box allows students to diagnose and address weaknesses.

Formative Assessment

You can use these pages to benchmark student progress.

📁 Standardized Test Practice, pp. 62–64

Answer Sheet Practice

Have students simulate taking a standardized test by record their answers on practice recording sheets.

Student Recording Sheet

McGraw-Hill
eAssessment Create practice tests that align to your state standards, the Common Core State Standards, and other national standards such as TIMSS and NAEP.

Homework Option

Get Ready for Chapter 2 Assign students the exercises on p. 59 as homework to assess whether they possess the prerequisite skills needed for the next chapter.

Lesson 1-2

36b. $5\left(2\frac{1}{2}\right) + 3\left(1\frac{1}{4}\right)$

$\qquad = 5\left(2 + \frac{1}{2}\right) + 3\left(1 + \frac{1}{4}\right)$ Definition of a mixed number

$\qquad = 5(2) + 5\left(\frac{1}{2}\right) + 3(1) + 3\left(\frac{1}{4}\right)$ Distributive Property

$\qquad = 10 + \frac{5}{2} + 3 + \frac{3}{4}$ Multiply.

$\qquad = 10 + 3 + \frac{5}{2} + \frac{3}{4}$ Commutative Property (+)

$\qquad = 13 + \frac{5}{2} + \frac{3}{4}$ Add.

$\qquad = 13 + \left(\frac{5}{2} + \frac{3}{4}\right)$ Associative Property (+)

$\qquad = 13 + 3\frac{1}{4}$ or $16\frac{1}{4}$ Add.

51a. Sample answer:

irrational	rational	integer	whole	natural
$-\sqrt{6}$, π	$3, \frac{-15}{3}, 4.1, 0,$ $\frac{3}{8}, \sqrt{36}$	$3, \frac{-15}{3}, 0,$ $\sqrt{36}$	$3, 0,$ $\sqrt{36}$	$3,$ $\sqrt{36}$

51b. $-\sqrt{6} \approx -2.449, 3 = 3.0, -\frac{15}{3} = -5, 4.1 = 4.1, \pi \approx 3.14,$

$0 = 0, \frac{3}{8} = 0.375, \sqrt{36} = 6; -\frac{15}{3}, -\sqrt{6}, 0, \frac{3}{8}, 3, \pi, 4.1,$

$\sqrt{36}$

51c.

Lesson 1-4

46. The 4 potential solutions are:
1. $(2x - 1) \geq 0$ and $(5 - x) \geq 0$
2. $(2x - 1) \geq 0$ and $(5 - x) < 0$
3. $(2x - 1) < 0$ and $(5 - x) \geq 0$
4. $(2x - 1) < 0$ and $(5 - x) < 0$

The resulting equations corresponding to these cases are:
1. $2x - 1 + 3 = 5 - x : x = 1$
2. $2x - 1 + 3 = x - 5 : x = -7$
3. $1 - 2x + 3 = 5 - x : x = -1$
4. $1 - 2x + 3 = x - 5 : x = 3$

The solutions from case 1 and case 3 work. The others are extraneous. The solution set is $\{-1, 1\}$.

47. Sometimes; this is only true for certain values of a. For example, it is true for $a = 8$; if $8 > 7$, then $11 > 10$. However it is not true for $a = -8$; if $8 > 7$, then $5 \not> 10$.

48. Always; if $|x| < 3$, then x is between -3 or 3. Adding 3 to the absolute value of any of the numbers in this set will produce a positive number.

49. Always; starting with numbers between 1 and 5 and subtracting 3 will produce numbers between -2 and 2. These all have an absolute value less than or equal to 2.

51. Sample answer: Symbols can be used as a shorthand way to represent ideas such as operations, equality, absolute value, and the empty set. For example, instead of writing 5 minus the absolute value of $2x$ equals 10, you could write $5 - |2x| = 10$.

Lesson 1-5

47. Using the Triangle Inequality Theorem, we know that the sum of the lengths of any 2 sides of a triangle must be greater than the length of the remaining side. This generates 3 inequalities to examine.

$3x + 4 + 2x + 5 > 4x$ $3x + 4 + 4x > 2x + 5$

$\qquad\qquad x > -9$ $x > 0.2$

$\qquad 2x + 5 + 4x > 3x + 4$

$\qquad\qquad x > -\frac{1}{3}$

In order for all 3 conditions to be true, x must be greater than 0.2.

48. Sample answer: $4x + 5 > 4(x + 1)$; This has a solution set of all real numbers because it simplifies to $4x + 5 > 4x + 4$ or $5 > 4$. This indicates that for any real value of x the inequality is equivalent to $1 > 0$, that is the left side will always be 1 greater than the right side.

49. Sample answer: When one number is greater than another number, it is either more positive or less negative than that number. When these numbers are multiplied by a negative value, their roles are reversed. That is, the number that was more positive is now more negative than the other number. Thus, it is now *less than* that number and the inequality symbol needs to be reversed.

Lesson 1-6 (Guided Practice)

1A. $\{x \mid -5 \leq x \leq 6\}$

1B. $\{z \mid -4 < z \leq -1\}$

2A. $\{j \mid j \geq 3 \text{ or } j \leq -7\}$

2B. $\{g \mid g > -5 \text{ or } g < -9.5\}$

Lesson 1-6

12. $\{v \mid 6 < v < 10\}$

13. $\{d \mid -1 \le d \le 0.5\}$

14. $\left\{r \mid r < -\frac{9}{4} \text{ or } r > 3\right\}$

15. $\{y \mid y < -4 \text{ or } y > 7\}$

16. $\{h \mid -2 < h < 2\}$

17. $\{k \mid -4 > k \text{ or } k > 4\}$

18. $\left\{x \mid x > \frac{14}{3} \text{ or } x < -2\right\}$

19. $\left\{t \mid -\frac{7}{8} \le t \le \frac{1}{8}\right\}$

20. $\left\{n \mid -1 < n < \frac{1}{3}\right\}$

21. $\left\{j \mid j \ge \frac{8}{5} \text{ or } j \le -\frac{16}{5}\right\}$

33. $\{k \mid 2 < k < 4\}$

34. $\{m \mid m > -5 \text{ or } m < -12\}$

35. $\{h \mid h < -15 \text{ or } h > 15\}$

36. $\varnothing$

37. $\{z \mid z < -1 \text{ or } z > 5\}$

38. $\left\{p \mid -\frac{9}{4} < p < \frac{5}{4}\right\}$

39. $\left\{f \mid f > \frac{26}{5} \text{ or } f < -\frac{22}{5}\right\}$

40. $\left\{w \mid w \le -\frac{23}{2} \text{ or } w \ge \frac{7}{2}\right\}$

	Diagnostic Assessment Quick Check			
	LESSON 2-1 45 min: 1 day / 90 min: 0.5 day	**EXTEND 2-1** 45 min: 0.5 day / 90 min: 0.25 day	**LESSON 2-2** 45 min: 1 day / 90 min: 0.5 day	**EXTEND 2-2** 45 min: 0.5 day / 90 min: 0.25 day
Title	Relations and Functions	Algebra Lab: Discrete and Continuous Functions	Linear Relations and Functions	Algebra Lab: Roots of Equations and Zeros of Functions
Objectives	▪ Analyze relations and functions. ▪ Use equations of relations and functions.	▪ Use discrete and continuous functions to solve real-world problems.	▪ Identify linear relations and functions. ▪ Write linear equations in standard form.	▪ Distinguish among roots, solutions, and zeros.
Key Vocabulary	one-to-one function onto function discrete relation continuous relation vertical line test independent variable dependent variable function notation		linear relations linear equation linear function standard form y-intercept x-intercept	
CCSS	F.IF.4, F.IF.5	F.IF.4	F.IF.4, F.IF.9	F.IF.4
Multiple Representations	⧉		⧉	
Lesson Resources	connectED.mcgraw-hill.com 🗀 Leveled Worksheets ᵃᵇᶜ Vocabulary PT Personal Tutor 🖱 Virtual Manipulatives ✓ Self-Check Quiz ▪ *5-Minute Check* ▪ *Study Notebook*	connectED.mcgraw-hill.com ▪ *Teaching Algebra with Manipulatives*	connectED.mcgraw-hill.com 🗀 Leveled Worksheets 🗀 Quiz 1 ᵃᵇᶜ Vocabulary PT Personal Tutor 🖱 Virtual Manipulatives ✓ Self-Check Quiz ▪ *5-Minute Check* ▪ *Study Notebook*	connectED.mcgraw-hill.com 🖱 Virtual Manipulatives ▪ *Teaching Algebra with Manipulatives*
Resources for Every Lesson	IWB eStudent Edition IWB Interactive Classroom	▪ eTeacher Edition ▪ eSolutions Manual ▪ McGraw-Hill ᵉAssessment		
Differentiated Instruction	pp. 64, 67		pp. 71, 74	

IWB All digital assets are Interactive Whiteboard ready.

Suggested Pacing			
Time Periods	Instruction	Review & Assess	Total
45-minute	11 days	2 days	13 days
90-minute	6 days	1 day	7 days

LESSON 2-3 45 min: 1 day 90 min: 0.5 day	**LESSON 2-4** 45 min: 1 day 90 min: 0.5 day	**EXTEND 2-4** 45 min: 0.5 day 90 min: 0.25 day	**LESSON 2-5** 45 min: 1 day 90 min: 0.75 day	**EXTEND 2-5** 45 min: 1 day 90 min: 0.5 day
Rate of Change and Slope	**Writing Linear Equations**	**Graphing Technology Lab: Direct Variation**	**Scatter Plots and Lines of Regression**	**Algebra Lab: Correlation and Causation**
■ Find rate of change. ■ Determine the slope of a line.	■ Write an equation of a line given the slope and a point on the line. ■ Write an equation of a line parallel or perpendicular to a given line.	■ Use functions to model direct variation.	■ Use scatter plots and prediction equations. ■ Model data using lines of regression.	■ Explore the difference between correlation and causation. ■ Identify lurking variables.
rate of change slope	slope-intercept form point-slope form parallel perpendicular		bivariate data scatter plot dot plot positive correlation negative correlation line of fit prediction equation regression line correlation coefficient	
F.IF.4, F.IF.6	A.SSE.1b, A.CED.2	F.IF.4	F.IF.4	F.IF.4
connectED.mcgraw-hill.com ▢ Leveled Worksheets Vocabulary **PT** Personal Tutor Virtual Manipulatives ☑ Self-Check Quiz ■ *5-Minute Check* ■ *Study Notebook*	connectED.mcgraw-hill.com ▢ Leveled Worksheets ▢ Quiz 2 Vocabulary **PT** Personal Tutor Virtual Manipulatives ☑ Self-Check Quiz ■ *5-Minute Check* ■ *Study Notebook*	connectED.mcgraw-hill.com **PT** Personal Tutor **Materials:** ■ TI-83/84 Plus or other graphing calculator	connectED.mcgraw-hill.com ▢ Leveled Worksheets Vocabulary 🏃 Animations **PT** Personal Tutor Virtual Manipulatives ☑ Self-Check Quiz ■ *5-Minute Check* ■ *Study Notebook*	connectED.mcgraw-hill.com **Materials:** ■ TI 83/84 Plus or other graphing calculator
IWB eStudent Edition **IWB** Interactive Classroom	■ eTeacher Edition ■ eSolutions Manual ■ *e*Assessment McGraw-Hill			
pp. 78, 82	pp. 84, 89		pp. 94, 98	

Formative Assessment
Mid-Chapter Quiz

	LESSON 2-6 — 45 min: 1 day / 90 min: 0.5 day	EXPLORE 2-7 — 45 min: 0.5 day / 90 min: 0.25 day	LESSON 2-7 — 45 min: 1 day / 90 min: 0.5 day	LESSON 2-8 — 45 min: 1 day / 90 min: 0.5 day	
Title	Special Functions	Graphing Technology Lab: Families of Lines	Parent Functions and Transformations	Graphing Linear and Absolute Value Inequalities	
Objectives	■ Write and graph piecewise-defined functions. ■ Write and graph step and absolute value functions.	■ Use a graphing calculator to determine how changing the parameters m and b affects the graphs of the functions.	■ Identify and use parent functions. ■ Describe transformations of functions.	■ Graph linear inequalities. ■ Graph absolute value inequalities.	
Key Vocabulary	piecewise-defined function piecewise-linear function step function greatest integer function absolute value function		family of graphs quadratic function parent graph translation parent function reflection constant function line of reflection identity function dilation	linear inequality boundary	
CCSS	F.IF.4, F.IF.7b	F.BF.3	F.IF.4, F.BF.3	A.CED.3	
Multiple Representations	🔧				
Lesson Resources	connectED.mcgraw-hill.com 📁 Leveled Worksheets 📁 Quiz 3 🔤 Vocabulary 🏃 Animations PT Personal Tutor ✓ Self-Check Quiz ■ *5-Minute Check* ■ *Study Notebook* ■ *Teaching Algebra with Manipulatives*	connectED.mcgraw-hill.com PT Personal Tutor **Materials:** ■ TI-83/84 Plus or other graphing calculator	connectED.mcgraw-hill.com 📁 Leveled Worksheets 🔤 Vocabulary PT Personal Tutor 🖐 Virtual Manipulatives ✓ Self-Check Quiz ■ *5-Minute Check* ■ *Study Notebook*	connectED.mcgraw-hill.com 📁 Leveled Worksheets 📁 Quiz 4 🔤 Vocabulary PT Personal Tutor 🖐 Virtual Manipulatives ✓ Self-Check Quiz ■ *5-Minute Check* ■ *Study Notebook*	
Resources for Every Lesson	IWB eStudent Edition IWB Interactive Classroom	■ eTeacher Edition ■ eSolutions Manual ■ McGraw-Hill eAssessment			
Differentiated Instruction	pp. 103, 104		pp. 111, 116	p. 121	
				Summative Assessment Study Guide and Review Practice Test	

IWB All digital assets are Interactive Whiteboard ready.

Project CRISSSM

STUDY SKILL

Students can create their own representation of the concepts they are learning by using a free-form map. Free-form mapping is done as a cooperative activity after students have read about and discussed a lesson or chapter. Give groups of students large sheets of paper and colored markers. Allow them to decide the important ideas of the lesson to represent on their map. Then present their ideas on paper with words, pictures, and diagrams. This free-form map represents the concepts in Lesson 2-7.

Parent Functions and Transformations		
Translation	**Reflection**	**Dilation**

• $f(x) \pm k$
  graph translated up/down
	• $-f(x)$
	  graph reflected in x-axis
		• $a \cdot f(x)$
		  graph stretched/compressed vertically

• $f(x \pm k)$
  graph translated left/right
	• $f(-x)$
	  graph reflected in y-axis.
		• $f(bx)$
		  graph stretched/compressed horizontally

Creating Independence through Student-owned Strategies

Assessment and Intervention

SE = Student Edition, **TE** = Teacher Edition, **CRM** = Chapter Resource Masters

Diagnosis	Prescription

Beginning Chapter 2

Get Ready for Chapter 2 **SE**	Response to Intervention **TE**

Beginning Every Lesson

Then, Now, Why? **SE** 5-Minute Checks	Chapter 0 **SE**

During/After Every Lesson

Guided Practice **SE**, every example Check Your Understanding **SE** H.O.T. Problems **SE** Spiral Review **SE** Additional Examples **TE** Watch Out! **TE** Step 4, Assess **TE** Chapter 2 Quizzes **CRM**, pp. 59–60 Self-Check Quizzes connectED.mcgraw-hill.com	TIER 1 Intervention Skills Practice **CRM**, Ch. 1–2 connectED.mcgraw-hill.com TIER 2 Intervention Differentiated Instruction **TE**; Study Guide and Intervention **CRM**, Ch. 1–2 TIER 3 Intervention *Math Triumphs, Alg. 2,* Ch. 1 and 3

Mid-Chapter

Mid-Chapter Quiz **SE** Mid-Chapter Test **CRM**, p. 61 eAssessment McGraw-Hill	TIER 1 Intervention Skills Practice **CRM**, Ch. 1–2 connectED.mcgraw-hill.com TIER 2 Intervention Study Guide and Intervention **CRM**, Ch. 1–2 TIER 3 Intervention *Math Triumphs, Alg. 2,* Ch. 1 and 3

Before Chapter Test

Chapter Study Guide and Review **SE** Practice Test **SE** Standardized Test Practice **SE** Chapter Test connectED.mcgraw-hill.com Standardized Test Practice connectED.mcgraw-hill.com Vocabulary Review connectED.mcgraw-hill.com eAssessment McGraw-Hill	TIER 1 Intervention Skills Practice **CRM**, Ch. 1–2 connectED.mcgraw-hill.com TIER 2 Intervention Study Guide and Intervention **CRM**, Ch. 1–2 TIER 3 Intervention *Math Triumphs, Alg. 2,* Chs. 1 and 3

After Chapter 2

Multiple-Choice Tests, Forms 1, 2A, 2B **CRM**, pp. 63–68 Free-Response Tests, Forms 2C, 2D, 3 **CRM**, pp. 69–74 Vocabulary Test **CRM**, p. 62 Extended Response Test **CRM**, p. 75 Standardized Test Practice **CRM**, pp. 76–78 eAssessment McGraw-Hill	Study Guide and Intervention **CRM**, Ch. 1–2 connectED.mcgraw-hill.com

Option 1 Reaching All Learners AL OL BL ELL

Kinesthetic Have students work in pairs to measure the length of their arms from shoulders to finger tips and their heights in centimeters. Record the measurements as an ordered pair (arm length, height) for each student. Use the class data to create a scatter plot and a line of fit, and describe the correlation. Then use a graphing calculator to find an equation for a line of regression.

Interpersonal As a class, create a word wall that contains a list of formulas and the purpose of each formula. As more and more formulas are added, ask questions about the variables such as, "Is m always used to represent slope?"

Option 2 Approaching Level AL

Provide each student with a ruler and a copy of a variety of graphs—some that are functions and some that are not. Ask students to draw multiple vertical lines on each graph, trying to intersect the graph in more than one point. Then have students indicate which graphs are graphs of functions.

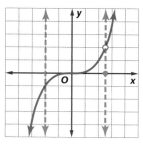

Option 3 English Learners ELL

Write the words *function, slope,* and *linear* on the board. For each one, have students suggest other words having similar or related meanings. Discuss when and how each of these words is used in real life versus how it is used in mathematics.

Option 4 Beyond Level BL

Create a set of vocabulary and/or formula cards from this chapter. Then ask the class to play a game of charades. One student draws a card and then must draw or act out what is on the card without uttering a word. The rest of the class tries to figure out the vocabulary word or formula written on the card.

Stretch students' understanding of data trends and linearity by analyzing real-world examples that display each trend. Ask students to create graphs that demonstrate each behavior:

- linear and continuous
- linear and discrete
- positively correlated linear trend
- negatively correlated linear trend
- linear trend based on an average increase or decrease

VerticalAlignment

Before Chapter 2

Related Topics from Algebra 1

- identify domains and ranges for given situations
- determine intercepts of the graphs of linear functions
- determine slopes from graphs, tables, and algebraic representations
- graph and write equations of lines
- use data to determine functional relationships between quantities
- formulate linear inequalities to solve problems and investigate methods for solving them

Chapter 2

- identify the mathematical domains and ranges of functions and determine reasonable domain and range values for continuous and discrete situations
- identify and sketch graphs of parent functions, including linear functions ($f(x) = x$) and absolute value functions ($f(x) = |x|$)
- collect and organize data, make and interpret scatter plots, fit the graph of a function to the data, interpret the results, and proceed to model, predict, and make decisions and critical judgments
- graph linear and absolute value inequalities

After Chapter 2

Preparation for Precalculus

- describe parent functions symbolically and graphically, including $f(x) = |x|$
- determine the domain and range of functions using graphs, tables, and symbols
- use regression to determine the appropriateness of a linear function to model real-life data

Lesson-by-LessonPreview

2-1 Relations and Functions

Recall that a relation is a set of ordered pairs, and a function is a relation in which each element of the domain is paired with exactly one element of the range. Functions can be *one-to-one*, *onto*, or *both one-to-one and onto* depending on the way the elements of the domain and range are paired.

Numerical and graphical means are used to compare functions with relations. Mappings are used to show how elements of the domain are paired with elements of the range. A mapping can be used to identify relations that are functions. The vertical line test can distinguish a relation from a function.

Relations and functions in which the domain is a set of individual points are said to be *discrete*. If the domain can be graphed with a line or smooth curve, then it is *continuous*.

2-2 Linear Relations and Functions

A *linear function* is a function whose ordered pairs satisfy a related linear equation. Linear functions can be written in slope-intercept form, $f(x) = mx + b$, where m and b are real numbers.

Linear equations can be written in slope-intercept form, $y = mx + b$, or in standard form, $Ax + By = C$. Since the graph of a linear function or equation is always a straight line, one way to graph a linear function or equation is to connect the coordinates of the x- and y-intercepts.

2-3 Rate of Change and Slope

The slope (or rate of change) of a line is the ratio of the change in the y-coordinates to the corresponding change in the x-coordinates. When two points on the line are known, the slope m of a line is given by the formula $m = \dfrac{y_2 - y_1}{x_2 - x_1}$, where $x_1 \neq x_2$.

2-4 Writing Linear Equations

When two points or a point and the slope of a line are known, an equation of the line can be found.

- *Slope-intercept form* of a linear equation: $y = mx + b$, m represents the slope of the graph of the equation and b represents the y-intercept.

- *Point-slope form:* $y - y_1 = m(x - x_1)$, m represents the slope of the graph of the equation and (x_1, y_1) represents a point on the line.

If you know the slopes of two lines, you can determine whether they are parallel or perpendicular.

- Two nonvertical lines are parallel if and only if the lines have the same slope. (All vertical lines are parallel.)

- Two nonvertical lines are perpendicular if and only if the product of their slopes is −1. (All vertical lines are perpendicular to horizontal lines.)

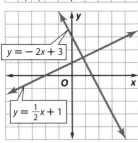

2-5 Scatter Plots and Lines of Regression

When *bivariate data* are graphed as ordered pairs in a coordinate plane, the graph is called a *scatter plot*. If a set of data exhibits a linear trend, an equation of a line of fit can be used to approximate the relation between domain values and range values of the data. This line of fit, also called a *prediction equation*, can be used to make predictions.

A *regression line,* a line of fit that is more precise than the one found by simply drawing a line that appears to represent the data in the scatter plot, can be determined through complex calculations. This line can be found easily using a graphing calculator.

2-6 Special Functions

This lesson examines several special functions. Special types of linear functions include the following.

- *Piecewise-defined functions* are usually written using two or more algebraic expressions.

- A common *piecewise-linear function*, called a *step function,* consists of a series of line segments that look like steps.

- The graph of an *absolute value function* is shaped like a V and is made up of portions of two lines.

2-7 Parent Functions and Transformations

A family of graphs is a group of graphs that display one or more similar characteristics. The parent graph, which is a graph of the parent function, is the simplest of the graphs in a family. This graph can be *transformed* to create other members in a family of graphs.

- A *translation* moves a figure up, down, left, or right. A constant added to or subtracted from a parent function results in a translation of the graph up or down. If a constant is added to or subtracted from x before evaluating a parent function, the result is a translation right or left.

- A *reflection* flips a figure over a line called the line of reflection. The reflection $-f(x)$ reflects the graph of $f(x)$ across the x-axis. The reflection $f(-x)$ reflects the graph of $f(x)$ across the y-axis.

- A *dilation* shrinks or enlarges a figure proportionally. When the graph of a parent function is multiplied by a nonzero number the result is a graph that is dilated.

2-8 Graphing Linear and Absolute Value Inequalities

The solution set of a linear inequality is the set of all ordered pairs that make the statement true. The graph of a linear inequality can be shown as a shaded region that is formed by a linear boundary that divides the coordinate plane into two regions. The boundary itself may or may not be included in the solution set depending on the particular inequality symbol used in the inequality. If the inequality symbol is $\geq$ or $\leq$, the boundary is included in the solution and a solid line is used in the graph. A dashed line is used when the inequality symbol is $<$ or $>$.

Mc Graw Hill Professional Development

Targeted professional development has been articulated throughout Algebra 2. More quality, customized professional development is available from McGraw-Hill Professional Development. Visit connectED.mcgraw-hill.com for details on each product.

- Online Lessons emphasize the strategies and techniques used to teach Algebra 2. Includes streaming video, interactive pages, and online tools.

- Video Workshops allow mentors, coaches, or leadership personnel to facilitate on-site workshops on educational strategies in mathematics and mathematical concepts.

- MHPD Online (www.mhpdonline.com) offers online professional development with video clips of instructional strategies, links, student activities, and news and issues in education.

- Teaching Today (teachingtoday.glencoe.com) gives secondary teachers practical strategies and materials that inspire excellence and innovation in teaching.

Study Guide and Intervention
AL OL ELL

NAME _____ DATE _____ PERIOD _____

2-1 Study Guide and Intervention
Relations and Functions

Relations and Functions A relation can be represented as a set of ordered pairs or as an equation; the relation is then the set of all ordered pairs (x, y) that make the equation true. A **function** is a relation in which each element of the domain is paired with exactly one element of the range.

One-to-One Function	Each element of the domain pairs to exactly one unique element of the range.	
Onto Function	Each element of the range also corresponds to an element of the domain.	
Both One-to-One and Onto	Each element of the domain is paired to exactly one element of the range and each element of the range	

Example State the domain and range of the relation. Does the relation represent a function?

The domain and range are both all real numbers. Each element of the domain corresponds with exactly one element of the range, so it is a function.

Exercises
State the domain and range of each relation. Then determine whether each relation is a *function*. If it is a function, determine if it is *one-to-one, onto, both,* or *neither*.

1. {(0.5, 3), (0.4, 2), (3.1, 1), (0.4, 0)} 2. {(–5, 2), (4, –2), (3, –13), (–7, 2)}

3. {(0.5, –3), (0.1, 2), (6, 8)} 4. {(–15, 12), (–14, 11), (–13, 10), (–12, 12)}

Practice
AL OL BL ELL

NAME _____ DATE _____ PERIOD _____

2-1 Practice
Relations and Functions

State the domain and range of each relation. Then determine whether each relation is a *function*. If it is a function, determine if it is *one-to-one, onto, both,* or *neither*.

1. 2.

3. 4.

Graph each equation and determine the domain and range. Determine whether the relation is a *function*, is *one-to-one, onto, both,* or *neither*. Then state whether it is *discrete* or *continuous*.

5. $x = -1$ 6. $y = 2x - 1$

Find each value if $f(x) = \frac{5}{x+2}$ and $g(x) = -2x + 3$.

7. $f(2)$ 8. $f(-4)$ 9. $g(\frac{1}{2})$

10. $f(-2)$ 11. $g(-6)$ 12. $f(m - 2)$

13. **MUSIC** The ordered pairs (1, 16), (2, 16), (3, 32), (4, 32), and (5, 48) represent the cost of buying various numbers of CDs through a music club. Identify the domain and range of the relation. Is the relation discrete or continuous? Is the relation a function?

14. **COMPUTING** If a computer can do one calculation in 0.00000015 second, then the function $T(n) = 0.00000015n$ gives the time required for the computer to do n calculations. How long would it take the computer to do 5 billion calculations?

Word Problem Practice
AL OL BL ELL

NAME _____ DATE _____ PERIOD _____

2-1 Word Problem Practice
Relations and Functions

1. **PLANETS** The table below gives the mean distance from the Sun and orbital period of the eight major planets in our Solar System. Think of the mean distance as the domain and the orbital period as the range of a relation. Is this relation a function? Explain.

Planet	Mean Distance from Sun (AU)	Orbital Period (years)
Mercury	0.387	0.241
Venus	0.723	0.615
Earth	1.0	1.0
Mars	1.524	1.881
Jupiter	5.204	11.75
Saturn	9.582	29.5
Uranus	19.201	84
Neptune	30.047	165

2. **PROBABILITY** Martha rolls a number cube several times and makes the frequency graph shown. Write a relation to represent this data.

3. **SCHOOL** The number of students N in Vanita's school is given by $N = 120 + 30G$, where G is the grade level. Is 285 in the range of this function?

4. **FLOWERS** Anthony decides to decorate a ballroom with $r = 3n + 20$ roses, where n is the number of dancers. It occurs to Anthony that the dancers always come in pairs. That is, $n = 2p$, where p is the number of pairs. What is r as a function of p?

5. **SALES** Cool Athletics introduced the new Power Sneaker in one of their stores. The table shows the sales for the first 6 weeks.

Week	1	2	3	4	5	6
Pairs Sold	8	10	15	22	31	44

a. Graph the data.

b. Identify the domain and range.

c. Is the relation a function? Explain.

B **33** **PODCASTS** Chaz has a collection of 15 podcasts downloaded on his digital audio player. He decides to download 3 more podcasts each month. The function $P(t) = 15 + 3t$ counts the number of podcasts $P(t)$ he has after t months. How many podcasts will he have after 8 months? **39**

C **34.** 🔁 **MULTIPLE REPRESENTATIONS** In this problem you will investigate one-to-one and onto functions. **34a–e. See Chapter 2 Answer Appendix.**

a. **Graphical** Graph each function on a separate graphing calculator screen.
$$f(x) = x^2 \qquad g(x) = 2^x \qquad h(x) = x^3 - 3x^2 - 5x + 6 \qquad j(x) = x^3$$

b. **Tabular** Use the graphs to create a table showing the number of times a horizontal line could intersect the graph of each function. List all possibilities.

c. **Analytical** For a function to be one-to-one, a horizontal line on the graph of the function can intersect the function at most once. Which functions meet this condition? Which do not? Explain your reasoning.

d. **Analytical** For a function to be onto, every possible horizontal line on the graph of the function must intersect the function at least once. Which functions meet this condition? Which do not? Explain your reasoning.

e. **Graphical** Create a table showing whether each function is one-to-one and/or onto.

35. Sample answer: Omar; Madison did not square the 3 before multiplying by –4.

H.O.T. Problems *Use Higher-Order Thinking Skills*

35. 🟦 **CRITIQUE** Omar and Madison are finding $f(3d)$ for the function $f(x) = -4x^2 - 2x + 1$. Is either of them correct? Explain your reasoning.

37. Never; if the graph crosses the y-axis twice, then there will be two separate y-values that correspond to $x = 0$, which violates the vertical line test.

Omar	Madison
$f(3d) = -4(3d)^2 - 2(3d) + 1$	$f(3d) = -4(3d)^2 - 2(3d) + 1$
$= -4(9d^2) - 6d + 1$	$= 12d^2 - 6d + 1$
$= -36d^2 - 6d + 1$	

39. Sample answer: False; a function is onto and not one-to-one if all of the elements of the domain correspond to an element of the range, but more than one element of the domain corresponds to the same element of the range.

36. CHALLENGE Consider the functions $f(x)$ and $g(x)$. $f(a) = 19$ and $g(a) = 33$, while $f(b) = 31$ and $g(b) = 51$. If $a = 5$ and $b = 8$, find two possible functions to represent $f(x)$ and $g(x)$. **Sample answer: $f(x) = 4x - 1$; $g(x) = 6x + 3$**

37. REASONING If the graph of a relation crosses the y-axis at more than one point, is the relation *sometimes, always,* or *never* a function? Explain your reasoning.

38. OPEN ENDED Graph a relation that can be used to represent each of the following.
a. the height of a baseball that is hit into the outfield **a–d. See Chapter 2 Answer Appendix.**
b. the speed of a car that travels to the store, stopping at two lights along the way
c. the height of a person from age 5 to age 80
d. the temperature on a typical day from 6 A.M. to 11 P.M.

39. REASONING Determine whether the following statement is *true* or *false*. Explain your reasoning.

If a function is onto, then it must be one-to-one as well.

40. WRITING IN MATH Explain why the vertical line test can determine if a relation is a function. **See margin.**

🔊 **66** | Lesson 2-1 | Relations and Functions

🗂 **Enrichment**
OL BL

NAME _____ DATE _____ PERIOD _____

2-1 Enrichment
Real Number Relations and Functions

It is interesting to think about one-to-one and onto correspondences between different subsets of the real numbers.

Example 1 Is there one-to-one and onto correspondence between natural numbers and integers?

Set the two groups up as follows:

Even though the natural numbers are a subset of the integers, there is one unique natural number for each integer. The two sets do have a one-to-one correspondence.

Example 2 Is there one-to-one and onto correspondence between integers and rational numbers?

🔁 **Multiple Representations**

In Exercise 34, students use a graphing calculator, a table of information, and visual inspection to identify properties of functions.

41. Patricia's swimming pool contains 19,500 gallons of water. She drains the pool at a rate of 6 gallons per minute. Which of these equations represents the number of gallons of water g remaining in the pool after m minutes? **A**

  A $g = 19,500 - 6m$
  B $g = 19,500 + 6m$
  C $g = \dfrac{19,500}{6m}$
  D $g = \dfrac{6m}{19,500}$

42. SHORT RESPONSE Look at the pattern below.

$$-\frac{5}{2}, -2, -\frac{3}{2}, -1, \ldots$$

If the pattern continues, what will the next term be? $-\dfrac{1}{2}$

43. GEOMETRY Which set of dimensions represents a triangle similar to the triangle shown below? **J**

  F 1 unit, 2 units, 3 units
  G 7 units, 11 units, 12 units
  H 10 units, 23 units, 24 units
  J 20 units, 48 units, 52 units

44. SAT/ACT If $g(x) = x^2$, which expression is equal to $g(x + 1)$? **C**
  A 1
  B $x^2 + 1$
  C $x^2 + 2x + 1$
  D $x^2 - x$
  E $x^2 + x + 1$

Solve each inequality. (Lesson 1-6)

45. $48 > 7y + 6 > 20$ $6 > y > 2$
46. $z + 12 > 18$ or $-2z + 16 > 12$ $z > 6$ or $z < 2$
47. $2|4x + 2| + 3 > 21$ $x > \dfrac{7}{4}$ or $x < -\dfrac{11}{4}$

48. CLUBS Mr. Willis is starting a chess club at his high school. He sent the advertisement at the right to all of the homerooms. Write an absolute value inequality representing the situation. (Lesson 1-6) $|x - 14| \le 8$

Chess Club!!
Mondays and Wednesdays after school in Mr. Willis's room. We are looking for **14** members, give or take **8**.

49. SALES Ling can spend no more than $120 at the summer sale of a department store. She wants to buy shirts on sale for $15 each. Write and solve an inequality to determine the number of shirts she can buy. (Lesson 1-5) $15x \le 120$; She can buy up to 8 shirts.

Solve each equation. Check your solutions. (Lesson 1-4)

50. $18 = 2|2a + 6| - 2$ 2 or -8
51. $2 = -3|4c - 5| + 8$ $\dfrac{3}{4}$ or $\dfrac{7}{4}$
52. $-5 = 2|3b + 4| - 9$ -2 or $-\dfrac{2}{3}$

Simplify each expression. (Lesson 1-2)

53. $6(3a - 2b) + 3(5a + 4b)$ $33a$
54. $-4(5x - 3y) + 2(y + 3x)$ $-14x + 14y$
55. $-7(2c - 4d) + 8(3c + d)$ $10c + 36d$

Solve each equation. Check your solutions.

56. $5x + 2 = 32$ 6
57. $6a - 3 = 21$ 4
58. $-2x + 5 = 5x + 19$ -2
59. $6b + 4 = -2b - 28$ -4
60. $2(x + 5) - 3(x - 4) = 19$ 3
61. $4(2y - 3) + 5(3y + 1) = -99$ -4
62. $5c - 8 + 2c = 4c + 10$ 6
63. $8d - 4 + 3d = 2d - 100 - 7d$ -6
64. $10y - 5 - 3y = 4(2y + 3) - 20$ 3

DifferentiatedInstruction (BL)

Extension Some relations (and functions) are neither discrete nor completely continuous. Ask students to consider the example below as a relation between two variables. Then ask them to determine whether each is a function, and whether the domain and/or the range are discrete.

the number of bacteria in a colony and time (Assume the colony starts with one cell and each cell divides into two new cells after exactly 30 minutes.) The number of bacteria is a function of time. The domain is continuous for non-negative values of time. The dependent variable (the number of bacteria) is discrete.

Exercise Alert
Graphing Calculator Exercise 34 requires use of a graphing calculator.

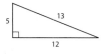 **Teaching the Mathematical Practices**

Critique Mathematically proficient students are also able to compare the effectiveness of two plausible arguments, distinguish correct logic or reasoning from that which is flawed, and—if there is a flaw in an argument—explain what it is.

WatchOut!
Error Analysis For Exercise 35, suggest that students rewrite the original function by substituting the expression ($3d$) for *each* occurrence of the variable x *before* they begin simplifying.

4 Assess

Crystal Ball In Lesson 2-2 students will be identifying linear relations and functions. Have students write about how they think today's lesson will connect with the theme of Lesson 2-2.

Tips for New Teachers
Intervention This lesson has a number of vocabulary words that may be new or challenging for some students. Make sure that all students are comfortable with the mathematical language of this lesson before they begin the next lesson.

Additional Answer

40. Sample answer: A relation is a function if each x-value only pairs with one y-value. If the vertical line test fails then there is an x-value that pairs with more than one y-value, so the relation is not a function.

1 Focus

Objective Distinguish between discrete and continuous functions to explore real-world problems.

Teaching Tip
Begin this activity by asking students to think of examples where only whole number values for the domain make sense.

Some examples might be:

- number of wheels on a car
- number of students absent from school

2 Teach

Working in Cooperative Groups
Divide the class into groups of two, mixing abilities. Have groups read the first four paragraphs.

Practice Have students complete Exercises 1–5.

3 Assess

Formative Assessment
Use Exercise 5 to assess whether students understand the difference between a discrete and a continuous function.

From Concrete to Abstract
Ask students to summarize what they learned about discrete and continuous functions. Then have them give an example of each.

A cup of frozen yogurt costs $2 at the Yogurt Shack. We might describe the cost of x cups of yogurt using the continuous function $y = 2x$, where y is the total cost in dollars. The graph of that function is shown at the right.

From the graph, you can see that 2 cups of yogurt cost $4, 3 cups cost $6, and so on. The graph also shows that 1.5 cups of yogurt cost 2(1.5) or $3. However, the Yogurt Shack probably will not sell partial cups of yogurt. This function is more accurately modeled with a discrete function.

Buying Frozen Yogurt

$y = 2x$

CCSS Common Core State Standards
Content Standards
F.IF.4 For a function that models a relationship between two quantities, interpret key features of graphs and tables in terms of the quantities, and sketch graphs showing key features given a verbal description of the relationship.

The graph of the discrete function at the right also models the cost of buying cups of frozen yogurt. The domain in this graph makes sense in this situation.

When choosing a discrete function or a continuous function to model a real-world situation, consider whether all real numbers make sense as part of the domain.

Buying Frozen Yogurt

Exercises

Determine whether each function is correctly modeled using a discrete or continuous function. Explain your reasoning.

1. **Converting Units**

Continuous; there can be measures of weight that are not integers.

2. **E-Mails Received**

Discrete; you cannot receive a fraction of an e-mail.

3. y represents the distance a car travels in x hours. **3–5. See Chapter 2 Answer Appendix.**

4. y represents the total number of riders who have ridden on a roller coaster after x rides.

5. **WRITING IN MATH** Give an example of a real-world function that is discrete and a real-world function that is continuous. Explain your reasoning.

68 | Extend 2-1 | Algebra Lab: Discrete and Continuous Functions

Extending the Concept
- Have students draw a graph of a real-life situation that is modeled by a continuous function.

- Have students draw a graph of a real-life situation that is modeled by a discrete function.

LESSON 2-2

Linear Relations and Functions

Then	Now	Why?
• You analyzed relations and functions.	**1** Identify linear relations and functions. **2** Write linear equations in standard form.	• Laura does yard work to earn money during the summer. She either cuts grass x or does general gardening y, and she schedules 5 jobs per day. The equation $x + y = 5$ can be used to relate how many of each task Laura can do in a day.

Yard Work

Gardening / Cuts Grass

$x + y = 5$

 NewVocabulary
linear relation
nonlinear relation
linear equation
linear function
standard form
y-intercept
x-intercept

Common Core State Standards

Content Standards

F.IF.4 For a function that models a relationship between two quantities, interpret key features of graphs and tables in terms of the quantities, and sketch graphs showing key features given a verbal description of the relationship.

F.IF.9 Compare properties of two functions each represented in a different way (algebraically, graphically, numerically in tables, or by verbal descriptions).

Mathematical Practices
3 Construct viable arguments and critique the reasoning of others.

1 **Linear Relations and Functions** The points on the graph above lie along a straight line. Relations that have straight line graphs are called **linear relations**. Relations that are not linear are called **nonlinear relations**.

An equation such as $x + y = 5$ is called a linear equation. A **linear equation** has no operations other than addition, subtraction, and multiplication of a variable by a constant. The variables may not be multiplied together or appear in a denominator. A linear equation does not contain variables with exponents other than 1. The graph of a linear equation is always a line.

Linear equations	Nonlinear equations
$4x - 5y = 16$	$2x + 6y^2 = -25$
$x = 10$	$y = \sqrt{x} + 2$
$y = -\frac{2}{3}x - 1$	$x + xy = -\frac{5}{8}$
$y = \frac{1}{2}x$	$y = \frac{1}{x}$

A **linear function** is a function with ordered pairs that satisfy a linear equation. Any linear function can be written in the form $f(x) = mx + b$, where m and b are real numbers.

Example 1 **Identify Linear Functions**

State whether each function is a linear function. Write *yes* or *no*. Explain.

a. $f(x) = 8 - \frac{3}{4}x$

Yes; it can be written as $f(x) = -\frac{3}{4}x + 8$.

$m = -\frac{3}{4}, b = 8$

b. $f(x) = \frac{2}{x}$

No; the expression includes division by the variable.

c. $g(x, y) = 3xy - 4$

No; the two variables are multiplied together.

GuidedPractice   1A. No; the variable appears in the denominator.

1A. $f(x) = \frac{5}{x + 6}$ **1B.** $g(x) = -\frac{3}{2}x + \frac{1}{3}$ yes; $m = -\frac{3}{2}, b = \frac{1}{3}$

You can evaluate linear functions by substituting values for x.

 connectED.mcgraw-hill.com   **69**

1 Focus

VerticalAlignment

Before Lesson 2-2 Analyze relations and functions.

Lesson 2-2 Identify linear relations and functions and write linear equations in standard form.

After Lesson 2-2 Identify and sketch graphs of parent functions.

2 Teach

Scaffolding Questions
Have students read the **Why?** section of the lesson.

Ask:
■ What does a value of zero for *x* mean in this situation? Laura does not cut any grass.

■ What does a negative value for *x* or *y* mean in this situation? It has no meaning. She cannot have a negative number of jobs.

Lesson 2-2 Resources

Resource	Approaching Level **AL**	On Level **OL**	Beyond Level **BL**	English Learners **ELL**
Teacher Edition	■ Differentiated Instruction, p. 71	■ Differentiated Instruction, pp. 71, 74	■ Differentiated Instruction, pp. 71, 74	■ Differentiated Instruction, p. 71
Chapter Resource Masters	■ Study Guide and Intervention, pp. 11–12 ■ Skills Practice, p. 13 ■ Practice, p. 14 ■ Word Problem Practice, p. 15	■ Study Guide and Intervention, pp. 11–12 ■ Skills Practice, p. 13 ■ Practice, p. 14 ■ Word Problem Practice, p. 15 ■ Enrichment, p. 16	■ Practice, p. 14 ■ Word Problem Practice, p. 15 ■ Enrichment, p. 16	■ Study Guide and Intervention, pp. 11–12 ■ Skills Practice, p. 13 ■ Practice, p. 14 ■ Word Problem Practice, p. 15
Other	■ 5-Minute Check 2-2 ■ Study Notebook	■ 5-Minute Check 2-2 ■ Study Notebook	■ 5-Minute Check 2-2 ■ Study Notebook	■ 5-Minute Check 2-2 ■ Study Notebook

1 Linear Relations and Functions

Example 1 illustrates how to determine whether a function is linear. **Example 2** shows how to evaluate a linear function by substitution.

Formative Assessment

Use the Guided Practice exercises after each example to determine students' understanding of concepts.

Additional Examples

1 State whether each function is a linear function. Write *yes* or *no*. Explain.

a. $g(x) = 2x - 5$
yes; $m = 2$, $b = -5$

b. $p(x) = x^3 + 2$ No; x has an exponent other than 1.

c. $t(x) = 4 + 7x$
yes; $m = 7$, $b = 4$

2 **METEOROLOGY** The linear function $f(C) = 1.8C + 32$ can be used to find the number of degrees Fahrenheit $f(C)$ that are equivalent to a given number of degrees Celsius C.

a. On the Celsius scale, normal body temperature is 37°C. What is it in degrees Fahrenheit? 98.6°F

b. There are 100 Celsius degrees between the freezing and boiling points of water and 180 Fahrenheit degrees between these two points. How many Fahrenheit degrees equal 1 Celsius degree? 1.8°F = 1°C

▶ **Additional Examples** also in Interactive Classroom PowerPoint® Presentations

IWB Interactive White Board READY

Real-WorldLink

The largest member of the grass family, bamboo, is capable of growing from 1 to 4 feet per day.

Source: Infoplease

2B. Sample answer: No; the grass could not support its own weight after a long period of time.

 70 | **Lesson 2-2** | Linear Relations and Functions

● Real-World Example 2 Evaluate a Linear Function

PLANTS The growth rate of a sample of Bermuda grass is given by the function $f(x) = 5.9x + 3.25$, where $f(x)$ is the total height in inches x days after an initial measurement.

a. How tall is the sample after 3 days?

$f(x) = 5.9x + 3.25$	Original function
$f(3) = 5.9(3) + 3.25$	Substitute 3 for x.
$= 20.95$	Simplify.

The height of the sample after 3 days is 20.95 inches.

b. The term 3.25 in the function represents the height of the grass when it was initially measured. The sample is how many times as tall after 3 days?

Divide the height after 3 days by the initial height. $\frac{20.95}{3.25} \approx 6.4$

The height after 3 days is about 6.4 times as great as the initial height.

▶ **Guided**Practice

2A. If the Bermuda grass is 50.45 inches tall, how many days has it been since it was last cut? **8 days**

2B. Is it reasonable to think that this rate of growth can be maintained for long periods of time? Explain.

---

2 **Standard Form** Any linear equation can be written in **standard form**, $Ax + By = C$, where A, B, and C are integers with a greatest common factor of 1.

⬥ KeyConcept Standard Form of a Linear Equation

Words	The standard form of a linear equation is $Ax + By = C$, where A, B, and C are integers with a greatest common factor of 1, $A \geq 0$, and A and B are not both zero.
Example	$3x + 5y = 12$; $A = 3$, $B = 5$, and $C = 12$

Example 3 Standard Form

Write $-\frac{3}{10}x = 8y - 15$ in standard form. Identify A, B, and C.

$-\frac{3}{10}x = 8y - 15$	Original equation
$-\frac{3}{10}x - 8y = -15$	Subtract $8y$ from each side.
$3x + 80y = 150$	Multiply each side by -10.
$A = 3$, $B = 80$, and $C = 150$	

▶ **Guided**Practice

Write each equation in standard form. Identify A, B, and C.

3A. $2y = 4x + 5$ $4x - 2y = -5$; 4, -2, -5 **3B.** $3x - 6y - 9 = 0$ $x - 2y = 3$; 1, -2, 3

Focus on Mathematical Content

Linear Equations When determining whether an equation is linear, it is helpful to examine the equation for certain characteristics. A linear equation has no operations other than addition, subtraction, and multiplication of a variable by a constant. The variables may not be multiplied together or appear in a denominator. A linear equation does not contain variables with exponents other than 1.

Teach with Tech

Document Camera Give students several equations and ask them to identify each as linear or nonlinear. If the equation is linear, have students write it in standard form. Choose several students to explain and share their work with the class.

Mike Powles/Photolibrary/Getty Images

Since two points determine a line, one way to graph a linear function is to find the points at which the graph intersects each axis and connect them with a line. The y-coordinate of the point at which a graph crosses the y-axis is called the **y-intercept**. Likewise, the x-coordinate of the point at which it crosses the x-axis is called the **x-intercept**.

StudyTip

Vertical and Horizontal Lines When C represents a constant, an equation of the form $x = C$ represents a vertical line with only an x-intercept. The equation $y = C$ represents a horizontal line with only a y-intercept.

Example 4 Use Intercepts to Graph a Line

Find the x-intercept and the y-intercept of the graph of $2x - 3y + 8 = 0$. Then graph the equation.

The x-intercept is the value of x when $y = 0$.

$2x - 3y + 8 = 0$	Original equation
$2x - 3(0) + 8 = 0$	Substitute 0 for y.
$2x = -8$	Subtract 8 from each side.
$x = -4$	Divide each side by 2.

The x-intercept is -4.

Likewise, the y-intercept is the value of y when $x = 0$.

$2x - 3y + 8 = 0$	Original equation
$2(0) - 3y + 8 = 0$	Substitute 0 for x.
$-3y = -8$	Subtract 8 from each side.
$y = \dfrac{8}{3}$	Divide each side by -3.

The y-intercept is $\dfrac{8}{3}$.

Use these ordered pairs to graph the equation.

▶ **Guided**Practice

4. Find the x-intercept and the y-intercept of the graph of $2x + 5y - 10 = 0$. Then graph the equation. **See margin.**

1. Yes; it can be written as $f(x) = \dfrac{x}{5} + \dfrac{12}{5}$.

Check Your Understanding   ● = **Step-by-Step Solutions begin on page R14.**

Example 1   State whether each function is a linear function. Write *yes* or *no*. Explain.

1. $f(x) = \dfrac{x + 12}{5}$ **2.** $g(x) = \dfrac{7 - x}{x}$ **3.** $p(x) = 3x^2 - 4$ **4.** $q(x) = -8x - 21$

   2. No; it cannot be written as $f(x) = mx + b$.   **3.** No; x has an exponent that is not 1.

Example 2   **5** **RECREATION** You want to make sure that you have enough music for a car trip. If each CD is an average of 45 minutes long, the linear function $m(x) = 0.75x$ could be used to find out how many CDs you need to bring.

4. Yes; it is written in $f(x) = mx + b$ form.

   a. How many hours of music are there on 4 CDs? **3 hours**

   b. If the trip you are taking is 6 hours, how many CDs should you bring? **8 CDs**

Example 3   **CCSS** **STRUCTURE** Write each equation in standard form. Identify A, B, and C. **6–11. See Chapter 2 Answer Appendix.**

   6. $y = -4x - 7$ **7.** $y = 6x + 5$ **8.** $3x = -2y - 1$

   9. $-8x = 9y - 6$ **10.** $12y = 4x + 8$ **11.** $4x - 6y = 24$

Example 4   Find the x-intercept and the y-intercept of the graph of each equation. Then graph the equation using the intercepts. **12–15. See Chapter 2 Answer Appendix.**

   12. $y = 5x + 12$ **13.** $y = 4x - 10$ **14.** $2x + 3y = 12$ **15.** $3x - 4y - 6 = 15$

DifferentiatedInstruction **AL** **OL** **BL** **ELL**

Verbal/Linguistic Learners Write a letter to another student in the class explaining how to write an equation in standard form. The letter should include an explanation and an example.

---

2 Standard Form

Example 3 shows how to write a linear equation in standard form. **Example 4** explains how to find the x- and y-intercepts of a graph and use the ordered pairs to graph the equation.

Additional Examples

3 Write $y = 3x - 9$ in standard form. Identify A, B, and C.
$3x - y = 9$;
$A = 3, B = -1, C = 9$

4 Find the x-intercept and the y-intercept of the graph of $-2x + y - 4 = 0$. Then graph the equation. x-intercept: -2; y-intercept: 4

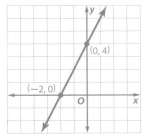

3 Practice

Formative Assessment
Use Exercises 1–15 to check for understanding.

Use the chart at the bottom of the next page to customize assignments for your students.

Additional Answer (Guided Practice)

4. 5; 2

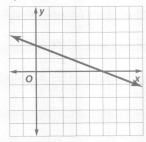

Tips for New Teachers

Graphing It may be helpful to have rulers and graph paper available for students to use as they work through the exercises.

Additional Answers

16. Yes; it can be written in $f(x) = mx + b$ form, where $m = \frac{4}{3}$ and $b = \frac{20}{3}$.

17. No; x has an exponent other than 1.

18. Yes, it can be written in $f(x) = mx + b$ form, where $m = 0$ and $b = 6$.

19. No; x has an exponent other than 1.

20. No; it cannot be written in $f(x) = mx + b$ form.

21. No; it cannot be written in $f(x) = mx + b$ form.

22. No; it cannot be written in $f(x) = mx + b$ form.

23. No; it cannot be written in $f(x) = mx + b$ form; There is an xy term.

24. Yes; it can be written in $f(x) = mx + b$ form, where $m = \frac{4}{5}$ and $b = \frac{8}{3}$.

25b. Kingda Ka; Sample answer: The Kingda Ka travels 847.5 meters in 25 seconds, so it travels a greater distance in the same amount of time.

26. $7x + 5y = -35$; $A = 7$, $B = 5$, $C = -35$

27. $8x + 3y = -6$; $A = 8$, $B = 3$, $C = -6$

28. $3x - 10y = -5$; $A = 3$, $B = -10$, $C = -5$

29. $2x + y = -11$; $A = 2$, $B = 1$, $C = -11$

30. $3x - y = 4$; $A = 3$, $B = -1$, $C = 4$

31. $6x + y = 0$; $A = 6$, $B = 1$, $C = 0$

32. $9x - 8y = 2$; $A = 9$, $B = -8$, $C = 2$

33. $5x + 32y = 160$; $A = 5$, $B = 32$, $C = 160$

34. $2x + 31y = 78$; $A = 2$, $B = 31$, $C = 78$

Example 1
16–24. See margin.

State whether each equation or function is a linear function. Write *yes* or *no*. Explain.

16. $3y - 4x = 20$
17. $y = x^2 - 6$
18. $h(x) = 6$
19. $j(x) = 2x^2 + 4x + 1$
20. $g(x) = 5 + \frac{6}{x}$
21. $f(x) = \sqrt{7 - x}$
22. $4x + \sqrt{y} = 12$
23. $\frac{1}{x} + \frac{1}{y} = 1$
24. $f(x) = \frac{4x}{5} + \frac{8}{3}$

Example 2

25. **ROLLER COASTERS** The speed of the Steel Dragon 2000 roller coaster in Mie Prefecture, Japan, can be modeled by $y = 10.4x$, where y is the distance traveled in meters in x seconds.
   a. How far does the coaster travel in 25 seconds? **260 m**
   b. The speed of the Kingda Ka roller coaster in Jackson, New Jersey, can be described by $y = 33.9x$. Which coaster travels faster? Explain your reasoning. **See margin.**

Example 3

Write each equation in standard form. Identify *A*, *B*, and *C*. **26–34. See margin.**

26. $-7x - 5y = 35$
27. $8x + 3y + 6 = 0$
28. $10y - 3x + 6 = 11$
29. $-6x - 3y - 12 = 21$
30. $3y = 9x - 12$
31. $2.4y = -14.4x$
32. $\frac{2}{3}y - \frac{3}{4}x + \frac{1}{6} = 0$
33. $\frac{4}{5}y + \frac{1}{8}x = 4$
34. $-0.08x = 1.24y - 3.12$

Example 4

Find the *x*-intercept and the *y*-intercept of the graph of each equation. Then graph the equation using the intercepts. **35–40. See Chapter 2 Answer Appendix.**

35. $y = -8x - 4$
36. $5y = 15x - 90$
37. $-4y + 6x = -42$
38. $-9x - 7y = -30$
39. $\frac{1}{3}x - \frac{2}{9}y = 4$
40. $\frac{3}{4}y - \frac{2}{3}x = 12$

 B

41b. See Chapter 2 Answer Appendix.

41. **CCSS MODELING** Latonya earns a commission of $1.75 for each magazine subscription that she sells and $1.50 for each newspaper subscription that she sells. Her goal is to earn a total of $525 in commissions in the next two weeks.
   a. Write an equation that is a model for the different numbers of magazine and newspaper subscriptions that can be sold to meet the goal. $1.75m + 1.5n = 525$
   b. Graph the equation. Does this equation represent a function? Explain.
   c. If Latonya sells 100 magazine subscriptions and 200 newspaper subscriptions, will she meet her goal? Explain. **No; the amount that Latonya will sell is $1.75 \cdot 100 + 1.5 \cdot 200$, which is $475.**

42. **SNAKES** Suppose the body length L in inches of a baby snake is given by $L(m) = 1.5 + 2m$, where m is the age of the snake in months until it becomes 12 months old.
   a. Find the length of an 8-month-old snake. **17.5 in.**
   b. Find the snake's age if the length of the snake is 25.5 inches. **12 mo**

43. **STATE FAIR** The Ohio State Fair charges $8 for admission and $5 for parking. After Joey pays for admission and parking, he plans to spend all of his remaining money at the ring game, which costs $3 per game.
   a. Write an equation representing the situation. $y = 3x + 13$
   b. How much did Joey spend at the fair if he paid $6 for food and drinks and played the ring game 4 times? **$31**

Differentiated Homework Options

Level	Assignment	Two-Day Option	
AL Basic	16–40, 53–76	17–39 odd, 57–60	16–40 even, 53–56, 61–76
OL Core	17–39 odd, 41–43, 45–49 odd, 50–51, 53–76	16–40, 57–60	41–51, 53–56, 61–76
BL Advanced	41–68, (optional: 69–76)		

44. $x + 6y =$
$7; A = 1, B =$
$6, C = 7$

45. $4x -$
$40y = -59;$
$A = 4,$
$B = -40,$
$C = -59$

51a. See
Chapter 2
Answer
Appendix.

56a. Sample
answer: When
the earnings
are determined
by a constant
hourly wage,
the total
earnings can
be represented
by $y = mx$
where m **is the**
hourly wage.

56b. Sample
answer: The
relationship
between the
cost and the
number of
gallons of
gasoline
purchased.

> **C** Write each equation in standard form. Identify A, B, and C. **46. See margin.**

44. $\dfrac{x + 5}{3} = -2y + 4$ **45.** $\dfrac{4x - 1}{5} = 8y - 12$ **46.** $\dfrac{-2x - 8}{3} = -12y + 18$

Find the x-intercept and the y-intercept of the graph of each equation.

47 $\dfrac{6x + 15}{4} = 3y - 12$ **48.** $\dfrac{-8x + 12}{3} = 16y + 24$ **49.** $\dfrac{15x + 20}{4} = \dfrac{3y + 6}{5}$ $-1\dfrac{1}{75}; 6\dfrac{1}{3}$

$-10.5; 5.25$ $-7.5; -1.25$

50. FUNDRAISING The Freshman Class Student Council wanted to raise money by giving car washes. The students spent $10 on supplies and charged $2 per car wash.

   a. Write an equation to model the situation. $E = 2c - 10$

   b. Graph the equation. **See Chapter 2 Answer Appendix.**

   c. How much money did they earn after 20 car washes? $30

   d. How many car washes are needed for them to earn $100? 55

51. 🔀 **MULTIPLE REPRESENTATIONS** Consider the following linear functions.

$$f(x) = -2x + 4 \qquad g(x) = 6 \qquad h(x) = \tfrac{1}{3}x + 5$$

   a. Graphical Graph the linear functions on separate graphs.

   b. Tabular Use the graphs to complete the table.

51c. No; horizontal lines are neither one-to-one nor onto because only one y-value is used and it is repeated for every x-value. Every other linear function is one-to-one and onto because every x-value has one unique y-value that is not used by any other x-element and every possible y-value is used.

Function	One-to-One	Onto
$f(x) = -2x + 4$	yes	yes
$g(x) = 6$	no	no
$h(x) = \tfrac{1}{3}x + 5$	yes	yes

   c. Verbal Are all linear functions one-to-one and/or onto? Explain your reasoning.

H.O.T. Problems   Use Higher-Order Thinking Skills

52. CHALLENGE Write a function with an x-intercept of $(a, 0)$ and a y-intercept of $(0, b)$.
   Sample answer: $f(x) = -\dfrac{bx}{a} + b$

53. OPEN ENDED Write an equation of a line with an x-intercept of 3.   Sample answer: $f(x) = 2(x - 3)$

54. REASONING Determine whether an equation of the form $x = a$, where a is a constant, is *sometimes*, *always*, or *never* a function. Explain your reasoning.   Sample answer: Never; the graph of $x = a$ is a vertical line.

55. **CCSS** **ARGUMENTS** Of the four equations shown, identify the one that does not belong. Explain your reasoning. $y = 2xy$; Sample answer: $y = 2xy$ is not a linear function.

$y = 2x + 3$

x	y
0	4
1	2
2	0
3	-2

$y = 2xy$

56. WRITING IN MATH Consider the graph of the relationship between hours worked and earnings.

   a. When would this graph represent a linear relationship? Explain your reasoning.

   b. Provide another example of a linear relationship in a real-world situation.

🔀 **Multiple Representations**

In Exercise 51, students use graphs, a table, and verbal reasoning to identify properties of linear functions.

Additional Answer

46. $x - 18y = -31; A = 1, B = -18,$
$C = -31$

Study Guide and Intervention

AL   OL   ELL

Practice

AL   OL   BL   ELL

Word Problem Practice

AL   OL   BL   ELL

Enrichment

OL   BL

Ticket Out the Door Ask each student to write a short explanation of how to determine if an equation is linear.

Formative Assessment

Check for student understanding of concepts in Lessons 2-1 and 2-2.

 Quiz 1, p. 59

Follow-up

Students have explored linear relations and functions.

Ask:

- What is a linear relationship?
  Sample answer: A constant relationship between two quantities.

- How can a linear relationship be represented? Sample answers: in words; using the graph of a line, as a relation, equation or function; using a table of values

---

Standardized Test Practice

57. Tom bought n DVDs for a total cost of $15n - 2$ dollars. Which expression represents the cost of each DVD? **C**

 A $n(15n - 2)$

 B $n + (15n - 2)$

 C $(15n - 2) \div n; n \neq 0$

 D $(15n - 2) - n$

58. SHORT RESPONSE What is the complete solution of the equation?

$$|9 - 3x| = 18 \quad -3, 9$$

59. NUMBER THEORY If a, b, c, and d are consecutive odd integers and $a < b < c < d$, how much greater is $c + d$ than $a + b$? **J**

 F 2   H 6

 G 4   J 8

60. SAT/ACT Which function is linear? **E**

 A $f(x) = x^2$ D $g(x) = \frac{2.7}{x}$

 B $g(x) = \sqrt{x - 1}$ E $f(x) = 2x$

 C $f(x) = \sqrt{9 - x^2}$

Spiral Review

State the domain and range of each relation. Then determine whether each relation is a *function*. If it is a function, determine if it is *one-to-one*, *onto*, *both*, or *neither*. (Lesson 2-1)

61.

D = {−4, −1, 8}, R = {3, 6, 9}; not a function

62.

D = {−2, 1, 2, 4}, R = {1, 3, 5}; function; onto

63.

x	y
−4	−2
−3	−1
−3	−1
7	9

D = {−4, −3, 7}, R = {−2, −1, 9}; function; both

64. SHOPPING Claudio is shopping for a new television. The average price of the televisions he likes is $800, and the actual prices differ from the average by up to $350. Write and solve an absolute value inequality to determine the price range of the televisions. (Lesson 1-6) $|x - 800| \leq 350; \$450 \leq x \leq \$1150$

Evaluate each expression if $a = -6$, $b = 5$, and $c = 3.6$. (Lesson 1-1)

65. $\frac{6a - 3c}{2ab}$ 0.78

66. $\frac{a + 7b}{4bc}$ $\frac{29}{72}$

67. $\frac{b - c}{a + c}$ $-\frac{7}{12}$

68. FOOD Brandi can order a small, medium, or large pizza with pepperoni, mushrooms, or sausage. How many different one-topping pizzas can she order? (Lesson 0-4) **9**

Skills Review

Evaluate each expression.

69. $\frac{12 - 8}{4 - (-2)}$ $\frac{2}{3}$

70. $\frac{5 - 9}{-3 - (-6)}$ $-\frac{4}{3}$

71. $\frac{-2 - 8}{3 - (-5)}$ $-\frac{5}{4}$

72. $\frac{-2 - (-6)}{-1 - (-8)}$ $\frac{4}{7}$

73. $\frac{-7 - (-11)}{-3 - 9}$ $-\frac{1}{3}$

74. $\frac{-1 - 8}{7 - (-3)}$ $-\frac{9}{10}$

75. $\frac{-12 - (-3)}{-6 - (-5)}$ **9**

76. $\frac{4 - 3}{2 - 5}$ $-\frac{1}{3}$

 74 | **Lesson 2-2** | Linear Relations and Functions

DifferentiatedInstruction OL BL

Extension Have students place a piece of spaghetti or a pencil on a large coordinate plane to model the graphs of equations such as: $x = 4$, $x = -2$, $y = 0$, $y = -3$, $x = y$, and $x = -y$.

EXTEND 2-2 Algebra Lab
Roots of Equations and Zeros of Functions

The *solution* of an equation is called the *root* of the equation.

CCSS Common Core State Standards
Content Standards
F.IF.4 For a function that models a relationship between two quantities, interpret key features of graphs and tables in terms of the quantities, and sketch graphs showing key features given a verbal description of the relationship.

Example   Determine Roots

Find the root of $0 = 5x - 10$.

$0 = 5x - 10$	Original equation
$10 = 5x$	Add 10 to each side.
$2 = x$	Divide each side by 5.

The root of the equation is 2.

You can also find the root of an equation by finding the *zero* of its related function. Values of x for which $f(x) = 0$ are called *zeros* of the function f.

Linear Equation

$0 = 5x - 10$

Related Linear Function

$f(x) = 5x - 10$ or $y = 5x - 10$

The zero of a function is the *x-intercept* of its graph. Since the graph of $y = 5x - 10$ intersects the *x*-axis at 2, the zero of the function is 2.

1. The root and solution of $0 = 4x + 10$ is −2.5, because it is the value of *x* that makes the equation a true statement. The zero of $f(x) = 4x + 10$ is −2.5 because it is the value of *x* for which $f(x) = 0$.

3. False; the *y*-intercept does not represent a zero of the function. The *x*-intercept −2 is the only zero.

Exercises

1. Use $0 = 4x + 10$ and $f(x) = 4x + 10$ to distinguish among roots, solutions, and zeros.

2. Relate solutions of equations and *x*-intercepts of graphs. The solution of an equation equals the *x*-intercept of the graph of the related function.

Determine whether each statement is *true* or *false*. Explain your reasoning.

3. The function graphed at the right has two zeros, −2 and −1.

4. The root of $6x + 9 = 0$ is −1.5. True; the root of an equation is its solution.

5. $f(0)$ is a zero of the function $f(x) = -\frac{2}{3}x + 12$. False; 18 is a zero of the function because it is the value of *x* for which $f(x) = 0$.

6. FUNDRAISERS The function $y = 2x - 150$ represents the money raised *y* when the Boosters sell *x* soft drinks at a basketball game. Find the zero and describe what it means in the context of this situation. Make a connection between the zero of the function and the root of $0 = 2x - 150$. 75; After selling 75 soft drinks, the Boosters have made 0 dollars. The zero of the function equals the root of the equation because it is the value of *x* when $y = 0$.

From Concrete to Abstract
Ask students to summarize what they have learned about the mathematical terms, such as roots, solutions, and zeros, in everyday words. Then have them cite examples.

1 Focus

Objective Distinguish among roots, solutions, and zeros.

Teaching Tip
Begin the lesson by asking students if they know of any words that have more than one meaning depending on how they are used. For example, the word *sentence* can mean a group of words that expresses a complete thought or a judgment by a court specifying the punishment for a crime committed.

2 Teach

Working in Cooperative Groups
Divide the class into groups of two or three, mixing abilities. Ask each group to read about roots of equations and zeros of functions.

Explain to students that if they can relate a word they are trying to learn to something familiar, it makes it easier to remember what the new word means. By relating mathematical terms to everyday things, students can recall their meanings more readily.

Practice Have students complete Exercises 1–5.

3 Assess

Formative Assessment
Use Exercise 6 to assess whether students can distinguish between roots of equations and zeros of functions.

LESSON 2-3

Rate of Change and Slope

:: Then	:: Now	:: Why?
• You graphed linear relations.	**1** Find rate of change. **2** Determine the slope of a line.	• The table shows the total distance a car traveled over various time intervals. The distance formula, $rt = d$ or $r = \frac{d}{t}$, relates time and distance.

Time (h)	Distance (mi)
1	68
2.5	170
3	204
4.5	306
5	340

1 Focus

VerticalAlignment

Before Lesson 2-3 Graph linear relations.

Lesson 2-3 Find rate of change and determine the slope of a line.

After Lesson 2-3 Identify and sketch graphs of parent functions, including quadratic.

2 Teach

Scaffolding Questions

Have students read the **Why?** section of the lesson.

Ask:

- What do the variables *r*, *d*, and *t* represent? *r* represents rate, *d* represents distance, and *t* represents time.

- How many miles did the car drive in the first three hours? 204 mi

- At what rate was the car traveling in the first three hours? 68 miles per hour

NewVocabulary
rate of change
slope

Common Core State Standards

Content Standards

F.IF.4 For a function that models a relationship between two quantities, interpret key features of graphs and tables in terms of the quantities, and sketch graphs showing key features given a verbal description of the relationship.

F.IF.6 Calculate and interpret the average rate of change of a function (presented symbolically or as a table) over a specified interval. Estimate the rate of change from a graph.

Mathematical Practices

8 Look for and express regularity in repeated reasoning.

1 Rate of Change **Rate of change** is a ratio that compares how much one quantity changes, on average, relative to the change in another quantity. If x is the independent variable and y is the dependent variable, then rate of change $= \frac{\text{change in } y}{\text{change in } x}$. This is sometimes referred to as $\frac{\Delta y}{\Delta x}$.

Real-World Example 1 Constant Rate of Change

CHEMISTRY The table shows the temperature of a solution after it has been removed from a heat source. Find the rate of change in temperature for the solution.

Use the ordered pairs (2, 139.4) and (5, 133.1).

$$\text{rate of change} = \frac{\text{change in } y}{\text{change in } x}$$
$$= \frac{\text{change in temperature}}{\text{change in time}}$$
$$= \frac{133.1 - 139.4}{5 - 2}$$
$$= \frac{-6.3}{3} \text{ or } -2.1$$

The rate of change is -2.1. This means that the temperature is decreasing by 2.1°C each minute.

Time (min)	Temperature (°C)
0	143.6
2	139.4
5	133.1
8	126.8
12	118.4

GuidedPractice

1. RECREATION The graph at the right shows the number of gallons of water in a swimming pool as it is being filled. At what rate is the pool being filled? 36 gal/min

Glen Allison/Photodisc/Getty Images

Lesson 2-3 Resources

Resource	Approaching Level **AL**	On Level **OL**	Beyond Level **BL**	English Learners **ELL**
Teacher Edition	• Differentiated Instruction, p. 78	• Differentiated Instruction, p. 78	• Differentiated Instruction, p. 82	• Differentiated Instruction, p. 78
Chapter Resource Masters	• Study Guide and Intervention, pp. 17–18 • Skills Practice, p. 19 • Practice, p. 20 • Word Problem Practice, p. 21	• Study Guide and Intervention, pp. 17–18 • Skills Practice, p. 19 • Practice, p. 20 • Word Problem Practice, p. 21 • Enrichment, p. 22	• Practice, p. 20 • Word Problem Practice, p. 21 • Enrichment, p. 22	• Study Guide and Intervention, pp. 17–18 • Skills Practice, p. 19 • Practice, p. 20 • Word Problem Practice, p. 21
Other	• 5-Minute Check 2-3 • Study Notebook	• 5-Minute Check 2-3 • Study Notebook	• 5-Minute Check 2-3 • Study Notebook	• 5-Minute Check 2-3 • Study Notebook

Rubberball/Getty Images

Up to this point, you have used rates of change that are constant. Many real-world situations involve rates of change that are not constant. These situations are often described using an average rate of change over a specified interval.

Real-World Example 2 Average Rate of Change

MUSIC Refer to the graph at the right. Find the average rate of change of the percent of total music sales for both CDs and downloads from 2001 to 2008. Compare the rates.

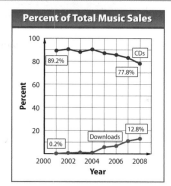

Percent of Total Music Sales

Source: Recording Industry Association of America

CDs:

$$\text{rate of change} = \frac{\text{change in } y}{\text{change in } x}$$

$$= \frac{\text{change in percent}}{\text{change in time}}$$

$$= \frac{77.8 - 89.2}{2008 - 2001}$$

$$= \frac{-11.4}{7} \text{ or } -1.63$$

Downloads:

$$\text{rate of change} = \frac{\text{change in } y}{\text{change in } x}$$

$$= \frac{\text{change in percent}}{\text{change in time}}$$

$$= \frac{12.8 - 0.2}{2008 - 2001}$$

$$= \frac{12.6}{7} \text{ or } 1.8$$

The percent of CD music sales declined at an average rate of 1.63% per year, while the percent of downloaded music sales increased at an average rate of 1.8% per year.

Real-WorldLink
As of April 2007, 36% of teens preferred to purchase CDs and 64% preferred to download music online.

Source: CNN

▶ **Guided**Practice   **State College: average increase about 803 applicants per year; Central University: average increase about 177 applicants per year**

2. **EDUCATION** In 2002, 23,142 students applied to State College, and 34,689 students applied to Central University. In 2010, 29,563 students applied to State College, and 36,107 applied to Central University. Determine the average rate of change in applicants for both schools from 2002 to 2010.

2 Slope The **slope** of a line is the ratio of the change in the *y*-coordinates to the corresponding change in the *x*-coordinates. The slope of a line is the same as its rate of change.

Suppose a line passes through points at (x_1, y_1) and (x_2, y_2).

$$\text{Slope} = \frac{\text{change in } y\text{-coordinates}}{\text{change in } x\text{-coordinates}} = \frac{y_2 - y_1}{x_2 - x_1}$$

Focus on Mathematical Content
Slope The steepness of a line is measured by its slope. The slope of a nonvertical line is found by identifying two points on the line and dividing the difference in their *y*-coordinates by the difference in their *x*-coordinates. The slope of a horizontal line is 0, and the slope of a vertical line is undefined.

1 **Rate of Change**
Examples 1 and 2 involve finding the rate of change in real-world situations.

Formative Assessment
Use the Guided Practice exercises after each example to determine students' understanding of concepts.

Additional Examples

1 **COLLEGE ADMISSIONS** In 2004, 56,878 students applied to UCLA. In 2006, 60,291 students applied. Find the rate of change in the number of students applying for admission from 2004 to 2006. 1706.5 students per year

2 **BUSINESS** Refer to the graph below, which shows data on the fastest-growing restaurant chain in the U.S. during the time period of the graph. Find the rate of change of the number of stores from 2001 to 2006.

Stores in U.S.

Between 2000 and 2006, the number of stores in the U.S. increased at an average rate of 5.4(1000), or 5400 stores per year.

▶ **Additional Examples** also in Interactive Classroom PowerPoint® Presentations

IWB **Interactive White Board** READY

2 Slope

Examples 3 and 4 ask students to find the slope of a line.

Additional Examples

3 Find the slope of the line that passes through $(-1, 4)$ and $(1, -2)$.
$-\dfrac{6}{2}$ or -3; The slope is -3.

4 Find the slope of the line shown below. $m = \dfrac{1}{2}$

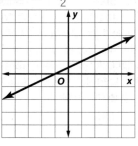

Teach with Tech

Blog Have students write a blog entry explaining how to find the slope of a line. Have them explain how the slope describes the appearance of the line.

StudyTip

Slope The formula for slope is often remembered as *rise over run*, where the rise is the difference in y-coordinates and the run is the difference in x-coordinates.

KeyConcept Slope of a Line

Words   The slope of a line is the ratio of the change in y-coordinates to the change in x-coordinates.

Symbols The slope m of a line passing through (x_1, y_1) and (x_2, y_2) is given by $m = \dfrac{y_2 - y_1}{x_2 - x_1}$, where $x_1 \neq x_2$.

Example 3 Find Slope Using Coordinates

Find the slope of the line that passes through $(-4, 3)$ and $(2, 5)$.

$m = \dfrac{y_2 - y_1}{x_2 - x_1}$ Slope Formula

$= \dfrac{5 - 3}{2 - (-4)}$ $(x_1, y_1) = (-4, 3), (x_2, y_2) = (2, 5)$

$= \dfrac{2}{6}$ or $\dfrac{1}{3}$ Simplify.

Guided Practice

Find the slope of the line that passes through each pair of points.

3A. $(1, -3)$ and $(3, 5)$ 4 **3B.** $(-8, 11)$ and $(24, -9)$ $-\dfrac{5}{8}$

You can choose any two points from the graph of a line to find the slope.

StudyTip

Slope is Constant The slope of a line is the same, no matter what two points on the line are used.

Example 4 Find Slope Using a Graph

Find the slope of the line shown at the right.

The line passes through $(-2, 0)$ and $(0, -3)$.

$m = \dfrac{y_2 - y_1}{x_2 - x_1}$ Slope Formula

$= \dfrac{-3 - 0}{0 - (-2)}$ $(x_1, y_1) = (-2, 0), (x_2, y_2) = (0, -3)$

$= \dfrac{-3}{2}$ or $-\dfrac{3}{2}$ Simplify.

Guided Practice

Find the slope of each line.

4A. $\dfrac{3}{8}$

4B. -4

DifferentiatedInstruction AL OL ELL

If some students have trouble remembering whether a vertical line or a horizontal line has a slope of 0,

Then they can remember that a horizontal line has a slope of 0 by associating the o in the word horizontal with 0.

Example 1   **CCSS REGULARITY**   Find the rate of change for each set of data.

1.

Time (min)	2	4	6	8	10
Distance (ft)	12	24	36	48	60

6 feet/min

2.

Time (sec)	5	10	15	20	25
Volume (cm³)	16	32	48	64	80

3.2 cm³/sec

Example 2

3. CAMERAS The graph shows the number of digital still cameras and film cameras sold by Yellow Camera Stores in recent years.

    a. Find the average rate of change of the number of digital cameras sold from 2004 to 2009.   about 11,000 per year

    b. Find the average rate of change of the number of film cameras sold from 2004 to 2009.   about −5000 per year

    c. What do the signs of each rate of change represent? The positive rate in part a represents an increase in sales of digital cameras. The negative rate in part b represents a decrease in sales of film cameras.

Digital Cameras vs. Film Cameras

Example 3   **Find the slope of the line that passes through each pair of points.**

    4. $(3, 2), (8, 12)$ 2   **5.** $(-1, 4), (3, -8)$ −3   **6.** $(-2, -5), (-7, 10)$ −3

Example 4   **Determine the rate of change of each graph.**

7. $\dfrac{3}{5}$

8.   5

Practice and Problem Solving   Extra Practice is on page R2.

Example 1   **Find the rate of change for each set of data.**

9.

Time (day)	3	6	9	12	15
Height (mm)	20	40	60	80	100

$\dfrac{20}{3}$ mm/day

10.

Weight (lb)	11	22	33	44	55
Cost ($)	8	16	24	32	40

$\dfrac{8}{11}$ $/lb

Differentiated Homework Options

Level	Assignment	Two-Day Option	
AL Basic	9–21, 36, 38–57	9–21 odd, 41–44	10–20 even, 36, 38–40, 45–57
OL Core	9–29 odd, 30, 31–35 odd, 36, 38–57	9–21, 41–44	22–36, 38–40, 45–57
BL Advanced	30–54, (optional: 55–57)		

Formative Assessment

Use Exercises 1–8 to check for understanding.

Use the chart on the bottom of this page to customize assignments for your students.

CCSS   **Teaching the Mathematical Practices**

Regularity Mathematically proficient students notice if calculations are repeated, and look both for general methods and for shortcuts. As they work to solve a problem, mathematically proficient students maintain oversight of the process, while attending to the details.

Study Guide and Intervention
AL OL ELL

Practice
AL OL BL ELL

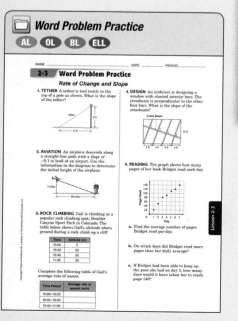

Word Problem Practice
AL OL BL ELL

Example 2

11. HEALTH The table below shows Lisa's temperature during an illness over a 3-day period.

Day	Monday		Tuesday		Wednesday	
Time	8:00 A.M.	8:00 P.M.	8:00 A.M.	8:00 P.M.	8:00 A.M.	8:00 P.M.
Temp (°F)	100.5	102.3	103.1	100.7	99.9	98.6

11b. −0.125°/h;
Yes; the number should be negative because her temperature is dropping.

a. What was the average rate of change in Lisa's temperature from 8:00 A.M. on Monday to 8:00 P.M. on Monday? 0.15°/h

b. What was the average rate of change in Lisa's temperature from 8:00 A.M. on Tuesday to 8:00 P.M. on Wednesday? Is your answer reasonable? What does the sign of the rate mean?

c. During which 12-hour period was the average rate of change in Lisa's temperature the greatest? Tuesday 8:00 A.M.–Tuesday 8:00 P.M.

Example 3   **Find the slope of the line that passes through each pair of points. Express as a fraction in simplest form.**

12. $(-2, 11)$, $(5, 6)$ $-\dfrac{5}{7}$

13. $(-9, -11)$, $(6, 3)$ $\dfrac{14}{15}$

14. $(-1.5, 3.5)$, $(4.5, 6)$ $\dfrac{5}{12}$

15. $(-4.5, 9.5)$, $(-1, 2.5)$ -2

16. $(-8, -0.5)$, $(-4, 5)$ $\dfrac{11}{8}$

17. $(-6, -2)$, $(-1.5, 5.5)$ $\dfrac{5}{3}$

Example 4   **Determine the rate of change of each graph.**

18. 2

19. 5

20. −6

21 −0.8

B ▶ **22. CCSS REASONING** The table shows your height on a water slide at various time intervals. **a. See margin.**

Time (s)	Height (ft)
0	120
1	90
2	60
3	30
4	0
5	0

a. Graph the height versus the time on the water slide.

b. Find the average rate of change in a rider's height between 1 and 3 seconds. −30 ft/s

c. Find the average rate of change in a rider's height between 0 and 5 seconds. −24 ft/s

d. What is another word for *rate of change* in this situation?   speed or velocity

Determine the rate of change for each equation.

23. $6y = 8x - 40$ $\dfrac{4}{3}$

24. $-2y - 16x = 41$ -8

25. $12x - 4y + 5 = 18$ 3

26. $20x + 85y = 120$ $-\dfrac{4}{17}$

27. $\dfrac{3}{2}x - \dfrac{5}{4}y = 15$ $\dfrac{6}{5}$

28. $\dfrac{1}{6}y + \dfrac{3}{8}x = 24$ $-\dfrac{9}{4}$

Enrichment
OL BL

2-3 Enrichment
The Increase in Greenhouse Gases

The atmosphere is composed of about 50% carbon dioxide, CO_2. The level of carbon dioxide is increasing due to increased fuel consumption and housing and commercial development. The concentration of a compound is measured in parts per million (ppm). For example, if there were 800 CO_2 molecules out of one million air particles, then the CO_2 level would be 500 ppm.

1. In 1965, the concentration of CO_2 was 320 ppm. In 2007, the concentration was 383 ppm. Determine the rate at which CO_2 increased in ppm per year.

2. Carbon dioxide concentration is related to human consumption of fossil fuels and the decrease of trees due to development; therefore an increase in human population will result in an increase in carbon dioxide. In 1980 the U.S. population was 225 million. The 2000 census reported 281 million. At what rate is the population increasing per year? What do you estimate the U.S. population to be in 2020?

Additional Answer

22a.

29. WASHINGTON MONUMENT The Washington Monument is 555 feet $5\frac{1}{8}$ inches tall and weighs 90,854 tons. The monument is topped by an aluminum square pyramid. The sides of the pyramid's base measure 5.6 inches, and the pyramid is 8.9 inches tall. Estimate the slope that a face of the pyramid makes with its base. **about 3.2**

 30. MARINE LIFE The illustrations show the growth of a starfish over time.

   a. Find the average rate of change in the measure over time. **2 in./yr**

   b. Predict the size of the starfish in 2014. **14 in.**

2012
10 in.
2008
2 in.

Find the value of r so that the line that passes through each pair of points has the given slope.

31. $(6, r), (3, 3), m = 2$ **9**

32. $(8, 1), (5, r), m = \frac{1}{3}$ **0**

33. $(10, r), (4, -3), m = \frac{4}{3}$ **5**

34. $(8, -2), (r, -6), m = -4$ **9**

35. **MULTIPLE REPRESENTATIONS** In this problem, you will explore the rate of change for the function $f(x) = x^2$.

   a. Graphical Graph $f(x) = x^2$. **See margin.**

   b. Tabular Copy and complete the table. To complete the slope row, find the slope of the line containing two consecutive points such as $(-4, 16)$ and $(-3, 9)$. The first one is completed for you.

x	−4	−3	−2	−1	0	1	2	3	4
f(x)	16	9	4	1	0	1	4	9	16
slope		−7	−5	−3	−1	1	3	5	7

   c. Verbal Describe what happens to the rate of change for $f(x) = x^2$ as x increases. **Sample answer: The rate of change is not constant. The rate of change increases as x approaches infinity.**

H.O.T. Problems   Use Higher-Order Thinking Skills

36. **CRITIQUE** Patty and Tim are asked to find the slope of the line passing through the points $(4, 3)$ and $(7, 9)$. Is either of them correct? Explain.

36. Patty; Tim calculated slope as the ratio of the change in x to the change in y.

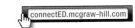
Patty
$m = \frac{9-3}{7-4}$
$= \frac{6}{3}$ or 2

Tim
$m = \frac{7-4}{9-3}$
$= \frac{3}{6}$ or $\frac{1}{2}$

37. CHALLENGE The graph of a line passes through the points $(2, 3)$ and $(5, 8)$. Explain how you would find the y-coordinate of the point $(11, y)$ on the same line. Then find y. **See margin.**

38. **WRITING IN MATH** In what ways can change be represented mathematically? **See margin.**

39. REASONING Determine whether the statement *A line has a slope that is a real number* is *sometimes, always,* or *never* true. Explain your reasoning. **Sometimes; the slope of a vertical line is undefined.**

40. WRITING IN MATH Describe the process of finding the rate of change for each.

   a. a table of values   **b.** a graph   **c.** an equation **40a–c. See margin.**

Comstock Images/PictureQuest

40a. Sample answer: Pick two sets of values in the table and determine the difference of the outputs (y) and the difference of the inputs (x). Divide the difference of the outputs by the difference of the inputs. This is the rate of change. Follow this process with the rest of the data to confirm the rate of change is constant and, thus, the function is linear.

40b. Sample answer: Pick two separate coordinates that fall on the graph of the straight line. Determine the difference of the outputs (y) and the difference of the inputs (x). Divide the difference of the outputs by the difference of the inputs. This is the rate of change.

40c. Sample answer: Pick two random x-values and substitute them into the equation to determine the corresponding y-values. Determine the difference of the outputs (y) and the difference of the inputs (x). Divide the difference of the outputs by the difference of the inputs. This is the rate of change.

Multiple Representations

In Exercise 35, students use a table of values and a graph to determine how the slope of a quadratic function changes as the input value increases.

CCSS **Teaching the Mathematical Practices**

Critique Mathematically proficient students are also able to compare the effectiveness of two plausible arguments, distinguish correct logic or reasoning from that which is flawed, and—if there is a flaw in an argument—explain what it is.

Additional Answers

35a.

37. Sample answer: Because the slope from $(2, 3)$ to $(5, 8)$ is the same as the slope from $(5, 8)$ to $(11, y)$, find the slope between each pair of points and set them equal to each other. Then solve for y.

$$\frac{8-3}{5-2} = \frac{y-8}{11-5}$$
$$\frac{5}{3} = \frac{y-8}{6}$$
$$30 = 3(y-8)$$
$$10 = y-8$$
$$18 = y$$

38. Sample answer: Change can be represented visually as a graph or as a pattern, numerically as a table of values, sequence of numbers, or ratio, or algebraically as an equation.

Pacing If there is any doubt whether your students thoroughly understand the concept of slope, consider spending an extra day on this lesson.

4 Assess

Name the Math Have students verbally describe the concept of slope and explain how to calculate it.

Standardized Test Practice

41. GRIDDED RESPONSE What is the slope of the line shown in the graph? **3/2 or 1.5**

42. SAT/ACT In the figure below, the large square contains two smaller squares. If the areas of the two smaller squares are 4 and 25, what is the sum of the perimeters of the two shaded rectangles? **D**

A 14   D 28

B 20   E 49

C 24

43. GEOMETRY In $\triangle ABC$ shown, $AC = 16$ and $m\angle DAB = 60°$. What is the measure of $\overline{BD}$? **H**

F $9\sqrt{2}$

G 9

H $4\sqrt{3}$

J 4

44. The table shows the cost of bananas depending on the amount purchased. Which conclusion can be made based on information in the table? **B**

Cost of Bananas	
Number of Pounds	Cost ($)
5	1.45
20	4.60
50	10.50
100	19.00

A The cost of 10 pounds of bananas would be more than $4.

B The cost of 200 pounds of bananas would be at most $38.

C The cost of bananas is always more than $0.20 per pound.

D The cost of bananas is always less than $0.28 per pound.

45. Yes; it can be written in $f(x) = mx + b$ form.

Spiral Review

State whether each equation or function is a linear function. Write *yes* or *no*. **Explain.** (Lesson 2-2)

45. $6y - 8x = 19$

46. $4x^2 = 2y - 9$
No; it cannot be written in $f(x) = mx + b$ form.

47. $18 = 2xy + 6$
No; it cannot be written in $f(x) = mx + b$ form.

Evaluate each function. (Lesson 2-1)

48. $f(-9)$ if $f(x) = -7x + 8$ **71**

49. $g(-4)$ if $g(x) = -3x^2 + 2$ **−46**

50. $h(12)$ if $h(x) = 4x^2 - 10x$ **456**

51. RACING There are 8 contestants in a 400-meter race. In how many different ways can the top three runners finish? (Lesson 0-4) **336**

Determine the quadrant of the coordinate plane where each point is located. (Lesson 0-1)

52. $(-4, -8)$ **III**

53. $(-2, 6)$ **II**

54. $(3, -1)$ **IV**

Skills Review

Solve each equation.

55. $8 = 4m - 6$ **3.5**

56. $-6 = 3(8) + b$ **−30**

57. $-2 = -3x + 5$ $\frac{7}{3}$

DifferentiatedInstruction **BL**

Extension Have students draw a line with a slope of 0. Let them discover that a horizontal line has a slope of 0. Ask them if all horizontal lines have a slope of 0.

LESSON 2-4 Writing Linear Equations

Then	Now	Why?
• You determined slopes of lines.	**1** Write an equation of a line given the slope and a point on the line. **2** Write an equation of a line parallel or perpendicular to a given line.	• Medical insurance companies often require their customers to make a co-payment for every doctor's office visit in addition to an annual insurance premium. If an insurance company charges $2280 annually and requires a copayment of $35 per doctor's office visit, then the linear equation $y = 35x + 2280$ represents the total annual cost y for x doctor's office visits.

 NewVocabulary
slope-intercept form
point-slope form
parallel
perpendicular

 Common Core State Standards

Content Standards
A.SSE.1.b Interpret complicated expressions by viewing one or more of their parts as a single entity.
A.CED.2 Create equations in two or more variables to represent relationships between quantities; graph equations on coordinate axes with labels and scales.

Mathematical Practices
2 Reason abstractly and quantitatively.

---

1 Forms of Equations Consider the line through $A(0, b)$ and $C(x, y)$. Notice that b is the y-intercept. You can use these two points to find the slope of $\overleftrightarrow{AC}$.

$$m = \frac{y_2 - y_1}{x_2 - x_1} \qquad \text{Slope Formula}$$

$$= \frac{y - b}{x - 0} \qquad (x_1, y_1) = (0, b), (x_2, y_2) = (x, y)$$

$$= \frac{y - b}{x} \qquad \text{Simplify.}$$

Now solve the equation for y.

$$mx = y - b \qquad \text{Multiply each side by } x.$$

$$mx + b = y \qquad \text{Add } b \text{ to each side.}$$

$$y = mx + b \qquad \text{Symmetric Property of Equality}$$

Equations written in this format are in **slope-intercept form**.

KeyConcept Slope-Intercept Form

Words	The slope-intercept form of the equation of a line is $y = mx + b$, where m is the slope and b is the y-intercept.	**Model**
Symbols	$y = mx + b$ slope ⟶ ⟵ y-intercept	

If you are given the slope and y-intercept of a line, you can find an equation of the line by substituting the values of m and b into the slope-intercept form.

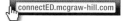 connectED.mcgraw-hill.com 83

Corbis

1 Focus

VerticalAlignment

Before Lesson 2-4 Determine slopes of lines.

Lesson 2-4 Write an equation of a line given the slope and a point on the line and write an equation of a line parallel or perpendicular to a given line.

After Lesson 2-4 Apply functions to problem situations.

2 Teach

Scaffolding Questions
Have students read the **Why?** section of the lesson.

Ask:

■ What does the variable x represent in this situation? the number of doctor visits

■ What does the variable y represent? the total annual cost

■ What would the total annual cost be for a person who visited the doctor nine times? $2595

Lesson 2-4 Resources

Resource	Approaching Level **AL**	On Level **OL**	Beyond Level **BL**	English Learners **ELL**
Teacher Edition	■ Differentiated Instruction, p. 84	■ Differentiated Instruction, p. 84	■ Differentiated Instruction, p. 89	■ Differentiated Instruction, p. 84
Chapter Resource Masters	■ Study Guide and Intervention, pp. 23–24 ■ Skills Practice, p. 25 ■ Practice, p. 26 ■ Word Problem Practice, p. 27	■ Study Guide and Intervention, pp. 23–24 ■ Skills Practice, p. 25 ■ Practice, p. 26 ■ Word Problem Practice, p. 27 ■ Enrichment, p. 28 ■ Spreadsheet Activity, p. 29	■ Practice, p. 26 ■ Word Problem Practice, p. 27 ■ Enrichment, p. 28	■ Study Guide and Intervention, pp. 23–24 ■ Skills Practice, p. 25 ■ Practice, p. 26 ■ Word Problem Practice, p. 27
Other	■ 5-Minute Check 2-4 ■ Study Notebook	■ 5-Minute Check 2-4 ■ Study Notebook	■ 5-Minute Check 2-4 ■ Study Notebook	■ 5-Minute Check 2-4 ■ Study Notebook

1 Forms of Equations

Examples 1–3 show how to write equations for lines in slope-intercept form.

Formative Assessment

Use the Guided Practice exercises after each example to determine students' understanding of concepts.

Additional Examples

1 Write an equation in slope-intercept form for the line.
$y = -\frac{1}{3}x + \frac{11}{3}$

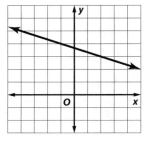

2 Write an equation of the line through $(5, -2)$ with a slope of $-\frac{3}{5}$. $y = -\frac{3}{5}x + 1$

▶ **Additional Examples** also in Interactive Classroom PowerPoint® Presentations

IWB Interactive White Board READY

CCSS **Teaching the Mathematical Practices**

Reasoning Mathematically proficient students make sense of quantities and their relationships in problem situations. Encourage students to further explain why the equation of a vertical line cannot be written in slope-intercept form.

WatchOut!

CCSS Reasoning The equation of a vertical line cannot be written in slope-intercept form because its slope is undefined.

StudyTip

Check Your Results You can check that your equation satisfies the conditions by graphing it.

Sometimes it is necessary to calculate the slope before you can write an equation.

Example 1 Write an Equation in Slope-Intercept Form

Write an equation in slope-intercept form for the line.

The graph intersects the y-axis at -2. So $b = -2$.

Step 1 Find the slope.

$$m = \frac{y_2 - y_1}{x_2 - x_1} \quad \text{Slope Formula}$$

$$= \frac{-2 - (-1)}{0 - (-4)} \quad (x_1, y_1) = (-4, -1), (x_2, y_2) = (0, -2)$$

$$= \frac{-1}{4} \text{ or } -\frac{1}{4} \quad \text{Simplify.}$$

Step 2 Substitute the values into the slope-intercept equation.

$$y = mx + b \quad \text{Slope-intercept form}$$

$$y = -\frac{1}{4}x - 2 \quad m = -\frac{1}{4}, b = -2$$

▶ **Guided**Practice **1A.** $y = \frac{4}{3}x + 4$ **1B.** $y = -4x - 6$

Write an equation in slope-intercept form for the line described.

1A. slope $\frac{4}{3}$, passes through $(0, 4)$ **1B.** passes through $(0, -6)$ and $(-4, 10)$

If you know the slope of a line and the coordinates of a point on the line, you can use the **point-slope form** to find an equation of the line.

KeyConcept Point-Slope Form

Words	The point-slope form of the equation of a line is $y - y_1 = m(x - x_1)$, where (x_1, y_1) are the coordinates of a point on the line and m is the slope of the line.	**Symbols** **slope**

coordinates of a point on the line

Example 2 Write an Equation Given Slope and One Point

Write an equation of the line through $(6, -2)$ with a slope of -4.

$$y - y_1 = m(x - x_1) \quad \text{Point-slope form}$$

$$y - (-2) = -4(x - 6) \quad (x_1, y_1) = (6, -2), m = -4$$

$$y + 2 = -4x + 24 \quad \text{Simplify.}$$

$$y = -4x + 22 \quad \text{Subtract 2 from each side.}$$

▶ **Guided**Practice **2A.** $y = \frac{1}{2}x + 2$ **2B.** $y = -3x - 7$

Write an equation in slope-intercept form for the line described.

2A. passes through $(2, 3)$; $m = \frac{1}{2}$ **2B.** passes through $(-2, -1)$; $m = -3$

 84 | Lesson 2-4 | Writing Linear Equations

DifferentiatedInstruction **AL** **OL** **ELL**

If ▶ students use m and b for the wrong attributes of a line,

Then ▶ make sure they understand that the letter m is always used for slope and b for the y-intercept in the slope-intercept form of an equation $y = mx + b$.

You can use any two points on a line to write an equation.

PT

Standardized Test Example 3 Write an Equation Given Two Points

Which is an equation of the line that passes through $(-2, 7)$ and $(3, -3)$?

A $y = -\frac{1}{2}x - \frac{3}{2}$ **C** $y = \frac{1}{2}x + 8$

B $y = -2x + 3$ **D** $y = 2x + 11$

Read the Test Item

You are given the coordinates of two points on the line.

Solve the Test Item

Step 1 Find the slope of the line.

$m = \dfrac{y_2 - y_1}{x_2 - x_1}$ Slope Formula

$= \dfrac{-3 - 7}{3 - (-2)}$ $(x_1, y_1) = (-2, 7),$
$(x_2, y_2) = (3, -3)$

$= -\dfrac{10}{5}$ Subtract.

$= -2$ Divide.

Step 2 Write an equation. Use either ordered pair for (x_1, y_1).

$y - y_1 = m(x - x_1)$ Point-slope form

$y - (-3) = -2(x - 3)$ $(x_1, y_1) = (3, -3)$ and $m = -2$

$y + 3 = -2x + 6$ Simplify.

$y = -2x + 3$ Subtract 3 from each side.

The answer is B.

> **GuidedPractice**

3. Which is an equation of the line that passes through $(4, -9)$ and $(2, -4)$? F

F $y = -\frac{5}{2}x + 1$ **H** $y = -\frac{2}{5}x + \frac{37}{5}$

G $y = -\frac{5}{2}x - 1$ **J** $y = -\frac{2}{5}x - \frac{37}{5}$

Test-TakingTip

Definitions Be certain to review key vocabulary, such as *y*-intercept, so that you understand what is being asked in a question.

2 Parallel and Perpendicular Lines Slopes can help you determine whether two lines are parallel or perpendicular.

KeyConcept Parallel and Perpendicular Lines

Parallel Lines	Perpendicular Lines
Two nonvertical lines are **parallel** if and only if they have the same slope. All vertical lines are parallel.	Two nonvertical lines are **perpendicular** if and only if the product of the slopes is -1. Vertical lines and horizontal lines are perpendicular.
$y = 2x + 5$ and $y = 2x - 1$	$y = 3x + 2$ and $y = -\frac{1}{3}x - 4$

Additional Example

3 STANDARDIZED TEST PRACTICE
What is an equation of the line through $(2, -3)$ and $(-3, 7)$? D

A $y = -2x - 1$

B $y = -\frac{1}{2}x + 1$

C $y = \frac{1}{2}x + 1$

D $y = -2x + 1$

Tips for New Teachers

Reasoning Point out that finding the slope in Example 3 eliminated three of the choices. Emphasize that eliminating answer choices helps you to use your time efficiently when taking a timed test.

Teach with Tech

Interactive Whiteboard Display a coordinate plane on the whiteboard. Plot two points on the plane and ask students to find the equation of the line that goes through these two points. Then drag the points to other locations on the plane and repeat.

Focus on Mathematical Content

Point-Slope Form The point-slope form of an equation of a line is closely related to the formula for calculating slope. As an alternative to memorizing point-slope form, replace the coordinates (x_2, y_2) with (x, y) in the slope formula, $m = \dfrac{y - y_1}{x - x_1}$, and multiply each side by $x - x_1$.

2 Parallel and Perpendicular Lines

Example 4 asks students to write an equation for a line passing through a given point that is perpendicular to a line with a given equation.

Additional Example

4 Write an equation in slope-intercept form for the line that passes through $(3, -2)$ and is perpendicular to the line with equation $y = -5x + 1$.
$y = \frac{1}{5}x - \frac{13}{5}$

Tips for New Teachers

Making Choices In Example 4, point out to students that they could choose to begin with slope-intercept form instead of point-slope form.

---

3 Practice

Formative Assessment

Use Exercises 1–7 to check for understanding.

Use the chart at the bottom of the next page to customize assignments for your students.

 Teaching the Mathematical Practices

Perseverance Mathematically proficient students start by explaining to themselves the meaning of a problem and looking for entry points to its solution. They analyze givens, constraints, relationships, and goals and make conjectures about the form and meaning of the solution and plan a solution pathway rather than simply jumping into a solution attempt.

Example 4 Write an Equation of a Parallel or Perpendicular Line

Write an equation in slope-intercept form for the line that passes through $(5, -6)$ and is perpendicular to the line with equation $y = -\frac{3}{2}x + 7$.

The slope of the given line is $-\frac{3}{2}$. Because the slopes of perpendicular lines are opposite reciprocals, the slope of the line perpendicular to the given line is $\frac{2}{3}$.

Use the point-slope form and the ordered pair $(5, -6)$.

$y - y_1 = m(x - x_1)$ \quad Point-slope form

$y - (-6) = \frac{2}{3}(x - 5)$ \quad $(x_1, y_1) = (5, -6)$ and $m = \frac{2}{3}$

$y + 6 = \frac{2}{3}x - \frac{10}{3}$ \quad Distributive Property

$y = \frac{2}{3}x - \frac{28}{3}$ \quad Subtract 6 from each side and simplify.

CHECK Graph both equations to verify the solution.

▶ **Guided Practice**

4. Write an equation in slope-intercept form for the line that passes through $(3, 7)$ and is parallel to the line with equation $y = \frac{3}{4}x - 5$. \quad $y = \frac{3}{4}x + \frac{19}{4}$

Check Your Understanding

 = Step-by-Step Solutions begin on page R14.

Example 1 Write an equation in slope-intercept form for the line described.

1. slope 1.5, passes through $(0, 5)$
$y = 1.5x + 5$

2. passes through $(-2, 3)$ and $(0, 1)$ \quad $y = -x + 1$

Example 2

3. passes through $(3, 5)$; $m = -2$
$y = -2x + 11$

4. passes through $(-8, -2)$; $m = \frac{5}{2}$ \quad $y = \frac{5}{2}x + 18$

Example 3

5. **MULTIPLE CHOICE** Which is an equation of the line? **A**

A $y = -4x - 25$

B $y = -\frac{2}{3}x - 5$

C $y = \frac{4}{5}x + \frac{29}{25}$

D $y = 6x + 35$

Example 4   **PERSEVERANCE** Write an equation in slope-intercept form for the line that satisfies each set of conditions.

6. passes through $(-9, -3)$, perpendicular to $y = -\frac{5}{3}x - 8$ \quad $y = 0.6x + 2.4$

7. passes through $(4, -10)$, parallel to $y = \frac{7}{8}x - 3$ \quad $y = \frac{7}{8}x - \frac{27}{2}$

Example 1

10. $y = -\frac{6}{5}x + 8$

8. slope 3, passes through $(0, -2)$ $y = 3x - 2$ **9.** slope $-\frac{1}{2}$, passes through $(0, 5)$ $y = -\frac{1}{2}x + 5$

10. slope $-\frac{6}{5}$, passes through $(0, 8)$ **11.** slope $\frac{9}{2}$, passes through $\left(0, -\frac{13}{2}\right)$ $y = 4.5x - 6.5$

Write an equation in slope-intercept form for the line described.

Example 2

12. $y = -2x + 8$

12. slope -2, passes through $(-3, 14)$ **13.** slope 4, passes through $(6, 9)$ $y = 4x - 15$

14. slope $\frac{3}{5}$, passes through $(-6, -8)$ $y = \frac{3}{5}x - \frac{22}{5}$ **15.** slope $-\frac{1}{4}$, passes through $(12, -4)$ $y = -\frac{1}{4}x - 1$

16. PART-TIME JOB Each week, Carmen earns a base pay of \$15 plus \$0.17 for every pamphlet that she delivers. Write an equation that can be used to find how much Carmen earns each week. How much will she earn the week that she delivers 300 pamphlets? $y = 0.17x + 15$; \$66

Example 3

Write an equation of the line passing through each pair of points. **17–22. See margin.**

17. $(-2, -6)$, $(4, 6)$ **18.** $(-8, -5)$, $(-3, 10)$ **19.** $(-4, 12)$, $(-2, -4)$

20. $(4.6, 3.4)$, $(2.2, 2.8)$ **21.** $(5.5, 0.6)$, $(1.1, 2.8)$ **22.** $(-25, -16)$, $(-29, 12)$

Example 4

CCSS PERSEVERANCE Write an equation in slope-intercept form for the line that satisfies each set of conditions.

23. passes through $(4, 2)$, perpendicular to $y = -2x + 3$ $y = \frac{1}{2}x$

24. passes through $(-6, -6)$, parallel to $y = \frac{4}{3}x + 8$ $y = \frac{4}{3}x + 2$

25. passes through $(12, 0)$, parallel to $y = -\frac{1}{2}x - 3$ $y = -\frac{1}{2}x + 6$

26. passes through $(10, 2)$, perpendicular to $y = 4x + 6$ $y = -0.25x + 4.5$

B **27. FINANCIAL LITERACY** Julio buys a used car for \$5900. Monthly expenses for the car—which include insurance, maintenance, and gas—average \$180 per month. Write an equation that represents the total cost of buying and owning the car for x months. **27–29. See margin.**

28. DELI The sales of a sandwich store increased approximately linearly from \$52,000 to \$116,000 during the first five years of business. Write an equation that models the sales y after x years. Determine what the sales will be at the end of 12 years if the pattern continues.

29. WHALES In 2009, it was estimated that there were 300 northern right whales in existence. The population of northern right whales is expected to decline by at least 25 whales each generation. Write an equation that represents the number of northern right whales that will be in existence in x generations.

Write an equation in slope-intercept form for each graph. **30–32. See margin.**

30.

31.

32.

Tips for New Teachers

Slope and Steepness Point out that when the units and/or scales differ on the two axes, you *cannot* estimate the slope of the graphed line by comparing it to the slope of $y = x$.

Additional Answers

17. $y = 2x - 2$

18. $y = 3x + 19$

19. $y = -8x - 20$

20. $y = 0.25x + 2.25$

21. $y = -0.5x + 3.35$

22. $y = -7x - 191$

27. $y = 180x + 5900$

28. $y = 16,000x + 36,000$; \$228,000

29. $y = -25x + 300$

30. $y = -\frac{5}{3}x + 12$

31. $y = \frac{2}{3}x + 6$

32. $y = 4x - 15$

Differentiated Homework Options

Level	Assignment	Two-Day Option	
AL Basic	8–26, 37, 39–65	9–25 odd, 43–46	8–26 even, 37, 39–42, 47–65
OL Core	9–25 odd, 27–29, 31, 33–37, 39–65	8–26, 43–46	30–37, 39–42, 47–65
BL Advanced	27–59 (optional: 60–65)		

Study Guide and Intervention
AL OL ELL

NAME _____ DATE _____ PERIOD _____

2-4 Study Guide and Intervention

Writing Linear Equations

Forms of Equations

Practice
AL OL BL ELL

NAME _____ DATE _____ PERIOD _____

2-4 Practice

Writing Linear Equations

Word Problem Practice
AL OL BL ELL

NAME _____ DATE _____ PERIOD _____

2-4 Word Problem Practice

Writing Linear Equations

33. ROSES Brad wants to send his girlfriend Kelli a dozen roses. He visits two stores. For what distance do the two stores charge the same amount to deliver a dozen roses? **10 mi**

Full Bloom	Flowers R US
Dozen roses $30	Dozen roses $40
Delivery: $3 per mile	Delivery: $2 per mile

34. TYPING The equation $y = 55(23 - x)$ can be used to model the number of words y you have left to type after x minutes.

  a. Write this equation in slope-intercept form. $y = 1265 - 55x$

  b. Identify the slope and y-intercept. $-55; 1265$

  c. Find the number of words you have left to type after 20 minutes. **165 words**

35. RECRUITING As an army recruiter, Ms. Cooper is paid a daily salary plus commission. When she recruits 10 people, she earns $100. When she recruits 14 people, she earns $120.

  a. Write a linear equation to model this situation. $y = 5x + 50$

  b. What is Ms. Cooper's daily salary? $50

  c. How much would Ms. Cooper earn in a day if she recruits 20 people? $150

36. MODELING Refer to the table at the right.

  a. Write and graph the linear equation that gives the distance y in kilometers in terms of the number x in miles. **See margin.**

Miles	Kilometers
100	161
50	80.5

  b. What distance in kilometers corresponds to 20 miles? **32.2 km**

  c. What number is the same in kilometers and miles? Explain your reasoning.
  0; because this is the point on the graph where the x-value and the y-value are the same

38. Sample answer: $y = \dfrac{ax}{d-b} - \dfrac{a^2}{d-b} + b$ or $y = \dfrac{ax}{d-b} + \dfrac{bd - b^2 - a^2}{d-b}$

H.O.T. Problems   Use Higher-Order Thinking Skills

37. Sample answer: Sometimes; while the two sets of parallel and perpendicular lines will always form a quadrilateral with four 90° angles, that figure will always be a rectangle, but not necessarily a square.

37. REASONING Determine whether the following statement is *always*, *sometimes*, or *never* true. Explain your reasoning.

    The quadrilateral formed by any two parallel lines and two lines perpendicular to those lines is a square.

38. CHALLENGE Given $\square ABCD$ with vertices $A(a, b)$, $B(c - a, d)$, $C(c + a, d)$, and $D(c, b)$, write an equation of a line perpendicular to diagonal $\overline{BD}$ that contains A.

39. REASONING Write $y = ax + b$ in point-slope form. Sample answer: $y - 0 = a\left(x + \dfrac{b}{a}\right)$

40. OPEN ENDED Write the equations of two parallel lines with negative slopes.
  Sample answer: $y = -2x + 3$ and $y = -2x - 1$

41. REASONING Write an equation in point-slope form of a line with an x-intercept of c and y-intercept of d. Sample answer: $y - d = -\dfrac{d}{c}(x - 0)$

42. WRITING IN MATH Why do we represent linear equations in more than one form? **See margin.**

Enrichment
OL BL

NAME _____ DATE _____ PERIOD _____

2-4 Enrichment

Two-Intercept Form of a Linear Equation

Teaching the Mathematical Practices

Modeling Mathematically proficient students can apply the mathematics they know to solve problems arising in everyday life, analyze relationships mathematically to draw conclusions, and interpret their mathematical results in the context of a situation.

Todd Gipstein/National Geographic/Getty Images

43. The total cost c in dollars to go to a water park and ride n water rides is given by the equation

$$c = 15 + 3n.$$

If the total cost was \$33, how many water rides were ridden? **A**

A 6   **B** 7   **C** 8   **D** 9

44. SHORT RESPONSE To raise money, the service club bought 1000 candy bars for \$0.60 each. If the club sells all of the candy bars for \$1 each, what will be their total profit? **\$400**

45. PROBABILITY A fair six-sided die is tossed. What is the probability that a number less than 3 will show on the face of the die? **G**

F $\frac{1}{6}$ **G** $\frac{1}{3}$ **H** $\frac{1}{2}$ **J** $\frac{2}{3}$

46. SAT/ACT What is an equation of the line through $\left(\frac{1}{2}, -\frac{3}{2}\right)$ and $\left(-\frac{1}{2}, \frac{1}{2}\right)$? **A**

A $y = -2x - \frac{1}{2}$ **D** $y = \frac{1}{2}x + 1$

B $y = -3x$ **E** $y = -2x - \frac{5}{2}$

C $y = 2x - 5$

Spiral Review

Determine the rate of change of each graph. (Lesson 2-3)

47. $-\frac{5}{3}$

48. $\frac{3}{4}$

49. $\frac{1}{5}$

50. RECREATION Scott is currently on page 210 of an epic novel that is 980 pages long. He plans to read 30 pages per day until he finishes the novel. Write and solve a linear equation to determine how many days it will take Scott to complete the novel. (Lesson 2-2) $30x + 210 = 980;\ 26$ **days**

Solve each inequality. (Lesson 1-5)

51. $-6x - 4 \leq 12 - 2x$ $x \geq -4$

52. $\frac{x+2}{5} > -3x + 1$ $x > \frac{3}{16}$

53. $\frac{5x+3}{3} \geq \frac{4x-2}{5}$ $x \geq -\frac{21}{13}$

Determine if the triangles with the following lengths are right triangles. (Lesson 0-8)

54. $5, 12, 13$ **yes**

55. $36, 48, 60$ **yes**

56. $7, 23, 25$ **no**

Multiply. (Lesson 0-2)

57. $(4c - 6)(2c + 5)$ $8c^2 + 8c - 30$

58. $(-3b + 2)(b + 3)$ $-3b^2 - 7b + 6$

59. $(2a - 5)(-3a - 4)$ $-6a^2 + 7a + 20$

Skills Review

Find the slope of the line that passes through each pair of points. Express as a fraction in simplest form.

60. $(4, 8), (-2, -6)$ $\frac{7}{3}$

61. $(-6, 3), (-2, 9)$ $\frac{3}{2}$

62. $(-4, -1), (-8, -8)$ $\frac{7}{4}$

63. $(12, 4), (42, 10)$ $\frac{1}{5}$

64. $(10.5, -3), (18, -8)$ $-\frac{2}{3}$

65. $(3.5, -2.5), (-1, -2)$ $-\frac{1}{9}$

DifferentiatedInstruction (BL)

Extension Draw a set of coordinate axes on the board and label each scale with only one number. Then draw several lines with arbitrary slopes (but include a horizontal line and a vertical line). Ask students to estimate values for the slopes and intercepts of the lines and to write a reasonable equation for each line. Try this when the scales on the two axes are the same. Then try it again when the scales differ on the two axes.

Tips for New Teachers

Looking Ahead Lesson 2-5 presents modeling real-world data using scatter plots. You may want to review how to find the median of a set of numbers.

4 Assess

Yesterday's News Have students explain how Lesson 2-3 on slope helped prepare them for this lesson's work on writing equations.

Formative Assessment

Check for students' understanding of concepts in Lessons 2-3 and 2-4.

📁 Quiz 2, p. 59

Additional Answers

36a.

$y = 1.61x$

42. Sample answer: Depending on what information is given and what the problem is, it might be easier to represent a linear equation in one form over another. For example, if you are given the slope and the y-intercept, you could represent the equation in slope-intercept form. If you are given a point and the slope, you could represent the equation in point-slope form. If you are trying to graph an equation using the x- and y-intercepts, you could represent the equation in standard form.

1 Focus

Objective Use functions to model direct variation.

Materials for Each Student

- TI-83/84 Plus or other graphing calculator

Teaching Tip

To help students understand direct variation, give them an example with which they are familiar. As a person drives a car at a constant rate, the distance increases as the driving time increases. This is an example of direct variation.

2 Teach

Working in Cooperative Groups

Divide the class into pairs. Give each pair a graphing calculator. Try to pair a student who is knowledgeable with a student who is less knowledgeable.

As a class, work through how to graph $y = 4x$, and discuss direct variation.

Then ask students to work with their partner to complete the Activity and Exercise 1.

Practice Have students complete Exercise 2.

3 Assess

Formative Assessment

Use Exercise 2 to assess whether students can graph direct variation equations.

From Concrete to Abstract

Ask students to give other examples of real-life situations that vary directly. For example: the pressure in your ears varies directly with the depth at which you are swimming.

EXTEND 2-4 Graphing Technology Lab
Direct Variation

An equation of a direct variation is a special case of a linear equation. A **direct variation** can be expressed in the form $y = kx$. This means that y is a multiple of x. The k in this equation is a constant and is called the **constant of variation**.

CCSS Common Core State Standards
Content Standards
F.IF.4 For a function that models a relationship between two quantities, interpret key features of graphs and tables in terms of the quantities, and sketch graphs showing key features given a verbal description of the relationship.

Notice that the graph of $y = 4x$ is a straight line through the origin. An equation of a direct variation is a special case of an equation written in slope-intercept form, $y = mx + b$. When $m = k$ and $b = 0$, $y = mx + b$ becomes $y = kx$. So the slope of a direct variation equation is its constant of variation.

To express a direct variation, we say that y varies directly as x. In other words, as x increases, y increases or decreases at a constant rate.

$y = 4x$

[−10, 10] scl: 1 by [−10, 10] scl: 1

> **KeyConcept** Direct Variation
>
> y varies directly as x if there is some nonzero constant k such that $y = kx$.
> k is called the *constant of variation*.

Activity

GOLD The karat rating r of a gold object varies directly as the percentage p of gold in the object. A 14-karat ring is 58.25% gold.

a. Write and graph a direct variation equation relating r and p.

Use the point (0.5825, 14) to find the constant of variation.

$y = kx$	Direct variation equation
$14 = k(0.5825)$	$x = 0.5825, y = 14$
$24.03 \approx k$	Divide each side by 0.5825.

The direct variation equation is $r = 24.03p$.

b. Find the karat rating of a ring that is 75% gold.

Use the calculator to find the karat rating.

KEYSTROKES: [2nd] [CALC] 0.75 [ENTER] 18.0225

The karat rating of a ring that is 75% gold is 18 karats.

[0, 1] scl: 0.1 by [0, 24] scl: 1

MENTAL CHECK 75% of $24 = \frac{3}{4}$ of 24 Think $75\% = \frac{3}{4}$.
$= 18$ ✓   Think $\frac{3}{4}$ of 24 is 18.

Exercises

1. **SWIMMING** When you swim under water, the pressure on your ears varies directly with the depth at which you are swimming. If you are swimming in 8 feet of water, the pressure on your ears is 3.44 pounds per square inch. Write and graph a direct variation equation relating pressure and depth. Then find the pressure at a depth of 65 feet.

2. Graph the direct variation equations $y = -4x$, $y = -2x$, $y = 4x$, and $y = 2x$. Compare and contrast the graphs of the equations. **1–2. See Chapter 2 Answer Appendix.**

90 | Extend 2-4 | Graphing Technology Lab: Direct Variation

Extending the Concept

Write the equation $y = kx^3$ on the board. Ask students to describe the kind of variation modeled by this equation. Have them describe what happens to the value of y when the value of x is doubled, tripled, halved, etc. For this equation, y varies directly as the cube of x. When x is tripled, y is multiplied by 27. When x is halved, y is divided by 8.

2 Mid-Chapter Quiz
Lessons 2-1 through 2-4

1. State the domain and range of the relation $\{(-3, 2), (4, 1), (0, 3), (5, -2), (2, 7)\}$. Then determine whether the relation is a function. (Lesson 2-1)
$D = \{-3, 0, 2, 4, 5\}; R = \{-2, 1, 2, 3, 7\};$ **function**

2. Graph $y = 2x - 3$ and determine whether the equation is a *function*, is *one-to-one, onto, both,* or *neither*. State whether it is *discrete* or *continuous*. (Lesson 2-1) **See margin.**

Given $f(x) = 3x^3 - 2x + 7$, find each value. (Lesson 2-1)

3. $f(-2)$ **−13**
4. $f(2y)$ **$24y^3 - 4y + 7$**
5. $f(1.4)$ **12.432**

6. State whether $f(x) = 2x^2 - 9$ is a linear function. Explain. (Lesson 2-2) **No, the variable is squared**

7. **MULTIPLE CHOICE** The daily pricing for renting a mid-sized car is given by the function $f(x) = 0.35x + 49$, where $f(x)$ is the total rental price for a car driven x miles. Find the rental cost for a car driven 250 miles. (Lesson 2-2) **C**

A $84
B $112.50
C $136.50
D $215

Write each equation in standard form. Identify *A, B,* and *C*. (Lesson 2-2)

8. $y = -6x + 5$ **$6x + y = 5;$ 6, 1, 5**
9. $y = 10x$ **$10x - y = 0;$ 10, −1, 0**
10. $-\frac{5}{8}x = 2y + 11$ **$5x + 16y = -88;$ 5, 16, −88**
11. $0.5x = 3$ **$x = 6;$ 1, 0, 6**

Find the *x*-intercept and the *y*-intercept of the graph of each equation. Then graph the equation using the intercepts. (Lesson 2-2)

12. $4x - 3y + 12 = 0$ **12, 13. See Chapter 2 Answer Appendix.**
13. $10 - x = 2y$

14. **SPEED** The table shows the distance traveled by a car after each time given in minutes. Find the rate of change in distance for the car. (Lesson 2-3) **1.33 mi/min**

Time (min)	Distance (mi)
15	20
30	40
45	60
60	80
75	100

Find the slope of the line that passes through each pair of points. Express as a fraction in simplest form. (Lesson 2-3)

15. $(-2, 6), (1, 15)$ **3**
16. $(3, 5), (7, 15)$ **$\frac{5}{2}$**
17. $(4, 8), (4, -3)$ **undefined**
18. $(-2.5, 4), (1.5, -2)$ **$-\frac{3}{2}$**
19. Find the slope of the line shown. (Lesson 2-3) **$-\frac{1}{2}$**

Write an equation for the line that satisfies each set of conditions. (Lesson 2-4)

20. slope $\frac{2}{3}$, passes through $(3, -4)$ **$2x - 3y = 18$**

21. slope -2.5, passes through $(1, 2)$ **$y = -2.5x + 4.5$**

Write an equation of the line through each set of points. (Lesson 2-4)

22. $(-2, 3), (4, 1)$
$y = -\frac{1}{3}x + \frac{7}{3}$
23. $(4.2, 3.6), (1.8, -1.2)$
$y = 2x - 4.8$

24. **MULTIPLE CHOICE** Each week, Jaya earns $32 plus $0.25 for each newspaper she delivers. Write an equation that can be used to determine how much Jaya earns each week. How much will she earn during a week in which she delivers 240 papers? (Lesson 2-4) **G**

F $75
G $92
H $148
J $212

25. **PART-TIME JOB** Jesse is a pizza delivery driver. Each day his employer gives him $20 plus $0.50 for every pizza that he delivers. (Lesson 2-4)   **$y = 0.5x + 20$**

a. Write an equation that can be used to determine how much Jesse earns each day if he delivers x pizzas.

b. How much will he earn the day he delivers 20 pizzas? **$30**

Formative Assessment
Use the Mid-Chapter Quiz to assess students' progress in the first half of the chapter.

For problems answered incorrectly, have students review the lessons indicated in parentheses.

Summative Assessment
📁 Mid-Chapter Test, p. 61

eAssessment **McGraw-Hill** Customize and create multiple versions of your Mid-Chapter Quiz and their answer keys.

FOLDABLES StudyOrganizer

Dinah Zike's Foldables®
Before students complete the Mid-Chapter Quiz, encourage them to review the information for Lessons 2-1 through 2-4 in their Foldables.

Additional Answer
2.

function, both, continuous

InterventionPlanner

TIER 1 On Level OL

If students miss about 25% of the exercises or less,

Then choose a resource:

SE Lessons 2-1, 2-2, 2-3, and 2-4

📁 Skills Practice, pp. 7, 13, 19, and 25

TE Chapter Project

 connectED.mcgraw-hill.com Self-Check Quiz

TIER 2 Strategic Intervention AL
approaching grade level

If students miss about 50% of the exercises,

Then choose a resource:

📁 Study Guide and Intervention, pp. 5, 11, 17, and 23

 connectED.mcgraw-hill.com Extra Examples, Personal Tutor, Homework Help

TIER 3 Intensive Intervention
2 or more grades below level

If students miss about 75% of the exercises,

Then use *Math Triumphs, Alg. 2,* Ch. 1 and 3

 connectED.mcgraw-hill.com Extra Examples, Personal Tutor, Homework Help, Review Vocabulary

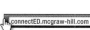

LESSON 2-5 Scatter Plots and Lines of Regression

1 Focus

VerticalAlignment

Before Lesson 2-5 Write linear equations.

Lesson 2-5 Use scatter plots and prediction equations. Model data using lines of regression.

After Lesson 2-5 Create models from data and use the models to make decisions and critical judgments.

2 Teach

Scaffolding Questions

Have students read the **Why?** section of the lesson.

Ask:

- About how many people visited the park in 2000? 21,000

- About how many people visited in 2005? 17,000

- What is a reasonable estimate for the number of people who visited this park in 2008? 16,000 or 17,000

:·Then
- You wrote linear equations.

:·Now
1. Use scatter plots and prediction equations.
2. Model data using lines of regression.

:·Why?
- The scatter plot shows the number of visitors to Isle Royale National Park in Michigan per year.

Visitors to Isle Royale National Park

Source: National Park Service

 NewVocabulary
bivariate data
scatter plot
dot plot
positive correlation
negative correlation
line of fit
prediction equation
regression line
correlation coefficient

 Common Core State Standards

Content Standards
F.IF.4 For a function that models a relationship between two quantities, interpret key features of graphs and tables in terms of the quantities, and sketch graphs showing key features given a verbal description of the relationship.

Mathematical Practices
4 Model with mathematics.
5 Use appropriate tools strategically.

1 Scatter Plots and Prediction Equations Data with two variables, such as year and number of visitors, are called **bivariate data**. A set of bivariate data graphed as ordered pairs in a coordinate plane is called a **scatter plot** or **dot plot**.

A scatter plot can show whether there is a positive, negative, or no correlation between two variables. Correlations are usually described as *strong* or *weak*. In a strong correlation, the points of the scatterplot are closer to the graph of a line than the points representing a weak correlation.

KeyConcept Scatter Plots

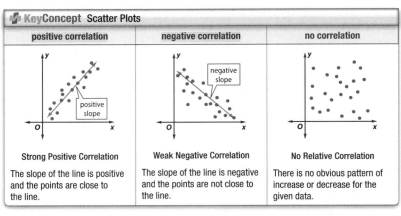

positive correlation	negative correlation	no correlation
Strong Positive Correlation	Weak Negative Correlation	No Relative Correlation
The slope of the line is positive and the points are close to the line.	The slope of the line is negative and the points are not close to the line.	There is no obvious pattern of increase or decrease for the given data.

When you find a line that closely approximates a set of data, you are finding a **line of fit** for the data. An equation of such a line is often called a **prediction equation** because it can be used to predict one of the variables given the other variable.

To find a line of fit and a prediction equation for a set of data, select two points that appear to represent the data well. This is a matter of personal judgment, so your line and prediction equation may differ from those of others.

 92 | Lesson 2-5

Per Brefelhagen/The Image Bank/Getty Images

Lesson 2-5 Resources

Resource	Approaching Level **AL**	On Level **OL**	Beyond Level **BL**	English Learners **ELL**
Teacher Edition	• Differentiated Instruction, p. 94	• Differentiated Instruction, p. 94	• Differentiated Instruction, pp. 94, 98	
Chapter Resource Masters	• Study Guide and Intervention, pp. 30–31 • Skills Practice, p. 32 • Practice, p. 33 • Word Problem Practice, p. 34	• Study Guide and Intervention, pp. 30–31 • Skills Practice, p. 32 • Practice, p. 33 • Word Problem Practice, p. 34 • Enrichment, p. 35	• Practice, p. 33 • Word Problem Practice, p. 34 • Enrichment, p. 35	• Study Guide and Intervention, pp. 30–31 • Skills Practice, p. 32 • Practice, p. 33 • Word Problem Practice, p. 34
Other	• 5-Minute Check 2-5 • Study Notebook	• 5-Minute Check 2-5 • Study Notebook	• 5-Minute Check 2-5 • Study Notebook	• 5-Minute Check 2-5 • Study Notebook

Jose Luis Pelaez Inc/Blend Images/Getty Images

Real-World Example 1 Use a Scatter Plot and Prediction Equation

TECHNOLOGY The table shows the percent of U.S. households with Internet access.

Year	1997	2000	2001	2003	2007
Percent	18.0	41.5	50.4	54.7	61.7

Source: U.S. Census Bureau

a. Make a scatter plot and a line of fit, and describe the correlation. Let x be the number of years since 1995.

Graph the data as ordered pairs with the number of years since 1995 on the horizontal axis and the percent of households on the vertical axis.

The points (5, 41.5) and (12, 61.7) appear to represent the data well. Draw a line through these two points. The data show a strong positive correlation.

Percent of Households with Internet Access

b. Use two ordered pairs to write a prediction equation.

Find an equation of the line through (5, 41.5) and (12, 61.7).

$m = \dfrac{y_2 - y_1}{x_2 - x_1}$ Slope Formula

$= \dfrac{61.7 - 41.5}{12 - 5}$ Substitute.

≈ 2.89 Simplify.

$y - y_1 = m(x - x_1)$ Point-Slope form

$y - 41.5 \approx 2.89(x - 5)$ Substitute.

$y - 41.5 \approx 2.89x - 14.45$ Distributive Property

$y \approx 2.89x + 27.05$ Simplify.

One prediction equation is $y = 2.89x + 27.05$.

c. Predict the percent of households with Internet access in 2020.

The year 2020 is 25 years after 1995, so find y when $x = 25$.

$y \approx 2.89x + 27.05$ Prediction equation

$\approx 2.89(25) + 27.05$ $x = 25$

≈ 99.3 Simplify.

The model predicts that 99.3% of U.S. households will have Internet access in 2020.

d. How accurate does your prediction appear to be?

Except for the outlier at (2, 18.0), the line fits the data well, so the prediction value should be fairly accurate.

ReviewVocabulary
outlier a data point that does not appear to belong to the rest of the set.

GuidedPractice

1. **HOUSING** The table shows the mean selling price of new, privately-owned, single-family homes for six consecutive years.

Year	0	1	2	3	4	5
Price($1000)	154.5	166.4	181.9	207.0	228.7	273.5

A. Make a scatter plot and a line of fit, and describe the correlation. **See margin.**

B. Write a prediction equation. **Sample answer:** $y = 23.4x + 135.1$

C. Predict the selling price of a new home for year 8.

D. How accurate does your prediction appear to be?

1C. Sample answer: $322,300

1D. Sample answer: The line fits the data fairly well, so the prediction is reasonably accurate.

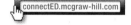

Real-World Link
According to the National Technology Scan, about 70% of U.S. households have computers connected to the Internet.

Source: *The National Technology Scan*

Additional Answer (Guided Practice)

1A.

Mean House Prices

Sample answer: The data show a strong positive correlation.

Focus on Mathematical Content

Correlation When a line is fit to a scatter plot, the type of correlation is related to the sign of the slope of the line. A positive slope indicates positive correlation, while a negative slope indicates negative correlation.

Tips for New Teachers

Scatter Plots Make sure students understand that data points must be clustered linearly to see a correlation.

1 Scatter Plots and Prediction Equations

Example 1 illustrates how to make a scatter plot from a table of data, draw a line of fit, and interpret the graph's meaning.

Formative Assessment

Use the Guided Practice exercises after each example to determine students' understanding of concepts.

Additional Example

1 **EDUCATION** The table below shows the approximate percent of students who sent applications to two colleges in various years since 1985.

Years Since 1985	Percent
0	20
3	18
6	15
9	15
12	14
15	13

Source: *U.S. News and World Report*

a. Make a scatter plot and a line of fit, and describe the correlation.

Percent of Students Applying to Two Colleges

The data show a strong negative correlation.

(continued on the next page)

Additional Example

b. Use two ordered pairs to write a prediction equation.

Using (3, 18) and (15, 13): $y = -\frac{5}{12}x + 19.25$; the slope indicates that the percent is falling by about 42% each year. The y-intercept indicates the percent in 1985 would have been about 19%.

c. Predict the percent of students who will send applications to two colleges in 2010. about 8.83%

d. How accurate does your prediction appear to be? The fit is only approximate, so the prediction may not be very accurate.

▶ **Additional Examples** also in Interactive Classroom PowerPoint® Presentations

IWB **Interactive White Board READY**

CCSS **Teaching the Mathematical Practices**

Modeling Mathematically proficient students can apply the mathematics they know to solve problems arising in everyday life, society, and the workplace. Encourage students to interpret their mathematical results in the context of the situation and reflect on whether the results make sense.

Additional Answer

1a.

Ocean Temperature

weak negative correlation

---

2 Lines of Regression

Another method for writing a line of fit is to use a line of regression. A **regression line** is determined through complex calculations to ensure that the distance of all data points to the line of fit are at a minimum. Most graphing calculators and spreadsheets can perform these calculations easily.

The **correlation coefficient** r, $-1 \le r \le 1$, is a measure that shows how well data are modeled by a linear equation.

• When r is close to -1, the data have a negative correlation.

• When $r = 0$, the data have no correlation.

• When r is close to 1, the data have a positive correlation.

Real-World Example 2   Regression Line

The table shows the life expectancy for people born in the United States.

Year of Birth	1980	1983	1990	1995	2000	2006
Life Expectancy (yr)	73.7	74.6	75.4	75.8	76.8	77.7

Source: U.S. CDC

Use a graphing calculator to make a scatter plot of the data. Find an equation for and graph a line of regression. Then use the equation to predict the life expectancy of a person born in 2025.

Step 1 Make a scatter plot.

• Enter the years of birth in **L1** and the ages in **L2**.

KEYSTROKES: [STAT] [ENTER] 1980 [ENTER] 1983 [ENTER] 1990 [ENTER] …

• Set the viewing window to fit the data.

KEYSTROKES: [WINDOW] 1975 [ENTER] 2010 [ENTER] 5 [ENTER] 70 [ENTER] 90 [ENTER] 2

[1975, 2010] scl: 5 by [70, 90] scl: 2

• Use **STAT PLOT** to graph the scatter plot.

KEYSTROKES: [2nd] [STAT PLOT] [ENTER] [ENTER] [GRAPH]

Step 2 Find the equation of the line of regression.

• Find the regression equation by selecting **LinReg(ax + b)** on the **STAT CALC** menu.

KEYSTROKES: [STAT] [▶] 4 [ENTER]

The regression equation is about $y = 0.14x - 211.43$. The slope indicates that the life expectancy increases at a rate of about 0.14 per year. The correlation coefficient r is about 0.99, which is very close to 1. So, the data fit the regression line very well.

Step 3 Graph the regression equation.

• Copy the equation to the **Y=** list and graph.

KEYSTROKES: [Y=] [VARS] 5 [▶] [▶] 1 [GRAPH]

Notice that the regression line comes close to all of the data points. As the correlation coefficient indicated, the line fits the data very well.

[1975, 2010] scl: 5 by [70, 90] scl: 2

StudyTip

CCSS Modeling
An alternate method for creating a model of the data is to let x represent the number of years since the first year in the data set. This simplifies the math involved in finding a function to model the data. But the rate of change and fit of the regression equation are the same. Only the y-intercept of the equation is changed.

DifferentiatedInstruction **AL** **OL** **BL**

Linguistic/Visual/Spatial Learners Newspapers and magazines can be good sources of real-world examples of mathematics, especially with regard to tables and graphs of data describing current issues of interest. Some publications even offer special news-related resources for teachers. Ask students to find a table of data in a newspaper or magazine to cut out and discuss with the class.

ReadingMath

Predictions

When you are predicting an x-value greater than or less than any in the data set, the process is known as **extrapolation**.

When you are predicting an x-value between the least and greatest in the data set, the process is known as **interpolation**.

Step 4 Predict using the function.

- Find y when $x = 2025$. Use **VALUE** on the **CALC** menu. Reset the window size to accommodate the x-value of 2025.

  KEYSTROKES: [2nd] [CALC] 1 2025 [ENTER]

  According to the function, the life expectancy of a person born in 2025 will be about 80.4 years.

[1975, 2025] scl: 5 by [70, 90] scl: 2

GuidedPractice

2. **MUSIC** The table at the right shows the percent of sales that were made in music stores in the United States for the period 1999–2008. Use a graphing calculator to make a scatter plot of the data. Find and graph a line of regression. Then use the function to predict the percent of sales made in a music store in 2018.

[1995, 2010] scl: 2 by [25, 50] scl: 5

$y = -1.4x + 2926.7$; 15.9%

Music Store Sales	
Year	Sales (percent)
1999	44.5
2000	42.4
2001	42.5
2002	36.8
2003	33.2
2004	32.5
2005	39.4
2006	35.4
2007	31.1
2008	30.0

Source: Recording Industry Association of America

Check Your Understanding   ⬤ = **Step-by-Step Solutions begin on page R14.**

Example 1   (1) **OCEANS** The table shows the temperature in the ocean at various depths.

Depth (in meters)	0	300	500	1000	2000	2500
Temp (°C)	22	20	13	7	6	?

Source: NDAA

1b. Sample answer using (0, 22) and (2000, 6): $y = -0.008x + 22$

a. Make a scatter plot and a line of fit, and describe the correlation. **See margin.**

b. Use two ordered pairs to write a prediction equation.

c. Use your prediction equation to predict the missing value. **Sample answer: 2°C**

Example 2   2. **(CCSS) TOOLS** The table shows the median income of families in North Carolina by family size in a recent year. Use a graphing calculator to make a scatter plot of the data. Find an equation for and graph a line of regression. Then use the equation to predict the median income of a North Carolina family of 9.

Sample answer: $y = 8455.1x + 25{,}837$; $101{,}932$; see margin for graph.

Family Size	Income ($)
1	33,265
2	44,625
3	50,528
4	59,481

Source: U.S. Department of Justice

 connectED.mcgraw-hill.com **95**

Additional Answer

2.

[0, 6] scl: 1 by [30,000, 60,000] scl: 5000

(CCSS) Teaching the Mathematical Practices

Tools When making mathematical models, mathematically proficient students know that technology can enable them to visualize the results of varying assumptions, explore consequences, and compare predictions with data.

---

2 Lines of Regression

Example 2 shows how to use a graphing calculator to make a scatter plot, graph a line of regression, and make a prediction.

Additional Example

2 **INCOME** The table shows the median income of U.S. families for the period 1970–2002.

Year	Income ($)
1970	9867
1980	21,023
1985	27,735
1990	35,353
1995	40,611
1998	46,737
2000	50,732
2002	51,680

Use a graphing calculator to make a scatter plot of the data. Find an equation for and graph a line of regression. Then use the equation to predict the median income in 2015.

[1965, 2015] scl: 5 by [0, 55,000] scl: 10,000

$y = 1349.87x - 2{,}650{,}768.34$; median family income in 2015 will be about $69,220.

Tips for New Teachers

Sense-Making Encourage students to think about why trends are occurring in Example 2 and Guided Practice 2. Then ask students how problems of this type could be important in other fields such as engineering, marketing, and statistics.

3 Practice

Formative Assessment

Use Exercises 1–2 to check for understanding.

Use the chart at the bottom of this page to customize assignments for your students.

Additional Answers

3a.

Compact Disc Sales

strong negative correlation

4a.

2006 Miami Heat

no correlation

5a.

Sunee's Homemade Ice Cream Sales

strong positive correlation

Example 1

For Exercises 3–6, complete parts a–c.

a. Make a scatter plot and a line of fit, and describe the correlation.

b. Use two ordered pairs to write a prediction equation.

c. Use your prediction equation to predict the missing value.

3. COMPACT DISC SALES The table shows the number of CDs sold in recent years at Jerome's House of Music. Let x be the number of years since 2000.

Year	2004	2005	2006	2007	2008	2017
Number of CDs sold	49,300	47,280	43,450	40,125	35,792	?

3a. See margin.
3b. Sample answer, using (4, 49,300) and (8, 35,792): $y = -3377x + 62,808$
3c. Sample answer: 5399 CDs

4. BASKETBALL The table shows the number of field goals and assists for some of the members of the Miami Heat in a recent NBA season. **4a.** See margin.

Field Goals	472	353	278	283	238	265	186	162	144
Assists	384	97	81	79	18	130	94	95	?

Source: NBA

4b. Sample answer: No equation can be written because there is no correlation.
4c. unpredictable

5 ICE CREAM The table shows the amount of ice cream Sunee's Homemade Ice Creams sold for eight months. Let $x = 1$ for January. **5a.** See margin.

Month	Jan	Feb	Mar	Apr	May	June	July	Aug	Sept
Gallons sold	37	44	72	80	105	110	119	131	?

5b. Sample answer using (1, 37) and (8, 131): $y = \frac{94}{7}x + \frac{165}{7}$

5c. Sample answer: about 144 gal

6. DRAMA CLUB The table shows the total revenue of all of Central High School's plays in recent school years. Let x be the number of years since 2003. **6a.** See margin.

School Year	2005	2006	2007	2008	2009	2016
Revenue ($)	603	666	643	721	771	?

6b. Sample answer using (2, 603) and (6, 771): $y = 42x + 519$
6c. Sample answer: $1065

Example 2

7. SALES The table shows the sales of Chayton's Computers. Let x be the number of years since 2002 and use a graphing calculator to make a scatter plot of the data. Find an equation for and graph a line of regression. Then use the function to predict the sales in 2018.

See Chapter 2 Answer Appendix for graph; $y = 61.9x + 530.2$; $1.52 million in sales

Year	Sales ($ thousands)
2004	640
2005	715
2006	791
2007	852
2008	910
2009	944

8. CCSS TOOLS The table shows the number of employees of a small company. Let x be the number of years since 2000 and use a graphing calculator to make a scatter plot of the data. Find an equation for and graph a line of regression. Then use the function to predict the number of employees in 2025.

See Chapter 2 Answer Appendix for graph; $y = 3.91x - 4.45$; about 93 employees

Year	Number of Employees
2002	4
2003	7
2004	11
2005	14
2006	20
2007	23

96 | Lesson 2-5 | Scatter Plots and Lines of Regression

Differentiated Homework Options

Level	Assignment	Two-Day Option	
AL Basic	3–8, 12, 14–32	3–7 odd, 17–20	4–8 even, 12, 14–16, 21–32
OL Core	3–7 odd, 9–12, 14–32	3–8, 17–20	9–12, 14–16, 21–32
BL Advanced	9–29, (optional: 30–32)		

9 BASEBALL The table at the right shows the total attendance and wins for the Cleveland Indians in some recent years.

9a. See Ch. 2 Answer Appendix.

a. Make a scatter plot of the years and attendance.

b. If a linear shape is apparent, find a regression equation to summarize the trend. $y = 40{,}634.9x - 79{,}543{,}385.8$

c. Interpret the slope of the regression line in the context of the data. **Attendance increases by about 40,634.9 each year.**

d. Is the relationship between years and attendance stronger than the one between wins and attendance? Explain. **See Ch. 2 Answer Appendix.**

Year	Wins	Attendance
2009	65	1,776,904
2008	81	2,182,087
2007	96	2,275,916
2006	78	1,997,936
2005	93	2,014,220
2004	80	1,814,401
2003	68	1,730,001

10. CLASS SIZE The table at the right shows the relationship between the number of students in a mathematics class and the average grade for each class.

10a. See Chapter 2 Answer Appendix.

a. Make a scatter plot and find a regression equation for the data. Then graph the regression line.

b. What is the correlation coefficient of the data? $r = -0.712$

c. Describe the correlation. How accurate is the regression equation?

Class Size	Class Average
16	81.2
19	80.5
24	82.5
25	79.9
27	78.6
29	79.3
32	77.7

10c. Sample answer: relatively accurate with a negative correlation

11. CCSS TOOLS Jocelyn is analyzing the sales of her company. The table at the right shows the total sales for each of six years. **11a.** $y = 3.1x - 6177; r = 0.63$

a. Find a regression equation and correlation coefficient for the data. Let x be the year.

b. Use the regression equation to predict the 2020 sales. **about $79.7 million**

c. Remove the outlier from the data set and find a new regression equation and correlation coefficient.

11c. $y = 2.6x - 5170; r = 0.986$

d. Use the new regression equation to predict the sales in 2020. **about $75.4 million**

e. Compare the correlation coefficients for the two equations. Which function fits the data better? Which prediction should Jocelyn expect to be more accurate? **See Chapter 2 Answer Appendix.**

Year	Sales ($ millions)
2003	31.2
2004	34.6
2005	18.6
2006	37.7
2007	41.3
2008	45.1

H.O.T. Problems Use Higher-Order Thinking Skills

12. REASONING What is the relevance of the correlation coefficient of a linear regression line? Explain your reasoning. **See Chapter 2 Answer Appendix.**

13. CHALLENGE If statements a and b have a positive correlation, b and c have a negative correlation, and c and d have a positive correlation, what can you determine about the correlation between statements a and d? Explain your reasoning. **See Chapter 2 Answer Appendix.**

14a. Sample answer: years and height of a teenager

14. OPEN ENDED Provide real-world quantities that represent each of the following.

a. positive correlation **b.** negative correlation **c.** no correlation

15. CHALLENGE Draw a scatter plot for the following data set.

x	1.0	1.5	2.0	2.8	3.2	4.0	4.8	5.8
y	3.5	4.7	5.1	6.8	7.1	7.5	8.8	10.3

14b. Sample answer: time and capacity of a standard battery

14c. Sample answer: a person's weight and his or her income

Which of the following best represents the correlation coefficient r for the data? Justify your answer. **See Chapter 2 Answer Appendix.**

a. 0.99 **b.** −0.98 **c.** 0.62 **d.** 0.08

16. WRITING IN MATH What are the strengths and weaknesses of using a regression equation to approximate data? **See Chapter 2 Answer Appendix.**

6a.

Drama Club Earnings

weak positive correlation

Study Guide and Intervention

AL OL ELL

Practice

AL OL BL ELL

Word Problem Practice

AL OL BL ELL

Enrichment

OL BL

Tips for New Teachers

Looking Ahead Lesson 2-6 presents the graphing of special functions, including absolute value functions. Exercises 30–32 should be used to determine your students' familiarity with solving absolute value equations.

4 Assess

Ticket Out the Door Give each student a small piece of paper. Have each student write three or four things to examine when reading a graph such as the one shown in the example on p. 93. Collect the papers as the students leave the room.

 Follow-up

Students have explored equations of linear functions and data that can be modeled by linear functions.

Ask:

- Why is it useful to be able to represent a relationship with a linear function? Sample answer: A linear function allows you to predict future behavior based on past relationships between quantities.

Standardized Test Practice

17. SHORT RESPONSE What is the value of the expression below? 2

$$17 - 3[-1 + 2(7 - 4)]$$

18. Anna took brownies to a club meeting. She gave half of her brownies to Selena. Selena gave a third of her brownies to Randall. Randall gave a fourth of his brownies to Trina. If Trina has 3 brownies, how many brownies did Anna bring to the meeting? C

A 12
B 36
C 72
D 144

19. GEOMETRY Which is always true? J

F A parallelogram is a square.
G A parallelogram is a rectangle.
H A quadrilateral is a trapezoid.
J A square is a rectangle.

20. SAT/ACT Which line best fits the data in the graph? D

A $y = x$
B $y = -0.5x + 4$
C $y = -0.5x - 4$
D $y = 0.5x + 0.5$
E $y = 1.5x - 1.5$

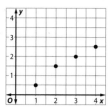

Spiral Review

Write an equation in slope-intercept form for each graph. (Lesson 2-4)

21.

$y = 2.5x - 6$

22.

$y = -\dfrac{2}{3}x + 8$

23.

$y = -3x - 6$

Find the rate of change for each set of data. (Lesson 2-3)

24.

Time (day)	3	6	9	12	15
Height (mm)	12	24	36	48	60

4 mm/day

25.

Time (h)	2	4	6	8
Distance (mi)	35	70	105	140

17.5 mi/hr

26.

Time (s)	12	16	20	24	28
Volume (cm³)	45	60	75	90	105

3.75 cm³/sec

27.

Force (N)	32	40	48	56	64
Work (J)	48	60	72	84	96

1.5 J/N

28. RECREATION Ramona estimates that she will need 50 tennis balls for every player that signs up for the tennis club and at least 150 more just in case. Write an inequality to express the situation. (Lesson 1-5) $t \geq 50p + 150$

29. DODGEBALL Six teams played in a dodgeball tournament. In how many ways can the top three teams finish? (Lesson 0-4) 120

Skills Review

Solve each equation.

30. $-4|x - 2| = -12$ 5, −1

31. $|3x + 4| = 21$ $\dfrac{17}{3}, -\dfrac{25}{3}$

32. $2|4x - 1| + 3 = 9$ $1, -\dfrac{1}{2}$

DifferentiatedInstruction BL

Extension Challenge students to research how the correlation coefficient is calculated. What mathematics is involved in the calculation? Why does it represent how well the regression equation fits the data?

EXTEND 2-5 Algebra Lab
Correlation and Causation

You have learned that the correlation coefficient measures how well an equation fits a set of data. A correlation coefficient close to 1 or −1 indicates a high correlation. However, this does not imply causation. When there is **causation**, one data set is the direct cause of the other data set. When there is a correlation between two sets of data, the data sets are related.

The news clip to the right uses correlation to imply causation. While a study may have found a high correlation between these two variables, this does not mean that cell phone use causes brain tumors. Other variables, such as an individual's family history, diet, and environment could also have an effect on the formation of brain tumors.

A **lurking variable** affects the relationship between two other variables, but is not included in the study that compares them.

> **NEWS**
>
> **Cell Phones Cause Brain Tumors!**
>
> Studies have shown that brain tumors are linked to cell phone use. One particular study showed a very strong correlation between the two events.

CCSS **Common Core State Standards**
Content Standards
F.IF.4 For a function that models a relationship between two quantities, interpret key features of graphs and tables in terms of the quantities, and sketch graphs showing key features given a verbal description of the relationship.

EXAMPLE 1

Determine whether the following correlations possibly show causation. Write *yes* or *no*. If not, identify other lurking variables.

a. Studies have shown that students are less energetic after they eat lunch.

No; students could have had gym class before lunch, or the class after lunch could be one that they do not find interesting.

b. If Kevin lifts weights, he will make the football team.

No; Kevin may need to maintain a certain grade-point average to make the team. He also needs to be talented enough to make the team.

c. When the Sun is visible, we have daylight.

Yes; no other variables cause daylight.

Exercises

Determine whether the following correlations possibly show causation. Write *yes* or *no*. If not, identify other lurking variables. 1–3. See margin.

1. If Allison studies, she will get an A.

2. When Lisa exercises, she is in a better mood.

3. The gun control laws have reduced violent crime.

4. Smoking causes lung cancer. **yes**

5. If we have a Level 2 snow emergency, we do not have school. **yes**

6. Reading more increases one's intelligence.

6. No; a person's experiences in school, at the workplace, and in everyday life could improve his or her intelligence.

7. Sample answer: A large number of unbiased studies need to be completed that show a direct link between the variables and all of the lurking variables need to be eliminated from the studies.

Think About It

7. What do you think must be done to show that a correlation between two variables actually shows causation?

(continued on the next page)

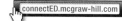
connectED.mcgraw-hill.com **99**

1 Focus

Objective Explore the difference between correlation and causation.

Materials for Each Group
- graphing calculator

Teaching Tip
This activity requires students to calculate and analyze the correlation coefficient. You may wish to review this concept as a class before you begin the Examples.

2 Teach

Working in Cooperative Groups
For each Example, each student should work with a partner.

Have the pairs discuss lurking variables and their effect on other variables. Have them think of three situations where lurking variables are influencing a relationship between two other variables that show a high correlation. Then have them try to think of two variables in which there is causation.

Practice Have students complete Exercises 1–7.

Additional Answers

1. No; she will have to complete her assignments with high quality work, participate in class, understand the material, score high on her quizzes and tests, and meet any other requirements designated by her instructor.

2. No; she could have an argument with her family, her friends could do something to upset her, or something could happen at work.

3. No; more violent criminals could already be in prison, rehabilitation could be working, unemployment could be falling, or the economy could be improving.

3 Assess

Formative Assessment
Use Exercise 7 to assess whether students understand the difference between correlation and causation.

From Concrete to Abstract
Ask students why they think there is a causation in Exercise 4. Discuss why this is accepted as fact in today's society.

Extending the Concept
Ask students to identify another situation in which causation is possible. Have them collect the data and calculate the correlation coefficient. Then ask them to say why they think the relationship shows causation.

Practice Have students complete Exercises 8–11.

Additional Answers

8. −0.73, very weak negative correlation; no causation is possible when the correlation is weak.

9. 0.89, strong positive correlation; no causation is possible because humidity, rain, time of day, and price of the ice cream are also factors.

10. 0.99, very strong positive correlation; no causation is possible because the popularity of the company, the products they sell, the economy, and the accessibility and user-friendliness of the company's Web site are also factors.

11. 1.00, very strong correlation, yes, causation exists because the age of a tree is a direct cause for the number of rings from pith to bark.

Algebra Lab
Correlation and Causation *Continued*

Statistics are often provided that show a correlation between two events and causation is implied, but not confirmed. The best method to find evidence that *x* causes *y* is to actually perform multiple experiments with large sample sizes. When an experiment is not available, the only option is to determine if causation is *possible*.

EXAMPLE 2

The table shows IQ scores for ten high school students and their corresponding grade-point averages.

a. **Calculate the correlation coefficient.**
The value of *r* is about 0.96.

b. **Describe the correlation.**
There is a strong positive correlation.

c. **Is there causation? Explain your reasoning.**
There is no causation. Many other factors determine grade-point averages, including effort, study habits, knowledge of the material in the particular classes, and extracurricular activities.

IQ	GPA	IQ	GPA
106	3.4	118	3.9
110	3.6	107	3.3
109	3.5	111	3.6
98	2.9	109	3.7
115	3.8	102	3.1

EXAMPLE 3

The table shows the approximate boiling temperatures of water at different altitudes.

a. **Calculate the correlation coefficient.**
The value of *r* is about −0.99.

b. **Describe the correlation.**
There is a strong negative correlation.

c. **Is there causation? Explain your reasoning.**
There appears to be causation. Altitude is a direct cause for the boiling temperature of water to decrease.

Altitude (ft)	Temp. (°F)
0	212
984	210
2000	208
3000	206
5000	203
7500	198
10,000	194
20,000	178
26,000	168

Exercises

Calculate the correlation coefficient and describe the correlation. Determine if causation is possible. Explain your reasoning. 8–11. See margin.

8.

Car Value ($)	Miles per Gallon	Car Value ($)	Miles per Gallon
14,000	55	25,000	33
16,000	48	29,000	29
18,500	37	32,000	19
20,000	35	35,000	26
22,500	29	40,000	33

9.

Temp. (°F)	Ice Cream Sales ($)	Temp. (°F)	Ice Cream Sales ($)
76	850	85	905
78	885	86	1005
80	875	87	1060
82	920	88	1105
84	955	89	1165

10.

Year	Sales via Internet ($)	Year	Sales via Internet ($)
2001	1923	2006	7067
2002	2806	2007	8568
2003	3467	2008	9991
2004	4848	2009	10,587
2005	5943	2010	12,695

11.

Number of Rings from Pith to Bark	Age of Tree (yr.)	Number of Rings from Pith to Bark	Age of Tree (yr.)
5	5	30	30
10	10	40	40
15	15	50	50
20	20	60	60

100 | Extend 2-5 | Algebra Lab: Correlation and Causation

LESSON 2-6 Special Functions

:: Then	:: Now	:: Why?
• You modeled data using lines of regression.	**1** Write and graph piecewise-defined functions. **2** Write and graph step and absolute value functions.	• The table shows a recent federal income tax rate schedule. The amount of federal income tax an individual is required to pay is a function of income.

Federal Tax Rate Schedule – Filing Single		
If taxable income is over	But not over	The tax is:
$0	$7,825	10% of the amount over $0
$7,825	$31,850	$782.50 plus 15% of the amount over $7,825
$31,850	$77,100	$4,386.25 plus 25% of the amount over $31,850
$77,100	$160,850	$15,698.75 plus 28% of the amount over $77,100
$160,850	$349,700	$39,148.75 plus 33% of the amount over $160,850
$349,700	no limit	$101,469.25 plus 35% of the amount over $349,700

Source: Internal Revenue Service

NewVocabulary
piecewise-defined function
piecewise-linear function
step function
greatest integer function
absolute value function

Common Core State Standards

Content Standards
F.IF.4 For a function that models a relationship between two quantities, interpret key features of graphs and tables in terms of the quantities, and sketch graphs showing key features given a verbal description of the relationship.

F.IF.7.b Graph square root, cube root, and piecewise-defined functions, including step functions and absolute value functions.

Mathematical Practices
1 Make sense of problems and persevere in solving them.

1 Piecewise-Defined Functions The function relating income and tax is not a linear function because each interval, or piece, of the function is defined by a different expression. A function that is written using two or more expressions is called a **piecewise-defined function.** On the graph of a piecewise-defined function, a dot indicates that the point is included in the graph. A circle indicates that the point is not included in the graph.

Example 1 Piecewise-Defined Function

Graph $f(x) = \begin{cases} x - 2 \text{ if } x < -1 \\ x + 3 \text{ if } x \geq -1 \end{cases}$. Identify the domain and range.

Step 1 Graph $f(x) = x - 2$ for $x < -1$.
$$f(x) = x - 2$$
$$= (-1) - 2$$
$$= -3$$
Because -1 does not satisfy the inequality, begin with a circle at $(-1, -3)$.

Step 2 Graph $f(x) = x + 3$ for $x \geq -1$.
$$f(x) = x + 3$$
$$= (-1) + 3$$
$$= 2$$
Because -1 satisfies the inequality, begin with a dot at $(-1, 2)$.

The function is defined for all values of x, so the domain is all real numbers.

The $f(x)$-coordinates of points on the graph are all real numbers less than -3 and all real numbers greater than or equal to 2, so the range is $\{f(x) | f(x) < -3 \text{ or } f(x) \geq 2\}$.

GuidedPractice

1. Graph $f(x) = \begin{cases} x + 2 \text{ if } x < 0 \\ x \quad\; \text{ if } x \geq 0 \end{cases}$. Identify the domain and range. See Chapter 2 Answer Appendix.

connectED.mcgraw-hill.com **101**

1 Focus

VerticalAlignment

▼

Before Lesson 2-6 Model data using lines of regression.

▼

Lesson 2-6 Write and graph piecewise-defined functions and write and graph step and absolute value functions.

▼

After Lesson 2-6 Identify and sketch graphs of parent functions, including quadratic functions.

2 Teach

Scaffolding Questions
Have students read the **Why?** section of the lesson.

Ask:
- What is the tax for a taxable income of $50,000? $8923.75
- What is the tax for a taxable income of $31,850? $4386.25

Lesson 2-6 Resources

Resource	Approaching Level **AL**	On Level **OL**	Beyond Level **BL**	English Learners **ELL**
Teacher Edition	• Differentiated Instruction, p. 103	• Differentiated Instruction, p. 103	• Differentiated Instruction, p. 104	• Differentiated Instruction, p. 103
Chapter Resource Masters	• Study Guide and Intervention, pp. 36–37 • Skills Practice, p. 38 • Practice, p. 39 • Word Problem Practice, p. 40	• Study Guide and Intervention, pp. 36–37 • Skills Practice, p. 38 • Practice, p. 39 • Word Problem Practice, p. 40 • Enrichment, p. 41 • Graphing Calculator, p. 42	• Practice, p. 39 • Word Problem Practice, p. 40 • Enrichment, p. 41	• Study Guide and Intervention, pp. 36–37 • Skills Practice, p. 38 • Practice, p. 39 • Word Problem Practice, p. 40
Other	• 5-Minute Check 2-6 • Study Notebook • Teaching Algebra with Manipulatives	• 5-Minute Check 2-6 • Study Notebook • Teaching Algebra with Manipulatives	• 5-Minute Check 2-6 • Study Notebook	• 5-Minute Check 2-6 • Study Notebook • Teaching Algebra with Manipulatives

1 Piecewise-Defined Functions

Example 1 shows how to graph a piecewise-defined function. **Example 2** shows how to write a piecewise-defined function.

Formative Assessment

Use the Guided Practice exercises after each example to determine students' understanding of concepts.

Additional Examples

1 Graph $f(x) = \begin{cases} x - 1 \text{ if } x \le 3 \\ -1 \text{ if } x > 3 \end{cases}$

Identify the domain and range.

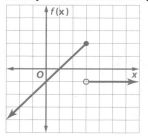

The domain is all real numbers. The range is $\{f(x)|f(x) \le 2\}$.

2 Write the piecewise-defined function shown in the graph.

$f(x) = \begin{cases} x - 4 \text{ if } x < 2 \\ 1 \text{ if } x \ge 2 \end{cases}$

Additional Examples also in Interactive Classroom PowerPoint® Presentations

IWB Interactive White Board READY

StudyTip

Graphs of Piecewise-Defined Functions The graphs of different parts of a piecewise-defined function may or may not connect. A graph may also stop at a given *x*-value and then begin at a different *y*-value for the same value of *x*.

Piecewise-defined functions are often defined by several linear functions.

Example 2 Write a Piecewise-Defined Function

Write the piecewise-defined function shown in the graph.

Examine and write a function for each portion of the graph.

The left portion of the graph is the graph of $f(x) = 2x + 3$. There is a circle at (1, 5), so the linear function is defined for $\{x|x < 1\}$.

The center portion of the graph is the graph of $f(x) = -x + 2$. There are dots at (1, 1) and (2, 0), so the linear function is defined for $\{x|1 \le x \le 2\}$.

The right portion of the graph is the constant function $f(x) = 3$. There is a circle at (2, 3), so the constant function is defined for $\{x|x > 2\}$.

Write the piecewise-defined function.

$f(x) = \begin{cases} 2x + 3 \text{ if } x < 1 \\ -x + 2 \text{ if } 1 \le x \le 2 \\ 3 \text{ if } x > 2 \end{cases}$

2A. $f(x) = \begin{cases} -\frac{1}{3}x - 2 \text{ if } x \le -3 \\ 2 \text{ if } -3 < x < 1 \\ 4x - 2 \text{ if } x \ge 1 \end{cases}$

CHECK The graph shows a portion of a line with positive slope for $x < 1$. The graph has negative slope for $1 \le x \le 2$ and constant slope for $x > 2$. The function is reasonable for the graph.

2B. $f(x) = \begin{cases} 3x \text{ if } x < -1 \\ -x \text{ if } -1 \le x < 3 \\ -x + 7 \text{ if } x \ge 3 \end{cases}$

▶ **Guided Practice**

Write the piecewise-defined function shown in each graph.

2A.   **2B.**

---

2 Step Functions and Absolute Value Functions

Unlike a piecewise-defined function, a **piecewise-linear function** contains a single expression. A common piecewise-linear function is the step function. The graph of a **step function** consists of line segments.

StudyTip

Greatest Integer Function Notice that the domain of this step function is all real numbers and the range is all integers.

The **greatest integer function**, written $f(x) = [\![x]\!]$, is one kind of step function. The symbol $[\![x]\!]$ means *the greatest integer less than or equal to x*. For example, $[\![3.25]\!] = 3$ and $[\![-4.6]\!] = -5$.

$f(x) = [\![x]\!]$

Focus on Mathematical Content

Piecewise-Defined Functions Piecewise-defined functions are sometimes called *multi-part functions* or *split-domain functions*. For example, the absolute value function $y = |x|$ is the same as $y = -x$ for $x < 0$, and $y = x$ for $x \ge 0$.

Real-World Example 3 Use a Step Function

BUSINESS An automotive repair center charges $50 for any part of the first hour of labor, and $35 for any part of each additional hour. Draw a graph that represents this situation.

Understand The total labor charge is $50 for the first hour plus $35 for each additional fraction of an hour, so the graph will be a step function.

Plan If the time spent on labor is greater than 0 hours, but less than or equal to 1 hour, then the labor charge is $50. If the time is greater than 1 hour but less than 2 hours, then the labor charge is $85, and so on.

Solve Use the pattern of times and costs to make a table, where x is the number of hours of labor and $T(x)$ is the total labor charge. Then graph.

x	T(x)
$0 < x \le 1$	$50
$1 < x \le 2$	$85
$2 < x \le 3$	$120
$3 < x \le 4$	$155
$4 < x \le 5$	$190

Check Since the repair center rounds any fraction of an hour up to the next whole number, each segment of the graph has a circle at the left endpoint and a dot at the right endpoint.

▶ **Guided**Practice

3. **RECYCLING** A recycling company pays $5 for every full box of newspaper. They do not give any money for partial boxes. Draw a graph that shows the amount of money $P(x)$ for the number of boxes x brought to the recycling center. **See margin.**

Another piecewise-linear function is the absolute value function. An **absolute value function** is a function that contains an algebraic expression within absolute value symbols.

KeyConcept Parent Function of Absolute Value Functions

Parent function: $f(x) = |x|$, defined as

$$f(x) = \begin{cases} x \text{ if } x > 0 \\ 0 \text{ if } x = 0 \\ -x \text{ if } x < 0 \end{cases}$$

Type of graph: V-shaped
Domain: all real numbers
Range: all nonnegative real numbers
Intercepts: $x = 0$, $f(x) = 0$
Not defined: $f(x) < 0$

Math History Link

Karl Weierstrass (1815–1897) At the wishes of his father, Weierstrass studied law, economics, and finance at the University of Bonn, but then dropped out to study his true interest, mathematics, at the University of Münster. In an 1841 essay, Weierstrass first used | | to denote absolute value.

Photo: The Granger Collection

The Granger Collection, New York - All rights reserved.

2 Step Functions and Absolute Value Functions

Example 3 shows how to graph and use a step function. **Example 4** shows how to graph an absolute value function and identify its domain and range.

Additional Example

3 **PSYCHOLOGY** One psychologist charges for counseling sessions at the rate of $85 per hour or any fraction thereof. Draw a graph that represents this situation.

Teach with Tech

Blog Give students several pictures of graphs of piecewise functions. Have students write a blog entry explaining how to interpret each of the graphs.

Additional Answer (Guided Practice)

3.

Differentiated Instruction AL OL ELL

Verbal/Linguistic Learners Have students verbally explain to a partner how to graph one of the piecewise-defined functions they learned about in this lesson.

Additional Example

4 Graph $y = |x| + 1$. Identify the domain and range. The domain is all real numbers. The range is $\{y|y \geq 1\}$.

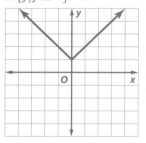

Additional Answers (Guided Practice)

4A.

$D = \{\text{all real numbers}\};$
$R = \{f(x) \mid f(x) \geq 0\}$

4B.

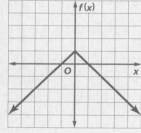

$D = \{\text{all real numbers}\};$
$R = \{f(x) \mid f(x) \leq 1\}$

Additional Answer

12.

$D = \{x \mid x \leq -4 \text{ or } 0 < x\};$
$R = \{f(x) \mid f(x) \geq 12, f(x) = 8, \text{ or }$
    $0 < f(x) \leq 3\}$

Example 4 Absolute Value Functions

Graph $f(x) = |2x| - 4$. Identify the domain and range.

Create a table of values.

x	$|2x| - 4$
-3	2
-2	0
-1	-2
0	-4
1	-2
2	0
3	2

Graph the points and connect them.

The domain is the set of all real numbers. The range is $\{f(x)|f(x) \geq -4\}$.

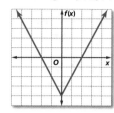

▶ **Guided**Practice **4A, 4B.** See margin.

Graph each function. Identify the domain and range.

4A. $f(x) = |x - 2|$ 　　　　　　　　**4B.** $f(x) = -|x| + 1$

Check Your Understanding

= Step-by-Step Solutions begin on page R14.

Example 1　Graph each function. Identify the domain and range. **1, 2.** See Chapter 2 Answer Appendix.

1. $g(x) = \begin{cases} -3 & \text{if } x \leq -4 \\ x & \text{if } -4 < x < 2 \\ -x + 6 & \text{if } x \geq 2 \end{cases}$
　　
2. $f(x) = \begin{cases} 8 & \text{if } x \leq -1 \\ 2x & \text{if } -1 < x < 4 \\ -4 - x & \text{if } x \geq 4 \end{cases}$

Example 2　Write the piecewise-defined function shown in each graph. **3, 4.** See Chapter 2 Answer Appendix.

3.

4.

Example 3　**5** **CCSS** **REASONING** Springfield High School's theater can hold 250 students. The drama club is performing a play in the theater. Draw a graph of a step function that shows the relationship between the number of tickets sold x and the minimum number of plays y that the drama club must perform. **See Chapter 2 Answer Appendix.**

Graph each function. Identify the domain and range. **6, 7.** See Chapter 2 Answer Appendix.

6. $g(x) = -2[\![x]\!]$ 　　　　　　　　**7.** $h(x) = [\![x - 5]\!]$

Example 4　Graph each function. Identify the domain and range. **8–11.** See Chapter 2 Answer Appendix.

8. $g(x) = |-3x|$ 　　　　　　　　**9.** $f(x) = 2|x|$

10. $h(x) = |x + 4|$ 　　　　　　　　**11.** $s(x) = |-2x| + 6$

104 | Lesson 2-6 | Special Functions

DifferentiatedInstruction **BL**

Extension Have students draw a large coordinate plane on a sheet of paper. Then have them use toothpicks (or other similar objects) to model the general shapes of step, piecewise-defined, and absolute value functions. Students should identify each type of graph as they model it.

Example 1　　Graph each function. Identify the domain and range. **12–15. See margin.**

12. $f(x) = \begin{cases} -3x \text{ if } x \le -4 \\ x \text{ if } 0 < x \le 3 \\ 8 \text{ if } x > 3 \end{cases}$　　**13.** $f(x) = \begin{cases} 2x \text{ if } x \le -6 \\ 5 \text{ if } -6 < x \le 2 \\ -2x + 1 \text{ if } x > 4 \end{cases}$

14. $g(x) = \begin{cases} 2x + 2 \text{ if } x < -6 \\ x \text{ if } -6 \le x \le 2 \\ -3 \text{ if } x > 2 \end{cases}$　　**15.** $g(x) = \begin{cases} -2 \text{ if } x < -4 \\ x - 3 \text{ if } -1 \le x \le 5 \\ 2x - 15 \text{ if } x > 7 \end{cases}$

Example 2　　Write the piecewise-defined function shown in each graph.

16–19. See Chapter 2 Answer Appendix.

16.

17.

18.

19.

Example 3　　Graph each function. Identify the domain and range. **20–23. See Chapter 2 Answer Appendix.**

20. $f(x) = [\![x]\!] - 6$　　　　　　　**21.** $h(x) = [\![3x]\!] - 8$

22. $f(x) = [\![3x + 2]\!]$　　　　　　**23.** $g(x) = 2[\![0.5x + 4]\!]$

Example 4　　Graph each function. Identify the domain and range. **24–29. See Chapter 2 Answer Appendix.**

24. $f(x) = |x - 5|$　　　　　　　**25.** $g(x) = |x + 2|$

26. $h(x) = |2x| - 8$　　　　　　**27.** $k(x) = |-3x| + 3$

28. $f(x) = 2|x - 4| + 6$　　　　**29.** $h(x) = -3|0.5x + 1| - 2$

30. **GIVING** Patrick is donating money and volunteering his time to an organization that restores homes for the needy. His employer will match his monetary donations up to $100.

　　a. Identify the type of function that models the total amount of money received by the charity when Patrick donates x dollars. **piecewise**

　　b. Write and graph a function for the situation. **See Chapter 2 Answer Appendix.**

31. **CCSS SENSE-MAKING** A car's speedometer reads 60 miles an hour.

　　a. Write an absolute value function for the difference between the car's actual speed a and the reading on the speedometer. $f(a) = |a - 60|$

　　b. What is an appropriate domain for the function? Explain your reasoning. $\{a | a \ge 0\}$

　　c. Use the domain to graph the function. **See Chapter 2 Answer Appendix.**

Differentiated Homework Options

Level	Assignment	Two-Day Option	
AL Basic	12–31, 40, 42–59	13–31 odd, 45–48	12–30 even, 40, 42–44, 49–59
OL Core	13–31 odd, 32, 33–37 odd, 39, 40, 42–59	12–31, 45–48	32–40, 42–44, 49–59
BL Advanced	32–56, (optional: 57–59)		

Formative Assessment

Use Exercises 1–11 to check for understanding.

Use the chart on the bottom of the page to customize assignments for your students.

CCSS **Teaching the Mathematical Practices**

Sense-Making Mathematically proficient students can explain correspondences between equations, verbal descriptions, tables, and graphs or draw diagrams of important features and relationships, graph data, and search for regularity or trends.

Additional Answers

13.

$D = \{x \mid x \le 2 \text{ or } x > 4\};$
$R = \{f(x) \mid f(x) < -7, \text{ or } f(x) = 5\}$

14.

$D = \{\text{all real numbers}\};$
$R = \{g(x) \mid g(x) < -10 \text{ or } -6 \le g(x) \le 2\}$

15.

$D = \{x \mid x < -4, -1 \le x \le 5, \text{ or } x > 7\};$
$R = \{g(x) \mid g(x) \ge -4\}$

Study Guide and Intervention
AL OL ELL

NAME _____ DATE _____ PERIOD _____

2-6 Study Guide and Intervention
Special Functions

Piecewise-Defined Functions A piecewise-defined function is written using two or more expressions. Its graph is often disjointed.

Example Graph $f(x) = \begin{cases} 2x & \text{if } x < 2 \\ x - 1 & \text{if } x \geq 2 \end{cases}$

First, graph the linear function $f(x) = 2x$ for $x < 2$. Since 2 does not satisfy this inequality, stop with a circle at (2, 4). Next, graph the linear function $f(x) = x - 1$ for $x \geq 2$. Since 2 does satisfy this inequality, begin with a dot at (2, 1).

Exercises

Graph each function. Identify the domain and range.

1. $f(x) = \begin{cases} x + 2 & \text{if } x < 0 \\ 2x + 5 & \text{if } 0 \leq x \leq 2 \\ -x + 1 & \text{if } x > 2 \end{cases}$

2. $f(x) = \begin{cases} -x - 4 & \text{if } x < -7 \\ 5x - 1 & \text{if } -7 \leq x \leq 0 \\ 2x + 1 & \text{if } x > 0 \end{cases}$

3. $h(x) = \begin{cases} \frac{x}{2} & \text{if } x \leq 0 \\ 2x - 6 & \text{if } 0 < x < 2 \\ 1 & \text{if } x \geq 2 \end{cases}$

Practice
AL OL BL ELL

NAME _____ DATE _____ PERIOD _____

2-6 Practice
Special Functions

Graph each function. Identify the domain and range.

1. $f(x) = \begin{cases} x + 2 & \text{if } x \leq -2 \\ 3x & \text{if } x > -2 \end{cases}$

2. $h(x) = \begin{cases} 4 - x & \text{if } x > 0 \\ -2x - 2 & \text{if } x < 0 \end{cases}$

3. $f(x) = [[0.5x]]$

4. $f(x) = |x| - 2$

5. $g(x) = -2|x|$

6. $f(x) = |x + 1|$

7. **BUSINESS** A Stitch in Time charges $40 per hour or any fraction thereof for labor. Draw a graph of the step function that represents this situation.

8. **BUSINESS** A wholesaler charges a store $3.00 per pound for less than 20 pounds of candy and $2.50 per pound for 20 or more pounds. Draw a graph of the function that represents this situation.

Word Problem Practice
AL OL BL ELL

NAME _____ DATE _____ PERIOD _____

2-6 Word Problem Practice
Special Functions

1. **SAVINGS** Nathan puts $200 into a checking account as soon as he gets his paycheck. The value of his checking account is modeled by the formula $200[[m]]$, where m is the number of months that Nathan has been working. After 105 days, how much money is in the account?

2. **FINANCE** A financial advisor handles the transactions for a customer. The median annual earnings for financial advisors is around $60,000. For every transaction, a certain financial advisor gets a 5% commission, regardless of whether the transaction is a deposit or withdrawal. Write a formula using the absolute value function for the advisor's commission. Let D represent the value of one transaction.

3. **ROUNDING** A science teacher instructs students to round their measurements as follows: If a number is less than 0.5 of a millimeter, students are instructed to round down. If a number is exactly 0.5 or greater, students are told to round up to the next millimeter. Write a formula that takes a measurement x millimeters and yields the rounded off number.

4. **ARCHITECTURE** The cross-section of a roof is shown in the figure. Write an absolute value function that models the shape of the roof.

5. **GAMES** Some young people are playing a game where a wooden plank is used as a target. It is marked off into 6 equal parts. A value is written in each section to represent the score earned if the dart lands in that section. Let x denote the horizontal position of a dart on the board, where the center of the board is the origin. Negative values correspond to the left half of the dart board, and positive values correspond to the right half. A player's score depends on the distance of the dart from the origin.

 | 1 | 2 | 3 | 2 | 1 |

a. Write a formula that gives the horizontal distance from the center of the dartboard.

b. Write a formula using the greatest integer function that can be used to find the person's score.

32. **RECREATION** The charge for renting a bicycle from a rental shop for different amounts of time is shown at the right.

Time	Price
$\frac{1}{2}$ hour	$6
1 hour	$10
2 hours	$16
Daily	$24

a. Identify the type of function that models this situation. **step function**

b. Write and graph a function for the situation. **See Chapter 2 Answer Appendix.**

Use each graph to write the absolute value function.

33. $f(x) = |0.5x|$

34. 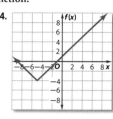 $f(x) = |x + 5| - 4$

C Graph each function. Identify the domain and range.

35. $f(x) = [[|0.5x|]]$

36. $g(x) = |[[2x]]|$

35–38. See Chapter 2 Answer Appendix.

37. $g(x) = \begin{cases} [[x]] & \text{if } x < -4 \\ x + 1 & \text{if } -4 \leq x \leq 5 \\ -|x| & \text{if } x > 3 \end{cases}$

38. $h(x) = \begin{cases} -|x| & \text{if } x < -6 \\ |x| & \text{if } -6 \leq x \leq 2 \\ |-x| & \text{if } x > 2 \end{cases}$

39. **MULTIPLE REPRESENTATIONS** Consider the following absolute value functions.

$$f(x) = |x| - 4 \qquad g(x) = |3x|$$

39a–c. See Chapter 2 Answer Appendix.

a. **Tabular** Use a graphing calculator to create a table of $f(x)$ and $g(x)$ values for $x = -4$ to $x = 4$.

b. **Graphical** Graph the functions on separate graphs.

c. **Numerical** Determine the slope between each two consecutive points in the table.

d. **Verbal** Describe how the slopes of the two sections of an absolute value graph are related. **The two sections of an absolute value graph have opposite slopes. The slope is constant for each section of the graph.**

H.O.T. Problems Use Higher-Order Thinking Skills

40. **OPEN ENDED** Write an absolute value relation in which the domain is all nonnegative numbers and the range is all real numbers. **Sample answer:** $|y| = x$

41. **CHALLENGE** Graph $|y| = 2|x + 3| - 5$. **See margin.**

42. **CCSS ARGUMENTS** Find a counterexample to the statement and explain your reasoning. **See margin.**

In order to find the greatest integer function of x when x is not an integer, round x to the nearest integer.

43. **OPEN ENDED** Write an absolute value function in which $f(5) = -3$. **Sample answer:** $f(x) = -|x - 2|$

44. **WRITING IN MATH** Explain how piecewise functions can be used to accurately represent real-world problems. **Sample answer: Piecewise functions can be used to represent the cost of items when purchased in quantities, such as a dozen eggs.**

106 | Lesson 2-6 | Special Functions

Ben Blankenburg/Corbis

Enrichment
OL BL

NAME _____ DATE _____ PERIOD _____

2-6 Enrichment

Graphing Greatest Integer Functions

Some equations involving the greatest integer function produce interesting graphs. It will be helpful to make a chart of values for each function and to use a colored pen or pencil.

Graph each function.

1. $y = 2x - [[x]]$

2. $y = \frac{[[x]]}{[[x]]}$

Multiple Representations

In Exercise 39, students use a table of values and a graphing calculator to compare the slopes of the two sections of an absolute-value function.

45. SHORT RESPONSE What expression gives the *n*th term of the linear pattern defined by the table? $3n + 1$

2	4	6	8	*n*
7	13	19	25	?

46. Solve: $5(x + 4) = x + 4$

Step 1: $5x + 20 = x + 4$

Step 2: $4x + 20 = 4$

Step 3: $\qquad 4x = 24$

Step 4: $\qquad x = 6$

Which is the first *incorrect* step in the solution shown above? **B**

A Step 4 C Step 2

B Step 3 D Step 1

47. NUMBER THEORY Twelve consecutive integers are arranged in order from least to greatest. If the sum of the first six integers is 381, what is the sum of the last six integers? **J**

F 345

G 381

H 387

J 417

48. SAT/ACT For which function does $f\left(-\frac{1}{2}\right) \neq -1$? **B**

A $f(x) = 2x$ D $f(x) = [\![2x]\!]$

B $f(x) = |-2x|$ E $f(x) = -|2x|$

C $f(x) = [\![x]\!]$

Spiral Review

49. FOOTBALL The table shows the relationship between the total number of male students per school and the number of students who tried out for the football team. (Lesson 2-5)

a. Find a regression equation for the data. $y = 0.10x + 30.34$

b. Determine the correlation coefficient. $r = 0.987$

c. Predict how many students will try out for football at a school with 800 male students. **about 110**

Number of Male Students	Number of Tryouts
180	46
212	51
274	62
401	75
513	81
589	90

Write an equation in slope-intercept form for the line described. (Lesson 2-4) **50. $y = 0.5x - 4.5$**

50. passes through $(-3, -6)$, perpendicular to $y = -2x + 1$

51. passes through $(4, 0)$, parallel to $3x + 2y = 6$ $\quad y = -\frac{3}{2}x + 6$

52. passes through the origin, perpendicular to $4x - 3y = 12$ $\quad y = -\frac{3}{4}x$

Find each value if $f(x) = -4x + 6$, $g(x) = -x^2$, and $h(x) = -2x^2 - 6x + 9$. (Lesson 2-1)

53. $f(2c)$ $-8c + 6$ **54.** $g(a + 1)$ $-a^2 - 2a - 1$ **55.** $h(6)$ -99

56. Determine whether the figures below are similar. (Lesson 0-7) **yes**

  33   26.4

12   9.6

Skills Review

Graph each equation. 57–59. See margin.

57. $y = -0.25x + 8$ **58.** $y = \frac{4}{3}x + 2$ **59.** $8x + 4y = 32$

CCSS **Teaching the Mathematical Practices**

Arguments Mathematically proficient students are able to analyze situations by breaking them into cases, and can recognize and use counterexamples. They justify their conclusions, communicate them to others.

59.

4 Assess

Crystal Ball In Lesson 2-7 students will learn about parent functions and transformations. Have students write about how they think today's lesson on special functions will help them with tomorrow's lesson.

Formative Assessment

Check for student understanding of concepts in Lessons 2-5 and 2-6.

Quiz 3, p. 60

Additional Answers

41.

42. Sample answer: 8.6; The greatest integer function asks for the greatest integer less than or equal to the given value; thus 8 is the greatest integer. If we were to round this value to the nearest integer, we would round up to 9.

57.

58.

1 Focus

Objective Use a graphing calculator to determine how changing the parameters m and b affects the graphs of the functions.

Materials for Each Student

- TI-83/84 Plus or other graphing calculator

Teaching Tip

Equations are entered as functions by first pressing the **Y=** button. For this reason, it is a good idea to encourage students to refer to this screen as the "Function editor." This is particularly helpful when calling up equations from the **VARS**, or Variables menu, since they are accessed via the **Function . . .** option under the **Y-VARS** submenu.

Note that the "standard viewing window" called for in Activity 1 does not show equal scales on the x- and y-axes. For example, the line $y = x$ will not make a 45° angle with the x-axis unless the squared-off viewing window **ZSquare** is chosen.

2 Teach

Working in Cooperative Groups

Have students work in pairs so they can help each other correct keystroke errors. Have them complete Activities 1 and 2 and Exercise 1.

- Point out to students that they can turn the graph of a particular function on or off by pressing $\boxed{\text{ENTER}}$ when the cursor is on the "=" in the function editor.

- By pressing the $\boxed{\text{TRACE}}$ button, students can view the equation associated with each line on the graph screen. The up and down arrows can be used to switch among the different functions and their graphs. If the equations don't appear in $\boxed{\text{TRACE}}$ mode, check the **FORMAT** screen to see if ExprOn is selected.

EXPLORE 2-7 Graphing Technology Lab
Families of Lines

The parent function of the family of linear functions is $f(x) = x$. You can use a graphing calculator to investigate how changing the parameters m and b in $f(x) = mx + b$ affects the graphs as compared to the parent function.

CCSS Common Core State Standards
Content Standards
F.BF.3 Identify the effect on the graph of replacing $f(x)$ by $f(x) + k$, $kf(x)$, $f(kx)$, and $f(x + k)$ for specific values of k (both positive and negative); find the value of k given the graphs. Experiment with cases and illustrate an explanation of the effects on the graph using technology.

PT 📱

Activity 1 b in $f(x) = mx + b$

Graph $f(x) = x$, $f(x) = x + 3$, and $f(x) = x - 5$ in the standard viewing window.

Enter the equations in the **Y=** list as **Y1**, **Y2**, and **Y3**. Then graph the equations.

KEYSTROKES: $\boxed{\text{Y=}}$ $\boxed{\text{X,T,θ,}n}$ $\boxed{\text{ENTER}}$ $\boxed{\text{X,T,θ,}n}$ $\boxed{+}$ 3
$\boxed{\text{ENTER}}$ $\boxed{\text{X,T,θ,}n}$ $\boxed{-}$ 5 $\boxed{\text{ENTER}}$ $\boxed{\text{ZOOM}}$ 6

$[-10, 10]$ scl: 1 by $[-10, 10]$ scl: 1

1A. Compare and contrast the graphs. The graphs have the same slope, but different y-intercepts.

1B. How would you obtain the graphs of $f(x) = x + 3$ and $f(x) = x - 5$ from the graph of $f(x) = x$? You would move the graph 3 units up; you would move the graph 5 units down.

The parameter m in $f(x) = mx + b$ affects the graphs in a different way than b.

Activity 2 m in $f(x) = mx + b$

Graph $f(x) = x$, $f(x) = 3x$, and $f(x) = \frac{1}{2}x$ in the standard viewing window.

Enter the equations in the **Y=** list and graph.

2A. How do the graphs compare? They have the same y-intercept.

2B. Which graph is steepest? Which graph is the least steep? $f(x) = 3x$; $f(x) = \frac{1}{2}x$

2C. Graph $f(x) = -x$, $f(x) = -3x$, and $f(x) = -\frac{1}{2}x$ in the standard viewing window. How do these graphs compare?

$[-10, 10]$ scl: 1 by $[-10, 10]$ scl: 1

2C. These graphs have the same y-intercept, but pass through quadrants II and IV instead of I and III.

Analyze the Results

Graph each set of equations on the same screen. Describe the similarities or differences among the graphs. 1–3. See Chapter 2 Answer Appendix.

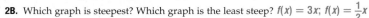

1. $f(x) = 3x$
$f(x) = 3x + 1$
$f(x) = 3x - 2$

2. $f(x) = x + 2$
$f(x) = 5x + 2$
$f(x) = \frac{1}{2}x + 2$

3. $f(x) = x - 3$
$f(x) = 2x - 3$
$f(x) = 0.75x - 3$

4. What do the graphs of equations of the form $f(x) = mx + b$ have in common? They are all lines.

5. What are the domain and range of functions of the form $f(x) = mx + b$, where $m \neq 0$? all real numbers

6. How do the values of b and m affect the graph of $f(x) = mx + b$ as compared to the parent function $f(x) = x$? The parameter b changes the y-intercept and the parameter m changes the slope.

7. Summarize your results. How can knowing about the effects of m and b help you sketch the graph of a function? Sample answer: Knowing b tells you how much to shift the parent graph vertically. The value of m tells you whether the graph is steeper or less steep than the parent graph.

108 | Explore 2-7 | Graphing Technology Lab: Families of Lines

---

Practice Have students complete Exercises 2–6.

3 Assess

Formative Assessment

Use Exercise 3 to assess students' ability to see similarities and differences in the graphs.

From Concrete to Abstract

Use Exercise 6 to bridge the algebraic idea of solving equations from the modeling activities.

Extending the Concept

In the general form of a linear function $f(x) = mx + b$, which constant would you change to move the graph up or down? b

LESSON 2-7 Parent Functions and Transformations

:: Then	:: Now	:: Why?
• You analyzed and used relations and functions.	**1** Identify and use parent functions. **2** Describe transformations of functions.	• Nick makes $8 an hour working at a pizza shop. The red line represents his wages. If he also delivers the pizzas, he is paid $2 more per hour. The blue line represents Nick's wages when he delivers the pizzas. These graphs are examples of transformations.

NewVocabulary
family of graphs
parent graph
parent function
constant function
identity function
quadratic function
translation
reflection
line of reflection
dilation

 Common Core State Standards

Content Standards
F.IF.4 For a function that models a relationship between two quantities, interpret key features of graphs and tables in terms of the quantities, and sketch graphs showing key features given a verbal description of the relationship.

F.BF.3 Identify the effect on the graph of replacing $f(x)$ by $f(x) + k$, $k\,f(x)$, $f(kx)$, and $f(x + k)$ for specific values of k (both positive and negative); find the value of k given the graphs. Experiment with cases and illustrate an explanation of the effects on the graph using technology.

Mathematical Practices
6 Attend to precision.

1 Parent Graphs A **family of graphs** is a group of graphs that display one or more similar characteristics. The **parent graph**, which is the graph of the **parent function**, is the simplest of the graphs in a family. This is the graph that is transformed to create other members in a family of graphs.

KeyConcept Parent Functions

Constant Function

(graph: $y = 1$)

The general equation of a **constant function** is $f(x) = a$, where a is any number. The domain is all real numbers, and the range consists of a single real number a.

Identity Function

(graph: $y = x$)

The **identity function** $f(x) = x$ passes through all points with coordinates (a, a). It is the parent function of most linear functions. Its domain and range are all real numbers.

Absolute Value Function

 (graph: $y = |x|$)

Recall that the parent function of absolute value functions is $f(x) = |x|$. The domain of $f(x) = |x|$ is the set of real numbers, and the range is the set of real numbers greater than or equal to 0.

Quadratic Function

 (graph: $y = x^2$)

The parent function of **quadratic functions** is $f(x) = x^2$. The domain of $f(x) = x^2$ is the set of real numbers, and the range is the set of real numbers greater than or equal to 0.

connectED.mcgraw-hill.com **109**

1 Focus

VerticalAlignment

Before Lesson 2-7 Analyze and use relations and functions.

Lesson 2-7 Identify and use parent functions and describe transformations of functions.

After Lesson 2-7 Analyze a situation modeled by a rational function, formulate an equation or inequality, and solve the problem.

2 Teach

Scaffolding Questions
Have students read the **Why?** section of the lesson.

Ask:
- If Nick works three hours at the pizza shop but does not deliver pizzas, how much will he make? $24

- How much per hour does Nick make working at the pizza shop and delivering pizzas? $10 an hour

- What are the slopes of the two lines? 8 and 10

Lesson 2-7 Resources

Resource	Approaching Level **AL**	On Level **OL**	Beyond Level **BL**	English Learners **ELL**
Teacher Edition	• Differentiated Instruction, p. 111	• Differentiated Instruction, p. 111	• Differentiated Instruction, pp. 116	• Differentiated Instruction, p. 111
Chapter Resource Masters	• Study Guide and Intervention, pp. 43–44 • Skills Practice, p. 45 • Practice, p. 46 • Word Problem Practice, p. 47	• Study Guide and Intervention, pp. 43–44 • Skills Practice, p. 45 • Practice, p. 46 • Word Problem Practice, p. 47 • Enrichment, p. 48 • Graphing Calculator Activity, p. 49	• Practice, p. 46 • Word Problem Practice, p. 47 • Enrichment, p. 48	• Study Guide and Intervention, pp. 43–44 • Skills Practice, p. 45 • Practice, p. 46 • Word Problem Practice, p. 47
Other	• 5-Minute Check 2-7 • Study Notebook	• 5-Minute Check 2-7 • Study Notebook	• 5-Minute Check 2-7 • Study Notebook	• 5-Minute Check 2-7 • Study Notebook

1 Parent Graphs

Example 1 shows how to identify a function when its graph is given.

Formative Assessment

Use the Guided Practice exercises after each example to determine students' understanding of concepts.

Additional Example

1 Identify the type of function represented by each graph.

a.

absolute value function

b.

quadratic function

Additional Examples also in Interactive Classroom PowerPoint® Presentations

IWB Interactive White Board READY

2 Transformations

Examples 2–4 show how to describe and graph translations, reflections, and dilations. **Example 5** shows how to identify transformations.

Additional Example

2 Describe the translation in $y = (x + 1)^2$. Then graph the function. Translates the graph $y = x^2$ left 1 unit.

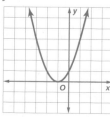

Example 1 Identify a Function Given the Graph

Identify the type of function represented by each graph.

a.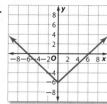

The graph is in the shape of a V. The graph represents an absolute value function.

b.

The graph is a horizontal line that crosses the *y*-axis at 4. The graph represents a constant function.

GuidedPractice

1A. quadratic

1B. 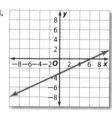 linear

---

2 Transformations

Transformations of a parent graph may appear in a different location, flip over an axis, or appear to have been stretched or compressed. The transformed graph may resemble the parent graph, or it may not.

ReadingMath

translation A translation is also called a *slide*, a *shift*, or a *glide*.

A **translation** moves a figure up, down, left, or right.

• When a constant *k* is added to or subtracted from a parent function, the result $f(x) \pm k$ is a translation of the graph up or down.

• When a constant *h* is added to or subtracted from *x* before evaluating a parent function, the result, $f(x \pm h)$, is a translation left or right.

Example 2 Describe and Graph Translations

Describe the translation in $y = |x| + 2$. Then graph the function.

The graph of $y = |x| + 2$ is a translation of the graph of $y = |x|$ up 2 units.

GuidedPractice   2A, 2B. See margin.

Describe the translation in each function. Then graph the function.

2A. $y = |x + 3|$ **2B.** $y = x^2 - 4$

Additional Answers (Guided Practice)

2A. The graph of $y = |x + 3|$ is a translation of the graph of $y = |x|$ left 3 units.

2B. The graph of $y = x^2 - 4$ is a translation of the graph of $y = x^2$ down 4 units.

A **reflection** flips a figure over a line called the **line of reflection**.

- When a parent function is multiplied by −1, the result −$f(x)$ is a reflection of the graph in the x-axis.

- When only the variable is multiplied by −1, the result $f(-x)$ is a reflection of the graph in the y-axis.

Example 3 Describe and Graph Reflections

Describe the reflection in $y = -x^2$. Then graph the function.

The graph of $y = -x^2$ is a reflection of the graph of $y = x^2$ in the x-axis.

▶ **Guided**Practice **3A, 3B. See margin.**

Describe the reflection in each function. Then graph the function.

3A. $y = -|x|$ **3B.** $y = -x$

A **dilation** shrinks or enlarges a figure proportionally. When the variable in a linear parent function is multiplied by a nonzero number, the slope of the graph changes.

- When a nonlinear parent function is multiplied by a nonzero number, the function is stretched or compressed vertically.

- Coefficients greater than 1 cause the graph to be stretched vertically, and coefficients between 0 and 1 cause the graph to be compressed vertically.

Example 4 Describe and Graph Dilations

Describe the dilation in $y = 4x$. Then graph the function.

The graph of $y = 4x$ is a dilation of the graph of $y = x$. The slope of the graph of $y = 4x$ is steeper than that of the graph of $y = x$.

▶ **Guided**Practice **4A, 4B. See Chapter 2 Answer Appendix.**

Describe the dilation in each function. Then graph the function.

4A. $y = 2x^2$ **4B.** $y = \left|\frac{1}{3}x\right|$

DifferentiatedInstruction AL OL ELL

If ▶ students have access to poster-making supplies,

Then ▶ ask them to create posters that display the four parent functions discussed in this lesson as well as the transformations. Display the posters in the room.

Additional Examples

3 Describe the reflection in $y = -|x|$. Then graph the function. It reflects the graph across the x-axis.

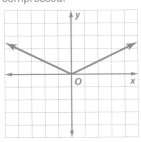

4 Describe the dilation on $y = \frac{1}{2}|x|$. Then graph the function. The graph is vertically compressed.

Additional Answers (Guided Practice)

3A. The graph of $y = -|x|$ is a reflection of the graph of $y = |x|$ across the x-axis.

3B. The graph of $y = -x$ is a reflection of the graph of $y = x$ across the x-axis, described by $-f(x)$, or it is a reflection of the graph of $y = x$ across the y-axis, described by $f(-x)$.

Ariel Skelley/Photographer's Choice/Getty Images

Additional Example

5 **ARCHWAYS** The function
$f(x) = -\frac{1}{2}(x - 5)^2 + 12.5$ can
be used to represent a parabolic
archway. Describe the
transformations in the function.
Then graph the function.
−5 translates $f(x) = x^2$ right
5 units. +12.5 translates
$f(x) = x^2$ up 12.5 units. $-\frac{1}{2}$
reflects $f(x) = x^2$ across the
x-axis and compresses the
graph vertically.

LANDSCAPING Ethan is going to add a brick
walkway around the perimeter of his
vegetable garden. The area of the walkway
can be represented by the function
$f(x) = 4(x + 2.5)^2 - 25$. Describe the
transformations in the function. Then
graph the function.

The graph of $f(x) = 4(x + 2.5)^2 - 25$ is a combination of transformations of the parent
graph $f(x) = x^2$. Determine how each transformation affects the parent graph.

$f(x) = 4(x + 2.5)^2 - 25$

+ 2.5 translates $f(x) = x^2$ left 2.5 units.

− 25 translates $f(x) = x^2$ down 25 units.

4 stretches $f(x) = x^2$ vertically.

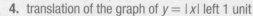

▶ **Guided**Practice

5. **SCIENCE** The function $C(x) = \frac{5}{9}(x - 32)$ can be used to determine the
temperature in degrees Celsius when given the temperature in degrees
Fahrenheit. Describe the transformations in the function. Then graph
the function. **See margin.**

Real-World Link
The average American
homeowner spends over
$350 each year on indoor
and outdoor plants.
Source: American Nursery &
Landscape Association

CCSS **Teaching the**
Mathematical Practices

Regularity Mathematically proficient
students notice if calculations are
repeated, and look both for general
methods and for shortcuts. As they
work to solve a problem,
mathematically proficient students
maintain oversight of the process, while
attending to the details. Encourage
students to identify the various values
involved in transformations as they
begin to analyze functions.

Additional Answer (Guided Practice)

5. The graph of $C(x) = \frac{5}{9}(x - 32)$ is
a translation of the parent graph
$C(x) = x$ right 32 units. The dilation
represents a slope that is less than
that of the parent graph.

StudyTip

CCSS Regularity
Ask yourself these questions
to help you identify
transformations.
1. What type of function
   is it?
2. Does the graph open
   up or down?
3. Does the vertex lie on
   an axis?

The table summarizes the changes to the parent graph under different transformations.

ConceptSummary Transformations of Functions	
Transformation	**Change to Parent Graph**
Translation	
$f(x + h), h > 0$	Translates graph h units left.
$f(x - h), h > 0$	Translates graph h units right.
$f(x) + k, k > 0$	Translates graph k units up.
$f(x) - k, k > 0$	Translates graph k units down.
Reflection	
$-f(x)$	Reflects graph in the x-axis.
$f(-x)$	Reflects graph in the y-axis.
Dilation	
$a \cdot f(x), \lvert a \rvert > 1$	Stretches graph vertically.
$a \cdot f(x), 0 < \lvert a \rvert < 1$	Compresses graph vertically
$f(bx), \lvert b \rvert > 1$	Compresses graph horizontally.
$f(bx), 0 < \lvert b \rvert < 1$	Stretches graph horizontally.

Additional Answers

3. translation of the graph of $y = x^2$ down 4 units

4. translation of the graph of $y = \lvert x \rvert$ left 1 unit

Example 1   Identify the type of function represented by each graph.

1. linear

2. absolute value

Example 2   **CCSS** SENSE-MAKING Describe the translation in each function. Then graph the function.

3, 4. See margin.

3. $y = x^2 - 4$ **4.** $y = |x + 1|$

Example 3   Describe the reflection in each function. Then graph the function. **5, 6.** See margin.

5. $y = -|x|$ **6.** $y = (-x)^2$

Example 4   Describe the dilation in each function. Then graph the function. **7, 8.** See margin.

7. $y = \frac{3}{5}x$ **8.** $y = 3x^2$

Example 5   **9. FOOD** The manager of a coffee shop is randomly checking coffee drinks prepared by employees to ensure that the correct amount of coffee is in each cup. Each 12-ounce drink should contain half coffee and half steamed milk. The amount of coffee by which each drink varies can be represented by $f(x) = \frac{1}{2}|x - 12|$. Describe the transformations in the function. Then graph the function. **See Chapter 2 Answer Appendix.**

Example 1   Identify the type of function represented by each graph.

10. constant

11 quadratic

12. absolute value

13. linear

Differentiated Homework Options

Level	Assignment	Two-Day Option	
AL Basic	10–32, 44–65	11–31 odd, 48–51	10–32 even, 44–47, 52–65
OL Core	11–31 odd, 32, 33–37 odd, 39–41, 44–65	10–32, 48–51	33–42, 44–47, 52–65
BL Advanced	33–62, (optional: 63–65)		

---

3 Practice

Formative Assessment

Use Exercises 1–9 to check for understanding.

Use the chart at the bottom of this page to customize assignments for your students.

Additional Answers

5. reflection of the graph of $y = |x|$ across the x-axis

6. reflection of the graph of $y = x^2$ across the y-axis

7. A vertical compression of the graph of $y = x$; the slope is not as steep as that of $y = x$.

8. The dilation stretches the graph of $y = x^2$ vertically.

Teaching the Mathematical Practices

Sense-Making Mathematically proficient students can explain correspondences between equations, verbal descriptions, tables, and graphs or draw diagrams of important features and relationship.

Additional Answers

14. translation of the graph of $y = x^2$ up 4 units

15. translation of the graph of $y = |x|$ down 3 units

16. translation of the graph of $y = x$ down 1 unit or right 1 unit

17. translation of the graph of $y = x$ up 2 units or left 2 units

18. translation of the graph of $y = x^2$ right 5 units

19. translation of the graph of $y = |x|$ left 6 units

Example 2   Describe the translation in each function. Then graph the function.   14–19. See margin.

14. $y = x^2 + 4$ **15.** $y = |x| - 3$ **16.** $y = x - 1$

17. $y = x + 2$ **18.** $y = (x - 5)^2$ **19.** $y = |x + 6|$

20–25. See Chapter 2 Answer Appendix.

Example 3   Describe the reflection in each function. Then graph the function.

20. $y = -x$ **21.** $y = -x^2$ **22.** $y = (-x)^2$

23. $y = |-x|$ **24.** $y = -|x|$ **25.** $y = (-x)$

26–31. See Chapter 2 Answer Appendix.

Example 4   Describe the dilation in each function. Then graph the function.

26. $y = (3x)^2$ **27.** $y = 6x$ **28.** $y = 4|x|$

29. $y = |2x|$ **30.** $y = \frac{2}{3}x$ **31.** $y = \frac{1}{2}x^2$

Example 5   **32.** **CCSS** SENSE-MAKING A non-impact workout can burn up to 7.5 Calories per minute. The equation to represent how many Calories a person burns after m minutes of the workout is $C(m) = 7.5m$. Identify the transformation in the function. Then graph the function. See Chapter 2 Answer Appendix.

B   Write an equation for each function.

33. $y = x^2 + 1$ **34.** $y = \frac{3}{16}x^2$

35. 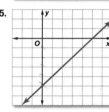 $y = x - 5$ **36.** $y = |x + 3|$

37. $y = (x - 2)^2$ **38.** $y = |x| - 4$

39 **BUSINESS** The graph of the cost of producing x widgets is represented by the blue line in the graph. After hiring a consultant, the cost of producing x widgets is represented by the red line in the graph. Write the equations of both lines and describe the transformation from the blue line to the red line.

Blue: $y = x + 4$; red: $y = x + 2$; the red line is a translation of the blue line 2 units down.

40. ROCKETRY Kenji launched a toy rocket from ground level. The height $h(t)$ of Kenji's rocket after t seconds is shown in blue. Emily believed that her rocket could fly higher and longer than Kenji's. The flight of Emily's rocket is shown in red.

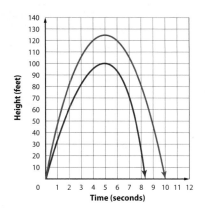

40c. about 25 ft higher

40d. a dilation in which the red graph is an expansion of the blue graph

a. Identify the type of function shown. quadratic

b. How much longer than Kenji's rocket did Emily's rocket stay in the air? about 1.5 seconds

c. How much higher than Kenji's rocket did Emily's rocket go?

d. Describe the type of transformation between the two graphs.

43. Sample answer: Since a vertical translation concerns only y-values and a horizontal translation concerns only x-values, order is irrelevant.

Write an equation for each function.

$y = (x + 4)^2 - 6$

42.

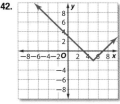

$y = |x - 5| - 2$

H.O.T. Problems Use Higher-Order Thinking Skills

43. CHALLENGE Explain why performing a horizontal translation followed by a vertical translation ends up being the same transformation as performing a vertical translation followed by a horizontal translation.

44. Kimi; sample answer: Each point in the table corresponds to a point on the graph. Therefore they represent the same relation.

44. CCSS CRITIQUE Kimi thinks that the graph and table below are representations of the same linear relation. Carla disagrees. Who is correct? Explain your reasoning.

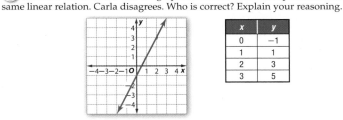

x	y
0	−1
1	1
2	3
3	5

46. Sample answer: The graph of $y = x^2$ is positive at its rightmost points and leftmost points.

45. OPEN ENDED Draw a figure in Quadrant II. Use any of the transformations you learned in this lesson to move your figure to Quadrant IV. Describe your transformation.
See Chapter 2 Answer Appendix.

46. REASONING Study the parent graphs at the beginning of this lesson. Select a parent graph with positive y-values at its leftmost points and positive y-values at its rightmost points.

47. WRITING IN MATH Explain why the reflection of the graph of $f(x) = x^2$ in the y-axis is the same as the graph of $f(x) = x^2$. Is this true for all reflections of quadratic equations? If not, describe a case when it is false. See margin.

Additional Answer

47. Sample answer: It is not always true. When the axis of symmetry of the parabola is not along the y-axis, the graphs of the preimage and image will be different.

Study Guide and Intervention
AL OL ELL

Practice
AL OL BL ELL

Word Problem Practice
AL OL BL ELL

Enrichment
OL BL

Ticket Out the Door Have students write the names of the parent functions on separate slips of paper. Then have them describe the characteristics of one function's graph.

Teach with Tech

Student Response System Show students a slide show of graphs of different types of functions, and ask them to identify what type of function is shown. Give students a key to use for their responses.

Additional Answers

52.

D = {all real numbers}
R = {f(x) | f(x) ≥ 0}

53.

D = {all real numbers}
R = {all integers}

54.

D = {all real numbers}
R = {f(x) | −2 < f(x) ≤ 1 or f(x) ≥ 4}

Standardized Test Practice

48. What is the solution set of the inequality? **D**

$$6 - |x + 7| \le -2$$

A $\{x | -15 \le x \le 1\}$

B $\{x | x \le -1 \text{ or } x \ge 3\}$

C $\{x | -1 \le x \le 3\}$

D $\{x | x \le -15 \text{ or } x \ge 1\}$

49. GEOMETRY The measures of two angles of a triangle are x and $4x$. Which of these expressions represents the measure of the third angle? **G**

F $180 + x + 4x$

G $180 - x - 4x$

H $180 - x + 4x$

J $180 + x - 4x$

50. GRIDDED RESPONSE Find the value of x that makes $\frac{1}{2} = \frac{x-2}{x+2}$ true. **6**

51. SAT/ACT Which could be the equation for the graph? **A**

A $y = 3x + 2$

B $y = 3x - 2$

C $y = -3x + 2$

D $y = -\frac{1}{3}x + 2$

E $y = \frac{1}{3}x + 2$

Spiral Review

Graph each function. Identify the domain and range. (Lesson 2-6) **52–54. See margin.**

52. $f(x) = |x - 3|$

53. $h(x) = [\![x]\!] - 5$

54. $f(x) = \begin{cases} -2x & \text{if } x \le -2 \\ x & \text{if } -2 < x \le 1 \\ 4 & \text{if } x > 1 \end{cases}$

55. ATTENDANCE The table shows the annual attendance to West High School's Summer Celebration. (Lesson 2-5)

Year	Attendance
2008	61
2009	83
2010	85
2011	92
2012	97
2013	106

a. Find a regression equation for the data. **$y = 7.83x - 15{,}620.70$**

b. Determine the correlation coefficient. **$r = 0.953$**

c. Predict how many people will attend the Summer Celebration in 2014. **about 118 people**

Solve each inequality. (Lesson 1-6)

56. $-12 \le 2x + 4 \le 8$ **$-8 \le x \le 2$**

57. $-4 < -3y + 2 < 11$ **$2 > y > -3$**

58. $|x - 3| > 7$ **$x < -4 \text{ or } x > 10$**

59. CARS Loren is buying her first car. She is considering 4 different models and 5 different colors. How many different cars could she buy? (Lesson 0-4) **20**

Determine if each relation is a function. (Lesson 0-1)

60. **no**

61. **yes**

62. **yes**

Skills Review

Evaluate each expression if $x = -4$ and $y = 6$.

63. $4x - 8y + 12$ **-52**

64. $5y + 3x - 8$ **10**

65. $-12x + 10y - 24$ **84**

 116 | Lesson 2-7 | Parent Functions and Transformations

DifferentiatedInstruction BL

Extension Ask students to describe how different parent functions could be combined to create more complex functions and to visualize what their graphs might look like. For instance, an absolute value function added to a constant function would exhibit a graph of the absolute value function shifted up or down.

LESSON 2-8 Graphing Linear and Absolute Value Inequalities

Large Pizza
$11.00

Soft Drinks
$2.25

::Then ::Now ::Why?

::Then	::Now	::Why?
● You described transformations of functions.	**1** Graph linear inequalities. **2** Graph absolute value inequalities.	● Randy is planning to treat his lacrosse team to a pizza party after the championship game, but he does not want to spend more than $200. Randy can use the inequality $11p + 2.25d \leq 200$, where p represents the number of pizzas and d represents the number of soft drinks, to check whether certain combinations of pizzas and drinks will fall within his budget.

 NewVocabulary
linear inequality
boundary

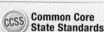 **Common Core State Standards**

Content Standards
A.CED.3 Represent constraints by equations or inequalities, and by systems of equations and/or inequalities, and interpret solutions as viable or nonviable options in a modeling context.

Mathematical Practices
1 Make sense of problems and persevere in solving them.

1 Graph Linear Inequalities A **linear inequality** resembles a linear equation, but with an inequality symbol instead of an equals symbol. For example, $y > -3x - 2$ is a linear inequality and $y = -3x - 2$ is the related linear equation.

The graph of the inequality $y > -3x - 2$ is shown at the right as a shaded region. Every point in the shaded region satisfies the inequality. The graph of $y = -3x - 2$ is the **boundary** of the region. It is drawn as a dashed line to show that points on the line do not satisfy the inequality. If the symbol was $\leq$ or $\geq$, then points on the boundary would satisfy the inequality, so the boundary would be drawn as a solid line.

Example 1 Dashed Boundary

Graph $x + 4y > 2$.

Step 1 The boundary of the graph is the graph of $x + 4y = 2$. Since the inequality symbol is $>$, the boundary will be dashed.

Step 2 Test the point $(0, 0)$ because it is not on the boundary.

$x + 4y > 2$	Original inequality
$0 + 4(0) \overset{?}{>} 2$	$(x, y) = (0, 0)$
$0 > 2$ ✗	False

The region that does *not* contain $(0, 0)$ is shaded.

CHECK The graph indicates that $(0, 3)$ is a solution.

$x + 4y > 2$	Original inequality
$0 + 4(3) \overset{?}{>} 2$	$(x, y) = (0, 3)$
$12 > 2$ ✓	True

The solution checks.

▶ **Guided**Practice **1A, 1B.** See Chapter 2 Answer Appendix.

1A. Graph $3x + \frac{1}{2}y < 2$.   **1B.** Graph $-x + 2y > 4$.

 connectED.mcgraw-hill.com **117**

1 Focus

VerticalAlignment

▼

Before Lesson 2-8 Describe transformations of functions.

▼

Lesson 2-8 Graph linear and absolute value inequalities.

▼

After Lesson 2-8 Formulate systems of inequalities.

2 Teach

Scaffolding Questions

Have students read the **Why?** section of the lesson.

Ask:

● Explain what the inequality $11p + 2.25d \leq 200$ means. The number of pizzas times $11 each plus the number of drinks times $2.25 each must be less than or equal to $200.

● Would Randy have enough money to buy 10 large pizzas and 20 drinks? yes

● If Randy bought 15 large pizzas, what is the maximum number of drinks he could buy without spending more than $200? 15 drinks

Lesson 2-8 Resources

Resource	Approaching Level **AL**	On Level **OL**	Beyond Level **BL**	English Learners **ELL**
Teacher Edition	▪ Differentiated Instruction, p. 121	▪ Differentiated Instruction, p. 121		▪ Differentiated Instruction, p. 121
Chapter Resource Masters	▪ Study Guide and Intervention, pp. 50–51 ▪ Skills Practice, p. 52 ▪ Practice, p. 53 ▪ Word Problem Practice, p. 54	▪ Study Guide and Intervention, pp. 50–51 ▪ Skills Practice, p. 52 ▪ Practice, p. 53 ▪ Word Problem Practice, p. 54 ▪ Enrichment, p. 55	▪ Practice, p. 53 ▪ Word Problem Practice, p. 54 ▪ Enrichment, p. 55	▪ Study Guide and Intervention, pp. 50–51 ▪ Skills Practice, p. 52 ▪ Practice, p. 53 ▪ Word Problem Practice, p. 54
Other	▪ 5-Minute Check 2-8 ▪ Study Notebook	▪ 5-Minute Check 2-8 ▪ Study Notebook	▪ 5-Minute Check 2-8 ▪ Study Notebook	▪ 5-Minute Check 2-8 ▪ Study Notebook

1 Graph Linear Inequalities

Example 1 shows how to graph a linear inequality. In **Example 2,** a linear inequality and its graph are used to model and analyze a real-world situation.

Formative Assessment

Use the Guided Practice exercises after each example to determine students' understanding of concepts.

Additional Examples

1 Graph $x - 2y < 4$.

$x - 2y = 4$

2 **EDUCATION** One tutoring company advertises that it specializes in helping students who have a combined SAT verbal and math score of 900 or less.

a. Write an inequality to describe the combined scores of students who are prospective tutoring clients. Let x represent the verbal score and y the math score. $x + y \leq 900$

b. Does a student with a verbal score of 480 and a math score of 410 fit the tutoring company's guidelines? **yes**

Additional Examples also in Interactive Classroom PowerPoint® Presentations

 IWB **Interactive White Board** READY

 The Math Forum

mathforum.org
The Math Forum
© Drexel University

Have your students refer to the Math Forum's Ask Dr. Math to learn more about absolute value inequalities. They can start their search at connectED.mcgraw-hill.com.

Real-WorldCareer

Recreation Workers
Recreation workers plan, organize, and manage recreational activities to meet the needs of a variety of people. Educational requirements range from a high school diploma to a graduate degree.

Teach with Tech

Portable Media Player Create a video showing your students how to graph absolute value inequalities. Load the video onto your students' portable media players for use as an additional reference outside of class.

RECREATION A recreation center offers various 30-minute and 60-minute art classes. The recreation director has allotted up to 20 hours per week for art classes.

a. Write an inequality to represent the number of classes that can be offered per week. Graph the inequality.

Let x represent the number of 30-minute or $\frac{1}{2}$-hour art classes, and let y represent the number of 60-minute or 1-hour art classes. Because the sum can equal the maximum, the inequality symbol is $\leq$, and the boundary is solid. The inequality is $\frac{1}{2}x + y \leq 20$.

Step 1 Graph the boundary $\frac{1}{2}x + y = 20$.

Step 2 Test the point $(0, 0)$.

$$\frac{1}{2}x + y \leq 20 \qquad \text{Original inequality}$$
$$\frac{1}{2}(0) + (0) \overset{?}{\leq} 20 \qquad (x, y) = (0, 0)$$
$$0 \leq 20 \checkmark \qquad \text{True}$$

The region that contains $(0, 0)$ is shaded.

$\frac{1}{2}x + y = 20$

b. Can the recreation director schedule 25 of the 30-minute classes and 15 of the 60-minute classes during a given week? Explain your reasoning.

The point $(25, 15)$ lies outside the shaded region, so it does not satisfy the inequality. Thus, the recreation director cannot schedule 25 30-minute and 15 60-minute classes.

▶ **Guided**Practice **2. See Chapter 2 Answer Appendix.**

2. Manuel has $15 to spend at the county fair. The fair costs $5 for admission, $0.75 for each ride ticket, and $0.25 for each game ticket. Write an inequality, and draw a graph that represents the number of r ride and g game tickets that Manuel can buy.

2 Graph Absolute Value Inequalities

Graphing absolute value inequalities is similar to graphing linear inequalities. First you graph the absolute value equation. Then you determine whether the boundary is dashed or solid and which region should be shaded.

Example 3 Absolute Value Inequality

Graph $y \geq |x| - 4$.

Since the inequality symbol is $\geq$, the boundary is solid. Graph the equation. Then test $(0, 0)$.

$$y \geq |x| - 4 \qquad \text{Original inequality}$$
$$0 \overset{?}{\geq} |0| - 4 \qquad (x, y) = (0, 0)$$
$$0 \geq -4 \checkmark \qquad \text{True}$$

The region that includes $(0, 0)$ is shaded.

$y = |x| - 4$

▶ **Guided**Practice **3A, 3B. See Chapter 2 Answer Appendix.**

3A. Graph $y \leq 2|x| + 3$.

3B. Graph $y \geq 3|x + 1|$.

 118 | Lesson 2-8 | Graphing Linear and Absolute Value Inequalities

Pauline St. Denis/Fancy/Corbis

Additional Answer

1.

Example 1 Graph each inequality. **1–4. See margin.**

1. $y \leq 4$

2. $x \geq -6$

3. $x + 4y \leq 2$

4. $3x + y > -8$

Example 2 **5.** (CCSS) **MODELING** Gregg needs to buy gas and oil for his car. Gas costs \$3.45 a gallon, and oil costs \$2.41 a quart. He has \$50 to spend.

5c. No; (10, 8) is not in the shaded region.

   a. Write an inequality to represent the situation, where g is the number of gallons of gas he buys and q is the number of quarts of oil. $3.45g + 2.41q \leq 50$

   b. Graph the inequality. **See Chapter 2 Answer Appendix.**

   c. Can Gregg buy 10 gallons of gasoline and 8 quarts of oil? Explain.

Example 3 Graph each inequality. **6, 7. See Chapter 2 Answer Appendix.**

6. $y \geq |x + 3|$

7. $y - 6 < |x|$

Practice and Problem Solving   Extra Practice is on page R2.

Example 1 Graph each inequality. **8–13. See Chapter 2 Answer Appendix.**

8. $x + 2y > 6$

9. $y \geq -3x - 2$

10. $2y + 3 \leq 11$

11. $4x - 3y > 12$

12. $6x + 4y \leq -24$

13. $y \geq \frac{3}{4}x + 6$

Example 2 **14. COLLEGE** April's guidance counselor says that she needs a combined score of at least 1700 on her college entrance exams to be eligible for the college of her choice. The highest possible score is 2400. There are 1200 possible points on the math portion and 1200 on the verbal portion.

   a. The inequality $x + y \geq 1700$ represents this situation, where x is the verbal score and y is the math score. Graph this inequality. **See Chapter 2 Answer Appendix.**

   b. Refer to your graph. If she scores a 680 on the math portion of the test and 910 on the verbal portion of the test, will April be eligible for the college of her choice? no

Example 3 Graph each inequality. **15–20. See Chapter 2 Answer Appendix.**

15. $y > |3x|$

16. $y + 4 \leq |x - 2|$

17. $y - 6 < |-2x|$

18. $y + 8 < 2\left|\frac{2}{3}x + 6\right|$

19. $2y > |4x - 5|$

20. $-y \leq |3x - 4|$

21. SCHOOL DANCE Carlos estimates that he will need to earn at least \$700 to take his girlfriend to the prom. Carlos works two jobs as shown in the table.

21b. See Chapter 2 Answer Appendix.

Job	Pay
Main St. Deli	\$8 per hour
Babysitting	\$6 per hour

   a. Write an inequality to represent this situation. $8a + 6b \geq 700$

   b. Graph the inequality.

   c. Will he make enough money if he works 50 hours at each job? yes

B Graph each inequality. **22–26. See Chapter 2 Answer Appendix.**

22. $y \geq |-2x - 6|$

23. $y \leq |x - 3| + 4$

24. $y - 3 > -2|x + 4|$

25. $|y| > |x|$

26. $|x - y| > 5$

27. $|x + 3y| \geq -2$

27. ordered pairs of all real numbers (The graph would be shaded everywhere.)

Differentiated Homework Options

Level	Assignment	Two-Day Option	
AL Basic	8–21, 33, 35–57	9–21 odd, 38–41	8–20 even, 33, 35–37, 42–57
OL Core	9–27 odd, 28, 29, 31, 33, 35–57	8–21, 38–41	22–33, 35–37, 42–57
BL Advanced	22–54, (optional: 55–57)		

---

2 Graph Absolute Value Inequalities

Example 3 illustrates how to graph an absolute value inequality.

Additional Example

3 Graph $y \geq |x| - 2$.

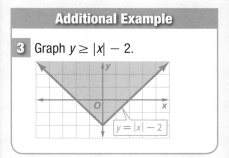

$y = |x| - 2$

---

3 Practice

Formative Assessment

Use Exercises 1–7 to check for understanding.

Use the chart on the bottom of this page to customize assignments for your students.

Additional Answers

2.

3.

4.

28. **CCSS MODELING** Mei is making necklaces and bracelets to sell at a craft show. She has enough beads to make 50 pieces. Let x represent the number of bracelets and y represent the number of necklaces.

a. Write an inequality that shows the possible number of necklaces and bracelets Mei can make. **$x + y \leq 50$**

b. Graph the inequality. **See Chapter 2 Answer Appendix.**

c. Give three possible solutions for the number of necklaces and bracelets that can be made. **Sample answers: 0 bracelets and 50 necklaces, 25 necklaces and 25 bracelets, or 30 bracelets and 20 necklaces**

29. GIFT CARDS Susan received a gift card from an electronics store for $400. She wants to spend the money on DVDs, which cost $20 each, and CDs, which cost $15 each.

a. Let d equal the number of DVDs, and let c equal the number of CDs. Write an inequality that shows the possible combinations of DVDs and CDs that Susan can purchase. **$20d + 15c \leq 400$**

b. Graph the inequality. **See Chapter 2 Answer Appendix.**

c. Give three possible solutions for the number of DVDs and CDs she can buy. **Sample answer: 18 CDs and 5 DVDs, 12 CDs and 10 DVDs, or 6 CDs and 15 DVDs**

▶ **C** Graph each inequality. **30–32. See Chapter 2 Answer Appendix.**

30. $y \geq [\![x]\!]$

31. $y < [\![x + 2]\!]$

32. $y \geq |[\![x]\!]|$

36. Sample answer: It will be possible when we have a situation where y and x are both inside an absolute value. An example is $|y| = |x|$. When this happens, positive and negative values of y will need to be considered, as well as the positive and negative values of x.

H.O.T. Problems   Use Higher-Order Thinking Skills

33. OPEN ENDED Create an absolute value inequality in which none of the possible solutions fall in the second or third quadrant. **Sample answer: $|y| < x$**

34. CHALLENGE Graph the following inequality. **See margin.**

$$g(x) > \begin{cases} |x + 1| \text{ if } x \leq -4 \\ -|x| \text{ if } -4 < x < 2 \\ |x - 4| \text{ if } x \geq 2 \end{cases}$$

35. ERROR ANALYSIS Paulo and Janette are graphing $x - y \geq 2$. Is either of them correct? Explain your reasoning. **Paulo; $x - y \geq 2$ can be written as $y \leq x - 2$.**

37. Sample answer: One possibility is when $|y| < 0$. In order for there to be a solution, the absolute value of y will need to be less than 0, and, by definition of absolute value, this is impossible.

Paulo

Janette

36. REASONING When will it be possible to shade two different areas when graphing a linear absolute value inequality? Explain your reasoning.

37. WRITING IN MATH Describe a situation in which there are no solutions to an absolute value inequality. Explain your reasoning.

Additional Answer

34.

Standardized Test Practice

38. EXTENDED RESPONSE Craig scored 85%, 96%, 79%, and 81% on his first four math tests. He hopes to score high enough on the final test to earn a 90% average. If the final test is worth twice as much as one of the other tests, determine if Craig can earn a 90% average. If so, what score does Craig need to get on the final test to accomplish this? Explain how you found your answer. **See margin.**

39. Which of the following sets of numbers represents an infinite set? **C**

A {2, 4, 6}

B {whole numbers between −50 and 50}

C {integers}

D $\left\{\frac{1}{2}, \frac{3}{4}, \frac{4}{5}, \frac{5}{6}\right\}$

40. SHORT RESPONSE Which theorem of congruence should be used to prove $\triangle ABC \cong \triangle XYZ$? **SAS**

41. SAT/ACT For which function is the range $\{f(x)\,|\,f(x) \leq 0\}$? **K**

F $f(x) = -x$ **J** $f(x) = |x|$

G $f(x) = [\![x]\!]$ **K** $f(x) = -|x|$

H $f(x) = [\![-x]\!]$

Spiral Review

Write an equation for each graph. (Lesson 2-7) **42.** $y = (x-3)^2 - 2$ **43.** $y = |x + 4| - 5$ **44.** $y = -|x - 2| + 4$

42.

43.

44.

Graph each function. (Lesson 2-6) **45–47. See Chapter 2 Answer Appendix.**

45. $f(x) = \begin{cases} x & \text{if } x < 1 \\ 3 & \text{if } 1 \leq x \leq 3 \\ -2x & \text{if } x > 3 \end{cases}$

46. $f(x) = \begin{cases} x + 3 & \text{if } x < -2 \\ 2x & \text{if } -2 \leq x \leq 2 \\ -3x & \text{if } x > 2 \end{cases}$

47. $f(x) = \begin{cases} -2x & \text{if } x \leq -2 \\ x + 1 & \text{if } 0 < x \leq 6 \\ x - 5 & \text{if } x > 6 \end{cases}$

Write each equation in standard form. Identify **A**, **B**, and **C**. (Lesson 2-2)

48. $-6y = 8x - 3$
$8x + 6y = 3$; $A = 8$, $B = 6$, $C = 3$

49. $12y + x = -3y + 5x - 6$
$4x - 15y = 6$; $A = 4$, $B = -15$, $C = 6$

50. $\frac{x+3}{4} + \frac{y-1}{2} = 3$
$x + 2y = 11$; $A = 1$, $B = 2$, $C = 11$

51. TENNIS Sixteen players signed up for tennis lessons. The instructor plans to use 50 tennis balls for every player and have 200 extra. How many tennis balls are needed for the lessons? (Lesson 1-3) **1000**

Multiply. (Lesson 0-2)

52. $(3x - 4)(2x + 1)$
$6x^2 - 5x - 4$

53. $(6x + 5)(-x - 3)$
$-6x^2 - 23x - 15$

54. $(5x + 2)(-2x + 3)$
$-10x^2 + 11x + 6$

Skills Review

Graph each linear equation. **55–57. See Chapter 2 Answer Appendix.**

55. $y = 2x - 8$

56. $y = -\frac{3}{4}x + 2$

57. $3y - 4x = 24$

DifferentiatedInstruction **AL OL ELL**

If students sometimes confuse equations and inequalities,

Then have them discuss the differences and similarities between solving an equation and solving an inequality.

Modeling Mathematically proficient students can apply the mathematics they know to solve problems arising in everyday life, analyze relationships mathematically to draw conclusions, and interpret their mathematical results in the context of a situation.

4 Assess

Name the Math Have students write how they know when the graph of an inequality has a dashed boundary and when it has a solid boundary.

Formative Assessment
Check for student understanding of concepts in Lessons 2-7 and 2-8.

☐ Quiz 4, p. 60

WatchOut!

Error Analysis In Exercise 35, students must decide which graph is shaded correctly. Tell students to solve the inequality for *y* before deciding who is correct.

Additional Answer

38. Sample answer: Craig can earn a 90% average by scoring a 99.5% or higher on his last test. This was determined by finding the sum of the first 4 tests. By making each original test worth 100 points, then so far he has 341 out of 400 points. If the last test is worth twice as much, then it is worth 200, so overall he has 600 possible points. In order to earn a 90%, he needs 0.90 · 600 or 540 points. So, Craig needs 540 − 341 or 199 out of 200 on the last test: 99.5%.

CHAPTER 2 Study Guide and Review

Formative Assessment

KeyVocabulary The page references after each word denote where that term was first introduced. If students have difficulty answering questions 1–6, remind them that they can use these page references to refresh their memories about the vocabulary.

Summative Assessment

📁 Vocabulary Test, p. 62

🔤 **Vocabulary Review**

Vocabulary Review provides students the opportunity to check their understanding of important concepts and terminology in an online game format.

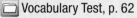 **StudyOrganizer**

Dinah Zike's Foldables®

Have students look through the chapter to make sure they have included examples in their Foldables. Suggest that students keep their Foldables handy while completing the Study Guide and Review pages. Point out that their Foldables can serve as a quick review tool when studying for the chapter test.

Study Guide

KeyConcepts

Relations and Functions (Lesson 2-1)

- A function is a relation where each member of the domain is paired with exactly one member of the range.

Linear Equations and Slope (Lessons 2-2 to 2-4)

- Standard Form: $Ax + By = C$, where A, B, and C are integers whose greatest common factor is 1, $A \geq 0$, and A and B are not both zero.
- Slope-Intercept Form: $y = mx + b$
- Point-Slope Form: $y - y_1 = m(x - x_1)$

Scatter Plots and Lines of Regression (Lesson 2-5)

- A prediction equation can be used to predict the value of one of the variables given the value of the other variable.
- A line of regression can be used to model data.

Special Functions and Parent Functions
(Lessons 2-6 and 2-7)

- A piecewise-defined function is made up of two or more expressions.
- Translations, reflections, and dilations of a parent graph form a family of graphs.

Graphing Linear and Absolute Value Inequalities
(Lesson 2-8)

- You can graph an inequality by following these steps.

  Step 1 Determine whether the boundary is solid or dashed. Graph the boundary.

  Step 2 Choose a point not on the boundary and test it in the inequality.

  Step 3 If a true inequality results, shade the region containing your test point. If a false inequality results, shade the other region.

FOLDABLES StudyOrganizer

Be sure the Key Concepts are noted in your Foldable.

KeyVocabulary

absolute value function (p. 103)	nonlinear relation (p. 69)
bivariate data (p. 92)	parent function (p. 109)
continuous relation (p. 62)	piecewise-defined function (p. 101)
correlation coefficient (p. 94)	
dependent variable (p. 64)	point-slope form (p. 84)
dilation (p. 111)	positive correlation (p. 92)
direct variation (p. 90)	prediction equation (p. 92)
discrete relation (p. 62)	quadratic function (p. 109)
family of graphs (p. 109)	rate of change (p. 76)
greatest integer function (p. 102)	reflection (p. 111)
	regression line (p. 94)
independent variable (p. 64)	scatter plot (p. 92)
linear equation (p. 69)	slope (p. 77)
linear function (p. 69)	slope-intercept form (p. 83)
linear inequality (p. 117)	standard form (p. 70)
line of fit (p. 92)	step function (p. 102)
negative correlation (p. 92)	translation (p. 110)
	vertical line test (p. 62)

VocabularyCheck

Choose the correct term to complete each sentence.

1. A function is (discrete, <u>one-to-one</u>) if each element of the domain is paired to exactly one unique element of the range.

2. The (<u>domain</u>, range) of a relation is the set of all first coordinates from the ordered pairs which determine the relation.

3. The (constant, <u>identity</u>) function is a linear function described by $f(x) = x$.

4. If you are given the coordinates of two points on a line, you can use the (slope-intercept, <u>point-slope</u>) form to find the equation of the line that passes through them.

5. A set of bivariate data graphed as ordered pairs in a coordinate plane is called a (<u>scatter plot</u>, line of fit).

6. A function that is written using two or more expressions is called a (linear, <u>piecewise-defined</u>) function.

 122 | **Chapter 2** | Study Guide and Review

Lesson-by-Lesson Review

2-1 Relations and Functions

State the domain and range of each relation. Then determine whether the relation is a function. If it is a function, determine if it is *one-to-one*, *onto*, *both*, or *neither*. **7–10. See margin.**

7. {(1, 2), (3, 4), (5, 6), (7, 8)}

8. {(−3, 0), (0, 2), (2, 4), (4, 5), (5, 2)}

9. {(−4, 1), (3, 3), (1, 1), (−2, 5), (3, −4)}

10. {(7, −4), (5, −2), (3, 0), (1, 2), (−1, 4)}

Find each value if $f(x) = -3x + 2$.

11. $f(4)$ -10 **12.** $f(-3)$ 11

13. $f(0)$ 2   **14.** $f(y)$ $-3y + 2$

15. $f(-a)$ $3a + 2$ **16.** $f(2w)$ $-6w + 2$

17. BOWLING A bowling alley charges $2.50 for shoe rental and $3.25 per game bowled. The amount a bowler is charged can be expressed as $y = 2.50 + 3.25x$, when $x \geq 1$, and is an integer. Find the domain and range. Then determine whether the equation is a function. Is the equation discrete or continuous? **See margin.**

Example 1

State the domain and range of the relation {(−4, 3), (−1, 0), (−2, 4), (3, −1), (2, 6)}. Then determine whether the relation is a function. If it is a function, determine if it is *one-to-one, onto, both,* or *neither*.

Domain: {−4, −2, −1, 2, 3}
Range: {−1, 0, 3, 4, 6}

Each element of the domain is paired with one element of the range, so the relation is a function. The function is both because each element of the domain is paired with a unique element of the range and each element of the range is paired with a unique element of the domain.

Example 2

Find $f(-2)$ if $f(x) = 4x - 3$.

$f(-2) = 4(-2) - 3$ Substitute −3 for *x*.

$ = -8 - 3$ Multiply.

$ = -11$ Simplify.

19. No; the variables have an exponent other than 1.

20. No; x^3 has an exponent other than 1.

23. No; *x* appears in a denominator.

25. $12x - y = 0$; 12, −1, 0

2-2 Linear Relations and Functions

State whether each function is a linear function. Write *yes* or *no*. Explain.

18. $3x + 4y = 12$ yes   **19.** $x^2 + y^2 = 4$

20. $y = x^3 - 6$ **21.** $y = 6x - 19$ yes

22. $f(x) = -2x + 9$ yes   **23.** $\frac{1}{x} + 3y = -5$

Write each equation in standard form. Identify *A, B,* and *C*.

24. $2x + 5y = 10$ 2, 5, 10   **25.** $y = 12x$

26. $-4y = 3x - 24$
$3x + 4y = 24$; 3, 4, 24   **27.** $4x = 8y - 12$
$x - 2y = -3$; 1, −2, −3

28. TRAVEL The distance the Green family traveled during their family vacation is given by the equation $y = 65x$, where *x* represents the number of hours spent driving. How far does the Green family travel in 8 hours? **520 miles**

Example 3

State whether $f(x) = 3x^2$ is a linear function. Write *yes* or *no*. Explain.

No, because the expression includes a variable raised to the second power.

Example 4

Write the equation $y = -5x + 8$ in standard form. Identify *A, B,* and *C*.

$y = -5x + 8$ Original equation

$5x + y = 8$ Add 5*x* to each side.

$A = 5, B = 1,$ and $C = 8$

Additional Answers

7. D = {1, 3, 5, 7}, R = {2, 4, 6, 8}; a function; both

8. D = {−3, 0, 2, 4, 5}, R = {0, 2, 4, 5}; a function; onto

9. D = {−4, −2, 1, 3}, R = {−4, 1, 3, 5}; not a function

10. D = {−1, 1, 3, 5, 7}, R = {−4, −2, 0, 2, 4}; a function; both

17. D = {1, 2, 3, 4, 5, . . .}, R = {5.75, 9, 12.25, 15.5, 18.75, . . .}; a function; discrete

Lesson-by-Lesson Review
Intervention If the given examples are not sufficient to review the topics covered by the questions, remind students that the lesson references tell them where to review that topic in their textbook.

Two-Day Option Have students complete the Lesson-by-Lesson Review. Then you can use McGraw-Hill eAssessment to customize another review worksheet that practices all the objectives of this chapter or only the objectives on which your students need more help.

Additional Answers

37. $y = 4x + 25$

38. $y = \frac{3}{5}x + \frac{22}{5}$

42.

Sample answer: using (1, 72) and (5, 224): $y = 38x + 34$

2-3 Rate of Change and Slope

29. RETAIL The table shows the number of DVDs sold each week at the Super Movie Store. Find the average rate of change of the number of DVDs sold from week 2 to week 5. **18**

Week	1	2	3	4	5
DVDs Sold	76	58	94	83	112

Find the slope of the line that passes through each pair of points.

30. (2, 5), (6, −3) **−2**

31. (8, 2), (2, 8) **−1**

32. Determine the rate of change of the graph. **2**

Example 5

Find the slope of the line that passes through each pair of points.

a. (−2, 9), (1, 4)

$m = \dfrac{y_2 - y_1}{x_2 - x_1}$ Slope Formula

$= \dfrac{4 - 9}{1 - (-2)}$ $(x_1, y_1) = (-2, 9), (x_2, y_2) = (1, 4)$

$= -\dfrac{5}{3}$ Simplify.

b. (−3, 6), (4, 6)

$m = \dfrac{y_2 - y_1}{x_2 - x_1}$ Slope Formula

$= \dfrac{6 - 6}{4 - (-3)}$ $(x_1, y_1) = (-3, 6), (x_2, y_2) = (4, 6)$

$= \dfrac{0}{7}$ or 0   Simplify.

2-4 Writing Linear Equations

Write an equation in slope-intercept form for the line that satisfies each set of conditions.

33. slope −2, passes through (−3, −5) $y = -2x - 11$

34. slope $\frac{2}{3}$, passes through (4, −1) $y = \frac{2}{3}x - \frac{11}{3}$

35. passes through (−2, 4) and (0, 8) $y = 2x + 8$

36. passes through (3, 5) and (−1, 5) $y = 5$

Write an equation of the line passing through each pair of points. **37, 38. See margin.**

37. (6, 1), (4, 9)   **38.** (−4, 2), (6, 8)

Write an equation in slope-intercept form for the line that satisfies each set of conditions.

39. through (1, 2), parallel to $y = 4x - 3$ $y = 4x - 2$

40. through (−3, 5), perpendicular to $y = \frac{2}{3}x - 8$

41. PETS Drew paid a $250 fee when he adopted a puppy. The average monthly cost of feeding and caring for the puppy is $32. Write an equation that represents the total cost of adopting and caring for the puppy for x months.

Example 6

Write an equation of the line through (−2, 5) and (0, −9).

Find the slope of the line.

$m = \dfrac{y_2 - y_1}{x_2 - x_1}$ Slope Formula

$= \dfrac{-9 - 5}{0 - (-2)}$ $(x_1, y_1) = (-2, 5),$ $(x_2, y_2) = (0, -9)$

$= \dfrac{-14}{2}$ or −7   Simplify.

Write an equation.

$y - y_1 = m(x - x_1)$ Point-slope form

$y - 5 = -7(x - (-2))$ Substitute.

$y - 5 = -7(x + 2)$ Simplify.

$y - 5 = -7x - 14$ Distributive Property

$y = -7x - 9$ Add 5 to each side.

The equation is $y = -7x - 9$.

40. $y = -\frac{3}{2}x + \frac{1}{2}$ **41.** $y = 32x + 250$

2-5 Scatter Plots and Lines of Regression

Make a scatter plot and a line of fit and describe the correlation for each set of data. Then, use two ordered pairs to write a prediction equation.

42. HEATING The table shows the monthly heating cost for a large home. **See margin.**

Month	Sep	Oct	Nov	Dec	Jan	Feb
Bill ($)	72	114	164	198	224	185

43. AMUSEMENT PARK The table shows the annual attendance in thousands at an amusement park during the last 5 years. **See margin.**

Year	1	2	3	4	5
People (thousands)	44	42	39	31	24

Example 7

SCHOOL ENROLLMENT The table shows the number of students each year at a school.

Year	'07	'08	'09	'10	'11	'12
Students	125	116	142	154	146	175

Use (2007, 125) and (2012, 175) to find a prediction equation.

$$m = \frac{y_2 - y_1}{x_2 - x_1}$$ Slope Formula

$$= \frac{175 - 125}{2012 - 2007}$$ Substitution

$$= \frac{50}{5} \text{ or } 10$$ Simplify.

$$y - y_1 = m(x - x_1)$$ Point-slope form

$$y - 125 = 10(x - 2007)$$ Substitution

$$y - 125 = 10x - 20{,}070$$ Distributive Property

$$y = 10x - 20{,}195$$ Add 125 to each side.

2-6 Special Functions

Graph each function. Identify the domain and range.

44–46. See margin.

44. $f(x) = \begin{cases} -2x & \text{if } x \le -1 \\ x + 1 & \text{if } -1 < x < 3 \\ x & \text{if } x \ge 3 \end{cases}$

45. $f(x) = \begin{cases} -3 & \text{if } x < -1 \\ 4x - 3 & \text{if } -1 \le x \le 3 \\ x & \text{if } x > 3 \end{cases}$

46. Write the piecewise-defined function shown in the graph.

47, 48. See margin.
Graph each function. Identify the domain and range.

47. $f(x) = [\![x]\!] + 2$ **48.** $f(x) = [\![x + 3]\!]$

Example 8

Write the piecewise-defined function shown in the graph.

The left portion of the graph is the graph of $f(x) = 3$. There is a circle at $(-2, 3)$, so the linear function is defined for $x < -2$.

The center portion of the graph is the graph of $f(x) = x - 1$. There is a dot at $(-2, -3)$ and a circle at $(2, 1)$, so the linear function is defined for $-2 \le x < 2$.

The right portion of the graph is the graph of $f(x) = 2x$. There is a dot at $(2, 4)$, so the linear function is defined for $x \ge 2$.

$$f(x) = \begin{cases} 3 & \text{if } x < -2 \\ x - 1 & \text{if } -2 \le x < 2 \\ 2x & \text{if } x \ge 2 \end{cases}$$

43.

Sample answer using (1, 44) and (5, 24):
$y = -5x + 49$

Additional Answers

44.

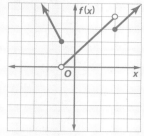

$D = \{\text{all real numbers}\};$
$R = \{f(x) \mid f(x) > 0\}$

45.

$D = \{\text{all real numbers}\};$
$R = \{f(x) \mid f(x) \ge -7\}$

46. $f(x) = \begin{cases} x - 1 & \text{if } x \le -2 \\ -2x & \text{if } -2 < x < 1 \\ 3 & \text{if } x \ge 1 \end{cases}$

47.

$D = \{\text{all real numbers}\};$
$R = \{\text{all integers}\}$

48.

$D = \{\text{all real numbers}\};$
$R = \{\text{all integers}\}$

Additional Answers

54.

55.

56.

57.

58.

2-7 Parent Functions and Transformations

Identify the type of function represented by each graph.

49.

quadratic

50.

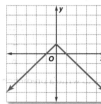

absolute value

51. $y = x^2$ shifted down 3 units

52. $y = x^2$ reflected over the x-axis

51. Describe the translation in $y = x^2 - 3$.

52. Describe the reflection in $y = -x^2$.

53. **CONSTRUCTION** A large arch is being constructed at the entrance of a new city hall building. The shape of the arch resembles the graph of the function $f(x) = -0.025x^2 + 3.64x - 0.038$. Describe the shape of the arch. **parabola**

Example 9

Identify the type of function represented by the graph.

The graph is in the shape of a V. The graph represents an absolute value function.

Example 10

Describe the translation in $y = |x + 6|$.

The graph of $y = |x + 6|$ is a translation of the graph of $y = |x|$ shifted left 6 units.

2-8 Graphing Linear and Absolute Value Inequalities

Graph each inequality. **54–61. See margin.**

54. $x - 3y < 6$

55. $y \geq 2x + 1$

56. $2x + 4y \leq 12$

57. $y < -3x - 5$

58. $y > |2x|$

59. $y \geq |2x - 2|$

60. $y + 3 < |x + 1|$

61. $2y \leq |x - 3|$

62. **BOOKS** Spencer has saved $96 for a trip to his favorite bookstore. Each paperback book costs $8 and each hardback book costs $12. Write and graph an inequality that shows the number of paperback books and hardback books Spencer can purchase.

$8x + 12y \leq 96$; See Chapter 2 Answer Appendix for graph.

Example 11

Graph $x - 2y > 6$.

Since the inequality symbol is $>$, the graph of the boundary line should be dashed. Graph $x - 2y = 6$.

Test $x - 2y > 6$ at $(0, 0)$.

$x - 2y > 6$

$0 - 2(0) \overset{?}{>} 6$

$0 > 6$ ✗

59.

60.

61.

CHAPTER 2 Practice Test

1. State the domain and range of the relation shown in the table. Then determine if it is a function. If it is a function, determine if it is *one-to-one*, *onto*, *both*, or *neither*.

x	y
−2	3
4	−1
3	2
6	3

   D = {−2, 3, 4, 6};
   R = {−1, 2, 3};
   onto

Find each value if $f(x) = -2x + 3$.

2. $f(-4)$ **11**

3. $f(3y)$ **$-6y + 3$**

4. Write $2y = -6x + 4$ in standard form. Identify A, B and C. **$3x + y = 2; 3, 1, 2$**

5. Find the x-intercept and the y-intercept for $3x - 4y = -24$. **(−8, 0), (0, 6)**

6. **MULTIPLE CHOICE** The cost of producing x pumpkin pies at a small bakery is given by $C(x) = 49 + 1.75x$. Find the cost of producing 25 pies. **C**

   A $74.00

   B $81.50

   C $92.75

   D $108.25

Find the slope of the line that passes through each pair of points.

7. (1, 6), (3, 10) **2**

8. (−2, 7), (3, −1) **$-\dfrac{8}{5}$**

9. **MULTIPLE CHOICE** Find the equation of the line that passes through (0, −3) and (4, 1). **H**

   F $y = -x + 3$

   G $y = -x - 3$

   H $y = x - 3$

   J $y = x + 3$

10. Write an equation in slope-intercept form for the line that has slope −2 and passes through the point (3, −4). **$y = -2x + 2$**

11. Write an equation of the line that passes through the points (2, −4) and (1, 6). **$y = -10x + 16$**

12. Write an equation in slope-intercept form for the line that passes through (−3, 5) and is parallel to $y = -6x + 1$. **$y = -6x - 13$**

13. **EMERGENCY ROOM** A hospital tracks the number of emergency room visits during the fall and winter months. **a. See Chapter 2 Answer Appendix.**

Month	Oct	Nov	Dec	Jan	Feb
Visits	124	163	155	171	192

    a. Make a scatter plot and describe the correlation.

    b. Use two ordered pairs to write a prediction equation. **$y = 17x + 107$**

    c. Use your prediction equation to predict the number of emergency room visits for March. **209**

14. Graph $f(x) = \begin{cases} -x & \text{if } x < -2 \\ x + 2 & \text{if } -2 \le x \le 2 \\ 5 & \text{if } x > 2 \end{cases}$. **See Chapter 2 Answer Appendix.**

15. Write the piecewise-defined function shown. **See margin.**

16. Identify the domain and range of $y = [\![x]\!] + 2$. **D = {all real numbers}; R = {all integers}**

17. Describe the translation to $y = x^2 + 5$. **$y = x^2$ shifted 5 units up**

18. Describe the reflection in $y = -|x|$. **over the x-axis**

Graph each inequality. **19, 20. See Chapter 2 Answer Appendix.**

19. $y \ge 4x - 1$

20. $2x + 6y < -12$

Summative Assessment

Use these alternate leveled chapter tests to differentiate assessment for your students.

Leveled Chapter 2 Tests

Form	Type	Level	📁 Page(s)
1	MC	AL	63–64
2A	MC	OL	65–66
2B	MC	OL	67–68
2C	FR	OL	69–70
2D	FR	OL	71–72
3	FR	BL	73–74
Vocabulary Test			62
Extended-Response Test			75

MC = multiple-choice questions
FR = free-response questions

McGraw-Hill eAssessment Customize and create multiple versions of your chapter tests and their answer keys. All of the questions from the leveled chapter tests in the *Chapter 2 Resource Masters* are also available on McGraw-Hill eAssessment.

Additional Answer

15. $y = \begin{cases} -2x & \text{if } x < -1 \\ -1 & \text{if } -1 \le x \le 3 \\ x + 1 & \text{if } x > 3 \end{cases}$

InterventionPlanner

TIER 1 On Level OL

If students miss about 25% of the exercises or less,

Then choose a resource:

SE Lessons 2-1, 2-2, 2-3, 2-4, 2-5, 2-6, 2-7, and 2-8

📁 Skills Practice, pp. 7, 13, 19, 25, 32, 38, 45, and 52

TE Chapter Project

connectED.mcgraw-hill.com Self-Check Quiz

TIER 2 Strategic Intervention AL
approaching grade level

If students miss about 50% of the exercises,

Then choose a resource:

📁 Study Guide and Intervention, pp. 5, 11, 17, 23, 30, 36, 43, and 50

connectED.mcgraw-hill.com Extra Examples, Personal Tutor, Homework Help

TIER 3 Intensive Intervention
2 or more grades below level

If students miss about 75% of the exercises,

Then use *Math Triumphs, Alg. 2,* Ch. 1 and 3

connectED.mcgraw-hill.com Extra Examples, Personal Tutor, Homework Help, Review Vocabulary

1 Focus

Objective Use the strategy of reading math problems to solve standardized test problems.

2 Teach

Scaffolding Questions
Ask:

- Have you ever had to reread a math problem to get the facts you need to answer the question? Answers will vary.

- Will you always use all the information given in a math problem to solve the problem? Answers will vary. Sample answer: No, there is often extra information given.

- How can you be sure you have the correct answer? Answers will vary. Sample answer: quickly estimate and compare your answer to the estimate; check that your answer has the correct units of measure.

CHAPTER 2

Preparing for Standardized Tests

Reading Math Problems

The first step to solving any math problem is to read the problem. When reading a math problem to get the information you need to solve, it is helpful to use special reading strategies.

Strategies for Reading Math Problems

Step 1

Read the problem quickly to gain a general understanding of it.

- **Ask yourself:** "What do I know?" "What do I need to find out?"

- **Think:** "Is there enough information to solve the problem? Is there extra information?"

- **Highlight:** If you are allowed to write in your test booklet, underline or highlight important information. Cross out any information you do not need.

Step 2

Reread the problem to identify relevant facts.

- **Analyze:** Determine how the facts are related.

- **Key Words:** Look for key words to solve the problem.

- **Vocabulary:** Identify mathematical terms. Think about the concepts and how they are related.

- **Plan:** Make a plan to solve the problem.

- **Estimate:** Quickly estimate the answer.

Step 3

Identify any obvious wrong answers.

- **Eliminate:** Eliminate any choices that are very different from your estimate.

- **Units of Measure:** Identify choices that are possible answers based on the units of measure in the question. For example, if the question asks for area, eliminate all the answer choices that are not in square units.

Step 4

Look back after solving the problem.
Check: Make sure you have answered the question.

Corbis

Standardized Test Example

Read the problem. Identify what you need to know. Then use the information in the problem to solve.

Sandy heated a solution over a burner and then removed it from the heat source. The temperature of the solution decreased linearly as it cooled. The temperatures after 0, 2, 5, and 9 minutes are shown in the table. What is the rate of change in the temperature of the solution as it cools?

Time (min)	Temperature (°C)
0	133.2
2	130.4
5	126.2
9	120.6

A −1.4 degrees per minute

B −0.8 degrees per minute

C 0.8 degrees per minute

D 1.4 degrees per minute

Read the problem carefully. There is extra information in the problem. To determine the slope, you only need information from two points on the linear function. Use two of the points to find the slope.

$$m = \frac{130.4 - 133.2}{2 - 0} = -1.4$$

The correct answer is A.

Exercises

Read each question. Then fill in the correct answer on the answer document provided by your teacher or on a sheet of paper.

1. The graph shows the cost of shipping packages. How much would it cost to ship a package that weighs 2 pounds 8 ounces? B

Shipping Costs

A $3.50

B $4.50

C $5.00

D $5.50

2. What is the slope of the line shown in the graph? J

F −2

G $-\frac{1}{2}$

H $\frac{1}{2}$

J 2

STANDARDIZED TEST PRACTICE

The number of cups of milk in a pitcher decreases linearly as it is poured. The graph below shows the number of cups of milk remaining in the pitcher as the milk is being poured. At what rate is the milk being poured? C

A 2 cups/min

B 2 sec/cup

C 5 cups/sec

D 10 cups/sec

3 Assess

Use Exercises 1 and 2 to assess students' understanding.

Diagnose Student Errors

Survey student responses for each item. Class trends may indicate common errors and misconceptions.

1. A these are the positive values
B these are the values in the range
C correct
D these are the negative values

2. F correct
G found the reciprocal of the slope
H found the inverse of the reciprocal of the slope
J simplified incorrectly

3. A incorrectly subtracted the years
B correct
C found only the difference in the values of the home
D guess

4. F guess
G did not identify 450 as a constant
H correct
J changed % to a decimal incorrectly

5. A did not solve for the case in which the value inside the absolute value symbols is negative
B correct
C incorrectly solved for the case in which the value inside the absolute value symbols is positive
D guess

6. F did not include the first 100 messages
G guess
H correct
J rounded the value up

7. A correct
B guess
C does not know which value is the value of the slope
D does not know that the slope of perpendicular lines are opposite reciprocals of each other

---

Standardized Test Practice
Cumulative, Chapters 1 through 2

Multiple Choice

Read each question. Then fill in the correct answer on the answer document provided by your teacher or on a sheet of paper.

1. What is the domain of the relation shown below? **C**

x	y
−3	4
1	−1
2	0
6	−3

A {0, 1, 2, 4, 6}

B {−3, −1, 0, 4}

C {−3, 1, 2, 6}

D {−3, −1}

2. What is the slope of the line? **F**

F −2

G $-\dfrac{1}{2}$

H $\dfrac{1}{2}$

J 2

3. The Robinson family bought their home in 1998 for $152,400. When they sold it in 2010, the value was $174,900. What was the annual rate of change in the value of the home? **B**

A $1225

B $1875

C $22,500

D $27,275

Test-Taking Tip

Question 2 Since the graph slopes downward from left to right, you know the slope is negative. So, answer choices H and J can be eliminated.

---

4. Carmen works for an electronics retailer. She earns a weekly salary of $450 plus a commission of 4.5% on her weekly sales. Write a linear equation for Carmen's weekly earnings E if she has d dollars in sales. **H**

F $E = (450 + 4.5)d$

G $E = (450 + 0.045)d$

H $E = 450 + 0.045d$

J $E = 450 + 4.5d$

5. Which of the graphs represents the solution set for $|x − 3| − 4 = 0$? **B**

A
$-4\ -3\ -2\ -1\ \ 0\ \ 1\ \ 2\ \ 3\ \ 4\ \ 5\ \ 6\ \ 7\ \ 8$

B
$-4\ -3\ -2\ -1\ \ 0\ \ 1\ \ 2\ \ 3\ \ 4\ \ 5\ \ 6\ \ 7\ \ 8$

C
$-4\ -3\ -2\ -1\ \ 0\ \ 1\ \ 2\ \ 3\ \ 4\ \ 5\ \ 6\ \ 7\ \ 8$

D
$-4\ -3\ -2\ -1\ \ 0\ \ 1\ \ 2\ \ 3\ \ 4\ \ 5\ \ 6\ \ 7\ \ 8$

6. Karissa has $10 per month to spend text messaging on her cell phone. The phone company charges $4.95 for the first 100 messages and $0.10 for each additional message. How many text messages can Karissa afford to send and receive each month? **H**

F 50

G 100

H 150

J 151

7. Given $y = 2.24x + 16.45$, which statement best describes the effect of decreasing the y-intercept by 20.25? **A**

A The x-intercept increases.

B The y-intercept increases.

C The new line has a greater rate of change.

D The new line is perpendicular to the original.

Short Response/Gridded Response

Record your answers on the answer sheet provided by your teacher or on a sheet of paper.

8. Write an equation for the piecewise-defined function shown in the graph below. **See margin.**

9. **GRIDDED RESPONSE** Evaluate the piecewise-defined function shown in Exercise 8 for $x = -3$. **1**

10. The graph of the absolute value parent function is shown below. **a. $y = |x|$**

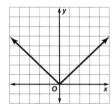

a. What is the equation of this parent function?

b. What equation would result in the graph of the parent function being reflected over the x-axis and shifted up 2 units? $y = -|x| + 2$

c. What equation would result in the graph of the parent function being shifted left 3 units and down 1 unit? $y = |x + 3| - 1$

Extended Response

Record your answers on a sheet of paper. Show your work.

11. The soccer team is having a bake sale this week to raise money for the program. For each cookie sold, the profit is \$0.45, and for each brownie sold, the profit is \$0.50. **a. $0.45x + 0.5y \geq 150$**

a. The team hopes to earn \$150 in profits from the bake sale. Let x represent the number of cookies sold and y the number of brownies sold. Write an inequality to model the situation.

b. Graph the inequality. **See margin.**

c. If the team sells 180 cookies and 160 brownies this week, will they meet their goal? Explain. **Yes, the profit earned will be \$161**

12. Use the scatter plot to answer each question.

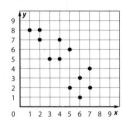

a. What type of correlation is shown by the data in the plot? **negative**

b. Determine a regression line for the data. **Sample answer: $y = -0.99x + 9.13$**

c. Use your regression line to predict the value of y when $x = 12$. **Sample answer: -2.75**

Need ExtraHelp?

If you missed Question...	1	2	3	4	5	6	7	8	9	10	11	12
Go to Lesson...	2-1	2-3	2-3	2-4	1-4	1-3	2-7	2-6	2-6	2-7	2-8	2-5

Additional Answers

8. $y = \begin{cases} 5 & \text{if } x < -4 \\ -x - 2 & \text{if } -4 \leq x \leq 4 \\ 2x - 12 & \text{if } x > 4 \end{cases}$

11b.

Formative Assessment

You can use these pages to benchmark student progress.

📁 Standardized Test Practice, pp. 76–78

Answer Sheet Practice

Have students simulate taking a standardized test by recording their answers on a practice recording sheet.

Student Recording Sheet

McGraw-Hill
Assessment Create practice tests that align to your state standards, the Common Core State Standards, and other national standards such as TIMSS and NAEP.

Homework Option

Get Ready for Chapter 3 Assign students the exercises on p. 133 as homework to assess whether they possess the prerequisite skills needed for the next chapter.

Lesson 2-1

5.

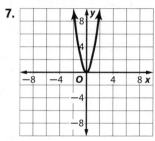

D = {all real numbers};
R = {all real numbers};
function; both; continuous

6.

D = {all real numbers};
R = {all real numbers};
function; both; continuous

7.

D = {all real numbers};
R = {y | y ≥ 0};
function; onto; continuous

8.

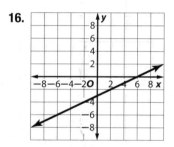

D = {7}; R = {all real numbers};
not a function; not continuous

15.

D = {all real numbers};
R = {all real numbers};
function; both; continuous

16.

D = {all real numbers};
R = {all real numbers};
function; both; continuous

17.

D = {all real numbers};
R = {y | y ≥ 0};
function; onto; continuous

18.

D = {all real numbers};
R = {y | y ≤ 0};
function; onto; continuous

19.

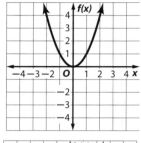

D = {all real numbers};
R = {y | y ≥ −8};
function; onto; continuous

20.

D = {all real numbers};
R = {all real numbers};
function; both; continuous

34a.

34b.

Function	Possible Intersection Points
$f(x) = x^2$	0, 1, 2
$g(x) = 2^x$	0, 1
$h(x) = x^3 - 3x^2 - 5x + 6$	1, 2, 3
$j(x) = x^3$	1

34c. $g(x)$ and $j(x)$ are one-to-one, and $f(x)$ and $h(x)$ are not.

34d. $h(x)$ and $j(x)$ are onto, and $f(x)$ and $g(x)$ are not.

34e.

Function	One-to-one	Onto
$f(x) = x^2$	no	no
$g(x) = 2^x$	yes	no
$h(x) = x^3 - 3x^2 - 5x + 6$	no	yes
$j(x) = x^3$	yes	yes

38a. Sample answer:

38b. Sample answer:

38c. Sample answer:

38d. Sample answer: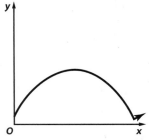

Extend 2-1

3. Continuous; the domain, the number of hours, is the set of values greater than or equal to 0, so the graph of the function is a straight line.

4. Discrete; the domain, the number of rides, is the set of whole numbers, so the graph of the function consists of individual points.

5. Sample answer: A function describing the cost y of x football tickets is discrete because only whole-number values make sense for the domain, the number of tickets. A function describing the cost y of x pounds of grapes is continuous because the domain, the pounds of grapes, is the set of values greater than or equal to 0.

Lesson 2-2

6. $4x + y = -7$; $A = 4, B = 1, C = -7$

7. $6x - y = -5$; $A = 6, B = -1, C = -5$

8. $3x + 2y = -1$; $A = 3, B = 2, C = -1$

9. $8x + 9y = 6$; $A = 8, B = 9, C = 6$

10. $x - 3y = -2$; $A = 1, B = -3, C = -2$

11. $2x - 3y = 12$; $A = 2, B = -3, C = 12$

12. $-\dfrac{12}{5}$; 12

13. $\dfrac{5}{2}$; -10

14. 6; 4

15. 7; $-\dfrac{21}{4}$

35. -0.5; -4

36. 6; -18

37. -7; 10.5

38. $\dfrac{10}{3}$; $\dfrac{30}{7}$

39. 12; -18

40. -18; 16

41b.

Yes; the graph passes the vertical line test.

50b.

51a.

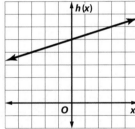

Extend 2-4

1. $p = 0.43d$; 27.95 lb/in^2

[0, 80] scl: 5 by [0, 50] scl: 5

2.

[−10, 10] scl: 1 by [−10, 10] scl: 1

The graphs of $y = -4x$ and $y = -2x$ have negative slope. As x increases, y decreases. The graphs of $y = 4x$ and $y = 2x$ have positive slope. As x increases, y increases. All of the graphs are lines, all have a constant rate of change, and all pass through the origin.

Mid-Chapter Quiz

12. $-3, 4$

13. $10, 5$

Lesson 2-5

7.

[0,12] scl: 2 by [0, 1000] scl: 100

8.

[0,15] scl: 2 by [0, 50] scl: 5

9a.

[2002, 2010] scl: 1 by
[1,600,000, 2,400,000] scl: 100,000

9d. The correlation is stronger between wins and attendance. The correlation coefficient between years and attendance is 0.42; the correlation coefficient between wins and attendance is about 0.79. The greater correlation coefficient implies a stronger correlation.

10a. $y = -0.21x + 85.1$

[10, 40] scl: 5 by [50, 100] scl: 10

11e. Sample answer: The new equation has a correlation coefficient, 0.986, that is extremely close to 1, so this equation should accurately represent the data.

12. Sample answer: The correlation coefficient is very valuable to a linear regression line because it determines how close the actual data points are to the regression line. The closer the points are to the line, or the closer the correlation coefficient is to 1 or -1, the more accurate the regression line is.

13. Sample answer: If a and b have a positive correlation, then they are both increasing. If b and c have a negative correlation and b is increasing, then c must be decreasing. If c and d have a positive correlation and c is decreasing, then d must be decreasing. If a is increasing and d is decreasing, then they must have a negative correlation.

15.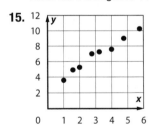

a; Sample answer: The data show a strong positive correlation which means that the correlation coefficient r should be close to 1.

16. A strength of using regression equations is that the regression equation can be used to make predictions when the values fall close to the domain of the original data set. A weakness of using regression equations is that they assume that a trend in the original data set will continue, and they are very sensitive to outliers. Both weaknesses can make predictions inaccurate.

Lesson 2-6 (Guided Practice)

1.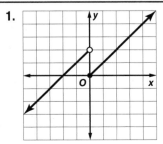

D = {all real numbers}, R = {all real numbers}

Lesson 2-6

1.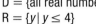

D = {all real numbers};
R = {$y \mid y \leq 4$}

2.

D = {all real numbers};
R = {$y \mid 8 \geq y > -2$ or $y \leq -8$}

3. $g(x) = \begin{cases} x + 4 & \text{if } x < -2 \\ -3 & \text{if } -2 \leq x \leq 3 \\ -2x + 12 & \text{if } x > 3 \end{cases}$

4. $g(x) = \begin{cases} 6 & \text{if } x \leq -5 \\ -x + 4 & \text{if } -5 < x < -2 \\ \frac{1}{2}x + 1 & \text{if } x \geq -2 \end{cases}$

5.

Tickets Sold

6.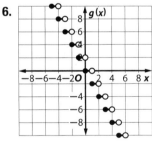

D = {all real numbers};
R = {all even integers}

7.

D = {all real numbers};
R = {all integers}

8.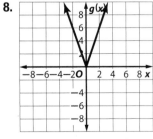

D = {all real numbers};
R = {$g(x) \mid g(x) \geq 0$}

9.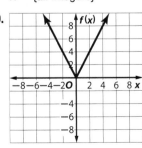

D = {all real numbers};
R = {$f(x) \mid f(x) \geq 0$}

10.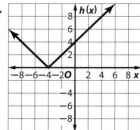

D = {all real numbers};
R = {h(x) | h(x) ≥ 0}

11.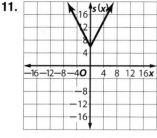

D = {all real numbers};
R = {s(x) | s(x) ≥ 6}

16. $g(x) = \begin{cases} -8 & \text{if } x \le -6 \\ 0.25x + 2 & \text{if } -4 \le x \le 4 \\ 4 & \text{if } x > 6 \end{cases}$

17. $g(x) = \begin{cases} -x - 4 & \text{if } x < -3 \\ x + 1 & \text{if } -3 \le x \le 1 \\ -6 & \text{if } x > 4 \end{cases}$

18. $g(x) = \begin{cases} -9 & \text{if } x < -5 \\ x + 4 & \text{if } 0 \le x \le 3 \\ x - 3 & \text{if } x > 7 \end{cases}$

19. $g(x) = \begin{cases} 8 & \text{if } x \le -1 \\ 2x & \text{if } 4 \le x \le 6 \\ 2x - 15 & \text{if } x > 7 \end{cases}$

20.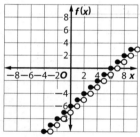

D = {all real numbers};
R = {all integers}

21.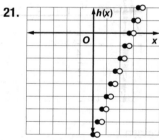

D = {all real numbers};
R = {all integers}

22.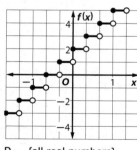

D = {all real numbers};
R = {all integers}

23.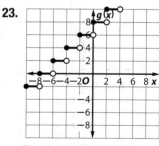

D = {all real numbers};
R = {all even integers}

24.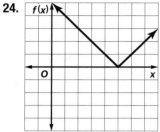

D = {all real numbers};
R = {f(x) | f(x) ≥ 0}

25.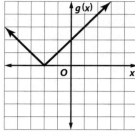

D = {all real numbers};
R = {g(x) | g(x) ≥ 0}

26.

D = {all real numbers};
R = {h(x) | h(x) ≥ −8}

27.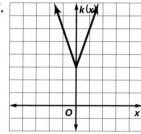

D = {all real numbers};
R = {k(x) | k(x) ≥ 3}

28.

D = {all real numbers};
R = {f(x) | f(x) ≥ 6}

29.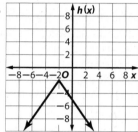

D = {all real numbers};
R = {h(x) | h(x) ≤ −2}

30b. $f(x) = \begin{cases} 2x & \text{if } 0 < x \le 100 \\ x + 100 & \text{if } x > 100 \end{cases}$

31c.

32b. $c(t) = \begin{cases} 6 & \text{if} \quad t \le \frac{1}{2} \\ 10 & \text{if} \quad \frac{1}{2} < t \le 1 \\ 16 & \text{if} \quad 1 < t \le 2 \\ 24 & \text{if} \quad 2 < t \le 24 \end{cases}$

35.

D = {all real numbers}; R = {all whole numbers}

36.

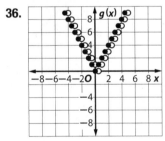

D = {all real numbers}; R = {non-negative integers}

37.

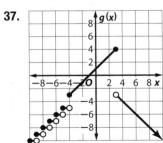

D = {all real numbers}; R = {$g(x) \mid g(x) \le 4$}

38.

D = {all real numbers}; R = {$h(x) \mid h(x) \le -6$ or $0 \le h(x)$}

39a.

x	−4	−3	−2	−1	0	1	2	3	4
f(x)	0	−1	−2	−3	−4	−3	−2	−1	0

x	−4	−3	−2	−1	0	1	2	3	4
g(x)	12	9	6	3	0	3	6	9	12

39b.

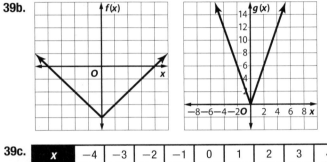

39c.

x	−4	−3	−2	−1	0	1	2	3	4
f(x)	0	−1	−2	−3	−4	−3	−2	−1	0
slope		−1	−1	−1	−1	1	1	1	1

x	−4	−3	−2	−1	0	1	2	3	4
g(x)	12	9	6	3	0	3	6	9	12
slope		−3	−3	−3	−3	3	3	3	3

Explore 2-7

1.

The slopes are the same, and the *y*-intercepts vary.

2.

The *y*-intercepts are the same, and the slopes vary.

3.

The *y*-intercepts are the same, and the slopes vary.

Lesson 2-7 (Guided Practice)

4A. The graph of $y = 2x^2$ is a dilation, stretching the graph of $y = x^2$ vertically.

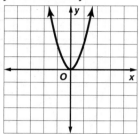

4B. The graph of $y = \left|\frac{1}{3}x\right|$ is a dilation, stretching the graph of $y = |x|$ horizontally.

Lesson 2-7

9. The function is a dilation and translation. The graph of $f(x) = \frac{1}{2}|x - 12|$ compresses the graph $f(x) = |x|$ vertically and translates it 12 units to the right.

20. reflection of the graph of $y = x$ across the *x*-axis

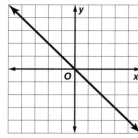

21. reflection of the graph of $y = x^2$ across the *x*-axis

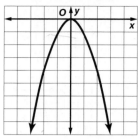

22. reflection of the graph of $y = x^2$ across the *y*-axis

23. reflection of the graph of $y = |x|$ across the *y*-axis

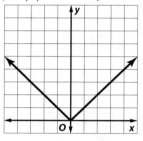

24. reflection of the graph of $y = |x|$ across the *x*-axis

25. reflection of the graph of $y = x$ across the *y*-axis

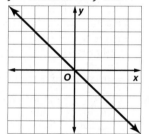

26. horizontal compression of the graph of $y = x^2$

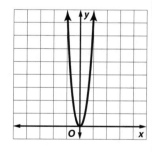

27. vertical expansion of the graph of $y = x$; The slope is steeper than that of $y = x$.

28. The dilation stetches the graph of $y = |x|$ vertically.

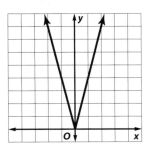

29. The dilation compresses the graph of $y = |x|$ horizontally.

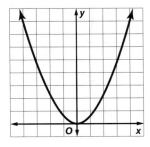

30. The dilation compresses the graph of $y = x$ vertically; the slope is not as steep as that of $y = x$.

31. vertical compression of the graph of $y = x^2$

32. The graph is a dilation of the graph of $y = x$. The dilation stretches the graph vertically.

45. Sample graph:

Sample answer: The figure in Quadrant II has been reflected and moved right 10 units.

Lesson 2-8 (Guided Practice)

1A.

1B.

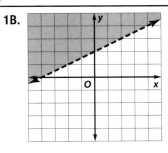

2. $g \leq -3r + 40$

3A.

3B.

Lesson 2-8

5b.

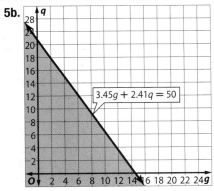

$3.45g + 2.41q = 50$

6.

7.

28b.

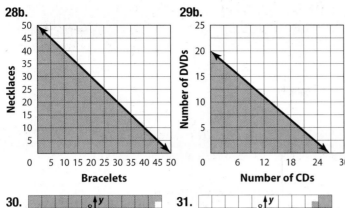

29b.

Number of CDs

Study Guide and Review

62.

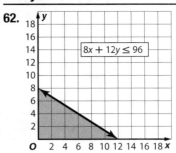
$8x + 12y \leq 96$

30.

31.

Practice Test

13a.

positive correlation

32.

14.

45.

46.

47.

55.

19.

20.

56.

57.

| | Diagnostic Assessment
Quick Check | | | |
	EXPLORE **3-1** 45 min: 0.5 day 90 min: 0.25 day	LESSON **3-1** 45 min: 1.5 days 90 min: 0.75 day	LESSON **3-2** 45 min: 1 day 90 min: 0.5 day	EXTEND **3-2** 45 min: 0.5 day 90 min: 0.25 day
Title	Graphing Technology Lab: Intersections of Graphs	Solving Systems of Equations	Solving Systems of Inequalities by Graphing	Graphing Technology Lab: Solving Systems of Linear Inequalities
Objectives	▪ Find intersections of graphs using a graphing calculator.	▪ Solve systems of linear equations graphically. ▪ Solve systems of linear equations algebraically.	▪ Solve systems of linear equations by graphing. ▪ Determine the coordinates of the vertices of a region formed by the graph of a system of inequalities.	▪ Use a graphing calculator to solve systems of linear inequalities.
Key Vocabulary		system of equations dependent break-even point substitution consistent method inconsistent elimination independent method	substitution method elimination method	
CCSS	A.CED.3, A.REI.11	A.CED.3	A.CED.3	
Multiple Representations			🔧	
Lesson Resources	connectED.mcgraw-hill.com PT Personal Tutor **Materials:** ▪ TI-83/84 Plus or other graphing calculator	connectED.mcgraw-hill.com 📁 Leveled Worksheets 🔤 Vocabulary PT Personal Tutor ✋ Virtual Manipulatives ✓ Self-Check Quiz ▪ *5-Minute Check* ▪ *Study Notebook*	connectED.mcgraw-hill.com 📁 Leveled Worksheets 📁 Quiz 1 🔤 Vocabulary PT Personal Tutor ✓ Self-Check Quiz ▪ *5-Minute Check* ▪ *Study Notebook* ▪ *Teaching Algebra with Manipulatives*	connectED.mcgraw-hill.com PT Personal Tutor **Materials:** ▪ TI-83/84 Plus or other graphing calculator
Resources for Every Lesson	IWB eStudent Edition IWB Interactive Classroom	▪ eTeacher Edition ▪ eSolutions Manual ▪ McGraw-Hill eAssessment		
Differentiated Instruction		pp. 137, 140, 145	pp. 148, 150, 152	

IWB All digital assets are Interactive Whiteboard ready.

Suggested Pacing			
Time Periods	Instruction	Review & Assess	Total
45-minute	13 days	2 days	15 days
90-minute	7 days	1 day	8 days

LESSON 3-3 45 min: 1.5 days 90 min: 0.75 day	**LESSON 3-4** 45 min: 2 days 90 min: 1 day	**EXPLORE 3-5** 45 min: 0.5 day 90 min: 0.25 day	**LESSON 3-5** 45 min: 1 day 90 min: 0.5 day	**LESSON 3-6** 45 min: 1 day 90 min: 0.5 day
Optimization with Linear Programming	**Systems of Equations in Three Variables**	**Spreadsheet Lab: Organizing Data with Matrices**	**Operations with Matrices**	**Multiplying Matrices**
▪ Find the maximum and minimum values of a function over a region. ▪ Solve real-world optimization problems using linear programming.	▪ Solve systems of linear equations in three variables. ▪ Solve real-world problems using systems of linear equations in three variables.	▪ Organize and display data using matrices and spreadsheets.	▪ Analyze data in matrices. ▪ Perform algebraic operations with matrices.	▪ Multiply matrices. ▪ Use the properties of matrix multiplication.
linear programming feasible region optimize unbounded	ordered triple	matrix row element cell dimensions spreatsheet	scalar scalar multiplication	
A.CED.3	A.CED.3			
connectED.mcgraw-hill.com 🗀 Leveled Worksheets abc Vocabulary 🏃 Animations PT Personal Tutor 🖐 Virtual Manipulatives ✓ Self-Check Quiz ▪ *5-Minute Check* ▪ *Study Notebook* ▪ *Teaching Algebra with Manipulatives*	connectED.mcgraw-hill.com 🗀 Leveled Worksheets 🗀 Quiz 2 abc Vocabulary 🏃 Animations PT Personal Tutor ✓ Self-Check Quiz ▪ *5-Minute Check* ▪ *Study Notebook* ▪ *Teaching Algebra with Manipulatives*	connectED.mcgraw-hill.com abc Vocabulary **Materials:** ▪ computer with spreadsheet program	connectED.mcgraw-hill.com 🗀 Leveled Worksheets abc Vocabulary PT Personal Tutor ✓ Self-Check Quiz ▪ *5-Minute Check* ▪ *Study Notebook*	connectED.mcgraw-hill.com 🗀 Leveled Worksheets 🗀 Quiz 3 🏃 Animations PT Personal Tutor ✓ Self-Check Quiz ▪ *5-Minute Check* ▪ *Study Notebook*
IWB eStudent Edition IWB Interactive Classroom	▪ eTeacher Edition ▪ eSolutions Manual ▪ eAssessment McGraw-Hill			
pp. 156, 160	pp. 164, 167		pp. 174, 178	pp. 180, 181
	Formative Assessment Mid-Chapter Quiz			

	EXTEND 3-6 — 45 min: 0.5 day / 90 min: 0.25 day	LESSON 3-7 — 45 min: 1 day / 90 min: 0.75 day	LESSON 3-8 — 45 min: 1.5 days / 90 min: 0.75 day	EXTEND 3-8 — 45 min: 0.5 day / 90 min: 0.5 day	
Title	Graphing Technology Lab: Operations with Matrices	Solving Systems of Equations Using Cramer's Rule	Solving Systems of Equations Using Inverse Matrices	Graphing Technology Lab: Augmented Matrices	
Objectives	▪ Use a graphing calculator to explore operations with matrices.	▪ Evaluate determinants. ▪ Solve systems of linear equations by using Cramer's Rule.	▪ Find the inverse of a 2×2 matrix. ▪ Write and solve matrix equations for a system of equations.	▪ Use a graphing calculator and the augmented matrix for a system of equations to solve the system.	
Key Vocabulary	zero matrix additive inverse	determinant second-order determinant third-order determinant diagonal rule Cramer's Rule coefficient matrix	identity matrix square matrix inverses matrix equation variable matrix constant matrix	augmented matrix	
CCSS		A.CED.3	A.CED.3		
Multiple Representations					
Lesson Resources	connectED.mcgraw-hill.com PT Personal Tutor abc Vocabulary **Materials:** ▪ TI-83/84 Plus or other graphing calculator	connectED.mcgraw-hill.com ☐ Leveled Worksheets abc Vocabulary PT Personal Tutor ✓ Self-Check Quiz ▪ 5-Minute Check ▪ Study Notebook ▪ Teaching Algebra with Manipulatives	connectED.mcgraw-hill.com ☐ Leveled Worksheets ☐ Quiz 4 abc Vocabulary PT Personal Tutor ✓ Self-Check Quiz ▪ 5-Minute Check ▪ Study Notebook	connectED.mcgraw-hill.com abc Vocabulary PT Personal Tutor **Materials:** ▪ TI-83/84 Plus or other graphing calculator	
Resources for Every Lesson	IWB eStudent Edition IWB Interactive Classroom	▪ eTeacher Edition ▪ eSolutions Manual ▪ McGraw-Hill eAssessment			
Differentiated Instruction		pp. 190, 193, 197	pp. 200, 204		
				Summative Assessment Study Guide and Review Practice Test	

IWB All digital assets are Interactive Whiteboard ready.

Project CRISS SM

STUDY SKILL

Students may find it helpful to organize information they are learning in a frame. A sample content frame showing major concepts about systems of linear equations is shown. Students can complete this sample and/or develop their own frames to explain other concepts from Chapter 3.

Term	Sample Graph	Description	Definition (own words)
consistent system		intersecting	a system of equations with 1 or more solutions
inconsistent		parallel	a system of equations with no solution

Creating Independence through Student-owned Strategies

Assessment and Intervention

SE = Student Edition, TE = Teacher Edition, CRM = Chapter Resource Masters

Diagnosis	Prescription
Beginning Chapter 3	
Get Ready for Chapter 3 **SE**	Response to Intervention **TE**
Beginning Every Lesson	
Then, Now, Why? **SE** 5-Minute Checks	Chapter 0 **SE**

DIAGNOSTIC ASSESSMENT

During/After Every Lesson	
Guided Practice **SE** Check Your Understanding **SE** H.O.T. Problems **SE** Spiral Review **SE** Additional Examples **TE** Watch Out! **TE** Step 4, Assess **TE** Chapter 3 Quizzes **CRM**, pp. 59–60 Self-Check Quizzes connectED.mcgraw-hill.com	<div>**TIER 1 Intervention** Skills Practice **CRM**, Ch. 1–3 connectED.mcgraw-hill.com</div> <div>**TIER 2 Intervention** Differentiated Instruction **TE**; Study Guide and Intervention **CRM**, Ch. 1–3</div> <div>**TIER 3 Intervention** *Math Triumphs, Alg. 2,* Ch. 3</div>
Mid-Chapter	
Mid-Chapter Quiz **SE** Mid-Chapter Test **CRM**, p. 61 McGraw-Hill ℮Assessment	<div>**TIER 1 Intervention** Skills Practice **CRM**, Ch. 1–3 connectED.mcgraw-hill.com</div> <div>**TIER 2 Intervention** Study Guide and Intervention **CRM**, Ch. 1–3</div> <div>**TIER 3 Intervention** *Math Triumphs, Alg. 2,* Ch. 3</div>
Before Chapter Test	
Chapter Study Guide and Review **SE** Practice Test **SE** Standardized Test Practice **SE** Chapter Test connectED.mcgraw-hill.com Standardized Test Practice connectED.mcgraw-hill.com Vocabulary Review connectED.mcgraw-hill.com McGraw-Hill ℮Assessment	<div>**TIER 1 Intervention** Skills Practice **CRM**, Ch. 1–3 connectED.mcgraw-hill.com</div> <div>**TIER 2 Intervention** Study Guide and Intervention **CRM**, Ch. 1–3</div> <div>**TIER 3 Intervention** *Math Triumphs, Alg. 2,* Ch. 3</div>

FORMATIVE ASSESSMENT

After Chapter 3	
Multiple-Choice Tests, Forms 1, 2A, 2B **CRM**, pp. 63–68 Free-Response Tests, Forms 2C, 2D, 3 **CRM**, pp. 69–74 Vocabulary Test **CRM**, p. 62 Extended Response Test **CRM**, p. 75 Standardized Test Practice **CRM**, pp. 76–78 McGraw-Hill ℮Assessment	Study Guide and Intervention **CRM**, Ch. 1–3 connectED.mcgraw-hill.com

SUMMATIVE ASSESSMENT

Option 1 Reaching All Learners

Kinesthetic Put students into pairs. Give each pair a system of equations that is consistent and independent. Ask pairs to graph each equation on a separate transparency coordinate grid. Then have students combine the graphs by overlapping them. Have each student tell one point that is a solution and one point that is not a solution.

Interpersonal Give pairs of students a system of equations. Ask them to explain which method they would use to solve the system, how they decided on that method, and what steps that method will involve when they are actually solving the system.

Logical Identify those students who know how to perform matrix operations using their graphing calculators. Ask these students to make a list of the series of keystrokes needed. Then have them demonstrate how to use the graphing calculator to perform matrix operations.

Option 2 Approaching Level **AL**

On a coordinate grid, draw a pair of intersecting lines. Discuss with the class how to find the slope, *y*-intercept, point of intersection, and equation for each line. Explain that since these equations have been graphed on the same coordinate grid, they are called a system of equations. Demonstrate how the point of intersection is a solution to both equations. Then, shade the graph above one line and below the other. Discuss with students how the shading changes the equals signs in the system of equations to inequality symbols. Discuss how the shading affects the solution set.

On the board, draw a table of data having at least three columns and two rows. As you tell students that the data in your table can be represented efficiently in a matrix, erase the column and row titles and draw matrix brackets around the data. Use this matrix to introduce the vocabulary pertaining to matrices: *element, dimensions, row, column, etc.*

Option 3 English Learners **ELL**

Create a set of flashcards to help students understand the vocabulary in the first lesson of this chapter. On one side of the card, draw a graph of intersecting lines, one line, or a pair of parallel lines. On the other side, write *consistent and independent, consistent and dependent,* or *inconsistent.*

Option 4 Beyond Level **BL**

Have students use three different colored index cards or sheets of lightweight cardboard to model the graphs of system of equations in three variables. Once a model is constructed, have students indicate the number of solutions their system has. Below are four such models.

Infinitely many solutions

No solution

One solution

No solution

Focus on Mathematical Content

VerticalAlignment

Before Chapter 3

Related Topics from Algebra 1

- graph equations of lines
- transform and solve equations

Chapter 3

Related Topics from Algebra 2

- analyze situations and formulate systems of equations in two or more unknowns to solve problems
- use algebraic methods, graphs, or tables to solve systems of equations or inequalities
- interpret and determine the reasonableness of solutions to systems of equations for given contexts

After Chapter 3

Preparation for Precalculus

- define functions, describe characteristics of functions, and translate among verbal, numerical, graphical, and symbolic representations of functions
- use functions and their properties to model and solve real-life problems

Lesson-by-LessonPreview

3-1 Solving Systems of Equations

Two or more equations with the same variables are called a *system of equations.* A system of equations can be solved by graphing each equation on the same coordinate plane. A solution to a system is an ordered pair that represents a point on both lines and satisfies both equations.

For any two lines in a plane, the lines must be intersecting, coincident, or parallel.

- **consistent and dependent**
  intersecting lines: The system has exactly one solution and is *consistent and independent.*
- **consistent and dependent**
  coincident lines: The system has an infinite number of solutions and is *consistent and dependent.*
- **inconsistent**
  parallel lines: The system has no solution and is inconsistent.

The methods of substitution and elimination can be used to solve systems of equations algebraically.

- To use the substitution method, solve one of the equations for one of the two variables. Then substitute the resulting expression into the other equation. This results in a single equation in one variable that can be solved. Then substitute to solve for the other variable.

- The elimination method is used when the coefficients of one of the variables in both equations are the same or additive inverses. Add or subtract the two equations. Solve the resulting equation, and then solve for the other variable.

- If the equations do not have one variable with coefficients that are the same or additive inverses, then multiplication is used to write equivalent equations so that the result is a system of equations in which one of the variables has coefficients that are the same or additive inverses.

3-2 Solving Systems of Inequalities by Graphing

The inequalities in the system are graphed on the same coordinate plane, and the ordered pairs that satisfy all of the inequalities in the system are in the region that is common to the inequalities. If the regions do not intersect, the solution is the empty set, and no solution exists.

If the graph of a system of three or more linear inequalities forms an enclosed bounded region, the vertices of the region can be found by determining the coordinates of where the boundary lines intersect.

Get educator support from the Math Forum's T2T. You may want to search, or ask, for teaching strategies on systems of equations. Start your search at connectED.mcgraw-hill.com.

mathforum.org
The Math Forum
@ Drexel University

3-3 Optimization with Linear Programming

Solving a linear programming problem builds on the skills learned in Lesson 3-2. After the system of inequalities has been graphed and the coordinates of the vertices of the feasible region determined, a given function must be evaluated using the vertex coordinates. The solution to the problem is the point that maximizes or minimizes the function.

A bounded feasible region is one whose outline is a polygon, and an unbounded feasible region is one whose boundary is not enclosed by the constraints.

3-4 Systems of Equations in Three Variables

The graph of a first-degree equation in three variables is a plane. There are several possibilities for the solution to a system of equations in three variables, since the three planes can intersect in a variety of ways.

- at a *point:* The ordered triple for that point is a single solution to the system.

- in a *line:* The system has an infinite number of solutions, any point on the line.

- in the same *plane:* This system would also have an infinite number of solutions, all points in the plane.

- *no points in common:* The solution is the empty set.

3-5 Operations with Matrices

To add or subtract two matrices, first check to see that the matrices have the same dimensions. Then calculate the sum or difference of corresponding elements. The sum or difference of the two matrices will have the same dimensions as the two matrices being added or subtracted.

The product of a scalar and a matrix is found by multiplying every element in the matrix by the scalar. The product is a matrix that has the same dimensions as the matrix being multiplied. For example,

$$2\begin{bmatrix} -2 & 1 & 0 \\ 4 & -1 & 3 \end{bmatrix} = \begin{bmatrix} 2(-2) & 2(1) & 2(0) \\ 2(4) & 2(-1) & 2(3) \end{bmatrix} = \begin{bmatrix} -4 & 2 & 0 \\ 8 & -2 & 6 \end{bmatrix}.$$

Properties of matrix operations are examined. Students look at examples that illustrate the Commutative and Associative Properties of matrix addition and the Distributive Property of scalar multiplication.

3-6 Multiplying Matrices

Two matrices can only be multiplied if the number of columns in the first matrix is equal to the number of rows in the second matrix. For example, the product of $A_{m \times n}$ and $B_{n \times t}$ is a matrix with dimensions $m \times t$.

In matrix multiplication, the elements of a row in one matrix are multiplied by the elements in a column of the second matrix. Like matrix addition and subtraction, the result of the combination is a single element. For example,

$$\begin{bmatrix} 3 & -2 \\ 1 & 4 \end{bmatrix} \cdot \begin{bmatrix} 2 & 4 \\ 5 & 1 \end{bmatrix} = \begin{bmatrix} 3(2) + (-2)(5) & 3(4) + (-2)(1) \\ 1(2) + 4(5) & 1(4) + 4(1) \end{bmatrix} = \begin{bmatrix} -4 & 10 \\ 22 & 8 \end{bmatrix}.$$

3-7 Solving Systems of Equations using Cramer's Rule

Every square matrix has a determinant. The determinant of a 2×2 matrix can be evaluated by calculating the difference of the products of the two diagonals. The determinant of a 3×3 matrix can be evaluated using the diagonal rule.

Cramer's Rule can be used to solve systems of equations. To use Cramer's Rule to solve a system of equations,

- write each equation in standard form,

- then set each variable equal to a fraction whose denominator and numerator are both determinants.

A system has no solution if the value of the determinant in the denominator is zero.

3-8 Solving Systems of Equations using Inverse Matrices

The identity matrix for multiplication is a square matrix with 1 for every element of the main diagonal and 0 in all other positions.

Two square matrices are inverses of each other if their product is the identity matrix. The multiplicative inverse for a 2×2 matrix is given by a product of two factors.

- One factor is a fraction whose numerator is 1 and whose denominator is the determinant of the given matrix.

- The second factor is the given matrix modified. The elements along the left-to-right diagonal are switched and the elements along the other diagonal are multiplied by -1.

A single matrix equation can be written to represent a system of equations. One side of the matrix equation shows the matrix of coefficients times a column matrix of variables. On the other side of the equation is a column matrix of the constants.

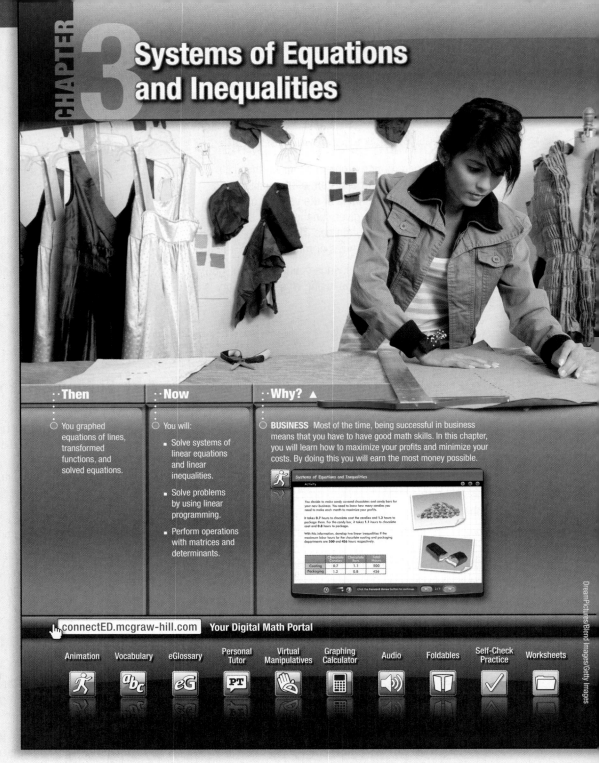

Systems of Equations and Inequalities

CHAPTER 3

Chapter Project

Operation Optimization

Students use what they have learned about linear programming to complete a project.

This chapter project addresses business literacy, as well as several specific skills identified as being essential to student success by the Framework for 21st Century Learning.

Visit connectED.mcgraw-hill.com for student and teacher handouts.

KeyVocabulary Introduce the key vocabulary in the chapter using the routine below.

<u>Define</u>: A consistent system of equations is one that has at least one solution.

<u>Example</u>: The system of equations shown has one solution, so it is consistent.

<u>Ask</u>: Would a system of equations that has infinitely many solutions be considered consistent? Explain.

$y = -\frac{1}{3}x + 4$

$y = 2x + 1$

Yes; because it would have more than one solution.

··Then

○ You graphed equations of lines, transformed functions, and solved equations.

··Now

○ You will:
- Solve systems of linear equations and linear inequalities.
- Solve problems by using linear programming.
- Perform operations with matrices and determinants.

··Why? ▲

○ **BUSINESS** Most of the time, being successful in business means that you have to have good math skills. In this chapter, you will learn how to maximize your profits and minimize your costs. By doing this you will earn the most money possible.

connectED.mcgraw-hill.com Your Digital Math Portal

| Animation | Vocabulary | eGlossary | Personal Tutor | Virtual Manipulatives | Graphing Calculator | Audio | Foldables | Self-Check Practice | Worksheets |

Get Ready for the Chapter

Diagnose Readiness | You have two options for checking prerequisite skills.

1 **Textbook Option** Take the Quick Check below. Refer to the Quick Review for help.

QuickCheck	QuickReview

Graph each equation. **1–6. See Chapter 3 Answer Appendix.**

1. $x = 4y$

2. $y = \frac{1}{3}x + 5$

3. $x + 2y = 4$

4. $y = -x + 6$

5. $3x + 5y = 15$

6. $3y - 2x = -12$

7. BUSINESS A museum charges $8.50 for adult tickets and $5.25 for children's tickets. On Friday they made $650.

  a. Write an equation that can be used to model the ticket sales. $8.50a + 5.25c = 650$

  b. Graph the equation. **See Chapter 3 Answer Appendix.**

Example 1 (Used in Lesson 3-1)

Graph $2y + 5x = -10$.

Find the x- and y-intercepts.

$$2(0) + 5x = -10 \qquad 2y + 5(0) = -10$$
$$5x = -10 \qquad\qquad 2y = -10$$
$$x = -2 \qquad\qquad\quad y = -5$$

The graph crosses the x-axis at $(-2, 0)$ and the y-axis at $(0, -5)$. Use these ordered pairs to graph the equation.

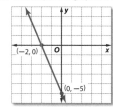

Graph each inequality.

8. $y < 3$

9. $x + y \geq 1$

10. $3x - y > 6$

11. $x + 2y \leq 5$

12. $y > 4x - 1$

13. $5x - 4y < 12$

8–13. See Chapter 3 Answer Appendix.

14. FUNDRAISER The student council is selling T-shirts for $15 and sweatshirts for $25. They must make $2500 to cover expenses. Write and graph an inequality to show the number of T-shirts and sweatshirts that they must sell. **See Chapter 3 Answer Appendix.**

Example 2 (Used in Lessons 3-2 and 3-3)

Graph $y \geq 3x - 2$.

The boundary is the graph of $y = 3x - 2$. Since the inequality symbol is $\geq$, the boundary will be solid.

Test the point $(0, 0)$.

$$0 \overset{?}{\geq} 3(0) - 2 \qquad (x, y) = (0, 0)$$
$$0 \geq -2 \;\checkmark$$

Shade the region that includes $(0, 0)$.

---

2 **Online Option** Take an online self-check Chapter Readiness Quiz at connectED.mcgraw-hill.com.

133

Get Ready for the Chapter

Response to Intervention (RtI) Use the *Quick Check* results and the Intervention Planner to help you determine your Response to Intervention. The If-Then statements in the chart below help you decide the appropriate tier of RtI and suggest intervention resources for each tier.

InterventionPlanner

TIER 1 **On Level** (OL)

If students miss about 25% of the exercises or less,

Then choose a resource:

SE Lessons 3-1 and 3-2

 Skills Practice, Chapter 3, pp. 7, 14

 Self-Check Quiz

TIER 2 **Strategic Intervention** (AL) approaching grade level

If students miss about 50% of the exercises,

Then choose a resource:

 Study Guide and Intervention, Chapter 3, pp. 5, 12

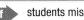 Extra Examples, Personal Tutor, Homework Help

TIER 3 **Intensive Intervention** 2 or more grades below level

If students miss about 75% of the exercises,

Then use *Math Triumphs Alg. 2*, Ch. 3

 Extra Examples, Personal Tutor, Homework Help, Review Vocabulary

---

E? **Essential Question**

• How can you find the solution to a math problem? Sample answers: Use a graph; analyze a table of values; solve an equation; guess and check.

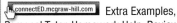

FOLDABLES® StudyOrganizer

Dinah Zike's Foldables®

Focus Students write notes about the vocabulary and concepts in the lessons of this chapter.

Teach Have students make and label their Foldables as illustrated. For each tab, have students develop their own example of the concept that is indicated. Above each example, encourage students to write a definition of the concept in their own words.

When to Use It Encourage students to add to their Foldables as they work through the chapter and to use them to review for the chapter test.

Differentiated Instruction

📁 Student-Built Glossary, pp. 1–2

Students should complete the chart by providing a definition of each term and an example as they progress through Chapter 3. This study tool can also be used to review for the chapter test.

Get Started on the Chapter

You will learn several new concepts, skills, and vocabulary terms as you study Chapter 3. To get ready, identify important terms and organize your resources. You may refer to Chapter 0 to review prerequisite skills.

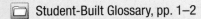

Systems of Equations and Inequalities Make this Foldable to help you organize your Chapter 3 notes about systems of equations and inequalities. Begin with a sheet of $8\frac{1}{2}''$ by $11''$ paper.

1 **Fold** in half along the height.

2 **Cut** along the fold.

3 **Fold** each sheet along the width into fourths.

4 **Tape** the ends of two sheets together.

5 **Label** the tabs with *Solve by Graphing, Substitution Method, Elimination Method, Optimization, Systems in Three Variables, Matrix Multiplication, Cramer's Rule,* and *Inverse Matrices.*

NewVocabulary

English		Español
break-even point	p. 136	punto de equilibrio
consistent	p. 137	consistente
inconsistent	p. 137	inconsistente
independent	p. 137	independiente
dependent	p. 137	dependiente
substitution method	p. 138	método de substitución
elimination method	p. 139	método de eliminación
feasible region	p. 154	región viable
bounded	p. 154	acotada
unbounded	p. 154	no acotado
optimize	p. 156	optimizer
dimensions	p. 169	tamaño
scalar	p. 173	escalar
determinant	p. 189	determinante
Cramer's Rule	p. 192	regula de Crámer
coefficient matrix	p. 192	matriz coeficiente
identity matrix	p. 198	matriz identidad
square matrix	p. 198	matriz cuadrada
inverse matrix	p. 198	matriz inversa
variable matrix	p. 200	matriz variables
constant matrix	p. 200	matriz constante

ReviewVocabulary

inequality *desigualdad*
an open sentence that contains the symbol $<$, $\leq$, $>$, or $\geq$

linear equation *ecuación lineal*
an equation that has no operations other than addition, subtraction, and multiplication of a variable by a constant

EXPLORE 3-1

Graphing Technology Lab
Intersections of Graphs

You can use a TI-83/84 Plus graphing calculator to find where two graphs intersect. You can use the Y= menu to graph each equation on the same set of axes.

CCSS Common Core State Standards
Content Standards
A.CED.3 Represent constraints by equations or inequalities, and by systems of equations and/or inequalities, and interpret solutions as viable or nonviable options in a modeling context.
A.REI.11 Explain why the x-coordinates of the points where the graphs of the equations $y = f(x)$ and $y = g(x)$ intersect are the solutions of the equation $f(x) = g(x)$; find the solutions approximately, e.g., using technology to graph the functions, make tables of values, or find successive approximations. Include cases where $f(x)$ and/or $g(x)$ are linear, polynomial, rational, absolute value, exponential, and logarithmic functions.

Example   Intersection of Two Graphs

Graph both equations in the standard viewing window.

$3x + y = 9$
$x - y = -1$

Step 1 Write each equation in the form $y = mx + b$.

$$3x + y = 9 \qquad\qquad x - y = 1$$
$$y = -3x + 9 \qquad\qquad -y = -x - 1$$
$$\qquad\qquad\qquad y = x + 1$$

Step 2 Enter $y = -3x + 9$ as **Y1** and $y = x + 1$ as **Y2**. Then graph the lines.

KEYSTROKES: Y= (−) 3 X,T,θ,n + 9 ENTER
X,T,θ,n + 1 ENTER ZOOM 6

Step 3 Find the intersection of the lines.

KEYSTROKES: 2nd [CALC] 5 ENTER ENTER ENTER

The intersection is at (2, 3).

[−10, 10] scl: 1 by [−10, 10] scl: 1

Exercises

Use a graphing calculator to find where each pair of graphs intersect.

1. $2x + 4y = 36$
$10y - 5x = 0$ **(9, 4.5)**

2. $2y - 3x = 7$
$5x = 4y - 12$ **(−2, 0.5)**

3. $4x - 2y = 16$
$7x + 3y = 15$ **(3, −2)**

4. $2x + 4y = 4$
$x + 3y = 13$ **(−20, 11)**

5. $5x + y = 13$
$3x = 15 - 3y$ **(2, 3)**

6. $4y - 5 = 20 - 3x$
$4x - 7y + 16 = 0$ **(3, 4)**

7. $\frac{1}{4}x + y = \frac{11}{4}$
$x - \frac{1}{2}y = 2$ **(3, 2)**

8. $3x + 2y = -3$
$x + \frac{1}{3}y = -4$ **(−7, 9)**

9. $3x - 6y = 6$
$2x - 4y = 4$ **infinite intersections**

10. $6x + 8y = -16$
$3x + 4y = 12$ **no intersection**

From Concrete to Abstract
For each exercise, ask students which equation represents each line. Then, ask them to identify, in the graph, the location of the intersection.

Extending the Concept
Ask:
- How many solutions of the system of equations are there when there is an intersection at one point? no intersections? infinite intersections? one solution; no solutions; infinitely many solutions

1 Focus

Objective Find intersections of graphs using a graphing calculator.

Materials for Each Student
- TI-83/84 Plus or other graphing calculator

Teaching Tip
Be sure students know how to find, use, and clear the Y= screen, and that they know how to graph the equation(s) entered.

2 Teach

Working in Cooperative Groups
Have students work in small groups. Then, for each exercise, have one student rewrite the system's equations in $y = mx + b$ form, have another student enter the equations in the Y= screen, and have a third student enter the keystrokes to find the point(s) of intersection.

Ask:
- How do you isolate y to rewrite an equation in $y = mx + b$ form? Isolate the y term on one side of the equation. Then divide both sides of the equation by the coefficient of y.

Practice Have students complete Exercises 1–10.

3 Assess

Formative Assessment
Use Exercises 1–10 to assess whether students can use a graphing calculator to find intersections of graphs.

1 Focus

VerticalAlignment

Before Lesson 3-1 Graph and solve linear equations.

Lesson 3-1 Solve systems of linear equations, graphically. Solve system of equations algebraically.

After Lesson 3-1 Solve systems of linear inequalities.

2 Teach

Scaffolding Questions

Have students read the **Why?** section of the lesson.

Ask:

- What is the *y*-intercept of the line representing Libby's cost? (0, 450)

- What does this point represent? Libby's start-up cost is $450. She will have to pay this amount even if she doesn't mow any lawns.

- What happens if Libby mows 17 lawns or more? She makes a profit.

LESSON 3-1 Solving Systems of Equations

Income Versus Cost

:: Then	:: Now	:: Why?
● You graphed and solved linear equations.	**1** Solve systems of linear equations graphically. **2** Solve systems of linear equations algebraically.	● Libby borrowed $450 to start a lawn-mowing business. She charges $35 per lawn and incurs $8 in operating costs per lawn. A system of equations can be used to determine the break-even point. The **break-even point** is the point at which income equals cost.

NewVocabulary
break-even point
system of equations
consistent
inconsistent
independent
dependent
substitution method
elimination method

CCSS Common Core State Standards

Content Standards
A.CED.3 Represent constraints by equations or inequalities, and by systems of equations and/or inequalities, and interpret solutions as viable or nonviable options in a modeling context.

Mathematical Practices
2 Reason abstractly and quantitatively.
6 Attend to precision.

1 Solve Systems Graphically A **system of equations** is two or more equations with the same variables. To solve a system of equations with two variables, find the ordered pair that satisfies all of the equations.

To solve a system of equations by using a table, first write each equation in slope-intercept form. Then substitute different values for *x* and solve for the corresponding *y*-values. For ease of use, choose 0 and 1 as your first *x*-values.

$y_1 = -2x + 8$
$y_2 = 4x - 7$

x	y_1	y_2	Difference
0	8	−7	15
1	6	−3	9
2	4	1	3
3	2	5	−3

Because the difference between the *y*-values is closer to 0 for *x* = 1 than *x* = 0, a value greater than 1 should be tried next.

Because the difference between the *y*-values changed signs from *x* = 2 to *x* = 3, a value between these should be tried next.

The solution is between 2 and 3.

Example 1 Solve by Using a Table

Solve the system of equations.

$3x + 2y = -2$
$-4x + 5y = -28$

Write each equation in slope-intercept form.

$3x + 2y = -2 \rightarrow y = -1.5x - 1$
$-4x + 5y = -28 \rightarrow y = 0.8x - 5.6$

Use a table to find the solution that satisfies both equations.

The solution of the system is (2, −4).

x	y_1	y_2	(x, y_1)	(x, y_2)
0	−1	−5.6	(0, −1)	(0, −5.6)
1	−2.5	−4.8	(1, −2.5)	(1, −4.8)
2	−4	−4	(2, −4)	(2, −4)

GuidedPractice

1A. $2x - 5y = 11$
$-3x + 4y = -13$ (3, −1)

1B. $4x + 3y = -17$
$-7x - 2y = 20$ (−2, −3)

 136 | Lesson 3-1

Lesson 3-1 Resources

Resource	Approaching Level **AL**	On Level **OL**	Beyond Level **BL**	English Learners **ELL**
Teacher Edition	▪ Differentiated Instruction, p. 140	▪ Differentiated Instruction, pp. 137, 140	▪ Differentiated Instruction, pp. 137, 145	
Chapter Resource Masters	▪ Study Guide and Intervention, pp. 5–6 ▪ Skills Practice, p. 7 ▪ Practice, p. 8 ▪ Word Problem Practice, p. 9	▪ Study Guide and Intervention, pp. 5–6 ▪ Skills Practice, p. 7 ▪ Practice, p. 8 ▪ Word Problem Practice, p. 9 ▪ Enrichment, p. 10 ▪ Spreadsheet Activity, p. 11	▪ Practice, p. 8 ▪ Word Problem Practice, p. 9 ▪ Enrichment, p. 10	▪ Study Guide and Intervention, pp. 5–6 ▪ Skills Practice, p. 7 ▪ Practice, p. 8 ▪ Word Problem Practice, p. 9
Other	▪ 5-Minute Check 3-1 ▪ Study Notebook	▪ 5-Minute Check 3-1 ▪ Study Notebook	▪ 5-Minute Check 3-1 ▪ Study Notebook	▪ 5-Minute Check 3-1 ▪ Study Notebook

Another method for solving a system of equations is to graph the equations on the same coordinate plane. The point of intersection represents the solution.

Example 2 Solve by Graphing

Solve the system of equations by graphing.

$$2x - y = -1 \qquad 2y + 5x = -16$$

Write each equation in slope-intercept form.

$$2x - y = -1 \quad \rightarrow \quad y = 2x + 1$$
$$2y + 5x = -16 \quad \rightarrow \quad y = -2.5x - 8$$

The graphs of the lines appear to intersect at $(-2, -3)$.

CHECK Substitute the coordinates into each original equation.

$$
\begin{array}{lll}
2x - y = -1 & 2y + 5x = -16 & \text{Original equations} \\
2(-2) - (-3) \stackrel{?}{=} -1 & 2(-3) + 5(-2) \stackrel{?}{=} -16 & x = -2 \text{ and } y = -3 \\
-1 = -1 \; \checkmark & -16 = -16 \; \checkmark & \text{Simplify.}
\end{array}
$$

The solution of the system is $(-2, -3)$.

> **GuidedPractice**
>
> **2A.** $4x + 3y = 12$
> $-6x + 4y = -1$ $(1.5, 2)$
>
> **2B.** $-3y + 8x = 36$
> $6x + y = -21$ $(-4, 3)$

StudyTip

Checking Solutions Always check to see if the values work for **both** of the original equations.

Systems of equations can be classified by the number of solutions. A system is **consistent** if it has at least one solution and **inconsistent** if it has no solutions. If it has exactly one solution, it is **independent**, and if it has an infinite number of solutions, it is **dependent**.

StudyTip

Slope and Classifying Systems If the equations have different slopes, then the system is consistent and independent.

Example 3 Classify Systems

Graph each system of equations and describe them as *consistent and independent*, *consistent and dependent*, or *inconsistent*.

a. $4x + 3y = 24$
$-3x + 5y = 30$

b. $-2x + 5y = 10$
$4x - 10y = -20$

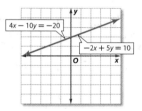

The graphs of the lines intersect at one point, so there is one solution. The system is consistent and independent.

Because the equations are equivalent, their graphs are the same line. The system is consistent and dependent.

> **GuidedPractice**   3A–3B. See Chapter 3 Answer Appendix.
>
> **3A.** $6x - 4y = 15$
> $-6x + 4y = 18$
>
> **3B.** $-4x + 5y = -17$
> $-4x - 2y = 15$

connectED.mcgraw-hill.com **137**

DifferentiatedInstruction OL BL

If students demonstrate an understanding of how to graph a system of equations and find the point of intersection,

Then have them create a list of tips to help someone draw the graphs for a system of equations and find the point of intersection easily and efficiently.

---

1 Solve Systems Graphically

Example 1 shows how to solve a system of equations by completing a table. **Example 2** shows how to solve a system of equations by graphing. **Example 3** shows how to classify systems of linear equations as *consistent and independent, consistent and dependent*, or *inconsistent*.

Formative Assessment

Use the Guided Practice exercises after each example to determine students' understanding of concepts.

Additional Examples

1 Solve the system of equations by completing a table. $(3, 0)$
$$x + y = 3$$
$$-2x + y = -6$$

2 Solve the system of equations by graphing. $(4, 2)$
$$x - 2y = 0$$
$$x + y = 6$$

3 Graph each system of equations and describe it as *consistent and independent, consistent and dependent*, or *inconsistent*.

a. $x - y = 5$
$x + 2y = -4$

consistent and independent

(continued on the next page)

> **Additional Examples** also in Interactive Classroom PowerPoint® Presentations

IWB Interactive White Board
READY

connectED.mcgraw-hill.com **137**

Additional Example

b. $15x - 6y = 0$
$\quad\ 5x - 2y = 10$

$15x - 6y = 0$

$5x - 2y = 10$

inconsistent

2 Solve Systems Algebraically

Example 4 shows how to solve a system of linear equations by using the algebraic method of substitution.

Examples 5–6 show how to solve systems of linear equations by using the algebraic method of elimination.

Additional Example

4 FURNITURE Lancaster Woodworkers Furniture Store builds two types of wooden outdoor chairs. A rocking chair sells for $265, and an Adirondack chair with footstool sells for $320. The books show that last month, the business earned $13,930 for the 48 outdoor chairs sold. How many rocking chairs were sold? **26**

IWB Interactive White Board
READY

The relationship between the graph and the solutions of a system is summarized below.

ConceptSummary Characteristics of Linear Systems

Consistent and Independent	Consistent and Dependent	Inconsistent
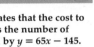		
intersecting lines; one solution	same line; infinitely many solutions	parallel lines; no solution

2 Solve Systems Algebraically Algebraic methods are used to find exact solutions of systems of equations. One algebraic method is called the **substitution method**.

KeyConcept Substitution Method

Step 1 Solve one equation for one of the variables.

Step 2 Substitute the resulting expression into the other equation to replace the variable. Then solve the equation.

Step 3 Substitute to solve for the other variable.

Systems of equations can be used to solve many real-world problems involving constraints modeled by two or more different functions.

Real-World Example 4 Use the Substitution Method

BUSINESS Alejandro has a computer support business. He estimates that the cost to run his business can be represented by $y = 48x + 500$, where x is the number of customers. He also estimates that his income can be represented by $y = 65x - 145$. How many customers will Alejandro need in order to break even? What will his profit be if he has 60 customers?

$y = 65x - 145$	Income equation
$48x + 500 = 65x - 145$	Substitute $48x + 500$ for y.
$500 = 17x - 145$	Subtract $48x$ from each side.
$645 = 17x$	Add 145 to each side.
$37.9 \approx x$	Divide each side by 17.

Alejandro needs 38 customers to break even. If he has 60 customers, his income will be $65(60) - 145$ or $3755, and his costs will be $48(60) + 500$ or $3380, so his profit will be $3755 - 3380$ or $375.

CHECK You can use a graphing calculator to check this solution. The break-even point is near (37.9, 2321.2). Use the **CALC** function to find the cost and income for 60 customers.

Intersection
X=37.941176 Y=2321.1765

[0, 65] scl: 5 by [0, 3000] scl: 300

 138 | Lesson 3-1 | Solving Systems of Equations

WatchOut!

Parallel Lines Remind students that the graphs of equations with the same slope and different intercepts are, by definition, parallel.

Teach with Tech

Student Response System Write a system of equations on the board. Ask students to graphically determine the number of solutions for this system. Have them reply with A for no solutions, B for one solution, or C for infinitely many solutions. Discuss students' responses with the class.

▶ GuidedPractice

Use substitution to solve each system of equations.

4A. $5x - 3y = 23$
$2x + y = 7$ $(4, -1)$

4B. $x - 7y = 11$
$5x + 4y = -23$ $(-3, -2)$

4C. $-6x - y = 27$
$3x + 8y = 9$ $(-5, 3)$

You can use the **elimination method** to solve a system when one of the variables has the same coefficient in both equations.

> **KeyConcept** Elimination Method
>
> **Step 1** Multiply one or both equations by a number to result in two equations that contain opposite terms.
>
> **Step 2** Add the equations, eliminating one variable. Then solve the equation.
>
> **Step 3** Substitute to solve for the other variable.

Variables can be eliminated by addition or subtraction.

StudyTip

CCSS Perseverance
Remember when you add or subtract one equation from another to add or subtract every term, including the constant on the other side of the equal sign.

Example 5 Solve by Using Elimination

Use the elimination method to solve the system of equations.

$5x + 3y = -19$
$8x + 3y = -25$ ◄ — Notice that solving by substitution would involve fractions.

Step 1 Multiply one equation by -1 so the equations contain $3y$ and $-3y$.

$8x + 3y = -25$ [Multiply by -1.] $-8x - 3y = 25$

Step 2 Add the equations to eliminate one variable.

$ 5x + 3y = -19$ Equation 1
$\underline{(+) -8x - 3y = 25}$ Equation 2 × (−1)
$ -3x = 6$ Add the equations.
$x = -2$ Divide each side by −3.

Step 3 Substitute -2 for x into either original equation.

$8x + 3y = -25$ Equation 2
$8(-2) + 3y = -25$ $x = -2$
$-16 + 3y = -25$ Multiply.
$3y = -9$ Add 16 to each side.
$y = -3$ Divide each side by 3.

The solution is $(-2, -3)$.

▶ GuidedPractice

5A. $4x - 3y = -22$
$2x + 3y = 16$ $(-1, 6)$

5B. $6x - 5y = -8$
$4x - 5y = -12$ $(2, 4)$

5C. $2x - 9y = 34$
$-2x + 6y = -28$ $(8, -2)$

ReviewVocabulary

Least Common Multiple
the least number that is a common multiple of two or more numbers

Sometimes, adding or subtracting equations will not eliminate either variable. You can use multiplication and least common multiples to find a common coefficient.

Additional Example

5 Use the elimination method to solve the system of equations.
$x + 2y = 10$
$x + y = 6$
$(2, 4)$

WatchOut!

Preventing Errors In Example 4, recommend to students that when they find a solution, they check their answers against the facts presented in the original problem.

CCSS **Teaching the Mathematical Practices**

Perseverance Mathematically proficient students monitor and evaluate their progress and change course if necessary. Remind students that they are adding or subtracting an *equation* rather than a single term.

Teach with Tech
Document Camera Choose two students to share their work with the class. Select students who solved the same system of equations in different ways (by solving for different variables).

Additional Example

6 **STANDARDIZED TEST PRACTICE**
Solve the system of equations.

$2x + 4y = 12$
$-5x - 10y = -30$

A $(2, 3)$

B $(6, 0)$

C no solution

D infinite solutions   D

Tips for New Teachers

Reasoning Remind students to take time to plan a problem-solving strategy before starting calculations. Advise students that they may find it easier to eliminate one of the variables rather than the other using this technique.

Focus on Mathematical Content

Solving Systems of Equations If the coefficients for one of the variables in the equation are already the same or opposite, that variable can be easily eliminated by subtracting or adding the given equations. If this is not the case, it may be a better choice to use the method of substitution. Substitution and elimination will both solve a system of equations. You will get the same solution no matter which method you choose.

Tips for New Teachers

Subtraction Remind students that when a variable has identical coefficients in two equations, it can be eliminated by subtraction.

Solve the system of equations.

$5x + 3y = 52$
$15x + 9y = 54$

A $(3, 1)$ **B** $(8, 4)$ **C** no solution   **D** infinite solutions

Read the Test Item

You are given a system of two linear equations and are asked to find the solution.

Solve the Test Item

Neither variable has a common coefficient. The coefficients of the y-variables are 3 and 9 and their least common multiple is 9, so multiply each equation by the value that will make the y-coefficient 9.

$5x + 3y = 52$ Multiply by 4. $15x + 9y = 156$
$15x + 9y = 54$ $(-)\ 15x + 9y = 54$
                                       $\overline{\qquad\qquad 0 = 102}$ Subtract the equations.

Because $0 = 102$ is not true, this system has no solution.

The correct answer is C.

> **StudyTip**
>
> **Adding and Subtracting Equations** If you add or subtract two equations in a system and the result is an equation that is never true, then the system is inconsistent. When you add or subtract two equations in a system and the result is an equation that is always true, then the system is dependent.

GuidedPractice

6. Solve the system of equations. $2x + 3y = 5$
                                        $6x + 9y = 15$ J

   F $(-2, 3)$ **G** $(7, 3)$ **H** no solution   **J** infinite solutions

The following summarizes the various methods for solving systems.

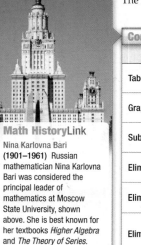

Math HistoryLink
Nina Karlovna Bari (1901–1961) Russian mathematician Nina Karlovna Bari was considered the principal leader of mathematics at Moscow State University, shown above. She is best known for her textbooks *Higher Algebra* and *The Theory of Series.*

ConceptSummary Solving Systems of Equations

Method	The Best Time to Use
Table	to estimate the solution, since a table may not provide an exact solution
Graphing	to estimate the solution, since graphing usually does not give an exact solution
Substitution	if one of the variables in either equation has a coefficient of 1 or −1
Elimination Using Addition	if one of the variables has opposite coefficients in the two equations
Elimination Using Subtraction	if one of the variables has the same coefficient in the two equations
Elimination Using Multiplication	if none of the coefficients are 1 or −1 and neither of the variables can be eliminated by simply adding or subtracting the equations

 140 | **Lesson 3-1** | Solving Systems of Equations

Brendan Hoffman/Alamy

Differentiated Instruction **AL** **OL**

Logical Learners Have students summarize the various algebraic methods for solving a system of equations using if-then statements and examples. Sample: If one of the equations has a variable with a coefficient of 1 (such as $x + 3y = 9$ or $5x + y = 13$), then consider the substitution method.

Example 1 Solve each system of equations by using a table.

1. $y = 3x - 4$
$y = -2x + 11$ (3, 5)

2. $4x - y = 1$
$5x + 2y = 24$ (2, 7)

Example 2 Solve each system of equations by graphing.

3. $y = -3x + 6$
$2y = 10x - 36$ (3, −3)

4. $y = -x - 9$
$3y = 5x + 5$ (−4, −5)

5. $y = 0.5x + 4$
$3y = 4x - 3$ (6, 7)

6. $-3y = 4x + 11$
$2x + 3y = -7$ (−2, −1)

7 $4x + 5y = -41$
$3y - 5x = 5$ (−4, −5)

8. $8x - y = 50$
$x + 4y = -2$ (6, −2)

9. **CCSS** **MODELING** Refer to the table at the right. **a.** $y = 0.15x + 2.70$, $y = 0.25x$

a. Write equations that represent the cost of printing digital photos at each lab.

b. Under what conditions is the cost to print digital photos the same at both stores? $6.75 for 27 photos

c. When is it best to use EZ Online Digital Photos and when is it best to use the local store?

9c. You should use EZ Online Digital Photos if you are printing more than 27 digital photos, and the local pharmacy if you are printing fewer than 27 photos.

Digital Photos

Online Store
$0.15 per photo + $2.70 shipping

Local Store
$0.25 per photo

Example 3 Graph each system of equations and describe it as *consistent and independent*, *consistent and dependent*, or *inconsistent*. **10–12.** See margin.

10. $y + 4x = 12$
$3y = 8 - 12x$

11. $-2x - 3y = 9$
$4x + 6y = -18$

12. $9x - 2y = 11$
$5x + 4y = 13$

Example 4 Solve each system of equations by using substitution.

13. $x + 5y = 3$
$3x - 2y = -8$ (−2, 1)

14. $y = 2x - 10$
$y = -4x + 8$ (3, −4)

15 $2a + 8b = -8$
$3a - 5b = 22$ (4, −2)

16. $a - 3b = -22$
$4a + 2b = -4$ (−4, 6)

17. $6x - 7y = 23$
$8x + 4y = 44$ (5, 1)

18. $9c - 3d = -33$
$6c + 5d = -8$ (−3, 2)

Examples 5–6 Solve each system of equations by using elimination.

19. $-6w - 8z = -44$
$3w + 6z = 36$ (−2, 7)

20. $4x - 3y = 29$
$4x + 3y = 35$ (8, 1)

21. $3a + 5b = -27$
$4a + 10b = -46$ (−4, −3)

22. $8a - 3b = -11$
$5a + 2b = -3$ (−1, 1)

23. $5a + 15b = -24$
$-2a - 6b = 28$ no solution

24. $6x - 4y = 30$
$12x + 5y = -18$ (1, −6)

25. **MULTIPLE CHOICE** What is the solution of the linear system? B

$4x + 3y = 2$
$4x - 2y = 12$

A (8, −10) **B** (2, −2) **C** (−10, 14) **D** no solution

connectED.mcgraw-hill.com **141** 🔊

Harald Hinze/age fotostock

Differentiated Homework Options

Level	Assignment	Two-Day Option	
AL Basic	26–58, 76, 78–97	27–57 odd, 81–84	26–58 even, 76, 78–80, 85–97
OL Core	27–67 odd, 68–76, 81–97	26–58, 81–84	59–76, 78–80, 85–97
BL Advanced	59–94, (optional: 95–97)		

---

3 Practice

Formative Assessment
Use Exercises 1–25 to check for understanding.

Use the chart on the bottom of the page to customize assignments for your students.

CCSS **Teaching the Mathematical Practices**

Modeling Mathematically proficient students can apply the mathematics they know to solve problems arising in everyday life, analyze relationships mathematically to draw conclusions, and interpret their mathematical results in the context of a situation.

Additional Answers

10.

inconsistent

11.

consistent, dependent

12.

consistent, independent

Follow-up

Students have explored solving systems of linear equations by graphing and using algebra.

Ask:

- What are the benefits of having different strategies for solving systems of equations? Sample answer: You can use the strategy that is most efficient. For example, if an estimate of the solution is sufficient, graphing can be used. You can use algebraic methods to find exact solutions.

Additional Answers

69a. Sample answer for men using (0, 10) and (44, 9.69):
$y_m = -0.00705x + 10$; sample answer for women using (0, 11.4) and (44, 10.78):
$y_w = -0.01409x + 11.4$

69b.

Years after 1964

Based on these data, the women's performance will catch up to the men's performance 198 years after 1964, or in the year 2162. The next Olympic year would be 2164; this prediction is not reasonable. It is unlikely that women's times will ever catch up to men's times because the times cannot continue to increase and decrease infinitely.

Practice and Problem Solving

Extra Practice is on page R3.

Example 1 Solve each system of equations by using a table.

26. $y = 5x + 3$
$y = x - 9$ $(-3, -12)$

27. $3x - 4y = 16$
$-6x + 5y = -29$ $(4, -1)$

28. $2x - 5 = y$
$-3x + 4y = 0$ $(4, 3)$

29. FUNDRAISER To raise money for new uniforms, the band boosters sell T-shirts and hats. The cost and sale price of each item is shown. The boosters spend a total of $2000 on T-shirts and hats. They sell all of the merchandise, and make $3375. How many T-shirts did they sell? 250 T-shirts

Item	Cost	Sale Price
T-Shirt	$6	$10
Hat	$4	$7

Example 2 Solve each system of equations by graphing. **33.** infinite solutions

30. $-3x + 2y = -6$
$-5x + 10y = 30$ $(6, 6)$

31. $4x + 3y = -24$
$8x - 2y = -16$ $(-3, -4)$

32. $6x - 5y = 17$
$6x + 2y = 31$ $(4.5, 2)$

33. $-3x - 8y = 12$
$12x + 32y = -48$

34. $y - 3x = -29$
$9x - 6y = 102$ $(8, -5)$

35. $-10x + 4y = 7$
$2x - 5y = 7$ $(-1.5, -2)$

36. CCSS MODELING Jerilyn has a $10 coupon and a 15% discount coupon for her favorite store. The store has a policy that only one coupon may be used per purchase. When is it best for Jerilyn to use the $10 coupon, and when is it best for her to use the 15% discount coupon? $10 coupon for a purchase less than $66.67 and 15% discount coupon for a purchase over $66.67

Example 3 Graph each system of equations and describe it as *consistent and independent, consistent and dependent,* or *inconsistent.* **37–42.** See Chapter 3 Answer Appendix.

37. $y = 3x - 4$
$y = 6x - 8$

38. $y = 2x - 1$
$y = 2x + 6$

39. $2x + 5y = 10$
$-4x - 10y = 20$

40. $x - 6y = 12$
$3x + 18y = 14$

41. $-5x - 6y = 13$
$12y + 10x = -26$

42. $8y - 3x = 15$
$-16y + 6x = -30$

Example 4 Solve each system of equations by using substitution.

43. $9y + 3x = 18$ infinite
$-3y - x = -6$ solutions

44. $5x - 20y = 70$
$6x + 5y = -32$ $(-2, -4)$

45. $-4x - 16y = -96$
$7x + 3y = 68$ $(8, 4)$

46. $-4a - 5b = 14$
$9a + 3b = -48$ $(-6, 2)$

47. $-9c - 4d = 31$
$6c + 6d = -24$ $(-3, -1)$

48. $8f + 3g = 12$
$-32f - 12g = 48$ no solution

49. TENNIS At a park, there are 38 people playing tennis. Some are playing doubles, and some are playing singles. There are 13 matches in progress. A doubles match requires 4 players, and a singles match requires 2 players.

   a. Write a system of two equations that represents the number of singles and doubles matches going on. $x + y = 13$ and $4x + 2y = 38$

   b. How many matches of each kind are in progress? 6 doubles matches and 7 singles matches

Examples 5–6 Solve each system of equations by using elimination. **53.** no solution

50. $8x + y = 27$
$-3x + 4y = 3$ $(3, 3)$

51. $2a - 5b = -20$
$2a + 5b = 20$ $(0, 4)$

52. $6j + 4k = -46$
$2j + 4k = -26$ $(-5, -4)$

53. $3x - 8y = 24$
$-12x + 32y = 96$

54. $5a - 2b = -19$
$8a + 5b = -55$ $(-5, -3)$

55. $r - 6t = 44$
$9r + 12t = 0$ $(8, -6)$

56. $6d + 5f = -32$
$5d - 9f = 26$ $(-2, -4)$

57. $11u = 5v + 35$
$8v = -6u + 62$ $(5, 4)$

58. $-1.2c + 3.4d = 6$ infinite
$6c = -30 + 17d$ solutions

B Use a graphing calculator to solve each system of equations. Round the coordinates of the intersection to the nearest hundredth.

59. $12y = 5x - 15$
$4.2y + 6.1x = 11$
$(2.07, -0.39)$

60. $-3.8x + 2.9y = 19$
$6.6x - 5.4y = -23$
$(-26.01, -27.54)$

61. $5.8x - 6.3y = 18$
$-4.3x + 8.8y = 32$ $(15.03, 10.98)$

Solve each system of equations.

62. $11p + 3q = 6$
$-0.75q - 2.75p = -1.5$
infinite solutions

63. $8r - 5t = -60$
$6r + 3t = -18$ $(-5, 4)$

64. $10t + 4v = 13$
$-4t - 7v = 11$ $(2.5, -3)$

65. $6w = 12 - 4x$

$6x = -9w + 18$
infinite solutions

66. $\frac{3}{2}y + z = 3$

$-y - \frac{2}{3}z = -2$
infinite solutions

67. $\frac{5}{2}a - \frac{3}{4}b = 46$

$-\frac{7}{8}a - 3b = 10$ $(16, -8)$

68. ROWING Allison can row a boat 1 mile upstream (against the current) in 24 minutes. She can row the same distance downstream in 13 minutes. Assume that both the rowing speed and the speed of the current are constant. **a. 3.56 mph; 1.06 mph**

a. Find the speed at which Allison is rowing and the speed of the current.

b. If Allison plans to meet her friends 3 miles upstream one hour from now, will she be on time? Explain. **No; she will be 12 minutes late.**

69. CCSS MODELING The table shows the winning times in seconds for the 100-meter dash at the Olympics between 1964 and 2008. **a–b. See margin.**

Years Since 1964, x	Men's Gold Medal Time	Women's Gold Medal Time
0	10.0	11.4
4	9.90	11.0
8	10.14	11.07
12	10.06	11.08
16	10.25	11.06
20	9.99	10.97
24	9.92	10.54
28	9.96	10.82
32	9.84	10.94
36	9.87	10.75
40	9.85	10.93
44	9.69	10.78

a. Write equations that represent the winning times for men and women since 1964. Assume that both times continue along the same trend.

b. Graph both equations. Estimate when the women's performance will catch up to the men's performance. Do you think that your prediction is reasonable? Explain.

70. JOBS Levi has a job offer in which he will receive $800 per month plus a commission of 2% of the total price of the cars that he sells. At his current job, he receives $1200 per month plus a commission of 1.5% of his total sales. How much must he sell per month to make the new job a better deal? **more than $80,000**

71 TRAVEL A youth group went on a trip to an amusement park, travelling in two vans. The number of people in each van and the total cost of admission are shown in the table. Find the adult price and student price of admission. **adult: $16; student: $9**

Van	Adults	Students	Total Cost
A	2	5	$77
B	2	7	$95

(t)Michael Steele/Getty Images Sport/Getty Images, (b)Purestock/Getty Images

📁 **Study Guide and Intervention**

AL OL ELL

NAME _____ DATE _____ PERIOD _____

3-1 Study Guide and Intervention
Solving Systems of Equations

Solve Systems Graphically A system of equations is two or more equations with the same variables. You can solve a system of linear equations by using a table or by graphing the equations on the same coordinate plane. If the lines intersect, the solution is that intersection point. The following chart summarizes the possibilities for graphs of two linear equations in two variables.

Graphs of Equations	Slopes of Lines	Classification of System	Number of Solutions
Lines intersect	Different slopes	Consistent and independent	One
Lines coincide (same line)	Same slope, same y-intercept	Consistent and dependent	Infinitely many
Lines are parallel	Same slope, different y-intercepts	Inconsistent	None

Example Graph the system of equations and describe it as *consistent and independent, consistent and dependent,* or *inconsistent.* $x - 3y = 6$ $2x - y = -3$

Write each equation in slope-intercept form.

$x - 3y = 6 \rightarrow y = \frac{1}{3}x - 2$
$2x - y = -3 \rightarrow y = 2x + 3$

The graphs intersect at $(-3, -3)$. Since there is one solution, the system is consistent and independent.

Exercises

Graph each system of equations and describe it as *consistent and independent, consistent and dependent,* or *inconsistent.*

📁 **Practice**

AL OL BL ELL

NAME _____ DATE _____ PERIOD _____

3-1 Practice
Solving Systems of Equations

Solve each system of equations by graphing.

Solve each system of equations.

16. SPORTS Last year the volleyball team paid $5 per pair for socks and $17 per pair for shorts on a total purchase of $315. This year they spent $342 to buy the same number of pairs of socks and shorts because the socks now cost 86 a pair and the shorts cost $18.

📁 **Word Problem Practice**

AL OL BL ELL

NAME _____ DATE _____ PERIOD _____

3-1 Word Problem Practice
Solving Systems of Equations

📁 **Enrichment**

OL BL

NAME _____ DATE _____ PERIOD _____

3-1 Enrichment

Solutions to Nonlinear Equations

Real-life situations are often not capable of being represented by a linear equation. Systems of nonlinear equations are often used in the study of population dynamics, modeling carbon monoxide exposure, and determining the height of an object in free fall. **Nonlinear equations** have one variable raised to a power other than one or multiplication of two or more variables.

Multiple Representations

In Exercise 75, students use a table of values and a graph in the coordinate plane to find solutions for a system of equations and interpret the results.

Teaching the Mathematical Practices

Critique Mathematically proficient students are also able to compare the effectiveness of two plausible arguments, distinguish correct logic or reasoning from that which is flawed, and—if there is a flaw in an argument—explain what it is.

Additional Answers

75a.

Equation 1		Equation 2	
x	y	x	y
0	$\frac{16}{3}$	0	−4
1	5	1	−2
2	$\frac{14}{3}$	2	0
3	$\frac{13}{3}$	3	2
4	4	4	4

Equation 3	
x	y
0	10
1	5
2	0
3	−5
4	−10

75b. Equations 1 and 2 intersect at (4, 4), equations 2 and 3 intersect at (2, 0), and equations 1 and 3 intersect at (1, 5); there is no solution that satisfies all three equations.

75c.

GEOMETRY Find the point at which the diagonals of the quadrilaterals intersect.

72. (6, 3.5)

73 $\left(\frac{53}{13}, \frac{153}{26}\right)$

74. ELECTIONS In the election for student council, Candidate A received 55% of the total votes, while Candidate B received 1541 votes. If Candidate C received 40% of the votes that Candidate A received, how many total votes were cast? **6700 votes**

75. ⚡ **MULTIPLE REPRESENTATIONS** In this problem, you will explore systems of three linear equations and two variables. **a–d. See margin.**

$$3y + x = 16$$
$$y - 2x = -4$$
$$y + 5x = 10$$

a. Tabular Make a table of x- and y-values for each equation.

b. Analytical Which values from the table indicate intersections? Is there a solution that satisfies all three equations?

c. Graphical Graph the three equations on a single coordinate plane.

d. Verbal What conditions must be met for a system of three equations with two variables to have a solution? What conditions result in no solution?

H.O.T. Problems — Use Higher-Order Thinking Skills

76. CCSS CRITIQUE Gloria and Syreeta are solving the system $6x - 4y = 26$ and $-3x + 4y = -17$. Is either of them correct? Explain your reasoning.

76. Sample answer: Gloria; Syreeta subtracted 26 from 17 instead of 17 from 26 and got $3x = -9$ instead of $3x = 9$. She proceeded to get a value of −11 for y. She would have found her error if she had substituted the solution into the original equations.

Gloria
$6x - 4y = 26$ $6(3) - 4y = 26$
$-3x + 4y = -17$ $18 - 4y = 26$
$3x = 9$ $-4y = 8$
$x = 3$ $y = -2$
The solution is $(3, -2)$.

Syreeta
$6x - 4y = 26$ $6(-3) - 4y = 26$
$-3x + 4y = -17$ $-18 - 4y = 26$
$3x = -9$ $-4y = 44$
$x = -3$ $y = -11$
The solution is $(-3, -11)$.

77. CHALLENGE Find values of a and b for which the following system has a solution of $(b - 1, b - 2)$. $a \neq 0, b = \pm 3$

$$-8ax + 4ay = -12a$$
$$2bx - by = 9$$

78. REASONING If a is consistent and dependent with b, b is inconsistent with c, and c is consistent and independent with d, then a will *sometimes*, *always*, or *never* be consistent and independent with d. Explain your reasoning.

78. Sample answer: Always; a and b are the same line. b is parallel to c, so a is also parallel to c. Since c and d are consistent and independent, then c is not parallel to d and, thus, intersects d. Since a and c are parallel, then a cannot be parallel to d, so, a must intersect d and must be consistent and independent with d.

79. OPEN ENDED Write a system of equations in which one equation needs to be multiplied by 3 and the other needs to be multiplied by 4 in order to solve the system with elimination. Then solve your system. **See Chapter 3 Answer Appendix.**

80. ✍ **WRITING IN MATH** Why is substitution sometimes more helpful than elimination, and vice versa? **See margin.**

 144 | Lesson 3-1 | Solving Systems of Equations

75d. If all three lines intersect at the same point, then the system has a solution. The system has no solution if the lines intersect at 3 different points, or if two or three lines are parallel.

80. Sample answer: It is more helpful to use substitution when one of the variables has a coefficient of 1 or if a coefficient can be reduced to 1 without turning other coefficients into fractions. Otherwise, elimination is more helpful because it will avoid the use of fractions in solving the system.

81. SHORT RESPONSE Simplify $3y(4x + 6y - 5)$.
$12xy + 18y^2 - 15y$

82. SAT/ACT Which of the following best describes the graph of the equations? **A**

$$4y = 3x + 8$$
$$-6x = -8y + 24$$

A The lines are parallel.
B The lines are perpendicular.
C The lines have the same x-intercept.
D The lines have the same y-intercept.
E The lines are the same.

83. GEOMETRY Which set of dimensions corresponds to a triangle similar to the one shown at the right? **J**

F 1 unit, 2 units, 3 units
G 7 units, 11 units, 12 units
H 10 units, 23 units, 24 units
J 20 units, 48 units, 52 units

84. Move-A-Lot Rentals will rent a moving truck for $100 plus $0.10 for every mile it is driven. Which equation can be used to find C, the cost of renting a moving truck, and driving it for m miles? **B**

A $C = 0.1(100 + m)$
B $C = 100 + 0.1m$
C $C = 100m + 0.1$
D $C = 100(m + 0.1)$

85. CRAFTS Priscilla sells stuffed animals at a local craft show. She charges $10 for the small ones and $15 for the large ones. To cover her expenses, she needs to sell at least $350 worth of animals. (Lesson 2-8)

a. Write an inequality for this situation. $10s + 15\ell \geq 350$

b. Graph the inequality. **See Chapter 3 Answer Appendix.**

c. If she sells 10 small and 15 large animals, will she cover her expenses? **no**

Write an equation for each function. (Lesson 2-7)

86.
$y = x^2 + 6$

87.
$y = -|x - 3|$

88.
$y = -(x - 4)^2$

Solve each equation. Check your solution. (Lesson 1-3)

89. $2p = 14$ **7**

90. $-14 + n = -6$ **8**

91. $7a - 3a + 2a - a = 16$ **3.2**

92. $x + 9x - 6x + 4x = 20$ **2.5**

93. $27 = -9(y + 5) + 6(y + 8)$ **−8**

94. $-7(p + 7) + 3(p - 4) = -17$ **−11**

Determine whether the given point satisfies each inequality.

95. $4x + 5y \leq 15$; $(2, -2)$ **yes**

96. $3x + 5y \geq 8$; $(1, 1)$ **yes**

97. $6x + 9y < -1$; $(0, 0)$ **no**

---

4 Assess

Name the Math Have students either write on a piece of paper or tell you whether elimination or substitution would be better to solve this system of equations.

$$2m - n = 2$$
$$2m + 3n = 22$$

DifferentiatedInstruction **BL**

Extension Give an example of a system of equations that has infinitely many solutions. Sample answer: $x + y = 3$ and $2x + 2y = 6$

1 Focus

VerticalAlignment

Before Lesson 3-2 Solve systems of linear equations using tables and graphs.

Lesson 3-2 Solve systems of inequalities by graphing. Determine the coordinates of the vertices of a region formed by the graph of a system of inequalities.

After Lesson 3-2 Solve real-world optimization problems using systems of inequalities.

2 Teach

Scaffolding Questions

Have students read the **Why?** section of the lesson.

Ask:

- If *x* represents outside air temperature in degrees Fahrenheit, what inequality represents the range of acceptable air temperatures for a space shuttle launch? $35 < x < 100$

- If *y* represents wind speed in knots, what inequality represents the range of acceptable wind speeds for a space shuttle launch? $y \leq 30$

- What are three possible temperature and wind speed pairs (x, y) that satisfy the given conditions? Sample answer: (40, 25), (60, 10), (95, 15)

LESSON 3-2 Solving Systems of Inequalities by Graphing

Then	**Now**	**Why?**
• You solved systems of linear equations graphically and algebraically.	**1** Solve systems of inequalities by graphing. **2** Determine the coordinates of the vertices of a region formed by the graph of a system of inequalities.	• Many weather conditions need to be met before a space shuttle can launch. The temperature must be greater than 35°F and less than 100°F, and the wind speed cannot exceed 30 knots. A system of inequalities can be used to show these three conditions.

NewVocabulary
system of inequalities

CCSS Common Core State Standards

Content Standards
A.CED.3 Represent constraints by equations or inequalities, and by systems of equations and/or inequalities, and interpret solutions as viable or nonviable options in a modeling context.

Mathematical Practices
1 Make sense of problems and persevere in solving them.

1 Systems of Inequalities Solving a **system of inequalities** means finding the ordered pairs that satisfy all of the inequalities in the system.

KeyConcept Solving Systems of Inequalities

Step 1 Graph each inequality, shading the correct area.

Step 2 Identify the region that is shaded for all of the inequalities. This is the solution of the system.

Example 1 Intersecting Regions

Solve the system of inequalities.
$y > 2x - 4$
$y \leq -0.5x + 3$

Solution of $y > 2x - 4 \rightarrow$ Regions 1 and 3

Solution of $y \leq -0.5x + 3 \rightarrow$ Regions 2 and 3

Region 3 is part of the solution of both inequalities, so it is the solution of the system.

CHECK Notice that the origin is part of the solution of the system. The origin can be used as a test point. You can test the solution by substituting (0, 0) for *x* and *y* in each equation.

$$y > 2x - 4 \qquad\qquad y \leq -0.5x + 3$$
$$0 \overset{?}{>} 2(0) - 4 \qquad\qquad 0 \overset{?}{\leq} -0.5(0) + 3$$
$$0 \overset{?}{>} 0 - 4 \qquad\qquad 0 \overset{?}{\leq} 0 + 3$$
$$0 > -4 \checkmark \qquad\qquad 0 \leq 3 \checkmark$$

GuidedPractice **1A, 1B.** See Chapter 3 Answer Appendix.

1A. $y \leq -2x + 5$
$y > -\frac{1}{4}x - 6$

1B. $y \geq |x|$
$y < \frac{4}{3}x + 5$

Purestock/SuperStock

Lesson 3-2 Resources

Resource	Approaching Level **AL**	On Level **OL**	Beyond Level **BL**	English Learners **ELL**
Teacher Edition		• Differentiated Instruction, p. 148	• Differentiated Instruction, pp. 148, 150, 152	
Chapter Resource Masters	• Study Guide and Intervention, pp. 12–13 • Skills Practice, p. 14 • Practice, p. 15 • Word Problem Practice, p. 16	• Study Guide and Intervention, pp. 12–13 • Skills Practice, p. 14 • Practice, p. 15 • Word Problem Practice, p. 16 • Enrichment, p. 17	• Practice, p. 15 • Word Problem Practice, p. 16 • Enrichment, p. 17	• Study Guide and Intervention, pp. 12–13 • Skills Practice, p. 14 • Practice, p. 15 • Word Problem Practice, p. 16
Other	• 5-Minute Check 3-2 • Study Notebook • Teaching Algebra with Manipulatives	• 5-Minute Check 3-2 • Study Notebook • Teaching Algebra with Manipulatives	• 5-Minute Check 3-2 • Study Notebook	• 5-Minute Check 3-2 • Study Notebook • Teaching Algebra with Manipulatives

BananaStock/Jupiterimages

It is possible that the regions do not intersect. When this occurs, there is no solution of the system or the solution set is the *empty set*.

Example 2 Separate Regions

Solve the system of inequalities by graphing.

$y \geq x + 5$
$y < x - 4$

Graph both inequalities.

Since the graphs of the inequalities do not overlap, there are no points in common and there is no solution to the system.

The solution set is the empty set.

▶ **Guided**Practice   **2A, 2B.** See Chapter 3 Answer Appendix.

2A. $y \geq -4x + 8$
  $y < -4x + 4$

2B. $y \geq |x|$
  $y < 2x - 24$

Real-World Example 3 Write and Use a System of Inequalities

TIME MANAGEMENT Chelsea has final exams in calculus, physics, and history. She has up to 25 hours to study for the exams. She plans to study history for 2 hours. She needs to spend at least 7 hours studying for calculus, but over 14 is too much. She hopes to spend between 8 and 12 hours on physics. Write and graph a system of inequalities to represent the situation.

Calculus: at least 7 hours, but no more than 14
$7 \leq c \leq 14$

Physics: at least 8 hours, but no more than 12
$8 \leq p \leq 12$

Chelsea has 25 hours available, and 2 of those will be spent on history. She has up to 23 hours left for calculus and physics.
$c + p \leq 23$

Graph all of the inequalities. Any ordered pair in the intersection is a solution of the system. One solution is 10 hours on physics and 12 hours on calculus.

Hours Spent on Physics

Real-WorldLink
Typical incoming freshmen will spend more than 3 times as many hours studying in college as in high school.

Source: *National Survey of Student Engagement*

▶ **Guided**Practice

3. TRAVEL Mr. and Mrs. Rodriguez are driving across the country with their two children. They plan on driving a maximum of 10 hours each day. Mr. Rodriguez wants to drive at least 4 hours a day but no more than 8 hours a day. Mrs. Rodriguez can drive between 2 and 5 hours per day. Write and graph a system of inequalities that represents this information. See margin.

Additional Answer (Guided Practice)

3. Let j = the number of hours Mr. Rodriguez drives, and let d = the number of hours Mrs. Rodriguez drives.

$j + d \leq 10$

$4 \leq j \leq 8$

$2 \leq d \leq 5$

Mrs. Rodriguez's Hours Driving

Mr. Rodriguez's Hours Driving

---

1 Systems of Inequalities

Examples 1 and **2** show how to solve systems of inequalities. **Example 3** shows how to write and use a system of inequalities in a real-world situation.

Formative Assessment

Use the Guided Practice exercises after each example to determine students' understanding of concepts.

Additional Examples

1 Solve the system of inequalities by graphing.

$y \geq 2x - 3$

$y < -x + 2$

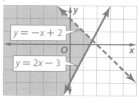

2 Solve the system of inequalities by graphing. Ø

$y \geq -\frac{3}{4}x + 1$

$y \leq -\frac{3}{4}x - 2$

▶ **Additional Examples** also in Interactive Classroom PowerPoint® Presentations

IWB Interactive White Board READY

Tips for New Teachers

Sense-Making Make sure that students understand that the solution of an inequality must be true for all of the inequalities of the system. Have them test the coordinates of points in different regions of the graph and analyze the results.

Additional Example

3 **MEDICINE** Medical professionals recommend that patients have a cholesterol level below 200 milligrams per deciliter (mg/dL) of blood and a triglyceride level below 150 mg/dL. Write and graph a system of inequalities that represents the range of cholesterol levels c and triglyceride levels t for patients.
$0 < c < 200, 0 < t < 150$

2 Find Vertices of an Enclosed Region

Example 4 shows how to find the vertices of the region formed by a system of inequalities.

Additional Example

4 Find the coordinates of the vertices of the triangle formed by $2x - y \geq -1$, $x + y \leq 4$, and $x + 4y \geq 4$.

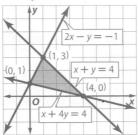

Additional Answers (Guided Practice)

4A. $(8, -4), (-3, 3), \left(\frac{4}{7}, -\frac{54}{7}\right)$

4B. $(-2, 1), (3, 3), (7, -1)$

StudyTip

Boundaries If the inequality that forms the boundary is $<$ or $>$, then the boundary is not included in the solution, and the line should be dashed.

2 **Find Vertices of an Enclosed Region** Sometimes the graph of a system of inequalities produces an enclosed region in the form of a polygon. To find the vertices of the region, determine the coordinates of the points at which the boundaries intersect.

Example 4 Find Vertices

Find the coordinates of the vertices of the triangle formed by $y \geq 2x - 8$, $y \leq -\frac{1}{4}x + 6$, and $4y \geq -15x - 32$.

Step 1 Graph each inequality.

The coordinates $(-4, 7)$ and $(0, -8)$ can be determined from the graph. To find the coordinates of the third vertex, solve the system of equations $y = 2x - 8$ and $y = -\frac{1}{4}x + 6$.

Step 2 Substitute for y in the second equation.

$2x - 8 = -\frac{1}{4}x + 6$ Replace y with $2x - 8$.

$2x = -\frac{1}{4}x + 14$ Add 8 to each side.

$\frac{9}{4}x = 14$ Add $\frac{1}{4}x$ to each side.

$x = \frac{56}{9}$ or $6\frac{2}{9}$ Divide each side by $\frac{9}{4}$.

Step 3 Find y.

$y = 2\left(6\frac{2}{9}\right) - 8$ Replace x with $6\frac{2}{9}$.

$= 12\frac{4}{9} - 8$ Distributive Property

$= 4\frac{4}{9}$ Simplify.

CHECK Compare the coordinates to the coordinates on the graph. The x-coordinate of the third vertex is between 6 and 7, so $6\frac{2}{9}$ is reasonable. The y-coordinate of the third vertex is between 4 and 5, so $4\frac{4}{9}$ is reasonable.

The vertices of the triangle are at $(-4, 7)$, $(0, -8)$, and $\left(6\frac{2}{9}, 4\frac{4}{9}\right)$.

GuidedPractice

Find the coordinates of the vertices of the triangle formed by each system of inequalities. **4A, 4B. See margin.**

4A. $y \geq -3x - 6$
$2y \geq x - 16$
$11y + 7x \leq 12$

4B. $5y \leq 2x + 9$
$y \leq -x + 6$
$9y \geq -2x + 5$

 148 | Lesson 3-2 | Solving Systems of Inequalities by Graphing

DifferentiatedInstruction **OL** **BL**

If students demonstrate an understanding of how to solve a system of inequalities by graphing,

Then have them write a list of tips to help someone draw the graphs of systems of inequalities and find the vertices easily and efficiently.

● = Step-by-Step Solutions begin on page R14. ✓

Examples 1–2 Solve each system of inequalities by graphing. **1–3. See margin.**

1. $y \leq 6$
$y > -3 + x$

2. $y \leq -3x + 4$
$y \geq 2x - 1$

3. $y > -2x + 4$
$y \leq -3x - 3$

Example 3

4. **CCSS REASONING** The most Kala can spend on hot dogs and buns for her cookout is $35. A package of 10 hot dogs costs $3.50. A package of buns costs $2.50 and contains 8 buns. She needs to buy at least 40 hot dogs and 40 buns. **a, b. See margin.**

   a. Graph the region that shows how many packages of each item she can purchase.

   b. Give an example of three different purchases she can make.

Example 4 Find the coordinates of the vertices of the triangle formed by each system of inequalities. **5, 6. See Chapter 3 Answer Appendix for graphs.**

5. $y \geq 2x + 1$
$y \leq 8$
$4x + 3y \geq 8$
$(3.5, 8), (-4, 8), (0.5, 2)$

6. $3y \geq -7x - 16$
$7y \leq x + 32$
$y \geq 15x - 40$
$(2, -10), (3, 5), (-4, 4)$

Practice and Problem Solving Extra Practice is on page R3.

Examples 1–2 Solve each system of inequalities by graphing. **7–15. See Chapter 3 Answer Appendix.**

7. $x < 3$
$y \geq -4$

8. $y > 3x - 5$
$y \leq 4$

9 $y < -3x + 4$
$3y + x > -6$

10. $y \geq 0$
$y < x$

11. $6x - 2y \geq 12$
$3x + 4y > 12$

12. $-8x > -2y - 1$
$-4y \geq 2x - 5$

13. $5y < 2x + 10$
$y - 4x > 8$

14. $3y - 2x \leq -24$
$y \geq \frac{2}{3}x - 1$

15. $y > -\frac{2}{5}x + 2$
$5y \leq -2x - 15$

Example 3

16. **RECORDING** Jane's band wants to spend no more than $575 recording their first CD. The studio charges at least $35 per hour to record. Graph a system of inequalities to represent this situation. **See Chapter 3 Answer Appendix.**

17. **SUMMER TRIP** Rondell has to save at least $925 to go to Rome with his Latin class in 8 weeks. He earns $9 an hour working at the Pizza Palace and $12 an hour working at a car wash. By law, he cannot work more than 25 hours per week. Graph two inequalities that Rondell can use to determine the number of hours he needs to work at each job if he wants to make the trip. **See Chapter 3 Answer Appendix.**

Example 4 Find the coordinates of the vertices of the triangle formed by each system of inequalities. **18–23. See Chapter 3 Answer Appendix for graphs.**

18. $x \geq 0$
$y \geq 0$
$x + 2y < 4$
$(0, 2), (4, 0), (0, 0)$

19. $y \geq 3x - 7$
$y \leq 8$
$x + y > 1$
$(2, -1), (5, 8), (-7, 8)$

20. $x \leq 4$
$y > -3x + 12$
$y \leq 9$
$(1, 9), (4, 0), (4, 9)$

21. $-3x + 4y \leq 15$
$2y + 5x > -12$
$10y + 60 \geq 27x$
$(-3, 1.5), (5, 7.5), (0, -6)$

22. $8y - 19x < 74$
$38y + 26x \leq 119$
$54y - 12x \geq -198$
$(-6, -5), (-2, 4.5), (7.5, -2)$

23. $6y - 24x \geq -168$
$8y + 7x > 10$
$20y - 2x \leq 64$
$(8, 4), (6, -4), (-2, 3)$

24. **BAKING** Rebecca wants to bake cookies and cupcakes for a bake sale. She can bake 15 cookies at a time and 12 cupcakes at a time. She needs to make at least 120 baked goods, but no more than 360, and she wants to have at least three times as many cookies as cupcakes. What combination of batches of each could Rebecca make?
Sample answer: 15 batches of cookies and 6 batches of cupcakes.

connectED.mcgraw-hill.com **149** 🔊

Differentiated Homework Options

Level	Assignment		Two-Day Option
AL Basic	7–25, 46, 48–68	7–25 odd, 51–54	8–24 even, 46, 48–50, 55–68
OL Core	7–25 odd, 29–35 odd, 41, 46, 48–68	7–25, 51–54	28–36, 40–42, 43–44, 46, 48–50, 55–68
BL Advanced	26–62, (optional: 63–68)		

---

3 Practice

Formative Assessment
Use Exercises 1–6 to check for understanding.

Use the chart on the bottom of this page to customize assignments for your students.

Additional Answers

1.

2.

3.

4a.

4b. Sample answer: 4 packages of hotdogs, 5 packages of buns; 5 packages of hotdogs, 6 packages of buns; 6 packages of hotdogs, 5 packages of buns

CCSS **Teaching the Mathematical Practices**

Reasoning Mathematically proficient students decontextualize information to represent it symbolically, manipulate the symbols, and then contextualize to interpret the solution.

Teach with Tech

Wiki Have students work in pairs to create a wiki describing how to graph a system of inequalities. Be sure they describe how to decide whether to use a solid or dashed line and how to decide which region to shade. Be sure they also explain how to interpret the graph to find solutions of the system of inequalities.

Additional Answers

26a. $h \geq 56$, $h \leq 72$, $c \geq 48$, $c \leq 60$

27a.

25 **CELL PHONES** Dale has a maximum of 800 minutes on his cell phone plan that he can use each month. Daytime minutes cost $0.15, and nighttime minutes cost $0.10. Dale plans to use at least twice as many daytime minutes as nighttime minutes. If Dale uses at least 200 nighttime minutes and does not go over his limit, what is his maximum bill? his minimum bill? **maximum = $110, minimum = $80**

B **26.** **TREES** Trees are divided into four categories according to height and trunk circumference. Data for the trees in a forest are described in the table.

Crown Class	dominant	co-dominant	intermediate	suppressed
Height (in feet)	over 72	56–72	40–55	under 39
Trunk Circumference (in inches)	over 60	48–60	34–48	under 33

Source: USDA Forest Service

a. Write and graph the system of inequalities that represents the range of heights h and circumferences c for a co-dominant tree. **See margin.**

b. Determine the crown class of a basswood that is 48 feet tall. Find the expected trunk circumference. **intermediate; 34–48 in.**

27. **CCSS** **REASONING** On a camping trip, Jessica needs at least 3 pounds of food and 0.5 gallon of water per day. Marc needs at least 5 pounds of food and 0.5 gallon of water per day. Jessica's equipment weighs 10 pounds, and Marc's equipment weighs 20 pounds. A gallon of water weighs approximately 8 pounds. Each of them carries their own supplies, and Jessica is capable of carrying 35 pounds while Marc can carry 50 pounds.

a. Graph the inequalities that represent how much they can carry. **See margin.**

b. How many days can they camp, assuming that they bring all their supplies in at once? $3\frac{1}{3}$ **days**

c. Who will run out of supplies first? **Marc; Jessica could last about a quarter of a day longer than Marc.**

Solve each system of inequalities by graphing. 28–36. See Chapter 3 Answer Appendix.

28. $y \geq |2x + 4| - 2$
$3y + x \leq 15$

29. $y \geq |6 - x|$
$|y| \leq 4$

30. $|y| \geq x$
$y < 2x$

31. $y > -3x + 1$
$4y \leq x - 8$
$3x - 5y < 20$

32. $6y + 2x \leq 9$
$2y + 18 \geq 5x$
$y > -4x - 9$

33. $|x| > y$
$y \leq 6$
$y \geq -2$

34. $2x + 3y \geq 6$
$y \leq |x - 6|$

35. $8x + 4y < 10$
$y > |2x - 1|$

36. $y \geq |x - 2| + 4$
$y \leq [\![x]\!] - 3$

37. **MUSIC** Steve is trying to decide what to put on his MP3 player. Audio books are 3 hours long and songs are 2.5 minutes long. Steve wants no more than 4 audio books on his MP3 player, but at least ten songs and one audio book. Each book costs $15.00 and each song costs $0.95. Steve has $63 to spend on books and music. Graph the inequalities to show possible combinations of books and songs that Steve can have. **See Chapter 3 Answer Appendix.**

38. **JOBS** Louie has two jobs and can work no more than 25 total hours per week. He wants to earn at least $150 per week. Graph the inequalities to show possible combinations of hours worked at each job that will help him reach his goal. **See Chapter 3 Answer Appendix.**

Job	Pay
Busboy	$6.50
Clerk	$8.00

 150 | Lesson 3-2 | Solving Systems of Inequalities by Graphing

DifferentiatedInstruction **BL**

Intrapersonal Learners For Exercises 25–27, 37, and 38, tell students that systems of inequalities are often used to solve real-world problems. Challenge students to search newspapers, magazines, and the Internet for real-world information and use that information to write a real-world problem that can be solved using a system of inequalities. This will help students to better understand the concept of solving systems of inequalities.

39. TIME MANAGEMENT Ramir uses his spare time to write a novel and to exercise. He has budgeted 35 hours per week. He wants to exercise at least 7 hours a week but no more than 15. He also hopes to write between 20 and 25 hours per week. Write and graph a system of inequalities that represents this situation. **See Chapter 3 Answer Appendix.**

Find the coordinates of the vertices of the figure formed by each system of inequalities. 40–42. See Chapter 3 Answer Appendix.

40. $y \geq 2x - 12$
$y \leq -4x + 20$
$4y - x \leq 8$
$y \geq -3x + 2$

41. $y \geq -x - 8$
$2y \geq 3x - 20$
$4y + x \leq 24$
$y \leq 4x + 22$

42. $2y - x \geq -20$
$y \geq -3x - 6$
$y \leq -2x + 2$
$y \leq 2x + 14$

44b. (6, 4), (6, 5), (7, 3), (7, 4), (7, 5), (7, 6), (8, 2), (8, 3), (8, 4), (8, 5), (8, 6), (8, 7), (9,1), (9, 2), (9, 3), (9, 4), (9, 5), (9, 6), (10, 0), (10, 1), (10, 2), (10, 3), (10, 4), (10, 5), (11, 0), (11, 1), (11, 2), (11, 3), (11, 4), (12, 0), (12, 1), (12, 2), (12, 3), (13, 0), (13, 1), (13, 2), (14, 0), (14, 1), (15, 0)

43. FINANCIAL LITERACY Mr. Hoffman is investing $10,000 in two funds. One fund will pay 6% interest, and a riskier second fund will pay 10% interest. What is the least amount he can invest in the risky fund and still earn at least $740 after one year? **$3500**

44. DODGEBALL A high school is selecting a dodgeball team to play in a fund-raising exhibition against their rival. There can be between 10 and 15 players on the team and there must be more girls than boys on the team.

44a. See Chapter 3 Answer Appendix.

a. Write and graph a system of inequalities to represent the situation.

b. List all of the possible combinations of boys and girls for the team.

c. Explain why there is not an infinite number of possibilities.
Sample answer: You cannot have a fraction of a person.

H.O.T. Problems Use Higher-Order Thinking Skills

45. CHALLENGE Find the area of the region defined by the following inequalities.

$y \geq -4x - 16$
$4y \leq 26 - x$
$3y + 6x \leq 30$
$4y - 2x \geq -10$

75 units²

47. Sample answer:
$y \geq 2x - 6$,
$y \leq -0.5x + 4$,
$y \geq -3x - 6$; 47

49. Sample answer: Shade each inequality in its standard way, by shading above the line if $y >$ and shading below the line if $y <$ (or you can use test points). Once you determine where to shade for each inequality, the area where *every* inequality needs to be shaded is the actual solution. This is only the shaded area.

46. OPEN ENDED Write a system of two inequalities in which the solution:

a. lies only in the third quadrant. **Sample answer: $y < -2$, $x < -1$**

b. does not exist. **Sample answer: $y > 2$, $y < -2$**

c. lies only on a line. **Sample answer: $y \geq x$, $y \leq x$**

d. lies on exactly one point. **Sample answer: $y \geq |x|$, $y < -|x|$; solution at (0, 0)**

47. CHALLENGE Write a system of inequalities to represent the solution shown at the right. How many points with integer coordinates are solutions of the system?

48. CCSS ARGUMENTS Determine whether the statement is *true* or *false*. If false, give a counterexample. **true**

A system of two linear inequalities has either no points or infinitely many points in its solution.

49. WRITING IN MATH Write a how-to manual for determining where to shade when graphing a system of inequalities.

50. WRITING IN MATH Explain how you would test to see whether $(-4, 6)$ is a solution of a system of inequalities. **See Chapter 3 Answer Appendix.**

connectED.mcgraw-hill.com **151**

CCSS Teaching the **Mathematical Practices**

Arguments Mathematically proficient students are able to analyze situations by breaking them into cases, and can recognize and use counterexamples. They justify their conclusions, communicate them to others.

Enrichment
OL BL

NAME _____ DATE _____ PERIOD _____

3-2 Enrichment

Creative Designs

A system of linear inequalities can be used to define the region bounded by a geometric shape graphed on a coordinate plane. For example, the rectangle shown can be defined by the system

$x \leq 4$,
$x \geq 0$,
$y \leq 3$, and
$y \geq 0$.

The triangle shown can be described using the inequalities

$x + 2y \leq 4$,
$x \geq 0$, and
$y \geq 1$.

Study Guide and Intervention
AL OL ELL

3-2 Study Guide and Intervention
Solving Systems of Inequalities by Graphing

Practice
AL OL BL ELL

3-2 Practice
Solving Systems of Inequalities by Graphing

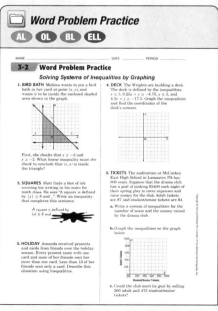

Word Problem Practice
AL OL BL ELL

3-2 Word Problem Practice
Solving Systems of Inequalities by Graphing

Crystal Ball Have students look ahead to Lesson 3-3. Have them write how they think what they learned today will connect with the theme in Lesson 3-3.

Formative Assessment
Check for student understanding of the concepts in Lessons 3-1 and 3-2.

 Quiz 1, p. 59

Standardized Test Practice

51. To be a member of the marching band, a student must have a grade-point average of at least 2.0 and must have attended at least five after-school practices. Choose the system of inequalities that best represents this situation. **A**

A $x \geq 2$ $y \geq 5$

C $x < 2$ $y < 5$

B $x \leq 2$ $y \leq 5$

D $x > 2$ $y > 5$

52. **SAT/ACT** The table at the right shows a relationship between x and y. Which equation represents this relationship? **G**

x	y
1	5
2	8
3	11
4	14
5	17
6	20

F $y = 3x - 2$

G $y = 3x + 2$

H $y = 4x + 1$

J $y = 4x + 2$

K $y = 4x - 1$

53. **SHORT RESPONSE** If $3x = 2y$ and $5y = 6z$, what is the value of x in terms of z? $\frac{4}{5}z$

54. **GEOMETRY** Look at the graph below. Which of these statements describes the relationship between the two lines? **D**

A They intersect at $(6, 2)$.

B They intersect at $(0, 2)$.

C They intersect at $(3.5, 0)$.

D They intersect at $(2, 6)$.

Spiral Review

55. **GEOMETRY** Find the coordinates of the vertices of the parallelogram with sides that are contained in the lines with equations $y = 3$, $y = 7$, $y = 2x$, and $y = 2x - 13$. (Lesson 3-1)
$(1.5, 3), (3.5, 7), (8, 3), (10, 7)$

Graph each inequality. (Lesson 2-8) **56–58. See Ch. 3 Answer Appendix.**

56. $x + y \geq 6$

57. $4x - 3y < 10$

58. $5x + 7y \geq -20$

Graph each function. Identify the domain and range. (Lesson 2-6) **59–61. See Chapter 3 Answer Appendix.**

59. $g(x) = \begin{cases} 0 \text{ if } x < 0 \\ -x + 2 \text{ if } x \geq 0 \end{cases}$

60. $h(x) = \begin{cases} x + 3 \text{ if } x \leq -1 \\ 2x \text{ if } x > -1 \end{cases}$

61. $h(x) = \begin{cases} -1 \text{ if } x < -2 \\ 1 \text{ if } x > 2 \end{cases}$

62. **BOOK CLUB** For each meeting of the Putnam High School book club, $25 is taken from the activities account to buy snacks and materials. After their sixth meeting, there will be $350 left in the activities account. (Lesson 2-4)

a. If no money is put back into the account, what equation can be used to show how much money is left in the activities account after having x number of meetings? $y = 500 - 25x$

b. How much money was originally in the account? **$500**

c. After how many meetings will there be no money left in the activities account? **20**

Skills Review

Find each value if $f(x) = 2x + 5$ **and** $g(x) = 3x - 4$.

63. $f(-3)$ -1

64. $g(-2)$ -10

65. $f(-1)$ 3

66. $g(-0.5)$ -5.5

67. $f(-0.25)$ 4.5

68. $g(-0.75)$ -6.25

 152 | **Lesson 3-2** | Solving Systems of Inequalities by Graphing

DifferentiatedInstruction BL

Extension Plot the points $(-2, 0)$, $(0, 2)$, and $(2, 0)$. Draw line segments connecting the points to form a triangular region. Write a system of inequalities that defines the region.
$y \geq 0, y \leq -x + 2, y \leq x + 2$

3-2
Graphing Technology Lab
Systems of Linear Inequalities

You can graph systems of linear inequalities with a graphing calculator by using the Y= menu. You can choose different graphing styles to shade above or below a line.

Example   Intersection of Two Graphs

Graph the system of inequalities in the standard viewing window.

$y \geq -3x + 4$
$y \leq 2x - 1$

Step 1 Enter $-3x + 4$ as **Y1**. Because y is greater than $-3x + 4$, shade above the line.

KEYSTROKES: [Y=] [◄] [◄] [ENTER] [ENTER] [►] [►]
[(−)] 3 [X,T,θ,n] [+] 4 [ENTER]

Step 2 Enter $2x - 1$ as **Y2**. Because y is less than $2x - 1$, shade below the line.

KEYSTROKES: [◄] [◄] [ENTER] [ENTER] [ENTER] [►] [►] 2
[X,T,θ,n] [−] 1 [ENTER]

Step 3 Display the graphs in the standard viewing window.

KEYSTROKES: [ZOOM] 6

$y = -3x + 4$
$y = 2x - 1$

[−10, 10] scl: 1 by [−10, 10] scl: 1

Notice the shading pattern above the line $y = -3x + 4$ and the shading pattern below the line $y = 2x - 1$. The intersection of the graphs is the region where the patterns overlap. This region includes all the points that satisfy the system $y \geq -3x + 4$ and $y \leq 2x - 1$.

Exercises

Use a graphing calculator to solve each system of inequalities. **1–9. See Chapter 3 Answer Appendix.**

1. $y \geq 3$
   $y \leq -x + 1$

2. $y \geq -4x$
   $y \leq -5$

3. $y \geq 2 - x$
   $y \leq x + 3$

4. $y \geq 2x + 1$
   $y \leq -x - 1$

5. $2y \geq 3x - 1$
   $3y \leq -x + 7$

6. $y + 5x \geq 12$
   $y - 3 \leq 10$

7. $5y + 3x \geq 11$
   $3y - x \leq -8$

8. $10y - 7x \geq -19$
   $7y - 5x \leq 11$

9. $\frac{1}{6}y - x \geq -3$
   $\frac{1}{5}y + x \leq 7$

---

1 Focus

Objective Use a graphing calculator to solve systems of linear inequalities.

Materials for Each Student

- TI-83/84 Plus or other graphing calculator

2 Teach

Working in Cooperative Groups

Put students into groups of two, mixing abilities. Then have each group complete the steps in the Example.

Explain to students that when they sketch the graphs from their calculator windows on paper, they need to draw the axes, the intercepts, and the points of intersection, as well as the lines.

Practice Have students complete Exercises 1–9.

3 Assess

Formative Assessment

Use Exercises 1–9 to assess whether students comprehend how to graph systems of linear inequalities with a graphing calculator.

From Concrete to Abstract

Ask the students how many points are in the solution set of a system of inequalities such as those in this activity. infinitely many

Extending the Concept

Ask:

- When is the point of intersection of the two boundary lines included in the solution set?
  The point of intersection is in the solution set when both boundary lines on the graph are solid. This occurs when the symbol in both inequalities is either $\geq$ or $\leq$.

1 Focus

VerticalAlignment

Before Lesson 3-3 Solve systems of linear inequalities by graphing.

Lesson 3-3 Find the maximum and minimum values of a function over a region. Solve real-world optimization problems using linear programming.

After Lesson 3-3 Formulate systems of equations in more than two unknowns.

2 Teach

Scaffolding Questions

Have students read the **Why?** section of the lesson.

Ask:

- How much does it cost to produce 1000 audio players? $55,000

- Is it cheaper to produce an audio player or a phone? an audio player

- Can the company produce 2000 audio players per shift? Explain. No; the maximum number of audio players that can be produced is 1500.

LESSON 3-3 Optimization with Linear Programming

| :: Then | :: Now | :: Why? |

:: Then
- You solved systems of linear inequalities by graphing.

:: Now
1. Find the maximum and minimum values of a function over a region.
2. Solve real-world optimization problems using linear programming.

:: Why?
An electronics company produces digital audio players and phones. A sign on the company bulletin board is shown.

If at least 2000 items must be produced per shift, how many of each type should be made to minimize costs?

The company is experiencing limitations, or constraints, on production caused by customer demand, shipping, and the productivity of their factory. A system of inequalities can be used to represent these constraints.

Keeping Costs Down: We Can Do It!

Our Goal: Production per Shift			
Unit	Minimum	Maximum	Cost per Unit
audio	600	1500	$55
phone	800	1700	$95

NewVocabulary
linear programming
feasible region
bounded
unbounded
optimize

Common Core State Standards

Content Standards
A.CED.3 Represent constraints by equations or inequalities, and by systems of equations and/or inequalities, and interpret solutions as viable or nonviable options in a modeling context.

Mathematical Practices
4 Model with mathematics.
8 Look for and express regularity in repeated reasoning.

1 Maximum and Minimum Values Situations often occur in business in which a company hopes to either maximize profits or minimize costs, and many constraints need to be considered. These issues can often be addressed by the use of systems of inequalities in linear programming.

Linear programming is a method for finding maximum or minimum values of a function over a given system of inequalities with each inequality representing a constraint. After the system is graphed and the vertices of the solution set, called the **feasible region**, are substituted into the function, you can determine the maximum or minimum value.

KeyConcept Feasible Regions

The feasible region is enclosed, or **bounded**, by the constraints. The maximum or minimum value of the related function *always* occurs at a vertex of the feasible region.

The feasible region is open and can go on forever. It is **unbounded**. Unbounded regions have either a maximum or a minimum.

Bloomberg/Getty Images

Lesson 3-3 Resources

Resource	Approaching Level (AL)	On Level (OL)	Beyond Level (BL)	English Learners (ELL)
Teacher Edition	▪ Differentiated Instruction, p. 156	▪ Differentiated Instruction, pp. 156, 160	▪ Differentiated Instruction, p. 160	
Chapter Resource Masters	▪ Study Guide and Intervention, pp. 18–19 ▪ Skills Practice, p. 20 ▪ Practice, p. 21 ▪ Word Problem Practice, p. 22	▪ Study Guide and Intervention, pp. 18–19 ▪ Skills Practice, p. 20 ▪ Practice, p. 21 ▪ Word Problem Practice, p. 22 ▪ Enrichment, p. 23 ▪ Graphing Calculator Activity, p. 24	▪ Practice, p. 21 ▪ Word Problem Practice, p. 22 ▪ Enrichment, p. 23	▪ Study Guide and Intervention, pp. 18–19 ▪ Skills Practice, p. 20 ▪ Practice, p. 21 ▪ Word Problem Practice, p. 22
Other	▪ 5-Minute Check 3-3 ▪ Study Notebook ▪ Teaching Algebra with Manipulatives	▪ 5-Minute Check 3-3 ▪ Study Notebook ▪ Teaching Algebra with Manipulatives	▪ 5-Minute Check 3-3 ▪ Study Notebook	▪ 5-Minute Check 3-3 ▪ Study Notebook ▪ Teaching Algebra with Manipulatives

Example 1 Bounded Region

Graph the system of inequalities. Name the coordinates of the vertices of the feasible region. Find the maximum and minimum values of the function for this region.

$3 \leq y \leq 6$
$y \leq 3x + 12$
$y \leq -2x + 6$
$f(x, y) = 4x - 2y$

ReadingMath

Function Notation
The notation $f(x, y)$ is used to represent a function with two variables, x and y. It is read f of x and y.

Step 1 Graph the inequalities and locate the vertices.

Step 2 Evaluate the function at each vertex.

(x, y)	$4x - 2y$	$f(x, y)$	
$(-3, 3)$	$4(-3) - 2(3)$	-18	
$(1.5, 3)$	$4(1.5) - 2(3)$	0	← maximum
$(0, 6)$	$4(0) - 2(6)$	-12	
$(-2, 6)$	$4(-2) - 2(6)$	-20	← minimum

The maximum value is 0 at (1.5, 3). The minimum value is -20 at $(-2, 6)$.

▶ **Guided**Practice   1A, 1B. See Chapter 3 Answer Appendix.

1A. $-2 \leq x \leq 6$
$1 \leq y \leq 5$
$y \leq x + 3$
$f(x, y) = -5x + 2y$

1B. $-6 \leq y \leq -2$
$y \leq -x + 2$
$y \leq 2x + 2$
$f(x, y) = 6x + 4y$

When a system of inequalities does not form a closed region, it is unbounded.

WatchOut!

CCSS **Precision** Do not assume that there is no maximum if the feasible region is unbounded above the vertices. Test points are needed to determine if there is a minimum or maximum.

Example 2 Unbounded Region

Graph the system of inequalities. Name the coordinates of the vertices of the feasible region. Find the maximum and minimum values of the function for this region.

$2y + 3x \geq -12$
$y \leq 3x + 12$
$y \geq 3x - 6$
$f(x, y) = 9x - 6y$

Evaluate the function at each vertex.

(x, y)	$9x - 6y$	$f(x, y)$
$(-4, 0)$	$9(-4) - 6(0)$	-36
$(0, -6)$	$9(0) - 6(-6)$	36

The maximum value is 36 at (0, -6). There is no minimum value. Notice that another point in the feasible region, (0, 8), yields a value of -48, which is less than -36.

▶ **Guided**Practice   2A, 2B. See Chapter 3 Answer Appendix.

2A. $y \leq 8$
$y \geq -x + 4$
$y \leq -x + 10$
$f(x, y) = -6x + 8y$

2B. $y \geq x - 9$
$y \leq -4x + 16$
$y \geq -4x - 4$
$f(x, y) = 10x + 7y$

Additional Answer (Additional Example)

2.

CCSS **Teaching the Mathematical Practices**

Precision Mathematically proficient students are careful about specifying units of measure, and labeling axes to clarify the correspondence with quantities in a problem. They calculate accurately and efficiently, express numerical answers with a degree of precision appropriate for the problem context. Encourage students to test all possible points before deciding on maximum or minimum values.

1 Maximum and Minimum Values

Examples 1 and **2** show how to find the maximum and minimum values of a function for a bounded and an unbounded region.

Formative Assessment

Use the Guided Practice exercises after each example to determine students' understanding of concepts.

Additional Examples

1 Graph the following system of inequalities. Name the coordinates of the vertices of the feasible region. Find the maximum and minimum values of the function for this region.

$x \leq 5$ $x + y \geq 2$
$y \leq 4$ $f(x, y) = 3x - 2y$

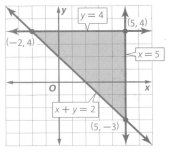

max at $(5, -3) = 21$;
min at $(-2, 4) = -14$

2 Graph the following system of inequalities. Name the coordinates of the vertices of the feasible region. Find the maximum and minimum values of the function for this region.

$-x + 2y \leq 2$
$x - 2y \leq 4$
$x + y \geq -2$
$f(x, y) = 2x + 3y$

See bottom margin for graph.
max: none; min at $(0, -2) = -6$

▶ **Additional Examples** also in Interactive Classroom PowerPoint® Presentations

IWB **Interactive White Board READY**

2 Optimization

Example 3 shows how to use linear programming to solve a real-world problem.

Additional Example

3 **LANDSCAPING** A landscaping company has crews who mow lawns and prune shrubbery. The company schedules one hour for mowing jobs and three hours for pruning jobs. Each crew is scheduled for no more than two pruning jobs per day. Each crew's schedule is set up for a maximum of nine hours per day. On average, the charge for mowing a lawn is $40, and the charge for pruning shrubbery is $120. Find a combination of mowing lawns and pruning shrubs that will maximize the income the company receives per day for one of its crews.

either three mowings and two prunings, or nine mowings and zero prunings

Focus on Mathematical Content

Linear Programming The process of finding maximum or minimum values of a function for a region defined by linear inequalities is called linear programming. Linear programming is a very useful tool for solving many real-world problems.

WatchOut!

Preventing Errors When using linear programming to solve real-world problems, it is important for students to follow the seven steps from p. 156. Encourage students to show their work and label each step as they work through the problem.

Real-WorldCareer

Operations Manager Operations management is an area of business that is concerned with the production of goods and services, and involves the responsibility of ensuring that business operations are efficient and effective. A master's degree in business and experience in operations are preferred.

StudyTip

Reasonableness Check your solutions for reasonableness by thinking of the context of the problem.

---

2 Optimization

To **optimize** means to seek the best price or amount to minimize costs or maximize profits. This is often obtained with the use of linear programming.

KeyConcept Optimization with Linear Programming

Step 1 Define the variables.
Step 2 Write a system of inequalities.
Step 3 Graph the system of inequalities.
Step 4 Find the coordinates of the vertices of the feasible region.
Step 5 Write a linear function to be maximized or minimized.
Step 6 Substitute the coordinates of the vertices into the function.
Step 7 Select the greatest or least result. Answer the problem.

When using a system of inequalities to describe constraints in real-world problems, often only whole-number solutions will make sense.

Real-World Example 3 Optimization with Linear Programming

BUSINESS Refer to the application at the beginning of the lesson. Determine how many of each type of device should be made per shift.

Step 1 Let a = number of audio players produced.
Let p = number of phones produced.

Step 2
$600 \leq a \leq 1500$
$800 \leq p \leq 1700$
$a + p \geq 2000$

Steps 3 and 4 The system is graphed at the right. Note the vertices of the feasible region.

Graph: Number of Phones (vertical axis) vs Number of Audio Players (horizontal axis). Vertices labeled (600, 1700), (1500, 1700), (600, 1400), (1200, 800), (1500, 800).

Step 5 The function to be minimized is $f(a, p) = 55a + 95p$.

Step 6

(a, p)	$55a + 95p$	$f(a, p)$	
(600, 1700)	55(600) + 95(1700)	194,500	
(600, 1400)	55(600) + 95(1400)	166,000	
(1500, 1700)	55(1500) + 95(1700)	244,000	← maximum
(1500, 800)	55(1500) + 95(800)	158,500	
(1200, 800)	55(1200) + 95(800)	142,000	← minimum

Step 7 Produce 1200 audio players and 800 phones to minimize costs.

GuidedPractice 40 earrings and 25 necklaces

3. **JEWELRY** Each week, Mackenzie can make 10 to 25 necklaces and 15 to 40 pairs of earrings. If she earns profits of $3 on each pair of earrings and $5 on each necklace, and she plans to sell at least 30 pieces of jewelry, how can she maximize profit?

 156 | Lesson 3-3 | Optimization with Linear Programming

DifferentiatedInstruction (AL) (OL)

If → students have trouble with the relationship between the different regions of a graph of a system of inequalities,

Then → have students use different colored pencils to shade the different regions of a graph defined by the inequalities in a linear programming problem. This should help students clarify the relationship between the various regions in these graphs.

ColorBlind Images/Blend Images/Getty Images

Examples 1–2 Graph each system of inequalities. Name the coordinates of the vertices of the feasible region. Find the maximum and minimum values of the given function for this region. **1–6. See margin.**

1. $y \le 5$
$x \le 4$
$y \ge -x$
$f(x, y) = 5x - 2y$

2. $y \ge -3x + 6$
$-y \le x$
$y \le 3$
$f(x, y) = 8x + 4y$

3. $y \ge -3x + 2$
$9x + 3y \le 24$
$y \ge -4$
$f(x, y) = 2x + 14y$

4. $-2 \le y \le 6$
$3y \le 4x + 26$
$y \le -2x + 2$
$f(x, y) = -3x - 6y$

5. $-3 \le y \le 7$
$4y \ge 4x - 8$
$6y + 3x \le 24$
$f(x, y) = -12x + 9y$

6. $y \le 2x + 6$
$y \ge 2x - 8$
$y \ge -2x - 18$
$f(x, y) = 5x - 4y$

Example 3 **7.** **CCSS** **PRECISION** The total number of workers' hours per day available for production in a skateboard factory is 85 hours. There are 40 hours available for finishing decks and quality control each day. The table shows the number of hours needed in each department for two different types of skateboards.

Skateboard Manufacturing Time

Board Type	Production Time	Deck Finishing/Quality control
Pro Boards	1.5 hours	2 hours
Specialty Boards	1 hour	0.5 hour

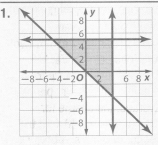

7a. $g \ge 0$, $c \ge 0$, $1.5g + c \le 85$, $2g + 0.5c \le 40$

a. Write a system of inequalities to represent the situation.

b. Draw the graph showing the feasible region. **See Chapter 3 Answer Appendix.**

c. List the coordinates of the vertices of the feasible region. **(0, 0), (0, 20), (80, 0)**

d. If the profit on a pro board is $50 and the profit on a specialty board is $65, write a function for the total profit on the skateboards. $f(c, g) = 65c + 50g$

e. Determine the number of each type of skateboard that needs to be made to have a maximum profit. What is the maximum profit? **80 specialty boards, 0 pro boards; $5200**

Practice and Problem Solving
Extra Practice is on page R3.

Examples 1–2 Graph each system of inequalities. Name the coordinates of the vertices of the feasible region. Find the maximum and minimum values of the given function for this region. **8–13. See Chapter 3 Answer Appendix.**

8. $1 \le y \le 4$
$4y - 6x \ge -32$
$2y \ge -x + 4$
$f(x, y) = -6x + 3y$

9. $2 \ge x \ge -3$
$y \ge -2x - 6$
$4y \le 2x + 32$
$f(x, y) = -4x - 9y$

10. $-2 \le x \le 4$
$5 \le y \le 8$
$2x + 3y \le 26$
$f(x, y) = 8x - 10y$

11. $-8 \le y \le -2$
$y \le x$
$y \le -3x + 10$
$f(x, y) = 5x + 14y$

12. $x + 4y \ge 2$
$2x + 4y \le 24$
$2 \le x \le 6$
$f(x, y) = 6x + 7y$

13. $3 \le y \le 7$
$2y + x \le 8$
$y - 2x \le 23$
$f(x, y) = -3x + 5y$

5.

$(4, 2), (-1, -3), (-6, 7)$; max does not exist; min $= -30$

6.

$(-6, -6), (-2.5, -13)$, no min, max $= 39.5$

Formative Assessment

Use Exercises 1–7 to check for understanding.

Use the chart on the bottom of the next page to customize assignments for your students.

Additional Answers

1.

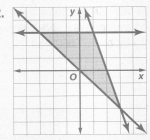

$(4, 5), (4, -4), (-5, 5)$; max $= 28$, min $= -35$

2.

$(1, 3), (3, -3), (-3, 3)$; max $= 20$, min $= -12$

3.

$(2, -4), (4, -4)$; max does not exist, min $= -52$

4.

$(2, -2), (-8, -2), (-2, 6)$; max $= 36$, min $= -30$

Problem Solving Tip Remind students
that neither the maximum nor minimum
automatically occurs at the smallest/
largest (x, y) coordinate. The max/min
depend on the $f(x, y)$ function in the
situation.

**Teaching the
Mathematical Practices**

Precision Mathematically proficient
students are careful about specifying
units of measure, and labeling axes to
clarify the correspondence with
quantities in a problem. They calculate
accurately and efficiently, express
numerical answers with a degree of
precision appropriate for the problem
context.

Additional Answers

14–22. See Ch. 3 Answer Appendix
for graphs.

14. $(-9, -9), (-4, -9), (-5, -5),$
$(-9, -5)$; max $= -140,$
min $= -252$

15. $(6, 3), (-8, 10), (-8, -18);$
max $= 42$, min $= -140$

16. $(2, 0), (5, 3), (-3, 8), (-6, 8);$
max $= -10$, min $= -105$

17. $(-6, 1), (6, -7), (-6, 5);$
max $= 48$, min $= 0$

18. $(5, -5), (8, -17), (-12, -17),$
$(-8, -5)$; max $= 115,$
min $= -49$

19. $(-8, 44), (16, 32), (-8, -26),$
$(16, 22)$; max $= 672$, min $= -486$

20. $(3, 7), (7, 3), (-3, -7), (-7, -3);$
max $= 43$, min $= -43$

21. $(5, -1), (1, 6), (-2, -8),$
$(-4, -8), (-4, 6)$; max $= 60,$
min $= -112$

22. $(-4, 6), (2, 4), (2, 1), (1, 0),$
$(-3, 0), (-6, 3), (-6, 6);$
max $= 26$, min $= -18$

Examples 1–2 Graph each system of inequalities. Name the coordinates of the vertices of the feasible region. Find the maximum and minimum values of the given function for this region. **14–22. See margin.**

14. $-9 \leq x \leq -3$
$-9 \leq y \leq -5$
$3y + 12x \leq -75$
$f(x, y) = 20x + 8y$

15. $x \geq -8$
$3x + 6y \leq 36$
$2y + 12 \geq 3x$
$f(x, y) = 10x - 6y$

16. $y \geq |x - 2|$
$y \leq 8$
$8y + 5x \leq 49$
$f(x, y) = -5x - 15y$

17. $x \geq -6$
$y + x \leq -1$
$2x + 3y \geq -9$
$f(x, y) = -10x - 12y$

18. $-5 \geq y \geq -17$
$y \leq 3x + 19$
$y \leq -4x + 15$
$f(x, y) = 8x - 3y$

19. $-8 \leq x \leq 16$
$y \geq 2x - 10$
$2y + x \leq 80$
$f(x, y) = 12x + 15y$

20. $y \leq x + 4$
$y \geq x - 4$
$y \leq -x + 10$
$y \geq -x - 10$
$f(x, y) = -10x + 9y$

21. $-4 \leq x \leq 8$
$-8 \leq y \leq 6$
$y \geq x - 6$
$4y + 7x \leq 31$
$f(x, y) = 12x + 8y$

22. $y \geq |x + 1| - 2$
$0 \leq y \leq 6$
$-6 \leq x \leq 2$
$x + 3y \leq 14$
$f(x, y) = 5x + 4y$

Example 3

23. **COOKING** Jenny's Bakery makes two types of birthday cakes: yellow cake, which sells for $25, and strawberry cake, which sells for $35. Both cakes are the same size, but the decorating and assembly time required for the yellow cake is 2 hours, while the time is 3 hours for the strawberry cake. There are 450 hours of labor available for production. How many of each type of cake should be made to maximize revenue? **225 yellow cakes, 0 strawberry cakes**

24. **BUSINESS** The manager of a travel agency is printing brochures and fliers to advertise special discounts on vacation spots during the summer months. Each brochure costs $0.08 to print, and each flier costs $0.04 to print. A brochure requires 3 pages, and a flier requires 2 pages. The manager does not want to use more than 600 pages, and she needs at least 50 brochures and 150 fliers. How many of each should she print to minimize the cost? **50 brochures, 150 fliers**

25. **CCSS PRECISION** Sean has 20 days to paint as many play houses and sheds as he is able. The sheds can be painted at a rate of 2.5 per day, and the play houses can be painted at a rate of 2 per day. He has 45 structures that need to be painted.

a. Write a system of inequalities to represent the possible ways Sean can paint the structures. $a \geq 0, b \geq 0, a + b \leq 45, 4a + 5b \leq 200$

b. Draw a graph showing the feasible region and list the coordinates of the vertices of the feasible region. **See Chapter 3 Answer Appendix.**

c. If the profit is $26 per shed and $30 per play house, how many of each should he paint? **25 sheds, 20 play houses**

d. What is the maximum profit? **$1250**

26. **MOVIES** Employees at a local movie theater work 8-hour shifts from noon to 8 P.M. or from 4 P.M. to midnight. The table below shows the number of employees needed and their corresponding pay. Find the numbers of day-shift workers and night-shift workers that should be scheduled to minimize the cost. What is the minimal cost? **8 day-shift and 6 night-shift workers; $776**

 158 | Lesson 3-3 | Optimization with Linear Programming

Time	noon to 4 P.M.	4 P.M. to 8 P.M.	8 P.M. to midnight
Number of Employees Needed	at least 5	at least 14	6
Rate per Hour	$5.50	$7.50	$7.50

Differentiated Homework Options

Level	Assignment	Two-Day Option	
AL Basic	8–24, 29–55	9–23 odd, 34–37	8–24 even, 29–33, 38–55
OL Core	9–23 odd, 25–29, 31–55	8–24, 34–37	25–29, 31–33, 38–55
BL Advanced	25–49, (optional: 50–55)		

27 **BUSINESS** Each car on a freight train can hold 4200 pounds of cargo and has a capacity of 480 cubic feet. The freight service handles two types of packages: small—which weigh 25 pounds and are 3 cubic feet each, and large—which are 50 pounds and are 5 cubic feet each. The freight service charges $5 for each small package and $8 for each large package.

a. Find the number of each type of package that should be placed on a train car to maximize revenue. **160 small packages, 0 large packages**

b. What is the maximum revenue per train car? **$800**

c. In this situation, is maximizing the revenue necessarily the best thing for the company to do? Explain. **See margin.**

28. RECYCLING A recycling plant processes used plastic into food or drink containers. The plant processes up to 1200 tons of plastic per week. At least 300 tons must be processed for food containers, while at least 450 tons must be processed for drink containers. The profit is $17.50 per ton for processing food containers and $20 per ton for processing drink containers. What is the profit if the plant maximizes processing? **$23,250**

H.O.T. Problems Use Higher-Order Thinking Skills

31. b; The feasible region of Graph b is unbounded while the other three are bounded.

29. OPEN ENDED Create a set of inequalities that forms a bounded region with an area of 20 units^2 and lies only in the fourth quadrant. **Sample answer:** $-2 \geq y \geq -6, 4 \leq x \leq 9$

30. CHALLENGE Find the area of the bounded region formed by the following constraints: $y \geq |x| - 3, y \leq -|x| + 3$, and $x \geq |y|$. **4.5 units^2**

31. CCSS ARGUMENTS Identify the system of inequalities that is not the same as the other three. Explain your reasoning.

32. Sample answer: Always; if a point on the unbounded region forms a minimum, then a maximum cannot also be formed because of the unbounded region. There will always be a value in the solution that will produce a higher value than any projected maximum.

a.

b.

c.

d.

33. Sample answer: Even though the region is bounded, multiple maximums occur at A and B and all of the points on the boundary of the feasible region containing both A and B. This happened because that boundary of the region has the same slope as the function.

32. REASONING Determine whether the following statement is *sometimes*, *always*, or *never* true. Explain your reasoning.

An unbounded region will not have both a maximum and minimum value.

33. WRITING IN MATH Upon determining a bounded feasible region, Ayumi noticed that vertices $A(-3, 4)$ and $B(5, 2)$ yielded the same maximum value for $f(x, y) = 16y + 4x$. Kelvin confirmed that his constraints were graphed correctly and his vertices were correct. Then he said that those two points were not the only maximum values in the feasible region. Explain how this could have happened.

Additional Answer

27c. No; if revenue is maximized, the company will not deliver any large packages, and customers with large packages to ship will probably choose another carrier.

Study Guide and Intervention
AL OL ELL

Practice
AL OL BL ELL

Word Problem Practice
AL OL BL ELL

Enrichment
OL BL

4 Assess

Yesterday's News Have students write how yesterday's lesson helped them with today's new material.

Additional Answers

38.

39.

40.

no solution

Standardized Test Practice

34. Kelsey worked 350 hours during the summer and earned $2978.50. She earned $6.85 per hour when she worked at a video store and $11 per hour as an architectural intern. Let x represent the number of hours she worked at the video store and y represent the number of hours that she interned. Which system of equations represents this situation? **B**

A $x + y = 350$
$11x + 6.85y = 2978.50$

B $x + y = 350$
$6.85x + 11y = 2978.50$

C $x + y = 2978.50$
$6.85x + 11y = 350$

D $x + y = 2978.50$
$11x + 6.85y = 350$

35. SHORT RESPONSE A family of four went out to dinner. Their bill, including tax, was $60. They left a 17% tip on the total cost of their bill. What is the total cost of the dinner including tip? $70.20

36. SAT/ACT For a game she is playing, Liz must draw a card from a deck of 26 cards, one with each letter of the alphabet on it, and roll a die. What is the probability that Liz will draw a letter in her name and roll an odd number? **H**

F $\frac{2}{3}$

G $\frac{1}{13}$

H $\frac{3}{52}$

J $\frac{1}{26}$

K $\frac{1}{52}$

37. GEOMETRY Which of the following best describes the graphs of $y = 3x - 5$ and $4y = 12x + 16$? **D**

A The lines have the same y-intercept.
B The lines have the same x-intercept.
C The lines are perpendicular.
D The lines are parallel.

Spiral Review

Solve each system of inequalities by graphing. (Lesson 3-2) **38–40. See margin.**

38. $3x + 2y \geq 6$
$4x - y \geq 2$

39. $4x - 3y < 7$
$2y - x < -6$

40. $3y \leq 2x - 8$
$y \geq \frac{2}{3}x - 1$

41. BUSINESS Last year the chess team paid $7 per hat and $15 per shirt for a total purchase of $330. This year they spent $360 to buy the same number of shirts and hats because the hats now cost $8 and the shirts cost $16. Write and solve a system of two equations that represents the number of hats and shirts bought each year. (Lesson 3-1) $7x + 15y = 330$, $8x + 16y = 360$; hats: 15, shirts: 15

Write an equation in slope-intercept form for the line that satisfies each set of conditions. (Lesson 2-4)

42. passes through $(5, 1)$ and $(8, -4)$ $y = -\frac{5}{3}x + \frac{28}{3}$

43. passes through $(-3, 5)$ and $(3, 2)$ $y = -\frac{1}{2}x + \frac{7}{2}$

Find the x-intercept and the y-intercept of the graph of each equation. Then graph the equation. (Lesson 2-2) **44–49. See Chapter 3 Answer Appendix for graphs.**

44. $5x + 3y = 15$ 3; 5

45. $2x - 6y = 12$ 6; -2

46. $3x - 4y - 10 = 0$ $\frac{10}{3}$; $-\frac{5}{2}$

47. $2x + 5y - 10 = 0$ 5; 2

48. $y = x$ 0; 0

49. $y = 4x - 2$ $\frac{1}{2}$; -2

Skills Review

Evaluate each expression if $x = -1$, $y = 3$, and $z = 7$.

50. $x + y + z$ 9

51. $2x - y + 2z$ 9

52. $-x + 4y - 3z$ -8

53. $4x + 2y - z$ -5

54. $5x - y + 4z$ 20

55. $-3x - 3y + 3z$ 15

 160 | Lesson 3-3 | Optimization with Linear Programming

DifferentiatedInstruction OL BL

Extension Linear programming is a great lens for looking at effective business practice. Ask students to create a business similar to those mentioned in the Exercises. Students must determine their costs and profits for the chosen business and analyze the information to determine maximum profit options. Make the task more authentic by having students pursue a business that is of personal interest to them and estimate profit margins using the Internet as a resource.

LESSON 3-4

Systems of Equations in Three Variables

Then	Now	Why?
• You solved systems of linear equations in two variables.	**1** Solve systems of linear equations in three variables. **2** Solve real-world problems using systems of linear equations in three variables.	• Seats closest to an amphitheater stage cost $30. The seats in the next section cost $25, and lawn seats are $20. There are twice as many seats in section B as in section A. When all 19,200 seats are sold, the amphitheater makes $456,000. A system of equations in three variables can be used to determine the number of seats in each section.

NewVocabulary
ordered triple

Common Core State Standards

Content Standards
A.CED.3 Represent constraints by equations or inequalities, and by systems of equations and/or inequalities, and interpret solutions as viable or nonviable options in a modeling context.

Mathematical Practices
3 Construct viable arguments and critique the reasoning of others.

1 Systems in Three Variables Like systems of equations in two variables, systems in three variables can have one solution, infinite solutions, or no solution. A solution of such a system is an **ordered triple** (x, y, z).

The graph of an equation in three variables is a three-dimensional graph in the shape of a plane. The graphs of a system of equations in three variables form a system of planes.

One Solution

The three individual planes intersect at a specific point.

(x, y, z)

Infinitely Many Solutions

The planes intersect in a line.

Every coordinate on the line represents a solution of the system.

The planes intersect in the same plane.

Every equation is equivalent.
Every coordinate in the plane represents a solution of the system.

No Solution There are no points in common with all three planes.

connectED.mcgraw-hill.com **161**

---

1 Focus

VerticalAlignment

Before Lesson 3-4 Solve systems of linear equations in two variables.

Lesson 3-4 Solve systems of linear equations in three variables. Solve real-world problems using systems of linear equations in three variables.

After Lesson 3-4 Use matrices to solve systems of equations.

2 Teach

Scaffolding Questions
Have students read the **Why?** section of the lesson.

Ask:
- How many seats are in the amphitheater? 19,200
- How much do the seats in section A cost? $30 in section B? $25
- Suppose the amphitheater sells 3100 seats in section A, 5050 seats in section B, and 7200 lawn seats. How much money is taken in? $363,250

Lesson 3-4 Resources

Resource	Approaching Level **AL**	On Level **OL**	Beyond Level **BL**	English Learners **ELL**
Teacher Edition	• Differentiated Instruction, p. 164	• Differentiated Instruction, p. 164, 167	• Differentiated Instruction, p. 167	• Differentiated Instruction, p. 164
Chapter Resource Masters	• Study Guide and Intervention, pp. 25–26 • Skills Practice, p. 27 • Practice, p. 28 • Word Problem Practice, p. 29	• Study Guide and Intervention, pp. 25–26 • Skills Practice, p. 27 • Practice, p. 28 • Word Problem Practice, p. 29 • Enrichment, p. 30	• Practice, p. 28 • Word Problem Practice, p. 29 • Enrichment, p. 30	• Study Guide and Intervention, pp. 25–26 • Skills Practice, p. 27 • Practice, p. 28 • Word Problem Practice, p. 29
Other	• 5-Minute Check 3-4 • Study Notebook • Teaching Algebra with Manipulatives	• 5-Minute Check 3-4 • Study Notebook • Teaching Algebra with Manipulatives	• 5-Minute Check 3-4 • Study Notebook	• 5-Minute Check 3-4 • Study Notebook • Teaching Algebra with Manipulatives

Examples 1 and **2** show how to solve systems of equations in three variables.

Formative Assessment

Use the Guided Practice exercises after each example to determine students' understanding of concepts.

Additional Example

1 Solve the system of equations.
$5x + 3y + 2z = 2$
$2x + y - z = 5$
$x + 4y + 2z = 16$
$(-2, 6, -3)$

▶ **Additional Examples** also in Interactive Classroom PowerPoint® Presentations

IWB Interactive White Board READY

WatchOut!

Preventing Errors Suggest that when solving systems of equations in three variables, students should take a few moments before beginning their calculations to examine the equations and make a plan for solving the system. Also, remind students that the three values of the solution represent a point in space that is contained in all three planes.

Tips for New Teachers

Elimination Ask students what happens to the graph of an equation when it is multiplied by a constant. The graph is unchanged because the new equation is equivalent to the original equation.

Solving systems of equations in three variables is similar to solving systems of equations in two variables. Use the strategies of substitution and elimination to find the ordered triple that represents the solution of the system.

Example 1 A System with One Solution

Solve the system of equations.

$3x - 2y + 4z = 35$
$-4x + y - 5z = -36$ ◀ The coefficient of 1 in the second equation makes y a good choice for elimination.
$5x - 3y + 3z = 31$

Step 1 Eliminate one variable by using two pairs of equations.

$3x - 2y + 4z = 35$
$-4x + y - 5z = -36$ → Multiply by 2. →

$\begin{array}{rl} 3x - 2y + 4z = & 35 \quad \text{Equation 1} \\ (+) -8x + 2y - 10z = & -72 \quad \text{Equation 2} \times 2 \\ \hline -5x \quad\quad - 6z = & -37 \end{array}$

$-4x + y - 5z = -36$ → Multiply by 3. →
$5x - 3y + 3z = 31$

$\begin{array}{rl} -12x + 3y - 15z = & -108 \quad \text{Equation 2} \times 3 \\ (+) 5x - 3y + 3z = & 31 \quad \text{Equation 3} \\ \hline -7x \quad\quad - 12z = & -77 \end{array}$

The y-terms in each equation have been eliminated. We now have a system of two equations and two variables, x and z.

Step 2 Solve the system of two equations.

$-5x - 6z = -37$ → Multiply by −2. →
$-7x - 12z = -77$

$\begin{array}{rll} 10x + 12z = & 74 \\ (+) -7x - 12z = & -77 \\ \hline 3x \quad\quad = & -3 & \text{Eliminate } z. \\ x = & -1 & \text{Divide by 3.} \end{array}$

Use substitution to solve for z.

$\begin{array}{rll} -5x - 6z = -37 & \text{Equation with two variables} \\ -5(-1) - 6z = -37 & \text{Substitution} \\ 5 - 6z = -37 & \text{Multiply.} \\ -6z = -42 & \text{Subtract 5 from each side.} \\ z = 7 & \text{Divide each side by } -6. \end{array}$

The result is $x = -1$ and $z = 7$.

Step 3 Substitute the two values into one of the original equations to find y.

$\begin{array}{rll} -4x + y - 5z = -36 & \text{Equation 2} \\ -4(-1) + y - 5(7) = -36 & \text{Substitution} \\ 4 + y - 35 = -36 & \text{Multiply.} \\ y = -5 & \text{Add 31 to each side.} \end{array}$

$\begin{array}{lrll} \text{CHECK} & -4x + y - 5z = -36 & \text{Equation 2} \\ & -4(-1) + (-5) - 5(7) \stackrel{?}{=} -36 & x = -1, y = -5, z = 7 \\ & 4 + (-5) - 35 \stackrel{?}{=} -36 & \text{Simplify.} \\ & -36 = -36 \checkmark \end{array}$

The solution is $(-1, -5, 7)$.

> **StudyTip**
> Checking Solutions Always substitute your answer into all of the original equations to confirm your answer.

▶ **Guided**Practice

1A. $2x + 4y - 5z = 18$
$-3x + 5y + 2z = -27$
$-5x + 3y - z = -17$ $(3, -2, -4)$

1B. $4x - 3y + 6z = 18$
$-x + 5y + 4z = 48$
$6x - 2y + 5z = 0$ $(-6, 2, 8)$

 162 | Lesson 3-4 | Systems of Equations in Three Variables

Teach with Tech

Video Recording Have students work in groups to create a video showing how to solve a system of equations in three variables. Be sure students show how to check the solution in the original system of equations. Share each group's video with the entire class.

Tips for New Teachers

Choosing Variables If one of the equations is missing a variable already, point out that students can use the other two equations and eliminate the same variable.

When solving a system of three linear equations with three variables, it is important to check your answer using all three of the original equations. This is necessary because it is possible for a solution to work for two of the equations but not the third.

Example 2 No Solution and Infinite Solutions

Solve each system of equations.

a. $5x + 4y - 5z = -10$
$-4x - 10y - 8z = -16$
$6x + 15y + 12z = 24$

Eliminate x in the second two equations.

$-4x - 10y - 8z = -16$ **Multiply by 3.** $-12x - 30y - 24z = -48$

$6x + 15y + 12z = 24$ **Multiply by 2.** $\underline{(+) \; 12x + 30y + 24z = 48}$
$0 = 0$

The equation $0 = 0$ is always true. This indicates that the last two equations represent the same plane. Check to see if this plane intersects the first plane.

$5x + 4y - 5z = -10$ **Multiply by 4.** $20x + 16y - 20z = -40$

$-4x - 10y - 8z = -16$ **Multiply by 5.** $\underline{(+) \; -20x - 50y - 40z = -80}$
$-34y - 60z = -120$

The planes intersect in a line. So, there are an infinite number of solutions.

b. $-6a + 9b - 12c = 21$
$-2a + 3b - 4c = 7$
$10a - 15b + 20c = -30$

Eliminate a in the first two equations.

$-6a + 9b - 12c = 21$ $-6a + 9b - 12c = 21$

$-2a + 3b - 4c = 7$ **Multiply by -3.** $\underline{(+) \; 6a - 9b + 12c = -21}$
$0 = 0$

The equation $0 = 0$ is always true. This indicates that the first two equations represent the same plane. Check to see if this plane intersects the last plane.

$-2a + 3b - 4c = 7$ **Multiply by 5.** $-10a + 15b - 20c = 35$

$10a - 15b + 20c = -30$ $\underline{(+) \; 10a - 15b + 20c = -30}$
$0 = 5$

The equation $0 = 5$ is never true. So, there is no solution of this system.

GuidedPractice

2A. $-4x - 2y - z = 15$ **no solution**
$12x + 6y + 3z = 45$
$2x + 5y + 7z = -29$

2B. $3x + 5y - 2z = 13$ **infinite solutions**
$-5x - 2y - 4z = 20$
$-14x - 17y + 2z = -19$

2 Real-World Problems When solving problems involving three variables, use the four-step plan to help organize the information. Identify the three variables and what they represent. Then use the information in the problem to form equations using the variables. Once you have three equations and all three variables are represented, you can solve the problem.

Additional Example

2 Solve each system of equations.

a. $2x + y - 3z = 5$
$x + 2y - 4z = 7$
$6x + 3y - 9z = 15$
There are an infinite number of solutions.

b. $3x - y - 2z = 4$
$6x - 2y - 4z = 11$
$9x - 3y - 6z = 12$
There is no solution to this system.

WatchOut!

Preventing Errors To help students visualize a system with no solutions, ask them to sketch a drawing of three planes whose equations form a system with no solutions.

WatchOut!

Preventing Errors Some students may think that *any* ordered triple will be a solution to a system in three variables where there are an infinite number of solutions (as in Example 2a). Ask students to discuss why only the infinite set of ordered triples that names points on the line (or plane) of intersection contains solutions of the system of equations.

StudyTip

Infinite and No Solutions When solving a system of more than two equations, $0 = 5$ will always yield no solution, while $0 = 0$ may not always yield infinite solutions.

2 Real-World Problems

Example 3 shows how to solve a real-world problem involving three variables.

Additional Example

3 **SPORTS** There are 49,000 seats in a sports stadium. Tickets for the seats in the upper level sell for $25, the ones in the middle level cost $30, and the ones in the bottom level are $35 each. The number of seats in the middle and bottom levels together equals the number of seats in the upper level. When all of the seats are sold for an event, the total revenue is $1,419,500. How many seats are there in each level? bottom level: 14,400 seats; middle level: 10,100 seats; upper level: 24,500 seats

---

Focus on Mathematical Content

Systems in Three Variables To graph equations in three variables, it is necessary to add a third dimension to the coordinate system. Ordered triples such as (x, y, z) represent a point in space just as ordered pairs such as (x, y) represent a point in a plane. The graph of an equation of the form $Ax + By + Cz = D$, where A, B, C, and D are not all equal to zero, is a plane.

Real-World Career

Music Management
Music management involves acting as the talent manager to the artists. Other duties include negotiating with record labels, music promoters, and tour promoters. Managers usually earn a percentage of the artist's income. A bachelor's degree is usually required.

Real-World Example 3 Write and Solve a System of Equations

CONCERTS Refer to the beginning of the lesson. Write and solve a system of equations to determine how many seats are in each section of the amphitheater.

Understand Define the variables. x = seats in section A
                                           y = seats in section B
                                           z = lawn seats

Plan There are 19,200 seats.
$x + y + z = 19{,}200$ Equation 1

The total revenue is $456,000.
$30x + 25y + 20z = 456{,}000$ Equation 2

There are twice as many seats in section B as in section A.
$y = 2x$ Equation 3

Solve Solve the system.

> **Step 1** Substitute $y = 2x$ in the first two equations.

$$x + y + z = 19{,}200 \qquad \text{Equation 1}$$
$$x + 2x + z = 19{,}200 \qquad y = 2x$$
$$3x + z = 19{,}200 \qquad \text{Add.}$$
$$30x + 25y + 20z = 456{,}000 \qquad \text{Equation 2}$$
$$30x + 25(2x) + 20z = 456{,}000 \qquad y = 2x$$
$$80x + 20z = 456{,}000 \qquad \text{Simplify.}$$

> **Step 2** Solve the system of two equations in two variables.

$$3x + z = 19{,}200 \quad \boxed{\text{Multiply by } -20.} \quad -60x - 20z = -384{,}000$$
$$80x + 20z = 456{,}000 \qquad\qquad\qquad\quad (+)\ 80x + 20z = 456{,}000$$
$$\overline{20x = 72{,}000}$$
$$x = 3600$$

> **Step 3** Substitute to find z.

$$3x + z = 19{,}200 \qquad \text{Remaining equation in two variables}$$
$$3(3600) + z = 19{,}200 \qquad x = 3600$$
$$10{,}800 + z = 19{,}200 \qquad \text{Multiply.}$$
$$z = 8400 \qquad \text{Simplify.}$$

> **Step 4** Substitute to find y.

$$y = 2x \qquad \text{Equation 3}$$
$$y = 2(3600) \text{ or } 7200 \qquad x = 3600$$

The solution is (3600, 7200, 8400). There are 3600 seats in section A, 7200 in section B, and 8400 lawn seats.

Check Substitute the values into either of the first two equations.

> **Guided Practice**

6000 at 10%; 18,000 at 8%; 26,000 at 12%

3. Ms. Garza invested $50,000 in three different accounts. She invested three times as much money in an account that paid 8% interest than an account that paid 10% interest. The third account earned 12% interest. If she earned a total of $5160 in interest in a year, how much did she invest in each account?

 164 | Lesson 3-4 | Systems of Equations in Three Variables

DifferentiatedInstruction AL OL ELL

If some students struggle with solving real-world problems using systems of equations in three variables,

Then pair those students with students that are having success. Encourage students to explain to partners their explorations and plans for solving real-world problems using systems of three equations in three variables. Suggest that the listening partner take notes about the speaking partner's strategies, asking questions as needed for clarification.

Examples 1–2 Solve each system of equations.

1. $(-2, -3, 5)$
2. $(4, -6, 1)$
3. $(-4, 3, 6)$

1. $-3a - 4b + 2c = 28$
$a + 3b - 4c = -31$
$2a + 3c = 11$

2. $3y - 5z = -23$
$4x + 2y + 3z = 7$
$-2x - y - z = -3$

3. $3x + 6y - 2z = -6$
$2x + y + 4z = 19$
$-5x - 2y + 8z = 62$

4. $(-2, 2, -5)$
5. infinite solutions
6. $(3, -4, 8)$

4. $-4r - s + 3t = -9$
$3r + 2s - t = 3$
$r + 3s - 5t = 29$

5. $3x + 5y - z = 12$
$-2x - 3y + 5z = 14$
$4x + 7y + 3z = 38$

6. $2a - 3b + 5c = 58$
$-5a + b - 4c = -51$
$-6a - 8b + c = 22$

Example 3

7. DOWNLOADING Heather downloaded some television shows. A sitcom uses 0.3 gigabyte of memory; a drama, 0.6 gigabyte; and a talk show, 0.6 gigabyte. She downloaded 7 programs totaling 3.6 gigabytes. There were twice as many episodes of the drama as the sitcom.

   a. Write a system of equations for the number of episodes of each type of show.

   b. How many episodes of each show did she download? **2 sitcoms, 4 dramas, 1 talk show**

   7a. $s + d + t = 7$, $d = 2s$, $0.3s + 0.6d + 0.6t = 3.6$

Practice and Problem Solving Extra Practice is on page R3.

Examples 1–2 Solve each system of equations.

8. $-5x + y - 4z = 60$
$2x + 4y + 3z = -12$
$6x - 3y - 2z = -52$
$(-8, 4, -4)$

9. $4a + 5b - 6c = 2$
$-3a - 2b + 7c = -15$
$-a + 4b + 2c = -13$
$(-3, -2, -4)$

10. $-2x + 5y + 3z = -25$ $(8, -3, 2)$
$-4x - 3y - 8z = -39$
$6x + 8y - 5z = 14$

11. $4r + 6s - t = -18$
$3r + 2s - 4t = -24$
$-5r + 3s + 2t = 15$
$(-2, -1, 4)$

12. $-2x + 15y + z = 44$
$4x + 3y + 3z = 18$
$-3x + 6y - z = 8$
no solution

13. $4x + 2y + 6z = 13$ **infinite solutions**
$-12x + 3y - 5z = 8$
$-4x + 7y + 7z = 34$

14. $8x + 3y + 6z = 43$
$-3x + 5y + 2z = 32$
$5x - 2y + 5z = 24$
$(-1, 3, 7)$

15. $-6x - 5y + 4z = 53$
$5x + 3y + 2z = -11$
$8x - 6y + 5z = 4$
$(-4, -1, 6)$

16. $-9a + 3b - 2c = 61$
$8a + 7b + 5c = -138$
$5a - 5b + 8c = -45$
$(-8, -7, -5)$

17. $2x - y + z = 1$
$x + 2y - 4z = 3$
$4x + 3y - 7z = -8$
no solution

18. $x + 2y = 12$
$3y - 4z = 25$
$x + 6y + z = 20$
$(6, 3, -4)$

19. $r - 3s + t = 4$
$3r - 6s + 9t = 5$
$4r - 9s + 10t = 9$
infinite solutions

20b. 7 swimmers placed third, 5 swimmers placed second, and 12 swimmers placed first.

Example 3

20c. The statement is false because when you solve for second place, you get a negative as an answer and you cannot have a negative person.

20. SENSE-MAKING A friend e-mails you the results of a recent high school swim meet. The e-mail states that 24 individuals placed, earning a combined total of 53 points. First place earned 3 points, second place earned 2 points, and third place earned 1 point. There were as many first-place finishers as second- and third-place finishers combined.

   a. Write a system of three equations that represents how many people finished in each place. $x + y + z = 24$, $3x + 2y + z = 53$, $x = y + z$

   b. How many swimmers finished in first place, in second place, and in third place?

   c. Suppose the e-mail had said that the athletes scored a combined total of 47 points. Explain why this statement is false and the solution is unreasonable.

B

21. AMUSEMENT PARKS Nick goes to the amusement park to ride roller coasters, bumper cars, and water slides. The wait for the roller coasters is 1 hour, the wait for the bumper cars is 20 minutes long, and the wait for the water slides is only 15 minutes long. Nick rode 10 total rides during his visit. Because he enjoys roller coasters the most, the number of times he rode the roller coasters was the sum of the times he rode the other two rides. If Nick waited in line for a total of 6 hours and 20 minutes, how many of each ride did he go on? **roller coasters: 5; bumper cars: 1; water slides: 4**

3 Practice

Formative Assessment

Use Exercises 1–7 to check for understanding.

Use the chart at the bottom of this page to customize assignments for your students.

CCSS **Teaching the Mathematical Practices**

Sense-Making Mathematically proficient students start by explaining to themselves the meaning of a problem and looking for entry points to its solution. They analyze givens, constraints, relationships, and goals, and make conjectures about the form and meaning of the solution and plan a solution pathway rather than simply jumping into a solution attempt.

Tips for New Teachers

Using Models Some students have difficulty visualizing situations in three dimensions. Encourage them to use paper, pencils, and other objects to help them model the situations they encounter in the lesson. If you or some of your students have access to computer software that draws models in three dimensions, encourage them to use this tool to demonstrate what happens visually with these systems.

Differentiated Homework Options

Level	Assignment	Two-Day Option	
AL Basic	8–20, 24, 26–42	9–19 odd, 30–33	8–20 even, 24, 26–29, 34–42
OL Core	9–19 odd, 21–24, 26–42	8–20, 30–33	21–24, 26–29, 34–42
BL Advanced	21–38, (optional: 39–42)		

Study Guide and Intervention
AL OL ELL

(Study Guide and Intervention worksheet 3-4: Systems of Equations in Three Variables)

Practice
AL OL BL ELL

(Practice worksheet 3-4: Systems of Equations in Three Variables)

Word Problem Practice
AL OL BL ELL

(Word Problem Practice worksheet 3-4: Systems of Equations in Three Variables)

---

22. BUSINESS Ramón usually gets one of the routine maintenance options at Annie's Garage. Today however, he needs a different combination of work than what is listed.

ANNIE'S GARAGE
ROUTINE MAINTENANCE OPTIONS
1. Oil Change & Radiator Flush — $29.99
2. Brake Pads & Oil Change — $39.99
3. Oil Change, Radiator Flush & Brake Pads — $49.99

a. Assume that the price of an option is the same price as purchasing each item separately. Find the prices for an oil change, a radiator flush, and a brake pad replacement.

oil change: $19.99; brake pad replacement: $20; radiator flush: $10

b. If Ramón wants his brake pads replaced and his radiator flushed, how much should he plan to spend? **$30**

23. FINANCIAL LITERACY Kate invested $100,000 in three different accounts. If she invested $30,000 more in account A than account C and is expected to earn $6300 in interest, how much did she invest in each account?

A: $55,000; B: $20,000; C: $25,000

Account	Expected Interest
A	4%
B	8%
C	10%

H.O.T. Problems   Use Higher-Order Thinking Skills

24. CCSS REASONING Write a system of equations to represent the three rows of figures below. Use the system to find the number of red triangles that will balance one green circle.

24. $t + c = s$, $p + t = c$, $2s = 3p$ where t represents triangle, c represents circle, s represents square, and p represents pentagon; 5 red triangles

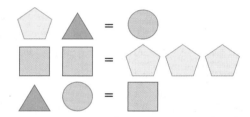

25. CHALLENGE The general form of an equation for a parabola is $y = ax^2 + bx + c$, where (x, y) is a point on the parabola. If three points on a parabola are $(2, -10)$, $(-5, -101)$, and $(6, -90)$, determine the values of a, b, and c and write the general form of the equation. $y = -3x^2 + 4x - 6$; $a = -3$, $b = 4$, $c = -6$

26. PROOF Consider the following system and prove that if $b = c = -a$, then $ty = a$.

$$rx + ty + vz = a$$
$$rx - ty + vz = b$$
$$rx + ty - vz = c$$

See Chapter 3 Answer Appendix.

27. OPEN ENDED Write a system of three linear equations that has a solution of $(-5, -2, 6)$. Show that the ordered triple satisfies all three equations.   See margin.

28. REASONING Use the diagrams of solutions of systems of equations on page 167 to consider a system of inequalities in three variables. Describe the solution of such a system.   See margin.

29. WRITING IN MATH Use your knowledge of solving a system of three linear equations with three variables to explain how to solve a system of four equations with four variables.   See Chapter 3 Answer Appendix.

Enrichment
OL BL

3-4 Enrichment

Homogenous Systems

A system of equations is called homogenous if it is of the form:
$$gx + hy + kz = 0$$
$$dx + ey + fz = 0$$
$$ax + by + cz = 0$$

Homogenous systems have some unique characteristics that set them apart from general systems of equations. The following exercises will explore some of these characteristics.

1. Evaluate the following statement. Is this statement *always, sometimes,* or *never* true? Explain your reasoning.
Every homogenous system of equations will have at least one trivial solution: (0, 0, 0).

2. Find a non-trivial solution to the following homogenous system of equations.

CCSS Teaching the Mathematical Practices

Reasoning Mathematically proficient students make sense of quantities and their relationships in problem situations. Quantitative reasoning entails habits of creating a coherent representation of the problem at hand; considering the units involved; attending to the meaning of quantities, not just how to compute them; and knowing and flexibly using different properties of operations and objects.

30. Solve the system of equations shown below. **B**

$$\begin{cases} x - y + z = 0 \\ -5x + 3y - 2z = -1 \\ 2x - y + 4z = 11 \end{cases}$$

 A $(0, 3, 3)$

 B $(2, 5, 3)$

 C no solution

 D infinitely many solutions

31. SAT/ACT The graph shows which system of equations? **A**

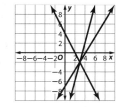

 A $y + 14 = 4x$
   $y = 4 - 2x$
   $-7 = y - \frac{5}{3}x$

 B $y - 14 = 4x$ **D** $y - 14x = 4$
   $y = 4 + 2x$ $2x = 4 + y$
   $-7 = y + \frac{5}{3}x$ $7 = y - \frac{5}{3}x$

 C $y + 14x = 4$ **E** $y - 4x = 14$
   $-2y = 4 + y$ $y = 2x + 4$
   $-7 = y - \frac{5}{3}x$ $7 = y + \frac{5}{3}x$

32. EXTENDED RESPONSE Use the graph to find the solution of the systems of equations. Describe one way to check the solution. **See margin.**

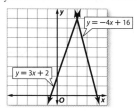

33. Which of the following represents a correct procedure for solving each equation? **J**

 F $-3(x - 7) = -16$ **H** $2(x - 4) = 20$
   $-3x - 21 = -16$ $2x - 8 = 20$
   $-3x = 5$ $2x = 12$
   $x = -\frac{5}{3}$ $x = 6$

 G $7 - 4x = 3x + 27$ **J** $6(2x + 1) = 30$
   $7 - 7x = 27$ $12x + 6 = 30$
   $-7x = -\frac{20}{7}$ $12x = 24$
   $x = 20$ $x = 2$

A feasible region has vertices at $(-3, 2)$, $(1, 3)$, $(6, 1)$, and $(2, -2)$. Find the maximum and minimum values of each function. (Lesson 3-3)

34. $f(x, y) = 2x - y$ **11; −8**   **35.** $f(x, y) = x + 5y$ **16; −8**   **36.** $f(x, y) = y - 4x$ **14; −23**   **37.** $f(x, y) = -x + 3y$ **9; −8**

38. SKI CLUB The ski club's budget for the year is $4250. They are able to find skis for $75 per pair and boots for $40 per pair. They know they should buy more boots than skis because the skis are adjustable to several sizes of boots. (Lesson 3-2)

 a. Give an example of three different purchases that the ski club can make.

 b. Suppose the ski club wants to spend all of its budget. What combination of skis and boots should they buy? Explain. **50 boots and 30 skis cost exactly $4250.**

38a. Sample answer: 40 boots, 35 skis; 45 boots, 32 skis; 50 boots, 30 skis

Solve each system of equations.

39. $x = y + 5$ **40.** $3x - 2y = 1$ **41.** $5x + 3y = 25$ **42.** $y = x - 7$
   $3x + y = 19$ **(6, 1)** $4x + 2y = 20$ **(3, 4)** $4x + 7y = -3$ **(8, −5)** $2x - 8y = 2$ **(9, 2)**

Differentiated Instruction OL BL

Extension The values in the table represent the sums of each row and column. Have students find the missing value. Then have students identify the values for each symbol or set of symbols. (*Hint:* Let each symbol or set of symbols represent one variable.)

💾	🖥	❖♦✈	⊙◎	19
⊙◎	💾	🖥	💾	15
⊙◎	🖥	🖥	⊙◎	14
💾	❖♦✈	❖♦✈	⊙◎	22
12	22	26	10	

💾 $= 4$, 🖥 $= 5$, ❖♦✈ $= 8$, ⊙◎ $= 2$

Name the Math Have students name the four steps they used in solving problems involving three variables.

Formative Assessment
Check for student understanding of the concepts in Lessons 3-3 and 3-4.

📁 Quiz 2, p. 59

Additional Answers

27. Sample answer:

$$3x + 4y + z = -17$$
$$2x - 5y - 3z = -18$$
$$-x + 3y + 8z = 47$$
$$3x + 4y + z = -17$$
$$3(-5) + 4(-2) + 6 = -17$$
$$-15 + (-8) + 6 = -17$$
$$-17 = -17 \checkmark$$
$$2x - 5y - 3z = -18$$
$$2(-5) - 5(-2) - 3(6) = -18$$
$$-10 + 10 - 18 = -18$$
$$-18 = -18 \checkmark$$
$$-x + 3y + 8z = 47$$
$$-(-5) + 3(-2) + 8(6) = 47$$
$$5 - 6 + 48 = 47$$
$$47 = 47 \checkmark$$

28. Sample answer: The solution of an inequality in 3 variables would be the region of space on one side or the other of a plane, with the plane included if the inequality is $\leq$ or $\geq$. The solution of a system of inequalities in 3 variables would be the intersection of the regions of space that are solutions to the individual inequalities in the system.

32. $(2, 8)$; Sample answer: You can substitute $(2, 8)$ into each of the equations and make sure the equations are true.

Formative Assessment

Use the Mid-Chapter Quiz to assess students' progress in the first half of the chapter.

For problems answered incorrectly, have students review the lessons indicated in parentheses.

Summative Assessment

📁 Mid-Chapter Test, p. 61

e✗Assessment Customize and create multiple versions of your Mid-Chapter Quiz and their answer keys.

FOLDABLES **StudyOrganizer**

Dinah Zike's Foldables®

Before students complete the Mid-Chapter Quiz, encourage them to review the information for Lessons 3-1 through 3-4 in their Foldables.

CHAPTER 3 Mid-Chapter Quiz
Lessons 3-1 through 3-4

Solve each system of equations. (Lesson 3-1)

1. $2x - 3y = 9$ **(3, −1)**
$4x + 3y = 9$

2. $x + 2y = 7$ **(1, 3)**
$y = 5x - 2$

3. $-x + y = 2$ **(3, 5)**
$4x - 3y = -3$

4. $\frac{1}{2}x + \frac{1}{3}y = 7$ **(10, 6)**
$\frac{1}{5}x - \frac{2}{3}y = -2$

5–8. See Chapter 3 Answer Appendix.
Solve each system of inequalities by graphing. (Lesson 3-2)

5. $x + y \le 4$
$y \ge x$

6. $2x + 3y > 12$
$3x - y < 21$

7. $x - y > 0$
$4 + y \le 2x$

8. $2y - 5x \le 6$
$4x + y < -4$

9. MULTIPLE CHOICE Which statement best describes the graphs of the two equations? (Lesson 3-1) **A**

$$x + 4y = 8$$
$$3x + 12y = 2$$

A The lines are parallel.

B The lines are the same.

C The lines intersect in only one point.

D The lines intersect in more than one point, but are not the same.

Solve each system of equations. (Lesson 3-4)

10. $x - 2y + 3z = 1$
$4y - 4z = 12$
$8y - 14z = 0$ **(3, 7, 4)**

11. $x + y + z = 4$
$x + 3y + 3z = 10$
$2x + y - z = 3$ **(1, 2, 1)**

12. $2x - y - 2z = 5$
$10x + 8z = -4$
$3x - y = 1$ **(2, 5, −3)**

13. $2x + 3y + z = 0$
$3x + y = 1$
$x - 2y + z = 9$
(1, −2, 4)

14. MULTIPLE CHOICE Seela rented a raft from River Rafter's Inc. She paid $100 to rent the raft and $25 for each hour. Martin rented a raft from Oscar's Outdoor Shop. He paid $50 to rent the raft and $35 per hour. For what number of hours will both rafting companies charge the same amount? (Lesson 3-1) **H**

F 0

G 4

H 5

J 10

---

15. CARPENTRY Cal's Carpentry makes tables and chairs. The process involves some carpentry time and some finishing time. The carpentry times and finishing times are listed in the table below.

Product	Carpentry Time (hr)	Finishing Time (hr)
chair	3	0.5
table	2	1

Cal's Carpentry can work for a maximum of 108 carpentry hours and 20 finishing hours per day. The profit is $35 for a table and $25 for a chair. How many tables and chairs should be made each day to maximize profit? (Lesson 3-3)

a. Using c for the number of chairs and t for the number of tables, write a system of inequalities to represent this situation. **a–c. See Chapter 3 Answer Appendix.**

b. Draw the graph showing the feasible region.

c. Determine the number of tables and chairs that need to be made to maximize profit. What is the maximum profit?

16. DRAMA On opening night of the drama club's play, they made $1366. They sold a total of 199 tickets. They charged $8.50 for each adult ticket and $5.00 for each child's ticket. Write a system of equations that can be used to find the number of adult tickets and the number of children's tickets sold. (Lesson 3-1) $a + c = 199$
$8.50a + 5.00c = 1366$

17, 18. See Chapter 3 Answer Appendix.
Graph each system of inequalities. Name the coordinates of the vertices of the feasible region. Find the maximum and the minimum values of the given function. (Lesson 3-3)

17. $5 \ge y \ge -3$
$4x + y \le 5$
$-2x + y \le 5$
$f(x, y) = 4x - 3y$

18. $x \ge -10$
$1 \ge y \ge -6$
$3x + 4y \le -8$
$2y \ge x - 10$
$f(x, y) = 2x + y$

19a. $a = 11$ in., $b = 10$ in., $c = 27$ in.
19. GEOMETRY An isosceles trapezoid has shorter base of measure a, longer base of measure c, and congruent legs of measure b. The perimeter of the trapezoid is 58 inches. The average of the bases is 19 inches and the longer base is twice the leg plus 7. (Lesson 3-4)

a. Find the lengths of the sides of the trapezoid.

b. Find the area of the trapezoid. **114 in²**

InterventionPlanner

TIER 1 On Level **OL**	**TIER 2** Strategic Intervention **AL** approaching grade level	**TIER 3** Intensive Intervention 2 or more grades below level
If students miss about 25% of the exercises or less,	**If** students miss about 50% of the exercises,	**If** students miss about 75% of the exercises,
Then choose a resource:	**Then** choose a resource:	**Then** use *Math Triumphs, Alg. 2,* Ch. 3
SE Lessons 3-1, 3-2, 3-3, and 3-4	📁 Study Guide and Intervention, pp. 5, 12, 18, and 25	connectED.mcgraw-hill.com Extra Examples, Personal Tutor, Homework Help, Review Vocabulary
📁 Skills Practice, pp. 7, 14, 20, and 27	connectED.mcgraw-hill.com Extra Examples, Personal Tutor, Homework Help	
TE Chapter Project		
connectED.mcgraw-hill.com Self-Check Quiz		

EXPLORE 3-5

Spreadsheet Lab
Organizing Data with Matrices

A **matrix** is a rectangular array of variables or constants in rows and columns, usually enclosed in brackets. In a matrix, the numbers or data are organized so that each position in the matrix has a purpose. Each value in the matrix is called an **element**. A matrix is usually named using an uppercase letter.

The element −1 is in Row 2, Column 1, depicted by a_{21}.

$$A = \begin{bmatrix} 8 & -2 & 5 & 6 \\ -1 & 3 & -3 & 6 \\ 7 & -8 & 1 & 4 \end{bmatrix} \Big\} \text{ 3 rows}$$

The element −8 is in Row 3, Column 2, depicted by a_{32}.

4 columns

A matrix can be described by its **dimensions**. A matrix with m rows and n columns is an $m \times n$ matrix (read "m by n"). Matrix A above is a 3×4 matrix because it has 3 rows and 4 columns. a_{12} refers to an element of A, whereas b_{12} refers to an element of B.

Activity 1 Organize Data in a Matrix

FOOTBALL The West High School football team used five running backs throughout its season. Coach Williams wanted to compare the statistics of each player.

Joey: 11 games, 72 attempts, 439 yards, 6.10 average, 8 TDs

DeShawn: 9 games, 143 attempts, 1024 yards, 7.16 average, 12 TDs

Dario: 11 games, 164 attempts, 885 yards, 5.40 average, 15 TDs

Leo: 11 games, 84 attempts, 542 yards, 6.45 average, 7 TDs

Alex: 10 games, 151 attempts, 966 yards, 6.40 average, 11 TDs

a. Organize the data in a matrix from greatest to least attempts.

Player	Games	Attempts	Yards	Average	TDs
Dario	11	164	885	5.40	15
Alex	10	151	966	6.40	11
DeShawn	9	143	1024	7.16	12
Leo	11	84	542	6.45	7
Joey	11	72	439	6.10	8

b. What are the dimensions of the matrix? What value is a_{34}?

There are five rows and five columns, so the dimensions are 5×5. The value a_{34}, which is in the third row and fourth column, is 7.16.

$$\begin{bmatrix} 11 & 164 & 885 & 5.40 & 15 \\ 10 & 151 & 966 & 6.40 & 11 \\ 9 & 143 & 1024 & \boxed{7.16} & 12 \\ 11 & 84 & 542 & 6.45 & 7 \\ 11 & 72 & 439 & 6.10 & 8 \end{bmatrix}$$

Exercise

1. Julie is shopping for a new smartphone and discovers that many different options are available.

   a. Organize the data in a matrix. List the options in descending order, and label the columns options, price, memory, color, and interface respectively. **See margin.**

   b. What are the dimensions of the matrix? What is the value of a_{23}? **4×4; 24**

Option 1
$420
Memory: 512, Color: 24,
Interface: infrared

Option 2
$399
Memory: 512, Color: 24,
Interface: Bluetooth

Option 3
$315
Memory: 256, Color: 24,
Interface: infrared

Option 4
$289
Memory: 128, Color: 18,
Interface: wi-fi

(continued on the next page)

---

1 Focus

Objective Organize and display data using matrices and spreadsheets.

Materials for Each Student
- computer with spreadsheet program

Teaching Tip
Explain to students that a matrix that only has one row is called a *row matrix* and a matrix that only has one column is called a *column matrix*. A *zero matrix* is a matrix in which every element is zero.

2 Teach

Working in Cooperative Groups
Have students work with partners, mixing abilities so that a student with more knowledge of spreadsheets is paired with one who has less experience.

Practice Have students complete Exercises 1–4.

Additional Answer
1a.

	Price	Memory	Color	Interface
Choice 1	$420	512	24	infrared
Choice 2	$399	512	24	Bluetooth
Choice 3	$315	256	24	infrared
Choice 4	$289	128	18	wi-fi

1 Analyze Data

Example 1 shows how to use matrices to analyze data using row and column operations.

Formative Assessment

Use the Guided Practice exercises after each example to determine students' understanding of concepts.

Additional Example

1 **COLLEGE** Kaitlin wants to attend one of three Iowa universities next year. She has gathered information about tuition (T), room and board (R/B), and enrollment (E) for the universities.

Iowa State University
T–$6160, R/B–$5958, E–26,160

University of Iowa
T–$6293, R/B–$7250, E–30,409

University of Northern Iowa
T–$5352, R/B–$6280, E–12,609

a. Organize the data as a matrix, with columns in the order of tuition, room and board, and enrollment.

$$\begin{array}{c} \\ \text{ISU} \\ \text{UI} \\ \text{UNI} \end{array} \begin{array}{ccc} \text{T} & \text{R/B} & \text{E} \end{array} \\ \begin{bmatrix} 6160 & 5958 & 26{,}160 \\ 6293 & 7250 & 30{,}409 \\ 5352 & 6280 & 12{,}609 \end{bmatrix}$$

b. Find the average of the elements in column 1, and interpret the result. The average tuition at the three universities is $5935.

c. Would adding the elements of the rows provide meaningful data? Explain. No, the first two elements of a row are in dollars and the third is in numbers of people.

d. Would adding the elements of the third column provide meaningful data? Yes, the sum of the elements of the third column would be the total enrollment of all three schools.

▶ **Additional Examples** also in Interactive Classroom PowerPoint® Presentations

IWB **Interactive White Board**
READY

---

2 Algebraic Operations Several algebraic operations can be performed on data that are organized in matrices. Matrices can be added or subtracted if and only if they have the same dimensions.

KeyConcept Adding and Subtracting Matrices

Words To add or subtract two matrices with the same dimensions, add or subtract their corresponding elements.

Symbols
$$A + B = A+B$$
$$\begin{bmatrix} a & b \\ c & d \end{bmatrix} + \begin{bmatrix} e & f \\ g & h \end{bmatrix} = \begin{bmatrix} a+e & b+f \\ c+g & d+h \end{bmatrix}$$

$$A - B = A-B$$
$$\begin{bmatrix} a & b \\ c & d \end{bmatrix} - \begin{bmatrix} e & f \\ g & h \end{bmatrix} = \begin{bmatrix} a-e & b-f \\ c-g & d-h \end{bmatrix}$$

Example
$$\begin{bmatrix} 3 & -5 \\ 1 & 7 \end{bmatrix} + \begin{bmatrix} 2 & 0 \\ -9 & 10 \end{bmatrix} = \begin{bmatrix} 3+2 & -5+0 \\ 1+(-9) & 7+10 \end{bmatrix}$$

Example 2 Add and Subtract Matrices

Find each of the following for $A = \begin{bmatrix} 16 & 2 \\ -9 & 8 \end{bmatrix}$, $B = \begin{bmatrix} -4 & -1 \\ -3 & -7 \end{bmatrix}$, and $C = \begin{bmatrix} 8 \\ 6 \end{bmatrix}$.

a. $A + B$

$$A + B = \begin{bmatrix} 16 & 2 \\ -9 & 8 \end{bmatrix} + \begin{bmatrix} -4 & -1 \\ -3 & -7 \end{bmatrix}$$ Substitution

$$= \begin{bmatrix} 16+(-4) & 2+(-1) \\ -9+(-3) & 8+(-7) \end{bmatrix}$$ Add corresponding elements.

$$= \begin{bmatrix} 12 & 1 \\ -12 & 1 \end{bmatrix}$$ Simplify.

b. $B - C$

$$B - C = \begin{bmatrix} -4 & -1 \\ -3 & -7 \end{bmatrix} - \begin{bmatrix} 8 \\ 6 \end{bmatrix}$$ Substitution

Since the dimensions of B and C are different, you cannot subtract the matrices.

c. $B - A$

$$B - A = \begin{bmatrix} -4 & -1 \\ -3 & -7 \end{bmatrix} - \begin{bmatrix} 16 & 2 \\ -9 & 8 \end{bmatrix}$$ Substitution

$$= \begin{bmatrix} -4-16 & -1-2 \\ -3-(-9) & -7-8 \end{bmatrix}$$ Subtract corresponding elements.

$$= \begin{bmatrix} -20 & -3 \\ 6 & -15 \end{bmatrix}$$ Simplify.

StudyTip
Corresponding Elements Elements are *corresponding* if they are in the exact same position in each matrix.

▶ **Guided Practice**

2A. $\begin{bmatrix} -3 & 4 \\ -9 & -5 \end{bmatrix} - \begin{bmatrix} -4 & 12 \\ 8 & -7 \end{bmatrix}$ $\begin{bmatrix} 1 & -8 \\ -17 & 2 \end{bmatrix}$

2B. $\begin{bmatrix} -9 & 8 & 3 \\ -2 & 4 & -7 \end{bmatrix} + \begin{bmatrix} -4 & -3 & 6 \\ -9 & -5 & 18 \end{bmatrix}$ $\begin{bmatrix} -13 & 5 & 9 \\ -11 & -1 & 11 \end{bmatrix}$

2C. $\begin{bmatrix} 8 & -3 \\ -2 & 0 \\ 1 & 7 \end{bmatrix} - \begin{bmatrix} 5 & 1 & -4 & 2 \\ 10 & -6 & 9 & 0 \end{bmatrix}$ impossible

Focus on Mathematical Content

Matrices A matrix provides a way to organize data. Each element in the matrix has a specific purpose. In order for matrices to be equal, they must have the same dimensions, and each element in one matrix must be equal to the corresponding element in the other matrix.

You can multiply any matrix by a constant called a **scalar**. When you do this, you multiply each individual element by the value of the scalar. This operation is called **scalar multiplication.**

ReadingMath

Scalar Think of a scalar as a coefficient for a variable, but instead it is for a matrix.

KeyConcept Multiplying by a Scalar

Words | To multiply a matrix by a scalar k, multiply each element by k.

$$k \cdot A = kA$$

Symbols | $k\begin{bmatrix} a & b \\ c & d \end{bmatrix} = \begin{bmatrix} ka & kb \\ kc & kd \end{bmatrix}$

Example | $-3\begin{bmatrix} 4 & 1 \\ 7 & -2 \end{bmatrix} = \begin{bmatrix} -3(4) & -3(1) \\ -3(7) & -3(-2) \end{bmatrix}$

StudyTip

Scalar Multiplication
The bracket of a matrix is treated just like a regular grouping symbol. So when multiplying by a scalar, distribute the same way as with a grouping symbol.

Example 3 Multiply a Matrix by a Scalar

If $R = \begin{bmatrix} -12 & 8 & 6 \\ -16 & 4 & 19 \end{bmatrix}$, find $5R$.

$5R = 5\begin{bmatrix} -12 & 8 & 6 \\ -16 & 4 & 19 \end{bmatrix}$ Substitution

$= \begin{bmatrix} 5(-12) & 5(8) & 5(6) \\ 5(-16) & 5(4) & 5(19) \end{bmatrix}$ Distribute the scalar.

$= \begin{bmatrix} -60 & 40 & 30 \\ -80 & 20 & 95 \end{bmatrix}$ Multiply.

▶ **Guided**Practice

3. If $T = \begin{bmatrix} 8 & 0 & 3 & -2 \\ -1 & -4 & -2 & 9 \end{bmatrix}$, find $-4T$. $\begin{bmatrix} -32 & 0 & -12 & 8 \\ 4 & 16 & 8 & -36 \end{bmatrix}$

Many properties of real numbers also hold true for matrices. A summary of these properties is listed below.

KeyConcept Properties of Matrix Operations

For any matrices A, B, and C for which the matrix sum and product are defined and any scalar k, the following properties are true.

Commutative Property of Addition $A + B = B + A$

Associative Property of Addition $(A + B) + C = A + (B + C)$

Left Scalar Distributive Property $k(A + B) = kA + kB$

Right Scalar Distributive Property $(A + B)k = kA + kB$

Multi-step operations can be performed on matrices. The order of these operations is the same as with real numbers.

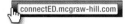
---

2 Algebraic Operations

Example 2 shows how to add or subtract two matrices. **Example 3** shows how to multiply a matrix by a scalar. **Examples 4 and 5** show how to perform a combination of matrix operations.

Additional Examples

2 **a.** Find $A + B$ if $A = \begin{bmatrix} 6 & 4 \\ -1 & 0 \end{bmatrix}$ and $B = \begin{bmatrix} -3 & 1 \\ 0 & 3 \end{bmatrix}$. $\begin{bmatrix} 3 & 5 \\ -1 & 3 \end{bmatrix}$

b. Find $A - B$ if $A = \begin{bmatrix} 4 & -2 & 0 \\ 1 & 5 & -1 \end{bmatrix}$ and $B = \begin{bmatrix} -6 & 7 \\ -9 & 3 \end{bmatrix}$. Since their dimensions are different, these matrices cannot be subtracted.

c. Find $B - A$ if $A = \begin{bmatrix} 2 & 0 \\ 3 & 4 \\ 1 & 5 \end{bmatrix}$ and $B = \begin{bmatrix} -4 & 8 \\ 2 & -1 \\ 0 & 5 \end{bmatrix}$. $\begin{bmatrix} 6 & -8 \\ 1 & 5 \\ 1 & 0 \end{bmatrix}$

3 If $A = \begin{bmatrix} 2 & 1 \\ -1 & 3 \\ 0 & 5 \end{bmatrix}$, find $2A$. $\begin{bmatrix} 4 & 2 \\ -2 & 6 \\ 0 & 10 \end{bmatrix}$

Tips for New Teachers

Building on Prior Knowledge
Point out that scalar multiplication of matrices is similar to using the Distributive Property to multiply the expression $3(x + y)$.

Focus on Mathematical Content

Adding and Subtracting Matrices In order to add or subtract matrices, they must have the same dimensions. To add or subtract two matrices, simply add or subtract the corresponding elements of the matrices. Both the Commutative and Associative Properties hold for matrix addition.

WatchOut!

Unlocking Misconceptions In order to help students see why matrices must have the same dimensions if they are to be added or subtracted, suggest that they try to add a 2×3 matrix to a 3×2 matrix.

Additional Examples

4 If $A = \begin{bmatrix} 2 & 3 \\ -1 & 0 \end{bmatrix}$ and

$B = \begin{bmatrix} -2 & 1 \\ 0 & -1 \end{bmatrix}$, find $4A - 3B$.

$\begin{bmatrix} 14 & 9 \\ -4 & 3 \end{bmatrix}$

5 **BUSINESS** A small company makes unfinished desks and cabinets. Each item requires different amounts of hardware as shown in the matrices.

	DESK		CABINET	
	short	long	short	long
nails	10	6	nails	4	8
screws	8	4	screws	3	4

nails $\begin{bmatrix} 10 & 6 \\ 8 & 4 \end{bmatrix}$ screws nails $\begin{bmatrix} 4 & 8 \\ 3 & 4 \end{bmatrix}$

The company has orders for 3 desks and 4 cabinets. Express the company's total needs for hardware in a single matrix.

$\begin{bmatrix} 46 & 50 \\ 36 & 28 \end{bmatrix}$

Tips for New Teachers

Reasoning Adding and subtracting matrices can be challenging for many students. Encourage students to develop strategies for keeping track of where they are in a matrix. Students may find it helpful to circle the elements they are adding or subtracting, use color coding, or point with their fingers to the two elements as they do the calculations.

Additional Answer (Guided Practice)

5.

	Wages	Sales
Entry	741,000	97,890,000
Assistant	1,560,000	131,430,000
Associate	1,638,000	156,130,000

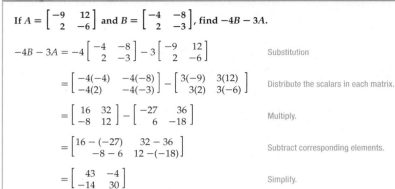

Example 4 Multi-Step Operations

If $A = \begin{bmatrix} -9 & 12 \\ 2 & -6 \end{bmatrix}$ and $B = \begin{bmatrix} -4 & -8 \\ 2 & -3 \end{bmatrix}$, find $-4B - 3A$.

$-4B - 3A = -4\begin{bmatrix} -4 & -8 \\ 2 & -3 \end{bmatrix} - 3\begin{bmatrix} -9 & 12 \\ 2 & -6 \end{bmatrix}$ — Substitution

$= \begin{bmatrix} -4(-4) & -4(-8) \\ -4(2) & -4(-3) \end{bmatrix} - \begin{bmatrix} 3(-9) & 3(12) \\ 3(2) & 3(-6) \end{bmatrix}$ — Distribute the scalars in each matrix.

$= \begin{bmatrix} 16 & 32 \\ -8 & 12 \end{bmatrix} - \begin{bmatrix} -27 & 36 \\ 6 & -18 \end{bmatrix}$ — Multiply.

$= \begin{bmatrix} 16-(-27) & 32-36 \\ -8-6 & 12-(-18) \end{bmatrix}$ — Subtract corresponding elements.

$= \begin{bmatrix} 43 & -4 \\ -14 & 30 \end{bmatrix}$ — Simplify.

▶ **Guided**Practice

4. If $A = \begin{bmatrix} -5 & 3 \\ 6 & -8 \\ 2 & 9 \end{bmatrix}$ and $B = \begin{bmatrix} 12 & 5 \\ 5 & -4 \\ 4 & -7 \end{bmatrix}$, find $-6B + 7A$. $\begin{bmatrix} -107 & -9 \\ 12 & -32 \\ -10 & 105 \end{bmatrix}$

Matrices can be used in many business applications.

Real-World Career

Financial Planner
Financial planners often use matrices to organize and describe the data they use. Financial planners need a bachelor's degree. Usually they have degrees in accounting, finance, economics, business, marketing, or commerce.

Real-World Example 5 Use Multi-Step Operations with Matrices

BUSINESS Refer to the application at the beginning of the lesson. Express the average wages and sales for the entire company for a 5-day week.

To calculate the 5-day sales for the entire company, each matrix needs to be multiplied by 5 and the totals added together.

$5\begin{bmatrix} 900 & 145,000 \\ 2400 & 225,000 \\ 2700 & 290,000 \end{bmatrix} + 5\begin{bmatrix} 900 & 122,000 \\ 1800 & 145,500 \\ 1800 & 160,000 \end{bmatrix} + 5\begin{bmatrix} 1050 & 109,500 \\ 1800 & 135,000 \\ 1800 & 150,500 \end{bmatrix}$ — Write matrices.

$= \begin{bmatrix} 4500 & 725,000 \\ 12,000 & 1,125,000 \\ 13,500 & 1,450,000 \end{bmatrix} + \begin{bmatrix} 4500 & 610,000 \\ 9000 & 727,500 \\ 9000 & 800,000 \end{bmatrix} + \begin{bmatrix} 5250 & 547,500 \\ 9000 & 675,000 \\ 9000 & 752,500 \end{bmatrix}$ — Multiply scalars.

	Wages	Sales
Entry	14,250	1,882,500
Assistant	30,000	2,527,500
Associate	31,500	3,002,500

$= \begin{matrix} \text{Entry} \\ \text{Assistant} \\ \text{Associate} \end{matrix} \begin{bmatrix} 14,250 & 1,882,500 \\ 30,000 & 2,527,500 \\ 31,500 & 3,002,500 \end{bmatrix}$ — Add matrices.

The final matrix indicates the average weekly sales and wages for all of the representatives of the company.

StudyTip

Corresponding Elements
When representing quantities with multiple matrices, make sure the corresponding elements represent corresponding quantities.

▶ **Guided**Practice

5. Use the data above to calculate the average yearly sales and wages for the company, assuming 260 working days. **See margin.**

 174 | Lesson 3-5 | Operations with Matrices

DifferentiatedInstruction **AL**

Verbal/Linguistic Learners Students may find it helpful to talk softly, or even silently, to themselves as they work with matrices. For example, they might recite the words "row by column" to remind themselves how to write the dimensions of a matrix. When multiplying by a scalar, students may find it helps to say, for example, "5 times 1 is 5, and 5 times negative 3 is negative 15." In this way, students use more than one of their senses to check their calculations.

Example 1

1. CCSS MODELING Use the table that shows the city and highway gas mileage of five different types of vehicles.

1c. Sample answer: The sums are 48, 45, 53, 91, and 131. These values are irrelevant since they are the sums of 2 different types of data.

Vehicle	SUV	Mini-van	Sedan	Compact	APV
City	23	21	21	42	61
Highway	25	24	32	49	70

Source: Auto Hoppers

$$\begin{bmatrix} 23 & 21 & 21 & 42 & 61 \\ 25 & 24 & 32 & 49 & 70 \end{bmatrix}$$

a. Organize the gas mileages in a matrix.
b. Add the elements of each row and interpret the results.
c. Add the elements of each column and interpret the results.

1b. Sample answer: City: The sum is 168. However, this value is irrelevant since it is the sum of 5 different types of data. Highway: The sum is 200. However, this value is irrelevant since it is the sum of 5 different types of data.

Example 2
Perform the indicated operations. If the matrix does not exist, write *impossible*.

2–5.
See margin.

2. $\begin{bmatrix} -8 & 2 & 6 \end{bmatrix} + \begin{bmatrix} 11 & -7 & 1 \end{bmatrix}$

3. $\begin{bmatrix} 9 & -8 & 4 \end{bmatrix} + \begin{bmatrix} 12 & 2 \end{bmatrix}$

4. $\begin{bmatrix} 7 & -12 \\ 15 & 4 \end{bmatrix} - \begin{bmatrix} 9 & 6 \\ 4 & -9 \end{bmatrix}$

5. $\begin{bmatrix} 5 & 13 & -6 \\ 3 & -17 & 2 \end{bmatrix} - \begin{bmatrix} -2 & -18 & 8 \\ 2 & -11 & 0 \end{bmatrix}$

Example 3
Perform the indicated operations. If the matrix does not exist, write *impossible*.

6, 7.
See margin.

6. $3\begin{bmatrix} 6 & 4 & 0 \\ -2 & 14 & -8 \\ -4 & -6 & 7 \end{bmatrix}$

7. $-6\begin{bmatrix} 15 & -9 & 2 & 3 \\ 6 & -11 & 14 & -2 \\ 4 & -8 & -10 & 27 \end{bmatrix}$

Example 4
Use matrices *A*, *B*, *C*, and *D* to find the following.

8–11.
See margin.

$A = \begin{bmatrix} 6 & -4 \\ 3 & -5 \end{bmatrix}$ $B = \begin{bmatrix} 8 & -1 \\ -2 & 7 \end{bmatrix}$ $C = \begin{bmatrix} -4 & -6 \\ 12 & -7 \end{bmatrix}$ $D = \begin{bmatrix} 9 & 6 & 0 \\ -2 & 8 & 0 \end{bmatrix}$

12b. $\begin{bmatrix} 157 \\ 149 \\ 179 \end{bmatrix}$

8. $4B - 2A$

9. $-8C + 3A$

10. $-5B - 2D$

11. $-4C - 5B$

12a. Test 1: $\begin{bmatrix} 85 \\ 75 \\ 96 \end{bmatrix}$ Test 2: $\begin{bmatrix} 72 \\ 74 \\ 83 \end{bmatrix}$

Example 5

12c. $\begin{bmatrix} 13 \\ 1 \\ 13 \end{bmatrix}$

12. GRADES Geraldo, Olivia, and Nikki have had two tests in their math class. The table shows the test grades for each student.

a. Write a matrix for the information from each test.
b. Find the sum of the scores from the two tests expressed as a matrix.
c. Express the difference in scores from test 1 to test 2 as a matrix.

Student	Test 1	Test 2
Geraldo	85	72
Olivia	75	74
Nikki	96	83

13b. Sample answer: Brand C; it was given the highest rating possible for cost and comfort, and a high rating for looks, and it will last a fairly long time.

Practice and Problem Solving
Extra Practice is on page R3.

Example 1

13. SHOES A consumer service company rated several pairs of shoes by cost, level of comfort, look, and longevity using a scale of 1–5, with 1 being low and 5 being high.

13a. See margin.
a. Write a 4 × 4 matrix to organize this information.
b. Which shoe would you buy based on this information, and why?
c. Would finding the sum of the rows or columns provide any useful information? Explain your reasoning.

Brand	Cost	Comfort	Look	Longevity
A	3	2	2	1
B	4	3	2	3
C	5	5	4	4
D	1	5	5	2

Example 2
Perform the indicated operations. If the matrix does not exist, write *impossible*. **14.** $\begin{bmatrix} 6 & 6 \\ -15 & -1 \end{bmatrix}$

14. $\begin{bmatrix} 12 & -5 \\ -8 & -3 \end{bmatrix} + \begin{bmatrix} -6 & 11 \\ -7 & 2 \end{bmatrix}$

15 $\begin{bmatrix} 9 & 5 \\ -2 & 16 \end{bmatrix} + \begin{bmatrix} -6 & -3 & 7 \\ 12 & 2 & -4 \end{bmatrix}$ impossible

13c. Sample answer: Yes; finding the sum of the rows and then calculating the average will provide an easy way to compare the data.

 connectED.mcgraw-hill.com **175**

---

3 Practice

Formative Assessment
Use Exercises 1–12 to check for understanding.

Use the chart at the bottom of this page to customize assignments for your students.

CCSS Teaching the Mathematical Practices

Modeling Mathematically proficient students can apply the mathematics they know to solve problems arising in everyday life, analyze relationships mathematically to draw conclusions, and interpret their mathematical results in the context of a situation.

Additional Answers

2. $\begin{bmatrix} 3 & -5 & 7 \end{bmatrix}$

3. impossible

4. $\begin{bmatrix} -2 & -18 \\ 11 & 13 \end{bmatrix}$ **5.** $\begin{bmatrix} 7 & 31 & -14 \\ 1 & -6 & 2 \end{bmatrix}$

6. $\begin{bmatrix} 18 & 12 & 0 \\ -6 & 42 & -24 \\ -12 & -18 & 21 \end{bmatrix}$

7. $\begin{bmatrix} -90 & 54 & -12 & -18 \\ -36 & 66 & -84 & 12 \\ -24 & 48 & 60 & -162 \end{bmatrix}$

8. $\begin{bmatrix} 20 & 4 \\ -14 & 38 \end{bmatrix}$ **9.** $\begin{bmatrix} 50 & 36 \\ -87 & 41 \end{bmatrix}$

10. impossible

11. $\begin{bmatrix} -24 & 29 \\ -38 & -7 \end{bmatrix}$

13a. $\begin{bmatrix} 3 & 2 & 2 & 1 \\ 4 & 3 & 2 & 3 \\ 5 & 5 & 4 & 4 \\ 1 & 5 & 5 & 2 \end{bmatrix}$

Differentiated Homework Options

Level	Assignment	Two-Day Option	
AL Basic	13–21, 31, 32, 34–52	13–21 odd, 37–40	14–20 even, 31, 32, 34–36, 41–52
OL Core	13–27 odd, 28–32, 34–52	13–21, 37–40	22–32, 34–36, 41–52
BL Advanced	22–49, (optional: 50–52)		

Additional Answers

16a.
$$\begin{bmatrix} \$0.95 & \$1.00 & \$1.05 \\ \$0.75 & \$0.80 & \$0.85 \\ \$0.75 & \$0.80 & \$0.85 \\ \$1.00 & \$1.10 & \$1.20 \end{bmatrix}$$

16b. 1.1

16c.
$$\begin{bmatrix} \$1.05 & \$1.10 & \$1.16 \\ \$0.83 & \$0.88 & \$0.94 \\ \$0.83 & \$0.88 & \$0.94 \\ \$1.10 & \$1.21 & \$1.32 \end{bmatrix}$$

16d.
$$\begin{bmatrix} \$0.10 & \$0.10 & \$0.11 \\ \$0.08 & \$0.08 & \$0.09 \\ \$0.08 & \$0.08 & \$0.09 \\ \$0.10 & \$0.11 & \$0.12 \end{bmatrix}$$

Sample answer: this matrix represents the price increases for each item.

17. $\begin{bmatrix} -33 & -26 \\ 25 & 75 \end{bmatrix}$

18. $\begin{bmatrix} 80 & 50 & -18 \\ -7 & -133 & 131 \end{bmatrix}$

19. impossible

20. $\begin{bmatrix} -22 & -57 \\ 62 & 118 \end{bmatrix}$

21a. Library A: $\begin{bmatrix} 10,000 \\ 5000 \\ 5000 \end{bmatrix}$;

Library B: $\begin{bmatrix} 15,000 \\ 10,000 \\ 2500 \end{bmatrix}$;

Library C: $\begin{bmatrix} 4000 \\ 700 \\ 800 \end{bmatrix}$

21b. $\begin{bmatrix} 29,000 \\ 15,700 \\ 8300 \end{bmatrix}$ **21c.** $\begin{bmatrix} 6000 \\ 4300 \\ 4200 \end{bmatrix}$

21d. $\begin{bmatrix} 25,000 \\ 15,000 \\ 7500 \end{bmatrix}$; The sum represents the combined size of the two libraries.

22. $\begin{bmatrix} 2 & 28.4 \\ -40 & 41.4 \end{bmatrix}$ **23.** $\begin{bmatrix} 5 & 24 \\ -\dfrac{167}{12} & -10 \end{bmatrix}$

24. $\begin{bmatrix} -54 & 18 & 24 \\ 15 & 9 & -36 \\ 0 & -9x & 3y \end{bmatrix}$

25. $\begin{bmatrix} -8a & 32b & 8c-8b \\ -104 & 80 & -40c \end{bmatrix}$

26. $\begin{bmatrix} -16 \\ 4x+14 \\ -75 \end{bmatrix}$ **27.** $\begin{bmatrix} -40 & 50 \\ -25 & 75 \end{bmatrix}$

28a.

	Jan	Feb	Mar
Grand Rapids	21.1	12.2	9.0
Boston	13.3	11.3	8.3
Buffalo	26.1	17.8	12.4
Pittsburgh	12.3	8.5	7.9

Grand Rapids $\begin{bmatrix} 15.4 & 33.6 & 13.6 \end{bmatrix}$
Boston $\begin{bmatrix} 1.0 & 4.6 & 10.2 \end{bmatrix}$
Buffalo $\begin{bmatrix} 15.5 & 33.5 & 5.4 \end{bmatrix}$
Pittsburgh $\begin{bmatrix} 11.3 & 14.0 & 9.3 \end{bmatrix}$

28b.
Grand Rapids $\begin{bmatrix} -5.7 & 21.4 & 4.6 \end{bmatrix}$
Boston $\begin{bmatrix} -12.3 & -6.7 & 1.9 \end{bmatrix}$
Buffalo $\begin{bmatrix} -10.6 & 15.7 & -7.0 \end{bmatrix}$
Pittsburgh $\begin{bmatrix} -1.0 & 5.5 & 1.4 \end{bmatrix}$

The matrix represents the difference from normal for each city and month.

Examples 3–4 **16. BUSINESS** The drink menu from a fast-food restaurant is shown at the right. The store owner has decided that all of the prices must be increased by 10%. **a–d. See margin.**

Drink	Small	Medium	Large
Soda	$0.95	$1.00	$1.05
Iced tea	$0.75	$0.80	$0.85
Lemonade	$0.75	$0.80	$0.85
Coffee	$1.00	$1.10	$1.20

  a. Write matrix C to represent the current prices.

  b. What scalar can be used to determine a matrix N to represent the new prices?

  c. Find N.

  d. What is $N - C$? What does this represent in this situation?

Use matrices A, B, C, and D to find the following.

17–20. See margin.
$A = \begin{bmatrix} 0 & -7 \\ 8 & 12 \end{bmatrix}$ $B = \begin{bmatrix} 11 & 4 \\ -3 & -17 \end{bmatrix}$ $C = \begin{bmatrix} 8 & 2 & -2 \\ 1 & -9 & 15 \end{bmatrix}$ $D = \begin{bmatrix} -2 & -8 & 0 \\ 4 & 13 & 1 \end{bmatrix}$

17. $-3B + 2A$

18. $9C - 4D$

19. $2C + 11A$

20. $7A - 2B$

Example 5 **21. CCSS MODELING** Library A has 10,000 novels, 5000 biographies, and 5000 children's books. Library B has 15,000 novels, 10,000 biographies, and 2500 children's books. Library C has 4000 novels, 700 biographies, and 800 children's books. **a–d. See margin.**

  a. Express each library's number of books as a matrix. Label the matrices A, B, and C.

  b. Find the total number of each type of book in all 3 libraries. Express as a matrix.

  c. How many more books of each type does Library A have than Library C?

  d. Find $A + B$. Does the matrix have meaning in this situation? Explain.

28c. A negative number means the city was below normal on snowfall for the month; a positive number means the city's snowfall was above normal for the month. Sample answer: All the cities had below normal snowfall in January. Grand Rapids and Buffalo had snowfall well above normal in February.

 Perform the indicated operations. If the matrix does not exist, write _impossible_. 22–27. See margin.

22. $-2\begin{bmatrix} -9.2 & -8.4 \\ 5.6 & -4.3 \end{bmatrix} - 4\begin{bmatrix} 4.1 & -2.9 \\ 7.2 & -8.2 \end{bmatrix}$

23. $-\dfrac{3}{4}\begin{bmatrix} 12 & -16 \\ 15 & 8 \end{bmatrix} + \dfrac{2}{3}\begin{bmatrix} 21 & 18 \\ -4 & -6 \end{bmatrix}$

24. $-3\begin{bmatrix} 18 & -6 & -8 \\ -5 & -3 & 12 \\ 0 & 3x & -y \end{bmatrix}$

25. $8\begin{bmatrix} -a & 4b & c-b \\ -13 & 10 & -5c \end{bmatrix}$

26. $-4\begin{bmatrix} -7 \\ 4 \\ -3 \end{bmatrix} + 3\begin{bmatrix} -8 \\ 3x \\ -9 \end{bmatrix} - 5\begin{bmatrix} 4 \\ x-6 \\ 12 \end{bmatrix}$

27 $-5\left(\begin{bmatrix} 4 & -8 \\ 8 & -9 \end{bmatrix} + \begin{bmatrix} 4 & -2 \\ -3 & -6 \end{bmatrix}\right)$

28. WEATHER The table shows snowfall in inches. **28a, b. See margin.**

  a. Express the normal snowfall data and the 2007 data in two 4×3 matrices.

  b. Subtract the matrix of normal data from the matrix of 2007 data. What does the difference represent in the context of the situation?

  c. Explain the meaning of positive and negative numbers in the difference matrix. What trends do you see in the data?

City	Normal Snowfall			2007 Snowfall		
	Jan	Feb	Mar	Jan	Feb	Mar
Grand Rapids, MI	21.1	12.2	9.0	15.4	33.6	13.6
Boston, MA	13.3	11.3	8.3	1.0	4.6	10.2
Buffalo, NY	26.1	17.8	12.4	15.5	33.5	5.4
Pittsburgh, PA	12.3	8.5	7.9	11.3	14.0	9.3

Source: National Weather Service

29b. In the 50 m, the fastest American time is 0.5 second behind the world record. In the 100 m, the fastest American time is 0.47 second behind the world record. In the 200 m, the fastest American time is 0.87 second behind the world record. In the 800 m, it was an American who set the world record.

(29) **MODELING** The table shows some of the world, Olympic, and American women's freestyle swimming records.

Distance (m)	World	Olympic	American
50	24.13 s	24.13 s	24.63 s
100	53.52 s	53.52 s	53.99 s
200	1:56.54 min	1:57.65 min	1:57.41 min
800	8:16.22 min	8:19.67 min	8:16.22 min

Source: USA Swimming

a. Find the difference between the American and World records expressed as a matrix.

$$\begin{bmatrix} 0.5\text{ s} \\ 0.47\text{ s} \\ 0.87\text{ s} \\ 0\text{ s} \end{bmatrix}$$

b. What is the meaning of each row in the column?

c. In which events were the fastest times set at the Olympics? **50 m and 100 m**

30. **MULTIPLE REPRESENTATIONS** In this problem, you will investigate using matrices to represent transformations.

a. Algebraic The matrix $\begin{bmatrix} -3 & -4 & 1 \\ 8 & 6 & 0 \end{bmatrix}$

represents a triangle with vertices at $(-3, 8)$, $(-4, 6)$, and $(1, 0)$. Write a matrix to represent $\triangle ABC$. **See Chapter 3 Answer Appendix.**

b. Geometric Multiply the matrix you wrote by 2. Then graph the figure represented by the new matrix. **See Chapter 3 Answer Appendix.**

c. Analytical How do the figures compare? Make a conjecture about the result of multiplying the matrix by 0.5. Verify your conjecture.
The new triangle is similar to the original, with sides twice as long as the original. The result of multiplying the matrix by 0.5 would be a figure similar to the original with sides half as long.

H.O.T. Problems *Use Higher-Order Thinking Skills*

31. PROOF Prove that matrix addition is commutative for 2×2 matrices.
See Chapter 3 Answer Appendix.

32. PROOF Prove that matrix addition is associative for 2×2 matrices.
See Chapter 3 Answer Appendix.

33. CHALLENGE Find the elements of C if:
$A = \begin{bmatrix} -3 & -4 \\ 8 & 6 \end{bmatrix}$, $B = \begin{bmatrix} 5 & -1 \\ 2 & -4 \end{bmatrix}$, and $3A - 4B + 6C = \begin{bmatrix} 13 & 22 \\ 10 & 4 \end{bmatrix}$. $\begin{bmatrix} 7 & 5 \\ -1 & -5 \end{bmatrix}$

34. REASONING Determine whether each statement is *sometimes*, *always*, or *never* true for matrices A and B. Explain your reasoning. **a–e. See Chapter 3 Answer Appendix.**

a. If $A + B$ exists, then $A - B$ exists.

b. If k is a real number, then kA and kB exist.

c. If $A - B$ does not exist, then $B - A$ does not exist.

d. If A and B have the same number of elements, then $A + B$ exists.

e. If kA exists and kB exists, then $kA + kB$ exists.

35. OPEN ENDED Give an example of matrices A and B if $4B - 3A = \begin{bmatrix} -6 & 5 \\ -2 & -1 \end{bmatrix}$.
See Chapter 3 Answer Appendix.

36. WRITING IN MATH Explain how to find $4D - 3C$ for two given matrices, C and D with the same dimensions. **See Chapter 3 Answer Appendix.**

Teach with Tech

Video Recording Separate the class into groups. Have each group create a video showing how to perform one of the operations with matrices: addition, subtraction, or scalar multiplication.

Study Guide and Intervention

Practice

Word Problem Practice

Enrichment

1 Multiply Matrices

Example 1 shows how to determine the dimensions of matrix products.
Example 2 shows how to find the product of two square matrices.
Example 3 shows how to solve a real-world problem by multiplying two matrices with different dimensions.

Formative Assessment

Use the Guided Practice exercises after each example to determine students' understanding of concepts.

Additional Examples

1 Determine whether each matrix product is defined. If so, state the dimensions of the product.

a. $A_{3 \times 4}$ and $B_{4 \times 2}$ The matrix product is defined. The dimensions are 3×2.

b. $A_{3 \times 2}$ and $B_{4 \times 3}$ The matrix product is not defined.

2 Find RS if $R = \begin{bmatrix} 3 & 2 \\ -1 & 0 \end{bmatrix}$ and $S = \begin{bmatrix} -2 & 1 \\ 1 & -1 \end{bmatrix}$. $\begin{bmatrix} -4 & 1 \\ 2 & -1 \end{bmatrix}$

Additional Examples also in Interactive Classroom PowerPoint® Presentations

IWB **Interactive White Board** READY

WatchOut!

Preventing Errors Encourage students to show all of their work while multiplying matrices. This will allow them to find computation errors if any are made.

The product of two matrices is found by multiplying columns and rows.

KeyConcept Multiplying Matrices

Words The element in the mth row and rth column of matrix AB is the sum of the products of the corresponding elements in row m of matrix A and column r of matrix B.

Symbols

$$\begin{array}{ccc} A & B & AB \\ \begin{bmatrix} a & b \\ c & d \end{bmatrix} \cdot \begin{bmatrix} e & f \\ g & h \end{bmatrix} & = & \begin{bmatrix} ae + bg & af + bh \\ ce + dg & cf + dh \end{bmatrix} \end{array}$$

Example 2 Multiply Square Matrices

Find XY if $X = \begin{bmatrix} 6 & -3 \\ -10 & -2 \end{bmatrix}$ and $Y = \begin{bmatrix} -5 & -4 \\ 3 & 3 \end{bmatrix}$.

$XY = \begin{bmatrix} 6 & -3 \\ -10 & -2 \end{bmatrix} \cdot \begin{bmatrix} -5 & -4 \\ 3 & 3 \end{bmatrix}$

WatchOut!

Saving Your Place It is easy to lose your place as you multiply matrices. It may help to cover rows or columns not being multiplied as you find elements of the product matrix.

Step 1 Multiply the numbers in the first row of X by the numbers in the first column of Y, add the products, and put the result in the first row, first column of XY.

$$\begin{bmatrix} 6 & -3 \\ -10 & -2 \end{bmatrix} \cdot \begin{bmatrix} -5 & -4 \\ 3 & 3 \end{bmatrix} = \begin{bmatrix} 6(-5) + (-3)(3) & \\ & \end{bmatrix}$$

Step 2 Follow the same procedure as in Step 1 using the first row and the second column numbers. Write the result in the first row, second column.

$$\begin{bmatrix} 6 & -3 \\ -10 & -2 \end{bmatrix} \cdot \begin{bmatrix} -5 & -4 \\ 3 & 3 \end{bmatrix} = \begin{bmatrix} 6(-5) + (-3)(3) & 6(-4) + (-3)(3) \\ & \end{bmatrix}$$

Step 3 Follow the same procedure with the second row and the first column numbers. Write the result in the second row, first column.

$$\begin{bmatrix} 6 & -3 \\ -10 & -2 \end{bmatrix} \cdot \begin{bmatrix} -5 & -4 \\ 3 & 3 \end{bmatrix} = \begin{bmatrix} 6(-5) + (-3)(3) & 6(-4) + (-3)(3) \\ -10(-5) + (-2)(3) & \end{bmatrix}$$

Step 4 The procedure is the same for the numbers in the second row, second column.

$$\begin{bmatrix} 6 & -3 \\ -10 & -2 \end{bmatrix} \cdot \begin{bmatrix} -5 & -4 \\ 3 & 3 \end{bmatrix} = \begin{bmatrix} 6(-5) + (-3)(3) & 6(-4) + (-3)(3) \\ -10(-5) + (-2)(3) & -10(-4) + (-2)(3) \end{bmatrix}$$

Step 5 Simplify the product matrix.

$$\begin{bmatrix} 6(-5) + (-3)(3) & 6(-4) + (-3)(3) \\ -10(-5) + (-2)(3) & -10(-4) + (-2)(3) \end{bmatrix} = \begin{bmatrix} -39 & -33 \\ 44 & 34 \end{bmatrix}$$

GuidedPractice

2. Find UV if $U = \begin{bmatrix} 5 & 9 \\ -3 & -2 \end{bmatrix}$ and $V = \begin{bmatrix} 2 & -1 \\ 6 & -5 \end{bmatrix}$. $\begin{bmatrix} 64 & -50 \\ -18 & 13 \end{bmatrix}$

 180 | Lesson 3-6 | Multiplying Matrices

DifferentiatedInstruction AL OL BL

If students need more practice with matrices,

Then have them use the class seating arrangement to form matrices, with each student's desk as an element. Have students make 24 large cards showing the values for the elements of two matrices. Form two matrices $A_{4 \times 3}$ and $B_{3 \times 4}$ with student seats and give each student sitting in these "matrices" a card to show the element in that position. Have students model the matrix multiplication $AB = C$. Begin by drawing a blank 4×4 matrix on the board. For each element on matrix C, have students walk out the products, compute the sums, and then write the results in the correct locations of the matrix on the board.

Real-World Example 3 Multiply Matrices

Real-WorldLink

Swim meets consist of racing and diving competitions. There are more than 241,000 high schools that participate each year.

Source: National Federation of State High School Associations

SWIM MEET At a particular swim meet, 7 points were awarded for each first-place finish, 4 points for second, and 2 points for third. Find the total number of points for each school. Which school won the meet?

School	First Place	Second Place	Third Place
Central	4	7	3
Franklin	8	9	1
Hayes	10	5	3
Lincoln	3	3	6

Understand The final scores can be found by multiplying the swim results for each school by the points awarded for each first-, second-, and third-place finish.

Plan Write the results of the races and the points awarded in matrix form. Set up the matrices so that the number of rows in the points matrix equals the number of columns in the results matrix.

$$
\begin{array}{cc}
\text{Results} & \text{Points} \\
R = \begin{bmatrix} 4 & 7 & 3 \\ 8 & 9 & 1 \\ 10 & 5 & 3 \\ 3 & 3 & 6 \end{bmatrix} & P = \begin{bmatrix} 7 \\ 4 \\ 2 \end{bmatrix}
\end{array}
$$

Solve Multiply the matrices.

$$
RP = \begin{bmatrix} 4 & 7 & 3 \\ 8 & 9 & 1 \\ 10 & 5 & 3 \\ 3 & 3 & 6 \end{bmatrix} \cdot \begin{bmatrix} 7 \\ 4 \\ 2 \end{bmatrix}
$$
Write an equation.

$$
= \begin{bmatrix} 4(7) + 7(4) + 3(2) \\ 8(7) + 9(4) + 1(2) \\ 10(7) + 5(4) + 3(2) \\ 3(7) + 3(4) + 6(2) \end{bmatrix}
$$
Multiply columns by rows.

$$
= \begin{bmatrix} 62 \\ 94 \\ 96 \\ 45 \end{bmatrix}
$$
Simplify.

The product matrix shows the scores for Central, Franklin, Hayes, and Lincoln, respectively. Hayes won the swim meet with a total of 96 points.

Check R is a 4×3 matrix and P is a 3×1 matrix, so their product should be a 4×1 matrix.

GuidedPractice

3. **BASKETBALL** Refer to the beginning of the lesson. Use matrix multiplication to determine in which season Lisa Leslie scored the most points. How many points did she score that season? **2006; 680 points**

2 **Multiplicative Properties** Recall that the properties of real numbers also held true for matrix addition. However, some of these properties do *not* always hold true for matrix multiplication.

  connectED.mcgraw-hill.com **181**

Michael Newman/PhotoEdit

Additional Example

3 **CHESS** Three teams competed in the final round of the Chess Club's championships. For each win, a team was awarded 3 points, and for each draw a team received 1 point. Which team won the tournament?

Team	Wins	Draws
Blue	5	4
Red	6	3
Green	4	5

The Red Team won the tournament with a total of 21 points.

Teach with Tech

Video Recording Have students write a rap or song to help them remember the steps for matrix multiplication. Then videotape their presentation and post to your favorite video sharing site. One such song can be sung to the tune of "My Darling Clementine."

Row by column, row by column
Multiply them line by line
Add them up to form a matrix
Now you're doing it just fine!

DifferentiatedInstruction BL

Extension Write the following problems on the board: $A = \begin{bmatrix} 1 & 0 \\ 0 & 1 \end{bmatrix} \cdot \begin{bmatrix} 2 & 6 \\ 1 & -2 \end{bmatrix}$ $B = \begin{bmatrix} 1 & 0 \\ 0 & 1 \end{bmatrix} \cdot \begin{bmatrix} -5 & 0 \\ 5 & 8 \end{bmatrix}$.

Have students find each product. Then predict the product $C = \begin{bmatrix} 1 & 0 \\ 0 & 1 \end{bmatrix} \cdot \begin{bmatrix} 6 & -3 \\ 5 & 2 \end{bmatrix}$. Have them multiply

to check their prediction. $A = \begin{bmatrix} 2 & 6 \\ 1 & -2 \end{bmatrix}$ $B = \begin{bmatrix} -5 & 0 \\ 5 & 8 \end{bmatrix}; \begin{bmatrix} 6 & -3 \\ 5 & 2 \end{bmatrix}$

Example 4 shows that the Commutative Property of Multiplication does not hold for matrix multiplication. **Example 5** shows how the Distributive Property is used with matrices.

Additional Examples

4 Find each product if

$K = \begin{bmatrix} -3 & 2 & 2 \\ -1 & -2 & 0 \end{bmatrix}$ and

$L = \begin{bmatrix} 1 & -2 \\ 4 & 3 \\ 0 & -1 \end{bmatrix}$.

a. KL $\begin{bmatrix} 5 & 10 \\ -9 & -4 \end{bmatrix}$

b. LK $\begin{bmatrix} -1 & 6 & 2 \\ -15 & 2 & 8 \\ 1 & 2 & 0 \end{bmatrix}$

5 Find each product if

$A = \begin{bmatrix} -1 & 2 \\ 2 & 1 \end{bmatrix}$, $B = \begin{bmatrix} 1 & 0 \\ 3 & -2 \end{bmatrix}$,

and $C = \begin{bmatrix} -3 & 1 \\ -1 & 0 \end{bmatrix}$.

a. $A(B + C)$ $\begin{bmatrix} 6 & -5 \\ -2 & 0 \end{bmatrix}$

b. $AB + AC$ $\begin{bmatrix} 6 & -5 \\ -2 & 0 \end{bmatrix}$

Tips for New Teachers

Building on Prior Knowledge Ask students to compare the process in Example 5 to the procedures they have used previously when applying the Distributive Property to real numbers and algebraic expressions.

---

4. $AB = \begin{bmatrix} -8 & 19 \\ -7 & 20 \end{bmatrix}$

and

$BA = \begin{bmatrix} 18 & -9 \\ 9 & -6 \end{bmatrix}$,

so $AB \neq BA$.

StudyTip

Proof and Counterexamples To show that a property is *not* always true, you need to find only one counterexample.

---

5. $(S + T)R = \begin{bmatrix} 15 & 38 \\ 1 & 45 \end{bmatrix}$ and

$SR + TR = \begin{bmatrix} 15 & 38 \\ 1 & 45 \end{bmatrix}$,

so $(S + T)R = SR + TR$.

Example 4 Test of the Commutative Property

Find each product if $G = \begin{bmatrix} 1 & 3 & -5 \\ 4 & -2 & 0 \end{bmatrix}$ and $H = \begin{bmatrix} 2 & 3 \\ -2 & -8 \\ 1 & 7 \end{bmatrix}$.

a. GH

$GH = \begin{bmatrix} 1 & 3 & -5 \\ 4 & -2 & 0 \end{bmatrix} \cdot \begin{bmatrix} 2 & 3 \\ -2 & -8 \\ 1 & 7 \end{bmatrix}$ Substitution

$= \begin{bmatrix} 2 - 6 - 5 & 3 - 24 - 35 \\ 8 + 4 + 0 & 12 + 16 + 0 \end{bmatrix}$ or $\begin{bmatrix} -9 & -56 \\ 12 & 28 \end{bmatrix}$

b. HG

$HG = \begin{bmatrix} 2 & 3 \\ -2 & -8 \\ 1 & 7 \end{bmatrix} \cdot \begin{bmatrix} 1 & 3 & -5 \\ 4 & -2 & 0 \end{bmatrix}$ Substitution

$= \begin{bmatrix} 2 + 12 & 6 - 6 & -10 + 0 \\ -2 - 32 & -6 + 16 & 10 + 0 \\ 1 + 28 & 3 - 14 & -5 + 0 \end{bmatrix}$ or $\begin{bmatrix} 14 & 0 & -10 \\ -34 & 10 & 10 \\ 29 & -11 & -5 \end{bmatrix}$ Notice that $GH \neq HG$.

▶ **GuidedPractice**

4. Determine if $AB = BA$ is true for $A = \begin{bmatrix} 4 & -1 \\ 5 & -2 \end{bmatrix}$ and $B = \begin{bmatrix} -3 & 6 \\ -4 & 5 \end{bmatrix}$.

Example 4 demonstrates that the Commutative Property of Multiplication does not hold for matrix multiplication. The order in which you multiply matrices is very important.

Example 5 Test of the Distributive Property

Find each product if $J = \begin{bmatrix} 2 & 4 \\ -5 & -2 \end{bmatrix}$, $K = \begin{bmatrix} 3 & 2 \\ -1 & 3 \end{bmatrix}$, and $L = \begin{bmatrix} -4 & -1 \\ 3 & 0 \end{bmatrix}$.

a. $J(K + L)$

$J(K + L) = \begin{bmatrix} 2 & 4 \\ -5 & -2 \end{bmatrix} \cdot \left(\begin{bmatrix} 3 & 2 \\ -1 & 3 \end{bmatrix} + \begin{bmatrix} -4 & -1 \\ 3 & 0 \end{bmatrix} \right)$ Substitution

$= \begin{bmatrix} 2 & 4 \\ -5 & -2 \end{bmatrix} \cdot \begin{bmatrix} -1 & 1 \\ 2 & 3 \end{bmatrix}$ Add.

$= \begin{bmatrix} -2 + 8 & 2 + 12 \\ 5 - 4 & -5 - 6 \end{bmatrix}$ or $\begin{bmatrix} 6 & 14 \\ 1 & -11 \end{bmatrix}$ Multiply.

b. $JK + JL$

$JK + JL = \begin{bmatrix} 2 & 4 \\ -5 & -2 \end{bmatrix} \cdot \begin{bmatrix} 3 & 2 \\ -1 & 3 \end{bmatrix} + \begin{bmatrix} 2 & 4 \\ -5 & -2 \end{bmatrix} \cdot \begin{bmatrix} -4 & -1 \\ 3 & 0 \end{bmatrix}$

$= \begin{bmatrix} 2(3) + 4(-1) & 2(2) + 4(3) \\ -5(3) + (-2)(-1) & -5(2) + (-2)(3) \end{bmatrix} + \begin{bmatrix} 2(-4) + 4(3) & 2(-1) + 4(0) \\ -5(-4) + (-2)(3) & -5(-1) + (-2)(0) \end{bmatrix}$

$= \begin{bmatrix} 2 & 16 \\ -13 & -16 \end{bmatrix} + \begin{bmatrix} 4 & -2 \\ 14 & 5 \end{bmatrix}$ or $\begin{bmatrix} 6 & 14 \\ 1 & -11 \end{bmatrix}$ Notice that $J(K + L) = JK + JL$.

▶ **GuidedPractice**

5. Use the matrices $R = \begin{bmatrix} 2 & -1 \\ 1 & 3 \end{bmatrix}$, $S = \begin{bmatrix} 4 & 6 \\ -2 & 5 \end{bmatrix}$, and $T = \begin{bmatrix} -3 & 7 \\ -4 & 8 \end{bmatrix}$ to determine if $(S + T)R = SR + TR$.

The previous example suggests that the Distributive Property is true for matrix multiplication. Some properties of matrix multiplication are shown below.

KeyConcept Properties of Matrix Multiplication

For any matrices A, B, and C for which the matrix product is defined and any scalar k, the following properties are true.

Associative Property of Matrix Multiplication	$(AB)C = A(BC)$
Associative Property of Scalar Multiplication	$k(AB) = (kA)B = A(kB)$
Left Distributive Property	$C(A + B) = CA + CB$
Right Distributive Property	$(A + B)C = AC + BC$

Check Your Understanding   = Step-by-Step Solutions begin on page R14.

Example 1
Determine whether each matrix product is defined. If so, state the dimensions of the product.

1. $A_{2 \times 4} \cdot B_{4 \times 3}$ 2×3 **2.** $C_{5 \times 4} \cdot D_{5 \times 4}$ undefined   **3.** $E_{8 \times 6} \cdot F_{6 \times 10}$ 8×10

Examples 2–3 Find each product, if possible.

4. $\begin{bmatrix} 2 & 1 \\ 7 & -5 \end{bmatrix} \cdot \begin{bmatrix} -6 & 3 \\ -2 & -4 \end{bmatrix}$ $\begin{bmatrix} -14 & 2 \\ -32 & 41 \end{bmatrix}$ **5.** $\begin{bmatrix} 10 & -2 \\ -7 & 3 \end{bmatrix} \cdot \begin{bmatrix} 1 & 4 \\ 5 & -2 \end{bmatrix}$ $\begin{bmatrix} 0 & 44 \\ 8 & -34 \end{bmatrix}$

6. $[9 \quad -2] \cdot \begin{bmatrix} -2 & 4 \\ 6 & -7 \end{bmatrix}$ $[-30 \quad 50]$ $\begin{bmatrix} -9 \\ 6 \end{bmatrix} \cdot [-1 \quad -10 \quad 1]$ $\begin{bmatrix} 9 & 90 & -9 \\ -6 & -60 & 6 \end{bmatrix}$

8. $\begin{bmatrix} -8 & 7 & 4 \\ -5 & -3 & 8 \end{bmatrix} \cdot \begin{bmatrix} 10 & 6 \\ 8 & 4 \end{bmatrix}$ undefined   **9.** $\begin{bmatrix} 2 & 8 \\ 3 & -1 \end{bmatrix} \cdot \begin{bmatrix} 6 \\ -7 \end{bmatrix}$ $\begin{bmatrix} -44 \\ 25 \end{bmatrix}$

10. $\begin{bmatrix} -4 & 3 & 2 \\ -1 & -5 & 4 \end{bmatrix} \cdot \begin{bmatrix} 2 & 1 & 6 \\ 8 & 4 & -1 \\ 5 & 3 & -2 \end{bmatrix}$ $\begin{bmatrix} 26 & 14 & -31 \\ -22 & -9 & -9 \end{bmatrix}$ **11.** $\begin{bmatrix} 2 & 5 & 3 & -1 \\ -3 & 1 & 8 & -3 \end{bmatrix} \cdot \begin{bmatrix} 6 & -3 \\ -7 & 1 \\ 2 & 0 \\ -1 & 0 \end{bmatrix}$ $\begin{bmatrix} -16 & -1 \\ -6 & 10 \end{bmatrix}$

12. (CCSS) **SENSE-MAKING** The table shows the number of people registered for aerobics for the first quarter.
a. See margin.
Quinn's Gym charges the following registration fees: class-by-class, $165; 11-class pass, $110; unlimited pass, $239.

Quinn's Gym		
Payment	Aerobics	Step Aerobics
class-by-class	35	28
11-class pass	32	17
unlimited pass	18	12

a. Write a matrix for the registration fees and a matrix for the number of students.

b. Find the total amount of money the gym received from aerobics and step aerobic registrations. $22,955

Examples 4–5 Use $X = \begin{bmatrix} -10 & -3 \\ 2 & -8 \end{bmatrix}$, $Y = \begin{bmatrix} -5 & 6 \\ -1 & 9 \end{bmatrix}$, and $Z = \begin{bmatrix} -5 & -1 \\ -8 & -4 \end{bmatrix}$ to determine whether the following equations are true for the given matrices. **13, 14.** See margin.

13. $XY = YX$ **14.** $X(YZ) = (XY)Z$

connectED.mcgraw-hill.com **183**

Focus on Mathematical Content
Multiplying Matrices Before you multiply two matrices, you must determine whether the matrix product is defined. Two matrices can be multiplied if and only if the number of columns in the first matrix is equal to the number of rows in the second matrix.

3 Practice

Formative Assessment
Use Exercises 1–14 to check for understanding.

Use the chart at the bottom of the next page to customize assignments for your students.

(CCSS) **Teaching the Mathematical Practices**

Sense-Making Mathematically proficient students start by explaining to themselves the meaning of a problem and looking for entry points to its solution. They analyze givens, constraints, relationships, and goals, and make conjectures about the form and meaning of the solution and plan a solution pathway rather than simply jumping into a solution attempt.

Additional Answers

12a. $[165 \quad 110 \quad 239] \begin{bmatrix} 35 & 28 \\ 32 & 17 \\ 18 & 12 \end{bmatrix}$

13. No; $\begin{bmatrix} 53 & -87 \\ -2 & -60 \end{bmatrix} \neq \begin{bmatrix} 62 & -33 \\ 28 & -69 \end{bmatrix}$.

14. Yes; $X(YZ) = \begin{bmatrix} 431 & 295 \\ 490 & 242 \end{bmatrix}$ and $(XY)Z = \begin{bmatrix} 431 & 295 \\ 490 & 242 \end{bmatrix}$.

WatchOut!

Preventing Errors In applications involving matrix products, even if the order of the matrices can be reversed, the product may only make sense in one direction.

CCSS Teaching the Mathematical Practices

Arguments Mathematically proficient students understand and use stated assumptions, definitions, and previously established results in constructing arguments. They justify their conclusions, communicate them to others.

Additional Answers

29a. $I = \begin{bmatrix} 3 & 2 & 2 \\ 2 & 3 & 1 \\ 4 & 3 & 0 \end{bmatrix}$, $C = \begin{bmatrix} 220 \\ 250 \\ 360 \end{bmatrix}$

30. Yes; $k(PQ) = \begin{bmatrix} 52 & 42 \\ 4 & -12 \end{bmatrix}$ and $P(kQ) = \begin{bmatrix} 52 & 42 \\ 4 & -12 \end{bmatrix}$.

31. No; $PQR = \begin{bmatrix} -22 & 240 \\ 44 & -12 \end{bmatrix}$ and $RQP = \begin{bmatrix} 34 & -40 \\ -220 & -44 \end{bmatrix}$.

32. Yes; $PR + QR = \begin{bmatrix} 22 & 72 \\ 14 & -18 \end{bmatrix}$ and $(P + Q)R = \begin{bmatrix} 22 & 72 \\ 14 & -18 \end{bmatrix}$.

33. No; $R(P + Q) = \begin{bmatrix} 34 & -6 \\ -64 & -30 \end{bmatrix}$ and $PR + QR = \begin{bmatrix} 22 & 72 \\ 14 & -18 \end{bmatrix}$.

Practice and Problem Solving

Extra Practice is on page R3.

Example 1   Determine whether each matrix product is defined. If so, state the dimensions of the product.

15. $P_{2 \times 3} \cdot Q_{3 \times 4}$ 2 × 4

16. $A_{5 \times 5} \cdot B_{5 \times 5}$ 5 × 5

17. $M_{3 \times 1} \cdot N_{2 \times 3}$ undefined

18. $X_{2 \times 6} \cdot Y_{6 \times 3}$ 2 × 3

19. $J_{2 \times 1} \cdot K_{2 \times 1}$ undefined

20. $S_{5 \times 2} \cdot T_{2 \times 4}$ 5 × 4

Examples 2–3 Find each product, if possible.

21. $[1 \quad 6] \cdot \begin{bmatrix} -10 \\ 6 \end{bmatrix}$ [26]

22. $\begin{bmatrix} 6 \\ -3 \end{bmatrix} \cdot [2 \quad -7]$ $\begin{bmatrix} 12 & -42 \\ -6 & 21 \end{bmatrix}$

23. $\begin{bmatrix} -3 & -7 \\ -2 & -1 \end{bmatrix} \cdot \begin{bmatrix} 4 & 4 \\ 9 & -3 \end{bmatrix}$ $\begin{bmatrix} -75 & 9 \\ -17 & -5 \end{bmatrix}$

24. $\begin{bmatrix} -1 & 0 \\ 5 & 2 \end{bmatrix} \cdot \begin{bmatrix} 6 & -3 \\ 7 & -2 \end{bmatrix}$ $\begin{bmatrix} -6 & 3 \\ 44 & -19 \end{bmatrix}$

25. $\begin{bmatrix} -1 & 0 & 6 \\ -4 & -10 & 4 \end{bmatrix} \cdot \begin{bmatrix} 5 & -7 \\ -2 & -9 \end{bmatrix}$ undefined

26. $\begin{bmatrix} -6 & 4 & -9 \\ 2 & 8 & 7 \end{bmatrix} \cdot \begin{bmatrix} 7 \\ 2 \\ 4 \end{bmatrix}$ $\begin{bmatrix} -70 \\ 58 \end{bmatrix}$

27. $\begin{bmatrix} 2 & 9 & -3 \\ 4 & -1 & 0 \end{bmatrix} \cdot \begin{bmatrix} 4 & 2 \\ -6 & 7 \\ -2 & 1 \end{bmatrix}$ $\begin{bmatrix} -40 & 64 \\ 22 & 1 \end{bmatrix}$

28. $\begin{bmatrix} -4 \\ 8 \end{bmatrix} \cdot [-3 \quad -1]$ $\begin{bmatrix} 12 & 4 \\ -24 & -8 \end{bmatrix}$

29. **TRAVEL** The Wolf family owns three bed and breakfasts in a vacation spot. A room with a single bed is \$220 per night, a room with two beds is \$250 per night, and a suite is \$360.

   a. Write a matrix for the number of each type of room at each bed and breakfast. Then write a room-cost matrix. **See margin.**

   b. Write a matrix for total daily income, assuming that all the rooms are rented.

29b. $\begin{bmatrix} \$1880 \\ \$1550 \\ \$1630 \end{bmatrix}$

   c. What is the total daily income from all three bed and breakfasts, assuming that all the rooms are rented? **\$5060**

Available Rooms at a Wolf Bed and Breakfast			
B & B	Single	Double	Suite
1	3	2	2
2	2	3	1
3	4	3	0

Examples 4–5 Use $P = \begin{bmatrix} 4 & -1 \\ 1 & 2 \end{bmatrix}$, $Q = \begin{bmatrix} 6 & 4 \\ -2 & -5 \end{bmatrix}$, $R = \begin{bmatrix} 4 & 6 \\ -6 & 4 \end{bmatrix}$, and $k = 2$ to determine whether the following equations are true for the given matrices. **30–33. See margin.**

30. $k(PQ) = P(kQ)$

31. $PQR = RQP$

32. $PR + QR = (P + Q)R$

33. $R(P + Q) = PR + QR$

34. **CCSS SENSE-MAKING** Student Council is selling flowers for Mother's Day. They bought 200 roses, 150 daffodils, and 100 orchids for the purchase prices shown. They sold all of the flowers for the sales prices shown.

   a. Organize the data in two matrices, and use matrix multiplication to find the total amount that was spent on the flowers. **\$747.50**

   b. Write two matrices, and use matrix multiplication to find the total amount the student council received for the flower sale. **\$1387.50**

   c. Use matrix operations to find how much money the student council made on their project. **\$640**

Flower	Purchase Price	Sales Price
rose	\$1.67	\$3.00
daffodil	\$1.03	\$2.25
orchid	\$2.59	\$4.50

Differentiated Homework Options

Level	Assignment	Two-Day Option	
AL Basic	15–33, 46–48, 50–65	15–33 odd, 51–54	16–32 even, 46–48, 50, 55–65
OL Core	15–33 odd, 34, 35–43 odd, 44–48, 50–65	15–33, 51–54	34–45, 46–48, 50, 55–65
BL Advanced	34–62, (optional: 63–65)		

35. AUTO SALES A car lot has four sales associates. At the end of the year, each sales associate gets a bonus of $1000 for every new car they have sold and $500 for every used car they have sold.

a. Use a matrix to determine which sales associate earned the most money. **See margin.**

b. What is the total amount of money the car lot spent on bonuses for the sales associates this year? **$187,500**

Cars Sold by Each Associate		
Sales Associate	New Cars	Used Cars
Mason	27	49
Westin	35	36
Gallagher	9	56
Stadler	15	62

Use matrices $X = \begin{bmatrix} 2 & -6 \\ 3y & -4.5 \end{bmatrix}$, $Y = \begin{bmatrix} -5 & -1.5 \\ x+2 & y \\ 13 & 1.2 \end{bmatrix}$, and $Z = \begin{bmatrix} -3 \\ x+y \end{bmatrix}$ to find each of the following. If the matrix does not exist, write *undefined*.

37. See margin.
39, 40, 43. See Chapter 3 Answer Appendix.

36. XY **undefined**   **37.** YX **38.** ZY **undefined**   **39.** YZ
40. $(YX)Z$ **41.** $(XZ)X$ **undefined**   **42.** $X(ZZ)$ **undefined**   **43.** $(XX)Z$

44. CAMERAS Prices of digital cameras depend on features like optical zoom, digital zoom, and megapixels.

a. The 10-mp cameras are on sale for 20% off, and the other models are 10% off. Write a cost matrix.

b. Write a new matrix allowing for a 6.25% sales tax on the discounted prices.

c. Describe what the differences in these two matrices represent. **sales tax**

Optical Zoom	6 MP	7 MP	10 MP
3 to 4	$189.99	$249.99	$349.99
5 to 6	$199.99	$289.99	$399.99
10 to 12	$299.99	$399.99	$499.99

a, b. See Chapter 3 Answer Appendix.

45. BUSINESS The Kangy Studio has packages available for senior portraits.

a. Use matrices to determine the total cost of each package.

b. The studio offers an early bird discount of 15% off any package. Find the early bird price for each package.
A: $357.85; B: $232.90;
C: $127.50; D: $57.80

Size (price)	Packages			
	A	B	C	D
4 × 5 ($7)	10	10	8	0
5 × 7 ($10)	4	4	4	4
8 × 10 ($14)	2	2	2	2
11 × 14 ($45)	1	1	0	0
16 × 20 ($95)	1	0	0	0
Wallets (8 for $13)	88	56	16	0

45a.
A: $421;
B: $274;
C: $150;
D: $68

H.O.T. Problems   Use Higher-Order Thinking Skills

46. REASONING If the product matrix AB has dimensions 5×8, and A has dimensions 5×6, what are the dimensions of matrix B? **6×8**

47. CCSS ARGUMENTS Show that each property of matrices is true for all 2×2 matrices.

a. Scalar Distributive Property **a–d. See Chapter 3 Answer Appendix.**

b. Matrix Distributive Property

c. Associative Property of Multiplication

d. Associative Property of Scalar Multiplication

48. OPEN ENDED Write two matrices A and B such that $AB = BA$.

49. CHALLENGE Find the missing values in $\begin{bmatrix} a & b \\ c & d \end{bmatrix} \cdot \begin{bmatrix} 4 & 3 \\ 2 & 5 \end{bmatrix} = \begin{bmatrix} 10 & 11 \\ 20 & 29 \end{bmatrix}$.
$a = 2$, $b = 1$, $c = 3$, $d = 4$

50. WRITING IN MATH Use the data on Lisa Leslie found at the beginning of the lesson to explain how matrices can be used in sports statistics. Describe a matrix that represents the total number of points she has scored during her career and an example of a sport in which different point values are used in scoring. **See Chapter 3 Answer Appendix.**

48. Sample answer:
$A = \begin{bmatrix} 2 & 2 \\ 2 & 2 \end{bmatrix}$,
$B = \begin{bmatrix} 3 & 3 \\ 3 & 3 \end{bmatrix}$

Additional Answers

35a.
$\begin{bmatrix} 27 & 49 \\ 35 & 36 \\ 9 & 56 \\ 15 & 62 \end{bmatrix} \cdot \begin{bmatrix} 1000 \\ 500 \end{bmatrix} = \begin{bmatrix} 51,500 \\ 53,000 \\ 37,000 \\ 46,000 \end{bmatrix}$;
Westin

37.
$\begin{bmatrix} -10 - 4.5y & 36.75 \\ 2x + 4 + 3y^2 & -6x - 4.5y - 12 \\ 3.6y + 26 & -83.4 \end{bmatrix}$

Study Guide and Intervention
AL OL ELL

NAME _____ DATE _____ PERIOD _____

3-6 Study Guide and Intervention
Multiplying Matrices

Multiply Matrices You can multiply two matrices if and only if the number of columns in the first matrix is equal to the number of rows in the second matrix.

Practice
AL OL BL ELL

NAME _____ DATE _____ PERIOD _____

3-6 Practice
Multiplying Matrices

Word Problem Practice
AL OL BL ELL

NAME _____ DATE _____ PERIOD _____

3-6 Word Problem Practice
Multiplying Matrices

Enrichment
OL BL

NAME _____ DATE _____ PERIOD _____

3-6 Enrichment
Properties of Matrices

connectED.mcgraw-hill.com   **185**

Ticket Out the Door Have students write on pieces of paper whether or not a 4 × 3 matrix can be multiplied by a 2 × 3 matrix. Ask them to explain their answers.

Formative Assessment

Check for student understanding of Lessons 3-5 and 3-6.

 Quiz 3, p. 60

Standardized Test Practice

51. GRIDDED RESPONSE The average (arithmetic mean) of r, w, x, and y is 8, and the average of x and y is 4. What is the average of r and w? **12**

52. Carla, Meiko, and Kayla went shopping to get ready for college. Their purchases and total amounts spent are shown in the table below.

Person	Shirts	Pants	Shoes	Total Spent
Carla	3	4	2	$149.79
Meiko	5	3	3	$183.19
Kayla	6	5	1	$181.14

Assume that all of the shirts were the same price, all of the pants were the same price, and all of the shoes were the same price. What was the price of each item? **A**

A shirt, $12.95; pants, $15.99; shoes, $23.49

B shirt, $15.99; pants, $12.95; shoes, $23.49

C shirt, $15.99; pants, $23.49; shoes, $12.95

D shirt, $23.49; pants, $15.99; shoes, $12.95

53. GEOMETRY Rectangle *LMNQ* has diagonals that intersect at point *P*.

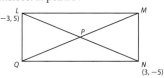

Which of the following represents point *P*? **H**

F (2, 2)   H (0, 0)

G (1, 1)   J (−1, −1)

54. SAT/ACT What are the dimensions of the matrix that results from the multiplication shown? **D**

$$\begin{bmatrix} a & b & c \\ d & e & f \\ g & h & i \\ j & k & l \end{bmatrix} \cdot \begin{bmatrix} 7 \\ 4 \\ 6 \end{bmatrix}$$

A 1×4 D 4×1

B 3×3 E 4×3

C 3×1

55–57. See Chapter 3 Answer Appendix.

Spiral Review

Perform the indicated operations. If the matrix does not exist, write *impossible*. (Lesson 3-5)

55. $4\begin{bmatrix} 8 & -1 \\ -3 & -4 \end{bmatrix} - 5\begin{bmatrix} -2 & 4 \\ 6 & 3 \end{bmatrix}$

56. $5\left(2\begin{bmatrix} -2 & -5 \\ -1 & 3 \end{bmatrix} - 3\begin{bmatrix} -1 & -2 \\ 6 & 4 \end{bmatrix}\right)$

57. $-4\left(\begin{bmatrix} 8 & 9 \\ -5 & 5 \end{bmatrix} - 2\begin{bmatrix} -6 & -1 \\ 6 & 3 \end{bmatrix}\right)$

Solve each system of equations. (Lesson 3-4)

58. $2x - 4y + 3z = -3$
$-7x + 5y - 4z = 11$
$x - y - 2z = -21$ **(−2, 5, 7)**

59. $-4x - 2y + 9z = -29$
$10x - 12y + 7z = 51$
$3x + 5y - 14z = 25$ **(8, 3, 1)**

60. $-7x + 8y - z = 43$
$3x - 2y + 5z = -43$
$2x - 4y + 6z = -50$ **(−3, 2, −6)**

61. MEDICINE The graph shows how much Americans spent on doctors' visits in some recent years and a prediction for 2014. (Lesson 2-5)

a. Find a regression equation for the data without the predicted value. **Sample answer: $y = 116.25x - 231,176.97$**

b. Use your equation to predict the expenditures for 2014. **Sample answer: $2950.53**

c. Compare your prediction to the one given in the graph. **The value predicted by the equation is significantly lower than the one given in the graph.**

62. How many different ways can the letters of the word *SAMPLING* be arranged? (Lesson 0-4) **40,320**

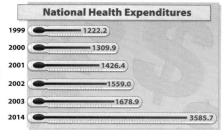

National Health Expenditures

Year	Expenditures (billions)
1999	1222.2
2000	1309.9
2001	1426.4
2002	1559.0
2003	1678.9
2014	3585.7

Expenditures (billions)

Skills Review

Describe the transformation in each function. Then graph the function.

63–65. See Chapter 3 Answer Appendix.

63. $f(x) = |x - 4| + 3$

64. $f(x) = 2|x + 3| - 5$

65. $f(x) = (x + 2)^2 - 6$

Follow-up

Students have explored matrices and matrix operations.

Ask:

- What are the advantages of using matrices to solve problems? Sample answer: They provide a convenient way to organize data; they can be used to shorten notation; technology can be used to quickly perform matrix operations.

EXTEND 3-6

Graphing Technology Lab
Operations with Matrices

A graphing calculator can be used to perform operations with matrices.

Activity 1 Perform Operations

Use A, B, and C to find the following.

$$A = \begin{bmatrix} -3.2 & 1.7 \\ 0.4 & -5.8 \end{bmatrix} \qquad B = \begin{bmatrix} 4.9 & 0.3 \\ -7.1 & 2.6 \end{bmatrix} \qquad C = \begin{bmatrix} 5.6 & -6.1 & 2.1 \\ -8.2 & 7.6 & 0.2 \end{bmatrix}$$

a. $3A + 2B$

Begin by entering matrix A into a graphing calculator.

KEYSTROKES: [2nd] [MATRIX] [▶] [▶] [ENTER] 2 [ENTER] 2
[ENTER] [(−)] 3.2 [ENTER] 1.7 [ENTER]
0.4 [ENTER] [(−)] 5.8 [ENTER] [2nd] [QUIT]

Enter matrix B into the graphing calculator using similar keystrokes. Then, perform the indicated operations.

KEYSTROKES: 3 [2nd] [MATRIX] [ENTER] + 2 [2nd] [MATRIX] [▼]
[ENTER] [ENTER]

$3A + 2B$ is equal to $\begin{bmatrix} 0.2 & 5.7 \\ -13 & -12.2 \end{bmatrix}$.

1. $\begin{bmatrix} 69.57 & -29.36 \\ 47.23 & 161.26 \end{bmatrix}$

b. $4C + 3A$

Enter matrix C into the graphing calculator. Perform the indicated operations. Notice that the calculator displays an error message when the dimensions do not allow the operations to be performed.

2. $\begin{bmatrix} 27.7 & -51.2 \\ -57.6 & 55.8 \end{bmatrix}$ 5. $\begin{bmatrix} -18.3 & -170.4 \\ -19.3 & 213.32 \end{bmatrix}$ 6. $\begin{bmatrix} 85.63 & -373.16 \\ -378.09 & -221.58 \end{bmatrix}$

Exercises

Use A, B, C, and D to find the following. If the matrix does not exist, write *impossible*.

$$A = \begin{bmatrix} 4.5 & -9.0 \\ -7.4 & 9.4 \end{bmatrix} \quad B = \begin{bmatrix} 1.9 & -5.9 \\ 2.9 & 5.0 \end{bmatrix} \quad C = \begin{bmatrix} 7.0 & 5.5 & -1.9 \\ 7.6 & -9.9 & 0.5 \end{bmatrix} \quad D = \begin{bmatrix} 8.5 & 8.0 \\ -1.2 & -5.9 \\ 0.7 & 8.9 \end{bmatrix}$$

1. $CD + 4A$

2. $-2B + 7A$

3. $4(DC)$ See margin.

4. $6B + DC$ impossible

5. $2(AB) - 3B + 5A$

6. $-3(CD) + 4(BA) + 7A$

(continued on the next page)

---

1 Focus

Objective Use a graphing calculator to explore operations with matrices.

Materials for Each Student
- TI-83/84 Plus or other graphing calculator

Teaching Tip
Prior to beginning the activity, ask students to explain the Associative, Commutative, Identity, and Inverse Properties of real numbers.

2 Teach

Working in Cooperative Groups
Have students work in groups of two or three to complete Activities 1 and 2 and Exercises 1–14.

- Before starting Activity 1, have each group explore the **MATRIX** menu on graphing calculators by pressing [2nd] [MATRIX].

- Have each group list the benefits and drawbacks to using a graphing calculator to perform operations with matrices.

Additional Answer

3. $\begin{bmatrix} 481.2 & -129.8 & -48.6 \\ -212.96 & 207.24 & -2.68 \\ 290.16 & -337.04 & 12.48 \end{bmatrix}$

EXTEND

3 Assess

Formative Assessment
Use Exercise 6 to assess whether students can perform operations with matrices using graphing calculators.

From Concrete to Abstract
After students complete Exercise 11, explain that there is an *identify matrix I* for multiplication such that for a matrix A, $AI = A$. Have students use graphing calculators to find a matrix I that satisfies $AI = A$.

Additional Answers

9. $(AB)C$ and $A(BC)$ are both equal to
$\begin{bmatrix} -498 & 136 \\ 197 & -52 \end{bmatrix}$; Associative Property of Matrix Multiplication.

10. $3(AB)$, $(3A)B$, and $A(3B)$ are all
equal to $\begin{bmatrix} -138 & 6 \\ 3 & 57 \end{bmatrix}$;
Associative Property of Scalar Multiplication.

11. For $A + O$ to be defined, O must be a 2×2 matrix. So,
$O = \begin{bmatrix} 0 & 0 \\ 0 & 0 \end{bmatrix}$. Enter O as matrix D in the graphing calculator. A and $A + D$ are both equal to
$\begin{bmatrix} -6 & 2 \\ 7 & 3 \end{bmatrix}$. Therefore, $A + O = A$.

12. $E = \begin{bmatrix} 6 & -2 \\ -7 & -3 \end{bmatrix}$; $A + E$ is equal to O. Therefore, A and E are additive inverses.

13. The additive inverse for matrix B is
$\begin{bmatrix} -5 & -1 \\ 8 & -4 \end{bmatrix}$ and for matrix C is
$\begin{bmatrix} -11 & 3 \\ -4 & 1 \end{bmatrix}$.

14. Sample answer: A matrix and its additive inverse have the same dimensions, but corresponding elements have opposite signs. The additive inverse of $\begin{bmatrix} w & -x \\ -y & z \end{bmatrix}$
is $\begin{bmatrix} -w & x \\ y & -z \end{bmatrix}$.

Activity 2 Explore Properties of Matrix Operations

Use A, B, and C to find the following.

$A = \begin{bmatrix} -6 & 2 \\ 7 & 3 \end{bmatrix}$
$B = \begin{bmatrix} 5 & 1 \\ -8 & 4 \end{bmatrix}$
$C = \begin{bmatrix} 11 & -3 \\ 4 & -1 \end{bmatrix}$

a. $A + B$ and $B + A$

Find $A + B$.
Then find $B + A$.

$A + B$ and $B + A$ are both equal to $\begin{bmatrix} -1 & 3 \\ -1 & 7 \end{bmatrix}$.

b. $(A + B) + C$ and $A + (B + C)$

Find $(A + B) + C$.
Then find $A + (B + C)$.

$(A + B) + C$ and $A + (B + C)$ are both equal to $\begin{bmatrix} 10 & 0 \\ 3 & 6 \end{bmatrix}$.

Analyze the Results

What property is illustrated in each part of Activity 2?

7. part a **Commutative Property of Addition**

8. part b **Associative Property of Addition**

Find each set of products. How are they similar or different? What property is illustrated? 9–14. See margin.

9. $(AB)C$, $A(BC)$

10. $3(AB)$, $(3A)B$, $A(3B)$

11. The zero matrix O is an $m \times n$ matrix with all elements equal to 0. If $A + O$ is defined, verify the Additive Identity Property for Matrices, $A + O = A$.

12. Two matrices are additive inverses if their sum is the zero matrix. Find matrix E so that $A + E$ are additive inverses. Then verify that $A + E = O$.

13. Find the additive inverses for matrix B and matrix C.

14. **CHALLENGE** What observations can be made about a matrix and its additive inverse?
Find the additive inverse for $\begin{bmatrix} w & -x \\ -y & z \end{bmatrix}$.

LESSON 3-7

Solving Systems of Equations Using Cramer's Rule

::Then	::Now	::Why?
● You solved systems of equations algebraically.	**1** Evaluate determinants. **2** Solve systems of linear equations by using Cramer's Rule.	● A zoologist tagged a tiger with a GPS tracker so that she could determine the tiger's territory. After several days, the zoologist determined that the tiger's territory was a triangular region. By using the coordinates of the vertices of this triangle, she could use matrices and determinants to determine the size of the tiger's territory.

 NewVocabulary
determinant
second-order determinant
third-order determinant
diagonal rule
Cramer's Rule
coefficient matrix

 Common Core State Standards

Content Standards
A.CED.3 Represent constraints by equations or inequalities, and by systems of equations and/or inequalities, and interpret solutions as viable or nonviable options in a modeling context

Mathematical Practices
7 Look for and make use of structure.

1 Determinants Every square matrix has a **determinant**. The determinant of a 2×2 matrix is called a **second-order determinant**.

KeyConcept Second-Order Determinant

Words	The value of a second-order determinant is the difference of the products of the two diagonals.
Symbols	$\det \begin{bmatrix} a & b \\ c & d \end{bmatrix} = \begin{vmatrix} a & b \\ c & d \end{vmatrix} = ad - bc$
Example	$\begin{vmatrix} 4 & 5 \\ -3 & 6 \end{vmatrix} = 4(6) - (5)(-3) = 39$

Example 1 Second-Order Determinant

Evaluate each determinant.

a. $\begin{vmatrix} 5 & -4 \\ 8 & 9 \end{vmatrix}$

$\begin{vmatrix} 5 & -4 \\ 8 & 9 \end{vmatrix} = 5(9) - (-4)(8)$ Definition of determinant

$= 45 + 32$ Simplify.

$= 77$

b. $\begin{vmatrix} 0 & 6 \\ 4 & -11 \end{vmatrix}$

$\begin{vmatrix} 0 & 6 \\ 4 & -11 \end{vmatrix} = 0(-11) - 6(4)$ Definition of determinant

$= 0 - 24$ Simplify.

$= -24$

GuidedPractice

1A. $\begin{vmatrix} -6 & -7 \\ 10 & 8 \end{vmatrix}$ 22

1B. $\begin{vmatrix} 7 & 5 \\ 9 & -4 \end{vmatrix}$ −73

  connectED.mcgraw-hill.com **189**

MORITSCH, MARC (VIA HORNOCKER, MAURICE)/National Geographic Stock

1 Focus

VerticalAlignment

Before Lesson 3-7 Solve systems of equations algebraically.

Lesson 3-7 Evaluate determinants. Solve systems of linear equations by using Cramer's Rule.

After Lesson 3-7 Use matrices to solve systems of equations.

2 Teach

Scaffolding Questions
Have students read the **Why?** section of the lesson.

Ask:
- How is GPS related to the use of coordinates? Sample answer: A GPS tracker shows the latitude and longitude coordinates of a point on Earth's surface.

- What method do you already know for finding the area of a triangle? the formula $A = \frac{1}{2}bh$

- Why would it be difficult to use this method in this situation? The height of the triangle would be difficult to find.

Lesson 3-7 Resources

Resource	Approaching Level **AL**	On Level **OL**	Beyond Level **BL**	English Learners **ELL**
Teacher Edition	■ Differentiated Instruction, p. 190	■ Differentiated Instruction, pp. 190, 193	■ Differentiated Instruction, pp. 193, 197	
Chapter Resource Masters	■ Study Guide and Intervention, pp. 43–44 ■ Skills Practice, p. 45 ■ Practice, p. 46 ■ Word Problem Practice, p. 47	■ Study Guide and Intervention, pp. 43–44 ■ Skills Practice, p. 45 ■ Practice, p. 46 ■ Word Problem Practice, p. 47 ■ Enrichment, p. 48 ■ Spreadsheet Activity, p. 49	■ Practice, p. 46 ■ Word Problem Practice, p. 47 ■ Enrichment, p. 48	■ Study Guide and Intervention, pp. 43–44 ■ Skills Practice, p. 45 ■ Practice, p. 46 ■ Word Problem Practice, p. 47
Other	■ 5-Minute Check 3-7 ■ Study Notebook ■ Teaching Algebra with Manipulatives	■ 5-Minute Check 3-7 ■ Study Notebook ■ Teaching Algebra with Manipulatives	■ 5-Minute Check 3-7 ■ Study Notebook	■ 5-Minute Check 3-7 ■ Study Notebook ■ Teaching Algebra with Manipulatives

1 Determinants

Example 1 shows how to find the value of a second-order determinant.
Example 2 shows how to use diagonals to evaluate a third-order determinant.
Example 3 shows how to use a determinant to evaluate the area of a triangle in a real-world situation.

Formative Assessment

Use the Guided Practice exercises after each example to determine students' understanding of concepts.

Additional Examples

1 Evaluate the determinant $\begin{vmatrix} 6 & 4 \\ -1 & 0 \end{vmatrix}$. 4

2 Evaluate $\begin{vmatrix} 3 & -2 & -1 \\ 2 & -1 & 0 \\ 1 & 2 & -3 \end{vmatrix}$ using diagonals. -8

Additional Examples also in Interactive Classroom PowerPoint® Presentations

 IWB Interactive White Board READY

Teach with Tech

Interactive Whiteboard Work through several examples with the class, and save your work as several notes pages. At the end of class, e-mail your notes to students or post them on a class Web page. This may help students to focus during the lesson rather than trying to copy the computations from the board.

StudyTip
Diagonal Rule The diagonal rule can only be used for 3 × 3 matrices.

Determinants of 3 × 3 matrices are called **third-order determinants**. They can be evaluated by using the **diagonal rule**.

KeyConcept Diagonal Rule

Step 1 Rewrite the first two columns to the right of the determinant.

Step 2 Draw diagonals, beginning with the upper left-hand element. Multiply the elements in each diagonal. Repeat the process, beginning with the upper right-hand element.

Step 3 Find the sum of the products of the elements in each set of diagonals.

Step 4 Subtract the second sum from the first sum.

Example 2 Use Diagonals

Evaluate $\begin{vmatrix} 4 & -8 & 3 \\ -3 & 2 & 6 \\ -4 & 5 & 9 \end{vmatrix}$ using diagonals.

Step 1 Rewrite the first two columns to the right of the determinant.

$$\begin{array}{rrr|rr} 4 & -8 & 3 & 4 & -8 \\ -3 & 2 & 6 & -3 & 2 \\ -4 & 5 & 9 & -4 & 5 \end{array}$$

Step 2 Find the products of the elements of the diagonals.

$4(2)(9) = 72$ $-4(2)(3) = -24$

$-8(6)(-4) = 192$ $5(6)(4) = 120$

$3(-3)(5) = -45$ $9(-3)(-8) = 216$

Step 3 Find the sum of each group.

$72 + 192 + (-45) = 219$ $-24 + 120 + 216 = 312$

Step 4 Subtract the sum of the second group from the sum of the first group.

$219 - 312 = -93$

The value of the determinant is -93.

▶ **Guided**Practice

Evaluate each determinant.

2A. $\begin{vmatrix} -5 & 9 & 4 \\ -2 & -1 & 5 \\ -4 & 6 & 2 \end{vmatrix}$ -48 **2B.** $\begin{vmatrix} -8 & -4 & 4 \\ 0 & -5 & -8 \\ 3 & 4 & 1 \end{vmatrix}$ -60

DifferentiatedInstruction AL OL

If ▶ students are using evaluation by diagonals for the first time,

Then ▶ urge them to write down each step in the procedure. Have them compare their work with classmates to find any errors in either their calculations or their use of the procedure.

Determinants can also be used to find the areas of triangles. If the coordinates of the vertices of the triangle are known, the formula below can be used to calculate the area of the triangle.

⚙ KeyConcept Area of a Triangle

Words

The area of a triangle with vertices (a, b), (c, d), and (e, f) is $|A|$, where

$$A = \frac{1}{2}\begin{vmatrix} a & b & 1 \\ c & d & 1 \\ e & f & 1 \end{vmatrix}.$$

$(-4, 3)$ $(3, 1)$ $(-2, -2)$

Example

$$A = \frac{1}{2}\begin{vmatrix} -4 & 3 & 1 \\ 3 & 1 & 1 \\ -2 & -2 & 1 \end{vmatrix}$$

🔵 Real-World Example 3 Use Determinants   **PT**

ZOOLOGY Refer to the application at the beginning of the lesson. The coordinates of the vertices of the tiger's territory are shown to the right. Use determinants to find the area of the tiger's territory.

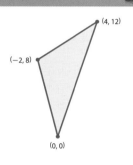

$(4, 12)$
$(-2, 8)$
$(0, 0)$

Real-World Link
Tigers are very territorial, solitary animals. Their territories can measure up to 100 square kilometers.

Source: *National Geographic*

$$A = \frac{1}{2}\begin{vmatrix} a & b & 1 \\ c & d & 1 \\ e & f & 1 \end{vmatrix}$$

$$= \frac{1}{2}\begin{vmatrix} 0 & 0 & 1 \\ 4 & 12 & 1 \\ -2 & 8 & 1 \end{vmatrix}$$
$(a, b) = (0, 0)$
$(c, d) = (4, 12)$
$(e, f) = (-2, 8)$

Diagonal Rule

$0 + 0 + 32 = 32 \qquad\qquad -24 + 0 + 0 = -24$ Sum of products of diagonals

$$A = \frac{1}{2}\begin{vmatrix} 0 & 0 & 1 \\ 4 & 12 & 1 \\ -2 & 8 & 1 \end{vmatrix}$$
Area of a Triangle

$$= \left(\frac{1}{2}\right)[32 - (-24)] \text{ or } 28$$ Simplify.

The area of the tiger's territory is 28 square kilometers.

▶ **Guided**Practice

3. CAR WASH To raise money for their rowing club, Hannah, Christina, and Dario are advertising a car wash at three different street corners in a neighborhood. On a map, the coordinates for the corners are $(3, 15)$, $(6, 4)$, and $(11, 9)$. Each unit represents 0.5 kilometer. What is the area of the neighborhood in which they are advertising? **8.75 km²**

Additional Example

3 SURVEYING A surveying crew located three points on a map that formed the vertices of a triangular area. A coordinate grid in which one unit corresponds to 10 miles is placed over the map so that the vertices are located at $(0, -1)$, $(-2, -6)$, and $(3, -2)$. Use a determinant to find the area of the triangle.
850 square miles

Focus on Mathematical Content
Determinants Every square matrix has a real number that is associated with it called the determinant of the matrix. The determinant of a 2×2 matrix is called a second-order determinant and is equal to the difference of the products of the elements on the two diagonals. The determinant of a 3×3 matrix is called a third-order determinant and can be evaluated using a method called the *diagonal rule*.

WatchOut!

Preventing Errors Have students make posters that show the process for finding the determinant of a 3×3 matrix, using colored markers to clearly identify the diagonals.

Digital Vision/Getty Images

Example 4 shows how to use Cramer's Rule to solve a system of two linear equations. **Example 5** shows how to use Cramer's Rule to solve a system of three linear equations.

Additional Example

4 Use Cramer's Rule to solve the system of equation.

$5x + 4y = 28$
$3x - 2y = 8$ $(4, 2)$

ReadingMath
Determinants The determinant is used to *determine* whether a system has a unique solution.

2 Cramer's Rule You can use determinants to solve systems of equations. If a determinant is nonzero, then the system has a unique solution. If a determinant is 0, then the system either has no solution or infinite solutions. A method called **Cramer's Rule** uses the coefficient matrix. The **coefficient matrix** is a matrix that contains only the coefficients of the system.

KeyConcept Cramer's Rule

Let C be the coefficient matrix of the system $\begin{matrix} ax + by = m \\ fx + gy = n \end{matrix} \rightarrow \begin{bmatrix} a & b \\ f & g \end{bmatrix}$.

The solution of this system is $x = \dfrac{\begin{vmatrix} m & b \\ n & g \end{vmatrix}}{|C|}$ and $y = \dfrac{\begin{vmatrix} a & m \\ f & n \end{vmatrix}}{|C|}$, if $|C| \neq 0$.

Test-TakingTip
Cramer's Rule When the determinant of the coefficient matrix C is 0, the system does not have a unique solution.

Example 4 Solve a System of Two Equations

Solve the system by using Cramer's Rule.

$5x - 6y = 15$
$3x + 4y = -29$

$x = \dfrac{\begin{vmatrix} m & b \\ n & g \end{vmatrix}}{|C|}$ Cramer's Rule $y = \dfrac{\begin{vmatrix} a & m \\ f & n \end{vmatrix}}{|C|}$

$= \dfrac{\begin{vmatrix} 15 & -6 \\ -29 & 4 \end{vmatrix}}{\begin{vmatrix} 5 & -6 \\ 3 & 4 \end{vmatrix}}$ Substitute values. $= \dfrac{\begin{vmatrix} 5 & 15 \\ 3 & -29 \end{vmatrix}}{\begin{vmatrix} 5 & -6 \\ 3 & 4 \end{vmatrix}}$

$= \dfrac{15(4) - (-29)(-6)}{5(4) - (3)(-6)}$ Evaluate. $= \dfrac{5(-29) - 3(15)}{5(4) - (3)(-6)}$

$= \dfrac{60 - 174}{20 + 18}$ Multiply. $= \dfrac{-145 - 45}{20 + 18}$

$= -\dfrac{114}{38}$ Add and subtract. $= -\dfrac{190}{38}$

$= -3$ Simplify. $= -5$

The solution of the system is $(-3, -5)$.

CHECK $5(-3) - 6(-5) \stackrel{?}{=} 15$ $x = -3, y = -5$

$-15 + 30 \stackrel{?}{=} 15$ Simplify.

$15 = 15$ ✓

$3(-3) + 4(-5) \stackrel{?}{=} -29$ $x = -3, y = -5$

$-9 - 20 \stackrel{?}{=} -29$ Simplify.

$-29 = -29$ ✓

▶ **GuidedPractice**

Solve each system using Cramer's Rule.

4A. $7x + 3y = 37$ $(4, 3)$
$\quad\,\, -5x - 7y = -41$

4B. $8x - 5y = 70$ $(5, -6)$
$\quad\,\, 9x + 7y = 3$

Cramer's Rule can also be used for systems of three equations.

KeyConcept Cramer's Rule for a System of Three Equations

Let C be the coefficient matrix of the system
$$\begin{aligned} ax + by + cz &= m \\ fx + gy + hz &= n \\ jx + ky + \ell z &= p \end{aligned} \rightarrow \begin{bmatrix} a & b & c \\ f & g & h \\ j & k & \ell \end{bmatrix}.$$

The solution of this system is $x = \dfrac{\begin{vmatrix} m & b & c \\ n & g & h \\ p & k & \ell \end{vmatrix}}{|C|}$, $y = \dfrac{\begin{vmatrix} a & m & c \\ f & n & h \\ j & p & \ell \end{vmatrix}}{|C|}$, and $z = \dfrac{\begin{vmatrix} a & b & m \\ f & g & n \\ j & k & p \end{vmatrix}}{|C|}$, if $|C| \neq 0$.

Example 5 Solve a System of Three Equations

Solve the system by using Cramer's Rule.
$$\begin{aligned} 4x + 5y - 6z &= -14 \\ 3x - 2y + 7z &= 47 \\ 7x - 6y - 8z &= 15 \end{aligned}$$

$$x = \dfrac{\begin{vmatrix} m & b & c \\ n & g & h \\ p & k & \ell \end{vmatrix}}{|C|} \qquad y = \dfrac{\begin{vmatrix} a & m & c \\ f & n & h \\ j & p & \ell \end{vmatrix}}{|C|} \qquad z = \dfrac{\begin{vmatrix} a & b & m \\ f & g & n \\ j & k & p \end{vmatrix}}{|C|}$$

$$= \dfrac{\begin{vmatrix} -14 & 5 & -6 \\ 47 & -2 & 7 \\ 15 & -6 & -8 \end{vmatrix}}{\begin{vmatrix} 4 & 5 & -6 \\ 3 & -2 & 7 \\ 7 & -6 & -8 \end{vmatrix}} \qquad = \dfrac{\begin{vmatrix} 4 & -14 & -6 \\ 3 & 47 & 7 \\ 7 & 15 & -8 \end{vmatrix}}{\begin{vmatrix} 4 & 5 & -6 \\ 3 & -2 & 7 \\ 7 & -6 & -8 \end{vmatrix}} \qquad = \dfrac{\begin{vmatrix} 4 & 5 & -14 \\ 3 & -2 & 47 \\ 7 & -6 & 15 \end{vmatrix}}{\begin{vmatrix} 4 & 5 & -6 \\ 3 & -2 & 7 \\ 7 & -6 & -8 \end{vmatrix}}$$

$$= \dfrac{3105}{621} \text{ or } 5 \qquad = -\dfrac{1242}{621} \text{ or } -2 \qquad = \dfrac{2484}{621} \text{ or } 4$$

The solution of the system is $(5, -2, 4)$.

CHECK $4(5) + 5(-2) - 6(4) \stackrel{?}{=} -14 \qquad 3(5) - 2(-2) + 7(4) \stackrel{?}{=} 47$
$$20 - 10 - 24 \stackrel{?}{=} -14 \qquad 15 + 4 + 28 \stackrel{?}{=} 47$$
$$-14 = -14 \checkmark \qquad 47 = 47 \checkmark$$

$$7(5) - 6(-2) - 8(4) \stackrel{?}{=} 15$$
$$35 + 12 - 32 \stackrel{?}{=} 15$$
$$15 = 15 \checkmark$$

StudyTip
Check for Accuracy
Always substitute your answers into the initial equations to confirm accuracy.

GuidedPractice

Solve each system using Cramer's Rule.
5A. $3x + 5y + 2z = -7 \quad \left(\dfrac{23}{22}, -\dfrac{57}{22}, \dfrac{31}{22}\right)$ **5B.** $6x + 5y + 2z = -1 \quad (-4, 5, -1)$
$\qquad -4x + 3y - 5z = -19 \qquad\qquad\qquad\qquad -x + 3y + 7z = 12$
$\qquad 5x + 4y - 7z = -15 \qquad\qquad\qquad\qquad\quad 5x - 7y - 3z = -52$

connectED.mcgraw-hill.com   **193**

Additional Example

5 Solve the system by using Cramer's Rule.
$$\begin{aligned} 2x + y - z &= -2 \\ -x + 2y + z &= -0.5 \\ x + y + 2z &= 3.5 \end{aligned}$$
$(0.5, -1, 2)$

WatchOut!
Preventing Errors When solving a system of three equations in three variables, it is important to point out that the numerator for x is found by replacing the coefficients of x with the constant terms of the system. Similarly, the numerators for y and z are found by replacing the coefficients of y and z, respectively, with the constant terms of the system.

Focus on Mathematical Content
Cramer's Rule One advantage of using Cramer's Rule to solve a system of three linear equations in three variables is that if the system has a unique solution, Cramer's Rule can give the value of any one of the variables without having to solve for the others. If the determinant is zero, then Cramer's Rule does not give the solution, but it does indicate that the system is either dependent or inconsistent.

Differentiated Instruction OL BL

If students wish to review the process of the manipulation of matrices,

Then ask students to write a brief reflection on their reactions to the various methods they have learned involving the manipulation of matrices. Ask them to comment on what aspects they found efficient and helpful, and what aspects they found to be difficult or confusing.

3 Practice

Formative Assessment

Use Exercises 1–25 to check for understanding.

Use the chart at the bottom of the next page to customize assignments for your students.

CCSS Teaching the Mathematical Practices

Perseverance Mathematically proficient students start by explaining to themselves the meaning of a problem and looking for entry points to its solution. They make conjectures about the form and meaning of the solution and plan a solution pathway rather than simply jumping into a solution attempt.

Tips for New Teachers

Sense-Making Remind students that Cramer's Rule does not apply when the determinant of the coefficient matrix is 0, because the solution would have a denominator of 0, which is undefined.

Check Your Understanding $\qquad$ ● = Step-by-Step Solutions begin on page R14.

Example 1 Evaluate each determinant.

1. $\begin{vmatrix} 8 & 6 \\ 5 & 7 \end{vmatrix}$ 26

2. $\begin{vmatrix} -6 & -6 \\ 8 & 10 \end{vmatrix}$ −12

3. $\begin{vmatrix} -4 & 12 \\ 9 & 5 \end{vmatrix}$ −128

4. $\begin{vmatrix} 16 & -10 \\ -8 & 5 \end{vmatrix}$ 0

Example 2 Evaluate each determinant using diagonals.

5. $\begin{vmatrix} 3 & -2 & 2 \\ -4 & 2 & -5 \\ -3 & 1 & 4 \end{vmatrix}$ −19

6. $\begin{vmatrix} 2 & -3 & 5 \\ -4 & 6 & -2 \\ 4 & -1 & -6 \end{vmatrix}$ −80

7. $\begin{vmatrix} 8 & 4 & 0 \\ -2 & -6 & -1 \\ 5 & -3 & 6 \end{vmatrix}$ −284

8. $\begin{vmatrix} -5 & -3 & 4 \\ -2 & -4 & -3 \\ 8 & -2 & 4 \end{vmatrix}$ 302

9. $\begin{vmatrix} 8 & 3 & 4 \\ 2 & 4 & 2 \\ 1 & 6 & 5 \end{vmatrix}$ 72

10. $\begin{vmatrix} -4 & 3 & 0 \\ 1 & 5 & -2 \\ -1 & -8 & -3 \end{vmatrix}$ 139

11. $\begin{vmatrix} 2 & -6 & -3 \\ 7 & 9 & -4 \\ -6 & 4 & 9 \end{vmatrix}$ 182

12. $\begin{vmatrix} -5 & -6 & 7 \\ 4 & 0 & 5 \\ -3 & 8 & 2 \end{vmatrix}$ 562

Example 4 Use Cramer's Rule to solve each system of equations.

13. $4x - 5y = 39$ (6, −3)
    $3x + 8y = -6$

14. $5x + 6y = 20$ (−2, 5)
    $-3x - 7y = -29$

15. $-8a - 5b = -27$ (4, −1)
    $7a + 6b = 22$

16. $10c - 7d = -59$ (−8, −3)
    $6c + 5d = -63$

Examples 3–5 17. **CCSS PERSEVERANCE** The "Bermuda Triangle" is an area located off the southeastern Atlantic coast of the United States, and is noted for reports of unexplained losses of ships, small boats, and aircraft.

   a. Find the area of the triangle on the map. **15.75 units²**
   b. Suppose each grid represents 175 miles. What is the area of the Bermuda Triangle? **482,343.75 mi²**

Use Cramer's Rule to solve each system of equations.

18. $4x - 2y + 7z = 26$ (−3, −5, 4)
    $5x + 3y - 5z = -50$
    $-7x - 8y - 3z = 49$

19. $-3x - 5y + 10z = -4$ $\left(\dfrac{66}{7}, -\dfrac{116}{7}, -\dfrac{41}{7}\right)$
    $-8x + 2y - 3z = -91$
    $6x + 8y - 7z = -35$

20. $6x - 5y + 2z = -49$ (−3, 7, 2)
    $-5x - 3y - 8z = -22$
    $-3x + 8y - 5z = 55$

21. $-9x + 5y + 3z = 50$ (−4, −2, 8)
    $7x + 8y - 2z = -60$
    $-5x + 7y + 5z = 46$

22. $x + 2y = 12$ (6, 3, −4)
    $3y - 4z = 25$
    $x + 6y + z = 20$

23. $9a + 7b = -30$ (−1, −3, 7)
    $8b + 5c = 11$
    $-3a + 10c = 73$

24. $2n + 3p - 4w = 20$ (5, 2, −1)
    $4n - p + 5w = 13$
    $3n + 2p + 4w = 15$

25. $x + y + z = 12$ (4, 0, 8)
    $6x - 2y - z = 16$
    $3x + 4y + 2z = 28$

 194 | Lesson 3-7 | Solving Systems of Equations Using Cramer's Rule

Examples 1–2 Evaluate each determinant.

26. $\begin{vmatrix} -7 & 12 \\ 5 & 6 \end{vmatrix}$ −102   27. $\begin{vmatrix} -8 & -9 \\ 11 & 12 \end{vmatrix}$ 3   28. $\begin{vmatrix} -5 & 8 \\ -6 & -7 \end{vmatrix}$ 83

29. $\begin{vmatrix} 3 & 5 & -2 \\ -1 & -4 & 6 \\ -6 & -2 & 5 \end{vmatrix}$ −135   30. $\begin{vmatrix} 2 & 0 & -6 \\ -3 & -4 & -5 \\ -2 & 5 & 8 \end{vmatrix}$ 124   31. $\begin{vmatrix} -5 & -1 & -2 \\ 1 & 8 & 4 \\ 0 & -6 & 9 \end{vmatrix}$ −459

32. $\begin{vmatrix} 6 & -3 & -5 \\ 0 & -7 & 0 \\ 3 & -6 & -4 \end{vmatrix}$ 63   33. $\begin{vmatrix} -8 & -3 & -9 \\ 0 & 0 & 0 \\ 8 & -2 & -4 \end{vmatrix}$ 0   34. $\begin{vmatrix} 1 & 6 & 7 \\ -2 & -5 & -8 \\ 4 & 4 & 9 \end{vmatrix}$ −13

35. $\begin{vmatrix} 1 & -8 & -9 \\ 6 & 5 & -6 \\ -2 & -8 & 10 \end{vmatrix}$ 728   36. $\begin{vmatrix} 5 & -5 & -5 \\ -8 & -3 & -2 \\ -2 & 4 & 6 \end{vmatrix}$ −120   37. $\begin{vmatrix} -4 & 1 & -2 \\ 10 & 12 & 9 \\ -6 & 0 & 13 \end{vmatrix}$ −952

38. **TRAVEL** Mr. Smith's art class took a bus trip to an art museum. The bus averaged 65 miles per hour on the highway and 25 miles per hour in the city. The art museum is 375 miles away from the school, and it took the class 7 hours to get there. Use Cramer's Rule to find how many hours the bus was on the highway and how many hours it was driving in the city. 5 hours on the highway, 2 hours in the city

Examples 4–5 Use Cramer's Rule to solve each system of equations.

39. $6x - 5y = 73$ (8, −5)
    $-7x + 3y = -71$

40. $10a - 3b = -34$ (−4, −2)
    $3a + 8b = -28$

41. $-4c - 5d = -39$ (6, 3)
    $5c + 8d = 54$

42. $-6f - 8g = -22$ (5, −1)
    $-11f + 5g = -60$

43. $9r + 4s = -55$ (−3, −7)
    $-5r - 3s = 36$

44. $-11u - 7v = 4$ (−8, 12)
    $9u + 4v = -24$

45. $5x - 4y + 6z = 58$ (4, −2, 5)
    $-4x + 6y + 3z = -13$
    $6x + 3y + 7z = 53$

46. $8x - 4y + 7z = 34$ (−3, −4, 6)
    $5x + 6y + 3z = -21$
    $3x + 7y - 8z = -85$

47. **DOUGHNUTS** Mi-Ling is ordering doughnuts for a class party. The box contains 2 dozen doughnuts, some of which are plain and some of which are jelly-filled. The plain doughnuts each cost $0.50, and the jelly-filled cost $0.60. If the total cost is $12.60, use Cramer's Rule to find how many jelly-filled doughnuts Mi-Ling ordered. 6

Doughnuts
Plain : $0.50
Jelly Filled : $0.60

Examples 3–4 48. **CCSS PERSEVERANCE** The salary for each of the stars of a new movie is $5 million, and the supporting actors each receive $1 million. The total amount spent for the salaries of the actors and actresses is $19 million. If the cast has 7 members, use Cramer's Rule to find the number of stars in the movie. 3

49. **ARCHAEOLOGY** Archaeologists found whale bones at coordinates (0, 3), (4, 7), and (5, 9). If the units of the coordinates are meters, find the area of the triangle formed by these finds. 2 m²

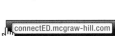

connectED.mcgraw-hill.com **195**

Differentiated Homework Options

Level	Assignment	Two-Day Option	
AL Basic	26–49, 56–58, 60–75	27–49 odd, 62–65	26–48 even, 56–58, 60–61, 66–75
OL Core	27–53 odd, 54–58, 60, 75	26–49, 62–65	50–58, 60–61, 66–75
BL Advanced	50–72, (optional: 73–75)		

WatchOut!

Error Analysis In Exercise 58, remind students that products along "upward-sloping" diagonals are subtracted when evaluating a determinant.

B Use Cramer's Rule to solve each system of equations.

50. $6a - 7b = -55$ $(-1, 7, -3)$
$2a + 4b - 3c = 35$
$-5a - 3b + 7c = -37$

51. $3a - 5b - 9c = 17$ $(4, 8, -5)$
$4a - 3c = 31$
$-5a - 4b - 2c = -42$

52. $4x - 5y = -2$ $(-8, -6, 3)$
$7x + 3z = -47$
$8y - 5z = -63$

53. $7x + 8y + 9z = -149$ $\left(-\dfrac{6187}{701}, -\dfrac{2904}{701}, -\dfrac{4212}{701}\right)$
$-6x + 7y - 5z = 54$
$4x + 5y - 2z = -44$

54. GARDENING Rob wants to build a triangular flower garden. To plan out his garden he uses a coordinate grid where each of the squares represents one square foot. The coordinates for the vertices of his garden are $(-1, 7)$, $(2, 6)$, and $(4, -3)$. Find the area of the garden. **12.5 ft²**

C **55** **FINANCIAL LITERACY** A vendor sells small drinks for \$1.15, medium drinks for \$1.75, and large drinks for \$2.25. During a week in which he sold twice as many small drinks as medium drinks, his total sales were \$2,238.75 for 1385 drinks. **small: 650; medium: 325; large: 410**

a. Use Cramer's Rule to determine how many of each drink were sold.

b. The vendor decided to increase the price for small drinks to \$1.25 the next week. The next week, he sold 140 fewer small drinks, 125 more medium drinks, and 35 more large drinks. Calculate his sales for that week. **\$2426.25**

c. Was raising the price of the small drink a good business move? Explain.

55c. It seems like it was a good move for the vendor. Although he sold fewer small drinks, he sold more medium and large drinks and on the whole, made more money this week than in the previous week.

H.O.T. Problems Use Higher-Order Thinking Skills

61. Sample answer: Given a 2 × 2 system of linear equations, if the determinant of the matrix of coefficients is 0, then the system does not have a unique solution. The system may have no solution and the graphical representation shows two parallel lines. The system may have infinitely many solutions in which the graphical representation will be the same line.

56. REASONING Some systems of equations cannot be solved by using Cramer's Rule.

a. Find the value of $\begin{vmatrix} a & b \\ f & g \end{vmatrix}$. When is the value 0? **a, b. See Chapter 3 Answer Appendix.**

b. Choose values for a, b, f, and g to make the determinant of the coefficient matrix 0. What type of system is formed?

57. REASONING What can you determine about the solution of a system of linear equations if the determinant of the coefficients is 0? **Sample answer: There is no unique solution of the system. There are either infinite or no solutions.**

58. **CCSS** **CRITIQUE** James and Amber are finding the value of $\begin{vmatrix} 8 & 3 \\ -5 & 2 \end{vmatrix}$. **James; because 3(−5) = −15**

Is either of them correct? Explain your reasoning.

59. CHALLENGE Find the determinant of a 3 × 3 matrix defined by

$$a_{mn} = \begin{cases} 0 & \text{if } m + n \text{ is even} \\ m + n & \text{if } m + n \text{ is odd} \end{cases}. \quad 0$$

60. OPEN ENDED Write a 2 × 2 matrix with each of the following characteristics.

a. The determinant equals 0. **a–c. See Chapter 3 Answer Appendix.**

b. The determinant equals 25.

c. The elements are all negative numbers and the determinant equals −32.

61. WRITING IN MATH Describe the possible graphical representations of a 2 × 2 system of linear equations if the determinant of the matrix of coefficients is 0.

196 | Lesson 3-7 | Solving Systems of Equations Using Cramer's Rule

Enrichment

OL BL

3-7 Enrichment

Matrix Transpose and Determinants

196 | Lesson 3-7 | Solving Systems of Equations Using Cramer's Rule

Teaching the Mathematical Practices

Critique Mathematically proficient students are also able to compare the effectiveness of two plausible arguments, distinguish correct logic or reasoning from that which is flawed, and—if there is a flaw in an argument—explain what it is.

62. Tyler paid $25.25 to play three games of miniature golf and two rides on go-karts. Brent paid $25.75 for four games of miniature golf and one ride on the go-karts. How much does one game of miniature golf cost? **C**

   A $4.25   **C** $5.25

   B $4.75   **D** $5.75

63. Use the table to determine the expression that best represents the number of faces of any prism having a base with n sides. **H**

Base	Sides of Base	Faces of Prisms
triangle	3	5
quadrilateral	4	6
pentagon	5	7
hexagon	6	8
heptagon	7	9
octagon	8	10

   F $2(n-1)$ **H** $n+2$

   G $2(n+1)$ **J** $2n$

64. SHORT RESPONSE A right circular cone has radius 4 inches and height 6 inches.

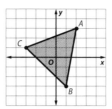

6 in.   4 in.

What is the lateral area of the cone? (lateral area of cone = $\pi r \ell$, where ℓ = slant height) $8\sqrt{13}\,\pi$ in^2

65. SAT/ACT Find the area of $\triangle ABC$. **D**

   A 10 units^2 **D** 14 units^2

   B 12 units^2 **E** 16 units^2

   C 13 units^2

Determine whether each matrix product is defined. If so, state the dimensions of the product. (Lesson 3-6)

66. $A_{4 \times 2} \cdot B_{2 \times 6}$ yes; 4×6 **67.** $C_{5 \times 4} \cdot D_{5 \times 3}$ no   **68.** $E_{2 \times 7} \cdot F_{7 \times 1}$ yes; 2×1

69. BUSINESS The table lists the prices at the Sandwich Shoppe. (Lesson 3-5) **a–c. See margin.**

   a. List the prices in a 4×3 matrix.

   b. The manager decides to cut the prices of every item by 20%. List this new set of data in a 4×3 matrix.

   c. Subtract the second matrix from the first and determine the savings to the customer for each sandwich.

Sandwich	Small	Medium	Large
ham	$4.50	$6.75	$9.50
salami	$4.50	$6.75	$9.50
veggie	$4.00	$6.25	$8.75
meatball	$4.75	$7.50	$10.25

Graph each function. (Lesson 2-6) **70–72. See Chapter 3 Answer Appendix.**

70. $f(x) = 2|x-3| - 4$ **71.** $f(x) = -3|2x| + 4$ **72.** $f(x) = |3x - 1| + 2$

Solve each system of equations.

73. $2x - 5y = -26$ $(-8, 2)$ **74.** $4y + 6x = 10$ $\left(\frac{79}{25}, -\frac{56}{25}\right)$ **75.** $-3x - 2y = 17$ $(-3, -4)$

   $5x + 3y = -34$ $2x - 7y = 22$ $-4x + 5y = -8$

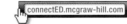
Vocabulary Point out to students that "lateral" means "side." Lateral area refers to the sum of the areas of all of the surfaces of the figure except the base or bases.

4 Assess

Name the Math Have each student tell a partner tips and techniques for remembering the current order of the calculations to be performed in the multi-step process discussed in this lesson.

Additional Answers

69a.
$$\begin{bmatrix} \$4.50 & \$6.75 & \$9.50 \\ \$4.50 & \$6.75 & \$9.50 \\ \$4.00 & \$6.25 & \$8.75 \\ \$4.75 & \$7.50 & \$10.25 \end{bmatrix}$$

69b.
$$\begin{bmatrix} \$3.60 & \$5.40 & \$7.60 \\ \$3.60 & \$5.40 & \$7.60 \\ \$3.20 & \$5.00 & \$7.00 \\ \$3.80 & \$6.00 & \$8.20 \end{bmatrix}$$

69c.
$$\begin{bmatrix} \$0.90 & \$1.35 & \$1.90 \\ \$0.90 & \$1.35 & \$1.90 \\ \$0.80 & \$1.25 & \$1.75 \\ \$0.95 & \$1.50 & \$2.05 \end{bmatrix}$$

DifferentiatedInstruction **BL**

Extension Write the following system on the board.
$$2x + 4y = 10$$
$$-x - 2y = 3$$

Ask students if they can use Cramer's Rule to solve the system. Have them explain why or why not.
Cramer's rule cannot be used because the determinant is zero.
Then have them graph each equation and explain why there is no solution to the system.
The graphs of the two equations are parallel.

Additional Example

2 Find the inverse of each matrix, if it exists.

a. $B = \begin{bmatrix} -1 & 0 \\ 8 & -2 \end{bmatrix}$ $\begin{bmatrix} -1 & 0 \\ -4 & -\frac{1}{2} \end{bmatrix}$

b. $T = \begin{bmatrix} -4 & 6 \\ -2 & 3 \end{bmatrix}$
No inverse exists.

2 Matrix Equations

Example 3 shows how to write a system of equations to model a real-world situation and then use a matrix equation for the system.

Math History Link

Seki Kowa **(1642–1708)**
Known as The Arithmetical Sage, Seki Kowa was the first to develop the theory of determinants.

Example 2 Find the Inverse of a Matrix

Find the inverse of each matrix, if it exists.

a. $P = \begin{bmatrix} 7 & -5 \\ 2 & -1 \end{bmatrix}$

$\begin{vmatrix} 7 & -5 \\ 2 & -1 \end{vmatrix} = -7 - (-10)$ or 3   Find the determinant.

Since the determinant does not equal 0, P^{-1} exists.

$P^{-1} = \frac{1}{ad - bc} \begin{bmatrix} d & -b \\ -c & a \end{bmatrix}$ Definition of inverse

$= \frac{1}{7(-1) - (-5)(2)} \begin{bmatrix} -1 & 5 \\ -2 & 7 \end{bmatrix}$ $a = 7, b = -5, c = 2, d = -1$

$= \frac{1}{3} \begin{bmatrix} -1 & 5 \\ -2 & 7 \end{bmatrix}$ or $\begin{bmatrix} -\frac{1}{3} & \frac{5}{3} \\ -\frac{2}{3} & \frac{7}{3} \end{bmatrix}$ Simplify.

CHECK Find the product of the matrices. If the product is I, then they are inverses.

$\begin{bmatrix} 7 & -5 \\ 2 & -1 \end{bmatrix} \cdot \begin{bmatrix} -\frac{1}{3} & \frac{5}{3} \\ -\frac{2}{3} & \frac{7}{3} \end{bmatrix} = \begin{bmatrix} -\frac{7}{3} + \frac{10}{3} & \frac{35}{3} - \frac{35}{3} \\ -\frac{2}{3} + \frac{2}{3} & \frac{10}{3} - \frac{7}{3} \end{bmatrix}$ or $\begin{bmatrix} 1 & 0 \\ 0 & 1 \end{bmatrix}$ ✓

b. $Q = \begin{bmatrix} -8 & -6 \\ 12 & 9 \end{bmatrix}$

$\begin{vmatrix} -8 & -6 \\ 12 & 9 \end{vmatrix} = -72 - (-72) = 0$ Find the determinant.

Since the determinant equals 0, Q^{-1} does not exist.

▶ **Guided**Practice

2A. $\begin{bmatrix} 3 & 7 \\ 1 & -4 \end{bmatrix}$ $\begin{bmatrix} \frac{4}{19} & \frac{7}{19} \\ \frac{1}{19} & -\frac{3}{19} \end{bmatrix}$ **2B.** $\begin{bmatrix} 2 & 1 \\ -4 & 3 \end{bmatrix}$ $\begin{bmatrix} \frac{3}{10} & \frac{-1}{10} \\ \frac{2}{5} & \frac{1}{5} \end{bmatrix}$

2 Matrix Equations Matrices can be used to represent and solve systems of equations. You can write a **matrix equation** to solve the system of equations below.

$\begin{array}{l} x + 2y = 9 \\ 3x - 6y = 3 \end{array} \rightarrow \begin{bmatrix} x + 2y \\ 3x - 6y \end{bmatrix} = \begin{bmatrix} 9 \\ 3 \end{bmatrix}$

Write the left side of the matrix equation as the product of the coefficient matrix and the variable matrix. Write the right side as a constant matrix.

$$\underset{\text{coefficient matrix}}{\overset{A}{\begin{bmatrix} 1 & 2 \\ 3 & -6 \end{bmatrix}}} \cdot \underset{\substack{\text{variable matrix} \\ \text{only the variables} \\ \text{of a system}}}{\overset{X}{\begin{bmatrix} x \\ y \end{bmatrix}}} = \underset{\substack{\text{constant matrix} \\ \text{only the constants} \\ \text{of a system}}}{\overset{B}{\begin{bmatrix} 9 \\ 3 \end{bmatrix}}}$$

 200 | Lesson 3-8 | Solving Systems of Equations Using Inverse Matrices

Courtesy Ichinoseki City Museum, Japan

DifferentiatedInstruction **OL**

Logical Learners Ask each student to write a comparison of the inverse of a matrix to the multiplicative and additive inverses of a number.

StudyTip

Inverse You can use this method to solve systems of equations only if A has an inverse. If A does not have an inverse, then the system either has no solution or infinitely many solutions.

Then solve the matrix equation in the same way that you would solve any other equation.

$ax = b$	Write the equation.	$AX = B$
$\left(\frac{1}{a}\right)ax = \left(\frac{1}{a}\right)b$	Multiply each side by the inverse of the coefficient, if it exists.	$A^{-1}AX = A^{-1}B$
$1x = \frac{b}{a}$	$\left(\frac{1}{a}\right)a = 1, A^{-1}A = I$	$IX = A^{-1}B$
$x = \frac{b}{a}$	$1x = x, IX = X$	$X = A^{-1}B$

Notice that the solution of the matrix equation is the product of the inverse of the coefficient matrix and the constant matrix.

Real-World Example 3 Solve a System of Equations

TRAVEL Helena stopped for gasoline twice during a car trip. The price of gasoline at the first station where she stopped was $3.75 per gallon. At the second station, the price was $3.50 per gallon. Helena bought a total of 24.2 gallons of gasoline and spent $88.05. How much gasoline did Helena buy at each gas station?

A system of equations to represent the situation is as follows.

$x + y = 24.2$
$3.75x + 3.50y = 88.05$

The matrix equation is $\begin{bmatrix} 1 & 1 \\ 3.75 & 3.50 \end{bmatrix} \cdot \begin{bmatrix} x \\ y \end{bmatrix} = \begin{bmatrix} 24.2 \\ 88.05 \end{bmatrix}$.

Step 1 Find the inverse of the coefficient matrix.

$A^{-1} = \frac{1}{3.50 - 3.75}\begin{bmatrix} 3.50 & -1 \\ -3.75 & 1 \end{bmatrix}$ or $-\frac{1}{0.25}\begin{bmatrix} 3.50 & -1 \\ -3.75 & 1 \end{bmatrix}$

Step 2 Multiply each side of the matrix equation by the inverse matrix.

$-\frac{1}{0.25}\begin{bmatrix} 3.50 & -1 \\ -3.75 & 1 \end{bmatrix} \cdot \begin{bmatrix} 1 & 1 \\ 3.75 & 3.50 \end{bmatrix} \cdot \begin{bmatrix} x \\ y \end{bmatrix} = -\frac{1}{0.25}\begin{bmatrix} 3.50 & -1 \\ -3.75 & 1 \end{bmatrix} \cdot \begin{bmatrix} 24.2 \\ 88.05 \end{bmatrix}$

$\begin{bmatrix} 1 & 0 \\ 0 & 1 \end{bmatrix} \cdot \begin{bmatrix} x \\ y \end{bmatrix} = -\frac{1}{0.25}\begin{bmatrix} -3.35 \\ -2.70 \end{bmatrix}$

$\begin{bmatrix} x \\ y \end{bmatrix} = \begin{bmatrix} 13.4 \\ 10.8 \end{bmatrix}$

The solution is (13.4, 10.8), where x represents the amount of gas Helena purchased at the first gas station, and y represents the amount purchased at the second gas station.

CHECK You can check your answer by using inverses.

Enter $\begin{bmatrix} 1 & 1 \\ 3.75 & 3.50 \end{bmatrix}$ as matrix A.

Enter $\begin{bmatrix} 24.2 \\ 88.05 \end{bmatrix}$ as matrix B.

Multiply the inverse of A by B.

Real-WorldLink

Average gas prices increased fivefold from $0.70 per gallon in 1977 to $3.50 per gallon in 2007.

Source: U.S. Department of Energy

GuidedPractice

3. **COMIC BOOKS** Dante and Erica just returned from a comic book store that sells new and used comic books. Dante spent $11.25 on 3 new and 4 old books, and Erica spent $15.75 on 10 used and 3 new ones. If comic books of one type are sold at the same price, what is the price in dollars of a new comic book? $2.75

Additional Example

3 **RENTAL COSTS** The Booster Club for North High School plans a picnic. The rental company charges $15 to rent a popcorn machine and $18 to rent a water cooler. The club spends $261 for a total of 15 items. How many of each do they rent? 3 popcorn machines, 12 water coolers

Focus on Mathematical Content

Solving a System of Equations Using A^{-1} If a system of equations has a unique solution, then the solution is given by $X = A^{-1}B$, where A is the coefficient matrix, B is the constant matrix, and X is the variable matrix. If there is no solution, or there are infinitely many solutions to the system, the coefficient matrix does not have an inverse, or is not invertible.

3 Practice

Formative Assessment

Use Exercises 1–12 to check for understanding.

Use the chart at the bottom of this page to customize assignments for your students.

 Teaching the Mathematical Practices

Perseverance Mathematically proficient students start by explaining to themselves the meaning of a problem and looking for entry points to its solution. They make conjectures about the form and meaning of the solution and plan a solution pathway rather than simply jumping into a solution attempt.

Additional Answers

5. $\begin{bmatrix} 0 & -1 \\ -\frac{1}{3} & -2 \end{bmatrix}$

6. $\begin{bmatrix} 0 & -\frac{1}{3} \\ -\frac{1}{4} & -\frac{1}{6} \end{bmatrix}$

7. $\begin{bmatrix} -\frac{1}{3} & 0 \\ \frac{5}{6} & \frac{1}{2} \end{bmatrix}$

8. does not exist

17. $\begin{bmatrix} \frac{1}{3} & 0 \\ 0 & \frac{1}{2} \end{bmatrix}$

18. $\begin{bmatrix} -\frac{2}{5} & \frac{3}{5} \\ \frac{3}{5} & -\frac{2}{5} \end{bmatrix}$

19. $\begin{bmatrix} \frac{1}{3} & 0 \\ -\frac{5}{3} & 1 \end{bmatrix}$

20. $\begin{bmatrix} \frac{1}{7} & -\frac{1}{7} \\ -\frac{6}{7} & -\frac{1}{7} \end{bmatrix}$

21. $\begin{bmatrix} \frac{1}{3} & \frac{2}{3} \\ -\frac{2}{3} & -\frac{5}{6} \end{bmatrix}$

22. $\begin{bmatrix} -2 & -\frac{9}{4} \\ -1 & -\frac{5}{4} \end{bmatrix}$

23. $\begin{bmatrix} \frac{9}{74} & \frac{5}{74} \\ -\frac{2}{37} & \frac{3}{37} \end{bmatrix}$

24. $\begin{bmatrix} -\frac{4}{9} & -\frac{1}{9} \\ \frac{7}{18} & \frac{2}{9} \end{bmatrix}$

25. $\begin{bmatrix} \frac{7}{22} & \frac{4}{11} \\ \frac{4}{11} & \frac{3}{11} \end{bmatrix}$

Example 1 Determine whether the matrices in each pair are inverses.

1. $A = \begin{bmatrix} 2 & 1 \\ -1 & 0 \end{bmatrix}$, $B = \begin{bmatrix} 1 & 2 \\ 2 & 1 \end{bmatrix}$ **no**   2. $C = \begin{bmatrix} 2 & 1 \\ 5 & 3 \end{bmatrix}$, $D = \begin{bmatrix} 2 & 1 \\ 5 & -3 \end{bmatrix}$ **no**

3. $F = \begin{bmatrix} -1 & 1 \\ 0 & -1 \end{bmatrix}$, $G = \begin{bmatrix} -1 & -1 \\ 0 & -1 \end{bmatrix}$ **yes**   4. $H = \begin{bmatrix} -3 & -1 \\ -4 & -2 \end{bmatrix}$, $J = \begin{bmatrix} -1 & 2 \\ 3 & -4 \end{bmatrix}$ **no**

Example 2 Find the inverse of each matrix, if it exists. **5–8. See margin.**

5. $\begin{bmatrix} 6 & -3 \\ -1 & 0 \end{bmatrix}$ 6. $\begin{bmatrix} 2 & -4 \\ -3 & 0 \end{bmatrix}$

7. $\begin{bmatrix} -3 & 0 \\ 5 & 2 \end{bmatrix}$ 8. $\begin{bmatrix} 2 & 4 \\ 1 & 2 \end{bmatrix}$

Example 3 Use a matrix equation to solve each system of equations.

9. $-2x + y = 9$ $(-2, 5)$ 10. $4x - 2y = 22$ $(4, -3)$ 11. $-2x + y = -4$ $(1, -2)$
   $x + y = 3$ $6x + 9y = -3$ $3x + y = 1$

12. **MONEY** Kevin had 25 quarters and dimes. The total value of all the coins was $4. How many quarters and dimes did Kevin have? **10 quarters and 15 dimes**

Practice and Problem Solving   Extra Practice is on page R3.

Example 1 Determine whether each pair of matrices are inverses of each other.

13. $K = \begin{bmatrix} 1 & 2 \\ 3 & 0 \end{bmatrix}$, $L = \begin{bmatrix} 0 & 1 \\ 2 & -1 \end{bmatrix}$ **no**   14. $M = \begin{bmatrix} 0 & 2 \\ 4 & 5 \end{bmatrix}$, $N = \begin{bmatrix} 1 & 1 \\ 0 & 0 \end{bmatrix}$ **no**

15. $P = \begin{bmatrix} 4 & 0 \\ 3 & 0 \end{bmatrix}$, $Q = \begin{bmatrix} -1 & -1 \\ \frac{2}{3} & 5 \end{bmatrix}$ **no**   16. $R = \begin{bmatrix} \frac{1}{2} & -\frac{1}{4} \\ \frac{1}{4} & -\frac{1}{2} \end{bmatrix}$, $S = \begin{bmatrix} 2 & 4 \\ 4 & 2 \end{bmatrix}$ **no**

Example 2 Find the inverse of each matrix, if it exists. **17–25. See margin.**

17. $\begin{bmatrix} 3 & 0 \\ 0 & 2 \end{bmatrix}$ 18. $\begin{bmatrix} 2 & 3 \\ 3 & 2 \end{bmatrix}$ 19. $\begin{bmatrix} 3 & 0 \\ 5 & 1 \end{bmatrix}$

20. $\begin{bmatrix} 1 & -1 \\ -6 & -1 \end{bmatrix}$ ㉑ $\begin{bmatrix} -5 & -4 \\ 4 & 2 \end{bmatrix}$ 22. $\begin{bmatrix} -5 & 9 \\ 4 & -8 \end{bmatrix}$

23. $\begin{bmatrix} 6 & -5 \\ 4 & 9 \end{bmatrix}$ 24. $\begin{bmatrix} -4 & -2 \\ 7 & 8 \end{bmatrix}$ 25. $\begin{bmatrix} -6 & 8 \\ 8 & -7 \end{bmatrix}$

Example 3

26. 6 mL of the red food coloring and 19 mL of the blue food coloring

26. **BAKING** Peggy is preparing a colored frosting for a cake. For the right shade of purple, she needs 25 milliliters of a 44% concentration food coloring. The store has a 25% red and a 50% blue concentration of food coloring. How many milliliters each of blue food coloring and red food coloring should be mixed to make the necessary amount of purple food coloring?

CCSS PERSEVERANCE Use a matrix equation to solve each system of equations.

27. $-x + y = 4$ **no solution**   28. $-x + y = 3$ $(-3, 0)$ 29. $x + y = 4$ $(-1, 5)$
    $-x + y = -4$ $-2x + y = 6$ $-4x + y = 9$

30. $3x + y = 3$ $\left(\frac{3}{4}, \frac{3}{4}\right)$ 31. $y - x = 5$ **no solution**   32. $4x + 2y = 6$ $(1.5, 0)$
    $5x + 3y = 6$ $2y - 2x = 8$ $6x - 3y = 9$

33. $1.6y - 0.2x = 1$ $(-5, 0)$ 34. $4y - x = -2$ $(-30, -8)$ 35. $2y - 4x = 3$ $\left(\frac{3}{4}, 3\right)$
    $0.4y - 0.1x = 0.5$ $3y - x = 6$ $4x - 3y = -6$

 202 | Lesson 3-8 | Solving Systems of Equations Using Inverse Matrices

Differentiated Homework Options

Level	Assignment	Two-Day Option	
AL Basic	13–36, 38, 40–56	13–35 odd, 43–46	14–38 even, 40–42, 47–56
OL Core	13–37 odd, 38, 40–56	13–35, 43–46	37, 38, 40–42, 47–56
BL Advanced	37–53, (optional: 54–56)		

36. POPULATIONS The diagram shows the annual percent migration between a city and its suburbs.

a. Write a matrix to represent the transitions in city population and suburb population. **See Ch. 3 Answer Appendix.**

b. There are currently 16,275 people living in the city and 17,552 people living in the suburbs. Assuming that the trends continue, predict the number of people who will live in the suburbs next year. **about 17,839**

c. Use the inverse of the matrix from part b to find the number of people who lived in the city last year. **about 16,587**

B 37 MUSIC The diagram shows the trends in digital audio player and portable CD player ownership over the past five years for Central City. Every person in Central City has either a digital audio player or a portable CD player. Central City has a stable population of 25,000 people, of whom 17,252 own digital audio players and 7748 own portable CD players.

39. The system would have to consist of two equations that are the same or one equation that is a multiple of the other.

a. Write a matrix to represent the transitions in player ownership. **See Ch. 3 Answer Appendix.**

b. Assume that the trends continue. Predict the number of people who will own digital audio players next year. **about 20,218**

c. Use the inverse of the matrix from part **b** to find the number of people who owned digital audio players last year. **about 4357**

42. Sample answer: Some systems in two variables can be easier to solve by using algebraic methods such as combination or elimination. More complex systems can be easier to solve by using matrices.

H.O.T. Problems Use Higher-Order Thinking Skills

41. Sample answer:
$$\begin{bmatrix} 2 & 3 \\ 4 & 6 \end{bmatrix} \cdot \begin{bmatrix} x \\ y \end{bmatrix} = \begin{bmatrix} 9 \\ 10 \end{bmatrix};$$ **any matrix that has a determinant equal to 0, such as**
$$\begin{bmatrix} 1 & 1 \\ 0 & 0 \end{bmatrix}$$

C 38. CCSS CRITIQUE Cody and Megan are setting up matrix equations for the system $5x + 7y = 19$ and $3y + 4x = 10$. Is either of them correct? Explain your reasoning.

Cody
$\begin{bmatrix} 5 & 7 \\ 3 & 4 \end{bmatrix}\begin{bmatrix} x \\ y \end{bmatrix} = \begin{bmatrix} 19 \\ 10 \end{bmatrix}$

Megan
$\begin{bmatrix} 5 & 7 \\ 4 & 3 \end{bmatrix}\begin{bmatrix} x \\ y \end{bmatrix} = \begin{bmatrix} 19 \\ 10 \end{bmatrix}$

Megan; Cody put 3 for x in the second equation instead of 4.

39. CHALLENGE Describe what a matrix equation with infinite solutions looks like.

40. REASONING Determine whether the following statement is *always*, *sometimes*, or *never* true. Explain your reasoning.

A square matrix has a multiplicative inverse.

Sometimes; sample answer: A square matrix has a multiplicative inverse if its determinant does not equal 0.

41. OPEN ENDED Write a matrix equation that does not have a solution.

42. WRITING IN MATH When would you prefer to solve a system of equations using algebraic methods, and when would you prefer to use matrices? Explain.

(l)Corbis/Punchstock, (r)Andrew Hetherington/Stone/Getty Images

---

CCSS Teaching the Mathematical Practices

Critique Mathematically proficient students are also able to compare the effectiveness of two plausible arguments, distinguish correct logic or reasoning from that which is flawed, and—if there is a flaw in an argument—explain what it is.

Study Guide and Intervention
AL OL ELL

Practice
AL OL BL ELL

Enrichment
OL BL

Word Problem Practice
AL OL BL ELL

Ticket Out the Door Have students describe how matrix equations can be used to solve systems of two equations in two variables.

Formative Assessment
Check for student understanding of concepts in Lessons 3-7 and 3-8.

 Quiz 4, p. 60

Standardized Test Practice

43. The Yogurt Shoppe sells cones in three sizes: small, $0.89; medium, $1.19; and large, $1.39. One day Santos sold 52 cones. He sold seven more medium cones than small cones. If he sold $58.98 in cones, how many medium cones did he sell? **C**

   A 11   **B** 17   **C** 24   **D** 36

44. The chart shows an expression evaluated for different values of x.

A student concludes that for all values of x, $x^2 + x + 1$ produces a prime number. Which value of x serves as a counterexample to prove this conclusion false? **J**

x	$x^2 + x + 1$
1	3
2	7
3	13
4	21

   F -4 **G** -3 **H** 2   **J** 4

45. **SHORT RESPONSE** What is the solution of the system of equations $6a + 8b = 5$ and $10a - 12b = 2$? $\left(\frac{1}{2}, \frac{1}{4}\right)$

46. **SAT/ACT** Each year at Capital High School the students vote to choose the theme of the homecoming dance. The theme "A Night Under the Stars" received 225 votes, and "The Time of My Life" received 480 votes. If 40% of girls voted for "A Night Under the Stars" and 75% of boys voted for "The Time of My Life," how many girls and boys voted? **C**

   A 176 boys and 351 girls

   B 395 boys and 310 girls

   C 380 boys and 325 girls

   D 705 boys and 325 girls

   E 854 boys and 176 girls

Spiral Review

Evaluate each determinant. (Lesson 3-7)

47. $\begin{vmatrix} 8 & -3 \\ 6 & -9 \end{vmatrix}$ -54

48. $\begin{vmatrix} 9 & -7 \\ -5 & -3 \end{vmatrix}$ -62

49. $\begin{vmatrix} 8 & 6 & -1 \\ -4 & 5 & 1 \\ -3 & -2 & 9 \end{vmatrix}$ 551

Find each product, if possible. (Lesson 3-6) **50.** $\begin{bmatrix} 34 & 10 \\ -21 & -5 \end{bmatrix}$ **51.** $\begin{bmatrix} -22 \\ -7 \end{bmatrix}$

50. $\begin{bmatrix} 4 & 2 \\ -1 & -3 \end{bmatrix} \cdot \begin{bmatrix} 6 & 2 \\ 5 & 1 \end{bmatrix}$

51. $\begin{bmatrix} 8 & -2 \\ -4 & -5 \end{bmatrix} \cdot \begin{bmatrix} -2 \\ 3 \end{bmatrix}$

52. $\begin{bmatrix} -3 \\ -4 \end{bmatrix} \cdot \begin{bmatrix} -6 & -8 \\ -4 & 5 \end{bmatrix}$ impossible

53. **MILK** The Yoder Family Dairy produces at most 200 gallons of skim and whole milk each day for delivery to large bakeries and restaurants. Regular customers require at least 15 gallons of skim and 21 gallons of whole milk each day. If the profit on a gallon of skim milk is $0.82 and the profit on a gallon of whole milk is $0.75, how many gallons of each type of milk should the dairy produce each day to maximize profits? (Lesson 3-4) **179 gal of skim and 21 gal of whole milk**

Skills Review

Identify the type of the function represented by each graph.

54.

quadratic

55.

absolute value

56. constant

DifferentiatedInstruction **BL**

Extension Write the following system on the board.

$$-3x + y = 5$$
$$-4x - 2y = 20$$

Have students use four different methods (graphing, by hand, Cramer's Rule, and inverse matrices) to solve the system. Then ask them to compare and contrast the methods using criteria such as ease of use and quickness. The solution to the system is $(-3, -4)$.

EXTEND 3-8 Graphing Technology Lab
Augmented Matrices

Using a TI-83/84 Plus graphing calculator, you can solve a system of linear equations using the MATRIX function. An **augmented matrix** contains the coefficient matrix with an extra column containing the constant terms. You can use a graphing calculator to reduce the augmented matrix so that the solution of the system of equations can be easily determined.

Example

Write an augmented matrix for the following system of equations. Then solve the system by using a graphing calculator.

$2x + y + z = 1$
$3x + 2y + 3z = 12$
$4x + y + 2z = -1$

Step 1 Write the augmented matrix and enter it into a calculator.

The augmented matrix $B = \begin{bmatrix} 2 & 1 & 1 & \vdots & 1 \\ 3 & 2 & 3 & \vdots & 12 \\ 4 & 1 & 2 & \vdots & -1 \end{bmatrix}$.

Begin by entering the matrix.

KEYSTROKES: [2nd] [MATRIX] [▶] [▶] [ENTER] 3 [ENTER] 4 [ENTER] 2 [ENTER] 1
[ENTER] 1 [ENTER] 1 [ENTER] 3 [ENTER] 2 [ENTER] 3 [ENTER] 12
[ENTER] 4 [ENTER] 1 [ENTER] 2 [ENTER] [(−)] 1 [ENTER]

Step 2 Find the reduced row echelon form (rref) using the graphing calculator.

KEYSTROKES: [2nd] [QUIT] [2nd] [MATRIX] [▶] [ALPHA] [B] [2nd] [MATRIX]
[ENTER] [ )] [ENTER]

Study the reduced echelon matrix. The first three columns are the same as a 3 × 3 identity matrix. The first row represents $x = -4$, the second row represents $y = 3$, and the third row represents $z = 6$. The solution is $(-4, 3, 6)$.

```
rref([A])
   [[1 0 0 -4]
    [0 1 0 3 ]
    [0 0 1 6 ]]
```

Exercises

Write an augmented matrix for each system of equations. Then solve with a graphing calculator. **1–6. See Chapter 3 Answer Appendix.**

1. $3x + 2y = -4$
$4x + 7y = 13$

2. $2x + y = 6$
$6x - 2y = 0$

3. $2x + 2y = -4$
$7x + 3y = 10$

4. $4x + 6y = 0$
$8x - 2y = 7$

5. $6x - 4y + 2z = -4$
$2x - 2y + 6z = 10$
$2x + 2y + 2z = -2$

6. $5x - 5y + 5z = 10$
$5x - 5z = 5$
$5y + 10z = 0$

---

1 Focus

Objective Use a graphing calculator and the augmented matrix for a system of equations to solve the system.

Materials
- TI-83/84 Plus or other graphing calculator

Teaching Tip
Point out that the TI-83/84 Plus allows you to perform row operations on matrices. These row operations are items C through F on the **[MATRIX] MATH** menu. Successive use of row operations allows you to transform a matrix to reduced row echelon form. The **rref(** function performs all the steps at once, thereby saving a great deal of time.

2 Teach

Working in Cooperative Groups
Put students in groups of two, mixing abilities. Then have groups use their calculators to complete Steps 1 and 2 of the Example and Exercise 1.

- Point out that if one of the variables is absent from an equation in a system of equations, then its coefficient is zero.

- In order to determine the correct augmented matrix in Exercise 6, students may find it helpful to rewrite the equation, showing 0 as the coefficient of the missing variables in the second and third equations.

Practice Have students complete Exercises 2–6.

3 Assess

Formative Assessment
Use Exercise 6 to assess whether students comprehend how to solve a system of equations using a graphing calculator.

From Concrete to Abstract
Ask students which method they prefer for solving systems of two equations in two variables, the graphing calculator method shown in this investigation, or the method presented in Lesson 3-8. Have them choose their preferred methods for solving systems of equations. Have them explain their choices.

Formative Assessment

Key Vocabulary The page references after each word denote where that term was first introduced. If students have difficulty answering questions 1–10, remind them that they can use these page references to refresh their memories about the vocabulary.

Summative Assessment
☐ Vocabulary Test, p. 62

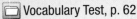 **Vocabulary Review**
Vocabulary Review provides students the opportunity to check their understanding of important concepts and terminology in an online game format.

FOLDABLES Study Organizer

Dinah Zike's Foldables®
Have students look through the chapter to make sure they have included examples in their Foldables. Suggest that students keep their Foldables handy while completing the Study Guide and Review pages. Point out that their Foldables can serve as a quick review tool when studying for the chapter test.

Study Guide

Key Concepts

Systems of Equations and Inequalities (Lessons 3-1 and 3-2)
• In the substitution method, one equation is solved for a variable and substituted to find the value of another variable. In the elimination method, one variable is eliminated by adding or subtracting the equations.
• The solution of a system of inequalities is found by graphing the inequalities and determining the intersection of the graphs.

Linear Programming (Lesson 3-3)
• Linear programming is a method for finding maximum or minimum values of a function over a given system of inequalities with each inequality representing a constraint.

Systems of Equations in Three Variables (Lesson 3-4)
• A system of equations in three variables can be solved algebraically by using the substitution method or the elimination method.

Operations with Matrices (Lessons 3-5 and 3-6)
• Matrices can be added or subtracted if they have the same dimensions. Add or subtract corresponding elements.
• Two matrices can be multiplied if and only if the number of columns in the first matrix is equal to the number of rows in the second matrix.

Solving Systems Using Cramer's Rule (Lesson 3-7)
• If a determinant is nonzero, then the system has a unique solution. If the determinant is 0, then the system either has no solution or infinite solutions.

Solving Systems with Inverse Matrices (Lesson 3-8)
• An identity matrix is a square matrix with ones on the main diagonal and zeros in the other positions.
• Two matrices are inverses of each other if their product is the identity matrix.
• To solve a matrix equation, find the inverse of the coefficient matrix. Then multiply each side of the equation by the inverse of the coefficient matrix.

FOLDABLES Study Organizer

Be sure the Key Concepts are noted in your Foldable.

 206 | Chapter 3 | Study Guide and Review

Key Vocabulary

bounded (p. 154)
break-even point (p. 136)
coefficient matrix (p. 192)
consistent (p. 137)
constant matrix (p. 200)
Cramer's Rule (p. 192)
dependent (p. 137)
determinant (p. 189)
diagonal rule (p. 190)
dimensions (p. 169)
elimination method (p. 139)
feasible region (p. 154)
identity matrix (p. 198)

inconsistent (p. 137)
independent (p. 137)
inverse matrices (p. 198)
matrix (p. 169)
matrix equation (p. 200)
optimize (p. 156)
ordered triple (p. 161)
scalar (p. 173)
scalar multiplication (p. 173)
substitution method (p. 138)
unbounded (p. 154)
variable matrix (p. 200)

Vocabulary Check

Choose the term from above to complete each sentence.

1. A feasible region that is open and can go on forever is called _____. **unbounded**

2. To _____ means to seek the best price or profit using linear programming. **optimize**

3. A matrix that contains the constants in a system of equations is called a(n) _____. **constant matrix**

4. A matrix can be multiplied by a constant called a(n) _____. **scalar**

5. The _____ of a matrix with 4 rows and 3 columns are 4 × 3. **dimensions**

6. A system of equations is _____ if it has at least one solution. **consistent**

7. The _____ matrix is a square matrix that, when multiplied by another matrix, equals that same matrix. **identity**

8. The _____ is the point at which income equals cost. **break-even point**

9. A system of equations is _____ if it has no solutions. **inconsistent**

10. If the product of two matrices is the identity matrix, they are _____. **inverses**

Lesson-by-Lesson Review

3-1 Solving Systems of Equations

Solve each system of equations by graphing.

11. $3x + 4y = 8$ $(0, 2)$
   $x - 3y = -6$

12. $x + \frac{8}{3}y = 12$
   $\frac{1}{2}x + \frac{4}{3}y = 6$

13. $y - 3x = 13$ $(-3, 4)$
   $y = \frac{1}{3}x + 5$

14. $6x - 14y = 5$
   $3x - 7y = 5$

15. LAWN CARE André and Paul each mow lawns. André charges a $30 service fee and $10 per hour. Paul charges a $10 service fee and $15 per hour. After how many hours will André and Paul charge the same amount? **4 hours**

Solve each system of equations by using either substitution or elimination. **18. (5.25, −1.75)**

16. $x + y = 6$ $(2, 4)$
   $3x - 2y = -2$

17. $5x - 2y = 4$ $(-2, -7)$
   $-2y + x = 12$

18. $x + y = 3.5$
   $x - y = 7$

19. $3y - 5x = 0$ $(3, 5)$
   $2y - 4x = -2$

20. SCHOOL SUPPLIES At an office supply store, Emilio bought 3 notebooks and 5 pens for $13.75. If a notebook costs $1.25 more than a pen, how much does a notebook cost? How much does a pen cost? **notebook: $2.50; pen: $1.25**

Example 1

Solve the system of equations by graphing.
$x + y = 4$
$x + 2y = 5$

Graph both equations on the coordinate plane.

The solution of the system is (3, 1).

12. infinitely many solutions

14. no solution

Example 2

Solve the system of equations by using either substitution or elimination.
$3x + 2y = 1$

$y = -x + 1$

Substitute $-x + 1$ for y in the first equation. Then solve for y.

$$3x + 2y = 1$$
$$3x + 2(-x + 1) = 1$$
$$3x - 2x + 2 = 1$$
$$x + 2 = 1$$
$$x = -1$$

$$y = -x + 1$$
$$= -(-1) + 1$$
$$= 2$$

The solution is $(-1, 2)$.

3-2 Solving Systems of Inequalities by Graphing

Solve each system of inequalities by graphing.

21. $y < 2x - 3$
   $y \geq 4$

22. $|y| > 2$
   $x > 3$

23. $y \geq x + 3$
   $2y \leq x - 5$

24. $y > x + 1$
   $x < -2$

25. JEWELRY Payton makes jewelry to sell at her mother's clothing store. She spends no more than 3 hours making jewelry on Saturdays. It takes her 15 minutes to set up her supplies and 25 minutes to make each bracelet. Draw a graph that represents this.
21–25. See margin.

Example 3

Solve the system of inequalities by graphing.
$y \geq \frac{3}{2}x - 3$
$y < 4 - 2x$

The solution of the system is the region that satisfies both inequalities. The solution of this system is the shaded region.

Intervention If the given examples are not sufficient to review the topics covered by the questions, remind students that the lesson references tell them where to review that topic in their textbooks.

Two-Day Option Have students complete the Lesson-by-Lesson Review. Then you can use McGraw-Hill eAssessment to customize another review worksheet that practices all the objectives of this chapter or only the objectives on which your students need more help.

Additional Answers

21.

22.

23.

24.

25.

CHAPTER 3

Study Guide and Review *Continued*

3-3 Optimization with Linear Programming

26. FLOWERS A florist can make a grand arrangement in 18 minutes or a simple arrangement in 10 minutes. The florist makes at least twice as many of the simple arrangements as the grand arrangements. The florist can work only 40 hours per week. The profit on the simple arrangements is $10 and the profit on the grand arrangements is $25. Find the number and type of arrangements that the florist should produce to maximize profit. **126 simple and 63 grand**

27. MANUFACTURING A shoe manufacturer makes outdoor and indoor soccer shoes. There is a two-step process for both kinds of shoes. Each pair of outdoor shoes requires 2 hours in step one and 1 hour in step two, and produces a profit of $20. Each pair of indoor shoes requires 1 hour in step one and 3 hours in step two, and produces a profit of $15. The company has 40 hours of labor available per day for step one and 60 hours available for step two. What is the manufacturer's maximum profit? What is the combination of shoes for this profit?
$480; 12 outdoor, 16 indoor

Example 4

A gardener is planting two types of herbs in a 5184-square-inch garden. Herb A requires 6 square inches of space, and herb B requires 24 square inches of space. The gardener will plant no more than 300 plants. If herb A can be sold for $8 and herb B can be sold for $12, how many of each herb should be sold to maximize income?

Let a = the number of herb A and b = the number of herb B.

$a \geq 0,\ b \geq 0,$
$6a + 24b \leq 5184,$
and $a + b \leq 300$

Graph the inequalities. The vertices of the feasible region are $(0, 0)$, $(300, 0)$, $(0, 216)$, and $(112, 188)$.

The profit function is $f(a, b) = 8a + 12b$.

The maximum value of $3152 occurs at $(112, 188)$. So the gardener should plant 112 of herb A and 188 of herb B.

3-4 Systems of Equations in Three Variables

Solve each system of equations.

28. $a - 4b + c = 3$
$b - 3c = 10$
$3b - 8c = 24$
$(-23, -8, -6)$

29. $2x - z = 14$
$3x - y + 5z = 0$
$4x + 2y + 3z = -2$
$(5, -5, -4)$

30. AMUSEMENT PARKS Dustin, Luis, and Marci went to an amusement park. They purchased snacks from the same vendor. Their snacks and how much they paid are listed in the table. How much did each snack cost?

Name	Hot Dogs	Popcorn	Soda	Price
Dustin	1	2	3	$15.25
Luis	2	0	3	$14.00
Marci	1	2	1	$10.25

hot dog: $3.25; popcorn: $2.25; soda: $2.50

Example 5

Solve the system of equations.

$x + y + 2z = 6$
$2x + 5z = 12$
$x + 2y + 3z = 9$

$2x + 2y + 4z = 12$	Equation 1 × 2
$(-)\ x + 2y + 3z = 9$	Equation 3
$x + z = 3$	Subtract.

Solve the system of two equations.

$2x + 5z = 12$	Equation 2
$(-)\ 2x + 2z = 6$	$2 \times (x + z = 3)$
$3z = 6$	Subtract.
$z = 2$	Divide each side by 3.

Substitute 2 for z in one of the equations with two variables, and solve for y. Then, substitute 2 for z and the value you got for y into an equation from the original system to solve for x.

The solution is $(1, 1, 2)$.

33a. buying price: $\begin{bmatrix} 15 \\ 25 \\ 30 \end{bmatrix}$;

33b. selling price: $\begin{bmatrix} 35 \\ 55 \\ 85 \end{bmatrix}$;

33c. $\begin{bmatrix} 35 \\ 55 \\ 85 \end{bmatrix} - \begin{bmatrix} 15 \\ 25 \\ 30 \end{bmatrix} = \begin{bmatrix} 20 \\ 30 \\ 55 \end{bmatrix}$

3-5 Operations with Matrices

Perform the indicated operations. If the matrix does not exist, write *impossible*.

31. $3\left(\begin{bmatrix} -2 & 0 \\ 6 & 8 \end{bmatrix} + \begin{bmatrix} 1 & 9 \\ -3 & -4 \end{bmatrix} \right)$ $\begin{bmatrix} -3 & 27 \\ 9 & 12 \end{bmatrix}$

32. $\begin{bmatrix} 2 \\ -6 \end{bmatrix} - \begin{bmatrix} -3 \\ 2 \end{bmatrix} + \begin{bmatrix} 6 \\ 0 \end{bmatrix}$ $\begin{bmatrix} 11 \\ -8 \end{bmatrix}$

33. RETAIL Current Fashions buys shirts, jeans, and shoes from a manufacturer, marks them up, and then sells them. The table shows the purchase price and the selling price.

Item	Purchase Price	Selling Price
shirts	$15	$35
jeans	$25	$55
shoes	$30	$85

a. Write a matrix for the purchase price. **a–c. See margin.**

b. Write a matrix for the selling price.

c. Use matrix operations to find the profit on 1 shirt, 1 pair of jeans, and 1 pair of shoes.

Example 6

Find $2A + 3B$ if $A = \begin{bmatrix} 9 & 1 \\ 1 & 2 \end{bmatrix}$ and $B = \begin{bmatrix} 1 & 4 \\ 3 & 7 \end{bmatrix}$.

$2B = 2\begin{bmatrix} 1 & 4 \\ 3 & 7 \end{bmatrix}$ or $\begin{bmatrix} 2 & 8 \\ 6 & 14 \end{bmatrix}$

$3A = 3\begin{bmatrix} 9 & 1 \\ 1 & 2 \end{bmatrix}$ or $\begin{bmatrix} 27 & 3 \\ 3 & 6 \end{bmatrix}$

$2B + 3A = \begin{bmatrix} 2 & 8 \\ 6 & 14 \end{bmatrix} + \begin{bmatrix} 27 & 3 \\ 3 & 6 \end{bmatrix}$ or $\begin{bmatrix} 29 & 11 \\ 9 & 20 \end{bmatrix}$

Example 7

Find $3C - 5D$ if $C = \begin{bmatrix} 3 \\ -7 \end{bmatrix}$ and $D = [9 \quad 8]$.

$3C - 5D = 3\begin{bmatrix} 3 \\ -7 \end{bmatrix} - 5[9 \quad 8]$.

Because the dimensions are different, you cannot subtract the matrices.

3-6 Multiplying Matrices

Find each product, if possible.

34. $[3 \quad -7] \cdot \begin{bmatrix} 9 \\ -5 \end{bmatrix}$ $[62]$

35. $\begin{bmatrix} -3 & 0 & 2 \\ 6 & -1 & 5 \end{bmatrix} \cdot \begin{bmatrix} 8 & -1 \\ -4 & 3 \\ 6 & 7 \end{bmatrix}$ $\begin{bmatrix} -12 & 17 \\ 82 & 26 \end{bmatrix}$

36. $\begin{bmatrix} 2 & 11 \\ 0 & -3 \\ -6 & 7 \end{bmatrix} \cdot \begin{bmatrix} 0 & 8 & -5 \\ 12 & 0 & 9 \\ 4 & -6 & 7 \end{bmatrix}$ undefined

37. GROCERIES Martin bought 1 gallon of milk, 2 apples, 4 frozen dinners, and 1 box of cereal. The following matrix shows the prices for each item, respectively.

$[\$2.59 \quad \$0.49 \quad \$5.25 \quad \$3.99]$

Use matrix multiplication to find the total amount of money Martin spent at the grocery store. **$28.56**

Example 8

Find XY if $X = \begin{bmatrix} 0 & -6 \\ 3 & 5 \end{bmatrix}$ and $Y = \begin{bmatrix} 8 \\ -1 \end{bmatrix}$.

$XY = \begin{bmatrix} 0 & -6 \\ 3 & 5 \end{bmatrix} \cdot \begin{bmatrix} 8 \\ -1 \end{bmatrix}$ Write an equation.

$= \begin{bmatrix} 0(8) + (-6)(-1) \\ 3(8) + 5(-1) \end{bmatrix}$ Multiply columns by rows.

$= \begin{bmatrix} 6 \\ 19 \end{bmatrix}$ Simplify.

Anticipation Guide

Have students complete the Chapter 3 Anticipation Guide and discuss how their responses have changed now that they have completed Chapter 3.

Additional Answers (Practice Test)

6.

7.

8.

9.

12. $\begin{bmatrix} -12a - 8 \\ 12 \\ 5 \end{bmatrix}$

13. $\begin{bmatrix} -6 & -12 \\ -28 & 4 \end{bmatrix}$

14. $\begin{bmatrix} 6 \\ -19 \\ -5 \end{bmatrix}$

15. impossible

3-7 Solving Systems of Equations Using Cramer's Rule

Evaluate each determinant.

38. $\begin{vmatrix} 2 & 4 \\ 7 & -3 \end{vmatrix}$ -34

39. $\begin{vmatrix} 2 & 3 & -1 \\ 0 & 2 & 4 \\ -2 & 5 & 6 \end{vmatrix}$ -44

Use Cramer's Rule to solve each system of equations.

40. $3x - y = 0$ (2, 6)
    $5x + 2y = 22$

41. $5x + 2y = 4$
    $3x + 4y + 2z = 6$
    $7x + 3y + 4z = 29$ (2, −3, 6)

42. **JEWELRY** Alana paid $98.25 for 3 necklaces and 2 pairs of earrings. Petra paid $133.50 for 2 necklaces and 4 pairs of earrings. Use Cramer's Rule to find out how much 1 necklace costs and how much 1 pair of earrings costs. **necklace: $15.75; pair of earrings: $25.50**

Example 9

Evaluate $\begin{vmatrix} 4 & -6 \\ 2 & 5 \end{vmatrix}$.

$\begin{vmatrix} 4 & -6 \\ 2 & 5 \end{vmatrix} = 4(5) - (-6)(2)$ Definition of determinant

$= 20 + 12$ or 32 Simplify.

Example 10

Use Cramer's Rule to solve $2a + 6b = -1$ and $a + 8b = 2$.

$a = \dfrac{\begin{vmatrix} -1 & 6 \\ 2 & 8 \end{vmatrix}}{\begin{vmatrix} 2 & 6 \\ 1 & 8 \end{vmatrix}}$ Cramer's Rule $b = \dfrac{\begin{vmatrix} 2 & -1 \\ 1 & 2 \end{vmatrix}}{\begin{vmatrix} 2 & 6 \\ 1 & 8 \end{vmatrix}}$

$= \dfrac{-8 - 12}{16 - 6}$ Evaluate each determinant. $= \dfrac{4 + 1}{16 - 6}$

$= \dfrac{-20}{10}$ or -2 Simplify. $= \dfrac{5}{10}$ or $\dfrac{1}{2}$

The solution is $\left(-2, \dfrac{1}{2}\right)$.

3-8 Solving Systems of Equations Using Inverse Matrices

Find the inverse of each matrix, if it exists.

43. $\begin{bmatrix} 7 & 4 \\ 3 & 2 \end{bmatrix}$ $\dfrac{1}{2}\begin{bmatrix} 2 & -4 \\ -3 & 7 \end{bmatrix}$

44. $\begin{bmatrix} 2 & 5 \\ -5 & -13 \end{bmatrix}$ $\begin{bmatrix} 13 & 5 \\ -5 & -2 \end{bmatrix}$

45. $\begin{bmatrix} 6 & -3 \\ -8 & 4 \end{bmatrix}$ No inverse exists.

Use a matrix equation to solve each system of equations.

46. $\begin{bmatrix} 5 & 3 \\ 3 & 2 \end{bmatrix} \cdot \begin{bmatrix} x \\ y \end{bmatrix} = \begin{bmatrix} 4 \\ 0 \end{bmatrix}$ (8, −12)

47. $\begin{bmatrix} 3 & -1 \\ 1 & 2 \end{bmatrix} \cdot \begin{bmatrix} a \\ b \end{bmatrix} = \begin{bmatrix} 5 \\ 4 \end{bmatrix}$ (2, 1)

48. **HEALTH FOOD** Heath sells nuts and raisins by the pound. Sonia bought 2 pounds of nuts and 2 pounds of raisins for $23.50. Drew bought 3 pounds of nuts and 1 pound of raisins for $22.25. What is the cost of 1 pound of nuts and 1 pound of raisins? **nuts: $5.25 per pound; raisins: $6.50 per pound**

Example 11

Solve $\begin{bmatrix} 2 & -5 \\ 3 & -6 \end{bmatrix} \cdot \begin{bmatrix} x \\ y \end{bmatrix} = \begin{bmatrix} 15 \\ 36 \end{bmatrix}$.

Step 1 Find the inverse of the coefficient matrix.

$A^{-1} = \dfrac{1}{-12 - (-15)}\begin{bmatrix} -6 & 5 \\ -3 & 2 \end{bmatrix}$ or $\dfrac{1}{3}\begin{bmatrix} -6 & 5 \\ -3 & 2 \end{bmatrix}$

Step 2 Multiply each side by the inverse matrix.

$\dfrac{1}{3}\begin{bmatrix} -6 & 5 \\ -3 & 2 \end{bmatrix} \cdot \begin{bmatrix} 2 & -5 \\ 3 & -6 \end{bmatrix} \cdot \begin{bmatrix} x \\ y \end{bmatrix} = \dfrac{1}{3}\begin{bmatrix} -6 & 5 \\ -3 & 2 \end{bmatrix} \cdot \begin{bmatrix} 15 \\ 36 \end{bmatrix}$

$\begin{bmatrix} 1 & 0 \\ 0 & 1 \end{bmatrix} \cdot \begin{bmatrix} x \\ y \end{bmatrix} = \dfrac{1}{3}\begin{bmatrix} 90 \\ 27 \end{bmatrix}$

$\begin{bmatrix} x \\ y \end{bmatrix} = \begin{bmatrix} 30 \\ 9 \end{bmatrix}$

CHAPTER 3 Practice Test

Solve each system of equations by using either substitution or elimination.

1. $y = x + 4$
$x + y = -12$ $(-8, -4)$

2. $3x + 5y = -7$
$6x - 4y = 0$ $\left(-\frac{2}{3}, -1\right)$

3. $5x + 2y = 4$
$3y - 4x = -40$ $(4, -8)$

4. $8x - 3y = -13$
$-3x + 5y = 1$ $(-2, -1)$

5. MULTIPLE CHOICE Which graph shows the solution of the system of inequalities? **B**

$$y \le 2x + 3$$
$$y < \frac{1}{3}x + 5$$

A

C

B

D

6–9. See margin.

Solve each system of inequalities by graphing.

6. $x + y > 6$
$x - y < 0$

7. $y \ge 2x - 5$
$y \le x + 4$

8. $3x + 4y \le 12$
$6x - 3y \ge 18$

9. $5y + 2x \le 20$
$4x + 3y > 12$

10. SALONS Sierra King is a nail technician. She allots 20 minutes for a manicure and 45 minutes for a pedicure in her 7-hour work day. No more than 5 pedicures can be scheduled each day. The prices are $18 for a manicure and $45 for a pedicure. How many manicures and pedicures should Ms. King schedule to maximize her daily income? What is her maximum daily income? **9 manicures and 5 pedicures; $387**

11. COLLEGE FOOTBALL Darren McFadden of Arkansas placed second overall in the Heisman Trophy voting. Players are given 3 points for every first-place vote, 2 points for every second-place vote, and 1 point for every third-place vote. McFadden received 490 total votes for first, second, and third place, for a total of 878 points. If he had 4 more than twice as many second-place votes as third-place votes, how many votes did he receive for each place? **45 first, 298 second, 147 third**

Perform the indicated operations. If the matrix does not exist, write *impossible*. **12–15. See margin.**

12. $-3\begin{bmatrix} 4a \\ 0 \\ -3 \end{bmatrix} + 4\begin{bmatrix} -2 \\ 3 \\ -1 \end{bmatrix}$

13. $\begin{bmatrix} -3 & 0 \\ 1 & 5 \end{bmatrix} \cdot \begin{bmatrix} 2 & 4 \\ -6 & 0 \end{bmatrix}$

14. $\begin{bmatrix} 2 & 0 \\ -3 & 5 \\ 1 & 4 \end{bmatrix} \cdot \begin{bmatrix} 3 \\ -2 \end{bmatrix}$

15. $\begin{bmatrix} -5 & 7 \\ 6 & 8 \end{bmatrix} - \begin{bmatrix} 4 & 0 & -2 \\ 9 & 0 & 1 \end{bmatrix}$

16. MULTIPLE CHOICE What is the value of

$$\begin{vmatrix} 2 & 3 & -1 \\ 0 & 2 & 4 \\ -2 & 5 & 6 \end{vmatrix}?$$ **F**

F -44

G $-\frac{1}{44}$

H $\frac{1}{44}$

J 44

Find the inverse of each matrix, if it exists.

17. $\begin{bmatrix} 5 & 0 \\ 0 & 1 \end{bmatrix}$ $\begin{bmatrix} \frac{1}{5} & 0 \\ 0 & 1 \end{bmatrix}$

18. $\begin{bmatrix} 1 & 2 \\ 2 & 1 \end{bmatrix}$ $\begin{bmatrix} -\frac{1}{3} & \frac{2}{3} \\ \frac{2}{3} & -\frac{1}{3} \end{bmatrix}$

19. $\begin{bmatrix} 6 & 3 \\ 8 & 4 \end{bmatrix}$ **No inverse exists.**

20. $\begin{bmatrix} -3 & -2 \\ 6 & 4 \end{bmatrix}$ **No inverse exists.**

Use Cramer's Rule to solve each system of equations.

21. $2x - y = -9$ $(-2, 5)$
$x + 2y = 8$

22. $x - y + 2z = 0$ $(4, 2, -1)$
$3x + z = 11$
$-x + 2y = 0$

connectED.mcgraw-hill.com **211** ✓

Summative Assessment
Use these alternate leveled chapter tests to differentiate assessment for your students.

Leveled Chapter 3 Tests

Form	Type	Level	📁 Page(s)
1	MC	AL	63–64
2A	MC	OL	65–66
2B	MC	OL	67–68
2C	FR	OL	69–70
2D	FR	OL	71–72
3	FR	BL	73–74
Vocabulary Test			62
Extended-Response Test			75

MC = multiple-choice questions
FR = free-response questions

eAssessment McGraw-Hill Customize and create multiple versions of your chapter tests and their answer keys. All of the questions from the leveled chapter tests in the *Chapter 3 Resource Masters* are also available on McGraw-Hill eAssessment.

InterventionPlanner

TIER 1 On Level OL

If students miss about 25% of the exercises or less,

Then choose a resource:

SE Lessons 3-1, 3-2, 3-3, 3-4, 3-5, 3-6, 3-7, and 3-8

Skills Practice, pp. 7, 14, 20, 27, 33, 39, 45, and 52

TE Chapter Project

connectED.mcgraw-hill.com Self-Check Quiz

TIER 2 Strategic Intervention AL approaching grade level

If students miss about 50% of the exercises,

Then choose a resource:

Study Guide and Intervention, pp. 5, 12, 18, 25, 31, 37, 43, and 50

connectED.mcgraw-hill.com Extra Examples, Personal Tutor, Homework Help

TIER 3 Intensive Intervention 2 or more grades below level

If students miss about 75% of the exercises,

Then use *Math Triumphs, Alg. 2*

connectED.mcgraw-hill.com Extra Examples, Personal Tutor, Homework Help, Review Vocabulary

1 Focus

Objective Use the strategy of solving short answer questions to solve standardized test problems.

2 Teach

Scaffolding Questions

Ask:

- What is the difference between a short answer question and a multiple choice question? Sample answer: A short answer question does not provide a list of possible answers to choose from. The answer must be written.

- Have you ever been given partial credit for an answer to a short answer question? If so, explain how the partial credit might be determined. Sample answer: giving the correct number but the wrong label, giving an answer but not showing an explanation or the steps in the solution

Short Answer Questions

Short answer questions require you to provide a solution to the problem, along with a method, explanation, and/or justification used to arrive at the solution.

Strategies for Solving Short Answer Questions

Short answer questions are typically graded using a **rubric**, or a scoring guide. The following is an example of a short answer question scoring rubric.

Scoring Rubric	
Criteria	Score
Full Credit: The answer is correct and a full explanation is provided that shows each step.	2
Partial Credit: • The answer is correct but the explanation is incomplete. • The answer is incorrect but the explanation is correct.	1
No Credit: Either an answer is not provided or the answer does not make sense.	0

In solving short answer questions, remember to…

- explain your reasoning or state your approach to solving the problem.
- show all of your work or steps.
- check your answer if time permits.

Standardized Test Example

Read the problem. Identify what you need to know. Then use the information in the problem to solve.

Company A charges a monthly fee of $14.50 plus $0.05 per minute for cell phone service. Company B charges $20.00 per month plus $0.04 per minute. For what number of minutes would the total monthly charge be the same with each company?

Read the problem carefully. You are given information about two different cell phone companies and their monthly charges. Since the situation involves a fixed amount and a variable rate, you can set up and solve a system of equations.

Example of a 2-point response:

Set up and solve a system of equations.

flat fee + rate × minutes = total charges
y = total charges, x = minutes used

$y = 14.5 + 0.05x$ (Company A)

$y = 20 + 0.04x$ (Company B)

Corbis

Solve the system by graphing.

The solution is (550, 42). So, with each company, if the customer uses 550 minutes the total monthly charge is $42.

The steps, calculations, and reasoning are clearly stated. The student also arrives at the correct answer. So, this response is worth the full 2 points.

3. Sample answer: $4.75b + 6.5p \leq 50$, $b \geq 2$, $p \geq 2$; See students' graphs; 4 brushes and 4 pencils, 5 brushes and 4 pencils, 6 brushes and 3 pencils

Exercises

Read each problem. Then use the information in the problem to solve.

1. Shawn and Jerome borrowed $1400 to start a lawn mowing business. They charge their customers $45 per lawn, and with each lawn that they mow, they incur $10.50 in operating expenses. How many lawns must they mow in order to start earning a profit? **41**

2. A circle of radius r is circumscribed about a square. What is the exact ratio of the area of the circle to the area of the square? $\dfrac{\pi}{2}$

3. Mr. Williams can spend no more than $50 on art supplies. Packages of paint brushes cost $4.75 each, and boxes of colored pencils cost $6.50 each. He wants to buy at least 2 packages of each supply. Write a system of inequalities and plot the feasible region on a coordinate grid. Give three different solutions to the system.

4. Marla sells engraved necklaces over the Internet. She purchases 50 necklaces for $400, and it costs her an additional $3 for each personalized engraving. If she charges $20 for each necklace, how many will she need to sell in order to make a profit of at least $225? **39**

5. An auto dealership sold 7378 cars during 2011. This was an 8.5% increase in the number of cars sold during 2010. What was the increase in the number of cars sold in 2011? **578**

6. The sides of two similar triangles are in a ratio of 3:5. If the area of the larger triangle is 600 square centimeters, what is the area of the smaller triangle? **216 cm^2**

7. Raul had $35 in a savings account and started adding $25 a week. At the same time, his sister Tina had $365 in her account and began spending $30 a week. After how many weeks will Raul and Tina have the same amount in their savings accounts? **6 weeks**

8. A city planner wishes to build a sidewalk diagonally across a rectangular-shaped park. The park measures 140 feet by 225 feet. It will cost $30 per foot to construct the sidewalk. What will be the total cost of the sidewalk? **$7950**

connectED.mcgraw-hill.com **213**

Additional Example

Samuel borrowed $35 to purchase a used graphing calculator. He charges $9 an hour to tutor others on how to use a graphing calculator. He tutors only one client for one hour per day at the community library. He spends $2 for parking at the library. How many hours must he tutor to make a profit?

Set up and solve a system of equations.

Let $x =$ the number of hours tutoring, and let $y =$ number of dollars.

Total Income   Total Cost
$9x = y$ $2x + 35 = y$

Solve the system by graphing:

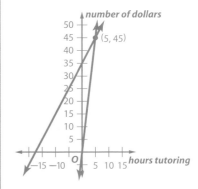

The solution is (5, 45). This is the break-even point. If Samuel tutors more than 5 hours, he will make a profit.

3 Assess

Use Exercises 1–8 to assess students' understanding.

Diagnose Student Errors

Survey student responses for each item. Class trends may indicate common errors and misconceptions.

1. A misinterpreted the key word for the operation
   B correct
   C misinterpreted the key word for the operation
   D reversed the two matrices

2. F added the elements instead of multiplying
   G correct
   H guess
   J did not understand when a matrix product is defined

3. A correct
   B did not use the Distributive Property correctly
   C did not use the Distributive Property correctly
   D guess

4. F incorrectly found the sum of the products of the diagonals
   G incorrectly found the sum of the products of the diagonals
   H incorrectly found the sum of the products of the diagonals
   J correct

5. A found the number of tomatoes sold
   B used an incorrect matrix equation
   C used an incorrect matrix equation
   D correct

6. F correct
   G reversed the rows with the columns
   H counted the elements instead of the columns
   J guess

7. A did not reverse the inequality sign when multiplying or dividing by a negative
   B guess
   C subtracted 6–7 before finding the value of absolute value expression
   D correct

Standardized Test Practice
Cumulative, Chapters 1 through 3

Multiple Choice

Read each question. Then fill in the correct answer on the answer document provided by your teacher or on a sheet of paper.

1. Matrix L shows the average low temperature, in degrees Fahrenheit, each month where Terrance lives. Matrix H shows the monthly average high temperature.

$$L = \begin{bmatrix} 24.1 & 27.7 & 35.9 \\ 44.1 & 53.6 & 62.2 \\ 66.4 & 64.9 & 57.9 \\ 46.4 & 37.3 & 28.4 \end{bmatrix}$$

$$H = \begin{bmatrix} 39.9 & 45.2 & 55.3 \\ 65.1 & 74.0 & 82.3 \\ 85.9 & 84.6 & 78.1 \\ 66.9 & 54.5 & 44.3 \end{bmatrix}$$

Which operation would you use to find the difference between the average high temperature and the average low temperature each month? **B**

A $L + H$ C $H \times L$

B $H - L$ D $L - H$

2. Find $[3 \quad 1] \cdot \begin{bmatrix} 2 \\ 5 \end{bmatrix}$, if possible. **G**

F $[-3]$ H $\begin{bmatrix} 8 & -4 \\ 12 & 6 \end{bmatrix}$

G $[11]$ J undefined

3. Which equation is equivalent to $4x - 3(2x + 7) = 5x$? **A**

A $-2x - 21 = 5x$ C $-2x + 21 = 5x$

B $-2x + 7 = 5x$ D $6x - 7 = 5x$

> **Test-Taking Tip**
> Question 2 The product of a 1-by-2 matrix and a 2-by-1 matrix is a 1-by-1 matrix. So, answer choices H and J can be eliminated.

4. Triangle DEF has vertices $D(-6, 2)$, $E(3, 5)$, and $F(8, -7)$. Evaluate the determinant below to find the area of the triangle. **J**

$$A = \frac{1}{2} \begin{vmatrix} -6 & 2 & 1 \\ 3 & 5 & 1 \\ 8 & -7 & 1 \end{vmatrix}$$

F 54.5 square units

G 58 square units

H 60 square units

J 61.5 square units

5. Suppose Kendall sells apples and tomatoes at a farmer's market. If he sold 280 items one morning and earned \$65.20, how many apples did he sell? **D**

Item	Cost
apple	\$0.25
tomato	\$0.20

A 96   C 168

B 126   D 184

6. What are the dimensions of $D = \begin{bmatrix} 4 & -6 \\ 9 & 2 \\ 1 & 0 \\ -3 & -5 \end{bmatrix}$? **F**

F 4×2

G 2×4

H 4×8

J 8×4

7. What is the solution set of $6 - |x + 7| \leq -2$? **D**

A $\{x \mid -15 \leq x \leq 1\}$

B $\{x \mid -1 \leq x \leq 3\}$

C $\{x \mid x \leq -1 \text{ or } x \geq 3\}$

D $\{x \mid x \leq -15 \text{ or } x \geq 1\}$

Short Response/Gridded Response

Record your answers on the answer sheet provided by your teacher or on a sheet of paper.

8. Does matrix B have an inverse? Explain why or why not.

Sample answer: No. The determinant of the matrix is 0, so it does not have an inverse.

$$B = \begin{bmatrix} 3 & -2 \\ -9 & 6 \end{bmatrix}$$

9. GRIDDED RESPONSE Evaluate the determinant of

$$W = \begin{bmatrix} 3 & 1 & 0 \\ 2 & 5 & -4 \\ 0 & -1 & 1 \end{bmatrix}.$$ 1

10. Valeria has 14 quarters and dimes. The total value of all the coins is $2.75. Use this information to answer each question.

  a. Let d represent the number of dimes that Valeria has, and let q represent the number of quarters. Write a system of equations to model the situation. $d + q = 14; 0.1d + 0.25q = 2.75$

  b. Write a matrix equation that can be used to solve for d and q. **See margin.**

  c. Solve your matrix equation using inverses. How many dimes and quarters does Valeria have? **5 dimes, 9 quarters**

11. GRIDDED RESPONSE Andrea is using a coordinate grid to design a new deck for her backyard. The deck is represented by the intersection of $y \le 20$, $x \le 16$, $y \ge 0$, $x \ge 0$ and $y \le -x + 32$. If each unit of the coordinate grid represents 1 foot, what is the area of the deck? Express your answer in square feet. **312**

12. What are the coordinates of the x- and y-intercepts of the graph of $2y = 4x + 3$?

$\left(-\dfrac{3}{4}, 0\right); \left(0, \dfrac{3}{2}\right)$

Extended Response

Record your answers on a sheet of paper. Show your work.

13. Suppose Tonya is baking cookies and muffins for a bake sale. Each tray of cookies uses 5 cups of flour and 2 cups of sugar. Each tray of muffins uses 5 cups of flour and 1 cup of sugar. She has 40 cups of flour and 15 cups of sugar available for baking. Tonya will make $12 profit for each tray of cookies that is sold and $8 profit for each tray of muffins sold.

  a. Let x represent the number of trays of cookies baked, and let y represent the number of trays of muffins baked. Write a system of inequalities to model the different number of trays Tonya can bake. $x \ge 0, y \ge 0, 5x + 5y \le 40, 2x + y \le 15$

  b. Graph the system of inequalities to show the feasible region. List the coordinates of the vertices of the feasible region. **See students' graphs; (0, 0), (0, 8), (7.5, 0), (7, 1).**

  c. Write a profit function for selling x trays of cookies and y trays of muffins. $P = 12x + 8y$

  d. How any trays of cookies and muffins should Tonya bake to maximize the profit? What will the total profit be? **7 trays of cookies and 1 tray of muffins; $92**

If you missed Question...	1	2	3	4	5	6	7	8	9	10	11	12	13
Go to Lesson...	3-5	3-6	1-2	3-7	3-1	3-5A	2-8	3-8	3-7	3-8	3-2	2-2	3-3

Need ExtraHelp?

Formative Assessment

You can use these pages to benchmark student progress.

📁 Standardized Test Practice, pp. 76–78

McGraw-Hill Assessment Create practice tests that align to your state standards, the Common Core State Standards, and other national standards such as TIMSS and NAEP.

Answer Sheet Practice

Have students simulate taking a standardized test by recording their answers on practice recording sheets.

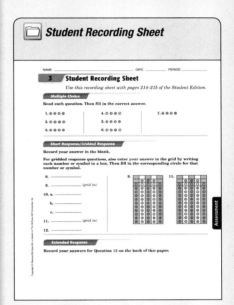

📁 **Student Recording Sheet**

Homework Option

Get Ready for Chapter 4 Assign students the exercises on p. 217 as homework to assess whether they possess the prerequisite skills needed for the next chapter.

Additional Answer

10b. 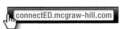 $\begin{bmatrix} 1 & 1 \\ 0.1 & 0.25 \end{bmatrix} \begin{bmatrix} d \\ q \end{bmatrix} = \begin{bmatrix} 14 \\ 2.75 \end{bmatrix}$

Get Ready for Chapter 3

1.

2.

3.

4.

5.

6.

7b.

8.

9.

10.

11.

12.

13.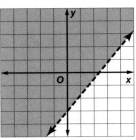

14. $15t + 25w \geq 2500$

Lesson 3-1 (Guided Practice)

3A.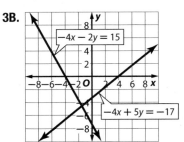

inconsistent

3B.

consistent, independent

37.

consistent and independent

38.

y = 2x + 6
y = 2x − 1

inconsistent

39.

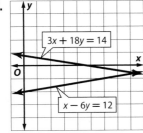

2x + 5y = 10
−4x − 10y = 20

inconsistent

40.

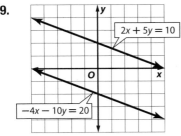

3x + 18y = 14
x − 6y = 12

consistent and independent

41.

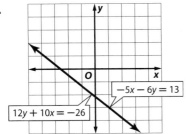

−5x − 6y = 13
12y + 10x = −26

consistent and dependent

42.

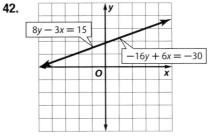

8y − 3x = 15
−16y + 6x = −30

consistent and dependent

79. Sample answer:

$$4x + 5y = 21 \quad \rightarrow \quad 3(4x + 5y = 21) \quad \rightarrow$$
$$3x − 2y = 10 \qquad\qquad 4(3x − 2y = 10)$$

$$12x + 15y = 63$$
$$(−)\ 12x − 8y = 40$$
$$\overline{23y = 23}$$
$$y = 1$$

$$4x + 5(1) = 21$$
$$4x + 5 = 21$$
$$4x = 16$$
$$x = 4 \qquad \text{The solution is (4, 1).}$$

85b.

10s + 15ℓ = 350

Lesson 3-2 (Guided Practice)

1A.

y = −2x + 5
y = −¼x − 6

1B.

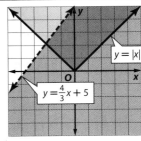

y = |x|
y = 4/3 x + 5

2A.

y = −4x + 4
y = −4x + 8

2B.

Lesson 3-2

5.

6.

7.

8.

9.

10.

11.

12.

13.

14.

15.

16.

17.

18.

19.

20.

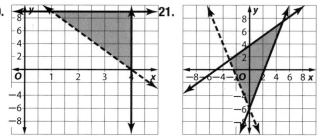

21.

215C │ Chapter 3 │ Answer Appendix
Chapter 3 Answer Appendix

22.

23.

28.

29.

30.

31.

32.

33.

34.

35.

36.

no solution

37.

38.

39. Let w = the number of hours writing, and let e = the number of hours exercising.

$w + e \le 35$

$7 \le e \le 15$

$20 \le w \le 25$

40. $(0, 2), \left(5\frac{1}{3}, -1\frac{1}{3}\right), \left(4\frac{4}{17}, 3\frac{1}{17}\right) (2.8, -6.4)$

41. $(-6, -2), \left(-3\frac{13}{17}, 6\frac{16}{17}\right), \left(9\frac{1}{7}, 3\frac{5}{7}\right), (0.8, -8.8)$

42. $(-4, 6), (-3, 8), (4.8, -7.6), \left(1\frac{1}{7}, -9\frac{3}{7}\right)$

44a. $10 \leq g + b \leq 15$; $g > b$; $g \geq 0$; $b \geq 0$;

50. Sample answer: Determine whether the point falls in the shaded area of the graphs and/or determine whether the values satisfy each inequality.

56.

$x + y = 6$

57.

$4x - 3y = 10$

58.

$5x + 7y = -20$

59. D = {all real numbers},
R = {$g(x) \mid g(x) \leq 2$}

60. D = {all real numbers},
R = {all real numbers}

61. D = {$x \mid x < -2$ or $x > 2$}, R = {-1, 1}

Extend 3-2

1.

$[-10, 10]$ scl: 1 by $[-10, 10]$ scl: 1

2.

$[-10, 10]$ scl: 1 by $[-10, 10]$ scl: 1

3.

$[-10, 10]$ scl: 1 by $[-10, 10]$ scl: 1

4.

$[-10, 10]$ scl: 1 by $[-10, 10]$ scl: 1

5.

$[-10, 10]$ scl: 1 by $[-10, 10]$ scl: 1

6.

$[-10, 10]$ scl: 1 by $[-5, 15]$ scl: 1

7.

$[-10, 10]$ scl: 1 by $[-10, 10]$ scl: 1

8.

$[-10, 10]$ scl: 1 by $[-10, 10]$ scl: 1

9.

$[-10, 10]$ scl: 1 by $[-10, 10]$ scl: 1

Lesson 3-3 (Guided Practice)

1A.
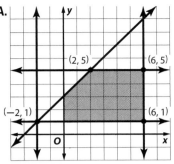

min at $(6, 1) = -28$;
max at $(0, 3) = 12$

1B.

min at $(-4, -6) = -48$;
max at $(8, -6) = 24$

2A.

2B.
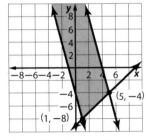

max at $(-4, 8) = 88$; no min

min at $(1, -8) = -46$; no max

Lesson 3-3

7b.

8.

$(8, 4)$, $(6, 1)$, $(2, 1)$, $(-4, 4)$;
max = 36, min = −36

9.

$(2, -10)$, $(-3, 0)$, $(-3, 6.5)$,
$(2, 9)$; max = 82, min = −89

10.
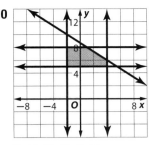

$(1, 8)$, $(4, 6)$, $(4, 5)$,
$(-2, 5)$, $(-2, 8)$;
max = −18, min = −96

11.

$(6, -8)$, $(4, -2)$,
$(-2, -2)$, $(-8, -8)$;
max = −8, min = −152

12.
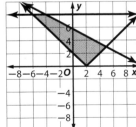

$(2, 0)$, $(6, -1)$, $(6, 3)$, $(2, 5)$;
max = 57, min = 12

13.
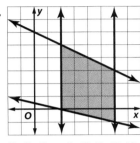

$(-10, 3)$, $(2, 3)$, $(-6, 7)$,
$(-8, 7)$; max = 59, min = 9

14.

15.

16.

17.

18.

19.

20.

21.

22.

25b.

44.

45.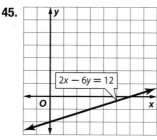

$5x + 3y = 15$

$2x - 6y = 12$

46.

47.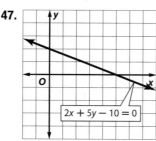

$3x - 4y - 10 = 0$

$2x + 5y - 10 = 0$

48.

49.

$y = x$

$y = 4x - 2$

Lesson 3-4

26. Sample answer:

$$a + b = (rx + ty + vz) + (rx - ty + vz)$$

Replace a with $rx + ty + vz$, and b with $rx - ty + vz$.

$a + b = 2rx + 2vz$	Simplify.
$a + (-a) = 2rx + 2vz$	Replace b with $-a$.
$0 = 2rx + 2vz$	Simplify.
$0 = rx + vz$	Divide each side by 2.
$rx + ty + vz = a$	Given
$ty + (rx + vz) = a$	Commutative and Associative Properties of Addition
$ty + 0 = a$	Substitution
$ty = a$	Simplify.

29. Sample answer: First, combine two of the original equations using elimination to form a new equation with three variables. Next, combine a different pair of the original equations using elimination to eliminate the same variable and form a second equation with three variables. Do the same thing with a third pair of the original equations. You now have a system of three equations with three variables. Follow the same procedure you learned in this section. Once you find the three variables, you need to use them to find the eliminated variable.

Mid-Chapter Quiz

5.

6.

7.

8.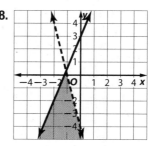

15a. $c \geq 0$; $t \geq 0$; $3c + 2t \leq 108$; $0.5c + t \leq 20$

15b.

15c. 34 chairs and 3 tables will give a maximum profit of $955

17. vertices: $(-4, -3)$, $(0, 5)$, $(2, -3)$; max: $f(2, -3) = 17$;
min: $f(0, 5) = -15$

18. vertices: $(-10, 1)$, $(-4, 1)$, $(2.4, -3.8)$, $(-2, -6)$, $(-10, -6)$;
max: $f(2.4, -3.8) = 1$; min: $f(-10, -6) = -26$

Lesson 3-5 (Guided Practice)

1A.

$$\begin{array}{c} \\ 0\text{–}19 \\ 20\text{–}39 \\ 40\text{–}59 \\ 60+ \end{array} \begin{matrix} \text{Male} & \text{Female} \\ \begin{bmatrix} 7.1 & 6.6 \\ 6.8 & 5.9 \\ 3.2 & 2.2 \\ 1.1 & 1.4 \end{bmatrix} \end{matrix}$$

1B. Column 1: 18.2 million Hispanic males in the U.S.; Column 2: 16.1 million Hispanic females in the U.S.

1C. Row 1: 13.7 million Hispanics aged 0–19 in the U.S.; Row 2: 12.7 million Hispanics aged 20–39 in the U.S.; Row 3: 5.4 million Hispanics aged 40–59 in the U.S.; Row 4: 2.5 million Hispanics aged 60+ in the U.S.

1D. Sample answer: The average of the columns or the rows would not be meaningful.

Lesson 3-5

30a. $\begin{bmatrix} 4 & -1 & -3 \\ 2 & -1 & 1 \end{bmatrix}$

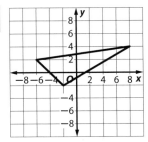

30b. $\begin{bmatrix} 8 & -2 & -6 \\ 4 & -2 & 2 \end{bmatrix}$

31. To show that the Commutative Property of Matrix Addition is true for 2×2 matrices, let $A = \begin{bmatrix} a & b \\ c & d \end{bmatrix}$ and $B = \begin{bmatrix} e & f \\ g & h \end{bmatrix}$.
Show that $A + B = B + A$.

$A + B = \begin{bmatrix} a & b \\ c & d \end{bmatrix} + \begin{bmatrix} e & f \\ g & h \end{bmatrix}$ Substitution

$= \begin{bmatrix} a+e & b+f \\ c+g & d+h \end{bmatrix}$ Definition of matrix addition

$= \begin{bmatrix} e+a & f+b \\ g+c & h+d \end{bmatrix}$ Commutative Property of Addition for Real Numbers

$= \begin{bmatrix} e & f \\ g & h \end{bmatrix} + \begin{bmatrix} a & b \\ c & d \end{bmatrix}$ Definition of matrix addition

$= B + A$ Substitution

32. $(A + B) + C = \left(\begin{bmatrix} a & b \\ c & d \end{bmatrix} + \begin{bmatrix} e & f \\ g & h \end{bmatrix} \right) + \begin{bmatrix} j & k \\ m & n \end{bmatrix}$ Substitution

$= \begin{bmatrix} a+e & b+f \\ c+g & d+h \end{bmatrix} + \begin{bmatrix} j & k \\ m & n \end{bmatrix}$ Definition of matrix addition

$= \begin{bmatrix} a+e+j & b+f+k \\ c+g+m & d+h+n \end{bmatrix}$ Definition of matrix addition

$= \begin{bmatrix} a+(e+j) & b+(f+k) \\ c+(g+m) & d+(h+n) \end{bmatrix}$ Associative Property of Addition

$= \begin{bmatrix} a & b \\ c & d \end{bmatrix} + \begin{bmatrix} e+j & f+k \\ g+m & h+n \end{bmatrix}$ Definition of matrix addition

$= \begin{bmatrix} a & b \\ c & d \end{bmatrix} + \left(\begin{bmatrix} e & f \\ g & h \end{bmatrix} + \begin{bmatrix} j & k \\ m & n \end{bmatrix} \right)$ Definition of matrix addition

$= A + (B + C)$ Substitution

34a. Always; if $A + B$ exists, A and B have the same dimensions. If A and B have the same dimensions, then $A - B$ exists.

34b. Always; if k is a real number, then by the definition of scalar multiplication, $kA = \begin{bmatrix} -3k & -4k \\ 8k & 6k \end{bmatrix}$ and $kB = \begin{bmatrix} 5k & -k \\ 2k & -4k \end{bmatrix}$.

34c. Always; if $A - B$ does not exist, then A and B must have different dimensions. If A and B have different dimensions, then $A + B$ does not exist.

34d. Sometimes; matrices must have the same dimensions for their sum to exist.

34e. Sometimes; matrices must have the same dimensions for their sum to exist.

35. Sample Answer: $A = \begin{bmatrix} 6 & 1 \\ 6 & 3 \end{bmatrix}$, $B = \begin{bmatrix} 3 & 2 \\ 4 & 2 \end{bmatrix}$

36. Sample answer: First, multiply every element in D by 4. Then, multiply every element in C by 3. Finally, subtract the elements in $3C$ from the corresponding elements in $4D$. The result is a matrix equivalent to $4D - 3C$.

45.

$x - 2y = -4$

$y = -2x - 3$

46.

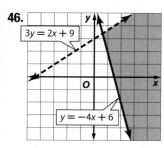

$3y = 2x + 9$

$y = -4x + 6$

47.

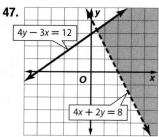

$4y - 3x = 12$

$4x + 2y = 8$

39. $\begin{bmatrix} -1.5x - 1.5y + 15 \\ y^2 + xy - 3x - 6 \\ 1.2x + 1.2y - 39 \end{bmatrix}$

40. $\begin{bmatrix} 36.75x + 50.25y + 30 \\ -6x^2 - 18x - 10.5xy - 13.5y^2 - 12y - 12 \\ -83.4x - 94.2y - 78 \end{bmatrix}$

43. $\begin{bmatrix} 15x + 69y - 12 \\ -18y^2 + 42.75y - 18xy + 20.25x \end{bmatrix}$

44a. $\begin{bmatrix} \$170.99 & \$224.99 & \$279.99 \\ \$179.99 & \$260.99 & \$319.99 \\ \$269.99 & \$359.99 & \$399.99 \end{bmatrix}$

44b. $\begin{bmatrix} \$181.68 & \$239.05 & \$297.49 \\ \$191.24 & \$277.30 & \$339.99 \\ \$286.86 & \$382.49 & \$424.99 \end{bmatrix}$

47a.

$c(A + B) = c\left(\begin{bmatrix} a & b \\ d & e \end{bmatrix} + \begin{bmatrix} w & x \\ y & z \end{bmatrix}\right)$ Substitution

$= c\begin{bmatrix} a + w & b + x \\ d + y & e + z \end{bmatrix}$ Definition of matrix addition

$= \begin{bmatrix} ca + cw & cb + cx \\ cd + cy & ce + cz \end{bmatrix}$ Definition of scalar multiplication

$= \begin{bmatrix} ca & cb \\ cd & ce \end{bmatrix} + \begin{bmatrix} cw & cx \\ cy & cz \end{bmatrix}$ Definition of matrix addition

$= cA + cB$ Substitution

47b.

$C(A + B) = \begin{bmatrix} a & b \\ c & d \end{bmatrix}\left(\begin{bmatrix} e & f \\ g & h \end{bmatrix} + \begin{bmatrix} j & k \\ m & n \end{bmatrix}\right)$ Substitution

$= \begin{bmatrix} a & b \\ c & d \end{bmatrix}\begin{bmatrix} e + j & f + k \\ g + m & h + n \end{bmatrix}$ Definition of matrix addition

$= \begin{bmatrix} a(e + j) + b(g + m) & a(f + k) + b(h + n) \\ c(e + j) + d(g + m) & c(f + k) + d(h + n) \end{bmatrix}$ Definition of matrix multiplication

$= \begin{bmatrix} ea + ja + gb + mb & fa + ka + hb + nb \\ ec + jc + gd + md & fc + kc + hd + nd \end{bmatrix}$ Distributive Property

$= \begin{bmatrix} ea + gb + ja + mb & fa + hb + ka + nb \\ ec + gd + jc + md & fc + hd + kc + nd \end{bmatrix}$ Commutative Property of Addition

$= \begin{bmatrix} ea + gb & fa + hb \\ ec + gd & fc + hd \end{bmatrix} + \begin{bmatrix} ja + mb & ka + nb \\ jc + md & kc + nd \end{bmatrix}$ Definition of matrix addition

$= CA + CB$ Definition of matrix multiplication

$$(A + B)C = \left(\begin{bmatrix} a_{11} & a_{12} \\ a_{21} & a_{22} \end{bmatrix} + \begin{bmatrix} b_{11} & b_{12} \\ b_{21} & b_{22} \end{bmatrix}\right)\begin{bmatrix} c_{11} & c_{12} \\ c_{21} & c_{22} \end{bmatrix}$$ Substitution

$$= \begin{bmatrix} a_{11} + b_{11} & a_{12} + b_{12} \\ a_{21} + b_{21} & a_{22} + b_{22} \end{bmatrix}\begin{bmatrix} c_{11} & c_{12} \\ c_{21} & c_{22} \end{bmatrix}$$ Definition of matrix addition

$$= \begin{bmatrix} (a_{11} + b_{11})c_{11} + (a_{12} + b_{12})c_{21} & (a_{11} + b_{11})c_{12} + (a_{12} + b_{12})c_{22} \\ (a_{21} + b_{21})c_{11} + (a_{22} + b_{22})c_{21} & (a_{21} + b_{21})c_{12} + (a_{22} + b_{22})c_{22} \end{bmatrix}$$ Definition of matrix multiplication

$$= \begin{bmatrix} a_{11}c_{11} + b_{11}c_{11} + a_{12}c_{21} + b_{12}c_{21} & a_{11}c_{12} + b_{11}c_{12} + a_{12}c_{22} + b_{12}c_{22} \\ a_{21}c_{11} + b_{21}c_{11} + a_{22}c_{21} + b_{22}c_{21} & a_{21}c_{12} + b_{21}c_{12} + a_{22}c_{22} + b_{22}c_{22} \end{bmatrix}$$ Distributive Property

$$= \begin{bmatrix} a_{11}c_{11} + a_{12}c_{21} + b_{11}c_{11} + b_{12}c_{21} & a_{11}c_{12} + a_{12}c_{22} + b_{11}c_{12} + b_{12}c_{22} \\ a_{21}c_{11} + a_{22}c_{21} + b_{21}c_{11} + b_{22}c_{21} & a_{21}c_{12} + a_{22}c_{22} + b_{21}c_{12} + b_{22}c_{22} \end{bmatrix}$$ Commutative Property of Addition

$$= \begin{bmatrix} a_{11}c_{11} + a_{12}c_{21} & a_{11}c_{12} + a_{12}c_{22} \\ a_{21}c_{11} + a_{22}c_{21} & a_{21}c_{12} + a_{22}c_{22} \end{bmatrix} + \begin{bmatrix} b_{11}c_{11} + b_{12}c_{21} & b_{11}c_{12} + b_{12}c_{22} \\ b_{21}c_{11} + b_{22}c_{21} & b_{21}c_{12} + b_{22}c_{22} \end{bmatrix}$$ Definition of matrix addition

$$= AC + BC$$ Definition of matrix multiplication

47c. $$(AB)C = \left(\begin{bmatrix} a_{11} & a_{12} \\ a_{21} & a_{22} \end{bmatrix}\begin{bmatrix} b_{11} & b_{12} \\ b_{21} & b_{22} \end{bmatrix}\right)\begin{bmatrix} c_{11} & c_{12} \\ c_{21} & c_{22} \end{bmatrix}$$ Substitution

$$= \begin{bmatrix} a_{11}b_{11} + a_{12}b_{21} & a_{11}b_{12} + a_{12}b_{22} \\ a_{21}b_{11} + a_{22}b_{21} & a_{21}b_{12} + a_{22}b_{22} \end{bmatrix}\begin{bmatrix} c_{11} & c_{12} \\ c_{21} & c_{22} \end{bmatrix}$$ Definition of matrix multiplication

$$= \begin{bmatrix} (a_{11}b_{11} + a_{12}b_{21})c_{11} + (a_{11}b_{12} + a_{12}b_{22})c_{21} & (a_{11}b_{11} + a_{12}b_{21})c_{12} + (a_{11}b_{12} + a_{12}b_{22})c_{22} \\ (a_{21}b_{11} + a_{22}b_{21})c_{11} + (a_{21}b_{12} + a_{22}b_{22})c_{21} & (a_{21}b_{11} + a_{22}b_{21})c_{12} + (a_{21}b_{12} + a_{22}b_{22})c_{22} \end{bmatrix}$$ Definition of matrix multiplication

$$= \begin{bmatrix} a_{11}b_{11}c_{11} + a_{12}b_{21}c_{11} + a_{11}b_{12}c_{21} + a_{12}b_{22}c_{21} & a_{11}b_{11}c_{12} + a_{12}b_{21}c_{12} + a_{11}b_{12}c_{22} + a_{12}b_{22}c_{22} \\ a_{21}b_{11}c_{11} + a_{22}b_{21}c_{11} + a_{21}b_{12}c_{21} + a_{22}b_{22}c_{21} & a_{21}b_{11}c_{12} + a_{22}b_{21}c_{12} + a_{21}b_{12}c_{22} + a_{22}b_{22}c_{22} \end{bmatrix}$$ Distributive Property

$$= \begin{bmatrix} a_{11}b_{11}c_{11} + a_{11}b_{12}c_{21} + a_{12}b_{21}c_{11} + a_{12}b_{22}c_{21} & a_{11}b_{11}c_{12} + a_{11}b_{12}c_{22} + a_{12}b_{21}c_{12} + a_{12}b_{22}c_{22} \\ a_{21}b_{11}c_{11} + a_{21}b_{12}c_{21} + a_{22}b_{21}c_{11} + a_{22}b_{22}c_{21} & a_{21}b_{11}c_{12} + a_{21}b_{12}c_{22} + a_{22}b_{21}c_{12} + a_{22}b_{22}c_{22} \end{bmatrix}$$ Commutative Property of Addition

$$= \begin{bmatrix} a_{11}(b_{11}c_{11} + b_{12}c_{21}) + a_{12}(b_{21}c_{11} + b_{22}c_{21}) & a_{11}(b_{11}c_{12} + b_{12}c_{22}) + a_{12}(b_{21}c_{12} + b_{22}c_{22}) \\ a_{21}(b_{11}c_{11} + b_{12}c_{21}) + a_{22}(b_{21}c_{11} + b_{22}c_{21}) & a_{21}(b_{11}c_{12} + b_{12}c_{22}) + a_{22}(b_{21}c_{12} + b_{22}c_{22}) \end{bmatrix}$$ Distributive Property

$$= \begin{bmatrix} a_{11} & a_{12} \\ a_{21} & a_{22} \end{bmatrix}\begin{bmatrix} b_{11}c_{11} + b_{12}c_{21} & b_{11}c_{12} + b_{12}c_{22} \\ b_{21}c_{11} + b_{22}c_{21} & b_{21}c_{12} + b_{22}c_{22} \end{bmatrix}$$ Definition of matrix multiplication

$$= \begin{bmatrix} a_{11} & a_{12} \\ a_{21} & a_{22} \end{bmatrix}\left(\begin{bmatrix} b_{11} & b_{12} \\ b_{21} & b_{22} \end{bmatrix}\begin{bmatrix} c_{11} & c_{12} \\ c_{21} & c_{22} \end{bmatrix}\right)$$ Definition of matrix multiplication

$$= A(BC)$$ Substitution

47d. $$c(AB) = c\left(\begin{bmatrix} a_{11} & a_{12} \\ a_{21} & a_{22} \end{bmatrix}\begin{bmatrix} b_{11} & b_{12} \\ b_{21} & b_{22} \end{bmatrix}\right)$$ Substitution

$$= c\begin{bmatrix} a_{11}b_{11} + a_{12}b_{21} & a_{11}b_{12} + a_{12}b_{22} \\ a_{21}b_{11} + a_{22}b_{21} & a_{21}b_{12} + a_{22}b_{22} \end{bmatrix}$$ Definition of matrix multiplication

$$= \begin{bmatrix} c(a_{11}b_{11} + a_{12}b_{21}) & c(a_{11}b_{12} + a_{12}b_{22}) \\ c(a_{21}b_{11} + a_{22}b_{21}) & c(a_{21}b_{12} + a_{22}b_{22}) \end{bmatrix}$$ Definition of scalar multiplication

$$= \begin{bmatrix} ca_{11}b_{11} + ca_{12}b_{21} & ca_{11}b_{12} + ca_{12}b_{22} \\ ca_{21}b_{11} + ca_{22}b_{21} & ca_{21}b_{12} + ca_{22}b_{22} \end{bmatrix}$$ Distributive Property

$$= \begin{bmatrix} ca_{11} & ca_{12} \\ ca_{21} & ca_{22} \end{bmatrix}\begin{bmatrix} b_{11} & b_{12} \\ b_{21} & b_{22} \end{bmatrix}$$ Definition of matrix multiplication

$$= c\begin{bmatrix} a_{11} & a_{12} \\ a_{21} & a_{22} \end{bmatrix}\begin{bmatrix} b_{11} & b_{12} \\ b_{21} & b_{22} \end{bmatrix}$$ Definition of scalar multiplication

$$= (cA)B$$ Substitution

$$c(AB) = c\left(\begin{bmatrix} a_{11} & a_{12} \\ a_{21} & a_{22} \end{bmatrix}\begin{bmatrix} b_{11} & b_{12} \\ b_{21} & b_{22} \end{bmatrix}\right)$$ Substitution

$$= c\begin{bmatrix} a_{11}b_{11} + a_{12}b_{21} & a_{11}b_{12} + a_{12}b_{22} \\ a_{21}b_{11} + a_{22}b_{21} & a_{21}b_{12} + a_{22}b_{22} \end{bmatrix}$$ Definition of matrix multiplication

$$= \begin{bmatrix} c(a_{11}b_{11} + a_{12}b_{21}) & c(a_{11}b_{12} + a_{12}b_{22}) \\ c(a_{21}b_{11} + a_{22}b_{21}) & c(a_{21}b_{12} + a_{22}b_{22}) \end{bmatrix}$$ Definition of scalar multiplication

$$= \begin{bmatrix} ca_{11}b_{11} + ca_{12}b_{21} & ca_{11}b_{12} + ca_{12}b_{22} \\ ca_{21}b_{11} + ca_{22}b_{21} & ca_{21}b_{12} + ca_{22}b_{22} \end{bmatrix}$$ Distributive Property

$$= \begin{bmatrix} a_{11}cb_{11} + a_{12}cb_{21} & a_{11}cb_{12} + a_{12}cb_{22} \\ a_{21}cb_{11} + a_{22}cb_{21} & a_{21}cb_{12} + a_{22}cb_{22} \end{bmatrix}$$ Commutative Property

$$= \begin{bmatrix} a_{11} & a_{12} \\ a_{21} & a_{22} \end{bmatrix}\begin{bmatrix} cb_{11} & cb_{12} \\ cb_{21} & cb_{22} \end{bmatrix}$$ Definition of matrix multiplication

$$= A(cB)$$ Substitution

50. Sports statistics are often listed in columns and matrices. In this case, you can find the total number of points scored by multiplying the point value matrix, which does not change, by the scoring matrix, which changes after each season. The total number of points for her career can be found by multiplying the scoring matrix S by the point matrix P. Basketball and wrestling use different point values in scoring.

55. $\begin{bmatrix} 42 & -24 \\ -42 & -31 \end{bmatrix}$

56. $\begin{bmatrix} -5 & -20 \\ -100 & -30 \end{bmatrix}$

57. $\begin{bmatrix} -80 & -44 \\ 68 & 4 \end{bmatrix}$

63. Translated 4 units to the right and 3 units up.

$f(x) = |x - 4| + 3$

64. Translated 3 units to the left and 5 units down and stretched vertically.

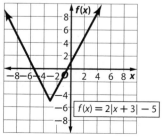

$f(x) = 2|x + 3| - 5$

65. Translated 2 units to the left and 6 units down.

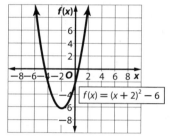

$f(x) = (x + 2)^2 - 6$

Lesson 3-7

56a. $\begin{vmatrix} a & b \\ f & g \end{vmatrix} = ag - bf$; If $ag - bf = 0$, then $ag = bf$.

56b. Sample answer: $x + 3y = 8$ and $2x + 6y = 12$; The system is dependent or inconsistent depending on the values of m and n.

60a. Sample answer: $\begin{bmatrix} 6 & 7 \\ 6 & 7 \end{bmatrix}$

60b. Sample answer: $\begin{bmatrix} 4 & 3 \\ -3 & 4 \end{bmatrix}$

60c. Sample answer: $\begin{bmatrix} -4 & -6 \\ -8 & -4 \end{bmatrix}$

70.

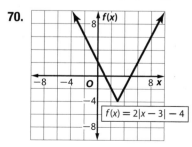

$f(x) = 2|x - 3| - 4$

71.

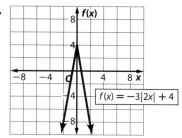

$f(x) = -3|2x| + 4$

72.

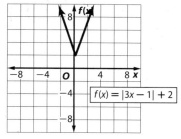

$f(x) = |3x - 1| + 2$

Lesson 3-8

36a.

$$\begin{array}{c} \text{From} \\ \begin{array}{cc} \text{city} & \text{suburbs} \end{array} \end{array}$$
$$\begin{array}{c} \text{To city} \\ \text{Suburbs} \end{array} \begin{bmatrix} 0.95 & 0.03 \\ 0.05 & 0.97 \end{bmatrix}$$

37a.

$$\begin{array}{c} \text{From} \\ \begin{array}{cc} \text{CD} & \text{MP3} \end{array} \end{array}$$
$$\begin{array}{c} \text{To CD} \\ \text{Mp3} \end{array} \begin{bmatrix} 0.35 & 0.12 \\ 0.65 & 0.88 \end{bmatrix}$$

Extend 3-8

1. $\begin{bmatrix} 3 & 2 & | & -4 \\ 4 & 7 & | & 13 \end{bmatrix}; \left(-\dfrac{54}{13}, \dfrac{55}{13} \right)$

2. $\begin{bmatrix} 2 & 1 & | & 6 \\ 6 & -2 & | & 0 \end{bmatrix}; (1.2, 3.6)$

3. $\begin{bmatrix} 2 & 2 & | & -4 \\ 7 & 3 & | & 10 \end{bmatrix}; (4, -6)$

4. $\begin{bmatrix} 4 & 6 & | & 0 \\ 8 & -2 & | & 7 \end{bmatrix}; \left(\dfrac{3}{4}, -\dfrac{1}{2} \right)$

5. $\begin{bmatrix} 6 & -4 & 2 & | & -4 \\ 2 & -2 & 6 & | & 10 \\ 2 & 2 & 2 & | & -2 \end{bmatrix}; (-2, -1, 2)$

6. $\begin{bmatrix} 5 & -5 & 5 & | & 10 \\ 5 & 0 & -5 & | & 5 \\ 0 & 5 & 10 & | & 0 \end{bmatrix}; (1.25, -0.5, 0.25)$

	Diagnostic Assessment Quick Check			
	LESSON 4-1 45 min: 1 day 90 min: 0.5 day	**EXTEND 4-1** 45 min: 0.5 day 90 min: 0.25 day	**LESSON 4-2** 45 min: 1 day 90 min: 0.5 day	**EXTEND 4-2** 45 min: 0.5 day 90 min: 0.25 day
Title	Graphing Quadratic Functions	Graphing Technology Lab: Modeling Real-World Data	Solving Quadratic Equations by Graphing	Graphing Technology Lab: Solving Quadratic Equations by Graphing
Objectives	▪ Graph quadratic functions. ▪ Find and interpret the maximum and minimum values of a quadratic function.	▪ Use a graphing calculator to model data points for which the curve of best fit is a quadratic function.	▪ Solve quadratic equations by graphing. ▪ Estimate solutions of quadratic equations by graphing.	▪ Use a graphing calculator to solve quadratic equations.
Key Vocabulary	quadratic function, quadratic, linear, constant terms, parabola, axis of symmetry, vertex, maximum value, minimum value		quadratic equations standard form root zero	
CCSS	A.SSE.1a, F.IF.9	F.IF.4	A.CED.2, F.IF.4	A.REI.11
Multiple Representations	🔧			
Lesson Resources	connectED.mcgraw-hill.com 📁 Leveled Worksheets 🔤 Vocabulary PT Personal Tutor ✋ Virtual Manipulatives ✓ Self-Check Quiz ▪ *5-Minute Check* ▪ *Study Notebook*	connectED.mcgraw-hill.com PT Personal Tutor **Materials:** ▪ TI-83/84 Plus or other graphing calculator	connectED.mcgraw-hill.com 📁 Leveled Worksheets 📁 Quiz 1 🔤 Vocabulary PT Personal Tutor ✋ Virtual Manipulatives ✓ Self-Check Quiz ▪ *5-Minute Check* ▪ *Study Notebook* ▪ *Teaching Algebra with Manipulatives*	connectED.mcgraw-hill.com PT Personal Tutor **Materials:** ▪ TI-83/84 Plus or other graphing calculator
Resources for Every Lesson	IWB eStudent Edition IWB Interactive Classroom	▪ eTeacher Edition ▪ eSolutions Manual ▪ eAssessment McGraw-Hill		
Differentiated Instruction	pp. 223, 227		pp. 232, 236	

IWB All digital assets are Interactive Whiteboard ready.

| Suggested Pacing | | | |
Time Periods	Instruction	Review & Assess	Total
45-minute	12 days	2 days	14 days
90-minute	6 days	1 day	7 days

LESSON 4-3 45 min: 1 day / 90 min: 0.5 day	**LESSON 4-4** 45 min: 1 day / 90 min: 0.5 day	**EXTEND 4-4** 45 min: 0.5 day / 90 min: 0.25 day	**LESSON 4-5** 45 min: 1 day / 90 min: 0.5 day	**EXTEND 4-5** 45 min: 1 day / 90 min: 0.5 day
Solving Quadratic Equations by Factoring	**Complex Numbers**	**Algebra Lab: The Complex Plane**	**Completing the Square**	**Graphing Technology Lab: Solving Quadratic Equations**
▪ Write quadratic equations in intercept form. ▪ Solve quadratic equations by factoring.	▪ Perform operations with pure imaginary numbers. ▪ Perform operations with complex numbers.	▪ Graph complex numbers in the complex plane and determine absolute value of complex numbers.	▪ Solve quadratic equations by using the Square Root Property. ▪ Solve quadratic equations by completing the square.	▪ Use a calculator containing a computer algebra system to solve quadratic equations.
factored form FOIL method	imaginary unit pure imaginary number complex number complex conjugates		completing the square	
A.SSE.2, F.IF.8a	N.CN.1, N.CN.2		N.CN.7, F.IF.8a	N.CN.7
connectED.mcgraw-hill.com	connectED.mcgraw-hill.com	connectED.mcgraw-hill.com	connectED.mcgraw-hill.com	connectED.mcgraw-hill.com
🗀 Leveled Worksheets 🔤 Vocabulary PT Personal Tutor ✓ Self-Check Quiz ▪ *5-Minute Check* ▪ *Study Notebook*	🗀 Leveled Worksheets 🗀 Quiz 2 🔤 Vocabulary 🏃 Animations PT Personal Tutor ✓ Self-Check Quiz ▪ *5-Minute Check* ▪ *Study Notebook* ▪ *Teaching Algebra with Manipulatives*		🗀 Leveled Worksheets 🔤 Vocabulary 🏃 Animations PT Personal Tutor ✓ Self-Check Quiz ▪ *5-Minute Check* ▪ *Study Notebook* ▪ *Teaching Algebra with Manipulatives*	PT Personal Tutor **Materials:** ▪ TI-Nspire CAS technology
IWB eStudent Edition IWB Interactive Classroom	▪ eTeacher Edition ▪ eSolutions Manual ▪ e*Assessment* McGraw-Hill			
pp. 239, 245	pp. 247, 252		pp. 259, 262	

Formative Assessment
Mid-Chapter Quiz

	LESSON 4-6 — 45 min: 1 day / 90 min: 0.5 day	EXPLORE 4-7 — 45 min: 0.5 day / 90 min: 0.25 day	LESSON 4-7 — 45 min: 1 day / 90 min: 0.5 day	
Title	The Quadratic Formula and the Discriminant	Graphing Technology Lab: Families of Parabolas	Transformations of Quadratic Graphs	
Objectives	▪ Solve quadratic equations by using the Quadratic Formula. ▪ Use the discriminant to determine the number and type of roots of a quadratic equation.	▪ Use a graphing calculator to investigate changes to parabolas.	▪ Write a quadratic function in the form $y = a(x - h)^2 + k$. ▪ Transform graphs of quadratic functions of the form $y = a(x - h)^2 + k$.	
Key Vocabulary	Quadratic Formula discriminant		vertex form	
CCSS	N.CN.7, A.SSE.1b	F.IF.4, F.BF.3	F.IF.8a, F.BF.3	
Multiple Representations				
Lesson Resources	connectED.mcgraw-hill.com 📁 Leveled Worksheets 📁 Quiz 3 ᵃᵇᶜ Vocabulary PT Personal Tutor ✓ Self-Check Quiz ▪ *5-Minute Check* ▪ *Study Notebook*	connectED.mcgraw-hill.com PT Personal Tutor **Materials:** ▪ TI-83/84 Plus or other graphing calculator	connectED.mcgraw-hill.com 📁 Leveled Worksheets ᵃᵇᶜ Vocabulary 🏃 Animations PT Personal Tutor ✓ Self-Check Quiz ▪ *5-Minute Check* ▪ *Study Notebook* ▪ *Teaching Algebra with Manipulatives*	
Resources for Every Lesson	IWB eStudent Edition IWB Interactive Classroom	▪ eTeacher Edition ▪ eSolutions Manual ▪ McGraw-Hill ᵉAssessment		
Differentiated Instruction	pp. 265, 268		pp. 277, 280	

IWB All digital assets are Interactive Whiteboard ready.

EXTEND 4-7 **45 min:** 0.5 day **90 min:** 0.25 day	LESSON 4-8 **45 min:** 1 day **90 min:** 0.5 day	EXTEND 4-8 **45 min:** 0.5 day **90 min:** 0.25 day
Algebra Lab: Quadratics and Rate of Change	**Quadratic Inequalities**	**Graphing Technology Lab: Modeling Motion**
▪ Investigate the rate of change of a quadratic function by examining first- and second-order differences.	▪ Graph quadratic inequalities in two variables. ▪ Solve quadratic inequalities in one variable.	▪ Use a data collection device to investigate the relationship between the time and the distance traveled by a car on a ramp.
	quadratic inequality	
F.IF.4, F.IF.6	A.CED.1, A.CED.3	

connectED.mcgraw-hill.com	connectED.mcgraw-hill.com	connectED.mcgraw-hill.com
▪ *Teaching Algebra with Manipulatives*	▸ Leveled Worksheets ▸ Quiz 4 🔤 Vocabulary 🏃 Animations **PT** Personal Tutor ✓ Self-Check Quiz ▪ *5-Minute Check* ▪ *Study Notebook*	**Materials:** ▪ TI-83/84 Plus or other graphing calculator ▪ long, flat, smooth-surface board ▪ stack of books ▪ data collection device (TI CBR Motion Detector) ▪ medium-size toy car

IWB eStudent Edition	▪ eTeacher Edition	
IWB Interactive Classroom	▪ eSolutions Manual	
	▪ McGraw-Hill ᵉAssessment	

	pp. 285, 288	
		Summative Assessment Study Guide and Review Practice Test

SE = Student Edition, **TE** = Teacher Edition, **CRM** = Chapter Resource Masters

Diagnosis	Prescription
Beginning Chapter 4	
Get Ready for Chapter 4 **SE**	Response to Intervention **TE**
Beginning Every Lesson	
Then, Now, Why? **SE** 5-Minute Checks	Chapter 0 **SE** Concepts and Skills Bank **SE**

DIAGNOSTIC ASSESSMENT

Diagnosis	Prescription
During/After Every Lesson	
Guided Practice **SE**, every example Check Your Understanding **SE** H.O.T. Problems **SE** Spiral Review **SE** Additional Examples **TE** Watch Out! **TE** Step 4, Assess **TE** Chapter 4 Quizzes **CRM**, pp. 59–60 Self-Check Quizzes connectED.mcgraw-hill.com	TIER 1 Intervention Skills Practice **CRM**, Ch. 1–4 connectED.mcgraw-hill.com TIER 2 Intervention Differentiated Instruction **TE**; Study Guide and Intervention **CRM**, Ch. 1–4 TIER 3 Intervention *Math Triumphs, Alg. 2,* Chs. 1, 3, and 4
Mid-Chapter	
Mid-Chapter Quiz **SE** Mid-Chapter Test **CRM**, p. 61 McGraw-Hill eAssessment	TIER 1 Intervention Skills Practice **CRM**, Ch. 1–4 connectED.mcgraw-hill.com TIER 2 Intervention Study Guide and Intervention **CRM**, Ch. 1–4 TIER 3 Intervention *Math Triumphs, Alg. 2,* Chs. 1, 3, and 4
Before Chapter Test	
Chapter Study Guide and Review **SE** Practice Test **SE** Standardized Test Practice **SE** Chapter Test connectED.mcgraw-hill.com Standardized Test Practice connectED.mcgraw-hill.com Vocabulary Review connectED.mcgraw-hill.com McGraw-Hill eAssessment	TIER 1 Intervention Skills Practice **CRM**, Ch. 1–4 connectED.mcgraw-hill.com TIER 2 Intervention Study Guide and Intervention **CRM**, Ch. 1–4 TIER 3 Intervention *Math Triumphs, Alg. 2,* Chs. 1, 3, and 4

FORMATIVE ASSESSMENT

Diagnosis	Prescription
After Chapter 4	
Multiple-Choice Tests, Forms 1, 2A, 2B **CRM**, pp. 63–68 Free-Response Tests, Forms 2C, 2D, 3 **CRM**, pp. 69–74 Vocabulary Test **CRM**, p. 62 Extended Response Test **CRM**, p. 75 Standardized Test Practice **CRM**, pp. 76–78 McGraw-Hill eAssessment	Study Guide and Intervention **CRM**, Ch. 1–4 connectED.mcgraw-hill.com

SUMMATIVE ASSESSMENT

Option 1 Reaching All Learners

Kinesthetic Have pairs of students work with algebra tiles to help them write five equations that can be solved by completing the square. Provide each pair with x^2, x, and unit tiles. Have pairs begin by creating a square arrangement of their tiles, then writing the expression modeled by the tiles. Then pairs replace the constant with c. Finally, have pairs trade their five expressions with another pair of students to solve for the value of c that will make the expression a perfect square.

x	1	1	1
x	1	1	1
x	1	1	1
x^2	x	x	x

$x^2 + 6x + 9$
$x^2 + 6x + c$

Visual Have students make a poster listing the four different numbers and types of roots that can result when solving a quadratic equation. Tell students to include a sample equation, a graph that results in each number and type of root, and an explanation of how the value of the discriminant is indicative of the number and root type.

Option 2 Approaching Level

Provide students with grid paper and tracing paper or a transparency. Have them place a piece of tracing paper over the grid paper and graph $y = x^2$. Then have students slide the tracing paper in order to translate the graph up, down, to the right, or to the left. Discuss the equation of each new graph.

Option 3 English Learners

Have students research the root words that form the words *quadratic* and *discriminant*. Discuss how the root words relate to the mathematical meanings of *quadratic* and *discriminant*.

Option 4 Beyond Level

Play a section of a song that rises to a crescendo and then gradually returns to its previous volume for the class. Ask students to describe how the song can be modeled by a parabola. Then have students write or find a music score that might model the path of a parabola. Ask students to indicate if the music has a maximum value or a minimum value.

VerticalAlignment

Before Chapter 4

Related Topics before Algebra 1

- approximate the value of irrational numbers

Related Topics from Algebra 1

- analyze graphs of quadratic functions and draw conclusions
- solve linear inequalities using graphs

Previous Topics from Algebra 2

- use tools including factoring to transform and solve equations

Chapter 4

Related Topics from Algebra 2

- determine reasonable domain and range values of quadratic functions
- analyze situations involving quadratic functions and formulate quadratic equations and inequalities to solve problems
- solve quadratic equations and inequalities using graphs, tables, and algebraic methods, including the Quadratic Formula
- use complex numbers to describe the solutions of quadratic equations
- determine a quadratic function from its zeros
- identify and sketch graphs of parent functions, including quadratic functions
- use the parent function to investigate, describe, and predict the effects of changes in a, h, and k on the graphs of the $y = a(x - h)^2 + k$ form of a function

After Chapter 4

Preparation for Precalculus

- determine the domain and range of functions using graphs, tables, and symbols
- recognize and use connections among significant values of a function, points on the graph of a function, and the symbolic representation of a function
- apply basic transformations to the parent functions

Lesson-by-LessonPreview

4-1 Graphing Quadratic Functions

The graph of a quadratic function is called a *parabola*. All parabolas have an axis of symmetry, a vertex, and a *y*-intercept. Some parabolas open up, and others open down.

axis of symmetry: $x = -\dfrac{b}{2a}$

y-intercept: c

vertex

One way to graph a quadratic function is to create a table of values. It is helpful to include the vertex of the graph and the *y*-intercept in that table. After the points in the table have been plotted, they should be connected with a smooth curve. The graph should have a U-shape and not a V-shape at the vertex of the graph. Unlike the letter U, however, the graph should become progressively wider and have arrows indicating that the graph continues to infinity.

4-2 Solving Quadratic Equations by Graphing

A quadratic equation in the form $ax^2 + bx + c = 0$ has a related function $f(x) = ax^2 + bx + c$. The *zeros* of the function are the *x*-intercepts of its graph. These *x*-values are the solutions or *roots* of the related quadratic equation. A quadratic equation can have *one real* solution, *two real* solutions, or *no real* solutions. Finding the *x*-intercepts (roots) from a graph without the use of a graphing calculator usually provides a rather imprecise estimate of the solutions. Solutions that appear to be integers can be verified by substituting them into the original equation.

4-3 Solving Quadratic Equations by Factoring

To solve a quadratic equation by factoring,

- make sure the equation is in the form, $ax^2 + bx + c = 0$,
- factor the polynomial expression,
- set each factor equal to zero,
- and then solve the resulting equations.

When factoring, a polynomial of degree two or more is rewritten as a product of polynomials each having a lesser degree. Techniques used to solve quadratic equations include the techniques for factoring general trinomials, perfect square trinomials, and a difference of squares.

4-4 Complex Numbers

Introduced in this lesson is not only a new symbol i, but a new number system called the *complex number system*. The real number system is a subset of this new number system. Complex numbers are numbers of the form $a + bi$, where a and b are real numbers and i is a number whose square is -1, that is $i = \sqrt{-1}$. A complex number of the form bi is called a *pure imaginary number*.

The complex number $a + bi$ can be treated as if it is a binomial, and operations on complex numbers follow properties for adding, subtracting, multiplying, and dividing binomials, with one exception. That exception is to replace i^2 with -1 whenever i^2 appears in an expression.

4-5 Completing the Square

The *Square Root Property* can be used to solve an equation of the form $ax^2 + bx + c = d$ when $ax^2 + bx + c$ is a perfect square. If the trinomial is not a perfect square trinomial, a method called *completing the square* can be used to rewrite the equation so that the trinomial is a perfect square.

To complete the square for any quadratic expression of the form $x^2 + bx$,

- find half of b and square it
- add this amount to $x^2 + bx$.

When you solve a quadratic equation by completing the square, add the value you used to complete the square to each side of the equation and solve the resulting equation by taking the square root of each side.

If the coefficient of x^2 is not 1, divide each term by this coefficient before completing the square and solving the equation.

4-6 The Quadratic Formula and the Discriminant

The *Quadratic Formula* can be used to solve any equation in the form $ax^2 + bx + c = 0$, where $a \neq 0$. To determine the roots of the equation, substitute the coefficients a and b and the constant c into the formula $x = \dfrac{-b \pm \sqrt{b^2 - 4ac}}{2a}$ and then simplify the resulting expression.

The expression in the Quadratic Formula that appears under the radical sign, $b^2 - 4ac$, is called the *discriminant*. This value can be used to determine the number and type of roots of the equation.

- If $b^2 - 4ac > 0$, there are two real roots.
- If $b^2 - 4ac = 0$, there is one real root.
- If $b^2 - 4ac < 0$, there are two complex roots.

Checking the discriminant can provide a quick check when solving quadratic equations.

4-7 Transformations of Quadratic Graphs

Quadratic equations can be written in *vertex form*, $y = a(x - h)^2 + k$, where the vertex of the graph of the equation is at (h, k) and the axis of symmetry is the line $x = h$. The value of k determines the graph's vertical translation. The value of h determines the graph's horizontal translation. The value of a determines the direction of opening of the graph and the shape of the parabola.

- If $a > 0$, the graph opens up.
- If $a < 0$, the graph opens down.

4-8 Quadratic Inequalities

The approach to graphing quadratic inequalities is similar to graphing linear inequalities. First, the related quadratic equation is graphed and then a point inside the parabola is checked. If the point is a solution to the inequality, then the region inside of the parabola is shaded. Otherwise, the region outside is shaded.

Solving a quadratic inequality in one variable algebraically is similar to solving a linear inequality. The difference lies in the fact that many quadratic inequalities have not one but two solutions. This means that the number line is divided in three intervals. Testing a value from each interval on the number line reveals which solution set or sets are correct. The solution set of a quadratic inequality will often be a compound inequality.

Mc Graw Hill Professional Development

Targeted professional development has been articulated throughout Algebra 2. More quality, customized professional development is available from McGraw-Hill Professional Development. Visit connectED.mcgraw-hill.com for details on each product.

- Online Lessons emphasize the strategies and techniques used to teach Algebra 2. Includes streaming video, interactive pages, and online tools.

- Video Workshops allow mentors, coaches, or leadership personnel to facilitate on-site workshops on educational strategies in mathematics and mathematical concepts.

- MHPD Online (www.mhpdonline.com) offers online professional development with video clips of instructional strategies, links, student activities, and news and issues in education.

- Teaching Today (teachingtoday.glencoe.com) gives secondary teachers practical strategies and materials that inspire excellence and innovation in teaching.

Chapter Project

Just Drop It

Students use what they have learned about quadratic functions to complete a project.

This chapter project addresses global awareness, as well as several specific skills identified as being essential to student success by the Framework for 21st Century Learning.

Visit connectED.mcgraw-hill.com for student and teacher handouts.

KeyVocabulary Introduce the key vocabulary in the chapter using the routine below.

Define: A quadratic inequality is an inequality in the form
$y > ax^2 + bx + c$;
$y \geq ax^2 + bx + c$;
$y < ax^2 + bx + c$;
$y \leq ax^2 + bx + c$

Example: The graph shows the inequality $y < x^2 - 5x + 4$.

Ask: What is the boundary of a quadratic inequality called?

a parabola

$y < x^2 - 5x + 4$

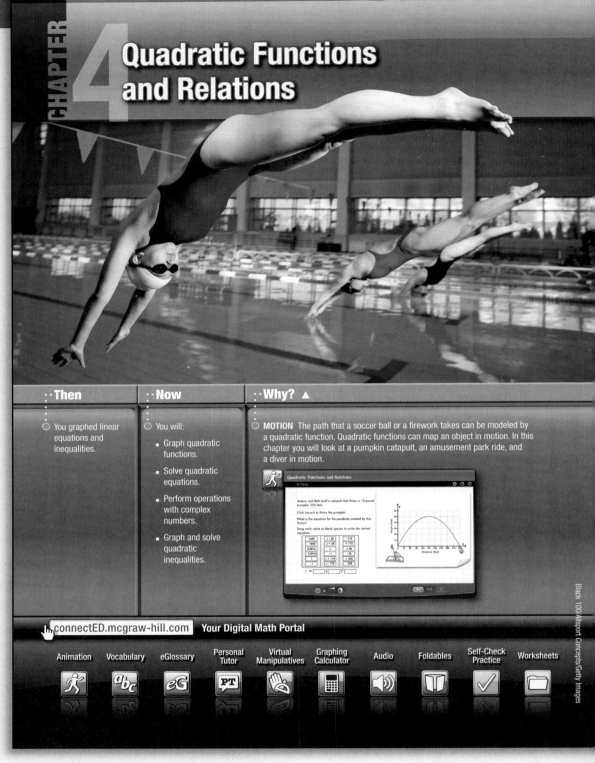

CHAPTER 4

Quadratic Functions and Relations

Then

You graphed linear equations and inequalities.

Now

You will:

- Graph quadratic functions.
- Solve quadratic equations.
- Perform operations with complex numbers.
- Graph and solve quadratic inequalities.

Why? ▲

MOTION The path that a soccer ball or a firework takes can be modeled by a quadratic function. Quadratic functions can map an object in motion. In this chapter you will look at a pumpkin catapult, an amusement park ride, and a diver in motion.

connectED.mcgraw-hill.com **Your Digital Math Portal**

Animation Vocabulary eGlossary Personal Tutor Virtual Manipulatives Graphing Calculator Audio Foldables Self-Check Practice Worksheets

Black 100/Allsport Concepts/Getty Images

Additional Answers (Get Ready for Chapter 4)

9. $(x + 8)(x + 5)$

10. $(x - 3)(x - 7)$

11. $(2x - 1)(x + 4)$

12. $(2x + 3)(x - 5)$

Get Ready for the Chapter

Diagnose Readiness | You have two options for checking prerequisite skills.

1 **Textbook Option** Take the Quick Check below. Refer to the Quick Review for help.

QuickCheck	QuickReview

QuickCheck

Given $f(x) = 2x^2 + 4$ and $g(x) = -x^2 - 2x + 3$, find each value.

1. $f(-1)$ **6**
2. $f(3)$ **22**
3. $f(0)$ **4**
4. $g(4)$ **−21**
5. $g(0)$ **3**
6. $g(-3)$ **0**

7. **FISH** Tuna swim at a steady rate of 9 miles per hour until they die, and they never stop moving.
   a. Write a function that is a model for the situation. $f(x) = 9x$
   b. Evaluate the function to estimate how far a 2-year old tuna has traveled. **157,680 mi**

8. **BUDGET** Marla has budgeted $65 per day on food during a business trip. Write a function that is a model for the situation and evaluate what she would spend on a 2 week business trip. $f(x) = 65x$; **$910**

Factor completely. If the polynomial is not factorable, write *prime*.

9. $x^2 + 13x + 40$
10. $x^2 - 10x + 21$ **9–12. See**
11. $2x^2 + 7x - 4$
12. $2x^2 - 7x - 15$ **margin.**
13. $x^2 - 11x + 15$ **prime**
14. $x^2 + 12x + 36$ $(x + 6)^2$

15. **FLOOR PLAN** The rectangular room pictured below has an area of $x^2 + 14x + 48$ square feet. If the width of the room is $(x + 6)$ feet, what is the length? $(x + 8)$ **feet**

$A = (x^2 + 14x + 48) \text{ ft}^2$ $(x + 6)$ ft

QuickReview

Example 1 (Used in Lessons 4-1 and 4-2)

Given $f(x) = -2x^2 + 3x - 1$ and $g(x) = 3x^2 - 5$, find each value.

a. $f(2)$

$f(x) = -2x^2 + 3x - 1$	Original function
$f(2) = -2(2)^2 + 3(2) - 1$	Substitute 2 for x.
$= -8 + 6 - 1$ or -3	Simplify.

b. $g(-2)$

$g(x) = 3x^2 - 5$	Original function
$g(-2) = 3(-2)^2 - 5$	Substitute -2 for x.
$= 12 - 5$ or 7	Simplify.

Example 2 (Used in Lessons 4-3 and 4-5)

Factor $2x^2 - x - 3$ completely. If the polynomial is not factorable, write *prime*.

To find the coefficients of the x-terms, you must find two numbers whose product is $2(-3)$ or -6, and whose sum is -1. The two coefficients must be 2 and -3 since $2(-3) = -6$ and $2 + (-3) = -1$. Rewrite the expression and factor by grouping.

$2x^2 - x - 3$

$= 2x^2 + 2x - 3x - 3$	Substitute $2x - 3x$ for $-x$.
$= (2x^2 + 2x) + (-3x - 3)$	Associative Property
$= 2x(x + 1) + -3(x + 1)$	Factor out the GCF.
$= (2x - 3)(x + 1)$	Distributive Property

2 **Online Option** Take an online self-check Chapter Readiness Quiz at connectED.mcgraw-hill.com.

217

Get Ready for the Chapter

Response to Intervention (RtI)
Use the *Quick Check* results and the Intervention Planner chart to help you determine your Response to Intervention. The If-Then statements in the chart below help you decide the appropriate tier of RtI and suggest intervention resources for each tier.

InterventionPlanner

TIER 1 **On Level** OL

If students miss about 25% of the exercises or less,

Then choose a resource:

SE Lessons 0-3, 2-1, and 2-4

Skills Practice, Chapter 2, pp. 7 and 19

 connectED.mcgraw-hill.com Self-Check Quiz

TIER 2 **Strategic Intervention** AL
approaching grade level

If students miss about 50% of the exercises,

Then choose a resource:

Study Guide and Intervention, Chapter 2, pp. 5–6 and 23–24

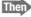 connectED.mcgraw-hill.com Extra Examples, Personal Tutor, Homework Help

TIER 3 **Intensive Intervention**
2 or more grades below level

If students miss about 75% of the exercises,

Then use *Math Triumphs, Alg. 2,* Chs. 1, 3, and 4

 connectED.mcgraw-hill.com Extra Examples, Personal Tutor, Homework Help, Review Vocabulary

Essential Question

- Why do we use different methods to solve math problems? Depending on the information given, one method may be easier to use than another. It also depends on whether you need an approximate or exact answer. For example, you could approximate the answer using a graph or mental math; you could find an exact answer using algebraic techniques.

FOLDABLES **Study**Organizer

Dinah Zike's Foldables®

Focus Students write about the different ways quadratic functions and relations can be solved as these methods are presented in the lessons of this chapter.

Teach Have students make and label their Foldables as illustrated. Students should fill in the appropriate sections with their notes, diagrams, and examples as they cover each lesson in this chapter.

Have students use the appropriate tabs as they cover each lesson in this chapter.

When to Use It Encourage students to add to their Foldables as they work through the chapter and to use them to review for the chapter test.

Differentiated Instruction

☐ Student-Built Glossary, p. 1

Students should complete the chart by providing a definition of each term and an example as they progress through Chapter 4. This study tool can also be used to review for the chapter test.

Additional Answers (Guided Practice)

1A.

1B.

Get Started on the Chapter

You will learn several new concepts, skills, and vocabulary terms as you study Chapter 4. To get ready, identify important terms and organize your resources. You may wish to refer to Chapter 0 to review prerequisite skills.

FOLDABLES **Study**Organizer

Quadratic Functions Make this Foldable to help you organize your Chapter 4 notes about quadratic functions and relations. Begin with one sheet of 11″ by 17″ paper.

1 **Fold** in half lengthwise.

2 **Fold** in fourths crosswise.

3 **Cut** along the middle fold from the edge to the last crease as shown.

4 **Refold** along the lengthwise fold and tape the uncut section at the top. Label each section with a lesson number and close to form a booklet.

NewVocabulary

English		Español
quadratic term	p. 219	término cuadrático
linear term	p. 219	término lineal
constant term	p. 219	término constante
vertex	p. 220	vértice
maximum value	p. 222	valor máximo
minimum value	p. 222	valor mínimo
quadratic equation	p. 229	ecuación cuadrática
standard form	p. 229	forma estándar
root	p. 229	raíz
zero	p. 229	cero
imaginary unit	p. 246	unidad imaginaria
pure imaginary number	p. 246	número imaginario puro
complex number	p. 247	número complejo
complex conjugates	p. 249	conjugados complejos
completing the square	p. 257	completar el cuadrado
Quadratic Formula	p. 264	fórmula cuadrática
discriminant	p. 267	discriminante
vertex form	p. 275	forma de vértice
quadratic inequality	p. 282	desigualdad cuadrática

ReviewVocabulary

domain *dominio* the set of all *x*-coordinates of the ordered pairs of a relation

function *función* a relation in which each *x*-coordinate is paired with exactly one *y*-coordinate

range *rango* the set of all *y*-coordinates of the ordered pairs of a relation

218 | **Chapter 4** | Quadratic Functions and Relations

LESSON 4-1 Graphing Quadratic Functions

:: Then	:: Now	:: Why?
● You identified and manipulated graphs of functions.	**1** Graph quadratic functions. **2** Find and interpret the maximum and minimum values of a quadratic function.	Eddie is organizing a charity tournament. He plans to charge a $20 entry fee for each of the 80 players. He recently decided to raise the entry fee by $5, and 5 fewer players entered with the increase. He used this information to determine how many fee increases will maximize the money raised.

Tournament Prize Pool

The quadratic function at the right represents this situation. The tournament prize pool increases when he first increases the fee, but eventually the pool starts to decrease as the fee gets even higher.

NewVocabulary
quadratic function
quadratic term
linear term
constant term
parabola
axis of symmetry
vertex
maximum value
minimum value

Common Core State Standards

Content Standards
A.SSE.1.a Interpret parts of an expression, such as terms, factors, and coefficients.

F.IF.9 Compare properties of two functions each represented in a different way (algebraically, graphically, numerically in tables, or by verbal descriptions).

Mathematical Practices
1 Make sense of problems and persevere in solving them.

1 Graph Quadratic Functions In a **quadratic function**, the greatest exponent is 2. These functions can have a **quadratic term**, a **linear term**, and a **constant term**. The general quadratic function is shown below.

$$f(x) = ax^2 + bx + c, \text{ where } a \neq 0$$

quadratic term | linear term | constant term

The graph of a quadratic function is called a **parabola**. To graph a quadratic function, graph ordered pairs that satisfy the function.

Example 1 Graph a Quadratic Function by Using a Table

Graph $f(x) = 3x^2 - 12x + 6$ by making a table of values.

Choose integer values for x, and evaluate the function for each value. Graph the resulting coordinate pairs, and connect the points with a smooth curve.

x	$3x^2 - 12x + 6$	$f(x)$	$(x, f(x))$
0	$3(0)^2 - 12(0) + 6$	6	$(0, 6)$
1	$3(1)^2 - 12(1) + 6$	-3	$(1, -3)$
2	$3(2)^2 - 12(2) + 6$	-6	$(2, -6)$
3	$3(3)^2 - 12(3) + 6$	-3	$(3, -3)$
4	$3(4)^2 - 12(4) + 6$	6	$(4, 6)$

GuidedPractice

Graph each function by making a table of values. **1A, 1B. See margin.**

1A. $g(x) = -2x^2 + 8x - 3$ **1B.** $h(x) = 4x^2 - 8x + 1$

connectED.mcgraw-hill.com   **219**

---

1 Focus

VerticalAlignment

Before Lesson 4-1 Indentify and manipulate graphs of functions.

Lesson 4-1 Graph quadratic functions. Find and interpret the maximum and minimum values of quadratic functions.

After Lesson 4-1 Solve quadratic equations using graphs.

2 Teach

Scaffolding Questions
Have students read the **Why?** section of the lesson.

Ask:
- What is the independent variable for this function? number of increases
- What is the dependent variable? pool size
- What is significant about the value of the expected pool size for 6 entry fee increases? The value of pool size is greatest for 6 increases.

Lesson 4-1 Resources

Resource	Approaching Level (AL)	On Level (OL)	Beyond Level (BL)	English Learners (ELL)
Teacher Edition	▪ Differentiated Instruction, p. 223	▪ Differentiated Instruction, pp. 223, 227	▪ Differentiated Instruction, pp. 223, 227	▪ Differentiated Instruction, p. 223
Chapter Resource Masters	▪ Study Guide and Intervention, pp. 5–6 ▪ Skills Practice, p. 7 ▪ Practice, p. 8 ▪ Word Problem Practice, p. 9	▪ Study Guide and Intervention, pp. 5–6 ▪ Skills Practice, p. 7 ▪ Practice, p. 8 ▪ Word Problem Practice, p. 9 ▪ Enrichment, p. 10	▪ Practice, p. 8 ▪ Word Problem Practice, p. 9 ▪ Enrichment, p. 10	▪ Study Guide and Intervention, pp. 5–6 ▪ Skills Practice, p. 7 ▪ Practice, p. 8
Other	▪ 5-Minute Check 4-1 ▪ Study Notebook	▪ 5-Minute Check 4-1 ▪ Study Notebook	▪ 5-Minute Check 4-1 ▪ Study Notebook	▪ 5-Minute Check 4-1 ▪ Study Notebook

1 Graph Quadratic Functions

Example 1 shows how to graph a quadratic function by making a table of values. **Example 2** shows how the axis of symmetry, the y-intercept, and the vertex can be used to graph a quadratic function.

Teach with Tech

Computer Projector Display a coordinate grid on the board. Use this throughout the lesson as you work through examples graphing quadratic functions.

Formative Assessment

Use the Guided Practice exercises after each example to determine students' understanding of concepts.

Additional Example

1 Graph $f(x) = x^2 + 3x - 1$ by making a table of values.

x	-3	-2	-1	0	1
$f(x)$	-1	-3	-3	-1	3

$f(x) = x^2 + 3x - 1$

Additional Examples also in Interactive Classroom PowerPoint® Presentations

 IWB **Interactive White Board** READY

ReviewVocabulary

Symmetry When something is symmetrical, its opposite sides are mirror images of each other.

Notice in Example 1 that there seemed to be a pattern in the values for $f(x)$. This is due to the axis of symmetry of parabolas. The **axis of symmetry** is a line through the graph of a parabola that divides the graph into two congruent halves. Each side of the parabola is a reflection of the other side.

The axis of symmetry will intersect a parabola at only one point, called the **vertex**. The vertex of the graph at the right is $A(3, -3)$.

Notice that the x-coordinates of points B and C are both 4 units away from the x-coordinate of the vertex, and they have the same y-coordinate. This is due to the symmetrical nature of the graph.

KeyConcept Graph of a Quadratic Function—Parabola

Words   Consider the graph of $y = ax^2 + bx + c$, where $a \neq 0$.

- The y-intercept is $a(0)^2 + b(0) + c$ or c.
- The equation of the axis of symmetry is $x = -\dfrac{b}{2a}$.
- The x-coordinate of the vertex is $-\dfrac{b}{2a}$.

Model

axis of symmetry: $x = -\dfrac{b}{2a}$

y-intercept: c

reflection of y-intercept

vertex

StudyTip

Plotting Reflections The reflection of a point is its mirror image on the other side of the axis of symmetry.

Now you can use the axis of symmetry to help plot points and graph a parabola. For $y = x^2 + 6x - 2$ below, the axis of symmetry is $x = -\dfrac{b}{2a} = -\dfrac{6}{2(1)}$ or $x = -3$.

Find the axis of symmetry and the vertex.

Find the y-intercept and its reflection.

Connect the points with a smooth curve.

Example 2 Axis of Symmetry, *y*-intercept, and Vertex

Consider $f(x) = x^2 + 4x - 3$.

a. Find the *y*-intercept, the equation of the axis of symmetry, and the *x*-coordinate of the vertex.

The function is of the form $f(x) = ax^2 + bx + c$, so we can identify *a*, *b*, and *c*.

$$f(x) = ax^2 + bx + c$$
$$\downarrow \quad \downarrow \quad \downarrow$$
$$f(x) = 1x^2 + 4x - 3 \quad \rightarrow \quad a = 1, b = 4, \text{ and } c = -3$$

The *y*-intercept is $c = -3$.

Use *a* and *b* to find the equation of the axis of symmetry.

$$x = -\frac{b}{2a} \qquad \text{Equation of the axis of symmetry}$$
$$= -\frac{4}{2(1)} \qquad a = 1 \text{ and } b = 4$$
$$= -2 \qquad \text{Simplify.}$$

The equation of the axis of symmetry is $x = -2$. Therefore, the *x*-coordinate of the vertex is -2.

b. Make a table of values that includes the vertex.

Select five specific points, with the vertex in the middle and two points on either side of the vertex, including the *y*-intercept and its reflection. Use symmetry to determine the *y*-values of the reflections.

x	$x^2 + 4x - 3$	f(x)	(x, f(x))
−6	$(-6)^2 + 4(-6) - 3$	9	(−6, 9)
−4	$(-4)^2 + 4(-4) - 3$	−3	(−4, −3)
−2	$(-2)^2 + 4(-2) - 3$	−7	(−2, −7)
0	$(0)^2 + 4(0) - 3$	−3	(0, −3)
2	$(2)^2 + 4(2) - 3$	9	(2, 9)

c. Use this information to graph the function.

Graph the points from the table and connect them with a smooth curve.

Draw the axis of symmetry, $x = -2$, as a dashed line. The graph should be symmetrical about this line.

Guided Practice

2. Consider $f(x) = -5x^2 - 10x + 6$.

A. Find the *y*-intercept, the equation of the axis of symmetry, and the *x*-coordinate of the vertex. *y*-intercept = 6; axis of symmetry: $x = -1$; *x*-coordinate = −1

B. Make a table of values that includes the vertex.

C. Use this information to graph the function. **See margin.**

connectED.mcgraw-hill.com **221**

StudyTip

Quadratic Form Make sure the function is in standard quadratic form, $y = ax^2 + bx + c$, before graphing.

StudyTip

Fractions When the *x*-coordinate of the vertex is a fraction, select the nearest integer for the next point to avoid using fractions and simplify the calculations.

2B.

x	f(x)
−3	−9
−2	6
−1	11
0	6
1	−9

Additional Example

2 Consider $f(x) = 2 - 4x + x^2$.

a. Find the *y*-intercept, the equation of the axis of symmetry, and the *x*-coordinate of the vertex. 2; $x = 2$; 2

b. Make a table of values that includes the vertex.

x	0	1	2	3	4
f(x)	2	−1	−2	−1	2

c. Use this information to graph the function.

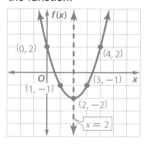

Tips for New Teachers

Extending the Concept Make sure students understand that the graph shows values for all the points that satisfy the function, even when the *x*-value is not an integer.

WatchOut!

Common Misconceptions When working through Example 2, make sure students realize that $f(x)$ and *y* can be used interchangeably, and also that the maximum or minimum value of the function is given by the *y*-coordinate of the vertex of the parabola.

Additional Answer (Guided Practice)

2C.

2 Maximum and Minimum Values

Example 3 shows how to determine whether a quadratic function has a maximum or minimum value and to state that value. **Example 4** shows how to find a maximum value for a real-world situation.

Additional Example

3 Consider $f(x) = -x^2 + 2x + 3$.

a. Determine whether the function has a maximum or minimum value. maximum

b. State the maximum or minimum value of the function. 4

c. State the domain and range of the function. D = {all real numbers}, R = {f(x) | f(x) ≤ 4}

WatchOut!

Preventing Errors Make sure students understand that a parabola which opens upward is the graph of a function with a minimum value, and that a parabola that opens downward is the graph of a function with a maximum value. Compare these parabolas to valleys (where the altitude of the valley floor is a minimum) and hills (where the peak of the hill is the maximum altitude).

WatchOut!

Maxima and Minima The terms *minimum point* and *minimum value* are not interchangeable. The minimum point on the graph of a quadratic function is the ordered pair that describes the location of the vertex. The minimum value of a function is the y-coordinate of the minimum point. It is the smallest value obtained when f(x) is evaluated for all values of x.

2 Maximum and Minimum Values

The y-coordinate of the vertex of a quadratic function is the **maximum value** or the **minimum value** of the function. These values represent the greatest or lowest possible value the function can reach.

KeyConcept Maximum and Minimum Value

Words   The graph of $f(x) = ax^2 + bx + c$, where $a \neq 0$,

- opens up and has a minimum value when $a > 0$, and
- opens down and has a maximum value when $a < 0$.

Model

a is positive.	a is negative.
The y-coordinate is the minimum value.	The y-coordinate is the maximum value.

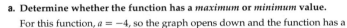

Example 3 Maximum or Minimum Values

Consider $f(x) = -4x^2 + 12x + 18$.

a. Determine whether the function has a *maximum* or *minimum* value.

For this function, $a = -4$, so the graph opens down and the function has a maximum value.

b. State the maximum or minimum value of the function.

The maximum value of the function is the y-coordinate of the vertex.

The x-coordinate of the vertex is $-\frac{12}{2(-4)}$ or 1.5.

Find the y-coordinate of the vertex by evaluating the function for $x = 1.5$.

$$f(x) = -4x^2 + 12x + 18 \qquad \text{Original function}$$
$$= -4(1.5)^2 + 12(1.5) + 18 \qquad x = 1.5$$
$$= -9 + 18 + 18 \text{ or } 27 \qquad \text{The maximum value of the function is 27.}$$

c. State the domain and range of the function.

The domain is all real numbers. The range is all real numbers less than or equal to the maximum value, or $\{f(x) \mid f(x) \leq 27\}$.

StudyTip
Domain and Range The domain of a quadratic function will always be all real numbers. The range will either be all real numbers less than or equal to the maximum or all real numbers greater than or equal to the minimum.

Guided Practice

3. Consider $f(x) = 4x^2 - 24x + 11$.

A. Determine whether the function has a maximum or minimum value. minimum

B. State the maximum or minimum value of the function. minimum = −25

C. State the domain and range of the function.
D = {all real numbers}; R = {f(x) | f(x) ≥ −25}

Focus on Mathematical Content

Graphs of Quadratic Functions Graphs of quadratic functions of the form $f(x) = ax^2 + bx + c$, where $a \neq 0$, are called parabolas. Except possibly for compression or stretching and the placement of the axes, the shape of the graph of a quadratic function will look like one of the parabolas at the top of the page.

Peter Dazeley/Getty Images

Real-World Example 4 Quadratic Equations in the Real World

CHARITY Refer to the beginning of the lesson.

a. How much should Eddie charge in order to maximize charity income?

| Words | Total | equals | fee | times | number of entrants. |

Variable Let x = the number of price increases.
Let $P(x)$ = the total pool as a function of x.

Equation $P(x)$ = $20 + 5x$ · $(80 - 5x)$

Solve for the x-value of the vertex.

$P(x) = (20 + 5x) \cdot (80 - 5x)$

$\quad = 20(80) + 20(-5x) + 5x(80) + 5x(-5x)$ Distribute.

$\quad = 1600 - 100x + 400x - 25x^2$ Multiply.

$\quad = 1600 + 300x - 25x^2$ Simplify.

$\quad = -25x^2 + 300x + 1600$ $ax^2 + bx + c$ form

Use the formula for the axis of symmetry, $x = -\dfrac{b}{2a}$, to find the x-coordinate.

$x = -\dfrac{300}{2(-25)}$ or 6 $a = -25$ and $b = 300$

Eddie needs to have 6 price increases, so he should charge $20 + 6(5)$ or \$50.

b. What will be the maximum value of the pool?

Find the maximum value of the quadratic function $P(x)$ by evaluating $P(6)$.

$P(x) = -25x^2 + 300x + 1600$ Total pool function

$P(6) = -25(6)^2 + 300(6) + 1600$ $x = 6$

$\quad = -900 + 1800 + 1600$ or 2500   Simplify.

Thus, the maximum prize pool is \$2500 after 6 price increases.

CHECK Graph the function on a graphing calculator and use the **CALC:maximum** function to confirm the solution.

Select a left bound of 0 and a right bound of 10. The calculator will display the coordinates of the maximum at the bottom of the screen.

[0, 10] scl: 1 by [0, 2500] scl: 100

The domain is $\{x \mid x \geq 0\}$ because there can be no negative increases in price. The range is $\{y \mid 0 \leq y \leq 2500\}$ because the prize pool cannot have a negative monetary value.

StudyTip

CCSS Modeling Use logic and the information from the problem to determine the domain and range that are reasonable in the situation.

▶ **Guided**Practice

4. Suppose a different tournament that Eddie organizes has 120 players and the entry fee is \$40. Each time he increases the fee by \$5, he loses 10 players. Determine what the entry fee should be to maximize the value of the pool. **\$50**

Real-WorldLink

As of 2006, there were approximately 900,000 public charities in the United States.

Source: National Center for Charitable Statistics

 connectED.mcgraw-hill.com **223**

Additional Example

4 ECONOMICS A souvenir shop sells about 200 coffee mugs each month for \$6 each. The shop owner estimates that for each \$0.50 increase in the price, he will sell about 10 fewer coffee mugs per month.

a. How much should the owner charge for each mug in order to maximize the monthly income from their sales? \$8

b. What is the maximum monthly income the owner can expect to make from the mugs? \$1280

WatchOut!

Preventing Errors Remind students to use the equation for the axis of symmetry to determine the x-coordinate of the vertex and then find the y-coordinate of the vertex.

CCSS **Teaching the Mathematical Practices**

Modeling Mathematically proficient students can apply the mathematics they know to solve problems arising in everyday life, society, and the workplace. Encourage students to interpret their mathematical results in the context of the situation and reflect on whether the results make sense.

DifferentiatedInstruction AL OL BL ELL

If students struggle to solve real-world problems involving maximum or minimum values,

Then encourage students to explain to partners their plans for solving. Suggest that the listening partner take notes about the speaking partner's strategies, asking questions as needed for clarification.

3 Practice

Formative Assessment

Use Exercises 1–11 to check for understanding.

Use the chart at the bottom of this page to customize assignments for your students.

 Teaching the Mathematical Practices

Modeling Mathematically proficient students can apply the mathematics they know to solve problems arising in everyday life, analyze relationships mathematically to draw conclusions, and interpret their mathematical results in the context of a situation.

Check Your Understanding

= Step-by-Step Solutions begin on page R14.

Examples 1–2 Complete parts a–c for each quadratic function. **1–6.** See Chapter 4 Answer Appendix.

a. Find the y-intercept, the equation of the axis of symmetry, and the x-coordinate of the vertex.

b. Make a table of values that includes the vertex.

c. Use this information to graph the function.

1. $f(x) = 3x^2$

2. $f(x) = -6x^2$

3. $f(x) = x^2 - 4x$

4. $f(x) = -x^2 - 3x + 4$

5. $f(x) = 4x^2 - 6x - 3$

6. $f(x) = 2x^2 - 8x + 5$

7. max = 8; D = {all real numbers}, R = {f(x)|f(x) ≤ 8}

8. min = −14.25; D = {all real numbers}, R = {f(x)|f(x) ≥ −14.25}

Example 3 Determine whether each function has a *maximum* or *minimum* value, and find that value. Then state the domain and range of the function.

7. $f(x) = -x^2 + 6x - 1$

8. $f(x) = x^2 + 3x - 12$

9. $f(x) = 3x^2 + 8x + 5$

10. $f(x) = -4x^2 + 10x - 6$

9. min = $-\frac{1}{3}$; D = {all real numbers}, R = $\left\{ f(x) \mid f(x) \geq -\frac{1}{3} \right\}$

Example 4 **11. BUSINESS** A store rents 1400 videos per week at $2.25 per video. The owner estimates that they will rent 100 fewer videos for each $0.25 increase in price. What price will maximize the income of the store? **$2.88**

10. max = 0.25; D = {all real numbers}, R = {f(x)|f(x) ≤ 0.25}

Practice and Problem Solving

Extra Practice is on page R4.

Examples 1–2 Complete parts a–c for each quadratic function. **12–21.** See Chapter 4 Answer Appendix.

a. Find the y-intercept, the equation of the axis of symmetry, and the x-coordinate of the vertex.

b. Make a table of values that includes the vertex.

c. Use this information to graph the function.

12. $f(x) = 4x^2$

13. $f(x) = -2x^2$

14. $f(x) = x^2 - 5$

15. $f(x) = x^2 + 3$

16. $f(x) = 4x^2 - 3$

17. $f(x) = -3x^2 + 5$

18. $f(x) = x^2 - 6x + 8$

 19. $f(x) = x^2 - 3x - 10$

20. $f(x) = -x^2 + 4x - 6$

21. $f(x) = -2x^2 + 3x + 9$

24. min = 0; D = {all real numbers}, R = {f(x)|f(x) ≥ 0}

25. max = 13.25; D = {all real numbers}, R = {f(x)|f(x) ≤ 13.25}

26. max = $\frac{22}{3}$; D = {all real numbers}, R = $\left\{ f(x)|f(x) \leq \frac{22}{3} \right\}$

22. min = 0; D = {all real numbers}, R = {f(x)|f(x) ≥ 0}

Example 3 Determine whether each function has a *maximum* or *minimum* value, and find that value. Then state the domain and range of the function.

22. $f(x) = 5x^2$

23. $f(x) = -x^2 - 12$

24. $f(x) = x^2 - 6x + 9$

25. $f(x) = -x^2 - 7x + 1$

26. $f(x) = 8x - 3x^2 + 2$

27. $f(x) = 5 - 4x - 2x^2$

28. $f(x) = 15 - 5x^2$

29. $f(x) = x^2 + 12x + 27$

30. $f(x) = -x^2 + 10x + 30$

31. $f(x) = 2x^2 - 16x - 42$

23. max = −12; D = {all real numbers}, R = {f(x)|f(x) ≤ −12}

27. max = 7; D = {all real numbers}, R = {f(x)|f(x) ≤ 7}

28. max = 15; D = {all real numbers}, R = {f(x)|f(x) ≤ 15}

29. min = −9; D = {all real numbers}, R = {f(x)|f(x) ≥ −9}

30. max = 55; D = {all real numbers}, R = {f(x)|f(x) ≤ 55}

Example 4 **32. CCSS MODELING** A financial analyst determined that the cost, in thousands of dollars, of producing bicycle frames is $C = 0.000025f^2 - 0.04f + 40$, where f is the number of frames produced.

a. Find the number of frames that minimizes cost. **800**

b. What is the total cost for that number of frames? **$24,000**

31. min = −74; D = {all real numbers}, R = {f(x)|f(x) ≥ −74}

Differentiated Homework Options

Level	Assignment		Two-Day Option
AL Basic	12–32, 61, 62, 64–82	13–31 odd, 66–69	12–32 even, 61, 62, 64, 65, 70–82
OL Core	13–39 odd, 42, 58–62, 64–82	12–32, 66–69	33–65, 70–82
BL Advanced	33–79, (optional: 80–82)		

B Complete parts a–c for each quadratic function. **33–40. See Chapter 4 Answer Appendix.**

a. Find the y-intercept, the equation of the axis of symmetry, and the x-coordinate of the vertex.

b. Make a table of values that includes the vertex.

c. Use this information to graph the function.

33. $f(x) = 2x^2 - 6x - 9$

34. $f(x) = -3x^2 - 9x + 2$

35. $f(x) = -4x^2 + 5x$

36. $f(x) = 2x^2 + 11x$

37. $f(x) = 0.25x^2 + 3x + 4$

38. $f(x) = -0.75x^2 + 4x + 6$

39. $f(x) = \frac{3}{2}x^2 + 4x - \frac{5}{2}$

40. $f(x) = \frac{2}{3}x^2 - \frac{7}{3}x + 9$

41c. $11; Because the function has a maximum at $x = 3$, it is in the domain. Therefore, three $0.50 increases is reasonable.

49. max = -7.2; D = {all real numbers}, R = {$f(x) | f(x) \leq -7.2$}

50. max = 2.5625; D = {all real numbers}, R = {$f(x) | f(x) \leq 2.5625$}

51. min = -9.375; D = {all real numbers}, R = {$f(x) | f(x) \geq -9.375$}

52. max = -4.75; D = {all real numbers}, R = {$f(x) | f(x) \leq -4.75$}

53. min = -23.5; D = {all real numbers}, R = {$f(x) | f(x) \geq -23.5$}

54. max = $-\frac{4}{3}$; D = {all real numbers}, R = $\left\{ f(x) | f(x) \leq -\frac{4}{3} \right\}$

41. FINANCIAL LITERACY A babysitting club sits for 50 different families. They would like to increase their current rate of $9.50 per hour. After surveying the families, the club finds that the number of families will decrease by about 2 for each $0.50 increase in the hourly rate. **b.** D = {$x | 0 \leq x \leq 25$}, R = {$y | 0 \leq y \leq 484$}

a. Write a quadratic function that models this situation. $y = -x^2 + 6x + 475$

b. State the domain and range of this function as it applies to the situation.

c. What hourly rate will maximize the club's income? Is this reasonable?

d. What is the maximum income the club can expect to make? **$484**

42. ACTIVITIES Last year, 300 people attended the Franklin High School Drama Club's winter play. The ticket price was $8. The advisor estimates that 20 fewer people would attend for each $1 increase in ticket price.

a. What ticket price would give the greatest income for the Drama Club? **$11.50**

b. If the Drama Club raised its tickets to this price, how much income should it expect to bring in? **$2645**

CCSS TOOLS Use a calculator to find the maximum or minimum of each function. Round to the nearest hundredth if necessary.

43. $f(x) = -9x^2 - 12x + 19$ max = 23

44. $f(x) = 12x^2 - 21x + 8$ min = -1.19

45. $f(x) = -8.3x^2 + 14x - 6$ max = -0.10

46. $f(x) = 9.7x^2 - 13x - 9$ min = -13.36

47. $f(x) = 28x - 15 - 18x^2$ max = -4.11

48. $f(x) = -16 - 14x - 12x^2$ max = -11.92

Determine whether each function has a *maximum* or *minimum* value, and find that value. Then state the domain and range of the function.

49 $f(x) = -5x^2 + 4x - 8$

50. $f(x) = -4x^2 - 3x + 2$

51. $f(x) = -9 + 3x + 6x^2$

52. $f(x) = 2x - 5 - 4x^2$

53. $f(x) = \frac{2}{3}x^2 + 6x - 10$

54. $f(x) = -\frac{3}{5}x^2 + 4x - 8$

55. $f(x) = x^2 - 4x - 5$

56. $f(x) = x^2 + 2x - 6$

57. $f(x) = x^2 - 6x + 8$

Determine the function represented by each graph.

55.

56.

57.

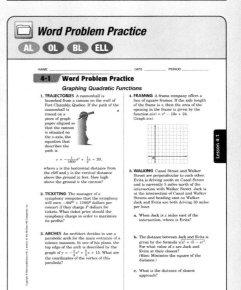

CCSS Teaching the Mathematical Practices

Tools When making mathematical models, mathematically proficient students know that technology can enable them to visualize the results of varying assumptions, explore consequences, and compare predictions with data. They are able to use technological tools to explore and deepen their understanding of concepts.

Enrichment
OL BL

Multiple Representations

In Exercise 58, students use a table of values, a graph, and analytical observation to show how the leading coefficient affects the shape of a graphed function.

 Teaching the Mathematical Practices

Critique Mathematically proficient students are also able to compare the effectiveness of two plausible arguments, distinguish correct logic or reasoning from that which is flawed, and—if there is a flaw in an argument—explain what it is.

Additional Answers

58a.

x	f(x)	g(x)
−4	40	88
−3	29	56
−2	20	32
−1	13	16
0	8	8
1	5	8
2	4	16
3	5	32
4	8	56

58b.

58c. Sample answer: $g(x)$ is much narrower than $f(x)$. The value of a changed from 1 to 4.

58d. Sample answer: The graph of $h(x)$ will be wider than $f(x)$.

59b. Omar can charge at 45 cents or 50 cents.

59c. $450 per week

---

 58. **MULTIPLE REPRESENTATIONS** Consider $f(x) = x^2 - 4x + 8$ and $g(x) = 4x^2 - 4x + 8$.

  a. **Tabular** Make a table of values for $f(x)$ and $g(x)$ if $-4 \le x \le 4$. **a–d. See margin.**
  b. **Graphical** Graph $f(x)$ and $g(x)$.
  c. **Verbal** Explain the difference in the shapes of the graphs of $f(x)$ and $g(x)$. What value was changed to cause this difference?
  d. **Analytical** Predict the appearance of the graph of $h(x) = 0.25x^2 - 4x + 8$. Confirm your prediction by graphing all three functions if $-10 \le x \le 10$.

59 **VENDING MACHINES** Omar owns a vending machine in a bowling alley. He currently sells 600 cans of soda per week at $0.65 per can. He estimates that he will lose 100 customers for every $0.05 increase in price and gain 100 customers for every $0.05 decrease in price. (*Hint:* The charge *must* be a multiple of 5.) **a–c. See margin.**

  a. Write and graph the related quadratic equation for a price increase.
  b. If Omar lowers the price, what should it be to maximize his income?
  c. What will be his income per week from the vending machine?

61. Madison; sample answer: $f(x)$ has a maximum of -2. $g(x)$ has a maximum of 1.

60. **BASEBALL** Lolita throws a baseball into the air and the height h of the ball in feet at a given time t in seconds after she releases the ball is given by the function $h(t) = -16t^2 + 30t + 5$.

  a. State the domain and range for this situation. $D = \{t | 0 < t < 2.09\}$, $R = \{h(t) | 0 \le h(t) \le 19.0625\}$
  b. Find the maximum height the ball will reach. **19.0625 ft**

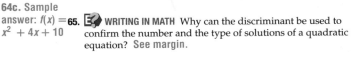

H.O.T. Problems   Use Higher-Order Thinking Skills

61. **CCSS** **CRITIQUE** Trent thinks that the function $f(x)$ graphed below, and the function $g(x)$ described next to it have the same maximum. Madison thinks that $g(x)$ has a greater maximum. Is either of them correct? Explain your reasoning.

$g(x)$ is a quadratic function with roots of 4 and 2 and a y-intercept of -8.

62. Sample answer: Always; the coordinates of a quadratic function are symmetrical, so x-coordinates equidistant from the vertex will have the same y-coordinate.

64a. Sample answer: $f(x) = -x^2 + 8$

64b. Sample answer: $f(x) = x^2 - 4$

64c. Sample answer: $f(x) = x^2 + 4x + 10$

62. **REASONING** Determine whether the following statement is *sometimes*, *always*, or *never* true. Explain your reasoning.

  In a quadratic function, if two x-coordinates are equidistant from the axis of symmetry, then they will have the same y-coordinate.

63. **CHALLENGE** The table at the right represents some points on the graph of a quadratic function.

  a. Find the values of a, b, c, and d. $a = 22$; $b = 26$; $c = -6$; $d = 2$
  b. What is the x-coordinate of the vertex? **0**
  c. Does the function have a maximum or a minimum? **maximum**

x	y
−20	−377
c	−13
−5	−2
−1	22
d − 1	a
5	a − 24
7	−b
15	−202
14 − c	−377

64. **OPEN ENDED** Give an example of a quadratic function with a
  a. maximum of 8.   b. minimum of -4.   c. vertex of $(-2, 6)$.

65. **WRITING IN MATH** Why can the discriminant be used to confirm the number and the type of solutions of a quadratic equation? **See margin.**

 226 | Lesson 4-1 | Graphing Quadratic Functions

59a. $f(x) = 39{,}000 - 3500x - 500x^2$

65. Sample answer: If the discriminant is positive, the Quadratic Formula will result in two real solutions because you are adding and subtracting the square root of a positive number in the numerator of the expression. If the discriminant is zero, there will be one real solution because you are adding and subtracting the square root of zero. If the discriminant is negative, there will be two complex solutions because you are adding and subtracting the square root of a negative number in the numerator of the expression.

Standardized Test Practice

66. Which expression is equivalent to $\frac{8!}{5!}$? **B**

A $\frac{8}{5}$

C 3!

B $8 \cdot 7 \cdot 6$

D $8 \cdot 7 \cdot 6 \cdot 5$

67. SAT/ACT The price of coffee beans is d dollars for 6 ounces, and each ounce makes c cups of coffee. In terms of c and d, what is the cost of the coffee beans required to make 1 cup of coffee? **K**

F $\frac{cd}{6}$

J $6cd$

G $\frac{6c}{d}$

K $\frac{d}{6c}$

H $\frac{6}{cd}$

68. SHORT RESPONSE Each side of the square base of a pyramid is 20 feet, and the pyramid's height is 90 feet. What is the volume of the pyramid?
12,000 ft³

69. Which ordered pair is the solution of the following system of equations? **C**

$$3x - 5y = 11$$
$$3x - 8y = 5$$

A $(2, 1)$

C $(7, 2)$

B $(7, -2)$

D $\left(\frac{1}{3}, -2\right)$

Name the Math Have students explain how to tell by examining a quadratic function whether its graph will have a maximum or minimum value. Then have them give examples of what such values might mean in a real-world problem.

Additional Answers

70. $\begin{bmatrix} -\frac{1}{5} & \frac{4}{5} \\ -\frac{2}{5} & \frac{3}{5} \end{bmatrix}$

76b.

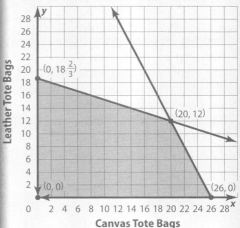

Spiral Review

Find the inverse of each matrix, if it exists. (Lesson 3-8)

70. $\begin{bmatrix} 3 & -4 \\ 2 & -1 \end{bmatrix}$ See margin.

71. $\begin{bmatrix} -4 & -1 \\ 0 & 6 \end{bmatrix}$ $\begin{bmatrix} -\frac{1}{4} & -\frac{1}{24} \\ 0 & \frac{1}{6} \end{bmatrix}$

72. $\begin{bmatrix} 2 & 8 \\ -3 & -5 \end{bmatrix}$ $\begin{bmatrix} -\frac{5}{14} & -\frac{4}{7} \\ \frac{3}{14} & \frac{1}{7} \end{bmatrix}$

Evaluate each determinant. (Lesson 3-7)

73. $\begin{vmatrix} 6 & -3 \\ -1 & 8 \end{vmatrix}$ **45**

74. $\begin{vmatrix} -3 & -5 \\ -1 & -9 \end{vmatrix}$ **22**

75. $\begin{vmatrix} 8 & 6 \\ 4 & 3 \end{vmatrix}$ **0**

76. MANUFACTURING The Community Service Committee is making canvas tote bags and leather tote bags for a fundraiser. They will line both types of bags with canvas and use leather handles on both. For the canvas bags, they need 4 yards of canvas and 1 yard of leather. For the leather bags, they need 3 yards of leather and 2 yards of canvas. The committee leader purchased 56 yards of leather and 104 yards of canvas. (Lesson 3-3)

   a. Let c represent the number of canvas bags, and let ℓ represent the number of leather bags. Write a system of inequalities for the number of bags that can be made. $c \geq 0$, $\ell \geq 0$, $c + 3\ell \leq 56$, $4c + 2\ell \leq 104$

   b. Draw the graph showing the feasible region. See margin.

   c. List the coordinates of the vertices of the feasible region. $(0, 0)$, $(26, 0)$, $(20, 12)$, $\left(0, 18\frac{2}{3}\right)$

   d. If the club plans to sell the canvas bags at a profit of \$20 each and the leather bags at a profit of \$35 each, write a function for the total profit on the bags. $f(c, \ell) = 20c + 35\ell$

   e. How can the club make the maximum profit? Make 20 canvas tote bags and 12 leather tote bags.

   f. What is the maximum profit? **\$820**

State whether each function is a linear function. Write *yes* or *no*. Explain. (Lesson 2-2)

77. $y = 4x^2 - 3x$

78. $y = -2x - 4$

79. $y = 4$

77. No; it cannot be written as $y = mx + b$.

78. Yes; it is written in $y = mx + b$ form.

79. Yes; it is written in $y = mx + b$ form, $m = 0$.

Skills Review

Evaluate each function for the given value.

80. $f(x) = 3x^2 - 4x + 6$, $x = -2$ **26**

81. $f(x) = -2x^2 + 6x - 5$, $x = 4$ **−13**

82. $f(x) = 6x^2 + 18$, $x = -5$ **168**

  227

DifferentiatedInstruction OL BL

Extension Write the following quadratic functions on the board. Have students write each function in the form $f(x) = ax^2 + bx + c$.

$f(x) = x(x - 2) - 1$ $f(x) = x^2 - 2x - 1$

$f(x) = 2x\left(\frac{1}{2}x + 2\right) - 3$ $f(x) = x^2 + 4x - 3$

$f(x) = 2x(x - 4)$ $f(x) = 2x^2 - 8x$

1 Focus

Objective Use a graphing calculator to model data points for which the curve of best fit is a quadratic function.

Materials for Each Group
- TI-83/84 Plus or other graphing calculator

Teaching Tip
In Step 2, the value of the coefficient *a* is displayed as 2.1035215E-4. Point out that this is how the calculator displays the scientific notation 2.1035215×10^{-4}.

When students use the procedure in Step 2 to copy the regression equation from Step 1 to the **Y=** list, the coefficients will have several more digits than the coefficients displayed on the home screen. The coefficients on the home screen are rounded versions of those in the **Y=** list.

2 Teach

Working in Cooperative Groups
Put students in groups of two or three, mixing abilities. Have groups work through the activity.

- Make sure students have cleared the **L1** and **L2** lists before entering new data. Also have them enter the **WINDOW** dimensions shown.

- For Step 1, point out that you can use the same keystrokes shown in Step 2, substituting 4 for the first 5, to select **LinReg.**

- If an error message appears in Step 2, have students clear the **Y=** list before trying Step 2 again.

Practice Have students complete Exercises 1–4.

EXTEND 4-1 Graphing Technology Lab
Modeling Real-World Data

You can use a TI-83/84 Plus graphing calculator to model data points for which a curve of best fit is a quadratic function.

WATER A bottle is filled with water. The water is allowed to drain from a hole made near the bottom of the bottle. The table shows the level of the water *y* measured in centimeters from the bottom of the bottle after *x* seconds.

CCSS Common Core State Standards
Content Standards
F.IF.4 For a function that models a relationship between two quantities, interpret key features of graphs and tables in terms of the quantities, and sketch graphs showing key features given a verbal description of the relationship.
Mathematical Practices
5 Use appropriate tools strategically.

Time (s)	0	20	40	60	80	100	120	140	160	180	200	220
Water level (cm)	42.6	40.7	38.9	37.2	35.8	34.3	33.3	32.3	31.5	30.8	30.4	30.1

Find and graph a linear regression equation and a quadratic regression equation. Determine which equation is a better fit for the data.

Activity

Step 1 Find and graph a linear regression equation.

- Enter the times in **L1** and the water levels in **L2**. Then find a linear regression equation.

  KEYSTROKES: *Refer to Lesson 2-5.*

- Use **STAT PLOT** to graph a scatter plot. Copy the equation to the **Y=** list and graph.

  KEYSTROKES: *Review statistical plots and graphing a regression equation in Lesson 2-5.*

[0, 260] scl: 20 by [25, 45] scl: 5

Step 2 Find and graph a quadratic regression equation.

- Find the quadratic regression equation. Then copy the equation to the **Y=** list and graph.

  KEYSTROKES: STAT ▶ 5 ENTER Y= VARS 5 ▶ ▶ ENTER GRAPH

[0, 260] scl: 20 by [25, 45] scl: 5

Notice that the graph of the linear regression equation appears to pass through just two data points. However, the graph of the quadratic regression equation fits the data very well.

Exercises

Refer to the table.

1–3. See Chapter 4 Answer Appendix.

1. Find and graph a linear regression equation and a quadratic regression equation for the data. Determine which equation is a better fit for the data.

2. Estimate the height of the player's feet after 1 second and 1.5 seconds. Use mental math to check the reasonableness of your estimates.

3. Compare and contrast the estimates you found in Exercise 2.

4. How might choosing a regression equation that does not fit the data well affect predictions made by using the equation? **It could give misleading predictions.**

Height of Player's Feet above Floor

Time (s)	Height (in.)
0.1	3.04
0.2	5.76
0.3	8.16
0.4	10.24
0.5	12
0.6	13.44
0.7	14.56

228 | Extend 4-1 | Graphing Technology Lab: Modeling Real-World Data

---

3 Assess

Formative Assessment
Use Exercise 4 to assess whether students understand that a regression equation that does not fit data well is a poor model of the given data and that it will most likely be a poor predictor.

From Concrete to Abstract
Have students explain how a scatter plot is useful for gaining insight into possible relationships between two variables. If the scatter plot shows that the data are in a straight line, a linear equation might be a good model. If they appear curved, a nonlinear model should be explored.

LESSON 4-2 Solving Quadratic Equations by Graphing

∴ Then	∴ Now	∴ Why?

- You solved systems of equations by graphing.

1 Solve quadratic equations by graphing.

2 Estimate solutions of quadratic equations by graphing.

Arielle works in the marketing department of a major retailer. Her job is to set prices for new products sold in the stores. Arielle determined that for a certain product, the function $f(p) = -6p^2 + 192p - 1440$ tells the profit $f(p)$ made at price p.

Arielle can determine the price range by finding the prices for which the profit is equal to $0. This can be done by finding the solutions of the related quadratic equation $0 = -6p^2 + 192p - 1440$.

The graph of the function indicates that the profit is zero at 12 and 20, so the profitable price range of the item is between $12 and $20.

Profit Function

(graph: y-axis labeled "Profit for Company" from 0 to 100 in increments of 10; x-axis labeled "Price of Item" from 4 to 40 in increments of 4)

NewVocabulary
quadratic equation
standard form
root
zero

Common Core State Standards

Content Standards

A.CED.2 Create equations in two or more variables to represent relationships between quantities; graph equations on coordinate axes with labels and scales.

F.IF.4 For a function that models a relationship between two quantities, interpret key features of graphs and tables in terms of the quantities, and sketch graphs showing key features given a verbal description of the relationship.

Mathematical Practices

3 Construct viable arguments and critique the reasoning of others.

Jack Sullivan/Alamy

1 **Solve Quadratic Equations** **Quadratic equations** are quadratic functions that are set equal to a value. The **standard form** of a quadratic equation is $ax^2 + bx + c = 0$, where $a \neq 0$ and a, b, and c are integers.

The solutions of a quadratic equation are called the **roots** of the equation. One method for finding the roots of a quadratic equation is to find the **zeros** of the related quadratic function.

The zeros of the function are the x-intercepts of its graph.

Quadratic Function

$$f(x) = x^2 - x - 6$$

$$f(-2) = (-2)^2 - (-2) - 6 \text{ or } 0$$
$$f(3) = 3^2 - 3 - 6 \text{ or } 0$$

-2 and 3 are zeros of the function.

Quadratic Equation

$$x^2 - x - 6 = 0$$

$$(-2)^2 - (-2) - 6 \text{ or } 0$$
$$3^2 - 3 - 6 \text{ or } 0$$

-2 and 3 are roots of the equation.

Graph of Function

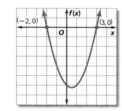

(graph with points $(-2, 0)$ and $(3, 0)$ marked, parabola $f(x)$ through origin region)

The x-intercepts are -2 and 3.

connectED.mcgraw-hill.com **229** 🔊

---

1 Focus

VerticalAlignment

▼

Before Lesson 4-2 Solve systems of equations by graphing.

▼

Lesson 4-2 Solve quadratic equations by graphing. Estimate solutions of quadratic equations by graphing.

▼

After Lesson 4-2 Solve quadratic equations using algebraic methods.

2 Teach

Scaffolding Questions

Have students read the Why? section of the lesson.

Ask:

- How can you tell by looking at the profit function's equation that the parabola opens downward? The coefficient of the p^2 term is negative.

- What item price yields the greatest profit? $16

- Explain how you can verify that a $12 item will give a $0 profit? Substitute 12 for p in the function $f(p) = -6p^2 + 192p - 1440$.

Lesson 4-2 Resources

Resource	Approaching Level AL	On Level OL	Beyond Level BL	English Learners ELL
Teacher Edition	▪ Differentiated Instruction, p. 232	▪ Differentiated Instruction, p. 236	▪ Differentiated Instruction, p. 236	▪ Differentiated Instruction, p. 232
Chapter Resource Masters	▪ Study Guide and Intervention, pp. 11–12 ▪ Skills Practice, p. 13 ▪ Practice, p. 14 ▪ Word Problem Practice, p. 15	▪ Study Guide and Intervention, pp. 11–12 ▪ Skills Practice, p. 13 ▪ Practice, p. 14 ▪ Word Problem Practice, p. 15 ▪ Enrichment, p. 16	▪ Practice, p. 14 ▪ Word Problem Practice, p. 15 ▪ Enrichment, p. 16	▪ Study Guide and Intervention, pp. 11–12 ▪ Skills Practice, p. 13 ▪ Practice, p. 14 ▪ Word Problem Practice, p. 15
Other	▪ 5-Minute Check 4-2 ▪ Study Notebook ▪ Teaching Algebra with Manipulatives	▪ 5-Minute Check 4-2 ▪ Study Notebook ▪ Teaching Algebra with Manipulatives	▪ 5-Minute Check 4-2 ▪ Study Notebook	▪ 5-Minute Check 4-2 ▪ Study Notebook ▪ Teaching Algebra with Manipulatives

1 Solve Quadratic Equations

Example 1 shows how to solve a quadratic equation with two real solutions by graphing. **Example 2** shows how to solve a quadratic equation with one real solution by graphing. **Example 3** shows how to solve a quadratic equation with no real solutions by graphing.

Formative Assessment

Use the Guided Practice exercises after each example to determine students' understanding of concepts.

Additional Examples

1 Solve $x^2 + 6x + 8 = 0$ by graphing. −4 and −2

2 Solve $x^2 - 4x = -4$ by graphing. 2

▶ **Additional Examples** also in Interactive Classroom PowerPoint® Presentations

IWB Interactive White Board READY

Example 1 Two Real Solutions

Solve $x^2 - 3x - 4 = 0$ by graphing.

Graph the related function, $f(x) = x^2 - 3x - 4$. The equation of the axis of symmetry is $x = -\dfrac{-3}{2(1)}$ or 1.5. Make a table using x-values around 1.5. Then graph each point.

x	−1	0	1	1.5	2	3	4
$f(x)$	0	−4	−6	−6.25	−6	−4	0

The zeros of the function are −1 and 4. Therefore, the solutions of the equation are −1 and 4 or $\{x \mid x = -1, 4\}$.

▶ **Guided**Practice

Solve each equation by graphing. 1A, 1B. See margin.

1A. $x^2 + 2x - 15 = 0$ **1B.** $x^2 - 8x = -12$

The graph of the related function in Example 1 has two zeros; therefore, the quadratic equation has two real solutions. This is one of the three possible outcomes when solving a quadratic equation.

KeyConcept Solutions of a Quadratic Equation

| Words | A quadratic equation can have one real solution, two real solutions, or no real solutions. |

Models

one real solution   two real solutions   no real solution

Example 2 One Real Solution

Solve $14 - x^2 = -6x + 23$ by graphing.

$14 - x^2 = -6x + 23$	Original equation
$14 = x^2 - 6x + 23$	Add x^2 to each side.
$0 = x^2 - 6x + 9$	Subtract 14.

Graph the related function $f(x) = x^2 - 6x + 9$.

x	1	2	3	4	5
$f(x)$	4	1	0	1	4

The function has only one zero, 3. Therefore, the solution is 3 or $\{x \mid x = 3\}$.

▶ **Guided**Practice **2A, 2B. See Chapter 4 Answer Appendix for graphs.**

Solve each equation by graphing.

2A. $x^2 + 5 = -8x - 11$ −4   **2B.** $12 - x^2 = 48 - 12x$ 6

StudyTip

Set-Builder Notation
In Lesson 1-5, you learned how to express the solution set of an inequality using set-builder notation. The solutions of an equation can also be expressed in set-builder notation. For example, the solutions of $x^2 = 25$ can be expressed as $\{x \mid x = -5, 5\}$.

StudyTip

Optional Graph
$f(x) = -x^2 + 6x - 9$ could also have been graphed for this example. The graph would appear different, but it would have the same solution.

Additional Answers (Guided Practice)

1A.

−5, 3

1B.

$-8-6-4-2$ O   2 4 6 8 x

6, 2

Example 3 No Real Solution

NUMBER THEORY Use a quadratic equation to find two real numbers with a sum of 15 and a product of 63.

Understand Let x represent one of the numbers. Then $15 - x$ is the other number.

Plan

$x(15 - x) = 63$	The product of the numbers is 63.
$15x - x^2 = 63$	Distributive Property
$-x^2 + 15x - 63 = 0$	Subtract 63.

Solve Graph the related function.

The graph has no x-intercepts. This means the original equation has no real solution. Thus, it is not possible for two real numbers to have a sum of 15 and a product of 63.

Check Try finding the product of several pairs of numbers with sums of 15. Is each product less than 63 as the graph suggests?

> **Guided**Practice
>
> **3.** Find two real numbers with a sum of 6 and a product of -55, or show that no such numbers exist. **11 and −5**

WatchOut!

Zeros You will see in later chapters that many zeros can appear within small intervals.

2 Estimate Solutions

Often exact roots cannot be found by graphing. You can estimate the solutions by stating the integers between which the roots are located.

When the value of the function is positive for one value and negative for a second value, then there is at least one zero between those two values.

x	-3	-2	-1	0	1	2	3
$f(x)$	12	3	-6	-2	4	8	14

zero   zero

Example 4 Estimate Roots

Solve $x^2 - 6x + 4 = 0$ by graphing. If exact roots cannot be found, state the consecutive integers between which the roots are located.

x	0	1	2	3	4	5	6
$f(x)$	4	-1	-4	-5	-4	-1	4

The x-intercepts of the graph indicate that one solution is between 0 and 1, and the other solution is between 5 and 6.

> **Guided**Practice
>
> **4.** Solve $x^2 - x - 10 = 0$ by graphing. If exact roots cannot be found, state the consecutive integers between which the roots are located.
> **between 3 and 4; between −3 and −2**

Additional Example

3 **NUMBER THEORY** Use a quadratic equation to find two real numbers with a sum of 4 and a product of 5.

The graph of the related function does not intersect the x-axis. Thus, no such numbers exist.

2 Estimate Solutions

Example 4 shows how to estimate the roots of a quadratic equation when the exact roots cannot be found. **Example 5** shows how to use a table to approximate the roots of a quadratic equation. **Example 6** shows how to use a graphing calculator to approximate the solutions of a quadratic equation in a real-world situation.

Additional Example

4 Solve $-x^2 + 4x - 1 = 0$ by graphing. If exact roots cannot be found, state the consecutive integers between which the roots are located. One solution is between 0 and 1; the other is between 3 and 4.

Focus on Mathematical Content

Solutions of a Quadratic Equation In a complex number system, a quadratic equation with real coefficients can have one real solution (a double root), two unequal real solutions, or two complex solutions. If the graph of the related quadratic function has one x-intercept, then there is one real solution (double root). If the graph of the related quadratic function has two x-intercepts, then there are two real solutions. If the graph of the related quadratic function does not intersect the x-axis, then there are no real solutions, but two imaginary solutions.

Additional Examples

5 Solve $x^2 + 5x - 7 = 0$.

To the nearest thousandth, the solutions are −6.140 and 1.140.

6 **ROYAL GORGE BRIDGE** The highest bridge in the U.S. is the Royal Gorge Bridge in Colorado. The deck is 1053 feet above the river. Suppose a marble is dropped over the railing from a height of 3 feet above the bridge deck. How long will it take the marble to reach the surface of the water, assuming there is no air resistance? Use the formula $h(t) = -16t^2 + h_0$, where t is time in seconds and h_0, is the initial height above the water in feet. about 8 s

Teach with Tech

Portable Media Player Load images of graphs of quadratic functions onto students' portable media players. For each graph, have students find the zeros of the function.

You can also use tables to solve quadratic equations. After entering the equation in your calculator, scroll through the table to locate the solutions.

Example 5 **Solve by Using a Table**

Solve $x^2 - 6x + 2 = 0$.

Enter **Y1** $= x^2 - 6x + 2$ in your graphing calculator. Use the **TABLE** window to find where the sign of **Y1** changes. Change $\triangle$**Tbl** to 0.1 and look again for the sign change. Repeat the process with 0.01 and 0.001 to get a more accurate location of the zero.

One solution is approximately 0.354.

GuidedPractice

5. Locate the second zero in the function above to the nearest thousandth. 5.646

Quadratic equations can be solved with a calculator as well. After entering the equation, use the **zero** feature in the **CALC** menu.

Real-World Example 6 **Solve by Using a Calculator**

RELIEF A package of supplies is tossed from a helicopter at an altitude of 200 feet. The package's height above the ground is modeled by $h(t) = -16t^2 + 28t + 200$, where t is the time in seconds after it is tossed. How long will it take the package to reach the ground?

We need to find t when $h(t)$ is 0. Solve $0 = -16t^2 + 28t + 200$. Then graph the related function $f(t) = -16t^2 + 28t + 200$ on a graphing calculator.

- Use the **zero** feature in the **CALC** menu to find the positive zero of the function, since time cannot be negative.

- Use the arrow keys to select a left bound and press [ENTER].

- Locate a right bound and press [ENTER] twice.

- The positive zero of the function is about 4.52. The package would take about 4.52 seconds to reach the ground.

[−10, 10] scl: 1 by [−200, 200] scl: 20

GuidedPractice

6. How long would it take to reach the ground if the height was modeled by $h(t) = -16t^2 + 48t + 400$? about 6.7 seconds

Real-WorldLink

Each year, the American Red Cross responds to over 70,000 disaster situations ranging from house fires to natural disasters such as hurricanes and earthquakes.

 232 | Lesson 4-2 | Solving Quadratic Equations by Graphing

Emmanuel Duparcq/AFP/Getty Images

DifferentiatedInstruction (AL) (ELL)

Visual/Spatial Learners Provide students with a variety of parabolas. Ask students to sort the parabolas into three piles: those that model a quadratic equation with one real solution, those that model a quadratic equation with two real solutions, and those that model a quadratic equation with no real solutions. Finally, ask students to name the real solutions for those in the piles having real solutions.

Example 1   Use the related graph of each equation to determine its solutions.

1. no real solution

1. $x^2 + 2x + 3 = 0$

2. $x^2 - 3x - 10 = 0$ $-2, 5$

3. $-x^2 - 8x - 16 = 0$ -4

Examples 2–5 (CCSS) **PRECISION** Solve each equation. If exact roots cannot be found, state the consecutive integers between which the roots are located. **4–11.** See margin.

4. $x^2 + 8x = 0$

5. $x^2 - 3x - 18 = 0$

6. $4x - x^2 + 8 = 0$

7. $-12 - 5x + 3x^2 = 0$

8. $x^2 - 6x + 4 = -8$

9. $9 - x^2 = 12$

10. $5x^2 + 10x - 4 = -6$

11. $x^2 - 20 = 2 + x$

12. **NUMBER THEORY** Use a quadratic equation to find two real numbers with a sum of 2 and a product of -24. 6 and -4

Example 6   **13.** **PHYSICS** How long will it take an object to fall from the roof of a building 400 feet above the ground? Use the formula $h(x) = -16t^2 + h_0$, where t is the time in seconds and the initial height h_0 is in feet. 5 seconds

Practice and Problem Solving   Extra Practice is on page R4.

Example 1   Use the related graph of each equation to determine its solutions.

15. no real solution

14. $x^2 + 4x = 0$ $-4, 0$

15. $-2x^2 - 4x - 5 = 0$

16. $0.5x^2 - 2x + 2 = 0$ 2

18. no real solution

17. $-0.25x^2 - x - 1 = 0$ -2

18. $x^2 - 6x + 11 = 0$

19. $-0.5x^2 + 0.5x + 6 = 0$ $-3, 4$

Differentiated Homework Options

Level	Assignment	Two-Day Option	
AL Basic	14–37, 52, 54–76	15–37 odd, 57–60	14–36 even, 52, 54–56, 61–76
OL Core	5–37 odd, 38–40, 41–47 odd, 49–52, 54–76	14–37, 57–60	38–52, 54–56, 61–76
BL Advanced	38–73, (optional: 74–76)		

3 Practice

Formative Assessment

Use Exercises 1–13 to check for understanding.

Use the chart on the bottom of this page to customize assignments for your students.

(CCSS) **Teaching the Mathematical Practices**

Precision Mathematically proficient students try to use clear definitions in their reasoning, calculate accurately and efficiently, and make explicit use of definitions.

Tips for New Teachers

Reasoning Some students may notice in Exercise 3 that $-x^2 - 8x - 16 = 0$ is equivalent to $x^2 + 8x + 16 = 0$. Although the two equations have the same solution of -4, the related functions $f(x) = -x^2 - 8x - 16$ and $f(x) = x^2 + 8x + 16$ are not equivalent. This can be seen by looking at their graphs, which open in opposite directions.

Additional Answers

4–11. See Ch. 4 Answer Appendix for graphs.

4. $0, -8$

5. $-3, 6$

6. between -2 and -1, between 5 and 6

7. between -2 and -1, 3

8. no real solution

9. no real solution

10. between -2 and -1, between -1 and 0

11. between -5 and -4, between 5 and 6

Study Guide and Intervention
AL OL ELL

Practice
AL OL BL ELL

Word Problem Practice
AL OL BL ELL

Examples 2–4 Solve each equation. If exact roots cannot be found, state the consecutive integers between which the roots are located. **20–29. See Chapter 4 Answer Appendix for graphs.**

20. $x^2 = 5x$ **0, 5**

21. $-2x^2 - 4x = 0$ **−2, 0**

22. $x^2 - 5x - 14 = 0$ **−2, 7**

23. $-x^2 + 2x + 24 = 0$ **−4, 6**

24. $x^2 - 18x = -81$ **9**

25. $2x^2 - 8x = -32$ **no real solution**

26. $2x^2 - 3x - 15 = 4$ **between −3 and −2 and between 3 and 4**

27. $-3x^2 - 7 + 2x = -11$ **between −1 and 0 and between 1 and 2**

28. $-0.5x^2 + 3 = -5x - 2$

29. $-2x + 12 = x^2 + 16$ **no real solution**

28. between −1 and 0 and between 10 and 11

Example 5 Use the tables to determine the location of the zeros of each quadratic function.

30.

x	−7	−6	−5	−4	−3	−2	−1	0
f(x)	−8	−1	4	4	−1	−8	−22	−48

between −6 and −5; between −4 and −3

31.

x	−2	−1	0	1	2	3	4	5
f(x)	32	14	2	−3	−3	2	14	32

between 0 and 1; between 2 and 3

32.

x	−6	−3	0	3	6	9	12	15
f(x)	−6	−1	3	5	3	−1	−6	−14

between −3 and 0; between 6 and 9

Example 6 **NUMBER THEORY** Use a quadratic equation to find two real numbers that satisfy each situation, or show that no such numbers exist.

33. Their sum is −15, and their product is −54. **3 and −18**

34. Their sum is 4, and their product is −117. **13 and −9**

35. Their sum is 12, and their product is −84. **about −5 and 17**

36. Their sum is −13, and their product is 42. **−6 and −7**

37. Their sum is −8 and their product is −209. **11 and −19**

B **CCSS MODELING** For Exercises 38–40, use the formula $h(t) = v_0 t - 16t^2$, where $h(t)$ is the height of an object in feet, v_0 is the object's initial velocity in feet per second, and t is the time in seconds.

38. **BASEBALL** A baseball is hit with an initial velocity of 80 feet per second. Ignoring the height of the baseball player, how long does it take for the ball to hit the ground? **5 seconds**

39. **CANNONS** A cannonball is shot with an initial velocity of 55 feet per second. Ignoring the height of the cannon, how long does it take for the cannonball to hit the ground? **about 3.4375 seconds**

40. **GOLF** A golf ball is hit with an initial velocity of 100 feet per second. How long will it take for it to hit the ground? **6.25 seconds**

45. between −1 and 0, between 4 and 5

47. between 3 and 4, between 8 and 9

Solve each equation. If exact roots cannot be found, state the consecutive integers between which the roots are located. **41–48. See Chapter 4 Answer Appendix for graphs.**

41. $2x^2 + x = 15$ **−3, between 2 and 3**

42. $-5x - 12 = -2x^2$ **$-\frac{3}{2}$, 4**

43. $4x^2 - 15 = -4x$ **between −3 and −2, between 1 and 2**

44. $-35 = -3x - 2x^2$ **−5, between 3 and 4**

45. $-3x^2 + 11x + 9 = 1$

46. $13 - 4x^2 = -3x$ **between −2 and −1, between 2 and 3**

47. $-0.5x^2 + 18 = -6x + 33$

48. $0.5x^2 + 0.75 = 0.25x$ **no real solution**

234 | Lesson 4-2 | Solving Quadratic Equations by Graphing

Enrichment
OL BL

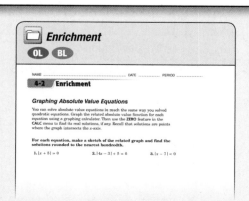

4-2 Enrichment

Graphing Absolute Value Equations

You can solve absolute value equations in much the same way you solved quadratic equations. Graph the related absolute value function for each equation using a graphing calculator. Then use the ZERO feature in the CALC menu to find its real solutions, if any. Recall that solutions are points where the graph intersects the x-axis.

For each equation, make a sketch of the related graph and find the solutions rounded to the nearest hundredth.

1. $|x + 5| = 0$ 2. $|4x - 3| + 5 = 0$ 3. $|x - 7| = 0$

CCSS **Teaching the Mathematical Practices**

Modeling Mathematically proficient students can apply the mathematics they know to solve problems arising in everyday life, analyze relationships mathematically to draw conclusions, and interpret their mathematical results in the context of a situation.

49 **WATER BALLOONS** Tony wants to drop a water balloon so that it splashes on his brother. Use the formula $h(t) = -16t^2 + h_0$, where t is the time in seconds and the initial height h_0 is in feet, to determine how far his brother should be from the target when Tony lets go of the balloon. **about 8.5 ft**

60 feet

4.4 ft/sec

50. **WATER HOSES** A water hose can spray water at an initial velocity of 40 feet per second. Use the formula $h(t) = v_0t - 16t^2$, where $h(t)$ is the height of the water in feet, v_0 is the initial velocity in feet per second, and t is the time in seconds.

  a. How long will it take the water to hit the nozzle on the way down? **2.5 seconds**

  b. Assuming the nozzle is 5 feet up, what is the maximum height of the water? **30 ft**

51. **SKYDIVING** In 2003, John Fleming and Dan Rossi became the first two blind skydivers to be in free fall together. They jumped from an altitude of 14,000 feet and free fell to an altitude of 4000 feet before their parachutes opened. Ignoring air resistance and using the formula $h(t) = -16t^2 + h_0$, where t is the time in seconds and the initial height h_0 is in feet, determine how long they were in free fall. **25 seconds**

Jump from plane at 14,000 ft

Free-fall

Open parachutes at 4,000 ft

H.O.T. Problems   Use Higher-Order Thinking Skills

52. **CRITIQUE** Hakeem and Tanya were asked to find the location of the roots of the quadratic function represented by the table. Is either of them correct? Explain.

x	−4	−2	0	2	4	6	8	10
f(x)	52	26	8	−2	−4	2	16	38

> **Hakeem**
> The roots are between 4 and 6 because $f(x)$ stops decreasing and begins to increase between $x = 4$ and $x = 6$.

> **Tanya**
> The roots are between −2 and 0 because x changes signs at that location.

Sample answer: No; roots are located where $f(x)$ changes signs.

56. Sample answer: Graph the function using the axis of symmetry. Determine where the graph intersects the x-axis. The x-coordinates of those points are solutions to the quadratic equation.

53. **CHALLENGE** Find the value of a positive integer k such that $f(x) = x^2 - 2kx + 55$ has roots at $k + 3$ and $k - 3$. **$k = 8$**

54. **REASONING** If a quadratic function has a minimum at $(-6, -14)$ and a root at $x = -17$, what is the other root? Explain your reasoning. **Sample answer: The other root is at $x = 5$ because the x-coordinates of the roots need to be equidistant from the x-value of the vertex.**

55. **OPEN ENDED** Write a quadratic function with a maximum at $(3, 125)$ and roots at -2 and 8. **$f(x) = -5x^2 + 30x + 80$**

56. **WRITING IN MATH** Explain how to solve a quadratic equation by graphing its related quadratic function.

Critique Mathematically proficient students are also able to compare the effectiveness of two plausible arguments, distinguish correct logic or reasoning from that which is flawed, and—if there is a flaw in an argument—explain what it is.

WatchOut!

Error Analysis For Exercise 52, have students think about the graph of a quadratic function, and contrast behavior near an intercept with behavior near a maximum or minimum.

Follow-up

- How can the graph of a quadratic function help you solve the corresponding quadratic equation?
  Sample answer: From the placement of the graph, you can see whether there are 0, 1, or 2 real solutions. The symmetry of the graph helps you identify a second solution when one solution is found.

4 Assess

Crystal Ball Have students write how what they learned today about solving quadratic equations by graphing will connect with tomorrow's theme of solving quadratic equations by factoring.

Formative Assessment

Check for student understanding of concepts in Lessons 4-1 and 4-2.

 Quiz 1, p. 59

Additional Answers

61. maximum, −12; D: {all real numbers}, R: $\{f(x) \mid f(x) \le -12\}$

62. minimum, −30; D: {all real numbers}, R: $\{f(x) \mid f(x) \ge -30)\}$

63. maximum, 15; D: {all real numbers}, R: $\{f(x) \mid f(x) \le 15\}$

67a. $\begin{bmatrix} 10 & 10 & 15 \\ 25 & 35 & 45 \end{bmatrix}$

67b. $2; \begin{bmatrix} 20 & 20 & 30 \\ 50 & 70 & 90 \end{bmatrix}$

67c. $\begin{bmatrix} 10 & 10 & 15 \\ 25 & 35 & 45 \end{bmatrix}$; the number of additional shirts that he needs to stock

---

Standardized Test Practice

Standardized Test Practice

57. SHORT RESPONSE A bag contains five different colored marbles. The colors of the marbles are black, silver, red, green, and blue. A student randomly chooses a marble. Then, without replacing it, he chooses a second marble. What is the probability that the student chooses the red and then the green marble? $\frac{1}{20}$

58. Which number would be closest to zero on the number line? **B**

    0

  A −0.6   **C** 0.5

  B $\frac{2}{5}$ **D** $\frac{\sqrt{2}}{2}$

59. SAT/ACT A salesman's monthly gross pay consists of $3500 plus 20 percent of the dollar amount of his sales. If his gross pay for one month was $15,500, what was the dollar amount of his sales for that month? **H**

  F $12,000   **J** $70,000

  G $16,000   **K** $77,500

  H $60,000

60. Find the next term in the sequence below. **A**

$$\frac{2x}{5}, \frac{3x}{5}, \frac{4x}{5}, \dots$$

  A x **C** $\frac{x}{5}$

  B $5x$ **D** $\frac{5x}{4}$

Spiral Review

Determine whether each function has a *maximum* or *minimum* value, and find that value. Then state the domain and range of the function. (Lesson 4-1) **61–63. See margin.**

61. $f(x) = -4x^2 + 8x - 16$ **62.** $f(x) = 3x^2 + 12x - 18$ **63.** $f(x) = 4x + 13 - 2x^2$

Determine whether each pair of matrices are inverses of each other. (Lesson 3-8)

64. $\begin{bmatrix} 4 & -3 \\ -1 & -6 \end{bmatrix}$ and $\begin{bmatrix} \frac{3}{13} & -\frac{1}{18} \\ -\frac{1}{26} & -\frac{2}{13} \end{bmatrix}$ **65.** $\begin{bmatrix} 6 & -3 \\ 4 & 8 \end{bmatrix}$ and $\begin{bmatrix} \frac{1}{10} & \frac{1}{20} \\ -\frac{1}{15} & \frac{2}{15} \end{bmatrix}$ **66.** $\begin{bmatrix} 2 & 4 \\ -3 & -2 \end{bmatrix}$ and $\begin{bmatrix} -\frac{1}{4} & -\frac{1}{2} \\ \frac{3}{8} & \frac{1}{4} \end{bmatrix}$

  no   no   yes

67. SALES Alex is in charge of stocking shirts for the concession stand at the high school football game. The numbers of shirts needed for a regular season game are listed in the matrix. Alex plans to double the number of shirts stocked for a playoff game. (Lesson 3-6)

Size	small	medium	large
Child	10	10	15
Adult	25	35	45

  a. Write a matrix A to represent the regular season stock. **a–c. See margin.**

  b. What scalar can be used to determine a matrix M to represent the new numbers? Find M.

  c. What is $M - A$? What does this represent in this situation?

Solve each system of equations. (Lesson 3-1)

68. $4x - 7y = -9$ $(-4, -1)$ **69.** $8y - 2x = 38$ $(-3, 4)$ **70.** $3x + 8y = 24$ **no solution**

    $5x + 2y = -22$ $5x - 3y = -27$ $-16y - 6x = 48$

Solve each inequality. (Lesson 1-5)

71. $3x - 6 \le -14$ $x \le -\frac{8}{3}$ **72.** $6 - 4x \le 2$ $x \ge 1$ **73.** $-6x + 3 \ge 3x - 16$ $x \le \frac{19}{9}$

Skills Review

Find the GCF of each set of numbers.

74. 16, 48, 128 **16**   **75.** 15, 21, 49 **1**   **76.** 12, 28, 36 **4**

DifferentiatedInstruction **BL** **OL**

Extension Tell students that they can solve absolute value equations by graphing just as they solve quadratic equations by graphing. Write the following equations on the board:

$$|x + 1| = 0 \qquad\qquad |x - 4| - 1 = 0$$

Have students use a graphing calculator to graph the related absolute value function for each equation. Then have them use the **ZERO** feature from the **CALC** menu to find its real solutions, if any, rounded to the nearest hundredth. **−1; 3, 5**

EXTEND 4-2 Graphing Technology Lab
Solving Quadratic Equations by Graphing

You can use a TI-83/84 Plus graphing calculator to solve quadratic equations.

CCSS Common Core State Standards
Content Standards
A.REI.11 Explain why the x-coordinates of the points where the graphs of the equations $y = f(x)$ and $y = g(x)$ intersect are the solutions of the equation $f(x) = g(x)$; find the solutions approximately, e.g., using technology to graph the functions, make tables of values, or find successive approximations. Include cases where $f(x)$ and/or $g(x)$ are linear, polynomial, rational, absolute value, exponential, and logarithmic functions.

Activity   Solving Quadratic Equations

Solve $x^2 - 8x + 15 = 0$.

Step 1 Let **Y1** $= x^2 - 8x + 15$ and **Y2** $= 0$.

Step 2 Graph **Y1** and **Y2** in the standard viewing window.

KEYSTROKES: Y= X,T,θ,n x² (−) 8 X,T,θ,n + 15 ENTER 0 ZOOM 6

[−10, 10] scl: 1 by [−10, 10] scl: 1

Step 3 To find the x-intercepts, determine the points where **Y1** = **Y2**.

KEYSTROKES: 2nd [CALC] 5

Press ENTER for the first equation. Press ENTER for the second equation.

Move the cursor as close to the left x-intercept and press ENTER.

[−10, 10] scl: 1 by [−10, 10] scl: 1

Find the second x-intercept.

KEYSTROKES: 2nd [CALC] 5

Press ENTER for the first equation. Press ENTER for the second equation.

Move the cursor as close to the right x-intercept and press ENTER.

The x-intercepts are 3 and 5, so $x = 3$ and $x = 5$.

[−10, 10] scl: 1 by [−10, 10] scl: 1

Exercises

Solve each equation. Round to the nearest tenth if necessary.

1. $x^2 - 7x + 12 = 0$ **3, 4**
2. $x^2 + 5x + 6 = 0$ **−3, −2**
3. $x^2 - 3 = 2x$ **−1, 3**
4. $x^2 + 5x + 6 = 12$ **−6, 1**
5. $x^2 + 5x = 0$ **0, −5**
6. $x^2 - 4 = 0$ **−2, 2**
7. $x^2 + 8x + 16 = 0$ **−4**
8. $x^2 - 10x = -25$ **5**
9. $9x^2 + 48x + 64 = 0$ **−2.7**
10. $2x^2 + 3x - 1 = 0$ **−1.8, 0.3**
11. $5x^2 - 7x = -2$ **0.4, 1**
12. $6x^2 + 2x + 1 = 0$ **no real solution**

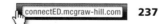 connectED.mcgraw-hill.com **237**

From Concrete to Abstract

Have students explain why the intercepts of the graph of a function can be used to identify the roots of an equation. The x-intercepts of a graph occur where $y = 0$. At these points, the equation of the function reduces to a single-variable equation in x.

Extending the Concept
Ask:
- The x-intercepts in the Activity are 3 and 5. Have students multiply $(x - 3)(x - 5)$ and ask them to compare the result to the equation they have solved.

1 Focus

Objective Use a graphing calculator to solve quadratic equations.

Materials for Each Group
- TI-83/84 Plus or other graphing calculator

Teaching Tips
In order to use a graph to solve the quadratic equation $ax^2 + bx + c = 0$, students must graph the related function $y = ax^2 + bx + c$.

2 Teach

Working in Cooperative Groups
Put students in groups of two or three, mixing abilities. Have each group work through the Activity, comparing screens with each other as they go.

- Steps 3 and 4 can only be done while the graph is showing on the screen. If students try to do these steps from the home screen, remind them to display the graph first.
- When using TRACE, the cursor's initial position is at the y-intercept, so it may not appear on the screen.

Practice Have students complete Exercises 1–12.

3 Assess

Formative Assessment
Use Exercise 3 to assess whether students understand that a quadratic equation should first be written in the form $ax^2 + bx + c = 0$ before solving.

1 Focus

VerticalAlignment

Before Lesson 4-3 Find the greatest common factors of sets of numbers.

Lesson 4-3 Write quadratic equations in intercept form. Solve quadratic equations by factoring.

After Lesson 4-3 Solve quadratic equations using the Quadratic Formula.

2 Teach

Scaffolding Questions
Have students read the **Why?** section of the lesson.

Ask:

- Does $x^2 - 8x + 12$ have a maximum or minimum value? minimum

- Solve $y = x^2 - 8x + 12 = 0$ by graphing. $2, 6$

- Compare the soultions to $x^2 - 8x + 12 = 0$ and $(x - 6)(x - 2) = 0$. The solutions are the same because the equations are equivalent.

LESSON 4-3 Solving Quadratic Equations by Factoring

::Then	::Now	::Why?
• You found the greatest common factors of sets of numbers.	**1** Write quadratic equations in standard form. **2** Solve quadratic equations by factoring.	The **factored form** of a quadratic equation is $0 = a(x - p)(x - q)$. In the equation, p and q represent the x-intercepts of the graph of the equation. The x-intercepts of the graph at the right are 2 and 6. In this lesson, you will learn how to change a quadratic equation in factored form into standard form and vice versa.

Related Graph
2 and 6 are
x-intercepts.

Standard Form   Factored Form

$$0 = x^2 - 8x + 12 \qquad 0 = (x - 6)(x - 2)$$

Factors

NewVocabulary
factored form
FOIL method

Common Core State Standards

Content Standards
A.SSE.2 Use the structure of an expression to identify ways to rewrite it.

F.IF.8.a Use the process of factoring and completing the square in a quadratic function to show zeros, extreme values, and symmetry of the graph, and interpret these in terms of a context.

Mathematical Practices
2 Reason abstractly and quantitatively.

1 Standard Form You can use the FOIL method to write a quadratic equation that is in factored form in standard form. The **FOIL method** uses the Distributive Property to multiply binomials.

KeyConcept FOIL Method for Multiplying Binomials

Words — To multiply two binomials, find the sum of the products of F the *First terms*, O the *Outer terms*, I the *Inner terms*, and L the *Last terms*.

Examples

Product of First Terms	Product of Outer Terms	Product of Inner Terms	Product of Last Terms
↓	↓	↓	↓

$$(x - 6)(x - 2) = (x)(x) + (x)(-2) + (-6)(x) + (-6)(-2)$$

$$= x^2 - 2x - 6x + 12 \text{ or } x^2 - 8x + 12$$

Example 1 Translate Sentences into Equations

Write a quadratic equation in standard form with $-\dfrac{1}{3}$ and 6 as its roots.

$$(x - p)(x - q) = 0 \qquad \text{Write the pattern.}$$

$$\left[x - \left(-\tfrac{1}{3}\right)\right](x - 6) = 0 \qquad \text{Replace } p \text{ with } -\tfrac{1}{3} \text{ and } q \text{ with 6.}$$

$$\left(x + \tfrac{1}{3}\right)(x - 6) = 0 \qquad \text{Simplify.}$$

$$x^2 - \tfrac{17}{3}x - 2 = 0 \qquad \text{Multiply.}$$

$$3x^2 - 17x - 6 = 0 \qquad \text{Multiply each side by 3 so that } b \text{ and } c \text{ are integers.}$$

> **Guided**Practice
> $4x^2 + 17x - 15 = 0$
> **1.** Write a quadratic equation in standard form with $\dfrac{3}{4}$ and -5 as its roots.

 238 | Lesson 4-3

Lesson 4-3 Resources

Resource	Approaching Level **AL**	On Level **OL**	Beyond Level **BL**	English Learners **ELL**
Teacher Edition	• Differentiated Instruction, p. 239	• Differentiated Instruction, pp. 239, 245	• Differentiated Instruction, pp. 239, 245	
Chapter Resource Masters	• Study Guide and Intervention, pp. 17–18 • Skills Practice, p. 19 • Practice, p. 20 • Word Problem Practice, p. 21	• Study Guide and Intervention, pp. 17–18 • Skills Practice, p. 19 • Practice, p. 20 • Word Problem Practice, p. 21 • Enrichment, p. 22 • Graphing Calculator Activity, p. 23	• Practice, p. 20 • Word Problem Practice, p. 21 • Enrichment, p. 22	• Study Guide and Intervention, pp. 17–18 • Skills Practice, p. 19 • Practice, p. 20 • Word Problem Practice, p. 21
Other	• 5-Minute Check 4-3 • Study Notebook	• 5-Minute Check 4-3 • Study Notebook	• 5-Minute Check 4-3 • Study Notebook	• 5-Minute Check 4-3 • Study Notebook

2 Solve Equations by Factoring
Solving quadratic equations by factoring is an application of the Zero Product Property.

KeyConcept Zero Product Property

Words	For any real numbers a and b, if $ab = 0$, then either $a = 0$, $b = 0$, or both a and b equal zero.
Example	If $(x + 3)(x - 5) = 0$, then $x + 3 = 0$ or $x - 5 = 0$.

Example 2 Factor the GCF

Solve $16x^2 + 8x = 0$.

$$16x^2 + 8x = 0 \qquad \text{Original equation.}$$
$$8x(2x) + 8x\,(1) = 0 \qquad \text{Factor the GCF.}$$
$$8x(2x + 1) = 0 \qquad \text{Distributive Property}$$
$$8x = 0 \text{ or } 2x + 1 = 0 \qquad \text{Zero Product Property}$$
$$x = 0 \qquad 2x = -1 \qquad \text{Solve both equations.}$$
$$x = -\frac{1}{2}$$

▶ **Guided**Practice Solve each equation.

2A. $20x^2 + 15x = 0$ $0, -\frac{3}{4}$ **2B.** $4y^2 + 16y = 0$ $0, -4$ **2C.** $6a^5 + 18a^4 = 0$ $0, -3$

ReviewVocabulary
perfect square a number with a positive square root that is a whole number

Trinomials and binomials that are perfect squares have special factoring rules. In order to use these rules, the first and last terms need to be perfect squares and the middle term needs to be twice the product of the square roots of the first and last terms.

Example 3 Perfect Squares and Differences of Squares

Solve each equation.

a. $x^2 + 16x + 64 = 0$

$$x^2 = (x)^2; 64 = (8)^2 \qquad \text{First and last terms are perfect squares.}$$
$$16x = 2(x)(8) \qquad \text{Middle term equals } 2ab.$$

$x^2 + 16x + 64$ is a perfect square trinomial.

$$x^2 + 16x + 64 = 0 \qquad \text{Original equation}$$
$$(x + 8)^2 = 0 \qquad \text{Factor using the pattern.}$$
$$x + 8 = 0 \qquad \text{Take the square root of each side.}$$
$$x = -8 \qquad \text{Solve.}$$

b. $x^2 = 64$

$$x^2 = 64 \qquad \text{Original equation}$$
$$x^2 - 64 = 0 \qquad \text{Subtract 64 from each side.}$$
$$x^2 - (8)^2 = 0 \qquad \text{Write in the form } a^2 - b^2.$$
$$(x + 8)(x - 8) = 0 \qquad \text{Factor the difference of squares.}$$
$$x + 8 = 0 \text{ or } x - 8 = 0 \qquad \text{Zero Product Property}$$
$$x = -8 \qquad x = 8 \qquad \text{Solve.}$$

StudyTip
Square Roots By inspection, notice that the square root of 64 is -8 and the 8. Also, for $x^2 = 4$, the solutions would be -2 and 2.

▶ **Guided**Practice

3A. $4x^2 - 12x + 9 = 0$ $\frac{2}{3}$.   **3B.** $81x^2 - 9x = 0$ $0, \frac{1}{9}$ **3C.** $6a^2 - 3a = 0$ $0, \frac{1}{2}$

connectED.mcgraw-hill.com **239**

---

1 Factored Form
Example 1 shows how to write a quadratic equation for a given pair of roots.

Formative Assessment
Use the Guided Practice exercises after each example to determine students' understanding of concepts.

Additional Example

1 Write a quadratic equation in standard form with $\frac{1}{2}$ and -5 as its roots. Sample answer: $2x^2 + 9x - 5 = 0$

▶ **Additional Examples** also in Interactive Classroom PowerPoint® Presentations

IWB Interactive White Board READY

2 Solve Equations by Factoring
Examples 2 and 3 show how to solve quadratic equations by factoring and by inspection. **Example 4** shows how to solve quadratic equations using a pattern. **Example 5** shows how to solve a real-world problem using factoring to solve a quadratic equation.

Additional Examples

2 Solve each equation.
a. $9y^2 + 3y = 0$
  $-\frac{1}{3}, 0$
b. $5a^2 - 20a = 0$
  $0, 4$

3 Solve each equation.
a. $x^2 - 6x + 9 = 0$
  3
b. $y^2 = 36$
  $-6, 6$

DifferentiatedInstruction ⒶⓁ ⓄⓁ ⒷⓁ

If ▶ students think that the steps in Example 1 provide the only possible equation for the given roots,

Then ▶ provide each student with a sheet of grid paper. Have students begin by drawing a coordinate grid with two points on the *x*-axis plotted as the roots of a quadratic equation. Ask students to draw several parabolas that might be the graphs of different equations having those two points as their solutions. Point out that this demonstrates that the steps shown in Example 1 yield just *one* of the possible equations having the given roots.

Additional Example

4 Solve each equation.

a. $x^2 - 2x - 15 = 0$
$-3, 5$

b. $5x^2 + 34x + 24 = 0$
$-\dfrac{4}{5}, -6$

CCSS Teaching the Mathematical Practices

Structure Mathematically proficient students look closely to discern a pattern or structure. Encourage students to look for patterns in factors and their sums, especially when factors have opposite signs.

StudyTip

CCSS Structure If values for m and p exist, then the trinomial can always be factored.

A special pattern is used when factoring trinomials of the form $ax^2 + bx + c$. First, multiply the values of a and c. Then, find two values, m and p, such that their product equals ac and their sum equals b.

Consider $6x^2 + 13x - 5$: $ac = 6(-5) = -30$.

Factors of −30	Sum	Factors of −30	Sum
1, −30	−29	−1, 30	29
2, −15	−13	−2, 15	13
3, −10	−7	−3, 10	7
5, −6	−1	−5, 6	1

Now the middle term, $13x$, can be rewritten as $-2x + 15x$.

This polynomial can now be factored by grouping.

$$
\begin{aligned}
6x^2 + 13x - 5 &= 6x^2 + mx + px - 5 & &\text{Write the pattern.}\\
&= 6x^2 - 2x + 15x - 5 & &m = -2 \text{ and } p = 15\\
&= (6x^2 - 2x) + (15x - 5) & &\text{Group terms.}\\
&= 2x(3x - 1) + 5(3x - 1) & &\text{Factor the GCF.}\\
&= (2x + 5)(3x - 1) & &\text{Distributive Property}
\end{aligned}
$$

Example 4 Factor Trinomials

Solve each equation.

a. $x^2 + 9x + 20 = 0$

$ac = 20 \qquad a = 1, c = 20$

Factors of 20	Sum	Factors of 20	Sum
1, 20	21	−1, −20	−21
2, 10	12	−2, −10	−12
4, 5	9	−4, −5	−9

StudyTip

Trinomials It does not matter if the values of m and p are switched when grouping.

$$
\begin{aligned}
x^2 + 9x + 20 &= 0 & &\text{Original expression}\\
x^2 + mx + px + 20 &= 0 & &\text{Write the pattern.}\\
x^2 + 4x + 5x + 20 &= 0 & &m = 4 \text{ and } p = 5\\
(x^2 + 4x) + (5x + 20) &= 0 & &\text{Group terms with common factors.}\\
x(x + 4) + 5(x + 4) &= 0 & &\text{Factor the GCF from each grouping.}\\
(x + 5)(x + 4) &= 0 & &\text{Distributive Property}
\end{aligned}
$$

$x + 5 = 0 \quad \text{or} \quad x + 4 = 0 \qquad$ Zero Product Property

$\qquad x = -5 \qquad\qquad x = -4 \qquad$ Solve each equation.

b. $6y^2 - 23y + 20 = 0$

$ac = 120 \qquad\qquad\qquad a = 6, c = 20$

$m = -8, p = -15 \qquad -8(-15) = 120; -8 + (-15) = -23$

$$
\begin{aligned}
6y^2 - 23y + 20 &= 0 & &\text{Original equation}\\
6y^2 + my + py + 20 &= 0 & &\text{Write the pattern.}\\
6y^2 - 8y - 15y + 20 &= 0 & &m = -8 \text{ and } p = -15\\
(6y^2 - 8y) + (-15y + 20) &= 0 & &\text{Group terms with common factors.}\\
2y(3y - 4) - 5(3y - 4) &= 0 & &\text{Factor the GCF from each grouping.}\\
(2y - 5)(3y - 4) &= 0 & &\text{Distributive Property}
\end{aligned}
$$

$2y - 5 = 0 \quad \text{or} \quad 3y - 4 = 0 \qquad$ Zero Product Property

$2y = 5 \qquad\qquad 3y = 4 \qquad$ Solve both equations.

$y = \dfrac{5}{2} \qquad\qquad y = \dfrac{4}{3}$

Focus on Mathematical Content

Solving Quadratics by Factoring Quadratic equations can be solved using several different methods. Factoring can be a quick method. Once a polynomial has been factored, the Zero Product Property may be used to find the roots of the equation. If the polynomial is difficult to factor or is not factorable, then other methods must be used.

Guided Practice

4A. $x^2 - 11x + 30 = 0$ 5, 6

4B. $x^2 - 4x - 21 = 0$ -3, 7

4C. $15x^2 - 8x + 1 = 0$ $\frac{1}{5}$, $\frac{1}{3}$

4D. $-12x^2 + 8x + 15 = 0$ $-\frac{5}{6}$, $\frac{3}{2}$

Real-World Example 5 Solve Equations by Factoring

TRACK AND FIELD The height of a javelin in feet is modeled by $h(t) = -16t^2 + 79t + 5$, where t is the time in seconds after the javelin is thrown. How long is it in the air?

To determine how long the javelin is in the air, we need to find when the height equals 0. We can do this by solving $-16t^2 + 79t + 5 = 0$.

$$-16t^2 + 79t + 5 = 0 \qquad \text{Original equation}$$

$$m = 80; p = -1 \qquad -16(5) = -80, 80 \cdot (-1) = -80, 80 + (-1) = 79$$

$$-16t^2 + 80t - t + 5 = 0 \qquad \text{Write the pattern.}$$

$$(-16t^2 + 80t) + (-t + 5) = 0 \qquad \text{Group terms with common factors.}$$

$$16t(-t + 5) + 1(-t + 5) = 0 \qquad \text{Factor GCF from each group.}$$

$$(16t + 1)(-t + 5) = 0 \qquad \text{Distributive Property}$$

$$16t + 1 = 0 \quad \text{or} \quad -t + 5 = 0 \qquad \text{Zero Product Property}$$

$$16t = -1 \qquad\quad -t = -5 \qquad \text{Solve both equations.}$$

$$t = -\frac{1}{16} \qquad\quad t = 5 \qquad \text{Solve.}$$

CHECK We have two solutions.

- The first solution is negative and since time cannot be negative, this solution can be eliminated.
- The second solution of 5 seconds seems reasonable for the time a javelin spends in the air.
- The answer can be confirmed by substituting back into the original equation.

$$-16t^2 + 79t + 5 = 0$$

$$-16(5)^2 + 79(5) + 5 \stackrel{?}{=} 0$$

$$-400 + 395 + 5 \stackrel{?}{=} 0$$

$$0 = 0 \checkmark$$

The javelin is in the air for 5 seconds.

Guided Practice

5. **BUNGEE JUMPING** Juan recorded his brother bungee jumping from a height of 1100 feet. At the time the cord lifted his brother back up, he was 76 feet above the ground. If Juan started recording as soon as his brother fell, how much time elapsed when the cord snapped back? Use $f(t) = -16t^2 + c$, where c is the height in feet. **8 seconds**

Real-World Link

Cuba's Osleidys Menendez broke the javelin world record in 2002 with a distance of 234 feet 8 inches.

Source: New York Times

Stu Forster/Getty Images Sport/Getty Images

Additional Example

5 **ARCHITECTURE** The entrance to an office building is an arch in the shape of a parabola whose vertex is the height of the arch. The height of the arch is given by $h = 9 - x^2$, where x is the horizontal distance from the center of the arch. Both h and x are measured in feet. How wide is the arch at ground level?   6 ft

WatchOut!

Common Misconceptions In Example 5, some students may suggest solving the equation by dividing both sides by t. Point out that this cannot be done because the value of t could be zero, and division by zero is undefined.

Teach with Tech

Document Camera Choose several students to share their work with the class and explain their answers. Have students check their work by substituting their solutions into the original equation.

3 Practice

Formative Assessment

Use Exercises 1–16 to check for understanding.

Use the chart at the bottom of this page to customize assignments for your students.

 Teaching the Mathematical Practices

Sense-Making Mathematically proficient students start by explaining to themselves the meaning of a problem and looking for entry points to its solution. They analyze givens, constraints, relationships, and goals. They check their answers to problems using a different method, and they continually ask themselves, "Does this make sense?"

Additional Answers

68a.

68c.

Check Your Understanding

○ = Step-by-Step Solutions begin on page R14.

Example 1 Write a quadratic equation in standard form with the given root(s).

1. $-8, 5$ $x^2 + 3x - 40 = 0$ **2.** $\frac{3}{2}, \frac{1}{4}$ $8x^2 - 14x + 3 = 0$ **3.** $-\frac{2}{3}, \frac{5}{2}$ $6x^2 - 11x - 10 = 0$

Examples 2–4 Factor each polynomial.   **5.** $(6x - 1)(3x + 4)$ **7.** $(x - 7)(x + 3)$ **8.** $(2x - 5)(x + 6)$

4. $35x^2 - 15x$ $5x(7x - 3)$ **5.** $18x^2 - 3x + 24x - 4$ **6.** $x^2 - 12x + 32$ $(x - 8)(x - 4)$

7. $x^2 - 4x - 21$ **8.** $2x^2 + 7x - 30$ **9.** $16x^2 - 16x + 3$ $(4x - 3)(4x - 1)$

Example 5 Solve each equation.

10. $x^2 - 36 = 0$ $-6, 6$ **11.** $12x^2 - 18x = 0$ $0, \frac{3}{2}$ **12.** $12x^2 - 2x - 2 = 0$ $-\frac{1}{3}, \frac{1}{2}$

13. $x^2 - 9x = 0$ $0, 9$ **14.** $x^2 - 3x - 28 = 0$ $-4, 7$ **15.** $2x^2 - 24x = -72$ 6

16. **CCSS SENSE-MAKING** Tamika wants to double the area of her garden by increasing the length and width by the same amount. What will be the dimensions of her garden then? **9 m by 12 m**

9 m / 6 m

Practice and Problem Solving

Extra Practice is on page R4.

Example 1 Write a quadratic equation in standard form with the given root(s).

17. 7 $x^2 - 14x + 49 = 0$ **18.** $-5, \frac{1}{2}$ $2x^2 + 9x - 5 = 0$ **19.** $\frac{1}{5}, 6$ $5x^2 - 31x + 6 = 0$

Examples 2–4 Factor each polynomial.

22. $(8x - 3a) \cdot (4y + 5b)$ **20.** $40a^2 - 32a$ $8a(5a - 4)$ **21.** $51c^3 - 34c$ $17c(3c^2 - 2)$ **22.** $32xy + 40bx - 12ay - 15ab$

23. $3x^2 - 12$ $3(x + 2)(x - 2)$ **24.** $15y^2 - 240$ **25.** $48cg + 36cf - 4dg - 3df$

24. $15(y + 4) \cdot (y - 4)$ **26.** $x^2 + 13x + 40$ **27.** $x^2 - 9x - 22$ **28.** $3x^2 + 12x - 36$

29. $15x^2 + 7x - 2$ **30.** $4x^2 + 29x + 30$ **31.** $18x^2 + 15x - 12$

25. $(12c - d) \cdot (4g + 3f)$ **32.** $8x^2z^2 - 4xz^2 - 12z^2$ $4z^2(2x - 3)(x + 1)$ **33.** $9x^2 - 25$ $(3x + 5)(3x - 5)$ **34.** $18x^2y^2 - 24xy^2 + 36y^2$ $6y^2(3x^2 - 4x + 6)$

Example 5 Solve each equation.

26. $(x + 8) \cdot (x + 5)$ **35.** $15x^2 - 84x - 36 = 0$ $-\frac{2}{5}, 6$ **36.** $12x^2 + 13x - 14 = 0$ $-\frac{7}{4}, \frac{2}{3}$ **37.** $12x^2 - 108x = 0$ $0, 9$

27. $(x - 11) \cdot (x + 2)$ **38.** $x^2 + 4x - 45 = 0$ $5, -9$ **39.** $x^2 - 5x - 24 = 0$ $8, -3$ **40.** $x^2 = 121$ $11, -11$

41. $x^2 + 13 = 17$ $2, -2$ **42.** $-3x^2 - 10x + 8 = 0$ $-4, \frac{2}{3}$ **43.** $-8x^2 + 46x - 30 = 0$ $5, \frac{3}{4}$

28. $3(x + 6) \cdot (x - 2)$ **44.** **GEOMETRY** The hypotenuse of a right triangle is 1 centimeter longer than one side and 4 centimeters longer than three times the other side. Find the dimensions of the triangle. **7 cm, 24 cm, 25 cm**

45. **NUMBER THEORY** Find two consecutive even integers with a product of 624. **24 and 26 or −24 and −26**

29. $(5x - 1) \cdot (3x + 2)$ **GEOMETRY** Find x and the dimensions of each rectangle.

30. $(4x + 5) \cdot (x + 6)$ **46.** $A = 96 \text{ ft}^2$ $x - 2$ ft / $x + 2$ ft $x = 10$; 8 ft by 12 ft

47. $A = 432 \text{ in}^2$ $x - 2$ in. / $x + 4$ in. $x = 20$; 24 in. by 18 in.

48. $x = 12$; 14 ft by 32 ft $A = 448 \text{ ft}^2$ $3x - 4$ ft / $x + 2$ ft

31. $3(2x - 1) \cdot (3x + 4)$

 242 | Lesson 4-3 | Solving Quadratic Equations by Factoring

Differentiated Homework Options

Level	Assignment	Two-Day Option	
AL Basic	17–48, 79, 82–100	17–47 odd, 87–90	18–48 even, 79, 82–85, 91–100
OL Core	17–43 odd, 44, 45–65 odd, 66–70, 71–79 odd, 82–100	17–48, 87–90	49–79, 91–100
BL Advanced	49–97, (optional: 98–100)		

B Solve each equation by factoring.

49. $12x^2 - 4x = 5$ $-\dfrac{1}{2}, \dfrac{5}{6}$

50. $5x^2 = 15x$ $0, 3$

51. $16x^2 + 36 = -48x$ $-\dfrac{3}{2}$

52. $75x^2 - 60x = -12$ $\dfrac{2}{5}$

53. $4x^2 - 144 = 0$ $6, -6$

54. $-7x + 6 = 20x^2$ $\dfrac{2}{5}, -\dfrac{3}{4}$

55. **MOVIE THEATER** A company plans to build a large multiplex theater. The financial analyst told her manager that the profit function for their theater was $P(x) = -x^2 + 48x - 512$, where x is the number of movie screens, and $P(x)$ is the profit earned in thousands of dollars. Determine the range of production of movie screens that will guarantee that the company will not lose money. **16 to 32 screens**

Write a quadratic equation in standard form with the given root(s).

56. $-\dfrac{4}{7}, \dfrac{3}{8}$
$56x^2 + 11x - 12 = 0$

57. $3.4, 0.6$
$25x^2 - 100x + 51 = 0$

58. $\dfrac{2}{11}, \dfrac{5}{9}$ $99x^2 - 73x + 10 = 0$

Solve each equation by factoring.

59. $10x^2 + 25x = 15$ $-3, \dfrac{1}{2}$

60. $27x^2 + 5 = 48x$ $\dfrac{5}{3}, \dfrac{1}{9}$

61. $x^2 + 0.25x = 1.25$ $1, -\dfrac{5}{4}$

62. $48x^2 - 15 = -22x$ $\dfrac{3}{8}, -\dfrac{5}{6}$

63. $3x^2 + 2x = 3.75$ $-\dfrac{3}{2}, \dfrac{5}{6}$

64. $-32x^2 + 56x = 12$ $\dfrac{1}{4}, \dfrac{3}{2}$

65. **DESIGN** A square is cut out of the figure at the right. Write an expression for the area of the figure that remains, and then factor the expression. $x^2 - 6^2; (x + 6)(x - 6)$

66. **CCSS PERSEVERANCE** After analyzing the market, a company that sells Web sites determined the profitability of their product was modeled by $P(x) = -16x^2 + 368x - 2035$, where x is the price of each Web site and $P(x)$ is the company's profit. Determine the price range of the Web sites that will be profitable for the company. **$9.25 to $13.75**

67. **PAINTINGS** Enola wants to add a border to her painting, distributed evenly, that has the same area as the painting itself. What are the dimensions of the painting with the border included? **20 in. by 15 in.**

15 in.

10 in.

68. **MULTIPLE REPRESENTATIONS** In this problem, you will consider $a(x - p)(x - q) = 0$.

a. Graphical Graph the related function for $a = 1$, $p = 2$, and $q = -3$. **See margin.**

b. Analytical What are the solutions of the equation? **2 and -3**

c. Graphical Graph the related functions for $a = 4, -3$, and $\dfrac{1}{2}$ on the same graph. **See margin.**

d. Verbal What are the similarities and differences between the graphs?

e. Verbal What conclusion can you make about the relationship between the factored form of a quadratic equation and its solutions?

69. **GEOMETRY** The area of the triangle is 26 square centimeters. Find the length of the base. **13 cm**

$x - 2$

$x + 7$

68d. Sample answer: They all have the same roots, p and q. Therefore, they all have the same solutions. The graphs are shaped differently due to the value of a. The graph with $a = -3$ is flipped due to the negative.

68e. When quadratic equations have the same factors, they will have the same solutions, regardless of the value of a, which only affects the shape of the graphs.

Multiple Representations
In Exercise 68, students use algebra and a graph in the coordinate plane to relate the factors of a quadratic equation to its solutions.

Study Guide and Intervention
AL OL ELL

Practice
AL OL BL ELL

Enrichment
OL BL

4-3 Enrichment

Using Patterns to Factor

Word Problem Practice
AL OL BL ELL

Teaching the Mathematical Practices

Arguments Mathematically proficient students understand and use stated assumptions, definitions, and previously established results in constructing arguments. They make conjectures and build a logical progression of statements to explore the truth of their conjectures. And they are able to analyze situations by breaking them into cases, and can recognize and use counterexamples.

WatchOut!

Error Analysis For Exercise 79 remind students that when a polynomial is subtracted, every term in the polynomial is subtracted. Always check answers by substitution in the original equation.

Additional Answers

83. Sample answer:

$(x - p)(x - q) = 0$
(Original equation)

$x^2 - px - qx + pq = 0$
(Multiply)

$x^2 - (p + q)x + pq = 0$
(Simplify)

$x = -\dfrac{b}{2a}$
(Formula for axis of symmetry)

$x = -\dfrac{-(p + q)}{2(1)}$

$a = 1$ and $b = -(p + q)$

$x = \dfrac{p + q}{2}$
(Simplify)

x is midway between p and q.
(Definition of midpoint)

86. Sample answer: In standard form, we have $ax^2 + bx + c$. Multiply a and c. Then find a pair of integers, g and h, that multiply to equal ac and add to equal b. Then write out the quadratic, turning the middle term, bx, into $gx + hx$. We now have $ax^2 + gx + hx + c$. Now factor the GCF from the first two terms and then factor the GCF from the second two terms. So we now

have $GCF(x - q) + GCF_2(x - q)$.
Simplifying, we get
$(GCF + GCF_2)(x - q)$ or
$(x - p)(x - q)$.

---

70. SOCCER When a ball is kicked in the air, its height in meters above the ground can be modeled by $h(t) = -4.9t^2 + 14.7t$ and the distance it travels can be modeled by $d(t) = 16t$, where t is the time in seconds.

  a. How long is the ball in the air? **3 seconds**

  b. How far does it travel before it hits the ground? (*Hint:* Ignore air resistance.) **48 m**

  c. What is the maximum height of the ball? **11.025 m**

Factor each polynomial.

71. $18a - 24ay + 48b - 64by$ $2(3 - 4y)(3a + 8b)$

72. $3x^2 + 2xy + 10y + 15x$ $(3x + 2y)(x + 5)$

73 $6a^2b^2 - 12ab^2 - 18b^3$ $6b^2(a^2 - 2a - 3b)$

74. $12a^2 - 18ab + 30ab^3$ $6a(2a - 3b + 5b^3)$

75. $32ax + 12bx - 48ay - 18by$ $2(2x - 3y)(8a + 3b)$

76. $30ac + 80bd + 40ad + 60bc$ $10(a + 2b)(3c + 4d)$

77. $5ax^2 - 2by^2 - 5ay^2 + 2bx^2$ $(x + y)(x - y)(5a + 2b)$

78. $12c^2x + 4d^2y - 3d^2x - 16c^2y$ $(2c + d)(2c - d)(3x - 4y)$

H.O.T. Problems   Use Higher-Order Thinking Skills

79. ERROR ANALYSIS Gwen and Morgan are solving $-12x^2 + 5x + 2 = 0$. Is either of them correct? Explain your reasoning.

Gwen	Morgan
$-12x^2 + 5x + 2 = 0$	$-12x^2 + 5x + 2 = 0$
$-12x^2 + 8x - 3x + 2 = 0$	$-12x^2 + 8x - 3x + 2 = 0$
$4x(-3x + 2) - (3x + 2) = 0$	$4x(-3x + 2) + (-3x + 2) = 0$
$(4x - 1)(3x + 2) = 0$	$(4x + 1)(-3x + 2) = 0$
$x = \frac{1}{4}$ or $-\frac{2}{3}$	$x = -\frac{1}{4}$ or $\frac{2}{3}$

79. Sample answer: Morgan; Gwen did not have like terms in the parentheses in the third line.

80. CHALLENGE Solve $3x^6 - 39x^4 + 108x^2 = 0$ by factoring. **0, 3, −3, 2, or −2**

81. CHALLENGE The rule for factoring a difference of cubes is shown below. Use this rule to factor $40x^5 - 135x^2y^3$. $5x^2(2x - 3y)(4x^2 + 6xy + 9y^2)$

$$a^3 - b^3 = (a - b)(a^2 + ab + b^2)$$

82. OPEN ENDED Choose two integers. Then write an equation in standard form with those roots. How would the equation change if the signs of the two roots were switched?

82. Sample answer: 3 and 6 → $x^2 - 9x + 18 = 0$. −3 and −6 → $x^2 + 9x + 18 = 0$. The linear term changes sign.

83. CHALLENGE For a quadratic equation of the form $(x - p)(x - q) = 0$, show that the axis of symmetry of the related quadratic function is located halfway between the x-intercepts p and q. **See margin.**

84. WRITE A QUESTION A classmate is using the guess-and-check strategy to factor trinomials of the form $x^2 + bx + c$. Write a question to help him think of a way to use that strategy for $ax^2 + bx + c$.

84. Sample answer: What do you know about $a \cdot c$ to use guess-and-check to factor trinomials of the form $ax^2 + bx + c$?

85. **ARGUMENTS** Determine whether the following statement is *sometimes, always,* or *never* true. Explain your reasoning.

In a quadratic equation in standard form where a, b, and c are integers, if b is odd, then the quadratic cannot be a perfect square trinomial.

85. Sample answer: Always; in order to factor using perfect square trinomials, the coefficient of the linear term, bx, must be a multiple of 2, or even.

86. WRITING IN MATH Explain how to factor a trinomial in standard form with $a > 1$. **See margin.**

 244 | **Lesson 4-3** | Solving Quadratic Equations by Factoring

87. SHORT RESPONSE If $ABCD$ is transformed by $(x, y) \rightarrow (3x, 4y)$, determine the area of $A'B'C'D'$. **192 square units**

88. For $y = 2|6 - 3x| + 4$, which set describes x when $y < 6$? **A**

A $\left\{ x \mid \frac{5}{3} < x < \frac{7}{3} \right\}$ C $\left\{ x \mid x < \frac{5}{3} \right\}$

B $\left\{ x \mid x < \frac{5}{3} \text{ or } x > \frac{7}{3} \right\}$ D $\left\{ x \mid x > \frac{7}{3} \right\}$

89. PROBABILITY A 5-character password can contain the numbers 0 through 9 and 26 letters of the alphabet. None of the characters can be repeated. What is the probability that the password begins with a consonant? **H**

F $\frac{21}{26}$ H $\frac{21}{36}$

G $\frac{21}{35}$ J $\frac{5}{36}$

90. SAT/ACT If $c = \frac{8a^3}{b}$, what happens to the value of c when both a and b are doubled? **D**

A c is unchanged.

B c is halved.

C c is doubled.

D c is multiplied by 4.

E c is multiplied by 8.

Spiral Review

Use the related graph of each equation to determine its solutions. (Lesson 4-2)

91. $x^2 - 2x - 8 = 0$ **−2, 4**

92. $x^2 + 4x = 12$ **−6, 2**

93. $x^2 + 4x + 4 = 0$ **−2**

Graph each function. (Lesson 4-1) **94–96. See margin.**

94. $f(x) = x^2 - 6x + 2$

95. $f(x) = -2x^2 + 4x + 1$

96. $f(x) = (x - 3)(x + 4)$

97. FUNDRAISING Lawrence High School sold wrapping paper and boxed cards for their fundraising event. The school gets $1.00 for each roll of wrapping paper sold and $0.50 for each box of cards sold. (Lesson 3-6)

a. Write a matrix that represents the amounts sold for each class and a matrix that represents the amount of money the school earns for each item sold. **See margin.**

b. Write a matrix that shows how much each class earned. **See margin.**

c. Which class earned the most money? **juniors**

d. What is the total amount of money the school made from the fundraiser? **$431**

Total Amounts for Each Class		
Class	Wrapping Paper	Cards
freshmen	72	49
sophomores	68	63
juniors	90	56
seniors	86	62

Skills Review

Simplify.

98. $\sqrt{5} \cdot \sqrt{15}$ **$5\sqrt{3}$**

99. $\sqrt{8} \cdot \sqrt{32}$ **16**

100. $2\sqrt{3} \cdot \sqrt{27}$ **18**

Name the Math Have students explain the Zero Product Property. Have them discuss why it is true and how it is used in finding the roots of a quadratic equation.

Additional Answers

94.

95.

96.

97a. $\begin{bmatrix} 72 & 49 \\ 68 & 63 \\ 90 & 56 \\ 86 & 62 \end{bmatrix}, \begin{bmatrix} 1.00 \\ 0.50 \end{bmatrix}$

97b. $\begin{bmatrix} 96.50 \\ 99.50 \\ 118 \\ 117 \end{bmatrix}$

DifferentiatedInstruction OL BL

Extension Pose the following question to students:

If the roots of a quadratic equation are 6 and −3, what is the equation of the axis of symmetry?

$x = \frac{3}{2}$

2 Operations with Complex Numbers

Example 4 shows how to equate complex numbers. **Example 5** shows how to add and subtract complex numbers. **Examples 6–7** show how to multiply and divide complex numbers.

Additional Examples

4 Find the values of x and y that make the equation
$2x + yi = -14 - 3i$ true.
$x = -7, y = -3$

5 Simplify.
  a. $(3 + 5i) + (2 - 4i)$ $5 + i$
  b. $(4 - 6i) - (3 - 7i)$ $1 + i$

WatchOut!

Preventing Errors Emphasize that two complex numbers are equal if and only if their real parts are equal and their imaginary parts are equal.

StudyTip
Complex Numbers Whereas all real numbers are also complex, the term *complex number* usually refers to a number that is not real.

The Venn diagram shows the set of complex numbers.

- If $b = 0$, the complex number is a real number.
- If $b \neq 0$, the complex number is imaginary.
- If $a = 0$, the complex number is a pure imaginary number.

Complex Numbers ($a + bi$)

Real Numbers $b = 0$	Imaginary Numbers $b \neq 0$
	Pure Imaginary Numbers $a = 0$

Two complex numbers are equal if and only if their real parts are equal and their imaginary parts are equal. That is, $a + bi = c + di$ if and only if $a = c$ and $b = d$.

Example 4   Equate Complex Numbers

Find the values of x and y that make $3x - 5 + (y - 3)i = 7 + 6i$ true.

Set the real parts equal to each other and the imaginary parts equal to each other.

$3x - 5 = 7$	Real parts		$y - 3 = 6$	Imaginary parts
$3x = 12$	Add 5 to each side.		$y = 9$	Add 3 to each side.
$x = 4$	Divide each side by 3.			

▸ **Guided**Practice   4. $x = -1, y = -9$

     4. Find the values of x and y that make $5x + 1 + (3 + 2y)i = 2x - 2 + (y - 6)i$ true.

The Commutative, Associative, and Distributive Properties of Multiplication and Addition hold true for complex numbers. To add or subtract complex numbers, combine like terms. That is, combine the real parts, and combine the imaginary parts.

Example 5   Add and Subtract Complex Numbers

Simplify.

a. $(5 - 7i) + (2 + 4i)$

$\quad (5 - 7i) + (2 + 4i) = (5 + 2) + (-7 + 4)i$ Commutative and Associative Properties

$\quad\quad\quad\quad\quad\quad\quad\quad\quad = 7 - 3i$ Simplify.

b. $(4 - 8i) - (3 - 6i)$

$\quad (4 - 8i) - (3 - 6i) = (4 - 3) + [-8 - (-6)]i$ Commutative and Associative Properties

$\quad\quad\quad\quad\quad\quad\quad\quad\quad = 1 - 2i$ Simplify.

▸ **Guided**Practice

     5A. $(-2 + 5i) + (1 - 7i)$ $-1 - 2i$ **5B.** $(4 + 6i) - (-1 + 2i)$ $5 + 4i$

StudyTip
Reading Math Electrical engineers use *j* as the imaginary unit to avoid confusion with the *i* for current.

Complex numbers are used with electricity. In these problems, *j* usually represents the imaginary unit. In a circuit with alternating current, the voltage, current, and impedance, or hindrance to current, can be represented by complex numbers. To multiply these numbers, use the FOIL method.

Real-World Example 6 Multiply Complex Numbers

ELECTRICITY In an AC circuit, the voltage V, current C, and impedance I are related by the formula $V = C \cdot I$. Find the voltage in a circuit with current $2 + 4j$ amps and impedance $9 - 3j$ ohms.

$V = C \cdot I$	Electricity formula
$= (2 + 4j) \cdot (9 - 3j)$	$C = 2 + 4j$ and $I = 9 - 3j$
$= 2(9) + 2(-3j) + 4j(9) + 4j(-3j)$	FOIL Method
$= 18 - 6j + 36j - 12j^2$	Multiply.
$= 18 + 30j - 12(-1)$	$j^2 = -1$
$= 30 + 30j$	Add.

The voltage is $30 + 30j$ volts.

▶ **Guided**Practice **6.** $-2 - 16j$ volts

6. Find the voltage in a circuit with current $2 - 4j$ amps and impedance $3 - 2j$ ohms.

Real-WorldLink

An example of a series circuit is a string of holiday lights. The number of bulbs on a circuit affects the strength of the current, which in turn affects the brightness of the lights.

Source: *Popular Science*

Two complex numbers of the form $a + bi$ and $a - bi$ are called **complex conjugates**. The product of complex conjugates is always a real number. You can use this fact to simplify the quotient of two complex numbers.

Example 7 Divide Complex Numbers

Simplify.

a. $\dfrac{2i}{3 + 6i}$

$\dfrac{2i}{3 + 6i} = \dfrac{2i}{3 + 6i} \cdot \dfrac{3 - 6i}{3 - 6i}$	$3 + 6i$ and $3 - 6i$ are complex conjugates.
$= \dfrac{6i - 12i^2}{9 - 36i^2}$	Multiply.
$= \dfrac{6i - 12(-1)}{9 - 36(-1)}$	$i^2 = -1$
$= \dfrac{6i + 12}{45}$	Simplify.
$= \dfrac{4}{15} + \dfrac{2}{15}i$	$a + bi$ form

b. $\dfrac{4 + i}{5i}$

$\dfrac{4 + i}{5i} = \dfrac{4 + i}{5i} \cdot \dfrac{i}{i}$	Multiply by $\dfrac{i}{i}$.
$= \dfrac{4i + i^2}{5i^2}$	Multiply.
$= \dfrac{4i - 1}{-5}$	$i^2 = -1$
$= \dfrac{1}{5} - \dfrac{4}{5}i$	$a + bi$ form

StudyTip

Technology Operations with complex numbers can be preformed with a TI-83/84 Plus graphing calculator. Use the [2nd] [i] function to enter the expression. Then press [MATH] [ENTER] [ENTER] to view the answer.

▶ **Guided**Practice

7A. $\dfrac{-2i}{3 + 5i}$ $-\dfrac{5}{17} - \dfrac{3}{17}i$

7B. $\dfrac{2 + i}{1 - i}$ $\dfrac{1}{2} + \dfrac{3}{2}i$

Design Pics Inc./Alamy

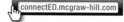 **249**

Teach with Tech

Interactive Whiteboard Work through an example multiplying two complex numbers. Write an example on the board using red for the imaginary part and blue for the real part of each complex number. In each calculation, continue to use red and blue to show the imaginary and real parts of the complex numbers.

Additional Examples

6 **ELECTRICITY** In an AC circuit, the voltage E, current I, and impedance Z are related by the formula $E = I \cdot Z$. Find the voltage in a circuit with current $1 + 4j$ amps and impedance $3 - 6j$ ohms. $27 + 6j$

7 Simplify.

a. $\dfrac{5i}{3 + 2i}$ $\dfrac{10}{13} + \dfrac{15}{13i}$

b. $\dfrac{5 + i}{2i}$ $\dfrac{1}{2} - \dfrac{5}{2i}$

Focus on Mathematical Content

Complex Numbers A complex number is any number that can be written in the form $a + bi$, where a and b are real numbers and i is the imaginary unit. If $b = 0$, the complex number is a real number. If $b \neq 0$, the complex number is imaginary. If $a = 0$, the complex number is a pure imaginary number. Pure imaginary and real numbers are both subsets of the set of complex numbers. Hence, every real number is complex, and every pure imaginary number is complex.

3 Practice

Formative Assessment

Use Exercises 1–17 to check for understanding.

Use the chart at the bottom of this page to customize assignments for your students.

 Teaching the Mathematical Practices

Structure Mathematically proficient students look closely to discern a pattern or structure.

⚙ Multiple Representations

In Exercise 65, students use graphs to analyze quadratic equations that have complex roots.

Additional Answers

65b.

$f(x) = x^2 + 9$

65d.

$f(x) = x^2 - 4x + 5$

65e. Sample answer: A quadratic equation will have only complex solutions when the graph of the related function has no x-intercepts.

 Check Your Understanding   ⬤ = Step-by-Step Solutions begin on page R14.

Examples 1–2 **Simplify.**

1. $\sqrt{-81}$ $9i$
2. $\sqrt{-32}$ $4i\sqrt{2}$
3. $(4i)(-3i)$ 12
4. $3\sqrt{-24} \cdot 2\sqrt{-18}$ $-72\sqrt{3}$
5. i^{40} 1
6. i^{63} $-i$

Example 3 **Solve each equation.**

7. $4x^2 + 32 = 0$ $\pm 2i\sqrt{2}$
8. $x^2 + 1 = 0$ $\pm i$

Example 4 **Find the values of a and b that make each equation true.**

9. $3a + (4b + 2)i = 9 - 6i$ $3, -2$
10. $4b - 5 + (-a - 3)i = 7 - 8i$ $5, 3$

Examples 5 and 7 **Simplify.**

11. $(-1 + 5i) + (-2 - 3i)$ $-3 + 2i$
12. $(7 + 4i) - (1 + 2i)$ $6 + 2i$
13. $(6 - 8i)(9 + 2i)$ $70 - 60i$
14. $(3 + 2i)(-2 + 4i)$ $-14 + 8i$
15. $\frac{3 - i}{4 + 2i}$ $\frac{1}{2} - \frac{1}{2}i$
16. $\frac{2 + i}{5 + 6i}$ $\frac{16}{61} - \frac{7}{61}i$

Example 6 17. **ELECTRICITY** The current in one part of a series circuit is $5 - 3j$ amps. The current in another part of the circuit is $7 + 9j$ amps. Add these complex numbers to find the total current in the circuit. $12 + 6j$ amps

---

Practice and Problem Solving   Extra Practice is on page R4.

Examples 1–2 ㏄ **STRUCTURE** **Simplify.**

18. $\sqrt{-121}$ $11i$
19. $\sqrt{-169}$ $13i$
20. $\sqrt{-100}$ $10i$
21. $\sqrt{-81}$ $9i$
22. $(-3i)(-7i)(2i)$ $-42i$
23. $4i(-6i)^2$ $-144i$
24. i^{11} $-i$
25. i^{25} i
26. $(10 - 7i) + (6 + 9i)$ $16 + 2i$
27. $(-3 + i) + (-4 - i)$ -7
28. $(12 + 5i) - (9 - 2i)$ $3 + 7i$
29. $(11 - 8i) - (2 - 8i)$ 9
30. $(1 + 2i)(1 - 2i)$ 5
31. $(3 + 5i)(5 - 3i)$ $30 + 16i$
32. $(4 - i)(6 - 6i)$ $18 - 30i$
33. $\frac{2i}{1 + i}$ $1 + i$
34. $\frac{5}{2 + 4i}$ $\frac{1}{2} - i$
35. $\frac{5 + i}{3i}$ $\frac{1}{3} - \frac{5}{3}i$

Example 3 **Solve each equation.**

36. $4x^2 + 4 = 0$ $\pm i$
37. $3x^2 + 48 = 0$ $\pm 4i$
38. $2x^2 + 50 = 0$ $\pm 5i$
39. $2x^2 + 10 = 0$ $\pm i\sqrt{5}$
40. $6x^2 + 108 = 0$ $\pm 3i\sqrt{2}$
41. $8x^2 + 128 = 0$ $\pm 4i$

Example 4 **Find the values of x and y that make each equation true.**

42. $9 + 12i = 3x + 4yi$ $3, 3$
43. $x + 1 + 2yi = 3 - 6i$ $2, -3$
44. $2x + 7 + (3 - y)i = -4 + 6i$ $-\frac{11}{2}, -3$
45. $5 + y + (3x - 7)i = 9 - 3i$ $\frac{4}{3}, 4$
46. $a + 3b + (3a - b)i = 6 + 6i$ $\frac{12}{5}, \frac{6}{5}$
47. $(2a - 4b)i + a + 5b = 15 + 58i$ $25, -2$

 250 | Lesson 4-4 | Complex Numbers

Differentiated Homework Options

Level	Assignment	Two-Day Option	
AL Basic	18–60, 66, 68–89	19–59 odd, 71–74	18–60 even, 66, 68–89
OL Core	19–59 odd, 61–89	18–60, 71–74	61–66, 68–70, 75–89
BL Advanced	61–83, (optional: 84–89)		

Simplify.

48. $\sqrt{-10} \cdot \sqrt{-24}$ $-4\sqrt{15}$ 49. $4i\left(\frac{1}{2}i\right)^2(-2i)^2$ $4i$ 50. i^{41} i

51. $(4 - 6i) + (4 + 6i)$ 8 52. $(8 - 5i) - (7 + i)$ $1 - 6i$ 53. $(-6 - i)(3 - 3i)$ $-21 + 15i$

54. $\dfrac{(5 + i)^2}{3 - i}$ $\dfrac{31}{5} + \dfrac{27}{5}i$ 55. $\dfrac{6 - i}{2 - 3i}$ $\dfrac{15}{13} + \dfrac{16}{13}i$ 56. $(-4 + 6i)(2 - i)(3 + 7i)$ $-118 + 34i$

57. $(1 + i)(2 + 3i)(4 - 3i)$
$11 + 23i$ 58. $\dfrac{4 - i\sqrt{2}}{4 + i\sqrt{2}}$ $\dfrac{7}{9} - \dfrac{4i\sqrt{2}}{9}$ 59. $\dfrac{2 - i\sqrt{3}}{2 + i\sqrt{3}}$ $\dfrac{1}{7} - \dfrac{4\sqrt{3}}{7}i$

Example 6

60. **ELECTRICITY** The impedance in one part of a series circuit is $7 + 8j$ ohms, and the impedance in another part of the circuit is $13 - 4j$ ohms. Add these complex numbers to find the total impedance in the circuit. $20 + 4j$ ohms

B **ELECTRICITY** Use the formula $V = C \cdot I$.   63. $(3 + i)x^2 + (-2 + i)x - 8i + 7$

61 The current in a circuit is $3 + 6j$ amps, and the impedance is $5 - j$ ohms. What is the voltage? $21 + 27j$ volts

62. The voltage in a circuit is $20 - 12j$ volts, and the impedance is $6 - 4j$ ohms. What is the current? $\dfrac{42}{13} + \dfrac{2}{13}j$ amps

63. Find the sum of $ix^2 - (4 + 5i)x + 7$ and $3x^2 + (2 + 6i)x - 8i$.

64. Simplify $[(2 + i)x^2 - ix + 5 + i] - [(-3 + 4i)x^2 + (5 - 5i)x - 6]$. $(5 - 3i)x^2 +$
$(-5 + 4i)x + i + 11$

65. **MULTIPLE REPRESENTATIONS** In this problem, you will explore quadratic equations that have complex roots.

   a. **Algebraic** Write a quadratic equation in standard form with $3i$ and $-3i$ as its roots. Sample answer: $x^2 + 9 = 0$

   b. **Graphical** Graph the quadratic equation found in part **a** by graphing its related function. See margin.

   c. **Algebraic** Write a quadratic equation in standard form with $2 + i$ and $2 - i$ as its roots. Sample answer: $x^2 - 4x + 5 = 0$

   d. **Graphical** Graph the quadratic equation found in part **c** by graphing its related function. See margin.

   e. **Analytical** How do you know when a quadratic equation will have only complex solutions? See margin.

68. Sample answer: Always; the value of 5 can be represented by $5 + 0i$, and the value of $3i$ can be represented by $0 + 3i$.

H.O.T. Problems Use Higher-Order Thinking Skills

66. **CCSS CRITIQUE** Joe and Sue are simplifying $(2i)(3i)(4i)$. Is either of them correct? Explain your reasoning. Sue; $i^3 = -i$, not -1.

Joe	Sue
$24i^3 = -24$	$24i^3 = -24i$

70. Some quadratic equations have complex solutions and cannot be solved using only the real numbers.

67. **CHALLENGE** Simplify $(1 + 2i)^3$. $-11 - 2i$

68. **REASONING** Determine whether the following statement is *always*, *sometimes*, or *never* true. Explain your reasoning.

   Every complex number has both a real part and an imaginary part.

69. **OPEN ENDED** Write two complex numbers with a product of 20. Sample answer: $(4 + 2i)(4 - 2i)$

70. **WRITING IN MATH** Explain how complex numbers are related to quadratic equations.

CCSS Teaching the Mathematical Practices

Critique Mathematically proficient students are also able to compare the effectiveness of two plausible arguments, distinguish correct logic or reasoning from that which is flawed, and—if there is a flaw in an argument—explain what it is.

4-4 Study Guide and Intervention
Complex Numbers

4-4 Practice
Complex Numbers

4-4 Word Problem Practice
Complex Numbers

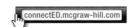

Enrichment
OL BL

4-4 Enrichment

Conjugates and Absolute Value

Ticket Out the Door Have students write two complex numbers that have a product of 10 on small pieces of paper and hand them to you as they leave the classroom. Sample answer: $1 + 3i$ and $1 - 3i$

Formative Assessment
Check for student understanding of concepts in Lessons 4-3 and 4-4.

📁 Quiz 2, p. 59

Additional Answers

71b. $\angle AED \cong \angle CEB$ (Vertical angles) $\overline{DE} \cong \overline{BE}$ (Both have length x.) $\angle ADE \cong \angle CBE$ (Given) Consecutive angles and the included side are all congruent, so the triangles are congruent by the ASA Property.

71c. $\overline{EC} \cong \overline{EA}$ by CPCTC (corresponding parts of congruent triangles are congruent). $EA = 7$, so $EC = 7$.

82a.
Residents:
$$\begin{array}{c}\text{Child}\quad\text{Adult}\\ \begin{bmatrix} 3.00 & 4.50 \\ 2.00 & 3.50 \end{bmatrix}\end{array},$$
Nonresidents:
$$\begin{array}{c}\text{Child}\quad\text{Adult}\\ \begin{bmatrix} 4.50 & 6.75 \\ 3.00 & 5.25 \end{bmatrix}\end{array}$$

82b.
$$\begin{array}{c}\quad\quad\text{Child}\quad\text{Adult}\\ \begin{array}{r}\text{Before 6}\\ \text{After 6}\end{array}\begin{bmatrix} 1.50 & 2.25 \\ 1.00 & 1.75 \end{bmatrix}\end{array}$$

82c.
$$\begin{array}{c}\quad\quad\quad\text{Child}\quad\text{Adult}\\ \begin{array}{r}\text{Resident}\\ \text{Nonresident}\end{array}\begin{bmatrix} 1.00 & 1.00 \\ 1.50 & 1.50 \end{bmatrix}\end{array}$$

83.

---

71. EXTENDED RESPONSE Refer to the figure to answer the following. **b, c. See margin.**

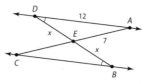

a. Name two congruent triangles with vertices in correct order. $\triangle CBE \cong \triangle ADE$

b. Explain why the triangles are congruent.

c. What is the length of $\overline{EC}$? Explain your procedure.

72. $(3 + 6)^2 =$ **B**

   A $2 \times 3 + 2 \times 6$ **C** $3^2 + 6^2$

   B 9^2 **D** $3^2 \times 6^2$

73. SAT/ACT A store charges $49 for a pair of pants. This price is 40% more than the amount it costs the store to buy the pants. After a sale, any employee is allowed to purchase any remaining pairs of pants at 30% off the store's cost. How much would it cost an employee to purchase the pants after the sale? **J**

   F $10.50   **J** $24.50

   G $12.50   **K** $35.00

   H $13.72

74. What are the values of x and y when $(5 + 4i) - (x + yi) = (-1 - 3i)$? **A**

   A $x = 6, y = 7$

   B $x = 4, y = i$

   C $x = 6, y = i$

   D $x = 4, y = 7$

Solve each equation by factoring. (Lesson 4-3)

75. $2x^2 + 7x = 15$ $-5, \dfrac{3}{2}$ **76.** $4x^2 - 12 = 22x$ $-\dfrac{1}{2}, 6$ **77.** $6x^2 = 5x + 4$ $-\dfrac{1}{2}, \dfrac{4}{3}$

NUMBER THEORY Use a quadratic equation to find two real numbers that satisfy each situation, or show that no such numbers exist. (Lesson 4-2)

78. Their sum is -3, and their product is -40. $-8, 5$ **79.** Their sum is 19, and their product is 48. $3, 16$

80. Their sum is -15, and their product is 56. -7 and -8 **81.** Their sum is -21, and their product is 108. $-9, -12$

82. RECREATION Refer to the table. (Lesson 3-5) **a–c. See margin.**

  a. Write a matrix that represents the cost of admission for residents and a matrix that represents the cost of admission for nonresidents.

  b. Write the matrix that represents the additional cost for nonresidents.

  c. Write a matrix that represents the difference in cost if a child or adult goes after 6:00 P.M. instead of before 6:00 P.M.

83. PART-TIME JOBS Terrell makes $10 per hour cutting grass and $12 per hour for raking leaves. He cannot work more than 15 hours per week. Graph two inequalities that Terrell can use to determine how many hours he needs to work at each job if he wants to earn at least $120 per week. (Lesson 3-2) **See margin.**

Daily Admission Fees		
Residents Time of day	**Child**	**Adult**
Before 6:00 p.m.	$3.00	$4.50
After 6:00 p.m.	$2.00	$3.50
Nonresidents Time of day	**Child**	**Adult**
Before 6:00 p.m.	$4.50	$6.75
After 6:00 p.m.	$3.00	$5.25

Determine whether each trinomial is a perfect square trinomial. Write *yes* or *no*.

84. $x^2 + 16x + 64$ **yes**   **85.** $x^2 - 12x + 36$ **yes**   **86.** $x^2 + 8x - 16$ **no**

87. $x^2 - 14x - 49$ **no**   **88.** $x^2 + x + 0.25$ **yes**   **89.** $x^2 + 5x + 6.25$ **yes**

DifferentiatedInstruction **BL** **OL**

Extension Tell students that you are thinking of two complex numbers that have a sum of $3 + i$ and a difference of $-5 + 7i$. Ask them to find the product of the two numbers. $8 + 19i$

EXTEND 4-4 Algebra Lab
The Complex Plane

A complex number $a + bi$ can be graphed in the **complex plane** by representing it with the point (a, b). Similar to a coordinate plane, the complex plane is comprised of two axes. The real component is plotted on the **real axis**, which is horizontal. The imaginary component is plotted on the **imaginary axis**, which is vertical. The complex plane may also be referred to as the **Argand (ar GON) plane**.

Example 1 Graph in the Complex Plane

Graph $z = 3 + 4i$ in the complex plane.

Step 1 Represent z with the point (a, b).

The real component a of z is 3.

The imaginary component bi of z is $4i$.

z can be represented by the point (a, b) or $(3, 4)$.

Step 2 Graph z in the complex plane.

Construct the complex plane and plot the point $(3, 4)$.

Recall that for a real number, the absolute value is its distance from zero on the number line. Similarly, the **absolute value of a complex number** is its distance from the origin in the complex plane. When $a + bi$ is graphed in the complex plane, the absolute value of $a + bi$ is the distance from (a, b) to the origin. This can be found by using the Distance Formula.

$$\sqrt{(a - 0)^2 + (b - 0)^2} \text{ or } \sqrt{a^2 + b^2}$$

KeyConcept Absolute Value of a Complex Number

The absolute value of the complex number $z = a + bi$ is

$$|z| = |a + bi| = \sqrt{a^2 + b^2}.$$

---

1 Focus

Objective Graph complex numbers in the complex plane and detemine absolute values of complex numbers.

Teaching Tip
Students should be familiar with the Distance Formula before attempting this lab.

2 Teach

Working in Cooperative Groups
Have students work in pairs, mixing abilities. Then have pairs work through the Examples.

Ask:
- Can any real number be written as a complex number? Explain. Yes, sample answer: Any real number can be written as $a + 0i$.

- How is the absolute value of a complex number related to the absolute value of a real number? Sample answer: Both the absolute value of a complex number and the absolute value of a real number are distances the numbers are from zero.

Practice Have students complete
Exercises 1–9.

3 Assess

Formative Assessment
Use Exercises 4–6 to assess whether
students can find the absolute value of
a complex number.

From Concrete to Abstract
Ask:
- What is the absolute value of
  $z = a + bi$ when $b = 0$? Explain.
  $|a|$; Sample answer: When $b = 0$,
  the absolute value of z is
  $|z| = \sqrt{a^2 + 0^2}$ or $\sqrt{a^2}$, $\sqrt{a^2}$ can
  be written as $|a|$.
- Can the point (x, y) in the coordinate
  plane be graphed in the complex
  plane? Explain. No, sample answer:
  The point (x, y) has x and y
  components. Points that can be
  graphed in the complex plane have
  real and imaginary components.

Algebra Lab
The Complex Plane *Continued*

Example 2 Absolute Value of a Complex Number

Find the absolute value of $z = -5 + 12i$.

Step 1 Determine values for a and b.

The real component a of z is -5. The imaginary component bi of z is $12i$.
Thus, $a = -5$ and $b = 12$.

Step 2 Find the absolute value of z.

$$|z| = \sqrt{a^2 + b^2} \qquad \text{Absolute value of a complex number}$$
$$= \sqrt{(-5)^2 + 12^2} \qquad a = -5 \text{ and } b = 12$$
$$= \sqrt{169} \text{ or } 13 \qquad \text{Simplify.}$$

The absolute value of $z = -5 + 12i$ is 13.

Addition and subtraction of complex numbers can be performed graphically.

Example 3 Simplify by Graphing

Simplify $(1 - 2i) - (-2 - 5i)$ by graphing.

Step 1 Write $(1 - 2i) - (-2 - 5i)$ as $(1 - 2i) + (2 + 5i)$.

Step 2 Graph $1 - 2i$ and $2 + 5i$ on the same complex plane. Connect each point
with the origin using a dashed segment.

Step 3 Complete the parallelogram that has the two segments as two of its sides.
Plot a point where the two additional sides meet. The solution of
$(1 - 2i) - (-2 - 5i)$ is $3 + 3i$.

Step 2   Step 3

Exercises

Graph each number in the complex plane. 1–3. See Chapter 4 Answer Appendix.

1. $z = 3 + i$ **2.** $z = -4 - 2i$ **3.** $z = 2 - 2i$

Find the absolute value of each complex number.

4. $z = -4 - 3i$ 5   **5.** $z = 7 - 2i$ $\sqrt{53}$ or about 7.28   **6.** $z = -6 - i$ $\sqrt{37}$ or about 6.08

Simplify by graphing. 7–9. See Chapter 4 Answer Appendix.

7. $(6 + 5i) + (-2 - 3i)$ **8.** $(8 - 2i) - (4 + 7i)$ **9.** $(5 + 6i) + (-4 + 3i)$

254 | Extend 4-4 | Algebra Lab: The Complex Plane

Mid-Chapter Quiz
Lessons 4-1 through 4-4

1. Find the y-intercept, the equation of the axis of symmetry, and the x-coordinate of the vertex for $f(x) = 2x^2 + 8x - 3$. Then graph the function by making a table of values. (Lesson 4-1) **See margin.**

2. **MULTIPLE CHOICE** For which equation is the axis of symmetry $x = 5$? (Lesson 4-1) **B**

   A $f(x) = x^2 - 5x + 3$

   B $f(x) = x^2 - 10x + 7$

   C $f(x) = x^2 + 10x - 3$

   D $f(x) = x^2 + 5x + 2$

3. Determine whether $f(x) = 5 - x^2 + 2x$ has a maximum or a minimum value. Then find this maximum or minimum value and state the domain and range of the function. (Lesson 4-1)
   max.; 6; D = {all real numbers}; R = {$f(x)$ | $f(x) \le 6$}

4. **PHYSICAL SCIENCE** From 4 feet above the ground, Maya throws a ball upward with a velocity of 18 feet per second. The height $h(t)$ of the ball t seconds after Maya throws the ball is given by $h(t) = -16t^2 + 18t + 4$. Find the maximum height reached by the ball and the time that this height is reached. (Lesson 4-1) **9.0625 feet at 0.5625 seconds**

5. Solve $3x^2 - 17x + 5 = 0$ by graphing. If exact roots cannot be found, state the consecutive integers between which the roots are located. (Lesson 4-2)
   between 0 and 1, and between 5 and 6

Use a quadratic equation to find two real numbers that satisfy each situation, or show that no such numbers exist. (Lesson 4-2)

6. Their sum is 15, and their product is 36. **3 and 12**

7. Their sum is 7, and their product is 15. **See margin.**

8. **MULTIPLE CHOICE** Using the graph of the function $f(x) = x^2 + 6x - 7$, what are the solutions to the equation $x^2 + 6x - 7 = 0$? (Lesson 4-2) **J**

   F $-1, 6$ H $-1, 7$

   G $1, -6$ J $1, -7$

9. **BASEBALL** A baseball is hit upward with a velocity of 40 feet per second. Ignoring the height of the baseball player, how long does it take for the ball to fall to the ground? Use the formula $h(t) = v_0 t - 16t^2$ where $h(t)$ is the height of an object in feet, v_0 is the object's initial velocity in feet per second, and t is the time in seconds. (Lesson 4-2)
   2.5 seconds

Solve each equation by factoring. (Lesson 4-3)

10. $x^2 - x - 12 = 0$ {$-3, 4$}

11. $3x^2 + 7x + 2 = 0$ $\left\{-2, -\dfrac{1}{3}\right\}$

12. $x^2 - 2x - 15 = 0$ {$-3, 5$}

13. $2x^2 + 5x - 3 = 0$ $\left\{-3, \dfrac{1}{2}\right\}$

14. Write a quadratic equation in standard form with roots -6 and $\dfrac{1}{4}$. (Lesson 4-3) **$0 = 4x^2 + 23x - 6$**

15. **TRIANGLES** Find the dimensions of a triangle if the base is $\dfrac{2}{3}$ the measure of the height and the area is 12 square centimeters. (Lesson 4-3) **base = 4 cm, height = 6 cm**

16. **PATIO** Eli is putting a cement slab in his backyard. The original slab was going to have dimensions of 8 feet by 6 feet. He decided to make the slab larger by adding x feet to each side. The area of the new slab is 120 square feet. (Lesson 4-3)

   a. Write a quadratic equation that represents the area of the new slab. **$120 = x^2 + 14x + 48$**

   b. Find the new dimensions of the slab. **12 feet by 10 feet**

Simplify. (Lesson 4-4) **19. $11 + 9i$**

17. $\sqrt{-81}$ $9i$

18. $\sqrt{-25x^4y^5}$ $5x^2y^2 i\sqrt{y}$

19. $(15 - 3i) - (4 - 12i)$

20. i^{37} i

21. $(5 - 3i)(5 + 3i)$ 34

22. $\dfrac{3 - i}{2 + 5i}$ $\dfrac{1}{29} - \dfrac{17}{29}i$

23. The impedance in one part of a series circuit is $3 + 4j$ ohms and the impedance in another part of the circuit is $6 - 7j$ ohms. Add these complex numbers to find the total impedance in the circuit. (Lesson 4-4) **$9 - 3j$ ohms**

Formative Assessment

Use the Mid-Chapter Quiz to assess students' progress in the first half of the chapter.

For problems answered incorrectly, have students review the lessons indicated in parentheses.

Summative Assessment

Mid-Chapter Test, p. 61

McGraw-Hill Assessment Customize and create multiple versions of your Mid-Chapter Quiz and their answer keys.

FOLDABLES StudyOrganizer

Dinah Zike's Foldables®
Before students complete the Mid-Chapter Quiz, encourage them to review the information for Lessons 4-1 through 4-4 in their Foldables.

Additional Answers

1. y-intercept $= -3$; axis of symmetry $x = -2$; x-coordinate of vertex $= -2$; See Ch. 4 Answer Appendix for graph.

7. Let x be the first number. Then $7 - x$ is the other number. $x(7 - x) = 15$; $-x^2 + 7x - 15 = 0$. Since the graph of the related function does not intersect the x-axis, this equation has no real solutions. Therefore, no such numbers exist.

InterventionPlanner

TIER 1 **On Level** OL	TIER 2 **Strategic Intervention** AL approaching grade level	TIER 3 **Intensive Intervention** 2 or more grades below level
If students miss about 25% of the exercises or less,	**If** students miss about 50% of the exercises,	**If** students miss about 75% of the exercises,
Then choose a resource:	**Then** choose a resource:	**Then** use *Math Triumphs, Alg. 2*, Chs. 1, 3, and 4
SE Lessons 4-1, 4-2, 4-3, and 4-4	Study Guide and Intervention, pp. 5, 11, 17, and 24	connectED.mcgraw-hill.com Extra Examples, Personal Tutor, Homework Help, Review Vocabulary
Skills Practice, pp. 7, 13, 19, and 26	connectED.mcgraw-hill.com Extra Examples, Personal Tutor, Homework Help	
TE Chapter Project		
connectED.mcgraw-hill.com Self-Check Quiz		

LESSON 4-5 Completing the Square

:: Then	:: Now	:: Why?
• You factored perfect square trinomials.	**1** Solve quadratic equations by using the Square Root Property. **2** Solve quadratic equations by completing the square.	• When going through a school zone, drivers must slow to a speed of 20 miles per hour. Once they are out of the school zone, the drivers can increase their speed. Suppose Arturo is leaving school to go home for lunch, and he lives 5000 feet from the school zone. If Arturo accelerates at a constant rate of 8 feet per second squared, the equation $t^2 + 2t + 8 = 16$ represents the time t it takes him to reach home. To solve this equation, you can use the Square Root Property.

1 Focus

VerticalAlignment

▼ **Before Lesson 4-5** Factor perfect square trinomials.

▼ **Lesson 4-5** Solve quadratic equations by using the Square Root Property. Solve quadratic equations by completing the square.

▼ **After Lesson 4-5** Solve quadratic equations using the Quadratic Formula.

2 Teach

Scaffolding Questions

Have students read the **Why?** section of the lesson.

Ask:

- How are the equations $t^2 + 2t + 8 = 16$ and $t^2 + 2t - 8 = 0$ related? They are equivalent equations.

- What is the factored form of $t^2 + 2t - 8$? $(t + 4)(t - 2)$

- Would you expect the solution to the equation to be positive or negative? Why? positive; time is positive

 NewVocabulary
completing the square

CCSS **Common Core State Standards**

Content Standards
N.CN.7 Solve quadratic equations with real coefficients that have complex solutions.

F.IF.8.a Use the process of factoring and completing the square in a quadratic function to show zeros, extreme values, and symmetry of the graph, and interpret these in terms of a context.

Mathematical Practices
7 Look for and make use of structure.

1 **Square Root Property** You have solved equations like $x^2 - 25 = 0$ by factoring. You have also used the Square Root Property to solve such equations. This method can be useful with equations like the one above that describes the car's speed.

Example 1 Equation with Rational Roots

Solve $x^2 + 6x + 9 = 36$ by using the Square Root Property.

$x^2 + 6x + 9 = 36$	Original equation
$(x + 3)^2 = 36$	Factor the perfect square trinomial.
$x + 3 = \pm\sqrt{36}$	Square Root Property
$x + 3 = \pm 6$	$\sqrt{36} = 6$
$x = -3 \pm 6$	Subtract 3 from each side.
$x = -3 + 6$ or $x = -3 - 6$	Write as two equations.
$= 3$ $= -9$	Simplify.

The solution set is $\{-9, 3\}$ or $\{x \mid x = -9, 3\}$.

CHECK Substitute both values into the original equation.

$x^2 + 6x + 9 = 36$	Original equation	$x^2 + 6x + 9 = 36$
$3^2 + 6(3) + 9 \stackrel{?}{=} 36$	Substitute 3 and −9.	$(-9)^2 + 6(-9) + 9 \stackrel{?}{=} 36$
$9 + 18 + 9 \stackrel{?}{=} 36$	Simplify.	$81 - 54 + 9 \stackrel{?}{=} 36$
$36 = 36 ✓$	Both solutions are correct.	$36 = 36 ✓$

▶ **GuidedPractice**

Solve each equation by using the Square Root Property.

1A. $x^2 - 12x + 36 = 25$ {1, 11}   **1B.** $x^2 - 16x + 64 = 49$ {1, 15}

256 | Lesson 4-5

Lesson 4-5 Resources

Resource	Approaching Level **AL**	On Level **OL**	Beyond Level **BL**	English Learners **ELL**
Teacher Edition	• Differentiated Instruction, p. 259	• Differentiated Instruction, pp. 259, 262	• Differentiated Instruction, p. 262	• Differentiated Instruction, p. 259
Chapter Resource Masters	• Study Guide and Intervention, pp. 30–31 • Skills Practice, p. 32 • Practice, p. 33 • Word Problem Practice, p. 34	• Study Guide and Intervention, pp. 30–31 • Skills Practice, p. 32 • Practice, p.33 • Word Problem Practice, p. 34 • Enrichment, p. 35	• Practice, p. 33 • Word Problem Practice, p. 34 • Enrichment, p. 35	• Study Guide and Intervention, pp. 30–31 • Skills Practice, p. 32 • Practice, p. 33 • Word Problem Practice, p. 34
Other	• 5-Minute Check 4-5 • Study Notebook • Teaching Algebra with Manipulatives	• 5-Minute Check 4-5 • Study Notebook • Teaching Algebra with Manipulatives	• 5-Minute Check 4-5 • Study Notebook	• 5-Minute Check 4-5 • Study Notebook • Teaching Algebra with Manipulatives

Arthur S. Aubry/Photodisc/Getty Images

Roots that are irrational numbers may be written as exact answers in radical form or as *approximate* answers in decimal form when a calculator is used.

Example 2 Equation with Irrational Roots

Solve $x^2 - 10x + 25 = 27$ by using the Square Root Property.

$x^2 - 10x + 25 = 27$	Original equation
$(x - 5)^2 = 27$	Factor the perfect square trinomial.
$x - 5 = \pm\sqrt{27}$	Square Root Property
$x = 5 \pm 3\sqrt{3}$	Add 5 to each side; $\sqrt{27} = 3\sqrt{3}$.
$x = 5 + 3\sqrt{3}$ or $x = 5 - 3\sqrt{3}$	Write as two equations.
≈ 10.2 ≈ -0.2	Use a calculator.

The exact solutions of this equation are $5 + 3\sqrt{3}$ and $5 - 3\sqrt{3}$. The approximate solutions are -0.2 and 10.2. Check these results by finding and graphing the related quadratic function.

$x^2 - 10x + 25 = 27$	Original equation
$x^2 - 10x - 2 = 0$	Subtract 27 from each side.
$y = x^2 - 10x - 2$	Related quadratic function

CHECK Use the **zero** function of a graphing calculator. The approximate zeros of the related function are -0.2 and 10.2.

[-5, 15] scl: 1 by [-30, 20] scl: 2

▶ **Guided**Practice

Solve each equation by using the Square Root Property.

2A. $x^2 + 8x + 16 = 20$ $\{0.47, -8.47\}$ **2B.** $x^2 - 6x + 9 = 32$ $\{-2.66, 8.66\}$

2 Complete the Square All quadratic equations can be solved using the Square Root Property by manipulating the equation until one side is a perfect square. This method is called **completing the square**.

Consider $x^2 + 16x = 9$. Remember to perform each operation on each side of the equation.

$x^2 + 16x + \blacksquare = 9$	What value is needed for the perfect square?
$x^2 + 16x + 64 = 9 + 64$	$\left(\frac{16}{2}\right)^2 = 64$; add 64 to each side.
$x^2 + 16x + 64 = 73$	Simplify.
$(x + 8)^2 = 73$	We can now use the Square Root Property.

Use this pattern of coefficients to complete the square of a quadratic expression.

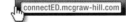

connectED.mcgraw-hill.com **257**

StudyTip

Plus or Minus When using the Square Root Property, remember to put a $\pm$ sign before the radical.

WatchOut!

Common Misconceptions Point out that the constant, 27, on the right side of the equation given in Example 2 is not a perfect square. Stress that this occurrence means the roots will be irrational numbers involving radicals. Also emphasize the use of the $\pm$ symbol in the step where the Square Root Property is utilized.

---

1 Square Root Property
Example 1 shows how to solve a quadratic equation using the Square Root Property. **Example 2** shows how to use the Square Root Property to solve a quadratic equation with irrational roots.

Formative Assessment
Use the Guided Practice exercises after each example to determine students' understanding of concepts.

Additional Examples

1 Solve $x^2 + 14x + 49 = 64$ by using the Square Root Property. $\{-15, 1\}$

2 Solve $x^2 - 4x + 4 = 13$ by using the Square Root Property. $\{2 \pm \sqrt{13}\}$

Additional Examples also in Interactive Classroom PowerPoint® Presentations

IWB **Interactive White Board** READY

WatchOut!

Preventing Errors In Example 1, point out that both constants in the equation, 9 and 36, are perfect squares.

connectED.mcgraw-hill.com **257**

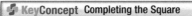

Focus on Mathematical Content

Completing the Square To complete the square for an expression of the form $x^2 + bx$, add the square of one half of the coefficient b to $x^2 + bx$. Written in symbols:

$$x^2 + bx + \left(\frac{b}{2}\right)^2 = \left(x + \frac{b}{2}\right)^2$$

Teach with Tech

Interactive Whiteboard Show a trinomial on the board where the constant term is c, and find the value that makes the trinomial a perfect square. Drag and copy the coefficient of x to a new line, and divide it by 2. Then drag and copy this quotient to a new line and square it. Take this result and drag it to replace c in the original trinomial.

WatchOut!

Preventing Errors When discussing the steps for completing the square, emphasize that the coefficient of the quadratic term must be 1.

Common Errors Remind students to think carefully about the difference between multiplying a quantity by 2 and squaring a quantity.

Additional Examples

3 Find the value of c that makes $x^2 + 12x + c$ a perfect square. Then write the trinomial as a perfect square. $36; (x + 6)^2$

4 Solve $x^2 + 4x - 12 = 0$ by completing the square. $\{-6, 2\}$

KeyConcept Completing the Square

Words	To complete the square for any quadratic expression of the form $x^2 + bx$, follow the steps below.
	Step 1 Find one half of b, the coefficient of x.
	Step 2 Square the result in Step 1.
	Step 3 Add the result of Step 2 to $x^2 + bx$.
Symbols	$x^2 + bx + \left(\frac{b}{2}\right)^2 = \left(x + \frac{b}{2}\right)^2$

 PT

Example 3 Complete the Square

Find the value of c that makes $x^2 + 16x + c$ a perfect square. Then write the trinomial as a perfect square.

Step 1	Find one half of 16.	$\frac{16}{2} = 8$
Step 2	Square the result in Step 1.	$8^2 = 64$
Step 3	Add the result of Step 2 to $x^2 + 16x$.	$x^2 + 16x + 64$

The trinomial $x^2 + 16x + 64$ can be written as $(x + 8)^2$.

▸ **Guided**Practice

3. Find the value of c that makes $x^2 - 14x + c$ a perfect square. Then write the trinomial as a perfect square. $49; (x - 7)^2$

You can solve any quadratic equation by completing the square. Because you are solving an equation, add the value you use to complete the square to each side.

 PT

Example 4 Solve an Equation by Completing the Square

Solve $x^2 + 10x - 11 = 0$ by completing the square.

$x^2 + 10x - 11 = 0$	Notice that $x^2 + 10x - 11$ is not a perfect square.
$x^2 + 10x = 11$	Rewrite so the left side is of the form $x^2 + bx$.
$x^2 + 10x + 25 = 11 + 25$	Since $\left(\frac{10}{2}\right)^2 = 25$, add 25 to each side.
$(x + 5)^2 = 36$	Write the left side as a perfect square.
$x + 5 = \pm 6$	Square Root Property
$x = -5 \pm 6$	Subtract 5 from each side.
$x = -5 + 6$ or $x = -5 - 6$	Write as two equations.
$= 1$ $= -11$	Simplify.

The solution set is $\{-11, 1\}$ or $\{x \mid x = -11, 1\}$. Check the result by using factoring.

▸ **Guided**Practice

Solve each equation by completing the square.

4A. $x^2 - 10x + 24 = 0$ $\{4, 6\}$ **4B.** $x^2 + 10x + 9 = 0$ $\{-9, -1\}$

WatchOut!

Each Side When solving equations by completing the square, don't forget to add $\left(\frac{b}{2}\right)^2$ to *each* side of the equation.

 258 | Lesson 4-5 | Completing the Square

When the coefficient of the quadratic term is not 1, you must divide the equation by that coefficient before completing the square.

Example 5 Equation with $a \neq 1$

Solve $2x^2 - 7x + 5 = 0$ by completing the square.

$2x^2 - 7x + 5 = 0$	Notice that $2x^2 - 7x + 5$ is not a perfect square.
$x^2 - \frac{7}{2}x + \frac{5}{2} = 0$	Divide by the coefficient of the quadratic term, 2.
$x^2 - \frac{7}{2}x = -\frac{5}{2}$	Subtract $\frac{5}{2}$ from each side.
$x^2 - \frac{7}{2}x + \frac{49}{16} = -\frac{5}{2} + \frac{49}{16}$	Since $\left(-\frac{7}{2} \div 2\right)^2 = \frac{49}{16}$, add $\frac{49}{16}$ to each side.
$\left(x - \frac{7}{4}\right)^2 = \frac{9}{16}$	Write the left side as a perfect square by factoring. Simplify the right side.
$x - \frac{7}{4} = \pm\frac{3}{4}$	Square Root Property
$x = \frac{7}{4} \pm \frac{3}{4}$	Add $\frac{7}{4}$ to each side.
$x = \frac{7}{4} + \frac{3}{4}$ or $x = \frac{7}{4} - \frac{3}{4}$	Write as two equations.
$= \frac{5}{2}$ $= 1$	

The solution set is $\left\{1, \frac{5}{2}\right\}$. or $\left\{x \mid x = 1, \frac{5}{2}\right\}$.

▶ **Guided**Practice

Solve each equation by completing the square.

5A. $3x^2 + 10x - 8 = 0$ $\left\{-4, \frac{2}{3}\right\}$ **5B.** $3x^2 + 14x - 16 = 0$ $\left\{2, \frac{8}{3}\right\}$

Not all solutions of quadratic equations are real numbers. In some cases, the solutions are complex numbers of the form $a + bi$, where $b \neq 0$.

StudyTip

CCSS Perseverance A graph of the related function shows that the equation has no real solutions since the graph has no x-intercepts. Imaginary solutions must be checked algebraically by substituting them in the original equation.

Example 6 Equation with Imaginary Solutions

Solve $x^2 + 8x + 22 = 0$ by completing the square.

$x^2 + 8x + 22 = 0$	Notice that $x^2 + 8x + 22$ is not a perfect square.
$x^2 + 8x = -22$	Rewrite so the left side is of the form $x^2 + bx$.
$x^2 + 8x + 16 = -22 + 16$	Since $\left(\frac{8}{2}\right)^2 = 16$, add 16 to each side.
$(x + 4)^2 = -6$	Write the left side as a perfect square.
$x + 4 = \pm\sqrt{-6}$	Square Root Property
$x + 4 = \pm i\sqrt{6}$	$\sqrt{-1} = i$
$x = -4 \pm i\sqrt{6}$	Subtract 4 from each side.

The solution set is $\{-4 + i\sqrt{6}, -4 - i\sqrt{6}\}$ or $\{x \mid x = -4 + i\sqrt{6}, -4 - i\sqrt{6}\}$.

▶ **Guided**Practice

Solve each equation by completing the square.

6A. $x^2 + 2x + 2 = 0$ $\{-1 + i, -1 - i\}$ **6B.** $x^2 - 6x + 25 = 0$ $\{3 + 4i, 3 - 4i\}$

Additional Examples

5 Solve $3x^2 - 2x - 1 = 0$ by completing the square. $\left\{-\frac{1}{3}, 1\right\}$

6 Solve $x^2 + 4x + 11 = 0$ by completing the square. $\{-2 \pm i\sqrt{7}\}$

WatchOut!

Preventing Misconceptions Some students may notice that the left side of the equation in Example 5 can be factored into the product of two binomials: $(2x - 5)(x - 1)$. Then the Zero Product Property can be used to obtain the same solutions. This is a good time to point out that more than one method of solution is often possible when solving a quadratic equation.

2 Complete the Square
Example 3 shows how to find a value for the constant in a quadratic expression so that the expression will be a perfect square. **Example 4** shows how to solve a quadratic equation by completing the square. **Example 5** shows how to solve a quadratic equation by completing the square when the coefficient of the quadratic term is not 1. **Example 6** shows how to solve a quadratic equation whose solutions are complex numbers of the form $a + bi$, where $b \neq 0$.

CCSS Teaching the Mathematical Practices

Perseverance Mathematically proficient students monitor and evaluate their progress and change course if necessary. Remind students that graphing a function can help them to verify *real* solutions only.

DifferentiatedInstruction **AL** **OL** **ELL**

Verbal/Linguistic Learners Have students solve the equation $x^2 + 6x - 40 = 0$ by completing the square. Then have them discuss with a partner as many ways as they can to check their solutions.

3 Practice

Formative Assessment

Use Exercises 1–13 to check for understanding.

Use the chart on the bottom of this page to customize assignments for your students.

CCSS Teaching the Mathematical Practices

Modeling Mathematically proficient students can apply the mathematics they know to solve problems arising in everyday life, analyze relationships mathematically to draw conclusions, and interpret their mathematical results in the context of a situation.

Exercise Alert

For Exercise 57, students will need a straightedge and a compass.

Multiple Representations

In Exercise 57, students use geometric constructions, algebra, a table of values, and logical analysis to investigate golden rectangles and the golden ratio.

Additional Answers

57c.

CQ	x
2	$1 + \sqrt{5}$
3	$\dfrac{3 + 3\sqrt{5}}{2}$
4	$2 + 2\sqrt{5}$

57d. Sample answer: the x-values are multiples of $\dfrac{1 + \sqrt{5}}{2}$;

$x = \dfrac{n(1 + \sqrt{5})}{2}$

Check Your Understanding

 Check Your Understanding   ● = Step-by-Step Solutions begin on page R14.

Examples 1–2 Solve each equation by using the Square Root Property. Round to the nearest hundredth if necessary.

1. $x^2 + 12x + 36 = 6$ $\{-8.45, -3.55\}$
2. $x^2 - 8x + 16 = 13$ $\{0.39, 7.61\}$
3. $x^2 + 18x + 81 = 15$ $\{-12.87, -5.13\}$
4. $9x^2 + 30x + 25 = 11$ $\{-2.77, -0.56\}$

5. **LASER LIGHT SHOW** The area A in square feet of a projected laser light show is given by $A = 0.16d^2$, where d is the distance from the laser to the screen in feet. At what distance will the projected laser light show have an area of 100 square feet? **25 ft**

Example 3 Find the value of c that makes each trinomial a perfect square. Then write the trinomial as a perfect square.

6. $x^2 - 10x + c$ $25; (x - 5)^2$
7. $x^2 - 5x + c$ $6.25; (x - 2.5)^2$

Examples 4–6 Solve each equation by completing the square. 9. $\{2 - i\sqrt{5}, 2 + i\sqrt{5}\}$

8. $x^2 + 2x - 8 = 0$ $\{-4, 2\}$
9. $x^2 - 4x + 9 = 0$
10. $2x^2 - 3x - 3 = 0$ $\{-0.69, 2.19\}$
11. $2x^2 + 6x - 12 = 0$ $\{-4.37, 1.37\}$
12. $x^2 + 4x + 6 = 0$ $\{-2 - i\sqrt{2}, -2 + i\sqrt{2}\}$
13. $x^2 + 8x + 10 = 0$ $\{-6.45, -1.55\}$

Practice and Problem Solving

Practice and Problem Solving   Extra Practice is on page R4.

Examples 1–2 Solve each equation by using the Square Root Property. Round to the nearest hundredth if necessary. 14. $\{-5.16, 1.16\}$ 15. $\{-1.47, 7.47\}$ 16. $\{-8.24, 0.24\}$ 17. $\{-7.65, -2.35\}$
18. $\{-8.24, -3.76\}$

14. $x^2 + 4x + 4 = 10$
15. $x^2 - 6x + 9 = 20$
16. $x^2 + 8x + 16 = 18$
17. $x^2 + 10x + 25 = 7$
18. $x^2 + 12x + 36 = 5$
19. $x^2 - 2x + 1 = 4$ $\{-1, 3\}$
20. $x^2 - 5x + 6.25 = 4$
21. $x^2 - 15x + 56.25 = 8$
22. $x^2 + 32x + 256 = 1$ $\{-17, -15\}$
23. $x^2 - 3x + \dfrac{9}{4} = 6$ $\{-0.95, 3.95\}$
24. $x^2 + 7x + \dfrac{49}{4} = 4$ $\{-5.5, -1.5\}$
25. $x^2 - 9x + \dfrac{81}{4} = \dfrac{1}{4}$ $\{4, 5\}$

20. $\{0.5, 4.5\}$ 21. $\{4.67, 10.33\}$

Example 3 Find the value of c that makes each trinomial a perfect square. Then write the trinomial as a perfect square.

26. $x^2 + 8x + c$ $16; (x + 4)^2$
27. $x^2 + 16x + c$ $64; (x + 8)^2$
28. $x^2 - 11x + c$ $\dfrac{121}{4}; \left(x - \dfrac{11}{2}\right)^2$
29. $x^2 + 9x + c$ $20.25; (x + 4.5)^2$

Examples 4–6 Solve each equation by completing the square. 30. $\{2 - 2i\sqrt{2}, 2 + 2i\sqrt{2}\}$ 31. $\{-4.61, 2.61\}$

30. $x^2 - 4x + 12 = 0$
31. $x^2 + 2x - 12 = 0$
32. $x^2 + 6x + 8 = 0$ $\{-4, -2\}$
33. $x^2 - 4x + 3 = 0$ $\{1, 3\}$
34. $2x^2 + x - 3 = 0$
35. $2x^2 - 3x + 5 = 0$
34. $\left\{-\dfrac{3}{2}, 1\right\}$
36. $2x^2 + 5x + 7 = 0$
(37) $3x^2 - 6x - 9 = 0$
38. $x^2 - 2x + 3 = 0$
39. $x^2 + 4x + 11 = 0$
40. $x^2 - 6x + 18 = 0$
41. $x^2 - 10x + 29 = 0$

35–45. See Chapter 4 Answer Appendix.

42. $3x^2 - 4x = 2$
43. $2x^2 - 7x = -12$
44. $x^2 - 2.4x = 2.2$
45. $x^2 - 5.3x = -8.6$
46. $x^2 - \dfrac{1}{5}x - \dfrac{11}{5} = 0$ $\{-1.39, 1.59\}$
47. $x^2 - \dfrac{9}{2}x - \dfrac{24}{5} = 0$ $\{-0.89, 5.39\}$

48. **CCSS MODELING** An architect's blueprints call for a dining room measuring 13 feet by 13 feet. The customer would like the dining room to be a square, but with an area of 250 square feet. How much will this add to the dimensions of the room? **about 2.81 ft**

B Solve each equation. Round to the nearest hundredth if necessary.

49. $4x^2 - 28x + 49 = 5$ $\{2.38, 4.62\}$
50. $9x^2 + 30x + 25 = 11$ $\{-2.77, -0.56\}$
51. $x^2 + x + \dfrac{1}{3} = \dfrac{2}{3}$ $\{-1.26, 0.26\}$
52. $x^2 + 1.2x + 0.56 = 0.91$ $\{-1.44, 0.24\}$

 260 | Lesson 4-5 | Completing the Square

Differentiated Homework Options

Level	Assignment	Two-Day Option	
AL Basic	14–48, 58, 60–83	15–47 odd, 63–66	14–48 even, 58, 60–62, 67–83
OL Core	15–57 odd, 58, 60–83	14–48, 63–66	49–58, 60–62, 67–83
BL Advanced	49–80, (optional: 81–83)		

53 **FIREWORKS** A firework's distance d meters from the ground is given by $d = -1.5t^2 + 25t$, where t is the number of seconds after the firework has been lit.

  a. How many seconds have passed since the firework was lit when the firework explodes if it explodes at the maximum height of its path? $8\frac{1}{3}$ seconds

  b. What is the height of the firework when it explodes? about 104.2 ft

▷ Find the value of c that makes each trinomial a perfect square. Then write the trinomial as a perfect square. **54.** 0.1225; $(x + 0.35)^2$ **55.** 2.56; $(x - 1.6)^2$ **56.** $\frac{81}{100}$; $\left(x - \frac{9}{10}\right)^2$

54. $x^2 + 0.7x + c$ **55.** $x^2 - 3.2x + c$ **56.** $x^2 - 1.8x + c$

57. **MULTIPLE REPRESENTATIONS** In this problem, you will use quadratic equations to investigate golden rectangles and the golden ratio.

  a. Geometric

  • Draw square $ABCD$.
  • Locate the midpoint of $\overline{CD}$. Label the midpoint P. Draw $\overline{PB}$.
  • Construct an arc with a radius of $\overline{PB}$ from B clockwise past the bottom of the square.
  • Extend $\overline{CD}$ until it intersects the arc. Label this point Q.
  • Construct rectangle $ARQD$.

  57a.

  b. Algebraic Let $AD = x$ and $CQ = 1$. Use completing the square to solve $\frac{DQ}{AD} = \frac{QR}{CQ}$ for x. $x = \frac{1 + \sqrt{5}}{2}$

  c. Tabular Make a table of x and values for $CQ = 2$, 3, and 4. See margin.

  d. Verbal What do you notice about the x-values? Write an equation you could use to determine x for $CQ = n$, where n is a nonzero real number. See margin.

H.O.T. Problems   Use Higher-Order Thinking Skills

58. **ERROR ANALYSIS** Alonso and Aida are solving $x^2 + 8x - 20 = 0$ by completing the square. Is either of them correct? Explain your reasoning.

Alonso
$$x^2 + 8x - 20 = 0$$
$$x^2 + 8x = 20$$
$$x^2 + 8x + 16 = 20 + 16$$
$$(x + 4)^2 = 36$$
$$x + 4 = \pm 6$$
$$x = -4 \pm 6$$

Aida
$$x^2 + 8x - 20 = 0$$
$$x^2 + 8x = 20$$
$$x^2 + 8x + 16 = 20$$
$$(x + 4)^2 = 20$$
$$x + 4 = \pm\sqrt{20}$$
$$x = -4 \pm \sqrt{20}$$

Alonso; Aida did not add 16 to each side; she added it only to the left side.

59. **CHALLENGE** Solve $x^2 + bx + c = 0$ by completing the square. Your answer will be an expression for x in terms of b and c. $x = \frac{-b}{2} \pm \sqrt{\frac{b^2}{4} - c}$

60. **CCSS** **ARGUMENTS** Without solving, determine how many unique solutions there are for each equation. Are they rational, real, or complex? Justify your reasoning. See Chapter 4 Answer Appendix.

  a. $(x + 2)^2 = 16$ **b.** $(x - 2)^2 = 16$ **c.** $-(x - 2)^2 = 16$
  d. $36 - (x - 2)^2 = 16$ **e.** $16(x + 2)^2 = 0$ **f.** $(x + 4)^2 = (x + 6)^2$

61. **OPEN ENDED** Write a perfect square trinomial equation in which the linear coefficient is negative and the constant term is a fraction. Then solve the equation.
  Sample answer: $x^2 - \frac{2}{3}x + \frac{1}{9} = \frac{1}{4}$; $\left\{\frac{5}{6}, -\frac{1}{6}\right\}$

62. **WRITING IN MATH** Explain what it means to complete the square. Describe each step.

62. Completing the square allows you to rewrite one side of a quadratic equation in the form of a perfect square. Once in this form, the equation can be solved by using the Square Root Property.

CCSS **Teaching the Mathematical Practices**

Arguments Mathematically proficient students understand and use stated assumptions, definitions, and previously established results in constructing arguments. They make conjectures and build a logical progression of statements to explore the truth of their conjectures. And they are able to analyze situations by breaking them into cases, and can recognize and use counterexamples.

📁 **Study Guide and Intervention**
AL OL ELL

📁 **Practice**
AL OL BL ELL

📁 **Word Problem Practice**
AL OL BL ELL

📁 **Enrichment**
OL BL

connectED.mcgraw-hill.com **261**

4 Assess

Name the Math Prepare two paper bags containing pieces of paper: one containing equations that can be solved using the Square Root Property, and the other with equations to be solved by completing the square. Have each student select an equation from each bag. Then have students tell how to solve the equations.

Additional Answers

75.

76.

77.

78. $y = \begin{cases} -4 \text{ if } x \leq -2 \\ -2x + 3 \text{ if } -2 < x < 1 \\ x - 5 \text{ if } x \geq 2 \end{cases}$

79. $y = \begin{cases} -8 \text{ if } x \leq -2 \\ 2x - 6 \text{ if } -2 < x \leq 5 \\ 2x - 18 \text{ if } x > 5 \end{cases}$

80. $y = \begin{cases} x + 12 \text{ if } x \leq -6 \\ 8 \text{ if } -6 < x < 2 \\ -2.5x + 15 \text{ if } x \geq 2 \end{cases}$

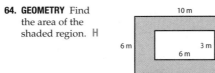

Standardized Test Practice

63. SAT/ACT If $x^2 + y^2 = 2xy$, then y must equal **E**

  A -1 **C** 1   **E** x

  B 0   **D** $-x$

64. GEOMETRY Find the area of the shaded region. **H**

  F 14 m^2 **G** 18 m^2 **H** 42 m^2 **J** 60 m^2

65. SHORT RESPONSE What value of c should be used to solve the following equation by completing the square? **125**

$$5x^2 - 50x + c = 12 + c$$

66. If $5 - 3i$ is a solution for $x^2 + ax + b = 0$, where a and b are real numbers, what is the value of b? **C**

  A 10   **C** 34

  B 14   **D** 40

Spiral Review

Simplify. (Lesson 4-4)

67. $(8 + 5i)^2$ $39 + 80i$

68. $4(3 - i) + 6(2 - 5i)$ $24 - 34i$

69. $\dfrac{5 - 2i}{6 + 9i}$ $\dfrac{4}{39} - \dfrac{19}{39}i$

Write a quadratic equation in standard form with the given root(s). (Lesson 4-3)

70. $\dfrac{4}{5}, \dfrac{3}{4}$ $20x^2 - 31x + 12 = 0$

71. $-\dfrac{2}{5}, 6$ $5x^2 - 28x - 12 = 0$

72. $-\dfrac{1}{4}, -\dfrac{6}{7}$ $28x^2 + 31x + 6 = 0$

73. TRAVEL Yoko is going with the Spanish Club to Costa Rica. She buys 10 traveler's checks in denominations of $20, $50, and $100, totaling $370. She has twice as many $20 checks as $50 checks. How many of each denomination of traveler's checks does she have? (Lesson 3-4) **1 $100, 3 $50, and 6 $20 checks**

74. SHOPPING Main St. Media sells all DVDs for one price and all books for another price. Alex bought 4 DVDs and 6 books for $170, while Matt bought 3 DVDs and 8 books for $180. What is the cost of a DVD and the cost of a book? (Lesson 3-1) **DVD: $20; book: $15**

Graph each inequality. (Lesson 2-8) **75–77. See margin.**

75. $y \geq 4x - 3$

76. $2x - 3y < 6$

77. $5x + 2y + 3 \leq 0$

Write the piecewise function shown in each graph. (Lesson 2-6) **78–80. See margin.**

78.

79.

80.

Skills Review

Evaluate $b^2 - 4ac$ **for the given values of** a, b, **and** c.

81. $a = 5, b = 6, c = 2$ -4

82. $a = -2, b = -7, c = 3$ 73

83. $a = -5, b = -8, c = -10$ -136

DifferentiatedInstruction (OL) (BL)

Extension Write $x^4 + 10x^2 + 25 = 49$ on the board. Explain that this is a quartic equation. Ask students to use what they have learned from this lesson to solve the equation for x.

Point out that a quadratic equation (degree 2) has at most 2 real solutions and a quartic equation (degree 4) has at most 4 real solutions.

$$x^4 + 10x^2 + 25 = 49$$
$$(x^2 + 5)^2 = 49$$
$$x^2 + 5 = \pm\sqrt{49}$$
$$x^2 + 5 = \pm 7$$
$$x^2 = -5 \pm 7$$
$$x^2 = -12 \text{ or } x^2 = 2$$
$$x = \pm\sqrt{-12} \text{ or } x = \pm\sqrt{2}$$
$$x = \pm 2i\sqrt{3}, x = \pm\sqrt{2}$$

EXTEND 4-5
Graphing Technology Lab
Solving Quadratic Equations

You can use a TI-Nspire™ CAS Technology to solve quadratic equations.

 Common Core State Standards
Content Standards
N.CN.7 Solve quadratic equations with real coefficients that have complex solutions.

Activity   Finding Roots

Solve each equation.

a. $3x^2 - 4x + 1 = 0$

Step 1 Add a new **Calculator** page.

Step 2 Select the **Solve** tool from the **Algebra** menu.

Step 3 Type $3x^2 - 4x + 1 = 0$ followed by a comma, x, and then **enter**. The solutions are $x = \frac{1}{3}$ or $x = 1$.

b. $6x^2 + 4x - 3 = 0$

Step 1 Select the **Solve** tool from the **Algebra** menu.

Step 2 Type $6x^2 - 4x - 3 = 0$ followed by a comma, x, and then **enter**. The solutions are $x = \frac{-2 \pm \sqrt{22}}{6}$.

c. $x^2 - 6x + 10 = 0$.

Step 1 Select the **Solve** tool from the **Algebra** menu.

Step 2 Type $x^2 - 6x + 10 = 0$ followed by a comma, x, and then **enter**. The calculator returns a value of *false*, meaning that there are no real solutions.

Step 3 Under menu, select **Algebra**, **Complex**, then **Solve**. Reenter the equation. The solutions are $x = 3 \pm i$.

Exercises

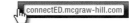

Solve each equation.

1. $x^2 - 2x - 24 = 0$ $-4, 6$

2. $-x^2 + 4x - 1 = 0$ $2 \pm \sqrt{3}$

3. $0 = -3x^2 - 6x + 9$ $-3, 1$

4. $x^2 - 2x + 5 = 0$ $1 \pm 2i$

5. $0 = 4x^2 - 8$ $\pm\sqrt{2}$

6. $0 = 2x^2 - 4x + 1$ $\frac{2 \pm \sqrt{2}}{2}$

7. $x^2 + 3x + 8 = 5$ $\frac{-3 \pm i\sqrt{3}}{2}$

8. $25 + 4x^2 = -20x$ $-\frac{5}{2}$

9. $x^2 - x = -6$ $\frac{1 \pm i\sqrt{23}}{2}$

---

1 Focus

Objective Use a calculator containing a computer algebra system to solve quadratic equations.

Materials for Each Group
- TI-Nspire CAS technology

Teaching Tip
The graphing calculator opens on the same screen as when it was turned off. Have students press the **home** key to begin the lab.

2 Teach

Working in Cooperative Groups
Have students work in groups of two or three, mixing abilities, to complete the Activity and Exercises 1–3.

- The **X**'s indicated in the **KEYSTROKES** steps refer to the letter **X** key. Make sure that students do not mistake these for multiplication symbols, which are not needed in these examples.

- Solutions are displayed in exact form. To convert to decimal form: Under **menu**, select **Number**, then select **Convert to Decimal.**

Practice Have students complete Exercises 4–9.

3 Assess

Formative Assessment
Use Exercise 1 to assess whether students can correctly use menu selections and then solve the quadratic equation.

From Concrete to Abstract
TI-Nspire CAS technology can be used to solve Exercises 7–9 without first writing equations in standard form $(ax^2 + bx + c = 0)$.

Extending the Concept
Ask:
- Do the signs of the numbers in a quadratic equation in standard form provide any clues as to whether its solutions are complex?
Sample answer: If the coefficient of x^2 and the constant term are the same sign, the solutions might be complex.

1 Focus

VerticalAlignment

Before Lesson 4-6 Solve equations by completing the square.

Lesson 4-6 Solve quadratic equations by using the Quadratic Formula. Use the discriminant to determine the number and type of roots of a quadratic equation.

After Lesson 4-6 Solve quadratic inequalities using graphs and algebraic methods.

2 Teach

Scaffolding Questions

Have students read the **Why?** section of the lesson.

Ask:

- As the value of *t* increases in this equation, what happens to the value of *h*? It increases then decreases.

- What shape does the graph of this equation have? a parabola

- Which way does the parabola open? downward

LESSON 4-6 The Quadratic Formula and the Discriminant

Then	Now	Why?
You solved equations by completing the square.	**1** Solve quadratic equations by using the Quadratic Formula. **2** Use the discriminant to determine the number and type of roots of a quadratic equation.	Pumpkin catapult is an event in which a contestant builds a catapult and launches a pumpkin at a target.

The path of the pumpkin can be modeled by the quadratic function $h = -4.9t^2 + 117t + 42$, where h is the height of the pumpkin and t is the number of seconds.

To predict when the pumpkin will hit the target, you can solve the equation $0 = -4.9t^2 + 117t + 42$. This equation would be difficult to solve using factoring, graphing, or completing the square.

 NewVocabulary
Quadratic Formula
discriminant

CCSS Common Core State Standards

Content Standards
N.CN.7 Solve quadratic equations with real coefficients that have complex solutions.
A.SSE.1.b Interpret complicated expressions by viewing one or more of their parts as a single entity.

Mathematical Practices
8 Look for and express regularity in repeated reasoning.

1 Quadratic Formula You have found solutions of some quadratic equations by graphing, by factoring, and by using the Square Root Property. There is also a formula that can be used to solve any quadratic equation. This formula can be derived by solving the standard form of a quadratic equation.

General Case		Specific Case
$ax^2 + bx + c = 0$	Standard quadratic equation	$2x^2 + 8x + 1 = 0$
$x^2 + \frac{b}{a}x + \frac{c}{a} = 0$	Divide each side by a.	$x^2 + 4x + \frac{1}{2} = 0$
$x^2 + \frac{b}{a}x = -\frac{c}{a}$	Subtract $\frac{c}{a}$ from each side.	$x^2 + 4x = -\frac{1}{2}$
$x^2 + \frac{b}{a}x + \frac{b^2}{4a^2} = -\frac{c}{a} + \frac{b^2}{4a^2}$	Complete the square.	$x^2 + 4x + \left(\frac{4}{2}\right)^2 = -\frac{1}{2} + \left(\frac{4}{2}\right)^2$
$\left(x + \frac{b}{2a}\right)^2 = -\frac{c}{a} + \frac{b^2}{4a^2}$	Factor the left side.	$(x + 2)^2 = -\frac{1}{2} + \left(\frac{4}{2}\right)^2$
$\left(x + \frac{b}{2a}\right)^2 = \frac{b^2 - 4ac}{4a^2}$	Simplify the right side.	$(x + 2)^2 = \frac{7}{2}$
$x + \frac{b}{2a} = \pm\frac{\sqrt{b^2 - 4ac}}{2a}$	Square Root Property	$x + 2 = \pm\sqrt{\frac{7}{2}}$
$x = -\frac{b}{2a} \pm \frac{\sqrt{b^2 - 4ac}}{2a}$	Subtract $\frac{b}{2a}$ from each side.	$x = -2 \pm \sqrt{\frac{7}{2}}$
$x = \frac{-b \pm \sqrt{b^2 - 4ac}}{2a}$	Simplify.	$x = \frac{-4 \pm \sqrt{14}}{2}$

The equation $x = \frac{-b \pm \sqrt{b^2 - 4ac}}{2a}$ is known as the **Quadratic Formula**.

 264 | Lesson 4-6

Izzy Schwartz/Photodisc/Getty Images

Lesson 4-6 Resources

Resource	Approaching Level **AL**	On Level **OL**	Beyond Level **BL**	English Learners **ELL**
Teacher Edition	▪ Differentiated Instruction, p. 265	▪ Differentiated Instruction, pp. 265, 268	▪ Differentiated Instruction, p. 268	▪ Differentiated Instruction, p. 265
Chapter Resource Masters	▪ Study Guide and Intervention, pp. 36–37 ▪ Skills Practice, p. 38 ▪ Practice, p. 39 ▪ Word Problem Practice, p. 40	▪ Study Guide and Intervention, pp. 36–37 ▪ Skills Practice, p. 38 ▪ Practice, p. 39 ▪ Word Problem Practice, p. 40 ▪ Enrichment, p. 41 ▪ Spreadsheet Activity, p. 42	▪ Practice, p. 39 ▪ Word Problem Practice, p. 40 ▪ Enrichment, p. 41	▪ Study Guide and Intervention, pp. 36–37 ▪ Skills Practice, p. 38 ▪ Practice, p. 39 ▪ Word Problem Practice, p. 40
Other	▪ 5-Minute Check 4-6 ▪ Study Notebook	▪ 5-Minute Check 4-6 ▪ Study Notebook	▪ 5-Minute Check 4-6 ▪ Study Notebook	▪ 5-Minute Check 4-6 ▪ Study Notebook

StudyTip

Quadratic Formula Although factoring may be an easier method to solve some of the equations, the Quadratic Formula can be used to solve any quadratic equation.

KeyConcept Quadratic Formula

Words	The solutions of a quadratic equation of the form $ax^2 + bx + c = 0$, where $a \neq 0$, are given by the following formula.
	$$x = \frac{-b \pm \sqrt{b^2 - 4ac}}{2a}$$
Example	$x^2 + 5x + 6 = 0 \rightarrow x = \frac{-5 \pm \sqrt{5^2 - 4(1)(6)}}{2(1)}$

Example 1 Two Rational Roots

Solve $x^2 - 10x = 11$ by using the Quadratic Formula.

First, write the equation in the form $ax^2 + bx + c = 0$ and identify a, b, and c.

$$\begin{array}{ccccc} ax^2 & + & bx & + & c = 0 \\ \downarrow & & \downarrow & & \downarrow \end{array}$$

$$x^2 - 10x = 11 \quad \rightarrow \quad 1x^2 - 10x - 11 = 0$$

Then, substitute these values into the Quadratic Formula.

$$x = \frac{-b \pm \sqrt{b^2 - 4ac}}{2a} \qquad \text{Quadratic Formula}$$

$$= \frac{-(-10) \pm \sqrt{(-10)^2 - 4(1)(-11)}}{2(1)} \qquad \text{Replace } a \text{ with 1, } b \text{ with } -10, \text{ and } c \text{ with } -11.$$

$$= \frac{10 \pm \sqrt{100 + 44}}{2} \qquad \text{Multiply.}$$

$$= \frac{10 \pm \sqrt{144}}{2} \qquad \text{Simplify.}$$

$$= \frac{10 \pm 12}{2} \qquad \sqrt{144} = 12$$

$$x = \frac{10 + 12}{2} \quad \text{or} \quad x = \frac{10 - 12}{2} \qquad \text{Write as two equations.}$$

$$= 11 \qquad\qquad = -1 \qquad \text{Simplify.}$$

The solutions are -1 and 11.

CHECK Substitute both values into the original equation.

$$x^2 - 10x = 11 \qquad\qquad x^2 - 10x = 11$$
$$(-1)^2 - 10(-1) \overset{?}{=} 11 \qquad (11)^2 - 10(11) \overset{?}{=} 11$$
$$1 + 10 \overset{?}{=} 11 \qquad\qquad 121 - 110 \overset{?}{=} 11$$
$$11 = 11 \checkmark \qquad\qquad\qquad 11 = 11 \checkmark$$

GuidedPractice

Solve each equation by using the Quadratic Formula.

1A. $x^2 + 6x = 16$ 2, -8 **1B.** $2x^2 + 25x + 33 = 0$ $-11, -\frac{3}{2}$

ReviewVocabulary

radicand the value underneath the radical symbol

When the value of the radicand in the Quadratic Formula is 0, the quadratic equation has exactly one rational root.

DifferentiatedInstruction AL OL ELL

If students substitute values into the Quadratic Formula incorrectly,

Then encourage students to write down the values of a, b, and c from the standard form of the quadratic equation before they begin substituting them into the formula.

---

1 Quadratic Formula

Example 1 shows how to solve a quadratic equation using the Quadratic Formula. **Example 2** shows how to solve a quadratic equation when the radicand in the Quadratic Formula simplifies to 0. **Example 3** shows how to express irrational roots of a quadratic equation by writing them in radical form. **Example 4** shows how to solve a quadratic equation for complex solutions when the radicand in the Quadratic Formula simplifies to a negative value.

Formative Assessment

Use the Guided Practice exercises after each example to determine students' understanding of concepts.

Additional Example

1 Solve $x^2 - 8x = 33$ by using the Quadratic Formula. $\{-3, 11\}$

  ▶ **Additional Examples** also in Interactive Classroom PowerPoint® Presentations

IWB **Interactive White Board** READY

Focus on Mathematical Content

The Quadratic Formula Any quadratic equation written in the form $ax^2 + bx + c = 0$, where $a \neq 0$, can be solved using the Quadratic Formula. Substitute the values for a, b, and c into the Quadratic Formula to find the value(s) for x. The Quadratic Formula is

$$x = \frac{-b \pm \sqrt{b^2 - 4ac}}{2a}$$ where $a \neq 0$.

WatchOut!

Common Misconceptions Some students may have commented that the equations in Examples 1 and 2 could have been solved by factoring. Before beginning Example 3, stress that many quadratic equations cannot be solved easily by factoring. State that the quadratic equation presented in Example 3 is such an equation. Emphasize that the Quadratic Formula provides a way to find the roots for *any* quadratic equation.

Additional Examples

2 Solve $x^2 - 34x + 289 = 0$ by using the Quadratic Formula. $\{17\}$

3 Solve $x^2 - 6x + 2 = 0$ by using the Quadratic Formula. $\{3 \pm \sqrt{7}\}$, or approximately 0.4 and 5.6

Math HistoryLink
Brahmagupta **(598–668)**
Indian mathematician
Brahmagupta offered the first general solution of the quadratic equation $ax^2 + bx = c$, now known as the Quadratic Formula.

Example 2 One Rational Root

Solve $x^2 + 8x + 16 = 0$ by using the Quadratic Formula.

Identify a, b, and c. Then, substitute these values into the Quadratic Formula.

$$x = \frac{-b \pm \sqrt{b^2 - 4ac}}{2a} \qquad \text{Quadratic Formula}$$

$$= \frac{-(8) \pm \sqrt{(8)^2 - 4(1)(16)}}{2(1)} \qquad \text{Replace } a \text{ with 1, } b \text{ with 8, and } c \text{ with 16.}$$

$$= \frac{-8 \pm \sqrt{0}}{2} \qquad \text{Simplify.}$$

$$= \frac{-8}{2} \text{ or } -4 \qquad \sqrt{0} = 0$$

The solution is -4.

CHECK A graph of the related function shows that there is one solution at $x = -4$.

[−10, 10] scl: 1 by [−10, 10] scl: 1

▶ **Guided**Practice

Solve each equation by using the Quadratic Formula.

2A. $x^2 - 16x + 64 = 0$ **8**   **2B.** $x^2 + 34x + 289 = 0$ **−17**

You can express irrational roots exactly by writing them in radical form.

Example 3 Irrational Roots

Solve $2x^2 + 6x - 7 = 0$ by using the Quadratic Formula.

$$x = \frac{-b \pm \sqrt{b^2 - 4ac}}{2a} \qquad \text{Quadratic Formula}$$

$$= \frac{-(6) \pm \sqrt{(6)^2 - 4(2)(-7)}}{2(2)} \qquad \text{Replace } a \text{ with 2, } b \text{ with 6, and } c \text{ with } -7.$$

$$= \frac{-6 \pm \sqrt{92}}{4} \qquad \text{Simplify.}$$

$$= \frac{-6 \pm 2\sqrt{23}}{4} \text{ or } \frac{-3 \pm \sqrt{23}}{2} \qquad \sqrt{92} = \sqrt{4 \cdot 23} \text{ or } 2\sqrt{23}$$

The approximate solutions are -3.9 and 0.9.

CHECK Check these results by graphing the related quadratic function, $y = 2x^2 + 6x - 7$. Using the **zero** function of a graphing calculator, the approximate zeros of the related function are -3.9 and 0.9.

[−10, 10] scl: 1 by [−10, 10] scl: 1

▶ **Guided**Practice

Solve each equation by using the Quadratic Formula.

3A. $3x^2 + 5x + 1 = 0$ $\dfrac{-5 \pm \sqrt{13}}{6}$ **3B.** $x^2 - 8x + 9 = 0$ $4 \pm \sqrt{7}$

Dinodia/age fotostock

StudyTip

Complex Numbers
Remember to write your solutions in the form $a + bi$, sometimes called the *standard form* of a complex number.

When using the Quadratic Formula, if the value of the radicand is negative, the solutions will be complex. Complex solutions always appear in conjugate pairs.

Example 4 Complex Roots

Solve $x^2 - 6x = -10$ by using the Quadratic Formula.

$$x = \frac{-b \pm \sqrt{b^2 - 4ac}}{2a}$$ Quadratic Formula

$$= \frac{-(-6) \pm \sqrt{(-6)^2 - 4(1)(10)}}{2(1)}$$ Replace a with 1, b with −6, and c with 10.

$$= \frac{6 \pm \sqrt{-4}}{2}$$ Simplify.

$$= \frac{6 \pm 2i}{2}$$ $\sqrt{-4} = \sqrt{4 \cdot (-1)}$ or $2i$

$$= 3 \pm i$$ Simplify.

The solutions are the complex numbers $3 + i$ and $3 - i$.

CHECK A graph of the related function shows that the solutions are complex, but it cannot help you find them. To check complex solutions, substitute them into the original equation.

[−10, 10] scl: 1 by [−10, 10] scl: 1

$$x^2 - 6x = -10$$ Original equation
$$(3 + i)^2 - 6(3 + i) \stackrel{?}{=} -10$$ $x = 3 + i$
$$9 + 6i + i^2 - 18 - 6i \stackrel{?}{=} -10$$ Square of a sum; Distributive Property
$$-9 + i^2 \stackrel{?}{=} -10$$ Simplify.
$$-9 - 1 = -10 \checkmark$$ $i^2 = -1$

$$x^2 - 6x = -10$$ Original equation
$$(3 - i)^2 - 6(3 - i) \stackrel{?}{=} -10$$ $x = 3 - i$
$$9 - 6i + i^2 - 18 + 6i \stackrel{?}{=} -10$$ Square of a sum; Distributive Property
$$-9 + i^2 \stackrel{?}{=} -10$$ Simplify.
$$-9 - 1 = -10 \checkmark$$ $i^2 = -1$

▶ **Guided**Practice

Solve each equation by using the Quadratic Formula.

4A. $3x^2 + 5x + 4 = 0$ $\dfrac{-5 \pm i\sqrt{23}}{6}$ **4B.** $x^2 - 4x = -13$ $2 \pm 3i$

2 Roots and the Discriminant In the previous examples, observe the relationship between the value of the expression under the radical and the roots of the quadratic equation. The expression $b^2 - 4ac$ is called the **discriminant**.

$$x = \frac{-b \pm \sqrt{b^2 - 4ac}}{2a} \quad \leftarrow \text{ discriminant}$$

The value of the discriminant can be used to determine the number and type of roots of a quadratic equation. The table on the following page summarizes the possible types of roots.

The discriminant can also be used to confirm the number and type of solutions after you solve the quadratic equation.

Additional Example

4 Solve $x^2 + 13 = 6x$ by using the Quadratic Formula. $\{3 \pm 2i\}$

WatchOut!

Preventing Errors Remind students that conjugate pairs are two complex numbers of the form $a + bi$ and $a - bi$.

2 Roots and the Discriminant

Example 5 shows how to find the value of the discriminant for a quadratic equation and use it to describe the number and type of roots for the equation.

Focus on Mathematical Content

Roots The value of the discriminant can be used to determine the number and type of roots of a quadratic equation. Consider a quadratic equation with rational coefficients. If the discriminant is a nonzero perfect square, there are 2 rational roots. If it is 0, there is 1 rational root. If it is positive, but not a perfect square, there are 2 irrational roots. If it is a negative number, there are 2 complex roots.

Additional Example

5 Find the value of the discriminant for each quadratic equation. Then describe the number and type of roots for the equation.

a. $x^2 + 3x + 5 = 0$ −11; two complex roots

b. $x^2 − 11x + 10 = 0$ 81; two rational roots

Teach with Tech

Student Response System Give students a quadratic equation, and ask them to use the discriminant to determine the number of real roots for the equation. Have them respond with A for two complex roots, B for one real root, and C for two real roots.

StudyTip

Roots Remember that the solutions of an equation are called *roots* or *zeros* and are the value(s) where the graph crosses the *x*-axis.

KeyConcept Discriminant

Consider $ax^2 + bx + c = 0$, where a, b, and c are rational numbers and $a \neq 0$.

Value of Discriminant	Type and Number of Roots	Example of Graph of Related Function
$b^2 − 4ac > 0$; $b^2 − 4ac$ is a perfect square.	2 real, rational roots	
$b^2 − 4ac > 0$; $b^2 − 4ac$ is *not* a perfect square.	2 real, irrational roots	
$b^2 − 4ac = 0$	1 real rational root	
$b^2 − 4ac < 0$	2 complex roots	

Example 5 Describe Roots

Find the value of the discriminant for each quadratic equation. Then describe the number and type of roots for the equation.

a. $7x^2 − 11x + 5 = 0$

$a = 7, b = −11, c = 5$

$b^2 − 4ac = (−11)^2 − 4(7)(5)$

$\qquad = 121 − 140$

$\qquad = −19$

The discriminant is negative, so there are two complex roots.

b. $x^2 + 22x + 121 = 0$

$a = 1, b = 22, c = 121$

$b^2 − 4ac = (22)^2 − 4(1)(121)$

$\qquad = 484 − 484$

$\qquad = 0$

The discriminant is 0, so there is one rational root.

▶ **Guided**Practice

5A. $−5x^2 + 8x − 1 = 0$ 44; 2 irrational roots

5B. $−7x + 15x^2 − 4 = 0$ 289; 2 rational roots

 268 | Lesson 4-6 | The Quadratic Formula and the Discriminant

---

DifferentiatedInstruction **OL** **BL**

Extension Write $x^3 − 8 = 0$ on the board. Explain that this is a cubic equation. Ask students to use what they have learned from this lesson to solve this equation for *x*.

Point out that a quadratic equation (degree 2) has at most 2 solutions, while a cubic equation (degree 3) has at most 3 solutions.

$$x^3 − 8 = 0$$
$$(x − 2)(x^2 + 2x + 4) = 0$$
$$x − 2 = 0 \text{ or } (x^2 + 2x + 4) = 0$$
$$x = 2 \text{ or } x = \frac{−2 \pm \sqrt{2^2 − 4(1)(4)}}{2(1)}$$
$$x = \frac{−2 \pm \sqrt{−12}}{2}$$
$$x = −1 \pm i\sqrt{3}$$

The solutions are $2, −1 + i\sqrt{3}, −1 − i\sqrt{3}$.

You have studied a variety of methods for solving quadratic equations. The table below summarizes these methods.

StudyTip
Study Notebook You may wish to copy this list of methods to your math notebook or Foldable to keep as a reference as you study.

ConceptSummary Solving Quadratic Equations

Method	Can be Used	When to Use
graphing	sometimes	Use only if an exact answer is not required. Best used to check the reasonableness of solutions found algebraically.
factoring	sometimes	Use if the constant term is 0 or if the factors are easily determined. **Example** $x^2 - 7x = 0$
Square Root Property	sometimes	Use for equations in which a perfect square is equal to a constant. **Example** $(x - 5)^2 = 18$
completing the square	always	Useful for equations of the form $x^2 + bx + c = 0$, where b is even. **Example** $x^2 + 6x - 14 = 0$
Quadratic Formula	always	Useful when other methods fail or are too tedious. **Example** $2.3x^2 - 1.8x + 9.7 = 0$

Check Your Understanding   ○ = Step-by-Step Solutions begin on page R14.

Examples 1–4 Solve each equation by using the Quadratic Formula.

 1. $x^2 + 12x - 9 = 0$ $-6 \pm 3\sqrt{5}$

2. $x^2 + 8x + 5 = 0$ $-4 \pm \sqrt{11}$

3. $4x^2 - 5x - 2 = 0$ $\dfrac{5 \pm \sqrt{57}}{8}$

4. $9x^2 + 6x - 4 = 0$ $\dfrac{-1 \pm \sqrt{5}}{3}$

5. $10x^2 - 3 = 13x$ $(1.5, -0.2)$

6. $22x = 12x^2 + 6$ $\left(\dfrac{3}{2}, \dfrac{1}{3}\right)$

7. $-3x^2 + 4x = -8$ $\dfrac{2 \pm 2\sqrt{7}}{3}$

8. $x^2 + 3 = -6x + 8$ $-3 \pm \sqrt{14}$

Examples 3–4 **9.** **MODELING** An amusement park ride takes riders to the top of a tower and drops them at speeds reaching 80 feet per second. A function that models this ride is $h = -16t^2 - 64t + 60$, where h is the height in feet and t is the time in seconds. About how many seconds does it take for riders to drop from 60 feet to 0 feet?   about 0.78 second

60 ft

Example 5 Complete parts a and b for each quadratic equation.
a. Find the value of the discriminant.
b. Describe the number and type of roots.

10a. 40
10b. 2 irrational roots
11a. −36
11b. 2 complex roots

10. $3x^2 + 8x + 2 = 0$

11. $2x^2 - 6x + 9 = 0$

12a. 0
12b. 1 rational root

12. $-16x^2 + 8x - 1 = 0$

13a. −76
13b. 2 complex roots

13. $5x^2 + 2x + 4 = 0$

3 Practice

Formative Assessment
Use Exercises 1–13 to check for understanding.

Use the chart on the bottom of the next page to customize assignments for your students.

CCSS **Teaching the Mathematical Practices**
Modeling Mathematically proficient students can apply the mathematics they know to solve problems arising in everyday life, analyze relationships mathematically to draw conclusions, and interpret their mathematical results in the context of a situation.

Tips for New Teachers
Sense-Making While the chart on this page offers suggestions for when to use each method to solve a quadratic equation, every student may not be able to analyze each equation and determine the best method to use to save time.

Teaching the Mathematical Practices

Sense-Making Mathematically proficient students start by explaining to themselves the meaning of a problem and looking for entry points to its solution. They analyze givens, constraints, relationships, and goals. They check their answers to problems using a different method, and they continually ask themselves, "Does this make sense?"

Additional Answers

21a. 33

21b. 2 irrational

21c. $\dfrac{-3 \pm \sqrt{33}}{4}$

22a. 4

22b. 2 rational

22c. $\dfrac{1}{2}$, 1

23a. 49

23b. 2 rational

23c. $\dfrac{1}{6}$, −1

24a. 121

24b. 2 rational

24c. 1, $-\dfrac{5}{6}$

25a. −87

25b. 2 complex

25c. $\dfrac{3 \pm i\sqrt{87}}{6}$

26a. −40

26b. 2 complex

26c. $\dfrac{-2 \pm i\sqrt{10}}{2}$

27a. 36

27b. 2 rational

27c. 1, $-\dfrac{1}{5}$

28a. 0

28b. 1 rational

28c. 3

29a. 1

29b. 2 rational

29c. −1, $-\dfrac{4}{3}$

30a. 176

30b. 2 irrational

30c. $\dfrac{1 \pm \sqrt{11}}{4}$

Examples 1–4 Solve each equation by using the Quadratic Formula.

14. $x^2 + 45x = -200$ −5, −40

15. $4x^2 - 6 = -12x$ $\dfrac{-3 \pm \sqrt{15}}{2}$

16. $3x^2 - 4x - 8 = -6$ $\dfrac{2 \pm \sqrt{10}}{3}$

17. $4x^2 - 9 = -7x - 4$ $\dfrac{-7 \pm \sqrt{129}}{8}$

18. $5x^2 - 9 = 11x$ $\dfrac{11 \pm \sqrt{301}}{10}$

19. $12x^2 + 9x - 2 = -17$ $\dfrac{-3 \pm i\sqrt{71}}{8}$

20. DIVING Competitors in the 10-meter platform diving competition jump upward and outward before diving into the pool below. The height h of a diver in meters above the pool after t seconds can be approximated by the equation $h = -4.9t^2 + 3t + 10$.

　　a. Determine a domain and range for which this function makes sense. $D = \{t \mid 0 \le t \le 2\}$, $R = \{h \mid 0 \le h \le 10\}$

　　b. When will the diver hit the water? about 1.77 seconds

Example 5　**Complete parts a–c for each quadratic equation.**

　　a. Find the value of the discriminant.

　　b. Describe the number and type of roots.

　　c. Find the exact solutions by using the Quadratic Formula. 21–32. See margin.

21 $2x^2 + 3x - 3 = 0$

22. $4x^2 - 6x + 2 = 0$

23. $6x^2 + 5x - 1 = 0$

24. $6x^2 - x - 5 = 0$

25. $3x^2 - 3x + 8 = 0$

26. $2x^2 + 4x + 7 = 0$

27. $-5x^2 + 4x + 1 = 0$

28. $x^2 - 6x = -9$

29. $-3x^2 - 7x + 2 = 6$

30. $-8x^2 + 5 = -4x$

31. $x^2 + 2x - 4 = -9$

32. $-6x^2 + 5 = -4x + 8$

33. VIDEO GAMES While Darnell is grounded his friend Jack brings him a video game. Darnell stands at his bedroom window, and Jack stands directly below the window. If Jack tosses a game cartridge to Darnell with an initial velocity of 35 feet per second, an equation for the height h feet of the cartridge after t seconds is $h = -16t^2 + 35t + 5$.

　　a. If the window is 25 feet above the ground, will Darnell have 0, 1, or 2 chances to catch the video game cartridge? 0

　　b. If Darnell is unable to catch the video game cartridge, when will it hit the ground? about 2.3 seconds

34. CCSS SENSE-MAKING Civil engineers are designing a section of road that is going to dip below sea level. The road's curve can be modeled by the equation $y = 0.00005x^2 - 0.06x$, where x is the horizontal distance in feet between the points where the road is at sea level and y is the elevation. The engineers want to put stop signs at the locations where the elevation of the road is equal to sea level. At what horizontal distances will they place the stop signs? 0 ft and 1200 ft

35a. 64
35b. 2 rational
35c. 0, $-\dfrac{8}{5}$
36a. 36
36b. 2 rational
36c. $\dfrac{1}{4}$, $-\dfrac{1}{2}$

Complete parts a–c for each quadratic equation.

　a. Find the value of the discriminant.

　b. Describe the number and type of roots.

　c. Find the exact solutions by using the Quadratic Formula.

35. $5x^2 + 8x = 0$

36. $8x^2 = -2x + 1$

37. $4x - 3 = -12x^2$

38. $0.8x^2 + 2.6x = -3.2$

39. $0.6x^2 + 1.4x = 4.8$

40. $-4x^2 + 12 = -6x - 8$

37a. 160
37b. 2 irrational
37c. $\dfrac{-1 \pm \sqrt{10}}{6}$
38a. −3.48
38b. imaginary
38c. $\dfrac{-1.3 \pm i\sqrt{0.87}}{0.8}$
39a. 13.48
39b. 2 irrational
39c. $\dfrac{-0.7 \pm \sqrt{3.37}}{0.6}$
40a. 356
40b. 2 irrational
40c. $\dfrac{3 \pm \sqrt{89}}{4}$

 270 | Lesson 4-6 | The Quadratic Formula and the Discriminant

---

Differentiated Homework Options

Level	Assignment		Two-Day Option
AL Basic	14–32, 43, 45, 46, 48–63	15–31 odd, 49–52	14–32 even, 43, 45, 46, 48, 53–63
OL Core	15–33 odd, 34, 35–39 odd, 41–43, 45, 46, 48–63	14–32, 49–52	33–43, 45, 46, 48, 53–63
BL Advanced	33–60, (optional: 61–63)		

41c. 2018; Sample answer: No; the death rate from cancer will never be 0 unless a cure is found. If and when a cure will be found cannot be predicted.

41 **SMOKING** A decrease in smoking in the United States has resulted in lower death rates caused by lung cancer. The number of deaths per 100,000 people y can be approximated by $y = -0.26x^2 - 0.55x + 91.81$, where x represents the number of years after 2000.

Year	Deaths per 100,000
2000	91.8
2002	89.7
2004	85.5
2010	60.3
2015	?
2017	?

a. Calculate the number of deaths per 100,000 people for 2015 and 2017. **25.1, 7.3**

b. Use the Quadratic Formula to solve for x when $y = 50$. **11.7**

c. According to the quadratic function, when will the death rate be 0 per 100,000? Do you think that this prediction is reasonable? Why or why not?

42. NUMBER THEORY The sum S of consecutive integers 1, 2, 3, …, n is given by the formula $S = \frac{1}{2}n(n + 1)$. How many consecutive integers, starting with 1, must be added to get a sum of 666? **36**

H.O.T. Problems Use Higher-Order Thinking Skills

43. CCSS CRITIQUE Tama and Jonathan are determining the number of solutions of $3x^2 - 5x = 7$. Is either of them correct? Explain your reasoning.

Jonathan; you must first write the equation in the form $ax^2 + bx + c = 0$ to determine the values of a, b, and c. Therefore, the value of c is -7, not 7.

Tama

$3x^2 - 5x = 7$

$b^2 - 4ac = (-5)^2 - 4(3)(7)$

$= -59$

Since the discriminant is negative, there are no real solutions.

Jonathan

$3x^2 - 5x = 7$

$3x^2 - 5x - 7 = 0$

$b^2 - 4ac = (-5)^2 - 4(3)(-7)$

$= 109$

Since the discriminant is positive, there are two real roots.

45a. Sample answer: Always; when a and c are opposite signs, then ac will always be negative and $-4ac$ will always be positive. Since b^2 will also always be positive, then $b^2 - 4ac$ represents the addition of two positive values, which will never be negative. Hence, the discriminant can never be negative and the solutions can never be imaginary.

44. CHALLENGE Find the solutions of $4ix^2 - 4ix + 5i = 0$ by using the Quadratic Formula. $\frac{1 \pm 2i}{2}$

45. REASONING Determine whether each statement is *sometimes*, *always*, or *never* true. Explain your reasoning.

a. In a quadratic equation in standard form, if a and c are different signs, then the solutions will be real.

b. If the discriminant of a quadratic equation is greater than 1, the two roots are real irrational numbers.

46. OPEN ENDED Sketch the corresponding graph and state the number and type of roots for each of the following. **a–e. See Chapter 4 Answer Appendix.**

a. $b^2 - 4ac = 0$

b. A quadratic function in which $f(x)$ never equals zero.

c. A quadratic function in which $f(a) = 0$ and $f(b) = 0; a \neq b$.

d. The discriminant is less than zero.

e. a and b are both solutions and can be represented as fractions.

45b. Sample answer: Sometimes; the roots will only be irrational if $b^2 - 4ac$ is not a perfect square.

47. CHALLENGE Find the value(s) of m in the quadratic equation $x^2 + x + m + 1 = 0$ such that it has one solution. -0.75

48. WRITING IN MATH Describe three different ways to solve $x^2 - 2x - 15 = 0$. Which method do you prefer, and why? **See Chapter 4 Answer Appendix.**

Additional Answers

31a. -16

31b. 2 complex

31c. $-1 \pm 2i$

32a. -56

32b. 2 complex

32c. $\frac{2 \pm i\sqrt{14}}{6}$

Study Guide and Intervention
AL OL ELL

NAME _____ DATE _____ PERIOD _____

4-6 Study Guide and Intervention
The Quadratic Formula and the Discriminant

Practice
AL OL BL ELL

NAME _____ DATE _____ PERIOD _____

4-6 Practice
The Quadratic Formula and the Discriminant

Word Problem Practice
AL OL BL ELL

NAME _____ DATE _____ PERIOD _____

4-6 Word Problem Practice
The Quadratic Formula and the Discriminant

Enrichment
OL BL

NAME _____ DATE _____ PERIOD _____

4-6 Enrichment

Sum and Product of Roots

Yesterday's News Ask students to write how an earlier lesson on simplifying radical expressions helped them with today's lesson.

Formative Assessment
Check for student understanding of concepts in Lessons 4-5 and 4-6.

 Quiz 3, p. 60

Additional Answers

60a. $I = \begin{bmatrix} 290 & 165 & 210 \\ 175 & 240 & 190 \\ 110 & 75 & 0 \end{bmatrix}$, $C = \begin{bmatrix} 22 \\ 25 \\ 18 \end{bmatrix}$

60b. $\begin{bmatrix} 14{,}285 \\ 13{,}270 \\ 4295 \end{bmatrix}$

Standardized Test Practice

49. A company determined that its monthly profit P is given by $P = -8x^2 + 165x - 100$, where x is the selling price for each unit of product. Which of the following is the best estimate of the maximum price per unit that the company can charge without losing money? **B**

   A $10   **B** $20   **C** $30   **D** $40

50. SAT/ACT For which of the following sets of numbers is the mean greater than the median? **H**

   F {4, 5, 6, 7, 8}   **J** {3, 5, 6, 7, 8}
   G {4, 6, 6, 6, 8}   **K** {2, 6, 6, 6, 6}
   H {4, 5, 6, 7, 9}

51. SHORT RESPONSE In the figure below, P is the center of the circle with radius 15 inches. What is the area of $\triangle APB$? **112.5 in²**

52. 75% of 88 is the same as 60% of what number? **D**

   A 100   **B** 101   **C** 108   **D** 110

Spiral Review

Find the value of c that makes each trinomial a perfect square. Then write the trinomial as a perfect square. (Lesson 4-5)

53. $x^2 + 13x + c$ **42.25; $(x + 6.5)^2$**   **54.** $x^2 + 2.4x + c$ **1.44; $(x + 1.2)^2$**   **55.** $x^2 + \frac{4}{5}x + c$ **$\frac{4}{25}$; $\left(x + \frac{2}{5}\right)^2$**

Simplify. (Lesson 4-4)

56. i^{26} **−1**   **57.** $\sqrt{-16}$ **$4i$**   **58.** $4\sqrt{-9} \cdot 2\sqrt{-25}$ **−120**

59. PILOT TRAINING Evita is training for her pilot's license. Flight instruction costs $105 per hour, and the simulator costs $45 per hour. She spent 4 more hours in airplane training than in the simulator. If Evita spent $3870, how much time did she spend training in an airplane and in a simulator? (Lesson 3-8)   **27 hours of flight instruction and 23 hours in the simulator**

60. BUSINESS Ms. Larson owns three fruit farms on which she grows apples, peaches, and apricots. She sells apples for $22 a case, peaches for $25 a case, and apricots for $18 a case. (Lesson 3-6)

   a. Write an inventory matrix for the number of cases for each type of fruit for each farm and a cost matrix for the price per case for each type of fruit. **a, b. See margin.**

   b. Find the total income of the three fruit farms expressed as a matrix.

   c. What is the total income from all three fruit farms? **$31,850**

| \multicolumn{4}{c}{Number of Cases in Stock of Each Type of Fruit} |
Fruit	Farm 1	Farm 2	Farm 3
apples	290	175	110
peaches	165	240	75
apricots	210	190	0

Skills Review

Write an equation for each graph.   **61.** $y = x^2 + 1$ **62.** $y = 0.25x^2$ **63.** $y = |x + 3|$

61.

62.

63.

 272 | **Lesson 4-6** | The Quadratic Formula and the Discriminant

Follow-up
Students have explored solving quadratic equations.

Ask:
- How do you know what method to use when solving a quadratic equation? Sample answer: If the equation has terms that you know are easily factorable, you could solve by factoring. If the equation has more complex terms, you could solve by using the Quadratic Formula, completing the square, or by graphing. You could also use one method to solve and a second method to check your answer.

EXPLORE 4-7

Graphing Technology Lab
Families of Parabolas

The general form of a quadratic function is $y = a(x - h)^2 + k$. Changing the values of a, h, and k results in a different parabola in the family of quadratic functions. You can use a TI-83/84 Plus graphing calculator to analyze the effects that result from changing each of these parameters.

CCSS Common Core State Standards
Content Standards

F.IF.4 For a function that models a relationship between two quantities, interpret key features of graphs and tables in terms of the quantities, and sketch graphs showing key features given a verbal description of the relationship.

F.BF.3 Identify the effect on the graph of replacing $f(x)$ by $f(x) + k$, $k f(x)$, $f(kx)$, and $f(x + k)$ for specific values of k (both positive and negative); find the value of k given the graphs. Experiment with cases and illustrate an explanation of the effects on the graph using technology.

Activity 1   Change in k

Graph each set of functions on the same screen in the standard viewing window. Describe any similarities and differences among the graphs.

$y = x^2$, $y = x^2 + 4$, $y = x^2 - 3$

The graphs have the same shape, and all open up. The vertex of each graph is on the y-axis. However, the graphs have different vertical positions.

[−10, 10] scl: 1 by [−10, 10] scl: 1

Activity 1 shows how changing the value of k in the function $y = a(x - h)^2 + k$ *translates* the parabola along the y-axis. If $k > 0$, the parabola is translated k units up, and if $k < 0$, it is translated $|k|$ units down.

How do you think changing the value of h will affect the graph of $y = a(x - h)^2 + k$?

Activity 2   Change in h

Graph each set of functions on the same screen in the standard viewing window. Describe any similarities and differences among the graphs.

$y = x^2$, $y = (x + 4)^2$, $y = (x - 3)^2$

These three graphs all open up and have the same shape. The vertex of each graph is on the x-axis. However, the graphs have different horizontal positions.

[−10, 10] scl: 1 by [−10, 10] scl: 1

Activity 2 shows how changing the value of h in the equation $y = a(x - h)^2 + k$ *translates* the graph horizontally. If $h > 0$, the graph translates to the right h units. If $h < 0$, the graph translates to the left $|h|$ units.

Activity 3   Change in h and k

Graph each set of functions on the same screen in the standard viewing window. Describe any similarities and differences among the graphs.

$y = x^2$, $y = (x + 6)^2 - 5$, $y = (x - 4)^2 + 6$

These three graphs all open up and have the same shape. However, the graphs have different horizontal and vertical positions.

[−10, 10] scl: 1 by [−10, 10] scl: 1

(continued on the next page)

connectED.mcgraw-hill.com   **273**

---

1 Focus

Objective Use a graphing calculator to investigate changes to parabolas.

Materials for Each Group

- TI-83/84 Plus or other graphing calculator

Teaching Tip

Students can use the calculator to confirm the location of the vertex of each parabola. A good way to do this is to change the window settings for the x-axis to [−9.4, 9.4]. Use the **TRACE** feature and symmetry properties of parabolas to check that the graph is symmetric with respect to the vertical line through the point that appears to be the vertex.

2 Teach

Working in Cooperative Groups
Put students in pairs, mixing abilities. Then have pairs work through Activities 1–4 and Exercises 1–3.

- Ask students to describe the three constants (a, h, and k) in the general form of a quadratic function $y = a(x - h)^2 + k$. Sample answer: a: coefficient of the squared quantity involving the variable x; h: value subtracted from x in the quantity being squared and then multiplied by a; k: value added at the end.

- Before discussing the examples, have students make conjectures about the effect of the value of each of the constants a, h, and k on the graph of the parabola.

- After completing the discussion of Activity 4, have students compare the conjectures they made at the beginning of the lab to the knowledge they gained during the discussions.

Practice Have students complete Exercises 4–15.

3 Assess

Formative Assessment
Use Exercise 15 to assess whether students comprehend how changing the constants a, h, and k affects the graph of $y = a(x - h)^2 + k$.

From Concrete to Abstract
Ask:
- In the general form of a quadratic function, which constant would you change to move the graph left or right? h

- Which constant would you change to move the graph up or down? k

- Which constant would you change to make the graph wider or narrower? a

Additional Answers

1. Changing the value of h moves the graph to the left and the right. If $h > 0$, the graph translates to the right, and if $h < 0$, it translates to the left. In $y = x^2$, the vertex is at $(0, 0)$ and in $y = (x - 2)^2$, the vertex is at $(2, 0)$. The graph has been translated to the right.

2. Changing the value of k moves the graph up and down. If $k > 0$, the graph translates upward, and if $k < 0$, it translates downward. In $y = x^2$, the vertex is at $(0, 0)$ and in $y = x^2 - 3$, the vertex is at $(0, -3)$. The graph has been translated downward.

3. Using $-a$ instead of a reflects the graph over the x-axis. The graph of $y = x^2$ opens upward, while the graph of $y = -x^2$ opens downward.

Graphing Technology Lab
Families of Parabolas *Continued*

How does the value of a affect the graph of $y = a(x - h)^2 + k$?

Activity 4   Change in a

Graph each set of functions on the same screen in the standard viewing window. Describe any similarities and differences among the graphs.

a. $y = x^2$, $y = -x^2$

The graphs have the same vertex and the same shape. However, the graph of $y = x^2$ opens up and the graph of $y = -x^2$ opens down.

[−10, 10] scl: 1 by [−10, 10] scl: 1

b. $y = x^2$, $y = 5x^2$, $y = \frac{1}{5}x^2$

The graphs have the same vertex, $(0, 0)$, but each has a different shape. The graph of $y = 5x^2$ is narrower than the graph of $y = x^2$. The graph of $y = \frac{1}{5}x^2$ is wider than the graph of $y = x^2$.

[−10, 10] scl: 1 by [−10, 10] scl: 1

Changing the value of a in the function $y = a(x - h)^2 + k$ can affect the direction of the opening and the shape of the graph. If $a > 0$, the graph opens up, and if $a < 0$, the graph opens down or is *reflected* over the x-axis. If $|a| > 1$, the graph is expanded vertically and is narrower than the graph of $y = x^2$. If $|a| < 1$, the graph is compressed vertically and is wider than the graph of $y = x^2$. Thus, a change in the absolute value of a results in a *dilation* of the graph of $y = x^2$.

Analyze the Results 1–3. See margin.
1. How does changing the value of h in $y = a(x - h)^2 + k$ affect the graph? Give an example.
2. How does changing the value of k in $y = a(x - h)^2 + k$ affect the graph? Give an example.
3. How does using $-a$ instead of a in $y = a(x - h)^2 + k$ affect the graph? Give an example.

Examine each pair of functions and predict the similarities and differences in their graphs. Use a graphing calculator to confirm your predictions. Write a sentence or two comparing the two graphs. 4–15. See Chapter 4 Answer Appendix.

4. $y = x^2$, $y = x^2 + 3.5$
5. $y = -x^2$, $y = x^2 - 7$
6. $y = x^2$, $y = 4x^2$
7. $y = x^2$, $y = -8x^2$
8. $y = x^2$, $y = (x + 2)^2$
9. $y = -\frac{1}{6}x^2$, $y = -\frac{1}{6}x^2 + 2$
10. $y = x^2$, $y = (x - 5)^2$
11. $y = x^2$, $y = 2(x + 3)^2 - 6$
12. $y = x^2$, $y = -\frac{1}{8}x^2 + 1$
13. $y = (x + 5)^2 - 4$, $y = (x + 5)^2 + 7$
14. $y = 2(x + 1)^2 - 4$, $y = 5(x + 3)^2 - 1$
15. $y = 5(x - 2)^2 - 3$, $y = \frac{1}{4}(x - 5)^2 - 6$

274 | **Explore 4-7** | Graphing Technology Lab: Families of Parabolas

LESSON 4-7 Transformations of Quadratic Graphs

Then	Now	Why?
● You transformed graphs of functions.	**1** Write a quadratic function in the form $y = a(x - h)^2 + k$. **2** Transform graphs of quadratic functions of the form $y = a(x - h)^2 + k$.	Recall that a family of graphs is a group of graphs that display one or more similar characteristics. The parent graph is the simplest graph in the family. For the family of quadratic functions, $y = x^2$ is the parent graph. Other graphs in the family of quadratic functions, such as $y = (x - 2)^2$ and $y = x^2 - 4$, can be drawn by transforming the graph of $y = x^2$.

 NewVocabulary
vertex form

CCSS Common Core State Standards

Content Standards
F.IF.8.a Use the process of factoring and completing the square in a quadratic function to show zeros, extreme values, and symmetry of the graph, and interpret these in terms of a context.

F.BF.3 Identify the effect on the graph of replacing $f(x)$ by $f(x) + k$, $k f(x)$, $f(kx)$, and $f(x + k)$ for specific values of k (both positive and negative); find the value of k given the graphs. Experiment with cases and illustrate an explanation of the effects on the graph using technology.

Mathematical Practices
7 Look for and make use of structure.

1 Write Quadratic Functions in Vertex Form Each function above is written in **vertex form**, $y = a(x - h)^2 + k$, where (h, k) is the vertex of the parabola, $x = h$ is the axis of symmetry, and a determines the shape of the parabola and the direction in which it opens.

When a quadratic function is in the form $y = ax^2 + bx + c$, you can complete the square to write the function in vertex form. If the coefficient of the quadratic term is not 1, then factor the coefficient from the quadratic and linear terms *before* completing the square. After completing the square and writing the function in vertex form, the value of k indicated a minimum value if $a < 0$ or a maximum value if $a > 0$.

Example 1 Write Functions in Vertex Form

Write each function in vertex form.

a. $y = x^2 + 6x - 5$

$y = x^2 + 6x - 5$	Original function
$y = (x^2 + 6x + 9) - 5 - 9$	Complete the square.
$y = (x + 3)^2 - 14$	Simplify.

b. $y = -2x^2 + 8x - 3$

$y = -2x^2 + 8x - 3$	Original function
$y = -2(x^2 - 4x) - 3$	Group $ax^2 + bx$ and factor, dividing by a.
$y = -2(x^2 - 4x + 4) - 3 - (-2)(4)$	Complete the square.
$y = -2(x - 2)^2 + 5$	Simplify.

GuidedPractice

1A. $y = x^2 + 4x + 6$ $y = (x + 2)^2 + 2$ **1B.** $y = 2x^2 - 12x + 17$ $y = 2(x - 3)^2 - 1$

 connectED.mcgraw-hill.com **275**

1 Focus

VerticalAlignment

Before Lesson 4-7 Transform graphs of functions.

Lesson 4-7 Write a quadratic function in the form $y = a(x - h)^2 + k$. Transform graphs of quadratic functions of the form $y = a(x - h)^2 + k$.

After Lesson 4-7 Solve quadratic inequalities using graphs and algebraic methods.

2 Teach

Scaffolding Questions
Have students read the **Why?** section of the lesson.

Ask:
- For the function $y = x^2$, what value of x makes y equal 0? $x = 0$
- What value of x makes y equal 0 if the function is $y = (x - 2)^2$? $x = 2$
- Compare the graph of $y = (x - 2)^2$ with $y = x^2$. What difference does subtracting 2 within the parentheses make? The graph of $y = x^2$ moves 2 units to the right.

Lesson 4-7 Resources

Resource	Approaching Level **AL**	On Level **OL**	Beyond Level **BL**	English Learners **ELL**
Teacher Edition	▪ Differentiated Instruction, p. 277	▪ Differentiated Instruction, p. 277	▪ Differentiated Instruction, pp. 277, 280	▪ Differentiated Instruction, p. 277
Chapter Resource Masters	▪ Study Guide and Intervention, pp. 43–44 ▪ Skills Practice, p. 45 ▪ Practice, p. 46 ▪ Word Problem Practice, p. 47	▪ Study Guide and Intervention, pp. 43–44 ▪ Skills Practice, p. 45 ▪ Practice, p. 46 ▪ Word Problem Practice, p. 47 ▪ Enrichment, p. 48	▪ Practice, p. 46 ▪ Word Problem Practice, p. 47 ▪ Enrichment, p. 48	▪ Study Guide and Intervention, pp. 43–44 ▪ Skills Practice, p. 45 ▪ Practice, p. 46 ▪ Word Problem Practice, p. 47
Other	▪ 5-Minute Check 4-7 ▪ Study Notebook ▪ Teaching Algebra with Manipulatives	▪ 5-Minute Check 4-7 ▪ Study Notebook ▪ Teaching Algebra with Manipulatives	▪ 5-Minute Check 4-7 ▪ Study Notebook	▪ 5-Minute Check 4-7 ▪ Study Notebook ▪ Teaching Algebra with Manipulatives

1 Write Quadratic Equations in Vertex Form

Example 1 shows how to write quadratic equations in vertex form.
Example 2 shows how the coefficient of the quadratic term in a function changes the width of the graph in a multiple-choice test format.

Formative Assessment

Use the Guided Practice exercises after each example to determine students' understanding of concepts.

Additional Example

1 Write each equation in vertex form.

a. $y = x^2 - 2x + 4$
$y = (x - 1)^2 + 3$

b. $y = -3x^2 - 18x + 10$
$y = -3(x + 3)^2 + 37$

▶ **Additional Examples** also in Interactive Classroom PowerPoint® Presentations

 Interactive White Board READY

WatchOut!

Preventing Errors As $|a|$ increases, the graph gets narrower, because a greater multiplier for the quantity $(x - h)^2$ will make the corresponding y-value greater. Greater values of y result in a steeper (and thus narrower) graph.

Teach with Tech

Interactive Whiteboard Show a coordinate grid on the board. Draw a graph of a quadratic function, give students the equation of the graph, and show students how to write the equation of the graph in vertex form. Drag the graph to move its vertex to other locations on the grid and have students find the equations of the new graphs. Discuss how the equations are similar and different.

If the vertex and one additional point on the graph of a parabola are known, you can write the equation of the parabola in vertex form.

Standardized Test Example 2 Write an Equation Given a Graph

Which is an equation of the function shown in the graph?

A $y = -4(x - 3)^2 + 2$

B $y = -\frac{1}{4}(x - 3)^2 + 2$

C $y = \frac{1}{4}(x + 3)^2 - 2$

D $y = 4(x + 3)^2 - 2$

Read the Test Item

You are given a graph of a parabola with the vertex and a point on the graph labeled. You need to find an equation of the parabola.

Solve the Test Item

The vertex of the parabola is at $(3, 2)$, so $h = 3$ and $k = 2$. Since $(-1, -2)$ is a point on the graph, let $x = -1$ and $y = -2$. Substitute these values into the vertex form of the equation and solve for a.

$y = a(x - h)^2 + k$	Vertex form
$-2 = a(-1 - 3)^2 + 2$	Substitute -2 for y, -1 for x, 3 for h and 2 for k.
$-2 = a(16) + 2$	Simplify.
$-4 = 16a$	Subtract 2 from each side.
$-\frac{1}{4} = a$	Divide each side by 16.

The equation of the parabola in vertex form is $y = -\frac{1}{4}(x - 3)^2 + 2$.

The answer is B.

Test-TakingTip

The Meaning of a The sign of a in vertex form does not determine the width of the parabola. The sign indicates whether the parabola opens up or down. The width of a parabola is determined by the absolute value of a.

▶ **Guided**Practice

2. Which is an equation of the function shown in the graph? **H**

F $y = \frac{9}{25}(x - 1)^2 + 2$

G $y = \frac{3}{5}(x + 1)^2 - 2$

H $y = \frac{5}{3}(x + 1)^2 - 2$

J $y = \frac{25}{9}(x - 1)^2 + 2$

2 Transformations of Quadratic Graphs In Lesson 2-7, you learned how different transformations affect the graphs of parent functions. The following summarizes these transformations for quadratic functions.

Focus on Mathematical Content

Quadratic Function in Vertex Form The values of a, k, and h, in the vertex form of a quadratic function, $y = a(x - h)^2 + k$, affect the graph of the parabola. The vertex is at (h, k). The value of a determines the direction in which the parabola opens and the width of the parabola. The value of h determines the direction in which the parabola is translated horizontally. The value of k determines the direction in which the parabola is translated vertically. The function must first be rewritten in vertex form before a, h, and k can be determined and analyzed.

ConceptSummary **Transformations of Quadratic Functions**

$$f(x) = a(x - h)^2 + k$$

h, Horizontal Translation
h units to the right if h is positive
|h| units to the left if h is negative

k, Vertical Translation
k units up if k is positive
|k| units down if k is negative

StudyTip

Absolute Value
$0 < |a| < 1$ means that a is a rational number between 0 and 1, such as $\frac{3}{4}$, or a rational number between -1 and 0, such as -0.3.

a, Reflection
If $a > 0$, the graph opens up.
If $a < 0$, the graph opens down.

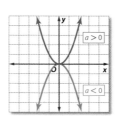

a, Dilation
If $|a| > 1$, the graph is stretched vertically. If $0 < |a| < 1$, the graph is compressed vertically.

Example 3 Graph Equations in Vertex Form

Graph $y = 4x^2 - 16x - 40$.

Step 1 Rewrite the equation in vertex form.

$y = 4x^2 - 16x - 40$	Original equation
$y = 4(x^2 - 4x) - 40$	Distributive Property
$y = 4(x^2 - 4x + 4) - 40 - 4(4)$	Complete the square.
$y = 4(x - 2)^2 - 56$	Simplify.

Step 2 The vertex is at $(2, -56)$. The axis of symmetry is $x = 2$. Because $a = 4$, the graph is narrower than the graph of $y = x^2$.

Step 3 Plot additional points to help you complete the graph.

GuidedPractice

3A, 3B. See Chapter 4 Answer Appendix.

3A. $y = (x - 3)^2 - 2$

3B. $y = 0.25(x + 1)^2$

Additional Example

2 **STANDARDIZED TEST PRACTICE**
Which is an equation of the function shown in the graph? B

a $y = -\frac{1}{2}(x - 2)^2 + 3$

B $y = \frac{1}{2}(x - 2)^2 - 3$

c $y = -2(x - 2)^2 + 3$

d $y = 2(x - 2)^2 - 3$

2 Transformations of Quadratic Graphs

Example 3 shows how to graph an equation after rewriting it in vertex form.

Additional Example

3 Graph $y = -2x^2 + 4x + 1$.

DifferentiatedInstruction **AL OL BL ELL**

Intrapersonal Learners Have students observe or research some natural events that can be modeled by parabolas. Students should report their observations and findings to the class. If students are able to determine a quadratic function that models the event, they should present the function and explain how the characteristics of the equation can be used to analyze its graph.

3 Practice

Formative Assessment

Use Exercises 1–7 to check for understanding.

Use the chart on the bottom of this page to customize assignments for your students.

 Teaching the Mathematical Practices

Modeling Mathematically proficient students can apply the mathematics they know to solve problems arising in everyday life, analyze relationships mathematically to draw conclusions, and interpret their mathematical results in the context of a situation.

Additional Answers

5.

6.

7.

20.

Example 1 Write each function in vertex form. **1.** $y = (x + 3)^2 - 7$ **2.** $y = -2(x - 2)^2 + 3$

1. $y = x^2 + 6x + 2$ **2.** $y = -2x^2 + 8x - 5$ **3.** $y = 4x^2 + 24x + 24$
                                                                             $y = 4(x + 3)^2 - 12$

Example 2 **4. MULTIPLE CHOICE** Which function is shown in the graph? **A**

A $y = -(x + 3)^2 + 6$

B $y = -(x - 3)^2 - 6$

C $y = -2(x + 3)^2 + 6$

D $y = -2(x - 3)^2 - 6$

Example 3 Graph each function. **5–7.** See margin.

5. $y = (x - 3)^2 - 4$ **6.** $y = -2x^2 + 5$ **7.** $y = \frac{1}{2}(x + 6)^2 - 8$

8. $y = \left(x + \frac{9}{2}\right)^2 - \frac{49}{4}$ **9.** $y = (x - 3)^2 - 6$ **10.** $y = -2\left(x - \frac{5}{4}\right)^2 + \frac{25}{8}$ **11.** $y = (x + 1)^2 + 6$

Practice and Problem Solving
Extra Practice is on page R4.

Example 1 Write each function in vertex form.   **15.** $y = 3\left(x + \frac{5}{3}\right)^2 - \frac{25}{3}$

8. $y = x^2 + 9x + 8$ **9.** $y = x^2 - 6x + 3$ **10.** $y = -2x^2 + 5x$

11 $y = x^2 + 2x + 7$ **12.** $y = -3x^2 + 12x - 10$ **13.** $y = x^2 + 8x + 16$ $y = (x + 4)^2$
                                   $y = -3(x - 2)^2 + 2$

14. $y = 2x^2 - 4x - 3$ **15.** $y = 3x^2 + 10x$ **16.** $y = x^2 - 4x + 9$ $y = (x - 2)^2 + 5$
   $y = 2(x - 1)^2 - 5$

17. $y = -4x^2 - 24x - 15$ **18.** $y = x^2 - 12x + 36$ **19.** $y = -x^2 - 4x - 1$
   $y = -4(x + 3)^2 + 21$ $y = (x - 6)^2$ $y = -(x + 2)^2 + 3$

Example 2 **20. FIREWORKS** During an Independence Day fireworks show, the height h in meters of a specific rocket after t seconds can be modeled by $h = -4.9(t - 4)^2 + 80$. Graph the function. **See margin.**

21. FINANCIAL LITERACY A bicycle rental shop rents an average of 120 bicycles per week and charges $25 per day. The manager estimates that there will be 15 additional bicycles rented for each $1 reduction in the rental price. The maximum income the manager can expect can be modeled by $y = -15x^2 + 255x + 3000$, where y is the weekly income and x is the number of bicycles rented. Write this function in vertex form. Then graph.
       $y = -15(x - 8.5)^2 + 4083.75$; See Chapter 4 Answer Appendix for graph.

Example 3 Graph each function. **22–33.** See Chapter 4 Answer Appendix.

22. $y = (x - 5)^2 + 3$ **23.** $y = 9x^2 - 8$ **24.** $y = -2(x - 5)^2$

25. $y = \frac{1}{10}(x + 6)^2 + 6$ **26.** $y = -3(x - 5)^2 - 2$ **27.** $y = -\frac{1}{4}x^2 - 5$

28. $y = 2x^2 + 10$ **29.** $y = -(x + 3)^2$ **30.** $y = \frac{1}{6}(x - 3)^2 - 10$

31. $y = (x - 9)^2 - 7$ **32.** $y = -\frac{5}{8}x^2 - 8$ **33.** $y = -4(x - 10)^2 - 10$

34. **CCSS MODELING** A sailboard manufacturer uses an automated process to manufacture the masts for its sailboards. The function $f(x) = \frac{1}{250}x^2 + \frac{3}{5}x$ is programmed into a computer to make one such mast. **a. See Chapter 4 Answer Appendix.**

    a. Write the quadratic function in vertex form. Then graph the function.

    b. Describe how the manufacturer can adjust the function to make its masts with a greater or smaller curve. **They can adjust the coefficient of x^2.**

Differentiated Homework Options

Level	Assignment		Two-Day Option
AL Basic	8–34, 48, 51–76	9–33 odd, 53–56	8–34 even, 48, 51, 52, 57–76
OL Core	9–33 odd, 34, 35–45 odd, 47, 48, 51–76	8–34, 53–56	35–48, 51, 52, 57–76
BL Advanced	35–73, (optional: 74–76)		

B Write an equation in vertex form for each parabola. 37. $y = -\frac{2}{3}(x-3)^2$

35.

(7, 10)
(6, 1)
$y = 9(x-6)^2 + 1$

36.

(−3, 6)
(−4, 3)
$y = 3(x+4)^2 + 3$

37.

(3, 0)
(6, −6)

38.

(5, 4)
(6, 1)
$y = -3(x-5)^2 + 4$

39.

(3, 8)
(0, 5)
$y = \frac{1}{3}x^2 + 5$

40.

(−1, 8)
(−3, 2)
$y = \frac{3}{2}(x+3)^2 + 2$

Write each function in vertex form. Then identify the vertex, axis of symmetry, and direction of opening. **41–46. See Chapter 4 Answer Appendix.**

41. $3x^2 - 4x = 2 + y$

42. $-2x^2 + 7x = y - 12$

43. $-x^2 - 4.7x = y - 2.8$

44. $x^2 + 1.4x - 1.2 = y$

45. $x^2 - \frac{2}{3}x - \frac{26}{9} = y$

46. $x^2 + 7x + \frac{49}{4} = y$

47 CARS The formula $S(t) = \frac{1}{2}at^2 + v_0 t$ can be used to determine the position $S(t)$ of an object after t seconds at a rate of acceleration a with initial velocity v_0. Valerie's car can accelerate 0.002 miles per second squared.

a. Express $S(t)$ in vertex form as she accelerates from 35 miles per hour to enter highway traffic. $S(t) = 0.001(t + 4.861)^2 - 0.024$

b. How long will it take Valerie to match the average speed of highway traffic of 68 miles per hour? (*Hint*: Use *acceleration · time = velocity*.) **4.58 seconds**

c. If the entrance ramp is $\frac{1}{8}$-mile long, will Valerie have sufficient time to match the average highway speed? Explain.

47c. Yes; if we substitute $\frac{1}{8}$ for $S(t)$ and solve for t we get 7.35 seconds. This is how long Valerie will be on the ramp. Since it will take her 4.58 seconds to accelerate to 68 mph, she will be on the ramp long enough to accelerate to match the average expressway speed.

H.O.T. Problems Use Higher-Order Thinking Skills

48. OPEN ENDED Write an equation for a parabola that has been translated, compressed, and reflected in the x-axis. Sample answer: $y = -\frac{1}{2}(x-4)^2$

49. CHALLENGE Explain how you can find an equation of a parabola using the coordinates of three points on the graph. **See Chapter 4 Answer Appendix.**

50. CHALLENGE Write the standard form of a quadratic function $ax^2 + bx + c = y$ in vertex form. Identify the vertex and the axis of symmetry. **See Chapter 4 Answer Appendix.**

51. REASONING Describe the graph of $f(x) = a(x - h)^2 + k$ when $a = 0$. Is the graph the same as that of $g(x) = ax^2 + bx + c$ when $a = 0$? Explain. **See Chapter 4 Answer Appendix.**

52. CCSS ARGUMENTS Explain how the graph of $y = x^2$ can be used to graph any quadratic function. Include a description of the effects produced by changing a, h, and k in the equation $y = a(x - h)^2 + k$, and a comparison of the graph of $y = x^2$ and the graph of $y = a(x - h)^2 + k$ using values you choose for a, h, and k. **See Chapter 4 Answer Appendix.**

connectED.mcgraw-hill.com **279**

CCSS Teaching the Mathematical Practices

Arguments Mathematically proficient students understand and use stated assumptions, definitions, and previously established results in constructing arguments. They make conjectures and build a logical progression of statements to explore the truth of their conjectures. And they are able to analyze situations by breaking them into cases, and can recognize and use counterexamples.

Enrichment
OL **BL**

NAME _____ DATE _____ PERIOD _____
4-7 **Enrichment**

A Shortcut to Complex Roots
When graphing a quadratic function, the real roots are shown in the graph. You have learned that quadratic graphs can also have imaginary roots that cannot be seen on the graph of the function. However, there is a way to graphically represent the complex roots of a quadratic function.

Example Find the complex roots of the quadratic function $y = x^2 - 4x + 5$.

Step 1 Graph the function.

Step 2 Reflect the graph over the horizontal line containing the vertex. In this example, the vertex is (2, 1).

Study Guide and Intervention
AL **OL** **ELL**

Practice
AL **OL** **BL** **ELL**

Word Problem Practice
AL **OL** **BL** **ELL**

connectED.mcgraw-hill.com **279**

Crystal Ball Ask students to write how today's lesson on analyzing and graphing quadratic functions will help them with tomorrow's lesson on graphing and solving quadratic inequalities.

Additional Answers

67.

68.

69.

70.

Standardized Test Practice

53. Flowering bushes need a mixture of 70% soil and 30% vermiculite. About how many buckets of vermiculite should you add to 20 buckets of soil? **B**

A 6.0   C 14.0

B 8.0   D 24.0

54. SAT/ACT The sum of the integers x and y is 495. The units digit of x is 0. If x is divided by 10, the result is equal to y. What is the value of x? **K**

F 40   J 250

G 45   K 450

H 245

55. What is the solution set of the inequality $|4x - 1| < 9$? **D**

A $\{x \mid 2.5 < x \text{ or } x < -2\}$

B $\{x \mid x < 2.5\}$

C $\{x \mid x > -2\}$

D $\{x \mid -2 < x < 2.5\}$

56. SHORT RESPONSE At your store, you buy wrenches for $30.00 a dozen and sell them for $3.50 each. What is the percent markup for the wrenches? **40%**

Spiral Review

Solve each equation by using the method of your choice. Find exact solutions. (Lesson 4-6)

57. $4x^2 + 15x = 21$ $\dfrac{-15 \pm \sqrt{561}}{8}$ **58.** $-3x^2 + 19 = 5x$ $\dfrac{-5 \pm \sqrt{253}}{6}$ **59.** $6x - 5x^2 + 9 = 3$ $\dfrac{3 \pm \sqrt{39}}{5}$

Find the value of c that makes each trinomial a perfect square. (Lesson 4-5)

60. $x^2 - 12x + c$ **36**   **61.** $x^2 + 0.1x + c$ **0.0025**   **62.** $x^2 - 0.45x + c$ **0.050625**

Determine whether each function has a maximum or minimum value, and find that value. (Lesson 4-1) **63.** minimum, $9\frac{1}{3}$ **64.** maximum, $-17\frac{15}{16}$

63. $f(x) = 6x^2 - 8x + 12$ **64.** $f(x) = -4x^2 + x - 18$ **65.** $f(x) = 3x^2 - 9 + 6x$ minimum, -12

66. ARCHAEOLOGY A coordinate grid is laid over an archaeology dig to identify the location of artifacts. Three corners of a building have been partially unearthed at $(-1, 6)$, $(4, 5)$, and $(-1, -2)$. If each square on the grid measures one square foot, estimate the area of the floor of the building. (Lesson 3-7) **about 20 ft²**

67. ART Rai can spend no more than $225 on the art club's supply of brushes and paint. She needs at least 20 brushes and 56 tubes of paint. Graph the region that shows how many boxes of each item can be purchased. (Lesson 3-2) **See margin.**

Solve each system of equations by graphing. (Lesson 3-1) **68–70. See margin.**

68. $y = 3x - 4$ **69.** $2x + 5y = 1$ **70.** $4x + 3y = -30$
      $y = -2x + 16$ $6y - 5x = 16$ $3x - 2y = 3$

Evaluate each function. (Lesson 2-1)

71. $f(3)$ if $f(x) = x^2 - 4x + 12$ **9**   **72.** $f(-2)$ if $f(x) = -4x^2 + x - 8$ **-26**   **73.** $f(4)$ if $f(x) = 3x^2 + x$ **52**

Skills Review

Determine whether the given value satisfies the inequality.

74. $3x^2 - 5 > 6$; $x = 2$ **yes**   **75.** $-2x^2 + x - 1 < 4$; $x = -2$ **yes**   **76.** $4x^2 + x - 3 \le 36$; $x = 3$ **yes**

DifferentiatedInstruction BL

Extension In this lesson on quadratic functions, only equations of parabolas that open upward or downward are analyzed and graphed. Ask students to explain why parabolas opening to the right or left are not included in this lesson. As an example, you can draw a graph of a parabola on the board with a vertex at $(-2, 0)$, axis of symmetry through $y = 0$, and through $(0, 2)$. Sample answer: This lesson is about quadratic functions. Parabolas opening right or left do not represent functions since 2 elements in the range are paired with one element of the domain, except at the vertex.

EXTEND 4-7

Algebra Lab
Quadratics and Rate of Change

You have learned that a linear function has a constant rate of change. In this lab, you will investigate the rate of change for quadratic functions.

CCSS **Common Core State Standards**
Content Standards
F.IF.4 For a function that models a relationship between two quantities, interpret key features of graphs and tables in terms of the quantities, and sketch graphs showing key features given a verbal description of the relationship.
F.IF.6 Calculate and interpret the average rate of change of a function (presented symbolically or as a table) over a specified interval. Estimate the rate of change from a graph.

Activity   Determine Rate of Change

Consider $f(x) = 0.1875x^2 - 3x + 12$. **Steps 1–5. See Chapter 4 Answer Appendix.**

Step 1 Make a table like the one below. Use values from 0 through 16 for x.

x	0	1	2	3	…	16
y	12	9.1875	6.75			
First-Order Differences						
Second-Order Differences						

Step 2 Find each y-value. For example, when $x = 1$, $y = 0.1875(1)^2 - 3(1) + 12$ or 9.1875.

Step 3 Graph the ordered pairs (x, y). Then connect the points with a smooth curve. Notice that the function *decreases* when $0 < x < 8$ and *increases* when $8 < x < 16$.

Step 4 The rate of change from one point to the next can be found by using the slope formula. From $(0, 12)$ to $(1, 9.1875)$, the slope is $\frac{9.1875 - 12}{1 - 0}$ or -2.8125. This is the first-order difference at $x = 1$. Complete the table for all the first-order differences. Describe any patterns in the differences.

Step 5 The second-order differences can be found by subtracting consecutive first-order differences. For example, the second-order difference at $x = 2$ is found by subtracting the first order difference at $x = 1$ from the first-order difference at $x = 2$. Describe any patterns in the differences.

Exercises

For each function make a table of values for the given x-values. Graph the function. Then determine the first-order and second-order differences. **1–3. See Chapter 4 Answer Appendix.**

1. $y = -x^2 + 2x - 1$ for $x = -3, -2, -1, 0, 1, 2, 3$

2. $y = 0.5x^2 + 2x - 2$ for $x = -5, -4, -3, -2, -1, 0, 1$

3. $y = -3x^2 - 18x - 26$ for $x = -6, -5, -4, -3, -2, -1, 0$

4. **MAKE A CONJECTURE** Repeat the activity for a cubic function. At what order difference would you expect $g(x) = x^4$ to be constant? $h(x) = x^n$? **fourth order; nth order**

---

1 Focus

Objective Investigate the rate of change of quadratic functions by examining first- and second-order differences.

Teaching Tip
Each student's table should have a column for every integer value of x from 0 through 16.

2 Teach

Working in Cooperative Groups
Have students work in pairs, mixing abilities. Then have the pairs complete the Activity.

Ask:
- Why is there no first difference written in the column labeled $x = 0$?
  Finding a difference requires two values of y.

- What is significant about the point where $x = 8$? $(8, 0)$ is the vertex of the parabola.

Practice Have students complete Exercises 1–4.

3 Assess

Formative Assessment
Use Exercise 1 to assess whether students are evaluating both the function and the rate of change correctly when x is negative.

From Concrete to Abstract
Exercise 4 asks students to observe and then generalize a pattern as the order of differences increases.

Teach with Tech

Blog Have students write a blog entry describing how graphing a quadratic inequality is similar to and different from graphing a linear inequality.

2 Solve Quadratic Inequalities continued

Example 3 shows how to solve a quadratic inequality of the form $ax^2 + bx + c \geq 0$ by graphing the related quadratic function. **Example 4** shows how to use the graph of a quadratic inequality to solve a real-world problem. **Example 5** shows how to solve a quadratic inequality algebraically.

Additional Examples

3 Solve $0 \leq -2x^2 - 6x + 1$ by graphing.

$\{x \mid -3.16 \leq x \leq 0.16\}$

4 **SPORTS** The height of a ball above the ground after it is thrown upwards at 40 feet per second can be modeled by the function $h(x) = 40x - 16$, where the height $h(x)$ is given in feet and the time x is in seconds. At what time in its flight is the ball within 15 feet of the ground? The ball is within 15 feet of the ground for the first 0.46 second of its flight and again after 2.04 seconds until the ball hits the ground at 2.5 seconds.

5 Solve $x^2 + x \leq 2$ algebraically. $\{x \mid -2 \leq x \leq 1\}$

Example 3 Solve $ax^2 + bx + c \geq 0$ by Graphing

Solve $2x^2 + 4x - 5 \geq 0$ by graphing.

The solution consists of x-values for which the graph of the related function lies *on and above* the x-axis. Begin by finding the roots of the related function.

$2x^2 + 4x - 5 = 0$	Related equation
$x = \dfrac{-b \pm \sqrt{b^2 - 4ac}}{2a}$	Use the Quadratic Formula
$x = \dfrac{-4 \pm \sqrt{4^2 - 4(2)(-5)}}{2(2)}$	Replace a with 4, b with 2, and c with −5.
$x = \dfrac{-4 + \sqrt{56}}{4}$ or $x = \dfrac{-4 - \sqrt{56}}{4}$	Simplify and write as two equations.
≈ 0.87 ≈ -2.87	Simplify.

Sketch the graph of a parabola with x-intercepts at −2.87 and 0.87. The graph opens up since $a > 0$. The graph lies on and above the x-axis at about $x \leq -2.87$ and $x \geq 0.87$. Therefore, the solution is approximately $\{x \mid x \leq -2.87 \text{ or } x \geq 0.87\}$ or $(-\infty, -2.87] \cup [0.87, \infty)$.

$y = 2x^2 + 4x - 5$

▶ **Guided**Practice

Solve each inequality by graphing.

3A. $x^2 - 6x + 2 > 0$

3B. $-4x^2 + 5x + 7 \geq 0$

3A. $\{x \mid x < 0.35 \text{ or } x > 5.65\}$

3B. $\{x \mid -0.84 \leq x \leq 2.09\}$

Real-world problems can be solved by graphing quadratic inequalities.

Real-World Example 4 Solve a Quadratic Inequality

WATER BALLOONS Refer to the beginning of the lesson. At what time will a water balloon be within 3 meters of the ground after it has been launched?

The function $h(t) = -4.9t^2 + 32t + 1.2$ describes the height of the water balloon. Therefore, you want to find the values of t for which $h(t) \leq 3$.

$h(t) \leq 3$	Original inequality
$-4.9t^2 + 32t + 1.2 \leq 3$	$h(t) = -4.9t^2 + 32t + 1.2$
$-4.9t^2 + 32t - 1.8 \leq 0$	Subtract 3 from each side.

Graph the related function $y = -4.9x^2 + 32x - 1.8$ using a graphing calculator. The zeros of the function are about 0.06 and 6.47, and the graph lies below the x-axis when $x < 0.06$ and $x > 6.47$.

So, the water balloon is within 3 meters of the ground during the first 0.06 second after being launched and again after about 6.47 seconds until it hits the ground.

[−1, 9] scl: 1 by [−5, 55] scl: 5

Real-World Link

It takes just milliseconds for a water balloon to break. A high-speed camera can capture the impact on the fluid before gravity makes it fall.

Source: NASA

▶ **Guided**Practice

4. **ROCKETS** The height $h(t)$ of a model rocket in feet t seconds after its launch can be represented by the function $h(t) = -16t^2 + 82t + 0.25$. During what interval is the rocket at least 100 feet above the ground? $\{t \mid 1.99 \leq t \leq 3.14\}$

Focus on Mathematical Content

Quadratic Inequalities Graph quadratic inequalities in two variables using the same techniques as when graphing linear inequalities in two variables. Graph the related function. Use a dashed line if the symbol is $>$ or $<$. Test a point inside the parabola. If it is a solution, shade inside the parabola. If it is not a solution, shade outside the parabola.

StudyTip
Solving Quadratic Inequalities Algebraically The solution set of a quadratic inequality is all real numbers when all three test points satisfy the inequality. It is the empty set when none of the test points satisfy the inequality.

Example 5 Solve a Quadratic Inequality Algebraically

Solve $x^2 - 3x \leq 18$ algebraically.

Step 1 Solve the related quadratic equation $x^2 - 3x = 18$.

$$x^2 - 3x = 18 \qquad \text{Related quadratic equation}$$
$$x^2 - 3x - 18 = 0 \qquad \text{Subtract 18 from each side.}$$
$$(x + 3)(x - 6) = 0 \qquad \text{Factor.}$$
$$x + 3 = 0 \quad \text{or} \quad x - 6 = 0 \qquad \text{Zero Product Property}$$
$$x = -3 \qquad x = 6 \qquad \text{Solve each equation.}$$

Step 2 Plot -3 and 6 on a number line. Use dots since these values are solutions of the original inequality. Notice that the number line is divided into three intervals.

Step 3 Test a value from each interval to see if it satisfies the original inequality.

$x \leq -3$	$-3 \leq x \leq 6$	$x \geq 6$
Test $x = -5$.	Test $x = 0$.	Test $x = 8$.
$x^2 - 3x \leq 18$	$x^2 - 3x \leq 18$	$x^2 - 3x \leq 18$
$(-5)^2 - 3(-5) \overset{?}{\leq} 18$	$(0)^2 - 3(0) \overset{?}{\leq} 18$	$(8)^2 - 3(8) \overset{?}{\leq} 18$
$40 \not\leq 18$	$0 \leq 18$	$40 \not\leq 18$

The solution set is $\{x \mid -3 \leq x \leq 6\}$ or $[-3, 6]$.

▶ **Guided Practice**

Solve each inequality algebraically.

5A. $x^2 + 5x < -6$ $\{x \mid -3 < x < -2\}$ **5B.** $x^2 + 11x + 30 \geq 0$
$\{x \mid x \leq -6 \text{ or } x \geq -5\}$

Check Your Understanding   ● = Step-by-Step Solutions begin on page R14.

Example 1 Graph each inequality. **1–3.** See margin.

1. $y \leq x^2 - 8x + 2$ **2.** $y > x^2 + 6x - 2$ **3.** $y \geq -x^2 + 4x + 1$

Examples 2–3 (CCSS) **SENSE-MAKING** Solve each inequality by graphing.

4. $0 < x^2 - 5x + 4$ $\{x \mid x < 1 \text{ or } x > 4\}$ **5.** $x^2 + 8x + 15 < 0$ $\{x \mid -5 < x < -3\}$

6. $-2x^2 - 2x + 12 \geq 0$ $\{x \mid -3 \leq x \leq 2\}$ **7.** $0 \geq 2x^2 - 4x + 1$ $\{x \mid 0.29 \leq x \leq 1.71\}$

Example 4 **8. SOCCER** A midfielder kicks a ball toward the goal during a match. The height of the ball in feet above the ground $h(t)$ at time t can be represented by $h(t) = -0.1t^2 + 2.4t + 1.5$. If the height of the goal is 8 feet, at what time during the kick will the ball be able to enter the goal? $\{t \mid 0 < t < 3.11\}$ or $\{t \mid 20.89 < t \leq 24.61\}$

Example 5 Solve each inequality algebraically.

9. $x^2 + 6x - 16 < 0$ $\{x \mid -8 < x < 2\}$ **10.** $x^2 - 14x > -49$ $\{x \mid x < 7 \text{ or } x > 7\}$

(11) $-x^2 + 12x \geq 28$ $\{x \mid 3.17 \leq x \leq 8.83\}$ **12.** $x^2 - 4x \leq 21$ $\{x \mid -3 \leq x \leq 7\}$

DifferentiatedInstruction (AL) (OL)

If ▶ students are having trouble making connections between the graph of a quadratic inequality and the inequality itself,

Then ▶ have students think about how the graph of a quadratic inequality helps them understand the meaning of the inequality. Ask them to explore whether the quadratic inequality itself or the graph of the inequality is more meaningful to them. Ask them to give explanations for their choices.

(CCSS) **Teaching the Mathematical Practices**

Sense-Making Mathematically proficient students start by explaining to themselves the meaning of a problem and looking for entry points to its solution. They analyze givens, constraints, relationships, and goals. They check their answers to problems using a different method, and they continually ask themselves, "Does this make sense?"

3 Practice

Formative Assessment
Use Exercises 1–12 to check for understanding.

Use the chart at the bottom of the next page to customize assignments for your students.

Additional Answers

1.

2.

3.

13.

14.

15.

16.

17.

18.

Practice and Problem Solving

Extra Practice is on page R4.

Example 1 Graph each inequality. **13–18. See margin.**

13. $y \geq x^2 + 5x + 6$ **14.** $x^2 - 2x - 8 < y$ **15.** $y \leq -x^2 - 7x + 8$

16. $-x^2 + 12x - 36 > y$ **17.** $y > 2x^2 - 2x - 3$ **18.** $y \geq -4x^2 + 12x - 7$

23. $\{x \mid x < -1.42 \text{ or } x > 8.42\}$

19. $\{x \mid 1.1 < x < 7.9\}$ **20.** $\{x \mid -4 \leq x \leq 6\}$

Examples 2–3 Solve each inequality by graphing. **21.** {all real numbers} **22.** $\{x \mid x < -5.45 \text{ or } x > -0.55\}$

24. {all real numbers}

19. $x^2 - 9x + 9 < 0$ **20.** $x^2 - 2x - 24 \leq 0$ **21.** $x^2 + 8x + 16 \geq 0$

26. $\{x \mid -2.30 < x < 1.30\}$

22. $x^2 + 6x + 3 > 0$ **23.** $0 > -x^2 + 7x + 12$ **24.** $-x^2 + 2x - 15 < 0$

27. $\{x \mid x < -0.73 \text{ or } x > 2.73\}$

25. $4x^2 + 12x + 10 \leq 0$ ∅   **26.** $-3x^2 - 3x + 9 > 0$ **27.** $0 > -2x^2 + 4x + 4$

28. $3x^2 + 12x + 36 \leq 0$ ∅   **29.** $0 \leq -4x^2 + 8x + 5$ $\{x \mid -0.5 \leq x \leq 2.5\}$ **30.** $-2x^2 + 3x + 3 \leq 0$ $\{x \mid x \leq -0.69 \text{ or } x \geq 2.19\}$

Example 4 **31** ARCHITECTURE An arched entry of a room is shaped like a parabola that can be represented by the equation $f(x) = -x^2 + 6x + 1$. How far from the sides of the arch is its height at least 7 feet? **about 1.26 ft to 4.73 ft**

32. MANUFACTURING A box is formed by cutting 4-inch squares from each corner of a square piece of cardboard and then folding the sides. If $V(x) = 4x^2 - 64x + 256$ represents the volume of the box, what should the dimensions of the original piece of cardboard be if the volume of the box cannot exceed 750 cubic inches? **greater than 8 in. but no more than 21.69 in.**

Example 5 Solve each inequality algebraically. **33–44. See Chapter 4 Answer Appendix.**

33. $x^2 - 9x < -20$ **34.** $x^2 + 7x \geq -10$ **35.** $2 > x^2 - x$

36. $-3 \leq -x^2 - 4x$ **37.** $-x^2 + 2x \leq -10$ **38.** $-6 > x^2 + 4x$

39. $2x^2 + 4 \geq 9$ **40.** $3x^2 + x \geq -3$ **41.** $-4x^2 + 2x < 3$

42. $-11 \geq -2x^2 - 5x$ **43.** $-12 < -5x^2 - 10x$ **44.** $-3x^2 - 10x > -1$

B **45.** (CCSS) PERSEVERANCE The Sanchez family is adding a deck along two sides of their swimming pool. The deck width will be the same on both sides and the total area of the pool and deck cannot exceed 750 square feet. **a. See Chapter 4 Answer Appendix.**

a. Graph the quadratic inequality.

b. Determine the possible widths of the deck.
greater than 0 ft but no more than 10.04 ft

Write a quadratic inequality for each graph.

46.

46. $y > x^2 - 4x - 6$

47. $y \leq -x^2 + 2x + 6$

48. $y > -0.25x^2 - 4x + 2$

47.

48.

286 | Lesson 4-8 | Quadratic Inequalities

Differentiated Homework Options

Level	Assignment	Two-Day Option	
AL Basic	13–44, 57–60, 62–82	13–43 odd, 63–66	14–44 even, 57–60, 62, 67–82
OL Core	13–55 odd, 56, 57–60, 62–82	13–44, 63–66	45–60, 62, 67–82
BL Advanced	45–76, (optional: 77–82)		

Solve each quadratic inequality by using a graph, a table, or algebraically.

49. $-2x^2 + 12x < -15$ **50.** $5x^2 + x + 3 \geq 0$ **51** $11 \leq 4x^2 + 7x$

52. $x^2 - 4x \leq -7$ **53.** $-3x^2 + 10x < 5$ **54.** $-1 \geq -x^2 - 5x$

49. $\{x \mid x < -1.06 \text{ or } x > 7.06\}$

50. {all real numbers}

51. $\{x \mid x \leq -2.75 \text{ or } x \geq 1\}$

52. $\varnothing$

53. $\{x \mid x < 0.61 \text{ or } x > 2.72\}$

54. $\{x \mid x \leq -5.19 \text{ or } x \geq 0.19\}$

55. BUSINESS An electronics manufacturer uses the function $P(x) = x(-27.5x + 3520) + 20,000$ to model their monthly profits when selling x thousand digital audio players.

  a. Graph the quadratic inequality for a monthly profit of at least $100,000. **See margin.**

  b. How many digital audio players must the manufacturer sell to earn a profit of at least $100,000 in a month? **from 30,000 to 98,000 digital audio players**

  c. Suppose the manufacturer has an additional monthly expense of $25,000. Explain how this affects the graph of the profit function. Then determine how many digital audio players the manufacturer needs to sell to have at least $100,000 in profits.

56b. 1.28 in. to 1.59 in.

56. UTILITIES A contractor is installing drain pipes for a shopping center's parking lot. The outer diameter of the pipe is to be 10 inches. The cross sectional area of the pipe must be at least 35 square inches and should not be more than 42 square inches.

  a. Graph the quadratic inequalities. **See Chapter 4 Answer Appendix.**

  b. What thickness of drain pipe can the contractor use?

55c. The graph is shifted down 25,000 units. The manufacturer must sell from 47,000 to 81,000 digital audio players.

H.O.T. Problems   Use Higher-Order Thinking Skills

57. OPEN ENDED Write a quadratic inequality for each condition.

  a. The solution set is all real numbers. **Sample answer:** $x^2 + 2x + 1 \geq 0$

  b. The solution set is the empty set. **Sample answer:** $x^2 - 4x + 6 < 0$

58. CCSS CRITIQUE Don and Diego used a graph to solve the quadratic inequality $x^2 - 2x - 8 > 0$. Is either of them correct? Explain.

58. Neither; Don graphed the inequality in two variables, and Diego graphed the wrong interval.

59–62. See Chapter 4 Answer Appendix.

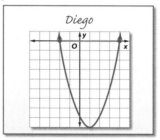

Don

Diego

59. REASONING Are the boundaries of the solution set of $x^2 + 4x - 12 \leq 0$ twice the value of the boundaries of $\frac{1}{2}x^2 + 2x - 6 \leq 0$? Explain.

60. REASONING Determine if the following statement is *sometimes*, *always*, or *never* true. Explain your reasoning.

    The intersection of $y \leq -ax^2 + c$ and $y \geq ax^2 - c$ is the empty set.

61. CHALLENGE Graph the intersection of the graphs of $y \leq -x^2 + 4$ and $y \geq x^2 - 4$.

62. WRITING IN MATH How are the techniques used when solving quadratic inequalities and quadratic equations similar? different?

Additional Answer

55a.

Study Guide and Intervention

AL   OL   ELL

NAME _____ DATE _____ PERIOD _____

4-8 Study Guide and Intervention
Quadratic Inequalities

Graph Quadratic Inequalities To graph a quadratic inequality in two variables, use the following steps:

1. Graph the related quadratic equation, $y = ax^2 + bx + c$. Use a dashed line for $<$ or $>$; use a solid line for $\leq$ or $\geq$.

2. Test a point inside the parabola. If it satisfies the inequality, shade the region inside the parabola; otherwise, shade the region outside the parabola.

Example Graph the inequality $y > x^2 + 6x + 7$.

Exercises

Graph each inequality.

1. $y > x^2 - 8x + 17$ 2. $y \leq x^2 + 6x + 4$ 3. $y \geq x^2 + 2x + 2$

4. $y < -x^2 + 4x - 6$ 5. $y \geq 2x^2 + 4x$ 6. $y > -2x^2 - 4x + 2$

Practice

AL   OL   BL   ELL

NAME _____ DATE _____ PERIOD _____

4-8 Practice
Quadratic Inequalities

Graph each inequality.

1. $y \leq x^2 + 4$ 2. $y > x^2 + 6x + 6$ 3. $y < 2x^2 - 4x - 2$

Solve each inequality.

4. $x^2 + 2x + 1 > 0$ 5. $x^2 - 3x + 2 \leq 0$ 6. $x^2 + 10x + 7 \geq 0$

7. $x^2 - x - 20 > 0$ 8. $x^2 - 10x + 16 < 0$ 9. $x^2 + 4x + 5 \leq 0$

10. $x^2 + 14x + 49 \geq 0$ 11. $x^2 - 5x > 14$ 12. $-x^2 - 15 \leq 8x$

13. $-x^2 + 5x - 7 \leq 0$ 14. $9x^2 + 36x + 36 \leq 0$ 15. $9x \leq 12x^2$

16. $4x^2 + 4x + 1 > 0$ 17. $5x^2 + 10 \geq 27x$ 18. $9x^2 + 31x + 12 \leq 0$

19. **FENCING** Vanessa has 180 feet of fencing that she intends to use to build a rectangular play area for her dog. She wants the play area to enclose at least 1800 square feet. What are the possible widths of the play area?

20. **BUSINESS** A bicycle maker sold 300 bicycles last year at a profit of $300 each. The maker wants to increase the profit margin this year, but predicts that each $20 increase in profit will reduce the number of bicycles sold by 10. How many $20 increases in profit can the maker add in and expect to make a total profit of at least $100,000?

Word Problem Practice

AL   OL   BL   ELL

NAME _____ DATE _____ PERIOD _____

4-8 Word Problem Practice
Quadratic Inequalities

1. **HUTS** The space inside a hut is shaded in the graph. The parabola is described by the equation $y = -\frac{4}{5}(x - 1)^2 + 4$.

Write an inequality that describes the shaded region.

2. **DISCRIMINANTS** Consider the equation $ax^2 + bx + c = 0$. Assume that the discriminant is zero and that a is positive. What are the solutions of the inequality $ax^2 + bx + c \leq 0$?

3. **KIOSKS** Caleb is designing a kiosk by wrapping a piece of sheet metal with dimensions $x + 5$ inches by $4x + 8$ inches into a cylindrical shape. Ignoring cost, Caleb would like a kiosk that's surface area of at least 4480 square inches. What values of x satisfy this condition?

4. **DAMS** The Hoover Dam is a concrete arch dam designed to hold the water of Lake Mead. At its center, the dam's height is approximately 725 feet, and the dam varies from 45 to 660 feet thick. The dark line on this sketch of the cross-section of the dam is a parabola.

a. Write an equation for the Hoover Dam parabola. Let the height be the y-value of the parabola and the thickness be the x-value of the parabola. (Hint: the equation will be in the form $y = \&x$ − maximum thickness) + maximum height.)

b. Using your equation, graph the parabola of the Hoover Dam for $45 \leq x \leq 660$.

c. Estimate to the nearest foot the thickness of the dam when the height is 200 feet.

Enrichment

OL   BL

NAME _____ DATE _____ PERIOD _____

4-8 Enrichment

Graphing Absolute Value Inequalities

You can solve absolute value inequalities by graphing in much the same manner you graphed quadratic inequalities. Graph the related absolute value function for each inequality by using a graphing calculator. For $>$ and $\geq$, identify the x-values, if any, for which the graph lies *below* the x-axis. For $<$ and $\leq$, identify the x values, if any, for which the graph lies *above* the x-axis.

For each inequality, make a sketch of the related graph and find the solutions rounded to the nearest hundredth.

1. $|x - 3| > 0$ 2. $|x| - 6 < 0$ 3. $-|x + 4| + 8 < 0$

WatchOut!

Error Analysis For an inequality containing x as the only variable, the solution set should refer only to values of x. Point out that in Exercise 58 Don has graphed an inequality in two variables.

4 Assess

Ticket Out the Door Write a quadratic inequality in one variable on the board. On an index card, have students solve the inequality graphically and algebraically.

Formative Assessment

Check for student understanding of concepts in Lessons 4-7 and 4-8.

☐ Quiz 4, p. 60

Standardized Test Practice

63. GRIDDED RESPONSE You need to seed an area that is 80 feet by 40 feet. Each bag of seed can cover 25 square yards of land. How many bags of seed will you need? **15**

64. SAT/ACT The product of two integers is between 107 and 116. Which of the following *cannot* be one of the integers? **D**

A 5   D 15
B 10   E 23
C 12

65 PROBABILITY Five students are to be arranged side by side with the tallest student in the center and the two shortest students on the ends. If no two students are the same height, how many different arrangements are possible? **G**

F 2   H 5
G 4   J 6

66. SHORT RESPONSE Simplify $\frac{5+i}{6-3i}$. $\frac{3}{5} + \frac{7}{15}i$

Spiral Review

Write an equation in vertex form for each parabola. (Lesson 4-7)

67.

$y = 2(x - 3)^2 - 4$

68.

$y = -3(x + 2)^2 + 1$

69.

$y = 0.25(x + 4)^2 + 3$

Complete parts a and b for each quadratic equation.
a. Find the value of the discriminant.
b. Describe the number and type of roots. (Lesson 4-6)

70. $4x^2 + 7x - 3 = 0$
97; 2 irrational roots

71. $-3x^2 + 2x - 4 = 9$
−152; 2 complex roots

72. $6x^2 + x - 4 = 12$
385; 2 irrational roots

Perform the indicated operation. If the matrix does not exist, write *impossible*. (Lesson 3-5)

73. $4\begin{bmatrix} 3 & -6 \\ -5 & 2 \end{bmatrix} - 3\begin{bmatrix} 4 & -1 \\ -2 & 8 \end{bmatrix}$ $\begin{bmatrix} 0 & -21 \\ -14 & -16 \end{bmatrix}$

74. $-2\begin{bmatrix} 5 & -9 \\ 5 & 11 \end{bmatrix} - 6\begin{bmatrix} 3 & -7 \\ -5 & 8 \end{bmatrix}$ $\begin{bmatrix} -28 & 60 \\ 20 & -70 \end{bmatrix}$

75. $\begin{bmatrix} 2 & -6 \\ -4 & 6 \end{bmatrix} \cdot \begin{bmatrix} 2 & -1 & 1 \\ -1 & 6 & 4 \end{bmatrix}$ $\begin{bmatrix} 10 & -38 & -22 \\ -14 & 40 & 20 \end{bmatrix}$

76. EXERCISE Refer to the graphic. (Lesson 3-1)

a. For each option, write an equation that represents the cost of belonging to the gym. $y = 400$; $y = 150 + 5x$

b. Graph the equations. Estimate the break-even point for the gym memberships. **See Chapter 4 Answer Appendix.**

c. Explain what the break-even point means.

d. If you plan to visit the gym at least once per week during the year, which option should you chose? **Option 1**

76c. It means that the options cost the same if you visit 50 times in a year.

OPTION 1:
$400/yr
Unlimited visits

OPTION 2:
$150/yr
$5 per visit

Skills Review

Use the Distributive Property to find each product.

77. $-6(x - 4)$ $-6x + 24$

78. $8(w + 3x)$ $8w + 24x$

79. $-4(-2y + 3z)$ $8y - 12z$

80. $-1(c - d)$ $d - c$

81. $0.5(5x + 6y)$ $2.5x + 3y$

82. $-3(-6y - 4z)$ $18y + 12z$

 288 | Lesson 4-8 | Quadratic Inequalities

DifferentiatedInstruction (OL) (BL)

Extension Write $\sqrt{x^2 - 2x - 3}$ on the board. Ask students to find the values for x for which this expression is a real number. Explain that the solution is all values of x such that the expression under the radical is greater than or equal to zero. $\{x \mid x \leq -1 \text{ or } x \geq 3\}$

EXTEND 4-8

Graphing Technology Lab
Modeling Motion

Set Up the Lab

- Place a board on a stack of books to create a ramp.

- Connect the data collection device to the graphing calculator. Place at the top of the ramp so that the data collection device can read the motion of the car on the ramp.

- Hold the car still about 6 inches up from the bottom of the ramp and zero the collection device.

CCSS Common Core State Standards
Mathematical Practices
4 Model with mathematics.

Activity

Step 1 One group member should press the button to start collecting data.

Step 2 Another group member places the car at the bottom of the ramp. After data collection begins, gently but quickly push the car so it travels up the ramp toward the motion detector.

Step 3 Stop collecting data when the car returns to the bottom of the ramp. Save the data as Trial 1.

Step 4 Remove one book from the stack. Then repeat the experiment. Save the data as Trial 2. For Trial 3, create a steeper ramp and repeat the experiment.

1. Sample answer: The graph resembles a parabola, so a quadratic function could be used to model the data.
5. Sample answer: Trial 3 is widest and Trial 2 is most narrow. The car moved more quickly on the steeper ramps. The absolute value of a is least in the widest graph.

Analyze the Results 2–4. See students' work.

1. What type of function could be used to represent the data? Justify your answer.

2. Use the **CALC** menu to find the vertex of the graph. Record the coordinates in a table like the one at the right.

3. Use the **TRACE** feature of the calculator to find the coordinates of another point on the graph. Then use the coordinates of the vertex and the point to find an equation of the graph.

Trial	Vertex (h, k)	Point (x, y)	Equation
1			
2			
3			

4. Find an equation for each of the graphs of Trials 2 and 3.

5. How do the equations for Trials 1, 2, and 3 compare? Which graph is widest and which is most narrow? Explain what this represents in the context of the situation. How is this represented in the equations?

6. What do the x-intercepts and vertex of each graph represent? the times at which the car passed the zero point; the time at which the car stopped its forward motion

7. Why were the values of h and k different in each trial? The time at which the car stopped is different.

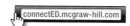

Ed-Imaging

---

1 Focus

Objective Use a data collection device to investigate the relationship between the time and the distance traveled by a car on a ramp.

Materials for Each Group

- long, flat, smooth-surface board
- stack of books
- data collection device (TI CBR Motion Detector)
- TI-83/84 Plus graphing calculator
- medium-size toy car (a 9-inch playground ball can also be used)

Teaching Tip

- Before starting the Activity, ask students to sketch prediction graphs of the distance vs. time as the car moves up and back down the ramp.

- You may want students to practice giving the car a quick but gentle push so it rolls halfway up the ramp. They must press **START** to begin collecting data as soon as they let go of the car. It is crucial that students get their hands out of the path of the data collection device quickly after releasing the car.

- The car should not come any closer than 1.5 feet (0.5 meters) to the collection device.

2 Teach

Working in Cooperative Groups

Put students in groups of three or four, mixing abilities. Then have groups complete the Activity and Exercises 1–3.

Ask:

- Ask students if the car covers more distance during the first two seconds of movement or the last two seconds.

Practice Have students complete Exercises 4–7.

---

3 Assess

Formative Assessment

Use Exercise 5 to assess whether students comprehend how changes in the car's movement affect the shape of the graph.

From Concrete to Abstract

Ask students what they think would happen to their graphs if the ramp got steeper and steeper. Then have them repeat the experiment to check their conjectures.

Extending the Concept

- Modify the activity by using a longer ramp.
- Modify the activity by adding weight to the car.

Formative Assessment
KeyVocabulary The page references after each word denote where that term was first introduced. If students have difficulty answering questions 1–8, remind them that they can use these page references to refresh their memories about the vocabulary.

Summative Assessment
Vocabulary Test, p. 62

Vocabulary Review
Vocabulary Review provides students the opportunity to check their understanding of important concepts and terminology in an online game format.

Additional Answers

9a. y-int: 12; $x = \dfrac{-5}{2}; \dfrac{-5}{2}$

9b.

x	$f(x)$
-3	6
$\dfrac{-5}{2}$	$\dfrac{23}{4}$
-2	6
-1	8
0	12

9c.

$f(x) = x^2 + 5x + 12$

10a. y-int: 15; $x = \dfrac{7}{2}; \dfrac{7}{2}$

10b.

x	$f(x)$
2	5
3	3
$\dfrac{7}{2}$	$\dfrac{11}{4}$
4	3
5	5

CHAPTER 4 Study Guide and Review

Study Guide

KeyConcepts

Graphing Quadratic Functions (Lesson 4-1)

- The graph of $y = ax^2 + bx + c$, $a \neq 0$, opens up, and the function has a minimum value when $a > 0$. The graph opens down, and the function has a maximum value when $a < 0$.

Solving Quadratic Equations (Lessons 4-2 and 4-3)

- Roots of a quadratic equation are the zeros of the related quadratic function. You can find the zeros of a quadratic function by finding the x-intercepts of the graph.

Complex Numbers (Lesson 4-4)

- i is the imaginary unit; $i^2 = -1$ and $i = \sqrt{-1}$.

Solving Quadratic Equations (Lessons 4-5 and 4-6)

- Completing the square: **Step 1** Find one half of b, the coefficient of x. **Step 2** Square the result in Step 1. **Step 3** Add the result of Step 2 to $x^2 + bx$.
- Quadratic Formula: $x = \dfrac{-b \pm \sqrt{b^2 - 4ac}}{2a}$

Transformations of Quadratic Graphs (Lesson 4-7)

- The graph of $y = (x - h)^2 + k$ is the graph of $y = x^2$ translated $|h|$ units left if h is negative or h units right if h is positive and k units up if k is positive or $|k|$ units down if k is negative.
- Consider $y = a(x - h)^2 + k$, $a \neq 0$. If $a > 0$, the graph opens up; if $a < 0$ the graph opens down. If $|a| > 1$, the graph is narrower than the graph of $y = x^2$. If $|a| < 1$, the graph is wider than the graph of $y = x^2$.

Quadratic Inequalities (Lesson 4-8)

- Graph the related function, test a point *not* on the parabola and determine if it is a solution, and shade accordingly.

FOLDABLES StudyOrganizer

Be sure the Key Concepts are noted in your Foldable.

 290 | Chapter 4 | Study Guide and Review

KeyVocabulary

axis of symmetry (p. 220)	pure imaginary number (p. 246)
completing the square (p. 257)	quadratic equation (p. 229)
complex conjugates (p. 249)	Quadratic Formula (p. 264)
complex number (p. 247)	quadratic function (p. 219)
constant term (p. 219)	quadratic inequality (p. 282)
discriminant (p. 267)	quadratic term (p. 219)
factored form (p. 238)	root (p. 229)
FOIL method (p. 238)	Square Root Property (p. 247)
imaginary unit (p. 246)	standard form (p. 229)
linear term (p. 219)	vertex (p. 220)
maximum value (p. 222)	vertex form (p. 275)
minimum value (p. 222)	zero (p. 229)
parabola (p. 219)	

VocabularyCheck

State whether each sentence is *true* or *false*. If *false*, replace the underlined term to make a true sentence.

1. The <u>factored form</u> of a quadratic equation is $ax^2 + bx + c = 0$, where $a \neq 0$ and a, b, and c are integers.
   false, standard form
2. The graph of a quadratic function is called a <u>parabola</u>. **true**
3. The <u>vertex form</u> of a quadratic function is $y = a(x - p)(x - q)$. **false, factored form**
4. The axis of symmetry will intersect a parabola in one point called the <u>vertex</u>. **true**
5. A method called <u>FOIL method</u> is used to make a quadratic expression a perfect square in order to solve the related equation. **false, completing the square**
6. The equation $x = \dfrac{-b \pm \sqrt{b^2 - 4ac}}{2a}$ is known as the <u>discriminant</u>. **false, Quadratic Formula**
7. The number $6i$ is called a <u>pure imaginary number</u>. **true**
8. The two numbers $2 + 3i$ and $2 - 3i$ are called <u>complex conjugates</u>. **true**

FOLDABLES StudyOrganizer

Dinah Zike's Foldables®
Have students look through the chapter to make sure they have included examples in their Foldables. Suggest that students keep their Foldables handy while completing the Study Guide and Review pages. Point out that their Foldables can serve as a quick review tool when studying for the Chapter Test.

Lesson-by-Lesson Review

4-1 Graphing Quadratic Functions

Complete parts **a**–**c** for each quadratic function.

a. Find the *y*-intercept, the equation of the axis of symmetry, and the *x*-coordinate of the vertex.

b. Make a table of values that includes the vertex.

c. Use this information to graph the function.

9. $f(x) = x^2 + 5x + 12$ **10.** $f(x) = x^2 - 7x + 15$

11. $f(x) = -2x^2 + 9x - 5$ **12.** $f(x) = -3x^2 + 12x - 1$

Determine whether each function has a maximum or minimum value and find the maximum or minimum value. Then state the domain and range of the function.

13. $f(x) = -x^2 + 3x - 1$ **14.** $f(x) = -3x^2 - 4x + 5$

15. BUSINESS Sal's Shirt Store sells 100 T-shirts per week at a rate of $10 per shirt. Sal estimates that he will sell 5 fewer shirts for each $1 increase in price. What price will maximize Sal's T-shirt income? **75 T-shirts at $15 each**

9–12. See margin.

14. max; $\dfrac{19}{3}$; D = all real numbers; R = $\left\{ f(x) \middle| f(x) \le \dfrac{19}{3} \right\}$

Example 1

Consider the quadratic function $f(x) = x^2 - 4x + 11$. Find the *y*-intercept, the equation for the axis of symmetry, and the *x*-coordinate of the vertex.

In the function, $a = 1$, $b = -4$, and $c = 11$. The *y*-intercept is $c = 11$.

Use *a* and *b* to find the equation of the axis of symmetry.

$$x = -\frac{b}{2a} \qquad \text{Equation of the axis of symmetry}$$

$$= -\frac{-4}{2(1)} \qquad a = 1 \text{ and } b = -4$$

$$= 2 \qquad \text{Simplify.}$$

The equation of the axis of symmetry is $x = 2$. Therefore, the *x*-coordinate of the vertex is 2.

13. max; 1.25; D = all real numbers; $R = \{f(x) \mid f(x) \le 1.25\}$

4-2 Solving Quadratic Equations by Graphing

Solve each equation by graphing. If exact roots cannot be found, state the consecutive integers between which the roots are located.

16. $x^2 - x - 20 = 0$ $\{-4, 5\}$

17. $2x^2 - x - 3 = 0$ $\left\{ -1, \dfrac{3}{2} \right\}$

18. $4x^2 - 6x - 15 = 0$ **18.** between −1 and −2; between 2 and 3

19. BASEBALL A baseball is hit upward at 120 feet per second. Use the formula $h(t) = v_0 t - 16t^2$, where $h(t)$ is the height of an object in feet, v_0 is the object's initial velocity in feet per second, and *t* is the time in seconds. Ignoring the height of the ball when it was hit, how long does it take for the ball to hit the ground? **7.5 seconds**

Example 2

Solve $2x^2 - 7x + 3 = 0$ by graphing.

The equation of the axis of symmetry is $-\dfrac{-7}{2(2)}$ or $x = \dfrac{7}{4}$.

x	0	1	$\frac{7}{4}$	2	3
f(x)	3	−2	$-2\frac{5}{8}$	−3	0

The zeros of the related function are $\frac{1}{2}$ and 3. Therefore, the solutions of the equation are $\frac{1}{2}$ and 3.

12a. *y*-int: −1; $x = 2$; 2

12b.

x	f(x)
0	−1
1	8
2	11
3	8
4	−1

12c.

Lesson-by-Lesson Review

Intervention If the given examples are not sufficient to review the topics covered by the questions, remind students that the lesson references tell them where to review that topic in their textbook.

Two-Day Option Have students complete the Lesson-by-Lesson Review. Then you can use McGraw-Hill eAssessment to customize another review worksheet that practices all the objectives of this chapter or only the objectives on which your students need more help.

Additional Answers

10c.

11a. *y*-int: −5; $x = \dfrac{9}{4}$; $\dfrac{9}{4}$

11b.

x	f(x)
1	2
2	5
$\frac{9}{4}$	$\frac{41}{8}$
3	4
4	−1

11c.

Additional Answers

20. $x^2 - 11x + 30 = 0$

21. $x^2 + 10x + 21 = 0$

22. $x^2 + 2x - 8 = 0$

23. $3x^2 - x - 2 = 0$

24. $6x^2 - 31x + 5 = 0$

25. $4x^2 + 5x + 1 = 0$

4-3 Solving Quadratic Equations by Factoring

Write a quadratic equation in standard form with the given roots. **20–25. See margin.**

20. 5, 6

21. −3, −7

22. −4, 2

23. $-\frac{2}{3}$, 1

24. $\frac{1}{6}$, 5

25. $-\frac{1}{4}$, −1

Solve each equation by factoring.

26. $2x^2 - 2x - 24 = 0$ {−3, 4}

27. $2x^2 - 5x - 3 = 0$ $\left\{-\frac{1}{2}, 3\right\}$

28. $3x^2 - 16x + 5 = 0$ $\left\{\frac{1}{3}, 5\right\}$

29. Find x and the dimensions of the rectangle below.

$A = 126\ \text{ft}^2$ | $x - 3$

$x + 2$

$x = 12$; 9 feet by 14 feet

Example 3

Write a quadratic equation in standard form with $-\frac{1}{2}$ and 4 as its roots.

$(x - p)(x - q) = 0$ — Write the pattern.

$\left[x - \left(-\frac{1}{2}\right)\right](x - 4) = 0$ — Replace p with $-\frac{1}{2}$ and q with 4.

$\left(x + \frac{1}{2}\right)(x - 4) = 0$ — Simplify.

$x^2 - \frac{7}{2}x - 2 = 0$ — Multiply.

$2x^2 - 7x - 4 = 0$ — Multiply each side by 2 so that b and c are integers.

Example 4

Solve $2x^2 - 3x - 5 = 0$ by factoring.

$2x^2 - 3x - 5 = 0$ — Original equation

$(2x - 5)(x + 1) = 0$ — Factor the trinomial.

$2x - 5 = 0$ or $x + 1 = 0$ — Zero Product Property

$x = \frac{5}{2}$ \qquad $x = -1$

The solution set is $\left\{-1, \frac{5}{2}\right\}$ or $\left\{x \mid x = -1, \frac{5}{2}\right\}$.

4-4 Complex Numbers

Simplify. **32.** $2 + 5i$

30. $\sqrt{-8}$ $2i\sqrt{2}$

31. $(2 - i) + (13 + 4i)$ $15 + 3i$

32. $(6 + 2i) - (4 - 3i)$

33. $(6 + 5i)(3 - 2i)$ $28 + 3i$

34. **ELECTRICITY** The impedance in one part of a series circuit is $3 + 2j$ ohms, and the impedance in the other part of the circuit is $4 - 3j$ ohms. Add these complex numbers to find the total impedance in the circuit. $7 - j$ ohms

Solve each equation.

35. $2x^2 + 50 = 0$ $\pm 5i$

36. $4x^2 + 16 = 0$ $\pm 2i$

37. $3x^2 + 15 = 0$ $\pm i\sqrt{5}$

38. $8x^2 + 16 = 0$ $\pm i\sqrt{2}$

39. $4x^2 + 1 = 0$ $\pm\frac{1}{2}i$

Example 5

Simplify $(12 + 3i) - (-5 + 2i)$.

$(12 + 3i) - (-5 + 2i)$

$= [12 - (-5)] + (3 - 2)i$ — Group the real and imaginary parts.

$= 17 + i$ — Simplify.

Example 6

Solve $3x^2 + 12 = 0$.

$3x^2 + 12 = 0$ — Original equation

$3x^2 = -12$ — Subtract 12 from each side.

$x^2 = -4$ — Divide each side by 3.

$x = \pm\sqrt{-4}$ — Square Root Property

$x = \pm 2i$ — $\sqrt{-4} = \sqrt{4} \cdot \sqrt{-1}$

292 | Chapter 4 | Study Guide and Review

4-5 Completing the Square

Find the value of *c* that makes each trinomial a perfect square. Then write the trinomial as a perfect square.

40. $x^2 + 18x + c$ **41.** $x^2 - 4x + c$ **40–45.**
                                      See margin.

42. $x^2 - 7x + c$ **43.** $x^2 + 2.4x + c$

44. $x^2 - \frac{1}{2}x + c$ **45.** $x^2 + \frac{6}{5}x + c$

Solve each equation by completing the square.

46. $x^2 - 6x - 7 = 0$ $\{-1, 7\}$

47. $x^2 - 2x + 8 = 0$ $\{1 \pm i\sqrt{7}\}$

48. $2x^2 + 4x - 3 = 0$ $\left\{\dfrac{-2 \pm \sqrt{10}}{2}\right\}$

49. $2x^2 + 3x - 5 = 0$ $\left\{1, -\dfrac{5}{2}\right\}$

50. FLOOR PLAN Mario's living room has a length 6 feet wider than the width. The area of the living room is 280 square feet. What are the dimensions of his living room?
                            20 feet by 14 feet

Example 7

Find the value of *c* that makes $x^2 + 14x + c$ a perfect square. Then write the trinomial as a perfect square.

Step 1 Find one half of 14.

Step 2 Square the result of Step 1.

Step 3 Add the result of Step 2 to $x^2 + 14x$.

The trinomial $x^2 + 14x + 49$ can be written as $(x + 7)^2$.

Example 8

Solve $x^2 + 12x - 13 = 0$ by completing the square.

$$x^2 + 12x - 13 = 0$$
$$x^2 + 12x = 13$$
$$x^2 + 12x + 36 = 13 + 36$$
$$(x + 6)^2 = 49$$
$$x + 6 = \pm 7$$
$$x + 6 = 7 \quad \text{or} \quad x + 6 = -7$$
$$x = 1 \qquad\qquad x = -13$$

The solution set is $\{-13, 1\}$ or $\{x \mid x = -13, 1\}$.

4-6 The Quadratic Formula and the Discriminant

Complete parts **a–c** for each quadratic equation.

a. Find the value of the discriminant.

b. Describe the number and type of roots.

c. Find the exact solutions by using the Quadratic Formula.

51. $x^2 - 10x + 25 = 0$ **51–57. See margin.**

52. $x^2 + 4x - 32 = 0$

53. $2x^2 + 3x - 18 = 0$

54. $2x^2 + 19x - 33 = 0$

55. $x^2 - 2x + 9 = 0$

56. $4x^2 - 4x + 1 = 0$

57. $2x^2 + 5x + 9 = 0$

58. PHYSICAL SCIENCE Lauren throws a ball with an initial velocity of 40 feet per second. The equation for the height of the ball is $h = -16t^2 + 40t + 5$, where *h* represents the height in feet and *t* represents the time in seconds. When will the ball hit the ground? **about 2.62 seconds**

Example 9

Solve $x^2 - 4x - 45 = 0$ by using the Quadratic Formula.

In $x^2 - 4x - 45 = 0$, $a = 1$, $b = -4$, and $c = -45$.

$$x = \frac{-b \pm \sqrt{b^2 - 4ac}}{2a} \qquad \text{Quadratic Formula}$$

$$= \frac{-(-4) \pm \sqrt{(-4)^2 - 4(1)(-45)}}{2(1)}$$

$$= \frac{4 \pm 14}{2}$$

Write as two equations.

$$x = \frac{4 + 14}{2} \quad \text{or} \quad x = \frac{4 - 14}{2}$$
$$= 9 \qquad\qquad = -5$$

The solution set is $\{-5, 9\}$ or $\{x \mid x = -5, 9\}$.

connectED.mcgraw-hill.com   **293**

Additional Answers

40. $81; (x + 9)^2$

41. $4; (x - 2)^2$

42. $\dfrac{49}{4}; \left(x - \dfrac{7}{2}\right)^2$

43. $1.44; (x + 1.2)^2$

44. $\dfrac{1}{16}; \left(x - \dfrac{1}{4}\right)^2$

45. $\dfrac{9}{25}; \left(x + \dfrac{3}{5}\right)^2$

51a. 0

51b. 1 real rational root

51c. $\{5\}$

52a. 144

52b. 2 rational real roots

52c. $\{-8, 4\}$

53a. 153

53b. 2 irrational real roots

53c. $\left\{\dfrac{-3 \pm 3\sqrt{17}}{4}\right\}$

54a. 625

54b. 2 real rational roots

54c. $\left\{-11, \dfrac{3}{2}\right\}$

55a. -32

55b. 2 complex roots

55c. $\{1 \pm 2i\sqrt{2}\}$

56a. 0

56b. 1 real rational root

56c. $\left\{\dfrac{1}{2}\right\}$

57a. -47

57b. 2 complex roots

57c. $\left\{\dfrac{-5 \pm i\sqrt{47}}{4}\right\}$

Anticipation Guide

Have students complete the Chapter 4 Anticipation Guide and discuss how their responses have changed now that they have completed Chapter 4.

Additional Answers

59. $y = -3(x - 1)^2 + 5$; $(1, 5)$; $x = 1$; opens down

60. $y = 2(x + 3)^2 - 26$; $(-3, -26)$; $x = -3$; opens up

61. $y = -\frac{1}{2}(x + 2)^2 + 14$; $(-2, 14)$;

$x = -2$; opens down

4-7 Transformations of Quadratic Graphs

Write each quadratic function in vertex form, if not already in that form. Then identify the vertex, axis of symmetry, and direction of opening. Then graph the function.
59–62. See margin.

59. $y = -3(x - 1)^2 + 5$ **60.** $y = 2x^2 + 12x - 8$

61. $y = -\frac{1}{2}x^2 - 2x + 12$ **62.** $y = 3x^2 + 36x + 25$

63. The graph at the right shows a product of 2 numbers with a sum of 10. Find a function that models this product and use it to determine the two numbers that would give a maximum product.
$f(x) = -x^2 + 10x$; 5 and 5

Example 10

Write the quadratic function $y = 3x^2 + 24x + 15$ in vertex form. Then identify the vertex, axis of symmetry, and direction of opening.

$y = 3x^2 + 24x + 15$	Original equation
$y = 3(x^2 + 8x) + 15$	Group and factor.
$y = 3(x^2 + 8x + 16) + 15 - 3(16)$	Complete the square.
$y = 3(x + 4)^2 - 33$	Rewrite $x^2 + 8x + 16$ as a perfect square.

So, $a = 3$, $h = -4$, and $k = -33$. The vertex is at $(-4, -33)$ and the axis of symmetry is $x = -4$. Since a is positive, the graph opens up.

64–67. See Chapter 4 Answer Appendix.

4-8 Quadratic Inequalities

Graph each quadratic inequality.

64. $y \geq x^2 + 5x + 4$ **65.** $y < -x^2 + 5x - 6$

66. $y > x^2 - 6x + 8$ **67.** $y \leq x^2 + 10x - 4$

68. Solomon wants to put a deck along two sides of his garden. The deck width will be the same on both sides and the total area of the garden and deck cannot exceed 500 square feet. How wide can the deck be?
between 0 and 5 ft

Solve each inequality using a graph or algebraically.

69. $x^2 + 8x + 12 > 0$ $\{x \mid x < -6 \text{ or } x > -2\}$

70. $6x + x^2 \geq -9$ $\{$all real numbers$\}$

71. $2x^2 + 3x - 20 > 0$ $\left\{x \mid x < -4 \text{ or } x > \frac{5}{2}\right\}$

72. $4x^2 - 3 < -5x$ $\{x \mid -1.69 < x < 0.44\}$

73. $3x^2 + 4 > 8x$ $\left\{x \mid x < \frac{2}{3} \text{ or } x > 2\right\}$

Example 11

Graph $y > x^2 + 3x + 2$.

Step 1 Graph the related function, $y > x^2 + 3x + 2$. Because the inequality symbol $>$ is used, the parabola should be dashed.

Step 2 Test a point not on the graph of the parabola such as $(0, 0)$.

$y > x^2 + 3x + 2$

$(0) \overset{?}{>} (0)^2 + 3(0) + 2$

$0 \not> 2$

So, $(0, 0)$ is not a solution of the inequality.

Step 3 Shade the region that does not contain the point $(0, 0)$.

62. $y = 3(x + 6)^2 - 83$; $(-6, -83)$; $x = 6$; opens up

Additional Answer (Practice Test)

14b. No; if you graph the function, the vertex in 202 units above the horizontal axis. So, the height will never be 250.

4 Practice Test

Complete parts *a–c* for each quadratic function.

a. Find the *y*-intercept, the equation of the axis of symmetry, and the *x*-coordinate of the vertex.

b. Make a table of values that includes the vertex.

c. Use this information to graph the function.

1. $f(x) = x^2 + 4x - 7$ **1–3. See Chapter 4 Answer Appendix.**

2. $f(x) = -2x^2 + 5x$

3. $f(x) = -x^2 - 6x - 9$

Determine whether each function has a maximum or minimum value. State the maximum or minimum value of each function.

4. $f(x) = x^2 + 10x + 25$ **min.; 0**

5. $f(x) = -x^2 + 6x$ **max.; 9**

Solve each equation using the method of your choice. Find exact solutions.

6. $x^2 - 8x - 9 = 0$ **−1, 9**

7. $-4.8x^2 + 1.6x + 24 = 0$ $\dfrac{1 \pm \sqrt{181}}{6}$

8. $12x^2 + 15x - 4 = 0$ $\dfrac{-15 \pm \sqrt{417}}{24}$

9. $x^2 - 7x - \dfrac{17}{4} = 0$ $\dfrac{7 \pm \sqrt{66}}{2}$

10. $4x^2 + x = 3$ $-1, \dfrac{3}{4}$

11. $-9x^2 + 40x + 84 = 0$ $-\dfrac{14}{9}, 6$

12. PHYSICAL SCIENCE Parker throws a ball off the top of a building. The building is 350 feet high and the initial velocity of the ball is 30 feet per second. Find out how long it will take the ball to hit the ground by solving the equation $-16t^2 - 30t + 350 = 0$. **about 3.83 seconds**

13. MULTIPLE CHOICE Which equation below has roots at -6 and $\dfrac{1}{5}$? **C**

A $0 = 5x^2 - 29x - 6$

B $0 = 5x^2 + 31x + 6$

C $0 = 5x^2 + 29x - 6$

D $0 = 5x^2 - 31x + 6$

14a. about 1.4 seconds and 5.6 seconds

14. PHYSICS A ball is thrown into the air vertically with a velocity of 112 feet per second. The ball was released 6 feet above the ground. The height above the ground *t* seconds after release is modeled by $h(t) = -16t^2 + 112t + 6$.

a. When will the ball reach 130 feet? **b. See margin.**

b. Will the ball ever reach 250 feet? Explain.

c. In how many seconds after its release will the ball hit the ground? **about 7 seconds**

15. The rectangle below has an area of 104 square inches. Find the value of *x* and the dimensions of the rectangle. **x = 9; 8 inches by 13 inches**

$A = 104 \text{ in}^2$ $x - 1$
$x + 4$

Simplify.

16. $(3 - 4i) - (9 - 5i)$ **−6 + i**

17. $\dfrac{4i}{4 - i}$ $-\dfrac{4}{17} + \dfrac{16}{17}i$

18. MULTIPLE CHOICE Which value of *c* makes the trinomial $x^2 - 12x + c$ a perfect square trinomial? **H**

F 6

G 12

H 36

J 144

Complete parts *a–c* for each quadratic equation.

a. Find the value of the discriminant.

b. Describe the number and type of roots.

c. Find the exact solution by using the Quadratic Formula. **19–21. See margin.**

19. $6x^2 + 7x = 0$

20. $5x^2 = -6x + 1$

21. $2x^2 + 5x - 8 = -13$

Write each quadratic function in vertex form. Then identify the vertex, axis of symmetry, and direction of opening. **22–24. See Chapter 4 Answer Appendix.**

22. $3x^2 + 6x = 2 + y$

23. $x^2 + 9x + \dfrac{81}{4} = y$

24. Graph the quadratic inequality $0 < -3x^2 + 4x + 10$.

Solve each inequality by using a graph or algebraically.

25. $x^2 + 6x > -5$ $\{x \mid x < -5 \text{ or } x > -1\}$

26. $4x^2 - 19x \le -12$ $\left\{x \mid \dfrac{3}{4} \le x \le 4\right\}$

 connectED.mcgraw-hill.com **295**

Summative Assessment

Use these alternate leveled chapter tests to differentiate assessment for your students.

Leveled **Chapter 4 Tests**

Form	Type	Level	🗀 Page(s)
1	MC	AL	63–64
2A	MC	OL	65–66
2B	MC	OL	67–68
2C	FR	OL	69–70
2D	FR	OL	71–72
3	FR	BL	73–74
Vocabulary Test			62
Extended-Response Test			75

MC = multiple-choice questions
FR = free-response questions

eAssessment Customize and create multiple versions of your chapter test and their answer keys. All of the questions from the leveled chapter tests in the *Chapter 4 Resource Masters* are also available on McGraw-Hill eAssessment.

19a. 49

19b. 2 rational roots

19c. $-\dfrac{7}{6}, 0$

20a. 56

20b. 2 irrational roots

20c. $\dfrac{-3 \pm \sqrt{14}}{5}$

21a. −15

21b. 2 complex roots

21c. $\dfrac{-5 \pm i\sqrt{15}}{4}$

InterventionPlanner

TIER 1 On Level **OL**	**TIER 2** Strategic Intervention **AL** approaching grade level	**TIER 3** Intensive Intervention 2 or more grades below level
If students miss about 25% of the exercises or less,	**If** students miss about 50% of the exercises,	**If** students miss about 75% of the exercises,
Then choose a resource:	**Then** choose a resource:	**Then** use *Math Triumphs, Alg. 2* Chs. 1, 3, and 4
SE Lessons 4-1, 4-2, 4-3, 4-4, 4-5, 4-6, and 4-8	🗀 Study Guide and Intervention, pp. 5, 11, 17, 24, 30, 36, and 49	Extra Examples, Personal Tutor, Homework Help, Review Vocabulary
🗀 Skills Practice, pp. 7, 13, 19, 26, 32, 38, and 51	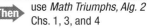 Extra Examples, Personal Tutor, Homework Help	
TE Chapter Project		
Self-Check Quiz		

1 Focus

Objective Use the strategy of using a graph to solve standardized test problems.

2 Teach

Scaffolding Questions

Ask:

- What are some parts of a graph?
  Sample answers: title, vertical and horizontal axes, scale

- What type of equation does the graph of a parabola represent? quadratic equation

- What do you know about each and every point on the graph of a parabola? Each point is a solution to the equation or function represented by the graph.

Use a Graph

Using a graph can help you solve many different kinds of problems on standardized tests. Graphs can help you solve equations, evaluate functions, and interpret solutions to real-world problems.

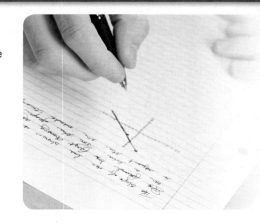

Strategies for Using a Graph

Step 1

Read the problem statement carefully.

Ask yourself:

- What am I being asked to solve?
- What information is given in the problem?
- How could a graph help me solve the problem?

Step 2

Create your graph.

- Sketch your graph on scrap paper if appropriate.
- If allowed, you can also use a graphing calculator to create the graph.

Step 3

Solve the problem.

- Use your graph to help you model and solve the problem.
- Check to be sure your answer makes sense.

Standardized Test Example

Read the problem. Identify what you need to know. Then use the information in the problem to solve.

The students in Mr. Himebaugh's physics class built a model rocket. The rocket is launched in a large field with an initial upward velocity of 128 feet per second. The function $h(t) = -16t^2 + 128t$ models the height of the rocket above the ground (in feet) t seconds after it is launched. How long will it take for the rocket to reach its maximum height?

A 4 seconds   **C** 6 seconds

B 5 seconds   **D** 8 seconds

JUPITERIMAGES/Polka Dot/Alamy

Graphing the quadratic function will allow you to determine the peak height of the rocket and when it occurs. A graphing calculator can help you quickly graph the function and analyze it.

KEYSTROKES: Y= (−) 16 X,T,θ,n x² + 128 X,T,θ,n GRAPH

After graphing the equation, use **maximum** under the **CALC** menu.

Press 2nd [CALC] 4. Then use ◄ to place the cursor to the left of the maximum point and press ENTER. Use ► to place the cursor to the right of the maximum point and press ENTER ENTER.

[0, 10] scl: 1 by [0, 300] scl: 50

The graph shows that the rocket takes 4 seconds to reach its maximum height of 256 feet. The correct answer is A.

Additional Example

STANDARDIZED TEST PRACTICE
Belinda throws a tennis ball upward at a velocity of 65 feet per second, releasing the ball when it is 4 feet above the ground. The height of the ball t seconds after being thrown is given by the formula
$h(t) = -16t^2 + 65t + 4$.
Find the time, to the nearest second, at which the tennis ball reaches its maximum height. A

A 2 seconds

B 4 seconds

C 70 seconds

D 140 seconds

3 Assess

Use Exercises 1–6 to assess students' understanding.

Exercises

Read each problem. Identify what you need to know. Then use the information in the problem to solve.

1. What are the roots of $y = 2x^2 + 10x - 48$? C

 A −5, 4

 B −6, 1

 C −8, 3

 D 2, 3

2. How many times does the graph of $f(x) = 2x^2 - 3x + 2$ cross the x-axis? F

 F 0   **H** 2

 G 1   **J** 3

3. Which statement best describes the graphs of the two equations? C

$$16x - 2y = 24$$
$$12x = 3y - 36$$

 A The lines are parallel.

 B The lines are the same.

 C The lines intersect in only one point.

 D The lines intersect in more than one point, but are not the same.

4. Adrian is using 120 feet of fencing to enclose a rectangular area for her puppy. One side of the enclosure will be her house.

(120 − 2x) ft

x ft   x ft

House

The function $f(x) = x(120 - 2x)$ represents the area of the enclosure. What is the greatest area Adrian can enclose with the fencing? G

 F 1650 ft²   **H** 1980 ft²

 G 1800 ft²   **J** 2140 ft²

5. For which equation is the x-coordinate of the vertex at 4? A

 A $f(x) = x^2 - 8x + 15$ **C** $f(x) = x^2 + 6x + 8$

 B $f(x) = -x^2 - 4x + 12$ **D** $f(x) = -x^2 - 2x + 2$

6. For what value of x does $f(x) = x^2 + 5x + 6$ reach its minimum value? H

 F −5   **H** $-\dfrac{5}{2}$

 G −3   **J** −2

Diagnose Student Errors

Survey student responses for each item. Class trends may indicate common errors and misconceptions.

1. A wrote an incorrect quadratic equation
B wrote an incorrect quadratic equation
C correct
D wrote an incorrect quadratic equation

2. F read the *x*-intercept values incorrectly
G read the *x*-intercept values incorrectly
H read the *x*-intercept values incorrectly
J correct

3. A correct
B made $10.50 the constant term
C did not represent a variable for the number of hours working
D guess

4. F guess
G guess
H correct
J combined like terms incorrectly

5. A guess
B correct
C factored incorrectly
D incorrectly used $8^2 = 16$ instead of $4^2 = 16$

6. F added 5 instead of subtracting 5
G correct
H subtracted 5 from the *x* term instead of the constant term
J added 5 to the *x* term instead of subtracting it from the constant term

7. A found a value that is a vertex
B correct
C found a value that is a vertex
D found a value that is a vertex

8. F guess
G guess
H misread the value of *x* on the graph when $P(t) = 200$
J correct

Standardized Test Practice
Cumulative, Chapters 1 through 4

Multiple Choice

Read each question. Then fill in the correct answer on the answer document provided by your teacher or on a sheet of paper.

1. A rental store charges $32 per day to rent a bicycle. At this rate, the store rents about 200 bikes per month. The owner of the store estimates that they will rent 5 fewer bikes per month for each $2 increase in the rental price. What price will maximize the income of the store? **C**

A $46 C $55
B $51 D $56

2. What are the solutions of the quadratic equation graphed below? **J**

F −4, −1 H −1, 4
G 1, 4 J −4, 1

3. Peyton works as a nanny. She charges at least $10 to drive to a home and $10.50 per hour. Which inequality best represents the relationship between the number of hours working *n* and the total charge *c*? **A**

A $c \geq 10 + 10.50n$
B $c \geq 10.50 + 10n$
C $c \leq 10.50 + 10$
D $c \leq 10n + 10.50n$

> **Test-Taking Tip**
> Question 1 Multiply the expressions for the new price and new number of customers after *x* price increases to write a quadratic equation.

4. Leo sells T-shirts at a local swim meet. It costs him $250 to set up the stand and rent the machine. It costs him an additional $5 to make each T-shirt. If he sells each T-shirt for $15, how many T-shirts does he have to sell before he can make a profit? **H**

F 10 H 25
G 15 J 50

5. Solve $x^2 - 2x = 15$ by completing the square. **B**

A −4, −1 C −2, 3
B −3, 5 D 5, 7

6. The graph of $g(x) = \frac{2}{5}x^2 - 4x + 2$ is translated down 5 units to produce the graph of the function $h(x)$. Which of the following could be $h(x)$? **G**

F $h(x) = \frac{2}{5}x^2 - 4x + 7$

G $h(x) = \frac{2}{5}x^2 - 4x - 3$

H $h(x) = \frac{2}{5}x^2 - 9x + 2$

J $h(x) = \frac{2}{5}x^2 + x + 2$

7. Which of the following points is *not* a vertex of the feasible region for the system of linear inequalities below? **B**

$$x \geq 0, y \geq 0$$
$$y \leq -2x + 6$$

A (0, 0) B (0, 3) C (0, 6) D (3, 0)

8. The function $P(t) = -0.068t^2 + 7.85t + 56$ can be used to approximate the population, in thousands, of Clarksville between 1960 and 2000. The domain *t* of the function is the number of years since 1960. According to the model, in what year did the population of Clarksville reach 200,000 people? **J**

F 1974 H 1981
G 1977 J 1983

Short Response/Gridded Response

Record your answers on the answer sheet provided by your teacher or on a sheet of paper.

9. **GRIDDED RESPONSE** What is the y-coordinate of the solution of the system of equations below? **1**

$$y = 4x - 7$$
$$y = -\frac{1}{2}x + 2$$

10. **GRIDDED RESPONSE** Simplify $-2i \cdot 5i$. **10**

11. Use the quadratic function below to answer each question.

a. How many real roots does the quadratic function have? **0**

b. How many complex roots does the function have? **2**

c. What do you know about the discriminant of the quadratic equation? Explain.
Sample answer: The discriminant is negative because the roots are complex.

12. If one of the roots of $x^2 + kx - 12 = 0$ is 4, what is the value of k? **−1**

13. Describe the translation of the graph of $y = (x + 5)^2 - 1$ to the graph of $y = (x - 1)^2 + 3$.
Sample answer: It is a translation 6 units right and 4 units up.

Extended Response

Record your answers on a sheet of paper. Show your work.

14. For a given quadratic equation $y = ax^2 + bx + c$, describe what the discriminant $b^2 - 4ac$ tells you about the roots of the equation. **See margin.**

15. Craig is checking in a shipment of ink jet printers that cost $200 each and laser printers that cost $500 each. There are 40 boxes in the shipment, and the invoice total is $11,600. **b. See margin.**

a. Write a system of equations to model the situation. Let x represent the number of ink jet printers, and let y represent the number of laser printers. $x + y = 40$; $200x + 500y = 11{,}600$

b. Write a matrix equation that can be used to solve the system of equations you wrote in part **a**.

c. Find the inverse of the coefficient matrix and solve the matrix equation. How many ink jet printers and laser printers were included in the shipment?
28 ink jet printers and 12 laser printers

16. If Robert kicks a football straight up into the air with an initial velocity of 100 feet per second, the function $h(t) = -16t^2 + 100t$ gives the height, in feet, of the ball after t seconds.

a. Graph the quadratic function on a coordinate grid. **See margin.**

b. What is the maximum height reached by the football? Round to the nearest foot. **156 ft**

c. How long is the football in the air before it hits the ground? **6.25 seconds**

Need Extra Help?

If you missed Question...	1	2	3	4	5	6	7	8	9	10	11	12	13	14	15	16
Go to Lesson...	4-1	4-2	2-4	1-3	4-5	4-7	3-3	4-6	3-1	4-4	4-2	4-2	4-7	4-6	3-8	4-1

Additional Answers

14. Sample answer: If negative, there are 2 complex roots; if 0, there is 1 real, rational root; if positive and a perfect square, there are 2 real, rational roots; if positive but not a perfect square, there are 2 real, irrational roots.

15b. $\begin{bmatrix} 1 & 1 \\ 200 & 500 \end{bmatrix} \cdot \begin{bmatrix} x \\ y \end{bmatrix} = \begin{bmatrix} 40 \\ 11{,}600 \end{bmatrix}$

16a.
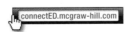

Formative Assessment

You can use these pages to benchmark student progress.

📁 Standardized Test Practice, pp. 76–78

McGraw-Hill
eAssessment Create practice tests that align to your state standards, the Common Core State Standards, and other national standards such as TIMSS and NAEP.

Answer Sheet Practice

Have students simulate taking a standardized test by recording their answers on practice recording sheets.

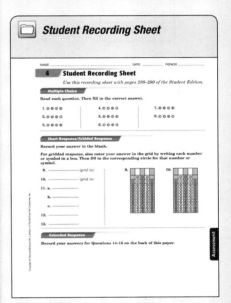

Homework Option

Get Ready for Chapter 5 Assign students the exercises on p. 301 as homework to assess whether they possess the prerequisite skills needed for the next chapter.

Lesson 4-1

1a. *y*-int = 0; axis of symmetry: *x* = 0; *x*-coordinate = 0

1b.

x	f(x)
−2	12
−1	3
0	0
1	3
2	12

1c.
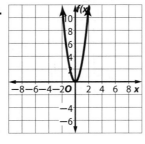

2a. *y*-int = 0; axis of symmetry: *x* = 0; *x*-coordinate = 0

2b.

x	f(x)
−2	−24
−1	−6
0	0
1	−6
2	−24

2c.
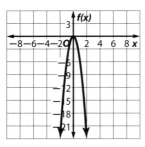

3a. *y*-int = 0; axis of symmetry: *x* = 2; *x*-coordinate = 2

3b.

x	f(x)
0	0
1	−3
2	−4
3	−3
4	0

3c.
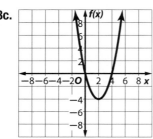

4a. *y*-int = 4; axis of symmetry: *x* = −1.5; *x*-coordinate = −1.5

4b.

x	f(x)
−3	4
−2	6
−1.5	6.25
−1	6
0	4

4c.
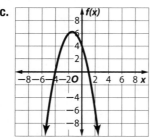

5a. *y*-int = −3; axis of symmetry: *x* = 0.75; *x*-coordinate = 0.75

5b.

x	f(x)
−1	7
0	−3
0.75	−5.25
1.5	−3
2.5	7

5c.

6a. *y*-int = 5; axis of symmetry: *x* = 2; *x*-coordinate = 2

6b.

x	f(x)
0	5
1	−1
2	−3
3	−1
4	5

6c.

12a. *y*-int = 0; axis of symmetry: *x* = 0; *x*-coordinate = 0

12b.

x	f(x)
−2	16
−1	4
0	0
1	4
2	16

12c.
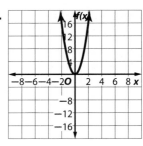

13a. *y*-int = 0; axis of symmetry: *x* = 0; *x*-coordinate = 0

13b.

x	f(x)
−2	−8
−1	−2
0	0
1	−2
2	−8

13c.
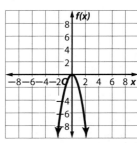

14a. *y*-int = −5; axis of symmetry: *x* = 0; *x*-coordinate = 0

14b.

x	f(x)
−2	−1
−1	−4
0	−5
1	−4
2	−1

14c.

15a. *y*-int = 3; axis of symmetry: *x* = 0; *x*-coordinate = 0

15b.

x	f(x)
−2	7
−1	4
0	3
1	4
2	7

15c.
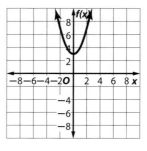

16a. y-int $= -3$; axis of symmetry: $x = 0$; x-coordinate $= 0$

16b.

x	$f(x)$
-2	13
-1	1
0	-3
1	1
2	13

16c.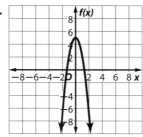

17a. y-int $= 5$; axis of symmetry: $x = 0$; x-coordinate $= 0$

17b.

x	$f(x)$
-2	-7
-1	2
0	5
1	2
2	-7

17c.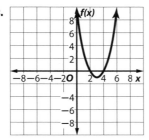

18a. y-int $= 8$; axis of symmetry: $x = 3$; x-coordinate $= 3$

18b.

x	$f(x)$
1	3
2	0
3	-1
4	0
5	3

18c.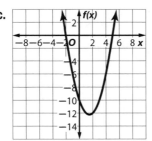

19a. y-int $= -10$; axis of symmetry: $x = 1.5$; x-coordinate $=1.5$

19b.

x	$f(x)$
0	-10
1	-12
1.5	-12.25
2	-12
3	-10

19c.

20a. y-int $= -6$; axis of symmetry: $x = 2$; x-coordinate $= 2$

20b.

x	$f(x)$
0	-6
1	-3
2	-2
3	-3
4	-6

20c.

21a. y-int $= 9$; axis of symmetry: $x = 0.75$; x-coordinate $= 0.75$

21b.

x	$f(x)$
-1	4
0	9
0.75	10.125
1.5	9
2.5	4

21c.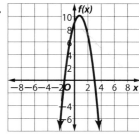

33a. y-int $= -9$; axis of symmetry: $x = 1.5$; x-coordinate of vertex $= 1.5$

33b.

x	$f(x)$
0	-9
1	-13
1.5	-13.5
2	-13
3	-9

33c.

34a. y-int $= 2$; axis of symmetry: $x = -1.5$; x-coordinate of vertex $= -1.5$

34b.

x	$f(x)$
-3	2
-2	8
-1.5	8.75
-1	8
0	2

34c.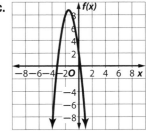

35a. y-int $= 0$; axis of symmetry: $x = \dfrac{5}{8}$; x-coordinate of vertex $= \dfrac{5}{8}$

35b.

x	$f(x)$
$-\dfrac{3}{4}$	-6
$\dfrac{1}{4}$	1
$\dfrac{5}{8}$	1.5625
1	1
2	-6

35c.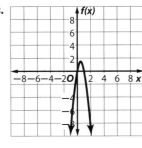

36a. y-int $= 0$; axis of symmetry: $x = -2.75$; x-coordinate of vertex $= -2.75$

36b.

x	$f(x)$
-4	-12
-3	-15
-2.75	-15.125
-2.5	-15
-1.5	-12

36c.

37a. y-int = 4; axis of symmetry: $x = -6$; x-coordinate of vertex = -6

37b.

x	$f(x)$
-10	-1
-8	-4
-6	-5
-4	-4
-2	-1

37c.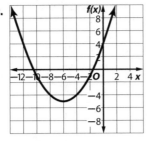

38a. y-int = 6; axis of symmetry: $x = \frac{8}{3}$; x-coordinate of vertex = $\frac{8}{3}$

38b.

x	$f(x)$
$\frac{4}{3}$	10
$\frac{7}{3}$	11.25
$\frac{8}{3}$	$11\frac{1}{3}$
3	11.25
4	10

38c.

39a. y-int = -2.5; axis of symmetry: $x = -\frac{4}{3}$; x-coordinate of vertex = $-\frac{4}{3}$

39b.

x	$f(x)$
$-\frac{11}{3}$	3
$-\frac{8}{3}$	-2.5
$-\frac{4}{3}$	$-5\frac{1}{6}$
0	-2.5
1	3

39c.

40a. y-int = 9; axis of symmetry: $x = 1.75$; x-coordinate of vertex = 1.75

40b.

x	$f(x)$
0.5	8
1.5	7
1.75	$6\frac{23}{24}$
2	7
3	8

40c.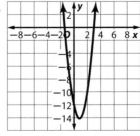

Extend 4-1

1. Linear regression graph:

[0, 1] scl: 0.1 by [0, 20] scl: 2
linear equation:
$h(t) = 19.2t + 1.92$

Quadratic regression graph:

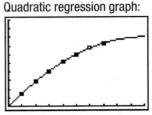

[0, 1] scl: 0.1 by [0, 20] scl: 2
quadratic equation:
$h(t) = -16t^2 + 32t$

The quadratic equation fits the data better.

2. Linear equation predictions: At 1 second, the height of the player will be 21.12 feet; at 1.5 seconds, the height of the player will be 30.72 feet. Quadratic equation predictions: At 1 second, the height of the player will be 16 feet; at 1.5 seconds, the height of the player will be 12 feet.

3. The linear equation predicts that the basketball player will continue to rise indefinitely, and the quadratic equation shows that the player will reach the highest point of his jump and then return to the ground.

Lesson 4-2 (Guided Practice)

2A.

2B.

Lesson 4-2

4.

5.

6.

7.

8.

9.

26.

27.

10.

11.

28.

29.

20.

21.

41.

42.

22.

23.

43.

44.

24.

45.

46.

25.

47.

48.

Extend 4-4

1.

2.

3.

7.

8.

9.

Mid-Chapter Quiz

1.

Lesson 4-5

35. $\left\{\dfrac{3 - i\sqrt{31}}{4}, \dfrac{3 + i\sqrt{31}}{4}\right\}$

36. $\left\{\dfrac{-5 - i\sqrt{31}}{4}, \dfrac{-5 + i\sqrt{31}}{4}\right\}$

37. $\{-1, 3\}$

38. $\left\{1 - i\sqrt{2}, 1 + i\sqrt{2}\right\}$

39. $\left\{-2 - i\sqrt{7}, -2 + i\sqrt{7}\right\}$

40. $\{3 - 3i, 3 + 3i\}$

41. $\{5 - 2i, 5 + 2i\}$

42. $\{-0.39, 1.72\}$

43. $\left\{\dfrac{7 - i\sqrt{47}}{4}, \dfrac{7 + i\sqrt{47}}{4}\right\}$

44. $\{-0.71, 3.11\}$

45. $\left\{2.65 - i\sqrt{1.5775}, 2.65 + i\sqrt{1.5775}\right\}$

60a. 2; rational; 16 is a perfect square so $x + 2$ and x are rational.

60b. 2; rational; 16 is a perfect square so $x - 2$ and x are rational.

60c. 2; complex; If the opposite of square is positive, the square is negative. The square root of a negative number is complex.

60d. 2; real; The square must equal 20. Since that is positive but not a perfect square, the solutions will be real but not rational.

60e. 1; rational; The expression must be equal to 0 and only -2 makes the expression equal to 0.

60f. 1; rational; The expressions $(x + 4)$ and $(x + 6)$ must either be equal or opposites. No value makes them equal, -5 makes them opposites. The only solution is -5.

Lesson 4-6

46a. 1 rational root

46b. 2 complex roots

46c. 2 real roots

46d. 2 complex roots

46e. 2 rational roots

Chapter 4 Answer Appendix

48. Sample answer: (1) Factor $x^2 - 2x - 15$ as $(x + 3)(x - 5)$. Then according to the Zero Product Property, either $x + 3 = 0$ or $x - 5 = 0$. Solving these equations, $x = -3$ or $x = 5$. (2) Rewrite the equation as $x^2 - 2x = 15$. Then add 1 to each side of the equation to complete the square on the left side. Then $(x - 1)^2 = 16$. Taking the square root of each side, $x - 1 = \pm 4$. Therefore, $x = 1 \pm 4$ and $x = -3$ or $x = 5$.

(3) Use the Quadratic Formula. Thus, $x = \dfrac{2 \pm \sqrt{2^2 - 4(1)(-15)}}{2(1)}$ or $x = \dfrac{2 \pm \sqrt{64}}{2}$. Simplifying the expression, $x = -3$ or $x = 5$. See students' preferences.

Explore 4-7

4.

[−10, 10] scl: 1 by [−10, 10] scl: 1

Both graphs have the same shape, but the graph of $y = x^2 + 3.5$ is 3.5 units above the graph of $y = x^2$.

5.

[−10, 10] scl: 1 by [−10, 10] scl: 1

Both graphs have the same shape, but the graph of $y = -x^2$ opens downward while the graph of $y = x^2 - 7$ opens upward and is 7 units lower than the graph of $y = x^2$.

6.

[−10, 10] scl: 1 by [−10, 10] scl: 1

The graph of $y = 4x^2$ is narrower than the graph of $y = x^2$.

7.

[−10, 10] scl: 1 by [−10, 10] scl: 1

The graph of $y = -8x^2$ opens downwards and is narrower than the graph of $y = x^2$.

8.

[−10, 10] scl: 1 by [−10, 10] scl: 1

The graphs have the same shape and open upward, but the graph of $y = (x + 2)^2$ is two units to the left of the graph of $y = x^2$.

9.
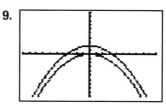
[−10, 10] scl: 1 by [−10, 10] scl: 1

The graphs have the same shape and open downward, but the graph of $y = -\frac{1}{6}x^2 + 2$ is two units above the graph of $y = -\frac{1}{6}x^2$.

10.

[−10, 10] scl: 1 by [−10, 10] scl: 1

The graphs have the same shape, but the graph of $y = (x - 5)^2$ is 5 units to the right of the graph of $y = x^2$.

11.

[−10, 10] scl: 1 by [−10, 10] scl: 1

The graph of $y = 2(x + 3)^2 - 6$ is 3 units to the left, 6 units below, and narrower than the graph of $y = x^2$.

12.

[−10, 10] scl: 1 by [−10, 10] scl: 1

The graph of $y = -\frac{1}{8}x^2 + 1$ opens downward, is wider than and 1 unit above the graph of $y = x^2$.

13.
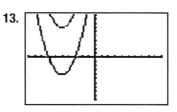
[−10, 10] scl: 1 by [−10, 10] scl: 1

The graphs have the same shape, but the graph of $y = (x + 5)^2 + 7$ is 11 units above the graph of $y = (x + 5)^2 - 4$.

14.

[−10, 10] scl: 1 by [−10, 10] scl: 1

The graph of $y = 5(x + 3)^2 - 1$ is narrower than the graph of $y = 2(x + 1)^2 - 4$.

15.

[−10, 10] scl: 1 by [−10, 10] scl: 1

The graph of $y = \frac{1}{4}(x - 5)^2 - 6$ is wider than the graph of $y = 5(x - 2)^2 - 3$, and its vertex is 3 units below and 3 units right of the vertex of $y = 5(x - 2)^2 - 3$.

Lesson 4-7 (Guided Practice)

3A.

3B.

Lesson 4-7

21.

22.

23.

24.

25.

26.

27.

28.

29.

30.

31.

32.

33.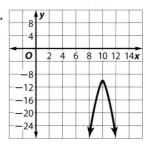

34a. $f(x) = \dfrac{1}{250}(x + 75)^2 - \dfrac{45}{2}$

41. $y = 3\left(x - \dfrac{2}{3}\right)^2 - \dfrac{10}{3}; \left(\dfrac{2}{3}, -\dfrac{10}{3}\right), x = \dfrac{2}{3}$, opens up

42. $y = -2\left(x - \dfrac{7}{4}\right)^2 + \dfrac{145}{8}; \left(\dfrac{7}{4}, \dfrac{145}{8}\right), x = \dfrac{7}{4}$, opens down

43. $y = -(x + 2.35)^2 + 8.3225; (-2.35, 8.3225), x = -2.35$, opens down

44. $y = (x + 0.7)^2 - 1.69; (-0.7, -1.69), x = -0.7$, opens up

45. $y = \left(x - \dfrac{1}{3}\right)^2 - 3; \left(\dfrac{1}{3}, -3\right), x = \dfrac{1}{3}$, opens up

46. $y = (x + 3.5)^2; (-3.5, 0), x = -3.5$, opens up

49. The equation of a parabola can be written in the form $y = ax^2 + bx + c$ with $a \neq 0$. For each of the three points, substitute the value of the *x*-coordinate for *x* in the equation and substitute the value of the *y*-coordinate for *y* in the equation. This will produce three equations in three variables *a, b,* and *c.* Solve the system of equations to find the values of *a, b,* and *c.* These values determine the quadratic equation.

50. $a\left(x + \dfrac{b}{2a}\right)^2 + \left(c - \dfrac{b^2}{4a}\right) = y; \left(\dfrac{-b}{2a}, c - \dfrac{b^2}{4a}\right); x = \dfrac{-b}{2a}$

51. Sample answer: The variable a represents different values for these functions, so making $a = 0$ will have a different effect on each function. For $f(x)$, when $a = 0$, the graph will be a horizontal line, $f(x) = k$. For $g(x)$, when $a = 0$, the graph will be linear, but not necessarily horizontal, $g(x) = bx + c$.

52. All quadratic functions are transformations of the parent graph $y = x^2$. By identifying these transformations when a quadratic function is written in vertex form, you can redraw the graph of $y = x^2$ with its vertex translated to (h, k), widened or narrowed as determined by a, opening downward if a is negative.

Extend 4-7

Activity Steps 1 and 2

x	0	1	2	3	4	5
y	12	9.1875	6.75	4.6875	3	1.6875
First-Order Differences						
Second-Order Differences						

x	6	7	8	9	10	11
y	0.75	0.1875	0	0.1875	0.75	1.6875
First-Order Differences						
Second-Order Differences						

x	12	13	14	15	16
y	3	4.6875	6.75	9.1875	12
First-Order Differences					
Second-Order Differences					

Step 3

Step 4

x	0	1	2	3	4	5
y	12	9.1875	6.75	4.6875	3	1.6875
First-Order Differences		−2.8125	−2.4375	−2.0625	−1.6875	−1.3125
Second-Order Differences						

x	6	7	8	9	10	11
y	0.75	0.1875	0	0.1875	0.75	1.6875
First-Order Differences	−0.9375	−0.5625	−0.1875	0.1875	0.5625	0.9375
Second-Order Differences						

x	12	13	14	15	16
y	3	4.6875	6.75	9.1875	12
First-Order Differences	1.3125	1.6875	2.0625	2.4375	2.8125
Second-Order Differences					

The differences decrease until $x = 8$, and then the differences increase.

Step 5

x	0	1	2	3	4	5
y	12	9.1875	6.75	4.6875	3	1.6875
First-Order Differences		−2.8125	−2.4375	−2.0625	−1.6875	−1.3125
Second-Order Differences			0.375	0.375	0.375	0.375

x	6	7	8	9	10	11
y	0.75	0.1875	0	0.1875	0.75	1.6875
First-Order Differences	−0.9375	−0.5625	−0.1875	0.1875	0.5625	0.9375
Second-Order Differences	0.375	0.375	0.375	0.375	0.375	0.375

x	12	13	14	15	16
y	3	4.6875	6.75	9.1875	12
First-Order Differences	1.3125	1.6875	2.0625	2.4375	2.8125
Second-Order Differences	0.375	0.375	0.375	0.375	

The second-order differences are all 0.375.

1.

x	−3	−2	−1	0	1	2	3
y	−16	−9	−4	−1	0	−1	−4
First-Order Differences		7	5	3	1	−1	−3
Second-Order Differences			−2	−2	−2	−2	−2

2.

x	−5	−4	−3	−2	−1	0	
y	0.5	−2	−3.5	−4	−3.5	−2	0.5
First-Order Differences		−2.5	−1.5	−0.5	0.5	1.5	2.5
Second-Order Differences			1	1	1	1	1

3.

x	−6	−5	−4	−3	−2	−1	0
y	−26	−11	−2	1	−2	−11	−26
First-Order Differences		15	9	3	−3	−9	−15
Second-Order Differences			−6	−6	−6	−6	−6

Lesson 4-8

33. $\{x \mid 4 < x < 5\}$

34. $\{x \mid x \leq -5 \text{ or } x \geq -2\}$

35. $\{x \mid -1 < x < 2\}$

36. $\{x \mid -4.65 \leq x \leq 0.65\}$

37. $\{x \mid x \leq -2.32 \text{ or } x \geq 4.32\}$

38. $\varnothing$

39. $\{x \mid x \leq -1.58 \text{ or } x \geq 1.58\}$

40. {all real numbers}

41. {all real numbers}

42. $\{x \mid x \leq -3.91 \text{ or } x \geq 1.41\}$

43. $\{x \mid -2.84 < x < 0.84\}$

44. $\{x \mid -3.43 < x < 0.10\}$

45a.

56a.

59. No; the graphs of the inequalities intersect the x-axis at the same points.

60. Sample answer: Sometimes; when a is positive and c is negative, there is no solution and when a is negative and c is positive there is a solution set.

61.

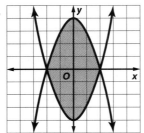

62. For both quadratic and linear inequalities, you must first graph the related equation. You use the inequality symbol to determine if the line is dashed or solid. Then you use test points to determine where to shade. One difference is that one related equation is a straight line while the other related equation is a curve.

76b.

Study Guide and Review

64.

$y \geq x^2 + 5x + 4$

65.

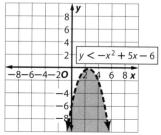

$y < -x^2 + 5x - 6$

66.

$y > x^2 - 6x + 8$

67.

$y \leq x^2 + 10x - 4$

Practice Test

1a. *y*-intercept : -7; axis of symmetry: $x = -2$;
x-coordinate of vertex: -2

1b.

x	f(x)
-4	-7
-3	-10
-2	-11
-1	-10
0	-7

1c.

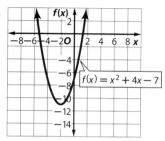

$f(x) = x^2 + 4x - 7$

2a. *y*-intercept: 0; axis of symmetry: $x = \dfrac{5}{4}$; *x*-coordinate of vertex: $\dfrac{5}{4}$

2b.

x	f(x)
0	0
1	3
$\dfrac{5}{4}$	$\dfrac{25}{8}$
2	2
3	-3

2c.

$f(x) = -2x^2 + 5x$

3a. *y*-intercept : -9; axis of symmetry: $x = -3$; *x*-coordinate of vertex: -3

3b.

x	f(x)
-5	-4
-4	-1
-3	0
-2	-1
-1	-4

3c.

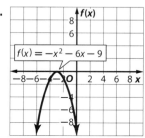

$f(x) = -x^2 - 6x - 9$

22. $y = 3(x + 1)^2 - 5$; vertex: $(-1, -5)$; axis of symmetry: $x = -1$; opens up

23. $y = \left(x + \dfrac{9}{2}\right)^2$; vertex: $\left(-\dfrac{9}{2}, 0\right)$; axis of symmetry: $x = -\dfrac{9}{2}$; opens up

24.

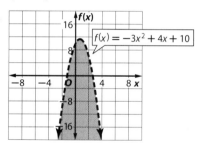

$f(x) = -3x^2 + 4x + 10$

	Diagnostic Assessment Quick Check			
	LESSON 5-1 45 min: 1 day 90 min: 0.5 day	**EXTEND 5-1** 45 min: 0.5 day 90 min: 0.25 day	**LESSON 5-2** 45 min: 1 day 90 min: 0.5 day	**EXTEND 5-2** 45 min: 0.5 day 90 min: 0.25 day
Title	**Operations with Polynomials**	**Algebra Lab: Dimensional Analysis**	**Dividing Polynomials**	**Graphing Technology Lab: Dividing Polynomials**
Objectives	▪ Multiply, divide, and simplify monomials and expressions involving powers. ▪ Add, subtract, and multiply polynomials.	▪ Use dimensional analysis to convert units and solve problems.	▪ Divide polynomials using long division. ▪ Divide polynomials using synthetic division.	▪ Use a graphing calculator with CAS to divide polynomials.
Key Vocabulary	simplify degree of a polynomial		synthetic division	
CCSS	A.APR.1		A.APR.6	A.APR.6
Multiple Representations	🔧		🔧	
Lesson Resources	[connectED.mcgraw-hill.com] 📁 Leveled Worksheets 🔤 Vocabulary 🏃 Animations [PT] Personal Tutor ✓ Self-Check Quiz ▪ *5-Minute Check* ▪ *Study Notebook*	▪ *Teaching Algebra with Manipulatives*	[connectED.mcgraw-hill.com] 📁 Leveled Worksheets 📁 Quiz 1 🔤 Vocabulary [PT] Personal Tutor ✓ Self-Check Quiz ▪ *5-Minute Check* ▪ *Study Notebook*	[connectED.mcgraw-hill.com] [PT] Personal Tutor **Materials:** ▪ TI-Nspire CAS technology
Resources for Every Lesson	[IWB] eStudent Edition [IWB] Interactive Classroom	▪ eTeacher Edition ▪ eSolutions Manual ▪ e*Assessment* McGraw-Hill		
Differentiated Instruction	pp. 304, 309		pp. 314, 317	

[IWB] All digital assets are Interactive Whiteboard ready.

Suggested Pacing			
Time Periods	Instruction	Review & Assess	Total
45-minute	12 days	2 days	14 days
90-minute	6 days	1 day	7 days

EXPLORE 5-3 45 min: 1 day / 90 min: 0.5 day	LESSON 5-3 45 min: 1 day / 90 min: 0.5 day	LESSON 5-4 45 min: 1 day / 90 min: 0.5 day	EXTEND 5-4 45 min: 0.5 day / 90 min: 0.25 day	EXPLORE 5-5 45 min: 0.5 day / 90 min: 0.25 day
Graphing Technology Lab: Power Functions	**Polynomial Functions**	**Analyzing Graphs of Polynomial Functions**	**Graphing Technology Lab: Modeling Data Using Polynomial Functions**	**Graphing Technology Lab: Solving Polynomial Equations by Graphing**
■ Use a graphing calculator to explore power functions.	■ Evaluate polynomial functions. ■ Identify general shapes of graphs of polynomial functions.	■ Graph polynomial functions and locate their zeros. ■ Find the relative maxima and minima of polynomial functions.	■ Use a graphing calculator to model data whose curve of best fit is a polynomial function.	■ Use a graphing calculator to find approximate solutions for polynomial equations.
	polynomial in one variable leading coefficient polynomial function power function end behavior quartic function quintic function	relative maximum relative minimum extrema turning points		
F.IF.4	F.IF.4, F.IF.7c	F.IF.4, F.IF.7c	F.IF.4, F.IF.7c	
Materials: ■ TI-83/84 Plus or other graphing calculator	connectED.mcgraw-hill.com 📁 Leveled Worksheets abc Vocabulary 🏃 Animations PT Personal Tutor ✋ Virtual Manipulatives ✓ Self-Check Quiz ■ *5-Minute Check* ■ *Study Notebook*	connectED.mcgraw-hill.com 📁 Leveled Worksheets 📁 Quiz 2 abc Vocabulary PT Personal Tutor ✋ Virtual Manipulatives ✓ Self-Check Quiz ■ *5-Minute Check* ■ *Study Notebook*	connectED.mcgraw-hill.com PT Personal Tutor **Materials:** ■ TI-83/84 Plus or other graphing calculator	connectED.mcgraw-hill.com PT Personal Tutor **Materials:** ■ TI-83/84 Plus or other graphing calculator
IWB eStudent Edition IWB Interactive Classroom	■ eTeacher Edition ■ eSolutions Manual ■ eAssessment McGraw-Hill			
	pp. 324, 329	pp. 333, 337	pp. 338	
			Formative Assessment Mid-Chapter Quiz	

	LESSON 5-5 — 45 min: 1 day / 90 min: 0.5 day	EXTEND 5-5 — 45 min: 0.5 day / 90 min: 0.25 day	LESSON 5-6 — 45 min: 1 day / 90 min: 0.5 day	
Title	**Solving Polynomial Equations**	**Graphing Technology Lab: Polynomial Identities**	**The Remainder and Factor Theorems**	
Objectives	■ Factor polynomials. ■ Solve polynomial equations by factoring.	■ Prove polynomial identities.	■ Evaluate functions by using synthetic substitution. ■ Determine whether a binomial is a factor of a polynomial by using synthetic substitution.	
Key Vocabulary	prime polynomials quadratic form	identity polynomial identity	synthetic substitution depressed polynomial	
CCSS	A.CED.1	A.APR.4	A.APR.2, F.IF.7c	
Multiple Representations			⚙	
Lesson Resources	connectED.mcgraw-hill.com 📁 Leveled Worksheets 🔤 Vocabulary **PT** Personal Tutor ✓ Self-Check Quiz ■ *5-Minute Check* ■ *Study Notebook*	connectED.mcgraw-hill.com 🔤 Vocabulary **PT** Personal Tutor **Materials:** ■ TI-Nspire CAS technology	connectED.mcgraw-hill.com 📁 Leveled Worksheets 📁 Quiz 3 🔤 Vocabulary **PT** Personal Tutor ✓ Self-Check Practice ■ *5-Minute Check* ■ *Study Notebook*	
Resources for Every Lesson	**IWB** eStudent Edition **IWB** Interactive Classroom	■ eTeacher Edition ■ eSolutions Manual ■ e**Assessment** McGraw-Hill		
Differentiated Instruction	pp. 344, 349		pp. 354, 357	

IWB All digital assets are Interactive Whiteboard ready.

LESSON 5-7	45 min: 1 day 90 min: 0.5 day	EXTEND 5-7	45 min: 0.5 day 90 min: 0.25 day	LESSON 5-8	45 min: 1 day 90 min: 0.5 day
Roots and Zeros		**Graphing Technology Lab: Analyzing Polynomial Functions**		**Rational Zero Theorem**	
▪ Determine the number and type of roots for a polynomial equation. ▪ Find the zeros of a polynomial function.		▪ Use a graphing calculator to analyze polynomial functions.		▪ Identify possible rational zeros of a polynomial function. ▪ Find all of the rational zeros of a polynomial function.	
N.CN.9, A.APR.3		A.REI.11			

connectED.mcgraw-hill.com

📁 Leveled Worksheets

🏃 Animations

PT Personal Tutor

✓ Self-Check Quiz

- *5-Minute Check*
- *Study Notebook*
- *Teaching Algebra with Manipulatives*

PT Personal Tutor

Materials:
- TI-83/84 Plus or other graphing calculator

connectED.mcgraw-hill.com

📁 Leveled Worksheets

📁 Quiz 4

PT Personal Tutor

✓ Self-Check Quiz

- *5-Minute Check*
- *Study Notebook*

IWB eStudent Edition
IWB Interactive Classroom

- eTeacher Edition
- eSolutions Manual
- **McGraw-Hill** *e*Assessment

---

pp. 361, 365

pp. 369, 372

Summative Assessment
Study Guide and Review
Practice Test

SE = Student Edition, **TE** = Teacher Edition, **CRM** = Chapter Resource Masters

Diagnosis	Prescription
Beginning Chapter 5	
Get Ready for Chapter 5 **SE**	Response to Intervention **TE**
Beginning Every Lesson	
Then, Now, Why? **SE** 5-Minute Checks	Chapter 5 **SE**
During/After Every Lesson	
Guided Practice **SE**, every example Check Your Understanding **SE** H.O.T. Problems **SE** Spiral Review **SE** Additional Examples **TE** Watch Out! **TE** Step 4, Assess **TE** Chapter 5 Quizzes **CRM**, pp. 57–58 Self-Check Quizzes connectED.mcgraw-hill.com	**TIER 1 Intervention** Skills Practice **CRM**, Ch. 1–5 connectED.mcgraw-hill.com **TIER 2 Intervention** Differentiated Instruction **TE** Study Guide and Intervention **CRM**, Ch. 1–5 **TIER 3 Intervention** Math Triumphs, Alg. 2, Chs. 1 and 4
Mid-Chapter	
Mid-Chapter Quiz **SE** Mid-Chapter Test **CRM**, p. 59 McGraw-Hill *eAssessment*	**TIER 1 Intervention** Skills Practice **CRM**, Ch. 1–5 connectED.mcgraw-hill.com **TIER 2 Intervention** Study Guide and Intervention **CRM**, Ch. 1–5 **TIER 3 Intervention** Math Triumphs, Alg. 2, Chs. 1 and 4
Before Chapter Test	
Chapter Study Guide and Review **SE** Practice Test **SE** Standardized Test Practice **SE** Chapter Test connectED.mcgraw-hill.com Standardized Test Practice connectED.mcgraw-hill.com Vocabulary Review connectED.mcgraw-hill.com McGraw-Hill *eAssessment*	**TIER 1 Intervention** Skills Practice **CRM**, Ch. 1–5 connectED.mcgraw-hill.com **TIER 2 Intervention** Study Guide and Intervention **CRM**, Ch. 1–5 **TIER 3 Intervention** Math Triumphs, Alg. 2, Chs. 1 and 4
After Chapter 5	
Multiple-Choice Tests, Forms 1, 2A, 2B **CRM**, pp. 61–66 Free-Response Tests, Forms 2C, 2D, 3 **CRM**, pp. 67–72 Vocabulary Test **CRM**, p. 60 Extended Response Test **CRM**, p. 73 Standardized Test Practice **CRM**, pp. 74–76 McGraw-Hill *eAssessment*	Study Guide and Intervention **CRM**, Ch. 1–5 connectED.mcgraw-hill.com

DIAGNOSTIC ASSESSMENT · FORMATIVE ASSESSMENT · SUMMATIVE ASSESSMENT

Option 1 Reaching All Learners

Logical On the board, show the steps in the long division algorithm of a problem, such as $2930 \div 25$. Then show the steps in the long division used to find $(x^2 + 2x - 29) \div (x - 4)$. Point out the similarities in the two divisions.

$$
\begin{array}{r}
117 \\
25\overline{)2930} \\
\underline{25} \\
43 \\
\underline{25} \\
180 \\
\underline{175} \\
5
\end{array}
\qquad
\begin{array}{r}
x + 6 \\
x - 4\overline{)x^2 + 2x - 29} \\
\underline{x^2 - 4x} \\
6x - 29 \\
\underline{6x - 24} \\
-5
\end{array}
$$

Interpersonal Put students in groups of three or four. Have each student in a group write a polynomial function. As a group, students decide whether the polynomial functions are odd-degree or even-degree functions. Then have students predict the end behavior and the number of zeros of each function. Finally, have students check their predictions by graphing the function on a graphing calculator.

Option 2 Approaching Level AL

Put students into groups of three or four. Instruct students to take turns explaining how to make a table of values for a polynomial function, how to plot several points to begin a graph of the function, how to locate the zeros of the function, and how to estimate the x-coordinates at which the relative maxima and relative minima of the function occur.

Option 3 English Learners ELL

Ask students to explain how to find all the possible rational zeros of a polynomial function. Ask them to demonstrate the technique using a polynomial function of degree 3 or higher while explaining the process.

Option 4 Beyond Level BL

Ask students to make up their own "Error Analysis" problems using concepts from this chapter. Have students write a problem, the incorrect steps used to "solve" it, and its incorrect solution on one side of an index card. On the flip side of the card, have students write the correct solution, the correct steps, and the correct solution to their problem. Then ask students to trade their cards with another student with the instructions to find the error in the problem, circle it in red, and then correct the error to find the correct solution.

Focus on Mathematical Content

VerticalAlignment

Before Chapter 5

Related Topics before Algebra 1

- simplify numerical expressions involving exponents

Related Topics from Algebra 1

- use the Distributive Property to simplify algebraic expressions
- find specific function values
- solve quadratic equations using graphs

Previous Topics from Algebra 2

- use complex numbers to describe the solutions of quadratic equations

Chapter 5

Related Topics from Algebra 2

- use tools including factoring and properties of exponents to simplify expressions and to transform and solve equations
- identify the mathematical domains and ranges of functions
- determine the reasonable domain and range values for continuous situations

After Chapter 5

Preparation for Precalculus

- recognize and use connections among significant values of a function, points on the graph of a function, and the symbolic representation of a function
- investigate properties of polynomial functions
- use functions such as a polynomial functions to model real-life data

Lesson-by-LessonPreview

5-1 Operations with Polynomials

To simplify an expression that contains powers means to rewrite the expression without parentheses or negative exponents. The rules for writing equivalent expressions include the *definition of negative exponents* and *properties of exponents.* The properties of exponents include finding the *product of powers, quotient of powers, power of a power, power of a product, power of a quotient,* and *zero power.*

- When adding (or subtracting) polynomials, rewrite the indicated sum (or difference) of polynomials as a sum of terms and then combine like terms.
- When multiplying a polynomial by a monomial, use the Distributive Property to rewrite the product as a single polynomial. When multiplying two binomials, use the Distributive Property and then combine like terms.

5-2 Dividing Polynomials

- To divide a polynomial by a monomial, use the Distributive Property. To divide a polynomial by a binomial (or by any polynomial), you can use a process and format analogous to the long division algorithm for whole numbers.
- An abbreviated form of the long division algorithm, called *synthetic division,* can be used when the divisor is a binomial of degree 1. If not, both the divisor and dividend must be rewritten so that it is.

5-3 Polynomial Functions

An expression made up of a sum of monomials that contains one variable is called a *polynomial in one variable.* The degree of a polynomial in one variable is the greatest exponent of its variable. The leading coefficient is the coefficient of the term with the highest degree. *A polynomial function* can be described by a polynomial equation in one variable.

If you know an element in the domain, x, of any polynomial function, you can find the corresponding value in the range, $f(x)$. Resulting sets of ordered pairs can be used to graph the functions. The maximum number of real zeros, where $f(x) = 0$, that a function will have is equal to the degree of the function. The degree and leading term of a polynomial function determine the graph's end behavior.

5-4 Analyzing Graphs of Polynomial Functions

Tables of values can be used to explore two types of changes in the values of a polynomial function.

Get educator support from the Math Forum's T2T. You may want to search, or ask, for teaching strategies on degrees of a polynomial. Start your search at connectED.mcgraw-hill.com.

mathforum.org
The Math Forum
@ Drexel University

- A change of signs in the value of $f(x)$ from one value of x to the next indicates that the graph of the function crosses the x-axis between the two x-values.

- A change between increasing values and decreasing values indicates that the graph is turning for that interval. A turning point on a graph is a relative maximum or minimum.

  For $f(x) = x^3 - 3x^2 + 5$, the table of values shown indicates the following graph.

x	$f(x)$
-2	-15
-1	1
0	5
1	3
2	1
3	5

zero between $x = -2$ and $x = -1$

←indicates a relative maximum

←indicates a relative minimum

$f(x) = x^3 - 3x^2 + 5$

5-5 Solving Polynomial Equations

Some methods for factoring polynomials were taught in Lesson 4-3; others are taught in this lesson.

- Binomials in the form of the difference of two squares or the sum or difference of two cubes can be rewritten as the product of two factors.

- A perfect square trinomial can be rewritten as the square of a binomial.

- Some other trinomials can be factored as the product of two binomials.

- Some polynomials with four or more terms can be regrouped to allow factoring.

- In some cases, higher-degree polynomials can be expressed in quadratic form and then factored.

A higher-degree polynomial equation can often be solved by writing the polynomial expression in quadratic form, factoring, and then using the Zero Product Property and/or the Quadratic Formula.

5-6 The Remainder and Factor Theorems

The *Remainder Theorem,* which says if a polynomial is divided by $x - a$, then the remainder k is equal to the value of the polynomial at a, or $f(a) = k$, allows for the use of synthetic division to evaluate a function. This method is called *synthetic substitution*.

The *Factor Theorem* states that $x - a$ is a factor of a polynomial if and only if $f(a) = 0$. This theorem can be used to help determine the factors of polynomials.

5-7 Roots and Zeros

According to the *Fundamental Theorem of Algebra*, every polynomial equation with a degree greater than 0 has at least one root in the set of complex numbers. And according to a corollary of that theorem, the number of complex roots of a polynomial equation is the same as the degree of the equation.

Descartes' Rule of Signs says that there is a relationship between the signs of the coefficients of a polynomial function and the number of positive and negative real zeros. By the *Complex Conjugates Theorem,* if an imaginary number is a zero of a function, its conjugate is also a zero.

5-8 Rational Zero Theorem

If a simplified fraction $\frac{p}{q}$ is a zero of a polynomial function with integral coefficients, then the fraction's numerator, p, must be a factor of the constant term of the function, and the denominator, q, must be a factor of the leading coefficient. To use this property to find all the rational zeros of a polynomial function,

- first, find all the factors, p, of the constant term and all the factors, q, of the leading coefficient and then list all the possible fractions, $\frac{p}{q}$.

- Then, test whether particular values are zeros. Once you find a zero, you can use the depressed polynomial to find other zeros.

Mc Graw Hill Professional Development

Targeted professional development has been articulated throughout Algebra 2. More quality, customized professional development is available from McGraw-Hill Professional Development. Visit connectED.mcgraw-hill.com for details on each product.

- Online Lessons emphasize the strategies and techniques used to teach Algebra 2. Includes streaming video, interactive pages, and online tools.

- Video Workshops allow mentors, coaches, or leadership personnel to facilitate on-site workshops on educational strategies in mathematics and mathematical concepts.

- MHPD Online (www.mhpdonline.com) offers online professional development with video clips of instructional strategies, links, student activities, and news and issues in education.

- Teaching Today (teachingtoday.glencoe.com) gives secondary teachers practical strategies and materials that inspire excellence and innovation in teaching.

Chapter Project

Here We Grow

Students use what they have learned about operations with polynomials and graphs of polynomial functions to complete a project.

This chapter project addresses global awareness, as well as several specific skills identified as being essential to student success by the Framework for 21st Century Learning.

Visit connectED.mcgraw-hill.com for student and teacher handouts.

KeyVocabulary Introduce the key vocabulary in the chapter using the routine below.

Define: The degree of a given polynomial is the greatest degree of any term in the polynomial.

Example: The degree of the polynomial
$\frac{2}{5}k^5 + 7m^3$ is 5.

Ask: What is the degree of the polynomial $9a^3 - 2b^8$? 8

CHAPTER 5

Polynomials and Polynomial Functions

·· Then

○ You graphed quadratic functions and solved quadratic equations.

·· Now

○ You will:

- Add, subtract, multiply, divide, and factor polynomials.

- Analyze and graph polynomial functions.

- Evaluate polynomial functions and solve polynomial equations.

- Find factors and zeros of polynomial functions.

·· Why? ▲

○ **TRANSPORTATION** Polynomial functions can be used to determine bus schedules, highway capacity, traffic patterns, average fuel costs, and the prices of new and used cars.

connectED.mcgraw-hill.com **Your Digital Math Portal**

Animation Vocabulary eGlossary Personal Tutor Virtual Manipulatives Graphing Calculator Audio Foldables Self-Check Practice Worksheets

Photodisc/Getty Images

Get Ready for the Chapter

Diagnose Readiness | You have two options for checking prerequisite skills.

1 Textbook Option Take the Quick Check below. Refer to the Quick Review for help.

QuickCheck	QuickReview
Rewrite each difference as a sum.	**Example 1** (Used in Lesson 5-2)

Rewrite each difference as a sum.

1. $-5 - 13$ $-5 + (-13)$ 2. $5 - 3y$ $5 + (-3y)$

3. $5mr - 7mp$ 4. $3x^2y - 14xy^2$
   $5mr + (-7mp)$ $3x^2y + (-14xy^2)$

5. **PARTIES** Twenty people attended a going away party for Zach. The guests left in groups of 2. By 9:00, x groups had left. Rewrite the number of guests remaining at 9:00 as a sum.
   $20 + (-2x)$

Example 1 (Used in Lesson 5-2)

Rewrite $2xy - 3 - z$ as a sum.

$2xy - 3 - z$ Original expression

$= 2xy + (-3) + (-z)$ Rewrite using addition.

Use the Distributive Property to rewrite each expression without parentheses.

6. $-4(a + 5)$ $-4a - 20$ 7. $-1(3b^2 + 2b - 1)$

8. $-\frac{1}{2}(2m - 5)$ $-m + \frac{5}{2}$ 9. $-\frac{3}{4}(3z + 5)$ $-\frac{9}{4}z - \frac{15}{4}$

   7. $-3b^2 - 2b + 1$

10. **MONEY** Mr. Chávez is buying pizza and soda for the members of the science club. A slice of pizza costs $2.25, and a soda costs $1.25. Write an expression to represent the amount that Mr. Chávez will spend on 15 students. Evaluate the expression by using the Distributive Property.
    $15(2.25 + 1.25)$ or $15(2.25) + 15(1.25)$; $52.50

Example 2 (Used in Lesson 5-2)

Use the Distributive Property to rewrite $-3(a + b - c)$.

$-3(a + b - c)$ Original expression

$= -3(a) + (-3)(b) + (-3)(-c)$ Distributive Property

$= -3a - 3b + 3c$ Simplify.

Solve each equation.

11. $x^2 + 2x - 8 = 0$ $-4, 2$ 12. $2x^2 + 7x + 3 = 0$ $-3, -\frac{1}{2}$

13. $6x^2 + 5x - 4 = 0$ $-\frac{4}{3}, \frac{1}{2}$ 14. $4x^2 - 2x - 1 = 0$ $\frac{1 \pm \sqrt{5}}{4}$

15. **PHYSICS** If an object is dropped from a height of 50 feet above the ground, then its height after t seconds is given by $h = -16t^2 + 50$. Use the equation $0 = -16t^2 + 50$ to find how long it will take until the ball reaches the ground.
    about 1.77 seconds

Example 3 (Used in Lesson 5-6)

Solve $2x^2 + 8x + 1 = 0$.

$x = \dfrac{-b \pm \sqrt{b^2 - 4ac}}{2a}$ Quadratic Formula

$= \dfrac{-8 \pm \sqrt{8^2 - 4(2)(1)}}{2(2)}$ $a = 2, b = 8, c = 1$

$= \dfrac{-8 \pm \sqrt{56}}{4}$ Simplify.

$= -2 \pm \dfrac{\sqrt{14}}{2}$ $\sqrt{56} = \sqrt{4 \cdot 14}$ or $2\sqrt{14}$

The exact solutions are $-2 + \dfrac{\sqrt{14}}{2}$ and $-2 - \dfrac{\sqrt{14}}{2}$.

The approximate solutions are -0.13 and -3.87.

---

2 Online Option Take an online self-check Chapter Readiness Quiz at connectED.mcgraw-hill.com.

301

Get Ready for the Chapter

Response to Intervention (RtI) Use the *Quick Check* results and the Intervention Planner to help you determine your Response to Intervention. The If-Then statements in the chart below help you decide the appropriate tier of RtI and suggest intervention resources for each tier.

InterventionPlanner

TIER 1 On Level OL

If students miss about 25% of the exercises or less,

Then choose a resource:

SE Lessons 5-2 and 5-6

Skills Practice, Chapter 5, pp. 13 and 37

connectED.mcgraw-hill.com Self-Check Quiz

TIER 2 Strategic Intervention AL approaching grade level

If students miss about 50% of the exercises,

Then choose a resource:

Study Guide and Intervention, Chapter 5, pp. 11 and 35

connectED.mcgraw-hill.com Extra Examples, Personal Tutor, Homework Help

TIER 3 Intensive Intervention 2 or more grades below level

If students miss about 75% of the exercises,

Then use *Math Triumphs, Alg. 2*, Chs. 1 and 4

connectED.mcgraw-hill.com Extra Examples, Personal Tutor, Homework Help, Review Vocabulary

---

Essential Question

- Why is math used to model real-world situations? Sample answer: in order to study trends, make predictions, understand phenomena in nature

FOLDABLES StudyOrganizer

Dinah Zike's Foldables®

Focus Students write notes as they explore polynomials and polynomial functions in this chapter.

Teach Have students make and label their Foldables as illustrated. As students work through the chapter, have them use index cards to write notes relating to the concepts on the tabs. Encourage students to write an example in their own words for each tab.

When to Use It Encourage students to add to their Foldables as they work through the chapter and to use them to review for the chapter test.

Differentiated Instruction

 Student-Built Glossary, pp. 1–2

Students should complete the chart by providing a definition of each term and an example as they progress through Chapter 5. This study tool can also be used to review for the chapter test.

Get Started on the Chapter

You will learn several new concepts, skills, and vocabulary terms as you study Chapter 5. To get ready, identify important terms and organize your resources. You may wish to refer to Chapter 0 to review prerequisite skills.

FOLDABLES StudyOrganizer

Polynomials and Polynomial Functions Make this Foldable to help you organize your Chapter 5 notes about polynomials and polynomial functions. Begin with one sheet of $8\frac{1}{2}$" by 14" paper.

1 **Fold** a 2" tab along the bottom of a long side.

2 **Fold** along the width into thirds.

3 **Staple** the outer edges of the tab.

4 **Label** the tabs *Polynomials*, *Polynomial Functions and Graphs*, and *Solving Polynomial Equations*.

NewVocabulary

English		Español
simplify	p. 303	reducer
degree of a polynomial	p. 305	grado de un polinomio
synthetic division	p. 312	división sintética
polynomial in one variable	p. 322	polinomio de una variable
leading coefficient	p. 322	coeficiente líder
polynomial function	p. 323	función polinomial
power function	p. 323	función potencia
end behavior	p. 324	comportamiento final
relative maximum	p. 331	máximo relativo
relative minimum	p. 331	mínimo relativo
extrema	p. 331	extrema
turning points	p. 331	momentos cruciales
prime polynomials	p. 342	polinomios primeros
quadratic form	p. 345	forma de ecuación cuadrática
synthetic substitution	p. 352	sustitución sintética
depressed polynomial	p. 354	polinomio reducido

ReviewVocabulary

factoring *factorización* to express a polynomial as the product of monomials and polynomials

function *función* a relation in which each element of the domain is paired with exactly one element in the range

| **Function** | **Not a Function** |

polynomial *polinomio* a monomial or sum of monomials

LESSON 5-1 Operations with Polynomials

:: Then	:: Now	:: Why?
● You evaluated powers.	**1** Multiply, divide, and simplify monomials and expressions involving powers. **2** Add, subtract, and multiply polynomials.	● The light from the Sun takes approximately 8 minutes to reach Earth. So if you are outside right now you are basking in sunlight that the Sun emitted approximately 8 minutes ago. Light travels very fast, at a speed of about 3×10^8 meters per second. How long would it take light to get here from the Andromeda galaxy, which is approximately 2.367×10^{21} meters away?

NewVocabulary
simplify
degree of a polynomial

CCSS Common Core State Standards

Content Standards
A.APR.1 Understand that polynomials form a system analogous to the integers, namely, they are closed under the operations of addition, subtraction, and multiplication; add, subtract, and multiply polynomials.

Mathematical Practices
2 Reason abstractly and quantitatively.

1 Multiply and Divide Monomials To **simplify** an expression containing powers means to rewrite the expression without parentheses or negative exponents. Negative exponents are a way of expressing the multiplicative inverse of a number. The following table summarizes the properties of exponents.

ConceptSummary Properties of Exponents

For any real numbers x and y, integers a and b:

Property	Definition	Examples
Product of Powers	$x^a \cdot x^b = x^{a+b}$	$3^2 \cdot 3^4 = 3^{2+4}$ or 3^6 $p^2 \cdot p^9 = p^{2+9}$ or p^{11}
Quotient of Powers	$\dfrac{x^a}{x^b} = x^{a-b}, x \neq 0$	$\dfrac{9^5}{9^2} = 9^{5-2}$ or 9^3 $\dfrac{b^6}{b^4} = b^{6-4}$ or b^2
Negative Exponent	$x^{-a} = \dfrac{1}{x^a}$ and $\dfrac{1}{x^{-a}} = x^a, x \neq 0$	$3^{-5} = \dfrac{1}{3^5}$ $\dfrac{1}{b^{-7}} = b^7$
Power of a Power	$(x^a)^b = x^{ab}$	$(3^3)^2 = 3^{3 \cdot 2}$ or 3^6 $(d^2)^4 = d^{2 \cdot 4}$ or d^8
Power of a Product	$(xy)^a = x^a y^a$	$(2k)^4 = 2^4 k^4$ or $16k^4$ $(ab)^3 = a^3 b^3$
Power of a Quotient	$\left(\dfrac{x}{y}\right)^a = \dfrac{x^a}{y^a}, y \neq 0$, and $\left(\dfrac{x}{y}\right)^{-a} = \left(\dfrac{y}{x}\right)^a$ or $\dfrac{y^a}{x^a}, x \neq 0, y \neq 0$	$\left(\dfrac{x}{y}\right)^2 = \dfrac{x^2}{y^2}$ $\left(\dfrac{a}{b}\right)^{-5} = \dfrac{b^5}{a^5}$
Zero Power	$x^0 = 1, x \neq 0$	$7^0 = 1$

Recall that a *monomial* is a number, a variable, or an expression that is the product of one or more variables with nonnegative integer exponents.

 connectED.mcgraw-hill.com **303**

---

1 Focus

VerticalAlignment

▼ **Before Lesson 5-1** Evaluate powers.

▼ **Lesson 5-1** Multiply, divide, and simplify monomials and expressions involving powers. Add, subtract, and multiply polynomials.

▼ **After Lesson 5-1** Use tools including factoring to transform and solve polynomial equations.

2 Teach

Scaffolding Questions
Have students read the **Why?** section of the lesson.

Ask:
- How is 3×10^8 written in standard form? 300,000,000

- How many seconds are there in 8 minutes? 480

- Using the given information, about how far is Earth from the Sun? 1.44×10^{11} meters

Lesson 5-1 Resources

Resource	Approaching Level **AL**	On Level **OL**	Beyond Level **BL**	English Learners **ELL**
Teacher Edition	▪ Differentiated Instruction, p. 304	▪ Differentiated Instruction, pp. 304, 309	▪ Differentiated Instruction, p. 309	▪ Differentiated Instruction, p. 304
Chapter Resource Masters	▪ Study Guide and Intervention, pp. 5–6 ▪ Skills Practice, p. 7 ▪ Practice, p. 8 ▪ Word Problem Practice, p. 9	▪ Study Guide and Intervention, pp. 5–6 ▪ Skills Practice, p. 7 ▪ Practice, p. 8 ▪ Word Problem Practice, p. 9 ▪ Enrichment, p. 10	▪ Practice, p. 8 ▪ Word Problem Practice, p. 9 ▪ Enrichment, p. 10	▪ Study Guide and Intervention, pp. 5–6 ▪ Skills Practice, p. 7 ▪ Practice, p. 8
Other	▪ 5-Minute Check 5-1 ▪ Study Notebook	▪ 5-Minute Check 5-1 ▪ Study Notebook	▪ 5-Minute Check 5-1 ▪ Study Notebook	▪ 5-Minute Check 5-1 ▪ Study Notebook

Corbis/Photolibrary

1 Multiply and Divide Monomials

Example 1 shows how to use properties of exponents to simplify expressions involving multiplication and division of monomials.

Formative Assessment

Use the Guided Practice exercises after each Example to determine students' understanding of concepts.

Additional Example

1 Simplify each expression. Assume that no variable equals 0.

a. $(a^{-3})(a^2b^4)(c^{-1})$ $\dfrac{b^4}{ac}$

b. $\dfrac{n^2}{n^{10}}$ $\dfrac{1}{n^8}$

c. $\left(\dfrac{3a^3}{b^4}\right)^2$ $\dfrac{9a^6}{b^8}$

Additional Examples also in Interactive Classroom PowerPoint® Presentations

IWB **Interactive White Board READY**

Focus on Mathematical Content

Properties of Exponents When multiplying or dividing powers of variables, be sure that the base is the same. Add the exponents if multiplying powers of the same variable, and subtract them if dividing powers.

WatchOut!

Common Errors When discussing Additional Example 1b, if any students get the incorrect answer $\dfrac{1}{n^5}$, lead them to understand that they divided the exponents instead of subtracting them.

When simplifying a monomial, check to be sure that it has been simplified fully.

KeyConcept Simplifying Monomials

A monomial expression is in simplified form when:
- there are no powers of powers,
- each base appears exactly once,
- all fractions are in simplest form, and
- there are no negative exponents.

Example 1 Simplify Expressions

Simplify each expression. Assume that no variable equals 0.

a. $(2a^{-2})(3a^3b^2)(c^{-2})$

$(2a^{-2})(3a^3b^2)(c^{-2})$	Original expression
$= 2\left(\dfrac{1}{a^2}\right)(3a^3b^2)\left(\dfrac{1}{c^2}\right)$	Definition of negative exponents
$= \left(\dfrac{2}{a \cdot a}\right)(3 \cdot a \cdot a \cdot a \cdot b \cdot b)\left(\dfrac{1}{c \cdot c}\right)$	Definition of exponents
$= \left(\dfrac{2}{\cancel{a} \cdot \cancel{a}}\right)(3 \cdot \cancel{a} \cdot \cancel{a} \cdot a \cdot b \cdot b)\left(\dfrac{1}{c \cdot c}\right)$	Divide out common factors.
$= \dfrac{6ab^2}{c^2}$	Simplify.

b. $\dfrac{q^2r^4}{q^7r^3}$

$\dfrac{q^2r^4}{q^7r^3} = q^{2-7} \cdot r^{4-3}$	Quotient of powers
$= q^{-5}r$	Subtract powers.
$= \dfrac{r}{q^5}$	Simplify.

Problem-SolvingTip

Check You can always check your answer using the definition of exponents.

$$\dfrac{q^2}{q^7} = \dfrac{q \cdot q}{q \cdot q \cdot q \cdot q \cdot q \cdot q \cdot q}$$
$$= \dfrac{1}{q^5}$$

c. $\left(\dfrac{-2a^4}{b^2}\right)^3$

$\left(\dfrac{-2a^4}{b^2}\right)^3 = \dfrac{(-2a^4)^3}{(b^2)^3}$	Power of a quotient
$= \dfrac{(-2)^3(a^4)^3}{(b^2)^3}$	Power of a product
$= \dfrac{-8a^{12}}{b^6}$	Power of a power

GuidedPractice

1A. $(2x^{-3}y^3)(-7x^5y^{-6})$ $\dfrac{-14x^2}{y^3}$

1B. $\dfrac{15c^5d^3}{-3c^2d^7}$ $\dfrac{-5c^3}{d^4}$

1C. $\left(\dfrac{a}{4}\right)^{-3}$ $\dfrac{64}{a^3}$

1D. $(-2x^3y^2)^5$ $-32x^{15}y^{10}$

DifferentiatedInstruction AL OL ELL

If students have difficulty describing or using properties of exponents,

Then have them write their own summary of the properties of powers, such as "to multiply expressions with exponents, you add the exponents; to divide, you subtract the exponents."

StudyTip

Power of 1 Remember that a variable with no exponent indicated can be written as a power of 1.

2 **Operations With Polynomials** The **degree of a polynomial** is the degree of the monomial with the greatest degree.

Example 2 Degree of a Polynomial

Determine whether each expression is a polynomial. If it is a polynomial, state the degree of the polynomial.

a. $\frac{1}{4}x^4y^3 - 8x^5$

This expression is a polynomial because each term is a monomial. The degree of the first term is $4 + 3$ or 7, and the degree of the second term is 5. The degree of the polynomial is 7.

b. $\sqrt{x} + x + 4$

This expression is not a polynomial because $\sqrt{x}$ is not a monomial.

c. $x^{-3} + 2x^{-2} + 6$

This expression is not a polynomial because x^{-3} and x^{-2} are not monomials: $x^{-3} = \frac{1}{x^3}$ and $x^{-2} = \frac{1}{x^2}$. Monomials cannot contain variables in the denominator.

GuidedPractice

2A. $\frac{x}{y} + 3x^2$ No; $\frac{x}{y}$ is not a monomial. **2B.** $x^5y + 9x^4y^3 - 2xy$ yes; 7

You can simplify a polynomial just like you simplify a monomial. Perform the operations indicated, and combine like terms.

StudyTip

Alternative Methods Notice that Example 3a uses a horizontal method, and Example 3b uses a vertical method to simplify. Either method will yield a correct solution.

Example 3 Simplify Polynomial Expressions

Simplify each expression.

a. $(4x^2 - 5x + 6) - (2x^2 + 3x - 1)$

Remove parentheses, and group like terms together.

$(4x^2 - 5x + 6) - (2x^2 + 3x - 1)$

$= 4x^2 - 5x + 6 - 2x^2 - 3x + 1$	Distribute the -1.
$= (4x^2 - 2x^2) + (-5x - 3x) + (6 + 1)$	Group like terms.
$= 2x^2 - 8x + 7$	Combine like terms.

b. $(6x^2 - 7x + 8) + (-4x^2 + 9x - 5)$

Align like terms vertically and add.

$$6x^2 - 7x + 8$$
$$\underline{(+) -4x^2 + 9x - 5}$$
$$2x^2 + 2x + 3$$

GuidedPractice **3A.** $-2x^2 - 5x - 1$

3A. $(-x^2 - 3x + 4) - (x^2 + 2x + 5)$ **3B.** $(3x^2 - 6) + (-x + 1)$ $3x^2 - x - 5$

Adding or subtracting integers results in an integer, so the set of integers is closed under addition and subtraction. Similarly, because adding or subtracting polynomials results in a polynomial, the set of polynomials is closed under addition and subtraction.

WatchOut!

Preventing Errors Suggest that students return to the basic definitions for exponents rather than just memorizing rules. For example, they can derive the rule for multiplying quantities such as $x^2 \cdot x^3$ by rewriting the problem as $x \cdot x \cdot x \cdot x \cdot x$.

2 **Operations with Polynomials**

Example 2 shows how to determine whether an expression is a polynomial and if it is, to state its degree. **Example 3** shows how to subtract and add polynomials. **Example 4** shows how to use the Distributive Property to multiply a polynomial by a monomial. **Example 5** shows how to represent a real-world situation using a polynomial. **Example 6** shows how to apply the Distributive Property two times to multiply a polynomial and a binomial.

Additional Examples

2 Determine whether each expression is a polynomial. If it is a polynomial, state the degree of the polynomial.

a. $c^4 - 4\sqrt{c} + 18$ no

b. $-16p^5 + \frac{3}{4}p^2t^7$ yes; 9

c. $x^2 - 3x^{-1} + 7$ no

3 Simplify each expression.

a. $(2a^3 + 5a - 7) - (a^3 - 3a + 2)$ $a^3 + 8a - 9$

b. $(4x^2 - 9x + 3) + (-2x^2 - 5x - 6)$ $2x^2 - 14x - 3$

Teach with Tech

Interactive Whiteboard Write an expression on the board to add or subtract two polynomials. Drag the like terms to group them together. Then combine like terms and simplify the expression.

Additional Examples

4 Find $-y(4y^2 + 2y - 3)$.
$-4y^3 - 2y^2 + 3y$

5 **E-SALES** A small online retailer estimates that the cost, in dollars, associated with selling x units of a particular product is given by the expression $0.001x^2 + 5x + 500$. The revenue from selling x units is given by $10x$. Write a polynomial to represent the profits generated by the product if profit = revenue − cost.
$-0.001x^2 + 5x - 500$

6 Find $(a^2 + 3a - 4)(a + 2)$.
$a^3 + 5a^2 + 2a - 8$

You can use the Distributive Property to multiply polynomials.

Example 4 Simplify by Using the Distributive Property

Find $3x(2x^2 - 4x + 6)$.

$$3x(2x^2 - 4x + 6) = 3x(2x^2) + 3x(-4x) + 3x(6) \qquad \text{Distributive Property}$$
$$= 6x^3 - 12x^2 + 18x \qquad \text{Multiply the monomials.}$$

GuidedPractice

Find each product.

4A. $\frac{4}{3}x^2(6x^2 + 9x - 12)$ $8x^4 + 12x^3 - 16x^2$

4B. $-2a(-3a^2 - 11a + 20)$ $6a^3 + 22a^2 - 40a$

Real-World Example 5 Write a Polynomial Expression

DRIVING The U.S. Department of Transportation limits the time a truck driver can work between periods of rest to ten hours. For the first part of his shift, Tom drives at a speed of 60 miles per hour, and for the second part of the shift, he drives at a speed of 70 miles per hour. Write a polynomial to represent the distance driven.

Words	60 mph for some time and 70 mph for the rest
Variable	Let $x =$ the number of hours he drives at 60 miles per hour.
Expression	60 $\quad$ x $\quad$ + $\quad$ 70 $\quad$ $(10 - x)$

$$60x + 70(10 - x) \qquad \text{Original expression}$$
$$= 60x + 700 - 70x \qquad \text{Distributive Property}$$
$$= 700 - 10x \qquad \text{Combine like terms.}$$

The polynomial is $700 - 10x$.

 GuidedPractice $\quad$ **5.** $37.8 - 0.024x$

5. Paul has \$900 to invest in a savings account that has an annual interest rate of 1.8%, and a money market account that pays 4.2% per year. Write a polynomial for the interest he will earn in one year if he invests x dollars in the savings account.

Real-World Career

Truck Driver Truck drivers are considered technical professionals because they are required to obtain specialized education and professional licensure. Although state motor vehicle departments administer the Commercial Driver's License program, federal law spells out the requirements to obtain one.

Like addition and subtraction, polynomials are closed under multiplication.

Example 6 Multiply Polynomials

Find $(n^2 + 4n - 6)(n + 2)$.

$$(n^2 + 4n - 6)(n + 2)$$
$$= n^2(n + 2) + 4n(n + 2) + (-6)(n + 2) \qquad \text{Distributive Property}$$
$$= n^2 \cdot n + n^2 \cdot 2 + 4n \cdot n + 4n \cdot 2 + (-6) \cdot n + (-6) \cdot 2 \qquad \text{Distributive Property}$$
$$= n^3 + 2n^2 + 4n^2 + 8n - 6n - 12 \qquad \text{Multiply monomials.}$$
$$= n^3 + 6n^2 + 2n - 12 \qquad \text{Combine like terms.}$$

GuidedPractice

Find each product. $\qquad$ **6B.** $6x^3 - 14x^2 + 19x - 5$

6A. $(x^2 + 4x + 16)(x - 4)$ $x^3 - 64$

6B. $(2x^2 - 4x + 5)(3x - 1)$

ColorBlind Images/Blend Images/Getty Images

Check Your Understanding

= Step-by-Step Solutions begin on page R14.

Example 1

Simplify. Assume that no variable equals 0.

1. $(2a^3b^{-2})(-4a^2b^4)$ $-8a^5b^2$

2. $\frac{12x^4y^2}{2xy^5}\cdot\frac{6x^3}{y^3}$

3. $\left(\frac{2a^2}{3b}\right)^3\cdot\frac{8a^6}{27b^3}$

4. $\left(6g^5h^{-4}\right)^3\cdot\frac{216g^{15}}{h^{12}}$

Example 2

Determine whether each expression is a polynomial. If it is a polynomial, state the degree of the polynomial.

5. $3x + 4y$ yes; 1

6. $\frac{1}{2}x^2 - 7y$ yes; 2

7. $x^2 + \sqrt{x}$ no

8. $\frac{ab^3 - 1}{az^4 + 3}$ no

Examples 3-4, and 6

Simplify. **9.** $-2x^2 - 6x + 3$

9. $(x^2 - 5x + 2) - (3x^2 + x - 1)$

10. $(3a + 4b) + (6a - 6b)$ $9a - 2b$

11. $2a(4b + 5)$ $8ab + 10a$

12. $3x^2(2xy - 3xy^2 + 4x^2y^3)$ $6x^3y - 9x^3y^2 + 12x^4y^3$

13. $(n - 9)(n + 7)$ $n^2 - 2n - 63$

14. $(a + 4)(a - 6)$ $a^2 - 2a - 24$

Example 5

15. EXERCISE Tara exercises 75 minutes a day. She does cardio, which burns an average of 10 Calories per minute, and weight training, which burns an average of 7.5 Calories per minute. Write a polynomial to represent the amount of Calories Tara burns in one day if she does x minutes of weight training. $750 - 2.5x$

Practice and Problem Solving

Extra Practice is on page R5.

Example 1

Simplify. Assume that no variable equals 0. **17.** $-8b^5c^3$

16. $\frac{20x^4}{y^2}$

16. $(5x^3y^{-5})(4xy^3)$

17. $(-2b^3c)(4b^2c^2)$

18. $\frac{a^3n^7}{an^4}$ a^2n^3

19. $\frac{-y^3z^5}{y^2z^3}$ $-yz^2$

20. $\frac{-7x^5y^5z^4}{21x^7y^5z^2}$ $\frac{z^2}{-3x^2}$

21. $\frac{9a^7b^5c^5}{18a^5b^9c^3}$ $\frac{a^2c^2}{2b^4}$

22. $(n^5)^4$ n^{20}

23. $(z^3)^6$ z^{18}

Example 2

Determine whether each expression is a polynomial. If it is a polynomial, state the degree of the polynomial.

24. $2x^2 - 3x + 5$ yes; 2

25. $a^3 - 11$ yes; 3

26. $\frac{5np}{n^2} - \frac{2g}{h}$ no

27. $\sqrt{m - 7}$ no

Examples 3-4, and 6

28. $2a^2 - a - 2$

29. $3b^2 + 6b - 5$

33. $a^4 + a^3b - 3a^2b - 4ab^2 - b^3$

REGULARITY Simplify. **32.** $x^3 + x^2y - xy^2 - y^3$

28. $(6a^2 + 5a + 10) - (4a^2 + 6a + 12)$

29. $(7b^2 + 6b - 7) - (4b^2 - 2)$

34. $-6a^3 + 4a^2 + 8a - 9$

30. $3p(np - z)$ $3np^2 - 3pz$

31. $4x(2x^2 + y)$ $8x^3 + 4xy$

35. $10c^3 - c^2 + 4c$

32. $(x - y)(x^2 + 2xy + y^2)$

33. $(a + b)(a^3 - 3ab - b^2)$

36. $10x^2y - 5xy^2 + 6x^2y^2 + 36y^2$

34. $4(a^2 + 5a - 6) - 3(2a^3 + 4a - 5)$

35. $5c(2c^2 - 3c + 4) + 2c(7c - 8)$

37. $12a^2b + 8a^2b^2 - 15ab^2 + 4b^2$

36. $5xy(2x - y) + 6y^2(x^2 + 6)$

37. $3ab(4a - 5b) + 4b^2(2a^2 + 1)$

38. $(x - y)(x + y)(2x + y)$ $2x^3 + x^2y - 2xy^2 - y^3$

39. $(a + b)(2a + 3b)(2x - y)$ $4a^2x - 2a^2y + 10abx - 5aby + 6b^2x - 3b^2y$

Example 5

40. PAINTING Connor has hired two painters to paint his house. The first painter charges $12 per hour and the second painter charges $11 per hour. It will take 15 hours of labor to paint the house.

a. Write a polynomial to represent the total cost of the job if the first painter does x hours of the labor. $x + 165$

b. Write a polynomial to represent the total cost of the job if the second painter does y hours of the labor. $180 - y$

Differentiated Homework Options

Level	Assignment	Two-Day Option	
AL Basic	16–50, 65, 67–96	17–49 odd, 70–73	16–50 even, 65, 67–69, 74–96
OL Core	17–61 odd, 64, 65, 67–96	16–50, 70–73	51, 61, 64, 65, 67–69, 74–96
BL Advanced	51–90, (optional: 91–96)		

Formative Assessment

Use Exercises 1–15 to check for understanding.

Use the chart at the bottom of this page to customize assignments for your students.

 Teaching the Mathematical Practices

Regularity Mathematically proficient students notice if calculations are repeated, and look both for general methods and for shortcuts. As they work to solve a problem, mathematically proficient students maintain oversight of the process, while attending to the details.

Study Guide and Intervention

AL OL ELL

Practice

AL OL BL ELL

Word Problem Practice

AL OL BL

Simplify. Assume that no variable equals 0.

41. $\left(\dfrac{8x^2y^3}{24x^3y^2}\right)^4$ $\dfrac{y^4}{81x^4}$

42. $\left(\dfrac{12a^3b^5}{4a^6b^3}\right)^3$ $\dfrac{27b^6}{a^9}$

43. $\left(\dfrac{4x^{-2}y^3}{xy^{-4}}\right)^{-2}$ $\dfrac{x^6}{16y^{14}}$

44. $\left(\dfrac{5a^{-7}b^2}{ab^{-6}}\right)^{-3}$ $\dfrac{a^{24}}{125b^{24}}$

45. $(a^2b^3)(ab)^{-2}$ b

46. $(-3x^3y)^2(4xy^2)$ $36x^7y^4$

47. $\dfrac{3c^2d(2c^3d^5)}{15c^4d^2}$ $\dfrac{2}{5}cd^4$

48. $\dfrac{-10g^6h^9(g^2h^3)}{30g^3h^3}$ $-\dfrac{1}{3}g^5h^9$

49. $\dfrac{5x^4y^2(2x^5y^6)}{20x^3y^5}$ $\dfrac{1}{2}x^6y^3$

50. $\dfrac{-12n^7p^5(n^2p^4)}{36n^6p^7}$ $-\dfrac{1}{3}n^3p^2$

51. **ASTRONOMY** Refer to the beginning of the lesson.

B

a. How long does it take light from Andromeda to reach Earth? 7.89×10^{12} s or about 250,190.26 yr

b. The average distance from the Sun to Mars is approximately 2.28×10^{11} meters. How long does it take light from the Sun to reach Mars? 760 seconds or about 12.67 min

Simplify. **52.** $2g^3 + 3g^2h - 4g^3h^2$ **53.** $2n^4 - 3n^3p + 6n^4p^4$ **54.** $x^2 - 3x + \dfrac{1}{x^3}$

52. $\dfrac{1}{4}g^2(8g + 12h - 16gh^2)$

53. $\dfrac{1}{3}n^3(6n - 9p + 18np^4)$

54. $x^{-2}(x^4 - 3x^3 + x^{-1})$

55. $a^{-3}b^2(ba^3 + b^{-1}a^2 + b^{-2}a)$

56. $(g^3 - h)(g^3 + h)$ $g^6 - h^2$

57. $(n^2 - 7)(2n^3 + 4)$

58. $(2x - 2y)^3$

59. $(4n - 5)^3$ $64n^3 - 240n^2 + 300n - 125$

60. $(3z - 2)^3$ $27z^3 - 54z^2 + 36z - 8$

61. **CCSS MODELING** The polynomials $0.108x^2 - 0.876x + 474.1$ and $0.047x^2 + 9.694x + 361.7$ approximate the number of bachelor's degrees, in thousands, earned by males and females, respectively, where x is the number of years after 1971.

a. Find the polynomial that represents the total number of bachelor's degrees (in thousands) earned by both men and women. $0.155x^2 + 8.818x + 835.8$

b. Find the polynomial that represents the difference between bachelor's degrees earned by men and by women. $0.061x^2 - 10.57x + 112.4$

55. $b^3 + \dfrac{b}{a} + \dfrac{1}{a^2}$

57. $2n^5 - 14n^3 + 4n^2 - 28$

58. $8x^3 - 24x^2y + 24xy^2 - 8y^3$

C **62.** If $5^{k+7} = 5^{2k-3}$, what is the value of k? 10

63. What value of k makes $q^{41} = q^{4k} \cdot q^5$ true? 9

64. **MULTIPLE REPRESENTATIONS** Use the model at the right that represents the product of $x + 3$ and $x + 4$.

a. **Geometric** The area of each rectangle is the product of its length and width. Use the model to find the product of $x + 3$ and $x + 4$. $x^2 + 7x + 12$

b. **Algebraic** Use FOIL to find the product of $x + 3$ and $x + 4$. $x^2 + 7x + 12$

c. **Verbal** Explain how each term of the product is represented in the model.

66. Sample answer: It approaches 0.

64c. Each term is represented by one or more rectangles with an area that represents the variable and the power in the term.

H.O.T. Problems Use Higher-Order Thinking Skills

65. **PROOF** Show how the property of negative exponents can be proven using the Quotient of Powers Property and the Zero Power Property. $\dfrac{1}{a^n} = \dfrac{a^0}{a^n} = a^{0-n} = a^{-n}$

67. Sample answer: We would have a 0 in the denominator, which makes the expression undefined.

66. **CHALLENGE** What happens to the quantity of x^{-y} as y increases, for $y > 0$ and $x > 1$?

67. **REASONING** Explain why the expression 0^{-2} is undefined.

68. **OPEN ENDED** Write three different expressions that are equivalent to x^{12}.

68. Sample answer: $x^9 \cdot x^3$, $\dfrac{x^{14}}{x^2}$, $(x^6)^2$

69. **WRITING IN MATH** Explain why properties of exponents are useful in astronomy. Include an explanation of how to find the amount of time it takes for light from a source to reach a planet. See margin.

Enrichment

OL BL

5-1 Enrichment

Polynomials with Fractional Coefficients

Polynomials may have fractional coefficients as long as there are no variables in the denominators. Computing with fractional coefficients is performed in the same way as computing with whole-number coefficients.

Simplify. Write all coefficients as fractions.

Additional Answer

69. Sample answer: Astronomy deals with very large numbers that are sometimes difficult to work with because they contain so many digits. Properties of exponents make very large or very small numbers more manageable. As long as you know how far away a planet is from a light source you can divide that distance by the speed of light to obtain how long it will take light to reach that planet.

70. SHORT RESPONSE Simplify $\frac{(2x^2)^3}{12x^4} \cdot \frac{2x^2}{3}$.

71. STATISTICS For the numbers a, b, and c, the average (arithmetic mean) is twice the median. If $a = 0$ and $a < b < c$, what is the value of $\frac{c}{b}$? **D**

 A 2   **C** 4
 B 3   **D** 5

72. Which is not a factor of $x^3 - x^2 - 2x$? **H**

 F x **H** $x - 1$
 G $x + 1$ **J** $x - 2$

73. SAT/ACT The expression $(-6 + i)^2$ is equivalent to which of the following expressions? **D**

 A 35
 B $-12i$
 C $-12 + i$
 D $35 - 12i$
 E $37 - 12i$

Spiral Review

Solve each inequality algebraically. (Lesson 4-8)

74. $x^2 - 6x \leq 16$ $-2 \leq x \leq 8$ **75.** $x^2 + 3x > 40$ $x > 5$ or $x < -8$ **76.** $2x^2 - 12 \leq -5x$ $-4 \leq x \leq 1.5$

Graph each function. (Lesson 4-7) **77–79. See margin.**

77. $y = 3(x - 2)^2 - 4$ **78.** $y = -2(x + 4)^2 + 3$ **79.** $y = \frac{1}{3}(x + 1)^2 + 6$

80. BASEBALL A baseball player hits a high pop-up with an initial upward velocity of 30 meters per second, 1.4 meters above the ground. The height $h(t)$ of the ball in meters t seconds after being hit is modeled by $h(t) = -4.9t^2 + 30t + 1.4$. How long does an opposing player have to get under the ball if he catches it 1.7 meters above the ground? Does your answer seem reasonable? Explain. (Lesson 4-3) **See margin.**

30 m/s

1.4 m

Evaluate each determinant. (Lesson 3-7)

81. $\begin{vmatrix} 3 & 0 & -2 \\ -1 & 4 & 3 \\ 5 & -2 & -1 \end{vmatrix}$ **42**   **82.** $\begin{vmatrix} -2 & -4 & -6 \\ 0 & 6 & -5 \\ -1 & 3 & -1 \end{vmatrix}$ **−74**   **83.** $\begin{vmatrix} -3 & -1 & -2 \\ -2 & 3 & 4 \\ 6 & 1 & 0 \end{vmatrix}$ **28**

84. FINANCIAL LITERACY A couple is planning to invest \$15,000 in certificates of deposit (CDs). For tax purposes, they want their total interest the first year to be \$800. They want to put \$1000 more in a 2-year CD than in a 1-year CD and then invest the rest in a 3-year CD. How much should they invest in each type of CD? (Lesson 3-4) **\$2500 in the 1-year; \$3500 in the 2-year; \$9000 in the 3-year**

Years	1	2	3
Rate	3.4%	5.0%	6.0%

Find the slope of the line that passes through each pair of points. (Lesson 2-3)

85. $(6, -2)$ and $(-2, -9)$ $\frac{7}{8}$ **86.** $(-4, -1)$ and $(3, 8)$ $\frac{9}{7}$ **87.** $(3, 0)$ and $(-7, -5)$ $\frac{1}{2}$

88. $\left(\frac{1}{2}, \frac{2}{3}\right)$ and $\left(\frac{1}{4}, \frac{1}{3}\right)$ $\frac{4}{3}$ **89.** $\left(\frac{2}{5}, \frac{1}{4}\right)$ and $\left(\frac{1}{10}, \frac{1}{12}\right)$ $\frac{5}{9}$ **90.** $(-4.5, 2.5)$ and $(-3, -1)$ $-\frac{7}{3}$

Skills Review

Factor each polynomial.

91. $12ax^3 + 20bx^2 + 32cx$ $4x(3ax^2 + 5bx + 8c)$ **92.** $x^2 + 2x + 6 + 3x$ $(x + 3)(x + 2)$ **93.** $12y^2 + 9y + 8y + 6$ $(3y + 2)(4y + 3)$

94. $2my + 7x + 7m + 2xy$ $(m + x)(2y + 7)$ **95.** $8ax - 6x - 12a + 9$ $(2x - 3)(4a - 3)$ **96.** $10x^2 - 14xy - 15x + 21y$ $(2x - 3)(5x - 7y)$

DifferentiatedInstruction **OL** **BL**

Extension Write 3^0, $(4y^3x)^0$, 0^0, and $\left(\frac{x^5}{y^2}\right)^0$ on the board. Ask students to simplify each expression. They all equal 1 except 0^0. Then, prompt students to use 0^{m-m} to explain why 0^0 does not equal 1 like the other three expressions. $0^{m-m} = \frac{0^m}{0^m}$. $0^m = 0$, $\frac{0^m}{0^m}$ implies division by 0, which is not defined. Therefore, 0^0 is not defined.

Multiple Representations

In Exercise 64, students use algebra tiles and algebraic operations to model multiplying binomials.

4 Assess

Crystal Ball Have students write how they think today's work with division of monomials will help them with dividing polynomials.

Additional Answers

77.

78.

79.

80. Sample answer: About 6.1 seconds; this answer seems reasonable. The equation has two solutions. The first solution, 0.01 second, is the time required for the ball to rise from 1.4 m to 1.7 m. 6.1 seconds is the time required for the ball to come back down to 1.7 m.

1 Focus

Objective Use dimensional analysis to convert units and solve problems.

Teaching Tip
Make sure that students have mastered multiplying fractions before attempting this lesson.

2 Teach

Working in Cooperative Groups
Have students work in pairs, mixing abilities. Have each pair work though the Example.

Ask:
- What number is the Multiplicative Identity? 1

- In what way can $\frac{5280 \text{ feet}}{1 \text{ mile}}$ be thought of as being equivalent to 1? Sample answer: One mile is the same distance as 5280 feet, so the numerator and denominator are equally-sized quantities.

Practice Have students complete Exercises 1—5.

3 Assess

Formative Assessment
Use Exercise 1 to assess whether students can set up a fraction that will change hours to minutes.

From Concrete to Abstract
In Exercise 5, ask students to give a specific example when they give their explanations.

EXTEND 5-1

Algebra Lab
Dimensional Analysis

Real-world problems often involve units of measure. Performing operations with units is called **dimensional analysis** or **unit analysis**. You can use dimensional analysis to convert units or to perform calculations.

Example

A car is traveling at 65 miles per hour. How fast is the car traveling in meters per second?

You want to find the speed in meters per second, so you need to change the unit of distance from miles to meters and the unit of time from hours to seconds. To make the conversion, use fractions that you can multiply.

Step 1 Change the units of length from miles to meters.
Use the relationships of miles to feet and feet to meters.

$$\frac{65 \text{ miles}}{1 \text{ hour}} \cdot \frac{5280 \text{ feet}}{1 \text{ mile}} \cdot \frac{1 \text{ meter}}{3.3 \text{ feet}}$$

Step 2 Change the units of time from hours to seconds.
Write fractions relating hours to minutes and minutes to seconds.

$$\frac{65 \text{ miles}}{1 \text{ hour}} \cdot \frac{5280 \text{ feet}}{1 \text{ mile}} \cdot \frac{1 \text{ meter}}{3.3 \text{ feet}} \cdot \frac{1 \text{ hour}}{60 \text{ minutes}} \cdot \frac{1 \text{ minute}}{60 \text{ seconds}}$$

Step 3 Simplify and check by canceling the units.

$$\frac{65 \text{ miles}}{1 \text{ hour}} \cdot \frac{5280 \text{ feet}}{1 \text{ mile}} \cdot \frac{1 \text{ meter}}{3.3 \text{ feet}} \cdot \frac{1 \text{ hour}}{60 \text{ minutes}} \cdot \frac{1 \text{ minute}}{60 \text{ seconds}}$$

$$= \frac{65 \cdot 5280}{3.3 \cdot 60 \cdot 60} \text{ m/s} \qquad \text{Simplify.}$$

$$\approx 28.9 \text{ m/s} \qquad \text{Use a calculator.}$$

So, 65 miles per hour is about 28.9 meters per second. This answer is reasonable because the final units are m/s, not m/hr, ft/s, or mi/hr.

5. Sample answer: If the units in the final answer are not correct, then a mistake was made in solving the problem. Dimensional analysis helps you catch numerical errors as well as errors in setting up the problem.

Exercises

Solve each problem by using dimensional analysis. Include the appropriate units with your answer.

1. A zebra can run 40 miles per hour. How far can a zebra run in 3 minutes? 2 mi

2. A cyclist traveled 43.2 miles at an average speed of 12 miles per hour. How long did the cyclist ride? 3.6 h

3. If you are driving 50 miles per hour, how many feet per second are you traveling? 73.33 ft/s

4. The equation $d = \frac{1}{2}(9.8 \text{ m/s}^2)(3.5 \text{ s})^2$ represents the distance d that a ball falls 3.5 seconds after it is dropped from a tower. Find the distance. 60.025 m

5. **WRITING IN MATH** Explain how dimensional analysis can be useful in checking the reasonableness of your answer.

310 | Extend 5-1 | Algebra Lab: Dimensional Analysis

LESSON 5-2 Dividing Polynomials

:: Then
- You divided monomials.

:: Now
1. Divide polynomials using long division.
2. Divide polynomials using synthetic division.

:: Why?
Arianna needed $140x^2 + 60x$ square inches of paper to make a book jacket $10x$ inches tall. In figuring the area, she allowed for a front and back flap. If the spine is $2x$ inches wide, and the front and back are $6x$ inches wide, how wide are the front and back flaps? You can use a quotient of polynomials to help you find the answer.

| f | 6x | 2x | 6x | f |

10x

f = flap width

 NewVocabulary
synthetic division

 Common Core State Standards

Content Standards
A.APR.6 Rewrite simple rational expressions in different forms; write $a(x)/b(x)$ in the form $q(x) + r(x)/b(x)$, where $a(x)$, $b(x)$, $q(x)$, and $r(x)$ are polynomials with the degree of $r(x)$ less than the degree of $b(x)$, using inspection, long division, or, for the more complicated examples, a computer algebra system.

Mathematical Practices
6 Attend to precision.

---

1 Long Division In Lesson 5-1, you learned how to divide monomials. You can divide a polynomial by a monomial by using those same skills.

Example 1 Divide a Polynomial by a Monomial

Simplify $\dfrac{6x^4y^3 + 12x^3y^2 - 18x^2y}{3xy}$.

$\dfrac{6x^4y^3 + 12x^3y^2 - 18x^2y}{3xy} = \dfrac{6x^4y^3}{3xy} + \dfrac{12x^3y^2}{3xy} - \dfrac{18x^2y}{3xy}$ Sum of quotients

$= \dfrac{6}{3} \cdot x^{4-1}y^{3-1} + \dfrac{12}{3} \cdot x^{3-1}y^{2-1} - \dfrac{18}{3} \cdot x^{2-1}y^{1-1}$ Divide.

$= 2x^3y^2 + 4x^2y - 6x$ $y^{1-1} = y^0$ or 1

▶ **Guided**Practice   Simplify.   **1A.** $5c^3d - 4f + 1$

1A. $(20c^4d^2f - 16cdf^2 + 4cdf) \div (4cdf)$ **1B.** $(18x^2y + 27x^3y^2z)(3xy)^{-1}$ $6x + 9x^2yz$

You can use a process similar to long division to divide a polynomial by a polynomial with more than one term. The process is known as the *division algorithm*.

Example 2 Division Algorithm

Use long division to find $(x^2 + 3x - 40) \div (x - 5)$.

$$
\begin{array}{r}
x + 8 \\
x - 5 \overline{\smash{)}\ x^2 + 3x - 40} \\
\underline{(-)\ x^2 - 5x} \\
8x - 40 \\
\underline{(-)\ 8x - 40} \\
0
\end{array}
$$

Multiply the divisor by x since $\frac{x^2}{x} = x$.

Subtract. Bring down the next term.

Multiply the divisor by 8 since $\frac{8x}{x} = 8$.

Subtract.

The quotient is $x + 8$. The remainder is 0.

▶ **Guided**Practice   Use long division to find each quotient.

2A. $(x^2 + 7x - 30) \div (x - 3)$ $x + 10$ **2B.** $(x^2 - 13x + 12) \div (x - 1)$ $x - 12$

 connectED.mcgraw-hill.com   **311**

---

1 Focus

VerticalAlignment

▼

Before Lesson 5-2 Divide monomials.

▼

Lesson 5-2 Divide polynomials using long division. Divide polynomials using synthetic division.

▼

After Lesson 5-2 Use tools including factoring to transform and solve polynomial equations.

2 Teach

Scaffolding Questions
Have students read the **Why?** section of the lesson.

Ask:
- What is the height of the book jacket? $10x$ inches
- What is the total length of the book jacket? $14x + 2f$ inches
- If you know the height of the book jacket is $10x$ inches and the area is $140x^2 + 60x$ square inches, how can you find the length of the jacket? Divide the area of the jacket by its height.

Lesson 5-2 Resources

Resource	Approaching Level **AL**	On Level **OL**	Beyond Level **BL**	English Learners **ELL**
Teacher Edition	• Differentiated Instruction, p. 314	• Differentiated Instruction, p. 314	• Differentiated Instruction, p. 317	
Chapter Resource Masters	• Study Guide and Intervention, pp. 11–12 • Skills Practice, p. 13 • Practice, p. 14 • Word Problem Practice, p. 15	• Study Guide and Intervention, pp. 11–12 • Skills Practice, p. 13 • Practice, p. 14 • Word Problem Practice, p. 15 • Enrichment, p. 16	• Practice, p. 14 • Word Problem Practice, p. 15 • Enrichment, p. 16	• Study Guide and Intervention, pp. 11–12 • Skills Practice, p. 13 • Practice, p. 14 • Word Problem Practice, p. 15
Other	• 5-Minute Check 5-2 • Study Notebook	• 5-Minute Check 5-2 • Study Notebook	• 5-Minute Check 5-2 • Study Notebook	• 5-Minute Check 5-2 • Study Notebook

1 Long Division

Example 1 shows how to divide a polynomial by a monomial. **Example 2** shows how to divide a polynomial by a polynomial using the division algorithm. **Example 3** shows how to solve a multiple choice test item involving division of polynomials resulting in a quotient with a remainder.

Formative Assessment

Use the Guided Practice exercises after each example to determine students' understanding of concepts.

Additional Examples

1 Simplify $\dfrac{5a^2b - 15ab^3 + 10a^3b^4}{5ab}$.

$a - 3b^2 + 2a^2b^3$

2 Use long division to find $(x^2 - 2x - 15) \div (x - 5)$.

$x + 3$

3 STANDARDIZED TEST PRACTICE

Which expression is equal to $(a^2 - 5a + 3)(2 - a)^{-1}$? D

A $a + 3$

B $-a + 3 + \dfrac{3}{2 - a}$

C $-a - 3 + \dfrac{3}{2 - a}$

D $-a + 3 - \dfrac{3}{2 - a}$

Additional Examples also in Interactive Classroom PowerPoint® Presentations

IWB Interactive White Board READY

Tips for New Teachers

Test Taking Strategy Point out that $3 - a$ in Example 3 is rewritten as $-a + 3$ before starting the division in order to have both the numerator and denominator written in descending order. The first step in the long division eliminates choice A, and students also can eliminate choice B by multiplying $-a + 10$ by $-a + 3$ and noting that the product is not $a^2 + 7a - 11$.

Just as with the division of whole numbers, the division of two polynomials may result in a quotient with a remainder. Remember that $11 \div 3 = 3 + R2$, which is often written as $3\frac{2}{3}$. The result of a division of polynomials with a remainder can be written in a similar manner.

Standardized Test Example 3 Divide Polynomials

Which expression is equal to $(a^2 + 7a - 11)(3 - a)^{-1}$?

A $a + 10 - \dfrac{19}{3 - a}$

C $-a - 10 + \dfrac{19}{3 - a}$

B $-a + 10$

D $-a - 10 - \dfrac{19}{3 - a}$

Read the Test Item

Since the second factor has an exponent of -1, this is a division problem.

$(a^2 + 7a - 11)(3 - a)^{-1} = \dfrac{a^2 + 7a - 11}{3 - a}$

Solve the Test Item

$$
\begin{array}{r}
-a - 10 \\
-a + 3 \overline{)a^2 + 7a - 11} \\
(-)\ a^2 - 3a \\
\hline
10a - 11 \\
(-)\ 10a - 30 \\
\hline
19
\end{array}
$$

For ease in dividing, rewrite $3 - a$ as $-a + 3$.

$-a(-a + 3) = a^2 - 3a$
$7a - (-3a) = 10a$

$-10(-a + 3) = 10a - 30$
$-11 - (-30) = 19$

The quotient is $-a - 10$, and the remainder is 19.

Therefore, $(a^2 + 7a - 11)(3 - a)^{-1} = -a - 10 + \dfrac{19}{3 - a}$. The answer is **C**.

▶ **Guided Practice**

3. Which expression is equal to $(r^2 + 5r + 7)(1 - r)^{-1}$? F

F $-r - 6 + \dfrac{13}{1 - r}$

H $r - 6 + \dfrac{13}{1 - r}$

G $r + 6$

J $r + 6 - \dfrac{13}{1 - r}$

2 Synthetic Division **Synthetic division** is a simpler process for dividing a polynomial by a binomial. Suppose you want to divide $2x^3 - 13x^2 + 26x - 24$ by $x - 4$ using long division. Compare the coefficients in this division with those in Example 4.

$$
\begin{array}{r}
2x^2 - 5x + 6 \\
x - 4 \overline{)2x^3 - 13x^2 + 26x - 24} \\
(-)\ 2x^3 - 8x^2 \\
\hline
-5x^2 + 26x \\
(-)\ -5x^2 + 20x \\
\hline
6x - 24 \\
(-)\ 6x - 24 \\
\hline
0
\end{array}
$$

When the polynomial in the dividend is missing a term, a zero must be used to represent the missing term. So, with a dividend of $2x^3 - 4x^2 + 6$, a 0 will be used as a placeholder for the x-term.

$\overline{)2x^3 - 4x^2 + 0x + 6}$

Test-Taking Tip

Multiple Choice You may be able to eliminate some of the answer choices by substituting the same value for a in the original expression and the answer choices, and then evaluate.

KeyConcept Synthetic Division

Step 1 Write the coefficients of the dividend so that the degrees of the terms are in descending order. Write the constant r of the divisor $x - r$ in the box. Bring the first coefficient down.

Step 2 Multiply the first coefficient by r, and write the product under the second coefficient.

Step 3 Add the product and the second coefficient.

Step 4 Repeat Steps 2 and 3 until you reach a sum in the last column. The numbers along the bottom row are the coefficients of the quotient. The power of the first term is one less than the degree of the dividend. The final number is the remainder.

Example 4 Synthetic Division

Use synthetic division to find $(2x^3 - 13x^2 + 26x - 24) \div (x - 4)$.

Step 1 Write the coefficients of the dividend. Write the constant r in the box. In this case, $r = 4$. Bring the first coefficient, 2, down.

$$\begin{array}{c|cccc} 4 & 2 & -13 & 26 & -24 \\ & \downarrow & & & \\ \hline & 2 & & & \mid \end{array}$$

Step 2 Multiply the first coefficient by r: $2 \cdot 4 = 8$. Write the product under the second coefficient.

$$\begin{array}{c|cccc} 4 & 2 & -13 & 26 & -24 \\ & & 8 & & \\ \hline & 2 \nearrow & & & \mid \end{array}$$

Step 3 Add the product and the second coefficient: $-13 + 8 = -5$.

$$\begin{array}{c|cccc} 4 & 2 & -13 & 26 & -24 \\ & & 8 & & \\ \hline & 2 & -5 & & \mid \end{array}$$

Step 4 Multiply the sum, -5, by r: $-5 \times 4 = -20$. Write the product under the next coefficient, and add: $26 + (-20) = 6$. Multiply the sum, 6, by r: $6 \cdot 4 = 24$. Write the product under the next coefficient and add: $-24 + 24 = 0$.

$$\begin{array}{c|cccc} 4 & 2 & -13 & 26 & -24 \\ & & 8 & -20 & 24 \\ \hline & 2 & -5 \nearrow & 6 \nearrow & \mid 0 \end{array}$$

CHECK Multiply the quotient by the divisor. The answer should be the dividend.

$$\begin{array}{r} 2x^2 - 5x + 6 \\ (\times) \quad x - 4 \\ \hline -8x^2 + 20x - 24 \\ 2x^3 - 5x^2 + 6x \\ \hline 2x^3 - 13x^2 + 26x - 24 \end{array}$$

The quotient is $2x^2 - 5x + 6$. The remainder is 0.

> **GuidedPractice** **4A.** $2x^2 - 3x + 5$ **4B.** $3x^2 - 2x + 7$
>
> **Use synthetic division to find each quotient.**
>
> **4A.** $(2x^3 + 3x^2 - 4x + 15) \div (x + 3)$ **4B.** $(3x^3 - 8x^2 + 11x - 14) \div (x - 2)$
>
> **4C.** $(4a^4 + 2a^2 - 4a + 12) \div (a + 2)$ **4D.** $(6b^4 - 8b^3 + 12b - 14) \div (b - 2)$

WatchOut!
Synthetic Division
Remember to *add* terms when performing synthetic division.

4C. $4a^3 - 8a^2 + 18a - 40 + \dfrac{92}{a + 2}$
4D. $6b^3 + 4b^2 + 8b + 28 + \dfrac{42}{b - 2}$

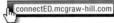

2 Synthetic Division
Example 4 shows how to use synthetic division to divide a polynomial by a binomial. **Example 5** shows how to use synthetic division to divide a polynomial by a binomial when the coefficient of x in the divisor is not 1.

Additional Example

4 Use synthetic division to find $(x^3 - 4x^2 + 6x - 4) \div (x - 2)$.
$x^2 - 2x + 2$

WatchOut!

Preventing Errors In the Guided Practice exercises, remind students to include a coefficient of 0 for any missing terms in the dividend.

Focus on Mathematical Content
Dividing Polynomials You can use synthetic division to divide a polynomial by a binomial. The terms in both the divisor and dividend must be in descending order of power, and a coefficient of 0 must be included for any term that is missing. If the divisor is not in the form $x - r$, divide every term in both the divisor and dividend by the coefficient of the first term of divisor and then use synthetic division.

Teach with Tech

Video Recording Have students work in groups, and assign a different polynomial division problem to each group. Have each group create a video showing how to use synthetic division to find the quotient. Post all of the videos to a video sharing Web site for students to view.

Additional Example

5 Use synthetic division to find
$(4y^3 - 6y^2 + 4y - 1) \div (2y - 1)$.
$2y^2 - 2y + 1$

(CCSS) Teaching the Mathematical Practices

Precision Mathematically proficient students calculate accurately and efficiently, express numerical answers with a degree of precision appropriate for the problem context. Encourage students to consider both long division and synthetic division and decide which method may be easier to use in a specific situation.

WatchOut!

CCSS Precision Remember to divide *all* terms in the numerator and denominator.

To use synthetic division, the divisor must be of the form $x - r$. If the coefficient of x in a divisor is not 1, you can rewrite the division expression so that you can use synthetic division.

Example 5 Divisor with First Coefficient Other than 1

Use synthetic division to find $(3x^4 - 5x^3 + x^2 + 7x) \div (3x + 1)$.

$$\frac{3x^4 - 5x^3 + x^2 + 7x}{3x + 1} = \frac{(3x^4 - 5x^3 + x^2 + 7x) \div 3}{(3x + 1) \div 3}$$

Rewrite the divisor with a leading coefficient of 1. Then divide the numerator and denominator by 3.

$$= \frac{x^4 - \frac{5}{3}x^3 + \frac{1}{3}x^2 + \frac{7}{3}x}{x + \frac{1}{3}}$$

Simplify the numerator and the denominator.

Since the numerator does not have a constant term, use a coefficient of 0 for the constant term.

$x - r = x + \frac{1}{3}$, so $r = -\frac{1}{3}$. →

$$-\frac{1}{3} \begin{array}{|ccccc} 1 & -\frac{5}{3} & \frac{1}{3} & \frac{7}{3} & 0 \\ & -\frac{1}{3} & \frac{2}{3} & -\frac{1}{3} & -\frac{2}{3} \\ \hline 1 & -2 & 1 & 2 & -\frac{2}{3} \end{array}$$

The result is $x^3 - 2x^2 + x + 2 - \dfrac{\frac{2}{3}}{x + \frac{1}{3}}$. Now simplify the fraction.

$$\frac{\frac{2}{3}}{x + \frac{1}{3}} = \frac{2}{3} \div \left(x + \frac{1}{3}\right)$$ *Rewrite as a division expression.*

$$= \frac{2}{3} \div \frac{3x + 1}{3}$$ $x + \frac{1}{3} = \frac{3x}{3} + \frac{1}{3} = \frac{3x + 1}{3}$

$$= \frac{2}{3} \cdot \frac{3}{3x + 1}$$ *Multiply by the reciprocal.*

$$= \frac{2}{3x + 1}$$ *Simplify.*

The solution is $x^3 - 2x^2 + x + 2 - \dfrac{2}{3x + 1}$.

CHECK Divide using long division.

$$
\begin{array}{r}
x^3 - 2x^2 + x + 2 \\
3x + 1 \overline{)3x^4 - 5x^3 + x^2 + 7x} \\
\underline{(-)\ 3x^4 + x^3} \\
-6x^3 + x^2 \\
\underline{(-)\ -6x^3 - 2x^2} \\
3x^2 + 7x \\
\underline{(-)\ 3x^2 + x} \\
6x + 0 \\
\underline{(-)\ 6x + 2} \\
-2
\end{array}
$$

The result is $x^3 - 2x^2 + x + 2 - \dfrac{2}{3x + 1}$. ✓

5A. $4x^3 - 2x^2 - x + 1 + \dfrac{3}{2x + 1}$

5B. $2y^4 - 4y - 1 + \dfrac{3}{4y - 1}$

▶ **Guided Practice**   **5C.** $3b^2 + 4b - 1 + \dfrac{2}{5b - 4}$ **5D.** $2c^2 - 3c - 2$

Use synthetic division to find each quotient.

5A. $(8x^4 - 4x^2 + x + 4) \div (2x + 1)$ **5B.** $(8y^5 - 2y^4 - 16y^2 + 4) \div (4y - 1)$

5C. $(15b^3 + 8b^2 - 21b + 6) \div (5b - 4)$ **5D.** $(6c^3 - 17c^2 + 6c + 8) \div (3c - 4)$

 314 | Lesson 5-2 | Dividing Polynomials

Differentiated Instruction (AL) (OL)

If ▶ some students have trouble keeping their concentration throughout the sequence of steps required in long division,

Then ▶ encourage them to compare intermediate results with a partner so they can ask questions and catch errors before completing the entire problem.

Examples 1, 2, and 4

Simplify.

5. $3z^3 - 15z^2 + 36z - 105 + \dfrac{309}{z+3}$　6. $y^4 + 2y^3 + 4y^2 + 5y + 10$

1. $\dfrac{4xy^2 - 2xy + 2x^2y}{xy}$　$4y + 2x - 2$

2. $(3a^2b - 6ab + 5ab^2)(ab)^{-1}$　$3a + 5b - 6$

3. $(x^2 - 6x - 20) \div (x + 2)$　$x - 8 - \dfrac{4}{x+2}$

4. $(2a^2 - 4a - 8) \div (a + 1)$　$2a - 6 - \dfrac{2}{a+1}$

5. $(3z^4 - 6z^3 - 9z^2 + 3z - 6) \div (z + 3)$

6. $(y^5 - 3y^2 - 20) \div (y - 2)$

Example 3

7. **MULTIPLE CHOICE** Which expression is equal to $(x^2 + 3x - 9)(4 - x)^{-1}$? **A**

　A $-x - 7 + \dfrac{19}{4-x}$　　**B** $-x - 7$　　**C** $x + 7 - \dfrac{19}{4-x}$　　**D** $-x - 7 - \dfrac{19}{4-x}$

Example 5

Simplify.　8. $2x + 1 + \dfrac{3}{x+1}$

8. $(10x^2 + 15x + 20) \div (5x + 5)$

9. $(18a^2 + 6a + 9) \div (3a - 2)$　$6a + 6 + \dfrac{21}{3a-2}$

10. $\dfrac{12b^2 + 23b + 15}{3b + 8}$　$4b - 3 + \dfrac{39}{3b+8}$

11. $\dfrac{27y^2 + 27y - 30}{9y - 6}$　$3y + 5$

Practice and Problem Solving　　　　　　　　　　　Extra Practice is on page R5.

Example 1

Simplify. 12. $3a^2b - 2ab^2$　13. $x + 3y - 2$　14. $7g^2h + 3g - 2h^2$

12. $\dfrac{24a^3b^2 - 16a^2b^3}{8ab}$

13. $\dfrac{5x^2y - 10xy + 15xy^2}{5xy}$

14. $\dfrac{7g^3h^2 + 3g^2h - 2gh^3}{gh}$

15. $\dfrac{4a^3b - 6ab + 2ab^2}{2ab}$　$2a^2 + b - 3$

16. $\dfrac{16c^4d^4 - 24c^2d^2}{4c^2d^2}$　$4c^2d^2 - 6$

17. $\dfrac{9n^3p^3 - 18n^2p^2 + 21n^2p^3}{3n^2p^2}$　$3np - 6 + 7p$

18. **ENERGY** Compact fluorescent light (CFL) bulbs reduce energy waste. The amount of energy waste that is reduced each day in a certain community can be estimated by $-b^2 + 8b$, where b is the number of bulbs. Divide by b to find the average amount of energy saved per CFL bulb. $-b + 8$

19. **BAKING** The number of cookies produced in a factory each day can be estimated by $-w^2 + 16w + 1000$, where w is the number of workers. Divide by w to find the average number of cookies produced per worker. $-w + 16 + \dfrac{1000}{w}$

Examples 2, 4, and 5

Simplify. 21. $b^2 - 5b + 6 - \dfrac{8}{b+1}$　23. $x^4 + 4x^3 + 12x^2 + 52x + 208 + \dfrac{832}{x-4}$

25. $g^3 + 2g^2 + g + 2 - \dfrac{14}{g-2}$

20. $(a^2 - 8a - 26) \div (a + 2)$　$a - 10 - \dfrac{6}{a+2}$

 $(b^3 - 4b^2 + b - 2) \div (b + 1)$

22. $(z^4 - 3z^3 + 2z^2 - 4z + 4)(z - 1)^{-1}$

23. $(x^5 - 4x^3 + 4x^2) \div (x - 4)$

26. $2a + \dfrac{1}{3} + \dfrac{29}{9a - 6}$

24. $\dfrac{y^3 + 11y^2 - 10y + 6}{y + 2}$

25. $(g^4 - 3g^2 - 18) \div (g - 2)$

22. $z^3 - 2z^2 - 4$

24. $y^2 + 9y - 28 + \dfrac{62}{y+2}$

27. $2x^4 + x^3 - x + \dfrac{2}{3} - \dfrac{2}{9x + 3}$

26. $(6a^2 - 3a + 9) \div (3a - 2)$

27. $\dfrac{6x^5 + 5x^4 + x^3 - 3x^2 + x}{3x + 1}$

28. $\dfrac{4g^4 - 6g^3 + 3g^2 - g + 12}{4g - 4}$

29. $(2b^3 - 6b^2 + 8b) \div (2b + 2)$　29. $b^2 - 4b + 8 - \dfrac{8}{b+1}$

28. $g^3 - \dfrac{1}{2}g^2 + \dfrac{1}{4}g + \dfrac{3}{g-1}$

30. $(6z^6 + 3z^4 - 9z^2)(3z - 6)^{-1}$

30. $2z^5 + 4z^4 + 9z^3 + 18z^2 + 33z + 66 + \dfrac{132}{z-2}$

31. $(10y^6 + 5y^5 + 10y^3 - 20y - 15)(5y + 5)^{-1}$　$2y^5 - y^4 + y^3 + y^2 - y - 3$

32. **CCSS REASONING** A rectangular box for a new product is designed in such a way that the three dimensions always have a particular relationship defined by the variable x. The volume of the box can be written as $6x^3 + 31x^2 + 53x + 30$, and the height is always $x + 2$. What are the width and length of the box? $2x + 3, 3x + 5$

33. **PHYSICS** The voltage V is related to current I and power P by the equation $V = \dfrac{P}{I}$. The power of a generator is modeled by $P(t) = t^3 + 9t^2 + 26t + 24$. If the current of the generator is $I = t + 4$, write an expression that represents the voltage. $V(t) = t^2 + 5t + 6$

 connectED.mcgraw-hill.com　**315**

Formative Assessment

Use Exercises 1–11 to check for understanding.

Use the chart at the bottom of this page to customize assignments for your students.

CCSS Teaching the Mathematical Practices

Reasoning Mathematically proficient students make sense of quantities and their relationships in problem situations. Quantitative reasoning entails habits of creating a coherent representation of the problem at hand; considering the units involved; attending to the meaning of quantities, not just how to compute them; and knowing and flexibly using different properties of operations and objects.

Tips for New Teachers

Reasoning When solving problems using the division algorithm, advise students to do a simple numeric division example such as 8 ÷ 5 to help them remember how to write the remainder as part of the quotient.

Differentiated Homework Options

Level	Assignment	Two-Day Option	
AL Basic	12–33, 43, 45–72	13–33 odd, 49–52	12–32 even, 43, 53–72
OL Core	13–33 odd, 34, 35, 37–41 odd, 42, 43, 45–72	12–33, 49–52	34–43, 45–48, 53–72
BL Advanced	34–66, (optional: 67–72)		

📁 **Study Guide and Intervention**
AL OL ELL

📁 **Practice**
AL OL BL ELL

📁 **Word Problem Practice**
AL OL BL

▶ **34. ENTERTAINMENT** A magician gives these instructions to a volunteer.

- Choose a number and multiply it by 4.
- Then add the sum of your number and 15 to the product you found.
- Now divide by the sum of your number and 3.

   a. What number will the volunteer always have at the end? **5**

   b. Explain the process you used to discover the answer. **See margin.**

35 BUSINESS The number of magazine subscriptions sold can be estimated by $n = \dfrac{3500a^2}{a^2 + 100}$, where a is the amount of money the company spent on advertising in hundreds of dollars and n is the number of subscriptions sold.

   a. Perform the division indicated by $\dfrac{3500a^2}{a^2 + 100}$. $3500 - \dfrac{350{,}000}{a^2 + 100}$

   b. About how many subscriptions will be sold if \$1500 is spent on advertising? **about 2423 subscriptions**

47. $\dfrac{5}{x^2}$ does not belong with the other three. The other three expressions are polynomials. Since the denominator of $\dfrac{5}{x^2}$ contains a variable, it is not a polynomial.

▶ **Simplify. 36–41. See margin.**

36. $(x^4 - y^4) \div (x - y)$

37. $(28c^3d^2 - 21cd^2) \div (14cd)$

38. $(a^3b^2 - a^2b + 2b)(-ab)^{-1}$

39. $\dfrac{n^3 + 3n^2 - 5n - 4}{n + 4}$

40. $\dfrac{p^3 + 2p^2 - 7p - 21}{p + 3}$

41. $\dfrac{3z^5 + 5z^4 + z + 5}{z + 2}$

42. 🔷 MULTIPLE REPRESENTATIONS Consider a rectangle with area $2x^2 + 7x + 3$ and length $2x + 1$. **a–c. See margin.**

   a. Concrete Use algebra tiles to represent this situation. Use the model to find the width.

   b. Symbolic Write an expression to represent the model.

   c. Numerical Solve this problem algebraically using synthetic or long division. Does your concrete model check with your algebraic model?

45. Sample answer: The degree of the quotient plus the degree of the divisor equals the degree of the dividend.

H.O.T. Problems   Use Higher-Order Thinking Skills

43. ERROR ANALYSIS Sharon and Jamal are dividing $2x^3 - 4x^2 + 3x - 1$ by $x - 3$. Sharon claims that the remainder is 26. Jamal argues that the remainder is -100. Is either of them correct? Explain your reasoning. **Sample answer: Sharon; Jamal actually divided by $x + 3$.**

48. Sample answer: By dividing $140x^2 + 60x$ by $10x$, the quotient of $14x + 6$ provides the length of the book jacket. Then subtracting $14x$, we are left with 6 inches. Half of this length is the width of each flap.

44. CHALLENGE If a polynomial is divided by a binomial and the remainder is 0, what does this tell you about the relationship between the binomial and the polynomial? **The binomial is a factor of the polynomial.**

45. REASONING Review any of the division problems in this lesson. What is the relationship between the degrees of the dividend, the divisor, and the quotient?

46. OPEN ENDED Write a quotient of two polynomials for which the remainder is 3.

46. Sample answer: $\dfrac{x^2 + 5x + 9}{x + 2}$

47. CCSS ARGUMENTS Identify the expression that does not belong with the other three. Explain your reasoning.

$3xy + 6x^2$	$\dfrac{5}{x^2}$	$x + 5$	$5b + 11c - 9ad^2$

48. WRITING IN MATH Use the information at the beginning of the lesson to write assembly instructions using the division of polynomials to make a paper cover for your textbook.

🔊 **316** | Lesson 5-2 | Dividing Polynomials

📁 **Enrichment**
OL BL

🔷 **Multiple Representations**

In Exercise 42, students use algebra tiles and algebraic operations to divide polynomials.

Additional Answer

34b. Sample answer: Let x be the number. Multiply x by 4 to get $4x$. Then add $x + 15$ to the product to get $5x + 15$. Divide the polynomial by $x + 3$. The quotient is 5.

49. An office employs x women and 3 men. What is the ratio of the total number of employees to the number of women? **A**

A $\dfrac{x+3}{x}$ C $\dfrac{3}{x}$

B $\dfrac{x}{x+3}$ D $\dfrac{x}{3}$

50. SAT/ACT Which polynomial has degree 3? **K**

F $x^3 + x^2 - 2x^4$ J $x^2 + x + 12^3$

G $-2x^2 - 3x + 4$ K $1 + x + x^3$

H $3x - 3$

51. GRIDDED RESPONSE In the figure below, $m + n + p = ?$ **360**

52. $(-4x^2 + 2x + 3) - 3(2x^2 - 5x + 1) =$ **C**

A $2x^2$ C $-10x^2 + 17x$

B $-10x^2$ D $2x^2 + 17x$

Spiral Review

Simplify. (Lesson 5-1) **53.** $3x^3 + 2x^2 + x + 4$ **54.** $2y^3 + 3y^2 - 9y + 8$ **56.** $2c^3 - 3c^2d - 2cd^2 + 3d^3$

53. $(5x^3 + 2x^2 - 3x + 4) - (2x^3 - 4x)$ **54.** $(2y^3 - 3y + 8) + (3y^2 - 6y)$ **55.** $4a(2a - 3) + 3a(5a - 4)$ $23a^2 - 24a$

56. $(c + d)(c - d)(2c - 3d)$ **57.** $(xy)^2(2xy^2z)^3$ $8x^5y^8z^3$ **58.** $(3ab^2)^{-2}(2a^2b)^2$ $\dfrac{4a^2}{9b^2}$

59. LANDSCAPING Amado wants to plant a garden and surround it with decorative stones. He has enough stones to enclose a rectangular garden with a perimeter of 68 feet, but he wants the garden to cover no more than 240 square feet. What could the width of his garden be? (Lesson 4-8) **0 to 10 ft or 24 to 34 ft**

Solve each equation by completing the square. (Lesson 4-5)

60. $x^2 + 6x + 2 = 0$ $-3 \pm \sqrt{7}$ **61.** $x^2 - 8x - 3 = 0$ $4 \pm \sqrt{19}$ **62.** $2x^2 + 6x + 5 = 0$ $-1.5 \pm \dfrac{i}{2}$

State the consecutive integers between which the zeros of each quadratic function are located. (Lesson 4-2)

63.

x	-7	-6	-5	-4	-3	-2	-1	0
$f(x)$	4	1	-3	-8	-1	2	8	16

between -6 and -5; between -3 and -2

64.

x	-2	-1	0	1	2	3	4	5
$f(x)$	-16	-7	-4	3	3	-4	-7	-16

between 0 and 1; between 2 and 3

65.

x	-2	-1	0	1	2	3	4	5
$f(x)$	6	1	-3	-5	-3	1	6	14

between -1 and 0; between 2 and 3

66. BUSINESS A landscaper can mow a lawn in 30 minutes and perform a small landscape job in 90 minutes. He works at most 10 hours per day, 5 days per week. He earns $35 per lawn and $125 per landscape job. He cannot do more than 3 landscape jobs per day. Find the combination of lawns mowed and completed landscape jobs per week that will maximize income. Then find the maximum income. (Lesson 3-3)
15 landscape jobs and 55 lawns; $3800

Skills Review

Find each value if $f(x) = 4x + 3$, $g(x) = -x^2$, and $h(x) = -2x^2 - 2x + 4$.

67. $f(-6)$ -21 **68.** $g(-8)$ -64 **69.** $h(3)$ -20

70. $f(c)$ $4c + 3$ **71.** $g(3d)$ $-9d^2$ **72.** $h(2b + 1)$ $-8b^2 - 12b$

DifferentiatedInstruction BL

Extension Dividing by polynomials that contain fractions or radicals requires attention to detail and careful calculations. Ask students to divide $x^4 - 6x^2 + 8$ by $(x - \sqrt{2})$. $x^3 + \sqrt{2}x^2 - 4x - 4\sqrt{2}$

CCSS Teaching the Mathematical Practices

Arguments Mathematically proficient students understand and use stated assumptions, definitions, and previously established results in constructing arguments. They make conjectures and build a logical progression of statements to explore the truth of their conjectures. And they are able to analyze situations by breaking them into cases, and can recognize and use counterexamples.

4 Assess

Name the Math Ask students to explain how to use synthetic division to divide a polynomial by a binomial. Ask them to demonstrate the technique using the polynomial $3x^3 - 4x^2 + 5x^2 - 7$ and the binomial $x - 2$.

Formative Assessment
Check for student understanding of concepts in Lessons 5-1 and 5-2.

▢ Quiz 1, p. 57

Additional Answers

36. $(x^2 + y^2)(x + y)$

37. $\dfrac{4c^2d - 3d}{2}$

38. $-a^2b + a - \dfrac{2}{a}$

39. $n^2 - n - 1$

40. $p^2 - p - 4 - \dfrac{9}{p+3}$

41. $3z^4 - z^3 + 2z^2 - 4z + 9 - \dfrac{13}{z+2}$

42a.

The width is $x + 3$.

42b. $2x^2 + 7x + 3 \div (2x + 1)$

42c.

$$-\dfrac{1}{2} \Big| \quad 1 \quad \dfrac{7}{2} \quad \dfrac{3}{2}$$
$$\quad\quad -\dfrac{1}{2} \quad -\dfrac{3}{2}$$
$$\overline{\quad 1 \quad 3 \quad | \quad 0}$$

yes

1 Focus

Objective Use a graphing calculator with CAS to divide polynomials.

Materials
- TI-Nspire CAS technology

Teaching Tip
- Sometimes long equations cannot be displayed on one screen. Use the right arrow to scroll the display to the right.

- Remind students that they cannot edit a line once they press **enter**. However, they can use **ctrl C** and **ctrl V** to copy and paste an equation, and then make edits.

- In Step 2 of Activity 3, students can use long division or the CAS to divide.

2 Teach

Working in Cooperative Groups
Divide the class into pairs. Work through Activities 1 and 2 as a class. Then ask students to work with their partners to complete Activity 3.

Practice Have students complete Exercises 1–5.

EXTEND **5-2** Graphing Technology Lab
Dividing Polynomials

Long division and synthetic division are two alternatives for dividing polynomials with linear divisors. You can use a graphing calculator with a computer algebra system (CAS) to divide polynomials with any divisor.

 Common Core State Standards
Content Standards
A.APR.6 Rewrite simple rational expressions in different forms; write $a(x)/b(x)$ in the form $q(x) + r(x)/b(x)$, where $a(x)$, $b(x)$, $q(x)$, and $r(x)$ are polynomials with the degree of $r(x)$ less than the degree of $b(x)$, using inspection, long division, or, for the more complicated examples, a computer algebra system.

Activity 1   Divide Polynomials Without Remainders

Use CAS to find $(x^4 + 3x^3 - x^2 - 5x + 2) \div (x^2 + 2x - 1)$.

Step 1 Add a new **Calculator** page on the TI-Nspire.

Step 2 From the menu, select **Algebra**, then **Polynomial Tools** and **Quotient of Polynomial**.

Step 3 Type the dividend, a comma, and the divisor.

The CAS indicates that $(x^4 + 3x^3 - x^2 - 5x + 2) \div (x^2 + 2x - 1)$ is $x^2 + x - 2$.

Step 4 To verify that there is no remainder, select **Remainder of a Polynomial** from the **Algebra, Polynomial Tools** menu then type the dividend, a comma, and the divisor.

In Activity 1, there was no remainder. But in many cases, there will be a remainder.

Activity 2   Divide Polynomials With Remainders

Use CAS to find $(4x^5 - 12x^4 - 7x^3 + 32x^2 + 3x + 20) \div (x^2 - 2x + 4)$.

Step 1 Add a new **Calculator** page on the TI-Nspire.

Step 2 From the menu, select **Algebra, Polynomial Tools,** and **Quotient of Polynomial**.

Step 3 Type the dividend, a comma, and the divisor.

The CAS indicates that $(4x^5 - 12x^4 - 7x^3 + 32x^2 + 3x + 20) \div (x^2 - 2x + 4)$ is $4x^3 - 4x^2 - 31x - 14$.

We need to determine whether there is a remainder.

Step 4 Use the **Remainder of a Polynomial** option from the **Algebra, Polynomial Tools** menu to determine the remainder. Then type the dividend, a comma, and the divisor.

The remainder is $99x + 76$.
Therefore, $(4x^5 - 12x^4 - 7x^3 + 32x^2 + 3x + 20) \div (x^2 - 2x + 4)$ is $4x^3 - 4x^2 - 31x - 14 + \dfrac{99x + 76}{x^2 - 2x + 4}$.

You can also use a graphing calculator to determine roots of a polynomial so you can divide with synthetic division.

Activity 3   Divide with Synthetic Division

Use synthetic division to find $(x^6 - 28x^4 + 14x^3 + 147x^2 - 14x - 120) \div (x^3 + 3x^2 - 18x - 40)$.

Step 1 Graph the dividend as $f1(x)$ and the divisor as $f2(x)$ on the same calculator page. Use the **Intersection Points** tool from **Points & Lines** menu to find where the graphs have the same x-intercepts.

Step 2 Use the roots from Step 1 as the divisors for synthetic division.

[−10, 10] scl: 1 by [−100, 100] scl: 10

Step 3 Use the **Expand** function to verify that $x^3 - 3x^2 - x + 3$ is the quotient when -5, -2, and 4 are roots.

Thus, $(x^6 - 28x^4 + 14x^3 + 147x^2 - 14x - 120) \div (x^3 + 3x^2 - 18x - 40)$ is $x^3 - 3x^2 - x + 3$.

$$5. \quad x^3 - 8 - \frac{24x}{2x^3 + 2x^2 - 4x - 2}$$

Exercises

Find each quotient.

1. $(2x^4 + x^3 - 8x^2 + 17x - 12) \div (x^2 + 2x - 3)$ $2x^2 - 3x + 4$

2. $(x^4 + 7x^3 + 8x^2 + x - 12) \div (x^2 + 3x - 4)$ $x^2 + 4x + \dfrac{17x - 12}{x^2 + 3x - 4}$

3. $(9x^5 - 9x^3 - 5x^2 + 5) \div (9x^3 - 5)$ $x^2 - 1$

4. $(x^5 - 8x^4 + 10x^3 + 14x^2 + 61x - 30) \div (x^2 - 5x + 3)$ $x^3 - 3x^2 - 8x - 17 + \dfrac{21}{x^2 - 5x + 3}$

5. $(2x^6 + 2x^5 - 4x^4 - 18x^3 - 16x^2 + 8x + 16) \div (2x^3 + 2x^2 - 4x - 2)$

6. $(6x^6 - 2x^5 - 14x^4 + 10x^3 - 4x^2 - 28x - 5) \div (3x^3 - x^2 - 7x - 1)$ $2x^3 + 4 - \dfrac{1}{3x^3 - x^2 - 7x - 1}$

7. Use synthetic division to find $(x^6 - 7x^5 - 21x^4 + 175x^3 + 56x^2 - 924x + 720) \div (x^3 - 5x^2 - 12x + 36)$.
   $x^3 - 2x^2 - 19x + 20$

connectED.mcgraw-hill.com **319**

3 Assess

Formative Assessment
Use Exercises 6 and 7 to assess each student's ability to divide polynomials.

From Concrete to Abstract
Ask students to summarize what they learned about dividing polynomials.

EXPLORE 5-3

Graphing Technology Lab
Power Functions

1 Focus

Objective Use a graphing calculator to explore power functions.

Materials

- TI-83/84 Plus or other graphing calculator

Teaching Tip

- Students may need to adjust the windows of their calculators to explore the graphs. Remind them that they can zoom in and out as well as use **zoomfit.**

- Remind students that they can use **zoomstandard** to restore the standard window.

2 Teach

Working in Cooperative Groups

Divide the class into pairs. Work through Activities 1 and 2 as a class. Then ask students to work with their partners to complete Activity 3.

Practice Have students complete Exercises 1–9.

A **power function** is any function of the form $f(x) = ax^n$, where a and n are nonzero constant real numbers. A power function in which n is a positive integer is called a **monomial function**.

CCSS Common Core State Standards
Content Standards
F.IF.4 For a function that models a relationship between two quantities, interpret key features of graphs and tables in terms of the quantities, and sketch graphs showing key features given a verbal description of the relationship.

Activity 1 $y = ax^4$

Graph $f(x) = 2x^4$. Analyze the graph.

Step 1 Graph the function.

$f(x) = 2x^4$

$[-5, 5]$ scl: 1 by $[-1, 9]$ scl: 1

Step 2 Analyze the graph.

The graph resembles the graph of $g(x) = x^2$, but it flattens out more at its vertex.

Analyze the Results

1. What assumptions can you make about the graph of $f(x) = 2x^4$ as x becomes more positive or more negative? Use the **TABLE** feature to evaluate the function for values of x to confirm this assumption, if necessary. **It appears that the value of $f(x)$ will continue to increase.**

2. State the domain and range of $f(x) = 2x^4$. Compare these values with the domain and range of $g(x) = x^2$. **D = {all real numbers}, R = {$f(x)$| $f(x) > 0$}; they are the same.**

3. What other characteristics does the graph of $f(x) = 2x^4$ share with the graph of $g(x) = x^2$? **The graphs have the same vertex, (0, 0); axis of symmetry, the line $x = 0$ (the y-axis); and y-intercept, 0.**

Activity 2 $y = ax^5$

Graph $f(x) = 3x^5$. Analyze the graph.

Step 1 Graph the function.

$f(x) = 3x^5$

$[-5, 5]$ scl: 1 by $[-5, 5]$ scl: 1

Step 2 Analyze the graph.

The graph resembles the graph of $f(x) = x^3$, but it flattens out more as the graph approaches the origin.

320 | **Explore 5-3** | Graphing Technology Lab: Power Functions

Analyze the Results

4. What assumptions can you make about the graph of $f(x) = 3x^5$ as x becomes more positive or more negative? Use the **TABLE** feature to evaluate the function for values of x to confirm this assumption, if necessary. **It appears that the value of $f(x)$ will continue to increase as x becomes more positive and decrease as x becomes more negative.**

5. State the domain and range of $f(x) = 3x^5$. Compare these values with the domain and range of $g(x) = x^3$. **D = {all real numbers}, R = {all real numbers}; they are the same.**

6. What other characteristics does the graph of $f(x) = 3x^5$ share with the graph of $g(x) = x^3$? **The graphs have the same y-intercept, 0, and the same shape.**

7. Compare the characteristics of the graph $h(x) = -3x^5$ with the graph of $f(x) = 3x^5$. What conclusions can you make about $f(x) = ax^5$ when a is positive and when a is negative? **See margin.**

Activity 3 $y = x^n$, where n is even

Graph $f(x) = x^2$, $g(x) = x^4$, and $h(x) = x^6$ on the same screen. Analyze the graphs.

Step 1 Graph the function.

[−5, 5] scl: 1 by [−1, 9] scl: 1

[−2, 2] scl: 1 by [−1, 2] scl: 1

Step 2 The graphs are all U-shaped, but the widths are different. The graphs also differ around the origin. If you zoom in around the origin, you can see this more clearly. You can do this by using the **ZOOM** feature or by adjusting the window manually.

Analyze the Results

8. The graphs will have the same appearance as $f(x) = x^2$, but they will get narrower and narrower as the exponent increases to greater even numbers.

8. What assumptions can you make about the graphs of $a(x) = x^8$, $b(x) = x^{10}$, and so on?

9. Identify the common characteristics of the graphs of power functions in which the power is an even number. **See margin.**

10. Graph $f(x) = x^3$, $g(x) = x^5$, and $h(x) = x^7$ on the same graph. **See margin.**

11. What assumptions can you make about the graphs of $a(x) = x^9$, $b(x) = x^{11}$, and so on? **The graphs will have the same appearance as $f(x) = x^3$, but they will get narrower and narrower as the exponent increases to greater odd numbers.**

12. Identify the common characteristics of the graphs of power functions in which the power is an odd number.

13. Graph $f(x) = x^4$ and $g(x) = -x^4$ on the same graph. **See margin.**

12. The graphs will appear to increase to the origin as x goes from negative infinity to zero, then cross the origin, then increase as x goes from 0 to infinity. The graphs will get narrower and narrower as the exponent increases to greater odd numbers.

14. Graph $f(x) = x^5$ and $g(x) = -x^5$ on the same graph. **See margin.**

15. What assumptions can you make about the effects of a negative value of a in $f(x) = ax^n$ when n is even? when n is odd? **When x is odd, changing a from positive to negative will reflect the graph in the origin. When x is even, changing a from positive to negative will reflect the graph in the x-axis.**

connectED.mcgraw-hill.com   **321**

---

3 Assess

Formative Assessment
Use Exercises 10–15 to assess each student's understanding of power functions.

From Concrete to Abstract
Ask students to summarize what they learned about how the graphs of power functions behave as x increases or decreases.

Additional Answers

7. The graphs are reflections in the origin. The graph of $f(x)$ increases as x increases while the graph of $h(x)$ decreases as x increases. When a is positive, the graph will increase from left to right, appearing to approach negative infinity as x approaches negative infinity, and appearing to approach positive infinity as x approaches positive infinity. When a is negative, the opposite is true.

9. The graphs will appear to decrease to the origin as x goes from negative infinity to zero, then cross the origin, and increase as x goes from 0 to infinity. The graphs will get narrower as the exponent increases to greater even numbers.

10.

[−5, 5] scl: 1 by [−50, 50] scl: 10

13.

[−5, 5] scl: 1 by [−5, 5] scl: 1

14.

[−5, 5] scl: 1 by [−5, 5] scl: 1

1 Focus

VerticalAlignment

Before Lesson 5-3 Analyze graphs of quadratic functions.

Lesson 5-3 Evaluate polynomial functions. Identify general shapes of graphs of polynomial functions.

After Lesson 5-3 Use tools including factoring to transform and solve polynomial equations.

2 Teach

Scaffolding Questions

Have students read the **Why?** section of the lesson.

Ask:

- What is a respiratory cycle? Sample answer: a sequence consisting of one inhalation and one exhalation

- Using the formula, what is $v(1)$? 0.288 liter

- What is the largest value of t that makes sense for this model? 5

LESSON 5-3 Polynomial Functions

:: Then	:: Now	:: Why?
• You analyzed graphs of quadratic functions.	**1** Evaluate polynomial functions. **2** Identify general shapes of graphs of polynomial functions.	• The volume of air in the lungs during a 5-second respiratory cycle can be modeled by $v(t) = -0.037t^3 + 0.152t^2 + 0.173t$, where v is the volume in liters and t is the time in seconds. This model is an example of a polynomial function.

NewVocabulary
polynomial in one variable
leading coefficient
polynomial function
power function
quartic function
quintic function
end behavior

(CCSS) Common Core State Standards

Content Standards
F.IF.4 For a function that models a relationship between two quantities, interpret key features of graphs and tables in terms of the quantities, and sketch graphs showing key features given a verbal description of the relationship.

F.IF.7.c Graph polynomial functions, identifying zeros when suitable factorizations are available, and showing end behavior.

Mathematical Practices
1 Make sense of problems and persevere in solving them.

1 Polynomial Functions A **polynomial in one variable** is an expression of the form $a_n x^n + a_{n-1} x^{n-1} + \cdots + a_2 x^2 + a_1 x + a_0$, where $a_n \neq 0$, a_{n-1}, a_2, a_1, and a_0 are real numbers, and n is a nonnegative integer.

The polynomial is written in standard form when the values of the exponents are in descending order. The degree of the polynomial is the value of the greatest exponent. The coefficient of the first term of a polynomial in standard form is called the **leading coefficient**.

Polynomial	Expression	Degree	Leading Coefficient
Constant	12	0	12
Linear	$4x - 9$	1	4
Quadratic	$5x^2 - 6x - 9$	2	5
Cubic	$8x^3 + 12x^2 - 3x + 1$	3	8
General	$a_n x^n + a_{n-1} x^{n-1} + \cdots + a_1 x + a_0$	n	a_n

Example 1 Degrees and Leading Coefficients

State the degree and leading coefficient of each polynomial in one variable. If it is not a polynomial in one variable, explain why.

a. $8x^5 - 4x^3 + 2x^2 - x - 3$

This is a polynomial in one variable. The greatest exponent is 5, so the degree is 5 and the leading coefficient is 8.

b. $12x^2 - 3xy + 8x$

This is not a polynomial in one variable. There are two variables, x and y.

c. $3x^4 + 6x^3 - 4x^8 + 2x$

This is a polynomial in one variable. The greatest exponent is 8, so the degree is 8 and the leading coefficient is -4.

▶ **Guided**Practice   1A–1C. See margin.

1A. $5x^3 - 4x^2 - 8x + \frac{4}{x}$ **1B.** $5x^6 - 3x^4 + 12x^3 - 14$ **1C.** $8x^4 - 2x^3 - x^6 + 3$

 322 | Lesson 5-3

Jamie Grill/Tetra Images RF/Getty Images

Lesson 5-3 Resources

Resource	Approaching Level **AL**	On Level **OL**	Beyond Level **BL**	English Learners **ELL**
Teacher Edition	• Differentiated Instruction, p. 324	• Differentiated Instruction, p. 324	• Differentiated Instruction, p. 329	
Chapter Resource Masters	• Study Guide and Intervention, pp. 17–18 • Skills Practice, p. 19 • Practice, p. 20 • Word Problem Practice, p. 21	• Study Guide and Intervention, pp. 17–18 • Skills Practice, p. 19 • Practice, p. 20 • Word Problem Practice, p. 21 • Enrichment, p. 22	• Practice, p. 20 • Word Problem Practice, p. 21 • Enrichment, p. 22	• Study Guide and Intervention, pp. 17–18 • Skills Practice, p. 19 • Practice, p. 20 • Word Problem Practice, p. 21
Other	• 5-Minute Check 5-3 • Study Notebook	• 5-Minute Check 5-3 • Study Notebook	• 5-Minute Check 5-3 • Study Notebook	• 5-Minute Check 5-3 • Study Notebook

A **polynomial function** is a continuous function that can be described by a polynomial equation in one variable. For example, $f(x) = 3x^3 - 4x + 6$ is a cubic polynomial function. The simplest polynomial functions of the form $f(x) = ax^b$ where a and b are nonzero real numbers are called **power functions**.

If you know an element in the domain of any polynomial function, you can find the corresponding value in the range.

PT

Real-World Example 2 Evaluate a Polynomial Function

RESPIRATION Refer to the beginning of the lesson. Find the volume of air in the lungs 2 seconds into the respiratory cycle.

By substituting 2 into the function we can find $v(2)$, the volume of air in the lungs 2 seconds into the respiratory cycle.

$v(t) = -0.037t^3 + 0.152t^2 + 0.173t$	Original function
$v(2) = -0.037(2)^3 + 0.152(2)^2 + 0.173(2)$	Replace t with 2.
$= -0.296 + 0.608 + 0.346$	Simplify.
$= 0.658$ L	Add.

▶ **Guided**Practice

2. Find the volume of air in the lungs 4 seconds into the respiratory cycle. **0.7332 L**

Real-World Link
Total lung capacity is approximately 6 liters in a healthy young adult.
Source: *Family Practice Notebook*

You can also evaluate functions for variables and algebraic expressions.

PT

Example 3 Function Values of Variables

Find $f(3c - 4) - 5f(c)$ if $f(x) = x^2 + 2x - 3$.

To evaluate $f(3c - 4)$, replace the x in $f(x)$ with $3c - 4$.

$f(x) = x^2 + 2x - 3$	Original function
$f(3c - 4) = (3c - 4)^2 + 2(3c - 4) - 3$	Replace x with $3c - 4$.
$= 9c^2 - 24c + 16 + 6c - 8 - 3$	Multiply.
$= 9c^2 - 18c + 5$	Simplify.

To evaluate $5f(c)$, replace x with c in $f(x)$, then multiply by 5.

$f(x) = x^2 + 2x - 3$	Original function
$5f(c) = 5(c^2 + 2c - 3)$	Replace x with c.
$= 5c^2 + 10c - 15$	Distributive Property

Now evaluate $f(3c - 4) - 5f(c)$.

$f(3c - 4) - 5f(c) = (9c^2 - 18c + 5) - (5c^2 + 10c - 15)$	
$= 9c^2 - 18c + 5 - 5c^2 - 10c + 15$	Distribute.
$= 4c^2 - 28c + 20$	Simplify.

▶ **Guided**Practice

3A. Find $g(5a - 2) + 3g(2a)$ if $g(x) = x^2 - 5x + 8$. **$37a^2 - 75a + 46$**

3B. Find $h(-4d + 3) - 0.5h(d)$ if $h(x) = 2x^2 + 5x + 3$. **$31d^2 - 70.5d + 34.5$**

connectED.mcgraw-hill.com **323**

Additional Answers (Guided Practice)

1A. Not a polynomial in one variable; $\frac{4}{x}$ has the variable with an exponent less than 0.

1B. degree $= 6$, leading coefficient $= 5$

1C. degree $= 6$, leading coefficient $= -1$

Teach with Tech

Document Camera Choose several students to show and explain to the class how to evaluate a polynomial function for a given value.

---

1 Polynomials Functions

Example 1 shows how to find the degree and leading coefficient of a polynomial in one variable. **Example 2** shows how to evaluate a polynomial function for a given value. **Example 3** shows how to evaluate a polynomial function for variables and algebraic expressions.

Formative Assessment

Use the Guided Practice exercises after each example to determine students' understanding of concepts.

Additional Examples

1 State the degree and leading coefficient of each polynomial in one variable. If it is not a polynomial in one variable, explain why.

a. $7z^3 - 4z^2 + z$ degree $= 3$, leading coefficient $= 7$

b. $6a^3 - 4a^2 + ab^2$ not in one variable because there are two variables, a and b

c. $3x^5 + 2x^2 - 4 - 8x^6$ degree $= 6$, leading coefficient $= -8$

2 **RESPIRATION** Refer to the application at the beginning of the lesson. Find the volume of air in the lungs 1.5 seconds into the respiratory cycle. **0.4766 L**

3 Find $b(2x - 1) - 3b(x)$ if $b(m) = 2m^2 + m - 1$. **$2x^2 - 9x + 3$**

▶ **Additional Examples** also in Interactive Classroom PowerPoint® Presentations

IWB Interactive White Board READY

WatchOut!

Preventing Errors Point out that the leading coefficient is not always the coefficient of the first term of a polynomial.

2 Graphs of Polynomial Functions

Example 4 shows how to use the degree and leading coefficient of a polynomial function to describe the end behavior, determine whether it is an odd-degree or even-degree polynomial function, and determine the maximum number of times the graph of the function intersects the *x*-axis.

Tips for New Teachers

Extending the Concept Since the graphs on this page show the *maximum* number of times each type of graph may intersect the *x*-axis, some students may ask about the minimum number of times each graph type may intersect the *x*-axis. Have students work in pairs using the given graphs to discuss this issue. Lead students to see that, for functions of degree 1 the minimum is 1 (the same as the maximum), for functions of degree 2 the minimum is 0, for functions of degree 4 the minimum is 0, and for functions of degree 5 the minimum is 1. Some students may notice the pattern for functions with odd and even degrees.

(CCSS) Teaching the Mathematical Practices

Sense-Making Mathematically proficient students analyze givens, constraints, relationships, and goals. They can explain correspondences between equations, verbal descriptions, tables, and graphs or draw diagrams of important features and relationships, graph data, and search for regularity or trends. Remind students that a polynomial function is a function whose rule is given by a polynomial in one variable. The leading term is the term with the variable that has the greatest exponent. The leading coefficient a_n is the coefficient of the leading term. The degree of the polynomial is the largest degree *n* of all of its terms.

---

2 Graphs of Polynomial Functions

The general shapes of the graphs of several polynomial functions show the *maximum* number of times the graph of each function may intersect the *x*-axis. This is the same number as the degree of the polynomial.

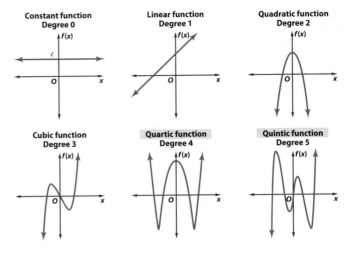

The domain of any polynomial function is all real numbers. The **end behavior** is the behavior of the graph of $f(x)$ as *x* approaches positive infinity ($x \rightarrow +\infty$) or negative infinity ($x \rightarrow -\infty$). The degree and leading coefficient of a polynomial function determine the end behavior of the graph and the range of the function.

Study Tip

(CCSS) Sense-Making The leading coefficient and degree are the sole determining factors for the end behavior of a polynomial function. With very large or very small numbers, the rest of the polynomial is insignificant in the appearance of the graph.

Review Vocabulary

infinity endless or boundless

KeyConcept End Behavior of a Polynomial Function

Degree: even **Leading Coefficient:** positive **End Behavior:** $f(x) \rightarrow +\infty$ as $x \rightarrow -\infty$; $f(x) \rightarrow +\infty$ as $x \rightarrow +\infty$; $y = x^2$; Domain: all real numbers; Range: all real numbers ≥ minimum	**Degree:** odd **Leading Coefficient:** positive **End Behavior:** $f(x) \rightarrow -\infty$ as $x \rightarrow -\infty$; $f(x) \rightarrow +\infty$ as $x \rightarrow +\infty$; $y = x^3$; Domain: all real numbers; Range: all real numbers
Degree: even **Leading Coefficient:** negative **End Behavior:** $f(x) \rightarrow -\infty$ as $x \rightarrow -\infty$; $f(x) \rightarrow -\infty$ as $x \rightarrow +\infty$; $y = -x^4$; Domain: all real numbers; Range: all real numbers ≤ maximum	**Degree:** odd **Leading Coefficient:** negative **End Behavior:** $f(x) \rightarrow +\infty$ as $x \rightarrow -\infty$; $f(x) \rightarrow -\infty$ as $x \rightarrow +\infty$; $y = -x^5$; Domain: all real numbers; Range: all real numbers

---

DifferentiatedInstruction (AL) (OL)

Logical Learners Power functions are single-term polynomial functions that can have any degree. Have students examine the graphs of a variety of power functions and describe their similarities and differences.

ReviewVocabulary

zero the *x*-coordinate of the point at which a graph intersects the *x*-axis

The number of real zeros of a polynomial function can be determined by examining its graph. Recall that real zeros occur at *x*-intercepts, so the number of times a graph crosses the *x*-axis equals the number of real zeros.

KeyConcept Zeros of Even- and Odd-Degree Functions

Odd-degree functions will always have an odd number of real zeros. Even-degree functions will always have an even number of real zeros or no real zeros at all.

Even-Degree Polynomials

Odd-Degree Polynomials

StudyTip

Double roots When a graph is tangent to the *x*-axis, there is a *double root*, which represents two of the same root.

Example 4 Graphs of Polynomial Functions

For each graph,
• describe the end behavior,
• determine whether it represents an odd-degree or an even-degree polynomial function, and
• state the number of real zeros.

a.

$f(x) \rightarrow -\infty$ as $x \rightarrow -\infty$.
$f(x) \rightarrow -\infty$ as $x \rightarrow +\infty$.

Since the end behavior is in the same direction, it is an even-degree function. The graph intersects the *x*-axis at two points, so there are two real zeros.

b.

$f(x) \rightarrow +\infty$ as $x \rightarrow -\infty$.
$f(x) \rightarrow -\infty$ as $x \rightarrow +\infty$.

Since the end behavior is in opposite directions, it is an odd-degree function. The graph intersects the *x*-axis at five points, so there are five real zeros.

GuidedPractice 4A, 4B. See margin.

4A.

4B.

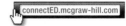
connectED.mcgraw-hill.com **325**

Additional Answers (Guided Practice)

4A. $f(x) \rightarrow +\infty$ as $x \rightarrow -\infty$. $f(x) \rightarrow -\infty$ as $x \rightarrow +\infty$. Since the end behavior is in opposite directions, it is an odd-degree function. The graph intersects the *x*-axis at three points, so there are three real zeros.

4B. $f(x) \rightarrow -\infty$ as $x \rightarrow -\infty$. $f(x) \rightarrow -\infty$ as $x \rightarrow +\infty$. Since the end behavior is in the same direction, it is an even-degree function. The graph intersects the *x*-axis at two points, so there are two real zeros.

Additional Example

4 For each graph,
• describe the end behavior,
• determine whether it represents an odd-degree or an even-degree polynomial function, and
• state the number of real zeros.

a.

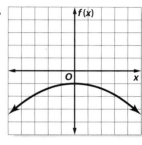

$f(x) \rightarrow -\infty$ as $x \rightarrow +\infty$.
$f(x) \rightarrow -\infty$ as $x \rightarrow -\infty$.
Since the end behavior is in the same direction, it is an even-degree polynomial function.

The graph does not intersect the *x*-axis, so the function has no real zeros.

b.

$f(x) \rightarrow +\infty$ as $x \rightarrow +\infty$.
$f(x) \rightarrow -\infty$ as $x \rightarrow -\infty$.
Since the end behavior is in opposite directions, it is an odd-degree polynomial function.

The graph intersects the *x*-axis at one point, so there is one real zero.

3 Practice

Formative Assessment

Use Exercises 1–12 to check for understanding.

Use the chart at the bottom of this page to customize assignments for your students.

Additional Answers

11a. $f(x) \to -\infty$ as $x \to -\infty$. $f(x) \to +\infty$ as $x \to +\infty$.

11b. Since the end behavior is in opposite directions, it is an odd-degree function.

11c. The graph intersects the x-axis at three points, so there are three real zeros.

12a. $f(x) \to -\infty$ as $x \to -\infty$. $f(x) \to -\infty$ as $x \to +\infty$.

12b. Since the end behavior is in the same direction, it is an even-degree function.

12c. The graph does not intersect the x-axis, so there are no real zeros.

13. not in one variable because there are two variables, x and y

14. not a polynomial because there is a negative exponent

15. degree = 6, leading coefficient = -12

16. degree = 7, leading coefficient = -21

17. degree = 4, leading coefficient = -5

18. degree = 5, leading coefficient = 3

19. degree = 2, leading coefficient = 3

20. degree = 2, leading coefficient = -6

21. degree = 9, leading coefficient = 2

22. degree = 8, leading coefficient = -2

23. $p(-6) = 1227$; $p(3) = 66$

24. $p(-6) = 546$; $p(3) = -93$

25. $p(-6) = -156$; $p(3) = 78$

26. $p(-6) = 2322$; $p(3) = 9$

27. $p(-6) = 319$; $p(3) = -5$

28. $p(-6) = 2232$; $p(3) = 153$

Example 1

State the degree and leading coefficient of each polynomial in one variable. If it is not a polynomial in one variable, explain why.

1. degree = 6, leading coefficient = 11   **2.** degree = 7, leading coefficient = -10

1. $11x^6 - 5x^5 + 4x^2$

2. $-10x^7 - 5x^3 + 4x - 22$

3. not in one variable because there are two variables, x and y

3. $14x^4 - 9x^3 + 3x - 4y$

4. not in one variable because there are two variables, x and y

4. $8x^5 - 3x^2 + 4xy - 5$

Example 2

Find $w(5)$ and $w(-4)$ for each function.

5. $w(x) = -2x^3 + 3x - 12$
$w(5) = -247$; $w(-4) = 104$

6. $w(x) = 2x^4 - 5x^3 + 3x^2 - 2x + 8$
$w(5) = 698$; $w(-4) = 896$

Example 3

If $c(x) = 4x^3 - 5x^2 + 2$ and $d(x) = 3x^2 + 6x - 10$, find each value.

7. $c(y^3)$ $4y^9 - 5y^6 + 2$

8. $-4[d(3z)]$ $-108z^2 - 72z + 40$

9. $6c(4a) + 2d(3a - 5)$ $1536a^3 - 426a^2 - 144a + 82$

10. $-3c(2b) + 6d(4b - 3)$ $-96b^3 + 348b^2 - 288b - 12$

Example 4

For each graph,
a. describe the end behavior,
b. determine whether it represents an odd-degree or an even-degree function, and
c. state the number of real zeros. **11, 12.** See margin.

11.

12.

Example 1

CCSS PERSEVERANCE State the degree and leading coefficient of each polynomial in one variable. If it is not a polynomial in one variable, explain why. **13–22.** See margin.

13. $-6x^6 - 4x^5 + 13xy$

14. $3a^7 - 4a^4 + \dfrac{3}{a}$

15. $8x^5 - 12x^6 + 14x^3 - 9$

16. $-12 - 8x^2 + 5x - 21x^7$

17. $15x - 4x^3 + 3x^2 - 5x^4$

18. $13b^3 - 9b + 3b^5 - 18$

19. $(d + 5)(3d - 4)$

20. $(5 - 2y)(4 + 3y)$

21. $6x^5 - 5x^4 + 2x^9 - 3x^2$

22. $7x^4 + 3x^7 - 2x^8 + 7$

Example 2

Find $p(-6)$ and $p(3)$ for each function. **23–28.** See margin.

23. $p(x) = x^4 - 2x^2 + 3$

24. $p(x) = -3x^3 - 2x^2 + 4x - 6$

25. $p(x) = 2x^3 + 6x^2 - 10x$

26. $p(x) = x^4 - 4x^3 + 3x^2 - 5x + 24$

(27) $p(x) = -x^3 + 3x^2 - 5$

28. $p(x) = 2x^4 + x^3 - 4x^2$

Example 3

If $c(x) = 2x^2 - 4x + 3$ and $d(x) = -x^3 + x + 1$, find each value.

29. $c(3a)$ $18a^2 - 12a + 3$

30. $5d(2a)$ $-40a^3 + 10a + 5$

31. $c(b^2)$ $2b^4 - 4b^2 + 3$

32. $d(4a^2)$ $-64a^6 + 4a^2 + 1$

33. $d(4y - 3)$ $-64y^3 + 144y^2 - 104y + 25$

34. $c(y^2 - 1)$ $2y^4 - 8y^2 + 9$

 326 | Lesson 5-3 | Polynomial Functions

Differentiated Homework Options

Level	Assignment	Two-Day Option	
AL Basic	13–42, 63, 66–85	13–41 odd, 69–72	14–42 even, 63, 66–68, 73–85
OL Core	13–41 odd, 43–55 odd, 56–63, 66–85	13–42, 69–72	55, 56, 63, 66–68, 73–85
BL Advanced	43–82, (optional: 83–85)		

Example 4 For each graph,

a. describe the end behavior,

b. determine whether it represents an odd-degree or an even-degree function, and

c. state the number of real zeros. **35–40. See Chapter 5 Answer Appendix.**

35.

36.

37.

38.

39.

40.

42b. Sample answer: [−500, 500], scl: 50 by [−10,000, 10,000], scl: 1000

42c. −41.0, 27.1, 228.2

42d. 28 to 228 microwaves

44. $p(-2) = -28$; $p(8) = -178$

45. $p(-2) = -0.5$; $p(8) = 3112$

46. $p(-2) = 1.5$; $p(8) = 304$

41 **PHYSICS** For a moving object with mass m in kilograms, the kinetic energy KE in joules is given by the function $KE(v) = 0.5mv^2$, where v represents the speed of the object in meters per second. Find the kinetic energy of an all-terrain vehicle with a mass of 171 kilograms moving at a speed of 11 meters/second. **10,345.5 joules**

42. **CCSS** **MODELING** A microwave manufacturing firm has determined that their profit function is $P(x) = -0.0014x^3 + 0.3x^2 + 6x - 355$, where x is the number of microwaves sold annually.

a. Graph the profit function using a calculator. **See margin.**

b. Determine a reasonable viewing window for the function.

c. Approximate all of the zeros of the function using the **CALC** menu.

d. What must be the range of microwaves sold in order for the firm to have a profit?

B Find $p(-2)$ and $p(8)$ for each function. **43.** $p(-2) = -16$; $p(8) = 1024$

43. $p(x) = \frac{1}{4}x^4 + \frac{1}{2}x^3 - 4x^2$

44. $p(x) = \frac{1}{8}x^4 - \frac{3}{4}x^3 + 12x - 18$

45. $p(x) = \frac{3}{4}x^4 - \frac{1}{8}x^2 + 6x$

46. $p(x) = \frac{5}{8}x^3 - \frac{1}{2}x^2 + \frac{3}{4}x + 10$

Use the degree and end behavior to match each polynomial to its graph.

A   **B**   **C**   **D**

51. $3a^3 - 24a^2 + 240a + 66$

52. $-4a^6 - 12a^4 - 36a^2 - 72a - 48$

47. $f(x) = x^3 + 3x^2 - 4x$ **D**

48. $f(x) = -2x^2 + 8x + 5$ **B**

49. $f(x) = x^4 - 3x^2 + 6x$ **A**

50. $f(x) = -4x^3 - 4x^2 + 8$ **C**

54. $6a^{12} + 18a^8 - 28a^6 + 6a^4 + 42a^3 - 62$

53. $5a^6 - 298a^2 + 1008a - 928$

If $c(x) = x^3 - 2x$ and $d(x) = 4x^2 - 6x + 8$, find each value.

51. $3c(a - 4) + 3d(a + 5)$

52. $-2d(2a + 3) - 4c(a^2 + 1)$

53. $5c(a^2) - 8d(6 - 3a)$

54. $-7d(a^3) + 6c(a^4 + 1)$

Additional Answer

42a.

Study Guide and Intervention
AL OL ELL

Practice
AL OL BL ELL

Enrichment
OL BL

Word Problem Practice
AL OL BL ELL

⚡ Multiple Representations

In Exercise 56, students use analysis, algebra, a table of values, and a graph to represent a higher-order function.

WatchOut!

Error Analysis In Exercise 63, point out that "double root" does not mean there are two different solutions to an equation, only that there are two identical factors in the solution process.

Additional Answers

55a.

x	p(x)
−7	−585
−6	0
−4	240
−3	135
−2	0
0	−144
1	−105
2	0
4	240
6	0
7	−585

55d. Sample answer: The negative values should not be considered because the company will not produce negative items.

64. $f(x)$; $f(x)$ has potential for 5 or more real roots and a degree of 5 or more. $g(x)$ has potential for 4 real roots and a degree of 4.

66.

---

55. BUSINESS A clothing manufacturer's profitability can be modeled by $p(x) = -x^4 + 40x^2 - 144$, where x is the number of items sold in thousands and $p(x)$ is the company's profit in thousands of dollars. **c. 2000 and 6000 items**

a. Use a table of values to sketch the function. **See margin.**

b. Determine the zeros of the function. **−6, −2, 2, 6**

c. Between what two values should the company sell in order to be profitable?

d. Explain why only two of the zeros are considered in part **c**. **See margin.**

56. 🔁 MULTIPLE REPRESENTATIONS Consider $g(x) = (x - 2)(x + 1)(x - 3)(x + 4)$.

a. **Analytical** Determine the x- and y-intercepts, roots, degree, and end behavior of $g(x)$. **See Chapter 5 Answer Appendix.**

b. **Algebraic** Write the function in standard form $g(x) = x^4 - 15x^2 + 10x + 24$

c. **Tabular** Make a table of values for the function. **See Chapter 5 Answer Appendix.**

d. **Graphical** Sketch a graph of the function by plotting points and connecting them with a smooth curve. **See Chapter 5 Answer Appendix.**

Describe the end behavior of the graph of each function.

57 $f(x) = -5x^4 + 3x^2$ **58.** $g(x) = 2x^5 + 6x^4$ **59.** $h(x) = -4x^7 + 8x^6 - 4x$

60. $f(x) = 6x - 7x^2$ **61.** $g(x) = 8x^4 + 5x^5$ **62.** $h(x) = 9x^6 - 5x^7 + 3x^2$

57. $f(x) \to -\infty$ as $x \to -\infty$; $f(x) \to -\infty$ as $x \to +\infty$

58. $g(x) \to -\infty$ as $x \to -\infty$; $g(x) \to +\infty$ as $x \to +\infty$

59. $h(x) \to +\infty$ as $x \to -\infty$; $h(x) \to -\infty$ as $x \to +\infty$

60. $f(x) \to -\infty$ as $x \to -\infty$; $f(x) \to -\infty$ as $x \to +\infty$

61. $g(x) \to -\infty$ as $x \to -\infty$; $g(x) \to +\infty$ as $x \to +\infty$

62. $h(x) \to +\infty$ as $x \to -\infty$; $h(x) \to +\infty$ as $x \to -\infty$

H.O.T. Problems   Use Higher-Order Thinking Skills

63. CCSS CRITIQUE Shenequa and Virginia are determining the number of zeros of the graph at the right. Is either of them correct? Explain your reasoning.

> **Shenequa**
> There are 7 zeros because the graph crosses the x-axis 7 times.

> **Virginia**
> There are 8 zeros because the graph crosses the x-axis 7 times, and there is a double root.

63. Sample answer: Virginia is correct; an even function will have an even number of zeros and the double root represents 2 zeros.

64. CHALLENGE Of $f(x)$ and $g(x)$, which function has more potential real roots? What is the degree of that function? **See margin.**

x	−24	−18	−12	−6	0	6	12	18	24
f(x)	−8	−1	3	−2	4	7	−1	−8	5

$g(x) = x^4 + x^3 - 13x^2 + x + 4$

65. CHALLENGE If $f(x)$ has a degree of 5 and a positive leading coefficient and $g(x)$ has a degree of 3 and a positive leading coefficient, determine the end behavior of $\frac{f(x)}{g(x)}$. Explain your reasoning.

66. OPEN ENDED Sketch the graph of an even-degree polynomial with 8 real roots, one of them a double root. **See margin.**

67. REASONING Determine whether the following statement is *always*, *sometimes*, or *never* true. Explain. **See margin.**

> A polynomial function that has four real roots is a fourth-degree polynomial.

68. WRITING IN MATH Describe what the end behavior of a polynomial function is and how to determine it. **See margin.**

65. Sample answer: $f(x) \to +\infty$ as $x \to -\infty$; $f(x) \to +\infty$ as $x \to +\infty$; $\frac{f(x)}{g(x)}$ will become a 2-degree function with a positive leading coefficient.

67. Sometimes; a polynomial function with four real roots may be a sixth-degree polynomial function with two imaginary roots. A polynomial function that has four real roots is at least a fourth-degree polynomial.

68. Sample answer: The end behavior of a polynomial function is what the graph does as the input value approaches negative and positive infinity. It can be determined by the leading coefficient and the degree of the polynomial.

69. SHORT RESPONSE Four students solved the same math problem. Each student's work is shown below. Who is correct? **Student A**

Student A	Student C
$x^2x^{-5} = \dfrac{x^2}{x^5}$	$x^2x^{-5} = \dfrac{x^2}{x^{-5}}$
$= \dfrac{1}{x^3}, x \neq 0$	$= x^7, x \neq 0$

Student B	Student D
$x^2x^{-5} = \dfrac{x^2}{x^{-5}}$	$x^2x^{-5} = \dfrac{x^2}{x^5}$
$= x^{-7}, x \neq 0$	$= x^3, x \neq 0$

70. SAT/ACT What is the remainder when $x^3 - 7x + 5$ is divided by $x + 3$? **B**

A -11 D 11

B -1 E 35

C 1

71. EXTENDED RESPONSE A company manufactures tables and chairs. It costs \$40 to make each table and \$25 to make each chair. There is \$1440 available to spend on manufacturing each week. Let t = the number of tables produced and c = the number of chairs produced. **See margin.**

a. The manufacturing equation is $40t + 25c = 1500$. Construct a graph of this equation.

b. The company always produces two chairs with each table. Write and graph an equation to represent this situation on the same graph as the one in part **a**.

c. Determine the number of tables and chairs that the company can produce each week.

d. Explain how to determine this answer using the graph.

72. If $i = \sqrt{-1}$, then $5i(7i) =$ **H**

F 70 H -35

G 35 J -70

Simplify. (Lesson 5-2) **73.** $2x^2y^2 + 4x^4y^4z^2$ **74.** $3b^3c^3 - 5a^3b^2 + 2a^4c$ **75.** $6c^3 - 1 + 4a^5cd^2$

73. $\dfrac{16x^4y^3 + 32x^6y^5z^2}{8x^2y}$

74. $\dfrac{18ab^4c^5 - 30a^4b^3c^2 + 12a^5bc^3}{6abc^2}$

75. $\dfrac{18c^5d^2 - 3c^2d^2 + 12a^5c^3d^4}{3c^2d^2}$

Determine whether each expression is a polynomial. If it is a polynomial, state the degree of the polynomial. (Lesson 5-1)

76. $8x^2 + 5xy^3 - 6x + 4$ **yes; 4**   **77.** $9x^4 + 12x^6 - 16$ **yes; 6**   **78.** $3x^4 + 2x^2 - x^{-1}$ **not a polynomial**

79. FOUNTAINS The height of a fountain's water stream can be modeled by a quadratic function. Suppose the water from a jet reaches a maximum height of 8 feet at a distance 1 foot away from the jet. (Lesson 4-7) **a, b. See margin.**

a. If the water lands 3 feet away from the jet, find a quadratic function that models the height $h(d)$ of the water at any given distance d feet from the jet. Then compare the graph of the function to the parent function.

b. Suppose a worker increases the water pressure so that the stream reaches a maximum height of 12.5 feet at a distance of 15 inches from the jet. The water now lands 3.75 feet from the jet. Write a new quadratic function for $h(d)$. How do the changes in h and k affect the shape of the graph?

1 ft

8 ft

3 ft

Solve each inequality. (Lesson 1-6)

80. $|2x + 4| \leq 8$ $-6 \leq x \leq 2$ **81.** $|-3x + 2| \geq 4$ $x \leq -\dfrac{2}{3}$ or $x \geq 2$ **82.** $|2x - 8| - 4 \leq -6$ **no solution**

Determine whether each function has a maximum or minimum value, and find that value.

83. $f(x) = 3x^2 - 8x + 4$ **minimum; $-\dfrac{4}{3}$**

84. $f(x) = -4x^2 + 2x - 10$ **maximum; -9.75**

85. $f(x) = -0.25x^2 + 4x - 5$ **maximum; 11**

DifferentiatedInstruction   **BL**

Extension Explain to students that functions can have symmetry with respect to the origin or to the y-axis. An even function is a function for which $f(x) = f(-x)$. Even functions are symmetrical with respect to the y-axis. An example of an even function is $f(x) = x^2$. An odd function is a function for which $f(x) = -f(-x)$. Odd functions are symmetrical with respect to the origin. An example of an odd function is $f(x) = x$. Ask students to determine if $f(x) = x^3 - 5x$ and $f(x) = x^4 - 3x^2$ are even or odd functions. odd, even   Ask students if it is possible for a function to have symmetry with respect to the x-axis.   Yes; $f(x) = 0$ is the only function that is symmetric with respect to the x-axis.

---

4 Assess

Ticket Out the Door Make several copies of five different polynomial functions. Give one function to each student. As students leave the room, ask them to tell you the values of the functions for a given numerical value, variable, or algebraic expression.

Additional Answers

71a.

71b. $t = 0.5c$;

71c. 16 tables and 32 chairs

71d. Sample answer: This can be determined by the intersection of the graphs. This point of intersection is the optimal amount of tables and chairs manufactured.

79a. $h(d) = -2d^2 + 4d + 6$; The graph opens downward and is narrower than the parent graph, and the vertex is at (1, 8).

79b. $h(d) = -2(d - 1.25)^2 + 12.5$; It shifted the graph up 4.5 ft and to the right 3 in.

1 Focus

VerticalAlignment

Before Lesson 5-4 Use maxima and minima and graphs of polynomials.

Lesson 5-4 Graph polynomial functions and locate their zeros. Find the relative maxima and minima of polynomial functions.

After Lesson 5-4 Use tools including factoring to transform and solve polynomial equations.

2 Teach

Scaffolding Questions

Have students read the **Why?** section of the lesson.

Ask:

- If you drew a graph with years on the *x*-axis and attendance on the *y*-axis, what would the *x*-intercept be? **1906**

- What can you say about the average slope of the graph from the 1920s to the 1970s? **It is negative.**

- Suppose the graph sloped upward to the right. What would that show about movie attendance? **that attendance was increasing**

:: Then
- You used maxima and minima and graphs of polynomials.

:: Now
1. Graph polynomial functions and locate their zeros.
2. Find the relative maxima and minima of polynomial functions.

:: Why?
- Annual attendance at the movies has fluctuated since the first movie theater, the Nickelodeon, opened in Pittsburgh in 1906. Overall attendance peaked during the 1920s, and it was at its lowest during the 1970s. A graph of the annual attendance to the movies can be represented by a polynomial function.

NewVocabulary
Location Principle
relative maximum
relative minimum
extrema
turning points

CCSS Common Core State Standards

Content Standards
F.IF.4 For a function that models a relationship between two quantities, interpret key features of graphs and tables in terms of the quantities, and sketch graphs showing key features given a verbal description of the relationship.

F.IF.7.c Graph polynomial functions, identifying zeros when suitable factorizations are available, and showing end behavior.

Mathematical Practices
3 Construct viable arguments and critique the reasoning of others.

1 Graphs of Polynomial Functions To graph a polynomial function, make a table of values to find several points and then connect them to make a smooth continuous curve. Knowing the end behavior of the graph will assist you in completing the graph.

Example 1 Graph of a Polynomial Function

Graph $f(x) = -x^4 + x^3 + 3x^2 + 2x$ by making a table of values.

x	f(x)	x	f(x)
−2.5	≈ −41	0.5	≈ 1.8
−2.0	−16	1.0	5.0
−1.5	≈ −4.7	1.5	≈ 8.1
−1.0	−1.0	2.0	8.0
−0.5	≈ −0.4	2.5	≈ 0.3
0.0	0.0	3.0	−21

This is an even-degree polynomial with a negative leading coefficient, so $f(x) \rightarrow -\infty$ as $x \rightarrow -\infty$ and $f(x) \rightarrow -\infty$ as $x \rightarrow +\infty$. Notice that the graph intersects the *x*-axis at two points, indicating there are two zeros for this function.

▶ **Guided**Practice **1. See Chapter 5 Answer Appendix.**

1. Graph $f(x) = x^4 - x^3 - 2x^2 + 4x - 6$ by making a table of values.

In Example 1, one of the zeros occurred at $x = 0$. Another zero occurred between $x = 2.5$ and $x = 3.0$. Because $f(x)$ is positive for $x = 2.5$ and negative for $x = 3.0$ and all polynomial functions are continuous, we know there is a zero between these two values.

(2.5, 0.3)
zero

So, if the value of $f(x)$ *changes signs* from one value of *x* to the next, then there is a zero between those two *x*-values. This idea is called the **Location Principle**.

George Marks/Hulton Archive/Getty Images

Lesson 5-4 Resources

Resource	Approaching Level **AL**	On Level **OL**	Beyond Level **BL**	English Learners **ELL**
Teacher Edition	• Differentiated Instruction, p. 333	• Differentiated Instruction, pp. 333, 337	• Differentiated Instruction, pp. 333, 337	• Differentiated Instruction, p. 333
Chapter Resource Masters	• Study Guide and Intervention, pp. 23–24 • Skills Practice, p. 25 • Practice, p. 26 • Word Problem Practice, p. 27	• Study Guide and Intervention, pp. 23–24 • Skills Practice, p. 25 • Practice, p. 27 • Word 26 Practice, p. 27 • Enrichment, p. 28	• Practice, p. 26 • Word Problem Practice, p. 27 • Enrichment, p. 28	• Study Guide and Intervention, pp. 23–24 • Skills Practice, p. 25 • Practice, p. 26 • Word Problem Practice, p. 27
Other	• 5-Minute Check 5-4 • Study Notebook	• 5-Minute Check 5-4 • Study Notebook	• 5-Minute Check 5-4 • Study Notebook	• 5-Minute Check 5-4 • Study Notebook

StudyTip

Degree Recall that the degree of the function is also the maximum number of zeros the function can have.

KeyConcept Location Principle

		Model
Words	Suppose $y = f(x)$ represents a polynomial function and a and b are two real numbers such that $f(a) < 0$ and $f(b) > 0$. Then the function has at least one real zero between a and b.	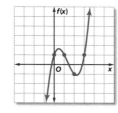

Example 2 Locate Zeros of a Function

Determine consecutive integer values of x between which each real zero of $f(x) = x^3 - 4x^2 + 3x + 1$ is located. Then draw the graph.

Make a table of values. Since $f(x)$ is a third-degree polynomial function, it will have either 3 or 1 real zeros. Look at the values of $f(x)$ to locate the zeros. Then use the points to sketch a graph of the function.

x	f(x)	
−2	−29	
−1	−7	
0	1	change in sign
1	1	
2	−1	change in sign
3	1	change in sign
4	13	

The changes in sign indicate that there are zeros between $x = -1$ and $x = 0$, between $x = 1$ and $x = 2$, and between $x = 2$ and $x = 3$.

> **Guided**Practice
> **2.** between 0 and 1 and between 3 and 4
> See margin for graph.
>
> **2.** Determine consecutive integer values of x between which each real zero of the function $f(x) = x^4 - 3x^3 - 2x^2 + x + 1$ is located. Then draw the graph.

2 Maximum and Minimum Points
The graph below shows the general shape of a third-degree polynomial function.

Point A on the graph is a **relative maximum** of the function since no other nearby points have a greater y-coordinate. The graph is increasing as it approaches A and decreasing as it moves from A.

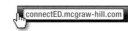

Likewise, point B is a **relative minimum** since no other nearby points have a lesser y-coordinate. The graph is decreasing as it approaches B and increasing as it moves from B. The maximum and minimum values of a function are called the **extrema**.

StudyTip

Odd Functions Some odd functions, like $f(x) = x^3$, have no turning points.

These points are often referred to as **turning points**. The graph of a polynomial function of degree n has at most $n - 1$ turning points.

connectED.mcgraw-hill.com **331**

---

1 Graphs of Polynomial Functions

Example 1 shows how to graph a polynomial function by making a table of values. **Example 2** shows how to graph a polynomial function by determining consecutive values of x between which each real zero of the function is located.

Formative Assessment
Use the Guided Practice exercises after each example to determine students' understanding of concepts.

Additional Examples

1 Graph $f(x) = -x^3 - 4x^2 + 5$ by making a table of values.

See bottom margin for graph.

2 Determine consecutive integer values of x between which each real zero of the function $f(x) = x^4 - x^3 - 4x^2 + 1$ is located. Then draw the graph.

There are zeros between $x = -2$ and $x = -1$, between $x = -1$ and $x = 0$, between $x = 0$ and $x = 1$, and between $x = 2$ and $x = 3$. See bottom margin for graph.

▶ **Additional Examples** also in Interactive Classroom PowerPoint® Presentations

IWB **Interactive White Board**
READY

Additional Answer (Guided Practice)

2.

Additional Answers (Additional Examples)

1.

$f(x) = -x^3 - 4x^2 + 5$

2.
$f(x) = x^4 - x^3 - 4x^2 + 1$

2 Maximum and Minimum Points

Example 3 shows how to graph a polynomial function and estimate the *x*-coordinates at which the relative maxima and relative minima occur.
Example 4 shows how to use the graph of a polynomial function to reveal trends in real-world data.

Additional Example

3 Graph $f(x) = x^3 - 3x^2 + 5$. Estimate the *x*-coordinates at which the relative maxima and relative minima occur.

relative maximum near $x = 0$, relative minimum near $x = 2$

Additional Answer (Guided Practice)

3.

relative maximum near $x = -1$, relative minimum near $x = 1$

StudyTip

Maximum and Minimum A polynomial with a degree greater than 3 may have more than one relative maximum or relative minimum.

Example 3   Maximum and Minimum Points

Graph $f(x) = x^3 - 4x^2 - 2x + 3$. **Estimate the *x*-coordinates at which the relative maxima and relative minima occur.**

Make a table of values and graph the function.

x	f(x)
−2	−17
−1	0
0	3
1	−2
2	−9
3	−12
4	−4
5	18

zero

indicates a relative maximum

indicates a relative minimum

zero between 4 and 5

Look at the table of values and the graph.

The value of $f(x)$ changes signs between $x = 4$ and $x = 5$, indicating a zero of the function.

The value of $f(x)$ at $x = 0$ is greater than the surrounding points, so there must be a relative maximum *near* $x = 0$.

The value of $f(x)$ at $x = 3$ is less than the surrounding points, so there must be a relative minimum *near* $x = 3$.

StudyTip

Rational Values Zeros and turning points will not always occur at integral values of *x*.

CHECK You can use a graphing calculator to find the relative maximum and relative minimum of a function and confirm your estimates.

Enter $y = x^3 - 4x^2 - 2x + 3$ in the **Y=** list and graph the function.

Use the **CALC** menu to find each maximum and minimum.

When selecting the left bound, move the cursor to the left of the maximum or minimum. When selecting the right bound, move the cursor to the right of the maximum or minimum.

Press **ENTER** twice.

[−10, 10] scl: 1 by [−15, 10] scl: 1

[−10, 10] scl: 1 by [−15, 10] scl: 1

The estimates for a relative maximum near $x = 0$ and a relative minimum near $x = 3$ are accurate.

▶ **Guided**Practice

3. Graph $f(x) = 2x^3 + x^2 - 4x - 2$. Estimate the *x*-coordinates at which the relative maxima and relative minima occur. **See margin.**

Teach with Tech

Student Response System Show a slideshow to students showing different graphs of polynomial functions. For each, ask students if this graph shows a polynomial function with an even or an odd degree. Have students reply with A for even and B for odd.

The graph of a polynomial function can reveal trends in real-world data. It is often helpful to note when the graph is increasing or decreasing.

Randy Faris/Corbis

Real-World Example 4 Graph a Polynomial Model

MOVIES Refer to the beginning of the lesson. Annual admissions to movies in the United States can be modeled by the function $f(x) = -0.0017x^4 + 0.31x^3 - 17.66x^2 + 277x + 3005$, where x is the number of years since 1926 and $f(x)$ is the annual admissions in millions.

a. Graph the function.

Make a table of values for the years 1926–2006. Plot the points and connect with a smooth curve. Finding and plotting the points for every tenth year gives a good approximation of the graph.

x	f(x)
0	3005
10	4302
20	3689
30	2414
40	1317
50	830
60	977
70	1374
80	1229

Real-WorldLink

Over 1.4 billion movie tickets were sold in the United States in 2006.

Source: CNN

b. Describe the turning points of the graph and its end behavior.

There are relative maxima near 1936 and 2000 and a relative minimum between 1976 and 1981. $f(x) \rightarrow -\infty$ as $x \rightarrow -\infty$ and $f(x) \rightarrow -\infty$ as $x \rightarrow \infty$.

c. What trends in movie admissions does the graph suggest? Is it reasonable that the trend will continue indefinitely?

Movie attendance peaked around 1936 and declined until about 1978. It then increased until 2000 and began a decline.

d. Is it reasonable that the trend will continue indefinitely?

This trend may continue for a couple of years, but the graph will soon become unreasonable as it predicts negative attendance for the future.

4B. The graph peaks around $x = 2$, then steadily decreases as the x increases.
4C. Fax machine sales peaked in 1992 and have decreased ever since. By 2006, fax machine sales hit zero.
4D. Sample answer: The graph is reasonable to suggest a sharp decline in sales, but not reasonable in suggesting 0 sales in 2006.

Guided Practice

4. FAX MACHINES The annual sales of fax machines for home use can be modeled by $f(x) = -0.17x^4 + 6.29x^3 - 77.65x^2 + 251x + 1100$, where x is the number of years after 1990 and $f(x)$ is the annual sales in millions of dollars.
   A. Graph the function. See margin.
   B. Describe the turning points of the graph and its end behavior.
   C. What trends in fax machine sales does the graph suggest?
   D. Is it reasonable that the trend will continue indefinitely?

  connectED.mcgraw-hill.com **333**

DifferentiatedInstruction AL OL BL ELL

If students ask how math functions can describe real-world situations,

Then have them discuss the appropriateness of describing real-world situations with mathematical functions. Help them to understand that a function is usually just an approximation of the real-world data, and is often only a reasonable model of a limited domain of values.

4 HEALTH The weight w, in pounds, of a patient during a 7-week illness is modeled by the cubic equation $w(n) = 0.1n^3 - 0.6n^2 + 110$, where n is the number of weeks since the patient became ill.

a. Graph the equation.

b. Describe the turning points of the graph and its end behavior. There is a relative minimum point at week 4. For the end behavior, $w(n)$ increases as n increases.

c. What trends in the patient's weight does the graph suggest? The patient lost weight for each of 4 weeks after becoming ill. After 4 weeks, the patient gained weight and continued to gain weight.

d. Is it reasonable to assume the trend will continue indefinitely? The trend may continue for a few weeks, but it is unlikely that the patient's weight will rise indefinitely.

Additional Answer (Guided Practice)

4A.

3 Practice

Formative Assessment

Use Exercises 1–13 to check for understanding.

Use the chart at the bottom of this page to customize assignments for your students.

Additional Answers

1.

2.

3.

4.

Check Your Understanding   ● = Step-by-Step Solutions begin on page R14.

Example 1   Graph each polynomial equation by making a table of values. **1–4. See margin.**

1. $f(x) = 2x^4 - 5x^3 + x^2 - 2x + 4$

2. $f(x) = -2x^4 + 4x^3 + 2x^2 + x - 3$

3. $f(x) = 3x^4 - 4x^3 - 2x^2 + x - 4$

4. $f(x) = -4x^4 + 5x^3 + 2x^2 + 3x + 1$

Example 2   Determine the consecutive integer values of x between which each real zero of each function is located. Then draw the graph. **5–8. See Chapter 5 Answer Appendix.**

5. $f(x) = x^3 - 2x^2 + 5$

6. $f(x) = -x^4 + x^3 + 2x^2 + x + 1$

7. $f(x) = -3x^4 + 5x^3 + 4x^2 + 4x - 8$

8. $f(x) = 2x^4 - x^3 - 3x^2 + 2x - 4$

Example 3   Graph each polynomial function. Estimate the x-coordinates at which the relative maxima and relative minima occur. State the domain and range for each function.   **9–12. See Chapter 5 Answer Appendix.**

9. $f(x) = x^3 + x^2 - 6x - 3$

10. $f(x) = 3x^3 - 6x^2 - 2x + 2$

11. $f(x) = -x^3 + 4x^2 - 2x - 1$

12. $f(x) = -x^3 + 2x^2 - 3x + 4$

Example 4   **13.** **CCSS** **SENSE-MAKING** Annual compact disc sales can be modeled by the quartic function $f(x) = 0.48x^4 - 9.6x^3 + 53x^2 - 49x + 599$, where x is the number of years after 1995 and $f(x)$ is annual sales in millions.

   a. Graph the function for $0 \leq x \leq 10$. **See Chapter 5 Answer Appendix.**

   b. Describe the turning points of the graph, its end behavior, and the intervals on which the graph is increasing or decreasing.

   c. Continue the graph for $x = 11$ and $x = 12$. What trends in compact disc sales does the graph suggest? **See Chapter 5 Answer Appendix.**   **See Chapter 5 Answer Appendix.**

   d. Is it reasonable that the trend will continue indefinitely? Explain.
   13b. Sample answer: Relative maximum at $x = 5$ and relative minimum at $x \approx 9.5$ $f(x) \rightarrow \infty$ as $x \rightarrow -\infty$ and $f(x) \rightarrow \infty$ as $x \rightarrow \infty$. The graph increases when $x < 5$ and $x > 9.5$ and decreases when $5 < x < 9.5$.

Practice and Problem Solving   Extra Practice is on page R5.

Examples 1–3   Complete each of the following. **14–21. See Chapter 5 Answer Appendix.**

   a. Graph each function by making a table of values.

   b. Determine the consecutive integer values of x between which each real zero is located.

   c. Estimate the x-coordinates at which the relative maxima and minima occur.

14. $f(x) = x^3 + 3x^2$

15 $f(x) = -x^3 + 2x^2 - 4$

16. $f(x) = x^3 + 4x^2 - 5x$

17. $f(x) = x^3 - 5x^2 + 3x + 1$

18. $f(x) = -2x^3 + 12x^2 - 8x$

19. $f(x) = 2x^3 - 4x^2 - 3x + 4$

20. $f(x) = x^4 + 2x - 1$

21. $f(x) = x^4 + 8x^2 - 12$

22b. Sample answer: The graph has a relative minimum at $x = 10$ and then increases as x increases.

Example 4   **22.** **FINANCIAL LITERACY** The average annual price of gasoline can be modeled by the cubic function $f(x) = 0.0007x^3 - 0.014x^2 + 0.08x + 0.96$, where x is the number of years after 1987 and $f(x)$ is the price in dollars. **a, d. See Chapter 5 Answer Appendix.**

   a. Graph the function for $0 \leq x \leq 30$.

   b. Describe the turning points of the graph and its end behavior.

   c. What trends in gasoline prices does the graph suggest?

   d. Is it reasonable that the trend will continue indefinitely? Explain.

22c. The graph suggests a fairly steep continuous increase and gas prices at $5 per gallon by 2012, which could be possible.

23. rel. max: $x = -2.73$; **rel. min:** $x = 0.73$

25. rel. max: $x = 1.34$; no rel. min

26. rel. max: $x = -1.87$; **rel. min:** $x = 1.52$

Use a graphing calculator to estimate the x-coordinates at which the maxima and minima of each function occur. Round to the nearest hundredth.

23. $f(x) = x^3 + 3x^2 - 6x - 6$

24. $f(x) = -2x^3 + 4x^2 - 5x + 8$ no relative max or min

25. $f(x) = -2x^4 + 5x^3 - 4x^2 + 3x - 7$

26. $f(x) = x^5 - 4x^3 + 3x^2 - 8x - 6$

 334 | Lesson 5-4 | Analyzing Graphs of Polynomial Functions

Differentiated Homework Options

Level	Assignment	Two-Day Option	
AL Basic	14–32, 47–70	15–31 odd, 54–57	14–32 even, 47–53, 58–70
OL Core	15–31 odd, 33, 40, 43, 45, 46–70	14–32, 54–57	33, 40, 41, 46–53, 58–70
BL Advanced	33–64, (optional: 65–70)		

Sketch the graph of polynomial functions with the following characteristics.

27. an odd function with zeros at -5, -3, 0, 2 and 4 **27–32.** See Chapter 5 Answer Appendix.

28. an even function with zeros at -2, 1, 3, and 5

29. a 4-degree function with a zero at -3, maximum at $x = 2$, and minimum at $x = -1$

30. a 5-degree function with zeros at -4, -1, and 3, maximum at $x = -2$

31. an odd function with zeros at -1, 2, and 5 and a negative leading coefficient

32. an even function with a minimum at $x = 3$ and a positive leading coefficient

33 DIVING The deflection d of a 10-foot-long diving board can be calculated using the function $d(x) = 0.015x^2 - 0.0005x^3$, where x is the distance between the diver and the stationary end of the board in feet.

   a. Make a table of values of the function for $0 \leq x \leq 10$. **See Chapter 5 Answer Appendix.**

   b. Graph the function. **See Chapter 5 Answer Appendix.**

   c. What does the end behavior of the graph suggest as x increases? $d(x)$ increases.

   d. Will this trend continue indefinitely? Explain your reasoning. Sample answer: no; x cannot be greater than 10

Complete each of the following. **34–39.** See Chapter 5 Answer Appendix.

a. Estimate the x-coordinate of every turning point and determine if those coordinates are relative maxima or relative minima.

b. Estimate the x-coordinate of every zero.

c. Determine the smallest possible degree of the function.

d. Determine the domain and range of the function.

40c. Sample answer: The graph suggests that the number of pager subscribers will increase dramatically and continue to increase.
40d. Sample answer: The graph is unreasonable for $x \geq 15$ since pager use is currently decreasing rapidly and pagers have been replaced by more efficient products.

34.

35.

36.

37.

38.

39.

40. CCSS REASONING The number of subscribers using pagers in the United States can be modeled by $f(x) = 0.015x^4 - 0.44x^3 + 3.46x^2 - 2.7x + 9.68$, where x is the number of years after 1990 and $f(x)$ is the number of subscribers in millions.

   a. Graph the function. **See margin.**

   b. Describe the end behavior of the graph. As x increases, $f(x)$ increases.

   c. What does the end behavior suggest about the number of pager subscribers?

   d. Will this trend continue indefinitely? Explain your reasoning.

connectED.mcgraw-hill.com **335**

Additional Answer

40a.

Study Guide and Intervention
AL OL ELL

Practice
AL OL BL ELL

Word Problem Practice
AL OL BL ELL

Enrichment
OL BL

NAME _____ DATE _____ PERIOD _____

5-4 Enrichment

Golden Rectangles

Use a straightedge, a compass, and the instructions below to construct a golden rectangle.

1. Construct square $ABCD$ with sides of 2 centimeters.

2. Construct the midpoint of $\overline{AB}$. Call the midpoint M.

3. Using M as the center, set your compass opening at MC. Construct an arc with center M that intersects $\overline{AB}$. Call the point of intersection P.

4. Construct a line through P that is perpendicular to $\overline{AB}$.

Multiple Representations

In Exercise 46, students use a table of values, analysis, and a graphing calculator to investigate the properties of a higher-order function.

Additional Answers

41.

$1.25

42a. zeros: $x = \pm 2$; x-intercepts: ± 2; y-intercept: 16; turning points: $x = -2, 0, 2$

42b. $x = 0$

42c. decreasing: $x < -2$ and $0 < x < 2$; increasing: $-2 < x < 0$ and $2 < x$

43a. zeros: $x \approx 1.75$; x-intercept: ≈ 1.75; y-intercept: -4; turning points: $x \approx -1.25, -0.5, 0.5, 1.25$

43b. no axis of symmetry

43c. decreasing: $-1.25 \le x \le -0.5$ and $0.5 \le x \le 1.25$; increasing: $x \le -1.25$, $-0.5 \le x \le 0.5$, and $x \ge 1.25$

44a. zeros: $x \approx -1$ and 0; x-intercept: ≈ -1 and 0; y-intercept: 0; turning point: $x \approx -0.5$

44b. no axis of symmetry

44c. decreasing: $x \ge -0.5$; increasing: $x \le -0.5$

45a. no zeros, no x-intercepts, y-intercept: 5; no turning points

45b. no axis of symmetry

45c. decreasing: $x \le -4$; constant: $-4 < x \le 0$; increasing $x > 0$

41. PRICING Jin's vending machines currently sell an average of 3500 beverages per week at a rate of $0.75 per can. She is considering increasing the price. Her weekly earnings can be represented by $f(x) = -5x^2 + 100x + 2625$, where x is the number of $0.05 increases. Graph the function and determine the most profitable price for Jin. **See margin.**

For each function, **42–45. See margin.**

a. determine the zeros, x- and y-intercepts, and turning points,

b. determine the axis of symmetry, and

c. determine the intervals for which it is increasing, decreasing, or constant.

42. $y = x^4 - 8x^2 + 16$

43 $y = x^5 - 3x^3 + 2x - 4$

44. $y = -2x^4 + 4x^3 - 5x$

45. $y = \begin{cases} x^2 \text{ if } x \le -4 \\ 5 \text{ if } -4 < x \le 0 \\ x^3 \text{ if } x > 0 \end{cases}$

46. **MULTIPLE REPRESENTATIONS** Consider the following function. **a–d. See Chapter 5 Answer Appendix.**

$$f(x) = x^4 - 8.65x^3 + 27.34x^2 - 37.2285x + 18.27$$

a. Analytical What are the degree, leading coefficient, and end behavior?

b. Tabular Make a table of integer values $f(x)$ if $-4 \le x \le 4$. How many zeros does the function appear to have from the table?

c. Graphical Graph the function by using a graphing calculator.

d. Graphical Change the viewing window to [0, 4] scl: 1 by [−0.4, 0.4] scl: 0.2. What conclusions can you make from this new view of the graph?

47. As the x-values approach large positive or negative numbers, the term with the largest degree becomes more and more dominant in determining the value of $f(x)$.

H.O.T. Problems   Use Higher-Order Thinking Skills

47. REASONING Explain why the leading coefficient and the degree are the only determining factors in the end behavior of a polynomial function.

48. REASONING The table below shows the values of $g(x)$, a cubic function. Could there be a zero between $x = 2$ and $x = 3$? Explain your reasoning.

x	−2	−1	0	1	2	3
g(x)	4	−2	−1	1	−2	−2

48. Sample answer: No; the cubic function is of degree 3 and cannot have any more than three zeros. Those zeros are located between −2 and −1, 0 and 1, and 1 and 2.

49. OPEN ENDED Sketch the graph of an odd polynomial function with 6 turning points and 2 double roots. **See margin.**

50. Sample answer: Always; the definition of a turning point of a graph is a point in which the graph stops increasing and begins to decrease, causing a maximum, or stops decreasing and begins to increase, causing a minimum.

50. ARGUMENTS Determine whether the following statement is *sometimes, always,* or *never* true. Explain your reasoning.

For any continuous polynomial function, the y-coordinate of a turning point is also either a relative maximum or relative minimum.

51. REASONING A function is said to be even if for every x in the domain of f, $f(x) = f(-x)$. Is every even-degree polynomial function also an even function? Explain.

52. REASONING A function is said to be *odd* if for every x in the domain, $-f(x) = f(-x)$. Is every odd-degree polynomial function also an odd function? Explain.

51. Sample answer: No; $f(x) = x^2 + x$ is an even degree, but $f(1) \ne f(-1)$.

52. Sample answer: No; $f(x) = x^3 + 2x^2$ is an odd degree, but $-f(1) \ne f(-1)$.

53. **WRITING IN MATH** How can you use the characteristics of a polynomial function to sketch its graph? **See margin.**

49. Sample answer:

53. Sample answer: From the degree, you can determine whether the graph is even or odd and the maximum number of zeros and turning points for the graph. You can create a table of values to help you find the approximate locations of turning points and zeros. The leading coefficient can be used to determine the end behavior of the graph, and, along with the degree, build the shape of the graph.

Standardized Test Practice

54. Which of the following is the factorization of $2x - 15 + x^2$? **B**

   A $(x - 3)(x - 5)$
   B $(x - 3)(x + 5)$
   C $(x + 3)(x - 5)$
   D $(x + 3)(x + 5)$

55. SHORT RESPONSE In the figure below, if $x = 35$ and $z = 50$, what is the value of y? **95**

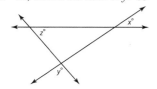

56. Which polynomial represents $(4x^2 + 5x - 3)(2x - 7)$? **H**

   F $8x^3 - 18x^2 - 41x - 21$
   G $8x^3 + 18x^2 + 29x - 21$
   H $8x^3 - 18x^2 - 41x + 21$
   J $8x^3 + 18x^2 - 29x + 21$

57. SAT/ACT The figure at the right shows the graph of a polynomial function $f(x)$. Which of the following could be the degree of $f(x)$? **C**

   A 2   **D** 5
   B 3   **E** 6
   C 4

Spiral Review

For each graph,
a. describe the end behavior,
b. determine whether it represents an odd-degree or an even-degree function, and
c. state the number of real zeros. (Lesson 5-3) **58–60. See margin.**

58.

59.

60.

Simplify. (Lesson 5-2)

61. $(x^3 + 2x^2 - 5x - 6) \div (x + 1)$
   $(x - 2)(x + 3)$

62. $(4y^3 + 18y^2 + 5y - 12) \div (y + 4)$
   $4y^2 + 2y - 3$

63. $(2a^3 - a^2 - 4a) \div (a - 1)$
   $2a^2 + a - 3 - \dfrac{3}{a - 1}$

64. CHEMISTRY Tanisha needs 200 milliliters of a 48% concentration acid solution. She has 60% and 40% concentration solutions in her lab. How many milliliters of 40% acid solution should be mixed with 60% acid solution to make the required amount of 48% acid solution? (Lesson 3-8) **80 mL of the 60% solution and 120 mL of the 40% solution**

Skills Review

Factor.

65. $x^2 + 6x + 3x + 18$ $(x + 6)(x + 3)$ **66.** $y^2 - 5y - 8y + 40$ $(y - 5)(y - 8)$ **67.** $a^2 + 6a - 16$ $(a + 8)(a - 2)$
68. $b^2 - 4b - 21$ $(b - 7)(b + 3)$ **69.** $6x^2 - 5x - 4$ $(3x - 4)(2x + 1)$ **70.** $4x^2 - 7x - 15$ $(4x + 5)(x - 3)$

DifferentiatedInstruction **OL** **BL**

Extension Real zeros of a function such as $f(x) = 4x^3 - x^2 - 11x + 3$ can be approximated using a method known as the *bisection method.* Confirm that this function has real zeros between -2 and -1, 0 and 1, and 1 and 2. $f(-2) = -11$, $f(-1) = 9$; $f(0) = 3$, $f(1) = -5$, $f(2) = 9$ Next, explain the following steps to students to approximate the zero between 0 and 1.

- Average (bisect) the interval from 0 to 1.
$$\frac{0 + 1}{2} = 0.5$$

- Evaluate the function for 0.5 to determine in which half of the interval the zero lies.
$f(0.5) = -2.25$

- Use the Location Principle to determine that the zero must lie between 0 and 0.5.

This procedure can be repeated until the desired accuracy of the zero is obtained.

---

4 Assess

Yesterday's News Have students write how yesterday's lesson on evaluating polynomial functions for given values helped them with graphing polynomial functions in today's lesson.

Formative Assessment
Check for student understanding of concepts from Lessons 5-3 and 5-4.

📁 Quiz 2, p. 57

📝 Follow-up
Students have explored polynomials and polynomial functions.

Ask:
- When would a nonlinear function be a good model? Sample answer: When the relationship between the quantities is not constant.

- What are the advantages of using polynomial functions for modeling? Sample answer: They have well-known and understood properties; there are multiple models that can be considered within the same family of functions; the computation that is done to make predictions is relatively easy to perform.

Additional Answers

58. $f(x) \to +\infty$ as $x \to -\infty$. $f(x) \to +\infty$ as $x \to +\infty$. Since the end behavior is in the same direction, it is an even-degree function. The graph intersects the x-axis at two points, so there are two real zeros.

59. $f(x) \to -\infty$ as $x \to -\infty$. $f(x) \to -\infty$ as $x \to +\infty$. Since the end behavior is in the same direction, it is an even-degree function. The graph intersects the x-axis at six points, so there are six real zeros.

60. $f(x) \to +\infty$ as $x \to -\infty$. $f(x) \to +\infty$ as $x \to +\infty$. Since the end behavior is in the same direction, it is an even-degree function. The graph intersects the x-axis at four points, so there are four real zeros.

1 Focus

Objective Use a graphing calculator to model data whose curve of best fit is a polynomial function.

Materials
- TI-83/84 Plus or other graphing calculator

Teaching Tip
Students should clear lists **L1** and **L2** before entering the data from the table in Step 1. This is a more reliable approach than simply "overwriting" old data with new data.

2 Teach

Working in Cooperative Groups
Put students in pairs, mixing abilities. Then have pairs complete all steps in the Example.

Ask:
- In Step 3, a quartic curve will fit the data best. An easy way to verify this is to find quadratic and cubic regression equations and then copy them to the **Y=** list along with the quartic regression equation. Change the window settings for the *x*-axis to [−0.2, 18.6]. Turn off the scatter plot and graph all three regression equations on the same screen. Then use **TRACE** to go to each *x* value in the first row of the table. While at each *x* value, use the up/down arrow keys to move between the curves, comparing the *y* values for each regression curve with the *y* value given in the table.

Practice Have students complete Exercises 1–3.

Modeling Data Using Polynomial Functions

You can use a TI-83/84 Plus graphing calculator to model data points when a curve of best fit is a polynomial function.

 Common Core State Standards
Content Standards
F.IF.4 For a function that models a relationship between two quantities, interpret key features of graphs and tables in terms of the quantities, and sketch graphs showing key features given a verbal description of the relationship.
F.IF.7.c Graph polynomial functions, identifying zeros when suitable factorizations are available, and showing end behavior.
Mathematical Practices
5 Use appropriate tools strategically.

Example

The table shows the distance a seismic wave produced by an earthquake travels from the epicenter. Draw a scatter plot and a curve of best fit to show how the distance is related to time. Then determine approximately how far away from the epicenter a seismic wave will be felt 8.5 minutes after an earthquake occurs.

Travel Time (min)	1	2	5	7	10	12	13
Distance (km)	400	800	2500	3900	6250	8400	10,000

Source: University of Arizona

Step 1 Enter time in **L1** and distance in **L2**.

KEYSTROKES: STAT 1 1 ENTER 2 ENTER 5 ENTER 7 ENTER 10 ENTER 12 ENTER 13 ENTER ▶

400 ENTER 800 ENTER 2500 ENTER 3900 ENTER 6250 ENTER 8400 ENTER 10000 ENTER

Step 2 Graph the scatter plot.

KEYSTROKES: 2nd [STAT PLOT] 1 ENTER ▼ ENTER ZOOM 9

[−0.2, 14.2] scl: 1
by [−1232, 11632] scl: 1000

Step 3 Determine and graph the equation for a curve of best fit. Use a quartic regression for the data.

KEYSTROKES: STAT ▶ 7 ENTER Y= VARS 5 ▶ ▶ 1 GRAPH

The equation is shown in the Y= screen. If rounded, the regression equation shown on the calculator can be written as the algebraic equation $y = 0.7x^4 − 17x^3 + 161x^2 − 21x + 293$.

Step 4 Use the [CALC] feature to find the value of the function for $x = 8.5$.

KEYSTROKES: 2nd [CALC] 1 8.5 ENTER

After 8.5 minutes, the wave could be expected to be felt approximately 4980 kilometers from the epicenter.

MENTAL CHECK The table gives the distance for 7 minutes as 3900 and the distance for 10 minutes as 6250. Since 8.5 is halfway between 7 and 10 a reasonable estimate for the distance is halfway between 3900 and 6250. ✓

(continued on the next page)

338 | Extend 5-4 | Graphing Technology Lab: Modeling Data Using Polynomial Functions

---

DifferentiatedInstruction (OL) (BL)

Extension The nature of polynomial functions makes them perfect for inventing funny stories about how things change over time. Challenge students to create stories about something that is growing or shrinking (or both) and display the growth graphically. Extend the problem by having them create scales for the graphs that match the context of their stories.

Exercises

The table shows how many minutes out of each eight-hour work day are used to pay one day's worth of taxes.

1. Draw a scatter plot of the data. Then graph several curves of best fit that relate the number of minutes to the number of years. Try LinReg, QuadReg, and CubicReg. **See margin.**

2. Write the equation for the curve that best fits the data.

3. Based on this equation, how many minutes should you expect to work each day in the year 2020 to pay one day's taxes? Use mental math to check the reasonableness of your estimate. **about 164 min**

2. Sample answer: $y = 0.000324x^3 - 1.94x^2 + 3861.99x - 2,566,944.04$

Year	Minutes
1930	56
1940	86
1950	119
1960	134
1970	144
1980	147
1990	148
2000	163
2005	151

Source: Tax Foundation

The table shows the estimated number of alternative-fueled vehicles in use in the United States per year from 1998 to 2007.

4. Draw a scatter plot of the data. Then graph several curves of best fit that relate the distance to the month. **See Chapter 5 Answer Appendix.**

5. Which curve best fits the data? Is that curve best for predicting future values? **See margin.**

6. Use the best-fit equation you think will give the most accurate prediction for how many alternative-fuel vehicles will be in use in 2018. Use mental math to check the reasonableness of your estimate. **See margin.**

Year	Number of Vehicles	Year	Number of Vehicles
1998	295,030	2003	533,999
1999	322,302	2004	565,492
2000	394,664	2005	592,122
2001	425,457	2006	634,559
2002	471,098	2007	695,763

Source: U.S. Department of Energy

The table shows the average distance from the Sun to Earth during each month of the year.

7. Draw a scatter plot of the data. Then graph several curves of best fit that relate the distance to the month. **See Chapter 5 Answer Appendix.**

8. Write the equation for the curve that best fits the data. **See margin.**

9. Based on your regression equation, what is the distance from the Sun to Earth halfway through September? **about 1.003**

10. Would you use this model to find the distance from the Sun to Earth in subsequent years? Explain your reasoning.
Sample answer: No; the distances should be cyclical, repeating from one year to the next.

Month	Distance (astronomical units)
January	0.9840
February	0.9888
March	0.9962
April	1.0050
May	1.0122
June	1.0163
July	1.0161
August	1.0116
September	1.0039
October	0.9954
November	0.9878
December	0.9837

Source: The Astronomy Cafe

Extension 11–13. See students' work.

11. Write a question that could be answered by examining data. For example, you might estimate the number of people living in your town 5 years from now or predict the future cost of a car.

12. Collect and organize the data you need to answer the question you wrote. You may need to research your topic on the Internet or conduct a survey to collect the data you need.

13. Make a scatter plot and find a regression equation for your data. Then use the regression equation to answer the question.

5. **Sample answer:** All three curves have strong correlation coefficients greater than 0.99. However, expanding the window for the graphs shows that the quadratic regression significantly decreases while the cubic regression significantly increases as time goes by. These are probably not realistic, so the linear regression is the best for predicting the future.

6. Using the linear model, about 1,175,301 alternative-fueled vehicles will be in use in 2018.

8. [0, 13] scl: 1 by [0.8, 1.1] scl: 0.05
Sample answer: $y = 0.0000065x^3 - 0.0012x^2 + 0.0147x + 0.967$

3 Assess

Formative Assessment
Use Exercises 4–10 to make sure students understand how to decide which of the several curves they graphed best fits the data. Ask students to justify their final choice.

In Exercises 3, 6, and 9, students' answers may vary slightly depending on their best-fit curves.

From Concrete to Abstract
Ask:
Given the description and command, what is the formula for each of the following?

Description	Command	Formula
Linear Regression	LinReg $(ax + b)$	$y = ax + b$
Quadratic Regression	QuadReg	$y = ax^2 + bx + c$
Cubic Regression	CubicReg	$y = ax^3 + bx^2 + cx + d$

Additional Answers

1. Sample graphs:

[1920, 2010] scl: 10 by [0, 200] scl: 20

[1920, 2010] scl: 10 by [0, 200] scl: 20

[1920, 2010] scl: 10 by [0, 200] scl: 20

Formative Assessment

Use the Mid-Chapter Quiz to assess students' progress in the first half of the chapter.

For problems answered incorrectly, have students review the lessons indicated in parentheses.

Summative Assessment

📁 Mid-Chapter Test, p. 59

McGraw-Hill
eAssessment Customize and create multiple versions of your Mid-Chapter Quiz and their answer keys.

FOLDABLES StudyOrganizer

Dinah Zike's Foldables®

Before students complete the Mid-Chapter Quiz, encourage them to review the information for Lessons 5-1 through 5-4 in their Foldables.

Additional Answer

15.

Simplify. Assume that no variable equals 0. (Lesson 5-1)

1. $(3x^2y^{-3})(-2x^3y^5)$ $-6x^5y^2$
2. $4t(3rt - r)$ $12rt^2 - 4rt$
3. $\dfrac{3a^4b^3c}{6a^2b^5c^3}$ $\dfrac{a^2}{2b^2c^2}$
4. $\left(\dfrac{p^2r^3}{pr^4}\right)^2$ $\dfrac{p^2}{r^2}$
5. $(4m^2 - 6m + 5) - (6m^2 + 3m - 1)$ $-2m^2 - 9m + 6$
6. $(x + y)(x^2 + 2xy - y^2)$ $x^3 + 3x^2y + xy^2 - y^3$

7. **MULTIPLE CHOICE** The volume of the rectangular prism is $6x^3 + 19x^2 + 2x - 3$. Which polynomial expression represents the area of the base? (Lesson 5-1) **C**

x + 3

A $6x^4 + 37x^3 + 59x^2 + 3x - 9$

B $6x^2 + x + 1$

C $6x^2 + x - 1$

D $6x + 1$

Simplify. (Lesson 5-2)

8. $(4r^3 - 8r^2 - 13r + 20) \div (2r - 5)$ $2r^2 + r - 4$

9. $\dfrac{3x^3 - 16x^2 + 9x - 24}{x - 5}$ $3x^2 - x + 4 - \dfrac{4}{x - 5}$

10. Describe the end behavior of the graph. Then determine whether it represents an odd-degree or an even-degree polynomial function and state the number of real zeros. (Lesson 5-3)

end behavior:
$f(x) \to \infty$ as
$x \to -\infty$ and
$f(x) \to -\infty$ as
$x \to \infty$; odd-degree function;
3 real zeros

11. **MULTIPLE CHOICE** Find $p(-3)$ if $p(x) = \frac{2}{3}x^3 + \frac{1}{3}x^2 - 5x$. (Lesson 5-3) **F**

F 0   H 30

G 11   J 36

---

12. **PENDULUMS** The formula $L(t) = \dfrac{8t^2}{\pi^2}$ can be used to find the length of a pendulum in feet when it swings back and forth in t seconds. Find the length of a pendulum that makes one complete swing in 4 seconds. (Lesson 5-3)
about 12.97 ft

13. **MULTIPLE CHOICE** Find $3f(a - 4) - 2h(a)$ if $f(x) = x^2 + 3x$ and $h(x) = 2x^2 - 3x + 5$. (Lesson 5-3) **D**

A $-a^2 + 15a - 74$

B $-a^2 - 2a - 1$

C $a^2 + 9a - 2$

D $-a^2 - 9a + 2$

14. **ENERGY** The power generated by a windmill is a function of the speed of the wind. The approximate power is given by the function $P(s) = \dfrac{s^3}{1000}$, where s represents the speed of the wind in kilometers per hour. Find the units of power $P(s)$ generated by a windmill when the wind speed is 18 kilometers per hour. (Lesson 5-3) **5.832 units**

Use $f(x) = x^3 - 2x^2 - 3x$ for Exercises 15–17. (Lesson 5-4)

15. Graph the function. **See margin.**

16. Estimate the x-coordinates at which the relative maxima and relative minima occur. $x = -0.5$ and 2

17. State the domain and range of the function.

18. Determine the consecutive integer values of x between which each real zero is located for $f(x) = 3x^2 - 3x - 1$. (Lesson 5-4)
between −1 and 0 and between 1 and 2

Refer to the graph. (Lesson 5-4)

17. D = {all real numbers},
R = {all real numbers}

maximum at $x \approx 1$; minima at $x \approx -1.5$ and $x \approx 4$

19. Estimate the x-coordinate of every turning point, and determine if those coordinates are relative maxima or relative minima.

20. Estimate the x-coordinate of every zero. $-2, -0.5, 2.5, 5$

21. What is the least possible degree of the function? **4**

InterventionPlanner

TIER 1 On Level **OL**	**TIER 2** Strategic Intervention **AL** approaching grade level	**TIER 3** Intensive Intervention 2 or more grades below level
If students miss about 25% of the exercises or less,	**If** students miss about 50% of the exercises,	**If** students miss about 75% of the exercises,
Then choose a resource: **SE** Lessons 5-1, 5-2, 5-3, and 5-4 📁 Skills Practice, pp. 7, 13, 19, and 25 **TE** Chapter Project connectED.mcgraw-hill.com Self-Check Quiz	**Then** choose a resource: 📁 Study Guide and Intervention, pp. 5, 11, 17, and 23 connectED.mcgraw-hill.com Extra Examples, Personal Tutor, Homework Help	**Then** use *Math Triumphs, Alg. 2,* Chs. 1 and 4 connectED.mcgraw-hill.com Extra Examples, Personal Tutor, Homework Help, Review Vocabulary

5-5
Graphing Technology Lab
Solving Polynomial Equations by Graphing

You can use a TI-83/84 Plus graphing calculator to solve polynomial equations.

Activity

Solve $x^4 + 2x^3 = 7$.

Method 1

Step 1 Graph each side of the equation separately in a standard viewing window.

Let **Y1** = $x^4 + 2x^3$ and **Y2** = 7.

KEYSTROKES: Y= X,T,θ,n ∧ 4 + 2 X,T,θ,n ∧ 3 ENTER 7 ZOOM 6

Step 2 Find the points of intersection.

KEYSTROKES: 2nd [CALC] 5

Use ◄ or ► to position the cursor on **Y1** near the first point of intersection.
Press ENTER ENTER ENTER.
Then use ► to position the cursor near the second intersection point.
Press ENTER ENTER ENTER.

Step 3 Examine the graphs.

Determine where the graph of $y = x^4 + 2x^3$ intersects $y = 7$.

The intersections of the graphs of **Y1** and **Y2** are approximately −2.47 and 1.29, so the solution is approximately −2.47 and 1.29

Method 2

Step 1 Rewrite the related equation so it is equal to 0.

$$x^4 + 2x^3 = 7$$
$$x^4 + 2x^3 - 7 = 0$$

Let **Y1** = $x^4 + 2x^3 - 7$ and **Y2** = 0.

KEYSTROKES: Y= X,T,θ,n ∧ 4 + 2 X,T,θ,n ∧ 3 − 7 ENTER 0 ZOOM 6

Step 2 Because **Y2** = 0, to find the intersection points of **Y1** and **Y2**, find where **Y1** crosses the x-axis.

KEYSTROKES: 2nd [CALC] 5

Use ◄ or ► to position the cursor on **Y1** near the first point of intersection.
Press ENTER ENTER ENTER.
Then use ► to position the cursor near the second intersection point.
Press ENTER ENTER ENTER.

Step 3 Examine the graphs.

Determine where the graph of $y = x^4 + 2x^3 - 7$ crosses the x-axis.

The intersections of the graphs of **Y1** and **Y2** are approximately −2.47 and 1.29, so the solution is approximately −2.47 and 1.29

[−10, 10] scl: 1 by [−10, 10] scl: 1

[−10, 10] scl: 1 by [−10, 10] scl: 1

[−10, 10] scl: 1 by [−10, 10] scl: 1

[−10, 10] scl: 1 by [−10, 10] scl: 1

Exercises

Solve each equation. Round to the nearest hundredth.

1. $\frac{2}{3}x^3 + x^2 - 5x = -9$ **−4.11**
2. $x^3 - 9x^2 + 27x = 20$ **1.09**
3. $x^3 + 1 = 4x^2$ **−0.47, 0.54, 3.94**
4. $x^6 - 15 = 5x^4 - x^2$ **−2.31, 2.31**
5. $\frac{1}{2}x^5 = \frac{1}{5}x^2 - 2$ **−1.27**
6. $x^8 = -x^7 + 3$ **−1.36, 1.06**
7. $x^4 - 15x^2 = -24$
8. $x^3 - 6x^2 + 4x = -6$
9. $x^4 - 15x^2 + x + 65 = 0$ **no solution**

7. **−3.63, −1.35, 1.35, 3.63** 8. **−0.69, 1.75, 4.95**

---

1 Focus

Objective Use a graphing calculator to find approximate solutions for polynomial equations.

Materials for Each Student
- TI-83/84 Plus or other graphing calculator

2 Teach

Working in Cooperative Groups
Have students work in pairs, mixing abilities. Have each pair work through the Activity.

- Make sure students understand that a calculator display of 1E−12 means 1×10^{-12}, and that such a small number can often be interpreted as essentially 0.

- Point out to students that they may have to experiment to find appropriate window settings in the Exercises.

Practice Have students complete Exercises 1–9.

3 Assess

Formative Assessment
Use Exercise 1 to assess whether students are locating x-intercepts correctly.

From Concrete to Abstract
What is the maximum number of x-intercepts that the graph of a polynomial function can have?
a number equal to the degree of the polynomial

1 Focus

VerticalAlignment

Before Lesson 5-5 Solve quadratic equations by factoring.

Lesson 5-5 Factor polynomials. Solve polynomial equations by factoring.

After Lesson 5-5 Use tools including factoring to transform and solve polynomial equations.

2 Teach

Scaffolding Questions

Have students read the **Why?** section of the lesson.

Ask:

- What is the formula for the volume of the cube? $V = x^3$

- What is the volume of the larger cube if $x = 5$ inches? 125 in^3

- What is the difference in the volume of the two cubes if $y = 2$ inches? 117 in^3

LESSON 5-5 Solving Polynomial Equations

:: Then	:: Now	:: Why?
● You solved quadratic functions by factoring.	● **1** Factor polynomials. **2** Solve polynomial equations by factoring.	● A small cube is cut out of a larger cube. The volume of the remaining figure is given and the dimensions of each cube need to be determined. This can be accomplished by factoring the cubic polynomial $x^3 - y^3$.

 NewVocabulary
prime polynomials
quadratic form

 Common Core State Standards

Content Standards
A.CED.1 Create equations and inequalities in one variable and use them to solve problems.

Mathematical Practices
4 Model with mathematics.

1 **Factor Polynomials** In Lesson 4-3, you learned that quadratics can be factored just like whole numbers. Their factors, however, are other polynomials. Like quadratics, some cubic polynomials can also be factored with special rules.

KeyConcept Sum and Difference of Cubes

Factoring Technique	General Case
Sum of Two Cubes	$a^3 + b^3 = (a + b)(a^2 - ab + b^2)$
Difference of Two Cubes	$a^3 - b^3 = (a - b)(a^2 + ab + b^2)$

Polynomials that cannot be factored are called **prime polynomials**.

Example 1 Sum and Difference of Cubes

Factor each polynomial. If the polynomial cannot be factored, write *prime*.

a. $16x^4 + 54xy^3$

$16x^4 + 54xy^3 = 2x(8x^3 + 27y^3)$ Factor out the GCF.

$8x^3$ and $27y^3$ are both perfect cubes, so we can factor the sum of two cubes.

$\begin{aligned} 8x^3 + 27y^3 &= (2x)^3 + (3y)^3 & (2x)^3 = 8x^3; (3y)^3 = 27y^3 \\ &= (2x + 3y)[(2x)^2 - (2x)(3y) + (3y)^2] & \text{Sum of two cubes} \\ &= (2x + 3y)(4x^2 - 6xy + 9y^2) & \text{Simplify.} \end{aligned}$

$16x^4 + 54xy^3 = 2x(2x + 3y)(4x^2 - 6xy + 9y^2)$ Replace the GCF.

b. $8y^3 + 5x^3$

The first term is a perfect cube, but the second term is not. So, the polynomial cannot be factored using the sum of two cubes pattern. The polynomial also cannot be factored using quadratic methods or the GCF. Therefore, it is a prime polynomial.

▶ **GuidedPractice**

1A. $5y^4 - 320yz^3$ **1A.** $5y(y - 4z)(y^2 + 4yz + 16z^2)$

1B. $-54w^4 - 250wz^3$ **1B.** $-2w(3w + 5z)(9w^2 - 15wz + 25z^2)$

Lesson 5-5 Resources

Resource	Approaching Level **AL**	On Level **OL**	Beyond Level **BL**	English Learners **ELL**
Teacher Edition	▪ Differentiated Instruction, p. 344	▪ Differentiated Instruction, p. 344	▪ Differentiated Instruction, p. 349	▪ Differentiated Instruction, p. 344
Chapter Resource Masters	▪ Study Guide and Intervention, pp. 29–30 ▪ Skills Practice, p. 31 ▪ Practice, p. 32 ▪ Word Problem Practice, p. 33	▪ Study Guide and Intervention, pp. 29–30 ▪ Skills Practice, p. 31 ▪ Practice, p.32 ▪ Word Problem Practice, p. 33 ▪ Enrichment, p. 34	▪ Practice, p. 32 ▪ Word Problem Practice, p. 33 ▪ Enrichment, p. 34	▪ Study Guide and Intervention, pp. 29–30 ▪ Skills Practice, p. 31 ▪ Practice, p. 32 ▪ Word Problem Practice, p. 33
Other	▪ 5-Minute Check 5-5 ▪ Study Notebook	▪ 5-Minute Check 5-5 ▪ Study Notebook	▪ 5-Minute Check 5-5 ▪ Study Notebook	▪ 5-Minute Check 5-5 ▪ Study Notebook

The table below summarizes the most common factoring techniques used with polynomials. Whenever you factor a polynomial, always look for a common factor first. Then determine whether the resulting polynomial factors can be factored again using one or more of the methods below.

ConceptSummary Factoring Techniques

Number of Terms	Factoring Technique	General Case
any number	Greatest Common Factor (GCF)	$4a^3b^2 - 8ab = 4ab(a^2b - 2)$
two	Difference of Two Squares	$a^2 - b^2 = (a + b)(a - b)$
	Sum of Two Cubes	$a^3 + b^3 = (a + b)(a^2 - ab + b^2)$
	Difference of Two Cubes	$a^3 - b^3 = (a - b)(a^2 + ab + b^2)$
three	Perfect Square Trinomials	$a^2 + 2ab + b^2 = (a + b)^2$ $a^2 - 2ab + b^2 = (a - b)^2$
	General Trinomials	$acx^2 + (ad + bc)x + bd$ $= (ax + b)(cx + d)$
four or more	Grouping	$ax + bx + ay + by$ $= x(a + b) + y(a + b)$ $= (a + b)(x + y)$

Math HistoryLink

Sophie Germain
(1776–1831)
Sophie Germain taught herself mathematics with books from her father's library during the French Revolution when she was confined for safety. Germain discovered the identity $x^4 + 4y^4 = (x^2 + 2y^2 + 2xy)(x^2 + 2y^2 - 2xy)$, which is named for her.

PT

Example 2 Factoring by Grouping

Factor each polynomial. If the polynomial cannot be factored, write *prime*.

a. $8ax + 4bx + 4cx + 6ay + 3by + 3cy$

$8ax + 4bx + 4cx + 6ay + 3by + 3cy$	Original expression
$= (8ax + 4bx + 4cx) + (6ay + 3by + 3cy)$	Group to find a GCF.
$= 4x(2a + b + c) + 3y(2a + b + c)$	Factor the GCF.
$= (4x + 3y)(2a + b + c)$	Distributive Property

b. $20fy - 16fz + 15gy + 8hz - 10hy - 12gz$

$20fy - 16fz + 15gy + 8hz - 10hy - 12gz$	Original expression
$= (20fy + 15gy - 10hy) + (-16fz - 12gz + 8hz)$	Group to find a GCF.
$= 5y(4f + 3g - 2h) - 4z(4f + 3g - 2h)$	Factor the GCF.
$= (5y - 4z)(4f + 3g - 2h)$	Distributive Property

StudyTip

Answer Checks
Multiply the factors to check your answer.

GuidedPractice

2A. $30ax - 24bx + 6cx - 5ay^2 + 4by^2 - cy^2$ $(6x - y^2)(5a - 4b + c)$

2B. $13ax + 18bz - 15by - 14az - 32bx + 9ay$ **prime**

Factoring by grouping is the only method that can be used to factor polynomials with four or more terms. For polynomials with two or three terms, it may be possible to factor according to one of the patterns listed above.

When factoring two terms in which the exponents are 6 or greater, look to factor perfect squares before factoring perfect cubes.

connectED.mcgraw-hill.com **343**

---

1 **Factor Polynomials**
Example 1 shows how to factor the sum or difference of two cubes.
Example 2 shows how to use grouping to factor a polynomial. **Example 3** shows how to factor combinations of cubes and squares.

Formative Assessment
Use the Guided Practice exercises after each example to determine students' understanding of concepts.

Additional Examples

1 Factor each polynomial. If the polynomial cannot be factored, write *prime*.

   a. $x^3 - 400$ The first term is a perfect cube, but the second term is not. It is a prime polynomial.

   b. $24x^5 + 3x^2y^3$ $3x^2(2x + y)(4x^2 - 2xy + y^2)$

2 **a.** $x^3 + 5x^2 - 2x - 10$ $(x + 5)(x^2 - 2)$

   b. $a^2 + 3ay + 2ay^2 + 6y^3$ $(a + 3y)(a + 2y^2)$

Additional Examples also in Interactive Classroom PowerPoint® Presentations

IWB **Interactive White Board**
READY

WatchOut!

Preventing Errors Point out to students that it is often difficult to recognize that grouping can be used to factor a polynomial. Stress that this technique should only be considered when trying to factor a polynomial with four or more terms.

Preventing Errors Emphasize the importance of checking each factor to make sure it is prime before deciding the final group of factors has been found.

Stock Montage/Archive Photos/Getty Images

Additional Example

3 Factor each polynomial. If the polynomial cannot be factored, write *prime*.

a. $x^2y^3 - 3xy^3 + 2y^3 + x^2z^3 - 3xz^3 + 2z^3$ $(x-2) \cdot (x-1)(y+z)(y^2 - yz + z^2)$

b. $64x^6 - y^6$ $(2x-y) \cdot (4x^2 + 2xy + y^2)(2x+y) \cdot (4x^2 - 2xy + y^2)$

2 Solve Polynomial Equations

Example 4 shows how to solve a polynomial equation by factoring. **Example 5** shows how some expressions can be written in quadratic form. **Example 6** shows how to rewrite higher-degree polynomial equations in order to solve the equation using the Zero Product Property.

Additional Example

4 **GEOMETRY** Refer to Example 4. Determine the dimensions of the cubes if the length of the smaller cube is one half the length of the larger cube, and the volume of the object is 23,625 cubic centimeters. larger cube: 30 cm; smaller cube: 15 cm

Focus on Mathematical Content

Solving Polynomial Equations If a polynomial can be written as a product of linear and quadratic factors, its zeros can be found. Set each of the factors equal to 0 and solve the resulting equations.

Example 3 Combine Cubes and Squares

Factor each polynomial. If the polynomial cannot be factored, write *prime*.

a. $x^6 - y^6$

This polynomial could be considered the difference of two squares or the difference of two cubes. The difference of two squares should always be done before the difference of two cubes for easier factoring.

$$x^6 - y^6 = (x^3 + y^3)(x^3 - y^3) \qquad \text{Difference of two squares}$$
$$= (x + y)(x^2 - xy + y^2)(x - y)(x^2 + xy + y^2) \qquad \text{Sum and difference of two cubes}$$

b. $a^3x^2 - 6a^3x + 9a^3 - b^3x^2 + 6b^3x - 9b^3$

With six terms, factor by grouping first.

$$a^3x^2 - 6a^3x + 9a^3 - b^3x^2 + 6b^3x - 9b^3$$
$$= (a^3x^2 - 6a^3x + 9a^3) + (-b^3x^2 + 6b^3x - 9b^3) \qquad \text{Group to find a GCF.}$$
$$= a^3(x^2 - 6x + 9) - b^3(x^2 - 6x + 9) \qquad \text{Factor the GCF.}$$
$$= (a^3 - b^3)(x^2 - 6x + 9) \qquad \text{Distributive Property}$$
$$= (a - b)(a^2 + ab + b^2)(x^2 - 6x + 9) \qquad \text{Difference of cubes}$$
$$= (a - b)(a^2 + ab + b^2)(x - 3)^2 \qquad \text{Perfect squares}$$

> **StudyTip**
> Grouping 6 or more terms Group the terms that have the *most* common values.

▶ **Guided**Practice

3A. $a^6 + b^6$ $(a^2 + b^2)(a^4 - a^2b^2 + b^4)$

3B. $x^5 + 4x^4 + 4x^3 + x^2y^3 + 4xy^3 + 4y^3$ $(x + y)(x^2 - xy + y^2)(x + 2)^2$

---

2 Solve Polynomial Equations In Chapter 4, you learned to solve quadratic equations by factoring and using the Zero Product Property. You can extend these techniques to solve higher-degree polynomial equations.

Real-World Example 4 Solve Polynomial Functions by Factoring

GEOMETRY Refer to the beginning of the lesson. If the small cube is half the length of the larger cube and the figure is 7000 cubic centimeters, what should be the dimensions of the cubes?

Since the length of the smaller cube is half the length of the larger cube, then their lengths can be represented by x and $2x$, respectively. The volume of the object equals the volume of the larger cube minus the volume of the smaller cube.

$$(2x)^3 - x^3 = 7000 \qquad \text{Volume of object}$$
$$8x^3 - x^3 = 7000 \qquad (2x)^3 = 8x^3$$
$$7x^3 = 7000 \qquad \text{Subtract.}$$
$$x^3 = 1000 \qquad \text{Divide.}$$
$$x^3 - 1000 = 0 \qquad \text{Subtract 1000 from each side.}$$
$$(x - 10)(x^2 + 10x + 100) = 0 \qquad \text{Difference of cubes}$$

$$x - 10 = 0 \quad \text{or} \quad x^2 + 10x + 100 = 0 \qquad \text{Zero Product Property}$$
$$x = 10 \qquad\qquad x = -5 \pm 5i\sqrt{3}$$

Since 10 is the only real solution, the lengths of the cubes are 10 cm and 20 cm.

▶ **Guided**Practice

4. Determine the dimensions of the cubes if the length of the smaller cube is one third of the length of the larger cube, and the volume of the object is 3250 cubic centimeters. **5 cm and 15 cm**

DifferentiatedInstruction AL OL ELL

If some students need a visual to help them solve problems like Example 4 and Additional Example 4,

Then it may help them to construct cardboard models of the figures.

In some cases, you can rewrite a polynomial in x in the form $au^2 + bu + c$. For example, by letting $u = x^2$, the expression $x^4 + 12x^2 + 32$ can be written as $(x^2)^2 + 12(x^2) + 32$ or $u^2 + 12u + 32$. This new, but equivalent, expression is said to be in **quadratic form**.

KeyConcept Quadratic Form

Words	An expression that is in quadratic form can be written as $au^2 + bu + c$ for any numbers a, b, and c, $a \neq 0$, where u is some expression in x. The expression $au^2 + bu + c$ is called the quadratic form of the original expression.
Example	$12x^6 + 8x^6 + 1 = 3(2x^3)^2 + 2(2x^3)^2 + 1$

StudyTip

Quadratic Form When writing a polynomial in quadratic form, choose the expression equal to u by examining the terms with variables. Pay special attention to the exponents in those terms. Not every polynomial can be written in quadratic form.

Example 5 Quadratic Form

Write each expression in quadratic form, if possible.

a. $150n^8 + 40n^4 - 15$

$150n^8 + 40n^4 - 15 = 6(5n^4)^2 + 8(5n^4) - 15$ $(5n^4)^2 = 25n^8$

b. $y^8 + 12y^3 + 8$

This cannot be written in quadratic form since $y^8 \neq (y^3)^2$.

▶ **Guided**Practice

5A. $x^4 + 5x + 6$
cannot be written in quadratic form

5B. $8x^4 + 12x^2 + 18$ $2(2x^2)^2 + 6(2x^2) + 18$

You can use quadratic form to solve equations with larger degrees.

Example 6 Solve Equations in Quadratic Form

Solve $18x^4 - 21x^2 + 3 = 0$.

$18x^4 - 21x^2 + 3 = 0$	Original equation
$2(3x^2)^2 - 7(3x^2) + 3 = 0$	$2(3x^2)^2 = 18x^4$
$2u^2 - 7u + 3 = 0$	Let $u = 3x^2$.
$(2u - 1)(u - 3) = 0$	Factor.

$u = \dfrac{1}{2}$ or $u = 3$ Zero Product Property

$3x^2 = \dfrac{1}{2}$ $3x^2 = 3$ Replace u with $3x^2$.

$x^2 = \dfrac{1}{6}$ $x^2 = 1$ Divide by 3.

$x = \pm\dfrac{\sqrt{6}}{6}$ $x = \pm 1$ Take the square root.

The solutions of the equation are 1, -1, $\dfrac{\sqrt{6}}{6}$, and $-\dfrac{\sqrt{6}}{6}$.

▶ **Guided**Practice   **6A.** $-\dfrac{\sqrt{2}}{2}, \dfrac{\sqrt{2}}{2}, -\dfrac{\sqrt{6}}{2}, \dfrac{\sqrt{6}}{2}$ **6B.** $-\dfrac{\sqrt{3}}{2}, \dfrac{\sqrt{3}}{2}, \sqrt{2}i, -\sqrt{2}i$

6A. $4x^4 - 8x^2 + 3 = 0$ **6B.** $8x^4 + 10x^2 - 12 = 0$

connectED.mcgraw-hill.com   **345**

Teach With Tech

Video Recording Record yourself as you work through an example showing how to solve a polynomial equation by first writing it in quadratic form. Post the video to a class Web site for students to use as a reference outside of class.

Additional Examples

5 Write each expression in quadratic form, if possible.
  a. $2x^6 - x^3 + 9$
     $2(x^3)^2 - (x^3) + 9$
  b. $x^4 - 2x^3 - 1$ This cannot be written in quadratic form since $x^4 \neq (x^3)^2$.

6 Solve $x^4 - 29x^2 + 100 = 0$
  $-5, -2, 2, 5$

WatchOut!

Common Misconceptions In Example 5, students may mistakenly conclude that variables must have even powers in order for the expression to be written in quadratic form. Clarify that the relationship between powers of two terms indicates whether an expression can be written in quadratic form. For example, $x - 9x^{\frac{1}{2}} + 3$ can be rewritten as $\left(x^{\frac{1}{2}}\right)^2 - 9\left(x^{\frac{1}{2}}\right) + 3$ since the power of the x term is twice the power of the $x^{\frac{1}{2}}$ term.

3 Practice

Formative Assessment

Use Exercises 1–19 to check for understanding.

Use the chart at the bottom of this page to customize assignments for your students.

Additional Answers

36. $(x^2)^2 + 12(x^2) - 8$

37. $-15(x^2)^2 + 18(x^2) - 4$

38. $2(2x^3)^2 + 3(2x^3) + 7$

39. not possible

40. $(3x^4)^2 - 7(3x^4) + 12$

41. $4(2x^5)^2 + 1(2x^5) + 6$

42. $\pm i\sqrt{5}, \pm i$

43. $\pm\sqrt{5}, \pm i\sqrt{2}$

44. $\pm\sqrt{2}, \pm\dfrac{\sqrt{6}}{2}$

45. $\pm\dfrac{2\sqrt{3}}{3}, \pm\dfrac{\sqrt{15}}{3}$

46. $\pm\sqrt{2}, \pm i\dfrac{\sqrt{3}}{2}$

47. $\pm\dfrac{\sqrt{6}}{6}, \pm i\dfrac{\sqrt{3}}{2}$

49. $(x^2 + 25)(x + 5)(x - 5)$

50. $(x + 2)(x^2 - 2x + 4)(x - 2)$ $(x^2 + 2x + 4)$

51. $x(x + 2)(x - 2)(x^2 + 4)$

52. $x^2 y^2 (2x - 3y)(4x^2 + 6xy + 9y^2)$

53. $(5x + 4y + 5z)(3a - 2b + c)$

54. $(6x^2 - 5y^3 + 2z^2)(a^2 - 4b^2 + 3c^2)$

55. $x(x + 3)(x - 3)(3x + 2)(2x - 5)$

56. $x^2(x + 5)(x - 5)(4x - 3)(5x + 2)$

58. $\pm\dfrac{1}{2}, \pm i\dfrac{\sqrt{6}}{2}$

59. $\pm\dfrac{2\sqrt{3}}{3}, \pm i\dfrac{\sqrt{2}}{2}$

60. $\pm\dfrac{3}{2}, \pm\dfrac{\sqrt{10}}{5}$

61. $\pm\dfrac{1}{3}, \pm i\dfrac{\sqrt{10}}{2}$

62. $\pm\dfrac{1}{2}, \pm\sqrt{2}$

63. $3, -3, \pm i\dfrac{\sqrt{15}}{3}$

64. $1, -2, \dfrac{-1 \pm i\sqrt{3}}{2}, 1 \pm i\sqrt{3}$

65. $-1, 3, \dfrac{-3 \pm 3i\sqrt{3}}{2}, \dfrac{1 \pm i\sqrt{3}}{2}$

Check Your Understanding

= Step-by-Step Solutions begin on page R14.

Examples 1–3 Factor completely. If the polynomial is not factorable, write *prime*.

1. $3ax + 2ay - az + 3bx + 2by - bz$ $(a + b)(3x + 2y - z)$

2. $2kx + 4mx - 2nx - 3ky - 6my + 3ny$ $(2x - 3y) \cdot (k + 2m - n)$

3. $2x^3 + 5y^3$ prime

4. $16g^3 + 2h^3$ $2(2g + h)(4g^2 - 2gh + h^2)$

5. $12qw^3 - 12q^4$ $12q(w - q)(w^2 + qw + q^2)$

6. $3w^2 + 5x^2 - 6y^2 + 2z^2 + 7a^2 - 9b^2$ prime

7. $a^6 x^2 - b^6 x^2$ $x^2(a - b)(a^2 + ab + b^2) \cdot (a + b)(a^2 - ab + b^2)$

8. $x^3 y^2 - 8x^3 y + 16x^3 + y^5 - 8y^4 + 16y^3$ $(x + y)(x^2 - xy + y^2) \cdot (y - 4)^2$

9. $8c^3 - 125d^3$ $(2c - 5d)(4c^2 + 10cd + 25d^2)$

10. $6bx + 12cx + 18dx - by - 2cy - 3dy$ $(6x - y)(b + 2c + 3d)$

Example 4   Solve each equation.

11. $x^4 - 19x^2 + 48 = 0$ $4, -4, \pm\sqrt{3}$

12. $x^3 - 64 = 0$ $4, -2 \pm 2i\sqrt{3}$

13. $x^3 + 27 = 0$ $-3, \dfrac{3 \pm 3i\sqrt{3}}{2}$

14. $x^4 - 33x^2 + 200 = 0$ $5, -5, \pm 2\sqrt{2}$

15. **CCSS PERSEVERANCE** A boardwalk that is x feet wide is built around a rectangular pond. The pond is 30 feet wide and 40 feet long. The combined area of the pond and the boardwalk is 2000 square feet. What is the width of the boardwalk?   5 ft

Example 5   Write each expression in quadratic form, if possible.

16. $4x^6 - 2x^3 + 8$ $(2x^3)^2 - 1(2x^3) + 8$

17. $25y^6 - 5y^2 + 20$ not possible

Example 6   Solve each equation.

18. $x^4 - 6x^2 + 8 = 0$ $2, -2, \sqrt{2}, -\sqrt{2}$

19. $y^4 - 18y^2 + 72 = 0$ $\sqrt{6}, -\sqrt{6}, 2\sqrt{3}, -2\sqrt{3}$

Practice and Problem Solving

Extra Practice is on page R5.

Examples 1–3 Factor completely. If the polynomial is not factorable, write *prime*.

20. $8c^3 - 27d^3$

21. $64x^4 + xy^3$

22. $a^8 - a^2 b^6$

20. $(2c - 3d) \cdot (4c^2 + 6cd + 9d^2)$

23. $x^6 y^3 + y^9$

24. $18x^6 + 5y^6$ prime

25. $w^3 - 2y^3$ prime

21. $x(4x + y) \cdot (16x^2 - 4xy + y^2)$

26. $gx^2 - 3hx^2 - 6fy^2 - gy^2 + 6fx^2 + 3hy^2$ $(x + y)(x - y)(6f + g - 3h)$

27. $12ax^2 - 20cy^2 - 18bx^2 - 10ay^2 + 15by^2 + 24cx^2$ $(6x^2 - 5y^2)(2a - 3b + 4c)$

22. $a^2(a - b)(a^2 + ab + b^2)(a + b) \cdot (a^2 - ab + b^2)$

28. $a^3 x^2 - 16a^3 x + 64a^3 - b^3 x^2 + 16b^3 x - 64b^3$ $(a - b)(a^2 + ab + b^2)(x - 8)^2$

(29) $8x^5 - 25y^3 + 80x^4 - x^2 y^3 + 200x^3 - 10xy^3$ $(2x - y)(4x^2 + 2xy + y^2)(x + 5)^2$

23. $y^3(x^2 + y^2) \cdot (x^4 - x^2 y^2 + y^4)$

Example 4   Solve each equation.   30. $3, -3, \pm i\sqrt{10}$ 31. $6, -6, \pm 2i\sqrt{5}$ 32. $\pm\sqrt{11}, \pm 2i$

30. $x^4 + x^2 - 90 = 0$

31. $x^4 - 16x^2 - 720 = 0$

32. $x^4 - 7x^2 - 44 = 0$

33. $x^4 + 6x^2 - 91 = 0$

34. $x^3 + 216 = 0$

35. $64x^3 + 1 = 0$ $-\dfrac{1}{4}, \dfrac{1 \pm i\sqrt{3}}{8}$

Example 5   Write each expression in quadratic form, if possible.   36–41. See margin.

36. $x^4 + 12x^2 - 8$

37. $-15x^4 + 18x^2 - 4$

38. $8x^6 + 6x^3 + 7$

33. $\pm\sqrt{7}, \pm i\sqrt{13}$

39. $5x^6 - 2x^2 + 8$

40. $9x^8 - 21x^4 + 12$

41. $16x^{10} + 2x^5 + 6$

34. $-6, 3 \pm 3i\sqrt{3}$

Example 6   Solve each equation.   42–47. See margin.

42. $x^4 + 6x^2 + 5 = 0$

43. $x^4 - 3x^2 - 10 = 0$

44. $4x^4 - 14x^2 + 12 = 0$

45. $9x^4 - 27x^2 + 20 = 0$

46. $4x^4 - 5x^2 - 6 = 0$

47. $24x^4 + 14x^2 - 3 = 0$

Differentiated Homework Options

Level	Assignment	Two-Day Option	
AL Basic	20–47, 81–99	21–47 odd, 84–87	20–46 even, 81–83, 88–99
OL Core	21–69 odd, 70–75, 77, 78, 81–99	20–47, 84–87	48, 57, 70–74, 78, 81–83, 88–99
BL Advanced	48–96, (optional: 97–99)		

B

48. ZOOLOGY A species of animal is introduced to a small island. Suppose the population of the species is represented by $P(t) = -t^4 + 9t^2 + 400$, where t is the time in years. Determine when the population becomes zero. **5 yr**

Factor completely. If the polynomial is not factorable, write *prime*. 49–56. See margin.

49. $x^4 - 625$ **50.** $x^6 - 64$ **51.** $x^5 - 16x$ **52.** $8x^5y^2 - 27x^2y^5$

53. $15ax - 10bx + 5cx + 12ay - 8by + 4cy + 15az - 10bz + 5cz$

54. $6a^2x^2 - 24b^2x^2 + 18c^2x^2 - 5a^2y^3 + 20b^2y^3 - 15c^2y^3 + 2a^2z^2 - 8b^2z^2 + 6c^2z^2$

55. $6x^5 - 11x^4 - 10x^3 - 54x^3 + 99x^2 + 90x$

56. $20x^6 - 7x^5 - 6x^4 - 500x^4 + 175x^3 + 150x^2$

57. GEOMETRY The volume of the figure at the right is 440 cubic centimeters. Find the value of x and the length, height, and width. **$x = 8$; 5, 8, 11**

Solve each equation. 58–65. See margin.

58. $8x^4 + 10x^2 - 3 = 0$ **59.** $6x^4 - 5x^2 - 4 = 0$

60. $20x^4 - 53x^2 + 18 = 0$ **61.** $18x^4 + 43x^2 - 5 = 0$

62. $8x^4 - 18x^2 + 4 = 0$ **63.** $3x^4 - 22x^2 - 45 = 0$

64. $x^6 + 7x^3 - 8 = 0$ **65** $x^6 - 26x^3 - 27 = 0$ **66.** $-5, \dfrac{1}{2}, \dfrac{-1 \pm i\sqrt{3}}{2}, \dfrac{5 \pm 5i\sqrt{3}}{2}$

66. $8x^6 + 999x^3 = 125$ **67.** $4x^4 - 4x^2 - x^2 + 1 = 0$ **$-1, 1, \pm\dfrac{1}{2}$**

68. $x^6 - 9x^4 - x^2 + 9 = 0$ **$\pm 3, \pm 1, \pm i$**   **69.** $x^4 + 8x^2 + 15 = 0$ **$\pm i\sqrt{5}, \pm i\sqrt{3}$**

70. CCSS SENSE-MAKING A rectangular prism with dimensions $x - 2$, $x - 4$, and $x - 6$ has a volume equal to $40x$ cubic units. **a.** $x^3 - 12x^2 + 44x - 48 = 40x$

a. Write a polynomial equation using the formula for volume.

b. Use factoring to solve for x. **12, $\pm 2i$**

c. Are any values for x unreasonable? Explain. Sample answer: $\pm 2i$ because they are imaginary numbers

d. What are the dimensions of the prism? **6, 8, and 10**

71. POOL DESIGN Andrea wants to build a pool following the diagram at the right. The pool will be surrounded by a sidewalk of a constant width.

a. If the total area of the pool itself is to be 336 ft², what is x? **2 ft**

b. If the value of x were doubled, what would be the new area of the pool? **176 ft²**

c. If the value of x were halved, what would be the new area of the pool? **428 ft²**

CCSS Teaching the Mathematical Practices

Perseverance Mathematically proficient students start by explaining to themselves the meaning of a problem and looking for entry points to its solution. They analyze givens, constraints, relationships, and goals and make conjectures about the form and meaning of the solution and plan a solution pathway rather than simply jumping into a solution attempt.

📁 **Study Guide and Intervention**

AL   OL   ELL

📁 **Practice**

AL   OL   BL   ELL

📁 **Enrichment**

OL   BL

📁 **Word Problem Practice**

AL   OL   BL   ELL

72b. Sample answer: Subtract 8000 from each side. Then multiply each side by 1000 to eliminate the decimals. Factor out $-12t$, then factor the remaining trinomial.

72c. Sample answer: 15 and 0 are reasonable, and -35 is unreasonable because time cannot be negative.

72d.

78a.

78c.

Tips for New Teachers

Sense-Making Ask students why the statement in Exercise 81, $a^2 + b^2 = (a + b)^2$, is not true. Show students that when you expand the expression $(a + b)^2$ the statement becomes $a^2 + b^2 = a^2 + 2ab + b^2$, which is not true.

---

72. BIOLOGY During an experiment, the number of cells of a virus can be modeled by $P(t) = -0.012t^3 - 0.24t^2 + 6.3t + 8000$, where t is the time in hours and P is the number of cells. Jack wants to determine the times at which there are 8000 cells.

  a. Solve for t by factoring. **15, 0, -35**

  b. What method did you use to factor? **b, c, d. See margin.**

  c. Which values for t are reasonable and which are unreasonable? Explain.

  d. Graph the function for $0 \le t \le 20$ using your calculator.

73 **HOME BUILDING** Alicia's parents want their basement home theater designed according to the diagram at the right.

  a. Write a function in terms of x for the area of the basement. $f(x) = 8x^2 + 34x + 24$

  b. If the basement is to be 1366 square feet, what is x? **11 ft**

74. BIOLOGY A population of parasites in an experiment can be modeled by $f(t) = t^3 + 5t^2 - 4t - 20$, where t is the time in days. **a. 2, -2, -5**

  a. Use factoring by grouping to determine the values of t for which $f(t) = 0$.

  b. At what times does the population reach zero? **2 days, -2 days, and -5 days**

  c. Are any of the values of t unreasonable? Explain. Sample answer: -2 and -5 are unreasonable because time cannot be negative.

Factor completely. If the polynomial is not factorable, write *prime*.

75. $x^6 - 4x^4 - 8x^4 + 32x^2 + 16x^2 - 64$ $(x + 2)^3(x - 2)^3$

76. $y^9 - y^6 - 2y^6 + 2y^3 + y^3 - 1$ $(y - 1)^3(y^2 + y + 1)^3$

77. $x^6 - 3x^4y^2 + 3x^2y^4 - y^6$ $(x + y)^3(x - y)^3$

78. **CCSS** **SENSE-MAKING** Fredo's corral, an enclosure for livestock, is currently 32 feet by 40 feet. He wants to enlarge the area to 4.5 times its current area by increasing the length and width by the same amount. **a, c. See margin.**

  a. Draw a diagram to represent the situation.

  b. Write a polynomial equation for the area of the new corral. Then solve the equation by factoring. $4x^2 + 144x + 1280 = 5760$; $x = -56$ or 20

  c. Graph the function.

  d. Which solution is irrelevant? Explain. Sample answer: -56 is irrelevant because length cannot be negative.

H.O.T. Problems Use Higher-Order Thinking Skills

79. CHALLENGE Factor $36x^{2n} + 12x^n + 1$. $(6x^n + 1)^2$

80. CHALLENGE Solve $6x - 11\sqrt{3x} + 12 = 0$. $\dfrac{16}{3}, \dfrac{3}{4}$

81. REASONING Find a counterexample to the statement $a^2 + b^2 = (a + b)^2$. Sample answer: $a = 1$, $b = -1$

82. OPEN ENDED The cubic form of an equation is $ax^3 + bx^2 + cx + d = 0$. Write an equation with degree 6 that can be written in *cubic* form. Sample answer: $12x^6 + 6x^4 + 8x^2 + 4 = 12(x^2)^3 + 6(x^2)^2 + 8(x^2) + 4$

83. WRITING IN MATH Explain how the graph of a polynomial function can help you factor the polynomial. Sample answer: The factors can be determined by the x-intercepts of the graph. An x-intercept of 5 represents a factor of $(x - 5)$.

CCSS **Teaching the Mathematical Practices**

Sense-Making Mathematically proficient students start by explaining to themselves the meaning of a problem and looking for entry points to its solution. They analyze givens, constraints, relationships, and goals. They check their answers to problems using a different method, and they continually ask themselves, "Does this make sense?"

84. SHORT RESPONSE Tiles numbered from 1 to 6 are placed in a bag and are drawn to determine which of six tasks will be assigned to six people. What is the probability that the tiles numbered 5 and 6 are the last two drawn? $\frac{1}{15}$

85. STATISTICS Which of the following represents a negative correlation? **D**

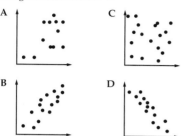

A   C

B   D

86. Which of the following most accurately describes the translation of the graph $y = (x + 4)^2 - 3$ to the graph of $y = (x - 1)^2 + 3$? **J**

F   down 1 and to the right 3

G   down 6 and to the left 5

H   up 1 and to the left 3

J   up 6 and to the right 5

87. SAT/ACT The positive difference between k and $\frac{1}{12}$ is the same as the positive difference between $\frac{1}{3}$ and $\frac{1}{5}$. Which of the following is the value of k? **D**

A $\frac{1}{60}$ D $\frac{13}{60}$

B $\frac{1}{20}$ E $\frac{37}{60}$

C $\frac{1}{15}$

Spiral Review

Graph each polynomial function. Estimate the x-coordinates at which the relative maxima and relative minima occur. (Lesson 5-4) **88–90. See margin.**

88. $f(x) = 2x^3 - 4x^2 + x + 8$ **89.** $f(x) = -3x^3 + 6x^2 + 2x - 1$ **90.** $f(x) = -x^3 + 3x^2 + 4x - 6$

State the degree and leading coefficient of each polynomial in one variable. If it is not a polynomial in one variable, explain why. (Lesson 5-3) **91–93. See margin.**

91. $f(x) = 4x^3 - 6x^2 + 5x^4 - 8x$ **92.** $f(x) = -2x^5 + 5x^4 + 3x^2 + 9$ **93.** $f(x) = -x^4 - 3x^3 + 2x^6 - x^7$

94. ELECTRICITY The impedance in one part of a series circuit is $3 + 4j$ ohms, and the impedance in another part of the circuit is $2 - 6j$ ohms. Add these complex numbers to find the total impedance of the circuit. (Lesson 4-4) **5 − 2j ohms**

95. SKIING All 28 members of a ski club went on a trip. The club paid a total of $478 for the equipment. How many skis and snowboards did they rent? (Lesson 3-2) **18 skis and 10 snowboards**

96. GEOMETRY The sides of an angle are parts of two lines whose equations are $2y + 3x = -7$ and $3y - 2x = 9$. The angle's vertex is the point where the two sides meet. Find the coordinates of the vertex of the angle. (Lesson 3-1) **(−3, 1)**

Ski Rental $16/day

or

Snowboard Rental $19/day

Skills Review

Divide.

97. $(x^2 + 6x - 2) \div (x + 4)$ **98.** $(2x^2 + 8x - 10) \div (2x + 1)$ **99.** $(8x^3 + 4x^2 + 6) \div (x + 2)$

97. $x + 2 - \dfrac{10}{x+4}$ **98.** $x + \dfrac{7}{2} - \dfrac{27}{4x+2}$ **99.** $8x^2 - 12x + 24 - \dfrac{42}{x+2}$

DifferentiatedInstruction **BL**

Extension Draw a right triangle on the board. Under the triangle write "Area: $6x^2 + zy + 2xz + 3xy$." Ask students to find binomials with whole number coefficients for the base and height of this triangle. (*Hint:* The formula for the area of a triangle is $A = \frac{1}{2}bh$.) There are two possible answers: base: $(6x + 2z)$, height: $(2x + y)$; or base: $(3x + z)$, height: $(4x + 2y)$.

---

4 Assess

Name the Math Prepare two paper bags containing small pieces of paper: one containing polynomial equations that can be solved by factoring, and the other containing polynomial equations that are in quadratic form. Have each student select one equation from each bag and solve both equations.

Additional Answers

88. rel. maximum at $x \approx 0.1$; rel. minimum at $x \approx 1.2$;

89. rel. maximum at $x \approx 1.5$, rel. minimum at $x \approx 0.1$;

90. rel. maximum at $x \approx 2.5$, rel. minimum at $x \approx -0.5$;

91. degree = 4; leading coefficient = 5

92. degree = 5; leading coefficient = −2

93. degree = 7; leading coefficient = −1

1 Focus

Objective Prove polynomial identities.

Materials

- TI-Nspire CAS technology

Teaching Tip

- A spreadsheet program can also be used to complete Activity 1.

- When formulas are entered into the TI-Nspire technology spreadsheet, it will prompt you to indicate whether x and y are column or variable references. Indicate that they are variable references.

- Sometimes the entire equation cannot be displayed on one CAS screen. Use the right arrow to scroll the display to the right.

2 Teach

Working in Cooperative Groups

Divide the class into pairs. Work through Activities 1 and 2 as a class. Then ask students to work with their partners to complete Activities 3 and 4 and Exercise 1.

Ask:

- After Activity 2, ask students if it matters which side of the equation they simplify. no

Practice Have students complete Exercises 2–5.

An **identity** is an equation that is satisfied by any numbers that replace the variables. Thus, a **polynomial identity** is a polynomial equation that is true for any values that are substituted for the variables.

CCSS Common Core State Standards
Content Standards
A.APR.4 Prove polynomial identities and use them to describe numerical relationships.

You can use a table or a spreadsheet on your graphing calculator to determine whether a polynomial equation may be an identity.

Activity 1 Use a Table

Determine whether $x^3 - y^3 = (x - y)(x^2 + xy + y^2)$ may be an identity.

Step 1 Add a new **Lists & Spreadsheets** page on the TI-Nspire. Label column A x and column B y. Type any values in columns A and B.

Step 2 Move the cursor to the formula row in column C and type $= x^3 - y^3$. In column D, type $= (x - y)(x^2 + xy + y^2)$.

No matter what values are entered for x and y in columns A and B, the values in columns C and D are the same. Thus, the equation may be an identity.

If you want to prove that an equation is an identity, you need to show that it is true for all values of the variables.

KeyConcept Verifying Identities by Transforming One Side

Step 1 Simplify one side of an equation until the two sides of the equation are the same. It is often easier to work with the more complicated side of the equation.

Step 2 Transform that expression into the form of the simpler side.

Activity 2 Transform One Side

Prove that $(x + y)^2 = x^2 + 2xy + y^2$ is an identity.

$$(x + y)^2 \stackrel{?}{=} x^2 + 2xy + y^2 \qquad \text{Original equation}$$
$$(x + y)(x + y) \stackrel{?}{=} x^2 + 2xy + y^2 \qquad \text{Write } (x + y)^2 \text{ as two factors.}$$
$$x^2 + xy + xy + y^2 \stackrel{?}{=} x^2 + 2xy + y^2 \qquad \text{FOIL Method}$$
$$x^2 + 2xy + y^2 = x^2 + 2xy + y^2 \checkmark \qquad \text{Simplify.}$$

Thus, the identity $(x + y)^2 = x^2 + 2xy + y^2$ is verified.

350 | Extend 5-5 | Graphing Technology Lab: Polynomial Identities

---

You can also use a TI-Nspire with a computer algebra system (CAS) to prove an identity.

Activity 3   Use CAS

Prove that $x^3 - y^3 = (x - y)(x^2 + xy + y^2)$ is an identity.

Step 1 Add a new **Calculator** page on the TI-Nspire CAS. Simplify the right side of the equation one step at a time.

Step 2 Enter the right side of the equation and then distribute.

Step 3 Multiply next. The CAS system will do the final simplification step.

The final step shown on the CAS screen is the results in $x^3 - y^3$. Thus, the identity has been proved.

You can also prove identities by transforming each side of the equation.

Activity 4   Use CAS to Transform Each Side

Prove that $(x + 2)^3(x - 1)^3 = (x^2 + x - 2)(x^4 + 2x^3 - 3x^2 - 4x + 4)$ is an identity.

Add a new **Calculator** page on the TI-Nspire. Simplify the left and the right sides of the equation simultaneously.

The CAS will indicate if the changes are true, otherwise it will simplify for you.

The CAS system will do the final simplification step. The identity $(x + 2)^3 (x - 1)^3 = (x^2 + x - 2)(x^4 + 2x^3 - 3x^2 - 4x + 4)$ has been proved.

Analyze

1. Determine whether $a^2 - b^2 = (a + b)(a - b)$ may be an identity.   **1–8. See Chapter 5 Answer Appendix.**

Exercises

Use CAS to prove each identity.

2. $x^3 + y^3 = (x + y)(x^2 - xy + y^2)$

3. $p^4 - q^4 = (p - q)(p + q)(p^2 + q^2)$

4. $a^5 - b^5 = (a - b)(a^4 + a^3b + a^2b^2 + ab^3 + b^4)$

5. $g^6 + h^6 = (g^2 + h^2)(g^4 - g^2h^2 + h^4)$

6. $a^5 + b^5 = (a + b)(a^4 - a^3b + a^2b^2 - ab^3 + b^4)$

7. $u^6 - w^6 = (u + w)(u - w)(u^2 + vw + w^2)(u^2 - vw + w^2)$

8. $(x + 1)^2(x - 4)^3 = (x^2 - 3x - 4)(x^3 - 7x^2 + 8x + 16)$

3 Assess

Formative Assessment
Use Exercises 6–8 to assess each student's ability to prove identities.

From Concrete to Abstract
Have students prove the identity $x^3 + y^3 = (x + y)(x^2 - xy + y^2)$ without the use of technology.

LESSON 5-6 The Remainder and Factor Theorems

blue jean images/Getty Images

1 Focus

VerticalAlignment

Before Lesson 5-6 Use the Distributive Property and factoring to simplify algebraic expressions.

Lesson 5-6 Evaluate functions by using synthetic substitution. Determine whether a binomial is a factor of a polynomial by using synthetic substitution.

After Lesson 5-6 Use tools including factoring to transform and solve equations.

2 Teach

Scaffolding Questions

Have students read the **Why?** section of the lesson.

Ask:

- What would be the value of x if you wanted to estimate the number of students in 1994? 1

- Would you expect the actual number of students in 2013 to match the number predicted by the function? Explain. No; sample answer: The equation is a model based on available data.

- Is the model more accurate for the years immediately following 1993 or for later years? Explain. Sample answer: The model closely matches actual data for the years immediately following 1993.

Then

- You used the Distributive Property and factoring to simplify algebraic expressions.

Now

1 Evaluate functions by using synthetic substitution.

2 Determine whether a binomial is a factor of a polynomial by using synthetic substitution.

Why?

- The number of college students from the United States who study abroad can be modeled by the function $S(x) = 0.02x^4 - 0.52x^3 + 4.03x^2 + 0.09x + 77.54$, where x is the number of years since 1993 and $S(x)$ is the number of students in thousands.

You can use this function to estimate the number of U.S. college students studying abroad in 2018 by evaluating the function for $x = 25$. Another method you can use is *synthetic substitution*.

NewVocabulary
synthetic substitution
depressed polynomial

CCSS Common Core State Standards

Content Standards
A.APR.2 Know and apply the Remainder Theorem: For a polynomial $p(x)$ and a number a, the remainder on division by $x - a$ is $p(a)$, so $p(a) = 0$ if and only if $(x - a)$ is a factor of $p(x)$.

F.IF.7.c Graph polynomial functions, identifying zeros when suitable factorizations are available, and showing end behavior.

Mathematical Practices
7 Look for and make use of structure.

---

1 Synthetic Substitution Synthetic division can be used to find the value of a function. Consider the polynomial function $f(x) = -3x^2 + 5x + 4$. Divide the polynomial by $x - 3$.

Method 1 Long Division

$$
\begin{array}{r}
-3x - 4 \\
x - 3 \overline{)-3x^2 + 5x + 4} \\
\underline{-3x^2 + 9x} \\
-4x + 4 \\
\underline{-4x + 12} \\
-8
\end{array}
$$

Method 2 Synthetic Division

$$
\begin{array}{r|rrr}
3 & -3 & 5 & 4 \\
  &   & -9 & -12 \\
\hline
  & -3 & -4 & -8
\end{array}
$$

Compare the remainder of -8 to $f(3)$.

$f(3) = -3(3)^2 + 5(3) + 4$ Replace x with 3.

$= -27 + 15 + 4$ Multiply.

$= -8$ Simplify.

Notice that the value of $f(3)$ is the same as the remainder when the polynomial is divided by $x - 3$. This illustrates the **Remainder Theorem**.

KeyConcept Remainder Theorem

Words   If a polynomial $P(x)$ is divided by $x - r$, the remainder is a constant $P(r)$, and

Dividend	equals	quotient	times	divisor	plus	remainder.
$P(x)$	$=$	$Q(x)$	$\bullet$	$(x - r)$	$+$	$P(r)$,

where $Q(x)$ is a polynomial with degree one less than $P(x)$.

Example $x^2 + 6x + 2 = (x - 4) \bullet (x + 10) + 42$

Applying the Remainder Theorem using synthetic division to evaluate a function is called **synthetic substitution**. It is a convenient way to find the value of a function, especially when the degree of the polynomial is greater than 2.

Lesson 5-6 Resources

Resource	Approaching Level **AL**	On Level **OL**	Beyond Level **BL**	English Learners **ELL**
Teacher Edition	▪ Differentiated Instruction, p. 354	▪ Differentiated Instruction, pp. 354, 357	▪ Differentiated Instruction, pp. 354, 357	▪ Differentiated Instruction, p. 354
Chapter Resource Masters	▪ Study Guide and Intervention, pp. 35–36 ▪ Skills Practice, p. 37 ▪ Practice, p. 38 ▪ Word Problem Practice, p. 39	▪ Study Guide and Intervention, pp. 35–36 ▪ Skills Practice, p. 37 ▪ Practice, p. 38 ▪ Word Problem Practice, p. 39 ▪ Enrichment, p. 40	▪ Practice, p. 38 ▪ Word Problem Practice, p. 39 ▪ Enrichment, p. 40	▪ Study Guide and Intervention, pp. 35–36 ▪ Skills Practice, p. 37 ▪ Practice, p. 38 ▪ Word Problem Practice, p. 39
Other	▪ 5-Minute Check 5-6 ▪ Study Notebook	▪ 5-Minute Check 5-6 ▪ Study Notebook	▪ 5-Minute Check 5-6 ▪ Study Notebook	▪ 5-Minute Check 5-6 ▪ Study Notebook

Example 1 Synthetic Substitution

If $f(x) = 3x^4 - 2x^3 + 5x + 2$, find $f(4)$.

Method 1 Synthetic Substitution

By the Remainder Theorem, $f(4)$ should be the remainder when the polynomial is divided by $x - 4$.

$$
\begin{array}{r|rrrrr}
4 & 3 & -2 & 0 & 5 & 2 \\
  &   & 12 & 40 & 160 & 660 \\
\hline
  & 3 & 10 & 40 & 165 & 662
\end{array}
$$

Because there is no x^2 term, a zero is placed in this position as a placeholder.

The remainder is 662. Therefore, by using synthetic substitution, $f(4) = 662$.

Method 2 Direct Substitution

Replace x with 4.

$f(x) = 3x^4 - 2x^3 + 5x + 2$ Original function

$f(4) = 3(4)^4 - 2(4)^3 + 5(4) + 2$ Replace x with 4.

$ = 768 - 128 + 20 + 2$ or 662 Simplify.

By using direct substitution, $f(4) = 662$. Both methods give the same result.

▶ **Guided**Practice

1A. If $f(x) = 3x^3 - 6x^2 + x - 11$, find $f(3)$. **19**

1B. If $g(x) = 4x^5 + 2x^3 + x^2 - 1$, find $f(-1)$. **−6**

Synthetic substitution can be used in situations in which direct substitution would involve cumbersome calculations.

Real-World Example 2 Find Function Values

COLLEGE Refer to the beginning of the lesson. How many U.S. college students will study abroad in 2018?

Use synthetic substitution to divide $0.02x^4 - 0.52x^3 + 4.03x^2 + 0.09x + 77.54$ by $x - 20$.

$$
\begin{array}{r|rrrrr}
25 & 0.02 & -0.52 & 4.03 & 0.09 & 77.54 \\
   &   & 0.5   & -0.5 & 88.25 & 2208.5 \\
\hline
   & 0.02 & -0.02 & 3.53 & 88.34 & 2286.04
\end{array}
$$

In 2018, there will be about 2,286,040 U.S. college students studying abroad.

▶ **Guided**Practice

2. **COLLEGE** The function $C(x) = 2.46x^3 - 22.37x^2 + 53.81x + 548.24$ can be used to approximate the number, in thousands, of international college students studying in the United States x years since 2000. How many international college students can be expected to study in the U.S. in 2015? **4,624,640**

Real-World Link
Some benefits of studying abroad include learning a new language, becoming more independent, and improving communication skills. Studying abroad also gives students a chance to experience different cultures and customs as well as try new foods.
Source: StudyAbroad

---

1 Synthetic Substitution

Example 1 shows how to use synthetic substitution to evaluate a function.
Example 2 shows how to use synthetic substitution to solve a real-world problem.

Formative Assessment
Use the Guided Practice exercises after each example to determine students' understanding of concepts.

Tips for New Teachers
Building on Prior Knowledge In Lesson 5-2, students learned synthetic division. In this lesson, students will use synthetic division to evaluate a function and to find factors of polynomials.

Additional Examples

1 If $f(x) = 2x^4 - 5x^2 + 8x - 7$, find $f(6)$. **2453**

2 **COLLEGE** Refer to the application at the beginning of the lesson. How many U.S. college students will study abroad in 2011? about **451,760**

▶ **Additional Examples** also in Interactive Classroom PowerPoint® Presentations

IWB **Interactive White Board** READY

Tips for New Teachers
Pacing Most of this lesson relies heavily on synthetic substitution. Before finishing your discussion of Example 1, be certain that students understand the method. Ask a student volunteer to demonstrate the steps of synthetic substitution shown in Method 1 of Example 1, and invite students to discuss any problems they have with the technique.

2 Factors of Polynomials

Example 3 shows how to use the Factor Theorem to show that a binomial is a factor of a polynomial and then to find the remaining factors of the polynomial.

Additional Example

3 Determine whether $x - 3$ is a factor of $x^3 + 4x^2 - 15x - 18$. Then find the remaining factors of the polynomial.

$$
\begin{array}{r|rrrr}
3 & 1 & 4 & -15 & -18 \\
  &   & 3 & 21 & 18 \\
\hline
  & 1 & 7 & 6 & 0
\end{array}
$$

So, $x^3 + 4x^2 - 15x - 18 = (x - 3)(x^2 + 7x + 6)$. Since $x^2 + 7x + 6 = (x + 1)(x + 6)$, $x^3 + 4x^2 - 15x - 18 = (x - 3)(x + 1)(x + 6)$.

Focus on Mathematical Content

Factor Theorem The Remainder Theorem says that the value of $f(a)$ is the same as the remainder when the polynomial is divided by $x - a$. The Factor Theorem is a special case of the Remainder Theorem. It says: If $f(a)$ has a value of 0, then $x - a$ is a factor of the polynomial.

WatchOut!

Preventing Errors Point out that the Factor Theorem does not say anything about which numbers to try. Techniques for identifying potential factors will be introduced later in the chapter.

Teach with Tech

Blog Have students write a blog entry explaining how the Remainder and Factor Theorems are related—specifically, have students describe how the Factor Theorem is a special case of the Remainder Theorem.

WatchOut!

Synthetic Substitution Remember that synthetic substitution is used to divide a polynomial by $(x - a)$. If the binomial is $(x - a)$, use a. If the binomial is $(x + a)$, use $-a$.

---

2 Factors of Polynomials

The synthetic division below shows that the quotient of $2x^3 - 3x^2 - 17x + 30$ and $x + 3$ is $2x^2 - 9x + 10$.

$$
\begin{array}{r|rrrr}
-3 & 2 & -3 & -17 & 30 \\
   &   & -6 & 27 & -30 \\
\hline
   & 2 & -9 & 10 & 0
\end{array}
$$

When you divide a polynomial by one of its binomial factors, the quotient is called a depressed polynomial. A **depressed polynomial** has a degree that is one less than the original polynomial. From the results of the division, and by using the Remainder Theorem, we can make the following statement.

$$
\underset{\text{Dividend}}{2x^3 - 3x^2 - 17x + 30} \; \underset{\text{equals}}{=} \; \underset{\text{quotient}}{(2x^2 - 9x + 10)} \; \underset{\text{times}}{\cdot} \; \underset{\text{divisor}}{(x + 3)} \; \underset{\text{plus}}{+} \; \underset{\text{remainder.}}{0}
$$

Since the remainder is 0, $f(-3) = 0$. This means that $x + 3$ is a factor of $2x^3 - 3x^2 - 17x + 30$. This illustrates the **Factor Theorem**, which is a special case of the Remainder Theorem.

KeyConcept Factor Theorem

The binomial $x - r$ is a factor of the polynomial $P(x)$ if and only if $P(r) = 0$.

The Factor Theorem can be used to determine whether a binomial is a factor of a polynomial. It can also be used to determine all of the factors of a polynomial.

Example 3 Use the Factor Theorem

Determine whether $x - 5$ is a factor of $x^3 - 7x^2 + 7x + 15$. Then find the remaining factors of the polynomial.

The binomial $x - 5$ is a factor of the polynomial if 5 is a zero of the related polynomial function. Use the Factor Theorem and synthetic division.

$$
\begin{array}{r|rrrr}
5 & 1 & -7 & 7 & 15 \\
  &   & 5 & -10 & -15 \\
\hline
  & 1 & -2 & -3 & 0
\end{array}
$$

Because the remainder is 0, $x - 5$ is a factor of the polynomial. The polynomial $x^3 - 7x^2 + 7x + 15$ can be factored as $(x - 5)(x^2 - 2x - 3)$. The polynomial $x^2 - 2x - 3$ is the depressed polynomial. Check to see if this polynomial can be factored.

$x^2 - 2x - 3 = (x + 1)(x - 3)$ Factor the trinomial.

So, $x^3 - 7x^2 + 7x + 15 = (x - 5)(x + 1)(x - 3)$.

You can check your answer by multiplying out the factors and seeing if you come up with the initial polynomial.

StudyTip

Factoring The factors of a polynomial do not have to be binomials. For example, the factors of $x^3 + x^2 - x + 15$ are $x + 3$ and $x^2 - 2x + 5$.

GuidedPractice

3. Show that $x - 2$ is a factor of $x^3 - 7x^2 + 4x + 12$. Then find the remaining factors of the polynomial. **See Chapter 5 Answer Appendix.**

 354 | Lesson 5-6 | The Remainder and Factor Theorems

DifferentiatedInstruction (AL) (OL) (BL) (ELL)

If students can describe two or three things about this lesson that they found difficult to understand,

Then have them address each item by writing explanations that will help them review the material later.

Example 1 Use synthetic substitution to find $f(4)$ and $f(-2)$ for each function.

1. $f(x) = 2x^3 - 5x^2 - x + 14$ **58; −20** **2.** $f(x) = x^4 + 8x^3 + x^2 - 4x - 10$ **758; −46**

Example 2 **3. NATURE** The approximate number of bald eagle nesting pairs in the United States can be modeled by the function $P(x) = -0.16x^3 + 15.83x^2 - 154.15x + 1147.97$, where x is the number of years since 1970. About how many nesting pairs of bald eagles can be expected in 2018? **12,526**

Example 3 **Given a polynomial and one of its factors, find the remaining factors of the polynomial.**

4. $x^3 - 6x^2 + 11x - 6; x - 1$ **$x - 2, x - 3$** **5.** $x^3 + x^2 - 16x - 16; x + 1$ **$x + 4, x - 4$**

6. $3x^3 + 10x^2 - x - 12; x - 1$ **$x + 3, 3x + 4$** **7.** $2x^3 - 5x^2 - 28x + 15; x + 3$ **$x - 5, 2x - 1$**

Practice and Problem Solving   Extra Practice is on page R5.

Example 1 Use synthetic substitution to find $f(-5)$ and $f(2)$ for each function.

8. $f(x) = x^3 + 2x^2 - 3x + 1$ **−59; 11** **9.** $f(x) = x^2 - 8x + 6$ **71; −6**

10. $f(x) = 3x^4 + x^3 - 2x^2 + x + 12$ **1707; 62** **11.** $f(x) = 2x^3 - 8x^2 - 2x + 5$ **−435; −15**

12. $f(x) = x^3 - 5x + 2$ **−98; 0** **13.** $f(x) = x^5 + 8x^3 + 2x - 15$ **−4150; 85**

14. $f(x) = x^6 - 4x^4 + 3x^2 - 10$ **13,190; 2** **15.** $f(x) = x^4 - 6x - 8$ **647; −4**

Example 2 **16. FINANCIAL LITERACY** A specific car's fuel economy in miles per gallon can be approximated by $f(x) = 0.00000056x^4 - 0.000018x^3 - 0.016x^2 + 1.38x - 0.38$, where x represents the car's speed in miles per hour. Determine the fuel economy when the car is traveling 40, 50 and 60 miles per hour. **29.5 mpg; 29.87 mpg; 28.19 mpg**

Example 3 **Given a polynomial and one of its factors, find the remaining factors of the polynomial.** **21. $x + 6, 2x + 7$** **23. $x + 1, x^2 + 2x + 3$**

17. $x^3 - 3x + 2; x + 2$ **$(x - 1)^2$** **18.** $x^4 + 2x^3 - 8x - 16; x + 2$ **$x - 2, x^2 + 2x + 4$**

19. $x^3 - x^2 - 10x - 8; x + 2$ **$x - 4, x + 1$** **20.** $x^3 - x^2 - 5x - 3; x - 3$ **$(x + 1)^2$**

21. $2x^3 + 17x^2 + 23x - 42; x - 1$ **22.** $2x^3 + 7x^2 - 53x - 28; x - 4$ **$x + 7, 2x + 1$**

B **23.** $x^4 + 2x^3 + 2x^2 - 2x - 3; x - 1$ **24.** $x^3 + 2x^2 - x - 2; x + 2$ **$x - 1, x + 1$**

25. $6x^3 - 25x^2 + 2x + 8; 2x + 1$ **26.** $16x^5 - 32x^4 - 81x + 162; 2x - 3$
   $x - 4, 3x - 2$ **$x - 2, 2x + 3, 4x^2 + 9$**

27 **BOATING** A motor boat traveling against waves accelerates from a resting position. Suppose the speed of the boat in feet per second is given by the function $f(t) = -0.04t^4 + 0.8t^3 + 0.5t^2 - t$, where t is the time in seconds.

a. Find the speed of the boat at 1, 2, and 3 seconds. **0.26 ft/s, 5.76 ft/s, 19.86 ft/s**

b. It takes 6 seconds for the boat to travel between two buoys while it is accelerating. Use synthetic substitution to find $f(6)$ and explain what this means. **132.96 ft/s; This means the boat is traveling at 132.96 ft/s when it passes the second buoy.**

28. CCSS REASONING A company's sales, in millions of dollars, of consumer electronics can be modeled by $S(x) = -1.2x^3 + 18x^2 + 26.4x + 678$, where x is the number of years since 2005. **a. $1513.2 million; $1074 million**

a. Use synthetic substitution to estimate the sales for 2017 and 2020.

b. Do you think this model is useful in estimating future sales? Explain. **See margin.**

 connectED.mcgraw-hill.com **355**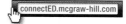

Differentiated Homework Options

Level	Assignment	Two-Day Option	
AL Basic	8–26, 36, 39, 41, 43, 44–63	8–26 odd, 44–47	8–26 even, 36, 39, 41, 43, 48–63
OL Core	9–35 odd, 36, 39, 41, 43, 44–63	8–26, 44–47	27–35, 36, 39, 41, 43, 48–63
BL Advanced	27–57, (optional: 58–63)		

3 Practice

Formative Assessment

Use Exercises 1–7 to check for understanding.

Use the chart at the bottom of this page to customize assignments for your students.

CCSS **Teaching the Mathematical Practices**

Reasoning Mathematically proficient students make sense of quantities and their relationships in problem situations. Quantitative reasoning entails habits of creating a coherent representation of the problem at hand; considering the units involved; attending to the meaning of quantities, not just how to compute them; and knowing and flexibly using different properties of operations and objects.

WatchOut!

Common Misconceptions Remind students that not all polynomials can be factored. Emphasize that the factors indicate where the graph of the function crosses the x-axis. If the graph of a polynomial function has no x-intercepts, then the polynomial cannot be factored. To see an example of a polynomial that cannot be factored, students can graph the function $f(x) = x^4 - x^3 - x^2 + 2$.

Additional Answer

28b. Sample answer: No; the graph of the function has a relative maximum at about $x = 8$ or the year 2008 and then the values for sales decrease. It is not likely that the sales would continually decrease.

Study Guide and Intervention
AL OL ELL

NAME _____ DATE _____ PERIOD _____

5-6 Study Guide and Intervention
The Remainder and Factor Theorems
Synthetic Substitution

Practice
AL OL BL ELL

NAME _____ DATE _____ PERIOD _____

5-6 Practice
The Remainder and Factor Theorems

Word Problem Practice
AL OL BL ELL

NAME _____ DATE _____ PERIOD _____

5-6 Word Problem Practice
The Remainder and Factor Theorems

40. If $x - a$ is a factor of $f(x)$, then $f(a)$ has a factor of $(a - a)$ or 0. Since a factor of $f(a)$ is 0, $f(a) = 0$. Now assume that $f(a) = 0$. If $f(a) = 0$, then the Remainder Theorem states that the remainder is 0 when $f(x)$ is divided by $x - a$. This means that $x - a$ is a factor of $f(x)$. This proves the Factor Theorem.

Use the graph to find all of the factors for each polynomial function.

29.

$f(x) = x^4 - 2x^3 - x^2 + 2x - 24$

$x + 2,$
$x - 3,$
$x^2 - x + 4$

30.

$f(x) = 20x^3 - 47x^2 + 8x + 12$

$x - 2,$
$4x - 3,$
$5x + 2$

31. 🔧 **MULTIPLE REPRESENTATIONS** In this problem, you will consider the function
$f(x) = -9x^5 + 104x^4 - 249x^3 - 456x^2 + 828x + 432.$ **31a–d. See Chapter 5 Answer Appendix.**

a. **Algebraic** If $x - 6$ is a factor of the function, find the depressed polynomial.

b. **Tabular** Make a table of values for $-5 \le x \le 6$ for the depressed polynomial.

c. **Analytical** What conclusions can you make about the locations of the other zeros based on the table? Explain your reasoning.

d. **Graphical** Graph the original function to confirm your conclusions.

CCSS PERSEVERANCE Find values of k so that each remainder is 3.

32. $(x^2 - x + k) \div (x - 1)$ 3

33 $(x^2 + kx - 17) \div (x - 2)$ 8

34. $(x^2 + 5x + 7) \div (x + k)$ 1, 4

35. $(x^3 + 4x^2 + x + k) \div (x + 2)$ −3

H.O.T. Problems Use Higher-Order Thinking Skills

43. Sample answer: A zero can be located using the Remainder Theorem and a table of values by determining when the output, or remainder, is equal to zero. For instance, if $f(6)$ leaves a remainder of 2 and $f(7)$ leaves a remainder of −1, then you know that there is a zero between $x = 6$ and $x = 7$.

36. OPEN ENDED Write a polynomial function that has a double root of 1 and a double root of −5. Graph the function. **See Chapter 5 Answer Appendix.**

CHALLENGE Find the solutions of each polynomial function.

37. $(x^2 - 4)^2 - (x^2 - 4) - 2 = 0$ $\pm\sqrt{6}, \pm\sqrt{3}$ **38.** $(x^2 + 3)^2 - 7(x^2 + 3) + 12 = 0$ −1, 0, 1

39. REASONING Polynomial $f(x)$ is divided by $x - c$. What can you conclude if:

a. the remainder is 0? $x - c$ is a factor of $f(x)$.

b. the remainder is 1? $x - c$ is not a factor of $f(x)$.

c. the quotient is 1, and the remainder is 0? $f(x) = x - c$

40. CHALLENGE Review the definition for the Factor Theorem. Provide a proof of the theorem.

41. OPEN ENDED Write a cubic function that has a remainder of 8 for $f(2)$ and a remainder of −5 for $f(3)$. **Sample answer:** $f(x) = -x^3 + x^2 + x + 10$

42. CHALLENGE Show that the quartic function $f(x) = ax^4 + bx^3 + cx^2 + dx + e$ will always have a rational zero when the numbers 1, −2, 3, 4, and −6 are randomly assigned to replace a through e, and all of the numbers are used. **See Chapter 5 Answer Appendix.**

43. WRITING IN MATH Explain how the zeros of a function can be located by using the Remainder Theorem and making a table of values for different input values and then comparing the remainders.

Enrichment
OL BL

NAME _____ DATE _____ PERIOD _____

5-6 Enrichment

Radical Notation

CCSS Teaching the Mathematical Practices

Perseverance Mathematically proficient students start by explaining to themselves the meaning of a problem and looking for entry points to its solution. They analyze givens, constraints, relationships, and goals and make conjectures about the form and meaning of the solution and plan a solution pathway rather than simply jumping into a solution attempt.

44. $27x^3 + y^3 =$ **B**

A $(3x + y)(3x + y)(3x + y)$

B $(3x + y)(9x^2 - 3xy + y^2)$

C $(3x - y)(9x^2 + 3xy + y^2)$

D $(3x - y)(9x^2 + 9xy + y^2)$

45. GRIDDED RESPONSE In the figure, a square with side length $2\sqrt{2}$ is inscribed in a circle. The area of the circle is $k\pi$. What is the exact value of k?   **4**

46. What is the product of the complex numbers $(4 + i)(4 - i)$?   **H**

F 15   H 17

G $16 - i$ J $17 - 8i$

47. SAT/ACT The measure of the largest angle of a triangle is 14 less than twice the measure of the smallest angle. The third angle measure is 2 more than the measure of the smallest angle. What is the measure of the smallest angle?   **B**

A 46   D 52

B 48   E 82

C 50

Solve each equation. (Lesson 5-5)

48. $x^4 - 4x^2 - 21 = 0$ $\pm\sqrt{7}, \pm i\sqrt{3}$ **49.** $x^4 - 6x^2 = 27$ $\pm 3, \pm i\sqrt{3}$ **50.** $4x^4 - 8x^2 - 96 = 0$ $\pm\sqrt{6}, \pm 2i$

Complete each of the following. (Lesson 5-4)

a. Estimate the x-coordinate of every turning point and determine if those coordinates are relative maxima or relative minima.

b. Estimate the x-coordinate of every zero.

c. Determine the smallest possible degree of the function.

d. Determine the domain and range of the function. **51–53. See margin.**

51.

52.

53.

54. HIGHWAY SAFETY Engineers can use the formula $d = 0.05v^2 + 1.1v$ to estimate the minimum stopping distance d in feet for a vehicle traveling v miles per hour. If a car is able to stop after 125 feet, what is the fastest it could have been traveling when the driver first applied the brakes? (Lesson 4-6)   **about 40.2 mph**

Solve by graphing. (Lesson 3-1) **55–57. See margin.**

55. $y = 3x - 1$
  $y = -2x + 4$

56. $3x + 2y = 8$
  $-4x + 6y = 11$

57. $5x - 2y = 6$
  $3x - 2y = 2$

If $c(x) = x^2 - 2x$ and $d(x) = 3x^2 - 6x + 4$, find each value. **58.** $-2a^2 + 32a - 76$ **59.** $4a^2 - 8a + 16$

58. $c(a + 2) - d(a - 4)$

59. $c(a - 3) + d(a + 1)$

60. $c(-3a) + d(a + 4)$ $12a^2 + 24a + 28$

61. $3d(3a) - 2c(-a)$
  $79a^2 - 58a + 12$

62. $c(a) + 5d(2a)$
  $61a^2 - 62a + 20$

63. $-2d(2a + 3) - 4c(a^2 + 1)$
  $-4a^4 - 24a^2 - 48a - 22$

DifferentiatedInstruction OL BL

Extension Explain to students that they already know how to use substitution to find $f(4)$ when $f(x) = x^3 - 6x^2 + 9x - 7$; $f(4) = 4^3 - 6(4)^2 + 9(4) - 7 = 64 - 96 + 36 - 7 = -3$. This method often requires a calculator. Tell students you will show them a method of evaluating a polynomial in which you do not have to compute powers using what is known as a recursive form of the polynomial. In this case, rewrite $x^3 - 6x^2 + 9x - 7$ as $[(x - 6)x + 9]x - 7$.

So $f(4) = [(4 - 6) \cdot 4 + 9] \cdot 4 - 7 = (-2 \cdot 4 + 9) \cdot 4 - 7 = 1 \cdot 4 - 7 = -3$.

---

4 Assess

Ticket out the Door Ask students to explain how synthetic substitution is related to synthetic division.

Formative Assessment
Check for student understanding of concepts in Lessons 5-5 and 5-6.

📁 Quiz 3, p. 58

Additional Answers

51a. -1.5(max), 0.5(min), 2.5(max)

51b. $-3.5, 3.75$ **51c.** 4

51d. D = {all real numbers};
  R = $\{y \mid y \le 4.5\}$

52a. -2(max), 0(min), 3(max), 3.5(min)

52b. $-3.75v, -1, 1.5, 3.25, 3.7$

52c. 5

52d. D = {all real numbers};
  R = {all real numbers}

53a. -3(min), -1(max), 1(min)

53b. $-0.25, 3$ **53c.** 4

53d. D = {all real numbers};
  R = $\{y \mid y \ge -4.5\}$

55.

56.

57.

1 Focus

VerticalAlignment

Before Lesson 5-7 Use complex numbers to describe the solutions of quadratic equations.

Lesson 5-7 Determine the number and type of roots for a polynomial equation. Find the zeros of a polynomial function.

After Lesson 5-7 Use tools including factoring to transform and solve equations.

2 Teach

Scaffolding Questions

Have students read the **Why?** section of the lesson.

Ask:

- According to the equation, what was the average price of a gallon of gas in 1990? about $1.37

- Would the given equation be valid for negative values of *x*? Possibly; negative values of *x* would correspond to years *before* 1990.

Roots and Zeros

:: Then	:: Now	:: Why?
● You used complex numbers to describe solutions of quadratic equations.	**1** Determine the number and type of roots for a polynomial equation. **2** Find the zeros of a polynomial function.	● The function $g(x) = 1.384x^4 - 0.003x^3 + 0.28x^2 - 0.078x + 1.365$ can be used to model the average price of a gallon of gasoline in a given year if *x* is the number of years since 1990. To find the average price of gasoline in a specific year, you can use the roots of the related polynomial equation.

Common Core State Standards

Content Standards
N.CN.9 Know the Fundamental Theorem of Algebra; show that it is true for quadratic polynomials.

A.APR.3 Identify zeros of polynomials when suitable factorizations are available, and use the zeros to construct a rough graph of the function defined by the polynomial.

Mathematical Practices
6 Attend to precision.

1 **Synthetic Types of Roots** Previously, you learned that a zero of a function $f(x)$ is any value c such that $f(c) = 0$. When the function is graphed, the real zeros of the function are the *x*-intercepts of the graph.

ConceptSummary Zeros, Factors, Roots, and Intercepts

Words	Let $P(x) = a_n x^n + \cdots + a_1 x + a_0$ be a polynomial function. Then the following statements are equivalent. • c is a zero of $P(x)$. • c is a root or solution of $P(x) = 0$. • $x - c$ is a factor of $a_n x^n + \cdots + a_1 x + a_0$. • If c is a real number, then $(c, 0)$ is an *x*-intercept of the graph of $P(x)$.
Example	Consider the polynomial function $P(x) = x^4 + 2x^3 - 7x^2 - 8x + 12$. The zeros of $P(x) = x^4 + 2x^3 - 7x^2 - 8x + 12$ are $-3, -2, 1,$ and 2. The roots of $x^4 + 2x^3 - 7x^2 - 8x + 12 = 0$ are $-3, -2, 1,$ and 2. The factors of $x^4 + 2x^3 - 7x^2 - 8x + 12$ are $(x + 3), (x + 2), (x - 1),$ and $(x - 2)$. The *x*-intercepts of the graph of $P(x) = x^4 + 2x^3 - 7x^2 - 8x + 12$ are $(-3, 0), (-2, 0), (1, 0),$ and $(2, 0)$.

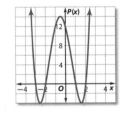

When solving a polynomial equation with degree greater than zero, there may be one or more real roots or no real roots (the roots are imaginary numbers). Since real numbers and imaginary numbers both belong to the set of complex numbers, all polynomial equations with degree greater than zero will have at least one root in the set of complex numbers. This is the **Fundamental Theorem of Algebra**.

KeyConcept Fundamental Theorem of Algebra

Every polynomial equation with degree greater than zero has at least one root in the set of complex numbers.

TORU YAMANAKA/AFP/Getty Images

Lesson 5-7 Resources

Resource	Approaching Level **AL**	On Level **OL**	Beyond Level **BL**	English Learners **ELL**
Teacher Edition	▪ Differentiated Instruction, p. 361	▪ Differentiated Instruction, pp. 361, 365	▪ Differentiated Instruction, pp. 361, 365	▪ Differentiated Instruction, p. 361
Chapter Resource Masters	▪ Study Guide and Intervention, pp. 41–42 ▪ Skills Practice, p. 43 ▪ Practice, p. 44 ▪ Word Problem Practice, p. 45	▪ Study Guide and Intervention, pp. 41–42 ▪ Skills Practice, p. 43 ▪ Practice, p. 44 ▪ Word Problem Practice, p. 45 ▪ Enrichment, p. 46	▪ Practice, p. 44 ▪ Word Problem Practice, p. 45 ▪ Enrichment, p. 46	▪ Study Guide and Intervention, pp. 41–42 ▪ Skills Practice, p. 43 ▪ Practice, p. 44 ▪ Word Problem Practice, p. 45
Other	▪ 5-Minute Check 5-7 ▪ Study Notebook ▪ Teaching Algebra with Manipulatives	▪ 5-Minute Check 5-7 ▪ Study Notebook ▪ Teaching Algebra with Manipulatives	▪ 5-Minute Check 5-7 ▪ Study Notebook	▪ 5-Minute Check 5-7 ▪ Study Notebook ▪ Teaching Algebra with Manipulatives

Example 1 Determine Number and Type of Roots

Solve each equation. State the number and type of roots.

a. $x^2 + 6x + 9 = 0$

$$x^2 + 6x + 9 = 0 \qquad \text{Original equation}$$
$$(x + 3)^2 = 0 \qquad \text{Factor.}$$
$$x + 3 = 0 \qquad \text{Take the root of each side.}$$
$$x = -3 \qquad \text{Solve for } x.$$

Because $(x + 3)$ is twice a factor of $x^2 + 6x + 9$, -3 is a double root. Thus, the equation has one real repeated root, -3.

CHECK The graph of the equation touches the x-axis at $x = -3$. Since -3 is a double root, the graph does not cross the axis. ✓

b. $x^3 + 25x = 0$

$$x^3 + 25x = 0 \qquad \text{Original equation}$$
$$x(x^2 + 25) = 0 \qquad \text{Factor.}$$
$$x = 0 \text{ or } x^2 + 25 = 0$$
$$x^2 = -25$$
$$x = \pm\sqrt{-25} \text{ or } \pm 5i$$

This equation has one real root, 0, and two imaginary roots, $5i$ and $-5i$.

CHECK The graph of this equation crosses the x-axis at only one place, $x = 0$. ✓

ReadingMath

Repeated Roots Polynomial equations can have double roots, triple roots, quadruple roots, and so on. In general, these are referred to as *multiple roots*.

1D. $\frac{1}{3}$, $i\sqrt{3}$, $-i\sqrt{3}$; 1 real, 2 imaginary

▶ **Guided Practice** **1A.** 0, $i\sqrt{2}$, $-i\sqrt{2}$; 1 real, 2 imaginary

1A. $x^3 + 2x = 0$ **1C.** $-1, 2, -5$; 3 real **1B.** $x^4 - 16 = 0$ $\quad$ 2, -2, $2i$, $-2i$; 2 real, 2 imaginary

1C. $x^3 + 4x^2 - 7x - 10 = 0$ **1D.** $3x^3 - x^2 + 9x - 3 = 0$

Examine the solutions for each equation in Example 1. Notice that the number of solutions for each equation is the same as the degree of each polynomial. The following corollary to the Fundamental Theorem of Algebra describes this relationship between the degree and the number of roots of a polynomial equation.

KeyConcept Corollary to the Fundamental Theorem of Algebra

Words	A polynomial equation of degree n has exactly n roots in the set of complex numbers, including repeated roots.
Example	$x^3 + 2x^2 + 6$ $\quad$ $4x^4 - 3x^3 + 5x - 6$ $\quad$ $-2x^5 - 3x^2 + 8$
	3 roots $\qquad$ 4 roots $\qquad$ 5 roots

Similarly, an nth degree polynomial function has exactly n zeros.

Additionally, French mathematician René Descartes discovered a relationship between the signs of the coefficients of a polynomial function and the number of positive and negative real zeros.

1 Synthetic Types of Roots
Example 1 shows how to determine the number and type of roots of a polynomial equation. **Example 2** shows how to determine the possible number of positive real zeros, negative real zeros, and imaginary zeros of a polynomial function.

Formative Assessment
Use the Guided Practice exercises after each example to determine students' understanding of concepts.

Additional Example

1 Solve each equation. State the number and type of roots.

a. $x^2 + 2x - 48 = 0$ This equation has two real roots, 6 and -8.

b. $y^4 - 256 = 0$ This equation has two real roots, 4 and -4, and two imaginary roots, $4i$ and $-4i$.

▶ **Additional Examples** also in Interactive Classroom PowerPoint® Presentations

IWB **Interactive White Board READY**

Focus on Mathematical Content
Zeros The real zeros of a polynomial function f are the x-intercepts of the graph of f. They are also the real solutions of the polynomial equation $f(x) = 0$. A polynomial function cannot have more zeros than its degree.

Teach with Tech

Interactive Whiteboard Write a polynomial function on the board. Show students how to apply Descartes' Rule of Signs by highlighting each change in signs in the coefficients.

Additional Example

2 State the possible number of positive real zeros, negative real zeros, and imaginary zeros of $p(x) = -x^6 + 4x^3 - 2x^2 - x - 1$. The function has either 2 or 0 positive zeros, 2 or 0 negative zeros, and 6, 4, or 2 imaginary zeros.

WatchOut!

Common Misconceptions Point out to students the method for determining the number of imaginary zeros for a polynomial function. In Example 2, the polynomial has degree 6, so it has a maximum of 6 real zeros. You find the numbers of positive and negative real zeros, and subtract the sum of these two numbers from 6 to find the number of imaginary zeros. Remind students that imaginary zeros come in conjugate pairs, so the number of imaginary zeros must be an even number.

StudyTip

Zero at the Origin If a zero of a function is at the origin, the sum of the number of positive real zeros, negative real zeros, and imaginary zeros is reduced by how many times 0 is a zero of the function.

KeyConcept Descartes' Rule of Signs

Let $P(x) = a_n x^n + \cdots + a_1 x + a_0$ be a polynomial function with real coefficients. Then

- the number of positive real zeros of $P(x)$ is the same as the number of changes in sign of the coefficients of the terms, or is less than this by an even number, and

- the number of negative real zeros of $P(x)$ is the same as the number of changes in sign of the coefficients of the terms of $P(-x)$, or is less than this by an even number.

Example 2 Find Numbers of Positive and Negative Zeros

State the possible number of positive real zeros, negative real zeros, and imaginary zeros of $f(x) = x^6 + 3x^5 - 4x^4 - 6x^3 + x^2 - 8x + 5$.

Because $f(x)$ has degree 6, it has six zeros, either real or imaginary. Use Descartes' Rule of Signs to determine the possible number and type of *real* zeros.

Count the number of changes in sign for the coefficients of $f(x)$.

$$f(x) = x^6 + 3x^5 - 4x^4 - 6x^3 + x^2 - 8x + 5$$

| no | yes | no | yes | yes | yes |
| + to + | + to − | − to − | − to + | + to − | − to + |

There are 4 sign changes, so there are 4, 2, or 0 positive real zeros.

Count the number of changes in sign for the coefficients of $f(-x)$.

$$f(-x) = (-x)^6 + 3(-x)^5 - 4(-x)^4 - 6(-x)^3 + (-x)^2 - 8(-x) + 5$$
$$= x^6 - 3x^5 - 4x^4 + 6x^3 + x^2 + 8x + 5$$

| yes | no | yes | no | no | no |
| + to − | − to − | − to + | + to + | + to + | + to + |

There are 2 sign changes, so there are 2, or 0 negative real zeros. Make a chart of the possible combinations of real and imaginary zeros.

Number of Positive Real Zeros	Number of Negative Real Zeros	Number of Imaginary Zeros	Total Number of Zeros
4	2	0	$4 + 2 + 0 = 6$
4	0	2	$4 + 0 + 2 = 6$
2	2	2	$2 + 2 + 2 = 6$
2	0	4	$2 + 0 + 4 = 6$
0	2	4	$0 + 2 + 4 = 6$
0	0	6	$0 + 0 + 6 = 6$

▶ **GuidedPractice**

2. State the possible number of positive real zeros, negative real zeros, and imaginary zeros of $h(x) = 2x^5 + x^4 + 3x^3 - 4x^2 - x + 9$. **2 or 0; 3 or 1; 4, 2, or 0**

2 Find Zeros You can use the various strategies and theorems you have learned to find all of the zeros of a function.

Example 3 Use Synthetic Substitution to Find Zeros

Find all of the zeros of $f(x) = x^4 - 18x^2 + 12x + 80$.

Step 1 Determine the total number of zeros.

Since $f(x)$ has degree 4, the function has 4 zeros.

Step 2 Determine the type of zeros.

Examine the number of sign changes for $f(x)$ and $f(-x)$.

$$f(x) = x^4 - 18x^2 + 12x + 80 \qquad\qquad f(-x) = x^4 - 18x^2 - 12x + 80$$

yes   yes   no   yes   no   yes

Because there are 2 sign changes for the coefficients of $f(x)$, the function has 2 or 0 positive real zeros. Because there are 2 sign changes for the coefficients of $f(-x)$, $f(x)$ has 2 or 0 negative real zeros. Thus, $f(x)$ has 4 real zeros, 2 real zeros and 2 imaginary zeros, or 4 imaginary zeros.

Step 3 Determine the real zeros.
List some possible values, and then use synthetic substitution to evaluate $f(x)$ for real values of x.

x	1	0	−18	12	80
−3	1	−3	−9	39	−37
−2	1	−2	−14	40	0
−1	1	−1	−17	29	51
0	1	0	−18	12	80
1	1	1	−17	−5	75
2	1	2	−14	−2	76

Each row shows the coefficients of the depressed polynomial and the remainder.

From the table, we can see that one zero occurs at $x = -2$. Since there are 2 negative real zeros, use synthetic substitution with the depressed polynomial function $f(x) = x^3 - 2x^2 - 14x + 40$ to find a second negative zero.

A second negative zero is at $x = -4$. Since the depressed polynomial $x^2 - 6x + 10$ is quadratic, use the Quadratic Formula to find the remaining zeros of $f(x) = x^2 - 6x + 10$.

x	1	−2	−14	40
−4	1	−6	10	0
−5	1	−7	21	−65
−6	1	−8	34	−164

$$x = \frac{-b \pm \sqrt{b^2 - 4ac}}{2a} \qquad\text{Quadratic Formula}$$

$$= \frac{-(-6) \pm \sqrt{(-6)^2 - 4(1)(10)}}{2(1)} \qquad\text{Replace } a \text{ with 1, } b \text{ with } -6, \text{ and } c \text{ with 10.}$$

$$= 3 \pm i \qquad\text{Simplify.}$$

The function has zeros at -4, -2, $3 + i$, and $3 - i$.

CHECK Graph the function on a graphing calculator. The graph crosses the x-axis two times, so there are two real zeros. Use the zero function under the **CALC** menu to locate each zero. The two real zeros are -4 and -2.

[−10, 10] scl: 1 by [−100, 100] scl: 10

[−10, 10] scl: 1 by [−100, 100] scl: 10

▸ **Guided**Practice

3. Find all of the zeros of $h(x) = x^3 + 2x^2 + 9x + 18$. $-2, 3i, -3i$

StudyTip

Testing for Zeros If a value is not a zero for a polynomial, then it will not be a zero for the depressed polynomial either, so it does not need to be checked again.

StudyTip

Locating Zeros Refer to Lesson 4-2 on how to use the CALC menu to locate a zero on your calculator.

2 Find Zeros
Example 3 shows how to find the zeros of a polynomial function using synthetic substitution. **Example 4** shows how to use zeros to write a polynomial function.

Additional Example

3 Find all of the zeros of $f(x) = x^3 - x^2 + 2x + 4$.
$-1, 1 + i\sqrt{3}, 1 - i\sqrt{3}$.

DifferentiatedInstruction AL OL BL ELL

If students sometimes make mistakes in mathematics exercises because they cannot read their own handwriting,

Then stress that throughout this course, students must work using neat and careful handwriting. It is extremely easy to misread coefficients and exponents, or misread *i* as the number 1.

Additional Example

4 Write a polynomial function of least degree with integral coefficients, the zeros of which include 4 and $4 - i$.

$f(x) = x^3 - 12x^2 + 49x - 68$

ReviewVocabulary

complex conjugates two complex numbers of the form $a + bi$ and $a - bi$

In Chapter 4, you learned that the product of complex conjugates is always a real number and that complex roots always come in conjugate pairs. For example, if one root of $x^2 - 8x + 52 = 0$ is $4 + 6i$, then the other root is $4 - 6i$.

This applies to the zeros of polynomial functions as well. For any polynomial function with real coefficients, if an imaginary number is a zero of that function, its conjugate is also a zero. This is called the **Complex Conjugates Theorem**.

KeyConcept Complex Conjugates Theorem

Words   Let a and b be real numbers, and $b \neq 0$. If $a + bi$ is a zero of a polynomial function with real coefficients, then $a - bi$ is also a zero of the function.

Example   If $3 + 4i$ is a zero of $f(x) = x^3 - 4x^2 + 13x + 50$, then $3 - 4i$ is also a zero of the function.

When you are given all of the zeros of a polynomial function and are asked to determine the function, convert the zeros to factors and then multiply all of the factors together. The result is the polynomial function.

Example 4 Use Zeros to Write a Polynomial Function

Write a polynomial function of least degree with integral coefficients, the zeros of which include -1 and $5 - i$.

Understand   If $5 - i$ is a zero, then $5 + i$ is also a zero according to the Complex Conjugates Theorem. So, $x + 1$, $x - (5 - i)$, and $x - (5 + i)$ are factors of the polynomial.

Plan   Write the polynomial function as a product of its factors.

$$P(x) = (x + 1)[x - (5 - i)][x - (5 + i)]$$

Solve   Multiply the factors to find the polynomial function.

$\begin{aligned}
P(x) &= (x + 1)\,[x - (5 - i)][x - (5 + i)] &&\text{Write the equation.}\\
&= (x + 1)\,[(x - 5) + i][(x - 5) - i] &&\text{Regroup terms.}\\
&= (x + 1)\,[(x - 5)^2 - i^2] &&\text{Difference of squares}\\
&= (x + 1)\,[(x^2 - 10x + 25 - (-1)] &&\text{Square terms.}\\
&= (x + 1)\,(x^2 - 10x + 26) &&\text{Simplify.}\\
&= x^3 - 10x^2 + 26x + x^2 - 10x + 26 &&\text{Multiply.}\\
&= x^3 - 9x^2 + 16x + 26 &&\text{Combine like terms.}
\end{aligned}$

Check   Because there are 3 zeros, the degree of the polynomial function must be 3, so $P(x) = x^3 - 9x^2 + 16x + 26$ is a polynomial function of least degree with integral coefficients and zeros of -1, $5 - i$, and $5 + i$.

GuidedPractice

4. Write a polynomial function of least degree with integral coefficients having zeros that include -1 and $1 + 2i$. $f(x) = x^3 - x^2 + 3x + 5$

Example 1 Solve each equation. State the number and type of roots.

1. $x^2 - 3x - 10 = 0$ $-2, 5$; 2 real

2. $x^3 + 12x^2 + 32x = 0$ $-8, -4, 0$; 3 real

3. $16x^4 - 81 = 0$ $-\dfrac{3}{2}, \dfrac{3}{2}, -\dfrac{3}{2}i, \dfrac{3}{2}i$; 2 real, 2 imaginary

4. $0 = x^3 - 8$ $2, -1 + i\sqrt{3}, -1 - i\sqrt{3}$; 1 real, 2 imaginary

Example 2 State the possible number of positive real zeros, negative real zeros, and imaginary zeros of each function. **6.** 1; 1 or 3; 0 or 2

5. $f(x) = x^3 - 2x^2 + 2x - 6$ 3 or 1; 0; 0 or 2

6. $f(x) = 6x^4 + 4x^3 - x^2 - 5x - 7$

7. $f(x) = 3x^5 - 8x^3 + 2x - 4$ 1 or 3; 0 or 2; 0, 2, or 4

8. $f(x) = -2x^4 - 3x^3 - 2x - 5$ 0; 0 or 2; 2 or 4

Example 3 Find all of the zeros of each function.

9. $f(x) = x^3 + 9x^2 + 6x - 16$ $-8, -2, 1$

10. $f(x) = x^3 + 7x^2 + 4x + 28$ $-7, -2i, 2i$

11. $f(x) = x^4 - 2x^3 - 8x^2 - 32x - 384$ $-4, 6, -4i, 4i$

12. $f(x) = x^4 - 6x^3 + 9x^2 + 6x - 10$ $-1, 1, 3 - i, 3 + i$

Example 4 Write a polynomial function of least degree with integral coefficients that have the given zeros.

13. $4, -1, 6$ $f(x) = x^3 - 9x^2 + 14x + 24$

14. $3, -1, 1, 2$ $f(x) = x^4 - 5x^3 + 5x^2 + 5x - 6$

15. $-2, 5, -3i$ $f(x) = x^4 - 3x^3 - x^2 - 27x - 90$

16. $-4, 4 + i$ $f(x) = x^3 - 4x^2 - 15x + 68$

Practice and Problem Solving Extra Practice is on page R5.

Example 1 Solve each equation. State the number and type of roots. 17–26. See margin.

17. $2x^2 + x - 6 = 0$

18. $4x^2 + 1 = 0$

19. $x^3 + 1 = 0$

20. $2x^2 - 5x + 14 = 0$

21. $-3x^2 - 5x + 8 = 0$

22. $8x^3 - 27 = 0$

23. $16x^4 - 625 = 0$

24. $x^3 - 6x^2 + 7x = 0$

25. $x^5 - 8x^3 + 16x = 0$

26. $x^5 + 2x^3 + x = 0$

Example 2 State the possible number of positive real zeros, negative real zeros, and imaginary zeros of each function.

(27) $f(x) = x^4 - 5x^3 + 2x^2 + 5x + 7$ 0 or 2; 0 or 2; 0, 2, or 4

28. $f(x) = 2x^3 - 7x^2 - 2x + 12$ 0 or 2; 1; 0 or 2

29. $f(x) = -3x^5 + 5x^4 + 4x^2 - 8$

30. $f(x) = x^4 - 2x^2 - 5x + 19$ 0 or 2; 0 or 2; 0, 2, or 4

31. $f(x) = 4x^6 - 5x^4 - x^2 + 24$ 0 or 2; 0 or 2; 2, 4, or 6

32. $f(x) = -x^5 + 14x^3 + 18x - 36$ 0 or 2; 1; 2 or 4

29. 0 or 2; 1; 2 or 4

Example 3 Find all of the zeros of each function.

33. $f(x) = x^3 + 7x^2 + 4x - 12$ $-6, -2, 1$

34. $f(x) = x^3 + x^2 - 17x + 15$ $-5, 1, 3$

35. $f(x) = x^4 - 3x^3 - 3x^2 - 75x - 700$ $-4, 7, -5i, 5i$

36. $f(x) = x^4 + 6x^3 + 73x^2 + 384x + 576$ $-3, -3, -8i, 8i$

37. $4, 4, -2i, 2i$ **37.** $f(x) = x^4 - 8x^3 + 20x^2 - 32x + 64$

38. $f(x) = x^5 - 8x^3 - 9x$ $-3, 0, 3, -i, i$

Example 4 Write a polynomial function of least degree with integral coefficients that have the given zeros. 39–44. See margin.

39. $5, -2, -1$

40. $-4, -3, 5$

41. $-1, -1, 2i$

42. $-3, 1, -3i$

43. $0, -5, 3 + i$

44. $-2, -3, 4 - 3i$

45b. Nonnegative roots represent numbers of computers produced per day which lead to no profit for the manufacturer.

B **45** **REASONING** A computer manufacturer determines that the profit for producing x computers per day is $P(x) = -0.006x^4 + 0.15x^3 - 0.05x^2 - 1.8x$.

a. How many positive real zeros, negative real zeros, and imaginary zeros exist? 2 or 0; 1; 1 or 3

b. What is the meaning of the zeros in this situation?

  363

Differentiated Homework Options

Level	Assignment	Two-Day Option	
AL Basic	17–48, 56, 58–76	17–47 odd, 61–64	18–48 even, 56, 58–60, 65–76
OL Core	17–49 odd, 53, 56, 58–76	17–48, 61–64	49–56, 58–60, 65–76
BL Advanced	49–73, (optional: 74–76)		

---

 3 Practice

Formative Assessment

Use Exercises 1–16 to check for understanding.

Use the chart at the bottom of this page to customize assignments for your students.

CCSS **Teaching the Mathematical Practices**

Reasoning Mathematically proficient students make sense of quantities and their relationships in problem situations. Quantitative reasoning entails habits of creating a coherent representation of the problem at hand; considering the units involved; attending to the meaning of quantities, not just how to compute them; and knowing and flexibly using different properties of operations and objects.

Additional Answers

17. $-2, \dfrac{3}{2}$; 2 real

18. $-\dfrac{1}{2}i, \dfrac{1}{2}i$; 2 imaginary

19. $-1, \dfrac{1 \pm i\sqrt{3}}{2}$; 1 real, 2 imaginary

20. $\dfrac{5 \pm i\sqrt{87}}{4}$; 2 imaginary

21. $-\dfrac{8}{3}, 1$; 2 real

22. $\dfrac{3}{2}, \dfrac{-3 \pm 3i\sqrt{3}}{4}$; 1 real, 2 imaginary

23. $-\dfrac{5}{2}, \dfrac{5}{2}, -\dfrac{5}{2}i, \dfrac{5}{2}i$; 2 real, 2 imaginary

24. $0, 3 + \sqrt{2}, 3 - \sqrt{2}$; 3 real

25. $-2, -2, 0, 2, 2$; 5 real

26. $0, -i, -i, i, i$; 1 real, 4 imaginary

39. $f(x) = x^3 - 2x^2 - 13x - 10$

40. $f(x) = x^3 + 2x^2 - 23x - 60$

41. $f(x) = x^4 + 2x^3 + 5x^2 + 8x + 4$

42. $f(x) = x^4 + 2x^3 + 6x^2 + 18x - 27$

43. $f(x) = x^4 - x^3 - 20x^2 + 50x$

44. $f(x) = x^4 - 3x^3 - 9x^2 + 77x + 150$

Study Guide and Intervention

AL OL ELL

Practice

AL OL BL ELL

Word Problem Practice

AL OL BL ELL

Enrichment

OL BL

Sketch the graph of each function using its zeros. See Chapter 5 Answer Appendix.

46. $f(x) = x^3 - 5x^2 - 2x + 24$

47. $f(x) = 4x^3 + 2x^2 - 4x - 2$

48. $f(x) = x^4 - 6x^3 + 7x^2 + 6x - 8$

49. $f(x) = x^4 - 6x^3 + 9x^2 + 4x - 12$

Match each graph to the given zeros.

a. $-3, 4, i, -i$ b. $-4, 3$ c. $-4, 3, i, -i$

50.   51.   52.

53c. 23.8; Sample answer: According to the model, the music hall will not earn any money after 2026.

53. **CONCERTS** The amount of money Hoshi's Music Hall took in from 2003 to 2010 can be modeled by $M(x) = -2.03x^3 + 50.1x^2 - 214x + 4020$, where x is the years since 2003.

a. How many positive real zeros, negative real zeros, and imaginary zeros exist? 3 or 1; 0; 2 or 0

b. Graph the function using your calculator. See Chapter 5 Answer Appendix.

c. Approximate all real zeros to the nearest tenth. What is the significance of each zero in the context of the situation?

Determine the number of positive real zeros, negative real zeros, and imaginary zeros for each function. Explain your reasoning.

54.

degree: 3

0 positive, 1 negative, 2 imaginary; Sample answer: The graph does not cross the positive x-axis, and crosses the negative x-axis once. Because the degree of the polynomial is 3, there are 3 − 1 or 2 imaginary zeros.

55.

degree: 5

1 positive, 2 negative, 2 imaginary; Sample answer: The graph crosses the positive x-axis once, and crosses the negative x-axis twice. Because the degree of the polynomial is 5, there are 5 − 3 or 2 imaginary zeros.

H.O.T. Problems Use Higher-Order Thinking Skills

a–c. See Chapter 5 Answer Appendix.

56. **OPEN ENDED** Sketch the graph of a polynomial function with:

a. 3 real, 2 imaginary zeros   b. 4 real zeros   c. 2 imaginary zeros

57. **CHALLENGE** Write an equation in factored form of a polynomial function of degree 5 with 2 imaginary zeros, 1 nonintegral zero, and 2 irrational zeros. Explain. See margin.

58. **CCSS ARGUMENTS** Determine which equation is not like the others. Explain. See margin.

$r^4 + 1 = 0$ $r^3 + 1 = 0$ $r^2 - 1 = 0$ $r^3 - 8 = 0$

59. **REASONING** Provide a counterexample for each statement.

a. All polynomial functions of degree greater than 2 have at least 1 negative real root. Sample answer: $f(x) = x^4 + 4x^2 + 4$

b. All polynomial functions of degree greater than 2 have at least 1 positive real root. Sample answer: $f(x) = x^3 + 6x^2 + 9x$

60. **WRITING IN MATH** Explain to a friend how you would use Descartes' Rule of Signs to determine the number of possible positive real roots and the number of possible negative roots of the polynomial function $f(x) = x^4 - 2x^3 + 6x^2 + 5x - 12$. See Chapter 5 Answer Appendix.

CCSS **Teaching the Mathematical Practices**

Arguments Mathematically proficient students understand and use stated assumptions, definitions, and previously established results in constructing arguments. They make conjectures and build a logical progression of statements to explore the truth of their conjectures. And they are able to analyze situations by breaking them into cases, and can recognize and use counterexamples.

61. Use the graph of the polynomial function below. Which is not a factor of the polynomial $x^5 + x^4 - 3x^3 - 3x^2 - 4x - 4$? **C**

    A $x - 2$
    B $x + 2$
    C $x - 1$
    D $x + 1$

62. **SHORT RESPONSE** A window is in the shape of an equilateral triangle. Each side of the triangle is 8 feet long. The window is divided in half by a support from one vertex to the midpoint of the side of the triangle opposite the vertex. Approximately how long is the support? **6.9 feet**

63. **GEOMETRY** In rectangle $ABCD$, $\overline{AD}$ is 8 units long. What is the length of $\overline{AB}$? **H**

    F 4 units
    G 8 units
    H $8\sqrt{3}$ units
    J 16 units

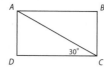

64. **SAT/ACT** The total area of a rectangle is $25a^4 - 16b^2$ square units. Which factors could represent the length and width? **B**

    A $(5a^2 + 4b)$ units and $(5a^2 + 4b)$ units
    B $(5a^2 + 4b)$ units and $(5a^2 - 4b)$ units
    C $(5a^2 - 4b)$ units and $(5a^2 - 4b)$ units
    D $(5a - 4b)$ units and $(5a - 4b)$ units
    E $(5a + 4b)$ units and $(5a - 4b)$ units

Spiral Review

Use synthetic substitution to find $f(-8)$ and $f(4)$ for each function. (Lesson 5-6) **65–67. See margin.**

65. $f(x) = 4x^3 + 6x^2 - 3x + 2$

66. $f(x) = 5x^4 - 2x^3 + 4x^2 - 6x$

67. $f(x) = 2x^5 - 3x^3 + x^2 - 4$

Factor completely. If the polynomial is not factorable, write *prime*. (Lesson 5-5)

68. $x^6 - y^6$ $(x + y)(x^2 - xy + y^2) \cdot$
    $(x - y)(x^2 + xy + y^2)$

69. $a^6 + b^6$ $(a^2 + b^2)(a^4 - a^2b^2 + b^4)$

70. $4x^2y + 8xy + 16y - 3x^2z - 6xz - 12z$
    $(x^2 + 2x + 4)(4y - 3z)$

71. $5a^3 - 30a^2 + 40a + 2a^2b - 12ab + 16b$
    $(a - 4)(a - 2)(5a + 2b)$

72. **BUSINESS** A mall owner has determined that the relationship between monthly rent charged for store space r (in dollars per square foot) and monthly profit $P(r)$ (in thousands of dollars) can be approximated by $P(r) = -8.1r^2 + 46.9r - 38.2$. Solve each quadratic equation or inequality. Explain what each answer tells about the relationship between monthly rent and profit for this mall. (Lesson 4-8) **a–d. See margin.**

    a. $-8.1r^2 + 46.9r - 38.2 = 0$
    b. $-8.1r^2 + 46.9r - 38.2 > 0$
    c. $-8.1r^2 + 46.9r - 38.2 > 10$
    d. $-8.1r^2 + 46.9r - 38.2 < 10$

73. **DIVING** To avoid hitting any rocks below, a cliff diver jumps up and out. The equation $h = -16t^2 + 4t + 26$ describes her height h in feet t seconds after jumping. Find the time at which she returns to a height of 26 feet. (Lesson 4-3) **0.25 s**

Skills Review

Find all of the possible values of $\pm\dfrac{b}{a}$ for each replacement set. **74–76. See margin.**

74. $a = \{1, 2, 4\}; b = \{1, 2, 3, 6\}$

75. $a = \{1, 5\}; b = \{1, 2, 4, 8\}$

76. $a = \{1, 2, 3, 6\}; b = \{1, 7\}$

DifferentiatedInstruction OL BL

Extension Explain to students that if $(x - r)^k$ is a factor of a polynomial $f(x)$, then r is a zero of the polynomial function $f(x)$ and k is the degree of the factor that produced r. From this, it is said that r has a *multiplicity of k*. When $k = 1$, we say that the zero has a multiplicity of 1 and is often called a *simple zero*. Use the following example to explain multiplicity and simple zeros: $f(x) = x^7 - 5x^6 + 6x^5 + 4x^4 - 8x^3$. When factored, $f(x) = x^3(x - 2)^3(x + 1)$. This polynomial function has 3 zeros: $x = -1$ (a simple zero), $x = 0$ with a multiplicity of 3, and $x = 2$ with multiplicity of 3. Challenge students to find the zeros and their multiplicities for the polynomial $f(x) = (x^6 - x^4)(x + 1)$. This polynomial has 3 zeros: $x = 1$, which is a simple zero, $x = 0$ with a multiplicity of 4, and $x = -1$ with multiplicity of 2.

---

4 Assess

Formative Assessment
Crystal Ball Have students write how what they learned today about the Fundamental Theorem of Algebra and Descartes' Rule of Signs will help them with identifying all the rational zeros of a polynomial function in tomorrow's lesson.

Additional Answers

57. Sample answer: $f(x) = (x + 2i)$
    $(x - 2i)(3x + 5)(x + \sqrt{5})$
    $(x - \sqrt{5})$ Use conjugates for the imaginary and irrational values.

58. $r^4 + 1 = 0$; Sample answer: The equation has imaginary solutions and all of the others have real solutions.

65. $f(-8) = -1638; f(4) = 342$

66. $f(-8) = 21,808; f(4) = 1192$

67. $f(-8) = -63,940; f(4) = 1868$

72a. 0.98, 4.81; The owner will break even if he charges $0.98 or $4.81 per square foot.

72b. $0.98 < r < 4.81$; The owner will make a profit if the rent per square foot is between $0.98 and $4.81.

72c. $1.34 < r < 4.45$; If rent is set between $1.34 and $4.45 per sq ft, the profit will be greater than $10,000.

72d. $r < 1.34$ or $r > 4.45$; If rent is set between $0 and $1.34 or above $4.45 per sq ft, the profit will be less than $10,000.

74. $\pm 1, \pm 2, \pm 3, \pm 6, \pm\dfrac{1}{2}, \pm\dfrac{3}{2},$
    $\pm\dfrac{1}{4}, \pm\dfrac{3}{4}$

75. $\pm 1, \pm 2, \pm 4, \pm 8, \pm\dfrac{1}{5}, \pm\dfrac{2}{5},$
    $\pm\dfrac{4}{5}, \pm\dfrac{8}{5}$

76. $\pm 1, \pm 7, \pm\dfrac{1}{2}, \pm\dfrac{7}{2}, \pm\dfrac{1}{3}, \pm\dfrac{7}{3},$
    $\pm\dfrac{1}{6}, \pm\dfrac{7}{6}$

1 Focus

Objective Use a graphing calculator to analyze polynomial functions.

Materials

- TI-83/84 Plus or other graphing calculator

Teaching Tip

- When finding zeros, students will have to perform Step 2 twice to find two zeros. In part **b,** they will need to perform Step 2 three times.

- Encourage students to set the window setting to match the samples in the Activity.

- Numbers have been rounded to two decimal places in this lab.

2 Teach

Working in Cooperative Groups

Divide the class into pairs. Ask students to work with their partners to complete the Activity.

Practice Have students complete Exercises 1–7.

Formative Assessment

Use Exercise 8 to assess each student's ability to analyze polynomials.

From Concrete to Abstract

Ask students to summarize what they learned about analyzing polynomial functions.

Graphing Technology Lab
Analyzing Polynomial Functions

You can use graphing technology to help you identify real zeros, maximum and minimum points, number and type of zeros, y-intercepts, and symmetry of polynomial functions.

CCSS Common Core State Standards
Content Standards
A.REI.11 Explain why the x-coordinates of the points where the graphs of the equations $y = f(x)$ and $y = g(x)$ intersect are the solutions of the equation $f(x) = g(x)$; find the solutions approximately, e.g., using technology to graph the functions, make tables of values, or find successive approximations. Include cases where $f(x)$ and/or $g(x)$ are linear, polynomial, rational, absolute value, exponential, and logarithmic functions.

Activity   Identify Polynomial Characteristics

Graph each function. Identify the real zeros, maximum and minimum points, number and type of zeros, y-intercepts, and symmetry.

a. $g(x) = 3x^4 - 15x^3 + 87x^2 - 375x + 300$

Step 1 Graph the equation.

Step 2 Use `2nd` [CALC] zero to find the real zeros at $x = 1$ and $x = 4$.

Step 3 Use `2nd` [CALC] minimum to find the relative minimum at $(2.68, -214.11)$. There is no relative maximum point.

Step 4 $g(x)$ has degree 4 and can have at most 4 zeros. Two real zeros were found through graphing. The other two zeros are either multiple zeros or imaginary zeros.

Step 5 Use `2nd` [CALC] value 0 to find the y-intercept, 300.

Step 6 The line of symmetry passes through the vertex. Its equation is $x = 2.68$.

$[-2, 8]$ scl: 1 by $[-300, 200]$ scl: 50

b. $f(x) = 2x^5 - 5x^4 - 3x^3 + 8x^2 + 4x$

Step 1 Graph the equation.

Step 2 Locate the real zeros at $x = -1$, $x = -\frac{1}{2}$, $x = 0$ and $x = 2$.

Step 3 Find the relative maxima at $(-0.81, 0.75)$ and $(1.04, 6.02)$ and the relative minima at $(-0.24, -0.48)$ and $(2, 0)$.

Step 4 $f(x)$ has degree 5 and can have at most 5 zeros. Four real zeros were found through graphing. The other zero is either a multiple zero or an imaginary zero. In this case, there is a double zero at $x = 2$.

Step 5 The y-intercept is 0 because the graph goes through the origin.

Step 6 There is no symmetry.

$[-4, 4]$ scl: 1 by $[-4, 8]$ scl: 2

Exercises

Graph each function. Identify the real zeros, maximum and minimum points, number and type of zeros, y-intercepts, and symmetry. **1–8. See Chapter 5 Answer Appendix.**

1. $f(x) = x^3 - 5x^2 + 6x$

2. $g(x) = x^4 - 3x^2 - 4$

3. $k(x) = -x^4 - x^3 + 2x^2$

4. $f(x) = -2x^3 - 4x^2 + 16x$

5. $g(x) = 3x^5 - 18x^4 + 27x^3$

6. $k(x) = x^4 - 8x^2 + 15$

7. $f(x) = -x^3 + 2x^2 + 8x$

8. $g(x) = x^5 + 3x^4 - 10x^2$

LESSON 5-8 Rational Zero Theorem

:: Then	:: Now	:: Why?
• You found zeros of quadratic functions of the form $f(x) = ax^2 + bx + c$.	**1** Identify possible rational zeros of a polynomial function. **2** Find all of the rational zeros of a polynomial function.	• Annual sales of recorded music in the United States can be approximated by $d(t) = 30x^3 - 478x^2 + 1758x + 10{,}092$, where $d(t)$ is the total sales in millions of dollars and t is the number of years since 2005. You can use this function to estimate when music sales will be $9 billion.

CCSS **Common Core State Standards**

Mathematical Practices
8 Look for and express regularity in repeated reasoning.

1 Identify Rational Zeros Usually it is not practical to test all possible zeros of a polynomial function using synthetic substitution. The **Rational Zero Theorem** can help you choose some possible zeros to test. If the leading coefficient is 1, the corollary applies.

> **KeyConcept** Rational Zero Theorem
>
> **Words**   If $P(x)$ is a polynomial function with integral coefficients, then every rational zero of $P(x) = 0$ is of the form $\frac{p}{q}$, a rational number in simplest form, where p is a factor of the constant term and q is a factor of the leading coefficient.
>
> **Example**   Let $f(x) = 6x^4 + 22x^3 + 11x^2 - 80x - 40$. If $\frac{4}{3}$ is a zero of $f(x)$, then 4 is a factor of -40, and 3 is a factor of 6.
>
> **Corollary to the Rational Zero Theorem**
>
> If $P(x)$ is a polynomial function with integral coefficients, a leading coefficient of 1, and a nonzero constant term, then any rational zeros of $P(x)$ must be factors of the constant term.

Example 1   Identify Possible Zeros

List all of the possible rational zeros of each function.

a. $f(x) = 4x^5 + x^4 - 2x^3 - 5x^2 + 8x + 16$

If $\frac{p}{q}$ is a rational zero, then p is a factor of 16 and q is a factor of 4.

p: $\pm1, \pm2, \pm4, \pm8, \pm16$ q: $\pm1, \pm2, \pm4$

Write the possible values of $\frac{p}{q}$ in simplest form.

$\frac{p}{q} = \pm1, \pm2, \pm4, \pm8, \pm16, \pm\frac{1}{2}, \pm\frac{1}{4}$

b. $f(x) = x^3 - 2x^2 + 5x + 12$

If $\frac{p}{q}$ is a rational zero, then p is a factor of 12 and q is a factor of 1.

p: $\pm1, \pm2, \pm3, \pm4, \pm6, \pm12$ q: ±1

So, $\frac{p}{q} = \pm1, \pm2, \pm3, \pm4, \pm6,$ and ±12

▶ **Guided**Practice   **1A, 1B.** See margin.

1A. $g(x) = 3x^3 - 4x + 10$ **1B.** $h(x) = x^3 + 11x^2 + 24$

connectED.mcgraw-hill.com   **367** 🔊

Javier Larrea/age fotostock

1 Focus

VerticalAlignment

▼

Before Lesson 5-8 Find zeros of quadratic functions of the form $f(x) = ax^2 + bx + c$.

Lesson 5-8 Identify possible rational zeros of a polynomial function. Find all of the rational zeros of a polynomial function.

▼

After Lesson 5-8 Use quotients of polynomials to describe the graphs of rational functions.

2 Teach

Scaffolding Questions
Have students read the **Why?** section of the lesson.

Ask:
- In what year was the total annual sales of recorded music about $10.092 billion? 2005

- What value of x would you use to approximate the total annual sales of recorded music for the year 2020? 15

Additional Answers (Guided Practice)

1A. $\pm\frac{10}{3}, \pm10, \pm\frac{5}{3}, \pm5, \pm\frac{2}{3}, \pm2,$ $\pm\frac{1}{3}, \pm1$

1B. $\pm24, \pm12, \pm8, \pm6, \pm4, \pm3, \pm2,$ ±1

Lesson 5-8 Resources

Resource	Approaching Level **AL**	On Level **OL**	Beyond Level **BL**	English Learners **ELL**
Teacher Edition	• Differentiated Instruction, p. 369	• Differentiated Instruction, pp. 369, 372	• Differentiated Instruction, pp. 369, 372	
Chapter Resource Masters	• Study Guide and Intervention, pp. 47–48 • Skills Practice, p. 49 • Practice, p. 50 • Word Problem Practice, p. 51	• Study Guide and Intervention, pp. 47–48 • Skills Practice, p. 49 • Practice, p. 50 • Word Problem Practice, p. 51 • Enrichment, p. 52 • Graphing Calculator Activity, p. 53	• Practice, p. 50 • Word Problem Practice, p. 51 • Enrichment, p. 52	• Study Guide and Intervention, pp. 47–48 • Skills Practice, p. 49 • Practice, p. 50 • Word Problem Practice, p. 51
Other	• 5-Minute Check 5-8 • Study Notebook	• 5-Minute Check 5-8 • Study Notebook	• 5-Minute Check 5-8 • Study Notebook	• 5-Minute Check 5-8 • Study Notebook

1 Identify Rational Zeros

Example 1 shows how to use the Rational Zero Theorem to identify all of the possible rational zeros of a polynomial function with integral coefficients.

Formative Assessment

Use the Guided Practice exercises after each example to determine students' understanding of concepts.

Additional Example

1 List all the possible rational zeros of each function.

a. $f(x) = 3x^4 - x^3 + 4$
$\pm 1, \pm 2, \pm 4, \pm \frac{1}{3}, \pm \frac{2}{3}, \pm \frac{4}{3}$

b. $f(x) = x^4 + 7x^3 - 15$
$\pm 1, \pm 3, \pm 5, \pm 15$

▶ **Additional Examples** also in Interactive Classroom PowerPoint® Presentations

 Interactive White Board READY

WatchOut!

Common Misconceptions While discussing Example 1, point out that ± 1 will always be possible rational zeros. Also make sure students clearly understand that these are just *possible* zeros. Until each potential zero has been tested by synthetic substitution, it should not be called a zero.

2 Find Rational Zeros

Example 2 shows how to identify the zeros once you have listed the possible zeros for a real-world problem.

Example 3 shows how to use one identified zero of a polynomial function to factor and find any other zeros.

---

2 Find Rational Zeros

Once you have written the possible rational zeros, you can test each number using synthetic substitution and use the other tools you have learned to determine the zeros of a function.

Real-World Example 2 Find Rational Zeros

WOODWORKING Adam is building a computer desk with a separate compartment for the computer. The compartment for the computer is a rectangular prism and will be 8019 cubic inches. The compartment will be 24 inches longer than it is wide and the height will be 18 inches greater than the width. Find the dimensions of the computer compartment.

Let x = width, $x + 24$ = length, and $x + 18$ = height.

Write an equation for the volume.

$$\ell wh = V \qquad \text{Formula for volume}$$
$$(x + 24)(x)(x + 18) = 8019 \qquad \text{Substitute.}$$
$$x^3 + 42x^2 + 432x = 8019 \qquad \text{Multiply.}$$
$$x^3 + 42x^2 + 432x - 8019 = 0 \qquad \text{Subtract 8019 from each side.}$$

The leading coefficient is 1, so the possible rational zeros are factors of 8019.

$\pm 1, \pm 3, \pm 9, \pm 11, \pm 27, \pm 33, \pm 81, \pm 99, \pm 243, \pm 297, \pm 729, \pm 891, \pm 2673,$ and ± 8019

Since length can only be positive, we only need to check positive values.

> **StudyTip**
>
> **CCSS Structure** Examine the signs of the coefficients of the equation. In this case, there is only one change of sign, so there is only one positive real zero.

There is one change of sign of the coefficients, so by Descartes' Rule of Signs, there is only one positive real zero. Make a table for synthetic division and test possible values.

p	1	42	432	-8019
1	1	43	475	-7544
2	1	45	567	-6318
9	1	51	891	0

One zero is 9. Since there is only one positive real zero, we do not have to test the other numbers. The other dimensions are $9 + 24$ or 33 inches, and $9 + 18$ or 27 inches.

CHECK Multiply the dimensions and see if they equal the volume of 8019 cubic inches.
$9 \times 33 \times 27 = 8019$ ✓

▶ **GuidedPractice**

2. GEOMETRY The volume of a rectangular prism is 1056 cubic centimeters. The length is 1 centimeter more than the width, and the height is 3 centimeters less than the width. Find the dimensions of the prism. **8 cm × 11 cm × 12 cm**

You usually do not need to test all of the possible zeros. Once you find a zero, you can try to factor the depressed polynomial to find any other zeros.

CCSS Teaching the Mathematical Practices

Structure Mathematically proficient students look closely to discern a pattern or structure. Encourage students to use Descartes' Rule of Signs to assist them in identifying zeros of a function.

Example 3 Find All Zeros

Find all of the zeros of $f(x) = 5x^4 - 8x^3 + 41x^2 - 72x - 36$.

From the corollary to the Fundamental Theorem of Algebra, there are exactly 4 complex zeros. According to Descartes' Rule of Signs, there are 3 or 1 positive real zeros and exactly 1 negative real zero. The possible rational zeros are ± 1, ± 2, ± 3, ± 4, ± 6, ± 9, ± 12, ± 18, ± 36, $\pm\frac{1}{5}$, $\pm\frac{2}{5}$, $\pm\frac{3}{5}$, $\pm\frac{4}{5}$, $\pm\frac{6}{5}$, $\pm\frac{9}{5}$, $\pm\frac{12}{5}$, $\pm\frac{18}{5}$, and $\pm\frac{36}{5}$.

Make a table and test some possible rational zeros.

$\frac{p}{q}$	5	−8	41	−72	−36
−1	5	−13	54	−126	90
1	5	−3	38	−34	−70
2	5	2	45	18	0

Because $f(2) = 0$, there is a zero at $x = 2$. Factor the depressed polynomial $5x^3 + 2x^2 + 45x + 18$.

$$5x^3 + 2x^2 + 45x + 18 = 0 \qquad \text{Write the depressed polynomial.}$$
$$(5x^3 + 2x^2) + (45x + 18) = 0 \qquad \text{Group terms.}$$
$$x^2(5x + 2) + 9(5x + 2) = 0 \qquad \text{Factor.}$$
$$(x^2 + 9)(5x + 2) = 0 \qquad \text{Distributive Property}$$
$$x^2 + 9 = 0 \quad \text{or} \quad 5x + 2 = 0 \qquad \text{Zero Product Property}$$
$$x^2 = -9 \qquad\qquad 5x = -2$$
$$x = \pm 3i \qquad\qquad x = -\frac{2}{5}$$

There is another real zero at $x = -\frac{2}{5}$ and two imaginary zeros at $x = 3i$ and $x = -3i$. The zeros of the function are $-\frac{2}{5}$, 2, $3i$, and $-3i$.

▶ **Guided**Practice

Find all of the zeros of each function.   **3B.** 2, $-\frac{1}{2}$, $\pm 3i$

3A. $h(x) = 9x^4 + 5x^2 - 4$ $\pm\frac{2}{3}$, $\pm i$ **3B.** $k(x) = 2x^4 - 5x^3 + 20x^2 - 45x + 18$

Check Your Understanding   ● = Step-by-Step Solutions begin on page R14.

Example 1   **List all of the possible rational zeros of each function.**

1. $f(x) = x^3 - 6x^2 - 8x + 24$
± 1, ± 2, ± 3, ± 4, ± 6, ± 8, ± 12, ± 24

2. $f(x) = 2x^4 + 3x^2 - x + 15$
± 1, ± 3, ± 5, ± 15, $\pm\frac{1}{2}$, $\pm\frac{3}{2}$, $\pm\frac{5}{2}$, $\pm\frac{15}{2}$

Example 2   **3.** **CCSS REASONING** The volume of the triangular pyramid is 210 cubic inches. Find the dimensions of the solid.
5 in. × 9 in. × 28 in.

$5x + 3$ in.
$2x - 1$ in.
x in.

Find all of the rational zeros of each function.

4. $f(x) = x^3 - 6x^2 - 13x + 42$ $-3, 2, 7$

5. $f(x) = 2x^4 + 11x^3 + 26x^2 + 29x + 12$ $-\frac{3}{2}, -1$

9. $-\frac{1}{2}, \frac{3}{2}, 1 + 2i, 1 - 2i$

Example 3   **Find all of the zeros of each function.**

8. $-4, \frac{3}{4}, -i, i$

6. $f(x) = 3x^3 - 2x^2 - 8x + 5$ $\frac{5}{3}, \frac{-1 \pm \sqrt{5}}{2}$

7. $f(x) = 8x^3 + 14x^2 + 11x + 3$ $-\frac{1}{2}, \frac{-5 \pm i\sqrt{23}}{8}$

8. $f(x) = 4x^4 + 13x^3 - 8x^2 + 13x - 12$

9. $f(x) = 4x^4 - 12x^3 + 25x^2 - 14x - 15$

DifferentiatedInstruction (AL) (OL) (BL)

Logical Learners Organize students into groups of four or five. Have the students in each group split the work shown in Example 3 into four or five steps. The students should give explanations to the group about their parts of the Example. In particular, students should explain any mathematical processes, the results of their steps, and how the results relate to the next step in the process.

Additional Examples

2 **GEOMETRY** The volume of a rectangular solid is 1120 cubic feet. The width is 2 feet less than the height, and the length is 4 feet more than the height. Find the dimensions of the solid.

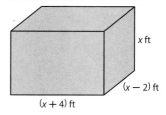
x ft
$(x - 2)$ ft
$(x + 4)$ ft

length: 14 ft, width: 8 ft, height: 10 ft

3 Find all of the zeros of $f(x) = x^4 + x^3 - 19x^2 + 11x + 30$.
$-5, -1, 2, 3$

---

3 Practice

Formative Assessment
Use Exercises 1−9 to check for understanding.

Use the chart at the bottom of the next page to customize assignments for your students.

CCSS **Teaching the Mathematical Practices**

Reasoning Mathematically proficient students make sense of quantities and their relationships in problem situations. Quantitative reasoning entails habits of creating a coherent representation of the problem at hand; considering the units involved; attending to the meaning of quantities, not just how to compute them; and knowing and flexibly using different properties of operations and objects.

Extra Practice is on page R5.

Example 1   List all of the possible rational zeros of each function. **10–17. See margin.**

10. $f(x) = x^4 + 8x - 32$ **11.** $f(x) = x^3 + x^2 - x - 56$

12. $f(x) = 2x^3 + 5x^2 - 8x - 10$ **⑬** $f(x) = 3x^6 - 4x^4 - x^2 - 35$

14. $f(x) = 6x^5 - x^4 + 2x^3 - 3x^2 + 2x - 18$ **15.** $f(x) = 8x^4 - 4x^3 - 4x^2 + x + 42$

16. $f(x) = 15x^3 + 6x^2 + x + 90$ **17.** $f(x) = 16x^4 - 5x^2 + 128$

Example 2   **18. MANUFACTURING** A box is to be constructed by cutting out equal squares from the corners of a square piece of cardboard and turning up the sides. **a. See margin.**

   a. Write a function $V(x)$ for the volume of the box.

   b. For what value of x will the volume of the box equal 1152 cubic centimeters? **2 or 8**

   c. What will be the volume of the box if $x = 6$ centimeters? **1536 cm³**

Find all of the rational zeros of each function.

19. $f(x) = x^3 + 10x^2 + 31x + 30$ **−5, −3, −2**   **20.** $f(x) = x^3 - 2x^2 - 56x + 192$ **−8, 4, 6**

21. $f(x) = 4x^3 - 3x^2 - 100x + 75$ **−5, $\frac{3}{4}$, 5**   **22.** $f(x) = 4x^4 + 12x^3 - 5x^2 - 21x + 10$ **$-\frac{5}{2}$, −2, $\frac{1}{2}$, 1**

23. $f(x) = x^4 + x^3 - 8x - 8$ **−1, 2**   **24.** $f(x) = 2x^4 - 3x^3 - 24x^2 + 4x + 48$ **−2, 4, $\frac{3}{2}$**

25. $f(x) = 4x^3 + x^2 + 16x + 4$ **$-\frac{1}{4}$**   **26.** $f(x) = 81x^4 - 256$ **$-\frac{4}{3}, \frac{4}{3}$**

Example 3   Find all of the zeros of each function.

27. $f(x) = x^3 + 3x^2 - 25x + 21$ **−7, 1, 3**   **28.** $f(x) = 6x^3 + 5x^2 - 9x + 2$ **$\frac{2}{3}, \frac{-3 \pm \sqrt{17}}{4}$**

29. $f(x) = x^4 - x^3 - x^2 - x - 2$ **2, −1, i, $-i$**   **30.** $f(x) = 10x^3 - 17x^2 - 7x + 2$ **30. $-\frac{1}{2}, \frac{1}{5}$, 2**

31. $f(x) = x^4 - 3x^3 + x^2 - 3x$ **0, 3, $-i$, i**   **32.** $f(x) = 6x^3 + 11x^2 - 3x - 2$ **$\frac{1}{2}, -\frac{1}{3}$, −2**

33. $f(x) = 6x^4 + 22x^3 + 11x^2 - 38x - 40$ **34.** $f(x) = 2x^3 - 7x^2 - 8x + 28$ **−2, 2, $\frac{7}{2}$**

35. $f(x) = 9x^5 - 94x^3 + 27x^2 + 40x - 12$ **36.** $f(x) = x^5 - 2x^4 - 12x^3 - 12x^2 - 13x - 10$

37. $f(x) = 48x^4 - 52x^3 + 13x - 3$ **38.** $f(x) = 5x^4 - 29x^3 + 55x^2 - 28x$ **$\frac{4}{5}$, 0, $\frac{5 \pm i\sqrt{3}}{2}$**

33. −2, $\frac{4}{3}, \frac{-3 \pm i}{2}$ **35.** 3, $\frac{2}{3}, -\frac{2}{3}, \frac{-3 \pm \sqrt{13}}{2}$ **36.** −1, −2, 5, i, $-i$

39. SWIMMING POOLS A diagram of the swimming pool at the Midtown Community Center is shown below. The pool can hold 9175 cubic feet of water.

37. $-\frac{1}{2}, \frac{1}{3}, \frac{1}{2}, \frac{3}{4}$

  a. $V(x) = 324x^3 + 54x^2 - 19x - 2$

  b. $\frac{-57 \pm i\sqrt{8987}}{36}$, 3; 3 is the only reasonable value for x. The other two values are imaginary.

  a. Write a polynomial function that represents the volume of the swimming pool.

  b. What are the possible values of x? Which of these values are reasonable?

40. CCSS MODELING A portion of the path of a certain roller coaster can be modeled by $f(t) = t^4 - 31t^3 + 308t^2 - 1100t + 1200$ where t represents the time in seconds and $f(t)$ represents the height of the roller coaster. Use the Rational Zero Theorem to determine the four times at which the roller coaster is at ground level. **2 s, 4 s, 10 s, and 15 s**

Differentiated Homework Options

Level	Assignment	Two-Day Option	
AL Basic	10–38, 46, 48, 49, 51–69	11–37 odd, 52–55	10–38 even, 46, 48, 49, 51, 56–69
OL Core	11–43 odd, 46, 48, 49, 51–69	10–38, 52–55	39–46, 48, 49, 51, 56–69
BL Advanced	39–63, (optional: 64–69)		

WatchOut!

Error Analysis For Exercise 46, remind students that the leading coefficient is the coefficient of the term with highest degree, not necessarily the coefficient of the term given first.

Additional Answers

10. ±1, ±2, ±4, ±8, ±16, ±32

11. ±1, ±2, ±4, ±7, ±8, ±14, ±28, ±56

12. ±1, ±2, ±5, ±10, $\pm\frac{1}{2}, \pm\frac{5}{2}$

13. ±1, ±5, ±7, ±35, $\pm\frac{1}{3}, \pm\frac{5}{3}, \pm\frac{7}{3}, \pm\frac{35}{3}$

14. ±1, ±2, ±3, ±6, ±9, ±18, $\pm\frac{1}{2}, \pm\frac{3}{2}, \pm\frac{9}{2}, \pm\frac{1}{3}, \pm\frac{2}{3}, \pm\frac{1}{6}$

15. ±1, ±2, ±3, ±6, ±7, ±14, ±21, ±42, $\pm\frac{1}{2}, \pm\frac{3}{2}, \pm\frac{7}{2}, \pm\frac{21}{2}, \pm\frac{1}{4}, \pm\frac{3}{4}, \pm\frac{7}{4}, \pm\frac{21}{4}, \pm\frac{1}{8}, \pm\frac{3}{8}, \pm\frac{7}{8}, \pm\frac{21}{8}$

16. ±1, ±2, ±3, ±5, ±6, ±9, ±10, ±15, ±18, ±30, ±45, ±90, $\pm\frac{1}{3}, \pm\frac{2}{3}, \pm\frac{5}{3}, \pm\frac{10}{3}, \pm\frac{1}{5}, \pm\frac{2}{5}, \pm\frac{3}{5}, \pm\frac{6}{5}, \pm\frac{9}{5}, \pm\frac{18}{5}, \pm\frac{1}{15}, \pm\frac{2}{15}$

17. ±1, ±2, ±4, ±8, ±16, ±32, ±64, ±128, $\pm\frac{1}{2}, \pm\frac{1}{4}, \pm\frac{1}{8}, \pm\frac{1}{16}$

18a. $V(x) = (28 - 2x)(28 - 2x)x = 4x^3 - 112xt^2 + 784x$

E **Follow-up**

Students have explored functions.

Ask:

▪ What are the limitations of mathematical modeling? Sample answer: Not all real-world situations can be modeled. For those that can be modeled, predictions made using the model may not be accurate when based on data values that are outside the range of data values used to create the model. Therefore, after a model is created, it should be carefully analyzed before it is used to make predictions or decisions.

(41) **FOOD** A restaurant orders spaghetti sauce in cylindrical metal cans. The volume of each can is about 160π cubic inches, and the height of the can is 6 inches more than the radius.

   a. Write a polynomial equation that represents the volume of a can. Use the formula for the volume of a cylinder, $V = \pi r^2 h$. $V = \pi r^3 + 6\pi r^2$

   b. What are the possible values of r? Which of these values are reasonable for this situation? $4, -5 \pm i\sqrt{15}; 4$

   c. Find the dimensions of the can. $r = 4$ in., $h = 10$ in.

42. Refer to the graph at the right.

   a. Find all of the zeros of $f(x) = 2x^3 + 7x^2 + 2x - 3$ and $g(x) = 2x^3 - 7x^2 + 2x + 3$. $-1, \frac{1}{2}, -3; -\frac{1}{2}, 1, 3$

   b. Determine which function, f or g, is shown in the graph at the right. g

43. **MUSIC SALES** Refer to the beginning of the lesson.

43b. 1, 2, 3, 4, 6, 7, 12, 13, 14, 21, 26, 28, 39, 42, 52, 78, 84, 91, 156, 182, 273, 364, 546, 1092

   a. Write a polynomial equation that could be used to determine the year in which music sales would be about \$9,000,000,000. $30x^3 - 478x^2 + 1758x + 1092 = 0$

   b. List the possible whole number solutions for your equation in part **a**.

   c. Determine the approximate year in which music sales will be 9,000,000,000.   2013

   d. Does the model represent a realistic estimate for all future music sales? Explain your reasoning.   No; Sample answer: Music sales fluctuate from 2005 to 2015, then increase indefinitely. It is not reasonable to expect sales to increase forever.

Find all of the zeros of each function.

44. $f(x) = x^5 + 3x^4 - 19x^3 - 43x^2 + 18x + 40$ $1, -1, -2, 4, -5$

45. $f(x) = x^5 - x^4 - 23x^3 + 33x^2 + 126x - 216$ $2, 3, 3, -3, -4$

H.O.T. Problems   Use Higher-Order Thinking Skills

48. Sample answer: Always; in order for the possible zeros of a polynomial function to be integers, the value of q must be 1 or -1. Otherwise, the possible zeros could be a fraction. In order for q to be 1 or -1, the leading coefficient of the polynomial must also be 1 or -1.

46. **CCSS CRITIQUE** Doug and Mika are listing all of the possible rational zeros for $f(x) = 4x^4 + 8x^5 + 10x^2 + 3x + 16$. Is either of them correct? Explain your reasoning.

Doug	Mika
$\pm 1, \pm 2, \pm 4, \pm 8, \pm 16, \pm \frac{1}{2}, \pm \frac{1}{4}$	$\pm 1, \pm 2, \pm 4, \pm 8, \pm 16, \pm \frac{1}{2}, \pm \frac{1}{4}, \pm \frac{1}{8}$

46. Mika; sample answer: The value of q is the leading coefficient, which is 8, not 4.

47. **CHALLENGE** Give a polynomial function that has zeros at $1 + \sqrt{3}$ and $5 + 2i$.   Sample answer: $f(x) = x^4 - 12x^3 + 47x^2 - 38x - 58$

48. **REASONING** Determine if the following statement is *sometimes, always,* or *never* true. Explain your reasoning.

   If all of the possible zeros of a polynomial function are integers, then the leading coefficient of the function is 1 or -1.

49. **OPEN ENDED** Write a function that has possible zeros of $\pm 18, \pm 9, \pm 6, \pm 3, \pm 2, \pm 1, \pm \frac{9}{4}$, $\pm \frac{9}{2}, \pm \frac{3}{2}, \pm \frac{3}{4}, \pm \frac{1}{2}$, and $\pm \frac{1}{4}$.   Sample answer: $f(x) = 4x^5 + 3x^3 + 8x + 18$

50. **CHALLENGE** The roots of $x^2 + bx + c = 0$ are M and N. If $|M - N| = 1$, express c in terms of b. $c = \dfrac{b^2 - 1}{4}$

51. **WRITING IN MATH** How can you find the zeros of a polynomial function?   See margin.

Additional Answer

51. Sample answer: You can start by using the Rational Zero Theorem to generate a list of possible zeros. Then, you could graph the function to narrow the list down. You could then perform polynomial division using the possible zeros in order to rewrite the polynomial as the product of linear expressions and a quadratic expression. You could then solve the quadratic expression to find the remaining zeros.

Study Guide and Intervention   AL OL ELL

Practice   AL OL BL ELL

Word Problem Practice   AL OL BL ELL

Enrichment   OL BL

4 Assess

Name the Math Ask students:
- What must be true about the coefficients of a polynomial function in order to apply the Rational Zero Theorem?

- How can the possible zeros of a polynomial function be identified?

Formative Assessment
Check for student understanding of concepts in Lessons 5-7 and 5-8.

 Quiz 4, p. 58

Standardized Test Practice

52. ALGEBRA Which of the following is a zero of the function $f(x) = 12x^5 - 5x^3 + 2x - 9$? **D**

A -6 C $\frac{3}{8}$

B $-\frac{2}{3}$ D 1

53. SAT/ACT How many negative real zeros does $f(x) = x^5 - 2x^4 - 4x^3 + 4x^2 - 5x + 6$ have? **J**

F 5   J 1

G 3   K 0

H 2

54. ALGEBRA For all nonnegative numbers n, let $\boxed{n}$ be defined by $\boxed{n} = \frac{\sqrt{n}}{2}$. If $\boxed{n} = 4$, what is the value of n? **D**

A 2   C 16

B 4   D 64

55. GRIDDED RESPONSE What is the y-intercept of a line that contains the point $(-1, 4)$ and has the same x-intercept as $x + 2y = -3$? **6**

Spiral Review

Write a polynomial function of least degree with integral coefficients that has the given zeros. (Lesson 5-7)

56. $6, -3, \sqrt{2}$
$f(x) = x^4 - 3x^3 - 20x^2 + 6x + 36$

57. $5, -1, 4i$
$f(x) = x^4 - 4x^3 + 11x^2 - 64x - 80$

58. $-4, -2, i\sqrt{2}$
$f(x) = x^4 + 6x^3 + 10x^2 + 12x + 16$

Given a polynomial and one of its factors, find the remaining factors of the polynomial. (Lesson 5-6)

59. $x^4 + 5x^3 + 5x^2 - 5x - 6; x + 3$
$(x - 1)(x + 2)(x + 1)$

60. $a^4 - 2a^3 - 17a^2 + 18a + 72; a - 3$
$(a + 3)(a - 4)(a + 2)$

61. $x^4 + x^3 - 11x^2 + x - 12; x + i$
$(x - 3)(x + 4)(x - i)$

62. BRIDGES The supporting cables of the Golden Gate Bridge approximate the shape of a parabola. The parabola can be modeled by the quadratic function $y = 0.00012x^2 + 6$, where x represents the distance from the axis of symmetry and y represents the height of the cables. The related quadratic equation is $0.00012x^2 + 6 = 0$. (Lesson 4-6)

a. Calculate the value of the discriminant. -0.00288

b. What does the discriminant tell you about the supporting cables of the Golden Gate Bridge? **Sample answer: This means that the cables do not touch the floor of the bridge, since the graph does not intersect the x-axis and the roots are imaginary.**

63. RIDES An amusement park ride carries riders to the top of a 225-foot tower. The riders then free-fall in their seats until they reach 30 feet above the ground. (Lesson 4-2)

a. Use the formula $h(t) = -16t^2 + h_0$, where the time t is in seconds and the initial height h_0, is in feet, to find how long the riders are in free-fall. **about 3.5 s**

b. Suppose the designer of the ride wants the riders to experience free-fall for 5 seconds before stopping 30 feet above the ground. What should be the height of the tower? **430 ft**

Skills Review

Simplify.

64. $(x - 4)(x + 3)$ $x^2 - x - 12$

65. $3x(x^2 + 4)$ $3x^3 + 12x$

66. $x^2(x - 2)(x + 1)$ $x^4 - x^3 - 2x^2$

Find each value if $f(x) = 6x + 2$ and $g(x) = -4x^2$.

67. $f(5)$ **32**

68. $g(-3)$ **−36**

69. $f(3c)$ **18c + 2**

DifferentiatedInstruction **OL** **BL**

Extension Ask students to factor $x^4 - 2x^2 - 3$ as the product of factors that are irreducible over rational numbers. $(x^2 - 3)(x^2 + 1)$

Then have students factor as the product of factors that are irreducible over real numbers. $(x - \sqrt{3})(x + \sqrt{3})(x^2 + 1)$

Finally, have students factor completely, including complex numbers. $(x - \sqrt{3})(x + \sqrt{3})(x + i)(x - i)$

Study Guide and Review

Study Guide

KeyConcepts

Operations with Polynomials (Lessons 5-1 and 5-2)
- To add or subtract: Combine like terms.
- To multiply: Use the Distributive Property.
- To divide: Use long division or synthetic division.

Polynomial Functions and Graphs (Lessons 5-3 and 5-4)
- Turning points of a function are called *relative maxima* and *relative minima*.

Solving Polynomial Equations (Lesson 5-5)
- You can factor polynomials by using the GCF, grouping, or quadratic techniques.

The Remainder and Factor Theorems (Lesson 5-6)
- Factor Theorem: The binomial $x - a$ is a factor of the polynomial $f(x)$ if and only if $f(a) = 0$.

Roots, Zeros, and the Rational Zero Theorem
(Lessons 5-7 and 5-8)
- Complex Conjugates Theorem: If $a + bi$ is a zero of a function, then $a - bi$ is also a zero.
- Integral Zero Theorem: If the coefficients of a polynomial function are integers such that $a_0 = 1$ and $a_n = 0$, any rational zeros of the function must be factors of a_n.
- Rational Zero Theorem: If $P(x)$ is a polynomial function with integral coefficients, then every rational zero of $P(x) = 0$ is of the form $\frac{p}{q}$, a rational number in simplest form, where p is a factor of the constant term and q is a factor of the leading coefficient.

FOLDABLES StudyOrganizer

Be sure the Key Concepts are noted in your Foldable.

KeyVocabulary

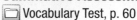

degree of a polynomial (p. 305)	power function (p. 323)
depressed polynomial (p. 354)	prime polynomials (p. 342)
end behavior (p. 324)	quadratic form (p. 345)
extrema (p. 331)	relative maximum (p. 331)
leading coefficient (p. 322)	relative minimum (p. 331)
Location principle (p. 330)	simplify (p. 303)
polynomial function (p. 323)	synthetic division (p. 312)
polynomial in one variable (p. 322)	synthetic substitution (p. 352)
	turning points (p. 331)

VocabularyCheck

State whether each sentence is *true* or *false*. If *false*, replace the underlined term to make a true sentence.

1. The coefficient of the first term of a polynomial in standard form is called the <u>leading coefficient</u>. **true**

2. Polynomials that cannot be factored are called <u>polynomials in one variable</u>. **false; prime polynomials**

3. A <u>prime polynomial</u> has a degree that is one less than the original polynomial. **false; depressed polynomial**

4. A point on the graph of a function where no other nearby point has a greater y-coordinate is called a <u>relative maximum</u>. **true**

5. A <u>polynomial function</u> is a continuous function that can be described by a polynomial equation in one variable. **true**

6. To <u>simplify</u> an expression containing powers means to rewrite the expression without parentheses or negative exponents. **true**

7. <u>Synthetic division</u> is a shortcut method for dividing a polynomial by a binomial. **true**

8. The relative maximum and relative minimum of a function are often referred to as <u>end behavior</u>. **false; turning points**

9. When a polynomial is divided by one of its binomial factors, the quotient is called a <u>depressed polynomial</u>. **true**

10. $(x^3)^2 + 3x^3 - 8 = 0$ is <u>a power function</u>.
10. false; written in quadratic form

Formative Assessment

KeyVocabulary The page references after each word denote where that term was first introduced. If students have difficulty answering questions 1–10, remind them that they can use these page references to refresh their memories about the vocabulary.

Summative Assessment

📁 Vocabulary Test, p. 60

Vocabulary Review provides students the opportunity to check their understanding of important concepts and terminology in an online game format.

FOLDABLES StudyOrganizer

Dinah Zike's Foldables®
Have students look through the chapter to make sure they have included examples in their Foldables. Suggest that students keep their Foldables handy while completing the Study Guide and Review pages. Point out that their Foldables can serve as a quick review tool when studying for the Chapter Test.

Lesson-by-Lesson Review

Intervention If the given examples are not sufficient to review the topics covered by the questions, remind students that the lesson references tell them where to review that topic in their textbook.

Two-Day Option Have students complete the Lesson-by-Lesson Review. Then you can use McGraw-Hill eAssessment to customize another review worksheet that practices all the objectives of this chapter or only the objectives on which your students need more help.

Additional Answers

22. degree: 6; leading coefficient: 5

23. This is not a polynomial in one variable. It has two variables, x and y.

24. degree: 8; leading coefficient: 6

25. $p(-2) = -3$; $p(x + h) = x^2 + 2xh + h^2 + 2x + 2h - 3$

26. $p(-2) = 14$; $p(x + h) = 3x^2 + 6xh + 3h^2 - x - h$

27. $p(-2) = -25$; $p(x + h) = 3 - 5x^2 - 10xh - 5h^2 + x^3 + 3hx^2 + 3h^2x + h^3$

Lesson-by-Lesson Review

5-1 Operations with Polynomials

Simplify. Assume that no variable equals 0.

11. $\dfrac{14x^4y}{2x^3y^5} \cdot \dfrac{7x}{y^4}$

12. $3t(tn - 5)$ $3t^2n - 15t$

13. $(4r^2 + 3r - 1) - (3r^2 - 5r + 4)$ $r^2 + 8r - 5$

14. $(x^4)^3$ x^{12}

15. $(m + p)(m^2 - 2mp + p^2)$ $m^3 - m^2p - mp^2 + p^3$

16. $3b(2b - 1) + 2b(b + 3)$ $8b^2 + 3b$

Example 1

Simplify each expression.

a. $(-4a^3b^5)(5ab^3)$

$(-4a^3b^5)(5ab^3) = (-4)(5)(a^3 \cdot a)(b^5 \cdot b^3)$ Product of Powers

$= -20a^4b^8$ Simplify.

b. $(2x^2 + 3x - 8) + (3x^2 - 5x - 7)$

$(2x^2 + 3x - 8) + (3x^2 - 5x - 7)$

$= (2x^2 + 3x^2) + (3x - 5x) + [-8 + (-7)]$

$= 5x^2 - 2x - 15$

5-2 Dividing Polynomials

Simplify.

17. $\dfrac{12x^4y^5 + 8x^3y^7 - 16x^2y^6}{4xy^5}$ $3x^3 + 2x^2y^2 - 4xy$

18. $(6y^3 + 13y^2 - 10y - 24) \div (y + 2)$ $6y^2 + y - 12$

19. $(a^4 + 5a^3 + 2a^2 - 6a + 4)(a + 2)^{-1}$ $a^3 + 3a^2 - 4a + 2$

20. $(4a^6 - 5a^4 + 3a^2 - a) \div (2a + 1)$

21. **GEOMETRY** The volume of the rectangular prism is $3x^3 + 11x^2 - 114x - 80$ cubic units. What is the area of the base? $x^2 + 3x - 40$ units^2

Example 2

Simplify $(6x^3 - 31x^2 - 34x + 22) \div (2x - 1)$.

$$
\begin{array}{r}
3x^2 - 14x - 24 \\
2x - 1 \overline{)6x^3 - 31x^2 - 34x + 22} \\
\underline{(-)\,6x^3 - 3x^2} \\
-28x^2 - 34x \\
\underline{(-)\,-28x^2 + 14x} \\
-48x + 22 \\
\underline{(-)\,-48x + 24} \\
-2
\end{array}
$$

20. $2a^5 - a^4 - 2a^3 + a^2 + a - 1 + \dfrac{1}{2a+1}$

The result is $3x^2 - 14x - 24 - \dfrac{2}{2x-1}$.

5-3 Polynomial Functions

State the degree and leading coefficient of each polynomial in one variable. If it is not a polynomial in one variable, explain why.

22. $5x^6 - 3x^4 + x^3 - 9x^2 + 1$ **22–24. See margin.**

23. $6xy^2 - xy + y^2$

24. $12x^3 - 5x^4 + 6x^8 - 3x - 3$

Find $p(-2)$ and $p(x + h)$ for each function.

25. $p(x) = x^2 + 2x - 3$ **25–27. See margin.**

26. $p(x) = 3x^2 - x$

27. $p(x) = 3 - 5x^2 + x^3$

Example 3

What are the degree and leading coefficient of $4x^3 + 3x^2 - 7x^7 + 4x - 1$?

The greatest exponent is 7, so the degree is 7. The leading coefficient is -7.

Example 4

Find $p(a - 2)$ if $p(x) = 3x + 2x^2 - x^3$.

$p(a - 2) = 3(a - 2) + 2(a - 2)^2 - (a - 2)^3$

$= 3a - 6 + 2a^2 - 8a + 8 - (a^3 - 6a^2 + 12a - 8)$

$= -a^3 + 8a^2 - 17a + 10$

374 | Chapter 5 | Study Guide and Review

5-4 Analyzing Graphs of Polynomial Functions

Complete each of the following.

a. Graph each function by making a table of values.

b. Determine the consecutive integer values of x between which each real zero is located.

c. Estimate the x-coordinates at which the relative maxima and minima occur.

28. $h(x) = x^3 - 4x^2 - 7x + 10$ **28–32. See margin.**

29. $g(x) = 4x^4 - 21x^2 + 5$

30. $f(x) = x^3 - 3x^2 - 4x + 12$

31. $h(x) = 4x^3 - 6x^2 + 1$

32. $p(x) = x^5 - x^4 + 1$
     2 relative maxima and 1 relative minima

33. **BUSINESS** Milo tracked the monthly profits for his sports store business for the first six months of the year. They can be modeled by using the following six points: (1, 675), (2, 950), (3, 550), (4, 250), (5, 600), and (6, 400). How many turning points would the graph of a polynomial function through these points have? Describe them.

Example 5

Graph $f(x) = x^3 + 3x^2 - 4$ by making a table of values.

Make a table of values for several values of x.

x	-3	-2	-1	0	1	2
$f(x)$	-4	0	-2	-4	0	16

Plot the points and connect the points with a smooth curve.

5-5 Solving Polynomial Equations

Factor completely. If the polynomial is not factorable, write *prime*.

34. $a^4 - 16$ $(a - 2)(a + 2)(a^2 + 4)$

35. $x^3 + 6y^3$ **prime**

36. $54x^3y - 16y^4$ $2y(3x - 2y)(9x^2 + 6xy + 4y^2)$

37. $6ay + 4by - 2cy + 3az + 2bz - cz$
     $(2y + z)(3a + 2b - c)$

Solve each equation.

38. $x^3 + 2x^2 - 35x = 0$ $-7, 0, 5$

39. $8x^4 - 10x^2 + 3 = 0$ $\pm\dfrac{\sqrt{3}}{2}, \pm\dfrac{\sqrt{2}}{2}$

40. **GEOMETRY** The volume of the prism is 315 cubic inches. Find the value of x and the length, height, and width.

$x = 6$, length = 9 in., height = 5 in., width = 7 in.

Example 6

Factor $r^7 + 64r$.

$r^7 + 64r = r(r^6 + 64)$ Factor by GCF.

$= r[(r^2)^3 + 4^3]$ Write as cubes.

$= r(r^2 + 4)(r^4 - 4r^2 + 16)$

Example 7

Solve $4x^4 - 25x^2 + 36 = 0$.

$(x^2 - 4)(4x^2 - 9) = 0$

$x^2 - 4 = 0$ or $4x^2 - 9 = 0$

$x^2 = 4$ $x^2 = \dfrac{9}{4}$

$x = \pm 2$ $x = \pm\dfrac{3}{2}$

The solutions are $-2, 2, -\dfrac{3}{2},$ and $\dfrac{3}{2}.$

Additional Answers

28a.

28b. The zeros are at -2, 1, and 5.

28c. rel. max: $x \approx -0.69$;
rel. min: $x \approx 3.36$

29a.

29b. between -3 and -2, between -1 and 0, between 0 and 1, between 2 and 3

29c. rel. max: $x \approx 0$; rel. min: $x \approx 1.62$ and $x \approx -1.62$

30a.

30b. zeros at -2, 2, and 3

30c. rel. max: $x \approx -0.53$;
rel. min: $x \approx 2.53$

31a.

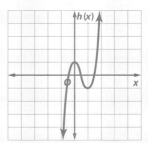

31b. between -1 and 0, between 0 and 1, and between 1 and 2

31c. rel. max: $x \approx 0$; rel. min: $x \approx 1$

32a.

32b. between -1 and 0

32c. rel. max: $x \approx 0$; rel. min: $x \approx 0.80$

Anticipation Guide
Have students complete the Chapter 5 Anticipation Guide and discuss how their responses have changed now that they have completed Chapter 5.

Additional Answers

48. positive real zeros: 3 or 1
negative real zeros: 0
imaginary zeros: 2 or 0

49. positive real zeros: 0
negative real zeros: 4, 2, or 0
imaginary zeros: 4, 2, or 0

50. positive real zeros: 3 or 1
negative real zeros: 1
imaginary zeros: 4 or 2

51. positive real zeros: 2 or 0
negative real zeros: 1
imaginary zeros: 4 or 2

52. positive real zeros: 2 or 0
negative real zeros: 2 or 0
imaginary zeros: 6, 4, or 2

Additional Answers (Practice Test)

9a.

9b. between -5 and -4

9c. rel. max: $x = -3$; rel. min: $x \approx 0.3$

10a.

10b. between -2 and -1, between -1 and 0, between 1 and 2, between 4 and 5

10c. rel max: $x \approx 0.5$; rel min: $x \approx -0.8$, $x \approx 3.3$

5-6 The Remainder and Factor Theorems

Use synthetic substitution to find $f(-2)$ and $f(4)$ for each function.

41. $f(x) = x^2 - 3$ $f(-2) = 1$; $f(4) = 13$

42. $f(x) = x^2 - 5x + 4$ $f(-2) = 18$; $f(4) = 0$

43. $f(x) = x^3 + 4x^2 - 3x + 2$ $f(-2) = 16$; $f(4) = 118$

44. $f(x) = 2x^4 - 3x^3 + 1$ $f(-2) = 57$; $f(4) = 321$

Given a polynomial and one of its factors, find the remaining factors of the polynomial.

45. $3x^3 + 20x^2 + 23x - 10$; $x + 5$ $x + 2, 3x - 1$

46. $2x^3 + 11x^2 + 17x + 5$; $2x + 5$ $x^2 + 3x + 1$

47. $x^3 + 2x^2 - 23x - 60$; $x - 5$ $x + 3, x + 4$

Example 8

Determine whether $x - 6$ is a factor of $x^3 - 2x^2 - 21x - 18$.

$$
\begin{array}{r|rrrr}
6 & 1 & -2 & -21 & -18 \\
  &   & 6 & 24 & 18 \\
\hline
  & 1 & 4 & 3   & 0
\end{array}
$$

$x - 6$ is a factor because $r = 0$.

$x^3 - 2x^2 - 21x - 18 = (x - 6)(x^2 + 4x + 3)$

5-7 Roots and Zeros

State the possible number of positive real zeros, negative real zeros, and imaginary zeros of each function.

48. $f(x) = -2x^3 + 11x^2 - 3x + 2$ **48–52. See margin.**

49. $f(x) = -4x^4 - 2x^3 - 12x^2 - x - 23$

50. $f(x) = x^6 - 5x^3 + x^2 + x - 6$

51. $f(x) = -2x^5 + 4x^4 + x^2 - 3$

52. $f(x) = -2x^6 + 4x^4 + x^2 - 3x - 3$

Example 9

State the possible number of positive real zeros, negative real zeros, and imaginary zeros of $f(x) = 3x^4 + 2x^3 - 2x^2 - 26x - 48$.

$f(x)$ has one sign change, so there is 1 positive real zero.

$f(-x)$ has 3 sign changes, so there are 3 or 1 negative real zeros.

There are 0 or 2 imaginary zeros.

5-8 Rational Zero Theorem

Find all of the zeros of each function.

53. $f(x) = x^3 + 4x^2 + 3x - 2$ $-2, -1 \pm \sqrt{2}$

54. $f(x) = 4x^3 + 4x^2 - x - 1$ $-1, -\dfrac{1}{2}, \dfrac{1}{2}$

55. $f(x) = x^3 + 2x^2 + 4x + 8$ $-2, \pm 2i$

56. STORAGE Melissa is building a storage box that is shaped like a rectangular prism. It will have a volume of 96 cubic feet. Using the diagram below, find the dimensions of the box.   width = 4 ft, length = 12 ft, height = 2 ft

Example 10

Find all of the zeros of $f(x) = x^3 + 4x^2 - 11x - 30$.

There are exactly 3 zeros.

There are 1 positive real zero and 2 negative real zeros. The possible rational zeros are $\pm 1, \pm 2, \pm 3, \pm 5, \pm 6, \pm 10, \pm 15, \pm 30$.

$$
\begin{array}{r|rrrr}
3 & 1 & 4 & -11 & -30 \\
  &   & 3 & 21 & 30 \\
\hline
  & 1 & 7 & 10 & 0
\end{array}
$$

$x^3 + 4x^2 - 11x - 30 = (x - 3)(x^2 + 7x + 10)$
$= (x - 3)(x + 2)(x + 5)$

Thus, the zeros are 3, -2, and -5.

CHAPTER 5 Practice Test

Simplify.

1. $(3a)^2(7b)^4$ $21,609a^2b^4$

2. $(7x - 2)(2x + 5)$ $14x^2 + 31x - 10$

3. $(2x^2 + 3x - 4) - (4x^2 - 7x + 1)$ $-2x^2 + 10x - 5$

4. $(4x^3 - x^2 + 5x - 4) + (5x - 10)$ $4x^3 - x^2 + 10x - 14$

5. $(x^4 + 5x^3 + 3x^2 - 8x + 3) \div (x + 3)$ $x^3 + 2x^2 - 3x + 1$

6. $(3x^3 - 5x^2 - 23x + 24) \div (x - 3)$
$$3x^2 + 4x - 11 - \frac{9}{x - 3}$$

7. **MULTIPLE CHOICE** How many unique real zeros does the graph have? **C**

A 0   C 3

B 2   D 5

8. If $c(x) = 3x^3 + 5x^2 - 4$, what is the value of $4c(3b)$?
$324b^3 + 180b^2 - 16$

Complete each of the following.

a. Graph each function by making a table of values.

b. Determine consecutive integer values of x between which each real zero is located.

c. Estimate the x-coordinates at which the relative maxima and relative minima occur.

9. $g(x) = x^3 + 4x^2 - 3x + 1$ **9, 10. See margin.**

10. $h(x) = x^4 - 4x^3 - 3x^2 + 6x + 2$

Factor completely. If the polynomial is not factorable, write *prime*.

11. $8y^4 + x^3y$ $y(2y + x)(4y^2 - 2xy + x^2)$

12. $2x^2 + 2x + 1$ **prime**

13. $a^2x + 3ax + 2x - a^2y - 3ay - 2y$ $(x - y)(a + 2)(a + 1)$

Solve each equation.

14. $8x^3 + 1 = 0$ $-\frac{1}{2}, \frac{1 \pm i\sqrt{3}}{4}$

15. $x^4 - 11x^2 + 28 = 0$ $\pm\sqrt{7}, \pm 2$

16. **FRAMING** The area of the picture and frame shown below is 168 square inches. What is the width of the frame? **2 in.**

17. **MULTIPLE CHOICE** Let $f(x) = x^4 - 3x^3 + 5x - 3$. Use synthetic substitution to find $f(-2)$. **G**

F 37   H −21

G 27   J −33

Given a polynomial and one of its factors, find the remaining factors of the polynomial.

18. $2x^3 + 15x^2 + 22x - 15; x + 5$ $2x - 1, x + 3$

19. $x^3 - 4x^2 + 10x - 12; x - 2$ $x^2 - 2x + 6$

State the possible number of positive real zeros, negative real zeros, and imaginary zeros of each function.

20. $p(x) = x^3 - x^2 - x - 3$ **1 positive, 2 or 0 negative, 2 or 0 imaginary**

21. $p(x) = 2x^6 + 5x^4 - x^3 - 5x - 1$
1 positive, 1 negative, 4 imaginary

Find all zeros of each function.

22. $p(x) = x^3 - 4x^2 + x + 6$ $-1, 2, 3$

23. $p(x) = x^3 + 2x^2 + 4x + 8$ $-2, \pm 2i$

24. **GEOMETRY** The volume of the rectangular prism shown is 612 cubic centimeters. Find the dimensions of the prism. **9 cm by 17 cm by 4 cm**

25. List all possible rational zeros of $f(x) = 2x^4 + 3x^2 - 12x + 8$. $\pm\frac{1}{2}, \pm 1, \pm 2, \pm 4, \pm 8$

Summative Assessment

Use these alternate leveled chapter tests to differentiate assessment for your students.

Leveled Chapter 5 Tests

Form	Type	Level	📁 Page(s)
1	MC	AL	61−62
2A	MC	OL	63−64
2B	MC	OL	65−66
2C	FR	OL	67−68
2D	FR	OL	69−70
3	FR	BL	71−72
Vocabulary Test			60
Extended-Response Test			73

MC = multiple-choice questions
FR = free-response questions

McGraw-Hill eAssessment Customize and create multiple versions of your chapter test and their answer keys. All of the questions from the leveled chapter tests in the *Chapter 5 Resource Masters* are also available on McGraw-Hill eAssessment.

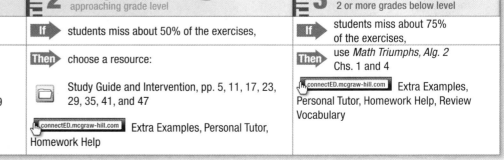

InterventionPlanner

TIER 1 On Level OL

If ▶ students miss about 25% of the exercises or less,

Then ▶ choose a resource:

SE Lessons 5-1 through 5-8

📁 Skills Practice, pp. 7, 13, 19, 25, 31, 37, 43, and 49

TE Chapter Project

connectED.mcgraw-hill.com Self-Check Quiz

TIER 2 Strategic Intervention AL
approaching grade level

If ▶ students miss about 50% of the exercises,

Then ▶ choose a resource:

📁 Study Guide and Intervention, pp. 5, 11, 17, 23, 29, 35, 41, and 47

connectED.mcgraw-hill.com Extra Examples, Personal Tutor, Homework Help

TIER 3 Intensive Intervention
2 or more grades below level

If ▶ students miss about 75% of the exercises,

Then ▶ use *Math Triumphs, Alg. 2* Chs. 1 and 4

connectED.mcgraw-hill.com Extra Examples, Personal Tutor, Homework Help, Review Vocabulary

1 Focus

Objective Use the strategy of drawing a picture to solve standardized test problems.

2 Teach

Scaffolding Questions

Ask:

- Have you ever needed directions to a location? Was it more helpful to get written directions or a hand-drawn map of the directions? Answers will vary.

- Was it enough that the map was hand drawn or did you need a professionally drawn map? Answers will vary.

- What information is necessary to make a drawing, whether a map or a diagram, most helpful? Sample answers: distances, lengths, points of interest, landmarks, starting point

---

Draw a Picture

Drawing a picture can be a helpful way for you to visualize how to solve a problem. Sketch your picture on scrap paper or in your test booklet (if allowed). Do not make any marks on your answer sheet other than your answers.

Strategies for Drawing a Picture

Step 1

Read the problem statement carefully.

    Ask yourself:

- What am I being asked to solve?
- What information is given in the problem?
- What are the unknowns that I need to model and solve for?

Step 2

Sketch and label your picture.

- Draw your picture as clearly and accurately as possible.
- Label the picture carefully. Be sure to include all of the information given in the problem statement.

Step 3

Solve the problem.

- Use your picture to help you model the problem situation with an equation. Solve the equation.
- Check to be sure your answer makes sense.

Standardized Test Example

Read the problem. Identify what you need to know. Then use the information in the problem to solve.

Mr. Nolan has a rectangular swimming pool that measures 25 feet by 14 feet. He wants to have a cement walkway installed around the perimeter of the pool. The combined area of the pool and walkway will be 672 square feet. What will be the width of the walkway?

A 2.75 ft   **C** 3.25 ft

B 3 ft   **D** 3.5 ft

378 | Chapter 5 | Preparing for Standardized Tests

Laurence Mouton/PhotoAlto/Getty Images

Draw a picture to help you visualize the problem situation. Let x represent the unknown width of the cement walkway.

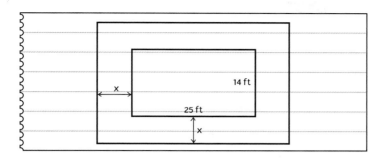

The width of the pool and walkway is $14 + 2x$, and the length is $25 + 2x$. Multiply these polynomial expressions and set the result equal to the combined area, 672 square feet. Then solve for x.

$(14 + 2x)(25 + 2x) = 672$

$350 + 78x + 4x^2 = 672$

$4x^2 + 78x - 322 = 0$

$x = -23$ or 3.5

Since the width cannot be negative, the width of the walkway will be 3.5 feet. The correct answer is D.

Exercises

Read each problem. Identify what you need to know. Then use the information in the problem to solve.

1. A farmer has 240 feet of fencing that he wants to use to enclose a rectangular area for his chickens. He plans to build the enclosure using the wall of his barn as one of the walls. What is the maximum amount of area he can enclose? A

   A 7200 ft^2

   B 4960 ft^2

   C 3600 ft^2

   D 3280 ft^2

2. Metal washers are made by cutting a hole in a circular piece of metal. Suppose a washer is made by removing the center of a piece of metal with a 1.8-inch diameter. What is the radius of the hole if the washer has an area of 0.65π square inches? H

   F 0.35 in.

   G 0.38 in.

   H 0.40 in.

   J 0.42 in.

STANDARDIZED TEST PRACTICE
Robin has glued a 3" by 5" index card to a sheet of red paper. She trims the red paper so that it creates an equal-width border around the index card. The index card and the red paper border have a combined area of 168 square inches. What is the width of the border? A

A 4.5 inches

B 8.5 inches

C 9 inches

D 14 inches

3 Assess

Use Exercises 1 and 2 to assess students' understanding.

Diagnose Student Errors

Survey student responses for each item. Class trends may indicate common errors and misconceptions.

1. A -1 was not distributed to the -5
   B correct
   C -1 was not distributed correctly
   D $2n^2$ was added to $5n^2$ instead of subtracted from it

2. F guess
   G did not understand that the coefficient of squared term affects the width of the graph
   H correct
   J did not understand that the value of h in $y = (x - h)^2 + k$ causes a shift left or right

3. A did not understand the key words indicating the operation to use
   B guess
   C correct
   D guess

4. F found a solution to the equation
   G found a solution to the equation
   H correct
   J found a solution to the equation

5. A did not square $2x + 1$
   B correct
   C confused the sign of the numbers
   D added the value of $(2x + 1)^2$ to 3

6. F correct
   G counted the x-and y-intercepts
   H guess
   J guess

7. A correct
   B did not understand that *between* means *not included*
   C did not add $100 to the low estimate
   D did not understand that *between* means *not included*

8. F simplified incorrectly
   G did not square 20 in the x^2 term
   H simplified incorrectly
   J correct

9. A incorrectly made a negative divided by a negative be a negative
   B made errors when combining like terms
   C made errors when combining like terms
   D correct

Standardized Test Practice
Cumulative, Chapters 1 through 5

Multiple Choice

Read each question. Then fill in the correct answer on the answer document provided by your teacher or on a sheet of paper.

1. Simplify the following expression.

$$(5n^2 + 11n - 6) - (2n^2 - 5)$$ **B**

   A $3n^2 + 11n - 11$ C $7n^2 + 11n - 11$

   B $3n^2 + 11n - 1$ D $7n^2 + 11n - 1$

2. What is the effect on the graph of the equation $y = x^2 + 4$ when it is changed to $y = x^2 - 3$? **H**

   F The slope of the graph changes.

   G The graph widens.

   H The graph is the same shape, and the vertex of the graph is moved down.

   J The graph is the same shape, and the vertex of the graph is shifted to the left.

3. Let p represent the price that Ella charges for a necklace. Let $f(x)$ represent the total amount of money that Ella makes for selling x necklaces. The function $f(x)$ is best represented by **C**

   A $f(x) = x + p$ C $f(x) = px$

   B $f(x) = xp^2$ D $f(x) = x^2 + p$

4. Which of the following is *not* a solution to the cubic equation below? **H**

$$x^3 - 37x - 84 = 0$$

   F -4 H 6

   G -3 J 7

5. What is the solution set for the equation $3(2x + 1)^2 = 27$? **B**

   A $\{-5, 4\}$ C $\{2, -1\}$

   B $\{-2, 1\}$ D $\{-3, 3\}$

Test-Taking Tip

Question 5 You can use substitution to check each possible solution and identify the one that does not result in a true number sentence.

---

6. How many real zeros does the polynomial function graphed below have? **F**

   F 2   H 4

   G 3   J 5

7. For Marla's vacation, it will cost $100 to drive her car, plus between $0.50 and $0.75 per mile. If she will drive her car 400 miles, what is a reasonable conclusion about c, the total cost to drive her car on the vacation? **A**

   A $300 < c < 400$

   B $200 < c \leq 400$

   C $100 < c < 400$

   D $200 \leq c \leq 300$

8. The function $P(x) = -0.000047x^2 + 0.027x + 3$ can be used to approximate the population of Ling's home country between 1960 and 2000. The domain of the function x represents the number of years since 1960, and P is given in millions of people. Evaluate $P(20)$ to estimate the population of the country in 1980. **J**

   F about 2 million people

   G about 2.5 million people

   H about 3 million people

   J about 3.5 million people

9. Solve $4x - 5 = 2x + 5 - 3x$ for x. **D**

   A -2

   B -1

   C 1

   D 2

Short Response/Gridded Response

Record your answers on the answer sheet provided by your teacher or on a sheet of paper.

10. A stone path that is x feet wide is built around a rectangular flower garden. The garden is 12 feet wide and 25 feet long as shown below.

If the combined area of the garden and the stone path is 558 square feet, what is the width of the walkway? Express your answer in feet. **3 ft**

11. Factor $64a^4 + ab^3$ completely. Show your work. $a(4a + b)(16a^2 - 4ab + b^2)$

12. Simplify $\dfrac{3x^3 - 4x^2 - 28x - 16}{x + 2}$. Give your answer in factored form. Show your work. $(3x + 2)(x - 4)$

13. GRIDDED RESPONSE What is the value of a in the matrix equation below? **7.5**

$$\begin{bmatrix} 4 & 3 \\ 2 & 2 \end{bmatrix} \cdot \begin{bmatrix} a \\ b \end{bmatrix} = \begin{bmatrix} 21 \\ 9 \end{bmatrix}$$

14. GRIDDED RESPONSE Matt has a cubic aquarium as shown below. He plans to fill it by emptying cans of water with the dimensions as shown.

About how many cylindrical cans will it take to fill the aquarium? **45**

15. What are the slope and y-intercept of the equation of the line graphed at the right? $m = \dfrac{1}{2}$; $b = 3$

Extended Response

Record your answers on a sheet of paper. Show your work. **b, c. See margin.**

16. The volume of a rectangular prism is 864 cubic centimeters. The length is 1 centimeter less than the height, and the width is 3 centimeters more than the height.

$$h^3 + 2h^2 - 3h = 864$$

a. Write a polynomial equation that can be used to solve for the height of the prism h.

b. How many possible roots are there for h in the polynomial equation you wrote? Explain.

c. Solve the equation from part **a** for all real roots h. What are the dimensions of the prism?

17. Scott launches a model rocket from ground level. The rocket's height h in meters is given by the equation $h = -4.9t^2 + 56t$, where t is the time in seconds after the launch.

See margin.

a. What is the maximum height the rocket will reach? Round to the nearest tenth of a meter. Show each step and explain your method.

b. How long after it is launched will the rocket reach its maximum height? Round to the nearest tenth of a second. **5.7 s**

Need ExtraHelp?

If you missed Question...	1	2	3	4	5	6	7	8	9	10	11	12	13	14	15	16	17
Go to Lesson...	5-2	4-7	2-4	5-5	4-5	5-7	3-2	5-3	1-3	5-5	5-5	5-2	3-8	5-7	2-3	5-8	4-7

Additional Answers

16b. Sample answer: The order of the equation is 3, so there are up to 3 possible solutions to the equation.

16c. $h = 9$, the prism is 9 centimeters tall, 8 centimeters long, and 12 centimeters wide.

17a. 160 m; Sample answer: the vertex form of the equation is $h = -4.9\left(t - \dfrac{40}{7}\right)^2 + 160$, so the vertex is at $\left(\dfrac{40}{7}\right)$, 160.

Formative Assessment

You can use these pages to benchmark student progress.

📁 Standardized Test Practice, pp. 70—72

McGraw-Hill
eAssessment Create practice tests that align to your state standards, the Common Core State Standards, and other national standards such as TIMSS and NAEP.

Answer Sheet Practice

Have students simulate taking a standardized test by recording their answers on practice recording sheets.

📁 **Student Recording Sheet**

Homework Option

Get Ready for Chapter 6 Assign students the exercises on p. 383 as homework to assess whether they possess the prerequisite skills needed for the next chapter.

Lesson 5-3

35a. $f(x) \to +\infty$ as $x \to -\infty$. $f(x) \to +\infty$ as $x \to +\infty$.

35b. Since the end behavior is in the same direction, it is an even-degree function.

35c. The graph intersects the x-axis at four points, so there are four real zeros.

36a. $f(x) \to +\infty$ as $x \to -\infty$. $f(x) \to -\infty$ as $x \to +\infty$.

36b. Since the end behavior is in opposite directions, it is an odd-degree function.

36c. The graph intersects the x-axis at one point, so there is one real zero.

37a. $f(x) \to -\infty$ as $x \to -\infty$. $f(x) \to +\infty$ as $x \to +\infty$.

37b. Since the end behavior is in opposite directions, it is an odd-degree function.

37c. The graph intersects the x-axis at one point, so there is one real zero.

38a. $f(x) \to +\infty$ as $x \to -\infty$. $f(x) \to +\infty$ as $x \to +\infty$.

38b. Since the end behavior is in the same direction, it is an even-degree function.

38c. The graph intersects the x-axis at no points, so there are no real zeros.

39a. $f(x) \to -\infty$ as $x \to -\infty$. $f(x) \to -\infty$ as $x \to +\infty$.

39b. Since the end behavior is in the same direction, it is an even-degree function.

39c. The graph intersects the x-axis at two points, so there are two real zeros.

40a. $f(x) \to -\infty$ as $x \to -\infty$. $f(x) \to -\infty$ as $x \to +\infty$.

40b. Since the end behavior is in the same direction, it is an even-degree function.

40c. The graph intersects the x-axis at two points, so there are two real zeros.

56a. degree: 4; x-intercepts: 2, -1, 3, -4; g-intercept: 24; roots: 2, -1, 3, -4, end behavior: as $x \to -\infty$, $g(x) \to +\infty$; as $x \to +\infty$, $g(x) \to +\infty$

56c.

x	$g(x)$
-5	224
-4	0
-3	-60
-2	-40
-1	0
0	24
1	20
2	0
3	0
4	80
5	324

56d.

Lesson 5-4 (Guided Practice)

1.

x	$f(x)$
-2.5	≈ 26
-2.0	2.0
-1.5	≈ -8.1
-1.0	-10
-0.5	≈ -8.3

x	$f(x)$
0.0	-6.0
0.5	≈ -4.6
1.0	-4.0
1.5	≈ -2.8
2.0	2.0

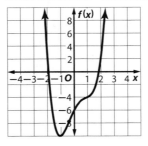

Lesson 5-4

5. between -2 and -1

6. at -1 and between 2 and 3

7. between 0 and 1 and between 2 and 3

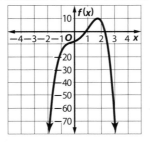

8. between -2 and -1 and between 1 and 2

9. rel. max at $x \approx -1.8$; rel. min at $x \approx 1.1$;
D = {all real numbers}, R = {all real numbers}

10. rel. max at $x \approx -0.1$; rel. min at $x \approx 1.5$;
D = {all real numbers}, R = {all reals numbers}

11. rel. max at $x \approx 2.4$; rel. min at $x \approx 0.3$; D = {all real numbers},
R = {all real numbers}

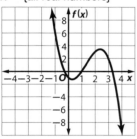

12. no relative max or min; D = {all real numbers},
R = {all real numbers}

13a.

Years Since 1995

13c.

Years Since 1995

Sample answer: This suggests a dramatic increase in sales.

13d. Sample answer: No; with so many other forms of media on the market today, CD sales will not increase dramatically. In fact, the sales will probably decrease. The function appears to be accurate only until about 2005.

14a.

x	f(x)
−4	−16
−3	0
−2	4
−1	2
0	0
1	4
2	20
3	54
4	112

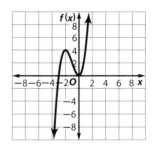

14b. at 0 and at −3

14c. rel. max: $x = -2$; rel. min: $x = 0$

15a.

x	f(x)
−4	92
−3	41
−2	12
−1	−1
0	−4
1	−3
2	−4
3	−13
4	−36

15b. between −2 and −1

15c. rel. min: $x = 0$; rel. max: $x = 1$

16a.

x	f(x)
−6	−42
−5	0
−4	20
−3	24
−2	18
−1	8
0	0
1	0
2	14
3	48

16b. −5, 0, and 1

16c. rel. max: $x = -3$; rel. min: between $x = 0$ and $x = 1$

17a.

x	f(x)
−4	−155
−3	−80
−2	−33
−1	−8
0	1
1	0
2	−5
3	−8
4	−3
5	16

17b. at $x = 1$, between −1 and 0, and between $x = 4$ and $x = 5$

17c. rel. max: $x \approx \frac{1}{3}$, rel. min: $x \approx 3$

18a.

x	f(x)
−1	22
0	0
1	2
2	16
3	30
4	32
5	10
6	−48
7	−154

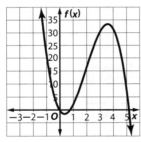

18b. at 0, between 0 and 1, and between 5 and 6

18c. rel. min: between $x = 0$ and $x = 1$; rel. max: near $x = 4$

19a.

x	f(x)
−4	−176
−3	−77
−2	−22
−1	1
0	4
1	−1
2	−2
3	13
4	56

19b. between $x = -2$ and $x = -1$, between $x = 0$ and $x = 1$, and between $x = 2$ and $x = 3$

19c. rel. max: near $x = -0.3$; rel. min: near $x = 1.6$

20a.

x	f(x)
−4	247
−3	74
−2	11
−1	−2
0	−1
1	2
2	19
3	86
4	263

20b. between $x = -2$ and $x = -1$ and between $x = 0$ and $x = 1$

20c. rel. min: near $x = -1$; no rel. max

21a.

x	f(x)
−4	372
−3	141
−2	36
−1	−3
0	−12
1	−3
2	36
3	141
4	372

21b. between $x = -2$ and $x = -1$ and between $x = 1$ and $x = 2$

21c. min: near $x = 0$

22a.

22d. Sample answer: While it is possible for gasoline prices to continue to soar at this rate, it is likely that alternate forms of transportation and fuel will slow down this rapid increase.

27.

28.

29.

30.

31.

32.

33a.

x	d(x)
0	0
1	0.0145
2	0.056
3	0.1215
4	0.208
5	0.3125
6	0.432
7	0.5635
8	0.704
9	0.8505
10	1

33b.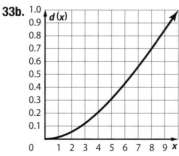

34a. −3.5(max), −2.5(min), −1(max), 2(min)

34b. −1.75, −0.25, 3.5

34c. 5

34d. D = {all real numbers}; R = {all real numbers}

35a. −2.5(min), −0.5(max), 1.5(min)

35b. −3.5, −1, 0, 3

35c. 4

35d. D = {all real numbers}; R = $\{y \mid y \geq -3.1\}$

36a. −2(max), 1(min), 2.5(max)

36b. −3.5, −0.5

36c. 4

36d. D = {all real numbers}; R = $\{y \mid y \leq 4.1\}$

37a. −3.5(min), −2.5(max), −2(min), −1(max), 1(min)

37b. −3.75, −3.25, −2, −1.75, −0.25, 2.9

37c. 6

37d. D = {all real numbers}; R = $\{y \mid y \geq -5\}$

38a. −3.5(max), −2.5(min), −1.75(max), −1(min), 1(max), 2(min)

38b. −4, −3, 0, 1.5, 2.75

38c. 7

38d. D = {all real numbers}; R = {all real numbers}

39a. −2(max), 1(min)

39b. −3, −0.5, 2

39c. 3

39d. D = {all real numbers}; R = {all real numbers}

46a. degree: 4; leading coefficient: 1; end behavior: as $x \to -\infty$, $f(x) \to +\infty$, as $x \to +\infty$, $f(x) \to +\infty$

46b.

x	f(x)
−4	1414
−3	690
−2	287
−1	92
0	18
1	0.7
2	−0.03
3	0.09
4	9.2

2 zeros

46c.

46d.

Sample answer: Sometimes it is necessary to have a more accurate viewing window or to change the interval values of the table function in order to assess the graph more accurately.

Extend 5-4

4. Sample graphs:

linear

[1995, 2075] scl: 10 by
[0, 2,000,000] sd: 200,000

quadratic

[1995, 2075] scl: 10 by
[0, 2,000,000] sd: 200,000

cubic

[1995, 2075] scl: 10 by
[0, 2,000,000] sd: 200,000

7. Sample graphs:

linear

[0, 13] scl: 1 by [0.8, 1.1] scl: 0.05

quadratic

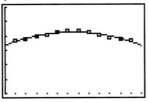

[0, 13] scl: 1 by [0.8, 1.1] scl: 0.05

cubic

[0, 13] scl: 1 by [0.8, 1.1] scl: 0.05

Extend 5-5

1. Sample answer:

2. Sample answer:

3. Sample answer:

4. Sample answer:

5. Sample answer:

6. Sample answer:

7. Sample answer:

8. Sample answer:

Lesson 5-6 (Guided Practice)

3.

$$
\begin{array}{r|rrrr}
2 & 1 & -7 & 4 & 12 \\
  &   & 2 & -10 & -12 \\
\hline
  & 1 & -5 & -6 & 0
\end{array}
$$

$(x - 2)(x - 6)(x + 1)$

Lesson 5-6

31a. $g(x) = -9x^4 + 50x^3 + 51x^2 - 150x - 72$

31b.

x	$g(x)$
-5	-9922
-4	-4160
-3	-1242
-2	-112
-1	70
0	-72
1	-130
2	88
3	558
4	1040
5	1078
6	0

31c. There is a zero between $x = -2$ and $x = -1$ because $g(x)$ changes sign between the two values. There are also zeros between $x = -1$ and 0 and between $x = 1$ and $x = 2$ because $g(x)$ changes sign between the two values. There is also a zero at $x = 6$.

31d.

36. $f(x) = 0.1(x - 1)^2(x + 5)^2$

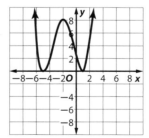

42. Sample answer: When $x = 1$, $f(1)$ is the sum of all of the coefficients and constants in $f(x)$, in this case, a, b, c, d, and e. The sum of a, b, c, d, and e is 0, so however the coefficients are arranged, $f(1)$ will always equal 0, and $f(x)$ will have a rational root.

46.

47.

48.

49.

53b.

[−10, 40] scl: 5 by [−4000, 13,200] scl: 100

56a.

56b.

56c.

60. Sample answer: To determine the number of positive real roots, determine how many time the signs change in the polynomial as you move from left to right. In this function there are 3 changes in sign. Therefore, there may be 3 or 1 positive real roots. To determine the number of negative real roots, I would first evaluate the polynomial for $-x$. All of the terms with an odd-degree variable would change signs. Then I would again count the number of sign changes as I move from left to right. There would be only one change. Therefore there may be 1 negative root.

1.

[−5, 5] scl: 1 by [−5, 5] scl: 1

$x = 0, 2, 3$; rel. max. at (0.78, 2.11), min. at (2.55, −0.63); 3 real zeros found; y-int. 0; no symmetry

2.

[−5, 5] scl: 1 by [−8, 2] scl: 1

$x = -2, 3$; rel. max. at (0, −4), min. at (−1.22, −6.25), (−1.22, −6.25); 2 real zeros found, two remaining are either multiple zeros or imaginary zeros; y-int. 0; line of symmetry $x = 0$

3.

[−5, 5] scl: 1 by [−5, 5] scl: 1

$x = -2, 0, 1$; max. at (−1.44, 2.8), rel. max. at (0.69, 0.40), min. at (0, 0); 3 real zeros found, one remaining is either multiple zero or imaginary zero; y-int. 0; no symmetry

4.

[−10, 10] scl: 1 by [−40, 20] scl: 4

$x = -4, 0, 3$; rel. max. at (1.10, 10.10), rel. min. at (−2.43, −33.8); 3 real zeros found; y-int. 0; no symmetry

5.

[−3, 5] scl: 1 by [−10, 30] scl: 4

$x = 0, 3$; rel. max. at (1.80, 25.19), rel. min. at (3, 0); 2 real zeros found, three remaining are either multiple zeros or imaginary zeros; y-int. 0; no symmetry

Chapter 5 Answer Appendix

6.

[−6, 6] scl: 1 by [−5, 20] scl: 2

$x = -2.23, -1.73, 1.73, 2.23$; rel. max. at $(0, 15)$, min. at $(-2, -1)$, $(2, -1)$; 4 real zeros found; y-int. 15; line of symmetry of $x = 0$

7.

[−4, 6] scl: 1 by [−5, 20] scl: 2

$x = -2, 0, 4$; rel. max. at $(2.43, 16.90)$, rel. min. at $(-1.10, -5.05)$; 3 real zeros found, one remaining is either multiple zero or imaginary zero; y-int. 0; no symmetry

8.

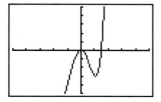

[−5, 5] scl: 1 by [−10, 10] scl: 2

$x = 0, 1.5$; rel. max. at $(0, 0)$, min. at $(1.07, -6.11)$; 2 real zeros found, three remaining are either multiple zeros or imaginary; y-int. 0; no symmetry

	Diagnostic Assessment Quick Check		
	LESSON 6-1 45 min: 1 day 90 min: 0.5 day	**LESSON 6-2** 45 min: 1 day 90 min: 0.5 day	**EXTEND 6-2** 45 min: 0.5 day 90 min: 0.5 day
Title	Operations on Functions	Inverse Functions and Relations	Graphing Technology Lab: Inverse Functions and Relations
Objectives	▪ Find the sum, difference, product, and quotient of functions. ▪ Find the composition of functions.	▪ Find the inverse of a function or relation. ▪ Determine whether two functions or relations are inverses.	▪ Compare a function and its inverse using a graphing calculator.
Key Vocabulary	composition of functions	inverse relation inverse function	
CCSS	F.IF.9, F.BF.1b	F.IF.4, F.BF.4a	
Multiple Representations	🔲	🔲	
Lesson Resources	connectED.mcgraw-hill.com 📁 Leveled Worksheets 🔤 Vocabulary 🏃 Animations PT Personal Tutor ✓ Self-Check Quiz ▪ *5-Minute Check* ▪ *Study Notebook* ▪ *Teaching Algebra with Manipulatives*	connectED.mcgraw-hill.com 📁 Leveled Worksheets 📁 Quiz 1 🔤 Vocabulary PT Personal Tutor ✓ Self-Check Quiz ▪ *5-Minute Check* ▪ *Study Notebook* ▪ *Teaching Algebra with Manipulatives*	connectED.mcgraw-hill.com PT Personal Tutor **Materials:** ▪ TI-83/84 Plus or other graphing calculator
Resources for Every Lesson	IWB eStudent Edition IWB Interactive Classroom	▪ eTeacher Edition ▪ eSolutions Manual ▪ eAssessment McGraw-Hill	
Differentiated Instruction	pp. 387, 388	pp. 395, 398	

IWB All digital assets are Interactive Whiteboard ready.

Suggested Pacing			
Time Periods	Instruction	Review & Assess	Total
45-minute	9 days	2 days	11 days
90-minute	5 days	1 day	6 days

LESSON 6-3 45 min: 1 day / 90 min: 0.5 day	**LESSON 6-4** 45 min: 1 day / 90 min: 0.5 day	**EXTEND 6-4** 45 min: 0.5 day / 90 min: 0.5 day	**LESSON 6-5** 45 min: 1 day / 90 min: 0.5 day
Square Root Functions and Inequalities	**_n_th Roots**	**Graphing Technology Lab: Graphing _n_th Root Functions**	**Operations with Radical Expressions**
■ Graph and analyze square root functions. ■ Graph square root inequalities.	■ Simplify radicals. ■ Use a calculator to approximate radicals.	■ Use a graphing calculator to graph _n_th root functions.	■ Simplify radical expressions. ■ Add, subtract, multiply, and divide radical expressions.
square root function radical function square root inequality	_n_th root radical sign index radicand principal root		rationalizing the denominator like radical expressions conjugate
F.IF.7b, F.BF.3	A.SSE.2	F.IF.7b, F.BF.3	A.SSE.2
connectED.mcgraw-hill.com	connectED.mcgraw-hill.com	connectED.mcgraw-hill.com	connectED.mcgraw-hill.com
📁 Leveled Worksheets 🔤 Vocabulary PT Personal Tutor ✓ Self-Check Quiz ■ _5-Minute Check_ ■ _Study Notebook_	📁 Leveled Worksheets 📁 Quiz 2 🔤 Vocabulary PT Personal Tutor ✓ Self-Check Quiz ■ _5-Minute Check_ ■ _Study Notebook_ ■ _Teaching Algebra with Manipulatives_	PT Personal Tutor **Materials:** ■ TI-83/84 Plus or other graphing calculator	📁 Leveled Worksheets 🔤 Vocabulary 🏃 Animations PT Personal Tutor ✓ Self-Check Quiz ■ _5-Minute Check_ ■ _Study Notebook_
IWB eStudent Edition IWB Interactive Classroom	■ eTeacher Edition ■ eSolutions Manual ■ eAssessment McGraw-Hill		
pp. 404, 406	pp. 409, 412		pp. 418, 421
		Formative Assessment Mid-Chapter Quiz	

	LESSON 6-6 — 45 min: 1 day / 90 min: 0.5 day	LESSON 6-7 — 45 min: 1 day / 90 min: 0.5 day	EXTEND 6-7 — 45 min: 1 day / 90 min: 0.5 day
Title	**Rational Exponents**	**Solving Radical Equations and Inequalities**	**Graphing Technology Lab: Solving Radical Equations and Inequalities**
Objectives	▪ Write expressions with rational exponents in radical form and vice versa. ▪ Simplify expressions in exponential or radical form.	▪ Solve equations containing radicals. ▪ Solve inequalities containing radicals.	▪ Use a graphing calculator to solve radical equations and inequalities.
Key Vocabulary		radical equation extraneous solution radical inequality	
CCSS		A.REI.2	A.REI.2, A.REI.11
Multiple Representations	⬚		
Lesson Resources	connectED.mcgraw-hill.com 📁 Leveled Worksheets 📁 Quiz 3 PT Personal Tutor ✓ Self-Check Quiz ▪ *5-Minute Check* ▪ *Study Notebook*	connectED.mcgraw-hill.com 📁 Leveled Worksheets 📁 Quiz 4 abc Vocabulary PT Personal Tutor ✓ Self-Check Quiz ▪ *5-Minute Check* ▪ *Study Notebook*	connectED.mcgraw-hill.com PT Personal Tutor **Materials:** ▪ TI-83/84 Plus or other graphing calculator
Resources for Every Lesson	IWB eStudent Edition IWB Interactive Classroom	▪ eTeacher Edition ▪ eSolutions Manual ▪ McGraw-Hill eAssessment	
Differentiated Instruction	pp. 423, 424	pp. 431, 435	
			Summative Assessment Study Guide and Review Practice Test

IWB All digital assets are Interactive Whiteboard ready.

What the Research Says...

Jitendra et al. (1999) found that each lesson should provide an adequate number of practice exercises on the new skill.

- After composition of functions is introduced for the first time in Lesson 6-1, a large number of exercises (both contextual and noncontextual) allow students to practice this new concept.

- Numerous practice exercises on properties and operations with radicals are provided in Lessons 6-4 through 6-7.

Source: Jitendra, A.K., Salmento, M.M., and Haydt, L.A. (1999). "Adherence to Important Instructional Design Criteria," *Learning Disabilities Research & Practice*, 14(2), pp. 69-79.

Teacher to Teacher

Melissa Williamson
Mathematics Department Head
Klein Collins High School
Spring, Texas

Use Before Lesson 6-5

When multiplying radical expressions, my students like the 'box method' best. For example, to multiply $(2\sqrt{3} + 5)(3\sqrt{3} - 2)$, they start with a 2×2 grid and write the terms along the top and left side. Then they fill in the grid, one row at a time. Simplify and combine like terms along the diagonal.

The product is $11\sqrt{3} + 8$. Students prefer this method because it is a good organizer.

STUDY SKILL

Writing out the steps in a mathematical process using a point-by-point format with an example can help students better understand that process. In order to explain a process, they must not only understand how to perform each step, but they must also understand the reasoning behind each step. Provide students with the description below as an example of using a point-by-point format to explain how to solve a radical equation.

Main Idea	Point-by-Point Steps	Example: $\sqrt{x-3} + 4 = 5$
Solving Radical Equations	1. Isolate the radical. 2. Raise each side of the equation to a power equal to the index of the radical. 3. Solve the resulting equation. 4. Verify solution and check for extraneous solutions.	1. $\sqrt{x-3} = 1$ 2. $x - 3 = 1$ 3. $x = 4$ 4. $\sqrt{4-3} + 4 = 5$ $\sqrt{1} + 4 = 5$ $1 + 4 = 5$ $5 = 5$

Creating Independence through Student-owned Strategies

SE = Student Edition, TE = Teacher Edition, CRM = Chapter Resource Masters

Diagnosis	Prescription
Beginning Chapter 6	
Get Ready for Chapter 6 **SE**	Response to Intervention **TE**
Beginning Every Lesson	
Then, Now, Why? **SE** 5-Minute Checks	Chapter 0 **SE**

FORMATIVE ASSESSMENT

During/After Every Lesson	
Guided Practice **SE**, every example Check Your Understanding **SE** H.O.T. Problems **SE** Spiral Review **SE** Additional Examples **TE** Watch Out! **TE** Step 4, Assess **TE** Chapter 6 Quizzes **CRM**, pp. 53–54 Self-Check Quizzes connectED.mcgraw-hill.com	TIER 1 Intervention Skills Practice **CRM**, Ch. 1–6 connectED.mcgraw-hill.com TIER 2 Intervention Differentiated Instruction **TE** Study Guide and Intervention **CRM**, Ch. 1–6 TIER 3 Intervention *Math Triumphs, Alg. 2,* Ch. 1
Mid-Chapter	
Mid-Chapter Quiz **SE** Mid-Chapter Test **CRM**, p. 55 ᵉAssessment **McGraw-Hill**	TIER 1 Intervention Skills Practice **CRM**, Ch. 1–6 connectED.mcgraw-hill.com TIER 2 Intervention Study Guide and Intervention **CRM**, Ch. 1–6 TIER 3 Intervention *Math Triumphs, Alg. 2,* Ch. 1
Before Chapter Test	
Chapter Study Guide and Review **SE** Practice Test **SE** Standardized Test Practice **SE** Chapter Test connectED.mcgraw-hill.com Standardized Test Practice connectED.mcgraw-hill.com Vocabulary Review connectED.mcgraw-hill.com ᵉAssessment **McGraw-Hill**	TIER 1 Intervention Skills Practice **CRM**, Ch. 1–6 connectED.mcgraw-hill.com TIER 2 Intervention Study Guide and Intervention **CRM**, Ch. 1–6 TIER 3 Intervention *Math Triumphs, Alg. 2,* Ch. 1

SUMMATIVE ASSESSMENT

After Chapter 6	
Multiple-Choice Tests, Forms 1, 2A, 2B **CRM**, pp. 57–62 Free-Response Tests, Forms 2C, 2D, 3 **CRM**, pp. 63–68 Vocabulary Test **CRM**, p. 56 Extended Response Test **CRM**, p. 69 Standardized Test Practice **CRM**, pp. 70–72 ᵉAssessment **McGraw-Hill**	Study Guide and Intervention **CRM**, Ch. 1–6 connectED.mcgraw-hill.com

Option 1 Reaching All Learners AL OL BL ELL

Auditory/Musical Divide the class into groups of three or four students. Challenge each group to give themselves a group name based on the vocabulary in a lesson, such as "The Radical Inequalities." Have students write a musical verse about one of the Key Concepts in the chapter. Have students share their verses with the entire class.

Visual/Spatial Have students use grid paper to draw rectangles of various sizes, and then cut them out. Then have them use the formula $d = \sqrt{\ell^2 + w^2}$ to find the length of the diagonal of each rectangle.

Option 2 Approaching Level AL

Have students read mathematical expressions and equations using correct mathematical language. For example, an exponent should be read in a way that distinguishes it from a coefficient or multiplier. Have students practice reading mathematical expressions and equations aloud. For example, they should read x^3 aloud as "x to the third power" or "x cubed."

Option 3 English Learners ELL

Whenever possible, write vocabulary words on the board, using diagrams that indicate the meaning of the words. For example, write the following diagram for nth roots on the board.

Ask students to explain each term in their own words.

Option 4 Beyond Level BL

After discussing how to determine if $f(x) = 5x + 10$ and $g(x) = \frac{1}{5}x - 2$ are inverses, challenge students to find two functions f and g such that $f(g(x)) \neq g(f(x))$, with one of the two compositions having a value of x. Allow students to work in groups to brainstorm as they attempt this puzzler.

VerticalAlignment

Before Chapter 6

Related Topics from Algebra 1

- simplify polynomial expressions
- transform and solve equations
- graph equations of lines
- use the Distributive Property to simplify algebraic expressions

Previous Topics from Algebra 2

- use the properties of exponents to simplify expressions

Chapter 6

Related Topics from Algebra 2

- relate representations of square root functions
- connect inverses of square root functions with quadratic functions
- determine solutions of square root equations and inequalities using graphs, tables, and algebraic methods
- determine the reasonable domain and range values of square root functions, and interpret and determine the reasonableness of solutions to square root equations and inequalities
- use the parent function to investigate, describe, and predict the effects of parameter changes on graphs of square root functions and describe limitations on the domains and ranges

After Chapter 6

Preparation for Precalculus

- define functions, describe characteristics of functions, and translate among verbal, numerical, graphical, and symbolic representations of functions, including power functions
- perform operations including composition of functions, finding inverses, and describing these procedures and results verbally, numerically, symbolically, and graphically

Lesson-by-LessonPreview

6-1 Operations on Functions

Any two functions, $f(x)$ and $g(x)$, can be added, subtracted, multiplied, and divided. If f and g are functions,

- their *sum* is the function defined by
  $(f + g)(x) = f(x) + g(x)$,
- their *difference* is the function defined by
  $(f - g)(x) = f(x) - g(x)$,
- their *product* is the function defined by
  $(f \cdot g)(x) = f(x) \cdot g(x)$, and
- their *quotient* is the function defined by
  $\left(\dfrac{f}{g}\right)(x) = \dfrac{f(x)}{g(x)}$, such that $g(x) \neq 0$.

The *composition of functions* provides another way of combining functions. Given two functions f and g, the composition, $f \circ g$, is described by $[f \circ g](x) = f[g(x)]$.

In most cases, $f \circ g \neq g \circ f$. Therefore, the order in which you compose functions is very important. It is also important to identify relationships between and restrictions on ranges and domains when composing functions.

6-2 Inverse Functions and Relations

The inverse of a function can be found by exchanging the domain and the range of the function. To find the inverse of $f(x)$:

- Rewrite $f(x)$ as an equation using x and y.
- Exchange x and y in the equation.
- Solve for y and then replace y with $f^{-1}(x)$.

To test whether two functions $f(x)$ and $g(x)$ are inverses, check that each of the two compositions $[f \circ g](x)$ and $[g \circ f](x)$ has the value x. When the inverse of a function is a function, then the original function is said to be *one-to-one*.

6-3 Square Root Functions and Inequalities

A function that contains a variable inside a square root symbol is called a *square root function.* The domains and ranges of these functions are limited to values for which the function is defined.

As with other functions, square root functions can be transformed.

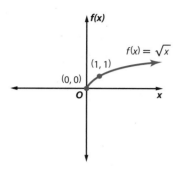

$f(x) = \sqrt{x}$

(1, 1)

(0, 0)

The parent function of the square root function is $f(x) = \sqrt{x}$, where the domain is $\{x \mid x \geq 0\}$ and the range is $\{f(x) \mid f(x) \geq 0\}$.

A *square root inequality* is an inequality involving square roots. To graph a square root inequality, first graph the related square root equation, forming two regions. Then select a point in one of the regions. If the point is a solution of the inequality, shade that region. If not, shade the other region.

6-4 *n*th Roots

A common definition of a square root, that a is a square root of b if $a^2 = b$, can be used to help students understand that a is an nth root of b if $a^n = b$. The symbol $\sqrt[n]{}$ indicates an nth root. In the symbol $\sqrt[n]{x}$, n is called the *index*, x is the *radicand*, and the symbol $\sqrt{}$ is called the *radical sign*.

The *principal root* of a number is always a nonnegative number. The value of a principal nth root depends on whether the *radicand* is positive or negative and whether its index is even or odd. When you find the nth root of an even power and the result is an odd power, you must use absolute value symbols around the result to ensure that the answer is nonnegative.

6-5 Operations with Radical Expressions

Radical expressions can be simplified using the *Product Property of Radicals* and the *Quotient Property of Radicals* as defined in this lesson. These properties are also used when multiplying or dividing radical expressions. In order to add or subtract radical expressions, the two radical expressions must be *like radical expressions*; that is, both the index and radicand are identical.

A process called *rationalizing the denominator* is used to eliminate radicals from a denominator or fractions from a radicand. To rationalize a denominator, multiply the numerator and denominator by a quantity so that the radicand has an exact root. A *conjugate* can be used to rationalize a denominator of the form $a\sqrt{b} \pm c\sqrt{d}$.

6-6 Rational Exponents

Exponents can be fractions. The rules for writing equivalent expressions for rational exponents include properties that describe how to translate between radical form and exponential form.

- For any real number b and for any positive integer n,
  $$b^{\frac{1}{n}} = \sqrt[n]{b}$$ except when $b < 0$ and n is even.

- For any non-zero real number b, and for any integers m and n, with $n > 1$, $b^{\frac{m}{n}} = \sqrt[n]{b^m} = \left(\sqrt[n]{b}\right)^m$, except when $b < 0$ and n is even.

An expression with rational exponents is in simplest form when all the exponents are positive and no fractional exponents are in the denominator. Also, the index used should be as small as possible.

6-7 Solving Radical Equations and Inequalities

Radical equations and radical inequalities are equations and inequalities that contain variables in the radicands. When solving a radical equation,

- first isolate the radical.
- Then raise each side of the equation to a power equal to the index of the radical. That is, to undo an nth root, you must raise each side of the equation to the nth power. This will eliminate the radical.
- Finally, solve the resulting equation.

Sometimes the process of raising both sides of a radical equation to a power does not produce an equivalent statement. For example, it is clear that $\sqrt{x} = -5$ has no real number solution. Squaring both sides results in $x = 25$. But $x = 25$ is not a solution to the original equation. Therefore, it is important to check solutions for *extraneous solutions*. When checking a solution to an inequality, test three numbers, one found within the solution interval, one less than, and one greater than the interval.

Professional Development

Targeted professional development has been articulated throughout Algebra 2. More quality, customized professional development is available from McGraw-Hill Professional Development. Visit connectED.mcgraw-hill.com for details on each product.

- Online Lessons emphasize the strategies and techniques used to teach Algebra 2. Includes streaming video, interactive pages, and online tools.
- Video Workshops allow mentors, coaches, or leadership personnel to facilitate on-site workshops on educational strategies in mathematics and mathematical concepts.
- MHPD Online (www.mhpdonline.com) offers online professional development with video clips of instructional strategies, links, student activities, and news and issues in education.
- Teaching Today (teachingtoday.glencoe.com) gives secondary teachers practical strategies and materials that inspire excellence and innovation in teaching.

FOLDABLES StudyOrganizer

Dinah Zike's Foldables®

Focus Students write notes as they study inverses and radical functions and relations in the lessons of this chapter.

Teach Have students make and label their Foldables as illustrated. Have students use the appropriate tab as they cover each lesson in this chapter. Encourage them to apply what they have learned by writing examples as well.

When to Use It Encourage students to add to their Foldables as they work through the chapter and to use them to review for the chapter test.

Differentiated Instruction

Student-Built Glossary, p. 1

Students should complete the chart by providing a definition of each term and an example as they progress through Chapter 6. This study tool can also be used to review for the chapter test.

Get Started on the Chapter

You will learn several new concepts, skills, and vocabulary terms as you study Chapter 6. To get ready, identify important terms and organize your resources. You may wish to refer to Chapter 0 to review prerequisite skills.

FOLDABLES StudyOrganizer

Inverses and Radical Functions Make this Foldable to help you organize your Chapter 6 notes about radical equations and inequalities. Begin with four sheets of notebook paper.

1 **Stack** four sheets of notebook paper so that each sheet is one inch higher than the sheet in front of it.

2 **Bring** the bottom of all the sheets upward and align the edges so that all of the layers or tabs are the same distance apart.

3 **When** all the tabs are an equal distance apart, fold the papers and crease well. Open the papers and staple them together along the valley or inner center fold. Label the pages with lesson titles.

NewVocabulary

English		Español
composition of functions	p. 387	composición de funciones
inverse relation	p. 393	relaciones inversas
inverse function	p. 393	función inversa
square root function	p. 400	función raíz cuadrada
radical function	p. 400	función radical
square root inequality	p. 402	desigualdad raíz cuadrada
nth root	p. 407	raíz enésima
radical sign	p. 407	signo radical
index	p. 407	índice
radicand	p. 407	radicando
principal root	p. 407	raíz principal
rationalizing the denominator	p. 416	racionalizar el denominador
conjugates	p. 418	conjugados
radical equation	p. 429	ecuación radical
extraneous solution	p. 429	solución extraña
radical inequality	p. 431	desigualdad radical

ReviewVocabulary

absolute value valor absoluto a number's distance from zero on the number line, represented by $|x|$

rational number número racional Any number $\frac{m}{n}$, where m and n are integers and n is not zero; the decimal form is either a terminating or repeating decimal.

relation relación a set of ordered pairs

LESSON 6-1

Operations on Functions

::Then	::Now	::Why?
• You performed operations on polynomials.	**1** Find the sum, difference, product, and quotient of functions. **2** Find the composition of functions.	• The graphs model the income for the Brooks family since 2000, where $m(x)$ represents Mr. Brooks' income and $f(x)$ represents Mrs. Brooks' income. The total household income for the Brooks household can be represented by $f(x) + m(x)$.

Brooks Household Income

NewVocabulary
composition of functions

Common Core State Standards

Content Standards
F.IF.9 Compare properties of two functions each represented in a different way (algebraically, graphically, numerically in tables, or by verbal descriptions).

F.BF.1.b Combine standard function types using arithmetic operations.

Mathematical Practices
2 Reason abstractly and quantitatively.

1 Arithmetic Operations In Chapter 6, you performed arithmetic operations with polynomials. You can also use addition, subtraction, multiplication, and division with functions.

You can perform arithmetic operations according to the following rules.

KeyConcept Operations on Functions

Operation	Definition	Example Let $f(x) = 2x$ and $g(x) = -x + 5$.
Addition	$(f + g)(x) = f(x) + g(x)$	$2x + (-x + 5) = x + 5$
Subtraction	$(f - g)(x) = f(x) - g(x)$	$2x - (-x + 5) = 3x - 5$
Multiplication	$(f \cdot g)(x) = f(x) \cdot g(x)$	$2x(-x + 5) = -2x^2 + 10x$
Division	$\left(\dfrac{f}{g}\right)(x) = \dfrac{f(x)}{g(x)}, g(x) \neq 0$	$\dfrac{2x}{-x + 5}, x \neq 5$

Example 1 Add and Subtract Functions

Given $f(x) = x^2 - 4$ and $g(x) = 2x + 1$, find each function. Indicate any restrictions in the domain or range.

a. $(f + g)(x)$

$(f + g)(x) = f(x) + g(x)$ Addition of functions

$= (x^2 - 4) + (2x + 1)$ $f(x) = x^2 - 4$ and $g(x) = 2x + 1$

$= x^2 + 2x - 3$ Simplify.

b. $(f - g)(x)$

$(f - g)(x) = f(x) - g(x)$ Subtraction of functions

$= (x^2 - 4) - (2x + 1)$ $f(x) = x^2 - 4$ and $g(x) = 2x + 1$

$= x^2 - 2x - 5$ Simplify.

GuidedPractice

Given $f(x) = x^2 + 5x - 2$ and $g(x) = 3x - 2$, find each function.

1A. $(f + g)(x)$ $(f + g)(x) = x^2 + 8x - 4$ **1B.** $(f - g)(x)$ $(f - g)(x) = x^2 + 2x$

connectED.mcgraw-hill.com   **385**

1 Focus

VerticalAlignment

Before Lesson 6-1 Perform operations on polynomials.

Lesson 6-1 Find the sum, difference, product, and quotient of functions. Find the composition of functions.

After Lesson 6-1 Describe and analyze the relationship between a function and its inverse.

2 Teach

Scaffolding Questions
Have students read the **Why?** section of the lesson.

Ask:

- What years are represented by $x = 0$, $x = 4$, and $x = 10$? 2000, 2004, 2010

- What are the approximate values of $f(0)$, $m(0)$, and $f(0) + m(0)$? $f(0) \approx 3,700$, $m(0) \approx 2,900$, $f(0) + m(0) \approx 6,600$

- Which graph, $m(x)$, $f(x)$, or $f(x) + m(x)$, has the greatest slope? Explain your answer. $f(x) + m(x)$; all three slopes are positive, and the slope of $f(x) + m(x)$ is the sum of the slopes of $f(x)$ and $m(x)$.

Lesson 6-1 Resources

Resource	Approaching Level **AL**	On Level **OL**	Beyond Level **BL**	English Learners **ELL**
Teacher Edition	• Differentiated Instruction, p. 388	• Differentiated Instruction, pp. 387, 388	• Differentiated Instruction, p. 387	• Differentiated Instruction, p. 388
Chapter Resource Masters	• Study Guide and Intervention, pp. 5–6 • Skills Practice, p. 7 • Practice, p. 8 • Word Problem Practice, p. 9	• Study Guide and Intervention, pp. 5–6 • Skills Practice, p. 7 • Practice, p. 8 • Word Problem Practice, p. 9 • Enrichment, p. 10 • Spreadsheet Activity, p. 11	• Practice, p. 8 • Word Problem Practice, p. 9 • Enrichment, p. 10	• Study Guide and Intervention, pp. 5–6 • Skills Practice, p. 7 • Practice, p. 8 • Word Problem Practice, p. 9
Other	• 5-Minute Check 6-1 • Study Notebook • Teaching Algebra with Manipulatives	• 5-Minute Check 6-1 • Study Notebook • Teaching Algebra with Manipulatives	• 5-Minute Check 6-1 • Study Notebook	• 5-Minute Check 6-1 • Study Notebook • Teaching Algebra with Manipulatives

1 Arithmetic Operations

Example 1 shows how to add and subtract functions. **Example 2** shows how to multiply and divide functions.

Formative Assessment

Use the Guided Practice exercises after each example to determine students' understanding of concepts.

Additional Examples

1 Given $f(x) = 3x^2 + 7x$ and $g(x) = 2x^2 - x - 1$, find each function. Indicate any restrictions in the domain or range.

a. $(f + g)(x)$
$(f + g)(x) = 5x^2 + 6x - 1$

b. $(f - g)(x)$
$(f - g)(x) = x^2 + 8x + 1$

2 Given $f(x) = 3x^2 - 2x + 1$ and $g(x) = x - 4$, find each function. Indicate any restrictions in the domain or range.

a. $(f \cdot g)(x)$
$(f \cdot g)(x) = 3x^3 - 14x^2 + 9x - 4$

b. $\left(\dfrac{f}{g}\right)(x)$

$\left(\dfrac{f}{g}\right)(x) = \dfrac{3x^2 - 2x + 1}{x - 4}, x \neq 4$

▶ **Additional Examples** also in Interactive Classroom PowerPoint® Presentations

IWB **Interactive White Board**
READY

Focus on Mathematical Content

Combining Functions Addition, subtraction, multiplication, and division can be used to combine functions. Composition of functions is a method of combining functions, and is fundamentally different from arithmetic operations on functions. In the composition $g \circ f$, the output $f(x)$ is used as input for g.

ReviewVocabulary
intersection the intersection of two sets is the set of elements common to both

You can graph sum and difference functions by graphing each function involved separately, then adding their corresponding functional values. Let $f(x) = x^2$ and $g(x) = x$. Examine the graphs of $f(x)$, $g(x)$, and their sum and difference.

Find $(f + g)(x)$.

x	$f(x) = x^2$	$g(x) = x$	$(f + g)(x) = x^2 + x$
−3	9	−3	$9 + (−3) = 6$
−2	4	−2	$4 + (−2) = 2$
−1	1	−1	$1 + (−1) = 0$
0	0	0	$0 + 0 = 0$
1	1	1	$1 + 1 = 2$
2	4	2	$4 + 2 = 6$
3	9	3	$9 + 3 = 12$

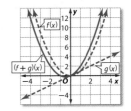

Find $(f - g)(x)$.

x	$f(x) = x^2$	$g(x) = x$	$(f - g)(x) = x^2 - x$
−3	9	−3	$9 - (−3) = 12$
−2	4	−2	$4 - (−2) = 6$
−1	1	−1	$1 - (−1) = 2$
0	0	0	$0 - 0 = 0$
1	1	1	$1 - 1 = 0$
2	4	2	$4 - 2 = 2$
3	9	3	$9 - 3 = 6$

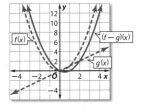

In Example 1, the functions $f(x)$ and $g(x)$ have the same domain of all real numbers. The functions $(f + g)(x)$ and $(f - g)(x)$ also have domains that include all real numbers. For each new function, the domain consists of the intersection of the domains of $f(x)$ and $g(x)$. Under division, the domain of the new function is restricted by excluded values that cause the denominator to equal zero.

PT

Example 2 **Multiply and Divide Functions**

Given $f(x) = x^2 + 7x + 12$ and $g(x) = 3x - 4$, find each function. Indicate any restrictions in the domain or range.

a. $(f \cdot g)(x)$

$(f \cdot g)(x) = f(x) \cdot g(x)$ Multiplication of functions

$= (x^2 + 7x + 12)(3x - 4)$ Substitution

$= 3x^3 + 21x^2 + 36x - 4x^2 - 28x - 48$ Distributive Property

$= 3x^3 + 17x^2 + 8x - 48$ Simplify.

b. $\left(\dfrac{f}{g}\right)(x)$

$\left(\dfrac{f}{g}\right)(x) = \dfrac{f(x)}{g(x)}$ Division of functions

$= \dfrac{x^2 + 7x + 12}{3x - 4}, x \neq \dfrac{4}{3}$ Substitution

Because $x = \dfrac{4}{3}$ makes the denominator $3x - 4 = 0$, $\dfrac{4}{3}$ is excluded from the domain of $\left(\dfrac{f}{g}\right)(x)$.

▶ **Guided Practice** Given $f(x) = x^2 - 7x + 2$ and $g(x) = x + 4$, find each function.

2A. $(f \cdot g)(x)$ $(f \cdot g)(x) = x^3 - 3x^2 - 26x + 8$ **2B.** $\left(\dfrac{f}{g}\right)(x)$ $\left(\dfrac{f}{g}\right)(x) = \dfrac{x^2 - 7x + 2}{x + 4}, x \neq -4$

 386 | Lesson 6-1 | Operations on Functions

WatchOut!

Student Misconceptions Some students may read $f \circ g$ as the word *fog*. Listen for students making this verbal error. Stress that the correct wording may help them understand the meaning. Lead students to understand the similarity in meaning between $f(x)$ (read "f of x") and $[f \circ g](x)$ (read "f of g of x"). Show how $f \circ g$ can also be written as $f[g(x)]$. You can also relate $f[g(x)]$ to an expression containing nested parentheses, such as $(1 + (3 \cdot 5(4)))$, in which the expressions in parentheses are evaluated from the innermost parentheses to the outermost.

---

2 Composition of Functions

2 Composition of Functions Another method used to combine functions is a composition of functions. In a **composition of functions**, the results of one function are used to evaluate a second function.

ReadingMath

Composition of Functions The composition of f and g, denoted by $f \circ g$ or $f[g(x)]$, is read f of g.

KeyConcept Composition of Functions

Words Suppose f and g are functions such that the range of g is a subset of the domain of f. Then the composition function $f \circ g$ can be described by

$$[f \circ g](x) = f[g(x)].$$

Model

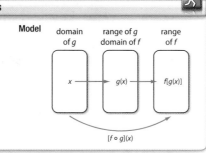

domain of g range of g domain of f range of f

$x \rightarrow g(x) \rightarrow f[g(x)]$

$[f \circ g](x)$

The composition of two functions may not exist. Given two functions f and g, $[f \circ g](x)$ is defined only if the range of $g(x)$ is a subset of the domain of f. Likewise, $[g \circ f](x)$ is defined only if the range of $f(x)$ is a subset of the domain of g.

StudyTip

Composition Be careful not to confuse a composition $f[g(x)]$ with multiplication of functions $(f \cdot g)(x)$.

Example 3 Compose Functions

For each pair of functions, find $[f \circ g](x)$ and $[g \circ f](x)$, if they exist. State the domain and range for each composed function.

a. $f = \{(1, 8), (0, 13), (15, 11), (14, 9)\}$, $g = \{(8, 15), (5, 1), (10, 14), (9, 0)\}$

To find $f \circ g$, evaluate $g(x)$ first. Then use the range to evaluate $f(x)$.

$f[g(8)] = f(15)$ or 11 $g(8) = 15$	$f[g(10)] = f(14)$ or 9 $g(10) = 14$
$f[g(5)] = f(1)$ or 8 $g(5) = 1$	$f[g(9)] = f(0)$ or 13 $g(9) = 0$

$f \circ g = \{(8, 11), (5, 8), (10, 9), (9, 13)\}$, D = $\{5, 8, 9, 10\}$, R = $\{8, 9, 11, 13\}$

To find $g \circ f$, evaluate $f(x)$ first. Then use the range to evaluate $g(x)$.

$g[f(1)] = g(8)$ or 15 $f(1) = 8$	$g[f(15)] = g(11)$ $g(11)$ is undefined.
$g[f(0)] = g(13)$ $g(13)$ is undefined.	$g[f(14)] = g(9)$ or 0 $f(14) = 0$

Because 11 and 13 are not in the domain of g, $g \circ f$ is undefined for $x = 11$ and $x = 13$. However, $g[f(1)] = 15$ and $g[f(14)] = 0$, so $g \circ f = \{(1, 15), (14, 0)\}$, D = $\{5, 8, 9, 10\}$, and R = $\{8, 9, 11, 13\}$.

b. $f(x) = 2a - 5$, $g(x) = 4a$

$[f \circ g](x) = f[g(x)]$	Composition of functions	$[g \circ f](x) = g[f(x)]$
$= f(4a)$	Substitute.	$= g(2a - 5)$
$= 2(4a) - 5$	Substitute again.	$= 4(2a - 5)$
$= 8a - 5$	Simplify.	$= 8a - 20$

For $[f \circ g](x)$, D = {all real numbers} and R = {all real numbers}, and for $[g \circ f](x)$, D = {all real numbers} and R = {all real numbers}.

> **GuidedPractice**
>
> **3A.** $f(x) = \{(3, -2), (-1, -5), (4, 7), (10, 8)\}$, $g(x) = \{(4, 3), (2, -1), (9, 4), (3, 10)\}$
>
> **3B.** $f(x) = x^2 + 2$ and $g(x) = x - 6$

3A. $f \circ g = \{(4, -2), (2, -5), (9, 7), (3, 8)\}$; D = $\{2, 3, 4, 9\}$; R = $\{-5, -2, 7, 8\}$; $g \circ f$ is undefined, D = $\varnothing$, R = $\varnothing$.

3B. $[f \circ g](x) = x^2 - 12x + 38$; D = {all real numbers}, R = $\{y \mid y \geq 2\}$; $[g \circ f](x) = x^2 - 4$; D = {all real numbers}, R = $\{y \mid y \geq -4\}$

---

2 Composition of Functions

Example 3 shows how to evaluate composition of functions. **Example 4** shows how to use composition of functions to solve a real-world problem.

Additional Example

3 For each pair of functions, find $[f \circ g](x)$ and $[g \circ f](x)$, if they exist. State the domain and range for each combined function.

a. $f = \{(2, 6), (9, 4), (7, 7), (0, -1)\}$ and $g = \{(7, 0), (-1, 7), (4, 9), (8, 2)\}$

$[f \circ g](x) = \{(7, -1), (-1, 7), (4, 4), (8, 6)\}$; D = $\{-1, 4, 7, 8\}$, R = $\{-1, 4, 6, 7\}$

$[g \circ f](x) = \{(9, 9), (7, 0), (0, 7)\}$; D = $\{0, 7, 9\}$, R = $\{0, 7, 9\}$

b. $f(x) = 3x^2 - x + 4$ and $g(x) = 2x - 1$

$[f \circ g](x) = 12x^2 - 14x + 8$; D = {all real numbers}, R = $\{y \mid y > 3.91\}$

$[g \circ f](x) = 6x^2 - 2x + 7$; D = {all real numbers}, R = $\{y \mid y > 6.33\}$

Teach with Tech

Interactive Whiteboard

Write two functions $f(x)$ and $g(x)$ on the board. To show students how to write $f(g(x))$, drag the expression for $g(x)$ and replace each x in $f(x)$. Repeat for $g(f(x))$.

DifferentiatedInstruction   OL   BL

Extension In some courses, a function $f(x)$ is modeled by a "machine" that accepts values for x as inputs and then outputs values for $f(x)$. Using this model, ask students to explain how the composition of two functions could be modeled by two such "machines" linked together. Also ask them to use the model to explain how the composition of two functions could be undefined for some initial input values.

4 TAXES Hector has $100 deducted from every paycheck for retirement. He can have this deduction taken before state taxes are applied, which reduces his taxable income. His state income tax rate is 4%. If Hector earns $1500 every pay period, find the difference in his net income if he has the retirement deduction taken before or after state taxes. *His net pay is $4 more by having his retirement deduction taken before state taxes.*

Additional Answers

1–15, 17–20. Domains and ranges are all real numbers unless otherwise specified.

1. $(f + g)(x) = 4x + 1$; $(f - g)(x) = -2x + 3$; $(f \cdot g)(x) = 3x^2 + 5x - 2$; $\left(\dfrac{f}{g}\right)(x) = \dfrac{x + 2}{3x - 1}, x \neq \dfrac{1}{3}$

2. $(f + g)(x) = x^2 - x + 3$; $(f - g)(x) = x^2 + x - 13$; $(f \cdot g)(x) = -x^3 + 8x^2 + 5x - 40$; $\left(\dfrac{f}{g}\right)(x) = \dfrac{x^2 - 5}{-x + 8}, x \neq 8$

3. $f \circ g$ is undefined, $D = \varnothing$, $R = \varnothing$; $g \circ f = \{(2, 8), (6, 13), (12, 11), (7, 15)\}$, $D = \{2, 6, 7, 12\}$, $R = \{8, 11, 13, 15\}$.

4. $f \circ g$ is undefined, $D = \varnothing$, $R = \varnothing$; $g \circ f = \{(0, 2)\}$, $D = \{0\}$, $R = \{2\}$.

5. $[f \circ g](x) = -15x + 18$, $R = \{$all multiples of 3$\}$ $[g \circ f](x) = -15x - 6$

6. $[f \circ g](x) = x^2 + 3x - 6$; $[g \circ f](x) = x^2 + 11x + 18$

7. Either way, she will have $228.95 taken from her paycheck. If she takes the college savings plan deduction before taxes, $76 will go to her college plan and $152.95 will go to taxes. If she takes the college savings plan deduction after taxes, only $62.70 will go to her college plan and $166.25 will go to taxes.

Real-WorldLink
Adjusted for inflation, the average price of a new car declined from $23,014 in 1995 to $22,013 in 2005.
Source: U.S. Department of Commerce

Notice that in most cases, $f \circ g \neq g \circ f$. Therefore, the order in which two functions are composed is important.

Real-World Example 4 Use Composition of Functions

SHOPPING A new car dealer is discounting all new cars by 12%. At the same time, the manufacturer is offering a $1500 rebate on all new cars. Mr. Navarro is buying a car that is priced $24,500. Will the final price be lower if the discount is applied before the rebate or if the rebate is applied before the discount?

Understand Let x represent the original price of a new car, $d(x)$ represent the price of a car after the discount, and $r(x)$ the price of the car after the rebate.

Plan Write equations for $d(x)$ and $r(x)$.

The original price is discounted by 12%. $\qquad d(x) = x - 0.12x$

There is a $1500 rebate on all new cars. $\qquad r(x) = x - 1500$

Solve If the discount is applied *before* the rebate, then the final price of Mr. Navarro's new car is represented by $[r \circ d](24,500)$.

$[r \circ d](x) = r[d(x)]$

$[r \circ d](24,500) = r[24,500 - 0.12(24,500)]$

$\qquad\qquad\qquad = r(24,500 - 2940)$

$\qquad\qquad\qquad = r(21,560)$

$\qquad\qquad\qquad = 21,560 - 1500$

$\qquad\qquad\qquad = 20,060$

If the rebate is given *before* the discount is applied, then the final price of Mr. Navarro's car is represented by $[d \circ r](24,500)$.

$[d \circ r](x) = d[r(x)]$

$[d \circ r](24,500) = d(24,500 - 1500)$

$\qquad\qquad\qquad = d(23,000)$

$\qquad\qquad\qquad = 23,000 - 0.12(23,000)$

$\qquad\qquad\qquad = 23,000 - 2760$

$\qquad\qquad\qquad = 20,240$

$[r \circ d](24,500) = 20,060$ and $[d \circ r](24,500) = 20,240$. So, the final price of the car is less when the discount is applied before the rebate.

Check The answer seems reasonable because the 12% discount is being applied to a greater amount. Thus, the dollar amount of the discount is greater.

GuidedPractice

4. SHOPPING Sounds-to-Go offers both an in-store $35 rebate and a 15% discount on a digital audio player that normally sells for $300. Which provides the better price: taking the discount before the rebate or taking the discount after the rebate? *It is better, by a difference of $5.25, to take the discount before the rebate.*

 388 | Lesson 6-1 | Operations on Functions

DifferentiatedInstruction ⒶⓁ ⓄⓁ ⒺⓁⓁ

Intrapersonal Learners Have students write expressions that involve compositions of functions and then annotate the expressions with notes to themselves such as, "$[f \circ g](x)$ is pronounced "f of g of x" and, "Start by finding the value $g(x)$; then use that value as the input for f."

Randy Faris/Corbis

Examples 1–2 Find $(f + g)(x)$, $(f - g)(x)$, $(f \cdot g)(x)$, and $\left(\dfrac{f}{g}\right)(x)$ for each $f(x)$ and $g(x)$. Indicate any restrictions in domain or range. **1, 2. See margin.**

1. $f(x) = x + 2$
$g(x) = 3x - 1$

2. $f(x) = x^2 - 5$
$g(x) = -x + 8$

Example 3 For each pair of functions, find $f \circ g$ and $g \circ f$, if they exist. State the domain and range for each composed function. **3, 4. See margin.**

3. $f = \{(2, 5), (6, 10), (12, 9), (7, 6)\}$
$g = \{(9, 11), (6, 15), (10, 13), (5, 8)\}$

4. $f = \{(-5, 4), (14, 8), (12, 1), (0, -3)\}$
$g = \{(-2, -4), (-3, 2), (-1, 4), (5, -6)\}$

Find $[f \circ g](x)$ and $[g \circ f](x)$, if they exist. State the domain and range for each composed function. **5, 6. See margin.**

5. $f(x) = -3x$
$g(x) = 5x - 6$

6. $f(x) = x + 4$
$g(x) = x^2 + 3x - 10$

Example 4 **7. CCSS MODELING** Dora has 8% of her earnings deducted from her paycheck for a college savings plan. She can choose to take the deduction either before taxes are withheld, which reduces her taxable income, or after taxes are withheld. Dora's tax rate is 17.5%. If her pay before taxes and deductions is \$950, will she save more money if the deductions are taken before or after taxes are withheld? Explain. **See margin.**

Practice and Problem Solving   Extra Practice is on page R6.

Examples 1–2 Find $(f + g)(x)$, $(f - g)(x)$, $(f \cdot g)(x)$, and $\left(\dfrac{f}{g}\right)(x)$ for each $f(x)$ and $g(x)$. Indicate any restrictions in domain or range. **8–15. See margin.**

8. $f(x) = 2x$
$g(x) = -4x + 5$

9. $f(x) = x - 1$
$g(x) = 5x - 2$

10. $f(x) = x^2$
$g(x) = -x + 1$

11 $f(x) = 3x$
$g(x) = -2x + 6$

12. $f(x) = x - 2$
$g(x) = 2x - 7$

13. $f(x) = x^2$
$g(x) = x - 5$

14. $f(x) = -x^2 + 6$
$g(x) = 2x^2 + 3x - 5$

15. $f(x) = 3x^2 - 4$
$g(x) = x^2 - 8x + 4$

16. POPULATION In a particular county, the population of the two largest cities can be modeled by $f(x) = 200x + 25$ and $g(x) = 175x - 15$, where x is the number of years since 2000 and the population is in thousands.

a. What is the population of the two cities combined after any number of years? $(f + g)(x) = 375x + 10$

b. What is the difference in the populations of the two cities? $(f - g)(x) = 25x + 40$

Example 3 For each pair of functions, find $f \circ g$ and $g \circ f$, if they exist. State the domain and range for each composed function. **17–20. See Chapter 6 Answer Appendix.**

17. $f = \{(-8, -4), (0, 4), (2, 6), (-6, -2)\}$
$g = \{(4, -4), (-2, -1), (-4, 0), (6, -5)\}$

18. $f = \{(-7, 0), (4, 5), (8, 12), (-3, 6)\}$
$g = \{(6, 8), (-12, -5), (0, 5), (5, 1)\}$

19. $f = \{(5, 13), (-4, -2), (-8, -11), (3, 1)\}$
$g = \{(-8, 2), (-4, 1), (3, -3), (5, 7)\}$

20. $f = \{(-4, -14), (0, -6), (-6, -18), (2, -2)\}$
$g = \{(-6, 1), (-18, 13), (-14, 9), (-2, -3)\}$

 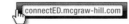

Differentiated Homework Options

Level	Assignment	Two-Day Option	
AL Basic	8–37, 59, 60, 62–86	9–37 odd, 64–67	8–36 even, 59, 60, 62, 63, 68–86
OL Core	9–49 odd, 50, 51–57 odd, 58–60, 62–86	8–37, 64–67	38–60, 62, 63, 68–86
BL Advanced	38–80, (optional: 81–86)		

$-(x^2 - 6)$

$(2x + 5)(x - 1)$

Formative Assessment

Use Exercises 1–7 to check for understanding.

Use the chart at the bottom of this page to customize assignments for your students.

Additional Answers

8. $(f + g)(x) = -2x + 5$; $(f - g)(x) = 6x - 5$; $(f \cdot g)(x) = -8x^2 + 10x$; $\left(\dfrac{f}{g}\right)(x) = \dfrac{2x}{-4x + 5}$, $x \neq \dfrac{5}{4}$

9. $(f + g)(x) = 6x - 3$; $(f - g)(x) = -4x + 1$; $(f \cdot g)(x) = 5x^2 - 7x + 2$; $\left(\dfrac{f}{g}\right)(x) = \dfrac{x - 1}{5x - 2}$, $x \neq \dfrac{2}{5}$

10. $(f + g)(x) = x^2 - x + 1$; $(f - g)(x) = x^2 + x - 1$; $(f \cdot g)(x) = -x^3 + x^2$; $\left(\dfrac{f}{g}\right)(x) = \dfrac{x^2}{-x + 1}$, $x \neq 1$

11. $(f + g)(x) = x + 6$; $(f - g)(x) = 5x - 6$; $(f \cdot g)(x) = -6x^2 + 18x$; $\left(\dfrac{f}{g}\right)(x) = \dfrac{3x}{-2x + 6}$, $x \neq 3$

12. $(f + g)(x) = 3x - 9$; $(f - g)(x) = -x + 5$; $(f \cdot g)(x) = 2x^2 - 11x + 14$; $\left(\dfrac{f}{g}\right)(x) = \dfrac{x - 2}{2x - 7}$, $x \neq \dfrac{7}{2}$

13. $(f + g)(x) = x^2 + x - 5$; $(f - g)(x) = x^2 - x + 5$; $(f \cdot g)(x) = x^3 - 5x^2$; $\left(\dfrac{f}{g}\right)(x) = \dfrac{x^2}{x - 5}$, $x \neq 5$

14. $(f + g)(x) = x^2 + 3x + 1$; $(f - g)(x) = -3x^2 - 3x + 11$; $(f \cdot g)(x) = -2x^4 - 3x^3 + 17x^2 + 18x - 30$; $\left(\dfrac{f}{g}\right)(x) = \dfrac{-x^2 + 6}{2x^2 + 3x - 5}$, $x \neq 1$ or $-\dfrac{5}{2}$

15. $(f + g)(x) = 4x^2 - 8x$; $(f - g)(x) = 2x^2 + 8x - 8$; $(f \cdot g)(x) = 3x^4 - 24x^3 + 8x^2 + 32x - 16$; $\left(\dfrac{f}{g}\right)(x) = \dfrac{3x^2 - 4}{x^2 - 8x + 4}$, $x \neq 4 \pm 2\sqrt{3}$

Study Guide and Intervention
AL OL ELL

Practice
AL OL BL ELL

Word Problem Practice
AL OL BL ELL

For each pair of functions, find $f \circ g$ and $g \circ f$, if they exist. State the domain and range for each composed function. **21–26. See margin.**

21. $f = \{(-15, -5), (-4, 12), (1, 7), (3, 9)\}$
$g = \{(3, -9), (7, 2), (8, -6), (12, 0)\}$

22. $f = \{(-1, 11), (2, -2), (5, -7), (4, -4)\}$
$g = \{(5, -4), (4, -3), (-1, 2), (2, 3)\}$

23. $f = \{(7, -3), (-10, -3), (-7, -8), (-3, 6)\}$
$g = \{(4, -3), (3, -7), (9, 8), (-4, -4)\}$

24. $f = \{(1, -1), (2, -2), (3, -3), (4, -4)\}$
$g = \{(1, -4), (2, -3), (3, -2), (4, -1)\}$

25. $f = \{(-4, -1), (-2, 6), (-1, 10), (4, 11)\}$
$g = \{(-1, 5), (3, -4), (6, 4), (10, 8)\}$

26. $f = \{(12, -3), (9, -2), (8, -1), (6, 3)\}$
$g = \{(-1, 5), (-2, 6), (-3, -1), (-4, 8)\}$

Find $[f \circ g](x)$ and $[g \circ f](x)$, if they exist. State the domain and range for each composed function. **27–35. See Chapter 6 Answer Appendix.**

27. $f(x) = 2x$
$g(x) = x + 5$

28. $f(x) = -3x$
$g(x) = -x + 8$

29. $f(x) = x + 5$
$g(x) = 3x - 7$

30. $f(x) = x - 4$
$g(x) = x^2 - 10$

31. $f(x) = x^2 + 6x - 2$
$g(x) = x - 6$

32. $f(x) = 2x^2 - x + 1$
$g(x) = 4x + 3$

33. $f(x) = 4x - 1$
$g(x) = x^3 + 2$

34. $f(x) = x^2 + 3x + 1$
$g(x) = x^2$

35. $f(x) = 2x^2$
$g(x) = 8x^2 + 3x$

36. FINANCE A ceramics store manufactures and sells coffee mugs. The revenue $r(x)$ from the sale of x coffee mugs is given by $r(x) = 6.5x$. Suppose the function for the cost of manufacturing x coffee mugs is $c(x) = 0.75x + 1850$.

a. Write the profit function. $P(x) = 5.75x - 1850$

b. Find the profit on 500, 1000, and 5000 coffee mugs. $P(500) = \$1025$; $P(1000) = \$3900$; $P(5000) = \$26,900$

37. CCSS SENSE-MAKING Ms. Smith wants to buy an HDTV, which is on sale for 35% off the original price of $2299. The sales tax is 6.25%.

37b. Since $[p \circ t](x) = [t \circ p](x)$, either function represents the price.

a. Write two functions representing the price after the discount $p(x)$ and the price after sales tax $t(x)$. $p(x) = 0.65x$; $t(x) = 1.0625x$

b. Which composition of functions represents the price of the HDTV, $[p \circ t](x)$ or $[t \circ p](x)$? Explain your reasoning.

c. How much will Ms. Smith pay for the HDTV? $\$1587.75$

B Perform each operation if $f(x) = x^2 + x - 12$ and $g(x) = x - 3$. State the domain of the resulting function. **39.** $2(g \cdot f)(x) = 2x^3 - 4x^2 - 30x + 72$; D = {all real numbers}

38. $(f - g)(x)$
$(f - g)(x) = x^2 - 9$; D = {all real numbers}

39. $2(g \cdot f)(x)$

40. $\left(\dfrac{f}{g}\right)(x)$ $\left(\dfrac{f}{g}\right)(x) = x + 4$; D = $\{x \mid x \neq 3\}$

If $f(x) = 5x$, $g(x) = -2x + 1$, and $h(x) = x^2 + 6x + 8$, find each value.

41. $f[g(-2)]$ 25

42. $g[h(3)]$ -69

43. $h[f(-5)]$ 483

44. $h[g(2)]$ -1

45. $f[h(-3)]$ -5

46. $h[f(9)]$ 2303

47. $f[g(3a)]$ $-30a + 5$

48. $f[h(a + 4)]$
$5a^2 + 70a + 240$

49. $g[f(a^2 - a)]$ $-10a^2 + 10a + 1$

50. MULTIPLE REPRESENTATIONS Let $f(x) = x^2$ and $g(x) = x$. **50a–d. See Chapter 6 Answer Appendix.**

a. Tabular Make a table showing values for $f(x)$, $g(x)$, $(f + g)(x)$, and $(f - g)(x)$.

b. Graphical Graph $f(x)$, $g(x)$, and $(f + g)(x)$ on the same coordinate grid.

c. Graphical Graph $f(x)$, $g(x)$, and $(f - g)(x)$ on the same coordinate grid.

d. Verbal Describe the relationship among the graphs of $f(x)$, $g(x)$, $(f + g)(x)$, and $(f - g)(x)$.

Enrichment
OL BL

Relative Maximum Values

The graph of $f(x) = x^3 - 6x - 9$ shows a relative maximum somewhere between $f(-2)$ and $f(-1)$. You can obtain a closer approximation by comparing values such as those shown in the table.

To the nearest tenth a relative maximum value for $f(x)$ is -3.3.

Using a calculator to find points, graph each function. To the nearest tenth, find a relative maximum value of the function.

Additional Answers

21. $f \circ g$ is undefined, D = ∅, R = ∅; $g \circ f = \{(-4, 0), (1, 2)\}$, D = $\{-4, 1\}$, R = $\{0, 2\}$.

22. $f \circ g = \{(-1, -2)\}$, D = $\{-1\}$, R = $\{-2\}$; $g \circ f$ is undefined, D = ∅, R = ∅.

23. $f \circ g = \{(4, 6), (3, -8)\}$, D = $\{3, 4\}$, R = $\{-8, 6\}$; $g \circ f$ is undefined, D = ∅, R = ∅.

51b. The function represents the difference in the number of men and women employed in the U.S.

(51) **EMPLOYMENT** The number of women and men age 16 and over employed each year in the United States can be modeled by the following equations, where x is the number of years since 1994 and y is the number of people in thousands.

women: $y = 1086.4x + 56,610$ men: $y = 999.2x + 66,450$

a. Write a function that models the total number of men and women employed in the United States during this time. $y = 2085.6x + 123,060$

b. If f is the function for the number of men, and g is the function for the number of women, what does $(f - g)(x)$ represent?

C If $f(x) = x + 2$, $g(x) = -4x + 3$, and $h(x) = x^2 - 2x + 1$, find each value.

52. $(f \cdot g \cdot h)(3)$ -180

53. $[(f + g) \cdot h](1)$ 0

54. $\left(\dfrac{h}{fg}\right)(-6)$ $-\dfrac{49}{108}$

55. $[f \circ (g \circ h)](2)$ 1

56. $[g \circ (h \circ f)](-4)$ -33

57. $[h \circ (f \circ g)](5)$ 256

58. 🔀 **MULTIPLE REPRESENTATIONS** You will explore $(f \cdot g)(x)$, $\left(\dfrac{f}{g}\right)(x)$, $[f \circ g](x)$, and $[g \circ f](x)$ if $f(x) = x^2 + 1$ and $g(x) = x - 3$. **58a–e. See Chapter 6 Answer Appendix.**

a. **Tabular** Make a table showing values for $(f \cdot g)(x)$, $\left(\dfrac{f}{g}\right)(x)$, $[f \circ g](x)$, and $[g \circ f](x)$.

b. **Graphical** Use a graphing calculator to graph $(f \cdot g)(x)$ and $\left(\dfrac{f}{g}\right)(x)$ on the same coordinate plane.

c. **Verbal** Explain the relationship between $(f \cdot g)(x)$ and $\left(\dfrac{f}{g}\right)(x)$.

d. **Graphical** Use a graphing calculator to graph $[f \circ g](x)$, and $[g \circ f](x)$ on the same coordinate plane.

e. **Verbal** Explain the relationship between $[f \circ g](x)$, and $[g \circ f](x)$.

H.O.T. Problems   Use Higher-Order Thinking Skills

59. **OPEN ENDED** Write two functions $f(x)$ and $g(x)$ such that $(f \circ g)(4) = 0$.   **Sample answer:** $f(x) = x - 9$, $g(x) = x + 5$

60. **CCSS CRITIQUE** Chris and Tobias are finding $(f \circ g)(x)$, where $f(x) = x^2 + 2x - 8$ and $g(x) = x^2 + 8$. Is either of them correct? Explain your reasoning.

Chris	Tobias
$(f \circ g)(x) = f[g(x)]$	$(f \circ g)(x) = f[g(x)]$
$= (x^2 + 8)^2 + 2x - 8$	$= (x^2 + 8)^2 + 2(x^2 + 8) - 8$
$= x^4 + 16x^2 + 64 + 2x - 8$	$= x^4 + 16x^2 + 64 + 2x^2 + 16 - 8$
$= x^4 + 16x^2 + 2x + 58$	$= x^4 + 18x^2 + 72$

Tobias; Chris did not substitute $g(x)$ for every x in $f(x)$.

62a. Always, since the range is dependent on the domain, the domain of $g[f(x)]$ is restricted by the domain of $f(x)$.

61. **CHALLENGE** Given $f(x) = \sqrt{x^3}$ and $g(x) = \sqrt{x^6}$, determine the domain for each of the following.

a. $g(x) \cdot g(x)$ $D = \{$all real numbers$\}$

b. $f(x) \cdot f(x)$ $D = \{x \mid x \geq 0\}$

62b. Sometimes; when $f(x) = 4x$ and $g(x) = \sqrt{x}$, $g[f(x)] = \sqrt{4x}$, $x \geq 0$. The domain of $g(x)$ restricts the domain of $g[f(x)]$. When $f(x) = 4x^2$ and $g(x) = \sqrt{x}$, $g[f(x)] = \sqrt{4x^2}$. In this case, the domain of $g(x)$ does not restrict the domain of $g[f(x)]$.

62. **REASONING** State whether each statement is *sometimes*, *always*, or *never* true. Explain.

a. The domain of two functions $f(x)$ and $g(x)$ that are composed $g[f(x)]$ is restricted by the domain of $f(x)$.

b. The domain of two functions $f(x)$ and $g(x)$ that are composed $g[f(x)]$ is restricted by the domain of $g(x)$.

63. 📝 **WRITING IN MATH** In the real world, why would you ever perform a composition of functions? See margin.

Javier Larrea/age fotostock

connectED.mcgraw-hill.com   **391**

DifferentiatedInstruction **AL** **OL** **BL** **ELL**

Extension Many states use function compositions to calculate speeding ticket amounts. They do this by calculating the number of miles over the speed limit a driver is traveling, multiplying this number by a per mile charge, and adding a flat rate. Ask students to create a composition of three different functions, substituting made-up or researched values for the posted speed limit, the per mile charge, and the flat rate. The function composition should be of the form $(f \circ g \circ h)(x)$ where x is the speed of the driver in miles per hour.

CCSS **Teaching the Mathematical Practices**

Critique Mathematically proficient students are also able to compare the effectiveness of two plausible arguments, distinguish correct logic or reasoning from that which is flawed, and—if there is a flaw in an argument—explain what it is.

WatchOut!

Error Analysis For Exercise 60, remind students that they will be finding $f(x^2 + 8)$. So every x in $x^2 + 2x - 8$ will be replaced with $x^2 + 8$.

🔀 **Multiple Representations**

In Exercise 58, students use a table of values, a graph, and analysis to compare and contrast the product, quotient, and composition of two functions.

Additional Answers

24. $f \circ g$ is undefined, $D = \varnothing$, $R = \varnothing$; $g \circ f$ is undefined, $D = \varnothing$, $R = \varnothing$.

25. $f \circ g = \{(3, -1), (6, 11)\}$, $D = \{3, 6\}$, $R = \{-1, 11\}$; $g \circ f = \{(-4, 5), (-2, 4), (-1, 8)\}$, $D = \{-4, -2, -1\}$, $R = \{4, 5, 8\}$

26. $f \circ g = \{(-2, 3), (-4, -1)\}$, $D = \{-4, -2\}$, $R = \{-1, 3\}$; $g \circ f = \{(12, -1), (9, 6), (8, 5)\}$, $D = \{8, 9, 12\}$, $R = \{-1, 5, 6\}$

63. Sample answer: Many situations in the real world involve complex calculations in which multiple functions are used. In order to solve some problems, a composition of those functions may need to be used. For example, the product of a manufacturing plant may have to go through several processes in a particular order, in which each process is described by a function. By finding the composition, only one calculation must be made to find the solution to the problem.

connectED.mcgraw-hill.com   **391**

Name the Math Have students tell a partner or write on a piece of paper the mathematical procedure they would use to determine their tax savings by having medical expenses deducted before they receive their pay. Some medical expenses can be paid from a special medical savings account that is not taxed. composition of functions

Standardized Test Practice

64. What is the value of x in the equation $7(x - 4) = 44 - 11x$? **D**

A 1
B 2
C 3
D 4

65. If $g(x) = x^2 + 9x + 21$ and $h(x) = 2(x + 5)^2$, which is an equivalent form of $h(x) - g(x)$? **G**

F $k(x) = -x^2 - 11x - 29$
G $k(x) = x^2 + 11x + 29$
H $k(x) = x + 4$
J $k(x) = x^2 + 7x + 11$

66. GRIDDED RESPONSE In his first three years of coaching basketball at North High School, Coach Lucas' team won 8 games the first year, 17 games the second year, and 6 games the third year. How many games does the team need to win in the fourth year so the coach's average will be 10 wins per year? **9**

67. SAT/ACT What is the value of $f[g(6)]$ if $f(x) = 2x + 4$ and $g(x) = x^2 + 5$? **D**

A 38   D 86
B 43   E 261
C 57

Spiral Review

Find all of the rational zeros of each function. (Lesson 5-8)

68. $f(x) = 2x^3 - 13x^2 + 17x + 12$ $-\frac{1}{2}, 3, 4$

69. $f(x) = x^3 - 3x^2 - 10x + 24$ $-3, 2, 4$

70. $f(x) = x^4 - 4x^3 - 7x^2 + 34x - 24$ $1, 2, 4, -3$

71. $f(x) = 2x^3 - 5x^2 - 28x + 15$ $-3, 5, \frac{1}{2}$

State the possible number of positive real zeros, negative real zeros, and imaginary zeros of each function. (Lesson 5-7)

72. $f(x) = 2x^4 - x^3 + 5x^2 + 3x - 9$ 3 or 1; 1; 2 or 0

73. $f(x) = -4x^4 - x^2 - x + 1$ 1; 1; 2

74. $f(x) = 3x^4 - x^3 + 8x^2 + x - 7$ 3 or 1; 1; 0 or 2

75. $f(x) = 2x^4 - 3x^3 - 2x^2 + 3$ 2 or 0; 2 or 0; 4, 2, or 0

76. MANUFACTURING A box measures 12 inches by 16 inches by 18 inches. The manufacturer will increase each dimension of the box by the same number of inches and have a new volume of 5985 cubic inches. How much should be added to each dimension? (Lesson 5-7) **3 in.**

Solve each system of equations. (Lesson 3-4)

77. $x + 4y - z = 6$ (1, 2, 3)
$3x + 2y + 3z = 16$
$2x - y + z = 3$

78. $2a + b - c = 5$ (4, -2, 1)
$a - b + 3c = 9$
$3a - 6c = 6$

79. $y + z = 4$ (3, -1, 5)
$2x + 4y - z = -3$
$3y = -3$

80. INTERNET A webmaster estimates that the time, in seconds, to connect to the server when n people are connecting is given by $t(n) = 0.005n + 0.3$. Estimate the time to connect when 50 people are connecting. (Lesson 2-2) **0.55 second**

81. $x = \dfrac{12 + 7y}{5}$ **82.** $y = \dfrac{1 - x^2}{-2x}$ **83.** $x = \dfrac{15 - 8yz}{4}$

Skills Review

Solve each equation or formula for the specified variable.   **86.** $y = \pm\sqrt{(x + 2)^2 - 4} - 5$

81. $5x - 7y = 12$, for x **82.** $3x^2 - 6xy + 1 = 4$, for y **83.** $4x + 8yz = 15$, for x

84. $D = mv$, for m $m = \dfrac{D}{v}$ **85.** $A = k^2 + b$, for k $k = \pm\sqrt{A - b}$ **86.** $(x + 2)^2 - (y + 5)^2 = 4$, for y

LESSON 6-2 Inverse Functions and Relations

·:Then	·:Now	·:Why?
● You transformed and solved equations for a specific variable.	**1** Find the inverse of a function or relation. **2** Determine whether two functions or relations are inverses.	● The table shows the value of $1 (U.S.) compared to Canadian dollars and Mexican pesos. The equation $p = 10.75d$ represents the number of pesos p you can receive for every U.S. dollar d. To determine how many U.S. dollars you can receive for one Mexican peso, solve the equation $p = 10.75d$ for d. The result, $d \approx 0.09p$, is the inverse function.

	U.S.	Canada	Mexico
U.S.		1.05	10.75
Canada	0.95		10.26
Mexico	0.09	0.10	

NewVocabulary
inverse relation
inverse function

CCSS Common Core State Standards

Content Standards
F.IF.4 For a function that models a relationship between two quantities, interpret key features of graphs and tables in terms of the quantities, and sketch graphs showing key features given a verbal description of the relationship.

F.BF.4.a Find inverse functions. - Solve an equation of the form $f(x) = c$ for a simple function f that has an inverse and write an expression for the inverse.

Mathematical Practices
7 Look for and make use of structure.
8 Look for and express regularity in repeated reasoning.

GeoStock/Photodisc/Getty Images

1 Find Inverses Recall that a relation is a set of ordered pairs. The **inverse relation** is the set of ordered pairs obtained by exchanging the coordinates of each ordered pair. The domain of a relation becomes the range of its inverse, and the range of the relation becomes the domain of its inverse.

KeyConcept Inverse Relations

Words	Two relations are inverse relations if and only if whenever one relation contains the element (a, b), the other relation contains the element (b, a).
Example	A and B are inverse relations.
	$A = \{(1, 5), (2, 6), (3, 7)\}$ $B = \{(5, 1), (6, 2), (7, 3)\}$

Example 1 Find an Inverse Relation

GEOMETRY The vertices of $\triangle ABC$ can be represented by the relation $\{(1, -2), (2, 5), (4, -1)\}$. Find the inverse of this relation. Describe the graph of the inverse.

Graph the relation. To find the inverse, exchange the coordinates of the ordered pairs. The inverse of the relation is $\{(-2, 1), (5, 2), (-1, 4)\}$.

Plotting these points shows that the ordered pairs describe the vertices of $\triangle A'B'C'$ as a reflection of $\triangle ABC$ in the line $y = x$.

▶ **Guided Practice** $\{(-3, -8), (-6, -8), (-6, -3)\}$; It is a reflection in the line $y = x$.

1. GEOMETRY The ordered pairs of the relation $\{(-8, -3), (-8, -6), (-3, -6)\}$ are the coordinates of the vertices of a right triangle. Find the inverse of this relation. Describe the graph of the inverse.

As with relations, the ordered pairs of **inverse functions** are also related. We can write the inverse of the function $f(x)$ as $f^{-1}(x)$.

 connectED.mcgraw-hill.com   **393**

1 Focus

Vertical Alignment

▼

Before Lesson 6-2 Transform and solve equations for a specific variable.

▼

Lesson 6-2 Find the inverse of a function or relation. Determine whether two functions or relations are inverses.

▼

After Lesson 6-2 Develop the definition of logarithm by exploring the relationship between exponential functions and their inverses.

2 Teach

Scaffolding Questions
Have students read the **Why?** section of the lesson.

Ask:
■ According to the table, how many Canadian dollars can you get for 5 U.S. dollars? for 10 U.S. dollars? 5.25, 10.50

■ How many U.S. dollars can you get for 1 Canadian dollar? for 5 Canadian dollars? $0.95; $4.75

■ What are the reciprocals of 1.05 and 10.75? Do the reciprocals appear in the table? $\frac{1}{1.05} \approx 0.95$; $\frac{1}{10.75} \approx 0.09$; the values 0.95 and 0.09 do appear in the table.

Lesson 6-2 Resources

Resource	Approaching Level **AL**	On Level **OL**	Beyond Level **BL**	English Learners **ELL**
Teacher Edition	▪ Differentiated Instruction, p. 395	▪ Differentiated Instruction, pp. 395, 398	▪ Differentiated Instruction, p. 398	
Chapter Resource Masters	▪ Study Guide and Intervention, pp. 12–13 ▪ Skills Practice, p. 14 ▪ Practice, p. 15 ▪ Word Problem Practice, p. 16	▪ Study Guide and Intervention, pp. 12–13 ▪ Skills Practice, p.14 ▪ Practice, p. 15 ▪ Word Problem Practice, p. 16 ▪ Enrichment, p. 17 ▪ Graphing Calculator Activity, p. 18	▪ Practice, p. 15 ▪ Word Problem Practice, p. 16 ▪ Enrichment, p. 17	▪ Study Guide and Intervention, pp. 12–13 ▪ Skills Practice, p. 14 ▪ Practice, p. 15 ▪ Word Problem Practice, p. 16
Other	▪ 5-Minute Check 6-2 ▪ Study Notebook ▪ Teaching Algebra with Manipulatives	▪ 5-Minute Check 6-2 ▪ Study Notebook ▪ Teaching Algebra with Manipulatives	▪ 5-Minute Check 6-2 ▪ Study Notebook	▪ 5-Minute Check 6-2 ▪ Study Notebook ▪ Teaching Algebra with Manipulatives

1 Find Inverses

Example 1 shows how to find the inverse relation for a set of ordered pairs. **Example 2** shows how to find the inverse of a function and graph it.

Formative Assessment

Use the Guided Practice exercises after each example to determine students' understanding of concepts.

Additional Examples

1 **GEOMETRY** The ordered pairs of the relation {(1, 3), (6, 3), (6, 0), (1, 0)} are the coordinates of the vertices of a rectangle. Find the inverse of this relation. Describe the graph of the inverse. The inverse of the relation is {(3, 1), (3, 6), (0, 6), (0, 1)}. The ordered pairs are the vertices of a rectangle; the vertices are the original vertices reflected across the line $y = x$.

2 Find the inverse of $f(x) = -\frac{1}{2}x + 1$. Then graph the function and its inverse.

$f^{-1}(x) = -2x + 2$

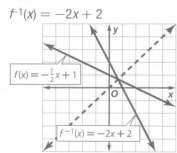

Additional Examples also in Interactive Classroom PowerPoint® Presentations

IWB **Interactive White Board READY**

ReadingMath

CCSS Sense-Making f^{-1} is read *f inverse* or *the inverse of f*. Note that −1 is *not* an exponent.

📑 KeyConcept Property of Inverses

Words If f and f^{-1} are inverses, then $f(a) = b$ if and only if $f^{-1}(b) = a$.

Example Let $f(x) = x - 4$ and represent its inverse as $f^{-1}(x) = x + 4$.

Evaluate $f(6)$.   Evaluate $f^{-1}(2)$.

$f(x) = x - 4$ $f^{-1}(x) = x + 4$

$f(6) = 6 - 4$ or 2 $f^{-1}(2) = 2 + 4$ or 6

Because $f(x)$ and $f^{-1}(x)$ are inverses, $f(6) = 2$ and $f^{-1}(2) = 6$.

When the inverse of a function is a function, the original function is one-to-one. Recall that the vertical line test can be used to determine whether a relation is a function. Similarly, the *horizontal line test* can be used to determine whether the inverse of a function is also a function.

No horizontal line can be drawn so that it passes through more than one point. The inverse of $y = f(x)$ is a function.

A horizontal line can be drawn that passes through more than one point. The inverse of $y = g(x)$ is not a function.

The inverse of a function can be found by exchanging the domain and the range.

Example 2 Find and Graph an Inverse

Find the inverse of each function. Then graph the function and its inverse.

a. $f(x) = 2x - 5$

Step 1 Rewrite the function as an equation relating x and y.

$f(x) = 2x - 5 \rightarrow y = 2x - 5$

Step 2 Exchange x and y in the equation. $x = 2y - 5$

Step 3 Solve the equation for y.

$x = 2y - 5$ Inverse of $y = 2x - 5$

$x + 5 = 2y$ Add 5 to each side.

$\frac{x + 5}{2} = y$ Divide each side by 2.

Step 4 Replace y with $f^{-1}(x)$.

$y = \frac{x + 5}{2} \rightarrow f^{-1}(x) = \frac{x + 5}{2}$

The inverse of $f(x) = 2x - 5$ is $f^{-1}(x) = \frac{x + 5}{2}$.

The graph of $f^{-1}(x) = \frac{x + 5}{2}$ is the reflection of the graph of $f(x) = 2x - 5$ in the line $y = x$.

Teach with Tech

Audio Recording Have students create audio recordings explaining how to find the inverse of a function. Also have students explain how they can check if two given functions are inverses of each other.

🗒 Follow-up

Students have explored inverse functions.

Ask:

▪ Why would you use the inverse of a function to model a real-world situation? Sample answer: You could use a function or the inverse of the function to model a relationship so that either quantity could be the independent variable. Then, you could use the functions to solve problems involving either quantity.

b. $f(x) = x^2 + 1$

Step 1 $f(x) = x^2 + 1 \rightarrow y = x^2 + 1$

Step 2 $x = y^2 + 1$

Step 3
$$x = y^2 + 1$$
$$x - 1 = y^2$$
$$\pm\sqrt{x - 1} = y \quad \text{Take the square root of each side.}$$

Step 4 $y = \pm\sqrt{x - 1}$

Graph $y = \pm\sqrt{x - 1}$ by reflecting the graph of $f(x) = x^2 + 1$ in the line $y = x$.

StudyTip
Functions The inverse of the function in part b is not a function since it does not pass the vertical line test.

▶ **Guided**Practice **2A, 2B.** See margin.

Find the inverse of each function. Then graph the function and its inverse.

2A. $f(x) = \dfrac{x - 3}{5}$

2B. $f(x) = 3x^2$

2 Verifying Inverses You can determine whether two functions are inverses by finding both of their compositions. If both compositions equal the identity function $I(x) = x$, then the functions are inverse functions.

KeyConcept Inverse Functions

Words	Two functions f and g are inverse functions if and only if both of their compositions are the identity function.
Symbols	$f(x)$ and $g(x)$ are inverses if and only if $[f \circ g](x) = x$ and $[g \circ f](x) = x$.

Example 3 Verify that Two Functions are Inverses

Determine whether each pair of functions are inverse functions. Explain your reasoning.

a. $f(x) = 3x + 9$ and $g(x) = \frac{1}{3}x - 3$

Verify that the compositions of $f(x)$ and $g(x)$ are identity functions.

$$[f \circ g](x) = f[g(x)] \qquad\qquad [g \circ f](x) = g[f(x)]$$
$$= f\left(\frac{1}{3}x - 3\right) \qquad\qquad = g(3x + 9)$$
$$= 3\left(\frac{1}{3}x - 3\right) + 9 \qquad\qquad = \frac{1}{3}(3x + 9) - 3$$
$$= x - 9 + 9 \text{ or } x \qquad\qquad = x + 3 - 3 \text{ or } x$$

The functions are inverses because $[f \circ g](x) = [g \circ f](x) = x$.

b. $f(x) = 4x^2$ and $g(x) = 2\sqrt{x}$
$$[f \circ g](x) = f(2\sqrt{x})$$
$$= 4(2\sqrt{x})^2$$
$$= 4(4x) \text{ or } 16x$$

Because $[f \circ g](x) \neq x$, $f(x)$ and $g(x)$ are not inverses.

▶ **Guided**Practice

3A. $f(x) = 3x - 3, g(x) = \frac{1}{3}x + 4$

3B. $f(x) = 2x^2 - 1, g(x) = \sqrt{\dfrac{x + 1}{2}}$

WatchOut!
Inverse Functions Be sure to check both $[f \circ g](x)$ and $[g \circ f](x)$ to verify that functions are inverses. By definition, both compositions must be the identity function.

3A. No; they are not inverse functions because $[f \circ g](x) = x + 9$ and $[g \circ f](x) = x + 3$.

3B. Yes; they are inverse functions because $[f \circ g](x) = x$ and $[g \circ f](x) = x$.

CCSS Teaching the Mathematical Practices

Sense-Making Mathematically proficient students start by explaining to themselves the meaning of a problem and looking for entry points to its solution. Explain to students that generally, f^{-1} refers to the inverse of the function f.

2 Verifying Inverses
Example 3 shows how to verify that two functions are inverses.

Additional Example

3 Determine whether $f(x) = \frac{3}{4}x - 6$ and $g(x) = \frac{4}{3}x + 8$ are inverse functions. Explain your reasoning. The functions are inverses since both $[f \circ g](x)$ and $[g \circ f](x)$ equal x.

Additional Answers (Guided Practice)

2A. $f^{-1}(x) = 5x + 3$

2B. $y = \pm\sqrt{\dfrac{1}{3}x}$

DifferentiatedInstruction AL OL

Kinesthetic Learners On a large coordinate grid, have students model the graph of the identity function $f(x) = x$ using a length of string, a piece of uncooked spaghetti, or something similar. Then place a second length of string to model the graph of $f(x) = 2x - 5$ from Example 2. Have students model the graph of the inverse of this function and relate the graphs of the function, and its inverse, to the graph of $f(x) = x$.

3 Practice

Formative Assessment

Use Exercises 1–8 to check for understanding.

Use the chart at the bottom of this page to customize assignments for your students.

Additional Answers

3. $f^{-1}(x) = -\frac{1}{3}x$

4. $g^{-1}(x) = \frac{x + 6}{4}$

5. $h^{-1}(x) = \pm\sqrt{x + 3}$

56. Sample answer: One of the functions carries out an operation on 5. Then the second function that is an inverse of the first function reverses the operation on 5. Thus, the result is 5.

Check Your Understanding

○ = Step-by-Step Solutions begin on page R14.

Example 1   Find the inverse of each relation.

1. $\{(-9, 10), (1, -3), (8, -5)\}$
$\{(10, -9), (-3, 1), (-5, 8)\}$

2. $\{(-2, 9), (4, -1), (-7, 9), (7, 0)\}$
$\{(9, -2), (-1, 4), (9, -7), (0, 7)\}$

Example 2   Find the inverse of each function. Then graph the function and its inverse. **3–5. See margin.**

3. $f(x) = -3x$ **4.** $g(x) = 4x - 6$ **5.** $h(x) = x^2 - 3$

Example 3   Determine whether each pair of functions are inverse functions. Write *yes* or *no*.

6. $f(x) = x - 7$ **7.** $f(x) = \frac{1}{2}x + \frac{3}{4}$ **8.** $f(x) = 2x^3$

$g(x) = x + 7$ yes $g(x) = 2x - \frac{4}{3}$ no $g(x) = \frac{1}{3}\sqrt{x}$ no

Practice and Problem Solving

Extra Practice is on page R6.

Example 1   Find the inverse of each relation. **9.** $\{(6, -8), (-2, 6), (-3, 7)\}$ **10.** $\{(7, 7), (9, 4), (-7, 3)\}$

11. $\{(-1, 8), (-1, -8), (-8, -2), (8, 2)\}$

9. $\{(-8, 6), (6, -2), (7, -3)\}$

10. $\{(7, 7), (4, 9), (3, -7)\}$

12. $\{(3, 4), (-4, -4), (-5, -3), (2, 5)\}$

11. $\{(8, -1), (-8, -1), (-2, -8), (2, 8)\}$

12. $\{(4, 3), (-4, -4), (-3, -5), (5, 2)\}$

13. $\{(1, -5), (2, 6), (3, -7), (4, 8), (5, -9)\}$
$\{(-5, 1), (6, 2), (-7, 3), (8, 4), (-9, 5)\}$

14. $\{(3, 0), (5, 4), (7, -8), (9, 12), (11, 16)\}$
$\{(0, 3), (4, 5), (-8, 7), (12, 9), (16, 11)\}$

Example 2   CCSS SENSE-MAKING Find the inverse of each function. Then graph the function and its inverse. **15–26. See Chapter 6 Answer Appendix.**

15. $f(x) = x + 2$ **16.** $g(x) = 5x$ **17.** $f(x) = -2x + 1$

18. $h(x) = \frac{x - 4}{3}$ **19.** $f(x) = -\frac{5}{3}x - 8$ **20.** $g(x) = x + 4$

21. $f(x) = 4x$ **22.** $f(x) = -8x + 9$ **23.** $f(x) = 5x^2$

24. $h(x) = x^2 + 4$ **25.** $f(x) = \frac{1}{2}x^2 - 1$ **26.** $f(x) = (x + 1)^2 + 3$

Example 3   Determine whether each pair of functions are inverse functions. Write *yes* or *no*.

27. $f(x) = 2x + 3$ **28.** $f(x) = 4x + 6$ **29.** $f(x) = -\frac{1}{3}x + 3$

$g(x) = 2x - 3$ no $g(x) = \frac{x - 6}{4}$ yes $g(x) = -3x + 9$ yes

30. $f(x) = -6x$ **31.** $f(x) = \frac{1}{2}x + 5$ **32.** $f(x) = \frac{x + 10}{8}$

$g(x) = \frac{1}{6}x$ no $g(x) = 2x - 10$ yes $g(x) = 8x - 10$ yes

33. $f(x) = 4x^2$ **34.** $f(x) = \frac{1}{3}x^2 + 1$ **35** $f(x) = x^2 - 9$

$g(x) = \frac{1}{2}\sqrt{x}$ yes $g(x) = \sqrt{3x - 3}$ yes $g(x) = x + 3$ no

36. $f(x) = \frac{2}{3}x^3$ **37.** $f(x) = (x + 6)^2$ **38.** $f(x) = 2\sqrt{x - 5}$

$g(x) = \sqrt{\frac{2}{3}x}$ no $g(x) = \sqrt{x} - 6$ yes $g(x) = \frac{1}{4}x^2 - 5$ no

39. FUEL The average miles traveled for every gallon g of gas consumed by Leroy's car is represented by the function $m(g) = 28g$.

   a. Find a function $c(g)$ to represent the cost per gallon of gasoline. $c(g) = 2.95g$

   b. Use inverses to determine the function used to represent the cost per mile traveled in Leroy's car. $c(m) \approx 0.105m$

Differentiated Homework Options

Level	Assignment	Two-Day Option	
AL Basic	9–39, 52, 53, 55–76	9–39 odd, 57–60	10–38 even, 52, 53, 55, 56, 61–76
OL Core	9–47 odd, 48–53, 55–76	9–39, 57–60	40–53, 55, 56, 61–76
BL Advanced	40–73, (optional: 74–76)		

B **40. SHOES** The shoe size for the average U.S. teen or adult male can be determined using the formula $M(x) = 3x - 22$, where x is length of a foot in measured inches. The shoe size for the average U.S. teen or adult female can be found by using the formula $F(x) = 3x - 21$.

   a. Find the inverse of each function. $M^{-1}(x) = \dfrac{x + 22}{3}$; $F^{-1}(x) = \dfrac{x + 21}{3}$

   b. If Lucy wears a size $7\frac{1}{2}$ shoe, how long are her feet? $9\frac{1}{2}$ in.

41 **GEOMETRY** The formula for the area of a circle is $A = \pi r^2$.

   a. Find the inverse of the function. $r = \sqrt{\dfrac{A}{\pi}}$

   b. Use the inverse to find the radius of a circle with an area of 36 square centimeters. $\approx$**3.39 cm**

Use the horizontal line test to determine whether the inverse of each function is also a function.

42. $f(x) = 2x^2$ **no**   **43.** $f(x) = x^3 - 8$ **yes**   **44.** $g(x) = x^4 - 6x^2 + 1$ **no**

45. $h(x) = -2x^4 - x - 2$ **no**   **46.** $g(x) = x^5 + x^2 - 4x$ **no**   **47.** $h(x) = x^3 + x^2 - 6x + 12$ **no**

48. SHOPPING Felipe bought a used car. The sales tax rate was 7.25% of the selling price, and he paid $350 in processing and registration fees. Find the selling price if Felipe paid a total of $8395.75. **$7501.86**

 49. TEMPERATURE A formula for converting degrees Celsius to Fahrenheit is $F(x) = \dfrac{9}{5}x + 32$.

   a. Find the inverse $F^{-1}(x)$. Show that $F(x)$ and $F^{-1}(x)$ are inverses. **See Chapter 6**

   b. Explain what purpose $F^{-1}(x)$ serves.   **Answer Appendix.**
      It can be used to convert Fahrenheit to Celsius.

50. MEASUREMENT There are approximately 1.852 kilometers in a nautical mile.

   a. Write a function that converts nautical miles to kilometers. $K(m) = 1.852m$

   b. Find the inverse of the function that converts kilometers back to nautical miles.

   c. Using composition of functions, verify that these two functions are inverses.

50c. $K[K^{-1}(m)] = m$ and $K^{-1}[K(m)] = m$, so the two functions are inverses of each other.

50b. $K^{-1}(m) = \dfrac{1}{1.852}m$

51. 📊 **MULTIPLE REPRESENTATIONS** Consider the functions $y = x^n$ for $n = 0, 1, 2, \dots$.

   a. Graphing Use a graphing calculator to graph $y = x^n$ for $n = 0, 1, 2, 3$, and 4. **51a. See Chapter 6 Answer Appendix.**

   b. Tabular For which values of n is the inverse a function? Record your results in a table. **See Chapter 6 Answer Appendix.**

   c. Analytical Make a conjecture about the values of n for which the inverse of $f(x) = x^n$ is a function. Assume that n is a whole number. *n is odd.*

52. Sample answer: Sometimes; $y = \pm\sqrt{x}$ **is an example of a relation that is not a function, with an inverse being a function. A circle is an example of a relation that is not a function with an inverse not being a function.**

H.O.T. Problems   *Use Higher-Order Thinking Skills*

52. REASONING If a relation is *not* a function, then its inverse is *sometimes*, *always*, or *never* a function. Explain your reasoning.

53. OPEN ENDED Give an example of a function and its inverse. Verify that the two functions are inverses. **Sample answer:** $f(x) = 2x$, $f^{-1}(x) = 0.5x$, $f[f^{-1}(x)] = f^{-1}[f(x)] = x$

54. CHALLENGE Give an example of a function that is its own inverse.
   Sample answer: $f(x) = x$ and $f^{-1}(x) = x$ or $f(x) = -x$ and $f^{-1}(x) = -x$

55. **CCSS ARGUMENTS** Show that the inverse of a linear function $y = mx + b$, where $m \neq 0$ and $x \neq b$, is also a linear function. $y = \dfrac{1}{m}x - \dfrac{b}{m}$

56. WRITING IN MATH Suppose you have a composition of two functions that are inverses. When you put in a value of 5 for x, why is the result always 5? **See margin.**

CCSS **Teaching the Mathematical Practices**

Arguments Mathematically proficient students understand and use stated assumptions, definitions, and previously established results in constructing arguments. They make conjectures and build a logical progression of statements to explore the truth of their conjectures. And they are able to analyze situations by breaking them into cases, and can recognize and use counterexamples.

📁 **Enrichment**
AL **OL**

NAME _____ DATE _____ PERIOD _____

6-2 **Enrichment**

Reading Algebra

In mathematics, the term *group* has a special meaning. The following numbered sentences discuss the idea of group and one interesting example of a group.

01 To be a group, a set of elements and a binary operation must satisfy four conditions: the set must be closed under the operation, the operation must be associative, there must be an identity element, and every element must have an inverse.

02 The following six functions form a group under the operation of composition of functions: $f_1(x) = x$, $f_2(x) = \frac{1}{x}$, $f_3(x) = 1 - x$,

$f_4(x) = \frac{(x-1)}{x}$, $f_5(x) = \frac{x}{(x-1)}$ and $f_6(x) = \frac{1}{(1-x)}$.

03 This group is an example of a noncommutative group. For example, $f_3 \circ f_2 = f_6$ but $f_2 \circ f_3 = f_4$.

04 Some experimentation with this group will show that the identity element is f_1.

05 Every element is its own inverse except for f_4 and f_6, each of which is the inverse of the other.

📁 **Study Guide and Intervention**
AL **OL** **ELL**

📁 **Practice**
AL **OL** **BL** **ELL**

📁 **Word Problem Practice**
AL **OL** **BL** **ELL**

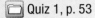

Yesterday's News Have students write how yesterday's concept of composition of functions helped them verify that two functions were inverses.

Formative Assessment

Check for student understanding of concepts in Lessons 6-1 and 6-2.

 Quiz 1, p. 53

Additional Answers

74.

75.

76.

57. SHORT RESPONSE If the length of a rectangular television screen is 24 inches and its height is 18 inches, what is the length of its diagonal in inches? **30 in.**

58. GEOMETRY If the base of a triangle is represented by $2x + 5$ and the height is represented by $4x$, which expression represents the area of the triangle? **D**

A $(2x + 5) + (4x)$

B $(2x + 5)(4x)$

C $\frac{1}{2}(2x + 5) + (4x)$

D $\frac{1}{2}(2x + 5)(4x)$

59. Which expression represents $f[g(x)]$ if $f(x) = x^2 + 3$ and $g(x) = -x + 1$? **J**

F $x^2 - x + 2$

G $-x^2 - 2$

H $-x^3 + x^2 - 3x + 3$

J $x^2 - 2x + 4$

60. SAT/ACT Which of the following is the inverse of $f(x) = \frac{3x - 5}{2}$? **A**

A $g(x) = \frac{2x + 5}{3}$

B $g(x) = \frac{2x - 5}{3}$

C $g(x) = \frac{3x + 5}{2}$

D $g(x) = 2x + 5$

E $g(x) = \frac{3x - 5}{2}$

Spiral Review

If $f(x) = 3x + 5$, $g(x) = x - 2$, and $h(x) = x^2 - 1$, find each value. (Lesson 6-1)

61. $g[f(3)]$ **12**

62. $f[h(-2)]$ **14**

63. $h[g(1)]$ **0**

64. CONSTRUCTION A picnic area has the shape of a trapezoid. The longer base is 8 more than 3 times the length of the shorter base, and the height is 1 more than 3 times the shorter base. What are the dimensions if the area is 4104 square feet? (Lesson 5-8) $b_1 = 25$ ft, $b_2 = 83$ ft, $h = 76$ ft

Find the value of c that makes each trinomial a perfect square. Then write the trinomial as a perfect square. (Lesson 4-5)

65. $x^2 + 34x + c$ **289;** $(x + 17)^2$

66. $x^2 - 11x + c$ **$\frac{121}{4}$;** $\left(x - \frac{11}{2}\right)^2$

Simplify. (Lesson 4-4)

67. $(3 + 4i)(5 - 2i)$ **$23 + 14i$**

68. $(\sqrt{6} + i)(\sqrt{6} - i)$ **7**

69. $\frac{1 + i}{1 - i}$ **i**

70. $\frac{4 - 3i}{1 + 2i}$ **$-\frac{2}{5} - \frac{11}{5}i$**

Determine the rate of change of each graph. (Lesson 2-3)

71.

$\frac{1}{2}$

72.
-3

73.

$\frac{3}{2}$

Skills Review

Graph each inequality. **74–76. See margin.**

74. $y > \frac{3}{4}x - 2$

75. $y \leq -3x + 2$

76. $y < -x - 4$

DifferentiatedInstruction (OL) (BL)

Extension The equation $f(x) = 2.3x + 61$ can be used to approximate height in centimeters when the length of a person's femur (thighbone) is known. Find $f^{-1}(x)$ and explain what it computes.

$f^{-1}(x) = \frac{x - 61}{2.3}$; it is used to find the approximate length of the femur when a person's height is known.

EXTEND 6-2

Graphing Technology Lab
Inverse Functions and Relations

You can use a TI-83/84 Plus graphing calculator to compare a function and its inverse using tables and graphs. Note that before you enter any values in the calculator, you should clear all lists.

Activity 1   Graph Inverses with Ordered Pairs

Graph $f(x) = \{(1, 2), (2, 4), (3, 6), (4, 8), (5, 10), (6, 12)\}$ and its inverse.

Step 1 Enter the x-values in **L1** and the y-values in **L2**. Then graph the function.

KEYSTROKES: [STAT] [ENTER] 1 [ENTER] 2 [ENTER] 3 [ENTER] 4 [ENTER]
5 [ENTER] 6 [ENTER] [▶] 2 [ENTER] 4 [ENTER] 6 [ENTER] 8 [ENTER]
10 [ENTER] 12 [ENTER] [2nd] [STAT PLOT] [ENTER] [ENTER] [GRAPH]

Adjust the window to reflect the domain and range.

Step 2 Define the inverse function by setting **Xlist** to **L2** and **Ylist** to **L1**. Then graph the inverse function.

KEYSTROKES: [2nd] [STAT PLOT] [▼] [ENTER] [ENTER] [▼] [▼] [2nd] [L2]
[▼] [2nd] [L1] [GRAPH]

Step 3 Graph the line $y = x$.

KEYSTROKES: [Y=] [X,T,θ,n] [GRAPH]

Step 2:

[0, 14] scl: 1 by [0, 14] scl: 1

Step 3:

[0, 14] scl: 1 by [0, 14] scl: 1

Activity 2   Graph Inverses with Function Notation

Graph $f(x) = 3x$ and its inverse $g(x) = \frac{x}{3}$.

Step 1 Clear the data from Activity 1.

KEYSTROKES: [2nd] [STAT PLOT] [ENTER] [▶] [ENTER] [▲] [▶] [ENTER]
[▶] [ENTER] [2nd] [QUIT]

Step 2 Enter $f(x)$ as **Y1**, $g(x)$ as **Y2**, and $y = x$ as **Y3**. Then graph.

KEYSTROKES: [Y=] 3 [X,T,θ,n] [ENTER] [X,T,θ,n] [÷] 3 [ENTER]
[X,T,θ,n] [GRAPH]

[0, 14] scl: 1 by [0, 14] scl: 1

1–6. See Chapter 6 Answer Appendix. 7. The graphs of a function and its inverse are reflections in the line $y = x$.

Exercises

Graph each function $f(x)$ and its inverse $g(x)$. Then graph $(f \circ g)(x)$.

1. $f(x) = 5x$
2. $f(x) = x - 3$
3. $f(x) = 2x + 1$
4. $f(x) = \frac{1}{2}x + 3$
5. $f(x) = x^2$
6. $f(x) = x^2 - 3$

7. What is the relationship between the graphs of a function and its inverse?

8. **MAKE A CONJECTURE** For any function $f(x)$ and its inverse $g(x)$, what is $(f \circ g)(x)$? $(f \circ g)(x) = x$

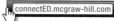 connectED.mcgraw-hill.com   **399**

From Concrete to Abstract

For each exercise, ask students to predict whether the graph passes through the origin. Then ask them to predict whether the graph of the inverse will pass through the origin. A function of the form $f(x) = ax^n$ contains the origin; if the graph of a function passes through the origin, its inverse will also pass through the origin.

Extending the Concept

Ask:
- If you have a set of ordered pairs that is a function, how can you generate the inverse of the function? Switch the values of x and y in each ordered pair.

1 Focus

Objective Compare a function and its inverse using a graphing calculator.

Materials for Each Student
- TI-83/84 Plus or other graphing calculator

Teaching Tip
Be sure students know how to enter data lists and switch between lists in their calculators.

2 Teach

Working in Cooperative Groups
Have students work in pairs, mixing abilities, to complete Activities 1 and 2. Students should confirm that their screens are identical at each step.

Ask:
- What is the difference between Activity 1 and Activity 2? Activity 1 compares discrete ordered pairs, while Activity 2 compares two continuous linear functions.

- How are the graphs of the function and its inverse related? They are reflections of each other across the line $y = x$.

Practice Have students complete Exercises 1–8.

3 Assess

Formative Assessment
Use Exercises 1, 2, and 5 to assess whether students can compare a function and its inverse using a graphing calculator.

Jose Luis Pelaez/Iconica/Getty Images

Additional Example

3 **PHYSICS** When an object is spinning in a circular path of radius 2 meters with velocity v, in meters per second, the centripetal acceleration a, in meters per second squared, is directed toward the center of the circle. The velocity v and acceleration a of the object are modeled by the function $v = \sqrt{2a}$.

a. Graph the function in the domain $\{a \mid a \geq 0\}$.

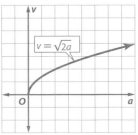

b. What would be the centripetal acceleration of an object spinning along the circular path with a velocity of 4 meters per second? 8 m/s²

2 Square Root Inequalities

Example 4 shows how to graph a square root inequality.

Additional Example

4 Graph $y > \sqrt{3x + 5}$.

Teach with Tech

Video Recording Have students create a video showing how to graph a square root function and explain how they determined the domain.

The Math Forum

The Math Forum
© Drexel University

Have your students refer to the Math Forum's Ask Dr. Math to learn more about square root functions. They can start their search at connectED.mcgraw-hill.com.

Real-WorldLink

On every string, the guitar player has an option of decreasing the length of the string in about 24 different ways. This will produce 24 different frequencies on each string.

Source: *Guitar World*

Problem-SolvingTip

CCSS Modeling Making a table is a good way to organize ordered pairs in order to see the general behavior of a graph.

Real-World Example 3 Use Graphs to Analyze Square Root Functions

MUSIC Refer to the application at the beginning of the lesson. The pitch, or frequency, measured in hertz (Hz) of a certain string can be modeled by $f(T) = \frac{1}{1.28}\sqrt{\dfrac{T}{0.0000708}}$, where T is tension in kilograms.

a. Graph the function for tension in the domain $\{T \mid 0 \leq T \leq 10\}$.

Make a table of values for $0 \leq T \leq 10$ and graph.

T	y(T)
0	0
1	92.8
2	131.3
3	160.8
4	185.7
5	207.6

T	f(T)
6	227.4
7	245.7
8	262.6
9	278.5
10	293.6

b. How much tension is needed for a pitch of over 200 Hz?

According to the graph and the table, more than 4.5 kilograms of tension is needed for a pitch of more than 200 hertz.

GuidedPractice See Chapter 6 Answer Appendix.

3. **MUSIC** The frequency of vibrations for a certain guitar string when it is plucked can be determined by $F = 200\sqrt{T}$, where F is the number of vibrations per second and T is the tension measured in pounds. Graph the function for $0 \leq T \leq 10$. Then determine the frequency for $T = 3$, 6, and 9 pounds.

---

2 **Square Root Inequalities** A **square root inequality** is an inequality involving square roots. They are graphed using the same method as other inequalities.

Example 4 Graph a Square Root Inequality

Graph $y < \sqrt{x - 4} - 6$.

Graph the boundary $y = \sqrt{x - 4} - 6$.

The domain is $\{x \mid x \geq 4\}$. Because y is *less than*, the shaded region should be *below* the boundary and within the domain.

CHECK Select a point in the shaded region, and verify that it is a solution of the inequality.

Test $(7, -5)$: $-5 \overset{?}{<} \sqrt{7 - 4} - 6$

$$-5 \overset{?}{<} \sqrt{3} - 6$$

$$-5 < -4.27 \checkmark$$

GuidedPractice **4A, 4B.** See Chapter 6 Answer Appendix.

4A. $f(x) \geq \sqrt{2x + 1}$

4B. $f(x) < -\sqrt{x + 2} - 4$

 402 | Lesson 6-3 | Square Root Functions and Inequalities

Follow-up

Students have explored square root functions.

Ask:

- Why would you choose a square root function to model a set of data instead of a polynomial function? Sample answer: The end behavior of a square root function might fit the data better than a polynomial function. Also, the domain of the square root function is $x \geq 0$, which may be applicable to many real-world situations that involve quantities such as time and distance.

Teaching the Mathematical Practices

CCSS

Modeling Mathematically proficient students are able to identify important quantities in a practical situation and map their relationships using such tools as diagrams, two-way tables, graphs, flowcharts and formulas. Encourage students to use various representations to better analyze functions.

Example 1 Identify the domain and range of each function. **2.** D = {x | x ≥ 5}; R = {f(x) | f(x) ≥ 0}

1. $f(x) = \sqrt{4x}$
   D = {x | x ≥ 0}; R = {f(x) | f(x) ≥ 0}

2. $f(x) = \sqrt{x-5}$

3. $f(x) = \sqrt{x+8} - 2$
   D = {x | x ≥ −8}; R = {f(x) | f(x) ≥ −2}

Example 2 Graph each function. State the domain and range. **4–7.** See margin.

4. $f(x) = \sqrt{x} - 2$

5. $f(x) = 3\sqrt{x-1}$

6. $f(x) = \frac{1}{2}\sqrt{x+4} - 1$

7. $f(x) = -\sqrt{3x-5} + 5$

Example 3 **8. OCEAN** The speed that a tsunami, or tidal wave, can travel is modeled by the equation $v = 356\sqrt{d}$, where v is the speed in kilometers per hour and d is the average depth of the water in kilometers. A tsunami is found to be traveling at 145 kilometers per hour. What is the average depth of the water? Round to the nearest hundredth of a kilometer. **0.17 km**

Example 4 Graph each inequality. **9–12. See Chapter 6 Answer Appendix.**

9. $f(x) \geq \sqrt{x} + 4$

10. $f(x) \leq \sqrt{x-6} + 2$

11. $f(x) < -2\sqrt{x+3}$

12. $f(x) > \sqrt{2x-1} - 3$

13. D = {x | x ≥ 0}; R = {f(x) | f(x) ≤ 2}   **14.** D = {x | x ≥ 0}; R = {f(x) | f(x) ≥ −6}

Practice and Problem Solving
Extra Practice is on page R6.

Example 1 Identify the domain and range of each function.   **15.** D = {x | x ≥ 2}; R = {f(x) | f(x) ≥ −8}

13. $f(x) = -\sqrt{2x} + 2$

14. $f(x) = \sqrt{x} - 6$

⑮ $f(x) = 4\sqrt{x-2} - 8$

16. $f(x) = \sqrt{x+2} + 5$

17. $f(x) = \sqrt{x-4} - 6$

18. $f(x) = -\sqrt{x-6} + 5$

Example 2 Graph each function. State the domain and range. **19–28. See Chapter 6 Answer Appendix.**

19. $f(x) = \sqrt{6x}$

20. $f(x) = -\sqrt{5x}$

16. D = {x | x ≥ −2};
    R = {f(x) | f(x) ≥ 5}

21. $f(x) = \sqrt{x-8}$

22. $f(x) = \sqrt{x+1}$

17. D = {x | x ≥ 4};
    R = {f(x) | f(x) ≥ −6}

23. $f(x) = \sqrt{x+3} + 2$

24. $f(x) = \sqrt{x-4} - 10$

18. D = {x | x ≥ 6};
    R = {f(x) | f(x) ≤ 5}

25. $f(x) = 2\sqrt{x-5} - 6$

26. $f(x) = \frac{3}{4}\sqrt{x+12} + 3$

27. $f(x) = -\frac{1}{5}\sqrt{x-1} - 4$

28. $f(x) = -3\sqrt{x+7} + 9$

Example 3 **29. SKYDIVING** The approximate time t in seconds that it takes an object to fall a distance of d feet is given by $t = \sqrt{\dfrac{d}{16}}$. Suppose a parachutist falls 11 seconds before the parachute opens. How far does the parachutist fall during this time? **1936 ft**

30. CCSS MODELING The velocity of a roller coaster as it moves down a hill is $V = \sqrt{v^2 + 64h}$, where v is the initial velocity in feet per second and h is the vertical drop in feet. The designer wants the coaster to have a velocity of 90 feet per second when it reaches the bottom of the hill.

  a. If the initial velocity of the coaster at the top of the hill is 10 feet per second, write an equation that models the situation. $90 = \sqrt{100 + 64h}$

  b. How high should the designer make the hill? **125 ft**

Differentiated Homework Options

Level	Assignment	Two-Day Option	
AL Basic	13–38, 48–66	13–37 odd, 54–57	14–38 even, 48–53, 58–66
OL Core	13–37 odd, 39–46, 48–66	13–38, 54–57	39–53, 58–66
BL Advanced	39–62, (optional: 63–66)		

Formative Assessment
Use Exercises 1–12 to check for understanding.

Use the chart at the bottom of this page to customize assignments for your students.

Additional Answers

4. D = {x | x ≥ 0};
    R = {f(x) | f(x) ≥ −2}

5. D = {x | x ≥ 1};
    R = {f(x) | f(x) ≥ 0}

6. D = {x | x ≥ −4};
    R = {f(x) | f(x) ≥ −1}

7. D = $\left\{ x \mid x \geq \dfrac{5}{3} \right\}$;
    R = {f(x) | f(x) ≤ 5}

Teaching the Mathematical Practices

CCSS

Reasoning Mathematically proficient students make sense of quantities and their relationships in problem situations. Quantitative reasoning entails habits of creating a coherent representation of the problem at hand; considering the units involved; attending to the meaning of quantities, not just how to compute them; and knowing and flexibly using different properties of operations and objects.

Multiple Representations

In Exercise 44, students use a graph, a table of values, and analysis to identify the effects of transformations on square-root functions.

WatchOut!

Error Analysis For Exercise 51, Remind students to graph $y = \sqrt{5x} + 10$ and then compare the graphs.

Additional Answers

45a.

45b. 1.57 s, 2.48 s, 3.14 s

46a.

[0, 400,000,000] scl: 50,000,000
by [0, 100,000] scl: 20,000

50. Sample answer: The domain is limited because square roots of negative numbers are imaginary. The range is limited due to the limitation of the domain.

Example 4   Graph each inequality. 31–38. See Chapter 6 Answer Appendix.

31. $y < \sqrt{x-5}$ **32.** $y > \sqrt{x+6}$

33 $y \geq -4\sqrt{x+3}$ **34.** $y \leq -2\sqrt{x-6}$

35. $y > 2\sqrt{x+7} - 5$ **36.** $y \geq 4\sqrt{x-2} - 12$

37. $y \leq 6 - 3\sqrt{x-4}$ **38.** $y < \sqrt{4x-12} + 8$

B **39. PHYSICS** The kinetic energy of an object is the energy produced due to its motion and mass. The formula for kinetic energy, measured in joules j, is $E = 0.5mv^2$, where m is the mass in kilograms and v is the velocity of the object in meters per second.

   a. Solve the above formula for v. $v = \sqrt{\dfrac{2E}{m}}$

   b. If a 1500-kilogram vehicle is generating 1 million joules of kinetic energy, how fast is it traveling? **about 36.5 m/s**

   c. *Escape velocity* is the minimum velocity at which an object must travel to escape the gravitational field of a planet or other object. Suppose a 100,000-kilogram ship must have a kinetic energy of 3.624×10^{14} joules to escape the gravitational field of Jupiter. Estimate the escape velocity of Jupiter. **about 85,135 m/s**

40. CCSS REASONING After an accident, police can determine how fast a car was traveling before the driver put on his or her brakes by using the equation $v = \sqrt{30fd}$. In this equation, v represents the speed in miles per hour, f represents the coefficient of friction, and d represents the length of the skid marks in feet. The coefficient of friction varies depending on road conditions. Assume that $f = 0.6$.

   a. Find the speed of a car that skids 25 feet. **about 21.2 mph**

   b. If your car is going 35 miles per hour, how many feet would it take you to stop? **about 68 ft**

   c. If the speed of a car is doubled, will the skid be twice as long? Explain.
   No; it is not a linear function. The skid will be 4 times as long.

Write the square root function represented by each graph.

41.

$y = \sqrt{x-4} - 6$

42.

$y = \sqrt{x+2} + 4$

43.

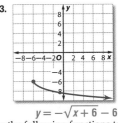

$y = -\sqrt{x+6} - 6$

C **44. MULTIPLE REPRESENTATIONS** In this problem, you will use the following functions to investigate transformations of square root functions.

$$f(x) = 4\sqrt{x-6} + 3 \qquad g(x) = \sqrt{16x+1} - 6 \qquad h(x) = \sqrt{x+3} + 2$$

   a. Graphical Graph each function on the same set of axes. **44a–e. See Chapter 6 Answer Appendix.**

   b. Analytical Identify the transformation on the graph of the parent function. What values caused each transformation?

   c. Analytical Which functions appear to be stretched or compressed vertically? Explain your reasoning.

   d. Verbal The two functions that are stretched appear to be stretched by the same magnitude. How is this possible?

   e. Tabular Make a table of the rate of change for all three functions between 8 and 12 as compared to 12 and 16. What generalization about rate of change in square root functions can be made as a result of your findings?

404 | Lesson 6-3 | Square Root Functions and Inequalities

DifferentiatedInstruction   AL   OL   ELL

If students continue to struggle with graphing square root inequalities,

Then have students work in pairs or small groups to discuss how to graph square root inequalities such as those in Exercises 31–38. Students should discuss how to determine the domain and range of an inequality, how to determine whether the boundary is solid or dashed, and how to determine where to shade the inequality.

45 **PENDULUMS** The period of a pendulum can be represented by $T = 2\pi\sqrt{\frac{L}{g}}$, where T is the time in seconds, L is the length in feet, and g is gravity, 32 feet per second squared.

a. See margin. Graph the function for $0 \leq L \leq 10$. **a, b. See margin.**

b. What is the period for lengths of 2, 5, and 8 feet?

46. PHYSICS Using the function $m = \dfrac{m_0}{\sqrt{1 - \left(\frac{v^2}{c^2}\right)}}$, Einstein's theory of relativity states that the apparent mass m of a particle depends on its velocity v. An object that is traveling extremely fast, close to the speed of light c, will appear to have more mass compared to its mass at rest, m_0.

a. Use a graphing calculator to graph the function for a 10,000-kilogram ship for the domain $0 \leq v \leq 300,000,000$. Use 300 million meters per second for the speed of light. **See margin.**

{0, 400,000,000} scl: 50,000,000 by {0, 100,000} scl: 20,000

b. What viewing window did you use to view the graph?

c. Determine the apparent mass m of the ship for speeds of 100 million, 200 million, and 299 million meters per second.
(100 million, 10,607)(200 million, 13,416)(299 million, 122,577)

H.O.T. Problems Use Higher-Order Thinking Skills

47. CHALLENGE Write an equation for a square root function with a domain of $\{x \mid x \geq -4\}$, a range of $\{y \mid y \leq 6\}$, and that passes through (5, 3). **Sample answer:** $y = -\sqrt{x + 4} + 6$

48. REASONING For what positive values of a are the domain and range of $f(x) = \sqrt[a]{x}$ the set of real numbers? **all positive odd numbers**

49. OPEN ENDED Write a square root function for which the domain is $\{x \mid x \geq 8\}$ and the range is $\{y \mid y \leq 14\}$. **Sample answer:** $y = -\sqrt{x - 8} + 14$

51. Molly; $y = \sqrt{5x + 10}$ has an x-intercept of -2 and would be to the right of the given graph.

50. WRITING IN MATH Explain why there are limitations on the domain and range of square root functions. **See margin.**

51. CCSS CRITIQUE Cleveland thinks that the graph and the equation represent the same function. Molly disagrees. Who is correct? Explain your reasoning.

$y = \sqrt{5x + 10}$

53a. Sample answer: The original is $y = x^2 + 2$ and inverse is $y = \pm\sqrt{x - 2}$.

53b. Sample answer: The original is $y = \pm\sqrt{x} + 4$ and inverse is $y = (x - 4)^2$.

52. WRITING IN MATH Explain why $y = \pm\sqrt{x}$ is not a function. **See margin.**

53. OPEN ENDED Write an equation of a relation that contains a radical and its inverse such that:

a. the original relation is a function, and its inverse is not a function.

b. the original relation is not a function, and its inverse is a function.

Additional Answer

52. To be a function, for every x-value there must be exactly one y-value. For every x in this equation there are two y-values, one that is negative and one that is positive. Also, the graph of $y = \pm\sqrt{x}$ does not pass the vertical line test.

Study Guide and Intervention
AL OL ELL

NAME _____ DATE _____ PERIOD _____

6-3 Study Guide and Intervention
Square Root Functions and Inequalities

Square Root Functions A function that contains the square root of a variable expression is a **square root function**. The domain of a square root function is those values for which the radicand is greater than or equal to 0.

Practice
AL OL BL ELL

NAME _____ DATE _____ PERIOD _____

6-3 Practice
Square Root Functions and Inequalities

Word Problem Practice
AL OL BL ELL

NAME _____ DATE _____ PERIOD _____

6-3 Word Problem Practice
Square Root Functions and Inequalities

Enrichment
OL BL

NAME _____ DATE _____ PERIOD _____

6-3 Enrichment

Reading Algebra

4 Assess

Crystal Ball In tomorrow's lesson, students will be simplifying radicals and using the calculator to approximate radicals. Have students write about how they think today's topic will connect with tomorrow's theme.

Additional Answers

62a.

62d. The number of cable TV systems rose steadily from 1985 to 2000. Then the number began to decline. The trend may continue for some years, but the number of cable TV systems cannot decline at this rate indefinitely. The number cannot fall below 0. It is not likely that the number would come close to 0 for the foreseeable future; there is no reason to believe that cable TV systems will not be in use.

Standardized Test Practice

54. The expression $-\dfrac{64x^6}{8x^3}$, $x \neq 0$, is equivalent to **D**

   A $8x^2$ **C** $-8x^2$

   B $8x^3$ **D** $-8x^3$

55. PROBABILITY For a game, Patricia must roll a standard die and draw a card from a deck of 26 cards, each card having a letter of the alphabet on it. What is the probability that Patricia will roll an odd number and draw a letter in her name? **G**

   F $\dfrac{2}{3}$ **H** $\dfrac{1}{13}$

   G $\dfrac{3}{26}$ **J** $\dfrac{1}{26}$

56. SHORT RESPONSE What is the product of $(d + 6)$ and $(d - 3)$? $d^2 + 3d - 18$

57. SAT/ACT Given the graph of the square root function below, which must be true? **E**

   I. The domain is all real numbers.
   II. The function is $y = \sqrt{x} + 3.5$.
   III. The range is $\{y \mid y \geq 3.5\}$.

   A I only   **D** II only

   B I, II, and III   **E** III only

   C II and III only

Spiral Review

Determine whether each pair of functions are inverse functions. Write *yes* or *no*. (Lesson 6-2)

58. $f(x) = 2x$ yes
   $g(x) = \dfrac{1}{2}x$

59. $f(x) = 3x - 7$ no
   $g(x) = \dfrac{1}{3}x - \dfrac{7}{16}$

60. $f(x) = \dfrac{3x + 2}{5}$ yes
   $g(x) = \dfrac{5x - 2}{3}$

61. TIME The formula $h = \dfrac{m}{60}$ converts minutes m to hours h, and $d = \dfrac{h}{24}$ converts hours h to days d. Write a function that converts minutes to days. (Lesson 6-1) $[d \circ h](m) = \dfrac{m}{1440}$

62. CABLE TV The number of households in the United States with cable TV after 1985 can be modeled by the function $C(t) = -43.2t^2 + 1343t + 790$, where t represents the number of years since 1985. (Lesson 5-4) **a, d. See margin.**

   a. Graph this equation for the years 1985 to 2005.

   b. Describe the turning points of the graph and its end behavior.

   c. What is the domain of the function? Use the graph to estimate the range for the function. $D = \{$all real numbers$\}$; $R = \{C(t) \mid C(t) \leq 11,225\}$

   d. What trends in households with cable TV does the graph suggest? Is it reasonable to assume that the trend will continue indefinitely?

   b. rel. max. between $t = 15$ and $t = 16$, and no rel. min.; $C(t) \rightarrow -\infty$ as $t \rightarrow -\infty$, $C(t) \rightarrow -\infty$ as $t \rightarrow +\infty$

Skills Review

Determine whether each number is *rational* or *irrational*.

63. 6.34   rational

64. 3.787887888…   irrational   **65.** 5.333…   rational

66. 1.25   rational

 406 | Lesson 6-3 | Square Root Functions and Inequalities

DifferentiatedInstruction   **OL**   **BL**

Extension How is the graph of $y = \sqrt{x - 3} - 5$ related to the graph of $y = \sqrt{x}$? It is translated 3 units right and 5 units down.

LESSON 6-4 *n*th Roots

:: Then	:: Now	:: Why?
• You worked with square root functions.	**1** Simplify radicals. **2** Use a calculator to approximate radicals.	• According to a world-wide injury prevention study, the number of collisions between bicycles and automobiles increased as the number of bicycles per intersection increased. The relationship can be expressed using the equation $c = \sqrt[5]{b^2}$, where b is the number of bicycles and c is the number of collisions.

NewVocabulary
*n*th root
radical sign
index
radicand
principal root

Common Core State Standards

Content Standards
A.SSE.2 Use the structure of an expression to identify ways to rewrite it.

Mathematical Practices
6 Attend to precision.

1 Simplify Radicals Finding the square root of a number and squaring a number are inverse operations. To find the square root of a number a, you must find a number with a square of a. Similarly, the inverse of raising a number to the nth power is finding the **nth root** of a number.

Powers	Factors	Words	Roots
$x^3 = 64$	$4 \cdot 4 \cdot 4 = 64$	4 is a cube root of 64.	$\sqrt[3]{64} = 4$
$x^4 = 625$	$5 \cdot 5 \cdot 5 \cdot 5 = 625$	5 is a fourth root of 625.	$\sqrt[4]{625} = 5$
$x^5 = 32$	$2 \cdot 2 \cdot 2 \cdot 2 \cdot 2 = 32$	2 is a fifth root of 32.	$\sqrt[5]{32} = 2$
$a^n = b$	$\underbrace{a \cdot a \cdot a \cdot \ldots \cdot a}_{n \text{ factors of } a} = b$	a is an nth root of b.	$\sqrt[n]{b} = a$

This pattern suggests the following formal definition of an nth root.

KeyConcept Definition of *n*th Root

Words	For any real numbers a and b, and any positive integer n, if $a^n = b$, then a is an nth root of b.
Example	Because $(-3)^4 = 81$, -3 is a fourth root of 81 and 3 is a principal root.

The symbol $\sqrt[n]{}$ indicates an nth root.

Some numbers have more than one real nth root. For example, 64 has two square roots, 8 and -8, since 8^2 and $(-8)^2$ both equal 64. When there is more than one real root and n is even, the nonnegative root is called the **principal root**.

Some examples of nth roots are listed below.

$\sqrt{25} = 5$ $\sqrt{25}$ indicates the principal square root of 25.

$-\sqrt{25} = -5$ $-\sqrt{25}$ indicates the opposite of the principal square root of 25.

$\pm\sqrt{25} = \pm5$ $\pm\sqrt{25}$ indicates both square roots of 25.

connectED.mcgraw-hill.com   **407**

Ingram Publishing/age fotostock

1 Focus

VerticalAlignment

Before Lesson 6-4 Work with square root functions.

Lesson 6-4 Simplify radicals. Use a calculator to approximate radicals.

After Lesson 6-4 Determine solutions of square root equations using algebraic methods.

2 Teach

Scaffolding Questions

Have students read the **Why?** section of the lesson.

Ask:

- What is the independent variable that is measured? number of bicycles

- What is the dependent variable that is calculated? number of collisions

- Do you think the number of collisions will increase or decrease as the number of bicycles increases? increase

Lesson 6-4 Resources

Resource	Approaching Level **AL**	On Level **OL**	Beyond Level **BL**	English Learners **ELL**
Teacher Edition	• Differentiated Instruction, p. 409	• Differentiated Instruction, pp. 409, 412	• Differentiated Instruction, p. 412	
Chapter Resource Masters	• Study Guide and Intervention, pp. 25–26 • Skills Practice, p. 27 • Practice, p. 28 • Word Problem Practice, p. 29	• Study Guide and Intervention, pp. 25–26 • Skills Practice, p. 27 • Practice, p. 28 • Word Problem Practice, p. 29 • Enrichment, p. 30	• Practice, p. 28 • Word Problem Practice, p. 29 • Enrichment, p. 30	• Study Guide and Intervention, pp. 25–26 • Skills Practice, p. 27 • Practice, p. 28 • Word Problem Practice, p. 29
Other	• 5-Minute Check 6-4 • Study Notebook • Teaching Algebra with Manipulatives	• 5-Minute Check 6-4 • Study Notebook • Teaching Algebra with Manipulatives	• 5-Minute Check 6-4 • Study Notebook	• 5-Minute Check 6-4 • Study Notebook • Teaching Algebra with Manipulatives

1 Simplify Radicals

Example 1 shows how to simplify *n*th root expressions. **Example 2** shows how to use the absolute value of the result to simplify the *n*th roots of expressions with even powers.

Formative Assessment
Use the Guided Practice exercises after each example to determine students' understanding of concepts.

Additional Example

1 **Simplify.**

a. $\pm\sqrt{16x^8}$ $\pm 4x^4$

b. $-\sqrt{(q^3+5)^4}$ $-(q^3+5)^2$

c. $\sqrt[5]{243a^{10}b^{15}}$ $3a^2b^3$

d. $\sqrt{-4}$ $2i$

Additional Examples also in Interactive Classroom PowerPoint® Presentations

IWB **Interactive White Board**
READY

WatchOut!

Common Error Be sure students understand that since $3^2 = 9$ and $(-3)^2 = 9$, the number 9 has two square roots, 3 and −3. However, the value of $\sqrt{9}$ is 3 only. To indicate both square roots and not just the principal root, the expression must be given as $\pm\sqrt{9}$.

Absolute values Remind students that adding absolute values only applies to expressions with single variables. Products or quotients of variables with odd powers could have domains that contain negative numbers and the simplified form would require absolute value bars.

Focus on Mathematical Content
Use of Absolute Value In general, if *n* is an integer greater than or equal to 2 and *a* is a real number, then $\sqrt[n]{a^n} = a$ if *n* is odd and $\sqrt[n]{a^n} = |a|$ if *n* is even.

ReviewVocabulary
pure imaginary numbers Square roots of negative real numbers; for any positive real number b, $\sqrt{-b^2} = \sqrt{b^2} \cdot \sqrt{-1}$, or bi, where i is the imaginary unit.

KeyConcept Real *n*th Roots

Suppose *n* is an integer greater than 1, and *a* is a real number.

a	*n* is even.	*n* is odd.
$a > 0$	1 unique positive and 1 unique negative real root: $\pm\sqrt[n]{a}$; positive root is principal root	1 unique positive and 0 negative real roots: $\sqrt[n]{a}$
$a < 0$	0 real roots	0 positive and 1 negative real root: $\sqrt[n]{a}$
$a = 0$	1 real root: $\sqrt[n]{0} = 0$	1 real root: $\sqrt[n]{0} = 0$

Example 1 Find Roots

Simplify.

a. $\pm\sqrt{16y^4}$

$\pm\sqrt{16y^4} = \pm\sqrt{(4y^2)^2}$

$\qquad = \pm 4y^2$

The square roots of $16y^4$ are $\pm 4y^2$.

b. $-\sqrt{(x^2-6)^8}$

$-\sqrt{(x^2-6)^8} = -\sqrt{[(x^2-6)^4]^2}$

$\qquad = -(x^2-6)^4$

The opposite of the principal square root of $(x^2-6)^8$ is $-(x^2-6)^4$.

c. $\sqrt[5]{243a^{20}b^{25}}$

$\sqrt[5]{243a^{20}b^{25}} = \sqrt[5]{(3a^4b^5)^5}$

$\qquad = 3a^4b^5$

The fifth root of $243a^{20}b^{25}$ is $3a^4b^5$

d. $\sqrt{-16x^4y^8}$

$\sqrt{-16x^4y^8}$

n is even. *b* is negative.

There are no real roots since $\sqrt{-16}$ is not a real number. However, there are two imaginary roots, $4ix^2y^4$ and $-4ix^2y^4$.

StudyTip
Odd Index If *n* is odd, there is only one real root. Therefore, there is no principal root when *n* is odd, and absolute value symbols are never needed.

GuidedPractice

1A. $\pm\sqrt{36x^{10}}$ $\pm 6x^5$

1B. $-\sqrt{(y+7)^{16}}$ $-(y+7)^8$

When you find an even root of an even power and the result is an odd power, you must use the absolute value of the result to ensure that the answer is nonnegative.

Example 2 Simplify Using Absolute Value

Simplify.

a. $\sqrt[4]{y^4}$

$\sqrt[4]{y^4} = |y|$

Since *y* could be negative, you must take the absolute value of *y* to identify the principal root.

b. $\sqrt[6]{64(x^2-3)^{18}}$

$\sqrt[6]{64(x^2-3)^{18}} = 2\left|(x^2-3)^3\right|$

Since the index 6 is even and the exponent 3 is odd, you must use absolute value.

GuidedPractice

2A. $\sqrt{36y^6}$ $6\left|y^3\right|$

2B. $\sqrt[4]{16(x-3)^{12}}$ $2\left|(x-3)^3\right|$

WatchOut!

Preventing Errors When discussing the information following Example 1, suggest that another way to simplify radicals that involve only numbers and no variables is to simplify the expression under the radical sign first. For example, $\sqrt{(-5)^2}$ could be rewritten by *first* simplifying under the radical sign to get $\sqrt{25}$, and then taking the principal root to get 5. Similarly, $\sqrt{(-2)^6}$ simplifies to $\sqrt{64}$, so the principal square root is 8.

2 Approximate Radicals with a Calculator Recall that real numbers that cannot be expressed as terminating or repeating decimals are irrational numbers. Approximations for irrational numbers are often used in real-world problems.

● Real-World Example 3 Approximate Radicals

INJURY PREVENTION Refer to the beginning of the lesson.

a. If $c = \sqrt[5]{b^2}$ represents the number of collisions and b represents the number of bicycle riders per intersection, estimate the number of collisions at an intersection that has 1000 bicycle riders per week.

Understand You want to find out how many collisions there are.

Plan Let 1000 be the number of bicycle riders. The number of collisions is c.

Solve
$$c = \sqrt[5]{b^2} \qquad \text{Original formula}$$
$$= \sqrt[5]{1000^2} \qquad b = 1000$$
$$\approx 15.85 \qquad \text{Use a calculator.}$$

There are about 16 collisions per week at the intersection.

Check
$$15.85 \stackrel{?}{=} \sqrt[5]{b^2} \qquad c = 15.85$$
$$15.85^5 \stackrel{?}{=} b^2 \qquad \text{Raise each side to the fifth power.}$$
$$1{,}000{,}337 \stackrel{?}{=} b^2 \qquad \text{Simplify.}$$
$$1000 \approx b \ \checkmark \qquad \text{Take the positive square root of each side.}$$

b. If the total number of collisions reported in one week is 21, estimate the number of bicycle riders that passed through that intersection.

$$c = \sqrt[5]{b^2} \qquad \text{Original formula}$$
$$21 = \sqrt[5]{b^2} \qquad c = 27$$
$$21^5 = b^2 \qquad \text{Raise each side to the fifth power.}$$
$$4{,}084{,}101 = b^2 \qquad \text{Simplify.}$$
$$2021 \approx b \qquad \text{Take the positive square root of each side.}$$

▸ **Guided**Practice

3A. The surface area of a sphere can be determined from the volume of the sphere using the formula $S = \sqrt[3]{36\pi V^2}$, where V is the volume. Determine the surface area of a sphere with a volume of 200 cubic inches. **about 165 in²**

3B. If the surface area of a sphere is about 214.5 square inches, determine the volume. **295.4 in²**

Real-World Link

77% of the employees in China commute by bicycle.

Source: International Bicycle Fund

Check Your Understanding ✓ ● = Step-by-Step Solutions begin on page R14.

Examples 1–2 Simplify.

1. $\pm\sqrt{100y^8}$ **±10y⁴**

2. $-\sqrt{49u^8v^{12}}$ **−7u⁴v⁶**

③ $\sqrt{(y-6)^8}$ **(y − 6)⁴**

4. $\sqrt[4]{16g^{16}h^{24}}$ **2g⁴h⁶**

5. $\sqrt{-16y^4}$ **±4iy²**

6. $\sqrt[6]{64(2y+1)^{18}}$ **2|(2y + 1)³|**

Example 3 Use a calculator to approximate each value to three decimal places.

7. $\sqrt{58}$ **7.616**

8. $-\sqrt{76}$ **−8.718**

9. $\sqrt[5]{-43}$ **−2.122**

10. $\sqrt[4]{71}$ **2.903**

11. **CCSS PERSEVERANCE** The radius r of the orbit of a television satellite is given by $\sqrt[3]{\dfrac{GMt^2}{4\pi^2}}$, where G is the universal gravitational constant, M is the mass of Earth, and t is the time it takes the satellite to complete one orbit. Find the radius of the satellite's orbit if G is $6.67 \times 10^{-11} N \cdot m^2/kg^2$, M is 5.98×10^{24} kg, and t is 2.6×10^6 seconds. **about 4.088 × 10⁸ m**

David C. Johnson

Differentiated Instruction ⒶⓁ ⓄⓁ

Logical Learners Some students tend to think that x must represent a positive number and $-x$ must represent a negative number. Reading $-x$ as "the opposite of x" should help them understand that $-x$ is 9 if $x = -9$. Also, explain that -9 has no square root that is a real number. That is, no real number can be squared to give -9. Remind students that $\sqrt{-9}$ is $3i$, an imaginary number.

Teach with Tech

Blog Have students write a blog entry explaining how to find the number of roots (and their signs) before performing any calculations.

Additional Example

2 Simplify.

a. $\sqrt[6]{t^6}$ **|t|**

b. $\sqrt[5]{243(x+2)^{15}}$ **3(x + 2)³**

---

2 Approximate Radicals With a Calculator

Example 3 shows students how to use a calculator to approximate the solution to a real-world situation.

Additional Example

3 **SPACE** Designers must create satellites that can resist damage from being struck by small particles of dust and rocks. A study showed that the diameter in millimeters d of the hole created in a solar cell by a dust particle traveling with energy k in joules is about $d = 0.926\sqrt[3]{k} + 0.169$.

a. Estimate the diameter of a hole created by a particle traveling with energy 3.5 joules. **1.237 mm**

b. If a hole has diameter 2.5 millimeters, estimate the energy with which the particle that made the hole was traveling. **23.9 joules**

CCSS **Teaching the Mathematical Practices**

Perseverance Mathematically proficient students start by explaining to themselves the meaning of a problem and looking for entry points to its solution. They analyze givens, constraints, relationships, and goals and make conjectures about the form and meaning of the solution and plan a solution pathway rather than simply jumping into a solution attempt.

Formative Assessment

Use Exercises 1–11 to check for understanding.

Use the chart at the bottom of this page to customize assignments for your students.

Multiple Representations

In Exercise 58, students use a table of values, a graph, and logical analysis to explore inverse functions.

(CCSS) Teaching the Mathematical Practices

Sense-Making Mathematically proficient students start by explaining to themselves the meaning of a problem and looking for entry points to its solution. They analyze givens, constraints, relationships, and goals. They check their answers to problems using a different method, and they continually ask themselves, "Does this make sense?"

Additional Answers

57. bald eagle: $\approx$ 226.5 Cal/d; golden retriever: $\approx$ 939.6 Cal/d; komodo dragon: $\approx$ 1811.8 Cal/d; bottlenose dolphin: $\approx$ 3235.5 Cal/d; Asian elephant: $\approx$ 24,344.4 Cal/d

64. Sample answer: They are needed to ensure that the answer is not a negative number. When we take any odd root of a number, we find that there is just one answer. If the number is positive, the root is positive. If the number is negative, the root is negative. Every positive real number has two *nth* roots when *n* is even; one of these roots is positive and one is negative. Negative real numbers do not have *nth* roots when *n* is even. Absolute value symbols are never needed when finding odd roots. When finding even *nth* roots, absolute value symbols are sometimes necessary, as with square roots.

Practice and Problem Solving Extra Practice is on page R6.

Examples 1–2 Simplify.

12. $\pm\sqrt{121x^4y^{16}}$ $\pm 11x^2y^8$
13. $\pm\sqrt{225a^{16}b^{36}}$ $\pm 15a^8b^{18}$
14. $\pm\sqrt{49x^4}$ $\pm 7x^2$

15. $-\sqrt{16c^4d^2}$ $-4c^2|d|$
16. $-\sqrt{81a^{16}b^{20}c^{12}}$ $-9a^8b^{10}c^6$
17. $-\sqrt{400x^{32}y^{40}}$ $-20x^{16}y^{20}$

18. $\sqrt{(x+15)^4}$ $(x+15)^2$
19. $\sqrt{(x^2+6)^{16}}$ $(x^2+6)^8$
20. $\sqrt{(a^2+4a)^{12}}$ $(a^2+4a)^6$

21. $\sqrt[3]{8a^6b^{12}}$ $2a^2b^4$
22. $\sqrt[6]{d^{24}x^{36}}$ d^4x^6
23. $\sqrt[3]{27b^{18}c^{12}}$ $3b^6c^4$

24. $-\sqrt{(2x+1)^6}$ $-|(2x+1)^3|$
25. $\sqrt[4]{-(x+2)^8}$ $i(x+2)^4$
26. $\sqrt[3]{-(y-9)^9}$ $-(y-9)^3$

27. $\sqrt[6]{x^{18}}$ $|x^3|$
28. $\sqrt[4]{a^{12}}$ $|a^3|$
29. $\sqrt[3]{a^{12}}$ a^4

30. $\sqrt[4]{81(x+4)^4}$ $3|(x+4)|$
31. $\sqrt[3]{(4x-7)^{24}}$ $(4x-7)^8$
32. $\sqrt[3]{(y^3+5)^{18}}$ $(y^3+5)^6$

33. $\sqrt[4]{256(5x-2)^{12}}$ $4|(5x-2)^3|$
34. $\sqrt[8]{x^{16}y^8}$ $x^2|y|$
35. $\sqrt[5]{32a^{15}b^{10}}$ $2a^3b^2$

Example 3 **36.** **SHIPPING** An online book store wants to increase the size of the boxes it uses to ship orders. The new volume *N* is equal to the old volume *V* times the scale factor *F* cubed, or $N = V \cdot F^3$. What is the scale factor if the old volume was 0.8 cubic feet and the new volume is 21.6 cubic feet? **3**

37. **GEOMETRY** The side length of a cube is determined by $r = \sqrt[3]{V}$, where *V* is the volume in cubic units. Determine the side length of a cube with a volume of 512 cm^3. **8 cm**

Use a calculator to approximate each value to three decimal places.

38. $\sqrt{92}$ 9.592
39. $-\sqrt{150}$ −12.247
40. $\sqrt{0.43}$ 0.656
41. $\sqrt{0.62}$ 0.787

42. $\sqrt[3]{168}$ 5.518
43. $\sqrt[5]{-4382}$ −5.350
44. $\sqrt[6]{(8912)^2}$ 20.733
45. $\sqrt[5]{(4756)^2}$ 29.573

46. **GEOMETRY** The radius *r* of a sphere with volume *V* can be found using the formula $r = \sqrt[3]{\dfrac{3V}{4\pi}}$.

a. Determine the radius for volumes of 1000 cm^3, 8000 cm^3, and 64,000 cm^3. (1000, 6.2), (8000, 12.4), (64,000, 24.8)

b. How does the volume of the sphere change if the radius is doubled? Explain. Sample answer: As *r* doubles, the volume increases by a factor of 2^3 or 8.

Simplify.
47. $\sqrt{196c^6d^4}$ $14|c^3|d^2$
48. $\sqrt{-64y^8z^6}$ $8iy^4|z^3|$
49. $\sqrt[3]{-27a^{15}b^9}$ $-3a^5b^3$
50. $\sqrt[4]{-16x^{16}y^8}$ $2x^4y^2\sqrt[4]{-1}$
51. $\sqrt{400x^{16}y^6}$ $20x^8|y^3|$
52. $\sqrt[3]{8c^3d^{12}}$ $2cd^4$
53. $\sqrt[3]{64(x+y)^6}$ $4(x+y)^2$
54. $\sqrt[5]{-(y-z)^{15}}$ $-(y-z)^3$

55. **PHYSICS** Johannes Kepler developed the formula $d = \sqrt[3]{6t^2}$, where *d* is the distance of a planet from the Sun in millions of miles and *t* is the number of Earth-days that it takes for the planet to orbit the Sun. If the length of a year on Mars is 687 Earth-days, how far from the Sun is Mars? **about 141 million mi**

56. **(CCSS) SENSE-MAKING** All matter is composed of atoms. The nucleus of an atom is the center portion of the atom that contains most of the mass of the atom. A theoretical formula for the radius *r* of the nucleus of an atom is $r = (1.3 \times 10^{-15})\sqrt[3]{A}$ meters, where *A* is the mass number of the nucleus. Find the radius of the nucleus for each atom in the table.

Atom	Mass Number
carbon	6
oxygen	8
sodium	11
aluminum	13
chlorine	17

carbon: $\approx 2.36 \times 10^{-15}$; oxygen: $= 2.6 \times 10^{-15}$; sodium: $\approx 2.9 \times 10^{-15}$; aluminum: $\approx 3.1 \times 10^{-15}$; chlorine: $\approx 3.34 \times 10^{-15}$

 410 | Lesson 6-4 | *nth* Roots

Differentiated Homework Options

Level	Assignment	Two-Day Option	
AL Basic	12–45, 59, 61, 63, 64, 70–92	13–45 odd, 70–73	12–44 even, 59, 61, 63, 64, 74–92
OL Core	13–45 odd, 46, 47–53 odd, 55–59, 61, 63, 64, 70–92	12–45, 70–73	46–59, 61, 63, 64, 74–92
BL Advanced	46–86, (optional: 87–92)		

57 **BIOLOGY** Kleiber's Law, $P = 73.3\sqrt[4]{m^3}$, shows the relationship between the mass m in kilograms of an organism and its metabolism P in Calories per day. Determine the metabolism for each of the animals listed at the right. **See margin.**

Animal	Mass (kg)
bald eagle	4.5
golden retriever	30
komodo dragon	72
bottlenose dolphin	156
Asian elephant	2300

58. **MULTIPLE REPRESENTATIONS** In this problem, you will use $f(x) = x^n$ and $g(x) = \sqrt[n]{x}$ to explore inverses. **a, b. See Chapter 6 Answer Appendix.**

   a. Tabular Make tables for $f(x)$ and $g(x)$ using $n = 3$ and $n = 4$.

   b. Graphical Graph the equations. **58c.** $f(x) = x^3$; $f(x) = x^4$; $g(x) = \sqrt[3]{x}$; $f(x) = x^3$ and $g(x) = \sqrt[3]{x}$

   c. Analytical Which equations are functions? Which functions are one-to-one?

   d. Analytical For what values of n are $g(x)$ and $f(x)$ inverses of each other? **positive odd values**

   e. Verbal What conclusions can you make about $g(x) = \sqrt[n]{x}$ and $f(x) = x^n$ for all positive even values of n? for odd values of n? **Sample answer: For all positive odd values of n, $f(x)$ and $g(x)$ are inverse functions. For all positive even values of n, $f(x)$ and $g(x)$ are inverse functions only if the range of $f(x)$ and the domain of $g(x)$ are restricted to positive values.**

H.O.T. Problems Use Higher-Order Thinking Skills

59. **CCSS CRITIQUE** Ashley and Kimi are simplifying $\sqrt[4]{16x^4y^8}$. Is either of them correct? Explain your reasoning.

Kimi; Ashley's error was keeping the y^2 inside the absolute value symbol.

Ashley	Kimi
$\sqrt[4]{16x^4y^8} = \sqrt[4]{(2xy^2)^4}$	$\sqrt[4]{16x^4y^8} = \sqrt[4]{(2xy^2)^4}$
$= 2\lvert xy^2 \rvert$	$= 2y^2\lvert x \rvert$

60. **CHALLENGE** Under what conditions is $\sqrt{x^2 + y^2} = x + y$ true? **when x or $y = 0$ and the other variable is ≥ 0**

61. **REASONING** Determine whether the statement $\sqrt[4]{(-x)^4} = x$ is *sometimes*, *always*, or *never* true. **Sample answer: Sometimes; when $x = -3$, $\sqrt[4]{(-x)^4} = \lvert(-x)\rvert$ or 3. When $x = 3$, $\sqrt[4]{(-x)^4} = \lvert 3 \rvert$ or 3.**

62. **CHALLENGE** For what real values of x is $\sqrt[3]{x} > x$? **$0 < x < 1$, $x < -1$**

63. **OPEN ENDED** Write a number for which the principal square root and cube root are both integers. **Sample answers: 1, 64**

64. **WRITING IN MATH** Explain when and why absolute value symbols are needed when taking an nth root. **See margin.**

65. **CHALLENGE** Write an equivalent expression for $\sqrt[3]{2x} \cdot \sqrt[3]{8y}$. Simplify the radical. **$2\sqrt[3]{2xy}$**

CHALLENGE Simplify each expression.

66. $\sqrt[4]{0.0016}$ **0.2**

67. $\sqrt[7]{-0.0000001}$ **−0.1**

68. $\dfrac{\sqrt[5]{-0.00032}}{\sqrt[3]{-0.027}}$ **$\dfrac{2}{3}$**

69. **CHALLENGE** Solve $-\dfrac{5}{\sqrt{a}} = -125$ for a. **$\dfrac{1}{625}$**

connectED.mcgraw-hill.com **411**

CCSS **Teaching the Mathematical Practices**

Critique Mathematically proficient students are also able to compare the effectiveness of two plausible arguments, distinguish correct logic or reasoning from that which is flawed, and—if there is a flaw in an argument—explain what it is.

4 Assess

4 Assess

Ticket Out the Door Make several copies of five different radicals similar to those in Example 1. Give one expression to each student. Have students simplify their radicals and, as the students leave the room, ask them to tell you the simplified expression.

Formative Assessment
Check for student understanding of concepts in Lessons 6-3 and 6-4.

 Quiz 2, p. 53

Additional Answers

74.

75.

76.

Standardized Test Practice

70. What is the value of w in the equation $\frac{1}{2}(4w + 36) = 3(4w - 3)$? **B**

A 2
B 2.7
C 27
D 36

71. What is the product of the complex numbers $(5 + i)$ and $(5 - i)$? **G**

F 24
G 26
H $25 - i$
J $26 - 10i$

72. EXTENDED RESPONSE A cylindrical cooler has a diameter of 9 inches and a height of 11 inches. Tate plans to use it for soda cans that have a diameter of 2.5 inches and a height of 4.75 inches.

a. Tate plans to place two layers consisting of 9 cans each into the cooler. What is the volume of the space that will not be filled with the cans? **280.1 in³**

b. Find the ratio of the volume of the cooler to the volume of the cans in part **a**. **about 1.7**

73. SAT/ACT Which of the following is closest to $\sqrt[3]{7.32}$? **B**

A 1.8   D 2.1
B 1.9   E 2.2
C 2.0

Spiral Review

Graph each function. (Lesson 6-3) **74–76. See margin.**

74. $y = \sqrt{x - 5}$ **75.** $y = \sqrt{x} - 2$ **76.** $y = 3\sqrt{x} + 4$

77. HEALTH The average weight of a baby born at a certain hospital is $7\frac{1}{2}$ pounds and the average length is 19.5 inches. One kilogram is about 2.2 pounds and 1 centimeter is about 0.3937 inches. Find the average weight in kilograms and the length in centimeters. (Lesson 6-2) **3.41 kg and 49.53 cm**

Simplify. (Lesson 5-1)

78. $(4c - 5) - (c + 11) + (-6c + 17)$ **$-3c + 1$**   **79.** $(11x^2 + 13x - 15) - (7x^2 - 9x + 19)$ **$4x^2 + 22x - 34$**

80. $(d - 5)(d + 3)$ **$d^2 - 2d - 15$**   **81.** $(2a^2 + 6)^2$ **$4a^4 + 24a^2 + 36$**

82. GAS MILEAGE The gas mileage y in miles per gallon for a certain vehicle is given by the equation $y = 10 + 0.9x - 0.01x^2$, where x is the speed of the vehicle between 10 and 75 miles per hour. Find the range of speeds that would give a gas mileage of at least 25 miles per gallon. (Lesson 4-8) **$22.087 \leq x \leq 67.91$ mph**

Write each equation in vertex form, if not already in that form. Identify the vertex, axis of symmetry, and direction of opening. Then graph the function. (Lesson 4-7)

83–86. See Chapter 6 Answer Appendix.

83. $y = -6(x + 2)^2 + 3$ **84.** $y = -\frac{1}{3}x^2 + 8x$ **85.** $y = (x - 2)^2 - 2$ **86.** $y = 2x^2 + 8x + 10$

Skills Review

Find each product.

87. $(x + 4)(x + 5)$ **$x^2 + 9x + 20$**   **88.** $(y - 3)(y + 4)$ **$y^2 + y - 12$**   **89.** $(a + 2)(a - 9)$ **$a^2 - 7a - 18$**

90. $(a - b)(a - 3b)$ **$a^2 - 4ab + 3b^2$**   **91.** $(x + 2y)(x - y)$ **$x^2 + xy - 2y^2$**   **92.** $2(w + z)(w - 4z)$ **$2w^2 - 6wz - 8z^2$**

 412 | Lesson 6-4 | nth Roots

DifferentiatedInstruction **OL** **BL**

Extension Under what conditions is it true that $\dfrac{\sqrt[n]{a}}{\sqrt[n]{b}} = \sqrt[n]{\dfrac{a}{b}}$? *$n$ must be a positive integer; if n is even then $a \geq 0$ and $b > 0$; if n is odd then $b \neq 0$.*

EXTEND 6-4

Graphing Technology Lab
Graphing *n*th Root Functions

You can use a graphing calculator to graph *n*th root functions.

CCSS Common Core State Standards
Content Standards
F.IF.7.b Graph square root, cube root, and piecewise-defined functions, including step functions and absolute value functions.
F.BF.3 Identify the effect on the graph of replacing $f(x)$ by $f(x) + k$, $k\,f(x)$, $f(kx)$, and $f(x + k)$ for specific values of k (both positive and negative); find the value of k given the graphs. Experiment with cases and illustrate an explanation of the effects on the graph using technology.
Mathematical Practices
5 Use appropriate tools strategically.

Example 1 Graph an *n*th Root Function

Graph $y = \sqrt[5]{x}$.

Enter the equation as **Y1** and graph.

KEYSTROKES: [Y=] 5 [MATH] 5 [X,T,θ,*n*] [GRAPH]

Another way to enter the equation is to use $y = x^{\frac{1}{5}}$. You will learn about this later in Chapter 6.

[−10, 10] scl: 1 by [−10, 10] scl: 1

Example 2 *n*th Root Functions with Different Roots

Graph and compare $y = \sqrt{x}$ **and** $y = \sqrt[4]{x}$.

Enter $y = \sqrt{x}$ as **Y1** and $y = \sqrt[4]{x}$ as **Y2**. Change the viewing window. Then graph.

KEYSTROKES: [Y=] [CLEAR] [2nd] [√] [X,T,θ,*n*] [ENTER] 4
[MATH] 5 [X,T,θ,*n*] [GRAPH]
[WINDOW] [(−)] 2 [ENTER] 10 [ENTER] 1 [ENTER]
[(−)] 2 [ENTER] 10 [ENTER] 1

[−2, 10] scl: 1 by [−2, 10] scl: 1

Example 3 *n*th Root Functions with Different Radicands

Graph and compare $y = \sqrt[3]{x}$, $y = \sqrt[3]{x + 4}$, **and** $y = \sqrt[3]{x} + 4$.

Enter $y = \sqrt[3]{x}$ as **Y1**, $y = \sqrt[3]{x + 4}$ as **Y2**, and $y = \sqrt[3]{x} + 4$ as **Y3**. Then graph in the standard viewing window.

KEYSTROKES: [Y=] [CLEAR] 3 [MATH] 5 [X,T,θ,*n*] [ENTER] [CLEAR] 3
[MATH] 5 [(] [X,T,θ,*n*] [+] 4 [)] [ENTER] 3 [MATH]
5 [X,T,θ,*n*] [+] 4 [ENTER] [ZOOM] 6

[−10, 10] scl: 1 by [−10, 10] scl: 1

Exercises

Graph each function. 1–6. See Chapter 6 Answer Appendix.

1. $y = \sqrt[4]{x}$

2. $y = \sqrt[4]{x + 2}$

3. $y = \sqrt[4]{x} + 2$

4. $y = \sqrt[5]{x}$

5. $y = \sqrt[5]{x - 5}$

6. $y = \sqrt[5]{x} - 5$

7. What is the effect of adding or subtracting a constant under the radical sign? **The graph shifts to the right or left.**

8. What is the effect of adding or subtracting a constant outside the radical sign? **The graph shifts up or down.**

connectED.mcgraw-hill.com **413**

---

1 Focus

Objective Use a graphing calculator to graph *n*th root functions.

Materials for Each Student

- TI-83/84 Plus or other graphing calculator

Teaching Tip

There is more than one way to enter the square root function into the Y= list. The function can be entered as $\sqrt{x}$ using the square root key, as $x \wedge (1/2)$ or as $x \wedge (0.5)$.

2 Teach

Working in Cooperative Groups

Have students work in pairs, mixing abilities, to complete the activity.

Ask:

- What values of *n* allow negative domain values for $\sqrt[n]{x}$? odd values of *n*

- Why is the domain restricted to nonnegative values for $\sqrt[n]{x}$ if *n* is even? An even root of a negative number is not a real number.

Practice Have students complete Exercises 1–8.

3 Assess

Formative Assessment

Use Exercises 1–3 to assess whether students can use a graphing calculator to graph *n*th root functions.

From Concrete to Abstract

Ask students to describe how to find the ordered pairs for the intersection of a graph and the *y*-axis, and the ordered pairs for the intersection of the graph and the *x*-axis. Let $x = 0$, then solve for *y*; let $y = 0$, then solve for *x*.

Formative Assessment

Use the Mid-Chapter Quiz to assess students' progress in the first half of the chapter.

For problems answered incorrectly, have students review the lessons indicated in parentheses.

Summative Assessment

📁 Mid-Chapter Test, p. 55

ᵉＡssessment Customize and create multiple versions of your Mid-Chapter Quiz and their answer keys.

FOLDABLES® StudyOrganizer

Dinah Zike's Foldables®

Before students complete the Mid-Chapter Quiz, encourage them to review the information for Lessons 6-1 through 6-4 in their Foldables.

Additional Answers

4. $\left(\frac{f}{g}\right)(x) = \frac{2x^2 + 4x - 3}{5x - 2}, x \neq \frac{2}{5}$

12. $h^{-1}(x) = \frac{5}{2}(x - 8)$

13. $f^{-1}(x) = \frac{9}{4}x + 3$

14. $h^{-1}(x) = -\frac{3}{10}x - 5$

15. $f^{-1}(x) = 7x - 12$

Mid-Chapter Quiz
Lessons 6-1 through 6-4

Given $f(x) = 2x^2 + 4x - 3$ and $g(x) = 5x - 2$, find each function. (Lesson 6-1)

1. $(f + g)(x)$ $(f + g)(x) = 2x^2 + 9x - 5$

2. $(f - g)(x)$ $(f - g)(x) = 2x^2 - x - 1$

3. $(f \cdot g)(x)$ $(f \cdot g)(x) = 10x^3 + 16x^2 - 23x + 6$

4. $\left(\frac{f}{g}\right)(x)$ See margin.

5. $[f \circ g](x)$ $(f \circ g)(x) = 50x^2 - 20x - 3$

6. $[g \circ f](x)$ $(g \circ f)(x) = 10x^2 + 20x - 17$

7. SHOPPING Mrs. Ross is shopping for her children's school clothes. She has a coupon for 25% off her total. The sales tax of 6% is added to the total after the coupon is applied. (Lesson 6-1)

   a. Express the total price after the discount and the total price after the tax using function notation. Let x represent the price of the clothing, $p(x)$ represent the price after the 25% discount, and $g(x)$ represent the price after the tax is added. $p(x) = 0.75x, g(x) = 1.06x$

   b. Which composition of functions represents the final price, $p[g(x)]$ or $g[p(x)]$? Explain your reasoning.
Since $p[g(x)] = g[p(x)]$, either function represents the price.

Determine whether each pair of functions are inverse functions. Write *yes* or *no*. (Lesson 6-2)

8. $f(x) = 2x + 16$
$g(x) = \frac{1}{2}x - 8$ **yes**

9. $g(x) = 4x + 15$
$h(x) = \frac{1}{4}x - 15$ **no**

10. $f(x) = x^2 - 5$
$g(x) = 5 + x^{-2}$ **no**

11. $g(x) = -6x + 8$
$h(x) = \frac{8 - x}{6}$ **yes**

Find the inverse of each function, if it exists. (Lesson 6-2)
12–15. See margin.

12. $h(x) = \frac{2}{5}x + 8$

13. $f(x) = \frac{4}{9}(x - 3)$

14. $h(x) = -\frac{10}{3}(x + 5)$

15. $f(x) = \frac{x + 12}{7}$

16. JOBS Louise runs a lawn care service. She charges $25 for supplies plus $15 per hour. The function $f(h) = 15h + 25$ gives the cost $f(h)$ for h hours of work. (Lesson 6-2)

   a. Find $f^{-1}(h)$. What is the significance of $f^{-1}(h)$?
See Chapter 6 Answer Appendix.

   b. If Louise charges a customer $85, how many hours did she work? **4 hours**

414 | Chapter 6 | Mid-Chapter Quiz

Graph each inequality. (Lesson 6-3)

17. $y < \sqrt{x - 5}$

18. $y \leq -2\sqrt{x}$

19. $y > \sqrt{x + 9} + 3$

20. $y \geq \sqrt{x + 4} - 5$

17–20. See Chapter 6 Answer Appendix.

Graph each function. State the domain and range of each function. (Lesson 6-3) **21, 22. See Chapter 6 Answer Appendix.**

21. $y = 2 + \sqrt{x}$

22. $y = \sqrt{x + 4} - 1$

23. MULTIPLE CHOICE What is the domain of $f(x) = \sqrt{2x + 5}$? (Lesson 6-3) **D**

A $\left\{x \mid x > \frac{5}{2}\right\}$

C $\left\{x \mid x \geq \frac{5}{2}\right\}$

B $\left\{x \mid x > -\frac{5}{2}\right\}$

D $\left\{x \mid x \geq -\frac{5}{2}\right\}$

Simplify. (Lesson 6-4)

24. $\pm\sqrt{121a^4b^{18}}$ $\pm 11a^2 \mid b^9 \mid$

25. $\sqrt{(x^4 + 3)^{12}}$ $(x^4 + 3)^6$

26. $\sqrt[3]{27(2x - 5)^{15}}$ $3(2x - 5)^5$

27. $\sqrt[5]{-(y - 6)^{20}}$ $-(y - 6)^4$

28. $\sqrt[3]{8(x + 4)^6}$ $2(x + 4)^2$

29. $\sqrt[4]{16(y + x)^8}$ $2(y + x)^2$

30. MULTIPLE CHOICE The radius of the cylinder below is equal to the height of the cylinder. The radius r can be found using the formula $r = \sqrt[3]{\frac{V}{\pi}}$. Find the radius of the cylinder if the volume is 500 cubic inches. (Lesson 6-4) **G**

F   2.53 inches

G   5.42 inches

H   7.94 inches

J   24.92 inches

31. PRODUCTION The cost in dollars of producing p cell phones in a factory is represented by $C(p) = 5p + 60$. The number of cell phones produced in h hours is represented by $P(h) = 40h$. (Lesson 6-1)

   a. Find the composition function. $C[P(h)] = 200h + 60$

   b. Determine the cost of producing cell phones for 8 hours. $1660

InterventionPlanner

TIER **1** On Level **OL**	TIER **2** Strategic Intervention **AL** approaching grade level	TIER **3** Intensive Intervention 2 or more grades below level
If students miss about 25% of the exercises or less,	**If** students miss about 50% of the exercises,	**If** students miss about 75% of the exercises,
Then choose a resource:	**Then** choose a resource:	**Then** use *Math Triumphs, Alg. 2,* Ch. 1
SE Lessons 6-1, 6-2, 6-3, and 6-4 📁 Skills Practice, pp. 7, 14, 21, and 27 **TE** Chapter Project 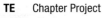 connectED.mcgraw-hill.com Self-Check Quiz	📁 Study Guide and Intervention, pp. 5, 12, 19 and 25 connectED.mcgraw-hill.com Extra Examples, Personal Tutor, Homework Help	connectED.mcgraw-hill.com Extra Examples, Personal Tutor, Homework Help, Review Vocabulary

LESSON 6-5 Operations with Radical Expressions

:: Then	:: Now	:: Why?
● You simplified expressions with nth roots.	**1** Simplify radical expressions. **2** Add, subtract, multiply, and divide radical expressions.	● Golden rectangles have been used by artists and architects to create beautiful designs. Many golden rectangles appear in the Parthenon in Athens, Greece. The ratio of the lengths of the sides of a golden rectangle is $\frac{2}{\sqrt{5}-1}$. In this lesson, you will learn to simplify radical expressions like $\frac{2}{\sqrt{5}-1}$.

 NewVocabulary
rationalizing the denominator
like radical expressions
conjugate

 Common Core State Standards

Content Standards
A.SSE.2 Use the structure of an expression to identify ways to rewrite it.

Mathematical Practices
1 Make sense of problems and persevere in solving them.

1 Simplify Radicals The properties you have used to simplify radical expressions involving square roots also hold true for expressions involving nth roots.

> **KeyConcept** Product Property of Radicals
>
> **Words** For any real numbers a and b and any integer $n > 1$, $\sqrt[n]{ab} = \sqrt[n]{a} \cdot \sqrt[n]{b}$, if n is even and a and b are both nonnegative or if n is odd.
>
> **Examples** $\sqrt{2} \cdot \sqrt{8} = \sqrt{16}$ or 4 and $\sqrt[3]{3} \cdot \sqrt[3]{9} = \sqrt[3]{27}$ or 3

In order for a radical to be in simplest form, the radicand must contain no factors that are nth powers of an integer or polynomial.

Example 1 Simplify Expressions with the Product Property

Simplify.

a. $\sqrt{32x^8}$

$\sqrt{32x^8} = \sqrt{4^2 \cdot 2 \cdot (x^4)^2}$ Factor into squares.

$= \sqrt{4^2} \cdot \sqrt{(x^4)^2} \cdot \sqrt{2}$ Product Property of Radicals

$= 4x^4\sqrt{2}$ Simplify.

b. $\sqrt[4]{16a^{24}b^{13}}$

$\sqrt[4]{16a^{24}b^{13}} = \sqrt[4]{2^4 \cdot (a^6)^4 (b^3)^4 \cdot b}$ Factor into squares.

$= \sqrt[4]{2^4} \cdot \sqrt[4]{(a^6)^4} \cdot \sqrt[4]{(b^3)^4} \cdot \sqrt[4]{b}$ Product Property of Radicals

$= 2a^6 |b^3| \sqrt[4]{b}$ Simplify.

In this case, the absolute value symbols are not necessary because in order for $\sqrt[4]{16a^{24}b^{13}}$ to be defined, b must be nonnegative.

Thus, $\sqrt[4]{16a^{24}b^{13}} = 2a^6b^3\sqrt[4]{b}$.

> **Guided** Practice
>
> **1A.** $\sqrt{12c^6d^3}$ $2c^3d\sqrt{3d}$ **1B.** $\sqrt[3]{27y^{12}z^7}$ $3y^4z^2\sqrt[3]{z}$

Photodisc/PunchStock

---

1 Focus

VerticalAlignment

▼ **Before Lesson 6-5** Simplify expressions with nth roots.

▼ **Lesson 6-5** Simplify radical expressions. Add, subtract, multiply, and divide radical expressions.

▼ **After Lesson 6-5** Determine solutions of square root equations using algebraic methods.

2 Teach

Scaffolding Questions
Have students read the **Why?** section of the lesson.

Ask:

■ Is the denominator, $\sqrt{5} - 1$, greater than 2 or less than 2? less than 2

■ Is the ratio $\frac{2}{\sqrt{5}-1}$ greater than 1 or less than 1? greater than 1

Lesson 6-5 Resources

Resource	Approaching Level **AL**	On Level **OL**	Beyond Level **BL**	English Learners **ELL**
Teacher Edition	■ Differentiated Instruction, p. 418	■ Differentiated Instruction, p. 418	■ Differentiated Instruction, p. 421	
Chapter Resource Masters	■ Study Guide and Intervention, pp. 31–32 ■ Skills Practice, p. 33 ■ Practice, p. 34 ■ Word Problem Practice, p. 35	■ Study Guide and Intervention, pp. 31–32 ■ Skills Practice, p. 33 ■ Practice, p. 34 ■ Word Problem Practice, p. 35 ■ Enrichment, p. 36 ■ Graphing Calculator Activity, p. 37	■ Practice, p. 34 ■ Word Problem Practice, p. 35 ■ Enrichment, p. 36	■ Study Guide and Intervention, pp. 31–32 ■ Skills Practice, p. 33 ■ Practice, p. 34 ■ Word Problem Practice, p. 35
Other	■ 5-Minute Check 6-5 ■ Study Notebook	■ 5-Minute Check 6-5 ■ Study Notebook	■ 5-Minute Check 6-5 ■ Study Notebook	■ 5-Minute Check 6-5 ■ Study Notebook

1 Simplify Radicals

Example 1 shows how to use the Product Property of Radicals to simplify a radical expression. **Example 2** shows how to use the Quotient Property of Radicals to simplify a radical expression.

Formative Assessment

Use the Guided Practice exercises after each example to determine students' understanding of concepts.

Additional Examples

1 Simplify.
  a. $\sqrt{25a^4b^9}$ $5a^2b^4\sqrt{b}$
  b. $\sqrt[3]{125m^{30}p^{20}}$ $5m^{10}p^6\sqrt[3]{p^2}$

2 Simplify.

  a. $\sqrt{\dfrac{y^8}{x^7}}$ $\dfrac{y^4\sqrt{x}}{x^4}$

  b. $\sqrt[3]{\dfrac{2}{9x}}$ $\dfrac{\sqrt[3]{6x^2}}{3x}$

▶ **Additional Examples** also in Interactive Classroom PowerPoint® Presentations

IWB Interactive White Board READY

Focus on Mathematical Content

Product Property of Radicals The Product Property of Radicals states that $\sqrt[n]{ab} = \sqrt[n]{a} \cdot \sqrt[n]{b}$. This statement is true for all real numbers a and b when n is an odd integer greater than 1. If n is even, then a and b must be nonnegative real numbers.

The Quotient Property of Radicals is another property used to simplify radicals.

KeyConcept Quotient Property of Radicals

Words	For any real numbers a and $b \neq 0$ and any integer $n > 1$, $\sqrt[n]{\dfrac{a}{b}} = \dfrac{\sqrt[n]{a}}{\sqrt[n]{b}}$, if all roots are defined.
Examples	$\dfrac{\sqrt{27}}{\sqrt{3}} = \sqrt{9}$ or 3 $\sqrt[3]{\dfrac{x^6}{8}} = \dfrac{\sqrt[3]{x^6}}{\sqrt[3]{8}} = \dfrac{x^2}{2}$ or $\dfrac{1}{2}x^2$

To eliminate radicals from a denominator or fractions from a radicand, you can use a process called **rationalizing the denominator**. To rationalize a denominator, multiply the numerator and denominator by a quantity so that the radicand has an exact root.

If the denominator is:	Multiply the numerator and denominator by:	Examples
$\sqrt{b}$	$\sqrt{b}$	$\dfrac{2}{\sqrt{3}} = \dfrac{2}{\sqrt{3}} \cdot \dfrac{\sqrt{3}}{\sqrt{3}}$ or $\dfrac{2\sqrt{3}}{3}$
$\sqrt[n]{b^x}$	$\sqrt[n]{b^{n-x}}$	$\dfrac{5}{\sqrt[3]{2}} = \dfrac{5}{\sqrt[3]{2}} \cdot \dfrac{\sqrt[3]{2^2}}{\sqrt[3]{2^2}}$ or $\dfrac{5\sqrt[3]{4}}{2}$

PT

Example 2 Simplify Expressions with the Quotient Property

Simplify.

a. $\sqrt{\dfrac{x^6}{y^7}}$

$\sqrt{\dfrac{x^6}{y^7}} = \dfrac{\sqrt{x^6}}{\sqrt{y^7}}$ Quotient Property

$= \dfrac{\sqrt{(x^3)^2}}{\sqrt{(y^3)^2 \cdot y}}$ Factor into squares.

$= \dfrac{\sqrt{(x^3)^2}}{\sqrt{(y^3)^2} \cdot \sqrt{y}}$ Product Property

$= \dfrac{|x^3|}{y^3\sqrt{y}}$ Simplify.

$= \dfrac{|x^3|}{y^3\sqrt{y}} \cdot \dfrac{\sqrt{y}}{\sqrt{y}}$ Rationalize the denominator.

$= \dfrac{|x^3|\sqrt{y}}{y^4}$ $\sqrt{y} \cdot \sqrt{y} = y$

b. $\sqrt[4]{\dfrac{6}{5x}}$

$\sqrt[4]{\dfrac{6}{5x}} = \dfrac{\sqrt[4]{6}}{\sqrt[4]{5x}}$ Quotient Property

$= \dfrac{\sqrt[4]{6}}{\sqrt[4]{5x}} \cdot \dfrac{\sqrt[4]{5^3x^3}}{\sqrt[4]{5^3x^3}}$ Rationalize the denominator.

$= \dfrac{\sqrt[4]{6 \cdot 5^3x^3}}{\sqrt[4]{5x \cdot 5^3x^3}}$ Product Property

$= \dfrac{\sqrt[4]{750x^3}}{\sqrt[4]{5^4x^4}}$ Multiply.

$= \dfrac{\sqrt[4]{750x^3}}{5x}$ $\sqrt[4]{5^4x^4} = 5x$

▶ **Guided Practice**

2A. $\dfrac{\sqrt{a^9}}{\sqrt{b^5}}$ $\dfrac{a^4\sqrt{ab}}{b^3}$

2B. $\sqrt[5]{\dfrac{3}{4y}}$ $\dfrac{\sqrt[5]{24y^4}}{2y}$

 416 | Lesson 6-5 | Operations with Radical Expressions

Tips for New Teachers

Product Property of Radicals When discussing the Product Property of Radicals, stress that a and b must both be nonnegative if n is even. This means that $\sqrt{-2}$ times $\sqrt{-8}$ *may not* be written as $\sqrt{16}$. This condition is necessary because $\sqrt{-2}$ and $\sqrt{-8}$ are *not* real numbers.

Here is a summary of the rules used to simplify radicals.

ConceptSummary Simplifying Radical Expressions

A radical expression is in simplified form when the following conditions are met.

- The index n is as small as possible.
- The radicand contains no factors (other than 1) that are nth powers of an integer or polynomial.
- The radicand contains no fractions.
- No radicals appear in a denominator.

2 **Operations with Radicals** You can use the Product and Quotient Properties to multiply and divide some radicals.

Example 3 Multiply Radicals

Simplify $5\sqrt[3]{-12ab^4} \cdot 3\sqrt[3]{18a^2b^2}$.

$$5\sqrt[3]{-12ab^4} \cdot 3\sqrt[3]{18a^2b^2} = 5 \cdot 3 \cdot \sqrt[3]{-12ab^4 \cdot 18a^2b^2}$$ Product Property of Radicals

$$= 15 \cdot \sqrt[3]{-2^2 \cdot 3 \cdot ab^4 \cdot 2 \cdot 3^2 \cdot a^2b^2}$$ Factor constants.

$$= 15 \cdot \sqrt[3]{-2^3 \cdot 3^3 \cdot a^3b^6}$$ Group into cubes if possible.

$$= 15 \cdot \sqrt[3]{-2^3} \cdot \sqrt[3]{3^3} \cdot \sqrt[3]{a^3} \cdot \sqrt[3]{b^6} \cdot$$ Product Property of Radicals

$$= 15 \cdot (-2) \cdot 3 \cdot a \cdot b^2$$ Simplify.

$$= -90ab^2$$ Multiply.

GuidedPractice

Simplify.

3A. $6\sqrt{8c^3d^5} \cdot 4\sqrt{2cd^3}$ $96c^2d^4$

3B. $2\sqrt[4]{8x^3y^2} \cdot 3\sqrt[4]{2x^5y^2}$ $12x^2|y|$

Radicals can be added and subtracted in the same manner as monomials. In order to add or subtract, the radicals must be like terms. Radicals are **like radical expressions** if *both* the index and the radicand are identical.

Like: $\sqrt{3b}$ and $4\sqrt{3b}$ Unlike: $\sqrt{3b}$ and $\sqrt[3]{3b}$ Unlike: $\sqrt{2b}$ and $\sqrt{3b}$

Example 4 Add and Subtract Radicals

Simplify $\sqrt{98} - 2\sqrt{32}$.

$$\sqrt{98} - 2\sqrt{32} = \sqrt{2 \cdot 7^2} - 2\sqrt{4^2 \cdot 2}$$ Factor using squares.

$$= \sqrt{7^2} \cdot \sqrt{2} - 2 \cdot \sqrt{4^2} \cdot \sqrt{2}$$ Product Property

$$= 7\sqrt{2} - 2 \cdot 4 \cdot \sqrt{2}$$ Simplify radicals.

$$= 7\sqrt{2} - 8\sqrt{2}$$ Multiply.

$$= -\sqrt{2}$$ $(7 - 8)\sqrt{2} = (-1)(\sqrt{2})$

GuidedPractice

4A. $4\sqrt{8} + 3\sqrt{50}$ $23\sqrt{2}$

4B. $5\sqrt{12} + 2\sqrt{27} - \sqrt{128}$ $16\sqrt{3} - 8\sqrt{2}$

StudyTip

CCSS Regularity Exact roots occur when the powers of the constants and variables are all identical to or multiples of the index.

For example,

$\sqrt[3]{2} \cdot \sqrt[3]{2^2} = \sqrt[3]{2^3}$ or 2.

StudyTip

Adding and Subtracting Radicals Simplify the individual radicals before attempting to combine like terms.

2 **Operations With Radicals**

Example 3 shows how to use the Product Property of Radicals to multiply radicals. **Example 4** shows how to add or subtract radicals. **Example 5** shows how to multiply binomials that contain radicals. **Example 6** shows how to use a conjugate to rationalize the denominator of a radical expression.

Additional Examples

3 Simplify $5\sqrt[3]{100a^2} \cdot \sqrt[3]{10a}$.
$50a$

4 Simplify
$3\sqrt{45} - 5\sqrt{80} + 4\sqrt{20}$.
$-3\sqrt{5}$

WatchOut!

Preventing Errors Urge students to verify that each of their final answers is in simplified form by testing it against the four conditions listed in the Concept Summary for simplifying radical expressions.

Teach with Tech

Interactive Whiteboard Write an expression on the board to add or subtract radical expressions. As you simplify the expression, drag the like terms to group them together. Then, combine like terms and simplify the expression.

CCSS **Teaching the Mathematical Practices**

Regularity Mathematically proficient students notice if calculations are repeated, and look both for general methods and for shortcuts. Encourage students to examine both the index and exponents when working with roots.

Additional Examples

5 Simplify
$(2\sqrt{3} + 3\sqrt{5})(3 - \sqrt{3})$.
$6\sqrt{3} - 6 + 9\sqrt{5} - 3\sqrt{15}$

6 **GEOMETRY** In a square with side a, the ratio of a side to the difference between the diagonal and a side is $\dfrac{a}{a\sqrt{2} - a}$. Use a conjugate to rationalize the denominator and simplify $\dfrac{a}{a\sqrt{2} - a}$.

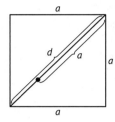

$\dfrac{\sqrt{2} + 1}{1}$ or $\sqrt{2} + 1$

Just as you can add and subtract radicals like monomials, you can multiply radicals using the FOIL method as you do when multiplying binomials.

Example 5 Multiply Radicals

Simplify $(4\sqrt{3} + 5\sqrt{2})(3\sqrt{2} - 6)$.

$(4\sqrt{3} + 5\sqrt{2})(3\sqrt{2} - 6) = 4\sqrt{3} \cdot 3\sqrt{2} + 4\sqrt{3} \cdot (-6) + 5\sqrt{2} \cdot 3\sqrt{2} + 5\sqrt{2} \cdot (-6)$

$\qquad = 12\sqrt{3 \cdot 2} - 24\sqrt{3} + 15\sqrt{2^2} - 30\sqrt{2}$ Product Property

$\qquad = 12\sqrt{6} - 24\sqrt{3} + 30 - 30\sqrt{2}$ Simplify.

▶ **Guided Practice**   **5A.** $12\sqrt{15} - 10\sqrt{5} + 24\sqrt{6} - 20\sqrt{2}$

Simplify.

5A. $(6\sqrt{3} - 5)(2\sqrt{5} + 4\sqrt{2})$ **5B.** $(7\sqrt{2} - 3\sqrt{3})(7\sqrt{2} + 3\sqrt{3})$ 71

StudyTip

Conjugates The product of conjugates is always a rational number.

Binomials of the form $a\sqrt{b} + c\sqrt{d}$ and $a\sqrt{b} - c\sqrt{d}$, where a, b, c, and d are rational numbers, are called **conjugates** of each other. You can use conjugates to rationalize denominators.

Real-World Example 6 Use a Conjugate to Rationalize a Denominator

ARCHITECTURE Refer to the beginning of the lesson. Use a conjugate to rationalize the denominator and simplify $\dfrac{2}{\sqrt{5} - 1}$.

$\dfrac{2}{\sqrt{5} - 1} = \dfrac{2}{\sqrt{5} - 1} \cdot \dfrac{\sqrt{5} + 1}{\sqrt{5} + 1}$ $\sqrt{5} + 1$ is the conjugate of $\sqrt{5} - 1$.

$\qquad = \dfrac{2\sqrt{5} + 2(1)}{(\sqrt{5})^2 + 1(\sqrt{5}) - 1(\sqrt{5}) - 1(1)}$ Multiply.

$\qquad = \dfrac{2\sqrt{5} + 2}{5 + \sqrt{5} - \sqrt{5} - 1}$ Simplify.

$\qquad = \dfrac{2\sqrt{5} + 2}{4}$ Subtract.

$\qquad = \dfrac{\sqrt{5} + 1}{2}$ Simplify.

Math HistoryLink

Theano (c. 5th century B.C.) Theano is believed to have been the wife of Pythagoras. It is also believed that she directed a famous mathematics academy and carried on the work of Pythagoras after his death. Her most important work was on the idea of the golden mean, which is the irrational number $\dfrac{1 + \sqrt{5}}{2}$.

Source: The Granger Collection, New York

▶ **Guided Practice**

6. GEOMETRY The area of the rectangle at the right is 900 ft². Write and simplify an equation for L in terms of x. $\dfrac{10{,}800\sqrt{3} - 900x}{432 - x^2}$

DifferentiatedInstruction AL OL

If ▶ when presented with a radical expression such as $11 + 6\sqrt{3}$, some students persist in trying to add the 11 and the 6,

Then ▶ to help them understand why this cannot be done, compare the radical expression $11 + 6\sqrt{3}$ to the expression $11 + 6x$. Stress that the radical $6\sqrt{3}$ is a multiplication expression just like $6x$. Remind students that the order of operations requires that multiplication be performed before addition. Students may find it helpful to rewrite $11 + 6\sqrt{3}$, as $11 + 6 \cdot \sqrt{3}$.

Examples 1–5 CCSS **PRECISION** Simplify.

1. $\sqrt{36ab^4c^5}$ $6b^2c^2\sqrt{ac}$
2. $\sqrt{144x^7y^5}$ $12x^3y^2\sqrt{xy}$
3. $\dfrac{\sqrt{c^5}}{\sqrt{d^9}}$ $\dfrac{c^2\sqrt{cd}}{d^5}$
4. $\sqrt[4]{\dfrac{5x}{8y}}$ $\dfrac{\sqrt[4]{10xy^3}}{2y}$

5. $5\sqrt{2x}\cdot3\sqrt{8x}$ $60x$
6. $4\sqrt{5a^5}\cdot\sqrt{125a^3}$ $100a^4$

7. $3\sqrt[3]{36xy}\cdot2\sqrt[3]{6x^2y^2}$ $36xy$
8. $\sqrt[4]{3x^3y^2}\cdot\sqrt[4]{27xy^2}$ $3x|y|$

9. $5\sqrt{32}+\sqrt{27}+2\sqrt{75}$ $20\sqrt{2}+13\sqrt{3}$
10. $4\sqrt{40}+3\sqrt{28}-\sqrt{200}$ $8\sqrt{10}+6\sqrt{7}-10\sqrt{2}$

11. $(4+2\sqrt{5})(3\sqrt{3}+4\sqrt{5})$ $12\sqrt{3}+16\sqrt{5}+40+6\sqrt{15}$
12. $(8\sqrt{3}-2\sqrt{2})(8\sqrt{3}+2\sqrt{2})$ 184

13. $\dfrac{5}{\sqrt{2}+3}$ $\dfrac{15-5\sqrt{2}}{7}$
14. $\dfrac{8}{\sqrt{6}-5}$ $\dfrac{-40-8\sqrt{6}}{19}$

15. $\dfrac{4+\sqrt{2}}{\sqrt{2}-3}$ $-2-\sqrt{2}$
16. $\dfrac{6-\sqrt{3}}{\sqrt{3}+4}$ $\dfrac{27-10\sqrt{3}}{13}$

Example 6

17. **GEOMETRY** Find the altitude of the triangle if the area is $189+4\sqrt{3}$ square centimeters. $32-2\sqrt{3}$ cm

$\leftarrow 12+\sqrt{3} \rightarrow$

18. $6a^4b^2\sqrt{2b}$
19. $3a^7b\sqrt{ab}$
20. $2a^8b^4\sqrt{6c}$
21. $3|a^3|bc^2\sqrt{2bc}$

Practice and Problem Solving   Extra Practice is on page R6.

Examples 1–4 Simplify.   **27.** $32a^5b^3\sqrt{b}$ **29.** $25x^6y^3\sqrt{2xy}$ **30.** $9\sqrt{10}+8\sqrt{5}+9\sqrt{2}$

18. $\sqrt{72a^8b^5}$
19. $\sqrt{9a^{15}b^3}$
20. $\sqrt{24a^{16}b^8c}$
21. $\sqrt{18a^6b^3c^5}$

22. $\dfrac{\sqrt{5a^5}}{\sqrt{b^{13}}}$ $\dfrac{a^2\sqrt{5ab}}{b^7}$
23. $\sqrt{\dfrac{7x}{10y^3}}$ $\dfrac{\sqrt{70xy}}{10y^2}$
24. $\dfrac{\sqrt[3]{6x^2}}{\sqrt[3]{5y}}$ $\dfrac{\sqrt[3]{150x^2y^2}}{5y}$
25. $\sqrt[4]{\dfrac{7x^3}{4b^2}}$ $\dfrac{\sqrt[4]{28b^2x^3}}{2|b|}$

26. $3\sqrt{5y}\cdot8\sqrt{10yz}$ $120y\sqrt{2z}$
27. $2\sqrt{32a^3b^5}\cdot\sqrt{8a^7b^2}$
28. $6\sqrt{3ab}\cdot4\sqrt{24ab^3}$ $144ab^2\sqrt{2}$

29. $5\sqrt{x^8y^3}\cdot5\sqrt{2x^5y^4}$
30. $3\sqrt{90}+4\sqrt{20}+\sqrt{162}$
31. $9\sqrt{12}+5\sqrt{32}-\sqrt{72}$ $18\sqrt{3}+14\sqrt{2}$

32. $4\sqrt{28}-8\sqrt{810}+\sqrt{44}$ $8\sqrt{7}-72\sqrt{10}+2\sqrt{11}$
33. $3\sqrt{54}+6\sqrt{288}-\sqrt{147}$ $9\sqrt{6}+72\sqrt{2}-7\sqrt{3}$

34. **GEOMETRY** Find the perimeter of the rectangle. $16+2\sqrt{3}+2\sqrt{6}$ ft $8+\sqrt{3}$ ft

35. **GEOMETRY** Find the area of the rectangle. $8\sqrt{6}+3\sqrt{2}$ ft^2

$\sqrt{6}$ ft

36. **GEOMETRY** Find the exact surface area of a sphere with radius of $4+\sqrt{5}$ inches. $(84+32\sqrt{5})\pi$ in^2

Examples 5–6 Simplify.   **37.** $56\sqrt{3}+42\sqrt{6}-36\sqrt{2}-54$ **40.** $36\sqrt{2}+36\sqrt{6}+20\sqrt{3}+60$

37. $(7\sqrt{2}-3\sqrt{3})(4\sqrt{6}+3\sqrt{12})$
38. $(8\sqrt{5}-6\sqrt{3})(8\sqrt{5}+6\sqrt{3})$ 212
42. $\dfrac{\sqrt{10}+\sqrt{6}}{2}$

39. $(12\sqrt{10}-6\sqrt{5})(12\sqrt{10}+6\sqrt{5})$ 1260
40. $(6\sqrt{3}+5\sqrt{2})(2\sqrt{6}+3\sqrt{8})$
43. $\dfrac{20-7\sqrt{3}}{11}$

41. $\dfrac{6}{\sqrt{3}-\sqrt{2}}$ $6\sqrt{3}+6\sqrt{2}$
42. $\dfrac{\sqrt{2}}{\sqrt{5}-\sqrt{3}}$
43. $\dfrac{9-2\sqrt{3}}{\sqrt{3}+6}$
44. $\dfrac{2\sqrt{2}+2\sqrt{5}}{\sqrt{5}+\sqrt{2}}$ 2

3 Practice

Formative Assessment

Use Exercises 1–17 to check for understanding.

Use the chart at the bottom of this page to customize assignments for your students.

CCSS **Teaching the Mathematical Practices**

Precision Mathematically proficient students try to use clear definitions in their reasoning, calculate accurately and efficiently, and make explicit use of definitions.

Differentiated Homework Options

Level	Assignment	Two-Day Option	
AL Basic	18–44, 60, 62, 64–88	19–43 odd, 66–69	18–44 even, 60, 62, 64, 65, 70–88
OL Core	19–53 odd, 54, 55, 57, 59, 60, 62, 64–88	18–44, 66–69	45–60, 62, 64, 65, 70–82
BL Advanced	45–82, (optional: 83–88)		

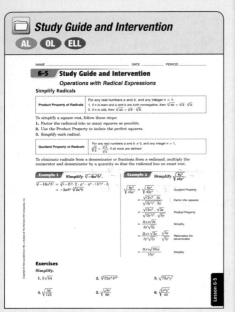

Study Guide and Intervention
AL OL ELL

NAME _____ DATE _____ PERIOD _____

6-5 **Study Guide and Intervention**
Operations with Radical Expressions

Simplify Radicals

Practice
AL OL BL ELL

NAME _____ DATE _____ PERIOD _____

6-5 **Practice**
Operations with Radical Expressions

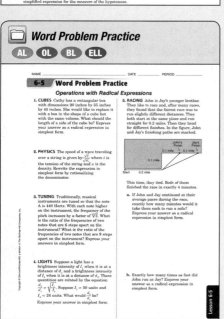

Word Problem Practice
AL OL BL ELL

NAME _____ DATE _____ PERIOD _____

6-5 **Word Problem Practice**
Operations with Radical Expressions

B Simplify.

49. $\dfrac{\sqrt[4]{1500a^2b^3x^3y^2}}{5|a|b}$

51. $\dfrac{(x+1)(\sqrt{x}+1)}{x-1}$ or $\dfrac{x\sqrt{x}+\sqrt{x}+x+1}{x-1}$

45. $\sqrt[3]{16y^4z^{12}}$ $2yz^4\sqrt[3]{2y}$

46. $\sqrt[3]{-54x^6y^{11}}$ $-3x^2y^3\sqrt[3]{2y^2}$

47. $\sqrt[4]{162a^6b^{13}c}$ $3|a|b^3\sqrt[4]{2a^2bc}$

48. $\sqrt[4]{48a^9b^3c^{16}}$ $2a^2c^4\sqrt[4]{3ab^3}$

49. $\sqrt[4]{\dfrac{12x^3y^2}{5a^2b}}$

50. $\dfrac{\sqrt[3]{36xy^2}}{\sqrt[3]{10xz}}$ $\dfrac{\sqrt[3]{450y^2z^2}}{5z}$

51. $\dfrac{x+1}{\sqrt{x}-1}$

52. $\dfrac{x-2}{\sqrt{x^2-4}}$ $\dfrac{\sqrt{x^2-4}}{x+2}$

53. $\dfrac{\sqrt{x}}{\sqrt{x^2-1}}$ $\dfrac{\sqrt{x^3-x}}{x^2-1}$

54. **APPLES** The diameter of an apple is related to its weight and can be modeled by the formula $d=\sqrt[3]{3w}$, where d is the diameter in inches and w is the weight in ounces. Find the diameter of an apple that weighs 6.47 ounces. **2.69 in.**

Simplify each expression if b is an even number.

55. $\sqrt[b]{a^b}$ $|a|$

56. $\sqrt[b]{a^{4b}}$ a^4

57. $\sqrt[b]{a^{2b}}$ a^2

58. $\sqrt[b]{a^{3b}}$ $|a^3|$

C 59. **MULTIPLE REPRESENTATIONS** In this problem, you will explore operations with like radicals. **a–e. See Chapter 6 Answer Appendix.**

    a. Numerical Copy the diagram at the right on dot paper. Use the Pythagorean Theorem to prove that the length of the red segment is $\sqrt{2}$ units.

    b. Graphical Extend the segment to represent $\sqrt{2}+\sqrt{2}$.

    c. Analytical Use your drawing to show that $\sqrt{2}+\sqrt{2}\neq\sqrt{2+2}$ or 2.

    d. Graphical Use the dot paper to draw a square with side lengths $\sqrt{2}$ units.

    e. Numerical Prove that the area of the square is $\sqrt{2}\cdot\sqrt{2}=2$ square units.

62. Sample answer: 0 is the only possible value for a since $\sqrt{a}$ is defined for $a\geq 0$, and $\sqrt{-a}$ is defined for $a\leq 0$.

H.O.T. Problems Use Higher-Order Thinking Skills

65. Sample answer: It is only necessary to use absolute values when it is possible that n could be odd or even and still be defined. It is when the radicand must be nonnegative in order for the root to be defined that the absolute values are not necessary.

60. **ERROR ANALYSIS** Twyla and Ben are simplifying $4\sqrt{32}+6\sqrt{18}$. Is either of them correct? Explain your reasoning.

Twyla; Ben's mistakes were multiplying the 4 by 16 instead of 4 and multiplying the 6 by 9 instead of 3.

Twyla	Ben
$4\sqrt{32}+6\sqrt{18}$	$4\sqrt{32}+6\sqrt{18}$
$=4\sqrt{4^2\cdot2}+6\sqrt{3^2\cdot2}$	$=4\sqrt{16\cdot2}+6\sqrt{9\cdot2}$
$=16\sqrt{2}+18\sqrt{2}$	$=64\sqrt{2}+54\sqrt{2}$
$=34\sqrt{2}$	$=118\sqrt{2}$

61. **CHALLENGE** Show that $\dfrac{-1-i\sqrt{3}}{2}$ is a cube root of 1. **See Chapter 6 Answer Appendix.**

62. **CCSS ARGUMENTS** For what values of a is $\sqrt{a}\cdot\sqrt{-a}$ a real number? Explain.

63. **CHALLENGE** Find four combinations of whole numbers that satisfy $\sqrt[b]{256}=b$.
$a=1, b=256$; $a=2, b=16$; $a=4, b=4$; $a=8, b=2$

64. **OPEN ENDED** Find a number other than 1 that has a positive whole number for a square root, cube root, and fourth root. **Sample answer: 4096**

65. **WRITING IN MATH** Explain why absolute values may be unnecessary when an nth root of an even power results in an odd power.

420 | Lesson 6-5 | Operations with Radical Expressions

Enrichment
OL BL

NAME _____ DATE _____ PERIOD _____

6-5 **Enrichment**

Special Products with Radicals

Notice that $(\sqrt{3}\times\sqrt{3})=3$, or $(\sqrt{3})^2=3$. In general, $(\sqrt{x})^2=x$ when $x\geq0$.
Also, notice that $(\sqrt{9}\times\sqrt{4})=\sqrt{36}$. In general, $(\sqrt{x}\times\sqrt{y})=\sqrt{xy}$ when x and y are not negative.
You can use these ideas to find the special products below.

$(\sqrt{a}+\sqrt{b})(\sqrt{a}-\sqrt{b})=(\sqrt{a})^2-(\sqrt{b})^2=a-b$
$(\sqrt{a}+\sqrt{b})^2=(\sqrt{a})^2+2\sqrt{ab}+(\sqrt{b})^2=a+2\sqrt{ab}+b$
$(\sqrt{a}-\sqrt{b})^2=(\sqrt{a})^2-2\sqrt{ab}+(\sqrt{b})^2=a-2\sqrt{ab}+b$

Example 1 Find the product: $(\sqrt{2}+\sqrt{5})(\sqrt{2}-\sqrt{5})$.

Example 2 Evaluate $(\sqrt{2}+\sqrt{5})^2$.

Multiple Representations
In Exercise 59, students use dot paper and the Pythagorean Theorem to explore operations with like radicals.

66. PROBABILITY A six-sided number cube has faces with the numbers 1 though 6 marked on it. What is the probability that a number less than 4 will occur on one toss of the number cube? **A**

A $\frac{1}{2}$ **C** $\frac{1}{4}$

B $\frac{1}{3}$ **D** $\frac{1}{5}$

67. When the number of a year is divisible by 4, the year is a leap year. However, when the year is divisible by 100, the year is not a leap year, unless the year is divisible by 400. Which is *not* a leap year? **G**

F 1884   **H** 1904
G 1900   **J** 1940

68. SHORT RESPONSE Which property is illustrated by $4x + 0 = 4x$? **Additive Identity Property**

69. SAT/ACT The expression $\sqrt{180a^2b^8}$ is equivalent to which of the following? **C**

A $3\sqrt{10}\,|a|b^4$

B $5\sqrt{6}\,|a|b^4$

C $6\sqrt{5}\,|a|b^4$

D $18\sqrt{10}\,|a|b^4$

E $36\sqrt{5}\,|a|b^4$

Simplify. (Lesson 6-4)

70. $\sqrt{81x^6}$ $9|x^3|$

71. $\sqrt[3]{729a^3b^9}$ $9ab^3$

72. $\sqrt{(g+5)^2}$ $|g+5|$

73. Graph $y \le \sqrt{x-2}$. (Lesson 6-3) **See margin.**

Solve each equation. (Lesson 5-5)

74. $x^4 - 34x^2 + 225 = 0$ $-5, -3, 3, 5$ **75.** $x^4 - 15x^2 - 16 = 0$ $-4, 4, -i, i$ **76.** $x^4 + 6x^2 - 27 = 0$ $-\sqrt{3}, \sqrt{3}, -3i, 3i$

77. $x^3 + 64 = 0$ $-4, 2 + 2i\sqrt{3}, 2 - 2i\sqrt{3}$

78. $27x^3 + 1 = 0$

79. $8x^3 - 27 = 0$ **78, 79. See margin.**

80. MODELS A model car builder is building a display table for model cars. He wants the perimeter of the table to be 26 feet, but he wants the area of the table to be no more than 30 square feet. What could be the width of the table? (Lesson 4-8) **between 0 and 3 ft or between 10 and 13 ft**

81. CONSTRUCTION Cho charges $1500 to build a small deck and $2500 to build a large deck. During the spring and summer, she built 5 more small decks than large decks. If she earned $23,500 how many of each type of deck did she build? (Lesson 3-8) **9 small, 4 large**

82. FOOD The Hot Dog Grille offers the lunch combinations shown. Assume that the price of a combo meal is the same price as purchasing each item separately. Find the prices for a hot dog, a soda, and a bag of potato chips. (Lesson 3-4)
hot dog, $1.95; soda, $1.50; potato chips, $0.90

Lunch Combo Meals
1. Two hot dogs, one soda $5.40
2. One hot dog, potato chips, one soda............... $4.35
3. Two hot dogs, two bags of chips $5.70

Evaluate each expression.

83. $2\left(\frac{1}{6}\right)$ $\frac{1}{3}$

84. $3\left(\frac{1}{8}\right)$ $\frac{3}{8}$

85. $\frac{1}{4} + \frac{1}{3}$ $\frac{7}{12}$

86. $\frac{1}{2} + \frac{3}{8}$ $\frac{7}{8}$

87. $\frac{2}{3} - \frac{1}{4}$ $\frac{5}{12}$

88. $\frac{5}{6} - \frac{2}{5}$ $\frac{13}{30}$

DifferentiatedInstruction BL

Extension Solve for *a* and *b*.

$$\sqrt{\frac{3^a}{7^b}} = \frac{3^2\sqrt{7}}{7^2} \quad a = 4, b = 3$$

WatchOut!

Error Analysis In Exercise 60, students should see that Ben made an error each time he "took out" a factor from the square root symbol. Explain to students that $4\sqrt{16 \cdot 2} = 4 \cdot 4\sqrt{2} = 16\sqrt{2}$ and $6\sqrt{9 \cdot 2} = 6 \cdot 3\sqrt{2} = 18\sqrt{2}$, as shown in Twyla's solution.

4 Assess

Name the Math Ask students to describe how combining radicals is the same as and different from combining expressions with variables.

WatchOut!

Common Misconceptions Students will need to simplify expressions involving radicals in much of their future work in algebra. Take time to help students uncover and correct their misconceptions by analyzing the errors they make.

Additional Answers

73.

78. $-\frac{1}{3}, \frac{1 + i\sqrt{3}}{6}, \frac{1 - i\sqrt{3}}{6}$

79. $\frac{3}{2}, \frac{-3 + 3i\sqrt{3}}{4}, \frac{-3 - 3i\sqrt{3}}{4}$

1 Focus

VerticalAlignment

Before Lesson 6-6 Use properties of exponents.

Lesson 6-6 Write expressions with rational exponents in radical form and vice versa. Simplify expressions in exponential or radical form.

After Lesson 6-6 Determine solutions of square root equations using algebraic methods.

2 Teach

Scaffolding Questions

Have students read the **Why?** section of the lesson.

Ask:

- In the formula $C = c(1 + r)^n$, what is the exponent of c? The exponent of c is 1.

- Which expression is raised to the exponent n? n is the exponent of the quantity $(1 + r)$.

- Why does $\frac{1}{2}$ appear as an exponent in the last sentence? n represents a number of years, and 6 months is half of a year.

LESSON 6-6 Rational Exponents

:: Then	:: Now	:: Why?
● You used properties of exponents.	**1** Write expressions with rational exponents in radical form and vice versa. **2** Simplify expressions in exponential or radical form.	● The formula $C = c(1 + r)^n$ can be used to estimate the future cost of an item due to inflation. C represents the future cost, c represents the current cost, r is the rate of inflation, and n is the number of years for the projection. For example, $C = c(1 + r)^{\frac{1}{2}}$ can be used to estimate the cost of a video game system in six months.

CCSS Common Core State Standards

Mathematical Practices
1 Make sense of problems and persevere in solving them.

1 **Rational Exponents and Radicals** You know that squaring a number and taking the square root of a number are inverse operations. But how would you evaluate an expression that contains a fractional exponent such as the one above? You can investigate such an expression by assuming that fractional exponents behave as integral exponents.

$$\left(b^{\frac{1}{2}}\right)^2 = b^{\frac{1}{2}} \cdot b^{\frac{1}{2}} \qquad \text{Write as a multiplication expression.}$$
$$= b^{\frac{1}{2} + \frac{1}{2}} \qquad \text{Add the exponents.}$$
$$= b^1 \text{ or } b \qquad \text{Simplify.}$$

Thus, $b^{\frac{1}{2}}$ is a number with a square equal to b. So $b^{\frac{1}{2}} = \sqrt{b}$.

🔑 KeyConcept $b^{\frac{1}{n}}$

Words	For any real number b and any positive integer n, $b^{\frac{1}{n}} = \sqrt[n]{b}$, except when $b < 0$ and n is even. When $b < 0$ and n is even, a complex root may exist.
Examples	$27^{\frac{1}{3}} = \sqrt[3]{27}$ or 3 $\qquad$ $(-16)^{\frac{1}{2}} = \sqrt{-16}$ or $4i$

Example 1 Radical and Exponential Forms

Simplify.

a. Write $x^{\frac{1}{6}}$ in radical form.

$x^{\frac{1}{6}} = \sqrt[6]{x}$ $\qquad$ Definition of $b^{\frac{1}{n}}$

b. Write $\sqrt[4]{z}$ in exponential form.

$\sqrt[4]{z} = z^{\frac{1}{4}}$ $\qquad$ Definition of $b^{\frac{1}{n}}$

▶ **Guided**Practice

1A. Write $a^{\frac{1}{5}}$ in radical form. $\sqrt[5]{a}$

1B. Write $\sqrt[8]{c}$ in exponential form. $c^{\frac{1}{8}}$

1C. Write $d^{\frac{7}{4}}$ in radical form. $\sqrt[4]{d^7}$

1D. Write $\sqrt[3]{c^{-5}}$ in exponential form. $c^{-\frac{5}{3}}$

 422 | Lesson 6-6

Lesson 6-6 Resources

Resource	Approaching Level **AL**	On Level **OL**	Beyond Level **BL**	English Learners **ELL**
Teacher Edition	▪ Differentiated Instruction, p. 423	▪ Differentiated Instruction, pp. 423, 424	▪ Differentiated Instruction, p. 424	▪ Differentiated Instruction, p. 423
Chapter Resource Masters	▪ Study Guide and Intervention, pp. 38–39 ▪ Skills Practice, p. 40 ▪ Practice, p. 41 ▪ Word Problem Practice, p. 42	▪ Study Guide and Intervention, pp. 38–39 ▪ Skills Practice, p. 40 ▪ Practice, p. 41 ▪ Word Problem Practice, p. 42 ▪ Enrichment, p. 43 ▪ Spreadsheet Activity, p. 44	▪ Practice, p. 41 ▪ Word Problem Practice, p. 42 ▪ Enrichment, p. 43	▪ Study Guide and Intervention, pp. 38–39 ▪ Skills Practice, p. 40 ▪ Practice, p. 41 ▪ Word Problem Practice, p. 42
Other	▪ 5-Minute Check 6-6 ▪ Study Notebook	▪ 5-Minute Check 6-6 ▪ Study Notebook	▪ 5-Minute Check 6-6 ▪ Study Notebook	▪ 5-Minute Check 6-6 ▪ Study Notebook

Ariel Skelley/Blend Images/JupiterImages

ReadingMath

Power Functions Square root functions are also *power functions* when the exponent is a fraction. In the case of square roots, the exponent is $\frac{1}{2}$.

The rules for negative exponents also apply to negative rational exponents.

Example 2 Evaluate Expressions with Rational Exponents

Evaluate each expression.

a. $81^{-\frac{1}{4}}$

$$81^{-\frac{1}{4}} = \frac{1}{81^{\frac{1}{4}}} \qquad b^{-n} = \frac{1}{b^n}$$

$$= \frac{1}{\sqrt[4]{81}} \qquad 81^{\frac{1}{4}} = \sqrt[4]{81}$$

$$= \frac{1}{\sqrt[4]{3^4}} \qquad 81 = 3^4$$

$$= \frac{1}{3} \qquad \text{Simplify.}$$

b. $216^{\frac{2}{3}}$

$$216^{\frac{2}{3}} = (6^3)^{\frac{2}{3}} \qquad 216 = 6^3$$

$$= 6^{3 \cdot \frac{2}{3}} \qquad \text{Power of a Power}$$

$$= 6^2 \qquad \text{Multiply exponents.}$$

$$= 36 \qquad \text{Simplify.}$$

▶ **Guided Practice**

2A. $-3125^{-\frac{1}{5}}$ $-\frac{1}{5}$

2B. $256^{\frac{3}{8}}$ 8

Examples **2a** and **2b** use the definition of $b^{\frac{1}{n}}$ and the properties of powers to evaluate an expression. Both methods suggest the following general definition of rational exponents

KeyConcept Rational Exponents

Words   For any real nonzero number b, and any integers x and y, with $y > 1$, $b^{\frac{x}{y}} = \sqrt[y]{b^x} = \left(\sqrt[y]{b}\right)^x$, except when $b < 0$ and y is even. When $b < 0$ and y is even, a complex root may exist.

Examples $27^{\frac{2}{3}} = \left(\sqrt[3]{27}\right)^2 = 3^2$ or 9 $(-16)^{\frac{3}{2}} = \left(\sqrt{-16}\right)^3 = (4i)^3$ or $-64i$

Real-World Example 3 Solve Equations with Rational Exponents

FINANCIAL LITERACY Refer to the beginning of the lesson. Suppose a video game system costs \$390 now. How much would the price increase in six months with an annual inflation rate of 5.3%?

$$C = c(1 + r)^n \qquad \text{Original formula}$$

$$= 390(1 + 0.053)^{\frac{1}{2}} \qquad c = 390, r = 0.053, \text{ and } n = \frac{6 \text{ months}}{12 \text{ months}} \text{ or } \frac{1}{2}$$

$$\approx 400.20 \qquad \text{Use a calculator.}$$

In six months the price of the video game system will be \$400.20 − \$390.00 or \$10.20 more than its current price.

Real-World Career

Economist Economists are employed in banking, finance, accounting, commerce, marketing, and business administration. Politicians often consult economists before enacting policy. Economists use shopping as one way of judging the economy. A bachelor's or master's degree in economics is required.

▶ **Guided Practice**

3. Suppose a gallon of milk costs \$2.99 now. How much would the price increase in 9 months with an annual inflation rate of 5.3%? **\$0.12**

  423

DifferentiatedInstruction AL OL ELL

 If students are not reading the exponent,

Then encourage students to read an exponent aloud in a way that distinguishes it from a coefficient or multiplier. Ask them to practice reading the exponent correctly, for example, reading x^3 as "x to the third power" or "x cubed." Also ask students to practice reading radical expressions correctly, for example, reading $\sqrt{y^3}$ as "the square root of y cubed" or "the square root of the third power of y."

Teach with Tech

Interactive Whiteboard Write an expression with a rational exponent on the board. Display a template such as $a^{\frac{\square}{\square}} = \sqrt[\square]{a^{\square}}$ and drag the numerator and denominator of the exponent to rewrite the expression as a radical.

1 **Rational Exponents and Radicals**

Example 1 shows how to write expressions in radical and exponential form. **Example 2** shows how to evaluate expressions with rational exponents. **Example 3** is a real-world example with a rational exponent.

Formative Assessment

Use the Guided Practice exercises after each example to determine students' understanding of concepts.

Additional Examples

1 **a.** Write $a^{\frac{1}{7}}$ in radical form. $\sqrt[7]{a}$

   b. Write $\sqrt{w}$ in exponential form. $w^{\frac{1}{2}}$

2 Evaluate each expression.

   a. $49^{-\frac{1}{2}}$ $\frac{1}{7}$ **b.** $32^{\frac{2}{5}}$ 4

3 **WEIGHTLIFTING** The formula $M = 512 - 146{,}230B^{-\frac{8}{5}}$ can be used to estimate the maximum total mass that a weightlifter of mass B kilograms can lift using the snatch and the clean and jerk. According to the formula, what is the maximum that a weightlifter weighing 168 kilograms can lift? The formula predicts that he can lift at most 472 kilograms.

▶ **Additional Examples** also in Interactive Classroom PowerPoint® Presentations

IWB **Interactive White Board** READY

Focus on Mathematical Content

Rational Exponents In the general definition of rational exponents, $b^{\frac{m}{n}} = \sqrt[n]{b^m} = \left(\sqrt[n]{b}\right)^m$, there are restrictions on the variables b, m, and n. The variable b must be *a* nonzero real number except when n is even, and then b must be *a* positive real number. The variable m must be an integer, and n must be a natural number. These restrictions on m and n do not decrease the generality of the definition since any rational number can be expressed as a quotient of an integer and a natural number.

2 Simplify Expressions

Example 4 shows how to simplify expressions with rational exponents. **Example 5** shows how to simplify radical expressions.

Additional Example

4 Simplify each expression.

a. $y^{\frac{1}{7}} \cdot y^{\frac{4}{7}} \quad y^{\frac{5}{7}}$

b. $x^{-\frac{2}{3}} \quad \dfrac{x^{\frac{1}{3}}}{x}$

WatchOut!

Preventing Errors If students are having difficulty remembering which part of the fractional exponent is the index, suggest that they recall the basic definition $b^{\frac{1}{2}} = \sqrt{b}$.

2 Simplify Expressions

All of the properties of powers you learned in Lesson 6-1 apply to rational exponents. Write each expression with all positive exponents. Also, any exponents in the denominator of a fraction must be positive *integers*. So, it may be necessary to rationalize a denominator.

Example 4 Simplify Expressions with Rational Exponents

> **StudyTip**
>
> **Simplifying Expressions** When simplifying expressions containing rational exponents, leave the exponent in rational form rather than writing the expression as a radical.

Simplify each expression.

a. $a^{\frac{2}{7}} \cdot a^{\frac{4}{7}}$

$a^{\frac{2}{7}} \cdot a^{\frac{4}{7}} = a^{\frac{2}{7} + \frac{4}{7}}$ Add powers.

$\qquad = a^{\frac{6}{7}}$ Add exponents.

b. $b^{-\frac{5}{6}}$

$b^{-\frac{5}{6}} = \dfrac{1}{b^{\frac{5}{6}}}$ $b^{-n} = \dfrac{1}{b^n}$

$\qquad = \dfrac{1}{b^{\frac{5}{6}}} \cdot \dfrac{b^{\frac{1}{6}}}{b^{\frac{1}{6}}}$ $\dfrac{b^{\frac{1}{6}}}{b^{\frac{1}{6}}} = 1$

$\qquad = \dfrac{b^{\frac{1}{6}}}{b^{\frac{6}{6}}}$ $b^{\frac{5}{6}} \cdot b^{\frac{1}{6}} = b^{\frac{5}{6} + \frac{1}{6}}$

$\qquad = \dfrac{b^{\frac{1}{6}}}{b}$ $b^{\frac{6}{6}} = b^1$ or b

> **Guided Practice**
>
> **4A.** $p^{\frac{1}{4}} \cdot p^{\frac{9}{4}} \quad p^{\frac{5}{2}}$
>
> **4B.** $r^{-\frac{4}{5}} \quad \dfrac{r^{\frac{1}{5}}}{r}$

When simplifying a radical expression, always use the least index possible. Using rational exponents makes this process easier, but the answer should be written in radical form.

Example 5 Simplify Radical Expressions

Simplify each expression.

a. $\dfrac{\sqrt[4]{27}}{\sqrt{3}}$

$\dfrac{\sqrt[4]{27}}{\sqrt{3}} = \dfrac{27^{\frac{1}{4}}}{3^{\frac{1}{2}}}$ Rational exponents

$\qquad = \dfrac{(3^3)^{\frac{1}{4}}}{3^{\frac{1}{2}}}$ $27 = 3^3$

$\qquad = \dfrac{3^{\frac{3}{4}}}{3^{\frac{1}{2}}}$ Power of a Power

$\qquad = 3^{\frac{3}{4} - \frac{1}{2}}$ Quotient of Powers

$\qquad = 3^{\frac{1}{4}}$ Simplify.

$\qquad = \sqrt[4]{3}$ Rewrite in radical form.

b. $\sqrt[3]{64z^6}$

$\sqrt[3]{64z^6} = (64z^6)^{\frac{1}{3}}$ Rational exponents

$\qquad = (8^2 \cdot z^6)^{\frac{1}{3}}$ $64 = 8^2$

$\qquad = 8^{\frac{2}{3}} \cdot z^{\frac{6}{3}}$ Power of a Power

$\qquad = 4z^2$ $8^{\frac{2}{3}} = 4$

DifferentiatedInstruction (OL) (BL)

Extension Assume that x and y are nonnegative real numbers.

Multiply $\left(x^{\frac{1}{2}} + y^{\frac{1}{2}}\right)\left(x^{\frac{1}{2}} - y^{\frac{1}{2}}\right)$. $x - y$

c. $\dfrac{x^{\frac{1}{2}} - 2}{3x^{\frac{1}{2}} + 2}$

$$\dfrac{x^{\frac{1}{2}} - 2}{3x^{\frac{1}{2}} + 2} = \dfrac{x^{\frac{1}{2}} - 2}{3x^{\frac{1}{2}} + 2} \cdot \dfrac{3x^{\frac{1}{2}} - 2}{3x^{\frac{1}{2}} - 2}$$ $3x^{\frac{1}{2}} - 2$ is the conjugate of $3x^{\frac{1}{2}} + 2$.

$$= \dfrac{3x^{\frac{2}{2}} - 8x^{\frac{1}{2}} + 4}{9x^{\frac{2}{2}} - 4}$$ Multiply.

$$= \dfrac{3x - 8x^{\frac{1}{2}} + 4}{9x - 4}$$ Simplify.

> **Guided**Practice
>
> **5A.** $\dfrac{\sqrt[4]{32}}{\sqrt[3]{2}}$ $\sqrt[12]{2^{11}}$ **5B.** $\sqrt[3]{16x^4}$ $2x\sqrt[3]{2x}$ **5C.** $\dfrac{y^{\frac{1}{2}} + 2}{y^{\frac{1}{2}} - 2}$ $\dfrac{y + 4y^{\frac{1}{2}} + 4}{y - 4}$

StudyTip

Radical Expressions Write the simplified expression in the same form as the beginning expression. When you start with a radical expression, end with a radical expression. When you start with an expression with rational exponents, end with an expression with rational exponents.

ConceptSummary **Expressions with Rational Exponents**

An expression with rational exponents is simplified when all of the following conditions are met.

- It has no negative exponents.
- It has no exponents that are not positive integers in the denominator.
- It is not a complex fraction.
- The index of any remaining radical is the least number possible.

Check Your Understanding

⬤ = Step-by-Step Solutions begin on page R14. ✓

Example 1   Write each expression in radical form, or write each radical in exponential form.

1. $10^{\frac{1}{4}}$ $\sqrt[4]{10}$ **2.** $x^{\frac{3}{5}}$ $\sqrt[5]{x^3}$ **3.** $\sqrt[3]{15}$ $15^{\frac{1}{3}}$ **4.** $\sqrt[4]{7x^6y^9}$ $7^{\frac{1}{4}}x^{\frac{3}{2}}y^{\frac{9}{4}}$

Example 2   Evaluate each expression.

5. $343^{\frac{1}{3}}$ 7   **6.** $32^{-\frac{1}{5}}$ $\dfrac{1}{2}$ **7.** $125^{\frac{2}{3}}$ 25   **8.** $\dfrac{24}{4^{\frac{3}{2}}}$ 3

Example 3   **9** GARDENING If the area A of a square is known, then the lengths of its sides ℓ can be computed using $\ell = A^{\frac{1}{2}}$. You have purchased a 169 ft² share in a community garden for the season. What is the length of one side of your square garden? **13 ft**

Examples 4–5 (CCSS) PRECISION Simplify each expression.

10. $a^{\frac{3}{4}} \cdot a^{\frac{1}{2}}$ $a^{\frac{5}{4}}$ **11.** $\dfrac{x^{\frac{4}{5}}}{x^{\frac{1}{5}}}$ $x^{\frac{3}{5}}$ **12.** $\dfrac{b^3}{c^{\frac{1}{2}}} \cdot \dfrac{c}{b^{\frac{1}{3}}}$ $b^{\frac{8}{3}}c^{\frac{1}{2}}$

13. $\sqrt[4]{9g^2}$ $\sqrt{3g}$ **14.** $\dfrac{\sqrt[5]{64}}{\sqrt[5]{4}}$ $2^{\frac{4}{5}}$ or $\sqrt[5]{16}$ **15.** $\dfrac{g^{\frac{1}{2}} - 1}{g^{\frac{1}{2}} + 1}$ $\dfrac{g - 2g^{\frac{1}{2}} + 1}{g - 1}$

connectED.mcgraw-hill.com **425**

Additional Example

5 Simplify each expression.

a. $\dfrac{\sqrt[6]{16}}{\sqrt[3]{2}}$ $2^{\frac{1}{3}}$ or $\sqrt[3]{2}$

b. $\sqrt[6]{4x^4}$ $\sqrt[3]{2x^2}$

c. $\dfrac{y^{\frac{1}{2}} + 1}{y^{\frac{1}{2}} - 1}$ $\dfrac{y + 2y^{\frac{1}{2}} + 1}{y - 1}$

Tips for New Teachers

Checking answers Encourage students to substitute values in both the original expression and their simplified expression. If the values of the expressions are equal, their answer is likely to be correct.

3 Practice

Formative Assessment

Use Exercises 1–15 to check for understanding.

Use the chart at the bottom of the next page to customize assignments for your students.

(CCSS) **Teaching the Mathematical Practices**

Precision Mathematically proficient students try to use clear definitions in their reasoning, calculate accurately and efficiently, and make explicit use of definitions.

52a. about 475

52b.

Year	0	1	2	3	4	5
Deer	100	141	200	282	400	565

52c.

52d. Sample answer: No; it is not reasonable to think that the population will continue to grow without bounds. This does not take into account the death rate of the deer.

65b.

67a. Sample answer:
$\sqrt[4]{(-16)^3} = \sqrt[4]{-4096}$;
there is no real number that when raised to the fourth power results in a negative number.

69. Sample answer: It may be easier to simplify an expression when it has rational exponents because all the properties of exponents apply. We do not have as many properties dealing directly with radicals. However, we can convert all radicals to rational exponents, and then use the properties of exponents to simplify.

Example 1 Write each expression in radical form, or write each radical in exponential form.

16. $8^{\frac{1}{5}}$ $\sqrt[5]{8}$

17. $4^{\frac{2}{7}}$ $\sqrt[7]{16}$

18. $a^{\frac{3}{4}}$ $\sqrt[4]{a^3}$

19. $(x^3)^{\frac{3}{2}}$ $\sqrt{x^9}$

20. $\sqrt{17}$ $17^{\frac{1}{2}}$

21. $\sqrt[4]{63}$ $63^{\frac{1}{4}}$

22. $\sqrt[3]{5xy^2}$ $5^{\frac{1}{3}}x^{\frac{1}{3}}y^{\frac{2}{3}}$

23. $\sqrt[4]{625x^2}$ $5x^{\frac{1}{2}}$

Example 2 Evaluate each expression.

24. $27^{\frac{1}{3}}$ 3

25. $256^{\frac{1}{4}}$ 4

26. $16^{-\frac{1}{2}}$ $\frac{1}{4}$

27. $81^{-\frac{1}{4}}$ $\frac{1}{3}$

Example 3

28. **CCSS** SENSE-MAKING A women's regulation-sized basketball is slightly smaller than a men's basketball. The radius r of the ball that holds V cubic units of air is $\left(\frac{3V}{4\pi}\right)^{\frac{1}{3}}$.

a. Find the radius of a women's basketball. about 4.62 in.

b. Find the radius of a men's basketball. about 4.77 in.

Men's 455 in³   Women's 413 in³

29. GEOMETRY The radius r of a sphere with volume V is given by $r = \left(\frac{3V}{4\pi}\right)^{\frac{1}{3}}$. Find the radius of a ball with a volume of 77 cm³. about 2.64 cm

Examples 4–5 Simplify each expression.

39. $\dfrac{x + 4x^{\frac{3}{4}} + 8x^{\frac{1}{2}} + 16x^{\frac{1}{4}} + 16}{x - 16}$

30. $x^{\frac{1}{3}} \cdot x^{\frac{2}{5}}$ $x^{\frac{11}{15}}$

31. $a^{\frac{4}{9}} \cdot a^{\frac{4}{4}}$ $a^{\frac{25}{36}}$

32. $b^{-\frac{3}{4}}$ $\dfrac{b^{\frac{1}{4}}}{b}$

33. $y^{-\frac{4}{5}}$ $\dfrac{y^{\frac{1}{5}}}{y}$

34. $\dfrac{\sqrt[8]{81}}{\sqrt[6]{3}}$ $\sqrt[3]{3}$

35. $\dfrac{\sqrt[4]{27}}{\sqrt[3]{3}}$ $\sqrt{3}$

36. $\sqrt[4]{25x^2}$ $\sqrt{5x}$

37. $\dfrac{\sqrt[6]{81g^3}}{\sqrt[3]{9} \cdot \sqrt{g}}$

38. $\dfrac{h^{\frac{1}{2}} + 1}{h^{\frac{1}{2}} - 1}$ $\dfrac{1}{h + 2h^{\frac{1}{2}} + 1}$ $\dfrac{h + 2h^{\frac{1}{2}} + 1}{h - 1}$

39. $\dfrac{x^{\frac{1}{4}} + 2}{x^{\frac{1}{4}} - 2}$

B GEOMETRY Find the area of each figure.

40.

$4r^{\frac{1}{4}}w^{\frac{1}{2}}$ $3r^{\frac{1}{2}}w^{\frac{1}{4}}$
$6r^{\frac{3}{4}}w^{\frac{3}{4}}$ units²

41.

$3x^{\frac{2}{3}}y^{\frac{1}{5}}z^2$
$28.27x^{\frac{4}{3}}y^{\frac{2}{5}}z^4$ units²

42. Find the simplified form of $18^{\frac{1}{2}} + 2^{\frac{1}{2}} - 32^{\frac{1}{2}}$. 0

43. What is the simplified form of $64^{\frac{1}{3}} - 32^{\frac{1}{3}} + 8^{\frac{1}{3}}$? $6 - 2 \cdot 4^{\frac{1}{3}}$

Simplify each expression.

44. $a^{\frac{7}{4}} \cdot a^{\frac{5}{4}}$ a^3

45. $x^{\frac{2}{3}} \cdot x^{\frac{8}{3}}$ $x^{\frac{10}{3}}$

46. $\left(b^{\frac{3}{4}}\right)^{\frac{1}{3}}$ $b^{\frac{1}{4}}$

47. $\left(y^{-\frac{3}{5}}\right)^{-\frac{1}{4}}$ $y^{\frac{3}{20}}$

48. $\sqrt[4]{64}$ $2\sqrt{2}$

49. $\sqrt[6]{216}$ $\sqrt{6}$

50. $d^{-\frac{5}{6}}$ $\dfrac{d^{\frac{1}{6}}}{d}$

51. $w^{-\frac{7}{8}}$ $\dfrac{w^{\frac{1}{8}}}{w}$

Differentiated Homework **Options**

Level	Assignment	Two-Day Option	
AL Basic	16–39, 66, 68–97	17–39 odd, 71–74	16–38 even, 66, 68–70, 75–97
OL Core	17–51 odd, 52, 53–65 odd, 66, 68–97	16–39, 71–74	40–66, 68–70, 75–97
BL Advanced	40–91, (optional: 92–97)		

52. WILDLIFE A population of 100 deer is reintroduced to a wildlife preserve. Suppose the population does extremely well and the deer population doubles in two years. Then the number D of deer after t years is given by $D = 100 \cdot 2^{\frac{t}{2}}$. **a–d. See margin.**

   a. How many deer will there be after $4\frac{1}{2}$ years?

   b. Make a table that charts the population of deer every year for the next five years.

   c. Make a graph using your table.

   d. Using your table and graph, decide whether this is a reasonable trend over the long term. Explain.

Simplify each expression. 57. $23\sqrt[6]{23}$ 63. $2\sqrt{6} - 5$

53. $\dfrac{f^{-\frac{1}{4}}}{4f^{\frac{1}{2}} \cdot f^{-\frac{1}{3}}}$ $\dfrac{f^{\frac{7}{12}}}{4f}$ **54.** $\dfrac{8^{\frac{5}{2}}}{g^{\frac{1}{2}} + 2}$ $\dfrac{g^3 - 2g^{\frac{5}{2}}}{g - 4}$ **55.** $\dfrac{c^{\frac{2}{3}}}{c^{-\frac{1}{6}}}$ $c^{\frac{1}{2}}$ **56.** $\dfrac{z^{\frac{4}{5}}}{z^{\frac{1}{2}}}$ $z^{\frac{3}{10}}$

57. $\sqrt{23} \cdot \sqrt[3]{23^2}$ **58.** $\sqrt[8]{36h^4j^4}$ $6^{\frac{1}{4}}h^{\frac{1}{2}}j^{\frac{1}{2}}$ **59.** $\sqrt{\sqrt{81}}$ 3 **60.** $\sqrt[4]{\sqrt{256}}$ 2

61. $\dfrac{ab}{\sqrt{c}}$ $\dfrac{ab\sqrt{c}}{c}$ **62.** $\dfrac{xy}{\sqrt[3]{z}}$ $\dfrac{xy\sqrt[3]{z^2}}{z}$ **63.** $\dfrac{8^{\frac{1}{6}} - 9^{\frac{1}{4}}}{\sqrt{3} + \sqrt{2}}$ **64.** $\dfrac{x^{\frac{5}{3}} - x^{\frac{1}{3}}z^{\frac{4}{3}}}{x^{\frac{2}{3}} + z^{\frac{2}{3}}}$ $x - x^{\frac{1}{3}}z^{\frac{2}{3}}$

65. MULTIPLE REPRESENTATIONS In this problem, you will explore the functions $f(x) = x^3$ and $g(x) = x^{\frac{1}{3}}$.

   a. Tabular Copy and complete the table to the right.

   b. Graphical Graph $f(x)$ and $g(x)$. **See margin.**

   c. Verbal Explain the transformation between $f(x)$ and $g(x)$. **It is a reflection in the line $y = x$.**

x	$f(x)$	$g(x)$
-2	-8	-1.26
-1	-1	-1
0	0	0
1	1	1
2	8	1.26

H.O.T. Problems   Use Higher-Order Thinking Skills

66. REASONING Determine whether $-x^{-2} = (-x)^{-2}$ is *always*, *sometimes*, or *never* true. Explain your reasoning.

66. Never; the quantities are not the same. When the negative is enclosed inside of the parentheses and the base is raised to an even power, the answer is positive. When the negative is not enclosed inside of the parentheses and the base is raised to an even power, the answer is negative.

67. CHALLENGE Consider $\sqrt[4]{(-16)^3}$.

   a. Explain why the expression is not a real number. **See margin.**

   b. Find n such that $n\sqrt[4]{(-16)^3}$ is a real number. **Sample answer:** $\sqrt[4]{-1}$

68. OPEN ENDED Find two different expressions that equal 2 in the form $x^{\frac{1}{a}}$.

68. Sample answer: $4^{\frac{1}{2}}$ and $16^{\frac{1}{4}}$

69. WRITING IN MATH Explain how it might be easier to simplify an expression using rational exponents rather than using radicals. **See margin.**

70. CRITIQUE Ayana and Kenji are simplifying $\dfrac{x^{\frac{3}{4}}}{x^{\frac{1}{2}}}$. Is either of them correct? Explain your reasoning.

Ayana
$$\dfrac{x^{\frac{3}{4}}}{x^{\frac{1}{2}}} = x^{\frac{3}{4} + \frac{1}{2}}$$
$$= x^{\frac{3}{4} + \frac{2}{4}}$$
$$= x^{\frac{5}{4}}$$

Kenji
$$\dfrac{x^{\frac{3}{4}}}{x^{\frac{1}{2}}} = x^{\frac{3}{4} \div \frac{1}{2}}$$
$$= x^{\frac{3}{4} \cdot \frac{2}{1}}$$
$$= x^{\frac{3}{2}}$$

No; Ayana added the exponents and Kenji divided the exponents. The exponents should have been subtracted.

connectED.mcgraw-hill.com **427**

Multiple Representations

In Exercise 65, students use a table of values and a graph to compare and contrast a cubic and a radical equation.

Study Guide and Intervention
AL OL ELL

NAME _____ DATE _____ PERIOD _____

6-6 Study Guide and Intervention
Rational Exponents
Rational Exponents and Radicals

Practice
AL OL BL ELL

NAME _____ DATE _____ PERIOD _____

6-6 Practice
Rational Exponents

Word Problem Practice
AL OL BL ELL

NAME _____ DATE _____ PERIOD _____

6-6 Word Problem Practice
Rational Exponents

Enrichment
OL BL

NAME _____ DATE _____ PERIOD _____

6-6 Enrichment

Lesser-Known Geometric Formulas
Many geometric formulas involve radical expressions.

Yesterday's News Have students write how yesterday's lesson on radical expressions helped them with writing and simplifying expressions with rational exponents in today's lesson.

Formative Assessment
Check for student understanding of concepts in Lessons 6-5 and 6-6.

📁 Quiz 3, p. 54

Additional Answers
83. $20; x^2 + 2xh + h^2 - x - h$

84. $-129; 2x^3 + 6x^2h + 6xh^2 + 2h^3 - 1$

Standardized Test Practice

71. The expression $\sqrt{56 - c}$ is equivalent to a positive integer when c is equal to **B**

  A -10 **B** -8 **C** 36 **D** 56

72. SAT/ACT Which of the following sentences is true about the graphs of $y = 2(x - 3)^2 + 1$ and $y = 2(x + 3)^2 + 1$? **G**

  F Their vertices are maximums.

  G The graphs have the same shape with different vertices.

  H The graphs have different shapes with the same vertices.

  J The graphs have different shapes with different vertices.

  K One graph has a vertex that is a maximum; the other has a vertex that is a minimum.

73. GEOMETRY What is the converse of the statement? *If it is summer, then it is hot outside.* **C**

  A If it is not hot outside, then it is not summer.

  B If it is not summer, then it is not hot outside.

  C If it is hot outside, then it is summer.

  D If it is hot outside, it is not summer.

74. SHORT RESPONSE If $3^5 \cdot p = 3^3$, then find p. **3^{-2}**

Spiral Review

Simplify. (Lesson 6-5)

75. $\sqrt{243}$ $9\sqrt{3}$

76. $\sqrt[3]{16y^3}$ $2y\sqrt[3]{2}$

77. $3\sqrt[3]{56y^6z^3}$ $6y^2z\sqrt[3]{7}$

78. PHYSICS The speed of sound in a liquid is $s = \sqrt{\dfrac{B}{d}}$, where B is the bulk modulus of the liquid and d is its density. For water, $B = 2.1 \times 10^9$ N/m^2 and $d = 10^3$ kg/m^3. Find the speed of sound in water to the nearest meter per second. (Lesson 6-4) **1449 m/s**

Find $p(-4)$ and $p(x + h)$ for each function. (Lesson 5-3)

79. $p(x) = x - 2$ $-6; x + h - 2$

80. $p(x) = -x + 4$ $8; -x - h + 4$

81. $p(x) = 6x + 3$ $-21; 6x + 6h + 3$

82. $p(x) = x^2 + 5$ $21; x^2 + 2xh + h^2 + 5$

83. $p(x) = x^2 - x$ **See margin.**

84. $p(x) = 2x^3 - 1$ **See margin.**

Solve each equation by factoring. (Lesson 4-3)

85. $x^2 - 11x = 0$ $\{0, 11\}$

86. $x^2 + 6x - 16 = 0$ $\{-8, 2\}$

87. $4x^2 - 13x = 12$ $\left\{-\dfrac{3}{4}, 4\right\}$

88. $x^2 - 14x = -49$ $\{7\}$

89. $x^2 + 9 = 6x$ $\{3\}$

90. $x^2 - 3x = -\dfrac{9}{4}$ $\left\{\dfrac{3}{2}\right\}$

91. GEOMETRY A rectangle is inscribed in an isosceles triangle as shown. Find the dimensions of the inscribed rectangle with maximum area. (*Hint*: Use similar triangles.) (Lesson 4-1) **5 in. by 4 in.**

8 in.

10 in.

Skills Review

Find each power.

92. $\left(\sqrt{x - 3}\right)^2$ $x - 3$

93. $\left(\sqrt[3]{3x - 4}\right)^3$ $3x - 4$

94. $\left(\sqrt[4]{7x - 1}\right)^4$ $7x - 1$

95. $\left(\sqrt{x} - 4\right)^2$ $x - 8\sqrt{x} + 16$

96. $\left(2\sqrt{x} - 5\right)^2$ $4x - 20\sqrt{x} + 25$

97. $\left(3\sqrt{x} + 1\right)^2$ $9x + 6\sqrt{x} + 1$

 428 | Lesson 6-6 | Rational Exponents

LESSON 6-7 Solving Radical Equations and Inequalities

Then	Now	Why?
• You solved polynomial equations.	**1** Solve equations containing radicals. **2** Solve inequalities containing radicals.	When you jump, the time that you are in the air is your hang time. Hang time can be calculated in seconds t if you know the height h of the jump in feet. The formula for hang time is $t = 0.5\sqrt{h}$. Michael Jordan had a hang time of about 0.98 second. How would you calculate the height of Jordan's jump?

 NewVocabulary
radical equation
extraneous solution
radical inequality

 Common Core State Standards

Content Standards
A.SSE.2 Use the structure of an expression to identify ways to rewrite it.

Mathematical Practices
4 Model with mathematics.

1 Solve Radical Equations **Radical equations** include radical expressions. You can solve a radical equation by raising each side of the equation to a power.

KeyConcept Solving Radical Equations

Step 1 Isolate the radical on one side of the equation.

Step 2 Raise each side of the equation to a power equal to the index of the radical to eliminate the radical.

Step 3 Solve the resulting polynomial equation. Check your results.

When solving radical equations, the result may be a number that does not satisfy the original equation. Such a number is called an **extraneous solution**.

Example 1 Solve Radical Equations

Solve each equation.

a. $\sqrt{x+2} + 4 = 7$

$\sqrt{x+2} + 4 = 7$	Original equation
$\sqrt{x+2} = 3$	Subtract 4 from each side to isolate the radical.
$\left(\sqrt{x+2}\right)^2 = 3^2$	Square each side to eliminate the radical.
$x + 2 = 9$	Find the squares.
$x = 7$	Subtract 2 from each side.

CHECK	$\sqrt{x+2} + 4 = 7$	Original equation
	$\sqrt{7+2} + 4 \overset{?}{=} 7$	Replace x with 7.
	$7 = 7$ ✓	Simplify.

b. $\sqrt{x-12} = 2 - \sqrt{x}$

$\sqrt{x-12} = 2 - \sqrt{x}$	Original equation
$\left(\sqrt{x-12}\right)^2 = \left(2 - \sqrt{x}\right)^2$	Square each side.
$x - 12 = 4 - 4\sqrt{x} + x$	Find the squares.
$-16 = -4\sqrt{x}$	Isolate the radical.
$4 = \sqrt{x}$	Divide each side by −4.
$16 = x$	Square each side.

 connectED.mcgraw-hill.com **429**

Streeter Lecka/Getty Images Sport/Getty Images

1 Focus

VerticalAlignment

Before Lesson 6-7 Solve polynomial equations.

Lesson 6-7 Solve equations containing radicals. Solve inequalities containing radicals.

After Lesson 6-7 Identify a conic section from a given equation.

2 Teach

Scaffolding Questions
Have students read the **Why?** section of the lesson.

Ask:

■ In the formula $t = 0.5\sqrt{h}$, which variable appears in the radical? *h*, the height of the jump in feet

■ What is the value of t if $h = 4$? Describe your results. If $h = 4$, then $t = 1$; a jump of height 4 feet results in a hang time of 1 second.

■ Looking at your answer to the previous question, how is t affected if h is multiplied by 4? t is doubled

Lesson 6-7 Resources

Resource	Approaching Level **AL**	On Level **OL**	Beyond Level **BL**	English Learners **ELL**
Teacher Edition	▪ Differentiated Instruction, p. 431	▪ Differentiated Instruction, pp. 431, 435	▪ Differentiated Instruction, pp. 431, 435	▪ Differentiated Instruction, p. 431
Chapter Resource Masters	▪ Study Guide and Intervention, pp. 45–46 ▪ Skills Practice, p. 47 ▪ Practice, p. 48 ▪ Word Problem Practice, p. 49	▪ Study Guide and Intervention, pp. 45–46 ▪ Skills Practice, p. 47 ▪ Practice, p. 48 ▪ Word Problem Practice, p. 49 ▪ Enrichment, p. 50	▪ Practice, p. 48 ▪ Word Problem Practice, p. 49 ▪ Enrichment, p. 50	▪ Study Guide and Intervention, pp. 45–46 ▪ Skills Practice, p. 47 ▪ Practice, p. 48 ▪ Word Problem Practice, p. 49
Other	▪ 5-Minute Check 6-7 ▪ Study Notebook	▪ 5-Minute Check 6-7 ▪ Study Notebook	▪ 5-Minute Check 6-7 ▪ Study Notebook	▪ 5-Minute Check 6-7 ▪ Study Notebook

1 Solve Radical Equations

Example 1 shows how to solve two radical equations, one with a solution and the other with an extraneous solution. **Example 2** shows how to solve a cube root equation. **Example 3** shows how to answer test questions involving radical equations.

Formative Assessment

Use the Guided Practice exercises after each example to determine students' understanding of concepts.

Additional Examples

1 Solve each equation.
   a. $\sqrt{y-2} - 1 = 5$ 38
   b. $\sqrt{x-12} = 2 - \sqrt{x}$
      no solution

2 Solve $(3y+1)^{\frac{1}{3}} + 5 = 0$. −42

> **Additional Examples** also in Interactive Classroom PowerPoint® Presentations

IWB **Interactive White Board**
READY

Focus on Mathematical Content

Extraneous Roots It is always important to check the solution to an equation or inequality in the *original* equation, but it is especially important when both sides of an equation are raised to a power. If a solution is an approximation, it is sometimes difficult to determine whether a discrepancy is due to rounding or if it is an incorrect solution. Students should check exact solutions whenever possible.

StudyTip
Solution Check You can use a graphing calculator to check solutions of equations. Graph each side of the original equation, and examine the intersection.

CHECK $\sqrt{x-12} = 2 - \sqrt{x}$

$\sqrt{16-12} \stackrel{?}{=} 2 - \sqrt{16}$

$\sqrt{4} \stackrel{?}{=} 2 - 4$

$2 \neq -2$ ✗

The solution does not check, so the equation has an extraneous solution. The graphs of $y = \sqrt{x-12}$ and $y = 2 - \sqrt{x}$ do not intersect, which confirms that there is no real solution.

[−10, 30] scl: 2 by [−5, 10] scl: 1

Guided Practice

1A. $5 = \sqrt{x-2} - 1$ 38   **1B.** $\sqrt{x+15} = 5 + \sqrt{x}$ no real solution

To undo a square root, you square the expression. To undo a cube root, you must raise the expression to the third power.

Example 2 Solve a Cube Root Equation

Solve $2(6x-3)^{\frac{1}{3}} - 4 = 0$.

In order to remove the $\frac{1}{3}$ power, or cube root, you must first isolate it and then raise each side of the equation to the third power.

$2(6x-3)^{\frac{1}{3}} - 4 = 0$ Original equation

$2(6x-3)^{\frac{1}{3}} = 4$ Add 4 to each side.

$(6x-3)^{\frac{1}{3}} = 2$ Divide each side by 2.

$\left[(6x-3)^{\frac{1}{3}}\right]^3 = 2^3$ Cube each side.

$6x-3 = 8$ Evaluate the cubes.

$6x = 11$ Add 3 to each side.

$x = \frac{11}{6}$ Divide each side by 6.

CHECK $2(6x-3)^{\frac{1}{3}} - 4 = 0$ Original equation

$2\left(6 \cdot \frac{11}{6} - 3\right)^{\frac{1}{3}} - 4 \stackrel{?}{=} 0$ Replace x with $\frac{11}{6}$.

$2(8)^{\frac{1}{3}} - 4 \stackrel{?}{=} 0$ Simplify.

$2(2) - 4 \stackrel{?}{=} 0$ The cube root of 8 is 2.

$0 = 0$ ✓   Subtract.

Guided Practice

Solve each equation.

2A. $(3n+2)^{\frac{1}{3}} + 1 = 0$ −1   **2B.** $3(5y-1)^{\frac{1}{3}} - 2 = 0$ $\frac{7}{27}$

WatchOut!

Preventing Errors Have a discussion with students about which operations may introduce extraneous solutions when solving an equation that contains a radical. Remind students that the square root sign in an equation means the principal root.

Teach with Tech

Blog Have students write a blog entry about extraneous solutions. Have them explain what extraneous solutions are and how they can check their solutions graphically and algebraically.

You can apply the methods used to solve square and cube root equations to solving equations with roots of any index. To undo an nth root, raise to the nth power.

Test-TakingTip
Substitute Values You could also solve the test question by substituting each answer for n in the equation to see if the solution is correct.

Standardized Test Example 3 Solve a Radical Equation

What is the solution of $3\left(\sqrt[4]{2n+6}\right) - 6 = 0$?

A -1 **B** 1 **C** 5 **D** 11

$3\left(\sqrt[4]{2n+6}\right) - 6 = 0$	Original equation
$3\left(\sqrt[4]{2n+6}\right) = 6$	Add 6 to each side.
$\sqrt[4]{2n+6} = 2$	Divide each side by 3.
$\left(\sqrt[4]{2n+6}\right)^4 = 2^4$	Raise each side to the fourth power.
$2n + 6 = 16$	Evaluate each side.
$2n = 10$	Subtract 6 from each side.
$n = 5$	The answer is C.

GuidedPractice

3. What is the solution of $4(3x+6)^{\frac{1}{4}} - 12 = 0$? **G**

F $x = 7$ **G** $x = 25$ **H** $x = 29$ **J** $x = 37$

2 **Solve Radical Inequalities** A **radical inequality** has a variable in the radicand. To solve radical inequalities, complete the following steps.

KeyConcept Solving Radical Inequalities

Step 1 If the index of the root is even, identify the values of the variable for which the radicand is nonnegative.

Step 2 Solve the inequality algebraically.

Step 3 Test values to check your solution.

StudyTip
Radical Inequalities
Since a principal square root is never negative, inequalities that simplify to the form $\sqrt{ax+b} \leq c$, where c is a negative number, have no solutions.

Example 4 Solve a Radical Inequality

Solve $3 + \sqrt{5x - 10} \leq 8$.

Step 1 Since the radicand of a square root must be greater than or equal to zero, first solve $5x - 10 \geq 0$ to identify the values of x for which the left side of the inequality is defined.

$5x - 10 \geq 0$	Set the radicand ≥ 0.
$5x \geq 10$	Add 10 to each side.
$x \geq 2$	Divide each side by 5.

Step 2 Solve $3 + \sqrt{5x - 10} \leq 8$.

$3 + \sqrt{5x - 10} \leq 8$	Original inequality
$\sqrt{5x - 10} \leq 5$	Isolate the radical.
$5x - 10 \leq 25$	Eliminate the radical.
$5x \leq 35$	Add 10 to each side.
$x \leq 7$	Divide each side by 5.

connectED.mcgraw-hill.com **431**

Additional Example

3 **STANDARDIZED TEST EXAMPLE**
What is the solution of
$7\left(\sqrt[6]{5m+4}\right) - 4 = 10$? C
A $m = -2$ **C** $m = 12$
B $m = 0$ **D** $m = 14$

2 Solve Radical Inequalities
Example 4 shows how to solve a radical inequality and confirm the solution.

Additional Example

4 Solve $\sqrt{3x - 6} + 4 \leq 7$.
$2 \leq x \leq 5$

DifferentiatedInstruction AL OL BL ELL

Interpersonal Learners Ask students to work in groups as they compare solving radical equations and inequalities to solving other types of equations and inequalities. Have them write or give a short presentation about the similarities and differences between the procedures used in the solution process.

WatchOut!

Preventing Errors Emphasize the importance of checking key test values in the appropriate ranges.

Step 3 It appears that $2 \le x \le 7$. You can test some x-values to confirm the solution. Use three test values: one less than 2, one between 2 and 7, and one greater than 7. Organize the test values in a table.

$x = 0$	$x = 4$	$x = 9$
$3 + \sqrt{5(0) - 10} \stackrel{?}{\le} 8$ $3 + \sqrt{-10} \le 8$ ✗	$3 + \sqrt{5(4) - 10} \stackrel{?}{\le} 8$ $6.16 \le 8$ ✓	$3 + \sqrt{5(9) - 10} \stackrel{?}{\le} 8$ $8.92 \le 8$ ✗
Since $\sqrt{-10}$ is not a real number, the inequality is not satisfied.	Since $6.16 \le 8$, the inequality is satisfied.	Since $8.92 \not\le 8$, the inequality is not satisfied.

The solution checks. Only values in the interval $2 \le x \le 7$ satisfy the inequality. You can summarize the solution with a number line.

GuidedPractice

Solve each inequality.

4A. $\sqrt{2x + 2} + 1 \ge 5$ $x \ge 7$

4B. $\sqrt{4x - 4} - 2 < 4$ $1 \le x < 10$

3 Practice

Formative Assessment

Use Exercises 1–22 to check for understanding.

Use the chart at the bottom of the next page to customize assignments for your students.

 Teaching the Mathematical Practices

Reasoning Mathematically proficient students make sense of quantities and their relationships in problem situations. Quantitative reasoning entails habits of creating a coherent representation of the problem at hand; considering the units involved; attending to the meaning of quantities, not just how to compute them; and knowing and flexibly using different properties of operations and objects.

WatchOut!

Preventing Errors Make sure that students know the constraints on the values of the variables in a radical equation so that the solutions are real numbers.

Check Your Understanding ● = Step-by-Step Solutions begin on page R14.

Examples 1–2 Solve each equation.

1. $\sqrt{x - 4} + 6 = 10$ 20
2. $\sqrt{x + 13} - 8 = -2$ 23
3. $8 - \sqrt{x + 12} = 3$ 13
4. $\sqrt{x - 8} + 5 = 7$ 12
5. $\sqrt[3]{x - 2} = 3$ 29
6. $(x - 5)^{\frac{1}{3}} - 4 = -2$ 13
7. $(4y)^{\frac{1}{3}} + 3 = 5$ 2
8. $\sqrt[3]{n + 8} - 6 = -3$ 19
9. $\sqrt{y} - 7 = 0$ 49
10. $2 + 4z^{\frac{1}{2}} = 0$ no solution
11. $5 + \sqrt{4y - 5} = 12$ $\frac{27}{2}$
12. $\sqrt{2t - 7} = \sqrt{t + 2}$ 9

13. **CCSS REASONING** The time T in seconds that it takes a pendulum to make a complete swing back and forth is given by the formula $T = 2\pi\sqrt{\frac{L}{g}}$, where L is the length of the pendulum in feet and g is the acceleration due to gravity, 32 feet per second squared.

13a. about 9.5 seconds

   a. In Tokyo, Japan, a huge pendulum in the Shinjuku building measures 73 feet 9.75 inches. How long does it take for the pendulum to make a complete swing?
   b. A clockmaker wants to build a pendulum that takes 20 seconds to swing back and forth. How long should the pendulum be? about 324 ft

Example 3

14. **MULTIPLE CHOICE** Solve $(2y + 6)^{\frac{1}{4}} - 2 = 0$. **B**

   A $y = 1$ **B** $y = 5$ **C** $y = 11$ **D** $y = 15$

Example 4 Solve each inequality.

15. $\sqrt{3x + 4} - 5 \le 4$ $-\frac{4}{3} \le x \le \frac{77}{3}$
16. $\sqrt{b - 7} + 6 \le 12$ $7 \le b \le 43$
17. $2 + \sqrt{4y - 4} \le 6$ $1 \le y \le 5$
18. $\sqrt{3a + 3} - 1 \le 2$ $-1 \le a < 2$
19. $1 + \sqrt{7x - 3} > 3$ $x > 1$
20. $\sqrt{3x + 6} + 2 \le 5$ $-2 \le x \le 1$
21. $-2 + \sqrt{9 - 5x} \ge 6$ $x \le -11$
22. $6 - \sqrt{2y + 1} < 3$ $y > 4$

Example 1   Solve each equation. Confirm by using a graphing calculator.

23. $\sqrt{2x+5} - 4 = 3$ **22**   24. $6 + \sqrt{3x+1} = 11$ **8**

25. $\sqrt{x+6} = 5 - \sqrt{x+1}$ **3**   26. $\sqrt{x-3} = \sqrt{x+4} - 1$ **12**

27. $\sqrt{x-15} = 3 - \sqrt{x}$ **no real solution**   28. $\sqrt{x-10} = 1 - \sqrt{x}$ **no real solution**

29. $6 + \sqrt{4x+8} = 9$ $\frac{1}{4}$ 30. $2 + \sqrt{3y-5} = 10$ **23**

31. $\sqrt{x-4} = \sqrt{2x-13}$ **9**   32. $\sqrt{7a-2} = \sqrt{a+3}$ $\frac{5}{6}$

33. $\sqrt{x-5} - \sqrt{x} = -2$ $\frac{81}{16}$ 34. $\sqrt{b-6} + \sqrt{b} = 3$ $\frac{25}{4}$

35. **CCSS SENSE-MAKING** Isabel accidentally dropped her keys from the top of a Ferris wheel. The formula $t = \frac{1}{4}\sqrt{d-h}$ describes the time t in seconds at which the keys are h meters above the ground and Isabel is d meters above the ground. If Isabel was 65 meters high when she dropped the keys, how many meters above the ground will the keys be after 2 seconds?   **1 m**

Example 2   Solve each equation.

36. $(5n-6)^{\frac{1}{3}} + 3 = 4$ $\frac{7}{5}$ 37. $(5p-7)^{\frac{1}{3}} + 3 = 5$ **3**

38. $(6q+1)^{\frac{1}{4}} + 2 = 5$ $\frac{40}{3}$ 39. $(3x+7)^{\frac{1}{4}} - 3 = 1$ **83**

40. $(3y-2)^{\frac{1}{5}} + 5 = 6$ **1**   41. $(4z-1)^{\frac{1}{5}} - 1 = 2$ **61**

42. $2(x-10)^{\frac{1}{3}} + 4 = 0$ **2**   43. $3(x+5)^{\frac{1}{3}} - 6 = 0$ **3**

44. $\sqrt[3]{5x+10} - 5 = 0$ **23**   45. $\sqrt[3]{4n-8} - 4 = 0$ **18**

46. $\frac{1}{7}(14a)^{\frac{1}{3}} = 1$ **24.5**   47. $\frac{1}{4}(32b)^{\frac{1}{3}} = 1$ **2**

Example 3   48. **MULTIPLE CHOICE** Solve $\sqrt[4]{y+2} + 9 = 14$.   **D**

A 23   B 53   C 123   D 623

49. **MULTIPLE CHOICE** Solve $(2x-1)^{\frac{1}{4}} - 2 = 1$.   **F**

F 41   G 28   H 13   J 1

Example 4   Solve each inequality.   53. no real solution   54. $x > 4$ 55. $d > -\frac{3}{4}$

56. $-3 \le x < 24$
57. $-\frac{5}{2} \le y \le 2$

50. $1 + \sqrt{5x-2} > 4$ $x > \frac{11}{5}$ 51. $\sqrt{2x+14} - 6 \ge 4$ $x \ge 43$ 52. $10 - \sqrt{2x+7} \le 3$ $x \ge 21$

53. $6 + \sqrt{3y+4} < 6$ 54. $\sqrt{2x+5} - \sqrt{9+x} > 0$ 55. $\sqrt{d+3} + \sqrt{d+7} > 4$

56. $\sqrt{3x+9} - 2 < 7$ 57. $\sqrt{2y+5} + 3 \le 6$ 58. $-2 + \sqrt{8-4z} \ge 8$ $z \le -23$

59. $-3 + \sqrt{6a+1} > 4$ $a > 8$ 60. $\sqrt{2} - \sqrt{b+6} \le -\sqrt{b}$ 61. $\sqrt{c+9} - \sqrt{}$ $c > \sqrt{3}$
$0 \le b \le 2$ $0 \le c < 3$

B 62. **PENDULUMS** The formula $s = 2\pi\sqrt{\frac{\ell}{32}}$ represents the swing of a pendulum, where s is the time in seconds to swing back and forth, and ℓ is the length of the pendulum in feet. Find the length of a pendulum that makes one swing in 1.5 seconds.   **about 1.82 ft**

$M = \left(\frac{L}{0.46}\right)^3$ 63. **FISH** The relationship between the length and mass of certain fish can be approximated by the equation $L = 0.46\sqrt[3]{M}$, where L is the length in meters and M is the mass in kilograms. Solve this equation for M.

 433

Differentiated Homework Options

Level	Assignment	Two-Day Option	
AL Basic	23–61, 67, 69–72, 74–98	23–61 odd, 76–79	24–60 even, 64–75, 80–98
OL Core	23–61 odd, 62–67, 69–72, 74–98	23–61, 76–79	62–67, 69–72, 74, 75, 80–98
BL Advanced	62–90, (optional: 91–98)		

CCSS **Teaching the Mathematical Practices**

Sense-Making Mathematically proficient students start by explaining to themselves the meaning of a problem and looking for entry points to its solution. They analyze givens, constraints, relationships, and goals. They check their answers to problems using a different method, and they continually ask themselves, "Does this make sense?"

Tips for New Teachers

Reasoning Advise students that they cannot use the Distributive Property in Exercises 46 and 47. Ask them why this is the case, reminding them that the order of operations requires operations with exponents to be performed before those involving multiplication.

Study Guide and Intervention
AL OL ELL

Practice
AL OL BL ELL

Word Problem Practice
AL OL BL ELL

 64. HANG TIME Refer to the information at the beginning of the lesson regarding hang time. Describe how the height of a jump is related to the amount of time in the air. Write a step-by-step explanation of how to determine the height of Jordan's 0.98-second jump. **See margin.**

65. CONCERTS The organizers of a concert are preparing for the arrival of 50,000 people in the open field where the concert will take place. Each person is allotted 5 square feet of space, so the organizers rope off a circular area of 250,000 square feet. Using the formula $A = \pi r^2$, where A represents the area of the circular region and r represents the radius of the region, find the radius of this region. **about 282 ft**

66. WEIGHTLIFTING The formula $M = 512 - 146{,}230B^{-\frac{8}{5}}$ can be used to estimate the maximum total mass that a weightlifter of mass B kilograms can lift using the snatch and the clean and jerk. According to the formula, how much does a person weigh who can lift at most 470 kilograms? **163 kg**

H.O.T. Problems   Use Higher-Order Thinking Skills

67. CCSS ARGUMENTS Which equation does not have a solution? $\sqrt{x+2} - 7 = -10$

$$\sqrt{x-1} + 3 = 4$$

$$\sqrt{x+1} + 3 = 4$$

$$\sqrt{x-2} + 7 = 10$$

$$\sqrt{x+2} - 7 = -10$$

68. Yes; since $\sqrt[4]{x+5} \geq 0$, the left side of the equation is nonnegative. Therefore, the left side of the equation cannot equal -4. Thus the equation has no solution.

68. CHALLENGE Lola is working to solve $(x+5)^{\frac{1}{4}} = -4$. She said that she could tell there was no real solution without even working the problem. Is Lola correct? Explain your reasoning.

69. REASONING Determine whether $\dfrac{\sqrt{(x^2)^2}}{-x} = x$ is *sometimes*, *always*, or *never* true when x is a real number. Explain your reasoning. **See margin.**

70. OPEN ENDED Select a whole number. Now work backward to write two radical equations that have that whole number as solutions. Write one square root equation and one cube root equation. You may need to experiment until you find a whole number you can easily use. **Sample answer using 6: $\sqrt{x-2} = 2$, $(x+21)^{\frac{1}{3}} = 3$**

71. WRITING IN MATH Explain the relationship between the index of the root of a variable in an equation and the power to which you raise each side of the equation to solve the equation. **They are the same number.**

72. OPEN ENDED Write an equation that can be solved by raising each side of the equation to the given power.

a. $\frac{3}{2}$ power   **b.** $\frac{5}{4}$ power   **c.** $\frac{7}{8}$ power **Sample answer: $10x^{\frac{8}{7}} = -1$**

Sample answer: $0 = 6x^{\frac{2}{3}} - 5$ Sample answer: $0 = x^{\frac{4}{5}} - 9$

73. CHALLENGE Solve $7^{3x-1} = 49^{x+1}$ for x. (*Hint:* $b^x = b^y$ if and only if $x = y$.) **3**

REASONING Determine whether the following statements are *sometimes*, *always*, or *never* true for $x^{\frac{1}{n}} = a$. Explain your reasoning.

74. If n is odd, there will be extraneous solutions. **Never; sample answer: The radicand can be negative.**

75. If n is even, there will be extraneous solutions. **Sometimes; sample answer: when the radicand is negative, then there will be extraneous roots.**

Enrichment
OL BL

Additional Answer

64. If the height of a person's jump and the amount of time he or she is in the air are related by an equation involving radicals, then the hang time associated with a given height can be found by solving a radical equation.

76. What is an equivalent form of $\frac{4}{5+i}$? **A**

A $\frac{10-2i}{13}$ **C** $\frac{6-i}{6}$

B $\frac{5-i}{6}$ **D** $\frac{6-i}{13}$

77. Which set of points describes a function? **G**

    F $\{(3, 0), (-2, 5), (2, -1), (2, 9)\}$

    G $\{(-3, 5), (-2, 3), (-1, 5), (0, 7)\}$

    H $\{(2, 5), (2, 4), (2, 3), (2, 2)\}$

    J $\{(3, 1), (-3, 2), (3, 3), (-3, 4)\}$

78. **SHORT RESPONSE** The perimeter of an isosceles triangle is 56 inches. If one leg is 20 inches long, what is the measure of the base of the triangle?

16 in.

79. **SAT/ACT** If $\sqrt{x+5} + 1 = 4$, what is the value of x? **A**

    A 4   **D** 12

    B 10   **E** 20

    C 11

Spiral Review

Evaluate. (Lesson 6-6)

80. $27^{-\frac{2}{3}}$ $\frac{1}{9}$

81. $9^{\frac{1}{3}} \cdot 9^{\frac{5}{3}}$ **81**

82. $\left(\frac{8}{27}\right)^{-\frac{2}{3}}$ $\frac{9}{4}$

83. **GEOMETRY** The measures of the legs of a right triangle can be represented by the expressions $4x^2y^2$ and $8x^2y^2$. Use the Pythagorean Theorem to find a simplified expression for the measure of the hypotenuse. (Lesson 6-5) $4x^2y^2\sqrt{5}$

Find the inverse of each function. (Lesson 6-2)

84. $y = 3x - 4$ $y^{-1} = \frac{x+4}{3}$

85. $y = -2x - 3$ $y^{-1} = \frac{-x-3}{2}$

86. $y = x^2$ $y^{-1} = \pm\sqrt{x}$

87. $y = (2x + 3)^2$ $y^{-1} = \pm\frac{1}{2}\sqrt{x} - \frac{3}{2}$

For each graph,

a. describe the end behavior,

b. determine whether it represents an odd-degree or an even-degree polynomial function, and

c. state the number of real zeros. (Lesson 5-3) **88–90. See margin.**

88.

89.

90.

Skills Review

Solve each equation. Write in simplest form.

91. $\frac{8}{5}x = \frac{4}{15}$ $\frac{1}{6}$

92. $\frac{27}{14}y = \frac{6}{7}$ $\frac{4}{9}$

93. $\frac{3}{10} = \frac{12}{25}a$ $\frac{5}{8}$

94. $\frac{6}{7} = 9m$ $\frac{2}{21}$

95. $\frac{9}{8}b = 18$ **16**

96. $\frac{6}{7}n = \frac{3}{4}$ $\frac{7}{8}$

97. $\frac{1}{3}p = \frac{5}{6}$ $2\frac{1}{2}$

98. $\frac{2}{3}q = 7$ $10\frac{1}{2}$

DifferentiatedInstruction OL BL

Extension Present the following problem to students: The radius r of the orbit of a satellite is given by the formula $r = \sqrt[3]{\dfrac{GMt^2}{4\pi^4}}$, where G is the universal gravitational constant, M is the mass of the central object, and t is the time it takes the satellite to complete one orbit. Solve this formula for t.

$t = 2\pi^2 r\sqrt{\dfrac{r}{GM}}$

---

4 Assess

Ticket Out the Door Make several copies each of five different radical equations or inequalities. Give one equation or inequality to each student. Have students solve their equations (or inequalities) and hand them to you as they leave the room.

Formative Assessment

Check for student understanding of concepts in Lesson 6-7.

Quiz 4, p. 54

Additional Answers

69. never;

$$\frac{\sqrt{(x^2)^2}}{-x} = x$$

$$\frac{\sqrt{(x^2)^2}}{-x} = x$$

$$\frac{x^2}{-x} = x$$

$$x^2 = (x)(-x)$$

$$x^2 \neq -x^2$$

88a. $f(x) \rightarrow -\infty$ as $x \rightarrow +\infty$, $f(x) \rightarrow +\infty$ as $x \rightarrow -\infty$;

88b. odd;

88c. 3

89a. $f(x) \rightarrow +\infty$ as $x \rightarrow +\infty$, $f(x) \rightarrow +\infty$ as $x \rightarrow -\infty$;

89b. even;

89c. 0

90a. $f(x) \rightarrow +\infty$ as $x \rightarrow +\infty$, $f(x) \rightarrow -\infty$ as $x \rightarrow -\infty$;

90b. odd;

90c. 1

1 Focus

Objective Use a graphing calculator to solve radical equations and inequalities.

Materials for Each Student
- TI-83/84 Plus or other graphing calculator

Teaching Tips
Explain to students the following information about their calculators.

- The TI-83/84 Plus automatically supplies a left parenthesis after each radical sign. When functions are entered on the Y= list, it is important to supply right parentheses as needed to ensure correct graphs and correct numerical results.

- To display the table in Step 2 on this page, students should check to be sure that the **AUTO** option has been selected on each of the last two lines of the **TABLE SETUP** screen.

- To determine exact solutions for repeating decimals, as in the approximate zero displayed on the screen shown in Step 4, go to the home screen immediately after Step 4 and use the keystrokes $\boxed{\text{X,T,θ,}n}$ $\boxed{\text{MATH}}$ 1 $\boxed{\text{ENTER}}$. The calculator will display a fraction for the exact solution, $\frac{49}{36}$. The calculator can be used to verify that this is indeed the exact solution of the equation.

You can use a TI-83/84 Plus graphing calculator to solve radical equations and inequalities. One way to do this is to rewrite the equation or inequality so that one side is 0. Then use the zero feature on the calculator.

CCSS Common Core State Standards
Content Standards
A.REI.2 Solve simple rational and radical equations in one variable, and give examples showing how extraneous solutions may arise.
A.REI.11 Explain why the x-coordinates of the points where the graphs of the equations $y = f(x)$ and $y = g(x)$ intersect are the solutions of the equation $f(x) = g(x)$; find the solutions approximately, e.g., using technology to graph the functions, make tables of values, or find successive approximations. Include cases where $f(x)$ and/or $g(x)$ are linear, polynomial, rational, absolute value, exponential, and logarithmic functions.

Example 1 Radical Equation

Solve $\sqrt{x} + \sqrt{x+2} = 3$.

Step 1 Rewrite the equation.

- Subtract 3 from each side of the equation to get $\sqrt{x} + \sqrt{x+2} - 3 = 0$.
- Enter the function $y = \sqrt{x} + \sqrt{x+2} - 3$ in the **Y=** list.

KEYSTROKES: $\boxed{\text{Y=}}$ $\boxed{\text{2nd}}$ [√] $\boxed{\text{X,T,θ,}n}$ $\boxed{)}$ $\boxed{+}$ $\boxed{\text{2nd}}$ [√] $\boxed{\text{X,T,θ,}n}$ $\boxed{+}$ 2 $\boxed{)}$ $\boxed{-}$ 3 $\boxed{\text{ENTER}}$

Step 2 Use a table.

- You can use the **TABLE** function to locate intervals where the solution(s) lie. First, enter the starting value and the interval for the table.

KEYSTROKES: $\boxed{\text{2nd}}$ [TBLSET] 0 $\boxed{\text{ENTER}}$ 1 $\boxed{\text{ENTER}}$

Step 3 Estimate the solution.

- Complete the table and estimate the solution(s).

KEYSTROKES: $\boxed{\text{2nd}}$ [TABLE]

Since the function changes sign from negative to positive between $x = 1$ and $x = 2$, there is a solution between 1 and 2.

Step 4 Use the **ZERO** feature.

- Graph the function in the standard viewing window; then select **ZERO** from the **CALC** menu.

KEYSTROKES: $\boxed{\text{ZOOM}}$ 6 $\boxed{\text{2nd}}$ [CALC] 2

[−10, 10] scl: 1 by [−10, 10] scl: 1

Place the cursor on a point at which $y < 0$ and press $\boxed{\text{ENTER}}$ for the **LEFT BOUND**. Then place the cursor on a point at which $y > 0$ and press $\boxed{\text{ENTER}}$ for the **RIGHT BOUND**. You can use the same point for the **GUESS**. The solution is about 1.36. This is consistent with the estimate made by using the **TABLE**.

Example 2 Radical Inequality

Solve $2\sqrt{x} > \sqrt{x+2} + 1$.

Step 1 Graph each side of the inequality and use the **TRACE** feature.

- In the **Y=** list, enter $y_1 = 2\sqrt{x}$ and $y_2 = \sqrt{x+2} + 1$. Then press $\boxed{\text{GRAPH}}$.

[−10, 10] scl: 1 by [−10, 10] scl: 1

- Press $\boxed{\text{TRACE}}$. You can use $\boxed{\blacktriangle}$ or $\boxed{\blacktriangledown}$ to switch the cursor between the two curves.

The calculator screen above shows that, for points to the left of where the curves cross, **Y1 < Y2** or $2\sqrt{x} < \sqrt{x+2} + 1$. To solve the original inequality, you must find points for which **Y1 > Y2**. These are the points to the right of where the curves cross.

Step 3 Use the **TABLE** feature to check your solution.

- Start the table at 2 and show x-values in increments of 0.1. Scroll through the table.

  KEYSTROKES: $\boxed{\text{2nd}}$ [TBLSET] 2 $\boxed{\text{ENTER}}$ .1 $\boxed{\text{ENTER}}$ $\boxed{\text{2nd}}$ [TABLE]

Notice that when x is less than or equal to 2.4, **Y1 < Y2**. This verifies the solution $\{x \mid x > 2.40\}$.

Step 2 Use the **intersect** feature.

- You can use the **intersect** feature on the **CALC** menu to approximate the x-coordinate of the point at which the curves cross.

  KEYSTROKES: $\boxed{\text{2nd}}$ [CALC] 5

- Press $\boxed{\text{ENTER}}$ for each of **FIRST CURVE?**, **SECOND CURVE?**, and **GUESS?**.

[−10, 10] scl: 1 by [−10, 10] scl: 1

The calculator screen shows that the x-coordinate of the point at which the curves cross is about 2.40. Therefore, the solution of the inequality is about $x > 2.40$. *Use the symbol $>$ in the solution because the symbol in the original inequality is $>$.*

Exercises

Use a graphical method to solve each equation or inequality. **8.** $0 \le x < 1$ **9.** about $1 \le x < 4.52$

1. $\sqrt{x+4} = 3$ 5   **2.** $\sqrt{3x-5} = 1$ 2   **3.** $\sqrt{x+5} = \sqrt{3x+4}$ 0.5

4. $\sqrt{x+3} + \sqrt{x-2} = 4$ about 3.89   **5.** $\sqrt{3x-7} = \sqrt{2x-2} - 1$ about 2.52   **6.** $\sqrt{x+8} - 1 = \sqrt{x+2}$ 4.25

7. $\sqrt{x-3} \ge 2$ $x \ge 7$ **8.** $\sqrt{x+3} > 2\sqrt{x}$ **9.** $\sqrt{x} + \sqrt{x-1} < 4$

10. WRITING IN MATH Explain how you could apply the technique in the first example to solving an inequality. **See margin.**

connectED.mcgraw-hill.com **437**

Additional Answer

10. Rewrite the inequality so that one side is 0. Then graph the other side and find the x-values for which the graph is above or below the x-axis, according to the inequality symbol. Use the zero feature to approximate the x-coordinate of the point at which the graph crosses the x-axis.

---

2 Teach

Working in Cooperative Groups

Put students in pairs, mixing abilities. Then have pairs complete both Activities and Exercises 1–3 and 7.

Activity 1

- Have students solve this radical equation again by treating each side as a separate function. Point out that the right side will simply be graphed as the function $y = 3$.

Activity 2

- Instead of treating each side of the inequality in Activity 2 as a separate function, have students solve it again by first subtracting $2\sqrt{x}$ from both sides and then graphing the function $y = \sqrt{x+2} + 1 - 2\sqrt{x}$. Point out that the portion of the graph below the x-axis shows the solution.

Practice Have students complete Exercises 4–6 and 8–9.

---

3 Assess

Formative Assessment

Use Exercises 5 and 9 to assess whether students comprehend how to solve a radical equation or inequality.

From Concrete to Abstract

Ask:

- When examining the table of values of the equation in Activity 1, how do you know a solution lies between the two x-values where the graphed function changes signs? Sample answer: Since the original equation was solved so that one side was zero, the function representing the other side has a value of 0 when the value of y is 0. That must occur for a value of x somewhere between two values of x whose corresponding values of y have different signs (because 0 lies between the positive numbers and negative numbers).

CHAPTER **6** **Study Guide and Review**

Formative Assessment

KeyVocabulary The page references after each word denote where that term was first introduced. If students have difficulty answering questions 1–9, remind them that they can use these page references to refresh their memories about the vocabulary.

Summative Assessment

☐ Vocabulary Test, p. 56

 Vocabulary Review

Vocabulary Review provides students the opportunity to check their understanding of important concepts and terminology in an online game format.

FOLDABLES StudyOrganizer

Dinah Zike's Foldables®

Have students look through the chapter to make sure they have included examples in their Foldables. Suggest that students keep their Foldables handy while completing the Study Guide and Review pages. Point out that their Foldables can serve as a quick review tool when studying for the chapter test.

Additional Answers

10. $[f \circ g](x) = 8x - 9$
    $[g \circ f](x) = 8x - 1$

11. $[f \circ g](x) = x^2 - 14x + 50$
    $[g \circ f](x) = x^2 - 6$

12. $[f \circ g](x) = 4x^2 - 4x + 5$
    $[g \circ f](x) = -2x^2 - 7$

13. $[f \circ g](x) = 20x - 4$
    $[g \circ f](x) = 20x - 1$

14. $[f \circ g](x) = x^3 - 3x^2 + 3x - 1$
    $[g \circ f](x) = x^3 - 1$

15. $[f \circ g](x) = x^2 + 4x$
    $[g \circ f](x) = x^2 + 2x - 2$

17. $f^{-1}(x) = \dfrac{x + 6}{5}$

Study Guide

KeyConcepts

Operations on Functions (Lesson 6-1)

Operation	Definition
Sum	$(f + g)(x) = f(x) + g(x)$
Difference	$(f - g)(x) = f(x) - g(x)$
Product	$(f \cdot g)(x) = f(x) \cdot g(x)$
Quotient	$\left(\dfrac{f}{g}\right)(x) = \dfrac{f(x)}{g(x)}, g(x) \neq 0$
Composition	$[f \circ g](x) = f[g(x)]$

Inverse and Square Root Functions (Lessons 6-2 and 6-3)

- Two functions are inverses if and only if both their compositions are the identity function.

nth Roots (Lesson 6-4)

	Real nth roots of b, $\sqrt[n]{b}$, or $-\sqrt[n]{b}$		
n	$\sqrt[n]{b}$ if $b > 0$	$\sqrt[n]{b}$ if $b < 0$	$\sqrt[n]{b}$ if $b = 0$
even	one positive root one negative root	no real roots	one real root, 0
odd	one positive root no negative roots	no positive roots one negative root	

Radicals (Lessons 6-5 through 6-7)

For any real numbers a and b and any integers n, x, and y, with $b \neq 0$, $n > 1$, and $y > 1$, the following are true.

- Product Property: $\sqrt[n]{ab} = \sqrt[n]{a} \cdot \sqrt[n]{b}$

- Quotient Property: $\sqrt[n]{\dfrac{a}{b}} = \dfrac{\sqrt[n]{a}}{\sqrt[n]{b}}$

- Rational Exponents: $b^{\frac{x}{y}} = \sqrt[y]{b^x} = \left(\sqrt[y]{b}\right)^x, b \geq 0$

FOLDABLES StudyOrganizer

Be sure the Key Concepts are noted in your Foldable.

KeyVocabulary

composition of functions (p. 387)

conjugates (p. 418)

extraneous solution (p. 429)

index (p. 407)

inverse function (p. 393)

inverse relation (p. 393)

like radical expressions (p. 417)

nth root (p. 407)

principal root (p. 407)

radical equation (p. 429)

radical function (p. 400)

radical inequality (p. 431)

radical sign (p. 407)

radicand (p. 407)

rationalizing the denominator (p. 416)

square root function (p. 400)

square root inequality (p. 402)

VocabularyCheck

Choose a word or term that best completes each statement.

1. If both compositions result in the _____, then the functions are inverse functions. **identity function**

2. Radicals are _____ if *both* the index and the radicand are identical. **like radical expressions**

3. In a(n) _____, the results of one function are used to evaluate a second function. **composition of functions**

4. When there is more than one real root, the nonnegative root is called the _____. **principal root**

5. To eliminate radicals from a denominator or fractions from a radicand, you use a process called _____. **rationalizing the denominator**

6. Equations with radicals that have variables in the radicands are called _____. **radical equations**

7. Two relations are _____ if and only if one relation contains the element (b, a) when the other relation contains the element (a, b). **inverse relations**

8. When solving a radical equation, sometimes you will obtain a number that does not satisfy the original equation. Such a number is called a(n) _____. **extraneous solution**

9. The square root function is a type of _____. **radical function**

Lesson-by-Lesson Review

6-1 Operations on Functions

Find $[f \circ g](x)$ and $[g \circ f](x)$. **10–15. See margin.**

10. $f(x) = 2x + 1$
$g(x) = 4x - 5$

11. $f(x) = x^2 + 1$
$g(x) = x - 7$

12. $f(x) = x^2 + 4$
$g(x) = -2x + 1$

13. $f(x) = 4x$
$g(x) = 5x - 1$

14. $f(x) = x^3$
$g(x) = x - 1$

15. $f(x) = x^2 + 2x - 3$
$g(x) = x + 1$

16. MEASUREMENT The formula $f = 3y$ converts yards y to feet f and $f = \frac{n}{12}$ converts inches n to feet f. Write a composition of functions that converts yards to inches.
$n = 36y$

Example 1

If $f(x) = x^2 + 3$ and $g(x) = 3x - 2$, find $g[f(x)]$ and $f[g(x)]$.

$g[f(x)] = 3(x^2 + 3) - 2$ Replace $f(x)$ with $x^2 + 3$.
$\quad\quad\;\; = 3x^2 + 9 - 2$ Distributive Property
$\quad\quad\;\; = 3x^2 + 7$ Simplify.

$f[g(x)] = (3x - 2)^2 + 3$ Replace $g(x)$ with $3x - 2$.
$\quad\quad\;\; = 9x^2 - 12x + 4 + 3$ Multiply.
$\quad\quad\;\; = 9x^2 - 12x + 7$ Simplify.

6-2 Inverse Functions and Relations

Find the inverse of each function. Then graph the function and its inverse. **17–22. See margin.**

17. $f(x) = 5x - 6$

18. $f(x) = -3x - 5$

19. $f(x) = \frac{1}{2}x + 3$

20. $f(x) = \frac{4x + 1}{5}$

21. $f(x) = x^2$

22. $f(x) = (2x + 1)^2$

23. SHOPPING Samuel bought a computer. The sales tax rate was 6% of the sale price, and he paid $50 for shipping. Find the sale price if Samuel paid a total of $1322. **$1200**

Use the horizontal line test to determine whether the inverse of each function is also a function. **26, 28. no**

24. $f(x) = 3x^2$ **no**

25. $h(x) = x^3 - 3$ **yes**

26. $g(x) = -3x^4 + 2x - 1$ **27.** $g(x) = 4x^3 - 5x$ **no**

28. $f(x) = -3x^5 + x^2 - 3$ **29.** $h(x) = 4x^4 + 7x$ **no**

30. FINANCIAL LITERACY During the last month, Jonathan has made two deposits of $45, made a deposit of double his original balance, and has withdrawn $35 five times. His balance is now $189. Write an equation that models this problem. How much money did Jonathan have in his account at the beginning of the month?
$x + 2(45) + 2x - 5(35) = 189$; about $91.33

Example 2

Find the inverse of $f(x) = -2x + 7$.

Rewrite $f(x)$ as $y = -2x + 7$. Then interchange the variables and solve for y.

$x = -2y + 7$ Interchange the variables.
$2y = -x + 7$ Solve for y.
$y = \frac{-x + 7}{2}$ Divide each side by 2.
$f^{-1}(x) = \frac{-x + 7}{2}$ Rewrite using function notation.

Example 3

Use the horizontal line test to determine whether the inverse of $f(x) = 2x^3 + 1$ is also a function.

Graph the function.

No horizontal line can be drawn so that it passes through more than one point. The inverse of this function is a function.

connectED.mcgraw-hill.com **439**

Lesson-by-Lesson Review

Intervention If the given examples are not sufficient to review the topics covered by the questions, remind students that the lesson references tell them where to review that topic in their textbook.

Two-Day Option Have students complete the Lesson-by-Lesson Review. Then you can use McGraw-Hill eAssessment to customize another review worksheet that practices all the objectives of this chapter or only the objectives on which your students need more help.

Additional Answers

18. $f^{-1}(x) = \frac{x + 5}{-3}$

19. $f^{-1}(x) = 2x - 6$

20. $f^{-1}(x) = \frac{5x - 1}{4}$

21. $f^{-1}(x) = \pm\sqrt{x}$

22. $f^{-1}(x) = \frac{-1 \pm \sqrt{x}}{2}$

Additional Answers

31.

$D = \{x \mid x \geq 0\}; R = \{f(x) \mid f(x) \geq 0\}$

32.

$D = \{x \mid x \geq 0\}; R = \{f(x) \mid f(x) \leq 0\}$

33.

$D = \{x \mid x \geq 7\}; R = \{f(x) \mid f(x) \geq 0\}$

34.

$D = \{x \mid x \geq -5\}; R = \{f(x) \mid f(x) \geq -3\}$

35.

$D = \{x \mid x \geq 1\}; R = \{f(x) \mid f(x) \geq 5\}$

36.

$D = \{x \mid x \geq -4\}; R = \{f(x) \mid f(x) \leq -1\}$

6-3 Square Root Functions and Inequalities

Graph each function. State the domain and range.

31. $f(x) = \sqrt{3x}$ 31–36. See margin.

32. $f(x) = -\sqrt{6x}$

33. $f(x) = \sqrt{x - 7}$

34. $f(x) = \sqrt{x + 5} - 3$

35. $f(x) = \frac{3}{4}\sqrt{x - 1} + 5$

36. $f(x) = -\frac{1}{3}\sqrt{x + 4} - 1$

37. GEOMETRY The area of a circle is given by the formula $A = \pi r^2$. What is the radius of a circle with an area of 300 square inches?   about 9.8 in.

Graph each inequality.   38–40. See margin.

38. $y \geq \sqrt{x} + 3$

39. $y < 2\sqrt{x - 5}$

40. $y > -\sqrt{x - 1} + 2$

Example 4

Graph $f(x) = \sqrt{x + 1} - 2$. State the domain and range.

Identify the domain.

$x + 1 \geq 0$ Write the radicand as greater than or equal to 0.

$x \geq -1$ Subtract 1 from each side.

Make a table of values for $x \geq -1$ and graph the function.

x	f(x)
−1	−2
0	−1
1	−0.59
2	−0.27
3	0
4	0.24
5	0.45

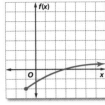

The domain is $\{x \mid x \geq -1\}$, and the range is $\{f(x) \mid f(x) \geq -2\}$.

6-4 nth Roots

Simplify.

41. $\pm\sqrt{121}$ ±11

42. $\sqrt[3]{-125}$ −5

43. $\sqrt{(-6)^2}$ 6

44. $\sqrt{-(x + 3)^4}$ $i(x + 3)^2$

45. $\sqrt[6]{(x^2 + 2)^{18}}$ $(x^2 + 2)^3$

46. $\sqrt[3]{27(x + 3)^3}$ $3(x + 3)$

47. $\sqrt[4]{a^8 b^{12}}$ $a^2 |b^3|$

48. $\sqrt[5]{243x^{10}y^{25}}$ $3x^2 y^5$

49. PHYSICS The velocity v of an object can be defined as $v = \sqrt{\dfrac{2K}{m}}$, where m is the mass of an object and K is the kinetic energy in joules. Find the velocity in meters per second of an object with a mass of 17 grams and a kinetic energy of 850 joules.   10 m/s

Example 5

Simplify $\sqrt{64x^6}$.

$\sqrt{64x^6} = \sqrt{(8x^3)^2}$ $64x^6 = (8x^3)^2$

$= 8|x^3|$ Simplify.

Use absolute value symbols because x could be negative.

Example 6

Simplify $\sqrt[6]{4096x^{12}y^{24}}$.

$\sqrt[6]{4096x^{12}y^{24}} = \sqrt[6]{(4x^2y^4)^6}$ $4096x^{12}y^{24} = (4x^2y^4)^6$

$= 4x^2y^4$ Simplify.

38.

6-5 Operations with Radical Expressions

Simplify.

50. $\sqrt[3]{54}$ $3\sqrt[3]{2}$

51. $\sqrt{144a^3b^5}$ $12ab^2\sqrt{ab}$

52. $4\sqrt{6y} \cdot 3\sqrt{7x^2y}$ $12|x|y\sqrt{42}$

53. $6\sqrt{72} + 7\sqrt{98} - \sqrt{50}$ $80\sqrt{2}$

54. $(6\sqrt{5} - 2\sqrt{2})(3\sqrt{5} + 4\sqrt{2})$ $74 + 18\sqrt{10}$

55. $\dfrac{\sqrt{6m^5}}{\sqrt{p^{11}}}$ $\dfrac{m^2\sqrt{6mp}}{p^6}$

56. $\dfrac{3}{5 + \sqrt{2}}$ $\dfrac{15 - 3\sqrt{2}}{23}$

57. $\dfrac{\sqrt{3}}{\sqrt{5} - \sqrt{6}}$ $-\sqrt{15} - 3\sqrt{2}$

58. GEOMETRY What are the perimeter and the area of the rectangle?

perimeter = $28 + 2\sqrt{3} - 2\sqrt{2}$ units;
area = $48 + 6\sqrt{3} - 8\sqrt{2} - \sqrt{6}$ units^2

Example 7

Simplify $2\sqrt[3]{18a^2b} \cdot 3\sqrt[3]{12ab^5}$.

$2\sqrt[3]{18a^2b} \cdot 3\sqrt[3]{12ab^5}$

$= (2 \cdot 3)\sqrt[3]{18a^2b \cdot 12ab^5}$ Product Property

$= 6\sqrt[3]{2^3 3^3 a^3 b^6}$ Factor.

$= 6 \cdot \sqrt[3]{2^3} \cdot \sqrt[3]{3^3} \cdot \sqrt[3]{a^3} \cdot \sqrt[3]{b^6}$ Product Property

$= 6 \cdot 2 \cdot 3 \cdot a \cdot b^2$ Find cube roots.

$= 36ab^2$ Simplify.

Example 8

Simplify $\sqrt{\dfrac{x^4}{y^5}}$.

$\sqrt{\dfrac{x^4}{y^5}} = \dfrac{\sqrt{x^4}}{\sqrt{y^5}}$ Quotient Property

$= \dfrac{\sqrt{(x^2)^2}}{\sqrt{(y^2)^2 \cdot \sqrt{y}}}$ Factor into squares.

$= \dfrac{x^2}{y^2\sqrt{y}} \cdot \dfrac{\sqrt{y}}{\sqrt{y}}$ Rationalize the denominator.

$= \dfrac{x^2\sqrt{y}}{y^3}$ $\sqrt{y} \cdot \sqrt{y} = y$

39.

40.

6-6 Rational Exponents

Simplify each expression.

59. $x^{\frac{1}{2}} \cdot x^{\frac{2}{3}}$ $x^{\frac{7}{6}}$ **60.** $\dfrac{m^{-\frac{3}{4}}}{m}$ $\dfrac{m^{\frac{1}{4}}}{m}$ **61.** $\dfrac{d^{\frac{1}{6}} d^{\frac{5}{12}}}{d^{\frac{3}{4}}}$ $\dfrac{1}{d}$

Simplify each expression.

62. $\dfrac{1}{y^{\frac{1}{4}}}$ $\dfrac{y^{\frac{3}{4}}}{y}$ **63.** $\sqrt[3]{\sqrt{729}}$ 3   **64.** $\dfrac{x^{\frac{2}{3}} - x^{\frac{1}{3}}y^{\frac{2}{3}}}{x^{\frac{1}{3}}}$ $x^{\frac{1}{3}} - y^{\frac{2}{3}}$

65. GEOMETRY What is the area of the circle?

$4a^{\frac{2}{3}}b^{\frac{4}{5}}c^2\pi$ units^2

Example 9

Simplify $a^{\frac{2}{3}} \cdot a^{\frac{1}{5}}$.

$a^{\frac{2}{3}} \cdot a^{\frac{1}{5}} = a^{\frac{2}{3} + \frac{1}{5}}$ Product of Powers

$= a^{\frac{13}{15}}$ Add.

Example 10

Simplify $\dfrac{2a}{\sqrt[3]{b}}$.

$\dfrac{2a}{\sqrt[3]{b}} = \dfrac{2a}{b^{\frac{1}{3}}}$ Rational exponents

$= \dfrac{2a}{b^{\frac{1}{3}}} \cdot \dfrac{b^{\frac{2}{3}}}{b^{\frac{2}{3}}}$ Rationalize the denominator.

$= \dfrac{2ab^{\frac{2}{3}}}{b}$ or $\dfrac{2a\sqrt[3]{b^2}}{b}$ Rewrite in radical form.

Anticipation Guide

Have students complete the Chapter 6 Anticipation Guide and discuss how their responses have changed now that they have completed Chapter 6.

Study Guide and Review *Continued*

6-7 Solving Radical Equations and Inequalities

Solve each equation.

66. $\sqrt{x-3}+5=15$ **103**

67. $-\sqrt{x-11}=3-\sqrt{x}$ $\frac{100}{9}$

68. $4+\sqrt{3x-1}=8$ $\frac{17}{3}$

69. $\sqrt{m+3}=\sqrt{2m+1}$ **2**

70. $\sqrt{2x+3}=3$ **3**

71. $(x+1)^{\frac{1}{4}}=-3$ **no solution**

72. $a^{\frac{1}{3}}-4=0$ **64**

73. $3(3x-1)^{\frac{1}{3}}-6=0$ **3**

74. PHYSICS The formula $t=2\pi\sqrt{\dfrac{\ell}{32}}$ represents the swing of a pendulum, where t is the time in seconds for the pendulum to swing back and forth and ℓ is the length of the pendulum in feet. Find the length of a pendulum that makes one swing in 2.75 seconds. **about 6.13 ft**

Solve each inequality.

75. $2+\sqrt{3x-1}<5$

76. $\sqrt{3x+13}-5\geq5$ $x\geq29$

77. $6-\sqrt{3x+5}\leq3$

78. $\sqrt{-3x+4}-5\geq3$ $x\leq-20$

79. $5+\sqrt{2y-7}<5$

80. $3+\sqrt{2x-3}\geq3$ $x\geq\frac{3}{2}$

81. $\sqrt{3x+1}-\sqrt{6+x}>0$ $x>\frac{5}{2}$

75. $\frac{1}{3}\leq x<\frac{10}{3}$

77. $x\geq\frac{4}{3}$

79. no solution

Example 11

Solve $\sqrt{2x+9}-2=5$.

$\sqrt{2x+9}-2=5$	Original equation
$\sqrt{2x+9}=7$	Add 2 to each side.
$(\sqrt{2x+9})^2=7^2$	Square each side.
$2x+9=49$	Evaluate the squares.
$2x=40$	Subtract 9 from each side.
$x=20$	Divide each side by 2.

Example 12

Solve $\sqrt{2x-5}+2>5$.

$\sqrt{2x-5}\geq0$	Radicand must be ≥0.
$2x-5\geq0$	Square each side.
$2x\geq5$	Add 5 to each side.
$x\geq2.5$	Divide each side by 2.

The solution must be greater than or equal to 2.5 to satisfy the domain restriction.

$\sqrt{2x-5}+2>5$	Original inequality
$\sqrt{2x-5}>3$	Subtract 2 from each side.
$(\sqrt{2x-5})^2>3^2$	Square each side.
$2x-5>9$	Evaluate the squares.
$2x>14$	Add 5 to each side.
$x>7$	Divide each side by 2.

Since $x\geq2.5$ contains $x>7$, the solution of the inequality is $x>7$.

CHAPTER 6 Practice Test

Determine whether each pair of functions are inverse functions. Write *yes* or *no*. Explain your reasoning.

1. $f(x) = 3x + 8$, $g(x) = \frac{x - 8}{3}$ **yes**

2. $f(x) = \frac{1}{3}x + 5$, $g(x) = 3x - 15$ **yes**

3. $f(x) = x + 7$, $g(x) = x - 7$ **yes**

4. $g(x) = 3x - 2$, $f(x) = \frac{x - 2}{3}$ **no**

5. **MULTIPLE CHOICE** Which inequality represents the graph below? **B**

A $y \geq \sqrt{x + 4}$

B $y \leq \sqrt{x + 4}$

C $y \geq \sqrt{x - 4}$

D $y \leq \sqrt{x - 4}$

If $f(x) = 3x + 2$ and $g(x) = x^2 - 2x + 1$, find each function. 7. $(f \cdot g)(x) = 3x^3 - 4x^2 - x + 2$

6. $(f + g)(x)$
$(f + g)(x) = x^2 + x + 3$

7. $(f \cdot g)(x)$

8. $(f - g)(x)$
$(f - g)(x) = -x^2 + 5x + 1$

9. $\left(\frac{f}{g}\right)(x)$

Solve each equation.
$\left(\frac{f}{g}\right)(x) = \frac{3x + 2}{x^2 - 2x + 1}; x \neq 1$

10. $\sqrt{a + 12} = \sqrt{5a - 4}$ **4**

11. $\sqrt{3x} = \sqrt{x - 2}$ **no solution**

12. $4(\sqrt[4]{3x + 1}) - 8 = 0$ **5**

13. $\sqrt[3]{5m + 6} + 15 = 21$ **42**

14. $\sqrt{3x + 21} = \sqrt{5x + 27}$ **−3**

15. $1 + \sqrt{x + 11} = \sqrt{2x + 15}$ **5**

16. $\sqrt{x - 5} = \sqrt{2x - 4}$ **no solution**

17. $\sqrt{x - 6} - \sqrt{x} = 3$ **no solution**

18. **MULTIPLE CHOICE** Which expression is equivalent to $125^{-\frac{1}{3}}$? **H**

F -5

G $-\frac{1}{5}$

H $\frac{1}{5}$

J 5

Simplify.

19. $(2 + \sqrt{5})(6 - 3\sqrt{5})$

20. $(3 - 2\sqrt{2})(-7 + \sqrt{2})$ $17\sqrt{2} - 25$

21. $\frac{12}{2 - \sqrt{3}}$ $12\sqrt{3} + 24$

22. $\frac{m^{\frac{1}{2}} - 1}{2m^{\frac{1}{2}} + 1}$ $\frac{2m - 3m^{\frac{1}{2}} + 1}{4m - 1}$

23. $4\sqrt{3} - 8\sqrt{48}$ $-28\sqrt{3}$

24. $5^{\frac{2}{3}} \cdot 5^{\frac{1}{2}} \cdot 5^{\frac{5}{6}}$ 5^2 or 25

25. $\sqrt[6]{729a^9b^{24}}$ $3ab^4\sqrt{a}$

26. $\sqrt[5]{32x^{15}y^{10}}$ $2x^3y^2$

27. $w^{-\frac{4}{5}} \cdot \frac{w^{\frac{1}{5}}}{w}$

28. $\frac{r^{\frac{2}{3}} \cdot r^{\frac{1}{2}}}{r^{\frac{1}{6}}}$

29. $\frac{a^{-\frac{1}{2}}}{6a^{\frac{1}{3}} \cdot a^{-\frac{1}{4}}}$ $\frac{a^{\frac{5}{12}}}{6a}$

30. $\frac{y^{\frac{3}{2}}}{y^{\frac{1}{2}} + 2} \cdot \frac{y^2 - 2y^{\frac{3}{2}}}{y - 4}$

31. **MULTIPLE CHOICE** What is the area of the rectangle? **A**

A $2\sqrt{3} + 3\sqrt{2}$ units^2

B $4 + 2\sqrt{6} + 2\sqrt{3}$ units^2

C $2\sqrt{3} + \sqrt{6}$ units^2

D $2\sqrt{3} + 3$ units^2

Solve each inequality.

32. $\sqrt{4x - 3} < 5\frac{3}{4}$ $\leq x < 7$

33. $-2 + \sqrt{3m - 1} < 4$

34. $2 + \sqrt{4x - 4} \leq 6$

35. $\sqrt{2x + 3} - 4 \leq 5$

36. $\sqrt{b + 12} - \sqrt{b} > 2$

37. $\sqrt{y - 7} + 5 \geq 10$ $y \geq 32$

38. $\sqrt{a - 5} - \sqrt{a + 7} \leq 4$ $a \geq 5$

39. $\sqrt{c + 5} + \sqrt{c + 10} > 2$ $c > -5$

33. $\frac{1}{3} \leq m < \frac{37}{3}$

34. $1 \leq x \leq 5$

35. $-\frac{3}{2} \leq x \leq 39$

36. $0 \leq b < 4$

40. **GEOMETRY** The area of a triangle with sides of length a, b, and c is given by $A = \sqrt{s(s - a)(s - b)(s - c)}$, where $s = \frac{1}{2}(a + b + c)$. What is the area of the triangle expressed in radical form? $2\sqrt{66}$ ft^2

Summative Assessment

Use these alternate leveled chapter tests to differentiate assessment for your students.

Leveled Chapter 6 Tests

Form	Type	Level	📁 Page(s)
1	MC	AL	57–58
2A	MC	OL	59–60
2B	MC	OL	61–62
2C	FR	OL	63–64
2D	FR	OL	65–66
3	FR	BL	67–68
Vocabulary Test			56
Extended-Response Test			69

MC = multiple-choice questions
FR = free-response questions

eAssessment McGraw-Hill Customize and create multiple versions of your chapter test and their answer keys. All of the questions from the leveled chapter tests in the *Chapter 6 Resource Masters* are also available on McGraw-Hill eAssessment.

InterventionPlanner

TIER 1 On Level OL

If ▶ students miss about 25% of the exercises or less,

Then ▶ choose a resource:

SE Lessons 6-1, 6-2, 6-3, 6-4, 6-5, 6-6, and 6-7

📁 Skills Practice, pp. 7, 14, 21, 27, 33, 40, and 47

TE Chapter Project

connectED.mcgraw-hill.com Self-Check Quiz

TIER 2 Strategic Intervention AL approaching grade level

If ▶ students miss about 50% of the exercises,

Then ▶ choose a resource:

📁 Study Guide and Intervention, pp. 5, 12, 19, 25, 31, 38, and 45

connectED.mcgraw-hill.com Extra Examples, Personal Tutor, Homework Help

TIER 3 Intensive Intervention 2 or more grades below level

If ▶ students miss about 75% of the exercises,

Then ▶ use *Math Triumphs, Alg. 2*, Ch. 1

connectED.mcgraw-hill.com Extra Examples, Personal Tutor, Homework Help, Review Vocabulary

CHAPTER 6

Preparing for Standardized Tests

1 Focus

Objective Use the strategy of working backward to solve standardized test problems.

2 Teach

Scaffolding Questions

Ask:

- Have you ever lost something and retraced your steps in order to find it? *Answers will vary.*

- Did you retrace your steps in reverse order or in any random order? *Answers will vary.*

- What is the reverse of adding? multiplying? *subtracting; dividing*

Work Backward

In certain math problems, you are given information about an end result, but you need to find out something that happened earlier. You can work backward to solve problems like this.

Strategies for Working Backward

Step 1

Read the problem statement carefully.

Ask yourself:

- What information am I given?
- What am I being asked to solve?
- Does any of the information given relate to an end result?
- Am I being asked to solve for a quantity that occurred "earlier" in the problem statement?
- What operations are being used in the problem?

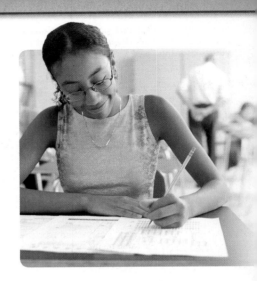

Step 2

Model the problem situation with an equation, an inequality, or a graph as appropriate. Then work backward to solve the problem.

- If needed, sketch a flow of events to show the sequence described in the problem statement.

- Use inverse operations to undo any operations while working backward until you arrive at your answer.

Step 3

Check by beginning with your answer and seeing if you arrive at the same result given in the problem statement.

Standardized Test Example

Read the problem. Identify what you need to know. Then use the information in the problem to solve.

Maria bought a used car. The sales tax rate was 6.75% of the selling price, and she had to pay $450 in processing, title, and registration fees. If Maria paid a total of $15,768.63, what was the sale price of the car? Show your work.

Read the problem carefully. You know the total amount that Maria paid for the car after sales tax was applied and after she paid all of the other fees. You need to find the sale price of the car before taxes and fees.

Sean Justice/Stone/Getty Images

Let x represent the sale price of the car and set up an equation. Use the work backward strategy to solve the problem.

Words	The sale price of the car plus the sales tax and other fees is equal to the final price.
Variable	Let x = sale price.
Equation	$1.0675x + 450 = 15,768.63$

Using the work backward strategy results in a simple equation. Use inverse operations to solve for x.

$1.0675x + 450 = 15,768.63$ Original equation

$1.0675x = 15,318.63$ Subtract 450 from each side.

$x \approx 14,350$ Divide each side by 1.0675.

Check your answer by working the problem forward. Begin with your answer and see if you get the same result as in the problem statement.

$14,350(1.0675) \approx 15,318.63$ Compute the sales tax.

$15,318.63 + 450 = 15,768.63$ Add the other fees.

$15,768.63 = 15,768.63$ The result is the same.

So, the sale price of the car was $14,350.

Exercises

Read the problem. Identify what you need to know. Then use the information in the problem to solve.

1. The equation $d = \dfrac{s^2}{30f}$ can be used to model the length of the skid marks left by a car when a driver applies the brakes to come to a sudden stop. In the equation, d is the length (in feet) of the skid marks left on the road, s is the speed of the car in miles per hour, and f is a coefficient of friction that describes the condition of the road. Suppose a car left skid marks that are 120 feet long. **a.** $s = \sqrt{30df}$

   a. Solve the equation for s, the speed of the car.

   b. If the coefficient of friction for the road is 0.75, about how fast was the car traveling?
   about 52 mph

   c. How fast was the car traveling if the coefficient of friction for the road is 1.1? **about 63 mph**

2. An object is shot straight upward into the air with an initial speed of 800 feet per second. The height h that the object will be after t seconds is given by the equation $h = -16t^2 + 800t$. When will the object reach a height of 10,000 feet? **B**

   A   10 seconds

   B   25 seconds

   C   100 seconds

   D   625 seconds

3. Pedro is creating a scale drawing of a car. He finds that the height of the car in the drawing is $\dfrac{1}{32}$ of the actual height of the car x. Which equation best represents this relationship? **H**

   F $y = x - \dfrac{1}{32}$ H $y = \dfrac{1}{32}x$

   G $y = -\dfrac{1}{32}x$ J $y = x + \dfrac{1}{32}$

Additional Example

STANDARDIZED TEST PRACTICE
Nate used a $15 gift card to partially pay for a dinner for two at his favorite restaurant. After a sales tax of 6.5% was added, a $5.25 tip was left, and he used the gift card, Nate paid a total of $18.21 for the cost of the dinners. What was the price of the two dinners before taxes? C

A $16.94

B $26.21

C $26.25

D $31.50

3 Assess

Use Exercises 1–3 to assess students' understanding.

Diagnose Student Errors
Survey student responses for each item. Class trends may indicate common errors and misconceptions.

1. A added the coupon to the original cost then multiplied by the percent to be paid
   B correct
   C subtracted the coupon from the original cost and then multiplied by the percent of discount
   D subtracted the coupon from the original cost and then multiplied by the percent to be paid

2. F guess
   G added the area of a circle to the area of the rectangle
   H did not find the correct value for the radius
   J correct

3. A did not follow the correct order of operations
   B correct
   C did not multiply correctly
   D did not multiply correctly

4. F used a tax rate of 9.6%
   G used a tax rate of 7.5%
   H correct
   J used a tax rate of 3.59%

5. A did not find the cube root of an exponent correctly
   B correct
   C did not find the cube root of −27 correctly
   D guess

6. F confused the constants that cause a shift up and down with those that shift the graph right and left
   G did not understand that the constant under the radical shifts the graph right or left
   H guess
   J correct

7. A did not find the least value for $\left|\frac{1}{a}\right|$

   B did not find the least value for $\left|\frac{1}{a}\right|$

   C correct

   D did not find the least value for $\left|\frac{1}{a}\right|$

Multiple Choice

Read each question. Then fill in the correct answer on the answer document provided by your teacher or on a sheet of paper.

1. A sporting goods store is discounting all camping equipment by 20% during the off-season. Charles also has a coupon good for $5.00 off his next purchase from the store. If the coupon is applied *after* the store discount, which of the following functions can be used to find the final price of a tent that originally cost *d* dollars? **B**

   A $P(d) = 0.8 \times (d + 5)$

   B $P(d) = (0.8 \times d) - 5$

   C $P(d) = 0.2 \times (d - 5)$

   D $P(d) = 0.8 \times (d - 5)$

2. Find the equation that can be used to determine the total area of the composite figure below. **J**

   F $A = \ell w + \frac{1}{2}\ell w$

   G $A = \ell w + \pi\left(\frac{1}{2}\ell\right)^2$

   H $A = \ell w + \frac{1}{2}\pi\ell^2$

   J $A = \ell w + \pi\left(\frac{1}{2}\ell\right)^2\left(\frac{1}{2}\right)$

3. Which expression is equivalent to $3a(2a + 1) - (2a - 2)(a + 3)$? **B**

   A $2a^2 + 6a + 7$

   B $4a^2 - a + 6$

   C $4a^2 + 6a - 6$

   D $4a^2 - 3a + 7$

---

4. Kay bought a used car. The sales tax rate was 6.5% of the selling price, and she also had to pay $325 in registration fees. Find the selling price if Kay spent a total of $15,501.25. **H**

   F $13,850   **H** $14,250

   G $14,120   **J** $14,650

5. Simplify $\sqrt[3]{-27b^6c^{12}}$. **B**

   A $-3b^3c^6$ **C** $3b^2c^4$

   B $-3b^2c^4$ **D** $3b^3c^6$

6. What is the equation of the square root function graphed at the right? **J**

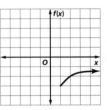

   F $f(x) = \sqrt{x - 3} - 1$

   G $f(x) = \sqrt{x + 1} - 3$

   H $f(x) = \sqrt{x + 3} + 1$

   J $f(x) = \sqrt{x - 1} - 3$

7. Which equation will produce the narrowest parabola when graphed? **C**

   A $y = 3x^2$ **C** $y = -6x^2$

   B $y = \frac{3}{4}x^2$ **D** $y = -\frac{3}{4}x^2$

8. Find the inverse of $f(x) = x - 5$. **F**

   F $f(x) = x + 5$ **H** $f(x) = \frac{x}{5}$

   G $f(x) = 5x$ **J** $f(x) = 5 - x$

9. The equations of two lines are $2x - y = 6$ and $4x - y = -2$. Which of the following describes their point of intersection? **C**

   A $(2, -2)$ **C** $(-4, -14)$

   B $(-8, -38)$ **D** no intersection

> **Test-TakingTip**
> **Question 4** You know the final price but need to know the sale price. Work backward to find the solution.

---

8. F correct
   G multiplied instead of adding
   H divided instead of adding
   J multiplied by −1 instead of exchanging variables and solving

9. A solved for values for the first equation only
   B guess
   C correct
   D simplified the system of equations incorrectly

Short Response/Gridded Response

Record your answers on the answer sheet provided by your teacher or on a sheet of paper.

10. Which of the following terms does *not* describe the system of equations graphed below: consistent, dependent, independent, or intersecting? **dependent**

11. Suppose a projectile is launched into the air from a platform. The formula $h = -16t^2 + 40t + 70$ relates the height h of the object (in feet) and the time t since it was launched (in seconds). What is the maximum height the object reaches? **95 ft**

12. The radius of a sphere with volume V can be found using the formula $r = \sqrt[3]{\dfrac{3V}{4\pi}}$.

a. What is the radius of the sphere at the right? Round to the nearest tenth. **12.7 in.**

$V = 8580$ cu. in.

b. Solve the formula for V to find the formula for the volume of a sphere given its radius. $V = \dfrac{4}{3}\pi r^3$

c. What is the volume of a basketball that has a diameter of 9 inches? Round to the nearest tenth. **381.7 in³**

13. GRIDDED RESPONSE The perimeter of the quadrilateral below is 160. What is the value of m? **4**

Extended Response

Record your answers on a sheet of paper. Show your work.

14. The amount that a retailer charges for shipping an electronics purchase is determined by the weight of the package. The charges for several different weights are given in the table.

Electronics Shipping Charges	
Weight (lb)	Shipping ($)
1	5.58
3	6.76
4	7.35
7	9.12
10	10.89
13	12.66
15	13.84

a. Find the rate of change of the shipping charge per pound. **$0.59**

b. Write an equation that could be used to find the shipping charge y for a package that weighs x pounds. $y = 0.59x + 4.99$

c. Find the shipping charge for a package that weighs 19 pounds. **$16.20**

15. Suppose $f(x)$ and $g(x)$ are inverse functions.

a. Describe how the graphs of $f(x)$ and $g(x)$ would appear on a coordinate grid. **a, b. See margin.**

b. What is the value of the composition $f[g(2)]$? Explain.

Need Extra Help?

If you missed Question...	1	2	3	4	5	6	7	8	9	10	11	12	13	14	15
Go to Lesson...	6-1	1-1	5-1	6-2	6-4	6-3	4-7	6-2	3-1	3-1	6-7	6-4	5-1	2-4	6-2

Formative Assessment

You can use these pages to benchmark student progress.

📁 *Chapter 6 Resource Masters*

McGraw-Hill
e**Assessment** Create practice tests that align to your state standards, the Common Core State Standards, and other national standards such as TIMSS and NAEP.

Answer Sheet Practice

Have students simulate taking a standardized test by recording their answers on practice recording sheets.

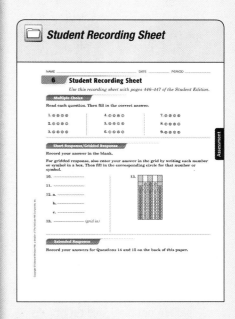

Homework Option

Get Ready for Chapter 7 Assign students the exercises on p. 449 as homework to assess whether they possess the prerequisite skills needed for the next chapter.

Additional Answers

15a. Sample answer: they would be reflections of each other across the line $y = x$

15b. Sample answer: The value is 2 because f and g are inverses. Whatever g does to 2, f undoes, so the output is the same as the input.

17. $f \circ g = \{(-4, 4)\}$, D = $\{-4\}$, R = $\{4\}$; $g \circ f = \{(-8, 0), (0, -4),$
$(2, -5), (-6, -1)\}$, D = $\{-6, 0, 2\}$, R = $\{-5, -4, -1, 0\}$

18. $f \circ g = \{(6, 12)\}$, D = $\{6\}$, R = $\{12\}$; $g \circ f = \{(-7, 5), (4, 1),$
$(-3, 8)\}$, D = $\{-7, -3, 4\}$, R = $\{1, 5, 8\}$

19. $f \circ g$ is undefined, D = $\varnothing$, R = $\varnothing$; $g \circ f$ is undefined, D = $\varnothing$,
R = $\varnothing$.

20. $f \circ g$ is undefined, D = $\varnothing$, R = $\varnothing$; $g \circ f = \{(-4, 9), (0, 1),$
$(-6, 13), (2, -3)\}$, D = $\{-6, -4, 0, 2\}$, R = $\{-3, 1, 9, 13\}$.

28–35. Domains and ranges are all real numbers unless
otherwise specified.

27. $[f \circ g](x) = 2x + 10$, D = {all real numbers}, R = {all even
numbers}; $[g \circ f](x) = 2x + 5$, D = {all real numbers}, R = {all
odd numbers}

28. $[f \circ g](x) = 3x - 24$; $[g \circ f](x) = 3x + 8$

29. $[f \circ g](x) = 3x - 2$; $[g \circ f](x) = 3x + 8$

30. $[f \circ g](x) = x^2 - 14$, R = $\{y \mid y \geq -14\}$; $[g \circ f](x) = x^2 - 8x + 6$,
R = $\{y \mid y \geq -10\}$

31. $[f \circ g](x) = x^2 - 6x - 2$, R = $\{y \mid y \geq -11\}$; $[g \circ f](x) = x^2 +$
$6x - 8$, R = $\{y \mid y \geq -17\}$

32. $[f \circ g](x) = 32x^2 + 44x + 16$, R = $\{y \mid y \geq 0.875\}$; $[g \circ f](x) =$
$8x^2 - 4x + 7$, R = $\{y \mid y \geq 6.5\}$

33. $[f \circ g](x) = 4x^3 + 7$; $[g \circ f](x) = 64x^3 - 48x^2 + 12x + 1$

34. $[f \circ g](x) = x^4 + 3x^2 + 1$, R = $\{y \mid y \geq 1\}$; $[g \circ f](x) = x^4 +$
$6x^3 + 11x^2 + 6x + 1$, R = $\{y \mid y \geq 0\}$

35. $[f \circ g](x) = 128x^4 + 96x^3 + 18x^2$, R = $\{y \mid y \geq 0\}$; $[g \circ f](x) =$
$32x^4 + 6x^2$, R = $\{y \mid y \geq 0\}$

50a.

x	$f(x) = x^2$	$g(x) = x$	$(f + g)(x)$ $= x^2 + x$	$(f - g)(x)$ $= x^2 - x$
−3	9	−3	6	12
−2	4	−2	2	6
−1	1	−1	0	2
0	0	0	0	0
1	1	1	2	0
2	4	2	6	2
3	9	3	12	6

50b.

50c.

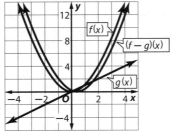

50d. Sample answer: For each value of x, the vertical distance
between the graph of $g(x)$ and the x-axis is the same as the
vertical distance between the graphs of $f(x)$ and $(f + g)(x)$ and
between $f(x)$ and $(f - g)(x)$.

58a.

x	$(f \cdot g)(x)$	$\left(\dfrac{f}{g}\right)(x)$	$[f \circ g](x)$	$[g \circ f](x)$
−3	−60	$-\dfrac{5}{3}$	37	7
−2	−25	−1	26	2
−1	−8	$-\dfrac{1}{2}$	17	−1
0	−3	$-\dfrac{1}{3}$	10	−2
1	−4	−1	5	−1
2	−5	−5	2	2
3	0	undef.	1	7

58b.

[−20, 20] scl: 2 by [−20, 20] scl: 2

58c. Sample answer: When x is 2 or 4, the functions are equal.

58d.

[−10, 10] scl: 1 by [−10, 10] scl: 1

58e. Sample answer: The functions are translations of the graph of
$y = x^2$.

Chapter 6 Answer Appendix

15. $f^{-1}(x) = x - 2$

16. $g^{-1}(x) = \frac{1}{5}x$

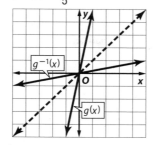

17. $f^{-1}(x) = -\frac{x}{2} + \frac{1}{2}$

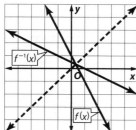

18. $h^{-1}(x) = 3x + 4$

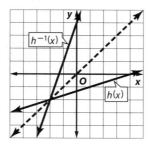

19. $f^{-1}(x) = -\frac{3}{5}(x + 8)$

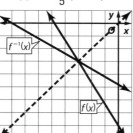

20. $g^{-1}(x) = x - 4$

21. $f^{-1}(x) = \frac{1}{4}x$

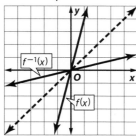

22. $f^{-1}(x) = -\frac{x}{8} + \frac{9}{8}$

23. $f^{-1}(x) = \pm\sqrt{\frac{1}{5}x}$

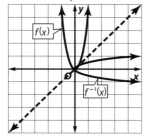

24. $h^{-1}(x) = \pm\sqrt{x - 4}$

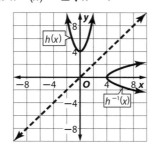

25. $f^{-1}(x) = \pm\sqrt{2x + 2}$

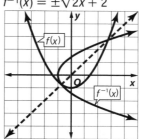

26. $f^{-1}(x) = \pm\sqrt{x - 3} - 1$

49a. $F^{-1}(x) = \frac{5}{9}(x - 32)$;

$$F[F^{-1}(x)] = \frac{9}{5}\left[\frac{5}{9}(x - 32)\right] + 32$$
$$= x - 32 + 32$$
$$= x$$

$$F^{-1}[F(x)] = \frac{5}{9}\left(\frac{9}{5}x + 32 - 32\right)$$
$$= \frac{5}{9}\left(\frac{9}{5}x + 0\right)$$
$$= x$$

51a.

[−10, 10] scl: 1 by [−10, 10] scl: 1

[−10, 10] scl: 1 by [−10, 10] scl: 1

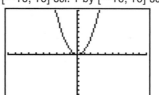

[−10, 10] scl: 1 by [−10, 10] scl: 1

[−10, 10] scl: 1 by [−10, 10] scl: 1

[−10, 10] scl: 1 by [−10, 10] scl: 1

[−10, 10] scl: 1 by [−10, 10] scl: 1

51b.

Function	Inverse a function?
$y = x^0$ or $y = 1$	no
$y = x^1$ or $y = x$	yes
$y = x^2$	no
$y = x^3$	yes
$y = x^4$	no

Extend 6-2

1.

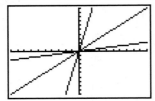

$[-10, 10]$ scl: 1 by $[-10, 10]$ scl: 1

2.

$[-10, 10]$ scl: 1 by $[-10, 10]$ scl: 1

3.

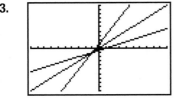

$[-10, 10]$ scl: 1 by $[-10, 10]$ scl: 1

4.

$[-10, 10]$ scl: 1 by $[-10, 10]$ scl: 1

5.

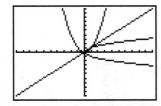

$[-10, 10]$ scl: 1 by $[-10, 10]$ scl: 1

6.

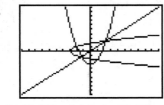

$[-10, 10]$ scl: 1 by $[-10, 10]$ scl: 1

Lesson 6-3 (Guided Practice)

3.

$(3, 346.4)$; $(6, 489.9)$; $(9, 600)$

4A.

4B.

Lesson 6-3

9.

10.

11.

12.

19.

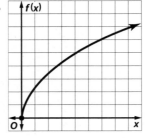

$D = \{x \mid x \geq 0\}$;
$R = \{f(x) \mid f(x) \geq 0\}$

20.

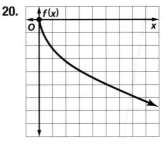

$D = \{x \mid x \geq 0\}$;
$R = \{f(x) \mid f(x) \leq 0\}$

21.

D = {x | x ≥ 8};
R = {f(x) | f(x) ≥ 0}

22.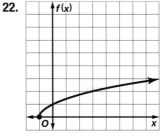

D = {x | x ≥ −1};
R = {f(x) | f(x) ≥ 0}

23.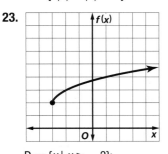

D = {x | x ≥ −3};
R = {f(x) | f(x) ≥ 2}

24.

D = {x | x ≥ 4};
R = {f(x) | f(x) ≥ −10}

25.

D = {x | x ≥ 5};
R = {f(x) | f(x) ≥ −6}

26.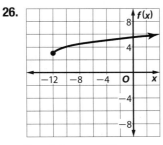

D = {x | x ≥ −12};
R = {f(x) | f(x) ≥ 3}

27.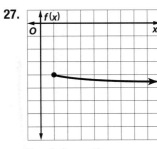

D = {x | x ≥ 1};
R = {f(x) | f(x) ≤ −4}

28.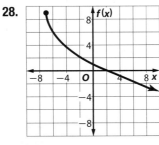

D = {x | x ≥ −7};
R = {f(x) | f(x) ≤ 9}

31.

32.

33.

34.

35.

36.

37.

38.

44a.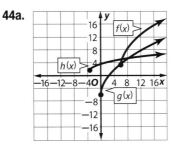

44b. f(x): 6 units to the right, 3 units up; g(x): $\frac{1}{16}$ to the left, 6 units down; h(x): 3 units to the left, 2 units up

44c. Sample answer: f(x) and g(x) appear to be stretched because the graph increases much more quickly than the parent graph.

44d. Sample answer: They are stretched by the same magnitude because $4 = \sqrt{16}$.

44e.

	x = 8	x = 12	x = 16	Rate of Change between 8 and 12	Rate of Change between 12 and 16
f(8) = 8.66	f(12) = 12.798	f(16) = 15.65		1.0345	0.713
g(8) = 5.36	g(12) = 7.89	g(16) = 10.03		0.6325	0.535
h(8) = 5.317	h(12) = 5.873	h(16) = 6.359		0.139	0.1215

The rate of change decreases as the x-values increase for square root functions of the form $a\sqrt{x - b} + c$, where a > 0.

Lesson 6-4

58a. $n = 3$

x	−5	−4	−3	−2	−1	0
f(x)	−125	−64	−27	−8	−1	0

x	1	2	3	4	5	
f(x)	1	8	27	64	125	

$n = 4$

x	−5	−4	−3	−2	−1	0
f(x)	625	256	81	16	1	0

x	1	2	3	4	5	
f(x)	1	16	81	256	625	

$n = 3$

x	−125	−64	−27	−8	−1	0
g(x)	−5	−4	−3	−2	−1	0

x	1	8	27	64	125	
g(x)	1	2	3	4	5	

$n = 4$

x	625	256	81	16	1	0
g(x)	imaginary	imaginary	imaginary	imaginary	imaginary	0

x	1	16	18	256	625	
g(x)	1	2	3	4	5	

58b. $n = 3$ $n = 4$

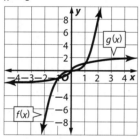

83. $(-2, 3)$; $x = -2$; down

84. $y = -\frac{1}{3}(x - 12)^2 + 48$; $(12, 48)$; $x = 12$; down

85. $(2, -2)$; $x = 2$; up

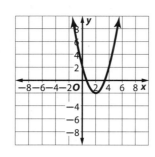

86. $y = 2(x + 2)^2 + 2$; $(-2, 2)$; $x = -2$; up

Extend 6-4

1.

$[-10, 10]$ scl: 1 by $[-10, 10]$ scl: 1

2.

$[-10, 10]$ scl: 1 by $[-10, 10]$ scl: 1

3.

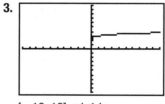

$[-10, 10]$ scl: 1 by $[-10, 10]$ scl: 1

4.

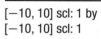

$[-10, 10]$ scl: 1 by $[-10, 10]$ scl: 1

5.

$[-10, 10]$ scl: 1 by $[-10, 10]$ scl: 1

6.

$[-10, 10]$ scl: 1 by $[-10, 10]$ scl: 1

Mid-Chapter Quiz

16a. $f^{-1}(h) = \frac{1}{15}h - \frac{5}{3}$; $f^{-1}(h)$ represents the number of hours worked

17.

18.

19.

20.

21.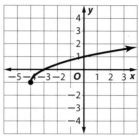

D = {x | x ≥ 0};
R = {y | y ≥ 2}

22.

D = {x | x ≥ −4};
R = {y | y ≥ −1}

Lesson 6-5

59a.
$$a^2 + b^2 = c^2$$
$$1^2 + 1^2 = c^2$$
$$2 = c^2$$
$$c = \sqrt{2}$$

59b.

59c. $\sqrt{2} + \sqrt{2}$ units is the length of the hypotenuse of an isosceles right triangle with legs of length 2 units.
Therefore, $\sqrt{2} + \sqrt{2} > 2$.

59d.

59e. The square creates 4 triangles with a base of 1 and a height of 1. Therefore the area of each triangle is $\frac{1}{2}bh = \frac{1}{2}(1)(1)$ or $\frac{1}{2} \cdot 4\left(\frac{1}{2}\right) = 2$. The area of the square is 2, so $\sqrt{2} \cdot \sqrt{2} = 2$.

61.
$$\left(\frac{-1 - i\sqrt{3}}{2}\right)^3 = \left(\frac{-1 - i\sqrt{3}}{2}\right) \cdot \left(\frac{-1 - i\sqrt{3}}{2}\right) \cdot \left(\frac{-1 - i\sqrt{3}}{2}\right)$$
$$= \frac{(-1 - i\sqrt{3})(-1 - i\sqrt{3})(-1 - i\sqrt{3})}{8}$$
$$= \frac{(1 + i\sqrt{3} + i\sqrt{3} + 3i^2)(-1 - i\sqrt{3})}{8}$$
$$= \frac{(2i\sqrt{3} - 2)(-1 - i\sqrt{3})}{8}$$
$$= \frac{-2i\sqrt{3} - 6i^2 + 2 + 2i\sqrt{3}}{8}$$
$$= \frac{-6i^2 + 2}{8} = \frac{8}{8} \text{ or } 1$$

	Diagnostic Assessment Quick Check			
	LESSON 7-1 — 45 min: 1 day / 90 min: 0.75 day	**EXPLORE 7-2** — 45 min: 0.5 day / 90 min: 0.25 day	**LESSON 7-2** — 45 min: 1.5 days / 90 min: 1.25 days	**LESSON 7-3** — 45 min: 1.5 days / 90 min: 1.5 days
Title	Graphing Exponential Functions	Graphing Technology Lab: Solving Exponential Equations and Inequalities	Solving Exponential Equations and Inequalities	Logarithms and Logarithmic Functions
Objectives	▪ Graph exponential growth functions. ▪ Graph exponential decay functions.	▪ Use a graphing calculator to solve exponential equations by graphing or by using the table feature.	▪ Solve exponential equations. ▪ Solve exponential inequalities.	▪ Evaluate logarithmic expressions. ▪ Graph logarithmic functions.
Key Vocabulary	exponential function exponential growth asymptote growth factor exponential decay decay factor		exponential equation compound interest exponential inequality	logarithm logarithmic function
CCSS	F.IF.7e, F.IF.8b	A.REI.11	A.CED.1, F.LE.4	F.IF.7e, F.BF.3
Multiple Representations	⟡		⟡	
Lesson Resources	🖱 connectED.mcgraw-hill.com 📁 Leveled Worksheets ᵃᵇᶜ Vocabulary 🏃 Animations PT Personal Tutor ✋ Virtual Manipulatives ✓ Self-Check Quiz ▪ *5-Minute Check* ▪ *Study Notebook* ▪ *Teaching Algebra with Manipulatives*	🖱 connectED.mcgraw-hill.com **Materials:** ▪ TI-83/84 Plus or other graphing calculator	🖱 connectED.mcgraw-hill.com 📁 Leveled Worksheets 📁 Quiz 1 ᵃᵇᶜ Vocabulary PT Personal Tutor ✓ Self-Check Quiz ▪ *5-Minute Check* ▪ *Study Notebook*	🖱 connectED.mcgraw-hill.com 📁 Leveled Worksheets ᵃᵇᶜ Vocabulary 🏃 Animations PT Personal Tutor ✓ Self-Check Quiz ▪ *5-Minute Check* ▪ *Study Notebook*
Resources for Every Lesson	IWB eStudent Edition IWB Interactive Classroom	▪ eTeacher Edition ▪ eSolutions Manual ▪ e𝒜ssessment McGraw-Hill		
Differentiated Instruction	pp. 453, 458		pp. 463, 465, 467	pp. 469, 475

IWB All digital assets are Interactive Whiteboard ready.

	Suggested Pacing		
Time Periods	Instruction	Review & Assess	Total
45-minute	12 days	2 days	14 days
90-minute	9 days	1 day	10 days

EXTEND 7-3 — 45 min: 0.5 day / 90 min: 0.25 day	LESSON 7-4 — 45 min: 1 day / 90 min: 0.75 day	LESSON 7-5 — 45 min: 1 day / 90 min: 0.75 day	LESSON 7-6 — 45 min: 1.5 days / 90 min: 1.25 days	EXTEND 7-6 — 45 min: 0.5 day / 90 min: 0.25 day
Graphing Technology Lab: Choosing the Best Model	**Solving Logarithmic Equations and Inequalities**	**Properties of Logarithms**	**Common Logarithms**	**Graphing Technology Lab: Solving Logarithmic Equations and Inequalities**
▪ Use a graphing calculator to find an equation of best fit for exponential and logarithmic functions.	▪ Solve logarithmic equations. ▪ Solve logarithmic inequalities.	▪ Simplify and evaluate expressions using the properties of logarithms. ▪ Solve logarithmic equations using the properties of logarithms.	▪ Solve exponential equations and inequalities using common logarithms. ▪ Evaluate logarithmic expressions using the Change of Base Formula.	▪ Use a graphing calculator to solve exponential and logarithmic equations and inequalities.
	logarithmic equation logarithmic inequality		common logarithm Change of Base Formula	
	A.SSE.2, A.CED.1	A.CED.1	A.CED.1	A.REI.11
connectED.mcgraw-hill.com Personal Tutor **Materials:** ▪ TI-83/84 Plus or other graphing calculator	**connectED.mcgraw-hill.com** Leveled Worksheets Quiz 2 Vocabulary Personal Tutor Self-Check Quiz ▪ 5-Minute Check ▪ Study Notebook	**connectED.mcgraw-hill.com** Leveled Worksheets Personal Tutor Self-Check Quiz ▪ 5-Minute Check ▪ Study Notebook	**connectED.mcgraw-hill.com** Leveled Worksheets Quiz 3 Vocabulary Animations Personal Tutor Self-Check Quiz ▪ 5-Minute Check ▪ Study Notebook	**connectED.mcgraw-hill.com** Personal Tutor **Materials:** ▪ TI-83/84 Plus or other graphing calculator
IWB eStudent Edition IWB Interactive Classroom	▪ eTeacher Edition ▪ eSolutions Manual ▪ McGraw-Hill eAssessment			
	pp. 479, 483	pp. 487, 491	pp. 494, 498	
	Formative Assessment Mid-Chapter Quiz			

	LESSON 7-7 — 45 min: 1 day / 90 min: 0.75 day	**EXPLORE 7-8** — 45 min: 0.5 day / 90 min: 0.25 day	**LESSON 7-8** — 45 min: 1 day / 90 min: 0.75 day	**EXTEND 7-8** — 45 min: 0.5 day / 90 min: 0.25 day	
Title	Base e and Natural Logarithms	Spreadsheet Lab: Compound Interest	Using Exponential and Logarithmic Functions	Graphing Technology Lab: Cooling	
Objectives	■ Evaluate expressions involving the natural base and natural logarithm. ■ Solve exponential equations and inequalities using natural logarithms.	■ Use a spreadsheet to display the growth of an investment over time.	■ Use logarithms to solve problems involving exponential growth and decay. ■ Use logarithms to solve problems involving logistic growth.	■ Use a data collection device to investigate the differences between types of insulated cups and cooling time.	
Key Vocabulary	natural base, e natural base exponential function natural logarithm		rate of continuous growth rate of continuous decay logistic growth model		
CCSS	A.SSE.2		F.IF.8b, F.LE.4	F.BF.1b	
Multiple Representations	⬛		⬛		
Lesson Resources	connectED.mcgraw-hill.com 📁 Leveled Worksheets 🔤 Vocabulary PT Personal Tutor ✅ Self-Check Quiz ■ *5-Minute Check* ■ *Study Notebook*	connectED.mcgraw-hill.com PT Personal Tutor ■ *Teaching Algebra with Manipulatives* **Materials:** ■ computer with spreadsheet program	connectED.mcgraw-hill.com 📁 Leveled Worksheets 📁 Quiz 4 🔤 Vocabulary PT Personal Tutor ✅ Self-Check Quiz ■ *5-Minute Check* ■ *Study Notebook*	connectED.mcgraw-hill.com **Materials:** ■ TI-83/84 Plus or other graphing calculator ■ data collection device and compatible temperature probe ■ variety of containers ■ very hot water	
Resources for Every Lesson	IWB eStudent Edition IWB Interactive Classroom	■ eTeacher Edition ■ eSolutions Manual ■ eAssessment *McGraw-Hill*			
Differentiated Instruction	pp. 503, 507		pp. 510, 511, 515		
				Summative Assessment Study Guide and Review Practice Test	

IWB All digital assets are Interactive Whiteboard ready.

STUDY SKILL

A comparison map can help students understand the differences in two or more similar concepts. Show students this sample map that compares exponential growth and exponential decay in Lesson 7-1. While studying Chapter 7, have students work in small cooperative groups to design comparison maps showing exponential and logarithmic functions, and base *e* exponential functions and natural logarithmic functions.

Exponential Functions	
Exponential Growth $A(t) = a(1 + r)^t$ • continuous, one-to-one, increasing • D = {all real numbers}, R = {positive real numbers} • horizontal asymptote	**Exponential Decay** $A(t) = a(1 - r)^t$ • continuous, one-to-one, decreasing • D = {all real numbers}, R = {positive real numbers} • horizontal asymptote

Creating Independence through Student-owned Strategies

SE = Student Edition, TE = Teacher Edition, CRM = Chapter Resource Masters

Diagnosis	Prescription
Beginning Chapter 7	
Get Ready for Chapter 7 **SE**	Response to Intervention **TE**
Beginning Every Lesson	
Then, Now, Why? **SE** 5-Minute Checks	Chapter 0 **SE**

DIAGNOSTIC ASSESSMENT

FORMATIVE ASSESSMENT

Diagnosis	Prescription
During/After Every Lesson	
Guided Practice **SE**, every example Check Your Understanding **SE** H.O.T. Problems **SE** Spiral Review **SE** Additional Examples **TE** Watch Out! **TE** Step 4, Assess **TE** Chapter 7 Quizzes **CRM**, pp. 63–64 Self-Check Quizzes connectED.mcgraw-hill.com	**TIER 1 Intervention** Skills Practice **CRM**, Ch. 1–7 connectED.mcgraw-hill.com **TIER 2 Intervention** Differentiated Instruction **TE** Study Guide and Intervention **CRM**, Ch. 1–7 **TIER 3 Intervention** _Math Triumphs, Alg. 2_
Mid-Chapter	
Mid-Chapter Quiz **SE** Mid-Chapter Test **CRM**, p. 65 **McGraw-Hill** *Assessment*	Skills Practice **CRM**, Ch. 1–7 connectED.mcgraw-hill.com **TIER 1 Intervention** Study Guide and Intervention **CRM**, Ch. 1–7 **TIER 2 Intervention** _Math Triumphs, Alg. 2_
Before Chapter Test	
Chapter Study Guide and Review **SE** Practice Test **SE** Standardized Test Practice **SE** Chapter Test connectED.mcgraw-hill.com Standardized Test Practice connectED.mcgraw-hill.com Vocabulary Review connectED.mcgraw-hill.com **McGraw-Hill** *Assessment*	**TIER 1 Intervention** Skills Practice **CRM**, Ch. 1–7 connectED.mcgraw-hill.com **TIER 2 Intervention** Study Guide and Intervention **CRM**, Ch. 1–7 **TIER 3 Intervention** _Math Triumphs, Alg. 2_

SUMMATIVE ASSESSMENT

Diagnosis	Prescription
After Chapter 7	
Multiple-Choice Tests, Forms 1, 2A, 2B **CRM**, pp. 67–72 Free-Response Tests, Forms 2C, 2D, 3 **CRM**, pp. 73–78 Vocabulary Test **CRM**, p. 66 Extended Response Test **CRM**, p. 79 Standardized Test Practice **CRM**, pp. 80–82 **McGraw-Hill** *Assessment*	Study Guide and Intervention **CRM**, Ch. 1–7 connectED.mcgraw-hill.com

Option 1 Reaching All Learners

Visual Because there are so many new properties taught in this chapter, suggest that students make up property posters. Encourage students to be as abbreviated as possible and to use color to help them see at a glance how the property works.

> **Properties of Logarithms**
>
> $\log_a(M \cdot N) = \log_a M + \log_a N$
>
> $\log_a(M/N) = \log_a M - \log_a N$
>
> $\log_a(M^P) = p \log_a M$

Logical Have pairs of students begin with $10, choose an interest rate that will be compounded continuously, and calculate how much they will have after 5, 10, 15, and 20 years. After each calculation, have students model the amount of money they have on a bar graph.

Option 2 Approaching Level

As a class, make a Venn diagram for the set of real numbers. Once the diagram is made, review with students what it means for a number to be a rational number, integer, whole number, natural number, or irrational number. Write an example of each type of real number on the diagram. Then discuss e as an irrational number.

Real Numbers

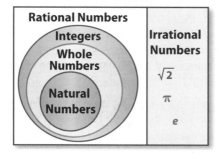

Option 3 English Learners

Write several exponential functions and their related logarithmic functions on the board. Ask students to read aloud each expression. For example, have students say *two to the third power is equal to eight* for $2^3 = 8$. Then have students say *the logarithm of eight with base two is equal to three* for $\log_2 8 = 3$.

Option 4 Beyond Level

Ask students to do a search for the tables of common logarithms of numbers either in the appendices of older algebra texts or online. As a class, learn to read and use these tables. Then ask students to use the tables to evaluate several common logarithms, such as $\log_{10} 125$. Then have students compare their results with what they get when they evaluate the same logarithms using their calculator.

VerticalAlignment

Before Chapter 7

Related Topics from Algebra 2

- use tools including the properties of exponents to simplify expressions

Chapter 7

- analyze a situation modeled by an exponential function, formulate an equation or inequality, and solve the problem

- develop the definition of logarithms by exploring and describing the relationships between exponential functions and their inverses

- use parent functions to investigate, describe, and predict the effects of parameter changes on the graphs of exponential and logarithmic functions, describe limitations on the domains and ranges, and examine asymptotic behavior

- determine solutions of exponential and logarithmic equations using graphs, tables, and algebraic methods

- interpret and determine the reasonableness of solutions to exponential and logarithmic equations and inequalities

After Chapter 7

Preparation for Precalculus

- describe parent functions symbolically and graphically, including $f(x) = \log_a x$ and $f(x) = \ln x$

- investigate the concepts of continuity, end behavior, and asymptotes, and connect these characteristics to functions represented graphically and numerically

- investigate logarithmic and exponential properties

Lesson-by-LessonPreview

7-1 Graphing Exponential Functions

An equation of the form $y = b^x$, where $b > 0$, and $b \neq 1$ is called an exponential function. The domain of an exponential function is the set of *all real numbers*. There are two types of exponential functions:

- *exponential growth*, where $b > 1$, and
- *exponential decay*, where $0 < b < 1$.

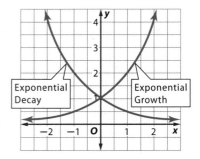

The graph of an exponential function has a horizontal asymptote. This characteristic means that the graph approaches a horizontal line. For the parent function $f(x) = b^x$, the horizontal asymptote is the x-axis.

The same transformations that have been performed on previously studied parent functions can be applied to the graphs of exponential functions.

7-2 Solving Exponential Equations and Inequalities

In an *exponential equation*, variables occur as exponents. The *Property of Equality for Exponential Functions* can be used to solve exponential equations. When the bases are the same, as in $3^x = 3^7$, the exponents must also be equal for the equation to be true. So, if $3^x = 3^7$, then $x = 7$. When the bases are not equal, as in $3^x = 9^4$, the expressions must be rewritten so that the bases are equal. In this case, 9 can be rewritten as 3^2, so that $9^4 = (3^2)^4 = 3^8$. Therefore $3^x = 3^8$, so $x = 8$.

In an *exponential inequality*, variables also occur as exponents. *The Property of Inequality for Exponential Functions* can be used to solve exponential inequalities. This property says that $b^x > b^y$ if and only if $x > y$, and $b^x < b^y$ if and only if $x < y$, for values of b greater than 1. This property also holds true for $\leq$ and $\geq$.

Get educator support from the Math Forum's T2T. You may want to search, or ask, for activities with logarithmic functions. Start your search at connectED.mcgraw-hill.com.

mathforum.org
The Math Forum
@ Drexel University

7-3 Logarithms and Logarithmic Functions

In general, the inverse of $y = b^x$ is $x = b^y$. In the equation $x = b^y$, b is the base and y is referred to as the *logarithm* of x. This relationship is usually written as $y = \log_b x$, where $b > 0$ and $b \neq 1$. It is read *y equals log base b of x*, and is called a *logarithmic function*.

The graph of $f(x) = \log_b x$ is continuous, one-to-one, and has a domain of all positive real numbers. The y-axis is a vertical asymptote of the graph. The same transformations that have been performed on previously studied parent functions can be applied to the graphs of logarithmic functions.

7-4 Solving Logarithmic Equations and Inequalities

When solving logarithmic equations and inequalities, it is important to remember that a defining characteristic of a logarithmic function is that its domain is the set of all *positive* numbers. This means that the logarithm of 0 or of a negative number for any base is undefined. It is very important to check possible solutions to logarithmic equations in the original equation to be sure that they do not result in taking the logarithm of 0 or a negative number. For logarithmic inequalities, this fact will exclude not just one value from the solution set, but a range of values.

7-5 Properties of Logarithms

Logarithms were invented to make computation easier. In fact, the word logarithm is a contraction of "<u>log</u>ical <u>arith</u>metic." According to the *Product and Quotient Properties of Logarithms*, when using logarithms, multiplication changes to addition and division changes to subtraction. For example, $\log_3 (5 \cdot 7)$ is equivalent to $\log_3 5 + \log_3 7$. The *Power Property of Logarithms* states that the logarithm of a power is the product of the logarithm and the exponent. For example, $\log_5 3^7 = 7 \log_5 3$. Using these properties of logarithms helps solve equations that involve logarithms.

7-6 Common Logarithms

Logarithms with a base of 10 are called *common logarithms*. The 10 is usually not written, that is, $\log_{10} x$ is written as $\log x$. Most calculators have a $\boxed{\text{LOG}}$ key for evaluating common logarithms.

Logarithms can be used to solve exponential equations and inequalities that cannot be easily written as powers of the same base. To find a solution to an equation such as $5^x = 41$, take the logarithm of both sides of the equation and then solve for x.

The *Change of Base Formula* makes it possible to evaluate a logarithmic expression of any base by translating the expression into one that involves common logarithms.

7-7 Base e and Natural Logarithms

An exponential function with base e is called a *natural base exponential function*. The number e is an irrational number with a value of 2.71828....

The logarithm with base e is called the *natural logarithm,* and is denoted by either $\log_e x$ or $\ln x$. It is the inverse function of the natural base exponential function. All the properties of logarithms from previous lessons apply to natural logarithms as well. Most calculators have an e^x function for evaluating natural base expressions and an $\boxed{\ln}$ key for finding the natural logarithm of a real number.

7-8 Using Exponential and Logarithmic Functions

The functions $f(x) = ae^{kt}$ and $f(x) = ae^{-kt}$, where a is the initial value, t is time in years, and k is a constant, that can be used to model exponential growth and decay, respectively. The constant k represents the rate of continuous growth in the growth model and the rate of continuous decay in the decay model.

When growth has a limiting factor, a logistic growth model, such as the logistic growth function $f(t) = \dfrac{c}{1 + ae^{-kt}}$ where t represents time, can be used.

Mc Graw Hill | **Professional Development**

Targeted professional development has been articulated throughout Algebra 2. More quality, customized professional development is available from McGraw-Hill Professional Development. Visit <u>connectED.mcgraw-hill.com</u> for details on each product.

- Online Lessons emphasize the strategies and techniques used to teach Algebra 2. Includes streaming video, interactive pages, and online tools.

- Video Workshops allow mentors, coaches, or leadership personnel to facilitate on-site workshops on educational strategies in mathematics and mathematical concepts.

- MHPD Online (<u>www.mhpdonline.com</u>) offers online professional development with video clips of instructional strategies, links, student activities, and news and issues in education.

- Teaching Today (<u>teachingtoday.glencoe.com</u>) gives secondary teachers practical strategies and materials that inspire excellence and innovation in teaching.

Chapter Project

The Population Puzzle

Students use what they have learned about exponential and logarithmic functions to complete a project.

This chapter project addresses environmental literacy, as well as several specific skills identified as being essential to student success by the Framework for 21st Century Learning.

Visit connectED.mcgraw-hill.com for student and teacher handouts.

KeyVocabulary Introduce the key vocabulary in the chapter using the routine below.

<u>Define</u>: In the function $x = b^y$, y is called the logarithm, base b, of x. Usually it is written as $y = \log_b x$ and is read "y equals log base b of x."

<u>Example</u>: $3 = \log_2 8$

<u>Ask</u>: How would you write the following in logarithmic form? $3^5 = 243$ $\log_3 243 = 5$

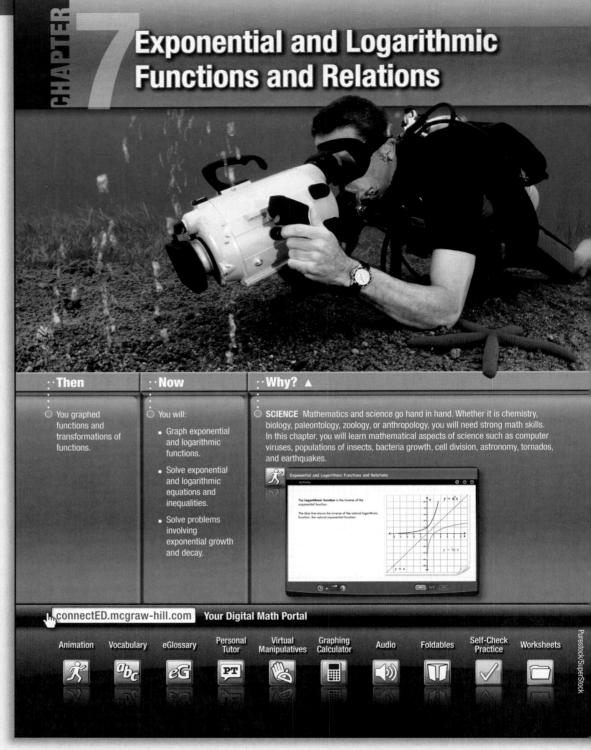

CHAPTER 7
Exponential and Logarithmic Functions and Relations

··Then

○ You graphed functions and transformations of functions.

··Now

○ You will:

- Graph exponential and logarithmic functions.

- Solve exponential and logarithmic equations and inequalities.

- Solve problems involving exponential growth and decay.

··Why? ▲

○ **SCIENCE** Mathematics and science go hand in hand. Whether it is chemistry, biology, paleontology, zoology, or anthropology, you will need strong math skills. In this chapter, you will learn mathematical aspects of science such as computer viruses, populations of insects, bacteria growth, cell division, astronomy, tornados, and earthquakes.

connectED.mcgraw-hill.com **Your Digital Math Portal**

Animation | Vocabulary | eGlossary | Personal Tutor | Virtual Manipulatives | Graphing Calculator | Audio | Foldables | Self-Check Practice | Worksheets

Purestock/SuperStock

Additional Answer (Get Ready for Chapter 7)

14. $f^{-1}(x) = \frac{1}{2}x - 6$; $f^{-1}(x)$ represents the number of toppings someone can have if they spend x dollars.

Get Ready for the Chapter

Diagnose Readiness | You have two options for checking prerequisite skills.

1 Textbook Option Take the Quick Check below. Refer to the Quick Review for help.

QuickCheck	QuickReview

QuickCheck

Simplify. Assume that no variable equals zero.

1. $a^4 a^3 a^5$ a^{12}

2. $(2xy^3z^2)^3$ $8x^3y^9z^6$

3. $\dfrac{-24x^8y^5z}{16x^2y^8z^6}$ $\dfrac{-3x^6}{2y^3z^5}$

4. $\left(\dfrac{-8r^2n}{36n^3t}\right)^2$ $\dfrac{4r^4}{81n^4t^2}$

5. **DENSITY** The density of an object is equal to the mass divided by the volume. An object has a mass of 7.5×10^3 grams and a volume of 1.5×10^3 cubic centimeters. What is the density of the object? **5 g/cm³**

Find the inverse of each function. Then graph the function and its inverse.

6. $f(x) = 2x + 5$ 7. $f(x) = x - 3$ **6–11. See Chapter 7 Answer Appendix.**

8. $f(x) = -4x$ 9. $f(x) = \frac{1}{4}x - 3$

10. $f(x) = \dfrac{x-1}{2}$ 11. $y = \frac{1}{3}x + 4$

Determine whether each pair of functions are inverse functions.

12. $f(x) = x - 6$ 13. $f(x) = 2x + 5$
    $g(x) = x + 6$ **yes** $g(x) = 2x - 5$ **no**

14. **FOOD** A pizzeria charges \$12 for a medium cheese pizza and \$2 for each additional topping. If $f(x) = 2x + 12$ represents the cost of a medium pizza with x toppings, find $f^{-1}(x)$ and explain its meaning. **See margin.**

QuickReview

Example 1 (Used in Lessons 7-1 through 7-3)

Simplify $\dfrac{(a^3bc^2)^2}{a^4a^2b^2bc^5c^3}$. Assume that no variable equals zero.

$$\dfrac{(a^3bc^2)^2}{a^4a^2b^2bc^5c^3}$$

$$= \dfrac{a^6b^2c^4}{a^6b^3c^8}$$

Simplify the numerator by using the Power of a Power Rule and the denominator by using the Product of Powers Rule.

$$= \dfrac{1}{bc^4} \text{ or } b^{-1}c^{-4}$$

Simplify by using the Quotient of Powers Rule.

Example 2 (Used in Lessons 7-2 and 7-3)

Find the inverse of $f(x) = 3x - 1$.

Step 1 Replace $f(x)$ with y in the original equation:
$f(x) = 3x - 1 \rightarrow y = 3x - 1$.

Step 2 Interchange x and y: $x = 3y - 1$.

Step 3 Solve for y.

$x = 3y - 1$	Inverse
$x + 1 = 3y$	Add 1 to each side.
$\dfrac{x+1}{3} = y$	Divide each side by 3.
$\frac{1}{3}x + \frac{1}{3} = y$	Simplify.

Step 4 Replace y with $f^{-1}(x)$.

$$y = \frac{1}{3}x + \frac{1}{3} \rightarrow f^{-1}(x) = \frac{1}{3}x + \frac{1}{3}$$

2 Online Option Take an online self-check Chapter Readiness Quiz at connectED.mcgraw-hill.com.

449

Essential Questions

- How can you make good decisions? Sample answer: Determine the available options, compare the advantages and disadvantages of each option, analyze the consequences, and choose the best option.

- What factors can affect good decision making? Sample answers: the amount of time that is available, the process used, the environment, the people that are involved, the available options

---

Get Ready for the Chapter

Response to Intervention (RtI)
Use the *Quick Check* results and the Intervention Planner chart to help you determine your Response to Intervention. The If-Then statements in the chart below help you decide the appropriate tier of RtI and suggest intervention resources for each tier.

 InterventionPlanner

TIER 1 On Level OL

If students miss about 25% of the exercises or less,

Then choose a resource:

SE Lessons 5-1, and 6-2

Skills Practice, Chapter 5, p. 7 Chapter 6, p. 14

connectED.mcgraw-hill.com Self-Check Quiz

TIER 2 Strategic Intervention AL approaching grade level

If students miss about 50% of the exercises,

Then choose a resource:

Study Guide and Intervention, Chapter 5, p. 5, Chapter 6, p. 12

connectED.mcgraw-hill.com Extra Examples, Personal Tutor, Homework Help

TIER 3 Intensive Intervention 2 or more grades below level

If students miss about 75% of the exercises,

Then use *Math Triumphs, Alg. 2*

connectED.mcgraw-hill.com Extra Examples, Personal Tutor, Homework Help, Review Vocabulary

FOLDABLES StudyOrganizer

Dinah Zike's Foldables®

Focus As students work through the lessons in this chapter, they write notes about exponential and logarithmic functions and relations.

Teach Have students make and label their Foldables as illustrated. Have students label one page of their Foldables for every two lessons in the chapter and use the appropriate pages as they cover the material. Have students list the Key Concepts and the vocabulary terms and their definitions in their Foldables. Point out that the Foldables can also be used to record positive and negative experiences during learning.

When to Use It Encourage students to add to their Foldables as they work through the chapter and to use them to review for the chapter test.

Differentiated Instruction

Student-Built Glossary, pp. 1–2

Students should complete the chart by providing a definition of each term and an example as they progress through Chapter 7. This study tool can also be used to review for the chapter test.

Get Started on the Chapter

You will learn several new concepts, skills, and vocabulary terms as you study Chapter 7. To get ready, identify important terms and organize your resources. You may wish to refer to Chapter 0 to review prerequisite skills.

FOLDABLES StudyOrganizer

Exponential and Logarithmic Functions and Relations
Make this Foldable to help you organize your Chapter 7 notes about exponential and logarithmic functions. Begin with two sheets of grid paper.

1 **Fold** in half along the width.

2 **On** the first sheet, cut 5 cm along the fold at the ends.

First Sheet

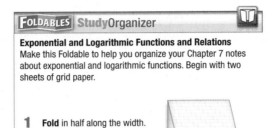

3 **On** the second sheet, cut in the center, stopping 5 cm from the ends.

Second Sheet

4 **Insert** the first sheet through the second sheet and align the folds. Label the pages with lesson numbers.

NewVocabulary

English		Español
exponential function	p. 451	función exponencial
exponential growth	p. 451	crecimiento exponencial
asymptote	p. 451	asíntota
growth factor	p. 453	factor de crecimiento
exponential decay	p. 453	desintegración exponencial
decay factor	p. 454	factor de desintegración
exponential equation	p. 461	ecuación exponencial
compound interest	p. 462	interés compuesto
exponential inequality	p. 463	desigualdad exponencial
logarithm	p. 468	logaritmo
logarithmic function	p. 469	función logarítmica
logarithmic equation	p. 478	ecuación logarítmica
logarithmic inequality	p. 479	desigualdad logarítmica
common logarithm	p. 492	logaritmos communes
Change of Base Formula	p. 494	fórmula del cambio de base
natural base, e	p. 501	e base natural
natural base exponential function	p. 501	base natural función exponencial
natural logarithm	p. 501	logaritmo natural

ReviewVocabulary

domain dominio the set of all x-coordinates of the ordered pairs of a relation

function función a relation in which each element of the domain is paired with exactly one element in the range

range rango the set of all y-coordinates of the ordered pairs of a relation

$\{(-3,1), (0, 2), (2, 4)\}$

Domain   Range

LESSON 7-1
Graphing Exponential Functions

·· Then	·· Now	·· Why?
● You graphed polynomial functions.	**1** Graph exponential growth functions. **2** Graph exponential decay functions.	Have you ever received an e-mail that tells you to forward it to 5 friends? If each of those 5 friends then forwards it to 5 of their friends, who each forward it to 5 of their friends, the number of people receiving the e-mail is growing exponentially. The equation $y = 5^x$ can be used to represent this situation, where x is the number of rounds that the e-mail has been forwarded.

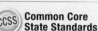
NewVocabulary
exponential function
exponential growth
asymptote
growth factor
exponential decay
decay factor

CCSS Common Core State Standards

Content Standards
F.IF.7.e Graph exponential and logarithmic functions, showing intercepts and end behavior, and trigonometric functions, showing period, midline, and amplitude.

F.IF.8.b Use the properties of exponents to interpret expressions for exponential functions.

Mathematical Practices
3 Construct viable arguments and critique the reasoning of others.

1 Exponential Growth A function like $y = 5^x$, where the base is a constant and the exponent is the independent variable, is an **exponential function**. One type of exponential function is exponential growth. An **exponential growth** function is a function of the form $f(x) = b^x$, where $b > 1$. The graph of an exponential function has an **asymptote**, which is a line that the graph of the function approaches.

KeyConcept Parent Function of Exponential Growth Functions

Parent Functions:	$f(x) = b^x$, $b > 1$
Type of graph:	continuous, one-to-one, and increasing
Domain:	all real numbers
Range:	all positive real numbers
Asymptote:	x-axis
Intercept:	$(0, 1)$

Example 1 Graph Exponential Growth Functions

Graph $y = 3^x$. State the domain and range.

Make a table of values. Then plot the points and sketch the graph.

x	-3	-2	$-\frac{1}{2}$	0
$y = 3^x$	$3^{-3} = \frac{1}{27}$	$3^{-2} = \frac{1}{9}$	$3^{-\frac{1}{2}} = \frac{\sqrt{3}}{3}$	$3^0 = 1$

x	1	$\frac{3}{2}$	2
$y = 3^x$	$3^1 = 3$	$3^{\frac{3}{2}} = \sqrt{27}$	$3^2 = 9$

The domain is all real numbers, and the range is all positive real numbers.

GuidedPractice

1. Graph $y = 4^x$. State the domain and range. **See Chapter 7 Answer Appendix.**

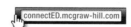
connectED.mcgraw-hill.com **451**

Stockbyte/Alamy

---

1 Focus

VerticalAlignment

Before Lesson 7-1 Graph polynomial functions.

Lesson 7-1 Graph exponential growth functions. Graph exponential decay functions.

After Lesson 7-1 Solve exponential equations and inequalities.

2 Teach

Scaffolding Questions
Have students read the **Why?** section of the lesson.

Ask:
■ How many people will receive the e-mail in the second round? 25

■ How many people will receive the e-mail in the fourth round? 625

■ Consider a scenario similar to the one in the **Why?** section, only this e-mail tells you to forward it on to 8 friends. Write an equation to represent this situation. $y = 8^x$

Lesson 7-1 Resources

Resource	Approaching Level **AL**	On Level **OL**	Beyond Level **BL**	English Learners **ELL**
Teacher Edition	■ Differentiated Instruction, p. 453	■ Differentiated Instruction, p. 453	■ Differentiated Instruction, p. 458	■ Differentiated Instruction, p. 453
Chapter Resource Masters	■ Study Guide and Intervention, pp. 5–6 ■ Skills Practice, p. 7 ■ Practice, p. 8 ■ Word Problem Practice, p. 9	■ Study Guide and Intervention, pp. 5–6 ■ Skills Practice, p. 7 ■ Practice, p. 8 ■ Word Problem Practice, p. 9 ■ Enrichment, p. 10 ■ Graphing Calculator Activity, p. 11	■ Practice, p. 8 ■ Word Problem Practice, p. 9 ■ Enrichment, p. 10	■ Study Guide and Intervention, pp. 5–6 ■ Skills Practice, p. 7 ■ Practice, p. 8 ■ Word Problem Practice, p. 9
Other	■ 5-Minute Check 7-1 ■ Study Notebook ■ Teaching Algebra with Manipulatives	■ 5-Minute Check 7-1 ■ Study Notebook ■ Teaching Algebra with Manipulatives	■ 5-Minute Check 7-1 ■ Study Notebook	■ 5-Minute Check 7-1 ■ Study Notebook ■ Teaching Algebra with Manipulatives

1 Exponential Growth

Example 1 shows how to graph an exponential growth function. **Example 2** shows how to graph transformations of exponential growth functions. **Example 3** shows how to graph an exponential function to model real-world growth.

Formative Assessment

Use the Guided Practice exercises after each example to determine students' understanding of concepts.

Additional Examples

1 Graph $y = 4^x$. State the domain and range.

D = {all real numbers};
R = {$y \mid y > 0$}

2 Graph each function. State the domain and range.

a. $y = 3^x - 2$

D = {all real numbers};
R = {$y \mid y > -2$}

b. $y = 2^{x-1}$

▶ **Additional Examples** also in Interactive Classroom PowerPoint® Presentations

IWB Interactive White Board READY

The graph of $f(x) = b^x$ represents a parent graph of the exponential functions. The same techniques used to transform the graphs of other functions you have studied can be applied to the graphs of exponential functions.

KeyConcept Transformations of Exponential Functions

$$f(x) = ab^{x-h} + k$$

h – Horizontal Translation	**k – Vertical Translation**
h units right if h is positive	k units up if k is positive
|h| units left if h is negative	|k| units down if k is negative

a – Orientation and Shape	
If $a < 0$, the graph is reflected in the x-axis.	If $\|a\| > 1$, the graph is stretched vertically. If $0 < \|a\| < 1$, the graph is compressed vertically.

StudyTip

CCSS Precision Remember that end behavior is the action of the graph as x approaches positive infinity or negative infinity. In Example 2a, as x approaches infinity, y approaches infinity. In Example 2b, as x approaches infinity, y approaches negative infinity.

Example 2 Graph Transformations

Graph each function. State the domain and range.

a. $y = 2^x + 1$

The equation represents a translation of the graph of $y = 2^x$ one unit up.

x	$y = 2^x + 1$
−3	$2^{-3} + 1 = 1.125$
−2	$2^{-2} + 1 = 1.25$
−1	$2^{-1} + 1 = 1.5$
0	$2^0 + 1 = 2$
1	$2^1 + 1 = 3$
2	$2^2 + 1 = 5$
3	$2^3 + 1 = 9$

Domain = {all real numbers}; Range = {$y \mid y > 1$}

b. $y = -\frac{1}{2} \cdot 5^{x-2}$

The equation represents a transformation of the graph of $y = 5^x$.

Graph $y = 5^x$ and transform the graph.

- $a = -\frac{1}{2}$: The graph is reflected in the x-axis and compressed vertically.
- $h = 2$: The graph is translated 2 units right.
- $k = 0$: The graph is not translated vertically.

Domain = {all real numbers}

Range = {$y \mid y < 0$}

▶ **Guided**Practice   **2A, 2B. See Chapter 7 Answer Appendix.**

2A. $y = 2^{x+3} - 5$ **2B.** $y = 0.1(6)^x - 3$

Additional Answer (Additional Example)

2b.

D = {all real numbers}; R = {$y \mid y > 0$}

CCSS **Teaching the Mathematical Practices**

Precision Mathematically proficient students try to communicate precisely, using clear definitions in discussion with others and in their own reasoning.

You can model exponential growth with a constant percent increase over specific time periods using the following function.

$$A(t) = a(1 + r)^t$$

The function can be used to find the amount $A(t)$ after t time periods, where a is the initial amount and r is the percent of increase per time period. Note that the base of the exponential expression, $1 + r$, is called the **growth factor**.

The exponential growth function is often used to model population growth.

Real-World Example 3 Graph Exponential Growth Functions

CENSUS The first U.S. census was conducted in 1790. At that time, the population was 3,929,214. Since then, the U.S. population has grown by approximately 2.03% annually. Draw a graph showing the population growth of the U.S. since 1790.

First, write an equation using $a = 3,929,214$, and $r = 0.0203$.

$$y = 3,929,214(1.0203)^t$$

Then graph the equation.

[0, 250] scl: 25 by [0, 400,000,000]
scl: 40,000,000

▶ **Guided**Practice   **3. See Chapter 7 Answer Appendix.**

3. FINANCIAL LITERACY Teen spending is expected to grow 3.5% annually from $79.7 billion in 2006. Draw a graph to show the spending growth.

2 Exponential Decay
The second type of exponential function is **exponential decay**.

KeyConcept Parent Function of Exponential Decay Functions

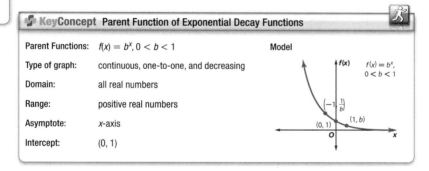

		Model
Parent Functions:	$f(x) = b^x, 0 < b < 1$	
Type of graph:	continuous, one-to-one, and decreasing	
Domain:	all real numbers	
Range:	positive real numbers	
Asymptote:	x-axis	
Intercept:	$(0, 1)$	

The graphs of exponential decay functions can be transformed in the same manner as those of exponential growth.

Creatas/SuperStock

Real-WorldLink
The U.S. Census Bureau's American Community Survey is mailed to approximately 1 out of every 480 households.
Source: Census Bureau

StudyTip
Interest The formula for simple interest, $i = prt$, illustrates linear growth over time. However, the formula for compound interest, $A(t) = a(1 + r)^t$, illustrates exponential growth over time. This is why investments with compound interest make more money.

---

3 INTERNET In 2006, there were 1,020,000,000 people worldwide using the Internet. At that time, the number of users was growing by 19.5% annually. Draw a graph showing how the number of users would grow from 2006 to 2016 if that rate continued.

Internet Usage

WatchOut!

Common Misconceptions
Be sure students do not confuse polynomial functions and exponential functions. While $y = x^2$ and $y = 2^x$ each have an exponent, $y = x^2$ is a polynomial function, and $y = 2^x$ is an exponential function.

Focus on Mathematical Content

Exponential Functions The function $f(x) = b^x$, where b is a positive real number and $b \neq 1$, is an exponential function. When $b > 1$, the function has no x-intercepts and one y-intercept. It is an increasing function with a horizontal asymptote (the x-axis). When $0 < b < 1$, the function has no x-intercepts and one y-intercept. It is a decreasing function with a horizontal asymptote (the x-axis).

DifferentiatedInstruction AL OL ELL

Verbal/Linguistic Learners Ask students where they have heard the term *exponential* before and what they think it might mean. Students may have heard terms like *exponential growth* on a television news program and they might think that exponential means "enormous." Use students' answers to introduce the concept of exponential functions.

2 Exponential Decay

Examples 4 and 5 show how to graph exponential decay functions.

Additional Example

4 Graph each function. State the domain and range.

a. $y = \left(\frac{1}{5}\right)^x$

$y = \left(\frac{1}{5}\right)^x$

D = {all real numbers};
R = {y | y > 0}

b. $y = -4\left(\frac{1}{2}\right)^{x-1} + 2$

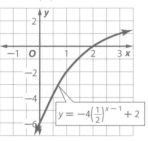

$y = -4\left(\frac{1}{2}\right)^{x-1} + 2$

D = {all real numbers};
R = {y | y < 2}

Additional Answers (Guided Practice)

4A.

D = {all real numbers};
R = {y | y < 2}

StudyTip

Exponential Decay Be sure not to confuse a dilation in which | a | < 1 with exponential decay in which 0 < b < 1.

Example 4 Graph Exponential Decay Functions

Graph each function. State the domain and range.

a. $y = \left(\frac{1}{3}\right)^x$

x	$y = \left(\frac{1}{3}\right)^x$
-3	$\left(\frac{1}{3}\right)^{-3} = 27$
-2	$\left(\frac{1}{3}\right)^{-2} = 9$
$-\frac{1}{2}$	$\left(\frac{1}{3}\right)^{-\frac{1}{2}} = \sqrt{3}$
0	$\left(\frac{1}{3}\right)^{0} = 1$
1	$\left(\frac{1}{3}\right)^{1} = \frac{1}{3}$
$\frac{3}{2}$	$\left(\frac{1}{3}\right)^{\frac{3}{2}} = \sqrt{\frac{1}{27}}$
2	$\left(\frac{1}{3}\right)^{2} = \frac{1}{9}$

The domain is all real numbers, and the range is all positive real numbers.

b. $y = 2\left(\frac{1}{4}\right)^{x+2} - 3$

The equation represents a transformation of the graph of $y = \left(\frac{1}{4}\right)^x$.

Examine each parameter.

- $a = 2$: The graph is stretched vertically.
- $h = -2$: The graph is translated 2 units left.
- $k = -3$: The graph is translated 3 units down.

$y = \left(\frac{1}{4}\right)^x$

$y = 2\left(\frac{1}{4}\right)^{x+2} - 3$

The domain is all real numbers, and the range is all real numbers greater than -3.

GuidedPractice **4A, 4B.** See margin.

4A. $y = -3\left(\frac{2}{5}\right)^{x-4} + 2$ **4B.** $y = \frac{3}{8}\left(\frac{5}{6}\right)^{x-1} + 1$

Similar to exponential growth, you can model exponential decay with a constant percent of decrease over specific time periods using the following function.

$$A(t) = a(1 - r)^t$$

The base of the exponential expression, $1 - r$, is called the **decay factor**.

4B.

D = {all real numbers}; R = {y | y > 1}

5.

[0, 10] scl: 1 by [0, 100] scl: 10

Real-WorldLink
After water, tea is the most consumed beverage in the U.S. It can be found in over 80% of American households. Just over half the American population drinks tea daily.

Source: The Tea Association of the USA

Real-World Example 5 Graph Exponential Decay Functions

TEA A cup of green tea contains 35 milligrams of caffeine. The average teen can eliminate approximately 12.5% of the caffeine from their system per hour.

a. Draw a graph to represent the amount of caffeine remaining after drinking a cup of green tea.

$$y = a(1 - r)^t$$
$$= 35(1 - 0.125)^t$$
$$= 35(0.875)^t$$

Graph the equation.

[0, 10] scl: 1 by [0, 50] scl: 5

b. Estimate the amount of caffeine in a teenager's body 3 hours after drinking a cup of green tea.

$$y = 35(0.875)^t$$ Equation from part a
$$= 35(0.875)^3$$ Replace t with 3.
$$\approx 23.45$$ Use a calculator.

The caffeine in a teenager will be about 23.45 milligrams after 3 hours.

▶ **Guided**Practice

5. A cup of black tea contains about 68 milligrams of caffeine. Draw a graph to represent the amount of caffeine remaining in the body of an average teen after drinking a cup of black tea. Estimate the amount of caffeine in the body 2 hours after drinking a cup of black tea. **See margin for graph; 52.06 mg**

Check Your Understanding
○ = Step-by-Step Solutions begin on page R14. ☑

Examples 1–2 Graph each function. State the domain and range. **1–6. See Chapter 7 Answer Appendix.**

1. $f(x) = 2^x$ **2.** $f(x) = 5^x$ **3** $f(x) = 3^{x-2} + 4$

4. $f(x) = 2^{x+1} + 3$ **5.** $f(x) = 0.25(4)^x - 6$ **6.** $f(x) = 3(2)^x + 8$

Example 3   **7.** **CCSS** SENSE-MAKING A virus spreads through a network of computers such that each minute, 25% more computers are infected. If the virus began at only one computer, graph the function for the first hour of the spread of the virus. **See margin.**

Example 4   Graph each function. State the domain and range. **8–11. See Chapter 7 Answer Appendix.**

8. $f(x) = 2\left(\frac{2}{3}\right)^{x-3} - 4$ **9.** $f(x) = -\frac{1}{2}\left(\frac{3}{4}\right)^{x+1} + 5$

10. $f(x) = -\frac{1}{3}\left(\frac{4}{5}\right)^{x-4} + 3$ **11.** $f(x) = \frac{1}{8}\left(\frac{1}{4}\right)^{x+6} + 7$

Example 5   **12.** **FINANCIAL LITERACY** A new SUV depreciates in value each year by a factor of 15%. Draw a graph of the SUV's value for the first 20 years after the initial purchase. **See margin.**

All New
Only $20,000

 connectED.mcgraw-hill.com **455**

Additional Answers

7.

12.

Core Imagery/Alamy

Additional Example

5 **AIR PRESSURE** The pressure of the atmosphere is 14.7 lb/in² at Earth's surface. It decreases by about 20% for each mile of altitude up to about 50 miles.

a. Draw a graph to represent atmospheric pressure for altitude from 0 to 50 miles.

$y = 14.7(.80)^x$

b. Estimate the atmospheric pressure at an altitude of 10 miles. **1.6 lb/in²**

---

3 Practice

Formative Assessment
Use Exercises 1–12 to check for understanding.

Use the chart at the bottom of the next page to customize assignments for your students.

Follow-Up
- How can being financially literate help you to make good decisions?

  Sample answer: If you are financially literate, you understand the vocabulary of financial terms and know how to analyze data and trends. Successfully applying these skills when considering your available options can help you to make good decisions in many real-world situations such as opening a bank account, applying for college loans, and buying a house.

Reasoning Mathematically proficient students make sense of quantities and their relationships in problem situations. Quantitative reasoning entails habits of creating a coherent representation of the problem at hand; considering the units involved; attending to the meaning of quantities, not just how to compute them; and knowing and flexibly using different properties of operations and objects.

Additional Answers

26.

27a. decay; 0.9

27b. The $P(x)$-intercept represents the number of pay phones in 1999. The asymptote is the x-axis. The number of pay phones can approach 0, but will never equal 0. This makes sense as there will probably always be a need for some pay phones.

28a. decay; 0.9

Practice and Problem Solving

Extra Practice is on page R7.

Examples 1–2 Graph each function. State the domain and range. **13–18. See Chapter 7 Answer Appendix.**

13. $f(x) = 2(3)^x$

14. $f(x) = -2(4)^x$

15. $f(x) = 4^{x+1} - 5$

16. $f(x) = 3^{2x} + 1$

17. $f(x) = -0.4(3)^{x+2} + 4$

18. $f(x) = 1.5(2)^x + 6$

Example 3

19. SCIENCE The population of a colony of beetles grows 30% each week for 10 weeks. If the initial population is 65 beetles, graph the function that represents the situation.
See Chapter 7 Answer Appendix.

Example 4 Graph each function. State the domain and range. **20–25. See Chapter 7 Answer Appendix.**

20. $f(x) = -4\left(\frac{3}{5}\right)^{x+4} + 3$

21. $f(x) = 3\left(\frac{2}{5}\right)^{x-3} - 6$

22. $f(x) = \frac{1}{2}\left(\frac{1}{5}\right)^{x+5} + 8$

23. $f(x) = \frac{3}{4}\left(\frac{2}{3}\right)^{x+4} - 2$

24. $f(x) = -\frac{1}{2}\left(\frac{3}{8}\right)^{x+2} + 9$

25. $f(x) = -\frac{5}{4}\left(\frac{4}{5}\right)^{x+4} + 2$

Example 5

26. ATTENDANCE The attendance for a basketball team declined at a rate of 5% per game throughout a losing season. Graph the function modeling the attendance if 15 home games were played and 23,500 people were at the first game. **See margin.**

27. PHONES The function $P(x) = 2.28(0.9^x)$ can be used to model the number of pay phones in millions x years since 1999.

a. Classify the function as either exponential *growth* or *decay*, and identify the growth or decay factor. Then graph the function. **See margin.**

b. Explain what the $P(x)$-intercept and the asymptote represent in this situation. **See margin.**

28. HEALTH Each day, 10% of a certain drug dissipates from the system.

a. Classify the function representing this situation as either exponential *growth* or *decay*, and identify the growth or decay factor. Then graph the function. **See margin.**

b. How much of the original amount remains in the system after 9 days? **a little less than 40%**

c. If a second dose should not be taken if more than 50% of the original amount is in the system, when should the label say it is safe to redose? Design the label and explain your reasoning. **Sample answer: The 7th day; see students' work.**

29. CCSS REASONING A sequence of numbers follows a pattern in which the next number is 125% of the previous number. The first number in the pattern is 18.

a. Write the function that represents the situation. $f(x) = 18(1.25)^{x-1}$

b. Classify the function as either exponential *growth* or *decay*, and identify the growth or decay factor. Then graph the function for the first 10 numbers. **See margin.**

c. What is the value of the tenth number? Round to the nearest whole number. **134**

C For each graph, $f(x)$ is the parent function and $g(x)$ is a transformation of $f(x)$. Use the graph to determine the equation of $g(x)$. **31.** $g(x) = 4(2)^{x-3}$ or $g(x) = \frac{1}{2}(2^x)$

30. $f(x) = 3^x$

31. $f(x) = 2^x$

32. $f(x) = 4^x$ $g(x) = -2(4)^{x+1} + 3$

$g(x) = 3^{x-4} + 5$

Differentiated Homework Options

Level	Assignment	Two-Day Option	
AL Basic	13–25, 34–35, 37–60	13–25 odd, 39–42	14–24 even, 34–35, 37–38, 43–60
OL Core	13–25 odd, 26–29, 31, 33, 34–35, 37–60	13–25, 39–42	26–35, 37–38, 43–60
BL Advanced	26–56, (optional: 57–60)		

33 **MULTIPLE REPRESENTATIONS** In this problem, you will use the tables below for exponential functions $f(x)$, $g(x)$, and $h(x)$. **a, b. See Chapter 7 Answer Appendix.**

x	−1	0	1	2	3	4	5
f(x)	2.5	2	1	−1	−5	−13	−29

x	−1	0	1	2	3	4	5
g(x)	5	11	23	47	95	191	383

x	−1	0	1	2	3	4	5
h(x)	3	2.5	2.25	2.125	2.0625	2.0313	2.0156

a. Graphical Graph the functions for $-1 \le x \le 5$ on separate graphs.

b. Verbal List any function with a negative coefficient. Explain your reasoning.

c. Analytical List any function with a graph that is translated to the left. **$g(x)$ and $h(x)$**

d. Analytical Determine which functions are growth models and which are decay models. **See Chapter 7 Answer Appendix.**

34a. Always; sample answer: The domain of exponential functions is all real numbers, so (0, y) always exists.
34b. Sometimes; sample answer: The graph of an exponential function crosses the x-axis when k < 0.

H.O.T. Problems   Use Higher-Order Thinking Skills

34. REASONING Determine whether each statement is *sometimes*, *always*, or *never* true. Explain your reasoning.

a. An exponential function of the form $y = ab^{x-h} + k$ has a y-intercept.

b. An exponential function of the form $y = ab^{x-h} + k$ has an x-intercept.

c. The function $f(x) = |b|^x$ is an exponential growth function if b is an integer.
Sometimes; Sample answer: The function is not exponential if $b = 1$ or -1.

35. **CCSS CRITIQUE** Vince and Grady were asked to graph the following functions. Vince thinks they are the same, but Grady disagrees. Who is correct? Explain your reasoning.

x	y
0	2
1	1
2	0.5
3	0.25
4	0.125
5	0.0625
6	0.03125

an exponential function with rate of decay of $\frac{1}{2}$ and an initial amount of 2

Vince; the graphs of the function would be the same.

36. CHALLENGE A substance decays 35% each day. After 8 days, there are 8 milligrams of the substance remaining. How many milligrams were there initially? **about 251 mg**

37. OPEN ENDED Give an example of a value of b for which $f(x) = \left(\frac{8}{b}\right)^x$ represents exponential decay. **Sample answer: 10**

38. WRITING IN MATH Write the procedure for transforming the graph of $g(x) = b^x$ to the graph of $f(x) = ab^{x-h} + k$. Justify each step.

38. Sample answer: The parent function, $g(x) = b^x$, is stretched if a is greater than 1 or compressed if a is less than 1. The parent function is translated up k units if k is positive and down $|k|$ units if k is negative. The parent function is translated h units to the right if h is positive and $|h|$ units to the left if h is negative.

Additional Answer

29b. growth; 1.25

Enrichment
OL BL

7-1 Enrichment

Families of Curves
Use these graphs for the problems below.

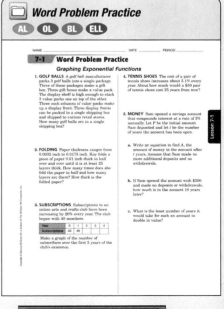

Ticket Out the Door Make several copies each of five exponential functions. Give one function to each student. As the students leave the room, ask them to tell you whether their functions represent exponential growth or decay.

Standardized Test Practice

39. GRIDDED RESPONSE In the figure, $\overline{PQ} \parallel \overline{RN}$, $QN = 12$, $MN = 6$, and $RN = 4$. What is the length of $\overline{PQ}$? **12**

40. Ivan has enough money to buy 12 used CDs. If the cost of each CD was $0.20 less, Ivan could buy 2 more CDs. How much money does Ivan have to spend on CDs? **A**

A $16.80   C $15.80

B $16.40   D $15.40

41. One hundred students will attend the fall dance if tickets cost $30 each. For each $5 increase in price, 10 fewer students will attend. What price will deliver the maximum dollar sales? **H**

F $30

G $35

H $40

J $45

42. SAT/ACT Javier mows a lawn in 2 hours. Tonya mows the same lawn in 1.5 hours. About how many minutes will it take to mow the lawn if Javier and Tonya work together? **C**

A 28 minutes   D 1.2 hours

B 42 minutes   E 1.4 hours

C 51 minutes

Spiral Review

Solve each equation or inequality. (Lesson 6-7)

43. $\sqrt{y + 5} = \sqrt{2y - 3}$ **8**

44. $\sqrt{y + 1} + \sqrt{y - 4} = 5$ **8**

45. $10 - \sqrt{2x + 7} \le 3$ **$x \ge 21$**

46. $6 + \sqrt{3y + 4} < 6$ **no solution**

47. $\sqrt{d + 3} + \sqrt{d + 7} > 4$ **$d > -\dfrac{3}{4}$**

48. $\sqrt{2x + 5} - \sqrt{9 + x} > 0$ **$x > 4$**

Simplify. (Lesson 6-6)

49. $\dfrac{1}{y^{\frac{2}{5}}} \cdot \dfrac{y^{\frac{3}{5}}}{y}$ **$y^{\frac{3}{5}}$**

50. $\dfrac{xy}{\sqrt[3]{z}} \cdot \dfrac{xyz^{\frac{2}{3}}}{z}$ **$\dfrac{2}{z}$**

51. $\dfrac{3x + 4x^2}{x^{-\frac{2}{3}}}$ **$3x^{\frac{5}{3}} + 4x^{\frac{8}{3}}$**

52. $\sqrt[6]{27x^3}$ **$\sqrt{3x}$**

53. $\dfrac{\sqrt[4]{27}}{\sqrt[4]{3}}$ **$\sqrt{3}$**

54. $\dfrac{a^{-\frac{1}{2}}}{6a^{\frac{1}{3}} \cdot a^{-\frac{1}{4}}}$ **$\dfrac{a^{\frac{5}{12}}}{6a}$**

55. FOOTBALL The path of a football thrown across a field is given by the equation $y = -0.005x^2 + x + 5$, where x represents the distance, in feet, the ball has traveled horizontally and y represents the height, in feet, of the ball above ground level. About how far has the ball traveled horizontally when it returns to ground level? (Lesson 4-6) **about 204.88 ft**

56. COMMUNITY SERVICE A drug awareness program is being presented at a theater that seats 300 people. Proceeds will be donated to a local drug information center. If every two adults must bring at least one student, what is the maximum amount of money that can be raised? (Lesson 3-3) **$500**

Skills Review

Simplify. Assume that no variable equals 0.

57. $f^{-7} \cdot f^4$ **$\dfrac{1}{f^3}$**

58. $(3x^2)^3$ **$27x^6$**

59. $(2y)(4xy^3)$ **$8xy^4$**

60. $\left(\dfrac{3}{5}c^2f\right)\left(\dfrac{4}{3}cd\right)^2$ **$\dfrac{16}{15}c^4d^2f$**

DifferentiatedInstruction BL

Extension Have students flip 50 pennies and count the number of heads. Then have students remove those pennies that landed on heads and repeat the activity. Students should record their results and make a plot of the trial number versus the number of heads counted in that trial. Have students graph their data and then explain why, in theory, their data should be modeled by the equation $y = \left(\dfrac{1}{2}\right)^x$.

EXPLORE

7-2 Graphing Technology Lab
Solving Exponential Equations and Inequalities

You can use a TI-83/84 Plus graphing calculator to solve exponential equations by graphing or by using the table feature. To do this, you will write the equations as systems of equations.

CCSS Common Core State Standards
Content Standards
A.REI.11 Explain why the x-coordinates of the points where the graphs of the equations $y = f(x)$ and $y = g(x)$ intersect are the solutions of the equation $f(x) = g(x)$; find the solutions approximately, e.g., using technology to graph the functions, make tables of values, or find successive approximations. Include cases where $f(x)$ and/or $g(x)$ are linear, polynomial, rational, absolute value, exponential, and logarithmic functions.

Activity 1

Solve $3^{x-4} = \frac{1}{9}$.

Step 1 Graph each side of the equation as a separate function. Enter 3^{x-4} as **Y1**. Be sure to include parentheses around the exponent. Enter $\frac{1}{9}$ as **Y2**. Then graph the two equations.

$[-10, 10]$ scl: 1 by $[-1, 1]$ scl: 0.1

Step 2 Use the **intersect** feature.

You can use the **intersect** feature on the **CALC** menu to approximate the ordered pair of the point at which the graphs cross.

The calculator screen shows that the x-coordinate of the point at which the curves cross is 2. Therefore, the solution of the equation is 2.

$[-10, 10]$ scl: 1 by $[-1, 1]$ scl: 0.1

Step 3 Use the **TABLE** feature.

You can also use the **TABLE** feature to locate the point at which the curves intersect.

The table displays x-values and corresponding y-values for each graph. Examine the table to find the x-value for which the y-values of the graphs are equal.

At $x = 2$, both functions have a y-value of $0.\overline{1}$ or $\frac{1}{9}$. Thus, the solution of the equation is 2.

CHECK Substitute 2 for x in the original equation.

$3^{x-4} \stackrel{?}{=} \frac{1}{9}$ Original equation

$3^{2-4} \stackrel{?}{=} \frac{1}{9}$ Substitute 2 for x.

$3^{-2} \stackrel{?}{=} \frac{1}{9}$ Simplify.

$\frac{1}{9} = \frac{1}{9}$ ✓   The solution checks.

A similar procedure can be used to solve exponential inequalities.

(continued on the next page)

1 Focus

Objective Use a graphing calculator to solve exponential equations by graphing or by using the table feature.

Materials
- TI-83/84 Plus or other graphing calculator

Teaching Tip
In Step 1 of Activity 1, remind students to include parentheses around the exponent.

2 Teach

Working in Cooperative Groups
Put students in groups of two or three, mixing abilities. Then have groups complete Activities 1 and 2 and Exercises 1 and 9.

Activity 1
- Before discussing Activity 1, use a simple equation such as $2x = 6$ to remind students how the equation can be solved by graphing. Graph the equations $y = 2x$ and $y = 6$ and then identify the point of intersection of the graphs.

- Ask students why it is necessary in Step 1 to enter the equation using parentheses around the exponent.

- Have students substitute the solution to Activity 1 into the original equation to verify that it is correct.

Activity 2

- In Activity 2, make sure students understand why the inequality needs to be rewritten as a system of inequalities.

Practice Have students complete Exercises 2–9.

3 Assess

Formative Assessment

Use Exercises 4 and 9 to assess whether students comprehend how to use a graphing calculator to solve exponential equations and inequalities.

From Concrete to Abstract

Have students explain how the solution set for Activity 2 would change if the inequality were $2^{x-2} \leq 0.5^{x-3}$.

Graphing Technology Lab
Solving Exponential Equations and Inequalities *Continued*

Activity 2

Solve $2^{x-2} \geq 0.5^{x-3}$.

Step 1 Enter the related inequalities.

Rewrite the problem as a system of inequalities.

The first inequality is $2^{x-2} \geq y$ or $y \leq 2^{x-2}$. Since this inequality includes the *less than or equal to* symbol, shade below the curve.

First enter the boundary, and then use the arrow and ENTER keys to choose the shade below icon, **▙**.

The second inequality is $y \geq 0.5^{x-3}$. Shade above the curve since this inequality contains *greater than or equal to*.

KEYSTROKES:

Step 2 Graph the system.

KEYSTROKES: GRAPH

The x-values of the points in the region where the shadings overlap is the solution set of the original inequality. Using the **intersect** feature, you can conclude that the solution set is $\{x \mid x \geq 2.5\}$.

[−10, 10] scl: 1 by [−10, 10] scl: 1

Step 3 Use the **TABLE** feature.

Verify using the **TABLE** feature. Set up the table to show x-values in increments of 0.5.

KEYSTROKES: 2nd [TBLSET] 0 ENTER .5 ENTER 2nd [TABLE]

Notice that for x-values greater than $x = 2.5$, **Y1 > Y2**. This confirms that the solution of the inequality is $\{x \mid x \geq 2.5\}$.

10. Because the system consists of the expressions on both sides of the original equation or inequality, whatever values are solutions of the system will satisfy the original equation or inequality.

Exercises

Solve each equation or inequality.

1. $9^{x-1} = \frac{1}{81}$ -1

2. $4^{x+3} = 2^{5x}$ 2

3. $5^{x-1} = 2^x$ 1.76

4. $3.5^{x+2} = 1.75^{x+3}$ -1.2

5. $-3^{x+4} = -0.5^{2x+3}$ -2.6

6. $6^{2-x} - 4 < -0.25^{x-2.5}$ $\{x \mid x > 1.8\}$

7. $16^{x-1} > 2^{2x+2}$ $\{x \mid x > 3\}$

8. $3^x - 4 \leq 5^{\frac{x}{2}}$ $\{x \mid x < 2\}$

9. $5^{x+3} \leq 2^{x+4}$ $\{x \mid x \leq -2.2\}$

10. WRITING IN MATH Explain why this technique of graphing a system of equations or inequalities works to solve exponential equations and inequalities.

460 | Explore 7-2 | Graphing Technology Lab: Solving Exponential Equations and Inequalities

LESSON 7-2

Solving Exponential Equations and Inequalities

:: Then	:: Now	:: Why?
• You graphed exponential functions.	**1** Solve exponential equations. **2** Solve exponential inequalities.	• Membership on Internet social networking sites tends to increase exponentially. The membership growth of one Web site can be modeled by the equation $y = 2.2(1.37)^x$, where x is the number of years since 2004 and y is the number of members in millions. You can use $y = 2.2(1.37)^x$ to determine how many members there will be in a given year, or to determine the year in which membership was at a certain level.

NewVocabulary
exponential equation
compound interest
exponential inequality

Common Core State Standards

Content Standards
A.CED.1 Create equations and inequalities in one variable and use them to solve problems.

F.LE.4 For exponential models, express as a logarithm the solution to $ab^{ct} = d$ where a, c, and d are numbers and the base b is 2, 10, or e; evaluate the logarithm using technology.

Mathematical Practices
2 Reason abstractly and quantitatively.

1 Solve Exponential Equations In an **exponential equation**, variables occur as exponents.

KeyConcept Property of Equality for Exponential Functions

Words	Let $b > 0$ and $b \neq 1$. Then $b^x = b^y$ if and only if $x = y$.
Example	If $3^x = 3^5$, then $x = 5$. If $x = 5$, then $3^x = 3^5$.

The Property of Equality can be used to solve exponential equations.

Example 1 Solve Exponential Equations

Solve each equation.

a. $2^x = 8^3$

$2^x = 8^3$	Original equation
$2^x = (2^3)^3$	Rewrite 8 as 2^3.
$2^x = 2^9$	Power of a Power
$x = 9$	Property of Equality for Exponential Functions

b. $9^{2x-1} = 3^{6x}$

$9^{2x-1} = 3^{6x}$	Original equation
$(3^2)^{2x-1} = 3^{6x}$	Rewrite 9 as 3^2.
$3^{4x-2} = 3^{6x}$	Power of a Power
$4x - 2 = 6x$	Property of Equality for Exponential Functions
$-2 = 2x$	Subtract $4x$ from each side.
$-1 = x$	Divide each side by 2.

▶ **GuidedPractice**

1A. $4^{2n-1} = 64$　**2**　　　　　　**1B.** $5^{5x} = 125^{x+2}$　**3**

connectED.mcgraw-hill.com　461

Susan Widers/UpperCut Images/Getty Images

1 Focus

VerticalAlignment

Before Lesson 7-2 Graph exponential functions.

Lesson 7-2 Solve exponential equations. Solve exponential inequalities.

After Lesson 7-2 Develop the definition of logarithms by exploring and describing the relationship between exponential functions and their inverses.

2 Teach

Scaffolding Questions
Have students read the **Why?** section of the lesson.

Ask:
- What value of x represents the year 2010? **6**
- How many members are represented by $y = 5.2$? **5,200,000**
- How many members will there be in 2012? **about 27,300,000**

Lesson 7-2 Resources

Resource	Approaching Level **AL**	On Level **OL**	Beyond Level **BL**	English Learners **ELL**
Teacher Edition	▪ Differentiated Instruction, p. 463	▪ Differentiated Instruction, pp. 463, 465, 467	▪ Differentiated Instruction, pp. 465, 467	
Chapter Resource Masters	▪ Study Guide and Intervention, pp. 12–13 ▪ Skills Practice, p. 14 ▪ Practice, p. 15 ▪ Word Problem Practice, p. 16	▪ Study Guide and Intervention, pp.12–13 ▪ Skills Practice, p. 14 ▪ Practice, p. 15 ▪ Word Problem Practice, p. 16 ▪ Enrichment, p. 17 ▪ Graphing Calculator Activity, p. 18	▪ Practice, p. 15 ▪ Word Problem Practice, p. 16 ▪ Enrichment, p. 17	▪ Study Guide and Intervention, pp. 12–13 ▪ Skills Practice, p. 14 ▪ Practice, p. 15 ▪ Word Problem Practice, p. 16
Other	▪ 5-Minute Check 7-2 ▪ Study Notebook	▪ 5-Minute Check 7-2 ▪ Study Notebook	▪ 5-Minute Check 7-2 ▪ Study Notebook	▪ 5-Minute Check 7-2 ▪ Study Notebook

1 Solve Exponential Equations

Example 1 shows how to solve an exponential equation. **Example 2** shows how to write an exponential function to model a real-world situation. **Example 3** shows how to determine compound interest using an exponential function.

Formative Assessment

Use the Guided Practice exercises after each example to determine students' understanding of concepts.

Additional Examples

1 Solve each equation.
   a. $3^x = 9^4$ 8
   b. $2^{5x} = 4^{2x-1}$ −2

2 **POPULATION** In 2000, the population of Phoenix was 1,321,045. By 2007, it was estimated at 1,512,986.

   a. Write an exponential function that could be used to model the population of Phoenix. Write x in terms of the number of years since 2000.
   $y = 1,321,045(1.0196)^x$

   b. Predict the population of Phoenix in 2013. 1,700,221

Additional Examples also in Interactive Classroom PowerPoint® Presentations

IWB **Interactive White Board** READY

You can use information about growth or decay to write the equation of an exponential function.

Real-World Example 2 Write an Exponential Function   [PT]

SCIENCE Kristin starts an experiment with 7500 bacteria cells. After 4 hours, there are 23,000 cells.

a. Write an exponential function that could be used to model the number of bacteria after x hours if the number of bacteria changes at the same rate.

At the beginning of the experiment, the time is 0 hours and there are 7500 bacteria cells. Thus, the y-intercept, and the value of a, is 7500.

When $x = 4$, the number of bacteria cells is 23,000. Substitute these values into an exponential function to determine the value of b.

$y = ab^x$	Exponential function
$23,000 = 7500 \cdot b^4$	Replace x with 4, y with 23,000, and a with 7500.
$3.067 \approx b^4$	Divide each side by 7500.
$\sqrt[4]{3.067} \approx b$	Take the 4th root of each side.
$1.323 \approx b$	Use a calculator.

An equation that models the number of bacteria is $y \approx 7500(1.323)^x$.

b. How many bacteria cells can be expected in the sample after 12 hours?

$y \approx 7500(1.323)^x$	Modeling equation
$\approx 7500(1.323)^{12}$	Replace x with 12.
$\approx 215,665$	Use a calculator.

There will be approximately 215,665 bacteria cells after 12 hours.

Guided Practice

2. **RECYCLING** A manufacturer distributed 3.2 million aluminum cans in 2005.

   A. In 2010, the manufacturer distributed 420,000 cans made from the recycled cans it had previously distributed. Assuming that the recycling rate continues, write an equation to model the distribution each year of cans that are made from recycled aluminum. $y = 3.2(0.67)^x$

   B. How many cans made from recycled aluminum can be expected in the year 2050? none

Exponential functions are used in situations involving compound interest. **Compound interest** is interest paid on the principal of an investment and any previously earned interest.

KeyConcept Compound Interest

You can calculate compound interest using the following formula.

$$A = P\left(1 + \frac{r}{n}\right)^{nt},$$

where A is the amount in the account after t years, P is the principal amount invested, r is the annual interest rate, and n is the number of compounding periods each year.

Real-WorldLink

In 2008, the U.S. recycling rate for metals of 35% prevented the release of approximately 25 million metric tons of carbon into the air—roughly the amount emitted annually by 4.5 million cars.

Source: Environmental Protection Agency

 462 | Lesson 7-2 | Solving Exponential Equations and Inequalities

---

Focus on Mathematical Content

Solving Exponential Equations Simple exponential equations can be solved by rewriting one or both sides of the equation so that the bases are the same. Once that has been achieved, the Property of Equality for Exponential Functions can be used to solve for the variable.

Nick Dolding/Cultura/Getty Images

Example 3 Compound Interest

PT

An investment account pays 4.2% annual interest compounded monthly. If $2500 is invested in this account, what will be the balance after 15 years?

Understand Find the total amount in the account after 15 years.

Plan Use the compound interest formula.
$$P = 2500, r = 0.042, n = 12, \text{ and } t = 15$$

Solve $A = P\left(1 + \frac{r}{n}\right)^{nt}$ Compound Interest Formula

$\quad\quad = 2500\left(1 + \frac{0.042}{12}\right)^{12 \cdot 15}$ $P = 2500, r = 0.042, n = 12, t = 15$

$\quad\quad \approx 4688.87$ Use a calculator.

Check Graph the corresponding equation $y = 2500(1.0035)^{12t}$. Use **CALC: value** to find y when $x = 15$.

The y-value 4688.8662 is very close to 4688.87, so the answer is reasonable.

[0, 20] scl: 1 by [0, 10,000] scl: 1000

WatchOut!

Percents Remember to convert all percents to decimal form; 4.2% is 0.042.

▶ **Guided**Practice

3. Find the account balance after 20 years if $100 is placed in an account that pays 1.2% interest compounded twice a month. **$127.12**

2 **Solve Exponential Inequalities** An **exponential inequality** is an inequality involving exponential functions.

KeyConcept Property of Inequality for Exponential Functions

Words   Let $b > 1$. Then $b^x > b^y$ if and only if $x > y$, and $b^x < b^y$ if and only if $x < y$.

Example   If $2^x > 2^6$, then $x > 6$. If $x > 6$, then $2^x > 2^6$.

This property also holds true for $\leq$ and $\geq$.

Example 4 Solve Exponential Inequalities

PT

Solve $16^{2x-3} < 8$.

$16^{2x-3} < 8$ Original inequality

$(2^4)^{2x-3} < 2^3$ Rewrite 16 as 2^4 and 8 as 2^3.

$2^{8x-12} < 2^3$ Power of a Power

$8x - 12 < 3$ Property of Inequality for Exponential Functions

$\quad 8x < 15$ Add 12 to each side.

$\quad\quad x < \frac{15}{8}$ Divide each side by 8.

▶ **Guided**Practice

Solve each inequality.

4A. $3^{2x-1} \geq \frac{1}{243}$ $\{x \mid x \geq -2\}$ **4B.** $2^{x+2} > \frac{1}{32}$ $\{x \mid x > -7\}$

Additional Example

3 An investment account pays 5.4% annual interest compounded quarterly. If $4000 is placed in this account, find the balance after 8 years. $6143.56

2 **Solve Exponential Inequalities**
Example 4 shows how to solve an exponential inequality.

Additional Example

4 Solve $5^{3-2x} > \frac{1}{625}$. $x < \frac{7}{2}$

Teach with Tech

Blog Have students write a blog entry explaining the properties of equality and inequality for exponential functions.

DifferentiatedInstruction   AL   OL

If ▶ you want students to check to see if their answer is correct,

Then ▶ remind students to choose any value in the solution interval and see if it satisfies the original inequality.

3 Practice

Formative Assessment
Use Exercises 1–8 to check for understanding.

Use the chart at the bottom of this page to customize assignments for your students.

 Teaching the Mathematical Practices

Modeling Mathematically proficient students can apply the mathematics they know to solve problems arising in everyday life, analyze relationships mathematically to draw conclusions, and interpret their mathematical results in the context of a situation.

Check Your Understanding   ◯ = Step-by-Step Solutions begin on page R14.

Example 1   Solve each equation.

1. $3^{5x} = 27^{2x-4}$ 12

2. $16^{2y-3} = 4^{y+1}$ $\frac{7}{3}$

3. $2^{6x} = 32^{x-2}$ −10

4. $49^{x+5} = 7^{8x-6}$ $\frac{8}{3}$

Example 2

5a. $c = 2^{\frac{t}{15}}$
5b. 16 cells

5. **SCIENCE** Mitosis is a process in which one cell divides into two. The *Escherichia coli* is one of the fastest growing bacteria. It can reproduce itself in 15 minutes.

   a. Write an exponential function to represent the number of cells c after t minutes.

   b. If you begin with one *Escherichia coli* cell, how many cells will there be in one hour?

Example 3

6. A certificate of deposit (CD) pays 2.25% annual interest compounded biweekly. If you deposit $500 into this CD, what will the balance be after 6 years? **$572.23**

Example 4   Solve each inequality.

7. $4^{2x+6} \le 64^{2x-4}$ $\{x \mid x \ge 4.5\}$

8. $25^{y-3} \le \left(\frac{1}{125}\right)^{y+2}$ $\{y \mid y \le 0\}$

Practice and Problem Solving   Extra Practice is on page R7.

Example 1   Solve each equation.

9. $8^{4x+2} = 64$ 0

10. $5^{x-6} = 125$ 9

 $81^{a+2} = 3^{3a+1}$ −7

12. $256^{b+2} = 4^{2-2b}$ −1

13. $9^{3c+1} = 27^{3c-1}$ $\frac{5}{3}$

14. $8^{2y+4} = 16^{y+1}$ −4

Example 2

15. **MODELING** In 2009, My-Lien received $10,000 from her grandmother. Her parents invested all of the money, and by 2021, the amount will have grown to $16,960.

   a. Write an exponential function that could be used to model the money y. Write the function in terms of x, the number of years since 2009. $y = 10{,}000(1.045)^x$

   b. Assume that the amount of money continues to grow at the same rate. What would be the balance in the account in 2031?   about **$26,336.52**

Write an exponential function for the graph that passes through the given points.

16. (0, 6.4) and (3, 100) $y = 6.4(2.5)^x$

17. (0, 256) and (4, 81) $y = 256(0.75)^x$

18. (0, 128) and (5, 371,293) $y = 128(4.926)^x$ 19. (0, 144), and (4, 21,609) $y = 144(3.5)^x$

Example 3

20. Find the balance of an account after 7 years if $700 is deposited into an account paying 4.3% interest compounded monthly. **$945.34**

21. Determine how much is in a retirement account after 20 years if $5000 was invested at 6.05% interest compounded weekly. **$16,755.63**

22. A savings account offers 0.7% interest compounded bimonthly. If $110 is deposited in this account, what will the balance be after 15 years? **$122.17**

23. A college savings account pays 13.2% annual interest compounded semiannually. What is the balance of an account after 12 years if $21,000 was initially deposited? **$97,362.61**

Example 4   Solve each inequality.

24. $625 \ge 5^{a+8}$ $\{a \mid a \le -4\}$

25. $10^{5b+2} > 1000$ $\left\{b \mid b > \frac{1}{5}\right\}$

26. $\left(\frac{1}{64}\right)^{c-2} < 32^{2c}$ $\left\{c \mid c > \frac{3}{4}\right\}$

27. $\left(\frac{1}{27}\right)^{2d-2} \le 81^{d+4}$ $\{d \mid d \ge -1\}$

28. $\left(\frac{1}{9}\right)^{3t+5} \ge \left(\frac{1}{243}\right)^{t-6}$ $\{t \mid t \le -40\}$

29. $\left(\frac{1}{36}\right)^{w+2} < \left(\frac{1}{216}\right)^{4w}$ $\left\{w \mid w < \frac{2}{5}\right\}$

🔊 **464** | Lesson 7-2 | Solving Exponential Equations and Inequalities

---

Differentiated Homework Options

Level	Assignment	Two-Day Option	
AL Basic	9–29, 42–43, 45–73	9–29 odd, 50–53	10–28 even, 42–43, 45–49, 54–73
OL Core	9–29 odd, 30, 31–37 odd, 38–43, 45–73	9–29, 50–53	30–43, 45–49, 54–73
BL Advanced	30–67, (optional: 68–73)		

30. SCIENCE A mug of hot chocolate is 90°C at time $t = 0$. It is surrounded by air at a constant temperature of 20°C. If stirred steadily, its temperature in Celsius after t minutes will be $y(t) = 20 + 70(1.071)^{-t}$.

   a. Find the temperature of the hot chocolate after 15 minutes. **45.02° C**

   b. Find the temperature of the hot chocolate after 30 minutes. **28.94° C**

   c. The optimum drinking temperature is 60°C. Will the mug of hot chocolate be at or below this temperature after 10 minutes? **below**

31. ANIMALS Studies show that an animal will defend a territory, with area in square yards, that is directly proportional to the 1.31 power of the animal's weight in pounds.

   a. If a 45-pound beaver will defend 170 square yards, write an equation for the area a defended by a beaver weighing w pounds. $a = 1.16w^{1.31}$

   b. Scientists believe that thousands of years ago, the beaver's ancestors were 11 feet long and weighed 430 pounds. Use your equation to determine the area defended by these animals. **about 3268 yd²**

Solve each equation.

32. $\left(\frac{1}{2}\right)^{4x+1} = 8^{2x+1}$ $-\frac{2}{5}$ **33.** $\left(\frac{1}{5}\right)^{x-5} = 25^{3x+2}$ $\frac{1}{7}$ **34.** $216 = \left(\frac{1}{6}\right)^{x+3}$ -6

35. $\left(\frac{1}{8}\right)^{3x+4} = \left(\frac{1}{4}\right)^{-2x+4}$ $-\frac{4}{13}$ **36.** $\left(\frac{2}{3}\right)^{5x+1} = \left(\frac{27}{8}\right)^{x-4}$ $\frac{11}{8}$ **37.** $\left(\frac{25}{81}\right)^{2x+1} = \left(\frac{729}{125}\right)^{-3x+1}$ 1

38. CCSS MODELING In 1950, the world population was about 2.556 billion. By 1980, it had increased to about 4.458 billion.

40a. $A = 5000$
$\left(\frac{4.065}{4}\right)^{4t}$;

$A = 5000$
$\left[\left(\frac{12.042}{12}\right)^{12t} +\right.$

$\left.\left(\frac{52.023}{52}\right)^{52t}\right]$

   a. Write an exponential function of the form $y = ab^x$ that could be used to model the world population y in billions for 1950 to 1980. Write the equation in terms of x, the number of years since 1950. (Round the value of b to the nearest ten-thousandth.) $y = 2.556(1.0187)^x$

   b. Suppose the population continued to grow at that rate. Estimate the population in 2000. **6.455 billion**

   c. In 2000, the population of the world was about 6.08 billion. Compare your estimate to the actual population. **The prediction was about 375 million greater than the actual.**

   d. Use the equation you wrote in Part **a** to estimate the world population in the year 2020. How accurate do you think the estimate is? Explain your reasoning. **See margin.**

39. TREES The diameter of the base of a tree trunk in centimeters varies directly with the $\frac{3}{2}$ power of its height in meters.

40c. Sample answer: During the first 22 years, Option B is the better choice because the total is greater than that of Option A. However, after about 22 years, the balance of Option A exceeds that of Option B, so Option A is the better choice.

   a. A young sequoia tree is 6 meters tall, and the diameter of its base is 19.1 centimeters. Use this information to write an equation for the diameter d of the base of a sequoia tree if its height is h meters high. $d = 1.30h^{\frac{3}{2}}$

   b. The General Sherman Tree in Sequoia National Park, California, is approximately 84 meters tall. Find the diameter of the General Sherman Tree at its base. **about 1001 cm**

40. FINANCIAL LITERACY Mrs. Jackson has two different retirement investment plans from which to choose.

Option A:	Option B:
6.5% annual rate compounded quarterly; minimum deposit $5000	4.2% annual rate compounded monthly; minimum deposit $5000
	PLUS
	2.3% annual rate compounded weekly; minimum deposit $5000

   a. Write equations for Option A and Option B given the minimum deposits.

   b. Draw a graph to show the balances for each investment option after t years. **See margin.**

   c. Explain whether Option A or Option B is the better investment choice.

connectED.mcgraw-hill.com   **465**

Additional Answers

38d. About 9.3498 billion; because the prediction for 2000 was greater than the actual population, this prediction is probably even higher than the actual population will be at the time.

40b.

Balance ($) vs. Years — Option A and Option B curves

DifferentiatedInstruction OL BL

Extension Allow students to develop their sense of consumerism by providing them with an initial deposit amount and having them shop around for the best interest rates. Students should record relevant information including bank name, account type, interest rate, how often interest is compounded, and restrictions on the account. Ask students to graph the growth of their initial deposit over time.

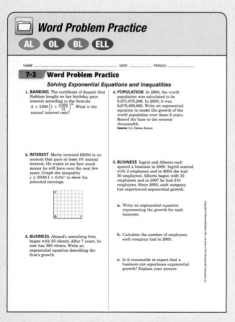

41. MULTIPLE REPRESENTATIONS In this problem, you will explore the rapid increase of an exponential function. A large sheet of paper is cut in half, and one of the resulting pieces is placed on top of the other. Then the pieces in the stack are cut in half and placed on top of each other. Suppose this procedure is repeated several times.

a. Concrete Perform this activity and count the number of sheets in the stack after the first cut. How many pieces will there be after the second cut? How many pieces after the third cut? How many pieces after the fourth cut? **2, 4, 8, 16**

b. Tabular Record your results in a table. **See Chapter 7 Answer Appendix.**

c. Symbolic Use the pattern in the table to write an equation for the number of pieces in the stack after x cuts. $y = 2^x$

d. Analytical The thickness of ordinary paper is about 0.003 inch. Write an equation for the thickness of the stack of paper after x cuts. $y = 0.003(2)^x$

e. Analytical How thick will the stack of paper be after 30 cuts? **about 3,221,225.47 in.**

H.O.T. Problems — Use Higher-Order Thinking Skills

46a. Always; 2^x will always be positive, and -8^{20x} will always be negative.

46b. Always; by definition the graph will always be increasing even if it is a small increase.

46c. Never; by definition the graph will always be decreasing even if it is a small decrease.

49. Sample answer: Divide the final amount by the initial amount. If n is the number of time intervals that pass, take the nth root of the answer.

42. WRITING IN MATH In a problem about compound interest, describe what happens as the compounding period becomes more frequent while the principal and overall time remain the same. **See Chapter 7 Answer Appendix.**

43. ERROR ANALYSIS Beth and Liz are solving $6^{x-3} > 36^{-x-1}$. Is either of them correct? Explain your reasoning.

Beth	Liz
$6^{x-3} > 36^{-x-1}$	$6^{x-3} > 36^{-x-1}$
$6^{x-3} > (6^2)^{-x-1}$	$6^{x-3} > (6^2)^{-x-1}$
$6^{x-3} > 6^{-2x-2}$	$6^{x-3} > 6^{-x+1}$
$x-3 > -2x-2$	$x-3 > -x+1$
$3x > 1$	$2x > 4$
$x > \frac{1}{3}$	$x > 2$

43. Sample answer: Beth; Liz added the exponents instead of multiplying them when taking the power of a power.

45. Reducing the term will be more beneficial. The multiplier is 1.3756 for the 4-year and 1.3828 for the 6.5%.

44. CHALLENGE Solve for x: $16^{18} + 16^{18} + 16^{18} + 16^{18} + 16^{18} = 4^x$. **37.1610**

45. OPEN ENDED What would be a more beneficial change to a 5-year loan at 8% interest compounded monthly: reducing the term to 4 years or reducing the interest rate to 6.5%?

46. ARGUMENTS Determine whether the following statements are *sometimes*, *always*, or *never* true. Explain your reasoning.

a. $2^x > -8^{20x}$ for all values of x.

b. The graph of an exponential growth equation is increasing.

c. The graph of an exponential decay equation is increasing.

47. OPEN ENDED Write an exponential inequality with a solution of $x \leq 2$. **Sample answer:** $4^x \leq 4^2$

48. PROOF Show that $27^{2x} \cdot 81^{x+1} = 3^{2x+2} \cdot 9^{4x+1}$. **See Chapter 7 Answer Appendix.**

49. WRITING IN MATH If you were given the initial and final amounts of a radioactive substance and the amount of time that passes, how would you determine the rate at which the amount was increasing or decreasing in order to write an equation?

Teaching the Mathematical Practices

Arguments Mathematically proficient students understand and use stated assumptions, definitions, and previously established results in constructing arguments. They make conjectures and build a logical progression of statements to explore the truth of their conjectures. And they are able to analyze situations by breaking them into cases, and can recognize and use counterexamples.

50. $3 \times 10^{-4} =$ **B**

   A 0.003   **C** 0.00003

   B 0.0003   **D** 0.000003

51. Which of the following could *not* be a solution to $5 - 3x < -3$? **F**

   F 2.5   **H** 3.5

   G 3   **J** 4

52. GRIDDED RESPONSE The three angles of a triangle are $3x$, $x + 10$, and $2x - 40$. Find the measure of the smallest angle in the triangle. **30**

53. SAT/ACT Which of the following is equivalent to $(x)(x)(x)(x)$ for all x? **E**

   A $x + 4$ **D** $4x^2$

   B $4x$ **E** x^4

   C $2x^2$

Graph each function. (Lesson 7-1) **54–56. See margin.**

54. $y = 2(3)^x$

55. $y = 5(2)^x$

56. $y = 4\left(\dfrac{1}{3}\right)^x$

Solve each equation. (Lesson 6-7)

57. $\sqrt{x + 5} - 3 = 0$ **4**

58. $\sqrt{3t - 5} - 3 = 4$ **18**

59. $\sqrt[4]{2x - 1} = 2$ **8.5**

60. $\sqrt{x - 6} - \sqrt{x} = 3$ **no solution**

61. $\sqrt[3]{5m + 2} = 3$ **5**

62. $(6n - 5)^{\frac{1}{3}} + 3 = -2$ **−20**

63. $(5x + 7)^{\frac{1}{5}} + 3 = 5$ **5**

64. $(3x - 2)^{\frac{1}{5}} + 6 = 5$ $\dfrac{1}{3}$

65. $(7x - 1)^{\frac{1}{3}} + 4 = 2$ **−1**

66. SALES A salesperson earns $10 per hour plus a 10% commission on sales. Write a function to describe the salesperson's income. If the salesperson wants to earn $1000 in a 40-hour week, what should his sales be? (Lesson 6-2) $I(m) = 400 + 0.1m$; $6000

67. STATE FAIR A dairy makes three types of cheese—cheddar, Monterey Jack, and Swiss—and sells the cheese in three booths at the state fair. At the beginning of one day, the first booth received x pounds of each type of cheese. The second booth received y pounds of each type of cheese, and the third booth received z pounds of each type of cheese. By the end of the day, the dairy had sold 131 pounds of cheddar, 291 pounds of Monterey Jack, and 232 pounds of Swiss. The table below shows the percent of the cheese delivered in the morning that was sold at each booth. How many pounds of cheddar cheese did each booth receive in the morning? (Lesson 3-4) **booth 1, 190 lb; booth 2, 150 lb; booth 3, 100 lb**

Type	Booth 1	Booth 2	Booth 3
Cheddar	40%	30%	10%
Monterey Jack	40%	90%	80%
Swiss	30%	70%	70%

Find $[g \circ h](x)$ and $[h \circ g](x)$.

68. $h(x) = 2x - 1$ $6x + 1$;
  $g(x) = 3x + 4$ $6x + 7$

69. $h(x) = x^2 + 2$
  $g(x) = x - 3$ $x^2 - 1$; $x^2 - 6x + 11$

70. $h(x) = x^2 + 1$ $-2x^2 - 1$;
  $g(x) = -2x + 1$ $4x^2 - 4x + 2$

71. $h(x) = -5x$
  $g(x) = 3x - 5$
  $-15x - 5$; $-15x + 25$

72. $h(x) = x^3$
  $g(x) = x - 2$
  $x^3 - 2$; $x^3 - 6x^2 + 12x - 8$

73. $h(x) = x + 4$ $|x + 4|$; $|x| + 4$
  $g(x) = |x|$

Extension Have students extend the solution to Example 3 for an increasing number of compounding periods. Try daily compounding ($n = 365$), and then explore what happens if n is varied up to tens of thousands of times per year. The final amount approaches an upper limit, which in this case is about $4694.03.

WatchOut!

Error Analysis In Exercise 43, students may fail to recognize the use of basic exponent rules when variables are involved. Remind them that $(a^m)^n = a^{mn}$.

4 Assess

Name the Math Have students describe the set of values for b that are possible in an exponential function of the form $y = b^x$.

Formative Assessment

Check for student understanding of the concepts in Lessons 7-1 and 7-2.

🗀 Quiz 1, p. 63

Additional Answers

54.

55.

56.

7-3 Logarithms and Logarithmic Functions

1 Focus

VerticalAlignment

Before Lesson 7-3 Find the inverse of a function.

Lesson 7-3 Evaluate logarithmic expressions. Graph logarithmic functions.

After Lesson 7-3 Determine the solutions of logarithmic equations using algebraic methods.

2 Teach

Scaffolding Questions

Have students read the **Why?** section of the lesson.

Ask:

- What kinds of objects are found near Earth? Sample answer: asteroids, meteors, comets

- What is meant by "likelihood of impact"? Sample answer: probability that an object hits Earth

:: Then	:: Now	:: Why?
● You found the inverse of a function.	**1** Evaluate logarithmic expressions. **2** Graph logarithmic functions.	● Many scientists believe the extinction of the dinosaurs was caused by an asteroid striking Earth. Astronomers use the Palermo scale to classify objects near Earth based on the likelihood of impact. To make comparing several objects easier, the scale was developed using *logarithms*. The Palermo scale value of any object can be found using the equation $PS = \log_{10} R$, where R is the relative risk posed by the object.

 NewVocabulary
logarithm
logarithmic function

CCSS Common Core State Standards

Content Standards
F.IF.7.e Graph exponential and logarithmic functions, showing intercepts and end behavior, and trigonometric functions, showing period, midline, and amplitude.
F.BF.3 Identify the effect on the graph of replacing $f(x)$ by $f(x) + k$, $kf(x)$, $f(kx)$, and $f(x + k)$ for specific values of k (both positive and negative); find the value of k given the graphs. Experiment with cases and illustrate an explanation of the effects on the graph using technology.

Mathematical Practices
6 Attend to precision.

1 Logarithmic Functions and Expressions Consider the exponential function $f(x) = 2^x$ and its inverse. Recall that you can graph an inverse function by interchanging the x- and y-values in the ordered pairs of the function.

$y = 2^x$	
x	y
-3	$\frac{1}{8}$
-2	$\frac{1}{4}$
-1	$\frac{1}{2}$
0	1
1	2
2	4
3	8

$x = 2^y$	
x	y
$\frac{1}{8}$	-3
$\frac{1}{4}$	-2
$\frac{1}{2}$	-1
1	0
2	1
4	2
8	3

As the value of y decreases, the value of x approaches 0.

The inverse of $y = 2^x$ can be defined as $x = 2^y$. In general, the inverse of $y = b^x$ is $x = b^y$. In $x = b^y$, the variable y is called the **logarithm** of x. This is usually written as $y = \log_b x$, which is read y *equals log base b of x.*

KeyConcept Logarithm with Base b

Words	Let b and x be positive numbers, $b \neq 1$. The *logarithm of x with base b* is denoted $\log_b x$ and is defined as the exponent y that makes the equation $b^y = x$ true.
Symbols	Suppose $b > 0$ and $b \neq 1$. For $x > 0$, there is a number y such that $$\log_b x = y \text{ if and only if } b^y = x.$$
Example	If $\log_3 27 = y$, then $3^y = 27$.

The definition of logarithms can be used to express logarithms in exponential form.

Example 1 Logarithmic to Exponential Form

Write each equation in exponential form.

a. $\log_2 8 = 3$

$\log_2 8 = 3 \rightarrow 8 = 2^3$

b. $\log_4 \frac{1}{256} = -4$

$\log_4 \frac{1}{256} = -4 \rightarrow \frac{1}{256} = 4^{-4}$

GuidedPractice

1A. $\log_4 16 = 2$ $16 = 4^2$

1B. $\log_3 729 = 6$ $729 = 3^6$

Dieter Spannknebel/The Image Bank/Getty Images

Lesson 7-3 Resources

Resource	Approaching Level **AL**	On Level **OL**	Beyond Level **BL**	English Learners **ELL**
Teacher Edition		■ Differentiated Instruction, pp. 469, 475	■ Differentiated Instruction, pp. 469, 475	
Chapter Resource Masters	■ Study Guide and Intervention, pp. 19–20 ■ Skills Practice, p. 21 ■ Practice, p. 22 ■ Word Problem Practice, p. 23	■ Study Guide and Intervention, pp. 19–20 ■ Skills Practice, p. 21 ■ Practice, p. 22 ■ Word Problem Practice, p. 23 ■ Enrichment, p. 24 ■ TI-Nspire Activity, p. 25	■ Practice, p.22 ■ Word Problem Practice, p. 23 ■ Enrichment, p. 24	■ Study Guide and Intervention, pp. 19–20 ■ Skills Practice, p. 21 ■ Practice, p. 22 ■ Word Problem Practice, p. 23
Other	■ 5-Minute Check 7-3 ■ Study Notebook	■ 5-Minute Check 7-3 ■ Study Notebook	■ 5-Minute Check 7-3 ■ Study Notebook	■ 5-Minute Check 7-3 ■ Study Notebook

The definition of logarithms can also be used to write exponential equations in logarithmic form.

Example 2 Exponential to Logarithmic Form

Write each equation in logarithmic form.

a. $15^3 = 3375$

$15^3 = 3375 \rightarrow \log_{15} 3375 = 3$

b. $4^{\frac{1}{2}} = 2$

$4^{\frac{1}{2}} = 2 \rightarrow \log_4 2 = \frac{1}{2}$

▶ **Guided**Practice

2A. $4^3 = 64$ $\log_4 64 = 3$

2B. $125^{\frac{1}{3}} = 5$ $\log_{125} 5 = \frac{1}{3}$

You can use the definition of a logarithm to evaluate a logarithmic expression.

> **WatchOut!**
>
> **Logarithmic Base** It is easy to get confused about which number is the base and which is the exponent in logarithmic equations. Consider highlighting each number as you solve to help organize your calculations.

Example 3 Evaluate Logarithmic Expressions

Evaluate $\log_{16} 4$.

$\log_{16} 4 = y$	Let the logarithm equal y.
$4 = 16^y$	Definition of logarithm
$4^1 = (4^2)^y$	$16 = 4^2$
$4^1 = 4^{2y}$	Power of a Power
$1 = 2y$	Property of Equality for Exponential Functions
$\frac{1}{2} = y$	Divide each side by 2.

Thus, $\log_{16} 4 = \frac{1}{2}$.

(handwritten notes: $128 = 4y$, $4^4 = 4^y$, $4 = 1y$, $y = 4$)

▶ **Guided**Practice

Evaluate each expression.

3A. $\log_3 81$ 4

3B. $\log_{\frac{1}{2}} 256$ -8

---

2 **Graphing Logarithmic Functions** The function $y = \log_b x$, where $b \neq 1$, is called a **logarithmic function**. The graph of $f(x) = \log_b x$ represents a parent graph of the logarithmic functions.

KeyConcept Parent Function of Logarithmic Functions

Parent function:	$f(x) = \log_b x$	Type of graph:	continuous, one-to-one
Domain:	all positive real numbers	Range:	all real numbers
Asymptote:	$f(x)$-axis	Intercept:	$(1, 0)$

Teach with Tech

Interactive Whiteboard As you introduce logarithms to your class, use a color code for each part of corresponding exponential and logarithmic functions. For example, use different colors for *x*, *y*, and *b* when showing that $\log_b x = y$ corresponds to $x = b^y$.

1 Logarithmic Functions and Expressions

Example 1 shows how to write logarithmic equations in exponential form. **Example 2** shows how to write exponential equations in logarithmic form. **Example 3** shows how to use the definition of a logarithm to evaluate a logarithmic expression.

Formative Assessment

Use the Guided Practice exercises after each example to determine students' understanding of concepts.

Additional Examples

1 Write each equation in exponential form.
a. $\log_3 9 = 2$ $9 = 3^2$
b. $\log_{10} \frac{1}{100} = -2$ $\frac{1}{100} = 10^{-2}$

2 Write each equation in logarithmic form.
a. $5^3 = 125$ $\log_5 125 = 3$
b. $27^{\frac{1}{3}} = 3$ $\log_{27} 3 = \frac{1}{3}$

3 Evaluate $\log_3 243$. 5

> **Additional Examples** also in Interactive Classroom PowerPoint® Presentations

IWB **Interactive White Board** READY

2 Graphing Logarithmic Functions

Example 4 shows how to graph a logarithmic function. **Example 5** shows how to use transformations to graph logarithmic functions. **Example 6** shows how to find the inverse of an exponential function to solve a real-world example.

DifferentiatedInstruction OL BL

Logical Learners After discussing the definition of logarithm, write $y = 2x$ on the board and ask students to rewrite the equation with x in terms of y. $x = \frac{1}{2}y$ Repeat for $y = x^2$. $x = \pm\sqrt{y}$ Now write $y = 2^x$ on the board and ask students to rewrite this equation with x in terms of y. This will likely have students baffled. Explain that the rewritten equation is $x = \log_2 y$. Stress that a logarithm is defined as the inverse of an exponential function.

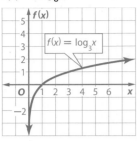

Additional Example

4 Graph each function.

a. $f(x) = \log_3 x$

b. $f(x) = \log_{\frac{1}{4}} x$

Tips for New Teachers

Pacing Students have not covered logarithmic functions before and are likely to find them confusing. Expect students to need extra time to absorb the material in this lesson before continuing with the rest of the chapter.

Additional Answers (Guided Practice)

4A.

4B.

Example 4 Graph Logarithmic Functions

Graph each function.

a. $f(x) = \log_5 x$

Step 1	Identify the base.

$b = 5$

Step 2	Determine points on the graph.

Because $5 > 1$, use the points $\left(\frac{1}{b}, -1\right)$, $(1, 0)$, and $(b, 1)$.

Step 3	Plot the points and sketch the graph.

$\left(\frac{1}{b}, -1\right) \rightarrow \left(\frac{1}{5}, -1\right)$

$(1, 0)$

$(b, 1) \rightarrow (5, 1)$

b. $f(x) = \log_{\frac{1}{3}} x$

Step 1	$b = \frac{1}{3}$

Step 2	$0 < \frac{1}{3} < 1$, so use the points

$\left(\frac{1}{3}, 1\right)$, $(1, 0)$ and $(3, -1)$.

Step 3	Sketch the graph.

> **GuidedPractice 4A, 4B.** See margin.
>
> **4A.** $f(x) = \log_2 x$ **4B.** $f(x) = \log_{\frac{1}{8}} x$

StudyTip

Zero Exponent Recall that for any $b \neq 0$, $b^0 = 1$. Therefore, $\log_2 0$ is undefined because $2^x \neq 0$ for any x-value.

The same techniques used to transform the graphs of other functions you have studied can be applied to the graphs of logarithmic functions.

KeyConcept Transformations of Logarithmic Functions	
$f(x) = a \log_b (x - h) + k$	
h – Horizontal Translation	**k – Vertical Translation**
h units right if h is positive $\lvert h \rvert$ units left if h is negative	k units up if k is positive $\lvert k \rvert$ units down if k is negative
a – Orientation and Shape	
If $a < 0$, the graph is reflected across the x-axis.	If $\lvert a \rvert > 1$, the graph is stretched vertically. If $0 < \lvert a \rvert < 1$, the graph is compressed vertically.

 470 | **Lesson 7-3** | Logarithms and Logarithmic Functions

Focus on Mathematical Content

Logarithms The equation $y = \log_b x$ is read "y equals the logarithm to the base b of the number x." The base b is always positive and $b \neq 1$. And since the equation $y = \log_b x$ is equivalent to the exponential equation $x = b^y$, a logarithm is an exponent. It is the exponent that the base b requires in order to equal the number x.

Winfield Parks/National Geographic/Getty Images

StudyTip

End Behavior In Example 5a, as x approaches infinity, $f(x)$ approaches infinity.

PT

Example 5 Graph Logarithmic Functions

Graph each function.

a. $f(x) = 3 \log_{10} x + 1$

This represents a transformation of the graph of $f(x) = \log_{10} x$.

- $|a| = 3$: The graph stretches vertically

- $h = 0$: There is no horizontal shift.

- $k = 1$: The graph is translated 1 unit up.

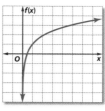

b. $f(x) = \frac{1}{2} \log_{\frac{1}{4}} (x - 3)$

This is a transformation of the graph of $f(x) = \log_{\frac{1}{4}} x$.

- $|a| = \frac{1}{2}$: The graph is compressed vertically.

- $h = 3$: The graph is translated 3 units to the right.

- $k = 0$: There is no vertical shift.

▶ **Guided**Practice **5A, 5B.** See margin.

Graph each function.

5A. $f(x) = 2 \log_3 (x - 2)$ **5B.** $f(x) = \frac{1}{4} \log_{\frac{1}{2}} (x + 1) - 5$

PT

● **Real-World Example 6** Find Inverses of Exponential Functions

EARTHQUAKES The Richter scale measures earthquake intensity. The increase in intensity between each number is 10 times. For example, an earthquake with a rating of 7 is 10 times more intense than one measuring 6. The intensity of an earthquake can be modeled by $y = 10^{x-1}$, where x is the Richter scale rating.

a. Use the information at the left to find the intensity of the strongest recorded earthquake in the United States.

$y = 10^{x-1}$	Original equation
$= 10^{9.2 - 1}$	Substitute 9.2 for x.
$= 10^{8.2}$	Simplify.
$= 158{,}489{,}319.2$	Use a calculator.

b. Write an equation of the form $y = \log_{10} x + c$ for the inverse of the function.

$y = 10^{x-1}$	Original equation
$x = 10^{y-1}$	Replace x with y, replace y with x, and solve for y.
$y - 1 = \log_{10} x$	Definition of logarithm
$y = \log_{10} x + 1$	Add 1 to each side.

▶ **Guided**Practice

6. Write an equation for the inverse of the function $y = 0.5^x$. $y = \log_{0.5} x$

Real-WorldLink

The largest recorded earthquake in the United States was a magnitude 9.2 that struck Prince William Sound, Alaska, on Good Friday, March 28, 1964.

Source: United States Geological Survey

 connectED.mcgraw-hill.com **471**

Additional Examples

5 Graph each function.

a. $f(x) = \frac{1}{3} \log_6 x - 1$

b. $f(x) = 4 \log_{\frac{1}{3}} (x + 2)$

6 **AIR PRESSURE** At Earth's surface, the air pressure is defined as 1 atmosphere. Pressure decreases by about 20% for each mile of altitude. Atmospheric pressure can be modeled by $P = 0.8^x$, where x measures altitude in miles.

a. Find the atmospheric pressure in atmospheres at an altitude of 8 miles.
0.168 atmospheres

b. Write an equation for the inverse of the function.
$x = \log_{0.8} P$

Additional Answers (Guided Practice)

5A.

5B.

Formative Assessment

Use Exercises 1–12 to check for understanding.

Use the chart at the bottom of this page to customize assignments for your students.

Teaching the Mathematical Practices

Precision Mathematically proficient students try to use clear definitions in their reasoning, calculate accurately and efficiently, and make explicit use of definitions.

Additional Answers

8.

9.

10.

11.

Check Your Understanding

 = Step-by-Step Solutions begin on page R14.

Example 1   Write each equation in exponential form.

1. $\log_8 512 = 3$ $8^3 = 512$ | **2.** $\log_5 625 = 4$ $5^4 = 625$

Example 2   Write each equation in logarithmic form.

3. $11^3 = 1331$ $\log_{11} 1331 = 3$ | **4.** $16^{\frac{3}{4}} = 8$ $\log_{16} 8 = \frac{3}{4}$

Example 3   Evaluate each expression.

5. $\log_{13} 169$ 2 | **6.** $\log_2 \frac{1}{128}$ -7 | **7.** $\log_6 1$ 0

Examples 4–5   Graph each function.   **8–11. See margin.**

8. $f(x) = \log_3 x$ | **9.** $f(x) = \log_{\frac{1}{6}} x$

10. $f(x) = 4 \log_4 (x - 6)$ | **11.** $f(x) = 2 \log_{\frac{1}{10}} x - 5$

Example 6   **12. SCIENCE** Use the information at the beginning of the lesson. The Palermo scale value of any object can be found using the equation $PS = \log_{10} R$, where R is the relative risk posed by the object. Write an equation in exponential form for the inverse of the function. $PS = 10^R$

Practice and Problem Solving

Extra Practice is on page R7.

Example 1   Write each equation in exponential form.

13. $\log_2 16 = 4$ $2^4 = 16$ | **14.** $\log_7 343 = 3$ $7^3 = 343$ | **15.** $\log_9 \frac{1}{81} = -2$ $9^{-2} = \frac{1}{81}$

16. $\log_3 \frac{1}{27} = -3$ $3^{-3} = \frac{1}{27}$ | **17.** $\log_{12} 144 = 2$ $12^2 = 144$ | **18.** $\log_9 1 = 0$ $9^0 = 1$

Example 2   Write each equation in logarithmic form.

19. $9^{-1} = \frac{1}{9}$ $\log_9 \frac{1}{9} = -1$ | **20.** $6^{-3} = \frac{1}{216}$ $\log_6 \frac{1}{216} = -3$ | **21.** $2^8 = 256$ $\log_2 256 = 8$

22. $4^6 = 4096$ $\log_4 4096 = 6$ | **23.** $27^{\frac{2}{3}} = 9$ $\log_{27} 9 = \frac{2}{3}$ | **24.** $25^{\frac{3}{2}} = 125$ $\log_{25} 125 = \frac{3}{2}$

Example 3   Evaluate each expression.

25. $\log_3 \frac{1}{9}$ -2 | **26.** $\log_4 \frac{1}{64}$ -3 | **27.** $\log_8 512$ 3 | **28.** $\log_6 216$ 3

29. $\log_{27} 3$ $\frac{1}{3}$ | **30.** $\log_{32} 2$ $\frac{1}{5}$ | **31.** $\log_9 3$ $\frac{1}{2}$ | **32.** $\log_{121} 11$ $\frac{1}{2}$

33. $\log_{\frac{1}{5}} 3125$ -5 | **34.** $\log_{\frac{1}{8}} 512$ -3 | **35.** $\log_{\frac{1}{3}} \frac{1}{81}$ 4 | **36.** $\log_{\frac{1}{6}} \frac{1}{216}$ 3

Examples 4–5   **PRECISION** Graph each function.   **37–48. See Chapter 7 Answer Appendix.**

37. $f(x) = \log_6 x$ | **38.** $f(x) = \log_{\frac{1}{5}} x$ | **39.** $f(x) = 4 \log_2 x + 6$

40. $f(x) = \log_{\frac{1}{9}} x$ | **41.** $f(x) = \log_{10} x$ | **42.** $f(x) = -3 \log_{\frac{1}{12}} x + 2$

43. $f(x) = 6 \log_{\frac{1}{8}} (x + 2)$ | **44.** $f(x) = -8 \log_3 (x - 4)$ | **45.** $f(x) = \log_{\frac{1}{4}} (x + 1) - 9$

46. $f(x) = \log_5 (x - 4) - 5$ | **47.** $f(x) = -\frac{1}{6} \log_8 (x - 3) + 4$ | **48.** $f(x) = -\frac{1}{3} \log_{\frac{1}{6}} (x + 2) - 5$

Differentiated Homework Options

Level	Assignment	Two-Day Option	
AL Basic	13–56, 60, 62–85	13–55 odd, 67–70	14–56 even, 60, 62–66, 71–85
OL Core	13–55 odd, 57–60, 62–85	13–56, 67–70	57–60, 62–66, 71–85
BL Advanced	57–81, (optional: 82–85)		

Example 6

49. PHOTOGRAPHY The formula $n = \log_2 \frac{1}{p}$ represents the change in the f-stop setting n to use in less light where p is the fraction of sunlight.

   a. Benito's camera is set up to take pictures in direct sunlight, but it is a cloudy day. If the amount of sunlight on a cloudy day is $\frac{1}{4}$ as bright as direct sunlight, how many f-stop settings should he move to accommodate less light? **2**

   b. Graph the function. **See margin.**

   c. Use the graph in part b to predict what fraction of daylight Benito is accommodating if he moves down 3 f-stop settings. Is he allowing more or less light into the camera?
$\frac{1}{8}$; **less light**

50. EDUCATION To measure a student's retention of knowledge, the student is tested after a given amount of time. A student's score on an Algebra 2 test t months after the school year is over can be approximated by $y(t) = 85 - 6 \log_2 (t + 1)$, where $y(t)$ is the student's score as a percent.

   a. What was the student's score at the time the school year ended ($t = 0$)? **85**

   b. What was the student's score after 3 months? **73**

   c. What was the student's score after 15 months? **61**

Graph each function. 51–56. See Chapter 7 Answer Appendix.

51 $f(x) = 4 \log_2 (2x - 4) + 6$ **52.** $f(x) = -3 \log_{12} (4x + 3) + 2$

53. $f(x) = 15 \log_{14} (x + 1) - 9$ **54.** $f(x) = 10 \log_5 (x - 4) - 5$

55. $f(x) = -\frac{1}{6} \log_8 (x - 3) + 4$ **56.** $f(x) = -\frac{1}{3} \log_6 (6x + 2) - 5$

B **57. CCSS MODELING** In general, the more money a company spends on advertising, the higher the sales. The amount of money in sales for a company, in thousands, can be modeled by the equation $S(a) = 10 + 20 \log_4(a + 1)$, where a is the amount of money spent on advertising in thousands, when $a \geq 0$.

57b. If $3000 is spent on advertising, $30,000 is returned in sales. If $15,000 is spent on advertising, $50,000 is returned in sales. If $63,000 is spent on advertising, $70,000 is returned in sales.

   a. The value of $S(0) \approx 10$, which means that if $10 is spent on advertising, $10,000 is returned in sales. Find the values of $S(3)$, $S(15)$, and $S(63)$. **$S(3) = 30$, $S(15) = 50$, $S(63) = 70$**

   b. Interpret the meaning of each function value in the context of the problem.

   c. Graph the function. **See Chapter 7 Answer Appendix.**

   d. Use the graph in part c and your answers from part a to explain why the money spent in advertising becomes less "efficient" as it is used in larger amounts.

57d. Because eventually the graph plateaus, and no matter how much money you spend you are still returning about the same in sales.

58. BIOLOGY The generation time for bacteria is the time that it takes for the population to double. The generation time G for a specific type of bacteria can be found using experimental data and the formula $G = \frac{t}{3.3 \log_b f}$, where t is the time period, b is the number of bacteria at the beginning of the experiment, and f is the number of bacteria at the end of the experiment.

   a. The generation time for mycobacterium tuberculosis is 16 hours. How long will it take four of these bacteria to multiply into 1024 bacteria? **264 h or 11 days**

   b. An experiment involving rats that had been exposed to salmonella showed that the generation time for the salmonella was 5 hours. After how long would 20 of these bacteria multiply into 8000? **49.5 h or 2 days 1.5 h**

   c. E. coli are fast growing bacteria. If 6 E. coli can grow to 1296 in 4.4 hours, what is the generation time of E. coli? **$\frac{1}{3}$ h or 20 min**

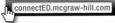 connectED.mcgraw-hill.com **473**

Additional Answer

49b.

Study Guide and Intervention
AL OL ELL

Practice
AL OL BL ELL

Word Problem Practice
AL OL BL ELL

Enrichment
OL BL

WatchOut!

Error Analysis For Exercise 62, point out that Betsy is thinking only of logarithmic functions of the form $f(x) = a \log_b x$.

Error Analysis For Exercise 65, point out that in the expression $a \log_b x$, b is the base of a power. Remind students that finding the logarithm of x with base b means finding what power of b equals x.

Additional Answers

59a.

64a–e. Sample answers given.

64a. $\log_2 33{,}554{,}432 = 25$

64b. $\log_4 \frac{1}{64} = -3$

64c. $\log_2 \sqrt{2} = \frac{1}{2}$

64d. $\log_7 1 = 0$

64e. There is no possible solution; this is the empty set.

65. No; Elisa was closer. She should have $-y = 2$ or $y = -2$ instead of $y = 2$. Matthew used the definition of logarithms incorrectly.

66. Sample answer: In $g(x) = a \log_{10}(x - h) + k$, the value of k is a vertical translation and the graph will shift up k units if k is positive and down $|k|$ units if k is negative. The value of h is a horizontal translation and the graph will shift h units to the right if h is positive and $|h|$ units to the left if h is negative. If $a < 0$, the graph will be reflected across the x-axis. If $|a| > 1$, the graph will be expanded vertically and if $0 < |a| < 1$, then the graph will be compressed vertically.

---

 59 **FINANCIAL LITERACY** Jacy has spent $2000 on a credit card. The credit card company charges 24% interest, compounded monthly. The credit card company uses $\log_{\left(1 + \frac{0.24}{12}\right)} \frac{A}{2000} = 12t$ to determine how much time it will be until Jacy's debt reaches a certain amount, if A is the amount of debt after a period of time, and t is time in years.

a. Graph the function for Jacy's debt. **See margin.**

b. Approximately how long will it take Jacy's debt to double? **≈ 3 years**

c. Approximately how long will it be until Jacy's debt triples? **≈ 4.5 years**

60. Sample answer: Exponential and logarithmic models can grow without bound, which is usually not the case of the situation that is being modeled. For instance, a population cannot grow without bound due to space and food constraints. Therefore, when using a model to make decisions, the situation that is being modeled should be carefully considered.

H.O.T. Problems   Use Higher-Order Thinking Skills

60. **WRITING IN MATH** What should you consider when using exponential and logarithmic models to make decisions?

61. **CCSS ARGUMENTS** Consider $y = \log_b x$ in which b, x, and y are real numbers. Zero can be in the domain *sometimes*, *always* or *never*. Justify your answer.

61. Never; if zero were in the domain, the equation would be $y = \log_b 0$. Then $b^y = 0$. However, for any real number b, there is no real power that would let $b^y = 0$.

62. **ERROR ANALYSIS** Betsy says that the graphs of all logarithmic functions cross the y-axis at $(0, 1)$ because any number to the zero power equals 1. Tyrone disagrees. Is either of them correct? Explain your reasoning.
Tyrone; sample answer: The graphs of logarithmic functions pass through $(1, 0)$, not $(0, 1)$.

63. **REASONING** Without using a calculator, compare $\log_7 51$, $\log_8 61$, and $\log_9 71$. Which of these is the greatest? Explain your reasoning.

64. **OPEN ENDED** Write a logarithmic equation of the form $y = \log_b x$ for each of the following conditions. **a–e. See margin.**

a. y is equal to 25.

b. y is negative.

c. y is between 0 and 1.

d. x is 1.

e. x is 0.

63. $\log_7 51$; sample answer: $\log_7 51$ equals a little more than 2. $\log_8 61$ equals a little less than 2. $\log_9 71$ equals a little less than 2. Therefore, $\log_7 51$ is the greatest.

65. **ERROR ANALYSIS** Elisa and Matthew are evaluating $\log_{\frac{1}{7}} 49$. Is either of them correct? Explain your reasoning. **See margin.**

Elisa

$\log_{\frac{1}{7}} 49 = y$

$\frac{1^y}{7} = 49$

$(7^{-1})^y = 7^2$

$(7)^{-y} = 7^2$

$y = 2$

Matthew

$\log_{\frac{1}{7}} 49 = y$

$49^y = \frac{1}{7}$

$(7^2)^y = (7)^{-1}$

$7^{2y} = (7)^{-1}$

$2y = -1$

$y = -\frac{1}{2}$

66. **WRITING IN MATH** A transformation of $\log_{10} x$ is $g(x) = a \log_{10}(x - h) + k$. Explain the process of graphing this transformation. **See margin.**

 474 | Lesson 7-3 | Logarithms and Logarithmic Functions

CCSS Teaching the Mathematical Practices

Arguments Mathematically proficient students understand and use stated assumptions, definitions, and previously established results in constructing arguments. They make conjectures and build a logical progression of statements to explore the truth of their conjectures. And they are able to analyze situations by breaking them into cases, and can recognize and use counterexamples.

67. A rectangle is twice as long as it is wide. If the width of the rectangle is 3 inches, what is the area of the rectangle in square inches? **D**

A 9

B 12

C 15

D 18

68. SAT/ACT Ichiro has some pizza. He sold 40% more slices than he ate. If he sold 70 slices of pizza, how many did he eat? **G**

F 25   J 98

G 50   K 100

H 75

69. SHORT RESPONSE In the figure $AB = BC$, $CD = BD$, and $m\angle CAD = 70°$. What is the measure of angle ADC? **80**

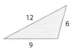

70. If $6x - 3y = 30$ and $4x = 2 - y$ then find $x + y$. **A**

A -4

B -2

C 2

D 4

Spiral Review

Solve each inequality. Check your solution. (Lesson 7-2)

71. $3^{n-2} > 27$ $\{n \mid n > 5\}$ **72.** $2^{2n} \leq \frac{1}{16}$ $\{n \mid n \leq -2\}$ **73.** $16^n < 8^{n+1}$ $\{n \mid n < 3\}$ **74.** $32^{5p} + 2 \geq 16^{5p}$

$\{p \mid p \geq -2\}$

Graph each function. (Lesson 7-1) **75–78. See margin.**

75. $y = -\left(\frac{1}{5}\right)^x$ **76.** $y = -2.5(5)^x$ **77.** $y = 30^{-x}$ **78.** $y = 0.2(5)^{-x}$

79. GEOMETRY The area of a triangle with sides of length a, b, and c is given by $\sqrt{s(s-a)(s-b)(s-c)}$, where $s = \frac{1}{2}(a + b + c)$. If the lengths of the sides of a triangle are 6, 9, and 12 feet, what is the area of the triangle expressed in radical form? (Lesson 6-5) $\frac{27\sqrt{15}}{4}$ ft²

80. GEOMETRY The volume of a rectangular box can be written as $6x^3 + 31x^2 + 53x + 30$ when the height is $x + 2$. (Lesson 5-5)

a. What are the width and length of the box? $2x + 3$ and $3x + 5$

b. Will the ratio of the dimensions of the box always be the same regardless of the value of x? Explain. No; for example, if $x = 1$, the ratio is 3:5:8, but if $x = 2$, the ratio is 4:7:11. The ratios are not equivalent.

81. AUTO MECHANICS Shandra is inventory manager for a local repair shop. She orders 6 batteries, 5 cases of spark plugs, and two dozen pairs of wiper blades and pays $830. She orders 3 batteries, 7 cases of spark plugs, and four dozen pairs of wiper blades and pays $820. The batteries are $22 less than twice the price of a dozen wiper blades. Use augmented matrices to determine what the cost of each item on her order is. (Lesson 3-8)
batteries, $74; spark plugs, $58; wiper blades, $48

Skills Review

Solve each equation or inequality. Check your solution.

82. $9^x = \frac{1}{81}$ -2 **83.** $2^{6x} = 4^{5x+2}$ -1 **84.** $49^{3p+1} = 7^{2p-5}$ $-\frac{7}{4}$ **85.** $9^{x^2} \leq 27^{x^2-2}$ $\{x \mid x \leq -\sqrt{6} \text{ or } x \geq \sqrt{6}\}$

DifferentiatedInstruction OL BL

Extension Point out that the division problem $\frac{32}{4} = 8$ can be written as $\frac{2^5}{2^2} = 2^3$. Ask students to identify the base 2 logarithm in the dividend, the divisor, and the resulting quotient. Then have them write an equation that relates the logarithms. $\log_2 32 = 5$; $\log_2 4 = 2$; $\log_2 8 = 3$; $\log_2 32 - \log_2 4 = \log_2 8$

4 Assess

Crystal Ball Ask students to discuss how they think today's work on logarithms will help them solve logarithmic equations in tomorrow's lesson.

Additional Answers

75.

76.

77.

78.

EXTEND 7-3

Graphing Technology Lab
Choosing the Best Model

Objective Use a graphing calculator to find an equation of best fit for exponential and logarithmic functions.

Materials

- TI-83/84 Plus or other graphing calculator

Teaching Tip

Before Step 1, students should use the keystrokes [2nd] **[STAT PLOT]** and check that both plot 2 and plot 3 are turned off.

Students should have their calculator set **DiagnosticOn**. To set the calculator for diagnostics, use [2nd] **[CATALOG]**, move the cursor down to **DiagnosticOn**, and press [ENTER] twice.

---

2 Teach

Working in Cooperative Groups

Put students in groups of two or three, mixing abilities. Then have groups complete the Activity and Exercises 1–3.

- Point out that the table of data in the example is arranged in two "double" columns.

- Suggest that students compare their graphs to the one shown.

- Have students estimate the population density in 2020 and 2050. How soon will the population density be twice what it was in 2000? about 2085

- When students begin the exercises, they should clear lists L1 and L2. They should also enter appropriate settings for the graphing window.

We can find exponential and logarithmic functions of best fit using a TI-83/84 Plus graphing calculator.

CCSS Common Core State Standards
Mathematical Practices
4 Model with mathematics.

Activity

The population per square mile in the United States has changed dramatically over a period of years. The table shows the number of people per square mile for several years.

a. Use a graphing calculator to enter the data. Then draw a scatter plot that shows how the number of people per square mile is related to the year.

Step 1 Enter the year into L1 and the people per square mile into L2.

KEYSTROKES: *See pages 94 and 95 to review how to enter lists.*

Be sure to clear the Y= list. Use the [▶] key to move the cursor from L1 to L2.

U.S. Population Density			
Year	People per square mile	Year	People per square mile
1790	4.5	1900	21.5
1800	6.1	1910	26.0
1810	4.3	1920	29.9
1820	5.5	1930	34.7
1830	7.4	1940	37.2
1840	9.8	1950	42.6
1850	7.9	1960	50.6
1860	10.6	1970	57.5
1870	10.9	1980	64.0
1880	14.2	1990	70.3
1890	17.8	2000	80.0

Source: Northeast-Midwest Institute

Step 2 Draw the scatter plot.

KEYSTROKES: *See pages 94 and 95 to review how to graph a scatter plot.*

Make sure that **Plot 1** is on, the scatter plot is chosen, **Xlist** is L1, and **Ylist** is L2.

[1780, 2020] scl: 10 by [0, 115] scl: 5

Step 3 Find a regression equation.

To find an equation that best fits the data, use the regression feature of the calculator. Examine various regressions to determine the best model.

Recall that the calculator returns the correlation coefficient r, which is used to indicate how well the model fits the data. The closer r is to 1 or −1, the better the fit.

Linear regression

KEYSTROKES: [STAT] [▶] 4 [ENTER]

Quadratic regression

KEYSTROKES: [STAT] [▶] 5 [ENTER]

$r^2 = 0.9974003374$

$r = \sqrt{0.9974003374}$

$r \approx 0.9986993228$

Exponential regression

KEYSTROKES: [STAT] [▶] 0 [ENTER]

Power regression

KEYSTROKES: [STAT] [▶] [ALPHA] [A] [ENTER]

Compare the *r*-values.

Linear: 0.945411996
Exponential: 0.991887235

Quadratic: 0.9986993228
Power: 0.9917543535

The *r*-value of the quadratic regression is closest to 1, so it appears to model the data best. You can examine the equation visually by graphing the regression equation with the scatter plot.

KEYSTROKES: [STAT] [▶] 5 [ENTER] [Y=] [VARS] 5 [▶] [▶] 1 [GRAPH]

b. **If this trend continues, what will be the population per square mile in 2020?**

To determine the population per square mile in 2020, find the value of *y* when *x* = 2020.

KEYSTROKES: [2nd] [CALC] [ENTER] 2020

If this trend continues, there will be approximately 94.9 people per square mile.

[1780, 2020] scl: 10 by [0, 115] scl: 5

[1780, 2020] scl: 10 by [0, 115] scl: 5

5. No; sample answer: The *r*-value of the exponential regression is 0.9936096203. The r-value of the quadratic regression is 0.9980751795, which is closer to 1.

Exercises

For Exercises 1–5, Jewel deposited $50 into an account, then forgot about it and made no further deposits or withdrawals. The table shows the account balance for several years. 1, 2. See margin.

1. Use a graphing calculator to draw a scatter plot of the data.

2. Calculate and graph a curve of fit using an exponential regression.

3. Write the equation of best fit. What are the domain and range?
   Sample answer: $y = 46.47(1.10^x)$; $D = \{x \mid x \geq 0\}$; $R = \{y \mid y \geq 4.6.47\}$

4. Based on the model, what will the account balance be after 25 years? $558.58

5. Is an exponential model the best fit for the data? Explain.

6. **YOUR TURN** Write a question that can be answered by examining the data of a logarithmic model. First choose a topic and then collect relevant data through Internet research or a survey. Next, make a scatter plot and find a regression equation for your data. Then answer your question. See students' work.

Elapsed Time (years)	Balance
0	$50.00
2	$55.80
4	$64.80
6	$83.09
8	$101.40
10	$123.14
12	$162.67

3 Assess

Formative Assessment
Use Exercise 2 to assess whether students can explain why their equations of best fit are good choices.

From Concrete to Abstract
When you discuss Exercise 5, you may want to ask for any ideas that students have about how to use the calculator to judge the relative merits of various models (quadratic, cubic, quartic, and exponential).

Additional Answers

1.

[0, 30] scl: 2 by [0, 250] scl: 25

2.

[0, 30] scl: 2 by [0, 250] scl: 25

LESSON 7-4 Solving Logarithmic Equations and Inequalities

1 Focus

VerticalAlignment

Before Lesson 7-4 Evaluate logarithmic expressions.

Lesson 7-4 Solve logarithmic equations. Solve logarithmic inequalities.

After Lesson 7-4 Determine the solutions of logarithmic equations using algebraic methods.

2 Teach

Scaffolding Questions

Have students read the **Why?** section of the lesson.

Ask:

- A tornado with a wind speed of 100 miles per hour is in which category? F-1

- How many category F-6 tornadoes are known to have occurred? 0

:·Then

- You evaluated logarithmic expressions.

:·Now

1. Solve logarithmic equations.
2. Solve logarithmic inequalities.

:·Why?

- Each year the National Weather Service documents about 1000 tornado touchdowns in the United States. The intensity of a tornado is measured on the Fujita scale. Tornados are divided into six categories according to their wind speed, path length, path width, and damage caused.

F-Scale	Wind Speed (mph)	Type of Damage
F-0 Gale	40-72	chimneys, branches
F-1 Moderate	73-112	mobile homes overturned
F-2 Significant	113-157	roof torn off
F-3 Severe	158-206	tree uprooted
F-4 Devastating	207-260	homes leveled, cars thrown
F-5 Incredible	261-318	homes thrown
F-6 Inconceivable	319-379	level has never been achieved

NewVocabulary
logarithmic equation
logarithmic inequality

Common Core State Standards

Content Standards
A.SSE.2 Use the structure of an expression to identify ways to rewrite it.
A.CED.1 Create equations and inequalities in one variable and use them to solve problems.

Mathematical Practices
4 Model with mathematics.

1 Solve Logarithmic Equations A **logarithmic equation** contains one or more logarithms. You can use the definition of a logarithm to help you solve logarithmic equations.

Example 1 Solve a Logarithmic Equation

Solve $\log_{36} x = \frac{3}{2}$.

$\log_{36} x = \frac{3}{2}$ Original equation

$x = 36^{\frac{3}{2}}$ Definition of logarithm

$x = (6^2)^{\frac{3}{2}}$ $36 = 6^2$

$x = 6^3$ or 216   Power of a Power

GuidedPractice

Solve each equation.

1A. $\log_9 x = \frac{3}{2}$ 27   **1B.** $\log_{16} x = \frac{5}{2}$ 1024

Use the following property to solve logarithmic equations that have logarithms with the same base on each side.

KeyConcept Property of Equality for Logarithmic Functions

Symbols   If b is a positive number other than 1, then $\log_b x = \log_b y$ if and only if $x = y$.

Example   If $\log_5 x = \log_5 8$, then $x = 8$. If $x = 8$, then $\log_5 x = \log_5 8$.

Comstock Images/Jupiterimages

Resource	Approaching Level AL	On Level OL	Beyond Level BL	English Learners ELL
Teacher Edition	• Differentiated Instruction, p. 479	• Differentiated Instruction, pp. 479, 483	• Differentiated Instruction, p. 483	
Chapter Resource Masters	• Study Guide and Intervention, pp. 26–27 • Skills Practice, p. 28 • Practice, p. 29 • Word Problem Practice, p. 30	• Study Guide and Intervention, pp. 26–27 • Skills Practice, p. 28 • Practice, p. 29 • Word Problem Practice, p. 30 • Enrichment, p. 31 • Graphing Calculator Activity, p. 32	• Practice, p.29 • Word Problem Practice, p. 23 • Enrichment, p. 24	• Study Guide and Intervention, pp. 26–27 • Skills Practice, p. 28 • Practice, p. 29 • Word Problem Practice, p. 30
Other	• 5-Minute Check 7-4 • Study Notebook	• 5-Minute Check 7-4 • Study Notebook	• 5-Minute Check 7-4 • Study Notebook	• 5-Minute Check 7-4 • Study Notebook

Math HistoryLink

Zhang Heng (A.D. 78–139)
The earliest known
seismograph was invented
by Zhang Heng in China
in 132 B.C. It was a large
brass vessel with a heavy
pendulum and several arms
that tripped when an
earthquake tremor was felt.
This helped determine the
direction of the quake.

Science & Society Picture Library/Getty Images

Test-TakingTip

Substitution To save time,
you can substitute each
answer choice in the original
equation to find the one that
results in a true statement.

Standardized Test Example 2 Solve a Logarithmic Equation

Solve $\log_2 (x^2 - 4) = \log_2 3x$.

A -2 B -1 C 2 D 4

Read the Test Item

You need to find x for the logarithmic equation.

Solve the Test Item

$\log_2 (x^2 - 4) = \log_2 3x$	Original equation
$x^2 - 4 = 3x$	Property of Equality for Logarithmic Functions
$x^2 - 3x - 4 = 0$	Subtract $3x$ from each side.
$(x - 4)(x + 1) = 0$	Factor.
$x - 4 = 0$ or $x + 1 = 0$	Zero Product Property
$x = 4$ $x = -1$	Solve each equation.

CHECK Substitute each value into the original equation.

$x = 4$ $x = -1$

$\log_2 (4^2 - 4) \stackrel{?}{=} \log_2 3(4)$ $\log_2 [(-1)^2 - 4] \stackrel{?}{=} \log_2 3(-1)$

$\log_2 12 = \log_2 12$ ✓ $\log_2 (-3) \stackrel{?}{=} \log_2 (-3)$ ✗

The domain of a logarithmic function cannot be 0, so $\log_2 (-3)$ is undefined and -1 is an extraneous solution. The answer is D.

GuidedPractice

2. Solve $\log_3 (x^2 - 15) = \log_3 2x$.   H

F -3 G -1 H 5 J 15

2 Solve Logarithmic Inequalities

A **logarithmic inequality** is an inequality that involves logarithms. The following property can be used to solve logarithmic inequalities.

KeyConcept Property of Inequality for Logarithmic Functions

If $b > 1$, $x > 0$, and $\log_b x > y$, then $x > b^y$.

If $b > 1$, $x > 0$, and $\log_b x < y$, then $0 < x < b^y$.

This property also holds true for $\leq$ and $\geq$.

Example 3 Solve a Logarithmic Inequality

Solve $\log_3 x > 4$.

$\log_3 x > 4$	Original inequality
$x > 3^4$	Property of Inequality for Logarithmic Functions
$x > 81$	Simplify.

GuidedPractice

Solve each inequality.

3A. $\log_4 x \geq 3$ $\{x \mid x \geq 64\}$ 3B. $\log_2 x < 4$ $\{x \mid 0 < x < 16\}$

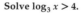

connectED.mcgraw-hill.com   479

---

1 Solve Logarithmic Equations

Example 1 shows how to solve logarithmic equations involving a single logarithm. **Example 2** shows how to solve an equation with logarithms on each side.

Formative Assessment

Use the Guided Practice exercises after each example to determine students' understanding of concepts.

Additional Examples

1 Solve $\log_8 x = \frac{4}{3}$.   16

2 **STANDARDIZED TEST EXAMPLE**
Solve $\log_4 x^2 = \log_4(-6x - 8)$.   C

A 4 C -4 and -2
B 2 D no solutions

Additional Examples also in Interactive Classroom PowerPoint® Presentations

IWB Interactive White Board READY

2 Solve Logarithmic Inequalties

Example 3 shows how to solve a logarithmic inequality. **Example 4** shows how to solve an inequality with logarithms on each side of the symbol.

Teach with Tech

Video Recording Have students work in pairs to create a video showing how to solve a logarithmic inequality. Be sure they explain each step of their work, specifically how to rewrite the logarithmic inequality as an exponential inequality.

Additional Example

3 Solve $\log_6 x > 3$. $\{x \mid x > 216\}$

DifferentiatedInstruction AL OL

If students need help visualizing the relative locations of the digits in equivalent logarithmic and exponential equations,

Then have students create colorful posters showing several equivalent exponential and logarithmic equations, such as $2^3 = 8$ and $3 = \log_2 8$. Suggest that students use a different color for each of the digits 2, 3, and 8.

Additional Example

4 Solve $\log_7 (2x + 8) > \log_7 (x + 5)$.
$\{x \mid x > -3\}$

3 Practice

Formative Assessment

Use Exercises 1–7 to check for understanding.

Use the chart at the bottom of this page to customize assignments for your students.

Additional Answers

36b. The graphs are reflections of each other over the x-axis.

36c. 1. The second graph is the same as the first, except it is shifted vertically up 2 units.

[−2, 8] scl: 1 by [−5, 5] scl: 1

2. The second graph is the same as the first, except it is shifted horizontally to the left 2 units.

[−4, 8] scl: 1 by [−5, 5] scl: 1

3. Each point on the second graph has a y-coordinate 3 times that of the corresponding point on the first graph.

[−2, 8] scl: 1 by [−5, 5] scl: 1

The following property can be used to solve logarithmic inequalities that have logarithms with the same base on each side. Exclude from your solution set values that would result in taking the logarithm of a number less than or equal to zero in the original inequality.

KeyConcept Property of Inequality for Logarithmic Functions

Symbols	If $b > 1$, then $\log_b x > \log_b y$ if and only if $x > y$, and $\log_b x < \log_b y$ if and only if $x < y$.
Example	If $\log_6 x > \log_6 35$, then $x > 35$

This property also holds true for $\le$ and $\ge$.

Example 4 Solve Inequalities with Logarithms on Each Side

Solve $\log_4 (x + 3) > \log_4 (2x + 1)$.

$\log_4 (x + 3) > \log_4 (2x + 1)$	Original inequality
$x + 3 > 2x + 1$	Property of Inequality for Logarithmic Functions
$2 > x$	Subtract $x + 1$ from each side.

Exclude all values of x for which $x + 3 \le 0$ or $2x + 1 \le 0$. So, $x > -3$, $x > -\frac{1}{2}$, and $x < 2$. The solution set is $\left\{x \mid -\frac{1}{2} < x < 2\right\}$ or $\left(-\frac{1}{2}, 2\right)$.

GuidedPractice

4. Solve $\log_5 (2x + 1) \le \log_5 (x + 4)$. Check your solution. $\left\{x \mid -\frac{1}{2} < x \le 3\right\}$

Check Your Understanding ◯ = Step-by-Step Solutions begin on page R14. ✓

Example 1 Solve each equation.

1. $\log_8 x = \frac{4}{3}$ 16

2. $\log_{16} x = \frac{3}{4}$ 8

Example 2 **3. MULTIPLE CHOICE** Solve $\log_5 (x^2 - 10) = \log_5 3x$. **C**

    A 10   **B** 2   **C** 5   **D** 2, 5

Example 3 Solve each inequality.

4. $\log_5 x > 3$ $\{x \mid x > 125\}$

5. $\log_8 x \le -2$ $\left\{x \mid 0 < x \le \frac{1}{64}\right\}$

6. $\log_4 (2x + 5) \le \log_4 (4x - 3)$ $\{x \mid x \ge 4\}$

7. $\log_8 (2x) > \log_8 (6x - 8)$ $\left\{x \mid 2 > x > \frac{4}{3}\right\}$

Practice and Problem Solving Extra Practice is on page R7.

Examples 1–2 CCSS **STRUCTURE** Solve each equation.

8. $\log_{81} x = \frac{3}{4}$ 27

9. $\log_{25} x = \frac{5}{2}$ 3125

10. $\log_8 \frac{1}{2} = x$ $-\frac{1}{3}$

11. $\log_6 \frac{1}{36} = x$ -2

12. $\log_x 32 = \frac{5}{2}$ 4

13. $\log_x 27 = \frac{3}{2}$ 9

14. $\log_3 (3x + 8) = \log_3 (x^2 + x)$ -2 or 4

15 $\log_{12} (x^2 - 7) = \log_{12} (x + 5)$ 4 or -3

16. $\log_6 (x^2 - 6x) = \log_6 (-8)$ no solution

17. $\log_9 (x^2 - 4x) = \log_9 (3x - 10)$ 5

18. $\log_4 (2x^2 + 1) = \log_4 (10x - 7)$ 1 or 4

19. $\log_7 (x^2 - 4) = \log_7 (-x + 2)$ -3

Differentiated Homework Options

Level	Assignment	Two-Day Option	
AL Basic	8–33, 38, 40–68	9–33 odd, 45–48	8–32 even, 38, 40–44, 49–68
OL Core	9–33 odd, 34–38, 40–68	9–33, 45–48	34–38, 40–44, 49–68
BL Advanced	34–62, (optional: 63–68)		

SCIENCE The equation for wind speed w, in miles per hour, near the center of a tornado is $w = 93 \log_{10} d + 65$, where d is the distance in miles that the tornado travels.

20. Write this equation in exponential form. $d = 10^{\frac{w-65}{93}}$

21. In May of 1999, a tornado devastated Oklahoma City with the fastest wind speed ever recorded. If the tornado traveled 525 miles, estimate the wind speed near the center of the tornado. **318 mph**

Solve each inequality.

Examples 3–4 **22.** $\log_6 x < -3$ $\left\{x \mid 0 < x < \frac{1}{216}\right\}$

23. $\log_4 x \geq 4$ $\{x \mid x \geq 256\}$

24. $\log_3 x \geq -4$ $\left\{x \mid x \geq \frac{1}{81}\right\}$

25 $\log_2 x \leq -2$ $\left\{x \mid 0 < x \leq \frac{1}{4}\right\}$

26. $\log_5 x > 2$ $\{x \mid x > 25\}$

27. $\log_7 x < -1$ $\left\{x \mid 0 < x < \frac{1}{7}\right\}$

28. $\log_2 (4x - 6) > \log_2 (2x + 8)$ $\{x \mid x > 7\}$

29. $\log_7 (x + 2) \geq \log_7 (6x - 3)$ $\left\{x \mid \frac{1}{2} < x \leq 1\right\}$

30. $\log_3 (7x - 6) < \log_3 (4x + 9)$ $\left\{x \mid \frac{6}{7} < x < 5\right\}$

31. $\log_5 (12x + 5) \leq \log_5 (8x + 9)$ $\left\{x \mid -\frac{5}{12} < x \leq 1\right\}$

32. $\log_{11} (3x - 24) \geq \log_{11} (-5x - 8)$ $\{x \mid x \geq 2\}$

33. $\log_9 (9x + 4) \leq \log_9 (11x - 12)$ $\{x \mid x \geq 8\}$

B **34.** **CCSS MODELING** The magnitude of an earthquake is measured on a logarithmic scale called the Richter scale. The magnitude M is given by $M = \log_{10} x$, where x represents the amplitude of the seismic wave causing ground motion.

34a.
10^3 or 1000 times as great

34b.
$10^{0.3}$ or about 2 times as great

  a. How many times as great as the amplitude caused by an earthquake with a Richter scale rating of 8 as an aftershock with a Richter scale rating of 5?

  b. In 1906, San Francisco was almost completely destroyed by a 7.8 magnitude earthquake. In 1911, an earthquake estimated at magnitude 8.1 occurred along the New Madrid fault in the Mississippi River Valley. How many times greater was the New Madrid earthquake than the San Francisco earthquake?

35. **MUSIC** The first key on a piano keyboard corresponds to a pitch with a frequency of 27.5 cycles per second. With every successive key, going up the black and white keys, the pitch multiplies by a constant. The formula for the frequency of the pitch sounded when the nth note up the keyboard is played is given by $n = 1 + 12 \log_2 \frac{f}{27.5}$.

  a. A note has a frequency of 220 cycles per second. How many notes up the piano keyboard is this? **37**

  b. Another pitch on the keyboard has a frequency of 880 cycles per second. After how many notes up the keyboard will this be found? **61**

36a. The shapes of the graphs are the same. The asymptote for each graph is the y-axis and the x-intercept for each graph is 1.

36. **MULTIPLE REPRESENTATIONS** In this problem, you will explore the graphs shown: $y = \log_4 x$ and $y = \log_{\frac{1}{4}} x$.

[−2, 8] scl: 1 by [−5, 5] scl: 1

  a. **Analytical** How do the shapes of the graphs compare? How do the asymptotes and the x-intercepts of the graphs compare?

  b. **Verbal** Describe the relationship between the graphs. **See margin.**

36d. The graphs are reflections of each other over the x-axis.
$D = \{x \mid x > 0\}$;
$R = \{$all real numbers$\}$

  c. **Graphical** Use what you know about transformations of graphs to compare and contrast the graph of each function and the graph of $y = \log_4 x$. **See margin.**

     1. $y = \log_4 x + 2$ 2. $y = \log_4 (x + 2)$ 3. $y = 3 \log_4 x$

  d. **Analytical** Describe the relationship between $y = \log_4 x$ and $y = -1(\log_4 x)$. What are a reasonable domain and range for each function?

  e. **Analytical** Write an equation for a function for which the graph is the graph of $y = \log_3 x$ translated 4 units left and 1 unit up. $y = \log_3 (x + 4) + 1$

CCSS **Teaching the Mathematical Practices**

Modeling Mathematically proficient students can apply the mathematics they know to solve problems arising in everyday life, analyze relationships mathematically to draw conclusions, and interpret their mathematical results in the context of a situation.

Study Guide and Intervention
AL OL ELL

Practice
AL OL BL ELL

Word Problem Practice
AL OL BL ELL

Enrichment
OL BL

Teaching the Mathematical Practices

Critique Mathematically proficient students are also able to compare the effectiveness of two plausible arguments, distinguish correct logic or reasoning from that which is flawed, and—if there is a flaw in an argument—explain what it is.

WatchOut!

Error Analysis For Exercise 38, remind students that raising a number to a negative power is not the same as multiplying by -1.

Additional Answers

37c. Sample answer: The power of the logarithm only changes by 2. The power is the answer to the logarithm. That 2 is multiplied by the 10 before the logarithm. So we expect the decibels to change by 20.

44. The y-intercept of the exponential function $y = b^x$ is $(0, 1)$. When the x and y coordinates are switched, the y-intercept is transformed to the x-intercept of $(1, 0)$. There was no x-intercept $(1, 0)$ in the exponential function of the form $y = b^x$. So when the x- and y-coordinates are switched there would be no point on the inverse of $(0, 1)$, and there is no y-intercept.

---

 37 SOUND The relationship between the intensity of sound I and the number of decibels β is $\beta = 10 \log_{10}\left(\frac{I}{10^{-12}}\right)$, where I is the intensity of sound in watts per square meter.

a. Find the number of decibels of a sound with an intensity of 1 watt per square meter. **120**

b. Find the number of decibels of sound with an intensity of 10^{-2} watts per square meter. **100**

c. The intensity of the sound of 1 watt per square meter is 100 times as much as the intensity of 10^{-2} watts per square meter. Why are the decibels of sound not 100 times as great? **See margin.**

Sound	Intensity	Decibels
pin drop	100	0
normal breathing	101	1
clothes dryer	106	6
subway train	1010	10
firecracker	1012	12

40. Sample answer: When $0 < b < 1$, $\log_b x > \log_b y$ if and only if $x < y$. The inequality symbol is switched because a fraction that is less than 1 becomes smaller when it is taken to a greater power.

H.O.T. Problems *Use Higher-Order Thinking Skills*

38. **CRITIQUE** Ryan and Heather are solving $\log_3 x \geq -3$. Is either of them correct? Explain your reasoning.

38. Sample answer: Ryan; Heather did not need to switch the inequality symbol when raising to a negative power.

> **Ryan**
> $\log_3 x \geq -3$
> $x \geq 3^{-3}$
> $x \geq \frac{1}{27}$

> **Heather**
> $\log_3 x \geq -3$
> $x \geq 3^{-3}$
> $0 < x \leq \frac{1}{27}$

39. CHALLENGE Find $\log_3 27 + \log_9 27 + \log_{27} 27 + \log_{81} 27 + \log_{243} 27$. $6\frac{17}{20}$

41. The logarithmic function of the form $y = \log_b x$ is the inverse of the exponential function of the form $y = b^x$. The domain of one of the two inverse functions is the range of the other. The range of one of the two inverse functions is the domain of the other.

40. REASONING The Property of Inequality for Logarithmic Functions states that when $b > 1$, $\log_b x > \log_b y$ if and only if $x > y$. What is the case for when $0 < b < 1$? Explain your reasoning.

41. WRITING IN MATH Explain how the domain and range of logarithmic functions are related to the domain and range of exponential functions.

42. OPEN ENDED Give an example of a logarithmic equation that has no solution. Sample answer: $\log_3 (x + 4) = \log_3 (2x + 12)$

43. REASONING Choose the appropriate term. Explain your reasoning. All logarithmic equations are of the form $y = \log_b x$.

a. If the base of a logarithmic equation is greater than 1 and the value of x is between 0 and 1, then the value for y is (*less than*, *greater than*, *equal to*) 0.

b. If the base of a logarithmic equation is between 0 and 1 and the value of x is greater than 1, then the value of y is (*less than*, *greater than*, *equal to*) 0.

c. There is/are (*no*, *one*, *infinitely many*) solution(s) for b in the equation $y = \log_b 0$.

d. There is/are (*no*, *one*, *infinitely many*) solution(s) for b in the equation $y = \log_b 1$.

44. WRITING IN MATH Explain why any logarithmic function of the form $y = \log_b x$ has an x-intercept of $(1, 0)$ and no y-intercept. **See margin.**

482 | Lesson 7-4 | Solving Logarithmic Equations and Inequalities

45. Find x if $\frac{6.4}{x} = \frac{4}{7}$. **C**

A 3.4
B 9.4
C 11.2
D 44.8

46. The monthly precipitation in Houston for part of a year is shown. **H**

Month	Precipitation (in.)
April	3.60
May	5.15
June	5.35
July	3.18
August	3.83

Find the median precipitation.

F 4.25 in.   H 3.83 in.
G 4.22 in.   J 3.60 in.

47. Clara received a 10% raise each year for 3 consecutive years. What was her salary after the three raises if her starting salary was $12,000 per year? **B**

A $14,520
B $15,972
C $16,248
D $16,410

48. SAT/ACT A vendor has 14 helium balloons for sale: 9 are yellow, 3 are red, and 2 are green. A balloon is selected at random and sold. If the balloon sold is yellow, what is the probability that the next balloon, selected at random, is also yellow? **J**

F $\frac{1}{9}$ H $\frac{36}{91}$ K $\frac{9}{14}$

G $\frac{1}{8}$ J $\frac{8}{13}$

4 Assess

Name the Math Have students write a step-by-step explanation of the procedure for solving a logarithmic equation such as $\log_8 n = \frac{7}{3}$.

Formative Assessment
Check for student understanding of Lessons 7-3 and 7-4.

Quiz 2, p. 63

Spiral Review

Evaluate each expression. (Lesson 7-3)

49. $\log_4 256$ **4**

50. $\log_2 \frac{1}{8}$ **−3**

51. $\log_6 216$ **3**

52. $\log_3 27$ **3**

53. $\log_5 \frac{1}{125}$ **−3**

54. $\log_7 2401$ **4**

Solve each equation or inequality. Check your solution. (Lesson 7-2)

55. $5^{2x+3} \le 125$ $\{x \mid x \le 0\}$

56. $3^{3x-2} > 81$ $\{x \mid x > 2\}$

57. $4^{4a+6} \le 16^a$ $\{a \mid a \le -3\}$

58. $11^{2x+1} = 121^{3x}$ $x = 0.25$

59. $3^{4x-7} = 27^{2x+3}$ $x = -8$

60. $8^{x-4} \le 2^{4-x}$ $\{x \mid x \le 4\}$

61. SHIPPING The height of a shipping cylinder is 4 feet more than the radius. If the volume of the cylinder is 5π cubic feet, how tall is it? Use the formula $V = \pi \cdot r^2 \cdot h$. (Lesson 5-8) **5 ft**

62. NUMBER THEORY Two complex conjugate numbers have a sum of 12 and a product of 40. Find the two numbers. (Lesson 4-4) $6 + 2i, 6 - 2i$

Skills Review

Simplify. Assume that no variable equals zero.

63. $x^5 \cdot x^3$ x^8

64. $a^2 \cdot a^6$ a^8

65. $(2p^2n)^3$ $8p^6n^3$

66. $(3b^3c^2)^2$ $9b^6c^4$

67. $\frac{x^4y^6}{xy^2}$ x^3y^4

68. $\left(\frac{c^9}{d^7}\right)^0$ **1**

DifferentiatedInstruction **OL** **BL**

Extension Ask students to evaluate $\log_3 9$ and $\log_3 27$. 2, 3 Then ask them to predict the value of $\log_3 (9 \cdot 27)$. After they have made their predictions, ask them to check to see if 3 raised to their predicted values is equal to $9 \cdot 27$, or 243. Have them predict the value of $\log_3(mn)$. $\log_3 m + \log_3 n$

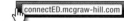

Formative Assessment

Use the Mid-Chapter Quiz to assess students' progress in the first half of the chapter.

For problems answered incorrectly, have students review the lessons indicated in parentheses.

Summative Assessment

📁 Mid-Chapter Test, p. 65

ℯAssessment McGraw-Hill Customize and create multiple versions of your Mid-Chapter Quiz and their answer keys.

FOLDABLES StudyOrganizer

Dinah Zike's Foldables®

Before students complete the Mid-Chapter Quiz, encourage them to review the information for Lessons 7-1 through 7-4 in their Foldables.

Mid-Chapter Quiz
Lessons 7-1 through 7-4

Graph each function. State the domain and range. (Lesson 7-1)

1. $f(x) = 3(4)^x$ **1–4. See Chapter 7 Answer Appendix.**

2. $f(x) = -(2)^x + 5$

3. $f(x) = -0.5(3)^{x+2} + 4$

4. $f(x) = -3\left(\frac{2}{3}\right)^{x-1} + 8$

5. **SCIENCE** You are studying a bacteria population. The population originally started with 6000 bacteria cells. After 2 hours, there were 28,000 bacteria cells. (Lesson 7-1)

  a. Write an exponential function that could be used to model the number of bacteria after x hours if the number of bacteria changes at the same rate. $f(x) = 6000(2.16025)^x$

  b. How many bacteria cells can be expected after 4 hours? **about 130,667**

6. **MULTIPLE CHOICE** Which exponential function has a graph that passes through the points at (0, 125) and (3, 1000)? (Lesson 7-1) **D**

  A $f(x) = 125(3)^x$

  B $f(x) = 1000(3)^x$

  C $f(x) = 125(1000)^x$

  D $f(x) = 125(2)^x$

7. **POPULATION** In 1995, a certain city had a population of 45,000. It increased to 68,000 by 2007. (Lesson 7-2)

  a. What is an exponential function that could be used to model the population of this city x years after 1995? $f(x) = 45,000(1.0350)^x$

  b. Use your model to estimate the population in 2020. **106,346**

8. **MULTIPLE CHOICE** Find the value of x for $\log_3 (x^2 + 2x) = \log_3 (x + 2)$. (Lesson 7-3) **H**

  F $x = -2, 1$

  G $x = -2$

  H $x = 1$

  J no solution

Graph each function. (Lesson 7-3) **9, 10. See Chapter 7 Answer Appendix.**

9. $f(x) = 3 \log_2 (x - 1)$

10. $f(x) = -4 \log_3 (x - 2) + 5$

11. **MULTIPLE CHOICE** Which graph below is the graph of the function $f(x) = \log_3 (x + 5) + 3$? (Lesson 7-3) **A**

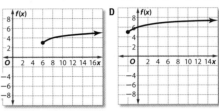

Evaluate each expression. (Lesson 7-3)

12. $\log_4 32$ $\frac{5}{2}$

13. $\log_5 5^{12}$ **12**

14. $\log_{16} 4$ $\frac{1}{2}$

15. Write $\log_9 729 = 3$ in exponential form. (Lesson 7-3) $9^3 = 729$

Solve each equation or inequality. Check your solution. (Lessons 7-2 and 7-4)

16. $3^x = 27^2$ **6**

17. $4^{3x-1} = 16^x$ **1**

18. $\frac{1}{9} = 243^{2x+1}$ $-\frac{7}{10}$

19. $16^{2x+3} < 64$ $\left\{x \mid x < -\frac{3}{4}\right\}$

20. $\left(\frac{1}{32}\right)^{x+3} \geq 16^{3x}$ $\left\{x \mid x \leq -\frac{15}{17}\right\}$

21. $\log_4 x = \frac{3}{2}$ **8**

22. $\log_7 (-x + 3) = \log_7 (6x + 5)$ $-\frac{2}{7}$

23. $\log_2 x < -3$ $\left\{x \mid 0 < x < \frac{1}{8}\right\}$

24. $\log_8 (3x + 7) = \log_8 (2x - 5)$ **no solution**

InterventionPlanner

TIER 1 On Level OL

If students miss about 25% of the exercises or less,

Then choose a resource:

SE Lessons 7-1, 7-2, 7-3, and 7-4

📁 Skills Practice, pp. 7, 14, 21, and 28

TE Chapter Project

🔗 connectED.mcgraw-hill.com Self-Check Quiz

TIER 2 Strategic Intervention AL approaching grade level

If students miss about 75% of the exercises,

Then choose a resource:

📁 Study Guide and Intervention, pp. 5, 12, 10, and 26

🔗 connectED.mcgraw-hill.com Extra Examples, Personal Tutor, Homework Help

TIER 3 Intensive Intervention 2 or more grades below level

If students miss about 75% of the exercises,

Then use *Math Triumphs, Alg. 2*

🔗 connectED.mcgraw-hill.com Extra Examples, Personal Tutor, Homework Help, Review Vocabulary

Ingram Publishing/SuperStock

LESSON 7-5
Properties of Logarithms

:: Then	:: Now	:: Why?
• You evaluated logarithmic expressions and solved logarithmic equations.	**1** Simplify and evaluate expressions using the properties of logarithms. **2** Solve logarithmic equations using the properties of logarithms.	• The level of acidity in food is important to some consumers with sensitive stomachs. Most of the foods that we consume are more acidic than basic. The pH scale measures acidity; a low pH indicates an acidic solution, and a high pH indicates a basic solution. It is another example of a logarithmic scale based on powers of ten. Black coffee has a pH of 5, while neutral water has a pH of 7. Black coffee is one hundred times as acidic as neutral water, because $10^{7-5} = 10^2$ or 100.

Product	pH Level
lemon juice	2.1
sauerkraut	3.5
tomatoes	4.2
black coffee	5.0
milk	6.4
pure water	7.0
eggs	7.8
milk of magnesia	10.0

Common Core State Standards

Content Standards
A.CED.1 Create equations and inequalities in one variable and use them to solve problems.

Mathematical Practices
8 Look for and express regularity in repeated reasoning.

1 Properties of Logarithms Because logarithms are exponents, the properties of logarithms can be derived from the properties of exponents. The Product Property of Logarithms can be derived from the Product of Powers Property of Exponents.

> **KeyConcept** Product Property of Logarithms
>
> | Words | The logarithm of a product is the sum of the logarithms of its factors. |
> |---|---|
> | Symbols | For all positive numbers a, b, and x, where $x \neq 1$, $\log_x ab = \log_x a + \log_x b$. |
> | Example | $\log_2 [(5)(6)] = \log_2 5 + \log_2 6$ |

To show that this property is true, let $b^x = a$ and $b^y = c$. Then, using the definition of logarithm, $x = \log_b a$ and $y = \log_b c$.

$b^x b^y = ac$	Substitution
$b^{x+y} = ac$	Product of Powers
$\log_b b^{x+y} = \log_b ac$	Property of Equality for Logarithmic Functions
$x + y = \log_b ac$	Inverse Property of Exponents and Logarithms
$\log_b a + \log_b c = \log_b ac$	Replace x with $\log_b a$ and y with $\log_b c$.

You can use the Product Property of Logarithms to approximate logarithmic expressions.

> **Example 1** Use the Product Property
>
> Use $\log_4 3 \approx 0.7925$ to approximate the value of $\log_4 192$.
>
> $\log_4 192 = \log_4 (4^3 \cdot 3)$ Replace 192 with 64 · 3 or $4^3 \cdot 3$.
> $= \log_4 4^3 + \log_4 3$ Product Property
> $= 3 + \log_4 3$ Inverse Property of Exponents and Logarithms
> $\approx 3 + 0.7925$ or 3.7925 Replace $\log_4 3$ with 0.7925.
>
> **Guided** Practice
>
> 1. Use $\log_4 2 = 0.5$ to approximate the value of $\log_4 32$. **2.5**

---

1 Focus

VerticalAlignment

Before Lesson 7-5 Evaluate logarithmic expressions and solve logarithmic equations.

Lesson 7-5 Simplify and evaluate expressions using the properties of logarithms. Solve logarithmic equations using the properties of logarithms.

After Lesson 7-5 Determine solutions of exponential equations using algebraic methods.

2 Teach

Scaffolding Questions
Have students read the **Why?** section of the lesson.

Ask:
- Is lemon juice acidic or basic? acidic
- Is milk of magnesia acidic or basic? basic
- Milk of magnesia is how many times as basic as neutral water? $10^{10-7} = 10^3 = 1000$

(handwritten notes)
$\log 48 = \log_5(24 \cdot 3)$
$= \log_5 24 + \log_5 3$
$4 + \log_5 3$
$4 + .6826$

Lesson 7-5 Resources

Resource	Approaching Level **AL**	On Level **OL**	Beyond Level **BL**	English Learners **ELL**
Teacher Edition	• Differentiated Instruction, p. 487	• Differentiated Instruction, p. 487	• Differentiated Instruction, pp. 487, 491	• Differentiated Instruction, p. 487
Chapter Resource Masters	• Study Guide and Intervention, pp. 34–34 • Skills Practice, p. 35 • Practice, p. 36 • Word Problem Practice, p. 37	• Study Guide and Intervention, pp. 34–34 • Skills Practice, p. 35 • Practice, p. 36 • Word Problem Practice, p. 37 • Enrichment, p. 38 • TI-Nspire Activity, p. 39	• Practice, p. 36 • Word Problem Practice, p. 37 • Enrichment, p. 38	• Study Guide and Intervention, pp. 34–34 • Skills Practice, p. 35 • Practice, p. 36 • Word Problem Practice, p. 37
Other	• 5-Minute Check 7-5 • Study Notebook	• 5-Minute Check 7-5 • Study Notebook	• 5-Minute Check 7-5 • Study Notebook	• 5-Minute Check 7-5 • Study Notebook

1 Properties of Logarithms

Example 1 shows how to use the Product Property to approximate the value of a logarithmic expression. **Example 2** shows how to use the Quotient Property to solve a real-world problem. **Example 3** shows how to use the Power Property of Logarithms to approximate the value of a logarithmic expression.

Additional Examples

1 Use $\log_5 2 \approx 0.4307$ to approximate the value of $\log_5 250$. 3.4307

2 **SCIENCE** Find the amount of hydrogen in a liter of acid rain that has a pH of 5.5. $10^{-5.5}$, or 0.0000032 moles

▶ **Additional Examples** also in Interactive Classroom PowerPoint® Presentations

 IWB Interactive White Board READY

Focus on Mathematical Content

Properties of Logarithms In addition to the Product, Quotient, and Power Properties of Logarithms, there are other properties such as the four listed below that can be helpful.

$$\log_b 1 = 0$$
$$\log_b b = 1$$
$$\log_b b^x = x$$
$$b^{\log_b x} = x$$

The first two of these properties become clear when the equations are written in exponential form. The third and fourth properties result from the fact that logarithmic and exponential functions are inverse functions.

Recall that the quotient of powers is found by subtracting exponents. The property for the logarithm of a quotient is similar. Let $b^x = a$ and $b^y = c$. Then $\log_b a = x$ and $\log_b c = y$

$$\frac{b^x}{b^y} = \frac{a}{c}$$

$b^{x-y} = \dfrac{a}{c}$	Quotient Property
$\log_b b^{x-y} = \log_b \dfrac{a}{c}$	Property of Equality for Logarithmic Equations
$x - y = \log_b \dfrac{a}{c}$	Inverse Property of Exponents and Logarithms
$\log_b a - \log_b c = \log_b \dfrac{a}{c}$	Replace x with $\log_b a$ and y with $\log_b c$.

🔑 KeyConcept Quotient Property of Logarithms

Words	The logarithm of a quotient is the difference of the logarithms of the numerator and the denominator.
Symbols	For all positive numbers a, b, and x, where $x \neq 1$, $\log_x \dfrac{a}{b} = \log_x a - \log_x b.$
Example	$\log_2 \dfrac{5}{6} = \log_2 5 - \log_2 6$

🔊 Real-World Example 2 Quotient Property

SCIENCE The pH of a substance is defined as the concentration of hydrogen ions $[H^+]$ in moles. It is given by the formula $\text{pH} = \log_{10} \dfrac{1}{H^+}$. Find the amount of hydrogen in a liter of acid rain that has a pH of 4.2.

Understand The formula for finding pH and the pH of the rain is given. You want to find the amount of hydrogen in a liter of this rain.

Plan Write the equation. Then, solve for $[H^+]$.

Solve

$\text{pH} = \log_{10} \dfrac{1}{H^+}$	Original equation
$4.2 = \log_{10} \dfrac{1}{H^+}$	Substitute 4.2 for pH.
$4.2 = \log_{10} 1 - \log_{10} H^+$	Quotient Property
$4.2 = 0 - \log_{10} H^+$	$\log_{10} 1 = 0$
$4.2 = -\log_{10} H^+$	Simplify.
$-4.2 = \log_{10} H^+$	Multiply each side by -1.
$10^{-4.2} = H^+$	Definition of logarithm

There are $10^{-4.2}$, or about 0.000063, mole of hydrogen in a liter of this rain.

Check

$4.2 = \log_{10} \dfrac{1}{H^+}$	pH = 4.2
$4.2 \stackrel{?}{=} \log_{10} \dfrac{1}{10^{-4.2}}$	$H^+ = 10^{-4.2}$
$4.2 \stackrel{?}{=} \log_{10} 1 - \log_{10} 10^{-4.2}$	Quotient Property
$4.2 \stackrel{?}{=} 0 - (-4.2)$	Simplify.
$4.2 = 4.2$ ✓	

▶ **Guided**Practice

2. **SOUND** The loudness L of a sound, measured in decibels, is given by $L = 10 \log_{10} R$, where R is the sound's relative intensity. Suppose one person talks with a relative intensity of 10^6 or 60 decibels. How much louder would 100 people be, talking at the same intensity? **20 decibels louder**

Real-WorldLink

Acid rain is more acidic than normal rain. Smoke and fumes from burning fossil fuels rise into the atmosphere and combine with the moisture in the air to form acid rain. Acid rain can be responsible for the erosion of statues, as in the photo above.

 486 | Lesson 7-5 | Properties of Logarithms

Tips for New Teachers

Problem Solving When discussing the Product Property of Logarithms, point out that the logarithms used in the example show that the property applies to all logarithms, not just those that can be simplified.

Teach with Tech

Interactive Whiteboard On the board, work through several examples that use the different properties of logarithms. Save each example as a notes page, and clearly label it with which property was used. Post your notes to a class Web page so students can use them as an additional reference outside of class.

Photoshot/age fotostock

Recall that the power of a power is found by multiplying exponents. The property for the logarithm of a power is similar.

> **KeyConcept** Power Property of Logarithms
>
> **Words** The logarithm of a power is the product of the logarithm and the exponent.
>
> **Symbols** For any real number p, and positive numbers m and b, where $b \neq 1$, $\log_b m^p = p \log_b m$.
>
> **Example** $\log_2 6^5 = 5 \log_2 6$

Additional Example

3 Given $\log_5 6 \approx 1.1133$, approximate the value of $\log_5 216$. 3.3399

StudyTip

CCSS Tools You can check this answer by evaluating $2^{4.6438}$ on a calculator. The calculator should give a result of about 25, since $\log_2 25 \approx 4.6438$ means $2^{4.6438} \approx 25$.

Example 3 Power Property of Logarithms

Given $\log_2 5 \approx 2.3219$, approximate the value of $\log_2 25$.

$$\log_2 25 = \log_2 5^2 \qquad \text{Replace 25 with } 5^2.$$
$$= 2 \log_2 5 \qquad \text{Power Property}$$
$$\approx 2(2.3219) \text{ or } 4.6438 \qquad \text{Replace } \log_2 5 \text{ with 2.3219.}$$

▶ **Guided**Practice

3. Given $\log_3 7 \approx 1.7712$, approximate the value of $\log_3 49$. ≈3.5424

CCSS **Teaching the Mathematical Practices**

Tools Mathematically proficient students consider the available tools when solving a mathematical problem. Encourage students to verify their solutions using a calculator.

2 Solve Logarithmic Equations
Example 4 shows how to solve logarithmic equations using properties of logarithms.

2 Solve Logarithmic Equations You can use the properties of logarithms to solve equations involving logarithms.

Example 4 Solve Equations Using Properties of Logarithms

Solve $\log_6 x + \log_6 (x - 9) = 2$.

$$\log_6 x + \log_6 (x - 9) = 2 \qquad \text{Original equation}$$
$$\log_6 x (x - 9) = 2 \qquad \text{Product Property}$$
$$x(x - 9) = 6^2 \qquad \text{Definition of logarithm}$$
$$x^2 - 9x - 36 = 0 \qquad \text{Subtract 36 from each side.}$$
$$(x - 12)(x + 3) = 0 \qquad \text{Factor.}$$
$$x - 12 = 0 \quad \text{or} \quad x + 3 = 0 \qquad \text{Zero Product Property}$$
$$x = 12 \qquad x = -3 \qquad \text{Solve each equation.}$$

CHECK $\log_6 x + \log_6 (x - 9) = 2$ $\log_6 x + \log_6 (x - 9) = 2$

$\log_6 12 + \log_6 (12 - 9) \stackrel{?}{=} 2$ $\log_6 (-3) + \log_6 (-3 - 9) \stackrel{?}{=} 2$

$\log_6 12 + \log_6 3 \stackrel{?}{=} 2$ $\log_6 (-3) + \log_6 (-12) \stackrel{?}{=} 2$

$\log_6 (12 \cdot 3) \stackrel{?}{=} 2$

$\log_6 36 \stackrel{?}{=} 2$ Because $\log_6 (-3)$ and $\log_6 (-12)$ are undefined, -3 is an extraneous solution.

$2 = 2 \checkmark$

The solution is $x = 12$.

▶ **Guided**Practice

4A. $2 \log_7 x = \log_7 27 + \log_7 3$ 9   **4B.** $\log_6 x + \log_6 (x + 5) = 2$ 4

Additional Example

4 Solve the equation.
$4 \log_2 x - \log_2 5 = \log_2 125$ 5

Tips for New Teachers
Sense-Making Encourage students to check their answers by estimating. For instance, in Example 3, $2^4 = 16$ and $2^5 = 32$. So, it is reasonable for $\log_2 25$ to be between 4 and 5.

DifferentiatedInstruction **AL** **OL** **BL** **ELL**

Interpersonal Learners Immediately after discussing Example 4, have pairs of students rework the Example together without looking at the solution in the text. Have the partners take turns explaining the solution steps to each other. Have them also discuss the reasonableness of their solutions.

3 Practice

Formative Assessment

Use Exercises 1–11 to check for understanding.

Use the chart at the bottom of this page to customize assignments for your students.

 Teaching the Mathematical Practices

Perseverance Mathematically proficient students start by explaining to themselves the meaning of a problem and looking for entry points to its solution. They analyze givens, constraints, relationships, and goals and make conjectures about the form and meaning of the solution and plan a solution pathway rather than simply jumping into a solution attempt.

Examples 1–2 Use $\log_4 3 \approx 0.7925$ and $\log_4 5 \approx 1.1610$ to approximate the value of each expression.

1. $\log_4 18$ 2.085

2. $\log_4 15$ 1.9535

3. $\log_4 \frac{5}{3}$ 0.3685

4. $\log_4 \frac{3}{4}$ −0.2075

5. Mt. Everest: 26,855.44 pascals;
Mt. Trisuli: 34,963.34 pascals;
Mt. Bonete: 36,028.42 pascals;
Mt. McKinley: 39,846.22 pascals;
Mt. Logan: 41,261.82 pascals

Example 2

5. **MOUNTAIN CLIMBING** As elevation increases, the atmospheric air pressure decreases. The formula for pressure based on elevation is $a = 15{,}500(5 - \log_{10} P)$, where a is the altitude in meters and P is the pressure in pascals (1 psi ≈ 6900 pascals). What is the air pressure at the summit in pascals for each mountain listed in the table at the right?

Mountain	Country	Height (m)
Everest	Nepal/Tibet	8850
Trisuli	India	7074
Bonete	Argentina/Chile	6872
McKinley	United States	6194
Logan	Canada	5959

Example 3 Given $\log_3 5 \approx 1.465$ and $\log_5 7 \approx 1.2091$, approximate the value of each expression.

6. $\log_3 25$ 2.93

7. $\log_5 49$ 2.4182

Example 4 Solve each equation. Check your solutions.

8. $\log_4 48 - \log_4 n = \log_4 6$ 8

9. $\log_3 2x + \log_3 7 = \log_3 28$ 2

10. $3 \log_2 x = \log_2 8$ 2

11. $\log_{10} a + \log_{10}(a - 6) = 2$ 13.4403

Practice and Problem Solving

Extra Practice is on page R7.

Examples 1–2 Use $\log_4 2 = 0.5$, $\log_4 3 \approx 0.7925$, and $\log_4 5 \approx 1.1610$ to approximate the value of each expression.

12. $\log_4 30$ 2.4535

13. $\log_4 20$ 2.1610

14. $\log_4 \frac{2}{3}$ −0.2925

(15) $\log_4 \frac{4}{3}$ 0.2075

16. $\log_4 9$ 1.5850

17. $\log_4 8$ 1.5

18a. $10^{2.7}$ or about 500 times as great

Example 2

18. **SCIENCE** In 2007, an earthquake near San Francisco registered approximately 5.6 on the Richter scale. The famous San Francisco earthquake of 1906 measured 8.3 in magnitude.

a. How much more intense was the 1906 earthquake than the 2007 earthquake?

b. Richter himself classified the 1906 earthquake as having a magnitude of 8.3. More recent research indicates it was most likely a 7.9. What is the difference in intensities?

Year	Location	Magnitude
1906	San Francisco	8.3
1923	Tokyo, Japan	8.3
1932	Gansu, China	7.6
1960	Chile	9.5
1964	Alaska	9.2
2007	San Francisco	5.6

Source: TLC

Richter thought the earthquake was $10^{0.4}$ or about $2\frac{1}{2}$ times greater than it actually was.

Example 3 Given $\log_6 8 \approx 1.1606$ and $\log_7 9 \approx 1.1292$, approximate the value of each expression.

19. $\log_6 48$ 2.1606

20. $\log_7 81$ 2.2584

21. $\log_6 512$ 3.4818

22. $\log_7 729$ 3.3876

Example 4 **PERSEVERANCE** Solve each equation. Check your solutions.

23. $\log_3 56 - \log_3 n = \log_3 7$ 8

24. $\log_2 (4x) + \log_2 5 = \log_2 40$ 2

25. $5 \log_2 x = \log_2 32$ 2

26. $\log_{10} a + \log_{10}(a + 21) = 2$ 4

Differentiated Homework Options

Level	Assignment	Two-Day Option	
AL Basic	12–26, 61–63, 65–88	13–25 odd, 69–72	12–26 even, 61–63, 73–88
OL Core	13–57 odd, 59–63, 65–88	13–25, 69–72	27–63, 65–68, 73–88
BL Advanced	27–82, (optional: 83–88)		

27 **PROBABILITY** In the 1930s, Dr. Frank Benford demonstrated a way to determine whether a set of numbers has been randomly chosen or manually chosen. If the sets of numbers were not randomly chosen, then the Benford formula, $P = \log_{10}\left(1 + \frac{1}{d}\right)$, predicts the probability of a digit d being the first digit of the set. For example, there is a 4.6% probability that the first digit is 9.

a. Rewrite the formula to solve for the digit if given the probability. $d = \dfrac{1}{10^P - 1}$

b. Find the digit that has a 9.7% probability of being selected. **4**

c. Find the probability that the first digit is 1 ($\log_{10} 2 \approx 0.30103$). **30.1%**

Use $\log_5 3 \approx 0.6826$ and $\log_5 4 \approx 0.8614$ to approximate the value of each expression.

28. $\log_5 40$ **2.2921**

29. $\log_5 30$ **2.1133**

30. $\log_5 \frac{3}{4}$ **−0.1788**

31. $\log_5 \frac{4}{3}$ **0.1788**

32. $\log_5 9$ **1.3652**

33. $\log_5 16$ **1.7228**

34. $\log_5 12$ **1.5440**

35. $\log_5 27$ **2.0478**

Solve each equation. Check your solutions.

36. $\log_3 6 + \log_3 x = \log_3 12$ **2**

37. $\log_4 a + \log_4 8 = \log_4 24$ **3**

38. $\log_{10} 18 - \log_{10} 3x = \log_{10} 2$ **3**

39. $\log_7 100 - \log_7 (y + 5) = \log_7 10$ **5**

40. $\log_2 n = \frac{1}{3}\log_2 27 + \log_2 36$ **108**

41. $3\log_{10} 8 - \frac{1}{2}\log_{10} 36 = \log_{10} x$ **$85\frac{1}{3}$**

Solve for n.

42. $\log_a 6n - 3\log_a x = \log_a x$ $\frac{x^4}{6}$

43. $2\log_b 16 + 6\log_b n = \log_b (x - 2)$ $\left(\frac{x-2}{256}\right)^{\frac{1}{6}}$

Solve each equation. Check your solutions. **46. no real solution**

44. $\log_{10} z + \log_{10} (z + 9) = 1$ **1**

45. $\log_3 (a^2 + 3) + \log_3 3 = 3$ $\sqrt{6}, -\sqrt{6}$

46. $\log_2 (15b - 15) - \log_2 (-b^2 + 1) = 1$

47. $\log_4 (2y + 2) - \log_4 (y - 2) = 1$ **5**

48. $\log_6 0.1 + 2\log_6 x = \log_6 2 + \log_6 5$ **10**

49. $\log_7 64 - \log_7 \frac{8}{3} + \log_7 2 = \log_7 4p$ **12**

50. **CCSS REASONING** The humpback whale is an endangered species. Suppose there are 5000 humpback whales in existence today, and the population decreases at a rate of 4% per year.

a. Write a logarithmic function for the time in years based upon population. $t = \log_{0.96}\left(\frac{p}{5000}\right)$

b. After how long will the population drop below 1000? Round your answer to the nearest year. **40 yr**

State whether each equation is *true* or *false*.

51. $\log_8 (x - 3) = \log_8 x - \log_8 3$ **false**

52. $\log_5 22x = \log_5 22 + \log_5 x$ **true**

53. $\log_{10} 19k = 19\log_{10} k$ **false**

54. $\log_2 y^5 = 5\log_2 y$ **true**

55. $\log_7 \frac{x}{3} = \log_7 x - \log_7 3$ **true**

56. $\log_4 (z + 2) = \log_4 z + \log_4 2$ **false**

57. $\log_8 p^4 = (\log_8 p)^4$ **false**

58. $\log_9 \frac{x^2 y^3}{z^4} = 2\log_9 x + 3\log_9 y - 4\log_9 z$ **true**

59. **PARADE** An equation for loudness L, in decibels, is $L = 10\log_{10} R$, where R is the relative intensity of the sound.

a. Solve $120 = 10\log_{10} R$ to find the relative intensity of the Macy's Thanksgiving Day Parade with a loudness of 120 decibels depending on how close you are. **10^{12}**

b. Some parents with young children want the decibel level lowered to 80. How many times less intense would this be? In other words, find the ratio of their intensities. **10^4 or about 10,000 times**

Teaching the Mathematical Practices

Reasoning Mathematically proficient students make sense of quantities and their relationships in problem situations. Quantitative reasoning entails habits of creating a coherent representation of the problem at hand; considering the units involved; attending to the meaning of quantities, not just how to compute them; and knowing and flexibly using different properties of operations and objects.

60. FINANCIAL LITERACY The average American carries a credit card debt of approximately $8600 with an annual percentage rate (APR) of 18.3%. The formula $m = \dfrac{b\left(\frac{r}{n}\right)}{1 - \left(1 + \frac{r}{n}\right)^{-nt}}$ can be used to compute the monthly payment m that is necessary to pay off a credit card balance b in a given number of years t, where r is the annual percentage rate and n is the number of payments per year.

a. What monthly payment should be made in order to pay off the debt in exactly three years? What is the total amount paid? **$312.21; $11,239.56**

b. The equation $t = \dfrac{\log\left(1 - \frac{br}{mn}\right)}{-n \log\left(1 + \frac{r}{n}\right)}$ can be used to calculate the number of years necessary for a given payment schedule. Copy and complete the table. **See margin.**

Payment (m)	Years (t)
$50	non-real
$100	non-real
$150	11.42
$200	5.87
$250	4.09
$300	3.16

c. Graph the information in the table from part b.

d. If you could only afford to pay $100 a month, will you be able to pay off the debt? If so, how long will it take? If not, why not? **No; the monthly interest is $131.15, so the payments do not even cover the interest.**

e. What is the minimum monthly payment that will work toward paying off the debt? **$131.16**

H.O.T. Problems   Use Higher-Order Thinking Skills

61. OPEN ENDED Write a logarithmic expression for each condition. Then write the expanded expression.

a. a product and a quotient   Sample answer: $\log_b \frac{xz}{5} = \log_b x + \log_b z - \log_b 5$

b. a product and a power   Sample answer: $\log_b m^4 p^6 = 4 \log_b m + 6 \log_b p$

c. a product, a quotient, and a power   Sample answer: $\log_b \frac{j^8 k}{h^5} = 8 \log_b j + \log_b k - 5 \log_b h$

62. CCSS ARGUMENTS Use the properties of exponents to prove the Power Property of Logarithms.   **See margin.**

63. WRITING IN MATH Explain why the following are true.   **63b.** $\log_b b = 1$, because $b^1 = b$.

a. $\log_b 1 = 0$ $\log_b 1 = 0$, because $b^0 = 1$.

b. $\log_b b = 1$

c. $\log_b b^x = x$ $\log_b b^x = x$, because $b^x = b^x$.

64. CHALLENGE Simplify $\log_{\sqrt{a}} (a^2)$ to find an exact numerical value.   **See margin.**

65. WHICH ONE DOESN'T BELONG? Find the expression that does not belong. Explain.

$\log_b 24 = \log_b 2 + \log_b 12$	$\log_b 24 = \log_b 20 + \log_b 4$
$\log_b 24 = \log_b 8 + \log_b 3$	$\log_b 24 = \log_b 4 + \log_b 6$

$\log_b 24 \neq \log_b 20 + \log_b 4$; all other choices are equal to $\log_b 24$.

66. REASONING Use the properties of logarithms to prove that $\log_a \frac{1}{x} = -\log_a x$.   **See Chapter 7 Answer Appendix.**

67. CHALLENGE Simplify $x^{3 \log_x 2 - \log_x 5}$ to find an exact numerical value.   **See Chapter 7 Answer Appendix.**

68. WRITING IN MATH Explain how the properties of exponents and logarithms are related. Include examples like the one shown at the beginning of the lesson illustrating the Product Property, but with the Quotient Property and Power Property of Logarithms.   **See Chapter 7 Answer Appendix.**

Teaching the Mathematical Practices

Arguments Mathematically proficient students understand and use stated assumptions, definitions, and previously established results in constructing arguments. They make conjectures and build a logical progression of statements to explore the truth of their conjectures. And they are able to analyze situations by breaking them into cases, and can recognize and use counterexamples.

69. Find the mode of the data.

22, 11, 12, 23, 7, 6, 17, 15, 21, 19 **D**

A 11 **C** 16

B 15 **D** There is no mode.

70. SAT/ACT What is the effect on the graph of $y = 4x^2$ when the equation is changed to $y = 2x^2$? **H**

F The graph is rotated 90 degrees about the origin.

G The graph is narrower.

H The graph is wider.

J The graph of $y = 2x^2$ is a reflection of the graph $y = 4x^2$ across the x-axis.

K The graph is unchanged.

71. SHORT RESPONSE In $y = 6.5(1.07)^x$, x represents the number of years since 2000, and y represents the approximate number of millions of Americans 7 years of age and older who went camping two or more times that year. Describe how the number of millions of Americans who go camping is changing over time.

growing exponentially

72. What are the x-intercepts of the graph of $y = 4x^2 - 3x - 1$? **D**

A $-\frac{1}{4}$ and $\frac{1}{4}$ **C** -1 and 1

B -1 and $\frac{1}{4}$ **D** 1 and $-\frac{1}{4}$

Spiral Review

Solve each equation. Check your solutions. (Lesson 7-4)

73. $\log_5 (3x - 1) = \log_5 (2x^2)$ $\frac{1}{2}$, 1 **74.** $\log_{10} (x^2 + 1) = 1$ ± 3 **75.** $\log_{10} (x^2 - 10x) = \log_{10} (-21)$

no solution

Evaluate each expression. (Lesson 7-3)

76. $\log_{10} 0.001$ -3 **77.** $\log_4 16^x$ **2x** **78.** $\log_3 27^x$ **3x**

79. ELECTRICITY The amount of current in amperes I that an appliance uses can be calculated using the formula $I = \left(\frac{P}{R}\right)^{\frac{1}{2}}$, where P is the power in watts and R is the resistance in ohms. How much current does an appliance use if $P = 120$ watts and $R = 3$ ohms? Round to the nearest tenth. (Lesson 6-6) **6.3**

Determine whether each pair of functions are inverse functions. Write *yes* or *no*. (Lesson 6-2)

80. $f(x) = x + 73$
$g(x) = x - 73$ **yes**

81. $g(x) = 7x - 11$
$h(x) = \frac{1}{7}x + 11$ **no**

82. SCULPTING Antonio is preparing to make an ice sculpture. He has a block of ice that he wants to reduce in size by shaving off the same amount from the length, width, and height. He wants to reduce the volume of the ice block to 24 cubic feet. (Lesson 5-7) **a.** $(3 - x)(4 - x)(5 - x) = 24$

a. Write a polynomial equation to model this situation.

b. How much should he take from each dimension? **1 ft**

Skills Review

Solve each equation or inequality. Check your solution.

83. $3^{4x} = 3^{3-x}$ $\frac{3}{5}$ **84.** $3^{2n} \le \frac{1}{9}$ $\{n \mid n \le -1\}$ **85.** $3^{5x} \cdot 81^{1-x} = 9^{x-3}$ **10**

86. $49^x = 7^{x^2 - 15}$ $-3, 5$ **87.** $\log_2 (x + 6) > 5$ $\{x \mid x > 26\}$ **88.** $\log_5 (4x - 1) = \log_5 (3x + 2)$ **3**

DifferentiatedInstruction **BL**

Extension Show students the following:

$\log_{10} 3 \approx 0.4771$
$\log_{10} 30 \approx 1.4771$
$\log_{10} 300 \approx 2.4771$
$\log_{10} 3000 \approx 3.4771$

Ask students to predict $\log_{10} 30{,}000$. 4.4771

Have students use properties of logarithms to explain this pattern. Sample explanation: 3, 30, 300, and 3000 can be written as 3×10^0, 3×10^1, 3×10^2, and 3×10^3 respectively. Then the base 10 logarithms of each can be rewritten as a sum of two logarithms. For example, $\log_{10} 3000$ can be written as $\log_{10} (3 \cdot 10^3)$. Then it follows that $\log_{10} (3 \cdot 10^3) = \log_{10} 3 + \log_{10} 10^3 = \log_{10} (3) + 3 = 3.4771$.

4 Assess

Crystal Ball Tell students that in the next lesson they will learn to solve equations using common (base 10) logarithms. Ask them to write how they think what they learned today will connect with the next lesson they will study.

Additional Answers

60c.

62.
$$m^p = m^p$$
$$\left(b^{\log_b m}\right)^p = b^{\log_b (m^p)}$$
$$b^{\log_b mp} = b^{\log_b (m^p)}$$
$$\log_b mp = \log_b (m^p)$$
$$p \log_b m = \log_b (m^p)$$

64.
$$\log_{\sqrt{a}} (a^2) = x$$
$$\left(\sqrt{a}\right)^x = a^2$$
$$\left(a^{\frac{1}{2}}\right)^x = a^2$$
$$a^{\frac{x}{2}} = a^2$$
$$\frac{x}{2} = 2$$
$$x = 4$$

LESSON 7-6 Common Logarithms

D. Falconer/PhotoLink/Getty Images

1 Focus

VerticalAlignment

Before Lesson 7-6 Simplify expressions and solve equations using properties of logarithms.

Lesson 7-6 Solve exponential equations and inequalities using common logarithms. Evaluate logarithmic expressions using the Change of Base Formula.

After Lesson 7-6 Analyze a situation modeled by an exponential function, and solve the problem.

2 Teach

Scaffolding Questions

Have students read the **Why?** section of the lesson.

Ask:

- How does the Richter number of a "Great" earthquake compare with the number for a "Light" earthquake? It is twice as large.

- How does the intensity of a "Great" earthquake compare to the intensity of a "Light" earthquake? It is 10,000 times as large.

- Where would an earthquake with an intensity half as large as 10^8 appear in the table? between "Major" and "Great"

::Then

- You simplified expressions and solved equations using properties of logarithms.

::Now

1. Solve exponential equations and inequalities using common logarithms.

2. Evaluate logarithmic expressions using the Change of Base Formula.

::Why?

- Seismologists use the Richter scale to measure the strength or magnitude of earthquakes. The magnitude of an earthquake is determined using the logarithm of the amplitude of waves recorded by seismographs.

The logarithmic scale used by the Richter scale is based on the powers of 10. For example, a magnitude 6.4 earthquake can be represented by $6.4 = \log_{10} x$.

Richter Number	Intensity
1	10^1 micro
2	10^2 minor
3	10^3 minor
4	10^4 light
5	10^5 moderate
6	10^6 strong
7	10^7 major
8	10^8 great

NewVocabulary
common logarithm
Change of Base Formula

Common Core State Standards

Content Standards
A.CED.1 Create equations and inequalities in one variable and use them to solve problems.

Mathematical Practices
4 Model with mathematics.

1 Common Logarithms

You have seen that the base 10 logarithm function, $y = \log_{10} x$, is used in many applications. Base 10 logarithms are called **common logarithms**. Common logarithms are usually written without the subscript 10.

$$\log_{10} x = \log x, x > 0$$

Most scientific calculators have a LOG key for evaluating common logarithms.

Example 1 Find Common Logarithms

Use a calculator to evaluate each expression to the nearest ten-thousandth.

a. log 5

KEYSTROKES: LOG 5 ENTER .6989700043

$\log 5 \approx 0.6990$

b. log 0.3

KEYSTROKES: LOG 0.3 ENTER −.5228787453

$\log 0.3 \approx -0.5229$

▶ **Guided**Practice

1A. log 7 **0.8451** **1B.** log 0.5 **−0.3010**

The common logarithms of numbers that differ by integral powers of ten are closely related. Remember that a logarithm is an exponent. For example, in the equation $y = \log x$, y is the power to which 10 is raised to obtain the value of x.

$\log x = y$	$\rightarrow$ means $\rightarrow$	$10^y = x$
$\log 1 = 0$	since	$10^0 = 1$
$\log 10 = 1$	since	$10^1 = 10$
$\log 10^m = m$	since	$10^m = 10^m$

Lesson 7-6 Resources

Resource	Approaching Level **AL**	On Level **OL**	Beyond Level **BL**	English Learners **ELL**
Teacher Edition		▪ Differentiated Instruction, pp. 494, 498	▪ Differentiated Instruction, pp. 494, 498	
Chapter Resource Masters	▪ Study Guide and Intervention, pp. 40–41 ▪ Skills Practice, p. 42 ▪ Practice, p. 43 ▪ Word Problem Practice, p. 44	▪ Study Guide and Intervention, pp. 40–41 ▪ Skills Practice, p. 42 ▪ Practice, p. 43 ▪ Word Problem Practice, p. 44 ▪ Enrichment, p. 45 ▪ Graphing Calculator Activity, p. 46	▪ Practice, p.43 ▪ Word Problem Practice, p. 44 ▪ Enrichment, p. 45	▪ Study Guide and Intervention, pp. 40–41 ▪ Skills Practice, p. 42 ▪ Practice, p. 43 ▪ Word Problem Practice, p. 44
Other	▪ 5-Minute Check 7-6 ▪ Study Notebook	▪ 5-Minute Check 7-6 ▪ Study Notebook	▪ 5-Minute Check 7-6 ▪ Study Notebook	▪ 5-Minute Check 7-6 ▪ Study Notebook

Common logarithms are used in the measurement of sound. Soft recorded music is about 36 decibels (dB).

Real-World Example 2 Solve Logarithmic Equations

ROCK CONCERT The loudness L, in decibels, of a sound is $L = 10 \log \frac{I}{m}$, where I is the intensity of the sound and m is the minimum intensity of sound detectable by the human ear. Residents living several miles from a concert venue can hear the music at an intensity of 66.6 decibels. How many times the minimum intensity of sound detectable by the human ear was this sound, if m is defined to be 1?

$L = 10 \log \frac{I}{m}$	Original equation
$66.6 = 10 \log \frac{I}{1}$	Replace L with 66.6 and m with 1.
$6.66 = \log I$	Divide each side by 10 and simplify.
$I = 10^{6.66}$	Exponential form
$I \approx 4{,}570{,}882$	Use a calculator.

The sound heard by the residents was approximately 4,570,000 times the minimum intensity of sound detectable by the human ear.

GuidedPractice

2. EARTHQUAKES The amount of energy E in ergs that an earthquake releases is related to its Richter scale magnitude M by the equation $\log E = 11.8 + 1.5M$. Use the equation to find the amount of energy released by the 2004 Sumatran earthquake, which measured 9.0 on the Richter scale and led to a tsunami. **about 2×10^{25} ergs**

Real-World Link
Acoustical Engineer
Acoustical engineers are concerned with reducing unwanted sounds, noise control, and making useful sounds. Examples of useful sounds are ultrasound, sonar, and sound reproduction. Employment in this field requires a minimum of a bachelor's degree.

If both sides of an exponential equation cannot easily be written as powers of the same base, you can solve by taking the logarithm of each side.

Example 3 Solve Exponential Equations Using Logarithms

Solve $4^x = 19$. Round to the nearest ten-thousandth.

$4^x = 19$	Original equation
$\log 4^x = \log 19$	Property of Equality for Logarithmic Functions
$x \log 4 = \log 19$	Power Property of Logarithms
$x = \frac{\log 19}{\log 4}$	Divide each side by log 4.
$x \approx 2.1240$	Use a calculator.

The solution is approximately 2.1240.

CHECK You can check this answer graphically by using a graphing calculator. Graph the line $y = 4^x$ and the line $y = 19$. Then use the **CALC** menu to find the intersection of the two graphs. The intersection is very close to the answer that was obtained algebraically. ✓

Intersection
X=2.1239638 Y=19

[−10, 10] scl: 1 by [−5, 25] scl: 1

GuidedPractice

3A. $3^x = 15$ ≈2.4650   **3B.** $6^x = 42$ ≈2.0860

---

1 Common Logarithms
Example 1 shows how to use a calculator to evaluate a common logarithm. **Example 2** shows how to use the inverse of logarithms, or exponentiation, to solve a logarithmic equation that models a real-world application. **Example 3** shows how to solve exponential equations using logarithms. **Example 4** shows how to use logarithms to solve exponential inequalities.

Formative Assessment
Use the Guided Practice Exercises after each example to determine students' understanding of concepts.

Additional Examples

1 Use a calculator to evaluate each expression to the nearest ten-thousandth.
   a. log 6 about 0.7782
   b. log 0.35 about −0.4559

2 **JET ENGINES** Refer to Example 2. The sound of a jet engine can reach a loudness of 125 decibels. How many times the minimum intensity of audible sound is this? $10^{12.5}$, or about 3×10^{12}, which is 3 trillion

3 Solve $5^x = 62$. about 2.5643

Additional Examples also in Interactive Classroom PowerPoint® Presentations

IWB **Interactive White Board**
READY

Hulton Archive/Getty Images

Additional Example

4 Solve $3^{7x} > 2^{5x-3}$. Round to the nearest ten-thousandth.

$\{x \mid x > -0.4922\}$

Focus on Mathematical Content

Common Logarithms Base 10 logarithms are called *common logarithms*. When the base of a logarithm is not shown, the base is assumed to be 10. When you have a logarithmic expression of any base, you can evaluate it using the Change of Base Formula to translate the expression into one that involves common logarithms.

2 Change of Base Formula

Example 5 shows how to express a logarithmic expression of any base in terms of common logarithms.

Tips for New Teachers

Reasoning As you discuss the Change of Base Formula, point out that the base *b* that students are changing to does not have to be 10. Any base could be used; however, 10 is the most common because this allows for the logarithms to be evaluated with a calculator. When the base of a logarithm is not shown, the base is assumed to be 10.

Teach with Tech

Interactive Whiteboard
Display a template such as $\log_\Box\Box = \dfrac{\log_b\Box}{\log_b\Box}$. Drag the values from the original logarithm to demonstrate the Change of Base Formula.

The same strategies that are used to solve exponential equations can be used to solve exponential inequalities.

StudyTip

Solving Inequalities Remember that the direction of an inequality must be switched if each side is multiplied or divided by a negative number. Since $5\log 3 - \log 7 > 0$, the inequality does not change.

Example 4 Solve Exponential Inequalities Using Logarithms

Solve $3^{5y} < 7^{y-2}$. Round to the nearest ten-thousandth.

$3^{5y} < 7^{y-2}$	Original inequality
$\log 3^{5y} < \log 7^{y-2}$	Property of Inequality for Logarithmic Functions
$5y \log 3 < (y-2) \log 7$	Power Property of Logarithms
$5y \log 3 < y \log 7 - 2 \log 7$	Distributive Property
$5y \log 3 - y \log 7 < -2 \log 7$	Subtract $y \log 7$ from each side.
$y(5 \log 3 - \log 7) < -2 \log 7$	Distributive Property
$y < \dfrac{-2 \log 7}{5 \log 3 - \log 7}$	Divide each side by $5 \log 3 - \log 7$.
$\{y \mid y < -1.0972\}$	Use a calculator.

CHECK Test $y = -2$.

$3^{5y} < 7^{y-2}$	Original inequality
$3^{5(-2)} \overset{?}{<} 7^{(-2)-2}$	Replace y with -2.
$3^{-10} \overset{?}{<} 7^{-4}$	Simplify.
$\dfrac{1}{59,049} < \dfrac{1}{2401}$ ✓	Negative Exponent Property

GuidedPractice

Solve each inequality. Round to the nearest ten-thousandth.

4A. $3^{2x} \geq 6^{x+1}$ $\{x \mid x \geq 4.4190\}$

4B. $4^y < 5^{2y+1}$ $\{y \mid y > -0.8782\}$

---

2 **Change of Base Formula** The **Change of Base Formula** allows you to write equivalent logarithmic expressions that have different bases.

KeyConcept Change of Base Formula

Symbols For all positive numbers a, b, and n, where $a \neq 1$ and $b \neq 1$,

$$\log_a n = \frac{\log_b n}{\log_b a}. \quad \begin{array}{l}\leftarrow \text{log base } b \text{ of original number} \\ \leftarrow \text{log base } b \text{ of old base}\end{array}$$

Example $\log_3 11 = \dfrac{\log_{10} 11}{\log_{10} 3}$

To prove this formula, let $\log_a n = x$.

$a^x = n$	Definition of logarithm
$\log_b a^x = \log_b n$	Property of Equality for Logarithmic Functions
$x \log_b a = \log_b n$	Power Property of Logarithms
$x = \dfrac{\log_b n}{\log_b a}$	Divide each side by $\log_b a$.
$\log_a n = \dfrac{\log_b n}{\log_b a}$	Replace x with $\log_a n$.

Math HistoryLink

John Napier (1550–1617)
John Napier was a Scottish mathematician and theologian who began the use of logarithms to aid in calculations. He is also known for popularizing the use of the decimal point.

DifferentiatedInstruction OL BL

Logical Learners Ask students to recall that an equation like $4^x = 19$ from Example 3 could be written in logarithmic form as $\log_4 19 = x$. Although this logarithm cannot be directly evaluated, the Change of Base Formula can be used to give the correct result of $x \approx 2.1234$.

The Change of Base Formula makes it possible to evaluate a logarithmic expression of any base by translating the expression into one that involves common logarithms.

Example 5 Change of Base Formula

Express $\log_3 20$ in terms of common logarithms. Then round to the nearest ten-thousandth.

$$\log_3 20 = \frac{\log_{10} 20}{\log_{10} 3} \qquad \text{Change of Base Formula}$$

$$\approx 2.7268 \qquad \text{Use a calculator.}$$

▶ Guided Practice $\dfrac{\log_{10} 8}{\log_{10} 6} \approx 1.1606$

5. Express $\log_6 8$ in terms of common logarithms. Then round to the nearest ten-thousandth.

Check Your Understanding

◯ = Step-by-Step Solutions begin on page R14.

Example 1   Use a calculator to evaluate each expression to the nearest ten-thousandth.

1. $\log 5$ 0.6990   **2.** $\log 21$ 1.3222   **3.** $\log 0.4$ −0.3979   **4.** $\log 0.7$ −0.1549

Example 2   **5. SCIENCE** The amount of energy E in ergs that an earthquake releases is related to its Richter scale magnitude M by the equation $\log E = 11.8 + 1.5M$. Use the equation to find the amount of energy released by the 1960 Chilean earthquake, which measured 8.5 on the Richter scale. 3.55×10^{24} ergs

Example 3   Solve each equation. Round to the nearest ten-thousandth.

6. $6^x = 40$ 2.0588   **7.** $2.1^{a+2} = 8.25$ 0.8442   **8.** $7^{x^2} = 20.42$ ±1.2451   **9** $11^{b-3} = 5^b$ 9.1237

Example 4   Solve each inequality. Round to the nearest ten-thousandth.

10. $5^{4n} > 33$ $\{n \mid n > 0.5431\}$ **11.** $6^{p-1} \le 4^p$ $\{p \mid p \le 4.4190\}$

Example 5   Express each logarithm in terms of common logarithms. Then approximate its value to the nearest ten-thousandth. **12–15. See margin.**

12. $\log_3 7$ **13.** $\log_4 23$ **14.** $\log_9 13$ **15.** $\log_2 5$

Practice and Problem Solving

Extra Practice is on page R7.

Example 1   Use a calculator to evaluate each expression to the nearest ten-thousandth.

16. $\log 3$ 0.4771   **17.** $\log 11$ 1.0414   **18.** $\log 3.2$ 0.5051

19. $\log 8.2$ 0.9138   **20.** $\log 0.9$ −0.0458   **21.** $\log 0.04$ −1.3979

Example 2   **22. CCSS SENSE-MAKING** Loretta had a new muffler installed on her car. The noise level of the engine dropped from 85 decibels to 73 decibels.
  a. How many times the minimum intensity of sound detectable by the human ear was the car with the old muffler, if m is defined to be 1? about 316,227,766 times
  b. How many times the minimum intensity of sound detectable by the human ear is the car with the new muffler? Find the percent of decrease of the intensity of the sound with the new muffler. about 19,952,623 times; about 93.7%

Differentiated Homework Options

Level	Assignment	Two-Day Option	
AL Basic	16–38, 68, 70–92	17–37 odd, 73–76	16–38 even, 68, 70–72, 77–92
OL Core	17–39 odd, 40, 41–65 odd, 66–68, 70–92	16–38, 73–76	39–68, 70–72, 77–92
BL Advanced	39–86, (optional: 87–92)		

---

5 Express $\log_5 140$ in terms of common logarithms. Then round to the nearest ten-thousandth.

$$\log_5 140 = \frac{\log_{10} 140}{\log_{10} 5};$$

$$\log_5 140 \approx 3.0704$$

3 Practice

Formative Assessment
Use Exercises 1–15 to check for understanding.

Use the chart at the bottom of this page to customize assignments for your students.

CCSS **Teaching the Mathematical Practices**

Sense-Making Mathematically proficient students start by explaining to themselves the meaning of a problem and looking for entry points to its solution. They analyze givens, constraints, relationships, and goals, and make conjectures about the form and meaning of the solution and plan a solution pathway rather than simply jumping into a solution attempt.

Additional Answers

12. $\dfrac{\log 7}{\log 3} \approx 1.7712$

13. $\dfrac{\log 23}{\log 4} \approx 2.2618$

14. $\dfrac{\log 13}{\log 9} \approx 1.1674$

15. $\dfrac{\log 5}{\log 2} \approx 2.3219$

Study Guide and Intervention

AL OL ELL

Practice

AL OL BL ELL

Word Problem Practice

AL OL BL ELL

Example 3   Solve each equation. Round to the nearest ten-thousandth.

23. $8^x = 40$ **1.7740**

24. $5^x = 55$ **2.4899**

25. $2.9^{a-4} = 8.1$ **5.9647**

26. $9^{b-1} = 7^b$ **8.7429**

27. $13^{x^2} = 33.3$ **±1.1691**

28. $15^{x^2} = 110$ **±1.3175**

Example 4   Solve each inequality. Round to the nearest ten-thousandth.

29. $6^{3n} > 36$ $\{n \mid n > 0.6667\}$

30. $2^{4x} \le 20$ $\{x \mid x \le 1.0805\}$

31. $3^{y-1} \le 4^y$ $\{y \mid y \ge -3.8188\}$

32. $5^{p-2} \ge 2^p$ $\{p \mid p \ge 3.5129\}$

Example 5   Express each logarithm in terms of common logarithms. Then approximate its value to the nearest ten-thousandth.

33. $\log_7 18$ $\dfrac{\log 18}{\log 7} \approx 1.4854$

34. $\log_5 31$ $\dfrac{\log 31}{\log 5} \approx 2.1337$

35. $\log_2 16$ $\dfrac{\log 16}{\log 2} = 4$

36. $\log_4 9$ $\dfrac{\log 9}{\log 4} \approx 1.5850$

37. $\log_3 11$ $\dfrac{\log 11}{\log 3} \approx 2.1827$

38. $\log_6 33$ $\dfrac{\log 33}{\log 6} \approx 1.9514$

B   **39**   **PETS** The number n of pet owners in thousands after t years can be modeled by $n = 35[\log_4 (t + 2)]$. Let $t = 0$ represent 2000. Use the Change of Base Formula to answer the following questions.

   a. How many pet owners were there in 2010?   **62,737 owners**

   b. How long until there are 80,000 pet owners? When will this occur?   **2022**

40. **CCSS PRECISION** Five years ago the grizzly bear population in a certain national park was 325. Today it is 450. Studies show that the park can support a population of 750.

   a. What is the average annual rate of growth in the population if the grizzly bears reproduce once a year?   **0.067 or 6.7%**

   b. How many years will it take to reach the maximum population if the population growth continues at the same average rate?   **8 yr**

Solve each equation or inequality. Round to the nearest ten-thousandth.

41. $3^x = 40$ **3.3578**

42. $5^{3p} = 15$ **0.5609**

43. $4^{n+2} = 14.5$ **−0.0710**

44. $8^{z-4} = 6.3$ **4.8851**

45. $7.4^{n-3} = 32.5$ **4.7393**

46. $3.1^{y-5} = 9.2$ **6.9615**

47. $5^x \ge 42$ $\{x \mid x \ge 2.3223\}$

48. $9^{2a} < 120$ $\{a \mid a < 1.0894\}$

49. $3^{4x} \le 72$ $\{x \mid x \le 0.9732\}$

50. $7^{2n} > 52^{4n+3}$ $\{n \mid n < -0.9950\}$

51. $6^p \le 13^{5-p}$ $\{p \mid p \le 2.9437\}$

52. $2^{y+3} \ge 8^{3y}$ $\{y \mid y \le 0.3750\}$

Express each logarithm in terms of common logarithms. Then approximate its value to the nearest ten-thousandth.

53. $\log_4 12$ $\dfrac{\log 12}{\log 4} \approx 1.7925$

54. $\log_3 21$ $\dfrac{\log 21}{\log 3} \approx 2.7712$

55. $\log_8 2$ $\dfrac{\log 2}{\log 8} \approx 0.3333$

56. $\log_6 7$ $\dfrac{\log 7}{\log 6} \approx 1.0860$

57. $\log_5 (2.7)^2$ $\dfrac{\log 7.29}{\log 5} \approx 1.2343$

58. $\log_7 \sqrt{5}$ $\dfrac{\log \sqrt{5}}{\log 7} \approx 0.4135$

59. **MUSIC** A musical cent is a unit in a logarithmic scale of relative pitch or intervals. One octave is equal to 1200 cents. The formula $n = 1200\left(\log_2 \dfrac{a}{b}\right)$ can be used to determine the difference in cents between two notes with frequencies a and b.

   a. Find the interval in cents when the frequency changes from 443 Hertz (Hz) to 415 Hz.   **113.03 cents**

   b. If the interval is 55 cents and the beginning frequency is 225 Hz, find the final frequency.   **about 218 Hz**

Enrichment

OL BL

CCSS **Teaching the Mathematical Practices**

Precision Mathematically proficient students try to use clear definitions in their reasoning, calculate accurately and efficiently, and make explicit use of definitions.

Solve each equation. Round to the nearest ten-thousandth. **61.** $\pm\sqrt{5} \approx \pm 2.2361$

60. $10^{x^2} = 60$ ± 1.3335 **61** $4^{x^2-3} = 16$ **62.** $9^{6y-2} = 3^{3y+1}$ 0.5556

63. $8^{2x-4} = 4^{x+1}$ 3.5   **64.** $16^x = \sqrt{4^{x+3}}$ 1   **65.** $2^y = \sqrt{3^{y-1}}$ −3.8188

 66. ENVIRONMENTAL SCIENCE An environmental engineer is testing drinking water wells in coastal communities for pollution, specifically unsafe levels of arsenic. The safe standard for arsenic is 0.025 parts per million (ppm). Also, the pH of the arsenic level should be less than 9.5. The formula for hydrogen ion concentration is pH $= -\log H$. (*Hint:* 1 kilogram of water occupies approximately 1 liter. 1 ppm = 1 mg/kg.)

  a. Suppose the hydrogen ion concentration of a well is 1.25×10^{-11}. Should the environmental engineer be worried about too high an arsenic content? **yes; 10.9 > 9.5**

  b. The environmental engineer finds 1 milligram of arsenic in a 3 liter sample, is the well safe? **no**

  c. What is the hydrogen ion concentration that meets the troublesome pH level of 9.5? 3.16×10^{-10}

67. MULTIPLE REPRESENTATIONS In this problem, you will solve the exponential equation $4^x = 13$. **a. The solution is between 1.8 and 1.9.**

  a. Tabular Enter the function $y = 4^x$ into a graphing calculator, create a table of values for the function, and scroll through the table to find x when $y = 13$.

  b. Graphical Graph $y = 4^x$ and $y = 13$ on the same screen. Use the **intersect** feature to find the point of intersection. **(1.85, 13)**

  c. Numerical Solve the equation algebraically. Do all of the methods produce the same result? Explain why or why not. **Yes; all methods produce the solution of 1.85. They all should produce the same result because you are starting with the same equation. If they do not, then an error was made.**

H.O.T. Problems   *Use Higher-Order Thinking Skills*

68. CCSS CRITIQUE Sam and Rosamaria are solving $4^{3p} = 10$. Is either of them correct? Explain your reasoning.

Rosamaria; Sam forgot to bring the 3 down from the exponent when he took the log of each side.

69. CHALLENGE Solve $\log_{\sqrt{a}} 3 = \log_a x$ for x and explain each step. **See margin.**

70. REASONING Write $\dfrac{\log_5 9}{\log_5 3}$ as a single logarithm. $\dfrac{\log_5 9}{\log_5 3} = \log_3 9$

71. PROOF Find the values of $\log_3 27$ and $\log_{27} 3$. Make and prove a conjecture about the relationship between $\log_a b$ and $\log_b a$. **See margin.**

72. WRITING IN MATH Explain how exponents and logarithms are related. Include examples like how to solve a logarithmic equation using exponents and how to solve an exponential equation using logarithms.

72. Logarithms are exponents. To solve logarithmic equations, write each side of the equation using exponents and solve by using the Inverse Property of Exponents and Logarithms. To solve exponential equations, use the Property of Equality for Logarithmic Functions and the Power Property of Logarithms.

71. $\log_3 27 = 3$ and $\log_{27} 3 = \dfrac{1}{3}$; Conjecture: $\log_a b = \dfrac{1}{\log_b a}$

Proof:

$\log_a b \overset{?}{=} \dfrac{1}{\log_b a}$ Original statement

$\dfrac{\log_b b}{\log_b a} \overset{?}{=} \dfrac{1}{\log_b a}$ Change of Base Formula

$\dfrac{1}{\log_b a} = \dfrac{1}{\log_b a}$ Inverse Property of Exponents and Logarithms

---

Multiple Representations

In Exercise 67, students use a graphing calculator and algebra to solve an exponential equation algebraically and graphically and compare the results.

CCSS Teaching the Mathematical Practices

Critique Mathematically proficient students are also able to compare the effectiveness of two plausible arguments, distinguish correct logic or reasoning from that which is flawed, and—if there is a flaw in an argument—explain what it is.

WatchOut!

Error Analysis For Exercise 68, point out to students that in the Power Property of Logarithms, $\log_b m^p = p \log_b m$, the p represents the entire exponent of m.

Additional Answers

69.

$\log_{\sqrt{a}} 3 = \log_a x$ Original equation

$\dfrac{\log_a 3}{\log_a \sqrt{a}} = \log_a x$ Change of Base Formula

$\dfrac{\log_a 3}{\frac{1}{2}} = \log_a x$ $\sqrt{a} = a^{\frac{1}{2}}$

$2 \log_a 3 = \log_a x$ Multiply numerator and denominator by 2.

$\log_a 3^2 = \log_a x$ Power Property of Logarithms

$3^2 = x$ Property of Equality for Logarithmic Functions

$9 = x$ Simplify.

4 Assess

Yesterday's News Have students write how knowing the properties of logarithms has helped them with solving the exponential equations and inequalities in today's lesson.

Formative Assessment

Check for student understanding of Lessons 7-5 and 7-6.

Quiz 3, p. 64

Additional Answers

74a. $3d + 2v + 2m = 16.29, d + 3v + 4m = 19.84, 2d + v + m = 9.14$

74b. documentaries: $1.99, video games: $2.79, movies: $2.37

Standardized Test Practice

73. Which expression represents $f[g(x)]$ if $f(x) = x^2 + 4x + 3$ and $g(x) = x - 5$? **B**

A $x^2 + 4x - 2$

B $x^2 - 6x + 8$

C $x^2 - 9x + 23$

D $x^2 - 14x + 6$

74. EXTENDED RESPONSE Colleen rented 3 documentaries, 2 video games, and 2 movies. The charge was $16.29. The next week, she rented 1 documentary, 3 video games, and 4 movies for a total charge of $19.84. The third week she rented 2 documentaries, 1 video game, and 1 movie for a total charge of $9.14.

a. Write a system of equations to determine the cost to rent each item. **a, b. See margin.**

b. What is the cost to rent each item?

75. GEOMETRY If the surface area of a cube is increased by a factor of 9, what is the change in the length of the sides of the cube? **G**

F The length is 2 times the original length.

G The length is 3 times the original length.

H The length is 6 times the original length.

J The length is 9 times the original length.

76. SAT/ACT Which of the following *most* accurately describes the translation of the graph $y = (x + 4)^2 - 3$ to the graph of $y = (x - 1)^2 + 3$? **E**

A down 1 and to the right 3

B down 6 and to the left 5

C up 1 and to the left 3

D up 1 and to the right 3

E up 6 and to the right 5

Spiral Review

Solve each equation. Check your solutions. (Lesson 7-5)

77. $\log_5 7 + \frac{1}{2}\log_5 4 = \log_5 x$ **14**

78. $2\log_2 x - \log_2 (x + 3) = 2$ **6**

79. $\log_6 48 - \log_6 \frac{16}{5} + \log_6 5 = \log_6 5x$ **15**

80. $\log_{10} a + \log_{10} (a + 21) = 2$ **4**

Solve each equation or inequality. (Lesson 7-4)

81. $\log_4 x = \frac{1}{2}$ **2**

82. $\log_{81} 729 = x$ **$\frac{3}{2}$**

83. $\log_8 (x^2 + x) = \log_8 12$ **−4, 3**

84. $\log_8 (3y - 1) < \log_8 (y + 5)$ **$\left\{y \mid \frac{1}{3} < y < 3\right\}$**

85. SAILING The area of a triangular sail is $16x^4 - 60x^3 - 28x^2 + 56x - 32$ square meters. The base of the triangle is $x - 4$ meters. What is the height of the sail? (Lesson 5-2) **$32x^3 + 8x^2 - 24x + 16$**

86. HOME REPAIR Mr. Turner is getting new locks installed. The locksmith charges $85 for the service call, $25 for each door, and $30 for each lock. (Lesson 2-4)

a. Write an equation that represents the cost for x number of doors. **$y = 85 + 55x$**

b. Mr. Turner wants the front, side, back, and garage door locks changed. How much will this cost? **$305**

Skills Review

Write an equivalent exponential equation.

87. $\log_2 5 = x$ **$2^x = 5$**

88. $\log_4 x = 3$ **$4^3 = x$**

89. $\log_5 25 = 2$ **$5^2 = 25$**

90. $\log_7 10 = x$ **$7^x = 10$**

91. $\log_6 x = 4$ **$6^4 = x$**

92. $\log_4 64 = 3$ **$4^3 = 64$**

 498 | Lesson 7-6 | Common Logarithms

DifferentiatedInstruction **OL** **BL**

Extension Remind students that the formula $A = P\left(1 + \frac{r}{n}\right)^{nt}$ can be used to find the final amount of an investment using compound interest. Have them use logarithms to find the number of years t that it will take for an investment of $5000 to grow to $8000 at an interest rate of 5% compounded monthly. about 9.4 years

7-6 Graphing Technology Lab
Solving Logarithmic Equations and Inequalities

You have solved logarithmic equations algebraically. You can also solve logarithmic equations by graphing or by using a table. The TI-83/84 Plus has $y = \log_{10} x$ as a built-in function. Enter $\boxed{Y=}$ $\boxed{LOG}$ $\boxed{X,T,\theta,n}$ $\boxed{GRAPH}$ to view this graph. To graph logarithmic functions with bases other than 10, you must use the Change of Base Formula, $\log_a n = \dfrac{\log_b n}{\log_b a}$.

$y = \log_{10} x$

[−2, 8] scl: 1 by [−10, 10] scl: 1

CCSS Common Core State Standards
Content Standards
A.REI.11 Explain why the x-coordinates of the points where the graphs of the equations $y = f(x)$ and $y = g(x)$ intersect are the solutions of the equation $f(x) = g(x)$; find the solutions approximately, e.g., using technology to graph the functions, make tables of values, or find successive approximations. Include cases where $f(x)$ and/or $g(x)$ are linear, polynomial, rational, absolute value, exponential, and logarithmic functions.

Activity 1

Solve $\log_2 (6x − 8) = \log_3 (20x + 1)$.

Step 1 Graph each side of the equation.

Graph each side of the equation as a separate function. Enter $\log_2 (6x − 8)$ as **Y1** and $\log_3 (20x + 1)$ as **Y2**. Then graph the two equations.

KEYSTROKES: $\boxed{Y=}$ $\boxed{LOG}$ 6 $\boxed{X,T,\theta,n}$ $\boxed{-}$ 8 $\boxed{)}$ $\boxed{\div}$ $\boxed{LOG}$ 2 $\boxed{)}$ $\boxed{ENTER}$ $\boxed{LOG}$ 20 $\boxed{X,T,\theta,n}$ $\boxed{+}$ 1 $\boxed{)}$ $\boxed{\div}$ $\boxed{LOG}$ 3 $\boxed{)}$ $\boxed{GRAPH}$

[−2, 8] scl: 1 by [−2, 8] scl: 1

Step 2 Use the **intersect** feature.

Use the **intersect** feature on the **CALC** menu to approximate the ordered pair of the point at which the curves intersect.

The calculator screen shows that the x-coordinate of the point at which the curves intersect is 4. Therefore, the solution of the equation is 4.

Intersection
X=4   Y=4

[−2, 8] scl: 1 by [−2, 8] scl: 1

Step 3 Use the **TABLE** feature.

Examine the table to find the x-value for which the y-values for the graphs are equal. At $x = 4$, both functions have a y-value of 4. Thus, the solution of the equation is 4.

You can use a similar procedure to solve logarithmic inequalities using a graphing calculator.

(continued on the next page)

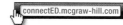

connectED.mcgraw-hill.com   **499**

1 Focus

Objective Use a graphing calculator to solve exponential and logarithmic equations and inequalities.

Materials for Each Group
- TI-83/84 Plus or other graphing calculator

Teaching Tip
Point out that the operation **log (** can be used with numbers or expressions. Remind students that it is always a good idea to use closing parentheses at the end of the expression.

2 Teach

Working in Cooperative Groups
Put students in groups of two or three, mixing abilities. Then have groups complete Activities 1 and 2 and Exercises 1–4.

Activity 1
- Before discussing the Activity, have students use the Change of Base Formula to express each side of the equation in terms of common logarithms.

- Be sure students understand how to read the table feature to find the x-value for which the y-values are equal.

- Have students substitute the solution into the original equation to verify that it is correct.

Activity 2

- Before discussing Activity 2, have students use the Change of Base Formula to express each side of the inequality in terms of common logarithms.

- Make sure students understand why the inequality needs to be rewritten as a system of inequalities.

- Be sure students understand how to read the table feature to find the x-value for which the y-values are undefined as well as the x-value for which the y-values are equal.

Practice Have students complete Exercises 5–8.

3 Assess

Formative Assessment

In Exercise 8, check that students record the inequalities in the solution set correctly. In particular, students must include the fact that x must be greater than 0.

From Concrete to Abstract

Have students explain how the solution set for Activity 2 would change if the inequality were $\log_4 (10x + 1) > \log_5 (16 + 6x)$.

Graphing Technology Lab
Solving Logarithmic Equations and Inequalities *Continued*

Activity 2

Solve $\log_4 (10x + 1) < \log_5 (16 + 6x)$.

Step 1 Enter the inequalities.

Rewrite the problem as a system of inequalities.

The first inequality is $\log_4 (10x + 1) < y$ or $y > \log_4 (10x + 1)$. Since this inequality includes the *greater than* symbol, shade above the curve.

First enter the boundary and then use the arrow and ENTER keys to choose the shade above icon, ▮.

The second inequality is $y < \log_5 (16 + 6x)$. Shade below the curve since this inequality contains *less than*.

KEYSTROKES: Y= ◄ ◄ ENTER ENTER ► ► LOG 10 X,T,θ,*n* + 1) ÷ LOG 4) ENTER ◄ ◄ ENTER ENTER ENTER ► ► LOG 16 + 6 X,T,θ,*n*) ÷ LOG 5)

Step 2 Graph the system.

KEYSTROKES: GRAPH

The left boundary of the solution set is where the first inequality is undefined. It is undefined for $10x + 1 \le 0$.

$$10x + 1 \le 0$$
$$10x \le -1$$
$$x \le -\frac{1}{10}$$

Use the calculator's **intersect** feature to find the right boundary. You can conclude that the solution set is $\{x \mid -0.1 < x < 1.5\}$.

[−2, 4] scl: 1 by [−2, 4] scl: 1

Step 3 Use the **TABLE** feature to check your solution.

Start the table at −0.1 and show x-values in increments of 0.1. Scroll through the table.

KEYSTROKES: 2nd [TBLSET] −0.1 ENTER 0.1 ENTER 2nd [TABLE]

The table confirms the solution of the inequality is $\{x \mid -0.1 < x < 1.5\}$.

Exercises

Solve each equation or inequality. Check your solution.

1. $\log_2 (3x + 2) = \log_3 (12x + 3)$ **0, 2**

2. $\log_6 (7x + 1) = \log_4 (4x - 4)$ **5**

3. $\log_2 3x = \log_3 (2x + 2)$ **about 0.7**

4. $\log_{10} (1 - x) = \log_5 (2x + 5)$ **about −1.5**

5. $\log_4 (9x + 1) > \log_3 (18x - 1)$ **$\{x \mid 0.06 < x < 0.17\}$**

6. $\log_3 (3x - 5) \ge \log_3 (x + 7)$ **$\{x \mid x \ge 6\}$**

7. $\log_5 (2x + 1) < \log_4 (3x - 2)$ **$\{x \mid x > 2\}$**

8. $\log_2 2x \le \log_4 (x + 3)$ **$\{x \mid 0 < x \le 1\}$**

500 | Extend 7-6 | Graphing Technology Lab: Solving Logarithmic Equations and Inequalities

LESSON 7-7 Base *e* and Natural Logarithms

::Then	::Now	::Why?
• You worked with common logarithms.	**1** Evaluate expressions involving the natural base and natural logarithm. **2** Solve exponential equations and inequalities using natural logarithms.	• The St. Louis Gateway Arch in Missouri is in the form of an inverted catenary curve. A catenary curve directs the force of its weight along itself, so that: • if a rope or chain is hanging, it is pulled into that shape, and, • if a catenary is standing upright, it can support itself. The equation for the catenary curve involves *e*, a special number that appears throughout mathematics and science.

NewVocabulary
natural base, *e*
natural base
  exponential function
natural logarithm

Common Core State Standards

Content Standards
A.SSE.2 Use the structure of an expression to identify ways to rewrite it.

Mathematical Practices
7 Look for and make use of structure.

---

1 Base *e* and Natural Logarithms Like π and $\sqrt{2}$, the number *e* is an irrational number. The value of *e* is 2.71828… . It is referred to as the **natural base, *e*.** An exponential function with base *e* is called a **natural base exponential function**.

> **KeyConcept** Natural Base Functions
>
> The function $f(x) = e^x$ is used to model continuous exponential growth.
> The function $f(x) = e^{-x}$ is used to model continuous exponential decay.
>
> The inverse of a natural base exponential function is called the **natural logarithm**. This logarithm can be written as $\log_e x$, but is more often abbreviated as ln *x*.
>
>
>
> **Exponential Growth** **Exponential Decay**

You can write an equivalent base *e* exponential equation for a natural logarithmic equation by using the fact that $\ln x = \log_e x$.

$$\ln 4 = x \quad \rightarrow \quad \log_e 4 = x \quad \rightarrow \quad e^x = 4$$

> **Example 1** Write Equivalent Expressions
>
> Write each exponential equation in logarithmic form.
>
> **a.** $e^x = 8$
> $$e^x = 8 \;\rightarrow\; \log_e 8 = x$$
> $$\ln 8 = x$$
>
> **b.** $e^5 = x$
> $$e^5 = x \;\rightarrow\; \log_e x = 5$$
> $$\ln x = 5$$
>
> **GuidedPractice**
>
> **1A.** $e^x = 9$ $\ln 9 = x$
> **1B.** $e^7 = x$ $\ln x = 7$

connectED.mcgraw-hill.com **501**

William Gutmann/Corbis

1 Focus

VerticalAlignment

Before Lesson 7-7 Work with common logarithms.

Lesson 7-7 Evaluate expressions involving the natural base and natural logarithm. Solve exponential equations and inequalities using natural logarithms.

After Lesson 7-7 Analyze a situation modeled by an exponential function, and solve the problem.

2 Teach

Scaffolding Questions
Have students read the **Why?** section of the lesson.

Ask:
• What other arch-shaped structures have you seen? Sample answers: arches in fast-food advertising, arch bridges

• What other special numbers have you studied? π

Lesson 7-7 Resources

Resource	Approaching Level AL	On Level OL	Beyond Level BL	English Learners ELL
Teacher Edition	• Differentiated Instruction, p. 503	• Differentiated Instruction, p. 507	• Differentiated Instruction, p. 507	
Chapter Resource Masters	• Study Guide and Intervention, pp. 47–48 • Skills Practice, p. 49 • Practice, p. 50 • Word Problem Practice, p. 51	• Study Guide and Intervention, pp. 47–48 • Skills Practice, p. 49 • Practice, p. 50 • Word Problem Practice, p. 51 • Enrichment, p. 52	• Practice, p. 50 • Word Problem Practice, p. 51 • Enrichment, p. 52	• Study Guide and Intervention, pp. 47–48 • Skills Practice, p. 49 • Practice, p. 50 • Word Problem Practice, p. 51
Other	• 5-Minute Check 7-7 • Study Notebook	• 5-Minute Check 7-7 • Study Notebook	• 5-Minute Check 7-7 • Study Notebook	• 5-Minute Check 7-7 • Study Notebook

1 Base *e* and Natural Logarithms

Example 1 shows how to write an equivalent natural logarithmic equation for an exponential equation. **Example 2** shows how to write an equivalent exponential equation for a natural logarithmic equation. **Example 3** shows how to simplify expressions containing natural logarithms.

Formative Assessment

Use the Guided Practice exercises after each example to determine students' understanding of concepts.

Additional Examples

1 Write each exponential equation in logarithmic form.

  a. $e^x = 23$ $\ln 23 = x$

  b. $e^4 = x$ $\ln x = 4$

2 Write each logarithmic equation in exponential form.

  a. $\ln x \approx 1.2528$ $x \approx e^{1.2528}$

  b. $\ln 25 = x$ $25 = e^x$

3 Write each expression as a single logarithm.

  a. $4 \ln 3 + \ln 6$ $\ln 486$

  b. $2 \ln 3 + \ln 4 + \ln y$ $\ln 36y$

▶ **Additional Examples** also in Interactive Classroom PowerPoint® Presentations

  Interactive White Board READY

WatchOut!

Common Misconceptions Stress that *e* is a constant like π and not a variable like *x* or *y*.

You can also write an equivalent natural logarithm equation for a natural base *e* exponential equation.

$$e^x = 12 \quad \rightarrow \quad \log_e 12 = x \quad \rightarrow \quad \ln 12 = x$$

Example 2 Write Equivalent Expressions

Write each logarithmic equation in exponential form.

a. $\ln x \approx 0.7741$

  $\ln x \approx 0.7741 \quad \rightarrow \quad \log_e x = 0.7741$
                                                   $x \approx e^{0.7741}$

b. $\ln 10 = x$

  $\ln 10 = x \quad \rightarrow \quad \log_e 10 = x$
                                $10 = e^x$

▶ **Guided**Practice

2A. $\ln x \approx 2.1438$ $x = e^{2.1438}$ **2B.** $\ln 18 = x$ $18 = e^x$

The properties of logarithms you learned in Lesson 7-5 also apply to the natural logarithms. The logarithmic expressions below can be simplified into a single logarithmic term.

Example 3 Simplify Expressions with *e* and the Natural Log

StudyTip

Simplifying When you simplify logarithmic expressions, verify that the logarithm contains no operations and no powers.

Write each expression as a single logarithm.

a. $3 \ln 10 - \ln 8$

$3 \ln 10 - \ln 8 = \ln 10^3 - \ln 8$	Power Property of Logarithms
$= \ln \dfrac{10^3}{8}$	Quotient Property of Logarithms
$= \ln 125$	Simplify.
$= \ln 5^3$	$5^3 = 125$
$= 3 \ln 5$	Power Property of Logarithms

  CHECK Use a calculator to verify the solution.

    KEYSTROKES: 3 [LN] 10 [)] [−] [LN] 8 [)] [ENTER] 4.828313737

    KEYSTROKES: 3 [LN] 5 [)] [ENTER] 4.828313737 ✓

b. $\ln 40 + 2 \ln \frac{1}{2} + \ln x$

$\ln 40 + 2 \ln \frac{1}{2} + \ln x = \ln 40 + \ln \frac{1}{4} + \ln x$	Power Property of Logarithms
$= \ln \left(40 \cdot \frac{1}{4} \cdot x \right)$	Product Property of Logarithms
$= \ln 10x$	Simplify.

▶ **Guided**Practice

3A. $6 \ln 8 - 2 \ln 4$ $14 \ln 2$ **3B.** $2 \ln 5 + 4 \ln 2 + \ln 5y$ $\ln 2000y$

Because the natural base and natural log are inverse functions, they can be used to *undo* or eliminate each other.

$$e^{\ln x} = x \qquad\qquad \ln e^x = x$$

Teach with Tech

Blog Have students write a blog entry describing how natural logarithms and common logarithms are similar and different.

2 Equations and Inequalities with *e* and ln Equations and inequalities involving base *e* are easier to solve by using natural logarithms rather than by using common logarithms, because ln *e* = 1.

Example 4 Solve Base *e* Equations

Solve $4e^{-2x} - 5 = 3$. Round to the nearest ten-thousandth.

$4e^{-2x} - 5 = 3$	Original equation
$4e^{-2x} = 8$	Add 5 to each side.
$e^{-2x} = 2$	Divide each side by 4.
$\ln e^{-2x} = \ln 2$	Property of Equality for Logarithms
$-2x = \ln 2$	$\ln e^x = x$
$x = \dfrac{\ln 2}{-2}$	Divide each side by -2.
$x \approx -0.3466$	Use a calculator.

KEYSTROKES: LN 2) ÷ −2 ENTER −.34657359

StudyTip

Calculators Most calculators have an e^x and LN key for evaluating natural base and natural log expressions.

▶ **Guided**Practice

Solve each equation. Round to the nearest ten-thousandth.

4A. $3e^{4x} - 12 = 15$ **0.5493**   **4B.** $4e^{-x} + 8 = 17$ **−0.8109**

Just like the natural logarithm can be used to eliminate e^x, the natural base exponential function can eliminate ln *x*.

Example 5 Solve Natural Log Equations and Inequalities

Solve each equation or inequality. Round to the nearest ten-thousandth.

a. $3 \ln 4x = 24$

$3 \ln 4x = 24$	Original equation
$\ln 4x = 8$	Divide each side by 3.
$e^{\ln 4x} = e^8$	Property of Equality for Exponential Functions
$4x = e^8$	$e^{\ln x} = x$
$x = \dfrac{e^8}{4}$	Divide each side by 4.
$x \approx 745.2395$	Use a calculator.

b. $\ln (x - 8)^4 < 4$

$\ln (x - 8)^4 < 4$	Original equation
$e^{\ln (x - 8)^4} < e^4$	Write each side using exponents and base *e*.
$(x - 8)^4 < e^4$	$e^{\ln x} = x$
$x - 8 < e$	Property of Equality for Exponential Functions
$x < e + 8$	Add 8 to each side.
$x < 10.7183$	Use a calculator.

▶ **Guided**Practice

Solve each equation or inequality. Round to the nearest ten-thousandth.

5A. $5 \ln 6x = 8$ **0.8255**   **5B.** $\ln (2x - 3)^3 > 6$ $\{x \mid x > 5.1945\}$

Focus on Mathematical Content

Base *e* A logarithm with base *e* is called a natural logarithm, written as $\log_e x$ or ln *x*. The natural logarithmic function, $y = \ln x$, is the inverse of the natural base exponential function $y = e^x$.

2 Equations and Inequalities with *e* and ln

Example 4 shows how to use natural logarithms to solve an equation involving base *e*. **Example 5** shows how to solve natural logarithmic equations and inequalities. **Example 6** shows how to use the formula $A = Pe^{rt}$ to solve a real-world problem involving interest that is compounded continuously.

Additional Examples

4 Solve $3e^{-2x} + 4 = 10$. Round to four decimal places.
$x \approx -0.3466$

5 Solve each equation or inequality. Round to the nearest ten-thousandth.

a. $2 \ln 5x = 6$ 4.0171

b. $\ln (3x + 1)^2 > 8$
$\{x \mid x > 17.8661\}$

DifferentiatedInstruction

If ▶ some students mistakenly think that an equation like $4e^{-2x} - 5 = 3$ contains two variables,

Then ▶ point out that the letter *e* represents a constant, just as π does. Both *e* and π are irrational numbers, which cannot be expressed exactly with numerals. To help students avoid this confusion, have them highlight the variables in the equation with a marker.

Additional Example

6 SAVINGS Suppose you deposit $700 into an account paying 3% annual interest, compounded continuously.

a. What is the balance after 8 years? $889.87

b. How long will it take for the balance in your account to reach at least $1200? about 18 years

c. How much would have to be deposited in order to reach a balance of $1500 after 12 years? $1046.51

CCSS Teaching the Mathematical Practices

Reasoning Mathematically proficient students make sense of quantities and their relationships in problem situations. Remind students that rounding throughout calculations can result in inaccurate results.

Interest that is compounded continuously can be found using e.

KeyConcept Continuously Compounded Interest

Calculate continuously compounded interest using the following formula:

$$A = Pe^{rt},$$

where A is the amount in the account after t years, P is the principal amount invested, and r is the annual interest rate.

● **Real-World Example 6** Solve Base e Inequalities

FINANCIAL LITERACY When Angelina was born, her grandparents deposited $3000 into a college savings account paying 4% interest compounded continuously.

a. Assuming there are no deposits or withdrawals from the account, what will the balance be after 10 years?

$A = Pe^{rt}$	Continuous Compounding Formula
$= 3000e^{(0.04)(10)}$	$P = 3000$, $r = 0.04$, and $t = 10$
$= 3000e^{0.4}$	Simplify.
≈ 4475.47	Use a calculator.

The balance will be $4475.47.

b. How long will it take the balance to reach at least $10,000?

$A < Pe^{rt}$	Continuous Compounding Formula
$10{,}000 < 3000e^{(0.04)t}$	$P = 3000$, $r = 0.04$, and $A = 10{,}000$
$\dfrac{10}{3} < e^{0.04t}$	Divide each side by 3000.
$\ln \dfrac{10}{3} < \ln e^{0.04t}$	Property of Equality of Logarithms
$\ln \dfrac{10}{3} < 0.04t$	$\ln e^x = x$
$\dfrac{\ln \frac{10}{3}}{0.04} < t$	Divide each side by 0.04.
$30.099 < t$	Use a calculator.

It will take about 30 years to reach at least $10,000.

c. If her grandparents want Angelina to have $10,000 after 18 years, how much would they need to invest?

$10{,}000 = Pe^{(0.04)18}$	$A = 10{,}000$, $r = 0.04$, and $t = 18$
$\dfrac{10{,}000}{e^{0.72}} = P$	Divide each side by $e^{0.72}$.
$4867.52 \approx P$	Use a calculator.

They need to invest $4867.52.

Real-WorldLink
The average cost of tuition, room, and board at four-year public colleges is about $12,804 per year.

Source: National Center for Education Statistics

StudyTip
CCSS Reasoning In order to avoid any errors due to rounding, do not round until the very end of your calculations.

▶ **Guided**Practice

6. Use the information in Example 6 to answer the following.

A. If they invested $8000 at 3.75% interest compounded continuously, how much money would be in the account in 30 years? **$24,641.73**

B. If they could only deposit $10,000 in the account above, at what rate would the account need to grow in order for Angelina to have $30,000 in 18 years? **about 6.1%**

C. If Angelina's grandparents found an account that paid 5% compounded continuously and wanted her to have $30,000 after 18 years, how much would they need to deposit? **$12,197.09**

 504 | Lesson 7-7 | Base e and Natural Logarithms

Comstock Images/Getty Images

Examples 1–2 Write an equivalent exponential or logarithmic function.

1. $e^x = 30$ $\ln 30 = x$

2. $\ln x = 42$ $e^{42} = x$

3. $e^3 = x$ $\ln x = 3$

4. $\ln 18 = x$ $e^x = 18$

Example 3 Write each as a single logarithm.

5. $3 \ln 2 + 2 \ln 4$ $7 \ln 2$

6. $5 \ln 3 - 2 \ln 9$ $\ln 3$

7. $3 \ln 6 + 2 \ln 9$ $\ln 17496$

8. $3 \ln 5 + 4 \ln x$ $\ln 125x^4$

Example 4 Solve each equation. Round to the nearest ten-thousandth.

9. $5e^x - 24 = 16$ 2.0794

10. $-3e^x + 9 = 4$ 0.5108

11. $3e^{-3x} + 4 = 6$ 0.1352

12. $2e^{-x} - 3 = 8$ -1.7047

Example 5 Solve each equation or inequality. Round to the nearest ten-thousandth.

13. $\ln 3x = 8$ 993.6527

14. $-4 \ln 2x = -26$ 332.5708

15. $\ln (x + 5)^2 < 6$

16. $\ln (x - 2)^3 > 15$ $\{x \mid x > 150.4132\}$

17. $e^x > 29$ $\{x \mid x > 3.3673\}$

18. $5 + e^{-x} > 14$ $\{x \mid x < -2.1972\}$

15. $\{x \mid -25.0855 < x < 15.0855, x \neq -5\}$

Example 6 19. **SCIENCE** A virus is spreading through a computer network according to the formula $v(t) = 30e^{0.1t}$, where v is the number of computers infected and t is the time in minutes. How long will it take the virus to infect 10,000 computers? about 58 min

Practice and Problem Solving Extra Practice is on page R7.

Examples 1–2 Write an equivalent exponential or logarithmic function.

22. $0.25 = e^x$ 23. $5.4 = e^x$

20. $e^{-x} = 8$ $\ln 8 = -x$

21. $e^{-5x} = 0.1$ $\ln 0.1 = -5x$

22. $\ln 0.25 = x$

23. $\ln 5.4 = x$

24. $e^{x-3} = 2$ $\ln 2 = x - 3$

25. $\ln (x + 4) = 36$ $e^{36} = x + 4$

26. $e^{-2} = x^6$ $-2 = 6 \ln x$

27. $\ln e^x = 7$ $e^7 = e^x$

Example 3 Write each as a single logarithm.

28. $\ln 125 - 2 \ln 5$ $\ln 5$

29. $3 \ln 10 + 2 \ln 100$ $7 \ln 10$

30. $4 \ln \frac{1}{3} - 6 \ln \frac{1}{9}$ $-8 \ln \frac{1}{3}$

31. $7 \ln \frac{1}{2} + 5 \ln 2$ $-2 \ln 2$

32. $8 \ln x - 4 \ln 5$ $\ln \frac{x^8}{625}$

33. $3 \ln x^2 + 4 \ln 3$ $\ln 81x^6$

Example 4 Solve each equation. Round to the nearest ten-thousandth.

34. $6e^x - 3 = 35$ 1.8458

35. $4e^x + 2 = 180$ 3.7955

36. $3e^{2x} - 5 = -4$ -0.5493

37. $-2e^{3x} + 19 = 3$ 0.6931

38. $6e^{4x} + 7 = 4$ no solution

39. $-4e^{-x} + 9 = 2$ -0.5596

Examples 5–6 40. **SENSE-MAKING** The value of a certain car depreciates according to $v(t) = 18500e^{-0.186t}$, where t is the number of years after the car is purchased new.

a. What will the car be worth in 18 months? $13,996

b. When will the car be worth half of its original value? about 3.73 yr

c. When will the car be worth less than $1000? about 15.69 yr

45. $\{x \mid x < -239.8802 \text{ or } x > 239.8802\}$

Solve each inequality. Round to the nearest ten-thousandth. 46. $\{x \mid 6 < x \leq 26.0855\}$

41. $e^x \leq 8.7$ $\{x \mid x \leq 2.1633\}$

42. $e^x \geq 42.1$ $\{x \mid x \geq 3.7400\}$

43. $\ln (3x + 4)^3 > 10$ $\{x \mid x > 8.0105\}$

44. $4 \ln x^2 < 72$

45. $\ln (8x^4) > 24$ $\{x \mid -8103.0839 < x < 8103.0839, x \neq 0\}$

46. $-2 [\ln (x - 6)^{-1}] \leq 6$

Formative Assessment

Use Exercises 1–19 to check for understanding.

Use the chart at the bottom of this page to customize assignments for your students.

Teaching the Mathematical Practices

Sense-Making Mathematically proficient students start by explaining to themselves the meaning of a problem and looking for entry points to its solution. They analyze givens, constraints, relationships, and goals, and make conjectures about the form and meaning of the solution and plan a solution pathway rather than simply jumping into a solution attempt.

Tips for New Teachers

Reasonableness Encourage students to evaluate each solution to logarithmic equations and inequalities for reasonableness.

Differentiated Homework Options

Level	Assignment	Two-Day Option	
AL Basic	20–46, 59–82	21–45 odd, 63–66	20–46 even, 59–62, 67–82
OL Core	21–55 odd, 56–57, 59–82	20–46, 63–66	47–57, 59–62, 67–82
BL Advanced	47–76, (optional: 77–82)		

Study Guide and Intervention

AL OL ELL

Practice

AL OL BL ELL

Word Problem Practice

AL OL BL ELL

B **47** **FINANCIAL LITERACY** Use the formula for continuously compounded interest.

a. If you deposited $800 in an account paying 4.5% interest compounded continuously, how much money would be in the account in 5 years? **$1001.86**

b. How long would it take you to double your money? **about 15.4 yr**

c. If you want to double your money in 9 years, what rate would you need? **about 7.7%**

d. If you want to open an account that pays 4.75% interest compounded continuously and have $10,000 in the account 12 years after your deposit, how much would you need to deposit? **about $5655.25**

Write the expression as a sum or difference of logarithms or multiples of logarithms.

48. $\ln 12x^2$ $\ln 12 + 2 \ln x$

49. $\ln \dfrac{16}{125}$ $4 \ln 2 - 3 \ln 5$

50. $\ln \sqrt[5]{x^3}$ $\dfrac{3}{5} \ln x$

51. $\ln xy^4z^{-3}$ $\ln x + 4 \ln y - 3 \ln z$

Use the natural logarithm to solve each equation.

52. $8^x = 24$ **1.5283** **53.** $3^x = 0.4$ **−0.8340** **54.** $2^{3x} = 18$ **1.3900** **55.** $5^{2x} = 38$ **1.1301**

C **56.** **CCSS MODELING** Newton's Law of Cooling, which can be used to determine how fast an object will cool in given surroundings, is represented by $T(t) = T_s + (T_0 - T_s)e^{-kt}$, where T_0 is the initial temperature of the object, T_s is the temperature of the surroundings, t is the time in minutes, and k is a constant value that depends on the type of object.

a. If a cup of coffee with an initial temperature of 180° is placed in a room with a temperature of 70° and the coffee cools to 140° after 10 minutes, find the value of k. **0.045**

b. Use this value of k to determine the temperature of the coffee after 20 minutes. **about 114.7°**

c. When will the temperature of the coffee reach 75°? **about 68 min**

57. **MULTIPLE REPRESENTATIONS** In this problem, you will use $f(x) = e^x$ and $g(x) = \ln x$.

a. **Graphical** Graph both functions and their axis of symmetry, $y = x$, for $-5 \leq x \leq 5$. Then graph $a(x) = e^{-x}$ on the same graph. **See margin.**

b. **Analytical** The graphs of $a(x)$ and $f(x)$ are reflections along which axis? What function would be a reflection of $f(x)$ along the other axis? *y*-axis; $a(x) = -e^x$

c. **Graphical** Determine the two functions that are reflections of $g(x)$. Graph these new functions. $\ln(-x)$ is a reflection across the *y*-axis. $-\ln x$ is a reflection across the *x*-axis.

d. **Verbal** We know that $f(x)$ and $g(x)$ are inverses. Are any of the other functions that we have graphed inverses as well? Explain your reasoning.
Sample answer: no; These functions are reflections along $y = -x$, which indicates that they are not inverses.

H.O.T. Problems Use Higher-Order Thinking Skills

58. **CHALLENGE** Solve $4^x - 2^{x+1} = 15$ for x. **2.3219**

59. **PROOF** Prove $\ln ab = \ln a + \ln b$ for natural logarithms. **See margin.**

60. **REASONING** Determine whether $x > \ln x$ is *sometimes, always,* or *never* true. Explain your reasoning. Sample answer: Always; the graph of $y = x$ is always greater than the graph of $y = \ln x$ and the graphs never intersect.

61. **OPEN ENDED** Express the value 3 using e^x and the natural log. Sample answer: $e^{\ln 3}$

62. **WRITING IN MATH** Explain how the natural log can be used to solve a natural base exponential function. **See margin.**

Enrichment

OL BL

Modeling Mathematically proficient students can apply the mathematics they know to solve problems arising in everyday life, analyze relationships mathematically to draw conclusions, and interpret their mathematical results in the context of a situation.

Teaching the Mathematical Practices

CCSS

63. Given the function $y = 2.34x + 11.33$, which statement best describes the effect of moving the graph down two units? **B**

   A The x-intercept decreases.

   B The y-intercept decreases.

   C The x-intercept remains the same.

   D The y-intercept remains the same.

64. **GRIDDED RESPONSE** Aidan sells wooden picture frames over the Internet. He purchases materials for $85 and pays $19.95 for his website. If he charges $15 each, how many frames will he need to sell in order to make a profit of at least $270? **25**

65. Solve $|2x - 5| = 17$. **G**

   F $-6, -11$

   G $-6, 11$

   H $6, -11$

   J $6, 11$

66. A local pet store sells rabbit food. The cost of two 5-pound bags is $7.99. The total cost c of purchasing n bags can be found by— **C**

   A multiplying n by c.

   B multiplying n by 5.

   C multiplying n by the cost of 1 bag.

   D dividing n by c.

Spiral Review

Solve each equation or inequality. Round to the nearest ten-thousandth. (Lesson 7-6)

67. $2^x = 53$ **5.7279**

68. $2.3^{x^2} = 66.6$ **±2.2452**

69. $3^{4x-7} < 4^{2x+3}$ **$x < 7.3059$**

70. $6^{3y} = 8^{y-1}$ **−0.6309**

71. $12^{x-5} \geq 9.32$ **$x \geq 5.8983$**

72. $2.1^{x-5} = 9.32$ **8.0086**

73. **SOUND** Use the formula $L = 10 \log_{10} R$, where L is the loudness of a sound and R is the sound's relative intensity. Suppose the sound of one alarm clock is 80 decibels. Find out how much louder 10 alarm clocks would be than one alarm clock. (Lesson 7-5) **10 decibels**

Given a polynomial and one of its factors, find the remaining factors of the polynomial. Some factors may not be binomials. (Lesson 5-6)

74. $x^3 + 5x^2 + 8x + 4; x + 1$ **$x + 2, x + 2$**

75. $x^3 + 4x^2 + 7x + 6; x + 2$ **$x^2 + 2x + 3$**

76. **CRAFTS** Mrs. Hall is selling crocheted items. She sells large afghans for $60, baby blankets for $40, doilies for $25, and pot holders for $5. She takes the following number of items to the fair: 12 afghans, 25 baby blankets, 45 doilies, and 50 pot holders. (Lesson 3-6)

   a. Write an inventory matrix for the number of each item and a cost matrix for the price of each item.

   b. Suppose Mrs. Hall sells all of the items. Find her total income as a matrix. **[3095]**

76a. $\begin{bmatrix} 12 & 25 & 45 & 50 \end{bmatrix}$
$\begin{bmatrix} 60 \\ 40 \\ 25 \\ 5 \end{bmatrix}$

Skills Review

Solve each equation.

77. $2^{3x+5} = 128$ **$\frac{2}{3}$**

78. $5^{n-3} = \frac{1}{25}$ **1**

79. $\left(\frac{1}{9}\right)^m = 81^{m+4}$ **$-\frac{8}{3}$**

80. $\left(\frac{1}{7}\right)^{y-3} = 343$ **0**

81. $10^{x-1} = 100^{2x-3}$ **$\frac{5}{3}$**

82. $36^{2p} = 216^{p-1}$ **−3**

DifferentiatedInstruction OL BL

Extension Explain to students that 2! means 2×1, 3! means $3 \times 2 \times 1$, 4! means $4 \times 3 \times 2 \times 1$, and so on. Ask students to use a calculator to find the value of the following series, for as many terms as they want:

$$1 + \frac{1}{1!} + \frac{1}{2!} + \frac{1}{3!} + \frac{1}{4!} + \dots.$$

Discuss with students how the value of this series approaches the value of e. Point out that the series expressed to about 20 terms will give the accuracy for e that Euler gave. Euler, a Swiss mathematician, studied the number e in the 1720s.

4 Assess

Name the Math Have each student write a natural logarithmic equation and a natural logarithmic inequality. Then have students write out all the steps for solving their problems.

Additional Answers

57a.

59. Let $p = \ln a$ and $q = \ln b$. That means that $e^p = a$ and $e^q = b$.

$$ab = e^p \times e^q$$
$$ab = e^{p+q}$$
$$\ln(ab) = (p+q)$$
$$\ln(ab) = \ln a + \ln b$$

62. Sample answer: The natural log and natural base are inverse functions, so taking the natural log of a natural base will *undo* the natural base and make the problem easier to solve.

1 Focus

Objective Use a spreadsheet to display the growth of an investment over time.

Materials for Each Student
- electronic spreadsheet

Teaching Tip
In Step 2 of the Activity, note that the dollar signs in cell C3 are required so that the rate in cell E2 is used for all cells in column C. This is called an "absolute cell reference," whereas all others are relative references.

2 Teach

Working in Cooperative Groups
Have students work in groups of two, mixing abilities, to complete the Activity and Exercise 1.

Ask:
- What formula appears in cell A4?
  $= A3 + 1$

- Why does each formula begin with "="? Cell entries without "=" are not calculated.

Practice Have students complete Exercises 2–6.

3 Assess

Formative Assessment
Use Exercise 4 to assess whether students understand how to find yearly balances after annual compounding with different starting values and interest rates.

Spreadsheet Lab
Compound Interest

You can use a spreadsheet to organize and display data. A spreadsheet is an easy way to track the amount of interest earned over a period of time.

Compound interest is earned not only on the original amount, but also on any interest that has been added to the principal.

CCSS Common Core State Standards
Mathematical Practices
5 Use appropriate tools strategically.

Activity

Find the total amount of money after 5 years if you deposit $100 at 7% compounded annually.

Step 1 Label your columns as shown. The period is one year. Enter the starting values and the rate.

Step 2 Each row will be generated using formulas. Enter the formulas as shown.

Savings Account

◇	A	B	C	D	E
1	End of Period	Principal	Interest	Balance	Rate per Period
2	0			$100.00	7%
3	=A2+1	=D2	=B3*E2	=C3+D2	

Sheet 1 / Sheet 2 / Sheet 3

Step 3 Use the **FILL DOWN** function to fill 4 additional rows.

Savings Account

◇	A	B	C	D	E
1	End of Period	Principal	Interest	Balance	Rate per Period
2	0			$100.00	7%
3	1	$100.00	$7.00	$107.00	
4	2	$107.00	$7.49	$114.49	
5	3	$114.49	$8.01	$122.50	
6	4	$122.50	$8.58	$131.08	
7	5	$131.08	$9.18	$140.26	

Sheet 1 / Sheet 2 / Sheet 3

If you deposit $100 at 7% annual interest for 5 years, you will have $140.26 at the end of the 5 years.

Exercises

Find the total balance for each situation.

1. deposit $500 for 7 years at 5% **$703.55**

2. deposit $1000 for 5 years at 6% **$1338.23**

3. deposit $200 for 2 years at 10% **$242.00**

4. deposit $800 for 3 years at 8% **$1007.77**

5. borrow $10,000 for 5 years at 5.05% **$12,793.23**

6. borrow $25,000 for 30 years at 8% **$251,566.42**

From Concrete to Abstract
Use Exercise 5 to see if students recognize that borrowed amounts will increase the same as deposits if allowed to accumulate.

Extending the Concept
Ask:
- How could the spreadsheet be changed to reflect monthly compounding? Divide the amount in cell E2 by 12.

- How would the yearly balances compare to those in the original spreadsheet? greater

Nik Wheeler/Corbis

LESSON 7-8 Using Exponential and Logarithmic Functions

::Then	::Now	::Why?
● You used exponential growth and decay formulas.	**1** Use logarithms to solve problems involving exponential growth and decay. **2** Use logarithms to solve problems involving logistic growth.	● The ancient footprints of Acahualinca, discovered in Managua, Nicaragua, are believed to be the oldest human footprints in the world. Using carbon dating, scientists estimate that these footprints are 6000 years old.

 NewVocabulary
rate of continuous growth
rate of continuous decay
logistic growth model

 Common Core State Standards

Content Standards
F.IF.8.b Use the properties of exponents to interpret expressions for exponential functions.

F.LE.4 For exponential models, express as a logarithm the solution to $ab^{ct} = d$ where a, c, and d are numbers and the base b is 2, 10, or e; evaluate the logarithm using technology.

Mathematical Practices
1 Make sense of problems and persevere in solving them.

1 Exponential Growth and Decay Scientists and researchers frequently use alternate forms of the growth and decay formulas that you learned in Lesson 7-1.

> **KeyConcept** Exponential Growth and Decay
>
> | Exponential Growth | Exponential Decay |
> |---|---|
> | Exponential growth can be modeled by the function

$$f(x) = ae^{kt},$$

where a is the initial value, t is time in years, and k is a constant representing the **rate of continuous growth**. | Exponential decay can be modeled by the function

$$f(x) = ae^{-kt},$$

where a is the initial value, t is time in years, and k is a constant representing the **rate of continuous decay**. |

> **Real-World Example 1** Exponential Decay
>
> **SCIENCE** The half-life of a radioactive substance is the time it takes for half of the atoms of the substance to disintegrate. The half-life of Carbon-14 is 5730 years. Determine the value of k and the equation of decay for Carbon-14.
>
> If a is the initial amount of the substance, then the amount y that remains after 5730 years can be represented by $\frac{1}{2}a$ or $0.5a$.
>
> | | |
> |---|---|
> | $y = ae^{-kt}$ | Exponential Decay Formula |
> | $0.5a = ae^{-k(5730)}$ | $y = 0.5a$ and $t = 5730$ |
> | $0.5 = e^{-5730k}$ | Divide each side by a. |
> | $\ln 0.5 = \ln e^{-5730k}$ | Property of Equality for Logarithmic Functions |
> | $\ln 0.5 = -5730k$ | $\ln e^x = x$ |
> | $\dfrac{\ln 0.5}{-5730} = k$ | Divide each side by -5730. |
> | $0.00012 \approx k$ | Use a calculator. |
>
> Thus, the equation for the decay of Carbon-14 is $y = ae^{-0.00012t}$.
>
> ▶ **GuidedPractice** 1. $k \approx 2.888 \cdot 10^{-5}$
>
> 1. The half-life of Plutonium-239 is 24,000 years. Determine the value of k.

 connectED.mcgraw-hill.com **509**

---

1 Focus

VerticalAlignment

▼

Before Lesson 7-8 Use exponential growth and decay formulas.

Lesson 7-8 Use logarithms to solve problems involving exponential growth and decay.
Use logarithms to solve problems involving logistic growth.

▼

After Lesson 7-8 Simplify rational expressions.

2 Teach

Scaffolding Questions
Have students read the **Why?** section of the lesson.

Ask:
■ In what region is Nicaragua located? Central America

■ Why can carbon be used for dating purposes? It decays over time in a measurable pattern.

Lesson 7-8 Resources

Resource	Approaching Level **AL**	On Level **OL**	Beyond Level **BL**	English Learners **ELL**
Teacher Edition	■ Differentiated Instruction, p. 510	■ Differentiated Instruction, pp. 510, 511, 515	■ Differentiated Instruction, pp. 510, 511, 515	■ Differentiated Instruction, p. 510
Chapter Resource Masters	■ Study Guide and Intervention, pp. 53–54 ■ Skills Practice, p. 55 ■ Practice, p. 56 ■ Word Problem Practice, p. 57	■ Study Guide and Intervention, pp. 53–54 ■ Skills Practice, p. 55 ■ Practice, p. 56 ■ Word Problem Practice, p. 57 ■ Enrichment, p. 58 ■ Spreadsheet Activity, p. 59	■ Practice, p. 56 ■ Word Problem Practice, p. 57 ■ Enrichment, p. 58	■ Study Guide and Intervention, pp. 53–54 ■ Skills Practice, p. 55 ■ Practice, p. 56 ■ Word Problem Practice, p. 57
Other	■ 5-Minute Check 7-8 ■ Study Notebook	■ 5-Minute Check 7-8 ■ Study Notebook	■ 5-Minute Check 7-8 ■ Study Notebook	■ 5-Minute Check 7-8 ■ Study Notebook

1 Exponential Growth and Decay

Examples 1 and 2 show how to use the algebraic model $y = ae^{-kt}$ to solve problems involving exponential decay.
Example 3 shows how to use the algebraic model $y = ae^{kt}$ to solve a real-world problem involving exponential growth.

Formative Assessment

Use the Guided Practice exercises after each example to determine students' understanding of concepts.

Additional Examples

1 **GEOLOGY** The half-life of Sodium-22 is 2.6 years. Determine the value of k and the equation of decay for Sodium-22.
$k \approx 0.2666$; $y = ae^{-0.2666t}$, where t is given in years

2 **GEOLOGY** A geologist examining a meteorite estimates that it contains only about 10% as much Sodium-22 as it would have contained when it reached Earth's surface. How long ago did the meteorite reach the surface of Earth? about 9 years ago

Additional Examples also in Interactive Classroom PowerPoint® Presentations

 Interactive White Board READY

Teach with Tech

Web Search Have students search the Web to find the half-life of different radioactive elements. Have them choose three elements and ask them to calculate the amount of a 500-gram sample that remains after 10 years.

Real-WorldLink

The oldest modern human fossil, found in Ethiopia, is approximately 160,000 years old.

Source: National Public Radio

Now that the value of k for Carbon-14 is known, it can be used to date fossils.

Real-World Example 2 Carbon Dating

SCIENCE A paleontologist examining the bones of a prehistoric animal estimates that they contain 2% as much Carbon-14 as they would have contained when the animal was alive.

a. How long ago did the animal live?

Understand The formula for the decay of Carbon-14 is $y = ae^{-0.00012t}$. You want to find out how long ago the animal lived.

Plan Let a be the initial amount of Carbon-14 in the animal's body. The amount y that remains after t years is 2% of a or $0.02a$.

Solve

$y = ae^{-0.00012t}$	Formula for the decay of Carbon-14
$0.02a = ae^{-0.00012t}$	$y = 0.02a$
$0.02 = e^{-0.00012t}$	Divide each side by a.
$\ln 0.02 = \ln e^{-0.00012t}$	Property of Equality for Logarithmic Functions
$\ln 0.02 = -0.00012t$	$\ln e^x = x$
$\dfrac{\ln 0.02}{-0.00012} = t$	Divide each side by -0.00012.
$32,600 \approx t$	Use a calculator.

The animal lived about 32,600 years ago.

Check Use the formula to find the amount of a sample remaining after 32,600 years. Use an original amount of 1.

$y = ae^{-0.00012t}$	Original equation
$= 1e^{-0.00012(32,600)}$	$a = 1$ and $t = 32,600$
≈ 0.02 or 2% ✓	Use a calculator.

> **StudyTip**
> Carbon Dating When given a percent or fraction of decay, use an original amount of 1 for a.

b. If prior research points to the animal being around 20,000 years old, how much Carbon-14 should be in the animal?

$y = ae^{-0.00012t}$	Formula for the decay of Carbon-14
$= 1e^{-0.00012(20,000)}$	$a = 1$ and $t = 20,000$
$= e^{-2.4}$	Simplify.
$= 0.09$ or 9%	Use a calculator.

GuidedPractice

2. Use the information in Example 2 to answer the following questions.

A. A specimen that originally contained 42 milligrams of Carbon-14 now contains 8 milligrams. How old is the fossil? about 13,819 yr

B. A wooly mammoth specimen was thought to be about 12,000 years old. How much Carbon-14 should be in the animal? about 23.7%

The exponential growth equation $y = ae^{kt}$ is identical to the continuously compounded interest formula you learned in Lesson 7-7.

Continuous Compounding	Population Growth
$A = Pe^{rt}$	$y = ae^{kt}$
$P =$ initial amount	$a =$ initial population
$A =$ amount at time t	$y =$ population at time t
$r =$ interest rate	$k =$ rate of continuous growth

Prehistoric/The Bridgeman Art Library/Getty Images

DifferentiatedInstruction AL OL BL ELL

Logical Learners Have students work in pairs or small groups. Ask them to examine the growth and decay formulas used in Examples 1–3 and to discuss how the equations are related. In particular, ask them to discuss how they can identify which equations are used for exponential decay situations (minus/negative sign) and which are used for exponential growth.

Focus on Mathematical Content

Exponential Growth and Decay

The exponential decay formulas are of the form $y = a(1 - r)^t$, or $y = ae^{-kt}$. The exponential growth formulas are of the form $y = a(1 + r)^t$, or $y = ae^{kt}$. Logarithms can be used to solve problems involving exponential growth and decay.

Additional Example

3 **POPULATION** In 2007, the population of China was 1.32 billion. In 2000, it was 1.26 billion.

a. Determine the value of k, China's relative rate of growth. **0.0066**

b. When will China's population reach 1.5 billion? **in 2026**

c. India's population in 2007 was 1.13 billion, and can be modeled by $1.13e^{0.015t}$. Determine when India's population will surpass China's. (Note: t represents years after 2007.) **after 18.5 years, or midway through the year 2025**

Real-World Example 3 Continuous Exponential Growth

PT

POPULATION In 2007, the population of the state of Georgia was 9.36 million people. In 2000, it was 8.18 million.

Problem-Solving Tip

Use a Formula When dealing with population, it is almost always necessary to use an exponential growth or decay formula.

a. Determine the value of k, Georgia's relative rate of growth.

$y = ae^{kt}$ — Formula for continuous exponential growth

$9.36 = 8.18e^{k(7)}$ — $y = 9.36$, $a = 8.18$, and $t = 2007 - 2000$ or 7

$\dfrac{9.36}{8.18} = e^{7k}$ — Divide each side by 8.18.

$\ln \dfrac{9.36}{8.18} = \ln e^{7k}$ — Property of Equality for Logarithmic Functions

$\ln \dfrac{9.36}{8.18} = 7k$ — $\ln e^x = x$

$\dfrac{\ln \frac{9.36}{8.18}}{7} = k$ — Divide each side by 7.

$0.01925 = k$ — Use a calculator.

Georgia's relative rate of growth is about 0.01925 or about 2%.

b. When will Georgia's population reach 12 million people?

$y = ae^{kt}$ — Formula for continuous exponential growth

$12 = 8.18e^{0.01925t}$ — $y = 10$, $a = 8.18$, and $k = 0.01925$

$1.4670 = e^{0.01925t}$ — Divide each side by 8.18.

$\ln 1.4670 = \ln e^{0.01925t}$ — Property of Equality for Logarithmic Functions

$1.4670 = 0.01925t$ — $\ln e^x = x$

$\dfrac{\ln 1.4670}{0.01925} = t$ — Divide each side by 0.01925.

$19.907 \approx t$ — Use a calculator.

Georgia's population will reach 12 million people by 2020.

c. Michigan's population in 2000 was 9.9 million and can be modeled by $y = 9.9e^{0.0028t}$. Determine when Georgia's population will surpass Michigan's.

$8.18e^{0.01925t} > 9.9e^{0.0028t}$ — Georgia's population will surpass Michigan's.

$\ln 8.18e^{0.01925t} > \ln 9.9e^{0.0028t}$ — Property of Inequality for Logarithms

$\ln 8.18 + \ln e^{0.01925t} > \ln 9.9 + \ln e^{0.0028t}$ — Product Property of Logarithms

$\ln 8.18 + 0.01925t > \ln 9.9 + 0.0028t$ — $\ln e^x = x$

$0.01645t > \ln 9.9 - \ln 8.18$ — Subtract (0.0028t + ln 8.18) from each side.

$t > \dfrac{\ln 9.9 - \ln 8.18}{0.01645}$ — Divide each side by 0.01645.

$t > 11.6$ — Use a calculator.

Georgia's population will surpass Michigan's during 2012.

Guided Practice

3. BIOLOGY A type of bacteria is growing exponentially according to the model $y = 1000e^{kt}$, where t is the time in minutes.

A. If there are 1000 cells initially and 1650 cells after 40 minutes, find the value of k for the bacteria. **$k \approx 0.0125$**

B. Suppose a second type of bacteria is growing exponentially according to the model $y = 50e^{0.0432t}$. Determine how long it will be before the number of cells of this bacteria exceed the number of cells in the other bacteria. **about 97.58 min**

Differentiated Instruction OL BL

Extension Mathematically and scientifically talented students can research the growth rates of different bacteria types. Students can explore how these growth rates are determined, environmental factors that cause them to thrive or inhibit their prosperity, and graph the growth of different types for comparison purposes.

2 Logistic Growth

Example 4 shows how to use a logistic growth function to model growth with a limiting factor.

Additional Example

4 A city's population in millions is modeled by

$$f(t) = \frac{1.432}{1 + 1.05e^{-0.32t}},$$ where t is the number of years since 2000.

a. Graph the function.

b. What is the horizontal asymptote? $f(t) = 1.432$

c. What will be the maximum population? 1,432,000

d. According to the function, when will the city's population reach 1 million? 2003

Additional Answer (Guided Practice)

4A.

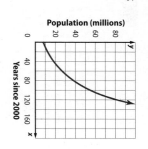

---

2 Logistic Growth

Logistic Growth Refer to the equation representing Georgia's population in Example 3. According to the graph at the right, Georgia's population will be about one billion by the year 2130. Does this seem logical?

Populations cannot grow infinitely large. There are limitations, such as food supplies, war, living space, diseases, available resources, and so on.

Exponential growth is unrestricted, meaning it will increase without bound. A **logistic growth model**, however, represents growth that has a limiting factor. Logistic models are the most accurate models for representing population growth.

KeyConcept Logistic Growth Function

Let a, b, and c be positive constants where $b < 1$. The logistic growth function is represented by $f(t) = \dfrac{c}{1 + ae^{-bt}}$, where t represents time.

Real-WorldLink
Phoenix is the fifth largest city in the country and has a population of 1.5 million.

Real-World Example 4 Logistic Growth

The population of Phoenix, Arizona, in millions can be modeled by the logistic function $f(t) = \dfrac{2.0666}{1 + 1.66e^{-0.048t}}$, where t is the number of years after 1980.

a. Graph the function for $0 \le t \le 500$.

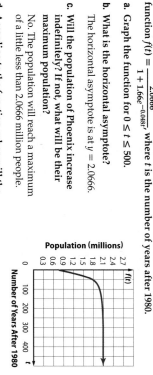

b. What is the horizontal asymptote?

The horizontal asymptote is at $y = 2.0666$.

c. Will the population of Phoenix increase indefinitely? If not, what will be their maximum population?

No. The population will reach a maximum of a little less than 2.0666 million people.

d. According to the function, when will the population of Phoenix reach 1.8 million people?

The graph indicates the population will reach 1.8 million people at $t \approx 50$. Replacing $f(t)$ with 1.8 and solving for t in the equation yields $t = 50.35$ years. So, the population of Phoenix will reach 1.8 million people by 2031.

StudyTip
Intersections To determine where the graph intersects 1.8 on the calculator, graph $y = 1.8$ on the same graph and select *intersection* in the CALC menu.

GuidedPractice

4. The population of a certain species of fish in a lake after t years can be modeled by the function $P(t) = \dfrac{1880}{1 + 1.42e^{-0.037t}}$, where $t \ge 0$.

A. Graph the function for $0 \le t \le 500$. See margin.

B. What is the horizontal asymptote? $y = 1880$

C. What is the maximum population of the fish in the lake? 1880

D. When will the population reach 1875? about 170 yr

Follow-Up

- How can mathematical models help you make good decisions? Sample answer: Mathematical models can be used to compare different options that are available, as well as to predict the impact of an option if it is chosen.

Lloyd Sutton/Masterfile

3 Practice

Formative Assessment

Use Exercises 1–3 to check for understanding.

Use the chart at the bottom of this page to customize assignments for your students.

CCSS Teaching the Mathematical Practices

Perseverance Mathematically proficient students start by explaining to themselves the meaning of a problem and looking for entry points to its solution. They analyze givens, constraints, relationships, and goals and make conjectures about the form and meaning of the solution and plan a solution pathway rather than simply jumping into a solution attempt.

Additional Answers

3a.

6a.

Check Your Understanding

 = Step-by-Step Solutions begin on page R14.

Examples 1–2

1. PALEONTOLOGY The half-life of Potassium-40 is about 1.25 billion years.

a. Determine the value of k and the equation of decay for Potassium-40. 5.545×10^{-10}

b. A specimen currently contains 36 milligrams of Potassium-40. How long will it take the specimen to decay to only 15 milligrams of Potassium-40? 1,578,843,530 yr

c. How many milligrams of Potassium-40 will be left after 300 million years? about 30.48 mg

d. How long will it take Potassium-40 to decay to one eighth of its original amount? 3,750,120,003 yr

Example 3

2. SCIENCE A certain food is dropped on the floor and is growing bacteria exponentially according to the model $y = 2e^{kt}$, where t is the time in seconds.

a. If there are 2 cells initially and 8 cells after 20 seconds, find the value of k for the bacteria. $k \approx 0.0693$

b. about 2.828 cells

b. The "5-second rule" says that if a person who drops food on the floor eats it within 5 seconds, there will be no harm. How much bacteria is on the food after 5 seconds?

c. Would you eat food that had been on the floor for 5 seconds? Why or why not? Do you think that the information you obtained in this exercise is reasonable? Explain.

2c. Sample answer: Yes; it has not even grown 1 cell in 5 seconds. There are many factors that affect this equation, such as how clean the floor is and what type of food was dropped.

Example 4

3. ZOOLOGY Suppose the red fox population in a restricted habitat follows the function
$$P(t) = \frac{16,500}{1 + 18e^{-0.085t}},$$ where t represents the time in years.

a. Graph the function for $0 \le t \le 200$. See margin.

b. What is the horizontal asymptote? $P(t) = 16,500$

c. What is the maximum population? 16,500

d. When does the population reach 16,450? about 102 years

Practice and Problem Solving

Extra Practice is on page R7.

Examples 1–2

4. CCSS PERSEVERANCE The half-life of Rubidium-87 is about 48.8 billion years.

a. Determine the value of k and the equation of decay for Rubidium-87. $k \approx 1.42 \times 10^{-11}$

b. A specimen currently contains 50 milligrams of Rubidium-87. How long will it take the specimen to decay to only 18 milligrams of Rubidium-87? 71,947,270,950 yr

c. How many milligrams of Rubidium-87 will be left after 800 million years? about 49.4 mg

d. How long will it take Rubidium-87 to decay to one-sixteenth its original amount? 195.3 billion yr

Example 3

5. BIOLOGY A certain bacteria is growing exponentially according to the model $y = 80e^{kt}$, where t is the time in minutes.

a. If there are 80 cells initially and 675 cells after 30 minutes, find the value of k for the bacteria. $k \approx 0.071$

b. When will the bacteria reach a population of 6000 cells? about 60.8 min

c. If a second type of bacteria is growing exponentially according to the model $y = 35e^{0.0978t}$, determine how long it will be before the number of cells of this bacteria exceed the number of cells in the other bacteria. about 30.85 min

Example 4

6. FORESTRY The population of trees in a certain forest follows the function
$$f(t) = \frac{18000}{1 + 16e^{-0.084t}},$$ where t is the time in years.

a. Graph the function for $0 \le t \le 100$. See margin.

b. When does the population reach 17500 trees? about 75.33 yr

---

Differentiated Homework Options

Level	Assignment	Two-Day Option	
AL Basic	4–6, 14, 16–34	5, 19–22	4, 6, 14, 16–18, 23–34
OL Core	5, 7–14, 16–34	4–6, 19–22	7–14, 16–18, 23–34
BL Advanced	7–26, (optional: 27–34)		

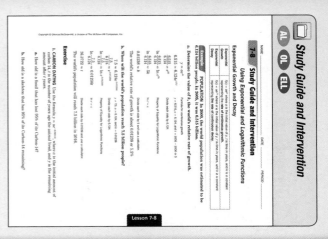

Study Guide and Intervention

Exponential Growth and Decay

Exercise

1. **CARBON DATING** Use the formula $y = ae^{-0.00012t}$, where t is the number of years ago, and y is the remaining amount of carbon-14, to answer the questions.

a. How old is a skeleton that has lost 95% of its Carbon-14?

Effective Annual Yield

Lesson 7-8

514 | Lesson 7-8 | Using Exponential and Logarithmic Functions

---

7. **PALEONTOLOGY** A paleontologist finds a human bone and determines that the Carbon-14 found in the bone is 85% of that found in living bone tissue. How old is the bone? **about 1354 yr old**

8. **ANTHROPOLOGY** An anthropologist has determined that a newly discovered human bone is 8000 years old. How much of the original amount of Carbon-14 is in the bone? **about 38%**

9. **RADIOACTIVE DECAY** 100 milligrams of Uranium-238 are stored in a container. If Uranium-238 has a half-life of about 4.47 billion years, after how many years will only 10 milligrams be present? **about 14.85 billion yr**

10. **POPULATION GROWTH** The population of the state of Oregon has grown from 3.4 million in 2000 to 3.7 million in 2006.

a. Write an exponential growth equation of the form $y = ae^{kt}$ for Oregon, where t is the number of years after 2000. $y = 3.4e^{0.014t}$

b. Use your equation to predict the population of Oregon in 2020. **4.5 million**

c. According to the equation, when will Oregon reach 6 million people? **about 2041**

11. **HALF-LIFE** A substance decays 99.9% of its total mass after 200 years. Determine the half-life of the substance. **about 20.1 yr**

12. **LOGISTIC GROWTH** The population in millions of the state of Ohio after 1900 can be modeled by $P(t) = \dfrac{12.95}{1 + 2.4e^{-t}}$, where t is the number of years after 1900 and k is a constant.

a. If Ohio had a population of 10 million in 1970, find the value of k. $k \approx 0.0299464$

b. According to the equation, when will the population of Ohio reach 12 million? **2014**

13. **MULTIPLE REPRESENTATIONS** In this problem, you will explore population growth. The population growth of a country follows the exponential function $f(t) = 8e^{0.0075t}$ or the logistic function $g(t) = \dfrac{400}{1 + 16e^{-0.025t}}$. The population is measured in millions and t is time in years.

a. **Graphical** Graph both functions for $0 \le t \le 100$. **See margin.**

b. **Analytical** Determine the intersection of the graphs. What is the significance of this intersection?

c. **Analytical** Which function is a more accurate estimate of the country's population 100 years from now? Explain your reasoning. **Sample answer: The logistic function $g(t)$ is a more accurate estimate of the country's population since $f(t)$ will continue to grow exponentially and $g(t)$ considers limitations on population growth such as food supply.**

H.O.T. Problems   Use Higher-Order Thinking Skills

14. **OPEN ENDED** Give an example of a quantity that grows or decays at a fixed rate. Write a real-world problem involving the rate and solve by using logarithms. **Sample answer: money in a bank; See students' work.**

15. **CHALLENGE** Solve $\dfrac{120,000}{1 + 48e^{-0.015t}} = 24e^{0.055t}$ for t. $t \approx 113.45$

16. **ARGUMENTS** Explain mathematically why $f(t) = \dfrac{c}{1 + 60e^{-0.5t}}$ approaches, but never reaches the value of c as $t \to +\infty$. **See margin.**

17. **OPEN ENDED** Give an example of a quantity that grows logistically and has limitations to growth. Explain why the quantity grows in this manner. **See margin.**

18. **WRITING IN MATH** How are exponential, continuous exponential, and logistic functions used to model different real-world situations? **See margin.**

CCSS

Teaching the Mathematical Practices

Arguments Mathematically proficient students understand and use stated assumptions, definitions, and previously established results in constructing arguments. They make conjectures and build a logical progression of statements to explore the truth of their conjectures. And they are able to analyze situations by breaking them into cases, and can recognize and use counterexamples.

Ticket Out the Door Ask students to write questions that can be solved using $y = ae^{kt}$ on one side of an index card. Then have students solve their problems on the reverse sides of the cards.

Formative Assessment

Check for student understanding of concepts in Lessons 7-7 and 7-8.

📋 Quiz 6, p. 64

Additional Answers

13a.

17. Sample answer: The spread of the flu throughout a small town. The growth of this is limited to the population of the town itself.

18. Sample answer: Exponential functions can be used to model situations that incorporate a percentage of growth or decay for a specific number of times per year. Continuous exponential functions can be used to model situations that incorporate a percentage of growth or decay continuously and consist of a limiting factor.

26c. Sample answer: Method 1: Use the Change of Base Formula and find that the value is about 0.4709. Method 2: Use the values $\log_7 2 \approx 0.3562$ and $\log_7 3 \approx 0.5646$. First, average the values and then guess and check by raising 7 to the various powers. Continue until you get the desired closeness to 2.5. To check find $7^{0.4709}$ which is very close to 2.5.

Standardized Test Practice

19. Kareem is making a circle graph showing the favorite ice cream flavors of customers at his store. The table summarizes the data. What central angle should Kareem use for the section representing chocolate? **C**

Flavor	Customers
chocolate	35
vanilla	42
strawberry	7
mint chip	12
butter pecan	4

A 35°
B 63°
C 126°
D 150°

20. PROBABILITY Lydia has 6 books on her bookshelf. Two are literature books, one is a science book, two are math books, and one is a dictionary. What is the probability that she randomly chooses a science book and the dictionary? **J**

F $\frac{1}{3}$
G $\frac{1}{4}$
H $\frac{1}{12}$
J $\frac{1}{15}$

21. SAT/ACT Peter has made a game for his daughter's birthday party. The playing board is a circle divided evenly into 8 sectors. If the circle has a radius of 18 inches, what is the approximate area of one of the sectors? **D**

A 4 in²
B 14 in²
C 32 in²
D 127 in²
E 254 in²

22. STATISTICS In a survey of 90 physical trainers, 15 said they went for a run at least 5 times per week. Of that group, 5 said they also swim during the week, and at least 25% of all trainers run and swim every week. Which conclusion is valid based on the information given? **H**

F The report is accurate because 15 out of 90 is 25%.
G The report is accurate because 5 out of 15 is 33%, which is at least 25%.
H The report is inaccurate because 5 out of 90 is only 5.6%.
J The report is inaccurate because no one knows if swimming is really exercising.

26a. Armenia and Yugoslavia, or Haiti and Yugoslavia; Morocco and Indonesia

Spiral Review

Write an equivalent exponential or logarithmic equation. (Lesson 7-7)

23. $e^7 = y$ $\ln y = 7$

24. $e^{2n-4} = 36$ $\ln 36 = 2n - 4$

25. $\ln 5 + 4 \ln x = 9$ $5x^4 = e^9$

26. EARTHQUAKES The table shows the magnitudes of some major earthquakes. (*Hint:* A magnitude x earthquake has an intensity of 10^x.) (Lessons 7-5 and 7-6)

Year	Location	Magnitude
1963	Yugoslavia	6.0
1970	Peru	7.8
1988	Armenia	7.0
2004	Morocco	6.4
2007	Indonesia	8.4
2010	Haiti	7.0

a. For which two earthquakes was the intensity of one 10 times that of the other? For which two was the intensity of one 100 times that of the other?

b. What would be the magnitude of an earthquake that is 1000 times as intense as the 1963 earthquake in Yugoslavia? **9.0**

c. Suppose you know that $\log_7 2 \approx 0.3562$ and $\log_7 3 \approx 0.5646$. Describe two different methods that you could use to approximate $\log_7 2.5$. (You may use a calculator, of course.) Then describe how you can check your result. **See margin.**

Skills Review

Solve each equation. Write in simplest form.

27. $\frac{8}{5}x = \frac{4}{15}$ $\frac{1}{6}$

28. $\frac{27}{14}n = \frac{6}{7}$ $\frac{4}{9}$

29. $\frac{3}{10} = \frac{12}{25}a$ $\frac{5}{8}$

30. $\frac{6}{7} = 9p$ $\frac{2}{21}$

31. $\frac{9}{8}b = 18$ 16

32. $\frac{6}{7}y = \frac{3}{4}$ $\frac{7}{8}$

33. $\frac{1}{3}z = \frac{5}{6}$ $2\frac{1}{2}$

34. $\frac{2}{3}q = 7$ $10\frac{1}{2}$

DifferentiatedInstruction OL BL

Extension *Catenary curves* are curves similar to those formed by a chain hanging between two hooks. Every segment of chain pulls on every other segment of chain, giving the hanging chain its curved shape. The Gateway Arch in St. Louis, Missouri is an example of an inverted catenary curve.

Ask students to use graphing calculators to graph $y = \dfrac{e^x + e^{-x}}{2}$ to see a model of a catenary curve.

13a. (graph) Population (millions) vs Time (yr), showing curves $f(t)$ and $g(t)$; y-axis 10–90, t-axis 10–90.

1 Focus

Objective Use a data collection device to investigate the differences between types of insulated cups and cooling time.

Materials

- TI-83/84 Plus graphing calculator
- data collection device and compatible temperature probe
- variety of containers
- very hot water

Teaching Tip

If the device being used is the Calculator-Based Laboratory 2, note the following: The CBL has three keys:

- **[TRANSFER]** begins transfer of programs of Calculator Software Applications between the CBL and an attached TI graphing calculator.
- **[QUICK SET-UP]** clears any data stored in CBLs memory.
- **[START/STOP]** begins sampling for Quick Set-Up. Sampling continues until the number of samples is collected or you press [START/STOP].

2 Teach

Working in Cooperative Groups

Put students in groups of three or four, mixing abilities. Then have groups complete the Activity and Exercises 1–4.

- To set up the calculator and CBL for data collection, select **Collect Data** from the Main Menu. Select **Time Graph** from the **Data Collection** menu. Enter 60 as the time between samples in seconds. Then enter 20 as the number of samples (the CBL will collect data for a total of 1200 seconds). Press **ENTER**, then select **Use Time Setup** to continue.

3 Assess

Formative Assessment

In Exercise 5, check that students have used different containers, but the same initial water temperature. Ask students why it is necessary to

- Discuss with students the **Ymin** and the **Ymax** and the **Yscl** that are appropriate for this lab.

In this lab, you will explore the type of equation that models the change in the temperature of water as it cools under various conditions.

EXTEND 7-8

Graphing Technology Lab
Cooling

CCSS Common Core State Standards
Content Standards
F.BF.1.b Combine standard function types using arithmetic operations.

Set Up the Lab

- Collect a variety of containers, such as a foam cup, a ceramic coffee mug, and an insulated cup.
- Boil water or collect hot water from a tap.
- Choose a container to test and fill with hot water. Place the temperature probe in the cup.
- Connect the temperature probe to your data collection device.

Activity

Step 1 Program the device to collect 20 or more samples in 1 minute intervals.

Step 2 Wait a few seconds for the probe to warm to the temperature of the water.

Step 3 Press the button to begin collecting data.

Analyze the Results

1. When the data collection is complete, graph the data in a scatter plot. Use time as the independent variable and temperature as the dependent variable. Write a sentence that describes the points on the graph. **Sample answer: It's an exponential pattern.**

2. Use the **STAT** menu to find an equation to model the data you collected. Try linear, quadratic, and exponential models. Which model appears to fit the data best? Explain.

   2. Sample answer: The exponential model fits best. Its correlation coefficient is closest to 1.

3. Would you expect the temperature of the water to drop below the temperature of the room? Explain your reasoning.

   3. Sample answer: No; the temperature will approach the temperature of the room but it will not drop below it.

4. Use the data collection device to find the temperature of the air in the room. Graph the function $y = t$, where t is the temperature of the room, along with the scatter plot and the model equation. Describe the relationship among the graphs. What is the meaning of the relationship in the context of the experiment?

   4. Sample answer: The linear graph of the room temperature is the asymptote for the exponential graph. The temperature of the water will approach the temperature of the room, but it will not go below that temperature.

Make a Conjecture

5. Do you think the results of the experiment would change if you used an insulated container for the water? What part of the function will change, the constant or the rate of decay? Repeat the experiment to verify your conjecture.

6. How might the results of the experiment change if you added ice to the water? What part of the function will change, the constant or the rate of decay? Repeat the experiment to verify your conjecture. **Sample answer: The ice will speed the rate of cooling. This affects the rate of decay in the function.**

From Concrete to Abstract

Have students describe the containers that make the best insulators.

Practice Have groups complete Exercises 5–6.

5. Sample answer: Yes; the thermal container will slow the rate of cooling. This affects the rate of decay in the function.

begin the second experiment with water at the same temperature as the first experiment.

Ed-Imaging

CHAPTER 7

Study Guide and Review

Study Guide

KeyConcepts

Exponential Functions (Lessons 7-1 and 7-2)

- An exponential function is in the form $y = ab^x$, where $a \neq 0$, $b > 0$ and $b \neq 1$.

- Property of Equality for Exponential Functions: If b is a positive number other than 1, then $b^x = b^y$ if and only if $x = y$.

- Property of Inequality for Exponential Functions: If $b > 1$, then $b^x > b^y$ if and only if $x > y$, and $b^x < b^y$ if and only if $x < y$.

Logarithms and Logarithmic Functions (Lessons 7-3 through 7-6)

- Suppose $b > 0$ and $b \neq 1$. For $x > 0$, there is a number y such that $\log_b x = y$ if and only if $b^y = x$.

- The logarithm of a product is the sum of the logarithms of its factors.

- The logarithm of a quotient is the difference of the logarithms of the numerator and the denominator.

- The logarithm of a power is the product of the logarithm and the exponent.

- The Change of Base Formula: $\log_a n = \dfrac{\log_b n}{\log_b a}$

Base e and Natural Logarithms (Lesson 7-7)

- Since the natural base function and the natural logarithmic function are inverses, these two can be used to "undo" each other.

Using Exponential and Logarithmic Functions (Lesson 7-8)

- Exponential growth can be modeled by the function $f(x) = ae^{kt}$, where k is a constant representing the rate of continuous growth.

- Exponential decay can be modeled by the function $f(x) = ae^{-kt}$, where k is a constant representing the rate of continuous decay.

FOLDABLES StudyOrganizer

Be sure the Key Concepts are noted in your Foldable.

KeyVocabulary

asymptote (p. 451)

Change of Base Formula (p. 494)

common logarithm (p. 492)

compound interest (p. 462)

decay factor (p. 454)

exponential decay (p. 453)

exponential equation (p. 461)

exponential function (p. 451)

exponential growth (p. 451)

exponential inequality (p. 463)

growth factor (p. 453)

logarithm (p. 468)

logarithmic equation (p. 478)

logarithmic function (p. 469)

common logarithm (p. 492)

logarithmic inequality (p. 479)

logistic growth model (p. 512)

natural base, e (p. 501)

natural base exponential function (p. 501)

natural logarithm (p. 501)

rate of continuous decay (p. 509)

rate of continuous growth (p. 509)

VocabularyCheck

Choose a word or term from the list above that best completes each statement or phrase.

1. A function of the form $f(x) = b^x$ where $b > 1$ is a(n) **exponential growth** function.

2. In $x = b^y$, the variable y is called the **logarithm** of x.

3. Base 10 logarithms are called **common logarithms**.

4. A(n) **exponential equation** is an equation in which variables occur as exponents.

5. The **change of base formula** allows you to write equivalent logarithmic expressions that have different bases.

6. The base of the exponential function, $A(t) = a(1 - r)^t$, $1 - r$ is called the **decay factor**.

7. The function $y = \log_b x$, where $b > 0$ and $b \neq 1$, is called a (n) **logarithmic function**.

8. An exponential function with base e is called the **natural base exponential function**.

9. The logarithm with base e is called the **natural logarithm**.

10. The number e is referred to as the **natural base**.

Formative Assessment

KeyVocabulary The page reference after each word denotes where that term was first introduced. If students have difficulty answering questions 1–10, remind them that they can use these page references to refresh their memories about the vocabulary terms.

Summative Assessment

Vocabulary Test, p. 66

Vocabulary Review

Vocabulary Review provides students the opportunity to check their understanding of important concepts and terminology in an online game format.

FOLDABLES StudyOrganizer

Dinah Zike's Foldables®

Have students look through the chapter to make sure they have included examples in their Foldables for each lesson of the chapter. Suggest that students keep their Foldables handy while completing the Study Guide and Review pages. Point out that their Foldables can serve as a quick review when studying for the chapter test.

Lesson-by-Lesson Review

Intervention If the given examples are not sufficient to review the topics covered by the questions, remind students that the lesson references tell them where to review that topic in their textbooks.

Two-Day Option Have students complete the Lesson-by-Lesson Review. Then you can use McGraw-Hill eAssessment to customize another review worksheet that practices all the objectives of this chapter or only the objectives on which your students need more help.

Additional Answers

11.

$f(x) = 3^x$

D = {all real numbers}
R = {f(x) | f(x) > 0}

12.

$f(x) = -5(2)^x$

D = {all real numbers}
R = {f(x) | f(x) < 0}

13.

$f(x) = 3(4)^x - 6$

D = {all real numbers}
R = {f(x) | f(x) > -6}

Study Guide and Review *Continued*

Lesson-by-Lesson Review

7-1 Graphing Exponential Functions

Graph each function. State the domain and range. **11–16.** *See margin.*

11. $f(x) = 3^x$

12. $f(x) = -5(2)^x$

13. $f(x) = 3(4)^x - 6$

14. $f(x) = 3(2)^x + 5$

15. $f(x) = 3\left(\frac{1}{4}\right)^{x+3} - 1$

16. $f(x) = \frac{3}{5}\left(\frac{2}{3}\right)^{x-2} + 3$

17. POPULATION A city with a population of 120,000 decreases at a rate of 3% annually.

a. Write the function that represents this situation.
$f(x) = 120,000(0.97)^x$

b. What will the population be in 10 years? **about 88,491**

Example 1

Graph $f(x) = -2(3)^x + 1$. State the domain and range.

The domain is all real numbers, and the range is all real numbers less than 1.

14.

$f(x) = 3^{2x} + 5$

D = {all real numbers}
R = {f(x) | f(x) > 5}

15.

$f(x) = 3\left(\frac{1}{4}\right)^{x+3} - 1$

D = {all real numbers}
R = {f(x) | f(x) > -1}

16.

$f(x) = \frac{3}{5}\left(\frac{2}{3}\right)^{x-2} + 3$

D = {all real numbers}
R = {f(x) | f(x) > 3}

7-2 Solving Exponential Equations and Inequalities

Solve each equation or inequality. **22.** $x > -\frac{2}{3}$

18. $16^x = \frac{1}{64} \quad -\frac{3}{2}$

19. $3^{4x} = 9^{3x+7} \quad -7$

20. $64^{3n} = 8^{2n-3} \quad -\frac{3}{4}$

21. $8^{3-y} = 256^{4y} \quad \frac{9}{41}$

22. $9^{x-2} > \left(\frac{1}{81}\right)^{x+2}$

23. $27^{3x} \leq 9^{2x-1} \quad x \leq -\frac{2}{5}$

24. BACTERIA A bacteria population started with 5000 bacteria. After 8 hours there were 28,000 in the sample.

a. Write an exponential function that could be used to model the number of bacteria after x hours if the number of bacteria changes at the same rate.
$y = 5000(1.240)^x$

b. How many bacteria can be expected in the sample after 32 hours? **about 4,880,496**

Example 2

Solve $4^{3x} = 32^{x-1}$ for x.

$4^{3x} = 32^{x-1}$	Original equation
$(2^2)^{3x} = (2^5)^{x-1}$	Rewrite so each side has the same base.
$2^{6x} = 2^{5x-5}$	Power of a Power
$6x = 5x - 5$	Property of Equality for Exponential Functions
$x = -5$	Subtract 5x from each side.

The solution is −5.

7-3 Logarithms and Logarithmic Functions

25. Write $\log_2 \frac{1}{16} = -4$ in exponential form. $2^{-4} = \frac{1}{16}$

26. Write $10^2 = 100$ in logarithmic form. $\log_{10} 100 = 2$

Evaluate each expression.

27. $\log_4 256$ **4**

28. $\log_2 \frac{1}{8}$ **−3**

Graph each function. **29, 30. See margin.**

29. $f(x) = 2\log_{10} x + 4$

30. $f(x) = \frac{1}{6}\log_3 (x - 2)$

Example 3

Evaluate $\log_2 64$.

$\log_2 64 = y$	Let the logarithm equal y.
$64 = 2^y$	Definition of logarithm
$2^6 = 2^y$	$64 = 2^6$
$6 = y$	Property of Equality for Exponential Functions

Additional Answers

29.

30.

7-4 Solving Logarithmic Equations and Inequalities

Solve each equation or inequality.

31. $\log_4 x = \dfrac{3}{2}$ 64

32. $\log_2 \dfrac{1}{64} = x$ -6

33. $\log_4 x < 3$ $\{x \mid 0 < x < 64\}$

34. $\log_5 x < -3$ $\left\{ x \mid 0 < x < \dfrac{1}{125} \right\}$

35. $\log_9 (3x - 1) = \log_9 (4x)$ no solution

36. $\log_2 (x^2 - 18) = \log_2 (-3x)$ -6

37. $\log_3 (3x + 4) \le \log_3 (x - 2)$ no solution

38. **EARTHQUAKE** The magnitude of an earthquake is measured on a logarithmic scale called the Richter scale. The magnitude M is given by $M = \log_{10} x$, where x represents the amplitude of the seismic wave causing ground motion. How many times as great is the amplitude caused by an earthquake with a Richter scale rating of 10 as an aftershock with a Richter scale rating of 7?   **1000**

Example 4

Solve $\log_{27} x < \dfrac{2}{3}$.

$\log_{27} x < \dfrac{2}{3}$	Original inequality
$x < 27^{\frac{2}{3}}$	Logarithmic to Exponential Inequality
$x < 9$	Simplify.

Example 5

Solve $\log_5 (p^2 - 2) = \log_5 p$.

$\log_5 (p^2 - 2) = \log_5 p$	Original equation
$p^2 - 2 = p$	Property of Equality
$p^2 - p - 2 = 0$	Subtract p from each side.
$(p - 2)(p + 1) = 0$	Factor.
$p - 2 = 0$ or $p + 1 = 0$	Zero Product Property
$p = 2$ $p = -1$	Solve each equation.

The solution is $p = 2$, since $\log_5 p$ is undefined for $p = -1$.

7-5 Properties of Logarithms

Use $\log_5 16 \approx 1.7227$ and $\log_5 2 \approx 0.4307$ to approximate the value of each expression.

39. $\log_5 8$ 1.2920

40. $\log_5 64$ 2.5841

41. $\log_5 4$ 0.8614

42. $\log_5 \dfrac{1}{8}$ -1.2921

43. $\log_5 \dfrac{1}{2}$ -0.4307

Solve each equation. Check your solution.

44. $\log_5 x - \log_5 2 = \log_5 15$ 30

45. $3 \log_4 a = \log_4 27$ 3

46. $2 \log_3 x + \log_3 3 = \log_3 36$ $2\sqrt{3}$

47. $\log_4 n + \log_4 (n - 4) = \log_4 5$ 5

48. **SOUND** Use the formula $L = 10 \log_{10} R$, where L is the loudness of a sound and R is the sound's relative intensity, to find out how much louder 20 people talking would be than one person talking. Suppose the sound of one person talking has a relative intensity of 80 decibels.   **361.6 times**

Example 6

Use $\log_5 16 \approx 1.7227$ and $\log_5 2 \approx 0.4307$ to approximate $\log_5 32$.

$\log_5 32 = \log_5 (16 \cdot 2)$	Replace 32 with 16.
$= \log_5 16 + \log_5 2$	Product Property
$\approx 1.7227 + 0.4307$	Use a calculator.
≈ 2.1534	

Example 7

Solve $\log_3 3x + \log_3 4 = \log_3 36$.

$\log_3 3x + \log_3 4 = \log_3 36$	Original equation
$\log_3 3x(4) = \log_3 36$	Product Property
$3x(4) = 36$	Definition of logarithm
$12x = 36$	Multiply.
$x = 3$	Divide each side by 12.

Anticipation Guide

Have students complete the Chapter 7 Anticipation Guide and discuss now how their responses have changed now that they have completed Chapter 7.

Additional Answers

56. 1.9459
57. −1.9459
58. 201.7144
59. $\{x \mid x > 1.9459\}$
60. $\{x \mid -3 < x < -0.2817\}$
61. $\{x \mid x < -2.8904\}$

CHAPTER 7

Study Guide and Review *Continued*

7-6 Common Logarithms

Solve each equation or inequality. Round to the nearest ten-thousandth.

49. $3^x = 15$ $x \approx 2.4650$
50. $6^{x^2} = 28$ $x \approx \pm 1.3637$
51. $8^{m+1} = 30$ $m \approx 0.6356$
52. $12^{r-1} = 7r$ $r \approx 4.6102$
53. $3^{5n} > 24$ $\{n \mid n > 0.5786\}$
54. $5^{x+2} \le 3^x$ $\{x \mid x \le -6.3013\}$
55. **SAVINGS** You deposited $1000 into an account that pays an annual interest rate r of 5% compounded quarterly. Use $A = P\left(1 + \dfrac{r}{n}\right)^{nt}$.
   a. How long will it take until you have $1500 in your account? about 8.2 years
   b. How long it will take for your money to double? about 13.9 years

Example 8

Solve $5^{3x} > 7^{x+1}$.

$5^{3x} > 7^{x+1}$	Original inequality
$\log 5^{3x} > \log 7^{x+1}$	Property of Inequality
$3x \log 5 > (x + 1) \log 7$	Power Property
$3x \log 5 > x \log 7 + \log 7$	Distributive Property
$x(3 \log 5 - x \log 7) > \log 7$	Subtract $x \log 7$.
$x > \dfrac{\log 7}{3 \log 5 - \log 7}$	Divide by $3 \log 5 - \log 7$.
$x > 0.6751$	Use a calculator.

The solution set is $\{x \mid x > 0.6751\}$.

7-7 Base *e* and Natural Logarithms

Solve each equation or inequality. Round to the nearest ten-thousandth. **56–61. See margin.**

56. $4e^x - 11 = 17$
57. $2e^{-x} + 1 = 15$
58. $\ln 2x = 6$
59. $2 + e^x > 9$
60. $\ln (x + 3)^5 < 5$
61. $e^{-x} > 18$
62. **SAVINGS** If you deposit $2000 in an account paying 6.4% interest compounded continuously, how long will it take for your money to triple? Use $A = Pe^{rt}$. about 17.2 years

Example 9

Solve $3e^{5x} + 1 = 10$. Round to the nearest ten-thousandth.

$3e^{5x} + 1 = 10$	Original equation
$3e^{5x} = 9$	Subtract 1 from each side.
$e^{5x} = 3$	Divide each side by 3.
$\ln e^{5x} = \ln 3$	Property of Equality
$5x = \ln 3$	$\ln e^x = x$
$x = \dfrac{\ln 3}{5}$	Divide each side by 5.
$x \approx 0.2197$	Use a calculator.

7-8 Using Exponential and Logarithmic Functions

63. **CARS** Abe bought a used car for $2500. It is expected to depreciate at a rate of 25% per year. What will be the value of the car in 3 years? $1054.69
64. **BIOLOGY** For a certain strain of bacteria, k is 0.728 when t is measured in days. Using the formula $y = ae^{kt}$, how long will it take 10 bacteria to increase to 675 bacteria? ≈5.8 days
65. **POPULATION** The population of a city 20 years ago was 24,330. Since then, the population has increased at a steady rate each year. If the population is currently 55,250, find the annual rate of growth for this city. about 4.1%

Example 10

A certain culture of bacteria will grow from 250 to 2000 bacteria in 1.5 hours. Find the constant k for the growth formula. Use $y = ae^{kt}$.

$y = ae^{kt}$	Exponential Growth Formula
$2000 = 250e^{k(1.5)}$	Replace y with 2000, a with 250, and t with 1.5.
$8 = e^{1.5k}$	Divide each side by 250.
$\ln 8 = \ln e^{1.5k}$	Property of Equality
$\ln 8 = 1.5k$	Inverse Property
$\dfrac{\ln 8}{1.5} = k$	Divide each side by 1.5.
$1.3863 \approx k$	Use a calculator.

Additional Answers (Practice Test)

1.

D = {all real numbers}
R = {f(x) | f(x) > 2}

2.

D = {all real numbers}
R = {f(x) | f(x) > −3}

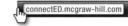
Study Guide and Intervention
AL **OL** **ELL**

NAME _____ DATE _____ PERIOD _____

7-8 Study Guide and Intervention
Using Exponential and Logarithmic Functions

Exponential Growth and Decay

Example POPULATION In 2000, the world population was estimated to be 6.124 billion people. In 2005, it was 6.515 billion.

Exercise

1. CARBON DATING Use the formula $y = ae^{-0.00012t}$, where a is the initial amount of carbon 14, t is the number of years ago the animal lived, and y is the remaining amount after t years.

   a. How old is a fossil that has lost 95% of its Carbon-14?

   b. How old is a skeleton that has 95% of its Carbon-14 remaining?

Practice
AL **OL** **BL** **ELL**

NAME _____ DATE _____ PERIOD _____

7-8 Practice
Using Exponential and Logarithmic Functions

1. BACTERIA How many hours will it take a culture of bacteria to increase from 20 to 2000? Use $k = 0.614$.

2. RADIOACTIVE DECAY A radioactive substance has a half-life of 32 years. Find the constant k in the decay formula for the substance.

3. RADIOACTIVE DECAY Cobalt, an element used to make alloys, has several isotopes. One of these, cobalt 60, is radioactive and has a half-life of 5.7 years. Cobalt 60 is used to trace the path of nonradioactive substances in a system. What is the value of k for cobalt 60?

4. WHALES Modern whales appeared 5–10 million years ago. The vertebrae of a whale discovered by paleontologists contain roughly 0.25% as much carbon-14 as they would have contained when the whale was alive. How long ago did the whale die? Use $k = 0.00012$.

5. POPULATION The population of rabbits in an area is modeled by the growth equation $P(t) = 8e^{0.26t}$, where P is in thousands and t is in years. How long will it take for the population to reach 25,000?

6. RADIOACTIVE DECAY A radioactive element decays exponentially. The decay model is given by the formula $A = A_0 e^{-0.048kt}$. A is the amount present after t days and A_0 is the amount present initially. Assume you are starting with 50g. How much of the element remains after 10 days? 30 days?

7. POPULATION A population is growing continuously at a rate of 3%. If the population is now 5 million, what will it be in 17 years' time?

8. BACTERIA A certain bacteria is growing exponentially according to the model $y = 80e^{kt}$. Using $k = 0.071$, find how many hours it will take for the bacteria to reach a population of 10,000 cells.

9. LOGISTIC GROWTH The population of a certain habitat follows the function $P(t) = \dfrac{16,300}{(1 + 17.5e^{-0.009t})}$.

   a. What is the maximum population?

   b. When does the population reach 16,200?

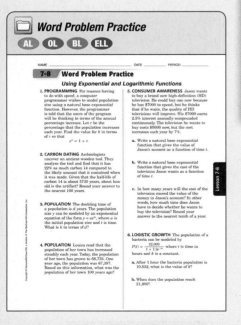

Word Problem Practice
AL **OL** **BL** **ELL**

NAME _____ DATE _____ PERIOD _____

7-8 Word Problem Practice
Using Exponential and Logarithmic Functions

1. PROGRAMMING For reasons having to do with speed, a computer programmer wishes to model population size using a natural base exponential function. However, the programmer is told that the users of the program will be thinking in terms of the annual percentage increase. Let r be the percentage that the population increases each year. Find the value for k in terms of r so that $e^k = 1 + r$.

2. CARBON DATING Archeologists uncover an ancient wooden tool. They analyze the tool and find that it has 22% as much carbon 14 as it would have in a similar new tool. Given that the half-life of carbon 14 is about 5730 years, about how old is the artifact? Round your answer to the nearest 100 years.

3. POPULATION The doubling time of a population is d years. The population size y can be modeled by an exponential equation of the form $y = ae^{kt}$, where a is the initial population size and t is time. What is k in terms of d?

4. POPULATION Louisa read that the population of her town has increased steadily each year. Today, the population of her town has grown to 68,735. One year ago, the population was 67,287. Based on this information, what was the population of her town 100 years ago?

5. CONSUMER AWARENESS Jason wants to buy a brand new high-definition (HD) television. He could buy one now because he has $7000 to spend, but he thinks that if he waits, the quality of HD televisions will improve. His $7000 earns 2.5% interest annually compounded continuously. The television he wants to buy costs $5000 now, but the cost increases each year by 7%.

   a. Write a natural base exponential function that gives the value of Jason's account as a function of time t.

   b. Write a natural base exponential function that gives the cost of the television Jason wants as a function of time t.

   c. In how many years will the cost of the television exceed the value of the money in Jason's account? In other words, how much time does Jason have to decide whether he wants to buy the television? Round your answer to the nearest tenth of a year.

6. LOGISTIC GROWTH The population of a bacteria can be modeled by $P(t) = \dfrac{22,000}{1 + 1.5e^{-kt}}$ where t is time in hours and k is a constant.

   a. After 1 hour the bacteria population is 10,532, what is the value of k?

   b. When does the population reach 21,900?

B **7** PALEONTOLOGY A paleontologist finds a human bone and determines that the Carbon-14 found in the bone is 85% of that found in living bone tissue. How old is the bone? **about 1354 yr old**

8. ANTHROPOLOGY An anthropologist has determined that a newly discovered human bone is 8000 years old. How much of the original amount of Carbon-14 is in the bone? **about 38%**

9. RADIOACTIVE DECAY 100 milligrams of Uranium-238 are stored in a container. If Uranium-238 has a half-life of about 4.47 billion years, after how many years will only 10 milligrams be present? **about 14.85 billion yr**

10. POPULATION GROWTH The population of the state of Oregon has grown from 3.4 million in 2000 to 3.7 million in 2006.

   a. Write an exponential growth equation of the form $y = ae^{kt}$ for Oregon, where t is the number of years after 2000. $y = 3.4e^{0.014t}$

   b. Use your equation to predict the population of Oregon in 2020. **4.5 million**

   c. According to the equation, when will Oregon reach 6 million people? **about 2041**

11. HALF-LIFE A substance decays 99.9% of its total mass after 200 years. Determine the half-life of the substance. **about 20.1 yr**

12. LOGISTIC GROWTH The population in millions of the state of Ohio after 1900 can be modeled by $P(t) = \dfrac{12.95}{1 + 2.4e^{-kt}}$, where t is the number of years after 1900 and k is a constant.

   a. If Ohio had a population of 10 million in 1970, find the value of k. $k \approx 0.0299464$

   b. According to the equation, when will the population of Ohio reach 12 million? **2014**

13. **MULTIPLE REPRESENTATIONS** In this problem, you will explore population growth. The population growth of a country follows the exponential function $f(t) = 8e^{0.075t}$ or the logistic function $g(t) = \dfrac{400}{1 + 16e^{-0.025t}}$. The population is measured in millions and t is time in years.

   a. **Graphical** Graph both functions for $0 \le t \le 100$. **See margin.**

   b. **Analytical** Determine the intersection of the graphs. What is the significance of this intersection?

   c. **Analytical** Which function is a more accurate estimate of the country's population 100 years from now? Explain your reasoning. **Sample answer: The logistic function $g(t)$ is a more accurate estimate of the country's population since $f(t)$ will continue to grow exponentially and $g(t)$ considers limitations on population growth such as food supply.**

13b. The graphs intersect at $t = 20.79$. Sample answer: This intersection indicates the point at which both functions determine the same population at the same time.

16. Sample answer: As $t \to +\infty$, $e^{-t} \to 0$. So, the denominator approaches $1 + 0$ or 1. As the denominator approaches 1, $f(r) \to \dfrac{c}{1}$ or c. However, since e^{-t} never reaches 0, $f(t)$ can never reach c.

H.O.T. Problems   Use Higher-Order Thinking Skills

14. OPEN ENDED Give an example of a quantity that grows or decays at a fixed rate. Write a real-world problem involving the rate and solve by using logarithms. **Sample answer: money in a bank; See students' work.**

15. CHALLENGE Solve $\dfrac{120,000}{1 + 48e^{-0.015t}} = 24e^{0.055t}$ for t. $t \approx 113.45$

16. CCSS ARGUMENTS Explain mathematically why $f(t) = \dfrac{c}{1 + 60e^{-0.5t}}$ approaches, but never reaches the value of c as $t \to +\infty$.

17. OPEN ENDED Give an example of a quantity that grows logistically and has limitations to growth. Explain why the quantity grows in this manner. **See margin.**

18. **WRITING IN MATH** How are exponential, continuous exponential, and logistic functions used to model different real-world situations? **See margin.**

Enrichment
OL **BL**

NAME _____ DATE _____ PERIOD _____

7-8 Enrichment

Effective Annual Yield

When interest is compounded more than once per year, the effective annual yield is higher than the annual interest rate. The effective annual yield, E, is the interest rate that would give the same amount of interest if the interest were compounded once per year. If P dollars are invested for one year, the value of the investment at the end of the year is $A = P(1 + E)$. If P dollars are invested for one year at a nominal rate r compounded n times per year, the value of the investment at the end of the year is $A = P\left(1 + \dfrac{r}{n}\right)^n$. Setting the amounts equal and solving for E will produce a formula for the effective annual yield.

$$P(1 + E) = P\left(1 + \dfrac{r}{n}\right)^n$$
$$1 + E = \left(1 + \dfrac{r}{n}\right)^n$$
$$E = \left(1 + \dfrac{r}{n}\right)^n - 1$$

If compounding is continuous, the value of the investment at the end of one year is $A = Pe^r$. Again set the amounts equal and solve for E. A formula for the effective annual yield under continuous compounding is obtained.

1 Focus

Objective Use a data collection device to investigate the differences between types of insulated cups and cooling time.

Materials
- TI-83/84 Plus graphing calculator
- data collection device and compatible temperature probe
- variety of containers
- very hot water

Teaching Tip
If the device being used is the Calculator-Based Laboratory 2, note the following: The CBL has three keys:

- [TRANSFER] begins transfer of programs of Calculator Software Applications between the CBL and an attached TI graphing calculator.

- [QUICK SET-UP] clears any data stored in CBLs memory.

- [START/STOP] begins sampling for Quick Set-Up. Sampling continues until the number of samples is collected or you press [START/STOP].

2 Teach

Working in Cooperative Groups
Put students in groups of three or four, mixing abilities. Then have groups complete the Activity and Exercises 1–4.

- To set up the calculator and CBL for data collection, select **Collect Data** from the Main Menu. Select **Time Graph** from the **Data Collection** menu. Enter 60 as the time between samples in seconds. Then enter 20 as the number of samples (the CBL will collect data for a total of 1200 seconds). Press ENTER , then select **Use Time Setup** to continue.

- Discuss with students the **Ymin** and the **Ymax** and the **Yscl** that are appropriate for this lab.

EXTEND 7-8 Graphing Technology Lab
Cooling

In this lab, you will explore the type of equation that models the change in the temperature of water as it cools under various conditions.

Set Up the Lab

CCSS Common Core State Standards
Content Standards
F.BF.1.b Combine standard function types using arithmetic operations.

- Collect a variety of containers, such as a foam cup, a ceramic coffee mug, and an insulated cup.

- Boil water or collect hot water from a tap.

- Choose a container to test and fill with hot water. Place the temperature probe in the cup.

- Connect the temperature probe to your data collection device.

Activity

Step 1 Program the device to collect 20 or more samples in 1 minute intervals.

Step 2 Wait a few seconds for the probe to warm to the temperature of the water.

Step 3 Press the button to begin collecting data.

> **2. Sample answer:** The exponential model fits best. Its correlation coefficient is closest to 1.
> **3. Sample answer:** No; the temperature will approach the temperature of the room but it will not drop below it.

Analyze the Results

1. When the data collection is complete, graph the data in a scatter plot. Use time as the independent variable and temperature as the dependent variable. Write a sentence that describes the points on the graph. **Sample answer: It's an exponential pattern.**

2. Use the **STAT** menu to find an equation to model the data you collected. Try linear, quadratic, and exponential models. Which model appears to fit the data best? Explain.

3. Would you expect the temperature of the water to drop below the temperature of the room? Explain your reasoning.

4. Use the data collection device to find the temperature of the air in the room. Graph the function $y = t$, where t is the temperature of the room, along with the scatter plot and the model equation. Describe the relationship among the graphs. What is the meaning of the relationship in the context of the experiment?

> **4. Sample answer:** The linear graph of the room temperature is the asymptote for the exponential graph. The temperature of the water will approach the temperature of the room, but it will not go below that temperature.

Make a Conjecture 5. **Sample answer: Yes; the thermal container will slow the rate of cooling. This affects the rate of decay in the function.**

5. Do you think the results of the experiment would change if you used an insulated container for the water? What part of the function will change, the constant or the rate of decay? Repeat the experiment to verify your conjecture.

6. How might the results of the experiment change if you added ice to the water? What part of the function will change, the constant or the rate of decay? Repeat the experiment to verify your conjecture. **Sample answer: The ice will speed the rate of cooling. This affects the rate of decay in the function.**

516 | Extend 7-8 | Graphing Technology Lab: Cooling

Practice Have groups complete Exercises 5–6.

3 Assess

Formative Assessment
In Exercise 5, check that students have used different containers, but the same initial water temperature. Ask students why it is necessary to

begin the second experiment with water at the same temperature as the first experiment.

From Concrete to Abstract
Have students describe the containers that make the best insulators.

Ed-Imaging

CHAPTER 7 Practice Test

Graph each function. State the domain and range.

1. $f(x) = 3^{x-3} + 2$ **1, 2. See margin.**

2. $f(x) = 2\left(\frac{3}{4}\right)^{x+1} - 3$

Solve each equation or inequality. Round to the nearest ten-thousandth if necessary.

3. $8^{c+1} = 16^{2c+3}$ $c = -\frac{9}{5}$

4. $9^{x-2} > \left(\frac{1}{27}\right)^x$ $\left\{x \mid x > \frac{4}{5}\right\}$

5. $2^{a+3} = 3^{2a-1}$ $a \approx 2.1130$

6. $\log_2 (x^2 - 7) = \log_2 6x$ $x = 7$

7. $\log_5 x > 2$ $\{x \mid x > 25\}$

8. $\log_3 x + \log_3 (x - 3) = \log_3 4$ $x = 4$

9. $6^{n-1} \le 11^n$ $\{n \mid n \le -2.9560\}$

10. $4e^{2x} - 1 = 5$ $x \approx 0.2027$

11. $\ln (x + 2)^2 > 2$ $\{x \mid x < -4.7183 \text{ or } x > 0.7183, x \ne -2\}$

Use $\log_5 11 \approx 1.4899$ and $\log_5 2 \approx 0.4307$ to approximate the value of each expression.

12. $\log_5 44$ **2.3513**

13. $\log_5 \frac{11}{2}$ **1.0592**

14. POPULATION The population of a city 10 years ago was 150,000. Since then, the population has increased at a steady rate each year. The population is currently 185,000. **a.** $y = 185,000(1.0212)^x$

  a. Write an exponential function that could be used to model the population after x years if the population changes at the same rate. **about**

  b. What will the population be in 25 years? **312,566**

15. Write $\log_9 27 = \frac{3}{2}$ in exponential form. $9^{\frac{3}{2}} = 27$

16. AGRICULTURE An equation that models the decline in the number of U.S. farms is $y = 3,962,520(0.98)^x$, where x is the number of years since 1960 and y is the number of farms. $\{b \mid b < 1\}$

  a. How can you tell that the number is declining?

  b. By what annual rate is the number declining? **2%**

  c. Predict when the number of farms will be less than 1 million. **in about 2028**

---

17. MULTIPLE CHOICE What is the value of $\log_4 \frac{1}{64}$? **A**

  A -3

  B $-\frac{1}{3}$

  C $\frac{1}{3}$

  D 3

18. SAVINGS You put $7500 in a savings account paying 3% interest compounded continuously.

  a. Assuming there are no deposits or withdrawals from the account, what is the balance after 5 years? **$8713.76**

  b. How long will it take your savings to double? **≈23.1 years**

  c. In how many years will you have $10,000 in your account? **about 9.6 years**

19. MULTIPLE CHOICE What is the solution of $\log_4 16 - \log_4 x = \log_4 8$? **G**

  F $\frac{1}{2}$

  G 2

  H 4

  J 8

20. MULTIPLE CHOICE Which function is graphed below? **C**

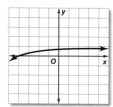

  A $y = \log_{10} (x - 5)$

  B $y = 5 \log_{10} x$

  C $y = \log_{10} (x + 5)$

  D $y = -5 \log_{10} x$

21. Write $2 \ln 6 + 3 \ln 4 - 5 \ln \left(\frac{1}{3}\right)$ as a single logarithm. $\ln \frac{(6^2)(4^3)}{\left(\frac{1}{3}\right)^5}$ or $\ln 559{,}872$

connectED.mcgraw-hill.com **521** ✓

Summative Assessment
Use these alternate leveled chapter tests to differentiate assessment for your students.

Leveled Chapter 7 Tests

Form	Type	Level	📁 Page(s)
1	MC	AL	67–68
2A	MC	OL	69–70
2B	MC	OL	71–72
2C	FR	OL	73–74
2D	FR	OL	75–76
3	FR	BL	77–78
Vocabulary Test			66
Extended-Response Test			79

MC = multiple-choice questions
FR = free-response questions

McGraw-Hill eAssessment Customize and create multiple versions of your chapter tests and their answer keys. All of the questions from the leveled chapter tests in the *Chapter 7 Resource Masters* are also available on McGraw-Hill eAssessment.

InterventionPlanner

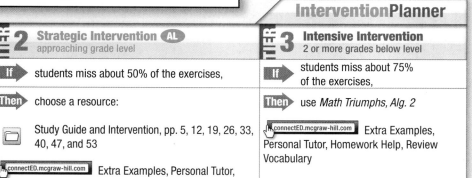

TIER 1 On Level OL

If students miss about 25% of the exercises or less,

Then choose a resource:

SE Lessons 7-1, 7-2, 7-3, 7-4, 7-5, 7-6, 7-7, and 7-8

📁 Skills Practice, pp. 7, 14, 21, 28, 35, 42, 49, and 55

TE Chapter Project

connectED.mcgraw-hill.com Self-Check Quiz

TIER 2 Strategic Intervention AL approaching grade level

If students miss about 50% of the exercises,

Then choose a resource:

📁 Study Guide and Intervention, pp. 5, 12, 19, 26, 33, 40, 47, and 53

connectED.mcgraw-hill.com Extra Examples, Personal Tutor, Homework Help

TIER 3 Intensive Intervention 2 or more grades below level

If students miss about 75% of the exercises,

Then use *Math Triumphs, Alg. 2*

connectED.mcgraw-hill.com Extra Examples, Personal Tutor, Homework Help, Review Vocabulary

1 Focus

Objective Use the strategy of using technology to solve standardized test problems.

2 Teach

Scaffolding Questions

Ask:

- What sort of tools have you used when solving math problems?
  Sample answer: calculator, protractor, compass, ruler, pencil, paper

- How do you know which tool to use?
  Sample answers: Use a calculator when graphing functions; use a protractor when measuring angles; use a compass when constructing angles; use a ruler when measuring; use pencil and paper when you do not have a calculator handy or when you cannot perform mental math.

Using Technology

Your calculator can be a useful tool in taking standardized tests. Some problems that you encounter might have steps or computations that require the use of a calculator. A calculator may also help you solve a problem more quickly.

Strategies for Using Technology

Step 1

A calculator is a useful tool, but typically it should be used sparingly. Standardized tests are designed to measure your ability to reason and solve problems, not to measure your ability to punch keys on a calculator.

Before using a calculator, ask yourself:

- How would I normally solve this type of problem?
- Are there any steps that I cannot perform mentally or by using paper and pencil?
- Is a calculator absolutely necessary to solve this problem?
- Would a calculator help me solve this problem more quickly or efficiently?

Step 2

When might a calculator come in handy?

- solving problems that involve large, complex computations
- solving certain problems that involve graphing functions, evaluating functions, solving equations, and so on
- checking solutions of problems

Standardized Test Example

Read the problem. Identify what you need to know. Then use the information in the problem to solve.

A certain can of soda contains 60 milligrams of caffeine. The caffeine is eliminated from the body at a rate of 15% per hour. What is the *half-life* of the caffeine? That is, how many hours does it take for half of the caffeine to be eliminated from the body?

A 4 hours   **C** 4.5 hours

B 4.25 hours   **D** 4.75 hours

522 | Chapter 7 | Preparing for Standardized Tests

Creativeact - Education series/Alamy

Read the problem carefully. The problem can be solved using an exponential function. Use the exponential decay formula to model the problem and solve for the half-life of caffeine.

$$y = a(1 - r)^t$$

$$y = 60(1 - 0.15)^t$$

Half of 60 milligrams is 30. So, let $y = 30$ and solve for t.

$$30 = 60(1 - 0.15)^t$$

$$0.5 = (0.85)^t$$

Take the log of each side and use the power property.

$$\log 0.5 = \log (0.85)^t$$

$$\log 0.5 = t \log 0.85$$

$$\frac{\log 0.5}{\log 0.85} = t$$

At this point, it is necessary to use a calculator to evaluate the logarithms and solve the problem. Doing so shows that $t \approx 4.265$. So, the half-life of caffeine is about 4.25 hours. The correct answer is B.

Exercises

Read each problem. Identify what you need to know. Then use the information in the problem to solve.

1. Jason recently purchased a new truck for $34,750. The value of the truck decreases by 12% each year. What will the approximate value of the truck be 7 years after Jason purchased it? **D**

   A $13,775

   B $13,890

   C $14,125

   D $14,200

2. A baseball is thrown upward at a velocity of 105 feet per second, and is released when it is 5 feet above the ground. The height of the baseball t seconds after being thrown is given by the formula $h(t) = -16t^2 + 105t + 5$. Find the time at which the baseball reaches its maximum height. **G**

F 1.0 s	H 6.6 s
G 3.3 s	J 177.3 s

3. Lucinda deposited $2500 in a CD with the terms described below.

   Super CD!

   Earn 4.25% interest compounded daily!

   (Minimum deposit of $1,000 over a period of at least 12 months.)

   Use the formula below to solve for t, the number of years needed to earn $250 in interest with the CD.

   $$2750 = 2500\left(1 + \frac{0.0425}{365}\right)^{365t}$$ **B**

   A about 2.15 years

   B about 2.24 years

   C about 2.35 years

   D about 2.46 years

Additional Example

STANDARDIZED TEST PRACTICE
A certain city had a population of 4590 in 1800. Since then, the population has grown by approximately 1.05% annually. In what year was the population of this city 37,075? D

A 1808

B 1820

C 1997

D 2000

3 Assess

Use Exercises 1–3 to assess students' understanding.

Diagnose Student Errors

Survey student responses for each item. Class trends may indicate common errors and misconceptions.

1. A correct
   B incorrectly simplified log
      $1 \div \log 4$
   C guess
   D incorrectly let $4^0 = 4$ instead of
      $4^0 = 1$

2. F incorrectly inputted values in the calculator
   G correct
   H incorrectly used the properties of logarithms
   J incorrectly used the properties of logarithms

3. A did not use a rate of 100%
   B used an incorrect formula for growth
   C incorrectly let $t = 5$ instead of 6
   D correct

4. F chose fixed cost
   G chose cost of m miles without fixed cost
   H guess
   J correct

5. A guess
   B guess
   C correct
   D did not understand that parallel lines have the same slope

6. F did not understand that no values for x can be ≤ 0
   G did not understand that no values for x can be ≤ 0
   H correct
   J did not understand that no values for x can be ≤ 0

7. A multiplied wrong quantities
   B incorrectly assumed the cost of 1 book is $5
   C correct
   D guess

Multiple Choice

Read each question. Then fill in the correct answer on the answer document provided by your teacher or on a sheet of paper.

1. What is the y-intercept of the exponential function below? **A**

$$y = 4^x - 1$$

   A 0   **B** 1   **C** 2   **D** 3

2. Suppose there are only 3500 birds of a particular endangered species left on an island and the population decreases at a rate of about 5% each year. The logarithmic function $t = \log_{0.95} \frac{p}{3500}$ predicts how many years t it will be for the population to decease to a number p. About how long will it take for the population to reach 3000 birds? **G**

   F 2 years   **H** 5 years
   G 3 years   **J** 8 years

3. Suppose a certain bacteria duplicates to reproduce itself every 20 minutes. If you begin with one cell of the bacteria, how many will there be after 2 hours? **D**

   A 2   **B** 6   **C** 32   **D** 64

4. Lucas determined that the total cost C to rent a car could be represented by the equation $C = 0.35m + 125$, where m is the number of miles that he drives. If the total cost to rent the car was $363, how many miles did he drive? **J**

   F 125   **H** 520
   G 238   **J** 680

5. Which of the following best describes the graph of $3y = 4x - 3$ and $8y = -6x - 5$? **C**

   A The lines have the same y-intercept.
   B The lines have the same x-intercept.
   C The lines are perpendicular.
   D The lines are parallel

Test-Taking Tip

Question 2 Use technology and the properties of logarithms to find t when $p = 3000$.

---

6. Graph $y = \log_5 x$. **H**

F

G

H

J
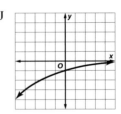

7. Ray's Book Store sells two used books for $7.99. The total cost c of purchasing n books can be found by— **C**

   A multiplying n by c.
   B multiplying n by 5.
   C multiplying n by the cost of 1 book.
   D dividing n by c.

Short Response/Gridded Response

Record your answers on the answer sheet provided by your teacher or on a sheet of paper.

8. The function $y = \left(\frac{1}{2}\right)^x$ is graphed below. What is the domain of the function? **all real numbers**

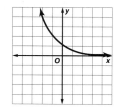

9. Suzanne bought a new car this year for $33,750. The value of the car is expected to decrease by 10.5% per year. What will be the *approximate* value of the car 6 years after Suzanne purchased it? Show your work. **about $17,347**

10. Julio is solving the system of equations $8x - 2y = 12$ and $-15x + 2y = -19$ by using elimination. His work is shown below.

$$\begin{array}{ll} 8x - 2y = 12 & 8x - 2y = 12 \\ \underline{-15x + 2y = -19} & 8(-1) - 2y = 12 \\ -7x = 7 & -8 - 2y = 12 \\ x = -1 & -2y = 20 \\ & y = -10 \end{array}$$

The solution is $(-1, -10)$.

a. What error does Julio make?

b. What is the correct solution of the system of equations? Show your work. **(1, −2)**

11. **GRIDDED RESPONSE** If $f(x) = 3x$ and $g(x) = x^2 - 1$, what is the value of $f(g(-3))$? **24**

12. Simplify $(-2a^{-2}b^{-6})(-3a^{-1}b^8)$. $\dfrac{6b^2}{a^3}$

10a. Sample answer: He has a sign error when he combines the equations; $12 + (-19) = -7$, not 7.

13. **GRIDDED RESPONSE** For what value of x would the rectangle below have an area of 48 square units? **12**

Extended Response

Record your answers on a sheet of paper. Show your work.

14. Suppose the number of whitetail deer in a particular region has increased at an annual rate of about 10% since 1995. There were 135,000 deer in 1995.

a. Write a function to model the number of whitetail deer after t years. **See margin.**

b. About how many whitetail deer inhabited the region in 2000? Round your answer to the nearest hundred deer. **about 217,400 deer**

15. Sandy inherited $250,000 from her aunt in 1998. She invested the money and increased it as shown in the table below.

Year	Amount
1998	$250,000
2006	$329,202
2011	$390,989

a. Write an exponential function that could be used to predict the amount of money A after investing for t years. $A(t) = 250{,}000(1.035)^t$

b. If the money continues to grow at the same rate, in what year will it be worth $500,000? **2018**

Need ExtraHelp?

If you missed Question...	1	2	3	4	5	6	7	8	9	10	11	12	13	14	15
Go to Lesson...	7-1	7-4	7-2	1-3	3-1	7-3	2-4	7-1	7-6	3-1	6-1	5-1	4-3	7-6	7-2

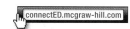

Additional Answer

14a. $N = 135{,}000(1.1)^t$, where N is the number of deer and t is the number of years since 1995.

Formative Assessment

You can use these pages to benchmark student progress.

📁 Standardized Test Practice, pp. 80–82

McGraw-Hill ℮Assessment Create practice tests that align to your state standards, the Common Core State Standards, and other national standards such as TIMSS and NAEP.

Answer Sheet Practice

Have students simulate taking a standardized test by recording their answers on a practice recording sheet.

Student Recording Sheet

Homework Option

Get Ready for Chapter 8 Assign students the exercises on p. 527 as homework to assess whether they possess the prerequisite skills needed for the next chapter.

Get Ready for Chapter 7

6. $f^{-1}(x) = \frac{1}{2}x - \frac{5}{2}$

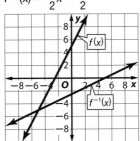

7. $f^{-1}(x) = x + 3$

8. $f^{-1}(x) = -\frac{1}{4}x$

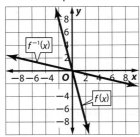

9. $f^{-1}(x) = 4x + 12$

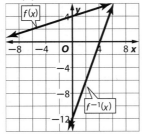

10. $f^{-1}(x) = 2x + 1$

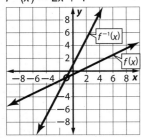

11. $f^{-1}(x) = 3x - 12$

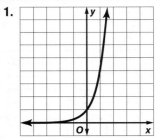

Lesson 7-1 (Guided Practice)

1.

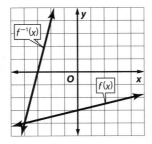

D = {all real numbers}; R = {y | y > 0}

2A.

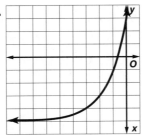

D = {all real numbers};
R = {y | y > −5}

2B.

D = {all real numbers};
R = {y | y > −3}

3.

[0, 15] scl: 1 by [70, 150] scl: 10

Lesson 7-1

1.

D = {all real numbers};
R = {f(x) | f(x) > 0}

2.

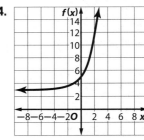

D = {all real numbers};
R = {f(x) | f(x) > 0}

3.

D = {all real numbers};
R = {f(x) | f(x) > 4}

4.

D = {all real numbers};
R = {f(x) | f(x) > 3}

5.

D = {all real numbers};
R = {f(x) | f(x) > −6}

6.

D = {all real numbers};
R = {f(x) | f(x) > 8}

8.

D = {all real numbers};
R = {f(x) | f(x) > −4}

9.

D = {all real numbers};
R = {f(x) | f(x) < 5}

17.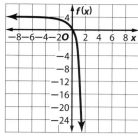

D = {all real numbers};
R = {f(x) | f(x) < 4}

18.

D = {all real numbers};
R = {f(x) | f(x) > 6}

10.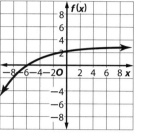

D = {all real numbers};
R = {f(x) | f(x) < 3}

11.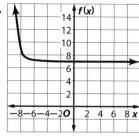

D = {all real numbers};
R = {f(x) | f(x) > 7}

19.

Number of Weeks

13.

D = {all real numbers};
R = {f(x) | f(x) > 0}

14.

D = {all real numbers};
R = {f(x) | f(x) < 0}

20.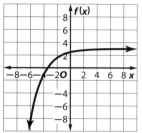

D = {all real numbers};
R = {f(x) | f(x) < 3}

21.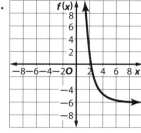

D = {all real numbers};
R = {f(x) | f(x) > −6}

15.

D = {all real numbers};
R = {f(x) | f(x) > −5}

16.

D = {all real numbers};
R = {f(x) | f(x) > 1}

22.

D = {all real numbers};
R = {f(x) | f(x) > 8}

23.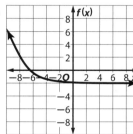

D = {all real numbers};
R = {f(x) | f(x) > −2}

24.

D = {all real numbers};
R = {f(x) | f(x) < 9}

25.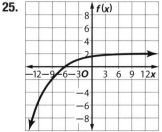

D = {all real numbers};
R = {f(x) | f(x) < 2}

33a.

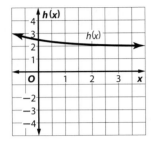

33b. Sample answer: $f(x)$; the graph of $f(x)$ is a reflection along the x-axis and the output values in the table are negative.

33d. Sample answer: $f(x)$ and $g(x)$ are growth and $h(x)$ is decay; The absolute value of the output is increasing for the growth functions and decreasing for the decay function.

Lesson 7-2

41b.

Cuts	Pieces
1	2
2	4
3	8
4	16

42. Sample answer: The more frequently interest is compounded, the higher the account balance becomes.

48.

$27^{2x} \cdot 81^{x+1} = 3^{2x+2} \cdot 9^{4x+1}$	Original equation
$(3^3)^{2x} \cdot (3^4)^{x+1} = 3^{2x+2} \cdot (3^2)^{4x+1}$	$3^2 = 9, 3^3 = 27,$ and $3^4 = 81$
$3^{6x} \cdot 3^{4x+4} = 3^{2x+2} \cdot 3^{8x+2}$	Power of a Power
$3^{10x+4} = 3^{10x+4}$	Product of Powers
$10x + 4 = 10x + 4$	Property of Equality for Exponential Functions
$10x = 10x$	Subtract 4 from each side.
$x = x$	Divide each side by 10.

Lesson 7-3

37.

38.

39.

40.

41.

42.

43.

44.

45.

46.

47.

48.

51.

52.

53.

54.

55.

56.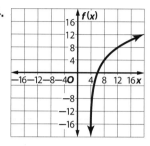

57c.

Sales versus Money Spent on Advertising

$S(a) = 10 + 20 \log_4 (a + 1)$

Mid-Chapter Quiz

1.

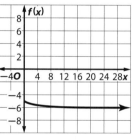

$f(x) = 3(4)^x$

D = {all real numbers}; R = {$f(x) \mid f(x) > 0$}

2.

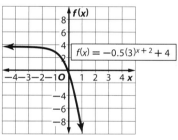

$f(x) = -(2)^x + 5$

D = {all real numbers}; R = {$f(x) \mid f(x) < 5$}

3.

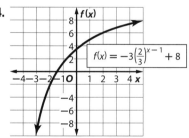

$f(x) = -0.5(3)^{x+2} + 4$

D = {all real numbers}; R = {$f(x) \mid f(x) < 4$}

4.

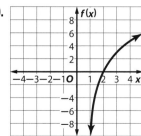

$f(x) = -3\left(\frac{2}{3}\right)^{x-1} + 8$

D = {all real numbers}; R = {$f(x) \mid f(x) < 8$}

9.

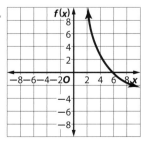

10.

66.

$\log_a \dfrac{1}{x} = -\log_a x$ Original equation

$\log_a x^{-1} = -\log_a x$ Definition of negative exponents

$\log_a x^{-1} = (-1)\log_a x$ Power Property of Logarithms

$\log_a \dfrac{1}{x} = -\log_a x$ Simplify.

67. $x^{3\log_x 2 - \log_x 5} = x^{\log_x 2^3 - \log_x 5}$

$\qquad\qquad\qquad = x^{\log_x 8 - \log_x 5}$

$\qquad\qquad\qquad = x^{\log_x \frac{8}{5}}$

$\qquad\qquad\qquad = \dfrac{8}{5}$

68. Since logarithms are exponents, the properties of logarithms are similar to the properties of exponents. The Product Property states that to multiply two powers that have the same base, add the exponents. Similarly, the logarithm of a product is the sum of the logarithms of its factors. The Quotient Property states that to divide two powers that have the same base, subtract their exponents. Similarly the logarithm of a quotient is the difference of the logarithms of the numerator and the denominator. The Power Property states that to find the power of a power, multiply the exponents. Similarly, the logarithm of a power is the product of the logarithm and the exponent. Answers should include the following.

- Quotient Property: $\log_2\left(\dfrac{32}{8}\right) = \log_2\left(\dfrac{2^5}{2^3}\right)$ Replace 32 with 2^5 and 8 with 2^8.

$\qquad\qquad\qquad\qquad = \log_2 2^{(5-3)}$ Quotient of Powers

$\qquad\qquad\qquad\qquad = 5 - 3$ or 2   Inverse Property of Exponents and Logarithms

$\log_2 32 - \log_2 8 = \log_2 2^5 - \log_2 2^3$ Replace 32 with 2^5 and 8 with 2^3.

$\qquad\qquad\qquad\quad = 5 - 3$ or 2   Inverse Property of Exponents and Logarithms

So, $\log_2\left(\dfrac{32}{8}\right) = \log_2 32 - \log_2 8$.

Power Property: $\log_3 9^4 = \log_3 (3^2)^4$ Replace 9 with 3^2.

$\qquad\qquad\qquad = \log_3 3^{(2\cdot4)}$ Power of a Power

$\qquad\qquad\qquad = 2 \cdot 4$ or 8   Inverse Property of Exponents and Logarithms

$4\log_3 9 = (\log_3 9) \cdot 4$ Commutative Property $(\times)$

$\qquad\quad = (\log_3 3^2) \cdot 4$ Replace 9 with 3^2.

$\qquad\quad = 2 \cdot 4$ or 8   Inverse Property of Exponents and Logarithms

So, $\log_3 9^4 = 4\log_3 9$.

- The Product of Powers Property and Product Property of Logarithms both involve the addition of exponents, since logarithms are exponents.

	Diagnostic Assessment Quick Check			
	LESSON 8-1 45 min: 1.5 days 90 min: 0.5 day	**LESSON 8-2** 45 min: 1.5 days 90 min: 0.5 day	**LESSON 8-3** 45 min: 1 day 90 min: 0.5 day	**LESSON 8-4** 45 min: 1.5 days 90 min: 0.5 day
Title	Multiplying and Dividing Rational Expressions	Adding and Subtracting Rational Expressions	Graphing Reciprocal Functions	Graphing Rational Functions
Objectives	■ Simplify rational expressions. ■ Simplify complex fractions.	■ Determine the LCM of polynomials. ■ Add and subtract rational expressions.	■ Determine properties of reciprocal functions. ■ Graph transformations of reciprocal functions.	■ Graph rational functions with vertical and horizontal asymptotes. ■ Graph rational functions with oblique asymptotes and point discontinuity.
Key Vocabulary	rational expression, complex fraction		reciprocal function, hyperbola, asymptote	rational function, point discontinuity
CCSS	A.APR.7	A.APR.7	A.CED.2, F.BF.3	A.CED.2, F.IF.9
Multiple Representations	⚙	⚙	⚙	
Lesson Resources	connectED.mcgraw-hill.com ▫ Leveled Worksheets ▫ Vocabulary ▫ Animations ▫ Personal Tutor ▫ Virtual Manipulatives ▫ Self-Check Quiz ■ 5-Minute Check ■ Study Notebook ■ Teaching Algebra with Manipulatives	connectED.mcgraw-hill.com ▫ Leveled Worksheets ▫ Quiz 1 ▫ Personal Tutor ▫ Self-Check Quiz ■ 5-Minute Check ■ Study Notebook	connectED.mcgraw-hill.com ▫ Leveled Worksheets ▫ Quiz 2 ▫ Vocabulary ▫ Personal Tutor ▫ Self-Check Quiz ■ 5-Minute Check ■ Study Notebook	connectED.mcgraw-hill.com ▫ Leveled Worksheets ▫ Vocabulary ▫ Animations ▫ Personal Tutor ▫ Virtual Manipulatives ▫ Self-Check Quiz ■ 5-Minute Check ■ Study Notebook
Resources for Every Lesson	[IWB] eStudent Edition [IWB] Interactive Classroom	■ eTeacher Edition ■ eSolutions Manual ■ eAssessment (McGraw-Hill)		
Differentiated Instruction	pp. 532, 537	pp. 540, 542	pp. 546, 548	pp. 555, 556, 560
			Formative Assessment Mid-Chapter Quiz	

[IWB] All digital assets are Interactive Whiteboard ready.

Suggested Pacing			
Time Periods	**Instruction**	**Review & Assess**	**Total**
45-minute	9 days	2 days	11 days
90-minute	4 days	1 day	5 days

EXTEND 8-4 — 45 min: 0.5 day / 90 min: 0.5 day	**LESSON 8-5** — 45 min: 1 day / 90 min: 0.5 day	**LESSON 8-6** — 45 min: 1.5 days / 90 min: 0.5 day	**EXTEND 8-6** — 45 min: 0.5 day / 90 min: 0.5 day
Graphing Technology Lab: Graphing Rational Functions	**Variation Functions**	**Solving Rational Equations and Inequalities**	**Graphing Technology Lab: Solving Rational Equations and Inequalities**
■ Use a graphing calculator to explore the graphs of rational functions.	■ Recognize and solve direct and joint variation problems. ■ Recognize and solve inverse and combined variation problems.	■ Solve rational equations. ■ Solve rational inequalities.	■ Use a graphing calculator to solve rational equations by graphing or by using the table feature.
	direct variation, constant of variation	rational equations, rational inequalities, weighted average	
	A.CED.2	A.CED.1, A.REI.2	A.REI.2, A.REI.11
connectED.mcgraw-hill.com PT Personal Tutor **Materials:** ■ TI-83/84 Plus or other graphing calculator	connectED.mcgraw-hill.com 📁 Leveled Worksheets 📁 Quiz 3 abc Vocabulary PT Personal Tutor ✓ Self-Check Quiz ■ *5-Minute Check* ■ *Study Notebook*	connectED.mcgraw-hill.com 📁 Leveled Worksheets 📁 Quiz 4 abc Vocabulary PT Personal Tutor ✓ Self-Check Quiz ■ *5-Minute Check* ■ *Study Notebook*	connectED.mcgraw-hill.com PT Personal Tutor **Materials:** ■ TI-83/84 Plus or other graphing calculator
IWB eStudent Edition IWB Interactive Classroom	■ eTeacher Edition ■ eSolutions Manual ■ eAssessment McGraw-Hill		
	pp. 567, 569	pp. 574, 578	
			Summative Assessment Study Guide and Review Practice Test

Assessment and Intervention

SE = Student Edition, **TE** = Teacher Edition, **CRM** = Chapter Resource Masters

Diagnosis	Prescription
Beginning Chapter 8	
Get Ready for Chapter 8 **SE**	Response to Intervention **TE**
Beginning Every Lesson	
Then, Now, Why? **SE** 5-Minute Checks	Chapter 0 **SE**
During/After Every Lesson	
Guided Practice **SE**, every example Check Your Understanding **SE** H.O.T. Problems **SE** Spiral Review **SE** Additional Examples **TE** Watch Out! **TE** Step 4, Assess **TE** Chapter 8 Quizzes **CRM**, pp. 47–48 Self-Check Quizzes connectED.mcgraw-hill.com	**TIER 1 Intervention** Skills Practice **CRM**, Ch. 1–8 connectED.mcgraw-hill.com **TIER 2 Intervention** Differentiated Instruction **TE** Study Guide and Intervention **CRM**, Ch. 1–8 **TIER 3 Intervention** *Math Triumphs, Alg. 2*, Ch. 2
Mid-Chapter	
Mid-Chapter Quiz **SE** Mid-Chapter Test **CRM**, p. 49 *e*Assessment McGraw-Hill	**TIER 1 Intervention** Skills Practice **CRM**, Ch. 1–8 connectED.mcgraw-hill.com **TIER 2 Intervention** Study Guide and Intervention **CRM**, Ch. 1–8 **TIER 3 Intervention** *Math Triumphs, Alg. 2*, Ch. 2
Before Chapter Test	
Chapter Study Guide and Review **SE** Practice Test **SE** Standardized Test Practice **SE** Chapter Test connectED.mcgraw-hill.com Standardized Test Practice connectED.mcgraw-hill.com Vocabulary Review connectED.mcgraw-hill.com *e*Assessment McGraw-Hill	**TIER 1 Intervention** Skills Practice **CRM**, Ch. 1–8 connectED.mcgraw-hill.com **TIER 2 Intervention** Study Guide and Intervention **CRM**, Ch. 1–8 **TIER 3 Intervention** *Math Triumphs, Alg. 2*, Ch. 2
After Chapter 8	
Multiple-Choice Tests, Forms 1, 2A, 2B **CRM**, pp. 51–56 Free-Response Tests, Forms 2C, 2D, 3 **CRM**, pp. 57–62 Vocabulary Test **CRM**, p. 50 Extended Response Test **CRM**, p. 63 Standardized Test Practice **CRM**, pp. 64–66 *e*Assessment McGraw-Hill	Study Guide and Intervention **CRM**, Ch. 1–8 connectED.mcgraw-hill.com

The left side bands are labeled: DIAGNOSTIC ASSESSMENT, FORMATIVE ASSESSMENT, SUMMATIVE ASSESSMENT.

Option 1 Reaching All Learners

Interpersonal Place students in groups of four. Since there are several tasks involved in graphing reciprocal functions of the form $f(x) = \dfrac{a}{x - h} + k$, have the group members decide which tasks they should each complete in order to graph a given function. For example, one member can be responsible for finding the a, h, and k values, another can identify the asymptotes, another can substitute values in order to determine points on the graph, and a fourth member can graph the points and draw the curve of the function.

Logical As you explain the various types of variation functions, have students copy and complete the following table to help them organize all the types studied.

Variation	Equation	Graph
Direct		
Joint		
Inverse		
Combined		

Option 2 Approaching Level AL

Write the following fraction and rational expression on the board.

$$\frac{32}{34} = \frac{2 \cdot 2 \cdot 2 \cdot 2 \cdot 2}{2 \cdot 17} = \frac{16}{17}$$

$$\frac{x^2 - 4x + 3}{x^2 - 6x + 5} = \frac{(x - 3)(x - 1)}{(x - 5)(x - 1)} = \frac{(x - 3)}{(x - 5)}$$

Review how to simplify fractions. Emphasize that you can only divide the numerator and denominator by common factors—common terms are not "fair game." Explain that the 3 in 32 and the 3 in 34 cannot be factored out because 3 is NOT a factor. As a class, simplify the rational expression. Explain that even though the x^2 term appears in both the numerator and the denominator, it cannot be factored out, since x^2 is a term in this expression and NOT a factor. Use colors to highlight the common factors.

Option 3 English Learners ELL

Have students think about the aspects of multiplying and dividing rational expressions that they find challenging. Then, have them write about or tell the procedures and cautions for multiplying and dividing rational expressions, demonstrating with examples.

Option 4 Beyond Level BL

Write a fraction and a rational expression on the board. Ask students to find two fractions whose sum is equal to the fraction you wrote. Then ask students to use similar thinking to find two rational expressions whose sum is equal to the expression you wrote. To make it a little more challenging, ask students to write two fractions whose denominators are different from the one you wrote, yet whose sum is equal to the fraction you wrote. Finally, challenge students to do the same with the rational expression you wrote.

Focus on Mathematical Content

VerticalAlignment

Before Chapter 8

Related Topics from Algebra 1

- use Commutative, Associative, and Distributive Properties to simplify algebraic expressions
- identify mathematical domains and ranges and determine reasonable domain and range values for given situations
- solve problems involving proportional change

Previous Topics from Algebra 2

- sketch graphs of parent functions, including linear functions
- use tools including factoring to transform and solve equations

Chapter 8

Topics from Algebra 2

- determine properties of reciprocal functions and graph their transformations
- use quotients of polynomials to describe the graphs of rational functions, describe limitations on the domains and ranges, and examine asymptotic behavior
- determine the reasonable domain and range values of rational functions and determine the reasonableness of solutions to rational equations and inequalities
- analyze a situation modeled by a rational function, formulate an equation composed of a linear or quadratic function, and solve the problem
- use functions to model and make predictions in problem situations involving direct and inverse variation

After Chapter 8

Preparation for Precalculus

- determine the domain and range of functions using graphs, tables, and symbols
- investigate the concepts of continuity, end behavior, and asymptotes, and connect these characteristics to functions represented graphically and numerically

Lesson-by-LessonPreview

8-1 Multiplying and Dividing Rational Expressions

In this lesson, students apply their prior knowledge of fractions to rational expressions (a ratio of two polynomial expressions).

- *A fraction is undefined if the denominator is zero.* Students examine polynomials in the denominators of rational expressions and identify values of variables for which the denominators are zero.
- *A fraction can be simplified by replacing a quotient of common factors in the numerator and denominator with 1.* In rational expressions, the common factors may be polynomials. Students should also recognize that the quotient of factors of the form $b - a$ and $a - b$ can be replaced with -1.
- *To multiply fractions, divide the product of the numerators by the product of the denominators.* Rational expressions are multiplied in the same way.
- *To divide by a fraction, multiply by its reciprocal.* Do the same to divide by a rational expression.
- *To simplify a complex fraction, rewrite it and treat it as a division expression.* The same is true when the complex fraction involves rational expressions.

8-2 Adding and Subtracting Rational Expressions

Adding and subtracting rational expressions is similar to adding and subtracting fractions. Least common multiples can be found by comparing prime factorizations. If two or more polynomials are factored, their least common multiple will include each factor with an exponent indicating the maximum number of times that factor appears in one of the polynomials.

To add and subtract rational expressions:

- Make sure the expressions have common denominators. If not, find the LCM of the given denominators and rewrite each rational expression as an equivalent expression having the LCM as its denominator.
- The sum (or difference) of rational expressions with common denominators is equal to an expression in which the denominator is the common denominator and the numerator is the sum (or difference) of the numerators.

8-3 Graphing Reciprocal Functions

A reciprocal function has an equation of the form $f(x) = \dfrac{1}{a(x)}$, where $a(x)$ is a linear function and $a(x) \neq 0$.

Reciprocal functions may have breaks in continuity for values that are excluded from the domain, and some may have an *asymptote,* a line that the graph of the function approaches.

- A *vertical asymptote* shows where a function is undefined. A reciprocal function is undefined for values of x that make $a(x) = 0$.

- A *horizontal asymptote* shows the end behavior of a graph.

The parent function of reciprocal functions is $f(x) = \frac{1}{x}$. The same techniques used to transform the graphs of other functions can be applied to the graphs of reciprocal functions.

8-4 Graphing Rational Functions

A rational function has an equation of the form $f(x) = \frac{a(x)}{b(x)}$, where $a(x)$ and $b(x)$ are polynomial functions and $b(x) \neq 0$. Some graphs of rational functions have breaks in continuity. A break in continuity may appear as:

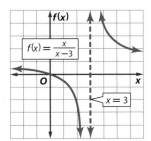

- a *vertical asymptote* whenever $b(x) = 0$.

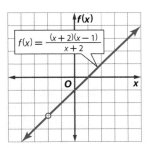

- *point discontinuity*, a hole in a graph. This type of discontinuity appears at $x = c$ whenever $x - c$ is a factor of both $a(x)$ and $b(x)$.

A rational function has at most one *horizontal asymptote*. It is:

- the line $y = 0$, whenever the degree of $a(x)$ is less than the degree of $b(x)$.

- the line $y = \dfrac{\text{leading coefficient of } a(x)}{\text{leading coefficient of } b(x)}$, whenever the degree of $a(x)$ equals the degree of $b(x)$.

- If the degree of $a(x)$ is greater than the degree of $b(x)$, there is no horizontal asymptote.

An *oblique asymptote,* also called a slant asymptote, occurs for rational functions that have a numerator polynomial that is one degree higher than the denominator polynomial.

Students can identify the asymptotes of a function and use them along with a table of values to graph the function.

8-5 Variation Functions

- A *direct variation* function is written as $y = kx$ (also written $\frac{y}{x} = k$), where k is a constant. Its graph is a *line* that passes through the origin and has slope k.

- A *joint variation* function occurs when y varies jointly as x and z and there is some nonzero constant k such that $y = kxz$.

- An *inverse variation* function is written as $y = \frac{k}{x}$ (or $xy = k$). Its graph is a *hyperbola.*

- A *combined variation* function occurs when one quantity varies directly and/or inversely as two or more other quantities.

8-6 Solving Rational Equations and Inequalities

Equations that contain one or more rational expressions are called *rational equations.* The same concepts and skills used to solve linear equations can be used to solve rational equations once the fractions are eliminated using the LCD. All solutions found should be checked in the original equation to identify extraneous solutions.

Rational equations can be written to solve *weighted average* problems. The weighted average of a set of data is the sum of the product of the number of units and the value per unit divided by the sum of the number of units. These types of problems include *mixture, work, distance,* and *interest* problems.

Mc Graw Hill **Professional Development**

Targeted professional development has been articulated throughout Algebra 2. More quality, customized professional development is available from McGraw-Hill Professional Development. Visit <u>connectED.mcgraw-hill.com</u> for details on each product.

- Online Lessons emphasize the strategies and techniques used to teach Algebra 2. Includes streaming video, interactive pages, and online tools.

- Video Workshops allow mentors, coaches, or leadership personnel to facilitate on-site workshops on educational strategies in mathematics and mathematical concepts.

- MHPD Online (<u>www.mhpdonline.com</u>) offers online professional development with video clips of instructional strategies, links, student activities, and news and issues in education.

- Teaching Today (<u>teachingtoday.glencoe.com</u>) gives secondary teachers practical strategies and materials that inspire excellence and innovation in teaching.

Chapter Project

Stop and Smell the Roses

Students use what they have learned about rational functions to complete a project.

This chapter project addresses financial literacy, as well as several specific skills identified as being essential to student success by the Framework for 21st Century Learning.

Visit connectED.mcgraw-hill.com for student and teacher handouts.

KeyVocabulary Introduce the key vocabulary in the chapter using the routine below.

<u>Define</u>: An asymptote is a line that a graph approaches.

<u>Example</u>:

The graph shows asymptotes at $x = -3$ and $f(x) = 2$.

$$f(x) = \frac{1}{x+3} + 2$$

<u>Ask</u>: What does a vertical asymptote show? *where a function is undefined*

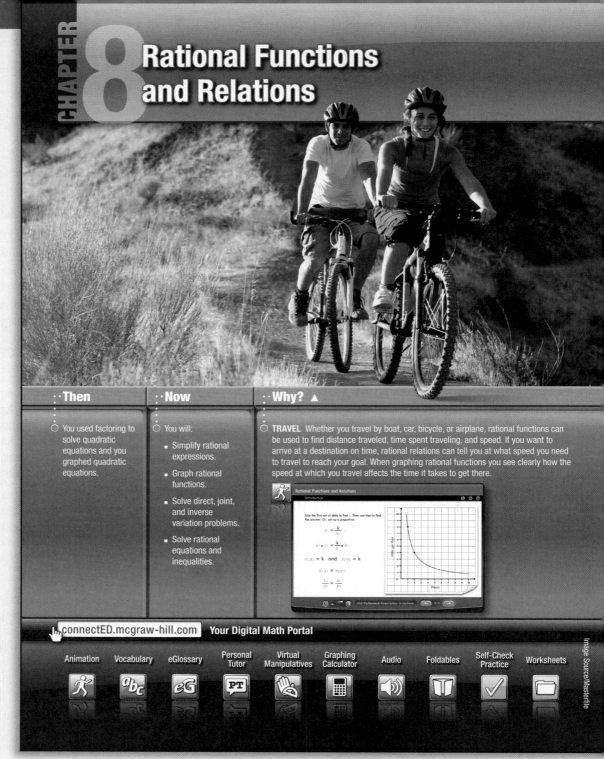

CHAPTER 8
Rational Functions and Relations

··Then

○ You used factoring to solve quadratic equations and you graphed quadratic equations.

··Now

○ You will:

- Simplify rational expressions.

- Graph rational functions.

- Solve direct, joint, and inverse variation problems.

- Solve rational equations and inequalities.

··Why? ▲

○ **TRAVEL** Whether you travel by boat, car, bicycle, or airplane, rational functions can be used to find distance traveled, time spent traveling, and speed. If you want to arrive at a destination on time, rational relations can tell you at what speed you need to travel to reach your goal. When graphing rational functions you see clearly how the speed at which you travel affects the time it takes to get there.

connectED.mcgraw-hill.com **Your Digital Math Portal**

| Animation | Vocabulary | eGlossary | Personal Tutor | Virtual Manipulatives | Graphing Calculator | Audio | Foldables | Self-Check Practice | Worksheets |

Image Source/Masterfile

Get Ready for the Chapter

Diagnose Readiness | You have two options for checking prerequisite skills.

1 Textbook Option Take the Quick Check below. Refer to the Quick Review for help.

QuickCheck	QuickReview

Solve each equation. Write in simplest form.

1. $\frac{5}{14} = \frac{1}{3}x$ $x = \frac{15}{14}$ **2.** $\frac{1}{8}m = \frac{7}{3}$ $m = \frac{56}{3}$

3. $\frac{8}{5} = \frac{1}{4}k$ $k = \frac{32}{5}$ **4.** $\frac{10}{9}p = 7$ $p = \frac{63}{10}$

5. TRUCKS Martin used $\frac{1}{3}$ of a tank of gas in his truck to get to work. He began with a full tank of gas. If he has 18 gallons of gas left, how many gallons does his tank hold? **27 gal**

Simplify each expression.

6. $\frac{3}{4} - \frac{7}{8}$ $-\frac{1}{8}$ **7.** $\frac{8}{9} - \frac{7}{6} + \frac{1}{3}$ $\frac{1}{18}$

8. $\frac{9}{10} - \frac{4}{15} + \frac{1}{3}$ $\frac{29}{30}$ **9.** $\frac{10}{3} + \frac{5}{6} + 3$ $\frac{43}{6}$

10. BAKING Annie baked cookies for a bake sale. She used $\frac{2}{3}$ cup of flour for one recipe and $4\frac{1}{2}$ cups of flour for the other recipe. How many cups did she use in all? $5\frac{1}{6}$ c

Solve each proportion.

11. $\frac{9}{12} = \frac{p}{36}$ $p = 27$

12. $\frac{9}{18} = \frac{6}{m}$ $m = 12$

13. $\frac{2}{7} = \frac{5}{k}$ $k = 17.5$

14. SALES TAX Kirsten pays $4.40 tax on $55 worth of clothes. What amount of tax will she pay on $35 worth of clothes? $2.80

Example 1 (Used in Lesson 8-1)

Solve $\frac{9}{11} = \frac{7}{8}r$. Write in simplest form.

$\frac{9}{11} = \frac{7}{8}r$

$\frac{72}{11} = 7r$ Multiply each side by 8.

$\frac{72}{77} = r$ Divide each side by 7.

Since the GCF of 72 and 77 is 1, the solution is in simplest form.

Example 2 (Used in Lesson 8-2)

Simplify $\frac{1}{3} + \frac{3}{4} - \frac{5}{6}$.

$\frac{1}{3} + \frac{3}{4} - \frac{5}{6}$

$= \frac{1}{3}\left(\frac{4}{4}\right) + \frac{3}{4}\left(\frac{3}{3}\right) - \frac{5}{6}\left(\frac{2}{2}\right)$ The GCF of 3, 4, and 6 is 12.

$= \frac{4}{12} + \frac{9}{12} - \frac{10}{12}$ Simplify.

$= \frac{3}{12}$ Add and subtract.

$= \frac{3 \div 3}{12 \div 3}$ or $\frac{1}{4}$ Simplify.

Example 3 (Used in Lesson 8-4)

Solve $\frac{5}{8} = \frac{u}{11}$.

$\frac{5}{8} = \frac{u}{11}$ Write the equation.

$5(11) = 8u$ Find the cross products.

$55 = 8u$ Simplify.

$\frac{55}{8} = u$ Divide each side by 8.

Since the GCF of 55 and 8 is 1, the answer is in simplified form.

$u = \frac{55}{8}$ or $6\frac{7}{8}$.

2 Online Option Take an online self-check Chapter Readiness Quiz at connectED.mcgraw-hill.com.

527

 Essential Question

- **Why are graphs useful?** Sample answer: Graphs are useful because they can help you visualize relationships between real-world quantities. They can also be used to estimate function values.

Get Ready for the Chapter

Response to Intervention (RtI)
Use the *Quick Check* results and the Intervention Planner to help you determine your Response to Intervention. The If-Then statements in the chart below help you decide the appropriate tier of RtI and suggest intervention resources for each tier.

 Intervention Planner

TIER 1 On Level OL

If students miss about 25% of the exercises or less,

Then choose a resource:

SE Lesson 1-3

 Skills Practice, Chapter 1, p. 19

 Self-Check Quiz

TIER 2 Strategic Intervention AL
approaching grade level

If students miss about 50% of the exercises,

Then choose a resource:

 Study Guide and Intervention, Chapter 1, p. 17

 Extra Examples, Personal Tutor, Homework Help

TIER 3 Intensive Intervention
2 or more grades below level

If students miss about 75% of the exercises,

Then use *Math Triumphs, Alg. 2*, Ch. 2

 Extra Examples, Personal Tutor, Homework Help, Review Vocabulary

FOLDABLES StudyOrganizer

Dinah Zike's Foldables®

Focus Students write notes about rational functions and relations in this chapter.

Teach Have students make and label their Foldables as illustrated. Students should use the appropriate tab to record their notes and examples for the concepts in each lesson of this chapter.

When to Use It Encourage students to add to their Foldables as they work through the chapter and to use them to review for the chapter test.

Differentiated Instruction

 Student-Built Glossary, pp. 1–2

Students should complete the chart by providing a definition of each term and an example as they progress through Chapter 8. This study tool can also be used to review for the chapter test.

Get Started on the Chapter

You will learn several new concepts, skills, and vocabulary terms as you study Chapter 8. To get ready, identify important terms and organize your resources. You may wish to refer to Chapter 0 to review prerequisite skills.

FOLDABLES StudyOrganizer

Rational Functions and Relations Make this Foldable to help you organize your Chapter 8 notes about rational functions and relations. Begin with an $8\frac{1}{2}'' \times 11''$ sheet of grid paper.

1 Fold in thirds along the height.

2 Fold the top edge down making a 2″ tab at the top. Cut along the folds.

3 Label the outside tabs *Expressions*, *Functions*, and *Equations*. Use the inside tabs for definitions and notes.

4 Write examples of each topic in the space below each tab.

NewVocabulary

English		Español
rational expression	p. 529	expresión racional
complex fraction	p. 532	fracción compleja
reciprocal function	p. 545	función recíproco
hyperbola	p. 545	hipérbola
rational function	p. 553	función racional
vertical asymptote	p. 553	asíntota vertical
horizontal asymptote	p. 553	asíntota horizontal
oblique asymptote	p. 555	asíntota oblicua
point discontinuity	p. 556	discontinuidad evitable
direct variation	p. 562	variación directa
constant of variation	p. 562	constante de variación
joint variation	p. 563	variación conjunta
inverse variation	p. 564	variación inversa
combined variation	p. 565	variación combinada
rational equation	p. 570	ecuación racional
weighted average	p. 572	media ponderada
rational inequality	p. 575	desigualdad racional

ReviewVocabulary

function *función* a relation in which each element of the domain is paired with exactly one element of the range

domain

$\{(1, -2), (3, -4), (5, -6), (7, -8), (9, -10)\}$

range

least common multiple *mínimo común múltiplo* the least number that is a common multiple of two or more numbers

rational number *número racional* a number expressed in the form $\frac{a}{b}$, where *a* and *b* are integers and $b \neq 0$

LESSON 8-1

Multiplying and Dividing Rational Expressions

∴Then	∴Now	∴Why?
● You factored polynomials.	**1** Simplify rational expressions. **2** Simplify complex fractions.	● If a scuba diver goes to depths greater than 33 feet, the rational function $T(d) = \frac{1700}{d-33}$ gives the maximum time a diver can remain at those depths and still surface at a steady rate with no stops. $T(d)$ represents the dive time in minutes, and d represents the depth in feet.

 NewVocabulary
rational expression
complex fraction

 Common Core State Standards

Content Standards
A.APR.7 Understand that rational expressions form a system analogous to the rational numbers, closed under addition, subtraction, multiplication, and division by a nonzero rational expression; add, subtract, multiply, and divide rational expressions.

Mathematical Practices
8 Look for and express regularity in repeated reasoning.

1 **Simplify Rational Expressions** A ratio of two polynomial expressions such as $\frac{1700}{d-33}$ is called a **rational expression**.

Because variables in algebra often represent real numbers, operations with rational numbers and rational expressions are similar. Just as with reducing fractions, to simplify a rational expression, you divide the numerator and denominator by their greatest common factor (GCF).

$$\frac{8}{12} = \frac{2 \cdot \overset{1}{\cancel{4}}}{3 \cdot \underset{1}{\cancel{4}}} = \frac{2}{3} \qquad \frac{x^2 - 4x + 3}{x^2 - 6x + 5} = \frac{(x-3)\overset{1}{\cancel{(x-1)}}}{(x-5)\underset{1}{\cancel{(x-1)}}} = \frac{(x-3)}{(x-5)}$$

$$\boxed{\text{GCF} = 4} \qquad \boxed{\text{GCF} = (x-1)}$$

Example 1 Simplify a Rational Expression

a. Simplify $\frac{5x(x^2 + 4x + 3)}{(x-6)(x^2-9)}$.

$$\frac{5x(x^2 + 4x + 3)}{(x-6)(x^2-9)} = \frac{5x(x+3)(x+1)}{(x-6)(x+3)(x-3)}$$ Factor numerator and denominator.

$$= \frac{5x(x+1)}{(x-6)(x-3)} \cdot \frac{\overset{1}{\cancel{(x+3)}}}{\underset{1}{\cancel{(x+3)}}}$$ Eliminate common factors.

$$= \frac{5x(x+1)}{(x-6)(x-3)}$$ Simplify.

b. Under what conditions is this expression undefined?

The original factored denominator is $(x-6)(x+3)(x-3)$.
Determine the values that would make the denominator equal to 0.
These values are 6, −3, or 3, so the expression is undefined when $x = 6$, −3 or 3.

▶ **Guided**Practice **1A.** $\frac{4(y+4)}{(y+2)}$; $y \neq 0$, −2, or 3   **1B.** $\frac{2z(z+4)}{(z-1)}$; $z \neq 1$, 2, or −5

Simplify each expression. Under what conditions is the expression undefined?

1A. $\frac{4y(y-3)(y+4)}{y(y^2-y-6)}$ **1B.** $\frac{2z(z+5)(z^2+2z-8)}{(z-1)(z+5)(z-2)}$

 connectED.mcgraw-hill.com   **529**

Pete Atkinson/Photographer's Choice/Getty Images

1 Focus

VerticalAlignment

▼ **Before Lesson 8-1** Factor polynomials.

▼ **Lesson 8-1** Simplify rational expressions. Simplify complex fractions.

▼ **After Lesson 8-1** Solve rational equations.

2 Teach

Scaffolding Questions
Have students read the **Why?** section of the lesson.

Ask:

■ How can the term *rational expression* help you identify what it means?
"Rational" contains the word "ratio."

■ What ratio is in the function $T(d) = \frac{1700}{d-33}$? $\frac{1700}{d-33}$

Lesson 8-1 Resources

Resource	Approaching Level **AL**	On Level **OL**	Beyond Level **BL**	English Learners **ELL**
Teacher Edition	■ Differentiated Instruction, p. 532	■ Differentiated Instruction, pp. 532, 537	■ Differentiated Instruction, p. 537	
Chapter Resource Masters	■ Study Guide and Intervention, pp. 5–6 ■ Skills Practice, p. 7 ■ Practice, p. 8 ■ Word Problem Practice, p. 9	■ Study Guide and Intervention, pp. 5–6 ■ Skills Practice, p. 7 ■ Practice, p. 8 ■ Word Problem Practice, p. 9 ■ Enrichment, p. 10	■ Practice, p. 8 ■ Word Problem Practice, p. 9 ■ Enrichment, p. 10	■ Study Guide and Intervention, pp. 5–6 ■ Skills Practice, p. 7 ■ Practice, p. 8 ■ Word Problem Practice, p. 9
Other	■ 5-Minute Check 8-1 ■ Study Notebook ■ Teaching Algebra with Manipulatives	■ 5-Minute Check 8-1 ■ Study Notebook ■ Teaching Algebra with Manipulatives	■ 5-Minute Check 8-1 ■ Study Notebook	■ 5-Minute Check 8-1 ■ Study Notebook ■ Teaching Algebra with Manipulatives

1 Simplify Rational Expressions

Example 1 shows how to simplify a rational expression and determine the excluded values. **Example 2** shows how to use elimination to answer a test item. **Example 3** shows how to factor out -1 in order to simplify a rational expression. **Example 4** shows how to multiply and divide rational expressions. **Example 5** shows how to simplify a rational expression containing polynomials.

Formative Assessment

Use the Guided Practice exercises after each Example to determine students' understanding of concepts.

Additional Examples

1 a. Simplify $\dfrac{3y(y + 7)}{(y + 7)(y^2 - 9)}$.

$\dfrac{3y}{y^2 - 9}$

b. Under what conditions is this expression undefined? when $y = -7$, $y = -3$, or $y = 3$

2 STANDARDIZED TEST EXAMPLE

For what value(s) is

$\dfrac{p^2 + 2p - 3}{p^2 - 2p - 15}$ undefined? B

A 5   C 3, -5

B $-3, 5$ D $5, 1, -3$

3 Simplify $\dfrac{a^4b - 2a^4}{2a^3 - a^3b}$. $-a$

▶ **Additional Examples** also in Interactive Classroom PowerPoint® Presentations

 Interactive White Board **READY**

For what value(s) is $\dfrac{x^2(x^2 - 5x - 14)}{4x(x^2 + 6x + 8)}$ undefined?

A $-2, -4$ C $0, -2, -4$

B $-2, 7$ D $0, -2, -4, 7$

Read the Test Item

You want to determine which values of x make the denominator equal to 0.

Solve the Test Item

With $4x$ in the denominator, x cannot equal 0. So, choices A and B can be eliminated. Next, factor the denominator.

$x^2 + 6x + 8 = (x + 2)(x + 4)$, so the denominator is $4x(x + 2)(x + 4)$.

Because the denominator equals 0 when $x = 0, -2$, and -4, the answer is C.

> **Test-Taking Tip**
>
> **Eliminating Choices** Sometimes you can save time by looking at the possible answers and eliminating choices.

▶ **Guided Practice**

2. For what value(s) of x is $\dfrac{x(x^2 + 8x + 12)}{-6(x^2 - 3x - 10)}$ undefined? G

   F $0, 5, -2$ G $5, -2$ H $0, -2, -6$ J $5, -2, -6$

Sometimes you can factor out -1 in the numerator or denominator to help simplify a rational expression.

[PT]

Example 3 Simplify Using -1

Simplify $\dfrac{(4w^2 - 3wy)(w + y)}{(3y - 4w)(5w + y)}$.

$\dfrac{(4w^2 - 3wy)(w + y)}{(3y - 4w)(5w + y)} = \dfrac{w(4w - 3y)(w + y)}{(3y - 4w)(5w + y)}$ Factor.

$= \dfrac{w(-1)(3y \overset{1}{\underset{}{-}} 4w)(w + y)}{(3y - 4w)(5w + y)}$ $4w - 3y = -1(3y - 4w)$

$= \dfrac{(-w)(w + y)}{5w + y}$ Simplify.

▶ **Guided Practice**

Simplify each expression.

3A. $\dfrac{(xz - 4z)}{z^2(4 - x)}$ $-\dfrac{1}{z}$ **3B.** $\dfrac{ab^2 - 5ab}{(5 + b)(5 - b)}$ $\dfrac{ab}{(5 + b)}$

The method for multiplying and dividing fractions also works with rational expressions. Remember that to multiply two fractions, you multiply the numerators and multiply the denominators. To divide two fractions, you multiply by the multiplicative inverse, or the reciprocal, of the divisor.

Multiplication

$\dfrac{2}{9} \cdot \dfrac{15}{4} = \dfrac{\overset{1}{2} \cdot \overset{1}{3} \cdot 5}{\underset{1}{3} \cdot 3 \cdot \underset{1}{2} \cdot 2} = \dfrac{5}{3 \cdot 2} = \dfrac{5}{6}$

Division

$\dfrac{3}{5} \div \dfrac{6}{35} = \dfrac{3}{5} \cdot \dfrac{35}{6} = \dfrac{\overset{1}{3} \cdot \overset{1}{5} \cdot 7}{\underset{1}{5} \cdot 2 \cdot \underset{1}{3}} = \dfrac{7}{2}$

 530 | Lesson 8-1 | Multiplying and Dividing Rational Expressions

---

> ### WatchOut!
>
> **Common Misconceptions** Point out that rational expressions are usually used without specifically excluding those values that make the expression undefined. It is understood that only those values for which the expression has meaning are included.

Teach with Tech

Interactive Whiteboard Use the board to help demonstrate canceling out common factors. Cross out the common factors as you simplify the expression, and drag the remaining factors to create the simplified expression.

The following table summarizes the rules for multiplying and dividing rational expressions.

KeyConcept

	Multiplying Rational Expressions
Words	To multiply rational expressions, multiply the numerators and multiply the denominators.
Symbols	For all rational expressions $\frac{a}{b}$ and $\frac{c}{d}$ with $b \neq 0$ and $d \neq 0$, $\frac{a}{b} \cdot \frac{c}{d} = \frac{ac}{bd}$.
	Dividing Rational Expressions
Words	To divide rational expressions, multiply by the reciprocal of the divisor.
Symbols	For all rational expressions $\frac{a}{b}$ and $\frac{c}{d}$ with $b \neq 0$, $c \neq 0$, and $d \neq 0$, $\frac{a}{b} \div \frac{c}{d} = \frac{a}{b} \cdot \frac{d}{c} = \frac{ad}{bc}$.

StudyTip
Eliminating Common Factors Be sure to eliminate factors from both the numerator and denominator.

Example 4 Multiply and Divide Rational Expressions

Simplify each expression.

a. $\dfrac{6c}{5d} \cdot \dfrac{15cd^2}{8a}$

$\dfrac{6c}{5d} \cdot \dfrac{15cd^2}{8a} = \dfrac{2 \cdot 3 \cdot c \cdot 5 \cdot 3 \cdot c \cdot d \cdot d}{5 \cdot d \cdot 2 \cdot 2 \cdot 2 \cdot a}$ Factor.

$= \dfrac{\overset{1}{2} \cdot 3 \cdot c \cdot \overset{1}{5} \cdot 3 \cdot c \cdot \overset{1}{d} \cdot d}{\underset{1}{5} \cdot \underset{1}{d} \cdot 2 \cdot \underset{1}{2} \cdot 2 \cdot a}$ Eliminate common factors.

$= \dfrac{3 \cdot 3 \cdot c \cdot c \cdot d}{2 \cdot 2 \cdot a}$ Simplify.

$= \dfrac{9c^2d}{4a}$ Simplify.

b. $\dfrac{18xy^3}{7a^2b^2} \div \dfrac{12x^2y}{35a^2b}$

$\dfrac{18xy^3}{7a^2b^2} \div \dfrac{12x^2y}{35a^2b} = \dfrac{18xy^3}{7a^2b^2} \cdot \dfrac{35a^2b}{12x^2y}$ Multiply by reciprocal of the divisor.

$= \dfrac{2 \cdot 3 \cdot 3 \cdot x \cdot y \cdot y \cdot y \cdot 5 \cdot 7 \cdot a \cdot a \cdot b}{7 \cdot a \cdot a \cdot b \cdot b \cdot 2 \cdot 2 \cdot 3 \cdot x \cdot x \cdot y}$ Factor.

$= \dfrac{\overset{1}{2} \cdot \overset{1}{3} \cdot 3 \cdot \overset{1}{x} \cdot \overset{1}{y} \cdot y \cdot y \cdot 5 \cdot \overset{1}{7} \cdot \overset{1}{a} \cdot \overset{1}{a} \cdot \overset{1}{b}}{\underset{1}{7} \cdot \underset{1}{a} \cdot \underset{1}{a} \cdot \underset{1}{b} \cdot b \cdot 2 \cdot 2 \cdot 3 \cdot x \cdot x \cdot y}$ Eliminate common factors.

$= \dfrac{3 \cdot 5 \cdot y \cdot y}{2 \cdot b \cdot x}$ Simplify.

$= \dfrac{15y^2}{2bx}$ Simplify.

GuidedPractice

4A. $\dfrac{12c^3d^2}{21ab} \cdot \dfrac{14a^2b}{8c^2d}$ acd

4B. $\dfrac{6xy}{15ab^2} \cdot \dfrac{21a^3}{18x^4y}$ $\dfrac{7a^2}{15b^2x^3}$

4C. $\dfrac{16mt^2}{21a^4b^3} \div \dfrac{24m^3}{7a^2b^2}$ $\dfrac{2t^2}{9a^2bm^2}$

4D. $\dfrac{12x^4y^2}{40a^4b^4} \div \dfrac{6x^2y^4}{16a^2x}$ $\dfrac{4x^3}{5a^2b^4y^2}$

Focus on Mathematical Content
Simplifying Rational Expressions
The primary skill needed to multiply and divide rational expressions is simplifying. After division is changed to multiplication by the reciprocal of the divisor, and numerators and denominators are multiplied, complete the problem by dividing by the common factors.

Additional Example

4 Simplify each expression.

a. $\dfrac{8x}{21y^3} \cdot \dfrac{7y^2}{16x^3}$ $\dfrac{1}{6x^2y}$

b. $\dfrac{10mk^2}{3c^2d} \div \dfrac{5m^5}{6c^2d^2}$ $\dfrac{4dk^2}{m^4}$

Tips for New Teachers
Building on Prior Knowledge To help students understand why division is equivalent to multiplying by the reciprocal, discuss simple examples such as this: dividing 18 marbles between two people means that each person gets one-half, or 9, of the marbles.

Focus on Mathematical Content
Division by Zero By definition, $\dfrac{a}{b} = c$ if $a = bc$. If students think $\dfrac{6}{0} = 0$, use the definition to show $\dfrac{6}{0} = 0$ if $6 = 0 \cdot 0$, which is false.

Additional Example

5 Simplify each expression.

a. $\dfrac{k-3}{k+1} \cdot \dfrac{1-k^2}{k^2-4k+3}$ -1

b. $\dfrac{2d+6}{d^2+d-2} \div \dfrac{d+3}{d^2+3d+2}$

$\dfrac{2(d+1)}{d-1}$

 Teaching the Mathematical Practices

Regularity Mathematically proficient students notice if calculations are repeated, and look both for general methods and for shortcuts. Encourage students to examine all of the factors involved, not just the factors in each expression individually.

2 Simplify Complex Fractions
Example 6 shows how to simplify a complex fraction by rewriting it as a division expression.

Sometimes you must factor the numerator and/or the denominator first before you can simplify a product or a quotient of rational expressions.

StudyTip

CCSS **Regularity** When simplifying rational expressions, factors in one polynomial will often reappear in other polynomials. In Example 5a, $x-8$ appears four times. Use this as a guide when factoring challenging polynomials.

Example 5 Polynomials in the Numerator and Denominator

Simplify each expression.

a. $\dfrac{x^2-6x-16}{x^2-16x+64} \cdot \dfrac{x-8}{x^2+5x+6}$

$\dfrac{x^2-6x-16}{x^2-16x+64} \cdot \dfrac{x-8}{x^2+5x+6} = \dfrac{(x-8)(x+2)}{(x-8)(x-8)} \cdot \dfrac{x-8}{(x+3)(x+2)}$ Factor.

$= \dfrac{\overset{1}{(x-8)}\overset{1}{(x+2)}}{(x-8)(x-8)} \cdot \dfrac{\overset{1}{x-8}}{(x+3)(x+2)}$ Eliminate common factors.

$= \dfrac{1}{x+3}$ Simplify.

b. $\dfrac{x^2-16}{12y+36} \div \dfrac{x^2-12x+32}{y^2-3y-18}$

$\dfrac{x^2-16}{12y+36} \div \dfrac{x^2-12x+32}{y^2-3y-18} = \dfrac{x^2-16}{12y+36} \cdot \dfrac{y^2-3y-18}{x^2-12x+32}$ Multiply by reciprocal.

$= \dfrac{(x+4)(x-4)}{12(y+3)} \cdot \dfrac{(y-6)(y+3)}{(x-4)(x-8)}$ Factor.

$= \dfrac{(x+4)\overset{1}{(x-4)}}{12\overset{}{(y+3)}} \cdot \dfrac{(y-6)\overset{}{(y+3)}}{\underset{1}{(x-4)}(x-8)}$ Eliminate common factors.

$= \dfrac{(x+4)(y-6)}{12(x-8)}$ Simplify.

Guided Practice

5A. $\dfrac{8x-20}{x^2+2x-35} \cdot \dfrac{x^2-7x+10}{4x^2-16}$ $\dfrac{2x-5}{(x+2)(x+7)}$

5B. $\dfrac{x^2-9x+20}{x^2+10x+21} \div \dfrac{x^2-x-12}{6x+42}$ $\dfrac{6x-30}{(x+3)^2}$

2 Simplify Complex Fractions A **complex fraction** is a rational expression with a numerator and/or denominator that is also a rational expression. The following expressions are complex fractions.

$$\dfrac{\tfrac{c}{6}}{5d} \qquad \dfrac{\tfrac{8}{x}}{x-2} \qquad \dfrac{x-3}{\tfrac{x-2}{x+4}} \qquad \dfrac{\tfrac{4}{a}+6}{\tfrac{12}{a}-3}$$

To simplify a complex fraction, first rewrite it as a division expression.

Example 6 Simplify Complex Fractions

Simplify each expression.

a. $\dfrac{\tfrac{a+b}{4}}{\tfrac{a^2+b^2}{4}}$

$\dfrac{\tfrac{a+b}{4}}{\tfrac{a^2+b^2}{4}} = \dfrac{a+b}{4} \div \dfrac{a^2+b^2}{4}$ Express as a division expression.

$= \dfrac{a+b}{4} \cdot \dfrac{4}{a^2+b^2}$ Multiply by the reciprocal.

$= \dfrac{a+b}{\underset{1}{4}} \cdot \dfrac{\overset{1}{4}}{a^2+b^2}$ or $\dfrac{a+b}{a^2+b^2}$ Simplify.

DifferentiatedInstruction (AL) (OL)

If ▶ students are having difficulty with these problems,

Then ▶ encourage them to use several steps, writing each one below the previous and keeping each line equivalent to the one above. Caution them to make only one change per step.

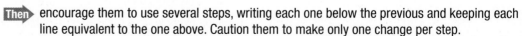

b. $\dfrac{\frac{x^2}{x^2-y^2}}{\frac{4x}{y-x}}$

$$\dfrac{\frac{x^2}{x^2-y^2}}{\frac{4x}{y-x}} = \frac{x^2}{x^2-y^2} \div \frac{4x}{y-x}$$ Express as a division expression.

$$= \frac{x^2}{x^2-y^2} \cdot \frac{y-x}{4x}$$ Multiply by the reciprocal.

$$= \frac{x \cdot x}{(x+y)(x-y)} \cdot \frac{(-1)(x-y)}{4x}$$ Factor.

$$= \frac{x \cdot \overset{1}{x}}{(x+y)(x-y)} \cdot \frac{(-1)(x-y)}{\underset{1}{4x}}$$ Eliminate Factors.

$$= \frac{-x}{4(x+y)}$$ Simplify.

▶ **Guided**Practice

Simplify each expression.

6A. $\dfrac{\frac{(x-2)^2}{2(x^2-5x+4)}}{\frac{x^2-4}{4x-10}}$ $\dfrac{(2x-5)(x-2)}{(x+2)(x-4)(x-1)}$

6B. $\dfrac{\frac{x^2-y^2}{y^2-49}}{\frac{y-x}{y+7}}$ $\dfrac{-x-y}{y-7}$

Check Your Understanding ⬤ = Step-by-Step Solutions begin on page R14. ☑

Example 1 | **Simplify each expression.**

1. $\dfrac{x^2-5x-24}{x^2-64} \cdot \dfrac{x+3}{x+8}$

2. $\dfrac{c+d}{3c^2-3d^2} \cdot \dfrac{1}{3(c-d)}$

Example 2

3. MULTIPLE CHOICE Identify all values of x for which $\dfrac{x+7}{x^2-3x-28}$ is undefined. **D**

   A $-7, 4$ **B** $7, 4$ **C** $4, -7, 7$ **D** $-4, 7$

Examples 3–6 **Simplify each expression.**

4. $\dfrac{y^2+3y-40}{25-y^2} \cdot \dfrac{y+8}{y+5}$

5 $\dfrac{a^2x-b^2x}{by-ay} \cdot \dfrac{-x(a+b)}{y}$

6. $\dfrac{27x^2y^4}{16yz^3} \cdot \dfrac{8z}{9xy^3} \cdot \dfrac{3x}{2z^2}$

7. $\dfrac{12x^3y}{13ab^2} \div \dfrac{36xy^3}{26b} \cdot \dfrac{2x^2}{3aby^2}$

8. $\dfrac{x^2-4x-21}{x^2-6x+8} \cdot \dfrac{x-4}{x^2-2x-35} \cdot \dfrac{x+3}{(x-2)(x+5)}$

9. $\dfrac{a^2-b^2}{3a^2-6a+3} \div \dfrac{4a+4b}{a^2-1} \cdot \dfrac{(a-b)(a+1)}{12(a-1)}$

10. $\dfrac{\frac{a^3b^3}{xy^4}}{\frac{a^2b}{x^2y}} \cdot \dfrac{ab^2x}{y^3}$

11. $\dfrac{\frac{4x}{x+6}}{\frac{x^2-3x}{x^2+3x-18}}$ 4

12. CCSS SENSE-MAKING The volume of a shipping container in the shape of a rectangular prism can be represented by the polynomial $6x^3+11x^2+4x$, where the height is x.

  a. Find the length and width of the container. $2x+1, 3x+4$

  b. Find the ratio of the three dimensions of the container when $x = 2$. $2 : 5 : 10$

  c. Will the ratio of the three dimensions be the same for all values of x? **no**

Volume = $6x^3 + 11x^2 + 4x$

Additional Example

6 Simplify $\dfrac{\frac{x^2}{9x^2-4y^2}}{\frac{x^3}{2y-3x}} \cdot \dfrac{-1}{3x^2+2xy}$

Tips for New Teachers

Reasoning Help students understand why the quotient of $(x-y)$ and $(y-x)$ is -1 by pointing out that these two expressions are opposites (or additive inverses), just like 2 and -2.

CCSS **Teaching the Mathematical Practices**

Sense-Making Mathematically proficient students start by explaining to themselves the meaning of a problem and looking for entry points to its solution. They analyze givens, constraints, relationships, and goals. They check their answers to problems using a different method, and they continually ask themselves, "Does this make sense?"

3 Practice

Formative Assessment

Use Exercises 1–12 to check for understanding.

Use the chart at the bottom of the next page to customize assignments for your students.

Teaching the Mathematical Practices

Reasoning Mathematically proficient students make sense of quantities and their relationships in problem situations. Quantitative reasoning entails habits of creating a coherent representation of the problem at hand; considering the units involved; attending to the meaning of quantities, not just how to compute them; and knowing and flexibly using different properties of operations and objects.

Practice and Problem Solving

Extra Practice is on page R8.

Example 1　Simplify each expression.

13. $\dfrac{x(x-3)(x+6)}{x^2+x-12}$　$\dfrac{x(x+6)}{x+4}$

14. $\dfrac{y^2(y^2+3y+2)}{2y(y-4)(y+2)}$　$\dfrac{y(y+1)}{2(y-4)}$

15. $\dfrac{(x^2-9)(x^2-z^2)}{4(x+z)(x-3)}$　$\dfrac{(x+3)(x-z)}{4}$

16. $\dfrac{(x^2-16x+64)(x+2)}{(x^2-64)(x^2-6x-16)}$　$\dfrac{1}{x+8}$

17. $\dfrac{x^2(x+2)(x-4)}{6x(x^2+x-20)}$　$\dfrac{x(x+2)}{6(x+5)}$

18. $\dfrac{3y(y-8)(y^2+2y-24)}{15y^2(y^2-12y+32)}$　$\dfrac{(y+6)}{5y}$

Example 2　19. **MULTIPLE CHOICE** Identify all values of x for which $\dfrac{(x-3)(x+6)}{(x^2-7x+12)(x^2-36)}$ is undefined.　**J**

F $3, -6$　　　G $4, 6$　　　H $-6, 6$　　　J $-6, 3, 4, 6$

Example 3　Simplify each expression.

20. $\dfrac{x^2-5x-14}{28+3x-x^2}$　$\dfrac{x+2}{x+4}$

21. $\dfrac{x^3-9x^2}{x^2-3x-54}$　$\dfrac{x^2}{x+6}$

22. $\dfrac{(x-4)(x^2+2x-48)}{(36-x^2)(x^2+4x-32)}$　$-\dfrac{1}{x+6}$

23. $\dfrac{16-c^2}{c^2+c-20}$　$-\dfrac{c+4}{c+5}$

24. **GEOMETRY** The cylinder at the right has a volume of $(x+3)(x^2-3x-18)\pi$ cubic centimeters. Find the height of the cylinder. $x-6$ cm

$2x+6$ cm

Examples 4–6　Simplify each expression.

25. $\dfrac{3ac^3f^3}{8a^2bcf^4}\cdot\dfrac{12ab^2c}{18ab^3c^2f}$　$\dfrac{c}{4ab^2f^2}$

26. $\dfrac{14xy^2z^3}{21w^4x^2yz}\cdot\dfrac{7wxyz}{12w^2y^3z}$　$\dfrac{7z^2}{18w^5y}$

27. $\dfrac{64a^2b^5}{35b^2c^3f^4}\div\dfrac{12a^4b^3c}{70abcf^2}$　$\dfrac{32b}{3ac^3f^2}$

28. $\dfrac{9x^2yz}{5z^4}\div\dfrac{12x^4y^2}{50xy^4z^2}$　$\dfrac{15y^3}{2xz}$

29. $\dfrac{15a^2b^2}{21ac}\cdot\dfrac{14a^4c^2}{6ab^3}$　$\dfrac{5a^4c}{3b}$

30. $\dfrac{14c^2f^5}{9a^2}\div\dfrac{35cf^4}{18ab^3}$　$\dfrac{4b^3cf}{5a}$

(31) $\dfrac{y^2+8y+15}{y-6}\cdot\dfrac{y^2-9y+18}{y^2-9}$　$y+5$

32. $\dfrac{c^2-6c-16}{c^2-d^2}\div\dfrac{c^2-8c}{c+d}$　$\dfrac{c+2}{c(c-d)}$

33. $\dfrac{x^2+9x+20}{8x+16}\cdot\dfrac{4x^2+16x+16}{x^2-25}$　$\dfrac{(x+4)(x+2)}{2(x-5)}$

34. $\dfrac{3a^2+6a+3}{a^2-3a-10}\div\dfrac{12a^2-12}{a^2-4}$　$\dfrac{(a+1)(a-2)}{4(a-5)(a-1)}$

35. $\dfrac{\dfrac{x^2-9}{6x-12}}{\dfrac{x^2+10x+21}{x^2-x-2}}$　$\dfrac{(x-3)(x+1)}{6(x+7)}$

36. $\dfrac{\dfrac{y-x}{z^3}}{\dfrac{x-y}{6z^2}}$　$-\dfrac{6}{z}$

37. $\dfrac{\dfrac{a^2-b^2}{b^3}}{\dfrac{b^2-ab}{a^2}}$　$-a^2(a+b)$ $\dfrac{}{b^4}$

38. $\dfrac{\dfrac{x-y}{a+b}}{\dfrac{x^2-y^2}{b^2-a^2}}$　$\dfrac{b-a}{x+y}$

39. **CCSS REASONING** At the end of her high school soccer career, Ashley had made 33 goals out of 121 attempts.

   a. Write a ratio to represent the ratio of the number of goals made to goals attempted by Ashley at the end of her high school career. $\dfrac{33}{121}$

   b. Suppose Ashley attempted a goals and made m goals during her first year at college. Write a rational expression to represent the ratio of the number of career goals made to the number of career goals attempted at the end of her first year in college. $\dfrac{33+m}{121+a}$

 534 | Lesson 8-1 | Multiplying and Dividing Rational Expressions

Differentiated Homework Options

Level	Assignment	Two-Day Option	
AL Basic	13–39, 59–60, 62–86	13–39 odd, 66–69	14–38 even, 59–60, 70–86
OL Core	13–39 odd, 40, 41–55 odd, 56–60, 62–86	13–39, 66–69	40–60, 62–65, 70–86
BL Advanced	40–80, (optional: 81–86)		

B **40.** **GEOMETRY** Parallelogram F has an area of $8x^2 + 10x - 3$ square meters and a height of $2x + 3$ meters. Parallelogram G has an area of $6x^2 + 13x - 5$ square meters and a height of $3x - 1$ meters. Find the area of right triangle H. $\frac{1}{2}(8x^2 + 18x - 5)$ m^2

41. **POLLUTION** The thickness of an oil spill from a ruptured pipe on a rig is modeled by the function $T(x) = \dfrac{0.4(x^2 - 2x)}{x^3 + x^2 - 6x}$, where T is the thickness of the oil slick in meters and x is the distance from the rupture in meters.

a. Simplify the function. $T(x) = \dfrac{0.4}{x + 3}$

b. How thick is the slick 100 meters from the rupture? about 3.9 mm thick

44. $\dfrac{(3 - x)(2x - 1)}{(x + 3)(2x + 1)}$

Simplify each expression.

45. $\dfrac{x(x + 2)(x - 1)}{(x + 3)(x - 7)}$

47. $\dfrac{15y^3}{4a^2cxz}$

42. $\dfrac{x^2 - 16}{3x^3 + 18x^2 + 24x} \cdot \dfrac{x^3 - 4x}{2x^2 - 7x - 4}$ $\dfrac{x - 2}{3(2x + 1)}$

43. $\dfrac{3x^2 - 17x - 6}{4x^2 - 20x - 24} \div \dfrac{6x^2 - 7x - 3}{2x^2 - x - 3}$ $\dfrac{1}{4}$

49. $\dfrac{2(4x + 1)(2x + 1)}{5(2x - 1)(x + 2)}$

44. $\dfrac{9 - x^2}{x^2 - 4x - 21} \cdot \left(\dfrac{2x^2 + 7x + 3}{2x^2 - 15x + 7}\right)^{-1}$

45. $\left(\dfrac{2x^2 + 2x - 12}{x^2 + 4x - 5}\right)^{-1} \cdot \dfrac{2x^3 - 8x}{x^2 - 2x - 35}$

46. $\left(\dfrac{3xy^3z}{2a^2bc^2}\right)^3 \cdot \dfrac{16a^4b^3c^5}{15x^7yz^3}$ $\dfrac{18y^8}{5a^2cx^4}$ **47** $\dfrac{20x^2y^6z^{-2}}{3a^3c^2} \cdot \left(\dfrac{16x^3y^3}{9acz}\right)^{-1}$

48. $\left(\dfrac{2xy^3}{3abc}\right)^{-2} \div \dfrac{6a^2b}{x^2y^4}$ $\dfrac{3bc^2}{8y^2}$

49. $\dfrac{\frac{8x^2 - 10x - 3}{10x^2 + 35x - 20}}{\frac{2x^2 + x - 6}{4x^2 + 18x + 8}}$

50. $\dfrac{\frac{2x^2 + 7x - 30}{-6x^2 + 13x + 5}}{\frac{4x^2 + 12x - 72}{3x^2 - 11x - 4}}$ $\dfrac{x - 4}{-4(x - 3)}$

51. $\dfrac{\frac{4x^2 - 1}{3x^3 - 6x^2 - 24x}}{\frac{12x^2 + 12x - 9}{-2x^2 + 5x + 12}}$ $\dfrac{2x + 1}{-9x(x + 2)}$

56b. $\dfrac{25}{27}$; Sample answer: The second airplane travels a bit farther than the first airplane.

56c. $x = -6$; Sample answer: When $x = -6$, the first airplane would travel for 0 hours and the second airplane would travel at a rate of 0 miles per hour.

52. **GEOMETRY** The area of the base of the rectangular prism at the right is 20 square centimeters.

a. Find the length of $\overline{BC}$ in terms of x. $\dfrac{20}{x}$

b. If $DC = 3BC$, determine the area of the shaded region in terms of x. $\dfrac{1200}{x^2}$

c. Determine the volume of the prism in terms of x. $\dfrac{1200}{x}$

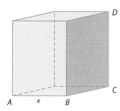

C **Simplify each expression.**

53. $\dfrac{x^2 + 4x - 32}{2x^2 + 9x - 5} \cdot \dfrac{3x^2 - 75}{3x^2 - 11x - 4} \div \dfrac{6x^2 - 18x - 60}{x^3 - 4x}$ $\dfrac{x(x - 2)(x + 8)}{2(2x - 1)(3x + 1)}$

54. $\dfrac{8x^2 + 10x - 3}{3x^2 - 12x - 36} \div \dfrac{2x^2 - 5x - 12}{3x^2 - 17x - 6} \cdot \dfrac{4x^2 + 3x - 1}{4x^2 - 40x + 24}$ $\dfrac{(4x - 1)^2(3x + 1)(x + 1)}{12(x + 2)(x - 4)(x^2 - 10x + 6)}$

55. $\dfrac{4x^2 - 9x - 9}{3x^2 + 6x - 18} \div \dfrac{-2x^2 + 5x + 3}{x^2 - 4x - 32} \div \dfrac{8x^2 + 10x + 3}{6x^2 - 6x - 12}$ $\dfrac{-2(x - 8)(x + 4)(x - 2)(x + 1)}{(2x + 1)^2(x^2 + 2x - 6)}$

56. **CCSS PERSEVERANCE** Use the formula $d = rt$ and the following information. An airplane is traveling at a rate r of 500 miles per hour for a time t of $(6 + x)$ hours. A second airplane travels at the rate r of $(540 + 90x)$ miles per hour for a time t of 6 hours.

a. Write a rational expression to represent the ratio of the distance d traveled by the first airplane to the distance d traveled by the second airplane. $\dfrac{(500)(6 + x)}{(540 + 90x)(6)}$

b. Simplify the rational expression. What does this expression tell you about the distances traveled by the two airplanes?

c. Under what condition is the rational expression undefined? Describe what this condition would tell you about the two airplanes.

Exercise Alert

Formula For Exercise 52, students will need to know the formula for the volume of a rectangular prism, $V = Bh$, where B is the area of the base, and h is the height. For Exercise 80, students will need to know the formula for the volume of a rectangular prism, $V = \ell wh$, where ℓ is the length, w is the width, and h is the height.

 Teaching the Mathematical Practices

Perseverance Mathematically proficient students start by explaining to themselves the meaning of a problem and looking for entry points to its solution. They analyze givens, constraints, relationships, and goals and make conjectures about the form and meaning of the solution and plan a solution pathway rather than simply jumping into a solution attempt.

Study Guide and Intervention

AL OL ELL

Practice

AL OL BL ELL

Word Problem Practice

AL OL BL ELL

57 **TRAINS** Trying to get into a train yard one evening, all of the trains are backed up for 2 miles along a system of tracks. Assume that each car occupies an average of 75 feet of space on a track and that the train yard has 5 tracks.

  a. Write an expression that could be used to determine the number of train cars involved in the backup.   **57a.** $5 \text{ tracks} \cdot \dfrac{2 \text{ miles}}{1 \text{ track}} \cdot \dfrac{5280 \text{ feet}}{1 \text{ mile}} \cdot \dfrac{1 \text{ car}}{75 \text{ feet}}$

  b. How many train cars are involved in the backup? **704**

  c. Suppose that there are 8 attendants doing safety checks on each car, and it takes each vehicle an average of 45 seconds for each check. Approximately how many hours will it take for all the vehicles in the backup to exit? **70.4 h**

58. 🔲 **MULTIPLE REPRESENTATIONS** In this problem, you will investigate the graph of a rational function. **b, d. See margin.**

  a. Algebraic Simplify $\dfrac{x^2 - 5x + 4}{x - 4}$. $x - 1$ **58c.** $f(4)$ results in an error because the function is undefined at $x = 4$. $g(4) = 3$

  b. Tabular Let $f(x) = \dfrac{x^2 - 5x + 4}{x - 4}$. Use the expression you wrote in part **a** to write the related function $g(x)$. Use a graphing calculator to make a table for both functions for $0 \le x \le 10$.

  c. Analytical What are $f(4)$ and $g(4)$? Explain the significance of these values.

  d. Graphical Graph the functions on the graphing calculator. Use the **TRACE** function to investigate each graph, using the ▲ and ▼ keys to switch from one graph to the other. Compare and contrast the graphs.

  e. Verbal What conclusions can you draw about the expressions and the functions? **The expressions and functions are equivalent except for $x = 4$.**

H.O.T. Problems   Use Higher-Order Thinking Skills

59. Sample answer: The two expressions are equivalent, except that the rational expression is undefined at $x = 3$.

59. REASONING Compare and contrast $\dfrac{(x-6)(x+2)(x+3)}{x+3}$ and $(x-6)(x+2)$.

60. **CCSS CRITIQUE** Troy and Beverly are simplifying $\dfrac{x+y}{x-y} \div \dfrac{4}{y-x}$. Is either of them correct? Explain your reasoning.

Troy
$$\dfrac{x+y}{x-y} \div \dfrac{4}{y-x} = \dfrac{x-y}{x+y} \cdot \dfrac{4}{y-x}$$
$$= \dfrac{-4}{x+y}$$

Beverly
$$\dfrac{x+y}{x-y} \div \dfrac{4}{y-x} = \dfrac{x+y}{x-y} \cdot \dfrac{y-x}{4}$$
$$= -\dfrac{x+y}{4}$$

61. CHALLENGE Find the expression that makes the following statement true. $x^2 + x - 6$

$$\dfrac{x-6}{x+3} \cdot \dfrac{?}{x-6} = x - 2$$

62. WHICH ONE DOESN'T BELONG? Identify the expression that does not belong with the other three. Explain your reasoning. **See margin.**

60. Sample answer: Beverly; Troy's mistake was multiplying by the reciprocal of the dividend instead of the divisor.

$$\boxed{\dfrac{1}{x-1}} \qquad \boxed{\dfrac{x^2 + 3x + 2}{x - 5}} \qquad \boxed{\dfrac{x+1}{\sqrt{x+3}}} \qquad \boxed{\dfrac{x^2 + 1}{3}}$$

63. REASONING Determine whether the following statement is *sometimes*, *always*, or *never* true. Explain your reasoning. **See margin.**

  A rational function that has a variable in the denominator is defined for all real values of x.

64. OPEN ENDED Write a rational expression that simplifies to $\dfrac{x-1}{x+4}$. **Sample answer:** $\dfrac{x^2 - 1}{x^2 + 5x + 4}$

65. WRITING IN MATH The rational expression $\dfrac{x^2 + 3x}{4x}$ is simplified to $\dfrac{x+3}{4}$. Explain why this new expression is not defined for all values of x. **See margin.**

Enrichment

OL BL

8-1 Enrichment

Dimensional Analysis

CCSS **Teaching the Mathematical Practices**

Critique Mathematically proficient students are also able to compare the effectiveness of two plausible arguments, distinguish correct logic or reasoning from that which is flawed, and—if there is a flaw in an argument—explain what it is.

66. SAT/ACT The Mason family wants to drive an average of 250 miles per day on their vacation. On the first five days, they travel 220 miles, 300 miles, 210 miles, 275 miles, and 240 miles. How many miles must they travel on the sixth day to meet their goal? **C**

A 235 miles   **D** 275 miles

B 251 miles   **E** 315 miles

C 255 miles

67. Which of the following equations gives the relationship between N and T in the table? **J**

N	1	2	3	4	5	6
T	1	4	7	10	13	16

F $T = 2 - N$ **H** $T = 3N + 1$

G $T = 4 - 3N$ **J** $T = 3N - 2$

68. A monthly cell phone plan costs $39.99 for up to 300 minutes and 20 cents per minute thereafter. Which of the following represents the total monthly bill (in dollars) to talk for x minutes if x is greater than 300? **B**

A $39.99 + 0.20(300 - x)$

B $39.99 + 0.20(x - 300)$

C $39.99 + 0.20x$

D $39.99 + 20x$

69. SHORT RESPONSE The area of a circle 6 meters in diameter exceeds the combined areas of a circle 4 meters in diameter and a circle 2 meters in diameter by how many square meters? **4π**

70. ANTHROPOLOGY An anthropologist studying the bones of a prehistoric person finds there is so little remaining Carbon-14 in the bones that instruments cannot measure it. This means that there is less than 0.5% of the amount of Carbon-14 the bones would have contained when the person was alive. The half-life of Carbon-14 is 5760 years. How long ago did the person die? (Lesson 7-8) **more than 44,000 years ago**

Solve each equation. Round to the nearest ten-thousandth. (Lesson 7-7)

71. $3e^x + 1 = 5$ **0.2877**   **72.** $2e^x - 1 = 0$ **−0.6931**   **73.** $-3e^{4x} + 11 = 2$ **0.2747**   **74.** $8 + 3e^{3x} = 26$ **0.5973**

75. NOISE ORDINANCE A proposed city ordinance will make it illegal in a residential area to create sound that exceeds 72 decibels during the day and 55 decibels during the night. How many times as intense is the noise level allowed during the day than at night? (Lesson 7-3) **$10^{1.7}$ or about 50 times**

Simplify. (Lesson 6-5)

76. $\sqrt{50x^4}$ **$5x^2\sqrt{2}$**   **77.** $\sqrt[3]{16y^3}$ **$2y\sqrt[3]{2}$**   **78.** $\sqrt{18x^2y^3}$ **$3|x|y\sqrt{2y}$**   **79.** $\sqrt{40a^3b^4}$ **$2ab^2\sqrt{10a}$**

80. AUTOMOBILES The length of the cargo space in a sport-utility vehicle is 4 inches greater than the height of the space. The width is 16 inches less than twice the height. The cargo space has a total volume of 55,296 cubic inches. (Lesson 5-8) **a. $V = 2h^3 - 8h^2 - 64h$**

a. Write a polynomial function that represents the volume of the cargo space.

b. Will a package 34 inches long, 44 inches wide, and 34 inches tall fit in the cargo space? Explain.
No; the dimensions of the space are $\ell = 36$ in., $w = 48$ in., $h = 32$ in., so the package is too tall to fit.

Simplify.

81. $(2a + 3b) + (8a - 5b)$ **$10a - 2b$**

82. $(x^2 - 4x + 3) - (4x^2 + 3x - 5)$ **$-3x^2 - 7x + 8$**

83. $(5y + 3y^2) + (-8y - 6y^2)$ **$-3y - 3y^2$**

84. $2x(3y + 9)$ **$6xy + 18x$**

85. $(x + 6)(x + 3)$ **$x^2 + 9x + 18$**

86. $(x + 1)(x^2 - 2x + 3)$ **$x^3 - x^2 + x + 3$**

DifferentiatedInstruction **OL** **BL**

Extension To prepare students for the next lesson and to build a strong base for future work with rational expressions, give them an expression like $\dfrac{5x^2(x^2 + 3)}{5x(x + 3)}$. Ask them to explain in detail, citing fundamentals from arithmetic, why the fives can be divided out but not the threes. Also explain why the first x^2 and x can be divided by x to simplify, but not those within parentheses. Students' explanations should mention that common factors of both the numerator and denominator can be divided out, but not terms that are parts of polynomials. Substitution of a number, like 2, for x may help some students reach this realization.

---

4 Assess

Crystal Ball Ask students to describe how they think their practice with multiplying and simplifying rational expressions will help them when they add or subtract rational expressions.

Additional Answers

58b.

x	0	1	2	3
f(x)	−1	0	1	2
g(x)	−1	0	1	2

x	4	5	6	7
f(x)	ERR	4	5	6
g(x)	3	4	5	6

x	8	9	10
f(x)	7	8	9
g(x)	7	8	9

58d.

The graphs appear to be the same on the graphing calculator. But $f(x)$ is undefined for $f(4)$ and $g(4) = 3$.

62. $\dfrac{x + 1}{\sqrt{x + 3}}$ does not belong with the other three. The other three expressions are rational expressions. Since the denominator of $\dfrac{x + 1}{\sqrt{x + 3}}$ is not a polynomial, $\dfrac{x + 1}{\sqrt{x + 3}}$ is not a rational expression.

63. Sample answer: Sometimes; with a denominator like $x^2 + 2$, in which the denominator cannot equal 0, the rational expression can be defined for all values of x.

65. Sample answer: When the original expression was simplified, a factor of x was taken out of the denominator. If x were to equal 0, then this expression would be undefined. So, the simplified expression is also undefined for x.

Adding and Subtracting Rational Expressions

1 Focus

VerticalAlignment

Before Lesson 8-2 Add and subtract polynomial expressions.

Lesson 8-2 Determine the LCM of polynomials. Add and subtract rational expressions.

After Lesson 8-2 Solve rational equations and inequalities.

:: Then	:: Now	:: Why?
• You added and subtracted polynomial expressions.	**1** Determine the LCM of polynomials. **2** Add and subtract rational expressions.	• As a fire engine moves toward a person, the pitch of the siren sounds higher to that person than it would if the fire engine were at rest. This is because the sound waves are compressed closer together, referred to as the *Doppler effect*. The Doppler effect can be represented by the rational expression $P_0\left(\dfrac{s_0}{s_0 - v}\right)$, where P_0 is the actual pitch of the siren, v is the speed of the fire truck, and s_0 is the speed of sound in air.

2 Teach

Scaffolding Questions

Have students read the **Why?** section of the lesson.

Ask:

■ Can $\dfrac{s_0}{s_0 - v}$ be factored to $s_0\left(\dfrac{1}{1 - v}\right)$? Explain. No; s_0 is not a common factor.

■ To multiply P_0 times $\dfrac{s_0}{s_0 - v}$, do you need a common denominator? no

1 LCM of Polynomials

Example 1 shows how to find the least common multiple of a set of polynomials.

Common Core State Standards

Content Standards
A.APR.7 Understand that rational expressions form a system analogous to the rational numbers, closed under addition, subtraction, multiplication, and division by a nonzero rational expression; add, subtract, multiply, and divide rational expressions.

Mathematical Practices
3 Construct viable arguments and critique the reasoning of others.

1 LCM of Polynomials
Just as with rational numbers in fractional form, to add or subtract two rational expressions that have unlike denominators, you must first find the least common denominator (LCD). The LCD is the least common multiple (LCM) of the denominators.

To find the LCM of two or more numbers or polynomials, factor them. The LCM contains each factor the greatest number of times it appears as a factor.

Numbers	Polynomials
$\dfrac{5}{6} + \dfrac{4}{9}$	$\dfrac{3}{x^2 - 3x + 2} + \dfrac{5}{2x^2 - 2}$
LCM of 6 and 9	**LCM of $x^2 - 3x + 2$ and $2x^2 - 2$**
$6 = 2 \cdot 3$	$x^2 - 3x + 2 = (x - 1)(x - 2)$
$9 = 3 \cdot 3$	$2x^2 - 2 = 2 \cdot (x - 1)(x + 1)$
LCM $= 2 \cdot 3 \cdot 3$ or 18	LCM $= 2(x - 1)(x - 2)(x + 1)$

Example 1 LCM of Monomials and Polynomials

Find the LCM of each set of polynomials.

a. $6xy$, $15x^2$, and $9xy^4$

$6xy = 2 \cdot 3 \cdot x \cdot y$	Factor the first monomial.
$15x^2 = 3 \cdot 5 \cdot x^2$	Factor the second monomial.
$9xy^4 = 3 \cdot 3 \cdot x \cdot y^4$	Factor the third monomial.
LCM $= 2 \cdot 3 \cdot 3 \cdot 5 \cdot x^2 \cdot y^4$	Use each factor the greatest number of times it appears.
$= 90x^2y^4$	Then simplify.

b. $y^4 + 8y^3 + 15y^2$ and $y^2 - 3y - 40$

$y^4 + 8y^3 + 15y^2 = y^2(y + 5)(y + 3)$	Factor the first polynomial.
$y^2 - 3y - 40 = (y + 5)(y - 8)$	Factor the second polynomial.
LCM $= y^2(y + 5)(y + 3)(y - 8)$	Use each factor the greatest number of times it appears as a factor.

▶ **Guided**Practice

1A. $12a^2b$, $15abc$, $8b^3c^4$ $120a^2b^3c^4$

1B. $4a^2 - 12a - 16$ and $a^3 - 9a^2 + 20a$ $4a(a - 4)(a - 5)(a + 1)$

Philip Rostron/Masterfile

Lesson 8-2 Resources

Resource	Approaching Level **AL**	On Level **OL**	Beyond Level **BL**	English Learners **ELL**
Teacher Edition	• Differentiated Instruction, p. 540	• Differentiated Instruction, p. 540	• Differentiated Instruction, p. 542	• Differentiated Instruction, p. 540
Chapter Resource Masters	• Study Guide and Intervention, pp. 11–12 • Skills Practice, p. 13 • Practice, p. 14 • Word Problem Practice, p. 15	• Study Guide and Intervention, pp. 11–12 • Skills Practice, p. 13 • Practice, p. 14 • Word Problem Practice, p. 15 • Enrichment, p. 16 • TI-Nspire Activity, p. 17	• Practice, p. 14 • Word Problem Practice, p. 15 • Enrichment, p. 16	• Study Guide and Intervention, pp. 11–12 • Skills Practice, p. 13 • Practice, p. 14 • Word Problem Practice, p. 15
Other	• 5-Minute Check 8-2 • Study Notebook	• 5-Minute Check 8-2 • Study Notebook	• 5-Minute Check 8-2 • Study Notebook	• 5-Minute Check 8-2 • Study Notebook

2 Add and Subtract Rational Expressions As with fractions, rational expressions must have common denominators in order to be added or subtracted.

KeyConcept	
Adding Rational Expressions	
Words	To add rational expressions, find the least common denominator (LCD). Rewrite each expression with the LCD. Then add.
Symbols	For all $\frac{a}{b}$ and $\frac{c}{d}$, with $b \neq 0$ and $d \neq 0$, $\frac{a}{b} + \frac{c}{d} = \frac{ad}{bd} + \frac{bc}{bd} = \frac{ad + bc}{bd}$.
Subtracting Rational Expressions	
Words	To subtract rational expressions, find the least common denominator (LCD). Rewrite each expression with the LCD. Then subtract.
Symbols	For all $\frac{a}{b}$ and $\frac{c}{d}$, with $b \neq 0$ and $d \neq 0$, $\frac{a}{b} - \frac{c}{d} = \frac{ad}{bd} - \frac{bc}{bd} = \frac{ad - bc}{bd}$.

Example 2 Monomial Denominators

Simplify $\frac{3y}{2x^3} + \frac{5z}{8xy^2}$.

$\frac{3y}{2x^3} + \frac{5z}{8xy^2} = \frac{3y}{2x^3} \cdot \frac{4y^2}{4y^2} + \frac{5z}{8xy^2} \cdot \frac{x^2}{x^2}$ The LCD is $8x^3y^2$.

$= \frac{12y^3}{8x^3y^2} + \frac{5x^2z}{8x^3y^2}$ Multiply fractions.

$= \frac{12y^3 + 5x^2z}{8x^3y^2}$ Add the numerators.

GuidedPractice

Simplify each expression.

2A. $\frac{4}{5a^3b^2} + \frac{9c}{10ab}$ $\frac{8 + 9a^2bc}{10a^3b^2}$ 2B. $\frac{3a^2}{16b^2} - \frac{8x}{5a^3b}$ $\frac{15a^5 - 128bx}{80a^3b^2}$

The LCD is also used to combine rational expressions with polynomial denominators.

Example 3 Polynomial Denominators

Simplify $\frac{5}{6x - 18} - \frac{x - 1}{4x^2 - 14x + 6}$.

$\frac{5}{6x - 18} - \frac{x - 1}{4x^2 - 14x + 6} = \frac{5}{6(x - 3)} - \frac{x - 1}{2(2x - 1)(x - 3)}$ Factor denominators.

$= \frac{5(2x - 1)}{6(x - 3)(2x - 1)} - \frac{(x - 1)(3)}{2(2x - 1)(x - 3)(3)}$ Multiply by missing factors.

$= \frac{10x - 5 - 3x + 3}{6(x - 3)(2x - 1)}$ Subtract numerators.

$= \frac{7x - 2}{6(x - 3)(2x - 1)}$ Simplify.

> **StudyTip**
> Simplifying Rational Expressions After you add or subtract rational expressions, it is possible that the resulting expression can be further simplified.

GuidedPractice

Simplify each expression.

3A. $\frac{x - 1}{x^2 - x - 6} - \frac{4}{5x + 10}$ $\frac{x + 7}{5(x + 2)(x - 3)}$ 3B. $\frac{x - 8}{4x^2 + 21x + 5} + \frac{6}{12x + 3}$ $\frac{3x + 2}{(4x + 1)(x + 5)}$

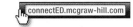 connectED.mcgraw-hill.com **539**

Formative Assessment

Use the Guided Practice exercises after each example to determine students' understanding of concepts.

Additional Example

1 Find the LCM of each set of polynomials.

   a. $15a^2bc^3$, $16b^5c^2$, and $20a^3c^6$ $240a^3b^5c^6$

   b. $x^3 - x^2 - 2x$ and $x^2 - 4x + 4$ $x(x + 1)(x - 2)^2$

Additional Examples also in Interactive Classroom PowerPoint® Presentations

IWB Interactive White Board READY

2 Add and Subtract Rational Expressions

Example 2 shows how to add two rational expressions with monomial denominators. **Example 3** shows how to subtract two rational expressions with polynomial denominators. **Examples 4 and 5** show how to simplify complex fractions.

Additional Examples

2 Simplify $\frac{5a^2}{6b} + \frac{9}{14a^2b^2}$.

   $\frac{35a^4b + 27}{42a^2b^2}$

3 Simplify $\frac{x + 10}{3x - 15} - \frac{3x + 15}{6x - 30}$.

   $-\frac{1}{6}$

Teach with Tech

Document Camera Choose several students to work through examples in front of the class. Be sure they clearly explain how to find the LCD.

> **Watch**Out!
>
> **Preventing Errors** Have students discuss the differences between procedures for adding and multiplying fractions. It is important that they see why common denominators are required for addition but not for multiplication.

Additional Examples

4 Simplify $\dfrac{\frac{1}{a} + \frac{1}{b}}{\frac{1}{b} - 1}$.

$\dfrac{a + b}{a(1 - b)}$ or $\dfrac{a + b}{a - ab}$

5 Simplify $\dfrac{\frac{2}{x} - 1}{\frac{1}{y} - \frac{3}{x}}$. $\dfrac{2y - xy}{x - 3y}$

Focus on Mathematical Content

Equivalent Expressions When rational expressions are given the same denominators in preparation for addition or subtraction, they are equivalent to the original expressions. Multiplying a rational expression by a form of 1 such as $\dfrac{6x}{6x}$ or $\dfrac{y - 3}{y - 3}$ does not change the value of the expression.

One way to simplify a complex fraction is to simplify the numerator and the denominator separately, and then simplify the resulting expressions.

StudyTip

Undefined Terms Remember that there are restrictions on variables in the denominator.

Example 4 Complex Fractions with Different LCDs

Simplify $\dfrac{1 + \frac{1}{x}}{1 - \frac{x}{y}}$.

$\dfrac{1 + \frac{1}{x}}{1 - \frac{x}{y}} = \dfrac{\frac{x}{x} + \frac{1}{x}}{\frac{y}{y} - \frac{x}{y}}$ The LCD of the numerator is x.
       The LCD of the denominator is y.

$= \dfrac{\frac{x + 1}{x}}{\frac{y - x}{y}}$ Simplify the numerator and denominator.

$= \dfrac{x + 1}{x} \div \dfrac{y - x}{y}$ Write as a division expression.

$= \dfrac{x + 1}{x} \cdot \dfrac{y}{y - x}$ Multiply by the reciprocal of the divisor.

$= \dfrac{xy + y}{xy - x^2}$ Simplify.

▶ **Guided**Practice

Simplify each expression.

4A. $\dfrac{1 - \frac{y}{x}}{\frac{1}{y} + \frac{1}{x}}$ $\dfrac{xy - y^2}{x + y}$ **4B.** $\dfrac{\frac{c}{d} - \frac{d}{c}}{\frac{d}{c} + 2}$ $\dfrac{c^2 - d^2}{d^2 + 2cd}$

Another method of simplifying complex fractions is to find the LCD of all of the denominators. Then, the denominators are all eliminated by multiplying by the LCD.

Example 5 Complex Fractions with Same LCDs

Simplify $\dfrac{1 + \frac{1}{x}}{1 - \frac{x}{y}}$.

$\dfrac{1 + \frac{1}{x}}{1 - \frac{x}{y}} = \dfrac{\left(1 + \frac{1}{x}\right)}{\left(1 - \frac{x}{y}\right)} \cdot \dfrac{xy}{xy}$ The LCD of all of the denominators is xy.
       Multiply by $\dfrac{xy}{xy}$.

$= \dfrac{xy + y}{xy - x^2}$ Distribute xy.

Notice that the same problem is solved in Examples 4 and 5 using different methods, but both produce the same answer. So, how you solve problems similar to these is left up to your own discretion.

▶ **Guided**Practice

Simplify each expression.

5A. $\dfrac{1 + \frac{2}{x}}{\frac{3}{y} - \frac{4}{x}}$ $\dfrac{xy + 2y}{3x - 4y}$ **5B.** $\dfrac{\frac{1}{d} - \frac{d}{c}}{\frac{1}{c} + 6}$ $\dfrac{c - d^2}{d + 6cd}$

5C. $\dfrac{\frac{1}{y} + \frac{1}{x}}{\frac{1}{y} - \frac{1}{x}}$ $\dfrac{x + y}{x - y}$ **5D.** $\dfrac{\frac{a}{b} + 1}{1 - \frac{b}{a}}$ $\dfrac{a(a + b)}{b(a - b)}$

 540 | Lesson 8-2 | Adding and Subtracting Rational Expressions

DifferentiatedInstruction   **AL**   **OL**   **ELL**

If ▶ students have difficulty adding and subtracting rational expressions,

Then ▶ have students work with a partner, one in the role of a coach, the other in the role of an athlete. The athlete works a problem, using steps and explaining the thinking while the coach listens and watches for errors, correcting as necessary. Then the partners exchange roles.

Example 1 Find the LCM of each set of polynomials.

1. $16x, 8x^2y^3, 5x^3y$　$80x^3y^3$

2. $7a^2, 9ab^3, 21abc^4$　$63a^2b^3c^4$

3. $3y^2 - 9y, y^2 - 8y + 15$　$3y(y - 3)(y - 5)$

4. $x^3 - 6x^2 - 16x, x^2 - 4$　$x(x + 2)(x - 2)(x - 8)$

Examples 2–3 Simplify each expression.

5. $\dfrac{12y}{5x} + \dfrac{5x}{4y^3}$　$\dfrac{48y^4 + 25x^2}{20xy^3}$

6. $\dfrac{5}{6ab} + \dfrac{3b^2}{14a^3}$　$\dfrac{35a^2 + 9b^3}{42a^3b}$

7. $\dfrac{7b}{12a} - \dfrac{1}{18ab^3}$　$\dfrac{21b^4 - 2}{36ab^3}$

8. $\dfrac{y^2}{8c^2d^2} - \dfrac{3x}{14c^4d}$　$\dfrac{7c^2y^2 - 12dx}{56c^4d^2}$

9. $\dfrac{4x}{x^2 + 9x + 18} + \dfrac{5}{x + 6}$　$\dfrac{9x + 15}{(x + 3)(x + 6)}$

10. $\dfrac{8}{y - 3} + \dfrac{2y - 5}{y^2 - 12y + 27}$　$\dfrac{10y - 77}{(y - 3)(y - 9)}$

11. $\dfrac{4}{3x + 6} - \dfrac{x + 1}{x^2 - 4}$　$\dfrac{x - 11}{3(x + 2)(x - 2)}$

12. $\dfrac{3a + 2}{a^2 - 16} - \dfrac{7}{6a + 24}$　$\dfrac{11a + 40}{6(a + 4)(a - 4)}$

13. **GEOMETRY** Find the perimeter of the rectangle.

$$\dfrac{14x - 10}{(x + 1)(x - 2)}$$

$\dfrac{3}{x - 2}$

$\dfrac{4}{x + 1}$

Examples 4–5 Simplify each expression.

14. $\dfrac{4 + \frac{2}{x}}{3 - \frac{2}{x}}$　$\dfrac{4x + 2}{3x - 2}$

15. $\dfrac{6 + \frac{4}{y}}{2 + \frac{6}{y}}$　$\dfrac{3y + 2}{y + 3}$

16. $\dfrac{\frac{3}{x} + \frac{2}{y}}{1 + \frac{4}{y}}$　$\dfrac{3y + 2x}{xy + 4x}$

17. $\dfrac{\frac{2}{b} + \frac{5}{a}}{\frac{3}{a} - \frac{8}{b}}$　$\dfrac{2a + 5b}{3b - 8a}$

Practice and Problem Solving　　　　　Extra Practice is on page R8.

Example 1 Find the LCM of each set of polynomials.　20. $(x - 4)(x - 5)(x + 6)$　21. $6(x + 4)(2x - 1)(2x + 3)$

18. $24cd, 40a^2c^3d^4, 15abd^3$　$120a^2bc^3d^4$

19. $4x^2y^3, 18xy^4, 10xz^2$　$180x^2y^4z^2$

20. $x^2 - 9x + 20, x^2 + x - 30$

21. $6x^2 + 21x - 12, 4x^2 + 22x + 24$

28. $\dfrac{17x + 58}{(x - 8)(x + 2)(x + 5)}$

Examples 2–3 (CCSS) **PERSEVERANCE** Simplify each expression.

22. $\dfrac{5a}{24cf^4} + \dfrac{a}{36bc^4f^3}$　$\dfrac{15abc^3 + 2af}{72bc^4f^4}$

23. $\dfrac{4b}{15x^3y^2} - \dfrac{3b}{35x^2y^4z}$　$\dfrac{28by^2z - 9bx}{105x^3y^4z}$

24. $\dfrac{5b}{6a} + \dfrac{3b}{10a^2} + \dfrac{2}{ab^2}$　$\dfrac{25ab^3 + 9b^3 + 60a}{30a^2b^2}$

25. $\dfrac{4}{3x} + \dfrac{8}{x^3} + \dfrac{2}{5xy}$　$\dfrac{20x^2y + 120y + 6x^2}{15x^3y}$

26. $\dfrac{8}{3y} + \dfrac{2}{9} - \dfrac{3}{10y^2}$　$\dfrac{240y + 20y^2 - 27}{90y^2}$

27. $\dfrac{1}{16a} + \dfrac{5}{12b} - \dfrac{9}{10b^3}$　$\dfrac{15b^3 + 100ab^2 - 216a}{240ab^3}$

28. $\dfrac{8}{x^2 - 6x - 16} + \dfrac{9}{x^2 - 3x - 40}$

29. $\dfrac{6}{y^2 - 2y - 35} + \dfrac{4}{y^2 + 9y + 20}$　$\dfrac{10y - 4}{(y - 7)(y + 5)(y + 4)}$

30. $\dfrac{12}{3y^2 - 10y - 8} - \dfrac{3}{y^2 - 6y + 8}$

31. $\dfrac{6}{2x^2 + 11x - 6} - \dfrac{8}{x^2 + 3x - 18}$　$\dfrac{-10x - 10}{(2x - 1)(x + 6)(x - 3)}$

32. $\dfrac{2x}{4x^2 + 9x + 2} + \dfrac{3}{2x^2 - 8x - 24}$

33. $\dfrac{4x}{3x^2 + 3x - 18} - \dfrac{2x}{2x^2 + 11x + 15}$　$\dfrac{2x^2 + 32x}{3(x - 2)(x + 3)(2x + 5)}$

34. **BIOLOGY** After a person eats something, the pH or acid level A of his or her mouth can be determined by the formula $A = \dfrac{20.4t}{t^2 + 36} + 6.5$, where t is the number of minutes that have elapsed since the food was eaten.

　a. Simplify the equation.　$A = \dfrac{6.5t^2 + 20.4t + 234}{t^2 + 36}$

　b. What would the acid level be after 30 minutes?　≈ 7.2

30. $\dfrac{3y - 30}{(3y + 2)(y - 4)(y - 2)}$

32. $\dfrac{4x^2 - 12x + 3}{2(x - 6)(4x + 1)(x + 2)}$

 connectED.mcgraw-hill.com　**541**

Formative Assessment
Use Exercises 1–17 to check for understanding.

Use the chart at the bottom of this page to customize assignments for your students.

(CCSS) **Teaching the Mathematical Practices**

Perseverance Mathematically proficient students start by explaining to themselves the meaning of a problem and looking for entry points to its solution. They analyze givens, constraints, relationships, and goals and make conjectures about the form and meaning of the solution and plan a solution pathway rather than simply jumping into a solution attempt.

Differentiated Homework Options

Level	Assignment	Two-Day Option	
AL Basic	18–39, 64–90	19–39 odd, 67–70	18–38 even, 64–66, 71–90
OL Core	19–39 odd, 40–57 odd, 59–62, 64–90	18–39, 67–70	40–62, 64–66, 71–90
BL Advanced	40–84, (optional: 85–90)		

Exercise Alerts

Formulas For Exercise 35, students will need to know the formula for the area of a triangle, $A = \frac{1}{2}bh$.

Grid Paper For Exercises 75–80, and 85–90, students will need grid paper.

 Teaching the Mathematical Practices

Modeling Mathematically proficient students can apply the mathematics they know to solve problems arising in everyday life, analyze relationships mathematically to draw conclusions, and interpret their mathematical results in the context of a situation.

Additional Answers

62b.

62c.

x	T(x)
20	4.5
50	4.2
100	4.1
200	4.05
400	4.025

64. Sample answer: The set of rational expressions is closed under all of these operations because the sum, difference, product, and quotient of two rational expressions is a rational expression.

66. Sample answer: First, factor the denominators of all of the expressions. Find the LCD of the denominators. Convert each expression so they all have the LCD. Add or subtract the numerators. Then simplify. It is the same.

35. GEOMETRY Both triangles in the figure at the right are equilateral. If the area of the smaller triangle is 200 square centimeters and the area of the larger triangle is 300 square centimeters, find the minimum distance from A to B in terms of x and y and simplify. $\dfrac{1000x + 800y}{x(x + 2y)}$

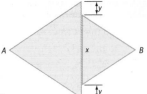

Examples 4–5 Simplify each expression.

36. $\dfrac{\frac{2}{x-3} + \frac{3x}{x^2-9}}{\frac{3}{x+3} - \frac{4x}{x^2-9}}$ $\dfrac{5x+6}{-x-9}$

37. $\dfrac{\frac{4}{x+5} + \frac{9}{x-6}}{\frac{5}{x-6} - \frac{8}{x+5}}$ $\dfrac{13x+21}{-3x+73}$

38. $\dfrac{\frac{5}{x+6} - \frac{2x}{2x-1}}{\frac{x}{2x-1} + \frac{4}{x+6}}$ $\dfrac{-2x^2 - 2x - 5}{x^2 + 14x - 4}$

39. $\dfrac{\frac{8}{x-9} - \frac{x}{3x+2}}{\frac{3}{3x+2} + \frac{4x}{x-9}}$ $\dfrac{-x^2 + 33x + 16}{12x^2 + 11x - 27}$

40. OIL PRODUCTION Managers of an oil company have estimated that oil will be pumped from a certain well at a rate based on the function $R(x) = \frac{20}{x} + \frac{200x}{3x^2 + 20}$, where $R(x)$ is the rate of production in thousands of barrels per year x years after pumping begins.

a. Simplify $R(x)$. $R(x) = \dfrac{260x^2 + 400}{3x^3 + 20x}$

b. At what rate will oil be pumping from the well in 50 years? **about 1730 barrels/yr**

Find the LCM of each set of polynomials.

41. $12xy^4, 14x^4y^2, 5xyz^3, 15x^5y^3$ $420x^5y^4z^3$

42. $-6abc^2, 18a^2b^2, 15a^4c, 8b^3$ $-360a^4b^3c^2$

43. $x^2 - 3x - 28, 2x^2 + 9x + 4, x^2 - 16$ $(x+4)(x-4)(2x+1)(x-7)$

44. $x^2 - 5x - 24, x^2 - 9, 3x^2 + 8x - 3$ $(x+3)(x-3)(x-8)(3x-1)$

Simplify each expression.

45. $\dfrac{1}{12a} + 6 - \dfrac{3}{5a^2}$ $\dfrac{360a^2 + 5a - 36}{60a^2}$

46. $\dfrac{5}{16y^2} - 4 - \dfrac{8}{3x^2y}$ $\dfrac{15x^2 - 192x^2y^2 - 128y}{48x^2y^2}$

47. $\dfrac{5}{6x^2 + 46x - 16} + \dfrac{2}{6x^2 + 57x + 72}$

48. $\dfrac{1}{8x^2 - 20x - 12} + \dfrac{4}{6x^2 + 27x + 12}$

49. $\dfrac{x^2 + y^2}{x^2 - y^2} + \dfrac{y}{x+y} - \dfrac{x}{x-y}$ 0

50. $\dfrac{x^2 + x}{x^2 - 9x + 8} + \dfrac{4}{x-1} - \dfrac{3}{x-8}$ $\dfrac{x^2 + 2x - 29}{x^2 - 9x + 8}$

51. $\dfrac{\frac{2}{a-1} + \frac{3}{a-4}}{\frac{6}{a^2 - 5a + 4}}$ $\dfrac{5a - 11}{6}$

52. $\dfrac{\frac{1}{x} + \frac{1}{y}}{\left(\frac{1}{x} - \frac{1}{y}\right)(x+y)}$ $\dfrac{1}{y-x}$

47. $\dfrac{42x + 41}{6(3x-1)(x+8)(2x+3)}$

48. $\dfrac{19x - 36}{12(2x+1)(x-3)(x+4)}$

53. GEOMETRY An expression for the length of one rectangle is $\frac{x^2 - 9}{x - 2}$. The length of a similar rectangle is expressed as $\frac{x + 3}{x^2 - 4}$. What is the scale factor of the lengths of the two rectangles? Write in simplest form. $(x-3)(x+2)$

54. MODELING Cameron is taking a 20-mile kayaking trip. He travels half the distance at one rate. The rest of the distance he travels 2 miles per hour slower.

a. If x represents the faster pace in miles per hour, write an expression that represents the time spent at that pace. $\dfrac{10}{x}$

b. Write an expression for the amount of time spent at the slower pace. $\dfrac{10}{x-2}$

c. Write an expression for the amount of time Cameron needed to complete the trip. $\dfrac{20(x-1)}{x(x-2)}$

Find the slope of the line that passes through each pair of points.

55. $A\left(\frac{2}{p}, \frac{1}{2}\right)$ and $B\left(\frac{1}{3}, \frac{3}{p}\right)$ $-\dfrac{3}{2}$

56. $C\left(\frac{1}{4}, \frac{4}{q}\right)$ and $D\left(\frac{5}{q}, \frac{1}{5}\right)$ $-\dfrac{4}{5}$

57. $E\left(\frac{7}{w}, \frac{1}{7}\right)$ and $F\left(\frac{1}{7}, \frac{7}{w}\right)$ -1

58. $G\left(\frac{6}{n}, \frac{1}{6}\right)$ and $H\left(\frac{1}{6}, \frac{6}{n}\right)$ -1

542 | Lesson 8-2 | Adding and Subtracting Rational Expressions

DifferentiatedInstruction BL

Extension *Partial fraction decomposition* is a useful algebraic skill in more advanced mathematics courses, including calculus. Give students a rational expression like $\dfrac{5x + 3}{x(x + 1)}$, and ask what kinds of simpler fractions might be added to yield this expression. If they can identify x and $x + 1$ as denominators for such "partial fractions," ask them to try and find values of A and B that will make the equation $\dfrac{A}{x} + \dfrac{B}{x + 1} = \dfrac{5x + 3}{x(x + 1)}$ a true statement for any value of x. $A = 3$ and $B = 2$

59. PHOTOGRAPHY The focal length of a lens establishes the field of view of the camera. The shorter the focal length is, the larger the field of view. For a camera with a fixed focal length of 70 mm to focus on an object x mm from the lens, the film must be placed a distance y from the lens. This is represented by $\frac{1}{x} + \frac{1}{y} = \frac{1}{70}$.

59b. Sample answer: When the object is 70 mm away, y needs to be 0, which is impossible.

a. Express y as a function of x. $y = \dfrac{70x}{x - 70}$

b. What happens to the focusing distance when the object is 70 mm away?

60. PHARMACOLOGY Two drugs are administered to a patient. The concentrations in the bloodstream of each are given by $f(t) = \dfrac{2t}{3t^2 + 9t + 6}$ and $g(t) = \dfrac{3t}{2t^2 + 6t + 4}$ where t is the time, in hours, after the drugs are administered.

a. Add the two functions together to determine a function for the total concentration of drugs in the patient's bloodstream. $h(t) = \dfrac{13t}{6t^2 + 18t + 12}$

b. What is the concentration of drugs after 8 hours? **about 0.19**

61a. $\dfrac{P_0 s_0 x - P_0 s_0 y}{(s_0 - x)(s_0 - y)}$

61 DOPPLER EFFECT Refer to the application at the beginning of the lesson. George is equidistant from two fire engines traveling toward him from opposite directions.

a. Let x be the speed of the faster fire engine and y be the speed of the slower fire engine. Write and simplify a rational expression representing the difference in pitch between the two sirens according to George.

b. If one is traveling at 45 meters per second and the other is traveling at 70 meters per second, what is the difference in their pitches according to George? The speed of sound in air is 332 meters per second, and both engines have a siren with a pitch of 500 Hz. **about 55.2 Hz**

62. RESEARCH A student studying learning behavior performed an experiment in which a rat was repeatedly sent through a maze. It was determined that the time it took the rat to complete the maze followed the rational function $T(x) = 4 + \dfrac{10}{x}$, where x represented the number of trials. **b, c. See margin.**

a. What is the domain of the function? $x \neq 0$

b. Graph the function for $0 \leq x \leq 10$.

c. Make a table of the function for $x = 20, 50, 100, 200,$ and 400.

d. If it were possible to have an infinite number of trials, what do you think would be the rat's best time? Explain your reasoning. **Sample answer: 4; The fraction approaches 0 as x approaches infinity; $4 + 0 = 4$.**

H.O.T. Problems   Use Higher-Order Thinking Skills

63. CHALLENGE Simplify $\dfrac{5x^{-2} - \dfrac{x+1}{x}}{\dfrac{4}{3 - x^{-1}} + 6x^{-1}} \cdot \dfrac{-3x^3 - 2x^2 + 16x - 5}{4x^3 + 18x^2 - 6x}$.

64. CCSS ARGUMENTS The sum of any two rational numbers is always a rational number. So, the set of rational numbers is said to be closed under addition. Determine whether the set of rational expressions is closed under addition, subtraction, multiplication, and division by a nonzero rational expression. Justify your reasoning. **See margin.**

65. OPEN ENDED Write three monomials with an LCM of $180a^4b^6c$.

Sample answer: $20a^4b^2c$, $15ab^6$, $9abc$

66. WRITING IN MATH Write a how-to manual for adding rational expressions that have unlike denominators. How does this compare to adding rational numbers? **See margin.**

CCSS Teaching the Mathematical Practices

Arguments Mathematically proficient students understand and use stated assumptions, definitions, and previously established results in constructing arguments. They make conjectures and build a logical progression of statements to explore the truth of their conjectures. And they are able to analyze situations by breaking them into cases, and can recognize and use counterexamples.

Study Guide and Intervention

AL OL ELL

8-2 Study Guide and Intervention

Adding and Subtracting Rational Expressions

Practice

AL OL BL ELL

8-2 Practice

Adding and Subtracting Rational Expressions

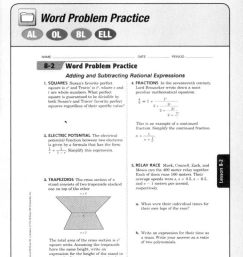

Word Problem Practice

AL OL BL ELL

8-2 Word Problem Practice

Adding and Subtracting Rational Expressions

Enrichment

OL BL

8-2 Enrichment

Zeno's Paradox

Yesterday's News Ask students to describe how yesterday's lesson on multiplication and division of rational expressions helped prepare them for today's lesson on addition and subtraction of rational expressions.

Formative Assessment

Check for student understanding of concepts in Lessons 8-1 and 8-2.

 Quiz 1, p. 47

Additional Answers

75. $D = \{x \mid x \geq -0.5\}, R = \{y \mid y \leq 0\}$

$y = -\sqrt{2x+1}$

76. $D = \{x \mid x \geq 0.6\}, R = \{y \mid y \geq 0\}$

$y = \sqrt{5x-3}$

77. $D = \{x \mid x \geq -6\}, R = \{y \mid y \geq -3\}$

$y = \sqrt{x+6} - 3$

78. $D = \{x \mid x \geq -4\}, R = \{y \mid y \leq 5\}$

$y = 5 - \sqrt{x+4}$

Standardized Test Practice

67. PROBABILITY A drawing is to be held to select the winner of a new bike. There are 100 seniors, 150 juniors, and 200 sophomores who had correct entries. The drawing will contain 3 tickets for each senior name, 2 for each junior, and 1 for each sophomore. What is the probability that a senior's ticket will be chosen? **D**

A $\frac{1}{8}$ **C** $\frac{2}{7}$

B $\frac{2}{9}$ **D** $\frac{3}{8}$

68. SHORT RESPONSE Find the area of the figure.

$24 + 12.5\pi \text{ cm}^2$

6 cm

8 cm

69. SAT/ACT If Mauricio receives b books in addition to the number of books he had, he will have t times as many as he had originally. In terms of b and t, how many books did Mauricio have at the beginning? **F**

F $\frac{b}{t-1}$ **J** $\frac{b}{t}$

G $\frac{b}{t+1}$ **K** $\frac{t}{b}$

H $\frac{t+1}{b}$

70. If $\frac{2a}{a} + \frac{1}{a} = 4$, then $a =$ ___. **B**

A 2   **C** $\frac{1}{8}$

B $\frac{1}{2}$ **D** $-\frac{1}{8}$

Spiral Review

Simplify each expression. (Lesson 8-1)

71. $\frac{-4ab}{21c} \cdot \frac{14c^2}{22a^2} - \frac{4bc}{33a}$

72. $\frac{x^2 - y^2}{6y} \div \frac{x+y}{36y^2}$ $6y(x-y)$

73. $\frac{n^2 - n - 12}{n+2} \div \frac{n-4}{n^2 - 4n - 12}$
$(n+3)(n-6)$

74. BIOLOGY Bacteria usually reproduce by a process known as *binary fission*. In this type of reproduction, one bacterium divides, forming two bacteria. Under ideal conditions, some bacteria reproduce every 20 minutes. (Lesson 7-8)

   a. Find the constant k for this type of bacteria under ideal conditions. **about 0.0347**

   b. Write the equation for modeling the exponential growth of this bacterium. $y = ae^{0.0347t}$

Graph each function. State the domain and range of each function. (Lesson 6-3) **75–80. See margin.**

75. $y = -\sqrt{2x+1}$ **76.** $y = \sqrt{5x-3}$ **77.** $y = \sqrt{x+6} - 3$

78. $y = 5 - \sqrt{x+4}$ **79.** $y = \sqrt{3x-6} + 4$ **80.** $y = 2\sqrt{3-4x} + 3$

Solve each equation. State the number and type of roots. (Lesson 5-7) **82.** $\frac{5 \pm i\sqrt{71}}{4}$; **2 imaginary**

81. $3x + 8 = 0$ **82.** $2x^2 - 5x + 12 = 0$ **83.** $x^3 + 9x = 0$ **84.** $x^4 - 81 = 0$
$-\frac{8}{3}$; **1 real**   **83.** 0, $3i$, $-3i$; **1 real, 2 imaginary**   **84.** 3, -3, $3i$, and $-3i$; **2 real, 2 imaginary**

Skills Review

Graph each function. 85–90. See Chapter 8 Answer Appendix.

85. $y = 4(x+3)^2 + 1$ **86.** $y = -(x-5)^2 - 3$ **87.** $y = \frac{1}{4}(x-2)^2 + 4$

88. $y = \frac{1}{2}(x-3)^2 - 5$ **89.** $y = x^2 + 6x + 2$ **90.** $y = x^2 - 8x + 18$

---

79. $D = \{x \mid x \geq 2\}, R = \{y \mid y \geq 4\}$

$y = \sqrt{3x-6} + 4$

80. $D = \{x \mid x \leq 0.75\}, R = \{y \mid y \geq 3\}$

$y = 2\sqrt{3-4x} + 3$

LESSON 8-3 Graphing Reciprocal Functions

::Then	::Now	::Why?
● You graphed polynomial functions.	**1** Determine properties of reciprocal functions. **2** Graph transformations of reciprocal functions.	The East High School Chorale wants to raise $5000 to fund a trip to a national competition in Tampa. They have decided to sell candy bars. They will make a $1 profit on each candy bar they sell, so they need to sell 5000 candy bars. If c represents the number of candy bars each student has to sell and n represents the number of students, then $c = \frac{5000}{n}$.

Number of Candy Bars

 NewVocabulary
reciprocal function
hyperbola

 Common Core State Standards

Content Standards
A.CED.2 Create equations in two or more variables to represent relationships between quantities; graph equations on coordinate axes with labels and scales.

F.BF.3 Identify the effect on the graph of replacing $f(x)$ by $f(x) + k$, $kf(x)$, $f(kx)$, and $f(x + k)$ for specific values of k (both positive and negative); find the value of k given the graphs. Experiment with cases and illustrate an explanation of the effects on the graph using technology.

Mathematical Practices
2 Reason abstractly and quantitatively.

1 **Vertical and Horizontal Asymptotes** The function $c = \frac{5000}{n}$ is a reciprocal function. A **reciprocal function** has an equation of the form $f(x) = \frac{1}{a(x)}$, where $a(x)$ is a linear function and $a(x) \neq 0$.

KeyConcept Parent Function of Reciprocal Functions

Parent function:	$f(x) = \frac{1}{x}$
Type of graph:	**hyperbola**
Domain and range:	all nonzero real numbers
Asymptotes:	$x = 0$ and $f(x) = 0$
Intercepts:	none
Not defined:	$x = 0$

The domain of a reciprocal function is limited to values for which the function is defined.

Functions: $\quad f(x) = \frac{-3}{x + 2} \qquad g(x) = \frac{4}{x - 5} \qquad h(x) = \frac{3}{x}$

Not defined at: $\quad x = -2 \qquad\qquad x = 5 \qquad\qquad x = 0$

Example 1 Limitations on Domain

Determine the value of x for which $f(x) = \frac{3}{2x + 5}$ is not defined.

Find the value for which the denominator of the expression equals 0.

$\frac{3}{2x + 5} \quad \rightarrow \quad 2x + 5 = 0$

$\qquad\qquad\qquad x = -\frac{5}{2} \qquad$ The function is undefined for $x = -\frac{5}{2}$.

GuidedPractice

Determine the value of x for which each function is not defined.

1A. $f(x) = \frac{2}{x - 1} \quad x = 1$ 　　　　　　**1B.** $f(x) = \frac{7}{3x + 2} \quad x = -\frac{2}{3}$

 connectED.mcgraw-hill.com **545**

---

1 Focus

VerticalAlignment

Before Lesson 8-3 Graph polynomial functions.

Lesson 8-3 Determine properties of reciprocal functions. Graph transformations of reciprocal functions.

After Lesson 8-3 Graph rational functions using asymptotes.

2 Teach

Scaffolding Questions
Have students read the **Why?** section of the lesson.

Ask:
- On what does the number of candy bars each student has to sell depend? the number of students participating

- What happens to the value of c as the value of n increases? It decreases.

- The variable n must be greater than or equal to what whole number? 1

- The variable n must be less than or equal to what number? the number of students in the chorale

Lesson 8-3 Resources

Resource	Approaching Level **AL**	On Level **OL**	Beyond Level **BL**	English Learners **ELL**
Teacher Edition	▪ Differentiated Instruction, p. 548	▪ Differentiated Instruction, pp. 546, 548	▪ Differentiated Instruction, p. 546	▪ Differentiated Instruction, p. 548
Chapter Resource Masters	▪ Study Guide and Intervention, pp. 18–19 ▪ Skills Practice, p. 20 ▪ Practice, p. 21 ▪ Word Problem Practice, p. 22	▪ Study Guide and Intervention, pp. 18–19 ▪ Skills Practice, p. 20 ▪ Practice, p. 21 ▪ Word Problem Practice, p. 22 ▪ Graphing Calculator, p. 24	▪ Practice, p. 21 ▪ Word Problem Practice, p. 22	▪ Study Guide and Intervention, pp. 18–19 ▪ Skills Practice, p. 20 ▪ Practice, p. 21 ▪ Word Problem Practice, p. 22
Other	▪ 5-Minute Check 8-3 ▪ Study Notebook	▪ 5-Minute Check 8-3 ▪ Study Notebook	▪ 5-Minute Check 8-3 ▪ Study Notebook	▪ 5-Minute Check 8-3 ▪ Study Notebook

1 Vertical and Horizontal Asymptotes

Example 1 shows how to find the limitations on the domain of a rational function. **Example 2** shows how to determine the properties of reciprocal functions.

Formative Assessment

Use the Guided Practice exercises after each example to determine students' understanding of concepts.

Additional Examples

1 Determine the values of x for which $f(x) = \dfrac{2}{x^2 + 5x - 24}$ is not defined. $x = -8$ and $x = 3$

2 Identify the asymptotes, domain, and range of each function.

a.

$f(x) = \dfrac{3}{x-2}$

asymptotes: $x = 2$, $y = 0$; domain: all real numbers not equal to 2; range: all real numbers not equal to 0

b.

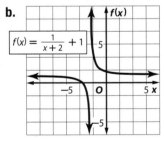

$f(x) = \dfrac{1}{x+2} + 1$

asymptotes: $x = -2$, $f(x) = 1$; domain: all real numbers not equal to -2; range: all real numbers not equal to 1

Additional Examples also in Interactive Classroom PowerPoint® Presentations

IWB Interactive White Board READY

The graphs of reciprocal functions may have breaks in continuity for excluded values. Some may have an asymptote, which is a line that the graph of the function approaches.

StudyTip

CCSS Structure Vertical asymptotes show where a function is undefined, while horizontal asymptotes show the end behavior of a graph.

Example 2 Determine Properties of Reciprocal Functions

Identify the asymptotes, domain, and range of each function.

a.

$f(x) = \dfrac{2}{x-3}$

Identify x-values for which $f(x)$ is undefined.

$x - 3 = 0$
$\quad\quad x = 3$

$f(x)$ is not defined when $x = 3$. So there is an asymptote at $x = 3$.

From $x = 3$, as x-values decrease, $f(x)$-values approach 0, and as x-values increase, $f(x)$-values approach 0. So there is an asymptote at $f(x) = 0$.

The domain is all real numbers not equal to 3 and the range is all real numbers not equal to 0.

b.

$g(x) = \dfrac{1}{x+2} - 1$

Identify x-values for which $g(x)$ is undefined.

$x + 2 = 0$
$\quad\quad x = -2$

$g(x)$ is not defined when $x = -2$. So there is an asymptote at $x = -2$.

From $x = -2$, as x-values decrease, $g(x)$-values approach -1, and as x-values increase, $g(x)$-values approach -1. So there is an asymptote at $g(x) = -1$.

The domain is all real numbers not equal to -2 and the range is all real numbers not equal to -1.

▶ **Guided**Practice

$x = 3$, $f(x) = -2$;
$D = \{x \mid x \neq 3\}$;
$R = \{f(x) \mid f(x) \neq -2\}$

$x = -1$, $g(x) = 5$;
$D = \{x \mid x \neq -1\}$;
$R = \{g(x) \mid g(x) \neq 5\}$

2A.

$f(x) = \dfrac{4}{x-3} - 2$

2B.

$g(x) = \dfrac{2}{x+1} + 5$

Differentiated Instruction ⬤OL ⬤BL

Extension Tell students the following:

- A function is called an **even** function if $f(-x) = f(x)$ for all values of x in its domain. The graph of an even function is *symmetric* with respect to the y-axis.

- A function is called an **odd** function if $f(-x) = -f(x)$ for all values of x in its domain. The graph of an odd function is *symmetric* with respect to the origin.

Ask students whether $f(x) = \dfrac{1}{x}$ is an odd or even function. odd

2 Transformations of Reciprocal Functions The same techniques used to transform the graphs of other functions you have studied can be applied to the graphs of reciprocal functions. In Example 2, note that the asymptotes have been moved along with the graphs of the functions.

StudyTip

Asymptotes The asymptotes of a reciprocal function move with the graph of the function and intersect at (h, k).

KeyConcept Transformations of Reciprocal Functions

$$f(x) = \frac{a}{x - h} + k$$

h – Horizontal Translation	k – Vertical Translation
h units right if h is positive	k units up if k is positive
$\lvert h \rvert$ units left if h is negative	$\lvert k \rvert$ units down if k is negative
The *vertical* asymptote is at $x = h$.	The *horizontal* asymptote is at $f(x) = k$.

a – Orientation and Shape	
If $a < 0$, the graph is reflected across the x-axis.	If $\lvert a \rvert > 1$, the graph is stretched vertically. If $0 < \lvert a \rvert < 1$, the graph is compressed vertically.

Example 3 Graph Transformations

Graph each function. State the domain and range.

a. $f(x) = \dfrac{2}{x - 4} + 2$

This represents a transformation of the graph of $f(x) = \frac{1}{x}$.

$a = 2$:	The graph is stretched vertically.
$h = 4$:	The graph is translated 4 units right. There is an asymptote at $x = 4$.
$k = 2$:	The graph is translated 2 units up. There is an asymptote at $f(x) = 2$.

Domain: $\{x \mid x \neq 4\}$ Range: $\{f(x) \mid f(x) \neq 2\}$

b. $f(x) = \dfrac{-3}{x + 1} - 4$

This represents a transformation of the graph of $f(x) = \frac{1}{x}$.

$a = -3$:	The graph is stretched vertically and reflected across the x-axis.
$h = -1$:	The graph is translated 1 unit left. There is an asymptote at $x = -1$.
$k = -4$:	The graph is translated 4 units down. There is an asymptote at $f(x) = -4$.

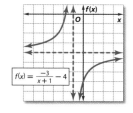

Domain: $\{x \mid x \neq -1\}$ Range: $\{f(x) \mid f(x) \neq -4\}$

GuidedPractice **3A, 3B.** See Chapter 8 Answer Appendix.

3A. $f(x) = \dfrac{-2}{x + 4} + 1$ **3B.** $g(x) = \dfrac{1}{3(x - 1)} - 2$

CCSS **Teaching the Mathematical Practices**

Structure Mathematically proficient students look closely to discern a pattern or structure. Encourage students to compare the equations of asymptotes to the continuity and behavior of a function.

2 Transformations of Reciprocal Functions

Example 3 illustrates graph transformations. **Example 4** is a real-world problem that shows that when a function is used to model a real situation, its meaningful domain may be smaller than its mathematical domain.

Additional Example

3 Graph each function. State the domain and range.

a. $f(x) = -\dfrac{1}{x + 1} + 3$

$D = \{x \mid x \neq -1\}$;
$R = \{f(x) \mid f(x) \neq 3\}$

b. $f(x) = \dfrac{-4}{x - 2} - 1$

$D = \{x \mid x \neq 2\}$;
$R = \{f(x) \mid f(x) \neq -1\}$

Teach with Tech

Interactive Whiteboard Use a graphing program to display a coordinate grid and the graph of a function $f(x)$. Choose several students to come to the board and draw what they think would be the graph of $\dfrac{1}{f(x)}$. Then use the graphing program to show the graph of $\dfrac{1}{f(x)}$, and compare it to students' graphs.

WatchOut!

Preventing Errors Suggest that students choose a large unit on their grid paper and estimate point coordinates to the nearest tenth. Point out that they may not be able to see the shape of the graph as a whole unless they use a graphing calculator or computer program.

DH Kong/Plush Studios/Blend Images/Getty Images

Additional Example

4 COMMUTING A commuter train has a nonstop service from one city to another, a distance of about 25 miles.

a. Write an equation to represent the travel time between these two cities as a function of rail speed. Then graph the equation. $t = \dfrac{25}{r}$

b. Explain any limitations to the range and domain in this situation. The range and domain are limited to all real numbers greater than 0 because negative values do not make sense. There will be further restrictions to the domain because the train has minimum and maximum speeds at which it can travel.

The Math Forum

mathforum.org
The Math Forum
© Drexel University

Have your students refer to the Math Forum's Ask Dr. Math to learn more about horizontal and vertical asymptotes. They can start their search at connectED.mcgraw-hill.com.

3 Practice

Formative Assessment

Use Exercises 1–6 to check for understanding.

Use the chart at the bottom of the next page to customize assignments for your students.

Real-World Example 4 Write Equations

TRAVEL An airline has a daily nonstop flight between Los Angeles, California, and Sydney, Australia. A one-way trip is about 7500 miles.

a. Write an equation to represent the travel time from Los Angeles to Sydney as a function of flight speed. Then graph the equation.

Solve $rt = d$ for t.

$rt = d$ — Original formula

$t = \dfrac{d}{r}$ — Divide each side by r.

$t = \dfrac{7500}{r}$ — $d = 7500$

Graph $t = \dfrac{7500}{r}$.

$t = \dfrac{7500}{r}$

b. Explain any limitations to the range or domain in this situation.

In this situation, the range and domain are limited to all real numbers greater than zero because negative values do not make sense. There will be further restrictions to the domain because the aircraft has minimum and maximum speeds at which it can travel.

▶ **Guided**Practice 4. See Chapter 8 Answer Appendix.

4. HOMECOMING DANCE The junior and senior class officers are sponsoring a homecoming dance. The total cost for the facilities and catering is $45 per person plus a $2500 deposit. Write and graph an equation to represent the average cost per person. Then explain any limitations to the domain and range.

Check Your Understanding

◯ = Step-by-Step Solutions begin on page R14.

Examples 1–2 Identify the asymptotes, domain, and range of each function.

1.

$f(x) = \dfrac{4}{x-1}$

$x = 1$, $f(x) = 0$;
$D = \{x \mid x \neq 1\}$;
$R = \{f(x) \mid f(x) \neq 0\}$

2.

$f(x) = \dfrac{3}{x+2} + 1$

$x = -2$, $f(x) = 1$;
$D = \{x \mid x \neq -2\}$;
$R = \{f(x) \mid f(x) \neq 1\}$

Example 3 Graph each function. State the domain and range. **3–5.** See margin.

3. $f(x) = \dfrac{5}{x}$

4. $f(x) = \dfrac{2}{x+3}$

5. $f(x) = \dfrac{-1}{x-2} + 4$

Example 4 **6. CCSS SENSE-MAKING** A group of friends plans to get their youth group leader a gift certificate for a day at a spa. The certificate costs $150.

a. If c represents the cost for each friend and f represents the number of friends, write an equation to represent the cost to each friend as a function of how many friends give. $c = \dfrac{150}{f}$

b. Graph the function. See margin.

c. Explain any limitations to the range or domain in this situation.

c. In this situation, the range and domain are limited to all real numbers greater than zero because negative values do not make sense.

548 | Lesson 8-3 | Graphing Reciprocal Functions

DifferentiatedInstruction
AL OL ELL

Visual/Spatial Learners Have students graph one of the functions from the lesson on a large sheet of poster board to clearly show how the graph approaches but never reaches an asymptote. Encourage students to use a variety of colored markers.

CCSS Teaching the Mathematical Practices

Sense-Making Mathematically proficient students start by explaining to themselves the meaning of a problem and looking for entry points to its solution. They analyze givens, constraints, relationships, and goals. They check their answers to problems using a different method, and they continually ask themselves, "Does this make sense?"

Real-World Career

Travel Agent Travel agents assess individual and business needs to help make the best possible travel arrangements. They may specialize by type of travel, such as leisure or business, or by destination, such as Europe or Africa. A high school diploma is required, and vocational training is preferred.

Examples 1–2 Identify the asymptotes, domain, and range of each function.

7.

$f(x) = \dfrac{5}{x+4}$

$x = -4$, $f(x) = 0$;
$D = \{x \mid x \neq -4\}$;
$R = \{f(x) \mid f(x) \neq 0\}$

8.

$f(x) = \dfrac{6}{x} - 3$

$x = 0$, $f(x) = -3$;
$D = \{x \mid x \neq 0\}$;
$R = \{f(x) \mid f(x) \neq -3\}$

9.

$f(x) = \dfrac{2}{x+6} - 2$

$x = -6$, $f(x) = -2$; $D = \{x \mid x \neq -6\}$;
$R = \{f(x) \mid f(x) \neq -2\}$

10.

$f(x) = \dfrac{-3}{x-1} + 5$

$x = 1$, $f(x) = 5$;
$D = \{x \mid x \neq 1\}$;
$R = \{f(x) \mid f(x) \neq 5\}$

Example 3 Graph each function. State the domain and range. **11–22.** See Chapter 8 Answer Appendix.

11. $f(x) = \dfrac{3}{x}$

12. $f(x) = \dfrac{-4}{x+2}$

13. $f(x) = \dfrac{2}{x-6}$

14. $f(x) = \dfrac{6}{x} - 5$

15 $f(x) = \dfrac{2}{x} + 3$

16. $f(x) = \dfrac{8}{x}$

17. $f(x) = \dfrac{-2}{x-5}$

18. $f(x) = \dfrac{3}{x-7} - 8$

19. $f(x) = \dfrac{9}{x+3} + 6$

20. $f(x) = \dfrac{8}{x+3}$

21. $f(x) = \dfrac{-6}{x+4} - 2$

22. $f(x) = \dfrac{-5}{x-2} + 2$

Example 4 **23. CYCLING** Marina's New Year's resolution is to ride her bike 5000 miles.

a. If m represents the mileage Marina rides each day and d represents the number of days, write an equation to represent the mileage each day as a function of the number of days that she rides. $m = \dfrac{5000}{d}$

b. Graph the function. **See Chapter 8 Answer Appendix.**

c. If she rides her bike every day of the year, how many miles should she ride each day to meet her goal? **13.7 mi**

24. CCSS MODELING Parker has 200 grams of an unknown liquid. Knowing the density will help him discover what type of liquid this is.

a. Density of a liquid is found by dividing the mass by the volume. Write an equation to represent the density of this unknown as a function of volume. $d = \dfrac{200}{v}$

b. Graph the function. **See Chapter 8 Answer Appendix.**

c. From the graph, identify the asymptotes, domain, and range of the function.
$v = 0$, $d = 0$; $D = \{v \mid v \neq 0\}$; $R = \{d \mid d \neq 0\}$

 Graph each function. State the domain and range. **25–28.** See Chapter 8 Answer Appendix.

25. $f(x) = \dfrac{3}{2x-4}$

26. $f(x) = \dfrac{5}{3x}$

27. $f(x) = \dfrac{2}{4x+1}$

28. $f(x) = \dfrac{1}{2x+3}$

 connectED.mcgraw-hill.com **549**

Differentiated Homework Options

Level	Assignment	Two-Day Option	
AL Basic	7–24, 38–40, 42–61	7–23 odd, 43–47	8–24 even, 38–40, 42, 48–61
OL Core	7–35 odd, 36–40, 42–61	7–24, 43–47	25–40, 42, 48–61
BL Advanced	25–59, (optional: 60–61)		

Exercise Alert

Grid Paper For Exercises 3–6, 11–39, 41, 52–55, and 60–61, students will need grid paper.

Additional Answers

3.

$f(x) = \dfrac{5}{x}$

$D = \{x \mid x \neq 0\}$;
$R = \{f(x) \mid f(x) \neq 0\}$

4.

$f(x) = \dfrac{2}{x+3}$

$D = \{x \mid x \neq -3\}$;
$R = \{f(x) \mid f(x) \neq 0\}$

5.

$f(x) = \dfrac{-1}{x-2} + 4$

$D = \{x \mid x \neq 2\}$;
$R = \{f(x) \mid f(x) \neq 4\}$

6b.

$c = \dfrac{150}{f}$

Lesson 8-3

29. BASEBALL The distance from the pitcher's mound to home plate is 60.5 feet.

a. If r represents the speed of the pitch and t represents the time it takes the ball to get to the plate, write an equation to represent the speed as a function of time. $r = \dfrac{60.5}{t}$

b. Graph the function. **See Chapter 8 Answer Appendix.**

c. If a two-seam fastball reaches the plate in 0.48 second, what was its speed? **126 ft/s**

30–35. See Chapter 8 Answer Appendix.

Graph each function. State the domain and range, and identify the asymptotes.

30. $f(x) = \dfrac{-3}{x + 7} - 1$

31. $f(x) = \dfrac{-4}{x + 2} - 5$

32. $f(x) = \dfrac{6}{x - 1} + 2$

33. $f(x) = \dfrac{2}{x - 4} + 3$

34. $f(x) = \dfrac{-7}{x - 8} - 9$

35. $f(x) = \dfrac{-6}{x - 7} - 8$

36. FINANCIAL LITERACY Lawanda's car went 440 miles on one tank of gas.

a. If g represents the number of miles to the gallon that the car gets and t represents the size of the gas tank, write an equation to represent the miles to the gallon as a function of tank size. $g = \dfrac{440}{t}$

b. Graph the function. **See Chapter 8 Answer Appendix.**

c. How many miles does the car get per gallon if it has a 15-gallon tank? $29\dfrac{1}{3}$ mi/gal

37. MULTIPLE REPRESENTATIONS In this problem you will investigate the similarities and differences between power functions with positive and negative exponents. **a–f. See Chapter 8 Answer Appendix.**

a. **TABULAR** Make a table of values for $a(x) = x^2$, $b(x) = x^{-2}$, $c(x) = x^3$, and $d(x) = x^{-3}$.

b. **GRAPHICAL** Graph $a(x)$ and $b(x)$ on the same coordinate plane.

c. **VERBAL** Compare the domain, range, end behavior, and behavior at $x = 0$ for $a(x)$ and $b(x)$.

d. **GRAPHICAL** Graph $c(x)$ and $d(x)$ on the same coordinate plane.

e. **VERBAL** Compare the domain, range, end behavior, and behavior at $x = 0$ for $c(x)$ and $d(x)$.

f. **ANALYTICAL** What conclusions can you make about the similarities and differences between power functions with positive and negative exponents?

H.O.T. Problems Use Higher-Order Thinking Skills

38. OPEN ENDED Write a reciprocal function for which the graph has a vertical asymptote at $x = -4$ and a horizontal asymptote at $f(x) = 6$. **Sample answer:** $f(x) = \dfrac{1}{x + 4} + 6$

39. REASONING Compare and contrast the graphs of each pair of equations.

a. $y = \dfrac{1}{x}$ and $y - 7 = \dfrac{1}{x}$

b. $y = \dfrac{1}{x}$ and $y = 4\left(\dfrac{1}{x}\right)$

c. $y = \dfrac{1}{x}$ and $y = \dfrac{1}{x + 5}$

d. Without making a table of values, use what you observed in parts **a–c** to sketch a graph of $y - 7 = 4\left(\dfrac{1}{x + 5}\right)$. **a–d. See Chapter 8 Answer Appendix.**

40. CCSS ARGUMENTS Find the function that does not belong. Explain.

$$f(x) = \dfrac{3}{x + 1}$$

$$g(x) = \dfrac{x + 2}{x^2 + 1}$$

$$h(x) = \dfrac{5}{x^2 + 2x + 1}$$

$$j(x) = \dfrac{20}{x - 7}$$

Sample answer: $g(x)$; all other choices have unknowns only in the denominator.

41. CHALLENGE Write two different reciprocal functions with graphs having the same vertical and horizontal asymptotes. Then graph the functions. **See Chapter 8 Answer Appendix.**

42. WRITING IN MATH Refer to the beginning of the lesson. Explain how rational functions can be used in fundraising. Explain why only part of the graph is meaningful in the context of the problem. **See Chapter 8 Answer Appendix.**

 550 | Lesson 8-3 | Graphing Reciprocal Functions

CCSS Teaching the Mathematical Practices

Arguments Mathematically proficient students understand and use stated assumptions, definitions, and previously established results in constructing arguments. They make conjectures and build a logical progression of statements to explore the truth of their conjectures. And they are able to analyze situations by breaking them into cases, and can recognize and use counterexamples.

43. SHORT RESPONSE What is the value of $(x + y)(x + y)$ if $xy = -3$ and $x^2 + y^2 = 10$? **4**

44. GRIDDED RESPONSE If $x = 2y$, $y = 4z$, $2z = w$, and $w \neq 0$, then $\frac{x}{w} = $ ___. **4**

45. If $c = 1 + \frac{1}{d}$ and $d > 1$, then c could equal ___. **B**

A $\frac{5}{7}$ **C** $\frac{15}{7}$

B $\frac{9}{7}$ **D** $\frac{19}{7}$

46. SAT/ACT A car travels m miles at the rate of t miles per hour. How many hours does the trip take? **F**

F $\frac{m}{t}$ **J** $\frac{t}{m}$

G $m - t$ **K** $t - m$

H mt

47. If $-1 < a < b < 0$, then which of the following has the greatest value? **B**

A $a - b$ **C** $a + b$

B $b - a$ **D** $2b - a$

Spiral Review

48. BUSINESS A small corporation decides that 8% of its profits will be divided among its six managers. There are two sales managers and four nonsales managers. Fifty percent will be split equally among all six managers. The other 50% will be split among the four nonsales managers. Let p represent the profits. (Lesson 8-2)

a. Write an expression to represent the share of the profits each nonsales manager will receive. $\frac{0.5(0.08p)}{6} + \frac{0.5(0.08p)}{4}$

b. Simplify this expression. $\frac{0.05p}{3}$

c. Write an expression in simplest form to represent the share of the profits each sales manager will receive. $\frac{0.02p}{3}$

Simplify each expression. (Lesson 8-1)

49. $\dfrac{\frac{p^3}{2n}}{\frac{p^2}{4n}}$ $-2p$

50. $\dfrac{\frac{m+q}{5}}{\frac{m^2+q^2}{5}}$ $\frac{m+q}{m^2+q^2}$

51. $\dfrac{\frac{x+y}{2x-y}}{\frac{x+y}{2x+y}}$ $\frac{2x+y}{2x-y}$

Graph each function. State the domain and range. (Lesson 7-1) **52–55. See margin.**

52. $y = 2(3)^x$ **53.** $y = 5(2)^x$ **54.** $y = 0.5(4)^x$ **55.** $y = 4\left(\frac{1}{3}\right)^x$

Find $(f + g)(x)$, $(f - g)(x)$, $(f \cdot g)(x)$, and $\left(\frac{f}{g}\right)(x)$ for each $f(x)$ and $g(x)$. (Lesson 6-1) **56–58. See margin.**

56. $f(x) = x + 9$
$g(x) = x - 9$

57. $f(x) = 2x - 3$
$g(x) = 4x + 9$

58. $f(x) = 2x^2$
$g(x) = 8 - x$

59. GEOMETRY The width of a rectangular prism is w centimeters. The height is 2 centimeters less than the width. The length is 4 centimeters more than the width. If the volume of the prism is 8 times the measure of the length, find the dimensions of the prism. (Lesson 5-5)
$w = 4$ cm, $\ell = 8$ cm, $h = 2$ cm

$w - 2$ $w + 4$

w

Skills Review

Graph each polynomial function. Estimate the x-coordinates at which the relative maxima and relative minima occur. State the domain and range for each function.

60, 61. See Chapter 8 Answer Appendix.

60. $f(x) = x^3 + 2x^2 - 3x - 5$ **61.** $f(x) = x^4 - 8x^2 + 10$

55. D = {all real numbers},
R = {y | y > 0}

56. $(f + g)(x) = 2x$; $(f - g)(x) = 18$;
$(f \cdot g)(x) = x^2 - 81$; $\left(\frac{f}{g}\right)(x) = \frac{x + 9}{x - 9}$,
$x \neq 9$

57. $(f + g)(x) = 6x + 6$; $(f - g)(x) = -2x - 12$; $(f \cdot g)(x) = 8x^2 + 6x - 27$;
$\left(\frac{f}{g}\right)(x) = \frac{2x - 3}{4x + 9}$, $x \neq -\frac{9}{4}$

58. $(f + g)(x) = 2x^2 - x + 8$; $(f - g)(x) = 2x^2 + x - 8$; $(f \cdot g)(x) = -2x^3 + 16x^2$;
$\left(\frac{f}{g}\right)(x) = \frac{2x^2}{8 - x}$, $x \neq 8$

4 Assess

Crystal Ball Ask students to tell how they think their practice with graphing reciprocal functions will help them when they graph rational functions.

Formative Assessment
Check for student understanding of concepts in Lesson 8-3.

 Quiz 2, p. 47

Additional Answers

52. D = {all real numbers},
R = {y | y > 0}

53. D = {all real numbers},
R = {y | y > 0}

54. D = {all real numbers},
R = {y | y > 0}

Formative Assessment

Use the Mid-Chapter Quiz to assess students' progress in the first half of the chapter.

For problems answered incorrectly, have students review the lessons indicated in parentheses.

Summative Assessment

☐ Mid-Chapter Test, p. 49

McGraw-Hill **eAssessment** Customize and create multiple versions of your Mid-Chapter Quiz and their answer keys.

FOLDABLES StudyOrganizer

Dinah Zike's Foldables®

Before students complete the Mid-Chapter Quiz, encourage them to review the information for Lessons 8-1 through 8-3 in their Foldables.

Additional Answer

25b.

Mid-Chapter Quiz
Lessons 8-1 through 8-3

Simplify each expression. *(Lesson 8-1)*

1. $\dfrac{2x^2 y^5}{7x^3 yz} \cdot \dfrac{14xyz^2}{18x^4 y}$ $\dfrac{2y^4 z}{9x^4}$

2. $\dfrac{24a^4 b^6}{35ab^3} \div \dfrac{12abc}{7a^2 c}$ $\dfrac{2a^4 b^2}{5}$

3. $\dfrac{3x - 3}{x^2 + x - 2} \cdot \dfrac{4x + 8}{6x + 18}$ $\dfrac{2}{x + 3}$

4. $\dfrac{m^2 + 3m + 2}{9} \div \dfrac{m + 1}{3m + 15}$ $\dfrac{(m+2)(m+5)}{3}$

5. $\dfrac{\frac{r^2 + 3r}{r + 1}}{\frac{3r}{3r + 3}}$ $r + 3$

6. $\dfrac{\frac{2y}{y^2 - 4}}{\frac{3}{y^2 - 4y + 4}}$ $\dfrac{2y(y - 2)}{3(y + 2)}$

7. MULTIPLE CHOICE For all $r \neq \pm 2$, $\dfrac{r^2 + 6r + 8}{r^2 - 4} =$ ___. *(Lesson 8-1)* **B**

A $\dfrac{r - 2}{r + 4}$ C $\dfrac{r + 2}{r - 4}$

B $\dfrac{r + 4}{r - 2}$ D $\dfrac{r + 4}{r + 2}$

8. MULTIPLE CHOICE Identify all values of x for which

$\dfrac{x^2 - 16}{(x^2 - 6x - 27)(x + 1)}$ is undefined. *(Lesson 8-1)* **H**

F $-3, -1$ H $-3, -1, 9$

G $3, 1, -9$ J -1

9. What is the LCM of $x^2 - x$ and $3 - 3x$? *(Lesson 8-2)*
$-3x(x - 1)$

Simplify each expression. *(Lesson 8-2)*

10. $\dfrac{2x}{4x^2 y} + \dfrac{x}{3xy^3}$ $\dfrac{6xy^2 + 4x^2}{12x^2 y^3}$

11. $\dfrac{3}{4m} + \dfrac{2}{3mn^2} - \dfrac{4}{n}$ $\dfrac{9n^2 + 8 - 48mn}{12mn^2}$

12. $\dfrac{6}{r^2 - 3r - 18} - \dfrac{1}{r^2 + r - 6}$ $\dfrac{5r - 6}{(r - 6)(r + 3)(r - 2)}$

13. $\dfrac{3x + 6}{x + y} + \dfrac{6}{-x - y} - \dfrac{3x}{x + y}$

14. $\dfrac{x - 4}{x^2 - 3x - 4} + \dfrac{x + 1}{2x - 8}$ $\dfrac{x^2 + 4x - 7}{2(x - 4)(x + 1)}$

15. Determine the perimeter of the rectangle. *(Lesson 8-2)*

$\dfrac{18x - 6}{(x - 3)(x + 1)}$

---

16. TRAVEL Lucita is going to a beach 100 miles away. She travels half the distance at one rate. The rest of the distance, she travels 15 miles per hour slower. *(Lesson 8-2)* **a.** $\dfrac{50}{x} h$

a. If x represents the faster pace in miles per hour, write an expression that represents the time spent at that pace.

b. Write an expression for the amount of time spent at the slower pace. $\dfrac{50}{x - 15} h$

c. Write an expression for the amount of time Lucita needs to complete the trip. $\dfrac{100(x - 7.5)}{x(x - 15)}$

Identify the asymptotes, domain, and range of each function. *(Lesson 8-3)*

17.

$x = -3; f(x) = 0;$
$D = \{x \mid x \neq -3\},$
$R = \{f(x) \mid f(x) \neq 0\}$

18.

$x = 6; f(x) = 4;$
$D = \{x \mid x \neq 6\},$
$R = \{f(x) \mid f(x) \neq 4\}$

Graph each reciprocal function. State the domain and range.
(Lesson 8-3) **19–24. See Chapter 8 Answer Appendix.**

19. $f(x) = \dfrac{4}{x}$ **20.** $f(x) = \dfrac{1}{3x}$

21. $f(x) = \dfrac{6}{x - 1}$ **22.** $f(x) = \dfrac{-2}{x} + 4$

23. $f(x) = \dfrac{3}{x + 2} - 5$ **24.** $f(x) = -\dfrac{1}{x - 3} + 2$

25. SANDWICHES A group makes 45 sandwiches to take on a picnic. The number of sandwiches a person can eat depends on how many people go on the trip. *(Lesson 8-3)*

a. Write a function to represent this situation. $f(x) = \dfrac{45}{x}$

b. Graph the function. **See margin.**

InterventionPlanner

TIER 1 On Level OL

If students miss about 25% of the exercises or less,

Then choose a resource:
- **SE** Lessons 8-1, 8-2, and 8-3
- ☐ Skills Practice, pp. 7, 13, and 20
- **TE** Chapter Project
- connectED.mcgraw-hill.com Self-Check Quiz

TIER 2 Strategic Intervention AL approaching grade level

If students miss about 50% of the exercises,

Then choose a resource:
- ☐ Study Guide and Intervention, pp. 5, 11, and 18
- *Quick Review Math Handbook*
- connectED.mcgraw-hill.com Extra Examples, Personal Tutor, Homework Help

TIER 3 Intensive Intervention 2 or more grades below level

If students miss about 75% of the exercises,

Then use *Math Triumphs, Alg. 2* Ch. 2
- connectED.mcgraw-hill.com Extra Examples, Personal Tutor, Homework Help, Review Vocabulary

LESSON 8-4 Graphing Rational Functions

: Then	: Now	: Why?
● You graphed reciprocal functions.	**1** Graph rational functions with vertical and horizontal asymptotes. **2** Graph rational functions with oblique asymptotes and point discontinuity.	● Regina bought a digital SLR camera and a photo printer for $350. The manufacturer claims that ink and photo paper cost $0.47 per photo. The rational function $C(p) = \dfrac{0.47p + 350}{p}$ can be used to determine the average cost $C(p)$ for printing p photos.

NewVocabulary
rational function
vertical asymptote
horizontal asymptote
oblique asymptote
point discontinuity

Common Core State Standards

Content Standards
A.CED.2 Create equations in two or more variables to represent relationships between quantities; graph equations on coordinate axes with labels and scales.

F.IF.9 Compare properties of two functions each represented in a different way (algebraically, graphically, numerically in tables, or by verbal descriptions).

Mathematical Practices
7 Look for and make use of structure.

1 Vertical and Horizontal Asymptotes A **rational function** has an equation of the form $f(x) = \dfrac{a(x)}{b(x)}$, where $a(x)$ and $b(x)$ are polynomial functions and $b(x) \neq 0$.

In order to graph a rational function, it is helpful to locate the zeros and asymptotes. A zero of a rational function $f(x) = \dfrac{a(x)}{b(x)}$ occurs at every value of x for which $a(x) = 0$.

KeyConcept Vertical and Horizontal Asymptotes

Words If $f(x) = \dfrac{a(x)}{b(x)}$, $a(x)$ and $b(x)$ are polynomial functions with no common factors other than 1, and $b(x) \neq 0$, then:

- $f(x)$ has a **vertical asymptote** whenever $b(x) = 0$.
- $f(x)$ has at most one **horizontal asymptote.**
  - If the degree of $a(x)$ is greater than the degree of $b(x)$, there is no horizontal asymptote.
  - If the degree of $a(x)$ is less than the degree of $b(x)$, the horizontal asymptote is the line $y = 0$.
  - If the degree of $a(x)$ equals the degree of $b(x)$, the horizontal asymptote is the line $y = \dfrac{\text{leading coefficient of } a(x)}{\text{leading coefficient of } b(x)}$.

Examples

No horizontal asymptote

$f(x) = \dfrac{x^2}{x+1}$

Vertical asymptote:
$x = -1$

One horizontal asymptote

$f(x) = \dfrac{3}{x^2 - 1}$

Vertical asymptotes:
$x = -1$, $x = 1$
Horizontal asymptote:
$f(x) = 0$

$f(x) = \dfrac{2x + 1}{x - 3}$

Vertical asymptote:
$x = 3$
Horizontal asymptote:
$f(x) = 2$

connectED.mcgraw-hill.com **553**

JGI/Blend Images/Getty Images

1 Focus

VerticalAlignment

Before Lesson 8-4 Graph reciprocal functions.

Lesson 8-4 Graph rational functions with vertical and horizontal asymptotes. Graph rational functions with oblique asymptotes and point discontinuity.

After Lesson 8-4 Solve rational equations by graphing.

2 Teach

Scaffolding Questions
Have students read the **Why?** section of the lesson.

Ask:

- On what does the average cost for printing photos depend? the number of photos printed

- What happens to the value of $C(p)$ as the value of p increases? It decreases.

- Can $p = 0$? no

- Can $C(p)$ ever equal 0? Yes, the function has a zero at $p = -\dfrac{350}{0.47}$, or about -744.68. However, negative values are not realistic to the context of the situation.

Lesson 8-4 Resources

Resource	Approaching Level **AL**	On Level **OL**	Beyond Level **BL**	English Learners **ELL**
Teacher Edition		▪ Differentiated Instruction, pp. 556, 560	▪ Differentiated Instruction, pp. 555, 560	▪ Differentiated Instruction, p. 556
Chapter Resource Masters	▪ Study Guide and Intervention, pp. 25–26 ▪ Skills Practice, p. 27 ▪ Practice, p. 28 ▪ Word Problem Practice, p. 29	▪ Study Guide and Intervention, pp. 25–26 ▪ Skills Practice, p. 27 ▪ Practice, p. 28 ▪ Word Problem Practice, p. 29 ▪ Enrichment, p. 30 ▪ Graphing Calculator Activity, p. 31	▪ Practice, p. 28 ▪ Word Problem Practice, p. 29 ▪ Enrichment, p. 30	▪ Study Guide and Intervention, pp. 25–26 ▪ Skills Practice, p. 27 ▪ Practice, p. 28 ▪ Word Problem Practice, p. 29
Other	▪ 5-Minute Check 8-4 ▪ Study Notebook	▪ 5-Minute Check 8-4 ▪ Study Notebook	▪ 5-Minute Check 8-4 ▪ Study Notebook	▪ 5-Minute Check 8-4 ▪ Study Notebook

1 Vertical and Horizontal Asymptotes

Example 1 shows how to identify and graph rational functions with no horizontal asymptotes. **Example 2** is a real-world problem that shows that when a function is used to model a real situation, its meaningful domain may be smaller than its mathematical domain.

Formative Assessment

Use the Guided Practice exercises after each Example to determine students' understanding of concepts.

Additional Example

1 Graph $f(x) = \dfrac{x^3}{x + 1}$.

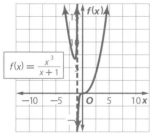

$f(x) = \dfrac{x^3}{x + 1}$

Additional Examples also in Interactive Classroom PowerPoint® Presentations

 IWB **Interactive White Board READY**

Additional Answer (Guided Practice)

1A.

$f(x) = \dfrac{x^2 - x - 6}{x + 1}$

WatchOut!
Zeros vs. Vertical Asymptotes Zeros of rational functions occur at the values that make the numerator equal to zero. Vertical asymptotes occur at the values that make the denominator equal to zero.

The asymptotes of a rational function can be used to draw the graph of the function. Additionally, the asymptotes can be used to divide a graph into regions to find ordered pairs on the graph.

Example 1 Graph with no Horizontal Asymptote

Graph $f(x) = \dfrac{x^3}{x - 1}$.

Step 1 Find the zeros.

$x^3 = 0$ Set $a(x) = 0$.

$x = 0$ Take the cube root of each side.

There is a zero at $x = 0$.

Step 2 Draw the asymptotes.

Find the vertical asymptote.

$x - 1 = 0$ Set $b(x) = 0$.

$x = 1$ Add 1 to each side.

There is a vertical asymptote at $x = 1$.

The degree of the numerator is greater than the degree of the denominator. So, there is no horizontal asymptote.

StudyTip
Graphing Calculator The TABLE feature of a graphing calculator can be used to calculate decimal values for x and y.

Step 3 Draw the graph.

Use a table to find ordered pairs on the graph. Then connect the points.

x	f(x)
−3	6.75
−2	2.67
−1	0.5
0	0
0.5	−0.25
1.5	6.75
2	8
3	13.5

Guided Practice

Graph each function.   **1A, 1B. See margin.**

1A. $f(x) = \dfrac{x^2 - x - 6}{x + 1}$ **1B.** $f(x) = \dfrac{(x + 1)^3}{(x + 2)^2}$

In the real world, sometimes values on the graph of a rational function are not meaningful. In the graph at the right, x-values such as time, distance, and number of people cannot be negative in the context of the problem. So, you do not even need to consider that portion of the graph.

 554 | Lesson 8-4 | Graphing Rational Functions

1B.

$f(x) = \dfrac{(x + 1)^3}{(x + 2)^2}$

Teach with Tech

Video Recording Have students create video recordings showing how to graph a rational function. Have students first describe how to find the horizontal and vertical asymptotes, and then use a table of values to draw the graph.

PA1 Tom Sperduto, US Coast Guard

● **Real-World Example 2** Use Graphs of Rational Functions

AVERAGE SPEED A boat traveled upstream at r_1 miles per hour. During the return trip to its original starting point, the boat traveled at r_2 miles per hour. The average speed for the entire trip R is given by the formula $R = \dfrac{2r_1 r_2}{r_1 + r_2}$.

a. Let r_1 be the independent variable, and let R be the dependent variable. Draw the graph if $r_2 = 10$ miles per hour.

The function is $R = \dfrac{2r_1(10)}{r_1 + (10)}$ or $R = \dfrac{20r_1}{r_1 + 10}$.

The vertical asymptote is $r_1 = -10$.

Graph the vertical asymptote and the function.

Notice that the horizontal asymptote is $R = 20$.

b. What is the R-intercept of the graph?

The R-intercept is 0.

c. What domain and range values are meaningful in the context of the problem?

In the problem context, speeds are nonnegative values. Therefore, only values of r_1 greater than or equal to 0 and values of R between 0 and 20 are meaningful.

▶ **Guided**Practice

2. SALARIES A company uses the formula $S(x) = \dfrac{45x + 25}{x + 1}$ to determine the salary in thousands of dollars of an employee during his xth year. Graph $S(x)$. What domain and range values are meaningful in the context of the problem? What is the meaning of the horizontal asymptote for the graph? **See margin.**

2 **Oblique Asymptotes and Point Discontinuity** An **oblique asymptote**, sometimes called a *slant asymptote*, is an asymptote that is neither horizontal nor vertical.

KeyConcept Oblique Asymptotes

Words   If $f(x) = \dfrac{a(x)}{b(x)}$, $a(x)$ and $b(x)$ are polynomial functions with no common factors other than 1 and $b(x) \neq 0$, then $f(x)$ has an oblique asymptote if the degree of $a(x)$ minus the degree of $b(x)$ equals 1. The equation of the asymptote is $f(x) = \dfrac{a(x)}{b(x)}$ with no remainder.

Example $f(x) = \dfrac{x^4 + 3x^3}{x^3 - 1}$

Vertical asymptote: $x = 1$

Oblique asymptote: $f(x) = x + 3$

Real-World Career

U.S. Coast Guard Boatswain's Mate
The most versatile member of the U.S. Coast Guard's operational team is the boatswain's mate. BMs are capable of performing almost any task. Training for BMs is accomplished through 12 weeks of intensive training.

Additional Example

2 **AVERAGE SPEED** Use the situation and formula given in Example 2.

a. Draw the graph if $r_2 = 15$ miles per hour.

b. What is the R-intercept of the graph? The R-intercept is 0.

c. What domain and range values are meaningful in the context of the problem? Values of r_1 greater than or equal to 0 and values of R between 0 and 30 are meaningful.

2 **Oblique Asymptotes and Point Discontinuity**
Example 3 shows how to graph a function with oblique asymptotes.
Example 4 shows a graph with point discontinuity.

Additional Answer (Guided Practice)

2. The number of years worked must be greater than or equal to 0, and salary values must be between 25 and 45. The asymptote represents a salary cap of $45,000.

DifferentiatedInstruction   **BL**

Extension Challenge students to explain the rules for finding horizontal and oblique asymptotes. While the chapter shows students how to find them, the explanation of why the rules work can be left to high ability students. Scaffold the task by having students examine graphs of varying degrees in the numerator and denominator and look for general patterns in the asymptotes.

Additional Example

3 Graph $f(x) = \dfrac{x^2}{x + 1}$.

Additional Answer (Guided Practice)

3A.

3B.

StudyTip

Oblique Asymptotes Oblique asymptotes occur for rational functions that have a numerator polynomial that is one degree higher than the denominator polynomial.

Example 3 Determine Oblique Asymptotes

Graph $f(x) = \dfrac{x^2 + 4x + 4}{2x - 1}$.

Step 1 Find the zeros.

$$x^2 + 4x + 4 = 0 \qquad \text{Set } a(x) = 0.$$
$$(x + 2)^2 = 0 \qquad \text{Factor.}$$
$$x + 2 = 0 \qquad \text{Take the square root of each side.}$$
$$x = -2 \qquad \text{Subtract 2 from each side.}$$

There is a zero at $x = -2$.

Step 2 Find the asymptotes.

$$2x - 1 = 0 \qquad \text{Set } b(x) = 0.$$
$$2x = 1 \qquad \text{Add 1 to each side.}$$
$$x = \frac{1}{2} \qquad \text{Divide each side by 2.}$$

There is a vertical asymptote at $x = \frac{1}{2}$.

The degree of the numerator is greater than the degree of the denominator, so there is no horizontal asymptote.

The difference between the degree of the numerator and the degree of the denominator is 1, so there is an oblique asymptote.

Divide the numerator by the denominator to determine the equation of the oblique asymptote.

The equation of the asymptote is the quotient excluding any remainder.

Thus, the oblique asymptote is the line $f(x) = \frac{1}{2}x + \frac{9}{4}$.

Step 3 Draw the asymptotes, and then use a table of values to graph the function.

GuidedPractice

Graph each function.   **3A, 3B. See margin.**

3A. $f(x) = \dfrac{x^2}{x - 2}$

3B. $f(x) = \dfrac{x^3 - 1}{x^2 - 4}$

In some cases, graphs of rational functions may have **point discontinuity**, which looks like a hole in the graph. This is because the function is undefined at that point.

 556 | Lesson 8-4 | Graphing Rational Functions

DifferentiatedInstruction OL ELL

Verbal/Linguistic Learners Have students write a list of tips to help someone draw the graphs of rational functions.

KeyConcept Point Discontinuity

Words If $f(x) = \dfrac{a(x)}{b(x)}$, $b(x) \neq 0$, and $x - c$ is a factor of both $a(x)$ and $b(x)$, then there is a point discontinuity at $x = c$.

Example $f(x) = \dfrac{(x + 2)(x + 1)}{x + 1}$
$= x + 2;\ x \neq -1$

$f(x) = \dfrac{(x + 2)(x + 1)}{x + 1}$

WatchOut!

Holes Remember that a common factor in the numerator and denominator can signal a hole.

Example 4 Graph with Point Discontinuity

Graph $f(x) = \dfrac{x^2 - 16}{x - 4}$.

Notice that $\dfrac{x^2 - 16}{x - 4} = \dfrac{(x + 4)(x - 4)}{x - 4}$ or $x + 4$.

Therefore, the graph of $f(x) = \dfrac{x^2 - 16}{x - 4}$ is the graph of $f(x) = x + 4$ with a hole at $x = 4$.

$f(x) = \dfrac{x^2 - 16}{x - 4}$

> **Guided**Practice   **4A, 4B.** See margin.

Graph each function.

4A. $f(x) = \dfrac{x^2 + 4x - 5}{x + 5}$

4B. $f(x) = \dfrac{x^3 + 2x^2 - 9x - 18}{x^2 - 9}$

Check Your Understanding

● = Step-by-Step Solutions begin on page R14.

Example 1 **Graph each function.** **1, 2. See Chapter 8 Answer Appendix.**

1. $f(x) = \dfrac{x^4 - 2}{x^2 - 1}$

2. $f(x) = \dfrac{x^3}{x + 2}$

Example 2 **3. CCSS REASONING** Eduardo is a kicker for his high school football team. So far this season, he has made 7 out of 11 field goals. He would like to improve his field goal percentage. If he can make x consecutive field goals, his field goal percentage can be determined using the function $P(x) = \dfrac{7 + x}{11 + x}$.

  a. Graph the function. **See Chapter 8 Answer Appendix.**
  b. What part of the graph is meaningful in the context of this problem? **the part in the first quadrant**
  c. Describe the meaning of the intercept of the vertical axis. **3c. It represents his original field goal percentage of 63.6%.**
  d. What is the equation of the horizontal asymptote? Explain its meaning with respect to Eduardo's field goal percentage. **$y = 1$; this represents 100% which he cannot achieve because he has already missed 4 field goals.**

Examples 3–4 **Graph each function.** **4–7. See Chapter 8 Answer Appendix.**

4. $f(x) = \dfrac{6x^2 - 3x + 2}{x}$

5. $f(x) = \dfrac{x^2 + 8x + 20}{x + 2}$

6. $f(x) = \dfrac{x^2 - 4x - 5}{x + 1}$

7. $f(x) = \dfrac{x^2 + x - 12}{x + 4}$

Additional Answers (Guided Practice)

4A.

$f(x) = \dfrac{x^2 + 4x - 5}{x + 5}$

4B.

$f(x) = \dfrac{x^3 + 2x^2 - 9x - 18}{x^2 - 9}$

Additional Example

4 Graph $f(x) = \dfrac{x^2 - 4}{x - 2}$.

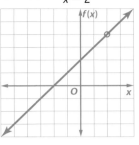

---

3 Practice

Formative Assessment

Use Exercises 1–7 to check for understanding.

Use the chart at the bottom of this page to customize assignments for your students.

(CCSS) Teaching the Mathematical Practices

Reasoning Mathematically proficient students make sense of quantities and their relationships in problem situations. Quantitative reasoning entails habits of creating a coherent representation of the problem at hand; considering the units involved; attending to the meaning of quantities, not just how to compute them; and knowing and flexibly using different properties of operations and objects.

Exercise Alert

Grid Paper For Exercises 1–42 and 51–53, students will need grid paper.

Follow-up

Students have explored rational expressions and functions.

Ask:

- How are the properties of a rational function reflected in its graph? Sample answer: Vertical asymptotes occur at values that make the denominator 0; horizontal asymptotes occur when the degree of the numerator is less than or equal to the degree of the denominator; oblique asymptotes occur when the degrees of the numerator and denominator differ by 1; holes occur when the numerator and denominator share a binomial factor.

8.

9.

10.

11.

12a. $c(t)$

43. Similarities: Both have vertical asymptotes at $x = 0$. Both approach 0 as x approaches $-\infty$ and approach 0 as x approaches ∞. Differences: $f(x)$ has holes at $x = 1$ and $x = -1$, while $g(x)$ has vertical asymptotes at $x = \sqrt{2}$ and $x = -\sqrt{2}$. $f(x)$ has no zeros, but $g(x)$ has at zeros at $x = 1$ and $x = -1$.

| **Practice and Problem Solving** | Extra Practice is on page R8. |

Example 1 Graph each function. **8–11.** See margin.

8. $f(x) = \dfrac{x^4}{6x + 12}$

9. $f(x) = \dfrac{x^3}{8x - 4}$

10. $f(x) = \dfrac{x^4 - 16}{x^2 - 1}$

11. $f(x) = \dfrac{x^3 + 64}{16x - 24}$

12a. $c(t) = \dfrac{9.5t - 75}{t - 15}$; See margin for graph.

Example 2 **12. SCHOOL SPIRIT** As president of Student Council, Brandy is getting T-shirts made for a pep rally. Each T-shirt costs \$9.50, and there is a set-up fee of \$75. The student council plans to sell the shirts, but each of the 15 council members will get one for free.

   a. Write a function for the average cost of a T-shirt to be sold. Graph the function.

   b. What is the average cost if 200 shirts are ordered? if 500 shirts are ordered? **\$10.68; \$9.95**

   c. How many T-shirts must be ordered to bring the average cost under \$9.75? **more than 885**

Examples 2–3 Graph each function. **13–26.** See Chapter 8 Answer Appendix.

13. $f(x) = \dfrac{x}{x + 2}$

14. $f(x) = \dfrac{5}{(x - 1)(x + 4)}$

(15) $f(x) = \dfrac{4}{(x - 2)^2}$

16. $f(x) = \dfrac{x - 3}{x + 1}$

17. $f(x) = \dfrac{1}{(x + 4)^2}$

18. $f(x) = \dfrac{2x}{(x + 2)(x - 5)}$

19. $f(x) = \dfrac{(x - 4)^2}{x + 2}$

20. $f(x) = \dfrac{(x + 3)^2}{x - 5}$

21. $f(x) = \dfrac{x^3 + 1}{x^2 - 4}$

22. $f(x) = \dfrac{4x^3}{2x^2 + x - 1}$

23. $f(x) = \dfrac{3x^2 + 8}{2x - 1}$

24. $f(x) = \dfrac{2x^2 + 5}{3x + 4}$

25. $f(x) = \dfrac{x^4 - 2x^2 + 1}{x^3 + 2}$

26. $f(x) = \dfrac{x^4 - x^2 - 12}{x^3 - 6}$

27. CCSS PERSEVERANCE The current in amperes in an electrical circuit with three resistors in a series is given by the equation $I = \dfrac{V}{R_1 + R_2 + R_3}$, where V is the voltage in volts in the circuit and $R_1, R_2,$ and R_3 are the resistances in ohms of the three resistors.

   a. Let R_1 be the independent variable, and let I be the dependent variable. Graph the function if $V = 120$ volts, $R_2 = 25$ ohms, and $R_3 = 75$ ohms. **See Chapter 8 Answer Appendix.**

   b. Give the equation of the vertical asymptote and the R_1- and I-intercepts of the graph. $R_1 = -100$; no R_1-intercept; 1.2

   c. Find the value of I when the value of R_1 is 140 ohms. **0.5 amperes**

   d. What domain and range values are meaningful in the context of the problem? $R_1 \geq 0$ and $0 < I \leq 1.2$

Example 4 Graph each function. **28–35.** See Chapter 8 Answer Appendix.

28. $f(x) = \dfrac{x^2 - 2x - 8}{x - 4}$

29. $f(x) = \dfrac{x^2 + 4x - 12}{x - 2}$

30. $f(x) = \dfrac{x^2 - 25}{x + 5}$

31. $f(x) = \dfrac{x^2 - 64}{x - 8}$

32. $f(x) = \dfrac{(x - 4)(x^2 - 4)}{x^2 - 6x + 8}$

33. $f(x) = \dfrac{(x + 5)(x^2 + 2x - 3)}{x^2 + 8x + 15}$

34. $f(x) = \dfrac{3x^4 + 6x^3 + 3x^2}{x^2 + 2x + 1}$

35. $f(x) = \dfrac{2x^4 + 10x^3 + 12x^2}{x^2 + 5x + 6}$

 558 | Lesson 8-4 | Graphing Rational Functions

Differentiated Homework Options

Level	Assignment	Two-Day Option	
AL Basic	8–35, 42, 44–61, 62	9–35 odd, 47–50	8–34 even, 42, 51–62
OL Core	9–35 odd, 36–39, 41–42, 44–62	8–35, 47–50	36–42, 44–46, 51–62
BL Advanced	36–61, (optional: 62)		

36. BUSINESS Liam purchased a snow plow for $4500 and plows the parking lots of local businesses. Each time he plows a parking lot, he incurs a cost of $50 for gas and maintenance.

a. Write and graph the rational function representing his average cost per customer as a function of the number of parking lots. **See Chapter 8 Answer Appendix.**

b. What are the asymptotes of the graph? $x = 0$ and $f(x) = 50$

c. Why is the first quadrant in the graph the only relevant quadrant?

d. How many total parking lots does Liam need to plow for his average cost per parking lot to be less than $80? **150**

37c. Sample answer: The number of months and the average cost cannot have negative values.

36c. Sample answer: The number of parking lots and the average cost cannot be negative.

37 FINANCIAL LITERACY Kristina bought a new cell phone with Internet access. The phone cost $150, and her monthly usage charge is $30 plus $10 for the Internet access.

a. Write and graph the rational function representing her average monthly cost as a function of the number of months Kristina uses the phone. **See Chapter 8 Answer Appendix.**

b. What are the asymptotes of the graph? $x = 0$ and $f(x) = 40$

c. Why is the first quadrant in the graph the only relevant quadrant?

d. After how many months will the average monthly charge be $45? **30**

38. CCSS SENSE-MAKING Alana plays softball for Centerville High School. So far this season she has gotten a hit 4 out of 12 times at bat. She is determined to improve her batting average. If she can get x consecutive hits, her batting average can be determined using $B(x) = \dfrac{4 + x}{12 + x}$. **c. It represents her original batting average of .333.**

a. Graph the function. **See Chapter 8 Answer Appendix.**

b. What part of the graph is meaningful in the context of the problem? **the part in the first quadrant**

c. Describe the meaning of the intercept of the vertical axis.

d. What is the equation of the horizontal asymptote? Explain its meaning with respect to Alana's batting average.

38d. $y = 1$; This represents 100%, which she can never achieve because she has already missed getting a hit 8 times.

Graph each function. 39–41. See Chapter 8 Answer Appendix.

39. $f(x) = \dfrac{x + 1}{x^2 + 6x + 5}$ **40.** $f(x) = \dfrac{x^2 - 10x - 24}{x + 2}$ **41.** $f(x) = \dfrac{6x^2 + 4x + 2}{x + 2}$

H.O.T. Problems   Use Higher-Order Thinking Skills

42. OPEN ENDED Sketch the graph of a rational function with a horizontal asymptote $y = 1$ and a vertical asymptote $x = -2$. **See Chapter 8 Answer Appendix.**

43. CHALLENGE Compare and contrast $g(x) = \dfrac{x^2 - 1}{x(x^2 - 2)}$ and $f(x)$ shown at the right. **See margin.**

44. REASONING What is the difference between the graphs of $f(x) = x - 2$ and $g(x) = \dfrac{(x + 3)(x - 2)}{x + 3}$? **The graph of $g(x)$ has a hole in it at -3.**

45. PROOF A rational function has an equation of the form $f(x) = \dfrac{a(x)}{b(x)}$, where $a(x)$ and $b(x)$ are polynomial functions and $b(x) \neq 0$. Show that $f(x) = \dfrac{x}{a - b} + c$ is a rational function.

45. $f(x) = \dfrac{x}{a - b} + \dfrac{c(a - b)}{(a - b)} \rightarrow \dfrac{x + ca - cb}{a - b}$

46. WRITING IN MATH How can factoring be used to determine the vertical asymptotes or point discontinuity of a rational function? **See margin.**

connectED.mcgraw-hill.com   **559**

Additional Answer

46. Sample answer: By factoring the denominator of a rational function and determining the values that cause each factor to equal zero you can determine the asymptotes of a rational function. After factoring the numerator and denominator of a rational function, if there is a common factor $x - c$, then there is point discontinuity at $x = c$.

Study Guide and Intervention
AL OL ELL

Practice
AL OL BL ELL

Word Problem Practice
AL OL BL ELL

Enrichment
OL BL

Name the Math Have students write their own examples of rational functions and graph them, showing any asymptotes or discontinuities.

Additional Answers

51.

$D = \{x \mid x \neq -2\}$,
$R = \{f(x) \mid f(x) \neq 0\}$

52.

$D = \{x \mid x \neq 1\}$,
$R = \{f(x) \mid f(x) \neq -3\}$

53.

$D = \{x \mid x \neq -6\}$,
$R = \{f(x) \mid f(x) \neq 1\}$

Standardized Test Practice

47. PROBABILITY Of the 6 courses offered by the music department at her school, Kaila must choose exactly 2 of them. How many different combinations of 2 courses are possible for Kaila if there are no restrictions on which 2 courses she can choose? **C**

A 48   **C** 15
B 18   **D** 12

48. The projected sales of a game cartridge is given by the function $S(p) = \dfrac{3000}{2p + a}$, where $S(p)$ is the number of cartridges sold, in thousands, p is the price per cartridge, in dollars, and a is a constant. If 100,000 cartridges are sold at \$10 per cartridge, how many cartridges will be sold at \$20 per cartridge? **H**

F 20,000   **H** 60,000
G 50,000   **J** 150,000

49. GRIDDED RESPONSE Five distinct points lie in a plane such that 3 of the points are on line ℓ and 3 of the points are on a different line m. What is the total number of lines that can be drawn so that each line passes through exactly 2 of these 5 points? **4**

50. GEOMETRY In the figure below, what is the value of $w + x + y + z$? **B**

A 140   **C** 320
B 280   **D** 360

Spiral Review

Graph each function. State the domain and range. (Lesson 8-3) **51–53. See margin.**

51. $f(x) = \dfrac{-5}{x + 2}$ **52.** $f(x) = \dfrac{4}{x - 1} - 3$ **53.** $f(x) = \dfrac{1}{x + 6} + 1$

Simplify each expression. (Lesson 8-2)

54. $\dfrac{m}{m^2 - 4} + \dfrac{2}{3m + 6}$ $\dfrac{5m - 4}{3(m + 2)(m - 2)}$

55. $\dfrac{y}{y + 3} - \dfrac{6y}{y^2 - 9}$ $\dfrac{y(y - 9)}{(y + 3)(y - 3)}$

56. $\dfrac{5}{x^2 - 3x - 28} + \dfrac{7}{2x - 14}$ $\dfrac{7x + 38}{2(x - 7)(x + 4)}$

57. $\dfrac{d - 4}{d^2 + 2d - 8} - \dfrac{d + 2}{d^2 - 16}$ $\dfrac{-8d + 20}{(d - 4)(d + 4)(d - 2)}$

Simplify each expression. (Lesson 6-6)

58. $y^{\frac{5}{3}} \cdot y^{\frac{7}{3}}$ y^4 **59.** $x^{\frac{3}{4}} \cdot x^{\frac{9}{4}}$ x^3 **60.** $\left(b^{\frac{1}{3}}\right)^{\frac{3}{5}}$ $b^{\frac{1}{5}}$ **61.** $\left(a^{-\frac{2}{3}}\right)^{-\frac{1}{6}}$ $a^{\frac{1}{9}}$

Skills Review

62. TRAVEL Mr. and Mrs. Wells are taking their daughter to college. The table shows their distances from home after various amounts of time.

a. Find the average rate of change in their distances from home between 1 and 3 hours after leaving home. **55 mph**

b. Find the average rate of change in their distances from home between 0 and 5 hours after leaving home. **45 mph**

Time (h)	Distance (mi)
0	0
1	55
2	110
3	165
4	165
5	225

DifferentiatedInstruction   **OL**   **BL**

Extension Draw a straight line on the board, but leave a hole in the line for some integral value of x. For example, draw the line representing $y = x - 2$, but leave a hole at the point $(1, -1)$. Ask students to write possible rational functions that could be described by the graph. Many answers are possible, as long as $x = 1$ is an excluded value and the simplified form of the function is $f(x) = x - 2$. Sample answer: $f(x) = \dfrac{(x - 1)(x - 2)}{x - 1}$ or $f(x) = \dfrac{x^2 - 3x + 2}{x - 1}$.

EXTEND 8-4

Graphing Technology Lab
Graphing Rational Functions

A TI-83/84 Plus graphing calculator can be used to explore graphs of rational functions. These graphs have some features that never appear in the graphs of polynomial functions.

Activity 1   Graph with Asymptotes

Graph $y = \dfrac{8x - 5}{2x}$ in the standard viewing window. Find the equations of any asymptotes. State the domain and range of the function.

Step 1   Enter the equation in the **Y=** list, and then graph.

KEYSTROKES: Y= ((8 X,T,θ,n — 5) ÷
(2 X,T,θ,n)) ZOOM 6

[−10, 10] scl: 1 by [−10, 10] scl: 1

Step 2   Examine the graph.

By looking at the equation, we can determine that if $x = 0$, the function is undefined. The equation of the vertical asymptote is $x = 0$. Notice what happens to the y-values as x grows larger and as x gets smaller. The y-values approach 4. So, the equation for the horizontal asymptote is $y = 4$. The domain is $\{x \mid x \neq 0\}$, and the range is all real numbers.

Activity 2   Graph with Point Discontinuity

Graph $y = \dfrac{x^2 - 16}{x + 4}$ in the window $[-5, 4.4]$ by $[-10, 2]$ with scale factors of 1.

Step 1   Because the function is not continuous, put the calculator in dot mode.

KEYSTROKES: MODE ▼ ▼ ▼ ▼ ▶ ENTER

[−5, 4.4] scl: 1 by [−20, 2] scl: 1

Step 2   Examine the graph.

This graph looks like a line with a break in continuity at $x = -4$. This happens because the denominator is 0 when $x = -4$. Therefore, the function is undefined when $x = -4$.

If you **TRACE** along the graph, when you come to $x = -4$, you will see that there is no corresponding y-value.

Exercises

Use a graphing calculator to graph each function. Write the x-coordinates of any points of discontinuity and/or the equations of any asymptotes. State the domain and range.

1. $f(x) = \dfrac{1}{x}$ $x = 0$, $y = 0$; $D = \{x \mid x \neq 0\}$, $R = \{y \mid y \neq 0\}$

2. $f(x) = \dfrac{x}{x + 2}$ $x = -2$, $y = 1$; $D = \{x \mid x \neq -2\}$, $R = \{y \mid y \neq 1\}$

3. $f(x) = \dfrac{2}{x - 4}$ $x = 4$, $y = 0$; $D = \{x \mid x \neq 4\}$, $R = \{y \mid y \neq 0\}$

4. $f(x) = \dfrac{2x}{3x - 6}$ $x = 2$, $y = \dfrac{2}{3}$; $D = \{x \mid x \neq 2\}$, $R = \left\{y \mid y \neq \dfrac{2}{3}\right\}$

5. $f(x) = \dfrac{4x + 2}{x - 1}$ $x = 1$, $y = 4$; $D = \{x \mid x \neq 1\}$, $R = \{y \mid y \neq 4\}$

6. $f(x) = \dfrac{x^2 - 9}{x + 3}$ point discontinuity at $x = -3$; $D = \{x \mid x \neq -3\}$, $R = \{\text{all real numbers}\}$

connectED.mcgraw-hill.com   **561**

From Concrete to Abstract

Exercise 5 asks students to apply their findings to sketching the graph of a general rational function.

Extending the Concept

In preparation for study of inverse variation, ask students to graph $y = \dfrac{1}{x}$, $y = \dfrac{2}{x}$, $y = \dfrac{3}{x}$, and $y = \dfrac{4}{x}$ on the same screen and describe their similarities. The coordinate axes are asymptotes for the graphs of all four functions.

---

1 Focus

Objective   Use a graphing calculator to explore the graphs of rational functions.

Materials for Each Group

- TI-83/84 Plus or other graphing calculator

Teaching Tip

Students may find it instructive to experiment with the graph style feature found to the left of the equation in the function editor (**Y=** list). They can begin by using the usual line style, indicated by a backslash icon. The path style slows down the graphing so students can follow it more easily. Its icon is calculator font, minus sign touching a zero.

2 Teach

Working in Cooperative Groups

Have students work in pairs to complete Activities 1 and 2.

- Students may need to trace a graph beyond the current window in order to identify asymptotes precisely.

- A calculator graphing in Connected mode may seem to graph a vertical asymptote when it is merely connecting two sequential pixels on a graph. Switching to Dot mode will eliminate this possibility.

Practice   Have students complete Exercises 1–6.

3 Practice

Formative Assessment

Use Exercises 1–3 to assess whether students can use a graphing calculator to explore the graphs of rational functions.

1 Focus

VerticalAlignment

Before Lesson 8-5 Write and graph linear equations.

Lesson 8-5 Recognize and solve direct and joint variation problems. Recognize and solve inverse and combined variation problems.

After Lesson 8-5 Graph hyperbolic functions.

2 Teach

Scaffolding Questions

Have students read the **Why?** section of the lesson.

Ask:

- If the height of the ramp increases, what happens to the length of the top of the ramp? It increases.

- If h decreases, what happens to ℓ? It decreases.

- Which value remains constant? the ratio $\frac{\ell}{h}$, which equals 1.5

LESSON 8-5 Variation Functions

::Then	::Now	::Why?
● You wrote and graphed linear equations.	**1** Recognize and solve direct and joint variation problems. **2** Recognize and solve inverse and combined variation problems.	● While building skateboard ramps, Yu determined that the best ramps were the ones in which the length of the top of the ramp was 1.5 times as long as the height of the ramp.

As shown in the table, the length of the top of the ramp depends on the height of a ramp. The length increases as the height increases, but the ratio remains the same, or is *constant*.

The equation $\frac{\ell}{h} = 1.5$ can be written as $\ell = 1.5h$.

The length *varies directly* with the height of the ramp.

Length (ℓ)	Height (h)	Ratio $\frac{\ell}{h}$
3	2	1.5
6	4	1.5
9	6	1.5
12	8	1.5

NewVocabulary
direct variation
constant of variation
joint variation
inverse variation
combined variation

CCSS Common Core State Standards

Content Standards
A.CED.2 Create equations in two or more variables to represent relationships between quantities; graph equations on coordinate axes with labels and scales.

Mathematical Practices
1 Make sense of problems and persevere in solving them.
4 Model with mathematics.

1 Direct Variation and Joint Variation The relationship given by $\ell = 1.5h$ is an example of direct variation. A **direct variation** can be expressed in the form $y = kx$. In this equation, k is called the **constant of variation**.

Notice that the graph of $\ell = 1.5h$ is a straight line through the origin. A direct variation is a special case of an equation written in slope-intercept form, $y = mx + b$. When $m = k$ and $b = 0$, $y = mx + b$ becomes $y = kx$. So the slope of a direct variation equation is its constant of variation.

To express a direct variation, we say that y varies directly as x. In other words, as x increases, y increases or decreases at a constant rate.

Height

KeyConcept Direct Variation

Words	y varies directly as x if there is some nonzero constant k such that $y = kx$. k is called the *constant of variation*.
Example	If $y = 3x$ and $x = 7$, then $y = 3(7)$ or 21.

If you know that y varies directly as x and one set of values, you can use a proportion to find the other set of corresponding values.

$$y_1 = kx_1 \qquad \text{and} \qquad y_2 = kx_2$$
$$\frac{y_1}{x_1} = k \qquad\qquad\qquad \frac{y_2}{x_2} = k \qquad \text{Therefore, } \frac{y_1}{x_1} = \frac{y_2}{x_2}.$$

Using the properties of equality, you can find many other proportions that relate these same x- and y-values.

562 | Lesson 8-5

Lesson 8-5 Resources

Resource	Approaching Level AL	On Level OL	Beyond Level BL	English Learners ELL
Teacher Edition	▪ Differentiated Instruction, p. 567	▪ Differentiated Instruction, pp. 567, 569	▪ Differentiated Instruction, p. 569	
Chapter Resource Masters	▪ Study Guide and Intervention, pp. 32–33 ▪ Skills Practice, p. 34 ▪ Practice, p. 35 ▪ Word Problem Practice, p. 36	▪ Study Guide and Intervention, pp. 32–33 ▪ Skills Practice, p. 34 ▪ Practice, p. 35 ▪ Word Problem Practice, p. 36 ▪ Enrichment, p. 37 ▪ Spreadsheet Activity, p. 38	▪ Practice, p. 35 ▪ Word Problem Practice, p. 36 ▪ Enrichment, p. 37	▪ Study Guide and Intervention, pp. 32–33 ▪ Skills Practice, p. 34 ▪ Practice, p. 35 ▪ Word Problem Practice, p. 36
Other	▪ 5-Minute Check 8-5 ▪ Study Notebook	▪ 5-Minute Check 8-5 ▪ Study Notebook	▪ 5-Minute Check 8-5 ▪ Study Notebook	▪ 5-Minute Check 8-5 ▪ Study Notebook

Example 1 Direct Variation

If y varies directly as x and $y = 15$ when $x = -5$, find y when $x = 7$.

Use a proportion that relates the values.

$$\frac{y_1}{x_1} = \frac{y_2}{x_2} \qquad \text{Direct variation}$$

$$\frac{15}{-5} = \frac{y_2}{7} \qquad y_1 = 15,\ x_1 = -5,\ \text{and}\ x_2 = 7$$

$$15(7) = -5(y_2) \qquad \text{Cross multiply.}$$

$$105 = -5y_2 \qquad \text{Simplify.}$$

$$-21 = y_2 \qquad \text{Divide each side by } -5.$$

> **Guided**Practice
>
> **1.** If r varies directly as t and $r = -20$ when $t = 4$, find r when $t = -6$. **30**

Another type of variation is joint variation. **Joint variation** occurs when one quantity varies directly as the product of two or more other quantities.

StudyTip

Joint Variation Some mathematicians consider joint variation a special type of combined variation.

KeyConcept Joint Variation

Words	y varies jointly as x and z if there is some nonzero constant k such that $y = kxz$.
Example	If $y = 5xz$, $x = 6$, and $z = -2$, then $y = 5(6)(-2)$ or -60.

If you know that y varies jointly as x and z and one set of values, you can use a proportion to find the other set of corresponding values.

$$y_1 = kx_1z_1 \qquad \text{and} \qquad y_2 = kx_2z_2$$

$$\frac{y_1}{x_1z_1} = k \qquad\qquad \frac{y_2}{x_2z_2} = k \qquad \text{Therefore, } \frac{y_1}{x_1z_1} = \frac{y_2}{x_2z_2}.$$

Example 2 Joint Variation

Suppose y varies jointly as x and z. Find y when $x = 9$ and $z = 2$, if $y = 20$ when $z = 3$ and $x = 5$.

Use a proportion that relates the values.

$$\frac{y_1}{x_1z_1} = \frac{y_2}{x_2z_2} \qquad \text{Joint variation}$$

$$\frac{20}{5(3)} = \frac{y_2}{9(2)} \qquad y_1 = 20,\ x_1 = 5,\ z_1 = 3,\ x_2 = 9,\ \text{and}\ z_2 = 2$$

$$20(9)(2) = 5(3)(y_2) \qquad \text{Cross multiply.}$$

$$360 = 15y_2 \qquad \text{Simplify.}$$

$$24 = y_2 \qquad \text{Divide each side by 15.}$$

> **Guided**Practice
>
> **2.** Suppose r varies jointly as v and t. Find r when $v = 2$ and $t = 8$, if $r = 70$ when $v = 10$ and $t = 4$. **28**

 connectED.mcgraw-hill.com **563**

1 Direct Variation and Joint Variation

Example 1 shows how to use a proportion to analyze a simple direct variation. **Example 2** shows how to solve a joint variation problem.

Formative Assessment

Use the Guided Practice exercises after each example to determine students' understanding of concepts.

Additional Examples

1 If y varies directly as x and $y = -15$ when $x = 5$, find y when $x = 3$. -9

2 Suppose y varies jointly as x and z. Find y when $x = 10$ and $z = 5$, if $y = 12$ when $z = 8$ and $x = 3$. 25

> **Additional Examples** also in Interactive Classroom PowerPoint® Presentations

IWB Interactive White Board READY

WatchOut!

Preventing Errors Discuss with students how to write variation equations that include a constant of variation.

Focus on Mathematical Content

Direct and Inverse Variation The type of variation present can sometimes be identified from a table of values for x and y. If the quotient $\frac{y}{x}$ has a constant value, y varies directly as x. If the product xy has a constant value, y varies inversely as x.

Example 3 shows how to use a proportion to solve an inverse variation problem. **Example 4** provides an example of a contextual application of inverse variation. **Example 5** shows how to solve a combined variation problem.

Additional Example

3 If r varies inversely as t and $r = -6$ when $t = 2$, find r when $t = -7$. $\dfrac{12}{7}$

Tips for New Teachers
Alternate Method In Example 3, students may wish to solve the problem by using the equation $a_1 b_1 = a_2 b_2$.

Tips for New Teachers
Sense-Making Help students understand the difference between the two types of variation by using the example of speed, distance, and driving time. When driving at a given rate, the distance increases as driving time increases (direct). However, for a given distance, the time needed decreases as speed increases (inverse).

---

2 **Inverse Variation and Combined Variation** Another type of variation is inverse variation. If two quantities x and y show **inverse variation**, their product is equal to a constant k.

Inverse variation is often described as one quantity increasing while the other quantity is decreasing. For example, speed and time for a fixed distance vary inversely with each other; the faster you go, the less time it takes you to get there.

🔑 KeyConcept Inverse Variation

Words	y varies inversely as x if there is some nonzero constant k such that $xy = k$ or $y = \dfrac{k}{x}$, where $x \neq 0$ and $y \neq 0$.
Example	If $xy = 2$, and $x = 6$, then $y = \dfrac{2}{6}$ or $\dfrac{1}{3}$.

StudyTip
Direct and Inverse Variation You can identify the type of variation by looking at a table of values for x and y. If the quotient $\dfrac{y}{x}$ has a constant value, y varies directly as x. If the product xy has a constant value, y varies inversely as x.

Suppose y varies inversely as x such that $xy = 6$ or $y = \dfrac{6}{x}$. The graph of this equation is shown at the right. Since k is a positive value, as the values of x increase, the values of y decrease.

Notice that the graph of an inverse variation is a reciprocal function.

A proportion can be used with inverse variation to solve problems in which some quantities are known. The following proportion is only one of several that can be formed.

$$x_1 y_1 = k \text{ and } x_2 y_2 = k$$

$$x_1 y_1 = x_2 y_2 \qquad \text{Substitution Property of Equality}$$

$$\frac{x_1}{y_2} = \frac{x_2}{y_1} \qquad \text{Divide each side by } y_1 y_2.$$

Example 3 Inverse Variation

If a varies inversely as b and $a = 28$ when $b = -2$, find a when $b = -10$.

Use a proportion that relates the values.

$$\frac{a_1}{b_2} = \frac{a_2}{b_1} \qquad \text{Inverse Variation}$$

$$\frac{28}{-10} = \frac{a_2}{-2} \qquad a_1 = 28, b_1 = -2, \text{ and } b_2 = -10$$

$$28(-2) = -10(a_2) \qquad \text{Cross multiply.}$$

$$-56 = -10(a_2) \qquad \text{Simplify.}$$

$$5\frac{3}{5} = a_2 \qquad \text{Divide each side by } -10.$$

▶ **Guided**Practice

3. If x varies inversely as y and $x = 24$ when $y = 4$, find x when $y = 12$. **8**

Inverse variation is often used in real-world situations.

RubberBall Productions

Real-World Link

When you pluck a string, it vibrates back and forth. This causes mechanical energy to travel through the air in waves. The number of times per second these waves hit our ear is called the *frequency*. The more waves per second, the higher the pitch.

StudyTip

Combined Variation
Quantities that vary directly appear in the numerator. Quantities that vary inversely appear in the denominator.

Real-World Example 4 Write and Solve an Inverse Variation

MUSIC The length of a violin string varies inversely as the frequency of its vibrations. A violin string 10 inches long vibrates at a frequency of 512 cycles per second. Find the frequency of an 8-inch violin string.

Let $v_1 = 10$, $f_1 = 512$, and $v_2 = 8$. Solve for f_2.

$$v_1 f_1 = v_2 f_2 \qquad \text{Original equation}$$
$$10 \cdot 512 = 8 \cdot f_2 \qquad v_1 = 10,\ f_1 = 512,\ \text{and } v_2 = 8$$
$$\frac{5120}{8} = f_2 \qquad \text{Divide each side by 8.}$$
$$640 = f_2 \qquad \text{Simplify.}$$

The 8-inch violin string vibrates at a frequency of 640 cycles per second.

GuidedPractice

4. The apparent length of an object is inversely proportional to one's distance from the object. Earth is about 93 million miles from the Sun. Jupiter is about 483.6 million miles from the Sun. Find how many times as large the diameter of the Sun would appear on Earth as on Jupiter. **5.2 times as large**

Another type of variation is combined variation. **Combined variation** occurs when one quantity varies directly and/or inversely as two or more other quantities.

If you know that y varies directly as x, y varies inversely as z, and one set of values, you can use a proportion to find the other set of corresponding values.

$$y_1 = \frac{kx_1}{z_1} \qquad \text{and} \qquad y_2 = \frac{kx_2}{z_2}$$
$$\frac{y_1 z_1}{x_1} = k \qquad\qquad\qquad \frac{y_2 z_2}{x_2} = k \qquad \text{Therefore, } \frac{y_1 z_1}{x_1} = \frac{y_2 z_2}{x_2}.$$

Example 5 Combined Variation

Suppose f varies directly as g, and f varies inversely as h. Find g when $f = 18$ and $h = -3$, if $g = 24$ when $h = 2$ and $f = 6$.

First set up a correct proportion for the information given.

$$f_1 = \frac{kg_1}{h_1} \text{ and } f_2 = \frac{kg_2}{h_2} \qquad \begin{array}{l}g \text{ varies directly as } f, \text{ so } g \text{ goes in the numerator. } h \text{ varies inversely as } f, \\ \text{so } h \text{ goes in the denominator.}\end{array}$$

$$k = \frac{f_1 h_1}{g_1} \text{ and } k = \frac{f_2 h_2}{g_2} \qquad \text{Solve for } k.$$

$$\frac{f_1 h_1}{g_1} = \frac{f_2 h_2}{g_2} \qquad \text{Set the two proportions equal to each other.}$$

$$\frac{6(2)}{24} = \frac{18(-3)}{g_2} \qquad f_1 = 6,\ g_1 = 24,\ h_1 = 2,\ f_2 = 18,\ \text{and } h_2 = -3$$

$$24(18)(-3) = 6(2)(g_2) \qquad \text{Cross multiply.}$$

$$-1296 = 12g_2 \qquad \text{Simplify.}$$

$$-108 = g_2 \qquad \text{Divide each side by 12.}$$

When $f = 18$ and $h = -3$, the value of g is -108.

GuidedPractice

5. Suppose p varies directly as r, and p varies inversely as t. Find t when $r = 10$ and $p = -5$, if $t = 20$ when $p = 4$ and $r = 2$. **−80**

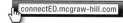

Additional Examples

4 **SPACE** Use the information from *Guided Practice* and the fact that Venus is about 67 million miles away from the Sun. How much larger would the diameter of the Sun appear on Venus than on Earth? about 1.39 times as large

5 Suppose f varies directly as g, and f varies inversely as h. Find g when $f = 6$ and $h = -5$, if $g = 18$ when $h = 3$ and $f = 5$. −36

Teach with Tech

Audio Recording Create a recording of several variation examples read out loud. Play the recordings for the class (or load them onto students' portable media players), and have students write and solve the variation problems.

3 Practice

Formative Assessment

Use Exercises 1–6 to check for understanding.

Use the chart at the bottom of this page to customize assignments for your students.

 Teaching the Mathematical Practices

Modeling Mathematically proficient students can apply the mathematics they know to solve problems arising in everyday life, analyze relationships mathematically to draw conclusions, and interpret their mathematical results in the context of a situation.

 Follow-up

Students have explored variation functions.

Ask:

- How can analyzing a rational function algebraically *and* graphically help you to see the "whole picture?" Sample answer: An algebraic analysis can help you to determine points of discontinuity that may not be clear or noticeable when viewing the graph of the function. A graphical analysis can help you to see the asymptotes and end behavior of the function.

Check Your Understanding

 = Step-by-Step Solutions begin on page R14.

Examples 1–3

1. If y varies directly as x and $y = 12$ when $x = 8$, find y when $x = 14$. **21**

2. Suppose y varies jointly as x and z. Find y when $x = 9$ and $z = -3$, if $y = -50$ when z is 5 and x is -10. **−27**

3. If y varies inversely as x and $y = -18$ when $x = 16$, find x when $y = 9$. **−32**

Example 4

4. TRAVEL A map of Illinois is scaled so that 2 inches represents 15 miles. How far apart are Chicago and Rockford if they are 12 inches apart on the map? **90 mi**

Example 5

5. Suppose a varies directly as b, and a varies inversely as c. Find b when $a = 8$ and $c = -3$, if $b = 16$ when $c = 2$ and $a = 4$. **−48**

6. Suppose d varies directly as f, and d varies inversely as g. Find g when $d = 6$ and $f = -7$, if $g = 12$ when $d = 9$ and $f = 3$. **−42**

Practice and Problem Solving

Extra Practice is on page R8.

Example 1

If x varies directly as y, find x when $y = 8$.

7. $x = 6$ when $y = 32$ **1.5**

8. $x = 11$ when $y = -3$ $-\dfrac{88}{3}$

9. $x = 14$ when $y = -2$ **−56**

10. $x = -4$ when $y = 10$ **−3.2**

11. MOON Astronaut Neil Armstrong, the first man on the Moon, weighed 360 pounds on Earth with all his equipment on, but weighed only 60 pounds on the Moon. Write an equation that relates weight on the Moon m with weight on Earth w. $m = \dfrac{1}{6}w$

Example 2

If a varies jointly as b and c, find a when $b = 4$ and $c = -3$.

12. $a = -96$ when $b = 3$ and $c = -8$ **−48**

13 $a = -60$ when $b = -5$ and $c = 4$ **−36**

14. $a = -108$ when $b = 2$ and $c = 9$ **72**

15. $a = 24$ when $b = 8$ and $c = 12$ **−3**

16. CCSS **MODELING** According to the A.C. Nielsen Company, the average American watches 4 hours of television a day.

a. Write an equation to represent the average number of hours spent watching television by m household members during a period of d days. $t = 4md$

b. Assume that members of your household watch the same amount of television each day as the average American. How many hours of television would the members of your household watch in a week? **Sample answer for four household members: 112 hours**

Example 3

If f varies inversely as g, find f when $g = -6$.

17. $f = 15$ when $g = 9$ **−22.5**

18. $f = 4$ when $g = 28$ $-\dfrac{56}{3}$

19. $f = -12$ when $g = 19$ **38**

20. $f = 0.6$ when $g = -21$ **2.1**

21. COMMUNITY SERVICE Every year students at West High School collect canned goods for a local food pantry. They plan to distribute flyers to homes in the community asking for donations. Last year, 12 students were able to distribute 1000 flyers in four hours.

a. Write an equation that relates the number of students s to the amount of time t it takes to distribute 1000 flyers. $s = \dfrac{48}{t}$

b. How long would it take 15 students to hand out the same number of flyers this year? **3.2 hours**

 566 | Lesson 8-5 | Variation Functions

Differentiated Homework Options

Level	Assignment	Two-Day Option	
AL Basic	7–24, 48, 50–72	7–23 odd, 53–56	8–24 even, 48, 50–52, 57–72
OL Core	7–43 odd, 44–48, 50–72	7–24, 53–56	25–48, 50–52, 57–72
BL Advanced	25–66, (optional: 67–72)		

Example 4

22. BIRDS When a group of snow geese migrate, the distance that they fly varies directly with the amount of time they are in the air.

    a. A group of snow geese migrated 375 miles in 7.5 hours. Write a direct variation equation that represents this situation. $d = 50t$

    b. Every year, geese migrate 3000 miles from their winter home in the southwest United States to their summer home in the Canadian Arctic. Estimate the number of hours of flying time that it takes for the geese to migrate. **60 hours**

Example 5

23. Suppose a varies directly as b, and a varies inversely as c. Find b when $a = 5$ and $c = -4$, if $b = 12$ when $c = 3$ and $a = 8$. **−10**

24. Suppose x varies directly as y, and x varies inversely as z. Find z when $x = 10$ and $y = -7$, if $z = 20$ when $x = 6$ and $y = 14$. **−6**

B Determine whether each relation shows *direct* or *inverse* variation, or *neither*.

25.

x	y
4	12
8	24
16	48
32	96

26.

x	y
8	2
4	4
−2	−8
−8	−2

27.

x	y
2	4
3	9
4	16
5	25

25. direct
26. inverse
27. neither

28. If y varies inversely as x and $y = 6$ when $x = 19$, find y when $x = 2$. **57**

29. If x varies inversely as y and $x = 16$ when $y = 5$, find x when $y = 20$. **4**

30. Suppose a varies directly as b, and a varies inversely as c. Find b when $a = 7$ and $c = -8$, if $b = 15$ when $c = 2$ and $a = 4$. **−105**

31. Suppose x varies directly as y, and x varies inversely as z. Find z when $x = 8$ and $y = -6$, if $z = 26$ when $x = 8$ and $y = 13$. **−12**

State whether each equation represents a *direct, joint, inverse,* or *combined* variation. Then name the constant of variation.

32. $\dfrac{x}{y} = 2.75$ **33.** $fg = -2$ **34.** $a = 3bc$ **35.** $10 = \dfrac{xy^2}{z}$

36. $y = -11x$ **37.** $\dfrac{n}{p} = 4$ **38.** $9n = pr$ **39.** $-2y = z$

40. $a = 27b$ **direct; 27**   **41.** $c = \dfrac{7}{d}$ **inverse; 7**   **42.** $-10 = gh$ **inverse; −10**   **43.** $m = 20cd$ **joint; 20**

32. direct; 2.75
33. inverse; −2
34. joint; 3
35. combined; 10
36. direct; −11
37. direct; 4
38. combined; 9
39. direct; −2

44. **PRECISION** The volume of a gas v varies inversely as the pressure p and directly as the temperature t.

    a. Write an equation to represent the volume of a gas in terms of pressure and temperature. Is your equation a *direct, joint, inverse,* or *combined* variation? $v = \dfrac{kt}{p}$; **combined**

    b. A certain gas has a volume of 8 liters, a temperature of 275 Kelvin, and a pressure of 1.25 atmospheres. If the gas is compressed to a volume of 6 liters and is heated to 300 Kelvin, what will the new pressure be? **approximately 1.82 atmospheres or $\dfrac{20}{11}$ atm**

    c. If the volume stays the same, but the pressure drops by half, then what must have happened to the temperature? **It dropped by half.**

45. VACATION The time it takes the Levensteins to reach Lake Tahoe varies inversely with their average rate of speed.

    a. If they are 800 miles away, write and graph an equation relating their travel time to their average rate of speed. **See margin.**

    b. What minimum average speed will allow them to arrive within 18 hours? **44.$\overline{4}$ mph**

Teaching the Mathematical Practices

Precision Mathematically proficient students try to use clear definitions in their reasoning, calculate accurately and efficiently, and make explicit use of definitions.

Additional Answer

45a.

Differentiated Instruction (AL) (OL)

If students have difficulty keeping the formulas for the different variation functions straight,

Then have students write the formulas using different colors for each variable and another color for k. For example: direct variation: $y = kx$; joint variation: $y = kxz$; and inverse variation: $y = \dfrac{k}{x}$.

Study Guide and Intervention
AL OL ELL

NAME _____ DATE _____ PERIOD _____

8-5 Study Guide and Intervention
Variation Functions
Direct Variation and Joint Variation

(Study guide content)

Practice
AL OL BL ELL

NAME _____ DATE _____ PERIOD _____

8-5 Practice
Variation Functions

Word Problem Practice
AL OL BL ELL

NAME _____ DATE _____ PERIOD _____

8-5 Word Problem Practice
Variation Functions

46. MUSIC The maximum number of songs that a digital audio player can hold depends on the lengths and the quality of the songs that are recorded. A song will take up more space on the player if it is recorded at a higher quality, like from a CD, than at a lower quality, like from the Internet.

a. If a certain player has 5400 megabytes of storage space, write a function that represents the number of songs the player can hold as a function of the average size of the songs. $f(x) = \dfrac{5400}{x}$

b. Is your function a *direct*, *joint*, *inverse*, or *combined* variation? **inverse**

c. Suppose the average file size for a high-quality song is 8 megabytes and the average size for a low-quality song is 5 megabytes. Determine how many more songs the player can hold if they are low quality than if they are high quality. **405**

47 GRAVITY According to the Law of Universal Gravitation, the attractive force F in newtons between any two bodies in the universe is directly proportional to the product of the masses m_1 and m_2 in kilograms of the two bodies and inversely proportional to the square of the distance d in meters between the bodies. That is,

$F = \dfrac{Gm_1m_2}{d^2}$. G is the universal gravitational constant. Its value is 6.67×10^{-11} Nm²/kg².

a. The distance between Earth and the Moon is about 3.84×10^8 meters. The mass of the Moon is 7.36×10^{22} kilograms. The mass of Earth is 5.97×10^{24} kilograms. What is the gravitational force that the Moon and Earth exert upon each other? **about 2×10^{20} newtons**

b. The distance between Earth and the Sun is about 1.5×10^{11} meters. The mass of the Sun is about 1.99×10^{30} kilograms. What is the gravitational force that the Sun and Earth exert upon each other? **about 3.5×10^{22} newtons**

c. Find the gravitational force exerted on each other by two 1000-kilogram iron balls at a distance of 0.1 meter apart. **6.67×10^{-3} newtons**

H.O.T. Problems Use Higher-Order Thinking Skills

48. CRITIQUE Jamil and Savannah are setting up a proportion to begin solving the combined variation in which z varies directly as x and z varies inversely as y. Who has set up the correct proportion? Explain your reasoning.

Jamil
$$z_1 = \dfrac{kx_1}{y_1} \text{ and } z_2 = \dfrac{kx_2}{y_2}$$
$$k = \dfrac{z_1y_1}{x_1} \text{ and } k = \dfrac{z_2y_2}{x_2}$$
$$\dfrac{z_1y_1}{x_1} = \dfrac{z_2y_2}{x_2}$$

Savannah
$$z_1 = \dfrac{kx_1}{y_1} \text{ and } z_2 = \dfrac{kx_2}{y_2}$$
$$k = \dfrac{z_1x_1}{y_1} \text{ and } k = \dfrac{z_2x_2}{y_2}$$
$$\dfrac{z_1x_1}{y_1} = \dfrac{z_2x_2}{y_2}$$

Jamil; Savannah multiplied when she should have divided and divided when she should have multiplied.

49. CHALLENGE If a varies inversely as b, c varies jointly as b and f, and f varies directly as g, how are a and g related? **a and g are directly related.**

50. REASONING Explain why some mathematicians consider every joint variation a combined variation, but not every combined variation a joint variation.

50. Sample answer: Every joint variation is a combined variation because there are two *combined* direct variations. However, a combined variation can have a combination of a direct and an inverse variation, so it cannot be considered as a joint variation.

51. OPEN ENDED Describe three real-life quantities that vary jointly with each other. **Sample answer: The force of an object varies jointly as its mass and acceleration.**

52. WRITING IN MATH Determine the type(s) of variation(s) for which 0 cannot be one of the values. Explain your reasoning.

52. Sample answer: Inverse and some types of combined variation functions cannot have a value of 0 in the domain because division by zero is undefined.

Enrichment
OL BL

NAME _____ DATE _____ PERIOD _____

8-5 Enrichment
Geosynchronous Satellites

Teaching the Mathematical Practices
CCSS

Critique Mathematically proficient students are also able to compare the effectiveness of two plausible arguments, distinguish correct logic or reasoning from that which is flawed, and—if there is a flaw in an argument—explain what it is.

53. SAT/ACT Rafael left the dorm and drove toward the cabin at an average speed of 40 km/h. Monica left some time later driving in the same direction at an average speed of 48 km/h. After driving for five hours, Monica caught up with Rafael. How long did Rafael drive before Monica caught up? **D**

A 1 hour   D 6 hours

B 2 hours   E 8 hours

C 4 hours

54. 75% of 88 is the same as 60% of what number? **J**

F 100

G 105

H 108

J 110

55. EXTENDED RESPONSE Audrey's hair is 7 inches long and is expected to grow at an average rate of 3 inches per year. **a. See margin.**

  a. Make a table that shows the expected length of Audrey's hair after each of the first 4 years.

  b. Write a function that can be used to determine the length of her hair after each year. $f(x) = 3x + 7$

  c. If she does not get a haircut, determine the length of her hair after 9 years. **34 in.**

56. Which of the following is equal to the sum of two consecutive even integers? **B**

A 144   C 147

B 146   D 148

Determine any vertical asymptotes and holes in the graph of each rational function. (Lesson 8-4)

57. $f(x) = \dfrac{1}{x^2 + 5x + 6}$ asymptotes: $x = -2, x = -3$

58. $f(x) = \dfrac{x + 2}{x^2 + 3x - 4}$ asymptotes: $x = -4, x = 1$

59. $f(x) = \dfrac{x^2 + 4x + 3}{x + 3}$ hole: $x = -3$

60. PHOTOGRAPHY The formula $\dfrac{1}{q} = \dfrac{1}{f} - \dfrac{1}{p}$ can be used to determine how far the film should be placed from the lens of a camera to create a perfect photograph. The variable q represents the distance from the lens to the film, f represents the focal length of the lens, and p represents the distance from the object to the lens. (Lesson 8-3)

  a. Solve the formula for $\dfrac{1}{p}$. $\dfrac{1}{p} = \dfrac{1}{f} - \dfrac{1}{q}$

  b. Write the expression containing f and q as a single rational expression. $\dfrac{q - f}{fq}$

  c. If a camera has a focal length of 8 centimeters and the lens is 10 centimeters from the film, how far should an object be from the lens so that the picture will be in focus? **40 cm**

Solve each equation. Check your solutions. (Lesson 7-5)

61. $\log_3 42 - \log_3 n = \log_3 7$ **6**

62. $\log_2 (3x) + \log_2 5 = \log_2 30$ **2**

63. $2 \log_5 x = \log_5 9$ **3**

64. $\log_{10} a + \log_{10} (a + 21) = 2$ **4**

Given a polynomial and one of its factors, find the remaining factors of the polynomial. Some factors may not be binomials. (Lesson 5-6)

65. $2x^3 - 5x^2 - 28x + 15;\ x - 5$
  $x + 3,\ x - \dfrac{1}{2},\ \text{or } 2x - 1$

66. $3x^3 + 10x^2 - x - 12;\ x + 3$
  $x - 1,\ x + \dfrac{4}{3},\ \text{or } 3x + 4$

Find the LCM of each set of polynomials.

67. $a, 2a, a + 1$ **$2a(a + 1)$**

68. $x, 4y, x - y$ **$4xy(x - y)$**

69. $8, 24x, 12$ **$24x$**

70. $x^4, 3x^2, 2xy$ **$6x^4y$**

71. $12a, 15, 4b^2$ **$60ab^2$**

72. $x + 2, x - 3, x^2 - x - 6$ **$x^2 - x - 6$**

DifferentiatedInstruction   OL   BL

Extension Write the equation $y = kx^3$ on the board. Ask students to describe the kind of variation modeled by this equation. Have them describe what happens to the value of y when the value of x is doubled, tripled, halved, etc. For this equation, y varies directly as the cube of x. When x is doubled, y is multiplied by 8. When x is tripled, y is multiplied by 27. When x is halved, y is divided by 8.

WatchOut!

Preventing Misconceptions Make sure that students understand the essential differences between direct and inverse variation. For example, ask them if the number of candles on a birthday cake varies directly or indirectly with the age of the birthday person. Then ask how the time the candle has burned varies with the length of the candle remaining.

4 Assess

Ticket Out the Door On a small slip of paper, have each student write an equation for some variation in daily life (time spent studying, hours of sleep, etc.). When leaving the classroom, have each student tell what kind of variation the equation exhibits.

Formative Assessment
Check for student understanding of concepts in Lessons 8-4 and 8-5.

📁 Quiz 3, p. 48

Additional Answer

55a.

Year	Length
1	10
2	13
3	16
4	19

1 Focus

VerticalAlignment

Before Lesson 8-6 Simplify rational expressions.

Lesson 8-6 Solve rational equations. Solve rational inequalities.

After Lesson 8-6 Solve equations with rational exponents.

2 Teach

Scaffolding Questions

Have students read the **Why?** section of the lesson.

Ask:

- What is the cost per visit if a member visits 10 times in a given month? $7

- If a member visits the club 10 times in a given month, how much does the monthly membership fee add to the cost of each visit? The fee adds $2 for each visit.

- If a member visits x times in a given month, how much does the monthly membership fee add to the cost of each visit? The fee adds $\frac{20}{x}$ dollars for each visit.

8-6 Solving Rational Equations and Inequalities

:: Then	:: Now	:: Why?
● You simplified rational expressions.	**1** Solve rational equations. **2** Solve rational inequalities.	● A gaming club charges $20 per month for membership. Members also have to pay $5 each time they visit the club. If a member visits the club x times in one month, then the charge for that month will be $20 + 5x$. The actual cost per visit will be $\frac{20 + 5x}{x}$. To determine how many visits are needed for the cost per visit to be $6, you would need to solve the equation $\frac{20 + 5x}{x} = 6$.

NewVocabulary
rational equation
weighted average
rational inequality

Common Core State Standards

Content Standards
A.CED.1 Create equations and inequalities in one variable and use them to solve problems.
A.REI.2 Solve simple rational and radical equations in one variable, and give examples showing how extraneous solutions may arise.

Mathematical Practices
6 Attend to precision.

1 Solve Rational Equations Equations that contain one or more rational expressions are called **rational equations**. These equations are often easier to solve once the fractions are eliminated. You can eliminate the fractions by multiplying each side by the least common denominator (LCD).

Example 1 Solve a Rational Equation

Solve $\frac{4}{x + 3} + \frac{5}{6} = \frac{23}{18}$. Check your solution.

The LCD for the terms is $18(x + 3)$.

$$\frac{4}{x + 3} + \frac{5}{6} = \frac{23}{18} \qquad \text{Original equation}$$

$$18(x + 3)\left(\frac{4}{x + 3}\right) + 18(x + 3)\left(\frac{5}{6}\right) = 18(x + 3)\frac{23}{18} \qquad \text{Multiply by LCD.}$$

$$\overset{1}{18(x + 3)}\left(\frac{4}{\underset{1}{x + 3}}\right) + \overset{3}{18(x + 3)}\left(\frac{5}{\underset{1}{6}}\right) = \overset{1}{18(x + 3)}\left(\frac{23}{\underset{1}{18}}\right) \qquad \text{Divide common factors.}$$

$$72 + 15x + 45 = 23x + 69 \qquad \text{Multiply.}$$

$$15x + 117 = 23x + 69 \qquad \text{Simplify.}$$

$$48 = 8x \qquad \text{Subtract 15}x \text{ and 69.}$$

$$6 = x \qquad \text{Divide.}$$

CHECK $\frac{4}{x + 3} + \frac{5}{6} = \frac{23}{18}$ Original equation

$\frac{4}{6 + 3} + \frac{5}{6} \overset{?}{=} \frac{23}{18}$ $x = 6$

$\frac{4}{9} + \frac{5}{6} \overset{?}{=} \frac{23}{18}$ Simplify.

$\frac{8}{18} + \frac{15}{18} \overset{?}{=} \frac{23}{18}$ Simplify.

$\frac{23}{18} = \frac{23}{18}$ ✓   Add.

GuidedPractice

Solve each equation. Check your solution.

1A. $\frac{2}{x + 3} + \frac{3}{2} = \frac{19}{10}$ 2

1B. $\frac{7}{12} + \frac{9}{x - 4} = \frac{55}{48}$ 20

Paul Costello/Stone/Getty Images

570 | Lesson 8-6

Lesson 8-6 Resources

Resource	Approaching Level **AL**	On Level **OL**	Beyond Level **BL**	English Learners **ELL**
Teacher Edition		▪ Differentiated Instruction, pp. 574, 578	▪ Differentiated Instruction, pp. 574, 578	▪ Differentiated Instruction, p. 574
Chapter Resource Masters	▪ Study Guide and Intervention, pp. 39–40 ▪ Skills Practice, p. 41 ▪ Practice, p. 42 ▪ Word Problem Practice, p. 43	▪ Study Guide and Intervention, pp. 39–40 ▪ Skills Practice, p. 41 ▪ Practice, p. 42 ▪ Word Problem Practice, p. 43 ▪ Enrichment, p. 44	▪ Practice, p. 42 ▪ Word Problem Practice, p. 43 ▪ Enrichment, p. 44	▪ Study Guide and Intervention, pp. 39–40 ▪ Skills Practice, p. 41 ▪ Practice, p. 42 ▪ Word Problem Practice, p. 43
Other	▪ 5-Minute Check 8-6 ▪ Study Notebook	▪ 5-Minute Check 8-6 ▪ Study Notebook	▪ 5-Minute Check 8-6 ▪ Study Notebook	▪ 5-Minute Check 8-6 ▪ Study Notebook

Multiplying each side of an equation by the LCD of rational expressions can yield results that are not solutions of the original equation. These are extraneous solutions.

Archive Photos/Getty Images

Math HistoryLink

Brook Taylor (1685–1731) English mathematician Taylor developed a theorem used in calculus known as Taylor's Theorem that relies on the remainders after computations with rational expressions.

Example 2 Solve a Rational Equation

Solve $\dfrac{2x}{x+5} - \dfrac{x^2 - x - 10}{x^2 + 8x + 15} = \dfrac{3}{x+3}$. **Check your solution.**

The LCD for the terms is $(x+3)(x+5)$.

$$\dfrac{2x}{x+5} - \dfrac{x^2 - x - 10}{x^2 + 8x + 15} = \dfrac{3}{x+3} \qquad \text{Original equation}$$

$$\dfrac{(x+3)(x+5)(2x)}{x+5} - \dfrac{(x+3)(x+5)(x^2 - x - 10)}{x^2 + 8x + 15} = \dfrac{(x+3)(x+5)3}{x+3} \qquad \substack{\text{Multiply by LCD.} \\ \text{Divide common factors.}}$$

$$\dfrac{(x+3)\cancel{(x+5)}(2x)}{\cancel{x+5}} - \dfrac{\cancel{(x+3)}\cancel{(x+5)}(x^2 - x - 10)}{\cancel{x^2 + 8x + 15}} = \dfrac{(x+5)\cancel{(x+3)}3}{\cancel{x+3}}$$

$$(x+3)(2x) - (x^2 - x - 10) = 3(x+5) \qquad \text{Simplify.}$$

$$2x^2 + 6x - x^2 + x + 10 = 3x + 15 \qquad \text{Distribute.}$$

$$x^2 + 7x + 10 = 3x + 15 \qquad \text{Simplify.}$$

$$x^2 + 4x - 5 = 0 \qquad \text{Subtract } 3x + 15.$$

$$(x+5)(x-1) = 0 \qquad \text{Factor.}$$

$$x+5=0 \quad \text{or} \quad x-1=0 \qquad \text{Zero Product Property}$$

$$x = -5 \qquad\qquad x = 1$$

CHECK Try $x = -5$.

$$\dfrac{2x}{x+5} - \dfrac{x^2 - x - 10}{x^2 + 8x + 15} = \dfrac{3}{x+3}$$

$$\dfrac{2(-5)}{-5+5} - \dfrac{(-5)^2 - (-5) - 10}{(-5)^2 + 8(-5) + 15} \overset{?}{=} \dfrac{3}{-5+3}$$

$$\dfrac{-10}{0} - \dfrac{25 + 5 - 10}{25 - 40 + 15} \neq -\dfrac{3}{2} \ \text{✗}$$

Try $x = 1$.

$$\dfrac{2x}{x+5} - \dfrac{x^2 - x - 10}{x^2 + 8x + 15} = \dfrac{3}{x+3}$$

$$\dfrac{2(1)}{1+5} - \dfrac{1^2 - 1 - 10}{1^2 + 8(1) + 15} \overset{?}{=} \dfrac{3}{1+3}$$

$$\dfrac{2}{6} - \dfrac{-10}{24} \overset{?}{=} \dfrac{3}{4}$$

$$\dfrac{8}{24} + \dfrac{10}{24} \overset{?}{=} \dfrac{3}{4}$$

$$\dfrac{3}{4} = \dfrac{3}{4} \ \text{✓}$$

When solving a rational equation, any possible solution that results in a zero in the denominator must be excluded from your list of solutions.

Since $x = -5$ results in a zero in the denominator, it is extraneous. Eliminate -5 from the list of solutions. The solution is 1.

ReviewVocabulary

extraneous solutions solutions that do not satisfy the original equation

▸ **Guided**Practice

2A. $\dfrac{5}{y-2} + 2 = \dfrac{17}{6}$ **8**

2B. $\dfrac{2}{z+1} - \dfrac{1}{z-1} = \dfrac{-2}{z^2 - 1}$ **no solutions**

2C. $\dfrac{7n}{3n+3} - \dfrac{5}{4n-4} = \dfrac{3n}{2n+2}$ $-\dfrac{1}{2}$, 3

2D. $\dfrac{1}{p-2} = \dfrac{2p+1}{p^2 + 2p - 8} + \dfrac{2}{p+4}$ $\dfrac{7}{3}$

connectED.mcgraw-hill.com **571**

---

1 Solve Rational Equations

Example 1 shows how to solve a rational equation. **Example 2** shows how to solve a rational equation and eliminate any extraneous solutions. **Examples 3–5** show how to use rational equations to solve mixture, distance, and work problems.

Formative Assessment

Use the Guided Practice exercises after each example to determine students' understanding of concepts.

Additional Examples

1 Solve $\dfrac{5}{24} + \dfrac{2}{3-x} = \dfrac{1}{4}$. Check your solution. $x = -45$

2 Solve $\dfrac{p^2 - p - 5}{p+1} = \dfrac{p^2 - 7}{p-1} + p$. Check your solution. $p = -3, -2, 2$

▸ **Additional Examples** also in Interactive Classroom PowerPoint® Presentations

IWB **Interactive White Board READY**

Focus on Mathematical Content

Solving Rational Equations During the solution process, a rational equation is usually transformed into another type of equation. Solving a rational equation may require solving a related linear, quadratic, or other type of equation.

WatchOut!

Preventing Misconceptions Remind students that a possible solution must always be checked in the original equation, rather than in any of the steps of the solution. However, it is not permissible simply to substitute the solution for the variable and then multiply both sides of the equation by the LCD, because the truth of the equation is being tested.

3 **BRINE** Aaron adds an 80% brine (salt and water) solution to 16 ounces of solution that is 10% brine. How much of the solution should be added to create a solution that is 50% brine? $21\frac{1}{3}$ ounces

(CCSS) **Teaching the Mathematical Practices**

Modeling Mathematically proficient students are able to identify important quantities in a practical situation and map their relationships using such tools as diagrams, two-way tables, graphs, flowcharts and formulas. Encourage students to use various representations to better analyze functions.

The **weighted average** is a method for finding the mean of a set of numbers in which some elements of the set carry more importance, or weight, than others. Many real-world problems involving mixtures, work, distance, and interest can be solved by using rational equations.

Real-World Example 3 Mixture Problem

CHEMISTRY Mia adds a 70% acid solution to 12 milliliters of a solution that is 15% acid. How much of the 70% acid solution should be added to create a solution that is 60% acid?

Understand Mia needs to know how much of a solution needs to be added to an original solution to create a new solution.

Plan Each solution has a certain percentage that is acid. The percentage of acid in the final solution must equal the amount of acid divided by the total solution.

	Original	Added	New
Amount of Acid	0.15(12)	0.7(x)	0.15(12) + 0.7x
Total Solution	12	x	12 + x

$$\text{Percentage of acid in solution} = \frac{\text{amount of acid}}{\text{total solution}}$$

Solve

$$\frac{\text{percent}}{100} = \frac{\text{amount of acid}}{\text{total solution}}$$ Write a proportion.

$$\frac{60}{100} = \frac{0.15(12) + 0.7x}{12 + x}$$ Substitute.

$$\frac{60}{100} = \frac{1.8 + 0.7x}{12 + x}$$ Simplify numerator.

$$100(12 + x)\frac{60}{100} = 100(12 + x)\frac{1.8 + 0.7x}{12 + x}$$ LCD is 100(12 + x). Multiply by LCD.

$$100(12 + x)\frac{60}{\underset{1}{100}} = 100(12 + x)\frac{1.8 + 0.7x}{\underset{1}{12 + x}}$$ Divide common factors.

$$(12 + x)60 = 100(1.8 + 0.7x)$$ Simplify.

$$720 + 60x = 180 + 70x$$ Distribute.

$$540 = 10x$$ Subtract 60x and 180.

$$54 = x$$ Divide by 10.

Check
$$\frac{60}{100} = \frac{0.15(12) + 0.7x}{12 + x}$$ Original equation

$$\frac{60}{100} \overset{?}{=} \frac{0.15(12) + 0.7(54)}{12 + 54}$$ x = 54

$$\frac{60}{100} \overset{?}{=} \frac{37.8}{66}$$ Simplify.

$$0.6 = 0.6 \checkmark$$ Simplify.

Mia needs to add 54 milliliters of the 70% acid solution.

GuidedPractice

3. Jimmy adds a 65% fruit juice solution to 15 milliliters of a drink that is 10% fruit juice. How much of the 65% fruit juice solution must be added to create a fruit punch that is 35% fruit juice? **12.5 mL**

StudyTip

(CCSS) Modeling Tables like the one in Example 3 are useful in organizing and solving mixture, work, weighted average, and distance problems.

The formula relating distance, rate, and time can also be used to solve rational equations. The most common use is $d = rt$. However, it can also be represented by $r = \frac{d}{t}$ and $t = \frac{d}{r}$.

Real-World Example 4 Distance Problem

ROWING Sandra is rowing a canoe on Stanhope Lake. Her rate in still water is 6 miles per hour. It takes Sandra 3 hours to travel 10 miles round trip. Assuming that Sandra rowed at a constant rate of speed, determine the rate of the current.

StudyTip

Distance Problems When distances involve round trips, the distance in one direction usually equals the distance in the other direction.

Understand We are given her speed in still water and the time it takes her to travel with the current and against it. We need to determine the speed of the current.

Plan She traveled 5 miles with the current and 5 miles against it. The formula that relates distance, rate, and time is $d = rt$, or $t = \frac{d}{r}$.

Time with the Current	Time Against the Current	Total Time
$\frac{5}{6+r}$	$\frac{5}{6-r}$	3 hours

Solve

$$\frac{5}{6+r} + \frac{5}{6-r} = 3 \qquad \text{Write the equation.}$$

$$(6+r)(6-r)\frac{5}{6+r} + (6+r)(6-r)\frac{5}{6-r} = (6+r)(6-r)3 \qquad \begin{array}{l}\text{LCD} = (6+r)(6-r)\\ \text{Multiply by LCD.}\end{array}$$

$$(6+r)(6-r)\frac{5}{6+r}\Big|_1 + (6+r)(6-r)\frac{5}{6-r}\Big|_1 = (6+r)(6-r)3 \qquad \text{Divide common factors.}$$

$$(6-r)5 + (6+r)5 = (36 - r^2)3 \qquad \text{Simplify.}$$

$$30 - 5r + 30 + 5r = 108 - 3r^2 \qquad \text{Distribute.}$$

$$60 = 108 - 3r^2 \qquad \text{Simplify.}$$

$$0 = -3r^2 + 48 \qquad \text{Subtract 10r.}$$

$$0 = -3(r+4)(r-4) \qquad \text{Factor.}$$

$$0 = (r+4)(r-4) \qquad \text{Divide each side by } -3.$$

$$r = 4 \text{ or } -4 \qquad \text{Zero Product Property}$$

Check

$$\frac{5}{6+r} + \frac{5}{6-r} = 3 \qquad \text{Original equation}$$

$$\frac{5}{6+4} + \frac{5}{6-4} \stackrel{?}{=} 3 \qquad r = 4$$

$$\frac{5}{10} + \frac{5}{2} \stackrel{?}{=} 3 \qquad \text{Simplify.}$$

$$\frac{1}{2} + \frac{5}{2} = \frac{6}{2} \checkmark \qquad \text{Simplify.}$$

Since speed cannot be negative, the speed of the current is 4 miles per hour.

GuidedPractice

4. **FLYING** The speed of the wind is 20 miles per hour. If it takes a plane 7 hours to fly 2368 miles round trip, determine the plane's speed in still air. **339.5 mph**

  connectED.mcgraw-hill.com **573**

Additional Example

4 **SWIMMING** Lilia swims for 5 hours in a stream that has a current of 1 mile per hour. She leaves her dock, swims upstream for 2 miles and then swims back to her dock. What is her swimming speed in still water? about 1.5 mi/h

WatchOut!

Avoiding Misconceptions In Example 4, make sure that students understand the difference between the solutions of the quadratic equation (of which there are two) and the solution of the problem (of which there is only one).

Preventing Errors When solving rational equations, students often forget that *all* terms in the equation must be multiplied by the LCD. Remind them to do so in order to obtain equivalent equations.

WatchOut!

Preventing Errors Suggest that when a problem involves completing part of a job while working together, students think about what part of the work gets done in one day, or one year, or one unit of the time.

Additional Example

5 **MOWING LAWNS** Wuyi and Uima mow lawns together. Wuyi working alone could complete a particular job in 4.5 hours, and Uima could complete it alone in 3.7 hours. How long does it take to complete the job when they work together? **about 2 hours**

Tips for New Teachers

Alternative Method The equation in Example 5 could also be solved by first subtracting $\frac{1}{24}$ from each side. Then find the cross products and solve.

Real-world problems that involve work can often be solved using rational equations.

Real-World Example 5 Work Problems

COMMUNITY SERVICE Every year, the junior and senior classes at Hillcrest High School build a house for the community. If it takes the senior class 24 days to complete a house and 18 days if they work with the junior class, how long would it take the junior class to complete a house if they worked alone?

Understand We are given how long it takes the senior class working alone and when the classes work together. We need to determine how long it would take the junior class by themselves.

Plan The senior class can complete 1 house in 24 days, so their rate is $\frac{1}{24}$ of a house per day.

The rate for the junior class is $\frac{1}{j}$.

The combined rate for both classes is $\frac{1}{18}$.

Senior Rate	Junior Rate	Combined Rate
$\frac{1}{24}$	$\frac{1}{j}$	$\frac{1}{18}$

Solve

$$\frac{1}{24} + \frac{1}{j} = \frac{1}{18} \qquad \text{Write the equation.}$$

$$72j\frac{1}{24} + 72j\frac{1}{j} = 72j\frac{1}{18} \qquad \begin{array}{l}\text{LCD} = 72j \\ \text{Multiply by LCD.}\end{array}$$

$$\overset{3}{72j}\frac{1}{24} + 72j\frac{1}{\underset{1}{j}} = \overset{4}{72j}\frac{1}{18} \qquad \text{Divide common factors.}$$

$$3j + 72 = 4j \qquad \text{Distribute.}$$

$$72 = j \qquad \text{Subtract } 3j.$$

Check Two methods are possible.

Method 1 Substitute values.

$$\frac{1}{24} + \frac{1}{j} = \frac{1}{18} \qquad \text{Original equation}$$

$$\frac{1}{24} + \frac{1}{72} \overset{?}{=} \frac{1}{18} \qquad j = 72$$

$$\frac{3}{72} + \frac{1}{72} \overset{?}{=} \frac{4}{72} \qquad \text{LCD} = 72$$

$$\frac{4}{72} = \frac{4}{72} \checkmark \qquad \text{Simplify.}$$

Method 2 Use a calculator.

It would take the junior class 72 days to complete the house by themselves.

Guided Practice

5A. It took Anthony and Travis 6 hours to rake the leaves together last year. The previous year it took Travis 10 hours to do it alone. How long will it take Anthony if he rakes them by himself this year? **15 hours**

5B. Noah and Owen paint houses together. If Noah can paint a particular house in 6 days and Owen can paint the same house in 5 days, how long would it take the two of them if they work together? **$2\frac{8}{11}$ days**

Real-World Link

Since 1997, students from Rock Point School in Burlington, Vermont, spend a week servicing communities throughout the world. While working with Habitat for Humanity, the students spent the time in rural Tennessee, starting and completing the roof of a Habitat home in one week.

Source: Vermont Community Work

Paul Burns/Photodisc/Getty Images

DifferentiatedInstruction OL BL ELL

Logical Learners Have students think about the difference between "pure" mathematics, such as solving an equation, and "applied" mathematics, such as solving a real-world problem. Ask them to list some ways in which these two are alike and some ways in which they are different.

2 Solve Rational Inequalities To solve **rational inequalities**, which are inequalities that contain one or more rational expressions, follow these steps.

KeyConcept Solving Rational Inequalities

Step 1 State the excluded values. These are the values for which the denominator is 0.

Step 2 Solve the related equation.

Step 3 Use the values determined from the previous steps to divide a number line into intervals.

Step 4 Test a value in each interval to determine which intervals contain values that satisfy the inequality.

Example 6 Solve a Rational Inequality

Solve $\dfrac{x}{3} - \dfrac{1}{x-2} < \dfrac{x+1}{4}$.

Step 1 The excluded value for this inequality is 2.

Step 2 Solve the related equation.

$$\dfrac{x}{3} - \dfrac{1}{x-2} = \dfrac{x+1}{4} \qquad \text{Related equation}$$

$$\overset{4}{\cancel{12}}(x-2)\dfrac{x}{\underset{1}{\cancel{3}}} - 12(x-2)\dfrac{1}{\cancel{x-2}} \;\underset{1}{=}\; \overset{3}{\cancel{12}}(x-2)\dfrac{x+1}{\underset{1}{\cancel{4}}} \qquad \begin{array}{l}\text{LCD is } 12(x-2).\\ \text{Multiply by LCD.}\end{array}$$

$$4x^2 - 8x - 12 = 3x^2 - 3x - 6 \qquad \text{Distribute.}$$

$$x^2 - 5x - 6 = 0 \qquad \text{Subtract } 3x^2 - 3x - 6.$$

$$(x-6)(x+1) = 0 \qquad \text{Factor.}$$

$$x = 6 \text{ or } -1 \qquad \text{Zero Product Property}$$

Step 3 Draw vertical lines at the excluded value and at the solutions to separate the number line into intervals.

StudyTip
Rational Inequalities It is possible that none or all of the intervals will produce a true statement.

Step 4 Now test a sample value in each interval to determine whether the values in the interval satisfy the inequality.

Test $x = -3$.

$$\dfrac{-3}{3} - \dfrac{1}{-3-2} \overset{?}{<} \dfrac{-3+1}{4}$$

$$-1 + \dfrac{1}{5} \overset{?}{<} -\dfrac{2}{4}$$

$$-\dfrac{4}{5} < -\dfrac{1}{2} \ \checkmark$$

Test $x = 0$.

$$\dfrac{0}{3} - \dfrac{1}{0-2} \overset{?}{<} \dfrac{0+1}{4}$$

$$0 + \dfrac{1}{2} \overset{?}{<} \dfrac{1}{4}$$

$$\dfrac{1}{2} \not< \dfrac{1}{4}$$

Test $x = 4$.

$$\dfrac{4}{3} - \dfrac{1}{4-2} \overset{?}{<} \dfrac{4+1}{4}$$

$$\dfrac{4}{3} - \dfrac{1}{2} \overset{?}{<} \dfrac{5}{4}$$

$$\dfrac{5}{6} < \dfrac{5}{4} \ \checkmark$$

Test $x = 8$.

$$\dfrac{8}{3} - \dfrac{1}{8-2} \overset{?}{<} \dfrac{8+1}{4}$$

$$\dfrac{32}{12} - \dfrac{2}{12} \overset{?}{<} \dfrac{27}{12}$$

$$\dfrac{30}{12} \not< \dfrac{27}{12}$$

The statement is true for $x = -3$ and $x = 4$. Therefore, the solution is $x < -1$ or $2 < x < 6$.

Guided Practice Solve each inequality.

6A. $\dfrac{5}{x} + \dfrac{6}{5x} > \dfrac{2}{3}$ $\quad 0 < x < 9.3$

6B. $\dfrac{4}{3x} + \dfrac{7}{x} < \dfrac{5}{9}$ $\quad x < 0$ or $x > 15$

2 Solve Rational Inequalities
Example 6 shows how to solve a rational inequality.

Additional Example

6 Solve $\dfrac{1}{3k} + \dfrac{2}{9k} < \dfrac{2}{3}$.

$k < 0$ or $k > \dfrac{5}{6}$

WatchOut!

Preventing Errors Suggest that students also verify whether the boundary indicated by the solution of the equation *is* or *is not* in the solution set of the inequality.

Teach with Tech

Document Camera Choose a student to show and explain how to solve a rational inequality. Be sure the student shows how to use test points and a sign chart to find the solution set.

3 Practice

Formative Assessment

Use Exercises 1–15 to check for understanding.

Use the chart at the bottom of this page to customize assignments for your students.

 CCSS **Teaching the Mathematical Practices**

Structure Mathematically proficient students look closely to discern a pattern or structure. They can see complicated things, such as some algebraic expressions, as single objects or as being composed of several objects.

Exercise Alert

Grid Paper For Exercises 33 and 47–49, students will need grid paper.

🖉 Multiple Representations

In Exercise 33, students will use the graphs of related functions to investigate solutions of rational equations.

Additional Answers

33b.

36. Sample answer:
$$\frac{4}{x+3} = \frac{x}{x-4} + \frac{7}{4}$$

38. Sample answer: The values -3 and 2 are undefined values. On the graph of $f(x)$ there would be vertical asymptotes at these values.

39. Sample answer: Multiplying each side of a rational equation or inequality by the LCD can result in extraneous solutions. Therefore, you should check all solutions to make sure that they satisfy the original equation or inequality.

Examples 1–2 Solve each equation. Check your solution.

1. $\frac{4}{7} + \frac{3}{x-3} = \frac{53}{56}$ 11

2. $\frac{7}{3} - \frac{3}{x-5} = \frac{19}{12}$ 9

3. $\frac{10}{2x+1} + \frac{4}{3} = 2$ 7

4. $\frac{11}{4} - \frac{5}{y+3} = \frac{23}{12}$ 3

5. $\frac{8}{x-5} - \frac{9}{x-4} = \frac{5}{x^2-9x+20}$ 8

6. $\frac{14}{x+3} + \frac{10}{x-2} = \frac{122}{x^2+x-6}$ 5

7. $\frac{14}{x-8} - \frac{5}{x-6} = \frac{82}{x^2-14x+48}$ 14

8. $\frac{5}{x+2} - \frac{3}{x-2} = \frac{12}{x^2-4}$ 14

Example 3

9. **CCSS STRUCTURE** Sara has 10 pounds of dried fruit selling for $6.25 per pound. She wants to know how many pounds of mixed nuts selling for $4.50 per pound she needs to make a trail mix selling for $5 per pound.

 a. Let m = the number of pounds of mixed nuts. Complete the following table.

	Pounds	Price per Pound	Total Price
Dried Fruit	10	$6.25	6.25(10)
Mixed Nuts	m	$4.50	4.5m
Trail Mix	$10 + m$	$5.00	5(10 + m)

 b. Write a rational equation using the last column of the table. $62.5 + 4.5m = 50 + 5m$

 c. Solve the equation to determine how many pounds of mixed nuts are needed. 25

Example 4

10. **DISTANCE** Alicia's average speed riding her bike is 11.5 miles per hour. She takes a round trip of 40 miles. It takes her 1 hour and 20 minutes with the wind and 2 hours and 30 minutes against the wind.

 a. Write an expression for Alicia's time with the wind. $\frac{20}{11.5+x}$

 b. Write an expression for Alicia's time against the wind. $\frac{20}{11.5-x}$

 c. How long does it take to complete the trip? **3 h and 50 min**

 d. Write and solve the rational equation to determine the speed of the wind. $\frac{20}{11.5+x} + \frac{20}{11.5-x} = \frac{23}{6};$ **3.5 mph**

Example 5

11. **WORK** Kendal and Chandi wax cars. Kendal can wax a particular car in 60 minutes and Chandi can wax the same car in 80 minutes. They plan on waxing the same car together and want to know how long it will take.

 a. How much will Kendal complete in 1 minute? $\frac{1}{60}$

 b. How much will Kendal complete in x minutes? $\frac{x}{60}$

 c. How much will Chandi complete in 1 minute? $\frac{1}{80}$

 d. How much will Chandi complete in x minutes? $\frac{x}{80}$

 e. Write a rational equation representing Kendal and Chandi working together on the car. $\frac{x}{60} + \frac{x}{80} = 1$

 f. Solve the equation to determine how long it will take them to finish the car. **about 34.3 min**

Example 6 Solve each inequality. Check your solutions.

12. $\frac{3}{5x} + \frac{1}{6x} > \frac{2}{3}$ $0 < x < 1.15$

13. $\frac{1}{4c} + \frac{1}{9c} < \frac{1}{2}$ $c < 0$, or $\frac{13}{18} < c$

14. $\frac{4}{3y} + \frac{2}{5y} < \frac{3}{2}$ $y > \frac{52}{45}$, or $y < 0$

15. $\frac{1}{3b} + \frac{1}{4b} < \frac{1}{5}$ $b < 0$, or $\frac{35}{12} < b$

Differentiated Homework Options

Level	Assignment	Two-Day Option	
AL Basic	16–30, 36, 38–54	17–29 odd, 40–43	16–30 even, 36, 38, 39, 44–54
OL Core	17–29 odd, 31–36, 38–54	16–30, 40–43	31–36, 38, 39, 44–54
BL Advanced	31–51, (optional: 52–54)		

Examples 1–2 Solve each equation. Check your solutions.

16. $\dfrac{9}{x-7} - \dfrac{7}{x-6} = \dfrac{13}{x^2 - 13x + 42}$ **9**

17. $\dfrac{13}{y+3} - \dfrac{12}{y+4} = \dfrac{18}{y^2 + 7y + 12}$ **2**

18. $\dfrac{14}{x-2} - \dfrac{18}{x+1} = \dfrac{22}{x^2 - x - 2}$ **7**

19. $\dfrac{11}{a+2} - \dfrac{10}{a+5} = \dfrac{36}{a^2 + 7a + 10}$ **1**

20. $\dfrac{x}{2x-1} + \dfrac{3}{x+4} = \dfrac{21}{2x^2 + 7x - 4}$ **−12, 2**

21. $\dfrac{2}{y-5} + \dfrac{y-1}{2y+1} = \dfrac{2}{2y^2 - 9y - 5}$ **∅**

Examples 3–5 **22.** **CHEMISTRY** How many milliliters of a 20% acid solution must be added to 40 milliliters of a 75% acid solution to create a 30% acid solution? **180 mL**

23 **GROCERIES** Ellen bought 3 pounds of bananas for $0.90 per pound. How many pounds of apples costing $1.25 per pound must she purchase so that the total cost for fruit is $1 per pound? **1.2 lb**

24. **BUILDING** Bryan's volunteer group can build a garage in 12 hours. Sequoia's group can build it in 16 hours. How long would it take them if they worked together? **about 6.86 hours**

Example 6 Solve each inequality. Check your solutions.

25. $3 - \dfrac{4}{x} > \dfrac{5}{4x}$ **$x < 0$ or $x > 1.75$**

26. $\dfrac{5}{3a} - \dfrac{3}{4a} > \dfrac{5}{6}$ **$0 < a < 1.1$**

27. $\dfrac{x-2}{x+2} + \dfrac{1}{x-2} > \dfrac{x-4}{x-2}$ **$x < -2$, or $2 < x < 14$**

28. $\dfrac{3}{4} - \dfrac{1}{x-3} > \dfrac{x}{x+4}$ **$x < -5$ or $-4 < x < 3$**

29. $\dfrac{x}{5} + \dfrac{2}{3} < \dfrac{3}{x-4}$ **$4 < x < \dfrac{17}{3}$**

30. $\dfrac{x}{x+2} + \dfrac{1}{x-1} < \dfrac{3}{2}$ **$2 < x$, $-2 < x < 1$, $x < -5$**

B **31.** **AIR TRAVEL** It takes a plane 20 hours to fly to its destination against the wind. The return trip takes 16 hours. If the plane's average speed in still air is 500 miles per hour, what is the average speed of the wind during the flight? **55.56 mph**

32. **FINANCIAL LITERACY** Judie wants to invest $10,000 in two different accounts. The risky account could earn 9% interest, while the other account earns 5% interest. She wants to earn $750 interest for the year. Of tables, graphs, or equations, choose the best representation needed and determine how much should be invested in each account.
$6250 at 9% and $3750 at 5%

33. **MULTIPLE REPRESENTATIONS** Consider $\dfrac{2}{x-3} + \dfrac{1}{x} = \dfrac{x-1}{x-3}$.

  a. Algebraic Solve the equation for x. Were any values of x extraneous? **1; yes; 3**

  b. Graphical Graph $y_1 = \dfrac{2}{x-3} + \dfrac{1}{x}$ and $y_2 = \dfrac{x-1}{x-3}$ on the same graph for $0 < x < 5$. **See margin.**

  c. Analytical For what value(s) of x do they intersect? Do they intersect where x is extraneous for the original equation? **1; no**

  d. Verbal Use this knowledge to describe how you can use a graph to determine whether an apparent solution of a rational equation is extraneous. **Graph both sides of the equation. Where the graphs intersect, there is a solution. If they do not, then the possible solution is extraneous.**

C Solve each equation. Check your solutions.

34. $\dfrac{2}{y+3} - \dfrac{3}{4-y} = \dfrac{2y-2}{y^2 - y - 12}$ **−1**

35. $\dfrac{2}{y+2} - \dfrac{y}{2-y} = \dfrac{y^2 + 4}{y^2 - 4}$ **∅**

H.O.T. Problems Use Higher-Order Thinking Skills

36. **OPEN ENDED** Give an example of a rational equation that can be solved by multiplying each side of the equation by $4(x+3)(x-4)$. **See margin.**

37. **CHALLENGE** Solve $\dfrac{1 + \dfrac{9}{x} + \dfrac{20}{x^2}}{1 - \dfrac{25}{x^2}} = \dfrac{x+4}{x-5}$. **all real numbers except 5, −5, 0**

38. **CCSS TOOLS** While using the table feature on the graphing calculator to explore $f(x) = \dfrac{1}{x^2 - x - 6}$, the values −2 and 3 say "**ERROR.**" Explain its meaning. **See margin.**

39. **WRITING IN MATH** Why should you check solutions of rational equations and inequalities? **See margin.**

Study Guide and Intervention
AL OL ELL

Practice
AL OL BL ELL

Word Problem Practice
AL OL BL ELL

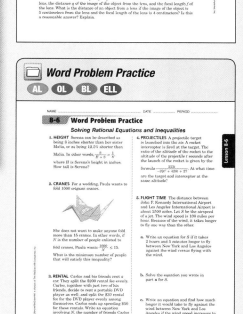

CCSS Teaching the Mathematical Practices

Tools When making mathematical models, mathematically proficient students know that technology can enable them to visualize the results of varying assumptions, explore consequences, and compare predictions with data. They are able to use technological tools to explore and deepen their understanding of concepts.

Enrichment
OL BL

Name the Math Have students write their own real-world problems similar to those they have seen in this lesson but using their own data. Have students solve the problems.

Formative Assessment
Check for student understanding of concepts in Lesson 8-6.

 Quiz 4, p. 48

Additional Answers

47.

$$f(x) = \frac{x+4}{x^2+7x+12}$$

48.

$$f(x) = \frac{x^2-5x-14}{x-7}$$

49.

$$f(x) = \frac{x^2+3x-6}{x-2}$$

Standardized Test Practice

40. Nine pounds of mixed nuts containing 55% peanuts were mixed with 6 pounds of another kind of mixed nuts that contain 40% peanuts. What percent of the new mixture is peanuts? **C**

   A 58%   **B** 51%   **C** 49%   **D** 47%

41. Working alone, Dato can dig a 10-foot by 10-foot hole in five hours. Pedro can dig the same hole in six hours. How long would it take them if they worked together? **J**

   F 1.5 hours   **H** 2.52 hours
   G 2.34 hours   **J** 2.73 hours

42. An aircraft carrier made a trip to Guam and back. The trip there took three hours and the trip back took four hours. It averaged 6 kilometers per hour on the return trip. Find the average speed of the trip to Guam. **B**

   A 6 km/h   **C** 10 km/h
   B 8 km/h   **D** 12 km/h

43. SHORT RESPONSE If a line ℓ is perpendicular to a segment CD at point F and $CF = FD$, how many points on line ℓ are the same distance from point C as from point D? **all of the points**

Spiral Review

Determine whether each relation shows *direct* or *inverse* variation, or *neither*. (Lesson 8-5)

44. **inverse**

x	y
14	3
28	1.5
56	0.75
112	0.375

45. **direct**

x	y
0.2	24
0.6	72
1.8	216
5.4	648

46. **neither**

x	y
12	18
24	36
36	18
72	9

Graph each function. (Lesson 8-4) **47–49. See margin.**

47. $f(x) = \dfrac{x+4}{x^2+7x+12}$ **48.** $f(x) = \dfrac{x^2-5x-14}{x-7}$ **49.** $f(x) = \dfrac{x^2+3x-6}{x-2}$

50. WEATHER The atmospheric pressure P, in bars, of a given height on Earth is given by the formula $P = a \cdot e^{-\frac{k}{H}}$. In the formula, a is the surface pressure on Earth, which is approximately 1 bar, k is the altitude for which you want to find the pressure in kilometers, and H is always 7 kilometers. (Lesson 7-7)

  a. Find the pressure for 2, 4, and 7 kilometers. **0.75 bars; 0.56 bars; 0.37 bars**

  b. What do you notice about the pressure as altitude increases? **The pressure decreases as the altitude increases.**

51. COMPUTERS Since computers have been invented, computational speed has multiplied by a factor of 4 about every three years. (Lesson 7-1)

  a. If a typical computer operates with a computational speed s today, write an expression for the speed at which you can expect an equivalent computer to operate after x three-year periods. $s \cdot 4^x$

  b. Suppose your computer operates with a processor speed of 2.8 gigahertz and you want a computer that can operate at 5.6 gigahertz. If a computer with that speed is currently unavailable for home use, how long can you expect to wait until you can buy such a computer? **0.5 three-yr periods or 1.5 yr**

Skills Review

Determine whether the following are possible lengths of the sides of a right triangle.

52. 5, 12, 13 **yes**   **53.** 60, 80, 100 **yes**   **54.** 7, 24, 25 **yes**

DifferentiatedInstruction OL BL

Extension The equation $\dfrac{1}{f} = \dfrac{1}{d} + \dfrac{1}{i}$ is sometimes called the "lens equation." It shows how the focal length f of a lens is related to the distances d and i from the lens to an object and to the object's image as seen through the lens, respectively. Ask students to find the focal length of a lens if, for a particular object, d is measured to be 20 cm and i is 12 cm. Ask them to solve the lens equation for f in terms of d and i. **7.5 cm; $f = \dfrac{di}{d+i}$**

EXTEND 8-6

Graphing Technology Lab
Solving Rational Equations and Inequalities

You can use a TI-83/84 Plus graphing calculator to solve rational equations by graphing or by using the table feature. Graph both sides of the equation, and locate the point(s) of intersection.

CCSS Common Core State Standards
Content Standards
A.REI.2 Solve simple rational and radical equations in one variable, and give examples showing how extraneous solutions may arise.
A.REI.11 Explain why the *x*-coordinates of the points where the graphs of the equations $y = f(x)$ and $y = g(x)$ intersect are the solutions of the equation $f(x) = g(x)$; find the solutions approximately, e.g., using technology to graph the functions, make tables of values, or find successive approximations. Include cases where $f(x)$ and/or $g(x)$ are linear, polynomial, rational, absolute value, exponential, and logarithmic functions.
Mathematical Practices
5 Use appropriate tools strategically.

Activity 1   Rational Equation

Solve $\dfrac{4}{x+1} = \dfrac{3}{2}$.

Step 1 Graph each side of the equation.

Graph each side of the equation as a separate function. Enter $\dfrac{4}{x+1}$ as **Y1** and $\dfrac{3}{2}$ as **Y2**. Then graph the two equations in the standard viewing window.

KEYSTROKES: Y= 4 ÷ (X,T,θ,*n* + 1)
ENTER 3 ÷ 2 ZOOM 6

[−10, 10] scl: 1 by [−10, 10] scl: 1

Because the calculator is in connected mode, a vertical line may appear connecting the two branches of the hyperbola. This line is not part of the graph.

Step 2 Use the **intersect** feature.

The **intersect** feature on the **CALC** menu allows you to approximate the ordered pair of the point at which the graphs cross.

KEYSTROKES: 2nd [CALC] 5

Select one graph and press ENTER. Select the other graph, press ENTER, and press ENTER again.

[−10, 10] scl: 1 by [−10, 10] scl: 1

The solution is $1\frac{2}{3}$.

Step 3 Use the **TABLE** feature.

Verify the solution using the **TABLE** feature. Set up the table to show *x*-values in increments of $\frac{1}{3}$.

KEYSTROKES: 2nd [TBLSET] 0 ENTER 1 ÷ 3 ENTER 2nd [TABLE]

The table displays *x*-values and corresponding *y*-values for each graph. At $x = 1\frac{2}{3}$, both functions have a *y*-value of 1.5. Thus, the solution of the equation is $1\frac{2}{3}$.

(continued on the next page)

---

1 Focus

Objective Use a graphing calculator to solve rational equations by graphing or by using the table feature.

Materials for Each Group
- TI-83/84 Plus or other graphing calculator

Teaching Tip
When students enter functions in the function editor, (**Y=** list), they should use parentheses around any numerator or denominator that is not a single number or variable.

2 Teach

Working in Cooperative Groups
Have students work in pairs so they can help each other correct keystroke errors. Have pairs complete Activities 1 and 2 and Exercises 1 and 6.

Activity 1
- When using the intersect feature in Step 2, students should press ENTER to select the graph of each function. They will then be prompted for a **Guess?**. They should move the cursor close to an estimated point of intersection before pressing ENTER a third time.

- As an alternative to using the table feature in Step 3, students can verify the solution for each function directly on the home screen. Entering **Y1** = (5/3) and **Y2** = (5/3) will both produce results of 1.5.

Activity 2

- In Activity 2, since the graph of the inequality $y > 9$ begins at the top of the standard window screen, students may want to change the window for y to $[-10, 20]$.

Practice Have students complete Exercises 2–5 and 7–9.

3 Assess

Formative Assessment

Use Exercise 2 to assess whether students comprehend how to find the intersection of the graphs of the left and right sides of a rational equation.

From Concrete to Abstract

Ask students to solve the equation $\frac{x}{2} = \frac{8}{x}$. They will find that the graphs of $y = \frac{x}{2}$ and $y = \frac{8}{x}$ intersect in two places. It is especially important in a case like this that they use the **intersect** feature correctly. In response to **Guess?**, they must first select one of the points of intersection, then repeat the **intersect** process and select the other point of intersection. The two solutions are $x = 4$ and $x = -4$.

Graphing Technology Lab
Solving Rational Equations and Inequalities *Continued*

You can use a similar procedure to solve rational inequalities using a graphing calculator.

Activity 2 Rational Inequality

Solve $\frac{3}{x} + \frac{7}{x} > 9$.

Step 1 Enter the inequalities.

Rewrite the problem as a system of inequalities.

The first inequality is $\frac{3}{x} + \frac{7}{x} > y$ or $y < \frac{3}{x} + \frac{7}{x}$. Since this inequality includes the *less than* symbol, shade below the curve. First enter the boundary and then use the arrow and ENTER keys to choose the shade below icon, ▙.

The second inequality is $y > 9$. Shade above the curve since this inequality contains *greater than*.

KEYSTROKES: Y= ◄ ◄ ENTER ENTER ENTER ► ► 3 ÷ X,T,θ,n + 7 ÷ X,T,θ,n ENTER ◄ ◄ ENTER ENTER ► ► 9

Step 2 Graph the system.

KEYSTROKES: GRAPH

[−10, 10] scl: 1 by [−10, 10] scl: 1

The solution set of the original inequality is the set of x-values of the points in the region where the shadings overlap. Using the calculator's **intersect** feature, you can conclude that the solution set is $\left\{x \,\middle|\, 0 < x < 1\frac{1}{9}\right\}$.

Step 3 Use the **TABLE** feature.

Verify using the **TABLE** feature. Set up the table to show x-values in increments of $\frac{1}{9}$.

KEYSTROKES: 2nd [TBLSET] 0 ENTER 1 ÷ 9 ENTER 2nd [TABLE]

Scroll through the table. Notice that for x-values greater than 0 and less than $1\frac{1}{9}$, Y1 > Y2. This confirms that the solution of the inequality is $\left\{x \,\middle|\, 0 < x < 1\frac{1}{9}\right\}$.

Exercises

Solve each equation or inequality. 4. $\{x \mid x \neq 1\}$ 5. ∅   7. $\left\{x \,\middle|\, x < 0 \text{ or } \frac{2}{3} < x < 1\right\}$

1. $\frac{1}{x} + \frac{1}{2} = \frac{2}{x}$ 2

2. $\frac{1}{x-4} = \frac{2}{x-2}$ 6

3. $\frac{4}{x} = \frac{6}{x^2}$ 1.5

4. $\frac{1}{1-x} = 1 - \frac{x}{x-1}$

5. $\frac{1}{x+4} = \frac{2}{x^2+3x-4} - \frac{1}{1-x}$

6. $\frac{1}{x} + \frac{1}{2x} > 5$ $\{x \mid 0 < x < 0.3\}$

7. $\frac{1}{x-1} + \frac{2}{x} < 0$

8. $1 + \frac{5}{x-1} \leq 0$ $\{x \mid -4 \leq x < 1\}$

9. $2 + \frac{1}{x-1} \geq 0$ $\{x \mid x \leq 0.5 \text{ or } x > 1\}$

Study Guide and Review

Study Guide

KeyConcepts

Rational Expressions (Lessons 8-1 and 8-2)

- Multiplying and dividing rational expressions is similar to multiplying and dividing fractions.

- To simplify complex fractions, simplify the numerator and the denominator separately, and then simplify the resulting expression.

Reciprocal and Rational Functions (Lessons 8-3 and 8-4)

- A reciprocal function is of the form $f(x) = \frac{1}{a(x)}$, where $a(x)$ is a linear function and $a(x) \neq 0$.

- A rational function is of the form $\frac{a(x)}{b(x)}$, where $a(x)$ and $b(x)$ are polynomial functions and $b(x) \neq 0$.

Direct, Joint, and Inverse Variation (Lesson 8-5)

- Direct Variation: There is a nonzero number k such that $y = kx$.

- Joint Variation: There is a nonzero number k such that $y = kxz$.

- Inverse Variation: There is a nonzero constant k such that $xy = k$ or $y = \frac{k}{x}$, where $x \neq 0$ and $y \neq 0$.

Rational Equations and Inequalities (Lesson 8-6)

- Eliminate fractions in rational equations by multiplying each side of the equation by the LCD.

- Possible solutions of a rational equation must exclude values that result in zero in the denominator.

FOLDABLES StudyOrganizer

Be sure the Key Concepts are noted in your Foldable.

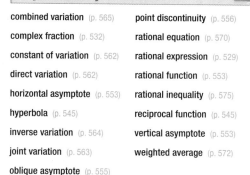

KeyVocabulary

combined variation (p. 565)	point discontinuity (p. 556)
complex fraction (p. 532)	rational equation (p. 570)
constant of variation (p. 562)	rational expression (p. 529)
direct variation (p. 562)	rational function (p. 553)
horizontal asymptote (p. 553)	rational inequality (p. 575)
hyperbola (p. 545)	reciprocal function (p. 545)
inverse variation (p. 564)	vertical asymptote (p. 553)
joint variation (p. 563)	weighted average (p. 572)
oblique asymptote (p. 555)	

VocabularyCheck

Choose a term from the list above that best completes each statement or phrase.

1. A(n) _____ is a rational expression whose numerator and/or denominator contains a rational expression.
   complex fraction

2. If two quantities show _____, their product is equal to a constant k. **inverse variation**

3. A(n) _____ asymptote is a linear asymptote that is neither horizontal nor vertical. **oblique**

4. A(n) _____ can be expressed in the form $y = kx$.
   direct variation

5. Equations that contain one or more rational expressions are called _____. **rational equations**

6. The graph of $y = \frac{x}{x+2}$ has a(n) _____ at $x = -2$.
   vertical asymptote

7. _____ occurs when one quantity varies directly as the product of two or more other quantities. **Joint variation**

8. A ratio of two polynomial expressions is called a(n) _____. **rational expression**

9. _____ looks like a hole in a graph because the graph is undefined at that point. **Point discontinuity**

10. _____ occurs when one quantity varies directly and/or inversely as two or more other quantities.
    Combined variation

 connectED.mcgraw-hill.com **581**

Formative Assessment

KeyVocabulary The page references after each word denote where that term was first introduced. If students have difficulty answering questions 1–10, remind them that they can use these page references to refresh their memories about the vocabulary.

Summative Assessment

Vocabulary Test, p. 50

Vocabulary Review

Vocabulary Review provides students the opportunity to check their understanding of important concepts and terminology in an online game format.

FOLDABLES StudyOrganizer

Dinah Zike's Foldables®

Have students look through the chapter to make sure they have included examples in their Foldables. Suggest that students keep their Foldables handy while completing the Study Guide and Review pages. Point out that their Foldables can serve as a quick review tool when studying for the chapter test.

Lesson-by-Lesson Review

Intervention If the given examples are not sufficient to review the topics covered by the questions, remind students that the lesson references tell them where to review that topic in their textbook.

Two-Day Option Have students complete the Lesson-by-Lesson Review. Then you can use McGraw-Hill eAssessment to customize another review worksheet that practices all the objectives of this chapter or only the objectives on which your students need more help.

Additional Answers

24.

$D = \{x \mid x \neq 0\}$,
$R = \{f(x) \mid f(x) \neq 0\}$

25.

$D = \{x \mid x \neq 0\}$,
$R = \{f(x) \mid f(x) \neq 2\}$

Study Guide and Review *Continued*

Lesson-by-Lesson Review

8-1 Multiplying and Dividing Rational Expressions

Simplify each expression.

11. $\dfrac{-16xy}{27z} \cdot \dfrac{15z^3}{8x^2}$ $-\dfrac{10yz^2}{9x}$

12. $\dfrac{x^2 - 2x - 8}{x^2 + x - 12} \cdot \dfrac{x^2 + 2x - 15}{x^2 + 7x + 10}$ $\dfrac{x-4}{x+4}$

13. $\dfrac{x^2 - 1}{x^2 - 4} \cdot \dfrac{x^2 - 5x - 14}{x^2 - 6x - 7}$ $\dfrac{x-1}{x-2}$

14. $\dfrac{x+y}{15x} \div \dfrac{x^2 - y^2}{3x^2}$ $\dfrac{x}{5(x-y)}$

15. $\dfrac{\frac{x^2 + 3x - 18}{x+4}}{\frac{x^2 + 7x + 6}{x+4}}$ $\dfrac{x-3}{x+1}$

16. GEOMETRY A triangle has an area of $3x^2 + 9x - 54$ square centimeters. If the height of the triangle is $x + 6$ centimeters, find the length of the base. $6x - 18$ cm

Example 1

Simplify $\dfrac{4a}{3b} \cdot \dfrac{9b^4}{2a^2}$.

$\dfrac{4a}{3b} \cdot \dfrac{9b^4}{2a^2} = \dfrac{2 \cdot 2 \cdot a \cdot 3 \cdot 3 \cdot b \cdot b \cdot b \cdot b}{3 \cdot b \cdot 2 \cdot a \cdot a}$

$= \dfrac{6b^3}{a}$

Example 2

Simplify $\dfrac{r^2 + 5r}{2r} \div \dfrac{r^2 - 25}{6r - 12}$.

$\dfrac{r^2 + 5r}{2r} \div \dfrac{r^2 - 25}{6r - 12} = \dfrac{r^2 + 5r}{2r} \cdot \dfrac{6r - 12}{r^2 - 25}$

$= \dfrac{r(r+5)}{2r} \cdot \dfrac{6(r-2)}{(r+5)(r-5)}$

$= \dfrac{3(r-2)}{r-5}$

8-2 Adding and Subtracting Rational Expressions

Simplify each expression.

17. $\dfrac{9}{4ab} + \dfrac{5a}{6b^2}$ $\dfrac{27b + 10a^2}{12ab^2}$

18. $\dfrac{3}{4x - 8} - \dfrac{x-1}{x^2 - 4}$ $\dfrac{-x + 10}{4(x-2)(x+2)}$

19. $\dfrac{y}{2x} + \dfrac{4y}{3x^2} - \dfrac{5}{6xy^2}$ $\dfrac{3xy^3 + 8y^3 - 5x}{6x^2y^2}$

20. $\dfrac{2}{x^2 - 3x - 10} - \dfrac{6}{x^2 - 8x + 15}$ $\dfrac{-4x - 18}{(x-5)(x+2)(x-3)}$

21. $\dfrac{3}{3x^2 + 2x - 8} + \dfrac{4x}{2x^2 + 6x + 4}$ $\dfrac{12x^2 - 10x + 6}{2(x+2)(3x-4)(x+1)}$

22. $\dfrac{\frac{3}{2x+3} - \frac{x}{x+1}}{\frac{2x}{x+1} + \frac{5}{2x+3}}$ $\dfrac{-2x^2 + 3}{4x^2 + 11x + 5}$

23. GEOMETRY What is the perimeter of the rectangle?

Example 3

Simplify $\dfrac{3a}{a^2 - 4} - \dfrac{2}{a - 2}$.

$\dfrac{3a}{a^2 - 4} - \dfrac{2}{a - 2} = \dfrac{3a}{(a-2)(a+2)} - \dfrac{2}{a-2}$

$= \dfrac{3a}{(a-2)(a+2)} - \dfrac{2(a+2)}{(a-2)(a+2)}$

$= \dfrac{3a - 2(a+2)}{(a-2)(a+2)}$ Subtract numerators.

$= \dfrac{3a - 2a - 4}{(a-2)(a+2)}$ Distributive Property

$= \dfrac{a - 4}{(a-2)(a+2)}$ Simplify.

582 | Chapter 8 | Study Guide and Review

26.

$D = \{x \mid x \neq -5\}$, $R = \{f(x) \mid f(x) \neq 0\}$

27.

$D = \{x \mid x \neq 9\}$, $R = \{f(x) \mid f(x) \neq 0\}$

8-3 Graphing Reciprocal Functions

Graph each function. State the domain and range.

24. $f(x) = \dfrac{10}{x}$

25. $f(x) = -\dfrac{12}{x} + 2$

26. $f(x) = \dfrac{3}{x + 5}$

27. $f(x) = \dfrac{6}{x - 9}$

28. $f(x) = \dfrac{7}{x - 2} + 3$

29. $f(x) = -\dfrac{4}{x + 4} - 8$

24–29. See margin.

30. CONSERVATION The student council is planting 28 trees for a service project. The number of trees each person plants depends on the number of student council members.

a. Write a function to represent this situation. $f(x) = \dfrac{28}{x}$

b. Graph the function. **See margin.**

Example 4

Graph $f(x) = \dfrac{3}{x + 2} - 1$. State the domain and range.

$a = 3$:   The graph is stretched vertically.

$h = -2$:   The graph is translated 2 units left. There is an asymptote at $x = -2$.

$k: = -1$:   The graph is translated 1 unit down. There is an asymptote at $f(x) = -1$.

Domain: $\{x \mid x \ne -2\}$,
Range: $\{f(x) \mid f(x) \ne -1\}$

8-4 Graphing Rational Functions

Determine the equations of any vertical asymptotes and the values of x for any holes in the graph of each rational function.

31. $f(x) = \dfrac{3}{x^2 + 4x}$ $x = -4$, $x = 0$

32. $f(x) = \dfrac{x + 2}{x^2 + 6x + 8}$ $x = -4$; hole: $x = -2$

33. $f(x) = \dfrac{x^2 - 9}{x^2 - 5x - 24}$ $x = 8$; hole: $x = -3$

Graph each rational function. **34–37. See margin.**

34. $f(x) = \dfrac{x + 2}{(x + 5)^2}$

35. $f(x) = \dfrac{x}{x + 1}$

36. $f(x) = \dfrac{x^2 + 4x + 4}{x + 2}$

37. $f(x) = \dfrac{x - 1}{x^2 + 5x + 6}$

38. SALES Aliyah is selling magazine subscriptions. Out of the first 15 houses, she sold subscriptions to 10 of them. Suppose Aliyah goes to x more houses and sells subscriptions to all of them. The percentage of houses that she sold to out of the total houses can be determined using $P(x) = \dfrac{10 + x}{15 + x}$.

a. Graph the function. **See margin of p. 584.**

b. What domain and range values are meaningful in the context of the problem?
$D = \{x \ge 0\}$, $R = \{0 \le P(x) \le 1.0\}$

Example 5

Determine the equation of any vertical asymptotes and the values of x for any holes in the graph of $f(x) = \dfrac{x^2 - 1}{x^2 + 2x - 3}$.

$\dfrac{x^2 - 1}{x^2 + 2x - 3} = \dfrac{(x - 1)(x + 1)}{(x - 1)(x + 3)}$

The function is undefined for $x = 1$ and $x = -3$.

Since $\dfrac{(x - 1)(x + 1)}{(x - 1)(x + 3)} = \dfrac{x + 1}{x + 3}$, $x = -3$ is a vertical asymptote, and $x = 1$ represents a hole in the graph.

Example 6

Graph $f(x) = \dfrac{1}{6x(x - 1)}$.

The function is undefined for $x = 0$ and $x = 1$. Because $\dfrac{1}{6x(x - 1)}$ is in simplest form, $x = 0$ and $x = 1$ are vertical asymptotes. Draw the two asymptotes and sketch the graph.

28.

$D = \{x \mid x \ne 2\}$,
$R = \{f(x) \mid f(x) \ne 3\}$

29.

$D = \{x \mid x \ne -4\}$,
$R = \{f(x) \mid f(x) \ne -8\}$

30b.

34.

35.

36.

37.

Anticipation Guide

Have students complete the Chapter 8 Anticipation Guide and discuss how their responses have changed now that they have completed Chapter 8.

Additional Answer

38a.

8-5 Variation Functions

39. If a varies directly as b and $b = 18$ when $a = 27$, find a when $b = 10$. **$a = 15$**

40. If y varies inversely as x and $y = 15$ when $x = 3.5$, find y when $x = -5$. **$y = -10.5$**

41. If y varies inversely as x and $y = -3$ when $x = 9$, find y when $x = 81$. **$y = -\dfrac{1}{3}$**

42. If y varies jointly as x and z, and $x = 8$ and $z = 3$ when $y = 72$, find y when $x = -2$ and $z = -5$. **$y = 30$**

43. If y varies jointly as x and z, and $y = 18$ when $x = 6$ and $z = 15$, find y when $x = 12$ and $z = 4$. **$y = \dfrac{48}{5}$**

44. JOBS Lisa's earnings vary directly with how many hours she babysits. If she earns $68 for 8 hours of babysitting, find her earnings after 5 hours of babysitting. **$42.50**

Example 7

If y varies inversely as x and $x = 24$ when $y = -8$, find x when $y = 15$.

$$\frac{x_1}{y_2} = \frac{x_2}{y_1} \qquad \text{Inverse variation}$$

$$\frac{24}{15} = \frac{x_2}{-8} \qquad x_1 = 24,\ y_1 = -8,\ y_2 = 15$$

$$24(-8) = 15(x_2) \qquad \text{Cross multiply.}$$

$$-192 = 15x_2 \qquad \text{Simplify.}$$

$$-12\frac{4}{5} = x_2 \qquad \text{Divide each side by 15.}$$

When $y = 15$, the value of x is $-12\frac{4}{5}$.

8-6 Solving Rational Equations and Inequalities

Solve each equation or inequality. Check your solutions.

45. $\dfrac{1}{3} + \dfrac{4}{x-2} = 6$ **$x = \dfrac{46}{17}$**

46. $\dfrac{6}{x+5} - \dfrac{3}{x-3} = \dfrac{6}{x^2 + 2x - 15}$ **$x = 13$**

47. $\dfrac{2}{x^2 - 9} = \dfrac{3}{x^2 - 2x - 3}$ **$x = -7$**

48. $\dfrac{4}{2x-3} + \dfrac{x}{x+1} = \dfrac{-8x}{2x^2 - x - 3}$ **$x = -\dfrac{1}{2}, -4$**

49. $\dfrac{x}{x+4} - \dfrac{28}{x^2 + x - 12} = \dfrac{1}{x-3}$ **$x = 8$**

50. $\dfrac{x}{2} + \dfrac{1}{x-1} < \dfrac{x}{4}$ **$x < 1$**

51. $\dfrac{1}{2x} - \dfrac{4}{5x} > \dfrac{1}{3}$ **$-\dfrac{9}{10} < x < 0$**

52. YARD WORK Lana can plant a garden in 3 hours. Milo can plant the same garden in 4 hours. How long will it take them if they work together? **$1\dfrac{5}{7}$ h**

Example 8

Solve $\dfrac{3}{x+2} + \dfrac{1}{x} = 0$.

The LCD is $x(x + 2)$.

$$\frac{3}{x+2} + \frac{1}{x} = 0$$

$$x(x+2)\left(\frac{3}{x+2} + \frac{1}{x}\right) = x(x+2)(0)$$

$$x(x+2)\left(\frac{3}{x+2}\right) + x(x+2)\left(\frac{1}{x}\right) = 0$$

$$3(x) + 1(x+2) = 0$$

$$3x + x + 2 = 0$$

$$4x + 2 = 0$$

$$4x = -2$$

$$x = -\frac{1}{2}$$

CHAPTER 8 Practice Test

Simplify each expression. 5–7. See margin.

1. $\frac{r^2 + rt}{2r} \div \frac{r + t}{16r^2}$ $8r^2$

2. $\frac{m^2 - 4}{3m^2} \cdot \frac{6m}{2 - m}$ $-\frac{2(m + 2)}{m}$

3. $\frac{m^2 + m - 6}{n^2 - 9} \div \frac{m - 2}{n + 3}$ $\frac{m + 3}{n - 3}$

4. $\frac{\frac{x^2 + 4x + 3}{x^2 - 2x - 15}}{\frac{x^2 - 1}{x^2 - x - 20}}$ $\frac{x + 4}{x - 1}$

5. $\frac{x + 4}{6x + 3} + \frac{1}{2x + 1}$

6. $\frac{x}{x^2 - 1} - \frac{3}{2x + 2}$

7. $\frac{1}{y} + \frac{2}{7} - \frac{3}{2y^2}$

8. $\frac{2 + \frac{1}{x}}{5 - \frac{1}{x}}$ $\frac{2x + 1}{5x - 1}$

9. Identify the asymptotes, domain, and range of the function graphed. **See margin.**

$f(x) = \frac{6}{x + 2} - 5$

10. **MULTIPLE CHOICE** What is the equation for the vertical asymptote of the rational function $f(x) = \frac{x + 1}{x^2 + 3x + 2}$? **A**

A $x = -2$

B $x = -1$

C $x = 1$

D $x = 2$

11–16. See Chapter 8 Answer Appendix.

Graph each function.

11. $f(x) = -\frac{8}{x} - 9$

12. $f(x) = \frac{2}{x + 4}$

13. $f(x) = \frac{3}{x - 1} + 8$

14. $f(x) = \frac{5x}{x + 1}$

15. $f(x) = \frac{x}{x - 5}$

16. $f(x) = \frac{x^2 + 5x - 6}{x - 1}$

17. Determine the equations of any vertical asymptotes and the values of x for any holes in the graph of the function $f(x) = \frac{x + 5}{x^2 - 2x - 35}$. **vertical asymptote:** $x = 7$; hole: $x = -5$

18. Determine the equations of any oblique asymptotes in the graph of the function $f(x) = \frac{x^2 + x - 5}{x + 3}$. $f(x) = x - 2$

Solve each equation or inequality.

19. $\frac{-1}{x + 4} = 6 - \frac{x}{x + 4}$ $x = -5$

20. $\frac{1}{3} = \frac{5}{m + 3} + \frac{8}{21}$ $m = -108$

21. $7 + \frac{2}{x} < -\frac{5}{x}$ $-1 < x < 0$

22. $r + \frac{6}{r} - 5 = 0$ $r = 2, 3$

23. $\frac{6}{7} - \frac{3m}{2m - 1} = \frac{11}{7}$ $m = \frac{5}{31}$

24. $\frac{r + 2}{3r} = \frac{r + 4}{r - 2} - \frac{2}{3}$ $r = -\frac{1}{4}$

25. If y varies inversely as x and $y = 18$ when $x = -\frac{1}{2}$, find x when $y = -10$. $\frac{9}{10}$

26. If m varies directly as n and $m = 24$ when $n = -3$, find n when $m = 30$. $-\frac{15}{4}$

27. Suppose r varies jointly as s and t. If $s = 20$ when $r = 140$ and $t = -5$, find s when $r = 7$ and $t = 2.5$. -2

28. **BICYCLING** When Susan rides her bike, the distance that she travels varies directly with the amount of time she is biking. Suppose she bikes 50 miles in 2.5 hours. At this rate, how many hours would it take her to bike 80 miles? **4 hours**

29. **PAINTING** Peter can paint a house in 10 hours. Melanie can paint the same house in 9 hours. How long would it take if they worked together? **about 4.7 hours**

30. **MULTIPLE CHOICE** How many liters of a 25% acid solution must be added to 30 liters of an 80% acid solution to create a 50% acid solution? **H**

F 18

G 30

H 36

J 66

31. What is the volume of the rectangular prism?

$\frac{1}{x + 1}$ **cubic units**

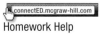

Summative Assessment

Use these alternate leveled chapter tests to differentiate assessment for your students.

Leveled Chapter 8 Tests

Form	Type	Level	🗀 Page(s)
1	MC	AL	51–52
2A	MC	OL	53–54
2B	MC	OL	55–56
2C	FR	OL	57–58
2D	FR	OL	59–60
3	FR	BL	61–62
Vocabulary Test			50
Extended-Response Test			63

MC = multiple-choice questions
FR = free-response questions

McGraw-Hill eAssessment Customize and create multiple versions of your chapter test and their answer keys. All of the questions from the leveled chapter tests in the *Chapter 8 Resource Masters* are also available on McGraw-Hill eAssessment.

Additional Answers

5. $\frac{x + 7}{3(2x + 1)}$

6. $\frac{-x + 3}{2(x - 1)(x + 1)}$

7. $\frac{4y^2 + 14y - 21}{14y^2}$

9. $x = -2$; $f(x) = -5$;
   $D = \{x \mid x \neq -2\}$,
   $R = \{f(x) \mid f(x) \neq -5\}$

InterventionPlanner

TIER 1 On Level OL

If students miss about 25% of the exercises or less,

Then choose a resource:

SE Lessons 8-1, 8-2, 8-3, 8-4, 8-5, and 8-6

🗀 Skills Practice, pp. 7, 13, 20, 27, 34, and 41

TE Chapter Project

connectED.mcgraw-hill.com Self-Check Quiz

TIER 2 Strategic Intervention AL
approaching grade level

If students miss about 50% of the exercises,

Then choose a resource:

🗀 Study Guide and Intervention, pp. 5, 11, 18, 25, 32, and 39

connectED.mcgraw-hill.com Extra Examples, Personal Tutor, Homework Help

TIER 3 Intensive Intervention
2 or more grades below level

If students miss about 75% of the exercises,

Then use *Math Triumphs, Alg. 2*, Ch. 2

connectED.mcgraw-hill.com Extra Examples, Personal Tutor, Homework Help, Review Vocabulary

1 Focus

Objective Use the strategy of guess and check to solve standardized test problems.

2 Teach

Scaffolding Questions

Ask:

- Why is guess and check a good strategy to use in standardized tests? Sample answer: You know that the correct answer is among the choices you are given. Once you make a reasonable guess, you can check it against the information in the problem to determine if it is correct or if you should go on to the next reasonable guess. Then you can be confident in your answer.

- When should you use guess and check strategy? Sample answer: You can use guess and check at any time in a standardized test. It is most helpful when the answer choices can replace information given in a problem, for example, when choices are values that can replace a variable in an expression or an equation.

Preparing for Standardized Tests

Guess and Check

It is very important to pace yourself and keep track of how much time you have when taking a standardized test. If time is running short, or if you are unsure how to solve a problem, the guess-and-check strategy may help you determine the correct answer quickly.

Strategies for Guessing and Checking

Step 1

Carefully look over each possible answer choice and evaluate for reasonableness. Eliminate unreasonable answers.

Ask yourself:

- Are there any answer choices that are clearly incorrect?
- Are there any answer choices that are not in the proper format?
- Are there any answer choices that do not have the proper units for the correct answer?

Step 2

For the remaining answer choices, use the guess-and-check method.

- **Equations:** If you are solving an equation, substitute the answer choice for the variable and see if this results in a true number sentence.
- **System of Equations:** For a system of equations, substitute the answer choice for all variables and make sure all equations result in a true number sentence.

Step 3

Choose an answer choice and see if it satisfies the constraints of the problem statement. Identify the correct answer.

- If the answer choice you are testing does not satisfy the problem, move on to the next reasonable guess and check it.
- When you find the correct answer choice, stop.

Standardized Test Example

Read the problem. Identify what you need to know. Then use the information in the problem to solve.

Solve: $\dfrac{2}{x-3} - \dfrac{4}{x+3} = \dfrac{8}{x^2-9}$.

A -1	**C** 5
B 1	**D** 7

Radius Images/Corbis

The solution of the rational equation will be a real number. Since all four answer choices are real numbers, they are all possible correct answers and must be checked. Begin with the first answer choice and check it in the rational equation. Continue until you find the answer choice that results in a true number sentence.

Guess: −1	**Check:**
	$\dfrac{2}{(-1)-3} - \dfrac{4}{(-1)+3} = \dfrac{8}{(-1)^2 - 9}$
	$-\dfrac{5}{2} \neq -1$ ✗

Guess: 1	**Check:**
	$\dfrac{2}{1-3} - \dfrac{4}{1+3} = \dfrac{8}{(1)^2 - 9}$
	$-2 \neq -1$ ✗

Guess: 5	**Check:**
	$\dfrac{2}{5-3} - \dfrac{4}{5+3} = \dfrac{8}{(5)^2 - 9}$
	$\dfrac{1}{2} = \dfrac{1}{2}$ ✓

If $x = 5$, the result is a true number sentence. So, the correct answer is C.

Exercises

Read each problem. Identify what you need to know. Then use the information in the problem to solve.

1. Solve: $\dfrac{2}{5x} - \dfrac{1}{2x} = -\dfrac{1}{2}$. **B**

   A $\dfrac{1}{10}$

   B $\dfrac{1}{5}$

   C $\dfrac{1}{4}$

   D $\dfrac{1}{2}$

2. The sum of Kevin's, Anna's, and Tia's ages is 40. Anna is 1 year more than twice as old as Tia. Kevin is 3 years older than Anna. How old is Anna? **H**

   F 7   **H** 15

   G 14   **J** 18

3. Determine the point(s) where the following rational function crosses the x-axis. **D**

$$f(x) = \dfrac{2}{x-1} - \dfrac{x+4}{3}$$

   A −5

   B 4

   C 2 or 3

   D −5 or 2

4. Rafael's Theatre Company sells tickets for $10. At this price, they sell 400 tickets. Rafael estimates that they would sell 40 fewer tickets for each $2 price increase. What charge would give the most income? **H**

   F 10   **H** 15

   G 13   **J** 20

Additional Example

STANDARDIZED TEST PRACTICE
Suppose y varies jointly as x and z. Find y when $x = 12$ and $z = 7$, if $y = 40$ when $z = 15$ and $x = 28$.
A

A 8

B 20

C 80

D 200

3 Assess

Use Exercises 1–4 to assess students' understanding.

Diagnose Student Errors
Survey student responses for each item. Class trends may indicate common errors and misconceptions.

1. A incorrectly wrote 1 hour 45 minutes as 1.45
   B mistakenly added denominators
   C correct
   D guess

2. F found the correct difference but added it when should have multiplied by it
   G correct
   H divided by the correct difference instead of multiplying by it
   J used multiplication but not the correct difference

3. A correct
   B incorrectly interpreted 2 as an *x*-intercept
   C did not use + sign for direction of shift
   D incorrectly interpreted 2 as an *x*-intercept

4. F ignored the word *not* in the question
   G ignored the word *not* in the question
   H ignored the word *not* in the question
   J correct

5. A did not factor $x^2 - 16$ correctly
   B simplified the fraction incorrectly
   C correct
   D mistakenly inverted the simplified fraction

6. F misinterpreted the graph
   G correct
   H misinterpreted the graph
   J misinterpreted the graph

7. A correct
   B in this function when *x* is negative, *y* is negative
   C in this function when *x* is negative, *y* is undefined
   D in this function when *x* is negative, *y* is negative

Standardized Test Practice
Cumulative, Chapters 1 through 8

Multiple Choice

Read each question. Then fill in the correct answer on the answer document provided by your teacher or on a sheet of paper.

1. Greg's father can mow the lawn on his riding mower in 45 minutes. It takes Greg 1 hour 45 minutes to mow the lawn with a push mower. Which of the following rational equations can be solved for the number of minutes *t* it would take them to mow the lawn working together? **C**

   A $\frac{t}{45} + \frac{t}{1.45} = 1$ C $\frac{t}{45} + \frac{t}{105} = 1$

   B $\frac{t}{150} = 1$ D $\frac{t + 45}{t + 105} = 1$

2. The total cost of reserving a campsite varies directly as the number of nights the site is rented, as shown in the table.

Days	Total Cost
1	$24
2	$48
3	$72
4	$96

Which equation represents the direct variation? **G**

   F $y = x + 24$ H $y = \frac{24}{x}$

   G $y = 24x$ J $y = 96x$

3. In which direction must the graph of $y = \frac{1}{x}$ be shifted to produce the graph of $y = \frac{1}{x} + 2$? **A**

   A up

   B down

   C right

   D left

> **Test-TakingTip**
> Question 2 Check your answer by substituting 1, 2, 3, and 4 for *x* and making sure the values in the table are produced.

---

4. Which of the following is *not* an asymptote of the rational function $f(x) = \frac{1}{x^2 - 49}$? **J**

   F $f(x) = 0$ H $x = 7$

   G $x = -7$ J $f(x) = 1$

5. Simplify the complex fraction. **C**

$$\frac{\frac{(x + 3)^2}{x^2 - 16}}{\frac{x + 3}{x + 4}}$$

   A $\frac{x + 3}{x + 4}$ C $\frac{x + 3}{x - 4}$

   B $\frac{1}{x - 4}$ D $\frac{x - 4}{x + 3}$

6. A ball was thrown upward with an initial velocity of 16 feet per second from the top of a building 128 feet high. Its height *h* in feet above the ground *t* seconds later will be $h = 128 + 16t - 16t^2$.

Which is the best conclusion about the ball's action? **G**

   F The ball stayed above 128 feet for more than 3 seconds.

   G The ball returned to the ground in less than 4 seconds.

   H The ball traveled a greater distance going up than it did going down.

   J The ball traveled less than 128 feet in 3.4 seconds.

7. Which of these equations describes a relationship in which every negative real number *x* corresponds to a nonnegative real number *y*? **A**

   A $y = -x$ C $y = \sqrt{x}$

   B $y = x$ D $y = x^3$

Short Response/Gridded Response

Record your answers on the answer sheet provided by your teacher or on a sheet of paper.

8. Martha is putting a stone walkway around the garden pictured at the right. About how many feet of stone are needed? **56.8 ft**

16 ft
10 ft
20 ft

9. **GRIDDED RESPONSE** Elaine had some money saved for a week-long vacation. The first day of the vacation she spent $125 on food and a hotel. On the second day, she was given $80 from her sister for expenses. Elaine then had $635 left for the rest of the vacation. How much money, in dollars, did she begin the vacation with? **680**

10. Carlos wants to print 800 one-page flyers for his landscaping business. He has a printer that is capable of printing 8 pages per minute. His business partner has another printer that prints 10 pages per minute.

    a. How long would it take Carlos' printer to print all the flyers? How long would it take his partner's printer? **100 min; 80 min**

    b. Set up a rational equation that can be used to find the number of minutes t it would take to print all 800 flyers if both printers are used simultaneously.

    c. Solve the equation you wrote in part **b**. How long would it take both printers to print all the flyers if they print simultaneously? Round to the nearest minute. **about 44 min**

11. **GRIDDED RESPONSE** The population of a country can be modeled by the equation $P(t) = 40e^{0.02t}$, where P is the population in millions and t is the number of years since 2005. When will the population be 400 million? **2120**

    10b. $\dfrac{t}{100} + \dfrac{t}{80} = 1$

12. What is the area of the shaded region of the rectangle expressed as a polynomial in simplest form? $3x^2 - 14x + 8$

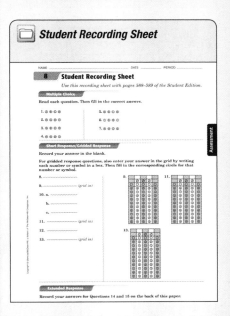
3x
2
x
4

13. **GRIDDED RESPONSE** Suppose y varies inversely as x and $y = 4$ when $x = 12$. What is y when x is 5? Round to the nearest tenth. **9.6**

Extended Response

Record your answers on a sheet of paper. Show your work. **14a, b. See margin.**

14. Use the graph of the rational function at the right to answer each question.

    a. Describe the vertical and horizontal asymptotes of the graph.

    b. Write the equation of the rational function. Explain how you found your answer.

15. Consider the polynomial function
    $$f(x) = 3x^4 + 19x^3 + 7x^2 - 11x - 2.$$

    a. What is the degree of the function? **4**

    b. What is the leading coefficient of the function? **3**

    c. Evaluate $f(1)$, $f(-2)$, and $f(2a)$. Show your work.
    16; -56; $48a^4 + 152a^3 + 28a^2 - 22a - 2$

Need ExtraHelp?

If you missed Question...	1	2	3	4	5	6	7	8	9	10	11	12	13	14	15
Go to Lesson...	8-5	8-4	8-3	8-3	8-1	8-7	2-4	1-4	3-1	8-5	7-8	5-1	8-4	8-3	6-1

Additional Answers

14a. Sample answer: vertical asymptotes at $x = -4$ and $x = 4$; horizontal asymptote at $y = 0$

14b. $y = \dfrac{1}{x^2 - 16}$

Formative Assessment

You can use these pages to benchmark student progress.

📁 Standardized Test Practice, pp. 64–66

McGraw-Hill
eAssessment Create practice tests that align to your state standards, the Common Core State Standards, and other national standards such as TIMSS and NAEP.

Answer Sheet Practice

Have students simulate taking a standardized test by recording their answers on practice recording sheets.

Student Recording Sheet

Homework Option

Get Ready for Chapter 9 Assign students the exercises on p. 591 as homework to assess whether they possess the prerequisite skills needed for the next chapter.

Lesson 8-2

85.

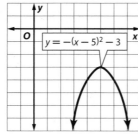

$y = 4(x + 3)^2 + 1$

86.

$y = -(x - 5)^2 - 3$

87.

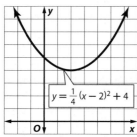

$y = \frac{1}{4}(x - 2)^2 + 4$

88.

$y = \frac{1}{2}(x - 3)^2 - 5$

89.

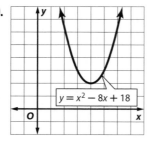

$y = x^2 + 6x + 2$

90.

$y = x^2 - 8x + 18$

Lesson 8-3 (Guided Practice)

3A.

$f(x) = \frac{-2}{x + 4} + 1$

$D = \{x \mid x \neq -4\}; R = \{f(x) \mid f(x) \neq 1\}$

3B.

$g(x) = \frac{1}{3(x - 1)} - 2$

$D = \{x \mid x \neq 1\}; R = \{g(x) \mid g(x) \neq -2\}$

4. $T = \dfrac{2500 + 45p}{p}$

$T = \dfrac{2500 + 45p}{p}$

The domain will never reach zero because someone has to be at the dance. The range will never reach zero because, even if there are thousands of people there, they will incur some cost. Neither value is negative because there cannot be a negative cost or a negative number of people.

Lesson 8-3

11.

$f(x) = \frac{3}{x}$

$D = \{x \mid x \neq 0\};$
$R = \{f(x) \mid f(x) \neq 0\}$

12.

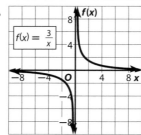

$f(x) = \frac{-4}{x + 2}$

$D = \{x \mid x \neq -2\};$
$R = \{f(x) \mid f(x) \neq 0\}$

13.

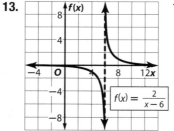

$f(x) = \frac{2}{x - 6}$

$D = \{x \mid x \neq 6\};$
$R = \{f(x) \mid f(x) \neq 0\}$

14.

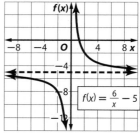

$f(x) = \frac{6}{x} - 5$

$D = \{x \mid x \neq 0\};$
$R = \{f(x) \mid f(x) \neq -5\}$

15.

$f(x) = \frac{2}{x} + 3$

$D = \{x \mid x \neq 0\};$
$R = \{f(x) \mid f(x) \neq 3\}$

16.

$f(x) = \frac{8}{x}$

$D = \{x \mid x \neq 0\};$
$R = \{f(x) \mid f(x) \neq 0\}$

17.

$f(x) = \dfrac{-2}{x-5}$

$D = \{x \mid x \neq 5\}$;
$R = \{f(x) \mid f(x) \neq 0\}$

18.

$f(x) = \dfrac{3}{x-7} - 8$

$D = \{x \mid x \neq 7\}$;
$R = \{f(x) \mid f(x) \neq -8\}$

19.

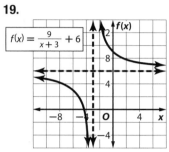

$f(x) = \dfrac{9}{x+3} + 6$

$D = \{x \mid x \neq -3\}$;
$R = \{f(x) \mid f(x) \neq 6\}$

20.

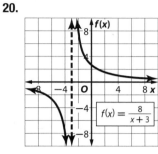

$f(x) = \dfrac{8}{x+3}$

$D = \{x \mid x \neq -3\}$;
$R = \{f(x) \mid f(x) \neq 0\}$

21.

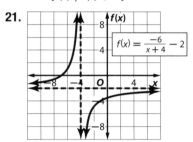

$f(x) = \dfrac{-6}{x+4} - 2$

$D = \{x \mid x \neq -4\}$; $R = \{f(x) \mid f(x) \neq -2\}$

22.

$f(x) = \dfrac{-5}{x-2} + 2$

$D = \{x \mid x \neq 2\}$; $R = \{f(x) \mid f(x) \neq 2\}$

23b.

$m = \dfrac{5000}{d}$

24b.

$d = \dfrac{200}{v}$

25.

$f(x) = \dfrac{3}{2x-4}$

$D = \{x \mid x \neq 2\}$;
$R = \{f(x) \mid f(x) \neq 0\}$

26.

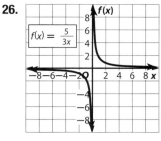

$f(x) = \dfrac{5}{3x}$

$D = \{x \mid x \neq 0\}$;
$R = \{f(x) \mid f(x) \neq 0\}$

27.

$f(x) = \dfrac{2}{4x+1}$

$D = \left\{x \mid x \neq -\dfrac{1}{4}\right\}$;

$R = \{f(x) \mid f(x) \neq 0\}$

28.

$f(x) = \dfrac{1}{2x+3}$

$D = \left\{x \mid x \neq -\dfrac{3}{2}\right\}$;

$R = \{f(x) \mid f(x) \neq 0\}$

29b.

$r = \dfrac{60.5}{t}$

30.

$f(x) = \dfrac{-3}{x+7} - 1$

$D = \{x \mid x \neq -7\}$; $R = \{f(x) \mid f(x) \neq -1\}$; $x = -7$, $f(x) = -1$

31.

$f(x) = \dfrac{-4}{x+2} - 5$

$D = \{x \mid x \neq -2\}$; $R = \{f(x) \mid f(x) \neq -5\}$; $x = -2$, $f(x) = -5$

32.

$$f(x) = \frac{6}{x-1} + 2$$

D = {$x \mid x \neq 1$}; R = {$f(x) \mid f(x) \neq 2$}; $x = 1$, $f(x) = 2$

33.

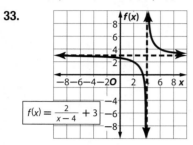

$$f(x) = \frac{2}{x-4} + 3$$

D = {$x \mid x \neq 4$}; R = {$f(x) \mid f(x) \neq 3$}; $x = 4$, $f(x) = 3$

34.

$$f(x) = \frac{-7}{x-8} - 9$$

D = {$x \mid x \neq 8$}; R = {$f(x) \mid f(x) \neq -9$}; $x = 8$, $f(x) = -9$

35.

$$f(x) = \frac{-6}{x-7} - 8$$

D = {$x \mid x \neq 7$}; R = {$f(x) \mid f(x) \neq -8$}; $x = 7$, $f(x) = -8$

36b.

$$g = \frac{440}{t}$$

37a.

x	$a(x) = x^2$	$b(x) = x^{-2}$	$c(x) = x^3$	$d(x) = x^{-3}$
-4	16	$\frac{1}{16}$	-64	$-\frac{1}{64}$
-3	9	$\frac{1}{9}$	-27	$-\frac{1}{27}$
-2	4	$\frac{1}{4}$	-8	$-\frac{1}{8}$
-1	1	1	-1	-1
0	0	undefined	0	undefined
1	1	1	1	1
2	4	$\frac{1}{4}$	8	$\frac{1}{8}$
3	9	$\frac{1}{9}$	27	$\frac{1}{27}$
4	16	$\frac{1}{16}$	64	$\frac{1}{64}$

37b.

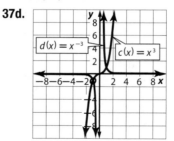

37c. $a(x)$: D = {all real numbers}, R = {$a(x) \mid a(x) \geq 0$}; as $x \rightarrow -\infty$, $a(x) \rightarrow \infty$, as $x \rightarrow \infty$, $a(x) \rightarrow \infty$; At $x = 0$, $a(x) = 0$, so there is a zero at $x = 0$.

$b(x)$: D = {$x \mid x \neq 0$}, R {$b(x) \mid b(x) > 0$}; as $x \rightarrow -\infty$, $a(x) \rightarrow 0$, as $x \rightarrow \infty$, $a(x) \rightarrow 0$; At $x = 0$, $b(x)$ is undefined, so there is an asymptote at $x = 0$.

37d.

(graph)

37e. $c(x)$: D = {all real numbers}, R = {all real numbers}; as $x \rightarrow -\infty$, $a(x) \rightarrow -\infty$, as $x \rightarrow \infty$, $a(x) \rightarrow \infty$; At $x = 0$, $a(x) = 0$, so there is a zero at $x = 0$.

$d(x)$: D = {$x \mid x \neq 0$}, R = {$b(x) \mid b(x) \neq 0$}; as $x \rightarrow -\infty$, $a(x) \rightarrow 0$, as $x \rightarrow \infty$, $a(x) \rightarrow 0$; At $x = 0$, $b(x)$ is undefined, so there is an asymptote at $x = 0$.

37f. For two power functions $f(x) = ax^n$ and $g(x) = ax^{-n}$, for every x, $f(x)$ and $g(x)$ are reciprocals. The domains are similar except that for $g(x)$, $x \neq 0$. Additionally, wherever $f(x)$ has a zero, $g(x)$ is undefined.

39a. The first graph has a vertical asymptote at $x = 0$ and a horizontal asymptote at $y = 0$. The second graph is translated 7 units up and has a vertical asymptote at $x = 0$ and a horizontal asymptote at $y = 7$.

39b. Both graphs have a vertical asymptote at $x = 0$ and a horizontal asymptote at $y = 0$. The second graph is stretched by a factor of 4.

39c. The first graph has a vertical asymptote at $x = 0$ and a horizontal asymptote at $y = 0$. The second graph is translated 5 units to the left and has a vertical asymptote at $x = -5$ and a horizontal asymptote at $y = 0$.

39d.

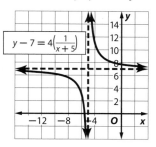

$y - 7 = 4\left(\dfrac{1}{x+5}\right)$

41. Sample answer = $f(x) = \dfrac{2}{x-3} + 4$ and $g(x) = \dfrac{5}{x-3} + 4$

$g(x) = \left(\dfrac{5}{x} - 3\right) + 4$

$f(x) = \left(\dfrac{2}{x} - 3\right) + 4$

42. A rational function can be used to determine how much each person needs to sell if the cost of the trip and the number of people going on the trip and selling the candy is s. The number of people selling and the number of bars sold per person must both be non-negative they also must both be integers. So only the first quadrant part of the graph is meaningful in the context of the problem.

60.

$f(x) = x^3 + 2x^2 - 3x - 5$

rel. max. at $x = -2$, rel. min. at $x = 0.5$; D = {all real numbers}, R = {all real numbers}

61.

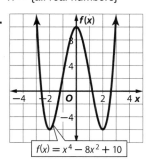

$f(x) = x^4 - 8x^2 + 10$

rel. max. at $x = 0$, rel. min. at $x = -2$ and at $x = 2$; D = {all real numbers}, R = $\{f(x) \mid f(x) \geq -6\}$

Mid-Chapter Quiz

19.

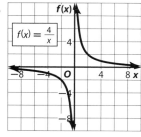

$f(x) = \dfrac{4}{x}$

D = $\{x \mid x \neq 0\}$, R = $\{f(x) \mid f(x) \neq 0\}$

20.

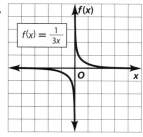

$f(x) = \dfrac{1}{3x}$

D = $\{x \mid x \neq 0\}$, R = $\{f(x) \mid f(x) \neq 0\}$

21.

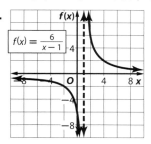

$f(x) = \dfrac{6}{x-1}$

D = $\{x \mid x \neq 1\}$, R = $\{f(x) \mid f(x) \neq 0\}$

22.

$f(x) = \dfrac{-2}{x} + 4$

D = $\{x \mid x \neq 0\}$, R = $\{f(x) \mid f(x) \neq 4\}$

23.

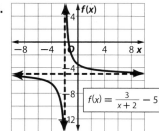

$f(x) = \dfrac{3}{x+2} - 5$

D = $\{x \mid x \neq -2\}$, R = $\{f(x) \mid f(x) \neq -5\}$

24.

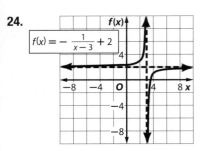

$$f(x) = -\frac{1}{x-3} + 2$$

$$D = \{x \mid x \neq 3\}, R = \{f(x) \mid f(x) \neq 2\}$$

Lesson 8-4

1.

$$f(x) = \frac{x^4 - 2}{x^2 - 1}$$

2.

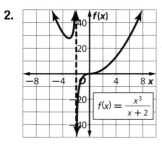

$$f(x) = \frac{x^3}{x + 2}$$

3a.

$$P(x) = \frac{7 + x}{11 + x}$$

4.

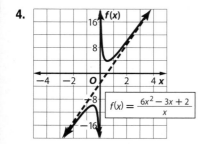

$$f(x) = \frac{6x^2 - 3x + 2}{x}$$

5.

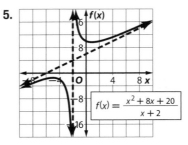

$$f(x) = \frac{x^2 + 8x + 20}{x + 2}$$

6.

$$f(x) = \frac{x^2 - 4x - 5}{x + 1}$$

7.

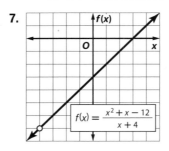

$$f(x) = \frac{x^2 + x - 12}{x + 4}$$

13.

$$f(x) = \frac{x}{x + 2}$$

14.

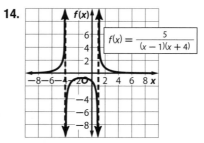

$$f(x) = \frac{5}{(x - 1)(x + 4)}$$

15.

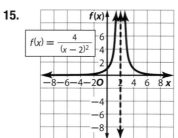

$$f(x) = \frac{4}{(x - 2)^2}$$

16.

$$f(x) = \frac{x-3}{x+1}$$

17.

$$f(x) = \frac{1}{(x+4)^2}$$

18.

$$f(x) = \frac{2x}{(x+2)(x-5)}$$

19.

$$f(x) = \frac{(x-4)^2}{x+2}$$

20.

$$f(x) = \frac{(x+3)^2}{x-5}$$

21.

$$f(x) = \frac{x^3+1}{x^2-4}$$

22.

$$f(x) = \frac{4x^3}{2x^2+x-1}$$

23.

$$f(x) = \frac{3x^2+8}{2x-1}$$

24.

$$f(x) = \frac{2x^2+5}{3x+4}$$

25.

$$f(x) = \frac{x^4-2x^2+1}{x^3+2}$$

26.

$$f(x) = \frac{x^4-x^2-12}{x^3-6}$$

27a.

28.

$$f(x) = \frac{x^2 - 2x - 8}{x - 4}$$

29.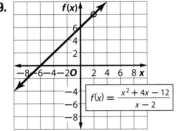

$$f(x) = \frac{x^2 + 4x - 12}{x - 2}$$

30.

$$f(x) = \frac{x^2 - 25}{x + 5}$$

31.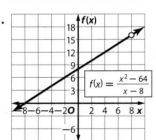

$$f(x) = \frac{x^2 - 64}{x - 8}$$

32.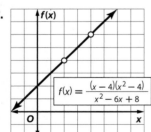

$$f(x) = \frac{(x - 4)(x^2 - 4)}{x^2 - 6x + 8}$$

33.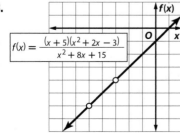

$$f(x) = \frac{(x + 5)(x^2 + 2x - 3)}{x^2 + 8x + 15}$$

34.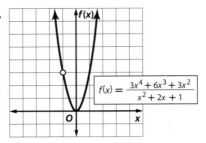

$$f(x) = \frac{3x^4 + 6x^3 + 3x^2}{x^2 + 2x + 1}$$

35.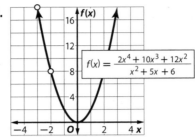

$$f(x) = \frac{2x^4 + 10x^3 + 12x^2}{x^2 + 5x + 6}$$

36a. $f(x) = \dfrac{4500 + 50x}{x}$

$$f(x) = \frac{4500 + 50x}{x}$$

Number of Parking Lots

37a. $f(x) = \dfrac{150 + 40x}{x}$

$$f(x) = \frac{150 + 40x}{x}$$

Number of Months

38a.

$$B(x) = \frac{4 + x}{12 + x}$$

39.

$$f(x) = \frac{x+1}{x^2+6x+5}$$

40.

$$f(x) = \frac{x^2-10x-24}{x+2}$$

41.

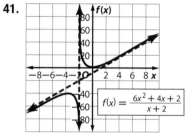

$$f(x) = \frac{6x^2+4x+2}{x+2}$$

42. Sample graph:

Practice Test

11.

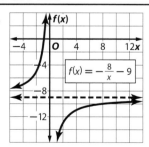

$$f(x) = -\frac{8}{x} - 9$$

12.

$$f(x) = \frac{2}{x+4}$$

13.

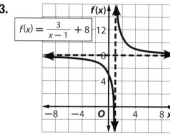

$$f(x) = \frac{3}{x-1} + 8$$

14.

$$f(x) = \frac{5x}{x+1}$$

15.

$$f(x) = \frac{x}{x-5}$$

16.

$$f(x) = \frac{x^2+5x-6}{x-1}$$

	Diagnostic Assessment Quick Check		
	LESSON 9-1 45 min: 1 day 90 min: 0.5 day	**LESSON 9-2** 45 min: 1 day 90 min: 0.5 day	**EXPLORE 9-3** 45 min: 0.5 day 90 min: 0.5 day
Title	Midpoint and Distance Formulas	Parabolas	Graphing Technology Lab: Equations of Circles
Objectives	■ Find the midpoint of a segment on the coordinate plane. ■ Find the distance between two points on the coordinate plane.	■ Write equations of parabolas in standard form. ■ Graph parabolas.	■ Use a graphing calculator to examine the characteristics of a circle and its equation.
Key Vocabulary		parabola focus directrix latus rectum standard form general form	
CCSS	A.CED.4	A.SSE.1b, A.CED.2	
Multiple Representations	🗗	🗗	
Lesson Resources	🖱 connectED.mcgraw-hill.com 📁 Leveled Worksheets **PT** Personal Tutor 🖐 Virtual Manipulatives ✅ Self-Check Quiz ■ *5-Minute Check* ■ *Study Notebook*	🖱 connectED.mcgraw-hill.com 📁 Leveled Worksheets 📁 Quiz 1 🔤 Vocabulary **PT** Personal Tutor 🖐 Virtual Manipulatives ✅ Self-Check Quiz ■ *5-Minute Check* ■ *Study Notebook* ■ *Teaching Algebra with Manipulatives*	🖱 connectED.mcgraw-hill.com **PT** Personal Tutor **Materials:** ■ TI-Nspire technology
Resources for Every Lesson	**IWB** eStudent Edition **IWB** Interactive Classroom	■ eTeacher Edition ■ eSolutions Manual ■ *e*Assessment McGraw-Hill	
Differentiated Instruction	pp. 594, 598	pp. 600, 605	

IWB All digital assets are Interactive Whiteboard ready.

Suggested Pacing			
Time Periods	**Instruction**	**Review & Assess**	**Total**
45-minute	10 days	2 days	12 days
90-minute	6 days	1 day	7 days

LESSON 9-3 45 min: 1 day / 90 min: 0.5 day	**EXPLORE 9-4** 45 min: 0.5 day / 90 min: 0.25 day	**LESSON 9-4** 45 min: 1 day / 90 min: 0.75 day	**LESSON 9-5** 45 min: 1.5 days / 90 min: 1 day
Circles	**Algebra Lab: Investigating Ellipses**	**Ellipses**	**Hyperbolas**
■ Write equations of circles. ■ Graph circles.	■ Determine how the graph of an ellipse is affected by changing the location of the foci.	■ Write equations of ellipses. ■ Graph ellipses.	■ Write equations of hyperbolas. ■ Graph hyperbolas.
circle center radius		ellipse foci major axis minor axis constant sum	hyperbola transverse axis conjugate axis vertices co-vertices constant difference
A.SSE.1b, A.CED.4		A.SSE.1b, A.CED.2	A.SSE.1b, A.CED.2
connectED.mcgraw-hill.com	connectED.mcgraw-hill.com	connectED.mcgraw-hill.com	connectED.mcgraw-hill.com
📁 Leveled Worksheets 🔤 Vocabulary PT Personal Tutor ✋ Virtual Manipulatives ✓ Self-Check Quiz ■ *5-Minute Check* ■ *Study Notebook*	■ *Teaching Algebra with Manipulatives*	📁 Leveled Worksheets 📁 Quiz 2 🔤 Vocabulary PT Personal Tutor ✓ Self-Check Quiz ■ *5-Minute Check* ■ *Study Notebook*	📁 Leveled Worksheets 🔤 Vocabulary PT Personal Tutor ✓ Self-Check Quiz ■ *5-Minute Check* ■ *Study Notebook* ■ *Teaching Algebra with Manipulatives*
IWB eStudent Edition IWB Interactive Classroom	■ eTeacher Edition ■ eSolutions Manual ■ McGraw-Hill eAssessment		
pp. 608, 610, 613		pp. 617, 618, 622	pp. 625, 631
		Formative Assessment Mid-Chapter Quiz	

	LESSON 9-6 — 45 min: 1.5 days / 90 min: 0.75 day	EXTEND 9-6 — 45 min: 0.5 day / 90 min: 0.25 day	EXPLORE 9-7 — 45 min: 0.5 day / 90 min: 0.5 day	LESSON 9-7 — 45 min: 1 day / 90 min: 0.5 day	
Title	**Identifying Conic Sections**	**Graphing Technology Lab: Analyzing Quadratic Relations**	**Graphing Technology Lab: Linear-Nonlinear Systems**	**Solving Linear-Nonlinear Systems**	
Objectives	■ Write equations of conic sections in standard form. ■ Identify conic sections from their equations.	■ Identify characteristics of quadratic relations. ■ Write equations of quadratic relations.	■ Solve linear-nonlinear systems using a graphing calculator.	■ Solve systems of linear and nonlinear equations algebraically and graphically. ■ Solve systems of linear and nonlinear inequalities graphically.	
Key Vocabulary					
CCSS	A.SSE.1b, F.IF.9			A.REI.11	
Multiple Representations	⟳				
Lesson Resources	connectED.mcgraw-hill.com ▭ Leveled Worksheets ▭ Quiz 3 **PT** Personal Tutor ☑ Self-Check Quiz ■ *5-Minute Check* ■ *Study Notebook* ■ *Teaching Algebra with Manipulatives*	connectED.mcgraw-hill.com **PT** Personal Tutor **Materials:** ■ TI-Nspire CAS technology	connectED.mcgraw-hill.com **PT** Personal Tutor **Materials:** ■ TI-83/84 Plus or other graphing calculator	connectED.mcgraw-hill.com ▭ Leveled Worksheets ▭ Quiz 4 🏃 Animations **PT** Personal Tutor ☑ Self-Check Quiz ■ *5-Minute Check* ■ *Study Notebook*	
Resources for Every Lesson	**IWB** eStudent Edition **IWB** Interactive Classroom	■ eTeacher Edition ■ eSolutions Manual ■ *e*Assessment McGraw-Hill			
Differentiated Instruction	pp. 633, 636			pp. 641, 645	
				Summative Assessment Study Guide and Review Practice Test	

IWB All digital assets are Interactive Whiteboard ready.

What the Research Says . . .

Students retain what they learn from their own efforts to address challenging problems that arise from situations that resonate with their own interests. (Steen and Forman, 1995)

- Examples in Chapter 9 include interesting contexts such as solar energy (Lesson 9-2) and astronomy (Lesson 9-5).

- Student-centered contexts such as baseball trajectories (Lesson 9-2), basketball free-throws (Lesson 9-5), and cell phones (Lesson 9-7).

Teacher to Teacher

Judie Campbell
Derry Area High School
Derry, PA

Use Before Lesson 9-6

As a final project, I have students make a poster with 6 different types of graphs, explaining how each was derived, and then have them label each part of the graph. An explanation of each graph is required..

Project CRISS^{SM}

STUDY SKILL

As you begin studying a new lesson or chapter, it may be helpful to students to complete a KWL frame. In a KWL frame, students record what they **k**now, what they **w**ant to know, and what they **l**earned. This completed sample frame shows information that a student may have recorded before and after studying Lesson 9-2 on parabolas.

What I **K**now	What I **W**ant To Know	What I **L**earned
• The graph of the quadratic parent function $f(x) = x^2$ is a parabola. • Changes to the parameters of a function result in translations, reflections, and/or dilations of the graphs of functions.	• How can you predict the shape, location, and orientation of the graph of a parabola by looking at the equation?	• The standard equation for a parabola opening up or down is $y = a(x - h)^2 + k$. The vertex is at (h, k). If $a > 0$ then the parabola opens up, and if $a < 0$ the parabola opens down. • The standard equation for a parabola opening left or right is $x = a(y - h)^2 + k$. The vertex is at (h, k). If $a > 0$ then the parabola opens right, and if $a < 0$ the parabola opens left.

Creating Independence through Student-owned Strategies

Assessment and Intervention

SE = Student Edition, **TE** = Teacher Edition, **CRM** = Chapter Resource Masters

Diagnosis	Prescription
Beginning Chapter 9	
Get Ready for Chapter 9 **SE**	Response to Intervention **TE**
Beginning Every Lesson	
Then, Now, Why? **SE** 5-Minute Checks	Chapter 0 **SE**

DIAGNOSTIC ASSESSMENT

Diagnosis	Prescription
During/After Every Lesson	
Guided Practice **SE**, every example Check Your Understanding **SE** H.O.T. Problems **SE** Spiral Review **SE** Additional Examples **TE** Watch Out! **TE** Step 4, Assess **TE** Chapter 9 Quizzes **CRM**, pp. 53–54 Self-Check Quizzes connectED.mcgraw-hill.com	**TIER 1 Intervention** Skills Practice **CRM**, Ch. 1–9 connectED.mcgraw-hill.com **TIER 2 Intervention** Differentiated Instruction **TE** Study Guide and Intervention **CRM**, Ch. 1–9 **TIER 3 Intervention** *Math Triumphs, Alg. 2*
Mid-Chapter	
Mid-Chapter Quiz **SE** Mid-Chapter Test **CRM**, p. 55 e*Assessment* McGraw-Hill	**TIER 1 Intervention** Skills Practice **CRM**, Ch. 1–9 connectED.mcgraw-hill.com **TIER 2 Intervention** Study Guide and Intervention **CRM**, Ch. 1–9 **TIER 3 Intervention** *Math Triumphs, Alg. 2*
Before Chapter Test	
Chapter Study Guide and Review **SE** Practice Test **SE** Standardized Test Practice **SE** Chapter Test connectED.mcgraw-hill.com Standardized Test Practice connectED.mcgraw-hill.com Vocabulary Review connectED.mcgraw-hill.com e*Assessment* McGraw-Hill	**TIER 1 Intervention** Skills Practice **CRM**, Ch. 1–9 connectED.mcgraw-hill.com **TIER 2 Intervention** Study Guide and Intervention **CRM**, Ch. 1–9 **TIER 3 Intervention** *Math Triumphs, Alg. 2*

FORMATIVE ASSESSMENT

Diagnosis	Prescription
After Chapter 9	
Multiple-Choice Tests, Forms 1, 2A, 2B **CRM**, pp. 57–62 Free-Response Tests, Forms 2C, 2D, 3 **CRM**, pp. 63–68 Vocabulary Test **CRM**, p. 56 Extended Response Test **CRM**, p. 69 Standardized Test Practice **CRM**, pp. 70–72 e*Assessment* McGraw-Hill	Study Guide and Intervention **CRM**, Ch. 1–9 connectED.mcgraw-hill.com

SUMMATIVE ASSESSMENT

Option 1 Reaching All Learners

Naturalist Ask each student to find a picture or photograph of an object in nature that is related to circles. Although circles in nature may not be mathematically perfect, they are seen often, as in the shape of a flower, the designs on an insect, or in the ripples on a pond after a stone hits the water. Have students draw a coordinate grid on the picture, find the ordered pair for the center point, a point on the graph, and write an equation for the graph.

Kinesthetic Write an equation of a parabola on the board. Have your class create a table of values and then graph this parabola using the tiles on the classroom floor as a coordinate grid. Masking tape can be used to mark the focus point and the directrix line. Mark the set of points on the graph with colored masking tape. Connect these points with a heavy rope to create a parabola.

Option 2 Approaching Level AL

As a class, make a set of double paper cone conic section models. Demonstrate that no matter where a cone is sliced horizontally, the result is a circle, as shown below. Demonstrate the other conic sections in a similar manner.

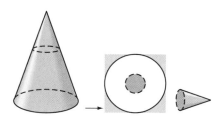

Option 3 English Learners ELL

Have students work in pairs. One student gives the other student two ordered pairs to graph on a coordinate grid. Then one student explains how to find the distance between them and the coordinates of the midpoint of the segment joining them. Then pairs switch roles and repeat the activity.

Option 4 Beyond Level BL

Have students research methods of drawing each conic section using only pushpins, string, and a pencil. For example, if the string is attached to the pushpins—representing the foci—an ellipse can be drawn.

Focus   Focus

Challenge students to develop three-dimensional models depicting the different types of conic sections or have them explore computer technology with the capacity to perform this task.

VerticalAlignment

Before Chapter 9

Related Topics from Algebra 1

- use tools including factoring and properties of exponents to simplify expressions
- graph equations of lines
- solve systems of linear equations using algebraic methods

Previous Topics from Algebra 2

- solve quadratic equations using algebraic methods
- determine solutions of square root equations using algebraic methods

Chapter 9

Related Topics from Algebra 2

- sketch graphs of conic sections to relate simple parameter changes in the equation to corresponding changes in the graph
- describe a conic section as the intersection of a plane and a cone
- identify symmetries from graphs of conic sections
- identify the conic section from a given equation
- use algebraic methods and graphs to solve systems of linear and nonlinear equations or inequalities

After Chapter 9

Preparation for Precalculus

- use conic sections to model motion, such as motion of the planets
- use conic sections to describe physical phenomena such as the reflective properties of light and sound

Lesson-by-LessonPreview

9-1   Midpoint and Distance Formulas

A *line segment* on a coordinate plane has two endpoints. The coordinates of these endpoints can be used to find the *midpoint* of the segment (Midpoint Formula) and the *length* of the segment (Distance Formula). The coordinates of the midpoint of a line segment are the arithmetic means of the corresponding coordinates of the endpoints. The Distance Formula can be derived from the Pythagorean Theorem.

9-2   Parabolas

A *parabola* is the set of all points in a plane that are the same distance from a given point called the *focus* and a given line called the *directrix.* This definition combined with the Distance Formula can be used to derive the equation $y = a(x - h)^2 + k$, which is the *standard form* of the equation of a parabola with vertex (h, k) and axis of symmetry $x = h$. The graph of $y = a(x - h)^2 + k$ is a transformation of the parent graph of $y = x^2$.

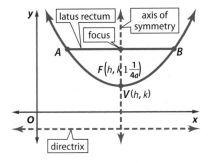

If a parabola has a vertical axis of symmetry, it is a function. If it has a horizontal axis of symmetry, its equation is of the form $x = a(y - k)^2 + h$, and it is not a function.

9-3   Circles

A *circle* is the set of all points in a plane that are a given distance (the radius) from a given point (the center) in the plane. Using (h, k) as the given point and r as the given radius, the equation of a circle is $\sqrt{(x - h)^2 + (y - k)^2} = r$. Squaring the equation, $(x - h)^2 + (y - k)^2 = r^2$. An equation for a circle that is not in standard form can be rewritten in standard form by completing the square (Lesson 5-5) for both the variables.

Get educator support from the Math Forum's T2T. You may want to search, or ask, for activities with conic sections. Start your search at connectED.mcgraw-hill.com.

mathforum.org
The Math Forum
@ Drexel University

9-4 Ellipses

An *ellipse* is the set of all points in a plane such that the sum of the distances from two fixed points is constant. The two fixed points are called the *foci* of the ellipse. If the orientation of an ellipse is horizontal and its center is the origin, the standard form of its equation is $\frac{x^2}{a^2} + \frac{y^2}{b^2} = 1$. If the orientation is vertical, the standard form of its equation is $\frac{y^2}{a^2} + \frac{x^2}{b^2} = 1$. Every ellipse has two axes of symmetry as shown below.

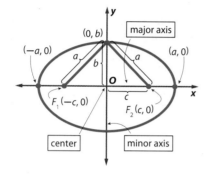

The values *a* and *b* determine the length of the major and minor axes and whether the major axis of the ellipse is horizontal or vertical. The equations $\frac{(x-h)^2}{a^2} + \frac{(y-k)^2}{b^2} = 1$ and $\frac{(y-k)^2}{a^2} + \frac{(x-h)^2}{b^2} = 1$ represent ellipses with centers translated to the point (h, k).

9-5 Hyperbolas

A *hyperbola* is the set of all points in a plane such that the absolute value of the difference of the distances from two fixed points (foci) is constant. The standard form of the equation for a hyperbola centered at the origin with a horizontal orientation is $\frac{x^2}{a^2} - \frac{y^2}{b^2} = 1$. If the orientation is vertical, the standard form of the equation is $\frac{y^2}{a^2} - \frac{x^2}{b^2} = 1$. A hyperbola has two axes of symmetry, the *transverse axis* and the *conjugate axis*. A hyperbola centered at the origin and having foci on the *x*-axis has asymptotes with the equations $y = \pm\frac{b}{a}x$. If the foci are on the *y*-axis, the equations of the asymptotes are $y = \pm\frac{a}{b}x$.

The equations $\frac{(x-h)^2}{a^2} - \frac{(y-k)^2}{b^2} = 1$ and $\frac{(y-k)^2}{a^2} - \frac{(x-h)^2}{b^2} = 1$ represent hyperbolas with centers translated to the point (h, k).

9-6 Identifying Conic Sections

The four curves (parabola, circle, ellipse, or hyperbola) are called *conic sections* because they are the cross-sections formed when a double cone is sliced by a plane.

The general equation for any conic section is $Ax^2 + Bxy + Cy^2 + Dx + Ey + F = 0$, where A, B, and C are not all zero. This equation is related to the standard forms of the equations of the four conic sections. If A, B, and C are not equal to zero, the expression $B^2 - 4AC$ (the discriminant) can be used to identify the conic.

9-7 Solving Linear-Nonlinear Systems

A system of a linear and a nonlinear equation can be solved algebraically or graphically.

- For a *linear-quadratic system,* a graph indicates whether the conic section and the line intersect in 0, 1, or 2 points. Substitution can be used as the first step in solving the system algebraically.

- For a *quadratic-quadratic system,* a graph indicates the number of solutions (0, 1, 2, 3, or 4). Elimination can be used to solve the system algebraically.

The solutions to systems of linear and nonlinear inequalities can be illustrated graphically and described algebraically.

Mc Graw Hill | **Professional Development**

Targeted professional development has been articulated throughout *Algebra 2*. More quality, customized professional development is available from McGraw-Hill Professional Development. Visit connectED.mcgraw-hill.com for details on each product.

- Online Lessons emphasize the strategies and techniques used to teach Algebra 2. Includes streaming video, interactive pages, and online tools.

- Video Workshops allow mentors, coaches, or leadership personnel to facilitate on-site workshops on educational strategies in mathematics and mathematical concepts.

- MHPD Online (www.mhpdonline.com) offers online professional development with video clips of instructional strategies, links, student activities, and news and issues in education.

- Teaching Today (teachingtoday.glencoe.com) gives secondary teachers practical strategies and materials that inspire excellence and innovation in teaching.

Chapter Project

Sound the Alarm

Students use what they have learned about conic sections to complete a project.

This chapter project addresses environmental literacy, as well as several specific skills identified as being essential to student success by the Framework for 21st Century Learning.

Visit connectED.mcgraw-hill.com for student and teacher handouts.

KeyVocabulary Introduce the key vocabulary in the chapter using the routine below.

Define: A parabola is the set of all points in a plane that are the same distance from a point, called the focus, and a line, called the directrix.

Example: The diagram shows a parabola.

Ask: What are the coordinates of the vertex of the parabola? (h, k)

CHAPTER 9 Conic Sections

··Then	··Now	··Why? ▲
○ You solved systems of linear equations algebraically and graphically.	○ You will: ■ Use the Midpoint and Distance Formulas. ■ Write and graph equations of parabolas, circles, ellipses, and hyperbolas. ■ Identify conic sections. ■ Solve systems of quadratic equations and inequalities.	**SPACE** Conic sections are evident in many aspects of space. Equations of circles are used to pilot spacecraft and satellites in circular orbits around Earth and the Moon. Planets travel in elliptical paths, not circular ones as previously thought. Comets travel along one branch of a hyperbola, which can help us to predict when they will appear again.

connectED.mcgraw-hill.com   **Your Digital Math Portal**

Animation   Vocabulary   eGlossary   Personal Tutor   Virtual Manipulatives   Graphing Calculator   Audio   Foldables   Self-Check Practice   Worksheets

Brand X Pictures/PunchStock

Get Ready for the Chapter

Diagnose Readiness | You have two options for checking prerequisite skills.

1 **Textbook Option** Take the Quick Check below. Refer to the Quick Review for help.

QuickCheck

Solve each equation by completing the square.

1. $x^2 + 8x + 7 = 0$ $\{-7, -1\}$ 2. $x^2 + 5x - 6 = 0$ $\{-6, 1\}$
3. $x^2 - 8x + 15 = 0$ $\{3, 5\}$ 4. $x^2 + 2x - 120 = 0$
$\{-12, 10\}$
5. $2x^2 + 7x - 15 = 0$ 6. $2x^2 + 3x - 5 = 0$
7. $x^2 - \frac{3}{2}x - \frac{23}{16} = 0$ 8. $3x^2 - 4x = 2$

5. $\left\{-5, \frac{3}{2}\right\}$ 6. $\left\{-\frac{5}{2}, 1\right\}$ 7. $\left\{\frac{3}{4} \pm \sqrt{2}\right\}$ 8. $\left\{\frac{2 \pm \sqrt{10}}{3}\right\}$

Solve each system of equations by using either substitution or elimination.

9. $y = x + 3$ $(2, 5)$
   $2x - y = -1$
10. $2x - 5y = -18$ $(1, 4)$
    $3x + 4y = 19$
11. $4y + 6x = -6$
    $5y - x = 35$ $(-5, 6)$
12. $x = y - 8$ $(-2, 6)$
    $4x + 2y = 4$

13. **MONEY** The student council paid $15 per registration for a conference. They also paid $10 for T-shirts for a total of $180. Last year, they spent $12 per registration and $9 per T-shirt for a total of $150 to buy the same number of registrations and T-shirts. Write and solve a system of two equations that represents the number of registrations and T-shirts bought each year. 13. $15x + 10y = 180$
    $12x + 9y = 150$
    number of registrations = 8
    number of T-shirts = 6

QuickReview

Example 1 (Used in Lessons 9-2 through 9-6)

Solve $x^2 + 6x - 16 = 0$ by completing the square.

$$x^2 + 6x = 16$$
$$x^2 + 6x + 9 = 16 + 9$$
$$(x + 3)^2 = 25$$
$$x + 3 = \pm 5$$
$$x + 3 = 5 \quad \text{or} \quad x + 3 = -5$$
$$x = 2 \qquad\qquad x = -8$$

Example 2 (Used in Lesson 9-7)

Solve the system of equations algebraically.
$3y = x - 9$
$4x + 5y = 2$

Since x has a coefficient of 1 in the first equation, use the substitution method. First, solve that equation for x.

$$3y = x - 9 \quad \rightarrow \quad x = 3y + 9$$

$4(3y + 9) + 5y = 2$ Substitute $3y + 9$ for x.
$12y + 36 + 5y = 2$ Distributive Property
$17y = -34$ Combine like terms.
$y = -2$ Divide each side by 17.

To find x, use $y = -2$ in the first equation.
$3(-2) = x - 9$ Substitute -2 for y.
$-6 = x - 9$ Multiply.
$3 = x$ Add 9 to each side.
The solution is $(3, -2)$.

2 **Online Option** Take an online self-check Chapter Readiness Quiz at <u>connectED.mcgraw-hill.com</u>.

591

Get Ready for the Chapter

Response to Intervention (RtI)
Use the *Quick Check* results and the Intervention Planner to help you determine your Response to Intervention. The If-Then statements in the chart below help you decide the appropriate tier of RtI and suggest intervention resources for each tier.

InterventionPlanner

TIER 1 **On Level** **OL**

If students miss about 25% of the exercises or less,

Then choose a resource:

SE Lessons 3-1 and 4-5

 Skills Practice, Chapter 3, p. 6, Chapter 4, p. 32

 Self-Check Quiz

TIER 2 **Strategic Intervention** **AL**
approaching grade level

If students miss about 50% of the exercises,

Then choose a resource:

 Study Guide and Intervention, Chapter 3, p. 5, Chapter 4, p. 30
Quick Review Math Handbook

 Extra Examples, Personal Tutor, Homework Help

TIER 3 **Intensive Intervention**
2 or more grades below level

If students miss about 75% of the exercises,

Then use *Math Triumphs, Alg. 2*

 Extra Examples, Personal Tutor, Homework Help, Review Vocabulary

---

Essential Questions

- How does mathematics help us to describe the physical world? Sample answer: Mathematics enables us to model real-world situations, which allows us to analyze and understand these situations better, and thus make better decisions.

- How are conics useful? Sample answer: Conics are useful because they have multiple forms that can be used to model various real-world situations, such as planetary orbits and projectile motion.

FOLDABLES StudyOrganizer

Dinah Zike's Foldables®

Focus Students write notes on what they learn about conic sections throughout this chapter.

Teach Have students make and label their Foldables as illustrated. Students should use the appropriate lines to record their notes and examples of the concepts in each lesson of this chapter.

When to Use It Encourage students to add to their Foldables as they work through the chapter and to use them to review conic sections for the chapter test.

Differentiated Instruction

 Student-Built Glossary, pp. 1–2

Students should complete the chart by providing a definition of each term and an example as they progress through Chapter 9. This study tool can also be used to review for the chapter test.

Get Started on the Chapter

You will learn several new concepts, skills, and vocabulary terms as you study Chapter 9. To get ready, identify important terms and organize your resources. You may wish to refer to Chapter 0 to review prerequisite skills.

FOLDABLES StudyOrganizer

Conic Sections Make this Foldable to help you organize your Chapter 9 notes about conic sections. Begin with eight sheets of grid paper.

1 Staple the stack of grid paper along the top to form a booklet.

2 Cut seven lines from the bottom of the top sheet, six lines from the second sheet, and so on.

3 Label with lesson numbers as shown.

NewVocabulary

English		Español
parabola	p. 599	parábola
focus	p. 599	foco
directrix	p. 599	directriz
circle	p. 607	círculo
center of a circle	p. 607	centro de un círculo
radius	p. 607	radio
ellipse	p. 615	elipse
foci	p. 615	focos
major axis	p. 615	eje mayor
minor axis	p. 615	eje menor
center of an ellipse	p. 615	centro de una elipse
vertices	p. 615	vértices
co-vertices	p. 615	co-vértices
constant sum	p. 616	suma constante
hyperbola	p. 624	hipérbola
transverse axis	p. 624	eje transversal
conjugate axis	p. 624	eje conjugado
constant difference	p. 627	diferencia constante

ReviewVocabulary

quadratic equation ecuación cuadrática **an equation of the form** $ax^2 + bx + c = 0$, where $a \neq 0$

system of equations sistema de ecuaciones **a set of equations with the same variables**

x- and *y*-intercepts intersecciones *x* y *y* the *x*- or *y*-coordinate of the point at which a graph crosses the *x*- or *y*-axis

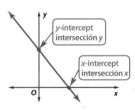

LESSON 9-1
Midpoint and Distance Formulas

::Then	::Now	::Why?
● You found the slope of a line passing through two points.	**1** Find the midpoint of a segment on the coordinate plane. **2** Find the distance between two points on the coordinate plane.	● The Zero Milestone in Washington, D.C., was established in 1919. It was intended to serve as the origin for all highway measures with highway markers across the United States displaying the distances from the Zero Milestone.

Common Core State Standards

Content Standards
A.CED.4 Rearrange formulas to highlight a quantity of interest, using the same reasoning as in solving equations.

Mathematical Practices
2 Reason abstractly and quantitatively.

1 The Midpoint Formula Recall that point M is the midpoint of segment PQ if M is between P and Q and $PM = MQ$. There is a formula for the coordinates of the midpoint of a segment in terms of the coordinates of the endpoints.

KeyConcept Midpoint Formula

Words If a line segment has endpoints $P(x_1, y_1)$ and $Q(x_2, y_2)$, then the midpoint of the segment has coordinates
$$M\left(\frac{x_1 + x_2}{2}, \frac{y_1 + y_2}{2}\right).$$

Model

Example 1 Find a Midpoint

Find the coordinates of M, the midpoint of $\overline{JK}$, for $J(-1, 2)$ and $K(6, 1)$.

Let J be (x_1, y_1) and K be (x_2, y_2).

$M\left(\dfrac{x_1 + x_2}{2}, \dfrac{y_1 + y_2}{2}\right)$ Midpoint Formula

$= M\left(\dfrac{-1 + 6}{2}, \dfrac{2 + 1}{2}\right)$ $(x_1, y_1) = (-1, 2), (x_2, y_2) = (6, 1)$

$= M\left(\dfrac{5}{2}, \dfrac{3}{2}\right)$ or $M\left(2\dfrac{1}{2}, 1\dfrac{1}{2}\right)$ Simplify.

> **GuidedPractice**
>
> **1A.** Find the coordinates of the midpoint of $\overline{AB}$ for $A(5, 12)$ and $B(-4, 8)$. $\left(\dfrac{1}{2}, 10\right)$
>
> **1B.** Find the coordinates of the midpoint of $\overline{CD}$ for $C(4, 5)$ and $D(14, 13)$.   **(9, 9)**

2 The Distance Formula The distance between two points, a and b, on a number line is $|a - b|$ or $|b - a|$. You can use this fact and the Pythagorean Theorem to derive a formula for the distance between two points on a coordinate plane.

  593

U.S. Department of Transportation

---

1 Focus

VerticalAlignment

Before Lesson 9-1 Find the slope of a line passing through two points.

Lesson 9-1 Find the midpoint of a segment on the coordinate plane. Find the distance between two points on the coordinate plane.

After Lesson 9-1 Describe a conic section as the intersection of a plane and a cone.

2 Teach

Scaffolding Questions
Have students read the **Why?** section of the lesson.

Ask:
- What does the word "origin" mean? Sample answer: the start or beginning

- If the Zero Milestone were graphed on a coordinate plane, what would its coordinates be? (0, 0)

- If a fountain were graphed on the same coordinate plane, at (0, 8), how far from the Zero Milestone would it be? 8 units

---

Lesson 9-1 Resources

Resource	Approaching Level **AL**	On Level **OL**	Beyond Level **BL**	English Learners **ELL**
Teacher Edition	▪ Differentiated Instruction, p. 594	▪ Differentiated Instruction, pp. 594, 598	▪ Differentiated Instruction, p. 598	
Chapter Resource Masters	▪ Study Guide and Intervention, pp. 5–6 ▪ Skills Practice, p. 7 ▪ Practice, p. 8 ▪ Word Problem Practice, p. 9	▪ Study Guide and Intervention, pp. 5–6 ▪ Skills Practice, p. 7 ▪ Practice, p. 8 ▪ Word Problem Practice, p. 9 ▪ Enrichment, p. 10	▪ Practice, p. 8 ▪ Word Problem Practice, p. 9 ▪ Enrichment, p. 10	▪ Study Guide and Intervention, pp. 5–6 ▪ Skills Practice, p. 7 ▪ Practice, p. 8 ▪ Word Problem Practice, p. 9
Other	▪ 5-Minute Check 9-1 ▪ Study Notebook	▪ 5-Minute Check 9-1 ▪ Study Notebook	▪ 5-Minute Check 9-1 ▪ Study Notebook	▪ 5-Minute Check 9-1 ▪ Study Notebook

1 The Midpoint Formula

Example 1 shows how to use the Midpoint Formula to find the coordinates of the midpoint of a segment given the coordinates of its endpoints.

Formative Assessment

Use the Guided Practice exercises after each example to determine students' understanding of concepts.

Additional Example

1 Find the coordinates of B, the midpoint of $\overline{AC}$, for $A(5, 2)$ and $C(7, 8)$. (6, 5)

▶ **Additional Examples** also in Interactive Classroom PowerPoint® Presentations

 IWB Interactive White Board READY

2 The Distance Formula

Example 2 shows how to use the Distance Formula to find the distance between two points. **Example 3** shows how to use both the Midpoint Formula and the Distance Formula to solve a real-world problem.

Additional Example

2 **DISC GOLF** Matt's disc is 3 feet short and 2 feet to the right of the basket. On his first putt the disc lands 4 feet beyond the basket and 1 foot to the left. If the disc went in a straight line, how far did it go? $\sqrt{58}$ feet or about 7.6 feet

---

StudyTip

Distance In mathematics, just as in real-world situations, distances are always nonnegative.

Let d represent the distance between (x_1, y_1) and (x_2, y_2).

$$c^2 = a^2 + b^2 \qquad \text{Pythagorean Theorem}$$

$$d^2 = |x_2 - x_1|^2 + |y_2 - y_1|^2 \qquad \text{Substitute.}$$

$$d^2 = (x_2 - x_1)^2 + (y_2 - y_1)^2 \qquad \begin{array}{l} |x_2 - x_1|^2 = (x_2 - x_1)^2, \\ |y_2 - y_1|^2 = (y_2 - y_1)^2 \end{array}$$

$$d = \sqrt{(x_2 - x_1)^2 + (y_2 - y_1)^2} \qquad \begin{array}{l} \text{Find the nonnegative} \\ \text{square root of each side.} \end{array}$$

🔑 KeyConcept Distance Formula

Words The distance between two points with coordinates (x_1, y_1) and (x_2, y_2) is given by $d = \sqrt{(x_2 - x_1)^2 + (y_2 - y_1)^2}$.

Model

⦿ Real-World Example 2 Find the Distance Between Two Points

DISC GOLF Troy's disc is 20 feet short and 8 feet to the right of the basket. On his first putt, the disc lands 2 feet to the left and 3 feet beyond the basket. If the disc went in a straight line, how far did it go?

Real-WorldLink

There are more than 700 disc golf courses in the U.S. These courses are permanent installations, usually located in public parks, where players actually "drive" and "putt" with specially-styled discs.

Model the situation. If the basket is at $(0, 0)$, then the location of the disc is $(8, -20)$. The location after the first putt is $(-2, 3)$.

$$d = \sqrt{(x_2 - x_1)^2 + (y_2 - y_1)^2} \qquad \text{Distance Formula}$$

$$= \sqrt{(-2 - 8)^2 + [3 - (-20)]^2} \qquad (x_1, y_1) = (8, -20) \text{ and } (x_2, y_2) = (-2, 3)$$

$$= \sqrt{(-10)^2 + 23^2} \qquad \text{Simplify.}$$

$$= \sqrt{629} \text{ or about } 25$$

The disc traveled about 25 feet on his first putt.

▶ **GuidedPractice**

2. Sharon hits a golf ball 12 feet above the hole and 3 feet to the left. Her first putt traveled to 2 feet above the cup and 1 foot to the right. How far did the ball travel on her first putt? $2\sqrt{29}$ or about 10.77 ft

---

DifferentiatedInstruction AL OL

Visual/Spatial Learners After finding the midpoints of line segments, suggest that students draw graphs to check the coordinates.

Rich Reed/National Geographic/Getty Images

StudyTip
Midpoints The coordinates of the midpoint are the means of the coordinates of the endpoints.

There most likely will be problems involving the Midpoint and Distance Formulas on standardized tests you will have to take.

Standardized Test Example 3   Find the Midpoint Between Coordinates

A coordinate grid is placed over a Florida map. St. Augustine is located at (3, 13), and Rockledge is located at (8, −1). If Port Orange is halfway between St. Augustine and Rockledge, which is closest to the distance in coordinate units from St. Augustine to Port Orange?

A 4.75   **B** 7.43   **C** 14.9   **D** 19

Read the Test Item

The question asks you to find the distance between one city and the midpoint. Find the midpoint, and then use the Distance Formula.

Solve the Test Item

Use the Midpoint Formula to find the coordinates of Port Orange.

$$\text{midpoint} = \left(\frac{3+8}{2}, \frac{13+(-1)}{2}\right) \qquad \text{Midpoint Formula}$$
$$= (5.5, 6) \qquad \text{Simplify.}$$

Use the Distance Formula to find the distance between St. Augustine (3, 13) and Port Orange (5.5, 6).

$$\text{distance} = \sqrt{(3-5.5)^2 + (13-6)^2} \qquad \text{Distance Formula}$$
$$= \sqrt{(-2.5)^2 + 7^2} \qquad \text{Evaluate exponents and add.}$$
$$= \sqrt{55.25} \text{ or about } 7.43 \qquad \text{Simplify.}$$

The answer is B.

Test-TakingTip
Answer Check To check your answer, find the distance between Port Orange and Rockledge. Since Port Orange is at the midpoint, these distances should be equal.

Guided Practice

3. The coordinates for points *A* and *B* are (−4, −5) and (10, −7), respectively. Find the distance between the midpoint of *A* and *B* and point *B*. **H**

F $\sqrt{10}$ units   **G** $5\sqrt{10}$ units   **H** $\sqrt{50}$ units   **J** $10\sqrt{5}$ units

---

Check Your Understanding   = Step-by-Step Solutions begin on page R14.

Example 1 **CCSS PRECISION** Find the midpoint of the line segment with endpoints at the given coordinates.

1. (−4, 7), (3, 9) $\left(-\frac{1}{2}, 8\right)$
2. (8, 2), (−1, −5) (3.5, −1.5)
3. (11, 6), (18, 13.5) (14.5, 9.75)
4. (−12, −2), (−10.5, −6) (−11.25, −4)

Example 2 Find the distance between each pair of points with the given coordinates.

5. (3, −5), (13, −11) 11.662 units
6. (8, 1), (−2, 9) 12.806 units
7. (0.25, 1.75), (3.5, 2.5) 3.335 units
8. (−4.5, 10.75), (−6.25, −7) 17.836 units

Example 3 **9. MULTIPLE CHOICE** The map of a mall is overlaid with a numeric grid. The kiosk for the cell phone store is halfway between The Ice Creamery and the See Clearly eyeglass store. If the ice cream store is at (2, 4) and the eyeglass store is at (78, 46), find the distance the kiosk is from the eyeglass store. **A**

A 43.4 units   **B** 47.2 units   **C** 62.4 units   **D** 94.3 units

---

CCSS Teaching the Mathematical Practices

Precision Mathematically proficient students try to use clear definitions in their reasoning, calculate accurately and efficiently, and make explicit use of definitions.

Teach with Tech

Interactive Whiteboard Display a coordinate plane on the board. Plot two points on the plane and have students find the midpoint and the distance of the segment that connects these two points. Then drag the points to other locations on the board and repeat.

Tips for New Teachers

Meaningful Solutions Have students refer to Example 3. Tell students that in radical form, $\sqrt{55.25}$ is an exact mathematical representation of the value of the distance between St. Augustine and Port Orange. However, since this value is being used to describe a distance, it is more useful to express the distance in decimal form by using a calculator to find an approximate value. For example, $\sqrt{55.25} \approx 7.43$. Also point out that writing the steps helps to prevent careless errors in calculations.

Additional Example

3 STANDARDIZED TEST EXAMPLE A coordinate grid is placed over a scale drawing of Nita's patio. A grill is located at (2, −3). A flowerpot is located at (−6, −1). A picnic table is halfway between the grill and the flowerpot. In coordinate units, about how far is it from the grill to the picnic table? **C**

A 0.2   **C** 4.1
B 2.24   **D** 5

---

3 Practice

Formative Assessment

Use Exercises 1–9 to check for understanding.

Use the chart at the bottom of the next page to customize assignments for your students.

Focus on Mathematical Content

The Distance Formula You can use the Distance Formula,

$$d = \sqrt{(x_2 - x_1)^2 + (y_2 - y_1)^2},$$ to

find the distance between two points, (x_1, y_1) and (x_2, y_2), on a coordinate plane. Since distance is always nonnegative, the equation must be solved for the nonnegative square root.

(CCSS) Teaching the Mathematical Practices

Modeling Mathematically proficient students can apply the mathematics they know to solve problems arising in everyday life, analyze relationships mathematically to draw conclusions, and interpret their mathematical results in the context of a situation.

Additional Answers

41a.

45. See students' graphs; the distance from A to B equals the distance from B to A. Using the Distance Formula, the solution is the same no matter which ordered pair is used first.

46. Most maps have a superimposed grid. Think of the grid as a coordinate system and assign approximate coordinates to the two locations. Then use the Midpoint Formula to find the midpoint between the points with those coordinates.

Practice and Problem Solving

Extra Practice is on page R9.

Example 1 Find the midpoint of the line segment with endpoints at the given coordinates.

10. (20, 3), (15, 5) **(17.5, 4)**   11. (−27, 4), (19, −6) **(−4, −1)** 12. (−0.4, 7), (11, −1.6) **(5.3, 2.7)**

13. (5.4, −8), (9.2, 10)   14. (−5.3, −8.6), (−18.7, 1)   15. (−6.4, −8.2), (−9.1, −0.8)
(7.3, 1)   **(−12, −3.8)**   **(−7.75, −4.5)**

Example 2 Find the distance between each pair of points with the given coordinates.

16. (1, 2), (6, 3) **5.099 units**

17. (3, −4), (0, 12) **16.279 units**

18. (−6, −7), (11, −12) **17.720 units**

19. (−10, 8), (−8, −8) **16.125 units**

20. (4, 0), (5, −6) **6.083 units**

21. (7, 9), (−2, −10) **21.024 units**

22. (−4, −5), (15, 17) **29.069 units**

23. (14, −20), (−18, 25) **55.218 units**

Example 3 24. **TRACK AND FIELD** A shot put is thrown from the inside of a circle. A coordinate grid is placed over the shot put circle. The toe board is located at the front of the circle at (−4, 1), and the back of the circle is at (5, 2). If the center of the circle is halfway between these two points, what is the distance from the toe board to the center of the circle? $\sqrt{20.5} \approx 4.528$ units

B Find the midpoint of the line segment with endpoints at the given coordinates. Then find the distance between the points.

25. (−93, 15), (90, −15)

26. (−22, 42), (57, 2)

27. (−70, −87), (59, −14)

28. (−98, 5), (−77, 64)

29. (41, −45), (−25, 75)

30. (90, 60), (−3, −2)

25. **(−1.5, 0); 185.443 units**
26. **(17.5, 22); 88.549 units**
27. **(−5.5, −50.5); 148.223 units**
28. **(−87.5, 34.5); 62.626 units**
29. **(8, 15); 136.953 units**
30. **(43.5, 29); 111.772 units**

31. (−1.2, 2.5), (0.34, −7)
(−0.43, −2.25); 9.624 units

32. (−7.54, 3.89), (4.04, −0.38)
(−1.75, 1.755); 12.342 units

33. $\left(-\dfrac{5}{12}, -\dfrac{1}{3}\right), \left(-\dfrac{17}{2}, -\dfrac{5}{3}\right)$
(−4.458, −1); 8.193 units

34. $\left(-\dfrac{5}{4}, -\dfrac{13}{2}\right), \left(-\dfrac{4}{3}, -\dfrac{5}{6}\right)$
(−1.292, −3.667); 5.667 units

35. $\left(-3\sqrt{2}, -4\sqrt{5}\right), \left(-3\sqrt{3}, 9\right)$
(−4.719, 0.028); 17.97 units

36. $\left(\dfrac{\sqrt{3}}{3}, \dfrac{\sqrt{2}}{4}\right), \left(\dfrac{-2\sqrt{3}}{3}, \dfrac{\sqrt{2}}{4}\right)$
(−0.289, 0.354); $\sqrt{3}$ units

37. **SPACE** Use the labeled points on the outline of the circular crater on Mars to estimate its diameter in kilometers. Assume each unit on the coordinate system is 1 kilometer. **14.53 km**

38c. Triangle *EFG* is also isosceles with two sides of measure $2\sqrt{5}$.
38d. The side lengths of △*EFG* are one half the side lengths of the sides of △*ABC*.

38. **(CCSS) MODELING** Triangle *ABC* has vertices *A*(2, 1), *B*(−6, 5), and *C*(−2, −3).

   a. An isosceles triangle has two sides with equal length. Is triangle *ABC* isosceles? Explain. **Yes; *AB* = *BC* = $4\sqrt{5}$.**

   b. An equilateral triangle has three sides of equal length. Is triangle *ABC* equilateral? Explain. **No; *AC* = $4\sqrt{2}$.**

   c. Triangle *EFG* is formed by joining the midpoints of the sides of triangle *ABC*. What type of triangle is *EFG*? Explain.

   d. Describe any relationship between the lengths of the sides of the two triangles.

 596 | Lesson 9-1 | Midpoint and Distance Formulas

JPL/NASA

Differentiated Homework Options

Level	Assignment	Two-Day Option	
AL Basic	10–24, 43–66	11–23 odd, 47–50	10–24 even, 43–46, 51–66
OL Core	11–37 odd, 38–41, 43–66	10–24, 47–50	25–41, 43–46, 51–66
BL Advanced	25–63, (optional: 64–66)		

39 **PACKAGE DELIVERY** To determine the mileage between cities for their overnight delivery service, a package delivery service superimposes a coordinate grid over the United States. Each side of a grid unit is equal to 0.316 mile. Suppose the locations of two distribution centers are at (132, 428) and (254, 105). Find the actual distance between these locations to the nearest mile. **109 mi**

40. **HIKING** Orlando wants to hike from his camp to a waterfall. The waterfall is 5 miles south and 8 miles east of his campsite.

  a. Use the Distance Formula to determine how far the waterfall is from the campsite. $\sqrt{89} \approx 9.4$ mi

  b. Verify your answer in part **a** by using the Pythagorean Theorem to determine the distance between the campsite and the waterfall.

  c. Orlando wants to stop for lunch halfway to the waterfall. If the camp is at the origin, where should he stop? $\left(4, -\dfrac{5}{2}\right)$

40b. $8^2 + 5^2 = c^2$
$64 + 25 = c^2$
$89 = c^2$
$c = \sqrt{89}$ or about 9.4 mi

41. **MULTIPLE REPRESENTATIONS** Triangle XYZ has vertices X(4, 9), Y(8, −9), and Z(−6, 5). **41b.** midpoint of $\overline{XY}$ = (6, 0); midpoint of $\overline{YZ}$ = (1, −2); midpoint of $\overline{XZ}$ = (−1, 7)

  a. **Concrete** Draw △XYZ on a coordinate plane. **See margin.**

  b. **Numerical** Find the coordinates of the midpoint of each side of the triangle.

  c. **Geometric** Find the perimeter of △XYZ and the perimeter of the triangle with vertices at the points found in part **b**.

  d. **Analytical** How do the perimeters in part **c** compare? **The perimeter of △XYZ is twice the perimeter of the smaller triangle.**

41c. The perimeter of △XYZ is $2\sqrt{29} + 14\sqrt{2} + 2\sqrt{85}$ units.
perimeter = $\sqrt{29} + 7\sqrt{2} + \sqrt{85}$

H.O.T. Problems Use Higher-Order Thinking Skills

42. **CHALLENGE** Find the coordinates of the point that is three fourths of the way from $P(-1, 12)$ to $Q(5, -10)$. $\left(\dfrac{7}{2}, -\dfrac{9}{2}\right)$

43. **REASONING** Identify all the points in a plane that are 3 units or less from the point (5, 6). What figure does this make? **a circle and its interior with center at (5, 6) and radius 3 units**

44. **CCSS ARGUMENTS** Triangle ABC is a right triangle.

  a. Find the midpoint of the hypotenuse. Call it point Q. $Q\left(\dfrac{9}{2}, 7\right)$

  b. Classify △BQC according to the lengths of its sides. Include sufficient evidence to support your conclusion.

  c. Classify △BQA according to its angles. **obtuse**

44b. $BC = 6$, $QC = \dfrac{\sqrt{205}}{2}$, $BQ = \dfrac{\sqrt{205}}{2}$; $\overline{QC}$ and $\overline{BQ}$ are the same length; therefore, △BQC is isosceles.

45. **OPEN ENDED** Plot two points, and find the distance between them. Does it matter which ordered pair is first when using the Distance Formula? Explain. **See margin.**

46. **WRITING IN MATH** Explain how the Midpoint Formula can be used to approximate the halfway point between two locations on a map. **See margin.**

CCSS **Teaching the Mathematical Practices**

Arguments Mathematically proficient students understand and use stated assumptions, definitions, and previously established results in constructing arguments. They make conjectures and build a logical progression of statements to explore the truth of their conjectures. And they are able to analyze situations by breaking them into cases, and can recognize and use counterexamples.

Enrichment
OL **BL**

NAME _____ DATE _____ PERIOD _____
9-1 **Enrichment**

Distance Between Points in Space

The Distance Formula and Midpoint Formula on the coordinate plane is derived from the Pythagorean Theorem $a^2 + b^2 = c^2$.

In three dimensions, the coordinate grid contains the z-axis and the y-axis, as in two-dimensional geometry, and also a z-axis. An example of a line segment drawn on a three-dimensional coordinate grid is shown at the right.

The three-dimensional distance formula is much like the one for two dimensions. The distance from A (x_1, y_1, z_1) to B (x_2, y_2, z_2) can be found using $d = \sqrt{(x_1 - x_2)^2 + (y_1 - y_2)^2 + (z_1 - z_2)^2}$.

Example Find the distance between the points (1, 3, 0) and (0, 1, 3).

$d = \sqrt{(1-0)^2 + (3-1)^2 + (0-3)^2}$
$d = \sqrt{1 + 4 + 9}$ or $\sqrt{14}$

NAME _____ DATE _____ PERIOD _____
9-1 **Study Guide and Intervention**
Midpoint and Distance Formulas

Practice
AL **OL** **BL** **ELL**

NAME _____ DATE _____ PERIOD _____
9-1 **Practice**
Midpoint and Distance Formulas

Word Problem Practice
AL **OL** **BL** **ELL**

NAME _____ DATE _____ PERIOD _____
9-1 **Word Problem Practice**
Midpoint and Distance Formulas

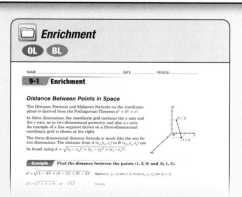

4 Assess

Name the Math Have students plot two points on a coordinate grid and explain how to find the distance between them and the coordinates of the midpoint of the segment joining them.

Additional Answer

54d.

Depth (ft)	Pressure (psi)
0	0
1	0.43
2	0.86
3	1.29
4	1.72

Standardized Test Practice

47. SHORT RESPONSE You currently earn $8.10 per hour and your boss gives you a 10% raise. What is your new hourly wage? **$8.91**

48. SAT/ACT A right circular cylinder has a radius of 3 and a height of 5. Which of the following dimensions of a rectangular solid will have a volume closest to that of the cylinder? **E**

- **A** 5, 5, 6
- **B** 5, 6, 6
- **C** 5, 5, 5
- **D** 4, 5, 6
- **E** 3, 5, 9

49. GEOMETRY If the sum of the lengths of the two legs of a right triangle is 49 inches and the hypotenuse is 41 inches, find the longer of the two legs. **G**

- **F** 9 in.
- **G** 40 in.
- **H** 42 in.
- **J** 49 in.

50. Five more than 3 times a number is 17. Find the number. **B**

- **A** 3
- **B** 4
- **C** 5
- **D** 6

Spiral Review

Solve each equation. Check your solutions. (Lesson 8-6)

51. $\frac{12}{v^2 - 16} - \frac{24}{v - 4} = 3$ $-6, -2$

52. $\frac{w}{w - 1} + w = \frac{4w - 3}{w - 1}$ 3

53. $\frac{4n^2}{n^2 - 9} - \frac{2n}{n + 3} = \frac{3}{n - 3}$ $\frac{3}{2}$

54. SWIMMING When a person swims under water, the pressure in his or her ears varies directly with the depth at which he or she is swimming. (Lesson 8-5)

a. Write a direct variation equation that represents this situation. $P = 0.43d$

b. Find the pressure at 60 feet. **25.8 psi**

c. It is unsafe for amateur divers to swim where the water pressure is more than 65 pounds per square inch. How deep can an amateur diver safely swim? **about 151 ft**

d. Make a table showing the number of pounds of pressure at various depths of water. Use the data to draw a graph of pressure versus depth. **See margin.**

4.3 pounds per square inch (psi)

10ft

Solve each equation or inequality. Round to the nearest ten-thousandth. (Lesson 7-6)

55. $9^{z - 4} = 6.28$ **4.8362**

56. $8.2^{n - 3} = 42.5$ **4.7820**

57. $2.1^{t - 5} = 9.32$ **8.0086**

58. $8^{2n} > 52^{4n + 3}$ $\{n \mid n < -1.0178\}$

59. $7^{p + 2} \leq 13^{5 - p}$ $\{p \mid p \leq 1.9803\}$

60. $3^{y + 2} \geq 8^{3y}$ $\{y \mid y \leq 0.4275\}$

Solve each equation. (Lesson 6-7)

61. $(6n - 5)^{\frac{1}{3}} + 3 = -2$ -20

62. $(5x + 7)^{\frac{1}{5}} + 3 = 5$ 5

63. $(3x - 2)^{\frac{1}{5}} + 6 = 5$ $\frac{1}{3}$

64. $y = -(x + 2)^2 + 12$; $(-2, 12)$; $x = -2$; down **65.** $y = (x - 3)^2 - 8$; $(3, -8)$; $x = 3$; up

Skills Review

Write each quadratic equation in vertex form. Then identify the vertex, axis of symmetry, and direction of opening. **66.** $y = -2(x - 5)^2 + 15$; $(5, 15)$; $x = 5$; down

64. $y = -x^2 - 4x + 8$

65. $y = x^2 - 6x + 1$

66. $y = -2x^2 + 20x - 35$

 598 | Lesson 9-1 | Midpoint and Distance Formulas

DifferentiatedInstruction **OL** **BL**

Extension Ask students how they would go about finding the coordinates of a point three-fourths of the distance from $A(1, -5)$ to $B(-4, -7)$. Then have students find the point. Sample answer: First find the midpoint of A and B, and call it C. Then find D, the midpoint of C and B. Point D is three-fourths of the distance from A to B. Since the midpoint is at $C\left(-\frac{3}{2}, -6\right)$, the point three-fourths of the distance is at $D\left(-2\frac{3}{4}, -6\frac{1}{2}\right)$.

LESSON 9-2 Parabolas

:: Then	:: Now	:: Why?
● You graphed quadratic functions.	**1** Write equations of parabolas in standard form. **2** Graph parabolas.	● Satellite dishes can be used to send and receive signals and can be seen attached to residential homes and businesses. A satellite dish is a type of antenna constructed to receive signals from orbiting satellites. The signals are reflected off of the dish's parabolic surface to a common collection point.

 NewVocabulary
parabola
focus
directrix
latus rectum
standard form
general form

 Common Core State Standards

Content Standards
A.SSE.1.b Interpret complicated expressions by viewing one or more of their parts as a single entity.
A.CED.2 Create equations in two or more variables to represent relationships between quantities; graph equations on coordinate axes with labels and scales.

Mathematical Practices
1 Make sense of problems and persevere in solving them.

1 Equations of Parabolas A **parabola** can be defined as the set of all points in a plane that are the same distance from a given point called the **focus** and a given line called the **directrix**.

The line segment through the focus of a parabola and perpendicular to the axis of symmetry is called the **latus rectum**. The endpoints of the latus rectum lie on the parabola.

KeyConcept Equations of Parabolas

Form of Equation	$y = a(x - h)^2 + k$	$x = a(y - k)^2 + h$
Direction of Opening	upward if $a > 0$, downward if $a < 0$	right if $a > 0$, left if $a < 0$
Vertex	(h, k)	(h, k)
Axis of Symmetry	$x = h$	$y = k$
Focus	$\left(h, k + \dfrac{1}{4a}\right)$	$\left(h + \dfrac{1}{4a}, k\right)$
Directrix	$y = k - \dfrac{1}{4a}$	$x = h - \dfrac{1}{4a}$
Length of Latus Rectum	$\left|\dfrac{1}{a}\right|$ units	$\left|\dfrac{1}{a}\right|$ units

The **standard form** of the equation of a parabola with vertex (h, k) and axis of symmetry $x = h$ is $y = a(x - h)^2 + k$.

• If $a > 0$, k is the minimum value of the related function and the parabola opens upward.

• If $a < 0$, k is the maximum value of the related function and the parabola opens downward.

An equation of a parabola in the form $y = ax^2 + bx + c$ is the **general form**. Any equation in general form can be written in standard form. The shape of a parabola and the distance between the focus and directrix depend on the value of a in the equation.

 connectED.mcgraw-hill.com **599**

1 Focus

VerticalAlignment

▼ **Before Lesson 9-2** Graph quadratic functions.

▼ **Lesson 9-2** Write equations of parabolas in standard form. Graph parabolas.

▼ **After Lesson 9-2** Identify the conic section from a given equation.

2 Teach

Scaffolding Questions
Have students read the **Why?** section of the lesson.

Ask:

▪ What does it mean to say that a satellite dish's surface is *parabolic*? cross-sections of its surface taken perpendicular to its base are parabolas

▪ Why is a parabola the perfect shape for a satellite dish? The parabola reflects all parallel incoming rays to the same point.

Lesson 9-2 Resources

Resource	Approaching Level **AL**	On Level **OL**	Beyond Level **BL**	English Learners **ELL**
Teacher Edition	▪ Differentiated Instruction, p. 600	▪ Differentiated Instruction, pp. 600, 605	▪ Differentiated Instruction, p. 605	
Chapter Resource Masters	▪ Study Guide and Intervention, pp. 11–12 ▪ Skills Practice, p. 13 ▪ Practice, p. 14 ▪ Word Problem Practice, p. 15	▪ Study Guide and Intervention, pp. 11–12 ▪ Skills Practice, p. 13 ▪ Practice, p. 14 ▪ Word Problem Practice, p. 15 ▪ Enrichment, p. 16 ▪ Spreadsheet Activity, p. 17	▪ Practice, p. 14 ▪ Word Problem Practice, p. 15 ▪ Enrichment, p. 16	▪ Study Guide and Intervention, pp. 11–12 ▪ Skills Practice, p. 13 ▪ Practice, p. 14 ▪ Word Problem Practice, p. 15
Other	▪ 5-Minute Check 9-2 ▪ Study Notebook ▪ Teaching Algebra with Manipulatives	▪ 5-Minute Check 9-2 ▪ Study Notebook ▪ Teaching Algebra with Manipulatives	▪ 5-Minute Check 9-2 ▪ Study Notebook	▪ 5-Minute Check 9-2 ▪ Study Notebook ▪ Teaching Algebra with Manipulatives

Sean Russell/ifStop/Getty Images

1 Equations of Parabolas

Example 1 shows how to write an equation of a parabola in standard form and then analyze the equation to identify the vertex, axis of symmetry, and direction of opening of the parabola.

Formative Assessment
Use the Guided Practice exercises after each example to determine students' understanding of concepts.

Additional Example

1 Write $y = -x^2 - 2x + 3$ in standard form. Identify the vertex, axis of symmetry, and direction of opening of the parabola. $y = -(x + 1)^2 + 4$; vertex: $(-1, 4)$; axis of symmetry: $x = -1$; opens downward

Additional Examples also in Interactive Classroom PowerPoint® Presentations

IWB Interactive White Board READY

2 Graph Parabolas

Example 2 shows how to use the parent function, $y = ax^2$, symmetry, and translations to graph the equation of a parabola with a vertical axis of symmetry. **Example 3** shows how to graph the equation of a parabola that is not in standard form and has a horizontal axis of symmetry. **Example 4** shows how to write and graph the equation of a parabola given the vertex and directrix. **Example 5** shows how to write and graph the equation of a parabola to solve a real-world problem.

CCSS Teaching the Mathematical Practices

Structure Mathematically proficient students look closely to discern a pattern or structure. Encourage students to identify all important values before beginning their graphs.

ReviewVocabulary

Completing the Square rewriting a quadratic expression as a perfect square trinomial

Example 1 Analyze the Equation of a Parabola

Write $y = 2x^2 - 12x + 6$ in standard form. Identify the vertex, axis of symmetry, and direction of opening of the parabola.

$y = 2x^2 - 12x + 6$	Original equation
$= 2(x^2 - 6x) + 6$	Factor 2 from the x- and x^2-terms.
$= 2(x^2 - 6x + \blacksquare) + 6 - 2(\blacksquare)$	Complete the square on the right side.
$= 2(x^2 - 6x + 9) + 6 - 2(9)$	The 9 added when you complete the square is multiplied by 2.
$= 2(x - 3)^2 - 12$	Factor.

The vertex of this parabola is located at $(3, -12)$, and the equation of the axis of symmetry is $x = 3$. The parabola opens upward.

> **Guided**Practice
>
> $y = 4(x + 2)^2 + 18$; vertex: $(h, k) = (-2, 18)$; axis of symmetry: $x = -2$; opens upward
>
> **1.** Write $y = 4x^2 + 16x + 34$ in standard form. Identify the vertex, axis of symmetry, and direction of opening of the parabola.

2 Graph Parabolas In Chapter 4, you learned that the graph of the quadratic equation $y = a(x - h)^2 + k$ is a transformation of the parent graph of $y = x^2$ translated h units horizontally and k units vertically, and reflected and/or dilated depending on the value of a.

WatchOut!

CCSS Structure Carefully examine the values for h and k before beginning to graph an equation.

- If h is positive, translate the graph h units to the right.
- If h is negative, translate the graph $|h|$ units to the left.
- If k is positive, translate the graph k units up.
- If k is negative, translate the graph $|k|$ units down.

Example 2 Graph Parabolas

Graph each equation.

a. $y = -3x^2$

For this equation, $h = 0$ and $k = 0$. The vertex is at the origin. Since the equation of the axis of symmetry is $x = 0$, substitute some small positive integers for x and find the corresponding y-values.

x	y
1	-3
2	-12
3	-27

Since the graph is symmetric about the y-axis, the points at $(-1, -3)$, $(-2, -12)$, and $(-3, -27)$ are also on the parabola. Use all of these points to draw the graph.

b. $y = -3(x - 4)^2 + 5$

The equation is of the form $y = a(x - h)^2 + k$, where $h = 4$ and $k = 5$. The graph of this equation is the graph of $y = -3x^2$ in part **a** translated 4 units to the right and up 5 units. The vertex is now at $(4, 5)$.

> **Guided**Practice
>
> 2A, 2B. See Chapter 9 Answer Appendix.
>
> **2A.** $y = 2x^2$ **2B.** $y = 2(x - 1)^2 - 4$

DifferentiatedInstruction AL OL

If some students think that any curve can be called a parabola,

Then explain that only curves with a certain well-defined shape meet the definition of a parabola. Have students use the Internet to research objects in the real world that are parabolas. Ask them to make a poster to display their findings.

StudyTip

Graphing When graphing these functions, it may be helpful to sketch the graph of the parent function.

Equations of parabolas with vertical axes of symmetry have the parent function $y = x^2$ and are of the form $y = a(x - h)^2 + k$. These are functions. Equations of parabolas with horizontal axes of symmetry are of the form $x = a(y - k)^2 + h$ and are not functions. The parent graph for these equations is $x = y^2$.

Example 3 Graph an Equation in General Form

Graph each equation.

a. $2x - y^2 = 4y + 10$

Step 1 Write the equation in the form $x = a(y - k)^2 + h$.

$2x - y^2 = 4y + 10$	Original equation
$2x = y^2 + 4y + 10$	Add y^2 to each side to isolate the x-term.
$2x = (y^2 + 4y + \blacksquare) + 10 - \blacksquare$	Complete the square.
$2x = (y^2 + 4y + 4) + 10 - 4$	Add and subtract 4, since $\left(\frac{4}{2}\right)^2 = 4$.
$2x = (y + 2)^2 + 6$	Factor and subtract.
$x = \frac{1}{2}(y + 2)^2 + 3$	$(h, k) = (3, -2)$

Step 2 Use the equation to find information about the graph. Then draw the graph based on the parent graph, $x = y^2$.

vertex: $(3, -2)$

axis of symmetry: $y = -2$

focus: $\left(3 + \dfrac{1}{4\left(\frac{1}{2}\right)}, -2\right)$ or $(3.5, -2)$

directrix: $x = 3 - \dfrac{1}{4\left(\frac{1}{2}\right)}$ or 2.5

direction of opening: right, since $a > 0$

length of latus rectum: $\left|\dfrac{1}{\left(\frac{1}{2}\right)}\right|$ or 2 units

ReadingMath

latus rectum from the Latin *latus*, meaning side, and *rectum*, meaning straight

b. $y + 2x^2 + 32 = -16x - 1$

Step 1

$y + 2x^2 + 32 = -16x - 1$	Original equation
$y = -2x^2 - 16x - 33$	Solve for y.
$y = -2(x^2 + 8x + \blacksquare) - 33 - \blacksquare$	Complete the square.
$y = -2(x^2 + 8x + 16) - 33 - (-32)$	Add and subtract -32.
$y = -2(x + 4)^2 - 1$	Factor and simplify.

Step 2 vertex: $(-4, -1)$

axis of symmetry: $x = -4$

focus: $\left(-4, -\dfrac{9}{8}\right)$

directrix: $y = -\dfrac{7}{8}$

length of latus rectum: $\dfrac{1}{2}$ unit

opens downward

▶ **Guided**Practice **3A, 3B.** See Chapter 9 Answer Appendix.

3A. $3x - y^2 = 4x + 25$

3B. $y = x^2 + 6x - 4$

Additional Examples

2 Graph each equation.

a. $y = 2x^2$

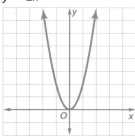

b. $y = 2(x - 1)^2 - 5$

3 Graph each equation.

a. $x + y^2 = 4y - 1$

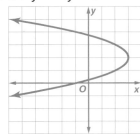

b. $y + 2x^2 + 10 = 8x + 5$

Teach with Tech

Video Recording Break the class into groups, and give each group an equation of a different parabola. Have the group create a video showing how to find all of the properties and information about the parabola. Share each group's video with the entire class.

Tips for New Teachers

Building on Prior Knowledge Remind students that the distance from a point to a line is measured on the perpendicular from the point to the line.

Additional Examples

4 Write an equation for a parabola with vertex at (8, 6) and focus (2, 6). Then graph the equation.

$x = -\frac{1}{24}(y - 6)^2 + 8$

5 **BRIDGES** The Hulme Arch Bridge in Manchester, England, is supported by cables suspended from a parabolic steel arch. The highest point of the arch is 25 meters above the bridge, and the focus of the arch is 18 meters above the bridge.

a. Let the bridge be the *x*–axis, and let the *y*-axis pass through the vertex of the arch. Write an equation that models the arch.

$y = -\frac{1}{28}x^2 + 25$

b. Graph the equation.

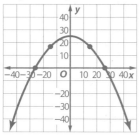

Additional Answers (Guided Practice)

4A. $y = \frac{1}{8}(x - 1)^2 + 3$

You can use specific information about a parabola to write an equation and draw a graph.

Example 4 Write an Equation of a Parabola

Write an equation for a parabola with vertex at (−2, −4) and directrix $y = 1$. Then graph the equation.

The directrix is a horizontal line, so the equation of the parabola is of the form $y = a(x - h)^2 + k$. Find *a*, *h*, and *k*.

- The vertex is at (−2, −4), so $h = -2$ and $k = -4$.
- Use the equation of the directrix to find *a*.

$y = k - \dfrac{1}{4a}$	Equation of directrix
$1 = -4 - \dfrac{1}{4a}$	Replace *y* with 1 and *k* with −4.
$5 = -\dfrac{1}{4a}$	Add 4 to each side.
$20a = -1$	Multiply each side by 4*a*.
$a = -\dfrac{1}{20}$	Divide each side by 20.

So, the equation of the parabola is $y = -\frac{1}{20}(x + 2)^2 - 4$.

▶ **Guided**Practice

4A, 4B. See margin.

Write an equation for each parabola described below. Then graph the equation.

4A. vertex (1, 3), focus (1, 5) **4B.** focus (5, 6), directrix $x = -2$

Parabolas are often used in the real world.

Real-World Example 5 Write an Equation for a Parabola

ENVIRONMENT Solar energy may be harnessed by using parabolic mirrors. The mirrors reflect the rays from the Sun to the focus of the parabola. The focus of each parabolic mirror at the facility described at the left is 6.25 feet above the vertex. The latus rectum is 25 feet long.

a. Assume that the focus is at the origin. Write an equation for the parabola formed by each mirror.

In order for the mirrors to collect the Sun's energy, the parabola must open upward. Therefore, the vertex must be below the focus.

focus: (0, 0)   vertex: (0, −6.25)

The measure of the latus rectum is 25. So $25 = \left|\frac{1}{a}\right|$, and $a = \frac{1}{25}$.

Using the form $y = a(x - h)^2 + k$, an equation for the parabola formed by each mirror is $y = \frac{1}{25}x^2 - 6.25$.

b. Graph the equation.

Now use all of the information to draw a graph.

▶ **Guided**Practice **5.** See margin.

5. Write and graph an equation for a parabolic mirror that has a focus 4.5 feet above the vertex and a latus rectum that is 18 feet long, when the focus is at the origin.

Real-World Link

In California's Mojave Desert, parabolic mirrors are used to heat oil that flows through tubes placed at the focus. The heated oil is used to produce electricity.

Source: Solel

4B. $x = \frac{1}{14}(y - 6)^2 + 1.5$

5. $y = \frac{1}{18}x^2 - 4.5$

Example 1 Write each equation in standard form. Identify the vertex, axis of symmetry, and direction of opening of the parabola. **1–4. See margin.**

1. $y = 2x^2 - 24x + 40$

2. $y = 3x^2 - 6x - 4$

3. $x = y^2 - 8y - 11$

4. $x + 3y^2 + 12y = 18$

Examples 2–3 Graph each equation. **5–8. See Chapter 9 Answer Appendix.**

5. $y = (x - 4)^2 - 6$

6. $y = 4(x + 5)^2 + 3$

7. $y = -3x^2 - 4x - 8$

8. $x = 3y^2 - 6y + 9$

9–12. See Chapter 9 Answer Appendix.

Example 4 Write an equation for each parabola described below. Then graph the equation.

9. vertex $(0, 2)$, focus $(0, 4)$

10. vertex $(-2, 4)$, directrix $x = -1$

11. focus $(3, 2)$, directrix $y = 8$

12. vertex $(-1, -5)$, focus $(-5, -5)$

Example 5 **13. ASTRONOMY** Consider a parabolic mercury mirror like the one described at the beginning of the lesson. The focus is 6 feet above the vertex and the latus rectum is 24 feet long.

    a. Assume that the focus is at the origin. Write an equation for the parabola formed by the parabolic microphone. $y = \frac{1}{24}x^2 - 6$

    b. Graph the equation. **See margin.**

Practice and Problem Solving   Extra Practice is on page R9.

Example 1 Write each equation in standard form. Identify the vertex, axis of symmetry, and direction of opening of the parabola. **14–19. See margin.**

14. $y = x^2 - 8x + 13$

15. $y = 3x^2 + 42x + 149$

16. $y = -6x^2 - 36x - 8$

17. $y = -3x^2 - 9x - 6$

18. $x = \frac{1}{3}y^2 - 3y + 4$

19. $x = \frac{2}{3}y^2 - 4y + 12$

Examples 2–3 Graph each equation. **20–25. See Chapter 9 Answer Appendix.**

20. $y = \frac{1}{3}x^2$

21. $y = -2x^2$

22. $y = -2(x - 2)^2 + 3$

23. $y = 3(x - 3)^2 - 5$

24. $x = \frac{1}{2}y^2$

25. $4x - y^2 = 2y + 13$

Example 4 Write an equation for each parabola described below. Then graph the equation.

26. vertex $(0, 1)$, focus $(0, 4)$

27. vertex $(1, 8)$, directrix $y = 3$ **26–31. See Chapter 9 Answer Appendix.**

28. focus $(-2, -4)$, directrix $x = -6$

29 focus $(2, 4)$, directrix $x = 10$

30. vertex $(-6, 0)$, directrix $x = 2$

31. vertex $(9, 6)$, focus $(9, 5)$

Example 5 **32. BASEBALL** When a ball is thrown, the path it travels is a parabola. Suppose a baseball is thrown from ground level, reaches a maximum height of 50 feet, and hits the ground 200 feet from where it was thrown. Assuming this situation could be modeled on a coordinate plane with the focus of the parabola at the origin, find the equation of the parabolic path of the ball. Assume the focus is on ground level. $y = -\frac{1}{200}x^2 + 50$

33. CCSS PERSEVERANCE Ground antennas and satellites are used to relay signals between the NASA Mission Operations Center and the spacecraft it controls. One such parabolic dish is 146 feet in diameter. Its focus is 48 feet from the vertex. **a. See Chapter 9 Answer Appendix.**

    a. Sketch two options for the dish, one that opens up and one that opens left.

    b. Write two equations that model the sketches in part **a.** $y = \frac{x^2}{192}$ and $x = \frac{y^2}{-192}$

    c. If you wanted to find the depth of the dish, does it matter which equation you use? Why or why not?   Sample answer: No; except for the direction in which the graphs are identical. they open,

  connectED.mcgraw-hill.com **603**

Differentiated Homework Options

Level	Assignment	Two-Day Option	
AL Basic	14–33, 37–61	15–33 odd, 41–44	14–32 even, 37–40, 45–61
OL Core	15–30 odd, 32–61	14–33, 41–44	34–40, 45–61
BL Advanced	34–57, (optional: 58–61)		

Formative Assessment

Use Exercises 1–13 to check for understanding.

Use the chart at the bottom of this page to customize assignments for your students.

Additional Answers

1. $y = 2(x - 6)^2 - 32$; vertex $(6, -32)$; axis of symmetry: $x = 6$; opens upward

2. $y = 3(x - 1)^2 - 7$; vertex $(1, -7)$; axis of symmetry: $x = 1$; opens upward

3. $x = (y - 4)^2 - 27$; vertex $(-27, 4)$; axis of symmetry: $y = 4$; opens right

4. $x = -3(y + 2)^2 + 30$; vertex $(30, -2)$; axis of symmetry: $y = -2$; opens left

13b.

14. $y = (x - 4)^2 - 3$; vertex $= (4, -3)$; axis of symmetry: $x = 4$; opens upward

15. $y = 3(x + 7)^2 + 2$; vertex $= (-7, 2)$; axis of symmetry: $x = -7$; opens upward

16. $y = -6(x + 3)^2 + 46$; vertex $= (-3, 46)$; axis of symmetry: $x = -3$; opens downward

17. $y = -3\left(x + \frac{3}{2}\right)^2 + \frac{3}{4}$; vertex $= \left(-\frac{3}{2}, \frac{3}{4}\right)$; axis of symmetry: $x = -\frac{3}{2}$; opens downward

18. $x = \frac{1}{3}(y - 4.5)^2 - 2.75$; vertex $= (-2.75, 4.5)$; axis of symmetry: $y = 4.5$; opens right

19. $x = \frac{2}{3}(y - 3)^2 + 6$; vertex $= (6, 3)$; axis of symmetry: $y = 3$; opens right

Study Guide and Intervention
AL OL ELL

NAME _____ DATE _____ PERIOD _____

9-2 Study Guide and Intervention
Parabolas

Equations of Parabolas A parabola is a curve consisting of all points in the coordinate plane that are the same distance from a given point (the *focus*) and a given line (the *directrix*). The following chart summarizes important information about parabolas.

Standard Form of Equation	$y = a(x - h)^2 + k$	$x = a(y - k)^2 + h$
Axis of Symmetry	$x = h$	$y = k$
Vertex	(h, k)	(h, k)
Focus	$\left(h, k + \frac{1}{4a}\right)$	$\left(h + \frac{1}{4a}, k\right)$
Directrix	$y = k - \frac{1}{4a}$	$x = h - \frac{1}{4a}$
Direction of Opening	upward if $a > 0$, downward if $a < 0$	right if $a > 0$, left if $a < 0$
Length of Latus Rectum	$\left|\frac{1}{a}\right|$ units	$\left|\frac{1}{a}\right|$ units

Example Write $y = 2x^2 - 12x - 25$ in standard form. Identify the vertex, axis of symmetry, and direction of opening of the parabola.

$y = 2x^2 - 12x - 25$ — Original equation
$y = 2(x^2 - 6x) - 25$ — Factor 2 from x terms.
$y = 2(x^2 - 6x + 9) - 25 - 2(9)$ — Complete the square on the right side.
$y = 2(x^2 - 6x + 9) - 25 - 2(9)$ — The 9 is added to complete the square is multiplied by 2.
$y = 2(x - 3)^2 - 43$ — Write in standard form.

The vertex of this parabola is located at $(3, -43)$, the equation of the axis of symmetry is $x = 3$, and the parabola opens upward.

Exercises
Write each equation in standard form. Identify the vertex, axis of symmetry, and direction of opening of the parabola.

1. $y = x^2 + 2x + 1$ 2. $y = -x^2$ 3. $y = 4x^2 + 4x - 15$

4. $y = x^2 + 6x - 4$ 5. $y = 8x - 2x^2 + 10$ 6. $x = y^2 - 8y + 6$

Practice
AL OL BL ELL

NAME _____ DATE _____ PERIOD _____

9-2 Practice
Parabolas

Write each equation in standard form. Identify the vertex, axis of symmetry, and direction of opening of the parabola.

1. $y = 2x^2 - 12x + 19$ 2. $y = \frac{1}{2}x^2 + 3x + \frac{3}{2}$ 3. $y = -8x^2 - 12x - 7$

Graph each equation.

4. $y = (x - 4)^2 + 3$ 5. $x = -\frac{1}{2}y^2 + 1$ 6. $x = 3(y + 1)^2 - 3$

Write an equation for each parabola described below. Then graph the equation.

7. vertex $(0, -4)$, focus $\left(0, -3\frac{7}{8}\right)$
8. vertex $(-1, 1)$, directrix $x = -3$
9. vertex $(1, 3)$, latus rectum: 2 units, $a < 0$, opens vertically

10. **TELEVISION** Write the equation in the form $y = ax^2$ for a satellite dish. Assume that the bottom of the upward-facing dish passes through $(0, 0)$ and that the distance from the bottom to the focus point is 8 inches.

Word Problem Practice
AL OL BL ELL

NAME _____ DATE _____ PERIOD _____

9-2 Word Problem Practice
Parabolas

1. **PROJECTILE** A projectile follows the graph of the parabola $y = -\frac{3}{2}x^2 + 6x$. Sketch the path of the projectile by graphing the equation.

2. **COMMUNICATION** David has just made a large parabolic dish whose cross section is based on the graph of the parabola $y = 0.25x^2$. Each unit represents one foot and the diameter of his dish is 4 feet. He wants to make a listening device by placing a microphone at the focus of the parabola. Where should the microphone be placed?

3. **BRIDGES** A bridge is in the shape of a parabola that opens downward. The equation of the parabola to model the arch of the bridge is given by $y = -\frac{x^2}{64} + \frac{2}{3}x + \frac{14}{3}$, where each unit is equivalent to 1 yard. The x-axis is the ground level. What is the maximum height of the bridge above the ground?

4. **TELESCOPES** An astronomer is working with a large reflecting telescope. The reflecting mirror in the telescope has the parabolic cross section shown in the graph whose equation is given by $y = \frac{1}{8}(x - 4)^2 + 2$. Each unit represents 1 meter. The astronomer is standing at the origin. How far from the vertex of the parabola is the point on the mirror directly over the astronomer's head?

5. **BRIDGES** Part of the Sydney Harbor Bridge in Sydney, Australia, can be modeled by a parabolic arch. If each unit corresponds to 10 meters, the arch would pass through the points at $(-25, 5)$, $(0, 10)$, and $(25, 5)$.

a. Write the equation of the parabola to model the arch.

b. Identify the coordinates of the focus of this parabola.

Enrichment
OL BL

NAME _____ DATE _____ PERIOD _____

9-2 Enrichment

Limits

Sequences of numbers with a rational expression for the general term often approach some number as a finite limit. For example, the reciprocals of the positive integers approach 0 as n gets larger and larger. This is written using the notation shown below. The symbol ∞ stands for infinity and $n \to \infty$ means that n is getting larger and larger, or "n goes to infinity."

$1, \frac{1}{2}, \frac{1}{3}, \frac{1}{4}, \ldots, \frac{1}{n}, \ldots$ $\lim_{n \to \infty} \frac{1}{n} = 0$

Example Find $\lim_{n \to \infty} \frac{n^2}{(n + 1)^2}$.

It is not immediately apparent whether the sequence approaches a limit or not. But notice what happens if we divide the numerator and denominator of the general term by n^2.

$\frac{n^2}{(n + 1)^2} = \frac{n^2}{n^2 + 2n + 1}$

---

34. **UMBRELLAS** A beach umbrella has an arch in the shape of a parabola that opens downward. The umbrella spans 6 feet across and is $1\frac{1}{2}$ feet high. Write an equation of a parabola to model the arch, assuming that the origin is at the point where the pole and umbrella meet at the vertex of the arch. $y = -\frac{1}{6}x^2 + \frac{3}{2}$

35. **AUTOMOBILES** An automobile headlight contains a parabolic reflector. The light coming from the source bounces off the parabolic reflector and shines out the front of the headlight. The equation of the cross section of the reflector is $y = \frac{1}{12}x^2$. How far from the vertex should the filament for the high beams be placed?   **3 units**

36. **MULTIPLE REPRESENTATIONS** Start with a sheet of wax paper that is about 15 inches long and 12 inches wide. **a–c. See students' work.**

a. **Concrete** Make a line that is perpendicular to the sides of the sheet by folding the sheet near one end. Open up the paper again. This line is the directrix. Mark a point about midway between the sides of the sheet so that the distance from the directrix is about 1 inch. This is the focus.

b. **Concrete** Start with a new sheet of wax paper. Form another outline of a parabola with a focus that is about 3 inches from the directrix.

c. **Concrete** On a new sheet of a wax paper, form a third outline of a parabola with a focus that is about 5 inches from the directrix.

d. **Verbal** Compare the shapes of the three parabolas. How does the distance between the focus and the directrix affect the shape of a parabola?
As the distance between the directrix and the focus increases, the parabola becomes wider.

H.O.T. Problems   Use Higher-Order Thinking Skills

37. **REASONING** How do you change the equation of the parent function $y = x^2$ to shift the graph to the right? **Rewrite it as $y = (x - h)^2$, where $h > 0$.**

38. **OPEN ENDED** Two different parabolas have their vertex at $(-3, 1)$ and contain the point with coordinates $(-1, 0)$. Write two possible equations for these parabolas.
Sample answers: $y = -\frac{1}{4}(x + 3)^2 + 1$ and $x = 2(y - 1)^2 - 3$

39. **CCSS CRITIQUE** Brianna and Russell are graphing $\frac{1}{4}y^2 + x = 0$. Is either of them correct? Explain your reasoning.

Brianna

Russell

39. Russell; the parabola should open to the left rather than to the right.

40. **WRITING IN MATH** Why are parabolic shapes used in the real world?   **See margin.**

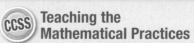

Teaching the Mathematical Practices
CCSS

Critique Mathematically proficient students are also able to compare the effectiveness of two plausible arguments, distinguish correct logic or reasoning from that which is flawed, and—if there is a flaw in an argument—explain what it is.

41. A gardener is placing a fence around a 1320-square-foot rectangular garden. He ordered 148 feet of fencing. If he uses all the fencing, what is the length of the longer side of the garden? **C**

A 30 ft   **C** 44 ft
B 34 ft   **D** 46 ft

42. SAT/ACT When a number is divided by 5, the result is 7 more than the number. Find the number. **F**

F $-\dfrac{35}{4}$ **J** $\dfrac{28}{4}$

G $-\dfrac{35}{6}$ **K** $\dfrac{35}{4}$

H $\dfrac{35}{6}$

43. GEOMETRY What is the area of the following square, if the length of $\overline{BD}$ is $2\sqrt{2}$? **D**

A 1
B 2
C 3
D 4

44. SHORT RESPONSE The measure of the smallest angle of a triangle is two thirds the measure of the middle angle. The measure of the middle angle is three sevenths of the measure of the largest angle. Find the largest angle's measure. **105°**

45. GEOMETRY Find the perimeter of a triangle with vertices at $(2, 4)$, $(-1, 3)$, and $(1, -3)$. (Lesson 9-1) $5\sqrt{2} + 3\sqrt{10}$ units

46. WORK A worker can powerwash a wall of a certain size in 5 hours. Another worker can do the same job in 4 hours. If the workers work together, how long would it take to do the job? Determine whether your answer is reasonable. (Lesson 8-6)

46. $2\dfrac{2}{9}$ h; The answer is reasonable. The time to complete the job when working together must be less than the time it would take either person working alone.

Solve each equation or inequality. Round to the nearest ten-thousandth. (Lesson 7-7)

47. $\ln (x + 1) = 1$ **1.7183**   **48.** $\ln (x - 7) = 2$ **14.3891**

49. $e^x > 1.6$ $x > 0.4700$ **50.** $e^{5x} \geq 25$ $x \geq 0.6438$

Simplify. (Lesson 6-4)

51. $\sqrt{0.25}$ **0.5**   **52.** $\sqrt[3]{-0.064}$ **−0.4**

53. $\sqrt[4]{z^8}$ z^2 **54.** $-\sqrt[6]{x^6}$ $-|x|$

List all of the possible rational zeros of each function. (Lesson 5-8)

55. $h(x) = x^3 + 8x + 6$
$\pm 1, \pm 2, \pm 3, \pm 6$

56. $p(x) = 3x^3 - 5x^2 - 11x + 3$
$\pm 1, \pm \dfrac{1}{3}, \pm 3$

57. $h(x) = 9x^6 - 5x^3 + 27$
$\pm 1, \pm \dfrac{1}{3}, \pm \dfrac{1}{9}, \pm 3, \pm 9, \pm 27$

Simplify each expression.

58. $\sqrt{24}$ $2\sqrt{6}$ **59.** $\sqrt{45}$ $3\sqrt{5}$ **60.** $\sqrt{252}$ $6\sqrt{7}$ **61.** $\sqrt{512}$ $16\sqrt{2}$

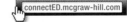

Wax Paper For Exercise 36, students will need three 15″ by 12″ sheets of wax paper.

WatchOut!

Error Analysis In Exercise 39, remind students that they must rewrite the equation in standard form before graphing.

4 Assess

Yesterday's News Have students write how understanding how to complete the square helped them in today's lesson with graphing equations of a parabola.

Formative Assessment
Check for student understanding of concepts in Lessons 9-1 and 9-2.

 Quiz 1, p. 53

Additional Answer

40. Sample answer: When rays which are parallel to the axis of symmetry of a parabolic mirror are reflected off the mirror, they are directed to the focus. This focuses all of the rays at one specific point. Also, a parabolic microphone can be used to make capturing sound more effective, because reflected sound waves are focused at the particular point where the microphone is located.

DifferentiatedInstruction **OL** **BL**

Extension Remind students that if they know two points on a line, they can write the equation of the line. Tell students that to write the equation of a parabola, they need three points on the graph. Ask students to write the equation for a parabola through $(0, 3)$, $(1, 4)$, and $(-2, 7)$. Using the standard quadratic equation $y = ax^2 + bx + c$, substitute each of the given values for x and y. Solve the system of three equations in a, b, and c. Substituting $(0, 3)$: $c = 3$; $(1, 4)$: $a + b + c = 4$; $(-2, 7)$: $4a - 2b + c = 7$. Solving simultaneously, $a = 1$, $b = 0$, and $c = 3$; the equation is $y = x^2 + 3$.

1 Focus

Objective Use a graphing calculator to examine the characteristics of a circle and its equation.

Materials for Each Group
- TI-Nspire technology

Teaching Tip
The calculator opens on the same screen as when it was turned off. Have students press the **home** key to begin the lab.

2 Teach

Working in Cooperative Groups
Have students of mixed abilities work in pairs to complete the Activity.

Ask:
- What do you know about all of the points on a circle? They are the same distance from a given point called the center.

- Is the center of a circle a part of the circle? no

- If the radius of a circle is increased, what happens to the circle? It becomes larger.

Practice Have students complete Exercises 1–4.

3 Assess

Formative Assessment
Use Exercise 3 to assess whether students can use a graphing calculator to examine the characteristics of a circle and its equation.

You can use TI-Nspire technology to examine characteristics of circles and the relationship with an equation of the circle.

Activity

Step 1 Draw a circle.

- Add a new **Graphs** page. Select **Window/Zoom** menu and use the **Windows Setting** tool to adjust the window size as shown. From the **View** menu, select **Show Grid**. Then from the **Shapes** menu, select **Circle**. Place the pointer at the point (2, 2) and press **enter** to set the center of the circle. Move the pointer out, creating a circle like the one shown.

- Use the **Point On** tool from the **Points & Line** menu to place a point on the circle.

- Use the **Segment** tool from the **Points & Line** menu to draw the radius.

Step 2 Add labels.

- From the **Actions** menu, select **Coordinates and Equations**. Use the pointer to select the center of the circle to display its coordinate. Then select the circle to display its equation. Move each display outside the circle.

- Use the **Length** tool from the **Measurement** menu to display the length of the radius.

Step 3 Change the radius.

Move the pointer so that a point on the circle is highlighted, then press and hold the center of the touchpad until it is selected. Examine the equation of the circle. Then move the edge of the circle in. Make note of changes in the equation.

Step 4 Move the center of the circle.

Move the pointer so that the center of the circle is highlighted, then press and hold the center of the touchpad until it is selected. Move the center of the circle. Again, examine the equation of the circle.

Analyze the Results

3. Sample answer: The equation is affected the same as when it was drawn centered in Quadrant I.

1. How does moving the edge of the circle in or out affect the equation of the circle? See margin.

2. What effect does moving the center of the circle have on the equation? See margin.

3. Repeat the activity by placing the center of a circle in Quadrant II. Move the center to each of the other two quadrants. How does the equation change?

4. **MAKE A CONJECTURE** Without graphing, write an equation of each circle.
   a. center: (4, 2), radius: 3
   b. center: (−1, 1), radius: 8
   c. center: (−6, −5), radius: 2.5
   d. center: (h, k), radius: r

4a. $(x − 4)^2 + (y − 2)^2 = 3^2$
4b. $(x + 1)^2 + (y − 1)^2 = 8^2$
4c. $(x + 6)^2 + (y + 5)^2 = 2.5^2$
4d. $(x − h)^2 + (y − k)^2 = r^2$

From Concrete to Abstract
Write the equation $x^2 + y^2 = 100$ on the board. Ask students to identify the center of the circle and the length of the radius without graphing.
center: (0, 0), radius: 10

Additional Answers

1. Sample answer: The squared constant term on the right side of the equation is the same as the length of the radius. It increases as the radius increases and decreases as the radius decreases.

2. Sample answer: The x-coordinate of the center is subtracted from x on the left side of the equation, and the y-coordinate of the center is subtracted from y in the equation.

LESSON 9-3 Circles

::Then	::Now	::Why?
• You graphed and wrote equations of parabolas.	**1** Write equations of circles. **2** Graph circles.	• When an object is thrown into water, ripples move out from the center forming concentric circles. If the point where the object entered the water is assigned coordinates, each ripple can be modeled by an equation of a circle.

 New Vocabulary
circle
center
radius

Common Core State Standards

Content Standards
A.SSE.1.b Interpret complicated expressions by viewing one or more of their parts as a single entity.

A.CED.4 Rearrange formulas to highlight a quantity of interest, using the same reasoning as in solving equations.

Mathematical Practices
4 Model with mathematics.

1 Equations of Circles A **circle** is the set of all points in a plane that are equidistant from a given point in the plane, called the **center**. Any segment with endpoints at the center and a point on the circle is a **radius** of the circle.

Assume that (x, y) are the coordinates of a point on the circle at the right. The center is at (h, k), and the radius is r. You can find an equation of the circle by using the Distance Formula.

$$\sqrt{(x_2 - x_1)^2 + (y_2 - y_1)^2} = d$$ Distance Formula
$(x_1, y_1) = (h, k)$,
$$\sqrt{(x - h)^2 + (y - k)^2} = r$$ $(x_2, y_2) = (x, y)$, $d = r$
$$(x - h)^2 + (y - k)^2 = r^2$$ Square each side.

KeyConcept Equations of Circles

Standard Form of Equation	$x^2 + y^2 = r^2$	$(x - h)^2 + (y - k)^2 = r^2$
Center	$(0, 0)$	(h, k)
Radius	r	r

You can use the standard form of the equation of a circle to write an equation for a circle given the center and the radius or diameter.

Real-World Example 1 Write an Equation Given the Radius

DELIVERY Appliances + More offers free delivery within 35 miles of the store. The Jacksonville store is located 100 miles north and 45 miles east of the corporate office. Write an equation to represent the delivery boundary of the Jacksonville store if the origin of the coordinate system is the corporate office.

Since the corporate office is at $(0, 0)$, the Jacksonville store is at $(45, 100)$. The boundary of the delivery region is the circle centered at $(45, 100)$ with radius 35 miles.

$(x - h)^2 + (y - k)^2 = r^2$ Equation of a circle
$(x - 45)^2 + (y - 100)^2 = 35^2$ $(h, k) = (45, 100)$ and $r = 35$
$(x - 45)^2 + (y - 100)^2 = 1225$ Simplify.

Guided Practice

1. $(x + 3)^2 + (y + 4)^2 = 900$

1. **WI-FI** A certain wi-fi phone has a range of 30 miles in any direction. If the phone is 4 miles south and 3 miles west of headquarters, write an equation to represent the area within which the phone can operate via the Wi-Fi system.

connectED.mcgraw-hill.com **607**

Photodisc/Getty Images

1 Focus

Vertical Alignment

Before Lesson 9-3 Graph and write equations of parabolas.

Lesson 9-3 Write equations of circles. Graph circles.

After Lesson 9-3 Write and graph equations of ellipses.

2 Teach

Scaffolding Questions
Have students read the **Why?** section of the lesson.

Ask:

• What are concentric circles? circles with a common center

• What part of a circle is the point where the rock hits the water? center

• Are all the circles similar? Explain. Yes; they are the same shape, but different sizes.

Lesson 9-3 Resources

Resource	Approaching Level **AL**	On Level **OL**	Beyond Level **BL**	English Learners **ELL**
Teacher Edition	• Differentiated Instruction, p. 608	• Differentiated Instruction, pp. 608, 613	• Differentiated Instruction, pp. 610, 613	
Chapter Resource Masters	• Study Guide and Intervention, pp. 18–19 • Skills Practice, p. 20 • Practice, p. 21 • Word Problem Practice, p. 22	• Study Guide and Intervention, pp. 18–19 • Skills Practice, p. 20 • Practice, p. 21 • Word Problem Practice, p. 22 • Enrichment, p. 23 • Graphing Calculator Activity, p. 24	• Practice, p. 21 • Word Problem Practice, p. 22 • Enrichment, p. 23	• Study Guide and Intervention, pp. 18–19 • Skills Practice, p. 20 • Practice, p. 21 • Word Problem Practice, p. 22
Other	• 5-Minute Check 9-3 • Study Notebook	• 5-Minute Check 9-3 • Study Notebook	• 5-Minute Check 9-3 • Study Notebook	• 5-Minute Check 9-3 • Study Notebook

1 Equations of Circles

Example 1 shows how to write an equation for a circle for a real-world problem given the center and radius. **Example 2** shows how to write an equation for a circle given the graph of the circle. **Example 3** shows how to write an equation for a circle given the endpoints of a diameter.

Formative Assessment

Use the Guided Practice exercises after each example to determine students' understanding of concepts.

Additional Examples

1 **LANDSCAPING** The plan for a park puts the center of a circular pond of radius 0.6 mile at 2.5 miles east and 3.8 miles south of the park headquarters. Write an equation to represent the border of the pond, using the headquarters as the origin.
$(x - 2.5)^2 + (y + 3.8)^2 = 0.36$

2 Write an equation for the graph.
$x^2 + (y - 3)^2 = 49$

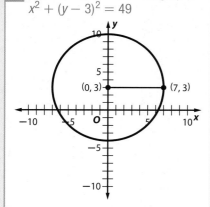

3 Write an equation for a circle if the endpoints of the diameter are at (2, 8) and (2, −2).
$(x - 2)^2 + (y - 3)^2 = 25$

Additional Examples also in Interactive Classroom PowerPoint® Presentations

IWB Interactive White Board READY

StudyTip

Center-Radius Form Standard form is sometimes referred to as *center-radius form* because the center and radius of the circle are apparent in the equation.

Example 2 Write an Equation from a Graph

Write an equation for the graph.

$(x - h)^2 + (y - k)^2 = r^2$	Standard form
$(2 + 3)^2 + (-1 - 1)^2 = r^2$	$x = 2, y = -1, h = -3, k = 1$
$(5)^2 + (-2)^2 = r^2$	Simplify.
$25 + 4 = r^2$	Evaluate the exponents.
$29 = r^2$	Add.

So, the equation of the circle is $(x + 3)^2 + (y - 1)^2 = 29$.

▶ **Guided**Practice   **2B.** $(x + 1)^2 + (y - 4)^2 = 13$

2A. ... $x^2 + y^2 = 26$ **2B.**

You can use the Midpoint and Distance Formulas when you know the endpoints of the radius or diameter of a circle.

Example 3 Write an Equation Given a Diameter

Write an equation for a circle if the endpoints of a diameter are at (7, 6) and (−1, −8).

Step 1 Find the center.

$(h, k) = \left(\dfrac{x_1 + x_2}{2}, \dfrac{y_1 + y_2}{2}\right)$	Midpoint Formula
$= \left(\dfrac{7 + (-1)}{2}, \dfrac{6 + (-8)}{2}\right)$	$(x_1, y_1) = (7, 6), (x_2, y_2) = (-1, -8)$
$= \left(\dfrac{6}{2}, \dfrac{-2}{2}\right)$	Add.
$= (3, -1)$	Simplify.

Step 2 Find the radius.

$r = \sqrt{(x_2 - x_1)^2 + (y_2 - y_1)^2}$	Distance Formula
$= \sqrt{(3 - 7)^2 + (-1 - 6)^2}$	$(x_1, y_1) = (7, 6), (x_2, y_2) = (3, -1)$
$= \sqrt{(-4)^2 + (-7)^2}$	Subtract.
$= \sqrt{65}$	Simplify.

The radius of the circle is $\sqrt{65}$ units, so $r^2 = 65$. Substitute h, k, and r^2 into the standard form of the equation of a circle. An equation of the circle is $(x - 3)^2 + (y + 1)^2 = 65$.

▶ **Guided**Practice   **3.** $(x - 2)^2 + (y - 1)^2 = 17$

3. Write an equation for a circle if the endpoints of a diameter are at (3, −3) and (1, 5).

DifferentiatedInstruction **AL** **OL**

If ▶ students need study tools,

Then ▶ have students write the Key Concepts for each conic section on a notecard.

StudyTip

Axis of Symmetry Every diameter in a circle is an axis of symmetry. There are infinitely many axes of symmetry in a circle.

2 Graph Circles You can use symmetry to help you graph circles.

Example 4 Graph an Equation in Standard Form

Find the center and radius of the circle with equation $x^2 + y^2 = 100$. Then graph the circle.

- The center of the circle is at $(0, 0)$, and the radius is 10.

- The table lists some integer values for x and y that satisfy the equation.

- Because the circle is centered at the origin, it is symmetric about the y-axis. Therefore, the points at $(-6, 8)$, $(-8, 6)$, and $(-10, 0)$ lie on the graph.

- The circle is also symmetric about the x-axis, so the points $(-6, -8)$, $(-8, -6)$, $(0, -10)$, $(6, -8)$, and $(8, -6)$ lie on the graph.

- Plot all of these points and draw the circle that passes through them.

x	y
0	10
6	8
8	6
10	0

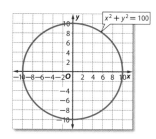

▶ **Guided**Practice

4. Find the center and radius of the circle with equation $x^2 + y^2 = 81$. Then graph the circle. **See margin.**

Circles with centers that are not $(0, 0)$ can be graphed by using translations. The graph of $(x - h)^2 + (y - k)^2 = r^2$ is the graph of $x^2 + y^2 = r^2$ translated h units horizontally and k units vertically.

Example 5 Graph an Equation Not in Standard Form

Find the center and radius of the circle with equation $x^2 + y^2 - 8x + 12y - 12 = 0$. Then graph the circle.

Complete the squares.

$$x^2 + y^2 - 8x + 12y - 12 = 0$$
$$x^2 - 8x + \blacksquare + y^2 + 12y + \blacksquare = 12 + \blacksquare + \blacksquare$$
$$x^2 - 8x + 16 + y^2 + 12y + 36 = 12 + 16 + 36$$
$$(x - 4)^2 + (y + 6)^2 = 64$$

The center of the circle is at $(4, -6)$, and the radius is 8. The graph of $(x - 4)^2 + (y + 6)^2 = 64$ is the same as $x^2 + y^2 = 64$ translated 4 units to the right and down 6 units.

▶ **Guided**Practice

5. Find the center and radius of the circle with equation $x^2 + y^2 + 4x - 10y - 7 = 0$. Then graph the circle. **See margin.**

Additional Answers (Guided Practice)

4. center: $(0, 0)$; radius: 9

5. center: $(-2, 5)$; radius: 6

Teach with Tech

Blog Divide the class into two groups. Have the first group write blog entries containing the coordinates of the center of a circle and its radius. Have the second group of students reply to the postings with the equation of the circle. Have students check each others' answers.

WatchOut!

Preventing Errors Suggest that students draw sketches showing the circle and the endpoints of the diameter to check their work.

2 Graph Circles

Example 4 shows how to find the center and radius of a circle and how to graph the circle given its equation in standard form. **Example 5** shows how to use translations to graph a circle with an equation not in standard form.

Additional Examples

4 Find the center and radius of the circle with equation $x^2 + y^2 = 16$. Then graph the circle. $(0, 0)$; 4

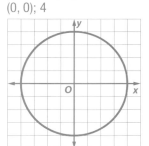

5 Find the center and radius of the circle with equation $x^2 + y^2 + 6x - 7 = 0$. Then graph the circle. $(-3, 0)$; 4

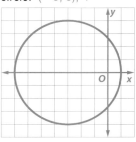

3 Practice

Formative Assessment

Use Exercises 1–11 to check for understanding.

Use the chart at the bottom of the next page to customize assignments for your students.

Additional Answers

8. center: (0, 0); radius: 4

$x^2 + y^2 = 16$

9. center: (0, 7); radius: 3

$x^2 + (y - 7)^2 = 9$

10. center: (4, 4); radius: 5

$(x - 4)^2 + (y - 4)^2 = 25$

11. center: (2, −4); radius: 5

$x^2 + y^2 - 4x + 8y - 5 = 0$

Example 1

1. WEATHER On average, the eye of a tornado is about 200 feet across. Suppose the center of the eye is at the point (72, 39). Write an equation to represent the boundary of the eye. $(x - 72)^2 + (y - 39)^2 = 10,000$

Write an equation for each circle given the center and radius.

2. center: (−2, −6), $r = 4$ units
$(x + 2)^2 + (y + 6)^2 = 16$

3. center: (1, −5), $r = 3$ units
$(x - 1)^2 + (y + 5)^2 = 9$

Example 2   **Write an equation for each graph.**

4.

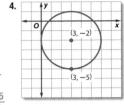

$(x - 3)^2 + (y + 2)^2 = 9$

5.

$(x + 5)^2 + (y + 3)^2 = 90$

6. $\left(x + \dfrac{1}{2}\right)^2 + \left(y + \dfrac{7}{2}\right)^2 = \dfrac{25}{2}$

Example 3   **Write an equation for each circle given the endpoints of a diameter.**

6. (−1, −7) and (0, 0)

7. (4, −2) and (−4, −6) $x^2 + (y + 4)^2 = 20$

Examples 4–5   **Find the center and radius of each circle. Then graph the circle. 8–11. See margin.**

8. $x^2 + y^2 = 16$

9. $x^2 + (y - 7)^2 = 9$

10. $(x - 4)^2 + (y - 4)^2 = 25$

11. $x^2 + y^2 - 4x + 8y - 5 = 0$

12. $(x - 4)^2 + (y - 9)^2 = 36$ **13.** $(x + 3)^2 + (y - 1)^2 = 16$ **14.** $(x + 7)^2 + (y + 3)^2 = 169$

15. $(x + 2)^2 + (y + 1)^2 = 81$ **16.** $(x - 1)^2 + y^2 = 15$ **17.** $x^2 + (y + 6)^2 = 35$

Practice and Problem Solving

Extra Practice is on page R9.

Example 1

Write an equation for each circle given the center and radius.

12. center: (4, 9), $r = 6$ **13.** center: (−3, 1), $r = 4$ **14.** center: (−7, −3), $r = 13$

15. center: (−2, −1), $r = 9$ **16.** center: (1, 0), $r = \sqrt{15}$ **⑰** center: (0, −6), $r = \sqrt{35}$

18. CCSS MODELING The radar for a county airport control tower is located at (5, 10) on a map. It can detect a plane up to 20 miles away. Write an equation for the outer limits of the detection area. $(x - 5)^2 + (y - 10)^2 = 400$

Example 2   **Write an equation for each graph.**

19.

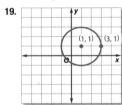

$(x - 1)^2 + (y - 1)^2 = 4$

20.

$(x - 6)^2 + (y + 3)^2 = 25$

21.

$x^2 + (y + 6)^2 = 53$

22.

$(x - 4)^2 + (y + 6)^2 = 202$

DifferentiatedInstruction **BL**

Extension Does your school have a track? Students can be challenged to develop a scale model of the track using equations and graph paper. Scaffold the task as needed using the following:

- What lines and/or curves make up a track?
- How can we locate the center so it can be placed at (0, 0)?
- What measurements do we need?
- What scale factor should be used to fit the model on a sheet of graph paper?

If there is no track at the school, the necessary information can be found using the Internet.

Example 3 Write an equation for each circle given the endpoints of a diameter.

23. (2, 1) and (2, −4) **24.** (−4, −10) and (4, −10) **25.** (5, −7) and (−2, −9)

26. (−6, 4) and (4, 8) **27.** (2, −5) and (6, 3) **28.** (18, 11) and (−19, −13)

29. **LAWN CARE** A sprinkler waters a circular section of lawn.

    a. Write an equation to represent the boundary of the sprinkler area if the endpoints of a diameter are at (−12, 16) and (12, −16). $x^2 + y^2 = 400$

    b. What is the area of the lawn that the sprinkler waters? approximately 1256.64 units^2

30. **SPACE** Apollo 8 was the first manned spacecraft to orbit the Moon at an average altitude of 185 kilometers above the Moon's surface. Write an equation to model a single circular orbit of the command module if the endpoints of a diameter of the Moon are at (1740, 0) and (−1740, 0). Let the center of the Moon be at the origin of the coordinate system measured in kilometers. $x^2 + y^2 = 3{,}705{,}625$

31–46. See Chapter 9 Answer Appendix.

Examples 4–5 Find the center and radius of each circle. Then graph the circle.

31. $x^2 + y^2 = 75$ **32.** $(x - 3)^2 + y^2 = 4$

33. $(x - 1)^2 + (y - 4)^2 = 34$ **34.** $x^2 + (y - 14)^2 = 144$

35. $(x - 5)^2 + (y + 2)^2 = 16$ **36.** $x^2 + y^2 = 256$

37. $(x - 4)^2 + y^2 = \frac{8}{9}$ **38.** $\left(x + \frac{2}{3}\right)^2 + \left(y - \frac{1}{2}\right)^2 = \frac{16}{25}$

39. $x^2 + y^2 + 4x = 9$ **40.** $x^2 + y^2 - 6y + 8x = 0$

41. $x^2 + y^2 + 2x + 4y = 9$ **42.** $x^2 + y^2 - 3x + 8y = 20$

43. $x^2 + y^2 + 6y = -50 - 14x$ **44.** $x^2 - 18x + 53 = 18y - y^2$

45. $2x^2 + 2y^2 - 4x + 8y = 32$ **46.** $3x^2 + 3y^2 - 6y + 12x = 24$

23. $(x - 2)^2 + \left(y + \frac{3}{2}\right)^2 = \frac{25}{4}$

24. $x^2 + (y + 10)^2 = 16$

25. $\left(x - \frac{3}{2}\right)^2 + (y + 8)^2 = \frac{53}{4}$

26. $(x + 1)^2 + (y - 6)^2 = 29$

27. $(x - 4)^2 + (y + 1)^2 = 20$

28. $\left(x + \frac{1}{2}\right)^2 + (y + 1)^2 = 486\frac{1}{4}$

B **47.** **SPACE** A satellite is in a circular orbit 25,000 miles above Earth.

    a. Write an equation for the orbit of this satellite if the origin is at the center of Earth. Use 8000 miles as the diameter of Earth. $x^2 + y^2 = 841{,}000{,}000$

    b. Draw a sketch of Earth and the orbit to scale. Label your sketch. See margin.

48. **CCSS SENSE-MAKING** Suppose an unobstructed radio station broadcast could travel 120 miles. Assume the station is centered at the origin.

    a. Write an equation to represent the boundary of the broadcast area with the origin as the center. $x^2 + y^2 = 14{,}400$

    b. If the transmission tower is relocated 40 miles east and 10 miles south of the current location, and an increased signal will transmit signals an additional 80 miles, what is an equation to represent the new broadcast area? $(x - 40)^2 + (y + 10)^2 = 40{,}000$

49a. $(x + 1)^2 + (y - 4)^2 = 36 + 16\sqrt{5}$

49. **GEOMETRY** Concentric circles are circles with the same center but different radii. Refer to the graph at the right where $\overline{AB}$ is a diameter of the circle.

    a. Write an equation of the circle concentric with the circle at the right, with radius 4 units greater.

    b. Write an equation of the circle concentric with the circle at the right, with radius 2 units less.

    c. Graph the circles from parts a and b on the same coordinate plane. See margin.

49b. $(x + 1)^2 + (y - 4)^2 = 24 - 8\sqrt{5}$ **50.** $(x + 35)^2 + (y - 40)^2 = 3025$

50. **EARTHQUAKES** The Rose Bowl is located about 35 miles west and 40 miles north of downtown Los Angeles. Suppose an earthquake occurs with its epicenter about 55 miles from the stadium. Assume that the origin of a coordinate plane is located at the center of downtown Los Angeles. Write an equation for the set of points that could be the epicenter of the earthquake.

CCSS **Teaching the Mathematical Practices**

Sense-Making Mathematically proficient students start by explaining to themselves the meaning of a problem and looking for entry points to its solution. They analyze givens, constraints, relationships, and goals. They check their answers to problems using a different method, and they continually ask themselves, "Does this make sense?"

Additional Answers

47b.

49c.

Differentiated Homework Options

Level	Assignment	Two-Day Option	
AL Basic	12–46, 62–88	13–45 odd, 68–71	12–46 even, 62–67, 72–88
OL Core	13–45 odd, 47–50, 51–61 odd, 62–88	12–46, 68–71	47–67, 72–88
BL Advanced	47–85, (optional: 86–88)		

Study Guide and Intervention
AL OL ELL

Practice
AL OL BL ELL

Word Problem Practice
AL OL BL ELL

 CCSS PRECISION Write an equation for the circle that satisfies each set of conditions.

51 center $(9, -8)$, passes through $(19, 22)$ $(x - 9)^2 + (y + 8)^2 = 1000$

52. center $\left(-\sqrt{15}, 30\right)$, passes through the origin $\left(x + \sqrt{15}\right)^2 + (y - 30)^2 = 915$

53. center at $(8, -9)$, tangent to y-axis $(x - 8)^2 + (y + 9)^2 = 64$

54. center at $(2, 4)$, tangent to x-axis $(x - 2)^2 + (y - 4)^2 = 16$

55. $(x - 2.5)^2 + (y - 2.5)^2 = 6.25$
56. $(x + 2)^2 + (y - 3)^2 = 4$

55. center in the first quadrant; tangent to $x = 5$, the x-axis, and the y-axis

56. center in the second quadrant; tangent to $y = 1$, $y = 5$, and the y-axis

57. **MULTIPLE REPRESENTATIONS** Graph $y = \sqrt{9 - x^2}$ and $y = -\sqrt{9 - x^2}$ on the same graphing calculator screen.

   a. **Verbal** Describe the graph formed by the union of these two graphs. **circle**

   b. **Algebraic** Write an equation for the union of the two graphs. **$x^2 + y^2 = 9$**

   c. **Verbal** Most graphing calculators cannot graph the equation $x^2 + y^2 = 49$ directly. Describe a way to use a graphing calculator to graph the equation. Then graph the equation. **See margin.**

   d. **Analytical** Solve $(x - 2)^2 + (y + 1)^2 = 4$ for y. Why do you need two equations to graph a circle on a graphing calculator?

   e. **Verbal** Do you think that it is easier to graph the equation in part **d** using graph paper and a pencil or using a graphing calculator? Explain. **See students' work.**

58–61. See Chapter 9

Find the center and radius of each circle. Then graph the circle. **Answer Appendix.**

58. $x^2 - 12x + 84 = -y^2 + 16y$ 59. $4x^2 + 4y^2 + 36y + 5 = 0$

60. $\left(x + \sqrt{5}\right)^2 + y^2 - 8y = 9$ 61. $x^2 + 2\sqrt{7}x + 7 + \left(y - \sqrt{11}\right)^2 = 11$

57d. $y = \pm\sqrt{4 - (x - 2)^2} - 1$; When you solve for y you must take the square root resulting in both a positive and negative answer, so you have to enter the positive equation as Y1 and the negative equation as Y2.

H.O.T. Problems   Use Higher-Order Thinking Skills

62. **ERROR ANALYSIS** Heather says that $(x - 2)^2 + (y + 3)^2 = 36$ and $(x - 2) + (y + 3) = 6$ are equivalent equations. Carlota says that the equations are *not* equivalent. Is either of them correct? Explain your reasoning.
   Carlota; the square root of $(x - 2)^2 + (y + 3)^2$ is not $(x - 2) + (y + 3)$.

63. **OPEN ENDED** Consider graphs with equations of the form $(x - 3)^2 + (y - a)^2 = 64$. Assign three different values for a, and graph each equation. Describe all graphs with equations of this form.
   See students' work; circles with a radius of 8 and centers on the graph of $x = 3$.

64. **REASONING** Explain why the phrase "in a plane" is included in the definition of a circle. What would be defined if the phrase were *not* included?
   If the phrase is not included, the figure would be a sphere.

65. **OPEN ENDED** Concentric circles have the same center, but most often, not the same radius. Write equations of two concentric circles. Then graph the circles. **See margin.**

66. **REASONING** Assume that (x, y) are the coordinates of a point on a circle. The center is at (h, k), and the radius is r. Find an equation of the circle by using the Distance Formula. **See margin.**

67. **WRITING IN MATH** The circle with equation $(x - a)^2 + (y - b)^2 = r^2$ lies in the first quadrant and is tangent to both the x-axis and the y-axis. Sketch the circle. Describe the possible values of a, b, and r. Do the same for a circle in Quadrants II, III, and IV. Discuss the similarities among the circles. **See Chapter 9 Answer Appendix.**

 612 | Lesson 9-3 | Circles

Enrichment
OL BL

CCSS **Teaching the Mathematical Practices**

Precision Mathematically proficient students try to use clear definitions in their reasoning, calculate accurately and efficiently, and make explicit use of definitions.

68. GRIDDED RESPONSE Two circles, both with a radius of 6, have exactly one point in common. If A is a point on one circle and B is a point on the other circle, what is the maximum possible length for the line segment $\overline{AB}$? **24**

69. In the senior class, there are 20% more girls than boys. If there are 180 girls, how many more girls than boys are there among the seniors? **A**

A 30
B 36
C 90
D 144

70. A $1000 deposit is made at a bank that pays 2% compounded weekly. How much will you have in your account at the end of 10 years? **H**

F $1200.00   H $1221.36
G $1218.99   J $1224.54

71. The mean of six numbers is 20. If one of the numbers is removed, the average of the remaining numbers is 15. What is the number that was removed? **C**

A 42   C 45
B 43   D 48

75. $\left(1, \frac{7}{22}\right)$; $\sqrt{65}$ units   **76.** $\left(-\frac{\sqrt{3}}{2}, 2\right)$; $\sqrt{271}$ units

Spiral Review

Graph each equation. (Lesson 9-2) **72–74. See Chapter 9 Answer Appendix.**

72. $y = -\frac{1}{2}(x - 1)^2 + 4$

73. $4(x - 2) = (y + 3)^2$

74. $(y - 8)^2 = -4(x - 4)$

Find the midpoint of the line segment with endpoints at the given coordinates. Then find the distance between the points. (Lesson 9-1)

75. $\left(-3, -\frac{2}{11}\right), \left(5, \frac{9}{11}\right)$

76. $\left(2\sqrt{3}, -5\right), \left(-3\sqrt{3}, 9\right)$

77. $(2.5, 4), (-2.5, 2)$ **(0, 3); $\sqrt{29}$ units**

78. If y varies directly as x and $y = 8$ when $x = 6$, find y when $x = 15$. (Lesson 8-5) **20**

79. If y varies jointly as x and z and $y = 80$ when $x = 5$ and $z = 8$, find y when $x = 16$ and $z = 2$. (Lesson 8-5) **64**

80. If y varies inversely as x and $y = 16$ when $x = 5$, find y when $x = 20$. (Lesson 8-5) **4**

Evaluate each expression. (Lesson 7-3)

81. $\log_9 243$ **$\frac{5}{2}$**

82. $\log_2 \frac{1}{32}$ **-5**

83. $\log_3 \frac{1}{81}$ **-4**

84. $\log_{10} 0.001$ **-3**

85. AMUSEMENT PARKS The velocity v in feet per second of a roller coaster at the bottom of a hill is related to the vertical drop h in feet and the velocity v_0 in feet per second of the coaster at the top of the hill by the formula $v_0 = \sqrt{v^2 - 64h}$. (Lesson 6-5)

a. Explain why $v_0 = v - 8\sqrt{h}$ is not equivalent to the given formula.

b. What velocity must the coaster have at the top of the hill to achieve a velocity of 125 feet per second at the bottom? **34.1 ft/s**

85a. The square root of a difference is not the difference of the square roots.

226 ft

Skills Review

Solve each equation by completing the square.

86. $x^2 + 3x - 18 = 0$ $\{-6, 3\}$

87. $2x^2 - 3x - 3 = 0$ $\left\{\frac{3 \pm \sqrt{33}}{4}\right\}$

88. $x^2 + 2x + 6 = 0$ $\{-1 \pm i\sqrt{5}\}$

connectED.mcgraw-hill.com   **613**

DifferentiatedInstruction OL BL

Extension Write $x + 2y = 2$ and $x^2 + y^2 = 25$ on the board. Ask students to determine how many points the graphs of these two equations have in common. Then ask students to explain whether all lines intersect a circle in two points. There are two points that the graphs of the equations have in common: $(-4, 3)$ and $(4.8, -1.4)$. Not all lines intersect a circle in two points. A tangent line intersects a circle at one point. Some lines do not intersect a circle at all. Hence, a line can intersect a circle in 0, 1, or 2 points.

WatchOut!

Error Analysis In Exercise 62, ask students if the square root of $(x - 2)^2 + (y + 3)^2$ is $(x - 2) + (y + 3)$. If they think it is, ask them to check to see if the square root of $4^2 + 3^2$ is equal to $4 + 3$.

4 Assess

Ticket Out the Door Write five different equations for circles on separate sheets of paper. Make several copies of each. Give one equation to each student. As students leave the room, ask them to tell you either the center or the radius of the circles formed by the equations.

Additional Answers

57c. Solve the equation for y: $y = \pm\sqrt{49 - x^2}$. Then graph the positive and negative answers.

65. Sample answer: $(x - 2)^2 + (y - 3)^2 = 25$ and $(x - 2)^2 + (y - 3)^2 = 36$

66. $\sqrt{(x_2 - x_1)^2 + (y_2 - y_1)^2} = d$

$\sqrt{(x - h)^2 + (y - k)^2} = r$

$(x - h)^2 + (y - k)^2 = r^2$

This is the standard form of the equation of a circle.

1 Focus

Objective Determine how the graph of an ellipse is affected by changing the location of the foci.

Materials for Each Student
- two thumbtacks
- cardboard
- string
- ruler
- grid paper

Easy to Make Manipulatives
Teaching Algebra with Manipulatives templates for:
- grid paper
- rulers

Teaching Tip
It may be easier to manipulate the string and pencil if students work in pairs. Make sure each member draws an ellipse.

2 Teach

Working in Cooperative Groups
Put students in pairs, mixing abilities. Then have pairs complete the Activity and Exercises 1–5.

- Ask each pair to stop at one point in the curve while drawing. Then find the total distance from the point to one thumbtack plus the distance from the same point to the other thumbtack. the length of the string minus the distance between the thumbtacks

- Ask students what the total distance to the tacks is for any point on the ellipse. the length of the string minus the distance between the thumbtacks

Practice Have students complete Exercises 6–12.

3 Assess

Formative Assessment
Use Exercises 6–10 to assess whether

EXPLORE 9-4 Algebra Lab
Investigating Ellipses

Follow the steps below to construct a type of conic section.

CCSS Common Core State Standards
Mathematical Practices
5 Use appropriate tools strategically.

Activity Make an Ellipse

Step 1 Place two thumbtacks in a piece of cardboard, about 1 foot apart.

Step 2 Tie a knot in a piece of string and loop it around the thumbtacks. Place your pencil in the string.

Step 3 Keep the string tight and draw a curve. Continue drawing until you return to your starting point.

The curve you have drawn is called an **ellipse**. The points where the thumbtacks are located are called the **foci** of the ellipse. *Foci* is the plural of *focus*.

Model and Analyze
Place a large piece of grid paper on a piece of cardboard.

1. Place the thumbtacks at (7, 0) and (−7, 0). Choose a string long enough to loop around both thumbtacks. Draw an ellipse. **See students' work.**

2. Repeat Exercise 1, but place the thumbtacks at (4, 0) and (−4, 0). Use the same loop of string and draw an ellipse. How does this ellipse compare to the one in Exercise 1? **See students' work; the ellipse is more circular.**

Place the thumbtacks at each set of points and draw an ellipse. You may change the length of the loop of string if you like. 3–5. See students' work.

3. (11, 0), (−11, 0)
4. (3, 0), (−3, 0)
5. (13, 3), (−9, 3)

Make a Conjecture
Describe what happens to the shape of an ellipse when each change is made.

6. The thumbtacks are moved closer together. The ellipse becomes more circular.

7. The thumbtacks are moved farther apart. The ellipse becomes more elongated.

8. The length of the loop of string is increased. The ellipse becomes larger.

9. The thumbtacks are arranged vertically. The ellipse is longer in the vertical direction than in the horizontal direction.

10. One thumbtack is removed, and the string is looped around the remaining thumbtack. The ellipse is a circle.

11. Pick a point on one of the ellipses you have drawn. Use a ruler to measure the distances from that point to the points where the thumbtacks were located. Add the distances. Repeat for other points on the same ellipse. What relationship do you notice? The sum of the distances is constant.

12. Could this activity be done with a rubber band instead of a piece of string? Explain. No; a rubber band might stretch so that the sum of the distances to the thumbtacks would not be constant.

614 | Explore 9-4 | Algebra Lab: Investigating Ellipses

Ed-Imaging

---

students comprehend the changes that result from varying the positions of the thumbtacks and the length of the string.

From Concrete to Abstract
Ask students to describe the set of points they studied in Lesson 9-3. Ask students to use what they have learned to describe the set of points called an ellipse. A circle is the set of points in a plane a constant distance from a given point in the plane, called the *center*. An ellipse is the set of all points in a plane such that the sum of the distances from two fixed points, the *foci*, is constant.

Extending the Concept
Ask:
As the coordinates of the foci of an ellipse come closer and closer together, what happens to the shape of the ellipse? It becomes more circular.

LESSON 9-4 Ellipses

Then	Now	Why?
• You graphed and wrote equations for circles.	**1** Write equations of ellipses. **2** Graph ellipses.	• Mercury, like all of the planets of our solar system, does not orbit the Sun in a perfect circular path. At its farthest point, Mercury is about 43 million miles from the Sun. At its closest point, it is only about 28.5 million miles from the Sun. This orbit is in the shape of an ellipse with the Sun at a focus.

 NewVocabulary
ellipse
foci
major axis
minor axis
center
vertices
co-vertices
constant sum

CCSS Common Core State Standards

Content Standards
A.SSE.1.b Interpret complicated expressions by viewing one or more of their parts as a single entity.
A.CED.2 Create equations in two or more variables to represent relationships between quantities; graph equations on coordinate axes with labels and scales.

Mathematical Practices
7 Look for and make use of structure.

1 Equations of Ellipses An **ellipse** is the set of all points in a plane such that the sum of the distances from two fixed points is constant. These two points are called the **foci** of the ellipse.

Every ellipse has two axes of symmetry, the **major axis** and the **minor axis**. The axes are perpendicular at the **center** of the ellipse.

The foci of an ellipse always lie on the major axis. The endpoints of the major axis are the **vertices** of the ellipse and the endpoints of the minor axis are the **co-vertices** of the ellipse.

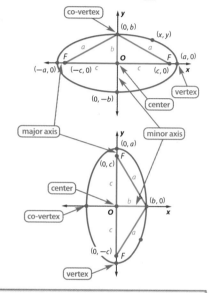

KeyConcept Equations of Ellipses Centered at the Origin

Standard Form	$\dfrac{x^2}{a^2} + \dfrac{y^2}{b^2} = 1$	$\dfrac{y^2}{a^2} + \dfrac{x^2}{b^2} = 1$
Orientation	horizontal	vertical
Foci	$(c, 0), (-c, 0)$	$(0, c), (0, -c)$
Length of Major Axis	$2a$ units	$2a$ units
Length of Minor Axis	$2b$ units	$2b$ units

There are several important relationships among the many parts of an ellipse.
• The length of the major axis, $2a$ units, equals the sum of the distances from the foci to any point on the ellipse.
• The values of a, b, and c are related by the equation $c^2 = a^2 - b^2$.
• The distance from a focus to either co-vertex is a units.

connectED.mcgraw-hill.com 615

1 Focus

VerticalAlignment

Before Lesson 9-4 Graph and write equations for circles.

Lesson 9-4 Write equations of ellipses. Graph ellipses.

After Lesson 9-4 Write and graph equations of hyperbolas.

2 Teach

Scaffolding Questions
Have students read the **Why?** section of the lesson.

Ask:
• What makes up our solar system? The Sun and the group of objects orbiting around the Sun.
• Is Earth always the same distance from the Sun? No; it is closest in January and farthest in July.
• What does it mean when it says that the Sun is at a focus? The Sun is not the center of the ellipse. It is at a point off to one side.

Lesson 9-4 Resources

Resource	Approaching Level **AL**	On Level **OL**	Beyond Level **BL**	English Learners **ELL**
Teacher Edition		• Differentiated Instruction, pp. 617, 618, 622	• Differentiated Instruction, pp. 617, 618, 622	• Differentiated Instruction, p. 618
Chapter Resource Masters	• Study Guide and Intervention, pp. 25–26 • Skills Practice, p. 27 • Practice, p. 28 • Word Problem Practice, p. 29	• Study Guide and Intervention, pp. 25–26 • Skills Practice, p. 27 • Practice, p. 28 • Word Problem Practice, p. 29 • Enrichment, p. 30 • Graphing Calculator Activity, p. 31	• Practice, p. 28 • Word Problem Practice, p. 29 • Enrichment, p. 30	• Study Guide and Intervention, pp. 25–26 • Skills Practice, p. 27 • Practice, p. 28 • Word Problem Practice, p. 29
Other	• 5-Minute Check 9-4 • Study Notebook	• 5-Minute Check 9-4 • Study Notebook	• 5-Minute Check 9-4 • Study Notebook	• 5-Minute Check 9-4 • Study Notebook

Document Camera Work on the document camera to show students how to construct an ellipse using two pins, string, and a pencil. Repeat several times to show how the shape of the ellipse changes when the foci are closer and farther away from each other.

1 Equation of Ellipses

Example 1 shows how to write an equation for an ellipse shown on a graph. **Example 2** shows how to write an equation for an ellipse when all four vertices are given. **Example 3** shows how to write the equation of an ellipse that models a real-world situation.

Formative Assessment

Use the Guided Practice exercises after each example to determine students' understanding of concepts.

Additional Example

1 Write an equation for the ellipse.

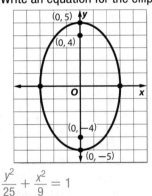

$$\frac{y^2}{25} + \frac{x^2}{9} = 1$$

Additional Examples also in Interactive Classroom PowerPoint® Presentations

IWB Interactive White Board READY

StudyTip

Major Axis In standard form, if the x^2-term has the greater denominator, then the major axis is horizontal. If the y^2-term has the greater denominator, then it is vertical.

The sum of the distances from the foci to any point on the ellipse, or the **constant sum**, must be greater than the distance between the foci.

Example 1 Write an Equation Given Vertices and Foci

Write an equation for the ellipse.

Step 1 Find the center.
The foci are equidistant from the center. The center is at $(0, 0)$.

Step 2 Find the value of a.
The vertices are $(0, 9)$ and $(0, -9)$, so the length of the major axis is 18. The value of a is $18 \div 2$ or 9, and $a^2 = 81$.

Step 3 Find the value of b.
We can use $c^2 = a^2 - b^2$ to find b. The foci are 7 units from the center, so $c = 7$.
$$c^2 = a^2 - b^2 \qquad \text{Equation relating } a, b, \text{ and } c$$
$$49 = 81 - b^2 \qquad a = 9 \text{ and } c = 7$$
$$b^2 = 32 \qquad \text{Solve for } b^2.$$

Step 4 Write the equation.
Because the major axis is vertical, a^2 goes with y and b^2 goes with x.
The equation for the ellipse is $\frac{y^2}{81} + \frac{x^2}{32} = 1$.

Guided Practice $\frac{x^2}{16} + \frac{y^2}{12} = 1$

1. Write an equation for an ellipse with vertices at $(-4, 0)$ and $(4, 0)$ and foci at $(2, 0)$ and $(-2, 0)$.

Like other graphs, the graph of an ellipse can be translated. When the graph is translated h units right and k units up, the center of the translation is (h, k). This is equivalent to replacing x with $x - h$ and replacing y with $y - k$ in the parent function.

KeyConcept Equations of Ellipses Centered at (h, k)

Standard Form	$\frac{(x-h)^2}{a^2} + \frac{(y-k)^2}{b^2} = 1$	$\frac{(y-k)^2}{a^2} + \frac{(x-h)^2}{b^2} = 1$
Orientation	horizontal	vertical
Foci	$(h \pm c, k)$	$(h, k \pm c)$
Vertices	$(h \pm a, k)$	$(h, k \pm a)$
Co-vertices	$(h, k \pm b)$	$(h \pm b, k)$

We can use this information to determine the equations for ellipses. The original ellipse at the right is horizontal and has a major axis of 10 units, so $a = 5$.

The length of the minor axis is 6 units, so $b = 3$.

The ellipse is translated 4 units right and 5 units down. So, the value of h is 4 and the value of k is -5.

The equation for the original ellipse is $\frac{x^2}{25} + \frac{y^2}{9} = 1$.

The equation for the translation is $\frac{(x-4)^2}{25} + \frac{(y+5)^2}{9} = 1$.

WatchOut!

Common Misconceptions Point out to students that when the Pythagorean Theorem is used to derive the standard form of the equation of an ellipse with its center at the origin and a horizontal major axis, it is a, not c, that is used to represent the length of the hypotenuse.

Focus on Mathematical Content

Ellipses Every ellipse has two axes of symmetry. The points at which the ellipse intersects its axes of symmetry determine two segments with endpoints on the ellipse. These segments are called the *major axis* and the *minor axis*. The intersection of the axes is the center of the ellipse.

You can also determine the equation for an ellipse if you are given all four vertices.

Example 2 Write an Equation Given the Lengths of the Axes

Write an equation for the ellipse with vertices at (6, −8) and (6, 4) and co-vertices at (3, −2) and (9, −2).

The *x*-coordinate is the same for both vertices, so the ellipse is vertical.

The center of the ellipse is at $\left(\frac{6 + 6}{2}, \frac{-8 + 4}{2}\right)$ or (6, −2).

The length of the major axis is 4 − (−8) or 12 units, so $a = 6$.

The length of the minor axis is 9 − 3 or 6 units, so $b = 3$.

The equation for the ellipse is $\frac{(y + 2)^2}{36} + \frac{(x - 6)^2}{9} = 1$. $a^2 = 36, b^2 = 9$

▸ **Guided Practice**

2. Write an equation for the ellipse with vertices at (−3, 8) and (9, 8) and co-vertices at (3, 12) and (3, 4). $\frac{(x - 3)^2}{36} + \frac{(y - 8)^2}{16} = 1$

Many real-world phenomena can be represented by ellipses.

Real-World Example 3 Write an Equation for an Ellipse

SPACE Refer to the application at the beginning of the lesson. Mercury's greatest distance from the Sun, or *aphelion*, is about 43 million miles. Mercury's closest distance, or *perihelion*, is about 28.5 million miles. The diameter of the Sun is about 870,000 miles. Use this information to determine an equation relating Mercury's elliptical orbit around the Sun in millions of miles.

Understand We need to determine an equation representing Mercury's orbit around the Sun.

Plan Including the diameter of the Sun, the sum of the perihelion and aphelion equals the length on the major axis of the ellipse. We can use this information to determine the values of *a*, *b*, and *c*.

Solve Find the value of *a*.
The value of *a* is one half the length of the major axis.
$a = 0.5(43 + 28.5 + 0.87)$ or 36.185

Find the value of *c*.
The value of *c* is the distance from the center of the ellipse to the focus. This distance is equal to *a* minus the perihelion and the radius of the Sun.
$c = 36.185 - 28.5 - 0.435$ or 7.25

(continued on the next page)

NASA-MSFC

Real-World Career

Aerospace Technician
Aerospace technicians work for NASA, helping engineers research and develop virtual reality and verbal communication between humans and computer systems. Although a bachelor's degree is desired, on-the-job training is available.
Source: NASA

Problem-Solving Tip

CCSS Sense-Making Draw a diagram when the problem situation involves spatial reasoning or geometric figures.

Additional Examples

2 Write an equation for the ellipse with vertices at (−6, −2) and (4, −2) and co-vertices at (−1, −4) and (−1, 0).
$\frac{(x + 1)^2}{25} + \frac{(y + 2)^2}{4} = 1$

3 **SOUND** A listener is standing in an elliptical room 150 feet wide and 320 feet long. When a speaker stands at one focus and whispers, the best place for the listener to stand is at the other focus. Write an equation to model this ellipse, assuming the major axis is horizontal and the center is at the origin.
$\frac{x^2}{25,600} + \frac{y^2}{5625} = 1$

Watch Out!

Preventing Errors Suggest that students make rough sketches of situations like the one in Example 3.

CCSS Teaching the Mathematical Practices

Sense-Making Mathematically proficient students analyze givens, constraints, relationships, and goals. They can explain correspondences between equations, verbal descriptions, tables, and graphs or draw diagrams of important features and relationships, graph data, and search for regularity or trends. Remind students that an ellipse has two axes of symmetry, the major axis and the minor axis. Point out that the major axis is the longer of the two axes.

Differentiated Instruction OL BL

Extension Ask students to create a town on graph paper. The town should include the following components:

- An elliptical border (give equation)
- Two police headquarters located at the foci of the ellipse (give coordinates)
- Five other locations of your choice (give coordinates)
- Create a series of problems about police responding to calls at various locations. Which police headquarters should respond to the call? What other factors must be considered?

2 Graph Ellipses

Example 4 shows how to graph an equation for an ellipse that is not in standard form.

Additional Example

4 Find the coordinates of the center and foci and the lengths of the major and minor axes of an ellipse with equation $x^2 + 4y^2 - 6x - 16y - 11 = 0$. Then graph the ellipse.

center: (3, 2); foci: $(3\sqrt{3} + 3, 2)$, $(-3\sqrt{3} + 3, 2)$: major axis 12; minor axis: 6

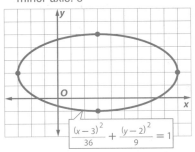

$$\frac{(x-3)^2}{36} + \frac{(y-2)^2}{9} = 1$$

CCSS Teaching the Mathematical Practices

Sense-Making Mathematically proficient students start by explaining to themselves the meaning of a problem and looking for entry points to its solution. They analyze givens, constraints, relationships, and goals. They check their answers to problems using a different method, and they continually ask themselves, "Does this make sense?"

Find the value of b.

$c^2 = a^2 - b^2$	Equation relating a, b, and c
$(7.25)^2 = (36.185)^2 - b^2$	$c = 7.25$ and $a = 36.185$
$52.5625 = 1309.3542 - b^2$	Simplify.
$b^2 = 1256.828$	Solve for b^2.
$b = 35.4518$	Take the square root of each side.

So, with the center of the orbit at the origin, the equation relating Mercury's orbit around the Sun can be modeled by

$$\frac{x^2}{1309.3542} + \frac{y^2}{1256.792} = 1.$$

Check Use your answer to recalculate a, b, and c. Then determine the aphelion and perihelion based on your answer. Compare to the actual values.

▶ **GuidedPractice**   3. $\dfrac{x^2}{13.47} + \dfrac{y^2}{12.42} = 1$

3. SPACE Pluto's distance from the Sun is 2.757 billion miles at perihelion and about 4.583 billion miles at aphelion. Determine an equation relating Pluto's orbit around the Sun in billions of miles with the center of the horizontal ellipse at the origin.

Real-WorldLink

Earth's orbit around the Sun is nearly circular, with only about a 3% difference between perihelion and aphelion.

Source: *The Astronomer*

2 Graph Ellipses

When you are given an equation for an ellipse that is not in standard form, you can write it in standard form by completing the square for both x and y. Once the equation is in standard form, you can use it to graph the ellipse.

Example 4   Graph an Ellipse

Find the coordinates of the center and foci, and the lengths of the major and minor axes of an ellipse with equation $25x^2 + 9y^2 + 250x - 36y + 436 = 0$. Then graph the ellipse.

Step 1 Write in standard form. Complete the square for each variable to write this equation in standard form.

$25x^2 + 9y^2 + 250x - 36y + 436 = 0$	Original equation
$25x^2 + 250x + 9y^2 - 36y = -436$	Associative Property
$25(x^2 + 10x) + 9(y^2 - 4y) = -436$	Distributive Property
$25(x^2 + 10x + \blacksquare) + 9(y^2 - 4y + \blacksquare) = -436 + 25(\blacksquare) + 9(\blacksquare)$	Complete the squares.
$25(x^2 + 10x + 25) + 9(y^2 - 4y + 4) = -436 + 25(25) + 9(4)$	$5^2 = 25$ and $(-2)^2 = 4$
$25(x + 5)^2 + 9(y - 2)^2 = 225$	Write as perfect squares.
$\dfrac{(x + 5)^2}{9} + \dfrac{(y - 2)^2}{25} = 1$	Divide each side by 225.

Step 2 Find the center.
$h = -5$ and $k = 2$, so the center of the ellipse is at $(-5, 2)$.

Step 3 Find the lengths of the axes and graph.
The ellipse is vertical.
$a^2 = 25$, so $a = 5$. $b^2 = 9$, so $b = 3$.
The length of the major axis is $2 \cdot 5$ or 10.
The length of the minor axis is $2 \cdot 3$ or 6.
The vertices are at $(-5, 7)$ and $(-5, -3)$.
The co-vertices are at $(-2, 2)$ and $(-8, 2)$.

DifferentiatedInstruction   OL   BL   ELL

Verbal/Linguistic Learners Remind students that the Moon travels around Earth in an elliptical orbit. The distance the Moon is from Earth varies from about 364,000 kilometers to about 405,000 kilometers. Ask students to research what shape the orbits are for other planets and their moons.

StockTrek/Photodisc/Getty Images

Step 4 Find the foci.
$c^2 = 25 - 9$ or 16, so $c = 4$.
The foci are at $(-5, 6)$ and $(-5, -2)$.

Step 5 Graph the ellipse.
Draw the ellipse that passes through the vertices and co-vertices.

▶ **Guided**Practice

4. Find the coordinates of the center and foci and the lengths of the major and minor axes of the ellipse with equation $x^2 + 4y^2 - 2x + 24y + 21 = 0$. Then graph the ellipse. **See margin.**

Check Your Understanding

⬤ = Step-by-Step Solutions begin on page R14. ✓

Example 1 Write an equation for each ellipse.

1.

$\dfrac{y^2}{25} + \dfrac{x^2}{9} = 1$

2.

$\dfrac{(x-4)^2}{49} + \dfrac{(y+6)^2}{36} = 1$

Example 2 Write an equation for an ellipse that satisfies each set of conditions.

3. vertices at $(-2, -6)$ and $(-2, 4)$, co-vertices at $(-5, -1)$ and $(1, -1)$ 3. $\dfrac{(y+1)^2}{25} + \dfrac{(x+2)^2}{9} = 1$

4. vertices at $(-2, 5)$ and $(14, 5)$, co-vertices at $(6, 1)$ and $(6, 9)$ $\dfrac{(x-6)^2}{64} + \dfrac{(y-5)^2}{16} = 1$

Example 3

5. **CCSS SENSE-MAKING** An architectural firm sent a proposal to a city for building a coliseum, shown at the right.

   a. Determine the values of a and b. $a = 240$, $b = 160$

   b. Assuming that the center is at the origin, write an equation to represent the ellipse. **5b.** $\dfrac{x^2}{57,600} + \dfrac{y^2}{25,600} = 1$

   c. Determine the coordinates of the foci. about $(179, 0)$ and $(-179, 0)$

320 m

480 m

6. **SPACE** Earth's orbit is about 91.4 million miles at perihelion and about 94.5 million miles at aphelion. Determine an equation relating Earth's orbit around the Sun in millions of miles with the center of the horizontal ellipse at the origin. $\dfrac{x^2}{8720.8} + \dfrac{y^2}{8718.4} = 1$

Example 4 Find the coordinates of the center and foci and the lengths of the major and minor axes for the ellipse with the given equation. Then graph the ellipse. **7–10. See margin.**

7. $\dfrac{(y+1)^2}{64} + \dfrac{(x-5)^2}{28} = 1$

8. $\dfrac{(x+2)^2}{48} + \dfrac{(y-1)^2}{20} = 1$

9. $4x^2 + y^2 - 32x - 4y + 52 = 0$

10. $9x^2 + 25y^2 + 72x - 150y + 144 = 0$

connectED.mcgraw-hill.com **619**

9. center $(4, 2)$; foci $(4, 5.46)$ and $(4, -1.46)$; major axis: 8; minor axis: 4

$4x^2 + y^2 - 32x - 4y + 52 = 0$

10. center $(-4, 3)$; foci $(0, 3)$ and $(-8, 3)$; major axis: 10; minor axis: 6

$9x^2 + 25y^2 + 72x - 150y + 144 = 0$

---

3 Practice

Formative Assessment
Use Exercises 1–10 to check for understanding.

Use the chart at the bottom of the next page to customize assignments for your students.

Additional Answers (Guided Practice)

4. center $(1, -3)$; foci $(4.46, -3)$ and $(-2.46, -3)$; major axis: 8; minor axis: 4

$x^2 + 4y^2 - 2x + 24y + 21 = 0$

Additional Answers

7. center $(5, -1)$; foci $(5, 5)$ and $(5, -7)$; major axis: 16; minor axis: ≈ 10.58

$\dfrac{(y+1)^2}{64} + \dfrac{(x-5)^2}{28} = 1$

8. center $(-2, 1)$; foci $(3.29, 1)$ and $(-7.29, 1)$; major axis: ≈ 13.86; minor axis: ≈ 8.94

$\dfrac{(x+2)^2}{48} + \dfrac{(y-1)^2}{20} = 1$

Teaching the Mathematical Practices

Modeling Mathematically proficient students can apply the mathematics they know to solve problems arising in everyday life, analyze relationships mathematically to draw conclusions, and interpret their mathematical results in the context of a situation.

Additional Answers

39a.

44. Sample answer: As an ellipse becomes more circular, the difference between a and b becomes smaller. This causes the value of c to become smaller since $c^2 = a^2 - b^2$. The value of $2c$ is the distance between the foci, so the foci get closer together.

45. For any point on an ellipse, the sum of the distances from that point to the foci is constant by the definition of an ellipse. So, if $(2, 14)$ is on the ellipse, then the sum of the distances from it to the foci will be a certain value consistent with every other point on the ellipse. The distance between $(-7, 2)$ and $(2, 14)$ is $\sqrt{(-7-2)^2 + (2-14)^2}$ or 15. The distance between $(18, 2)$ and $(2, 14)$ is $\sqrt{(18-2)^2 + (2-14)^2}$ or 20. The sum of these two distances is 35. The distance between $(-7, 2)$ and $(2, -10)$ is $\sqrt{(-7-2)^2 + [2-(-10)]^2}$ or 15. The distance between $(18, 2)$ and $(2, -10)$ is $\sqrt{(18-2)^2 + [2-(-10)]^2}$ or 20. The sum of these distances is also 35. Thus, $(2, -10)$ also lies on the ellipse.

620 | Lesson 9-4 | Ellipses

Example 1   Write an equation for each ellipse.

11.

12. $\dfrac{x^2}{81} + \dfrac{y^2}{16} = 1$

12.

13. $\dfrac{(x+5)^2}{49} + \dfrac{(y+4)^2}{25} = 1$

13.

11. $\dfrac{y^2}{100} + \dfrac{x^2}{36} = 1$

14.

15.

16. $\dfrac{(x-3)^2}{100} + \dfrac{(y+6)^2}{64} = 1$

16.

$\dfrac{(y+2)^2}{81} + \dfrac{(x-1)^2}{9} = 1$ $\dfrac{(y-1)^2}{64} + \dfrac{(x+5)^2}{16} = 1$

Example 2   Write an equation for an ellipse that satisfies each set of conditions.

17. vertices at $(-6, 4)$ and $(12, 4)$, co-vertices at $(3, 12)$ and $(3, -4)$

18. vertices at $(-1, 11)$ and $(-1, 1)$, co-vertices at $(-4, 6)$ and $(2, 6)$

19 center at $(-2, 6)$, vertex at $(-2, 16)$, co-vertex at $(1, 6)$

20. center at $(3, -4)$, vertex at $(8, -4)$, co-vertex at $(3, -2)$

21. vertices at $(4, 12)$ and $(4, -4)$, co-vertices at $(1, 4)$ and $(7, 4)$

22. vertices at $(-11, 2)$ and $(-1, 2)$, co-vertices at $(-6, 0)$ and $(-6, 4)$

17. $\dfrac{(x-3)^2}{81} + \dfrac{(y-4)^2}{64} = 1$

18. $\dfrac{(y-6)^2}{25} + \dfrac{(x+1)^2}{9} = 1$

19. $\dfrac{(y-6)^2}{100} + \dfrac{(x+2)^2}{9} = 1$

20. $\dfrac{(x-3)^2}{25} + \dfrac{(y+4)^2}{4} = 1$

21. $\dfrac{(y-4)^2}{64} + \dfrac{(x-4)^2}{9} = 1$

Example 3   **23.** **CCSS MODELING** The opening of a tunnel in the mountains can be modeled by semiellipses, or halves of ellipses. If the opening is 14.6 meters wide and 8.6 meters high, determine an equation to represent the opening with the center at the origin.
$\dfrac{y^2}{73.96} + \dfrac{x^2}{53.29} = 1$

Example 4   Find the coordinates of the center and foci and the lengths of the major and minor axes for the ellipse with the given equation. Then graph the ellipse.

24–31. See Chapter 9 Answer Appendix.

24. $\dfrac{(x-3)^2}{36} + \dfrac{(y-2)^2}{128} = 1$ **25.** $\dfrac{(x+6)^2}{50} + \dfrac{(y-3)^2}{72} = 1$

26. $\dfrac{x^2}{27} + \dfrac{(y-5)^2}{64} = 1$ **27.** $\dfrac{(x+4)^2}{16} + \dfrac{y^2}{75} = 1$

22. $\dfrac{(x+6)^2}{25} + \dfrac{(y-2)^2}{4} = 1$

28. $3x^2 + y^2 - 6x - 8y - 5 = 0$ **29.** $3x^2 + 4y^2 - 18x + 24y + 3 = 0$

30. $7x^2 + y^2 - 56x + 6y + 93 = 0$ **31.** $3x^2 + 2y^2 + 12x - 20y + 14 = 0$

32. **SPACE** Like the planets, Halley's Comet travels around the Sun in an elliptical orbit. The aphelion is 3282.9 million miles and the perihelion is 54.87 million miles. Determine an equation relating the comet's orbit around the Sun in millions of miles with the center of the horizontal ellipse at the origin.
$\dfrac{x^2}{2,786,629.3} + \dfrac{y^2}{181,584.8} = 1$

620 | Lesson 9-4 | Ellipses

JHP Travel/Alamy

Differentiated Homework Options

Level	Assignment	Two-Day Option	
AL Basic	11–32, 40, 41, 44–67	11–31 odd, 47–50	12–32 even, 40, 41, 44–46, 51–67
OL Core	11–31 odd, 33–41, 44–67	11–32, 47–50	33–41, 44–46, 51–67
BL Advanced	33–64, (optional: 65–67)		

B Write an equation for an ellipse that satisfies each set of conditions.

33. center at $(-5, -2)$, focus at $(-5, 2)$, co-vertex at $(-8, -2)$ $\dfrac{(y+2)^2}{25} + \dfrac{(x+5)^2}{9} = 1$

34. center at $(4, -3)$, focus at $(9, -3)$, co-vertex at $(4, -5)$

35. foci at $(-2, 8)$ and $(6, 8)$, co-vertex at $(2, 10)$ $\dfrac{(x-2)^2}{20} + \dfrac{(y-8)^2}{4} = 1$

36. foci at $(4, 4)$ and $(4, 14)$, co-vertex at $(0, 9)$

34. $\dfrac{(x-4)^2}{29} + \dfrac{(y+3)^2}{4} = 1$

36. $\dfrac{(y-9)^2}{41} + \dfrac{(x-4)^2}{16} = 1$

37. **GOVERNMENT** The Oval Office is located in the West Wing of the White House. It is an elliptical shaped room used as the main office by the President of the United States. The long axis is 10.9 meters long and the short axis is 8.8 meters long. Write an equation to represent the outer walls of the Oval Office. Assume that the center of the room is at the origin. $\dfrac{x^2}{29.7025} + \dfrac{y^2}{19.36} = 1$

38. **SOUND** A whispering gallery is an elliptical room in which a faint whisper at one focus cannot be heard by other people in the room, but can easily be heard by someone at the other focus. Suppose an ellipse is 400 feet long and 120 feet wide. What is the distance between the foci? about 381.58 ft

39. **MULTIPLE REPRESENTATIONS** The *eccentricity* of an ellipse measures how circular the ellipse is. **39b. Sample answer: The first graph is more circular than the second graph.**

   a. Graphical Graph $\dfrac{x^2}{81} + \dfrac{y^2}{36} = 1$ and $\dfrac{x^2}{81} + \dfrac{y^2}{9} = 1$ on the same graph. **See margin.**

   b. Verbal Describe the difference between the two graphs.

   c. Algebraic The eccentricity of an ellipse is $\dfrac{c}{a}$. Find the eccentricity for each.

   d. Analytical Make a conjecture about the relationship between the value of an ellipse's eccentricity and the shape of the ellipse as compared to a circle.
   39c. first graph: 0.745; second graph: 0.943
   39d. Sample answer: The closer the eccentricity is to 0, the more circular the ellipse.

H.O.T. Problems   *Use Higher-Order Thinking Skills*

C 40. **ERROR ANALYSIS** Serena and Karissa are determining the equation for an ellipse with foci at $(-4, -11)$ and $(-4, 5)$ and co-vertices at $(2, -3)$ and $(-10, -3)$. Is either of them correct? Explain your reasoning.

40. Sample answer: Neither; both are showing horizontal ellipses and answer is vertical.

Serena	Karissa
$\dfrac{(x-4)^2}{64} + \dfrac{(y+3)^2}{36} = 1$	$\dfrac{(x+4)^2}{100} + \dfrac{(y+3)^2}{36} = 1$

41. **OPEN ENDED** Write an equation for an ellipse with a focus at the origin.   **41. Sample answer:** $\dfrac{(x+4)^2}{40} + \dfrac{y^2}{24} = 1$

42. **CHALLENGE** When the values of a and b are equal, an ellipse is a circle. Use this information and your knowledge of ellipses to determine the formula for the area of an ellipse in terms of a and b. $A = \pi ab$

43. **CHALLENGE** Determine an equation for an ellipse with foci at $\left(2, \sqrt{6}\right)$ and $\left(2, -\sqrt{6}\right)$ that passes through $\left(3, \sqrt{6}\right)$. $\dfrac{y^2}{9} + \dfrac{(x-2)^2}{3} = 1$

44. **CCSS ARGUMENTS** What happens to the location of the foci as an ellipse becomes more circular? Explain your reasoning. **See margin.**

45. **REASONING** An ellipse has foci at $(-7, 2)$ and $(18, 2)$. If $(2, 14)$ is a point on the ellipse, show that $(2, -10)$ is also a point on the ellipse. **See margin.**

46. **WRITING IN MATH** Explain why the domain is $\{x \mid -a \le x \le a\}$ and the range is $\{y \mid -b \le y \le b\}$ for an ellipse with equation $\dfrac{x^2}{a^2} + \dfrac{y^2}{b^2} = 1$. **See Chapter 9 Answer Appendix.**

CCSS **Teaching the Mathematical Practices**

Arguments Mathematically proficient students understand and use stated assumptions, definitions, and previously established results in constructing arguments. They make conjectures and build a logical progression of statements to explore the truth of their conjectures. And they are able to analyze situations by breaking them into cases, and can recognize and use counterexamples.

Study Guide and Intervention
AL OL ELL

NAME _____ DATE _____ PERIOD _____

9-4 Study Guide and Intervention
Ellipses

Practice
AL OL BL ELL

NAME _____ DATE _____ PERIOD _____

9-4 Practice
Ellipses

Word Problem Practice
AL OL BL ELL

NAME _____ DATE _____ PERIOD _____

9-4 Word Problem Practice
Ellipses

Enrichment
OL BL

NAME _____ DATE _____ PERIOD _____

9-4 Enrichment

Eccentricity

In an ellipse, the ratio $\frac{c}{a}$ is called the **eccentricity** and is denoted by the letter e. Eccentricity measures the elongation of an ellipse. The closer e is to 0, the more an ellipse looks like a circle. The closer e is to 1, the more elongated it is. Recall that the equation of an ellipse is $\frac{x^2}{a^2} + \frac{y^2}{b^2} = 1$ or $\frac{x^2}{b^2} + \frac{y^2}{a^2} = 1$ where a is the length of the major axis, and that $c = \sqrt{a^2 - b^2}$.

Find the eccentricity of each ellipse rounded to the nearest hundredth.

1. $\frac{x^2}{9} + \frac{y^2}{36} = 1$ 2. $\frac{x^2}{81} + \frac{y^2}{9} = 1$ 3. $\frac{x^2}{4} + \frac{y^2}{9} = 1$

4 Assess

Yesterday's News Have students write how yesterday's lesson helped them with writing and graphing equations for ellipses in today's lesson.

Formative Assessment

Check for student understanding of concepts in Lessons 9-3 and 9-4.

☐ Quiz 2, p. 53

Additional Answers

54c.

$$y = \frac{1}{39}x^2 - 9.75$$

54d. Sample answer: The equation in feet is easier to graph because the numbers have fewer decimal places.

Standardized Test Practice

47. Multiply. **B**

$$(2 + 3i)(4 + 7i)$$

  A $8 + 21i$ **C** $-6 + 10i$

  B $-13 + 26i$ **D** $13 + 12i$

48. The average lifespan of American women has been tracked, and the model for the data is $y = 0.2t + 73$, where $t = 0$ corresponds to 1960. What is the meaning of the y-intercept? **H**

  F In 2007, the average lifespan was 60.

  G In 1960, the average lifespan was 58.

  H In 1960, the average lifespan was 73.

  J The lifespan is increasing 0.2 years every year.

49. GRIDDED RESPONSE If we decrease a number by 6 and then double the result, we get 5 less than the number. What is the number? **7**

50. SAT/ACT The length of a rectangular prism is one inch greater than its width. The height is three times the length. Find the volume of the prism. **C**

  A $3x^3 + x^2 + 3x$

  B $x^3 + x^2 + x$

  C $3x^3 + 6x^2 + 3x$

  D $3x^3 + 3x^2 + 3x$

  E $3x^3 + 3x^2$

Spiral Review

Write an equation for the circle that satisfies each set of conditions. (Lesson 9-3)

51. center $(8, -9)$, passes through $(21, 22)$ $(x - 8)^2 + (y + 9)^2 = 1130$

52. center at $(4, 2)$, tangent to x-axis $(x - 4)^2 + (y - 2)^2 = 4$

53. center in the second quadrant; tangent to $y = -1$, $y = 9$, and the y-axis $(x + 5)^2 + (y - 4)^2 = 25$

54. ENERGY A parabolic mirror is used to collect solar energy. The mirrors reflect the rays from the Sun to the focus of the parabola. The focus of a particular mirror is 9.75 feet above the vertex, and the latus rectum is 39 feet long. (Lesson 9-2)

  a. Assume that the focus is at the origin. Write an equation for the parabola formed by the mirror. $y = \frac{1}{39}x^2 - 9.75$

  b. One foot is exactly 0.3048 meter. Rewrite the equation for the mirror in meters. $y = \frac{1}{11.8872}x^2 - 2.9718$

  c. Graph one of the equations for the mirror. **See margin.**

  d. Which equation did you choose to graph? Explain why. **See margin.**

Simplify each expression. (Lesson 8-2)

55. $\dfrac{6}{d^2 + 4d + 4} + \dfrac{5}{d + 2}$ $\dfrac{5d + 16}{(d + 2)^2}$ **56.** $\dfrac{a}{a^2 - a - 20} + \dfrac{2}{a + 4}$ $\dfrac{3a - 10}{(a - 5)(a + 4)}$ **57.** $\dfrac{x}{x + 1} + \dfrac{3}{x^2 - 4x - 5}$ $\dfrac{x^2 - 5x + 3}{(x - 5)(x + 1)}$

Solve each equation. (Lesson 7-4)

58. $\log_{10}(x^2 + 1) = 1$ ± 3 **59.** $\log_b 64 = 3$ **4**   **60.** $\log_b 121 = 2$ **11**

Simplify. (Lesson 5-1)   **61.** $15a^3b^3 - 30a^4b^3 + 15a^5b^6$

61. $-5ab^2(-3a^2b + 6a^3b - 3a^4b^4)$

62. $2xy(3xy^3 - 4xy + 2y^4)$ $6x^2y^4 - 8x^2y^2 + 4xy^5$

63. $(4x^2 - 3y^2 + 5xy) - (8xy + 3y^2)$ $4x^2 - 3xy - 6y^2$

64. $(10x^2 - 3xy + 4y^2) - (3x^2 + 5xy)$ $7x^2 - 8xy + 4y^2$

Skills Review

Write an equation of the line passing through each pair of points.   **65.** $y = -\frac{4}{5}x + \frac{17}{5}$ **67.** $y = -\frac{3}{5}x + \frac{16}{5}$

65. $(-2, 5)$ and $(3, 1)$ **66.** $(7, 1)$ and $(7, 8)$ $x = 7$ **67.** $(-3, 5)$ and $(2, 2)$

 622 | Lesson 9-4 | Ellipses

DifferentiatedInstruction   **OL**   **BL**

Extension Tell students that the area of an ellipse can be found using the formula $A = \pi ab$. Have students find the approximate area of the first ellipse on page 616 by counting the number of squares on the grid inside the ellipse and multiplying by 4 since each square equals 4 square units. Next, have students find the area of the same ellipse to the nearest tenth using the formula. Discuss with students how this formula is similar to the formula for the area of a circle. Sample answer: $A =$ about 160 square units; answer: $A = 159.9$ square units

622 | Lesson 9-4 | Ellipses

Mid-Chapter Quiz
Lessons 9-1 through 9-4

CHAPTER 9

Find the midpoint of the line segment with endpoints at the given coordinates. (Lesson 9-1)

1. $(7, 4), (-1, -5)$ $\left(3, -\dfrac{1}{2}\right)$ **2.** $(-2, -9), (-6, 0)$

$\left(-4, -\dfrac{9}{2}\right)$

Find the distance between each pair of points with the given coordinates. (Lesson 9-1)

3. $(0, 6), (-2, 5)$ $\sqrt{5}$ units   **4.** $(10, 1), (0, -4)$

$5\sqrt{5}$ units

5. HIKING Carla and Lance left their campsite and hiked 6 miles directly north and then turned and hiked 7 miles east to view a waterfall. (Lesson 9-1) **a.** $\sqrt{85}$ or about 9.22 miles

a. How far is the waterfall from their campsite?

b. Let the campsite be located at the origin on a coordinate grid. At the waterfall they decide to head directly back to the campsite. If they stop halfway between the waterfall and the campsite for lunch, at what coordinates will they stop for lunch? $(3.5, 3)$

Write each equation in standard form. Identify the vertex, axis of symmetry, and direction of opening of the parabola. (Lesson 9-2) **6–9.** See margin.

6. $y = 3x^2 - 12x + 21$ **7.** $x - 2y^2 = 4y + 6$

8. $y = \dfrac{1}{2}x^2 + 12x - 8$ **9.** $x = 3y^2 + 5y - 9$

$y = \dfrac{2}{625}x^2$

10. BRIDGES Write an equation of a parabola to model the shape of the suspension cable of the bridge shown. Assume that the origin is at the lowest point of the cables. (Lesson 9-2)

500 m

200 m

Identify the coordinates of the vertex and focus, the equation of the axis of symmetry and directrix, and the direction of opening of the parabola with the given equation. Then find the length of the latus rectum. (Lesson 9-2) **11, 12.** See Chapter 9 Answer Appendix.

11. $y = x^2 + 6x + 5$

12. $x = -2y^2 + 4y + 1$

13. Find the center and radius of the circle with equation $(x - 1)^2 + y^2 = 9$. Then graph the circle. (Lesson 9-3)
See Chapter 9 Answer Appendix.

14. Write an equation for a circle that has center at $(3, -2)$ and passes through $(3, 4)$. (Lesson 9-3)
$(x - 3)^2 + (y + 2)^2 = 36$

15. Write an equation for a circle if the endpoints of a diameter are at $(8, 31)$ and $(32, 49)$. (Lesson 9-3)
$(x - 20)^2 + (y - 40)^2 = 225$

16. MULTIPLE CHOICE What is the radius of the circle with equation $x^2 + 2x + y^2 + 14y + 34 = 0$? (Lesson 9-3) **B**

A 2

B 4

C 8

D 16

Find the coordinates of the center and foci and the lengths of the major and minor axes of the ellipse with the given equation. Then graph the ellipse. (Lesson 9-4) **17–19.** See Chapter 9 Answer Appendix.

17. $\dfrac{(x + 4)^2}{16} + \dfrac{(y - 2)^2}{9} = 1$

18. $\dfrac{(x - 1)^2}{20} + \dfrac{(y + 2)^2}{4} = 1$

19. $4y^2 + 9x^2 + 16y - 90x + 205 = 0$

20. MULTIPLE CHOICE Which equation represents an ellipse with endpoints of the major axis at $(-4, 10)$ and $(-4, -6)$ and foci at about $(-4, 7.3)$ and $(-4, -3.3)$? (Lesson 9-4) **H**

F $\dfrac{(x - 2)^2}{36} + \dfrac{(y + 4)^2}{64} = 1$

G $\dfrac{(x + 4)^2}{64} + \dfrac{(y - 2)^2}{36} = 1$

H $\dfrac{(y - 2)^2}{64} + \dfrac{(x + 4)^2}{36} = 1$

J $\dfrac{(x - 2)^2}{64} + \dfrac{(y + 4)^2}{36} = 1$

Formative Assessment
Use the Mid-Chapter Quiz to assess students' progress in the first half of the chapter.

For problems answered incorrectly, have students review the lessons indicated in parentheses.

Summative Assessment
Mid-Chapter Test, p. 55

McGraw-Hill Assessment Customize and create multiple versions of your Mid-Chapter Quiz and their answer keys.

FOLDABLES StudyOrganizer

Dinah Zike's Foldables®
Before students complete the Mid-Chapter Quiz, encourage them to review the information for Lessons 9-1 through 9-4 in their Foldables.

Additional Answers

6. $y = 3(x - 2)^2 + 9$; $(2, 9)$; $x = 2$; opens up

7. $x = 2(y + 1)^2 + 4$; $(4, -1)$; $y = -1$; opens to the right

8. $y = \dfrac{1}{2}(x + 12)^2 - 80$; $(-12, -80)$; $x = -12$; opens up

9. $x = 3\left(y + \dfrac{5}{6}\right)^2 - 11\dfrac{1}{12}$; $\left(-\dfrac{133}{12}, -\dfrac{5}{6}\right)$; $y = -\dfrac{5}{6}$; opens to the right

---

InterventionPlanner

TIER **1** On Level **OL**	TIER **2** Strategic Intervention **AL** approaching grade level	TIER **3** Intensive Intervention 2 or more grades below level
If students miss about 25% of the exercises or less,	**If** students miss about 50% of the exercises,	**If** students miss about 75% of the exercises,
Then choose a resource:	**Then** choose a resource:	**Then** use *Math Triumphs, Alg. 2*
SE Lessons 9-1, 9-2, 9-3, and 9-4	Study Guide and Intervention, pp. 5, 11, 18, and 25	Extra Examples, Personal Tutor, Homework Help, Review Vocabulary
Skills Practice, pp. 7, 13, 20, and 27	*Quick Review Math Handbook*	
TE Chapter Project	connectED.mcgraw-hill.com Extra Examples, Personal Tutor, Homework Help	
connectED.mcgraw-hill.com Self-Check Quiz		

1 Focus

VerticalAlignment

Before Lesson 9-5 Graph and analyze equations of ellipses.

Lesson 9-5 Write equations of hyperbolas. Graph hyperbolas.

After Lesson 9-5 Identify the conic section from a given equation.

2 Teach

Scaffolding Questions

Have students read the **Why?** section of the lesson.

Ask:

- It takes about 76 years for Halley's comet to orbit the sun, and it was last seen in 1986. Approximately when will it appear again? 2062

- How is a hyperbola unlike other conic sections? A hyperbola has two branches.

- Of the conic sections studied so far, which is the only one that could be a function? a parabola with a parent function of $y = x^2$ that has a vertical axis of symmetry

- Could the graph of a hyperbola ever be a function? No.

LESSON 9-5 Hyperbolas

Then	Now	Why?
● You graphed and analyzed equations of ellipses.	**1** Write equations of hyperbolas. **2** Graph hyperbolas.	● Because Halley's Comet travels around the Sun in an elliptical path, it reappears in our sky. Other comets pass through our sky only once. Many of these comets travel in paths that resemble hyperbolas.

NewVocabulary
hyperbola
transverse axis
conjugate axis
foci
vertices
co-vertices
constant difference

CCSS Common Core State Standards

Content Standards
A.SSE.1.b Interpret complicated expressions by viewing one or more of their parts as a single entity.
A.CED.2 Create equations in two or more variables to represent relationships between quantities; graph equations on coordinate axes with labels and scales.

Mathematical Practices
6 Attend to precision.

 624 | Lesson 9-5

1 Equations of Hyperbolas Similar to an ellipse, a **hyperbola** is the set of all points in a plane such that the absolute value of the differences of the distances from the foci is constant.

Every hyperbola has two axes of symmetry, the **transverse axis** and the **conjugate axis**. The axes are perpendicular at the center of the hyperbola.

The **foci** of a hyperbola always lie on the transverse axis. The **vertices** are the endpoints of the transverse axis. The **co-vertices** are the endpoints of the conjugate axis.

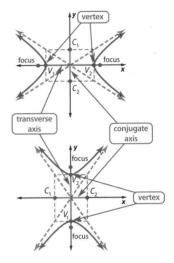

As a hyperbola recedes from the center, both halves approach asymptotes.

KeyConcept Equations of Hyperbolas Centered at the Origin

Standard Form	$\frac{x^2}{a^2} - \frac{y^2}{b^2} = 1$	$\frac{y^2}{a^2} - \frac{x^2}{b^2} = 1$
Orientation	horizontal	vertical
Foci	$(\pm c, 0)$	$(0, \pm c)$
Length of Transverse Axis	$2a$ units	$2a$ units
Length of Conjugate Axis	$2b$ units	$2b$ units
Equations of Asymptotes	$y = \pm\frac{b}{a}x$	$y = \pm\frac{a}{b}x$

As with ellipses, there are several important relationships among the parts of hyperbolas.

- There are two axes of symmetry.
- The values of a, b, and c are related by the equation $c^2 = a^2 + b^2$.

Lesson 9-5 Resources

Resource	Approaching Level **AL**	On Level **OL**	Beyond Level **BL**	English Learners **ELL**
Teacher Edition	▪ Differentiated Instruction, p. 625	▪ Differentiated Instruction, pp. 625, 631	▪ Differentiated Instruction, p. 631	
Chapter Resource Masters	▪ Study Guide and Intervention, pp. 32–33 ▪ Skills Practice, p. 34 ▪ Practice, p. 35 ▪ Word Problem Practice, p. 36	▪ Study Guide and Intervention, pp. 32–33 ▪ Skills Practice, p. 34 ▪ Practice, p. 35 ▪ Word Problem Practice, p. 36 ▪ Enrichment, p. 37 ▪ Graphing Calculator Activity, p. 38	▪ Practice, p. 35 ▪ Word Problem Practice, p. 36 ▪ Enrichment, p. 37	▪ Study Guide and Intervention, pp. 32–33 ▪ Skills Practice, p. 34 ▪ Practice, p. 35 ▪ Word Problem Practice, p. 36
Other	▪ 5-Minute Check 9-5 ▪ Study Notebook ▪ Teaching Algebra with Manipulatives	▪ 5-Minute Check 9-5 ▪ Study Notebook ▪ Teaching Algebra with Manipulatives	▪ 5-Minute Check 9-5 ▪ Study Notebook	▪ 5-Minute Check 9-5 ▪ Study Notebook ▪ Teaching Algebra with Manipulatives

Math HistoryLink

Hypatia (415–370 B.C.)
Hypatia was a mathematician, scientist, and philosopher in Alexandria, Egypt. She is considered the first woman to write on mathematical topics. Hypatia edited the book *On the Conics of Apollonius*, adding her own problems and examples to clarify the topic for her students. This book developed the ideas of hyperbolas, parabolas, and ellipses.

Example 1 Write an Equation Given Vertices and Foci

Write an equation for the hyperbola shown in the graph.

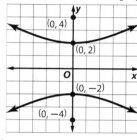

Step 1 Find the center.

The vertices are equidistant from the center.
The center is at $(0, 0)$.

Step 2 Find the values of a, b, and c.

The value of a is the distance between a vertex and the center, or 4 units.

The value of c is the distance between a focus and the center, or 5 units.

$c^2 = a^2 + b^2$ Equation relating a, b, and c for a hyperbola

$5^2 = 4^2 + b^2$ $c = 5$ and $a = 3$

$9 = b^2$ Subtract 4^2 from each side.

Step 3 Write the equation.

The transverse axis is horizontal, so the equation is $\frac{x^2}{16} - \frac{y^2}{9} = 1$.

▶ **Guided**Practice

1. Write an equation for a hyperbola with vertices at $(6, 0)$ and $(-6, 0)$ and foci at $(8, 0)$ and $(-8, 0)$. $\frac{x^2}{36} - \frac{y^2}{28} = 1$

Hyperbolas can also be determined using the equations of their asymptotes.

Example 2 Write an Equation Given Asymptotes

The asymptotes for a vertical hyperbola are $y = \frac{5}{3}x$ and $y = -\frac{5}{3}x$ and the vertices are at $(0, 5)$ and $(0, -5)$. Write the equation for the hyperbola.

Step 1 Find the center.

The vertices are equidistant from the center.
The center of the hyperbola is at $(0, 0)$.

Step 2 Find the values of a and b.

The hyperbola is vertical, so $a = 5$.
From the asymptotes, $b = 3$.
The value of c is not needed.

ReadingMath

Standard Form In the standard form of a hyperbola, the squared terms are subtracted. For an ellipse, they are added.

Step 3 Write the equation.

The equation for the hyperbola is $\frac{y^2}{25} - \frac{x^2}{9} = 1$.

▶ **Guided**Practice   **2.** $\frac{x^2}{81} - \frac{y^2}{49} = 1$

2. The asymptotes for a horizontal hyperbola are $y = \frac{7}{9}x$ and $y = -\frac{7}{9}x$. The vertices are $(9, 0)$ and $(-9, 0)$. Write an equation for the hyperbola.

Ancient Art & Architecture Collection Ltd/Alamy

1 **Equations of Hyperbolas**

Example 1 shows how to write an equation for a hyperbola given the graph of the hyperbola. **Example 2** shows how to write an equation for a hyperbola given the asymptotes and vertices.

Formative Assessment

Use the Guided Practice exercises after each example to determine students' understanding of concepts.

Additional Examples

1 Write an equation for the hyperbola shown in the graph.

$\frac{y^2}{4} - \frac{x^2}{12} = 1$

2 The asymptotes for a vertical hyperbola are $y = 3x$ and $y = -3x$ and the vertices are at $(0, 2\sqrt{10})$ and $(0, -2\sqrt{10})$. Write the equation for the hyperbola.

$\frac{y^2}{36} - \frac{x^2}{4} = 1$

▶ **Additional Examples** also in Interactive Classroom PowerPoint® Presentations

IWB **Interactive White Board**
READY

DifferentiatedInstruction   **AL**   **OL**

If ▶ some students think that a hyperbola has the shape of two parabolas,

Then ▶ explain that this is not true, and encourage students to draw parabolas on transparent paper and place them over hyperbolas to see that the shapes of the curves are different.

Focus on Mathematical Content

Hyperbolas Every hyperbola has *asymptotes*, or lines which the hyperbola branches approach. Every hyperbola has two axes of symmetry. The endpoints of the *transverse axis*, length 2a, are the vertices of the hyperbola. The *conjugate axis* of length 2b is perpendicular to the transverse axis and passes through the midpoint of the transverse axis (the center). The distance from the center to a focus is c. For a hyperbola, $a^2 + b^2 = c^2$.

2 Graphs of Hyperbolas

Example 3 shows how to use the asymptotes to graph a hyperbola from an equation in standard form.
Example 4 shows how to graph a hyperbola that models a real-world situation.

Additional Example

3 Graph $\dfrac{(x-3)^2}{4} - \dfrac{(y+5)^2}{9} = 1$.
Identify the vertices, foci, and asymptotes. vertices: $(5, -5)$, $(1, -5)$; foci: $(3 + \sqrt{13}, -5)$, $(3 - \sqrt{13}, -5)$; asymptotes: $y + 5 = \dfrac{3}{2}(x - 3)$, $y + 5 = -\dfrac{3}{2}(x - 3)$

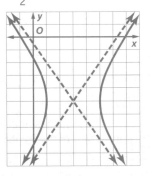

---

2 Graphs of Hyperbolas Hyperbolas can be translated in the same manner as the other conic sections.

KeyConcept Equations of Hyperbolas Centered at (h, k)

Standard Form	$\dfrac{(x-h)^2}{a^2} - \dfrac{(y-k)^2}{b^2} = 1$	$\dfrac{(y-k)^2}{a^2} - \dfrac{(x-h)^2}{b^2} = 1$
Orientation	horizontal	vertical
Foci	$(h \pm c, k)$	$(h, k \pm c)$
Vertices	$(h \pm a, k)$	$(h, k \pm a)$
Co-vertices	$(h, k \pm b)$	$(h \pm b, k)$
Equations of Asymptotes	$y - k = \pm\dfrac{b}{a}(x - h)$	$y - k = \pm\dfrac{a}{b}(x - h)$

StudyTip

Calculator You can graph a hyperbola on a graphing calculator by solving for *y*, and then graphing the two equations on the same screen.

Example 3 Graph a Hyperbola

Graph $\dfrac{(x-3)^2}{4} - \dfrac{(y+2)^2}{16} = 1$. Identify the vertices, foci, and asymptotes.

Step 1 Find the center. The center is at $(3, -2)$.

Step 2 Find *a*, *b*, and *c*. From the equation, $a^2 = 4$ and $b^2 = 16$, so $a = 2$ and $b = 4$.

$c^2 = a^2 + b^2$ Equation relating *a*, *b*, and *c* for a hyperbola
$c^2 = 2^2 + 4^2$ $a = 2$, $b = 4$
$c^2 = 20$ Simplify.
$c = \sqrt{20}$ or about 4.47   Take the square root of each side.

Step 3 Identify the vertices and foci. The hyperbola is horizontal and the vertices are 2 units from the center, so the vertices are at $(1, -2)$ and $(5, -2)$. The foci are about 4.47 units from the center. The foci are at $(-1.47, -2)$ and $(7.47, -2)$.

Step 4 Identify the asymptotes.

$y - k = \pm\dfrac{b}{a}(x - h)$ Equation for asymptotes of a horizontal hyperbola
$y - (-2) = \pm\dfrac{4}{2}(x - 3)$ $a = 2$, $b = 4$, $h = 3$, and $k = -2$

The equations for the asymptotes are $y = 2x - 8$ and $y = -2x + 4$.

Step 5 Graph the hyperbola. The hyperbola is symmetric about the transverse and conjugate axes. Use this symmetry to plot additional points for the hyperbola.

Use the asymptotes as a guide to draw the hyperbola that passes through the vertices and the other points.

GuidedPractice 3. See margin.

3. Graph $\dfrac{(y-4)^2}{9} - \dfrac{(x+3)^2}{25} = 1$. Identify the vertices, foci, and asymptotes.

Additional Answer (Guided Practice)

3.

$y = \dfrac{3}{5}x + \dfrac{29}{5}$

$y = -\dfrac{3}{5}x + \dfrac{11}{5}$

In the equation for any hyperbola, the value of $2a$ represents the **constant difference**. This is the absolute value of the difference between the distances from any point on the hyperbola to the foci of the hyperbola.

Any point on the hyperbola at the right will have the same constant difference, $|y - x|$ or $2a$.

Real-World Example 4 Write an Equation of a Hyperbola

SPACE Earth and the Sun are 146 million kilometers apart. A comet follows a path that is one branch of a hyperbola. Suppose the comet is 30 million miles farther from the Sun than from Earth. Determine the equation of the hyperbola centered at the origin for the path of the comet.

Understand We need to determine the equation for the hyperbola.

Plan Find the center and the values of a and b. Once we have this information, we can determine the equation.

Solve The foci are Earth and the Sun, with the origin between them.

The value of c is $146 \div 2$ or 73.

The difference of the distances from the comet to each body is 30. Therefore, a is $30 \div 2$ or 15 million miles.

$$c^2 = a^2 + b^2 \quad \text{Equation relating } a, b, \text{ and } c \text{ for a hyperbola}$$
$$73^2 = 15^2 + b^2 \quad a = 15 \text{ and } c = 73$$
$$5104 = b^2 \quad \text{Simplify.}$$

The equation of the hyperbola is $\dfrac{x^2}{225} - \dfrac{y^2}{5104} = 1$.

Since the comet is farther from the Sun, it is located on the branch of the hyperbola near Earth.

Check $(21, 70)$ is a point that satisfies the equation.

The distance between this point and the Sun $(-73, 0)$ is

$$\sqrt{[21 - (-73)]^2 + (70 - 0)^2} \text{ or } 117.2 \text{ million kilometers.}$$

The distance between this point and Earth $(73, 0)$ is

$$\sqrt{(21 - 73)^2 + (70 - 0)^2} \text{ or } 87.2 \text{ million kilometers.}$$

The difference between these distances is 30. ✓

GuidedPractice

4. **SEARCH AND RESCUE** Two receiving stations that are 150 miles apart receive a signal from a downed airplane. They determine that the airplane is 80 miles farther from station A than from station B. Determine the equation of the hyperbola centered at the origin on which the plane is located. $\dfrac{x^2}{1600} - \dfrac{y^2}{4025} = 1$

connectED.mcgraw-hill.com 627

StockTrek/Photodisc/Getty Images

Real-World Link

Halley's Comet becomes visible to the unaided eye about every 76 years as it nears the Sun.

Source: NASA

StudyTip

Exact Locations A third receiving station is necessary to determine the plane's exact location.

WatchOut!

Preventing Errors Tell students to make rough sketches for the situations in problems such as those in Example 4.

Additional Example

4 **NAVIGATION** The LORAN navigational system is based on hyperbolas. Two stations send out signals at the same time. A ship notes the difference in the times at which it receives the signals. The ship is on a hyperbola with the stations at the foci. Suppose a ship determines that the difference of its distances from two stations is 50 nautical miles. Write an equation for a hyperbola on which the ship lies if the stations are at $(-50, 0)$ and $(50, 0)$.

$$\dfrac{x^2}{625} - \dfrac{y^2}{1875} = 1$$

WatchOut!

Preventing Errors Suggest that students make lists of the values of a, b, c, h, and k. This will help avoid confusion about the signs of h and k.

Follow-ups

Students have explored ellipses, parabolas, and hyperbolas.

Ask:

- What are the similarities and differences between parabolas and ellipses? Sample answer: Similarities: Curved graphs; equations contain a variable raised to second power. Differences: Parabolas have one focus and one vertex; ellipses have two foci, 2 vertices, and 2 co-vertices.

- What are the similarities and differences between hyperbolas and the other conic sections? Sample answer: Similarities: Curved graphs; equations contain one or two variables raised to second power. Differences: Hyperbolas have two branches; other conic sections are continuous.

Teach with Tech

Web Page Have students search the Web to find applications and real-world situations involving hyperbolas. Have students create links on their social networking Web sites for pages with specific examples they find interesting.

3 Practice

Formative Assessment

Use Exercises 1–9 to check for understanding.

Use the chart at the bottom of this page to customize assignments for your students.

 Teaching the Mathematical Practices

Structure Mathematically proficient students look closely to discern a pattern or structure

Additional Answers

5.

6.

7.

8.

Check Your Understanding

= Step-by-Step Solutions begin on page R14.

Examples 1–2 Write an equation for each hyperbola.

1. $\dfrac{y^2}{36} - \dfrac{x^2}{28} = 1$

2. 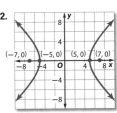 $\dfrac{x^2}{25} - \dfrac{y^2}{24} = 1$

3. $\dfrac{x^2}{64} - \dfrac{y^2}{25} = 1$

4. $\dfrac{y^2}{64} - \dfrac{x^2}{81} = 1$

Example 3 **CCSS STRUCTURE** Graph each hyperbola. Identify the vertices, foci, and asymptotes.

5–8. See margin.

5. $\dfrac{x^2}{64} - \dfrac{y^2}{49} = 1$

6. $\dfrac{y^2}{36} - \dfrac{x^2}{60} = 1$

7. $9y^2 + 18y - 16x^2 + 64x - 199 = 0$

8. $4x^2 + 24x - y^2 + 4y - 4 = 0$

Example 4 **9. NAVIGATION** A ship determines that the difference of its distances from two stations is 60 nautical miles. Write an equation for a hyperbola on which the ship lies if the stations are at $(-80, 0)$ and $(80, 0)$. $\dfrac{x^2}{900} - \dfrac{y^2}{5500} = 1$

Practice and Problem Solving

Extra Practice is on page R9.

Examples 1–2 Write an equation for each hyperbola.

10.

11.

10. $\dfrac{(x-3)^2}{25} - \dfrac{(y+6)^2}{11} = 1$

11. $\dfrac{(y-4)^2}{16} - \dfrac{(x+8)^2}{48} = 1$

12. $\dfrac{(y+5)^2}{16} - \dfrac{(x+4)^2}{36} = 1$

13. $\dfrac{(x+1)^2}{9} - \dfrac{(y-6)^2}{49} = 1$

12.

13.

 628 | Lesson 9-5 | Hyperbolas

Differentiated Homework Options

Level	Assignment	Two-Day Option	
AL Basic	10–24, 44, 46–67	11–23 odd, 50–53	10–24 even, 44, 46–49, 54–67
OL Core	11–31 odd, 32–44, 46–67	10–24, 50–53	25–44, 46–49, 54–67
BL Advanced	25–65, (optional: 66, 67)		

Example 3 | Graph each hyperbola. Identify the vertices, foci, and asymptotes. 14–23. See Chapter 9 Answer Appendix.

14. $\dfrac{x^2}{36} - \dfrac{y^2}{4} = 1$

15. $\dfrac{y^2}{9} - \dfrac{x^2}{49} = 1$

16. $\dfrac{y^2}{36} - \dfrac{x^2}{25} = 1$

17. $\dfrac{x^2}{16} - \dfrac{y^2}{16} = 1$

18. $\dfrac{(x-3)^2}{16} - \dfrac{(y+1)^2}{4} = 1$

19. $\dfrac{(y+5)^2}{16} - \dfrac{(x+2)^2}{36} = 1$

20. $9y^2 - 4x^2 - 54y + 32x - 19 = 0$

21. $16x^2 - 9y^2 + 128x + 36y + 76 = 0$

22. $25x^2 - 4y^2 - 100x + 48y - 144 = 0$

23. $81y^2 - 16x^2 - 810y + 96x + 585 = 0$

Example 4 | 24. **NAVIGATION** A ship determines that the difference of its distances from two stations is 80 nautical miles. Write an equation for a hyperbola on which the ship lies if the stations are at $(-100, 0)$ and $(100, 0)$. $\dfrac{x^2}{1600} - \dfrac{y^2}{8400} = 1$

 B Determine whether the following equations represent ellipses or hyperbolas.

25. $4x^2 = 5y^2 + 6$ **hyperbola**

26. $8x^2 - 2x = 8y - 3y^2$ **ellipse**

27. $-5x^2 + 4x = 6y + 3y^2$ **ellipse**

28. $7y - 2x^2 = 6x - 2y^2$ **hyperbola**

29. $6x - 7x^2 - 5y^2 = 2y$ **ellipse**

30. $4x + 6y + 2x^2 = -3y^2$ **ellipse**

31. **SPACE** Refer to the application at the beginning of the lesson. With the Sun as a focus and the center at the origin, a certain comet's path follows a branch of a hyperbola. If two of the coordinates of the path are $(10, 0)$ and $(30, 100)$ where the units are in millions of miles, determine the equation of the path. $\dfrac{x^2}{100} - \dfrac{y^2}{1250} = 1$

32. **COOLING** Natural draft cooling towers are shaped like hyperbolas for more efficient cooling of power plants. The hyperbola in the tower at the right can be modeled by $\dfrac{x^2}{16} - \dfrac{y^2}{225} = 1$, where the units are in meters. Find the width of the tower at the top and at its narrowest point in the middle. **8 m in the middle and 40.8 m at the top**

150 m

33. **MULTIPLE REPRESENTATIONS** Consider $xy = 16$. **a–c. See Chapter 9 Answer Appendix.**

   a. **Tabular** Make a table of values for the equation for $-12 \le x \le 12$.

   b. **Graphical** Graph the hyperbola represented by the equation.

   c. **Logical** Determine and graph the asymptotes for the hyperbola.

   d. **Analytical** What special property do you notice about the asymptotes? Hyperbolas that represent this property are called *rectangular hyperbolas*. **They are perpendicular.**

   e. **Analytical** Without any calculations, what do you think will be the coordinates of the vertices for $xy = 25$? for $xy = 36$? **For $xy = 25$, the vertices will be at $(5, 5)$ and $(-5, -5)$, and for $xy = 36$, they will be at $(-6, -6)$ and $(6, 6)$.**

35. $\dfrac{x^2}{2{,}722{,}500} - \dfrac{y^2}{1{,}277{,}500} = 1$

34. **CCSS MODELING** Two receiving stations that are 250 miles apart receive a signal from a downed airplane. They determine that the airplane is 70 miles farther from station B than from station A. Determine the equation of the horizontal hyperbola centered at the origin on which the plane is located. $\dfrac{x^2}{1225} - \dfrac{y^2}{14400} = 1$

35. **WEATHER** Luisa and Karl live exactly 4000 feet apart. While on the phone at their homes, Luisa hears thunder out of her window and Karl hears it 3 seconds later out of his. If sound travels 1100 feet per second, determine the equation for the horizontal hyperbola on which the lightning is located.

Multiple Representations

In Exercise 33, students use a table of values, a graph, and logical analysis to classify hyperbolas and predict their properties.

WatchOut!

Error Analysis For Exercise 44, remind students that they must use the standard form of the equation to determine the orientation of the hyperbola.

Additional Answers

44. Sample answer: Simon; The equation indicates a hyperbola that opens up and down. Gabriel drew a horizontal hyperbola.

46. Sample answer: The foci move farther away from the vertices. When a is only slightly smaller than b, the value of c is only fairly larger than a. However, as b becomes much larger than a, by way of $c^2 = a^2 + b^2$, c becomes much larger than a. When this happens, the distance between the foci c becomes much greater than the distance between the vertices a. Therefore, the foci are much farther away from the vertices.

49. Sample answer: Conic sections can be used to model phenomena that can't be modeled using functions. For example, parabolas can be used to model paths of comets and ellipses can be used to model planetary orbits.

56. (3, 0), 4 units

57. (0, 3), 5 units

58. $\left(-\dfrac{9}{2}, 4\right)$, $\dfrac{\sqrt{129}}{2}$ units

---

36. ARCHITECTURE Large pillars with cross sections in the shape of hyperbolas were popular in ancient Greece. The curves can be modeled by the equation $\dfrac{x^2}{0.16} - \dfrac{y^2}{4} = 1$, where the units are in feet. If the pillars are 9 feet tall, find the width of the top of each pillar and the width of each pillar at the narrowest point in the middle. Round to the nearest hundredth of a foot. $\approx$**1.96 ft, 0.8 ft**

Write an equation for the hyperbola that satisfies each set of conditions.

37. vertices $(-8, 0)$ and $(8, 0)$, conjugate axis of length 20 units $\dfrac{x^2}{64} - \dfrac{y^2}{100} = 1$

38. vertices $(0, -6)$ and $(0, 6)$, conjugate axis of length 24 units

38. $\dfrac{y^2}{36} - \dfrac{x^2}{144} = 1$

39 vertices $(6, -2)$ and $(-2, -2)$, foci $(10, -2)$ and $(-6, -2)$ $\dfrac{(x-2)^2}{16} - \dfrac{(y+2)^2}{48} = 1$

40. vertices $(-3, 4)$ and $(-3, -8)$, foci $(-3, 9)$ and $(-3, -13)$

40. $\dfrac{(y+2)^2}{36} - \dfrac{(x+3)^2}{85} = 1$

41. centered at the origin with a horizontal transverse axis of length 10 units and a conjugate axis of length 4 units $\dfrac{x^2}{25} - \dfrac{y^2}{4} = 1$

42. centered at the origin with a vertical transverse axis of length 16 units and a conjugate axis of length 12 units $\dfrac{y^2}{64} - \dfrac{x^2}{36} = 1$

43. TRIANGULATION While looking for their lost dog in the woods, Lae, Meg, and Cesar hear a bark. Meg hears it 2 seconds after Lae and Cesar hears it 3 seconds after Lae. With Lae at the origin, determine the exact location of their dog if sound travels 1100 feet per second. **(2308, 826)**

H.O.T. Problems Use Higher-Order Thinking Skills

44. **CCSS** **CRITIQUE** Simon and Gabriel are graphing $\dfrac{y^2}{25} - \dfrac{x^2}{4} = 1$. Is either of them correct? Explain your reasoning. **See margin.**

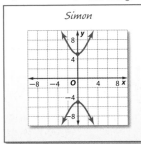

Simon   Gabriel

48. Sample answer:
$\dfrac{(x-5)^2}{9} - \dfrac{y^2}{16} = 1$

45. $\dfrac{(x-3)^2}{5} - \dfrac{(y+2)^2}{5} = 1$

45. CHALLENGE The origin lies on a horizontal hyperbola. The asymptotes for the hyperbola are $y = -x + 1$ and $y = x - 5$. Find the equation for the hyperbola.

46. REASONING What happens to the location of the foci of a hyperbola as the value of a becomes increasingly smaller than the value of b? Explain your reasoning. **See margin.**

47. REASONING Consider $\dfrac{y^2}{36} - \dfrac{x^2}{16} = 1$. Describe the change in the shape of the hyperbola and the locations of the vertices and foci if 36 is changed to 9. Explain why this happens. **See Chapter 9 Answer Appendix.**

48. OPEN ENDED Write an equation for a hyperbola with a focus at the origin.

49. WRITING IN MATH Why would you choose a conic section to model a set of data instead of a polynomial function? **See margin.**

CCSS **Teaching the Mathematical Practices**

Critique Mathematically proficient students are also able to compare the effectiveness of two plausible arguments, distinguish correct logic or reasoning from that which is flawed, and—if there is a flaw in an argument—explain what it is.

50. You have 6 more dimes than quarters. You have a total of $5.15. How many dimes do you have? **C**

    A 13   **C** 19
    B 16   **D** 25

51. How tall is a tree that is 15 feet shorter than a pole three times as tall as the tree? **J**

    F 24.5 ft
    G 22.5 ft
    H 21.5 ft
    J 7.5 ft

52. **SHORT RESPONSE** A rectangle is 8 feet long and 6 feet wide. If each dimension is increased by the same number of feet, the area of the new rectangle formed is 32 square feet more than the area of the original rectangle. By how many feet was each dimension increased? **2**

53. **SAT/ACT** When the equation $y = 4x^2 - 5$ is graphed in the coordinate plane, the graph is which of the following? **E**

    A line   **D** hyperbola
    B circle   **E** parabola
    C ellipse

Spiral Review

Write an equation for an ellipse that satisfies each set of conditions. (Lesson 9-4)

54. endpoints of major axis at (2, 2) and (2, −10), endpoints of minor axis at (0, −4) and (4, −4) $\dfrac{(y + 4)2}{36} + \dfrac{(x - 2)2}{4} = 1$

55. endpoints of major axis at (0, 10) and (0, −10), foci at (0, 8) and (0, −8) $\dfrac{y^2}{100} + \dfrac{x^2}{36} = 1$

Find the center and radius of the circle with the given equation. Then graph the circle. (Lesson 9-3) **56–58. See margin.**

56. $(x - 3)^2 + y^2 = 16$

57. $x^2 + y^2 - 6y - 16 = 0$

58. $x^2 + y^2 + 9x - 8y + 4 = 0$

59. **BASKETBALL** Zonta plays basketball for Centerville High School. So far this season, she has made 6 out of 10 free throws. She is determined to improve her free throw percentage. If she can make x consecutive free throws, her free throw percentage can be determined using $P(x) = \dfrac{6 + x}{10 + x}$. (Lesson 8-4)

    a. Graph the function. **See margin.**

    b. What part of the graph is meaningful in the context of the problem? **the part in the first quadrant**

    c. Describe the meaning of the y-intercept. **It represents her original free-throw percentage of 60%.**

    d. What is the equation of the horizontal asymptote? Explain its meaning with respect to Zonta's shooting percentage. **$P(x) = 1$; this represents 100%, which she cannot achieve because she has already missed 4 free throws.**

Solve each equation. (Lesson 7-2)

60. $\left(\dfrac{1}{7}\right)^{y - 3} = 343$ **0**

61. $10^{x - 1} = 100^{2x - 3}$ $\dfrac{5}{3}$

62. $36^{2p} = 216^{p - 1}$ **−3**

Graph each inequality. (Lesson 6-3) **63–65. See margin.**

63. $y \geq \sqrt{5x - 8}$

64. $y \geq \sqrt{x - 3} + 4$

65. $y < \sqrt{6x - 2} + 1$

Skills Review

66. Write an equation for a parabola with vertex at the origin that passes through (2, −8) $y = -2x^2$

67. Write an equation for a parabola with vertex at (−3, −4) that opens up and has y-intercept 8. $y = \dfrac{4}{3}(x + 3)^2 - 4$

DifferentiatedInstruction **OL** **BL**

Extension

- Use a compass to draw a circle on a piece of paper. Be sure not to take up the entire page.
- Use the sharp end of your compass to poke a small hole anywhere outside of your circle. Label the point where the hole is *F*.
- Fold the paper so that point *F* lands somewhere on the circle.
- Repeat the last step using point *F* and several other points on the circle. Predict what kind of curve these creases are forming. **The fixed point *F* is the focus of a hyperbola.**
- Discuss with students how the shape of the curve would change if they varied the location of point *F*. **As point *F* moves farther from the center of the circle, the branches of the hyperbola become wider.**

---

4 Assess

Name the Math Prepare two bags: one containing an ordered pair for the center of a hyperbola on each paper, the other with values for *a* and *b* on each paper. Have each student select both an ordered pair and values for *a* and *b*. Have students write the equations for the hyperbolas and tell as much about them as possible given the values they chose.

Additional Answers

59a.

$$P(x) = \dfrac{6 + x}{10 + x}$$

63.

$y = \sqrt{5x - 8}$

64.

$y = \sqrt{x - 3} + 4$

65.

$y = \sqrt{6x - 2} + 1$

1 Focus

VerticalAlignment

Before Lesson 9-6 Analyze different conic sections.

Lesson 9-6 Write equations of conic sections in standard form. Identify conic sections from their equations.

After Lesson 9-6 Use algebraic methods to solve systems of equations or inequalities of conic sections.

2 Teach

Scaffolding Questions

Have students read the **Why?** section of the lesson.

Ask:

- Describe the plane that forms a hyperbola. intersects both cones

- Describe the plane that forms a parabola. parallel to the slant height of a cone

- Describe the plane that forms a circle. parallel to the base of a cone

1 Conics in Standard Form

Example 1 shows how to rewrite an equation of a conic section in standard form and determine if the equation is that of a parabola, circle, ellipse, or hyperbola.

:: Then	:: Now	:: Why?
● You analyzed different conic sections.	**1** Write equations of conic sections in standard form. **2** Identify conic sections from their equations.	● Parabolas, circles, ellipses, and hyperbolas are called conic sections because they are the cross sections formed when a double cone is sliced by a plane.

Parabola   Circle and Ellipse   Hyperbola

Common Core State Standards

Content Standards

A.SSE.1.b Interpret complicated expressions by viewing one or more of their parts as a single entity.

F.IF.9 Compare properties of two functions each represented in a different way (algebraically, graphically, numerically in tables, or by verbal descriptions).

Mathematical Practices

3 Construct viable arguments and critique the reasoning of others.

8 Look for and express regularity in repeated reasoning.

1 Conics in Standard Form The equation for any conic section can be written in the form $Ax^2 + Bxy + Cy^2 + Dx + Ey + F = 0$, where A, B, and C are not all zero. This general form can be converted to the standard forms below by completing the square.

ConceptSummary **Standard Forms of Conic Sections**

Conic Section	Standard Form of Equation	
Circle	$(x - h)^2 + (y - k)^2 = r^2$	
	Horizontal Axis	**Vertical Axis**
Parabola	$y = a(x - h)^2 + k$	$x = a(y - k)^2 + h$
Ellipse	$\dfrac{(x - h)^2}{a^2} + \dfrac{(y - k)^2}{b^2} = 1$	$\dfrac{(y - k)^2}{a^2} + \dfrac{(x - h)^2}{b^2} = 1$
Hyperbola	$\dfrac{(x - h)^2}{a^2} - \dfrac{(y - k)^2}{b^2} = 1$	$\dfrac{(y - k)^2}{a^2} - \dfrac{(x - h)^2}{b^2} = 1$

Example 1 Rewrite an Equation of a Conic Section

Write $16x^2 - 25y^2 - 128x - 144 = 0$ in standard form. State whether the graph of the equation is a *parabola, circle, ellipse*, or *hyperbola*. Then graph the equation.

$16x^2 - 25y^2 - 128x - 144 = 0$	Original equation
$16(x^2 - 8x + \blacksquare) - 25y^2 = 144 + 16(\blacksquare)$	Isolate terms.
$16(x^2 - 8x + 16) - 25y^2 = 144 + 16(16)$	Complete the square.
$16(x - 4)^2 - 25y^2 = 400$	Perfect square
$\dfrac{(x - 4)^2}{25} - \dfrac{y^2}{16} = 1$	Divide each side by 400.

The graph is a hyperbola with its center at (4, 0).

▶ **Guided**Practice **1.** See Chapter 9 Answer Appendix.

1. Write $4x^2 + y^2 - 16x + 8y - 4 = 0$ in standard form. State whether the graph of the equation is a *parabola, circle, ellipse*, or *hyperbola*. Then graph the equation.

 632 | Lesson 9-6

Lesson 9-6 Resources

Resource	Approaching Level **AL**	On Level **OL**	Beyond Level **BL**	English Learners **ELL**
Teacher Edition	▪ Differentiated Instruction, p. 633	▪ Differentiated Instruction, pp. 633, 636	▪ Differentiated Instruction, p. 636	
Chapter Resource Masters	▪ Study Guide and Intervention, pp. 39–40 ▪ Skills Practice, p. 41 ▪ Practice, p. 42 ▪ Word Problem Practice, p. 43	▪ Study Guide and Intervention, pp. 39–40 ▪ Skills Practice, p. 41 ▪ Practice, p. 42 ▪ Word Problem Practice, p. 43 ▪ Enrichment, p. 44	▪ Practice, p. 42 ▪ Word Problem Practice, p. 43 ▪ Enrichment, p. 44	▪ Study Guide and Intervention, pp. 39–40 ▪ Skills Practice, p. 41 ▪ Practice, p. 42 ▪ Word Problem Practice, p. 43
Other	▪ 5-Minute Check 9-6 ▪ Study Notebook ▪ Teaching Algebra with Manipulatives	▪ 5-Minute Check 9-6 ▪ Study Notebook ▪ Teaching Algebra with Manipulatives	▪ 5-Minute Check 9-6 ▪ Study Notebook	▪ 5-Minute Check 9-6 ▪ Study Notebook ▪ Teaching Algebra with Manipulatives

ReviewVocabulary

discriminant the expression $b^2 - 4ac$ from the Quadratic Formula

2 Identify Conic Sections You can determine the type of conic without having to write $Ax^2 + Bxy + Cy^2 + Dx + Ey + F = 0$ in standard form. When there is an xy-term ($B \neq 0$), you can use the discriminant to identify the conic. $B^2 - 4AC$ is the discriminant of $Ax^2 + Bxy + Cy^2 + Dx + Ey + F = 0$.

ConceptSummary Classify Conics with the Discriminant

Discriminant	Conic Section
$B^2 - 4AC < 0$; $B = 0$ and $A = C$	circle
$B^2 - 4AC < 0$; either $B \neq 0$ or $A \neq C$	ellipse
$B^2 - 4AC = 0$	parabola
$B^2 - 4AC > 0$	hyperbola

When $B = 0$, the conic will be either vertical or horizontal. When $B \neq 0$, the conic will be neither vertical nor horizontal.

Horizontal Ellipse: $B = 0$

$x^2 + 4y^2 - 4 = 0$

Rotated Ellipse: $B \neq 0$

$7x^2 - 6\sqrt{3}xy + 13y^2 - 16 = 0$

StudyTip

Identifying Conics
When there is no xy-term ($B = 0$), use A and C.
Parabola: A or $C = 0$ but not both.
Circle: $A = C$
Ellipse: A and C have the same sign but are not equal.
Hyperbola: A and C have opposite signs.

Example 2 Analyze an Equation of a Conic Section

Without writing in standard form, state whether the graph of each equation is a *parabola, circle, ellipse,* or *hyperbola.*

a. $y^2 + 4x^2 - 3xy + 4x - 5y - 8 = 0$

$A = 4$, $B = -3$, and $C = 1$

The discriminant is $(-3)^2 - 4(4)(1)$ or -7.

Because the discriminant is less than 0 and $B \neq 0$, the conic is an ellipse.

b. $3x^2 - 6x + 4y - 5y^2 + 2xy - 4 = 0$

$A = 3$, $B = 2$, and $C = -5$

The discriminant is $2^2 - 4(3)(-5)$ or 64.

Because the discriminant is greater than 0, the conic is a hyperbola.

c. $4y^2 - 8x + 6y - 14 = 0$

$A = 0$, $B = 0$, and $C = 4$

The discriminant is $0^2 - 4(0)(4)$ or 0.

Because the discriminant equals 0, the conic is a parabola.

GuidedPractice

2A. $8y^2 - 6x^2 + 4xy - 6x + 2y - 4 = 0$ hyperbola

2B. $3xy + 4x^2 - 2y + 9x - 3 = 0$ hyperbola

2C. $3x^2 + 16x - 12y + 2y^2 - 6 = 0$ ellipse

DifferentiatedInstruction AL OL

If students need a visual aid in identifying conic sections,

Then divide students into groups of two or three. Assign each group a different conic section and have them make a poster describing how to identify the conic section from an equation in standard form and from an equation not in standard form.

Formative Assessment

Use the Guided Practice exercises after each example to determine students' understanding of concepts.

Additional Example

1 Write $y^2 = 18 - 2x^2$ in standard form. State whether the graph of the equation is a *parabola, circle, ellipse,* or *hyperbola.* Then graph the equation.

$\dfrac{x^2}{9} + \dfrac{y^2}{18} = 1$; ellipse

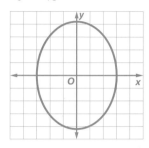

Additional Examples also in Interactive Classroom PowerPoint® Presentations

IWB Interactive White Board
READY

2 Identify Conic Sections

Example 2 shows how to determine if an equation is that of a parabola, circle, ellipse, or hyperbola without writing the equation in standard form.

Additional Example

2 Without writing the equation in standard form, state whether the graph of the equation is a *parabola, circle, ellipse,* or *hyperbola.*

a. $3y^2 - x^2 - 9 = 0$ hyperbola

b. $2x^2 + 2y^2 + 16x - 20y = -32$ circle

c. $y^2 - 2x - 4y + 10 = 0$ parabola

Follow-up

Students have explored the conic sections.

Ask:

- Why are parabolas, circles, ellipses, and hyperbolas called conic sections?

  Sample answer: They are the cross sections formed by a plane and a double napped cone.

3 Practice

Formative Assessment

Use Exercises 1–13 to check for understanding.

Use the chart at the bottom of this page to customize assignments for your students.

Additional Answers

34. hyperbola

43c.

44b.

Check Your Understanding

 = Step-by-Step Solutions begin on page R14.

Example 1 Write each equation in standard form. State whether the graph of the equation is a *parabola, circle, ellipse,* or *hyperbola.* Then graph the equation.

1–4. See Chapter 9 Answer Appendix.

1. $x^2 + 4y^2 - 6x + 16y - 11 = 0$ **2.** $x^2 + y^2 + 12x - 8y + 36 = 0$

3. $9y^2 - 16x^2 - 18y - 64x - 199 = 0$ **4.** $6y^2 - 24y + 28 - x = 0$

Example 2 Without writing in standard form, state whether the graph of each equation is a *parabola, circle, ellipse,* or *hyperbola.* **9. hyperbola**

5. $4x^2 + 6y^2 - 3x - 2y = 12$ **ellipse**   **6.** $5y^2 = 2x + 6y - 8 + 3x^2$ **hyperbola**

7. $8x^2 + 8y^2 + 16x + 24 = 0$ **circle**   **8.** $4x^2 - 6y = 8x + 2$ **parabola**

9. $4x^2 - 3y^2 + 8xy - 12 = 2x + 4y$ **10.** $5xy - 3x^2 + 6y^2 + 12y = 18$ **hyperbola**

11. $8x^2 + 12xy + 16y^2 + 4y - 3x = 12$ **12.** $16xy + 8x^2 + 8y^2 - 18x + 8y = 13$
ellipse   **parabola**

13. **CCSS MODELING** A military jet performs for an air show. The path of the plane during one maneuver can be modeled by a conic section with equation $24x^2 + 1000y - 31,680x - 45,600 = 0$, where distances are represented in feet.

  a. Identify the shape of the curved path of the jet. Write the equation in standard form. **parabola;** $y = -0.024(x - 660)^2 + 10,500$

  b. If the jet begins its path upward, or ascent, at $x = 0$, what is the horizontal distance traveled by the jet from the beginning of the ascent to the end of the descent? **about 1321 ft**

  c. What is the maximum height of the jet? **10,500 ft**

Practice and Problem Solving

Extra Practice is on page R9.

Example 1 Write each equation in standard form. State whether the graph of the equation is a *parabola, circle, ellipse,* or *hyperbola.* Then graph the equation.

14. $3x^2 - 2y^2 + 18x + 8y - 35 = 0$ **15.** $3x^2 + 24x + 4y^2 - 40y + 52 = 0$

16. $x^2 + y^2 = 16 + 6y$ **17.** $32x + 28 = y - 8x^2$

14–23. See Chapter 9 Answer Appendix.

18. $7x^2 - 8y = 84x - 2y^2 - 176$ **19.** $x^2 + 8y = 11 + 6x - y^2$

20. $4y^2 = 24y - x - 31$ **21.** $112y + 64x = 488 + 7y^2 - 8x^2$

22. $28x^2 + 9y^2 - 188 = 56x - 36y$ **23.** $25x^2 + 384y - 64y^2 + 200x = 1776$

Example 2 Without writing in standard form, state whether the graph of each equation is a *parabola, circle, ellipse,* or *hyperbola.* **29. hyperbola   31. hyperbola**

24. $4x^2 - 5y = 9x - 12$ **parabola**   **25.** $4x^2 - 12x = 18y - 4y^2$ **circle**

26. $9x^2 + 12y = 9y^2 + 18y - 16$ **hyperbola**   **27** $18x^2 - 16y = 12x - 4y^2 + 19$ **ellipse**

28. $12y^2 - 4xy + 9x^2 = 18x - 124$ **ellipse**   **29.** $5xy + 12x^2 - 16x = 5y + 3y^2 + 18$

30. $19x^2 + 14y = 6x - 19y^2 - 88$ **circle**   **31.** $8x^2 + 20xy + 18 = 4y^2 - 12 + 9x$

32. $5x - 12xy + 6x^2 = 8y^2 - 24y - 9$ **33.** $18x - 24y + 324xy = 27x^2 + 3y^2 - 5$
hyperbola   **hyperbola**

34. **LIGHT** A lamp standing near a wall throws an arc of light in the shape of a conic section. Suppose the edge of the light can be represented by the equation $3y^2 - 2y - 4x^2 + 2x - 8 = 0$. Identify the shape of the edge of the light and graph the equation. **See margin.**

 634 | Lesson 9-6 | Identifying Conic Sections

Differentiated Homework Options

Level	Assignment	Two-Day Option	
AL Basic	14–34, 45–62	15–33 odd, 48–51	14–34 even, 45–47, 52–62
OL Core	15–33 odd, 35–43, 45–62	14–34, 48–51	35–43, 45–47, 52–62
BL Advanced	35–59, (optional: 60–62)		

B Match each graph with its corresponding equation.

a. $x^2 + y^2 - 8x - 4y = -4$ b. $9x^2 - 16y^2 - 72x + 64y = 64$ c. $9x^2 + 16y^2 = 72x + 64y - 64$

35.
c

36.
a

37.
b

C For Exercises 38–41, match each situation with an equation that could be used to represent it.

a. $47.25x^2 - 9y^2 + 18y + 33.525 = 0$ b. $25x^2 + 100y^2 - 1900x - 2200y + 45,700 = 0$

c. $16x^2 - 90x + y - 0.25 = 0$ d. $x^2 + y^2 - 18x - 30y - 14,094 = 0$

38. COMPUTERS the boundary of a wireless network with a range of 120 feet **d**

39. FITNESS the oval path of your foot on an exercise machine **b**

40. COMMUNICATIONS the position of a cell phone between two cell towers **a**

41 SPORTS the height of a football above the ground after being kicked **c**

42. CCSS SENSE-MAKING The shape of the cables in a suspension bridge is approximately parabolic. If the towers for a planned bridge are 1000 meters apart and the lowest point of the suspension cables is 200 meters below the top of the towers, write the equation in standard form with the origin at the vertex. $y = 0.0008x^2$

43. MULTIPLE REPRESENTATIONS Consider an ellipse with center $(3, -2)$, vertex $M(-1, -2)$, and co-vertex $N(3, -4)$. **43a.** $\dfrac{(x-3)^2}{16} + \dfrac{(y+2)^2}{4} = 1$

a. **Analytical** Determine the standard form of the equation of the ellipse.

b. **Algebraic** Convert part **a** to $Ax^2 + Bxy + Cy^2 + Dx + Ey + F = 0$ form.
$x^2 + 4y^2 - 6x + 16y + 9 = 0$

c. **Graphical** Graph the ellipse. See margin.

d. **Analytical** If the ellipse is rotated such that M is moved to $(3, -6)$, determine the location of N and the angle of rotation. $N(5, -2)$; 90° counterclockwise

---

H.O.T. Problems   Use Higher-Order Thinking Skills

44. CHALLENGE When a plane passes through the vertex of a cone, a *degenerate* conic is formed.

a. Determine the type of conic represented by $4x^2 + 8y^2 = 0$. **ellipse**

b. Graph the conic. See margin.

c. Describe the difference between this degenerate conic and a standard conic of the same type with $A = 4$ and $B = 8$. **45. Sample answer: Always; when a conic is vertical, $B = 0$. When this is true and $A = C$, the conic is a circle.**

45. REASONING Is the following statement *sometimes*, *always*, or *never* true? Explain.

When a conic is vertical and $A = C$, it is a circle.

46. OPEN ENDED Write an equation of the form $Ax^2 + Bxy + Cy^2 + Dx + Ey + F = 0$, where $A = 9C$, that represents a parabola. **Sample answer: $9x^2 + 6xy + y^2 + 2x + 2y + 8 = 0$**

47. WRITING IN MATH Compare and contrast the graphs of the four types of conics and their corresponding equations. See Chapter 9 Answer Appendix.

44c. The graph of the standard conic is an ellipse, and the graph of the degenerate conic is a single point.

CCSS Teaching the Mathematical Practices

Sense-Making Mathematically proficient students start by explaining to themselves the meaning of a problem and looking for entry points to its solution. They analyze givens, constraints, relationships, and goals. They check their answers to problems using a different method, and they continually ask themselves, "Does this make sense?"

Enrichment
OL BL

NAME _____ DATE _____ PERIOD _____

9-6   Enrichment

Parabolic Football

A parabola is defined as all the points (x, y) in the plane whose distance from a fixed point, called the **focus** is the same as its distance from a fixed line, called the **directrix**. Examples of parabolas are the cables on a suspension bridge, satellite dishes, and the flight path of a football during a kick-off.

At the kick-off at the beginning of a football game the ball is placed on the 40-yard line of the kicking team. Suppose the ball lands on the goal line. Assume the 50-yard line has coordinates $(0, 0)$ and the 40-yard line of the kicking team has coordinates $(-10, 0)$.

Study Guide and Intervention
AL OL ELL

9-6   Study Guide and Intervention

Identifying Conic Sections

Conics in Standard Form Any conic section in the coordinate plane can be described by an equation of the form
$Ax^2 + Bxy + Cy^2 + Dx + Ey + F = 0$, where A, B, and C are not all zero.

One way to tell what kind of conic an equation represents is to rearrange terms and complete the square, if necessary, to get one of the standard forms from an earlier lesson. This method is especially useful if you are going to graph the equation.

Example Write the equation $3x^2 - 4y^2 - 30x - 8y + 59 = 0$ in standard form. State whether the graph of the equation is a *parabola*, *circle*, *ellipse*, or *hyperbola*.

Exercises

Write each equation in standard form. State whether the graph of the equation is a *parabola*, *circle*, *ellipse*, or *hyperbola*. Then graph the equation.

1. $4x^2 + 48x + y + 158 = 0$

2. $3x^2 + y^2 - 48x - 4y + 184 = 0$

3. $-3x^2 + 2y^2 - 18x + 20y + 5 = 0$

4. $x^2 + y^2 + 8x + 2y + 8 = 0$

Practice
AL OL BL ELL

NAME _____ DATE _____ PERIOD _____

9-6   Practice

Identifying Conic Sections

Write each equation in standard form. State whether the graph of the equation is a *parabola*, *circle*, *ellipse*, or *hyperbola*. Then graph the equation.

1. $y^2 = -3x$

2. $x^2 + y^2 + 6x = 7$

3. $5x^2 - 6y^2 - 30x - 12y = -9$

4. $196y^2 = 1225 - 100x^2$

5. $3x^2 = 9 - 3y^2 - 6y$

6. $9x^2 + y^2 + 54x - 6y = -81$

Without writing the equation in standard form, state whether the graph of each equation is a *parabola*, *circle*, *ellipse*, or *hyperbola*.

7. $6x^2 + 6y^2 = 36$

8. $4x^2 + 6xy - y^2 = 16$

9. $9x^2 + 16y^2 - 30x - 64y - 80 = 0$

10. $5x^2 + 5y^2 - 45 = 0$

11. $x^2 + 2x = y$

12. $4y^2 - 12xy - 36x^2 + 4x - 144 = 0$

13. **ASTRONOMY** A space probe flies by a planet in an hyperbolic orbit. It reaches the vertex of its orbit at $(5, 0)$ and then travels along a path that gets closer and closer to the line $y = \frac{2}{3}x$. Write an equation that describes the path of the space probe if the center of its hyperbolic orbit is at $(0, 0)$.

Word Problem Practice
AL OL BL ELL

NAME _____ DATE _____ PERIOD _____

9-6   Word Problem Practice

Identifying Conic Sections

1. **MISSING INFORMATION** Rick began reading a book on conic sections. He came to this passage and discovered an inkblot covering part of an equation.

$7x^2 - 12x + \blacksquare y^2 - 16y - 94 = 0$

Based on the information in the passage and your own knowledge of conic sections, what number is being covered by the inkblot?

2. **HEADLIGHTS** The light from the headlight of a car is in the shape of a cone. The axis of the cone is parallel to the ground. What shape does the edge of the lit region form on the road, assuming that the road is flat and level?

3. **GEOGRAPHY** Earth is shaped approximately like a sphere with a radius of 4000 miles. This means any straight path on Earth's surface is actually a conic section. Which types of conic sections cannot trace straight paths on Earth's surface?

4. **MIRROR** A painter used a can of spray paint to make an image. The boundary of the image is described by the equation
$4x^2 - 16x + y^2 - 6y + 21 = 0$.
Put this equation into standard form and describe whether the curve is a circle, ellipse, parabola, or hyperbola.

5. **NONSTANDARD EQUATIONS** Consider the equation $xy = 1$.

a. Are there any solutions of this equation that lie on the x- or y-axis?

b. Sketch a graph of the solutions of the equation.

c. Assuming that the equation represents a conic section, based on the graph, which type of conic section is it?

Crystal Ball Ask students to write how today's lesson will connect with solving linear-nonlinear systems.

Formative Assessment

Check for student understanding of concepts in Lessons 9-5 and 9-6.

 Quiz 3, p. 54

Additional Answers

52. $(0, \pm15)$; $(0, \pm25)$; $y = \pm\frac{3}{4}x$

53. $(0, 0)$; $(0, \pm3)$; $6\sqrt{2}$; 6

54. $(0, 0)$; $(\pm2, 0)$; $4\sqrt{2}$; 4

55. $(4, -2)$; $(4 \pm 2\sqrt{6}, -2)$; 10; 2

Standardized Test Practice

48. SAT/ACT A class of 25 students took a science test. Ten students had a mean score of 80. The other students had an average score of 60. What is the average score of the whole class? **B**

  A 66   **D** 72

  B 68   **E** 78

  C 70

49. Six times a number minus 11 is 43. What is the number? **J**

  F 12

  G 11

  H 10

  J 9

50. EXTENDED RESPONSE The amount of water remaining in a storage tank as it is drained can be represented by the equation $L = -4t^2 - 10t + 130$, where L represents the number of liters of water remaining and t represents the number of minutes since the drain was opened. How many liters of water were in the tank initially? Determine to the nearest tenth of a minute how long it will take for the tank to drain completely. **130 L; 4.6 min**

51. Ruben has a square piece of paper with sides 4 inches long. He rolled up the paper to form a cylinder. What is the volume of the cylinder? **B**

  A $\frac{4}{\pi}$ **C** 4π

  B $\frac{16}{\pi}$ **D** 16π

Spiral Review

52. ASTRONOMY Suppose a comet's path can be modeled by a branch of the hyperbola with equation $\frac{y^2}{225} - \frac{x^2}{400} = 1$. Find the coordinates of the vertices and foci and the equations of the asymptotes for the hyperbola. Then graph the hyperbola. (Lesson 9-5) **See margin.**

Find the coordinates of the center and foci and the lengths of the major and minor axes for the ellipse with the given equation. Then graph the ellipse. (Lesson 9-4) **53–55. See margin.**

53. $\frac{y^2}{18} + \frac{x^2}{9} = 1$ **54.** $4x^2 + 8y^2 = 32$ **55.** $x^2 + 25y^2 - 8x + 100y + 91 = 0$

Graph each function. (Lesson 8-3) **56–58. See Chapter 9 Answer Appendix.**

56. $f(x) = \frac{3}{x}$ **57.** $f(x) = \frac{-2}{x + 5}$ **58.** $f(x) = \frac{6}{x - 2} - 4$

59. SPACE A radioisotope is used as a power source for a satellite. The power output P (in watts) is given by $P = 50e^{-\frac{t}{250}}$, where t is the time in days. (Lesson 7-8)

  a. Is the formula for power output an example of exponential *growth* or *decay*? Explain your reasoning. **Decay; the exponent is negative.**

  b. Find the power available after 100 days. **about 33.5 watts**

  c. Ten watts of power are required to operate the equipment in the satellite. How long can the satellite continue to operate? **about 402 days**

Skills Review

Solve each system of equations.

60. $6g - 8h = 50$ $(7, -1)$
    $6h = 22 - 4g$

61. $3u + 5v = 6$ $\left(-\frac{1}{2}, \frac{3}{2}\right)$
    $2u - 4v = -7$

62. $10m - 9n = 15$ $(6, 5)$
    $5m - 4n = 10$

DifferentiatedInstruction   OL   BL

Extension Students have learned that a quadratic equation of the form $Ax^2 + Bxy + Cy^2 + Dx + Ey + F = 0$, where $B = 0$, is either a *parabola, circle, ellipse,* or a *hyperbola* when graphed. Ask students to determine the shape of the graph of the equations $x^2 + y^2 + 1 = 0$, $x^2 + y^2 = 0$ and $x^2 = 0$. **empty set; single point (0, 0); the line $x = 0$**

EXTEND 9-6

Graphing Technology Lab
Analyzing Quadratic Relations

You can use TI-Nspire technology to analyze quadratic relations.

Activity 1   Characteristics of a Parabolic Relation

Graph $f(x) = 9x^2 + 1$, $g(x) = -x^2 + 3x - 4$, and $(y - 3)^2 = -4(x - 2)$. Identify the maxima, minima, and axes of symmetry.

Step 1   Add a new **Graphs** and **Calculator** page.

Step 2   Enter $f(x)$ into **f1** and $g(x)$ into **f2**.

Step 3   On the **Calculator** page, use **solve** to solve $(y - 3)^2 = -4(x - 2)$ for y. Hold the **shift** and use ▶ to highlight one equation; then press **ctrl C**. Press **ctrl ▶** to go to the **Graphs** page; then press **tab ctrl V**. Repeat to copy the other equation into **f4**.

Step 4   Use **Analyze Graph**, **Maximum**, **Minimum**, and **Intersection** to find the coordinates of the extrema. For $f(x)$, the minimum is at $(0, 1)$; for $g(x)$, the maximum is at $(1.5, -1.75)$; for the relation, the vertex is at $(2, 3)$.

Step 5   You can use the coordinates of the extrema to find the equations of the axes of symmetry. For $f(x)$, the equation of the axis of symmetry is $x = 1.5$; for $g(x)$, the equation of the axis of symmetry is $x = 0$; for the relation, the equation of the axis of symmetry is $y = 3$.

$[-10, 10]$ scl: 1 by $[-10, 10]$ scl: 1

You can also use a graphing calculator to determine the equation of a parabola.

Activity 2   Write an Equation for a Parabola

Given that $f(x)$ has zeros at -2 and 4 and $f(x)$ opens downward, write an equation for the parabola.

Step 1   Add a new **Graphs** and **Calculator** page.

Step 2   Because the zeros are $x = -2$ and $x = 4$, the factors of the quadratic equation are $(x + 2)$ and $(x - 4)$.

Step 3   Use the **expand** command on the **Calculator** page to multiply $(x + 2)$ and $(x - 4)$. So, $y = x^2 - 2x - 8$.

Step 4   On the **Graphs** page, graph $y = x^2 - 2x - 8$. Verify the roots and direction of opening.

Step 5   The function in the graph opens upward, not downward, so multiply $x^2 - 2x - 8$ by -1. Thus, $y = -x^2 + 2x + 8$. Graph this function.

An equation for the parabola that has zeros at -2 and 4 and opens downward is $y = -x^2 + 2x + 8$.

$[-6, 6]$ scl: 0.5 by $[-12, 12]$ scl: 1

(continued on the next page)

connectED.mcgraw-hill.com   **637**

---

1 Focus

Objectives
- Identify characteristics of quadratic relations.
- Write equations of quadratic relations.

Materials
- TI-Nspire CAS technology

Teaching Tips
- To switch from one page to another, use **CTRL** with the left or right arrows.
- To use the **maximum** and **minimum** features, first click on the graph. Then indicate the lower and upper bounds, and make a guess by pressing the enter key between each prompt.
- To find the vertex of the relation, use the intersection feature. First click on the graph for **F3** and then click on the graph for **F4**. Then specify the left and right bound, and make a guess by pressing the enter key between each prompt.

2 Teach

Working in Cooperative Groups
Divide the class into pairs. Work through Activity 1 as a class. Then ask students to work with their partners to complete Activities 2 and 3.

Practice   Have students complete Exercises 1–5.

3 Assess

Formative Assessment
Use Exercise 6 to assess each student's understanding of how to write the equation of a quadratic relation.

From Concrete to Abstract
Ask students to summarize how to use technology to identify the characteristics of quadratic relations and how to write equations of quadratic relations.

Additional Answers

1.

For $f(x)$, max. at (3, 12) and axis of symmetry $x = 3$. For $g(x)$, min. at (−9.5, 12.3) and axis of symmetry $x = −9.5$. For relation, no extrema and axis of symmetry $y = −5$.

2.

For $f(x)$, max. at (−6, 3) and axis of symmetry $x = −6$. For $g(x)$, min. at (−0.5, 3.5) and axis of symmetry $x = −0.5$. For relation, no extrema and axis of symmetry $y = −1$.

5. $\dfrac{(x-1)^2}{25} + \dfrac{(y-4)^2}{16} = 1$

6. $\dfrac{(x+6)^2}{49} + \dfrac{(y-2)^2}{4} = 1$

Graphing Technology Lab
Analyzing Quadratic Relations *Continued*

You can use a graphing calculator to determine an equation of a quadratic relation.

Activity 3   Write an Equation for an Ellipse

Write an equation for an ellipse that has vertices at (−3, 3) and (−3, −7) and co-vertices at (0, −2) and (−6, −2).

Step 1 Add a new **Graphs** and **Calculator** page.

Step 2 Turn on the grid from **View, Show Axis.** Then use **Points, Points On** to graph the four points. Use **Actions, Coordinates and Equations** to show the coordinates of the points.

Step 3 Use the graph to identify the intersection point of the segment formed by the vertices and the segment formed by the co-vertices. The center is at (−3, −2).

Step 4 Identify other important characteristics. The ellipse is oriented vertically. The length of the major axis is 10, so $a = 5$. The length of the minor axis is 6, so $b = 3$.

Step 5 Write the equation in standard form.

$$\frac{[y - (-2)]^2}{5^2} + \frac{[x - (-3)]^2}{3^2} = 1 \text{ or}$$

$$\frac{(y + 2)^2}{25} + \frac{(x + 3)^2}{9} = 1$$

Step 6 Check the equation by using **Solve** under the **Algebra** menu on the **Calculator** page to solve for y. Copy and paste the two equations into the **Graph** page to graph the ellipse.

[−10, 10] scl: 1 by [−10, 10] scl: 1

Exercises

Graph each function and relation. Identify the maxima, minima, and axes of symmetry. 1–2. See margin.

1. $f(x) = -(x - 3)^2 + 12$, $g(x) = x^2 + x - 12$, and $(y + 5)^2 = -12(x - 2)$.

2. $f(x) = -0.25x^2 - 3x - 6$, $g(x) = 2x^2 + 2x + 4$, and $(y + 1)^2 = 2(x + 6)$.

3. Given that $f(x)$ has zeros at −1 and 3 and $f(x)$ opens upward, write an equation for the parabola. $y = x^2 - 2x - 3$

4. Given that $f(x)$ has zeros at −3 and −1 and $f(x)$ opens downward, write an equation for the parabola. $y = -x^2 - 4x - 3$

5. Write an equation for an ellipse that has vertices at (−6, 2) and (−6, −8) and co-vertices at (−3, −3) and (−9, −3). See margin.

6. Write an equation for an ellipse that has vertices at (−13, 2) and (1, 2) and co-vertices at (−6, 4) and (−6, 0). See margin.

EXPLORE 9-7
Graphing Technology Lab
Linear-Nonlinear Systems

You can use a TI-83/84 Plus application to solve linear-nonlinear systems by using the **Y=** menu to graph each equation on the same set of axes.

Example Linear-Quadratic System

Solve the system of equations.
$3y - 4x = -7$
$4x^2 + 3y^2 = 91$

Step 1 Solve each equation for y.

$3y - 4x = -7$ $\qquad$ $4x^2 + 3y^2 = 91$

$3y = 4x - 7$ $\qquad$ $3y^2 = 91 - 4x^2$

$y = \frac{4}{3}x - \frac{7}{3}$ $\qquad$ $y = \pm\sqrt{\frac{91 - 4x^2}{3}}$

Step 2 Enter $y = \frac{4}{3}x - \frac{7}{3}$ as **Y1**, $y = \sqrt{\frac{91 - 4x^2}{3}}$ as **Y2**, and $y = -\sqrt{\frac{91 - 4x^2}{3}}$ as **Y3**.
Then graph the equations in a standard viewing window.

KEYSTROKES: Y= 4 ÷ 3 X,T,θ,n — 7 ÷ 3 ENTER 2nd x² (91 — 4
X,T,θ,n x²) ÷ 3) ENTER (−) 2nd x² (91 — 4 X,T,θ,n x²
) ÷ 3) ENTER ZOOM 6

Step 3 Find the intersection of $y = \frac{4}{3}x - \frac{7}{3}$ with $y = \sqrt{\frac{91 - 4x^2}{3}}$.

KEYSTROKES: Press 2nd TRACE 5
ENTER ENTER ENTER. The two graphs
intersect at (4, 3).

Step 4 Find the intersection of $y = \frac{4}{3}x - \frac{7}{3}$ with $y = -\sqrt{\frac{91 - 4x^2}{3}}$.

KEYSTROKES: Press 2nd TRACE 5
ENTER ▼ ENTER. Then use ◄ to move
the cursor to the second intersection point.
Press ENTER. The two graphs intersect at $(-2, -5)$.
The solutions of the system are (4, 3) and $(-2, -5)$.

Exercises

Use a graphing calculator to solve each system of equations.

1. $x^2 + y^2 = 100$
   $x + y = 2$ $(-6, 8), (8, -6)$

2. $2y - x = 11$
   $5x^2 + 2y^2 = 407$ $(-9, 1), (7, 9)$

3. $21x + 9y = -36$
   $7x^2 + 9y^2 = 1152$ $(-6, 10), (3, -11)$

1 Focus

Objective Solve linear-nonlinear systems using a graphing calculator.

Materials for Each Group
- TI-83/84 Plus graphing calculator

Teaching Tip
Remind students that to return to a standard viewing window, they should press WINDOW and select the **ZStandard** feature.

2 Teach

Working in Cooperative Groups
Pair students with different abilities. Have students work through the Example Steps 1–4.

Ask:
- Can this system be solved algebraically? Explain. Yes; sample answer: Once the linear equation is solved for y, it can be substituted into the second equation and x can be found.

- How many possible solutions are there for linear-quadratic systems? Explain. 0, 1, or 2; Sample answer: Since one of the equations is linear, at most, it may intersect the quadratic equation twice.

Practice Have students complete Exercises 1–3.

3 Assess

Formative Assessment
Use Exercises 1–3 to assess whether students are able to use a graphing calculator to solve a linear-nonlinear system.

From Concrete to Abstract
Ask students if a system can ever have more than two solutions. Have them include drawings with their responses.

1 Focus

VerticalAlignment

Before Lesson 9-7 Solve systems of linear equations.

Lesson 9-7 Solve systems of linear and nonlinear equations algebraically and graphically. Solve systems of linear and nonlinear inequalities graphically.

2 Teach

Scaffolding Questions

Have students read the **Why?** section of the lesson.

Ask:

- Can a person be easily located using one cellular tower? Why or why not? No; the person could be anywhere in a circular path around the tower.

- If law enforcement only knew the distance a caller was from two cellular towers, what is the maximum number of locations they would have to search to find the caller? two

---

LESSON 9-7 Solving Linear-Nonlinear Systems

∴ Then	**∴ Now**	**∴ Why?**
● You solved systems of linear equations.	● **1** Solve systems of linear and nonlinear equations algebraically and graphically. **2** Solve systems of linear and nonlinear inequalities graphically.	● Have you ever wondered how law enforcement agencies can track a cell phone user's location? A person using a cell phone can be located in respect to three cellular towers. The respective coordinates and distances each tower is from the caller are used to pinpoint the caller's location. This is accomplished using a system of quadratic equations.

CCSS **Common Core State Standards**

Content Standards
A.REI.11 Explain why the x-coordinates of the points where the graphs of the equations $y = f(x)$ and $y = g(x)$ intersect are the solutions of the equation $f(x) = g(x)$; find the solutions approximately, e.g., using technology to graph the functions, make tables of values, or find successive approximations. Include cases where $f(x)$ and/or $g(x)$ are linear, polynomial, rational, absolute value, exponential, and logarithmic functions.

Mathematical Practices
6 Attend to precision.

1 Systems of Equations When a system of equations consists of a linear and a nonlinear equation, the system may have zero, one, or two solutions. Some of the possible solutions are shown below.

You can solve linear-quadratic systems by using graphical or algebraic methods.

Example 1 Linear-Quadratic System

Solve the system of equations.
$$9x^2 + 25y^2 = 225 \quad (1)$$
$$10y + 6x = 6 \quad (2)$$

Step 1 Solve the linear equation for y.
$$10y + 6x = 6 \qquad \text{Equation (2)}$$
$$y = -0.6x + 0.6 \qquad \text{Solve for } y.$$

Step 2 Substitute into the quadratic equation and solve for x.
$$9x^2 + 25y^2 = 225 \qquad \text{Quadratic equation}$$
$$9x^2 + 25(-0.6x + 0.6)^2 = 225 \qquad \text{Substitute } -0.6x + 0.6 \text{ for } y.$$
$$9x^2 + 25(0.36x^2 - 0.72x + 0.36) = 225 \qquad \text{Simplify.}$$
$$9x^2 + 9x^2 - 18x + 9 = 225 \qquad \text{Distribute.}$$
$$18x^2 - 18x - 216 = 0 \qquad \text{Simplify.}$$
$$x^2 - x - 12 = 0 \qquad \text{Divide each side by 18.}$$
$$(x - 4)(x + 3) = 0 \qquad \text{Factor.}$$
$$x = 4 \text{ or } -3 \qquad \text{Zero Product Property}$$

Step 3 Substitute x-values into the linear equation and solve for y.
$$y = -0.6x + 0.6 \qquad \text{Equation (2)} \qquad y = -0.6x + 0.6$$
$$= -0.6(4) + 0.6 \qquad \text{Substitute the } x\text{-values} \qquad = -0.6(-3) + 0.6$$
$$= -1.8 \qquad \text{Simplify.} \qquad = 2.4$$

The solutions of the system are $(4, -1.8)$ and $(-3, 2.4)$.

▶ **Guided**Practice   **1A.** $\left(-11, -49\frac{1}{3}\right)$ and $(5, 4)$

1A. $3y + x^2 - 4x - 17 = 0$ **1B.** $3(y - 4) - 2(x - 3) = -6$
$3y - 10x + 38 = 0$ $5x^2 + 2y^2 - 53 = 0$ $(3, 2)$ and $(-3, -2)$

 640 | Lesson 9-7

Lesson 9-7 Resources

Resource	Approaching Level **AL**	On Level **OL**	Beyond Level **BL**	English Learners **ELL**
Teacher Edition		■ Differentiated Instruction, pp. 641, 645	■ Differentiated Instruction, pp. 641, 645	
Chapter Resource Masters	■ Study Guide and Intervention, pp. 45–46 ■ Skills Practice, p. 47 ■ Practice, p. 48 ■ Word Problem Practice, p. 49	■ Study Guide and Intervention, pp. 45–46 ■ Skills Practice, p. 47 ■ Practice, p. 48 ■ Word Problem Practice, p. 49 ■ Enrichment, p. 50	■ Practice, p. 48 ■ Word Problem Practice, p. 49 ■ Enrichment, p. 50	■ Study Guide and Intervention, pp. 45–46 ■ Skills Practice, p. 47 ■ Practice, p. 48 ■ Word Problem Practice, p. 49
Other	■ 5-Minute Check 9-7 ■ Study Notebook	■ 5-Minute Check 9-7 ■ Study Notebook	■ 5-Minute Check 9-7 ■ Study Notebook	■ 5-Minute Check 9-7 ■ Study Notebook

If a quadratic system contains two conic sections, the system may have anywhere from zero to four solutions. Some graphical representations are shown below.

You can use elimination to solve quadratic-quadratic systems.

Example 2 Quadratic-Quadratic System

Solve the system of equations.

$$x^2 + y^2 = 45 \quad (1)$$
$$y^2 - x^2 = 27 \quad (2)$$

$y^2 + x^2 = 45$	Equation (1), Commutative Property
$(+) \ y^2 - x^2 = 27$	Equation (2)
$2y^2 = 72$	Add.
$y^2 = 36$	Divide each side by 2.
$y = \pm 6$	Take the square root of each side.

Substitute 6 and −6 into one of the original equations and solve for x.

$x^2 + y^2 = 45$	Equation (1)	$x^2 + y^2 = 45$
$x^2 + 6^2 = 45$	Substitute for y.	$x^2 + (-6)^2 = 45$
$x^2 = 9$	Subtract 36 from each side.	$x^2 = 9$
$x = \pm 3$	Take the square root of each side.	$x = \pm 3$

The solutions are $(-3, -6)$, $(-3, 6)$, $(3, -6)$, and $(3, 6)$.

▸ **Guided**Practice

2A. $x^2 + y^2 = 8$ $(-2, 2), (2, 2),$
    $x^2 + 3y = 10$ $(-\sqrt{7}, 1), (\sqrt{7}, 1)$

2B. $3x^2 + 4y^2 = 48$ $(-2, -3), (-2, 3),$
    $2x^2 - y^2 = -1$ $(2, -3), (2, 3)$

StudyTip

CCSS Tools If you use **ZSquare** on the **ZOOM** menu, the graph of the first equation will look like a circle.

---

2 Systems of Inequalities
Systems of quadratic inequalities can be solved by graphing.

Example 3 Quadratic Inequalities

Solve the system of inequalities by graphing.
$$x^2 + y^2 \le 49$$
$$x^2 - 4y^2 > 16$$

The intersection of the graphs, shaded green, represents the solution of the system.

CHECK $(6, 0)$ is in the shaded area. Use this point to check your solution.

$x^2 + y^2 \le 49$	$x^2 - 4y^2 > 16$
$6^2 + 0^2 \overset{?}{\le} 49$	$6^2 - 4(0)^2 \overset{?}{>} 16$
$36 \le 49 \ \checkmark$	$36 > 16 \ \checkmark$

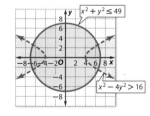

▸ **Guided**Practice

3A. $5x^2 + 2y^2 \le 10$
    $y \ge x^2 - 2x + 1$

3B. $x^2 - y^2 \le 8$
    $x^2 + y^2 \ge 120$

3A, 3B. See Chapter 9 Answer Appendix.

---

1 Systems of Equations
Example 1 shows how to solve a linear-quadratic system using substitution. **Example 2** shows how to solve a quadratic-quadratic system using elimination.

Formative Assessment
Use the Guided Practice exercises after each example to determine students' understanding of concepts.

Additional Examples

1 Solve the system of equations.
$$4x^2 - 16y^2 = 25$$
$$2y + x = 2 \quad \left(\frac{41}{16}, -\frac{9}{32}\right)$$

2 Solve the system of equations.
$$x^2 + y^2 = 16$$
$$4x^2 + y^2 = 23$$
$$\left(\frac{\sqrt{21}}{3}, \frac{\sqrt{123}}{3}\right),$$
$$\left(-\frac{\sqrt{21}}{3}, \frac{\sqrt{123}}{3}\right),$$
$$\left(\frac{\sqrt{21}}{3}, -\frac{\sqrt{123}}{3}\right),$$
$$\left(-\frac{\sqrt{21}}{3}, -\frac{\sqrt{123}}{3}\right)$$

▸ **Additional Examples** also in Interactive Classroom PowerPoint® Presentations

IWB **Interactive White Board** READY

Teaching the Mathematical Practices
Tools Mathematically proficient students consider the available tools when solving a mathematical problem. Encourage students to identify the types of graphs involved before setting their viewing windows.

DifferentiatedInstruction ⓄⓁ ⒷⓁ

Social Learners Point out that there are combinations of graphs other than those shown on page 639 just before Example 2, that are possible for each number of solutions. As a class, challenge students to sketch as many different possibilities as they can to add to the figures shown.

2 Systems of Inequalities

Example 3 shows how to solve a system of inequalities by graphing.
Example 4 shows how to solve a system of inequalities involving absolute value by graphing.

Additional Examples

3 Solve the system of inequalities by graphing.

$y > x^2 + 1$
$x^2 + y^2 \leq 9$

4 Solve the system of inequalities by graphing.

$x^2 + y^2 > 16$
$y > |x + 4|$

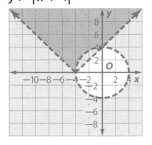

---

3 Practice

Formative Assessment

Use Exercises 1–13 to check for understanding.

Use the chart at the bottom of the next page to customize assignments for your students.

Systems involving absolute value can also be solved by graphing.

StudyTip
Graphing Calculator Like linear inequalities, systems of quadratic and absolute value inequalities can be checked with a graphing calculator.

Example 4 Quadratics with Absolute Value

Solve the system of inequalities by graphing.

$y \geq |2x - 4|$
$y \leq -x^2 + 4x + 2$

Graph the boundary equations. Then shade appropriately.

The intersection of the graphs, shaded green, represents the solution to the system.

CHECK (2, 4) is in the shaded area. Use the point to check your solution.

$y \geq |2x - 4|$	$y \leq -x^2 + 4x + 2$
$4 \stackrel{?}{\geq} |2(2) - 4|$	$4 \stackrel{?}{\leq} -(2)^2 + 4(2) + 2$
$4 \geq 0$ ✓	$4 \leq 6$ ✓

GuidedPractice **4A, 4B.** See Chapter 9 Answer Appendix.

4A. $y > |-0.5x + 2|$
$\dfrac{x^2}{16} + \dfrac{y^2}{36} \leq 1$

4B. $x^2 + y^2 \leq 49$
$y \geq |x^2 + 1|$

Check Your Understanding

○ = Step-by-Step Solutions begin on page R14.

Examples 1–2 Solve each system of equations.

1. $8y = -10x$
$y^2 = 2x^2 - 7$ (4, −5), (−4, 5)

2. $x^2 + y^2 = 68$
$5y = -3x + 34$ (−2, 8), (8, 2)

3. $y = 12x - 30$
$4x^2 - 3y = 18$ (3, 6), (6, 42)

4. $6y^2 - 27 = 3x$
$6y - x = 13$ (−7, 1), (−1, 2)

5. $x^2 + y^2 = 16$
$x^2 - y^2 = 20$ no solution

6. $y^2 - 2x^2 = 8$
$3y^2 + x^2 = 52$

7 $x^2 + 2y = 7$
$y^2 - x^2 = 8$

8. $4y^2 - 3x^2 = 11$
$3y^2 + 2x^2 = 21$

6. (2, 4), (−2, 4),
(−2, −4), (2, −4)
7. (−1, 3), (1, 3),
$(-\sqrt{17}, -5)$,
$(\sqrt{17}, -5)$
8. $(-\sqrt{3}, -\sqrt{5})$,
$(-\sqrt{3}, \sqrt{5})$,
$(\sqrt{3}, -\sqrt{5})$,
$(\sqrt{3}, \sqrt{5})$

9. **CCSS PERSEVERANCE** Refer to the beginning of the lesson. A person using a cell phone can be located with respect to three cellular towers. In a coordinate system where one unit represents one mile, the location of the caller is determined to be 50 miles from the tower at the origin. The person is also 40 miles from a tower at (0, 30) and 13 miles from a tower at (35, 18). Where is the caller? **(40, 30)**

Examples 3–4 Solve each system of inequalities by graphing. 10–13. See Chapter 9 Answer Appendix.

10. $6x^2 + 9(y - 2)^2 \leq 36$
$x^2 + (y + 3)^2 \leq 25$

11. $16x^2 + 4y^2 \leq 64$
$y \geq -x^2 + 2$

12. $4x^2 - 8y^2 \geq 32$
$y \geq |1.5x| - 8$

13. $x^2 + 8y^2 < 32$
$y < -|x - 2| + 2$

Teaching the Mathematical Practices

Perseverance Mathematically proficient students start by explaining to themselves the meaning of a problem and looking for entry points to its solution. They analyze givens, constraints, relationships, and goals and make conjectures about the form and meaning of the solution and plan a solution pathway rather than simply jumping into a solution attempt.

Examples 1–2 Solve each system of equations. **16–23, 25.** See margin.

14. $3x^2 - 2y^2 = -24$
$2y = -3x$ $(-4, 6), (4, -6)$

15. $5x^2 + 4y^2 = 20$
$5y = 7x + 35$ no solution

16. $x^2 + 3x = -4y - 2$
$y = -2x + 1$

17. $y = 2x$
$4x^2 - 2y^2 = -36$

18. $2y = x + 10$
$y^2 - 4y = 5x + 10$

19. $9y = 8x - 19$
$8x + 11 = 2y^2 + 5y$

20. $2y^2 + 5x^2 = 26$
$2x^2 - y^2 = 5$

21. $x^2 + y^2 = 16$
$x^2 - 4x + y^2 = 12$

22. $x^2 + y^2 = 8$
$5y^2 = 3x^2$

23. $y^2 - x^2 + 3y = 26$
$x^2 + 2y^2 = 34$

24. $x^2 - y^2 = 25$
$x^2 + y^2 + 7 = 0$ no solution

25. $x^2 - 10x + 2y^2 = 47$
$y^2 - 2x^2 = -14$

26. FIREWORKS Two fireworks are set off simultaneously but from different altitudes. The height y in feet of one is represented by $y = -16t^2 + 120t + 10$, where t is the time in seconds. The height of the other is represented by $y = -16t^2 + 60t + 310$.

　　a. After how many seconds are the fireworks the same height? **5 seconds**

　　b. What is that height? **210 ft**

Examples 3–4 **CCSS TOOLS** Solve each system of inequalities by graphing.

27–36. SEE CHAPTER 9 ANSWER APPENDIX.

27. $x^2 + y^2 \geq 36$
$x^2 + 9(y + 6)^2 \leq 36$

28. $-x > y^2$
$4x^2 + 14y^2 \leq 56$

29. $12x^2 - 4y^2 \geq 48$
$16(x - 4)^2 + 25y^2 < 400$

30. $8y^2 - 3x^2 \leq 24$
$2y > x^2 - 8x + 14$

31. $y > x^2 - 6x + 8$
$x \geq y^2 - 6y + 8$

32. $x^2 + y^2 \geq 9$
$25x^2 + 64y^2 \leq 1600$

33. $16(x - 3)^2 + 4y^2 \leq 64$
$y \leq -|x - 2| + 2$

34. $x^2 - 4x + y^2 + 6y \leq 23$
$y > |x - 2| - 6$

35. $2y - 4 \geq |x + 4|$
$12 - 2y > x^2 + 12x + 36$

36. $18y^2 - 3x^2 \leq 54$
$y \geq |2x| - 6$

37. $x^2 + y^2 < 16$
$y \geq |x - 2| + 6$ no solution

38. $x^2 < y - 2$
$y \leq |x + 8| - 7$ no solution

39. SPACE Two satellites are placed in orbit about Earth. The equations of the two orbits are $\frac{x^2}{(300)^2} + \frac{y^2}{(900)^2} = 1$ and $\frac{x^2}{(600)^2} + \frac{y^2}{(690)^2} = 1$, where distances are in kilometers and Earth is the center of each curve.

　　a. Solve each equation for y. **39a.** $y = \pm 900\sqrt{1 - \frac{x^2}{(300)^2}}$; $y = \pm 690\sqrt{1 - \frac{x^2}{(600)^2}}$

　　b. Use a graphing calculator to estimate the intersection points of the two orbits.

　　c. Compare the orbits of the two satellites. **39b.** Sample answer: $(209, 647)$, $(-209, 647)$, $(-209, -647)$, $(209, -647)$

39c. Sample answer: The orbit of the satellite modeled by the second equation is closer to a circle than the other orbit. The distance on the x-axis is twice as great for one satellite as the other.

40. PETS Taci's dog was missing one day. Fortunately, he was wearing an electronic monitoring device. If the dog is 10 units from the tree, 13 units from the tower, and 20 units from the house, determine the coordinates of his location. **(4, 5)**

41 BASEBALL In 1997, after Mark McGwire hit a home run, the claim was made that the ball would have traveled 538 feet if it had not landed in the stands. The path of the baseball can be modeled by $y = -0.0037x^2 + 1.77x - 1.72$ and the stands can be modeled by $y = \frac{3}{7}x - 128.6$. How far vertically and horizontally from home plate did the ball land in the stands? **440 ft from home plate, 60 ft above the playing surface**

Tower (−8, 10); (−12, −7); House; Tree (10, −3)

WatchOut!

Preventing Errors Remind students that it is always a good idea to check a point inside the solution region as well as one outside that region.

CCSS Teaching the Mathematical Practices

Tools Mathematically proficient students consider the available tools when solving a mathematical problem and are sufficiently familiar with tools appropriate for their grade or course to make sound decisions about when each of these tools might be helpful.

Additional Answers

16. $(3, -5), (2, -3)$

17. $(-3, -6), (3, 6)$

18. $(-2, 4), (10, 10)$

19. $(-1, -3), (8, 5)$

20. $(-2, -\sqrt{3}), (-2, \sqrt{3}),$
$(2, -\sqrt{3}), (2, \sqrt{3})$

21. $(1, -\sqrt{15}), (1, \sqrt{15})$

22. $(-\sqrt{5}, -\sqrt{3}), (-\sqrt{5}, \sqrt{3}),$
$(\sqrt{5}, -\sqrt{3}), (\sqrt{5}, \sqrt{3})$

23. $(-\sqrt{2}, 4), (\sqrt{2}, 4)$

25. $(5, -6), (5, 6), (-3, -2), (-3, 2)$

Differentiated Homework Options

Level	Assignment	Two-Day Option	
AL Basic	14–41, 52–74	15–41 odd, 55–58	14–40 even, 52–54, 59–74
OL Core	15–41 odd, 42–50, 52–74	14–41, 55–58	42–50, 52–54, 59–74
BL Advanced	42–70, (optional: 71–74)		

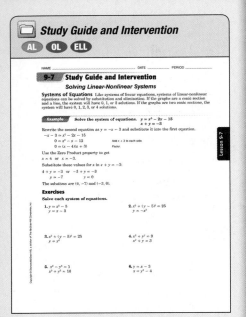

Study Guide and Intervention
AL OL ELL

Practice
AL OL BL ELL

Word Problem Practice
AL OL BL ELL

B **42. ADVERTISING** The corporate logo for an automobile manufacturer is shown at the right. Write a system of three equations to model this logo.

42. Sample answer: $x^2 + y^2 = 45$, $y = 3|x - 3| - 6$, $y = 3|x + 3| - 6$

Write a system of equations that satisfies each condition.

(43) a circle and an ellipse that intersect at one point

44. a parabola and an ellipse that intersect at two points

45. a hyperbola and a circle that do not intersect

46. an ellipse and a parabola that intersect at three points

47. an ellipse and a hyperbola that intersect at four points **43–47. See Chapter 9 Answer Appendix.**

48. FINANCIAL LITERACY Prices are often set on an equilibrium curve, where the supply of a certain product equals its corresponding demand by consumers. An economist represents the supply of a product with $y = p^2 + 10p$ and the corresponding demand with $y = -p^2 + 40p$, where p is the price. Determine the equilibrium price. **$15**

49. PAINTBALL The shape of a paintball field is modeled by $x^2 + 4y^2 = 10,000$ in yards where the center is at the origin. The teams are provided with short-range walkie-talkies with a maximum range of 80 yards. Are the teams capable of hearing each other anywhere on the field? Explain your reasoning graphically.

See Chapter 9 Answer Appendix.

C **50. MOVING** Lena is moving to a new city and needs for the location of her new home to satisfy the following conditions. **See Chapter 9 Answer Appendix.**

- It must be less than 10 miles from the office where she will work.

- Because of the terrible smell of the local paper mill, it must be at least 15 miles away from the mill.

If the paper mill is located 9.5 miles east and 6 miles north of Lena's office, write and graph a system of inequalities to represent the area(s) were she should look for a home.

54. Sample answer: By sketching a quick graph of a quadratic system, you can determine how many real solutions there are as well as estimate the values of the solutions. This can help you when confirming the answers that you get algebraically.

H.O.T. Problems Use Higher-Order Thinking Skills

51. CHALLENGE Find all values of k for which the following system of equations has two solutions. $k = a$ or $k = b$

$$\frac{x^2}{a^2} + \frac{y^2}{b^2} = 1 \qquad x^2 + y^2 = k^2$$

52. CCSS ARGUMENTS When the vertex of a parabola lies on an ellipse, how many solutions can the quadratic system represented by the two graphs have? Explain your reasoning using graphs. **See Chapter 9 Answer Appendix.**

53. OPEN ENDED Write a system of equations, one a hyperbola and the other an ellipse, for which a solution is $(-4, 8)$. **Sample answer:**

$$\frac{y^2}{128} + \frac{x^2}{32} = 1 \text{ and } \frac{x^2}{8} - \frac{y^2}{64} = 1$$

54. WRITING IN MATH Explain how sketching the graph of a quadratic system can help you solve it.

Enrichment
OL BL

CCSS **Teaching the Mathematical Practices**

Arguments Mathematically proficient students understand and use stated assumptions, definitions, and previously established results in constructing arguments. They make conjectures and build a logical progression of statements to explore the truth of their conjectures. And they are able to analyze situations by breaking them into cases, and can recognize and use counterexamples.

Mark Renders/Getty Images News/Getty Images

55. SHORT RESPONSE Solve.

$$4x - 3y = 0$$
$$x^2 + y^2 = 25 \quad (-3, -4), (3, 4)$$

56. You have 16 stamps. Some are postcard stamps that cost $0.23, and the rest cost $0.41. If you spent a total of $5.30 on the stamps, how many postcard stamps do you have? **A**

A 7
B 8
C 9
D 10

57. Ms. Talbot received a promotion and a 7.2% raise. Her new salary is $53,600 a year. What was her salary before the raise? **F**

F $50,000
G $53,600
H $55,000
J $57,500

58. SAT/ACT When a number is multiplied by $\frac{2}{3}$, the result is 188. Find the number. **B**

A 292
B 282
C 272
D 262
E $125\frac{1}{3}$

Spiral Review

Match each equation with the situation that it could represent. (Lesson 9-6)

a. $9x^2 + 4y^2 - 36 = 0$
b. $0.004x^2 - x + y - 3 = 0$
c. $x^2 + y^2 - 20x + 30y - 75 = 0$

59. SPORTS the flight of a baseball **b**

60. PHOTOGRAPHY the oval opening in a picture frame **a**

61. GEOGRAPHY the set of all points 20 miles from a landmark **c**

Find the coordinates of the vertices and foci and the equations of the asymptotes for the hyperbola with the given equation. Then graph the hyperbola. (Lesson 9-5) **62–64. See margin.**

62. $\dfrac{y^2}{16} - \dfrac{x^2}{25} = 1$

63. $\dfrac{(y-3)^2}{25} - \dfrac{(x-2)^2}{16} = 1$

64. $6y^2 = 2x^2 + 12$

Simplify each expression. (Lesson 8-1)

65. $\dfrac{12p^2 + 6p - 6}{4(p+1)^2} \div \dfrac{6p - 3}{2p + 10} \cdot \dfrac{p+5}{p+1}$

66. $\dfrac{x^2 + 6x + 9}{x^2 + 7x + 6} \div \dfrac{4x + 12}{3x + 3} \cdot \dfrac{3x + 9}{4x + 24}$

67. $\dfrac{r^2 + 2r - 8}{r^2 + 4r + 3} \div \dfrac{r - 2}{3r + 3} \cdot \dfrac{3(r+4)}{r+3}$

Graph each function. State the domain and range. (Lesson 7-1) **68–70. See Chapter 9 Answer Appendix.**

68. $f(x) = -\left(\dfrac{1}{5}\right)^x$

69. $y = -2.5(5)^x$

70. $f(x) = 2\left(\dfrac{1}{3}\right)^x$

Skills Review

Solve each equation or formula for the specified variable.

71. $d = rt$, for r $\quad \dfrac{d}{t} = r$

72. $x = \dfrac{-b}{2a}$, for a $\quad a = \dfrac{-b}{2x}$

73. $V = \dfrac{1}{3}\pi r^2 h$, for h $\quad \dfrac{3V}{\pi r^2} = h$

74. $A = \dfrac{1}{2}h(a + b)$, for b $\quad \dfrac{2A}{h} - a = b$

DifferentiatedInstruction **OL** **BL**

Extension Ask students to solve this system of inequalities by graphing.

$$y \geq x^2$$
$$x^2 + (y - 3)^2 \leq 4$$
$$y > -x + 2$$

Yesterday's News Ask students to write how yesterday's lesson on conic equations helped them with today's lesson.

Formative Assessment

Check for student understanding of concepts in Lesson 9-7.

📁 Quiz 4, p. 54

Additional Answers

62. $(0, \pm 4); (0, \pm\sqrt{41}); y = \pm\dfrac{4}{5}x$

63. $(2, -2), (2, 8); (2, 3 \pm \sqrt{41});$
$y - 3 = \pm\dfrac{5}{4}(x - 2)$

64. $(0, \pm\sqrt{2}); (0, \pm 2\sqrt{2});$
$y = \pm\dfrac{\sqrt{3}}{3}x,$

Formative Assessment

KeyVocabulary The page references after each word denote where that term was first introduced. If students have difficulty answering questions 1–10, remind them that they can use these page references to refresh their memories about the vocabulary.

Summative Assessment

📁 Vocabulary Test, p. 56

🔤 **Vocabulary Review**

Vocabulary Review provides students the opportunity to check their understanding of important concepts and terminology in an online game format.

FOLDABLES StudyOrganizer

Dinah Zike's Foldables®

Have students look through the chapter to make sure they have included examples in their Foldables. Suggest that students keep their Foldables handy while completing the Study Guide and Review pages. Point out that their Foldables can serve as a quick review tool when studying for the chapter test.

Study Guide

KeyConcepts

Midpoint and Distance Formulas (Lesson 9-1)

- $M = \left(\dfrac{x_1 + x_2}{2}, \dfrac{y_1 + y_2}{2} \right)$
- $d = \sqrt{(x_2 - x_1)^2 + (y_2 - y_1)^2}$

Parabolas (Lesson 9-2)

- Standard Form: $y = a(x - h)^2 + k$
  $x = a(y - k)^2 + h$

Circles (Lesson 9-3)

- The equation of a circle with center (h, k) and radius r can be written in the form $(x - h)^2 + (y - k)^2 = r^2$.

Ellipses (Lesson 9-4)

- Standard Form: horizontal $\dfrac{(x - h)^2}{a^2} + \dfrac{(y - k)^2}{b^2} = 1$

  vertical $\dfrac{(y - k)^2}{a^2} + \dfrac{(x - h)^2}{b^2} = 1$

Hyperbolas (Lesson 9-5)

- Standard Form: horizontal $\dfrac{(x - h)^2}{a^2} - \dfrac{(y - k)^2}{b^2} = 1$

  vertical $\dfrac{(y - k)^2}{a^2} - \dfrac{(x - h)^2}{b^2} = 1$

Solving Linear-Nonlinear Systems (Lesson 9-7)

- Systems of quadratic equations can be solved using substitution and elimination.
- A system of quadratic equations can have zero, one, two, three, or four solutions.

FOLDABLES StudyOrganizer

Be sure the Key Concepts are noted in your Foldable.

Conic Sections

KeyVocabulary 🔤

center (of a circle) (p. 607)

center (of an ellipse) (p. 615)

circle (p. 607)

conjugate axis (p. 624)

constant difference (p. 627)

constant sum (p. 616)

co-vertices (of a hyperbola) (p. 624)

co-vertices (of an ellipse) (p. 615)

directrix (p. 599)

ellipse (p. 615)

foci (of a hyperbola) (p. 624)

foci (of an ellipse) (p. 615)

focus (p. 599)

hyperbola (p. 624)

latus rectum (p. 599)

major axis (p. 615)

minor axis (p. 615)

parabola (p. 599)

radius (p. 607)

transverse axis (p. 624)

vertices (of a hyperbola) (p. 624)

vertices (of an ellipse) (p. 615)

VocabularyCheck

State whether each sentence is *true* or *false*. If *false*, replace the underlined term to make a true sentence.

false, center

1. The set of all points in a plane that are equidistant from a given point in the plane, called the <u>focus</u>, forms a circle.

2. A(n) <u>ellipse</u> is the set of all points in a plane such that the sum of the distances from the two fixed points is constant. **true**

3. The endpoints of the major axis of an ellipse are the <u>foci</u> of the ellipse. **false, vertices**

4. The <u>radius</u> is the distance from the center of a circle to any point on the circle. **true**

5. The line segment with endpoints on a parabola, through the focus of the parabola, and perpendicular to the axis of symmetry is called the <u>latus rectum</u>. **true**

6. Every hyperbola has two axes of symmetry, the transverse axis and the <u>major axis</u>. **false, conjugate axis**

7. A <u>directrix</u> is the set of all points in a plane that are equidistant from a given point in the plane, called the center. **false, circle**

8. A hyperbola is the set of all points in a plane such that the absolute value of the <u>sum</u> of the distances from any point on the hyperbola to two given points is constant. **false, difference**

9. A parabola can be defined as the set of all points in a plane that are the same distance from the focus and a given line called the <u>directrix</u>. **true**

10. The <u>major axis</u> is the longer of the two axes of symmetry of an ellipse. **true**

Lesson-by-Lesson Review

9-1 Midpoint and Distance Formulas

Find the midpoint of the line segment with endpoints at the given coordinates.

11. $(-8, 6), (3, 4)$ $\left(-\frac{5}{2}, 5\right)$ **12.** $(-6, 0), (-1, 4)$ $\left(-\frac{7}{2}, 2\right)$

13. $\left(\frac{3}{4}, \frac{2}{3}\right), \left(-\frac{1}{3}, \frac{1}{4}\right)$ **14.** $(15, 20), (18, 21)$ $\left(\frac{33}{2}, \frac{41}{2}\right)$
13. $\left(\frac{5}{24}, \frac{11}{24}\right)$

Find the distance between each pair of points with the given coordinates.

15. $(10, -3), (1, -5)$ $\sqrt{85}$ **16.** $(0, 6), (-9, 7)$ $\sqrt{82}$

17. $\left(\frac{1}{4}, \frac{1}{2}\right), \left(\frac{3}{2}, \frac{5}{4}\right)$ $\frac{\sqrt{34}}{4}$ **18.** $(5, -3), (7, -1)$ $2\sqrt{2}$

19. HIKING Marc wants to hike from his camp to a waterfall. The waterfall is 5 miles south and 8 miles east of his campsite.

   a. How far away is the waterfall? $\sqrt{89} \approx 9.4$ miles

   b. Marc wants to stop for lunch halfway to the waterfall. Where should he stop? $\left(4, -\frac{5}{2}\right)$

Example 1

Find the midpoint of a line segment whose endpoints are at $(-4, 8)$ and $(10, -1)$.

Let $(x_1, y_1) = (-4, 8)$ and $(x_2, y_2) = (10, -1)$.

$\left(\frac{x_1 + x_2}{2}, \frac{y_1 + y_2}{2}\right) = \left(\frac{-4 + 10}{2}, \frac{8 + (-1)}{2}\right)$

$= \left(\frac{6}{2}, \frac{7}{2}\right)$ or $\left(3, \frac{7}{2}\right)$

Example 2

Find the distance between $P(5, -3)$ and $Q(-1, 5)$.
Let $(x_1, y_1) = (5, -3)$ and $(x_2, y_2) = (-1, 5)$.

$d = \sqrt{(x_2 - x_1)^2 + (y_2 - y_1)^2}$ Distance Formula

$= \sqrt{(-1 - 5)^2 + [5 - (-3)]^2}$ Substitute.

$= \sqrt{36 + 64}$ Subtract.

$= \sqrt{100}$ or 10 units   Simplify.

9-2 Parabolas

Graph each equation. **20–23. See margin.**

20. $y = 3x^2 + 24x - 10$ **21.** $3y - x^2 = 8x - 11$

22. $x = \frac{1}{2}y^2 - 4y + 3$ **23.** $x = y^2 - 14y + 25$

24–27. See margin.

Write each equation in standard form. Identify the vertex, axis of symmetry, and direction of opening of the parabola.

24. $y = -\frac{1}{2}x^2$ **25.** $y = 4x^2 - 16x + 9$

26. $x - 6y = y^2 + 4$ **27.** $x = y^2 + 14y + 20$

28. SPORTS When a football is kicked, the path it travels is shaped like a parabola. Suppose a football is kicked from ground level, reaches a maximum height of 50 feet, and lands 200 feet away. Assuming the football was kicked at the origin, write an equation of the parabola that models the flight of the football. $y = -\frac{1}{200}(x - 100)^2 + 50$

Example 3

Write $3y - x^2 = 4x + 7$ in standard form. Identify the vertex, axis of symmetry, and direction of opening of the parabola.

Write the equation in the form $y = a(x - h)^2 + k$ by completing the square.

$3y = x^2 + 4x + 7$ Isolate the terms with x.

$3y = (x^2 + 4x + \blacksquare) + 7 - \blacksquare$ Complete the square.

$3y = (x^2 + 4x + 4) + 7 - 4$ $\left(\frac{4}{2}\right)^2 = 4$

$3y = (x + 2)^2 + 3$ $(x^2 + 4x + 4) = (x + 2)$

$y = \frac{1}{3}(x + 2)^2 + 1$ Divide each side by 3.

Vertex: $(-2, 1)$; axis of symmetry: $x = -2$; direction of opening: upward since $a > 0$.

connectED.mcgraw-hill.com **647**

23.

24. $y = -\frac{1}{2}x^2$; vertex: $(0, 0)$; axis of symmetry: $x = 0$; opens: upward

25. $y = 4(x - 2)^2 - 7$; vertex: $(2, -7)$; axis of symmetry: $x = 2$; opens: upward

26. $x = (y + 3)^2 - 5$; vertex: $(-5, -3)$; axis of symmetry: $y = -3$; opens to the right

27. $x = (y + 7)^2 - 29$; vertex $= (-29, -7)$; axis of symmetry $y = -7$; opens to the right

Lesson-by-Lesson Review

Intervention If the given examples are not sufficient to review the topics covered by the questions, remind students that the lesson references tell them where to review that topic in their textbook.

Two-Day Option Have students complete the Lesson-by-Lesson Review. Then you can use McGraw-Hill eAssessment to customize another review worksheet that practices all the objectives of this chapter or only the objectives on which your students need more help.

Additional Answers

20.

$y = 3x^2 + 24x - 10$

21.

$3y - x^2 = 8x - 11$

22.

$x = \frac{1}{2}y^2 - 4y + 3$

Additional Answers

32. $(-5, 0); r = 3$

33. $(3, -1); r = 5$

34. $(-2, 8); r = 1$

35. $(-2, 1); r = 4$

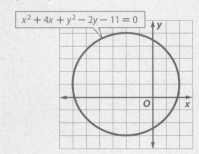

9-3 Circles

Write an equation for the circle that satisfies each set of conditions. **29.** $(x + 1)^2 + (y - 6)^2 = 9$

29. center $(-1, 6)$, radius 3 units

30. endpoints of a diameter $(2, 5)$ and $(0, 0)$

31. endpoints of a diameter $(4, -2)$ and $(-2, -6)$

Find the center and radius of each circle. Then graph the circle. **32–35. See margin.**

32. $(x + 5)^2 + y^2 = 9$

33. $(x - 3)^2 + (y + 1)^2 = 25$

34. $(x + 2)^2 + (y - 8)^2 = 1$

35. $x^2 + 4x + y^2 - 2y - 11 = 0$

36. SOUND A loudspeaker in a school is located at the point $(65, 40)$. The speaker can be heard in a circle with a radius of 100 feet. Write an equation to represent the possible boundary of the loudspeaker sound. $(x - 65)^2 + (y - 40)^2 = 100^2$

30. $(x - 1)^2 + (y - 2.5)^2 = \dfrac{29}{4}$ **31.** $(x - 1)^2 + (y + 4)^2 = 13$

Example 4

Find the center and radius of the circle with equation $x^2 - 2x + y^2 + 6y + 6 = 0$. Then graph the circle.

Complete the squares.

$$x^2 - 2x + y^2 + 6y + 6 = 0$$
$$(x^2 - 2x + \blacksquare) + (y^2 + 6y + \blacksquare) = -6 + \blacksquare + \blacksquare$$
$$(x^2 - 2x + 1) + (y^2 + 6y + 9) = -6 + 1 + 9$$
$$(x - 1)^2 + (y + 3)^2 = 4$$

The center of the circle is at $(1, -3)$ and the radius is 2.

9-4 Ellipses

Find the coordinates of the center and foci and the lengths of the major and minor axes for the ellipse with the given equation. Then graph the ellipse.

37. $\dfrac{x^2}{9} + \dfrac{y^2}{36} = 1$ **38.** $\dfrac{y^2}{10} + \dfrac{x^2}{5} = 1$

39. $\dfrac{x^2}{36} + \dfrac{(y - 4)^2}{4} = 1$ **40.** $27x^2 + 9y^2 = 81$

41. $\dfrac{(x + 1)^2}{25} + \dfrac{(y - 2)^2}{16} = 1$

42. $9x^2 + 4y^2 + 54x - 8y + 49 = 0$

43. $9x^2 + 25y^2 - 18x + 50y - 191 = 0$

44. $7x^2 + 3y^2 - 28x - 12y = -19$

37–44. See Chapter 9 Answer Appendix.

45. LANDSCAPING The Martins have a garden in their front yard that is shaped like an ellipse. The major axis is 16 feet and the minor axis is 10 feet. Write an equation to model the garden. Assume the origin is at the center of the garden and the major axis is horizontal.

$$\dfrac{x^2}{64} + \dfrac{y^2}{25} = 1$$

Example 5

Find the coordinates of the center and foci and the lengths of the major and minor axes for the ellipse with equation $9x^2 + 16y^2 - 54x + 32y - 47 = 0$. Then graph the ellipse.

First, convert to standard form.

$$9x^2 + 16y^2 - 54x + 32y - 47 = 0$$
$$9(x^2 - 6x + \blacksquare) + 16(y^2 + 2y + \blacksquare) = 47 + 9(\blacksquare) + 16(\blacksquare)$$
$$9(x^2 - 6x + 9) + 16(y^2 + 2y + 1) = 47 + 9(9) + 16(1)$$
$$9(x - 3)^2 + 16(y + 1)^2 = 144$$
$$\dfrac{(x - 3)^2}{16} + \dfrac{(y + 1)^2}{9} = 1$$

The center of the ellipse is $(3, -1)$. The ellipse is horizontal. $a^2 = 16$, so $a = 4$. $b^2 = 9$, so $b = 3$. The length of the major axis is $2 \cdot 4$ or 8. The length of the minor axis is $2 \cdot 3$ or 6. To find the foci: $c^2 = 16 - 9$ or 7, so $c = \sqrt{7}$. The foci are $\left(3 + \sqrt{7}, -1\right)$ and $\left(3 - \sqrt{7}, -1\right)$.

46. $(0, \pm 3); (0, \pm\sqrt{13}); y = \pm\dfrac{3}{2}x$

47. $(2, -2), (4, -2); (3 \pm \sqrt{5}, -2);$
$y + 2 = \pm 2(x - 3)$

9-5 Hyperbolas

Graph each hyperbola. Identify the vertices, foci, and asymptotes. **46–50. See margin.**

46. $\dfrac{y^2}{9} - \dfrac{x^2}{4} = 1$

47. $\dfrac{(x-3)^2}{1} - \dfrac{(y+2)^2}{4} = 1$

48. $\dfrac{(y+1)^2}{16} - \dfrac{(x-4)^2}{9} = 1$

49. $4x^2 - 9y^2 = 36$

50. $9y^2 - x^2 - 4x + 18y + 4 = 0$

51. MIRRORS A hyperbolic mirror is shaped like one branch of a hyperbola. It reflects light rays directed at one focus toward the other focus. Suppose a hyperbolic mirror is modeled by the upper branch of the hyperbola $\dfrac{y^2}{9} - \dfrac{x^2}{16} = 1$. A light source is located at $(-10, 0)$. Where should the light hit the mirror so that the light will be reflected to $(0, -5)$? **See margin.**

Example 6

Graph $9x^2 - 4y^2 - 36x - 8y - 4 = 0$. Identify the vertices, foci, and asymptotes.

Complete the square.
$$9x^2 - 4y^2 - 36x - 8y - 4 = 0$$
$$9(x^2 - 4x + \blacksquare) - 4(y^2 + 2y + \blacksquare) = 4 + 9(\blacksquare) - 4(\blacksquare)$$
$$9(x^2 - 4x + 4) - 4(y^2 + 2y + 1) = 4 + 9(4) - 4(1)$$
$$9(x-2)^2 - 4(y+1)^2 = 36$$
$$\dfrac{(x-2)^2}{4} - \dfrac{(y+1)^2}{9} = 1$$

The center is at $(2, -1)$. The vertices are at $(0, -1)$ and $(4, -1)$. The foci are at $\left(2 + \sqrt{13}, -1\right)$ and $\left(2 - \sqrt{13}, -1\right)$. The equations of the asymptotes are $y + 1 = \pm\dfrac{3}{2}(x - 2)$.

9-6 Identifying Conic Sections

Write each equation in standard form. State whether the graph of the equation is a *parabola, circle, ellipse,* or *hyperbola.* Then graph.

52. $3x^2 + 12x - y + 8 = 0$
53. $9x^2 + 16y^2 = 144$
54. $x^2 + y^2 - 8x - 2y + 8 = 0$
55. $-9x^2 + y^2 + 36x - 45 = 0$

52–55. See Chapter 9 Answer Appendix.

Without writing the equation in standard form, state whether the graph of the equation is a *parabola, circle, ellipse,* or *hyperbola.*

56. $7x^2 + 9y^2 = 63$ **ellipse**

57. $5y^2 + 2y + 4x - 13x^2 = 81$ **hyperbola**

58. $x^2 - 8x + 16 = 6y$ **parabola**

59. $x^2 + 4x + y^2 - 285 = 0$ **circle**

60. LIGHT Suppose the edge of a shadow can be represented by the equation $16x^2 + 25y^2 - 32x - 100y - 284 = 0$.

a. What is the shape of the shadow? **ellipse**

b. Graph the equation. **See margin.**

Example 7

Write $3x^2 + 3y^2 - 12x + 30y + 39 = 0$ in standard form. State whether the graph of the equation is a *parabola, circle, ellipse,* or *hyperbola.* Then graph the equation.

$$3x^2 + 3y^2 - 12x + 30y + 39 = 0$$
$$3(x^2 - 4x + \blacksquare) + 3(y^2 + 10y + \blacksquare) = -39 + 3(\blacksquare) + 3(\blacksquare)$$
$$3(x^2 - 4x + 4) + 3(y^2 + 10y + 25) = -39 + 3(4) + 3(25)$$
$$3(x-2)^2 + 3(y+5)^2 = 48$$
$$(x-2)^2 + (y+5)^2 = 16$$

In this equation $A = 3$ and $C = 3$. Since A and C are both positive and $A = C$, the graph is a circle. The center is at $(2, -5)$, and the radius is 4.

60b.

connectED.mcgraw-hill.com **649**

Additional Answers

48. $(4, -5), (4, 3), (4, -6), (4, 4);$
$y + 1 = \pm\dfrac{4}{3}(x - 4)$

49. $(\pm 3, 0); (\pm\sqrt{13}, 0); y = \pm\dfrac{2}{3}x$

50. $\left(-2, -\dfrac{2}{3}\right), \left(-2, -\dfrac{4}{3}\right);$
$\left(-2, -1 \pm \dfrac{\sqrt{10}}{3}\right);$
$y + 1 = \pm\dfrac{1}{3}(x + 2)$

51. $\left(\dfrac{40 - 24\sqrt{5}}{5}, \dfrac{45 - 12\sqrt{5}}{5}\right)$

Anticipation Guide

Have students complete the Chapter 9 Anticipation Guide and discuss how their responses have changed now that they have completed Chapter 9.

Additional Answers

69.

70.

71.

72.

9-7 Solving Linear-Nonlinear Systems

Solve each system of equations.

61. $x^2 + y^2 = 8$ $(2, -2),$
$x + y = 0$ $(-2, 2)$

62. $x - 2y = 2$ $(-2, -2)$
$y^2 - x^2 = 2x + 4$

63. $y + x^2 = 4x$ $(4, 0)$
$y + 4x = 16$

64. $3x^2 - y^2 = 11$
$x^2 + 4y^2 = 8$

65. $5x^2 + y^2 = 30$
$9x^2 - y^2 = -16$
$(1, \pm5), (-1, \pm5)$

66. $\dfrac{x^2}{30} + \dfrac{y^2}{6} = 1$
$x = y$
$(\sqrt{5}, \sqrt{5}), (-\sqrt{5}, -\sqrt{5})$

67. PHYSICAL SCIENCE Two balls are launched into the air at the same time. The heights they are launched from are different. The height y in feet of one is represented by $y = -16t^2 + 80t + 25$ where t is the time in seconds. The height of the other ball is represented by $y = -16t^2 + 30t + 100$.

  a. After how many seconds are the balls at the same height? **1.5 seconds**

  b. What is this height? **109 ft**

64. $(-2, -1), (-2, 1), (2, -1), (2, 1)$

68. ARCHITECTURE An architect is building the front entrance of a building in the shape of a parabola with the equation $y = -\dfrac{1}{10}(x - 10)^2 + 20$. While the entrance is being built, the construction team puts in two support beams with equations $y = -x + 10$ and $y = x - 10$. Where do the support beams meet the parabola? **(0, 10) and (20, 10)**

69–74. See margin.

Solve each system of inequalities by graphing.

69. $x^2 + y^2 < 64$
$x^2 + 16(y - 3)^2 < 16$

70. $x^2 + y^2 < 49$
$16x^2 - 9y^2 \geq 144$

71. $x + y < 4$
$9x^2 - 4y^2 \geq 36$

72. $x^2 + y^2 < 25$
$4x^2 - 9y^2 < 36$

73. $x^2 + y^2 < 36$
$4x^2 + 9y^2 > 36$

74. $y^2 < x$
$x^2 - 4y^2 < 16$

Example 8

Solve the system of equations.

$x^2 + y^2 = 100$
$3x - y = 10$

Use substitution to solve the system.
First, rewrite $3x - y = 10$ as $y = 3x - 10$.

$$x^2 + y^2 = 100$$
$$x^2 + (3x - 10)^2 = 100$$
$$x^2 + 9x^2 - 60x + 100 = 100$$
$$10x^2 - 60x + 100 = 100$$
$$10x^2 - 60x = 0$$
$$10x(x - 6) = 0$$

$10x = 0$ or $x - 6 = 0$
$x = 0$ $x = 6$

Now solve for y.

$y = 3x - 10$ $y = 3x - 10$
$= 3(0) - 10$ $= 3(6) - 10$
$= -10$ $= 8$

The solutions of the system are $(0, -10)$ and $(6, 8)$.

Example 9

Solve the system of inequalities by graphing.

$x^2 + y^2 \leq 9$
$2y \geq x^2 + 4$

The solution is the green shaded region.

73.

74.

CHAPTER 9 Practice Test

Find the midpoint of the line segment with endpoints at the given coordinates.

1. $(8, 3), (-4, 9)$ **(2, 6)**

2. $\left(\frac{3}{4}, 0\right), \left(\frac{1}{2}, -1\right)$ $\left(\frac{5}{8}, -\frac{1}{2}\right)$

3. $(-10, 0), (-2, 6)$ **(-6, 3)**

Find the distance between each pair of points with the given coordinates.

4. $(-5, 8), (4, 3)$ $\sqrt{106}$

5. $\left(\frac{1}{3}, \frac{2}{3}\right), \left(-\frac{5}{6}, -\frac{11}{6}\right)$ $\frac{\sqrt{274}}{6}$

6. $(4, -5), (4, 9)$ **14**

State whether the graph of each equation is a *parabola, circle, ellipse,* or *hyperbola.* Then graph the equation. 7–16. See Chapter 9 Answer Appendix.

7. $y^2 = 64 - x^2$

8. $4x^2 + y^2 = 16$

9. $4x^2 - 9y^2 + 8x + 36y = 68$

10. $\frac{1}{2}x^2 - 3 = y$

11. $y = -2x^2 - 5$

12. $16x^2 + 25y^2 = 400$

13. $x^2 + 6x + y^2 = 16$

14. $\frac{y^2}{4} - \frac{x^2}{16} = 1$

15. $(x + 2)^2 = 3(y - 1)$

16. $4x^2 + 16y^2 + 32x + 63 = 0$

17. **MULTIPLE CHOICE** Which equation represents a hyperbola that has vertices at $(-3, -3)$ and $(5, -3)$ and a conjugate axis of length 6 units? **B**

    A $\frac{(y - 1)^2}{16} - \frac{(x + 3)^2}{9} = 1$

    B $\frac{(x - 1)^2}{16} - \frac{(y + 3)^2}{9} = 1$

    C $\frac{(y + 1)^2}{16} - \frac{(x - 3)^2}{9} = 1$

    D $\frac{(x + 1)^2}{16} - \frac{(y - 3)^2}{9} = 1$

18. **CARPENTRY** Ellis built a window frame shaped like the top half of an ellipse. The window is 40 inches tall at its highest point and 160 inches wide at the bottom. What is the height of the window 20 inches from the center of the base?

18. $10\sqrt{15}$ or about 38.73 inches

Solve each system of equations.

19. $x^2 + y^2 = 100$ $(-8, 6), (6, -8)$
    $y = -x - 2$

20. $x^2 + 2y^2 = 11$ $(3, -1), \left(-\frac{1}{3}, \frac{7}{3}\right)$
    $x + y = 2$

21. $x^2 + y^2 = 34$ $\left(\pm\frac{5\sqrt{2}}{2}, \pm\frac{\sqrt{86}}{2}\right)$
    $y^2 - x^2 = 9$

22, 23. See Chapter 9 Answer Appendix.

Solve each system of inequalities.

22. $x^2 + y^2 \leq 9$
    $y > -x^2 + 2$

23. $\frac{(x - 2)^2}{4} - \frac{(y - 4)^2}{9} \geq 1$
    $x - 4y < 8$

24. **MULTIPLE CHOICE** Which is NOT the equation of a parabola? **J**

    F $y = 3x^2 + 5x - 3$

    G $2y + 3x^2 + x - 9 = 0$

    H $x = 3(y + 1)^2$

    J $x^2 + 2y^2 + 6x = 10$

25. **FORESTRY** A forest ranger at an outpost in the Sam Houston National Forest and another ranger at the primary station both heard an explosion. The outpost and the primary station are 6 kilometers apart.

    a, b. See Chapter 9 Answer Appendix.

    a. If one ranger heard the explosion 6 seconds before the other, write an equation that describes all the possible locations of the explosion. Place the two ranger stations on the *x*-axis with the midpoint between the stations at the origin. The transverse axis is horizontal. (*Hint:* The speed of sound is about 0.35 kilometer per second.)

    b. Draw a sketch of the possible locations of the explosion. Include the ranger stations in the drawing.

 651

Summative Assessment

Use these alternate leveled chapter tests to differentiate assessment for your students.

Leveled Chapter 9 Tests			
Form	Type	Level	Page(s)
1	MC	AL	57–58
2A	MC	OL	59–60
2B	MC	OL	61–62
2C	FR	OL	63–64
2D	FR	OL	65–66
3	FR	BL	67–68
Vocabulary Test			56
Extended-Response Test			69

MC = multiple-choice questions
FR = free-response questions

McGraw-Hill eAssessment Customize and create multiple versions of your chapter test and their answer keys. All of the questions from the leveled chapter tests in the *Chapter 9 Resource Masters* are also available on McGraw-Hill eAssessment.

InterventionPlanner

TIER 1 On Level OL	TIER 2 Strategic Intervention AL approaching grade level	TIER 3 Intensive Intervention 2 or more grades below level
If students miss about 25% of the exercises or less,	**If** students miss about 50% of the exercises,	**If** students miss about 75% of the exercises,
Then choose a resource:	**Then** choose a resource:	**Then** use *Math Triumphs, Alg. 2*
SE Lessons 9-1, 9-2, 9-3, 9-4, 9-5, 9-6, and 9-7	Study Guide and Intervention, pp. 5, 11, 18, 25, 32, 39 and 45	connectED.mcgraw-hill.com Extra Examples, Personal Tutor, Homework Help, Review Vocabulary
Skills Practice, pp. 7, 13, 20, 27, 34, 41, and 47	connectED.mcgraw-hill.com Extra Examples, Personal Tutor, Homework Help	
TE Chapter Project		
connectED.mcgraw-hill.com Self-Check Quiz		

1 Focus

Objective Use the strategy of using a formula to solve standardized test problems.

2 Teach

Scaffolding Questions
Ask:

- What are some types of problems you would use a formula to solve?
  Sample answers: finding a distance; finding the perimeter, area or volume of various geometric figures; solving a quadratic equation with the Quadratic Formula

- How do you use the formulas for the coordinates of the midpoint of a line segment if you are given both sets of coordinates of the endpoints?
  Sample answer: For the *x*-coordinate, add the endpoints' *x*-coordinates together and divide by 2. To find the *y*-coordinate, add the endpoints' *y*-coordinates, and then divide by 2.

Use a Formula

Sometimes it is necessary to use a formula to solve problems on standardized tests. In some cases you may even be given a sheet of formulas that you are permitted to reference while taking the test.

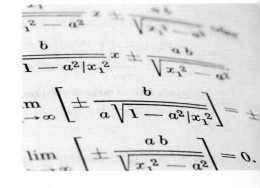

Strategies for Using a Formula

Step 1

Read the problem statement carefully.

Ask yourself:

- What am I being asked to solve?
- What information is given in the problem?
- Are there any formulas that I can use to help me solve the problem?

Step 2

Solve the problem and check your solution.

- Substitute the known quantities that are given in the problem statement into the formula.
- Simplify to solve for the unknown values in the formula.
- Check to make sure your answer makes sense. If time permits, check your answer.

Standardized Test Example

Read the problem. Identify what you need to know. Then use the information in the problem to solve. Show your work.

What is the distance between points *A* and *B* on the coordinate plane? Round your answer to the nearest tenth if necessary.

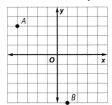

Scoring Rubric	
Criteria	Score
Full Credit: The answer is correct and a full explanation is provided that shows each step.	2
Partial Credit: • The answer is correct but the explanation is incomplete. • The answer is incorrect but the explanation is correct.	1
No Credit: Either an answer is not provided or the answer does not make sense.	0

pml/Alamy

Read the problem statement carefully. You are given the coordinates of two points on a coordinate plane and asked to find the distance between them. To solve this problem, you must use the **Distance Formula**.

Example of a 2-point response:

Use the Distance Formula to find the distance between points $A(-4, 3)$ and $B(1, -5)$.

$$d = \sqrt{(x_2 - x_1)^2 + (y_2 - y_1)^2}$$
$$= \sqrt{[1 - (-4)]^2 + [(-5) - 3]^2}$$
$$= \sqrt{5^2 + (-8)^2}$$
$$= \sqrt{25 + 64}$$
$$= \sqrt{89} \text{ or about } 9.4$$

The distance between points A and B is about 9.4 units.

The steps, calculations, and reasoning are clearly stated. The student also arrives at the correct answer. So, this response is worth the full 2 points.

Exercises

Read each problem. Identify what you need to know. Then use the information in the problem to solve. Show your work.

1. What is the midpoint of segment CD with endpoints $C(5, -12)$ and $D(-9, 4)$? $M(-2, -4)$

2. Katrina is making a map of her hometown on a coordinate plane. She plots the school at $S(7, 3)$ and the park at $P(-4, 12)$. If the scale of the map is 1 unit = 250 yards, what is the actual distance between the school and the park? Round to the nearest yard. about 3553 yd

3. Mr. Washington is making a concrete table for his backyard. The tabletop will be circular with a diameter of 6 feet and a depth of 6 inches. How much concrete will Mr. Washington need to make the top of the table? Round to the nearest cubic foot. 14 ft³

4. What is the equation, in standard form, of the hyperbola graphed below? $\dfrac{(y+1)^2}{9} - \dfrac{(x-2)^2}{16} = 1$

5. If the surface area of a cube is increased by a factor of 9, what is the change in the length of the sides of the cube? B

A The length is 2 times the original length.

B The length is 3 times the original length.

C The length is 6 times the original length.

D The length is 9 times the original length.

connectED.mcgraw-hill.com **653**

Additional Example

STANDARDIZED TEST PRACTICE
Find the exact solution for the following system of equations. D
$$y = 5 - x^2$$
$$y = 2x^2 + 2$$

A $(2, 2), (-2, 2)$

B $(1, 2), (-1, 2)$

C $(4, -1), (4, 1)$

D $(-1, 4), (1, 4)$

3 Assess

Use Exercises 1–5 to assess students' understanding.

Diagnose Student Errors

Survey student responses for each item. Class trends may indicate common errors and misconceptions.

1. A this line is *not* incorrect
   B correct
   C this is the *second* incorrect step
   D since there is an incorrect step, this is not true

2. F doesn't pass through (1, 9)
   G guess
   H correct
   J doesn't have (−3, −23) as focus

3. A these are the co-vertices
   B guess
   C correct
   D guess

4. F mistakenly divided 40 by 5
   G mistakenly added 14 and 5
   H correct
   J guess

5. A correct
   B guess
   C guess
   D guess

6. F correct
   G guess
   H solved $\sqrt{x + 5} = 4$
   J solved $\sqrt{x + 5} - 1 = 4$

7. A multiplied by *x* but not by 2
   B multiplied by 2 but not by *x*
   C correct
   D did not apply the Distributive Property consistently

8. F found $-\dfrac{1}{3}$ slope
   G found $\dfrac{3}{4}$ slope
   H found $-\dfrac{7}{3}$ slope
   J correct

Standardized Test Practice
Cumulative, Chapters 1 through 9

Multiple Choice

Read each question. Then fill in the correct answer on the answer document provided by your teacher or on a sheet of paper.

1. Which is the first *incorrect* step in simplifying $\log_3 \dfrac{3}{48}$? **B**

Step 1: $\log_3 \dfrac{3}{48} = \log_3 3 - \log_3 48$
Step 2: $= 1 - 16$
Step 3: $= -15$

A Step 1
B Step 2
C Step 3
D Each step is correct.

2. Which is the equation for the parabola that has vertex (−3, −23) and passes through the point (1, 9)? **H**

F $y = x^2 + 10x + 7$
G $y = x^2 - 6x + 19$
H $y = 2x^2 + 12x - 5$
J $y = 2x^2 - 3x + 10$

3. What are the vertices of the ellipse with equation $\dfrac{(x-3)^2}{36} + \dfrac{(y-2)^2}{144} = 1$? **C**

A (−3, 2) and (9, 2)
B (−2, 3) and (10, 3)
C (3, −10) and (3, 14)
D (2, −11) and (4, 13)

4. Hooke's Law states that the force needed to keep a spring stretched *x* units is directly proportional to *x*. If a force of 40 N is required to maintain a spring stretched to 5 centimeters, what force is needed to keep the spring stretched 14 centimeters? **H**

F 8 N
G 19 N
H 112 N
J 1600 N

Test-TakingTip

Question 2 You can check your answer by submitting 1 for *x* and making sure that the *y*-value is 9.

---

5. Angela is making a map of her backyard on a coordinate grid. She plots point $G(-4, -6)$ to represent her mom's garden and point $S(3, 7)$ to represent the rope swing hanging on an oak tree. If the scale of the map is 1 unit = 5 feet, what is the approximate distance between the garden and the rope swing? **A**

A 74 feet   C 82 feet
B 79 feet   D 90 feet

6. If $\sqrt{x + 5} + 1 = 4$, what is the value of *x*? **F**

F 4   G 10   H 11   J 20

7. The area of the base of a rectangular suitcase measures $3x^2 + 5x - 4$ square units. The height of the suitcase measures $2x$ units. Which polynomial expression represents the volume of the suitcase? **C**

A $3x^3 + 5x^2 - 4x$ C $6x^3 + 10x^2 - 8x$
B $6x^2 + 10x - 8$ D $3x^3 + 10x^2 - 4$

8. Malina was given this geoboard to model the slope $-\dfrac{3}{4}$.

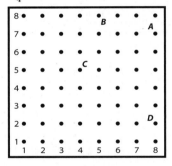

If the peg in the upper right-hand corner represents the origin on a coordinate plane, where could Malina place a rubber band to represent the given slope? **J**

F from peg *A* to peg *B*
G from peg *A* to peg *C*
H from peg *B* to peg *D*
J from peg *C* to peg *D*

Short Response/Gridded Response

Record your answers on the answer sheet provided by your teacher or on a sheet of paper.

9. A placekicker kicks a ball upward with a velocity of 32 feet per second. Ignoring the height of the kicking tee, how long after the football is kicked does it hit the ground? Use the formula $h(t) = v_0 t - 16t^2$, where $h(t)$ is the height of an object in feet, v_0 is the object's initial velocity in feet per second, and t is the time in seconds. **2 seconds**

10. GRIDDED RESPONSE What is the maximum number of solutions of a system of equations that consists of a circle and a hyperbola? **4**

11. Lupe is preparing boxes of assorted chocolates. Chocolate-covered peanuts cost $7 per pound. Chocolate-covered caramels cost $6.50 per pound. The boxes of assorted candies contain five more pounds of peanut candies than caramel candies. If the total amount sold was $575, how many pounds of each candy were needed to make the boxes? **40 lb caramel, 45 lb peanut**

12. GRIDDED RESPONSE What is the y-coordinate of the midpoint of segment AB with endpoints $A(0.8, 5.32)$ and $B(0.44, 2.2)$? **3.76**

13. Marc went shopping and bought two shirts, three pairs of pants, one belt, and two pairs of shoes. The following matrix shows the prices for each item respectively.

$$[\; \$20.15 \quad \$32 \quad \$15 \quad \$25.99 \;]$$

Use matrix multiplication to find the total amount of money Marc spent while shopping. **$203.28**

Extended Response

Record your answers on a sheet of paper. Show your work.

14. Clarence graphed the quadratic equation $h(t) = -16t^2 + 128t$ to model the flight of a firework. The parabola shows the height, in feet, of the firework t seconds after it was launched.

a. What is the vertex of the parabola? **(4, 256)**

b. What does the vertex of the parabola represent?

c. How long is the firework in the air before it lands? **8 seconds**

14b. the maximum height of the rocket

15. The Colonial High School Yearbook Staff is selling yearbooks and chrome picture frames engraved with the year. The number of yearbooks and frames sold to members of each grade is shown in the table. **15a. 82, 66**

Sales for Each Class		
Grade	Yearbooks	Frames
9th	423	256
10th	464	278
11th	546	344
12th	575	497

a. Find the difference in the sales of yearbooks and frames made to the 10th and 11th grade classes.

b. Find the total number of yearbooks and frames sold. **2008, 1375**

c. A yearbook costs $48 and a frame costs $18. Find the sales of yearbooks and frames for each class. **yearbooks: $20,304, $22,272, $26,208, $27,600; frames: $4608, $5004, $6192, $8946**

Need Extra Help?

If you missed Question...	1	2	3	4	5	6	7	8	9	10	11	12	13	14	15
Go to Lesson...	7-5	9-2	9-4	8-5	9-1	6-7	5-1	2-3	4-2	9-7	3-1	9-1	3-6	9-2	1-3

 connectED.mcgraw-hill.com **655**

Formative Assessment

You can use these pages to benchmark student progress.

📁 Standardized Test Practice, pp. 70–72

McGraw-Hill **eAssessment** Create practice tests that align to your state standards, the Common Core State Standards, and other national standards such as TIMSS and NAEP.

Answer Sheet Practice

Have students simulate taking a standardized test by recording their answers on practice recording sheets.

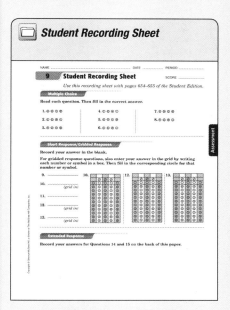

Student Recording Sheet

Homework Option

Get Ready for Chapter 10 Assign students the exercises on p. 657 as homework to assess whether they possess the prerequisite skills needed for the next chapter.

Lesson 9-2 (Guided Practice)

2A.

$y = 2x^2$

2B.

$y = 2(x - 1)^2 - 4$

3A. $x = \dfrac{1}{3}(y + 2)^2 + 7$

$x = \dfrac{1}{3}(y + 2)^2 + 7$

3B. $y = (x + 3)^2 - 13$

$y = (x + 3)^2 - 13$

Lesson 9-2

5.

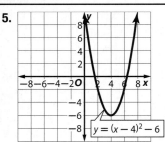

$y = (x - 4)^2 - 6$

6.

$y = 4(x + 5)^2 + 3$

7.

$y = -3x^2 - 4x - 8$

8.

$x = 3y^2 - 6y + 9$

9. $y = \dfrac{1}{8}x^2 + 2$

$y = \dfrac{1}{8}x^2 + 2$

10. $x = -\dfrac{1}{4}(y - 4)^2 - 2$

$x = -\dfrac{1}{4}(y - 4)^2 - 2$

11. $y = -\frac{1}{12}(x-3)^2 + 5$

12. $x = -\frac{1}{16}(y+5)^2 - 1$

20.

21.

22.

23.

24.

25.

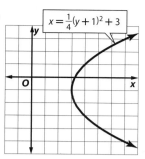

26. $y = \frac{1}{12}x^2 + 1$

27. $y = \frac{1}{20}(x-1)^2 + 8$

28. $x = \frac{1}{8}(y+4)^2 - 4$

29. $x = -\frac{1}{16}(y-4)^2 + 6$

30. $x = -\frac{1}{32}y^2 - 6$

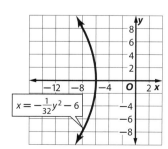

31. $y = -\frac{1}{4}(x-9)^2 + 6$

33a.

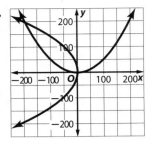

Lesson 9-3

31. center: (0,0); radius: $5\sqrt{3}$

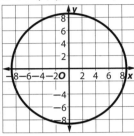

32. center: (3, 0); radius: 2

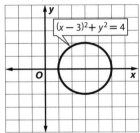

$(x-3)^2 + y^2 = 4$

33. center: (1, 4); radius: $\sqrt{34}$

$(x-1)^2 + (y-4)^2 = 34$

34. center: (0, 14); radius: 12

$x^2 + (y-14)^2 = 144$

35. center: (5, −2); radius: 4

$(x-5)^2 + (y+2)^2 = 16$

36. center: (0, 0); radius: 16

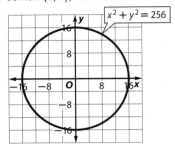

$x^2 + y^2 = 256$

37. center: (4, 0); radius: $\dfrac{\sqrt{8}}{3}$

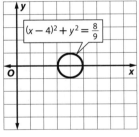

$(x-4)^2 + y^2 = \dfrac{8}{9}$

38. center: $\left(-\dfrac{2}{3}, \dfrac{1}{2}\right)$; radius: $\dfrac{4}{5}$

$\left(x + \dfrac{2}{3}\right)^2 + \left(y - \dfrac{1}{2}\right)^2 = \dfrac{16}{25}$

39. center: (−2, 0); radius: $\sqrt{13}$

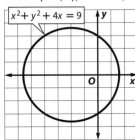

$x^2 + y^2 + 4x = 9$

40. center: $(-4, 3)$; radius: 5

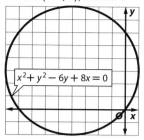
$x^2 + y^2 - 6y + 8x = 0$

41. center: $(-1, -2)$; radius: $\sqrt{14}$

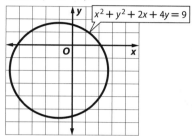
$x^2 + y^2 + 2x + 4y = 9$

42. center: $\left(\frac{3}{2}, -4\right)$; radius: $\frac{3\sqrt{17}}{2}$

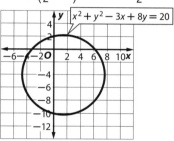
$x^2 + y^2 - 3x + 8y = 20$

43. center: $(-7, -3)$; radius: $2\sqrt{2}$ units

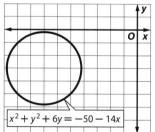
$x^2 + y^2 + 6y = -50 - 14x$

44. center: $(9, 9)$; radius: $\sqrt{109}$ units

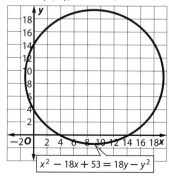
$x^2 - 18x + 53 = 18y - y^2$

45. center: $(1, -2)$; radius: $\sqrt{21}$

46. center: $(-2, 1)$; radius: $\sqrt{13}$

58. center: $(6, 8)$, radius: 4

$x^2 - 12x + 84 = -y^2 + 16y$

59. center: $\left(0, -\frac{9}{2}\right)$; radius: $\sqrt{19}$

$4x^2 + 4y^2 + 36y + 5 = 0$

60. center: $\left(-\sqrt{5}, 4\right)$; radius: 5

$(x + \sqrt{5})^2 + y^2 - 8y = 9$

61. center: $(-\sqrt{7}, \sqrt{11})$; radius: $\sqrt{11}$

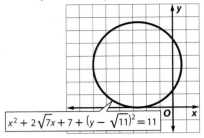

$x^2 + 2\sqrt{7}x + 7 + (y - \sqrt{11})^2 = 11$

67. Quadrant I
$a > 0, b > 0, a = b, r > 0$

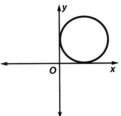

Quadrant II
$a < 0, b > 0, a = -b, r > 0$

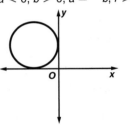

Quadrant III
$a < 0, b < 0, a = b, r > 0$

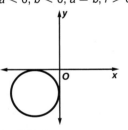

Quadrant IV
$a > 0, b < 0, a = -b, r > 0$

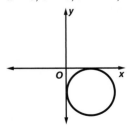

Sample answer: The circle is rotated 90° about the origin from one quadrant to the next.

72.

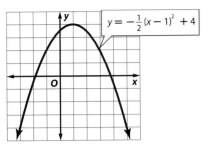

$y = -\frac{1}{2}(x - 1)^2 + 4$

73.

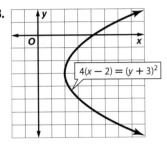

$4(x - 2) = (y + 3)^2$

74.

$(y - 8)^2 = -4(x - 4)$

Lesson 9-4

24. center $(3, 2)$; foci $(3, 11.59)$ and $(3, -7.59)$; major axis: ≈ 22.63; minor axis: 12

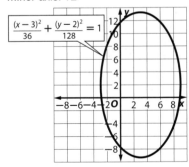

$\dfrac{(x - 3)^2}{36} + \dfrac{(y - 2)^2}{128} = 1$

25. center $(-6, 3)$; foci $(-6, 7.69)$ and $(-6, -1.69)$; major axis: ≈ 16.97; minor axis: ≈ 14.14

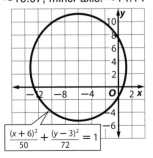

$\dfrac{(x + 6)^2}{50} + \dfrac{(y - 3)^2}{72} = 1$

26. center (0, 5); foci (0, 11.08) and (0, −1.08); major axis: 16; minor axis: ≈10.39

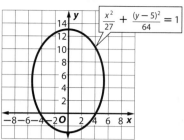

$$\frac{x^2}{27} + \frac{(y-5)^2}{64} = 1$$

27. center (−4, 0); foci (−4, 7.68) and (−4, −7.68); major axis: ≈17.32; minor axis: 8

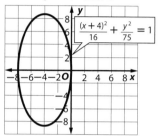

$$\frac{(x+4)^2}{16} + \frac{y^2}{75} = 1$$

28. center (1, 4); foci (1, 8) and (1, 0); major axis: ≈9.80; minor axis: ≈5.66

$$3x^2 + y^2 - 6x - 8y - 5 = 0$$

29. center (3, −3); foci (5.24, −3) and (0.76, −3); major axis: ≈8.94; minor axis: ≈7.75

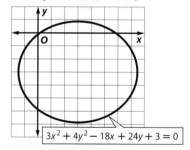

$$3x^2 + 4y^2 - 18x + 24y + 3 = 0$$

30. center (4, −3); foci (4, 1.90) and (4, −7.90); major axis: ≈10.58; minor axis: 4

$$7x^2 + y^2 - 56x + 6y + 93 = 0$$

31. center (−2, 5); foci (−2, 7.83) and (−2, 2.17); major axis: ≈9.80; minor axis: 8

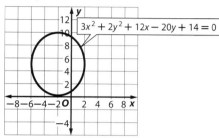

$$3x^2 + 2y^2 + 12x - 20y + 14 = 0$$

46. Sample answer: The domain is $\{x \mid -a \le x \le a\}$ because, if $|x| > a$, then $\frac{x^2}{a^2}$ would be greater than 1. This will force $\frac{y^2}{b^2}$ to be negative since $\frac{x^2}{a^2} + \frac{y^2}{b^2}$ must equal 1. In order for $\frac{y^2}{b^2}$ to be negative, either y^2 or b^2 must be negative, which cannot happen with real numbers. For any values of $\{x \mid -a \le x \le a\}$, $\frac{x^2}{a^2}$ will be between 0 and 1. The value of $\frac{y^2}{b^2}$ will also be between 0 and 1 for $\{y \mid -b \le y \le b\}$ and there are infinite combinations of $\frac{x^2}{a^2}$ and $\frac{y^2}{b^2}$ for which $\frac{x^2}{a^2} + \frac{y^2}{b^2} = 1$. Thus, the domain must be $\{x \mid -a \le x \le a\}$. The same method proves that the range must be $\{y \mid -b \le y \le b\}$.

Mid-Chapter Quiz

11. $(-3, -4)$; $\left(-3, -\frac{15}{4}\right)$; $x = -3$; $y = -\frac{17}{4}$; opens upward; 1 unit

12. $(3, 1)$; $\left(\frac{23}{8}, 1\right)$; $y = 1$; $x = \frac{25}{8}$; opens to the left; $\frac{1}{2}$ unit

13. $(1, 0)$; 3 units

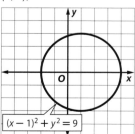

$$(x-1)^2 + y^2 = 9$$

17. $(-4, 2)$; $(-4 \pm \sqrt{7}, 2)$; 8; 6

$$\frac{(x+4)^2}{16} + \frac{(y-2)^2}{9} = 1$$

18. $(1, -2)$; $(5, -2)$, $(-3, -2)$; $4\sqrt{5}$; 4

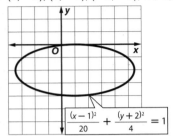

$$\frac{(x-1)^2}{20} + \frac{(y+2)^2}{4} = 1$$

19. $(5, -2)$; $(5, -2 \pm \sqrt{5})$; 6; 4

$$4y^2 + 9x^2 + 16y - 90x + 205 = 0$$

Lesson 9-5

14.

$y = \frac{1}{3}x$

$y = -\frac{1}{3}x$

15.

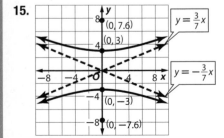

$y = \frac{3}{7}x$

$y = -\frac{3}{7}x$

16.

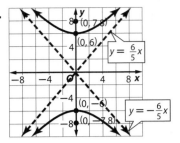

$y = \frac{6}{5}x$

$y = -\frac{6}{5}x$

17.

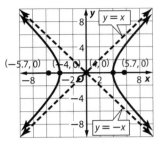

$y = x$

$y = -x$

18.

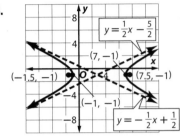

$y = \frac{1}{2}x - \frac{5}{2}$

$y = -\frac{1}{2}x + \frac{1}{2}$

19.

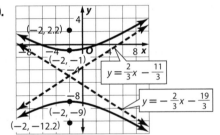

$y = \frac{2}{3}x - \frac{11}{3}$

$y = -\frac{2}{3}x - \frac{19}{3}$

20.

$y = -\frac{2}{3}x + \frac{17}{3}$

$y = \frac{2}{3}x + \frac{1}{3}$

21.

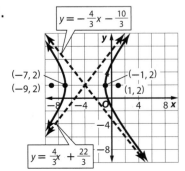

$y = -\frac{4}{3}x - \frac{10}{3}$

$y = \frac{4}{3}x + \frac{22}{3}$

22.

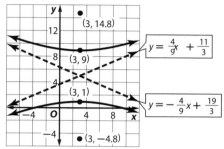

$y = \frac{5}{2}x + 1$

$(-3.4, 6)$ $(4, 6)$

$(0, 6)$ $(7.4, 6)$

$y = -\frac{5}{2}x + 11$

23.

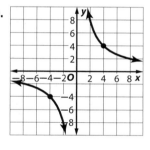

$(3, 14.8)$

$(3, 9)$

$y = \frac{4}{9}x + \frac{11}{3}$

$(3, 1)$

$y = -\frac{4}{9}x + \frac{19}{3}$

$(3, -4.8)$

33a.

x	y
−12	−1.33
−10	−1.6
−8	−2
−6	−2.67
−4	−4
−2	−8
0	undef
2	8
4	4
6	2.67
8	2
10	1.6
12	1.33

33b.

33c. The asymptotes are $y = 0$ and $x = 0$.

47. Sample answer: When 36 changes to 9, the vertical hyperbola widens (splits out from the y-axis faster). This is due to a smaller value of y being needed to produce the same value to x. The vertices are moved closer together due to the value of a decreasing from 6 to 3. The foci moved father from the vertices because the difference between c and a increased.

Lesson 9-6 (Guided Practice)

1. $\frac{(x-2)^2}{9} + \frac{(y+4)^2}{36} = 1$; ellipse

Lesson 9-6

1. $\frac{(x-3)^2}{36} + \frac{(y+2)^2}{9} = 1$; ellipse

2. $(x+6)^2 + (y-4)^2 = 16$; circle

3. $\frac{(y-1)^2}{16} - \frac{(x+2)^2}{9}$; hyperbola

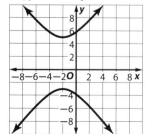

4. $x = 6(y - 2)^2 + 4$; parabola

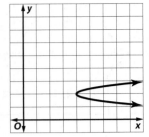

14. $\dfrac{(x + 3)^2}{18} - \dfrac{(y - 2)^2}{27} = 1$; hyperbola

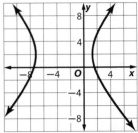

15. $\dfrac{(x + 4)^2}{32} + \dfrac{(y - 5)^2}{24} = 1$; ellipse

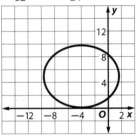

16. $x^2 + (y - 3)^2 = 25$; circle

17. $y = 8(x + 2)^2 - 4$; parabola

18. $\dfrac{(y - 2)^2}{42} + \dfrac{(x - 6)^2}{12} = 1$; ellipse

19. $(x - 3)^2 + (y + 4)^2 = 36$; circle

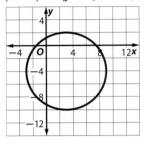

20. $x = -4(y - 3)^2 + 5$; parabola

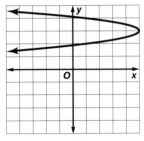

21. $\dfrac{(x + 4)^2}{21} - \dfrac{(y - 8)^2}{24} = 1$; hyperbola

22. $\dfrac{(y + 2)^2}{28} + \dfrac{(x - 1)^2}{9} = 1$; ellipse

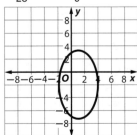

23. $\dfrac{(x+4)^2}{64} - \dfrac{(y-3)^2}{25} = 1$; hyperbola

47. Sample answer: An ellipse is a flattened circle. Both circles and ellipses are enclosed regions while hyperbolas and parabolas are not. A parabola has one branch, which is a smooth curve that never ends, and a hyperbola has two such branches that are reflections of each other. In standard form and when there is no *xy*-term: an equation for a parabola consists of only one squared term, an equation for a circle has values for A and C that are equal, an equation for an ellipse has values for A and C that are the same sign but not equal, and an equation for a hyperbola has values of A and C that have opposite signs.

56.

$f(x) = \dfrac{3}{x}$

57.

$f(x) = \dfrac{-2}{x+5}$

58.

$f(x) = \dfrac{6}{x-2} - 4$

Lesson 9-7 (Guided Practice)

3A.

3B.

4A.

4B.

Lesson 9-7

10.

11.

12.

13.

27.

28.

29.

30.

31.

32.

33.

34.

35.

36.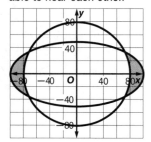

43. Sample answer: $\frac{x^2}{16} + \frac{y^2}{36} = 1$ and $(x + 10)^2 + y^2 = 36$

44. Sample answer: $y = x^2$ and $\frac{x^2}{16} + \frac{y^2}{36} = 1$

45. Sample answer: $x^2 + y^2 = 1$ and $\frac{x^2}{16} - \frac{y^2}{36} = 1$

46. Sample answer: $y = x^2$ and $\frac{x^2}{64} + \frac{(y - 4)^2}{16} = 1$

47. Sample answer: $\frac{x^2}{64} + \frac{y^2}{100} = 1$ and $x^2 - y^2 = 1$

49. Sample answer: No; if one player is in one of the shaded areas and the other player is in the other shaded area, they will not be able to hear each other.

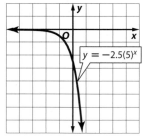

50. If we assume that Lena's office is located at the origin, then a system of equations describing this situation is $x^2 + y^2 < 100$ and $(x + 9.5)^2 + (y + 6)^2 \geq 225$.

52. 1, 2 or 3; the parabola could be tangent to the ellipse and have one solution. The parabola could intersect the ellipse at three points like this.

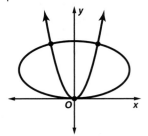

Or it could have two solutions like this.

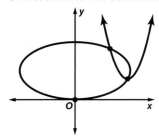

68. D = {all real numbers}, R = {$f(x) \mid f(x) < 0$}

$$f(x) = -\left(\frac{1}{5}\right)^x$$

69. D = {all real numbers}, R = {$y \mid y < 0$}

$$y = -2.5(5)^x$$

70. D = {all real numbers}, R = {$f(x) \mid f(x) > 0$}

Study Guide and Review

37. $(0, 0)$; $(0, \pm 3\sqrt{3})$; 12; 6

38. $(0, 0)$; $(0, \pm\sqrt{5})$; $2\sqrt{10}$; $2\sqrt{5}$

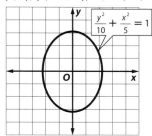

39. $(0, 4)$; $(\pm 4\sqrt{2}, 4)$; 12; 4

40. $(0, 0)$; $(0, \pm\sqrt{6})$; 6; $2\sqrt{3}$

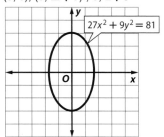

41. $(-1, 2)$; $(-4, 2), (2, 2)$; 10; 8

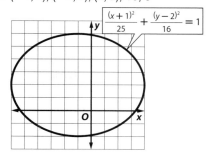

42. $(-3, 1)$; $(-3, 1 \pm \sqrt{5})$; 6; 4

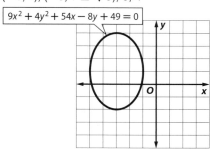

43. $(1, -1)$; $(-3, -1), (5, -1)$; 10; 6

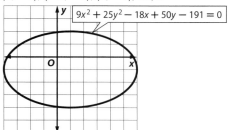

44. $(2, 2)$; $(2, 4), (2, 0)$; $2\sqrt{7}$; $2\sqrt{3}$

52. $y = 3(x + 2)^2 - 4$; parabola

53. $\dfrac{x^2}{16} + \dfrac{y^2}{9} = 1$; ellipse

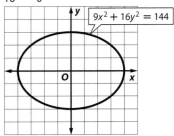

$9x^2 + 16y^2 = 144$

54. $(x - 4)^2 + (y - 1)^2 = 9$; circle

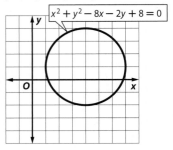

$x^2 + y^2 - 8x - 2y + 8 = 0$

55. $\dfrac{y^2}{9} - \dfrac{(x - 2)^2}{1} = 1$; hyperbola

$-9x^2 + y^2 + 36x - 45 = 0$

Practice Test

7. circle

$y^2 = 64 - x^2$

8. ellipse

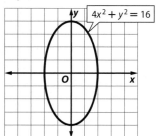

$4x^2 + y^2 = 16$

9. hyperbola

$4x^2 - 9y^2 + 8x + 36y = 68$

10. parabola

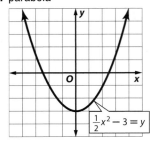

$\dfrac{1}{2}x^2 - 3 = y$

11. parabola

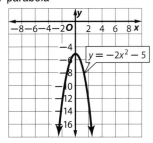

$y = -2x^2 - 5$

12. ellipse

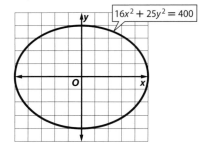

$16x^2 + 25y^2 = 400$

13. circle

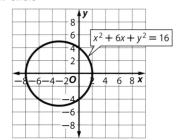

$x^2 + 6x + y^2 = 16$

14. hyperbola

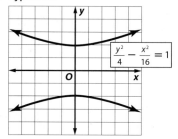

$$\frac{y^2}{4} - \frac{x^2}{16} = 1$$

15. parabola

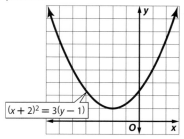

$(x + 2)^2 = 3(y - 1)$

16. ellipse

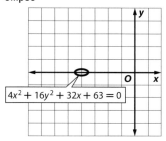

$4x^2 + 16y^2 + 32x + 63 = 0$

22.

23.

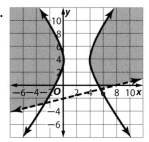

25a. $\dfrac{x^2}{1.1025} - \dfrac{y^2}{7.8975} = 1$

25b.

	Diagnostic Assessment Quick Check			
	LESSON 10-1 45 min: 1 day 90 min: 0.5 day	**LESSON 10-2** 45 min: 1.5 days 90 min: 0.75 day	**LESSON 10-3** 45 min: 1.5 days 90 min: 0.75 day	
Title	Sequences as Functions	Arithmetic Sequences and Series	Geometric Sequences and Series	
Objectives	▪ Relate arithmetic sequences to linear functions. ▪ Relate geometric sequences to exponential functions.	▪ Use arithmetic sequences. ▪ Find sums of arithmetic series.	▪ Use geometric sequences. ▪ Find sums of geometric series.	
Key Vocabulary	sequence term finite sequence infinite sequence arithmetic sequence common difference geometric sequence common ratio	arithmetic means series arithmetic series partial sum sigma notation	geometric means geometric series	
CCSS	F.IF.4	A.CED.4	A.SSE.4	
Multiple Representations		⚙		
Lesson Resources	connectED.mcgraw-hill.com 📁 Leveled Worksheets 🔤 Vocabulary 🏃 Animations PT Personal Tutor ✓ Self-Check Quiz ▪ *5-Minute Check* ▪ *Study Notebook* ▪ *Teaching Algebra with Manipulatives*	connectED.mcgraw-hill.com 📁 Leveled Worksheets 📁 Quiz 1 🔤 Vocabulary 🏃 Animations PT Personal Tutor ✓ Self-Check Quiz ▪ *5-Minute Check* ▪ *Study Notebook*	connectED.mcgraw-hill.com 📁 Leveled Worksheets 🔤 Vocabulary PT Personal Tutor ✓ Self-Check Quiz ▪ *5-Minute Check* ▪ *Study Notebook*	
Resources for Every Lesson	IWB eStudent Edition IWB Interactive Classroom	▪ eTeacher Edition ▪ eSolutions Manual ▪ *e*Assessment McGraw-Hill		
Differentiated Instruction	pp. 662, 665	pp. 668, 671	pp. 676, 680	

IWB All digital assets are Interactive Whiteboard ready.

	Suggested Pacing		
Time Periods	**Instruction**	**Review & Assess**	**Total**
45-minute	11 days	2 days	13 days
90-minute	6 days	1 day	7 days

EXPLORE **10-4**	45 min: 0.5 day 90 min: 0.25 day	LESSON **10-4**	45 min: 1 day 90 min: 0.5 day	EXTEND **10-4**	45 min: 0.5 day 90 min: 0.25 day	LESSON **10-5**	45 min: 1.5 days 90 min: 0.75 day
Algebra Lab: Area Under a Curve		**Infinite Geometric Series**		**Graphing Technology Lab: Limits**		**Recursion and Iteration**	
■ Approximate the area under a curve over a specified interval, using the sum of rectangular areas under the curve.		■ Find sums of infinite geometric series. ■ Write repeating decimals as fractions.		■ Use a graphing calculator to investigate limits of sequences.		■ Recognize and use special sequences. ■ Iterate functions.	
		convergent series divergent series infinity				Fibonacci sequence recursive sequence explicit formula recursive formula iteration	
connectED.mcgraw-hill.com 🖱 Virtual Manipulatives ■ *Teaching Algebra with Manipulatives* **Materials:** ■ grid paper		connectED.mcgraw-hill.com 📁 Leveled Worksheets 📁 Quiz 2 Vocabulary PT Personal Tutor ✓ Self-Check Quiz ■ *5-Minute Check* ■ *Study Notebook*		connectED.mcgraw-hill.com PT Personal Tutor **Materials:** ■ TI-83/84 Plus or other graphing calculator		connectED.mcgraw-hill.com 📁 Leveled Worksheets Vocabulary PT Personal Tutor ✓ Self-Check Quiz ■ *5-Minute Check* ■ *Study Notebook* ■ *Teaching Algebra with Manipulatives*	
IWB eStudent Edition IWB Interactive Classroom		■ eTeacher Edition ■ eSolutions Manual ■ McGraw-Hill eAssessment					
		pp. 685, 689				pp. 694, 697	
				Formative Assessment Mid-Chapter Quiz			

	EXTEND 10-5 45 min: 0.5 day / 90 min: 0.25 day	LESSON 10-6 45 min: 1.5 days / 90 min: 1 day	EXTEND 10-6 45 min: 0.5 day / 90 min: 0.25 day	LESSON 10-7 45 min: 1 day / 90 min: 0.75 day	
Title	Spreadsheet Lab: Amortizing Loans	The Binomial Theorem	Algebra Lab: Combinations and Pascal's Triangle	Proof by Mathematical Induction	
Objectives	▪ Use a spreadsheet to analyze the payments, interest, and balance on a loan.	▪ Use Pascal's triangle to expand powers of binomials. ▪ Use the Binomial Theorem to expand powers of binomials.	▪ Use combinations and Pascal's triangle to determine the number of ways the prizes of a game can be chosen.	▪ Prove statements by using mathematical induction. ▪ Disprove statements by finding a counterexample.	
Key Vocabulary		Pascal's triangle		mathematical induction induction hypothesis	
CCSS		A.APR.5	A.APR.5	A.SSE.1b, A.APR.5	
Multiple Representations					
Lesson Resources	connectED.mcgraw-hill.com 🖐 Virtual Manipulatives ▪ *Teaching Algebra with Manipulatives* **Materials:** ▪ computer and spreadsheet program	connectED.mcgraw-hill.com 📁 Leveled Worksheets 📁 Quiz 3 🔤 Vocabulary PT Personal Tutor ✓ Self-Check Quiz ▪ *5-Minute Check* ▪ *Study Notebook*	connectED.mcgraw-hill.com ▪ *Teaching Algebra with Manipulatives*	connectED.mcgraw-hill.com 📁 Leveled Worksheets 📁 Quiz 4 🔤 Vocabulary PT Personal Tutor ✓ Self-Check Quiz ▪ *5-Minute Check* ▪ *Study Notebook*	
Resources for Every Lesson	IWB eStudent Edition IWB Interactive Classroom	▪ eTeacher Edition ▪ eSolutions Manual ▪ e*Assessment* McGraw-Hill			
Differentiated Instruction		pp. 700, 703		pp. 706, 709	

Summative Assessment
Study Guide and Review
Practice Test

IWB All digital assets are Interactive Whiteboard ready.

What the Research Says . . .

Stohl Drier (2001) states that interactive spreadsheets can promote open-ended exploration of mathematical concepts, allow learners to extend beyond or significantly enhance what could be done using paper-and-pencil, and give teachers and students an opportunity to discover mathematical concepts in a laboratory-like setting.

- In Extend 10-5, students use a computer spreadsheet to investigate sequences involved in loan amortization.

- Suggest that students use a spreadsheet to explore answers to some of the financially-oriented exercises in Lesson 10-5, such as Exercises 7, 32, and 44.

Teacher to Teacher

Holly K. Plunkett
University High School, Morgantown, WV

 I have students investigate geometric sequences by bouncing rubber balls and using a CBL.

Project CRISS

STUDY SKILL

Students have used Venn diagrams before. Using Venn diagrams to compare and contrast concepts may be helpful as they assimilate new knowledge. Encourage students to look for the relationships between math concepts that are taught together. These relationships can be represented using Venn diagrams. Students can brainstorm ideas for Venn diagrams individually or as a cooperative learning activity. This Venn diagram represents the concepts of arithmetic and geometric sequences that students study in Lessons 10-2 and 10-3.

Arithmetic Sequences		Geometric Sequences
Differences	Similarities	Differences

Terms have a common difference To find the next term, add the common difference. The equation is linear.

list of numbers, contains terms, a discrete function

Terms have a common ratio To find the next term, multiply by the common ratio. The equation is exponential.

Creating Independence through Student-owned Strategies

Assessment and Intervention

SE = Student Edition, TE = Teacher Edition, CRM = Chapter Resource Masters

Diagnosis	Prescription
Beginning Chapter 10	
Get Ready for Chapter 10 **SE**	Response to Intervention **TE**
Beginning Every Lesson	
Then, Now, Why? **SE** 5-Minute Checks	Chapter 0 **SE**

Diagnosis	Prescription
During/After Every Lesson	
Guided Practice **SE**, every example Check Your Understanding **SE** H.O.T. Problems **SE** Spiral Review **SE** Additional Examples **TE** Watch Out! **TE** Step 4, Assess **TE** Chapter 10 Quizzes **CRM**, pp. 53–54 Self-Check Quizzes connectED.mcgraw-hill.com	**TIER 1 Intervention** Skills Practice **CRM**, Ch. 1–10 connectED.mcgraw-hill.com **TIER 2 Intervention** Differentiated Instruction **TE** Study Guide and Intervention **CRM**, Ch. 1–10 **TIER 3 Intervention** Math Triumphs, Alg. 2
Mid-Chapter	
Mid-Chapter Quiz **SE** Mid-Chapter Test **CRM**, p. 55 *e*Assessment McGraw-Hill	**TIER 1 Intervention** Skills Practice **CRM**, Ch. 1–10 connectED.mcgraw-hill.com **TIER 2 Intervention** Study Guide and Intervention **CRM**, Ch. 1–10 **TIER 3 Intervention** Math Triumphs, Alg. 2
Before Chapter Test	
Chapter Study Guide and Review **SE** Practice Test **SE** Standardized Test Practice **SE** Chapter Test connectED.mcgraw-hill.com Standardized Test Practice connectED.mcgraw-hill.com Vocabulary Review connectED.mcgraw-hill.com *e*Assessment McGraw-Hill	**TIER 1 Intervention** Skills Practice **CRM**, Ch. 1–10 connectED.mcgraw-hill.com **TIER 2 Intervention** Study Guide and Intervention **CRM**, Ch. 1–10 **TIER 3 Intervention** Math Triumphs, Alg. 2

Diagnosis	Prescription
After Chapter 10	
Multiple-Choice Tests, Forms 1, 2A, 2B **CRM**, pp. 57–62 Free-Response Tests, Forms 2C, 2D, 3 **CRM**, pp. 63–68 Vocabulary Test **CRM**, p. 56 Extended Response Test **CRM**, p. 69 Standardized Test Practice **CRM**, pp. 70–72 *e*Assessment McGraw-Hill	Study Guide and Intervention **CRM**, Ch. 1–10 connectED.mcgraw-hill.com

Sidebar labels: DIAGNOSTIC ASSESSMENT — FORMATIVE ASSESSMENT — SUMMATIVE ASSESSMENT

Option 1 Reaching All Learners

Logical Have students make a chart that compares and contrasts arithmetic and geometric sequences and series, explaining what the variables represent in each formula.

Interpersonal Create a friendly competition between two groups of students, the *Binomial Theorem Expanders* and the *Polynomial Multipliers.* Write an expression such as $(2x + y)^5$ on the board. Have each group expand the expression using the method for which their group is named. Compare the expanded expressions and the times the groups take to find them.

Option 2 Approaching Level AL

To explain conceptually how mathematical induction works, set up a series of 5 evenly spaced dominoes in a line, one behind the other. Knock the first domino over. Ask students to describe what they observe. Tell students that you think, based on what was just observed, that the 29th domino will get knocked over if you knock over the first domino. Set up 30 dominos to test and prove your assumption is true. Then ask the class what they think would happen to the 500th domino. Tell students that they just used inductive reasoning.

Explain that when they say that a mathematical statement is true for $n = 1$, it is the same as proving the first domino can be knocked down. Then proving it is true for $k + 1$ if it is true for k, means that you have proved that if any domino falls, the next one in line will fall. So, if #1 falls, #2 will fall, and if #2 falls, then #3 will fall, and so on.

Option 3 English Learners ELL

Have students create a list of the different mathematical notations (symbols) as they are presented in this chapter. As each notation is presented, ask students to record the notation, an example, its name, and why it is useful in their study of mathematics.

For example,

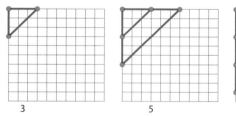

Notation	Name	Example	why it is useful
...	ellipsis	2, 4, 6, ...	indicates that a series is infinite
$\sum$	sigma	$\sum\limits_{k=1}^{\infty} 3k$	represents the sum of a series

Option 4 Beyond Level BL

With manipulatives or sketches, students use various geometric elements to model problems involving arithmetic or geometric sequences. Then have students write a general rule for the nth term of their sequence.

For example, the figures below show an arithmetic sequence in the number of red dots. The rule for the sequence is $a_n = 3 + 2(n - 1)$.

VerticalAlignment

Before Chapter 10

Related Topics before Algebra 1

- find and evaluate an algebraic expression to determine any term of an arithmetic sequence

Related Topics from Algebra 1

- look for patterns and represent generalizations algebraically
- transform and solve equations
- simplify polynomial expressions

Related Topics from Geometry

- use inductive reasoning to formulate a conjecture

Chapter 10

Related Topics from Algebra 2

- relate arithmetic sequences to linear functions and relate geometric sequences to exponential functions
- find specific terms and sums of arithmetic and geometric series
- find the sum of an infinite geometric series and write repeating decimals as fractions
- recognize and use special sequences and iterate functions
- use Pascal's triangle and the Binomial Theorem to expand powers of binomials
- use mathematical induction to prove statements

After Chapter 10

Preparation for Precalculus

- represent patterns using arithmetic and geometric sequences and series
- use arithmetic, geometric, and other sequences and series to solve real-life problems
- describe limits of sequences and apply their properties to investigate convergent and divergent series
- apply sequences and series to solve problems including sums and binomial expansion

Lesson-by-LessonPreview

10-1 Sequences as Functions

A set of numbers arranged in a particular order is called a sequence. Each number is a term of the sequence. A sequence is

- *finite* when there are a limited number of terms, or
- *infinite* when the terms continue without end.

A sequence is

- *arithmetic* when the terms are determined by adding a constant value (a common difference) to the previous term,
- *geometric* when each term is determined by multiplying a nonzero constant (the common ratio) by the previous term, or
- *neither* arithmetic nor geometric when there is no common difference or ratio.

Sequences are functions whose domains consist of natural numbers and whose ranges consist of real numbers. Arithmetic sequences are linear. Geometric sequences are exponential.

10-2 Arithmetic Sequences and Series

Any term a_n in an arithmetic sequence can be determined when you know the first term a_1 and the common difference d. The formula to use is $a_n = a_1 + (n - 1)d$, where n is any natural number.

An *arithmetic series* is formed when the terms of an arithmetic sequence are added. The formulas $S_n = n\left(\dfrac{a_1 + a_n}{2}\right)$ or $S_n = \dfrac{n}{2}[2a_1 + (n - 1)d]$ can be used to find the sum of the first n terms of any arithmetic series. Sigma notation can be used to indicate the sum of a series. For example, $\displaystyle\sum_{n=1}^{3}(n^2 + 2)$ represents a sum of three terms in which n is successively replaced by each of the values 1, 2, and 3.

10-3 Geometric Sequences and Series

The formula $a_n = a_1 \cdot r^{n-1}$ can be used to find any term a_n in a geometric sequence when you know the first term a_1 and the common ratio r.

A *geometric series* is formed when the terms of a geometric sequence are added. The formulas $S_n = \dfrac{a_1 + a_1 r^n}{1 - r}, r \neq 1$ or $S_n = \dfrac{a_1 + a_n r}{1 - r}, r \neq 1$ can be used to find the sum of the first n terms of a geometric sequence. Sigma notation can be used to represent the sum of a geometric series.

10-4 Infinite Geometric Series

If the sum of an infinite geometric series

- approaches a finite value, it is called a *convergent series* and $|r| < 1$. The formula $S = \dfrac{a_1}{1-r}$ can be used to find its sum.

- does not approach a finite value, it is called a *divergent series* and $|r| \geq 1$.

The formula for the sum of an infinite series can be used to convert a repeating decimal to a fraction as shown below.

$$0.353535... = 0.35 + 0.0035 + 0.000035 + ...$$

Common ratio: $r = \dfrac{0.0035}{0.35} = 0.01$

$$S = \dfrac{a_1}{1-r} = \dfrac{0.35}{1-0.01} = \dfrac{0.35}{0.99} = \dfrac{35}{99}$$

10-5 Recursion and Iteration

The formulas for terms of sequences that were investigated in previous lessons are called *explicit formulas*. They give a_n as a function of n. In *recursive formula*, terms are determined by a formula using one or more of the previous terms. For example, a recursive formula for the sequence 3, 7, 11, 15, 19, ... is

$$a_1 = 3, a_n = a_{n-1} + 4$$

Iteration is a process of repeatedly composing a function with itself. This process can be used to generate a sequence recursively. For example, how to find the first three iterates x_1, x_2, and x_3 of $f(x) = 2x + 1$ for an initial value $x_0 = 2$ is shown below.

$f(x) = 2x + 1$ when $x_0 = 2$

$x_1 = f(x_0)$	$x_2 = f(x_1)$	$x_3 = f(x_2)$
$= 2(2) + 1 = 5$	$= 2(5) + 1 = 11$	$= 2(11) + 1 = 23$

The first three iterates are 5, 11, and 23.

10-6 The Binomial Theorem

The pattern found in Pascal's triangle (shown below) can be used to determine the coefficients of an expanded binomial $(a + b)^n$.

$(a + b)^0$ 1

$(a + b)^1$ 1   1

$(a + b)^2$ 1   2   1

$(a + b)^3$ 1   3   3   1

$(a + b)^4$ 1   4   6   4   1

$(a + b)^5$ 1   5   10   10   5   1

Expanding the binomial $(a + b)^n$ for nonnegative values of n
requires finding the coefficient and the exponents for a and b in each term. Binomial expansions, such as in $(a + b)^4 = a^4 + 4a^3b + 6a^2b^2 + 4ab^3 + b^4$, contain many patterns.

- The sum of the exponents in each term is n.

- The exponent of a in the first term and b in the last term is n.

- The coefficients are the entries in Pascal's triangle.

- There are $n + 1$ terms.

The Binomial Theorem can be used to expand a binomial.

Written in sigma notation, $(a + b)^n = \displaystyle\sum_{k=0}^{n} \dfrac{n!}{k!\,(n-k)!} a^{n-k}b^k$.

10-7 Proof by Mathematical Induction

Mathematical induction is a powerful method of proving conjectures. It consists of two parts.

- First, a property is shown to be true for a particular number (often the number 1).

- The second, *inductive,* part involves demonstrating that if the property is true for *some* positive integer, it must be true for the *next* integer. As a result, the property is shown to be true for *all* integers, beginning with a certain value. Properties of series, divisibility, and proving a formula to be false by finding a counterexample are all examined in this lesson.

Mc Graw Hill Professional Development

Targeted professional development has been articulated throughout Algebra 2. More quality, customized professional development is available from McGraw-Hill Professional Development. Visit connectED.mcgraw-hill.com for details on each product.

- Online Lessons emphasize the strategies and techniques used to teach Algebra 2. Includes streaming video, interactive pages, and online tools.

- Video Workshops allow mentors, coaches, or leadership personnel to facilitate on-site workshops on educational strategies in mathematics and mathematical concepts.

- MHPD Online (www.mhpdonline.com) offers online professional development with video clips of instructional strategies, links, student activities, and news and issues in education.

- Teaching Today (teachingtoday.glencoe.com) gives secondary teachers practical strategies and materials that inspire excellence and innovation in teaching.

Chapter Project

Big Money

Students use what they have learned about sequences and series to complete a project.

This chapter project addresses financial literacy, as well as several specific skills identified as being essential to student success by the Framework for 21st Century Learning.

Visit connectED.mcgraw-hill.com for student and teacher handouts.

KeyVocabulary Introduce the key vocabulary in the chapter using the routine below.

<u>Define</u>: A term is any number in a sequence.

<u>Example</u>: The fourth term in the following sequence is 29. 2, 11, 20, 29, 39, . . .

<u>Ask</u>: What is the first term in the sequence above? 2

CHAPTER 10 Sequences and Series

Then

○ You simplified and evaluated algebraic expressions.

Now

○ You will:
- Use arithmetic and geometric sequences and series.
- Use special sequences and iterate functions.
- Expand powers by using the Binomial Theorem.
- Prove statements by using mathematical induction.

Why? ▲

○ **CONSERVATION AND NATURE** Mathematics occurs in aspects of nature in astonishing ways. The Fibonacci sequence manifests itself in seeds, flowers, pine cones, fruits, and vegetables. Sequences and series can further help us conserve our natural resources by making water filtration systems more efficient.

connectED.mcgraw-hill.com **Your Digital Math Portal**

Animation | Vocabulary | eGlossary | Personal Tutor | Virtual Manipulatives | Graphing Calculator | Audio | Foldables | Self-Check Practice | Worksheets

Martin Child/Photodisc/Getty Images

Additional Answers (Get Ready for Chapter 10)

6.

7.

8.

9.

Get Ready for the Chapter

Diagnose Readiness | You have two options for checking prerequisite skills.

1 Textbook Option Take the Quick Check below. Refer to the Quick Review for help.

QuickCheck	QuickReview

Solve each equation.

1. $-6 = 7x + 78$ $x = -12$

2. $768 = 3x^4$ $x = \pm 4$

3. $23 - 5x = 8$ $x = 3$

4. $2x^3 + 4 = -50$ $x = -3$

5. **PLANTS** Lauri has 48 plants for her two gardens. She plants 12 in the small garden. In the other garden she wants 4 plants in each row. How many rows will she have? **9 rows**

Example 1 (Used in Lessons 10-1 through 10-3)

Solve $25 = 3x^3 + 400$.

$25 = 3x^3 + 400$ Original equation

$-375 = 3x^3$ Subtract 400 from each side.

$-125 = x^3$ Divide each side by 3.

$\sqrt[3]{-125} = \sqrt[3]{x^3}$ Take the cube root of each side.

$-5 = x$ Simplify.

Graph each function. 6–9. See margin.

6. $\{(1, 3), (2, 5), (3, 7), (4, 9), (5, 11)\}$

7. $\{(1, -15), (2, -12), (3, -9), (4, -6), (5, -3)\}$

8. $\left\{(1, 27), (2, 9), (3, 3), (4, 1), \left(5, \frac{1}{3}\right)\right\}$

9. $\left\{(1, 1), (2, 2), \left(3, \frac{5}{2}\right), \left(4, \frac{11}{4}\right), \left(5, \frac{23}{8}\right)\right\}$

10. **DAY CARE** A child care center has expenses of $125 per day. They charge $50 per child per day. The function $P(c) = 50c - 125$ gives the amount of money the center makes when there are c children there. How much will they make if there are 8 children? **$275**

Example 2 (Used in Lessons 10-1 through 10-5)

Graph the function $\{(1, 1), (2, 4), (3, 9), (4, 16), (5, 25)\}$. State the domain and range.

The domain of a function is the set of all possible x-values. So, the domain of the function is $\{1, 2, 3, 4, 5\}$. The range of a function is the set of all possible y-values. So, the range of this function is $\{1, 4, 9, 16, 25\}$.

Evaluate each expression for the given value(s) of the variable(s).

11. $\frac{a}{3}(b + c)$ if $a = 9$, $b = -2$, and $c = -8$ -30

12. $r + (n - 2)t$ if $r = 15$, $n = 5$, and $t = -1$ 12

13. $x \cdot y^{z + 1}$ if $x = -2$, $y = \frac{1}{3}$, and $z = 5$ $-\frac{2}{729}$

14. $\frac{a(1 - bc)^2}{1 - b}$ if $a = -3$, $b = -4$, and $c = 1$ -15

Example 3 (Used in Lessons 10-1 through 10-5, and 10-8)

Evaluate $2 \cdot 3^{x + y}$ if $x = -2$ and $y = -3$.

$2 \cdot 3^{x + y} = 2 \cdot 3^{-2 + (-3)}$ Substitute.

$= 2 \cdot 3^{-5}$ Simplify.

$= \frac{2}{3^5}$ Rewrite with positive exponent.

$= \frac{2}{243}$ Evaluate the power.

2 Online Option Take an online self-check Chapter Readiness Quiz at connectED.mcgraw-hill.com.

657

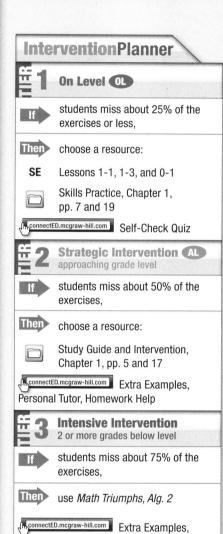

Get Ready for the Chapter

Response to Intervention (RtI)
Use the *Quick Check* results and the Intervention Planner to help you determine your Response to Intervention. The If-Then statements in the chart below help you decide the appropriate tier of RtI and suggest intervention resources for each tier.

InterventionPlanner

TIER 1 **On Level** **OL**

If students miss about 25% of the exercises or less,

Then choose a resource:

SE Lessons 1-1, 1-3, and 0-1

Skills Practice, Chapter 1, pp. 7 and 19

connectED.mcgraw-hill.com Self-Check Quiz

TIER 2 **Strategic Intervention** **AL**
approaching grade level

If students miss about 50% of the exercises,

Then choose a resource:

Study Guide and Intervention, Chapter 1, pp. 5 and 17

connectED.mcgraw-hill.com Extra Examples, Personal Tutor, Homework Help

TIER 3 **Intensive Intervention**
2 or more grades below level

If students miss about 75% of the exercises,

Then use *Math Triumphs, Alg. 2*

connectED.mcgraw-hill.com Extra Examples, Personal Tutor, Homework Help, Review Vocabulary

---

Essential Questions

- Where are patterns found in the real world? Sample answer: in nature, architecture, music, science, art

- How can recognizing patterns help you solve real-world problems? Sample answer: Recognizing a pattern can help you to predict future behavior.

Dinah Zike's Foldables®

Focus Students record notes and examples about sequences and series on index cards and place them in the appropriate pocket of their Foldables.

Teach Have students make and label their Foldables as illustrated. Students should use their Foldables to write notes, define terms, record concepts, and give examples of sequences and series on index cards. The cards should then be placed in the appropriate pocket.

When to Use It Encourage students to add to their Foldables as they work through the chapter and to use them to review for the chapter test.

Differentiated Instruction

☐ Student-Built Glossary, pp. 1–2

Students should complete the chart by providing a definition of each term and an example as they progress through Chapter 10. This study tool can also be used to review for the chapter test.

Get Started on the Chapter

You will learn several new concepts, skills, and vocabulary terms as you study Chapter 10. To get ready, identify important terms and organize your resources. You may wish to refer to Chapter 0 to review prerequisite skills.

FOLDABLES StudyOrganizer

Sequences and Series Make this Foldable to help you organize your Chapter 10 notes about sequences and series. Begin with one $8\frac{1}{2}''$ by $11''$ sheet of paper.

1 **Fold** in half, matching the short sides.

2 **Unfold** and fold the long side up 2 inches to form a pocket.

3 **Staple** or glue the outer edges to complete the pocket.

4 **Label** each side as shown. Use index cards to record notes and examples.

Sequences | Series

NewVocabulary

English		Español
sequence	p. 659	sucesión
finite sequence	p. 659	sucesión finita
infinite sequence	p. 659	sucesión infinita
arithmetic sequence	p. 659	sucesión aritmética
common difference	p. 659	diferencia común
geometric sequence	p. 661	sucesión geométrica
common ratio	p. 661	razón común
arithmetic means	p. 667	media aritmética
series	p. 668	serie
arithmetic series	p. 668	serie aritmética
partial sum	p. 668	suma parcial
geometric means	p. 675	media geométrica
geometric series	p. 676	serie geométrica
convergent series	p. 683	serie convergente
divergent series	p. 683	serie divergente
recursive sequence	p. 692	sucesión recursiva
iteration	p. 694	iteración
mathematical induction	p. 705	inducción matemática
induction hypothesis	p. 705	hipótesis inductiva

ReviewVocabulary

coefficient coeficiante **the numerical factor of a monomial**

coefficient
coeficiante

$$15x^3$$

formula fórmula **a mathematical sentence that expresses the relationship between certain quantities**

function función **a relation in which each element of the domain is paired with exactly one element in the range**

LESSON 10-1
Sequences as Functions

:: Then	:: Now	:: Why?
● You analyzed linear and exponential functions.	**1** Relate arithmetic sequences to linear functions. **2** Relate geometric sequences to exponential functions.	● During their routine, a high school marching band marches in rows. There is one performer in the first row, three performers in the next row, and five in the third row. This pattern continues for the rest of the rows.

NewVocabulary
sequence
term
finite sequence
infinite sequence
arithmetic sequence
common difference
geometric sequence
common ratio

Common Core State Standards

Content Standards
F.IF.4 For a function that models a relationship between two quantities, interpret key features of graphs and tables in terms of the quantities, and sketch graphs showing key features given a verbal description of the relationship.

Mathematical Practices
2 Reason abstractly and quantitatively.
7 Look for and make use of structure.

ThinkStock/SuperStock

1 Arithmetic Sequences A **sequence** is a set of numbers in a particular order or pattern. Each number in a sequence is called a **term**. A sequence may be a **finite sequence** containing a limited number of terms, such as {−2, 0, 2, 4, 6}, or an **infinite sequence** that continues without end, such as {0, 1, 2, 3, …}. The first term of a sequence is denoted a_1, the second term is denoted a_2, and so on.

KeyConcept Sequences as Functions

Words	A sequence is a function in which the domain consists of natural numbers, and the range consists of real numbers.
Symbols	Domain: 1 2 3 … n the position of a term Range: a_1 a_2 a_3 … a_n the terms of the sequence
Examples	Finite Sequence Infinite Sequence {3, 6, 9, 12, 15} {3, 6, 9, 12, 15, …} Domain: {1, 2, 3, 4, 5} Domain: {all natural numbers} Range: {3, 6, 9, 12, 15} Range: {$y \mid y$ is a multiple of 3, $y \geq 3$}

In an **arithmetic sequence**, each term is determined by adding a constant value to the previous term. This constant value is called the **common difference**.

Consider the sequence 3, 6, 9, 12, 15. This sequence is arithmetic because the terms share a common difference. Each term is 3 more than the previous term.

3   6   9   12   15
   +3   +3   +3   +3

Example 1 Identify Arithmetic Sequences

Determine whether each sequence is arithmetic.

a. 5, −6, −17, −28, …

5   −6   −17   −28
  −11   −11   −11

The common difference is −11.
The sequence is arithmetic.

b. −4, 12, 28, 42, …

−4   12   28   42
  +16   +16   +14

There is no common difference.
This is not an arithmetic sequence.

▶ **GuidedPractice**

1A. 7, 12, 16, 20, …   no

1B. −6, 3, 12, 21, …   yes

connectED.mcgraw-hill.com   **659**

---

1 Focus

VerticalAlignment

Before Lesson 10-1 Analyze linear and exponential functions.

Lesson 10-1 Relate arithmetic sequences to linear functions. Relate geometric sequences to exponential functions.

After Lesson 10-1 Use arithmetic, geometric, and other sequences and series to solve real-life problems.

2 Teach

Scaffolding Questions
Have students read the **Why?** section of the lesson.

Ask:

■ What pattern do the rows of marching performers make? 1, 3, 5, 7, . . .

■ How can you find the next four numbers in the pattern? Add 2 to each successive row to find the number of performers in that row.

■ What are the next four numbers? 9, 11, 13, 15

■ Would you expect this pattern to continue infinitely? No, there is a limit to the number of rows of performers.

Lesson 10-1 Resources

Resource	Approaching Level **AL**	On Level **OL**	Beyond Level **BL**	English Learners **ELL**
Teacher Edition	■ Differentiated Instruction, p. 662	■ Differentiated Instruction, p. 665	■ Differentiated Instruction, p. 665	■ Differentiated Instruction, p. 662
Chapter Resource Masters	■ Study Guide and Intervention, pp. 5–6 ■ Skills Practice, p. 7 ■ Practice, p. 8 ■ Word Problem Practice, p. 9	■ Study Guide and Intervention, pp. 5–6 ■ Skills Practice, p. 7 ■ Practice, p. 8 ■ Word Problem Practice, p. 9 ■ Enrichment, p. 10	■ Practice, p. 8 ■ Word Problem Practice, p. 9 ■ Enrichment, p. 10	■ Study Guide and Intervention, pp. 5–6 ■ Skills Practice, p. 7 ■ Practice, p. 8 ■ Word Problem Practice, p. 9
Other	■ 5-Minute Check 10-1 ■ Study Notebook ■ Teaching Algebra with Manipulatives	■ 5-Minute Check 10-1 ■ Study Notebook ■ Teaching Algebra with Manipulatives	■ 5-Minute Check 10-1 ■ Study Notebook	■ 5-Minute Check 10-1 ■ Study Notebook ■ Teaching Algebra with Manipulatives

PhotoLink/Photodisc/Getty Images

1 Arithmetic Sequences

Example 1 shows how to determine if there is a common difference for a sequence and, if there is one, how to identify the sequence as arithmetic. **Example 2** shows how to use the common difference to find terms of an arithmetic sequence and how to graph the terms of the sequence. **Example 3** shows how to find a term of a given arithmetic sequence to solve a real-world problem.

Formative Assessment

Use the Guided Practice exercises after each example to determine students' understanding of concepts.

Additional Examples

1 Determine whether each sequence is arithmetic. Write *yes* or *no*.
  a. $-3, -8, -13, -23, \ldots$ no
  b. $-8, -2, 4, 10, \ldots$ yes

2 Consider the arithmetic sequence $-8, -6, -4, \ldots$
  a. Find the next four terms of the sequence. $-2, 0, 2, 4$
  b. Graph the first seven terms of the sequence.

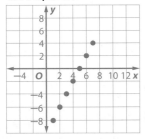

3 **MARCHING BAND** Use the information in Example 3 to determine how many performers will be in the 20th row during the routine. 39

▶ **Additional Examples** also in Interactive Classroom PowerPoint® Presentations

IWB **Interactive White Board** READY

You can use the common difference to find terms of an arithmetic sequence.

Example 2 Graph an Arithmetic Sequence

Consider the arithmetic sequence $18, 14, 10, \ldots$.

a. Find the next four terms of the sequence.

  Step 1 To determine the common difference, subtract any term from the term directly after it. The common difference is $10 - 14$ or -4.

  Step 2 To find the next term, add -4 to the last term.

  Continue to add -4 to find the following terms.

$$10 \quad 6 \quad 2 \quad -2 \quad -6$$
$$+(-4) \quad +(-4) \quad +(-4) \quad +(-4)$$

  The next four terms are $6, 2, -2$, and -6.

b. Graph the first seven terms of the sequence.
  The domain contains the terms $\{1, 2, 3, 4, 5, 6, 7\}$ and the range contains the terms $\{18, 14, 10, 6, 2, -2, -6\}$. So, graph the corresponding ordered pairs.

▶ **Guided**Practice

2. Find the next four terms of the arithmetic sequence $18, 11, 4, \ldots$. Then graph the first seven terms. **See margin.**

Notice that the graph of the terms of the arithmetic sequence lie on a line. An arithmetic sequence is a linear function in which the term number n is the independent variable, the term a_n is the dependent variable, and the common difference is the slope.

Real-World Example 3 Find a Term

MARCHING BANDS Refer to the beginning of the lesson. Suppose the director wants to determine how many performers will be in the 14th row during the routine.

Understand Because the difference between any two consecutive rows is 2, the common difference for the sequence is 2.

Plan Use point-slope form to write an equation for the sequence. Let $m = 2$ and $(x_1, y_1) = (3, 5)$. Then solve for $x = 14$.

Solve
$(y - y_1) = m(x - x_1)$	Point-slope form
$(y - 5) = 2(x - 3)$	$m = 2$ and $(x_1, y_1) = (3, 5)$
$y - 5 = 2x - 6$	Multiply.
$y = 2x - 1$	Add 5 to each side.
$y = 2(14) - 1$	Replace x with 14.
$y = 28 - 1$ or 27	Simplify.

Check You can find the terms of the sequence by adding 2, starting with row 1, until you reach row 14.

▶ **Guided**Practice

3. **MONEY** Geraldo's employer offers him a pay rate of $9 per hour with a $0.15 raise every three months. How much will Geraldo earn per hour after 3 years? **$10.80**

Real-World Link

Each year, about 100 bands compete in the Bands of America Grand National Championships.

Source: Bands of America

WatchOut!

Preventing Misconceptions Make sure all students understand that the subscript in a_n refers to a term, and that it is not an exponent.

Additional Answer (Guided Practice)

2. $-3, -10, -17, -24$

---

2 Geometric Sequences

Geometric Sequences Another type of sequence is a geometric sequence. In a **geometric sequence**, each term is determined by multiplying a nonzero constant by the previous term. This constant value is called the **common ratio**.

Consider the sequence $\frac{1}{16}$, $\frac{1}{4}$, 1, 4, 16. This sequence is geometric because the terms share a common ratio. Each term is 4 times as much as the previous term.

$$\frac{1}{16} \quad \frac{1}{4} \quad 1 \quad 4 \quad 16$$
$$\times 4 \quad \times 4 \quad \times 4 \quad \times 4$$

Example 4 Identify Geometric Sequences

Determine whether each sequence is geometric.

a. −2, 6, −18, 54, …

Find the ratios of the consecutive terms.

$$\frac{6}{-2} = -3 \qquad \frac{-18}{6} = -3 \qquad \frac{54}{-18} = -3$$

The ratios are the same, so the sequence is geometric.

b. 8, 16, 24, 32, …

$$\frac{16}{8} = 2 \qquad \frac{24}{16} = 1.5 \qquad \frac{32}{24} = 1.\overline{3}$$

The ratios are not the same, so the sequence is not geometric.

> **WatchOut!**
> **Ratios** If you find the ratio of a term to the previous term, set up the remaining ratios the same way.

GuidedPractice

4A. −8, 2, −0.5, 0.125, … **yes**   **4B.** 1, 3, 7, 15, … **no**

When given a set of information, you can create a problem that relates a story.

Example 5 Graph a Geometric Sequence

Consider the geometric sequence 32, 8, 2, … .

a. Find the next three terms of the sequence.

Step 1 Find the value of the common ratio: $\frac{2}{8}$ or $\frac{1}{4}$.

Step 2 To find the next term, multiply the previous term by $\frac{1}{4}$.

Continue multiplying by $\frac{1}{4}$ to find the following terms.

$$2 \quad \frac{1}{2} \quad \frac{1}{8} \quad \frac{1}{32}$$
$$\times \frac{1}{4} \quad \times \frac{1}{4} \quad \times \frac{1}{4}$$

The next three terms are $\frac{1}{2}$, $\frac{1}{8}$, and $\frac{1}{32}$.

b. Graph the first six terms of the sequence.

Domain: {1, 2, 3, 4, 5, 6}

Range: $\left\{ 32, 8, 2, \frac{1}{2}, \frac{1}{8}, \frac{1}{32} \right\}$

GuidedPractice

5. Find the next two terms of 7, 21, 63, … .
Then graph the first five terms. **See margin.**

Example 4 shows how to determine if a sequence has a common ratio, and, if it does, how to identify the sequence as geometric. **Example 5** shows how to use the common ratio to determine more terms of a geometric sequence and how to graph the terms of the sequence. **Example 6** shows how to determine whether a sequence is arithmetic, geometric, or neither.

Additional Examples

4 Determine whether each sequence is geometric. Write *yes* or *no*.

a. 8, 20, 50, 125, … yes

b. 19, 30, 41, 52, … no

5 Consider the geometric sequence 10, 15, 22.5, … .

a. Find the next three terms of the sequence. 33.75, 50.625, 75.9375

b. Graph the first six terms of the sequence.

Additional Answer (Guided Practice)

5. 189, 567

Follow-up

Students have explored sequences.

Ask:

- What types of patterns can be modeled mathematically? Sample answer: numerical patterns involving real number operations such as addition and multiplication

Additional Example

6 Determine whether each sequence is *arithmetic, geometric,* or *neither.* Explain your reasoning.

a. 13, 25, 37, 49, . . .
Arithmetic; the common difference is 12.

b. 2, 5, 9, 14, . . . Neither; there is no common difference or common ratio.

c. 6, −12, 24, −48, . . .
Geometric; the common ratio is −2.

Focus on Mathematical Content

Sequences Not all lists of numbers that follow a pattern are geometric or arithmetic. Knowing this will prepare students for Lesson 10-5 on special sequences called recursive sequences.

Teach with Tech

Instant Messaging Have students work in pairs. Each student will write an equation to create a sequence, and send a message with only the sequence to the other student. Each student should then find and reply with the equation for the sequence that he or she received.

Teaching the Mathematical Practices

Sense-Making Mathematically proficient students start by explaining to themselves the meaning of a problem and looking for entry points to its solution. They analyze givens, constraints, relationships, and goals. They check their answers to problems using a different method, and they continually ask themselves, "Does this make sense?"

Regularity Mathematically proficient students notice if calculations are repeated, and look both for general methods and for shortcuts. As they work to solve a problem, mathematically proficient students maintain oversight of the process, while attending to the details.

ReviewVocabulary
exponential function a function of the form $f(x) = b^x$, where $b > 0$ and $b \neq 1$

Examine the graph in Example 5. While the graph of an arithmetic sequence is linear, the graph of a geometric sequence is exponential and can be represented by $f(x) = r^x$, where r is the common ratio, $r > 0$, and $r \neq 1$.

Arithmetic

x	1	2	3	4	5	6	7	8	9	10
f(x)	4	8	12	16	20	24	28	32	36	40

Geometric

x	1	2	3	4	5	6
f(x)	2	4	8	16	32	64

Arithmetic and geometric sequences are functions in which the domain, defined by the term number n, contains the set of or subset of positive integers. The characteristics of arithmetic and geometric sequences can be used to classify sequences.

Example 6 Classify Sequences

Determine whether each sequence is *arithmetic, geometric,* or *neither.* Explain your reasoning.

a. 16, 24, 36, 54, ...
Check for a common difference.
$54 - 36 = 18 \qquad 36 - 24 = 12$ ✗
Check for a common ratio.
$\frac{54}{36} = \frac{3}{2} \qquad \frac{36}{24} = \frac{3}{2} \qquad \frac{24}{16} = \frac{3}{2}$ ✓
Because there is a common ratio, the sequence is geometric.

b. 1, 4, 9, 16, ...
Check for a common difference.
$16 - 9 = 7 \qquad 9 - 4 = 5$ ✗
Check for a common ratio.
$\frac{16}{9} = 1.\overline{7} \qquad \frac{9}{4} = 2.25$ ✗
Because there is no common difference or ratio, the sequence is neither arithmetic nor geometric.

6A. Arithmetic; the common difference is $\frac{1}{3}$.

6B. Geometric; the common ratio is $-\frac{3}{4}$.

6C. Neither; there is no common difference or ratio.

c. 23, 17, 11, 5, ...
Check for a common difference.
$5 - 11 = -6 \qquad 11 - 17 = -6 \qquad 17 - 23 = -6$ ✓
Because there is a common difference, the sequence is arithmetic.

▶ **Guided**Practice

6A. $\frac{5}{3}, 2, \frac{7}{3}, \frac{8}{3}, ...$ **6B.** $2, -\frac{3}{2}, \frac{9}{8}, -\frac{27}{32}, ...$ **6C.** $-4, 4, 5, -5, ...$

 662 | Lesson 10-1 | Sequences as Functions

Differentiated**Instruction** (AL) (ELL)

Interpersonal Learners Have students in small groups discuss any confusions they may have about the key concepts for arithmetic and geometric sequences. Suggest they help each other organize and complete their notes on these topics.

Check Your Understanding

● = Step-by-Step Solutions begin on page R14.

Example 1 Determine whether each sequence is arithmetic. Write *yes* or *no*.

1. 8, −2, −12, −22, yes

2. −19, −12, −5, 2, 9 yes

3. 1, 2, 4, 8, 16 no

4. 0.6, 0.9, 1.2, 1.8, … no

Example 2 Find the next four terms of each arithmetic sequence. Then graph the sequence.

5. 6, 18, 30, … **5–8. See margin.**

6. 15, 6, −3, …

7. −19, −11, −3, …

8. −26, −33, −40, …

Example 3 **9.** **FINANCIAL LITERACY** Kelly is saving her money to buy a car. She has $250, and she plans to save $75 per week from her job as a waitress.

  a. How much will Kelly have saved after 8 weeks? $850

  b. If the car costs $2000, how long will it take her to save enough money at this rate? 24 wk

Example 4 Determine whether each sequence is geometric. Write *yes* or *no*.

10. −8, −5, −1, 4, … no

11. 4, 12, 36, 108, … yes

12. 27, 9, 3, 1, … yes

13. 7, 14, 21, 28, … no

Example 5 Find the next three terms of each geometric sequence. Then graph the sequence.

14. 8, 12, 18, 27, …

15. 8, 16, 32, 64, …

16. 250, 50, 10, 2, …

17. 9, −3, 1, −$\frac{1}{3}$, …

14–17. See Chapter 10 Answer Appendix.

Example 6 Determine whether each sequence is *arithmetic, geometric,* or *neither*. Explain your reasoning. **18.** Neither; there is no common difference or ratio.

18. 5, 1, 7, 3, 9, …

19. 200, −100, 50, −25, …

20. 12, 16, 20, 24, …

19. Geometric; the common ratio is −$\frac{1}{2}$.

20. Arithmetic; the common difference is 4.

Practice and Problem Solving

Extra Practice is on page R10.

Example 1 Determine whether each sequence is arithmetic. Write *yes* or *no*.

21. $\frac{1}{2}$, $\frac{1}{3}$, $\frac{1}{4}$, $\frac{1}{5}$, … no

22. −9, −3, 0, 3, 9 no

23. 14, −5, −19, … no

24. $\frac{2}{9}$, $\frac{5}{9}$, $\frac{8}{9}$, $\frac{11}{9}$, … yes

Example 2 Find the next four terms of each arithmetic sequence. Then graph the sequence.

25. −4, −1, 2, 5, …

26. 10, 2, −6, −14, …

27. −5, −11, −17, −23, …

28. −19, −2, 15, …

29. $\frac{1}{5}$, $\frac{4}{5}$, $\frac{7}{5}$, …

30. $\frac{2}{3}$, −$\frac{1}{3}$, −$\frac{4}{3}$

25–30. See Chapter 10 Answer Appendix.

Example 3 **31** **THEATER** There are 28 seats in the front row of a theater. Each successive row contains two more seats than the previous row. If there are 24 rows, how many seats are in the last row of the theater? 74

32. **CCSS SENSE-MAKING** Mario began an exercise program to get back in shape. He plans to row 5 minutes on his rowing machine the first day and increase his rowing time by one minute and thirty seconds each day.

  a. How long will he row on the 18th day? 30 minutes and 30 seconds

  b. On what day will Mario first row an hour or more? on the 38th day

  c. Is it reasonable for this pattern to continue indefinitely? Explain.

32c. Sample answer: It is unreasonable because there are only so many hours in the day that can be dedicated to rowing.

connectED.mcgraw-hill.com **663**

Differentiated Homework Options

Level	Assignment	Two-Day Option	
AL Basic	21–50, 54, 55, 57–80	21–49 odd, 61–64	22–50 even, 54, 55, 57–60, 65–80
OL Core	21–49 odd, 51–53, 54, 55, 57–60, 61–80	21–50, 61–64	51–55, 57–60, 65–80
BL Advanced	51–74, (optional: 75–80)		

3 Practice

Formative Assessment
Use Exercises 1–20 to check for understanding.

Use the chart at the bottom of this page to customize assignments for your students.

Exercise Alert
Grid Paper For Exercises 5–8, 14–17, 25–30, 39–44, and 68–73, students will need grid paper.

Additional Answers

5. 42, 54, 66, 78

6. −12, −21, −30, −39

7. 5, 13, 21, 29

8. −47, −54, −61, −68

Example 4 Determine whether each sequence is geometric. Write *yes* or *no*.

33. 21, 14, 7, ... no **34.** 124, 186, 248, ... no **35** −27, 18, −12, ... yes

36. 162, 108, 72, ... yes **37.** $\frac{1}{2}$, $-\frac{1}{4}$, 1, $-\frac{1}{2}$, ... no **38.** −4, −2, 0, 2, ... no

39–44. See Chapter 10 Answer Appendix.

Example 5 Find the next three terms of the sequence. Then graph the sequence.

39. 0.125, −0.5, 2, ... **40.** 18, 12, 8, ... **41.** 64, 48, 36, ...

42. 81, 108, 144, ... **43.** $\frac{1}{3}$, 1, 3, 9, ... **44.** 1, 0.1, 0.01, 0.001, ...

45. Neither; there is no common difference or ratio.

Example 6 Determine whether each sequence is *arithmetic*, *geometric*, or *neither*. Explain your reasoning.

46. Arithmetic; the common difference is −3. **47.** Geometric; the common ratio is 3.

48. Geometric; the common ratio is $\frac{1}{5}$.

45. 3, 12, 27, 48, ... **46.** 1, −2, −5, −8, ... **47.** 12, 36, 108, 324, ...

49. Arithmetic; the common difference is $\frac{1}{2}$.

48. $-\frac{2}{5}$, $-\frac{2}{25}$, $-\frac{2}{125}$, $-\frac{2}{625}$, ... **49.** $\frac{5}{2}$, 3, $\frac{7}{2}$, 4, ... **50.** 6, 9, 14, 21, ...

50. Neither; there is no common difference or ratio.

B

51. **READING** Sareeta took an 800-page book on vacation. If she was already on page 112 and is going to be on vacation for 8 days, what is the minimum number of pages she needs to read per day to finish the book by the end of her vacation? **86 pg/day**

52. **DEPRECIATION** Tammy's car is expected to depreciate at a rate of 15% per year. If her car is currently valued at $24,000, to the nearest dollar, how much will it be worth in 6 years? **$9052**

53. **CCSS REGULARITY** When a piece of paper is folded onto itself, it doubles in thickness. If a piece of paper that is 0.1 mm thick could be folded 37 times, how thick would it be? **about 13,744 km**

H.O.T. Problems   Use Higher-Order Thinking Skills

C

54. **REASONING** Explain why the sequence 8, 10, 13, 17, 22 is not arithmetic.

54. Sample answer: The consecutive terms do not share a common difference. For instance, 22 − 17 = 5, while 17 − 13 = 4.

55. Sample answer: A babysitter earns $20 for cleaning the house and $8 extra for every hour she watches the children.

55. **OPEN ENDED** Describe a real-life situation that can be represented by an arithmetic sequence with a common difference of 8.

56. **CHALLENGE** The sum of three consecutive terms of an arithmetic sequence is 6. The product of the terms is −42. Find the terms. **−3, 2, 7**

57. Sample answer: Neither; the sequence is both arithmetic and geometric.

57. **ERROR ANALYSIS** Brody and Gen are determining whether the sequence 8, 8, 8, ... is *arithmetic*, *geometric*, *neither*, or *both*. Is either of them correct? Explain your reasoning.

> **Brody**
> The sequence has a common difference of 0. The sequence is arithmetic.

> **Gen**
> The sequence has a common ratio of 1. The sequence is geometric.

58. Sample answer: geometric: 3, 9, 27, 81, 243, ...
arithmetic: 3, 9, 15, 21, 27, ... neither geometric nor arithmetic: 3, 9, 21, 45, 93, ...

58. **OPEN ENDED** Find a geometric sequence, an arithmetic sequence, and a sequence that is neither geometric nor arithmetic that begins 3, 9, ...

59. **REASONING** If a geometric sequence has a ratio r such that $|r| < 1$, what happens to the terms as n increases? What would happen to the terms if $|r| \geq 1$? **See margin.**

60. **WRITING IN MATH** Describe what happens to the terms of a geometric sequence when the common ratio is doubled. What happens when it is halved? Explain your reasoning. **See Chapter 10 Answer Appendix.**

664 | Lesson 10-1 | Sequences as Functions

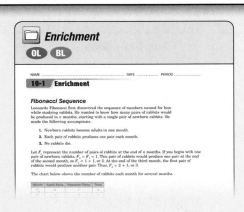

Additional Answer

59. Sample answer: If a geometric sequence has a ratio r such that $|r| < 1$, as n increases, the absolute value of the terms will decrease and approach zero because they are continuously being multiplied by a fraction. When $|r| \geq 1$, the absolute value of the terms will increase and approach infinity because they are continuously being multiplied by a value greater than 1.

61. SHORT RESPONSE Mrs. Aguilar's rectangular bedroom measures 13 feet by 11 feet. She wants to purchase carpet for the bedroom that costs $2.95 per square foot, including tax. How much will it cost to carpet her bedroom? **$421.85**

62. The pattern of filled circles and white circles below can be described by a relationship between two variables.

Which rule relates w, the number of white circles, to f, the number of dark circles? **C**

A $w = 3f$ **C** $w = 2f + 1$

B $f = \frac{1}{2}w - 1$ **D** $f = \frac{1}{3}w$

63. SAT/ACT Donna wanted to determine the average of her six test scores. She added the scores correctly to get T, but divided by 7 instead of 6. Her average was 12 less than the actual average. Which equation could be used to determine the value of T? **H**

F $6T + 12 = 7T$ **J** $\frac{T}{6} = \frac{T - 12}{7}$

G $\frac{T}{7} = \frac{T - 12}{6}$ **K** $\frac{T}{6} = 12 - \frac{T}{7}$

H $\frac{T}{7} + 12 = \frac{T}{6}$

64. Find the next term in the geometric sequence $8, 6, \frac{9}{2}, \frac{27}{8}, \dots$. **D**

A $\frac{11}{8}$ **C** $\frac{9}{4}$

B $\frac{27}{16}$ **D** $\frac{81}{32}$

Solve each system of equations. (Lesson 9-7)

65. $y = 5$ **($\pm$4, 5)**
$y^2 = x^2 + 9$

66. $y - x = 1$ **(−4, −3), (3, 4)**
$x^2 + y^2 = 25$

67. $3x = 8y^2$ **no solution**
$8y^2 - 2x^2 = 16$

Write each equation in standard form. State whether the graph of the equation is a *parabola*, *circle*, *ellipse*, or *hyperbola*. Then graph the equation. (Lesson 9-6) **68–70. See margin.**

68. $6x^2 + 6y^2 = 162$

69. $4y^2 - x^2 + 4 = 0$

70. $x^2 + y^2 + 6y + 13 = 40$

Graph each function. (Lesson 8-4) **71–73. See Chapter 10 Answer Appendix.**

71. $f(x) = \frac{6}{(x - 2)(x + 3)}$

72. $f(x) = \frac{-3}{(x - 2)^2}$

73. $f(x) = \frac{x^2 - 36}{x + 6}$

74. HEALTH A certain medication is eliminated from the bloodstream at a steady rate. It decays according to the equation $y = ae^{-0.1625t}$, where t is in hours. Find the half-life of this substance. (Lesson 7-8) **about 4.27 hours**

Write an equation of each line.

75. passes through $(6, 4)$, $m = 0.5$ $y = 0.5x + 1$

76. passes through $\left(2, \frac{1}{2}\right)$, $m = -\frac{3}{4}$ $y = -\frac{3}{4}x + 2$

77. passes through $(0, -6)$, $m = 3$ $y = 3x - 6$

78. passes through $(0, 4)$, $m = \frac{1}{4}$ $y = \frac{1}{4}x + 4$

79. passes through $(1, 3)$ and $\left(8, -\frac{1}{2}\right)$ $y = -\frac{1}{2}x + \frac{7}{2}$

80. passes through $(-5, 1)$ and $(5, 16)$ $y = \frac{3}{2}x + \frac{17}{2}$

DifferentiatedInstruction **OL** **BL**

Extension Ask students this question: "Is 1, 2, 4, 8, 16, . . . an arithmetic or geometric sequence? How do you know?" Geometric; the common ratio is 2. Then ask, "If each term in the sequence 1, 2, 4, 8, 16, . . . were replaced with its reciprocal, would the sequence still be a geometric sequence? How do you know?" Yes; the common ratio is $\frac{1}{2}$. Ask students to write their own geometric sequence and determine if its reciprocal counterpart is also a geometric sequence.

WatchOut!

Error Analysis For Exercise 57, ask students to find a sequence, other than 8, 8, 8, . . . , that is both arithmetic and geometric. Discuss how that sequence is similar to 8, 8, 8, ...

4 Assess

Name the Math Have students explain how to find any term of an arithmetic sequence or geometric sequence when the first few terms are given.

Additional Answers

68. circle; $x^2 + y^2 = 27$

69. hyperbola; $\frac{x^2}{4} - \frac{y^2}{1} = 1$

70. circle; $x^2 + (y + 3)^2 = 36$

1 Focus

VerticalAlignment

Before Lesson 10-2 Determine whether a sequence is arithmetic.

Lesson 10-2 Use arithmetic sequences. Find sums of arithmetic series.

After Lesson 10-2 Prove statements by using mathematical induction.

2 Teach

Scaffolding Questions

Have students read the **Why?** section of the lesson.

Ask:

- What type of sequence are the counting numbers, 1–100? arithmetic

- What is the value of the first term? 1

- How could you quickly determine the sum of the first 10 counting numbers? Sample answer: $5 \times 11 = 55$

$$1 + 2 + 3 + 4 + 5 + 6 + 7 + 8 + 9 + 10$$

:: Then	:: Now	:: Why?
● You determined whether a sequence was arithmetic.	**1** Use arithmetic sequences. **2** Find sums of arithmetic series.	● In the 18th century, a teacher asked his class of elementary students to find the sum of the counting numbers 1 through 100. A pupil named Karl Gauss correctly answered within seconds, astonishing the teacher. Gauss went on to become a great mathematician. He solved this problem by using an arithmetic series.

NewVocabulary
arithmetic means
series
arithmetic series
partial sum
sigma notation

CCSS Common Core State Standards

Content Standards
A.CED.4 Rearrange formulas to highlight a quantity of interest, using the same reasoning as in solving equations.

Mathematical Practices
8 Look for and express regularity in repeated reasoning.

1 Arithmetic Sequences In Lesson 10-1, you used the point-slope form to find a specific term of an arithmetic sequence. It is possible to develop an equation for any term of an arithmetic sequence using the same process.

Consider the arithmetic sequence $a_1, a_2, a_3, \ldots, a_n$ in which the common difference is d.

$$(y - y_1) = m(x - x_1) \qquad \text{Point-slope form}$$
$$(a_n - a_1) = d(n - 1) \qquad (x, y) = (n, a_n), (x_1, y_1) = (1, a_1), \text{ and } m = d$$
$$a_n = a_1 + d(n - 1) \qquad \text{Add } a_1 \text{ to each side.}$$

You can use this equation to find any term in an arithmetic sequence when you know the first term and the common difference.

> **KeyConcept** *n*th Term of an Arithmetic Sequence
>
> The *n*th term a_n of an arithmetic sequence in which the first term is a_1 and the common difference is d is given by the following formula, where n is any natural number.
>
> $$a_n = a_1 + (n - 1)d$$

You will prove this formula in Exercise 80.

Example 1 Find the *n*th Term

Find the 12th term of the arithmetic sequence 9, 16, 23, 30, … .

Step 1 Find the common difference.
$$16 - 9 = 7 \qquad 23 - 16 = 7 \qquad 30 - 23 = 7$$
So, $d = 7$.

Step 2 Find the 12th term.
$$a_n = a_1 + (n - 1)d \qquad \text{\textit{n}th term of an arithmetic sequence}$$
$$a_{12} = 9 + (12 - 1)(7) \qquad a_1 = 9, d = 7, \text{ and } n = 12$$
$$= 9 + 77 \text{ or } 86 \qquad \text{Simplify.}$$

▸ **GuidedPractice**

Find the indicated term of each arithmetic sequence.

1A. $a_1 = -4, d = 6, n = 9$ 44   **1B.** a_{20} for $a_1 = 15, d = -8$ −137

akg-images

Lesson 10-2 Resources

Resource	Approaching Level **AL**	On Level **OL**	Beyond Level **BL**	English Learners **ELL**
Teacher Edition		▪ Differentiated Instruction, pp. 668, 671	▪ Differentiated Instruction, p. 671	▪ Differentiated Instruction, p. 668
Chapter Resource Masters	▪ Study Guide and Intervention, pp. 11–12 ▪ Skills Practice, p. 13 ▪ Practice, p. 14 ▪ Word Problem Practice, p. 15	▪ Study Guide and Intervention, pp. 11–12 ▪ Skills Practice, p. 13 ▪ Practice, p. 14 ▪ Word Problem Practice, p. 15 ▪ Enrichment, p. 16 ▪ Spreadsheet Activity, p. 17 ▪ Graphing Calculator Activity, p. 18	▪ Practice, p. 14 ▪ Word Problem Practice, p. 15 ▪ Enrichment, p. 16	▪ Study Guide and Intervention, pp. 11–12 ▪ Skills Practice, p. 13 ▪ Practice, p. 14 ▪ Word Problem Practice, p. 15
Other	▪ 5-Minute Check 10-2 ▪ Study Notebook	▪ 5-Minute Check 10-2 ▪ Study Notebook	▪ 5-Minute Check 10-2 ▪ Study Notebook	▪ 5-Minute Check 10-2 ▪ Study Notebook

If you are given some terms of an arithmetic sequence, you can write an equation for the nth term of the sequence.

Example 2 Write Equations for the nth Term

Write an equation for the nth term of each arithmetic sequence.

a. 5, −13, −31, …

$d = -13 - 5$ or −18; 5 is the first term.

$a_n = a_1 + (n-1)d$	nth term of an arithmetic sequence
$a_n = 5 + (n-1)(-18)$	$a_1 = 5$ and $d = -18$
$a_n = 5 + (-18n + 18)$	Distributive Property
$a_n = -18n + 23$	Simplify.

b. $a_5 = 19$, $d = 6$

First, find a_1.

$a_n = a_1 + (n-1)d$	nth term of an arithmetic sequence
$19 = a_1 + (5-1)(6)$	$a_5 = 19$, $n = 5$, and $d = 6$
$19 = a_1 + 24$	Multiply.
$-5 = a_1$	Subtract 24 from each side.

Then write the equation.

$a_n = a_1 + (n-1)d$	nth term of an arithmetic sequence
$a_n = -5 + (n-1)(6)$	$a_1 = -5$ and $d = 6$
$a_n = -5 + (6n - 6)$	Distributive Property
$a_n = 6n - 11$	Simplify.

StudyTip
Checking Solutions Check your solution by using it to determine the first three terms of the sequence.

▶ **Guided**Practice

2A. 12, 3, −6, … $a_n = -9n + 21$ **2B.** $a_6 = 12$, $d = 8$ $a_n = 8n - 36$

Sometimes you are given two terms of a sequence, but they are not consecutive terms of the sequence. The terms between any two nonconsecutive terms of an arithmetic sequence, called **arithmetic means**, can be used to find missing terms of a sequence.

ReadingMath

arithmetic mean the average of two or more numbers

arithmetic means the terms between any two nonconsecutive terms of an arithmetic sequence

Example 3 Find Arithmetic Means

Find the arithmetic means in the sequence −8, _?_, _?_, _?_, _?_, 22, … .

Step 1 Since there are four terms between the first and last terms given, there are $4 + 2$ or 6 total terms, so $n = 6$.

Step 2 Find d.

$a_n = a_1 + (n-1)d$	nth term of an arithmetic sequence
$22 = -8 + (6-1)d$	$a_1 = -8$, $a_6 = 22$, and $n = 6$
$30 = 5d$	Distributive Property
$6 = d$	Divide each side by 5.

Step 3 Use d to find the four arithmetic means.

−8   −2   4   10   16   22

　+6　+6　+6　+6　+6

The arithmetic means are −2, 4, 10, and 16.

▶ **Guided**Practice

3. Find the five arithmetic means between −18 and 36. **−9, 0, 9, 18, 27**

---

1 Arithmetic Sequences

Example 1 shows how to find the nth term of an arithmetic sequence when you know the first term and the common difference. **Example 2** shows how to write an equation for the nth term of an arithmetic sequence when given some terms of the sequence. **Example 3** shows how to find the arithmetic means between two nonconsecutive terms of an arithmetic sequence.

Formative Assessment
Use the Guided Practice exercises after each example to determine students' understanding of concepts.

Additional Examples

1 Find the 20th term of the arithmetic sequence 3, 10, 17, 24, ... 136

2 Write an equation for the nth term of each arithmetic sequence.

a. −8, −6, −4, . . .
$a_n = 2n - 10$

b. $a_6 = 11$, $d = -11$
$a_n = -11n + 77$

3 Find the arithmetic means in the sequence 21, ___, ___, ___, 45, ... 27, 33, 39

▶ **Additional Examples** also in Interactive Classroom PowerPoint® Presentations

IWB **Interactive White Board** READY

Focus on Mathematical Content
Formula for an Arithmetic Sequence Any term of a sequence can be found from the value of the preceding term if the common difference is known, but a formula for the nth term requires knowledge of the first term in the sequence.

Example 4 shows how to use a formula to find the sum of an arithmetic series. **Example 5** shows how to use a sum formula to find terms in an arithmetic series. **Example 6** shows how to find the sum of a series written in sigma notation.

Additional Examples

4 Find the sum $8 + 12 + 16 + \ldots + 80$. 836

5 Find the first three terms of the arithmetic series. $S_n = 129$, $a_1 = 14$, and $a_n = 29$
14, 17, 20

Teach with Tech

Interactive Whiteboard Show an example on the board to help explain the formula for finding the sum of an arithmetic series. Write a sequence of numbers from 1 to 10, and drag the numbers to form the sums $1 + 10$, $2 + 9$, $3 + 8$, etc., and find the sum of the series. Then show how the formula simplifies this calculation.

---

2 Arithmetic Series A **series** is formed when the terms of a sequence are added. An **arithmetic series** is the sum of the terms of an arithmetic sequence. The sum of the first n terms is called the **partial sum** and is denoted S_n.

KeyConcept Partial Sum of an Arithmetic Series

Formula	Given	The sum S_n of the first n terms is:
General	a_1 and a_n	$S_n = n\left(\dfrac{a_1 + a_n}{2}\right)$
Alternate	a_1 and d	$S_n = \dfrac{n}{2}[2a_1 + (n-1)d]$

Sometimes a_1, a_n, or n must be determined before the sum of an arithmetic series can be found. When this occurs, use the formula for the nth term.

Example 4 Use the Sum Formulas

Find the sum of $12 + 19 + 26 + \ldots + 180$.

Step 1 $a_1 = 12$, $a_n = 180$, and $d = 19 - 12$ or 7.
We need to find n before we can use one of the formulas.

$a_n = a_1 + (n-1)d$	nth term of an arithmetic sequence
$180 = 12 + (n-1)(7)$	$a_n = 180$, $a_1 = 12$, and $d = 7$
$168 = 7n - 7$	Simplify.
$25 = n$	Solve for n.

Step 2 Use either formula to find S_n.

$S_n = \dfrac{n}{2}[2a_1 + (n-1)d]$	Sum formula
$S_{25} = \dfrac{25}{2}[2(12) + (25-1)(7)]$	$n = 25$, $a_1 = 12$, and $d = 7$
$S_{25} = 12.5(192)$ or 2400	Simplify.

▶ **Guided Practice**

Find the sum of each arithmetic series.

4A. $2 + 4 + 6 + \ldots + 100$ **2550**   **4B.** $n = 16$, $a_n = 240$, and $d = 8$. **2880**

You can use a sum formula to find terms of a series.

Example 5 Find the First Three Terms

Find the first three terms of the arithmetic series in which $a_1 = 7$, $a_n = 79$, and $S_n = 430$.

Step 1 Find n.

$S_n = n\left(\dfrac{a_1 + a_n}{2}\right)$	Sum formula
$430 = n\left(\dfrac{7 + 79}{2}\right)$	$S_n = 430$, $a_1 = 7$, and $a_n = 79$
$430 = n(43)$	Simplify.
$10 = n$	Divide each side by 43.

WatchOut!

Common Difference
Don't confuse the sign of the common difference in an arithmetic sequence. Check that the rule actually produces the terms of a sequence.

DifferentiatedInstruction OL ELL

Interpersonal Learners Discuss the difference between a sequence and a series, and ask students to suggest ways to remember which is which.

Step 2 Find d.

$$a_n = a_1 + (n - 1)d \qquad \text{\textit{n}th term of an arithmetic sequence}$$
$$79 = 7 + (10 - 1)d \qquad a_n = 79, a_1 = 7, \text{ and } n = 10$$
$$72 = 9d \qquad \text{Subtract 7 from each side.}$$
$$8 = d \qquad \text{Divide each side by 9.}$$

Step 3 Use d to determine a_2 and a_3.

$$a_2 = 7 + 8 \text{ or } 15 \qquad a_3 = 15 + 8 \text{ or } 23$$

The first three terms are 7, 15, and 23.

▶ **Guided**Practice

Find the first three terms of each arithmetic series. **5B.** $-24, -16, -8$

5A. $S_n = 120, n = 8, a_n = 36$ $-6, 0, 6$ **5B.** $a_1 = -24, a_n = 288, S_n = 5280$

The sum of a series can be written in shorthand by using **sigma notation**.

ReadingMath

Sigma Notation The name comes from the Greek letter sigma, which is used in the notation.

KeyConcept Sigma Notation

Symbols

$$\text{last value of } k \longrightarrow \quad \overset{n}{\underset{k=1}{\sum}} f(k) \longleftarrow \text{formula for the terms of the series}$$
$$\text{first value of } k \nearrow$$

Example $\overset{12}{\underset{k=1}{\sum}}(4k + 2) = [4(1) + 2] + [4(2) + 2] + [4(3) + 2] + \cdots + [4(12) + 2]$

$$= 6 + 10 + 14 + \cdots + 50$$

Standardized Test Example 6 Use Sigma Notation

Find $\overset{18}{\underset{k=4}{\sum}}(6k - 1)$.

A 846 **B** 910 **C** 975 **D** 1008

Read the Test Item

You need to find the sum of the series. Find a_1, a_n, and n.

Solve the Test Item

There are $18 - 4 + 1$ or 15 terms, so $n = 15$.

$$a_1 = 6(4) - 1 \text{ or } 23 \qquad a_n = 6(18) - 1 \text{ or } 107$$

Find the sum.

$$S_n = n\left(\frac{a_1 + a_n}{2}\right) \qquad \text{Sum formula}$$

$$S_{15} = 15\left(\frac{23 + 107}{2}\right) \qquad n = 15, a_1 = 23, \text{ and } a_n = 107$$

$$S_{15} = 15(65) \text{ or } 975 \qquad \text{The correct answer is C.}$$

Test-Taking Tip

CCSS Perseverance

Sometimes it is necessary to break a problem into parts, solve each part, then combine the solutions of the parts.

▶ **Guided**Practice

6. Find $\overset{21}{\underset{m=9}{\sum}}(5m + 6)$. **G**

F 972 **G** 1053 **H** 1281 **J** 1701

Focus on Mathematical Content

Finite Series Although most of the sequences studied in this lesson are infinite, the series in this lesson are not. The formula $S_n = \frac{n}{2}(a_1 + a_n)$ for the sum of an arithmetic series is valid only for a stated finite number of n terms.

Additional Example

6 STANDARDIZED TEST PRACTICE

Find $\overset{10}{\underset{k=3}{\sum}}(2k + 1)$. D

A 23

B 70

C 98

D 112

CCSS **Teaching the Mathematical Practices**

Perseverance Mathematically proficient students monitor and evaluate their progress and change course if necessary. Encourage students to examine expressions and list all important values before beginning their calculations.

 The Math Forum

Have your students refer to the Math Forum's Ask Dr. Math to learn more about real world applications of series and sequences. They can start their search at connectED. mcgraw-hill.com.

Check Your Understanding

Example 1 Find the indicated term of each arithmetic sequence.

1. $a_1 = 14, d = 9, n = 11$ **104**

2. a_{18} for 12, 25, 38, ... **233**

Example 2 Write an equation for the nth term of each arithmetic sequence.

3. 13, 19, 25, ... $a_n = 6n + 7$

4. $a_5 = -12, d = -4$ $a_n = -4n + 8$

Example 3 Find the arithmetic means in each sequence.

5. 6, _?_, _?_, _?_, 42 **15, 24, 33**

6. -4, _?_, _?_, _?_, 8 **-1, 2, 5**

Example 4 Find the sum of each arithmetic series.

7. the first 50 natural numbers **1275**

8. $4 + 8 + 12 + \cdots + 200$ **5100**

9. $a_1 = 12, a_n = 188, d = 4$ **4500**

10. $a_n = 145, d = 5, n = 21$ **1995**

Example 5 Find the first three terms of each arithmetic series.

11. $a_1 = 8, a_n = 100, S_n = 1296$ **8, 12, 16**

12. $n = 18, a_n = 112, S_n = 1098$ **10, 16, 22**

Example 6 **13. MULTIPLE CHOICE** Find $\displaystyle\sum_{k=1}^{12} (3k + 9)$. **C**

A 45

C 342

B 78

D 410

Practice and Problem Solving

Extra Practice is on page R10.

Example 1 Find the indicated term of each arithmetic sequence.

14. $a_1 = -18, d = 12, n = 16$ **162**

15. $a_1 = -12, n = 66, d = 4$ **248**

16. $a_1 = 9, n = 24, d = -6$ **-129**

17. a_{15} for $-5, -12, -19, \ldots$ **-103**

18. a_{10} for $-1, 1, 3, \ldots$ **17**

19. a_{24} for $8.25, 8.5, 8.75, \ldots$ **14**

20. $a_n = 11n + 13$
21. $a_n = -14n + 45$
22. $a_n = -3n + 72$
23. $a_n = 5n - 14$
24. $a_n = 0.25n + 11$
25. $a_n = 4.5n - 21$
26. $a_n = -7n + 16$
27. $a_n = 9n - 32$
28. $a_n = -2n + 8$
29. $a_n = \frac{2}{3}n - 3$
30. $a_n = -5n - 7$
31. $a_n = \frac{1}{2}n - \frac{23}{10}$

Example 2 Write an equation for the nth term of each arithmetic sequence.

20. 24, 35, 46, ...

21. 31, 17, 3, ...

22. $a_9 = 45, d = -3$

23. $a_7 = 21, d = 5$

24. $a_4 = 12, d = 0.25$

25. $a_5 = 1.5, d = 4.5$

26. 9, 2, -5, ...

27. $a_6 = 22, d = 9$

28. $a_8 = -8, d = -2$

29. $a_{15} = 7, d = \frac{2}{3}$

30. $-12, -17, -22, \ldots$

31. $a_3 = -\frac{4}{5}, d = \frac{1}{2}$

32. **CCSS STRUCTURE** José averaged 123 total pins per game in his bowling league this season. He is taking bowling lessons and hopes to bring his average up by 8 pins each new season. **b. 9th season**

a. Write an equation to represent the nth term of the sequence. $a_n = 115 + 8n$

b. If the pattern continues, during what season will José average 187 per game?

c. Is it reasonable for this pattern to continue indefinitely? Explain. **See margin.**

Example 3 Find the arithmetic means in each sequence.

33. 24, _?_, _?_, _?_, _?_, -1 **19, 14, 9, 4**

34. -6, _?_, _?_, _?_, _?_, 49 **5, 16, 27, 38**

35. -28, _?_, _?_, _?_, _?_, 7 **$-21, -14, -7, 0$**

36. 84, _?_, _?_, _?_, _?_, 39 **75, 66, 57, 48**

37. -12, _?_, _?_, _?_, _?_, _?_, -66
$-21, -30, -39, -48, -57$

38. 182, _?_, _?_, _?_, _?_, _?_, 104 **169, 156, 143, 130, 117**

WatchOut!

Preventing Errors Help students become comfortable with sigma notation by having them read aloud several expressions written in this notation. Explain that sigma is the uppercase letter S in the Greek alphabet. Ask them for other examples of mathematical notation that use Greek letters.
Sample answer: π

3 Practice

Formative Assessment

Use Exercises 1–13 to check for understanding.

Use the chart at the bottom of this page to customize assignments for your students.

CCSS Teaching the Mathematical Practices

Structure Mathematically proficient students look closely to discern a pattern or structure. They can see complicated things, such as some algebraic expressions, as single objects or as being composed of several objects.

Additional Answer

32c. Sample answer: No; there are a maximum of 300 points in a bowling game, so it would be impossible for the average to continue to climb indefinitely.

Differentiated Homework Options

Level	Assignment	Two-Day Option	
AL Basic	14–60, 74, 75, 78–82, 83–100	15–59 odd, 83–86	14–60 even, 74, 75, 78–82, 87–100
OL Core	15–59 odd, 61–71, 73–75, 78–100	14–60, 83–86	61–73, 74, 75, 78–82, 87–100
BL Advanced	61–96, (optional: 97–100)		

Example 4

Find the sum of each arithmetic series. 39. 10,100   40. 40,000   41. 10,000

39. the first 100 even natural numbers   40. the first 200 odd natural numbers

41. the first 100 odd natural numbers   42. the first 300 even natural numbers 90,300

43. $-18 + (-15) + (-12) + \cdots + 66$ **696**   44. $-24 + (-18) + (-12) + \cdots + 72$ **408**

45. $a_1 = -16, d = 6, n = 24$ **1272**   46. $n = 19, a_n = 154, d = 8$ **1558**

47. **CONTESTS** The prizes in a weekly radio contest began at $150 and increased by $50 for each week that the contest lasted. If the contest lasted for eleven weeks, how much was awarded in total? **$4400**

Example 5

Find the first three terms of each arithmetic series. 52. $-72, -57, -42$

48. $n = 32, a_n = -86, S_n = 224$ **100, 94, 88**   49. $a_1 = 48, a_n = 180, S_n = 1368$ **48, 60, 72**

50. $a_1 = 3, a_n = 66, S_n = 759$ **3, 6, 9**   �51 $n = 28, a_n = 228, S_n = 2982$ **-15, -6, 3**

52. $a_1 = -72, a_n = 453, S_n = 6858$ 53. $n = 30, a_n = 362, S_n = 4770$ **-44, -30, -16**

54. $a_1 = 19, n = 44, S_n = 9350$ **19, 28, 37**   55. $a_1 = -33, n = 36, S_n = 6372$ **-33, -21, -9**

56. **PRIZES** A radio station is offering a total of $8500 in prizes over ten hours. Each hour, the prize will increase by $100. Find the amounts of the first and last prize. **$400 and $1300**

Example 6

Find the sum of each arithmetic series.

57. $\displaystyle\sum_{k=1}^{16}(4k-2)$ **512**   58. $\displaystyle\sum_{k=4}^{13}(4k+1)$ **350**   59. $\displaystyle\sum_{k=5}^{16}(2k+6)$ **324**   60. $\displaystyle\sum_{k=0}^{12}(-3k+2)$ **-208**

B 61. **FINANCIAL LITERACY** Daniela borrowed some money from her parents. She agreed to pay $50 at the end of the first month and $25 more each additional month for 12 months. How much does she pay in total after the 12 months? **$2250**

62. **GRAVITY** When an object is in free fall and air resistance is ignored, it falls 16 feet in the first second, an additional 48 feet during the next second, and 80 feet during the third second. How many total feet will the object fall in 10 seconds? **1600 ft**

Use the given information to write an equation that represents the nth term in each arithmetic sequence 64. $a_n = -9n + 177$

63. The 100th term of the sequence is 245. The common difference is 13. $a_n = 13n - 1055$

64. The eleventh term of the sequence is 78. The common difference is -9.

65. The sixth term of the sequence is -34. The 23rd term is 119. $a_n = 9n - 88$

66. The 25th term of the sequence is 121. The 80th term is 506. $a_n = 7n - 54$

67. **CCSS MODELING** The rectangular tables in a reception hall are often placed end-to-end to form one long table. The diagrams below show the number of people who can sit at each of the table arrangements.

a. Make drawings to find the next three numbers as tables are added one at a time to the arrangement. **See margin.**

b. Write an equation representing the nth number in this pattern. $p_n = 4n + 2$

c. Is it possible to have seating for exactly 100 people with such an arrangement? Explain. **No; there is no whole number n for which $4n + 2 = 100$.**

connectED.mcgraw-hill.com **671**

CCSS **Teaching the Mathematical Practices**

Modeling Mathematically proficient students can apply the mathematics they know to solve problems arising in everyday life, analyze relationships mathematically to draw conclusions, and interpret their mathematical results in the context of a situation.

Exercise Alert

Grid Paper For Exercises 71, 90–92, and 94–96, students will need grid paper.

Additional Answer

67a. 14, 18, 22

DifferentiatedInstruction ㊦ ㊱

Extension Write an arithmetic series such as $5 + 9 + 13 + 17 + 21$ on the board and ask students to write the sum in terms of n using sigma notation with n starting at 1. $\displaystyle\sum_{n=1}^{5}(4n+1)$

Study Guide and Intervention
AL OL ELL

Practice
AL OL BL ELL

Word Problem Practice
AL OL BL ELL

68. PERFORMANCE A certain company pays its employees according to their performance. Belinda is paid a flat rate of $200 per week plus $24 for every unit she completes. If she earned $512 in one week, how many units did she complete? **13**

69. SALARY Terry currently earns $28,000 per year. If Terry expects a $4000 increase in salary every year, after how many years will he have a salary of $100,000 per year? **the 19th year**

70. SPORTS While training for cross country, Silvia plans to run 3 miles per day for the first week, and then increase the distance by a half mile each of the following weeks.

a. Write an equation to represent the nth term of the sequence. $a_n = 2.5 + 0.5n$

b. If the pattern continues, during which week will she be running 10 miles per day? **15th wk**

c. Is it reasonable for this pattern to continue indefinitely? Explain.

70c. Sample answer: No; eventually the number of miles per day will become unrealistic.

71. MULTIPLE REPRESENTATIONS Consider $\sum_{k=1}^{x} (2k + 2)$.

a–c. See Chapter 10 Answer Appendix.

a. Tabular Make a table of the partial sums of the series for $1 \le k \le 10$.

b. Graphical Graph $(k, \text{partial sum})$.

c. Graphical Graph $f(x) = x^2 + 3x$ on the same grid.

d. Verbal What do you notice about the two graphs?

71d. Sample answer: The graphs cover the same range. The domain of the series is the natural numbers, while the domain of the quadratic function is all real numbers, $0 \le x \le 10$.

e. Analytical What conclusions can you make about the relationship between quadratic functions and the sum of arithmetic series?

f. Algebraic Find the arithmetic series that relates to $g(x) = x^2 + 8x$. $\sum_{k=1}^{x} 2k + 7$

71e. Sample answer: For every partial sum of an arithmetic series, there is a corresponding quadratic function that shares the same range.

Find the value of x.

72. $\sum_{k=3}^{x} (6k - 5) = 928$ **18**

73. $\sum_{k=5}^{x} (8k + 2) = 1032$ **16**

H.O.T. Problems Use Higher-Order Thinking Skills

74. CRITIQUE Eric and Juana are determining the formula for the nth term for the sequence $-11, -2, 7, 16, \ldots$. Is either of them correct? Explain your reasoning.

74. Sample answer: Eric; Juana missed the step of multiplying d by $n - 1$.

Eric
$d = 16 - 7$ or 9, $a_1 = -11$
$a_n = -11 + (n - 1)9$
$= 9n - 20$

Juana
$d = 16 - 7$ or 9, $a_1 = -11$
$a_n = 9n - 11$

75. REASONING If a is the third term in an arithmetic sequence, b is the fifth term, and c is the eleventh term, express c in terms of a and b. $4b - 3a$

76. CHALLENGE There are three arithmetic means between a and b in an arithmetic sequence. The average of the arithmetic means is 16. What is the average of a and b? **16**

77. CHALLENGE Find S_n for $(x + y) + (x + 2y) + (x + 3y) + \ldots$.

77. $S_n = nx + y\left(\dfrac{n^2 + n}{2}\right)$

78. OPEN ENDED Write an arithmetic series with 8 terms and a sum of 324.

78. Sample answer: $9 + 18 + 27 + \cdots + 72$

79. WRITING IN MATH Compare and contrast arithmetic sequences and series. **See margin.**

80. PROOF Prove the formula for the nth term of an arithmetic sequence. **80–81. See Chapter 10 Answer Appendix.**

81. PROOF Derive a sum formula that does not include a_1.

82. PROOF Derive the Alternate Sum Formula using the General Sum Formula. **See margin.**

Enrichment
OL BL

10-2 **Enrichment**

Arithmetic Series in Computer Programming

Teaching the Mathematical Practices

Critique Mathematically proficient students are also able to compare the effectiveness of two plausible arguments, distinguish correct logic or reasoning from that which is flawed, and—if there is a flaw in an argument—explain what it is.

83. SAT/ACT The measures of the angles of a triangle form an arithmetic sequence. If the measure of the smallest angle is 36°, what is the measure of the largest angle? **C**

36°

A 54°
B 75°
C 84°
D 90°
E 97°

84. The area of a triangle is $\frac{1}{2}q^2 - 8$ and the height is $q + 4$. Which expression best describes the triangle's base? **J**

F $(q + 1)$ **H** $(q - 3)$
G $(q + 2)$ **J** $(q - 4)$.

85. The expression $1 + \sqrt{2} + \sqrt[3]{3}$ is equivalent to **A**

A $\sum_{k=1}^{3} k^{\frac{1}{k}}$ **C** $\sum_{k=1}^{3} k^{-k}$

B $\sum_{k=1}^{3} k^{k}$ **D** $\sum_{k=1}^{3} \sqrt{k}$

86. SHORT RESPONSE Trevor can type a 200-word essay in 6 hours. Minya can type the same essay in $4\frac{1}{2}$ hours. If they work together, how many hours will it take them to type the essay? $2\frac{4}{7}$

Determine whether each sequence is arithmetic. Write *yes* or *no*. (Lesson 10-1)

87. $-6, 4, 14, 24, \ldots$ **yes**

88. $2, \frac{7}{5}, \frac{4}{5}, \frac{1}{5}, \ldots$ **yes**

89. $10, 8, 5, 1, \ldots$ **no**

Solve each system of inequalities by graphing. (Lesson 9-7) **90–92. See Chapter 10 Answer Appendix.**

90. $x + 2y > 1$
$x^2 + y^2 \le 25$

91. $x + y \le 2$
$4x^2 - y^2 \ge 4$

92. $x^2 + y^2 \ge 4$
$4y^2 + 9x^2 \le 36$

93. PHYSICS The distance a spring stretches is related to the mass attached to the spring. This is represented by $d = km$, where d is the distance, m is the mass, and k is the spring constant. When two springs with spring constants k_1 and k_2 are attached in a series, the resulting spring constant k is found by the equation $\frac{1}{k} = \frac{1}{k_1} + \frac{1}{k_2}$. (Lesson 8-6)

Spring 1
$k_1 = 12$ cm/g

Spring 2
$k_2 = 8$ cm/g

Spring 1

d Spring 2

5 g

a. If one spring with constant of 12 centimeters per gram is attached in a series with another spring with constant of 8 centimeters per gram, find the resultant spring constant. **4.8 cm/g**

b. If a 5-gram object is hung from the series of springs, how far will the springs stretch? Is this answer reasonable in this context? **See margin.**

Graph each function. State the domain and range. (Lesson 7-1) **94–96. See Chapter 10 Answer Appendix.**

94. $f(x) = \frac{2}{3}(2^x)$

95. $f(x) = 4^x + 3$

96. $f(x) = 2\left(\frac{1}{3}\right)^x - 1$

Solve each equation. Round to the nearest ten-thousandth.

97. $5^x = 52$ **2.4550**

98. $4^{3p} = 10$ **0.5537**

99. $3^{n+2} = 14.5$ **0.4341**

100. $16^{d-4} = 3^{3-d}$ **3.7162**

82. $S_n = (a_1 + a_n) \cdot \left(\frac{n}{2}\right)$ General sum formula

$a_n = a_1 + (n - 1)d$ Formula for *nth* term

$S_n = [a_1 + a_1 + (n - 1)d] \cdot \left(\frac{n}{2}\right)$ Substitution

$S_n = [2a_1 + (n - 1)d] \cdot \left(\frac{n}{2}\right)$ Simplify.

93b. 24 cm; The answer is reasonable. The object would stretch the first spring 60 cm and would stretch the second spring 40 cm. The object would have to stretch the combined springs less than it would stretch either of the springs individually.

WatchOut!

Error Analysis For Exercise 74, tell students that to avoid errors such as the one Juana made, it is helpful to write the formula, then in the next line, show the substitution of known values.

Preventing Errors Make sure that all students can demonstrate understanding of sigma notation by asking them to write out the terms of a series described in sigma notation.

4 Assess

Yesterday's News Ask students to describe how their work with sequences in the previous lesson helped them with this lesson.

Formative Assessment
Check for student understanding of Lessons 10-1 and 10-2.

📁 Quiz 1, p. 53

Additional Answers

79. Sample answer: An arithmetic sequence is a list of terms such that any pair of successive terms has a common difference. An arithmetic series is the sum of the terms of an arithmetic sequence.

1 Focus

VerticalAlignment

Before Lesson 10-3 Determine whether a sequence is geometric.

Lesson 10-3 Use geometric sequences. Find the sums of geometric series.

After Lesson 10-3 Find the sum of an infinite geometric series.

2 Teach

Scaffolding Questions

Have students read the **Why?** section of the lesson.

Ask:

- What are the first 3 terms of the pattern? 5, 25, 125

- Why is this sequence not an arithmetic sequence? There is no common difference between the terms.

- What is the common ratio in the sequence? $r = 5$

- Compare a common difference with a common ratio. The first involves addition; the second involves multiplication.

LESSON 10-3 Geometric Sequences and Series

:: Then	:: Now	:: Why?
• You determined whether a sequence was geometric.	**1** Use geometric sequences. **2** Find sums of geometric series.	• Julian sees a new band at a concert. He e-mails a link for the band's Web site to five of his friends. They each forward the link to five of their friends. The link is forwarded again following the same pattern. How many people will receive the link on the eighth round of e-mails?

 NewVocabulary
geometric means
geometric series

 Common Core State Standards

Content Standards
A.SSE.4 Derive the formula for the sum of a finite geometric series (when the common ratio is not 1), and use the formula to solve problems.

Mathematical Practices
8 Look for and express regularity in repeated reasoning.

1 Geometric Sequences As with arithmetic sequences, there is a formula for the nth term of a geometric sequence. This formula can be used to determine any term of the sequence.

> **KeyConcept** nth Term of a Geometric Sequence
>
> The nth term a_n of a geometric sequence in which the first term is a_1 and the common ratio is r is given by the following formula, where n is any natural number.
>
> $$a_n = a_1 r^{n-1}$$

You will prove this formula in Exercise 68.

Real-World Example 1 Find the nth Term

MUSIC If the pattern continues, how many e-mails will be sent in the eighth round?

Understand We need to determine the number of forwarded e-mails on the eighth round. Five e-mails were sent on the first round. Each of the five recipients sent five e-mails on the second round, and so on.

Plan This is a geometric sequence, and the common ratio is 5. Use the formula for the nth term of a geometric sequence.

Solve $a_n = a_1 r^{n-1}$ nth term of a geometric sequence

$a_8 = 5(5)^{8-1}$ $a_1 = 5$, $r = 5$, and $n = 8$

$a_8 = 5(78{,}125)$ or $390{,}625$ $5^7 = 78{,}125$

Check Write out the first eight terms by multiplying by the common ratio.

5, 25, 125, 625, 3125, 15,625, 78,125, 390,625

There will be 390,625 e-mails sent on the 8th round.

▶ **GuidedPractice**

1. **E-MAILS** Shira receives a joke in an e-mail that asks her to forward it to four of her friends. She forwards it, then each of her friends forwards it to four of their friends, and so on. If the pattern continues, how many people will receive the e-mail on the ninth round of forwarding? **262,144**

 674 | Lesson 10-3

Lesson 10-3 Resources

Resource	Approaching Level **AL**	On Level **OL**	Beyond Level **BL**	English Learners **ELL**
Teacher Edition		▪ Differentiated Instruction, pp. 676, 680	▪ Differentiated Instruction, pp. 676, 680	▪ Differentiated Instruction, p. 676
Chapter Resource Masters	▪ Study Guide and Intervention, pp. 19–20 ▪ Skills Practice, p. 21 ▪ Practice, p. 22 ▪ Word Problem Practice, p. 23	▪ Study Guide and Intervention, pp. 19–20 ▪ Skills Practice, p. 21 ▪ Practice, p. 22 ▪ Word Problem Practice, p. 23 ▪ Enrichment, p. 24 ▪ Graphing Calculator Activity, p. 25	▪ Practice, p. 22 ▪ Word Problem Practice, p. 23 ▪ Enrichment, p. 24	▪ Study Guide and Intervention, pp. 19–20 ▪ Skills Practice, p. 21 ▪ Practice, p. 22 ▪ Word Problem Practice, p. 23
Other	▪ 5-Minute Check 10-3 ▪ Study Notebook	▪ 5-Minute Check 10-3 ▪ Study Notebook	▪ 5-Minute Check 10-3 ▪ Study Notebook	▪ 5-Minute Check 10-3 ▪ Study Notebook

Rubberball/Nicole Hill/Getty Images

If you are given some of the terms of a geometric sequence, you can determine an equation for finding the nth term of the sequence.

Example 2 Write an Equation for the nth Term

Write an equation for the nth term of each geometric sequence.

a. 0.5, 2, 8, 32, ...

$r = 8 \div 2$ or 4; 0.5 is the first term.
$a_n = a_1 r^{n-1}$ nth term of a geometric sequence
$a_n = 0.5(4)^{n-1}$ $a_1 = 0.5$ and $r = 4$

b. $a_4 = 5$ and $r = 6$

Step 1   Find a_1.
$a_n = a_1 r^{n-1}$ nth term of a geometric sequence
$5 = a_1(6^{4-1})$ $a_n = 5, r = 6$, and $n = 4$
$5 = a_1(216)$ Evaluate the power.
$\frac{5}{216} = a_1$ Divide each side by 216.

Step 2   Write the equation.
$a_n = a_1 r^{n-1}$ nth term of a geometric sequence
$a_n = \frac{5}{216}(6)^{n-1}$ $a_1 = \frac{5}{216}$ and $r = 6$

▶ **Guided**Practice   **2A.** $a_n = -0.25(-8)^{n-1}$

Write an equation for the nth term of each geometric sequence.

2A. $-0.25, 2, -16, 128, ...$ **2B.** $a_3 = 16, r = 4$ $a_n = 1(4)^{n-1}$

Like arithmetic means, **geometric means** are the terms between two nonconsecutive terms of a geometric sequence. The common ratio r can be used to find the geometric means.

ReadingMath

Geometric Means
A geometric mean can also be represented geometrically. In the figure below, h is the geometric mean between x and y.

Example 3 Find Geometric Means

Find three geometric means between 2 and 1250.

Step 1   Since there are three terms between the first and last term, there are $3 + 2$ or 5 total terms, so $n = 5$.

Step 2   Find r.
$a_n = a_1 r^{n-1}$ nth term of a geometric sequence
$1250 = 2r^{5-1}$ $a_n = 1250, a_1 = 2$, and $n = 5$
$625 = r^4$ Divide each side by 2.
$\pm 5 = r$ Take the 4th root of each side.

Step 3   Use r to find the four arithmetic means.

2   10   50   250   1250   or   2   −10   50   −250   1250
   × 5   × 5   × 5   × 5   × −5   × −5   × −5   × −5

The geometric means are 10, 50, and 250 or −10, 50, and −250.

▶ **Guided**Practice

3. Find four geometric means between 0.5 and 512.   **2, 8, 32, 128**

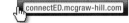 connectED.mcgraw-hill.com   **675**

Math History Link

Archytas (428–347 B.C.) Geometric sequences, or geometric progressions, were first studied by the Greek mathematician Archytas. His studies of these sequences came from his interest in music and octaves.

Hulton Archive/Getty Images

---

1 Geometric Sequences

Example 1 shows how to find a term in a geometric sequence when the first term and the common ratio are given. **Example 2** illustrates writing an equation for the nth term of a geometric sequence. **Example 3** shows how to find geometric means between two given numbers.

Formative Assessment

Use the Guided Practice exercises after each example to determine students' understanding of concepts.

Additional Examples

1 Find the sixth term of a geometric sequence for which $a_1 = -3$ and $r = -2$. $a_6 = 96$

2 Write an equation for the nth term of each geometric sequence.

a. 5, 10, 20, 40, ...
$a_n = 5 \cdot 2^{n-1}$

b. $a_5 = 4$ and $r = 3$
$a_n = \frac{4}{81}(3^{n-1})$

3 Find three geometric means between 3.12 and 49.92.
6.24, 12.48, 24.96, or −6.24, 12.48, −24.96

▶ **Additional Examples** also in Interactive Classroom PowerPoint® Presentations

IWB   **Interactive White Board** READY

WatchOut!

Preventing Errors Encourage students to begin a geometric sequence problem by writing the known values for each of the variables n, a, and r.

Focus on Mathematical Content

Increasing or Decreasing Geometric Sequences Many of the sequences in this lesson are either increasing or decreasing, and the common ratio for these sequences is positive. But if the common ratio in a geometric sequence is negative, the sequence will be neither increasing nor decreasing.

676 | Lesson 10-3 | Geometric Sequences and Series

Tips for New Teachers

Sense-Making You may wish to have students calculate one or more sums with and without using the formula in order to verify the formula. This will also demonstrate the formula's efficiency.

> ### WatchOut!
>
> **Preventing Errors** Emphasize the importance of writing every step of calculations as an equation, so that each numeric value found during the process is clearly identified.

2 Geometric Series

Example 4 illustrates finding the sum of the first n terms of a geometric series. **Example 5** shows how to evaluate a sum written in sigma notation. **Example 6** shows how to find the first term in a geometric series for which the sum of a specified number of terms and the common ratio are known.

Additional Examples

4 **MUSIC** Use the information in Example 4. How many e-mails will be sent after the sixth round? **19,530**

5 Find $\sum\limits_{n=1}^{12} 3 \cdot 2^{n-1}$. **12,285**

---

2 **Geometric Series** A **geometric series** is the sum of the terms of a geometric sequence. The sum of the first n terms of a series is denoted S_n. You can use either of the following formulas to find the partial sum S_n of the first n terms of a geometric series.

KeyConcept Partial Sum of a Geometric Series

Given	The sum S_n of the first n terms is:
a_1 and n	$S_n = \dfrac{a_1 - a_1 r^n}{1 - r}, r \neq 1$
a_1 and a_n	$S_n = \dfrac{a_1 - a_n r}{1 - r}, r \neq 1$

Real-World Example 4 Find the Sum of a Geometric Series

MUSIC Refer to the beginning of the lesson. If the pattern continues, what is the total number of e-mails sent in the eight rounds?

Five e-mails are sent in the first round and there are 8 rounds of e-mails.
So, $a_1 = 5$, $r = 5$ and $n = 8$.

$S_n = \dfrac{a_1 - a_1 r^n}{1 - r}$ ⠀⠀⠀ Sum formula

$S_8 = \dfrac{5 - 5 \cdot 5^8}{1 - 5}$ ⠀⠀⠀ $a_1 = 5$, $r = 5$, and $n = 8$

$S_8 = \dfrac{-1,953,120}{-4}$ ⠀⠀⠀ Simplify the numerator and denominator.

$S_8 = 488,280$ ⠀⠀⠀ Divide.

There will be 488,280 e-mails sent after 8 rounds.

> **Guided**Practice
>
> Find the sum of each geometric series.
>
> **4A.** $a_1 = 2$, $n = 10$, $r = 3$ ⠀ **59,048** ⠀⠀⠀ **4B.** $a_1 = 2000$, $a_n = 125$, $r = \frac{1}{2}$ ⠀ **3875**

As with arithmetic series, sigma notation can also be used to represent geometric series.

> ### WatchOut!
>
> **Sigma Notation** Notice in Example 5 that you are being asked to evaluate the sum from the 3rd term to the 10th term.

Example 5 Sum in Sigma Notation

Find $\sum\limits_{k=3}^{10} 4(2)^{k-1}$.

Find a_1, r, and n. In the first term, $k = 3$ and $a_1 = 4 \cdot 2^{3-1}$ or 16. The base of the exponential function is r, so $r = 2$. There are $10 - 3 + 1$ or 8 terms, so $n = 8$.

$S_n = \dfrac{a_1 - a_1 r^n}{1 - r}$ ⠀⠀⠀ Sum formula

$= \dfrac{16 - 16(2)^8}{1 - 2}$ ⠀⠀⠀ $a_1 = 16$, $r = 2$, and $n = 8$

$= 4080$ ⠀⠀⠀ Use a calculator.

> **Guided**Practice
>
> Find each sum.
>
> **5A.** $\sum\limits_{k=4}^{12} \frac{1}{4} \cdot 3^{k-1}$ ⠀ **66,426.75** ⠀⠀⠀ **5B.** $\sum\limits_{k=2}^{9} \frac{2}{3} \cdot 4^{k-1}$ ⠀ **58,253.333**

 676 | Lesson 10-3 | Geometric Sequences and Series

DifferentiatedInstruction ⓞⓛ ⓑⓛ ⓔⓛⓛ

If you think students might be interested in learning how this lesson applies to real-world situations,

Then have students research how biologists and ecologists use geometric series in their work to count and predict the population changes for various organisms.

You can use the formula for the sum of a geometric series to help find a particular term of the series.

Example 6 Find the First Term of a Series

Find a_1 in a geometric series for which $S_n = 13{,}116$, $n = 7$, and $r = 3$.

$$S_n = \frac{a_1 - a_1 r^n}{1 - r} \qquad \text{Sum formula}$$

$$13{,}116 = \frac{a_1 - a_1(3^7)}{1 - 3} \qquad S_n = 13{,}116,\ r = 3,\ \text{and } n = 7$$

$$13{,}116 = \frac{a_1(1 - 3^7)}{1 - 3} \qquad \text{Distributive Property}$$

$$13{,}116 = \frac{-2186a_1}{-2} \qquad \text{Subtract.}$$

$$13{,}116 = 1093a_1 \qquad \text{Simplify.}$$

$$12 = a_1 \qquad \text{Divide each side by 1093.}$$

▶ **Guided Practice**

6. Find a_1 in a geometric series for which $S_n = -26{,}240$, $n = 8$, and $r = -3$. **16**

Check Your Understanding   ○ **= Step-by-Step Solutions begin on page R14.**

Example 1

1. **CCSS REGULARITY** Dean is making a family tree for his grandfather. He was able to trace many generations. If Dean could trace his family back 10 generations, starting with his parents how many ancestors would there be? **2046**

Example 2

Write an equation for the nth term of each geometric sequence. 3. $a_n = 18\left(\frac{1}{3}\right)^{n-1}$

2. $2, 4, 8, \ldots$ $a_n = 2 \cdot 2^{n-1}$ 3. $18, 6, 2, \ldots$ 4. $-4, 16, -64, \ldots$ $a_n = -4(-4)^{n-1}$

⑤ $a_2 = 4, r = 3$ $a_n = \frac{4}{3}(3)^{n-1}$ 6. $a_6 = \frac{1}{8}, r = \frac{3}{4}$ $a_n = \frac{128}{243}\left(\frac{3}{4}\right)^{n-1}$ 7. $a_2 = -96, r = -8$ $a_n = 12(-8)^{n-1}$

Example 3

Find the geometric means of each sequence.

8. $0.25, \underline{\ ?\ }, \underline{\ ?\ }, \underline{\ ?\ }, 64$ **1, 4, 16 or -1, 4, -16**   9. $0.20, \underline{\ ?\ }, \underline{\ ?\ }, \underline{\ ?\ }, 125$ **1, 5, 25 or -1, 5, -25**

Example 4

10. **GAMES** Miranda arranges some rows of dominoes so that after she knocks over the first one, each domino knocks over two more dominoes when it falls. If there are ten rows, how many dominoes does Miranda use? **1023**

Example 5

Find the sum of each geometric series.

11. $\displaystyle\sum_{k=1}^{6} 3(4)^{k-1}$ **4095**   12. $\displaystyle\sum_{k=1}^{8} 4\left(\frac{1}{2}\right)^{k-1}$ **7.96875**

Example 6

Find a_1 for each geometric series described.

13. $S_n = 85\frac{5}{16}, r = 4, n = 6$ $\frac{1}{16}$ 14. $S_n = 91\frac{1}{12}, r = 3, n = 7$ $\frac{1}{12}$

15. $S_n = 1020, a_n = 4, r = \frac{1}{2}$ **512**   16. $S_n = 121\frac{1}{3}, a_n = \frac{1}{3}, r = \frac{1}{3}$ **81**

connectED.mcgraw-hill.com **677**

Additional Example

6 Find a_1 in a geometric series for which $S_8 = 765$, $n = 8$, and $r = 2$. 3

Focus on Mathematical Content

Two Formulas The formula for the sum of a geometric series $S_n = \dfrac{a_1 - a_n r}{1 - r}$ is used when the number of terms is not given. If n is known, the formula $S_n = \dfrac{a_1(1 - r^n)}{1 - r}$ can be used, since in this case it is not necessary to calculate a_n.

Teach with Tech

Document Camera Assign several exercises to the class, and give students time to work through them independently. Then choose students to share and explain their work to the class.

3 Practice

Formative Assessment

Use Exercises 1–16 to check for understanding.

Use the chart at the bottom of the next page to customize assignments for your students.

CCSS **Teaching the Mathematical Practices**

Regularity Mathematically proficient students notice if calculations are repeated, and look both for general methods and for shortcuts. As they work to solve a problem, mathematically proficient students maintain oversight of the process, while attending to the details.

Teaching the Mathematical Practices

Perseverance Mathematically proficient students start by explaining to themselves the meaning of a problem and looking for entry points to its solution. They analyze givens, constraints, relationships, and goals and make conjectures about the form and meaning of the solution and plan a solution pathway rather than simply jumping into a solution attempt.

Exercise Alert

Grid Paper For Exercises 81–83, students will need grid paper.

Additional Answers

64. $S_n = \dfrac{a_1 - a_1 r^n}{1-r}$

$= \dfrac{a_1 - a_1 r^{n-1} \cdot r}{1-r}$

$= \dfrac{a_1 - a_n r}{1-r}$

65. $S_n = \dfrac{a_1 - a_n r}{1-r}$

(Alternate sum formula)

$a_n = a_1 \cdot r^{n-1}$

(Formula for *n*th term)

$\dfrac{a_n}{r^{n-1}} = a_1$

(Divide each side by r^{n-1}.)

$S_n = \dfrac{\frac{a_n}{r^{n-1}} - a_n r}{1-r}$

(Substitution.)

$= \dfrac{\frac{a_n}{r^{n-1}} - \frac{a_n r \cdot r^{n-1}}{r^{n-1}}}{1-r}$

$\left(\text{Multiply by } \dfrac{\frac{r^{n-1}}{r^{n-1}}}{1}.\right)$

$= \dfrac{\frac{a_n(1-r^n)}{r^{n-1}}}{1-r}$

(Simplify.)

$= \dfrac{a_n(1-r^n)}{r^{n-1}(1-r)}$

(Divide by $(1-r)$.)

$= \dfrac{a_n(1-r^n)}{r^{n-1}-r^n}$

(Simplify.)

Practice and Problem Solving

Extra Practice is on page R10.

Example 1

17. **WEATHER** Heavy rain in Brieanne's town caused the river to rise. The river rose three inches the first day, and each day after rose twice as much as the previous day. How much did the river rise in five days? **93 in.**

Find a_n for each geometric sequence.

18. $a_1 = 2400, r = \frac{1}{4}, n = 7$ $7\frac{75}{128}$ or **0.5859375** **19.** $a_1 = 800, r = \frac{1}{2}, n = 6$ **25**

20. $a_1 = \frac{2}{9}, r = 3, n = 7$ **162** **21.** $a_1 = -4, r = -2, n = 8$ **512**

22. **BIOLOGY** A certain bacteria grows at a rate of 3 cells every 2 minutes. If there were 260 cells initially, how many are there after 21 minutes? **864,567**

Example 2 Write an equation for the *n*th term of each geometric sequence.

23. $-3, 6, -12, \ldots$ **24.** $288, -96, 32, \ldots$ **25.** $-1, 1, -1, \ldots$

26. $\frac{1}{3}, \frac{2}{9}, \frac{4}{27}, \ldots$ **27.** $8, 2, \frac{1}{2}, \ldots$ **28.** $12, -16, \frac{64}{3}, \ldots$

29. $a_3 = 28, r = 2$ **30.** $a_4 = -8, r = 0.5$ **31.** $a_6 = 0.5, r = 6$

32. $a_3 = 8, r = \frac{1}{2}$ **33.** $a_4 = 24, r = \frac{1}{3}$ **34.** $a_4 = 80, r = 4$

Example 3 Find the geometric means of each sequence.

35. $810, \underline{?}, \underline{?}, \underline{?}, 10$ **36.** $640, \underline{?}, \underline{?}, \underline{?}, 2.5$

37. $\frac{7}{2}, \underline{?}, \underline{?}, \underline{?}, \frac{56}{81}$ **38.** $\frac{729}{64}, \underline{?}, \underline{?}, \underline{?}, \frac{324}{9}$

39. Find two geometric means between 3 and 375. **15, 75**

40. Find two geometric means between 16 and -2. **-8, 4**

Example 4

41. **CCSS PERSEVERANCE** A certain water filtration system can remove 70% of the contaminants each time a sample of water is passed through it. If the same water is passed through the system four times, what percent of the original contaminants will be removed from the water sample? **99.99%**

Find the sum of each geometric series.

42. $a_1 = 36, r = \frac{1}{3}, n = 8$ **53.9918** **43.** $a_1 = 16, r = \frac{1}{2}, n = 9$ **31.9375**

44. $a_1 = 240, r = \frac{3}{4}, n = 7$ **831.855** **45.** $a_1 = 360, r = \frac{4}{3}, n = 8$ **9707.82**

46. **VACUUMS** A vacuum claims to pick up 80% of the dirt every time it is run over the carpet. Assuming this is true, what percent of the original amount of dirt is picked up after the seventh time the vacuum is run over the carpet? **99.99%**

Example 5 Find the sum of each geometric series.

47. $\sum_{k=1}^{7} 4(-3)^{k-1}$ **2188** **48.** $\sum_{k=1}^{8} (-3)(-2)^{k-1}$ **255** **49.** $\sum_{k=1}^{9} (-1)(4)^{k-1}$ **$-87,381$** **50.** $\sum_{k=1}^{10} 5(-1)^{k-1}$ **0**

Example 6 Find a_1 for each geometric series described.

51 $S_n = -2912, r = 3, n = 6$ **-8** **52.** $S_n = -10,922, r = 4, n = 7$ **-2**

53. $S_n = 1330, a_n = 486, r = \frac{3}{2}$ **64** **54.** $S_n = 4118, a_n = 128, r = \frac{2}{3}$ **1458**

55. $a_n = 1024, r = 8, n = 5$ **0.25** **56.** $a_n = 1875, r = 5, n = 7$ **$\frac{3}{25}$**

678 | Lesson 10-3 | Geometric Sequences and Series

---

23. $a_n = (-3)(-2)^{n-1}$

24. $a_n = 288\left(-\frac{1}{3}\right)^{n-1}$

25. $a_n = (-1)(-1)^{n-1}$

26. $a_n = \frac{1}{3}\left(\frac{2}{3}\right)^{n-1}$

27. $a_n = 8 \cdot \left(\frac{1}{4}\right)^{n-1}$

28. $a_n = 12\left(-\frac{4}{3}\right)^{n-1}$

29. $a_n = 7(2)^{n-1}$

30. $a_n = -64(0.5)^{n-1}$

31. $a_n = \frac{1}{15,552}(6)^{n-1}$

32. $a_n = 32\left(\frac{1}{2}\right)^{n-1}$

33. $a_n = 648\left(\frac{1}{3}\right)^{n-1}$

34. $a_n = \frac{5}{4}(4)^{n-1}$

35. 270, 90, 30 or -270, 90, -30

36. 160, 40, 10 or -160, 40, -10

37. $\frac{7}{3}, \frac{14}{9}, \frac{28}{27}$ or $-\frac{7}{3}, \frac{14}{9}, -\frac{28}{27}$

38. $\frac{243}{16}, \frac{81}{4}, 27$ or $-\frac{243}{16}, \frac{81}{4}, -27$

Differentiated Homework Options

Level	Assignment	Two-Day Option	
AL Basic	17–57, 64–68, 71–89	17–57 odd, 73–76	18–56 even, 64–68, 71, 72, 77–89
OL Core	17–57 odd, 58–68, 71–89	17–57, 73–76	58–68, 71, 72, 77–89
BL Advanced	58–85, (optional: 86–89)		

678 | Lesson 10-3 | Geometric Sequences and Series

57. SCIENCE One minute after it is released, a gas-filled balloon has risen 100 feet. In each succeeding minute, the balloon rises only 50% as far as it rose in the previous minute. How far will it rise in 5 minutes? **193.75 ft**

B 58. CHEMISTRY Radon has a half-life of about 4 days. This means that about every 4 days, half of the mass of radon decays into another element. How many grams of radon remain from an initial 60 grams after 4 weeks? **about 0.46875 g**

59. CCSS REASONING A virus goes through a computer, infecting the files. If one file was infected initially and the total number of files infected doubles every minute, how many files will be infected in 20 minutes? **524,288**

60. GEOMETRY In the figure, the sides of each equilateral triangle are twice the size of the sides of its inscribed triangle. If the pattern continues, find the sum of the perimeters of the first eight triangles. **about 119.5 cm**

— 20 cm —

61. PENDULUMS The first swing of a pendulum travels 30 centimeters. If each subsequent swing travels 95% as far as the previous swing, find the total distance traveled by the pendulum after the 30th swing. **about 471 cm**

C 62. PHONE CHAINS A school established a phone chain in which every staff member calls two other staff members to notify them when the school closes due to weather. The first round of calls begins with the superintendent calling both principals. If there are 94 total staff members and employees at the school, how many rounds of calls are there? **7**

63. TELEVISIONS High Tech Electronics advertises a weekly installment plan for the purchase of a popular brand of high definition television. The buyer pays $5 at the end of the first week, $5.50 at the end of the second week, $6.05 at the end of the third week, and so on for one year. (Assume that 1 year = 52 weeks.)

63a. $11.79, $30.58, $205.72

a. What will the payments be at the end of the 10th, 20th, and 40th weeks?

b. Find the total cost of the TV. **$7052.15**

c. Why is the cost found in part **b** not entirely accurate?

63c. Each payment made is rounded to the nearest penny, so the sum of the payments will actually be more than the sum found in part b.

H.O.T. Problems Use Higher-Order Thinking Skills

64. PROOF Derive the General Sum Formula using the Alternate Sum Formula. **See margin.**

65. PROOF Derive a sum formula that does not include a_1. **See margin.**

66. OPEN ENDED Write a geometric series for which $r = \frac{3}{4}$ and $n = 6$.

66. Sample answer:
$256 + 192 + 144 + 108 + 81 + \frac{243}{4}$

67. REASONING Explain how $\sum_{k=1}^{10} 3(2)^{k-1}$ needs to be altered to refer to the same series if $k = 1$ changes to $k = 0$. Explain your reasoning. **See Chapter 10 Answer Appendix.**

68. PROOF Prove the formula for the nth term of a geometric sequence. **See Chapter 10 Answer Appendix.**

69. CHALLENGE The fifth term of a geometric sequence is $\frac{1}{27}$th of the eighth term. If the ninth term is 702, what is the eighth term? **234**

70. CHALLENGE Use the fact that h is the geometric mean between x and y in the figure at the right to find h^4 in terms of x and y. x^2y^2

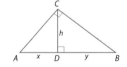

71. OPEN ENDED Write a geometric series with 6 terms and a sum of 252. **Sample answer:** $4 + 8 + 16 + 32 + 64 + 128$

72. WRITING IN MATH How can you classify a sequence? Explain your reasoning. **See margin.**

Additional Answer

72. Sample answer: A series is arithmetic if every pair of consecutive terms shares a common difference. A series is geometric if every pair of consecutive terms shares a common ratio. If the series displays both qualities, then it is both arithmetic and geometric. If the series displays neither quality, then it is neither geometric nor arithmetic.

Study Guide and Intervention
AL OL ELL

Practice
AL OL BL ELL

Word Problem Practice
AL OL BL ELL

Enrichment
OL BL

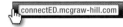

Crystal Ball Ask students to describe how they think their knowledge of sums of geometric series might help them learn about sums of infinite geometric series in the next lesson.

Additional Answers

81. $(3, 1)$, 5 units

82. $(-3, -7)$, 9 units

83. $(3, -7)$, $5\sqrt{2}$ units

Standardized Test Practice

73. Which of the following is closest to $\sqrt[3]{7.32}$? **B**

A 1.8
B 1.9
C 2.0
D 2.1

74. The first term of a geometric series is 5, and the common ratio is -2. How many terms are in the series if its sum is -6825? **J**

F 5
G 9
H 10
J 12

75. **SHORT RESPONSE** Danette has a savings account. She withdraws half of the contents every year. After 4 years, she has $2000 left. How much did she have in the savings account originally?
$32,000

76. **SAT/ ACT** The curve below could be part of the graph of which function? **E**

A $y = \sqrt{x}$
B $y = x^2 - 5x + 4$
C $y = -x + 20$
D $y = \log x$
E $xy = 4$

Spiral Review

77. **MONEY** Elena bought a high-definition LCD television at the electronics store. She paid $200 immediately and $75 each month for a year and a half. How much did Elena pay in total for the TV? (Lesson 10-2) **$1550**

79. Arithmetic; the common difference is $\frac{1}{50}$. **80.** Neither; there is no common difference or ratio.

Determine whether each sequence is *arithmetic*, *geometric*, or *neither*. Explain your reasoning. (Lesson 10-1) **78.** Neither; there is no common difference or ratio.

78. $\frac{1}{10}, \frac{3}{5}, \frac{7}{20}, \frac{17}{20}, \ldots$

79. $-\frac{7}{25}, -\frac{13}{50}, -\frac{6}{25}, -\frac{11}{50}, \ldots$

80. $-\frac{22}{3}, -\frac{68}{9}, -\frac{208}{27}, -\frac{632}{81}, \ldots$

Find the center and radius of each circle. Then graph the circle. (Lesson 9-3) **81–83. See margin.**

81. $(x - 3)^2 + (y - 1)^2 = 25$

82. $(x + 3)^2 + (y + 7)^2 = 81$

83. $(x - 3)^2 + (y + 7)^2 = 50$

84. Suppose y varies jointly as x and z. Find y when $x = 9$ and $z = -5$, if $y = -90$ when $z = 15$ and $x = -6$. (Lesson 8-5) **−45**

85. **SHOPPING** A certain store found that the number of customers who will attend a sale can be modeled by $N = 125\sqrt[3]{100Pt}$, where N is the number of customers expected, P is the percent of the sale discount, and t is the number of hours the sale will last. Find the number of customers the store should expect for a sale that is 50% off and will last four hours. (Lesson 6-4) **731 customers**

Skills Review

Evaluate each expression if $a = -2$, $b = \frac{1}{3}$, and $c = -12$.

86. $\frac{3ab}{c}$ $\frac{1}{6}$

87. $\frac{a - c}{a + c}$ $-\frac{5}{7}$

88. $\frac{a^3 - c}{b^2}$ 36

89. $\frac{c + 3}{ab}$ $\frac{27}{2}$

DifferentiatedInstruction **OL** **BL**

Extension Write the expression for the nth term of a geometric sequence, $a_1 r^{n-1}$, on the board. Ask students to write an expression for the *next* term, with term number $(n + 1)$, by replacing n with $(n + 1)$ in the expression $a_1 r^{n-1}$. Then have them show that $\frac{a_{n+1}}{a_n}$ is equal to the common ratio r. $\frac{a_1 r^{(n+1)-1}}{a_1 r^{n-1}} = \frac{r^n}{r^{n-1}} = r^{n-(n-1)} = r^1 = r$

EXPLORE 10-4

Algebra Lab
Area Under a Curve

The Morgans are renovating the outside of their house so that there is an archway above the front entrance. Mr. Morgan made a scale drawing of the archway in which each line on the grid paper represents one foot of the actual archway. Mrs. Morgan modeled the shape of the top with the quadratic equation $y = -0.25x^2 + 3x$.

4 ft

5 ft

8 ft

Activity

Find the area of the opening under the archway.

Method 1

Step 1 Make a table of values for $y = -0.25x^2 + 3x$. Then graph the equation.

x	0	1	2	3	4	5	6	7	8	9	10	11	12
y	0	2.75	5	6.75	8	8.75	9	8.75	8	6.75	5	2.75	0

Step 2 Divide the figure into regions.

To estimate the area inside the archway, you can divide the archway into rectangles as shown in blue.

Because the left and right sides of the archway are 5 feet high and $y = 5$ when $x = 2$ and when $x = 10$, the opening of the entrance extends from $x = 2$ to $x = 10$.

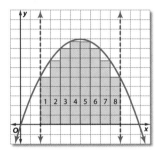

Step 3 Find the area of the regions.

Rectangle	1	2	3	4	5	6	7	8
Width (ft)	1	1	1	1	1	1	1	1
Height (ft)	5	6.75	8	8.75	8.75	8	6.75	5
Area (ft²)	5	6.75	8	8.75	8.75	8	6.75	5

The approximate area of the archway is the sum of the areas of the rectangles.

$5 + 6.75 + 8 + 8.75 + 8.75 + 8 + 6.75 + 5 = 57$ ft²

(continued on the next page)

1 Focus

Objective Approximate the area under a curve over a specified interval, using the sum of rectangular areas under the curve.

Materials for Each Group
- grid paper

Easy to Make Manipulatives
Teaching Algebra with Manipulatives
Template for:
- grid paper, p. 1

Teaching Tip
Students should be familiar with finding the area of irregular figures. In a sense, this lesson is also about finding the area of an irregular figure. Each rectangle the students divide the area into should have the same width. You might suggest that students number the rectangles they draw under the curve 1–8.

2 Teach

Working in Cooperative Groups
Pair students with different abilities. Have students work through the Steps for both Methods 1 and 2 of the Activity and then answer Exercises 1–4.

Ask:
- What does the width of each grid on your graph represent? 1 foot
- In Method 1, will your rectangles go beyond the graph of the quadratic equation? no
- How can you find the height of each rectangle? Use the values for x and y found in the table made in Step 1.

Practice Have students complete Exercises 5–7.

3 Assess

Formative Assessment
Use Exercise 5 to assess whether students can approximate the area under a curve for a specified interval.

From Concrete to Abstract
Ask students to repeat the Activity letting each grid represent 0.5 feet instead of 1 foot, thus using 16 rectangles instead of 8. Ask students if they think the results are closer to the actual area under the curve. Then ask how they might get even closer to the actual number of square feet found under the curve.

Additional Answers

4.

Rectangle	1	2	3	4
Width	2	2	2	2
Height	6.75	9	9	6.75
Area	13.5	18	18	13.5

Estimate of area: $2(13.5 + 18) = 63$ ft². The estimate will be greater than the actual area because there are 4 portions of the rectangles that lie outside the area under the curve and 2 portions of the area under the curve that are not in the rectangles. This estimate is probably more accurate than the other two estimates.

Algebra Lab
Area Under a Curve *Continued*

Method 2

Step 1 Draw a second graph of the equation and divide into regions. Divide the archway into rectangles as shown in blue.

Step 2 Find the area of the regions.

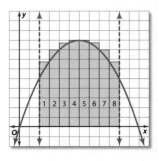

Rectangle	1	2	3	4	5	6	7	8
Width (ft)	1	1	1	1	1	1	1	1
Height (ft)	6.75	8	8.75	9	9	8.75	8	6.75
Area (ft²)	6.75	8	8.75	9	9	8.75	8	6.75

The approximate area of the archway is the sum of the areas of the rectangles.
$$6.75 + 8 + 8.75 + 9 + 9 + 8.75 + 8 + 6.75 = 65 \text{ ft}^2$$

Both Method 1 and Method 2 illustrate how to approximate the area under a curve over a specified interval.

Analyze the Results

1. Is the area of the regions calculated using Method 1 greater than or less than the actual area of the archway? Explain your reasoning.

2. Is the area of the regions calculated using Method 2 greater than or less than the actual area of the archway? Explain your reasoning.

3. Compare the area estimates for both methods. How could you find the best estimate for the area inside the archway? Explain your reasoning.

4. The diagram shows a third method for finding an estimate of the area of the archway. Is this estimate for the area greater than or less than the actual area? How does this estimate compare to the other two estimates of the area? **See margin.**

1. Less; sample answer: The rectangles are inside the curve leaving some area unaccounted for.
2. Greater; sample answer: The rectangles include area outside the curve.
3. Sample answer: Find the mean of the estimates: $(65 + 57) \div 2 = 61$.

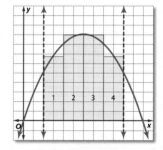

Exercises

Estimate the area described by any method. Make a table of values, draw graphs with rectangles, and make a table for the areas of the rectangles. Compare each estimate to the actual area. **5–7. See Chapter 10 Answer Appendix.**

5. the area under the curve for $y = -x^2 + 4$, from $x = -2$ to $x = 2$, and above the x–axis

6. the area under the curve for $y = x^3$, from $x = 0$ to $x = 4$, and above the x–axis

7. the area under the curve for $y = x^2$, from $x = -3$ to $x = 3$, and above the x–axis

LESSON 10-4 Infinite Geometric Series

Then	Now	Why?
● You found sums of finite geometric series.	**1** Find sums of infinite geometric series. **2** Write repeating decimals as fractions.	● With their opponent on the 10-yard line, the defense is penalized half the distance to the goal, placing the ball on the 5-yard line. If they continue to be penalized in this way, where will the ball eventually be placed? Will they ever reach the goal line? How many total penalty yards will the defense have incurred? These questions can be answered by looking at infinite geometric series.

NewVocabulary
infinite geometric series
convergent series
divergent series
infinity

Common Core State Standards

Mathematical Practices
6 Attend to precision.
8 Look for and express regularity in repeated reasoning.

1 **Infinite Geometric Series** An **infinite geometric series** has an infinite number of terms. A series that has a sum is a **convergent series**, because its sum converges to a specific value. A series that does not have a sum is a **divergent series**.

When you evaluated the sum S_n of an infinite geometric series for the first n terms, you were finding the partial sum of the series. It is also possible to find the sum of an entire series. In the application above, it seems that the ball will eventually reach the goal line, and the defense will be penalized a total of 10 yards. This value is the actual sum of the infinite series $5 + 2.5 + 1.25 + \ldots$. The graph of S_n for $1 \le n \le 10$ is shown on the left below. As n increases, S_n approaches 10.

KeyConcept Convergent and Divergent Series

Convergent Series	Divergent Series
Words The sum approaches a finite value.	**Words** The sum does not approach a finite value.
Ratio $|r| < 1$	**Ratio** $|r| \ge 1$
Example $5 + 2.5 + 1.25 + \ldots$	**Example** $\frac{1}{16} + \frac{1}{8} + \frac{1}{4} + \ldots$

Example 1 Convergent and Divergent Series

Determine whether each infinite geometric series is *convergent* or *divergent*.

a. $54 + 36 + 24 + \ldots$

Find the value of r.

$r = \frac{36}{54}$ or $\frac{2}{3}$; since $-1 < \frac{2}{3} < 1$, the series is convergent.

connectED.mcgraw-hill.com **683**

Darrin Klimek/Digital Vision/Getty Images

1 Focus

VerticalAlignment

Before Lesson 10-4 Find sums of finite geometric series.

Lesson 10-4 Find sums of infinite geometric series. Write repeating decimals as fractions.

After Lesson 10-4 Write recursive formulas for sequences.

2 Teach

Scaffolding Questions

Have students read the **Why?** section of the lesson.

Ask:

- To write a geometric series to model this situation, what would the first term be? 5

- What would the second term be? 2.5

- Write the first four terms of the series. $5 + 2.5 + 1.25 + 0.625 + \ldots$

- Why is this series called an infinite series? Sample answer: It goes on forever.

Lesson 10-4 Resources

Resource	Approaching Level AL	On Level OL	Beyond Level BL	English Learners ELL
Teacher Edition		● Differentiated Instruction, pp. 685, 689	● Differentiated Instruction, pp. 685, 689	
Chapter Resource Masters	● Study Guide and Intervention, pp. 26–27 ● Skills Practice, p. 28 ● Practice, p. 29 ● Word Problem Practice, p. 30	● Study Guide and Intervention, pp. 26–27 ● Skills Practice, p. 28 ● Practice, p. 29 ● Word Problem Practice, p. 30 ● Enrichment, p. 31	● Practice, p. 29 ● Word Problem Practice, p. 30 ● Enrichment, p. 31	● Study Guide and Intervention, pp. 26–27 ● Skills Practice, p. 28 ● Practice, p. 29 ● Word Problem Practice, p. 30
Other	● 5-Minute Check 10-4 ● Study Notebook	● 5-Minute Check 10-4 ● Study Notebook	● 5-Minute Check 10-4 ● Study Notebook	● 5-Minute Check 10-4 ● Study Notebook

Example 1 shows how to determine whether an infinite geometric series is convergent or divergent. **Example 2** shows how to find the sum of an infinite geometric series, while **Example 3** does the same for a series given in sigma notation.

Formative Assessment

Use the Guided Practice exercises after each example to determine students' understanding of concepts.

Additional Examples

1 Determine whether each infinite geometric series is *convergent* or *divergent*.

a. $729 + 243 + 81 + \ldots$
convergent

b. $2 + 5 + 12.5 + \ldots$
divergent

2 Find the sum of each infinite geometric series, if it exists.

a. $-\dfrac{4}{3} + 4 - 12 + 36 - 108 + \ldots$ no sum

b. $3 - \dfrac{3}{2} + \dfrac{3}{4} - \dfrac{3}{8} + \ldots$ 2

Additional Examples also in Interactive Classroom PowerPoint® Presentations

IWB Interactive White Board READY

Tips for New Teachers

Reasoning Make sure students can explain why $|r| < 1$ can also be written as $-1 < r < 1$. Graphing this inequality on a number line may help students understand what is meant by these two different mathematical notations.

---

b. $8 + 12 + 18 + \ldots$

$r = \dfrac{12}{8}$ or 1.5; since $1.5 > 1$, the series is divergent.

▶ **Guided**Practice

1A. $2 + 3 + 4.5 + \ldots$ **divergent**   **1B.** $100 + 50 + 25 + \ldots$ **convergent**

> **StudyTip**
> **Absolute Value** Recall that $|r| < 1$ means $-1 < r < 1$.

When $|r| < 1$, the value of r^n will approach 0 as n increases. Therefore, the partial sums of the infinite geometric series will approach $\dfrac{a_1 - a_1(0)}{1 - r}$ or $\dfrac{a_1}{1 - r}$.

> **KeyConcept** Sum of an Infinite Geometric Series
>
> The sum S of an infinite geometric series with $|r| < 1$ is given by
> $$S = \frac{a_1}{1 - r}.$$
> If $|r| \geq 1$, the series has no sum.

When an infinite geometric series is divergent, $|r| \geq 1$ and the series has no sum because the absolute value of r^n will increase infinitely as n increases.

The table at the right shows the partial sums for the divergent series $4 + 16 + 64 + \ldots$. As n increases, S_n increases rapidly without limit.

n	S_n
5	1364
10	1,398,100
15	1,431,655,764

> **Example 2** Sum of an Infinite Series

Find the sum of each infinite series, if it exists.

Determine whether each infinite geometric series is *convergent* or *divergent*.

a. $\dfrac{2}{3} + \dfrac{6}{15} + \dfrac{18}{75} + \ldots$

Step 1 Find the value of r to determine if the sum exists.

$r = \dfrac{6}{15} \div \dfrac{2}{3}$ or $\dfrac{3}{5}$ Divide consecutive terms.

Since $\left|\dfrac{3}{5}\right| < 1$, the sum exists.

Step 2 Use the formula to find the sum.

$S = \dfrac{a_1}{1 - r}$ Sum formula

$= \dfrac{\frac{2}{3}}{1 - \frac{3}{5}}$ $a_1 = \dfrac{2}{3}$ and $r = \dfrac{3}{5}$

$= \dfrac{2}{3} \div \dfrac{2}{5}$ or $\dfrac{5}{3}$ Simplify.

b. $6 + 9 + 13.5 + 20.25 + \ldots$

$r = \dfrac{9}{6}$ or 1.5; since $|1.5| \geq 1$, the series diverges and the sum does not exist.

> **StudyTip**
> **Convergence and Divergence** A series converges when the absolute value of a term is smaller than the absolute value of the previous term. An infinite arithmetic series will always be divergent.

▶ **Guided**Practice

2A. $4 - 2 + 1 - 0.5 + \ldots$ $\dfrac{8}{3}$ **2B.** $16 + 20 + 25 + \ldots$ **No sum exists.**

Focus on Mathematical Content

Sum of an Infinite Geometric Series Whether an infinite geometric series has a sum depends on the product of any term and the common ratio r. The absolute value of the product of any number and a fraction between -1 and 1 will always be less than the absolute value of the original number, so the series will have a sum if $|a_i| > |a_i r|$.

Sigma notation can be used to represent infinite series. If a sequence goes to **infinity**, it continues without end. The infinity symbol ∞ is placed above the $\sum$ to indicate that a series is infinite.

Example 3 Infinite Series in Sigma Notation

Find $\displaystyle\sum_{k=1}^{\infty} 18\left(\frac{4}{5}\right)^{k-1}$.

$$S = \frac{a_1}{1-r} \qquad \text{Sum formula}$$

$$= \frac{18}{1-\frac{4}{5}} \qquad a_1 = 18 \text{ and } r = \frac{4}{5}$$

$$= \frac{18}{\frac{1}{5}} \text{ or } 90 \qquad \text{Simplify.}$$

▶ **Guided**Practice

3. Find $\displaystyle\sum_{k=1}^{\infty} 12\left(\frac{3}{4}\right)^{k-1}$.   48

---

2 Repeating Decimals A repeating decimal is the sum of an infinite geometric series. For instance, $0.\overline{45} = 0.454545\ldots$ or $0.45 + 0.0045 + 0.000045 + \ldots$. The formula for the sum of an infinite series can be used to convert the decimal to a fraction.

Problem-Solving Tip

CCSS Sense-Making In many cases, it is possible to solve a problem in more than one way. Use the method with which you are most comfortable.

Example 4 Write a Repeating Decimal as a Fraction

Write $0.\overline{63}$ as a fraction.

Method 1 Use the sum of an infinite series.

$$0.\overline{63} = 0.63 + 0.0063 + \ldots$$

$$= \frac{63}{100} + \frac{63}{10,000} + \ldots$$

$$S = \frac{a_1}{1-r} \qquad \text{Sum formula}$$

$$= \frac{\frac{63}{100}}{1 - \frac{1}{100}} \qquad a_1 = \frac{63}{100} \text{ and } r = \frac{1}{100}$$

$$= \frac{63}{99} \text{ or } \frac{7}{11} \qquad \text{Simplify.}$$

Method 2 Use algebraic properties.

$$x = 0.\overline{63} \qquad\qquad \text{Let } x = 0.\overline{63}.$$

$$x = 0.636363\ldots \qquad \text{Write as a repeating decimal.}$$

$$100x = 63.636363\ldots \qquad \text{Multiply each side by 100.}$$

$$99x = 63 \qquad\qquad \text{Subtract } x \text{ from } 100x \text{ and } 0.\overline{63} \text{ from } 63.\overline{63}.$$

$$x = \frac{63}{99} \text{ or } \frac{7}{11} \qquad \text{Divide each side by 99.}$$

StudyTip

Repeating Decimals Every repeating decimal is a rational number and can be written as a fraction.

▶ **Guided**Practice

4. Write $0.\overline{21}$ as a fraction. $\frac{7}{33}$

Additional Example

3 Find $\displaystyle\sum_{k=1}^{\infty} 5\left(\frac{1}{2}\right)^{k-1}$.   10

WatchOut!

Preventing Errors Ask students to write a few terms of the series in Example 3 to make sure they know how to read the notation.

2 Repeating Decimals
Example 4 shows how to write a repeating decimal as a fraction.

Additional Example

4 Write $0.\overline{25}$ as a fraction. $\frac{25}{99}$

CCSS **Teaching the Mathematical Practices**

Sense-Making Mathematically proficient students start by explaining to themselves the meaning of a problem and looking for entry points to its solution. Encourage students to determine a method of solving a problem that makes sense to them.

Teach with Tech

Instant Messaging Have students work in pairs. The first student will send a message including a repeating decimal. The second student must then write the decimal as a fraction. Have students switch roles and repeat.

DifferentiatedInstruction **OL** **BL**

Logical Learners Have students research and read about the famous mathematical puzzle called Zeno's paradox. Have them discuss this story of the tortoise's race in terms of the content of this lesson.

3 Practice

Formative Assessment

Use Exercises 1–15 to check for understanding.

Use the chart on the bottom of this page to customize assignments for your students.

 Teaching the Mathematical Practices

Sense-Making Mathematically proficient students start by explaining to themselves the meaning of a problem and looking for entry points to its solution. They analyze givens, constraints, relationships, and goals. They check their answers to problems using a different method, and they continually ask themselves, "Does this make sense?"

Precision Mathematically proficient students try to use clear definitions in their reasoning, calculate accurately and efficiently, and make explicit use of definitions.

Check Your Understanding = Step-by-Step Solutions begin on page R14.

Example 1   Determine whether each infinite geometric series is *convergent* or *divergent*.

1. $16 - 8 + 4 - \ldots$ **convergent**   **2.** $32 - 48 + 72 - \ldots$ **divergent**

3. $0.5 + 0.7 + 0.98 + \ldots$ **divergent**   **4.** $1 + 1 + 1 + \ldots$ **divergent**

Example 2   Find the sum of each infinite series, if it exists.

5. $440 + 220 + 110 + \ldots$ **880**   **6.** $520 + 130 + 32.5 + \ldots$ **$693\frac{1}{3}$**

7. $\frac{1}{4} + \frac{3}{8} + \frac{9}{16} + \ldots$ **No sum exists.**   **8.** $\frac{32}{9} + \frac{16}{3} + 8 + \ldots$ **No sum exists.**

9. **CCSS SENSE-MAKING** A certain drug has a half-life of 8 hours after it is administered to a patient. What percent of the drug is still in the patient's system after 24 hours? **12.5%**

Example 3   Find the sum of each infinite series, if it exists.

10. $\sum_{k=1}^{\infty} 5 \cdot 4^{k-1}$ **No sum exists.**   **11.** $\sum_{k=1}^{\infty} (-2) \cdot (0.5)^{k-1}$ **−4**

12. $\sum_{k=1}^{\infty} 3 \cdot \left(\frac{4}{5}\right)^{k-1}$ **15**   **13.** $\sum_{k=1}^{\infty} \frac{1}{2} \cdot \left(\frac{3}{4}\right)^{k-1}$ **2**

Example 4   Write each repeating decimal as a fraction.

14. $0.\overline{35}$ **$\frac{35}{99}$**   **15.** $0.\overline{642}$ **$\frac{214}{333}$**

Practice and Problem Solving   Extra Practice is on page R10.

Example 1   Determine whether each infinite geometric series is *convergent* or *divergent*.

16. $21 + 63 + 189 + \ldots$ **divergent**   **17.** $480 + 360 + 270 + \ldots$ **convergent**

18. $\frac{3}{4} + \frac{9}{8} + \frac{27}{16} + \ldots$ **divergent**   **19.** $\frac{5}{6} + \frac{10}{9} + \frac{40}{27} + \ldots$ **divergent**

20. $0.1 + 0.01 + 0.001 + \ldots$ **convergent**   **21.** $0.008 + 0.08 + 0.8 + \ldots$ **divergent**

Example 2   Find the sum of each infinite series, if it exists.

22. $18 + 21.6 + 25.92 + \ldots$ **No sum exists.**   **23.** $-3 - 4.2 - 5.88 - \ldots$ **No sum exists.**

24. $\frac{1}{2} + \frac{1}{6} + \frac{1}{18} + \ldots$ **$\frac{3}{4}$**   **25** $\frac{12}{5} + \frac{6}{5} + \frac{3}{5} + \ldots$ **$\frac{24}{5}$**

26. $21 + 14 + \frac{28}{3} + \ldots$ **63**   **27.** $32 + 40 + 50 + \ldots$ **No sum exists.**

28. **SWINGS** If Kerry does not push any harder after his initial swing, the distance traveled per swing will decrease by 10% with each swing. If his initial swing traveled 6 feet, find the total distance traveled when he comes to rest. **60 ft**

Differentiated Homework Options

Level	Assignment	Two-Day Option	
AL Basic	16–40, 60, 61, 63–76	17–39 odd, 68–71	16–40 even, 60, 61, 63–67, 72–76
OL Core	17–39 odd, 41–43, 45–49 odd, 50–61, 63–76	16–40, 68–71	41–61, 63–67, 72–76
BL Advanced	41–73, (optional: 74–76)		

Example 3 Find the sum of each infinite series, if it exists. **29, 30, 32.** No sum exists.

29. $\sum_{k=1}^{\infty} \frac{4}{3} \cdot \left(\frac{5}{4}\right)^{k-1}$

30. $\sum_{k=1}^{\infty} \frac{1}{4} \cdot 3^{k-1}$

31. $\sum_{k=1}^{\infty} \frac{5}{3} \cdot \left(\frac{3}{7}\right)^{k-1}$ $\frac{35}{12}$

32. $\sum_{k=1}^{\infty} \frac{2}{3} \cdot \left(\frac{4}{3}\right)^{k-1}$

33. $\sum_{k=1}^{\infty} \frac{8}{3} \cdot \left(\frac{5}{6}\right)^{k-1}$ 16

34. $\sum_{k=1}^{\infty} \frac{1}{8} \cdot \left(\frac{1}{12}\right)^{k-1}$ $\frac{3}{22}$

Example 4 Write each repeating decimal as a fraction.

35 $0.3\overline{21}$ $\frac{53}{165}$

36. $0.1\overline{45}$ $\frac{8}{55}$

37. $2.\overline{18}$ $\frac{24}{11}$

38. $4.\overline{96}$ $\frac{164}{33}$

39. $0.12\overline{14}$ $\frac{601}{4950}$

40. $0.43\overline{36}$ $\frac{477}{1100}$

B **41. FANS** A fan is running at 10 revolutions per second. After it is turned off, its speed decreases at a rate of 75% per second. Determine the number of revolutions completed by the fan after it is turned off. $\frac{40}{3}$

42. CCSS PRECISION Kamiko deposited $5000 into an account at the beginning of the year. The account earns 8% interest each year.

   a. How much money will be in the account after 20 years? (*Hint:* Let $5000(1 + 0.08)^1$ represent the end of the first year.) **$23,304.79**

   b. Is this series *convergent* or *divergent*? Explain. It is a diverging series. The ratio is 1.08, which is greater than 1.

43. RECHARGEABLE BATTERIES A certain rechargeable battery is advertised to recharge back to 99.9% of its previous capacity with every charge. If its initial capacity is 8 hours of life, how many total hours should the battery last? **8000 hrs**

Find the sum of each infinite series, if it exists. **47, 48.** No sum exists.

44. $\frac{7}{5} + \frac{21}{20} + \frac{63}{80} + \dots$ $\frac{28}{5}$

45. $\frac{15}{4} + \frac{5}{2} + \frac{5}{3} + \dots$ $\frac{45}{4}$

46. $-\frac{16}{9} + \frac{4}{3} - 1 + \dots$ $-\frac{64}{63}$

47. $\frac{15}{8} + \frac{5}{2} + \frac{10}{3} + \dots$

48. $\frac{21}{16} + \frac{7}{4} + \frac{7}{3} + \dots$

49. $-\frac{18}{7} + \frac{12}{7} - \frac{8}{7} + \dots$ $-\frac{54}{35}$

50. **MULTIPLE REPRESENTATIONS** In this problem, you will use a square of paper that is at least 8 inches on a side.

50c. The original square has area 1 unit and the area of all the pieces cannot exceed 1.

   a. Concrete Let the square be one unit. Cut away one half of the square. Call this piece Term 1. Next, cut away one half of the remaining sheet of paper. Call this piece Term 2. Continue cutting the remaining paper in half and labeling the pieces with a term number as long as possible. List the fractions represented by the pieces. $\frac{1}{2}, \frac{1}{4}, \frac{1}{8}, \frac{1}{16}, \dots$

   b. Numerical If you could cut the squares indefinitely, you would have an infinite series. Find the sum of the series. **1**

   c. Verbal How does the sum of the series relate to the original square of paper?

C **51. PHYSICS** In a physics experiment, a steel ball on a flat track is accelerated, and then allowed to roll freely. After the first minute, the ball has rolled 120 feet. Each minute the ball travels only 40% as far as it did during the preceding minute. How far does the ball travel? **200 ft**

52. PENDULUMS A pendulum travels 12 centimeters on its first swing and 95% of that distance on each swing thereafter. Find the total distance traveled by the pendulum when it comes to rest. **240 cm**

53. TOYS If a rubber ball can bounce back to 95% of its original height, what is the total vertical distance that it will travel if it is dropped from an elevation of 30 feet? **1170 ft**

54. CARS During a maintenance inspection, a tire is removed from a car and spun on a diagnostic machine. When the machine is turned off, the spinning tire completes 20 revolutions the first second and 98% of the revolutions each additional second. How many revolutions does the tire complete before it stops spinning? **1000 revolutions**

🔌 Multiple Representations

In Exercise 50, students use a model, numerical analysis, and verbal description to represent an infinite series and its sum.

📁 **Study Guide and Intervention**
AL OL ELL

📁 **Practice**
AL OL BL ELL

📁 **Enrichment**
OL BL

📁 **Word Problem Practice**
AL OL BL ELL

Teaching the Mathematical Practices

Arguments Mathematically proficient students understand and use stated assumptions, definitions, and previously established results in constructing arguments. They make conjectures and build a logical progression of statements to explore the truth of their conjectures. And they are able to analyze situations by breaking them into cases, and can recognize and use counterexamples.

WatchOut!

Error Analysis For Exercise 60, remind students that the value of r determines whether the infinite geometric series is convergent or divergent.

Exercise Alert

Formula For Exercise 71, students may need to be reminded that the formula for the volume of a sphere is $V = \frac{4}{3}\pi r^3$.

Additional Answers

60. Sample answer: Austin; the common ratio of the series is $r = -1$, so the absolute value is $r = 1$ and the series diverges.

61. Sample answer: The sum of a geometric series is $S_n = \frac{a_1 - a_1 r^n}{1 - r}$. For an infinite series with $|r| < 1$, $r^n \rightarrow 0$ as $n \rightarrow \infty$. Thus, $S = \frac{a_1 - a_1(0)}{1 - r}$ or $\frac{a_1}{1 - r}$.

67. An arithmetic series has a common difference, so each term will eventually become more positive or more negative, but never approach 0. With the terms not approaching 0, the sum will never reach a limit and the series cannot converge.

55. ECONOMICS A state government decides to stimulate its economy by giving $500 to every adult. The government assumes that everyone who receives the money will spend 80% on consumer goods and that the producers of these goods will in turn spend 80% on consumer goods. How much money is generated for the economy for every $500 that the government provides? **$2500**

56. SCIENCE MUSEUM An exhibit at a science museum offers visitors the opportunity to experiment with the motion of an object on a spring. One visitor pulls the object down and lets it go. The object travels 1.2 feet upward before heading back the other way. Each time the object changes direction, it decreases its distance by 20% when compared to the previous direction. Find the total distance traveled by the object. **6 ft**

63. Sample answer: An infinite geometric series has a sum when the common ratio has an absolute value less than 1. When this occurs, the terms will approach 0 as n approaches infinity. With the future terms almost 0, the sum of the series will approach a limit. When the common ratio is 1 or greater, the terms will keep increasing and approach infinity as n approaches infinity and the sum of the series will have no limit.

Match each graph with its corresponding description.

57. b   **58.** d   **59.** a

a. converging geometric series   **b.** diverging geometric series
c. converging arithmetic series   **d.** diverging arithmetic series

H.O.T. Problems   Use Higher-Order Thinking Skills

60. ERROR ANALYSIS Emmitt and Austin are asked to find the sum of $1 - 1 + 1 - \ldots$. Is either of them correct? Explain your reasoning. **See margin.**

> **Emmitt**
> The sum is 0 because the sum of each pair of terms in the sequence is 0.

> **Austin**
> There is no sum because $|r| \geq 1$, and the series diverges.

61. PROOF Derive the formula for the sum of an infinite geometric series. **See margin.**

62. CHALLENGE For what values of b does $3 + 9b + 27b^2 + 81b^3 + \ldots$ have a sum? $-\frac{1}{3} < b < \frac{1}{3}$

63. REASONING When does an infinite geometric series have a sum, and when does it not have a sum? Explain your reasoning.

65. Sample answer:
$3 + 2 + \frac{4}{3} + \ldots$

66. $\displaystyle\sum_{k=1}^{\infty} 3(-2)^{k-1}$; $\displaystyle\sum_{k=0}^{\infty} 3(-2)^k$

64. ARGUMENTS Determine whether the following statement is *sometimes*, *always*, or *never* true. Explain your reasoning.

> If the absolute value of a term of any geometric series is greater than the absolute value of the previous term, then the series is divergent.

64. Sample answer: Sometimes; the statement is true for all *infinite* geometric series.

65. OPEN ENDED Write an infinite series with a sum that converges to 9.

66. OPEN ENDED Write $3 - 6 + 12 - \ldots$ using sigma notation in two different ways.

67. WRITING IN MATH Explain why an arithmetic series is always divergent. **See margin.**

68. SAT/ACT What is the sum of an infinite geometric series with a first term of 27 and a common ratio of $\frac{2}{3}$? **E**

A 18

B 34

C 41

D 65

E 81

69. Adelina, Michelle, Masao, and Brandon each simplified the same expression at the board. Each student's work is shown below. The teacher said that while two of them had a correct answer, only one of them had arrived at the correct conclusion using correct steps. **H**

Adelina's work

$x^2 x^{-5} = \dfrac{x^2}{x^{-5}}$

$= x^7, x \neq 0$

Masao's work

$x^2 x^{-5} = \dfrac{x^2}{x^5}$

$= \dfrac{1}{x^3}, x \neq 0$

Michelle's work

$x^2 x^{-5} = \dfrac{x^2}{x^{-5}}$

$= x^{-3}, x \neq 0$

Brandon's work

$x^2 x^{-5} = \dfrac{x^2}{x^5}$

$= x^3, x \neq 0$

Which is a completely accurate simplification?

F Adelina's work

G Michelle's work

H Masao's work

J Brandon's work

70. GRIDDED RESPONSE Evaluate $\log_8 60$ to the nearest hundredth. **1.97**

71. GEOMETRY The radius of a large sphere was multiplied by a factor of $\frac{1}{3}$ to produce a smaller sphere. **C**

Radius $= r$ Radius $= \frac{1}{3}r$

How does the volume of the smaller sphere compare to the volume of the larger sphere?

A The volume of the smaller sphere is $\frac{1}{9}$ as large.

B The volume of the smaller sphere is $\frac{1}{\pi^3}$ as large.

C The volume of the smaller sphere is $\frac{1}{27}$ as large.

D The volume of the smaller sphere is $\frac{1}{3}$ as large.

72. GAMES An audition is held for a TV game show. At the end of each round, one half of the prospective contestants are eliminated from the competition. On a particular day, 524 contestants begin the audition. (Lesson 10-3)

a. Write an equation for finding the number of contestants who are left after n rounds. $a_n = 524\left(\dfrac{1}{2}\right)^n$

b. Using this method, will the number of contestants who are to be eliminated always be a whole number? Explain. **No; at the beginning of the third round there will be 131 contestants and one-half of that is 65.5.**

73. CLUBS A quilting club consists of 9 members. Every week, each member must bring one completed quilt square. (Lesson 10-2)

a. Find the first eight terms of the sequence that describes the total number of squares that have been made after each meeting. **9, 18, 27, 36, 45, 54, 63, 72**

b. One particular quilt measures 72 inches by 84 inches and is being designed with 4-inch squares. After how many meetings will the quilt be complete? **42 meetings**

Find each function value.

74. $f(x) = 5x - 9, f(6)$ **21**

75. $g(x) = x^2 - x, g(4)$ **12**

76. $h(x) = x^2 - 2x - 1, h(3)$ **2**

Extension Have students work in pairs taking turns writing a repeating decimal and challenging their partners to express them as fractions.

4 Assess

Ticket Out the Door Make several copies of each of five different infinite geometric series—some that have sums and some that do not. Give a copy of one series to each student. As students leave the room, ask them to tell whether the series are convergent or divergent.

Formative Assessment
Check for student understanding of Lessons 10-3 and 10-4.

Quiz 2, p. 53

1 Focus

Objective Use a graphing calculator to investigate limits of sequences.

Materials

- TI-83/84 Plus or other graphing calculator

Teaching Tip

Stat plots for sequences are graphed in the same way as any other stat plot. It is essential that lists L_1 and L_2 contain the same number of elements.

2 Teach

Working in Cooperative Groups

Have students work in pairs so they can help each other correct keystroke errors. Then have the students complete Steps 1–2 of the Activity, and Exercises 1–2.

- If students are having trouble generating the sequence list, make sure they place the cursor on the *name* of the list, rather than on the first element in the list.

- Students may need to be reminded about how to read scientific notation as displayed on a calculator (e.g., 4.572E-4 represents the number 4.572×10^{-4}).

- Make sure students realize that for a sequence to have a limit, values of successive terms must continue to get closer and closer to the limiting value.

Practice Have students complete Exercises 3–6.

You may have noticed that in some geometric sequences, the later the term in the sequence, the closer the value is to 0. Another way to describe this is that as n increases, a_n approaches 0. The value that the terms of a sequence approach, in this case 0, is called the **limit** of the sequence. Other types of infinite sequences may also have limits. If the terms of a sequence do not approach a unique value, we say that the limit of the sequence does not exist.

You can use a TI-83/84 Plus graphing calculator to help find the limits of infinite sequences.

Activity

Find the limit of the geometric sequence $1, \frac{1}{4}, \frac{1}{16}, \dots$.

Step 1 Enter the sequence.

The formula for this sequence is $a_n = \left(\frac{1}{4}\right)^{n-1}$.

- Position the cursor on **L1** in the **STAT EDIT 1: Edit…** screen and enter the formula **seq(N,N,1,10,1)**. This generates the values 1, 2, …, 10 of the index **N**.

  KEYSTROKES: [STAT] [ENTER] [▲] [2nd] [STAT] [▶] 5 [X,T,θ,n] [,] [X,T,θ,n] [,] 1 [,] 10 [,] 1 [)] [ENTER]

- Position the cursor on **L2** and enter the formula **seq((1/4)^(N-1),N,1,10,1)**. This generates the first ten terms of the sequence.

  KEYSTROKES: [▶] [▲] [2nd] [STAT] [▶] 5 [(] 1 [÷] 4 [)] [^] [(] [X,T,θ,n] [−] 1 [)] [,] [X,T,θ,n] [,] 1 [,] 10 [,] 1 [)] [ENTER]

Notice that as n increases, the terms of the given sequence get closer and closer to 0. If you scroll down, you can see that for $n \geq 6$ the terms are so close to 0 that the calculator expresses them in scientific notation. This suggests that the limit of the sequence is 0.

Step 2 Graph the sequence.

Use **STAT PLOT** to graph the sequence. Use **L1** as the **Xlist** and **L2** as the **Ylist**.

The graph also shows that, as n increases, the terms approach 0. In fact, for $n \geq 3$, the marks appear to lie on the horizontal axis. This strongly suggests that the limit of the sequence is 0.

[0, 10] scl: 1 by [0, 1] scl: 0.1

Exercises

Find the limit of each sequence.

1. $a_n = \left(\frac{1}{3}\right)^n$ 0

2. $a_n = \left(-\frac{1}{3}\right)^n$ 0

3. $a_n = 5^n$ does not exist

4. $a_n = \frac{1}{n^2}$ 0

5. $a_n = \frac{3^n}{3^n + 1}$ 1

6. $a_n = \frac{n^2}{n + 2}$ does not exist

690 | Extend 10-4 | Graphing Technology Lab: Limits

---

3 Assess

Formative Assessment

Use Exercise 6 to assess whether students comprehend how to recognize when a sequence has no limit.

From Concrete to Abstract

Have students examine the sequence $a_n = \frac{3^n}{3^n + 1}$ in Exercise 5. Have them divide the numerator and denominator by 3^n and ask them to use the result to explain why the limit of the sequence must be 1.

10 Mid-Chapter Quiz
Lessons 10-1 through 10-4

Determine whether each sequence is *arithmetic*, *geometric*, or *neither*. Explain your reasoning. (Lesson 10-1)

1. $5, -3, -12, -22, -33\ldots$ **neither; no common ratio or difference**

2. $\frac{1}{5}, \frac{7}{10}, \frac{6}{5}, \frac{17}{10}, \frac{11}{5} \ldots$

arithmetic; common difference of $\frac{1}{2}$

3. HOUSING Laura is a real estate agent. She needs to sell 15 houses in 6 months. (Lesson 10-1)

a. By the end of the first 2 months she has sold 4 houses. If she sells 2 houses each month for the rest of the 6 months, will she meet her goal? Explain. **See margin.**

b. If she has sold 5 houses by the end of the first month, how many will she have to sell on average each month in order to meet her goal? **2 houses**

4. GEOMETRY The figures below show a pattern of filled squares and white squares. (Lesson 10-1)

Figure 1   Figure 2   Figure 3

a. Write an equation representing the nth number in this pattern where n is the number of white squares.
$a_n = 2n + 2$

b. Is it possible to have exactly 84 white squares in an arrangement? Explain. **See margin.**

Find the indicated term of each arithmetic sequence.
(Lesson 10-2)

5. $a_1 = 10, d = -5, n = 9$ **−30**

6. $a_1 = -8, d = 4, n = 99$ **384**

Find the sum of each arithmetic series. (Lesson 10-2)

7. $-15 + (-11) + (-7) + \cdots + 53$ **342**

8. $a_1 = -12, d = 8, n = 22$ **1584**

9. $\displaystyle\sum_{k=11}^{50} (-3k + 1)$ **−3620**

10. MULTIPLE CHOICE What is the sum of the first 50 odd numbers? (Lesson 10-2) **B**

A 2550

B 2500

C 2499

D 2401

Find the indicated term for each geometric sequence.
(Lesson 10-3)

11. $a_2 = 8, r = 2, a_8 = ?$ **512**

12. $a_3 = 0.5, r = 8, a_{10} = ?$ **1,048,576**

13. MULTIPLE CHOICE What are the geometric means of the sequence below? (Lesson 10-3) **J**

$$0.5, \underline{\hspace{1cm}}, \underline{\hspace{1cm}}, \underline{\hspace{1cm}}, 2048$$

F 512.375, 1024.25, 1536.125

G 683, 1365.5, 2048

H 2, 8, 32

J 4, 32, 256

14. INCOME Peter works for a house building company for 4 months per year. He starts out making $3000 per month. At the end of each month, his salary increases by 5%. How much money will he make in those 4 months? (Lesson 10-3)
$12,930.38

Evaluate the sum of each geometric series. (Lesson 10-3)

15. $\displaystyle\sum_{k=1}^{8} 3 \cdot 2^{k-1}$ **765**

16. $\displaystyle\sum_{k=1}^{9} 4 \cdot (-1)^{k-1}$ **4**

17. $\displaystyle\sum_{k=1}^{20} -2\left(\frac{2}{3}\right)^{k-1}$ **−5.998**

Find the sum of each infinite series, if it exists. (Lesson 10-4)

18. $\displaystyle\sum_{n=1}^{\infty} 9 \cdot 2^{n-1}$ **No sum exists.**

19. $\displaystyle\sum_{n=1}^{\infty} (4) \cdot (0.5)^{n-1}$ **8**

20. $\displaystyle\sum_{n=1}^{\infty} 12 \cdot \left(\frac{2}{3}\right)^{n-1}$ **36**

 connectED.mcgraw-hill.com **691**

Formative Assessment

Use the Mid-Chapter Quiz to assess students' progress in the first half of the chapter.

For problems answered incorrectly, have students review the lessons indicated in parentheses.

Summative Assessment

Mid-Chapter Test, p. 55

McGraw-Hill
Assessment Customize and create multiple versions of your Mid-Chapter Quiz and their answer keys.

FOLDABLES **Study Organizer**

Dinah Zike's Foldables®
Before students complete the Mid-Chapter Quiz, encourage them to review the information for Lessons 10-1 through 10-4 in their Foldables.

Additional Answers

3a. No, she will have 11 houses left to sell in 4 months. If she sold 2 houses per month for the remaining 4 months, she would only sell 8 more houses.

4b. Yes, when $2n + 2 = 84$, $n = 41$. In the 41st figure there will be 84 white squares.

InterventionPlanner

 1 **On Level** OL

If	students miss about 25% of the exercises or less,
Then	choose a resource:

SE Lessons 10-1, 10-2, 10-3, and 10-4

Skills Practice, pp. 7, 13, 21, and 28

TE Chapter Project

connectED.mcgraw-hill.com Self-Check Quiz

 2 **Strategic Intervention** AL
approaching grade level

If	students miss about 50% of the exercises,
Then	choose a resource:

Study Guide and Intervention, pp. 5, 11, 19, and 26

connectED.mcgraw-hill.com Extra Examples, Personal Tutor, Homework Help

3 **Intensive Intervention**
2 or more grades below level

If	students miss about 75% of the exercises,
Then	use *Math Triumphs, Alg. 2*

connectED.mcgraw-hill.com Extra Examples, Personal Tutor, Homework Help, Review Vocabulary

1 Focus

VerticalAlignment

Before Lesson 10-5 Explore compositions of functions.

Lesson 10-5 Recognize and use special sequences. Iterate functions.

After Lesson 10-5 Use mathematical induction to prove statements.

2 Teach

Scaffolding Questions

Have students read the **Why?** section of the lesson.

Ask:

- What number follows 8 in this sequence? 13

- Is this sequence an arithmetic sequence? no

- Is it a geometric sequence? no Explain. There is neither a common difference nor a common ratio.

1 Special Sequences

Example 1 shows how to use a recursive formula to find terms of a sequence. **Example 2** shows how to write a recursive formula. **Example 3** shows how to use a recursive formula to model a real-world situation.

	Then		Now		Why?
•	You explored compositions of functions.	**1**	Recognize and use special sequences.	•	The female honeybee is produced after the queen mates with a male, so the female has two parents, a male and a female. The male honeybee, however, is produced by the queen's unfertilized eggs and thus has only one parent, a female. The family tree for the honeybee follows a special sequence.
		2	Iterate functions.		

Generation	1	2	3	4	5	6
Ancestors	1	1	2	3	5	8

NewVocabulary
Fibonacci sequence
recursive sequence
explicit formula
recursive formula
iteration

CCSS Common Core State Standards

Mathematical Practices
6 Attend to precision.
8 Look for and express regularity in repeated reasoning.

1 Special Sequences Notice that every term in the list of ancestors is the sum of the previous two terms. This special sequence is called the **Fibonacci sequence**, and it is found in many places in nature. The Fibonacci sequence is an example of a **recursive sequence**. In a recursive sequence, each term is determined by one or more of the previous terms.

The formulas you have used for sequences thus far have been explicit formulas. An **explicit formula** gives a_n as a function of n, such as $a_n = 3n + 1$. The formula that describes the Fibonacci sequence, $a_n = a_{n-2} + a_{n-1}$, is a **recursive formula**, which means that every term will be determined by one or more of the previous terms. An initial term must be given in a recursive formula.

KeyConcept Recursive Formulas for Sequences

Arithmetic Sequence	$a_n = a_{n-1} + d$, where d is the common difference
Geometric Sequence	$a_n = r \cdot a_{n-1}$, where r is the common ratio

Example 1 Use a Recursive Formula

Find the first five terms of the sequence in which $a_1 = -3$ and $a_{n+1} = 4a_n - 2$, if $n \geq 1$.

$a_{n+1} = 4a_n - 2$ Recursive formula

$a_{1+1} = 4a_1 - 2$ $n = 1$

$a_2 = 4(-3) - 2$ or -14 $a_1 = -3$

$a_3 = 4(-14) - 2$ or -58 $a_2 = -14$

$a_4 = 4(-58) - 2$ or -234 $a_3 = -58$

$a_5 = 4(-234) - 2$ or -938 $a_4 = -234$

The first five terms of the sequence are -3, -14, -58, -234, and -938.

GuidedPractice

1. Find the first five terms of the sequence in which $a_1 = 8$ and $a_{n+1} = -3a_n + 6$, if $n \geq 1$. 8, −18, 60, −174, 528

 692 | Lesson 10-5

Gerry Ellis/Digital Vision/Getty Images

Lesson 10-5 Resources

Resource	Approaching Level **AL**	On Level **OL**	Beyond Level **BL**	English Learners **ELL**
Teacher Edition		• Differentiated Instruction, pp. 694, 697	• Differentiated Instruction, p. 697	
Chapter Resource Masters	• Study Guide and Intervention, pp. 32–33 • Skills Practice, p. 34 • Practice, p. 35 • Word Problem Practice, p. 36	• Study Guide and Intervention, pp. 32–33 • Skills Practice, p. 34 • Practice, p. 35 • Word Problem Practice, p. 36 • Enrichment, p. 37 • Graphing Calculator Activity, p. 38	• Practice, p. 35 • Word Problem Practice, p. 36 • Enrichment, p. 37	• Study Guide and Intervention, pp. 32–33 • Skills Practice, p. 34 • Practice, p. 35 • Word Problem Practice, p. 36
Other	• 5-Minute Check 10-5 • Study Notebook • Teaching Algebra with Manipulatives	• 5-Minute Check 10-5 • Study Notebook • Teaching Algebra with Manipulatives	• 5-Minute Check 10-5 • Study Notebook	• 5-Minute Check 10-5 • Study Notebook • Teaching Algebra with Manipulatives

In order to find a recursive formula, first determine the initial term. Then evaluate the pattern to generate the later terms. The recursive formula that generates a sequence does not include the value of the initial term.

StudyTip

Sequences and Recursive Formulas Like arithmetic and geometric sequences, recursive formulas define functions in which the domain is the set of positive integers, represented by the term number n.

Example 2 Write Recursive Formulas

Write a recursive formula for each sequence.

a. 2, 10, 18, 26, 34, …

Step 1 Determine whether the sequence is arithmetic or geometric. The sequence is arithmetic because each term after the first can be found by adding a common difference.

Step 2 Find the common difference.

$$d = 10 - 2 \text{ or } 8$$

Step 3 Write the recursive formula.

$a_n = a_{n-1} + d$ Recursive formula for arithmetic sequence
$a_n = a_{n-1} + 8$ $d = 8$

A recursive formula for the sequence is $a_n = a_{n-1} + 8$, $a_1 = 2$.

b. 16, 56, 196, 686, 2401, …

Step 1 Determine whether the sequence is arithmetic or geometric. The sequence is geometric because each term after the first can be found after multiplying by a common ratio.

Step 2 Find the common ratio.

$$r = \frac{56}{16} \text{ or } 3.5$$

Step 3 Write the recursive formula.

$a_n = r \cdot a_{n-1}$ Recursive formula for geometric sequence
$a_n = 3.5 a_{n-1}$ $r = 3.5$

A recursive formula for the sequence is $a_n = 3.5 a_{n-1}$, $a_1 = 16$.

c. $a_4 = 108$ and $r = 3$

Step 1 Determine whether the sequence is arithmetic or geometric. Because r is given, the sequence is geometric.

Step 2 Write the recursive formula.

$a_n = r \cdot a_{n-1}$ Recursive formula for geometric sequence
$a_n = 3 a_{n-1}$ $r = 3$

A recursive formula for the sequence is $a_n = 3a_{n-1}$, $a_1 = 4$.

GuidedPractice

Write a recursive formula for each sequence.

2A. 8, 20, 50, 125, 312.5, …   **2B.** 8, 17, 26, 35, 44, …   **2C.** $a_3 = 16$ and $r = 4$

2A. $a_n = 2.5 a_{n-1}$, $a_1 = 8$
2B. $a_n = a_{n-1} + 9$, $a_1 = 8$
2C. $a_n = 4 a_{n-1}$, $a_1 = 1$

   connectED.mcgraw-hill.com   **693**

Formative Assessment
Use the Guided Practice exercises after each example to determine students' understanding of concepts.

Additional Examples

1 Find the first five terms of the sequence in which $a_1 = 5$ and $a_{n+1} = 2a_n + 7$, $n \geq 1$.
5, 17, 41, 89, 185

2 Write a recursive formula for each sequence.

a. 3, 10, 17, 24, 31, …
$a_1 = 3$; $a_n = a_{n-1} + 7$

b. 5, 20, 80, 320, 1280, …
$a_1 = 5$; $a_n = 4 \cdot a_{n-1}$

c. $a_3 = 6$ and $d = 5$
$a_1 = -4$; $a_n = a_{n-1} + 5$

Additional Examples also in Interactive Classroom PowerPoint® Presentations

 Interactive White Board READY

WatchOut!

Preventing Misconceptions Make sure students understand that you use the value of one term to find the value of the next term when a sequence is given in recursive form. Point out that $a_n = a_1 + (n-1)d$ and $a_n = a_1 r^{n-1}$ are not recursive formulas. These *explicit* formulas determine each term by the number n of the term rather than by the value of the preceding term.

3 **BIOLOGY** Dr. Elliot is growing cells in lab dishes. She starts with 108 cells Monday morning and then removes 20 of these for her experiment. By Tuesday, the remaining cells have multiplied in number by 1.5. She again removes 20. This pattern repeats each day of the week. Write a recursive formula for the number of cells Dr. Elliot finds each day before she removes any. Then determine how many cells she will find on Friday morning.

$c_{n+1} = 1.5(c_n - 20)$ or
$c_{n+1} = 1.5c_n - 30$,
$c_1 = 108$; 303

Teach with Tech
Interactive Whiteboard
Show students how to calculate successive terms using a recursive formula. Use different colors and the highlight tool in your calculations to help show how each term is linked to the pervious terms.

2 Iteration
Example 4 shows how to find the first few iterates, or iterated values, of a function.

4 Find the first three iterates x_1, x_2, and x_3 of $f(x) = 3x - 1$ for an initial value of $x_0 = 5$.
14, 41, 122

FINANCIAL LITERACY Nate had $15,000 in credit card debt when he graduated from college. The balance increased by 2% each month due to interest, and Nate could only make payments of $400 per month. Write a recursive formula for the balance of his account each month. Then determine the balance after five months.

Step 1 Write the recursive formula.

Let a_n represent the balance of the account in the nth month. The initial balance a_1 is $15,000. After one month, interest is added and a payment is made.

initial balance	+	balance times 2%	−	monthly payment
$a_2 = $ a_1	+	$(a_1 \times 0.02)$	−	400

$a_2 = 1.02a_1 - 400$

The formula is $a_n = 1.02a_{n-1} - 400$.

Step 2 Find the next five terms.

$a_n = 1.02a_{n-1} - 400$ Recursive formula
$a_2 = (15,000 \times 1.02) - 400$ or **14,900** $a_1 = 15,000$
$a_3 = (14,900 \times 1.02) - 400$ or **14,798** $a_2 = 14,900$
$a_4 = (14,798 \times 1.02) - 400$ or **14,693.96** $a_3 = 14,798$
$a_5 = (14,693.96 \times 1.02) - 400$ or **14,587.84** $a_4 = 14,693.96$
$a_6 = (14,587.84 \times 1.02) - 400$ or **14,479.60** $a_5 = 14,587.84$

After the fifth month, the balance will be $14,479.60.

▶ **Guided**Practice
3. $a_n = 1.025a_{n-1} - 600$; $a_1 = 10,000$;
$10,000, $9650, $9291.25, $8923.53, $8546.62
3. Write a recursive formula for a $10,000 debt, at 2.5% interest per month, with a $600 monthly payment. Then find the first five balances.

ReviewVocabulary
Composition of functions A function is performed, and then a second function is performed on the result of the first function.

2 Iteration **Iteration** is the process of repeatedly composing a function with itself. Consider x_0. The first iterate is $f(x_0)$, the second iterate is $f(f(x_0))$, the third iterate is $f(f(f(x_0)))$, and so on.

Iteration can be used to recursively generate a sequence. Start with the initial value x_0. Let $x_1 = f(x_0)$, $x_2 = f(f(x_0))$, and so on.

Example 4 Iterate a Function

Find the first three iterates x_1, x_2, and x_3 of $f(x) = 5x + 4$ for an initial value of $x_0 = 2$.

$x_1 = f(x_0)$ Iterate the function.
$= 5(2) + 4$ or 14 $x_0 = 2$
$x_2 = f(x_1)$ Iterate the function.
$= 5(14) + 4$ or 74 $x_1 = 14$
$x_3 = f(x_2)$ Iterate the function.
$= 5(74) + 4$ or 374 $x_2 = 74$

The first three iterates are 14, 74, and 374.

▶ **Guided**Practice
4. Find the first three iterates x_1, x_2, and x_3 of $f(x) = -3x + 8$ for an initial value of $x_0 = 6$. **−10, 38, −106**

Corbis Premium RF/Alamy

Differentiated Instruction **OL**

Kinesthetic Learners Have students research and then play the game Tower of Hanoi. The object of the Tower of Hanoi game is to move a stack of 8 disks of graduated sizes from one of 3 pegs to a vacant peg in the fewest number of moves by following these rules:

- You may move only one disk at a time.
- A disk must be placed on top of another disk, not underneath.
- A smaller disk may be placed on top of a larger disk, but not vice versa.

Real-World Link
In 2008, the average credit card debt for college students was about $3173.
Source: USA Today

Check Your Understanding

⬤ = Step-by-Step Solutions begin on page R14.

Example 1   Find the first five terms of each sequence described.   4. $-4, -14, -34, -74, -154$

    1. $a_1 = 16, a_{n+1} = a_n + 4$ 16, 20, 24, 28, 32   **2.** $a_1 = -3, a_{n+1} = a_n + 8$ $-3, 5, 13, 21, 29$

    3. $a_1 = 5, a_{n+1} = 3a_n + 2$ 5, 17, 53, 161, 485   **4.** $a_1 = -4, a_{n+1} = 2a_n - 6$

Example 2   Write a recursive formula for each sequence.   **6.** $a_{n+1} = 3a_n - 1; a_1 = 5$

    5. $3, 8, 18, 38, 78, \ldots$ $a_{n+1} = 2a_n + 2; a_1 = 3$ **6.** $5, 14, 41, 122, 365, \ldots$

Example 3   **7. FINANCING** Ben financed a \$1500 rowing machine to help him train for the college rowing team. He could only make a \$100 payment each month, and his bill increased by 1% due to interest at the end of each month.   **a.** $a_n = 1.01a_{n-1} - 100, a_1 = 1500$

     a. Write a recursive formula for the balance owed at the end of each month.

     b. Find the balance owed after the first four months.   \$1415, \$1329.15, \$1242.44, \$1154.87

     c. How much interest has accumulated after the first six months?   \$77.08

Example 4   Find the first three iterates of each function for the given initial value.

    8. $f(x) = 5x + 2, x_0 = 8$ 42, 212, 1062   **9.** $f(x) = -4x + 2, x_0 = 5$ $-18, 74, -294$

    10. $f(x) = 6x + 3, x_0 = -4$ $-21, -123, -735$ **11.** $f(x) = 8x - 4, x_0 = -6$ $-52, -420, -3364$

   12. 10, 41, 165, 661, 2645   **13.** $-9, -10, -12, -16, -24$

Practice and Problem Solving

Extra Practice is on page R10.

Example 1   **CCSS PERSEVERANCE** Find the first five terms of each sequence described.

    12. $a_1 = 10, a_{n+1} = 4a_n + 1$ **13.** $a_1 = -9, a_{n+1} = 2a_n + 8$

    14. $a_1 = 12, a_{n+1} = a_n + n$ 12, 13, 15, 18, 22   **15.** $a_1 = -4, a_{n+1} = 2a_n + n$ $-4, -7, -12, -21, -38$

    16. $a_1 = 6, a_{n+1} = 3a_n - n$ 6, 17, 49, 144, 428   **17.** $a_1 = -2, a_{n+1} = 5a_n + 2n$ $-2, -8, -36, -174, -862$

    18. $a_1 = 7, a_2 = 10, a_{n+2} = 2a_n + a_{n+1}$ **19.** $a_1 = 4, a_2 = 5, a_{n+2} = 4a_n - 2a_{n+1}$ 4, 5, 6, 8, 8
       7, 10, 24, 44, 92

    20. $a_1 = 4, a_2 = 3x, a_n = a_{n-1} + 4a_{n-2}$ **21.** $a_1 = 3, a_2 = 2x, a_n = 4a_{n-1} - 3a_{n-2}$
       4, 3x, 3x + 16, 15x + 16, 27x + 80   3, 2x, 8x − 9, 26x − 36, 80x − 117

    22. $a_1 = 2, a_2 = x + 3, a_n = a_{n-1} + 6a_{n-2}$ **23.** $a_1 = 1, a_2 = x, a_n = 3a_{n-1} + 6a_{n-2}$
       2, x + 3, x + 15, 7x + 33, 13x + 123   1, x, 3x + 6, 15x + 18, 63x + 90

Example 2   Write a recursive formula for each sequence.   **24–31. See margin.**

    24. $16, 10, 7, 5.5, 4.75, \ldots$ **25.** $32, 12, 7, 5.75, \ldots$

    26. $4, 15, 224, 50,175, \ldots$ **27.** $1, 2, 9, 730, \ldots$

    28. $9, 33, 129, 513, \ldots$ **29.** $480, 128, 40, 18, \ldots$

    30. $393, 132, 45, 16, \ldots$ **31.** $68, 104, 176, 320, \ldots$

                          32a. $a_n = 1.08a_{n-1} + 20000$

Example 3   **32. FINANCIAL LITERACY** Mr. Edwards and his company deposit \$20,000 into his retirement account at the end of each year. The account earns 8% interest before each deposit.

     a. Write a recursive formula for the balance in the account at the end of each year.

     b. Determine how much is in the account at the end of each of the first 8 years.   See margin.

Example 4   Find the first three iterates of each function for the given initial value.

    33. $f(x) = 12x + 8, x_0 = 4$ 56, 680, 8168   **34.** $f(x) = -9x + 1, x_0 = -6$ 55, −494, 4447

    35. $f(x) = -6x + 3, x_0 = 8$ −45, 273, −1635   **36.** $f(x) = 8x + 3, x_0 = -4$ −29, −229, −1829

    37. $f(x) = -3x^2 + 9, x_0 = 2$ −3, −18, −963   **38.** $f(x) = 4x^2 + 5, x_0 = -2$ 21, 1769, 12,517,449

    39. $f(x) = 2x^2 - 5x + 1, x_0 = 6$ 43, 3484, 24,259,093   **40.** $f(x) = -0.25x^2 + x + 6, x_0 = 8$ −2, 3, 6.75

41. 4.25, 29.5625, 936.0664   **41.** $f(x) = x^2 + 2x + 3, x_0 = \frac{1}{2}$ **42.** $f(x) = 2x^2 + x + 1, x_0 = -\frac{1}{2}$ 1, 4, 37

 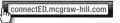

---

3 Practice

Formative Assessment

Use Exercises 1–11 to check for understanding.

Use the chart at the bottom of this page to customize assignments for your students.

CCSS   **Teaching the Mathematical Practices**

Perseverance Mathematically proficient students start by explaining to themselves the meaning of a problem and looking for entry points to its solution. They analyze givens, constraints, relationships, and goals and make conjectures about the form and meaning of the solution and plan a solution pathway rather than simply jumping into a solution attempt.

Additional Answers

24. $a_{n+1} = 0.5a_n + 2; a_1 = 16$

25. $a_{n+1} = 0.25a_n + 4; a_1 = 32$

26. $a_{n+1} = (a_n)^2 - 1; a_1 = 4$

27. $a_{n+1} = (a_n)^3 + 1; a_1 = 1$

28. $a_{n+1} = 4a_n - 3; a_1 = 9$

29. $a_{n+1} = 0.25a_n + 8; a_1 = 480$

30. $a_{n+1} = \frac{a_n}{3} + 1; a_1 = 393$

31. $a_{n+1} = 2a_n - 32; a_1 = 84$

32b. $a_1 = \$20,000,$
     $a_2 = \$41,600,$
     $a_3 = \$64,928,$
     $a_4 = \$90,122.24,$
     $a_5 = \$117,332.02,$
     $a_6 = \$146,718.58,$
     $a_7 = \$178,456.07,$
     $a_8 = \$212,732.56$

Differentiated Homework Options

Level	Assignment	Two-Day Option	
AL Basic	12–42, 49, 51–69	13–41 odd, 54–57	12–42 even, 49, 51–53, 58–69
OL Core	13–41 odd, 43–49, 51–69	12–42, 54–57	43–49, 51–53, 58–69
BL Advanced	43–63, (optional: 64–69)		

Study Guide and Intervention
AL OL ELL

NAME _____ DATE _____ PERIOD _____

10-5 Study Guide and Intervention
Recursion and Iteration

Practice
AL OL BL ELL

Word Problem Practice
AL OL BL ELL

B **43** **FRACTALS** Consider the figures at the right. The number of blue triangles increases according to a specific pattern.

a. Write a recursive formula for the number of blue triangles in the sequence of figures. $a_n = 3a_{n-1}, a_1 = 1$

b. How many blue triangles will be in the sixth figure? 243

44. $9815.15,
$9629.38,
$9442.67,
$9255.04,
$9066.46,
$8876.94,
$8686.48,
$8495.06

44. **FINANCIAL LITERACY** Miguel's monthly car payment is $234.85. The recursive formula $b_n = 1.005b_{n-1} - 234.85$ describes the balance left on the loan after n payments. Find the balance of the $10,000 loan after each of the first eight payments.

45. **CONSERVATION** Suppose a lake is populated with 10,000 fish. A year later, 80% of the fish have died or been caught, and the lake is replenished with 10,000 new fish. If the pattern continues, will the lake eventually run out of fish? If not, will the population of the lake converge to any particular value? Explain.

46. **GEOMETRY** Consider the pattern at the right.

C

45. No; the population of fish will reach 12,500. Each year, 20% of 12,500 or 2500 fish, plus 10,000 additional fish, yields 12,500 fish.

a. Write a sequence of the total number of triangles in the first six figures. 1, 4, 10, 19, 31, 46

b. Write a recursive formula for the number of triangles. $a_n = a_{n-1} + 3(n-1)$

c. How many triangles will be in the tenth figure? 136

47. **SPREADSHEETS** Consider the sequence with $x_0 = 20,000$ and $f(x) = 0.3x + 5000$.

a. Enter x_0 in cell A1 of your spreadsheet. Enter "= (0.3)*(A1) + 5000" in cell A2. What answer does it provide? 11,000

b. Copy cell A2, highlight cells A3 through A70, and paste. What do you notice about the sequence? It converges to 7142.857.

c. How do spreadsheets help analyze recursive sequences? See margin.

48. **VIDEO GAMES** The final monster in Helena's video game has 100 health points. During the final battle, the monster regains 10% of its health points after every 10 seconds. If Helena can inflict damage to the monster that takes away 10 health points every 10 seconds without getting hurt herself, will she ever kill the monster? If so, when? yes; between 250 and 260 seconds

H.O.T. Problems · Use Higher-Order Thinking Skills

49. **CCSS CRITIQUE** Marcus and Armando are finding the first three iterates of $f(x) = 5x - 3$ for an initial value of $x_0 = 4$. Is either of them correct? Explain.

49. Armando; Marcus included x_0 with the iterates and only showed the first 2 iterates.

52. Sample answer: $f(x) = 2x + 1$, $x_0 = 4$

Marcus	Armando
$f(4) = 5(4) - 3$ or 17	$f(4) = 5(4) - 3$ or 17
$f(17) = 5(17) - 3$ or 82	$f(17) = 5(17) - 3$ or 82
The first three iterates are 4, 17, and 82.	$f(82) = 5(82) - 3$ or 407
	The first three iterates are 17, 82, and 407.

50. **CHALLENGE** Find a recursive formula for 5, 23, 98, 401, ... $a_{n+1} = 4a_n + 3n, a_1 = 5$

51. **REASONING** Is the statement "If the first three terms of a sequence are identical, then the sequence is not recursive" sometimes, always, or never true? Explain your reasoning. See margin.

52. **OPEN ENDED** Write a function for which the first three iterates are 9, 19, and 39.

53. **WRITING IN MATH** Why is it useful to represent a sequence with an explicit or recursive formula? See margin.

Enrichment
OL BL

Additional Answers

47c. Sample answer: They make it easier to analyze recursive sequences because they can produce the first 100 terms instantaneously; it would take a long time to calculate the terms by hand.

51. Sample answer: Sometimes; the recursive formula could involve the first three terms. For example, 2, 2, 2, 8, 20, ... is recursive with $a_{n+3} = a_n + a_{n+1} + 2a_{n+2}$.

54. GEOMETRY In the figure shown, $a + b + c = ?$ **C**

A 180°
B 270°
C 360°
D 450°

55. EXTENDED RESPONSE Bill launches a model rocket from ground level. The rocket's height h in meters is given by the equation $h = -4.9t^2 + 56t$, where t is the time in seconds after the launch.

a. What is the maximum height the rocket will reach? **160 m**

b. How long after it is launched will the rocket reach its maximum height? Round to the nearest tenth of a second. **5.7 s**

c. How long after it is launched will the rocket land? Round to the nearest tenth of a second.

55c. **11.4 s**

56. Which of the following is true about the graphs of $y = 3(x - 4)^2 + 5$ and $y = 3(x + 4)^2 + 5$? **G**

F Their vertices are maximums.

G The graphs have the same shape with different vertices.

H The graphs have different shapes with different vertices.

J One graph has a vertex that is a maximum, while the other graph has a vertex that is a minimum.

57. Which factors could represent the length times the width? **C**

$A = 16x^4 - 25y^2$

A $(4x - 5y)(4x - 5y)$
B $(4x + 5y)(4x - 5y)$
C $(4x^2 - 5y)(4x^2 + 5y)$
D $(4x^2 + 5y)(4x^2 + 5y)$

Spiral Review

Write each repeating decimal as a fraction. (Lesson 10-4)

58. $0.\overline{7}$ $\dfrac{7}{9}$

59. $5.\overline{126}$ $5\dfrac{14}{111}$

60. $6.\overline{259}$ $6\dfrac{7}{27}$

61. SPORTS Adrahan is training for a marathon, about 26 miles. He begins by running 2 miles. Then, when he runs every other day, he runs one and a half times the distance he ran the time before. (Lesson 10-3)

a. Write the first five terms of a sequence describing his training schedule. **2, 3, 4.5, 6.75, 10.125**

b. When will he exceed 26 miles in one run? **the eighth session**

c. When will he have run 100 total miles? **during the ninth session**

State whether the events are *independent* or *dependent*. (Lesson 0-6)

62. tossing a penny and rolling a number cube **independent**

63. choosing first and second place in an academic competition **dependent**

Skills Review

Find each product.

64. $(y + 4)(y + 3)$ $y^2 + 7y + 12$

65. $(x - 2)(x + 6)$ $x^2 + 4x - 12$

66. $(a - 8)(a + 5)$ $a^2 - 3a - 40$

67. $(4h + 5)(h + 7)$ $4h^2 + 33h + 35$

68. $(9p - 1)(3p - 2)$ $27p^2 - 21p + 2$

69. $(2g + 7)(5g - 8)$ $10g^2 + 19g - 56$

 connectED.mcgraw-hill.com **697**

DifferentiatedInstruction OL BL

Extension Write a sequence like 2, 5, 7, 12, 19, 31, . . . , on the board and see if students can identify the pattern. Such a sequence, called a Lucas sequence, is similar to the Fibonacci sequence in that each term beginning with the third term is the sum of the two terms immediately preceding it. Have them write their own Lucas sequences with any two starting terms.

WatchOut!

Error Analysis For Exercise 49, remind students that $x_1 = f(x_0)$, where x_0 is the initial value and x_1 is the first iterate. Be sure students understand that the initial value does not count as an iterate. The first iterate is obtained by applying the function to the initial value.

Tips for New Teachers
Understanding Language Make sure that students understand the terminology used in this lesson, particularly *iteration* and *iterate*.

4 Assess

Name the Math Have students explain what it means to say that a formula or a function is *recursive*.

? Follow-up
Students have explored recursion and iteration.

Ask:

- Why is it helpful to represent a numeric pattern with a formula?
  Sample answer: By using a formula to represent a numeric pattern, you can find any term. If given a term, you can use the formula to determine the location of the term in the pattern. You can also determine other characteristics of the pattern such as whether the sequence representing the pattern is linear or whether the series representing the sequence is convergent or divergent.

Additional Answer

53. Sample answer: In a recursive sequence, each term is determined by one or more of the previous terms. A recursive formula is used to produce the terms of the recursive sequence.

1 Focus

Objective Use a spreadsheet to analyze the payments, interest, and balance on a loan.

Materials for Each Student
- computer spreadsheet (optional)

Teaching Tip
Suggest to students that they always type descriptive labels in cells adjacent to entered or computed values as well as at the tops of columns. This will make it easier to understand and interpret the results.

2 Teach

Working in Cooperative Groups
You may want to have students work in pairs of mixed abilities so that they can help each other with keystrokes and commands. Then have groups follow the Example and complete Exercises 1–2.

- Before students begin this investigation, point out that a spreadsheet can express a relationship in which the calculation of the value of the next term involves using the value of the previous term.

- Ask students why the balance is multiplied by 1.0075 rather than 0.0075. You need to find the balance plus the interest.

Practice Have students complete Exercises 3–6.

3 Assess

Formative Assessment
Use Exercise 1 to assess whether students comprehend how to relate the spreadsheet formula to the algebraic formulas they have studied.

EXTEND 10-5 Spreadsheet Lab
Amortizing Loans

When a payment is made on a loan, part of the payment is used to cover the interest that has accumulated since the last payment. The rest is used to reduce the *principal*, or original amount of the loan. This process is called *amortization*. You can use a spreadsheet to analyze the payments, interest, and balance on a loan. A table that shows this kind of information is called an *amortization schedule*.

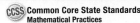
Common Core State Standards
Mathematical Practices
5 Use appropriate tools strategically.

Example

LOANS Gloria just bought a new computer for $695. The store is letting her make monthly payments of $60.78 at an interest rate of 9% for one year. How much will she still owe after six months?

Every month, the interest on the remaining balance will be $\frac{9\%}{12}$ or 0.75%. You can find the balance after a payment by multiplying the balance after the previous payment by $1 + 0.0075$ or 1.0075 and then subtracting 60.78.

In a spreadsheet, the column of numbers represents the number of payments, and Column B shows the balance. Enter the interest rate and monthly payment in cells in Column A so that they can be easily updated if the information changes.

The spreadsheet at the right shows the formulas for the balances after each of the first six payments. After six months, Gloria still owes $355.28.

Computer Loan

◇	A	B	C
1	Interest Rate	=695*(1+A2)−A5	
2	0.0075	=B1*(1+A2)−A5	
3		=B2*(1+A2)−A5	
4	Monthly payment	=B3*(1+A2)−A5	
5	60.78	=B4*(1+A2)−A5	
6		=B5*(1+A2)−A5	
7			

Sheet 1 Sheet 2 Sheet 3

Model and Analyze

1. Let b_n be the balance left on Gloria's loan after n months. Write an equation relating b_n and b_{n+1}. $b_{n+1} = 1.0075b_n − 60.78$, $b_1 = 695$

2. Payments at the beginning of a loan go more toward interest than payments at the end. What percent of Gloria's loan remains to be paid after half a year? about 51%

3. Extend the spreadsheet to the whole year. What is the balance after 12 payments? Why is it not 0? About −$0.01; the balance is not exactly 0 due to rounding.

4. Suppose Gloria decides to pay $70 every month. How long would it take her to pay off the loan? 11 mo

5. Suppose that, based on how much she can afford, Gloria will pay a variable amount each month in addition to the $60.78. Explain how the flexibility of a spreadsheet can be used to adapt to this situation. The variable amount can just be subtracted after A5.

6. Ethan has a three-year, $12,000 motorcycle loan. The annual interest rate is 6%, and his monthly payment is $365.06. After fifteen months, he receives an inheritance which he wants to use to pay off the loan. How much does he owe at that point? $7260.42

698 | Extend 10-5 | Spreadsheet Lab: Amortizing Loans

From Concrete to Abstract
Use Exercise 5 to see how well students can incorporate a variable payment amount into their spreadsheet models.

Extending the Concept
Ask students to try to create columns in their spreadsheets that will show just the interest paid each month.

10-6 The Binomial Theorem

LESSON

·:Then	·:Now	·:Why?

- You worked with combinations.

1 Use Pascal's triangle to expand powers of binomials.

2 Use the Binomial Theorem to expand powers of binomials.

- A manager plans to hire 8 new employees. Not wanting to appear biased, the manager wants to hire a combination of males and females that has at least a 10% chance of occurring randomly. If there are an equal number of male and female applicants, is the probability of randomly hiring 6 men and 2 women less than 10%?

NewVocabulary
Pascal's triangle

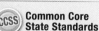
Common Core State Standards

Content Standards
A.APR.5 Know and apply the Binomial Theorem for the expansion of $(x + y)^n$ in powers of x and y for a positive integer n, where x and y are any numbers, with coefficients determined for example by Pascal's Triangle.

Mathematical Practices
4 Model with mathematics.

1 Pascal's Triangle In the 13th century, the Chinese discovered a pattern of numbers that would later be referred to as **Pascal's triangle**. This pattern can be used to determine the coefficients of an expanded binomial $(a + b)^n$.

$$
\begin{array}{c}
(a+b)^0 \quad\quad\quad\quad 1 \\
(a+b)^1 \quad\quad\quad 1 \quad\quad 1 \\
(a+b)^2 \quad\quad 1 \quad\quad 2 \quad\quad 1 \\
(a+b)^3 \quad 1 \quad\quad 3 \quad\quad 3 \quad\quad 1 \\
(a+b)^4 \quad 1 \quad 4 \quad 6 \quad 4 \quad 1 \\
(a+b)^5 \quad 1 \quad 5 \quad 10 \quad 10 \quad 5 \quad 1
\end{array}
$$

For example, the expanded form of
$(a + b)^5 = 1a^5 + 5a^4b + 10a^3b^2 + 10a^2b^3 + 5ab^4 + 1b^5$.

Real-World Example 1 Use Pascal's Triangle

Find the probability of hiring 6 men and 2 women by expanding $(m + f)^8$.

Write three more rows of Pascal's triangle and use the pattern to write the expansion.

$$
\begin{array}{c}
5 \quad\quad 1 \quad 5 \quad 10 \quad 10 \quad 5 \quad 1 \\
6 \quad\quad 1 \quad 6 \quad 15 \quad 20 \quad 15 \quad 6 \quad 1 \\
7 \quad 1 \quad 7 \quad 21 \quad 35 \quad 35 \quad 21 \quad 7 \quad 1 \\
8 \quad 1 \quad 8 \quad 28 \quad 56 \quad 70 \quad 56 \quad 28 \quad 8 \quad 1
\end{array}
$$

$(m + f)^8 = m^8 + 8m^7f + 28m^6f^2 + 56m^5f^3 + 70m^4f^4 + 56m^3f^5 + 28m^2f^6 + 8mf^7 + f^8$

By adding the coefficients of the polynomial, we determine that there are 256 combinations of males and females that could be hired.

$28m^6f^2$ represents the number of combinations with 6 males and 2 females. Therefore, there is a $\frac{28}{256}$ or about an 11% chance of randomly hiring 6 males and 2 females.

Guided Practice

1. Expand $(c + d)^9$. $c^9 + 9c^8d + 36c^7d^2 + 84c^6d^3 + 126c^5d^4 + 126c^4d^5 + 84c^3d^6 + 36c^2d^7 + 9cd^8 + d^9$

2 The Binomial Theorem Instead of writing out row after row of Pascal's triangle, you can use the **Binomial Theorem** to expand a binomial. Recall that $_nC_r = \frac{n!}{r!(n - r)!}$.

connectED.mcgraw-hill.com   **699**

---

1 Focus

VerticalAlignment

Before Lesson 10-6 Work with combinations.

Lesson 10-6 Use Pascal's triangle to expand powers of binomials. Use the Binomial Theorem to expand powers of binomials.

After Lesson 10-6 Find probabilities for binomial experiments.

2 Teach

Scaffolding Questions
Have students read the **Why?** section of the lesson.

Ask:
- Suppose only 2 men and 2 women are hired. How many sequences of 2 men and 2 women would there be? List them. 6; MMWW, MWMW, MWWM, WMMW, WMWM, WWMM

- Expand $(a + b)^4$. Which term has the coefficient 6? $a^4 + 4a^3b^1 + 6a^2b^2 + 4a^1b^3 + b^4$; $6a^2b^2$

- How many sequences of 3 men and 1 woman would there be? Use your expanded form of $(a + b)^4$. Then list them. 4; MMMW, MWMM, WMMM, MMWM

Lesson 10-6 Resources

Resource	Approaching Level **AL**	On Level **OL**	Beyond Level **BL**	English Learners **ELL**
Teacher Edition		• Differentiated Instruction, pp. 700, 703	• Differentiated Instruction, p. 703	• Differentiated Instruction, p. 700
Chapter Resource Masters	• Study Guide and Intervention, pp. 39–40 • Skills Practice, p. 41 • Practice, p. 42 • Word Problem Practice, p. 43	• Study Guide and Intervention, pp. 39–40 • Skills Practice, p. 41 • Practice, p. 42 • Word Problem Practice, p. 43 • Enrichment, p. 44	• Practice, p. 42 • Word Problem Practice, p. 43 • Enrichment, p. 44	• Study Guide and Intervention, pp. 39–40 • Skills Practice, p. 41 • Practice, p. 42 • Word Problem Practice, p. 43
Other	• 5-Minute Check 10-6 • Study Notebook	• 5-Minute Check 10-6 • Study Notebook	• 5-Minute Check 10-6 • Study Notebook	• 5-Minute Check 10-6 • Study Notebook

Somos/Veer/Getty Images

1 Pascal's Triangle

Example 1 shows how to use Pascal's triangle to write the expansion of a power of a binomial.

Formative Assessment

Use the Guided Practice exercises after each example to determine students' understanding of concepts.

Additional Example

1 Expand $(p + t)^5$. $p^5 + 5p^4t + 10p^3t^2 + 10p^2t^3 + 5pt^4 + t^5$

> **Additional Examples** also in Interactive Classroom PowerPoint® Presentations

IWB Interactive White Board READY

2 The Binomial Theorem

Example 2 shows how to use the Binomial Theorem to expand a power of a binomial. **Example 3** uses the factorial form of the Binomial Theorem to expand a power of a binomial with coefficients other than 1. **Example 4** shows how to use the Binomial Theorem to find a particular term of a binomial expansion.

Additional Examples

2 Expand $(t - w)^8$. $t^8 - 8t^7w + 28t^6w^2 - 56t^5w^3 + 70t^4w^4 - 56t^3w^5 + 28t^2w^6 - 8tw^7 + w^8$

3 Expand $(3x - y)^4$. $81x^4 - 108x^3y + 54x^2y^2 - 12xy^3 + y^4$

> **StudyTip**
> **Combinations** Recall that both $_nC_0$ and $_nC_n$ equal 1.

> **KeyConcept** Binomial Theorem
>
> If n is a natural number, then $(a + b)^n =$
> $$_nC_0\, a^nb^0 + {}_nC_1\, a^{n-1}b^1 + {}_nC_2\, a^{n-2}b^2 + \cdots + {}_nC_n\, a^0b^n = \sum_{k=0}^{n} \frac{n!}{k!(n-k)!}\, a^{n-k}b^k.$$

To use the theorem, replace n with the value of the exponent. Notice how the terms will follow the pattern of Pascal's triangle, and the coefficients will be symmetric.

Example 2 Use the Binomial Theorem

Expand $(a + b)^7$.

Method 1 Use combinations.

Replace n with 7 in the Binomial Theorem.

$(a + b)^7 = {}_7C_0\, a^7 + {}_7C_1\, a^6b + {}_7C_2\, a^5b^2 + {}_7C_3\, a^4b^3 + {}_7C_4\, a^3b^4 + {}_7C_5\, a^2b^5 + {}_7C_6\, ab^6 + {}_7C_7\, b^7$

$= a^7 + \frac{7!}{6!}a^6b + \frac{7!}{2!5!}a^5b^2 + \frac{7!}{3!4!}a^4b^3 + \frac{7!}{4!3!}a^3b^4 + \frac{7!}{5!2!}a^2b^5 + \frac{7!}{6!}ab^6 + b^7$

$= a^7 + 7a^6b + 21a^5b^2 + 35a^4b^3 + 35a^3b^4 + 21a^2b^5 + 7ab^6 + b^7$

Method 2 Use Pascal's triangle.

Use the Binomial Theorem to determine exponents, but instead of finding the coefficients by using combinations, look at the seventh row of Pascal's triangle.

6			1		6		15		20		15		6		1	
7		1		7		21		35		35		21		7		1
8	1		8		28		56		70		56		28		8		1

$(a + b)^7 = a^7 + 7a^6b + 21a^5b^2 + 35a^4b^3 + 35a^3b^4 + 21a^2b^5 + 7ab^6 + b^7$

> **GuidedPractice**
>
> **2.** Expand $(x + y)^{10}$.

2. $x^{10} + 10x^9y + 45x^8y^2 + 120x^7y^3 + 210x^6y^4 + 252x^5y^5 + 210x^4y^6 + 120x^3y^7 + 45x^2y^8 + 10xy^9 + y^{10}$

When the binomial to be expanded has coefficients other than 1, the coefficients will no longer be symmetric. In these cases, you may want to use the Binomial Theorem.

> **StudyTip**
> **Graphing Calculator**
> You can calculate $_nC_r$ by using a graphing calculator.
> Press [MATH] and choose PRB 3.

Example 3 Coefficients Other Than 1

Expand $(5a - 4b)^4$.

$(5a - 4b)^4$

$= {}_4C_0\, (5a)^4 + {}_4C_1\, (5a)^3(-4b) + {}_4C_2\, (5a)^2(-4b)^2 + {}_4C_3\, (5a)(-4b)^3 + {}_4C_4\, (-4b)^4$

$= 625a^4 + \frac{4!}{3!}(125a^3)(-4b) + \frac{4!}{2!2!}(25a^2)(16b^2) + \frac{4!}{3!}(5a)(-64b^3) + 256b^4$

$= 625a^4 - 2000a^3b + 2400a^2b^2 - 1280ab^3 + 256b^4$

> **GuidedPractice**
>
> **3.** Expand $(3x + 2y)^5$. $243x^5 + 810x^4y + 1080x^3y^2 + 720x^2y^3 + 240xy^4 + 32y^5$

Sometimes you may need to find only one term in a binomial expansion. To do this, you can use the summation formula for the Binomial Theorem, $\sum_{k=0}^{n} \frac{n!}{k!(n-k)!}\, a^{n-k}b^k$.

 700 | Lesson 10-6 | The Binomial Theorem

DifferentiatedInstruction **OL** **ELL**

If you want to give students the opportunity to be creative,

Then have pairs of students work together to make up a jingle or a poem that describes the patterns in the Binomial Theorem. Ask students to incorporate at least three of the five items listed in the Concept Summary on page 701 of the Student Edition.

Example 4 Determine a Single Term

Find the fifth term of $(y + z)^{11}$.

Step 1 Use the Binomial Theorem to write the expansion in sigma notation.

$$(y + z)^{11} = \sum_{k=0}^{11} \frac{11!}{k!(11 - k)!} y^{11 - k} z^k$$

Step 2 $\dfrac{11!}{k!(11 - k)!} y^{11 - k} z^k = \dfrac{11!}{4!(11 - 4)!} y^{11 - 4} z^4$ For the fifth term, $k = 4$.

$\qquad\qquad\qquad\qquad\qquad\quad = 330 y^7 z^4$ $C(11, 4) = 330$

▸ **Guided**Practice

4. Find the sixth term of $(c + d)^{10}$. $252 c^5 d^5$

ConceptSummary   **Binomial Expansion**

In a binomial expansion of $(a + b)^n$,

- there are $n + 1$ terms.
- n is the exponent of a in the first term and b in the last term.
- in successive terms, the exponent of a decreases by 1, and the exponent of b increases by 1.
- the sum of the exponents in each term is n.
- the coefficients are symmetric.

---

1. $c^5 + 5c^4 d + 10c^3 d^2 + 10c^2 d^3 + 5cd^4 + d^5$ 2. $g^7 + 7g^6 h + 21g^5 h^2 + 35g^4 h^3 + 35g^3 h^4 + 21g^2 h^5 + 7gh^6 + h^7$

Check Your Understanding

🔵 = Step-by-Step Solutions begin on page R14.

3. $x^6 - 24x^5 + 240x^4 - 1280x^3 + 3840x^2 - 6144x + 4096$
4. $32y^5 - 80y^4 z + 80y^3 z^2 - 40y^2 z^3 + 10yz^4 - z^5$

Examples 1–3   **Expand each binomial.**

1. $(c + d)^5$ **2.** $(g + h)^7$ **3.** $(x - 4)^6$

4. $(2y - z)^5$ **5.** $(x + 3)^5$ **6.** $(y - 4z)^4$

5. $x^5 + 15x^4 + 90x^3 + 270x^2 + 405x + 243$

7. GENETICS If a woman is equally as likely to have a baby boy or a baby girl, use binomial expansion to determine the probability that 5 of her 6 children are girls. Do not consider identical twins. $\dfrac{3}{32}$ or 0.09375

6. $y^4 - 16y^3 z + 96y^2 z^2 - 256yz^3 + 256z^4$

Example 4   **Find the indicated term of each expression.**

8. $84b^6 c^3$ **8.** fourth term of $(b + c)^9$ **9** fifth term of $(x + 3y)^8$ **10.** third term of $(a - 4b)^6$ $240a^4 b^2$

11. sixth term of $(2c - 3d)^8$ **12.** last term of $(5x + y)^5$ **13.** first term of $(3a + 8b)^5$ $243a^5$

9. $5670 x^4 y^4$
11. $-108{,}864 c^3 d^5$
12. y^5

14. CCSS MODELING The color of a particular flower is determined by the combination of two genes, also called *alleles*. If the flower has two red alleles r, the flower is red. If the flower has two white alleles w, the flower is white. If the flower has one allele of each color, the flower will be pink. In a lab, two pink flowers are mated and eventually produce 1000 offspring. How many of the 1000 offspring will be pink?   500

Differentiated Homework Options

Level	Assignment	Two-Day Option	
AL Basic	15–28, 35–52	15–27 odd, 39–42	16–28 even, 35–38, 43–52
OL Core	15–35 odd, 36–52	15–28, 39–42	29–33, 35–38, 43–52
BL Advanced	29–49, (optional: 50–52)		

Teach with Tech

Video Recording Have students work in groups to create videos about expanding powers of binomials. Have students explain how to use either Pascal's triangle or the Binomial Theorem to expand the power. Share each group's video with the entire class.

Focus on Mathematical Content

Patterns and the Binomial Theorem
For the expansion of $(a + b)^n$, the signs of the terms depend on the signs of a and b. The sign of each term is $+$ if both a and b are positive; the sign of each even-numbered term is $-$ if b alone is negative.

Additional Example

4 Find the fourth term of $(a + 3b)^4$.
$108ab^3$

3 Practice

Formative Assessment

Use Exercises 1–14 to check for understanding.

Use the chart at the bottom of this page to customize assignments for your students.

CCSS Teaching the Mathematical Practices

Modeling Mathematically proficient students can apply the mathematics they know to solve problems arising in everyday life, analyze relationships mathematically to draw conclusions, and interpret their mathematical results in the context of a situation.

Practice and Problem Solving

Extra Practice is on page R10.

Examples 1–3 **Expand each binomial.**

15. $(a - b)^6$ $a^6 - 6a^5b + 15a^4b^2 - 20a^3b^3 + 15a^2b^4 - 6ab^5 + b^6$

16. $(c - d)^7$ $c^7 - 7c^6d + 21c^5d^2 - 35c^4d^3 + 35c^3d^4 - 21c^2d^5 + 7cd^6 - d^7$

17. $(x + 6)^6$ $x^6 + 36x^5 + 540x^4 + 4320x^3 + 19{,}440x^2 + 46{,}656x + 46{,}656$

18. $(y - 5)^7$ $y^7 - 35y^6 + 525y^5 - 4375y^4 + 21{,}875y^3 - 65{,}625y^2 + 109{,}375y - 78{,}125$

19. $(2a + 4b)^4$ $16a^4 + 128a^3b + 384a^2b^2 + 512ab^3 + 256b^4$

20. $(3a - 4b)^5$ $243a^5 - 1620a^4b + 4320a^3b^2 - 5760a^2b^3 + 3840ab^4 - 1024b^5$

21. **COMMITTEES** If an equal number of men and women applied to be on a community planning committee and the committee needs a total of 10 people, find the probability that 7 of the members will be women. Assume that committee members will be chosen randomly. $\frac{120}{1024} = \frac{15}{128} \approx 0.117$

22. **BASEBALL** If a pitcher is just as likely to throw a ball as a strike, find the probability that 11 of his first 12 pitches are balls. $\frac{12}{4096} = \frac{3}{1024} \approx 0.00293$

Example 4 **Find the indicated term of each expression.**

23. third term of $(x + 2z)^7$ $84x^5z^2$

24. fourth term of $(y - 3x)^6$ $-540y^3x^3$

25. seventh term of $(2a - 2b)^8$ $7168a^2b^6$

26. sixth term of $(4x + 5y)^6$ $75{,}000xy^5$

27. fifth term of $(x - 4)^9$ $32{,}256x^5$

28. fourth term of $(c + 6)^8$ $12{,}096c^5$

B **Expand each binomial. 29–32. See margin.**

29. $\left(x + \frac{1}{2}\right)^5$

30. $\left(x - \frac{1}{3}\right)^4$

31. $\left(2b + \frac{1}{4}\right)^5$

32. $\left(3c + \frac{1}{3}\right)^5$

33. **CCSS SENSE-MAKING** In $\frac{n!}{k!(n-k)!} p^k q^{n-k}$, let p represent the likelihood of a success and q represent the likelihood of a failure.

  a. If a place-kicker makes 70% of his kicks within 40 yards, find the likelihood that he makes 9 of his next 10 attempts from within 40 yards. **0.121**

  b. If a quarterback completes 60% of his passes, find the likelihood that he completes 8 of his next 10 attempts. **0.121**

  c. If a team converts 30% of their two-point conversions, find the likelihood that they convert 2 of their next 5 conversions. **0.309**

H.O.T. Problems   Use Higher-Order Thinking Skills

34. **CHALLENGE** Find the sixth term of the expansion of $\left(\sqrt{a} + \sqrt{b}\right)^{12}$. Explain your reasoning. $792a^3b^2\sqrt{ab}$

35. **REASONING** Explain how the terms of $(x + y)^n$ and $(x - y)^n$ are the same and how they are different. **See margin.**

36. **REASONING** Determine whether the following statement is *true* or *false*. Explain your reasoning. **See margin.**

    The eighth and twelfth terms of $(x + y)^{20}$ have the same coefficients.

37. **OPEN ENDED** Write a power of a binomial for which the second term of the expansion is $6x^4y$. **Sample answer:** $\left(x + \frac{6}{5}y\right)^5$

38. **WRITING IN MATH** Explain how to write out the terms of Pascal's triangle. **See margin.**

Additional Answers

29. $x^5 + \frac{5}{2}x^4 + \frac{5}{2}x^3 + \frac{5}{4}x^2 + \frac{5}{16}x + \frac{1}{32}$

30. $x^4 - \frac{4}{3}x^3 + \frac{2}{3}x^2 - \frac{4}{27}x + \frac{1}{81}$

31. $32b^5 + 20b^4 + 5b^3 + \frac{5}{8}b^2 + \frac{5}{128}b + \frac{1}{1024}$

32. $243c^5 + 135c^4 + 30c^3 + \frac{10}{3}c^2 + \frac{5}{27}c + \frac{1}{243}$

39. PROBABILITY A desk drawer contains 7 sharpened red pencils, 5 sharpened yellow pencils, 3 unsharpened red pencils, and 5 unsharpened yellow pencils. If a pencil is taken from the drawer at random, what is the probability that it is yellow, given that it is one of the sharpened pencils? **A**

A $\frac{5}{12}$

B $\frac{7}{20}$

C $\frac{5}{8}$

D $\frac{1}{5}$

40. GRIDDED RESPONSE Two people are 17.5 miles apart. They begin to walk toward each other along a straight line at the same time. One walks at the rate of 4 miles per hour, and the other walks at the rate of 3 miles per hour. In how many hours will they meet? **2.5**

41. GEOMETRY Christie has a cylindrical block that she needs to paint for an art project.

20 in.

24 in.

What is the surface area of the cylinder in square inches rounded to the nearest square inch? **G**

F 1960   **H** 5127

G 2413   **J** 6635

42. Which of the following is a linear function? **C**

A $y = \frac{x+3}{x+2}$

B $y = (3x + 2)^2$

C $y = \frac{x+3}{2}$

D $y = |3x| + 2$

Find the first five terms of each sequence. (Lesson 10-5)

43. $a_1 = -2, a_{n+1} = a_n + 5$
$-2, 3, 8, 13, 18$

44. $a_1 = 3, a_{n+1} = 4a_n - 10$
$3, 2, -2, -18, -82$

45. $a_1 = 4, a_{n+1} = 3a_n - 6$
$4, 6, 12, 30, 84$

Find the sum of each infinite geometric series, if it exists. (Lesson 10-4)

46. $-6 + 3 - \frac{3}{2} + \ldots$ -4

47. $\frac{3}{4} + \frac{1}{4} + \frac{1}{12} + \ldots$ $1\frac{1}{8}$

48. $\sqrt{3} + 3 + \sqrt{27} + \ldots$ **No sum exists.**

49. TRAVEL A trip between two towns takes 4 hours under ideal conditions. The first 150 miles of the trip is on an interstate, and the last 130 miles is on a highway with a speed limit that is 10 miles per hour less than on the interstate. (Lesson 8-6)

a. If x represents the speed limit on the interstate, write expressions for the time spent at that speed and for the time spent on the other highway. $\frac{150}{x}, \frac{130}{x-10}$

b. Write and solve an equation to find the speed limits on the two highways. $\frac{150}{x} + \frac{130}{x-10} = 4$; 75 mph, 65 mph

State whether each statement is *true* or *false* when $n = 1$. Explain.

50. $\frac{(n+1)(n+1)}{2} = 2$

true; $\frac{(1+1)(1+1)}{2} = 2$

51. $3n + 5$ is even.

true; $3(1) + 5 = 8$, which is even

52. $n^2 - 1$ is odd.

false; $1^2 - 1 = 0$, which is not odd

DifferentiatedInstruction OL BL

Extension Write $(1.002)^5$ on the board and ask students how they could write this expression as the 5th power of a binomial. Then have them use the Binomial Theorem to find the decimal value of $(1.002)^5$ to five decimal places.   1.01004

Exercise Alert
Formula For Exercise 41, students will need to know the formula for the volume of a cylinder is $V = \pi r^2 h$.

4 Assess

Ticket Out the Door Give students small pieces of paper. Have them write the first seven rows of Pascal's triangle on the papers and hand them to you as they leave the room.

Formative Assessment
Check for student understanding of concepts in Lessons 10-5 and 10-6.

📁 Quiz 3, p. 54

Additional Answers

35. Sample answer: While they have the same terms, the signs for $(x + y)^n$ will all be positive, while the signs for $(x - y)^n$ will alternate.

36. Sample answer: False; a binomial to the 20th power will have 21 terms. The eleventh term will be in the middle and the rest of the terms will be symmetric. The tenth term corresponds with the twelfth term.

38. Sample answer: The first row is a 1. The second row is two 1s. Each new row begins and ends with 1. Each coefficient is the sum of the two coefficients above it in the previous row.

1 Focus

Objective Use combinations and Pascal's triangle to determine the number of ways the prizes of a game can be chosen.

Teaching Tip

The main difference between a permutation and a combination is whether order is considered. In a permutation, order matters. In a combination, order does not matter. To help students understand the difference, you may wish to show the number of combinations possible when choosing 2 snacks from a possible 6 different choices.

2 Teach

Working in Cooperative Groups

Pair students with different abilities. Have students work through the Activity and then complete Exercises 1–2.

Ask:

- What do you notice about $_5C_2$ and $_5C_3$? They are the same: 10.

- Can you ever have $_5C_6$? no

- How can you extend Pascal's triangle to include more rows? Sample answer: Use the patterns found in the triangle: the first and the last numbers are always 1, each next number is then the sum of the two numbers closest to it in the line above the number. Each line has one more number in it than the line above it.

Practice Have students complete Exercise 3.

EXTEND 10-6 Algebra Lab
Combinations and Pascal's Triangle

Recall that an arrangement or selection of objects in which order is not important is called a *combination*. For example, selecting 2 snacks from a choice of 6 is a combination of 6 objects taken 2 at a time and can be written $_6C_2$ or $C(6, 2)$.

CCSS **Common Core State Standards**
Content Standards
A.APR.5 Know and apply the Binomial Theorem for the expansion of $(x + y)^n$ in powers of x and y for a positive integer n, where x and y are any numbers, with coefficients determined for example by Pascal's Triangle.

Activity

A contestant on a game show has the opportunity to win up to five prizes, one for each of five rounds of the game. If the contestant wins a round, he or she may choose one prize. Determine the number of ways that prizes can be chosen.

Step 1 If a contestant does not win any rounds, he or she receives 0 prizes. This represents 5 items taken 0 at a time.

$$_nC_r = \frac{n!}{(n-r)!\,r!}$$ Definition of combination

$$_5C_0 = \frac{5!}{(5-0)!\,0!}$$ $n = 5$ and $r = 0$

$$= \frac{120}{120(1)} \text{ or } 1$$ $5! = 120$ and $0! = 1$

There is 1 way to receive 0 prizes.

If a contestant wins one round, any one of the prizes can be selected. If a contestant wins two rounds, two prizes can be chosen. If three rounds are won, three prizes can be chosen, and so on. In how many ways can 1 prize be chosen? 2 prizes? 3, 4, and 5 prizes? We can determine these answers by examining Pascal's triangle.

Step 2 Examine Pascal's triangle.

List Rows 0 through 5 of Pascal's triangle.

Row 0				1		
Row 1			1		1	
Row 2		1		2		1
Row 3	1		3		3		1
Row 4	1	4		6		4	1
Row 5	1	5	10	10	5	1

> The number of ways one prize can be chosen from 5 can be determined by looking at Row 5. The first number in Row 5 represents the number of ways to choose 0 prizes, the second number represents the number of ways to choose 1 prize, and so on.

Analyze the Result

> 1. Sample answer: There are n objects, so use Row n to find the number of choices. The first number is 0 objects selected, the second number is 1 object selected, and so on.

1. Make a conjecture about how the numbers in one of the rows can be used to find the number of ways that $0, 1, 2, 3, 4, \ldots, n$ objects can be selected from n objects.

2. Suppose the rules of the game are changed so that there are 6 rounds and 6 prizes from which to choose. Find the number of ways that $0, 1, 2, 3, 4, 5,$ or 6 prizes can be chosen. Which row of Pascal's triangle can be used to find the answers? 1, 6, 15, 20, 15, 6, 1; Row 6

3. Use Pascal's triangle to find $_8C_0, _8C_1, _8C_2, _8C_3, _8C_4, _8C_5, _8C_6, _8C_7,$ and $_8C_8$. State the row number that you used to find the answers. 1, 8, 28, 56, 70, 56, 28, 8, 1; Row 8

704 | Extend 10-6 | Algebra Lab: Combinations and Pascal's Triangle

---

3 Assess

Formative Assessment

Use Exercise 3 to assess whether students understand how to use combinations and Pascal's triangle to find the number of ways prizes can be chosen.

From Concrete to Abstract

Ask students to use Pascal's triangle to show that $_nC_k = {_nC_{n-k}}$. When Pascal's triangle is folded across a vertical center line, the numbers match. The corresponding numbers match k with $n - k$.

LESSON 10-7 Proof by Mathematical Induction

: Then	: Now	: Why?
● You have proved the sum of an arithmetic series.	**1** Prove statements by using mathematical induction. **2** Disprove statements by finding a counterexample.	● When dominoes are set up closely and the first domino is knocked down, the rest of the dominoes come tumbling down. All that is needed with this setup is for the first domino to fall, and the rest will follow. The same is true with mathematical induction.

 NewVocabulary
mathematical induction
induction hypothesis

 Common Core State Standards

Content Standards
A.SSE.1.b Interpret complicated expressions by viewing one or more of their parts as a single entity.

A.APR.5 Know and apply the Binomial Theorem for the expansion of $(x + y)^n$ in powers of x and y for a positive integer n, where x and y are any numbers, with coefficients determined for example by Pascal's Triangle.

Mathematical Practices
1 Make sense of problems and persevere in solving them.
3 Construct viable arguments and critique the reasoning of others.

1 Mathematical Induction **Mathematical induction** is a method of proving statements involving natural numbers.

KeyConcept Mathematical Induction

To prove that a statement is true for all natural numbers n,

Step 1 Show that the statement is true for $n = 1$.

Step 2 Assume that the statement is true for some natural number k. This assumption is called the **induction hypothesis**.

Step 3 Show that the statement is true for the next natural number $k + 1$.

Example 1 Prove Summation

Prove that $1^3 + 2^3 + 3^3 + \cdots + n^3 = \dfrac{n^2(n + 1)^2}{4}$.

Step 1 When $n = 1$, the left side of the equation is 1^3 or 1.

The right side is $\dfrac{1^2(1 + 1)^2}{4}$ or 1. Thus, the statement is true for $n = 1$.

Step 2 Assume that $1^3 + 2^3 + 3^3 + \cdots + k^3 = \dfrac{k^2(k + 1)^2}{4}$ for a natural number k.

Step 3 Show that the given statement is true for $n = k + 1$.

$$1^3 + 2^3 + 3^3 + \cdots + k^3 = \frac{k^2(k + 1)^2}{4} \qquad \text{Inductive hypothesis}$$

$$1^3 + 2^3 + \cdots + k^3 + (k + 1)^3 = \frac{k^2(k + 1)^2}{4} + (k + 1)^3 \qquad \text{Add } (k + 1)^3 \text{ to each side.}$$

$$= \frac{k^2(k + 1)^2 + 4(k + 1)^3}{4} \qquad \text{The LCD is 4.}$$

$$= \frac{(k + 1)^2[k^2 + 4(k + 1)]}{4} \qquad \text{Factor.}$$

$$= \frac{(k + 1)^2(k^2 + 4k + 4)}{4} \qquad \text{Simplify.}$$

$$= \frac{(k + 1)^2(k + 2)^2}{4} \qquad \text{Factor.}$$

The last expression is the statement to be proved, where n has been replaced by $k + 1$. This proves the conjecture.

▶ **GuidedPractice**   See Chapter 10 Answer Appendix.

1. Prove that $1^2 + 2^2 + 3^2 + \cdots + n^2 = \dfrac{n(n + 1)(2n + 1)}{6}$.

 connectED.mcgraw-hill.com   **705**

Comstock Images/Alamy

1 Focus

VerticalAlignment

▼ **Before Lesson 10-7** Prove the sum of an arithmetic series.

▼ **Lesson 10-7** Prove statements by using mathematical induction. Disprove statements by finding a counterexample.

▼ **After Lesson 10-7** Prove limits of sequences using mathematical induction.

2 Teach

Scaffolding Questions
Have students read the **Why?** section of the lesson.

Ask:
- Imagine the positive integers are assigned to each of the dominoes. What will make domino number 2 fall down? Domino number 1 must fall.

- What will make domino number 3 fall down? Domino number 2 must fall.

- If the domino n falls down, what happens to domino $n + 1$? It falls.

1 Mathematical Induction
Examples 1 and 2 illustrate the three-step procedure for proof by mathematical induction.

Lesson 10-7 Resources

Resource	Approaching Level **AL**	On Level **OL**	Beyond Level **BL**	English Learners **ELL**
Teacher Edition	▪ Differentiated Instruction, p. 706	▪ Differentiated Instruction, pp. 706, 709	▪ Differentiated Instruction, p. 709	
Chapter Resource Masters	▪ Study Guide and Intervention, pp. 45–46 ▪ Skills Practice, p. 47 ▪ Practice, p. 48 ▪ Word Problem Practice, p. 49	▪ Study Guide and Intervention, pp. 45–46 ▪ Skills Practice, p. 47 ▪ Practice, p. 48 ▪ Word Problem Practice, p. 49 ▪ Enrichment, p. 50	▪ Practice, p. 48 ▪ Word Problem Practice, p. 49 ▪ Enrichment, p. 50	▪ Study Guide and Intervention, pp. 45–46 ▪ Skills Practice, p. 47 ▪ Practice, p. 48 ▪ Word Problem Practice, p. 49
Other	▪ 5-Minute Check 10-7 ▪ Study Notebook	▪ 5-Minute Check 10-7 ▪ Study Notebook	▪ 5-Minute Check 10-7 ▪ Study Notebook	▪ 5-Minute Check 10-7 ▪ Study Notebook

Formative Assessment

Use the Guided Practice exercises after each example to determine students' understanding of concepts.

 Teaching the Mathematical Practices

Regularity Mathematically proficient students notice if calculations are repeated, and look both for general methods and for shortcuts. They continually evaluate the reasonableness of their intermediate results. Remind students that their claims must be true for all cases.

Additional Examples

1 Prove that $1 + 4 + 7 + \ldots + (3n - 2) = \dfrac{n(3n - 1)}{2}$.

Step 1: When $n = 1$, the left side of the given equation is 1. The right side is $\dfrac{1[3(1) - 1]}{2}$ or 1. Thus, the statement is true for $n = 1$.

Step 2: Assume $1 + 4 + 7 + \ldots + (3k - 2) = \dfrac{k(3k - 1)}{2}$ for a natural number k.

Step 3: $1 + 4 + 7 + \ldots + (3k - 2) + [3(k + 1) - 2]$

$= \dfrac{k(3k - 1)}{2} + [3(k + 1) - 2]$

$= \dfrac{3k^2 - k + 6k + 2}{2}$

$= \dfrac{3k^2 + 5k + 2}{2} = \dfrac{(k + 1)(3k + 2)}{2}$

$= \dfrac{(k + 1)[3(k + 1) - 1]}{2}$

The last expression is the statement to be proved, where n has been replaced by $k + 1$.

2 Prove that $6^n - 1$ is divisible by 5 for all positive integers n.
Proof uses steps similar to those in Example 2 in the Student Edition.

▶ **Additional Examples** also in Interactive Classroom PowerPoint® Presentations

IWB Interactive White Board READY

Example 2 Prove Divisibility

Prove that $8^n - 1$ is divisible by 7 for all natural numbers n.

Step 1 When $n = 1$, $8^n - 1 = 8^1 - 1$ or 7. Since 7 is divisible by 7, the statement is true for $n = 1$.

Step 2 Assume that $8^k - 1$ is divisible by 7 for some natural number k. This means that there is a natural number r such that $8^k - 1 = 7r$.

Step 3 Show that the statement is true for $n = k + 1$.

$8^k - 1 = 7r$	Inductive hypothesis
$8^k = 7r + 1$	Add 1 to each side.
$8(8^k) = 8(7r + 1)$	Multiply each side by 8.
$8^{k+1} = 56r + 8$	Simplify.
$8^{k+1} - 1 = 56r + 7$	Subtract 1 from each side.
$8^{k+1} - 1 = 7(8r + 1)$	Factor.

Since r is a natural number, $8r + 1$ is a natural number and $7(8r + 1)$ is divisible by 7. Therefore, $8^{k+1} - 1$ is divisible by 7.

This proves that $8^n - 1$ is divisible by 7 for all natural numbers n.

▶ **Guided**Practice   See Chapter 10 Answer Appendix.

2. Prove that $7^n - 1$ is divisible by 6 for all natural numbers n.

StudyTip

CCSS Regularity r is a whole number used to show divisibility in proof. When a value equals $4r$, then it must be divisible by 4.

ReviewVocabulary

counterexample One of the synonyms of *counter* is to *contradict*, so a counterexample is an example that contradicts a hypothesis.

2 Counterexamples Statements can be proved false by using mathematical induction. An easier method is by finding a counterexample, which is a specific case in which the statement is false.

Example 3 Use a Counterexample to Disprove

Find a counterexample to disprove the statement that $2^n + 2n^2$ is divisible by 4 for any natural number n.

Test different values of n.

n	$2^n + 2n^2$	Divisible by 4?
1	$2^1 + 2(1)^2 = 2 + 2$ or 4	yes
2	$2^2 + 2(2)^2 = 4 + 8$ or 12	yes
3	$2^3 + 2(3)^2 = 8 + 18$ or 26	no

The value $n = 3$ is a counterexample for the statement.

▶ **Guided**Practice

3. Find a counterexample to disprove $1^2 + 2^2 + 3^2 + \cdots + n^2 = \dfrac{n(3n - 1)}{2}$. $n = 3$

 706 | Lesson 10-7 | Proof by Mathematical Induction

DifferentiatedInstruction

Kinesthetic Learners Have students demonstrate proof by induction by organizing themselves into a line. Instruct each person to tell any message he hears to the person behind him. Tell the first person a message. Explain that the first person telling the next person assures that $n = 1$ is true. Your instructions to tell all messages heard to the next person assures that $k + 1$ is true.

Example 1 | **Prove that each statement is true for all natural numbers.**

1. $1 + 3 + 5 + \cdots + (2n - 1) = n^2$
See margin.

2. $1 + 2 + 3 + \cdots + n = \dfrac{n(n+1)}{2}$
See Chapter 10 Answer Appendix.

3. NUMBER THEORY A number is *triangular* if it can be represented visually by a triangular array. **c.** See Chapter 10 Answer Appendix.

10

   a. The first triangular number is 1. Find the next 5 triangular numbers. 3, 6, 10, 15, 21

   b. Write a formula for the *n*th triangular number. $a_n = \dfrac{n(n+1)}{2}$

   c. Prove that the sum of the first *n* triangular numbers equals $\dfrac{n(n+1)(n+2)}{6}$.

Example 2 | **Prove that each statement is true for all natural numbers.** 4, 5. See Chapter 10 Answer Appendix.

4. $10^n - 1$ is divisible by 9.

5. $4^n - 1$ is divisible by 3.

Example 3 | **Find a counterexample to disprove each statement.**

6. $3^n + 1$ is divisible by 4. $n = 2$

7. $2^n + 3^n$ is divisible by 4. $n = 1$

Practice and Problem Solving   Extra Practice is on page R10.

Example 1 | **CCSS ARGUMENTS Prove that each statement is true for all natural numbers.** 8–15. See Chapter 10 Answer Appendix.

8. $\dfrac{1}{2} + \dfrac{1}{2^2} + \dfrac{1}{2^3} + \cdots + \dfrac{1}{2^n} = 1 - \dfrac{1}{2^n}$

9. $2 + 5 + 8 + \cdots + (3n - 1) = \dfrac{n(3n + 1)}{2}$

10. $1 + 2 + 4 + \cdots + 2^{n-1} = 2^n - 1$

11. $1 + 5 + 9 + \cdots + (4n - 3) = n(2n - 1)$

12. $1 + 4 + 7 + \cdots + (3n - 2) = \dfrac{n(3n - 1)}{2}$

13 $3 + 7 + 11 + \cdots + (4n - 1) = 2n^2 + n$

14. $\dfrac{1}{2} + \dfrac{1}{6} + \dfrac{1}{12} + \cdots + \dfrac{1}{n(n+1)} = \dfrac{n}{n+1}$

15. $1^2 + 3^2 + 5^2 + \cdots + (2n - 1)^2 = \dfrac{n(2n-1)(2n+1)}{3}$

16. GEOMETRY According to the Interior Angle Sum Formula, if a convex polygon has *n* sides, then the sum of the measures of the interior angles of a polygon equals $180(n - 2)$. Prove this formula for $n \geq 3$ using mathematical induction and geometry.
See Chapter 10 Answer Appendix.

Example 2 | **Prove that each statement is true for all natural numbers.** 17–20. See Chapter 10 Answer Appendix.

17. $5^n + 3$ is divisible by 4.

18. $9^n - 1$ is divisible by 8.

19. $12^n + 10$ is divisible by 11.

20. $13^n + 11$ is divisible by 12.

Example 3 | **Find a counterexample to disprove each statement.**

21. $1 + 2 + 3 + \cdots + n = n^2$ $n = 2$

22. $1 + 8 + 27 + \cdots + n^3 = (2n + 2)^2$ $n = 3$

23. $n^2 - n + 15$ is prime. $n = 1$

24. $n^2 + n + 23$ is prime. $n = 1$

connectED.mcgraw-hill.com **707**

Differentiated Homework Options

Level	Assignment	Two-Day Option	
AL Basic	8–24, 34, 35, 37–62	9–23 odd, 41–45	8–24 even, 34, 35, 38–40, 45–62
OL Core	9–31 odd, 32, 34, 35, 37–62	8–24, 41–44	25–32, 34, 35, 37–40
BL Advanced	25–50, (optional: 51–62)		

---

2 Counterexamples

Example 3 illustrates disproving a formula by finding a counterexample.

Additional Example

3 Find a counterexample for the statement that $n^2 + n + 5$ is always a prime number for any positive integer *n*. Sample answer: $n = 4$

3 Practice

Formative Assessment
Use Exercises 1–7 to check for understanding.

Use the chart at the bottom of this page to customize assignments for your students.

Additional Answer

1. Step 1: When $n = 1$, the left side of the given equation is 1. The right side is 1^2 or 1, so the equation is true for $n = 1$.
Step 2: Assume that $1 + 3 + 5 + \ldots + (2k - 1) = k^2$ for some natural number *k*.
Step 3: $1 + 3 + 5 + \ldots + (2k - 1) + (2(k + 1) - 1)$
$= k^2 + (2(k + 1) - 1)$
$= k^2 + (2k + 2 - 1)$
$= k^2 + 2k + 1$
$= (k + 1)^2$
The last expression is the right side of the equation to be proved, where $n = k + 1$. Thus, the equation is true for $n = k + 1$. Therefore, $1 + 3 + 5 + \ldots + (2n - 1) = n^2$ for all natural numbers *n*.

Study Guide and Intervention
AL OL ELL

Practice
AL OL BL ELL

Word Problem Practice
AL OL BL ELL

 25 **NATURE** The terms of the Fibonacci sequence are found in many places in nature. The number of spirals of seeds in sunflowers is a Fibonacci number, as is the number of spirals of scales on a pinecone. The Fibonacci sequence begins 1, 1, 2, 3, 5, 8, …. Each element after the first two is found by adding the previous two terms. If f_n stands for the nth Fibonacci number, prove that $f_1 + f_2 + \ldots + f_n = f_{n+2} - 1$. **See margin.**

26, 27. See Chapter 10 Answer Appendix.
Prove that each statement is true for all natural numbers or find a counterexample.

26. $7^n + 5$ is divisible by 6.   **27.** $18^n - 1$ is divisible by 17.

28. $n^2 + 21n + 7$ is a prime number. $n = 6$ **29.** $n^2 + 3n + 3$ is a prime number. $n = 3$

30. $500 + 100 + 20 + \cdots + 4 \cdot 5^{4-n} = 625\left(1 - \dfrac{1}{5^n}\right)$ **See Chapter 10 Answer Appendix.**

31. $\dfrac{1}{1 \cdot 2 \cdot 3} + \dfrac{1}{2 \cdot 3 \cdot 4} + \dfrac{1}{3 \cdot 4 \cdot 5} + \cdots + \dfrac{1}{n(n+1)(n+2)} = \dfrac{n(n+3)}{4(n+1)(n+2)}$ **See Chapter 10 Answer Appendix.**

32. **CCSS PERSEVERANCE** Refer to the figures below.

34. Sample answer: False; Even if a counterexample cannot immediately be found, it does not mean that one doesn't exist. A statement can easily be proven false, but it is more difficult to prove that it is true. Statements need to be proven true by mathematical induction, geometrically, or by another method.

Figure 1   Figure 2   Figure 3

a. There is a total of 5 squares in the second figure. How many squares are there in the third figure? **14**

b. Write a sequence for the first five figures. **1, 5, 14, 30, 55**

c. How many squares are there in a standard 8 × 8 checkerboard? **204**

d. Write a formula to represent the number of squares in an $n \times n$ grid. $\dfrac{n(n+1)(2n+1)}{6}$

35. Sample answer: False; assume $k = 2$, just because a statement is true for $n = 2$ and $n = 3$ does not mean that it is true for $n = 1$.

H.O.T. Problems Use Higher-Order Thinking Skills

33. **CHALLENGE** Suggest a formula to represent $2 + 4 + 6 + \cdots + 2n$, and prove your hypothesis using mathematical induction. **See Chapter 10 Answer Appendix.**

REASONING Determine whether the following statements are *true* or *false*. Explain.

34. If you cannot find a counterexample to a statement, then it is true.

35. If a statement is true for $n = k$ and $n = k + 1$, then it is also true for $n = 1$.

36. **CHALLENGE** Prove $\displaystyle\sum_{k=1}^{n} k^3 = \left(\dfrac{n(n+1)}{2}\right)^2$. **See Chapter 10 Answer Appendix.**

37. **REASONING** Find a counterexample to $x^3 + 30 > x^2 + 20x$. $x = 3$

38. **OPEN ENDED** Write a sequence, the formula that produces it, and determine the formula for the sum of the terms of the sequence. Then prove the formula with mathematical induction. **See Chapter 10 Answer Appendix.**

39. **WRITING IN MATH** Explain how the concept of dominoes can help you understand the power of mathematical induction.

40. **WRITING IN MATH** Provide a real-world example other than dominoes that describes mathematical induction. **Sample answer: Climbing a ladder; each step leads to the next step.**

39. Sample answer: When dominoes are set up, after the first domino falls, the rest will fall as well. With induction, once it is proved that the statement is true for $n = 1$ (the first domino), $n = k$ (the second domino), and $n = k + 1$ (the next domino), it will be true for any integer value (any domino).

Enrichment
OL BL

10-7 Enrichment

More Proof by Induction

CCSS **Teaching the Mathematical Practices**

Perseverance Mathematically proficient students start by explaining to themselves the meaning of a problem and looking for entry points to its solution. They analyze givens, constraints, relationships, and goals and make conjectures about the form and meaning of the solution and plan a solution pathway rather than simply jumping into a solution attempt.

41. Which of the following is a counterexample to the statement below? **B**

$$n^2 + n - 11 \text{ is prime.}$$

A $n = -6$	**C** $n = 5$
B $n = 4$	**D** $n = 6$

42. PROBABILITY Latisha wants to create a 7-character password. She wants to use an arrangement of the first 3 letters of her first name (lat), followed by an arrangement of the 4 digits in 1986, the year she was born. How many possible passwords can she create in this way? **G**

F 72	**H** 288
G 144	**J** 576

43. GRIDDED RESPONSE A gear that is 8 inches in diameter turns a smaller gear that is 3 inches in diameter. If the larger gear makes 36 revolutions, how many revolutions does the smaller gear make in that time? **96**

44. SHORT RESPONSE Write an equation for the nth line. Show how it fits the pattern for each given line in the list. **See Ch. 10 Answer Appendix.**

Line 1: $1 \times 0 = 1 - 1$

Line 2: $2 \times 1 = 4 - 2$

Line 3: $3 \times 2 = 9 - 3$

Line 4: $4 \times 3 = 16 - 4$

Line 5: $5 \times 4 = 25 - 5$

Find the indicated term of each expansion. (Lesson 10-6)

45. fourth term of $(x + 2y)^6$ $160x^3y^3$

46. fifth term of $(a + b)^6$ $15a^2b^4$

47. fourth term of $(x - y)^9$ $-84x^6y^3$

48. BIOLOGY In a particular forest, scientists are interested in how the population of wolves will change over the next two years. One model for animal population is the Verhulst population model, $p_{n+1} = p_n + rp_n(1 - p_n)$, where n represents the number of time periods that have passed, p_n represents the percent of the maximum sustainable population that exists at time n, and r is the growth factor. (Lesson 10-5) **b. 0.45 represents the maximum sustainable population of 45% and 1.5 is the growth factor.**

   a. To find the population of the wolves after one year, evaluate $p_1 = 0.45 + 1.5(0.45)(1 - 0.45)$. **0.82125**

   b. Explain what each number in the expression in part **a** represents.

   c. The current population of wolves is 165. Find the new population by multiplying 165 by the value in part **a**. **about 136 wolves**

Find the exact solution(s) of each system of equations. (Lesson 9-7)

49. $x^2 + y^2 - 18x + 24y + 200 = 0$
$4x + 3y = 0$ **(6, −8), (12, −16)**

50. $4x^2 + y^2 = 16$ **(± 2, 0)**
$x^2 + 2y^2 = 4$

Evaluate each expression.

51. $P(8, 2)$ **56**

52. $P(9, 1)$ **9**

53. $P(12, 6)$ **665,280**

54. $C(5, 2)$ **10**

55. $C(8, 4)$ **70**

56. $C(20, 17)$ **1140**

57. $P(12, 2)$ **132**

58. $P(7, 2)$ **42**

59. $C(8, 6)$ **28**

60. $C(9, 4) \cdot C(5, 3)$ **1260**

61. $C(6, 1) \cdot C(4, 1)$ **24**

62. $C(10, 5) \cdot C(8, 4)$ **17,640**

DifferentiatedInstruction OL BL

Extension Tell students that the number $n!$ is defined as the product of all the natural numbers less than or equal to n. Then ask students to prove $1 \cdot 2 \cdot 3 \cdot \ldots \cdot (n - 1)(n) = n!$.

Step 1: When $n = 1$, the left side of the given equation is 1. The right side is $1!$ or 1. Thus, the statement is true for $n = 1$.

Step 2: Assume $1 \cdot 2 \cdot 3 \cdot \ldots \cdot (k - 1)(k) = k!$ for a natural number k.

Step 3: Show that the given statement is true for $n = k + 1$.
$1 \cdot 2 \cdot 3 \cdot \ldots \cdot (k - 1)(k)(k + 1) = k!(k + 1)$
$= [k(k - 1) \cdot \ldots \cdot 3 \cdot 2 \cdot 1](k + 1)$
$= (k + 1)k(k - 1) \cdot \ldots \cdot 3 \cdot 2 \cdot 1$ or $(k + 1)!$

Exercise Alert

Grid Paper For Exercise 32, students may benefit from drawing the first five figures on grid paper.

4 Assess

Name the Math Write an equation with one variable on the board that is not true for all values of the variable, and ask students to hand you pieces of paper on which they have written counterexamples to show that the equation is not always true.

Formative Assessment
Check for student understanding of concepts in Lesson 10-7.

📁 Quiz 4, p. 54

Additional Answer

25. Step 1: When $n = 1$, the left side of the given equation is f_1. The right side is $f_3 - 1$. Since $f_1 = 1$ and $f_3 = 2$ the equation becomes $1 = 2 - 1$ and is true for $n = 1$.

Step 2: Assume that $f_1 + f_2 + \ldots + f_k = f_{k+2} - 1$ for some natural number k.

Step 3: $f_1 + f_2 + \ldots + f_k + f_{k+1}$
$= f_{k+2} - 1 + f_{k+1}$
$= f_{k+1} + f_{k+2} - 1$
$= f_{k+3} - 1$, since Fibonacci numbers are produced by adding the two previous Fibonacci numbers.

The last expression is the right side of the equation to be proved, where $n = k + 1$. Thus, the equation is true for $n = k + 1$. Therefore, $f_1 + f_2 + \ldots + f_n = f_{n+2} - 1$ for all natural numbers n.

Formative Assessment

KeyVocabulary The page references after each word denote where that term was first introduced. If students have difficulty answering questions 1–10, remind them that they can use these page references to refresh their memories about the vocabulary.

Summative Assessment

☐ Vocabulary Test, p. 56

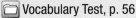 **Vocabulary Review**

Vocabulary Review provides students the opportunity to check their understanding of important concepts and terminology in an online game format.

FOLDABLES **StudyOrganizer**

Dinah Zike's Foldables®

Have students look through the chapter to make sure they have included examples in their Foldables. Suggest that students keep their Foldables handy while completing the Study Guide and Review pages. Point out that their Foldables can serve as a quick review tool when studying for the chapter test.

Study Guide

KeyConcepts

Arithmetic Sequences and Series (Lessons 10-1 and 10-2)

- The nth term a_n of an arithmetic sequence with first term a_1 and common difference d is given by $a_n = a_1 + (n - 1)d$, where n is any positive integer.

- The sum S_n of the first n terms of an arithmetic series is given by $S_n = \frac{n}{2}[2a_1 + (n - 1)d]$ or $S_n = \frac{n}{2}(a_1 + a_n)$.

Geometric Sequences and Series (Lessons 10-3 and 10-4)

- The nth term a_n of a geometric sequence with first term a_1 and common ratio r is given by $a_n = a_1 \cdot r^{n-1}$, where n is any positive integer.

- The sum S_n of the first n terms of a geometric series is given by $S_n = \frac{a_1(1 - r^n)}{1 - r}$ or $S_n = \frac{a_1 - a_1 r^n}{1 - r}$, where $r \neq 1$.

- The sum S of an infinite geometric series with $-1 < r < 1$ is given by $S_n = \frac{a_1}{1 - r}$.

Recursion and Iteration (Lesson 10-5)

- In a recursive formula, each term is formulated from one or more previous terms.

The Binomial Theorem (Lesson 10-6)

- The Binomial Theorem:
$$(a + b)^n = \sum_{k=0}^{n} \frac{n!}{(n - k)! \, k!} a^{n-k} b^k$$

Mathematical Induction (Lesson 10-7)

- Mathematical induction is a method of proof used to prove statements about the positive integers.

FOLDABLES **StudyOrganizer**

Be sure the Key Concepts are noted in your Foldable.

Sequences | Series

 710 | Chapter 10 | Study Guide and Review

KeyVocabulary

arithmetic means (p. 667)	induction hypothesis (p. 705)
arithmetic sequence (p. 659)	infinite geometric series (p. 683)
arithmetic series (p. 668)	infinite sequence (p. 659)
common difference (p. 659)	infinity (p. 685)
common ratio (p. 661)	iteration (p. 694)
convergent series (p. 683)	mathematical induction (p. 705)
divergent series (p. 683)	partial sum (p. 668)
explicit formula (p. 692)	Pascal's triangle (p. 699)
Fibonacci sequence (p. 692)	recursive formula (p. 692)
finite sequence (p. 659)	recursive sequence (p. 692)
geometric means (p. 675)	sequence (p. 659)
geometric sequence (p. 661)	series (p. 668)
geometric series (p. 676)	sigma notation (p. 669)
	term (p. 659)

4. false, sequence

VocabularyCheck

State whether each sentence is *true* or *false*. If *false*, replace the underlined term to make a true sentence.

1. An infinite geometric series that has a sum is called a <u>convergent series</u>. **true**

2. <u>Mathematical induction</u> is the process of repeatedly composing a function with itself. **false, iteration**

3. The <u>arithmetic means</u> of a sequence are the terms between any two non-successive terms of an arithmetic sequence. **true**

4. A <u>term</u> is a list of numbers in a particular order. **true**

5. The sum of the first n terms of a series is called the <u>partial sum</u>. **true**

6. The formula $a_n = a_{n-2} + a_{n-1}$ is a <u>recursive formula</u>. **true**

7. A <u>geometric sequence</u> is a sequence in which every term is determined by adding a constant value to the previous term. **false, arithmetic sequence**

8. An infinite geometric series that does not have a sum is called a <u>partial sum</u>. **false, divergent series**

9. Eleven and 17 are two <u>geometric means</u> between 5 and 23 in the sequence 5, 11, 17, 23. **false, arithmetic means**

10. Using the <u>Binomial Theorem</u>, $(x - 2)^4$ can be expanded to $x^4 - 8x^3 + 24x^2 - 32x + 16$. **true**

Lesson-by-Lesson Review

10-1 Sequences as Functions

Find the indicated term of each arithmetic sequence.

11. $a_1 = 9, d = 3, n = 14$ 48

12. $a_1 = -3, d = 6, n = 22$ 123

13. $a_1 = 10, d = -4, n = 9$ -22

14. $a_1 = -1, d = -5, n = 18$ -86

Example 1

Find the 11th term of an arithmetic sequence if $a_1 = -15$ and $d = 6$.

$a_n = a_1 + (n - 1)d$ Formula for the nth term

$a_{11} = -15 + (11 - 1)6$ $n = 11, a_1 = -15, d = 6$

$a_{11} = 45$ Simplify.

10-2 Arithmetic Sequences and Series

Find the arithmetic means in each sequence.

15. $-12, \_\_, \_\_, \_\_, 8$ $-7, -2, 3$

16. $15, \_\_, \_\_, 29$ $\dfrac{59}{3}, \dfrac{73}{3}$

17. $12, \_\_, \_\_, \_\_, \_\_, -8$ $8, 4, 0, -4$

18. $72, \_\_, \_\_, \_\_, 24$ $60, 48, 36$

19. BANKING Carson saves $40 every 2 months. If he saves at this rate for two years, how much will he have at the end of two years? $480

Find S_n for each arithmetic series.

20. $a_1 = 16, a_n = 48, n = 6$ 192

21. $a_1 = 8, a_n = 96, n = 20$ 1040

22. $9 + 14 + 19 + \cdots + 74$ 581

23. $16 + 7 + -2 + \cdots + -65$ -245

24. DRAMA Laura has a drama performance in 12 days. She plans to practice her lines each night. On the first night she rehearses her lines 2 times. The next night she rehearses her lines 4 times. The third night she rehearses her lines 6 times. On the eleventh night, how many times has she rehearsed her lines? 132

Find the sum of each arithmetic series.

25. $\displaystyle\sum_{k=5}^{21}(3k - 2)$ 629

26. $\displaystyle\sum_{k=0}^{10}(6k - 1)$ 319

27. $\displaystyle\sum_{k=4}^{12}(-2k + 5)$ -99

Example 2

Find the two arithmetic means between 3 and 39.

$a_n = a_1 + (n - 1)d$ Formula for the nth term

$a_4 = 3 + (4 - 1)d$ $n = 4, a_1 = 3$

$39 = 3 + 3d$ $a_4 = 39$

$12 = d$ Simplify.

The arithmetic means are $3 + 12$ or 15 and $15 + 12$ or 27.

Example 3

Find S_n for the arithmetic series with $a_1 = 18$, $a_n = 56$, and $n = 8$.

$S_n = \dfrac{n}{2}(a_1 + a_n)$ Sum formula

$S_8 = \dfrac{8}{2}(18 + 56)$ $n = 8, a_1 = 18, a_n = 56$

$= 296$ Simplify.

Example 4

Evaluate $\displaystyle\sum_{k=3}^{15} 5k + 1$.

Use the formula $S_n = \dfrac{n}{2}(a_1 + a_n)$. There are 13 terms, $a_1 = 5(3) + 1$ or 16, and $a_{13} = 5(15) + 1$ or 76.

$S_{13} = \dfrac{13}{2}(16 + 76)$

$= 598$

Lesson-by-Lesson Review

Intervention If the given examples are not sufficient to review the topics covered by the questions, remind students that the lesson references tell them where to review that topic in their textbook.

Two-Day Option Have students complete the Lesson-by-Lesson Review. Then you can use McGraw-Hill eAssessment to customize another review worksheet that practices all the objectives of this chapter or only the objectives on which your students need more help.

10-3 Geometric Sequences and Series

Find the indicated term for each geometric sequence.

28. $a_1 = 5, r = 2, n = 7$ **320**

29. $a_1 = 11, r = 3, n = 3$ **99**

30. $a_1 = 128, r = -\frac{1}{2}, n = 5$ **8**

31. a_8 for $\frac{1}{8}, \frac{3}{8}, \frac{9}{8} \cdots$ $\frac{2187}{8}$

Find the geometric means in each sequence.

32. $6, \_\_, \_\_, 162$ **18, 54**

33. $8, \_\_, \_\_, \_\_, 648$ **±24, 72, ±216**

34. $-4, \_\_, \_\_, 108$ **12, −36**

35. **SAVINGS** Nolan has a savings account with a current balance of \$1500. What would be Nolan's account balance after 4 years if he receives 5% interest annually? **\$1823.26**

Find S_n for each geometric series.

36. $a_1 = 15, r = 2, n = 4$ **225**

37. $a_1 = 9, r = 4, n = 6$ **12,285**

38. $5 - 10 + 20 - \cdots$ to 7 terms **215**

39. $243 + 81 + 27 + \cdots$ to 5 terms **363**

Evaluate the sum of each geometric series.

40. $\sum_{k=1}^{7} 3 \cdot (-2)^{k-1}$ **129** **41.** $\sum_{k=1}^{8} -1\left(\frac{2}{3}\right)^{k-1}$ $-\frac{6305}{2187}$

42. **ADVERTISING** Natalie is handing out fliers to advertise the next student council meeting. She hands out fliers to 4 people. Then, each of those 4 people hand out 4 fliers to 4 other people. Those 4 then hand out 4 fliers to 4 new people. If Natalie is considered the first round, how many people will have been given fliers after 4 rounds? **85**

Example 5

Find the sixth term of a geometric sequence for which $a_1 = 9$ and $r = 4$.

$a_n = a_1 \cdot r^{n-1}$ Formula for the *n*th term

$a_6 = 9 \cdot 4^{6-1}$ $n = 6, a_1 = 9, r = 4$

$a_6 = 9216$

The sixth term is 9216.

Example 6

Find two geometric means between 1 and 27.

$a_n = a_1 \cdot r^{n-1}$ Formula for the *n*th term

$a_4 = 1 \cdot r^{4-1}$ $n = 4$ and $a_1 = 1$

$27 = r^3$ $a_4 = 27$

$3 = r$ Simplify.

The geometric means are 1(3) or 3 and 3(3) or 9.

Example 7

Find the sum of a geometric series for which $a_1 = 3, r = 5$, and $n = 11$.

$S_n = \dfrac{a_1 - a_1 r^n}{1 - r}$ Sum formula

$S_{11} = \dfrac{3 - 3 \cdot 5^{11}}{1 - 5}$ $n = 11, a_1 = 3, r = 5$

$S_{11} = 36,621,093$ Use a calculator.

Example 8

Evaluate $\sum_{k=1}^{6} 2 \cdot (4)^{k-1}$.

$S_6 = \dfrac{2 - 2 \cdot 4^6}{1 - 4}$ $n = 6, a_1 = 2, r = 4$

$= \dfrac{-8190}{-3}$ Simplify.

$= 2730$ Simplify.

10-4 Infinite Geometric Series

Find the sum of each infinite series, if it exists.

43. $a_1 = 8, r = \frac{3}{4}$ **32**

44. $\frac{5}{6} - \frac{20}{18} + \frac{80}{54} - \frac{320}{162} + \cdots$ **does not exist**

45. $\sum\limits_{k=1}^{\infty} 3\left(\frac{1}{2}\right)^{k-1}$ **6**

46. PHYSICAL SCIENCE Maddy drops a ball off of a building that is 60 feet high. Each time the ball bounces, it bounces back to $\frac{2}{3}$ its previous height. If the ball continues to follow this pattern, what will be the total distance that the ball travels?   **300 ft**

Example 9

Find the sum of the infinite geometric series for which $a_1 = 15$ and $r = \frac{1}{3}$.

$$S = \frac{a_1}{1 - r} \qquad \text{Sum formula}$$

$$= \frac{15}{1 - \frac{1}{3}} \qquad a_1 = 15, r = \frac{1}{3}$$

$$= \frac{15}{\frac{2}{3}} \text{ or } 22.5 \qquad \text{Simplify.}$$

10-5 Recursion and Iteration

Find the first five terms of each sequence.

47. $a_1 = -3, a_{n+1} = a_n + 4$ **$-3, 1, 5, 9, 13$**

48. $a_1 = 5, a_{n+1} = 2a_n - 5$ **$5, 5, 5, 5, 5$**

49. $a_1 = 1, a_{n+1} = a_n + 5$ **$1, 6, 11, 16, 21$**

50. SAVINGS Sari has a savings account with a $12,000 balance. She has a 5% interest rate that is compounded monthly. Every month Sari adds $500 to the account. The recursive formula $b_n = 1.05b_{n-1} + 500$ describes the balance in Sari's savings account after n months. Find the balance of Sari's account after 3 months. Round your answer to the nearest penny.   **$15,467.75**

Find the first three iterates of each function for the given initial value.

51. $f(x) = 2x + 1, x_0 = 3$ **$7, 15, 31$**

52. $f(x) = 5x - 4, x_0 = 1$ **$1, 1, 1$**

53. $f(x) = 6x - 1, x_0 = 2$ **$11, 65, 389$**

54. $f(x) = 3x + 1, x_0 = 4$ **$13, 40, 121$**

Example 10

Find the first five terms of the sequence in which $a_1 = 1$, $a_{n+1} = 3a_n + 2$.

$a_{n+1} = 3a_n + 2$	Recursive formula
$a_{1+1} = 3a_1 + 2$	$n = 1$
$a_2 = 3(1) + 2$ or 5	$a_1 = 1$
$a_{2+1} = 3a_2 + 2$	$n = 2$
$a_3 = 3(5) + 2$ or 17	$a_2 = 5$
$a_{3+1} = 3a_3 + 2$	$n = 3$
$a_4 = 3(17) + 2$ or 53	$a_3 = 17$
$a_{4+1} = 3a_4 + 2$	$n = 4$
$a_5 = 3(53) + 2$ or 161	$a_4 = 53$

The first five terms of the sequence are 1, 5, 17, 53, and 161.

Example 11

Find the first three iterates of the function $f(x) = 3x - 2$ for the initial value of $x_0 = 2$.

$x_1 = f(x_0)$	$x_2 = f(x_1)$	$x_3 = f(x_2)$
$= f(2)$	$= f(4)$	$= f(10)$
$= 3(2) - 2$	$= 3(4) - 2$	$= 3(10) - 2$
$= 4$	$= 10$	$= 28$

The first three iterates are 4, 10, and 28.

Anticipation Guide

Have students complete the Chapter 10 Anticipation Guide and discuss how their responses have changed now that they have completed Chapter 10.

Additional Answers

56. $y^7 - 21y^6 + 189y^5 - 945y^4 + 2835y^3 - 5103y^2 + 5103y - 2187$

57. $-32z^5 + 240z^4 - 720z^3 + 1080z^2 - 810z + 243$

58. $256a^4 - 768a^3b + 864a^2b^2 - 432ab^3 + 81b^4$

59. $x^5 - \frac{5}{4}x^4 + \frac{5}{8}x^3 - \frac{5}{32}x^2 + \frac{5}{256}x - \frac{1}{1024}$

63. Step 1: When $n = 1$, the left side of the equation is equal to 2. The right side of the equation is also equal to 2. So the equation is true for $n = 1$.

Step 2: Assume that $2 + 6 + 12 + \ldots + k(k + 1) = \frac{k(k+1)(k+2)}{3}$ for some positive integer k.

Step 3: $1 \cdot 2 + 2 \cdot 3 + \ldots + k(k + 1) + (k + 1)(k + 2)$

$= \frac{k(k+1)(k+2)}{3} + (k+1)(k+2)$

$= \frac{k(k+1)(k+2)}{3} + \frac{3(k+1)(k+2)}{3}$

$= \frac{(k+1)[k(k+2) + 3(k+2)]}{3}$

$= \frac{(k+1)(k+2)(k+3)}{3}$

$= \frac{(k+1)[(k+1)+1][(k+1)+2]}{3}$

The last expression is the right side of the equation to be proved, where $n = k + 1$. Thus, the equation is true for $n = k + 1$. Therefore, $2 + 6 + 12 + \ldots + n(n + 1) = \frac{n(n+1)(n+2)}{3}$ for all positive integers n.

---

10 Study Guide and Review *Continued*

10-6 The Binomial Theorem

Expand each binomial. **56–59. See margin.**

55. $(a + b)^3$ $a^3 + 3a^2b + 3ab^2 + b^3$

56. $(y - 3)^7$

57. $(3 - 2z)^5$

58. $(4a - 3b)^4$

59. $\left(x - \frac{1}{4}\right)^5$

Find the indicated term of each expression.

60. third term of $(a + 2b)^8$ $112a^6b^2$

61. sixth term of $(3x + 4y)^7$ $193{,}536x^2y^5$

62. second term of $(4x - 5)^{10}$ $-13{,}107{,}200x^9$

Example 12

Expand $(x - 3y)^4$.

$(x - 3y)^4$

$= x^4 + {}_4C_1 x^3(-3y) + {}_4C_2 x^2(-3y)^2 + {}_4C_3(-3y)^4 + {}_4C_4(-3y)^4$

$= x^4 + \frac{4!}{3!}x^3(-3y) + \frac{4!}{2!2!}x^2(9y^2) + \frac{4!}{3!}x(-27y^3) + 81y^4$

$= x^4 + -12x^3y + 54x^2y^2 + -108xy^3 + 81y^4$

Example 13

Find the fourth term of $(x + y)^8$.

Use the Binomial Theorem to write the expansion in sigma notation.

$(x + y)^8 = \sum_{k=0}^{8} \frac{8!}{k!(8 - k)!} x^{8 - k} y^k$

For the fourth term, $k = 3$.

$\frac{8!}{k!(8 - k)!} x^{8 - k} y^k = \frac{8!}{3!(8 - 3)!} x^{8 - 3} y^3$

$= 56x^5 y^3$

10-7 Proof by Mathematical Induction

Prove that each statement is true for all positive integers.

63. $2 + 6 + 12 + \cdots + n(n + 1) = \frac{n(n+1)(n+2)}{3}$ **See margin.**

64. $7^n - 1$ is divisible by 6.

65. $5^n - 1$ is divisible by 4.
64, 65. See Chapter 10 Answer Appendix.

Find a counterexample for each statement.

66. $8^n + 3$ is divisible by 11. $n = 2$

67. $6^{n+1} - 2$ is divisible by 17. $n = 2$

68. $n^2 + 2n + 4$ is prime. $n = 2$

69. $n + 19$ is prime. $n = 1$

Example 14

Prove that $9^n + 3$ is divisible by 4.

Step 1 When $n = 1$, $9^n + 3 = 9^1 + 3$ or 12. Since 12 divided by 4 is 3, the statement is true for $n = 1$.

Step 2 Assume that $9^k + 3$ is divisible by 4 for some positive integer k. This means that $9^k + 3 = 4r$ for some whole number r.

Step 3
$9^k + 3 = 4r$
$9^k = 4r - 3$
$9^{k+1} = 36r - 27$
$9^{k+1} + 3 = 36r - 27 + 3$
$9^{k+1} + 3 = 36r - 24$
$9^{k+1} + 3 = 4(9r - 6)$

Since r is a whole number, $9r - 6$ is a whole number. Thus, $9^{k+1} + 3$ is divisible by 4, so the statement is true for $n = k + 1$.

Therefore, $9^n + 3$ is divisible by 4 for all positive integers n.

CHAPTER 10 Practice Test

1. Find the next 4 terms of the arithmetic sequence 81, 72, 63, … . **54, 45, 36, 27**

2. Find the 25th term of an arithmetic sequence for which $a_1 = 9$ and $d = 5$. **129**

3. **MULTIPLE CHOICE** What is the eighth term in the arithmetic sequence that begins 18, 20.2, 22.4, 24.6, …? **D**

  A 26.8

  B 29

  C 31.2

  D 33.4

4. Find the four arithmetic means between -9 and 11. **$-5, -1, 3, 7$**

5. Find the sum of the arithmetic series for which $a_1 = 11$, $n = 14$, and $a_n = 22$. **231**

6. **MULTIPLE CHOICE** What is the next term in the geometric sequence below? **H**

$$10, \frac{5}{2}, \frac{5}{8}, \frac{5}{32} \cdots$$

  F $\frac{5}{8}$

  G $\frac{5}{32}$

  H $\frac{5}{128}$

  J $\frac{5}{256}$

7. Find the three geometric means between 6 and 1536. **24, 96, 384**

8. Find the sum of the geometric series for which $a_1 = 15$, $r = \frac{2}{3}$, and $n = 5$. **$\frac{1055}{27}$**

Find the sum of each series, if it exists.

9. $\sum_{k=2}^{12} (3k - 1)$ **220**

10. $\sum_{k=1}^{\infty} \frac{1}{2}(3^k)$ **does not exist**

11. $45 + 37 + 29 + \cdots + -11$ **136**

12. $\frac{1}{8} + \frac{2}{24} + \frac{4}{72} + \cdots$ **$\frac{3}{8}$**

13. Write $0.\overline{65}$ as a fraction. **$\frac{65}{99}$**

Find the first five terms of each sequence.

14. $a_1 = -1$, $a_{n+1} = 3a_n + 5$ **$-1, 2, 11, 38, 119$**

15. $a_1 = 4$, $a_{n+1} = a_n + n$ **4, 5, 7, 10, 14**

16. **MULTIPLE CHOICE** What are the first 3 iterates of $f(x) = -5x + 4$ for an initial value of $x_0 = 3$? **B**

  A $3, -11, 59$

  B $-11, 59, -291$

  C $-1, -6, -11$

  D $59, -291, 1459$

17. Expand $(2a - 3b)^4$. **$16a^4 - 96a^3b + 216a^2b^2 - 216ab^3 + 81b^4$**

18. What is the coefficient of the fifth term of $(m + 3n)^6$? **1215**

19. Find the fourth term of the expansion of $(c + d)^9$. **$84c^6d^3$**

Prove that each statement is true for all positive integers. 20, 21. See Chapter 10 Answer Appendix.

20. $1 + 6 + 36 + \cdots + 6^{n-1} = \frac{1}{5}(6^n - 1)$.

21. $11^n - 1$ is divisible by 10.

22. Find a counterexample for the following statement.

$2^n + 4^n$ is divisible by 4. **$n = 1$**

23. **SCHOOL** There are an equal number of girls and boys in Mr. Marshall's science class. He needs to choose 8 students to represent his class at the science fair. What is the probability that 5 are boys? **about 21.9%**

24. **PENDULUM** Laurie swings a pendulum. The distance traveled per swing decreases by 15% with each swing. If the pendulum initially traveled 10 inches, find the total distance traveled when the pendulum comes to a rest. **about 66.7 inches**

10 in

 connectED.mcgraw-hill.com **715**

Summative Assessment
Use these alternate leveled chapter tests to differentiate assessment for your students.

Leveled Chapter 10 Tests

Form	Type	Level	📁 Page(s)
1	MC	AL	57–58
2A	MC	OL	59–60
2B	MC	OL	61–62
2C	FR	OL	63–64
2D	FR	OL	65–66
3	FR	BL	67–68
Vocabulary Test			56
Extended-Response Test			69

MC = multiple-choice questions
FR = free-response questions.

McGraw-Hill eAssessment Customize and create multiple versions of your chapter test and their answer keys. All of the questions from the leveled chapter tests in the *Chapter 10 Resource Masters* are also available on McGraw-Hill eAssessment .

Intervention Planner

TIER 1 On Level OL	**TIER 2** Strategic Intervention AL *approaching grade level*	**TIER 3** Intensive Intervention *2 or more grades below level*
If students miss about 25% of the exercises,	**If** students miss about 50% of the exercises,	**If** students miss about 75% of the exercises,
Then choose a resource:	**Then** choose a resource:	**Then** use *Math Triumphs, Alg. 2*
SE Lessons 10-1, 10-2, 10-3, 10-4, 10-5, 10-6, and 10-7	📁 Study Guide and Intervention, pp. 5, 11, 19, 26, 32, 39, and 45	connectED.mcgraw-hill.com Extra Examples, Personal Tutor, Homework Help, Review Vocabulary
📁 Skills Practice, pp. 7, 13, 21, 28, 34, 41, and 47	connectED.mcgraw-hill.com Extra Examples, Personal Tutor, Homework Help	
TE Chapter Project		
connectED.mcgraw-hill.com Self-Check Quiz		

1 Focus

Objective Use the strategy of looking for a pattern to solve standardized test problems.

2 Teach

Scaffolding Questions
Ask:

- For what types of questions do you think the strategy of looking for a pattern works best? *Possible answer: Questions such as, "What is the next term in the sequence?" or "What is an expression that models this pattern?"*

- How do you find a pattern? *Possible answer: See how the terms of the pattern are related and find common operations that express the pattern.*

Look For a Pattern

One of the most common problem-solving strategies is to look for a pattern. The ability to recognize patterns, model them algebraically, and extend them is a valuable problem-solving tool.

Strategies for Looking For a Pattern

Step 1

Identify the pattern.

- Compare the numbers, shapes, or graphs in the pattern.
- **Ask yourself:** How are the terms of the pattern related?
- **Ask yourself:** Are there any common operations that lead from one term to the next?

Step 2

Generalize the pattern.

- Write a rule using words to describe how the terms of the pattern are generated.
- Assign variables and write an algebraic expression to model the pattern if appropriate.

Step 3

Find missing terms, extend the pattern, and solve the problem.

- Use your pattern or your rule to finding missing terms and/or extend the pattern to solve the problem.
- Check your answer to make sure it makes sense.

Standardized Test Example

Read the problem. Identify what you need to know. Then use the information in the problem to solve.

Use the sequence of squares shown. How many squares will be needed to make the ninth figure of the sequence?

A 55   **C** 74

B 65   **D** 82

Figure 1   Figure 2   Figure 3

Digital Vision/Getty Images

Read the problem statement carefully. You are given three figures of a sequence and asked to find how many squares will be needed to make the ninth figure.

Look for a pattern in the figures of squares. Count the number of squares in each figure.

Write an expression to model this pattern.

Words	The number of squares is equal to the square of the figure number plus one.
Variable	Let n represent the figure number.
Equation	$a_n = n^2 + 1$

Use your expression to extend the pattern and find the number of squares in the ninth figure.

$a_9 = 9^2 + 1 = 82$

So, the ninth figure will have 82 squares. The correct answer is D.

Additional Example

STANDARDIZED TEST PRACTICE
Jonas has $2100 in a savings account. At the end of each month, the balance is increased by 0.3%. How much money will Jonas have in his savings account at the end of one year? B

A $2170.35

B $2176.86

C $2106.89

D $2994.09

3 Assess

Use Exercises 1 and 2 to assess students' understanding.

Exercises

Read each problem. Use a pattern to solve the problem.

1. The numbers below form a famous mathematical sequence of numbers known as the Fibonacci sequence. What is the next Fibonacci number in the sequence? B

$$1, 1, 2, 3, 5, 8, 13, 21, \ldots$$

A 36

B 34

C 31

D 29

2. What is the missing number in the table? H

n	a_n
1	0
2	2
3	6
4	12
5	??
6	30

F 17

G 18

H 20

J 21

Diagnose Student Errors

Survey student responses for each item. Class trends may indicate common errors and misconceptions.

1. A did not add the common difference and 31 correctly
B correct
C mistakenly found 7 as a common difference
D mistakenly found a pattern that every third term ends with 9

2. F guess
G only one dimension was taken into consideration
H mistakenly multiplied 4 by 2 instead of squaring 4
J correct

3. A only evaluated $9k - 1$ for $k = 15$
B found summation for $k = 1$ to 14
C correct
D found summation for $k = 1$ to 16

4. F guess
G guess
H mistakenly thought a lesser coefficient of x^2 causes the graph to become narrower
J correct

5. A guess
B correct
C transposed constant and base
D missed a_1

6. F confused the meanings of the variables
G correct
H did not recognize a geometric sequence
J did not recognize a geometric sequence

7. A found contaminants that remain after two times through the filter
B found contaminants that remain after three times through the filter
C found contaminants that were removed after four times through the filter
D correct

8. F transposed totals
G subtracted soda costs instead of added
H correct
J guess

Multiple Choice

Read each question. Then fill in the correct answer on the answer document provided by your teacher or on a sheet of paper.

1. Find the next term of the arithmetic sequence. **B**

$$7, 13, 19, 25, 31, \ldots$$

A 36   C 38

B 37   D 39

2. Lynette gets an enlargement of a 4-inch by 6-inch picture so that the new print has dimensions that are 4 times the dimensions of her original. How does the area of the enlargement compare to the area of the original picture? **J**

F The area is twice as large.

G The area is four times as large.

H The area is eight times as large.

J The area is sixteen times as large.

3. Evaluate $\displaystyle\sum_{k=1}^{15}(8k - 1)$. **C**

A 119   C 945

B 826   D 1072

4. What is the effect on the graph of the equation $y = 3x^2$ when the equation is changed to $y = 2x^2$? **J**

F The graph of $y = 2x^2$ is a reflection of the graph of $y = 3x^2$ across the y-axis.

G The graph is rotated 90 degrees about the origin.

H The graph is narrower.

J The graph is wider.

5. Write the formula for the nth term of the geometric sequence shown in the table. **B**

n	a_n
1	5
2	10
3	20
4	40
5	80

A $a_n = (5)^n$

B $a_n = 5(2)^{n-1}$

C $a_n = 2(5)^{n-1}$

D $a_n = 5(2)^n$

6. The table shows a dimension of a square tent and the number of people that the tent can fit.

Length of Tent (yards)	Number of People
2	7
5	28
6	39
8	67
12	147

Let ℓ represent the length of the tent and n represent the number of people that can fit in the tent. Identify the equation that best represents the relationship between the length of the tent and the number of people that can fit in the tent. **G**

F $\ell = n^2 + 3$ H $\ell = 3n + 1$

G $n = \ell^2 + 3$ J $n = 3\ell + 1$

7. An air filter claims to remove 90% of the contaminants in the air each time air is circulated through the filter. If the same volume of air is circulated through the filter three times, what percent of the original contaminants will be removed from the air? **D**

A 0.01%   B 0.1%   C 99.0%   D 99.9%

8. At the movies, the cost of 2 boxes of popcorn and 1 soft drink is $11.50. The cost of 3 boxes of popcorn and 4 soft drinks is $27.25. Which pair of equations can be used to determine p, the cost of a box of popcorn, and s, the cost of a soft drink? **H**

F $2p + s = 27.25$ H $2p + s = 11.50$
 $3p + 4s = 11.50$ $3p + 4s = 27.25$

G $2p - s = 11.50$ J $p + s = 11.50$
 $3p - 4s = 27.25$ $p + 4 = 27.25$

9. Which of the following geometric series does *not* converge to a sum? **B**

A $\displaystyle\sum_{k=1}^{\infty} 4 \cdot \left(\frac{9}{10}\right)^{k-1}$ C $\displaystyle\sum_{k=6}^{\infty} \frac{7}{6} \cdot \left(\frac{1}{3}\right)^{k-1}$

B $\displaystyle\sum_{k=1}^{\infty} \frac{1}{5} \cdot \left(\frac{3}{2}\right)^{k-1}$ D $\displaystyle\sum_{k=1}^{\infty} (-2) \cdot \left(\frac{5}{6}\right)^{k-1}$

Test-Taking Tip

Question 9 Understand the terms used in Algebra and how to apply them. A geometric series converges to a sum if the common ratio r has an absolute value less than 1.

9. A the absolute value of the common ratio will be less than one
B correct
C the absolute value of the common ratio will be less than one
D the absolute value of the common ratio will be less than one

Short Response/Gridded Response

Record your answers on the answer sheet provided by your teacher or on a sheet of paper.

10. What are the dimensions of the matrix that results from the multiplication shown? **4 × 1**

$$\begin{bmatrix} a & b & c \\ d & e & f \\ g & h & i \\ j & k & l \end{bmatrix} \cdot \begin{bmatrix} 7 \\ 4 \\ 6 \end{bmatrix}$$

11. **GRIDDED RESPONSE** Consider the pattern below. Into how many pieces will the sixth figure of the pattern be divided? **1024**

Figure 1	Figure 2	Figure 3
1 piece	4 pieces	16 pieces

12. Use the Binomial Theorem to expand the expression $(c + d)^6$. $c^6 + 6c^5d + 15c^4d^2 + 20c^3d^3 + 15c^2d^4 + 6cd^5 + d^6$

13. **GRIDDED RESPONSE** Kara has a cylindrical container that she needs to fill with dirt so she can plant some flowers.

20 in.

24 in.

What is the volume of the cylinder in cubic inches rounded to the nearest cubic inch? **9048**

14. Bacteria in a culture are growing exponentially with time, as shown in the table.

Hours	Bacteria
0	1000
1	2000
2	4000

Write an equation to express the number of bacteria, y, with respect to time, t. $y = (1000) \cdot (2)^t$

15. **GRIDDED RESPONSE** What is the value of $f[g(6)]$ if $f(x) = 2x + 4$ and $g(x) = x^2 + 5$? **86**

Extended Response

Record your answers on a sheet of paper. Show your work.

16. Prove that the sum of any two odd integers is even. **See margin.**

17. The endpoints of a diameter of a circle are at $(-1, 0)$ and $(5, -8)$. **a. (2, −4); The center is the midpoint of the diameter.**
  a. What are the coordinates of the center of the circle? Explain your method.
  b. Find the radius of the circle. Explain your method. **See margin.**
  c. Write an equation of the circle. $(x - 2)^2 + (y + 4)^2 = 25$

18. A cyclist travels from Centerville to Springfield in 2.5 hours. If she increases her speed, she can make the trip in 2 hours.
  a. Does this situation represent a direct or inverse variation? Explain your reasoning.
  b. If the trip from Centerville to Springfield takes 2.5 hours when traveling at 12 miles per hour, what must the speed be to make the trip in 2 hours? **15 mph**
    18a. Inverse; the slower speed takes more time to travel the same distance.

Need ExtraHelp?

If you missed Question...	1	2	3	4	5	6	7	8	9	10	11	12	13	14	15	16	17	18
Go to Lesson...	10-2	1-1	10-2	4-7	10-3	2-4	10-3	3-1	10-4	3-6	10-5	10-6	5-7	7-8	6-1	10-7	9-3	8-5

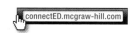
connectED.mcgraw-hill.com **719**

Additional Answers

16. Sample answer: Let $2n + 1$ and $2k + 1$ represent any two odd integers where n and k are both integers. Then their sum is $(2n + 1) + (2k + 1)$, or $2n + 2k + 2$. This can be written as $2(n + k + 1)$ which is the product of 2 and an integer, $n + k + 1$. So, the sum $2n + 2k + 2$ is even.

17b. 5; The radius is the distance between the center and any point on the circle.
$$r = \sqrt{(-1 - 2)^2 + (0 - (-4))^2}$$
$$= \sqrt{(-3)^2 + 4^2}$$
$$= 5$$

Formative Assessment

You can use these pages to benchmark student progress.

📁 Standardized Test Practice, pp. 70–72

McGraw-Hill
ᵉAssessment Create practice tests that align to your state standards, the Common Core State Standards, and other national standards such as TIMSS and NAEP.

Answer Sheet Practice

Have students simulate taking a standardized test by recording their answers on practice recording sheets.

📁 **Student Recording Sheet**

Homework Option

Get Ready for Chapter 11 Assign students the exercises on p. 721 as homework to assess whether they possess the prerequisite skills needed for the next chapter.

Lesson 10-1

14. 40.5, 60.75, 91.125

15. 128, 256, 512

16. $\dfrac{2}{5}, \dfrac{2}{25}, \dfrac{2}{125}$

17. $\dfrac{1}{9}, \dfrac{-1}{27}, \dfrac{1}{81}$

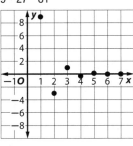

25. 8, 11, 14, 17

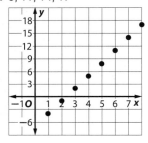

26. $-22, -30, -38, -46$

27. $-29, -35, -41, -47$

28. 32, 49, 66, 83

29. 2, $\dfrac{13}{5}, \dfrac{16}{5}, \dfrac{19}{5}$

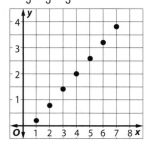

30. $-\dfrac{7}{3}, -\dfrac{10}{3}, -\dfrac{13}{3}, -\dfrac{16}{3}$

39. $-8, 32, -128$

40. $\dfrac{16}{3}, \dfrac{32}{9}, \dfrac{64}{27}$

41. 27, $\dfrac{81}{4}, \dfrac{243}{16}$

42. 192, 256, $\dfrac{1024}{3}$

43. 27, 81, 243

44. 0.0001, 0.00001, 0.000001

60. Sample answer: When the value of r is doubled, a_2 doubles, a_3 quadruples, a_4 is multiplied by 8, a_5 is multiplied by 2^4 or 16, and so on. So, the new terms are $a_n = a_n \cdot 2^{n-1}$. When the value of r is halved, the new terms are $a_n = a_n \cdot \left(\dfrac{1}{2}\right)^{n-1}$.

71.

$$f(x) = \dfrac{6}{(x-2)(x+3)}$$

72.

$$f(x) = \dfrac{-3}{(x-2)^2}$$

73.

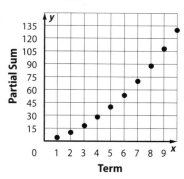

$$f(x) = \frac{x^2 - 36}{x + 6}$$

Lesson 10-2

71a.

n	S_n
1	4
2	10
3	18
4	28
5	40
6	54
7	70
8	88
9	108
10	130

71b.

71c.

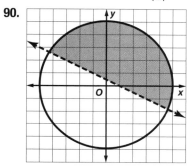

80. Sample answer:
Let a_n = the nth term of the sequence and d = the common difference

$a_2 = a_1 + d$	Definition of the second term of an arithmetic sequence
$a_3 = a_2 + d$	Definition of the third term of an arithmetic sequence
$a_3 = (a_1 + d) + d$	Substitution
$a_3 = a_1 + 2d$	Associative Property of Addition
$a_3 = a_1 + (3 - 1)d$	$3 - 1 = 2$
$a_n = a_1 + (n - 1)d$	$n = 3$

81.

$S_n = (a_1 + a_n) \cdot \left(\frac{n}{2}\right)$	General sum formula
$a_n = a_1 + (n - 1)d$	Formula for nth term
$a_n - (n - 1)d = a_1$	Subtract $(n - 1)d$ from both sides.
$S_n = [a_n - (n - 1)d + a_n] \cdot \left(\frac{n}{2}\right)$	Substitution
$S_n = [2a_n - (n - 1)d] \cdot \left(\frac{n}{2}\right)$	Simplify.

90.

91.

92.

94.

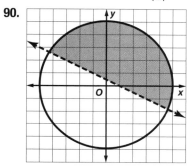

$f(x) = \frac{2}{3}(2^x)$

$D = \{$all real numbers$\}$, $R = \{f(x) \mid f(x) > 0\}$

95.

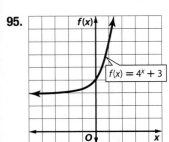

D = {all real numbers}, R = {$f(x) \mid f(x) > 3$}

96.

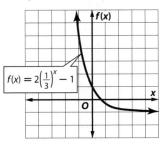

D = {all real numbers}, R = {$f(x) \mid f(x) > -1$}

Lesson 10-3

67. Sample answer: $n - 1$ needs to change to n, and the 10 needs to change to a 9. When this happens, the terms for both series will be identical (a_1 in the first series will equal a_0 in the second series, and so on), and the series will be equal to each other.

68. Sample answer:

Let $a_n =$ the nth term of the sequence and $r =$ the common ratio.

$a_2 = a_1 \cdot r$	Definition of the second term of a geometric sequence
$a_3 = a_2 \cdot r$	Definition of the third term of a geometric sequence
$a_3 = a_1 \cdot r \cdot r$	Substitution
$a_3 = a_1 \cdot r^2$	Associative Property of Multiplication
$a_3 = a_1 \cdot r^{3-1}$	$3 - 1 = 2$
$a_n = a_1 \cdot r^{n-1}$	$n = 3$

Explore 10-4

5. Sample answer:

x	−2	−1	0	1	2
y	0	3	4	3	0

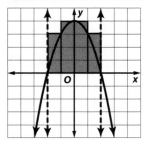

Rectangle	1	2	3	4
Width	1	1	1	1
Height	3	4	4	3
Area	3	4	4	3

14 units^2; The estimate will be greater than the actual area, because the rectangles extend outside the area under the curve.

6. Sample answer:

x	0	1	2	3	4
y	0	1	8	27	64

Rectangle	1	2	3
Width	1	1	1
Height	1	8	27
Area	1	8	27

36 units^2; The estimate will be less than the actual area, because rectangles are inside the curve leaving some area unaccounted for.

7. Sample answer:

x	−3	−2	−1	0	1	2	3
y	9	4	1	0	1	4	9

Rectangle	1	2	3	4	5	6
Width	1	1	1	1	1	1
Height	9	4	1	1	4	9
Area	9	4	1	1	4	9

28 units^2; The estimate will be greater than the actual area, because the rectangles extend outside the area under the curve.

Lesson 10-7 (Guided Practice)

1. Step 1: When $n = 1$, the left side of the given equation is 1^2 or 1.
   The right side is $\frac{1(1 + 1)(1 + 2)}{6}$ or 1, so the equation is true for $n = 1$.

   Step 2: Assume that $1^2 + 2^2 + 3^2 + \ldots + k^2 = \frac{k(k + 1)(2k + 1)}{6}$ for some natural number k.

   Step 3: $1^2 + 2^2 + 3^2 + \ldots + k^2 + (k + 1)^2$

   $= \frac{k(k + 1)(2k + 1)}{6} + (k + 1)^2$

   $= \frac{k(k + 1)(2k + 1)}{6} + \frac{6(k + 1)^2}{6}$

   $= \frac{(k + 1)[k(2k + 1) + 6(k + 1)]}{6}$

   $= \frac{(k + 1)(2k^2 + 7k + 6)}{6}$

   $= \frac{(k + 1)(k + 2)(2k + 3)}{6}$

   $= \frac{(k + 1)[(k + 1) + 1][2(k + 1) + 1]}{6}$

   The last expression is the right side of the equation to be proved, when $n = k + 1$. Thus, the equation is true for $n = k + 1$.

   Therefore, $1^2 + 2^2 + 3^2 + \ldots + n^2 = \frac{n(n + 1)(2n + 1)}{6}$ for all natural numbers n.

2. Step 1: $7^1 - 1 = 6$, which is divisible by 6. The statement is true for $n = 1$.

   Step 2: Assume that $7^k - 1$ is divisible by 6 for some natural number k. This means that $7^k - 1 = 6r$ for some natural number r.

   Step 3: $7^k - 1 = 6r$

   $7^k = 6r + 1$

   $7^{k + 1} = 7(6r + 1)$

   $7^{k + 1} - 1 = 42r + 6$

   $7^{k + 1} - 1 = 6(7r + 1)$

   Since r is a natural number, $7r + 1$ is a natural number. Thus, $7^{k + 1} - 1$ is divisible by 6, so the statement is true for $n = k + 1$. Therefore, $7^n - 1$ is divisible by 6 for all natural numbers n.

Lesson 10-7

2. Step 1: When $n = 1$, the left side of the given equation is 1. The right side is $\frac{1(1 + 1)}{2}$ or 1, so the equation is true for $n = 1$.

   Step 2: Assume that $1 + 2 + 3 + \ldots + k = \frac{k(k + 1)}{2}$ for some natural number k.

   Step 3: $1 + 2 + 3 + \ldots + k + (k + 1)$

   $= \frac{k(k + 1)}{2} + (k + 1)$

   $= \frac{k(k + 1) + 2(k + 1)}{2}$

   $= \frac{(k + 1)(k + 2)}{2}$

The last expression is the right side of the equation to be proved, where $n = k + 1$. Thus, the equation is true for $n = k + 1$.

Therefore, $1 + 2 + 3 + \ldots + n = \frac{n(n + 1)}{2}$ for all natural numbers n.

3c. Step 1: When $n = 1$, the left side of the given equation is $\frac{1(1 + 1)}{2}$ or 1. The right side is $\frac{1(1 + 1)(1 + 2)}{6}$ or 1, so the equation is true for $n = 1$.

   Step 2: Assume that $1 + 3 + 6 + \ldots + \frac{k(k + 1)}{2}$
   $= \frac{k(k + 1)(k + 2)}{6}$ for some natural number k.

   Step 3: $1 + 3 + 6 + \ldots + \frac{k(k + 1)}{2} + \frac{(k + 1)(k + 1 + 1)}{2}$

   $= \frac{k(k + 1)(k + 2)}{6} + \frac{(k + 1)(k + 1 + 1)}{2}$

   $= \frac{k(k + 1)(k + 2)}{6} + \frac{3(k + 1)(k + 2)}{6}$

   $= \frac{(k + 1)(k + 2)(k + 3)}{6}$

   $= \frac{(k + 1)[(k + 1) + 1][(k + 1) + 2]}{6}$

   The last expression is the right side of the equation to be proved, where $n = k + 1$. Thus, the equation is true for $n = k + 1$. Therefore, $1 + 3 + 6 + \ldots + \frac{n(n + 1)}{2}$
   $= \frac{n(n + 1)(n + 2)}{6}$ for all natural numbers n.

4. Step 1: $10^1 - 1 = 9$, which is divisible by 9. The statement is true for $n = 1$.

   Step 2: Assume $10^k - 1$ is divisible by 9 for some natural number k. This means that $10^k - 1 = 9r$ for some whole number r.

   Step 3: $10^k - 1 = 9r$

   $10^k = 9r + 1$

   $10^{k + 1} = 90r + 10$

   $10^{k + 1} - 1 = 90r + 9$

   $10^{k + 1} - 1 = 9(10r + 1)$

   Since r is a whole number, $10r + 1$ is a whole number. Thus, $10^{k + 1} - 1$ is divisible by 9, so the statement is true for $n = k + 1$. Therefore, $10^n - 1$ is divisible by 9 for all natural numbers n.

5. Step 1: $4^1 - 1 = 3$, which is divisible by 3. The statement is true for $n = 1$.

   Step 2: Assume that $4^k - 1$ is divisible by 3 for some natural number k. This means that $4^k - 1 = 3r$ for some whole number r.

   Step 3: $4^k - 1 = 3r$

   $4^k = 3r + 1$

   $4^{k + 1} = 12r + 4$

   $4^{k + 1} - 1 = 12r + 3$

   $4^{k + 1} - 1 = 3(4r + 1)$

   Since r is a whole number, $4r + 1$ is a whole number. Thus, $4^{k + 1} - 1$ is divisible by 3, so the statement is true for $n = k + 1$. Therefore, $4^n - 1$ is divisible by 3 for all natural numbers n.

8. Step 1: When $n = 1$, the left side of the given equation is $\frac{1}{2}$. The right side is $1 - \frac{1}{2}$ or $\frac{1}{2}$, so the equation is true for $n = 1$.

Step 2: Assume that $\frac{1}{2} + \frac{1}{2^2} + \frac{1}{2^3} + \ldots + \frac{1}{2^k} = 1 - \frac{1}{2^k}$ for some natural number k.

Step 3: $\frac{1}{2} + \frac{1}{2^2} + \frac{1}{2^3} + \ldots + \frac{1}{2^k} + \frac{1}{2^{k+1}}$

$= 1 - \frac{1}{2^k} + \frac{1}{2^{k+1}}$

$= 1 - \frac{2}{2^{k+1}} + \frac{1}{2^{k+1}}$

$= 1 - \frac{1}{2^{k+1}}$

The last expression is the right side of the equation to be proved, where $n = k + 1$. Thus, the equation is true for $n = k + 1$. Therefore, $\frac{1}{2} + \frac{1}{2^2} + \frac{1}{2^3} + \ldots + \frac{1}{2^n} = 1 - \frac{1}{2^n}$ for all natural numbers n.

9. Step 1: When $n = 1$, the left side of the given equation is 2. The right side is $\frac{1[3(1) + 1]}{2}$ or 2, so the equation is true for $n = 1$.

Step 2: Assume that $2 + 5 + 8 + \ldots + (3k - 1) = \frac{k(3k + 1)}{2}$ for some natural number k.

Step 3: $2 + 5 + 8 + \ldots + (3k - 1) + [3(k + 1) - 1]$

$= \frac{k(3k + 1)}{2} + [3(k + 1) - 1]$

$= \frac{k(3k + 1) + 2[3(k + 1) - 1]}{2}$

$= \frac{3k^2 + k + 6k + 6 - 2}{2}$

$= \frac{3k^2 + 7k + 4}{2}$

$= \frac{(k + 1)(3k + 4)}{2}$

$= \frac{(k + 1)[3(k + 1) + 1]}{2}$

The last expression is the right side of the equation to be proved, where $n = k + 1$. Thus, the equation is true for $n = k + 1$. Therefore, $2 + 5 + 8 + \ldots + (3n - 1) = \frac{n(3n + 1)}{2}$ for all natural numbers n.

10. Step 1: When $n = 1$, the left side of the given equation is 1. The right side is $2^1 - 1$ or 1, so the equation is true for $n = 1$.

Step 2: Assume that $1 + 2 + 4 + \ldots + 2^{k-1} = 2^k - 1$ for some natural number k.

Step 3: $1 + 2 + 4 + \ldots + 2^{k-1} + 2^{k-1+1}$

$= 2^k - 1 + 2^{k-1+1}$

$= 2^k + 2^k - 1$

$= 2(2^k) - 1$

$= 2^{k+1} - 1$

The last expression is the right side of the equation to be proved, where $n = k + 1$. Thus, the equation is true for $n = k + 1$. Therefore, $1 + 2 + 4 + \ldots + 2^{n-1} = 2^n - 1$ for all natural numbers n.

11. Step 1: When $n = 1$, the left side of the given equation is 1. The right side is $1[2(1) - 1]$ or 1, so the equation is true for $n = 1$.

Step 2: Assume that $1 + 5 + 9 + \ldots + (4k - 3) = k(2k - 1)$ for some natural number k.

Step 3: $1 + 5 + 9 + \ldots + (4k - 3) + [4(k + 1) - 3]$

$= k(2k - 1) + [4(k + 1) - 3]$

$= 2k^2 - k + 4k + 4 - 3$

$= 2k^2 + 3k + 1$

$= (k + 1)(2k + 1)$

$= (k + 1)[2(k + 1) - 1]$

The last expression is the right side of the equation to be proved, where $n = k + 1$. Thus, the equation is true for $n = k + 1$. Therefore, $1 + 5 + 9 + \ldots + (4n - 3) = n(2n - 1)$ for all natural numbers n.

12. Step 1: When $n = 1$, the left side of the given equation is 1. The right side is $\frac{1[3(1) - 1]}{2}$ or 1, so the equation is true for $n = 1$.

Step 2: Assume that $1 + 4 + 7 + \ldots + (3k - 2) = \frac{k(3k - 1)}{2}$ for some natural number k.

Step 3: $1 + 4 + 7 + \ldots + (3k - 2) + [3(k + 1) - 2]$

$= \frac{k(3k - 1)}{2} + [3(k + 1) - 2]$

$= \frac{k(3k - 1)}{2} + 3k + 1$

$= \frac{k(3k - 1)}{2} + \frac{6k + 2}{2}$

$= \frac{3k^2 + 5k + 2}{2}$

$= \frac{(k + 1)(3k + 2)}{2}$

$= \frac{(k + 1)[3(k + 1) - 1]}{2}$

The last expression is the right side of the equation to be proved, where $n = k + 1$. Thus, the equation is true for $n = k + 1$. Therefore, $1 + 4 + 7 + \ldots + (3n - 2) = \frac{n(3n - 1)}{2}$ for all natural numbers n.

13. Step 1: When $n = 1$, the left side of the given equation is $4(1) - 1$ or 3. The right side is $2(1)^2 + 1$ or 3, so the equation is true for $n = 1$.

Step 2: Assume that $3 + 7 + 11 + \ldots + (4k - 1) = 2k^2 + k$ for some natural number k.

Step 3: $3 + 7 + 11 + \ldots + (4k - 1) + [4(k + 1) - 1]$

$= 2k^2 + k + [4(k + 1) - 1]$

$= 2k^2 + k + 4k + 3$

$= 2k^2 + 5k + 3$

$= 2k^2 + 4k + 2 + k + 1$

$= [2(k + 1)^2] + (k + 1)$

The last expression is the right side of the equation to be proved, where $n = k + 1$. Thus, the equation is true for $n = k + 1$. Therefore, $3 + 7 + 11 + \ldots + (4n - 1) = 2n^2 + n$ for all natural numbers n.

14. Step 1: When $n = 1$, the left side of the given equation is $\frac{1}{1(1 + 1)}$ or $\frac{1}{2}$. The right side is $\frac{1}{1 + 1}$ or $\frac{1}{2}$, so the equation is true for $n = 1$.

Step 2: Assume that $\frac{1}{2} + \frac{1}{6} + \frac{1}{12} + \ldots + \frac{1}{k(k + 1)} = \frac{k}{k + 1}$ for some natural number k.

Step 3: $\frac{1}{2} + \frac{1}{6} + \frac{1}{12} + \ldots + \frac{1}{k(k+1)} + \frac{1}{(k+1)(k+1+1)}$

$= \frac{k}{k+1} + \frac{1}{(k+1)(k+1+1)}$

$= \frac{k(k+2)}{(k+1)(k+2)} + \frac{1}{(k+1)(k+2)}$

$= \frac{k^2 + 2k + 1}{(k+1)(k+2)}$

$= \frac{(k+1)(k+1)}{(k+1)(k+2)}$

$= \frac{k+1}{(k+1)+1}$

The last expression is the right side of the equation to be proved, where $n = k + 1$. Thus, the equation is true for $n = k + 1$. Therefore, $\frac{1}{2} + \frac{1}{6} + \frac{1}{12} + \ldots + \frac{1}{n(n+1)} = \frac{n}{n+1}$ for all natural numbers n.

15. Step 1: When $n = 1$, the left side of the given equation is 1^2 or 1. The right side is $\frac{1[2(1)-1][2(1)+1]}{3}$ or 1, so the equation is true for $n = 1$.

Step 2: Assume that $1^2 + 3^2 + 5^2 + \ldots + (2k-1)^2$
$= \frac{k(2k-1)(2k+1)}{3}$ for some natural number k.

Step 3: $1^2 + 3^2 + 5^2 + \ldots + (2k-1)^2 + [2(k+1)-1]^2$

$= \frac{k(2k-1)(2k+1)}{3} + [2(k+1)-1]^2$

$= \frac{k(2k-1)(2k+1) + 3(2k+1)^2}{3}$

$= \frac{(2k+1)[k(2k-1) + 3(2k+1)]}{3}$

$= \frac{(2k+1)(2k^2 - k + 6k + 3)}{3}$

$= \frac{(2k+1)(2k^2 + 5k + 3)}{3}$

$= \frac{(2k+1)(k+1)(2k+3)}{3}$

$= \frac{(k+1)[2(k+1)-1][2(k+1)+1]}{3}$

The last expression is the right side of the equation to be proved, where $n = k + 1$. Thus, the equation is true for $n = k + 1$.

Therefore, $1^2 + 3^2 + 5^2 + \ldots + (2n-1)^2 = \frac{n(2n-1)(2n+1)}{3}$ for all natural numbers n.

16. Step 1: When $n = 3$, $180(3 - 2) = 180$. When a polygon has 3 sides or 3 vertices, there are 180 degrees. The statement is true for $n = 3$.

Step 2: Assume that the statement is true for $n \geq 3$.

Step 3: Consider a convex polygon with $n + 1$ vertices. Since $n + 1 \geq 4$, if we take one vertex x there is another vertex y such that there is one vertex between x and y in one direction, and $n - 2$ in the other direction. Join x and y by a new edge, dividing our original polygon into two polygons. The new polygons' interior angles summed together make up the sum of the original polygon's interior angles. One of the new polygons is a triangle and the other a polygon of n vertices (all but the one isolated between x and y). The triangle has interior angle sum 180°, and by the inductive hypothesis the other polygon has interior angle sum $(n - 2) \cdot 180°$. Summing these we get $(n + 1 - 2) \cdot 180°$, and the theorem is proved.

17. Step 1: $5^1 + 3 = 8$, which is divisible by 4. The statement is true for $n = 1$.

Step 2: Assume $5^k + 3$ is divisible by 4 for some natural number k. This means that $5^k + 3 = 4r$ for some natural number r.

Step 3: $5^k + 3 = 4r$
$5^k = 4r - 3$
$5^{k+1} = 20r - 15$
$5^{k+1} + 3 = 20r - 12$
$5^{k+1} + 3 = 4(5r - 3)$

Since r is a natural number, $5r - 3$ is a natural number. Thus, $5^{k+1} + 3$ is divisible by 4, so the statement is true for $n = k + 1$. Therefore, $5^n + 3$ is divisible by 4 for all natural numbers n.

18. Step 1: $9^1 - 1 = 8$, which is divisible by 8. The statement is true for $n = 1$.

Step 2: Assume that $9^k - 1$ is divisible by 8 for some natural number k. This means that $9^k - 1 = 8r$ for some whole number r.

Step 3: $9^k - 1 = 8r$
$9^k = 8r + 1$
$9^{k+1} = 72r + 9$
$9^{k+1} - 1 = 72r + 8$
$9^{k+1} - 1 = 8(9r + 1)$

Since r is a whole number, $9r + 1$ is a whole number. Thus, $9^{k+1} - 1$ is divisible by 8, so the statement is true for $n = k + 1$. Therefore, $9^n - 1$ is divisible by 8 for all natural numbers n.

19. Step 1: $12^1 + 10 = 22$, which is divisible by 11. The statement is true for $n = 1$.

Step 2: Assume that $12^k + 10$ is divisible by 11 for some natural number k. This means that $12^k + 10 = 11r$ for some natural number r.

Step 3: $12^k + 10 = 11r$
$12^k = 11r - 10$
$12^{k+1} = 132r - 120$
$12^{k+1} + 10 = 132r - 110$
$12^{k+1} + 10 = 11(12r - 10)$

Since r is a natural number, $12r - 10$ is a natural number. Thus, $12^{k+1} + 10$ is divisible by 11, so the statement is true for $n = k + 1$. Therefore, $12^n + 10$ is divisible by 11 for all natural numbers n.

20. Step 1: $13^1 + 11 = 24$, which is divisible by 12. The statement is true for $n = 1$.

Step 2: Assume that $13^k + 11$ is divisible by 12 for some natural number k. This means that $13^k + 11 = 12r$ for some natural number r.

Step 3: $13^k + 11 = 12r$
$13^k = 12r - 11$
$13^{k+1} = 156r - 143$
$13^{k+1} + 11 = 156r - 132$
$13^{k+1} + 11 = 12(13r - 11)$

Since r is a natural number, $13r - 11$ is a natural number. Thus, $13^{k+1} + 11$ is divisible by 12, so the statement is true for $n = k + 1$. Therefore, $13^n + 11$ is divisible by 12 for all natural numbers n.

26. Step 1: $7^1 + 5 = 12$, which is divisible by 6. The statement is true for $n = 1$.

    Step 2: Assume that $7^k + 5$ is divisible by 6 for some natural number k. This means that $7^k + 5 = 6r$ for some whole number r.

    Step 3: $7^k + 5 = 6r$
    $$7^k = 6r - 5$$
    $$7^{k+1} = 7(6r - 5)$$
    $$7^{k+1} = 42r - 35$$
    $$7^{k+1} + 5 = 42r - 30$$
    $$7^{k+1} + 5 = 6(7r - 5)$$

    Since r is a natural number, $7r - 5$ is a natural number. Thus, $7^{k+1} + 5$ is divisible by 6, so the statement is true for $n = k + 1$. Therefore, $7^n + 5$ is divisible by 6 for all natural numbers n.

27. Step 1: $18^1 - 1 = 17$, which is divisible by 17. The statement is true for $n = 1$.

    Step 2: Assume that $18^k - 1$ is divisible by 17 for some natural number k. This means that $18^k - 1 = 17r$ for some natural number r.

    Step 3: $18^k - 1 = 17r$
    $$18^k = 17r + 1$$
    $$18^{k+1} = 18(17r + 1)$$
    $$18^{k+1} = 306r + 18$$
    $$18^{k+1} - 1 = 306r + 17$$
    $$18^{k+1} - 1 = 17(18r + 1)$$

    Since r is a natural number, $18r + 1$ is a natural number. Thus, $18^{k+1} - 1$ is divisible by 17, so the statement is true for $n = k + 1$. Therefore, $18^n - 1$ is divisible by 17 for all natural numbers n.

30. Step 1: When $n = 1$, the left side of the given equation is $4 \cdot 5^{4-1}$ or 500. The right side is $625\left(1 - \frac{1}{5^1}\right)$ or 500, so the equation is true for $n = 1$.

    Step 2: Assume that $500 + 100 + 20 + \ldots + 4 \cdot 5^{4-k} = 625\left(1 - \frac{1}{5^k}\right)$ for some natural number k.

    Step 3: $500 + 100 + 20 + \ldots + 4 \cdot 5^{4-k} + 4 \cdot 5^{4-(k+1)}$
    $$= 625\left(1 - \frac{1}{5^k}\right) + 4 \cdot 5^{4-(k+1)}$$
    $$= 625\left(\frac{5^k - 1}{5^k}\right) + 4 \cdot 5^{3-k}$$
    $$= 625\left(\frac{5^{k+1} - 5}{5^{k+1}}\right) + \frac{4 \cdot 5^4}{5^{k+1}}$$
    $$= \frac{5^4(5^{k+1} - 5) + 4 \cdot 5^4}{5^{k+1}}$$
    $$= \frac{5^4(5^{k+1} - 5 + 4)}{5^{k+1}}$$
    $$= 625\left(\frac{5^{k+1} - 1}{5^{k+1}}\right)$$
    $$= 625\left(1 - \frac{1}{5^{k+1}}\right)$$

The last expression is the right side of the equation to be proved, where $n = k + 1$. Thus, the equation is true for $n = k + 1$. Therefore, $500 + 100 + 20 + \ldots + 4 \cdot 5^{4-n} = 625\left(1 - \frac{1}{5^n}\right)$ for all natural numbers n.

31. Step 1: When $n = 1$, the left side of the given equation is $\frac{1}{1(1+1)(1+2)}$ or $\frac{1}{6}$. The right side is $\frac{1(1+3)}{4(1+1)(1+2)}$ or $\frac{1}{6}$, so the equation is true for $n = 1$.

    Step 2: Assume that $\frac{1}{1 \cdot 2 \cdot 3} + \frac{1}{2 \cdot 3 \cdot 4} + \frac{1}{3 \cdot 4 \cdot 5} + \ldots + \frac{1}{k(k+1)(k+2)} = \frac{k(k+3)}{4(k+1)(k+2)}$ for some natural number k.

    Step 3:
    $$\frac{1}{1 \cdot 2 \cdot 3} + \frac{1}{2 \cdot 3 \cdot 4} + \ldots + \frac{1}{k(k+1)(k+2)} + \frac{1}{(k+1)(k+2)(k+3)}$$
    $$= \frac{k(k+3)}{4(k+1)(k+2)} + \frac{1}{(k+1)(k+2)(k+3)}$$
    $$= \frac{k(k+3)(k+3)}{4(k+1)(k+2)(k+3)} + \frac{4}{4(k+1)(k+2)(k+3)}$$
    $$= \frac{k^3 + 6k^2 + 9k + 4}{4(k+1)(k+2)(k+3)}$$
    $$= \frac{(k+1)(k^2 + 5k + 4)}{4(k+1)(k+2)(k+3)}$$
    $$= \frac{(k+1)(k+4)}{4(k+2)(k+3)}$$
    $$= \frac{(k+1)[(k+1)+3]}{4[(k+1)+1][(k+1)+2]}$$

    The last expression is the right side of the equation to be proved, where $n = k + 1$. Thus, the equation is true for $n = k + 1$. Therefore, $\frac{1}{1 \cdot 2 \cdot 3} + \frac{1}{2 \cdot 3 \cdot 4} + \frac{1}{3 \cdot 4 \cdot 5} + \ldots + \frac{1}{n(n+1)(n+2)} = \frac{n(n+3)}{4(n+1)(n+2)}$ for all natural numbers n.

33. $n(n+1)$

    Step 1: When $n = 1$, the left side of the given equation is $2(1)$ or 2. The right side is $1(1+1)$ or 2, so the equation is true for $n = 1$.

    Step 2: Assume that $2 + 4 + 6 + \ldots + 2k = k(k+1)$ for some natural number k.

    Step 3: $2 + 4 + 6 + \ldots + 2k + 2(k+1)$
    $$= k(k+1) + 2(k+1)$$
    $$= (k+1)(k+2)$$
    $$= (k+1)[(k+1)+1]$$

    The last expression is the right side of the equation to be proved, where $n = k + 1$. Thus, the equation is true for $n = k + 1$. Therefore, $2 + 4 + 6 + \ldots + n^2 = n(n+1)$ for all natural numbers n.

36. Step 1: When $n = 1$, the left side of the given equation is 1^3 or 1. The right side is $\left(\frac{1(1+1)}{2}\right)^2$ or 1, so the equation is true for $n = 1$.

    Step 2: Assume that $1 + 8 + 27 + \ldots + k^3 = \left(\frac{k(k+1)}{2}\right)^2$ for some natural number k.

Step 3:
$$1 + 8 + \ldots + k^3 + (k+1)^3 = \left(\frac{k(k+1)}{2}\right)^2 + (k+1)^3$$
$$= \frac{k^2(k+1)^2}{4} + (k+1)^3$$
$$= \frac{k^2(k+1)^2}{4} + \frac{4(k+1)^3}{4}$$
$$= \frac{(k+1)^2[k^2 + 4(k+1)]}{4}$$
$$= \frac{(k+1)^2(k^2 + 4k + 4)}{4}$$
$$= \frac{(k+1)^2(k+2)^2}{2^2}$$
$$= \left(\frac{(k+1)[(k+1)+1]}{2}\right)^2$$

The last expression is the right side of the equation to be proved, where $n = k + 1$. Thus, the equation is true for $n = k + 1$.
Therefore, $1 + 8 + \ldots + n^3 = \left(\frac{n(n+1)}{2}\right)^2$ for all natural numbers n.

38. Sample answer: $6 + 10 + 14 + \ldots$. The sequence is produced by $a_n = 4n + 2$. The sum of the sequence is represented by $2n(n + 2)$.

Step 1: When $n = 1$, the left side of the given equation is $4(1) + 2$ or 6. The right side is $2(1)(1 + 2)$ or 6, so the equation is true for $n = 1$.

Step 2: Assume that $6 + 10 + 14 + \ldots 4k + 2 = 2k(k + 2)$ for some natural number k.

Step 3: $6 + 10 + 14 + \ldots 4k + 2 + 4(k + 1) + 2$
$$= 2k(k + 2) + 4(k + 1) + 2$$
$$= 2k^2 + 4k + 4k + 4 + 2$$
$$= 2k^2 + 8k + 6$$
$$= 2(k + 1)(k + 3)$$
$$= 2(k + 1)[(k + 1) + 2]$$

The last expression is the right side of the equation to be proved, where $n = k + 1$. Thus, the equation is true for $n = k + 1$. Therefore, $6 + 10 + 14 + \ldots (4n + 2) = 2n(n + 2)$ for all natural numbers n.

44. $n(n - 1) = n^2 - n$; Line 1: $1(1 - 1) = 1(0)$ or 0 and $1 - 1 = 0$; Line 2: $2(2 - 1) = 2(1)$ or 2 and $4 - 2 = 2$; Line 3: $3(3 - 1) = 3(2)$ or 6 and $9 - 3 = 6$; Line 4: $4(4 - 1) = 4(3)$ or 12 and $16 - 4 = 12$; Line 5: $5(5 - 1) = 5(4)$ or 20 and $25 - 5 = 20$

Study Guide and Review

64. Step 1: When $n = 1$, $7^1 - 1 = 7 - 1$ or 6. Since 6 divided by 6 is 1, the statement is true for $n = 1$.

Step 2: Assume that $7^k - 1$ is divisible by 6 for some positive integer k. This means that $7^k - 1 = 6r$ for some whole number r.

Step 3: $7^k - 1 = 6r$
$$7^k = 6r + 1$$
$$7^{k+1} = 42r + 7$$
$$7^{k+1} - 1 = 42r + 7 - 1$$
$$7^{k+1} - 1 = 42r + 6$$
$$7^{k+1} - 1 = 6(7r + 1)$$

Since r is a whole number, $7r + 1$ is a whole number. Thus, $7^{k+1} - 1$ is divisible by 6, so the statement is true for $n = k + 1$. Therefore, $7^n - 1$ is divisible by 6 for all positive integers n.

65. Step 1: When $n = 1$, $5^1 - 1 = 5 - 1$ or 4. Since 4 divided by 4 is 1, the statement is true for $n = 1$.

Step 2: Assume that $5^k - 1$ is divisible by 4 for some positive integer k. This means that $5^k - 1 = 4r$ for some whole number r.

Step 3:
$$5^k - 1 = 4r$$
$$5^k = 4r + 1$$
$$5^{k+1} = 20r + 5$$
$$5^{k+1} - 1 = 20r + 5 - 1$$
$$5^{k+1} - 1 = 20r + 4$$
$$5^{k+1} - 1 = 4(5r + 1)$$

Since r is a whole number, $5r + 1$ is a whole number. Thus, $5^{k+1} - 1$ is divisible by 4, so the statement is true for $n = k + 1$. Therefore, $5^n - 1$ is divisible by 4 for all positive integers n.

Practice Test

20. Step 1: When $n = 1$, the left side of the given equation is 1. The right side is 1 also, so the equation is true for $n = 1$.

Step 2: Assume $1 + 6 + 36 + \ldots + 6^{k-1} = \frac{1}{5}(6^k - 1)$ for some positive integer k.

Step 3: Show that the given equation is true for $n = k + 1$.
$$1 + 6 + 36 + \ldots + 6^{k-1} + 6^{k+1-1} = \frac{1}{5}(6^k - 1) + 6^{k+1-1}$$
$$= \frac{1}{5}(6^k - 1) + 6^k$$
$$= \frac{1}{5} \cdot 6^k - \frac{1}{5} + 6^k$$
$$= \frac{6}{5} \cdot 6^k - \frac{1}{5}$$
$$= \frac{1}{5} \cdot 6^{k+1} - \frac{1}{5}$$
$$= \frac{1}{5} \cdot (6^{k+1} - 1)$$

The last expression is the right side of the equation to be proved, where $n = k + 1$. Therefore, $1 + 6 + 36 + \ldots + 6^{n-1} = \frac{1}{5}(6^n - 1)$ for all positive integers n.

21. Step 1: $11^1 - 1 = 10$, which is divisible by 10. The statement is true for $n = 1$.

Step 2: Assume that $11^k - 1$ is divisible by 10 for some positive integer k. This means that $11^k - 1 = 10r$ for some whole number r.

Step 3: $11^k - 1 = 10r$
$$11^k = 10r + 1$$
$$11(11^k) = (10r + 1) 11$$
$$11^{k+1} = 110r + 11$$
$$11^{k+1} - 1 = 110r + 11 - 1$$
$$11^{k+1} - 1 = 110r + 10$$
$$11^{k+1} - 1 = 10(11r + 1)$$

Since r is a whole umber, $11r + 1$ is a whole number. Thus, $11^{k+1} - 1$ is divisible by 10, so the statement is true for $n = k + 1$. Therefore, $11^n - 1$ is divisible by 10 for all positive integers n.

	Diagnostic Assessment Quick Check			
	LESSON 11-1 45 min: 1 day / 90 min: 1 day	**EXTEND 11-1** 45 min: 0.5 day / 90 min: 0.5 day	**LESSON 11-2** 45 min: 1 day / 90 min: 1 day	**LESSON 11-3** 45 min: 1.5 days / 90 min: 1 day
Title	Designing a Study	Graphing Technology Lab: Simulations and Margin of Error	Distributions of Data	Probability Distributions
Objectives	■ Classify study types. ■ Design statistical studies.	■ Use a simulation to develop margins of error for various sizes of random samples.	■ Use the shapes of distributions to select appropriate statistics. ■ Use the shapes of distributions to compare data.	■ Construct a probability distribution. ■ Analyze a probability distribution and its summary statistics.
Key Vocabulary	parameter statistic bias random sample survey experiment observational study	margin of error	distribution negatively skewed distribution symmetric distribution positively skewed distribution	random variable discrete random variable continuous random variable probability distribution theoretical probability distribution experimental probability distribution Law of Large Numbers expected value
CCSS	S.IC.3, S.IC.5	S.IC.4, S.IC.6	S.IC.1	S.SMD.7
Multiple Representations	🔧			🔧
Lesson Resources	🖰 connectED.mcgraw-hill.com 📁 Leveled Worksheets 🔤 Vocabulary PT Personal Tutor ✅ Self-Check Quiz ■ *5-Minute Check* ■ *Study Notebook*	🖰 connectED.mcgraw-hill.com PT Personal Tutor **Materials:** ■ TI-83/84 Plus or other graphing calculator	🖰 connectED.mcgraw-hill.com 📁 Leveled Worksheets 📁 Quiz 1 🔤 Vocabulary PT Personal Tutor ✋ Virtual Manipulatives ✅ Self-Check Quiz ■ *5-Minute Check* ■ *Study Notebook*	🖰 connectED.mcgraw-hill.com 📁 Leveled Worksheets 📁 Quiz 2 🔤 Vocabulary PT Personal Tutor ✋ Virtual Manipulatives ✅ Self-Check Quiz ■ *5-Minute Check* ■ *Study Notebook*
Resources for Every Lesson	IWB eStudent Edition IWB Interactive Classroom	■ eTeacher Edition ■ eSolutions Manual ■ *e*Assessment McGraw-Hill		
Differentiated Instruction	pp. 725, 730		pp. 735, 741	pp. 745, 750
				Formative Assessment Mid-Chapter Quiz

IWB All digital assets are Interactive Whiteboard ready.

Suggested Pacing			
Time Periods	Instruction	Review & Assess	Total
45-minute	9 days	2 days	11 days
90-minute	7 days	1 day	8 days

LESSON 11-4 45 min: 1.5 days / 90 min: 1 day	**LESSON 11-5** 45 min: 1.5 days / 90 min: 1 day	**EXTEND 11-5** 45 min: 0.5 day / 90 min: 0.5 day	**LESSON 11-6** 45 min: 1.5 days / 90 min: 1 day
The Binomial Distribution	**The Normal Distribution**	**Spreadsheet Lab: Normal Approximation of Binomial Distributions**	**Confidence Intervals and Hypothesis Testing**
■ Identify and conduct a binomial experiment. ■ Find probabilities using binomial distributions.	■ Use the Empirical Rule to analyze normally distributed variables. ■ Apply the standard normal distribution and z-values.	■ Use a normal distribution to approximate a binomial distribution.	■ Find confidence intervals for normally distributed data. ■ Perform hypothesis tests on normally distributed data.
binomial experiment binomial distribution	normal distribution z-value standard normal distribution Empirical Rule		inferential statistics statistical inference confidence interval maximum error of estimate hypothesis test null hypothesis alternative hypothesis critical region left-tailed test two-tailed test right-tailed test
S.MD.6, S.MD.7	S.ID.4	S.ID.4	S.IC.1, S.IC.4
connectED.mcgraw-hill.com ▸ Leveled Worksheets ▸ Vocabulary ▸ Personal Tutor ▸ Self-Check Quiz ■ *5-Minute Check* ■ *Study Notebook*	connectED.mcgraw-hill.com ▸ Leveled Worksheets ▸ Quiz 3 ▸ Vocabulary ▸ Personal Tutor ▸ Virtual Manipulatives ▸ Self-Check Quiz ■ *5-Minute Check* ■ *Study Notebook*	connectED.mcgraw-hill.com ▸ Personal Tutor **Materials:** ■ Computers with spreadsheet program	connectED.mcgraw-hill.com ▸ Leveled Worksheets ▸ Quiz 4 ▸ Vocabulary ▸ Personal Tutor ▸ Self-Check Quiz ■ *5-Minute Check* ■ *Study Notebook*
▸ eStudent Edition ▸ Interactive Classroom	■ eTeacher Edition ■ eSolutions Manual ■ McGraw-Hill eAssessment		
pp. 755, 759	pp. 762, 766		pp. 771, 776
			Summative Assessment Study Guide and Review Practice Test

Assessment and Intervention

SE = Student Edition, TE = Teacher Edition, CRM = Chapter Resource Masters

Diagnosis	Prescription
Beginning Chapter 11	
Get Ready for Chapter 11 **SE**	Response to Intervention **TE**
Beginning Every Lesson	
Then, Now, Why? **SE** 5-Minute Checks	Chapter 0 **SE**

DIAGNOSTIC ASSESSMENT

Diagnosis	Prescription
During/After Every Lesson	
Guided Practice **SE**, every example Check Your Understanding **SE** H.O.T. Problems **SE** Spiral Review **SE** Additional Examples **TE** Watch Out! **TE** Step 4, Assess **TE** Chapter 11 Quizzes **CRM**, pp. 45–46 Self-Check Quizzes connectED.mcgraw-hill.com	**TIER 1 Intervention** Skills Practice **CRM**, Ch. 1–11 connectED.mcgraw-hill.com **TIER 2 Intervention** Differentiated Instruction **TE** Study Guide and Intervention **CRM**, Ch. 1–11 **TIER 3 Intervention** Math Triumphs, Alg. 2, Ch. 5
Mid-Chapter	
Mid-Chapter Quiz **SE** Mid-Chapter Test **CRM**, p. 47 McGraw-Hill eAssessment	**TIER 1 Intervention** Skills Practice **CRM**, Ch. 1–11 connectED.mcgraw-hill.com **TIER 2 Intervention** Study Guide and Intervention **CRM**, Ch. 1–11 **TIER 3 Intervention** Math Triumphs, Alg. 2, Ch. 5
Before Chapter Test	
Chapter Study Guide and Review **SE** Practice Test **SE** Standardized Test Practice **SE** Chapter Test connectED.mcgraw-hill.com Standardized Test Practice connectED.mcgraw-hill.com Vocabulary Review connectED.mcgraw-hill.com McGraw-Hill eAssessment	**TIER 1 Intervention** Skills Practice **CRM**, Ch. 1–11 connectED.mcgraw-hill.com **TIER 2 Intervention** Study Guide and Intervention **CRM**, Ch. 1–11 **TIER 3 Intervention** Math Triumphs, Alg. 2, Ch. 5

FORMATIVE ASSESSMENT

Diagnosis	Prescription
After Chapter 11	
Multiple-Choice Tests, Forms 1, 2A, 2B **CRM**, pp. 49–54 Free-Response Tests, Forms 2C, 2D, 3 **CRM**, pp. 55–60 Vocabulary Test **CRM**, p. 48 Extended Response Test **CRM**, p. 61 Standardized Test Practice **CRM**, pp. 62–64 McGraw-Hill eAssessment	Study Guide and Intervention **CRM**, Ch. 1–11 connectED.mcgraw-hill.com

SUMMATIVE ASSESSMENT

Option 1 Reaching All Learners AL OL BL ELL

Kinesthetic Ask each student to use a tape measure to measure the distance around the wrists of 15 classmates to the nearest tenth of a centimeter. Have students find the mean and standard deviation of their data. Then have them determine if their data appears to be normally distributed, positively skewed, or negatively skewed.

Visual/Spatial Have students work in pairs to make up a crossword puzzle using the terms studied in this chapter. Use either the definition or an example of the terms in the *across* and *down* hints. Make photocopies of the puzzle to distribute to the class. Ask students to keep the puzzles to use as a chapter review.

Option 2 Approaching Level AL

Have students work in groups of three or four. Ask each student to choose either the mean, median, or mode and to create a set of at least 15 data values that would be best represented using the chosen measure of central tendency. Each student, in turn, presents his/her set of data to the group, who then decides which measure to use. As a group, students actually find the measure of central tendency that best represents each set of data.

Option 3 English Learners ELL

Ask students to work with a partner. One partner writes an expression, such as $_{12}C_3$, and hands it to the other partner, who reads the notation aloud (for example, "the number of combinations of 12 items taken 3 at a time") and calculates the value. 220 The partners discuss and correct this value as necessary. Then they exchange roles.

Option 4 Beyond Level BL

Ask students to make a chart showing the relationship between the binomial coefficients and combinatorial notation. An example is shown below.

○
  Binomial notation:

○ $(x + y)^n = x^n + nx^{n-1}y + \frac{n(n-1)}{2!}x^{n-2}y^2$

  $+ \frac{n(n-1)(n-2)}{3!} x^{n-3}y^3 + \dots +$

  $\frac{n(n-1)(n-2) \dots (n-r+2)}{(r-1)!} x^{n-r+1}y^{r-1}$

  $+ \dots + nxy^{n-1} + y^n$

○ Combinatorial notation:

  $(x + y)^n = {}_nC_0 x^n + {}_nC_1 x^{n-1}y + {}_nC_2 x^{n-2}y^2$

  $+ {}_nC_3 x^{n-3}y^3 + \dots + {}_nC_{n-1} x^{n-r+1}y^{r-1}$

  $+ \dots + {}_nC_{n-1} xy^{n-1} + {}_nC_n y^n$

VerticalAlignment

Before Chapter 11

Related Topics from Algebra 1

- construct sample spaces for simple or composite experiments

- find the probabilities of dependent and independent events

- use theoretical probabilities and experimental results to make predictions and decisions

- select the appropriate measure of central tendency or range to describe a set of data

- select and use an appropriate representation for presenting and displaying relationships among collected data

- evaluate methods of sampling to determine validity of an inference made from a set of data

Previous Topics from Algebra 2

- simplify polynomial expressions

Chapter 11

Reinforcement of Topics Before Algebra 2

- use combinations and permutations to find probability and use them to solve problems

- find probabilities of two independent and two dependent events

- find the probability of mutually exclusive and inclusive events

Related Topics from Algebra 2

- use a measure of central tendency to represent a set of data and find measures of variation for a set of data

- determine whether a sample is biased and find margins of sampling error

- create and use graphs of probability distributions

- solve problems involving normally distributed data

- use binomial expressions to find probabilities

- use confidence intervals to estimate population parameters

After Chapter 11

This chapter reinforces skills needed in AP Statistics.

Lesson-by-LessonPreview

11-1 Designing a Study

A *parameter* is a measure that describes a characteristic of a population. A *statistic* is a measure that describes a characteristic of a sample.

Sample information can be collected using the following study types.
- *Survey*—Data are collected from responses given by the sample.
- *Experiment*—The sample is divided into an *experimental group* and a *control group*. The experimental group undergoes a change, and the effect is compared to the control group.
- *Observational study*—Data are collected by observing the sample.

Survey questions can introduce bias if they are confusing, cause a strong reaction, encourage a certain response, or address more than one issue. Experiments can also introduce bias if the sample is not randomly selected, or if the control and experimental groups are not similar.

11-2 Distributions of Data

A *distribution* of data shows the frequency of each possible data value. The shape of a distribution can be determined by looking at its histogram or box-and-whisker plot.

Negatively Skewed Distribution

Symmetric Distribution

Positively Skewed Distribution

When describing a distribution, use
- the mean and standard deviation if the distribution is symmetric, or
- the five-number summary if the distribution is skewed.

When comparing two sets of data, use
- the means and standard deviations if the distributions are both symmetric, or
- the five-number summaries if the distributions are both skewed, or if one distribution is symmetric and the other is skewed.

Get educator support from the Math Forum's T2T. You may want to search, or ask, for teaching strategies on the normal distribution. Start your search at connectED.mcgraw-hill.com.

mathforum.org
The Math Forum
@ Drexel University

11-3 Probability Distributions

A *random variable* can be either *discrete* or *continuous*. A discrete random variable is countable. A continuous random variable can take on any value.

A *probability distribution* is a table, equation, or graph that maps the sample space to the outcomes in the sample space.

A probability distribution for a random variable X has the following properties.
- can be created using theoretical or experimental probabilities
- can be discrete or continuous
- probability of each value of X must be between 0 and 1
- sum of all the probabilities of X must equal 1

The *expected value* of a discrete random variable X is the sum of the products of every possible value of X and its associated probability.

11-4 The Binomial Distribution

A *binomial experiment* satisfies the following conditions.
- There are n independent trials.
- Each trial has only two outcomes, success or failure.
- For each trial, the probability of success is p and the probability of failure q is $1 - p$.
- The random variable X is the number of successes in n trials.

In a *binomial distribution*, the probability of X successes in n independent trails can be found using the Binomial Probability Formula.

$$P(X) = {}_nC_X \, p^X q^{n-X}$$

11-5 The Normal Distribution

The *normal distribution* is a continuous, symmetric, bell-shaped distribution of a random variable. The mean, median, and mode are equal and located at the center of the curve. The curve approaches, but never touches, the x-axis. The total area under the curve is equal to 1.

The *Empirical Rule* can be used to determine the area under the normal curve at specific intervals.

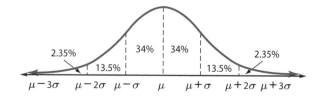

Data are standardized by converting them to z-values. The *z-value* represents the number of standard deviations that a given data value is from the mean. Once data are standardized, they can be compared using the *standard normal distribution*.

11-6 Confidence Intervals and Hypothesis Testing

A *confidence interval* is an estimate of a parameter stated as a range with a specific degree of certainty. A confidence interval for a population mean is given by

$$CI = \bar{x} \pm z \cdot \frac{s}{\sqrt{n}}$$

where $\bar{x}$ is the sample mean, z is the z-value that corresponds to a particular confidence level, s is the standard deviation of the sample, and n is the sample size.

A *hypothesis test* is used to assess a specific claim about the mean. The claim is either the null or the alternative hypothesis. The *null hypothesis* is a statement of equality. The *alternative hypothesis* is a statement of inequality that is the complement of the null hypothesis.

Mc Graw Hill Professional Development

Targeted professional development has been articulated throughout Algebra 2. More quality, customized professional development is available from McGraw-Hill Professional Development. Visit connectED.mcgraw-hill.com for details on each product.

- Online Lessons emphasize the strategies and techniques used to teach Algebra 2. Includes streaming video, interactive pages, and online tools.

- Video Workshops allow mentors, coaches, or leadership personnel to facilitate on-site workshops on educational strategies in mathematics and mathematical concepts.

- MHPD Online (www.mhpdonline.com) offers online professional development with video clips of instructional strategies, links, student activities, and news and issues in education.

- Teaching Today (teachingtoday.glencoe.com) gives secondary teachers practical strategies and materials that inspire excellence and innovation in teaching.

Chapter Project

Please Complete This Survey

Students use what they have learned about surveys, distributions of data, and normal distributions to complete a project.

This chapter project addresses business literacy, as well as several specific skills identified as being essential to student success by the Framework for 21st Century Learning.

Visit connectED.mcgraw-hill.com for student and teacher handouts.

KeyVocabulary Introduce the key vocabulary in the chapter using the routine below.

Define: Normal distribution is a frequency distribution that often occurs when there is a large number of values in a set of data: about 68% of the values are within one standard deviation of the mean, 95% of the values are within two standard deviations of the mean, and 99% of the values are within three standard deviations.

Example:

This diagram shows a normal distribution.

Normal Distribution

Ask: What do you notice about the normal distribution curve? It is shaped like a bell and is symmetrical.

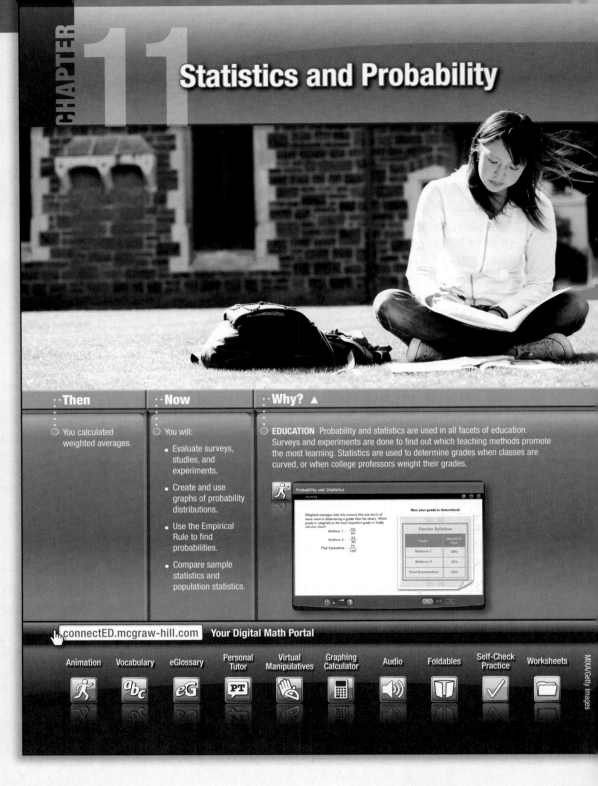

CHAPTER 11
Statistics and Probability

Then
○ You calculated weighted averages.

Now
○ You will:
- Evaluate surveys, studies, and experiments.
- Create and use graphs of probability distributions.
- Use the Empirical Rule to find probabilities.
- Compare sample statistics and population statistics.

Why? ▲
○ **EDUCATION** Probability and statistics are used in all facets of education. Surveys and experiments are done to find out which teaching methods promote the most learning. Statistics are used to determine grades when classes are curved, or when college professors weight their grades.

connectED.mcgraw-hill.com   **Your Digital Math Portal**

| Animation | Vocabulary | eGlossary | Personal Tutor | Virtual Manipulatives | Graphing Calculator | Audio | Foldables | Self-Check Practice | Worksheets |

MIXA/Getty Images

Get Ready for the Chapter

Diagnose Readiness | You have two options for checking prerequisite skills.

1 Textbook Option Take the Quick Check below. Refer to the Quick Review for help.

QuickCheck	QuickReview

Find the mean, median, and mode for each set of data.

1. number of customers at a store each day during the last two weeks:
78, 80, 101, 66, 73, 92, 97, 125, 110, 76, 89, 90, 82, 87
89 customers, 88 customers, no mode

2. a student's quiz scores for the first grading period:
88, 70, 85, 92, 88, 77, 98, 88, 70, 82 **83.8, 86.5, 88**

3. the number of touchdowns scored by a player over the last 10 years:
7, 5, 10, 12, 4, 10, 11, 6, 9, 3
7.7 touchdowns, 8 touchdowns, 10 touchdowns

Example 1 (Used in Lesson 11-2)

The number of days it rained in each month over the past year are shown below. Find the mean, median, and mode.

$$4, 2, 9, 16, 13, 9, 8, 9, 7, 6, 8, 5$$

Mean $\bar{x} = \dfrac{4 + 2 + 9 + 16 + 13 + 9 + 8 + 9 + 7 + 6 + 8 + 5}{12}$
or 8 days

Median 2, 4, 5, 6, 7, 8, 8, 9, 9, 9, 13, 16
$\dfrac{8 + 8}{2}$ or 8 days

Mode The value that occurs most often in the set is 9, so the mode of the data set is 9 days.

A die is rolled and a coin is tossed. Find each probability.

4. P (4, heads) $\dfrac{1}{12}$

5. P (odds, tails) $\dfrac{1}{4}$

6. P (2 or 4, heads) $\dfrac{1}{6}$

Example 2 (Used in Lesson 11-3)

A die is rolled and a coin is tossed. What is the probability that the die shows a 1 and the coin lands tails up?

$$P (1, \text{tails}) = \dfrac{1}{6} \cdot \dfrac{1}{2} \text{ or } \dfrac{1}{12}$$

Expand each binomial.

7. $(a - 2)^4$
8. $(m - a)^5$
9. $(2b - x)^4$
10. $(2a + b)^6$
11. $(3x - 2y)^5$
12. $(3x + 2y)^4$
13. $\left(\dfrac{a}{2} + 2\right)^5$
14. $\left(3 + \dfrac{m}{3}\right)^5$

7–14. See margin on page 722.

Example 3 (Used in Lesson 11-4)

Expand $(a + b)^4$.

Replace n with 4 in the Binomial Theorem.

$(a + b)^4$

$= {}_4C_0 a^4 + {}_4C_1 a^3 b + {}_4C_2 a^2 b^2 + {}_4C_3 a b^3 + {}_4C_4 b^4$

$= \dfrac{4!}{(4 - 0)! \cdot 0!} a^4 + \dfrac{4!}{(4 - 1)! \cdot 1!} a^3 b + \dfrac{4!}{(4 - 2)! \cdot 2!} a^2 b^2$
$\quad + \dfrac{4!}{(4 - 3)! \cdot 3!} a b^3 + \dfrac{4!}{(4 - 4)! \cdot 4!} b^4$

$= \dfrac{24}{24} a^4 + \dfrac{24}{6 \cdot 1} a^3 b + \dfrac{24}{2 \cdot 2} a^2 b^2 + \dfrac{24}{1 \cdot 6} a b^3 + \dfrac{24}{1 \cdot 4} b^4$

$= a^4 + 4a^3 b + 6a^2 b^2 + 4ab^3 + b^4$

2 Online Option Take an online self-check Chapter Readiness Quiz at <u>connectED.mcgraw-hill.com</u>.

721

Get Ready for the Chapter

Response to Intervention (RtI)
Use the *Quick Check* results and the Intervention Planner to help you determine your Response to Intervention. The If-Then statements in the chart below help you decide the appropriate tier of RtI and suggest intervention resources for each tier.

InterventionPlanner

TIER 1 On Level OL

If students miss about 25% of the exercises or less,

Then choose a resource:

SE Lessons 0-4 and 10-6

📁 Skills Practice, Chapter 10, p. 41

 connectED.mcgraw-hill.com Self-Check Quiz

TIER 2 Strategic Intervention AL
approaching grade level

If students miss about 50% of the exercises,

Then choose a resource:

📁 Study Guide and Intervention, Chapter 10, p. 39

connectED.mcgraw-hill.com Extra Examples, Personal Tutor, Homework Help

TIER 3 Intensive Intervention
2 or more grades below level

If students miss about 75% of the exercises,

Then use *Math Triumphs, Alg. 2*, Ch. 5

connectED.mcgraw-hill.com Extra Examples, Personal Tutor, Homework Help, Review Vocabulary

---

E? Essential Questions

- How can you effectively evaluate information? Sample answer: First, determine whether the information source is credible. Then critically analyze the information to determine whether it is useful for the given situation.

- How can you use information to make decisions? Sample answer: You can look for trends, and then make a decision based on what has happened in the past and/or is reflected in the information.

FOLDABLES StudyOrganizer

Dinah Zike's Foldables®

Focus Students write notes about probability and statistics.

Teach Have students make and label their Foldables as illustrated. At the end of each lesson, ask students to write about their experiences with the statistics and probability topics presented in the lessons. Encourage students to explain what they found interesting or challenging.

When to Use It Encourage students to add to their Foldables as they work through the chapter and to use them to review for the chapter test.

Differentiated Instruction

Student-Built Glossary, pp. 1–2

Students should complete the chart by providing a definition of each term and an example as they progress through Chapter 11. This study tool can also be used to review for the chapter test.

Additional Answers
(Get Ready for the Chapter)

7. $a^4 - 8a^3 + 24a^2 - 32a + 16$

8. $m^5 - 5m^4a + 10m^3a^2 - 10m^2a^3 + 5ma^4 - a^5$

9. $16b^4 - 32b^3x + 24b^2x^2 - 8bx^3 + x^4$

10. $64a^6 + 192a^5b + 240a^4b^2 + 160a^3b^3 + 60a^2b^4 + 12ab^5 + b^6$

11. $243x^5 - 810x^4y + 1080x^3y^2 - 720x^2y^3 + 240xy^4 - 32y^5$

12. $81x^4 + 216x^3y + 216x^2y^2 + 96xy^3 + 16y^4$

13. $\frac{a^5}{32} + \frac{5a^4}{8} + 5a^3 + 20a^2 + 40a + 32$

14. $243 + 135m + 30m^2 + \frac{10m^3}{3} + \frac{5m^4}{27} + \frac{m^5}{243}$

Get Started on the Chapter

You will learn several new concepts, skills, and vocabulary terms as you study Chapter 11. To get ready, identify important terms and organize your resources. You may wish to refer to Chapter 0 to review prerequisite skills.

FOLDABLES StudyOrganizer

Statistics and Probability Make this Foldable to help you organize your Chapter 11 notes about statistics and probability. Begin with a sheet of $8\frac{1}{2}$" by 11" paper.

1 **Fold** in half lengthwise.

2 **Fold** the top to the bottom.

3 **Open.** Cut along the second fold to make two tabs.

4 **Label** each tab as shown.

Statistics | Probability

NewVocabulary

English		Español
parameter	p. 723	parámetro
statistic	p. 723	estadística
survey	p. 723	exámenes
experiment	p. 723	experimento
observational study	p. 723	estudio de observación
random variable	p. 742	variable aleatoria
probability distribution	p. 743	distribución de probabilidad
expected value	p. 745	valor previsto
binomial experiment	p. 752	experimento binomio
binomial distribution	p. 754	distribución binomial
normal distribution	p. 760	distribución normal
z-value	p. 762	valor de z
confidence interval	p. 769	intervalo de la confianza
inferential statistics	p. 769	estadística deductiva
statistical inference	p. 769	inferencia estadística
hypothesis test	p. 771	prueba de hipótesis
null hypothesis	p. 771	hipótesis nula
alternative hypothesis	p. 771	hipótesis alternativa

ReviewVocabulary

combination combinación an arrangement or selection of objects in which order is not important

permutation permutación a group of objects or people arranged in a certain order

random arbitrario Unpredictable, or not based on any predetermined characteristics of the population; when a die is tossed, a coin is flipped, or a spinner is spun, the outcome is a random event.

11-1 Designing a Study

::Then	::Now	::Why?
• You identified various sampling techniques.	**1** Classify study types. **2** Design statistical studies.	• According to a recent study, 88% of teen cell phone users in the U.S. send text messages, and one in three teens sends more than 100 texts per day.

 NewVocabulary
parameter
statistic
bias
random sample
survey
experiment
observational study

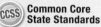 **Common Core State Standards**

Content Standards
S.IC.3 Recognize the purposes of and differences among sample surveys, experiments, and observational studies; explain how randomization relates to each.

S.IC.5 Use data from a randomized experiment to compare two treatments; use simulations to decide if differences between parameters are significant.

Mathematical Practices
3 Construct viable arguments and critique the reasoning of others.

1 Classifying Studies In a statistical study, data are collected and used to answer questions about a population characteristic or **parameter**. Due to time and money constraints, it may be impractical or impossible to collect data from each member of a population. Therefore, in many studies, a sample of the population is taken, and a measure called a **statistic** is calculated using the data. The sample statistic, such as the sample mean or sample standard deviation, is then used to make inferences about the population parameter.

The steps in a typical statistical study are shown below.

Identify the objective of the study.	→	Choose a sample, and collect data.	→	Organize the data, and calculate sample statistics.	→	Make inferences, and draw conclusions about the population.

To obtain good information and draw accurate conclusions about a population, it is important to select an *unbiased* sample. A **bias** is an error that results in a misrepresentation of members of a population. A poorly chosen sample can cause biased results. To reduce the possibility of selecting a biased sample, a **random sample** can be taken, in which members of the population are selected entirely by chance. You will review other sampling methods in Exercise 32.

The following study types can be used to collect sample information.

KeyConcept Study Types

Definition	Example
In a **survey**, data are collected from responses given by members of a population regarding their characteristics, behaviors, or opinions.	To determine whether the student body likes the new cafeteria menu, the student council asks a random sample of students for their opinion.
In an **experiment**, the sample is divided into two groups: • an *experimental group* that undergoes a change, and • a *control group* that does not undergo the change. The effect on the experimental group is then compared to the control group.	A restaurant is considering creating meals with chicken instead of beef. They randomly give half of a group of participants meals with chicken and the other half meals with beef. Then they ask how they like the meals.
In an **observational study**, members of a sample are measured or observed without being affected by the study.	Researchers at an electronics company observe a group of teenagers using different laptops and note their reactions.

connectED.mcgraw-hill.com **723**

Radius Images/Jupiterimages

---

1 Focus

VerticalAlignment

▼

Before Lesson 11-1 Identify sampling techniques.

Lesson 11-1 Classify study types. Design statistical studies.

After Lesson 11-1 Use the shapes of distributions to select appropriate statistics and compare data.

2 Teach

Scaffolding Questions
Have students read the **Why?** section of the lesson.

Ask:
- What do you think the purpose of the study was? Sample answer: to determine the number of teen cell phone users that send texts and the amount of texts that these users send per day

- How do you think the data could have been collected for this study? Sample answer: Teen cell phone users could have been asked if they send texts and, if so, how many texts they send per day.

- According to the study, if 50 teen cell phone users were randomly selected, how many would be texters? 44

Lesson 11-1 Resources

Resource	Approaching Level **AL**	On Level **OL**	Beyond Level **BL**	English Learners **ELL**
Teacher Edition		▪ Differentiated Instruction, pp. 725, 730	▪ Differentiated Instruction, p. 730	▪ Differentiated Instruction, p. 725
Chapter Resource Masters	▪ Study Guide and Intervention, pp. 5–6 ▪ Skills Practice, p. 7 ▪ Practice, p. 8 ▪ Word Problem Practice, p. 9	▪ Study Guide and Intervention, pp. 5–6 ▪ Skills Practice, p. 7 ▪ Practice, p. 8 ▪ Word Problem Practice, p. 9 ▪ Enrichment, p. 10	▪ Practice, p. 8 ▪ Word Problem Practice, p. 9 ▪ Enrichment, p. 10	▪ Study Guide and Intervention, pp. 5–6 ▪ Skills Practice, p. 7 ▪ Practice, p. 8 ▪ Word Problem Practice, p. 9
Other	▪ 5-Minute Check 11-1 ▪ Study Notebook	▪ 5-Minute Check 11-1 ▪ Study Notebook	▪ 5-Minute Check 11-1 ▪ Study Notebook	▪ 5-Minute Check 11-1 ▪ Study Notebook

1 Classifying Studies

Examples 1 shows how to classify study types. **Example 2** shows how to choose a study type.

Formative Assessment

Use the Guided Practice exercises after each example to determine students' understanding of concepts.

Additional Example

1 Determine whether each situation describes a *survey*, an *experiment*, or an *observational study*. Then identify the sample, and suggest a population from which it may have been selected.

a. **MOVIES** A retro movie theater wants to determine what genre of movies to play during the next year. They plan to poll 50 random area residents and ask them what their favorite movies are. survey; sample: the 50 residents that were polled; population: all potential movie-goers

b. **DRIVING** A driving school wants to determine the main issue drivers face while taking the driving test. They watch and record 30 random people taking the test. observational study; sample: the people observed while taking the test; population: potential driving school students

Additional Examples also in Interactive Classroom PowerPoint® Presentations

Interactive White Board
READY

Example 1 Classify Study Types

Determine whether each situation describes a *survey*, an *experiment*, or an *observational study*. Then identify the sample, and suggest a population from which it may have been selected.

a. **MUSIC** A record label wants to test three designs for an album cover. They randomly select 50 teenagers from local high schools to view the covers while they watch and record their reactions.

This is an observational study, because the company is going to observe the teens without them being affected by the study. The sample is the 50 teenagers selected, and the population is all potential purchasers of this album.

b. **RECYCLING** The city council wants to start a recycling program. They send out a questionnaire to 200 random citizens asking what items they would recycle.

This is a survey, because the data are collected from participants' responses in the questionnaire. The sample is the 200 people who received the questionnaire, and the population is all of the citizens of the city.

> **Guided**Practice

1A. experiment; sample: dogs given the heartworm medication; population: all dogs

1A. RESEARCH Scientists study the behavior of one group of dogs given a new heartworm treatment and another group of dogs given a false treatment or *placebo*.

1B. YEARBOOKS The yearbook committee conducts a study to determine whether students would prefer to have a print yearbook or both print and digital yearbooks.

To determine when to use a survey, experiment, or observational study, think about how the data will be obtained and whether or not the participants will be affected by the study.

Example 2 Choose a Study Type

Determine whether each situation calls for a *survey*, an *experiment*, or an *observational study*. Explain your reasoning.

a. **MEDICINE** A pharmaceutical company wants to test whether a new medicine is effective.

The treatment will need to be tested on a sample group, which means that the members of the sample will be affected by the study. Therefore, this situation calls for an experiment.

b. **ELECTIONS** A news organization wants to randomly call citizens to gauge opinions on a presidential election.

This situation calls for a survey because members of the sample population are asked for their opinion.

> **Guided**Practice

2A. RESEARCH A research company wants to study smokers and nonsmokers to determine whether 10 years of smoking affects lung capacity.

2B. PETS A national pet chain wants to know whether customers would pay a small annual fee to participate in a rewards program. They randomly select 200 customers and send them questionnaires.

724 | Lesson 11-1 | Designing a Study

StudyTip

Census A *census* is a survey in which each member of a population is questioned. Therefore, when a census is conducted, there is no sample.

1B. survey; sample: students participating in the survey; population: the student body

2A. Observational study; sample answer: The lung capacities of the participants are observed and compared without them being affected by the study.
2B. Survey; sample answer: The data is obtained from opinions given by members of the sample population.

Focus on Mathematical Content

Survey and Census When a census is conducted, data are collected from every member of a population. Therefore, the results are known to be correct. Since a survey investigates only part of a population, the results always involve some uncertainty.

3B. Biased; sample answer: The adjectives "exciting" and "dull" suggest that the respondent should agree that action movies are preferable to documentaries.

4. objective: to determine teachers' interest in teaching an online course; population: teachers at the school with at least five years of experience; sample survey questions: How long have you been teaching? If offered, would you be interested in teaching an online course?

2 Designing Studies The questions chosen for a survey or procedures used in an experiment can also introduce bias, and thus, affect the results of the study.

A survey question that is poorly written may result in a response that does not accurately reflect the opinion of the participant. Therefore, it is important to write questions that are clear and precise. Avoid survey questions that:

- are confusing or wordy
- encourage a certain response
- cause a strong reaction
- address more than one issue

Questions can also introduce bias if there is not enough information given for the participant to give an accurate response.

Example 3 Identify Bias in Survey Questions

Determine whether each survey question is *biased* or *unbiased*. If biased, explain your reasoning.

a. Don't you agree that the cafeteria should serve healthier food?

This question is biased because it encourages a certain response. The phrase "don't you agree" encourages you to agree that the cafeteria should serve healthier food.

b. How often do you exercise?

This question is unbiased because it is clearly stated and does not encourage a certain response.

▶ **Guided**Practice

3A. How many glasses of water do you drink a day? *unbiased*

3B. Do you prefer watching exciting action movies or boring documentaries?

When designing a survey, clearly state the objective, identify the population, and carefully choose unbiased survey questions.

Real-World Example 4 Design a Survey

TECHNOLOGY Jim is writing an article for his school newspaper about online courses. He wants to conduct a survey to determine how many students at his school would be interested in taking an online course from home. State the objective of the survey, suggest a population, and write two unbiased survey questions.

Step 1 State the objective of the survey.

The objective of the survey is to determine students' interest in taking an online course from home.

Step 2 Identify the population.

The population is the student body.

Step 3 Write unbiased survey questions.

Possible survey questions:
- "Do you have Internet access at home?"
- "If offered, would you take an online course?"

▶ **Guided**Practice

4. TECHNOLOGY In a follow-up article, Jim decides to conduct a survey to determine how many teachers from his school with at least five years of experience would be interested in teaching an online course. State the objective of the survey, suggest a population, and write two unbiased survey questions.

Real-World Link
Online Courses In 2009, about 1.2 million students took at least one online course.
Source: International Association for K-12 Online Learning

DifferentiatedInstruction OL ELL

Verbal/Linguistic Learners Divide students into small groups. Have each group design a survey question and practice asking it in such a way that there is bias built into the tone of voice and facial expression of the questioner. Then have them ask other groups the question and record their answers. As a class, discuss whether the answers corresponded to the bias that the question was designed to elicit.

---

2 Designing Studies

Example 3 shows how to identify biased survey questions. **Example 4** shows how to design a survey. **Example 5** shows how to identify flaws in experiments. **Example 6** shows how to design an experiment.

Additional Examples

2 Determine whether each situation calls for a *survey*, an *experiment*, or an *observational study*. Explain your reasoning.

a. VIDEO GAMES A gaming company plans to test whether a new controller is preferable to the old one. A group of teens will be observed while using the controllers to see which one they use the most. Observational study; sample answer: The teens will be observed without being affected by the study.

b. RESTAURANTS A restaurant wants to conduct an online study in which they will ask customers whether they were satisfied with their dining experience. Survey; sample answer: The members of the sample will be asked for their opinion.

3 Determine whether each survey question is *biased* or *unbiased*. If biased, explain your reasoning.

a. What is your favorite type of music? This question is unbiased because it is clearly stated and does not encourage a certain response.

b. Do you think that poisons, such as pesticides, should be sprayed on crops? This question is biased because the term "poison" could cause a strong reaction from the respondent.

4   **COLLEGE** A community college wants to determine whether college-bound students from local high schools would be interested in taking a class at the college. State the objective of the survey, suggest a population, and write two unbiased survey questions.   objective: to determine whether students that are planning on going to college would be interested in taking a class at the community college; population: all local high school students; sample survey questions: Are you planning on attending college after high school? Would you be interested in taking a course at the community college during high school?

5   Identify any flaws in the design of the experiment, and describe how they could be corrected.

Experiment: A research company wants to conduct a study to determine whether a new fishing reel is more effective than the old reel. The experimental procedure consists of using the new reel to catch fish in one lake, and using the old reel to catch fish in another nearby lake that is randomly chosen.

Results: The company concludes that the new reel is twice as effective as the old reel.   Sample answer: The flaw is that the fishing reels were used in two different lakes. The amount or type of fish in the two lakes may not be similar enough to perform a controlled experiment.

To avoid introducing bias in experiments, the experimental and control groups should be randomly selected and the experiment should be designed so that everything about the two groups is alike (except for the treatment or procedure).

Example 5 Identify Flaws in Experiments

Identify any flaws in the design of the experiment, and describe how they could be corrected.

Experiment: An electronics company wants to test whether using a new graphing calculator increases students' test scores. A random sample is taken. Calculus students in the experimental group are given the new calculator to use, and Algebra 2 students in the control group are asked to use their own calculator.

Results: When given the same test, the experimental group scored higher than the control group. The company concludes that the use of this calculator increases test scores.

Calculus students are more likely to score higher when given the same test as Algebra 2 students. Therefore, the flaw is that the experimental group consists of Calculus students and the control group consists of Algebra 2 students. This flaw could be corrected by selecting a random sample of all Calculus or all Algebra 2 students.

Guided Practice

5. **Experiment:** A research firm tests the effectiveness of a de-icer on car locks. They use a random sample of drivers in California and Minnesota for the control and experimental groups.

Results: They concluded that the de-icer is effective.

5. Sample answer: The flaw is that the control group is drivers in California while the experimental group is drivers in Minnesota. This flaw can be corrected by selecting both groups from a cold area like Minnesota where the de-icer is needed more.

When designing an experiment, clearly state the objective, identify the population, determine the experimental and control groups, and define the procedure.

Real-World Example 6 Design an Experiment

PLANTS A research company wants to test the claim of the advertisement shown at the right. State the objective of the experiment, suggest a population, determine the experimental and control groups, and describe a sample procedure.

Taller tomato plants in just 3 weeks!
Thrive

Step 1 State the objective, and identify the population.

The objective of the experiment is to determine whether tomato plants given the plant food grow taller in three weeks than tomato plants not given the food. The population is all tomato plants.

Step 2 Determine the experimental and control groups.

The experimental group is the tomato plants given the food, and the control group is the tomato plants not given the food.

Step 3 Describe a sample procedure.

Measure the heights of the plants in each group, and give the experimental group the plant food. Then, wait three weeks, measure the heights of the plants again, and compare the heights for each group to see if the claim was valid.

Guided Practice See margin.

6. SPORTS A company wants to determine whether wearing a new tennis shoe improves jogging time. State the objective of the experiment, suggest a population, determine the experimental and control groups, and describe a sample procedure.

 726 | Lesson 11-1 | Designing a Study

StudyTip

Bias in Experiments
An experiment is biased when the participants know which group they are in.

Additional Answer (Guided Practice)

6. objective: To determine whether wearing the tennis shoe improves jogging time; experimental group: participants that wear the new shoe; control group: participants that wear another type of shoe; sample procedure: The company could select participants that jog on a regular basis, are all healthy, and are close to the same age. The company could then randomly divide the sample into two groups: an experimental group of participants that wear the new tennis shoe and a control group of participants that wear another type of shoe. The company could record the jogging times of each group over a certain period of time and then determine if there were larger improvements in the experimental group compared to the control group.

Example 1
Determine whether each situation describes a *survey*, an *experiment*, or an *observational study*. Then identify the sample, and suggest a population from which it may have been selected.

1. **SCHOOL** A group of high school students is randomly selected and asked to complete the form shown.
survey; sample: the students in the study; population: the student body

2. **DESIGN** An advertising company wants to test a new logo design. They randomly select 20 participants and watch them discuss the logo. **See margin.**

> **Do you agree with the new lunch rules?**
> ☐ agree
> ☐ disagree
> ☐ don't care

Example 2
CCSS ARGUMENTS Determine whether each situation calls for a *survey*, an *experiment*, or an *observational study*. Explain your reasoning. 3–4. See Chapter 11 Answer Appendix.

3. **LITERACY** A literacy group wants to determine whether high school students that participated in a recent national reading program had higher standardized test scores than high school students that did not participate in the program.

4. **RETAIL** The research department of a retail company plans to conduct a study to determine whether a dye used on a new T-shirt will begin fading before 50 washes.

Example 3
Determine whether each survey question is *biased* or *unbiased*. If biased, explain your reasoning.

5. Which student council candidate's platform do you support? unbiased

6. How long have you lived at your current address? unbiased

Example 4
7. **HYBRIDS** A car manufacturer wants to determine what the demand in the U.S. is for hybrid vehicles. State the objective of the survey, suggest a population, and write two unbiased survey questions. 7–9. See Chapter 11 Answer Appendix.

Example 5
8. Identify any flaws in the experiment design, and describe how they could be corrected.

Experiment: A research company wants to determine whether a new vitamin boosts energy levels and decides to test the vitamin at a college campus. A random sample is taken. The experimental group consists of students who are given the vitamin, and the control group consists of instructors who are given a placebo.

Results: When given a physical test, the experimental group outperformed the control group. The company concludes that the vitamin is effective.

Example 6
9. **SPORTS** A research company wants to conduct an experiment to test the claim of the protein shake shown. State the objective of the experiment, suggest a population, determine the experimental and control groups, and describe a sample procedure.

Helps athletes recover from intense exercise!

Practice and Problem Solving Extra Practice is on page R11.

Example 1
Determine whether each situation describes a *survey*, an *experiment*, or an *observational study*. Then identify the sample, and suggest a population from which it may have been selected. 10–13. See Chapter 11 Answer Appendix.

10. **FOOD** A grocery store conducts an online study in which customers are randomly selected and asked to provide feedback on their shopping experience.

11. **GRADES** A research group randomly selects 80 college students, half of whom took a physics course in high school, and compares their grades in a college physics course.

12. **HEALTH** A research group randomly chooses 100 people to participate in a study to determine whether eating blueberries reduces the risk of heart disease for adults.

13. **TELEVISION** A television network mails a questionnaire to randomly selected people across the country to determine whether they prefer watching sitcoms or dramas.

Differentiated Homework Options

Level	Assignment	Two-Day Option	
AL Basic	10–24, 28–30, 32–61	11–23 odd, 33–36	10–24 even, 28–30, 32, 37–61
OL Core	11–23 odd, 25–61	10–24, 33–36	25–32, 37–61
BL Advanced	25–59, (optional: 60–61)		

Additional Example

6 **PETS** A research company wants to test a new food for overweight cats that promotes weight loss. State the objective of the experiment, suggest a population, determine the experimental and control groups, and describe a sample procedure.
objective: to determine whether overweight cats given a new weight loss food will lose weight; population: all overweight cats; experimental group: a group of overweight cats given the food; control group: a group of overweight cats given their regular food; sample procedure: The two groups of cats could be weighed at the start of the experiment and then the experimental group could be given the new food and the control group could be given their regular food. Then, after a certain period of time, the two groups could be weighed again, and the weights of the cats could be compared to determine any effect.

3 Practice

Formative Assessment
Use Exercises 1–9 to check for understanding.

Use the chart at the bottom of this page to customize assignments for your students.

CCSS Teaching the Mathematical Practices
Arguments Mathematically proficient students understand and use stated assumptions, definitions, and previously established results in constructing arguments. They make conjectures and build a logical progression of statements to explore the truth of their conjectures.

Additional Answer
2. observational study; sample: the participants in the study; population: potential customers

14. Survey; sample answer: The data will be obtained from opinions given by members of the sample population.

15. Survey; sample answer: The data will be obtained from opinions given by members of the sample population.

16. Observational study; sample answer: The eating habits of the participants will be observed and compared without them being affected by the study.

17. Experiment; sample answer: Metal samples will need to be tested, which means that the members of the sample will be affected by the study.

18. Biased; sample answer: The question is asking about two issues: whether the school needs a new gym *and* whether the school needs a new football field.

19. Biased; sample answer: The question only gives two options, and thus encourages a certain response.

20. unbiased

21. Biased; sample answer: The question encourages a certain response. The phrase "don't you agree" suggests that the people surveyed should agree.

22. objective: to determine the number of juniors in the district planning to attend college after high school; population: all juniors in the district; sample survey questions: What grade are you in? Do you plan on attending college after graduation?

23. Sample answer: The flaw is that the experimental group consists of stores in the Midwest, and the control group consists of stores in Arizona. On average, the temperature is higher in Arizona than in the Midwest, and people use more sunscreen. Therefore, the sunscreen sales in stores located in those regions would likely be different and should not be compared in an experiment.

Example 2 Determine whether each situation calls for a *survey*, an *experiment*, or an *observational study*. Explain your reasoning. **14–17. See margin.**

14. **FASHION** A fashion magazine plans to poll 100 people in the U.S. to determine whether they would be more likely to buy a subscription if given a free issue.

15. **TRAVEL** A travel agency randomly calls 250 U.S. citizens and asks them what their favorite vacation destination is.

16. **FOOD** Chee wants to examine the eating habits of 100 random students at lunch to determine how many students eat in the cafeteria.

17. **ENGINEERING** An engineer is planning to test 50 metal samples to determine whether a new titanium alloy has a higher strength than a different alloy.

Example 3 Determine whether each survey question is *biased* or *unbiased*. If biased, explain your reasoning. **18–21. See margin.**

18. Do you think that the school needs a new gym and football field?

19. Which is your favorite football team, the Dallas Cowboys or the Pittsburgh Steelers?

20. Do you play any extracurricular sports?

21. Don't you agree that students should carpool to school?

Example 4 22. **COLLEGE** A school district wants to conduct a survey to determine the number of juniors in the district who are planning to attend college after high school. State the objective of the survey, suggest a population, and write two unbiased survey questions. **See margin.**

Example 5 23. Identify any flaws in the experiment design, and describe how they could be corrected.

Experiment: A supermarket chain wants to determine whether shoppers are more likely to buy sunscreen if it is located near the checkout line. The experimental group consists of a group of stores in the midwest in which the sunscreen was moved next to the checkout line, and the control group consists of stores in Arizona in which the sunscreen was not moved. **See margin.**

Results: The Arizona stores sold more sunscreen than the midwest stores. The company concluded that moving the sunscreen closer to the checkout line did not increase sales.

Example 6 24. **CCSS ARGUMENTS** In chemistry class, Pedro learned that copper objects become dull over time because the copper reacts with air to form a layer of copper oxide. He plans to use the supplies shown below to determine whether a mixture of lemon juice and salt will remove copper oxide from pennies. **a–b. See Chapter 11 Answer Appendix.**

2 lemons   1 teaspoon   30 dull pennies   plastic bowl

a. State the objective of the experiment, suggest a population, determine the experimental and control groups, and describe a sample procedure.

b. What factors do you think should be considered when selecting pennies for the experiment? Explain your reasoning.

 25. **REPORTS** The graph shown is from a report on the average number of minutes 8- to 18-year-olds in the U.S. spend on cell phones each day. **a, c–d. See margin.**

a. Describe the sample and suggest a population.

b. What type of sample statistic do you think was calculated for this report? **average time**

c. Describe the results of the study for each age group.

d. Who do you think would be interested in this type of report? Explain your reasoning.

25a. sample: the 8- to 18-year-olds surveyed; population: all 8- to 18-year-olds in the U.S.

25c. Sample answer: The 8- to 10-year-old group talked for about 10 minutes a day and did not text at all. The 11- to 14-year-old group talked for about 30 minutes a day and texted for about 70 minutes a day. The 15- to 18-year-old group talked for about 40 minutes a day and texted for about 110 minutes a day.

25d. Sample answer: A cell phone company might use a report like this to determine which age group to target in their ads.

26a. survey; sample: the 2.4 million people polled, population: all U.S. citizens of voting age in 1936

26b. According to the predicted results, Landon should have won 57% of the popular vote. However, in the actual election, Roosevelt won 60.8% of the popular vote.

26. **CCSS** **PERSEVERANCE** In 1936, the *Literary Digest* reported the results of a statistical study used to predict whether Alf Landon or Franklin D. Roosevelt would win the presidential election that year. The sample consisted of 2.4 million Americans, including subscribers to the magazine, registered automobile owners, and telephone users. The results concluded that Landon would win 57% of the popular vote. The actual election results are shown.

ELECTORAL VOTE

POPULAR VOTE

Roosevelt	60.8%
Landon	36.5%

a–c. See margin.

a. Describe the type of study performed, the sample taken, and the population.

b. How do the predicted and actual results compare?

c. Do you think that the survey was biased? Explain your reasoning.

27. **MULTIPLE REPRESENTATIONS** The results of two experiments concluded that Product A is 70% effective and Product B is 80% effective. **a–d. See Chapter 11 Answer Appendix.**

a. **NUMERICAL** To simulate the experiment for Product A, use the random number generator on a graphing calculator to generate 30 integers between 0 and 9. Let 0–6 represent an effective outcome and 7–9 represent an ineffective outcome.

b. **TABULAR** Copy and complete the frequency table shown using the results from part **a.** Then use the data to calculate the probability that Product A was effective. Repeat to find the probability for Product B.

Product A	
Number	Frequency
0–6	
7–9	

c. **ANALYTICAL** Compare the probabilities that you found in part **b.** Do you think that the difference in the effectiveness of each product is significant enough to justify selecting one product over the other? Explain.

d. **LOGICAL** Suppose Product B costs twice as much as Product A. Do you think the probability of the product's effectiveness justifies the price difference to a consumer? Explain.

H.O.T. Problems   Use Higher-Order Thinking Skills

REASONING Determine whether each statement is *true* or *false*. If false, explain.
28, 30–32. See Chapter 11 Answer Appendix.

28. To save time and money, population parameters are used to estimate sample statistics.

29. Observational studies and experiments can both be used to study cause-and-effect relationships. **true**

30. **OPEN ENDED** Design an observational study. Identify the objective of the study, define the population and sample, collect and organize the data, and calculate a sample statistic.

31. **CHALLENGE** What factors should be considered when determining whether a given statistical study is reliable?

32. **WRITING IN MATH** Research each of the following sampling methods. Then describe each method and discuss whether using the method could result in bias.

a. convenience sample   **c.** stratified sample

b. self-selected sample   **d.** systematic sample

Additional Answer

26c. Yes; sample answer: The people sampled could afford magazine subscriptions, automobiles, and telephones, suggesting that they were wealthier than the average American citizen. The sampling method did not represent citizens that could not afford these things, and therefore, was not representative of the entire population.

Study Guide and Intervention
AL OL ELL

Practice
AL OL BL ELL

Word Problem Practice
AL OL BL ELL

Enrichment
OL BL

connectED.mcgraw-hill.com **729**

Ticket Out the Door Have each student research an experiment of their choosing, and identify the objective of the experiment, the population, the experimental and control groups, and the procedure.

Additional Answers

37. **Step 1:** $9^1 - 1 = 8$, which is divisible by 8. The statement is true for $n = 1$.

Step 2: Assume that $9^k - 1$ is divisible by 8 for some natural number k. This means that $9^k - 1 = 8r$ for some natural number r.

Step 3: $9^k - 1 = 8r$
$$9^k = 8r + 1$$
$$9^{k+1} = 72r + 9$$
$$9^{k+1} - 1 = 72r + 8$$
$$9^{k+1} - 1 = 8(9r + 1)$$

Since r is a natural number, $9r + 1$ is a natural number. Thus, $9^{k+1} - 1$ is divisible by 8, so the statement is true for $n = k + 1$. Therefore, $9^n - 1$ is divisible by 8 for all natural numbers n.

Standardized Test Practice

33. **GEOMETRY** In $\triangle ABC$, $BC > AB$. Which of the following must be true? **C**

   A $AB = BC$

   B $AC < AB$

   C $a > 60$

   D $a = b$

34. **SHORT RESPONSE** What is the solution set of $4^{4x^2 - 2x - 4} = 4^{-2}$? $\{1, -0.5\}$

35. **SAT/ACT** A pie is divided evenly between 3 boys and a girl. If one boy gives one half of his share to the girl and a second boy keeps two thirds of his share and gives the rest to the girl, what portion will the girl have in all? **G**

   F $\frac{5}{24}$ **H** $\frac{1}{2}$ **K** $\frac{13}{12}$

   G $\frac{11}{24}$ **J** $\frac{13}{24}$

36. Which equation represents a hyperbola? **D**

   A $y^2 = 49 - x^2$ **C** $y = 49x^2$

   B $y = 49 - x^2$ **D** $y = \frac{49}{x}$

42. $(0, 3)$, $\left(\pm\frac{\sqrt{23}}{2}, -\frac{11}{4}\right)$

Spiral Review

37. Prove that the statement $9^n - 1$ is divisible by 8 is true for all natural numbers. (Lesson 10-7) **See margin.**

38. **INTRAMURALS** Ofelia is taking ten shots in the intramural free-throw shooting competition. How many sequences of hits and misses are there that result in her making eight shots and missing two? (Lesson 10-6) **45**

Solve each system of equations. (Lesson 9-7)   40. $\left(-1 + \sqrt{17}, 1 + \sqrt{17}\right), \left(-1 - \sqrt{17}, 1 - \sqrt{17}\right)$

39. $y = x + 3$ $\left(\frac{3}{2}, \frac{9}{2}\right), (-1, 2)$ 40. $x^2 + y^2 = 36$ 41. $y^2 + x^2 = 9$ **no solution**
   $y = 2x^2$ $y = x + 2$ $y = 7 - x$

42. $y + x^2 = 3$ 43. $x^2 + y^2 = 64$ $(\pm 8, 0)$ 44. $y^2 = x^2 - 25$ **no solution**
   $x^2 + 4y^2 = 36$ $x^2 + 64y^2 = 64$ $x^2 - y^2 = 7$

Find the distance between each pair of points with the given coordinates. (Lesson 9-1)

45. $(9, -2), (12, -14)$ $3\sqrt{17}$ units   46. $(-4, -10), (-3, -11)$ $\sqrt{2}$ units   47. $(1, -14), (-6, 10)$ **25 units**

48. $(-4, 9), (1, -3)$ **13 units**   49. $(2.3, -1.2), (-4.5, 3.7)\sqrt{70.25}$ units   50. $(0.23, 0.4), (0.68, -0.2)$ **0.75 unit**

Simplify. Assume that no variable equals 0. (Lesson 5-1)

51. $(5cd^2)(-c^4d)$ $-5c^5d^3$ 52. $(7x^3y^{-5})(4xy^3)$ $\frac{28x^4}{y^2}$ 53. $\frac{a^2n^6}{an^5}$ an

54. $(n^4)^4$ n^{16} 55. $\frac{-y^5z^7}{y^2z^5}$ $-y^3z^2$ 56. $(-2r^2t)^3(3rt^2)$ $-24r^7t^5$

Write a quadratic equation with the given root(s). Write the equation in the form $ax^2 + bx + c = 0$, **where** a, b, **and** c **are integers.** (Lesson 4-3)

57. $-3, 9$ $x^2 - 6x - 27 = 0$ 58. $-\frac{1}{3}, -\frac{3}{4}$ $12x^2 + 13x + 3 = 0$ 59. $4, -5$ $x^2 + x - 20 = 0$

Skills Review

60. **TESTS** Ms. Bonilla's class of 30 students took a biology test. If 20 of her students had an average of 83 on the test and the other students had an average score of 74, what was the average score of the whole class? **80**

61. **DRIVING** During a 10-hour trip, Kwan drove 4 hours at 60 miles per hour and 6 hours at 65 miles per hour. What was her average rate, in miles per hour, for the entire trip? **63**

 730 | Lesson 11-1 | Designing a Study

DifferentiatedInstruction **OL** **BL**

Extension Have students investigate techniques that could be used to select random samples. For example, a random number table from the appendix of a statistics text or the random number generator of a spreadsheet or calculator could be used.

EXTEND 11-1

Graphing Calculator Lab
Simulations and Margin of Error

The Pew Research Center conducted a survey of a random sample of teens and concluded that 43% of all teens who take their cell phones to school text in class on a daily basis. How accurately did their random sample represent all teens?

As you learned in the previous lesson, a survey of a random sample is a valuable tool for generalizing information about a larger population. The program in the following activity makes use of a random number generator (randInt(a, b)) to simulate the results of a random sampling survey.

 Common Core State Standards
Content Standards
S.IC.4 Use data from a sample survey to estimate a population mean or proportion; develop a margin of error through the use of simulation models for random sampling.
S.IC.6 Evaluate reports based on data.
Mathematical Practices
5 Use appropriate tools strategically.

Activity 1 Random Sampling Simulation

Use the following program that simulates the texting survey to measure the percent of teens who text in class for random sample sizes of 20, 50, and 100 students.

Step 1 Input the following program into a graphing calculator.

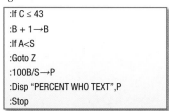

Program:SIMTEST	
:Input "SAMPLE SIZE ",S	:If C ≤ 43
:0→A	:B + 1→B
:0→B	:If A<S
:Lbl Z	:Goto Z
:A + 1→A	:100B/S→P
:randInt(1, 100) →C	:Disp "PERCENT WHO TEXT",P
	:Stop

Step 2 Run 10 trials of the program for each sample size of 20, 50, and 100. Press ENTER to run the program again each time.

Step 3 Record the percent who text for each trial in the table below.

Sample Size	1	2	3	4	5	6	7	8	9	10
20										
50										
100										

Analyze the Results

1. Discard the percent that is farthest from the Pew survey result of 43% for each sample size. What is the range of the remaining nine percents for each sample size? **See students' work.**

2. What is the farthest any of these remaining trials is from the 43% for each sample size? **See students' work.**

3. The positive or negative of the result found in Exercise 2 is known as the **margin of error**. For your results, which sample size had the smallest margin of error? **samples of size 100**

4. What would you expect to happen to the margin of error if we used a sample size of 500?

(continued on the next page)

Sample answer: The margin of error would decrease.

1 Focus

Objective Use a simulation to develop margins of error for various sizes of random samples.

Materials for Each Student
- TI-83/84 Plus or other graphing calculator

Teaching Tip
If students have limited programming experience on the calculator, review the list of commands by pressing PRGM when the calculator is in program editing mode. To save classroom time, the program can be downloaded from one calculator to the others.

2 Teach

Working in Cooperative Groups
Have students work in pairs, mixing abilities, to complete Activities 1–2 and Exercises 1–4.

Ask:
- What does the program line: randInt(1,100)→C do? It assigns a random integer from 1 to 100 to the variable C.

- What happens if C is not less than or equal to 43? The program will skip the next line (it will not add 1 to the value of B).

- What happens if a sample size of 500 or 1000 is entered into the program? The calculator will take several seconds longer to run the program.

3 Assess

Formative Assessment
Use Exercises 5 and 6 to assess whether students can correctly apply the margin of error formula.

From Concrete to Abstract
Ask:
- In Activity 1, what percent of the trials for each sample size did you use? 90%

- Why were all the trials not used to develop the margin of error? Sample answer: As seen in Exercise 9, a wide range of percents is possible, but not likely. By using 90% of the trials, a range with a smaller margin of error is obtained that more closely resembles our results.

Additional Answers

9. Yes; sample answer: The margin of error is a prediction of what is most likely to happen. It is possible though that a random sample could give results anywhere between 0% and 100%.

10. Sample answer: The margin of error predicts that using a sample size of 2500 should produce a percent that falls within the range 41%–45%. While it is possible that a random sample could give results anywhere between 0% and 100%, it is not likely that 10 trials all consisting of a sample size of 2500 would produce percents with a range of 19%–23%. Therefore, the validity of the model should be questioned.

Graphing Calculator Lab
Simulations and Margin of Error *Continued*

Statisticians have found that for large populations, the margin of error for a random sample of size n can be approximated by the following formula.

> **KeyConcept** Margin of Error Formula
>
> $$\text{Margin of error} = \pm \frac{1}{\sqrt{n}}(100)$$

Since n is in the denominator, the margin of error will decrease as the size of the random sample increases. This expression can also be used to determine the size of a random sample necessary to achieve a desired level of reliability.

Activity 2 Margin of Error and Sample Size

You are a member of a research team.

a. **You need to decide whether to conduct a survey with a margin of error of ±3% or ±2%. What sample size would be needed to achieve each goal?**

Set each percent equal to the margin of error formula and solve for n.

$\pm 3\% = \pm \frac{1}{\sqrt{n}}(100)$	Margin of error formula	$\pm 2\% = \pm \frac{1}{\sqrt{n}}(100)$
$0.03\sqrt{n} = 1$	Multiply by $\frac{\sqrt{n}}{100}$.	$0.02\sqrt{n} = 1$
$\sqrt{n} = 33.333$	Divide.	$\sqrt{n} = 50$
$n = 1111.11$	Square each side.	$n = 2500$

A random sample of about 1100 would have a margin of error of about ±3%, while a random sample of 2500 would have a margin of error of ±2%.

b. **Suppose the finance director would like to reduce the cost of the survey by using a random sample of 100. What would be the margin of error for this sample size?**

Substitute 100 for n in the margin of error formula.

$$\text{margin of error} = \pm \frac{1}{\sqrt{n}}(100) \qquad \text{Margin of error formula}$$

$$= \pm \frac{1}{\sqrt{100}}(100) \text{ or } \pm 10\% \qquad n = 100$$

A random sample of 100 would have a margin of error of ±10%.

Exercises

5. What random sample size would produce a margin of error of ±1%? **10,000**

6. What margin of error can be expected when using a sample size of 500? **about ±4.5%**

7. What are some reasons that a research center might decide that a survey with a margin of error of ±3% would be more desirable than one with a margin of error of ±2%?
Sample answer: The cost would be less, the survey would take less time, and it would require fewer people

8. What is the range for the percent of students that text in class that the research center can expect **to administer.** from any random survey they conduct with a sample size of 2500? **41%–45%**

9. If a survey with a random sample of 2500 students is conducted, is it possible that only 20% of the students could respond that they text during class? If so, how could this be possible? **See margin.**

10. If Step 2 from Activity 1 is repeated using a sample size of 2500 and the range for the percents is found to be 19%–23%, would this result cause you to question the model? **See margin.**

LESSON 11-2 Distributions of Data

:: Then	:: Now	:: Why?
● You calculated measures of central tendency and variation.	**1** Use the shapes of distributions to select appropriate statistics. **2** Use the shapes of distributions to compare data.	After four games as a reserve player, Craig joined the starting lineup and averaged 18 points per game over the *remaining* games. Craig's scoring average for the *entire* season was less than 18 points per game as a result of the lack of playing time in the first four games.

NewVocabulary
distribution
negatively skewed distribution
symmetric distribution
positively skewed distribution

Common Core State Standards

Content Standards
S.IC.1 Understand statistics as a process for making inferences about population parameters based on a random sample from that population.

Mathematical Practices
1 Make sense of problems and persevere in solving them.

1 Analyzing Distributions A **distribution** of data shows the observed or theoretical frequency of each possible data value. In Lesson 0-9, you described distributions of sample data using statistics. You used the mean or median to describe a distribution's center and standard deviation or quartiles to describe its spread. Analyzing the shape of a distribution can help you decide which measure of center or spread best describes a set of data.

The shape of the distribution for a set of data can be seen by drawing a curve over its histogram.

KeyConcept Symmetric and Skewed Distributions

Negatively Skewed Distribution	**Symmetric Distribution**	**Positively Skewed Distribution**
tail ↙ mean median		median mean ↘ tail
• The mean is less than the median. • The majority of the data are on the right of the mean.	• The mean and median are approximately equal. • The data are evenly distributed on both sides of the mean.	• The mean is greater than the median. • The majority of the data are on the left of the mean.

When a distribution is symmetric, the mean and standard deviation accurately reflect the center and spread of the data. However, when a distribution is skewed, these statistics are not as reliable. Recall that outliers have a strong effect on the mean of a data set, while the median is less affected. Similarly, when a distribution is skewed, the mean lies away from the majority of the data toward the tail. The median is less affected, so it stays near the majority of the data.

When choosing appropriate statistics to represent a set of data, first determine the skewness of the distribution.

• If the distribution is relatively symmetric, the mean and standard deviation can be used.
• If the distribution is skewed or has outliers, use the five-number summary to describe the center and spread of the data.

connectED.mcgraw-hill.com **733**

Rim Light Photography/Corbis

1 Focus

VerticalAlignment

Before Lesson 11-2 Calculate measures of central tendency and variation.

Lesson 11-2 Use the shapes of distributions to select appropriate statistics. Use the shapes of distributions to compare data.

After Lesson 11-2 Learn about probability distributions.

2 Teach

Scaffolding Questions
Have students read the **Why?** section of the lesson.

Ask:

■ Why is Craig's scoring average for the entire season less than 18 points per game? Sample answer: If Craig did not receive much playing time as a reserve player, he probably did not score many points during the first four games. As a result, these scores brought down his average.

■ Is Craig's scoring average for the entire season a good representation of his scoring average? Explain. Sample answer: No; his scoring average for the entire season is lower than it would have been had Craig been in the starting lineup for the entire season.

Lesson 11-2 Resources

Resource	Approaching Level **AL**	On Level **OL**	Beyond Level **BL**	English Learners **ELL**
Teacher Edition	■ Differentiated Instruction, p. 735	■ Differentiated Instruction, pp. 735, 741	■ Differentiated Instruction, pp. 735, 741	■ Differentiated Instruction, p. 735
Chapter Resource Masters	■ Study Guide and Intervention, pp. 11–12 ■ Skills Practice, p. 13 ■ Practice, p. 14 ■ Word Problem Practice, p. 15	■ Study Guide and Intervention, pp. 11–12 ■ Skills Practice, p. 13 ■ Practice, p. 14 ■ Word Problem Practice, p. 15 ■ Enrichment, p. 16	■ Practice, p. 14 ■ Word Problem Practice, p. 15 ■ Enrichment, p. 16	■ Study Guide and Intervention, pp. 11–12 ■ Skills Practice, p. 13 ■ Practice, p. 14 ■ Word Problem Practice, p. 15
Other	■ 5-Minute Check 11-2 ■ Study Notebook	■ 5-Minute Check 11-2 ■ Study Notebook	■ 5-Minute Check 11-2 ■ Study Notebook	■ 5-Minute Check 11-2 ■ Study Notebook

1 Analyzing Distributions

Example 1 shows how to choose appropriate statistics to describe a set of data using a histogram. **Example 2** shows how to choose appropriate statistics to describe a set of data using a box-and-whisker plot.

Formative Assessment
Use the Guided Practice exercises after each example to determine students' understanding of concepts.

Additional Example

1 **PRESENTATIONS** Ms. Shroyer's students each gave a presentation as part of their class project. The length of each presentation is shown in the table below.

Time (minutes)				
17	13	11	17	20
15	23	7	16	10
13	20	12	21	14
17	19	20	18	19

a. Use a graphing calculator to create a histogram. Then describe the shape of the distribution.

See bottom margin for graph; negatively skewed

b. Describe the center and spread of the data using either the mean and standard deviation or the five-number summary. Justify your choice.

Sample answer: The distribution is skewed, so use the five-number summary. The range is 7 to 23 minutes. The median is 17 minutes, and half of the times are between 13 and 19.5 minutes.

▶ **Additional Examples** also in Interactive Classroom PowerPoint® Presentations

IWB **Interactive White Board** **READY**

Real-World Link

The first portable computer, the Osborne I, was available for sale in 1981 for $1795. The computer weighed 24 pounds and included a 5-inch display. Laptops can now be purchased for as little as $250 and can weigh as little as 3 pounds.

Source: Computer History Museum

1A.

[15, 42] scl: 3 by [0, 8] scl: 1

symmetric

1B. Sample answer: The distribution is symmetric, so use the mean and standard deviation. The mean rainfall was 29 inches with standard deviation of 5.4 inches.

 734 | Lesson 11-2 | Distributions of Data

Additional Answer (Additional Example)

1a.

[3, 24] scl: 3 by [0, 8] scl: 1

Real-World Example 1 Describe a Distribution Using a Histogram

COMPUTERS The prices for a random sample of personal computers are shown.

Price (dollars)							
723	605	847	410	440	386	572	523
374	915	734	472	420	508	613	659
706	463	470	752	671	618	538	425
811	502	490	552	390	512	389	621

a. Use a graphing calculator to create a histogram. Then describe the shape of the distribution.

First, press ⎡STAT⎤ ⎡ENTER⎤ and enter each data value. Then, press ⎡2nd⎤ [STAT PLOT] ⎡ENTER⎤ ⎡ENTER⎤ and choose ⬚. Finally, adjust the window to the dimensions shown.

The majority of the computers cost between $400 and $700. Some of the computers are priced significantly higher, forming a tail for the distribution on the right. Therefore, the distribution is positively skewed.

[0, 1000] scl: 100 by [0, 10] scl: 1

b. Describe the center and spread of the data using either the mean and standard deviation or the five-number summary. Justify your choice.

The distribution is skewed, so use the five-number summary to describe the center and spread. Press ⎡STAT⎤ ▶ ⎡ENTER⎤ ⎡ENTER⎤ and scroll down to view the five-number summary.

The prices for this sample range from $374 to $915. The median price is $530.50, and half of the computers are priced between $451.50 and $665.

▶ **Guided**Practice

1. RAINFALL The annual rainfall for a region over a 24-year period is shown below.

A. Use a graphing calculator to create a histogram. Then describe the shape of the distribution.

B. Describe the center and spread of the data using either the mean and standard deviation or the five-number summary. Justify your choice.

Annual Rainfall (in.)					
27.2	30.2	35.8	26.1	39.3	20.6
28.9	23.0	32.7	26.8	22.7	25.4
29.6	36.8	33.4	28.4	21.9	20.8
24.7	30.6	27.7	31.4	34.9	37.1

A box-and-whisker plot can also be used to identify the shape of a distribution. The position of the line representing the median indicates the center of the data. The "whiskers" show the spread of the data. If one whisker is considerably longer than the other and the median is closer to the shorter whisker, then the distribution is skewed.

James Woodson/Digital Vision/Getty Images

Tips for New Teachers
Skewed Distributions Students may confuse negatively and positively skewed distributions. Remind them that when the tail is on the left of a distribution, the data appear to be going uphill, and since it is harder to go uphill, the distribution is negatively skewed. When the tail is on the right of a distribution, the data appear to be going downhill, and since it is easier to go downhill, the distribution is positively skewed.

KeyConcept Box-and-Whisker Plots as Distributions

Negatively Skewed	Symmetric	Positively Skewed
50% 50%	50% 50%	50% 50%
The data to the left of the median are distributed over a wider range than the data to the right. The data have a tail to the left.	The data are equally distributed to the left and right of the median.	The data to the right of the median are distributed over a wider range than the data to the left. The data have a tail to the right.

2A.

[500, 750] scl: 25 by [0, 5] scl: 1

negatively skewed

2B. Sample answer: The distribution is skewed, so use the five-number summary. Janet's used minutes range from 511 minutes to 695 minutes. The median is 655.5 minutes, and half of the data are between 616.5 and 670.5 minutes.

WatchOut!

Standard Deviation Recall from Lesson 0-9 that the formulas for standard deviation for a population σ and for a sample s are slightly different. In Example 2, times for all of the students in Mr. Fejis' class are being analyzed, so use the population standard deviation.

Example 2 Describe a Distribution Using a Box-and-Whisker Plot

HOMEWORK The students in Mr. Fejis' language arts class found the average number of minutes that they each spent on homework each night.

Minutes per Night					
62	53	46	66	38	45
52	46	73	39	42	56
64	54	48	59	70	60
49	54	48	57	70	33

a. Use a graphing calculator to create a box-and-whisker plot. Then describe the shape of the distribution.

Enter the data as **L1**. Press [2nd] [STAT PLOT] [ENTER] [ENTER] and choose ⊡‥. Adjust the window to the dimensions shown.

The lengths of the whiskers are approximately equal, and the median is in the middle of the data. This indicates that the data are equally distributed to the left and right of the median. Thus, the distribution is symmetric.

[30, 75] scl: 5 by [0, 5] scl: 1

b. Describe the center and spread of the data using either the mean and standard deviation or the five-number summary. Justify your choice.

The distribution is symmetric, so use the mean and standard deviation to describe the center and spread. The average number of minutes that a student spent on homework each night was 53.5 with standard deviation of about 10.5.

```
1-Var Stats
x̄=53.5
Σx=1284
Σx²=71320
Sx=10.68521937
σx=10.46024219
↓n=24
```

GuidedPractice

2. CELL PHONE Janet's parents have given her a prepaid cell phone. The number of minutes she used each month for the last two years are shown in the table.

Minutes Used per Month			
582	608	670	620
667	598	671	613
537	511	674	627
638	661	642	641
668	673	680	695
658	653	670	688

A. Use a graphing calculator to create a box-and-whisker plot. Then describe the shape of the distribution.

B. Describe the center and spread of the data using either the mean and standard deviation or the five-number summary. Justify your choice.

Additional Example

2 WAGES The hourly wages for a random sample of employees of a restaurant are shown in the table.

Wages ($)				
10.50	7.75	8.25	9.50	6.50
7.50	11.25	7.25	6.50	8.25
8.00	7.50	6.75	7.25	9.00
7.50	10.00	7.25	8.00	9.00

a. Use a graphing calculator to create a box-and-whisker plot. Then describe the shape of the distribution.

[6, 13] scl: 1 by [0, 8] scl: 1

positively skewed

b. Describe the center and spread of the data using either the mean and standard deviation or the five-number summary. Justify your choice.
Sample answer: The distribution is skewed, so use the five-number summary. The range is $6.50 to $11.25. The median is about $7.88, and half of the data are between $7.25 and $9.00.

DifferentiatedInstruction AL OL BL ELL

Interpersonal Learners Have students work in pairs to think of examples of data that may have distributions that are symmetric, negatively skewed, or positively skewed.

2 Comparing Distributions

Example 3 shows how to use histograms to compare the distributions of two data sets. **Example 4** shows how to use box-and-whisker plots to compare the distributions of two data sets.

Additional Example

3 **GAMES** Tyler and Jordan are working through several brainteasers on the computer. The time in minutes that it took to complete each game is shown.

Tyler (minutes)				
5.1	2.9	3.5	6.0	2.8
3.4	4.1	3.8	4.3	6.1
5.8	5.1	6.0	3.4	4.6
4.4	3.6	4.8	3.1	5.2

Jordan (minutes)				
4.1	5.4	3.8	2.9	6.4
3.1	3.7	5.3	4.5	2.7
3.4	4.3	4.2	5.8	4.7
6.1	5.9	5.4	4.4	3.9

a. Use a graphing calculator to create a histogram for each data set. Then describe the shape of each distribution. See bottom margin for graphs; Tyler, positively skewed; Jordan, symmetric

b. Compare the distributions using either the means and standard deviations or the five-number summaries. Justify your choice. Sample answer: One distribution is symmetric and the other is skewed, so use the five-number summaries. The median for both sets is 4.35 but 50% of Tyler's times occur between 3.45 and 5.15, while 50% of Jordan's times occur between 3.75 and 5.4. The smaller interquartile range for Jordan may suggest that she was slightly more consistent than Tyler.

3A.
3rd Period

[20, 50] scl: 4 by [0, 10] scl: 1

6th Period

[20, 50] scl: 4 by [0, 10] scl: 1

3rd period, symmetric; 6th period, positively skewed

StudyTip

CCSS Tools To compare two sets of data, enter one set as **L1** and the other as **L2**. In order to calculate statistics for a set of data in **L2**, press
[STAT] ▶ [ENTER]
[2nd] [L2] [ENTER].

3B. Sample answer: One distribution is symmetric and the other is skewed, so use the five-number summaries. The range for 3rd period is 23, and the range for 6th period is 25. However, the median for 3rd period is 33, and the median for 6th period is 27. The lower quartile for 3rd period is 28. Since this is greater than the median for 6th period, this means that 75% of the speeds for 3rd period are greater than 50% of the speeds for 6th period. Therefore, we can conclude that 3rd period had slightly better typing speeds overall.

 736 | Lesson 11-2 | Distributions of Data

---

2 Comparing Distributions

To compare two sets of data, first analyze the shape of each distribution. Use the mean and standard deviation to compare two symmetric distributions. Use the five-number summaries to compare two skewed distributions or a symmetric distribution and a skewed distribution.

Example 3 Compare Data Using Histograms

TEST SCORES Test scores from Mrs. Morash's class are shown below.

Chapter 3 Test Scores
81, 81, 92, 99, 61, 67, 86, 82, 76, 73, 62, 97, 97, 72, 72, 84, 77, 88, 92, 93, 76, 74, 66, 78, 76, 69, 84, 87, 83, 87, 92, 87, 82

Chapter 4 Test Scores
87, 73, 69, 83, 74, 86, 74, 69, 79, 84, 79, 74, 83, 74, 86, 69, 91, 73, 79, 83, 69, 79, 83, 74, 86, 79, 79, 78, 83, 79, 86, 79, 84

a. Use a graphing calculator to create a histogram for each data set. Then describe the shape of each distribution.

Chapter 3 Test Scores

[60, 100] scl: 5 by [0, 10] scl: 1

Chapter 4 Test Scores

[60, 100] scl: 5 by [0, 10] scl: 1

Both distributions are symmetric.

b. Compare the distributions using either the means and standard deviations or the five-number summaries. Justify your choice.

The distributions are symmetric, so use the means and standard deviations.

Chapter 3 Test Scores

1-Var Stats
x̄=81
Σx=2673
Σx²=219763
Sx=10.07782219
σx=9.923953269
↓n=33

Chapter 4 Test Scores

1-Var Stats
x̄=79
Σx=2607
Σx²=207085
Sx=5.947688627
σx=5.856878888
↓n=33

The Chapter 4 test scores, while lower in average, have a much smaller standard deviation, indicating that the scores are more closely grouped about the mean. Therefore, the mean for the Chapter 4 test scores is a better representation of the data than the mean for the Chapter 3 test scores.

▶ **Guided**Practice

3. **TYPING** The typing speeds of the students in two classes are shown below.

A. Use a graphing calculator to create a histogram for each data set. Then describe the shape of each distribution.

B. Compare the distributions using either the means and standard deviations or the five-number summaries. Justify your choice.

3rd Period (wpm)
23, 38, 27, 28, 40, 45, 32, 33, 34, 27, 40, 22, 26, 34, 29, 31, 35, 33, 37, 38, 28, 29, 39, 42

6th Period (wpm)
38, 26, 43, 46, 23, 24, 27, 36, 22, 21, 26, 27, 31, 32, 27, 25, 23, 22, 28, 29, 28, 33, 23, 24

Additional Answer (Additional Example)

3a.

Tyler

[1, 8] scl: 1 by [0, 8] scl: 1

Jordan

[1, 8] scl: 1 by [0, 8] scl: 1

CCSS **Teaching the Mathematical Practices**

Tools Mathematically proficient students consider the available tools when solving a mathematical problem. Remind students to select the correct data set when performing calculations.

Box-and-whisker plots can be displayed alongside one another, making them useful for side-by-side comparisons of data.

Example 4 Compare Data Using Box-and-Whisker Plots

POINTS The points scored per game by a professional football team for the 2008 and 2009 football seasons are shown.

2008							
7	51	24	27	17	35	27	33
28	30	27	21	24	30	14	20

2009							
20	9	3	10	6	14	3	10
3	37	7	21	13	41	20	23

a. Use a graphing calculator to create a box-and-whisker plot for each data set. Then describe the shape of each distribution.

Enter the 2008 scores as **L1**. Graph these data as **Plot1** by pressing [2nd] [STAT PLOT] [ENTER] [ENTER] and choosing ⊡⋯. Enter the 2009 scores as **L2**. Graph these data as **Plot2** by pressing [2nd] [STAT PLOT] [▼] [ENTER] [ENTER] and choosing ⊡⋯. For **Xlist**, enter **L2**. Adjust the window to the dimensions shown.

[0, 55] scl: 5 by [0, 5] scl: 1

For the 2008 scores, the left whisker is longer than the right and the median is closer to the right whisker. The distribution is negatively skewed.

For the 2009 scores, the right whisker is longer than the left and the median is closer to the left whisker. The distribution is positively skewed.

b. Compare the distributions using either the means and standard deviations or the five-number summaries. Justify your choice.

The distributions are skewed, so use the five-number summaries to compare the data.

The lower quartile for the 2008 season and the upper quartile for the 2009 season are both 20.5. This means that 75% of the scores from the 2008 season were greater than 20.5 and 75% of the scores from the 2009 season were less than 20.5.

The minimum of the 2008 season is approximately equal to the lower quartile for the 2009 season. This means that 25% of the scores from the 2009 season are lower than any score achieved in the 2008 season. Therefore, we can conclude that the team scored a significantly higher amount of points during the 2008 season than the 2009 season.

▶ **Guided**Practice

4. **GOLF** Robert recorded his golf scores for his sophomore and junior seasons.

A. Use a graphing calculator to create a box-and-whisker plot for each data set. Then describe the shape of each distribution.

B. Compare the distributions using either the means and standard deviations or the five-number summaries. Justify your choice.

Sophomore Season
42, 47, 43, 46, 50, 47, 52, 45, 53, 55, 48, 39, 40, 49, 47, 50

Junior Season
44, 38, 46, 48, 42, 41, 42, 46, 43, 40, 43, 43, 44, 45, 39, 44

connectED.mcgraw-hill.com **737**

StudyTip

Outliers Recall from Lesson 0-9 that outliers are data that are more than 1.5 times the interquartile range beyond the upper or lower quartile. All outliers should be plotted, but the whiskers should be drawn to the least and greatest values that are not outliers.

4A.

[35, 60] scl: 5 by [0, 5] scl: 1

both symmetric

4B. Sample answer: The distributions are symmetric, so use the means and standard deviations. The mean score for Robert's sophomore season is about 47.1 with standard deviation of about 4.4. The mean score for Robert's junior season is 43 with standard deviation of about 2.6. The lower mean and standard deviation from Robert's junior season indicates that he not only improved, but was also more consistent.

Additional Example

4 **TEMPERATURES** The daily high temperatures over a 20-day period for two cities are shown.

Clintonville (°F)				
60	55	68	63	70
59	63	61	54	68
66	57	72	65	67
65	62	58	72	58

Stockton (°F)				
61	63	62	61	64
63	64	60	65	63
63	65	62	60	64
64	63	66	62	66

a. Use a graphing calculator to create a box-and-whisker plot for each data set. Then describe the shape of each distribution.

[52, 76] scl: 4 by [0, 5] scl: 1

both symmetric

b. Compare the distributions using either the means and standard deviations or the five-number summaries. Justify your choice. Sample answer: The distributions are symmetric, so use the means and standard deviations. The mean temperature for Clintonville is about 63.15° with standard deviation of about 5.40°. The mean temperature for Stockton is about 63.05° with standard deviation of about 1.76°. The average temperatures for both cities are about the same, but the lower standard deviation for Stockton means that the temperatures there are more consistently near 63° than at Clintonville.

Teach with Tech

Graphing Calculator Have students enter the data points from Example 2 in a graphing calculator. Have students calculate the mean and median of the data set. Have students add outliers to the data set and then have them recalculate the mean and median. After they have reviewed their results, reiterate the concept of the median being less resistant to the effect of outliers.

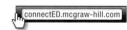

3 Practice

Formative Assessment

Use Exercises 1–4 to check for understanding.

Use the chart at the bottom of this page to customize assignments for your students.

Additional Answers

1a.

[4, 32] scl: 4 by [0, 8] scl: 1

negatively skewed

1b. Sample answer: The distribution is skewed, so use the five-number summary. The times range from 7 to 30 minutes. The median is 22.5 minutes, and half of the data are between 15.5 and 26 minutes.

2a.

[0, 25] scl: 5 by [0, 5] scl: 1

positively skewed

2b. Sample answer: The distribution is skewed, so use the five-number summary. The data range from 3 to 22 times. The median number is 7 times, and half of the data are between 5 and 13 times.

4a.

[0, 40] scl: 5 by [0, 5] scl: 1

junior class, positively skewed; senior class, symmetric

Example 1

1 **EXERCISE** The amount of time that James ran on a treadmill for the first 24 days of his workout is shown. **a–b. See margin.**

Time (minutes)											
23	10	18	24	13	27	19	7	25	30	15	22
10	28	23	16	29	26	26	22	12	23	16	27

a. Use a graphing calculator to create a histogram. Then describe the shape of the distribution.

b. Describe the center and spread of the data using either the mean and standard deviation or the five-number summary. Justify your choice.

Example 2

2. RESTAURANTS The total number of times that 20 random people either ate at a restaurant or bought fast food in a month are shown. **a–b. See margin.**

Restaurants or Fast Food									
4	7	5	13	3	22	13	6	5	10
7	18	4	16	8	5	15	3	12	6

a. Use a graphing calculator to create a box-and-whisker plot. Then describe the shape of the distribution.

b. Describe the center and spread of the data using either the mean and standard deviation or the five-number summary. Justify your choice.

Example 3

3. **TOOLS** The total fundraiser sales for the students in two classes at Cantonville High School are shown. **a–b. See Chapter 11 Answer Appendix.**

Mrs. Johnson's Class (dollars)					
6	14	17	12	38	15
11	12	23	6	14	28
16	13	27	34	25	32
21	24	21	17	16	

Mr. Edmunds' Class (dollars)					
29	38	21	28	24	33
14	19	28	15	30	6
31	23	33	12	38	28
18	34	26	34	24	37

a. Use a graphing calculator to create a histogram for each data set. Then describe the shape of each distribution.

b. Compare the distributions using either the means and standard deviations or the five-number summaries. Justify your choice.

Example 4

4. RECYCLING The weekly totals of recycled paper for the junior and senior classes are shown. **a–b. See margin.**

Junior Class (pounds)					
14	24	8	26	19	38
12	15	12	18	9	24
12	21	9	15	13	28

Senior Class (pounds)					
25	31	35	20	37	27
22	32	24	28	18	32
25	32	22	29	26	35

a. Use a graphing calculator to create a box-and-whisker plot for each data set. Then describe the shape of each distribution.

b. Compare the distributions using either the means and standard deviations or the five-number summaries. Justify your choice.

738 | Lesson 11-2 | Distributions of Data

Differentiated Homework Options

Level	Assignment	Two-Day Option	
AL Basic	5–10, 14–29	5–9 odd, 17–20	6–10 even, 14–16, 21–29
OL Core	5–12, 14–29	5–10, 17–20	11, 12, 14–16, 21–29
BL Advanced	11–25 (optional: 26–29)		

Examples 1–2 For Exercises 5 and 6, complete each step.

   a. Use a graphing calculator to create a histogram and a box-and-whisker plot. Then describe the shape of the distribution.

   b. Describe the center and spread of the data using either the mean and standard deviation or the five-number summary. Justify your choice.

5. FANTASY The weekly total points of Kevin's fantasy football team are shown. **a–b. See margin.**

Total Points							
165	140	88	158	101	137	112	127
53	151	120	156	142	179	162	79

6. MOVIES The students in one of Mr. Peterson's classes recorded the number of movies they saw over the past month. **a–b. See Chapter 11 Answer Appendix.**

Movies Seen											
14	11	17	9	6	11	7	8	12	13	10	9
5	11	7	13	9	12	10	9	15	11	13	15

Example 3   **CCSS MODELING** For Exercises 7 and 8, complete each step.

   a. Use a graphing calculator to create a histogram for each data set. Then describe the shape of each distribution.

   b. Compare the distributions using either the means and standard deviations or the five-number summaries. Justify your choice. **7–8. See Chapter 11 Answer Appendix.**

7. SAT A group of students took the SAT their sophomore year and again their junior year. Their scores are shown.

Sophomore Year Scores					
1327	1663	1708	1583	1406	1563
1637	1521	1282	1752	1628	1453
1368	1681	1506	1843	1472	1560

Junior Year Scores					
1728	1523	1857	1789	1668	1913
1834	1769	1655	1432	1885	1955
1569	1704	1833	2093	1608	1753

8. INCOME The total incomes for 18 households in two neighboring cities are shown.

Yorkshire (thousands of dollars)					
68	59	61	78	58	66
56	72	86	58	63	53
68	58	74	60	103	64

Applewood (thousands of dollars)					
52	55	60	61	55	65
65	60	45	37	41	71
50	61	65	66	87	55

Example 4   **9 TUITION** The annual tuitions for a sample of public colleges and a sample of private colleges are shown. Complete each step. **See Chapter 11 Answer Appendix.**

   a. Use a graphing calculator to create a box-and-whisker plot for each data set. Then describe the shape of each distribution.

   b. Compare the distributions using either the means and standard deviations or the five-number summaries. Justify your choice.

Public Colleges (dollars)					
3773	3992	3004	4223	4821	3880
3163	4416	5063	4937	3321	4308
4006	3508	4498	3471	4679	3612

Private Colleges (dollars)					
10,766	13,322	12,995	15,377	16,792	9147
15,976	11,084	17,868	7909	12,824	10,377
14,304	10,055	12,930	16,920	10,004	11,806

Additional Answer

4b. Sample answer: One distribution is symmetric and the other is skewed, so use the five-number summaries. The median for the junior class is 15, and the median for the senior class is 27.5. The minimum value for the senior class is 18. This means that all of the weekly totals for the senior class are greater than 50% of the weekly totals for the junior class. Therefore, we can conclude that the senior class's weekly totals were far greater than the junior class's weekly totals.

CCSS Teaching the Mathematical Practices

Modeling Mathematically proficient students can apply the mathematics they know to solve problems arising in everyday life, analyze relationships mathematically to draw conclusions, and interpret their mathematical results in the context of a situation.

WatchOut!

Standard Deviations For Exercises 6 and 7, students should calculate the population standard deviation. Every student in Mr. Peterson's class is included, and every student that was in the group of sophomores was included in the group of juniors. For Exercise 9, students should calculate the sample standard deviation since the tuitions for only a sample of colleges were collected.

Additional Answers

5a.

[50, 200] scl: 25 by [0, 8] scl: 1

[50, 200] scl: 25 by [0, 5] scl: 1

negatively skewed

5b. Sample answer: The distribution is skewed, so use the five-number summary. The points range from 53 to 179. The median is 138.5 points, and half of the data are between 106.5 and 157 points.

Study Guide and Intervention
AL OL ELL

Practice
AL OL BL ELL

Word Problem Practice
AL OL BL ELL

10. DANCE The total amount of money that a random sample of seniors spent on prom is shown. Complete each step. **See Chapter 11 Answer Appendix.**

  a. Use a graphing calculator to create a box-and-whisker plot for each data set. Then describe the shape of each distribution.

  b. Compare the distributions using either the means and standard deviations or the five-number summaries. Justify your choice.

Boys (dollars)					
253	288	304	283	348	276
322	368	247	404	450	341
291	260	394	302	297	272

Girls (dollars)					
682	533	602	504	635	541
489	703	453	521	472	368
562	426	382	668	352	587

11a–c. See margin.

11. BASKETBALL Refer to the beginning of the lesson. The points that Craig scored in the remaining games are shown.

  a. Use a graphing calculator to create a box-and-whisker plot. Describe the center and spread of the data.

  b. Craig scored 0, 2, 1, and 0 points in the first four games. Use a graphing calculator to create a box-and-whisker plot that includes the new data. Then find the mean and median of the new data set.

  c. What effect does adding the scores from the first four games have on the shape of the distribution and on how you should describe the center and spread?

Points Scored			
18	10	18	21
9	25	13	17
17	12	24	19
20	17	27	21

a–b. See Chapter 11 Answer Appendix.

12. SCORES Allison's quiz scores are shown.

  a. Use a graphing calculator to create a box-and-whisker plot. Describe the center and spread.

  b. Allison's teacher allows students to drop their two lowest quiz scores. Use a graphing calculator to create a box-and-whisker plot that reflects this change. Then describe the center and spread of the new data set.

Math Quiz Scores					
83	76	86	82	84	57
86	62	90	96	76	89
76	88	86	86	92	94

H.O.T. Problems   Use Higher-Order Thinking Skills

13. CHALLENGE Approximate the mean and median for each distribution of data. **a–c. See margin.**

a.

b.

c.

14. CCSS ARGUMENTS Distributions of data are not always symmetric or skewed. If a distribution has a gap in the middle, like the one shown, two separate clusters of data may result, forming a *bimodal distribution*. How can the center and spread of a bimodal distribution be described?

14. See Chapter 11 Answer Appendix.

15. OPEN ENDED Find a real-world data set that appears to represent a symmetric distribution and one that does not. Describe each distribution. Create a visual representation of each set of data. **See Chapter 11 Answer Appendix.**

16. WRITING IN MATH Explain the difference between positively skewed, negatively skewed, and symmetric sets of data, and give an example of each. **See Chapter 11 Answer Appendix.**

Enrichment
OL BL

11-2 Enrichment

The Harmonic Mean

The *harmonic mean*, H, is a useful measure of central tendency in special cases of averaging rates.

CCSS Teaching the Mathematical Practices

Arguments Mathematically proficient students understand and use stated assumptions, definitions, and previously established results in constructing arguments. They make conjectures and build a logical progression of statements to explore the truth of their conjectures.

17. DISTRIBUTIONS Which of the following is a characteristic of a negatively skewed distribution? **D**

   A The majority of the data are on the left of the mean.

   B The mean and median are approximately equal.

   C The mean is greater than the median.

   D The mean is less than the median.

18. SHORT RESPONSE The average test score of a class of c students is 80, and the average test score of a class of d students is 85. When the scores of both classes are combined, the average score is 82. What is the value of $\frac{c}{d}$? $\frac{3}{2}$

19. SAT/ACT What is the multiplicative inverse of $2i$? **H**

   F $-2i$ **J** $\frac{1}{2}$

   G -2 **K** $\frac{i}{2}$

   H $-\frac{i}{2}$

20. Which equation best represents the graph? **C**

   A $y = 4x$
   B $y = x^2 + 4$
   C $y = 4^{-x}$
   D $y = -4^x$

Spiral Review

Determine whether each survey question is biased. Explain your reasoning. (Lesson 11-1)

21. What toppings do you prefer on your pizza? **unbiased**

22. What is your favorite class, and what teacher gives the easiest homework?

> **22.** Biased; sample answer: The question combines two issues.

23. Don't you hate how high gas prices are?

> **23.** Biased; sample answer: The question encourages a certain response. The phrase "Don't you hate" encourages you to agree that gas prices are too high.

24. PARTIES Suppose each time a new guest arrives at a party, he or she shakes hands with each person already at the party. Prove that after n guests have arrived, a total of $\frac{n(n-1)}{2}$ handshakes have taken place. (Lesson 10-7) **See Chapter 11 Answer Appendix.**

25. ASTRONOMY The orbit of Pluto can be modeled by $\frac{x^2}{39.5^2} + \frac{y^2}{38.3^2} = 1$, where the units are astronomical units. Suppose a comet is following a path modeled by $x = y^2 + 20$. (Lesson 9-7)

   a. Find the point(s) of intersection of the orbits of Pluto and the comet. **(39.2, ± 4.4)**

   b. Will the comet necessarily hit Pluto? Explain.

   c. Where do the graphs of $y = 2x + 1$ and $2x^2 + y^2 = 11$ intersect?

> **25b.** No; the comet and Pluto may not be at either point of intersection at the same time.

   d. What are the coordinates of the points that lie on the graphs of both $x^2 + y^2 = 25$ and $2x^2 + 3y^2 = 66$? **(3, ± 4), (−3, ± 4)**   **25c.** $\left(-\frac{5}{3}, -\frac{7}{3}\right)$, **(1, 3)**

Skills Review

Determine whether each situation involves a *permutation* or a *combination*. Then find the number of possibilities.

26. the winner of the first, second, and third prizes in a contest with 8 finalists **permutation; 336**

27. selecting two of eight employees to attend a business seminar **combination; 28**

28. an arrangement of the letters in the word *MATH* **permutation; 24**

29. placing an algebra book, a geometry book, a chemistry book, an English book, and a health book on a shelf **permutation; 120**

DifferentiatedInstruction OL BL

Extension Have students complete Exercise 14. Examples of data that may have bimodal distributions are the annual tuitions for private and public schools, the selling prices for suburban and urban houses, and the heights of 6th grade and 12th grade students. Have students think of other examples that may result in bimodal distributions.

---

4 Assess

Ticket Out the Door Have students generate a set of data, create a histogram using the data, and describe the shape of the distribution. Then have the students describe the center and spread of the data using either the mean and standard deviation or the five-number summary.

Formative Assessment
Check for student understanding of Lessons 11-1 and 11-2.

📁 Quiz 1, p. 45

Additional Answers

11a.

[0, 30] sc: 3 by [0, 5] scl: 1

Sample answer: The distribution is symmetric, so use the mean and standard deviation. The mean of the data is 18 with sample standard deviation of about 5.2 points.

11b.

[0, 30] sc: 3 by [0, 5] scl: 1

mean: 14.6; median: 17

11c. Sample answer: Adding the scores from the first four games causes the shape of the distribution to go from being symmetric to being negatively skewed. Therefore, the center and spread should be described using the five-number summary.

13a. Sample answer: mean = 14; median = 10

13b. Sample answer: mean = 20; median = 24

13c. Sample answer: mean = 17; median = 17

1 Focus

VerticalAlignment

Before Lesson 11-3 Describe symmetrical and skewed distributions of data.

Lesson 11-3 Construct a probability distribution. Analyze a probability distribution and its summary statistics.

After Lesson 11-3 Understand the difference between continuous and discrete probability distributions.

Tips for New Teachers

Variables While both discrete and continuous variables can represent infinitely many values, it is important to distinguish between the two.

2 Teach

Scaffolding Questions

Have students read the **Why?** section of the lesson.

Ask:

- What do mutual funds offer investors? diversity

- Why is it beneficial to have an accurate analysis of the fund's performance? Sample answer: to determine if it is the right fund for the investor.

LESSON 11-3 Probability Distributions

Then	Now	Why?
• You used statistics to describe symmetrical and skewed distributions of data.	**1** Construct a probability distribution. **2** Analyze a probability distribution and its summary statistics.	• Mutual funds are professionally managed investments that offer diversity to investors. An accurate analysis of the fund's current and expected performance can help an investor determine if the fund suits their needs.

 NewVocabulary
random variable
discrete random variable
continuous random variable
probability distribution
theoretical probability distribution
experimental probability distribution
Law of Large Numbers
expected value

 Common Core State Standards

Content Standards
S.MD.7 Analyze decisions and strategies using probability concepts (e.g., product testing, medical testing, pulling a hockey goalie at the end of a game).

Mathematical Practices
2 Reason abstractly and quantitatively.

1 Construct a Probability Distribution A sample space is the set of all possible outcomes in a distribution. Consider a distribution of values represented by the sum of the values on two dice and a distribution of the miles per gallon for a sample of cars.

Sum of Two Dice   **Miles Per Gallon**

The sum of the values on the dice can be any integer from 2 to 12. So, the sample space is [2, 3, …, 11, 12]. This distribution is *discrete* because the number of possible values in the sample space can be counted.

The distribution of miles per gallon is *continuous*. While the sample space includes any positive value less than a certain maximum (around 100), the data can take on an infinite number of values within this range.

The value of a **random variable** is the numerical outcome of a random event. A random variable can be discrete or continuous. **Discrete random variables** represent countable values. **Continuous random variables** can take on any value.

Example 1 Identify and Classify Random Variables

Identify the random variable in each distribution, and classify it as *discrete* or *continuous*. Explain your reasoning.

a. the number of songs found on a random selection of mp3 players

The random variable X is the number of songs on any mp3 player in the random selection of players. The number of songs is countable, so X is discrete.

b. the weights of football helmets sent by a manufacturer

The random variable X is the weight of any particular helmet. The weight of any particular helmet can be anywhere within a certain range, typically 6 to 8 pounds. Therefore, X is continuous.

▶ **Guided**Practice **1A–1B.** See Chapter 11 Answer Appendix.

1A. the exact distances of a sample of discus throws

1B. the ages of counselors at a summer camp

742 | Lesson 11-3

Jack Hollingsworth/Corbis

Lesson 11-3 Resources

Resource	Approaching Level **AL**	On Level **OL**	Beyond Level **BL**	English Learners **ELL**
Teacher Edition	• Differentiated Instruction, p. 745	• Differentiated Instruction, p. 745	• Differentiated Instruction, pp. 745, 750	• Differentiated Instruction, p. 745
Chapter Resource Masters	• Study Guide and Intervention, pp. 18–19 • Skills Practice, p. 20 • Practice, p. 21 • Word Problem Practice, p. 22	• Study Guide and Intervention, pp. 18–19 • Skills Practice, p. 20 • Practice, p. 21 • Word Problem Practice, p. 22 • Enrichment, p. 23	• Practice, p. 21 • Word Problem Practice, p. 22 • Enrichment, p. 23	• Study Guide and Intervention, pp. 18–19 • Skills Practice, p. 20 • Practice, p. 21 • Word Problem Practice, p. 22
Other	• 5-Minute Check 11-3 • Study Notebook	• 5-Minute Check 11-3 • Study Notebook	• 5-Minute Check 11-3 • Study Notebook	• 5-Minute Check 11-3 • Study Notebook

StudyTip

Discrete vs. Continuous
Variables representing height, weight, and capacity will always be continuous variables because they can take on any positive value.

A **probability distribution** for a particular random variable is a function that maps the sample space to the probabilities of the outcomes in the sample space. Probability distributions can be represented by tables, equations, or graphs. In this lesson, we will focus on discrete probability distributions.

A probability distribution has the following properties.

KeyConcept Probability Distribution

- A probability distribution can be determined theoretically or experimentally.
- A probability distribution can be discrete or continuous.
- The probability of each value of X must be at least 0 and not greater than 1.
- The sum of all the probabilities for all of the possible values of X must equal 1. That is, $\Sigma P(X) = 1$.

Review Vocabulary

Theoretical and Experimental Probability Theoretical probability is based on assumptions, and experimental probability is based on experiments.
(Lesson 0-5)

A **theoretical probability distribution** is based on what is expected to happen. For example, the distribution for flipping a fair coin is $P(\text{heads}) = 0.5$, $P(\text{tails}) = 0.5$.

Example 2 Construct a Theoretical Probability Distribution

X represents the sum of the values on two dice.

a. Construct a relative-frequency table.

The theoretical probabilities associated with rolling two dice can be described using a relative-frequency table. When two dice are rolled, 36 total outcomes are possible. To determine the relative frequency, or theoretical probability, of each outcome, divide the frequency by 36.

Sum	2	3	4	5	6	7	8	9	10	11	12
Frequency	1	2	3	4	5	6	5	4	3	2	1
Relative Frequency	$\frac{1}{36}$	$\frac{1}{18}$	$\frac{1}{12}$	$\frac{1}{9}$	$\frac{5}{36}$	$\frac{1}{6}$	$\frac{5}{36}$	$\frac{1}{9}$	$\frac{1}{12}$	$\frac{1}{18}$	$\frac{1}{36}$

Sum: 36

b. Graph the theoretical probability distribution.

The graph shows the probability distribution for the sum of the values on two dice X. The bars are separated on the graph because the distribution is discrete (no other values of X are possible).

Each unique outcome of X is indicated on the horizontal axis, and the probability of each outcome occurring $P(X)$ is indicated on the vertical axis.

Sum of Two Dice

GuidedPractice

2. X represents the sum of the values of two spins of the wheel.

A. Construct a relative-frequency table.

B. Graph the theoretical probability distribution.

2A–2B. See Chapter 11 Answer Appendix.

1 Construct a Probability Distribution

Example 1 shows how to identify and classify random variables. **Example 2** shows how to use a relative frequency table to construct a probability distribution. **Example 3** shows how to use simulations to construct an experimental probability distribution.

Formative Assessment

Use the Guided Practice exercises after each example to determine students' understanding of concepts.

Additional Examples

1 Identify the random variable in each distribution, and classify it as *discrete* or *continuous*. Explain your reasoning.

a. the number of hits for the players of a baseball team The random variable X is the number of hits, which is finite and countable, so X is discrete.

b. the distances traveled by the tee shots in a golf tournament The random variable X is the distance traveled, which can take on any value in a certain range, so X is continuous.

2 X represents the sum of two cards drawn from a stack of cards numbered 1 through 8 with replacement.

a. Construct a relative-frequency table.

b. Graph the theoretical probability distribution.

Additional Answers (Additional Examples)

2a.

Sum	2	3	4	5	6	7	8	9
Relative Frequency	$\frac{1}{64}$	$\frac{1}{32}$	$\frac{3}{64}$	$\frac{1}{16}$	$\frac{5}{64}$	$\frac{3}{32}$	$\frac{7}{64}$	$\frac{1}{8}$
Sum	10	11	12	13	14	15	16	
Relative Frequency	$\frac{7}{64}$	$\frac{3}{32}$	$\frac{5}{64}$	$\frac{1}{16}$	$\frac{3}{64}$	$\frac{1}{32}$	$\frac{1}{64}$	

2b.

Additional Example

3 X represents the sum of two cards drawn from a stack of cards numbered 1 through 8 with replacement.

  a. Construct a relative frequency table for 100 trials.

  b. Graph the experimental probability distribution.

Teach with Tech

Interactive Whiteboard Choose two students to work through an example in front of the class. Have one student explain how to calculate a frequency distribution, and have the other student explain how to create a histogram from that distribution.

Tips for New Teachers

Law of Large Numbers Have students demonstrate the Law of Large Numbers by simulating the flipping of a coin for 10, 25, 50, 100, and 200 trials, and comparing the distributions. Show how the experimental probability more closely resembles the theoretical probability as the number of trials increases.

ReviewVocabulary

Simulations and Random Number Generators
For more practice on simulations and random number generators, see Extend 12-1.

StudyTip

Random Number Generators and Proportions When using a random number generator to simulate events with different probabilities, set up a proportion. For example, suppose there are 3 possible outcomes with probabilities of A: 0.25, B:0.35, and C:0.40. Random numbers 1–25 can represent A, 26–60 represent B, and 61–100 represent C.

An **experimental probability distribution** is a distribution of probabilities estimated from experiments. Simulations can be used to construct an experimental probability distribution. When constructing this type of distribution, use the frequency of occurrences of each observed value to compute its probability.

Example 3 Construct an Experimental Probability Distribution

X represents the sum of the values found by rolling two dice.

  a. Construct a relative-frequency table.

  Roll two dice 100 times or use a random number generator to complete the simulation and create a simulation tally sheet.

Sum	Tally	Frequency	Sum	Tally	Frequency
2	‖	2	8	𝍷𝍷 ‖	12
3	𝍷	5	9	𝍷𝍷 ∣	11
4	𝍷 ∣	6	10	𝍷 ‖‖	8
5	𝍷𝍷 ‖‖	13	11	‖‖	3
6	𝍷𝍷 ‖‖‖	14	12	‖‖‖‖	4
7	𝍷𝍷𝍷𝍷 ‖	22			

Calculate the experimental probability of each value by dividing its frequency by the total number of trials, 100.

Sum	2	3	4	5	6	7	8	9	10	11	12
Relative Frequency	0.02	0.05	0.06	0.13	0.14	0.22	0.12	0.11	0.08	0.03	0.04

  b. Graph the experimental probability distribution.

  The graph shows the discrete probability distribution for the sum of the values shown on two dice X.

GuidedPractice

3. X represents the sum of the values of two spins of the wheel.

  A. Construct a relative-frequency table for 100 trials.

  B. Graph the experimental probability distribution.

  3A–3B. See Chapter 11 Answer Appendix.

Notice that this graph is different from the theoretical graph in Example 2. With small sample sizes, experimental distributions may vary greatly from their associated theoretical distributions. However, as the sample size increases, experimental probabilities will more closely resemble their associated theoretical probabilities. This is due to the **Law of Large Numbers**, which states that the variation in a data set decreases as the sample size increases.

744 | Lesson 11-3 | Probability Distributions

Additional Answers (Additional Examples)

3a.

Sum	2	3	4	5	6	7	8	9	10	11	12	13	14	15	16
Frequency	1	3	6	5	6	8	12	10	4	2	1	2	2	1	1
Relative Frequency	$\frac{1}{64}$	$\frac{3}{64}$	$\frac{3}{32}$	$\frac{5}{64}$	$\frac{3}{32}$	$\frac{1}{8}$	$\frac{3}{16}$	$\frac{5}{32}$	$\frac{1}{16}$	$\frac{1}{32}$	$\frac{1}{64}$	$\frac{1}{32}$	$\frac{1}{32}$	$\frac{1}{64}$	$\frac{1}{64}$

3b.

2 Analyze a Probability Distribution

Analyze a Probability Distribution Probability distributions are often used to analyze financial data. The two most common statistics used to analyze a discrete probability distribution are the mean, or expected value, and the standard deviation. The **expected value** $E(X)$ of a discrete random variable of a probability distribution is the weighted average of the variable.

KeyConcept Expected Value of a Discrete Random Variable

Words The expected value of a discrete random variable is the weighted average of the values of the variable. It is calculated by finding the sum of the products of every possible value of X and its associated probability $P(X)$.

Symbols $E(X) = \Sigma\, [X \cdot P(X)]$

Real-World Example 4 Expected Value

GAME SHOWS A game-show contestant has won one spin of the wheel at the right. Find the expected value of his winnings.

Each prize value represents a value of X and each percent represents the corresponding probability $P(X)$. Find $E(X)$.

$E(X) = \Sigma[X \cdot P(X)]$

$= 0(0.20) + 25{,}000(0.08) + 15{,}000(0.14) + 10{,}000(0.22) + 5000(0.36)$

$= 0 + 2000 + 2100 + 2200 + 1800$

$= 8100$

The expected value of the contestant's winnings is $8100.

▶ **Guided**Practice

4. PRIZES Curt won a ticket for a prize. The distribution of the values of the tickets and their relative frequencies are shown. Find the expected value of his winnings. *about $8.48*

Value ($)	1	10	100	1000	5000	25,000
Frequency	5000	100	25	5	1	1

Sometimes the expected value does not provide enough information to fully analyze a probability distribution. For example, suppose two wheels had roughly the same expected value. Which one would you choose? Which one is *riskier*? The standard deviation can provide more insight into the expected value of a probability distribution.

The formula for calculating the standard deviation of a probability distribution is similar to the one used for a set of data.

KeyConcept Standard Deviation of a Probability Distribution

Words For each value of X, subtract the mean from X and square the difference. Then multiply by the probability of X. The sum of each of these products is the variance. The standard deviation is the square root of the variance.

Symbols Variance: $\sigma^2 = \Sigma[[X - E(X)]^2 \cdot P(X)]$

Standard Deviation: $\sigma = \sqrt{\sigma^2}$

---

akg-images

Math History Link

Christian Huygens (1629–1695) This Dutchman was the first to discuss games of chance. "Although in a pure game of chance the results are uncertain, the chance that one player has to win or to lose depends on a determined value." This became known as the *expected value*.

WatchOut!

Expected Value The expected value is what you *expect* to happen in the long run, not necessarily what *will* happen.

Focus on Mathematical Content

Symbols The expected value $E(X)$ of a discrete random variable is equal to the mean μ of the variable. Therefore, the symbols are interchangeable.

2 Analyze a Probability Distribution

Example 4 shows how to calculate the expected value. **Example 5** shows how to calculate and analyze the standard deviation.

Additional Example

4 A tetrahedral die has four sides numbered 1, 2, 3, and 4. Find the expected value of one roll of this die. 2.5

DifferentiatedInstruction **AL** **OL** **BL** **ELL**

Social Learners It is important to be aware that some students may have cultural or familial prohibitions against cards, dice, or gambling of any kind. Explain that, historically, the laws of probability were actually developed in the context of gambling but are now used in many other ways, including medicine and meteorology.

Additional Example

5 **RAFFLE** At a raffle, 400 tickets are sold for $1 each. One ticket wins $100, five tickets win $10, and ten tickets win $5. Calculate the expected value and standard deviation of the distribution of winnings for a $1 ticket.
—$0.50; 5.26

Real-WorldCareer

Mutual Fund Manager
Mutual fund managers buy and sell fund investments according to the investment objective of the fund. Investment management includes financial statement analysis, asset and stock selection, and monitoring of investments. Certification beyond a bachelor's degree is required.

Source: International Financial Services, London

5. Fund C: $E(X) = \$450$; $\sigma \approx 456.6$; Fund D: $E(X) = \$565$, $\sigma \approx 443$; sample answer: Funds C and D have roughly the same standard deviation, so they will have about the same risk. Therefore, Fund D will be the better investment because its expected value of $565 is greater than Fund C's expected value of $450.

StudyTip

Return on Investment When investing $1000 in a product that has a 6% expected return, the investor can expect a 0.06(1000) or $60 profit.

Real-World Example 5 Standard Deviation of a Distribution

DECISION MAKING Jimmy is thinking about investing $10,000 in two different investment funds. The expected rates of return and the corresponding probabilities for each fund are listed below.

Fund A	Fund B
50% chance of an $800 profit	30% chance of a $2400 profit
20% chance of a $1200 profit	10% chance of a $1900 profit
20% chance of a $600 profit	40% chance of a $200 loss
10% chance of a $100 loss	20% chance of a $400 loss

a. Find the expected value of each investment.

Fund A: $E(X) = 0.50(800) + 0.20(1200) + 0.20(600) + 0.10(-100)$ or 750
Fund B: $E(X) = 0.30(2400) + 0.10(1900) + 0.40(-200) + 0.20(-400)$ or 750
An investment of $10,000 in Fund A or Fund B will expect to yield $750.

b. Find each standard deviation.

Fund A:

Profit, X	P(X)	$[X - E(X)]^2$	$[X - E(X)]^2 \cdot P(X)$
800	0.50	$(800 - 750)^2 = 2500$	$2500 \cdot 0.50 = 1250$
1200	0.20	$(1200 - 750)^2 = 202{,}500$	$202{,}500 \cdot 0.20 = 40{,}500$
600	0.20	$(600 - 750)^2 = 22{,}500$	$22{,}500 \cdot 0.20 = 4500$
−100	0.10	$(-100 - 750)^2 = 722{,}500$	$722{,}500 \cdot 0.10 = 72{,}250$

$$\Sigma[[(X - E(X)]^2 \cdot P(X)] = 118{,}500$$
$$\sqrt{118{,}500} \approx 344.2$$

Fund B:

Profit, X	P(X)	$[X - E(X)]^2$	$[X - E(X)]^2 \cdot P(X)$
2400	0.30	$(2400 - 750)^2 = 2{,}722{,}500$	$2{,}722{,}500 \cdot 0.30 = 816{,}750$
1900	0.10	$(1900 - 750)^2 = 1{,}322{,}500$	$1{,}322{,}500 \cdot 0.10 = 132{,}250$
−200	0.40	$(-200 - 750)^2 = 902{,}500$	$902{,}500 \cdot 0.40 = 361{,}000$
−400	0.20	$(-400 - 750)^2 = 1{,}322{,}500$	$1{,}322{,}500 \cdot 0.20 = 264{,}500$

$$\Sigma[[(X - E(X)]^2 \cdot P(X)] = 1{,}574{,}500$$
$$\sqrt{1{,}574{,}500} \approx 1254.8$$

c. Which investment would you advise Jimmy to choose, and why?

Jimmy should choose Fund A. While the funds have identical expected values, the standard deviation of Fund B is almost four times the standard deviation for Fund A. This means that the expected value for Fund B will have about four times the variability than Fund A and will be riskier with a greater chance for gains and losses.

GuidedPractice

5. **DECISION MAKING** Compare a $10,000 investment in the two funds. Which investment would you recommend, and why?

Fund C	Fund D
30% chance of a $1000 profit	40% chance of a $1000 profit
40% chance of a $500 profit	30% chance of a $600 profit
20% chance of a $100 loss	15% chance of a $100 profit
10% chance of a $300 loss	15% chance of a $200 loss

 746 | Lesson 11-3 | Probability Distributions

Tetra Images/Getty Images

Example 1 Identify the random variable in each distribution, and classify it as *discrete* or *continuous*. Explain your reasoning. **1–4. See margin.**

1. the number of pages linked to a Web page

2. the number of stations in a cable package

3. the amount of precipitation in a city per month

4. the number of cars passing through an intersection in a given time interval

Examples 2–5 5. X represents the sum of the values of two spins of the wheel.

a. Construct a relative-frequency table showing the theoretical probabilities.

b. Graph the theoretical probability distribution.

c. Construct a relative-frequency table for 100 trials.

d. Graph the experimental probability distribution.

e. Find the expected value for the sum of two spins of the wheel. **13.5**

f. Find the standard deviation for the sum of two spins of the wheel. **4.29**

5a–d. See Chapter 11 Answer Appendix.

Practice and Problem Solving Extra Practice is on page R11.

Example 1 Identify the random variable in each distribution, and classify it as *discrete* or *continuous*. Explain your reasoning. **6–9. See margin.**

6. the number of texts received per week

7. the number of diggs (or "likes") for a Web page

8. the height of a plant after a specific amount of time

9. the number of files infected by a computer virus

10a–d. See Chapter 11 Answer Appendix.

Examples 2–5 10. **CCSS PERSEVERANCE** A contestant has won a prize on a game show. The frequency table at the right shows the number of winners for 3200 hypothetical players.

a. Construct a relative-frequency table showing the theoretical probability.

b. Graph the theoretical probability distribution.

c. Construct a relative-frequency table for 50 trials.

d. Graph the experimental probability distribution.

e. Find the expected value. **$922.50**

f. Find the standard deviation. **1711.91**

Prize, X	Winners
$100	1120
$250	800
$500	480
$1000	320
$2500	256
$5000	128
$7500	64
$10,000	32

B 11 **SNOW DAYS** The following probability distribution lists the probable number of snow days per school year at North High School. Use this information to determine the expected number of snow days per year. **3.34**

Number of Snow Days Per Year									
Days	0	1	2	3	4	5	6	7	8
Probability	0.1	0.1	0.15	0.15	0.25	0.1	0.08	0.05	0.02

12. **CARDS** In a standard deck of 52 cards, there are 4 different suits.

a. If jacks = 11, queens = 12, kings = 13, and aces = 1, what is the expected value of a card that is drawn from a standard deck? **7**

b. If you are dealt 7 cards with replacement, what is the expected number of spades? **1.75**

Differentiated Homework Options

Level	Assignment	Two-Day Option	
AL Basic	6–10, 19–44	7, 9, 25–28	6–10 even, 20–24, 29–44
OL Core	7, 9, 11–44	6–10, 25–28	11–24, 29–44
BL Advanced	11–41, (optional: 42–44)		

Formative Assessment

Use Exercises 1–5 to check for understanding.

Use the chart on the bottom of this page to customize assignments for your students.

CCSS Teaching the Mathematical Practices

Perseverance Mathematically proficient students start by explaining to themselves the meaning of a problem and looking for entry points to its solution. They analyze givens, constraints, relationships, and goals and make conjectures about the form and meaning of the solution and plan a solution pathway rather than simply jumping into a solution attempt.

Additional Answers

1. The random variable X is the number of pages linked to a Web page. The pages are countable, so X is discrete.

2. The random variable X is the number of stations in a cable package. The stations are countable, so X is discrete.

3. The random variable X is the amount of precipitation in a city per month. Precipitation can be anywhere within a certain range. Therefore, X is continuous.

4. The random variable X is the number of cars passing through an intersection. The cars are countable, so X is discrete.

6. The random variable X is the number of texts per week. The texts are countable, so X is discrete.

7. The random variable X is the number of diggs for a web page. The diggs are countable, so X is discrete.

8. The random variable X is the height of a plant. Height can be anywhere within a certain range. Therefore, X is continuous.

9. The random variable X is the number of files infected by a computer virus. The files are countable, so X is discrete.

Study Guide and Intervention
AL OL ELL

Practice
AL OL BL ELL

Word Problem Practice
AL OL BL ELL

13. RAFFLES The table shows the probability distribution for a raffle if 100 tickets are sold for $1 each. There is 1 prize for $20, 5 prizes for $10, and 10 prizes for $5.

Distribution of Prizes				
Prize	no prize	$20	$10	$5
Probability	0.84	0.01	0.05	0.10

a. Graph the theoretical probability distribution.   **See Chapter 11 Answer Appendix.**

b. Find the expected value. **$1.20**

c. Interpret the results you found in part **b**. What can you conclude about the raffle? **See margin.**

14a–c. See Chapter 11 Answer Appendix.

14. CCSS TOOLS Based on previous data, the probability distribution of the number of students running for class president is shown.

Number of Students Running

a. Determine the expected number of students who will run. Interpret your results.

b. Construct a relative-frequency table for 50 trials.

c. Graph the experimental probability distribution.

15c–d. See Chapter 11 Answer Appendix.

15. BASKETBALL The distribution below lists the probability of the number of major upsets in the first round of a basketball tournament each year.

Number of Upsets Per Year									
Upsets	0	1	2	3	4	5	6	7	8
Probability	$\frac{1}{32}$	$\frac{1}{16}$	$\frac{3}{32}$	$\frac{1}{8}$	$\frac{1}{8}$	$\frac{5}{16}$	$\frac{1}{8}$	$\frac{3}{32}$	$\frac{1}{32}$

a. Determine the expected number of upsets. Interpret your results.

b. Find the standard deviation. **1.90**

c. Construct a relative-frequency table for 50 trials.

d. Graph the experimental probability distribution.

15a. 4.34; Sample answer: The expected number is 4.34, so we can expect there to be 4 upsets. We cannot have 0.34 upsets, so we round to the nearest whole number.

16. RAFFLES The French Club sold 500 raffle tickets for $1 each. The first prize ticket will win $100, 2 second prize tickets will each win $10, and 5 third prize tickets each win $5.

a. What is the expected value of a single ticket? **−$0.71**

b. Calculate the standard deviation of the probability distribution. **4.53**

c. **DECISION MAKING** The Glee Club is offering a raffle with a similar expected value and a standard deviation of 2.2. In which raffle should you participate? Explain your reasoning. **See margin.**

17 DECISION MAKING Carmen is thinking about investing $10,000 in two different investment funds. The expected rates of return and the corresponding probabilities for each fund are listed below. Compare the two investments using the expected value and standard deviation. Which investment would you advise Carmen to choose, and why? **See margin.**

Fund A
30% chance of a $1900 profit
30% chance of a $600 profit
15% chance of a $200 loss
25% chance of a $500 loss

Fund B
40% chance of a $1600 profit
10% chance of a $900 profit
10% chance of a $300 loss
40% chance of a $400 loss

Enrichment
OL BL

11-3 Enrichment

Geometric Probability

Additional Answer

13c. Sample answer: The expected value is positive, so a person buying a ticket can expect to win $0.20 even after the cost of the ticket is considered. Thus, a person would want to participate in this raffle. On the other hand, this raffle is guaranteed to lose money for the organizers and they should change the distribution of prizes or not do the raffle.

18. **MULTIPLE REPRESENTATIONS** In this problem, you will investigate geometric probability.

   a. Tabular The spinner shown has a radius of 2.5 inches. Copy and complete the table below.

Color	Probability	Sector Area	Total Area	Sector Area / Total Area
red	$\frac{1}{6}$	3.27 in²	19.63 in²	0.166
orange	$\frac{1}{6}$	3.27 in²	19.63 in²	0.166
yellow	$\frac{1}{6}$	3.27 in²	19.63 in²	0.166
green	$\frac{1}{4}$	4.91 in²	19.63 in²	0.25
blue	$\frac{1}{4}$	4.91 in²	19.63 in²	0.25

18b. Sample answer: The probability is equal to the ratio of the sector area to the total area.

18c. green: $\frac{1}{9}$; yellow: $\frac{1}{3}$; blue: $\frac{5}{9}$

   b. Verbal Make a conjecture about the relationship between the ratio of the area of the sector to the total area and the probability of the spinner landing on each color.

   c. Analytical Consider the dartboard shown. Predict the probability of a dart landing in each area of the board. Assume that any dart thrown will land on the board and is equally likely to land at any point on the board.

   d. Tabular Construct a relative-frequency table for throwing 100 darts.

   e. Graphical Graph the experimental probability distribution.
   18d–e. See Chapter 11 Answer Appendix.

H.O.T. Problems   *Use Higher-Order Thinking Skills*

19. **CCSS CRITIQUE** Liana and Shannon each created a probability distribution for the sum of two spins on the spinner at the right. Is either of them correct? Explain your reasoning.

19. Sample answer: Liana; Shannon didn't consider every scenario in determining the total probability. For example, in calculating the probability of a sum of 5, she considered spinning a 3 then a 2, but not a 2 then a 3.

20. Sample answer: False; the expected value is 3.5 which is not a possible outcome of a single roll.

20. **REASONING** Determine whether the following statement is *true* or *false*. Explain.
   If you roll a die 10 times, you will roll the expected value at least twice.

21. **OPEN ENDED** Create a discrete probability distribution that shows five different outcomes and their associated probabilities. **See Chapter 11 Answer Appendix.**

22. **REASONING** Determine whether the following statement is *true* or *false*. Explain.
   Random variables that can take on an infinite number of values are continuous.
     22. See Chapter 11 Answer Appendix.

23. **OPEN ENDED** Provide examples of a discrete probability distribution and a continuous probability distribution. Describe the differences between them. **See Chapter 11 Answer Appendix.**

24. **WRITING IN MATH** Compare and contrast two investments that have identical expected values and significantly different standard deviations. **See Chapter 11 Answer Appendix**

Multiple Representations
In Exercise 18, students use a diagram, information organized in a table, symbolic formulas, and verbal analysis to determine geometric probabilities.

CCSS **Teaching the Mathematical Practices**

Critique Mathematically proficient students are also able to compare the effectiveness of two plausible arguments, distinguish correct logic or reasoning from that which is flawed, and—if there is a flaw in an argument—explain what it is.

WatchOut!
Error Analysis For Exercise 19, one common error would be considering spinning a "2" and a "3", but not considering a "3" and a "2".

Additional Answers

16c. Sample answer: The standard deviation of the probability distribution for the Glee Club raffle is about half the standard deviation for the French Club raffle, so the Glee Club raffle is less risky. Since they have similar expected values, the riskier raffle will also have the potential to win more, so both raffles have good and bad qualities. It is up to the individual participant to decide which one to choose.

17. Sample answer: The expected value of Funds A and B is $595 and $540, respectively. The standard deviation for Fund A is about 951.6, while the standard deviation for Fund B is about 941.5. Since the standard deviations are about the same, the funds have about the same amount of risk. Therefore, with a higher expected value, Fund A is the better investment.

 Follow-up Students have explored probability.

Ask:

- **How can probability be used in decision making?** Sample answer: You can use probability to predict the most likely outcomes, and then make a decision based on those findings.

4 Assess

Crystal Ball Ask students how they think today's study of probability distributions will help them with their study of the binomial distribution in the next lesson.

Formative Assessment
Check for student understanding of Lesson 11-3.

📁 Quiz 2, p. 45

Additional Answers

29a.

Peter's Articles

[0, 70] scl: 5 by [0, 10] scl: 1

Paul's Articles

[0, 80] scl: 5 by [0, 10] scl: 1

Peter's articles, positively skewed; Paul's articles, symmetric

29b. Sample answer: One of the distributions is symmetric and the other is skewed, so use the five-number summaries. The range for Peter's articles is 64, and the range for Paul's articles is 53. However, the upper quartile for Peter's is 33, while the lower quartile for Paul's is 34. This means that 75% of Paul's articles have more likes (and are more popular) than 75% of Peter's articles. Therefore, we can conclude that Paul's articles are more popular overall.

Standardized Test Practice

25. GRIDDED RESPONSE The height $f(x)$ of a bouncing ball after x bounces is represented by $f(x) = 140(0.8)^x$. How many times higher is the first bounce than the fifth bounce? **2.4**

26. PROBABILITY Andres has a bag that contains 4 red, 6 yellow, 2 blue, and 4 green marbles. If he reaches into the bag and removes a marble without looking, what is the probability that it will not be yellow? **D**

A $\frac{1}{8}$ **C** $\frac{3}{8}$

B $\frac{1}{4}$ **D** $\frac{5}{8}$

27. GEOMETRY Find the area of the shaded portion of the figure to the nearest square inch. **H**
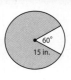

F 79   **H** 589
G 94   **J** 707

28. SAT/ACT If x and y are positive integers, which of the following expressions is equivalent to $\frac{(5^x)^y}{5^x}$? **E**

A 1^y **D** 5^{xy-1}
B ± 1 **E** 5^{xy-x}
C 5^y

Spiral Review

29. ARTICLES Peter and Paul each write articles for an online magazine. Their employer keeps track of the number of *likes* received by each article. (Lesson 11-2) **29a–b. See margin.**

 a. Use a graphing calculator to create a histogram for each data set. Then describe the shape of each distribution.

 b. Compare the distributions using either the means and standard deviations or the five-number summaries. Justify your choice.

Peter's Articles
16, 22, 19, 31, 24, 8, 40, 19, 33, 18, 36, 21, 55, 3, 16, 44, 22, 39, 12, 18, 13, 20, 67, 31, 13, 38, 31, 22, 26, 28

Paul's Articles
41, 38, 29, 33, 36, 55, 51, 19, 49, 56, 28, 52, 49, 19, 38, 33, 42, 61, 72, 55, 48, 39, 37, 43, 48, 45, 52, 43, 34, 29

30. Sample answer: This situation calls for an experiment because the treatment will be tested on a sample group, which means that the members of the sample will be affected by the study.

Determine whether the situation calls for a *survey,* **an** *observational study,* **or an** *experiment.* **Explain your reasoning.** (Lesson 11-1)

30. You want to test a drug that reverses male pattern baldness.

31. You want to find voters' opinions on recent legislation.

31. Sample answer: This situation calls for a survey because the data will be collected from responses from members of a sample of the population.

32. 0.125, 0.1875, 0.28125, 0.421875, 0.6328125   **33.** 0.5, 1.25, 3.125, 7.8125, 19.53125

Find the first five terms of each geometric sequence described. (Lesson 10-3)

32. $a_1 = 0.125, r = 1.5$ **33.** $a_1 = 0.5, r = 2.5$ **34.** $a_1 = 4, r = 0.5$ 4, 2, 1, 0.5, 0.25

35. $a_1 = 12, r = \frac{1}{3}$ 12, 4, $\frac{4}{3}, \frac{4}{9}, \frac{4}{27}$ **36.** $a_1 = 21, r = \frac{2}{3}$ 21, 14, $\frac{28}{3}, \frac{56}{9}, \frac{112}{27}$ **37.** $a_1 = 80, r = \frac{5}{4}$ 80, 100, 125, $\frac{625}{4}, \frac{3125}{16}$

38. COMMUNICATION A microphone is placed at the focus of a parabolic reflector to collect sound for the television broadcast of a football game. Write an equation for the cross section, assuming that the focus is at the origin, the focus is 6 inches from the vertex, and the parabola opens to the right. (Lesson 9-2) $x = \frac{1}{24}y^2 - 6$

Solve each equation. Check your solutions. (Lesson 7-4)

39. $\log_9 x = \frac{3}{2}$ 27   **40.** $\log_{\frac{1}{10}} x = -3$ 1000   **41.** $\log_b 9 = 2$ 3

Skills Review

Expand each power. **43.** $m^4 + 4m^3n + 6m^2n^2 + 4mn^3 + n^4$

44. $r^8 + 8r^7n + 28r^6n^2 + 56r^5n^3 + 70r^4n^4 + 56r^3n^5 + 28r^2n^6 + 8rn^7 + n^8$

42. $(a - b)^3$ $a^3 - 3a^2b + 3ab^2 - b^3$ **43.** $(m + n)^4$ **44.** $(r + n)^8$

DifferentiatedInstruction (BL)

Extension Write the numbers 1, 1, 1, 2, 2, 3, 4, 4 on eight slips of paper and mix them up in a paper bag. Have students construct a probability distribution for the number that results when one slip is drawn at random from the bag. Then have them take turns drawing (with replacement) and compare the relative frequencies with the theoretical probability distribution.

The theoretical probabilities are $P(1) = \frac{3}{8}$, $P(2) = \frac{1}{4}$, $P(3) = \frac{1}{8}$, and $P(4) = \frac{1}{4}$.

CHAPTER 11 Mid-Chapter Quiz
Lessons 11-1 through 11-3

Determine whether each situation describes a *survey*, an *experiment*, or an *observational study*. Then identify the sample, and suggest a population from which it may have been selected. (Lesson 11-1)

1. A high school principal wants to test five ideas for a new school mascot. He randomly selects 15 high school students to view pictures of the ideas while he watches and records their reactions. **observational study; sample: the 15 high school students selected; population: the student body**

2. Half of the employees of a grocery store are randomly chosen for an extra hour lunch break. The managers then compare their attitudes with their co-workers. **experiment; sample: the employees that were given an extra hour lunch break; population: all grocery store employees**

3. Students want to create a school yearbook. They send out a questionnaire to 100 students asking what they would like to showcase in the yearbook. **survey; sample: the 100 students who received the questionnaire; population: school body**

4. The producers of a sitcom want to determine if a new character that they are planning to introduce will be well received. They show a clip of the show with the new character to 50 randomly chosen participants and then record the participants' reactions. **observational study; sample: the 50 participants; population: all potential viewers**

5. **MULTIPLE CHOICE** Which survey question is *unbiased*? (Lesson 11-1) **D**

   A Do you like days like today?

   B Which is your favorite theme park, Park A or Park B?

   C Don't you think that carrots taste better than celery?

   D How often do you go to the movies?

6a–b. See Chapter 11 Answer Appendix.

6. **PARENTS** The table below shows the ages of parents who volunteered to assist in a neighborhood bake sale. (Lesson 11-2)

Ages of Parents (years)				
28	34	33	45	31
33	41	34	36	42
30	29	32	40	36
29	33	29	28	44
47	31	28	27	29

   a. Use a graphing calculator to create a box-and-whisker plot. Then describe the shape of the distribution.

   b. Describe the center and spread of the data using either the mean and standard deviation or the five-number summary. Justify your choice.

7. **TRAINING** Aiden and Mark's training times for the 40-meter dash are shown. (Lesson 11-2)
   a–b. See Chapter 11 Answer Appendix.

Aiden's 40-Meter Dash Times (seconds)					
4.84	4.94	4.87	4.78	5.04	4.98
4.83	5.03	4.74	5.15	4.82	4.91
4.62	4.83	4.76	4.93	4.85	4.82
4.76	4.98	4.94	5.05	4.94	5.04
4.86	4.85	4.71	4.66	4.91	4.82

Mark's 40-Meter Dash Times (seconds)					
5.03	4.76	4.69	4.52	4.81	4.78
4.65	4.66	4.83	4.95	4.64	4.76
4.43	4.64	4.50	4.58	4.68	4.65
4.83	4.78	4.71	4.81	4.76	4.84
4.61	4.63	4.33	4.46	4.74	4.63

   a. Use a graphing calculator to create a histogram for each data set. Then describe the shape of each distribution.

   b. Compare the distributions using either the means and standard deviations or the five-number summaries. Justify your choice.

8. **MULTIPLE CHOICE** Find the expected value of winning one of the following prizes. (Lesson 11-3) **J**

   F $1950

   G $2100

   H $3000

   J $3450

$10,000 15%
$1000 35%
$5000 20%
$2000 30%

9–12. See Chapter 11 Answer Appendix.

Identify the random variable in each distribution, and classify it as *discrete* or *continuous*. Explain your reasoning. (Lesson 11-3)

9. the number of calls received by an operator

10. the number of books sold at a yard sale

11. the height of students in a gym class

12. the weight of animals on a farm

connectED.mcgraw-hill.com **751**

Formative Assessment
Use the Mid-Chapter Quiz to assess students' progress in the first half of the chapter.

For problems answered incorrectly, have students review the lessons indicated in parentheses.

Summative Assessment
📁 Mid-Chapter Test, p. 47

McGraw-Hill eAssessment Customize and create multiple versions of your Mid-Chapter Quiz and their answer keys.

FOLDABLES StudyOrganizer

Dinah Zike's Foldables®
Before students complete the Mid-Chapter Quiz, encourage them to review the information for Lessons 11-1 through 11-3 in their Foldables.

InterventionPlanner

TIER 1 On Level OL	TIER 2 Strategic Intervention AL approaching grade level	TIER 3 Intensive Intervention 2 or more grades below level
If students miss about 25% of the exercises or less,	**If** students miss about 50% of the exercises,	**If** students miss about 75% of the exercises,
Then choose a resource: **SE** Lessons 11-1, 11-2, and 11-3 📁 Skills Practice, pp. 7, 13, and 20 **TE** Chapter Project connectED.mcgraw-hill.com Self-Check Quiz	**Then** choose a resource: 📁 Study Guide and Intervention, pp. 5, 11, and 18 connectED.mcgraw-hill.com Extra Examples, Personal Tutor, Homework Help	**Then** use *Math Triumphs, Alg. 2*, Ch. 5 connectED.mcgraw-hill.com Extra Examples, Personal Tutor, Homework Help, Review Vocabulary

1 Focus

VerticalAlignment

Before Lesson 11-4 Use the Binomial Theorem.

Lesson 11-4 Identify and conduct a binomial experiment. Find probabilities using binomial distribution.

After Lesson 11-4 Use the normal distribution to find probabilities.

2 Teach

Scaffolding Questions

Have students read the **Why?** section of the lesson.

Ask:

- What is the probability that Jessica answers a question correctly? incorrectly? $\frac{1}{4}, \frac{3}{4}$

- Are answering two successive questions independent events? yes

- If answering successive questions are independent events, what is the probability that Jessica correctly answers the first two questions? $\frac{1}{16}$ or 0.0625

:: Then

- You used the Binomial Theorem.

:: Now

1. Identify and conduct a binomial experiment.
2. Find probabilities using binomial distributions.

:: Why?

- Jessica forgot to study for her civics quiz. The quiz consists of five multiple-choice questions with each question having four answer choices. Jessica randomly circles an answer for each question. In order to pass, she needs to answer at least four questions correctly.

 NewVocabulary
binomial experiment
binomial distribution

Common Core State Standards

Content Standards
S.MD.6 Use probabilities to make fair decisions (e.g., drawing by lots, using a random number generator).
S.MD.7 Analyze decisions and strategies using probability concepts (e.g., product testing, medical testing, pulling a hockey goalie at the end of a game).

Mathematical Practices
4 Model with mathematics.

1 Binomial Experiments Each question on a multiple-choice quiz, like the one described above, can be thought of as a trial with two possible outcomes, correct or incorrect. If Jessica guesses on each question, the probability that she answers a question correctly is the same for all five questions.

Jessica's guessing on each question is an example of a binomial experiment. A **binomial experiment** is a probability experiment that satisfies the following conditions.

KeyConcept Binomial Experiments

- There is a fixed number of independent trials n.
- Each trial has only two possible outcomes, success or failure.
- The probability of success p is the same in every trial. The probability of failure q is $1 - p$.
- The random variable X is the number of successes in n trials.

Many probability experiments are or can be reduced to binomial experiments.

Example 1 Identify a Binomial Experiment

Determine whether each experiment is a binomial experiment or can be reduced to a binomial experiment. If so, describe a trial, determine the random variable, and state n, p, and q.

a. **The spinner at the right is spun 20 times to see how many times it lands on red.**

This experiment can be reduced to a binomial experiment with success being that the spinner lands on red and failure being any other outcome. Thus, a trial is a spin, and the random variable X represents the number of reds spun. The number of trials n is 20, the probability of success p is $\frac{1}{4}$ or 0.25, and the probability of failure q is $1 - 0.25$ or 0.75.

b. **One hundred students are randomly asked their favorite food.**

This is not a binomial experiment because there are many possible outcomes.

▶ **Guided**Practice **1A–1B. See Chapter 11 Answer Appendix.**

1A. Seventy-five students are randomly asked if they own a car.

1B. Four cards are removed from a deck to see how many aces are selected.

🔊 **752** | Lesson 11-4

Fancy Photography/Veer

Lesson 11-4 Resources

Resource	Approaching Level **AL**	On Level **OL**	Beyond Level **BL**	English Learners **ELL**
Teacher Edition	• Differentiated Instruction, p. 755	• Differentiated Instruction, pp. 755, 759	• Differentiated Instruction, pp. 755, 759	• Differentiated Instruction, p. 755
Chapter Resource Masters	• Study Guide and Intervention, pp. 24–25 • Skills Practice, p. 26 • Practice, p. 27 • Word Problem Practice, p. 28	• Study Guide and Intervention, pp. 24–25 • Skills Practice, p. 26 • Practice, p. 27 • Word Problem Practice, p. 28 • Enrichment, p. 29	• Practice, p. 27 • Word Problem Practice, p. 28 • Enrichment, p. 29	• Study Guide and Intervention, pp. 24–25 • Skills Practice, p. 26 • Practice, p. 27 • Word Problem Practice, p. 28
Other	• 5-Minute Check 11-4 • Study Notebook	• 5-Minute Check 11-4 • Study Notebook	• 5-Minute Check 11-4 • Study Notebook	• 5-Minute Check 11-4 • Study Notebook

Use the following guidelines when conducting a binomial experiment.

KeyConcept Conducting Binomial Experiments

Step 1 Describe a trial for the situation, and determine the number of trials to be conducted.

Step 2 Define a success, and calculate the theoretical probabilities of success and failure.

Step 3 Describe the random variable X.

Step 4 Design and conduct a simulation to determine the experimental probability.

A binomial experiment can be conducted to compare experimental and theoretical probabilities.

Example 2 Design a Binomial Experiment

Conduct a binomial experiment to determine the probability of drawing an odd-numbered card from a deck of cards. Then compare the experimental and theoretical probabilities of the experiment.

Step 1 A trial is drawing a card from a deck. The number of trials conducted can be any number greater than 0. We will use 52.

Step 2 A success is drawing an odd-numbered card. The odd-numbered cards in a deck are 3, 5, 7, and 9, and they occur once in each of the four suits. Therefore, there are $4 \cdot 4$ or 16 odd-numbered cards in the deck. The probability of drawing an odd-numbered card, or the probability of success, is $\frac{16}{52}$ or $\frac{4}{13}$. The probability of failure is $1 - \frac{4}{13}$ or $\frac{9}{13}$.

Step 3 The random variable X represents the number of odd-numbered cards drawn in 52 trials.

Step 4 Use the random number generator on a calculator to create a simulation. Assign the integers 0–12 to accurately represent the probability data.

Odd-numbered cards   0, 1, 2, 3
Other cards   4, 5, 6, 7, 8, 9, 10, 11, 12

Make a frequency table and record the results as you run the generator.

Outcome	Tally	Frequency
Odd-Numbered Card	ⲨⲎ ⲨⲎ ‖	12
Other Cards	ⲨⲎ ⲨⲎ ⲨⲎ ⲨⲎ ⲨⲎ ⲨⲎ ⲨⲎ ⲨⲎ	40

An odd-numbered card was drawn 12 times, so the experimental probability is $\frac{12}{52}$ or about 23.1%. This is less than the theoretical probability of $\frac{16}{52}$ or about 30.8%.

GuidedPractice

2. Conduct a binomial experiment to determine the probability of drawing an even-numbered card from a deck of cards. Then compare the experimental and theoretical probabilities of the experiment. **See Chapter 11 Answer Appendix.**

StudyTip
Random Number Generator
To generate all 52 random numbers using a graphing calculator, press MATH ◄ 5 and then enter the desired range followed by the number of trials. For example, type (0, 12, 52) for Example 2.

 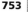
1 Binomial Experiments
Example 1 shows how to identify a binomial experiment. **Example 2** shows how to design a binomial experiment.

Formative Assessment
Use the Guided Practice exercises after each example to determine students' understanding of concepts.

Additional Example

1 Determine whether each experiment is a binomial experiment or can be reduced to a binomial experiment. If so, describe a trial, determine the random variable, and state n, p, and q.

a. In a class that consists of 12 male and 13 female students, a group of 5 students is randomly chosen to see how many males are selected. This experiment cannot be reduced to a binomial experiment because the events are not independent. The probability of choosing a male student changes after each selection.

b. A school study found that 65% of students have seen an episode of a specific show. Seven students are randomly asked if they have seen an episode of the show. This experiment can be reduced to a binomial experiment. Success is a yes, failure is a no, a trial is asking a student, and the random variable is the number of yeses; $n = 7$, $p = 0.65$, $q = 0.35$.

Additional Examples also in Interactive Classroom PowerPoint® Presentations

IWB Interactive White Board READY

2 Binomial Distribution

Example 3 shows how to find a probability by using the Binomial Probability Formula. **Example 4** shows how to create a probability distribution and find expected value.

Additional Examples

2 Conduct a binomial experiment to determine the probability of two fair coins both landing on heads. Then compare the experimental and theoretical probabilities of the experiment. *See margin.*

3 **CANDY** A candy company produces strawberry and vanilla flavored candies, 65% of which are strawberry on average. The production line mixes the candies randomly and packages 15 per bag. What is the probability that 7 candies in a bag are strawberry? *0.071 or 7.1%*

Additional Answer
(Additional Example)

2. Sample answer:

Step 1 A trial is tossing two coins. The simulation will consist of 20 trials.

Step 2 A success is both coins landing on heads. The probability of success is $\frac{1}{4}$ and the probability of failure is $\frac{3}{4}$.

Step 3 The random variable X represents the number of times both coins land on heads in 20 trials.

Step 4 Use a random number generator. Let 0 represent both coins landing on heads. Let 1–3 represent all other outcomes. Make a frequency table and record the results as you run the generator.

Outcome	Tally	Frequency
Two Heads	||||		6
Other Outcomes	|||| |||| ||||	14

The experimental probability is $\frac{6}{20}$ or 30%. This is greater than the theoretical probability of $\frac{1}{4}$ or 25%.

---

2 Binomial Distribution

In the binomial experiment in Example 2, there were 12 successes in 52 trials. If you conducted that same experiment again, there may be any number of successes from 0 to 52. This situation can be represented by a binomial distribution. A **binomial distribution** is a frequency distribution of the probability of each value of X, where the random variable X represents the number of successes in n trials. Because X is a discrete random variable, a binomial distribution is a *discrete probability distribution*.

The probabilities in a binomial distribution can be calculated using the following formula.

StudyTip
Binomial Probability Formula
In the Binomial Probability Formula, X represents the number of successes in n trials. Thus, the exponent for q, $n - X$, represents the number of failures in n trials.

> **KeyConcept** Binomial Probability Formula
>
> The probability of X successes in n independent trials is
> $$P(X) = {}_nC_X\, p^X q^{n-X},$$
> where p is the probability of success of an individual trial and q is the probability of failure on that same individual trial ($q = 1 - p$).

Notice that the Binomial Probability Formula is an adaptation of the Binomial Theorem you have already studied. The expression ${}_nC_X\, p^X q^{n-X}$ represents the $p^X q^{n-X}$ term in the binomial expansion of $(p + q)^n$.

> **Standardized Test Example 3** Find a Probability
>
> Garrett is selling items from a catalog to raise money for school. He has a 40% chance of making a sale each time he solicits a potential customer. Garrett asks 10 people to purchase an item. Find the probability that 6 people make a purchase.
>
> **A** 8.6%   **B** 11.2%   **C** 24%   **D** 40%
>
> **Read the Test Item**
>
> We need to find the probability that 6 people purchase an item. A success is making a sale, so $p = 0.4$, $q = 1 - 0.4$ or 0.6, and $n = 10$.
>
> **Solve the Test Item**
>
> $P(X) = {}_nC_X\, p^X q^{n-X}$ Binomial Probability Formula
>
> $P(6) = {}_{10}C_6\, (0.4)^6\, (0.6)^{10-6}$ $n = 10$, $X = 6$, $p = 0.4$, and $q = 0.6$
>
> ≈ 0.111 Simplify.
>
> The probability of Garrett making six sales is about 0.111 or 11.1%. So, the correct answer is B.

Test-TakingTip
CCSS Precision A common error when using the Binomial Probability Formula is to focus on only the successes and to forget the failures. Notice that $P(6) \neq (0.4)^6$ and $P(6) \neq {}_{10}C_6(0.4)^6$.

▸ **GuidedPractice**

3. TELEMARKETING At Jenny's telemarketing job, 15% of the calls that she makes to potential customers result in a sale. She makes 20 calls in a given hour. What is the probability that 5 calls result in a sale? **H**

F 6.7%   **G** 8.3%   **H** 10.3%   **J** 11.9%

If, on average, 40% of the people Garrett solicits make a purchase and he solicits 10 people, he can probably expect to make 10(0.40) or 4 sales. This value represents the mean of the binomial distribution. In general, the mean of a binomial distribution can be calculated by the following formula.

> **KeyConcept** Mean of a Binomial Distribution
>
> The mean μ of a binomial distribution is given by $\mu = np$, where n is the number of trials and p is the probability of success.

754 | Lesson 11-4 | The Binomial Distribution

CCSS **Teaching the Mathematical Practices**

Precision Mathematically proficient students use clear definitions in discussion with others and in their own reasoning. They have learned to examine claims and make explicit use of definitions. Encourage students to examine all aspects of a problem situation.

You can find the probability distribution for a binomial experiment by fully expanding the binomial $(p + q)^n$. A probability distribution can be helpful when solving for problems that allow multiple numbers of successes.

Real-World Example 4 Full Probability Distribution

TEST TAKING Refer to the beginning of the lesson.

a. Determine the probabilities associated with the number of questions Jessica answered correctly by calculating the probability distribution.

If there are four answer choices for each question, then the probability that Jessica guesses and answers a question correctly is $\frac{1}{4}$ or 0.25. In this binomial experiment, $n = 5$, $p = 0.25$, and $q = 1 - 0.25$ or 0.75. Expand the binomial $(p + q)^n$.

$(p + q)^n$

$= 1p^5 + 5p^4q + 10p^3q^2 + 10p^2q^3 + 5pq^4 + 1q^5$

$= (0.25)^5 + 5(0.25)^4(0.75) + 10(0.25)^3(0.75)^2 + 10(0.25)^2(0.75)^3 + 5(0.25)(0.75)^4 + (0.75)^5$

$\approx 0.001 + 0.015 + 0.089 + 0.264 + 0.396 + 0.237$

| 0.1% | 1.5% | 8.9% | 26.4% | 39.6% | 23.7% |
| 5 correct | 4 correct | 3 correct | 2 correct | 1 correct | 0 correct |

The graph shows the binomial probability distribution for the number of questions that Jessica answered correctly.

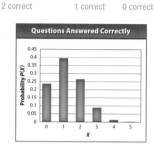

b. What is the probability that Jessica passes the quiz?

Jessica must answer at least four questions correctly to pass the quiz. The probability that Jessica answers *at least* four correct is the sum of the probabilities that she answers four or five correct and is about 1.5% + 0.1% or 1.6%. So, Jessica has about a 1.6% chance of passing, which is not likely.

c. How many questions should Jessica expect to answer correctly?

Find the mean.

$\mu = np$ Mean of a Binomial Distribution
$= 5(0.25)$ or 1.25 $n = 5$ and $p = 0.25$

The mean of the distribution is 1.25. On average, Jessica should expect to answer one question correctly when she guesses on five.

GuidedPractice

4A. 0 questions, 0.031 or 3.1%; 1 question, 0.156 or 15.6%; 2 questions, 0.313 or 31.3%; 3 questions, 0.313 or 31.3%; 4 questions, 0.156 or 15.6%; 5 questions, 0.031 or 3.1%

4. TEST TAKING Suppose Jessica's civics quiz consisted of five true-or-false questions instead of multiple-choice questions.

A. Determine the probabilities associated with the number of answers Jessica answered correctly by calculating the probability distribution.

B. What is the probability that Jessica passes the quiz? **0.187 or 18.7%**

C. How many questions should Jessica expect to answer correctly? **3**

Real-World Link

ACT The math portion of the ACT college entrance exam includes 60 multiple-choice questions that each have five answer choices.

Source: ACT

StudyTip

Mean and Expected Value The mean of a binomial distribution can be any positive rational number. The expected value of a binomial distribution, however, should be rounded to the nearest whole number since a fraction of a success is not possible.

Additional Example

4 FAMILY A family has 4 children. Assume that when a child is born, there is a 50% chance that the child is female.

a. Determine the probabilities associated with the number of daughters in the family by calculating the probability distribution. 0 daughters, 0.063 or 6.3%; 1 daughter, 0.25 or 25%; 2 daughters, 0.375 or 37.5%; 3 daughters, 0.25 or 25%; 4 daughters, 0.063 or 6.3%

b. What is the probability that the family has at least 3 daughters? 0.313 or 31.3%

c. How many daughters should the family expect to have? 2

Focus on Mathematical Content

Binomial Distributions A binomial distribution is called a *two-parameter* distribution, with the parameters being the number of trials n and the probability of one of the outcomes p. Probabilities for a binomial distribution are determined once the parameters are specified.

Teach with Tech

Web Search Have students search the Web for binomial distribution applets. Allow students to create and explore binomial probability histograms, and search for additional examples of binomial experiments.

DifferentiatedInstruction AL OL BL ELL

Kinesthetic Learners Have students in small groups do a binomial experiment by tossing a ball into a wastebasket about 20 times to establish the probability of scoring a goal. Then have them find the probability that they will score exactly 4 goals in 8 tries.

Photodisc/Getty Images

3 Practice

Formative Assessment

Use Exercises 1–6 to check for understanding.

Use the chart at the bottom of this page to customize assignments for your students.

Additional Answers

1. This experiment cannot be reduced to a binomial experiment because there are more than two possible outcomes.

2. This experiment cannot be reduced to a binomial experiment because there are more than two possible outcomes.

3. This experiment can be reduced to a binomial experiment. Success is yes, failure is no, a trial is asking a student, and the random variable is the number of yeses; $n = 30$, $p = 0.72$, $q = 0.28$.

4. Sample answer:

Step 1 A trial is drawing a card from a deck. The simulation will consist of 20 trials.

Step 2 A success is drawing an ace or a king. The probability of success is $\frac{2}{13}$ and the probability of failure is $\frac{11}{13}$.

Step 3 The random variable X represents the number of aces or kings drawn in 20 trials.

Step 4 Use a random number generator. Let 0–1 represent drawing an ace or a king. Let 2–12 represent drawing all other cards. Make a frequency table and record the results as you run the generator.

Outcome	Tally	Frequency
Ace or King	ⅢⅠ	6
Other Outcomes	Ⅲ Ⅲ Ⅲ	14

The experimental probability is $\frac{6}{20}$ or 30%. This is greater than the theoretical probability of $\frac{2}{13}$ or about 15.4%.

Check Your Understanding

= Step-by-Step Solutions begin on page R14.

1–4. See margin.

Example 1 Determine whether each experiment is a binomial experiment or can be reduced to a binomial experiment. If so, describe a trial, determine the random variable, and state *n*, *p*, and *q*.

1. A study finds that 58% of people have pets. You ask 100 people how many pets they have.

2. You roll a die 15 times and find the sum of all of the rolls.

3. A poll found that 72% of students plan on going to the homecoming dance. You ask 30 students if they are going to the homecoming dance.

Example 2
4. Conduct a binomial experiment to determine the probability of drawing an ace or a king from a deck of cards. Then compare the experimental and theoretical probabilities of the experiment.

Example 3
⑤ **GAMES** Aiden has earned five spins of the wheel on the right. He will receive a prize each time the spinner lands on WIN. What is the probability that he receives three prizes? **D**

A 4.2%   C 7.1%
B 5.8%   D 8.8%

Example 4
6. **CCSS PRECISION** A poll at Steve's high school was taken to see if students are in favor of spending class money to expand the junior-senior parking lot. Steve surveyed 6 random students from the population.

6a. See margin.

Expand the Parking Lot	
favor	85%
oppose	15%

a. Determine the probabilities associated with the number of students that Steve asked who are in favor of expanding the parking lot by calculating the probability distribution.

b. What is the probability that no more than 2 people are in favor of expanding the parking lot? **0.00589 or 0.589%**

c. How many students should Steve expect to find who are in favor of expanding the parking lot? **5**

Practice and Problem Solving

Extra Practice is on page R11.

Example 1 Determine whether each experiment is a binomial experiment or can be reduced to a binomial experiment. If so, describe a trial, determine the random variable, and state *n*, *p*, and *q*.

**7–9.
See margin.**

7. There is a 35% chance that it rains each day in a given month. You record the number of days that it rains for that month.

8. A survey found that on a scale of 1 to 10, a movie received a 7.8 rating. A movie theater employee asks 200 patrons to rate the movie on a scale of 1 to 10.

9. A ball is hidden under one of the hats shown below. A hat is chosen, one at a time, until the ball is found.

Example 2
10. **DICE** Conduct a binomial experiment to determine the probability of rolling a 7 with two dice. Then compare the experimental and theoretical probabilities of the experiment.

10–11. See Chapter 11 Answer Appendix.
11. **MARBLES** Conduct a binomial experiment to determine the probability of pulling a red marble from the bag. Then compare the experimental and theoretical probabilities of the experiment.

Differentiated Homework Options

Level	Assignment	Two-Day Option	
AL Basic	7–23, 43–71	7–23 odd, 47–50	8–22 even, 43–46, 51–71
OL Core	7–41 odd, 43–71	7–23, 47–50	24–41, 43–46, 51–71
BL Advanced	24–70, (optional: 71)		

12. SPINNER Conduct a binomial experiment to determine the probability of the spinner stopping on an even number. Then compare the experimental and theoretical probabilities of the experiment.

12–13. See Chapter 11 Answer Appendix.

13. CARDS Conduct a binomial experiment to determine the probability of drawing a face card out of a standard deck of cards. Then compare the experimental and theoretical probabilities of the experiment.

Example 3

14. PERSONAL MEDIA PLAYERS According to a recent survey, 85% of high school students own a personal media player. What is the probability that 6 out of 10 random high school students own a personal media player? **0.04 or 4%**

15. CARS According to a recent survey, 92% of high school seniors drive their own car. What is the probability that 10 out of 12 random high school students drive their own car? **0.183 or 18.3%**

16. SENIOR PROM According to a recent survey, 25% of high school upperclassmen think that the junior-senior prom is the most important event of the school year. What is the probability that 3 out of 15 random high school upperclassmen think this way? **0.225 or 22.5%**

17. FOOTBALL A certain football team has won 75.7% of their games. Find the probability that they win 7 of their next 12 games. **0.096 or 9.6%**

18. GARDENING Peter is planting 24 irises in his front yard. The flowers he bought were a combination of two varieties, blue and white. The flowers are not blooming yet, but Peter knows that the probability of having a blue flower is 75%. What is the probability that 20 of the flowers will be blue? **0.132 or 13.2%**

(19) FOOTBALL A field goal kicker is accurate 75% of the time from within 35 yards. What is the probability that he makes exactly 7 of his next 10 kicks from within 35 yards? **0.25 or 25%**

Range (yd)	Accuracy (%)
0–35	75
35–45	62
45+	20

20. BABIES Mr. and Mrs. Davis are planning to have 3 children. The probability of each child being a boy is 50%. What is the probability that they will have 2 boys? **0.375 or 37.5%**

Example 4

21. CCSS SENSE-MAKING According to a recent survey, 52% of high school students own a laptop. Ten random students are chosen.

a. Determine the probabilities associated with the number of students that own a laptop by calculating the probability distribution. **See margin.**

b. What is the probability that at least 8 of the 10 students own a laptop? **about 7%**

c. How many students should you expect to own a laptop? **5**

22. ATHLETICS A survey was taken to see the percent of students that participate in sports for their school. Six random students are chosen.

Student Athletics	
0 sports	20%
1 sport	55%
2 sports	20%
3+ sports	5%

a. Determine the probabilities associated with the number of students playing in at least one sport by calculating the probability distribution. **See margin.**

b. What is the probability that no more than 2 of the students participated in a sport? **0.01696 or 1.696%**

c. How many students should you expect to have participated in at least one sport? **5**

22a. 0 play at least 1 sport, 0.00006 or 0.006%; 1 plays at least 1 sport, 0.00154 or 0.154%; 2 play at least 1 sport, 0.01536 or 1.536%; 3 play at least 1 sport, 0.08192 or 8.192%; 4 play at least 1 sport, 0.24576 or 24.576%; 5 play at least 1 sport, 0.39322 or 39.322%; 6 play at least 1 sport, 0.26214 or 26.214%

CCSS Teaching the Mathematical Practices

Sense-Making Mathematically proficient students start by explaining to themselves the meaning of a problem and looking for entry points to its solution. They analyze givens, constraints, relationships, and goals. They check their answers to problems using a different method, and they continually ask themselves, "Does this make sense?"

Additional Answers

6a. 0 in favor, 0.00001 or 0.001%;
1 in favor, 0.00039 or 0.039%;
2 in favor, 0.00549 or 0.549%;
3 in favor, 0.04145 or 4.145%;
4 in favor, 0.17618 or 17.618%;
5 in favor, 0.39933 or 39.933%;
6 in favor, 0.37715 or 37.715%

7. This experiment can be reduced to a binomial experiment. Success is a day that it rains, failure is a day it does not rain, a trial is a day, and the random variable X is the number of days it rains; $n =$ the number of days in the month, $p = 0.35$, $q = 0.65$.

8. This experiment cannot be reduced to a binomial experiment because there are more than two possible outcomes.

9. This experiment cannot be reduced to a binomial experiment because the events are not independent. The probability of choosing the hat that covers the ball changes after each selection.

21a. 0 own a laptop, 0.0006 or 0.06%;
1 owns a laptop, 0.007 or 0.7%,
2 own a laptop, 0.0343 or 3.43%;
3 own a laptop, 0.0991 or 9.91%;
4 own a laptop, 0.1878 or 18.78%; 5 own a laptop, 0.2441 or 24.41%; 6 own a laptop, 0.2204 or 22.04%; 7 own a laptop, 0.1364 or 13.64%; 8 own a laptop, 0.0554 or 5.54%; 9 own a laptop, 0.0133 or 1.33%; 10 own a laptop, 0.0014 or 0.14%

Study Guide and Intervention
AL OL ELL

Practice
AL OL BL ELL

Word Problem Practice
AL OL BL ELL

23. CCSS MODELING An online poll showed that 57% of adults still own vinyl records. Moe surveyed 8 random adults from the population.

   a. Determine the probabilities associated with the number of adults that still own vinyl records by calculating the probability distribution. **See margin.**

   b. What is the probability that no less than 6 of the people surveyed still own vinyl records? **0.256 or 25.6%**

   c. How many people should Moe expect to still own vinyl records? **5**

A binomial distribution has a 60% rate of success. There are 18 trials.

24. What is the probability that there will be at least 12 successes? **0.374 or 37.4%**

25. What is the probability that there will be 12 failures? **0.015 or 1.5%**

26. What is the expected number of successes? **11**

27. DECISION MAKING Six roommates randomly select someone to wash the dishes each day.

   a. What is the probability that the same person has to wash the dishes 3 times in a given week? **about 7.8%**   **b. Sample answer: They can roll a six-sided die.**

   b. What method can the roommates use to select who washes the dishes each day?

28. DECISION MAKING A committee of five people randomly selects someone to take the notes of each meeting. **b. Sample answer: They can roll a six-sided die. If a six is rolled, they roll again.**

   a. What is the probability that a person takes notes less than twice in 10 meetings? **about 37.6%**

   b. What method can the committee use to select the notetaker each meeting?

   c. If the method described in part **b** results in the same person being notetaker for nine straight meetings, would this result cause you to question the method? **See margin.**

Each binomial distribution has n trials and p probability of success. Determine the most likely number of successes.

29. $n = 8, p = 0.6$ **5**   **30.** $n = 10, p = 0.4$ **4**   **31.** $n = 6, p = 0.8$ **5**

32. $n = 12, p = 0.55$ **7**   **33.** $n = 9, p = 0.75$ **7**   **34.** $n = 11, p = 0.35$ **4**

35. SWEEPSTAKES A beverage company is having a sweepstakes. The probabilities of winning selected prizes are shown at the right. If Ernesto purchases 8 beverages, what is the probability that he wins at least one prize? **0.603 or 60.3%**

Odds of Winning	
beverage	1 in 10
CD	1 in 200
hat	1 in 250
MP3 player	1 in 20,000
car	1 in 25,000,000

Each binomial distribution has n trials and p probability of success. Determine the probability of s successes.

36. $n = 8, p = 0.3, s \geq 2$ **0.745 or 74.5%**   **37.** $n = 10, p = 0.2, s > 2$ **0.322 or 32.2%**   **38.** $n = 6, p = 0.6, s \leq 4$ **0.767 or 76.7%**

39. $n = 9, p = 0.25, s \leq 5$ **0.99 or 99%**   **40.** $n = 10, p = 0.75, s \geq 8$ **0.526 or 52.6%**   **41.** $n = 12, p = 0.1, s < 3$ **0.889 or 88.9%**

H.O.T. Problems   Use Higher-Order Thinking Skills

42. CHALLENGE A poll of students determined that 88% wanted to go to college. Eight random students are chosen. The probability that at least x students want to go to college is about 0.752 or 75.2%. Solve for x. **7**

43–46. See margin.

43. WRITING IN MATH What should you consider when using a binomial distribution to make a decision?

44. OPEN ENDED Describe a real-world setting within your school or community activities that seems to fit a binomial distribution. Identify the key components of your setting that connect to binomial distributions.

45. WRITING IN MATH Describe how binomial distributions are connected to Pascal's triangle.

46. WRITING IN MATH Explain the relationship between a binomial experiment and a binomial distribution.

Enrichment
OL BL

Additional Answer

23a. 0 own vinyl records, 0.001 or 0.1%;
1 owns vinyl records, 0.012 or 1.2%;
2 own vinyl records, 0.058 or 5.8%;
3 own vinyl records, 0.152 or 15.2%;
4 own vinyl records, 0.253 or 25.3%;
5 own vinyl records, 0.268 or 26.8%;
6 own vinyl records, 0.178 or 17.8%;
7 own vinyl records, 0.067 or 6.7%;
8 own vinyl records, 0.011 or 1.1%

47. EXTENDED RESPONSE Carly is taking a 10 question multiple-choice test, in which each question has four choices. If she guesses on each question, what is the probability that she will get

   a. 7 questions correct? **0.003 or 0.3%**

   b. 9 questions correct? **0.00003 or 0.003%**

   c. 0 questions correct? **0.056 or 5.6%**

   d. 3 questions correct? **0.25 or 25%**

48. What is the maximum point of the graph of the equation $y = -2x^2 + 16x + 5$? **C**

   A $(-4, -59)$ **C** $(4, 37)$

   B $(-4, -91)$ **D** $(4, 101)$

49. GEOMETRY On a number line, point X has coordinate -8 and point Y has coordinate 4. Point P is $\frac{2}{3}$ of the way from X to Y. What is the coordinate of P? **H**

   F -4 **H** 0

   G -2 **J** 2

50. SAT/ACT The cost of 4 CDs is d dollars. At this rate, what is the cost, in dollars, of 36 CDs? **A**

   A $9d$ **D** $\frac{d}{36}$

   B $144d$ **E** $\frac{36}{d}$

   C $\frac{9d}{4}$

Spiral Review

Identify the random variable in each distribution, and classify it as *discrete* or *continuous*. Explain your reasoning. (Lesson 11-3) **51–54. See Chapter 11 Answer Appendix.**

51. the number of customers at an amusement park

52. the running time of a movie

53. the number of hot dogs sold at a sporting event

54. the distance between two cities

55. FINANCIAL LITERACY The prices of entrees offered by a restaurant are shown. (Lesson 11-2) **See Chapter 11 Answer Appendix.**

Prices (dollars)			
11.25	14.75	9.00	17.25
19.75	9.75	20.25	15.50
16.50	21.50	10.25	22.75
12.75	18.50	23.00	13.50

   a. Use a graphing calculator to create a box-and-whisker plot. Then describe the shape of the distribution.

   b. Describe the center and spread of the data using either the mean and standard deviation or the five-number summary. Justify your choice.

Find the missing value for each arithmetic sequence. (Lesson 10-1)

56. $a_5 = 12, a_{16} = 133, d = ?$ **11**

57. $a_9 = -34, a_{22} = 44, d = ?$ **6**

58. $a_4 = 18, a_n = 95, d = 7, n = ?$ **15**

59. $a_8 = ?, a_{19} = 31, d = 8$ **−57**

60. $a_6 = ?, a_{20} = 64, d = 7$ **−34**

61. $a_7 = -28, a_n = 76, d = 8, n = ?$ **20**

62. ASTRONOMY The table at the right shows the closest and farthest distances of Venus and Jupiter from the Sun in millions of miles. (Lesson 9-4)

Planet	Closest	Farthest
Venus	66.8	67.7
Jupiter	460.1	507.4

   a. Write an equation for the orbit of each planet. Assume that the center of the orbit is the origin and the center of the Sun is a focus that lies on the x-axis.

   b. Which planet has an orbit that is closer to a circle? **Venus**

62a. Venus: $\dfrac{x^2}{4522.5625} + \dfrac{y^2}{4522.36} = 1$;

Jupiter: $\dfrac{x^2}{234{,}014.06} + \dfrac{y^2}{233{,}454.74} = 1$

Write an equivalent exponential or logarithmic function. (Lesson 7-7)

63. $e^{-x} = 5$ $-x = \ln 5$

64. $e^2 = 6x$ $2 = \ln 6x$

65. $\ln e = 1$ $e^1 = e$

66. $\ln 5.2 = x$ $e^x = 5.2$

67. $e^{x+1} = 9$ $x + 1 = \ln 9$

68. $e^{-1} = x^2$ $-1 = \ln x^2$

69. $\ln \frac{7}{3} = 2x$ $e^{2x} = \frac{7}{3}$

70. $\ln e^x = 3$ $e^3 = e^x$

Skills Review

71. MUSIC Tina owns 11 pop, 6 country, 16 rock, and 7 rap CDs. Find each probability if she randomly selects 4 CDs.

   a. $P(2 \text{ rock})$ **0.36**

   b. $P(1 \text{ rap})$ **0.42**

   c. $P(1 \text{ rock and 2 country})$ **0.05**

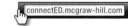

DifferentiatedInstruction OL BL

Extension Have students look at a histogram of a binomial distribution created on a graphing calculator for $n = 20$ and $p = 0.5$. Point out that they can use the histogram to answer a question like "What is the probability that x is at least 12?" by adding the heights of all the bars beginning with $x = 12$. Then have them find the probability that x is at least 12. 0.252 or 25.2%

Name the Math Have each student use his or her own family or the family of a friend (for example, 2 boys and a girl), and find the probabilities for that particular group of siblings. Then have students explain the steps they used.

Additional Answers

28c. Sample answer: While it is possible for the same person to be chosen nine straight times, it is rather unlikely. The fairness of the die should be questioned.

43. Sample answer: You should consider the type of situation for which the binomial distribution is being used. For example, if a binomial distribution is being used to predict outcomes regarding an athletic event, the probabilities of success and failure could change due to other variables such as weather conditions or player health. So, binomial distributions should be used cautiously when making decisions involving events that are not completely random.

44. Sample answer: During May and June, lunches are held outside, weather permitting. Also during this time, there has historically been a 15% chance of rain. So, to determine the probability of not having rain for at least 24 of these 28 days, the binomial distribution would use $p = 0.85$, $q = 0.15$, and $n = 28$.

45. Sample answer: A full binomial distribution can be determined by expanding the binomial, which itself utilizes Pascal's triangle.

46. Sample answer: A binomial distribution shows the probabilities of the outcomes of a binomial experiment.

The Normal Distribution

1 Focus

VerticalAlignment

Before Lesson 11-5 Analyze probability distributions of data.

Lesson 11-5 Use the Empirical Rule to analyze normally distributed variables.

After Lesson 11-5 Understand how to apply the standard normal distribution.

2 Teach

Scaffolding Questions

Have students read the **Why?** section of the lesson.

Ask:

- What dates are one standard deviation from the mean? April 11 and May 1

- What do you think the likelihood is of a Swiss cherry tree flowering before April 1? about 2.5%

:Then	:Now	:Why?
● You constructed and analyzed discrete probability distributions.	**1** Use the Empirical Rule to analyze normally distributed variables. **2** Apply the standard normal distribution and z-values.	● Extensive observations of Swiss cherry trees found that the mean flowering date is April 21 with a standard deviation of about 10 days. Therefore, 95% of the time, a Swiss cherry tree will have a flowering date between April 1 and May 3.

 NewVocabulary
normal distribution
Empirical Rule
z-value
standard normal distribution

Common Core State Standards

Content Standards
S.ID.4 Use the mean and standard deviation of a data set to fit it to a normal distribution and to estimate population percentages. Recognize that there are data sets for which such a procedure is not appropriate. Use calculators, spreadsheets, and tables to estimate areas under the normal curve.

Mathematical Practices
6 Attend to precision.
8 Look for and express regularity in repeated reasoning.

1 The Normal Distribution Distributions of mileages of different sample sizes of cars are shown below. As the sample size increases, the distributions become more and more symmetrical and resemble the curve at the right, due to the Law of Large Numbers.

$n = 100$

$n = 10,000$

$n =$ population

The curve at the right is a **normal distribution**, a continuous, symmetric, bell-shaped distribution of a random variable. It is the most common *continuous probability distribution*. The characteristics of the normal distribution are as follows.

KeyConcept The Normal Distribution

- The graph of the curve is continuous, bell-shaped, and symmetric with respect to the mean.
- The mean, median, and mode are equal and located at the center.
- The curve approaches, but never touches, the x-axis.
- The total area under the curve is equal to 1 or 100%.

The area under the normal curve represents the amount of data within a certain interval or the probability that a random data value falls within that interval. The **Empirical Rule** can be used to determine the area under the normal curve at specific intervals.

KeyConcept The Empirical Rule

In a normal distribution with mean μ and standard deviation σ,

- approximately 68% of the data fall within 1σ of the mean,
- approximately 95% of the data fall within 2σ of the mean, and
- approximately 99.7% of the data fall within 3σ of the mean.

Frank Krahmer/Photographer's Choice RF/Getty Images

Lesson 11-5 Resources

Resource	Approaching Level **AL**	On Level **OL**	Beyond Level **BL**	English Learners **ELL**
Teacher Edition	■ Differentiated Instruction, p. 762	■ Differentiated Instruction, p. 762	■ Differentiated Instruction, p. 766	
Chapter Resource Masters	■ Study Guide and Intervention, pp. 30–31 ■ Skills Practice, p. 32 ■ Practice, p. 33 ■ Word Problem Practice, p. 34	■ Study Guide and Intervention, pp. 30–31 ■ Skills Practice, p. 32 ■ Practice, p. 33 ■ Word Problem Practice, p. 34 ■ Enrichment, p. 35	■ Practice, p. 33 ■ Word Problem Practice, p. 34 ■ Enrichment, p. 35	■ Study Guide and Intervention, pp. 30–31 ■ Skills Practice, p. 32 ■ Practice, p. 33 ■ Word Problem Practice, p. 34
Other	■ 5-Minute Check 11-5 ■ Study Notebook	■ 5-Minute Check 11-5 ■ Study Notebook	■ 5-Minute Check 11-5 ■ Study Notebook	■ 5-Minute Check 11-5 ■ Study Notebook

StudyTip

Normal Distributions In all of these cases, the number of data values must be large for the distribution to be approximately normal.

Example 1 Use the Empirical Rule to Analyze Data

A normal distribution has a mean of 21 and a standard deviation of 4.

a. Find the range of values that represent the middle 68% of the distribution.

The middle 68% of data in a normal distribution is the range from $\mu - \sigma$ to $\mu + \sigma$. Therefore, the range of values in the middle 68% is $17 < X < 25$.

b. What percent of the data will be greater than 29?

29 is 2σ more than μ. 95% of the data fall between $\mu - 2\sigma$ and $\mu + 2\sigma$, so the remaining data values represented by the two tails covers 5% of the distribution. We are only concerned with the upper tail, so 2.5% of the data will be greater than 29.

▶ **Guided**Practice

1. A normal distribution has a mean of 8.2 and a standard deviation of 1.3.
   A. Find the range of values that represent the middle 95% of the distribution. $5.6 < X < 10.8$
   B. What percent of the data will be less than 4.3? **0.15%**

Real-World Example 2 Use the Empirical Rule to Analyze a Distribution

HEIGHTS The heights of 1800 adults are normally distributed with a mean of 70 inches and a standard deviation of 2 inches.

a. About how many adults are between 66 and 74 inches?

66 and 74 are 2σ away from the mean. Therefore, about 95% of the data are between 66 and 74.

Since $1800 \times 95\% = 1710$, we know that about 1710 of the adults are between 66 and 74 inches tall.

b. What is the probability that a random adult is more than 72 inches tall?

From the curve, values greater than 72 are more than 1σ from the mean. 13.5% are between 1σ and 2σ, 2.35% are between 2σ and 3σ, and 0.15% are greater than 3σ.

So, the probability that an adult selected at random has a height greater than 72 inches is $13.5 + 2.35 + 0.15$ or 16%.

▶ **Guided**Practice

2. **NETWORKING SITES** The number of friends per member in a sample of 820 members is normally distributed with a mean of 38 and a standard deviation of 12.
   A. About how many members have between 26 and 50 friends? **558**
   B. What is the probability that a random member will have more than 14 friends? **97.5%**

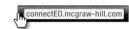

Real-World Link

While the average adult American male is 5 feet 10 inches, the average height of adult males in the Netherlands is the highest worldwide, at almost 6 feet 1 inch.

Source: Eurostats Statistical Yearbook

Image Source/Getty Images

connectED.mcgraw-hill.com **761**

---

1 The Normal Distribution

Example 1 shows how to use the Empirical Rule to analyze data.
Example 2 shows how to use the Empirical Rule to analyze a distribution.

Formative Assessment

Use the Guided Practice exercises after each example to determine students' understanding of concepts.

Additional Examples

1 A normal distribution has a mean of 45.1 and a standard deviation of 9.6.

  a. Find the values that represent the middle 99.7% of the distribution. $16.3 < X < 73.9$

  b. What percent of the data will be greater than 54.7? 16%

2 **PACKAGING** Students counted the number of candies in 100 small packages. They found that the number of candies per package was normally distributed with a mean of 23 candies per package and a standard deviation of 1 piece of candy.

  a. About how many packages have between 22 and 24 candies? about 68 packages

  b. What is the probability that a package selected at random has more than 25 candies? about 2.5%

Teach with Tech

Instant Messaging Have students work in pairs and send each other questions about a normal distribution (for example: "What percent of the values are within two standard deviations above the mean?") Students should reply with the answers and check each others' work.

Focus on Mathematical Content

Normal Distributions The graphs of all normally distributed variables have essentially the same shape. With appropriate labeling of the mean and the points that are one standard deviation from the mean, the same normal curve can represent any normal distribution.

2 The Standard Normal Distribution

Example 3 shows how to use *z*-values to locate position. **Example 4** shows how to use *z*-values and the standard normal distribution to find probabilities.

Additional Example

3 Find σ if $X = 28.3$, $\mu = 24.6$, and $z = 0.63$. Indicate the position of X in the distribution. **5.87; 0.63 standard deviations greater than the mean**

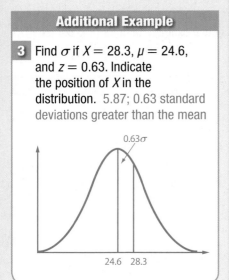

24.6 28.3

Tips for New Teachers

Relative Position Like percentiles, *z*-values can be used to compare the relative positions of two values in two different sets of data.

---

2 Standard Normal Distribution

The Empirical Rule is only useful for evaluating specific values, such as $\mu + \sigma$. Once the data set is *standardized*, however, any data value can be evaluated. Data are standardized by converting them to *z*-values, also known as *z*-scores. The **z-value** represents the number of standard deviations that a given data value is from the mean. Therefore, *z*-values can be used to determine the position of any data value within a set of data.

KeyConcept Formula for *z*-Values

The *z*-value for a data value X in a set of normally distributed data is given by $z = \dfrac{X - \mu}{\sigma}$, where μ is the mean and σ is the standard deviation.

> **StudyTip**
> **Symmetry** The normal distribution is symmetrical, so when you are asked for the middle or outside set of data, the *z*-values will be opposites.

Example 3 Use *z*-Values to Locate Position

Find z if $X = 18$, $\mu = 22$, and $\sigma = 3.1$. Indicate the position of X in the distribution.

$$z = \frac{X - \mu}{\sigma} \qquad \text{Formula for } z\text{-values}$$

$$= \frac{18 - 22}{3.1} \qquad X = 18, \mu = 22, \sigma = 3.1$$

$$\approx -1.29 \qquad \text{Simplify.}$$

The *z*-value that corresponds to $X = 18$ is approximately -1.29. Therefore, 18 is about 1.29 standard deviations less than the mean of the distribution.

> **GuidedPractice**
>
> **3.** Find X if $\mu = 39$, $\sigma = 8.2$, and $z = 0.73$. Indicate the position of X in the distribution. **See margin.**

Any combination of mean and standard deviation is possible for a normally distributed set of data. As a result, there are infinitely many normal probability distributions. This makes comparing two individual distributions difficult. Different distributions *can* be compared, however, once they are standardized using *z*-values. The **standard normal distribution** is a normal distribution with a mean of 0 and a standard deviation of 1.

> **StudyTip**
> **Standard Normal Distribution** The standard normal distribution is the set of all *z*-values.

KeyConcept Characteristics of the Standard Normal Distribution

- The total area under the curve is equal to 1 or 100%.
- Almost all of the area is between $z = -3$ and $z = 3$.
- The distribution is symmetric.
- The mean is 0, and the standard deviation is 1.

The standard normal distribution allows us to assign actual areas to the intervals created by *z*-values. The area under the normal curve corresponds to the proportion of data values in an interval as well as the probability of a random data value falling within the interval. For example, the area between $z = 0$ and $z = 1$ is 0.34. Therefore, the probability of a *z*-value being in this interval is 34%.

762 | Lesson 11-5 | The Normal Distribution

DifferentiatedInstruction AL OL

If students need an aid in drawing a normal curve,

Then it may be useful to know that the concave side of the curve switches from facing downward to facing upward at points that are one standard deviation from the mean. Drawing a normal curve for each problem can help students with their estimates.

Brand X Pictures/Jupiterimages

Real-World Example 4 Find Probabilities

VIDEOS The number of videos uploaded daily to a video sharing site is normally distributed with $\mu = 181,099$ videos and $\sigma = 35,644$ videos. Find each probability. Then use a graphing calculator to sketch the corresponding area under the curve.

a. $P(180,000 < X < 200,000)$

The question is asking for the percent of days when between 180,000 and 200,000 videos are uploaded. First, find the corresponding z-values for $X = 180,000$ and $X = 200,000$.

$z = \dfrac{X - \mu}{\sigma}$ Formula for z-values

$= \dfrac{180,000 - 181,099}{35,644}$ or about -0.03 $X = 180,000$, $\mu = 181,099$, and $\sigma = 35,644$

Use 200,000 to find the other z-value.

$z = \dfrac{X - \mu}{\sigma}$ Formula for z-values

$= \dfrac{200,000 - 181,099}{35,644}$ or about 0.53 $X = 200,000$, $\mu = 181,099$, and $\sigma = 35,644$

The range of z-values that corresponds to $180,000 < X < 200,000$ is $-0.03 < z < 0.53$. Find the area under the normal curve within this interval.

You can use a graphing calculator to display the area that corresponds to any z-value by selecting 2nd [DISTR]. Then, under the **DRAW** menu, select **ShadeNorm(lower z value, upper z value)**. The area between $z = -0.03$ and $z = 0.53$ is about 0.21 as shown in the graph.

[−4, 4] scl: 1 by [0, 0.5] scl: 0.125

Therefore, about 21% of the time, there will be between 180,000 and 200,000 video uploads on a given day.

b. $P(X > 250,000)$

$z = \dfrac{X - \mu}{\sigma}$ Formula for z-values

$= \dfrac{250,000 - 181,099}{35,644}$ or about 1.93 $X = 250,000$, $\mu = 181,099$, and $\sigma = 35,644$

Using a graphing calculator, you can find the area between $z = 1.93$ and $z = 4$ to be about 0.027.

Therefore, the probability that more than 250,000 videos will be uploaded is about 2.7%.

[−4, 4] scl: 1 by [0, 0.5] scl: 0.125

Guided Practice

4. TIRES The life spans of a certain tread of tire are normally distributed with $\mu = 31,066$ miles and $\sigma = 1644$ miles. Find each probability. Then use a graphing calculator to sketch the corresponding area under the curve.

A. $P(30,000 < X < 32,000)$ **B.** $P(X > 35,000)$ **4A–B.** See margin.

Another method for calculating the area between two z-values is 2nd [DISTR] **normalcdf**(*lower z value, upper z value*).

Real-WorldLink

Video Uploading According to a recent study, 52% of people who said they upload videos to the Web do it through sites such as Facebook and MySpace. The rest use video-sharing sites like YouTube and Google Video.

Source: Pew Internet and American Life Project

StudyTip

Range of z-values The majority of data values are within $\pm 4\sigma$ of the mean, so setting the maximum value of z equal to 4 is sufficient in part **b**. Use the window [−4, 4] by [0, 0.5] when using **ShadeNorm**.

Additional Example

4 **HEALTH** The cholesterol levels for adult males of a specific racial group are normally distributed with a mean of 158.3 and a standard deviation of 6.6. Find each probability. Then use a graphing calculator to sketch the corresponding area under the curve.

a. $P(X > 150)$ 0.90

b. $P(145 < X < 165)$ 0.82

a.

[−4, 4] scl: 1 by [0, 0.5] scl: 0.125

b.

[−4, 4] scl: 1 by [0, 0.5] scl: 0.125

Additional Answers (Guided Practice)

4A. 0.46

[−4, 4] scl: 1 by [0, 0.5] scl: 0.125

4B. 0.008

[−4, 4] scl: 1 by [0, 0.5] scl: 0.125

Additional Answer (Guided Practice)

3. $X \approx 45.0$

0.73σ

39   45.0

Tips for New Teachers

ClrDraw Have students select 2nd [DRAW] 1 to clear any previous drawings.

Tips for New Teachers

Sign Usage In a continuous distribution, there is no difference between $P(X \geq c)$ and $P(X > c)$ because the probability that x is equal to c is zero.

3 Practice

Formative Assessment
Use Exercises 1–8 to check for understanding.

Use the chart on the bottom of this page to customize assignments for your students.

 Teaching the Mathematical Practices

Tools When making mathematical models, mathematically proficient students know that technology can enable them to visualize the results of varying assumptions, explore consequences, and compare predictions with data. They are able to use technological tools to explore and deepen their understanding of concepts.

Additional Answers

4. −0.70; 0.70 standard deviations less than the mean

5. 2.05; 2.05 standard deviations greater than the mean

6. 59.8; 1.38 standard deviations less than the mean

7. 3.08; 2.40 standard deviations less than the mean

8a.

[−4, 4] scl: 1 by [0, .5] scl: 0.125

8b.

[−4, 4] scl: 1 by [0, .5] scl: 0.125

12. 2.5; 2.5 standard deviations greater than the mean

13. −1.33; 1.33 standard deviations less than the mean

14. 51.8; 0.92 standard deviations less than the mean

15. 177.7; 1.73 standard deviations greater than the mean

Check Your Understanding

= Step-by-Step Solutions begin on page R14.

Example 1

A normal distribution has a mean of 416 and a standard deviation of 55.

1. Find the range of values that represent the middle 99.7% of the distribution. $251 < X < 581$

2. What percent of the data will be less than 361? **16%**

Example 2

3. **CCSS TOOLS** The number of texts sent per day by a sample of 811 teens is normally distributed with a mean of 38 and a standard deviation of 7.

  a. About how many teens sent between 24 and 38 texts? **about 386 teens**

  b. What is the probability that a teen selected at random sent less than 45 texts? **84%**

Example 3

Find the missing variable. Indicate the position of X in the distribution. **4–7. See margin.**

4. z if $\mu = 89$, $X = 81$, and $\sigma = 11.5$

5. z if $\mu = 13.3$, $X = 17.2$, and $\sigma = 1.9$

6. X if $z = -1.38$, $\mu = 68.9$, and $\sigma = 6.6$

7. σ if $\mu = 21.1$, $X = 13.7$, and $z = -2.40$

Example 4

8. **CONCERTS** The number of concerts attended per year by a sample of 925 teens is normally distributed with a mean of 1.8 and a standard deviation of 0.5. Find each probability. Then use a graphing calculator to sketch the area under each curve.

  a. $P(X < 2)$ **65.5%**

  b. $P(1 < X < 3)$ **93.7%**

  a–b. See margin for graphs.

Practice and Problem Solving

Extra Practice is on page R11.

Example 1

A normal distribution has a mean of 29.3 and a standard deviation of 6.7.

9. Find the range of values that represent the outside 5% of the distribution. $X < 15.9$ or $X > 42.7$

10. What percent of the data will be between 22.6 and 42.7? **81.5%**

Example 2

11. **GYMS** The number of visits to a gym per year by a sample of 522 members is normally distributed with a mean of 88 and a standard deviation of 19.

  a. About how many members went to the gym at least 50 times? **about 509 members**

  b. What is the probability that a member selected at random went to the gym more than 145 times? **0.15%**

Example 3

Find the missing variable. Indicate the position of X in the distribution. **12–15. See margin.**

12. z if $\mu = 3.3$, $X = 3.8$, and $\sigma = 0.2$

13. z if $\mu = 19.9$, $X = 18.7$, and $\sigma = 0.9$

14. μ if $z = -0.92$, $X = 44.2$, and $\sigma = 8.3$

15. X if $\mu = 138.8$, $\sigma = 22.5$, and $z = 1.73$

Example 4

16. **VENDING** A vending machine dispenses about 8.2 ounces of coffee. The amount varies and is normally distributed with a standard deviation of 0.3 ounce. Find each probability. Then use a graphing calculator to sketch the corresponding area under the curve.

  a. $P(X < 8)$ **25.2%**

  b. $P(X > 7.5)$ **99.0%**

  16a–b. See Chapter 11 Answer Appendix for graphs.

17 **CAR BATTERIES** The useful life of a certain car battery is normally distributed with a mean of 113,627 miles and a standard deviation of 14,266 miles. The company makes 20,000 batteries a month.

  a. About how many batteries will last between 90,000 and 110,000 miles? **about 7000 batteries**

  b. About how many batteries will last more than 125,000 miles? **about 4200 batteries**

  c. What is the probability that if you buy a car battery at random, it will last less than 100,000 miles? **17.0%**

18. **FOOD** The shelf life of a particular snack chip is normally distributed with a mean of 173.3 days and a standard deviation of 23.6 days.

  a. About what percent of the product lasts between 150 and 200 days? **71.0%**

  b. About what percent of the product lasts more than 225 days? **1.4%**

  c. What range of values represents the outside 5% of the distribution? $X > 220.5$ or $X < 126.1$

 764 | Lesson 11-5 | The Normal Distribution

Differentiated Homework Options

Level	Assignment	Two-Day Option	
AL Basic	9–16, 21, 23–38	9–15 odd, 27, 29	10–16 even, 21, 23–26, 31–38
OL Core	9–15 odd, 17–21, 23–38	9–16, 27, 29	17–21, 23–38
BL Advanced	17–37 (optional: 38)		

19 FINANCIAL LITERACY The insurance industry uses various factors including age, type of car driven, and driving record to determine an individual's insurance rate. Suppose insurance rates for a sample population are normally distributed.

a. If the mean annual cost per person is $829 and the standard deviation is $115, what is the range of rates you would expect the middle 68% of the population to pay annually? **between $714 and $944**

b. If 900 people were sampled, how many would you expect to pay more than $1000 annually? **62**

c. Where on the distribution would you expect a person with several traffic citations to lie? Explain your reasoning. **c–d. See Chapter 11 Answer Appendix.**

d. How do you think auto insurance companies use each factor to calculate an individual's insurance rate?

20. STANDARDIZED TESTS Nikki took three national standardized tests and scored an 86 on all three. The table shows the mean and standard deviation of each test.

	Math	Science	Social Studies
μ	76	81	72
σ	9.7	6.2	11.6

a. Calculate the z-values that correspond to her score on each test.

b. What is the probability of a student scoring an 86 or *lower* on each test?

c. On which test was Nikki's standardized score the highest? Explain your reasoning.
20a–c. See Chapter 11 Answer Appendix.

H.O.T. Problems — Use Higher-Order Thinking Skills

21. **CCSS CRITIQUE** A set of normally distributed tree diameters have mean 11.5 centimeters, standard deviation 2.5, and range from 3.6 to 19.8. Monica and Hiroko are to find the range that represents the middle 68% of the data. Is either of them correct? Explain.
Sample answer: Hiroko; Monica's solution would work with a uniform distribution.

Monica	Hiroko
The data span 16.2 cm. 68% of 16.2 is about 11 cm. Center this 11-cm range around the mean of 11.5 cm. This 68% group will range from about 6 cm to about 17 cm.	The middle 68% span from $\mu + \sigma$ to $\mu - \sigma$. So we move 2.5 cm below 11.5 and then 2.5 cm above 11.5. The 68% group will range from 9 cm to 14 cm.

22. **CHALLENGE** A case of portable media players has an average battery life of 8.2 hours with a standard deviation of 0.7 hour. Eight of the players have a battery life greater than 9.3 hours. If the sample is normally distributed, how many portable media players are in the case? **about 138**

23. **REASONING** The term *six sigma process* comes from the notion that if one has six standard deviations between the mean of a process and the nearest specification limit, there will be practically no items that fail to meet the specifications. Is this a true assumption? Explain. **See margin.**

24. **REASONING** *True* or *false*: According to the Empirical Rule, in a normal distribution, most of the data will fall within one standard deviation of the mean. Explain. **See margin.**

25. **OPEN ENDED** Find a set of real-world data that appears to be normally distributed. Calculate the range of values that represent the middle 68%, the middle 95%, and the middle 99.7% of the distribution. **See Chapter 11 Answer Appendix.**

26. **WRITING IN MATH** Describe the relationship between the z-value, the position of an interval of X in the normal distribution, the area under the normal curve, and the probability of the interval occurring. Use an example to explain your reasoning. **See Chapter 11 Answer Appendix.**

Additional Answers

23. Sample answer: True; according to the Empirical Rule, 99% of the data lie within 3 standard deviations of the mean. Therefore, only 1% will fall outside of three sigma. An infinitesimally small amount will fall outside of six-sigma.

24. Sample answer: True; according to the Empirical Rule, 68% of the data lie within 1 standard deviation of the mean.

Study Guide and Intervention
AL OL ELL

NAME _____ DATE _____ PERIOD _____

11-5 Study Guide and Intervention
The Normal Distribution
The Empirical Rule

Example The heights of players in a basketball league are normally distributed with a mean of 6 feet 1 inch and a standard deviation of 2 inches.

Practice
AL OL BL ELL

NAME _____ DATE _____ PERIOD _____

11-5 Practice
The Normal Distribution

A normal distribution has a mean of 186.4 and a standard deviation of 48.9.

Word Problem Practice
AL OL BL ELL

NAME _____ DATE _____ PERIOD _____

11-5 Word Problem Practice
The Normal Distribution

Enrichment
OL BL

NAME _____ DATE _____ PERIOD _____

11-5 Enrichment
Working Backward with Normal Distribution

Name the Math Ask students to explain the difference between a normal distribution and the standard normal distribution.

WatchOut!

Error Analysis For Exercise 21, remind students that the middle 68% of any normal distribution is within one standard deviation of the mean.

Formative Assessment

Check for student understanding of Lessons 11-4 and 11-5.

📁 Quiz 3, p. 46

E? Follow-up

Students have explored binomial distributions.

Ask:

- **Can statistics lie?** Sample answer: Statistics can "lie" when they are manipulated and then used to influence the intended audiences' beliefs and behaviors.

Additional Answers

32. The random variable X is the number of stations in a cable package. The stations are finite and countable, so X is discrete.

33. The random variable X is the amount of precipitation in a city per month. Precipitation can be anywhere within a certain range. Therefore, X is continuous.

Standardized Test Practice

27. The lifetimes of 10,000 light bulbs are normally distributed. The mean lifetime is 300 days, and the standard deviation is 40 days. How many light bulbs will last between 260 and 340 days? **D**

  A 2500   **C** 5000

  B 3400   **D** 6800

28. Which description best represents the graph? **J**

  F negatively skewed   **H** normal distribution

  G no correlation   **J** positively skewed

29. **SHORT RESPONSE** In the figure below, $RT = TS$ and $QR = QT$. What is the value of x? **32.5**

30. **SAT/ACT** The integer 99 can be expressed as a sum of n consecutive positive integers. The value of n could be which of the following? **E**

    I.   2

    II.   3

    III. 6

  A I only   **D** I and II

  B II only   **E** I, II, and III

  C III

Spiral Review

31. **SNOW** There is a 25% chance that it snows each day during a given week. Find the probability that it snows 3 out of the next 7 days. (Lesson 11-4) **17.3%**

Identify the random variable in each distribution, and classify it as *discrete* or *continuous*. Explain your reasoning. (Lesson 11-3) **32–33. See margin.**

32. the number of pages in a newspaper

33. the amount of precipitation in a city per month

34. **BRIDGES** The Sydney Harbour Bridge connects the central business district to northern metropolitan Sydney. It has an arch in the shape of a parabola that opens downward. Write an equation of a parabola to model the arch, assuming that the origin is at the surface of the water, beneath the vertex of the arch. (Lesson 9-2)
about $y = -0.00065x^2 + 440$

Identify the type of function represented by each graph. (Lesson 2-7)

35.

greatest integer

36.

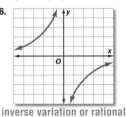

inverse variation or rational

37.

constant

Skills Review

38. Calculate the standard deviation of the population of data. **5.28**

13	18	17	21	16	9	11	28	8	10
7	19	16	16	12	19	21	11	8	13

 766 | Lesson 11-5 | The Normal Distribution

DifferentiatedInstruction **BL**

Extension Explain to students that they can use a graphing calculator to identify the range of z-values that corresponds with any area under the normal curve. For example, to find the range of z-values that corresponds to the middle 50% of data, they need to find the z-values that correspond to the data between 25% and 75% in the distribution. Select 2nd **[DISTR] 3** for the **InvNorm(** feature. Then enter 0.25 `)`. Do the same for 0.75 `)`.

Alberto Coto/Photodisc/Getty Images

EXTEND 11-5

Normal Approximation of Binomial Distributions

In Lesson 11-4, you used a binomial expansion to find a full probability distribution. A spreadsheet can be used to quickly find and graph a full distribution for any number of trials.

CCSS Common Core State Standards
Content Standards
S.ID.4 Use the mean and standard deviation of a data set to fit it to a normal distribution and to estimate population percentages. Recognize that there are data sets for which such a procedure is not appropriate. Use calculators, spreadsheets, and tables to estimate areas under the normal curve.

Activity 1 Full Probability Distribution

PLAYING CARDS Tom randomly selects a card from a deck of 52 playing cards, records its suit, and replaces it. Use a spreadsheet to construct and graph a full probability distribution for *X*, the number of hearts that Tom selects if he chooses 4, 20, or 100 cards.

Since one fourth of the cards in a standard deck are hearts, the probability of success is 25% or 0.25, and the probability of failure is 75% or 0.75.

Step 1 Enter the numbers 0 to 4 in column A. In B1, enter the binomial probability formula as =COMBIN(4,A1)*(0.25)^A1* (0.75)^(4−A1). Copy and paste this formula in cells B2:B5.

Step 2 Select cells B1:B5, and insert a clustered column bar graph. Use the values in column A for Category (*X*) axis labels.

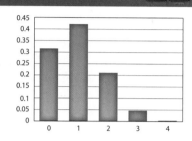

Step 3 Select cells A1:A5, and autofill the through A21. In C1, enter the formula =COMBIN(20,A1)*(0.25)^A1* (0.75)^(20−A1). Copy this formula in cells C2:C21. Repeat Step 2 to graph this distribution.

Step 4 Autofill column A through A101. In D1, enter the formula =COMBIN(100,A1)*(0.25)^A1* (0.75)^(100−A1). Copy this formula in cells D2:C101. Repeat Step 2 to graph this distribution.

Notice that the number of trials increases, the shape of the graph of the distribution becomes more symmetric with its center at the mean, $\mu = np$. If the number of trials becomes large enough, the shape of the distribution approaches a normal curve. Therefore, a normal distribution can be used to approximate the binomial distribution.

KeyConcept Normal Approximation of a Binomial Distribution

In a binomial distribution with *n* trials, a probability of success *p*, and a probability of failure *q*, such that $np \geq 5$ and $nq \geq 5$, a binomial distribution can be approximated by a normal distribution with $\mu = np$ and $\sigma = \sqrt{npq}$.

(continued on the next page)

---

1 Focus

Objective Use a normal distribution to approximate a binomial distribution.

Materials for Each Student
- computer with spreadsheet program

Teaching Tip
Have students review the Binomial Probability Formula and Example 4 in Lesson 11-4 before completing Activity 1.

Before Activity 2, remind students that in Lesson 11-5, they used *z*-values and graphing calculators to find probabilities associated with normally distributed data.

From Concrete to Abstract
Have students refer to Activity 1.

Ask:
- Find the probability that Tom selects no more than 30 hearts using the normal approximation if he chooses 100 cards. 87.5%

- Find the sum of the probabilities for 0 to 30 hearts by highlighting cells D1:D31 on the spreadsheet from Step 4 and applying the AutoSum function. 89.6%

- Explain why the probabilities found using these two methods are not the same. Sample answer: Using the normal distribution is an approximation. The larger the number of trials, the closer the answers should get to each other.

2 Teach

Working in Cooperative Groups

Have students work in pairs to complete Activities 1 and 2 and Exercise 1. Mix abilities so that a student with more knowledge of spreadsheets is partnered with one who has less experience.

3 Assess

Formative Assessment

Use Exercise 2 to assess whether students can use a normal distribution to approximate a binomial distribution.

Additional Answer

1.

Spreadsheet Lab
Normal Approximation of Binomial Distributions *Continued*

Once the mean and standard deviation are calculated, a *z*-value can be determined, and the corresponding probability can be found as demonstrated in the next activity.

Activity 2   Normal Approximation of a Binomial Distribution

JURY DUTY According to a poll, 60% of the registered voters in a city have never been called for jury duty. Mariah conducts a random survey of 300 registered voters. What is the probability that at least 170 of those voters have never been called for jury duty?

This is a binomial experiment with $n = 300$, $p = 0.6$, and $q = 0.4$. Since $np = 300(0.6)$ or 180 and $nq = 300(0.4)$ or 120 are both greater than 5, the normal distribution can be used to approximate the binomial distribution.

Step 1 The mean μ is np or 180. Find the standard deviation σ.

$$\sigma = \sqrt{npq} \qquad \text{Standard deviation of a binomial distribution}$$
$$= \sqrt{300(0.6)(0.4)} \qquad n = 300, p = 0.6, \text{ and } q = 0.4$$
$$\approx 8.49 \qquad \text{Simplify.}$$

Step 2 Write the problem in probability notation using X. The probability that at least 170 people have never been called for jury duty is $P(X \geq 170)$.

Step 3 Find the corresponding *z*-value for X.

$$z = \frac{X - \mu}{\sigma} \qquad \text{Formula for } z\text{-values}$$
$$= \frac{170 - 180}{8.49} \qquad X = 170, \mu = 180, \text{ and } \sigma = 8.49$$
$$\approx -1.18 \qquad \text{Simplify.}$$

Step 4 Use a calculator to find the area under the normal curve to the right of *z*.

KEYSTROKES: 2nd [DISTR] 2 (—) 1.18 , 4) ENTER

The approximate area to the right of *z* is 0.881. Therefore, the probability that at least 170 of the registered voters have never been called for jury duty is about 88.1%.

Exercises

1. Use a spreadsheet to construct the graph of the full probability distribution for X, the number of times a 3 is rolled from rolling a die 25 times. **See margin.**

2. **SUMMER JOBS** According to an online poll, 80% of high school upperclassmen have summer jobs. Tadeo thinks the number should be lower so he conducts a survey of 480 random upperclassmen. What is the probability that no more than 380 of the surveyed upperclassmen have summer jobs? **about 32.3%**

3. **WORK** According to an online poll, 28% of adults feel that the standard 40-hour work week should be increased. Sheila interviews 300 adults at the mall. What is the probability that more than 80 but fewer than 100 of those surveyed will say that the work week should be increased? **about 67.7%**

LESSON 11-6 Confidence Intervals and Hypothesis Testing

:: Then	:: Now	:: Why?
• You applied the standard normal distribution and z-values.	**1** Find confidence intervals for normally distributed data. **2** Perform hypothesis tests on normally distributed data.	• In a recent Gallup Poll, 1514 teens who owned a portable media player had an average of 1033 songs. The poll had the following disclaimer: "For results based on the total sample of national teens, one can say with 95% confidence that the margin of sampling error is ±31 songs."

NewVocabulary
inferential statistics
statistical inference
confidence interval
maximum error of
  estimate
hypothesis test
null hypothesis
alternative
  hypothesis
critical region
left-tailed test
two-tailed test
right-tailed test

Common Core State Standards

Content Standards
S.IC.1 Understand statistics as a process for making inferences about population parameters based on a random sample from that population.

S.IC.4 Use data from a sample survey to estimate a population mean or proportion; develop a margin of error through the use of simulation models for random sampling.

Mathematical Practices
7 Look for and make use of structure.

1 Confidence Intervals **Inferential statistics** are used to draw conclusions or **statistical inferences** about a population using a sample. For example, the sample mean of 1033 songs per portable media player can be used to estimate the population mean.

A **confidence interval** is an estimate of a population parameter stated as a range with a specific degree of certainty. Typically, statisticians use 90%, 95%, and 99% confidence intervals, but any percentage can be considered. In the opening example, we are 95% confident that the population mean is within 31 songs of 1033.

The confidence interval for a normal distribution is equivalent to the area under the standard normal curve between $-z$ and z, as shown. A 95% confidence interval for a population mean implies that we are 95% sure that the mean will fall within the range of z-values.

Suppose you want 95% confidence when conducting an experiment. The corresponding z-value is 1.960, where 2.5% of the area lies to the left of $-z$ and 2.5% lies to the right of z.

To find the confidence interval, use the maximum error of estimate and the sample mean. The **maximum error of estimate** E is the maximum difference between the estimate of the population mean μ and its actual value.

KeyConcept Maximum Error of Estimate

The maximum error of estimate E for a population mean is given by

$$E = z \cdot \frac{s}{\sqrt{n}},$$

where z is the z-value that corresponds to a particular confidence level, s is the standard deviation of the sample, and n is the sample size; $n \geq 30$.

---

connectED.mcgraw-hill.com **769**

Fancy Photography/Veer

---

1 Focus

VerticalAlignment

Before Lesson 11-6 Apply the standard normal distribution.

Lesson 11-6 Find confidence intervals and perform hypothesis tests on normally distributed data.

After Lesson 11-6 Understand how to draw conclusions about a population using a sample.

2 Teach

Scaffolding Questions
Have students read the **Why?** section of the lesson.

Ask:
- To what do you think the margin of sampling error refers? Sample answer: The average obtained using the sample is within 31 songs of the population average.

- Why do you think they are only 95% confident? Sample answer: The sample is only an estimate of the population. The entire population would need to be used in order to be 100% confident.

Lesson 11-6 Resources

Resource	Approaching Level **AL**	On Level **OL**	Beyond Level **BL**	English Learners **ELL**
Teacher Edition		• Differentiated Instruction, p. 771	• Differentiated Instruction, p. 776	• Differentiated Instruction, p. 771
Chapter Resource Masters	• Study Guide and Intervention, pp. 36–37 • Skills Practice, p. 38 • Practice, p. 39 • Word Problem Practice, p. 40	• Study Guide and Intervention, pp. 36–37 • Skills Practice, p. 38 • Practice, p. 39 • Word Problem Practice, p. 40 • Enrichment, p. 41	• Practice, p. 39 • Word Problem Practice, p. 40 • Enrichment, p. 41	• Study Guide and Intervention, pp. 36–37 • Skills Practice, p. 38 • Practice, p. 39 • Word Problem Practice, p. 40
Other	• 5-Minute Check 11-6 • Study Notebook	• 5-Minute Check 11-6 • Study Notebook	• 5-Minute Check 11-6 • Study Notebook	• 5-Minute Check 11-6 • Study Notebook

1 Confidence Intervals

Example 1 shows how to find a maximum error of estimate.

Example 2 shows how to find a confidence interval.

Formative Assessment

Use the Guided Practice exercises after each example to determine students' understanding of concepts.

Additional Examples

1 **ORAL HYGIENE** A poll of 422 randomly selected adults showed that they brushed their teeth an average of 11.4 times per week with a standard deviation of 1.6. Use a 99% confidence interval to find the maximum error of estimate for the number of times per week adults brush their teeth. 0.20

2 **PACKAGING** A poll of 156 randomly selected members of a golf course showed that they play an average of 4.6 times every summer with a standard deviation of 1.1. Determine a 95% confidence interval for the population mean. $4.43 \leq \mu \leq 4.77$

Additional Examples also in Interactive Classroom PowerPoint® Presentations

IWB **Interactive White Board READY**

CCSS Teaching the Mathematical Practices

Perseverance Mathematically proficient students start by explaining to themselves the meaning of a problem and looking for entry points to its solution. Mathematically proficient students can search for regularity or trends. Encourage student to become familiar with commonly used values.

Example 1 Maximum Error of Estimate

SOCIAL NETWORKING A poll of 218 randomly selected members of a social networking Web site showed that they spent an average of 14 minutes per day on the site with a standard deviation of 3.1 minutes. Use a 95% confidence interval to find the maximum error of estimate for the time spent on the site.

In a 95% confidence interval, 2.5% of the area lies in each tail. The corresponding z-value is 1.960.

$$E = z \cdot \frac{s}{\sqrt{n}} \qquad \text{Maximum Error of Estimate}$$

$$= 1.960 \cdot \frac{3.1}{\sqrt{218}} \qquad z = 1.960, s = 3.1, \text{ and } n = 218$$

$$\approx 0.41 \qquad \text{Simplify.}$$

This means that you can be 95% confident that the population mean time spent on the site will be within 0.41 minute of the sample mean of 14 minutes.

StudyTip

CCSS Perseverance The z-value that corresponds to a particular confidence level is known as the *critical value*. The most commonly used levels and their corresponding z-value are shown below.

Confidence Level	z-Value
90%	1.645
95%	1.960
99%	2.576

Guided Practice

1. TEACHING A poll of 184 randomly selected high school teachers showed that they spend an average of 16.8 hours per week grading, planning, and preparing for class. The standard deviation is 2.9 hours. Use a 90% confidence interval to find the maximum error of estimate for the amount of time spent per week. **0.35**

Once the maximum error of estimate E is found, a confidence interval CI for the population mean can be determined by adding $\pm E$ to the sample mean.

KeyConcept Confidence Interval for the Population Mean

A confidence interval for a population mean is given by

$$CI = \overline{x} \pm E \text{ or } \overline{x} \pm z \cdot \frac{s}{\sqrt{n}},$$

where $\overline{x}$ is the sample mean and E is the maximum error of estimate.

Real-WorldLink

On average, students sleep 8.1 hours per weekday and spend 7.5 hours on educational activities, such as attending class or doing homework.

Source: Bureau of Labor Statistics

Real-World Example 2 Confidence Interval

SCHOOL WORK A sample of 200 students was asked the average time they spent on homework each weeknight. The mean time was 52.5 minutes with a standard deviation of 5.1 minutes. Determine a 99% confidence interval for the population mean.

$$CI = \overline{x} \pm z \cdot \frac{s}{\sqrt{n}} \qquad \text{Confidence Interval for Population Mean}$$

$$= 52.5 \pm 2.576 \cdot \frac{5.1}{\sqrt{200}} \qquad \overline{x} = 52.5, z = 2.576, s = 5.1, \text{ and } n = 200$$

$$\approx 52.5 \pm 0.93 \qquad \text{Use a calculator.}$$

The 99% confidence interval is $51.57 \leq \mu \leq 53.43$. Therefore, we are 99% confident that the population mean time is between 51.57 and 53.43 minutes.

Guided Practice

2. SCHOOL ATHLETICS A sample of 224 students showed that they attend an average of 2.6 school athletic events per year with a standard deviation of 0.8. Determine a 90% confidence interval for the population mean. $2.51 \leq \mu \leq 2.69$

BananaStock/Jupiterimages/PictureQuest

2 Hypothesis Testing

Hypothesis Testing While a confidence interval provides an estimate of the population mean, a **hypothesis test** is used to assess a specific claim about the mean. Typical claims are that the mean is equal to, is greater than, or is less than a specific value.

There are two parts to a hypothesis test: the null hypothesis and the alternative hypothesis. The **null hypothesis** H_0 is a statement of equality to be tested. The **alternative hypothesis** H_a is a statement of inequality that is the complement of the null hypothesis. A *claim* can be part of the null or alternative hypothesis.

Example 3 Claims and Hypotheses

Identify the null and alternative hypotheses for each statement. Then identify the statement that represents the claim.

a. A school administrator thinks it takes less than 3 minutes to evacuate the entire building for a fire drill.

less than 3 minutes	not less than three minutes
$\mu < 3$	$\mu \geq 3$

The claim is $\mu < 3$, and it is the alternative hypothesis because it does not include equality. The null hypothesis is $\mu \geq 3$, which is the complement of $\mu < 3$.

$$H_0: \mu \geq 3 \qquad H_a: \mu < 3 \text{ (claim)}$$

b. The owner of a deli says that there are 2 ounces of ham in a sandwich.

$H_0: \mu = 2$ (claim); $H_a: \mu \neq 2$

The claim is $\mu = 2$, and it is the null hypothesis because it is an equality. The alternative hypothesis is $\mu \neq 2$, which is the complement of $\mu = 2$.

▶ **Guided**Practice

3A. Jill thinks it takes longer than 10 minutes for the school bus to reach school from her stop. $H_0: \mu \leq 10$; $H_a: \mu > 10$ (claim)

3B. Mindy thinks there aren't 35 peanuts in every package. $H_0: \mu = 35$; $H_a: \mu \neq 35$ (claim)

ReadingMath

Significance Level
A significance level of 5% means that we need the data to provide evidence against H_0 so strong that the data would occur no more than 5% of the time when H_0 is true.

Once a claim is made, a sample is collected and analyzed, and the null hypothesis is either *rejected* or *not rejected*. From this information, the claim is then rejected or not rejected. This is determined by whether the sample mean falls within the critical region. The **critical region** is the range of values that suggests a significant enough difference to reject the null hypothesis. The critical region is determined by the significance level α, most commonly 1%, 5%, or 10%, and the inequality sign of the alternate hypothesis.

ConceptSummary Tests of Significance

If $H_a: \mu < k$, the hypothesis test is a **left-tailed test**.	If $H_a: \mu \neq k$, the hypothesis test is a **two-tailed test**.	If $H_a: \mu > k$, the hypothesis test is a **right-tailed test**.

The value k represents the claim about the population mean μ.

StudyTip

z-Values and z-Statistics In hypothesis testing, a *z*-value is used to determine the critical region. A *z*-statistic is calculated to see if the sample mean falls within that critical region, and thus determines whether or not to reject the null hypothesis.

The type of hypothesis test that we are using in this lesson is known as a *z*-test. Once the area corresponding to the significance level is determined, a *z*-statistic for the sample mean is calculated. The *z*-statistic is the *z*-value for the sample. When testing at 5% significance, it is only 5% likely for the *z*-statistic to fall within the critical region *and* for H_0 to be true. Therefore, when the *z*-statistic falls within the critical region, H_0 is rejected.

---

2 Hypothesis Testing

Example 3 shows how to identify claims and hypotheses. **Example 4** shows how to perform a one-sided hypothesis test. **Example 5** shows how to perform a two-sided hypothesis test.

Additional Example

3 Identify the null and alternative hypotheses for each statement. Then identify which statement represents the claim.

a. Cindy thinks her employees spend over 60 minutes per day surfing the Internet at work. $H_0: \mu \leq 60$; $H_a: \mu > 60$ (claim)

b. Mr. Bacon thinks that he tells at least five jokes every class period. $H_0: \mu \geq 5$ (claim); $H_a: \mu < 5$

Differentiated Instruction OL ELL

Verbal/Linguistic Learners The formal conclusion of a hypothesis test is stated as "Reject the null hypothesis" or "Do not reject the null hypothesis." Students should always restate conclusions in their own words.

Additional Example

4 MAGAZINES Jenny thinks there are more than 10 pages of advertisements in every issue of a particular magazine. Using a sample of 42 magazines, she calculated a mean of 10.5 pages with a standard deviation of 1.6. Test the hypothesis at 1% significance.

H_0 is not rejected. There is not enough evidence to support Jenny's claim that there are more than 10 pages of advertisements in every issue.

Tip for New Teachers

Failing to Reject Versus Accepting the Null Hypothesis When the null hypothesis is not rejected, it cannot be accepted as true. There is merely not enough evidence to say that it is false.

Teach with Tech

Blog Have students write a blog entry to describe hypothesis testing in their own words. Check students' entries to be sure they understand the concept and importance of hypothesis testing.

Focus on Mathematical Content

Tests and Confidence Intervals
While a confidence interval is used to estimate the value of a quantity, a hypothesis test is performed in order to answer a *yes* or *no* question about whether a null hypothesis is true. A confidence interval can be used to help answer that question.

Real-World Career

Politician Politicians are elected officials in the federal, state, and local governments. They influence public decision making and make laws that affect everyone. Many politicians have business, teaching, or legal experience prior to getting involved in politics.

StudyTip

Rejecting the Null Hypothesis When the null hypothesis is rejected, it is implied that the alternative hypothesis is accepted.

4. H_0: $\mu \geq 60$; H_a: $\mu < 60$ (claim); H_0 is rejected. Therefore, there is not enough evidence to reject Julian's claim that it takes less than 60 minutes for people to get in a full workout.

Use these steps to test a hypothesis.

KeyConcept Steps for Hypothesis Testing

Step 1 State the hypotheses and identify the claim.

Step 2 Determine the critical value(s) and critical region.

Step 3 Calculate the z-statistic using $z = \dfrac{\bar{x} - \mu}{\frac{s}{\sqrt{n}}}$.

Step 4 Reject or fail to reject the null hypothesis.

Step 5 Make a conclusion about the claim.

Real-World Example 4 One-Sided Hypothesis Test

STUDENT COUNCIL Lindsey, the president of the student council, has heard complaints that the cafeteria lunch line moves too slowly. The dining services coordinator assures her that the average wait time is 6 minutes or less. Using a sample of 65 customers, Lindsey calculated a mean wait time of 6.3 minutes and a standard deviation of 1.1 minutes. Test the hypothesis at 5% significance.

Step 1 State the claim and hypotheses. H_0: $\mu \leq 6$ (claim) and H_a: $\mu > 6$.

Step 2 Determine the critical region.

The alternative hypothesis is $\mu > 6$, so this is a right-tailed test. We are testing at 5% significance, so we need to identify the z-value that corresponds with the upper 5% of the distribution.

```
invNorm(.95)
         1.644853626
```

KEYSTROKES: 2nd [DISTR] 3 .95 ENTER

Critical region

+ 1.645

Step 3 Calculate the z-statistic for the sample data.

$z = \dfrac{\bar{x} - \mu}{\frac{s}{\sqrt{n}}}$ Formula for z-statistic

$= \dfrac{6.3 - 6}{\frac{1.1}{\sqrt{65}}}$ $\bar{x} = 6.3$, $\mu = 6$, $s = 1.1$, and $n = 65$

≈ 2.20 Simplify.

Step 4 Decide whether to reject the null hypothesis.

H_0 is rejected because the z-statistic for the sample falls within the critical region.

Critical region
$z = 2.20$
+ 1.645

Step 5 Make a conclusion about the claim.

Therefore, there is enough evidence to reject the claim that the average wait time is 6 minutes or less.

GuidedPractice

4. **EXERCISE** Julian thinks it takes less than 60 minutes for people to get in a full workout at the gym. Using a sample of 128 members, he calculated a mean workout time of 58.4 minutes with a standard deviation of 8.9 minutes. Test the hypothesis at 10% significance.

Yellow Dog Productions/Lifesize/Getty Images

For a two-sided test, the significance level α must be divided by 2 in order to determine the critical value at each tail.

Example 5 Two-Sided Hypothesis Test

ADVERTISEMENTS Jerrod wants to determine if the advertisement shown is accurate. Using a sample of 42 pizzas, Jerrod calculates a mean of 99.8 pieces and a standard deviation of 0.8. Test the hypothesis at 5% significance.

100 Pepperoni on Every Large Pizza!

Step 1 State the claim and hypotheses.

$H_0: \mu = 100$ (claim) and $H_a: \mu \neq 100$.

Step 2 Determine the critical region.

The alternative hypothesis is $\mu \neq 100$, so this is a two-tailed test. We are testing at 5% significance, so we need to identify the z-value that corresponds with upper and lower 2.5% of the distribution.

KEYSTROKES: 2nd [DISTR] 3 0.25 ENTER and 2nd [DISTR] 3 .975 ENTER

```
invNorm(.025)
    -1.959963986
invNorm(.975)
    1.959963986
```

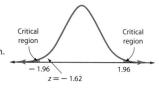

Critical region   Critical region

−1.96   1.96

StudyTip

z-Test The *z*-test can be used when σ is known or when $n \geq 30$.

Step 3 Calculate the z-statistic for the sample data.

$z = \dfrac{\bar{x} - \mu}{\frac{s}{\sqrt{n}}}$ Formula for z-statistic

$= \dfrac{99.8 - 100}{\frac{0.8}{\sqrt{42}}}$ $\bar{x} = 99.8$, $\mu = 100$, $s = 0.8$, and $n = 42$

≈ -1.62 Simplify.

Step 4 Reject or fail to reject the null hypothesis.

H_0 is not rejected because the z-value for the sample does not fall within the critical region.

Critical region   Critical region

−1.96   1.96
$z = -1.62$

WatchOut!

Inferential Statistics
Inferential statistics are not designed to definitely prove a hypothesis. They depend on probability statements and therefore, may not be correct.

Step 5 Make a conclusion about the claim.

Therefore, there is not enough evidence to reject the claim that there are 100 pieces of pepperoni on every large pizza.

5. $H_0: \mu = 12$ (claim); $H_a: \mu \neq 12$; H_0 is rejected. Therefore, the claim that there are 12 ounces in every can is rejected.

GuidedPractice

5. **JUICE** A manufacturer claims there are 12 ounces of juice in every can. Using a sample of 68 cans, Susan calculated a mean of 11.95 ounces with a standard deviation of 0.15 ounce. Test the hypothesis at 1% significance.

connectED.mcgraw-hill.com   **773**

Additional Example

5 CALORIES A label on a package of food states that there are 120 Calories per serving. Using a sample of 38 servings, Libby calculated a mean of 120.6 Calories with a standard deviation of 2.2. Test the hypothesis at 5% significance.
H_0 is not rejected. There is not enough evidence to reject the label's claim that there are 120 Calories per serving.

Tip for New Teachers

***t*-Test** When σ is unknown and $n < 30$, the *z*-test is no longer accurate. A different hypothesis test called the *t*-test needs to be performed.

3 Practice

Formative Assessment

Use Exercises 1–8 to check for understanding.

Use the chart on the bottom of this page to customize assignments for your students.

Additional Answers

17. $H_0: \mu = 12$; $H_a: \mu \neq 12$; The mean of the sample data is 12.9 with a standard deviation of about 1.08. The z-statistic is about 5.27, which falls in the critical region at 1% significance. Therefore, the null hypothesis is rejected and the company should not make the claim on the label.

18. $H_0: \mu = 20$; $H_a: \mu \neq 20$; The mean of the sample data is 19.175 with a standard deviation of 1.375. The z-statistic is about -3.79, which falls in the critical region at 1% significance. Therefore, the null hypothesis is rejected and the company should not make the claim on the label.

20. Neither; Sample answer: The null hypothesis is a statement of equality, so Tim is incorrect. The claim is that the delivery time is less than or equal to 45, so Judie is also incorrect.

24. Sample answer: A pizza company claims to put 100 pepperonis on its large pizza. $H_0: \mu = 100$ and H_1: $\mu > 100$; data collected: 100, 102, 101, 101, 100, 99, 99, 103, 102, 103, 103, 101, 102, 99, 105, 103, 102, 100, 101, 104; $\bar{x} = 101.5$, $s = 1.70$; a 95% confidence interval is $100.74 < \bar{x} < 102.26$. H_0 does not fall within the confidence interval, so we reject the null hypothesis and accept the alternative hypothesis.

Check Your Understanding

 = Step-by-Step Solutions begin on page R14.

Example 1
1. **LUNCH** A sample of 145 high school seniors was asked how many times they go out for lunch per week. The mean number of times was 2.4 with a standard deviation of 0.7. Use a 90% confidence level to calculate the maximum error of estimate. **0.096**

Example 2
2. **PRACTICE** A poll of 233 randomly chosen high school athletes showed that they spend an average of 1.6 hours practicing their sport during the off-season. The standard deviation is 0.5 hour. Determine a 99% confidence interval for the population mean. **$1.52 \leq \mu \leq 1.68$**

Example 3
Identify the null and alternative hypotheses for each statement. Then identify the statement that represents the claim. 5. $H_0: \mu \geq 20$ (claim); $H_a: \mu < 20$

3. Lori thinks it takes a fast-food restaurant less than 2 minutes to serve her meal after she orders it. $H_0: \mu \geq 2$; $H_a: \mu < 2$ (claim)

4. A snack label states that one serving contains one gram of fat. $H_0: \mu = 1$ (claim); $H_a: \mu \neq 1$

5. Mrs. Hart's review game takes at least 20 minutes to complete.

6. The tellers at a bank can complete no more than 18 transactions per hour. $H_0: \mu \leq 18$ (claim); $H_a: \mu > 18$

CCSS REASONING Identify the hypotheses and claim, decide whether to reject the null hypothesis, and make a conclusion about the claim.

Example 4
7. **COMPACT DISCS** A manufacturer of blank compact discs claims that each disc can hold at least 84 minutes of music. Using a sample of 219 compact discs, Cayla calculated a mean time of 84.1 minutes per disc with a standard deviation of 1.9 minutes. Test the hypothesis at 5% significance.

7. $H_0: \mu \geq 84$ (claim), $H_a: \mu < 84$; Do not reject H_0; The manufacturer's claim that the discs can hold at least 84 minutes cannot be rejected.

Example 5
8. **GOLF TEES** A company claims that each golf tee they produce is 5 centimeters in length. Using a sample of 168 tees, Angelene calculated a mean of 5.1 centimeters with a standard deviation of 0.3. Test the hypothesis at 10% significance. $H_0: \mu = 5$ (claim), $H_a: \mu \neq 5$; Reject H_0; The company's claim that each tee is 5 centimeters is rejected.

Practice and Problem Solving

Extra Practice is on page R11.

Example 1
(9) **MUSIC** A sample of 76 albums had a mean run time of 61.3 minutes with a standard deviation of 5.2 minutes. Use a 95% confidence level to calculate the maximum error of estimate. **1.17**

Example 2
10. **COLLEGE** A poll of 218 students at a university showed that they spend 11.8 hours per week studying. The standard deviation is 3.7 hours. Determine a 90% confidence interval for the population mean. **$11.39 \leq \mu \leq 12.21$**

12. $H_0: \mu = 27$ (claim); $H_a: \mu \neq 27$

Example 3
Identify the null and alternative hypotheses for each statement. Then identify the statement that represents the claim. 11. $H_0: \mu \geq 6$ (claim); $H_a: \mu < 6$

11. Julian sends at least six text messages to his best friend every day.

15. $H_0: \mu \geq 30$; H_a: $\mu < 30$ (claim); Do not reject H_0; There is not enough evidence to support the pizza chain's claim of a delivery time of less than 30 minutes cannot be rejected.

12. A car company states that one of their vehicles gets 27 miles per gallon.

13. A company advertisement states that it takes no more than 2 hours to paint a 200-square-foot room. $H_0: \mu \leq 2$ (claim); $H_a: \mu > 2$

14. A singer plays at least 18 songs at every concert. $H_0: \mu \geq 18$ (claim); $H_a: \mu < 18$

Identify the hypotheses and claim, decide whether to reject the null hypothesis, and make a conclusion about the claim.

Example 4
15. **PIZZA** A pizza chain promises a delivery time of less than 30 minutes. Using a sample of 38 deliveries, Chelsea calculated a mean delivery time of 29.6 minutes with a standard deviation of 3.9 minutes. Test the hypothesis at 1% significance.

 774 | Lesson 11-6 | Confidence Intervals and Hypothesis Testing

---

Differentiated Homework Options

Level	Assignment	Two-Day Option	
AL Basic	9–16, 20, 22–42	9–15 odd, 25–28	2–16 even, 20, 22–24, 29–42
OL Core	9–15 odd, 9–20, 22–42	9–16, 25–28	17–20, 22–42
BL Advanced	17–39, (optional; 40–42)		

Example 5

16. H_0: $\mu = 24$ (claim), H_a: $\mu \neq 24$; Do not reject H_0; The company's claim that each package contains 24 slices is not rejected.

16. CHEESE A company claims that each package of cheese contains exactly 24 slices. Using a sample of 93 packages, Mr. Matthews calculated a mean of 24.1 slices with a standard deviation of 0.5. Test the hypothesis at 5% significance.

17. DECISION MAKING The number of peaches in 40 random cans is shown below. Should the manufacturer place a label on the can promising exactly 12 peaches in every can? Explain your reasoning. **See margin.**

13, 14, 13, 14, 12, 12, 12, 11, 15, 12, 13, 13, 14, 13, 14, 12, 15, 11, 11, 14,
13, 14, 14, 13, 12, 12, 12, 12, 13, 13, 11, 14, 14, 13, 14, 13, 13, 14, 12, 12

18. CCSS ARGUMENTS The number of chocolate chips in 40 random cookies is shown below. Should the manufacturer place a label on the package promising exactly 20 chips on every cookie? Explain your reasoning. **See margin.**

21, 19, 20, 20, 19, 19, 18, 21, 19, 17, 19, 18, 18, 20, 20, 19, 18, 20, 19, 20,
21, 21, 19, 17, 17, 18, 19, 19, 20, 17, 22, 21, 21, 20, 19, 18, 19, 17, 17, 21

19. MULTIPLE REPRESENTATIONS In this problem, you will explore how the confidence interval is affected by the sample size and the confidence level. Consider a sample of data where $\bar{x} = 25$ and $s = 3$.

a. GRAPHICAL Graph the 90% confidence interval for $n = 50, 100,$ and 250 on a number line.

b. ANALYTICAL How does the sample size affect the confidence interval?

c. GRAPHICAL Graph the 90%, 95%, and 99% confidence intervals for $n = 150$.

d. ANALYTICAL How does the confidence level affect the confidence interval?

e. ANALYTICAL How does decreasing the size of the confidence interval affect the accuracy of the confidence interval?
a–e. See Chapter 11 Answer Appendix.

H.O.T. Problems Use Higher-Order Thinking Skills

20. ERROR ANALYSIS Tim and Judie want to test whether a delivery service meets their promised time of 45 minutes or less. Their hypotheses are shown below. Is either of them correct? **See margin.**

Tim	Judie
H_0: $\mu < 45$ (claim)	H_0: $\mu \leq 45$
H_a: $\mu \geq 45$	H_a: $\mu > 45$ (claim)

21. CHALLENGE A 95% confidence interval for the mean weight of a 20-ounce box of cereal was $19.932 \leq \mu \leq 20.008$ with a sample standard deviation of 0.128 ounces. Determine the sample size that led to this interval. **44**

22. Sample answer: Always; if the null hypothesis value of μ falls within the confidence interval, then it is not rejected.

22. REASONING Determine whether the following statement is *sometimes, always,* or *never* true. Explain your reasoning.

If a confidence interval contains the H_0 value of μ, then it is not rejected.

23. WRITING IN MATH How can a statistical test be used in a decision-making process? **Sample answer: You can use a statistical test to help you to determine the strength of your decision.**

24. OPEN ENDED Design and conduct your own research study, and draw conclusions based on the results of a hypothesis test. Write a brief summary of your findings. **See margin.**

CCSS **Teaching the Mathematical Practices**

Arguments Mathematically proficient students understand and use stated assumptions, definitions, and previously established results in constructing arguments. They make conjectures and build a logical progression of statements to explore the truth of their conjectures.

Study Guide and Intervention
AL OL ELL

Practice
AL OL BL ELL

Word Problem Practice
AL OL BL ELL

Enrichment
OL BL

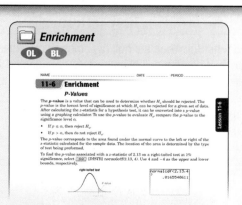

Yesterday's News Ask students to describe how their study of normal distributions relates to their study of confidence intervals and hypothesis testing.

Formative Assessment
Check for student understanding of Lesson 11-6.

 Quiz 4, p. 46

Additional Answers

34. ellipse

35. parabola

36. hyperbola

Standardized Test Practice

25. GEOMETRY In the graph below, line ℓ passes through the origin. What is the value of $\frac{a}{b}$? **C**

A -4

B $-\frac{1}{4}$

C $\frac{1}{4}$

D 4

26. SAT/ACT If $5 + i$ and $5 - i$ are the roots of the equation $x^2 - 10x + c = 0$, what is the value of c? **J**

F -26 G -25 H 25 J 26

27. The Service Club at Jake's school was founded 8 years ago. The number of members of the club by year is shown in the table. Which linear equation best models the data? **B**

A $y = 1.4x$
B $y = 1.4x + 10.4$
C $y = 1.6x$
D $y = 1.6x + 11.1$

Year	Participation
0	11
2	13
4	15
6	19
8	22

28. SHORT RESPONSE Solve for x: $\log_2 (x - 6) = 3$. **14**

Spiral Review

29. HEALTH The heights of students at Madison High School are normally distributed with a mean of 66 inches and a standard deviation of 2 inches. Of the 1080 students in the school, how many would you expect to be less than 62 inches tall? (Lesson 11-5) **27 students**

30. CAR WASH The Spanish Club is washing cars to raise money. They have determined that 65% of the customers donate more than the minimum amount for the car wash. What is the probability that at least 4 of the next 5 customers will donate more than the minimum? (Lesson 11-4) **42.8%**

Find a_n for each geometric sequence. (Lesson 10-3)

31. $a_1 = \frac{1}{3}, r = 3, n = 8$ **729**

32. $a_1 = \frac{1}{64}, r = 4, n = 9$ **1024**

33. $a_4 = 16, r = 0.5, n = 8$ **1**

Write each equation in standard form. State whether the graph of the equation is a *parabola, circle, ellipse,* or *hyperbola.* Then graph the equation. (Lesson 9-6) **34–36. See margin.**

34. $4x^2 + 2y^2 = 8$

35. $x^2 = 8y$

36. $(x - 1)^2 - 9(y - 4)^2 = 36$

Write an equation in slope-intercept form for each graph. (Lesson 2-4)

37.

$y = 0.8x$

38.

$y = -\frac{5}{3}x + \frac{29}{3}$

39.

$y = -4$

Skills Review

Find each missing measure. Round to the nearest tenth, if necessary.

40. **17.5**

9 cm
15 cm

41. **25**
24 cm
7 cm

42. **13.7**

17 cm
10 cm

 776 | **Lesson 11-6** | Confidence Intervals and Hypothesis Testing

---

DifferentiatedInstruction **BL**

Extension When the null hypothesis is rejected when it is actually true, the result is called a Type I error. When the null hypothesis is not rejected when it is actually false, the result is called a Type II error. Have students explore the risks of making Type I and Type II errors with different significant levels. What conclusions can they make? The risk of making a Type I error is identical to the significance level. Reducing the risk of a Type I error increases the risk of a Type II error by widening the acceptance level.

Study Guide and Review

Study Guide

KeyConcepts

Designing a Study (Lesson 11-1)

- A survey, an experiment, or an observational study can be used to collect information.
- A bias is an error that results in a misrepresentation of members of a population.

Distributions of Data (Lesson 11-2)

- Use the mean and standard deviation to describe a symmetric distribution.
- Use the five-number summary to describe a skewed distribution.

Probability Distributions (Lesson 11-3)

- A theoretical probability distribution is based on what is expected to happen. An experimental probability distribution is a distribution of probabilities estimated from experiments.
- The expected value of a discrete random variable is the weighted average of the values of the variable.

Binomial Distributions (Lesson 11-4)

- A binomial experiment has a fixed number of independent trials, only two possible outcomes for each trial, and the same probability of success for every trial.
- A binomial distribution is a frequency distribution of the probability of each value of a random variable.

Normal Distributions (Lesson 11-5)

- The graph of a normal distribution is bell-shaped.
- The z-value represents the number of standard deviations that a given data value is from the mean.

Hypothesis Testing (Lesson 11-6)

- A confidence interval for a population mean μ is given by $CI = \bar{x} \pm E$ or $\bar{x} \pm z \cdot \dfrac{s}{\sqrt{n}}$, where $\bar{x}$ is the sample mean and E is the maximum error of estimate.
- A hypothesis test is used to assess a specific claim.

FOLDABLES StudyOrganizer

Be sure the following Key Concepts are noted in your Foldable.

KeyVocabulary

alternative hypothesis H_a (p. 771)	normal distribution (p. 760)
bias (p. 723)	null hypothesis H_0 (p. 771)
binomial distribution (p. 754)	observational study (p. 723)
confidence interval (p. 769)	parameter (p. 723)
continuous random variable (p. 742)	positively skewed distribution (p. 733)
discrete random variable (p. 742)	probability distribution (p. 743)
Empirical Rule (p. 760)	random variable (p. 742)
expected value $E(X)$ (p. 745)	standard normal distribution (p. 762)
experiment (p. 723)	statistic (p. 723)
experimental probability distribution (p. 744)	statistical inference (p. 769)
hypothesis test (p. 771)	survey (p. 723)
inferential statistics (p. 769)	symmetric distribution (p. 733)
maximum error of estimate (p. 769)	theoretical probability distribution (p. 743)
negatively skewed distribution (p. 733)	z-value (p. 762)

VocabularyCheck

Choose a term from the list above that best completes each statement.

1. A(n) _____ for a particular random variable is a function that maps the sample space to the probabilities of the outcomes of the sample space. **probability distribution**

2. A(n) _____ is an error that results in a misrepresentation of members of a population. **bias**

3. In a statistical study, data are collected and used to answer questions about a population characteristic or _____. **parameter**

4. The _____ can be used to determine the area under the normal curve at specific intervals. **Empirical Rule**

5. In a(n) _____, members of a sample are measured or observed without being affected by the study. **observational study**

6. A _____ is an estimate of a population parameter stated as a range with a specific degree of certainty. **confidence interval**

Formative Assessment

KeyVocabulary The page references after each word denote where that term was first introduced. If students have difficulty answering questions 1–6, remind them that they can use these page references to refresh their memories about the vocabulary.

Summative Assessment

Vocabulary Test, p. 48

Vocabulary Review

Vocabulary Review provides students the opportunity to check their understanding of important concepts and terminology in an online game format.

FOLDABLES StudyOrganizer

Dinah Zike's Foldables®
Have students look through the chapter to make sure they have included examples in their Foldables. Suggest that students keep their Foldables handy while completing the Study Guide and Review pages. Point out that their Foldables can serve as a quick review tool when studying for the chapter test.

Lesson-by-Lesson Review

Intervention If the given examples are not sufficient to review the topics covered by the questions, remind students that the lesson references tell them where to review that topic in their textbook.

Two-Day Option Have students complete the Lesson-by-Lesson Review. Then you can use McGraw-Hill eAssessment to customize another review worksheet that practices all the objectives of this chapter or only the objectives on which your students need more help.

Additional Answers

7. survey; sample: every tenth shopper; population: all potential shoppers

8. observational study; sample: the 25 customers; population: all potential customers

9. survey; sample: every fifth person; population: student body

10a.

[8, 18] scl: 1 by [0, 10] scl: 1

positively skewed

10b. Sample answer: The distribution is skewed, so use the five-number summary. The values range from 8.7 to 18.1 days. The median is 11.55 days, and half of the data are between 9.55 and 13.3 days.

11a.

[300, 330] scl: 1 by [0, 5] scl: 1

positively skewed

11b. Sample answer: The distribution is skewed, so use the five-number summary. Kelly's times range from 301 to 329 seconds. The median is 311, and half of the data are between 307 and 316 seconds.

Lesson-by-Lesson Review

11-1 Designing a Study

Determine whether each situation describes a *survey*, an *experiment*, or an *observational study*. Then identify the sample, and suggest a population from which it may have been selected. **7–9. See margin.**

7. **SHOPPING** Every tenth shopper coming out of a store is asked questions about his or her satisfaction with the store.

8. **MILK SHAKE** A fast food restaurant gives 25 of their customers a sample of a new milk shake and employees monitor their reactions as they taste it.

9. **SCHOOL** Every fifth person coming out of a high school is asked what their favorite class is.

Example 1

A dealership wants to test different promotions. They randomly select 100 customers and ask them which promotion they prefer. Does this situation describe a *survey*, an *experiment*, or an *observational study*? Identify the sample, and suggest a population from which it may have been selected.

This is a survey, because the data are collected from participants' responses. The sample is the 100 customers that were selected, and the population is all potential customers.

11-2 Distributions of Data

10. **DOGSLED** The Iditarod is a race across Alaska. The table shows the winning times, in days, for recent years.

Iditarod Winning Times
9.1, 9.4, 10.3, 9.3, 9.6, 8.7, 9.5, 9.4, 9.2, 17.3, 15.4, 15.5, 14.2, 12.0, 16.6, 13.5, 13.0, 18.1, 12.4, 11.6, 11.5, 11.3, 11.3, 13.1, 11.2, 11.6, 11.6, 9.7
a–b. See margin.

a. Use a graphing calculator to create a histogram. Then describe the shape of the distribution.

b. Describe the center and spread of the data using either the mean and standard deviation or the five-number summary. Justify your choice.

11. **SWIMMING** Kelly's practice times in the 400-meter individual medley are shown in the table. **a–b. See margin.**

Times in Seconds
301, 311, 320, 308, 312, 307, 303, 305, 309, 308, 304, 302, 311, 313, 313, 316, 314, 306, 329, 326, 319, 310, 306, 309, 320, 318, 315, 318, 314, 309

a. Use a graphing calculator to create a box-and-whisker plot. Then describe the shape of the distribution.

b. Describe the center and spread of the data using either the mean and standard deviation or the five-number summary. Justify your choice.

Example 2

Data collected from a group of third graders is shown.

Number of Years Playing an Instrument
2.5, 2.4, 3.1, 2.9, 4.2, 1.3, 2.6, 2.4, 3.3, 1.9, 3.4, 4.8, 2.3, 1.7, 3.2, 2.3, 3.5, 2.2, 3.6, 1.2, 4.4, 2.1, 3.4, 4.5, 1.9, 1.5, 1.4, 0.7, 1.2, 2.5, 1.9, 2.0, 2.4, 2.5, 3.4

a. Use a graphing calculator to create a histogram. Then describe the shape of the distribution.

[0, 5] scl: 1 by [0, 15] scl: 1

The distribution is symmetric.

b. Describe the center and spread of the data using either the mean and standard deviation or the five-number summary. Justify your choice.

The distribution is symmetric, so use the mean and standard deviation. The mean number of years is about 2.6 with standard deviation of about 1 year.

11-3 Probability Distributions

Identify the random variable in each distribution, and classify it as *discrete* or *continuous*. Explain your reasoning.

12. the number of ice cream sandwiches sold at an ice cream shop **12–13. See margin.**

13. the time it takes to run a 5-kilometer race

14. **DANCE RECITALS** The probability distribution lists the probable number of dance recitals per year at Rena's Dance School. Determine the expected number of dance recitals per year. **1.51**

Number of Dance Recitals Per Year					
Recitals	0	1	2	3	4
Probability	0.3	0.3	0.13	0.13	0.14

15. **SNOW DAYS** The distribution lists the number of snow days per year at Washington Elementary over the past 26 years. Determine the expected number of snow days this year. **1.88 snow days**

Number of Snow Days Per Year					
Snow Days	0	1	2	3	4
Frequency	4	8	6	3	5

Example 3

Identify the random variable in a distribution of the number of DVDs on display at a store. Classify it as *discrete* or *continuous*. Explain your reasoning.

The random variable X is the number of DVDs on display. The number of DVDs is countable, so X is discrete.

Example 4

MEDICINE The probability distribution lists the probable number of drops of medicine that a veterinarian administers to her sick patients. Find the expected number of drops of medicine.

Number of Drops of Medicine				
Drops	1	2	3	4
Probability	0.5	0.3	0.1	0.1

Each drop of medicine represents a value of X, and each decimal represents the corresponding probability $P(X)$.

$E(X) = \Sigma[X \cdot P(X)]$

$= 1(0.5) + 2(0.3) + 3(0.1) + 4(0.1)$

$= 0.5 + 0.6 + 0.3 + 0.4$ or 1.8

The expected number of drops of medicine is 1.8.

11-4 The Binomial Distribution

Determine whether each experiment is a binomial experiment or can be reduced to a binomial experiment. If so, describe a trial, determine the random variable, and state *n, p,* and *q.*

16. A survey found that 30% of adults like chocolate ice cream more than any other flavor. You ask 35 adults if they prefer chocolate ice cream more than any other flavor.

17. Thirty random guests from Sheila's birthday party are asked their favorite song. **16–17. See margin.**

18. **WATCHES** According to an online poll, 74% of adults wear watches. Timmy surveyed 25 random adults. What is the probability that 20 of the adults surveyed wear a watch? **15.3%**

19. **SEASONS** Of 1108 people surveyed, 68% say that summer is their favorite season. What is the probability that at least 15 of 20 randomly selected people will prefer summer? **34.3%**

Example 5

WORK According to an online poll, 40% of adults feel that the standard 40-hour workweek should be increased. Sheila conducts a survey of 10 random adults. What is the probability that 3 of the surveyed adults feel that the standard 40-hour workweek should be increased?

A success is an adult agreeing that the 40-hour workweek should be increased, so $p = 0.4$, $X = 3$, $q = 1 - 0.4$ or 0.6, and $n = 10$.

$P(X) = {}_nC_X p^X q^{n-X}$

$P(6) = {}_{10}C_3(0.4)^3(0.6)^{10-3}$

≈ 0.215

The probability that three surveyed adults will feel that the standard 40-hour workweek should be increased is about 0.215 or 21.5%.

Additional Answers

12. The random variable X is the number of ice cream sandwiches sold. The number of ice cream sandwiches countable, so X is discrete.

13. The random variable X is the time it takes to run the race. Time can be anywhere within a certain range. Therefore, X is continuous.

16. This experiment can be reduced to a binomial experiment. Success is yes, failure is no, a trial is asking an adult, and the random variable is the number of yeses; $n = 35$, $p = 0.30$, $q = 0.70$.

17. This experiment cannot be reduced to a binomial experiment because there are more than two possible outcomes.

Anticipation Guide

Have students complete the Chapter 11 Anticipation Guide and discuss how their responses have changed now that they have completed Chapter 11.

Additional Answers (Practice Test)

2a.

[59, 75] scl: 1 by [0, 5] scl: 1

positively skewed

4. This experiment can be reduced to a binomial experiment. Success is yes, failure is no, a trial is asking a teacher, and the random variable is the number of yeses; $n = 15$, $p = 0.65$, $q = 0.35$.

5. This experiment cannot be reduced to a binomial experiment because there are more than two possible outcomes.

6. This experiment cannot be reduced to a binomial experiment because there are more than two possible outcomes.

7. The random variable X is the number of laps that Jesse swims. The laps are countable, so X is discrete.

8. The random variable X is the body temperatures of patients in a hospital. Temperature can be anywhere within a certain range. Therefore, X is continuous.

9. The random variable X is the weights of pets in a pet store. Weight can be anywhere within a certain range. Therefore, X is continuous.

11-5 The Normal Distribution

20. RUNNING TIMES The times in the 40-meter dash for a select group of professional football players are normally distributed with a mean of 4.74 seconds and a standard deviation of 0.13 second.

   a. About what percent of players have times between 4.6 and 4.8 seconds? **53.7%**

   b. About how many of a sample of 800 players will have times below 4.5 seconds? **about 26 players**

21. ATTENDANCE The number of tickets sold at high school basketball games in a particular conference are normally distributed with a mean of 68.7 and a standard deviation of 13.1.

   a. About what percent of the games sell fewer than 75 tickets? **68.5%**

   b. About how many of a sample of 200 games will sell more than 100 tickets? **about 2 games**

22. COMMUTING The number of minutes it takes Phil to commute to work each day are normally distributed with a mean of 18.6 and a standard deviation of 3.5.

   a. About what percent of the time will it take Phil more than 20 minutes to commute to work? **34.5%**

   b. About how many of a sample of 50 days will it take Phil less than 15 minutes to commute to work? **about 8 days**

Example 6

TEST SCORES The midterm test scores for the students in Mrs. Hendrix's classes are normally distributed with a mean of 73.2 and standard deviation of 7.8. About how many test scores are between 70 and 80?

Find the corresponding z-values for $X = 70$ and 80.

$$z = \frac{X - \mu}{\sigma}$$
$$= \frac{70 - 73.2}{7.8}$$
$$\approx -0.41$$

$$z = \frac{X - \mu}{\sigma}$$
$$= \frac{80 - 73.2}{7.8}$$
$$\approx 0.87$$

Using a graphing calculator, you can find the area between $-0.41 < z < 0.87$ to be about 0.47.

[−4, 4] scl: 1 by [0, 0.5] scl: 0.125

11-6 Confidence Intervals and Hypothesis Testing

23. INTERNET A sample of 300 students was asked the average amount of time they spend online during a weeknight. The mean time was 64.3 minutes with a standard deviation of 7.3 minutes. Determine a 95% confidence interval. $63.5 \leq \bar{x} \leq 65.1$

Identify the null and alternative hypotheses for each statement. Then identify the statement that represents the claim.

24. A cupcake shop owner says that they sell 200 cupcakes everyday. $H_0: \mu = 200$ (claim); $H_a: \mu \neq 200$

25. A technician says that it takes at least 45 minutes to diagnose a computer problem.
$H_0: \mu \geq 45$ (claim) ; $H_a: \mu < 45$

26. A member of a swim team says that she practices no more than 35 minutes on Mondays.
$H_0: \mu \leq 35$ (claim); $H_a: \mu > 35$

Example 7

TELEVISION WATCHING A sample of 100 teenagers was asked the average amount of time they spent watching television on the weekends. The mean time was 65 minutes with a standard deviation of 1.6 minutes. Determine a 95% confidence interval for the population.

$$CI = \bar{x} \pm 1.96 \cdot \frac{s}{\sqrt{n}} \quad \text{Confidence Interval Formula}$$
$$= 65 \pm 1.96 \cdot \frac{1.6}{\sqrt{100}} \quad \bar{x} = 65, s = 1.6, n = 100$$
$$= 65 \pm 0.31 \quad \text{Use a calculator.}$$

The 95% confidence interval is $64.69 \leq \mu \leq 65.31$. Therefore, we are 95% confident that the population mean time is between 64.69 and 65.31 minutes.

780 | Chapter 11 | Study Guide and Review

CHAPTER 11 Practice Test

1. **BUTTERFLIES** Students in a biology class are learning about the monarch butterfly's life cycle. Each student is given a caterpillar. When a caterpillar turns into chrysalis, it is placed in a glass enclosure with food and a heat lamp and examined.

   a. Determine whether the situation describes a *survey*, an *experiment* or an *observational study*. **observational study**

   b. Identify the sample, and suggest a population from which it was selected. **sample: the monarch butterflies observed in the study; population: all monarch butterflies**

2. **HEIGHTS** The heights of Ms. Joy's dance students are shown.

   . The distribution **skewed, so use** **e five-number** **mmary. The** **lues range from** **to 74. The median** **64, and half of the** **ta are between 63** **d 67.**

Height (Inches)				
60	64	62	69	64
63	65	64	66	73
74	63	62	65	64
68	70	66	63	61

   a. Use a graphing calculator to create a box-and-whisker plot. Then describe the shape of the distribution. **See margin.**

   b. Describe the center and spread of the data using either the mean and standard deviation or the five-number summary. Justify your choice.

3. A binomial distribution has a 65% rate of success. There are 15 trials.

   a. What is the probability that there will be exactly 12 successes? **about 11.1%**

   b. What is the probability that there will be at least 10 successes? **about 56.5%**

Determine whether each experiment is a binomial experiment or can be reduced to a binomial experiment. If so, describe a trial, determine the random variable, and state *n*, *p*, and *q*. 4–6. See margin.

4. A poll found that 65% of high school teachers own a pet. You ask 15 high school teachers if they own a pet.

5. A study finds that 20% of families in a town have a land line telephone. You ask 55 families how many land line telephones they have.

6. A survey found that on a scale of 1 to 5, a sandwich received a 3.5 rating. A restaurant manager asks 150 customers to rate the sandwich on a scale of 1 to 5.

Identify the random variable in each distribution, and classify it as *discrete* or *continuous*. Explain your reasoning. 7–9. See margin.

7. the number of laps that Jesse swims

8. the body temperatures of patients in a hospital

9. the weights of pets in a pet store

10. **MULTIPLE CHOICE** The table shows the number of gift cards previously won in a mall contest. What is the expected value of the gift card that is won? **C**

    A $250.00

    B $223.15

    C $143.25

    D $100.23

Amount, X	Winners
$100	495
$125	405
$150	285
$200	180
$250	90
$300	45

11. **WEIGHTS** The weights of 1500 bodybuilders are normally distributed with a mean of 190.6 pounds and a standard deviation of 5.8 pounds.

    a. About how many bodybuilders are between 180 and 190 pounds? **about 638 bodybuilders**

    b. What is the probability that a bodybuilder selected at random has a weight greater than 195 pounds? **22.4%**

A normal distribution has a mean of 16.4 and a standard deviation of 2.6.

12. Find the range of values that represent the middle 95% of the distribution. **11.2 < X < 21.6**

13. What percent of the data will be less than 19? **84%**

14. **MULTIPLE CHOICE** A survey of 300 randomly selected members of a fitness club showed that they spent an average of 25.3 minutes per visit in the gym with a standard deviation of 8.6 minutes. Which is the maximum error of estimate for the time spent in the gym using a 95% confidence interval? **G**

    F 0.82   G 0.97   H 1.28   J 2.86

Identify the null and alternative hypotheses for each statement. Then identify the statement that represents the claim.

15. A restaurant owner says that each milk shake contains 5 strawberries. H_0: $\mu = 5$ (claim); H_a: $\mu \neq 5$

16. A valet says that no more than 56 cars can be parked in the parking lot at one time. H_0: $\mu \leq 56$ (claim); H_a: $\mu > 56$

 connectED.mcgraw-hill.com **781** ✓

Summative Assessment

Use these alternate leveled chapter tests to differentiate assessment for your students.

Leveled Chapter 11 Tests

Form	Type	Level	📁 Page(s)
1	MC	AL	49–50
2A	MC	OL	51–52
2B	MC	OL	53–54
2C	FR	OL	55–56
2D	FR	OL	57–58
3	FR	BL	59–60
Vocabulary Test			48
Extended-Response Test			61

MC = multiple-choice questions
FR = free-response questions

McGraw-Hill eAssessment Customize and create multiple versions of your chapter test and their answer keys. All of the questions from the leveled chapter tests in the *Chapter 11 Resource Masters* are also available on McGraw-Hill eAssessment.

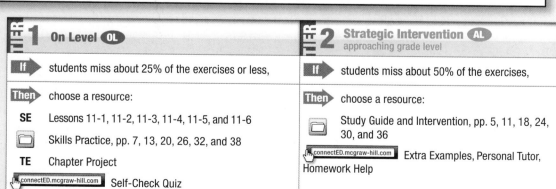

InterventionPlanner

TIER 1 On Level OL

If ▶ students miss about 25% of the exercises or less,

Then ▶ choose a resource:

SE Lessons 11-1, 11-2, 11-3, 11-4, 11-5, and 11-6

📁 Skills Practice, pp. 7, 13, 20, 26, 32, and 38

TE Chapter Project

connectED.mcgraw-hill.com Self-Check Quiz

TIER 2 Strategic Intervention AL
approaching grade level

If ▶ students miss about 50% of the exercises,

Then ▶ choose a resource:

📁 Study Guide and Intervention, pp. 5, 11, 18, 24, 30, and 36

connectED.mcgraw-hill.com Extra Examples, Personal Tutor, Homework Help

TIER 3 Intensive Intervention
2 or more grades below level

If ▶ students miss about 75% of the exercises,

Then ▶ use *Math Triumphs, Alg. 2, Ch. 5*

connectED.mcgraw-hill.com Extra Examples, Personal Tutor, Homework Help, Review Vocabulary

1 Focus

Objective Use the strategy for solving multi-step problems to solve standardized test problems.

2 Teach

Scaffolding Questions

Ask:

- What is another question to ask yourself when you think that the problem might require multiple steps to solve it? Sample answer: What information am I missing?

- How do you know if you have completed all the steps needed to solve the problem? Sample answer: Go back and check the steps you originally listed. Reread the original problem and see if your solution fits.

Solve Multi-Step Problems

Some problems that you will encounter on standardized tests require you to solve multiple parts in order to come up with the final solution. Use this lesson to practice these types of problems.

Strategies for Solving Multi-Step Problems

Step 1

Read the problem statement carefully.

Ask yourself:

- What am I being asked to solve? What information is given?
- Are there any intermediate steps that need to be completed before I can solve the problem?

Step 2

Organize your approach.

- List the steps you will need to complete in order to solve the problem.
- Remember that there may be more than one possible way to solve the problem.

Step 3

Solve and check.

- Work as efficiently as possible to complete each step and solve.
- If time permits, check your answer.

Standardized Test Example

Read the problem. Identify what you need to know. Then use the information in the problem to solve.

There are 15 boys and 12 girls in Mrs. Lawrence's homeroom. Suppose a committee is to be made up of 6 randomly selected students. What is the probability that the committee will contain 3 boys and 3 girls? Round your answer to the nearest tenth of a percent.

A 27.2%   C 31.5%

B 29.6%   D 33.8%

Comstock Images/Jupiterimages/PictureQuest

Read the problem statement carefully. You are asked to find the probability that a committee will be made up of 3 boys and 3 girls. Finding this probability involves successfully completing several steps.

Step 1 Find the number of possible successes.

There are C(15, 3) ways to choose 3 boys from 15, and there are C(12, 3) to choose 3 girls from 12. Use the Fundamental Counting Principle to find s, the number of possible successes.

$$s = C(15, 3) \times C(12, 3) = \frac{15!}{12!3!} \times \frac{12!}{9!3!} \text{ or } 100,100$$

Step 2 Find the total number of possible outcomes.

Compute the number of ways 6 people can be chosen from a group of 27 students.

$$C(27, 6) = 296,010$$

Step 3 Compute the probability.

Find the probability by comparing the number of successes to the number of possible outcomes.

$$P(3 \text{ boys, 3 girls}) = \frac{100,100}{296,010} \approx 0.33816$$

So, there is about a 33.8% chance of selecting 3 boys and 3 girls for the committee. The answer is D.

Additional Example

STANDARDIZED TEST PRACTICE

A bakery has 8 fudge cakes and 10 yellow cakes, each in a separate, identical box. Anna needs to ice 4 cakes. What is the probability that she will randomly choose 2 fudge cakes and 2 yellow cakes to ice? B

A   about 9.7%

B   about 25.7%

C   about 41.2%

D   about 50%

3 Assess

Use Exercises 1 and 2 to assess students' understanding.

Exercises

Read the problem. Identify what you need to know. Then use the information in the problem to solve.

1. There are 52 cards in a standard deck. Of these, 4 of the cards are aces. What is the probability of a randomly dealt 5-card hand containing a pair of aces? Round your answer to the nearest whole percent. A

A   4%

B   5%

C   6%

D   7%

2. According to the table, what is the probability that a randomly selected camper went on the horse ride, given that the camper is an 8th grader? H

Camp Activities			
Grade	Canoe Trip	Horse Ride	Nature Hike
6th	8	6	3
7th	5	4	7
8th	11	9	6

F   0.731   H   0.346

G   0.441   J   0.153

 connectED.mcgraw-hill.com   **783**

Diagnose Student Errors

Survey student responses for each item. Class trends may indicate common errors and misconceptions.

1. A found the probability that the score is more than 2 or more standard deviations above the mean
   B found the probability that the score is more than 2 standard deviations above the mean
   C correct
   D found the probability that the score is above the mean, but no more than 1 standard deviation above the mean

2. F found one fourth of diameter
   G correct
   H guess
   J found diameter

3. A correct
   B transposed x and y
   C found inverse of slope
   D found equation that works for only (9, 10.80)

4. F thinks a greater coefficient gives a wider graph
   G thinks a lessor coefficient gives a wider graph
   H correct
   J thinks a coefficient with a greater absolute value gives a wider graph

5. A found σx instead of Sx
   B correct
   C guess
   D eliminated duplicate grades

6. F missed multiplying $-\frac{1}{4}$
   G ignored $-\frac{1}{4}$
   H multiplied $12b$ by $\frac{1}{4}$ instead of $-\frac{1}{4}$
   J correct

7. A took cube root of 27 and divided 6 by 3 instead of dividing 3 by 6
   B took cube root of 27 and misapplied laws of exponents
   C failed to include x in square root
   D correct

8. F correct
   G subtracted incorrectly
   H used exterior angle measure instead of its supplement
   J used exterior angle measure instead of its supplement

9. A thinks coefficient determines decay
   B correct
   C thinks base greater than 1 determines decay
   D thinks fractional base determines decay

10. F found equation that works for only (1, 6)
   G found equation that works for only (1, 6)
   H found incorrect initial term
   J correct

Multiple Choice

Read each question. Then fill in the correct answer on the answer document provided by your teacher or on a sheet of paper.

1. Suppose the test scores on a final exam are normally distributed with a mean of 74 and a standard deviation of 3. What is the probability that a randomly selected test has a score higher than 77? **C**

   A 2.5%   C 16%
   B 13.5%   D 34%

2. A diameter of a circle has endpoints $A(4, 6)$ and $B(-3, -1)$. Find the approximate length of the radius. **G**

   F 2.5 units   H 5.1 units
   G 4.9 units   J 9.9 units

3. An equation can be used to find the total cost of a pizza with a certain diameter. Using the table below, find the equation that best represents y, the total cost, as a function of x, the diameter in inches. **A**

Diameter, x (in.)	Total Cost, y
9	$10.80
12	$14.40
20	$24.00

   A $y = 1.2x$ C $y = 0.83x$
   B $x = 1.2y$ D $y = x + 1.80$

4. Which shows the functions correctly listed in order from widest to narrowest graph? **H**

   F $y = 8x^2, y = 2x^2, y = \frac{1}{2}x^2, y = -\frac{4}{5}x^2$
   G $y = -\frac{4}{5}x^2, y = \frac{1}{2}x^2, y = 2x^2, y = 8x^2$
   H $y = \frac{1}{2}x^2, y = -\frac{4}{5}x^2, y = 2x^2, y = 8x^2$
   J $y = 8x^2, y = 2x^2, y = -\frac{4}{5}x^2, y = \frac{1}{2}x^2$

> **Test-TakingTip**
> Question 5 You can use a scientific calculator to find the standard deviation. Enter the data values as a list and calculate the 1-Var statistics.

5. The table at the right shows the grades earned by students on a science test. Calculate the standard deviation of the test scores. **B**

76	84	91	75	83
82	65	94	90	71
92	84	83	88	80
78	84	89	95	93

   A 7.82   B 8.03   C 8.23   D 8.75

6. Which expression is equivalent to $(6a - 2b) - \frac{1}{4}(4a + 12b)$? **J**

   F $5a + 10b$ H $5a + b$
   G $10a + 10b$ J $5a - 5b$

7. Simplify $\sqrt[6]{27x^3}$. **D**

   A $3x^2$ B $3x$ C $\sqrt{3}x$ D $\sqrt{3x}$

8. In the figure below, lines a and b are parallel. What are the measures of the angles in the triangle? **F**

   F 42, 48, 90   H 48, 52, 90
   G 42, 90, 132   J 48, 90, 132

9. Which of the following functions represents exponential decay? **B**

   A $y = 0.2(7)^x$ C $y = 4(9)^x$
   B $y = (0.5)^x$ D $y = 5\left(\frac{4}{3}\right)^x$

10. Using the table below, which expression can be used to determine the nth term of the sequence? **J**

n	1	2	3	4
y	6	10	14	18

   F $y = 6n$ H $y = 2n + 1$
   G $y = n + 5$ J $y = 2(2n + 1)$

Short Response/Gridded Response

Record your answers on the answer sheet provided by your teacher or on a sheet of paper.

11. Determine whether each of the following situations calls for a *survey*, an *observational study*, or an *experiment*. Explain the process. **a. experiment**

    a. Caroline wants to find out if a particular plant food helps plants to grow faster than just water.

    b. Allison wants to find opinions on favorite candidates in the upcoming student counsel elections. **survey**

    c. Manuel wants to find out if people who get regular exercise sleep better at night.
        observational study

12. GRIDDED RESPONSE Carla received a map of some walking paths through her college campus. Paths A, B, and C are parallel. What is the length *x* to the nearest tenth of a foot? **17.1**

13. Alex wants to find the area of a triangle. He draws the triangle on a coordinate plane and finds that it has vertices at (2, 1), (3, 4), and (1, 4). Find the area of the triangle using determinants. **3 units²**

14. GRIDDED RESPONSE Perry drove to the gym at an average rate of 30 miles per hour. It took him 45 minutes. Going home, he took the same route, but drove at a rate of 45 miles per hour. How many miles is it to his house from the gym? **22.5**

Extended Response

Record your answers on a sheet of paper. Show your work.

15. Martin is taking a multiple choice test that has 8 questions. Each question has four possible answers: A, B, C, or D. Martin forgot to study for the test, so he must guess at each answer.

    a. What is the probability of guessing a correct answer on the test? **0.25**

    b. What is the expected number of correct answers if Martin guesses at each question? **2**

    c. What is the probability that Martin will get at least half of the questions correct? Round your answer to the nearest tenth of a percent.
        about 11.4%

16. Christine had one dress and three sweaters cleaned at the dry cleaner and the charge was $19.50. The next week, she had two dresses and two sweaters cleaned for a total charge of $23.00.

    a. Let *d* represent the price of cleaning a dress and *s* represent the price of cleaning a sweater. Write a system of linear equations to represent the prices of cleaning each item. $d + 3s = 19.5$; $2d + 2s = 23$

    b. Solve the system of equations using substitution or elimination. Explain your choice of method.

    c. What will the charge be if Christine takes two dresses and four sweaters to be cleaned? **$31.00**

16b. $d = 7.50; $s = 4.00; substitution, because the variable *d* in the first equation has a coefficient of 1.

Need ExtraHelp?

If you missed Question...	1	2	3	4	5	6	7	8	9	10	11	12	13	14	15	16
Go to Lesson...	11-5	9-1	2-4	4-7	0-9	5-1	6-6	3-1	7-1	10-2	11-1	1-4	3-8	8-5	11-4	3-1

Formative Assessment

You can use these pages to benchmark student progress.

📁 Standardized Test Practice, pp. 62–64

McGraw-Hill
ᵉAssessment   Create practice tests that align to your state standards, the Common Core State Standards, and other national standards such as TIMSS and NAEP.

Answer Sheet Practice

Have students simulate taking a standardized test by recording their answers on practice recording sheets.

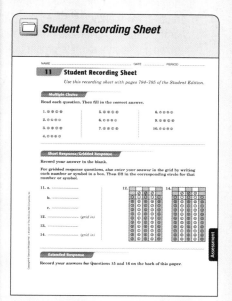

Homework Option

Get Ready for Chapter 12 Assign students the exercises on p. 787 as homework to assess whether they possess the prerequisite skills needed for the next chapter.

Chapter 11 Answer Appendix

3. Observation study; sample answer: The scores of the participants are observed and compared without them being affected by the study.

4. Experiment; sample answer: A sample of dyed shirts will need to be tested, which means that the members of the sample will be affected by the study.

7. objective: to determine how many people in the U.S. are interested in purchasing a hybrid vehicle; population: the people surveyed; sample survey questions: Do you currently own a hybrid vehicle? Are you planning on purchasing a hybrid vehicle?

8. Sample answer: The flaw is that the experimental group consists of students, and the control group consists of instructors. On average, college students are younger than their instructors, and thus, are more likely to score higher on a physical test whether given a vitamin or not.

9. objective: to determine whether the protein shake helps athletes recover from exercise; population: all athletes; experiment group: athletes given the protein shake; control group: athletes given a placebo; sample procedure: The researchers could randomly divide the athletes into two groups: an experimental group given the protein shake and a control group given the placebo. Next, they could have the athletes exercise and then drink the protein shake or placebo. Later, the researchers could interview the athletes to see how they feel.

10. survey; sample: customers that take the online survey; population: all customers

11. observational study; sample: physics students selected; population: all college students that take a physics course

12. experiment; sample: adults participating in the study; population: all adults

13. survey; sample: people that receive the questionnaire; population: all viewers

24a. objective: to determine whether a mixture of salt and lemon juice will remove copper oxide from copper objects; population: all copper objects; experimental group: pennies that are submerged in the mixture; control group: pennies that are submerged in a placebo mixture; sample procedure: Pedro could randomly assign the pennies into two groups, and create the lemon and salt mixture. Next, he could submerge the experimental group of pennies in the mixture for a certain period of time, remove them, and then visually compare the two groups of pennies.

24b. Sample answer: The pennies should be chosen so that they have roughly the same amount of copper oxide and are from the same time period. Using these guidelines could eliminate bias regarding the initial condition and composition, so that the conditions of the experimental and control groups are exactly the same at the start of the experiment.

27a. See students' work.

27b. Sample answer for Product A: $\approx 63.3\%$

Product A	
Number	**Frequency**
0–6	IIII IIII IIII IIII
7–9	IIII IIII I

Sample answer for Product B: $\approx 76.7\%$

Product B	
Number	**Frequency**
0–7	IIII IIII IIII IIII III
8–9	IIII II

27c. Sample answer: Yes; the probability that Product B is effective is 14.4% higher than that of Product A.

27d. Sample answer: It depends on what the product is and how it is being used. For example, if the product is a pencil sharpener, then the lower price may be more important than the effectiveness, and therefore, might not justify the price difference. However, if the product is a life-saving medicine, the effectiveness may be more important than the price, and therefore, might justify the price difference.

28. False; sample answer: A sample statistic is used to estimate a population parameter.

30. Sample answer:

objective: Determine the average amount of time that students spend studying at the library.

population: All students that study at the library.

sample: 30 randomly selected students studying at the library during a given week.

Study Time (minutes)				
38	16	45	41	63
18	20	17	8	15
41	28	55	19	15
30	11	20	79	24
78	24	26	32	19

mean: ≈ 26.1 min

31. Sample answer: the sampling method used, the type of sample that was selected, the type of study performed, the survey question(s) that were asked or procedures that were used

32a. Sample answer: In a convenience sample, members are selected based on the convenience of the researcher. One example is handing a survey to shoppers as they walk out of the mall. This method could result in bias if the members of the population who are readily available to be sampled are not representative of the entire population.

32b. Sample answer: In a self-selected sample, members volunteer to be in the sample. This method could result in bias if certain groups of people in the population choose not to volunteer.

32c. Sample answer: In a stratified sample, the population is first divided into similar, nonoverlapping groups, and members are then randomly selected from each group. This method could result in bias if the entire population is not represented when divided into groups or if the members are not randomly selected from each group.

32d. Sample answer: In a systematic sample, a rule is used to select the members. This method could result in bias if the rule does not include everyone in the population.

Lesson 11-2

3a.

Mrs. Johnson's Class

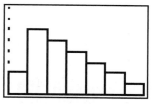

[5, 40] scl: 5 by [0, 8] scl: 1

Mr. Edmunds' Class

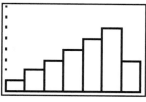

[5, 40] scl: 5 by [0, 8] scl: 1

Mrs. Johnson's class, positively skewed; Mr. Edmunds' class, negatively skewed

3b. Sample answer: The distributions are skewed, so use the five-number summaries. The range for both classes is the same. However, the median for Mrs. Johnson's class is 17 and the median for Mr. Edmunds' class is 28. The lower quartile for Mr. Edmunds' class is 20. Since this is greater than the median for Mrs. Johnson's class, this means that 75% of the data from Mr. Edmunds' class is greater than 50% of the data from Mrs. Johnson's class. Therefore, we can conclude that the students in Mr. Edmunds' class had slightly higher sales overall than the students in Mrs. Johnson's class.

6a.

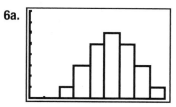

[0, 18] scl: 2 by [0, 8] scl: 1

[0, 18] scl: 2 by [0, 5] scl: 1

symmetric

6b. Sample answer: The distribution is symmetric, so use the mean and standard deviation. The mean number of movies watched was about 10.7 with standard deviation of about 3 movies.

7a.

Sophomore Year

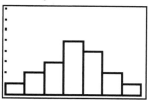

[1200, 1900] scl: 100 by [0, 8] scl: 1

Junior Year

[1300, 2200] scl: 100 by [0, 8] scl: 1

both symmetric

7b. Sample answer: The distributions are symmetric, so use the means and standard deviations. The mean score for sophomore year is about 1552.9 with standard deviation of about 147.2. The mean score for junior year is about 1753.8 with standard deviation of about 159.1. We can conclude that the scores and the variation of the scores from the mean both increased from sophomore year to junior year.

8a.

Yorkshire

[50, 105] scl: 5 by [0, 6] scl: 1

Applewood

[35, 90] scl: 5 by [0, 6] scl: 1

Yorkshire, positively skewed; Applewood, negatively skewed

8b. Sample answer: The distributions are skewed, so use the five-number summaries. The median for Yorkshire is 63.5, and the median for Applewood is 60. The lower quartile for Applewood is 52, while the minimum for Yorkshire is 53. This means that 25% of the incomes for Applewood are lower than any of the incomes for Yorkshire. Also, the upper 25% of incomes for Yorkshire is between 72 and 103, while the upper 25% of incomes for Applewood is between 65 and 87. We can conclude that the incomes for the households in Yorkshire are greater than the incomes for the households in Applewood.

9a.

[3000, 18,000] scl: 1000 by [0, 5] scl: 1

both symmetric

9b. Sample answer: The distributions are symmetric, so use the means and standard deviations. The mean for the public colleges is $4037.50 with standard deviation of about $621.93. The mean for private colleges is about $12,803.11 with standard deviation of about $2915.20. We can conclude that not only is the average cost of private schools far greater than the average cost of public schools, but the variation of the costs from the mean is also much greater.

10a.

[225, 775] scl: 50 by [0, 5] scl: 1

boys, positively skewed; girls, symmetric

10b. Sample answer: One distribution is skewed and the other is symmetric, so use the five-number summaries. The maximum value for the boys is 450. The lower quartile for the girls is 453. This means that 75% of the data for the girls is higher than any data for the boys. We can conclude that 75% of the girls spent more money on prom than any of the boys.

12a.

[50, 100] scl: 5 by [0, 5] scl: 1

Sample answer: The distribution is negatively skewed, so use the five-number summary. The scores range from 57 to 96. The median is 86, and half of the data are between 76 and 89.

12b.

[50, 100] scl: 5 by [0, 5] scl: 1

Sample answer: The distribution is symmetric, so use the mean and standard deviation. The mean is about 85.6 with standard deviation of about 5.9.

14. Sample answer: Since the distribution has two clusters, an overall summary of center and spread would give an inaccurate depiction of the data. Instead, summarize the center and spread of each cluster individually using its mean and standard deviation.

15. Sample answer: The heights of the players on the Pittsburgh Steelers roster appear to represent a normal distribution.

Heights of the Players on the 2009 Pittsburgh Steelers Roster (inches)							
75	74	71	70	74	75	77	72
71	72	70	70	75	78	71	75
77	71	69	70	77	75	74	73
77	71	73	76	76	74	72	75
75	70	70	74	73	76	79	73
71	69	70	77	77	80	75	77
67	74	69	76	77	76		

The mean of the data is about 73.61 in. or 6 ft 1.61 in. The standard deviation is about 2.97 in.

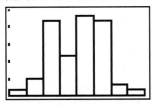

[66, 82] scl: 2 by [0, 15] scl: 3

The birth months of the players do not display central tendency.

Birth Months of the Players on the 2009 Pittsburgh Steelers Roster							
1	12	10	3	11	1	10	5
4	8	9	11	1	1	11	5
8	6	11	4	3	4	8	5
3	7	2	1	11	4	3	2
1	1	6	1	6	8	11	9
3	3	1	6	9	1	9	9
6	5	10	11	11	12		

[0, 12] scl: 2 by [0, 15] scl: 3

16. Sample answer: The distribution for a data set is positively skewed when the majority of the data is on the left of the mean and a tail appears to the right of the mean. An example is when the data set includes the height of everyone in an elementary school, most of the data will be on the left side (the students), while a comparatively small amount will be on the right (the teachers and staff). The distribution for a data set is negatively skewed when the majority of the data is on the right of the mean and a tail appears to the left of the mean. An example is when the batting averages of a baseball lineup are listed, most of the data will be at a certain level while the pitcher will typically be much lower. The distribution for a data set is symmetric when the data are evenly distributed on both sides of the mean. An example is when test scores are calculated for an entire state, most of the students will place in the middle, while some will place above or below.

24. Step 1: After the first guest has arrived, no handshakes have taken place. $\frac{1(1-1)}{2} = 0$, so the formula is correct for $n = 1$.

Step 2: Assume that after k guests have arrived, a total of $\frac{k(k-1)}{2}$ handshakes have taken place, for some positive integer k.

Step 3: When the $(k + 1)$st guest arrives, he or she shakes hands with the k guests already there, so the total number of handshakes that have then taken place is $\frac{k(k-1)}{2} + k$.

$$\frac{k(k-1)}{2} + k = \frac{k(k-1) + 2k}{2}$$
$$= \frac{k[(k-1) + 2]}{2}$$
$$= \frac{k(k+1)}{2} \text{ or } \frac{(k+1)k}{2}$$

The last expression is the formula to be proved, where $n = k + 1$. Thus, the formula is true for $n = k + 1$.

Therefore, the total number of handshakes is $\frac{n(n-1)}{2}$ for all positive integers n.

Lesson 11-3 (Guided Practice)

1A. The random variable X is the distance traveled for each throw. Distance can be anywhere within a certain range. Therefore, X is continuous.

1B. The random variable X is the ages of the counselors. The ages are countable, so X is discrete.

2A.

Sum	2	3	4	5	6	7	8	9	10	11	12	13	14	15	16	18	20
Frequency	1	2	3	4	5	6	5	6	5	6	5	4	4	2	3	2	1
Relative Frequency	$\frac{1}{64}$	$\frac{1}{32}$	$\frac{3}{64}$	$\frac{1}{16}$	$\frac{5}{64}$	$\frac{3}{32}$	$\frac{5}{64}$	$\frac{3}{32}$	$\frac{5}{64}$	$\frac{3}{32}$	$\frac{5}{64}$	$\frac{1}{16}$	$\frac{1}{16}$	$\frac{1}{32}$	$\frac{3}{64}$	$\frac{1}{32}$	$\frac{1}{64}$

2B.

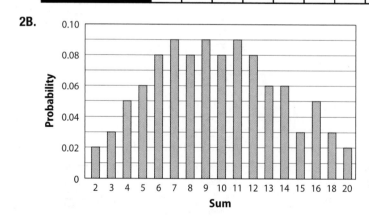

3A.

Sum	Frequency	Relative Frequency
2	0	0
3	8	0.08
4	4	0.04
5	5	0.05
6	5	0.05
7	7	0.07
8	8	0.08
9	7	0.07
10	10	0.10
11	7	0.07
12	10	0.10
13	9	0.09
14	5	0.05
15	4	0.04
16	2	0.02
18	7	0.07
20	2	0.02

3B.

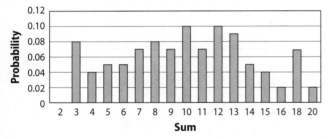

Sum of Two Spins

Lesson 11-3

5a.

Sum	Frequency	Relative Frequency
4	1	$\frac{1}{64}$
6	2	$\frac{1}{32}$
7	2	$\frac{1}{32}$
8	3	$\frac{3}{64}$
9	4	$\frac{1}{16}$
10	5	$\frac{5}{64}$
11	4	$\frac{1}{16}$
12	7	$\frac{7}{64}$
13	4	$\frac{1}{16}$
14	7	$\frac{7}{64}$
15	4	$\frac{1}{16}$
16	5	$\frac{5}{64}$
17	4	$\frac{1}{16}$
18	4	$\frac{1}{16}$
19	2	$\frac{1}{32}$
20	3	$\frac{3}{64}$
22	2	$\frac{1}{32}$
24	1	$\frac{1}{64}$

5b.

Sum of Two Spins

5c.

Sum	Frequency	Relative Frequency
4	1	$\frac{1}{64}$
6	1	$\frac{1}{64}$
7	3	$\frac{3}{64}$
8	3	$\frac{3}{64}$
9	2	$\frac{1}{32}$
10	6	$\frac{3}{32}$
11	5	$\frac{5}{64}$
12	7	$\frac{7}{64}$
13	5	$\frac{5}{64}$
14	8	$\frac{1}{8}$
15	3	$\frac{3}{64}$
16	3	$\frac{3}{64}$
17	4	$\frac{1}{16}$
18	5	$\frac{5}{64}$
19	2	$\frac{1}{32}$
20	3	$\frac{3}{64}$
22	2	$\frac{1}{32}$
24	1	$\frac{1}{64}$

5d.

Sum of Two Spins

10a.

Prize, X	P(X)
$100	0.35
$250	0.25
$500	0.15
$1000	0.10
$2500	0.08
$5000	0.04
$7500	0.02
$10,000	0.01

10b.

Prizes Won

10c.

Prize, X	Frequency	Relative Frequency
$100	23	0.46
$250	8	0.16
$500	6	0.12
$1000	5	0.10
$2500	3	0.06
$5000	2	0.04
$7500	1	0.02
$10,000	2	0.04

10d.

Prizes Won

13a.

Probability to Win Each Prize

14a. 4.2; Sample answer: The expected number is 4.2, so we can expect there to be 4 students running. We cannot have 0.2 people, so we round to the nearest whole number.

14b. Sample answer:

Number of Students, X	Frequency	Relative Frequency
1	5	0.10
2	8	0.16
3	7	0.14
4	1	0.02
5	20	0.40
6	9	0.18

14c. Sample answer:

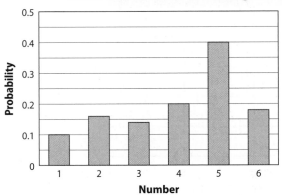

Number of Students Running

15c.

Number of Upsets, X	Frequency	Relative Frequency
0	1	0.02
1	2	0.04
2	7	0.14
3	4	0.08
4	9	0.18
5	17	0.34
6	4	0.08
7	6	0.12
8	0	0

15d.

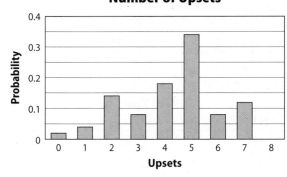

Number of Upsets

18d.

Color	Frequency	Relative Frequency
red	8	0.08
yellow	27	0.27
blue	65	0.65

18e.

Dart Landings

21. Sample answer: A spinner with 5 equal-sided areas shaded red, blue, yellow, green, and brown.

Color	red	blue	yellow	green	brown
Probability	0.2	0.2	0.2	0.2	0.2

22. Sample answer: False; a random variable X representing the number of Web sites on the Internet at any given time is infinite and countable. However, there cannot be 1.3 Web sites, so it is also discrete.

23. Sample answer: A discrete probability distribution can be the uniform distribution of the roll of a die. In this type of distribution, there are only a finite number of possibilities. A continuous probability distribution can be the distribution of the lives of 400 batteries. In this distribution, there are an infinite number of possibilities.

24. Sample answer: Since the investments have identical expected values, an investor would expect to earn the same amount of money on each investment. However, since they have significantly different standard deviations, the investment with the higher standard deviation is much riskier than the other investment. The greater standard deviation indicates a greater variability, so the riskier investment will have an opportunity to earn more money than the other investment, but it will also have an opportunity to lose more as well.

Mid-Chapter Quiz

6a.

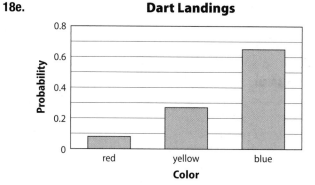

[25, 50] scl: 1 by [0, 5] scl: 1

positively skewed

6b. Sample answer: The distribution is skewed, so use the five-number summary. The ages range from 27 to 47 years. The median is 33 years, and half of the data are between 29 and 38 years.

7a.

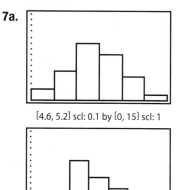

[4.6, 5.2] scl: 0.1 by [0, 15] scl: 1

[4.4, 5.1] scl: 0.1 by [0, 15] scl: 1

both symmetric

7b. Sample answer: The distributions are symmetric, so use the means and standard deviations. The mean time for Aiden is about 4.88 with standard deviation of about 0.12. The mean time for Mark is about 4.69 with standard deviation of about 0.15. We can conclude that Aiden's times were higher than Mark's and the variation of Mark's times from the mean is greater than Aiden's.

9. The random variable X is the number of calls. The calls are countable, so X is discrete.

10. The random variable X is the number of books sold. The books are countable, so X is discrete.

11. The random variable X is the height of students in gym class. Height can be anywhere within a certain range. Therefore, X is continuous.

12. The random variable X is the weight of the animals. Weight can be anywhere within a certain range. Therefore, X is continuous.

Lesson 11-4 (Guided Practice)

1A. This experiment can be reduced to a binomial experiment. Success is yes, failure is no, a trial is asking a student, and the random variable is the number of yeses; $n = 75$, $p = 0.34$, $q = 0.66$.

1B. This experiment cannot be reduced to a binomial experiment because the events are not independent. The probability of choosing an ace changes after each selection.

2. Sample answer:

Step 1 A trial is drawing a card from a deck. The simulation will consist of 26 trials.

Step 2 A success is drawing an even-numbered card. The probability of success is $\frac{5}{13}$ and the probability of failure is $\frac{8}{13}$.

Step 3 The random variable X represents the number of even-numbered cards drawn in 20 trials.

Step 4 Use a random number generator. Let 0–4 represent drawing an even-numbered card. Let 5–12 represent drawing all other cards. Make a frequency table and record the results as you run the generator.

Outcome	Tally	Frequency
Even-Numbered Card	~~IIII~~ ~~IIII~~	10
Other Cards	~~IIII~~ ~~IIII~~	10

The experimental probability is $\frac{10}{20}$ or 50%. This is greater than the theoretical probability of $\frac{5}{13}$ or about 38.5%.

Lesson 11-4

10. Sample answer:

Step 1 A trial is rolling two dice. The simulation will consist of 25 trials.

Step 2 A success is rolling a 7. The probability of success is $\frac{1}{6}$ and the probability of failure is $\frac{5}{6}$.

Step 3 The random variable X represents the number of times a 7 is rolled in 25 trials.

Step 4 Use a random number generator. Let 0 represent rolling a 7. Let 1–5 represent all other outcomes. Make a frequency table and record the results as you run the generator.

Outcome	Tally	Frequency
Rolling a 7	IIII	4
Other Outcomes	~~IIII~~ ~~IIII~~ ~~IIII~~ ~~IIII~~ I	21

The experimental probability is $\frac{4}{25}$ or 16%. This is approximately equal to the theoretical probability of $\frac{1}{6}$ or about 16.7%.

11. Sample answer:

Step 1 A trial is pulling out a marble. The simulation will consist of 20 trials.

Step 2 A success is pulling out a red marble. The probability of success is $\frac{5}{12}$ and the probability of failure is $\frac{7}{12}$.

Step 3 The random variable X represents the number of red marbles pulled out in 20 trials.

Step 4 Use a random number generator. Let 0–4 represent pulling out a red marble. Let 5–11 represent all other outcomes. Make a frequency table and record the results as you run the generator.

Outcome	Tally	Frequency
Red Marble	~~IIII~~ ~~IIII~~	10
Other Outcomes	~~IIII~~ ~~IIII~~	10

The experimental probability is $\frac{10}{20}$ or 50%. This is greater than the theoretical probability of $\frac{5}{12}$ or about 41.7%.

12. Sample answer:

Step 1 A trial is spinning the spinner. The simulation will consist of 25 trials.

Step 2 A success is the spinner landing on an even number. The probability of success is $\frac{2}{5}$ and the probability of failure is $\frac{3}{5}$.

Step 3 The random variable X represents the number of times the spinner stops on an even number in 25 trials.

Step 4 Use a random number generator. Let 0–1 represent the spinner stopping on an even number. Let 2–4 represent all other outcomes. Make a frequency table and record the results as you run the generator.

Outcome	Tally	Frequency
Even Number	~~IIII~~ ~~IIII~~ I	11
Other Outcomes	~~IIII~~ ~~IIII~~ IIII	14

The experimental probability is $\frac{11}{25}$ or 44%. This is slightly greater than the theoretical probability of $\frac{2}{5}$ or 40%.

13. Sample answer:

Step 1 A trial is drawing a card from a deck. The simulation will consist of 20 trials.

Step 2 A success is drawing a face card. The probability of success is $\frac{3}{13}$ and the probability of failure is $\frac{10}{13}$.

Step 3 The random variable X represents the number of face cards drawn in 20 trials.

Step 4 Use a random number generator. Let 0–2 represent drawing a face card. Let 3–12 represent all other outcomes. Make a frequency table and record the results as you run the generator.

Outcome	Tally	Frequency
Face Card	II	2
Other Cards	~~IIII~~ ~~IIII~~ ~~IIII~~ III	18

The experimental probability is $\frac{2}{20}$ or 10%. This is less than the theoretical probability of $\frac{3}{13}$ or about 23.1%.

51. The random variable X is the number of customers at an amusement park. The customers are finite and countable, so X is discrete.

52. The random variable X is the running time of a movie. The time can be anywhere within a certain range, so X is continuous.

53. The random variable X is the number of hot dogs sold at a sporting event. The hot dogs are finite and countable, so X is discrete.

54. The random variable X is the distance between two cities. The distance can be anywhere within a certain range, so X is continuous.

55a.

[5, 25] scl: 2 by [0, 5] scl: 1

symmetric

55b. Sample answer: The distribution is symmetric, so use the mean and standard deviation. The mean is about $16.02 with standard deviation of about $4.52.

Lesson 11-5

16a.

[−4, 4] scl: 1 by [0, .5] scl: 0.125

16b.

[−4, 4] scl: 1 by [0, .5] scl: 0.125

19c. Sample answer: I would expect people with several traffic citations to lie to the far right of the distribution where insurance costs are highest, because I think insurance companies would charge them more.

19d. Sample answer: As the probability of an accident occurring increases, the more an auto insurance company is going to charge. I think auto insurance companies would charge younger people more than older people because they have not been driving as long. I think they would charge more for expensive cars and sports cars and less for cars that have good safety ratings. I think they would charge a person less if they have a good driving record and more if they have had tickets and accidents.

20a. Math: 1.03; Science: 0.81; Social Studies: 1.21

20b. Math: 84.8%; Science: 79.1%; Social Studies: 88.7%

20c. Social Studies; Sample answer: When the distributions are standardized, Nikki's relative scores on each test are 84.8%, 79.1%, and 88.7%. These standardized scores are Nikki's scores in relation to the population of scores for each individual test. Therefore, her 86% on the Social Studies test was better than 88.7% of the other test takers' scores on that test. This is the highest percentage of the three tests, so Nikki performed the best in Social Studies.

25. Sample answer: The scores per team in each game of the first round of the 2010 NBA playoffs. The mean is 96.56 and the standard deviation is 11.06. The middle 68% of the distribution is $85.50 < X < 107.62$. The middle 95% is $74.44 < X < 118.68$. The middle 99.7% is $63.38 < X < 129.74$.

26. Sample answer: The z-value represents the position of a value X in a normal distribution. A z-value of 1.43 means that the corresponding data value X is 1.43 standard deviations to the right of the mean in the distribution. An interval of all of the values greater than X in the distribution will be represented by the area under the curve from $z = 1.43$ to $z = 4$. This area is equivalent to the probability of the interval occurring (a random data value falling within the interval).

Lesson 11-6

19a.

19b. Sample answer: With everything else held constant, increasing the same size will decrease the size of the confidence interval.

19c.

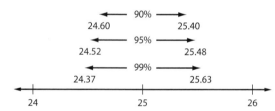

19d. Sample answer: With everything else held constant, increasing the confidence level will increase the size of the confidence interval.

19e. Sample answer: Expanding the confidence interval reduces the accuracy of the estimate. So decreasing the size of the confidence interval increases the accuracy of the estimate.

Chapter Planner

	Diagnostic Assessment Quick Check			
	EXPLORE 12-1 45 min: 0.5 day 90 min: 0.25 day	**LESSON 12-1** 45 min: 1.5 days 90 min: 0.75 day	**LESSON 12-2** 45 min: 1 day 90 min: 0.5 day	**EXTEND 12-2** 45 min: 0.5 day 90 min: 0.25 day
Title	Spreadsheet Lab: Investigating Special Right Triangles	Trigonometric Functions in Right Triangles	Angles and Angle Measure	Geometry Lab: Areas of Parallelograms
Objectives	■ Use a spreadsheet to investigate ratios of side lengths in special right triangles.	■ Find values of trigonometric functions. ■ Use trigonometric functions to find side lengths and angle measures of right triangles.	■ Draw and find angles in standard position. ■ Convert between degree measures and radian measures.	■ Use the sine ratio to find the area of a parallelogram.
Key Vocabulary		trigonometry trigonometric ratio trigonometric function sine cosine tangent cosecant secant cotangent reciprocal functions angle of elevation angle of depression	standard position initial side terminal side coterminal angles radian central angle	
CCSS			F.TF.1	
Multiple Representations			⟳	
Lesson Resources	connectED.mcgraw-hill.com **PT** Personal Tutor **Materials:** ■ spreadsheet program and computer	connectED.mcgraw-hill.com ☐ Leveled Worksheets ᵃᵇᶜ Vocabulary **PT** Personal Tutor ✋ Virtual Manipulatives ☑ Self-Check Quiz ■ *5-Minute Check* ■ *Study Notebook* ■ *Teaching Algebra with Manipulatives*	connectED.mcgraw-hill.com ☐ Leveled Worksheets ☐ Quiz 1 ᵃᵇᶜ Vocabulary 🏃 Animations **PT** Personal Tutor ☑ Self-Check Quiz ■ *5-Minute Check* ■ *Study Notebook*	connectED.mcgraw-hill.com ✋ Virtual Manipulatives ■ *Teaching Algebra with Manipulatives*
Resources for Every Lesson	IWB eStudent Edition IWB Interactive Classroom	■ eTeacher Edition ■ eSolutions Manual ■ McGraw-Hill *eAssessment*		
Differentiated Instruction		pp. 793, 798	pp. 801, 805	

IWB All digital assets are Interactive Whiteboard ready.

Suggested Pacing			
Time Periods	**Instruction**	**Review & Assess**	**Total**
45-minute	13 days	2 days	15 days
90-minute	7 days	1 day	8 days

LESSON 12-3 45 min: 1.5 days / 90 min: 0.75 day	**LESSON 12-4** 45 min: 1.5 days / 90 min: 0.75 day	**EXTEND 12-4** 45 min: 0.5 day / 90 min: 0.25 day	**LESSON 12-5** 45 min: 1 day / 90 min: 0.5 day	**LESSON 12-6** 45 min: 1 day / 90 min: 0.75 day
Trigonometric Functions of General Angles	**Law of Sines**	**Geometry Lab: Regular Polygons**	**Law of Cosines**	**Circular and Periodic Functions**
▪ Find values of trigonometric functions for general angles. ▪ Find values of trigonometric functions by using reference angles.	▪ Find the area of a triangle using two sides and an included angle. ▪ Use the Law of Sines to solve triangles.	▪ Investigate measures in regular polygons using trigonometry.	▪ Use the Law of Cosines to solve triangles. ▪ Choose methods to solve triangles.	▪ Find values of trigonometric functions based on the unit circle. ▪ Use the properties of periodic functions to evaluate trigonometric functions.
quadrantal angle reference angle	Law of Sines solving a triangle ambiguous case		Law of Cosines	unit circle circular function periodic function cycle period
				F.TF.1, F.TF.2
connectED.mcgraw-hill.com ▢ Leveled Worksheets ▢ Quiz 1 *abc* Vocabulary PT Personal Tutor ✓ Self-Check Quiz ▪ *5-Minute Check* ▪ *Study Notebook*	connectED.mcgraw-hill.com ▢ Leveled Worksheets *abc* Vocabulary PT Personal Tutor Virtual Manipulatives ✓ Self-Check Quiz ▪ *5-Minute Check* ▪ *Study Notebook*	connectED.mcgraw-hill.com Virtual Manipulatives ▪ *Teaching Algebra with Manipulatives* **Materials:** ▪ compass ▪ straightedge ▪ protractor	connectED.mcgraw-hill.com ▢ Leveled Worksheets ▢ Quiz 2 *abc* Vocabulary PT Personal Tutor Virtual Manipulatives ✓ Self-Check Quiz ▪ *5-Minute Check* ▪ *Study Notebook*	connectED.mcgraw-hill.com ▢ Leveled Worksheets *abc* Vocabulary 🏃 Animations PT Personal Tutor Virtual Manipulatives ✓ Self-Check Quiz ▪ *5-Minute Check* ▪ *Study Notebook*
IWB eStudent Edition IWB Interactive Classroom	▪ eTeacher Edition ▪ eSolutions Manual ▪ *e*Assessment McGraw-Hill			
pp. 810, 813	pp. 817, 821		pp. 825, 828	pp. 832, 836

Formative Assessment
Mid-Chapter Quiz

Title	LESSON **12-7** 45 min: 1 day / 90 min: 0.75 day	EXPLORE **12-8** 45 min: 0.5 day / 90 min: 0.25 day	LESSON **12-8** 45 min: 1.5 days / 90 min: 0.75 day	LESSON **12-9** 45 min: 1 day / 90 min: 0.5 day	
Title	**Graphing Trigonometric Functions**	**Graphing Technology Lab: Trigonometric Graphs**	**Translations of Trigonometric Graphs**	**Inverse Trigonometric Functions**	
Objectives	■ Describe and graph the sine, cosine, and tangent functions. ■ Describe and graph other trigonometric functions.	■ Use a graphing calculator to explore transformations of the graphs of trigonometric functions.	■ Graph horizontal translations of trigonometric graphs and find phase shifts. ■ Graph vertical translations of trigonometric graphs.	■ Find values of inverse trigonometric functions. ■ Solve equations by using inverse trigonometric functions.	
Key Vocabulary	amplitude frequency		phase shift vertical shift midline	principal values Arcsine function Arccosine function Arctangent function	
CCSS	F.IF.7e, F.TF.5	F.BF.3	F.IF.7e, F.BF.3	A.CED.2	
Multiple Representations				🔧	
Lesson Resources	connectED.mcgraw-hill.com 📁 Leveled Worksheets 📁 Quiz 3 🔤 Vocabulary 🏃 Animations PT Personal Tutor ✓ Self-Check Quiz ■ *5-Minute Check* ■ *Study Notebook*	connectED.mcgraw-hill.com PT Personal Tutor **Materials:** ■ TI-83/84 Plus or other graphing calculator	connectED.mcgraw-hill.com 📁 Leveled Worksheets 🔤 Vocabulary PT Personal Tutor ✓ Self-Check Practice ■ *5-Minute Check* ■ *Study Notebook*	connectED.mcgraw-hill.com 📁 Leveled Worksheets 📁 Quiz 4 🔤 Vocabulary PT Personal Tutor ✓ Self-Check Practice ■ *5-Minute Check* ■ *Study Notebook*	
Resources for Every Lesson	IWB eStudent Edition IWB Interactive Classroom	■ eTeacher Edition ■ eSolutions Manual ■ McGraw-Hill *e*Assessment			
Differentiated Instruction	pp. 838, 843		pp. 846, 852	pp. 855, 858	
				Summative Assessment Study Guide and Review Practice Test	

IWB All digital assets are Interactive Whiteboard ready.

STUDY SKILL

Students may find that making study cards is helpful when they are learning many new terms and concepts. Making the cards aids in learning because it requires that they summarize the information. Then the cards can be used as students review when preparing for tests. These cards describe some of the concepts of right triangle trigonometry that students study in Lesson 12-1. A name and sketch are on the front of the card and the relationship is on the back of the card.

Front

Back

Creating Independence through Student-owned Strategies

SE = Student Edition, TE = Teacher Edition, CRM = Chapter Resource Masters

Diagnosis	Prescription
Beginning Chapter 12	
Get Ready for Chapter 12 **SE**	Response to Intervention **TE**
Beginning Every Lesson	
Then, Now, Why? **SE** 5-Minute Checks	Chapter 0 **SE**

DIAGNOSTIC ASSESSMENT

Diagnosis	Prescription
During/After Every Lesson	
Guided Practice **SE**, every example Check Your Understanding **SE** H.O.T. Problems **SE** Spiral Review **SE** Additional Examples **TE** Watch Out! **TE** Step 4, Assess **TE** Chapter 12 Quizzes **CRM**, pp. 61–62 Self-Check Quizzes connectED.mcgraw-hill.com	`TIER 1 Intervention` Skills Practice **CRM**, Ch. 1–12 connectED.mcgraw-hill.com `TIER 2 Intervention` Differentiated Instruction **TE** Study Guide and Intervention **CRM**, Ch. 1–12 `TIER 3 Intervention` Math Triumphs, Alg. 2, Ch. 6
Mid-Chapter	
Mid-Chapter Quiz **SE** Mid-Chapter Test **CRM**, p. 63 McGraw-Hill eAssessment	`TIER 1 Intervention` Skills Practice **CRM**, Ch. 1–12 connectED.mcgraw-hill.com `TIER 2 Intervention` Study Guide and Intervention **CRM**, Ch. 1–12 `TIER 3 Intervention` Math Triumphs, Alg. 2, Ch. 6
Before Chapter Test	
Chapter Study Guide and Review **SE** Practice Test **SE** Standardized Test Practice **SE** Chapter Test connectED.mcgraw-hill.com Standardized Test Practice connectED.mcgraw-hill.com Vocabulary Review connectED.mcgraw-hill.com McGraw-Hill eAssessment	`TIER 1 Intervention` Skills Practice **CRM**, Ch. 1–12 connectED.mcgraw-hill.com `TIER 2 Intervention` Study Guide and Intervention **CRM**, Ch. 1–12 `TIER 3 Intervention` Math Triumphs, Alg. 2, Ch. 6

FORMATIVE ASSESSMENT

Diagnosis	Prescription
After Chapter 12	
Multiple-Choice Tests, Forms 1, 2A, 2B **CRM**, pp. 65–70 Free-Response Tests, Forms 2C, 2D, 3 **CRM**, pp. 71–76 Vocabulary Test **CRM**, p. 64 Extended Response Test **CRM**, p. 77 Standardized Test Practice **CRM**, pp. 78–80 McGraw-Hill eAssessment	Study Guide and Intervention **CRM**, Ch. 1–12 connectED.mcgraw-hill.com

SUMMATIVE ASSESSMENT

Option 1 Reaching All Learners AL OL BL ELL

Intrapersonal Have students choose an angle measuring greater than 90°, draw an angle with that measure in standard position, find the reference angle, and give the values of all 6 of the trigonometric functions in both degrees and radians.

Kinesthetic Have students work in groups of three or four. Provide each group with drinking straws, scissors, and glue. Ask students to make a poster of the Key Concepts in Lesson 12-4 and Lesson 12-5, using straws to form the triangles.

Option 2 Approaching Level AL

Working in groups of three or four, students draw a variety of right triangles, each having one angle that measures 30°. Have students measure the lengths of the hypotenuse and the leg opposite the 30° angle to the nearest millimeter. Then ask students to find the sine of 30° in each triangle to determine that sin 30° is the same for each triangle, 0.5. Discuss results with students. Then draw a right triangle with a 30° angle on the board. Label the hypotenuse 8 cm. Ask students how they could find the length of the leg opposite the 30° angle.

$$\sin 30° = \frac{1}{2} = \frac{? \text{ cm}}{8 \text{ cm}}$$

Option 3 English Learners ELL

Have students work in small groups to decide on an informal explanation of what a radian is and what coterminal angles are. Then have a reporter from each group share that explanation with the whole class.

Option 4 Beyond Level BL

Tell students that you want them to draw a triangle and determine its area using each of the following three different ways:

- using the basic formula for the area of a triangle $A = \frac{1}{2} bh$
  The lengths of the base and height are needed.

- using Heron's formula The lengths of all three sides are needed.

- using the formula $A = \frac{1}{2} bc \sin A$ The lengths of two sides and the sine of their included angle are needed.

Students may use a protractor and a ruler as needed. Then have students tell what information about the triangle they needed to know when using each method.

VerticalAlignment

Before Chapter 12

Related Topics before Algebra 1

- use the Pythagorean Theorem
- apply formulas to solve problems

Previous Topics from Algebra 2

- introduce new notations for inverse functions when logarithmic functions are explored
- restrict domains when inverses for functions, such as $y = x^2$, are found

Chapter 12

Related Topics from Algebra 2

- explore trigonometric functions, first in acute angles in standard form, and also for points on the unit circle
- derive and use the Law of Sines and the Law of Cosines as applications of trigonometric functions
- develop inverses for the sine, cosine, and tangent functions
- use trigonometric functions to explore amplitude and period
- investigate phase shifts and vertical shifts in the graphs of trigonometric functions

After Chapter 12

Preparation for Precalculus

- continue the study of amplitude, period, and frequency for trigonometric and other periodic functions
- explore trigonometric functions and periodic functions
- explore amplitude and frequency for periodic functions
- look at translations of graphs
- use trigonometric functions for sum and difference of angles
- solve equations involving trigonometric functions

Lesson-by-LessonPreview

12-1 Trigonometric Functions in Right Triangles

A rule given by a ratio that compares the lengths of the sides of a right triangle is called a *trigonometric function.* The hypotenuse, the leg opposite θ, and the leg adjacent to θ are used to define the six trigonometric functions.

12-2 Angles and Angle Measure

An angle on a coordinate plane is in *standard position* if one ray of the angle (*initial side*) is placed on the positive *x*-axis and the other ray (*terminal side*) rotates about the origin. The terminal side rotates counterclockwise to show an angle with a positive measure and rotates clockwise to show an angle with a negative measure. It is important to note that the measure of a −210° angle is not less than the measure of a 210° angle.

Coterminal angles are angles in standard position that have the same terminal side. To find angles that are coterminal to another angle, add or subtract a multiple of 360°.

Angles can be measured in degrees or in *radians*—units that are based on arc length. The fact that π radians = 180° can be used to convert between the two units.

12-3 Trigonometric Functions of General Angles

When a point $P(x, y)$ on the terminal side of the angle θ is known, the value of the six trigonometric functions can be found. Draw a segment from the point perpendicular to the *x*-axis, forming a right triangle with a leg measuring *x* units, a leg measuring *y* units, and a hypotenuse measuring *r* units, where $r = \sqrt{x^2 + y^2}$. Thus, $\sin \theta = \frac{y}{r}$ and $\cos \theta = \frac{x}{r}$.

To determine the values of trigonometric functions for any nonquadrantal angle θ, find the value of the function for the *reference angle* θ that is formed by the terminal side and the *x*-axis. The sign of the function is determined by the quadrant in which the terminal side of θ lies.

12-4 Law of Sines and 12-5 Law of Cosines

The *Law of Sines* can be used to solve a triangle if you know

- the measures of two angles and any side (AAS or ASA) or
- the measures of two sides and the angle opposite one of the sides (SSA). When solving a triangle in the SSA case, zero, one, or two solutions are possible.

The *Law of Cosines* can be used to solve a triangle if you know
- the measures of two sides and the included angle, (SAS) or
- the measures of the three sides (SSS).

12-6 Circular and Periodic Functions

A unit circle—a circle centered at the origin with a radius of 1 unit—can be used to generalize the sine and cosine functions. If the terminal side of an angle θ in standard position intersects the unit circle at a point P with coordinates (x, y), then $\cos \theta = x$ and $\sin \theta = y$.

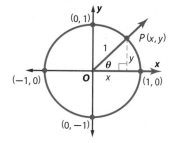

The exact values of $\cos \theta$ and $\sin \theta$ for special angles are shown on the unit circle below.

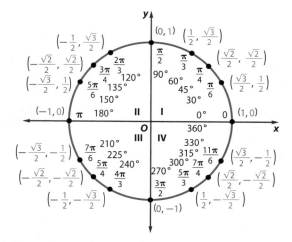

Cycles of the sine and cosine functions repeat every 360°, so they are *periodic functions*.

12-7 Graphing Trigonometric Functions

The graphs of sine, cosine, and tangent functions have repeating patterns (cycles). The horizontal length of each cycle is its *period*. For the sine and cosine functions, the period is 2π radians or 360°. For the tangent function, the period is π radians or 180°. For sine and cosine functions, half the difference between the maximum or minimum value is called the *amplitude* of the graph.

Algebraically, for the sine function $y = a \sin b\theta$ and the cosine function $y = a \cos b\theta$, the period is $360° \div |b|$ and the amplitude is $|a|$. For the tangent function $y = a \tan b\theta$, the period is $180° \div |b|$. The graphs of the cosecant, secant, and cotangent functions are related to the graphs of the sine, cosine, and tangent functions, respectively.

12-8 Translations of Trigonometric Graphs

The graphs of the functions $y = a \sin b(\theta - h) + k$, $y = a \cos b(\theta - h) + k$, and $y = a \tan b(\theta - h) + k$ are affected by changing the values of a, b, h, and k.

- A *horizontal translation* is a *phase shift*. If h is positive, the entire graph is shifted to the right; if h is negative, the entire graph is shifted to the left.
- A *vertical translation* is a *vertical shift* of the *horizontal midline*. If k is positive, the entire graph shifts upward; if k is negative, the entire graph shifts downward.
- The *amplitude* is determined by the value of $|a|$, so the maximum values of the function are $|a|$ above k and the minimum values of the function are $|a|$ below k.
- The *period* of a function can also change. As the value of $|b|$ increases, the period of the function decreases.

12-9 Inverse Trigonometric Functions

The section of the sine and tangent graphs between $-90° \left(-\dfrac{\pi}{2}\right)$ and $90° \left(\dfrac{\pi}{2}\right)$ and of the cosine graph between 0° (0) and 180° (π) pass the horizontal line test, so inverse functions can be identified for those domains. The functions $y = \sin x$, $y = \cos x$, and $y = \tan x$ are the restricted-domain functions. Their inverses are $y = \sin^{-1} x$ or $y = \arcsin x$, $y = \cos^{-1} x$ or $y = \arccos x$, and $y = \tan^{-1} x$ or $y = \arctan x$.

Mc Graw Hill **Professional Development**

Targeted professional development has been articulated throughout Algebra 2. More quality, customized professional development is available from McGraw-Hill Professional Development. Visit <u>connectED.mcgraw-hill.com</u> for details on each product.

- Online Lessons emphasize the strategies and techniques used to teach Algebra 2. Includes streaming video, interactive pages, and online tools.

- Video Workshops allow mentors, coaches, or leadership personnel to facilitate on-site workshops on educational strategies in mathematics and mathematical concepts.

- MHPD Online (<u>www.mhpdonline.com</u>) offers online professional development with video clips of instructional strategies, links, student activities, and news and issues in education.

- Teaching Today (<u>teachingtoday.glencoe.com</u>) gives secondary teachers practical strategies and materials that inspire excellence and innovation in teaching.

CHAPTER 12 Trigonometric Functions

Chapter Project

To Bee Or Not To Bee
Students use what they have learned about trigonometry to complete a project.

This chapter project addresses environmental awareness, as well as several specific skills identified as being essential to student success by the Framework for 21st Century Learning.

Visit connectED.mcgraw-hill.com for student and teacher handouts.

KeyVocabulary Introduce the key vocabulary in the chapter using the routine below.

<u>Define</u>: One radian is the measure of an angle θ in standard position whose rays intercept an arc of length 1 unit on the unit circle.

<u>Ask</u>: What does a radian measure? an angle

<u>Example</u>:

$\theta = 1$ radian

··Then
○ You have graphed and analyzed functions.

··Now
○ You will:
- Find values of trigonometric functions.
- Solve problems by using right triangle trigonometry.
- Solve triangles by using the Law of Sines and Law of Cosines.
- Graph trigonometric functions.

··Why? ▲
WATER SPORTS Knowing trigonometric functions has practical applications in water sports. For instance, you can use right triangle trigonometry to find the distance a kayak has traveled down river. If you are familiar with angles and angle measures, then you have a better understanding of how impressive it is to be able to do a 540° rotation on a wakeboard.

connectED.mcgraw-hill.com **Your Digital Math Portal**

| Animation | Vocabulary | eGlossary | Personal Tutor | Virtual Manipulatives | Graphing Calculator | Audio | Foldables | Self-Check Practice | Worksheets |

Dennis Welsh/UpperCut Images/Getty Images

Get Ready for the Chapter

Diagnose Readiness | You have two options for checking prerequisite skills.

1 **Textbook Option** Take the Quick Check below. Refer to the Quick Review for help.

QuickCheck	QuickReview

QuickCheck

Find the value of x to the nearest tenth.

1. **11.7**

2. **15**
3. **20.5**

4. Laura has a rectangular garden in her backyard that measures 12 feet by 15 feet. She wants to put a rock walkway on the diagonal. How long will the walkway be? Round to the nearest tenth of a foot. **19.2 feet**

Find each missing measure. Write all radicals in simplest form.

5. $x = 9, y = 9\sqrt{2}$

6.
   $x = 16, y = 8\sqrt{3}$

7. $x = 12,$ $y = 12\sqrt{3}$

8. A ladder leans against a wall at a 45° angle. If the ladder is 12 feet long, how far up the wall does the ladder reach?
   $6\sqrt{2}$ ft or about 8.5 ft

QuickReview

Example 1 (Used in Lessons 12-1 and 12-3)

Find the missing measure of the right triangle.

$c^2 = a^2 + b^2$	Pythagorean Theorem
$18^2 = a^2 + 5^2$	Replace c with 18 and b with 5.
$324 = a^2 + 25$	Simplify.
$299 = a^2$	Subtract 25 from each side.
$17.3 \approx a$	Take the positive square root of each side.

Example 2 (Used in Lesson 12-1)

Find the missing measures. Write all radicals in simplest form.

$x^2 + x^2 = 18^2$	Pythagorean Theorem
$2x^2 = 18^2$	Combine like terms.
$2x^2 = 324$	Simplify.
$x^2 = 162$	Divide each side by 2.
$x = \sqrt{162}$	Take the positive square root of each side.
$x = 9\sqrt{2}$	Simplify.

2 **Online Option** Take an online self-check Chapter Readiness Quiz at connectED.mcgraw-hill.com.

787

Get Ready for the Chapter

Response to Intervention (RtI)
Use the *Quick Check* results and the Intervention Planner to help you determine your Response to Intervention. The If-Then statements in the chart below help you decide the appropriate tier of RtI and suggest intervention resources for each tier.

InterventionPlanner

TIER 1 **On Level** **OL**

If students miss about 25% of the exercises or less,

Then choose a resource:

SE Lesson 0-7

connectED.mcgraw-hill.com Self-Check Quiz

TIER 2 **Strategic Intervention** **AL**
approaching grade level

If students miss about 50% of the exercises,

Then choose a resource:

connectED.mcgraw-hill.com Extra Examples, Personal Tutor, Homework Help

TIER 3 **Intensive Intervention**
2 or more grades below level

If students miss about 75% of the exercises,

Then use *Math Triumphs, Alg. 2,* Ch. 6

connectED.mcgraw-hill.com Extra Examples, Personal Tutor, Homework Help, Review Vocabulary

---

E? **Essential Question**

Ask:

- What types of real-world problems can be modeled and solved using trigonometry? Sample answer: periodic functions and physical situations involving triangles such as construction and surveying problems

FOLDABLES StudyOrganizer

Dinah Zike's Foldables®

Focus Students write notes as they explore trigonometric functions in the lessons of this chapter.

Teach Have students make and label their Foldables as illustrated. Have students use the appropriate tab as they cover each lesson in this chapter. Lessons 12-8 and 12-9 should go on the last tab. Encourage students to apply what they have learned by writing their own examples as well.

When to Use It Encourage students to add to their Foldables as they work through the chapter and to use them to review for the chapter test.

Differentiated Instruction

 Student-Built Glossary, pp. 1–2

Students should complete the chart by providing a definition of each term and an example as they progress through Chapter 12. This study tool can also be used to review for the chapter test.

Get Started on the Chapter

You will learn several new concepts, skills, and vocabulary terms as you study Chapter 12. To get ready, identify important terms and organize your resources. You may wish to refer to Chapter 0 to review prerequisite skills.

FOLDABLES StudyOrganizer

Trigonometric Functions Make this Foldable to help you organize your Chapter 12 notes about trigonometric functions. Begin with four pieces of grid paper.

1 **Stack** paper together and measure 2.5 inches from the bottom.

2 **Fold** on the diagonal.

3 **Staple** along the diagonal to form a book.

4 **Label** the edge as Trigonometric Functions.

NewVocabulary

English		Español
trigonometry	p. 790	trigonometría
sine	p. 790	seno
cosine	p. 790	coseno
tangent	p. 790	tangente
cosecant	p. 790	cosecante
secant	p. 790	secante
cotangent	p. 790	cotangente
angle of elevation	p. 794	ángulo de depresión
angle of depression	p. 794	ángulo de elevación
standard position	p. 799	posición estándar
radian	p. 801	radián
Law of Sines	p. 815	Ley de los senos
ambiguous case	p. 816	caso ambiguo
Law of Cosines	p. 823	Ley de los cosenos
unit circle	p. 830	círculo unitario
circular function	p. 830	funciones circulares
periodic function	p. 831	función periódica
cycle	p. 831	ciclo
period	p. 831	período
amplitude	p. 837	amplitud
frequency	p. 838	frecuencia

ReviewVocabulary

function función a relation in which each element of the domain is paired with exactly one element in the range
inverse function función inversa two functions *f* and *g* are inverse functions if and only if both of their compositions are the identity function

---

Additional Answer (Explore 12-1)

1.

	A	B	C	D	E	F
1	a	b	c	b/a	b/c	a/c
2	1	1.732050808	2	1.732050808	0.866025404	0.5
3	2	3.464101615	4	1.732050808	0.866025404	0.5
4	3	5.196152423	6	1.732050808	0.866025404	0.5
5	4	6.92820323	8	1.732050808	0.866025404	0.5
6	5	8.660254038	10	1.732050808	0.866025404	0.5
7	6	10.39230485	12	1.732050808	0.866025404	0.5

EXPLORE 12-1

Spreadsheet Lab
Investigating Special Right Triangles

You can use a spreadsheet to investigate side measures of special right triangles.

CCSS Common Core State Standards
Mathematical Practices
8 Look for and express regularity in repeated reasoning.

Activity 45°-45°-90° Triangle

The legs of a 45°-45°-90° triangle, a and b, are equal in measure. What patterns do you observe in the ratios of the side measures of these triangles?

Step 1 Enter the indicated formulas in the spreadsheet. The formula uses the Pythagorean Theorem in the form $c = \sqrt{a^2 + b^2}$.

=SQRT(A2^2+B2^2)   =B2/A2   =B2/C2   =A2/C2

45-45-90 triangles

	A	B	C	D	E	F
1	a	b	c	b/a	b/c	a/c
2	1	1	1.414213562	1	0.707106781	0.707106781
3	2	2	2.828427125	1	0.707106781	0.707106781
4	3	3	4.242640687	1	0.707106781	0.707106781
5	4	4	5.656854249	1	0.707106781	0.707106781

Sheet 1   Sheet 2   Sheet 3

Step 2 Examine the results. Because 45°-45°-90° triangles share the same angle measures, these triangles are all similar. The ratios of the sides of these triangles are all the same. The ratios of side b to side a are 1. The ratios of side b to side c and of side a to side c are approximately 0.71.

Model and Analyze

Use the spreadsheet below for 30°-60°-90° triangles.

30-60-90 triangles

	A	B	C	D	E	F
1	a	b	c	b/a	b/c	a/c
2	1		2			
3	2		4			
4	3		6			
5	4		8			

Sheet 1   Sheet 2   Sheet 3

1. Copy and complete the spreadsheet above. See margin.

2. Describe the relationship among the 30°-60°-90° triangles with the dimensions given.
The triangles are all similar.

3. What patterns do you observe in the ratios of the side measures of these triangles?

3. All of the ratios of side b to side a are approximately 1.73. All of the ratios of side b to side c are approximately 0.87. All of the ratios of side a to side c are 0.5.

connectED.mcgraw-hill.com   789

Extending the Concept
Ask:
- Ask students what formula could have been used in column B for the 45°-45°-90° triangle instead of entering the data? in column C for the 30°-60°-90° triangle? $b = a$; $c = 2a$

- Ask students to compare the 30°-60°-90° triangles. What is the same? What is different? The angle measures are all the same, but the side measures are different. The triangles are similar but not congruent.

---

1 Focus

Objective Use a spreadsheet to investigate ratios of side lengths in special right triangles.

Materials for Each Student
- spreadsheet program and computer

Teaching Tip
Have students practice their skills in using a spreadsheet by entering the data in the example. Have students format the cells before they start to enter the data by using the Format Cells command. Format columns C, E, and F for numbers with 9 decimal places. Format columns A, B, and D for integers.

2 Teach

Working in Cooperative Groups
Have students work with partners, mixing abilities so that a student with more knowledge of spreadsheets is paired with one who has less experience.

Practice Have students complete the Activity and Exercises 1–3.

3 Assess

Formative Assessment
Use Exercise 1 to assess whether students comprehend how to enter data into a spreadsheet.

From Concrete to Abstract
Ask students to describe the rule, pattern, or formula for various cells of the spreadsheet.

1 Focus

VerticalAlignment

Before Lesson 12-1 Use the Pythagorean Theorem to find side lengths of right triangles.

Lesson 12-1 Find values of trigonometric functions. Use trigonometric functions to find side lengths and angle measures of right triangles.

After Lesson 12-1 Investigate graphs of the sine and cosine functions.

2 Teach

Scaffolding Questions

Have students read the **Why?** section of the lesson.

Ask:

- What kind of angle is formed by the tow rope and the horizontal? acute

- Which side of the triangle is opposite the right angle? Which side is opposite the *x*° angle? the hypotenuse ℓ; altitude

- If length ℓ is constant and *x* increases, how does the altitude change? It increases.

Trigonometric Functions in Right Triangles

Then	**Now**	**Why?**
● You used the Pythagorean Theorem to find side lengths of right triangles.	**1** Find values of trigonometric functions for acute angles. **2** Use trigonometric functions to find side lengths and angle measures of right triangles.	● The altitude of a person parasailing depends on the length of the tow rope ℓ and the angle the rope makes with the horizontal *x*°. If you know these two values, you can use a ratio to find the altitude of the person parasailing.

altitude

NewVocabulary
trigonometry
trigonometric ratio
trigonometric function
sine
cosine
tangent
cosecant
secant
cotangent
reciprocal functions
inverse sine
inverse cosine
inverse tangent
angle of elevation
angle of depression

CCSS Common Core State Standards
Mathematical Practices
6 Attend to precision.

1 Trigonometric Functions for Acute Angles **Trigonometry** is the study of relationships among the angles and sides of a right triangle. A **trigonometric ratio** compares the side lengths of a right triangle. A **trigonometric function** has a rule given by a trigonometric ratio.

The Greek letter *theta* θ is often used to represent the measure of an acute angle in a right triangle. The *hypotenuse*, the *leg opposite* θ, and the *leg adjacent to* θ are used to define the six trigonometric functions.

KeyConcept Trigonometric Functions in Right Triangles

Words If θ is the measure of an acute angle of a right triangle, then the following trigonometric functions involving the opposite side *opp*, the adjacent side *adj*, and the hypotenuse *hyp* are true.

Symbols

$\sin(\text{sine})\ \theta = \dfrac{opp}{hyp}$ $\csc(\text{cosecant})\ \theta = \dfrac{hyp}{opp}$

$\cos(\text{cosine})\ \theta = \dfrac{adj}{hyp}$ $\sec(\text{secant})\ \theta = \dfrac{hyp}{adj}$

$\tan(\text{tangent})\ \theta = \dfrac{opp}{adj}$ $\cot(\text{cotangent})\ \theta = \dfrac{adj}{opp}$

Examples

$\sin\theta = \dfrac{4}{5}$ $\cos\theta = \dfrac{3}{5}$ $\tan\theta = \dfrac{4}{3}$

$\csc\theta = \dfrac{5}{4}$ $\sec\theta = \dfrac{5}{3}$ $\cot\theta = \dfrac{3}{4}$

Example 1 Evaluate Trigonometric Functions

Find the values of the six trigonometric functions for angle θ.

leg opposite θ: *BC* = 8   leg adjacent θ: *AC* = 15   hypotenuse: *AB* = 17

$\sin\theta = \dfrac{opp}{hyp} = \dfrac{8}{17}$ $\cos\theta = \dfrac{adj}{hyp} = \dfrac{15}{17}$ $\tan\theta = \dfrac{opp}{adj} = \dfrac{8}{15}$

$\csc\theta = \dfrac{hyp}{opp} = \dfrac{17}{8}$ $\sec\theta = \dfrac{hyp}{adj} = \dfrac{17}{15}$ $\cot\theta = \dfrac{adj}{opp} = \dfrac{15}{8}$

> **GuidedPractice**
> 1. Find the values of the six trigonometric functions for angle *B*.

See Chapter 12
Answer Appendix.

 790 | Lesson 12-1

isifa Image Service s.r.o./Alamy

Lesson 12-1 Resources

Resource	Approaching Level **AL**	On Level **OL**	Beyond Level **BL**	English Learners **ELL**
Teacher Edition	▪ Differentiated Instruction, p. 793	▪ Differentiated Instruction, pp. 793, 798	▪ Differentiated Instruction, p. 798	
Chapter Resource Masters	▪ Study Guide and Intervention, pp. 5–6 ▪ Skills Practice, p. 7 ▪ Practice, p. 8 ▪ Word Problem Practice, p. 9	▪ Study Guide and Intervention, pp. 5–6 ▪ Skills Practice, p. 7 ▪ Practice, p. 8 ▪ Word Problem Practice, p. 9 ▪ Enrichment, p. 10	▪ Practice, p. 8 ▪ Word Problem Practice, p. 9 ▪ Enrichment, p. 10	▪ Study Guide and Intervention, pp. 5–6 ▪ Skills Practice, p. 7 ▪ Practice, p. 8 ▪ Word Problem Practice, p. 9
Other	▪ 5-Minute Check 12-1 ▪ Study Notebook ▪ Teaching Algebra with Manipulatives	▪ 5-Minute Check 12-1 ▪ Study Notebook ▪ Teaching Algebra with Manipulatives	▪ 5-Minute Check 12-1 ▪ Study Notebook	▪ 5-Minute Check 12-1 ▪ Study Notebook ▪ Teaching Algebra with Manipulatives

StudyTip

Memorize Trigonometric Ratios SOH-CAH-TOA is a mnemonic device for remembering the first letter of each word in the ratios for sine, cosine, and tangent.

$\sin \theta = \frac{\text{opp}}{\text{hyp}}$

$\cos \theta = \frac{\text{adj}}{\text{hyp}}$

$\tan \theta = \frac{\text{opp}}{\text{adj}}$

Notice that the cosecant, secant, and cotangent ratios are reciprocals of the sine, cosine, and tangent ratios, respectively. These are called the **reciprocal functions**.

$$\csc \theta = \frac{1}{\sin \theta} \qquad \sec \theta = \frac{1}{\cos \theta} \qquad \cot \theta = \frac{1}{\tan \theta}$$

The domain of any trigonometric function is the set of all acute angles θ of a right triangle. So, trigonometric functions depend only on the measures of the acute angles, not on the side lengths of a right triangle.

Example 2 Find Trigonometric Ratios

If $\sin B = \frac{5}{8}$, find the exact values of the five remaining trigonometric functions for B.

Step 1 Draw a right triangle and label one acute angle B. Label the opposite side 5 and the hypotenuse 8.

Step 2 Use the Pythagorean Theorem to find a.

$a^2 + b^2 = c^2$	Pythagorean Theorem
$a^2 + 5^2 = 8^2$	$b = 5$ and $c = 8$
$a^2 + 25 = 64$	Simplify.
$a^2 = 39$	Subtract 25 from each side.
$a = \pm\sqrt{39}$	Take the square root of each side.
$a = \sqrt{39}$	Length cannot be negative.

Step 3 Find the other values.

Since $\sin B = \frac{5}{8}$, $\csc B = \frac{\text{hyp}}{\text{opp}}$ or $\frac{8}{5}$.

$\cos B = \frac{\text{adj}}{\text{hyp}} = \frac{\sqrt{39}}{8}$ $\qquad\qquad$ $\sec B = \frac{\text{hyp}}{\text{adj}} = \frac{8}{\sqrt{39}}$ or $\frac{8\sqrt{39}}{39}$

$\tan B = \frac{\text{opp}}{\text{adj}} = \frac{5}{\sqrt{39}}$ or $\frac{5\sqrt{39}}{39}$ $\qquad\qquad$ $\cot B = \frac{\text{adj}}{\text{opp}} = \frac{\sqrt{39}}{5}$

GuidedPractice

2. $\sin B = \frac{3\sqrt{58}}{58}$, $\sec B = \frac{\sqrt{58}}{7}$, $\cot B = \frac{7}{3}$, $\csc B = \frac{\sqrt{58}}{3}$, $\cos B = \frac{7\sqrt{58}}{58}$

2. If $\tan B = \frac{3}{7}$, find exact values of the remaining trigonometric fuctions for B.

ReadingMath

Labeling Triangles Throughout this chapter, a capital letter is used to represent both a vertex of a triangle and the measure of the angle at that vertex. The same letter in lowercase is used to represent both the side opposite that angle and the length of the side.

Angles that measure 30°, 45°, and 60° occur frequently in trigonometry.

KeyConcept Trigonometric Values for Special Angles

30°-60°-90°

$\sin 30° = \frac{1}{2}$ $\qquad$ $\cos 30° = \frac{\sqrt{3}}{2}$ $\qquad$ $\tan 30° = \frac{\sqrt{3}}{3}$

$\sin 60° = \frac{\sqrt{3}}{2}$ $\qquad$ $\cos 60° = \frac{1}{2}$ $\qquad$ $\tan 60° = \sqrt{3}$

45°-45°-90°

$\sin 45° = \frac{\sqrt{2}}{2}$ $\qquad$ $\cos 45° = \frac{\sqrt{2}}{2}$ $\qquad$ $\tan 45° = 1$

connectED.mcgraw-hill.com **791**

---

1 Trigonometric Functions for Acute Angles

Example 1 shows how to find the values of the six trigonometric functions for an angle. **Example 2** shows how to use one trigonometric ratio for an angle to find another.

Formative Assessment

Use the Guided Practice exercises after each example to determine students' understanding of concepts.

Additional Examples

1 Find the values of the six trigonometric functions for angle G.

$\sin G = \frac{3}{5}$, $\cos G = \frac{4}{5}$,

$\tan G = \frac{3}{4}$, $\cot G = \frac{4}{3}$,

$\sec G = \frac{5}{4}$, $\csc G = \frac{5}{3}$

2 If $\tan A = \frac{5}{3}$, find the exact values of the five remaining trigonometric functions for A.

$\cos A = \frac{3\sqrt{34}}{34}$, $\csc A = \frac{\sqrt{34}}{5}$,

$\cot A = \frac{3}{5}$, $\sec A = \frac{\sqrt{34}}{3}$,

$\sin A = \frac{5\sqrt{34}}{34}$

Additional Examples also in Interactive Classroom PowerPoint® Presentations

IWB Interactive White Board READY

Teach with Tech

Interactive Whiteboard Draw a right triangle on the board and label each of the side lengths. To write the trigonometric ratios, drag the measurement from the side of the triangle to the ratio.

Tips for New Teachers

Reasoning The relative lengths of the sides of the two special triangles in the Key Concept box on this page are critical. Have students learn how to recreate these triangles themselves from memory. From the triangles, they can generate the values shown in the Key Concept box.

2 Use Trigonometric Functions

Examples 3 and 4 show how to use a trigonometric ratio to find a missing side length of a right triangle. **Example 5** shows how to find missing angle measures of a right triangle. **Example 6** shows how to use angles of elevation and depression to solve real-world problems.

Additional Examples

3 Use a trigonometric function to find the value of x. Round to the nearest tenth if necessary. **10.4**

4 **BUILDINGS** In Example 4, what is the distance from the top of the building to Joel's eye? **about 827 ft**

 Use Trigonometric Functions You can use trigonometric functions to find missing side lengths and missing angle measures of right triangles.

Example 3 Find a Missing Side Length

Use a trigonometric function to find the value of x. Round to the nearest tenth if necessary.

The length of the hypotenuse is 8. The missing measure is for the side adjacent to the 30° angle. Use the cosine function to find x.

$$\cos \theta = \frac{\text{adj}}{\text{hyp}} \qquad \text{Cosine function}$$

$$\cos 30° = \frac{x}{8} \qquad \text{Replace } \theta \text{ with 30°, adj with } x, \text{ and hyp with 8.}$$

$$\frac{\sqrt{3}}{2} = \frac{x}{8} \qquad \cos 30° = \frac{\sqrt{3}}{2}$$

$$\frac{8\sqrt{3}}{2} = x \qquad \text{Multiply each side by 8.}$$

$$6.9 \approx x \qquad \text{Use a calculator.}$$

GuidedPractice

3A. 7

3B. 14.1

You can use a calculator to find the missing side lengths of triangles that do not have 30°, 45°, or 60° angles.

Real-World Example 4 Find a Missing Side Length

BUILDINGS To calculate the height of a building, Joel walked 200 feet from the base of the building and used an inclinometer to measure the angle from his eye to the top of the building. If his eye level is at 6 feet, how tall is the building?

The measured angle is 76°. The side adjacent to the angle is 200 feet. The missing measure is the side opposite the angle. Use the tangent function to find d.

$$\tan \theta = \frac{\text{opp}}{\text{adj}} \qquad \text{Tangent function}$$

$$\tan 76° = \frac{d}{200} \qquad \text{Replace } \theta \text{ with 76°, opp with } d, \text{ and adj with 200.}$$

$$200 \tan 76° = d \qquad \text{Multiply each side by 200.}$$

$$802 \approx d \qquad \text{Use a calculator to simplify: 200 } \boxed{\text{TAN}} \text{ 76 } \boxed{\text{ENTER}}.$$

Because the inclinometer was 6 feet above the ground, the height of the building is approximately 808 feet.

GuidedPractice

4. Use a trigonometric function to find the value of x. Round to the nearest tenth if necessary. **27.0**

StudyTip

Choose a Function If the length of the hypotenuse is unknown, then either the sine or cosine function must be used to find the missing measure.

 Real-WorldLink

Inclinometers measure the angle of Earth's magnetic field as well as the pitch and roll of vehicles, sailboats, and airplanes. They are also used for monitoring volcanoes and well drilling.

Source: Science Magazine

792 | Lesson 12-1 | Trigonometric Functions in Right Triangles

---

Focus on Mathematical Content

Trigonometric Functions Students may be familiar with finding values of trigonometric functions of any angle. However, at this point, the functions have been defined in terms of ratios of the sides of a right triangle. This means that the definitions are only valid for acute angles. The more general case will be considered later when reference angles are introduced.

792 | **Lesson 12-1** | Trigonometric Functions in Right Triangles

When solving equations like $3x = -27$, you use the inverse of multiplication to find x. You also can find angle measures by using the inverse of sine, cosine, or tangent.

ReadingMath

CCSS Precision The expression $\sin^{-1} x$ is read *the inverse sine of x* and is interpreted as the angle whose sine is x. Be careful not to confuse this notation with the notation for negative exponents; $\sin^{-1} x \neq \dfrac{1}{\sin x}$. Instead, this notation is similar to the notation for an inverse function, $f^{-1}(x)$.

KeyConcept Inverse Trigonometric Ratios

Words	If $\angle A$ is an acute angle and the sine of A is x, then the **inverse sine** of x is the measure of $\angle A$.
Symbols	If $\sin A = x$, then $\sin^{-1} x = m\angle A$.
Example	$\sin A = \dfrac{1}{2} \rightarrow \sin^{-1} \dfrac{1}{2} = m\angle A \rightarrow m\angle A = 30°$
Words	If $\angle A$ is an acute angle and the cosine of A is x, then the **inverse cosine** of x is the measure of $\angle A$.
Symbols	If $\cos A = x$, then $\cos^{-1} x = m\angle A$.
Example	$\cos A = \dfrac{\sqrt{2}}{2} \rightarrow \cos^{-1} \dfrac{\sqrt{2}}{2} = m\angle A \rightarrow m\angle A = 45°$
Words	If $\angle A$ is an acute angle and the tangent of A is x, then the **inverse tangent** of x is the measure of $\angle A$.
Symbols	If $\tan A = x$, then $\tan^{-1} x = m\angle A$.
Example	$\tan A = \sqrt{3} \rightarrow \tan^{-1} \sqrt{3} = m\angle A \rightarrow m\angle A = 60°$

If you know the sine, cosine, or tangent of an acute angle, you can use a calculator to find the measure of the angle, which is the inverse of the trigonometric ratio.

Example 5 Find a Missing Angle Measure

Find the measure of each angle. Round to the nearest tenth if necessary.

a. $\angle N$

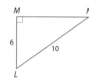

You know the measure of the side opposite $\angle N$ and the measure of the hypotenuse. Use the sine function.

$$\sin N = \dfrac{6}{10} \qquad \sin \theta = \dfrac{\text{opp}}{\text{hyp}}$$
$$\sin^{-1} \dfrac{6}{10} = m\angle N \qquad \text{Inverse sine}$$
$$36.9° \approx m\angle N \qquad \text{Use a calculator.}$$

b. $\angle B$

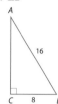

Use the cosine function.

$$\cos B = \dfrac{8}{16} \qquad \cos \theta = \dfrac{\text{adj}}{\text{hyp}}$$
$$\cos^{-1} \dfrac{8}{16} = m\angle B \qquad \text{Inverse cosine}$$
$$60° = m\angle B \qquad \text{Use a calculator.}$$

Guided Practice Find x. Round to the nearest tenth if necessary.

5A.

28.1

5B.

56.3

CCSS **Teaching the Mathematical Practices**

Precision Mathematically proficient students use clear definitions in discussion with others and in their own reasoning. They have learned to examine claims and make explicit use of definitions. Remind students that the notation for inverse trigonometric ratios is similar to the notation used for inverse functions.

Additional Example

5 Find the measure of each angle. Round to the nearest tenth if necessary.

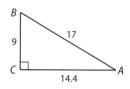

a. $\angle A$ 32°

b. $\angle B$ 58°

DifferentiatedInstruction AL OL

Visual/Spatial Learners Have students use a stack of books and a notebook to model a ramp and investigate how steep the ramp needs to be for a toy car to roll down it without being pushed. Have them report their results in terms of the trigonometric functions of a right triangle.

STAN HONDA/AFP/Getty Images

Additional Example

6 a. GOLF From a camera in a blimp, the apparent distance between the golfer and the hole in Example 6a is the horizontal distance. Find the horizontal distance.
169.4 yd

b. ROLLER COASTER From a blimp, the apparent distance traveled by the roller coaster in Example 6b is the horizontal distance from the top of the hill to the bottom. Find the horizontal distance. 112.6 ft

StudyTip
Angles of Elevation and Depression The angle of elevation and the angle of depression are congruent since they are alternate interior angles of parallel lines.

In the figure at the right, the angle formed by the line of sight from the swimmer and a line parallel to the horizon is called the **angle of elevation**. The angle formed by the line of sight from the lifeguard and a line parallel to the horizon is called the **angle of depression**.

Real-World Example 6 Use Angles of Elevation and Depression

a. GOLF A golfer is standing at the tee, looking up to the green on a hill. If the tee is 36 yards lower than the green and the angle of elevation from the tee to the hole is 12°, find the distance from the tee to the hole.

Write an equation using a trigonometric function that involves the ratio of the vertical rise (side opposite the 12° angle) and the distance from the tee to the hole (hypotenuse).

$\sin 12° = \dfrac{36}{x}$ $\sin \theta = \dfrac{\text{opp}}{\text{hyp}}$

$x \sin 12° = 36$ Multiply each side by x.

$x = \dfrac{36}{\sin 12°}$ Divide each side by sin 12°.

$x \approx 173.2$ Use a calculator.

So, the distance from the tee to the hole is about 173.2 yards.

Real-WorldLink
The steepest roller coasters in the world have angles of descent that are close to 90°.
Source: Ultimate Roller Coaster

b. ROLLER COASTER The hill of the roller coaster has an *angle of descent*, or an angle of depression, of 60°. Its vertical drop is 195 feet. Estimate the length of the hill.

Write an equation using a trigonometric function that involves the ratio of the vertical drop (side opposite the 60° angle) and the length of the hill (hypotenuse).

$\sin 60° = \dfrac{195}{x}$ $\sin \theta = \dfrac{\text{opp}}{\text{hyp}}$

$x \sin 60° = 195$ Multiply each side by x.

$x = \dfrac{195}{\sin 60°}$ Divide each side by sin 60°.

$x \approx 225.2$ Use a calculator.

So, the length of the hill is about 225.2 feet.

GuidedPractice

6A. MOVING A ramp for unloading a moving truck has an angle of elevation of 32°. If the top of the ramp is 4 feet above the ground, estimate the length of the ramp. about 7.5 ft

6B. LADDERS A 14-ft long ladder is placed against a house at an angle of elevation of 72°. How high above the ground is the top of the ladder? about 13.3 ft

 794 | Lesson 12-1 | Trigonometric Functions in Right Triangles

---

Additional Answers

1. $\sin B = \dfrac{4}{5}$; $\cos B = \dfrac{3}{5}$; $\tan B = \dfrac{4}{3}$;

$\csc B = \dfrac{5}{4}$; $\sec B = \dfrac{5}{3}$; $\cot B = \dfrac{3}{4}$

2. $\sin \theta = \dfrac{\sqrt{7}}{4}$; $\cos \theta = \dfrac{3}{4}$;

$\tan \theta = \dfrac{\sqrt{7}}{3}$; $\csc \theta = \dfrac{4\sqrt{7}}{7}$;

$\sec \theta = \dfrac{4}{3}$; $\cot \theta = \dfrac{3\sqrt{7}}{7}$

3. $\sin A = \dfrac{\sqrt{33}}{7}$, $\tan A = \dfrac{\sqrt{33}}{4}$,

$\csc A = \dfrac{7\sqrt{33}}{33}$, $\sec A = \dfrac{7}{4}$,

$\cot A = \dfrac{4\sqrt{33}}{33}$

4. $\sin A = \dfrac{20}{29}$, $\cos A = \dfrac{21}{29}$,

$\csc A = \dfrac{29}{20}$, $\sec A = \dfrac{29}{21}$,

$\cot A = \dfrac{21}{20}$

Check Your Understanding

= Step-by-Step Solutions begin on page R14.

Example 1 Find the values of the six trigonometric functions for angle θ. **1–2. See margin.**

1.

2.

Example 2 In a right triangle, $\angle A$ is acute. Find the values of the five remaining trigonometric funtions. **3–4. See margin.**

3. $\cos A = \frac{4}{7}$

4. $\tan A = \frac{20}{21}$

Examples 3–4 Use a trigonometric function to find the value of x. Round to the nearest tenth.

5 25.4

6. 7.7

7. 8.3

Example 5 Find the value of x. Round to the nearest tenth.

8. 61.9

9. 25.4

10. 68.0

Example 6 **11. CCSS SENSE-MAKING** Christian found two trees directly across from each other in a canyon. When he moved 100 feet from the tree on his side (parallel to the edge of the canyon), the angle formed by the tree on his side and the tree on the other side was 70°. Find the distance across the canyon. **about 274.7 ft**

12. LADDERS The recommended angle of elevation for a ladder used in fire fighting is 75°. At what height on a building does a 21-foot ladder reach if the recommended angle of elevation is used? Round to the nearest tenth. **20.3 ft**

Practice and Problem Solving

Extra Practice is on page R12.

Example 1 Find the values of the six trigonometric functions for angle θ. **13–16. See margin.**

13.

14.

15.

16.

Example 2 In a right triangle, $\angle A$ and $\angle B$ are acute. Find the values of the five remaining trigonometric funtions. **17–20. See margin.**

17. $\tan A = \frac{8}{15}$

18. $\cos A = \frac{3}{10}$

19. $\tan B = 3$

20. $\sin B = \frac{4}{9}$

connectED.mcgraw-hill.com **795**

Differentiated Homework Options

Level	Assignment	Two-Day Option	
AL Basic	13–36, 55–72	13–35 odd, 58–61	14–36 even, 55–57, 62–72
OL Core	13–35 odd, 37–45 odd, 46–53, 55–72	13–36, 58–61	37–53, 55–57, 62–72
BL Advanced	37–66, (optional: 67–72)		

Formative Assessment
Use Exercises 1–12 to check for understanding.

Use the chart at the bottom of this page to customize assignments for your students.

Additional Answers

13. $\sin \theta = \frac{12}{13}$; $\cos \theta = \frac{5}{13}$; $\tan \theta = \frac{12}{5}$; $\csc \theta = \frac{13}{12}$; $\sec \theta = \frac{13}{5}$; $\cot \theta = \frac{5}{12}$

14. $\sin \theta = \frac{9}{41}$; $\cos \theta = \frac{40}{41}$; $\tan \theta = \frac{9}{40}$; $\csc \theta = \frac{41}{9}$; $\sec \theta = \frac{41}{40}$; $\cot \theta = \frac{40}{9}$

15. $\sin \theta = \frac{\sqrt{51}}{10}$; $\cos \theta = \frac{7}{10}$; $\tan \theta = \frac{\sqrt{51}}{7}$; $\csc \theta = \frac{10\sqrt{51}}{51}$; $\sec \theta = \frac{10}{7}$; $\cot \theta = \frac{7\sqrt{51}}{51}$

16. $\sin \theta = \frac{2\sqrt{13}}{13}$; $\cos \theta = \frac{3\sqrt{13}}{13}$; $\tan \theta = \frac{2}{3}$; $\csc \theta = \frac{\sqrt{13}}{2}$; $\sec \theta = \frac{\sqrt{13}}{3}$; $\cot \theta = \frac{3}{2}$

17. $\sin A = \frac{8}{17}$, $\cos A = \frac{15}{17}$, $\csc A = \frac{17}{8}$, $\sec A = \frac{17}{15}$, $\cot A = \frac{15}{8}$

18. $\sin A = \frac{\sqrt{91}}{10}$, $\tan A = \frac{\sqrt{91}}{3}$, $\csc A = \frac{10\sqrt{91}}{91}$, $\sec A = \frac{10}{3}$, $\cot A = \frac{3\sqrt{91}}{91}$

19. $\sin B = \frac{3\sqrt{10}}{10}$, $\cos B = \frac{\sqrt{10}}{10}$, $\csc B = \frac{\sqrt{10}}{3}$, $\sec B = \sqrt{10}$, $\cot B = \frac{1}{3}$

20. $\cos B = \frac{\sqrt{65}}{9}$, $\tan B = \frac{4\sqrt{65}}{65}$, $\csc B = \frac{9}{4}$, $\sec B = \frac{9\sqrt{65}}{65}$, $\cot B = \frac{\sqrt{65}}{4}$

Tips for New Teachers

Calculators On a scientific calculator (in contrast to a graphing calculator), the sequence for finding the sine, cosine, or tangent of an angle may be to enter the angle measure, such as 20, first, and then press the SIN, COS, or TAN key.

Additional Answers

46a. Sample answer:

x ft
50°
24 ft

55. True; $\sin\theta = \dfrac{\text{opp}}{\text{hyp}}$ and the values of the opposite side and the hypotenuse of an acute triangle are positive, so the value of the sine function is positive.

56. The value of sin A equals sin C, so $\dfrac{\text{side opp } A}{\text{hyp}} = \dfrac{\text{side opp } C}{\text{hyp}}$. Since the hypotenuse is the same, the length of the side opposite angle A must equal the length of the side opposite angle C. Since the two sides have the same measure, the triangle is isosceles.

57. Sample answer: The slope describes the ratio of the vertical rise to the horizontal run of the roof. The vertical rise is opposite the angle that the roof makes with the horizontal. The horizontal run is the adjacent side. So, the tangent of the angle of elevation equals the ratio of the rise to the run, or the slope of the roof; $\theta = 33.7°$.

Examples 3–4 Use a trigonometric function to find each value of *x*. Round to the nearest tenth.

21. 12.7

22. 3.6

23 10.4

24. 32.9

25. 8.7

26. 5.1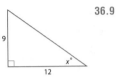

27. PARASAILING Refer to the beginning of the lesson and the figure at the right. Find *a*, the altitude of a person parasailing, if the tow rope is 250 feet long and the angle formed is 32°. Round to the nearest tenth. **132.5 ft**

28. CCSS MODELING Devon wants to build a rope bridge between his treehouse and Cheng's treehouse. Suppose Devon's treehouse is directly behind Cheng's treehouse. At a distance of 20 meters to the left of Devon's treehouse, an angle of 52° is measured between the two treehouses. Find the length of the rope. **25.6 m**

Example 5 Find the value of *x*. Round to the nearest tenth.

29. 30

30. 67.2

31. 36.9

32. 55.2

33. 32.5

34. 23.6

Example 6

35. SQUIRRELS Adult flying squirrels can make glides of up to 160 feet. If a flying squirrel glides a horizontal distance of 160 feet and the angle of descent is 9°, find its change in height. **25.3 ft**

36. HANG GLIDING A hang glider climbs at a 20° angle of elevation. Find the change in altitude of the hang glider when it has flown a horizontal distance of 60 feet. **21.8 ft**

B Use trigonometric functions to find the values of *x* and *y*. Round to the nearest tenth.

37.

38.

39.

37. *x* = 21.9, *y* = 20.8
38. *x* = 93.7, *y* = 60.2
39. *x* = 19.3, *y* = 70.7

Solve each equation.

40. $\cos A = \dfrac{3}{19}$ 80.9

41. $\sin N = \dfrac{9}{11}$ 54.9

42. $\tan X = 15$ 86.2

43. $\sin T = 0.35$ 20.5

44. $\tan G = 0.125$ 7.1

45. $\cos Z = 0.98$ 11.5

CCSS **Teaching the Mathematical Practices**

Modeling Mathematically proficient students can apply the mathematics they know to solve problems arising in everyday life, analyze relationships mathematically to draw conclusions, and interpret their mathematical results in the context of a situation.

46. MONUMENTS A monument casts a shadow 24 feet long. The angle of elevation from the end of the shadow to the top of the monument is 50°.

  a. Draw and label a right triangle to represent this situation. **See margin.**

  b. Write a trigonometric function that can be used to find the height of the monument. $\tan 50° = \dfrac{x}{24}$

  c. Find the value of the function to determine the height of the monument to the nearest tenth. **28.6 ft**

C **47** **NESTS** Tabitha's eyes are 5 feet above the ground as she looks up to a bird's nest in a tree. If the angle of elevation is 74.5° and she is standing 12 feet from the tree's base, what is the height of the bird's nest? Round to the nearest foot. **48 ft**

48. RAMPS Two bicycle ramps each cover a horizontal distance of 8 feet. One ramp has a 20° angle of elevation, and the other ramp has a 35° angle of elevation, as shown at the right.

  a. How much taller is the second ramp than the first? Round to the nearest tenth. **2.7 ft**

  b. How much longer is the second ramp than the first? Round to the nearest tenth. **1.3 ft**

49. FALCONS A falcon at a height of 200 feet sees two mice A and B, as shown in the diagram.

  a. What is the approximate distance z between the falcon and mouse B? **about 647.2 ft**

  b. How far apart are the two mice? **about 239.4 ft**

In $\triangle ABC$, $\angle C$ is a right angle. Use the given measurements to find the missing side lengths and missing angle measures of $\triangle ABC$. Round to the nearest tenth if necessary. **50.** $m\angle B = 54°$, $b = 16.5$, $c = 20.4$ **51.** $m\angle A = 59°$, $a = 31.6$, $c = 36.9$

50. $m\angle A = 36°$, $a = 12$

51. $m\angle B = 31°$, $b = 19$

52. $a = 8$, $c = 17$
$m\angle A = 28.1°$, $m\angle B = 61.9°$, $b = 15$

53. $\tan A = \dfrac{4}{5}$, $a = 6$
$m\angle A = 38.7°$, $m\angle B = 51.3°$, $b = 7.5$, $c = 9.6$

H.O.T. Problems Use Higher-Order Thinking Skills

54. CHALLENGE A line segment has endpoints $A(2, 0)$ and $B(6, 5)$, as shown in the figure at the right. What is the measure of the acute angle θ formed by the line segment and the x-axis? Explain how you found the measure.

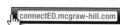

55. CCSS ARGUMENTS Determine whether the following statement is *true* or *false*. Explain your reasoning. **See margin.**

  For any acute angle, the sine function will never have a negative value.

56. OPEN ENDED In right triangle ABC, $\sin A = \sin C$. What can you conclude about $\triangle ABC$? Justify your reasoning. **See margin.**

57. WRITING IN MATH A roof has a slope of $\dfrac{2}{3}$. Describe the connection between the slope and the angle of elevation θ that the roof makes with the horizontal. Then use an inverse trigonometric function to find θ. **See margin.**

54. About 51.3°; if a right triangle is drawn with $\overline{AB}$ as the hypotenuse, then the side opposite angle θ is 5 and the side adjacent angle θ is 4. $\tan A = \dfrac{5}{4}$, so $A \approx 51.3°$.

Study Guide and Intervention
AL OL ELL

Practice
AL OL BL ELL

Word Problem Practice
AL OL BL ELL

CCSS Teaching the Mathematical Practices

Arguments Mathematically proficient students understand and use stated assumptions, definitions, and previously established results in constructing arguments. They make conjectures and build a logical progression of statements to explore the truth of their conjectures. And they are able to analyze situations by breaking them into cases, and can recognize and use counterexamples.

Enrichment
OL BL

Standardized Test Practice

58. EXTENDED RESPONSE Your school needs 5 cases of yearbooks. Neighborhood Yearbooks lists a case of yearbooks at $153.85 with a 10% discount on an order of 5 cases. Yearbooks R Us lists a case of yearbooks at $157.36 with a 15% discount on 5 cases. **a. Yearbooks R Us**

a. Which company would you choose?

b. What is the least amount that you would have to spend for the yearbooks? **$668.78**

59. SHORT RESPONSE As a fundraiser, the marching band sold T-shirts and hats. They sold a total of 105 items and raised $1170. If the cost of a hat was $10 and the cost of a T-shirt was $15, how many T-shirts were sold? **24**

60. A hot dog stand charges price x for a hot dog and price y for a drink. Two hot dogs and one drink cost $4.50. Three hot dogs and two drinks cost $7.25. Which matrix could be multiplied by $\begin{bmatrix} 4.50 \\ 7.25 \end{bmatrix}$ to find x and y? **B**

A $\begin{bmatrix} -1 & 1 \\ 2 & -1 \end{bmatrix}$ C $\begin{bmatrix} 1 & 2 \\ -1 & 3 \end{bmatrix}$

B $\begin{bmatrix} 2 & -1 \\ -3 & 2 \end{bmatrix}$ D $\begin{bmatrix} 1 & -1 \\ -1 & 2 \end{bmatrix}$

61. SAT/ACT The length and width of a rectangle are in the ratio of 5:12. If the rectangle has an area of 240 square centimeters, what is the length, in centimeters, of its diagonal? **G**

F 24   H 28   K 32

G 26   J 30

Spiral Review

Identify the null and alternative hypotheses for each statement. Then identify the statement that represents the claim. (Lesson 11-6)

62. Jack thinks that it takes less than 10 minutes to ride his bike from his home to the store. $H_0: \mu \geq 10$; $H_a: \mu < 10$ (claim

63. A deli sign says that one 12-inch turkey sandwich contains three ounces of meat. $H_0: \mu = 3$ (claim); $H_a: \mu \neq 3$

64. Mrs. Thomas takes at least 15 minutes to prepare a cake. $H_0: \mu \geq 15$ (claim); $H_a: \mu < 15$

65. SWIMMING POOL The number of visits to a community swimming pool per year by a sample of 425 members is normally distributed with a mean of 90 and a standard deviation of 15. (Lesson 11-5)

a. About what percent of the members went to the pool at least 45 times? **99.85%**

b. What is the probability that a member selected at random went to the pool more than 120 times? **2.5%**

c. What percent of the members went to the pool between 75 and 105 times? **64%**

66. POLLS A polling company wants to estimate how many people are in favor of a new environmental law. The polling company polls 20 people. The probability that a person is in favor of the law is 0.5. (Lesson 11-4)

a. What is the probability that exactly 12 people are in favor of the new law? **0.12**

b. What is the expected number of people in favor of the law? **10**

Skills Review

Find each product. Include the appropriate units with your answer.   **70. 86.4 in³   71. 216$\frac{2}{3}$ cm**

67. 4.3 miles$\left(\dfrac{5280 \text{ feet}}{1 \text{ mile}}\right)$ **22,704 ft**   **68.** 8 gallons$\left(\dfrac{8 \text{ pints}}{1 \text{ gallon}}\right)$ **64 pt**   **69.** $\left(\dfrac{5 \text{ dollars}}{3 \text{ meters}}\right)$21 meters **35 dollars**

70. $\left(\dfrac{18 \text{ cubic inches}}{5 \text{ seconds}}\right)$24 seconds   **71.** 65 degrees$\left(\dfrac{10 \text{ centimeters}}{3 \text{ degrees}}\right)$ **72.** $\left(\dfrac{7 \text{ liters}}{30 \text{ minutes}}\right)$10 minutes **2$\frac{1}{3}$ L**

 798 | **Lesson 12-1** | Trigonometric Functions in Right Triangles

DifferentiatedInstruction OL BL

Extension Have students compute the value of $\sin^2 x + \cos^2 x$ for some values of x. Have them make a conjecture about the value of the expression and use the definitions of the trigonometric functions to prove their conjecture. The value is always 1.

$$\sin^2 x + \cos^2 x \overset{?}{=} 1$$

$$\left(\frac{\text{opp}}{\text{hyp}}\right)^2 + \left(\frac{\text{adj}}{\text{hyp}}\right)^2 \overset{?}{=} 1$$

$$\text{opp}^2 + \text{adj}^2 = \text{hyp}^2, \text{ which is true by the Pythagorean Theorem}$$

Nick Koudis/Photodisc/Getty Images

LESSON 12-2 — Angles and Angle Measure

:: Then	:: Now	:: Why?
● You used angles with degree measures.	**1** Draw and find angles in standard position. **2** Convert between degree measures and radian measures.	● A sundial is an instrument that indicates the time of day by the shadow that it casts on a surface marked to show hours or fractions of hours. The shadow moves around the dial 15° every hour.

 NewVocabulary
standard position
initial side
terminal side
coterminal angles
radian
central angle
arc length

 Common Core State Standards

Content Standards
F.TF.1 Understand radian measure of an angle as the length of the arc on the unit circle subtended by the angle.

Mathematical Practices
2 Reason abstractly and quantitatively.

1 Angles in Standard Position An angle on the coordinate plane is in **standard position** if the vertex is at the origin and one ray is on the positive x-axis.

- The ray on the x-axis is called the **initial side** of the angle.
- The ray that rotates about the center is called the **terminal side**.

KeyConcept Angle Measures

If the measure of an angle is positive, the terminal side is rotated counterclockwise.

If the measure of an angle is negative, the terminal side is rotated clockwise.

Example 1 Draw an Angle in Standard Position [PT]

Draw an angle with the given measure in standard position.

a. 215° $215° = 180° + 35°$

Draw the terminal side of the angle 35° counterclockwise past the negative x-axis.

b. −40°

The angle is negative. Draw the terminal side of the angle 40° clockwise from the positive x-axis.

GuidedPractice   1A, 1B. See Chapter 12 Answer Appendix.
1A. 80°   **1B.** −105°

 connectED.mcgraw-hill.com   **799**

---

1 Focus

VerticalAlignment

Before Lesson 12-2 Use angles with degree measures.

Lesson 12-2 Draw and find angles in standard position. Convert between degree measures and radian measures.

After Lesson 12-2 Investigate graphs of the sine and cosine functions.

2 Teach

Scaffolding Questions
Have students read the **Why?** section of the lesson.

Ask:
- In what direction does the shadow point in the morning? west
- At what time do you get the shortest shadow? noon
- Explain why the shadow moves 15° every hour. The shadow "returns" to any particular position every 24 hours, and the entire circle, divided by 24, is 360° ÷ 24, or 15°.

Lesson 12-2 Resources

Resource	Approaching Level (AL)	On Level (OL)	Beyond Level (BL)	English Learners (ELL)
Teacher Edition	▪ Differentiated Instruction, p. 801	▪ Differentiated Instruction, p. 805	▪ Differentiated Instruction, p. 805	
Chapter Resource Masters	▪ Study Guide and Intervention, pp. 11–12 ▪ Skills Practice, p. 13 ▪ Practice, p. 14 ▪ Word Problem Practice, p. 15	▪ Study Guide and Intervention, pp. 11–12 ▪ Skills Practice, p. 13 ▪ Practice, p. 14 ▪ Word Problem Practice, p. 15 ▪ Enrichment, p. 16	▪ Practice, p. 14 ▪ Word Problem Practice, p. 15 ▪ Enrichment, p. 16	▪ Study Guide and Intervention, pp. 11–12 ▪ Skills Practice, p. 13 ▪ Practice, p. 14 ▪ Word Problem Practice, p. 15
Other	▪ 5-Minute Check 12-2 ▪ Study Notebook	▪ 5-Minute Check 12-2 ▪ Study Notebook	▪ 5-Minute Check 12-2 ▪ Study Notebook	▪ 5-Minute Check 12-2 ▪ Study Notebook

1 Angles in Standard Position

Example 1 shows how to draw an angle in standard position with a given measure. **Example 2** shows how to represent an angle greater than 360°. **Example 3** shows how to represent coterminal angles with a positive or a negative measure.

Formative Assessment

Use the Guided Practice exercises after each example to determine students' understanding of concepts.

Additional Examples

1 Draw an angle with the given measure in standard position.

a. 210°

b. −45°

2 **DIVING** In a springboard diving competition, a diver made a 900-degree rotation before slicing into the water. Draw an angle in standard position that measures 900°.

The terminal side of an angle can make more than one complete rotation. For example, a complete rotation of 360° plus a rotation of 120° forms an angle that measures 360° + 120° or 480°.

480° angle

Real-WorldLink

Wakeboarding is one of the fastest-growing water sports in the United States. Participation has increased more than 100% in recent years.

Source: King of Wake

● Real-World Example 2 Draw an Angle in Standard Position

WAKEBOARDING *Wakeboarding* is a combination of surfing, skateboarding, snowboarding, and water skiing. One maneuver involves a 540-degree rotation in the air. Draw an angle in standard position that measures 540°.

$540° = 360° + 180°$

Draw the terminal side of the angle 180° past the positive *x*-axis.

▶ **Guided**Practice

2. Draw an angle in standard position that measures 600°. **See margin.**

Two or more angles in standard position with the same terminal side are called **coterminal angles**. For example, angles that measure 60°, 420°, and −300° are coterminal, as shown in the figure at the right.

An angle that is coterminal with another angle can be found by adding or subtracting a multiple of 360°.

- $60° + 360° = 420°$
- $60° − 360° = −300°$

ReadingMath

Angle of Rotation
In trigonometry, an angle is sometimes referred to as an *angle of rotation*.

Example 3 Find Coterminal Angles

Find an angle with a positive measure and an angle with a negative measure that are coterminal with each angle.

a. 130°

positive angle: $130° + 360° = 490°$ Add 360°.
negative angle: $130° − 360° = −230°$ Subtract 360°.

b. −200°

positive angle: $−200° + 360° = 160°$ Add 360°.
negative angle: $−200° − 360° = −560°$ Subtract 360°.

▶ **Guided**Practice

3A. 15°   375°, −345°   **3B.** −45°   315°, −405°

Tips for New Teachers

Drawing Angles Discuss with students the importance of marking their drawings of angles with directional arrows, and labeling the angle measures.

Additional Answer (Guided Practice)

2.

600°

Paul Bradbury/Digital Vision/Getty Images

StudyTip

CCSS **Structure** As with degrees, radians measure the amount of rotation from the initial side to the terminal side.

- The measure of an angle in radians is positive if its rotation is counterclockwise.
- The measure is negative if the rotation is clockwise.

2 Convert Between Degrees and Radians Angles can also be measured in units that are based on arc length. One **radian** is the measure of an angle θ in standard position with a terminal side that intercepts an arc with the same length as the radius of the circle.

$\theta = 1$ radian

The circumference of a circle is $2\pi r$. So, one complete revolution around a circle is 2π radians. Since 2π radians $= 360°$, degree measure and radian measure are related by the following equations.

$$2\pi \text{ radians} = 360° \qquad \pi \text{ radians} = 180°$$

KeyConcept Convert Between Degrees and Radians

Degrees to Radians	Radians to Degrees
To convert from degrees to radians, multiply the number of degrees by $\dfrac{\pi \text{ radians}}{180°}$.	To convert from radians to degrees, multiply the number of radians by $\dfrac{180°}{\pi \text{ radians}}$.

ReadingMath

Radian Measures
The word *radian* is usually omitted when angles are expressed in radian measure. Thus, when no units are given for an angle measure, radian measure is implied.

Example 4 Convert Between Degrees and Radians

Rewrite the degree measure in radians and the radian measure in degrees.

a. $-30°$

$-30° = -30° \cdot \dfrac{\pi \text{ radians}}{180°}$

$= \dfrac{-30\pi}{180}$ or $-\dfrac{\pi}{6}$ radians

b. $\dfrac{5\pi}{2}$

$\dfrac{5\pi}{2} = \dfrac{5\pi}{2} \text{ radians} \cdot \dfrac{180°}{\pi \text{ radians}}$

$= \dfrac{900°}{2}$ or $450°$

Guided Practice

4A. $120°$ $\dfrac{2\pi}{3}$

4B. $-\dfrac{3\pi}{8}$ $-67.5°$

ConceptSummary Degrees and Radians

The diagram shows equivalent degree and radian measures for special angles.

You may find it helpful to memorize the following equivalent degree and radian measures. The other special angles are multiples of these angles.

$30° = \dfrac{\pi}{6}$ $\qquad 45° = \dfrac{\pi}{4}$

$60° = \dfrac{\pi}{3}$ $\qquad 90° = \dfrac{\pi}{2}$

DifferentiatedInstruction AL

If students need further practice with coterminal angles,

Then have them work with a partner so that one person models an angle with two pencils or yardsticks. The other partner then names a positive and negative angle, less than or more than a full circle, that are each coterminal with the modeled angle.

Additional Example

3 Find an angle with a positive measure and an angle with a negative measure that are coterminal with each angle.

a. $210°$ Sample answers: $570°, -150°$

b. $-120°$ Sample answers: $240°, -480°$

Additional Examples also in Interactive Classroom PowerPoint® Presentations

IWB Interactive White Board READY

Teach with Tech

Instant Messaging Have students work in pairs. One student should send a message with the measure of an angle. The other student should reply with the measures of two angles coterminal with the given angle. Students should then switch roles and repeat.

2 Convert Between Degrees and Radians

Example 4 shows how to convert between degrees and radians. **Example 5** shows how to convert and use angle measures in a real-world situation.

CCSS **Teaching the Mathematical Practices**

Structure Mathematically proficient students look closely to discern a pattern or structure. Refer students to the Concept Summary and point out that angle measures are arranged counterclockwise.

Additional Example

4 Rewrite the degree measure in radians and the radian measure in degrees.

a. $30°$ $\dfrac{\pi}{6}$

b. $-\dfrac{5\pi}{3}$ $-300°$

Additional Example

5 TRUCKS The steering wheel on a monster truck has a radius of 11 inches. How far does a point on the steering wheel travel if the wheel makes four fifths of a rotation? **about 55.3 in.**

Additional Answers

1.

2.

3.

11.

12.

13.

A **central angle** of a circle is an angle with a vertex at the center of the circle. If you know the measure of a central angle and the radius of the circle, you can find the length of the arc that is intercepted by the angle.

KeyConcept Arc Length

Words	For a circle with radius r and central angle θ (in radians), the **arc length** s equals the product of r and θ.	Model
Symbols	$s = r\theta$	

You will justify this formula in Exercise 52

PT

Real-World Example 5 Find Arc Length

TRUCKS Monster truck tires have a radius of 33 inches. How far does a monster truck travel in feet after just three fourths of a tire rotation?

Step 1 Find the central angle in radians.

$\theta = \dfrac{3}{4} \cdot 2\pi$ or $\dfrac{3\pi}{2}$ The angle is $\frac{3}{4}$ of a complete rotation.

Step 2 Use the radius and central angle to find the arc length.

$s = r\theta$ Write the formula for arc length.

$= 33 \cdot \dfrac{3\pi}{2}$ Replace r with 33 and θ with $\frac{3\pi}{2}$.

≈ 155.5 in.   Use a calculator to simplify.

≈ 13.0 ft   Divide by 12 to convert to feet.

So, the truck travels about 13 feet after three fourths of a tire rotation.

> **WatchOut!**
> **Arc Length** Remember to write the angle measure in radians, not degrees, when finding arc length. Also, recall that the number of radians in a complete rotation is 2π.

▶ **Guided**Practice

5. A circle has a diameter of 9 centimeters. Find the arc length if the central angle is 60°. Round to the nearest tenth. **4.7 cm**

Check Your Understanding

 = Step-by-Step Solutions begin on page R14.

Examples 1–2 Draw an angle with the given measure in standard position. **1–3. See margin.**

1. 140°   **2.** −60°   **3.** 390°

Example 3 Find an angle with a positive measure and an angle with a negative measure that are coterminal with each angle. **4. Sample answer: 385°, −335°**

4. 25°   **5** 175°   **6.** −100°

5. Sample answer: 535°, −185°
6. Sample answer: 260°, −460°

Example 4 Rewrite each degree measure in radians and each radian measure in degrees.

7. $\dfrac{\pi}{4}$ **45°**   **8.** 225° $\dfrac{5\pi}{4}$ **9.** −40° $-\dfrac{2\pi}{9}$

Example 5 **10. CCSS REASONING** A tennis player's swing moves along the path of an arc. If the radius of the arc's circle is 4 feet and the angle of rotation is 100°, what is the length of the arc? Round to the nearest tenth. **7.0 ft**

🔊 **802** | Lesson 12-2 | Angles and Angle Measure

CCSS Teaching the Mathematical Practices

Reasoning Mathematically proficient students make sense of quantities and their relationships in problem situations. Quantitative reasoning entails habits of creating a coherent representation of the problem at hand; considering the units involved; attending to the meaning of quantities, not just how to compute them; and knowing and flexibly using different properties of operations and objects.

Examples 1–2 Draw an angle with the given measure in standard position. **11–16. See margin.**

11. 75°   **12.** 160°   **13.** −90°

14. −120°   **15.** 295°   **16.** 510°

17. **GYMNASTICS** A gymnast on the uneven bars swings to make a 240° angle of rotation.

18. **FOOD** The lid on a jar of pasta sauce is turned 420° before it comes off. **17, 18. See margin.**

Example 3 Find an angle with a positive measure and an angle with a negative measure that are coterminal with each angle. **19–24. Sample answers given.**

19. 50° 410°, −310°   **20.** 95° 455°, −265°   **21.** 205° 565°, −155°

22. 350° 710°, −10°   **23.** −80° 280°, −440°   **24.** −195° 165°, −555°

Example 4 Rewrite each degree measure in radians and each radian measure in degrees.

25 330° $\frac{11\pi}{6}$ **26.** $\frac{5\pi}{6}$ 150°   **27.** $-\frac{\pi}{3}$ −60°

28. −50° $-\frac{5\pi}{18}$ **29.** 190° $\frac{19\pi}{18}$ **30.** $-\frac{7\pi}{3}$ −420°

Example 5

31. **SKATEBOARDING** The skateboard ramp at the right is called a *quarter pipe*. The curved surface is determined by the radius of a circle. Find the length of the curved part of the ramp. **about 12.6 ft**

32. **RIVERBOATS** The paddlewheel of a riverboat has a diameter of 24 feet. Find the arc length of the circle made when the paddlewheel rotates 300°. **about 62.8 ft**

Find the length of each arc. Round to the nearest tenth.

33.   6.7 cm   **34.**   94.2 m

35. **CLOCKS** How long does it take for the minute hand on a clock to pass through 2.5π radians? **1 h 15 min**

36. **CCSS** **PERSEVERANCE** Refer to the beginning of the lesson. A shadow moves around a sundial 15° every hour.

a. After how many hours is the angle of rotation of the shadow $\frac{8\pi}{5}$ radians? **19.2 h**

b. What is the angle of rotation in radians after 5 hours? $\frac{5\pi}{12}$

c. A sundial has a radius of 8 inches. What is the arc formed by a shadow after 14 hours? Round to the nearest tenth. **29.3 in.**

Find an angle with a positive measure and an angle with a negative measure that are coterminal with each angle. **37–40. Sample answers given.**

37. 620° 260°, −100° **38.** −400° 320°, −40° **39.** $-\frac{3\pi}{4}$ $\frac{5\pi}{4}$, $-\frac{11\pi}{4}$ **40.** $\frac{19\pi}{6}$ $\frac{7\pi}{6}$, $-\frac{5\pi}{6}$

Differentiated Homework Options

Level	Assignment	Two-Day Option	
AL Basic	11–34, 48, 50–67	11–33 odd, 54–57	12–34 even, 48, 50–53, 58–67
OL Core	11–39 odd, 41–48, 50–67	11–34, 54–57	35–48, 50–53, 58–67
BL Advanced	35–64, (optional: 65–67)		

Formative Assessment

Use Exercises 1–10 to check for understanding.

Use the chart at the bottom of this page to customize assignments for your students.

Additional Answers

14.

15.

16.

17.

18.

Study Guide and Intervention

AL OL ELL

NAME _____ DATE _____ PERIOD _____

12-2 Study Guide and Intervention

Angles and Angle Measure

Angles in Standard Position An angle is determined by two rays. The degree measure of an angle in standard position is described by the amount and direction of rotation from the **initial side**, which lies along the positive *x*-axis, to the **terminal side**. A counterclockwise rotation is associated with positive angle measure and a clockwise rotation has the same terminal side are called **coterminal angles**.

Example 1 Draw an angle with measure 290° in standard position.

The negative *y*-axis represents a positive rotation of 270°. To generate an angle of 290°, rotate the terminal side 20° more in the counterclockwise direction

Example 2 Find an angle with a positive measure and an angle with a negative measure that are coterminal with each angle.

a. 250°
A positive angle is 250° + 360° or 610°.   Add 360°
A negative angle is 250° − 360° or −110°.   Subtract 360°

b. −140°
A positive angle is −140° + 360° or 220°.   Add 360°
A negative angle is −140° − 360° or −500°.   Subtract 360°

Exercises

Draw an angle with the given measure in standard position.

1. 160°   2. 280°   3. 400°

Find an angle with a positive measure and an angle with a negative measure that are coterminal with each angle.

4. 65°   5. −75°   6. 230°   7. 420°

Practice

AL OL BL ELL

NAME _____ DATE _____ PERIOD _____

12-2 Practice

Angles and Angle Measure

Draw an angle with the given measure in standard position.

1. 210°   2. 305°   3. 580°

4. 135°   5. −450°   6. −560°

Find an angle with a positive measure and an angle with a negative measure that are coterminal with each angle.

7. 65°   8. 80°   9. 110°

10. $\frac{2\pi}{5}$ 11. $\frac{5\pi}{4}$ 12. $-\frac{3\pi}{2}$

Rewrite each degree measure in radians and each radian measure in degrees.

13. 18°   14. 6°   15. −72°   16. −820°

17. 4π   18. $\frac{5\pi}{2}$ 19. $-\frac{8\pi}{3}$ 20. $-\frac{7\pi}{12}$

Find the length of each arc. Round to the nearest tenth.

21.   22.   23.

24. **TIME** Find both the degree and radian measures of the angle through which the hour hand on a clock rotates from 5 A.M. to 10 P.M.

25. **ROTATION** A truck with 16-inch radius wheels is driven at 77 feet per second (52.5 miles per hour). Find the measure of the angle through which a point on the outside of the wheel travels each second. Round to the nearest degree and nearest radian.

Word Problem Practice

AL OL BL ELL

NAME _____ DATE _____ PERIOD _____

12-2 Word Problem Practice

Angles and Angle Measure

1. **AMUSEMENT PARKS** The carousel at an amusement park has 20 horses spaced evenly around its circumference. The horses are numbered consecutively from 1 to 20. The carousel completes one rotation about its axis every 40 seconds.

a. What is the central angle, in degrees, formed by horse #1 and horse #8?

b. What is the speed of the carousel in rotations per minute?

c. What is the speed of the carousel in radians per minute?

d. A child rides the carousel for 6 minutes. Through how many radians will the child pass in the course of the carousel ride?

2. **TIME** Through what angle, in degrees and radians, does the hour hand on a clock rotate between 4 P.M. and 7 P.M.? Assuming the length of the hour hand is 6 inches, find the arc length of the circle made by the hour hand during that time.

3. **TIME** Through what angle, in degrees and radians, does the minute hand rotate between 4 P.M. and 7 P.M.?

4. **PLANETS** Earth makes one full rotation on its axis every 24 hours. How long does it take Earth to rotate through 150°? Neptune makes one full rotation on its axis every 16 hours. How long does it take Neptune to rotate through 150°?

41. **SWINGS** A swing has a 165° angle of rotation.

a. Draw the angle in standard position. **See margin.**

b. Write the angle measure in radians. $\frac{11\pi}{12}$

c. If the chains of the swing are $6\frac{1}{2}$ feet long, what is the length of the arc that the swing makes? Round to the nearest tenth. **18.7 ft**

d. Describe how the arc length would change if the lengths of the chains of the swing were doubled.

41d. The arc length would double. Since $s = r\theta$, if r is doubled and θ remains unchanged, then the value of s is also doubled.

42b.
$\tan \angle BEA = -\frac{3}{2}$,
$\tan \angle DEC = \frac{4}{3}$

42d. Sample answer: In the coordinate plane, the tangent of the angle in standard position equals the slope of the terminal side of the angle.

42. **MULTIPLE REPRESENTATIONS** Consider $A(-4, 0)$, $B(-4, 6)$, $C(6, 0)$, and $D(6, 8)$.

a. **Geometric** Draw $\triangle EAB$ and $\triangle ECD$ with E at the origin. **See margin.**

b. **Algebraic** Find the values of the tangent of $\angle BEA$ and the tangent of $\angle DEC$.

c. **Algebraic** Find the slope of $\overline{BE}$ and $\overline{ED}$. slope of $\overline{BE} = -\frac{3}{2}$, slope of $\overline{ED} = \frac{4}{3}$

d. **Verbal** What conclusions can you make about the relationship between slope and tangent?

Rewrite each degree measure in radians and each radian measure in degrees.

43. $\frac{21\pi}{8}$ **472.5°**   44. 124° $\frac{31\pi}{45}$ 45. −200° $-\frac{10\pi}{9}$ 46. 5 $\frac{900}{\pi} \approx 286.5°$

47. **CAROUSELS** A carousel makes 5 revolutions per minute. The circle formed by riders sitting in the outside row has a radius of 17.2 feet. The circle formed by riders sitting in the inside row has a radius of 13.1 feet.

a. Find the angle θ in radians through which the carousel rotates in one second. $\frac{\pi}{6}$

b. In one second, what is the difference in arc lengths between the riders sitting in the outside row and the riders sitting in the inside row? **2.1 ft**

H.O.T. Problems Use Higher-Order Thinking Skills

48. **CCSS CRITIQUE** Tarshia and Alan are writing an expression for the measure of an angle coterminal with the angle shown at the right. Is either of them correct? Explain your reasoning.

Tarshia	Alan
The measure of a coterminal angle is $(x - 360)°$.	The measure of a coterminal angle is $(360 - x)°$.

49. **CHALLENGE** A line makes an angle of $\frac{\pi}{2}$ radians with the positive *x*-axis at the point (2, 0). Find an equation for this line. $x = 2$

50. **REASONING** Express $\frac{1}{8}$ of a revolution in degrees and in radians. Explain your reasoning. 45°, $\frac{\pi}{4}$; $\frac{1}{8}$ of 360° is 45° and $\frac{1}{8}$ of 2π is $\frac{\pi}{4}$.

51. **OPEN ENDED** Draw and label an acute angle in standard position. Find two angles, one positive and one negative, that are coterminal with the angle. **See margin.**

52. **REASONING** Justify the formula for the length of an arc. **See Chapter 12 Answer Appendix.**

53. **WRITING IN MATH** Use a circle with radius *r* to describe what one degree and one radian represent. Then explain how to convert between the measures. **See Chapter 12 Answer Appendix.**

48. Tarshia; a coterminal can be found by adding a multiple of 360° or by subtracting a multiple of 360°. Alan incorrectly subtracted the original angle measure from 360°.

 804 | Lesson 12-2 | Angles and Angle Measure

Enrichment

OL BL

NAME _____ DATE _____ PERIOD _____

12-2 Enrichment

Making and Using a Hypsometer

A **hypsometer** is a device that can be used to measure the height of an object. To construct your own hypsometer, you will need a rectangular piece of heavy cardboard that is at least 7 cm by 10 cm, a straw, transparent tape, a string about 20 cm long, and a small weight that can be attached to the string.

Mark off 1-cm increments along one short side and one long side of the cardboard. Tape the straw to the other short side. Then attach the weight to one end of the string, and attach the other end of the string to one corner of the cardboard, as shown in the figure below. The diagram below shows how your hypsometer should look.

Multiple Representations

In Exercise 42, students use coordinate geometry, trigonometry, algebra, and analysis to relate the slope of a line to the tangent of the angle it makes with the *x*-axis.

54. SHORT RESPONSE If $(x + 6)(x + 8) -$ $(x - 7)(x - 5) = 0$, find x. $-\dfrac{1}{2}$

55. Which of the following represents an inverse variation? **A**

A
x	2	5	10	20	25	50
y	50	20	10	5	4	2

B
x	2	4	6	8	10	12
y	−4	−8	−12	−16	−20	−24

C
x	1	2	3	4	5	6
y	5	10	15	20	25	30

D
x	10	9	8	7	6	5
y	5	6	7	8	9	10

56. GEOMETRY If the area of the figure is 60 square units, what is the length of side $\overline{XZ}$? **G**

F $2\sqrt{34}$ H $4\sqrt{34}$

G $2\sqrt{109}$ J $4\sqrt{109}$

57. SAT/ACT The first term of a sequence is −6, and every term after the first is 8 more than the term immediately preceding it. What is the value of the 101st term? **B**

A 788 D 806

B 794 E 814

C 802

Find the values of the six trigonometric functions for angle θ. (Lesson 12-1) **58–60. See Chapter 12 Answer Appendix.**

58.

59.

60.

Identify the null and alternative hypotheses for each statement. Then identify the statement that represents the claim. (Lesson 11-6)

61. Tom drinks at least eight glasses of water every day. $H_0: \mu \geq 8$ (claim); $H_a: \mu < 8$

62. Juanita says that she has two umbrellas in her car. $H_0: \mu = 2$ (claim); $H_a: \mu \neq 2$

63. MANUFACTURING The sizes of CDs made by a company are normally distributed with a standard deviation of 1 millimeter. The CDs are supposed to be 120 millimeters in diameter, and they are made for drives that are 122 millimeters wide. (Lesson 11-5)
 a. What percent of the CDs would you expect to be greater than 120 millimeters? **50%**
 b. If the company manufactures 1000 CDs per hour, how many of the CDs made in one hour would you expect to be between 119 and 122 millimeters? **815**
 c. About how many CDs per hour will be too large to fit in the drives? **25**

64. FINANCIAL LITERACY If the rate of inflation is 2%, the cost of an item in future years can be found by iterating the function $c(x) = 1.02x$. Find the cost of a $70 digital audio player in four years if the rate of inflation remains constant. (Lesson 10-5) **$75.77**

Use the Pythagorean Theorem to find the length of the hypotenuse for each right triangle with the given side lengths.

65. $a = 12, b = 15$ $3\sqrt{41}$ **66.** $a = 8, b = 17$ $\sqrt{353}$ **67.** $a = 14, b = 11$ $\sqrt{317}$

DifferentiatedInstruction OL BL

Extension There are infinitely many angles coterminal with a given angle. Have students write an expression that gives the angle measure of all angles coterminal with an angle of 50°.
$50° + k \cdot 360°$, where k is any integer.

WatchOut!

Error Analysis For Exercise 48, if students think that Alan is correct, point out that if the angle shown is measured clockwise, then the angle measure x is a negative quantity and $(360 - x)°$ will be greater, not less, than 360°.

4 Assess

Yesterday's News Ask students to describe how yesterday's topic of angles in right triangles helped them understand and illustrate today's topic of angles in standard position.

Additional Answers

41a.

42a.

51. Sample answer:
440° and −280°

1 Focus

Objective Use the sine ratio to find the area of a parallelogram.

Teaching Tip

If students use calculators to approximate areas, remind them to make sure that their calculators are in degree mode.

2 Teach

Working in Cooperative Groups

Have students work in pairs, mixing abilities, to complete the Activity.

Have each student in the pair draw a different diagonal of the parallelogram. Then have them compare their area calculations to check their work.

Ask:

- What part of the parallelogram is represented by $AB \sin A$?

$\sin A = \dfrac{h}{AB}$, so $AB \sin A$ is the height h of the parallelogram.

Practice Have students complete Exercise 1–3.

3 Assess

Formative Assessment

Use Exercise 2 to assess whether students can calculate the area of a parallelogram given two consecutive sides and the measure of the included angle.

The area of any triangle can be found using the sine ratios in the triangle. A similar process can be used to find the area of a parallelogram.

Activity

Find the area of parallelogram $ABCD$.

Step 1 Draw diagonal $\overline{BD}$.

$\overline{BD}$ divides the parallelogram into two congruent triangles, $\triangle ABD$ and $\triangle CDB$.

Step 2 Find the area of $\triangle ABD$.

Use the sine ratio to determine the height h from B to $\overline{AD}$.

$\sin \theta = \dfrac{\text{opp}}{\text{hyp}}$ Definition of sine

$\sin \theta = \dfrac{h}{AB}$ $h = \text{opp}, AB = \text{hyp}$

$AB \sin \theta = h$ Solve for h.

So, $h = AB \sin \theta$.

$\text{Area} = \frac{1}{2}bh$	Area of a triangle
$= \frac{1}{2}(AD)(AB) \sin A$	$b = AD, h = AB \sin A$
$= \frac{1}{2}(28)(16) \sin 60°$	$AD = 28, AB = 16, A = 60°$
$= 224\left[\dfrac{\sqrt{3}}{2}\right]$	Multiply and evaluate $\sin 60°$.
$= 112\sqrt{3}$	Simplify.

Step 3 Find the area of $\square ABCD$.

The area of $\square ABCD$ is equal to the sum of the areas of $\triangle ABD$ and $\triangle CDB$. Because $\triangle ABD \cong \triangle CDB$, the areas of $\triangle ABD$ and $\triangle CDB$ are equal. So, the area of $\square ABCD$ equals twice the area of $\triangle ABD$.

$2 \cdot 112\sqrt{3} = 224\sqrt{3}$ or about 387.98 square inches.

Exercises

For each of the following,

a. find the area of each parallelogram.

b. find the area of each parallelogram when the included angle is half the given measure.

c. find the area of each parallelogram when the included angle is twice the given measure.

1a. 106.07 m²   **1b.** 57.40 m²   **1c.** 150 m²   **2a.** 22.5 in²   **2b.** 11.65 in²
2c. 38.97 in²   **3a.** 19,318.52 ft²   **3b.** 12,175.23 ft²   **3c.** 10,000 ft²

1.   **2.**   **3.**

From Concrete to Abstract

Ask students to develop a formula for the area of a parallelogram if the measures of two consecutive sides are a and b, and one of the angles of the parallelogram is θ. $A = ab \sin \theta$

Additional Answer (Guided Practice)

1. $\sin \theta = \dfrac{\sqrt{10}}{10}$, $\cos \theta = -\dfrac{3\sqrt{10}}{10}$, $\tan \theta = -\dfrac{1}{3}$,

$\csc \theta = \sqrt{10}$, $\sec \theta = -\dfrac{\sqrt{10}}{3}$, $\cot \theta = -3$

12-3 Trigonometric Functions of General Angles

Then
- You found values of trigonometric functions for acute angles.

Now
1. Find values of trigonometric functions for general angles.
2. Find values of trigonometric functions by using reference angles.

Why?
- In the ride at the right, the cars rotate back and forth about a central point. The positions of the arms supporting the cars can be described using trigonometric angles in standard position, with the central point of the ride at the origin of a coordinate plane.

NewVocabulary
quadrantal angle
reference angle

Common Core State Standards

Mathematical Practices
6 Attend to precision.

---

1 Trigonometric Functions for General Angles
You can find values of trigonometric functions for angles greater than 90° or less than 0°.

KeyConcept Trigonometric Functions of General Angles

Let θ be an angle in standard position and let $P(x, y)$ be a point on its terminal side. Using the Pythagorean Theorem, $r = \sqrt{x^2 + y^2}$. The six trigonometric functions of θ are defined below.

$$\sin \theta = \frac{y}{r} \qquad \cos \theta = \frac{x}{r} \qquad \tan \theta = \frac{y}{x}, x \neq 0$$

$$\csc \theta = \frac{r}{y}, y \neq 0 \qquad \sec \theta = \frac{r}{x}, x \neq 0 \qquad \cot \theta = \frac{x}{y}, y \neq 0$$

Example 1 Evaluate Trigonometric Functions Given a Point

The terminal side of θ in standard position contains the point at $(-3, -4)$. Find the exact values of the six trigonometric functions of θ.

Step 1 Draw the angle, and find the value of r.

$$r = \sqrt{x^2 + y^2}$$
$$= \sqrt{(-3)^2 + (-4)^2}$$
$$= \sqrt{25} \text{ or } 5$$

Step 2 Use $x = -3$, $y = -4$, and $r = 5$ to write the six trigonometric ratios.

$$\sin \theta = \frac{y}{r} = \frac{-4}{5} \text{ or } -\frac{4}{5} \qquad \cos \theta = \frac{x}{r} = \frac{-3}{5} \text{ or } -\frac{3}{5} \qquad \tan \theta = \frac{y}{x} = \frac{-4}{-3} \text{ or } \frac{4}{3}$$

$$\csc \theta = \frac{r}{y} = \frac{5}{-4} \text{ or } -\frac{5}{4} \qquad \sec \theta = \frac{r}{x} = \frac{5}{-3} \text{ or } -\frac{5}{3} \qquad \cot \theta = \frac{x}{y} = \frac{-3}{-4} \text{ or } \frac{3}{4}$$

GuidedPractice

1. The terminal side of θ in standard position contains the point at $(-6, 2)$. Find the exact values of the six trigonometric functions of θ. **See margin.**

  807

---

1 Focus

VerticalAlignment

Before Lesson 12-3 Find values of trigonometric functions for acute angles.

Lesson 12-3 Find values of trigonometric functions for general angles. Find values of trigonometric functions by using reference angles.

After Lesson 12-3 Investigate graphs of the tangent, cotangent, secant, and cosecant functions.

---

2 Teach

Scaffolding Questions

Have students read the **Why?** section of the lesson and tell them that the positions of two of the arms are 220° clockwise and 50° counterclockwise from the standard position.

Ask:
- Which quadrant is indicated by "20° clockwise"? Quadrant IV
- Which is indicated by "200° counterclockwise"? Quadrant III
- How can you describe the position "20° clockwise" in terms of a counterclockwise rotation? 340° counterclockwise

---

Lesson 12-3 Resources

Resource	Approaching Level **AL**	On Level **OL**	Beyond Level **BL**	English Learners **ELL**
Teacher Edition	▪ Differentiated Instruction, p. 810	▪ Differentiated Instruction, pp. 810, 813	▪ Differentiated Instruction, p. 813	
Chapter Resource Masters	▪ Study Guide and Intervention, pp. 17–18 ▪ Skills Practice, p. 19 ▪ Practice, p. 20 ▪ Word Problem Practice, p. 21	▪ Study Guide and Intervention, pp. 17–18 ▪ Skills Practice, p. 19 ▪ Practice, p. 20 ▪ Word Problem Practice, p. 21 ▪ Enrichment, p. 22	▪ Practice, p. 20 ▪ Word Problem Practice, p. 21 ▪ Enrichment, p. 22	▪ Study Guide and Intervention, pp. 17–18 ▪ Skills Practice, p. 19 ▪ Practice, p. 20 ▪ Word Problem Practice, p. 21
Other	▪ 5-Minute Check 12-3 ▪ Study Notebook	▪ 5-Minute Check 12-3 ▪ Study Notebook	▪ 5-Minute Check 12-3 ▪ Study Notebook	▪ 5-Minute Check 12-3 ▪ Study Notebook

©Visions of America, LLC/Alamy

1 Trigonometric Functions for General Angles

Example 1 shows how to evaluate trigonometric functions for a given point. **Example 2** shows how to find the values of the six trigonometric functions for a quadrantal angle.

Formative Assessment

Use the Guided Practice exercises after each example to determine students' understanding of concepts.

Additional Examples

1 The terminal side of θ in standard position contains the point $(8, -15)$. Find the exact values of the six trigonometric functions of θ.

$\sin \theta = -\dfrac{15}{17}$; $\cos \theta = \dfrac{8}{17}$;

$\tan \theta = -\dfrac{15}{8}$; $\sec \theta = \dfrac{17}{8}$;

$\csc \theta = -\dfrac{17}{15}$; $\cot \theta = -\dfrac{8}{15}$

2 The terminal side of θ in standard position contains the point at $(0, -2)$. Find the values of the six trigonometric functions of θ. $\sin \theta = -1$; $\cos \theta = 0$; $\tan \theta$ is undefined; $\csc \theta = -1$; $\sec \theta$ is undefined; $\cot \theta = 0$

Additional Examples also in Interactive Classroom PowerPoint® Presentations

IWB Interactive White Board READY

If the terminal side of angle θ in standard position lies on the x- or y-axis, the angle is called a **quadrantal angle**.

StudyTip
Quadrantal Angles The measure of a quadrantal angle is a multiple of 90° or $\dfrac{\pi}{2}$.

KeyConcept Quadrantal Angles

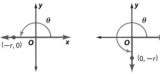

$\theta = 0°$ or 0 radians $\quad$ $\theta = 90°$ or $\dfrac{\pi}{2}$ radians $\quad$ $\theta = 180°$ or π radians $\quad$ $\theta = 270°$ or $\dfrac{3\pi}{2}$ radians

Example 2 Quadrantal Angles

The terminal side of θ in standard position contains the point at $(0, 6)$. Find the values of the six trigonometric functions of θ.

The point at $(0, 6)$ lies on the positive y-axis, so the quadrantal angle θ is 90°. Use $x = 0$, $y = 6$, and $r = 6$ to write the trigonometric functions.

$\sin \theta = \dfrac{y}{r} = \dfrac{6}{6}$ or 1 $\qquad$ $\cos \theta = \dfrac{x}{r} = \dfrac{0}{6}$ or 0 $\qquad$ $\tan \theta = \dfrac{y}{x} = \dfrac{6}{0}$ undefined

$\csc \theta = \dfrac{r}{y} = \dfrac{6}{6}$ or 1 $\qquad$ $\sec \theta = \dfrac{r}{x} = \dfrac{6}{0}$ undefined $\qquad$ $\cot \theta = \dfrac{x}{y} = \dfrac{0}{6}$ or 0

GuidedPractice

2. The terminal side of θ in standard position contains the point at $(-2, 0)$. Find the values of the six trigonometric functions of θ.

---

2. $\sin \theta = 0$ $\cos \theta = -1$, $\tan \theta = 0$, $\csc \theta =$ undefined, $\sec \theta = -1$, $\cot \theta =$ undefined

ReadingMath
Theta Prime θ' is read *theta prime*.

2 Trigonometric Functions with Reference Angles

If θ is a nonquadrantal angle in standard position, its **reference angle** θ' is the acute angle formed by the terminal side of θ and the x-axis. The rules for finding the measures of reference angles for $0° < \theta < 360°$ or $0° < \theta < 2\pi$ are shown below.

KeyConcept Reference Angles

Quadrant I	Quadrant II	Quadrant III	Quadrant IV
$\theta' = \theta$	$\theta' = 180° - \theta$ $\theta' = \pi - \theta$	$\theta' = \theta - 180°$ $\theta' = \theta - \pi$	$\theta' = 360° - \theta$ $\theta' = 2\pi - \theta$

 808 | Lesson 12-3 | Trigonometric Functions of General Angles

---

WatchOut!

Unlocking Misconceptions In Example 1, watch for students who think that r must be a negative number any time either x or y has a negative value.

Teach with Tech

Student Response System Present students with a number of examples (such as cos 120°, csc (–35°), and tan (–165°)). Ask students if the value of each trigonometric function is positive or negative. Have students respond with A for positive and B for negative.

If the measure of θ is greater than 360° or less than 0°, then use a coterminal angle with a positive measure between 0° and 360° to find the reference angle.

StudyTip

Graphing Angles You can refer to the diagram in the Lesson 12-2 Concept Summary to help you sketch angles.

Example 3 Find Reference Angles

Sketch each angle. Then find its reference angle.

a. 210°

The terminal side of 210° lies in Quadrant III.

$\theta' = \theta - 180°$

$= 210° - 180°$ or $30°$

b. $-\dfrac{5\pi}{4}$

coterminal angle: $-\dfrac{5\pi}{4} + 2\pi = \dfrac{3\pi}{4}$

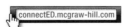

The terminal side of $\dfrac{3\pi}{4}$ lies in Quadrant III.

$\theta' = \pi - \theta$

$= \pi - \dfrac{3\pi}{4}$ or $\dfrac{\pi}{4}$

▶ **Guided**Practice   3A, 3B. See margin.

3A. −110°   **3B.** $\dfrac{2\pi}{3}$

You can use reference angles to evaluate trigonometric functions for any angle θ. The sign of a function is determined by the quadrant in which the terminal side of θ lies. Use these steps to evaluate a trigonometric function for any angle θ.

KeyConcept Evaluate Trigonometric Functions

Step 1 Find the measure of the reference angle θ'.

Step 2 Evaluate the trigonometric function for θ'.

Step 3 Determine the sign of the trigonometric function value. Use the quadrant in which the terminal side of θ lies.

	Quadrant II	Quadrant I
	$\sin \theta$, csc θ: +	$\sin \theta$, csc θ: +
	$\cos \theta$, sec θ: −	$\cos \theta$, sec θ: +
	$\tan \theta$, cot θ: −	$\tan \theta$, cot θ: +
	Quadrant III	Quadrant IV
	$\sin \theta$, csc θ: −	$\sin \theta$, csc θ: −
	$\cos \theta$, sec θ: −	$\cos \theta$, sec θ: +
	$\tan \theta$, cot θ: +	$\tan \theta$, cot θ: −

You can use the trigonometric values of angles measuring 30°, 45°, and 60° that you learned in Lesson 12-1.

Trigonometric Values for Special Angles					
Sine	**Cosine**	**Tangent**	**Cosecant**	**Secant**	**Cotangent**
$\sin 30° = \dfrac{1}{2}$	$\cos 30° = \dfrac{\sqrt{3}}{2}$	$\tan 30° = \dfrac{\sqrt{3}}{3}$	$\csc 30° = 2$	$\sec 30° = \dfrac{2\sqrt{3}}{3}$	$\cot 30° = \sqrt{3}$
$\sin 45° = \dfrac{\sqrt{2}}{2}$	$\cos 45° = \dfrac{\sqrt{2}}{2}$	$\tan 45° = 1$	$\csc 45° = \sqrt{2}$	$\sec 45° = \sqrt{2}$	$\cot 45° = 1$
$\sin 60° = \dfrac{\sqrt{3}}{2}$	$\cos 60° = \dfrac{1}{2}$	$\tan 60° = \sqrt{3}$	$\csc 60° = \dfrac{2\sqrt{3}}{3}$	$\sec 60° = 2$	$\cot 60° = \dfrac{\sqrt{3}}{3}$

connectED.mcgraw-hill.com   **809**

---

2 Trigonometric Functions with Reference Angles

Example 3 shows how to find the reference angle for a given angle. **Example 4** shows how to use a reference angle to find a trigonometric value. **Example 5** shows how to solve a real-world problem involving reference angles.

Additional Example

3 Sketch each angle. Then find its reference angle.

a. 330° 30°

b. $-\dfrac{5\pi}{6}$ $\dfrac{\pi}{6}$

Additional Answers (Guided Practice)

3A. 70°

3B. $\dfrac{\pi}{3}$

WatchOut!

Preventing Errors When finding trigonometric values for an angle, students may find it helpful to sketch an angle in standard position and drop a perpendicular to the x-axis to form a right triangle. Then they can see that, as the angle increases from 90° to 180°, the y value approaches 0, and the values of the other two sides approach each other.

Additional Examples

4 Find the exact value of each trigonometric function.

a. sin 135° $\dfrac{\sqrt{2}}{2}$

b. cot $\dfrac{7\pi}{3}$ $\dfrac{\sqrt{3}}{3}$

5 **RIDES** Suppose a ride similar to the one in Example 5 has a swing arm 89 feet long. The height of the axis is 99 feet, and the angles are the same. What is the total height of the new ride at the peak of the arc?
$y = 89 \sin 20° = 30.4$ ft; $30.4 + 99 = 129.4$ ft

CCSS **Teaching the Mathematical Practices**

Reasoning Mathematically proficient students make sense of quantities and their relationships in problem situations. Quantitative reasoning entails habits of creating a coherent representation of the problem at hand; considering the units involved; attending to the meaning of quantities, not just how to compute them; and knowing and flexibly using different properties of operations and objects.

Additional Answers

1. $\sin \theta = \dfrac{2\sqrt{5}}{5}$, $\cos \theta = \dfrac{\sqrt{5}}{5}$, $\tan \theta = 2$, $\csc \theta = \dfrac{\sqrt{5}}{2}$, $\sec \theta = \sqrt{5}$, $\cot \theta = \dfrac{1}{2}$

2. $\sin \theta = -\dfrac{15}{17}$, $\cos \theta = -\dfrac{8}{17}$, $\tan \theta = \dfrac{15}{8}$, $\csc \theta = -\dfrac{17}{15}$, $\sec \theta = -\dfrac{17}{8}$, $\cot \theta = \dfrac{8}{15}$

3. $\sin \theta = -1$, $\cos \theta = 0$, $\tan \theta =$ undefined, $\csc \theta = -1$, $\sec \theta =$ undefined, $\cot \theta = 0$

4. 60°

Example 4 Use a Reference Angle to Find a Trigonometric Value

Find the exact value of each trigonometric function.

a. cos 240°

The terminal side of 240° lies in Quadrant III.

$\theta' = \theta - 180°$ Find the measure of the reference angle.
$= 240° - 180°$ or 60° $\theta = 240°$

$\cos 240° = -\cos 60°$ or $-\dfrac{1}{2}$ The cosine function is negative in Quadrant III.

b. csc $\dfrac{5\pi}{6}$

The terminal side of $\dfrac{5\pi}{6}$ lies in Quadrant II.

$\theta' = \pi - \theta$ Find the measure of the reference angle.
$= \pi - \dfrac{5\pi}{6}$ or $\dfrac{\pi}{6}$ $\theta = \dfrac{5\pi}{6}$

$\csc \dfrac{5\pi}{6} = \csc \dfrac{\pi}{6}$ The cosecant function is positive in Quadrant II.

$= \csc 30°$ $\dfrac{\pi}{6}$ radians $= 30°$

$= 2$ $\csc 30° = \dfrac{1}{\sin 30}$

GuidedPractice

4A. cos 135° $-\dfrac{\sqrt{2}}{2}$ **4B.** tan $\dfrac{5\pi}{6}$ $-\dfrac{\sqrt{3}}{3}$

 Real-World Example 5 Use Trigonometric Functions

RIDES The swing arms of the ride at the right are 84 feet long and the height of the axis from which the arms swing is 97 feet. What is the total height of the ride at the peak of the arc?

coterminal angle: $-200° + 360° = 160°$

reference angle: $180° - 160° = 20°$

$\sin \theta = \dfrac{y}{r}$ Sine function
$\sin 20° = \dfrac{y}{84}$ $\theta = 20°$ and $r = 84$
$84 \sin 20° = y$ Multiply each side by 84.
$28.7 \approx y$ Use a calculator to solve for y.

Since y is approximately 28.7 feet, the total height of the ride at its peak is $28.7 + 97$ or about 125.7 feet.

GuidedPractice   about 106.6 ft

5. RIDES A similar ride that is smaller has swing arms that are 72 feet long. The height of the axis from which the arms swing is 88 feet, and the angle of rotation from the standard position is $-195°$. What is the total height of the ride at the peak of the arc?

Real-WorldLink

On a swing ride, riders experience weightlessness just like the drop side of a roller coaster. The ride lasts one minute and reaches speeds of 60 miles per hour in both directions.
Source: Cedar Point

---

810 | Lesson 12-3 | Trigonometric Functions of General Angles

Differentiated Instruction **AL** **OL**

Auditory/Musical Learners Have students work in small groups to create a jingle, song, rap, or short poem to help them remember the trigonometric values for special angles.

courtesy of Cedar Point

Examples 1–2 The terminal side of θ in standard position contains each point. Find the exact values of the six trigonometric functions of θ. **1–3. See margin.**

1. $(1, 2)$ **2.** $(-8, -15)$ **3.** $(0, -4)$

Example 3 Sketch each angle. Then find its reference angle. **4–6. See margin.**

4. $300°$ **5.** $115°$ **6.** $-\dfrac{3\pi}{4}$

Example 4 Find the exact value of each trigonometric function.

7. $\sin \dfrac{3\pi}{4}$ $\dfrac{\sqrt{2}}{2}$ **8.** $\tan \dfrac{5\pi}{3}$ $-\sqrt{3}$ **9.** $\sec 120°$ -2 **10.** $\sin 300°$ $-\dfrac{\sqrt{3}}{2}$

Example 5 **11. ENTERTAINMENT** Alejandra opens her portable DVD player so that it forms a $125°$ angle. The screen is $5\frac{1}{2}$ inches long.

    a. Redraw the diagram so that the angle is in standard position on the coordinate plane. **See Chapter 12 Answer Appendix.**

    b. Find the reference angle. Then write a trigonometric function that can be used to find the distance to the wall d that she can place the DVD player. $55°$; $\cos 55° = \dfrac{d}{5\frac{1}{2}}$

    c. Use the function to find the distance. Round to the nearest tenth. **3.2 in.**

Practice and Problem Solving   Extra Practice is on page R12.

Examples 1–2 The terminal side of θ in standard position contains each point. Find the exact values of the six trigonometric functions of θ. **12–17. See margin.**

12. $(5, 12)$ **13** $(-6, 8)$ **14.** $(3, 0)$

15. $(0, -7)$ **16.** $(4, -2)$ **17.** $(-9, -3)$

Example 3 Sketch each angle. Then find its reference angle. **18–23. See Chapter 12 Answer Appendix.**

18. $195°$ **19.** $285°$ **20.** $-250°$

21. $\dfrac{7\pi}{4}$ **22.** $-\dfrac{\pi}{4}$ **23.** $400°$

Example 4 Find the exact value of each trigonometric function.

24. $\sin 210°$ $-\dfrac{1}{2}$ **25.** $\tan 315°$ -1 **26.** $\cos 150°$ $-\dfrac{\sqrt{3}}{2}$ **27.** $\csc 225°$ $-\sqrt{2}$

28. $\sin \dfrac{4\pi}{3}$ $-\dfrac{\sqrt{3}}{2}$ **29.** $\cos \dfrac{5\pi}{3}$ $\dfrac{1}{2}$ **30.** $\cot \dfrac{5\pi}{4}$ 1 **31.** $\sec \dfrac{11\pi}{6}$ $\dfrac{2\sqrt{3}}{3}$

Example 5 **32. CCSS REASONING** A soccer player x feet from the goalie kicks the ball toward the goal, as shown in the figure. The goalie jumps up and catches the ball 7 feet in the air.

    a. Find the reference angle. Then write a trigonometric function that can be used to find how far from the goalie the soccer player was when he kicked the ball. $26°$; $\tan 26° = \dfrac{7}{x}$

    b. About how far away from the goalie was the soccer player? **about 14.4 ft**

Differentiated Homework Options

Level	Assignment	Two-Day Option	
AL Basic	12–32, 48–78	13–31 odd, 52–55	12–32 even, 48–51, 56–78
OL Core	13–31 odd, 33–36, 37–45 odd, 48–78	12–32, 52–55	33–46, 48–51, 56–78
BL Advanced	33–75, (optional: 76–78)		

Formative Assessment

Use Exercises 1–11 to check for understanding.

Use the chart at the bottom of this page to customize assignments for your students.

Additional Answers

5. $65°$

6. $\dfrac{\pi}{4}$

12. $\sin \theta = \dfrac{12}{13}$, $\cos \theta = \dfrac{5}{13}$, $\tan \theta = \dfrac{12}{5}$, $\csc \theta = \dfrac{13}{12}$, $\sec \theta = \dfrac{13}{5}$, $\cot \theta = \dfrac{5}{12}$

13. $\sin \theta = \dfrac{4}{5}$, $\cos \theta = -\dfrac{3}{5}$, $\tan \theta = -\dfrac{4}{3}$, $\csc \theta = \dfrac{5}{4}$, $\sec \theta = -\dfrac{5}{3}$, $\cot \theta = -\dfrac{3}{4}$

14. $\sin \theta = 0$, $\cos \theta = 1$, $\tan \theta = 0$, $\csc \theta = $ undefined, $\sec \theta = 1$, $\cot \theta = $ undefined

15. $\sin \theta = -1$, $\cos \theta = 0$, $\tan \theta = $ undefined, $\csc \theta = -1$, $\sec \theta = $ undefined, $\cot \theta = 0$

16. $\sin \theta = -\dfrac{\sqrt{5}}{5}$, $\cos \theta = \dfrac{2\sqrt{5}}{5}$, $\tan \theta = -\dfrac{1}{2}$, $\csc \theta = -\sqrt{5}$, $\sec \theta = \dfrac{\sqrt{5}}{2}$, $\cot \theta = -2$

17. $\sin \theta = -\dfrac{\sqrt{10}}{10}$, $\cos \theta = -\dfrac{3\sqrt{10}}{10}$, $\tan \theta = \dfrac{1}{3}$, $\csc \theta = -\sqrt{10}$, $\sec \theta = -\dfrac{\sqrt{10}}{3}$, $\cot \theta = 3$

NAME _____ DATE _____ PERIOD _____
12-3 Study Guide and Intervention
Trigonometric Functions of General Angles

Practice
AL OL BL ELL

Word Problem Practice
AL OL BL ELL

B (33) **SPRINKLER** A sprinkler rotating back and forth shoots water out a distance of 10 feet. From the horizontal position, it rotates 145° before reversing its direction. At a 145° angle, about how far to the left of the sprinkler does the water reach? **about 8.2 ft**

34. **BASKETBALL** The formula $R = \dfrac{v_0^2 \sin 2\theta}{32}$ gives the distance of a basketball shot with an initial velocity of v_0 feet per second at an angle θ with the ground.

   a. If the basketball was shot with an initial velocity of 24 feet per second at an angle of 75°, how far will the basketball travel? **9 ft**

   b. If the basketball was shot at an angle of 65° and traveled 10 feet, what was its initial velocity? **about 20.4 feet per second**

   c. If the basketball was shot with an initial velocity of 30 feet per second and traveled 12 feet, at what angle was it shot? **about 12.6°**

35. **PHYSICS** A rock is shot off the edge of a ravine with a slingshot at an angle of 65° and with an initial velocity of 6 meters per second. The equation that represents the horizontal distance of the rock x is $x = v_0 (\cos \theta)t$, where v_0 is the initial velocity, θ is the angle at which it is shot, and t is the time in seconds. About how far does the rock travel after 4 seconds? **about 10.1 m**

36. **FERRIS WHEELS** The Wonder Wheel Ferris wheel at Coney Island has a radius of about 68 feet and is 15 feet off the ground. After a person gets on the bottom car, the Ferris wheel rotates 202.5° counterclockwise before stopping. How high above the ground is this car when it has stopped? **145.8 ft**

C Suppose θ is an angle in standard position whose terminal side is in the given quadrant. For each function, find the exact values of the remaining five trigonometric functions of θ. **37–40. See margin.**

37. $\sin \theta = \dfrac{4}{5}$, Quadrant II

38. $\tan \theta = -\dfrac{2}{3}$, Quadrant IV

39. $\cos \theta = -\dfrac{8}{17}$, Quadrant III

40. $\cot \theta = -\dfrac{12}{5}$, Quadrant IV

Find the exact value of each trigonometric function.

41. $\cot 270°$ **0**

42. $\csc 180°$ **undefined**

43. $\sin 570°$ **$-\dfrac{1}{2}$**

44. $\tan\left(-\dfrac{7\pi}{6}\right)$ **$-\dfrac{\sqrt{3}}{3}$**

45. $\cos\left(-\dfrac{11\pi}{6}\right)$ **$\dfrac{\sqrt{3}}{2}$**

46. $\cot \dfrac{9\pi}{4}$ **1**

47. **No; for $\sin \theta = \dfrac{\sqrt{2}}{2}$ and $\tan \theta = -1$, the reference angle is 45°. However, for $\sin \theta$ to be positive and $\tan \theta$ to be negative, the reference angle must be in the second quadrant. So, the value of θ must be 135° or an angle coterminal with 135°.**

H.O.T. Problems   Use Higher-Order Thinking Skills

47. **CHALLENGE** For an angle θ in standard position, $\sin \theta = \dfrac{\sqrt{2}}{2}$ and $\tan \theta = -1$. Can the value of θ be 225°? Justify your reasoning.

48. **CCSS ARGUMENTS** Determine whether $3 \sin 60° = \sin 180°$ is *true* or *false*. Explain your reasoning.

49. **REASONING** Use the sine and cosine functions to explain why $\cot 180°$ is undefined. **See margin.**

50. **OPEN ENDED** Give an example of a negative angle θ for which $\sin \theta > 0$ and $\cos \theta < 0$.

51. **WRITING IN MATH** Describe the steps for evaluating a trigonometric function for an angle θ that is greater than 90°. Include a description of a reference angle. **See margin.**

48. False; $3 \sin 60° = 3 \cdot \dfrac{\sqrt{3}}{2}$ or $\dfrac{3\sqrt{3}}{2}$ and $\sin 180° = 0.0$

50. Sample answer: $\theta = -200°$

Enrichment
OL BL

NAME _____ DATE _____ PERIOD _____
12-3 Enrichment

Areas of Polygons and Circles

A regular polygon has sides of equal length and angles of equal measure. A regular polygon can be inscribed in or circumscribed about a circle. For n-sided regular polygons, the following area formulas can be used.

Area of circle $\quad A_c = \pi r^2$

Area of inscribed polygon $\quad A_i = \dfrac{nr^2}{2} \times \sin \dfrac{360°}{n}$

Area of circumscribed polygon $\quad A_c = nr^2 \times \tan \dfrac{180°}{n}$

Use a calculator to complete the chart below for a unit circle (a circle of radius 1).

Additional Answers

37. $\cos \theta = -\dfrac{3}{5}$, $\tan \theta = -\dfrac{4}{3}$, $\csc \theta = \dfrac{5}{4}$, $\sec \theta = -\dfrac{5}{3}$, $\cot \theta = -\dfrac{3}{4}$

38. $\sin \theta = -\dfrac{2\sqrt{13}}{13}$, $\cos \theta = \dfrac{3\sqrt{13}}{13}$, $\csc \theta = -\dfrac{\sqrt{13}}{2}$, $\sec \theta = \dfrac{\sqrt{13}}{3}$, $\cot \theta = -\dfrac{3}{2}$

52. GRIDDED RESPONSE If the sum of two numbers is 21 and their difference is 3, what is their product? **108**

53. GEOMETRY D is the midpoint of $\overline{BC}$, and A and E are the midpoints of $\overline{BD}$ and $\overline{DC}$, respectively. If the length of $\overline{AE}$ is 12, what is the length of $\overline{BC}$? **C**

A 6   **C** 24
B 12   **D** 48

54. The expression $(-6 + i)^2$ is equivalent to which of the following expressions? **J**

F $-12i$ **H** $36 - 12i$
G $36 - i$ **J** $35 - 12i$

55. SAT/ACT Of the following, which is least? **E**

A $1 + \frac{1}{4}$ **D** $1 \times \frac{1}{4}$

B $1 - \frac{1}{4}$ **E** $\frac{1}{4} - 1$

C $1 \div \frac{1}{4}$

Spiral Review

Rewrite each radian measure in degrees. (Lesson 12-2)

56. $\frac{4}{3}\pi$ **240°**

57. $\frac{11}{6}\pi$ **330°**

58. $-\frac{17}{4}\pi$ **−765°**

Solve each equation. (Lesson 12-1)

59. $\cos a = \frac{13}{17}$ **40.1°**

60. $\sin 30 = \frac{b}{6}$ **3**

61. $\tan c = \frac{9}{4}$ **66.0°**

62. ARCHITECTURE A memorial being constructed in a city park will be a brick wall, with a top row of six gold-plated bricks engraved with the names of six local war veterans. Each row has two more bricks than the row above it. Prove that the number of bricks in the top n rows is $n^2 + 5n$. (Lesson 10-7) **See Chapter 12 Answer Appendix.**

63. LEGENDS There is a legend of a king who wanted to reward a boy for a good deed. The king gave the boy a choice. He could have $1,000,000 at once, or he could be rewarded daily for a 30-day month, with one penny on the first day, two pennies on the second day, and so on, receiving twice as many pennies each day as the previous day. How much would the second option be worth? (Lesson 10-3) **$10,737,418.23**

Write an equation for each circle given the endpoints of a diameter. (Lesson 9-3)

64. $(2, -4)$, $(10, 2)$

65. $(-1, -10)$, $(-7, 6)$

66. $(9, 0)$, $(4, -7)$

64. $(x - 6)^2 + (y + 1)^2 = 25$
65. $(x + 4)^2 + (y + 2)^2 = 73$
66. $(x - 6.5)^2 + (y + 3.5)^2 = 18.5$

Simplify each expression. (Lesson 8-2) **67–69. See margin.**

67. $\frac{5}{x^2 + 6x + 8} + \frac{x}{x^2 - 3x - 28}$

68. $\frac{3x}{x^2 + 8x - 20} - \frac{6}{x^2 + 7x - 18}$

69. $\frac{4}{3x^2 + 12x} + \frac{2x}{x^2 - 2x - 24}$

Solve each equation or inequality. Round to the nearest ten-thousandth. (Lesson 7-6)

70. $8^x = 30$ **1.6356**

71. $5^x = 64$ **2.5841**

72. $3^{x + 2} = 41$ **1.3802**

Evaluate each expression. (Lesson 6-6)

73. $16^{-\frac{1}{4}}$ $\frac{1}{2}$

74. $27^{\frac{4}{3}}$ **81**

75. $25^{-\frac{5}{2}}$ $\frac{1}{3125}$

Skills Review

Solve for x.

76. $\frac{x + 2}{18} = \frac{x - 2}{9}$ **6**

77. $\frac{x + 5}{x - 1} = \frac{7}{4}$ **9**

78. $\frac{5}{x + 8} = \frac{15}{2x + 20}$ **−4**

DifferentiatedInstruction **OL** **BL**

Extension For an angle with a measure between 0° and 90°, the angle is its own reference angle. Have students write expressions for the measure of the reference angle of θ,

if $90° < \theta < 180°$. $180° - \theta$

if $180° < \theta < 270°$. $\theta - 180°$

if $270° < \theta < 360°$. $360° - \theta$

---

4 Assess

Crystal Ball Have students look ahead to Lesson 12-4. Have them write how they think what they learned today will connect with the theme in Lesson 12-4.

Formative Assessment
Check for student understanding of concepts in Lessons 12-1, 12-2, and 12-3.

📁 Quiz 1, p. 61

Additional Answers

39. $\sin \theta = -\frac{15}{17}$, $\tan \theta = \frac{15}{8}$, $\csc \theta = -\frac{17}{15}$, $\sec \theta = -\frac{17}{8}$, $\cot \theta = \frac{8}{15}$

40. $\sin \theta = -\frac{5}{13}$, $\cos \theta = \frac{12}{13}$, $\csc \theta = -\frac{13}{5}$, $\sec \theta = \frac{13}{12}$, $\tan \theta = -\frac{5}{12}$

49. Sample answer: We know that $\cot \theta = \frac{x}{y}$, $\sin \theta = \frac{y}{r}$, and $\cos \theta = \frac{x}{r}$. Since $\sin 180° = 0$, it must be true that $y = 0$. Thus $\cot 180° = \frac{x}{0}$, which is undefined.

51. Sample answer: First, sketch the angle and determine in which quadrant it is located. Then use the appropriate rule for finding its reference angle θ'. A reference angle is the acute angle formed by the terminal side of θ and the x-axis. Next, find the value of the trigonometric function for θ'. Finally, use the quadrant location to determine the sign of the trigonometric function value of θ.

67. $\frac{x^2 + 7x - 35}{(x + 2)(x + 4)(x - 7)}$

68. $\frac{3(x^2 + 7x - 20)}{(x + 10)(x + 9)(x - 2)}$

69. $\frac{2(3x^2 + 2x - 12)}{3x(x + 4)(x - 6)}$

1 Focus

VerticalAlignment

▼

Before Lesson 12-4 Find side lengths and angle measures of right triangles.

▼

Lesson 12-4 Find the area of a triangle using two sides and an included angle. Use the Law of Sines to solve triangles.

▼

After Lesson 12-4 Use the Law of Cosines to solve problems.

2 Teach

Scaffolding Questions

Have students read the **Why?** section of the lesson.

Ask:

- What is the size of the angle opposite the side connecting Wahoo and Naukan? Wahoo and Wabash?
  102°; 23°

- What is the measure of the angle opposite the side connecting Wabash and Naukan? 55°

- What is the distance between Wahoo and Wabash? 1.2 km

- Which crater is at the vertex of the angle across from the longest side?
  Wabash

LESSON 12-4 Law of Sines

::Then	::Now	::Why?
● You found side lengths and angle measures of right triangles.	**1** Find the area of a triangle using two sides and an included angle. **2** Use the Law of Sines to solve triangles.	● Mars has hundreds of thousands of craters. These craters are named after famous scientists, science fiction authors, and towns on Earth. The craters named Wahoo, Wabash, and Naukan are shown in the figure. You can use trigonometry to find the distance between Wahoo and Naukan.

 NewVocabulary
Law of Sines
solving a triangle
ambiguous case

 Common Core State Standards

Mathematical Practices
1 Make sense of problems and persevere in solving them.
3 Construct viable arguments and critique the reasoning of others.

1 Find the Area of a Triangle

In the triangle at the right, $\sin A = \frac{h}{c}$, or $h = c \sin A$.

Area $= \frac{1}{2}bh$ Formula for area of a triangle

Area $= \frac{1}{2}b(c \sin A)$ Replace h with $c \sin A$.

Area $= \frac{1}{2}bc \sin A$ Simplify.

You can use this formula or two other formulas to find the area of a triangle if you know the lengths of two sides and the measure of the included angle.

KeyConcept Area of a Triangle

Words	The area of a triangle is one half the product of the lengths of two sides and the sine of their included angle.

Symbols   Area $= \frac{1}{2}bc \sin A = \frac{1}{2}ac \sin B = \frac{1}{2}ab \sin C$

Example 1 Find the Area of a Triangle

Find the area of △ABC to the nearest tenth.

In △ABC, $a = 8$, $b = 9$, and $C = 104°$.

Area $= \frac{1}{2}ab \sin C$ Based on the known measures, use the third area formula.

$= \frac{1}{2}(8)(9) \sin 104°$ Substitution

≈ 34.9 cm²   Simplify.

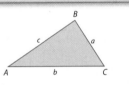

MENTAL CHECK   Round the sin 104° to sin 90° because the sin of 90° is 1.

$\frac{1}{2}(8)(9)\sin 90° = \frac{1}{2}(8)(9)(1) = 36$

This is close to the answer of 34.9 square centimeters.

▶ **GuidedPractice**

1. Find the area of △ABC to the nearest tenth if $A = 31°$, $b = 18$ meters, and $c = 22$ meters. **102.0 m²**

 814 | Lesson 12-4

JPL/NASA

Lesson 12-4 Resources

Resource	Approaching Level (AL)	On Level (OL)	Beyond Level (BL)	English Learners (ELL)
Teacher Edition	▪ Differentiated Instruction, p. 817	▪ Differentiated Instruction, pp. 817, 821	▪ Differentiated Instruction, pp. 817, 821	▪ Differentiated Instruction, p. 817
Chapter Resource Masters	▪ Study Guide and Intervention, pp. 23–24 ▪ Skills Practice, p. 25 ▪ Practice, p. 26 ▪ Word Problem Practice, p. 27	▪ Study Guide and Intervention, pp. 23–24 ▪ Skills Practice, p. 25 ▪ Practice, p. 26 ▪ Word Problem Practice, p. 27 ▪ Enrichment, p. 28	▪ Practice, p. 26 ▪ Word Problem Practice, p. 27 ▪ Enrichment, p. 28	▪ Study Guide and Intervention, pp. 23–24 ▪ Skills Practice, p. 25 ▪ Practice, p. 26 ▪ Word Problem Practice, p. 27
Other	▪ 5-Minute Check 12-4 ▪ Study Notebook	▪ 5-Minute Check 12-4 ▪ Study Notebook	▪ 5-Minute Check 12-4 ▪ Study Notebook	▪ 5-Minute Check 12-4 ▪ Study Notebook

Ree Collection courtesy of The Bancroft Library, University of California, Berkeley

Math HistoryLink

Pauline Sperry (1885–1967) During the 1920s, Pauline Sperry wrote two textbooks, *Short Course in Spherical Trigonometry* and *Plane Trigonometry*. In 1923, she became the first woman to be promoted to assistant professor in the mathematics department at the University of California, Berkeley.

2 Use the Law of Sines to Solve Triangles You can use the area formulas to derive the **Law of Sines**, which shows the relationships between side lengths of a triangle and the sines of the angles opposite them.

$\frac{1}{2}bc \sin A = \frac{1}{2}ac \sin B = \frac{1}{2}ab \sin C$ Set the area formulas equal to each other.

$bc \sin A = ac \sin B = ab \sin C$ Multiply each expression by 2.

$\frac{bc \sin A}{abc} = \frac{ac \sin B}{abc} = \frac{ab \sin C}{abc}$ Divide each expression by *abc*.

$\frac{\sin A}{a} = \frac{\sin B}{b} = \frac{\sin C}{c}$ Simplify.

KeyConcept   Law of Sines

In $\triangle ABC$, if sides with lengths *a*, *b*, and *c* are opposite angles with measures *A*, *B*, and *C*, respectively, then the following is true.

$$\frac{\sin A}{a} = \frac{\sin B}{b} = \frac{\sin C}{c}$$

You can use the Law of Sines to solve a triangle if you know either one of the following.

- the measures of two angles and any side (angle-angle-side AAS or angle-side-angle ASA cases)

- the measures of two sides and the angle opposite one of the sides (side-side-angle SSA case)

Using given measures to find all unknown side lengths and angle measures of a triangle is called **solving a triangle**.

StudyTip

CCSS Reasoning The Law of Sines may also be written as $\frac{a}{\sin A} = \frac{b}{\sin B} = \frac{c}{\sin C}$. So, the expressions below could also be used to solve the triangle in Example 2.

- $\frac{a}{\sin 55°} = \frac{3}{\sin 80°}$
- $\frac{b}{\sin 45°} = \frac{3}{\sin 80°}$

Example 2   Solve a Triangle Given Two Angles and a Side   [PT]

Solve $\triangle ABC$. **Round to the nearest tenth if necessary.**

Step 1 Find the measure of the third angle.
$m\angle A = 180 - (80 + 45)$ or 55°

Step 2 Use the Law of Sines to find side lengths *a* and *b*. Write an equation to find each variable.

$\frac{\sin A}{a} = \frac{\sin C}{c}$ Law of Sines $\frac{\sin B}{b} = \frac{\sin C}{c}$

$\frac{\sin 55°}{a} = \frac{\sin 80°}{3}$ Substitution $\frac{\sin 45°}{b} = \frac{\sin 80°}{3}$

$a = \frac{3 \sin 55°}{\sin 80°}$ Solve for each variable. $b = \frac{3 \sin 45°}{\sin 80°}$

$a \approx 2.5$ Use a calculator. $b \approx 2.2$

So, $A = 55°$, $a \approx 2.5$, and $b \approx 2.2$.

GuidedPractice

2. Solve $\triangle NPQ$ if $P = 42°$, $Q = 65°$, and $n = 5$.   **$N = 73°$, $p \approx 3.5$, $q \approx 4.7$**

---

1 Find the Area of a Triangle

Example 1 shows how to use the sine of an angle to find the area of a triangle.

Formative Assessment

Use the Guided Practice exercises after each example to determine students' understanding of concepts.

Additional Example

1 Find the area of $\triangle ABC$ to the nearest tenth.   3.8 cm²

Additional Examples also in Interactive Classroom PowerPoint® Presentations

[IWB] Interactive White Board READY

2 Use the Law of Sines to Solve Triangles

Example 2 shows how to use the Law of Sines to solve a triangle when two angles and a side are given. **Example 3** shows how to use the Law of Sines to solve a triangle that has zero, one, or two solutions. **Example 4** shows how to use the Law of Sines to solve a real-world problem.

Additional Example

2 Solve $\triangle ABC$. Round to the nearest tenth if necessary.

$A = 27°$, $a \approx 5.1$, $c \approx 11.1$

WatchOut!

Unlocking Misconceptions Students may think that the Law of Sines only works for right trangles. Clarify that this formula works for any triangle, as does the Law of Cosines explored in Lesson 12-5.

CCSS **Teaching the Mathematical Practices**

Reasoning Mathematically proficient students make sense of quantities and their relationships in problem situations. Remind students that ratios and proportions can be written in many forms.

3 Determine whether each triangle has *no* solution, *one* solution, or *two* solutions. Then solve the triangle. Round side lengths to the nearest tenth and angle measures to the nearest degree.

a. In $\triangle NPQ$, $Q = 110°$, $q = 11$, and $n = 8$. $P = 27°$, $N = 43°$, $p = 5.3$

b. In $\triangle DEF$, $E = 52°$, $e = 5$, and $f = 9$. no solution

c. In $\triangle XYZ$, $X = 28°$, $z = 15$, and $x = 9$. One solution is $Z = 51°$, $Y = 101°$, and $y = 18.8$; another solution is $Z = 129°$, $Y = 23°$, and $y = 7.5$.

Teach with Tech

Interactive Whiteboard Use the board to demonstrate the ambiguous case with the Law of Sines. Use a drawing program to accurately draw the two given sides of the triangle and the given angle. Draw one of the possible third sides of the triangle. Then drag the segment to show how to construct the other possible triangle.

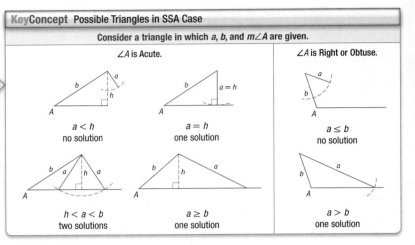

StudyTip

Two Solutions A situation in which two solutions for a triangle exist is called the *ambiguous case.*

StudyTip

A is Acute In the figures at the right, the altitude *h* is compared to *a* because *h* is the minimum distance from *C* to $\overline{AB}$ when *A* is acute.

$\sin A = \dfrac{\text{opp}}{\text{hyp}}$

$\sin A = \dfrac{h}{b}$

If you are given the measures of two angles and a side, exactly one triangle is possible. However, if you are given the measures of two sides and the angle opposite one of them, zero, one, or two triangles may be possible. This is known as the **ambiguous case**. So, when solving a triangle using the SSA case, zero, one, or two solutions are possible.

KeyConcept Possible Triangles in SSA Case

Consider a triangle in which *a*, *b*, and *m*∠*A* are given.

∠A is Acute.

$a < h$ — no solution

$a = h$ — one solution

$h < a < b$ — two solutions

$a \geq b$ — one solution

∠A is Right or Obtuse.

$a \leq b$ — no solution

$a > b$ — one solution

Since $\sin A = \dfrac{h}{b}$, you can use $h = b \sin A$ to find *h* in acute triangles.

Example 3 Solve a Triangle Given Two Sides and an Angle

Determine whether each triangle has *no* solution, *one* solution, or *two* solutions. Then solve the triangle. Round side lengths to the nearest tenth and angle measures to the nearest degree.

a. In $\triangle RST$, $R = 105°$, $r = 9$, and $s = 6$.

Because ∠*R* is obtuse and $9 > 6$, you know that one solution exists.

Step 1 Use the Law of Sines to find $m\angle S$.

$\dfrac{\sin S}{6} = \dfrac{\sin 105°}{9}$ Law of Sines

$\sin S = \dfrac{6 \sin 105°}{9}$ Multiply each side by 6.

$\sin S \approx 0.6440$ Use a calculator.

$S \approx 40°$ Use the $\sin^{-1}$ function.

Step 2 Find $m\angle T$.

$m\angle T \approx 180 - (105 + 40)$ or $35°$

Step 3 Use the Law of Sines to find *t*.

$\dfrac{\sin 35°}{t} \approx \dfrac{\sin 105°}{9}$ Law of Sines

$t \approx \dfrac{9 \sin 35°}{\sin 105°}$ Solve for *t*.

$t \approx 5.3$ Use a calculator.

So, $S \approx 40°$, $T \approx 35°$, and $t \approx 5.3$.

---

b. In $\triangle ABC$, $A = 54°$, $a = 6$, and $b = 8$.

Since $\angle A$ is acute and $6 < 8$, find h and compare it to a.

$b \sin A = 8 \sin 54°$ $b = 8$ and $A = 54°$

≈ 6.5 Use a calculator.

Since $6 \leq 6.5$ or $a \leq h$, there is no solution.

c. In $\triangle ABC$, $A = 35°$, $a = 17$, and $b = 20$.

Since $\angle A$ is acute and $17 < 20$, find h and compare it to a.

$b \sin A = 20 \sin 35°$ $b = 20$ and $A = 35°$

≈ 11.5 Use a calculator.

Since $11.5 < 17 < 20$ or $h < a < b$, there are two solutions. So, there are two triangles to be solved.

Case 1 $\angle B$ is acute.

Step 1 Find $m\angle B$.

$\dfrac{\sin B}{20} = \dfrac{\sin 35°}{17}$ Law of Sines

$\sin B = \dfrac{20 \sin 35°}{17}$ Solve for $\sin B$.

$\sin B \approx 0.6748$ Use a calculator.

$B \approx 42°$ Find $\sin^{-1} 0.6748$.

Step 2 Find $m\angle C$.

$m\angle C \approx 180 - (35 + 42)$ or $103°$

Step 3 Find c.

$\dfrac{\sin 103°}{c} = \dfrac{\sin 35°}{17}$ Law of Sines

$c = \dfrac{17 \sin 103°}{\sin 35°}$ Solve for c.

$c \approx 28.9$ Simplify.

Case 2 $\angle B$ is obtuse.

Step 1 Find $m\angle B$.

The sine function also has a positive value in Quadrant II. So, find an obtuse angle B for which $\sin B \approx 0.6748$.

$m\angle B \approx 180° - 42°$ or $138°$

Step 2 Find $m\angle C$.

$m\angle C \approx 180 - (35 + 138)$ or $7°$

Step 3 Find c.

$\dfrac{\sin 7°}{c} \approx \dfrac{\sin 35°}{17}$ Law of Sines

$c \approx \dfrac{17 \sin 7°}{\sin 35°}$ Solve for c.

$c \approx 3.6$ Simplify.

So, one solution is $B \approx 42°$, $C \approx 103°$, and $c \approx 28.9$, and another solution is $B \approx 138°$, $C \approx 7°$, and $c \approx 3.6$.

> **StudyTip**
> Reference Angle In the triangle in Case 2, you are using the reference angle 42° to find the other value of B.

Guided Practice

Determine whether each triangle has *no* solution, *one* solution, or *two* solutions. Then solve the triangle. Round side lengths to the nearest tenth and angle measures to the nearest degree.

3A. In $\triangle RST$, $R = 95°$, $r = 10$, and $s = 12$.   **no solution**

3B. In $\triangle MNP$, $N = 32°$, $n = 7$, and $p = 4$.   **one; $P \approx 18°$, $M \approx 130°$, $m \approx 10.1$**

3C. In $\triangle ABC$, $A = 47°$, $a = 15$, and $b = 18$.

3C. two; $B \approx 61°$, $C \approx 72°$, $c \approx 19.5$; $B \approx 119°$, $C \approx 14°$, $c \approx 5.0$

Differentiated Instruction   AL OL BL ELL

Intrapersonal Learners Have students write a journal entry about which example they found the most challenging and why. Ask them to include any questions they still have about the lesson.

Focus on Mathematical Content
Law of Sines In one case for which the Law of Sines may be used, more than one solution of the triangle is possible. For this ambiguous case, the process of solving the triangle must be performed twice. However, there may still be only one solution for the triangle.

Additional Example

4 **BASEBALL** In Example 4, suppose the angles at second base and third base are 58° and 41°, respectively. How far away from second base was the ball caught? about 59.8 ft

3 Practice

Formative Assessment

Use Exercises 1–12 to check for understanding.

Use the chart at the bottom of the next page to customize assignments for your students.

 Teaching the Mathematical Practices

Perseverance Mathematically proficient students start by explaining to themselves the meaning of a problem and looking for entry points to its solution. They analyze givens, constraints, relationships, and goals and make conjectures about the form and meaning of the solution and plan a solution pathway rather than simply jumping into a solution attempt.

Real-World Example 4 Use the Law of Sines to Solve a Problem

BASEBALL A baseball is hit between second and third bases and is caught at point *B*, as shown in the figure. How far away from second base was the ball caught?

$$\frac{\sin 72°}{90} = \frac{\sin 43°}{x}$$ Law of Sines

$x \sin 72° = 90 \sin 43°$ Cross products

$x = \dfrac{90 \sin 43°}{\sin 72°}$ Solve for *x*.

$x \approx 64.5$ Use a calculator.

So, the distance is about 64.5 feet.

> **GuidedPractice**
>
> **4.** How far away from third base was the ball caught? 85.8 ft

Real-WorldLink

High school and college baseball fields share the same infield dimensions as professional baseball fields. The outfield dimensions vary greatly.

Source: *Baseball Digest Magazine*

Check Your Understanding ⬤ = **Step-by-Step Solutions begin on page R14.**

Example 1 Find the area of △*ABC* to the nearest tenth, if necessary.

1. 27.9 mm²

2. 3 yd²

3 *A* = 40°, *b* = 11 cm, *c* = 6 cm 21.2 cm²   **4.** *B* = 103°, *a* = 20 in., *c* = 18 in. 175.4 in²

Example 2 Solve each triangle. Round side lengths to the nearest tenth and angle measures to the nearest degree.

5.

6.

5. *E* = 107°, *d* ≈ 7.9, *f* ≈ 7.0
6. *C* = 33°, *a* ≈ 6.9, *c* ≈ 4.9

7. Solve △*FGH* if *G* = 80°, *H* = 40°, and *g* = 14. *F* = 60°, *f* ≈ 12.3, *h* ≈ 9.1

Example 3 **PERSEVERANCE** Determine whether each △*ABC* has *no* solution, *one* solution, or *two* solutions. Then solve the triangle. Round side lengths to the nearest tenth and angle measures to the nearest degree. **8.** one; *B* ≈ 39°, *C* ≈ 46°, *c* ≈ 13.7

8. *A* = 95°, *a* = 19, *b* = 12

9. *A* = 60°, *a* = 15, *b* = 24 no solution

10. *A* = 34°, *a* = 8, *b* = 13

11. *A* = 30°, *a* = 3, *b* = 6

10. two; *B* ≈ 65°, *C* ≈ 81°, *c* ≈ 14.1; *B* ≈ 115°, *C* ≈ 31°, *c* ≈ 7.4
11. one; *B* = 90°, *C* = 60°, *c* ≈ 5.2

Example 4 **12. SPACE** Refer to the beginning of the lesson. Find the distance between the Wahoo Crater and the Naukan Crater on Mars. 3 kilometers

 818 | **Lesson 12-4** | Law of Sines

(t)Rich Pilling/MLB Photos via Getty Images, (b)JPL/NASA

Example 1

Find the area of △ABC to the nearest tenth.

13. 10.6 km²

14. 126.1 ft²

15. 36.8 m²

16. 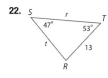 74.1 cm²

17. C = 25°, a = 4 ft, b = 7 ft 5.9 ft²

18. A = 138°, b = 10 in., c = 20 in. 66.9 in²

19. B = 92°, a = 14.5 m, c = 9 m 65.2 m²

20. C = 116°, a = 2.7 cm, b = 4.6 cm 5.6 cm²

Example 2

CCSS REASONING Solve each triangle. Round side lengths to the nearest tenth and angle measures to the nearest degree.

21.

22.

23.

24.

21. C = 30°, b ≈ 11.1, c ≈ 5.8
22. R = 80°, r ≈ 17.5, t ≈ 14.2
23. L = 74°, m ≈ 4.9, n ≈ 3.1
24. B = 38°, a ≈ 19.5, c ≈ 36.1

25. Solve △HJK if H = 53°, J = 20°, and h = 31. K = 107°, j ≈ 13.3, k ≈ 37.1

26. Solve △NPQ if P = 109°, Q = 57°, and n = 22. N = 14°, p ≈ 86.0, q ≈ 76.3

27. Solve △ABC if A = 50°, a = 2.5, and C = 67°. B = 63°, b ≈ 2.9, c ≈ 3.0

28. Solve △ABC if B = 18°, C = 142°, and b = 20. A = 20°, a ≈ 22.1, c ≈ 39.8

Example 3

Determine whether each △ABC has *no* solution, *one* solution, or *two* solutions. Then solve the triangle. Round side lengths to the nearest tenth and angle measures to the nearest degree.

29. A = 100°, a = 7, b = 3

30. A = 75°, a = 14, b = 11

31. A = 38°, a = 21, b = 18

32. A = 52°, a = 9, b = 20 no solution

33. A = 42°, a = 5, b = 6

34. A = 44°, a = 14, b = 19

35. A = 131°, a = 15, b = 32 no solution

36. A = 30°, a = 17, b = 34

29. one; B ≈ 25°, C ≈ 55°, c ≈ 5.8
30. one; B ≈ 49°, C ≈ 56°, c ≈ 12.0
31. one; B ≈ 32°, C ≈ 110°, c ≈ 32.1
33. two; B ≈ 53°, C ≈ 85°, c ≈ 7.4; B ≈ 127°, C ≈ 11°, c ≈ 1.4
34. two; B ≈ 71°, C ≈ 65°, c ≈ 18.3; B ≈ 109°, C ≈ 27°, c ≈ 9.1
36. one; B ≈ 90°, C = 60°, c ≈ 29.4

CCSS **Teaching the Mathematical Practices**

Reasoning Mathematically proficient students make sense of quantities and their relationships in problem situations. Quantitative reasoning entails habits of creating a coherent representation of the problem at hand; considering the units involved; attending to the meaning of quantities, not just how to compute them; and knowing and flexibly using different properties of operations and objects.

Differentiated Homework Options

Level	Assignment	Two-Day Option	
AL Basic	13–38, 43, 46–69	13–37 odd, 49–52	14–38 even, 43, 46–48, 53–69
OL Core	13–37 odd, 39–43, 46–69	13–38, 49–52	39–43, 46–48, 53–69
BL Advanced	39–66, (optional: 67–69)		

Example 4 **GEOGRAPHY** In Hawaii, the distance from Hilo to Kailua is 57 miles, and the distance from Hilo to Captain Cook is 55 miles.

37. What is the measure of the angle formed at Hilo? **about 28°**

38. What is the distance between Kailua and Captain Cook? **about 27.2 mi**

B **39** **TORNADOES** Tornado sirens A, B, and C form a triangular region in one area of a city. Sirens A and B are 8 miles apart. The angle formed at siren A is 112°, and the angle formed at siren B is 40°. How far apart are sirens B and C? **about 15.8 mi**

40. **MYSTERIES** The Bermuda Triangle is a region of the Atlantic Ocean between Bermuda, Miami, Florida, and San Juan, Puerto Rico. It is an area where ships and airplanes have been rumored to mysteriously disappear.

41b. Sample answer:
$$\frac{\sin 66°}{a} = \frac{\sin 64°}{4},$$
$$\frac{\sin 50°}{b} = \frac{\sin 64°}{4}$$

a. What is the distance between Miami and Bermuda? **about 1120.3 mi**

b. What is the approximate area of the Bermuda Triangle? **about 464,366.1 mi²**

41. **BICYCLING** One side of a triangular cycling path is 4 miles long. The angle opposite this side is 64°. Another angle formed by the triangular path measures 66°.

46. Sample answer:
$a = 12$, $b \approx 14.2$, $c \approx 5.0$; $a = 6$, $b \approx 7.1$, $c \approx 2.5$

a. Sketch a drawing of the situation. Label the missing sides a and b. **See margin.**

b. Write equations that could be used to find the lengths of the missing sides.

c. What is the perimeter of the path? **about 11.5 mi**

42. **ROCK CLIMBING** Savannah S and Leon L are standing 8 feet apart in front of a rock climbing wall, as shown at the right. What is the height of the wall? Round to the nearest tenth. **18.9 ft**

H.O.T. Problems   Use Higher-Order Thinking Skills

C **43.** **CCSS CRITIQUE** In $\triangle RST$, $R = 56°$, $r = 24$, and $t = 12$. Cameron and Gabriela are using the Law of Sines to find T. Is either of them correct? Explain your reasoning.
Cameron; R is acute and $r > t$, so there is one solution.

Cameron	Gabriela
$\dfrac{\sin T}{12} = \dfrac{\sin 56°}{24}$ $\sin T \approx 0.4145$ $T \approx 24.5°$	Since $r > t$, there is no solution.

44. **OPEN ENDED** Create an application problem involving right triangles and the Law of Sines. Then solve your problem, drawing diagrams if necessary. **See students' work.**

45. **CHALLENGE** Using the figure at the right, derive the formula Area $= \frac{1}{2}bc \sin A$. **See Chapter 12 Answer Appendix.**

46. **REASONING** Find the side lengths of two different triangles ABC that can be formed if $A = 55°$ and $C = 20°$.

47. **WRITING IN MATH** Use the Law of Sines to explain why a and b do not have unique values in the figure shown. **See margin.**

48. **OPEN ENDED** Given that $E = 62°$ and $d = 38$, find a value for e such that no triangle DEF can exist. Explain your reasoning. **See margin.**

Enrichment

OL BL

12-4 **Enrichment**

Navigation

CCSS **Teaching the Mathematical Practices**

Critique Mathematically proficient students are also able to compare the effectiveness of two plausible arguments, distinguish correct logic or reasoning from that which is flawed, and—if there is a flaw in an argument—explain what it is.

49. SHORT RESPONSE Given the graphs of $f(x)$ and $g(x)$, what is the value of $f(g(4))$? **2**

50. STATISTICS If the average of seven consecutive odd integers is n, what is the median of these seven integers? **C**

A 0 C n
B 7 D $n - 2$

51. One zero of $f(x) = x^3 - 7x^2 - 6x + 72$ is 4. What is the factored form of the expression $x^3 - 7x^2 - 6x + 72$? **G**

F $(x - 6)(x + 3)(x + 4)$
G $(x - 6)(x + 3)(x - 4)$
H $(x + 6)(x + 3)(x - 4)$
J $(x + 12)(x - 1)(x - 4)$

52. SAT/ACT Three people are splitting \$48,000 using the ratio $5 : 4 : 3$. What is the amount of the greatest share? **C**

A \$12,000 D \$24,000
B \$16,000 E \$30,000
C \$20,000

Find the exact value of each trigonometric function. (Lesson 12-3)

53. $\sin 210°$ $-\dfrac{1}{2}$

54. $\cos \dfrac{3}{4}\pi$ $-\dfrac{\sqrt{2}}{2}$

55. $\cot 60°$ $\dfrac{\sqrt{3}}{3}$

Find an angle with a positive measure and an angle with a negative measure that are coterminal with each angle. (Lesson 12-2)

56. $125°$ **485°, −235°**

57. $-32°$ **328°, −392°**

58. $\dfrac{2}{3}\pi$ $\dfrac{8}{3}\pi, -\dfrac{4}{3}\pi$

59. CLOCKS Jun's grandfather clock is broken. When she sets the pendulum in motion by holding it against the side of the clock and letting it go, it swings 24 centimeters to the other side, then 18 centimeters back, then 13.5 centimeters, and so on. What is the total distance that the pendulum swings before it stops? (Lesson 10-5) **96 cm**

Find the sum of each infinite series, if it exists. (Lesson 10-4)

60. $64 + 48 + 36 + \dots$ **256**

61. $27 + 36 + 48 + \dots$ **No sum exists.**

62. $\displaystyle\sum_{n=1}^{\infty} 0.5(1.1)^n$ **No sum exists.**

63. ASTRONOMY At its closest point, Earth is 91.8 million miles from the center of the Sun. At its farthest point, Earth is 94.9 million miles from the center of the Sun. Write an equation for the orbit of Earth, assuming that the center of the orbit is the origin and the Sun lies on the x-axis. (Lesson 9-4) $\dfrac{x^2}{8.714 \times 10^{15}} + \dfrac{y^2}{8.712 \times 10^{15}} = 1$

Simplify. (Lesson 6-4)

64. $\sqrt{(x-4)^2}$ $|x - 4|$

65. $\sqrt{(y+2)^4}$ $(y+2)^2$

66. $\sqrt[3]{(a-b)^6}$ $(a-b)^2$

Evaluate each expression if $w = 6$, $x = -4$, $y = 1.5$, and $z = \dfrac{3}{4}$.

67. $w^2 + y^2 - 6xz$ **56.25**

68. $x^2 + z^2 + 5wy$ $61\dfrac{9}{16}$

69. $wy + xz + w^2 - x^2$ **26**

DifferentiatedInstruction OL BL

Extension Have students compute the area of a 3-4-5 right triangle using the area formula presented in this lesson. Have them use the formula for all three angles of the triangle. 6; 6; 6

WatchOut!

Error Analysis For Exercise 43, students should see that Gabriela is incorrect because this is not a "no solution" situation. Explain to students that the "no solution" situation is when the side opposite the given angle θ is less than $\sin \theta$ times the length of the side adjacent to θ.

4 Assess

Name the Math Ask students to state the steps that are necessary for solving a triangle when one side and two angle measures are given.

Additional Answers

41a. Sample answer:

47. Sample answer: In the triangle, $B = 115°$. Using the Law of Sines, $\dfrac{\sin 50°}{a} = \dfrac{\sin 115°}{b}$. This equation cannot be solved because there are two unknowns. To solve a triangle using the Law of Sines, two sides and an angle must be given or two angles and a side opposite one of the angles must be given.

48. Sample answer: $e = 30$; for no triangle to exist, the length of the side opposite angle E must be less than 33.6 to satisfy the Law of Sines.

1 Focus

Objective Investigate measures in regular polygons using trigonometry.

Materials for Each Student
- compass
- straightedge
- protractor

Teaching Tip
Have students draw a diameter for each circle. This will make it easier to use the protractor to draw the angles in each circle.

2 Teach

Working in Cooperative Groups
Have students work with partners, mixing abilities so that a student with more experience using drawing tools is paired with one who has less experience. Have students complete the Activity.

Practice Have students complete Exercises 1–7.

3 Assess

Formative Assessment
Use Exercise 1 to assess whether students understand how to inscribe regular polygons in a circle.

From Concrete to Abstract
Exercises 5–7 require students to determine formulas that express relationships between the number of sides of a regular polygon and the length of the apothem. Students should use the results of Exercise 1 to determine the formulas.

You can use central angles of circles to investigate characteristics of regular polygons inscribed in a circle. Recall that a regular polygon is inscribed in a circle if each of its vertices lies on the circle.

Activity   Collect the Data

Step 1 Use a compass to draw a circle with a radius of one inch.

Step 2 Inscribe an equilateral triangle inside the circle. To do this, use a protractor to measure three angles of 120° at the center of the circle, since $\frac{360°}{3} = 120°$.

Then connect the points where the sides of the angles intersect the circle using a straightedge.

Step 3 The **apothem** of a regular polygon is a segment that is drawn from the center of the polygon perpendicular to a side of the polygon. Use the cosine of angle θ to find the length of an apothem, labeled a in the diagram.

Model and Analyze

1. Make a table like the one shown below and record the length of the apothem of the equilateral triangle. Inscribe each regular polygon named in the table in a circle with radius one inch. Copy and complete the table.

Number of Sides, n	θ	a	Number of Sides, n	θ	a
3	60	0.50	7	25.7	0.90
4	45	0.71	8	22.5	0.92
5	36	0.81	9	20	0.94
6	30	0.87	10	18	0.95

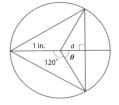

2. What do you notice about the measure of θ as the number of sides of the inscribed polygon increases? **The measure of θ decreases.**

3. What do you notice about the value of a? **The length of the apothem increases as the number of sides increases, approaching 1.**

4. **MAKE A CONJECTURE** Suppose you inscribe a 30-sided regular polygon inside a circle. Find the measure of angle θ. **6°**

5. Write a formula that gives the measure of angle θ for a polygon with n sides. **$\theta = 360 \div 2n$ or $\theta = 180 \div n$**

6. Write a formula that gives the length of the apothem of a regular polygon inscribed in a circle with radius one inch. **$a = \cos\theta$**

7. How would the formula you wrote in Exercise 5 change if the apothem of the circle was not one inch? **See margin.**

Extending the Concept

Ask:
- What happens to the area of the inscribed regular polygons as n increases? The area increases.

- What value does that area approach? the area of the circle

Additional Answer

7. To find the length of the apothem, you need to write the equation $\cos\theta = \dfrac{a}{\text{length of radius}}$.
If the radius is 1, then $\cos\theta = a$. If the radius is not 1, then $a = \text{length of radius} \cdot \cos\theta$.

Ed-Imaging

LESSON 12-5 Law of Cosines

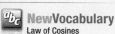

::·Then	::·Now	::·Why?
● You solved triangles by using the Law of Sines.	**1** Use the Law of Cosines to solve triangles. **2** Choose methods to solve triangles.	● A *submersible* is an underwater vessel used for exploring the depths of the ocean. You can use trigonometry to find the distance from a ship used to lower a submersible into the ocean and a shipwreck spotted by the submersible on the ocean floor.

520 m

70°

?

338 m

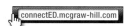

NewVocabulary
Law of Cosines

CCSS Common Core State Standards

Mathematical Practices
1 Make sense of problems and persevere in solving them.
3 Construct viable arguments and critique the reasoning of others.

1 **Use Law of Cosines to Solve Triangles** You cannot use the Law of Sines to solve a triangle like the one shown above. You can use the **Law of Cosines** if:

- the measures of two sides and the included angle are known (side-angle-side case).
- the measures of three sides are known (side-side-side case).

KeyConcept Law of Cosines

In $\triangle ABC$, if sides with lengths a, b, and c are opposite angles with measures A, B, and C, respectively, then the following are true.

$$a^2 = b^2 + c^2 - 2bc \cos A$$
$$b^2 = a^2 + c^2 - 2ac \cos B$$
$$c^2 = a^2 + b^2 - 2ab \cos C$$

Example 1 Solve a Triangle Given Two Sides and the Included Angle

Solve $\triangle ABC$.

Step 1 Use the Law of Cosines to find the missing side length.

$b^2 = a^2 + c^2 - 2ac \cos B$ — Law of Cosines
$b^2 = 7^2 + 5^2 - 2(7)(5) \cos 36°$ — $a = 7, c = 5, B = 36°$
$b^2 \approx 17.4$ — Use a calculator to simplify.
$b \approx 4.2$ — Take the positive square root of each side.

Step 2 Use the Law of Sines to find a missing angle measure.

$\dfrac{\sin A}{7} \approx \dfrac{\sin 36°}{4.2}$ $\dfrac{\sin A}{a} = \dfrac{\sin B}{b}$

$\sin A \approx \dfrac{7 \sin 36°}{4.2}$ — Multiply each side by 7.

$A \approx 78°$ — Use the $\sin^{-1}$ function.

Step 3 Find the measure of the other angle.
$m\angle C \approx 180° - (36° + 78°)$ or $66°$

So, $b \approx 4.2$, $A \approx 78°$, and $C \approx 66°$.

GuidedPractice

1. Solve $\triangle FGH$ if $G = 82°$, $f = 6$, and $h = 4$. $H \approx 36°$, $F = 62°$, $g = 6.7$

connectED.mcgraw-hill.com   **823**

---

1 Focus

VerticalAlignment

Before Lesson 12-5 Solve triangles by using the Law of Sines.

Lesson 12-5 Use the Law of Cosines to solve triangles. Choose methods to solve triangles.

After Lesson 12-5 Use the Law of Sines and the Law of Cosines to solve problems.

2 Teach

Scaffolding Questions
Have students read the **Why?** section of the lesson.

Ask:

- Is the triangle in the diagram acute, right, or obtuse? acute

- Explain why the angle with vertex at the shipwreck has greater measure than the angle with vertex at the ship. The side opposite the angle with vertex at the shipwreck is longer.

- Explain why the distance between the ship and the shipwreck must be less than 858 meters. Triangle Inequality tells us that the third side of the triangle must be less than 520 + 338 = 858 m.

Lesson 12-5 Resources

Resource	Approaching Level **AL**	On Level **OL**	Beyond Level **BL**	English Learners **ELL**
Teacher Edition	▪ Differentiated Instruction, p. 825	▪ Differentiated Instruction, pp. 825, 828	▪ Differentiated Instruction, p. 828	▪ Differentiated Instruction, p. 825
Chapter Resource Masters	▪ Study Guide and Intervention, pp. 29–30 ▪ Skills Practice, p. 31 ▪ Practice, p. 32 ▪ Word Problem Practice, p. 33	▪ Study Guide and Intervention, pp. 29–30 ▪ Skills Practice, p. 31 ▪ Practice, p. 32 ▪ Word Problem Practice, p. 33 ▪ Enrichment, p. 34	▪ Practice, p. 32 ▪ Word Problem Practice, p. 33 ▪ Enrichment, p. 34	▪ Study Guide and Intervention, pp. 29–30 ▪ Skills Practice, p. 31 ▪ Practice, p. 32 ▪ Word Problem Practice, p. 33
Other	▪ 5-Minute Check 12-5 ▪ Study Notebook	▪ 5-Minute Check 12-5 ▪ Study Notebook	▪ 5-Minute Check 12-5 ▪ Study Notebook	▪ 5-Minute Check 12-5 ▪ Study Notebook

1 Use Law of Cosines to Solve Triangles

Example 1 shows how to solve a triangle given two sides and the included angle. **Example 2** shows how to solve a triangle given three sides.

Formative Assessment

Use the Guided Practice exercises after each example to determine students' understanding of concepts.

Additional Examples

1 Solve △*ABC*.

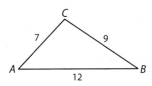

$A \approx 40°$; $B \approx 67°$; $c \approx 10.4$

2 Solve △*ABC*.

$A \approx 48°$; $B \approx 35°$; $C \approx 97°$

> **Additional Examples** also in Interactive Classroom PowerPoint® Presentations

IWB **Interactive White Board** READY

When you are only given the three side lengths of a triangle, you can solve it by using the Law of Cosines. The first step is to find the measure of the largest angle. This is done to ensure the other two angles are acute when using the Law of Sines.

Example 2 Solve a Triangle Given Three Sides

Solve △*ABC*.

Step 1 Use the Law of Cosines to find the measure of the largest angle, $\angle A$.

$a^2 = b^2 + c^2 - 2bc \cos A$	Law of Cosines
$16^2 = 9^2 + 10^2 - 2(9)(10) \cos A$	$a = 16$, $b = 9$, and $c = 10$
$16^2 - 9^2 - 10^2 = -2(9)(10) \cos A$	Subtract 9^2 and 10^2 from each side.
$\dfrac{16^2 - 9^2 - 10^2}{-2(9)(10)} = \cos A$	Divide each side by $-2(9)(10)$.
$-0.4167 \approx \cos A$	Use a calculator to simplify.
$115° \approx A$	Use the $\cos^{-1}$ function.

StudyTip
Alternative Method After finding $m\angle A$ in Step 1, the Law of Cosines could be used again to find the measure of a second angle.

Step 2 Use the Law of Sines to find the measure of $\angle B$.

$\dfrac{\sin B}{9} \approx \dfrac{\sin 115°}{16}$ $\qquad$ $\dfrac{\sin B}{b} = \dfrac{\sin A}{a}$

$\sin B \approx \dfrac{9 \sin 115°}{16}$ $\qquad$ Multiply each side by 9.

$\sin B \approx 0.5098$ $\qquad$ Use a calculator.

$B \approx 31°$ $\qquad$ Use the $\sin^{-1}$ function.

Step 3 Find the measure of $\angle C$.

$m\angle C \approx 180° - (115° + 31°)$ or about $34°$

So, $A \approx 115°$, $B \approx 31°$, and $C \approx 34°$.

GuidedPractice

2. Solve △*ABC* if $a = 5$, $b = 11$, and $c = 8$. **$A = 24.6°$, $B = 113.6°$, $C = 41.8°$**

ReviewVocabulary
oblique a triangle that has no right angle

2 Choose a Method to Solve Triangles

You can use the Law of Sines and the Law of Cosines to solve problems involving oblique triangles. You need to know the measure of at least one side and any two other parts. If the triangle has a solution, you must decide whether to use the Law of Sines or the Law of Cosines to begin solving it.

ConceptSummary Solving Oblique Triangles

Given	Begin by Using
two angles and any sides	Law of Sines
two sides and an angle opposite one of them	Law of Sines
two sides and their included angle	Law of Cosines
three sides	Law of Cosines

 824 | Lesson 12-5 | Law of Cosines

Focus on Mathematical Content

Law of Cosines The Law of Cosines can be used to solve a triangle if all three sides of the triangle are given or if two sides and the included angle are given. Unlike the Law of Sines cases, if a solution exists for cases in which the Law of Cosines can be used, the solution is unique.

Real-WorldLink
The record for the deepest seawater scuba dive was 1044 feet, made by a scuba diver in the Red Sea.
Source: *Guinness Book of World Records*

Real-World Example 3 Use the Law of Cosines

SCUBA DIVING A scuba diver looks up 20° and sees a turtle 9 feet away. She looks down 40° and sees a blue parrotfish 12 feet away. How far apart are the turtle and the blue parrotfish?

Understand You know the angles formed when the scuba diver looks up and when she looks down. You also know how far away the turtle and the blue parrotfish are from the scuba diver.

Plan Use the information to draw and label a diagram. Since two sides and the included angle of a triangle are given, you can use the Law of Cosines to solve the problem.

Solve $a^2 = b^2 + c^2 - 2bc \cos A$ Law of Cosines
$a^2 = 12^2 + 9^2 - 2(12)(9) \cos 60$ $b = 12$, $c = 9$, and $A = 60$
$a^2 = 117$ Use a calculator.
$a \approx 10.8$ Find the positive value of a.

So, the turtle and the blue parrotfish are about 10.8 feet apart.

Check Using the Law of Sines, you can find that $B \approx 74°$ and $C \approx 46°$. Since $C < A < B$ and $c < a < b$, the solution is reasonable.

▶ **Guided**Practice

3. MARATHONS Amelia ran 6 miles in one direction. She then turned 79° and ran 7 miles. At the end of the run, how far was Amelia from her starting point? **about 8.3 mi**

Check Your Understanding
⬤ = Step-by-Step Solutions begin on page R14. ☑

Examples 1–2 Solve each triangle. Round side lengths to the nearest tenth and angle measures to the nearest degree.

1.

1. $A \approx 36°$, $C \approx 52°$, $b \approx 5.1$
2. $A \approx 112°$, $B \approx 40°$, $C \approx 28°$

2.

3. $A \approx 18°$, $B \approx 29°$, $C \approx 133°$
4. $A \approx 48°$, $C \approx 22°$, $b \approx 7.6$

3. $a = 5$, $b = 8$, $c = 12$

4. $B = 110°$, $a = 6$, $c = 3$

Example 3 **CCSS PRECISION** Determine whether each triangle should be solved by beginning with the Law of *Sines* or the Law of *Cosines*. Then solve the triangle.

5.

Sines; $B \approx 40°$, $C \approx 33°$, $c \approx 6.8$

6.

Cosines; $A \approx 48°$, $C \approx 36°$, $b \approx 6.7$

7 In $\triangle RST$, $R = 35°$, $s = 16$, and $t = 9$. Cosines; $S \approx 114°$, $T \approx 31°$, $r \approx 10.1$

8. FOOTBALL In a football game, the quarterback is 20 yards from Receiver A. He turns 40° to see Receiver B, who is 16 yards away. How far apart are the two receivers? **about 12.9 yd**

DifferentiatedInstruction (AL) (OL) (ELL)

If ▶ students struggle with any of the methods shown to solve a triangle,

Then ▶ have students discuss in small groups how to choose which method to use when solving a triangle. Have them compare their approaches and develop a brief explanation to help others decide. Then have each group share their conclusions with the class.

Federico Cabello/SuperStock

---

2 Choose a Method to Solve Triangles

Example 3 shows how to apply the Law of Cosines to solve a real-world problem.

Additional Example

3 AIRPORT Two pilots in a stationary airplane look 38° to the left of their runway and see a bus 75 feet away. They look 28° to the right of their runway and see a truck 110 feet away. How far apart are the bus and the truck? about 105 ft

Teach with Tech

Document Camera Assign several problems to the class, and give students time to work through them. Then choose several students to share and explain their work to the class. Be sure the students sketch a diagram and explain how they decided whether to use the Law of Sines or the Law of Cosines to solve the problem.

WatchOut!

Preventing Errors When students must decide which method to use to solve, watch for students who think the angle is included. Review the definition of included angles with them.

CCSS Teaching the Mathematical Practices

Precision Mathematically proficient students try to use clear definitions in their reasoning, calculate accurately and efficiently, and make explicit use of definitions.

3 Practice

Formative Assessment

Use Exercises 1–8 to check for understanding.

Use the chart at the bottom of this page to customize assignments for your students.

4 Assess

Yesterday's News Ask students to describe how yesterday's topic, using the Law of Sines, was similar to and was different from today's topic, using the Law of Cosines.

Formative Assessment

Check for student understanding of concepts in Lessons 12-4 and 12-5.

 Quiz 2, p. 61

Tips for New Teachers

For Exercise 28, remind students that sometimes rounding can lead to ambiguous answers, such as a triangle appearing to be 181 degrees.

Additional Answers

29a. Sample answer:

34. $a^2 = (b - x)^2 + h^2$ (Use the
Pythagorean Theorem for $\triangle DBC$.)
$a^2 = b^2 - 2bx + x^2 + h^2$
(Expand $(b - x)^2$.)
$a^2 = b^2 - 2bx + c^2$
(In $\triangle ADB$, $c^2 = x^2 + h^2$.)
$a^2 = b^2 - 2b(c \cos A) + c^2$
($\cos A = \frac{x}{c}$, so $x = c \cos A$.)
$a^2 = b^2 + c^2 - 2bc \cos A$
(Commutative Property)

Practice and Problem Solving   Extra Practice is on page R12.

Examples 1–2 Solve each triangle. Round side lengths to the nearest tenth and angle measures to the nearest degree.

9.

10.

11.

12.

13. $A = 116°$, $b = 5$, $c = 3$

14. $C = 80°$, $a = 9$, $b = 2$

15. $f = 10$, $g = 11$, $h = 4$

16. $w = 20$, $x = 13$, $y = 12$

9. $A \approx 70°$, $B \approx 40°$, $c \approx 3.0$
10. $A \approx 48°$, $C \approx 40°$, $b \approx 18.8$
11. $A \approx 31°$, $B \approx 108°$, $C \approx 41°$
12. $A \approx 102°$, $B \approx 44°$, $C \approx 34°$
13. $a \approx 6.9$, $B \approx 41°$, $C \approx 23°$
14. $c \approx 8.9$, $A \approx 87°$, $B \approx 13°$
15. $F \approx 65°$, $G \approx 94°$, $H \approx 21°$

Example 3 Determine whether each triangle should be solved by beginning with the Law of *Sines* or the Law of *Cosines*. Then solve the triangle.

17.

18.

19.

20.

16. $W \approx 106°$, $X \approx 39°$, $Y \approx 35°$
17. Sines; $C \approx 45°$, $A \approx 85°$, $a \approx 18.2$
18. Cosines; $s \approx 28.9$, $R \approx 42°$, $T \approx 32°$
19. Cosines; $A \approx 27°$, $B \approx 115°$, $C \approx 38°$
20. Sines; $N \approx 53°$, $p \approx 38.2$, $m \approx 28.4$

21. Sines; $A \approx 17°$, $B \approx 79°$, $b \approx 6.9$
22. Cosines; $H \approx 48°$, $J = 25°$, $K = 107°$

21. In $\triangle ABC$, $C = 84°$, $c = 7$, and $a = 2$.
22. In $\triangle HJK$, $h = 18$, $j = 10$, and $k = 23$.

23. **EXPLORATION** Find the distance between the ship and the shipwreck shown in the diagram. Round to the nearest tenth. **514.2 m**

24. **GEOMETRY** A parallelogram has side lengths 8 centimeters and 12 centimeters. One angle between them measures 42°. To the nearest tenth, what is the length of the shorter diagonal? **8.1 cm**

25. **RACING** A triangular cross-country course has side lengths 1.8 kilometers, 2 kilometers, and 1.2 kilometers. What are the angles formed between each pair of sides? **81°, 36°, 63°**

26. **CCSS MODELING** A triangular plot of farm land measures 0.9 by 0.5 by 1.25 miles.
   a. If the plot of land is fenced on the border, what will be the angles at which the fences of the three sides meet? Round to the nearest degree. **19°, 37°, 124°**
   b. What is the area of the plot of land? **about 0.19 mi²**

27. **LAND** Some land is in the shape of a triangle. The distances between each vertex of the triangle are 140 yd, 210 yd and 300 yd, respectively. Use the Law of Cosines to find the area of the land to the nearest square yard. **about 13,148 yd²**

Differentiated Homework Options

Level	Assignment	Two-Day Option	
AL Basic	9–24, 35–55	9–23 odd, 38–41	10–24 even, 35–37, 42–55
OL Core	9–23 odd, 25–31, 33, 35–55	9–24, 38–41	25–33, 35–37, 42–55
BL Advanced	25–52, (optional: 53–55)		

28. RIDES Two bumper cars at an amusement park ride collide as shown below.

7 ft
118°
5.5 ft
d

29b. Sample answer: Use the Law of Cosines to find the measure of ∠A. Then use the formula Area $= \frac{1}{2}bc \sin A$.

   a. How far apart *d* were the two cars before they collided? **about 10.7 ft**

   b. Before the collision, a third car was 10 feet from the blue car and 13 feet from the orange car. Describe the angles formed by the three cars before the collision. **78°, 49°, 53°**

 29. PICNICS A triangular picnic area is 11 yards by 14 yards by 10 yards.

   a. Sketch and label a drawing to represent the picnic area. **See margin.**

   b. Describe how you could find the area of the picnic area.

   c. What is the area? Round to the nearest tenth. **54.6 yd²**

30. WATERSPORTS A person on a personal watercraft makes a trip from point *A* to point *B* to point *C* traveling 28 miles per hour. She then returns from point *C* back to her starting point traveling 35 miles per hour. How many minutes did the entire trip take? Round to the nearest tenth. **1.5 min**

B
0.15 mi
130°
0.25 mi
C
A

31. $B \approx 39°$, $C \approx 37°$, $c \approx 7.7$
32. $R \approx 107°$, $S \approx 48°$, $q \approx 16.0$
33. $F \approx 42°$, $G \approx 72°$, $H \approx 66°$

Solve each triangle. Round side lengths to the nearest tenth and angle measures to the nearest degree.

31.

B
12.4
c
104°
A   8.1   C

32.

Q   28   R
25°
36.2   *q*
S

33.

F   20.8   G
21.6   15.2
H

H.O.T. Problems   Use Higher-Order Thinking Skills

34. CHALLENGE Use the figure and the Pythagorean Theorem to derive the Law of Cosines. Use the hints below. **See margin.**

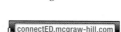

B
c   *h*   *a*
A   *x*   D   *b − x*   C
b

35. The longest side is 14.5 centimeters. Use the Law of Cosines to find the measure of the angle opposite the longest side; 102°.

   • First, use the Pythagorean Theorem for △DBC.

   • In △ADB, $c^2 = x^2 + h^2$.

   • $\cos A = \dfrac{x}{c}$

35. CCSS ARGUMENTS Three sides of a triangle measure 10.6 centimeters, 8 centimeters, and 14.5 centimeters. Explain how to find the measure of the largest angle. Then find the measure of the angle to the nearest degree.

36. OPEN ENDED Create an application problem involving right triangles and the Law of Cosines. Then solve your problem, drawing diagrams if necessary. **See students' work.**

37. WRITING IN MATH How do you know which method to use when solving a triangle? **See Chapter 12 Answer Appendix.**

CCSS **Teaching the Mathematical Practices**

Arguments Mathematically proficient students understand and use stated assumptions, definitions, and previously established results in constructing arguments. They make conjectures and build a logical progression of statements to explore the truth of their conjectures. And they are able to analyze situations by breaking them into cases, and can recognize and use counterexamples.

Study Guide and Intervention
AL OL ELL

Practice
AL OL BL ELL

Word Problem Practice
AL OL BL ELL

Enrichment
OL BL
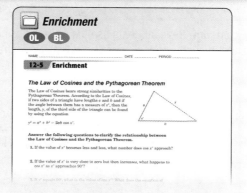

connectED.mcgraw-hill.com **827**

Additional Answers

45. $\sin \theta = \dfrac{5\sqrt{89}}{89}, \cos \theta = \dfrac{8\sqrt{89}}{89},$

$\tan \theta = \dfrac{5}{8}, \csc \theta = \dfrac{\sqrt{89}}{5},$

$\sec \theta = \dfrac{\sqrt{89}}{8}, \cot \theta = \dfrac{8}{5}$

46. $\sin \theta = \dfrac{-\sqrt{5}}{5}, \cos \theta = \dfrac{-2\sqrt{5}}{5},$

$\tan \theta = 0.5, \csc \theta = -\sqrt{5},$

$\sec \theta = \dfrac{-\sqrt{5}}{2}, \cot \theta = 2$

47. $\sin \theta = -\dfrac{3\sqrt{13}}{13}, \cos \theta = \dfrac{2\sqrt{13}}{13},$

$\tan \theta = -1.5, \csc \theta = \dfrac{-\sqrt{13}}{3},$

$\sec \theta = \dfrac{\sqrt{13}}{2}, \cot \theta = -\dfrac{2}{3}$

48a.

[50, 450] scl: 50 by [0, 5] scl: 1

The distribution is positively skewed.

53.

54.

55.

Standardized Test Practice

38. SAT/ACT If c and d are different positive integers and $4c + d = 26$, what is the sum of all possible values of c? **D**

  A 6   **D** 21

  B 10   **E** 28

  C 15

39. If $6^y = 21$, what is y? **G**

  F $\log 12 - \log 6$ **H** $\dfrac{\log 6}{\log 21}$

  G $\dfrac{\log 21}{\log 6}$ **J** $\log\left(\dfrac{6}{21}\right)$

40. GEOMETRY Find the perimeter of the figure. **C**

  A 24   **B** 30   **C** 36   **D** 48

41. SHORT RESPONSE Solve the equation below for x.

$4, \dfrac{23}{15}$ $\dfrac{1}{x-1} + \dfrac{5}{8} = \dfrac{23}{6x}$

Spiral Review

Find the area of $\triangle ABC$ to the nearest tenth. (Lesson 12-4)

42.

65.2 cm²

43. 7.5 yd²

44. 35.1 ft²

The terminal side of θ in standard position contains each point. Find the exact values of the six trigonometric functions of θ. (Lesson 12-3) **45–47. See margin.**

45. $(8, 5)$ **46.** $(-4, -2)$ **47.** $(6, -9)$

48. ATHLETIC SHOES The prices for a random sample of athletic shoes are shown. (Lesson 11-2)

Price (dollars)				
70	300	400	250	250
150	120	250	100	70
150	160	200	170	300

  a. Use a graphing calculator to create a box-and-whisker plot. Then describe the shape of the distribution. **See margin.**

  b. Describe the center and spread of the data using either the mean and standard deviation or the five-number summary. Justify your choice.

48b. Sample answe The distribution is positively skewed, use the five-numbe summary. The ran₣ is $70 to $400. Th₣ median is $190, a₣ half of the data are between $120 and $250.

49. BUSINESS During the month of June, MediaWorld had revenue of $2700 from sales of a certain DVD box set. During the July Blowout Sale, the set was on sale for $10 off. Revenue from the set was $3750 in July with 30 more sets sold than were sold in June. Find the price of the DVD set for June and the price for July. (Lesson 9-7) **$60, $50**

Without writing the equation in standard form, state whether the graph of each equation is a *parabola*, *circle*, *ellipse*, or *hyperbola*. (Lesson 9-6)

50. $x^2 + y^2 - 8x - 6y + 5 = 0$ **51.** $3x^2 - 2y^2 + 32y - 134 = 0$ **52.** $y^2 + 18y - 2x = -84$
    circle   hyperbola   parabola

Skills Review

Sketch each angle. Then find its reference angle. **53–55. See margin.**

53. $245°$ **54.** $-15°$ **55.** $\dfrac{5}{4}\pi$

 828 | Lesson 12-5 | Law of Cosines

DifferentiatedInstruction ⓄⓁ ⒷⓁ

Extension Have students use the Law of Cosines to attempt to solve a "triangle" with sides 5, 12, and 18 (such a triangle does not exist). Have them explain what they discover and what it means. Students will get an error when attempting to find the inverse cosine because the value is not between −1 and 1. This means that no such triangle exists.

12 Mid-Chapter Quiz
Lessons 12-1 through 12-5

Solve △XYZ by using the given measurements. Round measures of sides to the nearest tenth and measures of angles to the nearest degree. (Lesson 12-1)

$X = 25°$,
$y = 34.3$,
$z = 37.9$

$Y = 65°$,
$y = 17.2$,
$z = 18.9$

1. $Y = 65°$, $x = 16$ **2.** $X = 25°$, $x = 8$

3. Find the values of the six trigonometric functions for angle θ. (Lesson 12-1) **See margin.**

4. Draw an angle measuring $-80°$ in standard position. (Lesson 12-2) **See margin.**

Rewrite each degree measure in radians and each radian measure in degrees. (Lesson 12-2)

5. $215°$ $\dfrac{43\pi}{36}$ **6.** $-350°$ $-\dfrac{35\pi}{18}$

7. $\dfrac{8\pi}{5}$ $288°$ **8.** $\dfrac{9\pi}{2}$ $810°$

9. MULTIPLE CHOICE What is the length of the arc below rounded to the nearest tenth? (Lesson 12-2) **C**

A 4.2 cm

B 17.1 cm

C 53.9 cm

D 2638.9 cm

Find the exact value of each trigonometric function. (Lesson 12-3)

10. $\tan \pi$ 0 **11.** $\cos \dfrac{3\pi}{4}$ $-\dfrac{\sqrt{2}}{2}$

The terminal side of θ in standard position contains each point. Find the exact values of the six trigonometric functions of θ. (Lesson 12-3) **12, 13. See Chapter 12 Answer Appendix.**

12. $(0, -5)$ **13.** $(6, 8)$

---

14. MULTIPLE CHOICE Suppose θ is an angle in standard position with $\cos \theta > 0$. In which quadrant(s) does the terminal side of θ lie? (Lesson 12-3) **J**

F I

G II

H III

J I and IV

15. GARDEN Lana has a garden in the shape of a triangle as pictured below. She wants to fill the garden with top soil. What is the area of the triangle? (Lesson 12-4) **about 38.8 m²**

Determine whether each triangle has *no* solution, *one* solution, or *two* solutions. Then solve the triangle. Round side lengths to the nearest tenth and angle measures to the nearest degree. (Lesson 12-4)

16. $A = 38°$, $a = 18$, $c = 25$ **See Chapter 12 Answer Appendix.**

17. $A = 65°$, $a = 5$, $b = 7$ **no solution**

18. $A = 115°$, $a = 12$, $b = 8$ **one solution:** $B = 37°$, $C = 28°$, $c = 6.2$

Solve each triangle. Round side lengths to the nearest tenth and angle measures to the nearest degree. (Lesson 12-5)

19.   **20.**

$A = 50°$, $B = 87°$, $C = 43°$ $A = 40°$, $C = 35°$, $c = 10.7$

21. Eric and Zach are camping. Erik leaves Zach at the campsite and walks 4.5 miles. He then turns at a 120° angle and walks another 2.5 miles. If Eric were to walk directly back to Zach, how far would he walk? (Lesson 12-5) **about 6.1 mi**

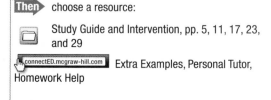

Formative Assessment
Use the Mid-Chapter Quiz to assess students' progress in the first half of the chapter.

For problems answered incorrectly, have students review the lessons indicated in parentheses.

Summative Assessment
Mid-Chapter Test, p. 63

McGraw-Hill
eAssessment Customize and create multiple versions of your Mid-Chapter Test and their answer keys.

FOLDABLES StudyOrganizer

Before students complete the Mid-Chapter Test, encourage them to review the information for Lessons 12-1 through 12-5 in their Foldables.

Additional Answers

3. $\sin \theta = \dfrac{\sqrt{7}}{4}$, $\cos \theta = \dfrac{3}{4}$,
$\tan \theta = \dfrac{\sqrt{7}}{3}$, $\csc \theta = \dfrac{4\sqrt{7}}{7}$,
$\sec \theta = \dfrac{4}{3}$, $\cot \theta = \dfrac{3\sqrt{7}}{7}$

4.

InterventionPlanner

TIER 1 On Level OL	TIER 2 Strategic Intervention AL approaching grade level	TIER 3 Intensive Intervention 2 or more grades below level
If students miss about 25% of the exercises or less,	**If** students miss about 50% of the exercises,	**If** students miss about 75% of the exercises,
Then choose a resource:	**Then** choose a resource:	**Then** use *Math Triumphs, Alg. 2*, Ch. 6
SE Lessons 12-1, 12-2, 12-3, 12-4, and 12-5	Study Guide and Intervention, pp. 5, 11, 17, 23, and 29	connectED.mcgraw-hill.com Extra Examples, Personal Tutor, Homework Help, Review Vocabulary
Skills Practice, pp. 7, 13, 19, 25, and 31	connectED.mcgraw-hill.com Extra Examples, Personal Tutor, Homework Help	
TE Chapter Project		
connectED.mcgraw-hill.com Self-Check Quiz		

LESSON 12-6 Circular and Periodic Functions

1 Focus

VerticalAlignment

Before Lesson 12-6 Evaluate trigonometric functions using reference angles.

Lesson 12-6 Find values of trigonometric functions based on the unit circle. Use the properties of periodic functions to evaluate trigonometric functions.

After Lesson 12-6 Sketch and interpret graphs of the sine and cosine functions.

2 Teach

Scaffolding Questions

Have students read the **Why?** section of the lesson.

Ask:

- Use the maximum and minimum heights of the pedal to find the diameter of the circle. diameter = maximum − minimum = 18 − 4 = 14 in.

- Use the height of the circle's center to find the circle's diameter. The radius is 18 − 11 = 7 in., or 11 − 4 = 7 in., so the diameter is 2 · 7 = 14 in.

- What is the starting position of the pedal? Explain. The starting position would correspond with $t = 0$ seconds; at that time the pedal is at the top of the diagram.

:: Then

- You evaluated trigonometric functions using reference angles.

:: Now

1 Find values of trigonometric functions based on the unit circle.

2 Use the properties of periodic functions to evaluate trigonometric functions.

:: Why?

- The pedals on a bicycle rotate as the bike is being ridden. The height of a pedal is a function of time, as shown in the figure at the right.

Notice that the pedal makes one complete rotation every two seconds.

NewVocabulary
unit circle
circular function
periodic function
cycle
period

CCSS Common Core State Standards

Content Standards
F.TF.1 Understand radian measure of an angle as the length of the arc on the unit circle subtended by the angle.

F.TF.2 Explain how the unit circle in the coordinate plane enables the extension of trigonometric functions to all real numbers, interpreted as radian measures of angles traversed counterclockwise around the unit circle.

Mathematical Practices
7 Look for and make use of structure.

---

1 Circular Functions A **unit circle** is a circle with a radius of 1 unit centered at the origin on the coordinate plane. You can use a point P on the unit circle to generalize sine and cosine functions.

$$\sin \theta = \frac{y}{r} = \frac{y}{1} \text{ or } y \qquad \cos \theta = \frac{x}{r} = \frac{x}{1} \text{ or } x$$

So, the values of $\sin \theta$ and $\cos \theta$ are the y-coordinate and x-coordinate, respectively, of the point where the terminal side of θ intersects the unit circle.

KeyConcept Functions on a Unit Circle

Words	If the terminal side of an angle θ in standard position intersects the unit circle at $P(x, y)$, then $\cos \theta = x$ and $\sin \theta = y$.
Symbols	$P(x, y) = P(\cos \theta, \sin \theta)$
Example	If $\theta = 120°$, $P(x, y) = P(\cos 120°, \sin 120°)$.

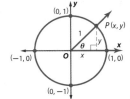

Model

Both $\cos \theta = x$ and $\sin \theta = y$ are functions of θ. Because they are defined using a unit circle, they are called **circular functions**.

Example 1 Find Sine and Cosine Given a Point on the Unit Circle

The terminal side of angle θ in standard position intersects the unit circle at $P\left(\frac{1}{2}, \frac{\sqrt{3}}{2}\right)$. Find $\cos \theta$ and $\sin \theta$.

$$P\left(\frac{1}{2}, \frac{\sqrt{3}}{2}\right) = P(\cos \theta, \sin \theta)$$

$$\cos \theta = \frac{1}{2} \qquad \sin \theta = \frac{\sqrt{3}}{2}$$

GuidedPractice

1. The terminal side of angle θ in standard position intersects the unit circle at $P\left(\frac{3}{5}, -\frac{4}{5}\right)$. Find $\cos \theta$ and $\sin \theta$. $\cos \theta = \frac{3}{5}$, $\sin \theta = -\frac{4}{5}$

830 | Lesson 12-6

Lesson 12-6 Resources

Resource	Approaching Level **AL**	On Level **OL**	Beyond Level **BL**	English Learners **ELL**
Teacher Edition	• Differentiated Instruction, p. 832	• Differentiated Instruction, pp. 832, 836	• Differentiated Instruction, p. 836	
Chapter Resource Masters	• Study Guide and Intervention, pp. 35–36 • Skills Practice, p. 37 • Practice, p. 38 • Word Problem Practice, p. 39	• Study Guide and Intervention, pp. 35–36 • Skills Practice, p. 37 • Practice, p. 38 • Word Problem Practice, p. 39 • Enrichment, p. 40	• Practice, p. 38 • Word Problem Practice, p. 39 • Enrichment, p. 40	• Study Guide and Intervention, pp. 35–36 • Skills Practice, p. 37 • Practice, p. 38 • Word Problem Practice, p. 39
Other	• 5-Minute Check 12-6 • Study Notebook	• 5-Minute Check 12-6 • Study Notebook	• 5-Minute Check 12-6 • Study Notebook	• 5-Minute Check 12-6 • Study Notebook

John P. Kelly/Photodisc/Getty Images

StudyTip

Cycles A cycle can begin at any point on the graph of a periodic function. In Example 2, if the beginning of the cycle is at $\frac{\pi}{2}$, then the pattern repeats at $\frac{3\pi}{2}$.

The period is $\frac{3\pi}{2} - \frac{\pi}{2}$ or π.

2 Periodic Functions

A **periodic function** has y-values that repeat at regular intervals. One complete pattern is a **cycle**, and the horizontal length of one cycle is a **period**.

θ	y
0°	1
180°	−1
360°	1
540°	−1
720°	1

The cycle repeats every 360°.

Example 2 Identify the Period

Determine the period of the function.

The pattern repeats at π, 2π, and so on. So, the period is π.

▶ **Guided**Practice

2. Graph a function with a period of 4. **See margin.**

The rotations of wheels, pedals, carousels, and objects in space are all periodic.

Real-World Example 3 Use Trigonometric Functions

CYCLING Refer to the beginning of the lesson. The height of a bicycle pedal varies periodically as a function of time, as shown in the figure.

a. Make a table showing the height of a bicycle pedal at 0, 0.5, 1.0, 1.5, 2.0, 2.5, and 3.0 seconds.

At 0 seconds, the pedal is 18 inches high. At 0.5 second, the pedal is 11 inches high. At 1.0 second, the pedal is 4 inches high, and so on.

Time (s)	Height (in.)
0	18
0.5	11
1.0	4
1.5	11
2.0	18
2.5	11
3.0	4

b. Identify the period of the function.

The period is the time it takes to complete one rotation. So, the period is 2 seconds.

c. Graph the function. Let the horizontal axis represent the time t and the vertical axis represent the height h in inches that the pedal is from the ground.

The maximum height of the pedal is 18 inches, and the minimum height is 4 inches. Because the period of the function is 2 seconds, the pattern of the graph repeats in intervals of 2 seconds.

▶ **Guided**Practice

3. CYCLING Another cyclist pedals the same bike at a rate of 1 revolution per second.

A. Make a table showing the height of a bicycle pedal at times 0, 0.5, 1.0, 1.5, 2.0, 2.5, and 3.0 seconds.

B. Identify the period and graph the function. **1; See margin for graph.**

Real-WorldLink

Most competitive cyclists pedal at rates of more than 200 rotations per minute. Most other people pedal at between 90 and 120 rotations per minute.

Source: SpringerLink

3A.

Time (s)	Height of Pedal (in.)
0	18
0.5	4
1.0	18
1.5	4
2.0	18
2.5	4
3.0	18

---

1 Circular Functions

Example 1 shows how to find the sine and cosine of a point on the unit circle.

Formative Assessment

Use the Guided Practice exercises after each example to determine students' understanding of concepts.

Additional Example

1 The terminal side of angle θ in standard position intersects the unit circle at $P\left(\dfrac{\sqrt{7}}{4}, \dfrac{3}{4}\right)$. Find $\cos \theta$ and $\sin \theta$.

$\sin \theta = \dfrac{3}{4}$, $\cos \theta = \dfrac{\sqrt{7}}{4}$

▶ **Additional Examples** also in Interactive Classroom PowerPoint® Presentations

 Interactive White Board READY

2 Periodic Functions

Example 2 shows how to use a graph to identify the period of the function.
Example 3 shows how to solve a real-world problem using the periodic nature of trigonometric functions. **Example 4** shows how to find the value of a trigonometric function using the period of the function.

Additional Example

2 Determine the period of the function. $\dfrac{\pi}{3}$

Additional Answers (Guided Practice)

2.

3B.

Additional Examples

3 **CYCLING** Refer to the application at the beginning of the lesson. The height of a bicycle pedal varies periodically as a function of the time, as shown in the figure.

a. Make a table showing the height of a bicycle pedal at 3.0, 3.5, 4.0, 4.5, and 5.0 seconds.

Time (s)	Height (in.)
3.0	4
3.5	11
4.0	18
4.5	11
5.0	4

b. Identify the period of the function. **2 seconds**

c. Graph the function. Let the horizontal axis represent the time t and the vertical axis represent the height h in inches that the pedal is from the ground.

4 Find the exact value of each function.

a. $\cos 690°$ $\dfrac{\sqrt{3}}{2}$

b. $\sin\left(-\dfrac{3\pi}{4}\right)$ $\dfrac{\sqrt{2}}{2}$

Teach with Tech

Audio Recording Have students create audio recordings explaining how to find the exact value of a trigonometric function for an angle greater than 360° (or less than −360°). Give students specific examples to explain.

StudyTip

Sine and Cosine To help you remember that for (x, y) on a unit circle, $x = \cos \theta$ and $y = \sin \theta$, notice that alphabetically x comes before y and *cosine* comes before *sine*.

The exact values of $\cos \theta$ and $\sin \theta$ for special angles are shown on the unit circle at the right. The cosine values are the x-coordinates of the points on the unit circle, and the sine values are the y-coordinates.

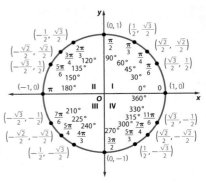

You can use this information to graph the sine and cosine functions. Let the horizontal axis represent the values of θ and the vertical axis represent the values of $\sin \theta$ or $\cos \theta$.

The cycles of the sine and cosine functions repeat every 360°. So, they are periodic functions. The period of each function is 360° or 2π.

Consider the points on the unit circle for $\theta = 45°$, $\theta = 150°$, and $\theta = 270°$.

$(\cos 45°, \sin 45°) = \left(\dfrac{\sqrt{2}}{2}, \dfrac{\sqrt{2}}{2}\right)$

$(\cos 150°, \sin 150°) = \left(-\dfrac{\sqrt{3}}{2}, \dfrac{1}{2}\right)$

$(\cos 270°, \sin 270°) = (0, -1)$

StudyTip

Radians The sine and cosine functions can also be graphed using radians as the units on the θ-axis.

These points can also be shown on the graphs of the sine and cosine functions.

Since the period of the sine and cosine functions is 360°, the values repeat every 360°. So, $\sin (x + 360°) = \sin x$, and $\cos (x + 360°) = \cos x$.

Example 4 Evaluate Trigonometric Expressions

Find the exact value of each expression.

a. $\cos 480°$

$\cos 480° = \cos (120° + 360°)$

$= \cos 120°$

$= -\dfrac{1}{2}$

b. $\sin \dfrac{11\pi}{4}$

$\sin \dfrac{11\pi}{4} = \sin \left(\dfrac{3\pi}{4} + \dfrac{8\pi}{4}\right)$

$= \sin \dfrac{3\pi}{4}$

$= \dfrac{\sqrt{2}}{2}$

▶ **Guided Practice**

4A. $\cos\left(-\dfrac{3\pi}{4}\right)$ $-\dfrac{\sqrt{2}}{2}$

4B. $\sin 420°$ $\dfrac{\sqrt{3}}{2}$

 832 | **Lesson 12-6** | Circular and Periodic Functions

DifferentiatedInstruction **AL** **OL**

Naturalist Learners Have students research various kinds of circular calendars, such as those used by the Maya, to predict the weather and determine the best time for planting crops.

Example 1   **CCSS STRUCTURE** The terminal side of angle θ in standard position intersects the unit circle at each point P. Find $\cos \theta$ and $\sin \theta$.

1. $P\left(\frac{15}{17}, \frac{8}{17}\right)$ $\cos \theta = \frac{15}{17}$, $\sin \theta = \frac{8}{17}$

2. $P\left(-\frac{\sqrt{2}}{2}, \frac{\sqrt{2}}{2}\right)$ $\cos \theta = -\frac{\sqrt{2}}{2}$, $\sin \theta = \frac{\sqrt{2}}{2}$

Example 2   Determine the period of each function.

3.   2

4. 4π

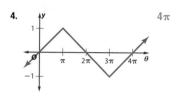

Example 3   **5. SWINGS** The height of a swing varies periodically as the function of time. The swing goes forward and reaches its high point of 6 feet. It then goes backward and reaches 6 feet again. Its lowest point is 2 feet. The time it takes to swing from its high point to its low point is 1 second.

   a. How long does it take for the swing to go forward and back one time?   **4 seconds**

   b. Graph the height of the swing h as a function of time t.   **See margin.**

Example 4   Find the exact value of each expression.

6. $\sin \frac{13\pi}{6}$ $\frac{1}{2}$

7. $\sin (-60°)$ $-\frac{\sqrt{3}}{2}$

8. $\cos 540°$ -1

Practice and Problem Solving   Extra Practice is on page R12.

Example 1   The terminal side of angle θ in standard position intersects the unit circle at each point P. Find $\cos \theta$ and $\sin \theta$.

9. $P\left(\frac{6}{10}, -\frac{8}{10}\right)$ $\cos \theta = \frac{3}{5}$, $\sin \theta = -\frac{4}{5}$

10. $P\left(-\frac{10}{26}, -\frac{24}{26}\right)$ $\cos \theta = -\frac{5}{13}$, $\sin \theta = -\frac{12}{13}$

11 $P\left(\frac{\sqrt{3}}{2}, \frac{1}{2}\right)$ $\cos \theta = \frac{\sqrt{3}}{2}$, $\sin \theta = \frac{1}{2}$

12. $P\left(\frac{\sqrt{6}}{5}, \frac{\sqrt{19}}{5}\right)$ $\cos \theta = \frac{\sqrt{6}}{5}$, $\sin \theta = \frac{\sqrt{19}}{5}$

Example 2   Determine the period of each function.

13.   3

14.   8

15.   12

16.   6

Differentiated Homework Options

Level	Assignment	Two-Day Option	
AL Basic	9–25, 37, 39–62	9–25 odd, 42–46	10–24 even, 37, 39–41, 47–62
OL Core	9–25 odd, 26–30, 31–39 odd, 40–62	9–25, 42–46	26–37, 39–41, 47–62
BL Advanced	26–59, (optional: 60–62)		

   The Math Forum

Have your students refer to the Math Forum's Ask Dr. Math to learn more about how degrees are related to radians. They can start their search at connectED. mcgraw-hill.com.

3 Practice

Formative Assessment
Use Exercises 1–8 to check for understanding.

Then use the chart at the bottom of this page to customize assignments for your students.

CCSS **Teaching the Mathematical Practices**

Structure Mathematically proficient students look closely to discern a pattern or structure.

Additional Answer

5b. Sample answer:

Teaching the Mathematical Practices

Sense-Making Mathematically proficient students start by explaining to themselves the meaning of a problem and looking for entry points to its solution. They analyze givens, constraints, relationships, and goals. They check their answers to problems using a different method, and they continually ask themselves, "Does this make sense?"

Multiple Representations
In Exercise 29, students draw a diagram and use a table of values and analysis to investigate the relationship between the terminal side of an angle and the slope of the line representing it.

Additional Answers

19a.

26b. Sample answer:

27b. Sample answer:

28. Sample answer:

Determine the period of each function.

17. 180°

18. 2π

Example 3

19. WEATHER In a city, the average high temperature for each month is shown in the table.

a. Sketch a graph of the function representing this situation. **See margin.**

b. Describe the period of the function. **12 mo or 1 yr**

Average High Temperatures			
Month	**Temperature (°F)**	**Month**	**Temperature (°F)**
Jan	36	July	85
Feb.	41	Aug.	84
Mar.	52	Sept.	78
Apr.	64	Oct.	66
May	74	Nov.	52
Jun.	82	Dec.	41

Source: The Weather Channel

Example 4

Find the exact value of each expression.

20. $\sin \frac{7\pi}{3}$ $\frac{\sqrt{3}}{2}$

21. $\cos(-60°)$ $\frac{1}{2}$

22. $\cos 450°$ **0**

23. $\sin \frac{11\pi}{4}$ $\frac{\sqrt{2}}{2}$

24. $\sin(-45°)$ $-\frac{\sqrt{2}}{2}$

25. $\cos 570°$ $-\frac{\sqrt{3}}{2}$

26. **CCSS SENSE-MAKING** In the engine at the right, the distance d from the piston to the center of the circle, called the *crankshaft*, is a function of the speed of the piston rod. Point R on the piston rod rotates 150 times per second.

a. Identify the period of the function as a fraction of a second. $\frac{1}{150}$

b. The shortest distance d is 0.5 inch, and the longest distance is 3.5 inches. Sketch a graph of the function. Let the horizontal axis represent the time t. Let the vertical axis represent the distance d. **See margin.**

27. TORNADOES A tornado siren makes 2.5 rotations per minute and the beam of sound has a radius of 1 mile. Ms. Miller's house is 1 mile from the siren. The distance of the sound beam from her house varies periodically as a function of time.

a. Identify the period of the function in seconds. **24 seconds**

b. Sketch a graph of the function. Let the horizontal axis represent the time t from 0 seconds to 60 seconds. Let the vertical axis represent the distance d the sound beam is from Ms. Miller's house at time t. **See margin.**

28. FERRIS WHEEL A Ferris wheel in China has a diameter of approximately 520 feet. The height of a compartment h is a function of time t. It takes about 30 seconds to make one complete revolution. Let the height at the center of the wheel represent the height at time 0. Sketch a graph of the function. **See margin.**

834 | Lesson 12-6 | Circular and Periodic Functions

30b.

40. Sample answer:

period: 2

29. **MULTIPLE REPRESENTATIONS** The terminal side of an angle in standard position intersects the unit circle at P, as shown in the figure.

a. Geometric Copy the figure. Draw lines representing 30°, 60°, 150°, 210°, and 315°.

b. Tabular Use a table of values to show the slope of each line to the nearest tenth.

c. Analytical What conclusions can you make about the relationship between the terminal side of the angle and the slope? Explain your reasoning. **29a–c. See Chapter 12 Answer Appendix.**

30. POGO STICK A person is jumping up and down on a pogo stick at a constant rate. The difference between his highest and lowest points is 2 feet. He jumps 50 times per minute.

a. Describe the independent variable and dependent variable of the periodic function that represents this situation. Then state the period of the function in seconds.

b. Sketch a graph of the jumper's change in height in relation to his starting point. Assume that his starting point is halfway between his highest and lowest points. Let the horizontal axis represent the time t in seconds. Let the vertical axis represent the height h. **See margin.**

30a. Sample answer: independent variable: time t in seconds; dependent variable: height h in feet; period: 1.2 seconds

Find the exact value of each expression.

31. $\cos 45° - \cos 30°$ $\dfrac{\sqrt{2}-\sqrt{3}}{2}$

32. $6(\sin 30°)(\sin 60°)$ $\dfrac{3\sqrt{3}}{2}$

33. $2 \sin \dfrac{4\pi}{3} - 3 \cos \dfrac{11\pi}{6}$ $-\dfrac{5\sqrt{3}}{2}$

34. $\cos\left(-\dfrac{2\pi}{3}\right) + \dfrac{1}{3} \sin 3\pi$ $-\dfrac{1}{2}$

35. $(\sin 45°)^2 + (\cos 45°)^2$ 1

36. $\dfrac{(\cos 30°)(\cos 150°)}{\sin 315°}$ $\dfrac{3\sqrt{2}}{4}$

37. Benita; Francis incorrectly wrote $\cos \dfrac{-\pi}{3} = -\cos \dfrac{\pi}{3}$.

H.O.T. Problems   Use Higher-Order Thinking Skills

37. **CRITIQUE** Francis and Benita are finding the exact value of $\cos \dfrac{-\pi}{3}$. Is either of them correct? Explain your reasoning.

39. Sometimes; the period of a sine curve could be $\dfrac{\pi}{2}$, which is not a multiple of π.

Francis	Benita
$\cos \dfrac{-\pi}{3} = -\cos \dfrac{\pi}{3}$ $= -0.5$	$\cos \dfrac{-\pi}{3} = \cos\left(-\dfrac{\pi}{3} + 2\pi\right)$ $= \cos \dfrac{5\pi}{3}$ $= 0.5$

41. The period of a periodic function is the horizontal distance of the part of the graph that is nonrepeating. Each nonrepeating part of the graph is one cycle.

38. CHALLENGE A ray has its endpoint at the origin of the coordinate plane, and point $P\left(\dfrac{1}{2}, -\dfrac{\sqrt{3}}{2}\right)$ lies on the ray. Find the angle θ formed by the positive x-axis and the ray. **−60°**

39. REASONING Is the period of a sine curve *sometimes*, *always*, or *never* a multiple of π? Justify your reasoning.

40. OPEN ENDED Draw the graph of a periodic function that has a maximum value of 10 and a minimum value of −10. Describe the period of the function. **See margin.**

41. WRITING IN MATH Explain how to determine the period of a periodic function from its graph. Include a description of a cycle.

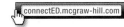

Teaching the Mathematical Practices

Critique Mathematically proficient students are also able to compare the effectiveness of two plausible arguments, distinguish correct logic or reasoning from that which is flawed, and—if there is a flaw in an argument—explain what it is.

📁 **Study Guide and Intervention**
AL OL ELL

📁 **Practice**
AL OL BL ELL

📁 **Word Problem Practice**
AL OL BL ELL

📁 **Enrichment**
OL BL

Name the Math Ask students to describe how they can find points on the unit circle in each of the quadrants and on each of the axes.

Additional Answer

42. Sample answer: Move the graph of $f(x)$ 4 units to the left and 3 units down to obtain the graph of $g(x)$.

Standardized Test Practice

42. SHORT RESPONSE Describe the translation of the graph of $f(x) = x^2$ to the graph of $g(x) = (x + 4)^2 - 3$. **See margin.**

43. The rate of population decline of Hampton Cove is modeled by $P(t) = 24{,}000e^{-0.0064t}$, where t is time in years from this year and 24,000 is the current population. In how many years will the population be 10,000? **C**

A 14   **B** 104   **C** 137   **D** 375

44. SAT/ACT If $d^2 + 8 = 21$, then $d^2 - 8 =$ **G**

   F 0   **H** 13   **K** 161

   G 5   **J** 31

45. STATISTICS If the average of three different positive integers is 65, what is the greatest possible value of one of the integers? **A**

   A 192   **B** 193   **C** 194   **D** 195

46. GRIDDED RESPONSE If $8xy + 3 = 3$, what is the value of xy? **0**

Spiral Review

Solve each triangle. Round side lengths to the nearest tenth and angle measures to the nearest degree. (Lesson 12-5)

47.

$A \approx 34°,\ C \approx 64°,\ c \approx 12.7$

48.
$a \approx 16.1,\ B \approx 49°,\ C \approx 21°$

49.
$B \approx 33°,\ C \approx 29°,\ c \approx 9.9$

Determine whether each triangle has *no* solution, *one* solution, or *two* solutions. Then solve the triangle. Round side lengths to the nearest tenth and angle measures to the nearest degree. (Lesson 12-4)

50. $A = 72°, a = 6, b = 11$
no solution

51. $A = 46°, a = 10, b = 8$

52. $A = 110°, a = 9, b = 5$

51. one solution; $B \approx 35°$, $C \approx 99°, c \approx 13.7$

52. one solution; $B \approx 31°$, $C \approx 39°, c \approx 6.0$

A binomial distribution has a 70% rate of success. There are 10 trials. (Lesson 11-4)

53. What is the probability that there will be 3 failures? **0.267**

54. What is the probability that there will be at least 7 successes? **0.6496**

55. What is the expected number of successes? **7**

56. GAMES The diagram shows the board for a game in which spheres are dropped down a chute. A pattern of nails and dividers causes the spheres to take various paths to the sections at the bottom. For each section, how many paths through the board lead to that section? (Lesson 10-6) **1, 4, 6, 4, 1**

57. SALARIES Phillip's current salary is $40,000 per year. His annual pay raise is always a percent of his salary at the time. What would his salary be if he got four consecutive 4% increases? (Lesson 10-2) **$46,794.34**

Find the exact solution(s) of each system of equations. (Lesson 9-7)

58. $y = x + 2$ **(2, 4), (−1, 1)**
$y = x^2$

59. $4x + y^2 = 20$ **(5, 0), (−4, ±6)**
$4x^2 + y^2 = 100$

Skills Review

Simplify each expression.

60. $\dfrac{240}{\left|1 - \frac{5}{4}\right|}$ **960**

61. $\dfrac{180}{\left|2 - \frac{1}{3}\right|}$ **108**

62. $\dfrac{90}{\left|2 - \frac{11}{4}\right|}$ **120**

DifferentiatedInstruction   OL   BL

Extension Show students the graphs of $y = \cos x$, $y = \cos 2x$, and $y = \cos 3x$. Ask them to state the period of each function. $360°, 180°, 120°$ Ask them to determine the period of the function $y = \cos kx$. $\dfrac{360°}{k}$

LESSON 12-7 Graphing Trigonometric Functions

Then	Now	Why?
● You examined periodic functions.	**1** Describe and graph the sine, cosine, and tangent functions. **2** Describe and graph other trigonometric functions.	● Visible light waves have different wavelengths or periods. Red has the longest wavelength and violet has the shortest wavelength.

 NewVocabulary
amplitude
frequency

 Common Core State Standards

Content Standards
F.IF.7.e Graph exponential and logarithmic functions, showing intercepts and end behavior, and trigonometric functions, showing period, midline, and amplitude.

F.TF.5 Choose trigonometric functions to model periodic phenomena with specified amplitude, frequency, and midline.

Mathematical Practices
1 Make sense of problems and persevere in solving them.

1 Sine, Cosine, and Tangent Functions Trigonometric functions can also be graphed on the coordinate plane. Recall that graphs of periodic functions have repeating patterns, or *cycles*. The horizontal length of each cycle is the *period*. The **amplitude** of the graph of a sine or cosine function equals half the difference between the maximum and minimum values of the function.

KeyConcept Sine and Cosine Functions

Parent Function	$y = \sin \theta$	$y = \cos \theta$
Graph		
Domain	{all real numbers}	{all real numbers}
Range	$\{y \mid -1 \leq y \leq 1\}$	$\{y \mid -1 \leq y \leq 1\}$
Amplitude	1	1
Period	360°	360°

As with other functions, trigonometric functions can be transformed. For the graphs of $y = a \sin b\theta$ and $y = a \cos b\theta$, the amplitude $= |a|$ and the period $= \frac{360°}{|b|}$.

Example 1 Find Amplitude and Period

Find the amplitude and period of $y = 4 \cos 3\theta$.

amplitude: $|a| = |4|$ or 4

period: $\frac{360°}{|b|} = \frac{360°}{|3|}$ or 120°

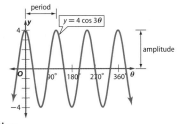

GuidedPractice

Find the amplitude and period of each function.

1A. $y = \cos \frac{1}{2}\theta$ amplitude: 1; period: 720°   **1B.** $y = 3 \sin 5\theta$ amplitude: 3; period: 72°

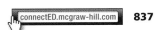

connectED.mcgraw-hill.com   **837**

1 Focus

VerticalAlignment

Before Lesson 12-7 Examine periodic functions.

Lesson 12-7 Describe and graph the sine, cosine, and tangent functions. Describe and graph other trigonometric functions.

After Lesson 12-7 Graph functions and interpret them in terms of their amplitude, frequency, period, and phase shift.

2 Teach

Scaffolding Questions
Have students read the **Why?** section of the lesson.

Ask:

- How are wavelengths measured in the diagram? In the diagram, wavelengths are measured between the highest points of two sequential waves.

- What is another way to measure a wavelength? The wavelengths could be measured as the distance between any two corresponding points of sequential waves.

- In the diagram, how many violet wavelengths are equivalent to one red wavelength? 3

Lesson 12-7 Resources

Resource	Approaching Level **AL**	On Level **OL**	Beyond Level **BL**	English Learners **ELL**
Teacher Edition	▪ Differentiated Instruction, p. 838	▪ Differentiated Instruction, pp. 838, 843	▪ Differentiated Instruction, p. 843	
Chapter Resource Masters	▪ Study Guide and Intervention, pp. 41–42 ▪ Skills Practice, p. 43 ▪ Practice, p. 44 ▪ Word Problem Practice, p. 45	▪ Study Guide and Intervention, pp. 41–42 ▪ Skills Practice, p. 43 ▪ Practice, p. 44 ▪ Word Problem Practice, p. 45 ▪ Enrichment, p. 46	▪ Practice, p. 44 ▪ Word Problem Practice, p. 45 ▪ Enrichment, p. 46	▪ Study Guide and Intervention, pp. 41–42 ▪ Skills Practice, p. 43 ▪ Practice, p. 44 ▪ Word Problem Practice, p. 45
Other	▪ 5-Minute Check 12-7 ▪ Study Notebook	▪ 5-Minute Check 12-7 ▪ Study Notebook	▪ 5-Minute Check 12-7 ▪ Study Notebook	▪ 5-Minute Check 12-7 ▪ Study Notebook

1 Sine, Cosine, and Tangent Functions

Example 1 shows how to find the amplitude and period of a trigonometric function. **Example 2** shows how to use the amplitude and period to graph a trigonometric function. **Example 3** shows how to write a trigonometric function to model a real-world situation. **Example 4** shows how to use the period and asymptotes of a tangent function to graph the function.

Formative Assessment

Use the Guided Practice exercises after each example to determine students' understanding of concepts.

Additional Examples

1 Find the amplitude and period of $y = \sin \frac{1}{3}\theta$. amplitude: 1; period: 1080°

2 Graph each function.

a. $y = \sin 3\theta$

b. $y = \frac{1}{2} \cos \theta$

Additional Examples also in Interactive Classroom PowerPoint® Presentations

IWB Interactive White Board READY

StudyTip

Periods In $y = a \sin b\theta$ and $y = a \cos b\theta$, b represents the number of cycles in 360°. In Example 1, the 3 in $y = 4 \cos 3\theta$ indicates that there are three cycles in 360°. So, there is one cycle in 120°.

StudyTip

Amplitude The graphs of $y = a \sin b\theta$ and $y = a \cos b\theta$ with amplitude of $|a|$ have maxima at $y = a$ and minima at $y = -a$.

Use the graphs of the parent functions to graph $y = a \sin b\theta$ and $y = a \cos b\theta$. Then use the amplitude and period to draw the appropriate sine and cosine curves. You can also use θ-intercepts to help you graph the functions.

The θ-intercepts of $y = a \sin b\theta$ and $y = a \cos b\theta$ in one cycle are as follows.

$y = a \sin b\theta$	$y = a \cos b\theta$
$(0, 0), \left(\frac{1}{2} \cdot \frac{360°}{b}, 0\right) \left(\frac{360°}{b}, 0\right)$	$\left(\frac{1}{4} \cdot \frac{360°}{b}, 0\right), \left(\frac{3}{4} \cdot \frac{360°}{b}, 0\right)$

Example 2 Graph Sine and Cosine Functions

Graph each function.

a. $y = 2 \sin \theta$

Find the amplitude, the period, and the x-intercepts: $a = 2$ and $b = 1$.

amplitude: $|a| = |2|$ or 2   →   The graph is stretched vertically so that the maximum value is 2 and the minimum value is -2.

period: $\frac{360°}{|b|} = \frac{360°}{|1|}$ or 360°   →   One cycle has a length of 360°.

x-intercepts: $(0, 0)$

$$\left(\frac{1}{2} \cdot \frac{360°}{b}, 0\right) = (180°, 0)$$

$$\left(\frac{360°}{b}, 0\right) = (360°, 0)$$

b. $y = \cos 4\theta$

amplitude: $|a| = |1|$ or 1

period: $\frac{360°}{|b|} = \frac{360°}{|4|}$ or 90°

x-intercepts: $\left(\frac{1}{4} \cdot \frac{360°}{b}, 0\right) = (22.5°, 0)$

$$\left(\frac{3}{4} \cdot \frac{360°}{b}, 0\right) = (67.5°, 0)$$

GuidedPractice   2A, 2B. See margin.

2A. $y = 3 \cos \theta$ **2B.** $y = \frac{1}{2} \sin 2\theta$

Trigonometric functions are useful for modeling real-world periodic motion such as electromagnetic waves or sound waves. Often these waves are described using *frequency*. **Frequency** is the number of cycles in a given unit of time.

The frequency of the graph of a function is the reciprocal of the period of the function. So, if the period of a function is $\frac{1}{100}$ second, then the frequency is 100 cycles per second.

DifferentiatedInstruction   AL   OL

Visual/Spatial Learners Have groups of students make posters showing sketches of the graphs of the six trigonometric functions. Encourage students to color-code the key features of all the graphs, such as period, amplitude, asymptotes, and so on.

Digital Vision/PunchStock

● Real-World Example 3 Model Periodic Situations

SOUND Sound that has a frequency below the human range is known as *infrasound*. Elephants can hear sounds in the infrasound range, with frequencies as low as 5 hertz (Hz), or 5 cycles per second.

a. Find the period of the function that models the sound waves.

There are 5 cycles per second, and the period is the time it takes for one cycle. So, the period is $\frac{1}{5}$ or 0.2 second.

Real-World Link
Elephants are able to hear sound coming from up to 5 miles away. Humans can hear sounds with frequencies between 20 Hz and 20,000 Hz.
Source: School for Champions

b. Let the amplitude equal 1 unit. Write a sine equation to represent the sound wave y as a function of time t. Then graph the equation.

$period = \frac{2\pi}{|b|}$ Write the relationship between the period and b.

$0.2 = \frac{2\pi}{|b|}$ Substitution

$0.2|b| = 2\pi$ Multiply each side by $|b|$.

$b = 10\pi$ Multiply each side by 5; b is positive.

$y = a \sin b\theta$ Write the general equation for the sine function.

$y = 1 \sin 10\pi t$ $a = 1$, $b = 10\pi$, and $\theta = t$

$y = \sin 10\pi t$ Simplify.

$y = \sin 10\pi t$

Study Tip

Amplitude and Period Note that the amplitude affects the graph along the vertical axis, and the period affects it along the horizontal axis.

▶ **Guided Practice**

3. SOUND Humans can hear sounds with frequencies as low as 20 hertz.

A. Find the period of the function. $\frac{1}{20}$ or 0.05 second

B. Let the amplitude equal 1 unit. Write a cosine equation to model the sound waves. Then graph the equation. $y = \cos 40\pi t$; See Chapter 12 Answer Appendix for graph.

The tangent function is one of the trigonometric functions whose graphs have asymptotes.

KeyConcept Tangent Functions

Parent Function	$y = \tan \theta$	Graph
Domain	$\{\theta \mid \theta \neq 90 + 180n,$ n is an integer$\}$	
Range	{all real numbers}	
Amplitude	undefined	
Period	$180°$	
θ intercepts in one cycle	$(0, 0), \left(\frac{1}{2} \cdot \frac{360°}{b}, 0\right), \left(\frac{360°}{b}, 0\right)$	

For the graph of $y = a \tan b\theta$, the period $= \frac{180°}{|b|}$, there is no amplitude, and the asymptotes are odd multiples of $\frac{180°}{2|b|}$.

connectED.mcgraw-hill.com   **839**

Additional Example

3 **SOUND** Humans can hear sounds with a frequency of 40 Hz.

a. Find the period of the function that models the sound waves. $\frac{1}{40}$ or 0.025 second

b. Let the amplitude equal 1 unit. Write a sine equation to represent the sound wave y as a function of time t. Then graph the equation. $y = \sin 80\pi t$

$y = \sin 80\pi t$

Focus on Mathematical Content

Amplitude and Period The graphs of the sine and cosine functions have an amplitude. The graphs of the other trigonometric functions do not have an amplitude because there is no maximum or minimum value. The period is the distance along the horizontal axis required for the graph to complete one cycle. The period is easy to determine from the graph.

Tips for New Teachers

Graphing You might wish to have students draw their own graphs of the sine, cosine, and tangent functions. Their graphs can be compared to those shown on pp. 838 and 840. This immediate feedback can be beneficial in helping students acquire the skills for graphing trigonometric functions, which are more complicated than most of the graphing students have done to this point.

Additional Answers (Guided Practice)

2A.

$y = 3 \cos \theta$

2B.

$y = \frac{1}{2} \sin 2\theta$

Additional Example

4 Find the period of $y = \tan \frac{1}{2}\theta$.

Then graph the function.
period: 360°

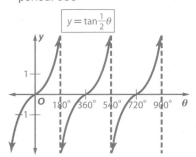

2 Graphs of Other Trigonometric Functions

Example 5 shows how to use the period of one of the three basic functions to graph another trigonometric function.

Additional Example

5 Find the period of $y = 3 \csc \theta$.

Then graph the function.
period: 360°

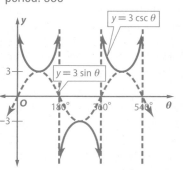

Teach with Tech

Document Camera Choose students to show and explain to the class how to sketch the graph of a trigonometric function. Be sure each student explains how to create a table of values, choose a scale for the graph, and plot the points to draw the graph.

StudyTip

Tangent The tangent function does not have an amplitude because it has no maximum or minimum values.

Example 4 Graph Tangent Functions

Find the period of $y = \tan 2\theta$. Then graph the function.

period: $\dfrac{180°}{|b|} = \dfrac{180°}{|2|}$ or 90°

asymptotes: $\dfrac{180°}{2|b|} = \dfrac{180°}{2|2|}$ or 45°

Sketch asymptotes at $-1 \cdot 45°$ or $-45°$, $1 \cdot 45°$ or $45°$, $3 \cdot 45°$ or $135°$, and so on.

Use $y = \tan \theta$, but draw one cycle every 90°.

▶ **Guided Practice**

4. Find the period of $y = \frac{1}{2} \tan \theta$. Then graph the function. **See margin.**

2 Graphs of Other Trigonometric Functions
The graphs of the cosecant, secant, and cotangent functions are related to the graphs of the sine, cosine, and tangent functions.

KeyConcept Cosecant, Secant, and Cotangent Functions			
Parent Function	$y = \csc \theta$	$y = \sec \theta$	$y = \cot \theta$
Graph			
Domain	$\{\theta \mid \theta \neq 180n,$ n is an integer$\}$	$\{\theta \mid \theta \neq 90 + 180n,$ n is an integer$\}$	$\{\theta \mid \theta \neq 180n,$ n is an integer$\}$
Range	$\{y \mid -1 > y$ or $y > 1\}$	$\{y \mid -1 > y$ or $y > 1\}$	{all real numbers}
Amplitude	undefined	undefined	undefined
Period	360°	360°	180°

StudyTip

Reciprocal Functions You can use the graphs of $y = \sin \theta$, $y = \cos \theta$, and $y = \tan \theta$ to graph the reciprocal functions, but these graphs are not part of the graphs of the cosecant, secant, and cotangent functions.

Example 5 Graph Other Trigonometric Functions

Find the period of $y = 2 \sec \theta$. Then graph the function.

Since $2 \sec \theta$ is a reciprocal of $2 \cos \theta$, the graphs have the same period, 360°. The vertical asymptotes occur at the points where $2 \cos \theta = 0$. So, the asymptotes are at $\theta = 90°$ and $\theta = 270°$.

Sketch $y = 2 \cos \theta$ and use it to graph $y = 2 \sec \theta$.

▶ **Guided Practice**

5. Find the period of $y = \csc 2\theta$. Then graph the function. **See margin.**

840 | Lesson 12-7 | Graphing Trigonometric Functions

Additional Answers (Guided Practice)

4. period: 180°

5. period: 180°

Examples 1–2 Find the amplitude and period of each function. Then graph the function. **1–4. See margin.**

1. $y = 4 \sin \theta$ 2. $y = \sin 3\theta$

3. $y = \cos 2\theta$ 4. $y = \frac{1}{2} \cos 3\theta$

Example 3 **5. SPIDERS** When an insect gets caught in a spider web, the web vibrates with a frequency of 14 hertz.

    a. Find the period of the function. $\frac{1}{14}$ **or about 0.07 second**

    b. Let the amplitude equal 1 unit. Write a sine equation to represent the vibration of the web y as a function of time t. Then graph the equation. **See margin.**

Examples 4–5 Find the period of each function. Then graph the function. **6–8. See Chapter 12 Answer Appendix.**

6. $y = 3 \tan \theta$ 7. $y = 2 \csc \theta$ 8. $y = \cot 2\theta$

Practice and Problem Solving   Extra Practice is on page R12.

Examples 1–2 Find the amplitude and period of each function. Then graph the function.

9. $y = 2 \cos \theta$ 10. $y = 3 \sin \theta$ 11. $y = \sin 2\theta$

12. $y = \cos 3\theta$ 13. $y = \cos \frac{1}{2}\theta$ 14. $y = \sin 4\theta$

15. $y = \frac{3}{4} \cos \theta$ 16. $y = \frac{3}{2} \sin \theta$ (17) $y = \frac{1}{2} \sin 2\theta$

18. $y = 4 \cos 2\theta$ 19. $y = 3 \cos 2\theta$ 20. $y = 5 \sin \frac{2}{3}\theta$

9–20. See Chapter 12 Answer Appendix.

Example 3 **21. CCSS REASONING** A boat on a lake bobs up and down with the waves. The difference between the lowest and highest points of the boat is 8 inches. The boat is at *equilibrium* when it is halfway between the lowest and highest points. Each cycle of the periodic motion lasts 3 seconds.

    a. Write an equation for the motion of the boat. Let h represent the height in inches and let t represent the time in seconds. Assume that the boat is at equilibrium at $t = 0$ seconds. $h = 4 \sin \frac{2}{3}\pi t$

    b. Draw a graph showing the height of the boat as a function of time. **See Chapter 12 Answer Appendix.**

22. ELECTRICITY The voltage supplied by an electrical outlet is a periodic function that *oscillates*, or goes up and down, between −165 volts and 165 volts with a frequency of 50 cycles per second.

    a. Write an equation for the voltage V as a function of time t. Assume that at $t = 0$ seconds, the current is 165 volts. $V = 165 \cos 100\pi t$

    b. Graph the function. **See Chapter 12 Answer Appendix.**

Examples 4–5 Find the period of each function. Then graph the function.

23–28. See Chapter 12 Answer Appendix.

23. $y = \tan \frac{1}{2}\theta$ 24. $y = 3 \sec \theta$ 25. $y = 2 \cot \theta$

26. $y = \csc \frac{1}{2}\theta$ 27. $y = 2 \tan \theta$ 28. $y = \sec \frac{1}{3}\theta$

Differentiated Homework **Options**

Level	Assignment	Two-Day Option	
AL Basic	9–28, 42–63	9–27 odd, 45–48	10–28 even, 42–44, 49–63
OL Core	9–27 odd, 29–31, 33–39 odd, 42–63	9–28, 45–48	29–40, 42–44, 49–63
BL Advanced	29–60, (optional: 61–63)		

---

3 Practice

Formative Assessment
Use Exercises 1–8 to check for understanding.

Use the chart at the bottom of this page to customize assignments for your students.

Additional Answers

1. amplitude: 4; period: 360°

2. amplitude: 1; period: 120°

3. amplitude: 1; period: 180°

4. amplitude: $\frac{1}{2}$; period: 120°

5b. $y = \sin 28\pi t$

Study Guide and Intervention
AL OL ELL

Practice
AL OL BL ELL

Word Problem Practice
AL OL BL ELL

B 29 **EARTHQUAKES** A seismic station detects an earthquake wave that has a frequency of 0.5 hertz and an amplitude of 1 meter. **b. See margin.**

a. Write an equation involving sine to represent the height of the wave h as a function of time t. Assume that the equilibrium point of the wave, $h = 0$, is halfway between the lowest and highest points. $h = \sin \pi t$

b. Graph the function. Then determine the height of the wave after 20.5 seconds.

30. **CCSS PERSEVERANCE** An object is attached to a spring as shown at the right. It oscillates according to the equation $y = 20 \cos \pi t$, where y is the distance in centimeters from its equilibrium position at time t.

30a. amplitude: 20 cm; frequency: 0.5 vibrations per second; period: 2 seconds

a. Describe the motion of the object by finding the following: the amplitude in centimeters, the frequency in vibrations per second, and the period in seconds.

b. Find the distance of the object from its equilibrium position at $t = \frac{1}{4}$ second. **about 14.1 cm**

c. The equation $v = (-20 \text{ cm})(\pi \text{ rad/s}) \cdot \sin (\pi \text{ rad/s} \cdot t)$ represents the velocity v of the object at time t. Find the velocity at $t = \frac{1}{4}$ second. **about −44.4 cm/s**

31. **PIANOS** A piano string vibrates at a frequency of 130 hertz.

C

a. Write and graph an equation using cosine to model the vibration of the string y as a function of time t. Let the amplitude equal 1 unit.

31a. $y = \cos 260\pi t$; See Chapter 12 Answer Appendix for graph.

b. Suppose the frequency of the vibration doubles. Do the amplitude and period increase, decrease, or remain the same? Explain.

31b. The amplitude remains the same. The period decreases because it is the reciprocal of the frequency.

Find the amplitude, if it exists, and period of each function. Then graph the function.

32. $y = 3 \sin \frac{2}{3}\theta$

33. $y = \frac{1}{2} \cos \frac{3}{4}\theta$

34. $y = 2 \tan \frac{1}{2}\theta$

35. $y = 2 \sec \frac{4}{5}\theta$

36. $y = 5 \csc 3\theta$

37. $y = 2 \cot 6\theta$

32–37. See Chapter 12 Answer Appendix.

Identify the period of the graph and write an equation for each function.

38.

$360°$; $y = \frac{3}{2} \cos \theta$

39.

$180°$; $y = 5 \sin 2\theta$

40.

$1800°$; $y = 2 \sin \frac{1}{5}\theta$

41. The domain of $y = a \cos \theta$ is the set of all real numbers. The domain of $y = a \sec \theta$ is the set of all real numbers except the values for which $\cos \theta = 0$. The range of $y = a \cos \theta$ is $-a \leq y \leq a$. The range of $y = a \sec \theta$ is $y \leq -a$ and $y \geq a$.

42. The graph of $y = \frac{1}{2} \sin \theta$ has an amplitude of $\frac{1}{2}$ and a period of $360°$. The graph of $y = \sin \frac{1}{2}\theta$ has an amplitude of 1 and a period of $720°$.

H.O.T. Problems — Use Higher-Order Thinking Skills

41. CHALLENGE Describe the domain and range of $y = a \cos \theta$ and $y = a \sec \theta$, where a is any positive real number.

42. REASONING Compare and contrast the graphs of $y = \frac{1}{2} \sin \theta$ and $y = \sin \frac{1}{2}\theta$.

43. OPEN ENDED Write a trigonometric function that has an amplitude of 3 and a period of $180°$. Then graph the function. **See margin.**

44. **WRITING IN MATH** How can you use the characteristics of a trigonometric function to sketch its graph? **See Chapter 12 Answer Appendix.**

Enrichment
OL BL

Additional Answer

29b.

$h = \sin \pi t$

Standardized Test Practice

45. SHORT RESPONSE Find the 100,001st term of the sequence. **700,013**

$$13, 20, 27, 34, 41, \ldots$$

46. STATISTICS You bowled five games and had the following scores: 143, 171, 167, 133, and 156. What was your average? **C**

    A 147   **B** 153   **C** 154   **D** 156

47. Your city had a population of 312,430 ten years ago. If its current population is 418,270, by what percentage has it grown over the past 10 years? **G**

    F 25%   **G** 34%   **H** 66%   **J** 75%

48. SAT/ACT If $h + 4 = b - 3$, then $(h - 2)^2 = $ **C**

    A $h^2 + 4$ **D** $b^2 - 14b + 49$

    B $b^2 - 6b + 3$ **E** $b^2 - 10b + 25$

    C $b^2 - 18b + 81$

Spiral Review

Find the exact value of each expression. (Lesson 12-6)

49. $\cos 120° - \sin 30°$ -1

50. $3(\sin 45°)(\sin 60°)$ $\dfrac{3\sqrt{6}}{4}$

51. $4 \sin \dfrac{4\pi}{3} - 2 \cos \dfrac{\pi}{6}$ $-3\sqrt{3}$

Solve each triangle. Round side lengths to the nearest tenth and angle measures to the nearest degree. (Lesson 12-5)

52.

$B \approx 17°$,
$A \approx 139°$,
$c \approx 7.2$

53.

$R \approx 140°$,
$S \approx 9°$,
$q \approx 48.6$

54.

$F \approx 49°$,
$G \approx 62°$,
$H \approx 69°$

A binomial distribution has a 40% rate of success. There are 12 trials. (Lesson 11-4)

55. What is the probability that there will be exactly 5 failures? **10.1%**

56. What is the probability that there will be at least 8 successes? **5.7%**

57. What is the expected number of successes? **5**

58. BANKING Rita has deposited $1000 in a bank account. At the end of each year, the bank posts interest to her account in the amount of 3% of the balance, but then takes out a $10 annual fee. (Lesson 10-6)

    a. Let b_0 be the amount Rita deposited. Write a recursive equation for the balance b_n in her account at the end of n years. $b_n = 1.03 b_{n-1} - 10$

    b. Find the balance in the account after four years.   **$1083.67**

Write an equation for an ellipse that satisfies each set of conditions. (Lesson 9-4)

59. center at $(6, 3)$, focus at $(2, 3)$, co-vertex at $(6, 1)$ $\dfrac{(x-6)^2}{20} + \dfrac{(y-3)^2}{4} = 1$

60. foci at $(2, 1)$ and $(2, 13)$, co-vertex at $(5, 7)$ $\dfrac{(x-2)^2}{9} + \dfrac{(y-7)^2}{45} = 1$

Skills Review

Graph each function. **61–63. See margin.**

61. $y = 2(x - 3)^2 - 4$

62. $y = \dfrac{1}{3}(x + 5)^2 + 2$

63. $y = -3(x + 6)^2 + 7$

DifferentiatedInstruction   OL   BL

Extension Ask students to graph $y = \sin(\theta - \pi)$, $y = \sin\left(\theta - \dfrac{\pi}{2}\right)$, and $y = \sin\left(\theta - \dfrac{\pi}{4}\right)$. Ask them to describe the effect of the term being subtracted from θ. The value being subtracted from θ determines how far to the right the graph of $y = \sin \theta$ is translated.

---

4 Assess

Ticket Out the Door Have students write a function of the form $y = a \sin b\theta$ or $y = a \cos b\theta$. Then have them state the amplitude and period of the graph.

Formative Assessment
Check student understanding of concepts in Lessons 12-6 and 12-7.

Quiz 3, p. 62

Additional Answers

43. Sample answer: $y = 3 \sin 2\theta$

61.

62.

63.

1 Focus

Objective Use a graphing calculator to explore transformations of the graphs of trigonometric functions.

Materials for Each Student
- TI-83/84 Plus or other graphing calculator

Teaching Tip
To set the calculator for degrees, press MODE and move the cursor to highlight **DEGREE** and press ENTER. Also, be sure to have students clear the Y= lists before beginning Exercise 1.

2 Teach

Working in Cooperative Groups
Have students work in pairs, mixing abilities, to complete the Activities and Exercises 1 and 2.

Ask:
- How do parentheses change the meanings of the functions in Steps 1 and 2? In Step 1, a number is added to the value of sin θ, while in Step 2, a number is added to θ before finding the sine.

- How do the parentheses change the graphs in Steps 1 and 2? The graphs in Step 1 "move" up or down, while the graphs in Step 2 "move" right or left.

Practice Have students complete Exercises 3–7.

12-8 Graphing Technology Lab
Trigonometric Graphs

You can use a TI-83/84 Plus graphing calculator to explore transformations of the graphs of trigonometric functions.

 Common Core State Standards
Content Standards
F.BF.3 Identify the effect on the graph of replacing $f(x)$ by $f(x) + k$, $k\,f(x)$, $f(kx)$, and $f(x + k)$ for specific values of k (both positive and negative); find the value of k given the graphs. Experiment with cases and illustrate an explanation of the effects on the graph using technology.
Mathematical Practices
5 Use appropriate tools strategically.

Activity 1 k in $y = \sin \theta + k$

Graph $y = \sin \theta$, $y = \sin \theta + 2$, and $y = \sin \theta - 3$ on the same coordinate plane. Describe any similarities and differences among the graphs.

Set the viewing window to match the window shown at the right. Let **Y1** $= \sin \theta$, **Y2** $= \sin \theta + 2$, and **Y3** $= \sin \theta - 3$.

KEYSTROKES: Y= SIN X,T,θ,n) ENTER
SIN X,T,θ,n) + 2 ENTER
SIN X,T,θ,n) – 3 GRAPH

[−360, 360] scl: 90 by [−5, 5] scl: 1

The graphs have the same shape, but different vertical positions.

Activity 2 h in $y = \sin (\theta - h)$

Graph $y = \sin \theta$, $y = \sin (\theta + 45°)$, and $y = \sin (\theta - 90°)$ on the same coordinate plane. Describe any similarities and differences among the graphs.

Let **Y1** $= \sin \theta$, **Y2** $= \sin (\theta + 45)$, and **Y3** $= \sin (\theta - 90)$.
Be sure to clear the entries from Activity 1.

KEYSTROKES: Y= SIN X,T,θ,n) ENTER
SIN X,T,θ,n + 45) ENTER
SIN X,T,θ,n – 90) GRAPH

[−360, 360] scl: 90 by [−5, 5] scl: 1

The graphs have the same shape, but different horizontal positions.

Model and Analyze

Repeat the activities for the cosine and tangent functions.

2. Sample answer: Adding a constant to a trigonometric function translates the graph of the function vertically.

1. What are the domain and range of the functions in Activities 1 and 2? **See Chapter 12 Answer Appendix.**

2. What is the effect of adding a constant to a trigonometric function?

3. What is the effect of adding a constant to θ in a trigonometric function? **Sample answer: Adding a constant to θ translates the graph of the function horizontally.**

Repeat the activities for each of the following. Describe the relationship between each pair of graphs.

4–7. See Chapter 12 Answer Appendix.

4. $y = \sin \theta + 4$
   $y = \sin (2\theta) + 4$

5. $y = \cos \left(\frac{1}{2}\theta\right)$
   $y = \cos \frac{1}{2}(\theta + 45°)$

6. $y = 2 \sin \theta$
   $y = 2 \sin \theta - 1$

7. $y = \cos \theta - 3$
   $y = \cos (\theta - 90°) - 3$

8. Write a general equation for the sine, cosine, and tangent functions after changes in amplitude a, period b, horizontal position h, and vertical position k. $y = a \sin (b\theta - h) + k$

---

3 Assess

Formative Assessment
Use Exercises 1 and 2 to assess whether students can describe the effect of adding a constant to a trigonometric function or to the value of θ in a trigonometric function.

From Concrete to Abstract
Ask students to describe how the value of n affects the graphs of $y = \sin \theta + n$ and $y = \sin(\theta + n)$ if n is a positive rational number. The graph of $y = \sin \theta + n$ is n units above $y = \sin \theta$; the graph of $y = \sin(\theta + n)$ is n units left of $y = \sin \theta$.

LESSON 12-8 Translations of Trigonometric Graphs

:: Then	:: Now	:: Why?
● You translated exponential functions.	**1** Graph horizontal translations of trigonometric graphs and find phase shifts. **2** Graph vertical translations of trigonometric graphs.	● The graphs at the right represent the waves in a bay during high and low tides. Notice that the shape of the waves does not change.

NewVocabulary
phase shift
vertical shift
midline

Common Core State Standards

Content Standards
F.IF.7.e Graph exponential and logarithmic functions, showing intercepts and end behavior, and trigonometric functions, showing period, midline, and amplitude.

F.BF.3 Identify the effect on the graph of replacing $f(x)$ by $f(x) + k$, $k\,f(x)$, $f(kx)$, and $f(x + k)$ for specific values of k (both positive and negative); find the value of k given the graphs. Experiment with cases and illustrate an explanation of the effects on the graph using technology.

Mathematical Practices
4 Model with mathematics.

1 Horizontal Translations Recall that a *translation* occurs when a figure is moved from one location to another on the coordinate plane without changing its orientation. A horizontal translation of a periodic function is called a **phase shift**.

KeyConcept Phase Shift

Words The phase shift of the functions $y = a \sin b(\theta - h)$, $y = a \cos b(\theta - h)$, and $y = a \tan b(\theta - h)$ is h, where $b > 0$.

Models

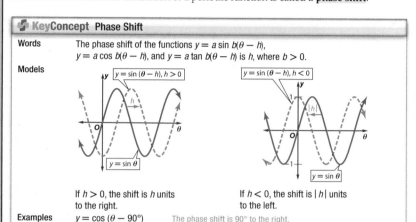

If $h > 0$, the shift is h units to the right.

If $h < 0$, the shift is $|h|$ units to the left.

Examples $y = \cos(\theta - 90°)$ The phase shift is 90° to the right.
$y = \tan(\theta + 30°)$ The phase shift is 30° to the left.

The secant, cosecant, and cotangent can be graphed using the same rules.

Example 1 Graph Phase Shifts

State the amplitude, period, and phase shift for $y = \sin(\theta - 90°)$. Then graph the function.

amplitude: $a = 1$

period: $\dfrac{360°}{|b|} = \dfrac{360°}{1}$ or $360°$

phase shift: $h = 90°$

Graph $y = \sin \theta$ shifted 90° to the right.

GuidedPractice

1. State the amplitude, period, and phase shift for $y = 2\cos(\theta + 45°)$. Then graph the function. **2; 360°; $h = -45°$; See Chapter 12 Answer Appendix for graph.**

845

1 Focus

VerticalAlignment

Before Lesson 12-8 Translate exponential functions.

Lesson 12-8 Graph horizontal translations of trigonometric graphs and find phase shifts. Graph vertical translations of trigonometric graphs.

After Lesson 12-8 Use trigonometry in a variety of applications and word problems.

2 Teach

Scaffolding Questions
Have students read the **Why?** section of the lesson.

Ask:

■ Compare the waves for high and low tides on their periods and amplitudes. The two waves have the same period and amplitude.

■ What is represented by the horizontal axis? The horizontal axis represents time measured in seconds.

■ What is represented by the vertical axis? The vertical axis shows the height above the bottom of the bay, measured in inches.

■ According to the diagram, is the greatest level at low tide greater than the lowest level at high tide? no

Lesson 12-8 Resources

Resource	Approaching Level **AL**	On Level **OL**	Beyond Level **BL**	English Learners **ELL**
Teacher Edition	■ Differentiated Instruction, p. 846	■ Differentiated Instruction, p. 852	■ Differentiated Instruction, p. 852	
Chapter Resource Masters	■ Study Guide and Intervention, pp. 47–48 ■ Skills Practice, p. 49 ■ Practice, p. 50 ■ Word Problem Practice, p. 51	■ Study Guide and Intervention, pp. 47–48 ■ Skills Practice, p. 49 ■ Practice, p. 50 ■ Word Problem Practice, p. 51 ■ Enrichment, p. 52	■ Practice, p. 50 ■ Word Problem Practice, p. 51 ■ Enrichment, p. 52	■ Study Guide and Intervention, pp. 47–48 ■ Skills Practice, p. 49 ■ Practice, p. 50 ■ Word Problem Practice, p. 51
Other	■ 5-Minute Check 12-8 ■ Study Notebook	■ 5-Minute Check 12-8 ■ Study Notebook	■ 5-Minute Check 12-8 ■ Study Notebook	■ 5-Minute Check 12-8 ■ Study Notebook

1 Horizontal Translations

Example 1 shows how to graph phase shifts of trigonometric functions.

Formative Assessment

Use the Guided Practice exercises after each example to determine students' understanding of concepts.

Additional Example

1 State the amplitude, period, and phase shift for $y = 2 \sin(\theta + 20°)$. Then graph the function.

amplitude: 2; period: 360°; phase shift: 20° left

Additional Examples also in Interactive Classroom PowerPoint® Presentations

IWB Interactive White Board READY

2 Vertical Translations

Example 2 shows how to graph vertical translations of trigonometric functions. **Example 3** shows how to graph a function that combines a horizontal and a vertical translation of a trigonometric function. **Example 4** shows how to solve a real-world problem using translations.

Additional Answer (Guided Practice)

2. no amplitude; 180°; $k = 3$; $y = 3$

---

2 Vertical Translations

Recall that the graph of $y = x^2 + 5$ is the graph of the parent function $y = x^2$ shifted up 5 units. Similarly, graphs of trigonometric functions can be translated vertically through a **vertical shift**.

StudyTip

Notation Note that $\sin(\theta + x) \neq \sin\theta + x$. The first expression indicates a phase shift. The second expression indicates a vertical shift.

KeyConcept Vertical Shift

Words The vertical shift of the functions $y = a\sin b\theta + k$, $y = a\cos b\theta + k$, and $y = a\tan b\theta + k$ is k.

Models

If $k > 0$, the shift is k units up.   If $k < 0$, the shift is $|k|$ units down.

Examples $y = \sin\theta + 4$ The vertical shift is 4 units up.
$y = \tan\theta - 3$ The vertical shift is 3 units down.

The secant, cosecant, and cotangent can be graphed using the same rules.

When a trigonometric function is shifted vertically k units, the line $y = k$ is the new horizontal axis about which the graph oscillates. This line is called the **midline**, and it can be used to help draw vertical translations.

StudyTip

Using Color It may be helpful to first graph the parent function in one color. Next, apply the vertical shift and graph the function in another color. Then apply the change in amplitude and graph the function in the final color.

Example 2 Graph Vertical Translations

State the amplitude, period, vertical shift, and equation of the midline for $y = \frac{1}{2}\cos\theta - 2$. Then graph the function.

amplitude: $|a| = \frac{1}{2}$

period: $\frac{2\pi}{|b|} = \frac{2\pi}{|1|}$ or 2π

vertical shift: $k = -2$

midline: $y = -2$

To graph $y = \frac{1}{2}\cos\theta - 2$, first draw the midline. Then use it to graph $y = \frac{1}{2}\cos\theta$ shifted 2 units down.

GuidedPractice

2. State the amplitude, period, vertical shift, and equation of the midline for $y = \tan\theta + 3$. Then graph the function. **See margin.**

DifferentiatedInstruction AL

If students struggle with translations of trigonometric graphs,

Then make coordinate axes with masking tape on the classroom floor. Give students at least 15 feet of rope and have them stand along the *x*-axis, positioning the rope to model the graph of $y = \sin x$. As you call out equations of functions with graphs being horizontal phase shifts of the graph of $y = \sin x$, students can step left or right to model the translated graph. Similarly, call out functions with graphs being vertical shifts of the graph $y = \sin x$.

You can use the following steps to graph trigonometric functions involving phase shifts and vertical shifts.

ConceptSummary Graph Trigonometric Functions

$$y = a \sin b(\theta - h) + k$$

amplitude ↓ | period ↓ | phase shift ↑ | vertical shift ↑

Step 1 Determine the vertical shift, and graph the midline.

Step 2 Determine the amplitude, if it exists. Use dashed lines to indicate the maximum and minimum values of the function.

Step 3 Determine the period of the function, and graph the appropriate function.

Step 4 Determine the phase shift, and translate the graph accordingly.

Example 3 Graph Transformations

State the amplitude, period, phase shift, and vertical shift for $y = 3 \sin \frac{2}{3}(\theta - \pi) + 4$. Then graph the function.

amplitude: $|a| = 3$

period: $\frac{2\pi}{|b|} = \frac{2\pi}{\left|\frac{2}{3}\right|}$ or 3π — The period indicates that the graph will be stretched.

phase shift: $h = \pi$ — The graph will shift π to the right.

vertical shift: $k = 4$ — The graph will shift 4 units up.

midline: $y = 4$ — The graph will oscillate around the line $y = 4$.

Step 1 Graph the midline.

Step 2 Since the amplitude is 3, draw dashed lines 3 units above and 3 units below the midline.

Step 3 Graph $y = 3 \sin \frac{2}{3}\theta + 4$ using the midline as a reference.

Step 4 Shift the graph π units to the right.

CHECK You can check the accuracy of your transformation by evaluating the function for various values of θ and confirming their location on the graph.

StudyTip

Verifying a Graph After drawing the graph of a trigonometric function, select values of θ and evaluate them in the equation to verify your graph.

GuidedPractice

3. State the amplitude, period, phase shift, and vertical shift for $y = 2 \cos \frac{1}{2}\left(\theta + \frac{\pi}{2}\right) - 2$. Then graph the function. **See margin.**

Additional Examples

2 State the amplitude, period, vertical shift, and equation of the midline for $y = \frac{1}{2}\cos \theta + 3$. Then graph the function.

amplitude: $\frac{1}{2}$; period: 360° or 2π; vertical shift: 3 units up; midline: $y = 3$

3 State the amplitude, period, phase shift, and vertical shift for $y = 3 \sin \left[2\left(\theta - \frac{\pi}{2}\right)\right] + 4$. Then graph the function.

amplitude: 3; period: π; phase shift: $\frac{\pi}{2}$ to the right; vertical shift: 4 units up

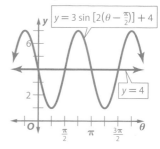

Additional Answer (Guided Practice)

3. 2; 4π; $h = -\frac{\pi}{2}$; $k = -2$

Tips for New Teachers

Reading Math Make sure students are clear about the difference between the terms *vertical* and *horizontal*. Students can use the words *vertigo* (a sense of dizziness some people experience when looking down from a great height) and *horizon* to help link vertical and horizontal to their relative direction.

Teach with Tech

Interactive Whiteboard Show a coordinate grid on the board. Draw a sine or cosine graph on the grid and have students find the equation of the graph. Drag the graph up or down along the y-axis and left or right along the x-axis. After each time you move the graph, have students find its equation. Discuss how the equations are similar and different.

Additional Example

4 WAVE POOL In Example 4, suppose the height of water oscillates between a maximum of 10 feet and a minimum of 6 feet, and the wave generator pumps 3 waves per minute. Write a sine function that represents the height of the water at time *t* seconds. Then graph the function.

$$h = 2 \sin \frac{\pi}{10} t + 8$$

$$h = 2 \sin \frac{\pi}{10}t + 8$$

Follow-up

Students have explored transformations of trigonometric functions.

Ask:

- How are transformations of trigonometric functions similar to transformations of other functions you have studied? Sample answer: Adding or subtracting to any function translates the graph; multiplying a function dilates the graph; multiplying a function by a negative number reflects the graph.

Additional Answer (Guided Practice)

4. $h = 4 \cos \frac{\pi}{6} t + 10$

The sine wave occurs often in physics, signal processing, music, electrical engineering, and many other fields.

● Real-World Example 4 Represent Periodic Functions

WAVE POOL The height of water in a wave pool oscillates between a maximum of 13 feet and a minimum of 5 feet. The wave generator pumps 6 waves per minute. Write a sine function that represents the height of the water at time *t* seconds. Then graph the function.

Step 1 Write the equation for the midline, and determine the vertical shift.

$$y = \frac{13 + 5}{2} \text{ or } 9$$ The midline lies halfway between the maximum and minimum values.

Since the midline is $y = 9$, the vertical shift is $k = 9$.

Step 2 Find the amplitude.

$$|a| = |13 - 9| \text{ or } 4$$ Find the difference between the midline value and the maximum value.

So, $a = 4$.

Step 3 Find the period.

Since there are 6 waves per minute, there is 1 wave every 10 seconds. So, the period is 10 seconds.

$$10 = \frac{2\pi}{|b|}$$ period $= \frac{2\pi}{|b|}$

$$|b| = \frac{2\pi}{10}$$ Solve for $|b|$.

$$b = \pm\frac{\pi}{5}$$ Simplify.

WatchOut!

Parent Functions Often the graph of a trigonometric function can be represented by more than one equation. For example, the graphs of $y = \cos \theta$ and $y = \sin(\theta + 90°)$ are the same.

Step 4 Write an equation for the function.

$$h = a \sin b(t - h) + k$$ Write the equation for sine relating height *h* and time *t*.

$$= 4 \sin \frac{\pi}{5}(t - 0) + 9$$ Substitution: $a = 4$, $b = \frac{\pi}{5}$, $h = 0$, $k = 9$

$$= 4 \sin \frac{\pi}{5}t + 9$$ Simplify.

Then graph the function.

$h = 4 \sin \frac{\pi}{5}t + 9$

▶ **Guided Practice**

4. WAVE POOL The height of water in a wave pool oscillates between a maximum of 14 feet and a minimum of 6 feet. The wave generator pumps 5 waves per minute. Write a cosine function that represents the height of water at time *t* seconds. Then graph the function. **See margin.**

Real-WorldLink

In some wave pools, surfers can ride waves up to 70 meters.

Source: Orlando Wave Pool

Additional Answers

1. 1; 360°; $h = 180°$

$y = \sin(\theta - 180°)$

2. no amplitude; 180°; $h = \frac{\pi}{4}$

$y = \tan\left(\theta - \frac{\pi}{4}\right)$

Stephen Frink/Photographer's Choice RF/Getty Images

Check Your Understanding

 = Step-by-Step Solutions begin on page R14.

Example 1 State the amplitude, period, and phase shift for each function. Then graph the function. **1–4. See margin.**

1. $y = \sin(\theta - 180°)$

2. $y = \tan\left(\theta - \frac{\pi}{4}\right)$

3. $y = \sin\left(\theta - \frac{\pi}{2}\right)$

4. $y = \frac{1}{2}\cos(\theta + 90°)$

Example 2 State the amplitude, period, vertical shift, and equation of the midline for each function. Then graph the function. **5–8. See margin.**

5. $y = \cos\theta + 4$

6. $y = \sin\theta - 2$

7. $y = \frac{1}{2}\tan\theta + 1$

8. $y = \sec\theta - 5$

Example 3 **CCSS REGULARITY** State the amplitude, period, phase shift, and vertical shift for each function. Then graph the function. **9–12. See Chapter 12 Answer Appendix.**

9. $y = 2\sin(\theta + 45°) + 1$

10. $y = \cos 3(\theta - \pi) - 4$

11. $y = \frac{1}{4}\tan 2(\theta + 30°) + 3$

12. $y = 4\sin\frac{1}{2}\left(\theta - \frac{\pi}{2}\right) + 5$

Example 4 **13. EXERCISE** While doing some moderate physical activity, a person's blood pressure oscillates between a maximum of 130 and a minimum of 90. The person's heart rate is 90 beats per minute. Write a sine function that represents the person's blood pressure P at time t seconds. Then graph the function. **See Chapter 12 Answer Appendix.**

Practice and Problem Solving

Extra Practice is on page R12.

Example 1 State the amplitude, period, and phase shift for each function. Then graph the function. **14–19. See Chapter 12 Answer Appendix.**

14. $y = \cos(\theta + 180°)$

15. $y = \tan(\theta - 90°)$

16. $y = \sin(\theta + \pi)$

17. $y = 2\sin\left(\theta + \frac{\pi}{2}\right)$

18. $y = \tan\frac{1}{2}(\theta + 30°)$

19. $y = 3\cos\left(\theta - \frac{\pi}{3}\right)$

Example 2 State the amplitude, period, vertical shift, and equation of the midline for each function. Then graph the function. **20–25. See Chapter 12 Answer Appendix.**

20. $y = \cos\theta + 3$

21. $y = \tan\theta - 1$

22. $y = \tan\theta + \frac{1}{2}$

23 $y = 2\cos\theta - 5$

24. $y = 2\sin\theta - 4$

25. $y = \frac{1}{3}\sin\theta + 7$

Example 3 State the amplitude, period, phase shift, and vertical shift for each function. Then graph the function. **26–33. See Chapter 12 Answer Appendix.**

26. $y = 4\sin(\theta - 60°) - 1$

27. $y = \cos\frac{1}{2}(\theta - 90°) + 2$

28. $y = \tan(\theta + 30°) - 2$

29. $y = 2\tan 2\left(\theta + \frac{\pi}{4}\right) - 5$

30. $y = \frac{1}{2}\sin\left(\theta - \frac{\pi}{2}\right) + 4$

31. $y = \cos 3(\theta - 45°) + \frac{1}{2}$

32. $y = 3 + 5\sin 2(\theta - \pi)$

33. $y = -2 + 3\sin\frac{1}{3}\left(\theta - \frac{\pi}{2}\right)$

Example 4 **34. TIDES** The height of the water in a harbor rose to a maximum height of 15 feet at 6:00 P.M. and then dropped to a minimum level of 3 feet by 3:00 A.M. The water level can be modeled by the sine function. Write an equation that represents the height h of the water t hours after noon on the first day. $h = 9 + 6\sin\left[\frac{\pi}{9}(t - 1.5)\right]$

 connectED.mcgraw-hill.com **849**

7. no amplitude; 180°; $k = 1$; $y = 1$

8. $k = -5$; $y = -5$; no amplitude; 360°

3 Practice

Formative Assessment
Use Exercises 1–13 to check for understanding.

Use the chart at the bottom of the next page to customize assignments for your students.

Additional Answers

3. $1; 2\pi; h = \frac{\pi}{2}$

4. $\frac{1}{2}$; 360°; $h = -90°$

5. 1; 360°; $k = 4$; $y = 4$

6. 1; 360°; $k = -2$; $y = -2$

Additional Answers

35.

$d = 1.8 \sin \frac{3\pi}{4}t + 12$

min: 10.2 ft; max: 13.8 ft

36. $h = 260 \cos \frac{\pi}{15}t + 265$

$h = 260 \cos \frac{\pi}{15}t + 265$

41b.

$d = 9 \sin \frac{2\pi}{7}t + 46$

42c.

$y = 5 \cos \frac{\pi}{2}d + 45$

60. Sometimes; if the function is shifted vertically, then you also need to know the value of the midline. The maximum value is the value of the midline plus the amplitude. The minimum value is the midline value minus the amplitude.

---

35. LAKES A buoy marking the swimming area in a lake oscillates each time a speed boat goes by. Its distance d in feet from the bottom of the lake is given by

$d = 1.8 \sin \frac{3\pi}{4}t + 12$, where t is the time in seconds. Graph the function. Describe the minimum and maximum distances of the buoy from the bottom of the lake when a boat passes by. **See margin.**

36. FERRIS WHEEL Suppose a Ferris wheel has a diameter of approximately 520 feet and makes one complete revolution in 30 minutes. Suppose the lowest car on the Ferris wheel is 5 feet from the ground. Let the height at the top of the wheel represent the height at time 0. Write an equation for the height of a car h as a function of time t minutes. Then graph the function. **See margin.**

Write an equation for each translation.

37. $y = \sin x$, 4 units to the right and 3 units up $y = \sin (x - 4) + 3$

38. $y = \cos x$, 5 units to the left and 2 units down $y = \cos (x + 5) - 2$

39. $y = \tan x$, π units to the right and 2.5 units up $y = \tan (x - \pi) + 2.5$

40a. At 1.25 seconds, the height of the rope is 68 inches; at 2.75 seconds, the height of the rope is 2 inches.

 40. JUMP ROPE The graph at the right approximates the height of a jump rope h in inches as a function of time t in seconds. A maximum point on the graph is (1.25, 68), and a minimum point is (2.75, 2).

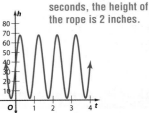

a. Describe what the maximum and minimum points mean in the context of the situation.

b. What is the equation for the midline, the amplitude, and the period of the function? $y = 35; 33, 1$

c. Write an equation for the function. $h = 33 \sin 2\pi t + 35$

41. CAROUSEL A horse on a carousel goes up and down 3 times as the carousel makes one complete rotation. The maximum height of the horse is 55 inches, and the minimum height is 37 inches. The carousel rotates once every 21 seconds. Assume that the horse starts and stops at its median height. **a.** $h = 9 \sin \frac{2\pi}{7}t + 46$

a. Write an equation to represent the height of the horse h as a function of time t seconds.

b. Graph the function. **See margin.**

c. Use your graph to estimate the height of the horse after 8 seconds. Then use a calculator to find the height to the nearest tenth. **Sample answer: 8 s; 53.0 in.**

42. CCSS REASONING During one month, the outside temperature fluctuates between 40°F and 50°F. A cosine curve approximates the change in temperature, with a high of 50°F being reached every four days.

a. Describe the amplitude, period, and midline of the function that approximates the temperature y on day d. **5; 4; $y = 45$**

b. Write a cosine function to estimate the temperature y on day d. $y = 5 \cos \frac{\pi}{2}d + 45$

c. Sketch a graph of the function. **See margin.**

d. Estimate the temperature on the 7th day of the month. **about 45°F**

Find a coordinate that represents a maximum for each graph.

43. $y = -2 \cos \left(x - \frac{\pi}{2}\right)$ $\left(\frac{3\pi}{2}, 2\right)$

44. $y = 4 \sin \left(x + \frac{\pi}{3}\right)$ $\left(\frac{\pi}{6}, 4\right)$

45. $y = 3 \tan \left(x + \frac{\pi}{2}\right) + 2$ **no maximum values**

46. $y = -3 \sin \left(x - \frac{\pi}{4}\right) - 4$ $\left(\frac{7\pi}{4}, -1\right)$

Differentiated Homework Options

Level	Assignment	Two-Day Option	
AL Basic	14–39, 61–85	15–39 odd, 65–68	16–38 even, 61–64, 69–85
OL Core	15–39 odd, 40–42, 43–59 odd, 61–85	15–39, 65–68	40–59, 61–64, 69–85
BL Advanced	40–79, (optional: 80–85)		

Compare each pair of graphs.

47. $y = -\cos 3\theta$ and $y = \sin 3(\theta - 90°)$ The graphs are reflections of each other over the x-axis.

48. $y = 2 + 0.5 \tan \theta$ and $y = 2 + 0.5 \tan(\theta + \pi)$ The graphs are identical.

49. $y = 2\sin\left(\theta - \frac{\pi}{6}\right)$ and $y = -2\sin\left(\theta + \frac{5\pi}{6}\right)$ The graphs are identical.

C Identify the period of each function. Then write an equation for the graph using the given trigonometric function. **50.** 360°; Sample answer: $y = \sin \theta - 5$

51. 360°; Sample answer: $y = 2\cos(\theta + 90°)$
52. 360°; Sample answer: $y = 4\cos\theta + 1$
53. 180°; Sample answer: $y = \sin 2(\theta - 45°) + 3$

50. sine

51. cosine

52. cosine

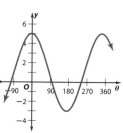

61. The graph of $y = 3\sin 2\theta + 1$ has an amplitude of 3 rather than an amplitude of 1. It is shifted up 1 unit from the parent graph and is compressed so that it has a period of 180°.

53 sine

State the period, phase shift, and vertical shift. Then graph the function.

54. $y = \csc(\theta + \pi)$

55. $y = \cot\theta + 6$

56. $y = \cot\left(\theta - \frac{\pi}{6}\right) - 2$

57. $y = \frac{1}{2}\csc 3(\theta - 45°) + 1$

58. $y = 2\sec\frac{1}{2}(\theta - 90°)$

59. $y = 4\sec 2\left(\theta + \frac{\pi}{2}\right) - 3$

54–59. See Chapter 12 Answer Appendix.

62. Sample answer: a phase shift 90° left, $y = \sin(\theta + 90°)$; a phase shift 270° right, $y = \sin(\theta - 270°)$

H.O.T. Problems Use Higher-Order Thinking Skills

64. Sample answer: Infinitely many; any change in amplitude will create a different graph that has the same θ-intercepts.

60. **CCSS ARGUMENTS** If you are given the amplitude and period of a cosine function, is it *sometimes, always,* or *never* possible to find the maximum and minimum values of the function? Explain your reasoning. **See margin.**

61. **REASONING** Describe how the graph of $y = 3\sin 2\theta + 1$ is different from $y = \sin\theta$.

62. **WRITING IN MATH** Describe two different phase shifts that will translate the sine curve onto the cosine curve shown at the right. Then write an equation for the new sine curve using each phase shift.

63. **OPEN ENDED** Write a periodic function that has an amplitude of 2 and midline at $y = -3$. Then graph the function. **See margin.**

64. **REASONING** How many different sine graphs pass through the origin $(n\pi, 0)$? Explain your reasoning.

Additional Answer

63. Sample answer: $y = 2\sin\theta - 3$

Enrichment
OL BL

Name the Math Have students write a function for which the graph involves both a horizontal and a vertical translation of a trigonometric function. Then have them describe the translations.

Additional Answers

69. amplitude: 2; period: 360°

70. amplitude: 3; period: 360°

71. amplitude: 1; period: 180°

Standardized Test Practice

65. GRIDDED RESPONSE The expression $\frac{3x-1}{4} + \frac{x+6}{4}$ is how much greater than x? **1.25**

66. Expand $(a - b)^4$. **D**

  A $a^4 - b^4$

  B $a^4 - 4ab + b^4$

  C $a^4 + 4a^3b + 6a^2b^2 + 4ab^3 + b^4$

  D $a^4 - 4a^3b + 6a^2b^2 - 4ab^3 + b^4$

67. Solve $\sqrt{x - 3} + \sqrt{x + 2} = 5$. **F**

  F 7   **H** 7, 13

  G 0, 7   **J** no solution

68. GEOMETRY Using the figures below, what is the average of a, b, c, d, and f? **D**

  A 21   **B** 45   **C** 50   **D** 54

77. observational study; sample: 100 students selected; population: all students that have parttime jobs

Spiral Review

Find the amplitude and period of each function. Then graph the function. (Lesson 12-7) **69–71. See margin.**

69. $y = 2 \cos \theta$ **70.** $y = 3 \sin \theta$ **71.** $y = \sin 2\theta$

Find the exact value of each expression. (Lesson 12-6)

72. $\sin \frac{4\pi}{3}$ $-\frac{\sqrt{3}}{2}$ **73.** $\sin (-30°)$ $-\frac{1}{2}$ **74.** $\cos 405°$ $\frac{\sqrt{2}}{2}$

Determine whether each situation describes a *survey*, an *experiment*, or an *observational study*. Then identify the sample, and suggest a population from which it may have been selected. (Lesson 11-1) **75. experiment; sample: people that exercise for an hour a day; population: all adults**

75. A group of 220 adults is randomly split into two groups. One group exercises for an hour a day and the other group does not. The body mass indexes are then compared.

76. A soccer coach randomly selects some of his players and gives them a questionnaire asking about their daily sleeping habits.

77. A teacher randomly selects 100 students who have part-time jobs and compares their grades.

76. survey; sample: players that received the questionnaire; population: all soccer players

78. GEOMETRY Equilateral triangle ABC has a perimeter of 39 centimeters. If the midpoints of the sides are connected, a smaller equilateral triangle results. Suppose the process of connecting midpoints of sides and drawing new triangles is continued indefinitely. (Lesson 10-4)

  a. Write an infinite geometric series to represent the sum of the perimeters of all of the triangles. **39 + 19.5 + 9.75 + …**

  b. Find the sum of the perimeters of all of the triangles. **78 cm**

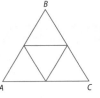

79. CONSTRUCTION A construction company will be fined for each day it is late completing a bridge. The daily fine will be $4000 for the first day and will increase by $1000 each day. Based on its budget, the company can only afford $60,000 in total fines. What is the maximum number of days it can be late? (Lesson 10-3) **8 days**

Skills Review

Find each value of θ. Round to the nearest degree.

80. $\sin \theta = \frac{7}{8}$ 61°   **81.** $\tan \theta = \frac{9}{10}$ 42°   **82.** $\cos \theta = \frac{1}{4}$ 76°

83. $\cos \theta = \frac{4}{5}$ 37°   **84.** $\sin \theta = \frac{5}{6}$ 56°   **85.** $\tan \theta = \frac{2}{7}$ 16°

 852 | Lesson 12-8 | Translations of Trigonometric Graphs

DifferentiatedInstruction   **OL** **BL**

Extension Have students choose a vertical shift, amplitude, period, and phase shift and write a function with these characteristics.

LESSON 12-9 Inverse Trigonometric Functions

:: Then	:: Now	:: Why?
• You graphed trigonometric functions.	**1** Find values of inverse trigonometric functions. **2** Solve equations by using inverse trigonometric functions.	• The leaning bookshelf at the right is 15 inches from the wall and reaches a height of 75 inches. In Lesson 13-1, you learned how to use the inverse of a trigonometric function to find the measure of acute angle θ. $\tan \theta = \frac{15}{75}$ or 0.2 Use the tangent function. Find an angle that has a tangent of 0.2. [2nd] [TAN⁻¹] .2 [ENTER] 11.30993247 So, the measure of θ is about 11°.

75 in.

15 in.

NewVocabulary

principal values
Arcsine function
Arccosine function
Arctangent function

Common Core State Standards

Content Standards
A.CED.2 Create equations in two or more variables to represent relationships between quantities; graph equations on coordinate axes with labels and scales.

Mathematical Practices
7 Look for and make use of structure.

1 Inverse Trigonometric Functions If you know the value of a trigonometric function for an angle, you can use the *inverse* to find the angle. Recall that an inverse function is the relation in which all values of x and y are reversed. The inverse of $y = \sin x$, $x = \sin y$, is graphed at the right.

Notice that the inverse is not a function because there are many values of y for each value of x. If you restrict the domain of the sine function so that $-\frac{\pi}{2} \le x \le \frac{\pi}{2}$, then the inverse is a function.

$x = \sin y$

The values in this restricted domain are called **principal values**. Trigonometric functions with restricted domains are indicated with capital letters.

• $y = \text{Sin } x$ if and only if $y = \sin x$ and $-\frac{\pi}{2} \le x \le \frac{\pi}{2}$.

• $y = \text{Cos } x$ if and only if $y = \cos x$ and $0 \le x \le \pi$.

• $y = \text{Tan } x$ if and only if $y = \tan x$ and $-\frac{\pi}{2} \le x \le \frac{\pi}{2}$.

You can use functions with restricted domains to define inverse trigonometric functions. The inverses of the sine, cosine, and tangent functions are the **Arcsine**, **Arccosine**, and **Arctangent** functions, respectively.

KeyConcept Inverse Trigonometric Functions

Inverse Function	Symbols	Domain	Range	Model
Arcsine	$y = \text{Arcsin } x$ $y = \text{Sin}^{-1} x$	$-1 \le x \le 1$	$-\frac{\pi}{2} \le y \le \frac{\pi}{2}$ $-90° \le y \le 90°$	
Arccosine	$y = \text{Arccos } x$ $y = \text{Cos}^{-1} x$	$-1 \le x \le 1$	$0 \le y \le \pi$ $0° \le y \le 180°$	
Arctangent	$y = \text{Arctan } x$ $y = \text{Tan}^{-1} x$	all real numbers	$-\frac{\pi}{2} \le y \le \frac{\pi}{2}$ $-90° \le y \le 90°$	

$y = \text{Sin}^{-1} x$

connectED.mcgraw-hill.com   **853**

Pieter Estersohn/Beateworks/Corbis

1 Focus

VerticalAlignment

Before Lesson 12-9 Graph trigonometric functions.

Lesson 12-9 Find values of inverse trigonometric functions. Solve equations by using inverse trigonometric functions.

After Lesson 12-9 Define and graph the inverse trigonometric functions.

2 Teach

Scaffolding Questions

Have students read the **Why?** section of the lesson.

Ask:

■ How can you find the length of the leaning bookshelf? Find that length.
Use the Pythagorean Theorem; $s^2 = 15^2 + 75^2$; $s^2 = 5850$; $s \approx 76.485$ in.

■ What ratio represents $\sin \theta$? $\frac{15}{76.485}$

■ What ratio represents $\cos \theta$? $\frac{75}{76.485}$

■ Use your values for $\sin \theta$ and $\cos \theta$ to find θ. Do you get the same value?
The same value for θ, 11.3099°, is obtained using $\sin \theta$ and $\cos \theta$.

Lesson 12-9 Resources

Resource	Approaching Level **AL**	On Level **OL**	Beyond Level **BL**	English Learners **ELL**
Teacher Edition		■ Differentiated Instruction, pp. 855, 858	■ Differentiated Instruction, pp. 855, 858	
Chapter Resource Masters	■ Study Guide and Intervention, pp. 53–54 ■ Skills Practice, p. 55 ■ Practice, p. 56 ■ Word Problem Practice, p. 57	■ Study Guide and Intervention, pp. 53–54 ■ Skills Practice, p. 55 ■ Practice, p. 56 ■ Word Problem Practice, p. 57 ■ Enrichment, p. 58	■ Practice, p. 56 ■ Word Problem Practice, p. 57 ■ Enrichment, p. 58	■ Study Guide and Intervention, pp. 53–54 ■ Skills Practice, p. 55 ■ Practice, p. 56 ■ Word Problem Practice, p. 57
Other	■ 5-Minute Check 12-9 ■ Study Notebook	■ 5-Minute Check 12-9 ■ Study Notebook	■ 5-Minute Check 12-9 ■ Study Notebook	■ 5-Minute Check 12-9 ■ Study Notebook

1 Inverse Trigonometric Functions

Example 1 shows how to evaluate inverse trigonometric functions for angles measured in degrees or radians.
Example 2 shows how to use a calculator to find values of expressions involving inverse trigonometric functions.

Formative Assessment

Use the Guided Practice exercises after each example to determine students' understanding of concepts.

Additional Examples

1 Find each value. Write angle measures in degrees and radians.

a. $\mathrm{Sin}^{-1}\left(\dfrac{\sqrt{2}}{2}\right)$ $45°; \dfrac{\pi}{4}$

b. Arcsin (-1) $-90°; -\dfrac{\pi}{2}$

2 Find $\tan\left(\mathrm{Cos}^{-1}\dfrac{4}{7}\right)$. Round to the nearest hundredth. 1.44

Additional Examples also in Interactive Classroom PowerPoint® Presentations

IWB Interactive White Board READY

Tips for New Teachers

Sense-Making It may help students to understand the nature of inverse functions if they read Arcsin x and $\mathrm{Sin}^{-1} x$ as "the angle with sine of x."

Teach with Tech

Video Recording Create a video explaining how to find the values of inverse trigonometric functions. Post the video to a class Web site so students can use it as an additional reference outside of class.

ReviewVocabulary

inverse functions If f and f^{-1} are inverse functions, then $f(a) = b$ if and only if $f^{-1}(b) = a$.

In the relation $y = \cos^{-1} x$, if $x = \frac{1}{2}$, $y = 60°$, $300°$, and all angles that are coterminal with those angles. In the function $y = \mathrm{Cos}^{-1} x$, if $x = \frac{1}{2}$, $y = 60°$ only.

Example 1 Evaluate Inverse Trigonometric Functions

Find each value. Write angle measures in degrees and radians.

a. $\mathrm{Cos}^{-1}\left(-\dfrac{1}{2}\right)$

Find the angle θ for $0° \le \theta \le 180°$ that has a cosine value of $-\dfrac{1}{2}$.

Method 1 Use a unit circle.

Find a point on the unit circle that has an x-coordinate of $-\dfrac{1}{2}$.

When $\theta = 120°$, $\cos\theta = -\dfrac{1}{2}$.

So, $\mathrm{Cos}^{-1}\left(-\dfrac{1}{2}\right) = 120°$ or $\dfrac{2\pi}{3}$.

StudyTip
Angle Measure Remember that when evaluating an inverse trigonometric function, the result is an angle measure.

Method 2 Use a calculator.

KEYSTROKES: [2nd] [COS⁻¹] [(−)] 1 [÷] 2 [)] [ENTER] 120

Therefore, $\mathrm{Cos}^{-1}\left(-\dfrac{1}{2}\right) = 120°$ or $\dfrac{2\pi}{3}$.

b. Arctan 1

Find the angle θ for $-90° \le \theta \le 90°$ that has a tangent value of 1.

KEYSTROKES: [2nd] [TAN⁻¹] 1 [ENTER] 45   Therefore, Arctan $1 = 45°$ or $\dfrac{\pi}{4}$.

▶ **Guided**Practice

1A. $\mathrm{Cos}^{-1} 0$ $90; \dfrac{\pi}{2}$ **1B.** Arcsin $\left(-\dfrac{\sqrt{2}}{2}\right)$ $-45°; -\dfrac{\pi}{4}$

When finding a value when there are multiple trigonometric functions involved, use the order of operations to solve.

Example 2 Find a Trigonometric Value

Find $\tan\left(\mathrm{Cos}^{-1}\dfrac{1}{2}\right)$. Round to the nearest hundredth.

Use a calculator.

KEYSTROKES: [TAN] [2nd] [COS⁻¹] 1 [÷] 2 [)] [)] [ENTER] 1.732050808

So, $\tan\left(\mathrm{Cos}^{-1}\dfrac{1}{2}\right) \approx 1.73$.

CHECK $\mathrm{Cos}^{-1}\dfrac{1}{2} = 60°$ and $\tan 60° \approx 1.73$. So, the answer is correct.

▶ **Guided**Practice

Find each value. Round to the nearest hundredth.

2A. $\sin\left(\mathrm{Tan}^{-1}\dfrac{3}{8}\right)$ 0.35   **2B.** $\cos\left(\mathrm{Arccos} -\dfrac{\sqrt{2}}{2}\right)$ -0.71

 854 | Lesson 12-9 | Inverse Trigonometric Functions

Focus on Mathematical Content

Inverse Trigonometric Functions Because the trigonometric functions are periodic, many angles correspond to the same function value. Therefore, the inverse of any of the trigonometric functions is not a function (since the value of the inverse would not be unique). However, if the domain is restricted to an appropriate interval, the inverse is a function. While there are infinitely many intervals that would work, standard intervals (centered at 0 or with 0 as an endpoint) have been chosen.

2 Solve Equations by Using Inverses

You can rewrite trigonometric equations to solve for the measure of an angle.

Standardized Test Example 3 Find an Angle Measure

If Sin $\theta = -0.35$, find θ.

A $-20.5°$ **B** $-0.6°$ **C** $0.6°$ **D** $20.5°$

Read the Test Item

The sine of angle θ is -0.35. This can be written as Arcsin $(-0.35) = \theta$.

Solve the Test Item

Use a calculator.

KEYSTROKES: 2nd [SIN⁻¹] (–) .35 ENTER -20.48731511

So, $\theta \approx -20.5°$. The answer is A.

> **GuidedPractice**
>
> **3.** If Tan $\theta = 1.8$, find θ. **H**
>
>   **F** $0.03°$ **G** $29.1°$ **H** $60.9°$ **J** no solution

Test-TakingTip

Eliminate Possibilities
The function Sin restricts the possible angle measures to Quadrants I or IV. Because -0.35 is negative, look for an angle measure in Quadrant IV.

Inverse trigonometric functions can be used to determine angles of inclination, depression, and elevation.

Real-World Example 4 Use Inverse Trigonometric Functions

WATER SKIING A water ski ramp is 6 feet tall and 9 feet long, as shown at the right. Write an inverse trigonometric function that can be used to find θ, the angle the ramp makes with the water. Then find the measure of the angle. Round to the nearest tenth.

Because the measures of the opposite side and the hypotenuse are known, you can use the sine function.

$\sin \theta = \dfrac{6}{9}$ Sine function

$\theta = \text{Sin}^{-1} \dfrac{6}{9}$ Inverse sine function

$\theta \approx 41.8°$ Use a calculator.

So, the angle of the ramp is about $41.8°$.

CHECK Using your calculator, $\sin 41.8 \approx 0.66653 \approx \dfrac{6}{9}$. So, the answer is correct.

> **GuidedPractice**
>
> **4.** **SKIING** A ski trail is shown at the right. Write an inverse trigonometric function that can be used to find θ, the angle the trail makes with the ground in the valley. Then find the angle. Round to the nearest tenth. $\theta = \text{Tan}^{-1} \dfrac{5}{12}$; $22.6°$

Real-WorldCareer

Sport Science Administrator
A sport science administrator provides sport science information to players, coaches, and parents. He or she implements testing, training, and treatment programs for athletes. A master's degree in sport science or a related area is recommended.

---

2 Solve Equations by Using Inverses

Example 3 shows how to solve an equation by using an inverse trigonometric function. **Example 4** shows how to apply an inverse trigonometric function to solve a real-world problem.

Additional Examples

3 **STANDARDIZED TEST EXAMPLE**
If Cos $\theta = -0.86$, find θ. D

A $-149.3°$ **C** $59.3°$

B $-59.3°$ **D** $149.3°$

4 **WATER SKIING** In Example 4, suppose the ramp is 5.5 feet tall and 12 feet long. Write an inverse trigonometric function that can be used to find θ. Then find the measure of θ. Round to the nearest tenth. $\sin \theta = \dfrac{5.5}{12}$; $\theta = \text{Sin}^{-1} \dfrac{5.5}{12} \approx 27.3°$

DifferentiatedInstruction OL BL

Visual/Spatial Learners Ask students to find Arcsin 2. If they use a calculator, suggest that they study the graph of $y = \sin x$ to explain why an error message was the result. The graph of $y = \sin x$ has no y-values greater than 1 or less than -1.

Formative Assessment

Use Exercises 1–11 to check for understanding.

Use the chart at the bottom of this page to customize assignments for your students.

CCSS **Teaching the Mathematical Practices**

Sense-Making Mathematically proficient students start by explaining to themselves the meaning of a problem and looking for entry points to its solution. They analyze givens, constraints, relationships, and goals. They check their answers to problems using a different method, and they continually ask themselves, "Does this make sense?"

Multiple Representations

In Exercise 40, students use a graph, symbolic notation, and numeric evaluation to investigate the inverse cosine function.

Additional Answers

40a.

domain: $-1 \le x \le 1$; range: $0 \le y \le \pi$

40d. Sample answer: The graph of $y = \cos x$ has a domain of all real numbers and a range from -1 to 1. The graph of $y = \text{Cos}^{-1} x$ has a domain from -1 to 1 and a range of 0 to 180°.

45. Sample answer: $y = \tan^{-1} x$ is a relation that has a domain of all real numbers and a range of all real numbers except odd multiples of $\frac{\pi}{2}$. The relation is not a function. $y = \text{Tan}^{-1} x$ is a function that has a domain of all real numbers and a range of $-\frac{\pi}{2} \le y \le \frac{\pi}{2}$.

Example 1   **Find each value. Write angle measures in degrees and radians.**

1. $\text{Sin}^{-1} \frac{1}{2}$ 30°; $\frac{\pi}{6}$ **2.** $\text{Arctan} (-\sqrt{3})$ $-60°$; $-\frac{\pi}{3}$ **3.** $\text{Arccos} (-1)$ 180°; π

Example 2   **Find each value. Round to the nearest hundredth if necessary.**

4. $\cos \left(\text{Arcsin} \frac{4}{5} \right)$ 0.6   **5.** $\tan (\text{Cos}^{-1} 1)$ 0   **6.** $\sin \left(\text{Sin}^{-1} \frac{\sqrt{3}}{2} \right)$ 0.87

Example 3   **7. MULTIPLE CHOICE** If $\text{Sin}\, \theta = 0.422$, find θ. **A**

   A 25°   **B** 42°   **C** 48°   **D** 65°

Solve each equation. Round to the nearest tenth if necessary.

8. $\text{Cos}\, \theta = 0.9$ 25.8°   **9.** $\text{Sin}\, \theta = -0.46$ $-27.4°$ **10.** $\text{Tan}\, \theta = 2.1$ 64.5°

Example 4   **11. SNOWBOARDING** A cross section of a superpipe for snowboarders is shown at the right. Write an inverse trigonometric function that can be used to find θ, the angle that describes the steepness of the superpipe. Then find the angle to the nearest degree. $\text{Arctan} \frac{6.2}{18}$; 19°

18 ft

6.2 ft

Example 1   **Find each value. Write angle measures in degrees and radians.**

12. $\text{Arcsin} \left(\frac{\sqrt{3}}{2} \right)$ 60°; $\frac{\pi}{3}$ **13** $\text{Arccos} \left(\frac{\sqrt{3}}{2} \right)$ 30°; $\frac{\pi}{6}$ **14.** $\text{Sin}^{-1} (-1)$ $-90°$; $-\frac{\pi}{2}$

15. $\text{Tan}^{-1} \sqrt{3}$ 60°; $\frac{\pi}{3}$ **16.** $\text{Cos}^{-1} \left(-\frac{\sqrt{3}}{2} \right)$ 150°; $\frac{5\pi}{6}$ **17.** $\text{Arctan} \left(-\frac{\sqrt{3}}{3} \right)$ $-30°$; $-\frac{\pi}{6}$

Example 2   **Find each value. Round to the nearest hundredth if necessary.**

18. $\tan (\text{Cos}^{-1} 1)$ 0   **19.** $\tan \left[\text{Arcsin} \left(-\frac{1}{2} \right) \right]$ -0.58 **20.** $\cos \left(\text{Tan}^{-1} \frac{3}{5} \right)$ 0.86

21. $\sin \left(\text{Arctan} \sqrt{3} \right)$ 0.87   **22.** $\cos \left(\text{Sin}^{-1} \frac{4}{9} \right)$ 0.90   **23.** $\sin \left[\text{Cos}^{-1} \left(-\frac{\sqrt{2}}{2} \right) \right]$ 0.71

Example 3   **Solve each equation. Round to the nearest tenth if necessary.**

24. $\text{Tan}\, \theta = 3.8$ 75.3°   **25.** $\text{Sin}\, \theta = 0.9$ 64.2°   **26.** $\text{Sin}\, \theta = -2.5$ no solution

27. $\text{Cos}\, \theta = -0.25$ 104.5°   **28.** $\text{Cos}\, \theta = 0.56$ 55.9°   **29.** $\text{Tan}\, \theta = -0.2$ $-11.3°$

Example 4   **30.** CCSS **SENSE-MAKING** A boat is traveling west to cross a river that is 190 meters wide. Because of the current, the boat lands at point Q, which is 59 meters from its original destination point P. Write an inverse trigonometric function that can be used to find θ, the angle at which the boat veered south of the horizontal line. Then find the measure of the angle to the nearest tenth.

$\text{Arctan} \frac{59}{190} = \theta$; 17.3°

 856 | Lesson 12-9 | Inverse Trigonometric Functions

Differentiated Homework Options

Level	Assignment	Two-Day Option	
AL Basic	12–31, 42–60	13–31 odd, 47–50	12–30 even, 42–46, 51–60
OL Core	13–39 odd, 40, 42–60	12–31, 47–50	32–40, 42–46, 51–60
BL Advanced	32–56, (optional: 57–60)		

31. **TREES** A 24-foot tree is leaning 2.5 feet left of vertical, as shown in the figure. Write an inverse trigonometric function that can be used to find θ, the angle at which the tree is leaning. Then find the measure of the angle to the nearest degree. $\text{Arcsin } \frac{2.5}{24}$; 6°

2.5 ft

24 ft θ

B 32. **DRIVING** An expressway off-ramp curve has a radius of 52 meters and is designed for vehicles to safely travel at speeds up to 45 kilometers per hour (or 12.5 meters per second). The equation below represents the angle θ of the curve. What is the measure of the angle to the nearest degree? 17°

$$\tan \theta = \frac{(12.5 \text{ m/s})^2}{(52 \text{ m})(9.8 \text{ m/s}^2)}$$

(33) **TRACK AND FIELD** A shot-putter throws the shot with an initial speed of 15 meters per second. The expression $\frac{15 \text{ m/s (sin } x)}{9.8 \text{ m/s}^2}$ represents the time in seconds at which the shot reached its maximum height. In the expression, x is the angle at which the shot was thrown. If the maximum height of the shot was reached in 1.0 second, at what angle was it thrown? Round to the nearest tenth. 40.8°

42. Sample answer: Neither; cosine is not positive in the second quadrant.

Solve each equation for $0 \le \theta \le 2\pi$.

34. $\csc \theta = 1$ $\frac{\pi}{2}$

35. $\sec \theta = -1$ π

36. $\sec \theta = 1$ $0, 2\pi$

43. The domain of $y = \text{Sin}^{-1} x$ is $-1 \le x \le 1$. This is the same as the range of $y = \text{Sin } x$.

37. $\csc \theta = \frac{1}{2}$ no solution

38. $\cot \theta = 1$ $\frac{\pi}{4}, \frac{5\pi}{4}$

39. $\sec \theta = 2$ $\frac{\pi}{3}, \frac{5\pi}{3}$

40. **MULTIPLE REPRESENTATIONS** Consider $y = \text{Cos}^{-1} x$. **a. See margin.**

  a. Graphical Sketch a graph of the function. Describe the domain and the range.

  b. Symbolic Write the function using different notation. $y = \text{Arccos } x$

  c. Numerical Choose a value for x between -1 and 0. Then evaluate the inverse cosine function. Round to the nearest tenth. **Sample answer: $x = -0.2$; $y = 101.5$**

  d. Analytical Compare the graphs of $y = \cos x$ and $y = \text{Cos}^{-1} x$. **See margin.**

H.O.T. Problems   Use Higher-Order Thinking Skills

C 41. **CHALLENGE** Determine whether $cos (Arccos x) = x$ for all values of x is *true* or *false*. If false, give a counterexample. **false; $x = 2\pi$**

44. Sample answer: $\text{Arcsin } \frac{1}{2} = 30°$; $\frac{1}{2} = \text{Sin } 30°$

46. Sample answer: The range of $y = \text{Sin } x$ and $y = \text{Cos } x$ is $-1 \le x \le 1$. The range of $y = \text{Tan}^{-1} x$ is all real numbers.

42. **CCSS CRITIQUE** Desiree and Oscar are solving $\cos \theta = 0.3$ where $90 < \theta < 180$. Is either of them correct? Explain your reasoning.

Desiree	Oscar
$cos \, \theta = 0.3$	$cos \, \theta = 0.3$
$cos^{-1} \, 0.3 = 162.5°$	$cos^{-1} \, 0.3 = 72.5°$

43. **REASONING** Explain how the domain of $y = \text{Sin}^{-1} x$ is related to the range of $y = \text{Sin } x$.

44. **OPEN ENDED** Write an equation with an Arcsine function and an equation with a Sine function that both involve the same angle measure.

45. **WRITING IN MATH** Compare and contrast the relations $y = \tan^{-1} x$ and $y = \text{Tan}^{-1} x$. Include information about the domains and ranges. **See margin.**

46. **REASONING** Explain how $\text{Sin}^{-1} 8$ and $\text{Cos}^{-1} 8$ are undefined while $\text{Tan}^{-1} 8$ is defined.

connectED.mcgraw-hill.com **857**

CCSS **Teaching the Mathematical Practices**

Critique Mathematically proficient students are also able to compare the effectiveness of two plausible arguments, distinguish correct logic or reasoning from that which is flawed, and—if there is a flaw in an argument—explain what it is.

Enrichment
OL BL

12-9 Enrichment

Snell's Law

Snell's Law describes what happens to a ray of light that passes from air into water or some other substance. In the figure, the ray starts at the left and makes an angle of incidence θ with the surface.

Part of the ray is reflected, creating an angle of reflection θ. The rest of the ray is bent, or refracted, as it passes through the other medium. This creates angle θ'.

The angle of incidence equals the angle of reflection.

The angles of incidence and refraction are related by Snell's Law:

$\sin \theta = k \sin \theta'$

The constant k is called the index of refraction.

Study Guide and Intervention
AL OL ELL

Practice
AL OL BL ELL

Word Problem Practice
AL OL BL ELL

connectED.mcgraw-hill.com **857**

4 Assess

Ticket Out the Door Have students write the domain and range for the Arcsin, Arccos, and Arctan functions.

Formative Assessment
Check for student understanding of concepts in Lessons 12-8 and 12-9.

 Quiz 4, p. 62

E **Follow-up**
Students have explored transformations of trigonometric functions.

Ask:
- How are inverses of trigonometric functions similar to inverses of other functions you have studied? **Sample answer: The composition of a trigonometric function and its inverse is the identity function; the graphs are symmetric with respect to the line $y = x$; the domain of an inverse trigonometric function must be restricted in order to be a function.**

Standardized Test Practice

47. Simplify $\dfrac{\frac{2}{x} + 2}{\frac{2}{x} - 2}$. **A**

A $\dfrac{1 + x}{1 - x}$ **C** $\dfrac{1 - x}{1 + x}$

B $\dfrac{2}{x}$ **D** $-x$

48. SHORT RESPONSE What is the equation of the graph below? $(x - 3)^2 + (y + 4)^2 = 25$

49. If $f(x) = 2x^2 - 3x$ and $g(x) = 4 - 2x$, what is $g[f(x)]$? **G**

   F $g[f(x)] = 4 + 6x - 8x^2$
   G $g[f(x)] = 4 + 6x - 4x^2$
   H $g[f(x)] = 20 - 26x + 8x^2$
   J $g[f(x)] = 44 - 38x + 8x^2$

50. If g is a positive number, which of the following is equal to $12g$? **D**

   A $\sqrt{144g}$

   B $\sqrt{12g^2}$

   C $\sqrt{24g^2}$

   D $6\sqrt{4g^2}$

Spiral Review

51. RIDES The Cosmoclock 21 is a huge Ferris wheel in Japan. The diameter is 328 feet. Suppose a rider enters the ride at 0 feet, and then rotates in 90° increments counterclockwise. The table shows the angle measures of rotation and the height in feet above the ground of the rider. (Lesson 12-8)

   a. A function that models the data is $y = 164 \cdot [\sin(x - 90°)] + 164$. Identify the vertical shift, amplitude, period, and phase shift of the graph. **164; 164; 360°, 90°**

   b. Write an equation using the sine that models the position of a rider on the Vienna Giant Ferris Wheel in Austria, with a diameter of 200 feet. Check by plotting the points and your equation with a graphing calculator. $y = 100 [\sin(x - 90°)] + 100$

Angle	Height	Angle	Height
0°	0	450°	164
90°	164	540°	328
180°	328	630°	164
270°	164	720°	0
360°	0		

52. TIDES The world's record for the highest tide is held by the Minas Basin in Nova Scotia, Canada, with a tidal range of 54.6 feet. A tide is at equilibrium when it is at its normal level halfway between its highest and lowest points. Write an equation to represent the height h of the tide. Assume that the tide is at equilibrium at $t = 0$, that the high tide is beginning, and that the tide completes one cycle in 12 hours. (Lesson 12-7) $h = 27.3 \sin \dfrac{\pi}{6} t$

Solve each equation. (Lesson 7-4)

53. $\log_3 5 + \log_3 x = \log_3 10$ **2**

54. $\log_4 a + \log_4 9 = \log_4 27$ **3**

55. $\log_{10} 16 - \log_{10} 2t = \log_{10} 2$ **4**

56. $\log_7 24 - \log_7 (y + 5) = \log_3 8$ **−2**

Skills Review

Find the exact value of each trigonometric function.

57. $\cos 3\pi$ **−1**   **58.** $\tan 120°$ **$-\sqrt{3}$**   **59.** $\sin 300°$ **$-\dfrac{\sqrt{3}}{2}$**   **60.** $\sec \dfrac{7\pi}{6}$ **$-\dfrac{2\sqrt{3}}{3}$**

DifferentiatedInstruction **OL** **BL**

Extension Have students consider the function $y = \sin^{-1} x + \cos^{-1} x$ and complete the table.

x	0	$\dfrac{1}{2}$	$\dfrac{\sqrt{2}}{2}$	$\dfrac{\sqrt{3}}{2}$	1	$-\dfrac{1}{2}$	$-\dfrac{\sqrt{2}}{2}$	$-\dfrac{\sqrt{3}}{2}$	-1
y	$\dfrac{\pi}{2}$	$\dfrac{\pi}{2}$	$\dfrac{\pi}{2}$	$\dfrac{\pi}{2}$	$\dfrac{\pi}{2}$	$\dfrac{\pi}{2}$	$\dfrac{\pi}{2}$	$\dfrac{\pi}{2}$	$\dfrac{\pi}{2}$

Have them make a conjecture about the function $y = \sin^{-1} x + \cos^{-1} x$. $\sin^{-1} x + \cos^{-1} x = \dfrac{\pi}{2}$ for all values of x. Have them make a conjecture about the functions $y = \tan^{-1} x + \cot^{-1} x$ and $y = \sec^{-1} x + \csc^{-1} x$. Both functions are equal to 1 for all values of x.

12 Study Guide and Review
CHAPTER

Study Guide

KeyConcepts

Right Triangle Trigonometry (Lesson 12-1)

- $\sin \theta = \frac{opp}{hyp}$, $\cos \theta = \frac{adj}{hyp}$, $\tan \theta = \frac{opp}{adj}$,

  $\csc \theta = \frac{hyp}{opp}$, $\sec \theta = \frac{hyp}{adj}$, $\cot \theta = \frac{adj}{opp}$

Angle Measures and Trigonometric Functions of General Angles (Lessons 12-2 and 12-3)

- The measure of an angle is determined by the amount of rotation from the initial side to the terminal side.

- You can find the exact values of the six trigonometric functions of θ, given the coordinates of a point $P(x, y)$ on the terminal side of the angle.

Law of Sines and Law of Cosines (Lessons 12-4 and 12-5)

- $\frac{\sin A}{a} = \frac{\sin B}{b} = \frac{\sin C}{c}$

- $a^2 = b^2 + c^2 - 2bc \cos A$

  $b^2 = a^2 + c^2 - 2ac \cos B$

  $c^2 = a^2 + b^2 - 2ab \cos C$

Circular and Inverse Trigonometric Functions
(Lessons 12-6 and 12-9)

- If the terminal side of an angle θ in standard position intersects the unit circle at $P(x, y)$, then $\cos \theta = x$ and $\sin \theta = y$.

- $y = \text{Sin } x$ if $y = \sin x$ and $-\frac{\pi}{2} \le x \le \frac{\pi}{2}$

Graphing Trigonometric Functions (Lesson 12-7)

- For trigonometric functions of the form $y = a \sin b\theta$ and $y = a \cos b\theta$, the amplitude is $|a|$, and the period is $\frac{360°}{|b|}$ or $\frac{2\pi}{b}$.

- The period of $y = a \tan b\theta$ is $\frac{180°}{|b|}$ or $\frac{\pi}{|b|}$.

FOLDABLES StudyOrganizer

Be sure the Key Concepts are noted in your Foldable.

KeyVocabulary

ambiguous case (p. 816)	period (p. 831)
amplitude (p. 837)	periodic function (p. 831)
angle of depression (p. 794)	phase shift (p. 845)
angle of elevation (p. 794)	principal values (p. 853)
Arccosine function (p. 853)	quadrantal angle (p. 808)
Arcsine function (p. 853)	radian (p. 801)
Arctangent function (p. 853)	reference angle (p. 808)
central angle (p. 802)	secant (p. 790)
circular function (p. 830)	sine (p. 790)
cosecant (p. 790)	solving a triangle (p. 815)
cosine (p. 790)	standard position (p. 799)
cotangent (p. 790)	tangent (p. 790)
coterminal angles (p. 800)	terminal side (p. 799)
cycle (p. 831)	trigonometric function (p. 790)
frequency (p. 838)	trigonometric ratio (p. 790)
initial side (p. 799)	trigonometry (p. 790)
Law of Cosines (p. 823)	unit circle (p. 830)
Law of Sines (p. 815)	vertical shift (p. 846)
midline (p. 846)	

VocabularyCheck

State whether each sentence is *true* or *false*. If *false*, replace the underlined term to make a true sentence.

1. The <u>Law of Cosines</u> is used to solve a triangle when two angles and any sides are known. **false, Law of Sines**

2. An angle on the coordinate plane is in <u>standard position</u> if the vertex is at the origin and one ray is on the positive *x*-axis. **true**

3. <u>Coterminal angles</u> are angles in standard position that have the same terminal side. **true**

4. A horizontal translation of a periodic function is called a <u>phase shift</u>. **true**

5. The inverse of the sine function is the <u>cosecant function</u>. **false, Arcsine function**

6. The <u>cycle</u> of the graph of a sine or cosine function equals half the difference between the maximum and minimum values of the function. **false, amplitude**

  connectED.mcgraw-hill.com **859**

Formative Assessment
KeyVocabulary The page references after each word denote where that term was first introduced. If students have difficulty answering questions 1–6, remind them that they can use these page references to refresh their memories about the vocabulary.

Summative Assessment
Vocabulary Test, p. 64

Vocabulary Review
Vocabulary Review provides students the opportunity to check their understanding of important concepts and terminology in an online game format.

FOLDABLES StudyOrganizer

Dinah Zike's Foldables®
Have students look through the chapter to make sure they have included examples in their Foldables. Suggest that students keep their Foldables handy while completing the Study Guide and Review pages. Point out that their Foldables can serve as a quick review tool when studying for the chapter test.

Lesson-by-Lesson Review

Intervention If the given examples are not sufficient to review the topics covered by the questions, remind students that the lesson references tell them where to review that topic in their textbook.

Two-Day Option Have students complete the Lesson-by-Lesson Review. Then you can use McGraw-Hill eAssessment to customize another review worksheet that practices all the objectives of this chapter or only the objectives on which your students need more help.

Study Guide and Review *Continued*

Lesson-by-Lesson Review

12-1 Trigonometric Functions in Right Triangles

Solve $\triangle ABC$ by using the given measurements. Round measures of sides to the nearest tenth and measures of angles to the nearest degree.

7. $c = 12$, $b = 5$

8. $a = 10$, $B = 55°$

9. $B = 75°$, $b = 15$ $A = 15°$; $a = 4.0$; $c = 15.5$

10. $B = 45°$, $c = 16$ $a = 11.3$; $b = 11.3$; $A = 45°$

11. $A = 35°$, $c = 22$ $B = 55°$; $a = 12.6$; $b = 18.0$

12. $\sin A = \frac{2}{3}$, $a = 6$ $A = 42°$; $B = 48°$; $b = 6.7$; $c = 9.0$

13. **TRUCK** The back of a moving truck is 3 feet off of the ground. What length does a ramp off the back of the truck need to be in order for the angle of elevation of the ramp to be 20°? **about 8.8 ft**

7. $a = 10.9$; $A = 65°$; $B = 25°$
8. $A = 35°$; $c = 17.4$; $b = 14.3$

Example 1

Solve $\triangle ABC$ by using the given measurements. Round measures of sides to the nearest tenth and measures of angles to the nearest degree.

Find b. $a^2 + b^2 = c^2$
 $9^2 + b^2 = 16^2$
 $b = \sqrt{16^2 - 9^2}$
 $b \approx 13.2$

Find A. $\sin A = \frac{9}{16}$
 Use a calculator.
 To the nearest degree, $A = 34°$.

Find B. $34° + B \approx 90°$
 $B \approx 56°$

Therefore, $b \approx 13.2$, $A \approx 34°$, and $B \approx 56°$.

12-2 Angles and Angle Measure

Rewrite each degree measure in radians and each radian measure in degrees.

14. $215°$ $\frac{43\pi}{36}$

15. $\frac{5\pi}{2}$ $450°$

16. -3π $-540°$

17. $-315°$ $-\frac{7\pi}{4}$

Find one angle with positive measure and one angle with negative measure coterminal with each angle.

18. $265°$ $625°$, $-95°$

19. $-65°$ $295°$, $-425°$

20. $\frac{7\pi}{2}$ $\frac{11\pi}{2}$, $-\frac{\pi}{2}$

21. **BICYCLE** A bicycle tire makes 8 revolutions in one second. The tire has a radius of 15 inches. Find the angle θ in radians through which the tire rotates in one second. $\frac{4\pi}{15}$

Example 2

Rewrite $160°$ in radians.

$160° = 160°\left(\dfrac{\pi \text{ radians}}{180°}\right)$

$ = \dfrac{160\pi}{180}$ radians or $\dfrac{8\pi}{9}$

Example 3

Find one angle with positive measure and one angle with negative measure coterminal with $150°$.

positive angle:

$150° + 360° = 510°$ Add $360°$.

negative angle:

$150° - 360° = -210°$ Subtract $360°$.

12-3 Trigonometric Functions of General Angles

Find the exact value of each trigonometric function.

22. $\cos 135°$ $-\dfrac{\sqrt{2}}{2}$ **23.** $\tan 150°$ $-\dfrac{\sqrt{3}}{3}$

24. $\sin 2\pi$ 0 **25.** $\cos \dfrac{3\pi}{2}$ 0

The terminal side of θ in standard position contains each point. Find the exact values of the six trigonometric functions of θ. **26–28. See margin.**

26. $P(-4, 3)$

27. $P(5, 12)$

28. $P(16, -12)$

29. BALL A ball is thrown off the edge of a building at an angle of 70° and with an initial velocity of 5 meters per second. The equation that represents the horizontal distance of the ball x is $x = v_0 (\cos \theta)t$, where v_0 is the initial velocity, θ is the angle at which it is thrown, and t is the time in seconds. About how far will the ball travel in 10 seconds? **about 17.1 meters**

Example 4

Find the exact value of $\sin 120°$.

Because the terminal side of 120° lies in Quadrant II, the reference angle θ' is 180° − 120° or 60°. The sine function is positive in Quadrant II, so $\sin 120° = \sin 60°$ or $\dfrac{\sqrt{3}}{2}$.

Example 5

The terminal side of θ in standard position contains the point (6, 5). Find the exact values of the six trigonometric functions of θ.

$\sin \theta = \dfrac{y}{r}$ or $\dfrac{5\sqrt{61}}{61}$ $\csc \theta = \dfrac{r}{y}$ or $\dfrac{\sqrt{61}}{5}$

$\cos \theta = \dfrac{x}{r}$ or $\dfrac{6\sqrt{61}}{61}$ $\sec \theta = \dfrac{r}{x}$ or $\dfrac{\sqrt{61}}{6}$

$\tan \theta = \dfrac{y}{x}$ or $\dfrac{5}{6}$ $\cot \theta = \dfrac{x}{y}$ or $\dfrac{6}{5}$

12-4 Law of Sines

Determine whether each triangle has *no* solution, *one* solution, or *two* solutions. Then solve each triangle. Round measures of sides to the nearest tenth and measures of angles to the nearest degree.

30. $C = 118°$, $c = 10$, $a = 4$ **See margin.**

31. $A = 25°$, $a = 15$, $c = 18$ **See margin.**

32. $A = 70°$, $a = 5$, $c = 16$ **no solution**

33. BOAT Kira and Mallory are standing on opposite sides of a river. How far is Kira from the boat? Round to the nearest tenth if necessary. **98.9 ft**

Example 6

Solve $\triangle ABC$.

First, find the measure of the third angle.

$60° + 70° + a = 180°$
$A = 50°$

Now use the Law of Sines to find a and c. Write two equations, each with one variable.

$\dfrac{\sin B}{b} = \dfrac{\sin C}{c}$ $\dfrac{\sin B}{b} = \dfrac{\sin A}{a}$

$\dfrac{\sin 60°}{8} = \dfrac{\sin 70°}{c}$ $\dfrac{\sin 60°}{8} = \dfrac{\sin 50°}{a}$

$c = \dfrac{8 \sin 70°}{\sin 60°}$ $a = \dfrac{8 \sin 50°}{\sin 60°}$

$c \approx 8.7$ $a \approx 7.1$

Therefore, $A = 50°$, $c \approx 8.7$, and $a \approx 7.1$.

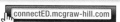

Additional Answers

26. $\sin \theta = \dfrac{3}{5}$, $\cos \theta = -\dfrac{4}{5}$,
$\tan \theta = -\dfrac{3}{4}$, $\csc \theta = \dfrac{5}{3}$,
$\sec \theta = -\dfrac{5}{4}$, $\cot \theta = -\dfrac{4}{3}$,

27. $\sin \theta = \dfrac{12}{13}$, $\cos \theta = \dfrac{5}{13}$,
$\tan \theta = \dfrac{12}{5}$, $\csc \theta = \dfrac{13}{12}$,
$\sec \theta = \dfrac{13}{5}$, $\cot \theta = \dfrac{5}{12}$,

28. $\sin \theta = -\dfrac{3}{5}$, $\cos \theta = \dfrac{4}{5}$,
$\tan \theta = -\dfrac{3}{4}$, $\csc \theta = -\dfrac{5}{3}$,
$\sec \theta = \dfrac{5}{4}$, $\cot \theta = -\dfrac{4}{3}$,

30. one solution; $A \approx 21°$, $B \approx 41°$, $b \approx 7.4$

31. two solutions; first solution: $C = 30°$, $B = 125°$, $b = 29.1$; second solution: $C = 150°$, $B = 5°$, $b = 3.1$

Additional Answers

34. Cosines; $A \approx 46°$, $B \approx 85°$, $C \approx 49°$

35. Sines; $B \approx 52°$, $C \approx 48°$, $c \approx 11.3$

36. Cosines; $A \approx 40°$, $B \approx 65°$, $c \approx 7.5$

37. Sines; $B \approx 75°$, $C \approx 63°$, $c \approx 12.0$ or $B \approx 105°$, $C \approx 33°$, $c \approx 7.3$

38. Cosines; $a \approx 9.9$, $B \approx 28°$, $C \approx 117°$

46. amplitude: 4, period: 180°

$y = 4 \sin 2\theta$

47. amplitude: 1, period: 720°

$y = \cos \frac{1}{2}\theta$

48. amplitude: not defined, period: 360°

$y = 3 \csc \theta$

12-5 Law of Cosines

Determine whether each triangle should be solved by beginning with the Law of *Sines* or Law of *Cosines*. Then solve each triangle. Round measures of sides to the nearest tenth and measures of angles to the nearest degree.

34–38. See margin.

34.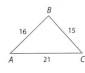

35.

36. $C = 75°$, $a = 5$, $b = 7$

37. $A = 42°$, $a = 9$, $b = 13$

38. $b = 8.2$, $c = 15.4$, $A = 35°$

39. FARMING A farmer wants to fence a piece of his land. Two sides of the triangular field have lengths of 120 feet and 325 feet. The measure of the angle between those sides is 70°. How much fencing will the farmer need?
about 750.5 ft

Example 7

Solve $\triangle ABC$ for $C = 55°$, $b = 11$, and $a = 18$.

You are given the measure of two sides and the included angle. Begin by drawing a diagram and using the Law of Cosines to determine c.

$c^2 = a^2 + b^2 - 2ab \cos C$

$c^2 = 18^2 + 11^2 - 2(18)(11) \cos 55°$

$c^2 \approx 217.9$

$c \approx 14.8$

Next, you can use the Law of Sines to find the measure of angle A.

$\dfrac{\sin A}{18} \approx \dfrac{\sin 55°}{14.8}$

$\sin A \approx \dfrac{18 \sin 55°}{14.8}$ or A is about 85.0°

The measure of the angle B is approximately $180 - (85.0 + 55)$ or 40.0°.

Therefore, $c \approx 14.8$, $A \approx 85.0°$, and $B \approx 40.0°$.

12-6 Circular and Periodic Functions

Find the exact value of each function.

40. $\cos(-210°)$ $\quad -\dfrac{\sqrt{3}}{2}$

41. $(\cos 45°)(\cos 210°)$ $\quad -\dfrac{\sqrt{6}}{4}$

42. $\sin -\dfrac{7\pi}{4}$ $\quad \dfrac{\sqrt{2}}{2}$

43. $\left(\cos \dfrac{\pi}{2}\right)\left(\sin \dfrac{\pi}{2}\right)$ $\quad 0$

44. Determine the period of the function.

6

45. A wheel with a diameter of 18 inches completes 4 revolutions in 1 minute. What is the period of the function that describes the height of one spot on the outside edge of the wheel as a function of time?
15 seconds

Example 8

Find the exact value of $\sin 510°$.

$\sin 510° = \sin(360° + 150°)$

$= \sin 150°$

$= \dfrac{1}{2}$

Example 9

Determine the period of the function below.

The pattern repeats itself at $\dfrac{\pi}{2}$, π, and so on. So, the period is $\dfrac{\pi}{2}$.

---

49. amplitude: not defined, period: 360°

50. amplitude: not defined, period: 90°

12-7 Graphing Trigonometric Functions

Find the amplitude, if it exists, and period of each function. Then graph the function. **46–51. See margin.**

46. $y = 4 \sin 2\theta$

47. $y = \cos \frac{1}{2}\theta$

48. $y = 3 \csc \theta$

49. $y = 3 \sec \theta$

50. $y = \tan 2\theta$

51. $y = 2 \csc \frac{1}{2}\theta$

52. When Lauren jumps on a trampoline it vibrates with a frequency of 10 hertz. Let the amplitude equal 5 feet. Write a sine equation to represent the vibration of the trampoline y as a function of time t. $y = 5 \sin 20\pi t$

Example 10

Find the amplitude and period of $y = 2 \cos 4\theta$. Then graph the function.

amplitude: $|a| = |2|$ or 2. The graph is stretched vertically so that the maximum value is 2 and the minimum value is -2.

period:
$\frac{360°}{|b|} = \frac{360°}{|4|}$ or $90°$

12-8 Translations of Trigonometric Graphs

State the vertical shift, amplitude, period, and phase shift of each function. Then graph the function. **53–57. See margin.**

53. $y = 3 \sin [2(\theta - 90°)] + 1$

54. $y = \frac{1}{2} \tan [2(\theta - 30°)] - 3$

55. $y = 2 \sec \left[3\left(\theta - \frac{\pi}{2}\right)\right] + 2$

56. $y = \frac{1}{2} \cos \left[\frac{1}{4}\left(\theta + \frac{\pi}{4}\right)\right] - 1$

57. $y = \frac{1}{3} \sin \left[\frac{1}{3}(\theta - 90°)\right] + 2$

58. The graph below approximates the height y of a rope that two people are twirling as a function of time t in seconds. Write an equation for the function. $y = 4 \sin 360t + 4$

Example 11

State the vertical shift, amplitude, period, and phase shift of $y = 2 \sin \left[3\left(\theta + \frac{\pi}{2}\right)\right] + 4$. Then graph the function.

Identify the values of k, a, b, and h.

$k = 4$, so the vertical shift is 4.

$a = 2$, so the amplitude is 2.

$b = 3$, so the period is $\frac{2\pi}{|3|}$ or $\frac{2\pi}{3}$.

$h = -\frac{\pi}{2}$, so the phase shift is $\frac{\pi}{2}$ to the left.

56. vertical shift: down 1, amplitude: $\frac{1}{2}$, period: 4π, phase shift: $\frac{\pi}{4}$ left

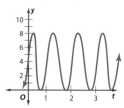

57. vertical shift: up 2, amplitude: $\frac{1}{3}$, period: 1080°, phase shift: 90° right

Additional Answers

51. amplitude: not defined, period: 720°

53. vertical shift: up 1, amplitude: 3, period: 180°, phase shift: 90° right

54. vertical shift: down 3, amplitude: undefined, period: 90°, phase shift: 30° right

55. vertical shift: up 2, amplitude: not defined, period: $\frac{2\pi}{3}$, phase shift: $\frac{\pi}{2}$ right

Anticipation Guide

Have students complete the Chapter 12 Anticipation Guide and discuss how their responses have changed now that they have completed Chapter 12.

Additional Answers (Practice Test)

19.

$y = 2\sin 3\theta$

20.

$y = \frac{1}{2}\cos 2\theta$

23.

$y = \cos(\theta + 180)$

12-9 Inverse Trigonometric Functions

Evaluate each inverse trigonometric function. Write angle measures in degrees and radians.

59. $\sin^{-1}(1)$ $90°, \frac{\pi}{2}$

60. $\arctan(0)$ $0°, 0$

61. $\arcsin \frac{\sqrt{3}}{2}$ $60°, \frac{\pi}{3}$

62. $\cos^{-1} \frac{\sqrt{2}}{2}$ $45°, \frac{\pi}{4}$

63. $\tan^{-1} 1$ $45°, \frac{\pi}{4}$

64. $\arccos 0$ $90°, \frac{\pi}{2}$

65. RAMPS A bicycle ramp is 5 feet tall and 10 feet long, as shown below. Write an inverse trigonometric function that can be used to find θ, the angle the ramp makes with the ground. Then find the angle.

$\sin^{-1} \frac{5}{10} = \theta$; $30°$

10 ft   5 ft θ

Evaluate each inverse trigonometric function. Round to the nearest hundredth if necessary.

66. $\tan\left(\cos^{-1} \frac{1}{3}\right)$ 2.83

67. $\sin\left(\arcsin -\frac{\sqrt{2}}{2}\right)$ -0.71

68. $\sin(\tan^{-1} 0)$ 0

Solve each equation. Round to the nearest tenth if necessary.

69. $\tan \theta = -1.43$ $-55.0°$

70. $\sin \theta = 0.8$ $53.1°$

71. $\cos \theta = 0.41$ $65.8°$

Example 12

Evaluate $\cos^{-1} \frac{1}{2}$. Write angle measures in degrees and radians.

Find the angle θ for $0° \le \theta \le 180°$ that has a cosine value of $\frac{1}{2}$.

Use a unit circle.

$\left(\frac{1}{2}, \frac{\sqrt{3}}{2}\right)$ $60° = \frac{\pi}{3}$

Find a point on the unit circle that has an x-coordinate of $\frac{1}{2}$.

When $\theta = 60°$, $\cos \theta = \frac{1}{2}$.

So, $\cos^{-1} = 60°$ or $\frac{\pi}{3}$.

Example 13

Evaluate $\sin\left(\tan^{-1} \frac{1}{2}\right)$. Round to the nearest hundredth.

Use a calculator.

KEYSTROKES: [SIN] [2nd] [TAN^{-1}] 1 [÷] 2 [)] [)]

[ENTER] 0.4472135955

So, $\sin\left(\tan^{-1} \frac{1}{2}\right) \approx 0.45$.

Example 14

If $\cos \theta = 0.72$, find θ.

Use a calculator.

KEYSTROKES: [2nd] [COS^{-1}] .72 [)] [ENTER] 43.9455195623

So, $\theta \approx 43.9°$.

24.

$y = \frac{1}{2}\tan\left(\theta - \frac{\pi}{2}\right)$

25.

$h = 10\cos\frac{2\pi}{45}t + 10$

Practice Test

Solve △ABC by using the given measurements. Round measures of sides to the nearest tenth and measures of angles to the nearest degree.

1. $A = 36°, c = 9$ $B = 54°, a = 5.3, b = 7.3$
2. $a = 12, A = 58°$ $B = 32°, c = 14.2, b = 7.5$
3. $B = 85°, b = 8$ $A = 5°, c = 8.0, a = 0.7$
4. $a = 9, c = 12$ $b = 7.9, B = 41°, A = 49°$

Rewrite each degree measure in radians and each radian measure in degrees.

5. $325°$ $\frac{65\pi}{36}$
6. $-175°$ $-\frac{35\pi}{36}$
7. $\frac{9\pi}{4}$ $405°$
8. $-\frac{5\pi}{6}$ $-150°$

9. Determine whether △ABC, with $A = 110°, a = 16,$ and $b = 21$, has *no* solution, *one* solution, or *two* solutions. Then solve the triangle, if possible. Round measures of sides to the nearest tenth and measures of angles to the nearest degree.
   no solution

Find the exact value of each function. Write angle measures in degrees.

10. $\cos(-90°)$ 0
11. $\sin 585°$ $-\frac{\sqrt{2}}{2}$
12. $\cot \frac{4\pi}{3}$ $\frac{\sqrt{3}}{3}$
13. $\sec\left(-\frac{9\pi}{4}\right)$ $\sqrt{2}$
14. $\tan\left(\cos^{-1}\frac{4}{5}\right)$ $\frac{3}{4}$
15. $\text{Arccos } \frac{1}{2}$ $60°$

16. The terminal side of angle θ in standard position intersects the unit circle at point $P\left(\frac{1}{2}, \frac{\sqrt{3}}{2}\right)$. Find $\cos \theta$ and $\sin \theta$. $\cos \theta = \frac{1}{2}, \sin \theta = \frac{\sqrt{3}}{2}$

17. **MULTIPLE CHOICE** What angle has a tangent and sine that are both negative? **D**

   A $65°$

   B $120°$

   C $265°$

   D $310°$

18. **NAVIGATION** Airplanes and ships measure distance in nautical miles. The formula 1 nautical mile = $6077 - 31 \cos 2\theta$ feet, where θ is the latitude in degrees, can be used to find the approximate length of a nautical mile at a certain latitude. Find the length of a nautical mile when the latitude is 120°.
   6092.5 ft

Find the amplitude and period of each function. Then graph the function. **19–20. See margin for graphs.**

19. $y = 2 \sin 3\theta$ $2, 120°$
20. $y = \frac{1}{2} \cos 2\theta$ $\frac{1}{2}, 180°$

21. **MULTIPLE CHOICE** What is the period of the function $y = 3 \cot \theta$? **G**

   F $120°$

   G $180°$

   H $360°$

   J $1080°$

22. Determine whether △XYZ, with $y = 15, z = 9,$ and $X = 105°$, should be solved by beginning with the Law of *Sines* or Law of *Cosines*. Then solve the triangle. Round measures of sides to the nearest tenth and measures of angles to the nearest degree. **Cosines; $Y \approx 48°, x \approx 19.4, Z \approx 27°$**

23–24. See margin for graphs.
State the amplitude, period, and phase shift for each function. Then graph the function.

23. $y = \cos(\theta + 180)$ $1, 360°, -180°$
24. $y = \frac{1}{2} \tan\left(\theta - \frac{\pi}{2}\right)$ **does not exist, $\pi, \frac{\pi}{2}$**

25. **WHEELS** A water wheel has a diameter of 20 feet. It makes one complete revolution in 45 seconds. Let the height at the top of the wheel represent the height at time 0. Write an equation for the height of point h in the diagram below as a function of time t. Then graph the function. $h = 10 \cos \frac{2\pi}{45} t + 10$; **See margin for graph.**

Summative Assessment
Use these alternate leveled chapter tests to differentiate assessment for your students.

Leveled Chapter 12 Tests

Form	Type	Level	📁 Page(s)
1	MC	**AL**	65–66
2A	MC	**OL**	67–68
2B	MC	**OL**	69–70
2C	FR	**OL**	71–72
2D	FR	**OL**	73–74
3	FR	**BL**	75–76
Vocabulary Test			64
Extended-Response Test			77

MC = multiple-choice questions
FR = free-response questions

McGraw-Hill eAssessment Customize and create multiple versions of your chapter tests and their answer keys. All of the questions from the leveled chapter tests in the *Chapter 12 Resource Masters* are also available on McGraw-Hill eAssessment.

InterventionPlanner

TIER 1 On Level **OL**

If students miss about 25% of the exercises or less,

Then choose a resource:

SE Lessons 12-1, 12-2, 12-3, 12-4, 12-5, 12-6, 12-7, 12-8, and 12-9

Skills Practice, pp. 7, 13, 19, 25, 31, 37, 44, 49, and 55

TE Chapter Project

📁 connectED.mcgraw-hill.com Self-Check Quiz

TIER 2 Strategic Intervention **AL** approaching grade level

If students miss about 50% of the exercises,

Then choose a resource:

📁 Study Guide and Intervention, pp. 5, 11, 17, 23, 29, 35, 41, 47, and 53

connectED.mcgraw-hill.com Extra Examples, Personal Tutor, Homework Help

TIER 3 Intensive Intervention 2 or more grades below level

If students miss about 75% of the exercises,

Then use *Math Triumphs, Alg. 2* Ch. 6

connectED.mcgraw-hill.com Extra Examples, Personal Tutor, Homework Help, Review Vocabulary

1 Focus

Objective Use the strategy of using a scientific calculator to solve standardized test problems.

2 Teach

Scaffolding Questions

Ask:

- Should you use a scientific calculator for all problems on a standardized test? Sample answer: No, some problems are more efficiently solved mentally or by hand.

- With what types of problems would the use of a scientific calculator be most helpful? Sample answer: Those problems that involve complicated calculations, logarithmic functions, exponential functions, trigonometric functions, or square roots are among those problems with which the use of a scientific calculator would be most helpful.

Using a Scientific Calculator

Scientific calculators and graphing calculators are powerful problem-solving tools. As you have likely seen, some test problems that you encounter have steps or computations that require the use of a scientific calculator.

Strategies for Using a Scientific Calculator

Step 1

Familiarize yourself with the various functions of a scientific calculator as well as when they should be used.

- **Scientific notation**—for calculating large numbers

- **Logarithmic and exponential functions**—growth and decay problems, compound interest

- **Trigonometric functions**—problems involving angles, triangle problems, indirect measurement problems

- **Square roots and *n*th roots**—distance on a coordinate plane, Pythagorean Theorem

Step 2

Use your scientific or graphing calculator to solve the problem.

- Remember to work as efficiently as possible. Some steps may be done mentally or by hand, while others must be done using your calculator.

- If time permits, check your answer.

Standardized Test Example

Read the problem. Identify what you need to know. Then use the information in the problem to solve.

When Molly stands at a distance of 18 feet from the base of a tree, she forms an angle of 57° with the top of the tree. What is the height of the tree to the nearest tenth?

A 27.7 ft

B 28.5 ft

C 29.2 ft

D 30.1 ft

McGraw-Hill Companies, Inc.

Read the problem carefully. You are given some measurements and asked to find the height of a tree. It may be helpful to first sketch a model of the problem.

Use a trigonometric function to relate the lengths and the angle measure in the right triangle.

$\text{tangent } \theta = \dfrac{\text{opposite}}{\text{adjacent}}$ Definition of tangent ratio

$\tan 57° = \dfrac{h}{18}$ Substitute.

You need to evaluate $\tan 57°$ to solve for the height of the tree h. Use a scientific calculator.

$1.53986 \approx \dfrac{h}{18}$ Use a calculator.

$27.71748 \approx h$ Multiply each side by 18.

The height of the tree is about 27.7 feet. The correct answer is A.

Exercises

Read each problem. Identify what you need to know. Then use the information in the problem to solve.

1. An airplane takes off and climbs at a constant rate. After traveling 800 yards horizontally, the plane has climbed 285 yards vertically. What is the plane's angle of elevation during the takeoff and initial climb? **C**

   A 15.6°

   B 18.4°

   C 19.6°

   D 22.3°

2. What is the angle of the bike ramp below? **J**

   F 26.3°

   G 28.5°

   H 30.4°

   J 33.6°

Additional Example

STANDARDIZED TEST PRACTICE
The sum of three numbers is 35. The sum of the first two numbers is five greater than the third number. Twice the second number is equal to three times the first number. What are the three numbers? C

   A 6, 9, 20

   B 6, 10, 21

   C 8, 12, 15

   D 8, 10, 17

3 Assess

Use Exercises 1 and 2 to assess students' understanding.

Diagnose Student Errors
Survey student responses for each item. Class trends may indicate common errors and misconceptions.

1. A used cosine instead of tangent
B guess
C rounded answer incorrectly
D correct

2. F guess
G guess
H correct
J guess

3. A correct
B guess
C guess
D used diameter instead of radius in circumference formula

4. F unfamiliar with point-slope form
G unfamiliar with point-slope form
H unfamiliar with point-slope form
J correct

5. A correct
B found cos 240°
C guess
D found sin 120°

6. F did not calculate *x* and *y* correctly
G correct
H guess
J guess

7. A correct
B guess
C used wrong angle measures with side measures
D used cosine instead of sine

8. F this is a valid conclusion but the problem asked for a conclusion that is not valid
G this is a valid conclusion but the problem asked for a conclusion that is not valid
H this is a valid conclusion but the problem asked for a conclusion that is not valid
J correct

Standardized Test Practice
Cumulative, Chapters 1 through 12

Multiple Choice

Read each question. Then fill in the correct answer on the answer document provided by your teacher or on a sheet of paper.

1. What is the value of *x*? Round to the nearest tenth if necessary. **D**

A 6.5
B 6.9
C 7.1
D 7.3

2. Marvin rides his bike at a speed of 21 miles per hour and can ride his training loop 10 times in the time that it takes his younger brother to complete the training loop 8 times. Which is a reasonable estimate for Marvin's younger brother's speed? **H**

F between 14 mph and 15 mph
G between 15 mph and 16 mph
H between 16 mph and 17 mph
J between 17 mph and 18 mph

3. Suppose a Ferris wheel has a diameter of 68 feet. The wheel rotates 12° each time a new passenger is picked up. How far would you travel when the wheel rotates 12°? Round to the nearest tenth if necessary. **A**

A 7.1 ft   C 7.8 ft
B 7.5 ft   D 14.2 ft

4. What is the slope of a line parallel to $y - 2 = 4(x + 1)$? **J**

F -4 H $\frac{1}{4}$
G $-\frac{1}{4}$ J 4

5. What is the exact value of sin 240°? **A**

A $-\frac{\sqrt{3}}{2}$ C $\frac{\sqrt{2}}{3}$
B $-\frac{1}{2}$ D $\frac{\sqrt{3}}{2}$

6. What is the solution of the system of equations shown below? **G**

$$\begin{cases} x - y + z = 0 \\ -5x + 3y - 2z = -1 \\ 2x - y + 4z = 11 \end{cases}$$

F (0, 3, 3)   H no solution
G (2, 5, 3)   J infinitely many solutions

7. Find *m* in triangle *MNO* if $n = 12.4$ centimeters, $M = 35°$, and $N = 74°$. Round to the nearest tenth. **A**

A 7.4 cm   C 14.6 cm
B 8.5 cm   D 35.9 cm

8. The results of a recent poll are organized in the matrix.

	For	Against
Proposition 1	1553	771
Proposition 2	689	1633
Proposition 3	2088	229

Based on these results, which conclusion is NOT valid? **J**

F There were 771 votes cast against Proposition 1.
G More people voted against Proposition 1 than voted for Proposition 2.
H Proposition 2 has little chance of passing.
J More people voted for Proposition 1 than for Proposition 3.

9. The graph of which of the following equations is symmetrical about the *y*-axis? **C**

A $y = x^2 + 3x - 1$ C $y = 6x^2 + 9$
B $y = -x^2 + x$ D $y = 3x^2 - 3x + 1$

10. What is the remainder when $x^3 - 7x + 5$ is divided by $x + 3$? **G**

F -11 G -1 H 1   J 11

Test-TakingTip
Question 7 Use the Law of Sines to solve the triangle.

9. A the coefficient of the *x* term prevents the graph from being symmetrical about the *y*-axis
B the coefficient of the *x* term prevents the graph from being symmetrical about the *y*-axis
C correct
D the coefficient of the *x* term prevents the graph from being symmetrical about the *y*-axis

10. F error in final subtraction of the division algorithm
G correct
H error in final subtraction of the division algorithm
J error in final subtraction of the division algorithm

Short Response/Gridded Response

Record your answers on the answer sheet provided by your teacher or on a sheet of paper.

11. The speed a tsunami, or tidal wave, can travel is modeled by the equation $s = 356\sqrt{d}$, where s is the speed in kilometers per hour and d is the average depth of the water in kilometers. A tsunami is found to be traveling at 145 kilometers per hour. What is the average depth of the water? Round to the nearest hundredth. **0.17 km**

12. **GRIDDED RESPONSE** Suppose you deposit $500 in an account paying 4.5% interest compounded semiannually. Find the dollar value of the account rounded to the nearest penny after 10 years. **780.25**

13. In order to remain healthy, a horse requires 10 pounds of hay per day.

    a. Write an equation to represent the amount of hay needed to sustain x horses for d days. **$h = 10xd$**

    b. Is your equation a direct, joint, or inverse variation? Explain. **See margin.**

    c. How much hay do three horses need for the month of July? **930 lb**

14. **GRIDDED RESPONSE** What is the radius of the circle with equation $x^2 + y^2 + 8x + 8y + 28 = 0$? **2**

15. Anna is training to run a 10-kilometer race. The table below lists the times she received in several 1-kilometer races. The times are listed in minutes. Describe the center and spread of the data using either the mean and standard deviation or the five-number summary. Justify your choice. **See margin.**

7.25	8.10
7.40	6.75
7.20	7.35
7.10	7.25
8.00	7.45

16. **GRIDDED RESPONSE** The pattern of squares below continues infinitely, with more squares being added at each step. How many squares are in the tenth step? **22**

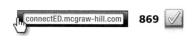

Step 1   Step 2   Step 3

Extended Response

Record your answers on a sheet of paper. Show your work.

17. Amanda's hours at her summer job for one week are listed in the table below. She earns $6 per hour.

Amanda's Work Hours	
Sunday	0
Monday	6
Tuesday	4
Wednesday	0
Thursday	2
Friday	6
Saturday	8

    a. Write an expression for Amanda's total weekly earnings. **$6(0 + 6 + 4 + 0 + 2 + 6 + 8)$**

    b. Evaluate the expression from part **a** by using the Distributive Property. **$156**

    c. Michael works with Amanda and also earns $6 per hour. If Michael's earnings were $192 this week, write and solve an equation to find how many more hours Michael worked than Amanda. **$6h + 156 = 192$; 6 hours**

Need ExtraHelp?

If you missed Question...	1	2	3	4	5	6	7	8	9	10	11	12	13	14	15	16	17
Go to Lesson...	12-1	1-3	12-2	2-4	12-3	3-4	12-4	3-5	4-1	5-2	6-3	7-2	8-5	9-3	11-2	10-2	1-3

connectED.mcgraw-hill.com **869** ✓

Additional Answer

13b. joint; Sample answer: The amount of hay needed is dependent upon both the number of days and the number of horses.

15. The distribution is skewed, so use the five-number summary. The median time is 7.3 minutes, the range is 6.75 to 8.10 minutes, and half of the data are between 7.2 and 7.45 minutes.

Formative Assessment

You can use these pages to benchmark student progress.

📁 Standardized Test Practice, pp. 78–80

McGraw-Hill
ᵉAssessment Create practice tests that align to your state standards, the Common Core State Standards, and other national standards such as TIMSS and NAEP.

Answer Sheet Practice

Have students simulate taking a standardized test by recording their answers on practice recording sheets.

📁 *Student Recording Sheet*

Homework Option

Get Ready for Chapter 13 Assign students the exercises on p. 871 as homework to assess whether they possess the prerequisite skills needed for the next chapter.

Lesson 12-1 (Guided Practice)

1. $\sin B = \frac{15}{17}$, $\cos B = \frac{8}{17}$, $\tan B = \frac{15}{8}$, $\csc B = \frac{17}{15}$, $\sec B = \frac{17}{8}$, $\cot B = \frac{8}{15}$

Lesson 12-2 (Guided Practice)

1A.

1B.

Lesson 12-2

52. Use a proportion.

$$\frac{\text{measure of the central angle}}{\text{measure of an entire circle}} = \frac{\text{the length of the arc}}{\text{the circumference}}$$

$\frac{\theta}{2\pi} = \frac{s}{2\pi r}$ Substitute.

$2\pi r\theta = 2\pi s$ Find the cross products.

$r\theta = s$ Divide each side by 2π.

53. One degree represents an angle measure that equals $\frac{1}{360}$ rotation around a circle. One radian represents the measure of an angle in standard position that intercepts an arc of length r. To change from degrees to radians, multiply the number of degrees by $\frac{\pi \text{ radians}}{180°}$. To change from radians to degrees, multiply the number of radians by $\frac{180°}{\pi \text{ radians}}$.

58. $\sin \theta = \frac{3}{\sqrt{205}}$ or $\frac{3\sqrt{205}}{205}$, $\cos \theta = \frac{14}{\sqrt{205}}$ or $\frac{14\sqrt{205}}{205}$, $\tan \theta = \frac{3}{14}$, $\csc \theta = \frac{\sqrt{205}}{3}$, $\sec \theta = \frac{\sqrt{205}}{14}$, $\cot \theta = \frac{14}{3}$

59. $\sin \theta = \frac{\sqrt{259}}{22}$, $\cos \theta = \frac{15}{22}$, $\tan \theta = \frac{\sqrt{259}}{15}$, $\csc \theta = \frac{22}{\sqrt{259}}$ or $\frac{22\sqrt{259}}{259}$, $\sec \theta = \frac{22}{15}$, $\cot \theta = \frac{15}{\sqrt{259}}$ or $\frac{15\sqrt{259}}{259}$

60. $\sin \theta = \frac{11}{\sqrt{290}}$ or $\frac{11\sqrt{290}}{290}$, $\cos \theta = \frac{13}{\sqrt{290}}$ or $\frac{13\sqrt{290}}{290}$, $\tan \theta = \frac{11}{13}$, $\csc \theta = \frac{\sqrt{290}}{11}$, $\sec \theta = \frac{\sqrt{290}}{13}$, $\cot \theta = \frac{13}{11}$

Lesson 12-3

11a.

18. 15°

19. 75°

20. 70°

21. $\frac{\pi}{4}$

22. $\frac{\pi}{4}$

23. 40°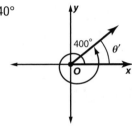

62. Sample answer:

Step 1: There are 6 bricks in the top row, and $1^2 + 5(1) = 6$, so the formula is true for $n = 1$.

Step 2: Assume that there are $k^2 + 5k$ bricks in the top k rows for some positive integer k.

Step 3: Since each row has 2 more bricks than the one above, the numbers of bricks in the rows form an arithmetic sequence. The number of bricks in the $(k + 1)$st row is $6 + [(k + 1) - 1](2)$ or $2k + 6$. Then the number of bricks in the top $k + 1$ rows is $k^2 + 5k + (2k + 6)$ or $k^2 + 7k + 6$. $k^2 + 7k + 6 = (k + 1)^2 + 5(k + 1)$, which is the formula to be proved, where $n = k + 1$. Thus, the formula is true for $n = k + 1$. Therefore, the number of bricks in the top n rows in $n^2 + 5n$ for all positive integers n.

Lesson 12-4

45. Sample answer:

$\sin A = \frac{\text{opposite}}{\text{hypotenuse}}$	Definition of sine
$\sin A = \frac{h}{c}$	$h = $ opposite side, $c = $ hypotenuse
$c \sin A = h$	Multiply both sides by c.
$\text{Area} = \frac{1}{2} \cdot \text{base} \cdot \text{height}$	Area of a triangle
$\text{Area} = \frac{1}{2} bh$	$b = $ base, $h = $ height
$\text{Area} = \frac{1}{2} bc \sin A$	Substitution

Lesson 12-5

37. Sample answer: To solve a right triangle, you can use the Pythagorean Theorem to find side lengths and trigonometric ratios to find angle measures and side lengths. To solve a nonright triangle, you can use the Law of Sines or the Law of Cosines, depending on what information is given. When two angles and a side are given or when two sides and an angle opposite one of the sides are given, you can use the Law of Sines. When two sides and an included angle or three sides are given, you can use the Law of Cosines.

Mid-Chapter Quiz

12. $\sin \theta = -1$, $\cos \theta = 0$, $\tan \theta = $ undefined, $\csc \theta = -1$, $\sec \theta = $ undefined, $\cot \theta = 0$

13. $\sin \theta = \frac{4}{5}$, $\cos \theta = \frac{3}{5}$, $\tan \theta = \frac{4}{3}$, $\csc \theta = \frac{5}{4}$, $\sec \theta = \frac{5}{3}$, $\cot \theta = \frac{3}{4}$

16. two solutions: $C = 59°$, $B = 83°$, $b = 29.0$ or $C = 121°$, $B = 21°$, $b = 10.5$

Lesson 12-6

29a.

29b.

Angle	Slope
30	0.6
60	1.7
120	−1.7
150	−0.6
210	0.6
315	−1

29c. Sample answer: The slope corresponds to the tangent of the angle. For $\theta = 120°$, the x-coordinate of P is $-\frac{1}{2}$ and the y-coordinate is $= \frac{\sqrt{3}}{2}$; slope $= \frac{\text{change in } y}{\text{change in } x}$.
Since change in $x = -\frac{1}{2}$ and change in $y = \frac{\sqrt{3}}{2}$,
slope $= \frac{\sqrt{3}}{2} \div \left(-\frac{1}{2}\right) = -\sqrt{3}$ or about -1.7.

Lesson 12-7 (Guided Practice)

3B.

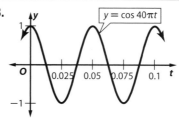

Lesson 12-7

6. period: 180°

7. period: 360°

8. period: 90°

9. amplitude: 2; period: 360°

10. amplitude: 3; period: 360°

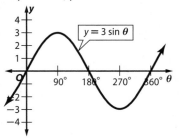

11. amplitude: 1; period: 180°

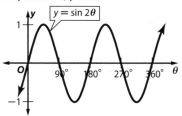

12. amplitude: 1; period: 120°

13. amplitude: 1; period: 720°

14. amplitude: 1; period: 90°

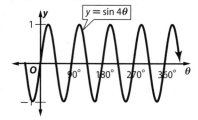

15. amplitude: $\frac{3}{4}$; period: 360°

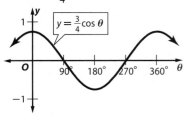

16. amplitude: $\frac{3}{2}$; period: 360°

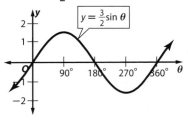

17. amplitude: $\frac{1}{2}$; period: 180°

18. amplitude: 4; period: 180°

19. amplitude: 3; period: 180°

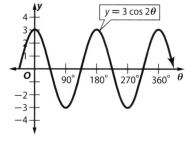

20. amplitude: 5; period: 540°

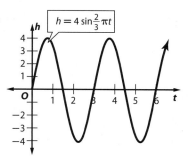

$y = 5 \sin \frac{2}{3}\theta$

21b.

$h = 4 \sin \frac{2}{3}\pi t$

22b.

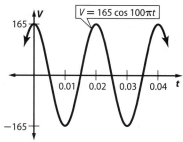

$V = 165 \cos 100\pi t$

23. period: 360°

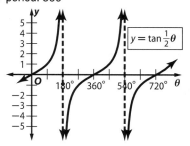

$y = \tan \frac{1}{2}\theta$

24. period: 360°

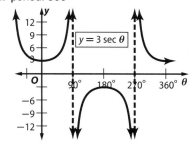

$y = 3 \sec \theta$

25. period: 180°

$y = 2 \cot \theta$

26. period: 720°

$y = \csc \frac{1}{2}\theta$

27. period: 180°

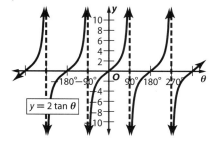

$y = 2 \tan \theta$

28. period: 1080°

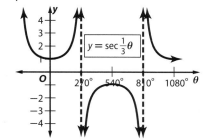

$y = \sec \frac{1}{3}\theta$

31a.

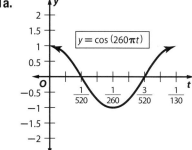

$y = \cos(260\pi t)$

32. amplitude: 3; period: 540°

$y = 3 \sin \frac{2}{3}\theta$

33. amplitude: $\frac{1}{2}$; period: 480°

$y = \frac{1}{2} \cos \frac{3}{4}\theta$

34. amplitude: does not exist; period: 360°

$y = 2 \tan \frac{1}{2}\theta$

35. amplitude: does not exist; period: 450°

$y = 2 \sec \frac{4}{5}\theta$

36. amplitude: does not exist; period: 120°

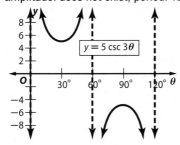
$y = 5 \csc 3\theta$

37. amplitude: does not exist; period: 30°

$y = 2 \cot 6\theta$

44. Sample answer: Determine the amplitude and period of the function; find and graph any x-intercepts, extrema, and asymptotes; use the parent function to sketch the graph.

Explore 12-8

1. **Y1:** D = {all real numbers}, R = {y | −1 ≤ y ≤ 1}, **Y2:** D = {all real numbers}, R = {y | 1 ≤ y ≤ 3}, **Y3:** D = {all real numbers}, R = {y | −4 ≤ y ≤ 2}; **Y1 − Y3:** D = {all real numbers}, R = {y | −1 ≤ y ≤ 1}

4. Sample answer: The graph of $y = \sin (2\theta) + 4$ goes through two cycles in 360°, and the graph of $y = \sin \theta + 4$ goes through one cycle in 360°.

5. Sample answer: The graph of $y = \cos \frac{1}{2} (\theta + 45°)$ has been translated 45° left from the graph of $y = \cos \left(\frac{1}{2}\theta\right)$.

6. Sample answer: The graph of $y = 2 \sin \theta − 1$ has been translated down 1 unit from the graph of $y = 2 \sin \theta$.

7. Sample answer: The graph of $y = \cos (\theta − 90°) − 3$ has been translated 90° right from the graph of $y = \cos \theta − 3$.

Lesson 12-8 (Guided Practice)

1.

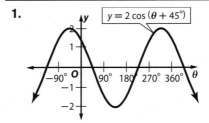
$y = 2 \cos (\theta + 45°)$

Lesson 12-8

9. 2; 360°; $h = −45°$; $k = 1$

$y = 2 \sin (\theta + 45°) + 1$

10. 1; 120°; $h = \pi$; $k = -4$

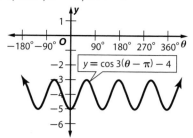

$y = \cos 3(\theta - \pi) - 4$

11. no amplitude; 90°; $h = -30°$; $k = 3$

$y = \frac{1}{4}\tan 2(\theta + 30°) + 3$

12. 4; 4π; $h = \frac{\pi}{2}$; $k = 5$

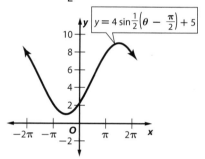

$y = 4 \sin \frac{1}{2}\left(\theta - \frac{\pi}{2}\right) + 5$

13. $P = 20 \sin 3\pi t + 110$

$P = 20 \sin 3\pi t + 110$

14. 1; 360°; $h = -180°$

$y = \cos(\theta + 180°)$

15. no amplitude; 180°; $h = 90°$

$y = \tan(\theta - 90°)$

16. 1; 2π; $h = -\pi$

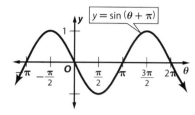

$y = \sin(\theta + \pi)$

17. 2; 2π; $h = -\frac{\pi}{2}$

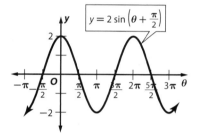

$y = 2 \sin\left(\theta + \frac{\pi}{2}\right)$

18. no amplitude; 360°; $h = -30°$

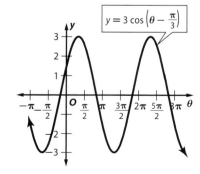

$y = \tan \frac{1}{2}(\theta + 30°)$

19. 3; 2π; $h = \frac{\pi}{3}$

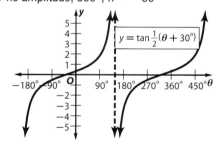

$y = 3 \cos\left(\theta - \frac{\pi}{3}\right)$

Chapter 12 Answer Appendix

20. 1; 360°; $k = 3$; $y = 3$

21. no amplitude; 180°; $k = -1$; $y = -1$

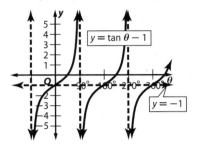

22. no amplitude; 180°; $k = \frac{1}{2}$; $y = \frac{1}{2}$

23. 2; 360°; $k = -5$; $y = -5$

24. 2; 360°; $k = -4$; $y = -4$

25. $\frac{1}{3}$; 360°; $k = 7$; $y = 7$

26. 4; 360°; $h = 60°$; $k = -1$

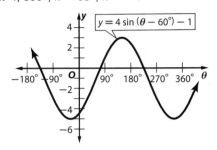

27. 1; 720°; $h = 90°$; $k = 2$

28. no amplitude; 180°; $h = -30°$; $k = -2$

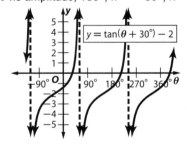

29. no amplitude; $\frac{\pi}{2}$; $h = -\frac{\pi}{4}$; $k = -5$

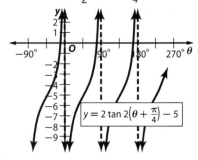

30. $\frac{1}{2}$; 2π; $h = \frac{\pi}{2}$; $k = 4$

$y = \frac{1}{2}\sin\left(\theta - \frac{\pi}{2}\right) + 4$

31. 1; 120°; $h = 45°$; $k = \frac{1}{2}$

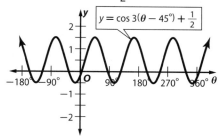

$y = \cos 3(\theta - 45°) + \frac{1}{2}$

32. 5; π; $h = \pi$; $k = 3$

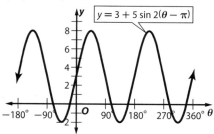

$y = 3 + 5\sin 2(\theta - \pi)$

33. 3; 6π; $h = \frac{\pi}{2}$; $k = -2$

$y = -2 + 3\sin\frac{1}{3}\left(\theta - \frac{\pi}{2}\right)$

54. 360°; $h = -\pi$; no vertical shift

$y = \csc(\theta + \pi)$

55. 180°; no phase shift; $k = 6$

$y = \cot\theta + 6$

56. π; $h = \frac{\pi}{6}$; $k = -2$

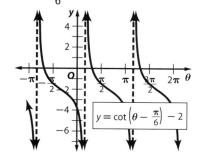

$y = \cot\left(\theta - \frac{\pi}{6}\right) - 2$

57. 120°; $h = 45°$; $k = 1$

$y = \frac{1}{2}\csc 3(\theta - 45°) + 1$

58. 720°; $h = 90°$; no vertical shift

$y = 2\sec\frac{1}{2}(\theta - 90°)$

59. π; $h = -\frac{\pi}{2}$; $k = -3$

$y = 4\sec 2\left(\theta + \frac{\pi}{2}\right) - 3$

	Diagnostic Assessment Quick Check		
	LESSON 13-1 45 min: 1 day / 90 min: 0.5 day	**LESSON 13-2** 45 min: 1 day / 90 min: 0.5 day	**LESSON 13-3** 45 min: 1 day / 90 min: 0.5 day
Title	Trigonometric Identities	Verifying Trigonometric Identities	Sum and Difference of Angles Identities
Objectives	■ Use trigonometric identities to find trigonometric values. ■ Use trigonometric identities to simplify expressions.	■ Verify trigonometric identities by transforming one side of an equation into the form of the other side. ■ Verify trigonometric identities by transforming each side of the equation into the same form.	■ Find values of sine and cosine by using sum and difference identities. ■ Verify trigonometric identities by using sum and difference identities.
Key Vocabulary	trigonometric identity		
CCSS	F.TF.8	F.TF.8	F.TF.8
Multiple Representations	🔧		🔧
Lesson Resources	connectED.mcgraw-hill.com 📁 Leveled Worksheets 🔤 Vocabulary PT Personal Tutor ☑ Self-Check Quiz ■ *5-Minute Check* ■ *Study Notebook*	connectED.mcgraw-hill.com 📁 Leveled Worksheets 📁 Quiz 1 PT Personal Tutor ☑ Self-Check Quiz ■ *5-Minute Check* ■ *Study Notebook*	connectED.mcgraw-hill.com 📁 Leveled Worksheets 📁 Quiz 2 PT Personal Tutor ☑ Self-Check Quiz ■ *5-Minute Check* ■ *Study Notebook*
Resources for Every Lesson	IWB eStudent Edition IWB Interactive Classroom	■ eTeacher Edition ■ eSolutions Manual ■ McGraw-Hill *e*Assessment	
Differentiated Instruction	pp. 874, 875, 879	pp. 881, 885	pp. 887, 891
			Formative Assessment Mid-Chapter Quiz

IWB All digital assets are Interactive Whiteboard ready.

Suggested Pacing			
Time Periods	Instruction	Review & Assess	Total
45-minute	6 days	2 days	8 days
90-minute	3 days	1 day	4 days

LESSON **13-4** 45 min: 1 day / 90 min: 0.5 day	EXPLORE **13-5** 45 min: 0.5 day / 90 min: 0.25 day	LESSON **13-5** 45 min: 1.5 days / 90 min: 0.75 day
Double-Angle and Half-Angle Identities	**Graphing Technology Lab: Solving Trigonometric Equations**	**Solving Trigonometric Equations**
■ Find values of sine and cosine by using double-angle identities. ■ Find values of sine and cosine by using half-angle identities.	■ Use a graphing calculator to solve trigonometric equations.	■ Solve trigonometric equations. ■ Find extraneous solutions from trigonometric equations.
		trigonometric equations
F.TF.8		F.TF.8
		🔗
connectED.mcgraw-hill.com	connectED.mcgraw-hill.com	connectED.mcgraw-hill.com
🗀 Leveled Worksheets 🗀 Quiz 3 PT Personal Tutor ✓ Self-Check Quiz ■ *5-Minute Check* ■ *Study Notebook*	PT Personal Tutor **Materials:** ■ TI-83/84 Plus or other graphing calculator	🗀 Leveled Worksheets 🗀 Quiz 4 ꞯᵇᶜ Vocabulary PT Personal Tutor ✓ Self-Check Quiz ■ *5-Minute Check* ■ *Study Notebook* ■ *Teaching Algebra with Manipulatives*
IWB eStudent Edition IWB Interactive Classroom	■ eTeacher Edition ■ eSolutions Manual ■ ᵉ McGraw-Hill Assessment	
pp. 896, 899		pp. 903, 907
		Summative Assessment Study Guide and Review Practice Test

SE = Student Edition, TE = Teacher Edition, CRM = Chapter Resource Masters

Diagnosis	Prescription
Beginning Chapter 13	
Get Ready for Chapter 13 **SE**	Response to Intervention **TE**
Beginning Every Lesson	
Then, Now, Why? **SE** 5-Minute Checks	Chapter 0 **SE**

DIAGNOSTIC ASSESSMENT

Diagnosis	Prescription
During/After Every Lesson	
Guided Practice **SE**, every example Check Your Understanding **SE** H.O.T. Problems **SE** Spiral Review **SE** Additional Examples **TE** Watch Out! **TE** Step 4, Assess **TE** Chapter 13 Quizzes **CRM**, pp. 37–38 Self-Check Quizzes connectED.mcgraw-hill.com	**TIER 1 Intervention** Skills Practice **CRM**, Ch. 1–13 connectED.mcgraw-hill.com **TIER 2 Intervention** Differentiated Instruction **TE** Study Guide and Intervention **CRM**, Ch. 1–13 **TIER 3 Intervention** *Math Triumphs, Alg. 2*
Mid-Chapter	
Mid-Chapter Quiz **SE** Mid-Chapter Test **CRM**, p. 39 McGraw-Hill eAssessment	**TIER 1 Intervention** Skills Practice **CRM**, Ch. 1–13 connectED.mcgraw-hill.com **TIER 2 Intervention** Study Guide and Intervention **CRM**, Ch. 1–13 **TIER 3 Intervention** *Math Triumphs, Alg. 2*
Before Chapter Test	
Chapter Study Guide and Review **SE** Practice Test **SE** Standardized Test Practice **SE** Chapter Test connectED.mcgraw-hill.com Standardized Test Practice connectED.mcgraw-hill.com Vocabulary Review connectED.mcgraw-hill.com McGraw-Hill eAssessment	**TIER 1 Intervention** Skills Practice **CRM**, Ch. 1–13 connectED.mcgraw-hill.com **TIER 2 Intervention** Study Guide and Intervention **CRM**, Ch. 1–13 **TIER 3 Intervention** *Math Triumphs, Alg. 2*

FORMATIVE ASSESSMENT

Diagnosis	Prescription
After Chapter 13	
Multiple-Choice Tests, Forms 1, 2A, 2B **CRM**, pp. 41–46 Free-Response Tests, Forms 2C, 2D, 3 **CRM**, pp. 47–52 Vocabulary Test **CRM**, p. 40 Extended Response Test **CRM**, p. 53 Standardized Test Practice **CRM**, pp. 54–56 McGraw-Hill eAssessment	Study Guide and Intervention **CRM**, Ch. 1–13 connectED.mcgraw-hill.com

SUMMATIVE ASSESSMENT

Option 1 Reaching All Learners

Kinesthetic Put students into pairs. Have each pair of students make up a set of Basic Trigonometric Identities cards. Pairs should make up two cards for every type of identity within each category (quotient, reciprocal, Pythagorean, cofunction, and negative angle). For example, for the quotient identities, four cards will be made: $\tan \theta$, $\dfrac{\sin \theta}{\cos \theta}$, $\cot \theta$, and $\dfrac{\cos \theta}{\sin \theta}$. A total of 28 should be made. Then, pairs will play a game of "Memory." The first partner will turn over two cards. The two cards are removed if they represent equal basic trigonometric values. If not, both are turned back over. The other partner then takes a turn.

$\cot \theta$	$\dfrac{\cos \theta}{\sin \theta}$				

Logical Have students work in pairs to create their own Trigonometric Identity equations. Ask students to start with a true statement such as $\cos \theta = \cos \theta$. Then, transform each side by substituting equivalent expressions. For example, $\cos \theta = \cos \theta$ can be transformed using the quotient identities to give the following result, $\dfrac{\sin \theta}{\tan \theta} + \dfrac{\cos \theta}{\sin \theta} = \dfrac{1}{\sec \theta} + \cot \theta$. Then have pairs trade and try to verify that each other's equation is an identity.

Option 2 Approaching Level AL

Work with students to create and complete a table like the one below that they can use as a reference for knowing the sign of each function of θ in each quadrant.

Quadrant	I	II	III	IV
Sine	+	+	−	−
Cosine	+	−	−	+
Tangent	+	−	+	−

Option 3 English Learners ELL

Have students make up a half-angle identity. Then ask them to verify that it actually is an identity by explaining in words or in writing each step needed to do so.

Option 4 Beyond Level BL

Have students make up a crossword puzzle using any of the terms related to trigonometry used in this chapter. Use either the definition or an example of the terms in the *across* and *down* hints. Have each student make photocopies of his/her puzzles to distribute to other class members. Ask students to complete the puzzles.

Focus on Mathematical Content

VerticalAlignment

Before Chapter 13

Previous Topics from Algebra 2

- investigate the six trigonometric functions
- work with angles measured in degrees or radians
- investigate angles as rays in standard position and as points on the unit circle
- explore periodicity and inverse trigonometric functions

Chapter 13

Related Topics from Algebra 2

- learn how to verify and use trigonometric identities
- prove trigonometric identities
- solve trigonometric equations using the ideas of factoring, the zero product property, trigonometric inverses, and periodic behavior

After Chapter 13

Preparation for Precalculus

- use trigonometric formulas and identities while analyzing vertical and horizontal translations of graphs and how these translations are related to changes in the algebraic description of the graphs

Lesson-by-LessonPreview

13-1 Trigonometric Identities

An equation that involves trigonometric functions that is true for all values for which every expression in the equation is defined is called a *trigonometric identity*. There are five categories of identities: Quotient, Reciprocal, Pythagorean, Cofunction, and Negative Angle Identities. These are shown in the table below.

Basic Trigonometric Identities		
Quotient Identities (no denominator equals 0)		
$\tan \theta = \dfrac{\sin \theta}{\cos \theta}$		$\cot \theta = \dfrac{\cos \theta}{\sin \theta}$
Reciprocal Identities (no denominator equals 0)		
$\csc \theta = \dfrac{1}{\sin \theta}$	$\sec \theta = \dfrac{1}{\cos \theta}$	$\cot \theta = \dfrac{1}{\tan \theta}$
Pythagorean Identities		
$\cos^2 \theta + \sin^2 \theta = 1$	$\tan^2 \theta + 1 = \sec^2 \theta$	$\cot^2 \theta + 1 = \csc^2 \theta$
Cofunction Identities		
$\sin \left(\dfrac{\pi}{2} - \theta\right) = \cos \theta$	$\cos \left(\dfrac{\pi}{2} - \theta\right) = \sin \theta$	$\tan \left(\dfrac{\pi}{2} - \theta\right) = \cot \theta$
Negative Angle Identities		
$\sin (-\theta) = -\sin \theta$	$\cos (-\theta) = \cos \theta$	$\tan (-\theta) = -\tan \theta$

These identities are helpful when simplifying trigonometric expressions and solving real-world problems.

13-2 Verifying Trigonometric Identities

Basic trigonometric identities and definitions can be used to verify identities. To verify an identity, the goal is to either transform one side of the equation, replacing expressions with equivalent expressions, until the two sides are identical or transform both sides of an equation separately into a common form. There are several approaches for writing equivalent expressions.

- Make substitutions using the Pythagorean Identities.
- Use the Distributive Property to factor an expression or to collect like terms.
- Transform a term by multiplying the term by an expression equivalent to 1.
- Rewrite all the trigonometric functions in terms of $\sin \theta$ or $\cos \theta$ by using the Quotient and Reciprocal Identities.

> **Warning:** Do not perform operations to each side of the unverified identity because the properties of equality do not apply to identities as they do to equations.

The Math Forum @ Drexel University mathforum.org

Get educator support from the Math Forum's T2T. You may want to search for, or ask about, trigonometry projects. Start your search at connectED.mcgraw-hill.com.

13-3 Sum and Difference of Angles Identities

The identities for the sum and difference of two angles, $\sin(a \pm b)$, $\cos(a \pm b)$, and $\tan(a \pm b)$, can be used to find values of sine and cosine. They can also be used to help verify identities such as $\sin(180° + \theta) = -\sin\theta$.

The sum and difference of angles identities are as follows.

- $\sin(a + b) = \sin a \cos b + \cos a \sin b$
- $\sin(a - b) = \sin a \cos b - \cos a \sin b$
- $\cos(a + b) = \cos a \cos b - \sin a \sin b$
- $\cos(a - b) = \cos a \cos b + \sin a \sin b$
- $\tan(a + b) = \dfrac{\tan a + \tan b}{1 - \tan a \tan b}$
- $\tan(a - b) = \dfrac{\tan a - \tan b}{1 + \tan a \tan b}$

13-4 Double-Angle and Half-Angle Identities

Using the identities for $\sin(a + b)$, $\cos(a + b)$, and $\tan(a + b)$, and replacing both a and b with θ, results in identities called the Double-Angle Identities. These identities are:

- $\sin 2\theta = 2 \sin\theta \cos\theta$
- $\cos 2\theta = 2 \cos^2\theta - 1$ or
  $\cos 2\theta = \cos^2\theta - \sin^2\theta$ or
  $\cos 2\theta = 1 - 2 \sin^2\theta$
- $\tan 2\theta = \dfrac{2 \tan\theta}{1 - \tan^2\theta}$

Three identities called Half-Angle Identities can be derived from the Double-Angle Identities. These identities are:

- $\sin \dfrac{a}{2} = \pm\sqrt{\dfrac{1 - \cos a}{2}}$
- $\cos \dfrac{a}{2} = \pm\sqrt{\dfrac{1 + \cos a}{2}}$
- $\cos \dfrac{a}{2} = \pm\sqrt{\dfrac{1 - \cos a}{1 + \cos a}}$, $\cos a \neq -1$

The Half-Angle and Double-Angle Identities, along with other identities, can be used to find exact values for particular trigonometric expressions. They can also be used to verify trigonometric identities.

13-5 Solving Trigonometric Equations

Trigonometric equations are true for only certain values of the variable. Solving a trigonometric equation is similar to solving an algebraic equation.

- The first step in solving a trigonometric equation is to use factoring, the Zero Product Property, and/or identities to rewrite a complicated equation as a string of simpler trigonometric equations.

- The second step is to use trigonometric inverses to isolate the variable; that is, to solve an equation such as $\cos\theta = -1$, for θ.

- The third step is to use ideas of periodicity to include all the occurrences of that value.

Some equations have infinitely many solutions and some equations have no solution. Hence, it is necessary to check your solutions in the original equation.

McGraw Hill Professional Development

Targeted professional development has been articulated throughout Algebra 2. More quality, customized professional development is available from McGraw-Hill Professional Development. Visit connectED.mcgraw-hill.com for details on each product.

- Online Lessons emphasize the strategies and techniques used to teach Algebra 2. Includes streaming video, interactive pages, and online tools.

- Video Workshops allow mentors, coaches, or leadership personnel to facilitate on-site workshops on educational strategies in mathematics and mathematical concepts.

- MHPD Online (www.mhpdonline.com) offers online professional development with video clips of instructional strategies, links, student activities, and news and issues in education.

- Teaching Today (teachingtoday.glencoe.com) gives secondary teachers practical strategies and materials that inspire excellence and innovation in teaching.

Chapter Project

Spin the Wheel

Students use what they have learned about trigonometric identities to complete a project.

This chapter project addresses business literacy, as well as several specific skills identified as being essential to student success by the Framework for 21st Century Learning.

Visit connectED.mcgraw-hill.com for student and teacher handouts.

KeyVocabulary Introduce the key vocabulary in the chapter using the routine below.

<u>Define</u>: A trigonometric identity is an equation involving a trigonometric function that is true for all values of the variable.

<u>Example</u>: The identity $\sin(-\theta) = -\sin\theta$ is a Negative Angle Identity.

<u>Ask</u>: What other identities have you learned about this year? Answers may vary; students may mention identity matrices or the identity function.

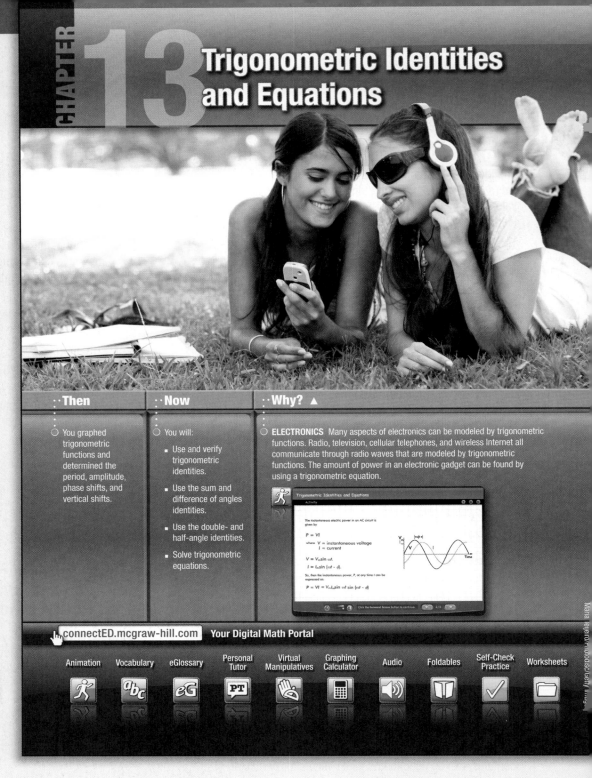

CHAPTER 13 Trigonometric Identities and Equations

Then
- You graphed trigonometric functions and determined the period, amplitude, phase shifts, and vertical shifts.

Now
- You will:
  - Use and verify trigonometric identities.
  - Use the sum and difference of angles identities.
  - Use the double- and half-angle identities.
  - Solve trigonometric equations.

Why? ▲
ELECTRONICS Many aspects of electronics can be modeled by trigonometric functions. Radio, television, cellular telephones, and wireless Internet all communicate through radio waves that are modeled by trigonometric functions. The amount of power in an electronic gadget can be found by using a trigonometric equation.

connectED.mcgraw-hill.com **Your Digital Math Portal**

| Animation | Vocabulary | eGlossary | Personal Tutor | Virtual Manipulatives | Graphing Calculator | Audio | Foldables | Self-Check Practice | Worksheets |

Get Ready for the Chapter

Diagnose Readiness | You have two options for checking prerequisite skills.

1 **Textbook Option** Take the Quick Check below. Refer to the Quick Review for help.

QuickCheck	QuickReview

Factor completely. If the polynomial is not factorable, write *prime*.

1. $-16a^2 + 4a$ $-4a(4a - 1)$ **2.** $5x^2 - 20$ $5(x + 2)(x - 2)$

3. $x^3 + 9$ prime **4.** $2y^2 - y - 15$ $(2y + 5)(y - 3)$

5. GEOMETRY The area of a rectangular piece of cardboard is $x^2 + 6x + 8$ square inches. If the cardboard has a length of $(x + 4)$ inches, what is the width? $(x + 2)$ in.

Example 1 (Used in Lesson 13-2)

Factor $x^3 + 2x^2 - 24x$ completely.

$x^3 + 2x^2 - 24x = x(x^2 + 2x - 24)$

The product of the coefficients of the x terms must be -24, and their sum must be 2. The product of 6 and -4 is -24 and their sum is 2.

$x(x^2 + 2x - 24) = x(x + 6)(x - 4)$

Solve each equation by factoring.

6. $x^2 + 6x = 0$ $\{-6, 0\}$ **7.** $x^2 + 2x - 35 = 0$ $\{-7, 5\}$

8. $x^2 - 9 = 0$ $\{-3, 3\}$ **9.** $x^2 - 7x + 12 = 0$ $\{3, 4\}$

10. GARDENING Peyton is building a flower bed in her back yard. The area of the flower bed will be 42 square feet. Find the possible values for x. 6

x ft

$x + 1$ ft

Example 2 (Used in Lesson 13-7)

Solve $x^2 + 6x + 5 = 0$ by factoring.

$x^2 + 6x + 5 = 0$ Original equation

$(x + 5)(x + 1) = 0$ Factor.

$x + 5 = 0$ or $x + 1 = 0$

$x = -5$ $x = -1$

The solution set is $\{-5, -1\}$.

Find the exact value of each trigonometric function.

11. $\sin 45°$ $\dfrac{\sqrt{2}}{2}$ **12.** $\cos 225°$ $-\dfrac{\sqrt{2}}{2}$

13. $\tan 150°$ $-\dfrac{\sqrt{3}}{3}$ **14.** $\sin 120°$ $\dfrac{\sqrt{3}}{2}$

15. RIDES The distance from the highest point of a Ferris wheel to the ground can be found by multiplying 90 feet by $\sin 90°$. What is the height of the Ferris wheel when it is halfway between the tallest point and the ground? 45 ft

Example 3 (Used in Lesson 13-3)

Find the exact value of $\cos 135°$.

The reference angle is $180° - 135°$ or $45°$.

$\cos 45°$ is $\dfrac{\sqrt{2}}{2}$. Since $135°$ is in the second quadrant, $\cos 135° = -\dfrac{\sqrt{2}}{2}$.

2 **Online Option** Take an online self-check Chapter Readiness Quiz at connectED.mcgraw-hill.com.

871

Essential Question

- **How can representing the same mathematical concept in different ways be helpful?** Sample answer: Depending on the situation, it might be more helpful to use a visual representation such as a graph or diagram. In other situations, it might be more helpful to use a numerical or algebraic representation such as a table of values or equation.

Get Ready for the Chapter

Response to Intervention (RtI)
Use the *Quick Check* results and the Intervention Planner to help you determine your Response to Intervention. The If-Then statements in the chart below help you decide the appropriate tier of RtI and suggest intervention resources for each tier.

Intervention Planner

TIER 1 **On Level** **OL**

 If students miss about 25% of the exercises or less,

Then choose a resource:

SE Lessons 5-5, 4-3, and 12-3

Skills Practice, Chapter 5, p. 37; Chapter 4, p. 19 and Chapter 12, p. 19

connectED.mcgraw-hill.com Self-Check Quiz

TIER 2 **Strategic Intervention** **AL** approaching grade level

 If students miss about 50% of the exercises,

Then choose a resource:

Study Guide and Intervention, Chapter 5, pp. 33–36, Chapter 4, pp. 17–18, Chapter 12, pp. 17–18

connectED.mcgraw-hill.com Extra Examples, Personal Tutor, Homework Help

TIER 3 **Intensive Intervention** 2 or more grades below level

 If students miss about 75% of the exercises,

Then use *Math Triumphs, Alg. 2*

connectED.mcgraw-hill.com Extra Examples, Personal Tutor, Homework Help, Review Vocabulary

FOLDABLES StudyOrganizer

Dinah Zike's Foldables®

Focus Students write notes as they explore trigonometric identities and equations in the lessons of this chapter.

Teach Have students make and label their Foldables as illustrated. Have students use the appropriate tab as they cover each lesson in this chapter. Encourage students to apply what they have learned about trigonometric identities by writing their own examples as well.

When to Use It Suggest that students add to their Foldables as they work through the chapter and to use them to review concepts for the chapter test.

Differentiated Instruction

📁 Student-Built Glossary, pp. 1–2

Students should complete the chart by providing a definition of each term and an example as they progress through Chapter 13. This study tool can also be used to review for the chapter test.

Get Started on the Chapter

You will learn several new concepts, skills, and vocabulary terms as you study Chapter 13. To get ready, identify important terms and organize your resources. You may wish to refer to Chapter 0 to review prerequisite skills.

FOLDABLES StudyOrganizer

Trigonometric Identities and Equations Make this Foldable to help you organize your Chapter 13 notes about trigonometric identities and equations. Begin with one sheet of 11″ × 17″ paper and four sheets of grid paper.

1 **Fold** the short sides of the 11″ × 17″ paper to meet in the middle.

2 **Cut** each tab in half as shown.

3 **Cut** four sheets of grid paper in half and fold the half-sheets in half.

4 **Insert** two folded half-sheets under each of the four tabs and staple along the fold. Label each tab as shown.

NewVocabulary

English		Español
trigonometric identity	p. 873	identidad trigonométrica
quotient identity	p. 873	identidad de cociente
reciprocal identity	p. 873	identidad recíproca
Pythagorean identity	p. 873	identidad Pitagórica
cofunction identity	p. 873	identidad de función conjunta
negative angle identity	p. 873	identidad negativa de ángulo
trigonometric equation	p. 901	ecuación trigonométrica

ReviewVocabulary

formula *fórmula* a mathematical sentence that expresses the relationship between certain quantities

identity *identidad* an equality that remains true regardless of the values of any variables that are in it

trigonometric functions *funciones rigonométricas* For any angle, with measure θ, a point $P(x, y)$ on its terminal side, $r = \sqrt{x^2 + y^2}$, the trigonometric functions of θ are as follows.

$$\sin \theta = \frac{y}{r} \qquad \cos \theta = \frac{x}{r} \qquad \tan \theta = \frac{y}{x}$$

$$\csc \theta = \frac{r}{y} \qquad \sec \theta = \frac{r}{x} \qquad \cot \theta = \frac{x}{y}$$

McGraw-Hill Companies, Inc.

LESSON 13-1
Trigonometric Identities

:: Then	:: Now	:: Why?
• You evaluated trigonometric functions.	**1** Use trigonometric identities to find trigonometric values. **2** Use trigonometric identities to simplify expressions.	• The amount of light that a source provides to a surface is called the *illuminance*. The illuminance E in foot candles on a surface is related to the distance R in feet from the light source. The formula $\sec \theta = \frac{I}{ER^2}$, where I is the intensity of the light source measured in candles and θ is the angle between the light beam and a line perpendicular to the surface, can be used in situations in which lighting is important, as in photography.

New Vocabulary
trigonometric identity

Common Core State Standards

Content Standards
F.T.F.8 Prove the Pythagorean identity $\sin^2(\theta) + \cos^2(\theta) = 1$ and use it to find $\sin(\theta)$, $\cos(\theta)$, or $\tan(\theta)$ given $\sin(\theta)$, $\cos(\theta)$, or $\tan(\theta)$ and the quadrant of the angle.

Mathematical Practices
2 Reason abstractly and quantitatively.
7 Look for and make use of structure.

1 Find Trigonometric Values The equation above can also be written as $E = \frac{I \cos \theta}{R^2}$. This is an example of a trigonometric identity. A **trigonometric identity** is an equation involving trigonometric functions that is true for all values for which every expression in the equation is defined.

If you can show that a specific value of the variable in an equation makes the equation false, then you have produced a *counterexample*. It only takes one counterexample to prove that an equation is not an identity.

KeyConcept Basic Trigonometric Identities

Quotient Identities

$$\tan \theta = \frac{\sin \theta}{\cos \theta}, \qquad\qquad \cot \theta = \frac{\cos \theta}{\sin \theta},$$
$$\cos \theta \neq 0 \qquad\qquad\qquad \sin \theta \neq 0$$

Reciprocal Identities

$$\sin \theta = \frac{1}{\csc \theta}, \csc \theta \neq 0 \qquad\qquad \csc \theta = \frac{1}{\sin \theta}, \sin \theta \neq 0$$
$$\cos \theta = \frac{1}{\sec \theta}, \sec \theta \neq 0 \qquad\qquad \sec \theta = \frac{1}{\cos \theta}, \cos \theta \neq 0$$
$$\tan \theta = \frac{1}{\cot \theta}, \cot \theta \neq 0 \qquad\qquad \cot \theta = \frac{1}{\tan \theta}, \tan \theta \neq 0$$

Pythagorean Identities

$$\cos^2 \theta + \sin^2 \theta = 1 \qquad \tan^2 \theta + 1 = \sec^2 \theta \qquad \cot^2 \theta + 1 = \csc^2 \theta$$

Cofunction Identities

$$\sin\left(\frac{\pi}{2} - \theta\right) = \cos \theta \qquad \cos\left(\frac{\pi}{2} - \theta\right) = \sin \theta \qquad \tan\left(\frac{\pi}{2} - \theta\right) = \cot \theta$$

Negative Angle Identities

$$\sin(-\theta) = -\sin \theta \qquad \cos(-\theta) = \cos \theta \qquad \tan(-\theta) = -\tan \theta$$

The negative angle identities are sometimes called *odd-even* identities.

The identity $\tan \theta = \frac{\sin \theta}{\cos \theta}$ is true except for angle measures such as $90°$, $270°$, … , $90° + k180°$, where k is an integer. The cosine of each of these angle measures is 0, so $\tan \theta$ is not defined when $\cos \theta = 0$. An identity similar to this is $\cot \theta = \frac{\cos \theta}{\sin \theta}$.

connectED.mcgraw-hill.com **873**

---

1 Focus

Vertical Alignment

Before Lesson 13-1 Evaluate trigonometric functions.

Lesson 13-1 Use trigonometric identities to find trigonometric values. Use trigonometric identities to simplify expressions.

After Lesson 13-1 Use trigonometry to solve real-world problems.

2 Teach

Scaffolding Questions
Have students read the **Why?** section of the lesson.

Ask:
- On the right-hand side of the illuminance formula, which variables appear in the numerator? In the denominator? intensity; illuminance and distance

- In a right triangle, what is the ratio for $\sec \theta$? $\sec \theta = \dfrac{\text{hypotenuse}}{\text{adjacent side}}$

- What is the reciprocal identity of $\sec \theta$? $\dfrac{1}{\cos \theta}, \cos \theta \neq 0$

Lesson 13-1 Resources

Resource	Approaching Level **AL**	On Level **OL**	Beyond Level **BL**	English Learners **ELL**
Teacher Edition	• Differentiated Instruction, p. 874	• Differentiated Instruction, pp. 874, 879	• Differentiated Instruction, pp. 875, 879	
Chapter Resource Masters	• Study Guide and Intervention, pp. 5–6 • Skills Practice, p. 7 • Practice, p. 8 • Word Problem Practice, p. 9	• Study Guide and Intervention, pp. 5–6 • Skills Practice, p. 7 • Practice, p. 8 • Word Problem Practice, p. 9 • Enrichment, p. 10	• Practice, p. 8 • Word Problem Practice, p. 9 • Enrichment, p. 10	• Study Guide and Intervention, pp. 5–6 • Skills Practice, p. 7 • Practice, p. 8 • Word Problem Practice, p. 9
Other	• 5-Minute Check 13-1 • Study Notebook	• 5-Minute Check 13-1 • Study Notebook	• 5-Minute Check 13-1 • Study Notebook	• 5-Minute Check 13-1 • Study Notebook

Example 1 shows how to find values of
a trigonometric function for an angle in
a specified quadrant.

Formative Assessment

Use the Guided Practice exercises after
each example to determine students'
understanding of concepts.

Additional Example

1 **a.** Find the exact value of tan θ
if sec $\theta = -2$ and $180° <$
$\theta < 270°$. tan $\theta = \sqrt{3}$

b. Find the exact value of sin θ
if cos $\theta = -\frac{1}{2}$ and $90° <$
$\theta < 180°$. sin $\theta = \frac{\sqrt{3}}{2}$

Additional Examples also in
Interactive Classroom
PowerPoint® Presentations

 **Interactive White Board
READY**

You can use trigonometric identities to find exact values of trigonometric functions. You
can find approximate values by using a graphing calculator.

Example 1 Use Trigonometric Identities

a. **Find the exact value of cos θ if sin $\theta = \frac{1}{4}$ and $90° < \theta < 180°$.**

$\cos^2 \theta + \sin^2 \theta = 1$	Pythagorean identity
$\cos^2 \theta = 1 - \sin^2 \theta$	Subtract $\sin^2 \theta$ from each side.
$\cos^2 \theta = 1 - \left(\frac{1}{4}\right)^2$	Substitute $\frac{1}{4}$ for sin θ.
$\cos^2 \theta = 1 - \frac{1}{16}$	Square $\frac{1}{4}$.
$\cos^2 \theta = \frac{15}{16}$	Subtract: $\frac{16}{16} - \frac{1}{16} = \frac{15}{16}$.
$\cos \theta = \pm\frac{\sqrt{15}}{4}$	Take the square root of each side.

Since θ is in the second quadrant, cos θ is negative. Thus, cos $\theta = -\frac{\sqrt{15}}{4}$.

CHECK Use a calculator to find an approximate answer.

Step 1 Find Arcsin $\frac{1}{4}$.

$\sin^{-1}\frac{1}{4} \approx 14.48°$ Use a calculator.

Because $90° < \theta < 180°$, $\theta \approx 180° - 14.48°$ or about $165.52°$.

Step 2 Find cos θ.

Replace θ with $165.52°$.

$\cos 165.52° \approx -0.97$

Step 3 Compare with the exact value.

$-\frac{\sqrt{15}}{4} \stackrel{?}{\approx} 0.97$

$-0.968 \approx 0.97$ ✓

b. **Find the exact value of csc θ if cot $\theta = -\frac{3}{5}$ and $270° < \theta < 360°$.**

$\cot^2 \theta + 1 = \csc^2 \theta$	Pythagorean identity
$\left(-\frac{3}{5}\right)^2 + 1 = \csc^2 \theta$	Substitute $-\frac{3}{5}$ for cot θ.
$\frac{9}{25} + 1 = \csc^2 \theta$	Square $-\frac{3}{5}$.
$\frac{34}{25} = \csc^2 \theta$	Add: $\frac{9}{25} + \frac{25}{25} = \frac{34}{25}$.
$\pm\frac{\sqrt{34}}{5} = \csc \theta$	Take the square root of each side.

Since θ is in the fourth quadrant, csc θ is negative. Thus, csc $\theta = -\frac{\sqrt{34}}{5}$.

Guided Practice

1A. Find sin θ if cos $\theta = \frac{1}{3}$ and $270° < \theta < 360°$. $-\frac{2\sqrt{2}}{3}$

1B. Find sec θ if sin $\theta = -\frac{2}{7}$ and $180° < \theta < 270°$. $-\frac{7\sqrt{5}}{15}$

2 Simplify Expressions Simplifying an expression that contains trigonometric
functions means that the expression is written as a numerical value or in terms
of a single trigonometric function, if possible.

StudyTip

Quadrants Here is a table
to help you remember which
ratios are positive and
which are negative in
each quadrant.

Function	+	−
sin θ	1, 2	3, 4
cos θ	1, 4	2, 3
tan θ	1, 3	2, 4
csc θ	1, 2	3, 4
sec θ	1, 4	2, 3
cot θ	1, 3	2, 4

 874 | Lesson 13-1 | Trigonometric Identities

DifferentiatedInstruction **AL** **OL**

If students have difficulty with trigonometric identities,

Then have students work in groups of three. Ask each group to choose one of the identities in the
Key Concept box on p. 873 and work together to demonstrate that it is true. Students should
verify their results using the definitions of sine, cosine, and tangent in terms of the sides of
a right triangle.

StudyTip
Simplifying It is often easiest to write all expressions in terms of sine and/or cosine.

Example 2 Simplify an Expression

Simplify $\dfrac{\sin \theta \csc \theta}{\cot \theta}$.

$$\dfrac{\sin \theta \csc \theta}{\cot \theta} = \dfrac{\sin \theta \left(\dfrac{1}{\sin \theta}\right)}{\dfrac{1}{\tan \theta}} \qquad \csc \theta = \dfrac{1}{\sin \theta} \text{ and } \cot \theta = \dfrac{1}{\tan \theta}$$

$$= \dfrac{1}{\dfrac{1}{\tan \theta}} \qquad \dfrac{\sin \theta}{\sin \theta} = 1$$

$$= \dfrac{1}{1} \cdot \dfrac{\tan \theta}{1} \text{ or } \tan \theta \qquad \dfrac{a}{b} \div \dfrac{c}{d} = \dfrac{a}{b} \cdot \dfrac{d}{c}$$

▶ **Guided**Practice

Simplify each expression.

2A. $\dfrac{\tan^2 \theta \csc^2 \theta - 1}{\sec^2 \theta}$ $\sin^2 \theta$

2B. $\dfrac{\sec \theta}{\sin \theta}(1 - \cos^2 \theta)$ $\tan \theta$

Simplifying trigonometric expressions can be helpful when solving real-world problems.

Real-World Example 3 Simplify and Use an Expression

LIGHTING Refer to the beginning of the lesson.

a. Solve the formula in terms of E.

$$\sec \theta = \dfrac{I}{ER^2} \qquad \text{Original equation}$$

$$ER^2 \sec \theta = I \qquad \text{Multiply each side by } ER^2.$$

$$ER^2 \dfrac{1}{\cos \theta} = I \qquad \dfrac{1}{\cos \theta} = \sec \theta$$

$$\dfrac{E}{\cos \theta} = \dfrac{I}{R^2} \qquad \text{Divide each side by } R^2.$$

$$E = \dfrac{I \cos \theta}{R^2} \qquad \text{Multiply each side by } \cos \theta.$$

b. Is the equation in part **a** equivalent to $R^2 = \dfrac{I \tan \theta \cos \theta}{E}$? **Explain.**

$$R^2 = \dfrac{I \tan \theta \cos \theta}{E} \qquad \text{Original equation}$$

$$ER^2 = I \tan \theta \cos \theta \qquad \text{Multiply each side by } E.$$

$$E = \dfrac{I \tan \theta \cos \theta}{R^2} \qquad \text{Divide each side by } R^2.$$

$$E = \dfrac{I \left(\dfrac{\sin \theta}{\cos \theta}\right) \cos \theta}{R^2} \qquad \tan \theta = \dfrac{\sin \theta}{\cos \theta}$$

$$E = \dfrac{I \sin \theta}{R^2} \qquad \text{Simplify.}$$

No; the equations are not equivalent. $R^2 = \dfrac{I \tan \theta \cos \theta}{E}$ simplifies to $E = \dfrac{I \sin \theta}{R^2}$.

▶ **Guided**Practice

3. Rewrite $\cot^2 \theta - \tan^2 \theta$ in terms of $\sin \theta$. $\dfrac{1 - 2 \sin^2 \theta}{\sin^2 \theta - \sin^4 \theta}$

Math HistoryLink
Aryabhatta (476–550 A.D.) Among Indian mathematicians, Aryabhatta is probably the most famous. His name is closely associated with trigonometry. He was the first to introduce inverse trigonometric functions and spherical trigonometry. Aryabhatta also calculated approximations for pi and trigonometric functions.

Dinodia Photo

connectED.mcgraw-hill.com   **875**

---

2 Simplify Expressions
Example 2 shows how to simplify an expression by writing it in terms of one trigonometric function. **Example 3** illustrates simplifying a real-world expression involving trigonometric functions.

Additional Examples

2 Simplify $\sin \theta (\csc \theta - \sin \theta)$. $\cos^2 \theta$

3 **LIGHTING** Refer to the beginning of the lesson.
   a. Solve the illuminance formula in terms of R. $R = \sqrt{\dfrac{I \cos \theta}{E}}$
   b. Is the equation in part **a** equivalent to $\dfrac{1}{R^2} = \dfrac{E}{I \sec \theta}$?   no

Focus on Mathematical Content
Identities Students must have good recall of the quotient, reciprocal, and Pythagorean identities. While it is helpful to remember other identities, the other identities can be derived from the basic identities.

Teach with Tech
Web Page Create a class Web page that includes all of the trigonometric identities and formulas from this chapter. Have students subscribe to an RSS feed so that they can easily tell when you update the page.

DifferentiatedInstruction **BL**

Extension Beyond-level students are often interested in learning how certain mathematical ideas came into existence. Provide an extension to the math history link about the Indian mathematician, Aryabhatta. Students can investigate the prior mathematical discoveries of Aryabhatta that led to the need for inverse trigonometric functions in the context of his work.

3 Practice

Formative Assessment

Use Exercises 1–8 to check for understanding.

Use the chart at the bottom of this page to customize assignments for your students.

 Teaching the Mathematical Practices

Perseverance Mathematically proficient students start by explaining to themselves the meaning of a problem and looking for entry points to its solution. They analyze givens, constraints, relationships, and goals and make conjectures about the form and meaning of the solution and plan a solution pathway rather than simply jumping into a solution attempt.

Tips for New Teachers

Trigonometric Ratios You can use the familiar definitions of sine, cosine, and tangent as the ratios of the opposite side, adjacent side, and hypotenuse of a right triangle to show why $\dfrac{\sin \theta}{\cos \theta} = \tan \theta$.

Example 1 Find the exact value of each expression if $0° < \theta < 90°$.

1. If $\cot \theta = 2$, find $\tan \theta$. $\dfrac{1}{2}$

2. If $\sin \theta = \dfrac{4}{5}$, find $\cos \theta$. $\dfrac{3}{5}$

3. If $\cos \theta = \dfrac{2}{3}$, find $\sin \theta$. $\dfrac{\sqrt{5}}{3}$

4. If $\cos \theta = \dfrac{2}{3}$, find $\csc \theta$. $\dfrac{3\sqrt{5}}{5}$

Example 2 Simplify each expression.

5. $\tan \theta \cos^2 \theta$ $\sin \theta \cos \theta$

6. $\csc^2 \theta - \cot^2 \theta$ 1

7. $\dfrac{\cos \theta \csc \theta}{\tan \theta}$ $\cot^2 \theta$

Example 3 **8.** **CCSS PERSEVERANCE** When unpolarized light passes through polarized sunglass lenses, the intensity of the light is cut in half. If the light then passes through another polarized lens with its axis at an angle of θ to the first, the intensity of the light is again diminished. The intensity of the emerging light can be found by using the formula $I = I_0 - \dfrac{I_0}{\csc^2 \theta}$, where I_0 is the intensity of the light incoming to the second polarized lens, I is the intensity of the emerging light, and θ is the angle between the axes of polarization.

Axis 2
Axis 1
Unpolarized light

a. Simplify the formula in terms of $\cos \theta$. $I = I_0 \cos^2 \theta$

b. Use the simplified formula to determine the intensity of light that passes through a second polarizing lens with axis at 30° to the original. $I = \dfrac{3}{4}I_0$; The light has three fourths the intensity it had before passing through the second polarizing lens.

Practice and Problem Solving   Extra Practice is on page R13.

Example 1 Find the exact value of each expression if $0° < \theta < 90°$.

9. If $\cos \theta = \dfrac{3}{5}$, find $\csc \theta$. $\dfrac{5}{4}$

10. If $\sin \theta = \dfrac{1}{2}$, find $\tan \theta$. $\dfrac{\sqrt{3}}{3}$

11. If $\sin \theta = \dfrac{3}{5}$, find $\cos \theta$. $\dfrac{4}{5}$

12. If $\tan \theta = 2$, find $\sec \theta$. $\sqrt{5}$

Find the exact value of each expression if $180° < \theta < 270°$.

13. If $\cos \theta = -\dfrac{3}{5}$, find $\csc \theta$. $-\dfrac{5}{4}$

14. If $\sec \theta = -3$, find $\tan \theta$. $2\sqrt{2}$

15 If $\cot \theta = \dfrac{1}{4}$, find $\csc \theta$. $\dfrac{-\sqrt{17}}{4}$

16. If $\sin \theta = -\dfrac{1}{2}$, find $\cos \theta$. $-\dfrac{\sqrt{3}}{2}$

Find the exact value of each expression if $270° < \theta < 360°$.

17. If $\cos \theta = \dfrac{5}{13}$, find $\sin \theta$. $-\dfrac{12}{13}$

18. If $\tan \theta = -1$, find $\sec \theta$. $\sqrt{2}$

19. If $\sec \theta = \dfrac{5}{3}$, find $\cos \theta$. $\dfrac{3}{5}$

20. If $\csc \theta = -\dfrac{5}{3}$, find $\cos \theta$. $\dfrac{4}{5}$

Example 2 Simplify each expression.

21. $\sec \theta \tan^2 \theta + \sec \theta$ $\sec^3 \theta$

22. $\cos \left(\dfrac{\pi}{2} - \theta\right)\cot \theta$ $\cos \theta$

23. $\cot \theta \sec \theta$ $\csc \theta$

24. $\sin \theta (1 + \cot^2 \theta)$ $\csc \theta$

25. $\sin \left(\dfrac{\pi}{2} - \theta\right)\sec \theta$ 1

26. $\dfrac{\cos (-\theta)}{\sin (-\theta)}$ $-\cot \theta$

Differentiated Homework Options

Level	Assignment	Two-Day Option	
AL Basic	9–27, 42, 44–48, 50–67	9–27 odd, 51–54	10–26 even, 42, 44–48, 50, 55–67
OL Core	9–33 odd, 34–42, 44–48, 50–67	9–27, 51–54	28–42, 44–48, 50, 55–67
BL Advanced	28–64, (optional: 65–67)		

Example 3

27 **ELECTRONICS** When there is a current in a wire in a magnetic field, such as in a hairdryer, a force acts on the wire. The strength of the magnetic field can be determined using the formula $B = \frac{F \csc \theta}{I\ell}$, where F is the force on the wire, I is the current in the wire, ℓ is the length of the wire, and θ is the angle the wire makes with the magnetic field. Rewrite the equation in terms of $\sin \theta$. (*Hint:* Solve for F.) $F = I\ell B \sin \theta$

 Simplify each expression.

28. $\frac{1 - \sin^2 \theta}{\sin^2 \theta} \cot^2 \theta$ **29.** $\tan \theta \csc \theta$ $\sec \theta$ **30.** $\frac{1}{\sin^2 \theta} - \frac{\cos^2 \theta}{\sin^2 \theta}$ 1

31. $2(\csc^2 \theta - \cot^2 \theta)$ 2 **32.** $(1 + \sin \theta)(1 - \sin \theta)$ $\cos^2 \theta$ **33.** $2 - 2 \sin^2 \theta$ $2 \cos^2 \theta$

34. **SUN** The ability of an object to absorb energy is related to a factor called the emissivity e of the object. The emissivity can be calculated by using the formula $e = \frac{W \sec \theta}{AS}$, where W is the rate at which a person's skin absorbs energy from the Sun, S is the energy from the Sun in watts per square meter, A is the surface area exposed to the Sun, and θ is the angle between the Sun's rays and a line perpendicular to the body. **a.** $W = eAS \cos \theta$

a. Solve the equation for W. Write your answer using only $\sin \theta$ or $\cos \theta$.

b. Find W if $e = 0.80$, $\theta = 40°$, $A = 0.75$ m², and $S = 1000$ W/m². Round to the nearest hundredth. **459.63 W**

35. **CCSS** **MODELING** The map shows some of the buildings in Maria's neighborhood that she visits on a regular basis. The sine of the angle θ formed by the roads connecting the dance studio, the school, and Maria's house is $\frac{4}{9}$.

a. What is the cosine of the angle? $\frac{\sqrt{65}}{9}$

b. What is the tangent of the angle? $\frac{4\sqrt{65}}{65}$

35c. $\frac{4}{9}$, $-\frac{\sqrt{65}}{9}$, $-\frac{4\sqrt{65}}{65}$

c. What are the sine, cosine, and tangent of the angle formed by the roads connecting the piano teacher's house, the school, and Maria's house?

36. **MULTIPLE REPRESENTATIONS** In this problem, you will use a graphing calculator to determine whether an equation may be a trigonometric identity. Consider the trigonometric identity $\tan^2 \theta - \sin^2 \theta = \tan^2 \theta \sin^2 \theta$.

a. Tabular Copy and complete the table below.

θ	0°	30°	45°	60°
$\tan^2 \theta - \sin^2 \theta$	0	$\frac{1}{12}$	$\frac{1}{2}$	$\frac{9}{4}$
$\tan^2 \theta \sin^2 \theta$	0	$\frac{1}{12}$	$\frac{1}{2}$	$\frac{9}{4}$

b. Graphical Use a graphing calculator to graph $\tan^2 \theta - \sin^2 \theta = \tan^2 \theta \sin^2 \theta$ as two separate functions. Sketch the graph. **See margin.**

c. Analytical If the graphs of the two functions do not match, then the equation is not an identity. Do the graphs coincide? **yes**

d. Analytical Use a graphing calculator to determine whether the equation $\sec^2 x - 1 = \sin^2 x \sec^2 x$ may be an identity. (Be sure your calculator is in degree mode.) **yes**

 connectED.mcgraw-hill.com **877**

CCSS **Teaching the Mathematical Practices**

Modeling Mathematically proficient students can apply the mathematics they know to solve problems arising in everyday life, analyze relationships mathematically to draw conclusions, and interpret their mathematical results in the context of a situation.

Multiple Representations
In Exercise 36, students use a table of values and a graphing calculator to determine whether an equation is a trigonometric identity.

Additional Answer
36b.

[−540, 540] scl: 90 by [−10, 10] scl: 1

Study Guide and Intervention
AL OL ELL

Practice
AL OL BL ELL

Word Problem Practice
AL OL BL ELL

---

37. SKIING A skier of mass m descends a θ-degree hill at a constant speed. When Newton's laws are applied to the situation, the following system of equations is produced: $F_n - mg \cos \theta = 0$ and $mg \sin \theta - \mu_k F_n = 0$, where g is the acceleration due to gravity, F_n is the normal force exerted on the skier, and μ_k is the coefficient of friction. Use the system to define μ_k as a function of θ. $\mu_k = \tan \theta$

Simplify each expression.

38. $\dfrac{\tan\left(\frac{\pi}{2} - \theta\right)\sec \theta}{1 - \csc^2 \theta}$ $-\tan \theta \sec \theta$

39 $\dfrac{\cos\left(\frac{\pi}{2} - \theta\right) - 1}{1 + \sin(-\theta)}$ -1

40. $\dfrac{\sec \theta \sin \theta + \cos\left(\frac{\pi}{2} - \theta\right)}{1 + \sec \theta}$ $\sin \theta$

41. $\dfrac{\cot \theta \cos \theta}{\tan(-\theta)\sin\left(\frac{\pi}{2} - \theta\right)}$ $-\cot^2 \theta$

H.O.T. Problems   Use Higher-Order Thinking Skills

42. CCSS CRITIQUE Clyde and Rosalina are debating whether an equation from their homework assignment is an identity. Clyde says that since he has tried ten specific values for the variable and all of them worked, it must be an identity. Rosalina argues that specific values could only be used as counterexamples to prove that an equation is not an identity. Is either of them correct? Explain your reasoning. **Rosalina; there may be other values for which the equation is not true.**

43. CHALLENGE Find a counterexample to show that $1 - \sin x = \cos x$ is *not* an identity. **Sample answer: $x = 45°$**

45. Sample answer: The functions $\cos \theta$ and $\sin \theta$ can be thought of as the lengths of the legs of a right triangle, and the number 1 can be thought of as the measure of the corresponding hypotenuse.

44. REASONING Demonstrate how the formula about illuminance from the beginning of the lesson can be rewritten to show that $\cos \theta = \dfrac{ER^2}{I}$. **See margin.**

45. WRITING IN MATH Pythagoras is most famous for the Pythagorean Theorem. The identity $\cos^2 \theta + \sin^2 \theta = 1$ is an example of a Pythagorean identity. Why do you think that this identity is classified in this way?

46. PROOF Prove that $\tan(-a) = -\tan a$ by using the quotient and negative angle identities. **See margin.**

47. OPEN ENDED Write two expressions that are equivalent to $\tan \theta \sin \theta$. **Sample answer: $\dfrac{\sin \theta}{\cos \theta} \cdot \sin \theta$ and $\dfrac{\sin^2 \theta}{\cos \theta}$**

48. REASONING Explain how you can use division to rewrite $\sin^2 \theta + \cos^2 \theta = 1$ as $1 + \cot^2 \theta = \csc^2 \theta$. **Divide all of the terms by $\sin^2 \theta$.**

49. CHALLENGE Find $\cot \theta$ if $\sin \theta = \dfrac{3}{5}$ and $90° \le \theta < 180°$. $-\dfrac{4}{3}$

50. ERROR ANALYSIS Jordan and Ebony are simplifying $\dfrac{\sin^2 \theta}{\cos^2 \theta + \sin^2 \theta}$. Is either of them correct? Explain your reasoning.

50. Ebony; Jordan did not use the identity that $\sin^2 \theta + \cos^2 \theta = 1$ and made an error adding rational expressions.

Jordan
$$\frac{\sin^2 \theta}{\cos^2 \theta + \sin^2 \theta} = \frac{\sin^2 \theta}{\cos^2 \theta} + \frac{\sin^2 \theta}{\sin^2 \theta}$$
$$= \tan^2 \theta + 1$$
$$= \sec^2 \theta$$

Ebony
$$\frac{\sin^2 \theta}{\cos^2 \theta + \sin^2 \theta} = \frac{\sin^2 \theta}{1}$$
$$= \sin^2 \theta$$

 878 | Lesson 13-1 | Trigonometric Identities

Enrichment
OL BL

13-1 Enrichment

Planetary Orbits
The orbit of a planet around the Sun is an ellipse with the Sun at one focus. Let the pole of a polar coordinate system be that focus and the polar axis be toward the other focus. The polar equation of an ellipse is

---

Additional Answer

44. $\sec \theta = \dfrac{I}{ER^2}$

$\dfrac{1}{\cos \theta} = \dfrac{I}{ER^2}$ $\sec \theta = \dfrac{1}{\cos \theta}$

$ER^2 = I \cos \theta$ Cross multiply.

$\cos \theta = \dfrac{ER^2}{I}$ Divide each side by I.

51. Refer to the figure below. If $\cos D = 0.8$, what is the length of $\overline{DF}$? **A**

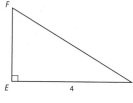

A 5   **C** 3.2
B 4   **D** $\frac{4}{5}$

52. PROBABILITY There are 16 green marbles, 2 red marbles, and 6 yellow marbles in a jar. How many yellow marbles need to be added to the jar in order to double the probability of selecting a yellow marble? **J**

F 4   **H** 8
G 6   **J** 12

53. SAT/ACT Ella is 6 years younger than Amanda. Zoe is twice as old as Amanda. The total of their ages is 54. Which equation can be used to find Amanda's age? **D**

A $x + (x - 6) + 2(x - 6) = 54$
B $x - 6x + (x + 2) = 54$
C $x - 6 + 2x = 54$
D $x + (x - 6) + 2x = 54$
E $2(x + 6) + (x + 6) + x = 54$

54. Which of the following functions represents exponential growth? **G**

F $y = (0.3)^x$
G $y = (1.3)^x$
H $y = x^3$
J $y = x^{\frac{1}{3}}$

Spiral Review

Find each value. Write angle measures in radians. Round to the nearest hundredth. (Lesson 12-9)

55. $\text{Cos}^{-1}\left(-\frac{1}{2}\right)$ **2.09**

56. $\text{Sin}^{-1}\frac{\pi}{2}$ **does not exist**

57. $\text{Arctan}\frac{\sqrt{3}}{3}$ **0.52**

58. $\tan\left(\text{Cos}^{-1}\frac{6}{7}\right)$ **0.60**

59. $\sin\left(\text{Arctan}\frac{\sqrt{3}}{3}\right)$ **0.5**

60. $\cos\left(\text{Arcsin}\frac{3}{5}\right)$ **0.8**

61. PHYSICS A weight is attached to a spring and suspended from the ceiling. At equilibrium, the weight is located 4 feet above the floor. The weight is pulled down 1 foot and released. Write the equation for the distance d of the weight above the floor as a function of time t seconds assuming the weight returns to its lowest position every 4 seconds. (Lesson 12-8) $d = 4 - \cos\frac{\pi}{2}t$ or $d = 4 - \cos 90°t$

Evaluate the sum of each geometric series. (Lesson 10-3)

62. $\sum_{k=1}^{5} \frac{1}{4} \cdot 2^{k-1}$ $\frac{31}{4}$

63. $\sum_{k=1}^{7} 81\left(\frac{1}{3}\right)^{k-1}$ $\frac{1093}{9}$

64. $\sum_{k=1}^{8} \frac{1}{3} \cdot 5^{k-1}$ **32,552**

Skills Review

Solve each equation.

65. $a + 1 = \frac{6}{a}$ $-3, 2$

66. $\frac{9}{t-3} = \frac{t-4}{t-3} + \frac{1}{4}$ **11**

67. $\frac{5}{x+1} - \frac{1}{3} = \frac{x+2}{x+1}$ **2**

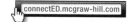

WatchOut!

Error Analysis For Exercise 42, students should see that Rosalina is correct. Explain to students that inductive reasoning (generalizing from several examples) can never prove that an identity is true, but that any specific counterexample is enough to show that an equation is not an identity. For Exercise 50, students should see that Ebony is correct (because $\cos^2 \theta + \sin^2 \theta = 1$) and that Jordan is not correct $\left(\text{because } \frac{a}{b+c} \neq \frac{a}{b} + \frac{a}{c}\right)$. Explain to students that when they simplify a trigonometric expression, they use the same properties as when they simplify any rational expression.

4 Assess

Crystal Ball Tell students to look ahead to Lesson 13-2. Ask them to write how they think what they learned today will help them with Lesson 13-2.

Additional Answer

46. $\tan(-A) = \dfrac{\sin(-A)}{\cos(-A)}$
$= \dfrac{-\sin A}{\cos A}$
$= -\dfrac{\sin A}{\cos A}$
$= -\tan A$

DifferentiatedInstruction OL BL

Extension Ask students if it is always possible to simplify a trigonometric expression by writing it in terms of one trigonometric function. Ask them to provide an example if it is not possible or explain why it is possible.

1 Focus

VerticalAlignment

Before Lesson 13-2 Use identities to find trigonometric values and simplify expressions.

Lesson 13-2 Verify trigonometric identities by transforming one side of an equation into the form of the other side. Verify trigonometric identities by transforming each side of the equation into the same form.

After Lesson 13-2 Use sum and difference identities to prove or simplify other trigonometric identities.

2 Teach

Scaffolding Questions

Have students read the **Why?** section of the lesson.

Ask:

- On the right-hand side of the *angle of incline* equation, which variables appear in the numerator? In the denominator? *v*; *g* and *R*

- How can you express tan θ in terms of sin θ and cos θ? $\tan \theta = \dfrac{\sin \theta}{\cos \theta}$

- Does $\dfrac{\sin \theta}{\cos \theta}$ equal $\dfrac{v^2}{gR}$ or $\dfrac{gR}{v^2}$? $\dfrac{v^2}{gR}$

	::Then	::Now	::Why?

- You used identities to find trigonometric values and simplify expressions.

1. Verify trigonometric identities by transforming one side of an equation into the form of the other side.

2. Verify trigonometric identities by transforming each side of the equation into the same form.

- While running on a circular track, Lamont notices that his body is not perpendicular to the ground. Instead, it leans away from a vertical position. The nonnegative acute angle θ that Lamont's body makes with the vertical is called the *angle of incline* and is described by the equation $\tan \theta = \dfrac{v^2}{gR}$.

This is not the only equation that describes the angle of incline in terms of trigonometric functions. Another such equation is $\sin \theta = \cos \dfrac{v^2}{gR} \theta$, where $0 \leq \theta \leq 90°$.

Are these two equations completely independent of one another or are they merely different versions of the same relationship?

Nick Wilson/Getty Images Sport/Getty Images

(CCSS) **Common Core State Standards**

Content Standards
F.TF.8 Prove the Pythagorean identity $\sin^2 (\theta) + \cos^2 (\theta) = 1$ and use it to find $\sin (\theta)$, $\cos (\theta)$, or $\tan (\theta)$ given $\sin (\theta)$, $\cos (\theta)$, or $\tan (\theta)$ and the quadrant of the angle.

Mathematical Practices
1 Make sense of problems and persevere in solving them.
8 Look for and express regularity in repeated reasoning.

---

1 Transform One Side of an Equation You can use the basic trigonometric identities along with the definitions of the trigonometric functions to verify identities. If you wish to show an identity, you need to show that it is true for all values of θ.

KeyConcept Verifying Identities by Transforming One Side

Step 1 Simplify one side of an equation until the two sides of the equation are the same. It is often easier to work with the more complicated side of the equation.

Step 2 Transform that expression into the form of the simpler side.

Example 1 Transform One Side of an Equation

Verify that $\dfrac{\sin^2 \theta}{1 - \cos \theta} = 1 + \cos \theta$ is an identity.

$$\dfrac{\sin^2 \theta}{1 - \cos \theta} \stackrel{?}{=} 1 + \cos \theta \qquad \text{Original equation}$$

$$\dfrac{1 + \cos \theta}{1 + \cos \theta} \cdot \dfrac{\sin^2 \theta}{1 - \cos \theta} \stackrel{?}{=} 1 + \cos \theta \qquad \text{Multiply the numerator and denominator by } 1 + \cos \theta.$$

$$\dfrac{\sin^2 \theta (1 + \cos \theta)}{1 - \cos^2 \theta} \stackrel{?}{=} 1 + \cos \theta \qquad (1 + \cos \theta)(1 - \cos \theta) = 1 - \cos^2 \theta$$

$$\dfrac{\sin^2 \theta (1 + \cos \theta)}{\sin^2 \theta} \stackrel{?}{=} 1 + \cos \theta \qquad \sin^2 \theta = 1 - \cos^2 \theta$$

$$1 + \cos \theta = 1 + \cos \theta \checkmark \qquad \text{Divide the numerator and denominator by } \sin^2 \theta.$$

▶ **Guided**Practice

1. Verify that $\cot^2 \theta - \cos^2 \theta = \cot^2 \theta \cos^2 \theta$ is an identity. **See margin.**

Lesson 13-2 Resources

Resource	Approaching Level **AL**	On Level **OL**	Beyond Level **BL**	English Learners **ELL**
Teacher Edition	▪ Differentiated Instruction, p. 881	▪ Differentiated Instruction, p. 881	▪ Differentiated Instruction, pp. 881, 885	▪ Differentiated Instruction, p. 881
Chapter Resource Masters	▪ Study Guide and Intervention, pp. 11–12 ▪ Skills Practice, p. 13 ▪ Practice, p. 14 ▪ Word Problem Practice, p. 15	▪ Study Guide and Intervention, pp. 11–12 ▪ Skills Practice, p. 13 ▪ Practice, p. 14 ▪ Word Problem Practice, p. 15 ▪ Enrichment, p. 16	▪ Practice, p. 14 ▪ Word Problem Practice, p. 15 ▪ Enrichment, p. 16	▪ Study Guide and Intervention, pp. 11–12 ▪ Skills Practice, p. 13 ▪ Practice, p. 14 ▪ Word Problem Practice, p. 15
Other	▪ 5-Minute Check 13-2 ▪ Study Notebook	▪ 5-Minute Check 13-2 ▪ Study Notebook	▪ 5-Minute Check 13-2 ▪ Study Notebook	▪ 5-Minute Check 13-2 ▪ Study Notebook

When you verify a trigonometric identity, you are really working backward. In Example 1, consider the last step $1 + \cos \theta = 1 + \cos \theta$. Since that step is clearly true, you can conclude that the next-to-last step is also true, and so on, all the way back to the original equation.

Standardized Test Example 2 Simplify an Expression

$$\frac{\cos \theta \, \csc \theta}{\tan \theta} =$$

A $\cot \theta$ **B** $\csc \theta$ **C** $\cot^2 \theta$ **D** $\csc^2 \theta$

Read the Test Item

Find an expression that is always equal to the given expression. Notice that all of the answer choices involve either $\cot \theta$ or $\csc \theta$. So work toward eliminating the other trigonometric functions.

Solve the Test Item

Transform the given expression to match one of the choices.

$$\frac{\cos \theta \, \csc \theta}{\tan \theta} = \frac{\cos \theta \left(\frac{1}{\sin \theta}\right)}{\frac{\sin \theta}{\cos \theta}} \qquad \csc \theta = \frac{1}{\sin \theta} \text{ and } \tan \theta = \frac{\sin \theta}{\cos \theta}$$

$$= \frac{\frac{\cos \theta}{\sin \theta}}{\frac{\sin \theta}{\cos \theta}} \qquad \text{Multiply.}$$

$$= \frac{\cos \theta}{\sin \theta} \cdot \frac{\cos \theta}{\sin \theta} \qquad \text{Invert the denominator and multiply.}$$

$$= \cot \theta \cdot \cot \theta \qquad \cot \theta = \frac{\cos \theta}{\sin \theta}$$

$$= \cot^2 \theta \qquad \text{Multiply.}$$

The answer is C.

> **WatchOut!**
> **CCSS** Perseverance
> Verifying an identity is like checking the solution of an equation. You must simplify one or both sides separately until they are the same.

> **Test-TakingTip**
> Checking Answers Verify your answer by choosing values for θ. Then evaluate the original expression and compare to your answer choice.

GuidedPractice

2. $\tan^2 \theta (\cot^2 \theta - \cos^2 \theta) =$ H

   F $\cot^2 \theta$ **G** $\tan^2 \theta$ **H** $\cos^2 \theta$ **J** $\sin^2 \theta$

2 **Transform Each Side of an Equation** Sometimes it is easier to transform each side of an equation separately into a common form. The following suggestions may be helpful as you verify trigonometric identities.

KeyConcept Suggestions for Verifying Identities

- Substitute one or more basic trigonometric identities to simplify the expression.
- Factor or multiply as necessary. You may have to multiply both the numerator and denominator by the same trigonometric expression.
- Write each side of the identity in terms of sine and cosine only. Then simplify each side as much as possible.
- The properties of equality do not apply to identities as with equations. Do not perform operations to the quantities on each side of an unverified identity.

DifferentiatedInstruction   AL   OL   BL   ELL

Interpersonal Learners Have groups or pairs of students work together to verify some of the identities in Exercises 8–17. Have students record the techniques they found helpful. Ask students to compare their list of techniques to the list of suggestions given at the bottom of p. 881. Also have them discuss failed strategies. What did and did not work, and why?

---

1 Transform One Side of an Equation

Example 1 shows how to verify a trigonometric identity by transforming one side of an equation. **Example 2** shows how to find an expression equivalent to a given trigonometric expression.

CCSS Teaching the Mathematical Practices

Perseverance Mathematically proficient students monitor and evaluate their progress and change course if necessary. Remind students that expressions are not equal until they are in the same form.

Formative Assessment

Use the Guided Practice exercises after each example to determine students' understanding of concepts.

Additional Examples

1 Verify that $\csc \theta \cos \theta \tan \theta = 1$ is an identity.

$$\csc \theta \cos \theta \tan \theta \overset{?}{=} 1$$

$$\frac{1}{\sin \theta} \cdot \cos \theta \cdot \frac{\sin \theta}{\cos \theta} \overset{?}{=} 1$$

$$1 = 1 \checkmark$$

2 **STANDARDIZED TEST PRACTICE**

$$\frac{\csc \theta}{\cos \theta} - \tan \theta = \text{A}$$

A $\cot \theta$ **C** 0

B $\frac{1 - \sin \theta}{\cos^2 \theta}$ **D** $\cos^2 \theta$

Additional Examples also in Interactive Classroom PowerPoint® Presentations

IWB Interactive White Board READY

Additional Answer (Guided Practice)

1. $\cot^2 \theta - \cos^2 \theta \overset{?}{=} \cot^2 \theta \cos^2 \theta$

$$\frac{\cos^2 \theta}{\sin^2 \theta} - \cos^2 \theta \overset{?}{=} \cot^2 \theta \cos^2 \theta$$

$$\cos^2 \theta \left(\frac{1}{\sin^2 \theta} - 1\right) \overset{?}{=} \cot^2 \theta \cos^2 \theta$$

$$\cos^2 \theta (\csc^2 \theta - 1) \overset{?}{=} \cot^2 \theta \cos^2 \theta$$

$$\cot^2 \theta \cos^2 \theta = \cot^2 \theta \cos^2 \theta \checkmark$$

2 Transform Each Side of an Equation

Example 3 shows how to verify a trigonometric identity by transforming both sides of an equation.

Additional Example

3 Verify that $\csc \theta + \sec \theta = \dfrac{1 + \cot \theta}{\cos \theta}$ is an identity.

$$\csc \theta + \sec \theta \stackrel{?}{=} \frac{1 + \cot \theta}{\cos \theta}$$

$$\frac{1}{\sin \theta} + \frac{1}{\cos \theta} \stackrel{?}{=} \frac{1 + \frac{\cos \theta}{\sin \theta}}{\cos \theta}$$

$$\frac{\cos \theta + \sin \theta}{\sin \theta \cos \theta} \stackrel{?}{=} \frac{\sin \theta \left(1 + \frac{\cos \theta}{\sin \theta}\right)}{\sin \theta \cos \theta}$$

$$\frac{\cos \theta + \sin \theta}{\sin \theta \cos \theta} = \frac{\sin \theta + \cos \theta}{\sin \theta \cos \theta} \; \checkmark$$

---

3 Practice

Formative Assessment

Use Exercises 1–7 to check for understanding.

Use the chart at the bottom of this page to customize assignments for your students.

 Teaching the Mathematical Practices

Precision Mathematically proficient students try to use clear definitions in their reasoning, calculate accurately and efficiently, and make explicit use of definitions.

Additional Answer (Guided Practice)

3. $\csc^2 \theta - \cot^2 \theta \stackrel{?}{=} \cot \theta \tan \theta$

$$\frac{1}{\sin^2 \theta} - \frac{\cos^2 \theta}{\sin^2 \theta} \stackrel{?}{=} \frac{\cos \theta}{\sin \theta} \cdot \frac{\sin \theta}{\cos \theta}$$

$$\frac{1 - \cos^2 \theta}{\sin^2 \theta} \stackrel{?}{=} 1$$

$$\frac{\sin^2 \theta}{\sin^2 \theta} \stackrel{?}{=} 1$$

$$1 = 1 \; \checkmark$$

 Example 3 Verify by Transforming Each Side

Verify that $1 - \tan^4 \theta = 2 \sec^2 \theta - \sec^4 \theta$ is an identity.

$1 - \tan^4 \theta \stackrel{?}{=} 2 \sec^2 \theta - \sec^4 \theta$	Original equation
$(1 - \tan^2 \theta)(1 + \tan^2 \theta) \stackrel{?}{=} \sec^2 \theta \,(2 - \sec^2 \theta)$	Factor each side.
$[1 - (\sec^2 \theta - 1)] \sec^2 \theta \stackrel{?}{=} (2 - \sec^2 \theta) \sec^2 \theta$	$1 + \tan^2 \theta = \sec^2 \theta$
$(2 - \sec^2 \theta) \sec^2 \theta = (2 - \sec^2 \theta) \sec^2 \theta \; \checkmark$	Simplify.

▶ **Guided**Practice

3. Verify that $\csc^2 \theta - \cot^2 \theta = \cot \theta \tan \theta$ is an identity. **See margin.**

Check Your Understanding

◯ = **Step-by-Step Solutions begin on page R14.**

Examples 1–3 CCSS PRECISION Verify that each equation is an identity. **1–6. See Chapter 13 Answer Appendix.**

1. $\cot \theta + \tan \theta = \dfrac{\sec^2 \theta}{\tan \theta}$

2. $\cos^2 \theta = (1 + \sin \theta)(1 - \sin \theta)$

3. $\sin \theta = \dfrac{\sec \theta}{\tan \theta + \cot \theta}$

4. $\tan^2 \theta = \dfrac{1 - \cos^2 \theta}{\cos^2 \theta}$

5. $\tan^2 \theta \csc^2 \theta = 1 + \tan^2 \theta$

6. $\tan^2 \theta = (\sec \theta + 1)(\sec \theta - 1)$

Example 2 **7** **MULTIPLE CHOICE** Which expression can be used to form an identity with $\dfrac{\tan^2 \theta + 1}{\tan^2 \theta}$? **D**

A $\sin^2 \theta$ **B** $\cos^2 \theta$ **C** $\tan^2 \theta$ **D** $\csc^2 \theta$

Practice and Problem Solving

Extra Practice is on page R13.

Example 1 Verify that each equation is an identity. **8–17. See Chapter 13 Answer Appendix.**

8. $\cos^2 \theta + \tan^2 \theta \cos^2 \theta = 1$

9. $\cot \theta (\cot \theta + \tan \theta) = \csc^2 \theta$

10. $1 + \sec^2 \theta \sin^2 \theta = \sec^2 \theta$

11. $\sin \theta \sec \theta \cot \theta = 1$

12. $\dfrac{1 - \cos \theta}{1 + \cos \theta} = (\csc \theta - \cot \theta)^2$

13. $\dfrac{1 - 2 \cos^2 \theta}{\sin \theta \cos \theta} = \tan \theta - \cot \theta$

14. $\tan \theta = \dfrac{\sec \theta}{\csc \theta}$

15. $\cos \theta = \sin \theta \cot \theta$

16. $(\sin \theta - 1)(\tan \theta + \sec \theta) = -\cos \theta$

17. $\cos \theta \cos (-\theta) - \sin \theta \sin (-\theta) = 1$

Example 2 **18. LADDER** Some students derived an expression for the length of a ladder that, when carried flat, could fit around a corner from a 5-foot-wide hallway into a 7-foot-wide hallway, as shown. They determined that the maximum length ℓ of a ladder that would fit was given by $\ell(\theta) = \dfrac{7 \sin \theta + 5 \cos \theta}{\sin \theta \cos \theta}$. When their teacher worked the problem, she concluded that $\ell(\theta) = 7 \sec \theta + 5 \csc \theta$. Are the two expressions equivalent? **yes**

7 ft

5 ft

Differentiated Homework Options

Level	Assignment	Two-Day Option	
AL Basic	8–32, 52, 54–57, 59–76	9–31 odd, 60–63	8–32 even, 52, 54–57, 59, 64–76
OL Core	9–31 odd, 33–34, 35–49 odd, 50–52, 54–57, 59–76	8–32, 60–63	33–52, 54–57, 59, 64–76
BL Advanced	33–72, (optional: 73–76)		

Example 3

Verify that each equation is an identity. **19–32.** See Chapter 13 Answer Appendix.

19. $\sec \theta - \tan \theta = \dfrac{1 - \sin \theta}{\cos \theta}$

20. $\dfrac{1 + \tan \theta}{\sin \theta + \cos \theta} = \sec \theta$

21. $\sec \theta \csc \theta = \tan \theta + \cot \theta$

22. $\sin \theta + \cos \theta = \dfrac{2 \sin^2 \theta - 1}{\sin \theta - \cos \theta}$

23. $(\sin \theta + \cos \theta)^2 = \dfrac{2 + \sec \theta \csc \theta}{\sec \theta \csc \theta}$

24. $\dfrac{\cos \theta}{1 - \sin \theta} = \dfrac{1 + \sin \theta}{\cos \theta}$

25. $\csc \theta - 1 = \dfrac{\cot^2 \theta}{\csc \theta + 1}$

26. $\cos \theta \cot \theta = \csc \theta - \sin \theta$

27. $\sin \theta \cos \theta \tan \theta + \cos^2 \theta = 1$

28. $(\csc \theta - \cot \theta)^2 = \dfrac{1 - \cos \theta}{1 + \cos \theta}$

29. $\csc^2 \theta = \cot^2 \theta + \sin \theta \csc \theta$

30. $\dfrac{\sec \theta - \csc \theta}{\csc \theta \sec \theta} = \sin \theta - \cos \theta$

31. $\sin^2 \theta + \cos^2 \theta = \sec^2 \theta - \tan^2 \theta$

32. $\sec \theta - \cos \theta = \tan \theta \sin \theta$

 33. **SENSE-MAKING** The diagram at the right represents a game of tetherball. As the ball rotates around the pole, a conical surface is swept out by the line segment $\overline{SP}$. A formula for the relationship between the length L of the string and the angle θ that the string makes with the pole is given by the equation $L = \dfrac{g \sec \theta}{\omega^2}$. Is $L = \dfrac{g \tan \theta}{\omega^2 \sin \theta}$ also an equation for the relationship between L and θ? **yes**

34. RUNNING A portion of a racetrack has the shape of a circular arc with a radius of 16.7 meters. As a runner races along the arc, the sine of her angle of incline θ is found to be $\frac{1}{4}$. Find the speed of the runner. Use the Angle of Incline Formula given at the beginning of the lesson, $\tan \theta = \dfrac{v^2}{gR}$, where $g = 9.8$ and R is the radius. (*Hint:* Find $\cos \theta$ first.) **6.5 m/s**

When simplified, would the expression be equal to 1 or −1?

35. $\cot (-\theta) \tan(-\theta)$ **1**

36. $\sin \theta \csc (-\theta)$ **−1**

37. $\sin^2 (-\theta) + \cos^2 (-\theta)$ **1**

38. $\sec (-\theta) \cos (-\theta)$ **1**

39. $\sec^2 (-\theta) - \tan^2 (-\theta)$ **1**

40. $\cot (-\theta) \cot \left(\dfrac{\pi}{2} - \theta \right)$ **−1**

Simplify the expression to either a constant or a basic trigonometric function.

41. $\dfrac{\tan \left(\frac{\pi}{2} - \theta \right) \csc \theta}{\csc^2 \theta}$ **cos θ**

42. $\dfrac{1 + \tan \theta}{1 + \cot \theta}$ **tan θ**

43. $(\sec^2 \theta + \csc^2 \theta) - (\tan^2 \theta + \cot^2 \theta)$ **2**

44. $\dfrac{\sec^2 \theta - \tan^2 \theta}{\cos^2 x + \sin^2 x}$ **1**

45. $\tan \theta \cos \theta$ **sin θ**

46. $\cot \theta \tan \theta$ **1**

47. $\sec \theta \sin \left(\dfrac{\pi}{2} - \theta \right)$ **1**

48. $\dfrac{1 + \tan^2 \theta}{\csc^2 \theta}$ **tan$^2 \theta$**

49. PHYSICS When a firework is fired from the ground, its height y and horizontal displacement x are related by the equation $y = \dfrac{-gx^2}{2v_0^2 \cos^2 \theta} + \dfrac{x \sin \theta}{\cos \theta}$, where v_0 is the initial velocity of the projectile, θ is the angle at which it was fired, and g is the acceleration due to gravity. Rewrite this equation so that $\tan \theta$ is the only trigonometric function that appears in the equation. $\quad y = -\dfrac{gx^2}{2v_0^2}(1 + \tan^2 \theta) + x \tan \theta$

Teaching the Mathematical Practices

Sense-Making Mathematically proficient students start by explaining to themselves the meaning of a problem and looking for entry points to its solution. They analyze givens, constraints, relationships, and goals. They check their answers to problems using a different method, and they continually ask themselves, "Does this make sense?"

Arguments Mathematically proficient students understand and use stated assumptions, definitions, and previously established results in constructing arguments. They make conjectures and build a logical progression of statements to explore the truth of their conjectures. And they are able to analyze situations by breaking them into cases, and can recognize and use counterexamples.

Follow-up

Students have explored and verified trigonometric identities.

Ask:

- Why are trigonometric identities useful? Sample answer: Trigonometric identities provide a way to simplify complex trigonometric functions by rewriting the expressions found within the functions in equivalent, but more convenient forms.

Focus on Mathematical Content

Transform One or Two Sides An identity can be verified by transforming one or both sides of an equation at the same time. Some students may prefer to work on only one side at a time to avoid confusion.

Study Guide and Intervention
AL OL ELL

Practice
AL OL BL ELL

Word Problem Practice
AL OL BL ELL

Enrichment
OL BL

 50. ELECTRONICS When an alternating current of frequency f and peak current I_0 passes through a resistance R, the power delivered to the resistance at time t seconds is $P = I_0^2 R \sin^2 2\pi f t$.

  a. Write an expression for the power in terms of $\cos^2 2\pi f t$. $P = I_0^2 R(1 - \cos^2 2\pi f t)$

  b. Write an expression for the power in terms of $\csc^2 2\pi f t$. $P = \dfrac{I_0^2 R}{\csc^2 2\pi f t}$

51 THROWING A BALL In this problem, you will investigate the path of a ball represented by the equation $h = \dfrac{v_0^2 \sin^2 \theta}{2g}$, where θ is the measure of the angle between the ground and the path of the ball, v_0 is its initial velocity in meters per second, and g is the acceleration due to gravity. The value of g is 9.8 m/s^2.

  a. If the initial velocity of the ball is 47 meters per second, find the height of the ball at $30°$, $45°$, $60°$, and $90°$. Round to the nearest tenth. **See Chapter 13 Answer Appendix.**

  b. Graph the equation on a graphing calculator. **See Chapter 13 Answer Appendix.**

  c. Show that the formula $h = \dfrac{v_0^2 \tan^2 \theta}{2g \sec^2 \theta}$ is equivalent to the one given above. **See Chapter 13 Answer Appendix.**

H.O.T. Problems   Use Higher-Order Thinking Skills

52. CCSS ARGUMENTS Identify the equation that does not belong with the other three. Explain your reasoning.

52. $\sin^2 \theta - \cos^2 \theta = 2\sin^2 \theta$; the other three are Pythagorean identities, but this is not.

$\boxed{\sin^2 \theta + \cos^2 \theta = 1}$ $\boxed{1 + \cot^2 \theta = \csc^2 \theta}$

$\boxed{\sin^2 \theta - \cos^2 \theta = 2\sin^2 \theta}$ $\boxed{\tan^2 \theta + 1 = \sec^2 \theta}$

53. $\tan^2 \theta = \dfrac{\sin^2 \theta}{\cos^2 \theta}$
$= \dfrac{1 - \cos^2 \theta}{\cos^2 \theta}$
$= \dfrac{1}{\cos^2 \theta} - \dfrac{\cos^2 \theta}{\cos^2 \theta}$
$= \sec^2 \theta - 1$

53. CHALLENGE Transform the right side of $\tan^2 \theta = \dfrac{\sin^2 \theta}{\cos^2 \theta}$ to show that $\tan^2 \theta = \sec^2 \theta - 1$.

54. WRITING IN MATH Explain why you cannot square each side of an equation when verifying a trigonometric identity. **The properties of equality do not apply to identities as they do with equations. Do not perform operations to the quantities from each side of an unverified identity.**

55. REASONING Explain why $\sin^2 \theta + \cos^2 \theta = 1$ is an identity, but $\sin \theta = \sqrt{1 - \cos \theta}$ is not. **Sample answer: counterexample $45°$, $30°$**

56. WRITE A QUESTION A classmate is having trouble trying to verify a trigonometric identity involving multiple trigonometric functions to multiple degrees. Write a question to help her work through the problem. **Sample answer: Have you tried using the most common identity, $\sin^2 \sigma + \cos^2 \sigma = 1$, to simplify?**

57. ⬛ **WRITING IN MATH** Why do you think expressions in trigonometric identities are often rewritten in terms of sine and cosine? **See margin.**

58. CHALLENGE Let $x = \dfrac{1}{2}\tan \theta$, where $-\dfrac{\pi}{2} < \theta < \dfrac{\pi}{2}$. Write $f(x) = \dfrac{x}{\sqrt{1 + 4x^2}}$ in terms of a single trigonometric function of θ. $f(\theta) = \dfrac{1}{2}\sin \theta$

59. REASONING Justify the three basic Pythagorean identities. **See Chapter 13 Answer Appendix.**

🔊 **884** | Lesson 13-2 | Verifying Trigonometric Identities

Additional Answer

57. Sample answer: Sine and cosine are the trigonometric functions with which most people are familiar, and all trigonometric expressions can be written in terms of sine and cosine. Also, by rewriting complex trigonometric expressions in terms of sine and cosine it may be easier to perform operations and to apply trigonometric properties.

60. SAT/ACT A small business owner must hire seasonal workers as the need arises. The following list shows the number of employees hired monthly for a 5-month period.

$$5, 14, 6, 8, 12$$

If the mean of these data is 9, what is the population standard deviation for these data? (Round your answer to the nearest tenth.) **A**

A 3.5   D 8.6
B 3.9   E 12.3
C 5.7

61. Find the center and radius of the circle with equation $(x - 4)^2 + y^2 - 16 = 0$. **H**

F $C(-4, 0); r = 4$ units
G $C(-4, 0); r = 16$ units
H $C(4, 0); r = 4$ units
J $C(4, 0); r = 16$ units

62. GEOMETRY The perimeter of a right triangle is 36 inches. Twice the length of the longer leg minus twice the length of the shorter leg is 6 inches. What are the lengths of all three sides? **C**

A 3 in., 4 in., 5 in.
B 6 in., 8 in., 10 in.
C 9 in., 12 in., 15 in.
D 12 in., 16 in., 20 in.

63. Simplify $128^{\frac{1}{4}}$. **G**

F $2\sqrt[4]{2}$
G $2\sqrt[4]{8}$
H 4
J $4\sqrt[4]{2}$

Find the exact value of each expression. (Lesson 13-1)

64. $\tan \theta$, if $\cot \theta = 2; 0° \le \theta < 90°$ $\dfrac{1}{2}$

65. $\sin \theta$, if $\cos \theta = \dfrac{2}{3}; 0° \le \theta < 90°$ $\dfrac{\sqrt{5}}{3}$

66. $\csc \theta$, if $\cos \theta = -\dfrac{3}{5}; 90° < \theta < 180°$ $\dfrac{5}{4}$

67. $\cos \theta$, if $\sec \theta = \dfrac{5}{3}; 270° < \theta < 360°$ $\dfrac{3}{5}$

68. ARCHITECTURE The support for a roof is shaped like two right triangles, as shown at the right. Find θ. (Lesson 12-9) **30°**

69. FAST FOOD The table shows the probability distribution for value meals ordered at a fast food restaurant on Saturday mornings. Use this information to determine the expected value of the meals ordered. (Lesson 11-3) **$4**

Value Meals Ordered			
Meals	$3	$4	$5	$6
Probability	0.5	0.2	0.1	0.2

Find the coordinates of the vertices and foci and the equations of the asymptotes for the hyperbolas with the given equations. Then graph the hyperbola. (Lesson 9-5) **70–72. See margin.**

70. $\dfrac{y^2}{18} - \dfrac{x^2}{20} = 1$

71. $\dfrac{(y + 6)^2}{20} - \dfrac{(x - 1)^2}{25} = 1$

72. $x^2 - 36y^2 = 36$

Simplify.

73. $\dfrac{2 + \sqrt{2}}{5 - \sqrt{2}}$ $\dfrac{12 + 7\sqrt{2}}{23}$

74. $\dfrac{x + 1}{\sqrt{x^2 - 1}}$ $\dfrac{\sqrt{x^2 - 1}}{x - 1}$

75. $\dfrac{x - 1}{\sqrt{x} - 1}$ $\sqrt{x} + 1$

76. $\dfrac{-2 - \sqrt{3}}{1 + \sqrt{3}}$ $\dfrac{-1 - \sqrt{3}}{2}$

DifferentiatedInstruction BL

Extension $\sin x + \cos x = 1$ is not an identity, meaning that it is not true for all real values of x. Find the values of x for which the equation is true. $0° + k \cdot 360°$ or $90° + k \cdot 360°$, where k is any integer

---

4 Assess

Yesterday's News Have students write how learning basic trigonometric identities in Lesson 13-1 has helped them with verifying the more complex identities in today's lesson.

Formative Assessment

Check for student understanding of concepts in Lessons 13-1 and 13-2.

▢ Quiz 1, p. 37

Additional Answers

70. $(0, \pm 3\sqrt{2}); (0, \pm\sqrt{38});$

$$y = \pm\dfrac{3\sqrt{10}}{10} x$$

71. $(1, -6 \pm 2\sqrt{5}); (1, -6 \pm 3\sqrt{5});$

$$y + 6 = \pm\dfrac{2\sqrt{5}}{5}(x - 1)$$

72. $(\pm 6, 0); (\pm\sqrt{37}, 0); y = \pm\dfrac{1}{6}x$

1 Focus

VerticalAlignment

Before Lesson 13-3 Find values of trigonometric functions for general angles.

Lesson 13-3 Find values of sine and cosine by using sum and difference identities. Verify trigonometric identities by using sum and difference identities.

After Lesson 13-3 Use double-angle and half-angle identities.

2 Teach

Scaffolding Questions

Have students read the **Why?** section of the lesson.

Ask:

- Where else have you heard the term *interference*? Sample answer: television and radio

- At its peak, how does the amplitude of the combined wave compare to the amplitude of the initial two waves? The amplitude of the combined wave is the sum of the amplitudes of the two initial waves.

(continued on the next page)

Then	Now	Why?
● You found values of trigonometric functions for general angles.	**1** Find values of sine and cosine by using sum and difference identities. **2** Verify trigonometric identities by using sum and difference identities.	● Have you ever been using a wireless Internet provider and temporarily lost the signal? Waves that pass through the same place at the same time cause interference. Interference occurs when two waves combine to have a greater, or smaller, amplitude than either of the component waves.

$y = \sin A$
$y = \sin (A - B)$
$y = 2 \sin \left[\frac{1}{2}(2A + B)\right]$

CCSS **Common Core State Standards**

Content Standards
F.TF.8 Prove the Pythagorean identity $\sin^2(\theta) + \cos^2(\theta) = 1$ and use it to find $\sin(\theta)$, $\cos(\theta)$, or $\tan(\theta)$ given $\sin(\theta)$, $\cos(\theta)$, or $\tan(\theta)$ and the quadrant of the angle.

Mathematical Practices
3 Construct viable arguments and critique the reasoning of others.
6 Attend to precision.

1 Sum and Difference Identities Notice that the third equation shown above involves the sum of A and B. It is often helpful to use formulas for the trigonometric values of the difference or sum of two angles. For example, you could find the exact value of $\sin 15°$ by evaluating $\sin(60° - 45°)$. Formulas exist that can be used to evaluate expressions like $\sin(A - B)$ or $\cos(A + B)$.

KeyConcept Sum and Difference Identities

Sum Identities
- $\sin(A + B) = \sin A \cos B + \cos A \sin B$
- $\cos(A + B) = \cos A \cos B - \sin A \sin B$
- $\tan(A + B) = \dfrac{\tan A + \tan B}{1 - \tan A \tan B}$

Difference Identities
- $\sin(A - B) = \sin A \cos B - \cos A \sin B$
- $\cos(A - B) = \cos A \cos B + \sin A \sin B$
- $\tan(A - B) = \dfrac{\tan A - \tan B}{1 + \tan A \tan B}$

Example 1 Find Trigonometric Values

Find the exact value of each expression.

a. sin 105°

Use the identity $\sin(A + B) = \sin A \cos B + \cos A \sin B$.

$\sin 105° = \sin(60° + 45°)$ $A = 60°$ and $B = 45°$

$= \sin 60° \cos 45° + \cos 60° \sin 45°$ Sum identity

$= \left(\dfrac{\sqrt{3}}{2} \cdot \dfrac{\sqrt{2}}{2}\right) + \left(\dfrac{1}{2} \cdot \dfrac{\sqrt{2}}{2}\right)$ Evaluate each expression.

$= \dfrac{\sqrt{6}}{4} + \dfrac{\sqrt{2}}{4}$ or $\dfrac{\sqrt{6} + \sqrt{2}}{4}$ Multiply.

b. cos (−120°)

Use the identity $\cos(A - B) = \cos A \cos B + \sin A \sin B$.

$\cos(-120) = \cos(60° - 180°)$ $A = 60°$ and $B = 180°$

$= \cos 60° \cos 180° + \sin 60° \sin 180°$ Difference identity

$= \dfrac{1}{2} \cdot (-1) + \dfrac{\sqrt{3}}{2} \cdot 0$ Evaluate each expression.

$= -\dfrac{1}{2}$ Multiply.

GuidedPractice

1A. $\sin 15°$ $\dfrac{\sqrt{6} - \sqrt{2}}{4}$ **1B.** $\cos(-15°)$ $\dfrac{\sqrt{6} + \sqrt{2}}{4}$

Lesson 13-3 Resources

Resource	Approaching Level **AL**	On Level **OL**	Beyond Level **BL**	English Learners **ELL**
Teacher Edition		▪ Differentiated Instruction, pp. 887, 891	▪ Differentiated Instruction, pp. 887, 891	▪ Differentiated Instruction, p. 887
Chapter Resource Masters	▪ Study Guide and Intervention, pp. 17–18 ▪ Skills Practice, p. 19 ▪ Practice, p. 20 ▪ Word Problem Practice, p. 21	▪ Study Guide and Intervention, pp. 17–18 ▪ Skills Practice, p. 19 ▪ Practice, p. 20 ▪ Word Problem Practice, p. 21 ▪ Enrichment, p. 22	▪ Practice, p. 20 ▪ Word Problem Practice, p. 21 ▪ Enrichment, p. 22	▪ Study Guide and Intervention, pp. 17–18 ▪ Skills Practice, p. 19 ▪ Practice, p. 20 ▪ Word Problem Practice, p. 21
Other	▪ 5-Minute Check 13-3 ▪ Study Notebook	▪ 5-Minute Check 13-3 ▪ Study Notebook	▪ 5-Minute Check 13-3 ▪ Study Notebook	▪ 5-Minute Check 13-3 ▪ Study Notebook

You can use the sum and difference of angles identities to solve real-world applications.

Real-World Example 2 Sum and Difference of Angles Identities

A geologist measures the angle between one side of a rectangular lot and the line from her position to the opposite corner of the lot as 30°. She then measures the angle between that line and the line to the point on the property where a river crosses as 45°. She stands 100 yards from the opposite corner of the property. How far is she from the point at which the river crosses the property line?

Understand The question asks for the distance between the geologist and the point where the river crosses the property line, or y.

Plan Draw a picture that labels all the things that you know from the information given.

Solve Solve for x.

$\sin 30° = \dfrac{x}{100}$ Definition of sine

$x = 100 \sin 30°$

$x = 50$ Since the lot is rectangular, opposite sides are equal.

Now look at the triangle on the far left and solve for y.

$\cos 15° = \dfrac{50}{y}$ Definition of cosine

$\cos(45° - 30°) = \dfrac{50}{y}$ $15 = 45 - 30$

$\cos 45° \cos 30° + \sin 45° \sin 30° = \dfrac{50}{y}$ Difference identity

$\dfrac{\sqrt{2}}{2} \cdot \dfrac{\sqrt{3}}{2} + \dfrac{\sqrt{2}}{2} \cdot \dfrac{1}{2} = \dfrac{50}{y}$ Evaluate.

$\dfrac{\sqrt{6} + \sqrt{2}}{4} = \dfrac{50}{y}$ Simplify.

$(\sqrt{6} + \sqrt{2})y = 200$ Cross products

$y = \dfrac{200}{(\sqrt{6} + \sqrt{2})} \cdot \dfrac{(\sqrt{6} - \sqrt{2})}{(\sqrt{6} - \sqrt{2})}$

$y = 50(\sqrt{6} - \sqrt{2})$

$y = 50\sqrt{6} - 50\sqrt{2}$ or about 51.8

The geologist is about 51.8 yards from the point where the river crosses the property line.

Check Use a calculator to find the Arccos of $\dfrac{50}{51.8} \approx 15°$. ✓

GuidedPractice

2. The harmonic motion of an object can be described by $x = 4 \cos\left(2\pi t - \dfrac{\pi}{4}\right)$, where x is distance from the equilibrium point in inches and t is time in minutes. Find the exact distance from the equilibrium point at 45 seconds. $2\sqrt{2}$ inches below

Problem-Solving Tip

Make a Model Make a model to visualize a problem situation. A model can be a drawing or a figure made of different objects, such as algebra tiles or folded paper.

- Why does the combined wave cross the x-axis at a point where neither of the two initial waves are crossing the axis? The combined wave is the sum of the other two waves. It crosses the x-axis at points where one of the initial waves is above the x-axis and the other wave is an equal distance below the x-axis.

1 Sum and Difference Identities
Example 1 shows how to use the sum and difference identities to find exact values of trigonometric expressions. **Example 2** shows how to use a difference identity to solve a real-world problem.

Formative Assessment
Use the Guided Practice exercises after each example to determine students' understanding of concepts.

Additional Examples

1 Find the exact value of each expression.

a. $\sin 75°$ $\dfrac{\sqrt{2} + \sqrt{6}}{4}$

b. $\cos(-75°)$ $\dfrac{\sqrt{6} - \sqrt{2}}{4}$

2 **DISTANCE** Refer to Example 2. If z represents the distance between the upper left corner of the property and the point where the river crosses the top boundary, then $\tan 15° = \dfrac{z}{50}$ or $\tan(45° - 30°) = \dfrac{z}{50}$. Use the identity for $\tan(A - B)$ to find an exact value for z.

$z = \dfrac{50\left(1 - \dfrac{1}{\sqrt{3}}\right)}{1 + \dfrac{1}{\sqrt{3}}}$, which can be simplified to $z = 100 - 50\sqrt{3}$.

Additional Examples also in Interactive Classroom PowerPoint® Presentations

IWB **Interactive White Board** READY

Differentiated Instruction OL BL ELL

Verbal/Linguistic Learners Ask students to identify patterns, similarities, and differences in the sum and difference identities. Then ask the students to write short sentences describing what they have identified.

Teach with Tech

Document Camera Choose several students to work through examples and explain how to apply the sum and difference of angles formulas.

2 Verify Trigonometric Identities

Example 3 shows how to use the sum and difference identities to verify trigonometric identities.

 Teaching the Mathematical Practices

Sense-Making Mathematically proficient students analyze givens, constraints, relationships, and goals. Encourage students to become familiar with angle measures for which identities can be easily applied.

Modeling Mathematically proficient students can apply the mathematics they know to solve problems arising in everyday life, analyze relationships mathematically to draw conclusions, and interpret their mathematical results in the context of a situation.

Additional Example

3 Verify that each equation is an identity.

a. $\cos(360° - \theta) = \cos\theta$

$$\cos(360° - \theta) \stackrel{?}{=} \cos\theta$$
$$\cos 360° \cos\theta + \sin 360° \sin\theta \stackrel{?}{=} \cos\theta$$
$$1 \cdot \cos\theta + 0 \cdot \sin\theta \stackrel{?}{=} \cos\theta$$
$$\cos\theta = \cos\theta \checkmark$$

b. $\cos(\pi - \theta) = -\cos\theta$

$$\cos(\pi - \theta) \stackrel{?}{=} -\cos\theta$$
$$\cos\pi\cos\theta + \sin\pi\sin\theta \stackrel{?}{=} -\cos\theta$$
$$-1 \cdot \cos\theta + 0 \cdot \sin\theta \stackrel{?}{=} -\cos\theta$$
$$-\cos\theta = -\cos\theta \checkmark$$

Formative Assessment

Use Exercises 1–11 to check for understanding.

Use the chart at the bottom of the next page to customize assignments for your students.

StudyTip

CCSS Sense-Making Make a list of the trigonometric values for the angles between 0° and 360° for which the sum and difference identities can be easily used. Use your list as a reference.

3A. $\sin(90° - \theta) \stackrel{?}{=} \cos\theta$;
$\sin 90° \cos\theta - \cos 90° \sin\theta \stackrel{?}{=} \cos\theta$;
$1\cos\theta - 0\sin\theta \stackrel{?}{=} \cos\theta$;
$\cos\theta = \cos\theta$

3B. $\cos(90° + \theta) \stackrel{?}{=} -\sin\theta$;
$\cos 90° \cos\theta - \sin 90° \sin\theta \stackrel{?}{=} -\sin\theta$;
$0\cos\theta - 1\sin\theta \stackrel{?}{=} -\sin\theta$;
$-\sin\theta = -\sin\theta$

---

2 **Verify Trigonometric Identities** You can also use the sum and difference identities to verify identities.

Example 3 Verify Trigonometric Identities

Verify that each equation is an identity.

a. $\cos(90° - \theta) = \sin\theta$

$\cos(90° - \theta) \stackrel{?}{=} \sin\theta$	Original equation
$\cos 90° \cos\theta + \sin 90° \sin\theta \stackrel{?}{=} \sin\theta$	Sum identity
$0 \cdot \cos\theta + 1 \cdot \sin\theta \stackrel{?}{=} \sin\theta$	Evaluate each expression.
$\sin\theta = \sin\theta \checkmark$	Simplify.

b. $\sin\left(\theta + \dfrac{\pi}{2}\right) = \cos\theta$

$\sin\left(\theta + \dfrac{\pi}{2}\right) \stackrel{?}{=} \cos\theta$	Original equation
$\sin\theta\cos\dfrac{\pi}{2} + \cos\theta\sin\dfrac{\pi}{2} \stackrel{?}{=} \cos\theta$	Sum identity
$\sin\theta \cdot 0 + \cos\theta \cdot 1 \stackrel{?}{=} \cos\theta$	Evaluate each expression.
$\cos\theta = \cos\theta \checkmark$	Simplify.

▶ **Guided** Practice

3A. $\sin(90° - \theta) = \cos\theta$ **3B.** $\cos(90° + \theta) = -\sin\theta$

Check Your Understanding

◯ **= Step-by-Step Solutions begin on page R14.**

Example 1   **Find the exact value of each expression.**

1 $\cos 165°$ $-\dfrac{\sqrt{2}+\sqrt{6}}{4}$ **2.** $\cos 105°$ $\dfrac{\sqrt{2}-\sqrt{6}}{4}$ **3.** $\cos 75°$ $\dfrac{\sqrt{6}-\sqrt{2}}{4}$

4. $\sin(-30°)$ $-\dfrac{1}{2}$ **5.** $\sin 135°$ $\dfrac{\sqrt{2}}{2}$ **6.** $\sin(-210°)$ $\dfrac{1}{2}$

Example 2   **7.** **CCSS** **MODELING** Refer to the beginning of the lesson. *Constructive interference* occurs when two waves combine to have a greater amplitude than either of the component waves. *Destructive interference* occurs when the component waves combine to have a smaller amplitude. The first signal can be modeled by the equation $y = 20\sin(3\theta + 45°)$. The second signal can be modeled by the equation $y = 20\sin(3\theta + 225°)$.

    a. Find the sum of the two functions.   **0**

    b. What type of interference results when signals modeled by the two equations are combined?   **The interference is destructive. The signals cancel each other completely.**

Example 3   **Verify that each equation is an identity.**   **8–11. See margin.**

8. $\sin(90° + \theta) = \cos\theta$ **9.** $\cos\left(\dfrac{3\pi}{2} - \theta\right) = -\sin\theta$

10. $\tan\left(\theta + \dfrac{\pi}{2}\right) = -\cot\theta$ **11.** $\sin(\theta + \pi) = -\sin\theta$

 888 | Lesson 13-3 | Sum and Difference of Angles Identities

Additional Answers

8.
$$\sin(90° + \theta) \stackrel{?}{=} \cos\theta$$
$$\sin 90° \cos\theta + \cos 90° \sin\theta \stackrel{?}{=} \cos\theta$$
$$1 \cdot \cos\theta + 0 \cdot \sin\theta \stackrel{?}{=} \cos\theta$$
$$\cos\theta = \cos\theta \checkmark$$

9.
$$\cos\left(\dfrac{3\pi}{2} - \theta\right) \stackrel{?}{=} -\sin\theta$$
$$\cos\dfrac{3\pi}{2}\cos\theta + \sin\dfrac{3\pi}{2}\sin\theta \stackrel{?}{=} -\sin\theta$$
$$0 \cdot \cos\theta - 1 \cdot \sin\theta \stackrel{?}{=} -\sin\theta$$
$$-\sin\theta = -\sin\theta \checkmark$$

Example 1

Find the exact value of each expression.

12. $\sin 165°$ $\dfrac{\sqrt{6} - \sqrt{2}}{4}$ **13.** $\cos 135°$ $-\dfrac{\sqrt{2}}{2}$ **14.** $\cos \dfrac{7\pi}{12}$ $\dfrac{\sqrt{2} - \sqrt{6}}{4}$

15. $\sin \dfrac{\pi}{12}$ $\dfrac{\sqrt{6} - \sqrt{2}}{4}$ **16.** $\tan 195°$ $2 - \sqrt{3}$ **17.** $\cos\left(-\dfrac{\pi}{12}\right)$ $\dfrac{\sqrt{2} + \sqrt{6}}{4}$

Example 2

18. ELECTRONICS In a certain circuit carrying alternating current, the formula $c = 2 \sin(120t)$ can be used to find the current c in amperes after t seconds.

    a. Rewrite the formula using the sum of two angles. Sample answer: $c = 2 \sin(90t + 30t)$

    b. Use the sum of angles formula to find the exact current at $t = 1$ second.
    $\sqrt{3}$ amperes

Example 3

Verify that each equation is an identity. **19–22. See margin.**

19. $\cos\left(\dfrac{\pi}{2} + \theta\right) = -\sin\theta$ **20.** $\cos(60° + \theta) = \sin(30° - \theta)$

21. $\cos(180° + \theta) = -\cos\theta$ **22.** $\tan(\theta + 45°) = \dfrac{1 + \tan\theta}{1 - \tan\theta}$

B **23.** **CCSS REASONING** The monthly high temperatures for Minneapolis, Minnesota, can be modeled by the equation $y = 31.65 \sin\left(\dfrac{\pi}{6}x - 2.09\right) + 52.35$, where the months x are represented by January = 1, February = 2, and so on. The monthly low temperatures for Minneapolis can be modeled by the equation $y = 30.15 \sin\left(\dfrac{\pi}{6}x - 2.09\right) + 32.95$.

    a. Write a new function by adding the expressions on the right side of each equation and dividing the result by 2. $y = 30.9 \sin\left(\dfrac{\pi}{6}x - 2.09\right) + 42.65$

    b. What is the meaning of the function you wrote in part **a**?
    The new function represents the average of the high and low temperatures for each month.

Find the exact value of each expression.

24. $\tan 165°$ $-2 + \sqrt{3}$ **25.** $\sec 1275°$ $\sqrt{2} - \sqrt{6}$ **26.** $\sin 735°$ $\dfrac{\sqrt{6} - \sqrt{2}}{4}$

27. $\tan \dfrac{23\pi}{12}$ $-2 + \sqrt{3}$ **28.** $\csc \dfrac{5\pi}{12}$ $\sqrt{6} - \sqrt{2}$ **29.** $\cot \dfrac{113\pi}{12}$ $2 - \sqrt{3}$

30. FORCE In the figure at the right, the effort F necessary to hold a safe in position on a ramp is given by $F = \dfrac{W(\sin A + \mu \cos A)}{\cos A - \mu \sin A}$, where W is the weight of the safe and $\mu = \tan\theta$. Show that $F = W \tan(A + \theta)$.
See Chapter 13 Answer Appendix.

31 **QUILTING** As part of a quilt that is being made, the quilter places two right triangular swatches together to make a new triangular piece. One swatch has sides 6 inches, 8 inches, and 10 inches long. The other swatch has sides 8 inches, $8\sqrt{3}$ inches, and 16 inches long. The pieces are placed with the sides of eight inches against each other, as shown in the figure, to form triangle ABC.

    a. What is the exact value of the sine of angle BAC? $\dfrac{3 + 4\sqrt{3}}{10}$

    b. What is the exact value of the cosine of angle BAC? $\dfrac{4 - 3\sqrt{3}}{10}$

    c. What is the measure of angle BAC? $96.9°$

    d. Is the new triangle formed from the two triangles also a right triangle? no

Differentiated Homework Options

Level	Assignment	Two-Day Option	
AL Basic	12–22, 38, 39, 41–57	13–21 odd, 43–46	12–22 even, 38, 39, 41, 42, 47–57
OL Core	13–29 odd, 30–33, 35, 37–39, 41–57	12–22, 43–46	23–39, 41, 42, 47–57
BL Advanced	23–54, (optional: 55–57)		

Additional Answers

10.
$$\tan\left(\theta + \dfrac{\pi}{2}\right) \overset{?}{=} -\cot\theta$$
$$\dfrac{\sin\left(\theta + \dfrac{\pi}{2}\right)}{\cos\left(\theta + \dfrac{\pi}{2}\right)} \overset{?}{=} -\cot\theta$$
$$\dfrac{\sin\theta\cos\dfrac{\pi}{2} + \cos\theta\sin\dfrac{\pi}{2}}{\cos\theta\cos\dfrac{\pi}{2} - \sin\theta\sin\dfrac{\pi}{2}} \overset{?}{=} -\cot\theta$$
$$\dfrac{(\sin\theta)\cdot 0 + (\cos\theta)\cdot 1}{(\cos\theta)\cdot 0 - (\sin\theta)\cdot 1} \overset{?}{=} -\cot\theta$$
$$-\dfrac{\cos\theta}{\sin\theta} \overset{?}{=} -\cot\theta$$
$$-\cot\theta = -\cot\theta \checkmark$$

11.
$$\sin(\theta + \pi) \overset{?}{=} -\sin\theta$$
$$\sin\theta\cos\pi + \cos\theta\sin\pi \overset{?}{=} -\sin\theta$$
$$(\sin\theta)(-1) + (\cos\theta)(0) \overset{?}{=} -\sin\theta$$
$$-\sin\theta = -\sin\theta \checkmark$$

19.
$$\cos\left(\dfrac{\pi}{2} + \theta\right) \overset{?}{=} -\sin\theta$$
$$\cos\dfrac{\pi}{2}\cos\theta - \sin\dfrac{\pi}{2}\sin\theta \overset{?}{=} -\sin\theta$$
$$(0)(\cos\theta) - (1)(\sin\theta) \overset{?}{=} -\sin\theta$$
$$-\sin\theta = -\sin\theta \checkmark$$

20.
$$\cos(60° + \theta) \overset{?}{=} \sin(30° - \theta)$$
$$\cos 60°\cos\theta - \sin 60°\sin\theta \overset{?}{=} \sin 30°\cos\theta - \cos 30°\sin\theta$$
$$\dfrac{1}{2}\cos\theta - \dfrac{\sqrt{3}}{2}\sin\theta = \dfrac{1}{2}\cos\theta - \dfrac{\sqrt{3}}{2}\sin\theta \checkmark$$

21.
$$\cos(180° + \theta) \overset{?}{=} -\cos\theta$$
$$\cos 180°\cos\theta - \sin 180°\sin\theta \overset{?}{=} -\cos\theta$$
$$-1\cdot\cos\theta - 0\cdot\sin\theta \overset{?}{=} -\cos\theta$$
$$-\cos\theta = -\cos\theta \checkmark$$

22.
$$\tan(\theta + 45°) \overset{?}{=} \dfrac{1 + \tan\theta}{1 - \tan\theta}$$
$$\dfrac{\tan\theta + \tan 45°}{1 - \tan\theta\tan 45°} \overset{?}{=} \dfrac{1 + \tan\theta}{1 - \tan\theta}$$
$$\dfrac{\tan\theta + 1}{1 - (\tan\theta)(1)} \overset{?}{=} \dfrac{1 + \tan\theta}{1 - \tan\theta}$$
$$\dfrac{1 + \tan\theta}{1 - \tan\theta} = \dfrac{1 + \tan\theta}{1 - \tan\theta} \checkmark$$

Study Guide and Intervention
AL OL ELL

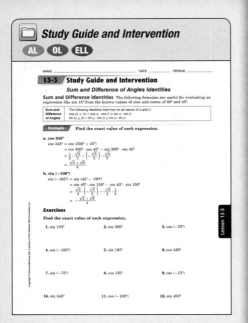

Practice
AL OL BL ELL

Word Problem Practice
AL OL BL ELL

32. OPTICS When light passes symmetrically through a prism, the index of refraction n of the glass with respect to air is 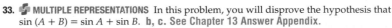 $n = \dfrac{\sin\left[\frac{1}{2}(a + b)\right]}{\sin \frac{b}{2}}$, where a is the measure of the deviation angle and b is the measure of the prism apex angle.

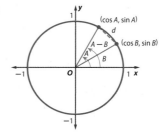

prism apex

a. Show that for the prism shown, $n = \sqrt{3} \sin \frac{a}{2} + \cos \frac{a}{2}$. See margin.

b. Find n for the prism shown. $\sqrt{3}$

33. 📊 **MULTIPLE REPRESENTATIONS** In this problem, you will disprove the hypothesis that $\sin (A + B) = \sin A + \sin B$. **b, c. See Chapter 13 Answer Appendix.**

a. Tabular Copy and complete the table.

b. Graphical Assume that B is always 15° less than A. Use a graphing calculator to graph $y = \sin (x + x - 15)$ and $y = \sin x + \sin (x - 15)$ on the same screen.

c. Analytical Determine whether $\cos (A + B) = \cos A + \cos B$ is an identity. Explain your reasoning.

A	B	sin A	sin B	sin (A + B)	sin A + sin B
30°	90°	$\frac{1}{2}$	1	$\frac{\sqrt{3}}{2}$	$\frac{3}{2}$
45°	60°	$\frac{\sqrt{2}}{2}$	$\frac{\sqrt{3}}{2}$	$\frac{\sqrt{2} + \sqrt{6}}{4}$	$\frac{\sqrt{2} + \sqrt{3}}{2}$
60°	45°	$\frac{\sqrt{3}}{2}$	$\frac{\sqrt{2}}{2}$	$\frac{\sqrt{2} + \sqrt{6}}{4}$	$\frac{\sqrt{2} + \sqrt{3}}{2}$
90°	30°	1	$\frac{1}{2}$	$\frac{\sqrt{3}}{2}$	$\frac{3}{2}$

Verify that each equation is an identity. 34–37. See Chapter 13 Answer Appendix.

34. $\sin (A + B) = \dfrac{\tan A + \tan B}{\sec A \sec B}$

35 $\cos (A + B) = \dfrac{1 - \tan A \tan B}{\sec A \sec B}$

36. $\sec (A - B) = \dfrac{\sec A \sec B}{1 + \tan A \tan B}$

37. $\sin (A + B) \sin (A - B) = \sin^2 A - \sin^2 B$

42. Sample answer: $A = 35°$, $B = 60°$, $C = 85°$; $0.7002 + 1.7321 + 11.4301 \overset{?}{=} (0.7002)(1.7321)(11.4301)$; $13.86 = 13.86$ ✓

H.O.T. Problems — Use Higher-Order Thinking Skills

38. REASONING Simplify the following expression without expanding any of the sums or differences. $\sin (-2\theta)$

$$\sin \left(\frac{\pi}{3} - \theta\right) \cos \left(\frac{\pi}{3} + \theta\right) - \cos \left(\frac{\pi}{3} - \theta\right) \sin \left(\frac{\pi}{3} + \theta\right)$$

39. WRITING IN MATH Use the information at the beginning of the lesson and in Exercise 7 to explain how the sum and difference identities are used to describe wireless Internet interference. Include an explanation of the difference between constructive and destructive interference.

39. Sample answer: To determine wireless Internet interference, you need to determine the sine or cosine of the sum or difference of two angles. Interference occurs when waves pass through the same space at the same time. When the combined waves have a greater amplitude, constructive interference results. When the combined waves have a smaller amplitude, destructive interference results.

40. CHALLENGE Derive an identity for $\cot (A + B)$ in terms of $\cot A$ and $\cot B$. **See Chapter 13 Answer Appendix.**

41. (CCSS) **ARGUMENTS** The figure shows two angles A and B in standard position on the unit circle. Use the Distance Formula to find d, where $(x_1, y_1) = (\cos B, \sin B)$ and $(x_2, y_2) = (\cos A, \sin A)$. **See Chapter 13 Answer Appendix.**

42. OPEN ENDED Consider the following theorem. *If A, B, and C are the angles of an oblique triangle, then* $\tan A + \tan B + \tan C = \tan A \tan B \tan C$. Choose values for A, B, and C. Verify that the conclusion is true for your specific values.

🔊 **890** | Lesson 13-3 | Sum and Difference of Angles Identities

Enrichment
OL BL

📊 Multiple Representations

In Exercise 33, students use information organized in a table and a graphing calculator to disprove a hypothesis about trigonometric operations.

43. GRIDDED RESPONSE The mean of seven numbers is 0. The sum of three of the numbers is −9. What is the sum of the remaining four numbers? **9**

44. The variables a, b, c, d, and f are integers in a sequence, where $a = 2$ and $b = 12$. To find the next term, double the last term and add that result to one less than the next-to-last term. For example, $c = 25$, because $2(12) = 24$, $2 − 1 = 1$, and $24 + 1 = 25$. What is the value of f? **C**

 A 74

 B 144

 C 146

 D 256

45. SAT/ACT Solve $x^2 − 5x < 14$. **H**

 F $\{x \,|\, {-7} < x < 2\}$

 G $\{x \,|\, x < {-7} \text{ or } x > 2\}$

 H $\{x \,|\, {-2} < x < 7\}$

 J $\{x \,|\, x < {-2} \text{ or } x > 7\}$

 K $\{x \,|\, x > {-2} \text{ and } x < 7\}$

46. PROBABILITY A math teacher is randomly distributing 15 yellow pencils and 10 green pencils. What is the probability that the first pencil she hands out will be yellow and the second pencil will be green? **B**

 A $\frac{1}{24}$ **C** $\frac{2}{5}$

 B $\frac{1}{4}$ **D** $\frac{23}{25}$

Verify that each equation is an identity. (Lesson 13-2)   **47, 48. See margin.**

47. $\dfrac{\sin \theta}{\tan \theta} + \dfrac{\cos \theta}{\cot \theta} = \cos \theta + \sin \theta$

48. $\sec \theta (\sec \theta − \cos \theta) = \tan^2 \theta$

Simplify each expression. (Lesson 13-1)

49. $\sin \theta \csc \theta − \cos^2 \theta$ $\sin^2 \theta$ **50.** $\cos^2 \theta \sec \theta \csc \theta \cot \theta$ **51.** $\cos \theta + \sin \theta \tan \theta$ $\sec \theta$

52. GUITAR When a guitar string is plucked, it is displaced from a fixed point in the middle of the string and vibrates back and forth, producing a musical tone. The exact tone depends on the frequency, or number of cycles per second, that the string vibrates. To produce an A, the frequency is 440 cycles per second, or 440 hertz (Hz). (Lesson 12-6)

 a. Find the period of this function. $\frac{1}{440}$ second

 b. Graph the height of the fixed point on the string from its resting position as a function of time. Let the maximum distance above the resting position have a value of 1 unit, and let the minimum distance below this position have a value of 1 unit. **See Chapter 13 Answer Appendix.**

Prove that each statement is true for all positive integers. (Lesson 10-7)   **53, 54. See Chapter 13 Answer Appendix.**

53. $4^n − 1$ is divisible by 3.

54. $5^n + 3$ is divisible by 4.

Solve each equation.

55. $7 + \sqrt{4x + 8} = 9$ −1   **56.** $\sqrt{y + 21} − 1 = \sqrt{y + 12}$ 4   **57.** $\sqrt{4z + 1} = 3 + \sqrt{4z − 2}$ **no solution**

DifferentiatedInstruction OL BL

Extension Tell students that sin 20° is approximately 0.3420. Tell them to use this information to find sin 65° and cos 65°. 0.9063, 0.4226

Ticket Out the Door Have students make a list of angles between 0° and 360° for which the sum and difference formulas can easily be used. Then have them tell what the angles have in common.

Formative Assessment
Check for student understanding of concepts in Lesson 13-3.

📁 Quiz 2, p. 37

Additional Answers

32a.
$$\frac{\sin \left[\frac{1}{2}(a + b) \right]}{\sin \frac{b}{2}}$$
$$= \frac{\sin \left[\frac{1}{2}(a + 60°) \right]}{\sin \frac{60°}{2}}$$
$$= \frac{\sin \left(\frac{a}{2} + 30° \right)}{\sin 30°}$$
$$= \frac{\sin \frac{a}{2} \cos 30° + \cos \frac{a}{2} \sin 30°}{\sin 30°}$$
$$= \frac{\left(\sin \frac{a}{2} \right) \left(\frac{\sqrt{3}}{2} \right) + \left(\cos \frac{a}{2} \right) \left(\frac{1}{2} \right)}{\frac{1}{2}}$$
$$= \sqrt{3} \sin \frac{a}{2} + \cos \frac{a}{2}$$

47. $\dfrac{\sin \theta}{\tan \theta} + \dfrac{\cos \theta}{\cot \theta} \stackrel{?}{=} \cos \theta + \sin \theta$

$\dfrac{\sin \theta}{\frac{\sin \theta}{\cos \theta}} + \dfrac{\cos \theta}{\frac{\cos \theta}{\sin \theta}} \stackrel{?}{=} \cos \theta + \sin \theta$

$\sin \theta \cdot \dfrac{\cos \theta}{\sin \theta} + \cos \theta \cdot \dfrac{\sin \theta}{\cos \theta} \stackrel{?}{=}$
$\cos \theta + \sin \theta$

$\cos \theta + \sin \theta = \cos \theta + \sin \theta \checkmark$

48. $\sec \theta (\sec \theta − \cos \theta) \stackrel{?}{=} \tan^2 \theta$

$\dfrac{1}{\cos \theta} \left(\dfrac{1}{\cos \theta} − \cos \theta \right) \stackrel{?}{=} \tan^2 \theta$

$\dfrac{1}{\cos^2 \theta} − 1 \stackrel{?}{=} \tan^2 \theta$

$\sec^2 \theta − 1 \stackrel{?}{=} \tan^2 \theta$

$\tan^2 \theta = \tan^2 \theta \checkmark$

Formative Assessment

Use the Mid-Chapter Quiz to assess students' progress in the first half of the chapter.

For problems answered incorrectly, have students review the lessons indicated in parentheses.

Summative Assessment
📁 Mid-Chapter Test, p. 39

McGraw-Hill *e*Assessment Customize and create multiple versions of your Mid-Chapter Quiz and their answer keys.

FOLDABLES **StudyOrganizer**

Dinah Zike's Foldables®
Before students complete the Mid-Chapter Quiz, encourage them to review the information for Lessons 13-1 through 13-3 in their Foldables.

Additional Answer

5. $\dfrac{31.5\sqrt{3593.25}}{3593.25}$

Simplify each expression. (Lesson 13-1)

1. $\cot \theta \sec \theta$ $\csc \theta$

2. $\dfrac{1 - \cos^2 \theta}{\sin^2 \theta}$ 1

3. $\dfrac{1}{\cos \theta} - \dfrac{\sin^2 \theta}{\cos \theta}$ $\cos \theta$

4. $\cos\left(\dfrac{\pi}{2} - \theta\right) \csc \theta$ 1

5. **HISTORY** In 1861, the United States 34-star flag was adopted. For this flag, $\tan \theta = \dfrac{31.5}{51}$. Find $\sin \theta$.
See margin.

Find the value of each expression. (Lesson 13-1)

6. $\sin \theta$, if $\cos \theta = \dfrac{3}{5}$; $0° < \theta < 90°$ $\dfrac{4}{5}$

7. $\csc \theta$, if $\cot \theta = \dfrac{1}{2}$; $270° < \theta < 360°$ $-\dfrac{\sqrt{5}}{2}$

8. $\tan \theta$, if $\sec \theta = \dfrac{4}{3}$; $0° < \theta < 90°$ $\dfrac{\sqrt{7}}{3}$

9. **MULTIPLE CHOICE** Which of the following is equivalent to $\dfrac{\cos \theta}{1 - \sin^2 \theta}$? (Lesson 13-1) **D**

A $\cos \theta$

B $\csc \theta$

C $\tan \theta$

D $\sec \theta$

10. **AMUSEMENT PARKS** Suppose a child on a merry-go-round is seated on an outside horse. The diameter of the merry-go-round is 16 meters. The angle of inclination is represented by the equation $\tan \theta = \dfrac{v^2}{gR}$, where R is the radius of the circular path, v is the speed in meters per second, and g is 9.8 meters per second squared. (Lesson 13-1)

a. If the sine of the angle of inclination of the child is $\dfrac{1}{5}$, what is the angle of inclination made by the child?
about 11.5°

b. What is the velocity of the merry-go-round? **about 4 m/s**

c. If the speed of the merry-go-round is 3.6 meters per second, what is the value of the angle of inclination of a rider? **about 9.4°**

Verify that each of the following is an identity. (Lesson 13-2)

11. $\cot^2 \theta + 1 = \dfrac{\cot \theta}{\cos \theta \cdot \sin \theta}$

12. $\dfrac{\cos \theta \csc \theta}{\cot \theta} = 1$

13. $\dfrac{\sin \theta \tan \theta}{1 - \cos \theta} = (1 + \cos \theta) \sec \theta$

11–14. See Chapter 13 Answer Appendix.

14. $\tan \theta(1 - \sin \theta) = \dfrac{\cos \theta \sin \theta}{1 + \sin \theta}$

15. **COMPUTER** The front of a computer monitor is usually measured along the diagonal of the screen as shown below. (Lesson 13-2)

a. Find h. **9**

b. Using the diagram shown, show that $\cot \theta = \dfrac{\cos \theta}{\sin \theta}$.
See Chapter 13 Answer Appendix.

Verify that each of the following is an identity. (Lesson 13-2)

16. $\tan^2 \theta + 1 = \dfrac{\tan \theta}{\cos \theta \cdot \sin \theta}$ **16–19. See Chapter 13 Answer Appendix.**

17. $\dfrac{\sin \theta \cdot \sec \theta}{\sec \theta - 1} = (\sec \theta + 1) \cot \theta$

18. $\sin^2 \theta \cdot \tan^2 \theta = \tan^2 \theta - \sin^2 \theta$

19. $\cot \theta(1 - \cos \theta) = \dfrac{\cos \theta \cdot \sin \theta}{1 + \cos \theta}$

Find the exact value of each expression. (Lesson 13-3)

20. $\cos 105°$ $\dfrac{\sqrt{2} - \sqrt{6}}{4}$

21. $\sin(-135°)$ $\dfrac{-\sqrt{2}}{2}$

22. $\tan 15°$ $2 - \sqrt{3}$

23. $\cot 75°$ $2 - \sqrt{3}$

24. **MULTIPLE CHOICE** What is the exact value of $\cos \dfrac{5\pi}{12}$? (Lesson 13-3) **H**

F $\sqrt{2}$

G $\dfrac{\sqrt{6} + \sqrt{2}}{2}$

H $\dfrac{\sqrt{6} - \sqrt{2}}{4}$

J $\dfrac{\sqrt{6} + \sqrt{2}}{4}$

25. Verify that $\cos 30° \cos \theta + \sin 30° \sin \theta = \sin 60° \cos \theta + \cos 60° \sin \theta$ is an identity. (Lesson 13-3)
See Chapter 13 Answer Appendix.

InterventionPlanner

TIER 1 **On Level** OL	TIER 2 **Strategic Intervention** AL approaching grade level	TIER 3 **Intensive Intervention** 2 or more grades below level
If students miss about 25% of the exercises or less,	**If** students miss about 50% of the exercises,	**If** students miss about 75% of the exercises,
Then choose a resource:	**Then** choose a resource:	**Then** use *Math Triumphs, Alg. 2*
SE Lessons 13-1, 13-2, and 13-3	📁 Study Guide and Intervention, pp. 5, 11, and 17	connectED.mcgraw-hill.com Extra Examples, Personal Tutor, Homework Help, Review Vocabulary
📁 Skills Practice, pp. 7, 13, and 19	connectED.mcgraw-hill.com Extra Examples, Personal Tutor, Homework Help	
TE Chapter Project		
connectED.mcgraw-hill.com Self-Check Quiz		

LESSON 13-4 Double-Angle and Half-Angle Identities

·:Then	·:Now	·:Why?
• You found values of sine and cosine by using sum and difference identities.	**1** Find values of sine and cosine by using double-angle identities. **2** Find values of sine and cosine by using half-angle identities.	• Chicago's Buckingham Fountain contains jets placed at specific angles that shoot water into the air to create arcs. When a stream of water shoots into the air with velocity v at an angle of θ with the horizontal, the model predicts that the water will travel a horizontal distance of $D = \frac{v^2}{g}\sin 2\theta$ and reach a maximum height of $H = \frac{v^2}{2g}\sin^2\theta$. The ratio of H to D helps determine the total height and width of the fountain. Express $\frac{H}{D}$ as a function of θ.

Common Core State Standards

Content Standards

F.T.F.8 Prove the Pythagorean identity $\sin^2(\theta) + \cos^2(\theta) = 1$ and use it to find $\sin(\theta)$, $\cos(\theta)$, or $\tan(\theta)$ given $\sin(\theta)$, $\cos(\theta)$, or $\tan(\theta)$ and the quadrant of the angle.

Mathematical Practices
3 Construct viable arguments and critique the reasoning of others.
6 Attend to precision.

1 Double-Angle Identities It is sometimes useful to have identities to find the value of a function of twice an angle or half an angle.

KeyConcept Double-Angle Identities

The following identities hold true for all values of θ.

$$\cos 2\theta = \cos^2\theta - \sin^2\theta$$

$$\sin 2\theta = 2\sin\theta\cos\theta \qquad \cos 2\theta = 2\cos^2\theta - 1 \qquad \tan 2\theta = \frac{2\tan\theta}{1-\tan^2\theta}$$

$$\cos 2\theta = 1 - 2\sin^2\theta$$

Example 1 Double-Angle Identities

Find the exact value of $\sin 2\theta$ if $\sin\theta = \frac{2}{3}$ and θ is between 0° and 90°.

Step 1 Use the identity $\sin 2\theta = 2\sin\theta\cos\theta$ to find the value of $\cos\theta$.

$\cos^2\theta = 1 - \sin^2\theta \qquad \cos^2\theta + \sin^2\theta = 1$

$\cos^2\theta = 1 - \left(\frac{2}{3}\right)^2 \qquad \sin\theta = \frac{2}{3}$

$\cos^2\theta = \frac{5}{9} \qquad$ Subtract.

$\cos\theta = \pm\frac{\sqrt{5}}{3} \qquad$ Take the square root of each side.

Since θ is in the first quadrant, cosine is positive. Thus, $\cos\theta = \frac{\sqrt{5}}{3}$.

Step 2 Find $\sin 2\theta$.

$\sin 2\theta = 2\sin\theta\cos\theta \qquad$ Double-angle identity

$\quad = 2\left(\frac{2}{3}\right)\left(\frac{\sqrt{5}}{3}\right) \qquad \sin\theta = \frac{2}{3}$ and $\cos\theta = \frac{\sqrt{5}}{3}$

$\quad = \frac{4\sqrt{5}}{9} \qquad$ Multiply.

▶ **Guided**Practice

1. Find the exact value of $\sin 2\theta$ if $\cos\theta = -\frac{1}{3}$ and $90° < \theta < 180°$. $\quad -\frac{4\sqrt{2}}{9}$

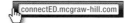
connectED.mcgraw-hill.com **893**

Brand X Pictures/PunchStock

1 Focus

VerticalAlignment

Before Lesson 13-4 Find values of sine and cosine by using sum and difference identities.

Lesson 13-4 Find values of sine and cosine by using double-angle identities. Find values of sine and cosine using half-angle identities.

After Lesson 13-4 Solve trigonometric equations.

2 Teach

Scaffolding Questions
Have students read the **Why?** section of the lesson.

Ask:
- What is the difference between $\sin 2\theta$ and $\sin^2\theta$? Explain. $\sin 2\theta$ represents the sine of the angle that is two times θ; $\sin^2\theta$ represents the square of the value of $\sin\theta$.

- Will expressing $\frac{H}{D}$ as a function of θ include the variable v? No, when simplifying, $\frac{v^2}{v^2} = 1$.

- Will it include g? Explain. No, $\frac{1}{2g} \div \frac{1}{g}$ is $\frac{1}{2}$, so it will not include the variable g.

Lesson 13-4 Resources

Resource	Approaching Level **AL**	On Level **OL**	Beyond Level **BL**	English Learners **ELL**
Teacher Edition	▪ Differentiated Instruction, p. 896	▪ Differentiated Instruction, p. 896	▪ Differentiated Instruction, p. 899	
Chapter Resource Masters	▪ Study Guide and Intervention, pp. 23–24 ▪ Skills Practice, p. 25 ▪ Practice, p. 26 ▪ Word Problem Practice, p. 27	▪ Study Guide and Intervention, pp. 23–24 ▪ Skills Practice, p. 25 ▪ Practice, p. 26 ▪ Word Problem Practice, p. 27 ▪ Enrichment, p. 28	▪ Practice, p. 26 ▪ Word Problem Practice, p. 27 ▪ Enrichment, p. 28	▪ Study Guide and Intervention, pp. 23–24 ▪ Skills Practice, p. 25 ▪ Practice, p. 26 ▪ Word Problem Practice, p. 27
Other	▪ 5-Minute Check 13-4 ▪ Study Notebook	▪ 5-Minute Check 13-4 ▪ Study Notebook	▪ 5-Minute Check 13-4 ▪ Study Notebook	▪ 5-Minute Check 13-4 ▪ Study Notebook

1 Double-Angle Identities

Examples 1 and 2 show how to find the exact value of an expression using the double-angle identities.

Formative Assessment

Use the Guided Practice exercises after each example to determine students' understanding of concepts.

Additional Examples

1 Find the exact value of $\cos 2\theta$ if $\sin \theta = \frac{3}{4}$ and θ is between $0°$ and $90°$. -0.125

2 Find the exact value of each expression if $\cos \theta = \frac{4}{5}$ and θ is between $0°$ and $90°$.

a. $\tan 2\theta$ $\frac{24}{7}$

b. $\sin 2\theta$ $\frac{24}{25}$

> **Additional Examples** also in Interactive Classroom PowerPoint® Presentations

IWB **Interactive White Board** READY

Tips for New Teachers

Sense-Making Remind students of the identity $\sin^2\theta + \cos^2\theta = 1$ and its two variations that occur by subtracting either $\sin^2\theta$ or $\cos^2\theta$ from both sides. Explain that there are three versions of the formula for $\cos 2\theta$ because of the three variations of the identity $\sin^2\theta + \cos^2\theta = 1$.

StudyTip

Deriving Formulas You can use the identity for $\sin(A + B)$ to find the sine of twice an angle θ, $\sin 2\theta$, and the identity for $\cos(A + B)$ to find the cosine of twice an angle θ, $\cos 2\theta$.

Real-World Career

Electrician An electrician specializes in the wiring of electrical components. Electricians serve an apprenticeship lasting 3–5 years. Schooling in electrical theory and building codes is required. Certification requires work experience and a passing score on a written test.

Example 2 Double-Angle Identities

Find the exact value of each expression if $\sin \theta = \frac{2}{3}$ and θ is between $0°$ and $90°$.

a. $\cos 2\theta$

Since we know the values of $\cos \theta$ and $\sin \theta$, we can use any of the double-angle identities for cosine. We will use the identity $\cos 2\theta = 1 - 2\sin^2 \theta$.

$$\cos 2\theta = 1 - 2\sin^2 \theta \qquad \text{Double-angle identity}$$

$$= 1 - 2\left(\frac{2}{3}\right)^2 \text{ or } \frac{1}{9} \qquad \sin \theta = \frac{2}{3}$$

b. $\tan 2\theta$

Step 1 Find $\tan \theta$ to use the double-angle identity for $\tan 2\theta$.

$$\tan \theta = \frac{\sin \theta}{\cos \theta} \qquad \text{Definition of tangent}$$

$$= \frac{\frac{2}{3}}{\frac{\sqrt{5}}{3}} \qquad \sin \theta = \frac{2}{3} \text{ and } \cos \theta = \frac{\sqrt{5}}{3}$$

$$= \frac{2}{\sqrt{5}} \text{ or } \frac{2\sqrt{5}}{5} \qquad \text{Rationalize the denominator.}$$

Step 2 Find $\tan 2\theta$.

$$\tan 2\theta = \frac{2\tan \theta}{1 - \tan^2 \theta} \qquad \text{Double-angle identity}$$

$$= \frac{2\left(\frac{2\sqrt{5}}{5}\right)}{1 - \left(\frac{2\sqrt{5}}{5}\right)^2} \qquad \tan \theta = \frac{2\sqrt{5}}{5}$$

$$= \frac{2\left(\frac{2\sqrt{5}}{5}\right)}{\frac{25}{25} - \frac{20}{25}} \qquad \text{Square the denominator.}$$

$$= \frac{\frac{4\sqrt{5}}{5}}{\frac{1}{5}} \qquad \text{Simplify.}$$

$$= \frac{4\sqrt{5}}{5} \cdot \frac{5}{1} \text{ or } 4\sqrt{5} \qquad \frac{a}{b} \div \frac{c}{d} = \frac{a}{b} \cdot \frac{d}{c}$$

Guided Practice

Find the exact value of each expression if $\cos \theta = -\frac{1}{3}$ and $90° < \theta < 180°$.

2A. $\cos 2\theta$ $-\frac{7}{9}$

2B. $\tan 2\theta$ $\frac{4\sqrt{2}}{7}$

2 Half-Angle Identities

It is sometimes useful to have identities to find the value of a function of half an angle.

KeyConcept Half-Angle Identities

The following identities hold true for all values of θ.

$$\sin \frac{\theta}{2} = \pm\sqrt{\frac{1 - \cos \theta}{2}} \qquad \cos \frac{\theta}{2} = \pm\sqrt{\frac{1 + \cos \theta}{2}} \qquad \tan \frac{\theta}{2} = \pm\sqrt{\frac{1 - \cos \theta}{1 + \cos \theta}}, \cos \theta \neq -1$$

 894 | Lesson 13-4 | Double-Angle and Half-Angle Identities

JGI/Blend Images/Getty Images

StudyTip

Choosing the Sign In the first step of the solution, you may want to determine the quadrant in which the terminal side of $\frac{\theta}{2}$ will lie. Then you can use the correct sign from that point on.

ReadingMath

Plus or Minus The first sign of the half-angle identity is read *plus or minus*. Unlike with the double-angle identities, you must determine the sign.

Example 3 Half-Angle Identities

a. Find the exact value of $\cos\frac{\theta}{2}$ if $\sin\theta = -\frac{4}{5}$ and θ is in the third quadrant.

$\cos^2\theta = 1 - \sin^2\theta$ Use a Pythagorean identity to find $\cos\theta$.

$\cos^2\theta = 1 - \left(-\frac{4}{5}\right)^2$ $\sin\theta = -\frac{4}{5}$

$\cos^2\theta = 1 - \frac{16}{25}$ Evaluate exponent.

$\cos^2\theta = \frac{9}{25}$ Subtract.

$\cos\theta = \pm\frac{3}{5}$ Take the square root of each side.

Since θ is in the third quadrant, $\cos\theta = -\frac{3}{5}$.

$\cos\frac{\theta}{2} = \pm\sqrt{\frac{1 + \cos\theta}{2}}$ Half-angle identity

$= \pm\sqrt{\frac{1 - \frac{3}{5}}{2}}$ $\cos\theta = -\frac{3}{5}$

$= \pm\sqrt{\frac{1}{5}}$ Simplify.

$= \pm\frac{1}{\sqrt{5}} \cdot \frac{\sqrt{5}}{\sqrt{5}}$ or $\pm\frac{\sqrt{5}}{5}$ Rationalize the denominator.

If θ is between 180° and 270°, $\frac{\theta}{2}$ is between 90° and 135°. So, $\cos\frac{\theta}{2}$ is $-\frac{\sqrt{5}}{5}$.

b. Find the exact value of $\cos 67.5°$.

$\cos 67.5° = \cos\frac{135°}{2}$ $67.5° = \frac{135°}{2}$

$= \sqrt{\frac{1 + \cos 135°}{2}}$ $\cos\frac{\theta}{2} = \pm\sqrt{\frac{1 + \cos\theta}{2}}$

$= \sqrt{\frac{1 - \frac{\sqrt{2}}{2}}{2}}$ 67.5° is in Quadrant I; the value is positive.

$= \sqrt{\frac{\frac{2}{2} - \frac{\sqrt{2}}{2}}{2}}$ $1 = \frac{2}{2}$

$= \sqrt{\frac{\frac{2 - \sqrt{2}}{2}}{2}}$ Subtract fractions.

$= \sqrt{\frac{2 - \sqrt{2}}{2} \cdot \frac{1}{2}}$ $\frac{a}{b} \div \frac{c}{d} = \frac{a}{b} \cdot \frac{d}{c}$

$= \sqrt{\frac{2 - \sqrt{2}}{4}}$ Multiply.

$= \frac{\sqrt{2 - \sqrt{2}}}{\sqrt{4}}$ $\sqrt{\frac{a}{b}} = \frac{\sqrt{a}}{\sqrt{b}}$

$= \frac{\sqrt{2 - \sqrt{2}}}{2}$ Simplify.

GuidedPractice

3. Find the exact value of $\sin\frac{\theta}{2}$ if $\sin\theta = \frac{2}{3}$ and θ is in the second quadrant. $\frac{\sqrt{18 + 6\sqrt{5}}}{6}$

---

2 Half-Angle Identities

Example 3 shows how to use the half-angle identities to find the exact value of a trigonometric function for an angle in a given quadrant. **Example 4** shows how to simplify an equation that contains trigonometric expressions. **Example 5** shows how to verify trigonometric identities using double-angle identities.

Additional Example

3 a. Find the exact value of $\cos\frac{\theta}{2}$ if $\sin\theta = \frac{3}{5}$ and θ is in the second quadrant.

$\sqrt{\frac{1}{10}}$ or $\frac{\sqrt{10}}{10}$

b. Find the exact value of $\sin 165°$. $\frac{\sqrt{2 - \sqrt{3}}}{2}$

WatchOut!

Preventing Errors Stress the Study Tip provided in the margin next to Example 3. Determining the proper sign for the answer at the beginning of the computation will help some students avoid forgetting this step at the end of their computations.

Teach with Tech

Video Recording Split the class into groups, and assign each group a double-angle or half-angle problem. Have the group create a video recording showing how to apply the appropriate formula to solve the problem. Try to assign as many different types of problems as possible to the class.

Focus on Mathematical Content

Deriving Formulas The Double-Angle Formulas are special cases of the Sum of Angles Formulas studied in the previous lesson. The Double-Angle Formulas are obtained by setting A and B equal to each other in the Sum of Angles Formulas.

Additional Examples

4 **FOUNTAIN** Refer to the beginning of the lesson. Find $\frac{D}{H}$.

$4 \cot \theta$ or $\frac{4}{\tan \theta}$

5 Verify that $\sin \theta(\cos^2 \theta - \cos 2\theta) = \sin^3 \theta$ is an identity.

$\sin \theta (\cos^2 \theta - \cos 2\theta) \stackrel{?}{=} \sin^3 \theta$

$\sin \theta [\cos^2 \theta - (\cos^2 \theta - \sin^2 \theta)] \stackrel{?}{=} \sin^3 \theta$

$\sin \theta (\cos^2 \theta - \cos^2 \theta + \sin^2 \theta) \stackrel{?}{=} \sin^3 \theta$

$\sin \theta (\sin^2 \theta) \stackrel{?}{=} \sin^3 \theta$

$\sin^3 \theta = \sin^3 \theta$ ✓

 Follow-up

Students have explored trigonometric identities and equations.

Ask:

- How do you decide what techniques to use when verifying a trigonometric identity? Sample answer: If possible, simplify the most complex side of the identity by substituting basic trigonometric identities. When dealing with a more complex identity, work each side separately to obtain a common expression.

Additional Answer (Guided Practice)

5. $4 \cos^2 x - \sin^2 2x \stackrel{?}{=} 4 \cos^4 x$

$4 \cos^2 x - 4 \sin^2 x \cos^2 x \stackrel{?}{=} 4 \cos^4 x$

$4 \cos^2 x(1 - \sin^2 x) \stackrel{?}{=} 4 \cos^4 x$

$4 \cos^2 x \cos^2 x \stackrel{?}{=} 4 \cos^4 x$

$4 \cos^4 x = 4 \cos^4 x$ ✓

Real-World Link

The City Hall Park Fountain in New York City is located in the heart of Manhattan in front of City Hall.

Source: Fodor's

Real-World Example 4 Simplify Using Double-Angle Identities

FOUNTAIN Refer to the beginning of the lesson. Find $\frac{H}{D}$.

$\frac{H}{D} = \dfrac{\frac{v^2}{2g} \sin^2 \theta}{\frac{v^2}{g} \sin 2\theta}$ Original equation

$= \dfrac{\frac{v^2 \sin^2 \theta}{2g}}{\frac{v^2 \sin 2\theta}{g}}$ Simplify the numerator and denominator.

$= \dfrac{v^2 \sin^2 \theta}{2g} \cdot \dfrac{g}{v^2 \sin 2\theta}$ $\frac{a}{b} \div \frac{c}{d} = \frac{a}{b} \cdot \frac{d}{c}$

$= \dfrac{\sin^2 \theta}{2 \sin 2\theta}$ Simplify.

$= \dfrac{\sin^2 \theta}{4 \sin \theta \cos \theta}$ $\sin 2\theta = 2 \sin \theta \cos \theta$

$= \dfrac{1}{4} \cdot \dfrac{\sin \theta}{\cos \theta}$ Simplify.

$= \dfrac{1}{4} \tan \theta$ $\frac{\sin \theta}{\cos \theta} = \tan \theta$

▶ **Guided**Practice

Find each value.

4A. $\sin 135°$ $\dfrac{\sqrt{2}}{2}$

4B. $\cos \dfrac{7\pi}{8}$ $-\dfrac{\sqrt{2 + \sqrt{2}}}{2}$

Recall that you can use the sum and difference identities to verify identities. Double- and half-angle identities can also be used to verify identities.

Example 5 Verify Identities

Verify that $\dfrac{\cos 2\theta}{1 + \sin 2\theta} = \dfrac{\cot \theta - 1}{\cot \theta + 1}$ is an identity.

$\dfrac{\cos 2\theta}{1 + \sin 2\theta} \stackrel{?}{=} \dfrac{\cot \theta - 1}{\cot \theta + 1}$ Original equation

$\dfrac{\cos 2\theta}{1 + \sin 2\theta} \stackrel{?}{=} \dfrac{\frac{\cos \theta}{\sin \theta} - 1}{\frac{\cos \theta}{\sin \theta} + 1}$ $\cot \theta = \frac{\cos \theta}{\sin \theta}$

$\dfrac{\cos 2\theta}{1 + \sin 2\theta} \stackrel{?}{=} \dfrac{\cos \theta - \sin \theta}{\cos \theta + \sin \theta}$ Multiply numerator and denominator by $\sin \theta$.

$\dfrac{\cos 2\theta}{1 + \sin 2\theta} \stackrel{?}{=} \dfrac{\cos \theta - \sin \theta}{\cos \theta + \sin \theta} \cdot \dfrac{\cos \theta + \sin \theta}{\cos \theta + \sin \theta}$ Multiply the right side by 1.

$\dfrac{\cos 2\theta}{1 + \sin 2\theta} \stackrel{?}{=} \dfrac{\cos^2 \theta - \sin^2 \theta}{\cos^2 \theta + 2 \cos \theta \sin \theta + \sin^2 \theta}$ Multiply.

$\dfrac{\cos 2\theta}{1 + \sin 2\theta} \stackrel{?}{=} \dfrac{\cos^2 \theta - \sin^2 \theta}{1 + 2 \cos \theta \sin \theta}$ Simplify.

$\dfrac{\cos 2\theta}{1 + \sin 2\theta} = \dfrac{\cos 2\theta}{1 + \sin 2\theta}$ ✓ $\cos^2 \theta - \sin^2 \theta = \cos 2\theta$; $2 \cos \theta \sin \theta = \sin 2\theta$

▶ **Guided**Practice

5. Verify that $4 \cos^2 x - \sin^2 2x = 4 \cos^4 x$. **See margin.**

 896 | Lesson 13-4 | Double-Angle and Half-Angle Identities

Differentiated Instruction   **AL**   **OL**

Auditory/Musical Learners If possible, ask a music teacher at your school to talk to students about harmonics. Students playing stringed instruments may also be willing to share what they have learned about harmonics and waves. If a music teacher is not available, a physics teacher may also be able to demonstrate harmonics or bring a device that creates standing waves in class.

Paul Cocoa/Painet Inc.

Examples 1–3 (CCSS) **PRECISION** Find the exact values of $\sin 2\theta$, $\cos 2\theta$, $\sin \frac{\theta}{2}$, and $\cos \frac{\theta}{2}$.

1. $\frac{\sqrt{15}}{8}, \frac{7}{8},$

$\frac{\sqrt{8 - 2\sqrt{15}}}{4},$

$\frac{\sqrt{8 + 2\sqrt{15}}}{4}$

1. $\sin \theta = \frac{1}{4}; 0° < \theta < 90°$

2. $\sin \theta = \frac{4}{5}; 90° < \theta < 180°$

2. $-\frac{24}{25}, -\frac{7}{25}, \frac{2\sqrt{5}}{5}, \frac{\sqrt{5}}{5}$

3. $\cos \theta = -\frac{5}{13}; \frac{\pi}{2} < \theta < \pi$

4. $\cos \theta = \frac{3}{5}; 270° < \theta < 360°$

3. $\frac{120}{169}, \frac{119}{169}, \frac{3\sqrt{13}}{13}, \frac{2\sqrt{13}}{13}$

5. $\tan \theta = -\frac{8}{15}; 90° < \theta < 180°$

6. $\tan \theta = \frac{5}{12}; \pi < \theta < \frac{3\pi}{2}$

4. $-\frac{24}{25}, \frac{7}{25}, \frac{\sqrt{5}}{5}, -\frac{2\sqrt{5}}{5}$

Find the exact value of each expression.

5. $-\frac{240}{289}, \frac{161}{289}, \frac{4\sqrt{17}}{17}, \frac{\sqrt{17}}{17}$

7. $\sin \frac{\pi}{8}$ $\frac{\sqrt{2 - \sqrt{2}}}{2}$

8. $\cos 15°$ $\frac{\sqrt{2 + \sqrt{3}}}{2}$

6. $\frac{120}{169}, \frac{119}{169}, \frac{5\sqrt{26}}{26}, -\frac{\sqrt{26}}{26}$

Example 4

9. **SOCCER** A soccer player kicks a ball at an angle of 37° with the ground with an initial velocity of 52 feet per second. The distance d that the ball will go in the air if it is not blocked is given by $d = \frac{2v^2 \sin \theta \cos \theta}{g}$. In this formula, g is the acceleration due to gravity and is equal to 32 feet per second squared, and v is the initial velocity.

a. Simplify this formula by using a double-angle identity. $d = \frac{v^2 \sin 2\theta}{g}$

b. Using the simplified formula, how far will this ball go? ≈81 ft

Example 5 Verify that each equation is an identity. **10, 11. See margin.**

10. $\tan \theta = \frac{1 - \cos 2\theta}{\sin 2\theta}$

11. $(\sin \theta + \cos \theta)^2 = 1 + 2 \sin \theta \cos \theta$

12. $-\frac{4\sqrt{5}}{9}, \frac{1}{9}, \frac{\sqrt{6} \cdot \sqrt{3 + \sqrt{5}}}{6}, \frac{\sqrt{6} \cdot \sqrt{3 - \sqrt{5}}}{6}$

13. $\frac{240}{289}, \frac{161}{189}, \frac{5\sqrt{34}}{34}, \frac{3\sqrt{34}}{34}$

Practice and Problem Solving Extra Practice is on page R13.

Examples 1–3 Find the exact values of $\sin 2\theta$, $\cos 2\theta$, $\sin \frac{\theta}{2}$, and $\cos \frac{\theta}{2}$.

14. $-\frac{24}{25}, -\frac{7}{25}, \frac{\sqrt{5}}{5}, \frac{2\sqrt{5}}{5}$

12. $\sin \theta = \frac{2}{3}; 90° < \theta < 180°$

13. $\sin \theta = -\frac{15}{17}; \pi < \theta < \frac{3\pi}{2}$

14. $\cos \theta = \frac{3}{5}; \frac{3\pi}{2} < \theta < 2\pi$

15. $\cos \theta = \frac{1}{5}; 270° < \theta < 360°$

15. $-\frac{4\sqrt{6}}{25}, -\frac{23}{25}, \frac{\sqrt{10}}{5}, -\frac{\sqrt{15}}{5}$

16. $\tan \theta = \frac{4}{3}; 180° < \theta < 270°$

17. $\tan \theta = -2; \frac{\pi}{2} < \theta < \pi$

16. $\frac{24}{25}, -\frac{7}{25}, \frac{2\sqrt{5}}{5}, -\frac{\sqrt{5}}{5}$

Find the exact value of each expression.

17. $-\frac{4}{5}, -\frac{3}{5}, \sqrt{\frac{\sqrt{5} + 1}{2\sqrt{5}}}, \sqrt{\frac{\sqrt{5} - 1}{2\sqrt{5}}}$

18. $\sin 75°$ $\frac{\sqrt{2 + \sqrt{3}}}{2}$

19. $\sin \frac{3\pi}{8}$ $\frac{\sqrt{2 + \sqrt{2}}}{2}$

20. $\cos \frac{7\pi}{12}$ $-\frac{\sqrt{2 - \sqrt{3}}}{2}$

21. $\tan 165°$ $\sqrt{3} - 2$

22. $\tan \frac{5\pi}{12}$ $2 + \sqrt{3}$

23. $\tan 22.5°$ $\sqrt{2} - 1$

24. **GEOGRAPHY** The Mercator projection of the globe is a projection on which the distance between the lines of latitude increases with their distance from the equator. The calculation of the location of a point on this projection involves the expression $\tan\left(45° + \frac{L}{2}\right)$, where L is the latitude of the point.

a. Write this expression in terms of a trigonometric function of L. **See margin.**

b. The latitude of Tallahassee, Florida, is 30° north. Find the value of the expression if $L = 30°$. $\sqrt{3}$

connectED.mcgraw-hill.com **897**

---

3 Practice

Formative Assessment

Use Exercises 1–11 to check for understanding.

Then use the chart at the bottom of this page to customize assignments for your students.

(CCSS) **Teaching the Mathematical Practices**

Precision Mathematically proficient students try to use clear definitions in their reasoning, calculate accurately and efficiently, and make explicit use of definitions.

Additional Answers

10. $\tan \theta \stackrel{?}{=} \frac{1 - \cos 2\theta}{\sin 2\theta}$

$\stackrel{?}{=} \frac{1 - (1 - 2 \sin^2 \theta)}{2 \sin \theta \cos \theta}$

$\stackrel{?}{=} \frac{2 \sin^2 \theta}{2 \sin \theta \cos \theta}$

$\stackrel{?}{=} \frac{\sin \theta}{\cos \theta}$

$= \tan \theta$ ✓

11. $(\sin \theta + \cos \theta)^2 \stackrel{?}{=} 1 + 2 \sin \theta \cos \theta$

$(\sin \theta + \cos \theta)(\sin \theta + \cos \theta) \stackrel{?}{=} 1 + 2 \sin \theta \cos \theta$

$\sin^2 \theta + 2 \sin \theta \cos \theta + \cos^2 \theta \stackrel{?}{=} 1 + 2 \sin \theta \cos \theta$

$1 + 2 \sin \theta \cos \theta = 1 + 2 \sin \theta \cos \theta$ ✓

24a. $\frac{1 \pm \sqrt{\frac{1 - \cos L}{1 + \cos L}}}{1 \mp \sqrt{\frac{1 - \cos L}{1 + \cos L}}}$

Differentiated Homework Options

Level	Assignment	Two-Day Option	
AL Basic	12–29, 37, 39–60	13–29 odd, 43–46	12–28 even, 37, 39–42, 47–60
OL Core	13–29 odd, 30, 31–37 odd, 39–60	12–29, 43–46	30–37, 39–42, 47–60
BL Advanced	30–57, (optional: 58–60)		

Study Guide and Intervention

AL OL ELL

Practice

AL OL BL ELL

Word Problem Practice

AL OL BL ELL

Example 4 **25 ELECTRONICS** Consider an AC circuit consisting of a power supply and a resistor. If the current I_0 in the circuit at time t is $I_0 \sin t\theta$, then the power delivered to the resistor is $P = I_0^2 R \sin^2 t\theta$, where R is the resistance. Express the power in terms of $\cos 2t\theta$. $\quad P = \frac{1}{2} I_0^2 R - \frac{1}{2} I_0^2 R \cos 2t\theta$

Example 5 Verify that each equation is an identity. **26–29. See Chapter 13 Answer Appendix.**

26. $\tan 2\theta = \dfrac{2}{\cot \theta - \tan \theta}$

27. $1 + \dfrac{1}{2} \sin 2\theta = \dfrac{\sec \theta + \sin \theta}{\sec \theta}$

28. $\sin \dfrac{\theta}{2} \cos \dfrac{\theta}{2} = \dfrac{\sin \theta}{2}$

29. $\tan \dfrac{\theta}{2} = \dfrac{\sin \theta}{1 + \cos \theta}$

B **30. FOOTBALL** Suppose a place kicker consistently kicks a football with an initial velocity of 95 feet per second. Prove that the horizontal distance the ball travels in the air will be the same for $\theta = 45° + A$ as for $\theta = 45° - A$. Use the formula given in Exercise 9. **See Chapter 13 Answer Appendix.**

Find the exact values of $\sin 2\theta$, $\cos 2\theta$, and $\tan 2\theta$.

31. $\cos \theta = \dfrac{4}{5}; 0° < \theta < 90° \quad \dfrac{24}{25}, \dfrac{7}{25}, \dfrac{24}{7}$

32. $\sin \theta = \dfrac{1}{3}; 0 < \theta < \dfrac{\pi}{2} \quad \dfrac{4\sqrt{2}}{9}, \dfrac{7}{9}, \dfrac{4\sqrt{2}}{7}$

33. $\tan \theta = -3; 90° < \theta < 180° \quad -\dfrac{3}{5}, -\dfrac{4}{5}, \dfrac{3}{4}$

34. $\sec \theta = -\dfrac{4}{3}; 90° < \theta < 180° \quad -\dfrac{3\sqrt{7}}{8}, \dfrac{1}{8}, -3\sqrt{7}$

35. $\csc \theta = -\dfrac{5}{2}; \dfrac{3\pi}{2} < \theta < 2\pi \quad -\dfrac{4\sqrt{21}}{25}, \dfrac{17}{25}, -\dfrac{4\sqrt{21}}{17}$

36. $\cot \theta = \dfrac{3}{2}; 180° < \theta < 270° \quad \dfrac{12}{13}, \dfrac{5}{13}, \dfrac{12}{5}$

H.O.T. Problems Use Higher-Order Thinking Skills

C **37. CCSS CRITIQUE** Teresa and Nathan are calculating the exact value of $\sin 15°$. Is either of them correct? Explain your reasoning. **See Chapter 13 Answer Appendix.**

38. $\angle PBD$ is an inscribed angle that subtends the same arc as the central angle $\angle POD$, so $m\angle PBD = \dfrac{1}{2}\theta$.

By right triangle trigonometry,

$\tan \dfrac{1}{2}\theta = \dfrac{PA}{BA} = \dfrac{PA}{1 + OA} = \dfrac{\sin \theta}{1 + \cos \theta}$.

Teresa

$\sin (A - B) = \sin A \cos B - \cos A \sin B$

$\sin (45 - 30) = \sin 45 \cos 30 - \cos 45 \sin 30$

$= \dfrac{\sqrt{2}}{2} \cdot \dfrac{\sqrt{3}}{2} - \dfrac{\sqrt{2}}{2} \cdot \dfrac{1}{2}$

$= \dfrac{\sqrt{4}}{4}$

Nathan

$\sin \dfrac{A}{2} = \pm\sqrt{\dfrac{1 - \cos A}{2}}$

$\sin \dfrac{30}{2} = \pm\sqrt{\dfrac{1 - \frac{1}{2}}{2}}$

$= 0.5$

38. CHALLENGE Circle O is a unit circle. Use the figure to prove that $\tan \dfrac{1}{2}\theta = \dfrac{\sin \theta}{1 + \cos \theta}$.

39. WRITING IN MATH Write a short paragraph about the conditions under which you would use each of the three identities for $\cos 2\theta$. **See Chapter 13 Answer Appendix.**

40. PROOF Use the formula for $\sin (A + B)$ to derive the formula for $\sin 2\theta$, and use the formula for $\cos (A + B)$ to derive the formula for $\cos 2\theta$. **See Chapter 13 Answer Appendix.**

41. REASONING Derive the half-angle identities from the double-angle identities. **See Chapter 13 Answer Appendix.**

42. OPEN ENDED Suppose a golfer consistently hits the ball so that it leaves the tee with an initial velocity of 115 feet per second and $d = \dfrac{2v^2 \sin \theta \cos \theta}{g}$. Explain why the maximum distance is attained when $\theta = 45°$. **Sample answer:** Since $d = \dfrac{v^2 \sin 2\theta}{g}$, d is at a maximum when $\sin 2\theta = 1$, that is, when $2\theta = 90°$ or $\theta = 45°$.

Enrichment

OL BL

13-4 Enrichment

Alternating Current

The figure at the right represents an alternating current generator. A rectangular coil of wire is suspended between the poles of a magnet. As the coil of wire is rotated, it passes through the magnetic field and generates current.

CCSS Teaching the Mathematical Practices

Critique Mathematically proficient students are also able to compare the effectiveness of two plausible arguments, distinguish correct logic or reasoning from that which is flawed, and—if there is a flaw in an argument—explain what it is.

43. SHORT RESPONSE Angles C and D are supplementary. The measure of angle C is seven times the measure of angle D. Find the measure of angle D in degrees. **22.5**

44. SAT/ACT Ms. Romero has a list of the yearly salaries of the staff members in her department. Which measure of data describes the middle income value of the salaries? **B**

   A mean
   B median
   C mode
   D range
   E standard deviation

45. Identify the domain and range of the function $f(x) = |4x + 1| - 8$. **G**

   F $D = \{x \mid -3 \le x \le 1\}$, $R = \{y \mid y \ge -8\}$
   G $D = \{$all real numbers$\}$, $R = \{y \mid y \ge -8\}$
   H $D = \{x \mid -3 \le x \le 1\}$,
      $R = \{$all real numbers$\}$
   J $D = \{$all real numbers$\}$,
      $R = \{$all real numbers$\}$

46. GEOMETRY Angel is putting a stone walkway around a circular pond. He has enough stones to make a walkway 144 feet long. If he uses all of the stones to surround the pond, what is the radius of the pond? **B**

   A $\frac{12}{\pi}$ ft
   B $\frac{72}{\pi}$ ft
   C 72π ft
   D 144π ft

Spiral Review

Find the exact value of each expression. (Lesson 13-3) **48.** $\frac{\sqrt{2} - \sqrt{6}}{4}$

47. $\sin 135°$ $\frac{\sqrt{2}}{2}$ **48.** $\cos 105°$ **49.** $\sin 285°$ $\frac{-\sqrt{6} - \sqrt{2}}{4}$

50. $\cos(-30°)$ $\frac{\sqrt{3}}{2}$ **51.** $\sin(-240°)$ $\frac{\sqrt{3}}{2}$ **52.** $\cos(-120°)$ $-\frac{1}{2}$

Verify that each equation is an identity. (Lesson 13-2) **53, 54. See margin.**

53. $\cot \theta + \sec \theta = \frac{\cos^2 \theta + \sin \theta}{\sin \theta \cos \theta}$ **54.** $\sin^2 \theta + \tan^2 \theta = (1 - \cos^2 \theta) + \frac{\sec^2 \theta}{\csc^2 \theta}$

Determine whether each triangle should be solved by beginning with the Law of *Sines* or Law of *Cosines*. Then solve each triangle. Round measures of sides to the nearest tenth and measures of angles to the nearest degree. (Lesson 12-5)

55.
sines; $B \approx 102°$, $C \approx 44°$, $b \approx 21.0$ or $B \approx 10°$, $C \approx 136°$, $b \approx 3.7$

56.
cosines; $A \approx 40°$, $B \approx 98°$, $c \approx 11.5$

57.
sines; $A = 80°$, $a \approx 10.9$, $c \approx 5.4$

Skills Review

Solve each equation by factoring.

58. $x^2 + 5x - 24 = 0$ $\{-8, 3\}$ **59.** $x^2 - 3x - 28 = 0$ $\{-4, 7\}$ **60.** $x^2 - 4x = 21$ $\{-3, 7\}$

DifferentiatedInstruction BL

Extension Ask students to write an expression for $\sin 4\theta$ in terms of $\sin \theta$ and $\cos \theta$.

Sample answer: $\sin 4\theta = 4 \sin \theta \cos \theta - 8 \sin^3 \theta \cos \theta$

WatchOut!

Error Analysis For Exercise 37, students should see that Teresa incorrectly replaced $\frac{\sqrt{6}}{4} - \frac{\sqrt{2}}{4}$ with $\frac{\sqrt{4}}{4}$, while Nathan incorrectly replaced $\cos 30°$ with $\frac{1}{2}$ rather than with $\frac{\sqrt{3}}{2}$. Explain to students that both first steps, $\sin 15° = \sin(45° - 30°)$ and $\sin 15° = \sin \frac{30°}{2}$, are valid first steps. However, Teresa's fourth step should be $\frac{\sqrt{6} - \sqrt{2}}{4}$ and Nathan's steps, after the first line, should be $\sin \frac{30°}{2} = \sqrt{\frac{1 - \frac{\sqrt{3}}{2}}{2}}$
$= \sqrt{\frac{2 - \sqrt{3}}{4}} = \frac{\sqrt{2 - \sqrt{3}}}{2}$.

4 Assess

Name the Math Have students tell how they determine whether a problem involves a double-angle or half-angle identity and how they use each type of equation.

Formative Assessment
Check for students' understanding of concepts in Lesson 13-4.

📁 Quiz 3, p. 38

Additional Answers

53. $\cot \theta + \sec \theta \stackrel{?}{=} \frac{\cos^2 \theta + \sin \theta}{\sin \theta \cos \theta}$

$\cot \theta + \sec \theta \stackrel{?}{=} \frac{\cos^2 \theta}{\sin \theta \cos \theta} + \frac{\sin \theta}{\sin \theta \cos \theta}$

$\cot \theta + \sec \theta \stackrel{?}{=} \frac{\cos \theta}{\sin \theta} + \frac{1}{\cos \theta}$

$\cot \theta + \sec \theta = \cot \theta + \sec \theta$ ✓

54. $\sin^2 \theta + \tan^2 \theta \stackrel{?}{=} (1 - \cos^2 \theta) + \frac{\sec^2 \theta}{\csc^2 \theta}$

$\sin^2 \theta + \tan^2 \theta \stackrel{?}{=} \sin^2 \theta + \frac{\sec^2 \theta}{\csc^2 \theta}$

$\sin^2 \theta + \tan^2 \theta \stackrel{?}{=} \sin^2 \theta + \frac{1}{\cos^2 \theta} \div \frac{1}{\sin^2 \theta}$

$\sin^2 \theta + \tan^2 \theta \stackrel{?}{=} \sin^2 \theta + \frac{\sin^2 \theta}{\cos^2 \theta}$

$\sin^2 \theta + \tan^2 \theta = \sin^2 \theta + \tan^2 \theta$ ✓

1 Focus

Objective Use a graphing calculator to solve trigonometric equations.

Materials for Each Student
- TI-83/84 Plus or other graphing calculator

Teaching Tip
In Activity 1, approximate solutions can also be found by using the **TRACE** feature. In most situations, however, the **intersect** feature will give more accurate solutions.

Remind students to use appropriate windows for their graphs.

2 Teach

Working in Cooperative Groups
Have students work in pairs so they can help each other correct keystroke errors. Have them complete Activities 1 and 2 and Exercise 1.

Ask:
- How are the solutions of the equations related to the points where the graphs intersect? The solutions are the x values of the points of intersection.

- How can you tell if an equation has no solution? The graphs of Y1 and Y2 do not intersect.

Practice Have students complete Exercises 2–6.

3 Assess

Formative Assessment
Use Exercise 6 to assess whether students understand how the specified interval affects the solutions.

Graphing Technology Lab

Solving Trigonometric Equations

The graph of a trigonometric function is made up of points that represent all values that satisfy the function. To solve a trigonometric equation, you need to find all values of the variable that satisfy the equation. You can use a TI-83/84 Plus graphing calculator to solve trigonometric equations by graphing each side of the equation as a function and then locating the points of intersection.

CCSS Common Core State Standards
Mathematical Practices
5 Use appropriate tools strategically.

Activity 1 Real Solutions

Use a graphing calculator to solve $\sin x = 0.4$ if $0° \leq x < 360°$.

Step 1 Enter and graph related equations. Rewrite the equation as two equations, **Y1** = sin x and **Y2** = 0.4. Then graph the two equations. Because the interval is in degrees, set your calculator to degree mode.

KEYSTROKES: MODE ▼ ▼ ▶ ENTER
Y= SIN X,T,θ,n)
ENTER 0.4 ENTER GRAPH

[0, 360] scl: 90 by [−1, 1] scl: 0.1

Step 2 Approximate the solutions. Based on the graph, you can see that there are two points of intersection in the interval $0° \leq x < 360°$. Use the **CALC** feature to determine the x-values at which the two graphs intersect.

The solutions are $x \approx 23.57°$ and $x \approx 156.4°$.

Activity 2 No Real Solutions

Use a graphing calculator to solve $\tan^2 x \cos x + 3 \cos x = 0$ if $0° \leq x < 360°$.

Step 1 Enter and graph related equations. The related equations to be graphed are **Y1** = $\tan^2 x \cos x + 3 \cos x$ and **Y2** = 0.

KEYSTROKES: Y= TAN X,T,θ,n) x² COS
X,T,θ,n) + 3 COS X,T,θ,n
) ENTER 0 ENTER

[0, 360] scl: 90 by [−15, 15] scl: 1

Step 2 These two functions do not intersect.

Therefore, the equation $\tan^2 x \cos x + 3 \cos x = 0$ has no real solutions.

Exercises

Use a graphing calculator to solve each equation for the values of x indicated.

1. $\sin x = 0.7$; $0° \leq x < 360°$ **44.4°, 135.6°**

2. $\tan x = \cos x$; $0° \leq x < 360°$ **38.17°, 141.8°**

3. $3 \cos x + 4 = 0.5$; $0° \leq x < 360°$ **no real solution**

4. $0.25 \cos x = 3.4$; $−720° \leq x < 720°$ **no real solution**

5. $\sin 2x = \sin x$; $0° \leq x < 360°$ **0°, 60°, 180°, 300°**

6. $\sin 2x − 3 \sin x = 0$ if $−360° \leq x < 360°$
−360°, −180°, 0°, 180°

From Concrete to Abstract
Have students consider Exercises 1–6 with a different interval or no interval and explain how this affects the solutions.

13-5 LESSON

Solving Trigonometric Equations

:: Then	:: Now	:: Why?
● You verified trigonometric identities.	**1** Solve trigonometric equations. **2** Find extraneous solutions from trigonometric equations.	● When you ride a Ferris wheel that has a diameter of 40 meters and turns at a rate of 1.5 revolutions per minute, the height above the ground, in meters, of your seat after t minutes can be modeled by the equation $$h = 21 - 20 \cos 3\pi t.$$ After the ride begins, how long is it before your seat is 31 meters above the ground for the first time?

 NewVocabulary
trigonometric equations

 Common Core State Standards

Content Standards
F.TF.8 Prove the Pythagorean identity $\sin^2(\theta) + \cos^2(\theta) = 1$ and use it to find $\sin(\theta)$, $\cos(\theta)$, or $\tan(\theta)$ given $\sin(\theta)$, $\cos(\theta)$, or $\tan(\theta)$ and the quadrant of the angle.

Mathematical Practices
4 Model with mathematics.
6 Attend to precision.

1 **Solve Trigonometric Equations** So far in this chapter, we have studied a special type of trigonometric equation called an identity. Trigonometric identities are equations that are true for all values of the variable for which both sides are defined. In this lesson, we will examine **trigonometric equations** that are true only for certain values of the variable. Solving these equations resembles solving algebraic equations.

Example 1 Solve Equations for a Given Interval

Solve $\sin \theta \cos \theta - \frac{1}{2} \cos \theta = 0$ if $0 \le \theta \le 180°$.

$\sin \theta \cos \theta - \frac{1}{2} \cos \theta = 0$ Original equation

$\cos \theta \left(\sin \theta - \frac{1}{2} \right) = 0$ Factor.

$\cos \theta = 0$ or $\sin \theta - \frac{1}{2} = 0$ Zero Product Property

$\theta = 90°$ $\sin \theta = \frac{1}{2}$

                  $\theta = 30°$ or $150°$

The solutions are 30°, 90°, and 150°.

CHECK You can check the answer by graphing $y = \sin \theta \cos \theta$ and $y = \frac{1}{2} \cos \theta$ in the same coordinate plane on a graphing calculator. Then find the points where the graphs intersect. You can see that there are infinitely many such points, but we are only interested in the points between 0° and 180°.

[0, 720] scl: 90 by [−1, 1] scl: 0.5

▶ **GuidedPractice**

1. Find all solutions of $\sin 2\theta = \cos \theta$ if $0 \le \theta \le 2\pi$. $\frac{\pi}{6}, \frac{\pi}{2}, \frac{5\pi}{6}, \frac{3\pi}{2}$

Trigonometric equations are usually solved for values of the variable between 0° and 360° or between 0 radians and 2π radians. There are solutions outside that interval. These other solutions differ by integral multiples of the period of the function.

 connectED.mcgraw-hill.com **901**

Hans-Jürgen Hermann/age fotostock

---

1 Focus

VerticalAlignment

▼

Before Lesson 13-5 Verify trigonometric identities.

Lesson 13-5 Solve trigonometric equations. Find extraneous solutions from trigonometric equations.

▼

After Lesson 13-5 Use trigonometry to solve real-world problems.

2 Teach

Scaffolding Questions
Have students read the **Why?** section of the lesson.

Ask:
- Take any point on the Ferris wheel. How far does that point travel in 1 revolution? 40π or about 125.7 meters

- How far does a position on the Ferris wheel travel in 1 minute? 60π or about 188.5 meters

- When $t = 0$, what is the value of $20 \cos 3\pi t$? 20

Lesson 13-5 Resources

Resource	Approaching Level **AL**	On Level **OL**	Beyond Level **BL**	English Learners **ELL**
Teacher Edition	▪ Differentiated Instruction, p. 903	▪ Differentiated Instruction, p. 903	▪ Differentiated Instruction, p. 907	
Chapter Resource Masters	▪ Study Guide and Intervention, pp. 29–30 ▪ Skills Practice, p. 31 ▪ Practice, p. 32 ▪ Word Problem Practice, p. 33	▪ Study Guide and Intervention, pp. 29–30 ▪ Skills Practice, p. 31 ▪ Practice, p. 32 ▪ Word Problem Practice, p. 33 ▪ Enrichment, p. 34	▪ Practice, p. 32 ▪ Word Problem Practice, p. 33 ▪ Enrichment, p. 34	▪ Study Guide and Intervention, pp. 29–30 ▪ Skills Practice, p. 31 ▪ Practice, p. 32 ▪ Word Problem Practice, p. 33
Other	▪ 5-Minute Check 13-5 ▪ Study Notebook ▪ Teaching Algebra with Manipulatives	▪ 5-Minute Check 13-5 ▪ Study Notebook ▪ Teaching Algebra with Manipulatives	▪ 5-Minute Check 13-5 ▪ Study Notebook	▪ 5-Minute Check 13-5 ▪ Study Notebook ▪ Teaching Algebra with Manipulatives

1 Solve Trigonometric Equations

Example 1 shows how to solve trigonometric equations for a given interval. **Example 2** shows how to solve trigonometric equations for angles measured in radians. **Example 3** shows how to solve real-world problems involving trigonometric equations.

Formative Assessment

Use the Guided Practice exercises after each example to determine students' understanding of concepts.

Additional Examples

1 Solve $2\cos^2\theta - 1 = \sin\theta$ if $0° \le \theta \le 180°$. $30°, 150°$

2 a. Solve $\cos\theta + \frac{1}{4} = \sin^2\theta$ for all values of θ if θ is measured in degrees. $60° + k \cdot 360°, 300° + k \cdot 360°$, where k is any integer

b. Solve $2\cos\theta = -1$ for all values of θ if θ is measured in radians. $\frac{2\pi}{3} + 2k\pi$ and $\frac{4\pi}{3} + 2k\pi$, where k is any integer

3 **AMUSEMENT PARKS** Refer to the beginning of the lesson. How long after the Ferris wheel starts will your seat first be $(10\sqrt{2} + 21)$ meters above the ground?

$10\sqrt{2} + 21 = 21 - 20\cos 3\pi t,$
$10\sqrt{2} = -20\cos 3\pi t,$
$-\frac{\sqrt{2}}{2} = \cos 3\pi t,$
$\cos^{-1}\left(-\frac{\sqrt{2}}{2}\right) = 3\pi t,$
$\frac{3}{4}\pi = 3\pi t$
$\frac{1}{4} = t$

$\frac{1}{4}$ min or 15 sec

StudyTip

Expressing Solutions as Multiples The expression $\pi + 2k\pi$ includes 3π and its multiples, so it is not necessary to list them separately.

2A. $90° + k \cdot 180°$, $120° + k \cdot 360°$, and $240° + k \cdot 360°$

2B. $\frac{7\pi}{6} + 2k\pi$ and $\frac{11\pi}{6} + 2k\pi$

Example 2 Infinitely Many Solutions

Solve $\cos\theta + 1 = 0$ for all values of θ if θ is measured in radians.

$\cos\theta + 1 = 0$
$\cos\theta = -1$

Look at the graph of $y = \cos\theta$ to find solutions of $\cos\theta = -1$.

The solutions are $\pi, 3\pi, 5\pi$, and so on, and $-\pi, -3\pi, -5\pi$, and so on. The only solution in the interval 0 radians to 2π radians is π. The period of the cosine function is 2π radians. So the solutions can be written as $\pi + 2k\pi$, where k is any integer.

▶ **Guided**Practice

2A. Solve $\cos 2\theta + \cos\theta + 1 = 0$ for all values of θ if θ is measured in degrees.

2B. Solve $2\sin\theta = -1$ for all values of θ if θ is measured in radians.

Trigonometric equations are often used to solve real-world problems.

Real-World Example 3 Solve Trigonometric Equations

AMUSEMENT PARKS Refer to the beginning of the lesson. How long after the Ferris wheel starts will your seat first be 31 meters above the ground?

$h = 21 - 20\cos 3\pi t$	Original equation
$31 = 21 - 20\cos 3\pi t$	Replace h with 31.
$10 = -20\cos 3\pi t$	Subtract 21 from each side.
$-\frac{1}{2} = \cos 3\pi t$	Divide each side by −20.
$\cos^{-1}\left(-\frac{1}{2}\right) = 3\pi t$	Take the Arccosine.
$\frac{2\pi}{3} = 3\pi t$ or $\frac{4\pi}{3} = 3\pi t$	The Arccosine of $-\frac{1}{2}$ is $\frac{2\pi}{3}$ or $\frac{4\pi}{3}$.
$\frac{2\pi}{3} + 2\pi k = 3\pi t$ or $\frac{4\pi}{3} + 2\pi k = 3\pi t$	k is any integer.
$\frac{2}{9} + \frac{2}{3}k = t$ $\frac{4}{9} + \frac{2}{3}k = t$	Divide each term by 3π.

The least positive value for t is obtained by letting $k = 0$ in the first expression. Therefore, $t = \frac{2}{9}$ of a minute or about 13 seconds.

▶ **Guided**Practice

3. How long after the Ferris wheel starts will your seat first be 41 meters above the ground? **about 20 seconds**

Tips for New Teachers

Reasoning In Example 2, show students how to look for patterns in the solutions. Students should look for pairs of solutions that differ by exactly π or 2π.

2 Extraneous Solutions Some trigonometric equations have no solution. For example, the equation $\cos \theta = 4$ has no solution because all values of $\cos \theta$ are between -1 and 1, inclusive. Thus, the solution set for $\cos \theta = 4$ is empty.

Example 4 Determine Whether a Solution Exists

Solve each equation.

a. $2 \sin^2 \theta - 3 \sin \theta - 2 = 0$ if $0 \le \theta \le 2\pi$

$2 \sin^2 \theta - 3 \sin \theta - 2 = 0$	Original equation
$(\sin \theta - 2)(2 \sin \theta + 1) = 0$	Factor.
$\sin \theta - 2 = 0 \quad$ or $\quad 2 \sin \theta + 1 = 0$	Zero Product Property

$$\sin \theta = 2 \qquad\qquad 2 \sin \theta = -1$$

This is not a solution since all values of $\sin \theta$ are between -1 and 1, inclusive.

$$\sin \theta = -\frac{1}{2}$$

$$\theta = \frac{7\pi}{6} \text{ or } \frac{11\pi}{6}$$

The solutions are $\frac{7\pi}{6}$ or $\frac{11\pi}{6}$.

CHECK $\quad 2 \sin^2 \theta - 3 \sin \theta - 2 = 0 \qquad\qquad 2 \sin^2 \theta - 3 \sin \theta - 2 = 0$

$$2 \sin^2 \left(\frac{7\pi}{6}\right) - 3 \sin \left(\frac{7\pi}{6}\right) - 2 \stackrel{?}{=} 0 \qquad 2 \sin^2 \left(\frac{11\pi}{6}\right) - 3 \sin \left(\frac{11\pi}{6}\right) - 2 \stackrel{?}{=} 0$$

$$2\left(\frac{1}{4}\right) - 3\left(-\frac{1}{2}\right) - 2 \stackrel{?}{=} 0 \qquad\qquad 2\left(\frac{1}{4}\right) - 3\left(-\frac{1}{2}\right) - 2 \stackrel{?}{=} 0$$

$$\frac{1}{2} + \frac{3}{2} - 2 \stackrel{?}{=} 0 \qquad\qquad\qquad \frac{1}{2} + \frac{3}{2} - 2 \stackrel{?}{=} 0$$

$$0 = 0 \ \checkmark \qquad\qquad\qquad\qquad 0 = 0 \ \checkmark$$

b. $\sin \theta = 1 + \cos \theta$ if $0° \le \theta < 360°$

$\sin \theta = 1 + \cos \theta$	Original equation
$\sin^2 \theta = (1 + \cos \theta)^2$	Square each side.
$1 - \cos^2 \theta = 1 + 2 \cos \theta + \cos^2 \theta$	$\sin^2 \theta = 1 - \cos^2 \theta$
$0 = 2 \cos \theta + 2 \cos^2 \theta$	Set the left side equal to 0.
$0 = 2 \cos \theta (1 + \cos \theta)$	Factor.
$1 + \cos \theta = 0 \quad$ or $\quad 2 \cos \theta = 0$	Zero Product Property

$$\cos \theta = -1 \qquad\qquad \cos \theta = 0$$

$$\theta = 180 \qquad\qquad \theta = 90° \text{ or } 270°$$

CHECK $\quad \sin \theta = 1 + \cos \theta \qquad\qquad \sin \theta = 1 + \cos \theta$

$$\sin 90° \stackrel{?}{=} 1 + \cos 90° \qquad\qquad \sin 180° \stackrel{?}{=} 1 + \cos 180°$$

$$1 \stackrel{?}{=} 1 + 0 \qquad\qquad\qquad\quad 0 \stackrel{?}{=} 1 + (-1)$$

$$1 = 1 \ \checkmark \qquad\qquad\qquad\quad 0 = 0 \ \checkmark$$

$$\sin \theta = 1 + \cos \theta$$

$$\sin 270° \stackrel{?}{=} 1 + \cos 270°$$

$$-1 \stackrel{?}{=} 1 + 0$$

$$-1 \ne 1 \ ✗$$

The solutions are 90° and 180°.

▸ **Guided Practice**

4A. $\sin^2 \theta + 2 \cos^2 \theta = 4$ **no solution**

4B. $\cos^2 \theta + 3 = 4 - \sin^2 \theta$

4B. identity; therefore, infinitely many solutions

If an equation cannot be solved easily by factoring, try rewriting the expression using trigonometric identities. However, using identities and some algebraic operations, such as squaring, may result in extraneous solutions. So, it is necessary to check your solutions using the original equation.

   903

Problem-Solving Tip

CCSS Regularity Look for patterns in your solutions. Look for pairs of solutions that differ by exactly π or 2π and write your solutions with the simplest possible pattern.

DifferentiatedInstruction ⒶⓁ ⒪Ⓛ

Interpersonal Learners As students work through this lesson, have them create a class list on the board that identifies common errors they made. Encourage students to add suggestions for how to avoid their errors. For example, one common error is having one's calculator set to degrees when it needs to be set to radians for a problem, and vice versa.

---

2 Extraneous Solutions

Example 4 shows how to solve a trigonometric equation and test for extraneous solutions. **Example 5** shows how to use identities to solve a trigonometric equation.

Additional Example

4 Solve each equation.

a. $\sin \theta \cos \theta = \cos^2 \theta$ if $0 \le \theta \le 2\pi$ $\quad \frac{\pi}{4}, \frac{\pi}{2}, \frac{5\pi}{4}, \frac{3\pi}{2}$

b. $\cos \theta = 1 - \sin \theta$ if $0° \le \theta \le 360°$ $\quad 0°, 90°, 360°$

CCSS **Teaching the Mathematical Practices**

Regularity Mathematically proficient students notice if calculations are repeated, and look both for general methods and for shortcuts. They continually evaluate the reasonableness of their intermediate results. Encourage students to examine their solutions closely, and to write them in the simplest possible form.

Focus on Mathematical Content

Infinitely Many Solutions Many trigonometric equations have an infinite number of solutions. If no interval is specified to restrict the number of solutions, the infinite number of solutions must be specified by using an expression rather than by simply using a number. There is a solution that occurs for each full rotation about the origin, a solution has the form $a° + k \cdot 360°$, where k is any integer.

Teach with Tech

Blog Have students write blog entries about how to solve trigonometric equations. Have them describe how the process is similar to and different from solving other types of equations.

Additional Example

5 Solve $\tan^4 \theta - 4\sec^2 \theta = -7$ for all values of θ if θ is measured in degrees.

$\theta = 60° + 180°k$, $120° + 180°k$, or $45° + 90° \cdot k$, where k is any integer.

3 Practice

Formative Assessment

Use Exercises 1–29 to check for understanding.

Then use the chart at the bottom of this page to customize assignments for your students.

 Teaching the Mathematical Practices

Regularity Mathematically proficient students notice if calculations are repeated, and look both for general methods and for shortcuts. As they work to solve a problem, mathematically proficient students maintain oversight of the process, while attending to the details and continually evaluate the reasonableness of their intermediate results.

Tips for New Teachers

Notation In all answers in this lesson involving k, k is any integer.

Additional Answer

21b. Every day from February 19 to October 20; sample explanation: Since the longest day of the year occurs around June 22, the days between February 19 and October 20 must increase in length until June 22 and then decrease in length until October 20.

StudyTip

Solving Trigonometric Equations Remember that *solving a trigonometric equation* means solving for all values of the variable.

Example 5 Solve Trigonometric Equations by Using Identities

Solve $2\sec^2 \theta - \tan^4 \theta = -1$ for all values of θ if θ is measured in degrees.

$2\sec^2 \theta - \tan^4 \theta = -1$	Original equation
$2(1 + \tan^2 \theta) - \tan^4 \theta = -1$	$\sec^2 \theta = 1 + \tan^2 \theta$
$2 + 2\tan^2 \theta - \tan^4 \theta = -1$	Distributive Property
$\tan^4 \theta - 2\tan^2 \theta - 3 = 0$	Set one side of the equation equal to 0.
$(\tan^2 \theta - 3)(\tan^2 \theta + 1) = 0$	Factor.

$\tan^2 \theta - 3 = 0$ or $\tan^2 \theta + 1 = 0$ Zero Product Property

$\tan^2 \theta = 3$ $\tan^2 \theta = -1$

$\tan \theta = \pm\sqrt{3}$ This part gives no solutions since $\tan^2 \theta$ is never negative.

$\theta = 60° + 180°k$ and $\theta = -60° + 180°k$, where k is any integer. The solutions are $60° + 180°k$ and $-60° + 180°k$.

Guided Practice

Solve each equation.

5A. $\sin \theta \cot \theta - \cos^2 \theta = 0$

5B. $\dfrac{\cos \theta}{\cot \theta} + 2\sin^2 \theta = 0$

5A. $90° + k \cdot 180°$

5B. $\dfrac{7\pi}{6} + 2k \cdot \pi$, and $\dfrac{11\pi}{6} + 2k \cdot \pi$

Check Your Understanding   ◯ = Step-by-Step Solutions begin on page R14.   ☑

Example 1   **CCSS REGULARITY** Solve each equation if $0° \le \theta \le 360°$.

1. $2\sin \theta + 1 = 0$ $210°, 330°$

2. $\cos^2 \theta + 2\cos \theta + 1 = 0$ $180°$

3. $\cos 2\theta + \cos \theta = 0$ $60°, 180°, 300°$

4. $2\cos \theta = 1$ $60°, 300°$

5. $\cos \theta = -\dfrac{\sqrt{3}}{2}$ $150°, 210°$

6. $\sin 2\theta = -\dfrac{\sqrt{3}}{2}$ $120°, 150°, 300°, 330°$

7. $\cos 2\theta = 8 - 15\sin \theta$ $30°, 150°$

8. $\sin \theta + \cos \theta = 1$ $0°, 90°, 360°$

Example 2   Solve each equation for all values of θ if θ is measured in radians.

9. $4\sin^2 \theta - 1 = 0$

9. $\pm\dfrac{\pi}{6} + 2k\pi$ or $\pm\dfrac{5\pi}{6} + 2k\pi$

10. $2\cos^2 \theta = 1$ $\dfrac{\pi}{4} + \dfrac{k}{2}\pi$

11. $\cos 2\theta \sin \theta = 1$ $\dfrac{3\pi}{2} + 2k\pi$

12. $\sin \dfrac{\theta}{2} + \cos \dfrac{\theta}{2} = \sqrt{2}$ $\dfrac{\pi}{2} + 4\pi k$

13. $\cos 2\theta + 4\cos \theta = -3$ $\pi + 2k\pi$

14. $\sin \dfrac{\theta}{2} + \cos \theta = 1$

14. $0 + 2k\pi$; $\dfrac{\pi}{3} + 4k\pi$; $\dfrac{5\pi}{3} + 4k\pi$

Solve each equation for all values of θ if θ is measured in degrees.

15. $\cos 2\theta - \sin^2 \theta + 2 = 0$

16. $\sin^2 \theta - \sin \theta = 0$

17. $2\sin^2 \theta - 1 = 0$ $45° + k \cdot 90°$

18. $\cos \theta - 2\cos \theta \sin \theta = 0$

19. $\cos 2\theta \sin \theta = 1$ $270° + k \cdot 360°$

20. $\sin \theta \tan \theta - \tan \theta = 0$ $0° + k \cdot 180°, 90° + k \cdot 360°$

15. $90° + k \cdot 180°$

16. $0° + k \cdot 180°, 90° + k \cdot 360°$

18. $30° + k \cdot 360°, 150° + k \cdot 360°, 90° + k \cdot 180°$

Example 3   **21. LIGHT** The number of hours of daylight d in Hartford, Connecticut, may be approximated by the equation $d = 3\sin \dfrac{2\pi}{365}t + 12$, where t is the number of days after March 21.

a. On what days will Hartford have exactly $10\frac{1}{2}$ hours of daylight?

b. Using the results in part **a**, tell what days of the year have at least $10\frac{1}{2}$ hours of daylight. Explain how you know. **See margin.**

21a. There will be $10\frac{1}{2}$ hours of daylight 213 and 335 days after March 21; that is, on October 20 and February 19.

904 | Lesson 13-5 | Solving Trigonometric Equations

Differentiated Homework Options

Level	Assignment	Two-Day Option	
AL Basic	30–48, 58, 60–62, 64–83	31–47 odd, 64–67	30–48 even, 58, 60–62, 68–83
OL Core	31–55 odd, 56–58, 60–62, 64–83	30–48, 64–67	49–58, 60–62, 68–83
BL Advanced	49–80, (optional: 81–83)		

Examples 4–5 Solve each equation. **25.** $\frac{\pi}{2} + \pi k, \frac{\pi}{6} + 2\pi k, \frac{5\pi}{6} + 2\pi k$

22. $\sin^2 2\theta + \cos^2 \theta = 0$ $\frac{\pi}{2} + \pi k$

23. $\tan^2 \theta + 2 \tan \theta + 1 = 0$ $\frac{3\pi}{4} + \pi k$

24. $\cos^2 \theta + 3 \cos \theta = -2$ $\pi + 2\pi k$

25. $\sin 2\theta - \cos \theta = 0$

26. $\tan \theta = 1$ $45° + k \cdot 180°$ or $\frac{\pi}{4} + k \cdot \pi$

27. $\cos 8\theta = 1$ $0° + k \cdot 45°$ or $0 + k \cdot \frac{\pi}{4}$

28. $\sin \theta + 1 = \cos 2\theta$
$0° + k \cdot 180°, 210° + k \cdot 360°, 330° + k \cdot 360°$

29. $2 \cos^2 \theta = \cos \theta$ $\frac{\pi}{3} + 2k\pi, \frac{\pi}{2} + k\pi, \frac{5\pi}{3} + 2k\pi$

Practice and Problem Solving

Extra Practice is on page R13.

Example 1 Solve each equation for the given interval.

30. $\cos^2 \theta = \frac{1}{4}; 0° \le \theta \le 360°$

31. $2 \sin^2 \theta = 1; 90° < \theta < 270°$ $135°, 225°$

32. $\sin 2\theta - \cos \theta = 0; 0 \le \theta \le 2\pi$

33. $3 \sin^2 \theta = \cos^2 \theta; 0 \le \theta \le \frac{\pi}{2}$ $\frac{\pi}{6}$

34. $2 \sin \theta + \sqrt{3} = 0; 180° < \theta < 360°$
$240°, 300°$

35. $4 \sin^2 \theta - 1 = 0; 180° < \theta < 360°$ $210°, 330°$

30. $60°, 120°, 240°, 300°$ **32.** $\frac{\pi}{6}, \frac{\pi}{2}, \frac{5\pi}{6}, \frac{3\pi}{2}$

Example 2 Solve each equation for all values of θ if θ is measured in radians.

36. $\cos 2\theta + 3 \cos \theta = 1$

37 $2 \sin^2 \theta = \cos \theta + 1$

38. $\cos^2 \theta - \frac{3}{2} = \frac{5}{2} \cos \theta$

39. $3 \cos \theta - \cos \theta = 2$ $0 + 2k\pi$

36. $\frac{\pi}{3} + 2k\pi, \frac{5\pi}{3} + 2k\pi$ **37.** $\pi + 2k\pi, \frac{\pi}{3} + 2k\pi, \frac{5\pi}{3} + 2k\pi$ **38.** $\frac{2\pi}{3} + 2k\pi, \frac{4\pi}{3} + 2k\pi$

45. $\frac{7\pi}{6} + 2k\pi, \frac{11\pi}{6} + 2k\pi$ or $210° + k \cdot 360°, 330° + k \cdot 360°$

Solve each equation for all values of θ if θ is measured in degrees.

40. $\sin \theta - \cos \theta = 0$ $45° + k \cdot 180°$

41. $\tan \theta - \sin \theta = 0$ $0° + k \cdot 180°$

42. $\sin^2 \theta = 2 \sin \theta + 3$ $270° + k \cdot 360°$

43. $4 \sin^2 \theta = 4 \sin \theta - 1$ $30° + k \cdot 360°, 150° + k \cdot 360°$

Example 3

44. **ELECTRONICS** One of the tallest structures in the world is a television transmitting tower located near Fargo, North Dakota, with a height of 2064 feet. What is the measure of θ if the length of the shadow is 1 mile? **about 21°**

Examples 4–5 Solve each equation.

45. $2 \sin^2 \theta = 3 \sin \theta + 2$

46. $2 \cos^2 \theta + 3 \sin \theta = 3$

47. $\sin^2 \theta + \cos 2\theta = \cos \theta$
$0 + 2k\pi, \frac{\pi}{2} + k\pi$ or $0° + k \cdot 360°, 90° + k \cdot 180°$

48. $2 \cos^2 \theta = -\cos \theta$

46. $\frac{\pi}{6} + 2k\pi, \frac{5\pi}{6} + 2k\pi, \frac{\pi}{2} + 2k\pi$ or $30° + k \cdot 360°, 150° + k \cdot 360°, 90° + k \cdot 360°$

B 49. **CCSS SENSE-MAKING** Due to ocean tides, the depth y in meters of the River Thames in London varies as a sine function of x, the hour of the day. On a certain day that function was $y = 3 \sin \left[\frac{\pi}{6}(x - 4) \right] + 8$, where $x = 0, 1, 2, \ldots, 24$ corresponds to 12:00 midnight, 1:00 A.M., 2:00 A.M., …, 12:00 midnight the next night.

a. What is the maximum depth of the River Thames on that day? **11 m**

b. At what times does the maximum depth occur?
7:00 A.M. and 7:00 P.M.

48. $\frac{\pi}{2} + k\pi, \frac{2\pi}{3} + 2k\pi, \frac{4\pi}{3} + 2k\pi$ or $90° + k \cdot 180°, 120° + k \cdot 360°, 240° + k \cdot 360°$

Solve each equation if θ is measured in radians.

50. $(\cos \theta)(\sin 2\theta) - 2 \sin \theta + 2 = 0$ $\frac{\pi}{2} + 2\pi k$

51. $2 \sin^2 \theta + (\sqrt{2} - 1) \sin \theta = \frac{\sqrt{2}}{2}$
$\frac{\pi}{6} + 2\pi k, \frac{5\pi}{6} + 2\pi k, \frac{5\pi}{4} + 2\pi k, \frac{7\pi}{4} + 2\pi k$

Solve each equation if θ is measured in degrees.

52. $\sin 2\theta + \frac{\sqrt{3}}{2} = \sqrt{3} \sin \theta + \cos \theta$
$30° + 360°k, 150° + 360°k, 330° + 360°k$

53. $1 - \sin^2 \theta - \cos \theta = \frac{3}{4}$ $120° + 360°k, 240° + 360°k$

CCSS Teaching the Mathematical Practices

Sense-Making Mathematically proficient students start by explaining to themselves the meaning of a problem and looking for entry points to its solution. They analyze givens, constraints, relationships, and goals. They check their answers to problems using a different method, and they continually ask themselves, "Does this make sense?"

60. Each type of equation may require adding, subtracting, multiplying, or dividing each side by the same number. Quadratic and trigonometric equations can often be solved by factoring. Linear and quadratic equations do not require identities. All linear and quadratic equations can be solved algebraically, whereas some trigonometric equations may be solved more easily by using a graphing calculator. A linear equation has at most one solution. A quadratic equation has at most two solutions. A trigonometric equation usually has infinitely many solutions, unless the values of the variable are restricted.

61. Sample answer: All trigonometric functions are periodic. Therefore, once one or more solutions are found for a certain interval, there will be additional solutions that can be found by adding integral multiples of the period of the function to those solutions.

72. $\sin(270° - \theta) \overset{?}{=} -\cos\theta$

$\sin 270° \cos\theta - \cos 270° \sin\theta \overset{?}{=} -\cos\theta$

$-1 \cos\theta - 0 \overset{?}{=} -\cos\theta$

$-\cos\theta = -\cos\theta$ ✓

73. $\cos(90° + \theta) \overset{?}{=} -\sin\theta$

$\cos 90° \cos\theta - \sin 90° \sin\theta \overset{?}{=} -\sin\theta$

$0 - 1 \sin\theta \overset{?}{=} -\sin\theta$

$-\sin\theta = -\sin\theta$ ✓

74. $\cos(90° - \theta) \overset{?}{=} \sin\theta$

$\cos 90° \cos\theta + \sin 90° \sin\theta \overset{?}{=} \sin\theta$

$0 \cdot \cos\theta + 1 \cdot \sin\theta \overset{?}{=} \sin\theta$

$\sin\theta = \sin\theta$ ✓

Solve each equation.

54. $2\sin\theta = \sin 2\theta$ πk

55. $\cos\theta\tan\theta - 2\cos^2\theta = -1$ $\frac{\pi}{6} + 2\pi k, \frac{5\pi}{6} + 2\pi k$

56. DIAMONDS According to Snell's Law, $n_1 \sin i = n_2 \sin r$, where n_1 is the index of refraction of the medium the light is exiting, n_2 is the index of refraction of the medium the light is entering, i is the degree measure of the angle of incidence, and r is the degree measure of the angle of refraction.

a. The index of refraction of a diamond is 2.42, and the index of refraction of air is 1.00. If a beam of light strikes a diamond at an angle of 35°, what is the angle of refraction?   **13.71°**

b. Explain how a gemologist might use Snell's Law to determine whether a diamond is genuine.   **Measure the angles of incidence and refraction to determine the index of refraction. If the index is 2.42, the diamond is genuine.**

57. **CCSS PERSEVERANCE** A wave traveling in a guitar string can be modeled by the equation $D = 0.5 \sin(6.5x) \sin(2500t)$, where D is the displacement in millimeters at the position x millimeters from the left end of the string at time t seconds. Find the first positive time when the point 0.5 meter from the left end has a displacement of 0.01 millimeter.   **0.0026 second**

58. MULTIPLE REPRESENTATIONS Consider the trigonometric inequality $\sin\theta \geq \frac{1}{2}$.

a. Tabular Construct a table of values for $0° \leq \theta \leq 360°$. For what values of θ is $\sin\theta \geq \frac{1}{2}$?   **a–d. See Chapter 13 Answer Appendix.**

b. Graphical Graph $y = \sin\theta$ and $y = \frac{1}{2}$ on the same graph for $0° \leq \theta \leq 360°$. For what values of θ is the graph of $y = \sin\theta$ above the graph of $y = \frac{1}{2}$?

c. Analytic Based on your answers for parts **a** and **b**, solve $\sin\theta \geq \frac{1}{2}$ for all values of θ.

d. Algebraic Solve each inequality if $0 \leq \theta \leq 360°$. Then solve each for all values of θ.

i. $\cos\theta \geq \frac{\sqrt{2}}{2}$ **ii.** $2\sin\theta \leq \sqrt{3}$

iii. $-\sin\theta \geq 0$ **iv.** $\cos\theta - 1 < -\frac{1}{2}$

H.O.T. Problems   Use Higher-Order Thinking Skills

59. CHALLENGE Solve $\sin 2x < \sin x$ for $0 \leq x \leq 2\pi$ without a calculator. $\frac{\pi}{3} < x < \pi$ or $\frac{5\pi}{3} < x < 2\pi$

60. REASONING Compare and contrast solving trigonometric equations with solving linear and quadratic equations. What techniques are the same? What techniques are different? How many solutions do you expect?   **See margin.**

61. **WRITING IN MATH** Why do trigonometric equations often have infinitely many solutions?   **See margin.**

62. OPEN ENDED Write an example of a trigonometric equation that has exactly two solutions if $0° \leq \theta \leq 360°$.   **Sample answer:** $2\cos\theta = 0$, 90° and 270°

63. CHALLENGE How many solutions in the interval $0° \leq \theta \leq 360°$ should you expect for $a\sin(b\theta + c) = d$, if $a \neq 0$ and b is a positive integer?   **0, b, or 2b**

 906 | Lesson 13-5 | Solving Trigonometric Equations

75. $\sin(90° - \theta) \overset{?}{=} \cos\theta$

$\sin 90° \cos\theta - \cos 90° \sin\theta \overset{?}{=} \cos\theta$

$1 \cdot \cos\theta - 0 \cdot \sin\theta \overset{?}{=} \cos\theta$

$\cos\theta - 0 \overset{?}{=} \cos\theta$

$\cos\theta = \cos\theta$ ✓

76b.

64. EXTENDED RESPONSE Charles received $2500 for a graduation gift. He put it into a savings account in which the interest rate was 5.5% per year.

  a. How much did he have in his savings account after 5 years if he made no deposits or withdrawals? **$3267.40**

  b. After how many years will the amount in his savings account have doubled? **about 13 yr**

65. PROBABILITY Find the probability of rolling three 3s if a number cube is rolled three times. **A**

  A $\frac{1}{216}$ **C** $\frac{1}{6}$

  B $\frac{1}{36}$ **D** $\frac{1}{4}$

66. Use synthetic substitution to find $f(-2)$ for the function below. **F**

$$f(x) = x^4 + 10x^2 + x + 8$$

  F 62 **H** 30
  G 38 **J** 8

67. SAT/ACT The pattern of dots below continues infinitely, with more dots being added at each step.

Step 1   Step 2   Step 3

Which expression can be used to determine the number of dots in the nth step? **E**

  A $2n$ **D** $2(n + 2)$
  B $n(n + 2)$ **E** $2(n + 1)$
  C $n(n + 1)$

Find the exact value of each expression. (Lesson 13-4)

68. $\cos 165°$ $-\dfrac{\sqrt{2 + \sqrt{3}}}{2}$

69. $\sin 22\frac{1}{2}°$ $\dfrac{\sqrt{2 - \sqrt{2}}}{2}$

70. $\sin \dfrac{7\pi}{8}$ $\dfrac{\sqrt{2 - \sqrt{2}}}{2}$

71. $\cos \dfrac{7\pi}{12}$ $-\dfrac{\sqrt{2 - \sqrt{3}}}{2}$

Verify that each equation is an identity. (Lesson 13-3) **72–75. See margin.**

72. $\sin (270° - \theta) = -\cos \theta$

73. $\cos (90° + \theta) = -\sin \theta$

74. $\cos (90° - \theta) = \sin \theta$

75. $\sin (90° - \theta) = \cos \theta$

76. WATER SAFETY A harbor buoy bobs up and down with the waves. The distance between the highest and lowest points is 4 feet. The buoy moves from its highest point to its lowest point and back to its highest point every 10 seconds. (Lesson 12-7)

  a. Write an equation for the motion of the buoy. Assume that it is at equilibrium at $t = 0$ and that it is on the way up from the normal water level. $y = 2 \sin \dfrac{\pi}{5}t$

  b. Draw a graph showing the height of the buoy as a function of time. **See margin.**

  c. What is the height of the buoy after 12 seconds? **about 1.9 ft**

4 ft

Find the first three terms of each arithmetic series described. (Lesson 10-2)

77. $a_1 = 17, a_n = 197, S_n = 2247$ **17, 26, 35**

78. $a_1 = -13, a_n = 427, S_n = 18,423$ **-13, -8, -3**

79. $n = 31, a_n = 78, S_n = 1023$ **-12, -9, -6**

80. $n = 19, a_n = 103, S_n = 1102$ **13, 18, 23**

Graph each rational function. **81–83. See margin.**

81. $f(x) = \dfrac{1}{(x + 3)^2}$

82. $f(x) = \dfrac{x + 4}{x - 1}$

83. $f(x) = \dfrac{x + 2}{x^2 - x - 6}$

DifferentiatedInstruction BL

Extension Ask students to explore the solutions of the equation $\sin x = \dfrac{x}{k}$, where k is a positive integer. Ask them to determine how the number of solutions to the equation changes as k changes. What value of x is a solution of the equation for every value of k? **0**

CCSS Teaching the Mathematical Practices

Perseverance Mathematically proficient students start by explaining to themselves the meaning of a problem and looking for entry points to its solution. They analyze givens, constraints, relationships, and goals and make conjectures about the form and meaning of the solution and plan a solution pathway rather than simply jumping into a solution attempt.

4 Assess

Ticket Out the Door Have students write an equation involving $\sin^2 \theta$ that has exactly one solution for the interval $90° < \theta < 270°$.

Formative Assessment
Check for students' understanding of concepts in Lesson 13-5.

Quiz 4, p. 38

Additional Answers

81.

$f(x) = \dfrac{1}{(x + 3)^2}$

82.

$f(x) = \dfrac{x + 4}{x - 1}$

83.

$f(x) = \dfrac{x + 2}{x^2 - x - 6}$

Formative Assessment

KeyVocabulary The page references after each word denote where that term was first introduced. If students have difficulty answering questions 1–9, remind them that they can use these page references to refresh their memories about the vocabulary.

Summative Assessment

Vocabulary Test, p. 40

Vocabulary Review

Vocabulary Review provides students the opportunity to check their understanding of important concepts and terminology in an online game format.

FOLDABLES StudyOrganizer

Dinah Zike's Foldables®

Have students look through the chapter to make sure they have included examples in their Foldables. Suggest that students keep their Foldables handy while completing the Study Guide and Review pages. Point out that their Foldables can serve as a quick review tool when studying for the chapter test.

Study Guide

KeyConcepts

Trigonometric Identities (Lessons 13-1, 13-2, and 13-5)

- Trigonometric identities describe the relationships between trigonometric functions.

- Trigonometric identities can be used to simplify, verify, and solve trigonometric equations and expressions.

Sum and Difference of Angles Identities (Lesson 13-3)

- For all values of A and B:

$$\cos (A \pm B) = \cos A \cos B \mp \sin A \sin B$$
$$\sin (A \pm B) = \sin A \cos B \pm \cos A \sin B$$
$$\tan (A \pm B) = \frac{\tan A \pm \tan B}{1 \mp \tan A \tan B}$$

Double-Angle and Half-Angle Identities (Lesson 13-4)

- Double-angle identities:

$$\sin 2\theta = 2 \sin \theta \cos \theta$$
$$\cos 2\theta = \cos^2 \theta - \sin^2 \theta$$
$$\cos 2\theta = 1 - 2 \sin^2 \theta$$
$$\cos 2\theta = 2 \cos^2 \theta - 1$$
$$\tan 2\theta = \frac{2 \tan \theta}{1 - \tan^2 \theta}$$

- Half-angle identities:

$$\sin \frac{\theta}{2} = \pm\sqrt{\frac{1 - \cos \theta}{2}}$$

$$\cos \frac{\theta}{2} = \pm\sqrt{\frac{1 + \cos \theta}{2}}$$

$$\tan \frac{\theta}{2} = \pm\sqrt{\frac{1 - \cos \theta}{1 + \cos \theta}}, \cos \theta \neq -1$$

FOLDABLES StudyOrganizer

Be sure the Key Concepts are noted in your Foldable.

 908 | Chapter 13 | Study Guide and Review

KeyVocabulary

cofunction identity (p. 873)

negative angle identity (p. 873)

Pythagorean identity (p. 873)

quotient identity (p. 873)

reciprocal identity (p. 873)

trigonometric equation (p. 901)

trigonometric identity (p. 873)

VocabularyCheck

Choose the correct term to complete each sentence.

1. The _____ can be used to find the sine or cosine of 75° if the sine and cosine of 90° and 15° are known.
difference of angles identity

2. The identities $\tan \theta = \frac{\sin \theta}{\cos \theta}$ and $\cot \theta = \frac{\cos \theta}{\sin \theta}$ are examples of _____. **quotient identities**

3. A _____ is an equation involving trigonometric functions that is true for all values for which every expression in the equation is defined. **trigonometric identity**

4. The _____ can be used to find sin 60° using 30° as a reference. **double-angle identity**

5. A _____ is true for only certain values of the variable. **trigonometric equation**

6. The _____ identity can be used to find $\cos 22\frac{1}{2}°$. **half-angle**

7. The identities $\csc \theta = \frac{1}{\sin \theta}$ and $\sec \theta = \frac{1}{\cos \theta}$ are examples of _____. **reciprocal identities**

8. The _____ can be used to find the sine or cosine of 120° if the sine and cosine of 90° and 30° are known. **sum of angles identity**

9. $\cos^2 \theta + \sin^2 \theta = 1$ is an example of a _____. **Pythagorean identity**

Lesson-by-Lesson Review

13-1 Trigonometric Identities

Find the value of each expression.

10. $\sin \theta$, if $\cos \theta = \frac{\sqrt{2}}{2}$ and $270° < \theta < 360°$ $-\frac{\sqrt{2}}{2}$

11. $\sec \theta$, if $\cot \theta = \frac{\sqrt{2}}{2}$ and $90° < \theta < 180°$ $-\sqrt{3}$

12. $\tan \theta$, if $\cot \theta = 2$ and $0° < \theta < 90°$ $\frac{1}{2}$

13. $\cos \theta$, if $\sin \theta = -\frac{3}{5}$ and $180° < \theta < 270°$ $-\frac{4}{5}$

14. $\csc \theta$, if $\cot \theta = -\frac{4}{5}$ and $270° < \theta < 360°$ $-\frac{\sqrt{41}}{5}$

15. **SOCCER** For international matches, the maximum dimensions of a soccer field are 110 meters by 75 meters. Find $\sin \theta$.

                       See margin.

75 m

110 m

Simplify each expression. **16.** $\cos^2 \theta$ **18.** $\csc \theta$

16. $1 - \tan \theta \sin \theta \cos \theta$ 17. $\tan \theta \csc \theta$ $\sec \theta$

18. $\sin \theta + \cos \theta \cot \theta$ 19. $\cos \theta (1 + \tan^2 \theta)$ $\sec \theta$

Example 1

Find $\sin \theta$ if $\cos \theta = \frac{3}{4}$ and $0° < \theta < 90°$.

$\cos^2 \theta + \sin^2 \theta = 1$ Trigonometric identity

$\sin^2 \theta = 1 - \cos^2 \theta$ Subtract $\cos^2 \theta$ from each side.

$\sin^2 \theta = 1 - \left(\frac{3}{4}\right)^2$ Substitute $\frac{3}{4}$ for $\cos \theta$.

$\sin^2 \theta = 1 - \frac{9}{16}$ Square $\frac{3}{4}$.

$\sin^2 \theta = \frac{7}{16}$ Subtract.

$\sin \theta = \pm\frac{\sqrt{7}}{4}$ Take the square root of each side.

Because θ is in the first quadrant, $\sin \theta$ is positive.

Thus, $\sin \theta = \frac{\sqrt{7}}{4}$.

Example 2

Simplify $\cos \theta \sec \theta \cot \theta$.

$\cos \theta \sec \theta \cot \theta = \cos \theta \left(\frac{1}{\cos \theta}\right)\left(\frac{\cos \theta}{\sin \theta}\right)$

$= \cot \theta$

13-2 Verifying Trigonometric Identities

Verify that each of the following is an identity.

20. $\tan \theta \cos \theta + \cot \theta \sin \theta = \sin \theta + \cos \theta$

21. $\frac{\cos \theta}{\cot \theta} + \frac{\sin \theta}{\tan \theta} = \sin \theta + \cos \theta$

22. $\sec^2 \theta - 1 = \frac{\sin^2 \theta}{1 - \sin^2 \theta}$ **20–23. See margin.**

23. **GEOMETRY** The right triangle shown at the right is used in a special quilt. Use the measures of the sides of the triangle to show that $\tan^2 \theta + 1 = \sec^2 \theta$.

3   4

$\sqrt{7}$

Example 3

Verify that $\frac{\cos \theta + 1}{\sin \theta} = \cot \theta + \csc \theta$ is an identity.

$\frac{\cos \theta + 1}{\sin \theta} \stackrel{?}{=} \cot \theta + \csc \theta$ Original equation

$\frac{\cos \theta}{\sin \theta} + \frac{1}{\sin \theta} \stackrel{?}{=} \cot \theta + \csc \theta$ Simplify.

$\cot \theta + \csc \theta = \cot \theta + \csc \theta \checkmark$ Simplify.

Intervention If the given examples are not sufficient to review the topics covered by the questions, remind students that the lesson references tell them where to review that topic in their textbook.

eAssessment (McGraw-Hill) Have students complete the Lesson-by-Lesson Review. Then you can use McGraw-Hill eAssessment to customize another review worksheet that practices all the objectives of this chapter or only the objectives on which your students need more help.

Additional Answers

15. $\frac{15\sqrt{709}}{709}$

20. $\tan \theta \cos \theta + \cot \theta \sin \theta \stackrel{?}{=}$
   $\sin \theta + \cos \theta$
   $\frac{\sin \theta}{\cos \theta} \cdot \cos \theta + \frac{\cos \theta}{\sin \theta} \cdot \sin \theta \stackrel{?}{=}$
   $\sin \theta + \cos \theta$
   $\sin \theta + \cos \theta = \sin \theta + \cos \theta \checkmark$

21. $\frac{\cos \theta}{\cot \theta} + \frac{\sin \theta}{\tan \theta} \stackrel{?}{=} \sin \theta + \cos \theta$
   $\cos \theta \div \frac{\cos \theta}{\sin \theta} + \sin \theta \div \frac{\sin \theta}{\cos \theta} \stackrel{?}{=}$
   $\sin \theta + \cos \theta$
   $\cos \theta \cdot \frac{\sin \theta}{\cos \theta} + \sin \theta \cdot \frac{\cos \theta}{\sin \theta} \stackrel{?}{=}$
   $\sin \theta + \cos \theta$
   $\sin \theta + \cos \theta = \sin \theta + \cos \theta \checkmark$

---

22. $\sec^2 \theta - 1 \stackrel{?}{=} \frac{\sin^2 \theta}{1 - \sin^2 \theta}$

   $\sec^2 \theta - 1 \stackrel{?}{=} \frac{\sin^2 \theta}{\cos^2 \theta}$

   $\sec^2 \theta - 1 \stackrel{?}{=} \tan^2 \theta$

   $\sec^2 \theta - 1 = \sec^2 \theta - 1 \checkmark$

23. $\tan^2 \theta + 1 = \left(\frac{\sqrt{7}}{3}\right)^2 + 1 = \frac{7}{9} + 1$

   $= \frac{7}{9} + \frac{9}{9} = \frac{16}{9}$;

   $\sec^2 \theta = \left(\frac{4}{3}\right)^2 = \frac{16}{9}$

Additional Answers (Practice Test)

2. $\cos(30° - \theta) \stackrel{?}{=} \sin(60° + \theta)$

$\cos 30° \cos \theta + \sin 30° \sin \theta \stackrel{?}{=}$
$\sin 60° \cos \theta + \cos 60° \sin \theta$

$\dfrac{\sqrt{3}}{2}\cos \theta + \dfrac{1}{2}\sin \theta =$

$\dfrac{\sqrt{3}}{2}\cos \theta + \dfrac{1}{2}\sin \theta \checkmark$

3. $\cos(\theta - \pi) \stackrel{?}{=} -\cos \theta$

$\cos \theta \cos \pi + \sin \theta \sin \pi \stackrel{?}{=} -\cos \theta$

$(\cos \theta)(-1) + (\sin \theta)(0) \stackrel{?}{=} -\cos \theta$

$-\cos \theta = -\cos \theta \checkmark$

10. $\sin \theta (\cot \theta + \tan \theta) \stackrel{?}{=} \sec \theta$

$\sin \theta \left(\dfrac{\cos \theta}{\sin \theta} + \dfrac{\sin \theta}{\cos \theta}\right) \stackrel{?}{=} \sec \theta$

$\cos \theta + \dfrac{\sin^2 \theta}{\cos \theta} \stackrel{?}{=} \sec \theta$

$\dfrac{\cos^2 \theta + \sin^2 \theta}{\cos \theta} \stackrel{?}{=} \sec \theta$

$\dfrac{1}{\cos \theta} \stackrel{?}{=} \sec \theta$

$\sec \theta = \sec \theta \checkmark$

11. $\dfrac{\cos^2 \theta}{1 - \sin \theta} \stackrel{?}{=} \dfrac{\cos \theta}{\sec \theta - \tan \theta}$

$\dfrac{\cos^2 \theta}{1 - \sin \theta} \stackrel{?}{=} \dfrac{\cos \theta}{\dfrac{1}{\cos \theta} - \dfrac{\sin \theta}{\cos \theta}}$

$\dfrac{\cos^2 \theta}{1 - \sin \theta} \stackrel{?}{=} \dfrac{\cos \theta}{\dfrac{1 - \sin \theta}{\cos \theta}}$

$\dfrac{\cos^2 \theta}{1 - \sin \theta} \stackrel{?}{=} \cos \theta \cdot \dfrac{\cos \theta}{1 - \sin \theta}$

$\dfrac{\cos^2 \theta}{1 - \sin \theta} = \dfrac{\cos^2 \theta}{1 - \sin \theta} \checkmark$

12. $(\tan \theta + \cot \theta)^2 \stackrel{?}{=} \csc^2 \theta \sec^2 \theta$

$\left(\dfrac{\sin \theta}{\cos \theta} + \dfrac{\cos \theta}{\sin \theta}\right)^2 \stackrel{?}{=} \csc^2 \theta \sec^2 \theta$

$\left(\dfrac{\sin^2 \theta + \cos^2 \theta}{\cos \theta \sin \theta}\right)^2 \stackrel{?}{=} \csc^2 \theta \sec^2 \theta$

$\left(\dfrac{1}{\cos \theta \sin \theta}\right)^2 \stackrel{?}{=} \csc^2 \theta \sec^2 \theta$

$\dfrac{1}{\cos^2 \theta} \cdot \dfrac{1}{\sin^2 \theta} \stackrel{?}{=} \csc^2 \theta \sec^2 \theta$

$\sec^2 \theta \csc^2 \theta = \csc^2 \theta \sec^2 \theta \checkmark$

13-3 Sum and Difference of Angles Identities

Find the exact value of each expression.

24. $\cos(-135°)$ $-\dfrac{\sqrt{2}}{2}$

25. $\cos 15°$ $\dfrac{\sqrt{6} + \sqrt{2}}{4}$

26. $\sin 210°$ $-\dfrac{1}{2}$

27. $\sin 105°$ $\dfrac{\sqrt{6} + \sqrt{2}}{4}$

28. $\tan 75°$ $\sqrt{3} + 2$

29. $\cos 105°$ $\dfrac{-\sqrt{6} + \sqrt{2}}{4}$

Verify that each of the following is an identity.

30. $\sin(\theta + 90) = \cos \theta$ **30–32. See Chapter 13 Answer Appendix.**

31. $\sin\left(\dfrac{3\pi}{2} - \theta\right) = -\cos \theta$

32. $\tan(\theta - \pi) = \tan \theta$

Example 4

Find the exact value of $\sin 75°$.

Use $\sin(A + B) = \sin A \cos B + \cos A \sin B$.

$\sin 75° = \sin(30° + 45°)$

$= \sin 30° \cos 45° + \cos 30° \sin 45°$

$= \left(\dfrac{1}{2}\right)\left(\dfrac{\sqrt{2}}{2}\right) + \left(\dfrac{\sqrt{3}}{2}\right)\left(\dfrac{\sqrt{2}}{2}\right)$

$= \dfrac{\sqrt{2}}{4} + \dfrac{\sqrt{6}}{4}$ or $\dfrac{\sqrt{2} + \sqrt{6}}{4}$

13-4 Double-Angle and Half-Angle Identities

Find the exact values of $\sin 2\theta$, $\cos 2\theta$, $\sin \dfrac{\theta}{2}$, and $\cos \dfrac{\theta}{2}$ for each of the following.

33. $\cos \theta = \dfrac{4}{5}; 0° < \theta < 90°$ **33–36. See Chapter 13 Answer Appendix.**

34. $\sin \theta = -\dfrac{1}{4}; 180° < \theta < 270°$

35. $\cos \theta = -\dfrac{2}{3}; \dfrac{\pi}{2} < \theta < \pi$

36. BASEBALL The infield of a baseball diamond is a square with side length 90 feet.

  a. Find the length of the diagonal.

  b. Write the ratio for $\sin 45°$ using the lengths of the baseball diamond.

  c. Use the formula $\sin \dfrac{\theta}{2} = \pm\sqrt{\dfrac{1 - \cos \theta}{2}}$ to verify the ratio you wrote in part **b**.

Example 5

Find the exact value of $\sin \dfrac{\theta}{2}$ if $\cos \theta = -\dfrac{3}{5}$ and θ is in the second quadrant.

$\sin \dfrac{\theta}{2} = \pm\sqrt{\dfrac{1 - \cos \theta}{2}}$ Half-angle identity

$= \pm\sqrt{\dfrac{1 - \left(-\dfrac{3}{5}\right)}{2}}$ $\cos \theta = -\dfrac{3}{5}$

$= \pm\sqrt{\dfrac{\dfrac{8}{5}}{2}}$ Subtract.

$= \pm\sqrt{\dfrac{4}{5}}$ Divide.

$= \pm\dfrac{2\sqrt{5}}{5}$ Simplify.

Since θ is in the second quadrant, $\sin \dfrac{\theta}{2} = \dfrac{2\sqrt{5}}{5}$.

13-5 Solving Trigonometric Equations

Find all solutions of each equation for the given interval.

37. $2 \cos \theta - 1 = 0; 0° \le \theta < 360°$ $60°, 300°$

38. $4 \cos^2 \theta - 1 = 0; 0 \le \theta < 2\pi$ $\dfrac{\pi}{3}, \dfrac{2\pi}{3}, \dfrac{4\pi}{3}, \dfrac{5\pi}{3}$

39. $\sin 2\theta + \cos \theta = 0; 0° \le \theta < 360°$ $90°, 210°, 270°, 330°$

40. $\sin^2 \theta = 2 \sin \theta + 3; 0° \le \theta < 360°$ $270°$

41. $4 \cos^2 \theta - 4 \cos \theta + 1 = 0; 0 \le \theta < 2\pi$ $\dfrac{\pi}{3}, \dfrac{5\pi}{3}$

Example 6

Find all solutions of $\sin 2\theta - \cos \theta = 0$ if $0 \le \theta < 2\pi$.

$\sin 2\theta - \cos \theta = 0$ Original equation

$2 \sin \theta \cos \theta - \cos \theta = 0$ Double-angle identity

$\cos \theta (2 \sin \theta - 1) = 0$ Factor.

$\cos \theta = 0$ or $2 \sin \theta - 1 = 0$

$\theta = \dfrac{\pi}{2}, \dfrac{3\pi}{2}$ $\sin \theta = \dfrac{1}{2}; \theta = \dfrac{\pi}{6}, \dfrac{5\pi}{6}$

13. $\dfrac{1 + \sec \theta}{\sec \theta} \stackrel{?}{=} \dfrac{\sin^2 \theta}{1 - \cos \theta}$

$\dfrac{1}{\sec \theta} + \dfrac{\sec \theta}{\sec \theta} \stackrel{?}{=} \dfrac{\sin^2 \theta}{1 - \cos \theta} \cdot \dfrac{1 + \cos \theta}{1 + \cos \theta}$

$\cos \theta + 1 \stackrel{?}{=} \dfrac{\sin^2 \theta (1 + \cos \theta)}{1 - \cos^2 \theta}$

$\cos \theta + 1 \stackrel{?}{=} \dfrac{\sin^2 \theta (1 + \cos \theta)}{\sin^2 \theta}$

$\cos \theta + 1 = 1 + \cos \theta \checkmark$

CHAPTER 13 Practice Test

1. MULTIPLE CHOICE Which expression is equivalent to $\sin\theta + \cos\theta\cot\theta$? **D**

A $\cot\theta$ C $\sec\theta$

B $\tan\theta$ D $\csc\theta$

2. Verify that $\cos(30° - \theta) = \sin(60° + \theta)$ is an identity. **See margin.**

3. Verify that $\cos(\theta - \pi) = -\cos\theta$. **See margin.**

4. MULTIPLE CHOICE What is the exact value of $\sin\theta$, if $\cos\theta = -\frac{3}{5}$ and $90° < \theta < 180°$? **J**

F $\frac{5}{3}$

G $\frac{\sqrt{34}}{8}$

H $-\frac{4}{5}$

J $\frac{4}{5}$

Find the value of each expression.

5. $\cot\theta$, if $\sec\theta = \frac{4}{3}$; $270° < \theta < 360°$ $-\frac{3\sqrt{7}}{7}$

6. $\tan\theta$, if $\cos\theta = -\frac{1}{2}$; $90° < \theta < 180°$ $-\sqrt{3}$

7. $\sec\theta$, if $\csc\theta = -2$; $180° < \theta < 270°$ $-\frac{2\sqrt{3}}{3}$

8. $\cot\theta$, if $\csc\theta = -\frac{5}{3}$; $270° < \theta < 360°$ $-\frac{4}{3}$

9. $\sec\theta$, if $\sin\theta = \frac{1}{2}$; $0° \le \theta < 90°$ $\frac{2\sqrt{3}}{3}$

Verify that each of the following is an identity.

10. $\sin\theta(\cot\theta + \tan\theta) = \sec\theta$ **10–13. See margin.**

11. $\frac{\cos^2\theta}{1-\sin\theta} = \frac{\cos\theta}{\sec\theta - \tan\theta}$

12. $(\tan\theta + \cot\theta)^2 = \csc^2\theta\sec^2\theta$

13. $\frac{1+\sec\theta}{\sec\theta} = \frac{\sin^2\theta}{1-\cos\theta}$

14. $\frac{\sin\theta}{1-\cos\theta} = \csc\theta + \cot\theta$

See Chapter 13 Answer Appendix.

15. MULTIPLE CHOICE What is the exact value of $\tan\frac{\pi}{8}$? **B**

A $\frac{\sqrt{2-\sqrt{3}}}{2}$

B $\sqrt{2}-1$

C $1-\sqrt{2}$

D $-\frac{\sqrt{2-\sqrt{3}}}{2}$

16. HISTORY Some researchers believe that the builders of ancient pyramids, such as the Great Pyramid of Khufu, may have tried to build the faces as equilateral triangles. Later they had to change to other types of triangles. Suppose a pyramid is built such that a face is an equilateral triangle of side length 18 feet. **a, b. See Chapter 13 Answer Appendix.**

18 ft

a. Find the height of the equilateral triangle.

b. Use the formula $\sin 2\theta = 2\sin\theta\cos\theta$ and the measures of the equilateral triangle and its height to show that $\sin 2(30°) = \sin 60°$. Find the exact values.

Find the exact value of each expression.

17. $\cos(-225°)$ $-\frac{\sqrt{2}}{2}$ **18.** $\sin 480°$ $\frac{\sqrt{3}}{2}$

19. $\cos 75°$ $\frac{\sqrt{6}-\sqrt{2}}{4}$ **20.** $\sin 165°$ $\frac{\sqrt{6}-\sqrt{2}}{4}$

21. ROCKETS A model rocket is launched with an initial velocity of 20 meters per second. The range of a projectile is given by the formula $R = \frac{v^2}{g}\sin 2\theta$, where R is the range, v is the initial velocity, g is acceleration due to gravity or 9.8 meters per second squared, and θ is the launch angle. What angle is needed in order for the rocket to reach a range of 25 meters? **See Chapter 13 Answer Appendix.**

Solve each equation for all values of θ if θ is measured in radians.

22. $2\cos^2\theta - 3\cos\theta - 2 = 0$ $\frac{2\pi}{3} + 2\pi k, \frac{4\pi}{3} + 2\pi k$

23. $2\sin 3\theta - 1 = 0$ $\frac{\pi}{18} + \frac{2k\pi}{3}, \frac{5\pi}{18} + \frac{2k\pi}{3}$

Solve each equation for $0° \le \theta \le 360°$ if θ is measured in degrees.

24. $\cos 2\theta + \cos\theta = 2$ $0°, 360°$

25. $\sin\theta\cos\theta - \frac{1}{2}\sin\theta = 0$ $0°, 60°, 180°, 300°, 360°$

 911

Summative Assessment
Use these alternate leveled chapter tests to differentiate assessment for your students.

Leveled Chapter 13 Tests

Form	Type	Level	📁 Page(s)
1	MC	AL	41–42
2A	MC	OL	43–44
2B	MC	OL	45–46
2C	FR	OL	47–48
2D	FR	OL	49–50
3	FR	BL	51–52
Vocabulary Test			40
Extended-Response Test			53

MC = multiple-choice questions
FR = free-response questions

eAssessment McGraw-Hill Customize and create multiple versions of your chapter test and their answer keys. All of the questions from the leveled chapter tests in the *Chapter 13 Resource Masters* are also available on McGraw-Hill eAssessment.

InterventionPlanner

TIER 1 — On Level OL

If students miss about 25% of the exercises or less,

Then choose a resource:

SE Lessons 13-1, 13-2, 13-3, 13-4, and 13-5

📁 Skills Practice, pp. 7, 13, 19, 25, and 31

TE Chapter Project

connectED.mcgraw-hill.com Self-Check Quiz

TIER 2 — Strategic Intervention AL approaching grade level

If students miss about 50% of the exercises,

Then choose a resource:

📁 Study Guide and Intervention, pp. 5, 11, 17, 23, and 29

connectED.mcgraw-hill.com Extra Examples, Personal Tutor, Homework Help

TIER 3 — Intensive Intervention 2 or more grades below level

If students miss about 75% of the exercises,

Then use *Math Triumphs, Alg. 2*

connectED.mcgraw-hill.com Extra Examples, Personal Tutor, Homework Help, Review Vocabulary

1 Focus

Objective Use the strategy of simplifying expressions to solve standardized test problems.

2 Teach

Scaffolding Questions

Ask:

- What kinds of problems might you have to simplify in order to solve?
  Possible answer: equations or inequalities where one or both sides need to be simplified before the equation or inequality is solved, trigonometric functions that need to be simplified.

- Why check if a problem could be simplified before using a calculator?
  Possible answer: if you start to use a calculator before simplifying you have more of a chance to enter information incorrectly. Also, you may find that simplifying the problem first gives you an answer without having to use the calculator.

CHAPTER 13 Preparing for Standardized Tests

Simplify Expressions

Some standardized test questions will require you to use the properties of algebra to simplify expressions. Follow the steps below to help prepare to solve these kinds of problems.

Strategies for Simplifying Expressions

Step 1

Study the expression that you are being asked to simplify.

Ask yourself:

- Are there any mathematical operations I can apply to help simplify the expression?
- Are there any laws or identities I can apply to help simplify the expression?

Step 2

Solve the problem and check your solution.

- Use the order of operations.
- Combine terms and factor as appropriate.
- Apply laws and identities.

Step 3

Check your solution if time permits.

- Retrace the steps in your work to make sure you answered the question thoroughly and accurately.
- If needed, sometimes you can use your scientific calculator to help you check your solution. Evaluate the original expression and your answer for some value and make sure they are the same.

Standardized Test Example

Solve the problem below. Responses will be graded using the short-response scoring rubric shown.

Simplify the trigonometric expression shown below by writing it in terms of $\sin \theta$. Show your work to receive full credit.

$$\frac{\cos \theta}{\sec \theta + \tan \theta}$$

Scoring Rubric	
Criteria	Score
Full Credit: The answer is correct and a full explanation is provided that shows each step.	2
Partial Credit: • The answer is correct, but the explanation is incomplete. • The answer is incorrect, but the explanation is correct.	1
No Credit: Either an answer is not provided or the answer does not make sense.	0

Corbis

Read the problem statement carefully. You are given a trigonometric expression and asked to simplify it by writing it in terms of sin θ. So, your final answer must contain only numbers and terms involving sin θ. Show your work to receive full credit.

Example of a 2-point response:

Use trigonometric identities to simplify the expression.

$$\frac{\cos \theta}{\sec \theta + \tan \theta} = \frac{\cos \theta}{\dfrac{1}{\cos \theta} + \dfrac{\sin \theta}{\cos \theta}}$$ Definition of sec θ and tan θ

$$= \frac{\cos \theta}{\dfrac{1 + \sin \theta}{\cos \theta}}$$ Simplify the denominator.

$$= \frac{\cos^2 \theta}{1 + \sin \theta}$$ Simplify the complex fraction.

$$= \frac{1 - \sin^2 \theta}{1 + \sin \theta}$$ Pythagorean identity

$$= \frac{(1 + \sin \theta)(1 - \sin \theta)}{1 + \sin \theta}$$ Factor.

$$= 1 - \sin \theta$$ Simplify.

The simplified expression is $1 - \sin \theta$.

The steps, calculations, and reasoning are clearly stated. The student also arrives at the correct answer. So this response is worth the full 2 points.

Additional Example

STANDARDIZED TEST PRACTICE
Simplify the expression
$\sec \theta \cot \theta \cos \theta$. B

A $\csc \theta$

B $\cot \theta$

C $\cos \theta$

D $\sin \theta$

3 Assess

Use Exercises 1–7 to assess students' understanding.

Exercises

Solve each problem. Show your work. Responses will be graded using the short-response scoring rubric given at the beginning of the lesson.

1. Simplify $\dfrac{\sec \theta}{\cot \theta + \tan \theta}$ by writing it in terms of sin θ.

$\sin \theta$

2. What is $\dfrac{10a^{-3}}{29b^4} \div \dfrac{5a^{-5}}{16b^{-7}}$? $\dfrac{32a^2}{29b^{11}}$

3. Write $\dfrac{y + 1}{y - 1} + \dfrac{y + 2}{y - 2} + \dfrac{y}{y^2 - 3y + 2}$ in simplest form. $\dfrac{2y^2 + y - 4}{(y - 1)(y - 2)}$

4. Simplify $\dfrac{\cot^2 \theta - \csc^2 \theta}{\tan^2 \theta - \sec^2 \theta}$ by writing it as a constant. 1

5. Multiply $(-5 + 2i)(6 - i)(4 + 3i)$. $-163 - 16i$

6. Simplify $(\cot \theta + 1)^2 - 2 \cot \theta$ by writing it in terms of csc θ. $\csc^2 \theta$

7. Express $\dfrac{4 - \sqrt{7}}{3 + \sqrt{7}}$ in simplest form. $\dfrac{19 - 7\sqrt{7}}{2}$

connectED.mcgraw-hill.com   **913**

Diagnose Student Errors

Survey student responses for each item. Class trends may indicate common errors and misconceptions.

1. A chose an amount not between $650 and $925
   B chose an amount not between $650 and $925
   C correct
   D chose an amount not between $650 and $925

2. F mistakenly reversed the cosine and sine values and subtracted in the formula
   G incorrect denominator, 2, found when multiplying values
   H correct
   J mistakenly subtracted when using the addition of angles formula

3. A did not understand that n represents the number of sides
   B did not evalute expression with data in the table
   C correct
   D only used data from the first line of the table to determine the expression

4. F mistakenly identified the slope as the y-intercept
   G mistakenly identified the slope as the x-intercept
   H does not understand that the slopes of two lines must be additive inverses in order for the lines to be perpendicular
   J correct

5. A product should be a 1×2 matrix not 2×1 matrix
   B correct
   C product should be a 1×2 matrix not 3×2
   D the first matrix is 1×3 and the second is 3×2 so the product of a 1×2 matrix is defined

6. F added instead of multiplied to find x^2 coefficient
   G added instead of multiplied to find x^2 coefficient
   H placed incorrect signs on x-term and constant
   J correct

Standardized Test Practice
Cumulative, Chapters 1 through 13

Multiple Choice

Read each question. Then fill in the correct answer on the answer document provided by your teacher or on a sheet of paper.

1. The profit p that Selena's Shirt Store makes in a day can be represented by the inequality $10t + 200 < p < 15t + 250$, where t represents the number of shirts sold. If the store sold 45 shirts on Friday, which of the following is a reasonable amount that the store made? **C**

A $200   B $625   C $850   D $950

2. Use a sum or difference of angles identity to find the exact value of cos 75°. **H**

F $\dfrac{\sqrt{2} - \sqrt{6}}{4}$ H $\dfrac{\sqrt{6} - \sqrt{2}}{4}$

G $\dfrac{\sqrt{2} + \sqrt{6}}{2}$ J $\dfrac{\sqrt{6} + \sqrt{2}}{4}$

3. Use the table to determine the expression that best represents the degree measure of an interior angle of a regular polygon with n sides. **C**

Polygon	Number of Sides	Angle Measure
triangle	3	60
quadrilateral	4	90
pentagon	5	108
hexagon	6	120
heptagon	7	128.5
octagon	8	135

A $(180 + n) \div n$
B $\dfrac{180}{n}$
C $[180(n - 2)] \div n$
D $30(n - 1)$

4. Which of the following best describes the graphs of $y = 3x - 5$ and $4y = 12x + 16$? **J**

F The lines have the same y-intercept.
G The lines have the same x-intercept.
H The lines are perpendicular.
J The lines are parallel.

> **Test-Taking Tip**
> Question 2 You can check your answer using a scientific calculator. Find cos 75° and compare it to the value of your answer.

---

5. What is the product of $[5 \ \ -2 \ \ 3]$ and $\begin{bmatrix} 1 & -2 \\ 0 & 3 \\ 2 & 5 \end{bmatrix}$? **B**

A $\begin{bmatrix} 11 \\ -1 \end{bmatrix}$ C $\begin{bmatrix} 5 & -10 \\ 0 & -6 \\ 6 & -15 \end{bmatrix}$

B $[\ 11 \ \ -1 \ ]$ D undefined

6. Which quadratic equation has roots $\frac{1}{2}$ and $\frac{1}{3}$? **J**

F $5x^2 - 5x - 2 = 0$
G $5x^2 - 5x + 1 = 0$
H $6x^2 + 5x - 1 = 0$
J $6x^2 - 5x + 1 = 0$

7. How can you express $\cos \theta \csc \theta \cot \theta$ in terms of $\sin \theta$? **A**

A $\dfrac{1 - \sin^2 \theta}{\sin^2 \theta}$ C $\dfrac{\sin^2 \theta}{2}$

B $\dfrac{1 + \sin^2 \theta}{\sin^2 \theta}$ D $\dfrac{1 - \sin^2 \theta}{\sin \theta}$

8. The area of a rectangle is $25a^4 - 16b^2$. Which factors could represent the length times width? **G**

F $(5a^2 + 4b)(5a^2 + 4b)$ H $(5a - 4b)(5a - 4b)$
G $(5a^2 + 4b)(5a^2 - 4b)$ J $(5a + 4b)(5a - 4b)$

9. What is the domain of $f(x) = \sqrt{5x - 3}$? **C**

A $\left\{ x \mid x > \frac{3}{5} \right\}$ C $\left\{ x \mid x \geq \frac{3}{5} \right\}$

B $\left\{ x \mid x > -\frac{3}{5} \right\}$ D $\left\{ x \mid x \geq -\frac{3}{5} \right\}$

10. If the equation $y = 3^x$ is graphed, which of the following values of x would produce a point closest to the x-axis? **J**

F $\dfrac{3}{4}$ H 0

G $\dfrac{1}{4}$ J $-\dfrac{3}{4}$

---

7. A correct
   B used $\cos^2 \theta = 1 + \sin^2 \theta$ instead of $\cos^2 \theta = 1 - \sin^2 \theta$
   C incorrectly simplified fraction
   D incorrectly multiplied denominator

8. F does not recognize the difference of squares but pays attention to exponents of terms
   G correct
   H does not recognize the difference of squares
   J does recognize the difference of squares but does not pay attention to exponents of terms

9. A did not include $\frac{3}{5}$ in domain
   B guess
   C correct
   D incorrectly solved $0 \leq 5x - 3$

10. F doesn't understand that exponential functions are everywhere increasing
   G doesn't understand that exponential functions are everywhere increasing
   H doesn't understand that exponential functions are everywhere increasing
   J correct

Short Response/Gridded Response

Record your answers on the answer sheet provided by your teacher or on a sheet of paper.

11. Use right triangle LMN at the right to show that $\sin 2N = \frac{2n\ell}{m^2}$.

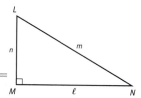

Sample answer:
$\sin 2N = 2 \sin N \cos N = 2\frac{n}{m} \cdot \frac{\ell}{m} = \frac{2n\ell}{m^2}$

12. GRIDDED RESPONSE Solve the trigonometric equation below in the interval from 0 to 2π. Round your answer to the nearest hundredth if necessary. **2.52**

$$3 \cos \frac{t}{3} = 2$$

13. Kelly is designing a 12-inch by 12-inch scrapbook page. She cuts one picture that is 4 inches by 6 inches. She decides that she wants the next picture to be 75% as large as the first picture and the third picture to be 150% as large as the second picture. What are the approximate dimensions of the third picture? **4.5 in. by 6.75 in.**

14. Identify the amplitude and period of the function graphed below. Then write an equation for the function.

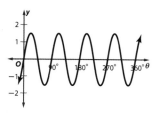

amplitude: 1.5, period: 90°, equation: $y = 1.5 \sin 4\theta$

15. GRIDDED RESPONSE A coordinate grid is placed over a map. Emilee's house is located at $(-3, -2)$, and Oliver's house is located at $(2, 5)$. A side of each square represents one block. What is the approximate distance between Emilee's house and Oliver's house? **8.6**

Extended Response

Record your answers on a sheet of paper. Show your work.

16. Kyla's annual salary is $50,000. Each year she gets a 6% raise.

a. To the nearest dollar, what will her salary be in four years? **$63,124**

b. To the nearest dollar, what will her salary be in 10 years? **$89,542**

17. COMPACT DISCS According to a recent survey, 91% of high school students do not buy compact discs. 8 random students are chosen.

a. Determine the probabilities associated with the number of students that do not buy compact discs by calculating the probability distribution. See margin.

b. What is the probability that at least 7 of the 8 students do not buy compact discs? **84.2%**

c. How many students should you expect to not buy compact discs? **8**

Formative Assessment

You can use these two pages to benchmark student progress.

📁 Standardized Test Practice, pp. 54–56

McGraw-Hill
ᵉAssessment   Create practice tests that align to your state standards, the Common Core State Standards, and other national standards such as TIMSS and NAEP.

Answer Sheet Practice

Have students simulate taking a standardized test by recording their answers on practice recording sheets.

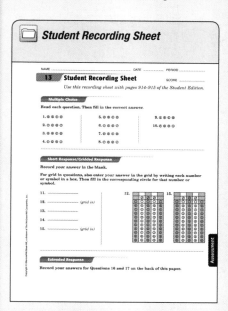

Need Extra Help?

If you missed Question...	1	2	3	4	5	6	7	8	9	10	11	12	13	14	15	16	17
Go to Lesson...	1-5	13-3	2-4	3-1	3-6	4-3	13-1	5-2	6-3	7-1	13-4	13-5	8-5	12-8	9-1	10-2	11-4

Additional Answer

17a.

High School Students Who Do Not Buy Compact Discs									
Students	0	1	2	3	4	5	6	7	8
Probability	0%	0%	0%	0%	0.3%	2.5%	12.9%	37.2%	47.0%

1. $\cot\theta + \tan\theta \overset{?}{=} \dfrac{\sec^2\theta}{\tan\theta}$

$\cot\theta + \tan\theta \overset{?}{=} \dfrac{\tan^2\theta + 1}{\tan\theta}$

$\cot\theta + \tan\theta \overset{?}{=} \dfrac{\tan^2\theta}{\tan\theta} + \dfrac{1}{\tan\theta}$

$\cot\theta + \tan\theta = \tan\theta + \cot\theta$ ✓

2. $\cos^2\theta \overset{?}{=} (1 + \sin\theta)(1 - \sin\theta)$

$\cos^2\theta \overset{?}{=} 1 - \sin^2\theta$

$\cos^2\theta = \cos^2\theta$ ✓

3. $\sin\theta \overset{?}{=} \dfrac{\sec\theta}{\tan\theta + \cot\theta}$

$\sin\theta \overset{?}{=} \dfrac{\dfrac{1}{\cos\theta}}{\dfrac{\sin\theta}{\cos\theta} + \dfrac{\cos\theta}{\sin\theta}}$

$\sin\theta \overset{?}{=} \dfrac{\dfrac{1}{\cos\theta}}{\dfrac{\sin^2\theta + \cos^2\theta}{\cos\theta\sin\theta}}$

$\sin\theta \overset{?}{=} \dfrac{\dfrac{1}{\cos\theta}}{\dfrac{1}{\cos\theta\sin\theta}}$

$\sin\theta \overset{?}{=} \dfrac{1}{\cos\theta} \cdot \dfrac{\cos\theta\sin\theta}{1}$

$\sin\theta = \sin\theta$ ✓

4. $\tan^2\theta \overset{?}{=} \dfrac{1 - \cos^2\theta}{\cos^2\theta}$

$\tan^2\theta \overset{?}{=} \dfrac{\sin^2\theta}{\cos^2\theta}$

$\tan^2\theta = \tan^2\theta$ ✓

5. $\tan^2\theta\csc^2\theta \overset{?}{=} 1 + \tan^2\theta$

$\dfrac{\sin^2\theta}{\cos^2\theta} \cdot \dfrac{1}{\sin^2\theta} \overset{?}{=} \sec^2\theta$

$\dfrac{1}{\cos^2\theta} \overset{?}{=} \sec^2\theta$

$\sec^2\theta = \sec^2\theta$ ✓

6. $\tan^2\theta \overset{?}{=} (\sec\theta + 1)(\sec\theta - 1)$

$\tan^2\theta \overset{?}{=} \sin^2\theta - 1$

$\tan^2\theta = \tan^2\theta$ ✓

8. $\cos^2\theta + \tan^2\theta\cos^2\theta \overset{?}{=} 1$

$\cos^2\theta + \dfrac{\sin^2\theta}{\cos^2\theta} \cdot \cos^2\theta \overset{?}{=} 1$

$\cos^2\theta + \sin^2\theta \overset{?}{=} 1$

$1 = 1$ ✓

9. $\cot\theta(\cot\theta + \tan\theta) \overset{?}{=} \csc^2\theta$

$\cot^2\theta + \cot\theta\tan\theta \overset{?}{=} \csc^2\theta$

$\cot^2\theta + \dfrac{\sin\theta}{\cos\theta} \cdot \dfrac{\cos\theta}{\sin\theta} \overset{?}{=} \csc^2\theta$

$\cot^2\theta + 1 \overset{?}{=} \csc^2\theta$

$\csc^2\theta = \csc^2\theta$ ✓

10. $1 + \sec^2\theta\sin^2\theta \overset{?}{=} \sec^2\theta$

$1 + \dfrac{1}{\cos^2\theta} \cdot \sin^2\theta \overset{?}{=} \sec^2\theta$

$1 + \tan^2\theta \overset{?}{=} \sec^2\theta$

$\sec^2\theta = \sec^2\theta$ ✓

11. $\sin\theta\sec\theta\cot\theta \overset{?}{=} 1$

$\sin\theta \cdot \dfrac{1}{\cos\theta} \cdot \dfrac{\cos\theta}{\sin\theta} \overset{?}{=} 1$

$1 = 1$ ✓

12. $\dfrac{1 - \cos\theta}{1 + \cos\theta} \overset{?}{=} (\csc\theta - \cot\theta)^2$

$\dfrac{1 - \cos\theta}{1 + \cos\theta} \overset{?}{=} \csc^2\theta - 2\cot\theta\csc\theta + \cot^2\theta$

$\dfrac{1 - \cos\theta}{1 + \cos\theta} \overset{?}{=} \dfrac{1}{\sin^2\theta} - 2 \cdot \dfrac{\cos\theta}{\sin\theta} \cdot \dfrac{1}{\sin\theta} + \dfrac{\cos^2\theta}{\sin^2\theta}$

$\dfrac{1 - \cos\theta}{1 + \cos\theta} \overset{?}{=} \dfrac{1}{\sin^2\theta} - \dfrac{2\cos\theta}{\sin^2\theta} + \dfrac{\cos^2\theta}{\sin^2\theta}$

$\dfrac{1 - \cos\theta}{1 + \cos\theta} \overset{?}{=} \dfrac{1 - 2\cos\theta + \cos^2\theta}{\sin^2\theta}$

$\dfrac{1 - \cos\theta}{1 + \cos\theta} \overset{?}{=} \dfrac{(1 - \cos\theta)(1 - \cos\theta)}{1 - \cos^2\theta}$

$\dfrac{1 - \cos\theta}{1 + \cos\theta} \overset{?}{=} \dfrac{(1 - \cos\theta)(1 - \cos\theta)}{(1 - \cos\theta)(1 + \cos\theta)}$

$\dfrac{1 - \cos\theta}{1 + \cos\theta} = \dfrac{1 - \cos\theta}{1 + \cos\theta}$ ✓

13. $\dfrac{1 - 2\cos^2\theta}{\sin\theta\cos\theta} \overset{?}{=} \tan\theta - \cot\theta$

$\dfrac{(1 - \cos^2\theta) - \cos^2\theta}{\sin\theta\cos\theta} \overset{?}{=} \tan\theta - \cot\theta$

$\dfrac{\sin^2\theta - \cos^2\theta}{\sin\theta\cos\theta} \overset{?}{=} \tan\theta - \cot\theta$

$\dfrac{\sin^2\theta}{\sin\theta\cos\theta} - \dfrac{\cos^2\theta}{\sin\theta\cos\theta} \overset{?}{=} \tan\theta - \cot\theta$

$\dfrac{\sin\theta}{\cos\theta} - \dfrac{\cos\theta}{\sin\theta} \overset{?}{=} \tan\theta - \cot\theta$

$\tan\theta - \cot\theta = \tan\theta - \cot\theta$ ✓

14. $\tan\theta \overset{?}{=} \dfrac{\sec\theta}{\csc\theta}$

$\tan\theta \overset{?}{=} \dfrac{\dfrac{1}{\cos\theta}}{\dfrac{1}{\sin\theta}}$

$\tan\theta \overset{?}{=} \dfrac{\sin\theta}{\cos\theta}$

$\tan\theta = \tan\theta$ ✓

15. $\cos\theta \overset{?}{=} \sin\theta\cot\theta$

$\cos\theta \overset{?}{=} \sin\theta\left(\dfrac{\cos\theta}{\sin\theta}\right)$

$\cos\theta = \cos\theta$ ✓

16.

$$(\sin \theta - 1)(\tan \theta + \sec \theta) \overset{?}{=} -\cos \theta$$

$$\sin \theta \tan \theta + \sin \theta \sec \theta - \tan \theta - \sec \theta \overset{?}{=} -\cos \theta$$

$$\frac{\sin^2 \theta}{\cos \theta} + \frac{\sin \theta}{\cos \theta} - \frac{\sin \theta}{\cos \theta} - \frac{1}{\cos \theta} \overset{?}{=} -\cos \theta$$

$$\frac{\sin^2 \theta}{\cos \theta} - \frac{1}{\cos \theta} \overset{?}{=} -\cos \theta$$

$$\frac{\sin^2 \theta - 1}{\cos \theta} \overset{?}{=} -\cos \theta$$

$$\frac{-\cos^2 \theta}{\cos \theta} \overset{?}{=} -\cos \theta$$

$$-\cos \theta = -\cos \theta \checkmark$$

17. $\cos \theta \cos(-\theta) - \sin \theta \sin(-\theta) \overset{?}{=} 1$

$$\cos \theta \cos \theta - \sin \theta (-\sin \theta) \overset{?}{=} 1$$

$$\cos^2 \theta + \sin^2 \theta \overset{?}{=} 1$$

$$1 = 1 \checkmark$$

19. $\sec \theta - \tan \theta \overset{?}{=} \dfrac{1 - \sin \theta}{\cos \theta}$

$$\frac{1}{\cos \theta} - \frac{\sin \theta}{\cos \theta} \overset{?}{=} \frac{1 - \sin \theta}{\cos \theta}$$

$$\frac{1 - \sin \theta}{\cos \theta} = \frac{1 - \sin \theta}{\cos \theta} \checkmark$$

20.

$$\frac{1 + \tan \theta}{\sin \theta + \cos \theta} \overset{?}{=} \sec \theta$$

$$\frac{1 + \dfrac{\sin \theta}{\cos \theta}}{\sin \theta + \cos \theta} \overset{?}{=} \sec \theta$$

$$\frac{\dfrac{\cos \theta + \sin \theta}{\cos \theta}}{\sin \theta + \cos \theta} \overset{?}{=} \sec \theta$$

$$\frac{\cos \theta + \sin \theta}{\cos \theta} \cdot \frac{1}{\sin \theta + \cos \theta} \overset{?}{=} \sec \theta$$

$$\frac{1}{\cos \theta} \overset{?}{=} \sec \theta$$

$$\sec \theta = \sec \theta \checkmark$$

21. $\sec \theta \csc \theta \overset{?}{=} \tan \theta + \cot \theta$

$$\frac{1}{\cos \theta} \cdot \frac{1}{\sin \theta} \overset{?}{=} \frac{\sin \theta}{\cos \theta} + \frac{\cos \theta}{\sin \theta}$$

$$\frac{1}{\cos \theta \sin \theta} \overset{?}{=} \frac{\sin^2 \theta}{\sin \theta \cos \theta} + \frac{\cos^2 \theta}{\sin \theta \cos \theta}$$

$$\frac{1}{\cos \theta \sin \theta} \overset{?}{=} \frac{\sin^2 \theta + \cos^2 \theta}{\sin \theta \cos \theta}$$

$$\frac{1}{\cos \theta \sin \theta} = \frac{1}{\cos \theta \sin \theta} \checkmark$$

22. $\sin \theta + \cos \theta \overset{?}{=} \dfrac{2 \sin^2 \theta - 1}{\sin \theta - \cos \theta}$

$$\sin \theta + \cos \theta \overset{?}{=} \frac{2 \sin^2 \theta - (\sin^2 \theta + \cos^2 \theta)}{\sin \theta - \cos \theta}$$

$$\sin \theta + \cos \theta \overset{?}{=} \frac{\sin^2 \theta - \cos^2 \theta}{\sin \theta - \cos \theta}$$

$$\sin \theta + \cos \theta \overset{?}{=} \frac{(\sin \theta - \cos \theta)(\sin \theta - \cos \theta)}{\sin \theta - \cos \theta}$$

$$\sin \theta + \cos \theta = \sin \theta + \cos \theta \checkmark$$

23. $(\sin \theta + \cos \theta)^2 \overset{?}{=} \dfrac{2 + \sec \theta \csc \theta}{\sec \theta \csc \theta}$

$$(\sin \theta + \cos \theta)^2 \overset{?}{=} \frac{2 + \dfrac{1}{\cos \theta} \cdot \dfrac{1}{\sin \theta}}{\dfrac{1}{\cos \theta} \cdot \dfrac{1}{\sin \theta}}$$

$$(\sin \theta + \cos \theta)^2 \overset{?}{=} \left(2 + \frac{1}{\cos \theta \sin \theta}\right) \cdot \frac{\cos \theta \sin \theta}{1}$$

$$(\sin \theta + \cos \theta)^2 \overset{?}{=} 2 \cos \theta \sin \theta + 1$$

$$(\sin \theta + \cos \theta)^2 \overset{?}{=} 2 \cos \theta \sin \theta + \cos^2 \theta + \sin^2 \theta$$

$$(\sin \theta + \cos \theta)^2 = (\sin \theta + \cos \theta)^2 \checkmark$$

24. $\dfrac{\cos \theta}{1 - \sin \theta} \overset{?}{=} \dfrac{1 + \sin \theta}{\cos \theta}$

$$\frac{\cos \theta}{1 - \sin \theta} \overset{?}{=} \frac{1 + \sin \theta}{\cos \theta} \cdot \frac{1 - \sin \theta}{1 - \sin \theta}$$

$$\frac{\cos \theta}{1 - \sin \theta} \overset{?}{=} \frac{1 - \sin^2 \theta}{\cos \theta (1 - \sin \theta)}$$

$$\frac{\cos \theta}{1 - \sin \theta} \overset{?}{=} \frac{\cos^2 \theta}{\cos \theta (1 - \sin \theta)}$$

$$\frac{\cos \theta}{1 - \sin \theta} = \frac{\cos \theta}{1 - \sin \theta} \checkmark$$

25. $\csc \theta - 1 \overset{?}{=} \dfrac{\cot^2 \theta}{\csc \theta + 1}$

$$\csc \theta - 1 \overset{?}{=} \frac{\csc^2 \theta - 1}{\csc \theta + 1}$$

$$\csc \theta - 1 \overset{?}{=} \frac{(\csc \theta - 1)(\csc \theta + 1)}{\csc \theta + 1}$$

$$\csc \theta - 1 = \csc \theta - 1 \checkmark$$

26. $\cos \theta \cot \theta \overset{?}{=} \csc \theta - \sin \theta$

$$(\cos \theta) \frac{\cos \theta}{\sin \theta} \overset{?}{=} \frac{1}{\sin \theta} - \sin \theta$$

$$\frac{\cos^2 \theta}{\sin \theta} \overset{?}{=} \frac{1}{\sin \theta} - \frac{\sin \theta}{1} \cdot \frac{\sin \theta}{\sin \theta}$$

$$\frac{\cos^2 \theta}{\sin \theta} \overset{?}{=} \frac{1}{\sin \theta} - \frac{\sin^2 \theta}{\sin \theta}$$

$$\frac{\cos^2 \theta}{\sin \theta} \overset{?}{=} \frac{1 - \sin^2 \theta}{\sin \theta}$$

$$\frac{\cos^2 \theta}{\sin \theta} = \frac{\cos^2 \theta}{\sin \theta} \checkmark$$

27. $\sin \theta \cos \theta \tan \theta + \cos^2 \theta \overset{?}{=} 1$

$$\sin \theta \cos \theta \cdot \frac{\sin \theta}{\cos \theta} + \cos^2 \theta \overset{?}{=} 1$$

$$\sin^2 \theta + \cos^2 \theta \overset{?}{=} 1$$

$$1 = 1 \checkmark$$

28. $(\csc \theta - \cot \theta)^2 \overset{?}{=} \dfrac{1 - \cos \theta}{1 + \cos \theta}$

$\left(\dfrac{1}{\sin \theta} - \dfrac{\cos \theta}{\sin \theta}\right)^2 \overset{?}{=} \dfrac{1 - \cos \theta}{1 + \cos \theta}$

$\left(\dfrac{1 - \cos \theta}{\sin \theta}\right)^2 \overset{?}{=} \dfrac{1 - \cos \theta}{1 + \cos \theta}$

$\dfrac{(1 - \cos \theta)^2}{\sin^2 \theta} \overset{?}{=} \dfrac{1 - \cos \theta}{1 + \cos \theta}$

$\dfrac{(1 - \cos \theta)(1 - \cos \theta)}{\sin^2 \theta} \overset{?}{=} \dfrac{1 - \cos \theta}{1 + \cos \theta}$

$\dfrac{(1 - \cos \theta)(1 - \cos \theta)}{1 - \cos^2 \theta} \overset{?}{=} \dfrac{1 - \cos \theta}{1 + \cos \theta}$

$\dfrac{(1 - \cos \theta)(1 - \cos \theta)}{(1 - \cos \theta)(1 + \cos \theta)} \overset{?}{=} \dfrac{1 - \cos \theta}{1 + \cos \theta}$

$\dfrac{1 - \cos \theta}{1 + \cos \theta} = \dfrac{1 - \cos \theta}{1 + \cos \theta}$ ✓

29. $\csc^2 \theta \overset{?}{=} \cot^2 \theta + \sin \theta \csc \theta$

$\csc^2 \theta \overset{?}{=} \cot^2 \theta + \sin \theta \cdot \dfrac{1}{\sin \theta}$

$\csc^2 \theta \overset{?}{=} \cot^2 \theta + 1$

$\csc^2 \theta = \csc^2 \theta$ ✓

30. $\dfrac{\sec \theta - \csc \theta}{\csc \theta \sec \theta} \overset{?}{=} \sin \theta - \cos \theta$

$\dfrac{\sec \theta}{\csc \theta \sec \theta} - \dfrac{\csc \theta}{\csc \theta \sec \theta} \overset{?}{=} \sin \theta - \cos \theta$

$\dfrac{1}{\csc \theta} - \dfrac{1}{\sec \theta} \overset{?}{=} \sin \theta - \cos \theta$

$\sin \theta - \cos \theta = \sin \theta - \cos \theta$ ✓

31. $\sin^2 \theta + \cos^2 \theta \overset{?}{=} \sec^2 \theta - \tan^2 \theta$

$1 \overset{?}{=} \tan^2 \theta + 1 - \tan^2 \theta$

$1 = 1$ ✓

32. $\sec \theta - \cos \theta \overset{?}{=} \tan \theta \sin \theta$

$\dfrac{1}{\cos \theta} - \cos \theta \overset{?}{=} \tan \theta \sin \theta$

$\dfrac{1}{\cos \theta} - \dfrac{\cos^2 \theta}{\cos \theta} \overset{?}{=} \tan \theta \sin \theta$

$\dfrac{1 - \cos^2 \theta}{\cos \theta} \overset{?}{=} \tan \theta \sin \theta$

$\dfrac{\sin^2 \theta}{\cos \theta} \overset{?}{=} \tan \theta \sin \theta$

$\left(\dfrac{\sin \theta}{\cos \theta}\right) \sin \theta \overset{?}{=} \tan \theta \sin \theta$

$\tan \theta \sin \theta = \tan \theta \sin \theta$ ✓

51a.

Angle measure	Height
30°	28.2 m
45°	56.4 m
60°	84.5 m
90°	112.7 m

51b.

[0, 180] scl: 10 by [0, 150] scl: 10

51c. $\dfrac{v_0^2 \tan^2 \theta}{2g \sec^2 \theta} \overset{?}{=} \dfrac{v_0^2 \sin^2 \theta}{2g}$

$\dfrac{v_0^2 \left(\dfrac{\sin^2 \theta}{\cos^2 \theta}\right)}{2g \left(\dfrac{1}{\cos^2 \theta}\right)} \overset{?}{=} \dfrac{v_0^2 \sin^2 \theta}{2g}$

$\dfrac{v_0^2 \sin^2 \theta}{2g} = \dfrac{v_0^2 \sin^2 \theta}{2g}$ ✓

59. Using the unit circle and the Pythagorean Theorem, we can justify $\cos^2 \theta + \sin^2 \theta = 1$.

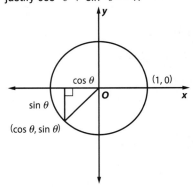

If we divide each term of the identity $\cos^2 \theta + \sin^2 \theta = 1$ by $\cos^2 \theta$, we can justify $1 + \tan^2 \theta = \sec^2 \theta$.

$\dfrac{\cos^2 \theta}{\cos^2 \theta} + \dfrac{\sin^2 \theta}{\cos^2 \theta} = \dfrac{1}{\cos^2 \theta}$

$1 + \tan^2 \theta = \sec^2 \theta$

If we divide each term of the identity $\cos^2 \theta + \sin^2 \theta = 1$ by $\sin^2 \theta$, we can justify $\cot^2 \theta + 1 = \csc^2 \theta$.

$\dfrac{\cos^2 \theta}{\sin^2 \theta} + \dfrac{\sin^2 \theta}{\sin^2 \theta} = \dfrac{1}{\sin^2 \theta}$

$\cot^2 \theta + 1 = \csc^2 \theta$

Lesson 13-3

30. $F = \dfrac{W(\sin A + \mu \cos A)}{\cos A - \mu \sin A}$

$= \dfrac{W(\sin A + \tan \theta \cos A)}{\cos A - \tan \theta \sin A}$

$= \dfrac{W\left(\dfrac{\sin A}{\cos A} + \dfrac{\tan \theta \cos A}{\cos A}\right)}{\dfrac{\cos A}{\cos A} - \dfrac{\tan \theta \sin A}{\cos A}}$

$= \dfrac{W(\tan A + \tan \theta)}{1 - \tan A \tan \theta}$

$= W \tan (A + \theta)$

33b.

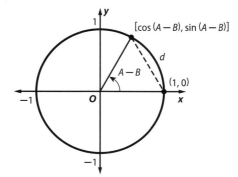

33c. No; a counterexample is: $\cos(30° + 45°) = \cos 30° + \cos 45°$, which equals $\dfrac{\sqrt{3}}{2} + \dfrac{\sqrt{2}}{2}$ or about 1.5731. Since a cosine value cannot be greater than 1, this statement must be false.

34. $\sin(A + B) \overset{?}{=} \dfrac{\tan A + \tan B}{\sec A \sec B}$

$\sin(A + B) \overset{?}{=} \dfrac{\dfrac{\sin A}{\cos A} + \dfrac{\sin B}{\cos B}}{\dfrac{1}{\cos A} \cdot \dfrac{1}{\cos B}}$

$\sin(A + B) \overset{?}{=} \dfrac{\dfrac{\sin A}{\cos A} + \dfrac{\sin B}{\cos B}}{\dfrac{1}{\cos A} \cdot \dfrac{1}{\cos B}} \cdot \dfrac{\cos A \cos B}{\cos A \cos B}$

$\sin(A + B) \overset{?}{=} \dfrac{\sin A \cos B + \cos A \sin B}{1}$

$\sin(A + B) = \sin(A + B) \checkmark$

35. $\cos(A + B) \overset{?}{=} \dfrac{1 - \tan A \tan B}{\sec A \sec B}$

$\cos(A + B) \overset{?}{=} \dfrac{1 - \dfrac{\sin A}{\cos A} \cdot \dfrac{\sin B}{\cos B}}{\dfrac{1}{\cos A} \cdot \dfrac{1}{\cos B}}$

$\cos(A + B) \overset{?}{=} \dfrac{1 - \dfrac{\sin A}{\cos A} \cdot \dfrac{\sin B}{\cos B}}{\dfrac{1}{\cos A} \cdot \dfrac{1}{\cos B}} \cdot \dfrac{\cos A \cos B}{\cos A \cos B}$

$\cos(A + B) \overset{?}{=} \dfrac{\cos A \cos B - \sin A \sin B}{1}$

$\cos(A + B) = \cos(A + B) \checkmark$

36. $\sec(A - B) \overset{?}{=} \dfrac{\sec A \sec B}{1 + \tan A \tan B}$

$\sec(A - B) \overset{?}{=} \dfrac{\dfrac{1}{\cos A} \cdot \dfrac{1}{\cos B}}{1 + \dfrac{\sin A}{\cos A} \cdot \dfrac{\sin B}{\cos B}}$

$\sec(A - B) \overset{?}{=} \dfrac{\dfrac{1}{\cos A} \cdot \dfrac{1}{\cos B}}{1 + \dfrac{\sin A}{\cos A} \cdot \dfrac{\sin B}{\cos B}} \cdot \dfrac{\cos A \cos B}{\cos A \cos B}$

$\sec(A - B) \overset{?}{=} \dfrac{1}{\cos A \cos B + \sin A \sin B}$

$\sec(A - B) \overset{?}{=} \dfrac{1}{\cos(A - B)}$

$\sec(A - B) = \sec(A - B) \checkmark$

37. $\sin(A + B)\sin(A - B) \overset{?}{=} \sin^2 A - \sin^2 B$

$(\sin A \cos B + \cos A \sin B)$
$\quad(\sin A \cos B - \cos A \sin B) \overset{?}{=} \sin^2 A - \sin^2 B$

$(\sin A \cos B)^2 - (\cos A \sin B)^2 \overset{?}{=} \sin^2 A - \sin^2 B$

$\sin^2 B \cos^2 B - \cos^2 A \sin^2 B \overset{?}{=} \sin^2 A - \sin^2 B$

$\sin^2 A \cos^2 B + \sin^2 A \sin^2 B -$
$\quad \sin^2 A \sin^2 B - \cos^2 A \sin^2 B \overset{?}{=} \sin^2 A - \sin^2 B$

$\sin^2 A(\cos^2 B + \sin^2 B) - \sin^2 B(\sin^2 A + \cos^2 A) \overset{?}{=} \sin^2 A - \sin^2 B$

$(\sin^2 A)(1) - (\sin^2 B)(1) \overset{?}{=} \sin^2 A - \sin^2 B$

$\sin^2 A - \sin^2 B = \sin^2 A - \sin^2 B \checkmark$

40. $\cot(A + B) = \dfrac{1}{\tan(A + B)}$

$= \dfrac{1}{\dfrac{\tan A + \tan B}{1 - \tan A \tan B}}$

$= \dfrac{1 - \tan A \tan B}{\tan A + \tan B}$

$= \dfrac{1 - \dfrac{1}{\cot A} \cdot \dfrac{1}{\cot B}}{\dfrac{1}{\cot A} + \dfrac{1}{\cot B}} \cdot \dfrac{\cot A \cot B}{\cot A \cot B}$

$= \dfrac{\cot A \cot B - 1}{\cot A + \cot B}$

41. $d = \sqrt{(\cos A - \cos B)^2 + (\sin A - \sin B)^2}$

$d^2 = (\cos A - \cos B)^2 + (\sin A - \sin B)^2$

$d^2 = (\cos^2 A - 2\cos A \cos B + \cos^2 B) + (\sin^2 A - 2\sin A \sin B + \sin^2 B)$

$d^2 = \cos^2 A + \sin^2 A + \cos^2 B + \sin^2 B - 2\cos A \cos B - 2\sin A \sin B$

$d^2 = 1 + 1 - 2\cos A \cos B - 2\sin A \sin B$

$d^2 = 2 - 2\cos A \cos B - 2\sin A \sin B$

Now find the value of d^2 when the angle having measure $A - B$ is in standard position on the unit circle, as shown in the figure above.

$$d = \sqrt{[\cos(A-B)-1]^2 + [\sin(A-B)-0]^2}$$

$$d^2 = [\cos(A-B)-1]^2 + [\sin(A-B)-0]^2$$

$$= [\cos^2(A-B) - 2\cos(A-B) + 1] + \sin^2(A-B)$$

$$= \cos^2(A-B) + \sin^2(A-B) - 2\cos(A-B) + 1$$

$$= 1 - 2\cos(A-B) + 1$$

$$= 2 - 2\cos(A-B)$$

52b.

53. Step 1: $4^1 - 1 = 3$, which is divisible by 3. The statement is true for $n = 1$.

Step 2: Assume that $4^k - 1$ is divisible by 3 for some positive integer k. This means that $4^k - 1 = 3r$ for some whole number r.

Step 3: $4^k - 1 = 3r$

$$4^k = 3r + 1$$

$$4^{k+1} = 12r + 4$$

$$4^{k+1} - 1 = 12r + 3$$

$$4^{k+1} - 1 = 3(4r + 1)$$

Since r is a whole number, $4r + 1$ is a whole number. Thus, $4^{k+1} - 1$ is divisible by 3, so the statement is true for $n = k + 1$. Therefore, $4^n - 1$ is divisible by 3 for all positive integers n.

54. Step 1: $5^1 + 3 = 8$, which is divisible by 4. The statement is true for $n = 1$.

Step 2: Assume that $5^k + 3$ is divisible by 4 for some positive integer k. This means that $5^k + 3 = 4r$ for some positive integer r.

Step 3: $5^k + 3 = 4r$

$$5^k = 4r - 3$$

$$5^{k+1} = 20r - 15$$

$$5^{k+1} + 3 = 20r - 12$$

$$5^{k+1} + 3 = 4(5r - 3)$$

Since r is a positive integer, $5r - 3$ is a positive integer. Thus, $5^{k+1} + 3$ is divisible by 4, so the statement is true for $n = k + 1$. Therefore, $5^n + 3$ is divisible by 4 for all positive integers n.

Mid-Chapter Quiz

11. $\sec^2\theta + 1 \overset{?}{=} \dfrac{\cot\theta}{\cos\theta \cdot \sin\theta}$

$\csc^2\theta \overset{?}{=} \dfrac{\cos\theta}{\cos\theta \cdot \sin\theta \cdot \sin\theta}$

$\csc^2\theta \overset{?}{=} \dfrac{1}{\sin^2\theta}$

$\csc^2\theta = \csc^2\theta$ ✓

12. $\dfrac{\cos\theta \csc\theta}{\cot\theta} \overset{?}{=} 1$

$\dfrac{\cos\theta}{\sin\theta} \cdot \tan\theta \overset{?}{=} 1$

$\dfrac{\cos\theta}{\sin\theta} \cdot \dfrac{\sin\theta}{\cos\theta} \overset{?}{=} 1$

$1 = 1$ ✓

13. $\dfrac{\sin\theta \tan\theta}{1 - \cos\theta} \overset{?}{=} (1 + \cos\theta)\sec\theta$

$\dfrac{\sin\theta\left(\dfrac{\sin\theta}{\cos\theta}\right)}{1 - \cos\theta} \overset{?}{=} (1 + \cos\theta)\dfrac{1}{\cos\theta}$

$\dfrac{\dfrac{\sin^2\theta}{\cos\theta}}{1 - \cos\theta} \overset{?}{=} \dfrac{1}{\cos\theta} + 1$

$\dfrac{\sin^2\theta}{\cos\theta} \cdot \dfrac{1}{1 - \cos\theta} \overset{?}{=} \dfrac{1}{\cos\theta} + 1$

$\dfrac{\sin^2\theta}{\cos\theta(1 - \cos\theta)} \overset{?}{=} \dfrac{1}{\cos\theta} + 1$

$\dfrac{1 - \cos^2\theta}{\cos\theta(1 - \cos\theta)} \overset{?}{=} \dfrac{1}{\cos\theta} + 1$

$\dfrac{(1 - \cos\theta)(1 + \cos\theta)}{\cos\theta(1 - \cos\theta)} \overset{?}{=} \dfrac{1}{\cos\theta} + 1$

$\dfrac{1 + \cos\theta}{\cos\theta} \overset{?}{=} \dfrac{1}{\cos\theta} + 1$

$\dfrac{1}{\cos\theta} + 1 = \dfrac{1}{\cos\theta} + 1$ ✓

14. $\tan\theta(1 - \sin\theta) \overset{?}{=} \dfrac{\cos\theta \sin\theta}{1 + \sin\theta}$

$\tan\theta(1 - \sin\theta) \overset{?}{=} \dfrac{\cos\theta \sin\theta}{1 + \sin\theta} \cdot \dfrac{1 - \sin\theta}{1 - \sin\theta}$

$\tan\theta(1 - \sin\theta) \overset{?}{=} \dfrac{\cos\theta \sin\theta(1 - \sin\theta)}{1 - \sin^2\theta}$

$\tan\theta(1 - \sin\theta) \overset{?}{=} \dfrac{\cos\theta \sin\theta(1 - \sin\theta)}{\cos^2\theta}$

$\tan\theta(1 - \sin\theta) \overset{?}{=} \dfrac{\sin\theta(1 - \sin\theta)}{\cos\theta}$

$\tan\theta(1 - \sin\theta) \overset{?}{=} \dfrac{\sin\theta}{\cos\theta} \cdot (1 - \sin\theta)$

$\tan\theta(1 - \sin\theta) = \tan\theta(1 - \sin\theta)$ ✓

15b. $\cot\theta = \dfrac{12}{9}$; $\dfrac{\cos\theta}{\sin\theta} = \dfrac{\frac{12}{15}}{\frac{9}{15}} = \dfrac{12}{9}$, so $\dfrac{12}{9} = \dfrac{12}{9}$

16. $\tan^2\theta + 1 \overset{?}{=} \dfrac{\tan\theta}{\cos\theta \cdot \sin\theta}$

$\sec^2\theta \overset{?}{=} \dfrac{\tan\theta}{\cos\theta \cdot \sin\theta}$

$\sec^2\theta \overset{?}{=} \dfrac{\frac{\sin\theta}{\cos\theta}}{\cos\theta \cdot \sin\theta}$

$\sec^2\theta \overset{?}{=} \dfrac{\sin\theta}{\cos^2\theta \cdot \sin\theta}$

$\sec^2\theta \overset{?}{=} \dfrac{1}{\cos^2\theta}$

$\sec^2\theta = \sec^2\theta$ ✓

17.

$$\frac{\sin\theta \cdot \sec\theta}{\sec\theta - 1} \overset{?}{=} (\sec\theta + 1)\cot\theta$$

$$\frac{\sec\theta + 1}{\sec\theta + 1} \cdot \frac{\sin\theta \cdot \sec\theta}{\sec\theta - 1} \overset{?}{=} (\sec\theta + 1)\cot\theta$$

$$\frac{\sin\theta \cdot \sec\theta(\sec\theta + 1)}{\sec^2\theta - 1} \overset{?}{=} (\sec\theta + 1)\cot\theta$$

$$\frac{\sin\theta \cdot \dfrac{1}{\cos\theta}(\sec\theta + 1)}{\tan^2\theta} \overset{?}{=} (\sec\theta + 1)\cot\theta$$

$$\frac{\dfrac{\sin\theta}{\cos\theta}(\sec\theta + 1)}{\tan^2\theta} \overset{?}{=} (\sec\theta + 1)\cot\theta$$

$$\frac{\tan\theta(\sec\theta + 1)}{\tan^2\theta} \overset{?}{=} (\sec\theta + 1)\cot\theta$$

$$\frac{\sec\theta + 1}{\tan\theta} \overset{?}{=} (\sec\theta + 1)\cot\theta$$

$$\frac{\sec\theta + 1}{1} \cdot \frac{1}{\tan\theta} \overset{?}{=} (\sec\theta + 1)\cot\theta$$

$$(\sec\theta + 1)\cot\theta = (\sec\theta + 1)\cot\theta \checkmark$$

18. $\sin^2\theta \cdot \tan^2\theta \overset{?}{=} \tan^2\theta - \sin^2\theta$

$$\sin^2\theta \cdot \tan^2\theta \overset{?}{=} \frac{\sin^2\theta}{\cos^2\theta} - \sin^2\theta$$

$$\sin^2\theta \cdot \tan^2\theta \overset{?}{=} \frac{\sin^2\theta - \sin^2\theta\cos^2\theta}{\cos^2\theta}$$

$$\sin^2\theta \cdot \tan^2\theta \overset{?}{=} \frac{\sin^2\theta(1 - \cos^2\theta)}{\cos^2\theta}$$

$$\sin^2\theta \cdot \tan^2\theta \overset{?}{=} \frac{\sin^2\theta(\sin^2\theta)}{\cos^2\theta}$$

$$\sin^2\theta \cdot \tan^2\theta \overset{?}{=} \sin^2\theta \cdot \frac{\sin^2\theta}{\cos^2\theta}$$

$$\sin^2\theta \cdot \tan^2\theta = \sin^2\theta \cdot \tan^2\theta \checkmark$$

19. $\cot\theta(1 - \cos\theta) \overset{?}{=} \dfrac{\cos\theta \cdot \sin\theta}{1 + \cos\theta}$

$$\cot\theta(1 - \cos\theta) \overset{?}{=} \frac{\cos\theta \cdot \sin\theta}{1 + \cos\theta} \cdot \frac{1 - \cos\theta}{1 - \cos\theta}$$

$$\cot\theta(1 - \cos\theta) \overset{?}{=} \frac{\cos\theta \cdot \sin\theta(1 - \cos\theta)}{1 - \cos^2\theta}$$

$$\cot\theta(1 - \cos\theta) \overset{?}{=} \frac{\cos\theta \cdot \sin\theta(1 - \cos\theta)}{\sin^2\theta}$$

$$\cot\theta(1 - \cos\theta) \overset{?}{=} \frac{\cos\theta(1 - \cos\theta)}{\sin\theta}$$

$$\cot\theta(1 - \cos\theta) \overset{?}{=} \frac{\cos\theta}{\sin\theta}(1 - \cos\theta)$$

$$\cot\theta(1 - \cos\theta) = \cot\theta(1 - \cos\theta) \checkmark$$

25. $\cos 30°\cos\theta + \sin 30°\sin\theta \overset{?}{=} \sin 60°\cos\theta + \cos 60°\sin\theta$

$$\frac{\sqrt{3}}{2}\cos\theta + \frac{1}{2}\sin\theta = \frac{\sqrt{3}}{2}\cos\theta + \frac{1}{2}\sin\theta \checkmark$$

26. $\tan 2\theta \overset{?}{=} \dfrac{2}{\cot\theta - \tan\theta}$

$$\tan 2\theta \overset{?}{=} \frac{2}{\cot\theta - \tan\theta} \cdot \frac{\tan\theta}{\tan\theta}$$

$$\tan 2\theta \overset{?}{=} \frac{2\tan\theta}{\cot\theta\tan\theta - \tan^2\theta}$$

$$\tan 2\theta \overset{?}{=} \frac{2\tan\theta}{1 - \tan^2\theta}$$

$$\tan 2\theta = \tan 2\theta \checkmark$$

27. $1 + \dfrac{1}{2}\sin 2\theta \overset{?}{=} \dfrac{\sec\theta + \sin\theta}{\sec\theta}$

$$\overset{?}{=} \frac{\dfrac{1}{\cos\theta} + \sin\theta}{\dfrac{1}{\cos\theta}}$$

$$\overset{?}{=} \frac{\dfrac{1}{\cos\theta} + \sin\theta}{\dfrac{1}{\cos\theta}} \cdot \frac{\cos\theta}{\cos\theta}$$

$$\overset{?}{=} 1 + \frac{1}{2} \cdot 2\sin\theta\cos\theta$$

$$= 1 + \frac{1}{2}\sin 2\theta \checkmark$$

28. $\sin\dfrac{\theta}{2}\cos\dfrac{\theta}{2} \overset{?}{=} \dfrac{\sin\theta}{2}$

$$\frac{2\sin\dfrac{\theta}{2}\cos\dfrac{\theta}{2}}{2} \overset{?}{=} \frac{\sin\theta}{2}$$

$$\frac{\sin 2\left(\dfrac{\theta}{2}\right)}{2} \overset{?}{=} \frac{\sin\theta}{2}$$

$$\frac{\sin\theta}{2} = \frac{\sin\theta}{2} \checkmark$$

29. $\tan\dfrac{\theta}{2} \overset{?}{=} \dfrac{\sin\theta}{1 + \cos\theta}$

$$\tan\frac{\theta}{2} \overset{?}{=} \frac{\sin 2\left(\dfrac{\theta}{2}\right)}{1 + \cos 2\left(\dfrac{\theta}{2}\right)}$$

$$\tan\frac{\theta}{2} \overset{?}{=} \frac{2\sin\dfrac{\theta}{2}\cos\dfrac{\theta}{2}}{1 + 2\cos^2\dfrac{\theta}{2} - 1}$$

$$\tan\frac{\theta}{2} \overset{?}{=} \frac{2\sin\dfrac{\theta}{2}\cos\dfrac{\theta}{2}}{2\cos^2\dfrac{\theta}{2}}$$

$$\tan\frac{\theta}{2} \overset{?}{=} \frac{\sin\dfrac{\theta}{2}}{\cos\dfrac{\theta}{2}}$$

$$\tan\frac{\theta}{2} = \tan\frac{\theta}{2} \checkmark$$

30. For $\theta = 45° + \alpha$,

$d = \dfrac{v^2 \sin 2\,(45° + \alpha)}{g}$

$= \dfrac{v^2 \sin (90° + 2\alpha)}{g}$

$= \dfrac{v^2(\sin 90° \cos 2\alpha + \cos 90° \sin 2\alpha)}{g}$

$= \dfrac{v^2(1 \cdot \cos 2\alpha + 0 \cdot \sin 2\alpha)}{g}$

$= \dfrac{v^2 \cos 2\alpha}{g}$

For $\theta = 45° - \alpha$,

$d = \dfrac{v^2 \sin 2\,(45° - \alpha)}{g}$

$= \dfrac{v^2 \sin (90° - 2\alpha)}{g}$

$= \dfrac{v^2(\sin 90° \cos 2\alpha - \cos 90° \sin 2\alpha)}{g}$

$= \dfrac{v^2(1 \cdot \cos 2\alpha - 0 \cdot \sin 2\alpha)}{g}$

$= \dfrac{v^2 \cos 2\alpha}{g}$

37. No; Teresa incorrectly added the square roots, and Nathan used the half-angle identity incorrectly. He used sin 30° in the formula instead of first finding the cosine.

39. If you are only given the value of cos θ, then $\cos 2\theta = 2 \cos^2 \theta - 1$ is the best identity to use. If you are only given the value of sin θ, then $\cos 2\theta = 1 - 2 \sin^2 \theta$ is the best identity to use. If you are given the values of both cos θ and sin θ, then $\cos 2\theta = \cos^2 \theta - \sin^2 \theta$ works just as well as the other two.

40. $\sin 2\theta = \sin(\theta + \theta)$
$\qquad = \sin \theta \cos \theta + \cos \theta \sin \theta$
$\qquad = 2 \sin \theta \cos \theta$
$\cos 2\theta = \cos(\theta + \theta)$
$\qquad = \cos \theta \cos \theta - \sin \theta \sin \theta$
$\qquad = \cos^2 \theta - \sin^2 \theta$

You can find alternate forms for cos 2θ by making substitutions into the expression $\cos^2 \theta - \sin^2 \theta$.

$\cos^2 \theta - \sin^2 \theta = (1 - \sin^2 \theta) - \sin^2 \theta$
$\qquad\qquad\quad = 1 - 2 \sin^2 \theta$
Substitute $1 - \sin^2 \theta$ for $\cos^2 \theta$.
Simplify.

$\cos^2 \theta - \sin^2 \theta = \cos^2 \theta - (1 - \cos^2 \theta)$
$\qquad\qquad\quad = 2 \cos^2 \theta - 1$
Substitute $1 - \cos^2 \theta$ for $\sin^2 \theta$.
Simplify.

41.

$1 - 2 \sin^2 \theta = \cos 2\theta$	Double-angle identity
$1 - 2 \sin^2 \dfrac{A}{2} = \cos A$	Substitute $\dfrac{A}{2}$ for θ and A for 2θ.
$\sin^2 \dfrac{A}{2} = \dfrac{1 - \cos A}{2}$	Solve for $\sin^2 \dfrac{A}{2}$.
$\sin \dfrac{A}{2} = \pm \sqrt{\dfrac{1 - \cos A}{2}}$	Take the square root of each side.

Find $\cos \dfrac{A}{2}$.

$2 \cos^2 \theta - 1 = \cos 2\theta$	Double-angle identity
$2 \cos^2 \dfrac{A}{2} - 1 = \cos A$	Substitute $\dfrac{A}{2}$ for θ and A for 2θ.
$\cos^2 \dfrac{A}{2} = \dfrac{1 + \cos A}{2}$	Solve for $\cos^2 \dfrac{A}{2}$.
$\cos \dfrac{A}{2} = \pm \sqrt{\dfrac{1 + \cos A}{2}}$	Take the square root of each side.

Lesson 13-5

58a.

θ	$\sin \theta$	θ	$\sin \theta$
0°	1	210°	$-\dfrac{1}{2}$
30°	$\dfrac{1}{2}$	225°	$-\dfrac{\sqrt{2}}{2}$
45°	$\dfrac{\sqrt{2}}{2}$	240°	$-\dfrac{\sqrt{3}}{2}$
60°	$\dfrac{\sqrt{3}}{2}$	270°	-1
90°	1	300°	$-\dfrac{\sqrt{3}}{2}$
120°	$\dfrac{\sqrt{3}}{2}$	315°	$-\dfrac{\sqrt{2}}{2}$
135°	$\dfrac{\sqrt{2}}{2}$	330°	$-\dfrac{1}{2}$
150°	$\dfrac{1}{2}$	360°	0
180°	0		

58b. The graph of $y = \sin \theta$ is above the graph of $y = \dfrac{1}{2}$ for $30° < \theta < 150°$.

58c. The graph of $y = \sin \theta$ is above the graph of $y = \dfrac{1}{2}$ for $30° < \theta < 150°$ and the period of sin θ is $[-360°, 360°]$, so the solutions of $\sin \theta > \dfrac{1}{2}$ are $30° + k \cdot 360° < \theta < 150° + k \cdot 360°$.

Chapter 13 Answer Appendix

58d. i. $0° \leq \theta \leq 45°$ and $315° \leq \theta \leq 360°$; $0° + k \cdot 360° \leq \theta \leq 45° + k \cdot 360°$ and $315° + k \cdot 360° \leq \theta \leq 360° + k \cdot 360°$; ii. $0° \leq \theta \leq 60°$ and $120° \leq \theta \leq 360°$; $0° + k \cdot 360° \leq \theta \leq 60° + k \cdot 360°$ and $120° + k \cdot 360° \leq \theta \leq 360° + k \cdot 360°$; iii. $180° \leq \theta \leq 360°$; $180° + k \cdot 360° \leq \theta \leq 360° + k \cdot 360°$; iv. $60° \leq \theta \leq 300°$; $60° + k \cdot 360° \leq \theta \leq 300° + k \cdot 360°$

Study Guide and Review

30.
$$\sin(\theta + 90) \stackrel{?}{=} \cos\theta$$
$$\sin\theta\cos 90° + \cos\theta\sin 90° \stackrel{?}{=} \cos\theta$$
$$(\sin\theta)(0) + (\cos\theta)(1) \stackrel{?}{=} \cos\theta$$
$$\cos\theta = \cos\theta \checkmark$$

31.
$$\sin\left(\frac{3\pi}{2} - \theta\right) \stackrel{?}{=} -\cos\theta$$
$$\sin\frac{3\pi}{2}\cos\theta - \cos\frac{3\pi}{2}\sin\theta \stackrel{?}{=} -\cos\theta$$
$$(-1)\cos\theta - (0)\sin\theta \stackrel{?}{=} -\cos\theta$$
$$-\cos\theta = -\cos\theta \checkmark$$

32.
$$\tan(\theta - \pi) \stackrel{?}{=} \tan\theta$$
$$\frac{\tan\theta - \tan\pi}{1 + \tan\theta\tan\pi} \stackrel{?}{=} \tan\theta$$
$$\frac{\tan\theta - 0}{1 + (\tan\theta)(0)} \stackrel{?}{=} \tan\theta$$
$$\frac{\tan\theta}{1} \stackrel{?}{=} \tan\theta$$
$$\tan\theta = \tan\theta \checkmark$$

33. $\sin 2\theta = \frac{24}{25}$, $\cos 2\theta = \frac{7}{25}$, $\sin\frac{\theta}{2} = \frac{\sqrt{10}}{10}$, and $\cos\frac{\theta}{2} = \frac{3\sqrt{10}}{10}$

34. $\sin 2\theta = \frac{\sqrt{15}}{8}$, $\cos 2\theta = \frac{7}{8}$, $\sin\frac{\theta}{2} = \frac{\sqrt{2}\sqrt{4 + \sqrt{15}}}{4}$, and $\cos\frac{\theta}{2} = -\frac{\sqrt{2}\sqrt{4 - \sqrt{15}}}{4}$

35. $\sin 2\theta = -\frac{4\sqrt{5}}{9}$, $\cos 2\theta = -\frac{1}{9}$, $\sin\frac{\theta}{2} = \frac{\sqrt{30}}{6}$, and $\cos\frac{\theta}{2} = \frac{\sqrt{6}}{6}$

36a. $c^2 = 90^2 + 90^2$; $c^2 = 8100 + 8100$; $c^2 = 16{,}200$; $c = 90\sqrt{2}$

36b. $\sin 45° = \frac{90}{90\sqrt{2}} = \frac{1}{\sqrt{2}} = \frac{\sqrt{2}}{2}$

36c. $\sin\frac{\theta}{2} = \pm\sqrt{\frac{1 - \cos\theta}{2}}$; $\sin\frac{90}{2} = \pm\sqrt{\frac{1 - \cos 90}{2}}$;
$$\sin\frac{90}{2} = \pm\sqrt{\frac{1 - 0}{2}} = \frac{1}{\sqrt{2}} = \frac{\sqrt{2}}{2}$$

Practice Test

14.
$$\frac{\sin\theta}{1 - \cos\theta} \stackrel{?}{=} \cos\theta + \cot\theta$$
$$\frac{\sin\theta}{1 - \cos\theta} \stackrel{?}{=} \frac{1}{\sin\theta} + \frac{\cos\theta}{\sin\theta}$$
$$\frac{\sin\theta}{1 - \cos\theta} \stackrel{?}{=} \frac{1 + \cos\theta}{\sin\theta}$$
$$\frac{\sin\theta}{1 - \cos\theta} \stackrel{?}{=} \frac{1 + \cos\theta}{\sin\theta} \cdot \frac{1 - \cos\theta}{1 - \cos\theta}$$
$$\frac{\sin\theta}{1 - \cos\theta} \stackrel{?}{=} \frac{1 - \cos^2\theta}{\sin\theta(1 - \cos\theta)}$$
$$\frac{\sin\theta}{1 - \cos\theta} \stackrel{?}{=} \frac{\sin^2\theta}{\sin\theta(1 - \cos\theta)}$$
$$\frac{\sin\theta}{1 - \cos\theta} \stackrel{?}{=} \frac{\sin\theta}{1 - \cos\theta} \checkmark$$

16a. $18^2 = 9^2 + a^2$; $324 = 81 + a^2$; $243 = a^2$; $a = 9\sqrt{3}$

16b. $\sin 2\theta = 2\sin\theta\cos\theta$;
$$\sin 2(30°) = 2\sin 30°\cos 30°;$$
$$\sin 60° = 2\left(\frac{9}{18}\right)\left(\frac{9\sqrt{3}}{18}\right) = \frac{162\sqrt{3}}{324} = \frac{\sqrt{3}}{2};$$
$$\sin 60° = 9\frac{\sqrt{3}}{18} = \frac{\sqrt{3}}{2}$$

21. $R = \frac{v^2}{g}\sin 2\theta$; $25 = \frac{20^2}{9.8}\sin 2\theta$; $0.6125 = \sin 2\theta$; $\theta \approx 18.9°$

Student Handbook

This **Student Handbook** can help you answer these questions.

What if I Need More Practice?

Extra Practice   R1

The **Extra Practice** section provides additional problems for each lesson so you have ample opportunity to practice new skills.

What if I Need to Check a Homework Answer?

Selected Answers and Solutions   R14

The answers to odd-numbered problems are included in **Selected Answers and Solutions**.

What if I Forget a Vocabulary Word?

Glossary/Glosario   R124

The **English-Spanish Glossary** provides definitions and page numbers of important or difficult words used throughout the textbook.

What if I Need to Find Something Quickly?

Index   R149

The **Index** alphabetically lists the subjects covered throughout the entire textbook and the pages on which each subject can be found.

What if I Forget a Formula?

Formulas,   **Inside Back Cover**
Symbols, and Parent Functions

Inside the back cover of your math book is a list of **Formulas and Symbols** that are used in the book.

Rubberball/Getty Images

Extra Practice

CHAPTER 1   Equations and Inequalities

Evaluate each expression if $x = 2.2$, $y = -4$, and $z = 5$. (Lesson 1-1)

1. $5x + z(2y - 3)$ -44

2. $(6 - y)^2 - 3$ 97

3. $z^3 - 7x(y + 6)$ 94.2

4. $3y(2z + 5x) - 7$ -259

Evaluate each expression if $a = \dfrac{3}{5}$, $b = -1.4$, and $c = 3$. (Lesson 1-1)

5. $\dfrac{a - b}{a + b}$ -2.5

6. $\dfrac{b^2 + a}{ac} \approx 1.42$

7. INTEREST The formula $A = P(1 + r)^t$ can be used to calculate the value of an investment A after t years that starts with a principal p and has an interest rate r compounded annually. If Jack deposits $3500 into an investment with an interest rate of 5%, how much interest will he earn in 3 years? (Lesson 1-1) $551.69

Simplify each expression. (Lesson 1-2)

8. $7(3x - 4y) + 2(4x + 3y)$ $29x - 22y$

9. $-2(5a + 3b) - 8(2a - b)$ $2b - 26a$

10. $3(-6c - 5d) - 4(9c - 8d)$ $17d - 54c$

11. FOOD The menu for a fast food restaurant is shown in the table below. (Lesson 1-2)

Item	Cost
cheeseburger	$1.19
French fries	$1.79
milkshake	$2.09

Meiko buys 2 cheeseburgers, 2 fries, and 2 milkshakes. (Lesson 1-2)

a. Illustrate the Distributive Property by writing two expressions to represent the cost of the food. $2(1.19 + 1.79 + 2.09)$ or $2(1.19) + 2(1.79) + 2(2.09)$

b. Use the Distributive Property to find out how much the restaurant receives from selling this food. $10.14

Solve each equation. Check your solution. (Lesson 1-3)

12. $a - 21 = 42$ 63

13. $2x + 5 = -13$ -9

14. $4(c - 5) - 2(3c + 5) = 6$ -18

Solve each equation. Check your solutions. (Lesson 1-4)

15. $|x + 11| = 21$ $\{-32, 10\}$

16. $4|y - 7| = -3$ $\varnothing$

17. $-3|2a - 5| = -9$ $\{1, 4\}$

Solve each inequality. Then graph the solution set on a number line. (Lesson 1-5)

18–21. See Extra Practice Answer Appendix for graphs.

18. $d - 6 > -3$ $d > 3$

19. $2c + 7 \geq 15$ $c \geq 4$

20. $3x + 8 \leq 5$ $x \leq -1$

21. $21 < -3(2y + 1)$ $y < -4$

22. TAXI A taxi's rates are shown below. (Lesson 1-5)

Initial fare	$2.00
Mileage charge	$1.75

a. Shannon wants to keep his fare under $35. How many miles can he travel in the taxi? Shannon can travel no more than 18.8 miles.

b. Samantha has $21 for her taxi fare. If she plans on traveling 20 miles in the taxi, does she have enough money to cover the fare? Explain your answer. Samantha's fare would be $37, so she does not have enough money to cover the fare.

Solve each inequality. Graph the solution set on a number line. (Lesson 1-6)

23. $-1 < d + 9 < 7$

24. $|y| < -5$ $\varnothing$

23. $|d| - 10 < d < -2)$

23–24. See Extra Practice Answer Appendix for graphs.

25. PARTY Mrs. Jackson is buying pizzas for her son's birthday party. She wants to spend at least $40 but no more than $50 for the pizzas. The pizza prices are shown below. (Lesson 1-6)

Size	Cost
medium	$8.50
large	$11.50

Mrs. Jackson can buy 2 or 3 medium pizzas.

If she buys 2 large pizzas, how many medium pizzas can she buy to stay within her range?

connectED.mcgraw-hill.com

Extra Practice

Extra Practice

CHAPTER 2 — Linear Relations and Functions

State the domain and range of each relation. Then determine whether each relation is a function. If it is a function, determine if it is one-to-one, onto, both, or neither. (Lesson 2-1)

1.
x	y
-2	4
-1	1
2	4
2	6

2. (mapping diagram)
0.5, 1, 2, 2.5 → 2, 3, 5, 6

1. D = {-2, -1, 2}, R = {4, 1, 6}, not a function
2. D = {0.5, 1, 2, 2.5}, R = {2, 3, 5, 6}; function; both

Find the x-intercept and y-intercept of the graph of each equation. Then graph the equation using the intercepts. (Lesson 2-2)

3. $y = 3x - 9$ 3; -9
4. $2y - 5x = 10$ -2; 5
5. $4x + 5y = -20$ -5; -4
6. $-2y + 3x = -12$ -4; 6

3–6. See Extra Practice Answer Appendix for graphs.

7. **REPAIR** An auto mechanic charges an initial fee of $25 plus an hourly fee of $35. (Lesson 2-2)
   a. Write an equation to represent the situation. $y = 35x + 25$
   b. How much did it cost Stacy if the mechanic fixed her car in 3.5 hours? $147.50

Find the slope of the line that passes through each pair of points. (Lesson 2-3)

8. (1, -7), (-3, 5) -3
9. (-1, 8), (3, -12) -5
10. (-2, -15), (4, 9) 4

11. **ANIMALS** The table shows the weight in pounds of a Chihuahua for the first four weeks after its birth. (Lesson 2-3)

Time (wk)	Weight (lb)
0	2.0
1	2.5
2	3.25
3	4.7
4	5.6

   a. Find the average rate of change in the weight of the puppy between weeks 1 and 3. 1.1 lb/wk
   b. Find the average rate of change in the weight of the puppy between weeks 0 and 4. 0.9 lb/wk

Write an equation of the line passing through each pair of points. (Lesson 2-4)

12. (-3, -14), (1, -2) $y = 3x - 5$
13. (-2, 11), (2, 3) $y = -2x + 7$
14. (-4, 2), (2, 5) $y = 0.5x + 4$

15. **TELEVISION** The table shows the number of LCD televisions sold in recent years at an electronics store. (Lesson 2-5)

Year	Number of TVs Sold
2006	733
2007	838
2008	943
2009	1048
2015	?

   a. Using x as the number of years since 2000, make a scatter plot, graph a line of fit, and describe the correlation. See Extra Practice Answer Appendix.
   b. Use two ordered pairs to write a prediction equation. Sample answer using (7, 838) and (9, 1048); $y = 105x + 103$
   c. Use your prediction equation to predict the missing value. 1678

16–18. See Extra Practice Answer Appendix for graphs. **Graph each function. Identify the domain and range.** (Lesson 2-6)

16. $g(x) = |-2x|$ 16. D = {all real numbers}; R = {g(x) | g(x) ≥ 0}
17. $f(x) = [[x - 3]]$ 17. D = {all real numbers}; R = {all integers}
18. $h(x) = 3|x| + 2$ 18. D = {all real numbers}; R = {h(x) | h(x) ≥ 2}

Describe the translation in each function. Then graph the function. (Lesson 2-7)

19. $y = x^2 - 3$
20. $y = |x| + 5$
21. $y = (x + 2)^2$
22. $y = |x - 4|$

19–22. See Extra Practice Answer Appendix.

Graph each inequality. (Lesson 2-8)

23. $y > -3$ 23–25. See Extra Practice Answer Appendix.
24. $y < -2$
25. $2x - 5y \geq 10$

CHAPTER 3 — Systems of Equations and Inequalities

Solve each system of equations by using substitution or elimination. (Lesson 3-1)

1. $y = 3x + 4$
   $y = -2x - 6$ (-2, -2)
2. $y = -4x - 5$
   $y - 5x = -14$ (1, -9)
3. $2a - 3b = -5$
   $2a + 3b = 13$ (2, 3)
4. $-5a + 2b = 16$
   $4a - 5b = -6$ (-4, -2)

Solve each system of inequalities by graphing. (Lesson 3-2) 5–6. See Extra Practice Answer Appendix.

5. $y > 3x + 1$
   $y < -2x - 1$
6. $y \leq -x - 2$
   $y \leq 4x + 1$

7. **COLLEGE** The total score on a college entrance exam for the math and verbal sections is 2000. The college requires a math score of at least 875 and a verbal score of at least 800. (Lesson 3-2)
   a. Write and graph a system of inequalities to represent this situation. See Extra Practice Answer Appendix.
   b. Give two examples of acceptable math and verbal scores for entrance to the college. Sample answer: (1000, 900) or (1100, 850)

Graph each system of inequalities. Name the coordinates of the vertices of the feasible region. Find the maximum and minimum value of the given function for this region. (Lesson 3-3) 8–9. See Extra Practice Answer Appendix.

8. $x \geq -2$
   $y \leq 4$
   $y \geq x - 1$
   $f(x, y) = 2x - y$
9. $y \geq -3$
   $x \leq 5$
   $y \leq 2x + 4$
   $f(x, y) = -4x + 5y$

Solve each system of equations. (Lesson 3-4)

10. $2x - 3y + z = -4$
    $x + 2y - 3z = -9$
    $4x - y + 2z = -3$ (-2, 1, 3)
11. $-3x + 4y - 2z = -14$
    $2x - 3y - 4z = 6$
    $-5x - 2y + 3z = 35$ (-4, -6; 1)

12. **BUSINESS** A car dealership is combining two of their lots. Each entry in the matrices represents the number of new car models of each color. Express the total number of new car models of each color after the lots are combined as a matrix. (Lesson 3-5) See Extra Practice Answer Appendix.

Lot 1 = $\begin{bmatrix} 1 & 5 & 7 \\ 8 & 4 & 3 \\ 6 & 5 & 2 \end{bmatrix}$, Lot 2 = $\begin{bmatrix} 7 & 5 & 3 \\ 6 & 2 & 5 \\ 6 & 7 & 5 \end{bmatrix}$

Find each product, if possible. (Lesson 3-6)

$A = \begin{bmatrix} -2 & 4 \\ 3 & -5 \end{bmatrix}$ $B = \begin{bmatrix} -1 & 2 & 6 \\ -4 & -2 & 5 \end{bmatrix}$

$C = \begin{bmatrix} -7 & 8 & 4 \\ -2 & -5 & 6 \\ 4 & 9 & 1 \end{bmatrix}$ $D = \begin{bmatrix} 1 & 6 \\ -7 & 5 \\ -5 & 4 \\ 2 & -2 \end{bmatrix}$

13. AD undefined
14. CD undefined
15. AB $\begin{bmatrix} -14 & -12 & 8 \\ 17 & 16 & -7 \end{bmatrix}$
16. BC $\begin{bmatrix} 27 & 36 & 14 \\ 52 & 23 & -23 \end{bmatrix}$

Evaluate each determinant. (Lesson 3-7)

17. $\begin{vmatrix} 3 & 4 \\ -5 & -3 \end{vmatrix}$ 11
18. $\begin{vmatrix} -2 & 6 & 0 \\ -1 & 3 & 2 \\ 4 & -1 & 5 \end{vmatrix}$ 44

Use Cramer's Rule to solve each system of equations. (Lesson 3-7)

19. $2x - 3y = 18$
    $5x + 7y = -13$ (3, -4)
20. $-3x - 2y = -5$
    $-4x + 6y = 28$ (-1, 4)
21. $x + 5y - z = 14$
    $2x - y + z = -8$
    $3x + 2y + z = -1$ (-2, 3, -1)
22. $x + y + z = -3$
    $3x - 2y = 16$
    $2x - y + 3z = -5$ (4, -2, -5)

23. **EVENT** At a high school football game, the concession stand sells hot chocolate for $1.75 and sodas for $1.25. At the last game, 175 drinks were sold for a total of $243.75. Use Cramer's Rule to determine how many of each type of drink were sold. (Lesson 3-7) soda: 125, hot chocolate: 50

Use a matrix equation to solve each system of equations. (Lesson 3-8)

24. $5x - 3y = -9$
    $-4x - 6y = 24$ (-3, -2)
25. $-2x + 4y = 4$
    $6x - 5y = -19$ (-4, -1)
26. $2x + 3y = 6$
    $6x + 9y = -10$ no solution

CHAPTER 4 — Quadratic Functions and Relations

Complete parts a–c for each quadratic function.
a. Find the *y*-intercept, the equation of the axis of symmetry, and the *x*-coordinate of the vertex.
b. Make a table of values that includes the vertex.
c. Use this information to graph the function.
(Lesson 4-1) 1–2. See Extra Practice Answer Appendix.

1. $f(x) = x^2 - 3$
2. $f(x) = 2x^2 - 5x$

Use the related graph of each equation to determine its solutions. (Lesson 4-2)

3. $x^2 + x - 6 = 0$ $-3, 2$
4. $x^2 - 2x - 3 = 0$ $-1, 3$

Solve each equation by factoring. (Lesson 4-3)

6. $2x^2 + x - 10 = 0$ $-\frac{5}{2}, 2$
7. $2x^2 + x = 28$ $-4, \frac{7}{2}$

8. **GAMES** Julio constructed a platform for a bean bag toss game. The plans for the original platform had dimensions of 3 feet by 5 feet. He made his platform larger by adding *x* feet to each side. The area of the new platform is 35 square feet. (Lesson 4-3)

a. Write a quadratic equation that represents the area of his platform. $x^2 + 8x + 15 = 35$

b. Find the dimensions of the platform Julio made. 5 ft by 7 ft

Simplify. (Lesson 4-4)

9. $(10 + 3i) + (3 - 7i)$ $13 - 4i$
10. $(2 + i)(2 - i)$ 5
11. $\dfrac{5}{1 + 3i}$ $\dfrac{1}{2} - \dfrac{3}{2}i$

Solve each equation by completing the square. (Lesson 4-5)

12. $x^2 - 2x - 24 = 0$ $-4, 6$
13. $x^2 - 2x = 35$ $-5, 7$
14. $5x^2 - 4x + 1 = 0$ $\dfrac{2}{5} + \dfrac{2}{5}i, \dfrac{2}{5} - \dfrac{2}{5}i$
15. $3x^2 + 6x + 10 = 0$ $\dfrac{-3 + i\sqrt{21}}{3}, \dfrac{-3 - i\sqrt{21}}{3}$

Solve each equation by using the Quadratic Formula. (Lesson 4-6)

16. $2x^2 - 3x - 9 = 0$ $-\dfrac{3}{2}, 3$
17. $4x^2 + 2x - 1 = 0$ $\dfrac{-1 \pm \sqrt{5}}{4}$
18. $3x^2 + 4x - 2 = 0$ $\dfrac{-2 \pm \sqrt{10}}{3}$
19. $3x^2 - 5x - 10 = 0$ $\dfrac{5 \pm \sqrt{145}}{6}$

Write an equation in vertex form for each parabola. (Lesson 4-7)

20. $y = (x - 3)^2 + 2$

21. $y = \dfrac{1}{2}(x + 1)^2 - 3$

22. **VIDEO GAMES** The daily sales of a video game company can be modeled by the function $S(p) = -2p^2 + 80p + 1250$, where $S(p)$ is the amount of sales and *p* is the price per game. (Lesson 4-7)

a. Write the function in vertex form. $S(p) = -2(p - 20)^2 + 2050$

b. What is the maximum amount of daily sales the company can expect to make? $2050

Solve each inequality algebraically. (Lesson 4-8)

23. $x^2 + 3x < 18$ $|x| -6 < x < 3|$
24. $x^2 - 3x \le 28$ $|x| -4 \le x \le 7|$
25. $2x^2 - 13x \ge -20$ $\left\{x \mid x \le \dfrac{5}{2} \text{ or } x \ge 4\right\}$

5. ROCKET A science teacher is conducting an experiment with a model rocket. The function $h = -16t^2 + 64t$ models the height of the rocket in feet, where *h* is the height of the rocket in feet and *t* is the time in seconds after it is launched. Determine how many seconds the rocket is in the air. (Lesson 4-2) 4 seconds

CHAPTER 5 — Polynomials and Polynomial Functions

Simplify. (Lesson 5-1)

1. $(5x^2 + 3x - 7) - (3x^2 - x + 4)$ $2x^2 + 4x - 11$
2. $(2y^2 - 4y - 5) + (3y^2 - 2y + 1)$ $5y^2 - 6y - 4$
3. $2cd(3c - 4d) + 4d(c + 2d)$ $6c^2d - 8cd^2 + 4cd + 8d^2$
4. $(r - s)(r + s)(3r - 2s)$ $3r^3 - 2r^2s - 3rs^2 + 2s^3$

Simplify. (Lesson 5-2)

5. $\dfrac{4x^3y^4 - 10xy^5}{2xy}$ $2x^2y^3 - 5y^4$
6. $(x^3 + 4x^2 - 4x - 7) \div (x + 1)$ $x^2 + 3x - 7$

7. **SHOES** The number of shoes sold by a store can be modeled by $-s^2 + 12s$, where *s* is the number of employees. Find the average number of shoes sold by each worker. (Lesson 5-2) $-s + 12$

Find $p(-2)$ and $p(4)$ for each function. (Lesson 5-3)

8. $p(x) = -2x^2 + 5x - 3$ $-21, -15$
9. $p(x) = -3x^3 - x^2 + 2x - 8$ $8, -208$
10. $p(x) = x^4 - 5x^3 + 4x + 6$ $54, -42$

11. **CELL PHONES** The yearly sales of cell phones can be modeled by the function $s(t) = 0.45t^4 + 5.4t^3 - 15t^2 - 300t + 1000$, where $s(t)$ is the annual sales in millions of dollars and *t* is the number of years after 2005. (Lesson 5-4) a–d. See Extra Practice Answer Appendix.

a. Graph the function for $0 \le t \le 10$.

b. Describe the turning points of the graph and its end behavior.

c. What trends in cell phone sales does this graph suggest?

d. Is it reasonable that the trend will continue indefinitely? Explain.

Factor completely. If the polynomial is not factorable, write *prime*. (Lesson 5-5)

12. $6x^4 - 5y^7$ prime
13. $27n^3 - 125d^3$ $(3a - 5d)(9a^2 + 15ad + 25d^2)$
14. $2ax^2 - 3bx^2 + cx^2 - 2ay^2 + 3by^2 - cy^2$ $(x - y)(x + y)(2a - 3b + c)$

Given a polynomial and one of its factors, find the remaining factors of the polynomial. (Lesson 5-6)

15. $x^3 - 2x^2 - 5x + 6$; $x - 1$ $(x + 2)(x - 3)$
16. $2x^3 - x^2 - 25x - 12$; $x + 3$ $(2x + 1)(x - 4)$
17. $6x^3 + 11x^2 + x - 4$; $2x - 1$ $(3x + 4)(x + 1)$

Find all the zeros of each function. (Lesson 5-7)

18. $f(x) = x^3 - 4x^2 - 7x + 10$ $-2, 1, 5$
19. $f(x) = x^4 - 8x^2 - 9$ $-3, 3, i, -i$
20. $f(x) = x^4 + 3x^2 - 4$ $-1, 1, 2i, -2i$

21. **BUSINESS** The profit in hundreds of dollars for selling *c* calculators per day can be modeled by $R(c) = -0.005c^4 + 0.25c^3 + 0.01c^2 - 2.5c + 100$. (Lesson 5-7)

a. How many positive real zeros, negative real zeros, and imaginary zeros exist? 3 or 1, 1, 0 or 2

b. What is the meaning of the zeros in this situation? Sample answer: Nonnegative zeros represent the number of calculators that need to be sold to break even.

Find all the rational zeros of each function. (Lesson 5-8)

22. $f(x) = 20x^3 - 29x^2 - 25x + 6$ $-\dfrac{3}{4}, \dfrac{1}{5}, 2$
23. $f(x) = 16x^3 - 44x^2 + 4x + 15$ $-\dfrac{1}{2}, \dfrac{5}{2}, \dfrac{3}{4}$
24. $f(x) = 4x^4 - 81x^2 - 110x + 75$ $-3, -\dfrac{5}{2}, \dfrac{5}{2}$

25. Refer to the graph below. Find all the zeros of $f(x) = x^3 - 5x^2 - 12x + 36$. (Lesson 5-8) $-3, 2, 6$

CHAPTER 6   Inverses and Radical Functions and Relations

Find $(f + g)(x)$, $(f - g)(x)$, $(f \cdot g)(x)$, and $\left(\dfrac{f}{g}\right)(x)$ for each $f(x)$ and $g(x)$. (Lesson 6-1)

1. $f(x) = x - 4$
$g(x) = 2x + 3$

2. $f(x) = -x + 3$
$g(x) = x^2 - 4$

1–2. See Extra Practice Answer Appendix.

Find $[f \circ g](x)$ and $[g \circ f](x)$, if they exist.

3. $f(x) = 3x + 2$
$g(x) = x - 3$

4. $f(x) = 3x^2$
$g(x) = x^2 + 5$

3. $[f \circ g](x) = 3x - 7$; $[g \circ f](x) = 3x - 1$
4. $[f \circ g](x) = 3x^4 + 30x^2 + 75$; $[g \circ f](x) = 9x^4 + 5$
Determine whether each pair of functions are inverse functions. Write yes or no. (Lesson 6-2)

5. $f(x) = 2x - 8$
$g(x) = \dfrac{1}{2}x + 4$ **yes**

6. $f(x) = 3x - 4$
$g(x) = 3x + 7$ **no**

7. GEOMETRY The formula for the volume of a sphere is $V = \dfrac{4}{3}\pi r^3$, where r is the radius of the sphere. (Lesson 6-2)

a. Find the inverse of the function. $\sqrt[3]{\dfrac{3V}{4\pi}}$

b. Explain what purpose $V^{-1}(x)$ serves. **The inverse function can be used to find the radius when the volume is known.**

Identify the domain and range of each function. (Lesson 6-3)

8. $f(x) = \sqrt{2x + 4}$ **D** $= \{x \mid x \geq 0\}$; **R** $= \{f(x) \mid f(x) \geq 4\}$

9. $f(x) = \sqrt{x - 1} - 2$
D $= \{x \mid x \geq 1\}$; **R** $= \{f(x) \mid f(x) \geq -2\}$

10. $f(x) = -3\sqrt{x + 2}$
D $= \{x \mid x \geq -2\}$; **R** $= \{f(x) \mid f(x) \leq 0\}$

Simplify. (Lesson 6-4)

11. $\sqrt{36a^2b^4}$ $6|a|b^2$

12. $\sqrt[3]{343x^6y^9}$ $7x^2y^3$

13. $\sqrt[5]{s^{10}t^{15}}$ s^2t^3

Simplify. (Lesson 6-5)

14. $4\sqrt{3x} \cdot 5\sqrt{12x}$ $120|x|$

15. $\sqrt{36} + 2\sqrt{32} - \sqrt{50}$ $3\sqrt{2} + 6$

16. $\dfrac{2}{\sqrt{5} - 3}$ $\dfrac{-\sqrt{5} - 3}{2}$

---

17. GEOMETRY Find the exact perimeter or circumference of each shape. (Lesson 6-5)

a.
$\sqrt{32}$ $14\sqrt{2}$
$\sqrt{18}$

b.
$\sqrt{48}$ $8\pi\sqrt{3}$

Simplify each expression. (Lesson 6-6)

18. $a^{\frac{1}{4}} \cdot a^{\frac{2}{3}}$ $a^{\frac{11}{12}}$

19. $\sqrt[6]{125x^9}$ $x\sqrt[6]{125x^3}$

20. $\dfrac{a^{\frac{1}{2}} - 3}{a^{\frac{1}{2}} + 3}$ $\dfrac{a - 6a^{\frac{1}{2}} + 9}{a - 9}$

21. HERON'S FORMULA The area A of a triangle with sides a, b, and c is given by $A = \sqrt{s(s - a)(s - b)(s - c)}$ where s is the semiperimeter of the triangle. Determine the exact area of the triangle below. (Lesson 6-6)

$8\sqrt{182}$ in^2

22 in.   18 in.   12 in.

Solve each equation. (Lesson 6-7)

22. $\sqrt{5x + 14} - 2 = 6$ 10

23. $\sqrt{4x + 15} = \sqrt{2x + 9}$ -3

24. $\sqrt{x + 7} = \sqrt{x} - 1$ **no real solution**

25. WIND The wind speed s in miles per hour can be modeled by the formula $s = 150\sqrt{\dfrac{P}{a}}$, where a is the area of a sail in square feet and p is the pressure of the wind on the sail in pounds per square feet. Determine the area of a sail if the wind speed is 75 mph and the pressure is 20 pounds per square feet. (Lesson 6-7) **80 ft^2**

CHAPTER 7   Exponential and Logarithmic Functions and Relations

Graph each function. State the domain and range. (Lesson 7-1) **1–2. See Extra Practice Answer Appendix for graphs.**

1. $f(x) = 3(4)^x$ **D** $= \{$all real numbers$\}$;
R $= \{f(x) \mid f(x) > 0\}$

2. $f(x) = 2^{3x} - 3$ **D** $= \{$all real numbers$\}$;
R $= \{f(x) \mid f(x) > -3\}$

Solve each equation. (Lesson 7-2)

3. $2^{x-1} = 8^{x+3}$ -5

4. $5^{2x+12} = 25^{10x-12}$ 2

5. BIOLOGY During an experiment, the number of cells of a virus can be modeled by $f(t) = 2^{t-2}$, where t is the time in days and $f(t)$ is the number of cells. Determine how many days have passed if there are 64 virus cells. (Lesson 7-2) **8 days**

Evaluate each expression. (Lesson 7-3)

6. $\log_8 64$ 2

7. $\log_7 1$ 0

8. $\log_5 \dfrac{1}{25}$ -2

9. DOLPHINS The number of dolphins living in an ocean region after t months can be approximated by $n(t) = 920 \log_{10}(t - 1)$. (Lesson 7-3)

a. How many dolphins are in the ocean region after 4 months? ≈ 439

b. How many dolphins are in the ocean region after 2 years? ≈ 1253

Solve each equation. (Lesson 7-4)

10. $\log_{25} x = \dfrac{3}{2}$ 125

11. $\log_9 81 = x$ 2

12. $\log_6 (5x - 3) = \log_6 (x + 9)$ 3

13. $\log_8 (x^2 + 6) = \log_8 (5x)$ 2 or 3

Solve each equation. Check your solution. (Lesson 7-5)

14. $\log_4 3 + \log_4 x = \log_4 12$ 4

15. $\log_7 6 - \log_7 2 = \log_7 x$ 3

16. $4 \log_2 x = \log_2 81$ 3

17. $\log_6 x + \log_6 (x + 5) = 2$ 4

---

18. POPULATION The wolf population per year at a national park is listed in the table below. (Lesson 7-5)

Year	Population
2005	420
2006	399
2007	379
2008	360
2009	342
2020	?

a. Determine the annual percent of decrease of the population. **5%**

b. Write a logarithmic function for the time in years based upon population from 2005.

c. How many wolves will there be in the national park in 2020? ≈ 195 **wolves**

18b. $T = \log_{0.95}\left(\dfrac{p}{420}\right)$

Solve each equation. Round to the nearest ten-thousandth. (Lesson 7-6)

19. $5^x = 60$ 2.5440

20. $4^{x^2} = 21$ 1.4819

21. $3^{x-2} = 4^x$ -7.6377

Solve each equation. Round to the nearest ten-thousandth. (Lesson 7-7)

22. $4e^x - 6 = 11$ 1.4469

23. $3e^{-x} + 4 = 17$ -1.4663

24. $2e^{3x} - 8 = 21$ 0.8914

25. INTEREST Stefanie deposited $2000 into a savings account paying 2.5% interest compounded continuously. (Lesson 7-8)

a. Determine how long it will take Stefanie to have a balance of $2400. **7.3 yr**

b. How long will it take Stefanie to double her investment? **27.7 yr**

c. If Stefanie wanted to have $5000 after 10 years, how much should she have invested? **$3894**

connectED.mcgraw-hill.com

CHAPTER 9 Conic Sections

Find the distance between each pair of points with the given coordinates. (Lesson 9-1)

1. (1, 3), (−4, 5)
5.385 units

2. (−2, 4), (6, 1)
8.544 units

3. (−2½, −2), (−6, 8)
10.770 units

4. (−3, −3), (4, −5)
7.280 units

Write each equation in standard form. Identify the vertex, axis of symmetry, and direction of opening of the parabola. (Lesson 9-2)

5. $y = x^2 + 6x + 7$

6. $y = 2x^2 - 8x + 12$
5–6. See Extra Practice Answer Appendix.

7. TRACK A long jumper makes a leap of 24 feet, and his feet reach a maximum height of 6 feet. A parabola on the coordinate plane with the focus at the origin models the event. What is the equation of the parabolic path of the jump? (Lesson 9-2) $y = -\frac{1}{24}x^2 + 6$

Write an equation for each circle given the endpoints of a diameter. (Lesson 9-3)

8. (1, 3), (−4, 5) $(x + 1.5)^2 + (y - 4)^2 = 7.25$

9. (−2, 4), (6, 1) $(x - 2)^2 + (y - 2.5)^2 = 18.25$

10. GEARS The endpoints of the diameter of a gear of a bicycle are (−2, 5) and (2, −5). Write an equation to represent a two-dimensional sketch of the gear. (Lesson 9-3) $x^2 + y^2 = 29$

Write an equation for each ellipse. (Lesson 9-4)

11.
$\frac{x^2}{16} + \frac{y^2}{81} = 1$

12.
$\frac{(x - 3)^2}{25} + \frac{(y + 2)^2}{9} = 1$

Write an equation for each hyperbola. (Lesson 9-5)

13.
$\frac{y^2}{9} - \frac{x^2}{16} = 1$

14.
$\frac{(x - 1)^2}{81} - \frac{(y + 3)^2}{144} = 1$

Without writing in standard form, state whether the graph of each equation is a *parabola, circle, ellipse,* or *hyperbola.* (Lesson 9-6)

15. $8y^2 - 16x + 12y - 20 = 0$ parabola

16. $9x^2 - 18x + 12y - 15y^2 + 10xy + 12 = 0$ hyperbola

17. $2y^2 + 8x^2 - 6xy + 8x - 10y - 12 = 0$ ellipse

18. TUNNEL A tunnel through a mountain has an elliptical opening that is 100 feet wide and 48 feet high. Write the equation of the ellipse in standard form. (Lesson 9-6) $\frac{x^2}{2500} + \frac{y^2}{2576} = 1$

Solve each system of equations. (Lesson 9-7)

19. $x^2 - y^2 = 21$
$y^2 + x^2 = 29$ (−5, −2), (−5, 2), (5, −2), (5, 2)

20. $y = -3x + 1$
$y - x^2 = 23 - 8x$ no solution

21. $y = x^2 - 2x - 3$
$y = 2x - 3$ (0, −3), (4, 5)

connectED.mcgraw-hill.com

CHAPTER 8 Rational Functions and Relations

Simplify each expression. (Lesson 8-1)

1. $\frac{(x - 2)(x + 5)}{x^2 + 2x - 8}$ $\frac{x + 5}{x + 4}$

2. $\frac{(x + 3)(x - 4)}{x^2 + x - 20}$ $\frac{x + 3}{x + 5}$

3. $\frac{x^2 - 16}{x^2 - 8x + 12} \cdot \frac{x^2 + 3x - 18}{x^2 - 7x + 12}$ $\frac{(x + 4)(x + 6)}{(x - 6)(x - 2)}$

4. $\frac{x^2 + 10x + 21}{x^2 - 5x + 6} \div \frac{x^2 + 3x - 28}{x^2 - 8x + 15}$ $\frac{(x + 3)(x - 5)}{(x - 2)(x - 4)}$

5. BUSINESS The average hourly income I of a college student x years after graduation can be modeled by the function $I(x) = \frac{11x^2 + 33x + 22}{x + 2}$. Simplify the function. (Lesson 8-1) $11(x + 1)$

Simplify each expression. (Lesson 8-2)

6. $\frac{2}{x^2 - 7x + 12} + \frac{6}{x^2 + x - 20}$ $\frac{8(x - 1)}{(x - 3)(x - 4)(x + 5)}$

7. $\frac{4}{x^2 - 11x + 30} - \frac{3}{x^2 - 13x + 40}$ $\frac{x - 14}{(x - 5)(x - 6)(x - 8)}$

Identify the asymptotes, domain, and range of each function. (Lesson 8-3)

8.
$x = -2, y = 1;$
$D = \{x \mid x \neq -2\};$
$R = \{y \mid y \neq 1\}$

9.
$x = 0, y = 2;$
$D = \{x \mid x \neq 0\};$
$R = \{y \mid y \neq 2\}$

Identify the asymptotes, domain, and range of each function. (Lesson 8-3)

10.
$x = -4, y = -4;$
$D = \{x \mid x \neq -4\};$
$R = \{y \mid y \neq -4\}$

11b. In this situation, the range and domain are limited to real numbers greater than zero because negative numbers do not make sense.

11. FIELD TRIP The American history class is planning a field trip to Washington, D.C. It costs $300 to rent the bus, and each ticket to the museum costs $12.

a. If C represents the cost for each student and n represents the number of students, write an equation to represent the cost to each student as a function of how many students go on the field trip. $C(n) = \frac{12n + 300}{n}$

b. Are there any limitations to the range or domain in this situation? Explain.

Graph each function. (Lesson 8-4)

12. $f(x) = \frac{2x}{x - 3}$

13. $f(x) = \frac{x - 4}{x + 2}$
12–13. See Extra Practice Answer Appendix.
If a varies jointly as b and c, find a when $b = 2$ and $c = -3$, given each of the following. (Lesson 8-5)

14. $a = -12$ when $b = 4$ and $c = 6$ 3

15. $a = -36$ when $b = 12$ and $c = -4$ −4.5

Solve each equation. Check your solution. (Lesson 8-6)

16. $\frac{3}{x + 4} + \frac{1}{x - 1} = \frac{17}{24}$ 4, $\frac{23}{17}$

17. $\frac{12}{x + 4} - \frac{24}{x - 3} = \frac{24}{x^2 + x - 12}$ 8

18. $\frac{14}{x - 5} - \frac{7}{x + 5} = \frac{-7}{x^2 - 25}$ −16

19. EXERCISE Natalie can bike 10 kilometers per hour faster than Aaron. By the time Natalie travels 60 kilometers, Aaron has gone 40 kilometers. (Lesson 8-6)

a. Write an expression for Natalie's time. $t = \frac{60}{r + 10}$

b. Write an expression for Aaron's time. $t = \frac{40}{r}$

c. Write and solve the rational equation to determine the speed of each biker.
$\frac{60}{r + 10} = \frac{40}{r}$; Natalie: 30 km/h, Aaron 20 km/h

connectED.mcgraw-hill.com R8–R9

Extra Practice

CHAPTER 10 Sequences and Series

Determine whether each sequence is *arithmetic*, *geometric*, or *neither*. Explain your reasoning. (Lesson 10-1)

1. 3, 6, 9, 12, ... Arithmetic; the common difference is 3.

2. 4, 12, 36, 108, ... Geometric; the common ratio is 3.

3. −1, 5, −25, 125, ... Geometric; the common ratio is −5.

4. 2, 5, 9, 12, ... Neither; there is no common difference or ratio.

Write an equation for the nth term of each arithmetic sequence. (Lesson 10-2)

5. 3, −12, −27, ... $a_n = -15n + 18$

6. 6, 13, 20, ... $a_n = 7n − 1$

7. GAMES Susie is playing a game in which she has to stack cups as shown in the diagram below. (Lesson 10-2) **36**

How many cups does she need to be able to stack the cups 8 rows high?

Write an equation for the nth term of each geometric sequence. (Lesson 10-3)

8. 0.75, 3, 12, ... $a_n = 0.75(4)^{n−1}$

9. 0.4, −2, 10, ... $a_n = 0.4(−5)^{n−1}$

Find the sum of each infinite series, if it exists. (Lesson 10-4)

10. 8 + 14 + 24.5 + ... No sum exists.

11. $\frac{1}{3} + \frac{1}{9} + \frac{1}{27} + \dots \frac{1}{2}$

12. 20 + 15 + 11.25 + ... 80

13. −5 + (−12) + (−28.8) + ... No sum exists.

R10 | Extra Practice

14. MONEY Dustin is saving money from his summer job to buy a gaming system. He saved one dollar the first week, and each following week he saved twice as much as the week before. Dustin saved enough money to buy the gaming system after 8 weeks. How much did the system cost? (Lesson 10-3) **$255**

Find the first five terms of each sequence described. (Lesson 10-5)

15. $a_1 = 6, a_{n+1} = 4a_n + 3$ 6, 27, 111, 447, 1791

16. $a_1 = 11, a_{n+1} = 3a_n − n$ 11, 32, 94, 279, 833

17. $a_1 = 4, a_2 = x, a_n = 2a_{n−2} − 3a_{n−1}$
4, x, 8 − 3x, 11x − 24, −39x + 88

Write a recursive formula for each sequence. (Lesson 10-5)

18. 3, 7, 11, 15, ... $a_n = a_{n−1} + 4, a_1 = 3$

19. $a_3 = 96$ and $r = 2$ $a_n = 2a_{n−1}, a_1 = 24$

20. 8, 16, 32, 64, ... $a_n = 2a_{n−1}, a_1 = 8$

21. $a_5 = −4$ and $d = −2$ $a_n = a_{n−1} − 2, a_1 = 4$

Find the indicated term of each expression. (Lesson 10-6)

22. third term of $(a + 2b)^6$ $60a^4b^2$

23. fourth term of $(y − 4x)^7$ $−2240x^3y^4$

24. GEOMETRY Suppose each of the dimensions of the cube shown below is increased by 0.25 centimeters. (Lesson 10-6)

a. Write a binomial expression to represent the volume of the new cube. $(s + 0.25)^3$

b. Expand the binomial expression.
$s^3 + 0.75s^2 + 0.1875s + 0.015625$

Find a counterexample to disprove each statement. (Lesson 10-7)

25. $n^2 + 4n − 3$ is prime. $n = 2$

26. $2n^2 + 3n − 3$ is prime. $n = 3$

CHAPTER 11 Statistics and Probability

Determine whether each situation describes a *survey*, an *experiment*, or an *observational study*. Then identify the sample, and suggest a population from which it may have been selected. (Lesson 11-1)

1. Every fifth person coming out of a football stadium is asked how often they go to a football game. survey; sample: every fifth person; population: everyone

2. A company posts a controversial advertisement in their store and records the reactions of the customers. observational study; sample: the participants in the study; population; potential customers

3. A randomly selected group of people are chosen to test the effects of a new drug on energy levels. See Extra Practice Answer Appendix.

4. A randomly selected group of people are asked their opinions on political topics. survey; sample: the participants in the survey; population: eligible voters

5. COLLEGE The SAT scores of Mr. Williams' calculus class are given below. (Lesson 11-2)

		SAT Scores		
2020	1250	1500	1700	1600
1830	1230	1330	1450	1250
1560	1450	1350	1430	1640
2210	1550	2200	2050	1750
2400	1900	2140	1280	1800

a. Use a graphing calculator to create a histogram. Then describe the shape of the distribution. See Extra Practice Answer Appendix. positively skewed.

b. Describe the center and spread of the data using either the mean and standard deviation or the five-number summary. Sample answer: The distribution is skewed, so use the five-number summary. The median is 1600, and half of the scores are between 1390 and 1960. The range is 1230 to 2400. **Identify the random variable in each distribution, and classify it as *discrete* or *continuous*. Explain your reasoning.** (Lesson 11-3)

6. the number of golf balls hit into a lake 6–9. See Extra Practice Answer Appendix.

7. the life of a battery

8. the amount of gasoline in a gas tank

9. the number of stamps in a collection

10. DOGS According to a recent survey, 34% of homeowners own a dog. What is the probability that out of 10 randomly selected homeowners, more than 5 own a dog? (Lesson 11-4) **22.7%**

11. ASSEMBLIES The number of assemblies at West High School each year is normally distributed with a mean of 12.4 and a standard deviation of 1.6. (Lesson 11-5)

a. What is the probability that there will be more than 10 assemblies in a given year? 93.3%

b. If the school has existed for 30 years, in how many of those years were there between 11 and 13 assemblies? about 14 yr

12. STUDENT COUNCIL The number of students that run for student council each year is normally distributed with a mean of 16.8 students and a standard deviation of 3.7. (Lesson 11-5)

a. What is the probability that fewer than 10 students run in a given year? 3.3%

b. If the school has kept records for 20 years, in how many of those years were there between 15 and 20 students who ran for student council? about 10 yr

13. CLOTHES A sample of 225 high school students spent a mean of $125 on clothes for the new school year. Calculate the maximum error of estimate for a 90% confidence level if the standard deviation was $10.40. (Lesson 11-6) ≈$1.14

14. CEREAL A cereal manufacturer claims that every box of cereal contains 100 raisins. Jackie thought there were fewer than 100, so she tested 30 random boxes of cereal. Test the hypothesis at 5% significance. Identify the hypotheses and claim, decide whether to reject the null hypothesis, and make a conclusion about the claim. (Lesson 11-6) See Extra Practice Answer Appendix.

Number of Raisins				
98	97	99	95	98	99
100	98	99	99	96	98
102	97	100	100	98	99
100	98	99	97	99	100
101	97	100	99	101	99

CHAPTER 12 Trigonometric Functions

Use a trigonometric function to find the value of x. Round to the nearest tenth. (Lesson 12-1)

1.

2. 6.6   6.1

3. **FLAGPOLE** Tina is standing away from a flagpole. The angle of elevation from the ground to the top of the flagpole is 15°. The flagpole is 50 feet high. Determine how far Tina is standing from the flagpole. (Lesson 12-1)
187 ft

Rewrite each degree measure in radians and each radian measure in degrees. (Lesson 12-2)

4. 300° $\frac{5\pi}{3}$

5. −80° $-\frac{4\pi}{9}$

6. $-\frac{2\pi}{3}$ −120°

7. $-\frac{\pi}{6}$ −30°

Find the exact value of each trigonometric function. (Lesson 12-3)

8. sin 120° $\frac{\sqrt{3}}{2}$

9. csc $\frac{3\pi}{4}$ $\sqrt{2}$

Solve each triangle. Round side lengths to the nearest tenth and angle measures to the nearest degree. (Lesson 12-4)

10.
$B = 88°$, $a \approx 8.8$, $b \approx 10.5$

11.
$Y = 94°$, $x \approx 7.3$, $y \approx 10.4$

12. **SAILS** One side of a triangular sail of a sailboat is 25 feet long. The angle opposite this side is 40°. Another angle formed by the sail measures 55°. What is the perimeter of the sail to the nearest tenth? (Lesson 12-4) 95.6 ft

Solve each triangle. Round side lengths to the nearest tenth and angle measures to the nearest degree. (Lesson 12-5)

13.
$y \approx 9.5$, $X \approx 39°$, $Z \approx 47°$

14.
$A \approx 135°$, $B \approx 15°$, $C \approx 30°$

15. **GARDENING** A triangular flower bed has sides measuring 4.5 feet, 6 feet, and 7.5 feet. Find the measure of the smallest angle. (Lesson 12-5) 37°

Find the exact value of each function. (Lesson 12-6)

16. cos (−60°) $\frac{1}{2}$

17. sin $\left(\frac{3\pi}{4}\right)$ $\frac{\sqrt{2}}{2}$

18–19. See Extra Practice Answer Appendix for graphs.
Find the amplitude and period of each function. Then graph the function. (Lesson 12-7)

18. $y = 3 \cos \theta$ amplitude 3; period 360°

19. $y = \frac{1}{2} \sin 3\theta$ amplitude $\frac{1}{2}$; period 120°

State the amplitude, period, phase shift, and vertical shift for each function. Then graph the function. (Lesson 12-8) **20–21. See Extra Practice Answer Appendix for graphs.**

20. $y = 2 \tan (\theta + 30°) + 3$ 2; 360°; $h = -30°$; $k = 3$

21. $y = 4 \cos \left(\theta - \frac{\pi}{2}\right) - 2$ 4; 360°; $h = \frac{\pi}{2}$; $k = -2$

Solve each equation. Round to the nearest tenth if necessary. (Lesson 12-9)

22. Sin $\theta = -0.58$ −35.5°

23. Cos $\theta = 0.32$ 71.3°

24. Tan $\theta = 2.7$ 69.7°

CHAPTER 13 Trigonometric Identities and Equations

Find the exact value of each expression if $0° < \theta < 90°$. (Lesson 13-1)

1. If cos $\theta = \frac{3}{5}$, find sin θ. $\frac{4}{5}$

2. If tan $\theta = 2$, find cot θ. $\frac{1}{2}$

3. If sin $\theta = \frac{\sqrt{5}}{3}$, find cos θ. $\frac{2}{3}$

4. If csc $\theta = \frac{3\sqrt{5}}{5}$, find tan θ. $\frac{\sqrt{5}}{2}$

5–8. See Extra Practice Answer Appendix.
Verify that each equation is an identity. (Lesson 13-2)

5. $\frac{\sin^2 \theta}{\cos^2 \theta} \cdot \csc^2 \theta = 1 + \tan^2 \theta$

6. sec θ cot $\theta = \csc \theta$

7. sin θ cot $\theta = \cos \theta$

8. $\frac{\sec^2 \theta}{\csc^2 \theta} = \tan^2 \theta$

9. **CONSTRUCTION** A window has the dimensions shown below. Use the measures of the sides of the triangle to show that $\sin^2\theta + \cos^2\theta = 1$. (Lesson 13-2)

$$\sin^2 \theta + \cos^2 \theta$$
$$= \left[\frac{3\sqrt{5}}{9}\right]^2 + \left[\frac{6}{9}\right]^2$$
$$= \frac{5}{9} + \frac{36}{81}$$
$$= \frac{5}{9} + \frac{4}{9}$$
$$= 1$$

Find the exact value of each expression. (Lesson 13-3)

10. cos (−30°) $\frac{\sqrt{3}}{2}$

11. sin (−120°) $-\frac{\sqrt{3}}{2}$

12. sin 135° $\frac{\sqrt{2}}{2}$

13. cos 225° $-\frac{\sqrt{2}}{2}$

Find the exact value of each expression. (Lesson 13-4)

14. sin 15° $\frac{\sqrt{2}-\sqrt{3}}{2}$

15. cos $\frac{\pi}{12}$ $\frac{\sqrt{2+\sqrt{3}}}{2}$

16. tan 67.5° $\sqrt{3} + 2\sqrt{2}$

17. sin 165° $\frac{\sqrt{2-\sqrt{3}}}{2}$

Solve each equation if $0° \le \theta \le 360°$. (Lesson 13-5)

18. 2 sin $\theta = 1$ 30°, 150°

19. 2 cos $\theta + 1 = 0$ 120°, 240°

20. 4 cos² $\theta - 1 = 0$ 60°, 120°, 240°, 300°

21. **TOYS** A toy's height h in inches can be modeled by the equation $h = 4 \sin \frac{11\pi}{12} t$, where t represents time in seconds. (Lesson 13-5)
   a. At what point will the toy be 4 inches above its resting position for the first time? $\frac{6}{11} \approx 0.55$ second
   b. At what point will the toy be 2 inches above its resting position for the first time? $\frac{2}{11} \approx 0.18$ second

22. **BUOYS** As waves move past a buoy, the height in feet of the buoy oscillates according to the equation $y = 6 \cos 30t$, where t is the time in seconds. If the buoy is at the top of the wave at 0 seconds, during what time intervals will the buoy be 3 feet below the midline? (Lesson 13-5)
$4 + 12n$ and $6 + 12n$ seconds where n is an integer

lowest point (no wave)   midline

connectED.mcgraw-hill.com

Extra Practice Answer Appendix

Chapter 1

18.

19.

20.

21.

23.

24.

Chapter 2

3.

4.

5.

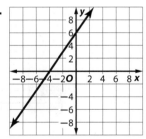

6.

15a.

[0, 10] scl: 1 by [0, 1200] scl: 100

strong positive correlation

16.

17.

18.

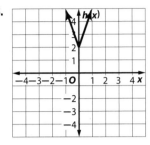

19. translation of the graph of $y = x^2$ down 3 units

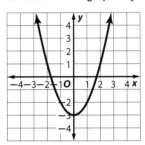

20. translation of the graph of $y = |x|$ up 5 units

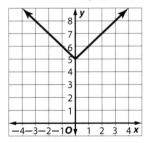

21. translation of the graph of $y = x^2$ left 2 units

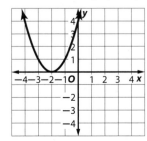

22. translation of the graph of $y = |x|$ right 4 units

23.

24.

25.

Chapter 3

5.

6.

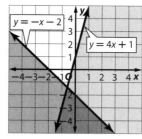

7a. $m + v \le 2000$
$m \ge 875$
$v \ge 800$

8.

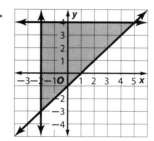

$(-2, 4), (-2, -3), (5, 4)$; max. $= 6$, min. $= -8$

9.

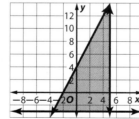

$(5, 14), (5, -3), (-3.5, -3)$; max. $= 50$, min. $= -35$

12. $\begin{bmatrix} 8 & 10 & 10 \\ 14 & 6 & 8 \\ 12 & 12 & 7 \end{bmatrix}$

Chapter 4

1a. -3; $x = 0$; 0

1b.

x	f(x)
-2	1
-1	-2
0	-3
1	-2
2	1

1c.

2a. 0; $x = \dfrac{5}{4}$; $\dfrac{5}{4}$

2b.

x	f(x)
0	0
1	-3
1.25	-3.125
2	-2
3	3

2c.

Chapter 5

11a.

11b. The graph has a relative minimum between $x = 4$ and $x = 5$, then steadily increases as the x increases.

11c. Cell phone sales will continue to increase and will total over 6 billion dollars in 2015.

11d. Sample answer: Every year new phones come out, so it is reasonable that the graph suggests sales will continue to rise until a different type of communication device is developed.

Chapter 6

1. $(f + g)(x) = 3x - 1$, $(f - g)(x) = -x - 7$, $(f \cdot g)(x) = 2x^2 - 5x - 12$, and $\left(\dfrac{f}{g}\right)(x) = \dfrac{x - 4}{2x + 3}$, $x \neq -\dfrac{3}{2}$

2. $(f + g)(x) = x^2 - x - 1$, $(f - g)(x) = -x^2 - x + 7$, $(f \cdot g)(x) = -x^3 + 3x^2 + 4x - 12$, and $\left(\dfrac{f}{g}\right)(x) = \dfrac{-x + 3}{x^2 - 4}$, $x \neq -2$ or 2

Chapter 7

1.

2.

Chapter 8

12.

13.

Chapter 9

5. $y = (x + 3)^2 - 2$; $(-3, -2)$; $x = -3$; opens upward

6. $y = 2(x - 2)^2 + 4$; $(2, 4)$; $x = 2$; opens upward

Chapter 11

3. experiment; sample: the participants in the experiment; population: the population of people that could take the drug

5a.

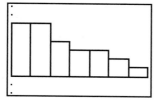

[1200, 2600] scl: 200 by [−2, 8] scl: 1

6. the number of golf balls hit into a lake; discrete

7. the life of the battery, which can be anywhere within a certain range; continuous

8. the amount of gasoline in a gas tank, which can be anywhere within a certain range; continuous

9. the number of stamps in the collection; discrete

14. $H_0: \mu = 100$ (claim), $H_a: \mu \neq 100$; Reject H_0; The manufacturer's claim that each box contains 100 raisins is rejected.

Chapter 12

18.

19.

20.

21.

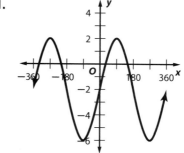

Chapter 13

5. $\dfrac{\sin^2\theta}{\cos^2\theta} \cdot \csc^2\theta \overset{?}{=} 1 + \tan^2\theta$

$\dfrac{\sin^2\theta}{\cos^2\theta} \cdot \dfrac{1}{\sin^2\theta} \overset{?}{=} \sec^2\theta$

$\dfrac{1}{\cos^2\theta} \overset{?}{=} \sec^2\theta$

$\sec^2\theta = \sec^2\theta$ ✔

6. $\sec\theta \cot\theta \overset{?}{=} \csc\theta$

$\dfrac{1}{\cos\theta} \cdot \dfrac{\cos\theta}{\sin\theta} \overset{?}{=} \csc\theta$

$\dfrac{1}{\sin\theta} \overset{?}{=} \csc\theta$

$\csc\theta = \csc\theta$ ✔

7. $\sin\theta \cot\theta \overset{?}{=} \cos\theta$

$\sin\theta \cdot \dfrac{\cos\theta}{\sin\theta} \overset{?}{=} \cos\theta$

$\cos\theta = \cos\theta$ ✔

8. $\dfrac{\sec^2\theta}{\csc^2\theta} \overset{?}{=} \tan^2\theta$

$\dfrac{\dfrac{1}{\cos^2\theta}}{\dfrac{1}{\sin^2\theta}} \overset{?}{=} \tan^2\theta$

$\dfrac{\sin^2\theta}{\cos^2\theta} \overset{?}{=} \tan^2\theta$

$\tan^2\theta = \tan^2\theta$ ✔

Selected Answers and Solutions

CHAPTER 0
Preparing for Advanced Algebra

Lesson 0-1

1. $D = \{1, 2, 3\}$, $R = \{6, 7, 10\}$; yes **3.** $D = \{1, 2\}$, $R = \{5, 7, 9\}$; no **5.** $D = \{-2, -1, 0, 3\}$, $R = \{-3, -2, 2\}$; yes **7.** $D = \{-1, 0, 1, 2, 3\}$, $R = \{-3, -2, -1, 2, 3, 4\}$; no **9.** I **11.** none

Lesson 0-2

1. $a^2 + 6a + 8$ **3.** $b^2 - 16$ **5.** $b^2 + b - 12$ **7.** $r^2 - 5r - 24$ **9.** $p^2 + 16p + 64$ **11.** $2c^2 - 9c - 5$ **13.** $6m^2 - 7m - 20$ **15.** $2t^2 - 13t - 34$ **17a.** $n - 7$, $n + 2$ **17b.** $n^2 - 5n - 14$

Lesson 0-3

1. $4x(3x + 1)$ **3.** $4ab(2b - 3)$ **5.** $(y + 3)(y + 9)$ **7.** $(3y + 1)(y + 4)$ **9.** $(3x + 4)(x + 8)$ **11.** $(y - 4)(y - 1)$ **13.** $2(3a - b)(a - 8b)$ **15.** $(2x - 3y)(9x - 2y)$ **17.** $(3x - 4)^2$ **19.** $(x + 12)(x - 12)$ **21.** $(4y + 1)(4y - 1)$ **23.** $4(3y + 2)(3y - 2)$

Lesson 0-4

1. 60 **3.** 18 **5.** 120 **7.** 6 **9.** 6 **11.** permutations, 720 **13.** permutations, 5040 **15.** combinations, 15 **17.** permutations, 3024 **19a.** 2,176,782,336; **19b.** 308,915,776; 712,882,560; The password with one digit is more secure, because the chance of someone guessing this password at random is $\frac{1}{712,882,560}$, which is less than the chance of someone guessing a 6-character password that contains only letters, $\frac{1}{308,915,776}$.

Lesson 0-5

1a.

Color	Frequency	Experimental Probability
red	6	0.12
blue	7	0.14
yellow	9	0.18
orange	12	0.24
purple	6	0.10
green	11	0.22

1b.

Spinner Experimental Probabilities

1c.

Color	Frequency	Experimental Probability	Theoretical Probability
red	6	0.12	$0.1\overline{6}$
blue	7	0.14	$0.1\overline{6}$
yellow	9	0.18	$0.1\overline{6}$
orange	12	0.24	$0.1\overline{6}$
purple	6	0.10	$0.1\overline{6}$
green	11	0.22	$0.1\overline{6}$

1d.

Spinner Theoretical Probabilities

1e. Sample answer: Since all bars in the graph of the theoretical probabilities are the same height, the graph represents a uniform distribution. This means that in theory, the chance of landing on any one of the colors is equally likely. The graph of the experimental probabilities indicates that in practice, it is more likely that the spinner will land on orange or green than on any of the other colors, since the heights of those bars are taller than any others in the graph. **3a.** mutually exclusive, $\frac{1}{2}$ **3b.** not mutually exclusive, $\frac{4}{13}$ **3c.** not mutually exclusive, $\frac{7}{13}$ **5a.** mutually exclusive, $\frac{11}{20}$ **5b.** not mutually exclusive, $\frac{29}{40}$ **5c.** not mutually exclusive, $\frac{2}{5}$

7. When events are mutually exclusive, $P(A \text{ and } B)$ will always equal 0, so the probability will simplify to $P(A \text{ or } B) = P(A) + P(B)$. **9a.** 1 to 3 **9b.** 1 to 1

Lesson 0-6

1. independent; $\frac{1}{36}$ **3.** $\frac{1}{36}$ **5.** $\frac{25}{36}$ **7.** $\frac{5}{18}$ **9.** $\frac{5}{17}$ **11.** $\frac{1}{8}$ **13.** 16% **15a.** $\frac{32}{41}$ **15b.** $\frac{2}{5}$ **15c.** $\frac{3}{7}$ **17a.** $\frac{91}{115}$ or about 79.1% **17b.** $\frac{239}{308}$ or about 77.6%

Lesson 0-7

1. similar **3.** neither **5.** similar **7.** 8; 21 **9.** 10.2; 13.6 **11.** $4\frac{1}{2}$ in.

Lesson 0-8

1. 39 ft **3.** 8.3 cm **5.** 5.5 **7.** 9.2 **9.** 8.5 **11.** yes **13.** no **15.** yes **17.** about 2.66 m

Lesson 0-9

1. 451.8 min, 399 min, no mode **3.** ≈ 34.4 text messages, 35 text messages, 35 text messages **5.** Sample; Walk A: 47, ≈ 242.0, ≈ 15.6; Walk B: 92, ≈ 1115.4, ≈ 33.4; since the sample standard deviation of Walk B is greater than that of Walk A, there is more variability in the number of sponsors obtained by participants in Walk B than in Walk A. **7.** 18, 23, 25, 27, 29; Sample answer: There are 18 students in the smallest math class at Central High and 29 students in the largest class. 25% of the classes have less than 23 students, 50% of the classes have less than 25 students, and 75% of the classes have less than 27 students. **9.** 13; Sample answer: The interval beyond which any outliers would lie is 14.25 < X < 60.25. Since 13 < 14.25, it is an outlier.

Removing the outlier did not affect the mode. However, the removal did affect the mean, median, standard deviation, and range. The mean and median increased, and the standard deviation and range decreased.

Data Set	Mean	Median	Mode	Range	Standard Deviation
with outlier	≈ 35.8	36	36	38	≈ 9.3
without outlier	≈ 37.3	36	36	29	≈ 7.5

11a. Yes; sample answer: The interval would lie is 15.35 < X < 17.35. Since 14.9 < 15.35, it is an outlier. **11b.** No; sample answer: The new interval beyond which any outliers would lie would be 15.4 < X < 17.4. Since 17.4 > 17.35, it would not be an outlier. **11c.** Sample answers: data recording errors, manufacturing errors

CHAPTER 1
Equations and Inequalities

Chapter 1   Get Ready

1. 12.25 **3.** −66.15 **5.** $1\frac{13}{15}$ **7.** $-1\frac{1}{3}$ **9.** $10\frac{1}{2}$ yd **11.** −64 **13.** 15.625 **15.** $\frac{2401}{81}$ **17.** $-\frac{3375}{8}$ **19.** true **21.** false **23.** yes

Lesson 1-1

1. 4.6 **3.** 18.4 **5.** 11.6 **7.** 0.96875 **9.** 0.6 **11.** $6\frac{4}{15}$ **13.** 28 **15.** −13.4 **17a.** 1524.6 mi **17b.** 720 mi **19.** 20

21 $\dfrac{b^2 d^2}{ad} = \dfrac{(-0.8)^2(5)^2}{(-4)(\frac{1}{5})}$ Evaluate the numerator and denominator separately.

  $= \dfrac{(0.64)(25)}{(-0.8)}$ Simplify the numerator.

  $= \dfrac{16}{-0.8}$ Simplify the fraction.

  $= -20$

23. 3.71 **25.** $\frac{1}{2}(x + 7)(2x)$ **27a.** 584,336,233.6 mi **27b.** 8761 h **27c.** yes; $\frac{8761}{24} = 365$ days or 1 year

29. 544 **31.** 13.8 **33.** 131.25 **35.** $6\pi x^3$

37 $t = 50 + \dfrac{4}{n - 40}$ Write the formula.

  $n = 120$

  $= 50 + \dfrac{4}{120 - 40}$

  $= 50 + \dfrac{80}{4}$

  $= 50 + 20$

  $= 70$ Evaluate the numerator of the fraction. Simplify the fraction. Simplify.

If the number of chirps is 120, then the temperature is 70°F.

39a. $3.91; $5.36; $7.31 **39b.** $4.42; $6.62; $11.62; Sample answer: the average prices found in part b become increasingly higher with time.

41 $y = \sqrt{b^2\left(1 - \dfrac{x^3}{a^2}\right)}$ Write the equation.

  $= \sqrt{8^2\left(1 - \dfrac{3^3}{6^2}\right)}$ $a = 6$, $b = 8$, $x = 3$

  $= \sqrt{64\left(1 - \dfrac{9}{36}\right)}$

  $= \sqrt{64\left(\dfrac{27}{36}\right)}$

  $= \sqrt{48}$ Simplify inside the parentheses. Simplify.

  ≈ 6.9 Use a calculator.

43. Lauren; −12 − 20 = −32. **45.** Sample answer: b; Since t is time, it must be nonnegative. So $-2t(2t + 1)$ will be negative for all values of t other than 0. The maximum value of $-2t(2t + 1)$ is 0, which occurs when $t = 0$. Thus, the maximum value of $-2t(2t + 1) + 6$ is 6. **47.** Sample answer: $y\left(\dfrac{-4z}{x^2} - x\right) + z$ **49.** A table of on-base percentages is limited to those situations listed, while a formula can be used to find any on-base percentage. **51.** 9 mo **53.** B **55.** 10 cm **57.** 6x(x + 2) **59.** 3 and 11 **61.** 5 **63.** 11 **65.** −4 **67.** $\frac{5}{8}$

connectED.mcgraw-hill.com

R16 (left page)

Lesson 1-2

1. N, W, Z, Q, R   3. I, R   5. Associative Property ($\times$)
7. Commutative Property ($+$)   9. $7, -\frac{1}{7}$ 11. $-3.8; \frac{1}{3.8}$
13a. $22(2 + 4 + 3 + 1 + 5 + 6 + 7)$ or $22(2) + 22(4) + 22(3) + 22(1) + 22(5) + 22(6) + 22(7)$ 13b. $616   13c. If she continues to mow the same number of lawns, at the end of next week she will have the money. This may not be reasonable because not all the lawns she mowed this week may need to be mowed again next week.
15. $24b + 9b$ 17. $-16x + 22y$ 19. Q, R   21. Q, R
23. $-\sqrt{144} = -12$ belongs to the set of integers (Z), the set of rationals (Q), and the set of reals (R).
25. I, R   27. Distributive Property   29. Inverse Property ($\times$)   31. $-12.1; \frac{1}{12.1}$ 33. $\frac{6}{13}, \frac{13}{6}$
35. $-\sqrt{15}, \frac{1}{\sqrt{15}}$ 37. $12b + 6c$ 39. $40x - 20y$
41. $28g - 48k$ 43. $53(60 + 60); 53(60) + 53(60); 6360$ yd²
45a. $(2 + 1)4.50 = (3)4.50 = \$13.50$ or 6.50.
45b. The amount left over is $20 - 13.50 or $6.50. Because Billie cannot buy part of a sandwich, she can buy 3 cold sandwiches.
45c. In two weeks, or ten school days, Billie buys a hot lunch 3 times and buys a cold sandwich 3 times. She has to pack lunch $10 - (3 + 3) = 10 - 6$ or 4 times.
47. $\frac{27}{5}c - \frac{199}{20}d$ 49. $-42x - 72y - 30z$
51a. $-\sqrt{6}$ is an irrational number because the square root of 6 is not a perfect square.

irrational	rational	integer	whole	natural
$-\sqrt{6}, \pi$	$3, \frac{-15}{3}, 4.1,$ $0, \frac{3}{8}, \sqrt{36}$	$3, \frac{-15}{3},$ $0, \sqrt{36}$	$3.0,$ $\sqrt{36}$	$3,$ $\sqrt{36}$

$\frac{3}{8}$ is a rational number.
$\sqrt{36}$, or 6, is a rational number, integer, whole number, and natural number.
π is an irrational number.
$0,$ or $\frac{0}{1}, \frac{0}{2}, \dots$, is a rational number, integer, and whole number.
$\frac{-15}{3}$, or $\frac{41}{10}$, is a rational number.
4.1, or $\frac{41}{10}$, is a rational number.
$-\frac{15}{3}$, or -5, is a rational number and integer.
3, or $\frac{3}{1}$, is a rational number, integer, whole number, and natural number.

51b. Use a calculator to find the decimal form of $-\sqrt{6}$ and π; $-\sqrt{6} \approx -2.449$, $3 = 3.0$, $\frac{-15}{3} = -5$, $4.1 = 4.1$, $\pi \approx 3.14$, $0 = 0$, $\frac{3}{8} = 0.375$, $\sqrt{36} = 6$. Since $-5 < -2.449 < 0 < 0.375 < 3.14 < 4.1 < 6$, the numbers from least to greatest are $\frac{-15}{3}, -\sqrt{6}, 0, \frac{3}{8}, \pi,$ 4.1, $\sqrt{36}$.

c. Draw a number line with tick marks at integers from -6 to 6. Then use the decimal forms in part b to graph each number.

d. Sample answer: By converting the real numbers into decimal form, the decimal points can be easily lined up and the numbers compared.
53. $\sqrt{81}$; It is a rational number, while the other three are irrational numbers.   55. No; Luna did not distribute the negative sign to the second term and Sophia switched the a and b terms because usually a comes first. The correct answer is $32a - 46b$.
57. Sample answer: $\sqrt{5} \cdot \sqrt{5} = \sqrt{25}$ or 5, which is not irrational   59. Sample answer: (a) 3.2 and (b) $\sqrt{10}$
61. Sample answer: The Commutative Property does not hold for subtraction or division because order matters with these two operations. In addition or multiplication, the order does not matter. For example, $2 + 4 \neq 4 + 2$ and $2 \cdot 4 \neq 4 \cdot 2$. However, with subtraction, $2 - 4 \neq 4 - 2$, and with division, $2 \div 4 \neq 4 \div 2$.   63. C   65. B   67. 24   69. about 2.66 m
71. $3(3x^2 - x + 6)$ 73. $10x(x - 2)$ 75. $6(2x^2 - 3x - 4)$
77. $y^2 + y - 2$ 79. $b^2 - 10b + 21$ 81. $p^2 - 8p - 9$
83. $\frac{10}{9}$ 85. 8   87. ≈ -1.176 89. -1.7

Lesson 1-3

1. $12[x + (-3)]$ 3. The sum of five times a number and 7 equals 18.   5. The difference between five times a number and the cube of that number is 12.
7. Reflexive Property   9. 53   11. -8 13. -6 15. 3
17. 4   19. $q = \frac{8r - 3}{5}$ 21. B   23. $8x^2$ 25. $\frac{x}{4} + 5$
27. The quotient of the sum of 3 and a number and 4 is 5.
29. Let $n = $ the number of home runs that Jacobs hit. Then $n + 6 = $ the number of home runs that Cabrera hit.
$n + (n + 6) = 46$ Cabrera and Jacobs hit a combined total of 46 home runs.
$2n + 6 = 46$ Simplify.
$2n = 40$ Subtract 6 from each side.
$n = 20$ Divide each side by 2.
So, Jacobs hit 20 home runs and Cabrera hit $n + 6 = 20 + 6$ or 26 home runs.
31. Substitution   33. Multiplication ($=$)   35. 5   37. -3
39. $5(-2x - 4) - 3(4x + 5) = 97$ Original equation
$-10x - 20 - 12x - 15 = 97$ Apply the Distributive Property.
$-22x - 35 = 97$ Simplify the left side.
$-22x = 132$ Add 35 to each side.
$x = -6$ Divide each side by -22.
41. -3 43. $s = $ length of a side; $5s = 100$; 20 in.
45. $m = \frac{E}{c^2}$ 47. $h = \frac{z}{\pi q^3}$ 49. $x = \frac{y - bx - c}{x^2}$
51a. $V = \pi \cdot r \cdot r \cdot h$ 51b. $h = \frac{V}{\pi r^2}$
53. -2 55. -4

R17 (right page)

$|x - 100| = 245$ means $x - 100 = 245$ or $x - 100 = -245$.

Case 1	Case 2
$x - 100 = 245$	$x - 100 = -245$
$x - 100 + 100 = 245 + 100$	$x - 100 + 100 = -245 + 100$
$x = 345$	$x = -145$

The solutions are 345 and -145. This means the maximum is 345 ft above sea level; the minimum is -145 ft or 145 ft below sea level. The maximum is reasonable, but the minimum is not. Florida's lowest point should be at sea level where Florida meets the Atlantic Ocean and the Gulf of Mexico.
45. Ling; Ana included an extraneous solution. She would have caught this error if she had checked to see if her answers were correct by substituting the values into the original equation.   47. Sometimes; this is only true for certain values of a. For example, it is true for $a = 8$; if $8 > 7$, then $11 > 10$. However, it is not true for $a = -8$; if $8 > 7$, then $5 \not> 10$.
49. Always; starting with numbers between 1 and 5 and subtracting 3 will produce numbers between -2 and 2. These all have an absolute value less than or equal to 2.   51. Sample answer: First, isolate the absolute value symbol by subtracting each side by c, and then dividing each side by a. You then have $|x - b|$ equals a mathematical expression. Take away the absolute value symbol, and form two new equations by setting $x - b$ equal to both the positive and negative values of the expression. Solve each equation for x. Then substitute each solution into the original equation, and confirm whether they are correct.   53. $\frac{5}{8}$ 55. E   57. -2 59a. $6800
59b. $535.83   59c. 1 mo   61. Distributive   63. $10x + 2y$
65. $11n + 10a$ 67. $32c - 46d$ 69. 2   71. -8 73. $\frac{4}{7}$

Lesson 1-5

1. $b < 8$
2. $x \leq -6$
5. $w < 2$
7.
$s \geq s + 6$ Original inequality
$5s \geq s + 6$ Multiply each side by 5.
$4s \geq 6$ Subtract s from each side.
$s \geq 1.5$ Divide each side by 4.
The solution set is $\{s \mid s \geq 1.5\}$.
9. 40 bags
11. $n \leq -3$
13. $t \leq \frac{1}{2}$
15. $k < 27$

57. $-\frac{117}{11}$ 59. $x = $ the cost of rent each month; $622 + 428 + 240 + 144 + 12x = 10{,}734$; $775 per month
61a. The integers from -5 to 5 are $-5, -4, -3, -2, -1, 0, 1, 2, 3, 4,$ and 5. Draw a number line and plot a point at each integer.
61b. -5 and 5 are 5 units from zero, -4 and 4 are 4 units from zero, and so on.
61c. The points $(x, y) = $ (integer, distance from zero) are $(-5, 5), (-4, 4), (-3, 3), (-2, 2), (-1, 1), (0, 0), (1, 1), (2, 2), (3, 3), (4, 4),$ and $(5, 5)$.

Integer	Distance from Zero
-5	5
-4	4
-3	3
-2	2
-1	1
0	0
1	1
2	2
3	3
4	4
5	5

d. For positive integers, the distance from zero is the same as the integer. For negative integers, the distance is the integer with the opposite sign because distance is always positive.
63. $y_1 = y_2 = \sqrt{d^2 - (x_2 - x_1)^2}$ 65. Sample answer: $3(x - 4) = 3x + 5$; $2(3x - 1) = 6x - 2$ 67. D   69. A
71. $-3x + 6y + 6z$ 73. 605 ft   75. $4\frac{4}{5}$ 77. $2x$
79. $-3\frac{2}{3}$ 81. $-5x$

Lesson 1-4

1. 12   3. -108 5a. $|x - 78| = 2$ 5b. least: 76°F; greatest: 80°F   5c. 77°F; This would ensure a minimum temperature of 76°F.   7. $\{15, -7\}$ 9. $\varnothing$ 11. $\left\{\frac{6}{5}, -\frac{4}{5}\right\}$
13. $\{2\}$ 15. 25   17. 9.2   19. 49.2   21. -63 23. $\{34, -8\}$
25. $\{4, -14\}$ 27. $\{-2, -10\}$ 29. $\{2\}$
31. $2|3x - 4| + 8 = 6$ Original equation
$2|3x - 4| + 8 - 8 = 6 - 8$ Subtract 8 from each side.
$2|3x - 4| = -2$ Simplify.
$|3x - 4| = -1$ Divide each side by 2.
Because the absolute value of a number is always positive or zero, this sentence is never true. So, there is no solution.
33. $\varnothing$ 35. $|x - 5.67| = 0.02$; heaviest: 5.69 g; lightest: 5.65 g   37. 28   39. $\left\{1, \frac{1}{5}\right\}$ 41. $\left\{\frac{8}{3}\right\}$
43. The average altitude c is 100 ft. Since the altitude can be plus or minus 245 feet, the range r is 245.
$|x - c| = r$ Absolute value equation
$|x - 100| = 245$ $c = 100$ and $r = 245$

Selected Answers and Solutions

Selected Answers and Solutions

17. $z < 3$

19. $12 < -4(3c - 6)$ Original inequality
$-3 > 3c$ Divide each side by -4, reversing the inequality symbol.
$3 > 3c$ Add 6 to each side.
$1 > c$ Divide each side by 3.
The solution set is $\{c \mid c < 1\}$.

21. $z < 3$

23. $3x - 12 < 21; x < 11$ **25.** $5x - 6 > x; x > 1.5$ **27.** 8 hours

29. $x > -\frac{3}{4}$

31. $y > 18.75$

33. $v > -4.5$

35. $r > \frac{3}{4}$

37a. $250 + 0.03(500a) \geq 700$ **37b.** $a \geq 30$; He must sell at least 30 advertisements.   **39.** $\frac{x}{3} + 4 \leq 2x + 12$; $x \geq -4.8$

41. **a.** Let d = the number of miles by which Jamie should increase her average daily run. Then $5 + d$ = her average daily distance after the increase.

	average daily distance	is at least	length of a marathon
3 times	3 ·		

So, the inequality is $3(5 + d) \geq 26.2$.
b. $3(5 + d) \geq 26.2$ Original inequality
$5 + d \geq 8.73$ Divide each side by 3. Round to the nearest hundredth.
$d \geq 3.73$ Subtract 5 from each side.
In order to have enough endurance to run a marathon, Jamie should increase the distance of her average daily run by at least 3.73 miles.

43a. Sample answer:

Point	Resulting Statement	True or False
(0, 0)	$0 \geq 3$	False
(1, 1)	$1 \geq \frac{5}{2}$	False
(2, 2)	$2 \geq 2$	True
(3, 3)	$3 \geq \frac{3}{2}$	True
(4, 4)	$4 \geq 1$	True

43b. Sample answer:

43c. Sample answer: The points on or above the line result in true statements, and the points below the line result in false statements. This is true for all points

---

on the coordinate plane.   **45.** No; Sample answer: Madlynn reversed the inequality sign when she added 1 to each side. Emilie did not reverse the inequality sign at all.   **47.** Using the Triangle Inequality Theorem, we know that the sum of the lengths of any 2 sides of a triangle must be greater than the length of the remaining side. This generates 3 inequalities to examine.
$3x + 4 + 2x + 5 > 4x$ $3x + 4 + 4x > 2x + 5$
   $x > -9$ $x > 0.2$
$2x + 5 + 4x > 3x + 4$
   $x > -\frac{1}{3}$

In order for all 3 conditions to be true, x must be greater than 0.2.   **49.** Sample answer: When one number is greater than another number, it is either more positive or less negative than that number. When these numbers are multiplied by a negative value, their roles are reversed. That is, the number that was more positive is now more negative than the other number. Thus, it is now *less than* that number and the inequality symbol needs to be reversed.
51. A   **53.** D   **55.** $\left\{-\frac{1}{3}, 3\right\}$ **57.** $-3647.5 \mid = 891.5$ **59a.** $SA = 2\pi r(r + h)$ **59b.** $78\pi \text{ cm}^2$ **59c.** Sample answer: The formula in part b is quicker.   **61.** $(-9, 9)$ **63.** $\left\{\frac{1}{2}, 7\right\}$ **65.** $(-6, 2)$

Lesson 1-6

1. $\{g \mid -12 < g < -2\}$

3. $\{z \mid z > -3 \text{ or } z < -6\}$

5. $\{c \mid c \geq 8 \text{ or } c \leq -8\}$

7. $\{z \mid -6 < z < 6\}$

9. $\left\{v \mid v > 3 \text{ or } v < -\frac{19}{3}\right\}$

11. $43.96 \leq c \leq 77.94$; between \$43.96 and \$77.94

13. $\{d \mid -1 \leq d \leq 0.5\}$

15. $\{y \mid y < -4 \text{ or } y > 7\}$

17. $\{k \mid -4 > k \text{ or } k > 4\}$

---

19. $|8t + 3| \leq 4$ is equivalent to $-4 \leq 8t + 3 \leq 4$.
$-4 \leq 8t + 3 \leq 4$
$-4 - 3 \leq 8t + 3 - 3 \leq 4 - 3$
$\frac{-7}{8} \leq \frac{8t}{8} \leq \frac{1}{8}$
$\frac{-7}{8} \leq t \leq \frac{1}{8}$
The solution set is $\left\{t \mid -\frac{7}{8} \leq t \leq \frac{1}{8}\right\}$ or $\left[-\frac{7}{8}, \frac{1}{8}\right]$.

21. $\left\{j \mid j \geq \frac{8}{5} \text{ or } j \leq -\frac{16}{5}\right\}$

23. $\{x \mid -1 \leq x \leq 5\}$ **25.** $\{x \mid x + 9 \leq 3\}$ **27.** $|x - 2| \geq 10$
29. $|x + 3| > 1$

31. A healthy weight w for a fully grown female Labrador retriever is 55 pounds to 70 pounds. This can be represented by $55 \leq w \leq 70$.

33. $\{k \mid 2 < k < 4\}$

35. $\{h \mid h < -15 \text{ or } h > 15\}$

37. $\{z \mid z < -1 \text{ or } z > 5\}$

39. $\left\{f \mid f > \frac{26}{5} \text{ or } f < -\frac{22}{5}\right\}$

41. $\{x \mid x + 5 \geq 4$ **43.** $6 \leq |x - 2| \leq 10$

45. $\{n \mid -7 < n < 1\}$

47. $\varnothing$

49. $\left\{g \mid g \geq -\frac{2}{3}\right\}$

51. $|s - 88| > 38$; $\{s \mid s > 126 \text{ or } s < 50\}$ **53.** Sample answer: David; when Sarah converted the absolute value into two inequalities, she mistakenly switched the inequality symbols.   **55.** False; sample answer: the graph of $x > 2$ and $x > 5$ is a ray bounded only on one end.   **57.** true   **59.** Sample answer: The graph on the left indicates a solution set from -3 to 5 for the inequality $|x - 1| \leq 4$. The graph on the right indicates a solution set of all numbers less than or equal to -3 or greater than or equal to 5 for the inequality $|x - 1| \geq 4$.   **61.** Each of these has a non-empty solution set except for $x > 5$ and $x < 1$. There are no values of x that are simultaneously

---

greater than 5 and less than 1.   **63.** C   **65.** 60   **67a.** $750 \leq x \leq 990$ **67b.** 110 g   **69.** $[0, 10]$ **71.** $\varnothing$
73. Transitive (=)

Chapter 1   **Study Guide and Review**

1. false; nonnegative   **3.** true   **5.** true   **7.** false; or
9. true   **11.** 3   **13.** 10   **15.** 21   **17.** $\approx 169.65 \text{ in}^3$
19. N, W, Z, Q, R   **21.** $11x + 2y$ **23.** $5m + 41n$
25. -7 **27.** $\frac{3}{2}$ **29.** 58   **31.** $m = \frac{r + 5}{pm}$ **33.** about 8 in.

35. $[2, 10]$ **37.** $[14]$

39. $a \geq -6$

41. $x \leq -\frac{2}{9}$

43. 3 or fewer slices each

45. $\{x \mid -2 < x < 4\}$

47. $\left\{m \mid \frac{13}{5} \leq m < \frac{24}{5}\right\}$

49. $\{p \mid -5 \leq p \leq 33\}$

51. $\varnothing$ **53.** $20 \leq 2.50(3) + 1.25b \leq 30$; $10 \leq b \leq 18$

Linear Relations and Functions

CHAPTER 2

Chapter 2   **Get Ready**

1. (4, 1); 1   **3.** (0, 0); origin   **5.** $(-4, -4)$; III
7. -15 **9.** 10   **11.** \$40   **13.** $b = \frac{a}{3} - 3$ **15.** $x = \frac{8}{3} + \frac{4}{3}y$

Lesson 2-1

1. $D = \{5, 6, -2\}$, $R = \{3, -8, 1\}$; function; both

3. The domain is the set of x-values: $\{-2, 1, 4, 8\}$; the range is the set of y-values: $\{-4, -2, 6\}$. Since each element of the domain is paired with exactly one element of the range, the relation is a function. Since each element of the range corresponds to an element of the domain, the relation is an onto function.

$D = \{$all real numbers$\}$,
$R = \{$all real numbers$\}$;
function; both; continuous

5.

connectED.mcgraw-hill.com

Selected Answers and Solutions

$\dfrac{6x + 15}{4} = 3y - 12$ Original equation

$\dfrac{6(0) + 15}{4} = 3y - 12$ Substitute 0 for x.

$\dfrac{15}{4} = 3y - 12$ Simplify.

$\dfrac{63}{4} = 3y$ Add 12 to each side.

$5.25 = y$ Divide each side by 3.

The y-intercept is 5.25.

49. $-1\frac{1}{75}; 6\frac{1}{3}$ **51a.**

51b.

Function	One-to-One	Onto
$f(x) = -2x + 4$	yes	yes
$g(x) = 6$	no	no
$h(x) = \frac{1}{3}x + 5$	yes	yes

51c. No; horizontal lines are neither one-to-one nor onto because only one y-value is used and it is repeated for every x-value. Every other linear function is one-to-one and onto because every x-value has one unique y-value that is not used by any other x-element and every possible y-value is used. **53.** Sample answer: $f(x) = 2(x - 3)$ **55.** $y = 2xy$; Sample answer: $y = 2xy$ is not a linear function. **57.** C **59.** J **61.** D = {8, −4, −1}, R = {6, 3, 9}; not a function; **63.** D = {−3, −4, 7}, R = {−1, −2, 9}; function; both **65.** 0.78 **67.** about −0.583 **69.** $\frac{2}{3}$ **71.** $-\frac{5}{4}$ **73.** $-\frac{1}{3}$ **75.** 9

Lesson 2-3

1. 6 feet/min **3a.** about 11,000 per year **3b.** about −5000 per year **3c.** The positive rate in part **a** represents an increase in the sales of digital cameras. The negative rate in part **b** represents a decrease in sales of film cameras. **5.** −3 **7.** $\frac{3}{5}$

9 Use the ordered pairs (3, 20) and (6, 40).

$$\text{rate of change} = \dfrac{\text{change in } y}{\text{change in } x}$$

$B = 3, C = -6$ **29.** $2x + y = -11; A = 2, B = 1, C = -11$

31
$\quad 2.4y = -14.4x$ Original equation
$14.4x + 2.4y = 0$ Add 14.4x to each side.
$144x + 24y = 0$ Multiply each side by 10.
$6x + y = 0$ Divide each side by 24.
$A = 6, B = 1, C = 0.$

33. $5x + 32y = 160; A = 5, B = 32, C = 160$

35. −0.5; −4

37. −7; 10.5

39. 12; −18

41a. $1.75m + 1.5n = 525$
41b.

$1.75m + 1.5n = 525$

Yes; the graph passes the vertical line test.

41c. No; the amount that Latonya will sell is $1.75 \cdot 100 + 1.5 \cdot 200$, which is $475. **43a.** $y = 3x + 13$ **43b.** $31 **45.** $4x - 40y = -59$; $A = 4, B = -40, C = -59$

47 The x-intercept is the value of x when y = 0.
$\dfrac{6x + 15}{4} = 3y - 12$ Original equation
$\dfrac{6x + 15}{4} = 3(0) - 12$ Substitute 0 for y.
$\dfrac{6x + 15}{4} = -12$ Simplify.
$6x + 15 = -48$ Multiply each side by 4.
$6x = -63$ Subtract 15 from each side.
$x = \dfrac{-63}{6}$ Divide each side by 6.
$x = -10.5$ Simplify.
The x-intercept is −10.5.
The y-intercept is the value of y when x = 0.

23c. D = $\{x \mid x \geq 0\}$, R = $\{y \mid y \geq 1\}$; continuous **23d.** Yes; each domain value is paired with only one range value so the relation is a function. **25.** 29 **27.** −72 **29.** −267 **31.** −4.5

33 $P(t) = 15 + 3t$ Original function
$P(8) = 15 + 3(8)$ Substitute 8 for each t.
$= 15 + 24$ Multiply.
$= 39$ Simplify.
After 8 months, Chaz will have 39 podcasts.

35. Sample answer: Omar; Madison did not square the 3 before multiplying by −4. **37.** Never; if the graph crosses the y-axis twice, then there will be two separate y-values that correspond to $x = 0$, which violates the vertical line test. **39.** Sample answer: False; a function is onto and not one-to-one if all of the elements of the domain correspond to an element of the range, but more than one element of the domain corresponds to the same element of the range. **41.** A **43.** J **45.** $6 > y > 2$ **47.** $x > \frac{7}{4}$ or $x < -\frac{11}{4}$ **49.** $15x \leq 120$; She can buy up to 8 shirts. **51.** $\frac{3}{4}$ or $\frac{7}{4}$ **53.** 33a **55.** $10 + 36d$ **57.** 4 **59.** −4 **61.** −4 **63.** −6

Lesson 2-2

1. Yes; it can be written as $f(x) = \frac{x}{5} + \frac{12}{5}$ **3.** No; x has an exponent that is not 1.

5 a. $m(x) = 0.75x$ Original function
$m(4) = 0.75(4)$ Substitute 4 for x.
$= 3$ Simplify.
If you have 4 CDs, you have 3 hours of music.
b. $m(x) = 0.75x$ Original function
$6 = 0.75x$ Substitute 6 for m(x).
$8 = x$ Divide each side by 0.75.
If the trip is 6 hours long, you should bring 8 CDs.
7. $6x - y = -5; A = 6, B = -1, C = -5$ **9.** $8x + 9y = 6; A = 8, B = 9, C = 6$ **11.** $2x - 3y = 12; A = 2, B = -3, C = 12$
13. $5\frac{1}{2}; -10$
15. 7; $-\frac{21}{4}$

17. No; x has an exponent other than 1. **19.** No; x has an exponent other than 1. **21.** No; it cannot be written in $f(x) = mx + b$ form. **23.** No; it cannot be written in $f(x) = mx + b$ form; There is an xy term. **25a.** 260 m **25b.** Kingda Ka; Sample answer: The Kingda Ka travels 847.5 meters in 25 seconds, so it travels a greater distance in the same amount of time. **27.** $8x + 3y = -6; A = 8,$

---

7. D = {all real numbers}, R = $\{y \mid y \geq 0\}$; function; onto; continuous
9. 4 **11.** D = {−0.3, 0.4, 1.2}, R = {−6, −3, −1, 4}; not a function **13.** D = {−3, −1, 3, 5}, R = {−4, 0, 3}; function; onto

15. D = {all real numbers}, R = {all real numbers}; function; both; continuous

17. D = {all real numbers}, R = $\{y \mid y \geq 0\}$; function; onto; continuous

19. D = {all real numbers}, R = $\{y \mid y \geq 0\}$; function; onto; continuous

21 $f(x) = 5x^3 + 1$ Original function
$f(-8) = 5(-8)^3 + 1$ Substitute −8 for each x.
$= 5(-512) + 1$ Evaluate $(-8)^3$.
$= -2560 + 1$ Multiply.
$= -2559$ Simplify.

23a. {(0, 1), (20, 1.6), (40, 2.2), (60, 2.8), (80, 3.4), (100, 4)}
23b.

Diving Pressure

(y-axis: Pressure (atmospheres), 1–10)
(x-axis: Depth (feet), 0 10 20 30 40 50 60 70 80 90 100)

Left column (R22)

$$\frac{\text{change in height}}{\text{change in time}} \leftarrow \text{mm} \leftarrow \text{days}$$

$$= \frac{40 - 20}{6 - 3}$$

$$= \frac{20}{3}$$

The rate of change is $\frac{20}{3}$ mm/day.

11a. $0.15°/h$ **11b.** $-0.125°/h$; Yes; the number should be negative because her temperature is dropping.

11c. Monday 8:00 A.M.–Monday 8:00 P.M. **13.** $\frac{14}{15}$

15. -2 **17.** $\frac{5}{3}$ **19.5**

21 The line passes through (0, 16) and (10, 8).

$m = \frac{y_2 - y_1}{x_2 - x_1}$ Slope Formula

$= \frac{8 - 16}{10 - 0}$ $(x_1, y_1) = (0, 16), (x_2, y_2) = (10, 8)$

$= \frac{-8}{10}$ or -0.8 Simplify.

23. $\frac{4}{3}$ **25.3** **27.** $\frac{6}{5}$

29 slope $= \frac{\text{change in } y}{\text{change in } x}$

$= \frac{\text{change in vertical distance}}{\text{change in horizontal distance}}$

$= \frac{8.9}{2.8}$

≈ 3.2

31.9 **33.5** **35a.**

35b.

x	−4	−3	−2	−1	0	1	2	3	4
f(x)	16	9	4	1	0	1	4	9	16
slope		−7	−5	−3	−1	1	3	5	7

35c. Sample answer: The rate of change is not constant. The rate of change decreases as x approaches zero and then increases as x approaches infinity. **37.** Sample answer: Because the slope from (2, 3) to (5, 8) is the same as the slope from (5, 8) to (11, y), find the slope between each pair of points and set them equal to each other. Then solve for y.

$\frac{8 - 3}{5 - 2} = \frac{y - 8}{11 - 5}$

$\frac{5}{3} = \frac{y - 8}{6}$

$30 = 3(y - 8)$

$10 = y - 8$

$18 = y$

39. Sometimes; the slope of a vertical line is undefined.

41. $\frac{3}{2}$ **43.** H **45.** Yes; it can be written in $f(x) = mx + b$ form. **47.** No; it cannot be written in $f(x) = mx + b$ form. **49.** -46 **51.** 336 **53.** 11

55. 3.5 **57.** $\frac{7}{3}$

R22 | Selected Answers

Lesson 2-4

1. $y = 1.5x + 5$ **3.** $y = -2x + 11$ **5.** A

7 The slope of the given line is $\frac{7}{8}$. Lines that are parallel have the same slope, so the slope of the line parallel to the given line is $\frac{7}{8}$.

$y - y_1 = m(x - x_1)$ Point-slope form

$y - (-10) = \frac{7}{8}(x - 4)$ $(x_1, y_1) = (4, -10)$ and $m = \frac{7}{8}$

$y + 10 = \frac{7}{8}x - \frac{7}{2}$ Distributive Property

$y = \frac{7}{8}x - \frac{27}{2}$ Subtract 10 from each side.

9. $y = -\frac{1}{2}x + 5$ **11.** $y = 4.5x - 6.5$ **13.** $y = 4x - 15$

15. $y = -\frac{1}{4}x - 1$ **17.** $y = 2x - 2$ **19.** $y = -8x - 20$

21. $y = -0.5x + 3.35$ **23.** $y = \frac{1}{2}x$ **25.** $y = -\frac{1}{2}x + 6$

27. $y = 180x + 5900$ **29.** $y = -25x + 300$

31 First, find the slope. The line passes through (−6, 2) and (0, 6).

$m = \frac{y_2 - y_1}{x_2 - x_1}$ Slope Formula

$= \frac{6 - 2}{0 - (-6)}$ $(x_1, y_1) = (-6, 2), (x_2, y_2) = (0, 6)$

$= \frac{4}{6}$ or $\frac{2}{3}$ Simplify.

The graph intersects the y-axis at 6. So, $b = 6$. Substitute the values into the slope-intercept equation.

$y = mx + b$ Slope-intercept form

$y = \frac{2}{3}x + 6$ $m = \frac{2}{3}, b = 6$

33. 10 mi

35 **a.** Let x be the number of people Ms. Cooper recruits and let y be the amount of money she earns. Use the points (10, 100) and (14, 120) to represent this situation.

$m = \frac{y_2 - y_1}{x_2 - x_1}$ Slope Formula

$= \frac{120 - 100}{14 - 10}$ $(x_1, y_1) = (10, 100), (x_2, y_2) = (14, 120)$

$= \frac{20}{4}$ or 5   Simplify.

Use the slope and either of the given points with the point-slope form to write the equation.

$y - y_1 = m(x - x_1)$ Point-slope form

$y - 100 = 5(x - 10)$ $(x_1, y_1) = (10, 100), m = 5$

$y - 100 = 5x - 50$ Distributive Property

$y = 5x + 50$ Add 100 to each side.

b. The y-intercept of the graph of $y = 5x + 50$ is 50. This represents the money Ms. Cooper would make if she had no recruits. So, $50 is her daily salary.

c. Find the value of y when $x = 20$.

$y = 5x + 50$ Use the equation you found in part a.

$y = 5(20) + 50$ Replace x with 20.

$y = 150$ Simplify.

So, Ms. Cooper would earn $150 in a day if she recruits 20 people.

37. Sample answer: Sometimes; while the two sets of parallel and perpendicular lines will always

Continued middle column (top)

form a quadrilateral with four 90° angles, that figure will always be a rectangle, but not necessarily a square. **39.** Sample answer: $y - 0 = a\left(x + \frac{b}{a}\right)$

41. Sample answer: $y - d = -\frac{d}{c}(x - 0)$ **43.** A

45. G **47.** $-\frac{5}{3}$ **49.** $\frac{1}{5}$ **51.** $x \geq -4$ **53.** $x \geq -\frac{21}{13}$

55. yes **57.** $8c^2 + 8c - 30$ **59.** $-6t^2 + 7t + 20$

61. $\frac{3}{2}$ **63.** $\frac{1}{5}$ **65.** $-\frac{9}{9}$

Lesson 2-5

1 **a.** Graph the data as ordered pairs with the depth on the horizontal axis and the temperature on the vertical axis. Draw a line through two points that appear to represent the data well, such as (0, 22) and (2000, 6).

The data show a weak negative correlation.

b. Sample answer: Use (0, 22) and (2000, 6) to find an equation. First, find the slope of the line through (0, 22) and (2000, 6).

$m = \frac{y_2 - y_1}{x_2 - x_1}$ Slope Formula

$= \frac{6 - 22}{2000 - 0}$ $(x_1, y_1) = (0, 22), (x_2, y_2) = (2000, 6)$

≈ -0.008 Simplify.

Then write the equation.

$y - y_1 = m(x - x_1)$ Point-slope form

$y - 22 = -0.008(x - 0)$ $(x_1, y_1) = (0, 22), m = -0.008$

$y - 22 = -0.008x$ Simplify.

$y = -0.008x + 22$ Add 22 to each side.

c. Sample answer: Find y when $x = 2500$.

$y = -0.008x + 22$ Prediction equation

$y = -0.008(2500) + 22$ $x = 2500$

$= -20 + 22$ or 2   Simplify.

So, at a depth of 2500 m, the temperature in the ocean is about 2°C.

3a.

strong negative correlation

Right column (R23)

3b. Sample answer, using (4, 49,300) and (8, 20,193): $y = -7276.75x + 78,407$ **3c.** Sample answer: 12,916 CDs

5 **a.** Graph the data as ordered pairs with the month on the horizontal axis and the number of gallons sold on the vertical axis. Let 1 represent January, 2 represent February, and so on. Draw a line through two points that appear to represent the data well, such as (1, 37) and (8, 131).

The data show a strong positive correlation.

b. Sample answer: Use (1, 37) and (8, 131) to find an equation. First, find the slope of the line through (1, 37) and (8, 131).

$m = \frac{y_2 - y_1}{x_2 - x_1}$ Slope Formula

$= \frac{131 - 37}{8 - 1}$ $(x_1, y_1) = (1, 37), (x_2, y_2) = (8, 131)$

$= \frac{94}{7}$ Simplify.

Then write the equation.

$y - y_1 = m(x - x_1)$ Point-slope form

$y - 37 = \frac{94}{7}(x - 1)$ $(x_1, y_1) = (1, 37), m = \frac{94}{7}$

$y - 37 = \frac{94}{7}x - \frac{94}{7}$ Distributive Property

$y = \frac{94}{7}x + \frac{165}{7}$ Add 37 to each side.

c. Sample answer: Find y when $x = 9$.

$y = \frac{94}{7}x + \frac{165}{7}$ Prediction equation

$y = \frac{94}{7}(9) + \frac{165}{7}$ $x = 9$

$= \frac{1011}{7}$ Simplify.

≈ 144.4 Use a calculator.

So, about 144 gallons of ice cream are sold in September.

7.

[0, 12] scl: 1 by [0, 1000] scl: 100

$y = 61.9x + 530.2$
(x is the number of years after 2002);
$1.149 million in sales

9 **a.** Graph the data as ordered pairs with the year on the horizontal axis and the attendance on the vertical axis.

connectED.mcgraw-hill.com

37.

D = {all real numbers};
R = {g(x) | g(x) ≤ 4}

35.

39a.

x	−4	−3	−2	−1	0	1	2	3	4
f(x)	0	−1	−2	−3	−4	−3	−2	−1	0

D = {all real numbers};
R = {all whole numbers}

x	−4	−3	−2	−1	0	1	2	3	4
g(x)	4	3	2	1	0	3	6	9	12

39b.

39c.

x	−4	−3	−2	−1	0	1	2	3	4
f(x)	0	−1	−2	−3	−4	−3	−2	−1	0
slope		−1	−1	−1		3	3	3	

x	−4	−3	−2	−1	0	1	2	3	4
g(x)	12	9	6	3	0	3	6	9	12
slope		−3	−3	−3		3	3	3	

39d. The two sections of an absolute value graph have opposite slopes. The slope is constant for each section of the graph.

41.

43. Sample answer: f(x) = −|x − 2| **45.** 3n + 1
47. J **49a.** y = 0.10x + 30.34 **49b.** r = 0.987
49c. about 110 **51.** y = −\frac{3}{2}x + 6 **53.** −8c + 6
55. −99

---

g(x) = −6. There is a circle at (4, −6), so the linear function is defined for {x | x > 4}. Write the piecewise-defined function.

$$g(x) = \begin{cases} -x - 4 & \text{if } x < -3 \\ x + 1 & \text{if } -3 \le x \le 1 \\ -6 & \text{if } x > 4 \end{cases}$$

19. g(x) = $\begin{cases} 8 & \text{if } x \le -1 \\ 2x & \text{if } 4 \le x \le 6 \\ 2x - 15 & \text{if } x > 7 \end{cases}$

21.

23.

D = {all real numbers};
R = {all integers}

25.

D = {all real numbers};
R = {g(x) | g(x) ≥ 0}

27.

D = {all real numbers};
R = {k(x) | k(x) ≥ 3}
D = {all even integers}

29.

D = {all real numbers};
R = {h(x) | h(x) ≥ 0}

D = {all real numbers};
R = {h(x) | h(x) ≤ −2}

31a. f(a) = |a − 60|
31b. {a | a ≥ 0}

31c.

33. f(x) = 0.5x if x > 0, f(x) = 0 if x = 0, and
f(x) = −0.5x if x < 0. So, according to the definition of absolute value, f(x) = |0.5x|.

---

5. If the number of tickets sold is greater than 0 but less than or equal to 250, then the drama club must do 1 performance. If the number of tickets sold is greater than 250 but less than or equal to 500, then the drama club must do 2 performances, and so on. You can use the pattern to make a table, where x is the number of tickets sold and P(x) is the number of performances. Then graph.

x	P(x)
0 < x ≤ 250	1
250 < x ≤ 500	2
500 < x ≤ 750	3
750 < x ≤ 1000	4
1000 < x ≤ 1250	5

Tickets Sold

7.

D = {all real numbers};
R = {all integers}

9.

D = {all real numbers};
R = {f(x) | f(x) ≥ 0}

11.

D = {all real numbers};
R = {s(x) | s(x) ≥ 6}

13.

D = {x | x ≤ 2 or x > 4};
R = {f(x) | f(x) < −7,
or f(x) = 5}

15.

D = {all real numbers};
R = {g(x) | g(x) ≥ −4}

17. The left portion of the graph is the graph of
g(x) = −x − 4. There is a circle at (−3, −1), so the linear function is defined for {x | x < −3}. The middle portion of the graph is the graph of
g(x) = x + 1. There are dots at (−3, −2) and (1, 2), so the linear function is defined for {x | −3 ≤ x ≤ 1}. The right portion of the graph is the graph of

---

b. Use a graphing calculator to find a regression equation for the data. Enter the years in **L1** and the attendance in **L2**. Then select **LinReg(ax + b)** on the **STAT CALC** menu. The regression equation is approximately y = 40,634.9x − 79,543,385.8.

c. The slope of the regression line indicates that the attendance increases by about 40,634.9 each year. **d.** The correlation is stronger between wins and attendance. The correlation coefficient between years and attendance is 0.42; the correlation coefficient between wins and attendance is about 0.79. The greater correlation coefficient implies a stronger correlation.

11a. y = 3.1x − 6177; r = 0.63 **11b.** about $64.2 million
11c. y = 2.6x − 5170; r = 0.986 **11d.** about $62.4 million
11e. Sample answer: The new equation has a correlation coefficient, 0.986, that is extremely close to 1, so this equation should accurately represent the data. **13.** Sample answer: If a and b have a positive correlation, then they are both increasing. If b and c have a negative correlation and b is increasing, then c must be decreasing. If c and d have a positive correlation and c is decreasing, then d must be decreasing. If a is increasing and d is decreasing, then they must have a negative correlation.

15. a; Sample answer: The data show a strong positive correlation which means that the correlation coefficient r should be close to 1.

17. 2 **19.** J **21.** y = 2.5x − 6 **23.** y = −3x − 6
25. 17.5 mi/hr **27.** 1.5 J/N **29.** 120 **31.** $\frac{17}{3}$, $\frac{25}{3}$

1.

D = {all real numbers};
R = {y | y ≤ 4}

3. g(x) = $\begin{cases} x + 4 & \text{if } x < -2 \\ -3 & \text{if } -2 \le x \le 3 \\ -2x + 12 & \text{if } x > 3 \end{cases}$

[2002, 2010] scl:1 by[1,600,000, 2,400,000] scl:100,000

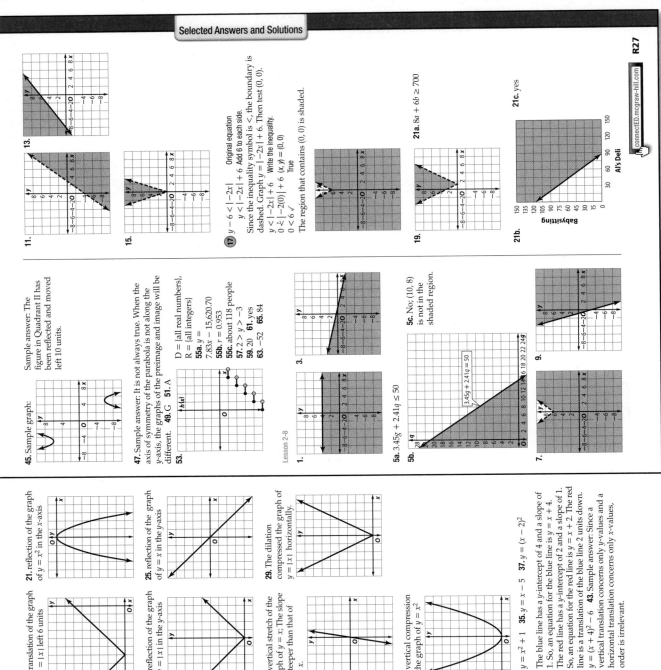

13.

11.

15.

17. $y - 6 < |-2x|$ Original equation
$y < |-2x| + 6$ Add 6 to each side.
Since the inequality symbol is <, the boundary is dashed. Graph $y = |-2x| + 6$. Then test $(0, 0)$.
$y < |-2x| + 6$ Write the inequality.
$0 \stackrel{?}{<} |-2(0)| + 6$ $(x, y) = (0, 0)$
$0 < 6 \checkmark$ True
The region that contains $(0, 0)$ is shaded.

19.

21a. $8a + 6b \geq 700$

21b.

Babysitting

Al's Deli

21c. yes

---

45. Sample graph:

Sample answer: The figure in Quadrant II has been reflected and moved left 10 units.

47. Sample answer: It is not always true. When the axis of symmetry of the parabola is not along the y-axis, the graphs of the preimage and image will be different. **49.** G   **51.** A
53.
$D = \{all\ real\ numbers\}$,
$R = \{all\ integers\}$
55a. $y =$
55b. $r = 0.953$
$7.83x - 15,620.70$
55c. about 118 people
57. $2 > y, y > -3$
59. 20   **61.** yes
63. -52 **65.** 84

Lesson 2-8

1.

3.

5a. $3.45g + 2.41q \leq 50$

5b.

$3.45g + 2.41q = 50$

5c. No; $(10, 8)$ is not in the shaded region.

7.

9.

---

19. translation of the graph of $y = |x|$ left 6 units

21. reflection of the graph of $y = x^2$ in the x-axis

23. reflection of the graph of $y = |x|$ in the y-axis

25. reflection of $y = x$ in the y-axis

27. vertical stretch of the graph of $y = x$; The slope is steeper than that of $y = x$.

29. The dilation compressed the graph of $y = |x|$ horizontally.

31. vertical compression of the graph of $y = x^2$

33. $y = x^2 + 1$ **35.** $y = x - 5$ **37.** $y = (x - 2)^2$

39. The blue line has a y-intercept of 4 and a slope of 1. So, an equation for the blue line is $y = x + 4$. The red line has a y-intercept of 2 and a slope of 1. So, an equation for the red line is $y = x + 2$. The red line is a translation of the blue line 2 units down.

41. $y = (x + 4)^2 - 6$ **43.** Sample answer: Since a vertical translation concerns only y-values and a horizontal translation concerns only x-values, order is irrelevant.

57.

59.

Lesson 2-7

1. linear

3. translation of the graph of $y = x^2$ down 4 units

5. reflection of the graph of $y = |x|$ in the x-axis

7. A vertical compression of the graph of $y = x$; the slope is not as steep as that of $y = x$.

9. The function is a dilation and translation. The graph of $f(x) = \frac{1}{2}|x - 12|$ compresses the graph $f(x) = |x|$ vertically and translates it 12 units to the right.

11. The graph is a curve that appears symmetrical. The graph represents a quadratic function.

13. linear

15. translation of the graph of $y = |x|$ down 3 units

17. translation of the graph of $y = x$ up 2 units or left 2 units

m.connectED.mcgraw-hill.com

13.

Lesson 3-1

1. (3, 5) **3.** (3, −3) **5.** (6, 7)

7 Write each equation in slope-intercept form. Then graph.

$4x + 5y = −41 \quad \rightarrow \quad y = −\frac{4}{5}x − \frac{41}{5}$

$3y − 5x = 5 \quad \rightarrow \quad y = \frac{5}{3}x + \frac{5}{3}$

The graphs appear to intersect at (−4, −5). So, the solution is (−4, −5).

9a. $y = 0.15x + 2.70, y = 0.25x$
9b. $6.75 for 27 photos
9c. You should use EZ Online photos if you are printing more than 27 digital photos, and the local pharmacy if you are printing fewer than 27 photos.

11.

consistent and dependent

13. (−2, 1)

15 $2a + 8b = −8$ Multiply by 5. → $10a + 40b = −40$
$3a − 5b = 22$ Multiply by 8. → $24a − 40b = 176$
$10a + 40b = −40$ Equation 1 × 5
$(+) 24a − 40b = 176$ Equation 2 × 8
$34a \qquad = 136$ Add the equations.
$a = 4$ Divide each side by 34.

Substitute 4 for a into either original equation.
$2a + 8b = −8$ Equation 1
$2(4) + 8b = −8$ $a = 4$
$8 + 8b = −8$ Multiply.
$8b = −16$ Subtract 8 from each side.
$b = −2$ Divide each side by 8.

The solution is (4, −2).

59.

61.

CHAPTER 3
Systems of Equations and Inequalities

Chapter 3 Get Ready

1.

3.

5.

7a. $8.50a + 5.25c = 650$

7b.

Adult Tickets

9.

11.

45.

47.

49. $4x − 15y = 6; A = 4, B = −15, C = 6$ **51.** 1000
53. $−6x^2 − 23x − 15$

55.

57.

Chapter 2 Study Guide and Review

1. one-to-one **3.** identity **5.** scatter plot
7. D = {1, 3, 5, 7}, R = {2, 4, 6, 8}; function; both
9. D = {−4, −2, 1, 3}, R = {−4, 1, 3, 5}; not a function
11. −10 **13.** 2 **15.** $3a + 2$ **17.** D = {1, 2, 3, 4, 5, ...}, R = {5.75, 9, 12.25, 15.5, 18.75, ...}; a function; discrete **19.** No; the variables have an exponent other than 1. **21.** yes **23.** No; x appears in a denominator. **25.** $12x − y = 0; 12, −1, 0$
27. $x − 2y = −3; 1, −2, −3$ **29.** 18 **31.** −1
33. $y = −2x − 11$ **35.** $y = 2x + 8$ **37.** $y = −4x + 25$
39. $y = 4x − 2$ **41.** $y = 32x + 250$ and (5, 24): $y = −5x + 49$ using (1, 44) and (5, 24) **43.** Sample answer

45.

47.

D = {all real numbers},
R = {all integers}

49. quadratic **51.** $y = x^2$ shifted down 3 units
53. parabola

55.

57.

23.

25.

27. all real numbers (The graph would be shaded everywhere.)

29 a. 20*d* represents the cost of *d* DVDs and 15*c* represents the cost of *c* CDs. Because the sum can equal the maximum $400, the inequality symbol is ≤. The inequality is $20d + 15c ≤ 400$.
b. Graph the boundary $20d + 15c = 400$. Since the inequality symbol is ≤, the boundary is solid. Then test the point (0, 0).

$20d + 15c ≤ 400$ Original inequality
$20(0) + 15(0) \overset{?}{≤} 400$ $(c, d) = (0, 0)$
$0 ≤ 400 ✓$ True

The region that contains (0, 0) is shaded.

c. Sample answer: Since the points (18, 5), (12, 10), and (6, 15) lie inside the shaded region, they satisfy the inequality. This means that Susan can buy 18 CDs and 5 DVDs, 12 CDs and 10 DVDs, or 6 CDs and 15 DVDs.

31.

33. Sample answer: $|y| < x$ **35.** Paulo; $x − y ≥ 2$ can be written as $y ≤ x − 2$. **37.** Sample answer: One possibility is when $|y| < 0$. In order for there to be a solution, the absolute value of y will need to be less than 0, and, by definition of absolute value, this is impossible. **39.** C **41.** K
43. $y = |x + 4| − 5$

R28 | Selected Answers

connectED.mcgraw-hill.com

R31

81. $12xy + 18y^2 - 15y$ 83. J   85a. $10s + 15\ell \geq 350$

85b.

$[10x + 15\ell = 350]$
45
40
35
30
25
20
15
10
5
O   5 10 15 20 25 30 35 40 45 s

85c. no   87. $-|x - 3|$ 89. 7   91. 3.2   93. -8

95. yes   97. no

Lesson 3-2

1.

$[y = 6]$
$[y = -3 + x]$

3.

$[y = 3x - 3]$
$[y = -2x + 4]$

5.

6
2
$-8 -6 -4 -2$ O   2 4 6 8 x
-6
-8
-10
-12

7.

9. The solution of $y < -3x + 4$ is the region to the left of the boundary. The solution of $3y + x > -6$ is the region above the boundary. The intersection of the two regions is the solution of the system.

(3.5, 8), (−4, 8), (0.5, 2)

connectED.mcgraw-hill.com

Slope:
$$m = \frac{y_2 - y_1}{x_2 - x_1}$$
$$= \frac{18 - 4}{11 - 3}$$
$$= \frac{14}{8} \text{ or } \frac{7}{4}$$

Equation:
$$y - y_1 = m(x - x_1)$$
$$y - 4 = \frac{7}{4}(x - 3)$$
$$y - 4 = \frac{7}{4}x - \frac{21}{4}$$
$$y = \frac{7}{4}x - \frac{5}{4}$$

Find the point of intersection of the diagonals.

$y = -\frac{3}{2}x + 12$ Equation 1

$\frac{7}{4}x - \frac{5}{4} = -\frac{3}{2}x + 12$ Substitute $\frac{7}{4}x - \frac{5}{4}$ for y.

$\frac{13}{4}x - \frac{5}{4} = 12$ Add $\frac{3}{2}x$ to each side.

$\frac{13}{4}x = \frac{53}{4}$ Add $\frac{5}{4}$ to each side.

$x = \frac{53}{13}$ Multiply each side by $\frac{4}{13}$.

$y = -\frac{3}{2}x + 12$ Equation 1

$y = -\frac{3}{2}\left(\frac{53}{13}\right) + 12$ $x = \frac{53}{13}$

$y = -\frac{159}{26} + 12$ Multiply.

$y = \frac{153}{26}$ Simplify.

The diagonals intersect at $\left(\frac{53}{13}, \frac{153}{26}\right)$.

75a.

Equation 1		Equation 2		Equation 3	
x	y	x	y	x	y
0	$\frac{16}{3}$	0	−4	0	10
1	5	1	−2	1	5
2	$\frac{14}{3}$	2	0	2	0
3	$\frac{13}{3}$	3	2	3	−5
4	4	4	4	4	−10

75b. Equations 1 and 2 intersect at (4, 4), equations 2 and 3 intersect at (2, 0), and equations 1 and 3 intersect at (1, 5); there is no solution that satisfies all three equations.

75c.
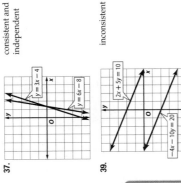
$[Equation 2]$ $[Equation 1]$ $[Equation 3]$

75d. If all three lines intersect at the same point, then the system has a solution. The system has no solution if the lines intersect at 3 different points, or if two or three lines are parallel.

77. $a \neq 0$, $b = 3$

79. Sample answer:
$4x + 5y = 21 \rightarrow 3(4x + 5y = 21)$
$3x - 2y = 10 \rightarrow 4(3x - 2y = 10)$

$12x + 15y = 63$
$(-)\ 12x - 8y = 40$
$23y = 23$
$y = 1$

$4x + 5(1) = 21$
$4x + 5 = 21$
$4x = 16$
$x = 4$

The solution is (4, 1).

$m = \frac{y_2 - y_1}{x_2 - x_1}$ Slope Formula

$= \frac{10.78 - 11.4}{44 - 0}$ $(x_1, y_1) = (0, 11.4)$, $(x_2, y_2) = (44, 10.78)$

$= -0.01409$ Simplify.

At $x = 0$, $y = 11.4$. So, the y-intercept is 11.4.

$y = mx + b$ Slope-intercept form

$y_w = -0.01409x + 11.4$ $m = -0.01409$, $b = 11.4$

b.

20
18
16
14
12
10
8
6
4
2
0

Time (seconds)
$[y_w = -0.01409x + 11.4]$
$[y_m = -0.00705x + 10]$

Years after 1964

The graphs appear to intersect at (198, 8.6). This means that 198 years after 1964, the winning time for both men and women will be about 8.6 seconds. So, based on these data, the women's performance will catch up to the men's performance 198 years after 1964, or in the year 2162. The next Olympic year would be 2164. This prediction is not reasonable. It is unlikely that women's times will ever catch up to men's times because the times cannot continue to increase and decrease infinitely.

71. Let x represent the cost of an adult and y represent the cost of a student.

Van A: $2x + 5y = 77$
Van B: $2x + 7y = 95$

$2x + 5y = 77$ Multiply by −1. $\rightarrow -2x - 5y = -77$
  Equation 1 × (−1)

$(+)\ 2x + 7y = 95$ Equation 2
$2y = 18$ Add the equations.
$y = 9$ Divide each side by 2.

$2x + 7y = 95$ Equation 2
$2x + 7(9) = 95$ $y = 9$
$2x + 63 = 95$ Multiply.
$2x = 32$ Subtract 63 from each side.
$x = 16$ Divide each side by 2.

The solution is (16, 9). So, the cost of an adult is $16 and the cost of a student is $9.

73. Find an equation for the diagonal that goes through (6, 3) and (2, 9).

Slope:
$$m = \frac{y_2 - y_1}{x_2 - x_1}$$
$$= \frac{9 - 3}{2 - 6}$$
$$= \frac{6}{-4} \text{ or } -\frac{3}{2}$$

Equation:
$$y - y_1 = m(x - x_1)$$
$$y - 3 = -\frac{3}{2}(x - 6)$$
$$y - 3 = -\frac{3}{2}x + 9$$
$$y = -\frac{3}{2}x + 12$$

Find an equation for the diagonal that goes through (3, 4) and (11, 18).

---

17. (5, 1)   19. (−2, 7)   21. (−4, −3)   23. no solution

25. B   27. (4, −1)   29. 250 T-shirts   31. (−3, −4)

33. infinite solutions   35. (−1.5, −2)

37.
consistent and independent
$[y = 3x - 4]$
$[y = 6x - 8]$

39.
inconsistent
$[2x + 5y = 10]$

41. Write each equation in slope-intercept form. Then graph.

$-5x - 6y = 13 \rightarrow y = -\frac{5}{6}x - \frac{13}{6}$

$12y + 10x = -26 \rightarrow y = -\frac{5}{6}x - \frac{13}{6}$

$[-5x - 6y = 13]$
$[12y + 10x = -26]$

Because the equations are equivalent, their graphs are the same line. The system is consistent and dependent.

43. infinite solutions   45. (8, 4)   47. (−3, −1)

49a. $x + y = 13$ and $4x + 2y = 38$ 49b. 6 doubles matches and 7 singles matches   51. (0, 4)   53. no solution   55. (8, −6)   57. (5, 4)   59. (2.07, −0.39)

61. (15.03, 10.98)   63. (−5, 4)   65. infinite solutions

67. (16, −8)

69. a. Sample answer: For men, you can use (0, 10) and (44, 9.69) to write an equation.

$m = \frac{y_2 - y_1}{x_2 - x_1}$ Slope Formula

$= \frac{9.69 - 10}{44 - 0}$ $(x_1, y_1) = (0, 10)$, $(x_2, y_2) = (44, 9.69)$

$= -0.00705$ Simplify.

At $x = 0$, $y = 10.0$. So, the y-intercept is 10.

$y = mx + b$ Slope-intercept form

$y_m = -0.00705x + 10$ $m = -0.00705$, $b = 10$

For women, you can use (0, 11.4) and (44, 10.78) to write an equation.

45.

$(-3, -4), (-6, 8), (2, 6), (5, 0)$; 75 units^2

47. Sample answer:
$y \geq 2x - 6$,
$y \leq -0.5x + 4$,
$y \geq -3x - 6$; 47

49. Sample answer: Shade each inequality in their standard way, by shading above the line if $y >$ and shading below the line if $y <$ (or you can use test points). Once you determine where to shade for each inequality, the area where *every* inequality needs to be shaded is the actual solution. This is only the shaded area. **51.** A **53.** $\frac{4}{5}z$

55.

$x + y = 6$

57.

$5x + 7y = -20$

59.

$g(x)$

61.

$h(x)$

$D = \{x \mid x < -2 \text{ or } x > 2\}$, $R = [-1, 1]$

$D = \{\text{all real numbers}\}$,
$R = \{g(x) \mid g(x) \leq 2\}$

63. −1 **65.** 3 **67.** 4.5

Lesson 3-3

1.

$(2, -4), (4, -4)$; max does not exist, min = −52

3.

$(4, 5), (4, -4), (-5, 5)$; max = 28, min = −35

33.

35.

37.

Number of Books

39. Let $w =$ the number of hours writing, and let $e =$ the number of hours exercising.
$w + e \leq 35$
$7 \leq e \leq 15$
$20 \leq w \leq 25$

Number of Hours Writing
Number of Hours Exercising

41. $(-6, -2), \left(-3\frac{13}{17}, 6\frac{16}{17}\right), \left(9\frac{1}{7}, 3\frac{5}{7}\right), (0.8, -8.8)$

43. Let x represent the amount in the fund that pays 6% interest and y represent the amount in the fund that pays 10% interest. Write a system of inequalities and then graph.

$x + y \leq 10{,}000$ Total amount invested is up to $10,000.
$0.06x + 0.10y \geq 740$ Total amount earned is at least $740.

The least amount Mr. Hoffman can invest in the risky fund, or the 10% interest fund, is $3500.

6% Interest Fund

---

25. Let d represent the number of daytime minutes and n represent the number of nighttime minutes. Write a system of inequalities and then graph.

$d + n \leq 800$ Maximum number of minutes is 800.
$d \geq 2n$ At least twice as many daytime minutes as nighttime minutes
$n \geq 200$ At least 200 nighttime minutes

The intersection points are at (400, 200), (533.3, 266.7), and (600, 200).

Daytime Minutes

$0.15d + 0.1n = 0.15(400) + 0.1(200)$
$= 60 + 20$ or 80 400 daytime min., 200 nighttime min.

$0.15d + 0.1n = 0.15(533.3) + 0.1(266.7)$
$\approx 80.0 + 26.7$ or 106.70 533.3 daytime min., 266.7 nighttime min. Simplify.

$0.15d + 0.1n = 0.15(600) + 0.1(200)$
$= 90 + 20$ or 110 600 daytime min., 200 nighttime min. Simplify.

So, his maximum bill is $110 and his minimum bill is $80.

27a.

Days

27b. $3\frac{1}{3}$ days **27c.** Marc; Jessica could last about a quarter of a day longer than Marc.

29.

31.

11.

13.

15.

17.

$x + y = 200$
$9x + 12y = 925$

19.

$(2, -1), (5, 8), (-7, 8)$

21.

$(-3, 1.5), (5, 7.5), (0, -6)$

23.

$(8, 4), (6, -4), (-2, 3)$

connectED.mcgraw-hill.com

Selected Answers and Solutions

45. 6; −2

2x − 6y = 12

47. 5; 2

2x + 5y − 10 = 0

49. $\frac{1}{2}$; −2

y = 4x − 2

51. 9   **53.** −5   **55.** 15

Lesson 3-4

1. (−2, −3, 5)   **3.** (−4, 3, 6)   **5.** infinite solutions
7a. $s + d + t = 7$, $d = 2s$, $0.3s + 0.6d + 0.6t = 3.6$
7b. 2 sitcoms, 4 dramas, 1 talk show

9 −a + 4b + 2c = −13   Multiply by 4.   →   −4a + 16b + 8c = −52   Equation 3 (× 4)

$$\begin{array}{rcl} -4a + 16b + 8c &=& -52 \quad \text{Equation 3 (×4)} \\ (+)\, 4a + 5b - 6c &=& 2 \quad \text{Equation 1} \\ \hline 21b + 2c &=& -50 \end{array}$$

−a + 4b + 2c = −13   Multiply by −3.   →   3a − 12b − 6c = 39   Equation 2

$$\begin{array}{rcl} 3a - 12b - 6c &=& 39 \quad \text{Equation 3 × (−3)} \\ (+)\, -3a - 2b + 7c &=& -15 \quad \text{Equation 2} \\ \hline -14b + c &=& 24 \end{array}$$

The resulting system of two equations and two variables is shown below.

21b + 2c = −50   New Equation 1
−14b + c = 24   New Equation 2

−14b + c = 24   Multiply by −2.   →   28b − 2c = −48   New Equation 1

$$\begin{array}{rcl} 21b + 2c &=& -50 \quad \text{New Equation 1} \\ (+)\, 28b - 2c &=& -48 \quad \text{New Equation 2 × (−2)} \\ \hline 49b &=& -98 \quad \text{Add the equations.} \end{array}$$

b = −2   Divide each side by 49.

−14b + c = 24   New Equation 2
−14(−2) + c = 24   Replace b with −2.
28 + c = 24   Multiply.
c = −4   Subtract 28 from each side.

−a + 4b + 2c = −13   Original Equation 3
−a + 4(−2) + 2(−4) = −13   b = −2 and c = −4
−a − 8 − 8 = −13   Multiply.
−a − 8 = −13   Add 16 to each side.
−a = 3   Multiply each side by −1.
a = 3

The solution is (3, −2, −4).

11. (−2, −1, 4)   **13.** infinite solutions   **15.** (−4, −1, 6)
17. no solution   **19.** infinite solutions   **21.** roller
coasters: 5; bumper cars: 1; water slides: 4

---

27 a. Let x represent the number of small packages and y represent the number of large packages. Write a system of inequalities. Then graph the inequalities and locate the vertices.

$x \geq 0$ number of small packages ≥ 0
$y \geq 0$ number of large packages ≥ 0
$25x + 50y \leq 4200$ weight of packages ≤ 4200 lb
$3x + 5y \leq 480$ capacity of packages ≤ 480 cu ft

The vertices are at (0, 84), (120, 24), and (160, 0). Evaluate the function f(x, y) = 5x + 8y at each vertex.

(x, y)	5x + 8y	f(x, y)
(0, 84)	5(0) + 8(84)	672
(120, 124)	5(120) + 8(24)	792
(160, 0)	5(160) + 8(0)	800

To maximize revenue, 160 small packages and 0 large packages should be placed on a train car. **b.** The maximum revenue per train car is $800. **c.** No; if revenue is maximized, the company will not deliver any large packages, and customers with large packages to ship will probably choose another carrier.

29. Sample answer: −2 ≥ y ≥ −6, 4 ≤ x ≤ 9
31. b; The feasible region of Graph b is unbounded while the other three are bounded.
33. Sample answer: Even though the region is bounded, multiple maximums occur at A and B and all of the points on the boundary of the feasible region containing both A and B. This happened because that boundary of the region has the same slope as the function.   **35.** $70.20   **37.** D

39.

41. 7x + 15y = 330, 8x + 16y = 360; hats: 15, shirts: 15
43. $y = -\frac{1}{2}x + \frac{7}{2}$

5.

(4, 2), (−1, −3), (−6, 7); max does not exist; min = −30

7a. $g \geq 0$, $c \geq 0$, $1.5g + c \leq 85$, $2g + 0.5c \leq 40$
7b.

7c. (0, 0), (0, 20), (80, 0)   **7d.** $f(c, g) = 65c + 50g$
7e. 80 specialty boards, 0 pro boards; $5200

9 Graph the inequalities and locate the vertices.

The vertices are at (2, −10), (−3, 0), (−3, 6.5), and (2, 9). Evaluate the function at each vertex.

(x, y)	−4x − 9y	f(x, y)
(2, −10)	−4(2) − 9(−10)	82
(−3, 0)	−4(−3) − 9(−0)	12
(−3, 6.5)	−4(−3) − 9(6.5)	−46.5
(2, 9)	−4(2) − 9(9)	−89

The maximum value is 82 at (2, −10). The minimum value is −89 at (2, 9).

11.

(6, −8), (4, −2), (−2, −2), (−8, −8); max = 59, min = −8

13.

(−10, 3), (2, 3), (−6, 7), (−8, 7); max = 59, min = 9

---

15 Graph the inequalities and locate the vertices.

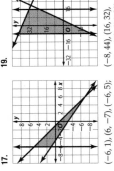

The vertices are at (6, 3), (−8, 10), and (−8, −18). Evaluate the function at each vertex.

(x, y)	10x − 6y	f(x, y)
(6, 3)	10(6) − 6(3)	42
(−8, 10)	10(−8) − 6(10)	−140
(−8, −18)	10(−8) − 6(−18)	28

The maximum value is 42 at (6, 3). The minimum value is −140 at (−8, 10).

17.

(−8, 44), (16, 32), (−2, −8), (−8, −26), (16, 22); max = 672, min = −486

19.

(5, −1), (1, 6), (−2, −8), (−4, −8), (−4, 6); max = 60, min = −112

21.

(−6, 1), (6, −7), (−6, 5); max = 48, min = 0

23. 225 yellow cakes, 0 strawberry cakes
25a. $a \geq 0$, $b \geq 0$, $a + b \leq 45$, $4a + 5b \leq 200$
25b.

(0, 0), (0, 40), (25, 20), (45, 0)

25c. 25 sheds, 20 play houses
25d. $1250

43. $(4, -3, -2)$

45.

47.

$y = -4x + 6$

$3y = 2x + 9$

$x - 2y = 8$

$y = 2x - 3$

49. $350,349 - x = 15,991; 334,358$

51. $-10a - b$

Lesson 3-6

1. 2×3 **3.** 8×10 **5.** $\begin{bmatrix} 0 & 44 \\ 8 & -34 \end{bmatrix}$

7. The product of a 2×1 matrix and a 1×3 matrix is a 2×3 matrix.

$\begin{bmatrix} -9 \\ 6 \end{bmatrix} \cdot \begin{bmatrix} -1 & -10 & 1 \end{bmatrix}$

$= \begin{bmatrix} -9(-1) & -9(-10) & -9(1) \\ 6(-1) & 6(-10) & 6(1) \end{bmatrix}$ Multiply the column by the row.

$= \begin{bmatrix} 9 & 90 & -9 \\ -6 & -60 & 6 \end{bmatrix}$ Simplify.

9. $\begin{bmatrix} -44 \\ 25 \end{bmatrix}$ **11.** $\begin{bmatrix} -16 & -1 \\ -6 & 10 \end{bmatrix}$

13. No; $\begin{bmatrix} 53 & -87 \\ -2 & -60 \end{bmatrix} \neq \begin{bmatrix} 62 & -33 \\ 28 & -69 \end{bmatrix}$.

15. 2×4 **17.** undefined   **19.** undefined   **21.** $\begin{bmatrix} 26 \end{bmatrix}$

23. $\begin{bmatrix} -3 & -7 \\ -2 & -1 \end{bmatrix} \cdot \begin{bmatrix} 4 & 4 \\ 9 & -3 \end{bmatrix}$ Multiply the 1st row in the 1st matrix by each column in the 2nd matrix.

$= \begin{bmatrix} -3(4) + (-7)(9) & -3(4) + (-7)(-3) \\ -2(4) + (-1)(9) & -2(4) + (-1)(-3) \end{bmatrix}$ Multiply the 2nd row in the 1st matrix by each column in the 2nd matrix.

$= \begin{bmatrix} -12 + (-63) & -12 + 21 \\ -8 + (-9) & -8 + 3 \end{bmatrix}$ Multiply.

$= \begin{bmatrix} -75 & 9 \\ -17 & -5 \end{bmatrix}$ Add.

25. undefined   **27.** $\begin{bmatrix} -40 & 64 \\ 22 & 1 \end{bmatrix}$

29a. $I = \begin{bmatrix} 3 & 2 \\ 4 & 3 \end{bmatrix}$, $C = \begin{bmatrix} 220 \\ 360 \end{bmatrix}$ **29b.** $\begin{bmatrix} \$1880 \\ \$1550 \\ \$1630 \end{bmatrix}$

29c. $\$5060$ **31.** $PQR = \begin{bmatrix} -22 & 240 \\ 44 & -12 \end{bmatrix}$ and

$RQP = \begin{bmatrix} 34 & -40 \\ -220 & -44 \end{bmatrix}$

33. No; $R(P + Q) = \begin{bmatrix} 34 & -6 \\ -64 & -30 \end{bmatrix}$ and

$PR + QR = \begin{bmatrix} 22 & 72 \\ 14 & -18 \end{bmatrix}$.

---

27. $-5 \left(\begin{bmatrix} 4 & -8 \\ 8 & -9 \end{bmatrix} + \begin{bmatrix} 4 & -2 \\ -3 & -6 \end{bmatrix} \right)$

$= -5 \begin{bmatrix} 4 + 4 & -8 + (-2) \\ 8 + (-3) & -9 + (-6) \end{bmatrix}$ Add corresponding elements.

$= -5 \begin{bmatrix} 8 & -10 \\ 5 & -15 \end{bmatrix}$ Simplify.

$= \begin{bmatrix} -5(8) & -5(-10) \\ -5(5) & -5(-15) \end{bmatrix}$ Distribute the scalar.

$= \begin{bmatrix} -40 & 50 \\ -25 & 75 \end{bmatrix}$ Multiply.

29a. Write a matrix to show the American records and a matrix to show the world records. Then subtract.

$\begin{bmatrix} 24.63 \text{ s} \\ 53.99 \text{ s} \\ 1:57.41 \text{ min} \\ 8:16.22 \text{ min} \end{bmatrix} - \begin{bmatrix} 24.13 \text{ s} \\ 53.52 \text{ s} \\ 1:56.54 \text{ min} \\ 8:16.22 \text{ min} \end{bmatrix}$

$= \begin{bmatrix} 24.63 \text{ s} - 24.13 \text{ s} \\ 53.99 \text{ s} - 53.52 \text{ s} \\ 1:57.41 \text{ min} - 1:56.54 \text{ min} \\ 8:16.22 \text{ min} - 8:16.22 \text{ min} \end{bmatrix}$ Subtract corresponding elements.

$= \begin{bmatrix} 0.5 \text{ s} \\ 0.47 \text{ s} \\ 0.87 \text{ s} \\ 0 \text{ s} \end{bmatrix}$ Simplify.

b. In the 50-meter, the fastest American time is 0.5 second behind the world record. In the 100 m, the fastest American time is 0.47 second behind the world record. In the 200 m, the fastest American time is 0.87 second behind the world record. In the 800 m, the American and world records are the same. So, it was an American who set the world record.

c. In the 50-meter and 100-meter events, the fastest times were set at the Olympics. These times became world records.

31. To show the Commutative Property of Matrix Addition is true for 2×2 matrices,

let $A = \begin{bmatrix} a & b \\ c & d \end{bmatrix}$ and $B = \begin{bmatrix} e & f \\ g & h \end{bmatrix}$. Show that

$A + B = B + A$.

$A + B = \begin{bmatrix} a & b \\ c & d \end{bmatrix} + \begin{bmatrix} e & f \\ g & h \end{bmatrix}$ Substitution

$= \begin{bmatrix} a + e & b + f \\ c + g & d + h \end{bmatrix}$ Definition of matrix addition

$= \begin{bmatrix} e + a & f + b \\ g + c & h + d \end{bmatrix}$ Commutative Property of Addition for Real Numbers

$= \begin{bmatrix} e & f \\ g & h \end{bmatrix} + \begin{bmatrix} a & b \\ c & d \end{bmatrix}$ Definition of matrix addition

$= B + A$ Substitution

33. $\begin{bmatrix} 7 & 5 \\ -1 & 2 \end{bmatrix}$ **35.** Sample answer: $A = \begin{bmatrix} 6 & 1 \\ 6 & 3 \end{bmatrix}$

and $B = \begin{bmatrix} 3 & 2 \\ 4 & 2 \end{bmatrix}$ **37.** C   **39.** F   **41.** $(3, -1, 4)$

connectED.mcgraw-hill.com

R37

---

23. a = the amount invested in account A
b = the amount invested in account B
c = the amount invested in account C

$a + b + c = 100,000$ She invested a total of $100,000.
$a = c + 30,000$ She invested $30,000 more in account A than account C.
$0.04a + 0.08b + 0.1c = 6300$ The expected interest earned is $6300.

Substitute $a = c + 30,000$ in Equations 1 and 3.

$a + b + c = 100,000$ Equation 1
$c + 30,000 + b + c = 100,000$ $a = c + 30,000$
$30,000 + b + 2c = 100,000$ Add.
$b + 2c = 70,000$ Simplify.

$0.04a + 0.08b + 0.1c = 6300$ Equation 3
$0.04(c + 30,000) + 0.08b + 0.1c = 6300$ $a = c + 30,000$
$0.04c + 1200 + 0.08b + 0.1c = 6300$ Distribute.
$1200 + 0.08b + 0.14c = 6300$ Add.
$0.08b + 0.14c = 5100$ Simplify.

Solve the system of two equations in two variables.
$b + 2c = 70,000$ Multiply by -0.08. $\rightarrow$ $-0.08b + 0.08b + 0.14c =$

$-0.08b - 0.16c = -5600$
$(+) \; 0.08b + 0.14c = 5100$
$-0.02c = -500$
$c = 25,000$

Substitute to find b.
$b + 2c = 70,000$ Remaining equation in two variables
$b + 2(25,000) = 70,000$ $c = 25,000$
$b + 50,000 = 70,000$ Distribute.
$b = 20,000$ Simplify.

Substitute to find a.
$a + b + c = 100,000$ Equation 1
$a + 20,000 + 25,000 = 100,000$ $b = 20,000, c = 25,000$
$a + 45,000 = 100,000$ Add.
$a = 55,000$ Simplify.

The solution is $(55,000, 20,000, 25,000)$. She invested $55,000 in account A, $20,000 in account B, and $25,000 in account C.

25. $y = -3x^2 + 4x - 6; a = -3, b = 4, c = -6$

27. Sample answer:
$3x + 4y + z = -17$
$2x - 5y - 3z = -18;$
$-x + 3y + 8z = 47$

$3x + 4y + z = -17$
$3(-5) + 4(-2) + 6 = -17$
$-15 + (-8) + 6 = -17$
$-17 = -17$ ✓

$2x - 5y - 3z = -18$
$2(-5) - 5(-2) - 3(6) = -18$
$-10 + 10 - 18 = -18$
$-18 = -18$ ✓

$-x + 3y + 8z = 47$
$-(-5) + 3(-2) + 8(6) = 47$
$5 - 6 + 48 = 47$
$47 = 47$ ✓

29. Sample answer: First, combine two of the original equations using elimination to form a new equation

with three variables. Next, combine a different pair of the original equations using elimination to eliminate the same variable and form a second equation with three variables. Do the same thing with a third pair of the original equations. You now have a system of three equations with three variables. Follow the same procedure you learned in this section. Once you find the three variables, you need to use them to find the eliminated variable.   **31.** J   **33.** A   **35.** 16; -8

37. 9; -8 **39.** $(6, 1)$ **41.** $(8, -5)$

Lesson 3-5

1a. $\begin{bmatrix} 23 & 21 & 21 & 42 & 61 \\ 25 & 24 & 32 & 49 & 70 \end{bmatrix}$ **1b.** Sample answer:

City: The sum is 168. However, this value is irrelevant since it is the sum of 5 different types of data. Highway: The sum is 200. However, this value is irrelevant since it is the sum of 5 different types of data.   **1c.** Sample answer: The sums are 48, 45, 53, 91, and 131. These values are irrelevant since they are the sums of 2 different types of data.   **3.** impossible

5. $\begin{bmatrix} 7 & 31 & -14 \\ 1 & -6 & 2 \end{bmatrix}$ **7.** $\begin{bmatrix} -90 & 54 & -12 & -18 \\ -36 & 66 & -84 & 12 \\ -24 & 48 & 60 & -162 \end{bmatrix}$

9. $\begin{bmatrix} 50 & 36 \\ -87 & 41 \end{bmatrix}$ **11.** $\begin{bmatrix} -24 & 29 \\ -38 & -7 \end{bmatrix}$

13a. $\begin{bmatrix} 3 & 2 & 1 \\ 4 & 3 & 2 \\ 5 & 5 & 4 \\ 1 & 5 & 2 \end{bmatrix}$ **13b.** Sample answer:

Brand C; it was given the highest rating possible for cost and comfort, and a high rating for looks, and it will last a fairly long time.   **13c.** Sample answer: Yes; finding the sum of the rows and then calculating the average will provide an easy way to compare the data.

15. To find the sum of two matrices, they must have the same dimensions. Since the dimensions of the matrices are 2×2 and 2×3, it is impossible to find the sum.

17. $\begin{bmatrix} -33 & -26 \\ 25 & 75 \end{bmatrix}$ **19.** impossible

21a. Library A: $\begin{bmatrix} 10,000 \\ 5000 \\ 5000 \end{bmatrix}$; Library B: $\begin{bmatrix} 15,000 \\ 10,000 \\ 2500 \end{bmatrix}$;

Library C: $\begin{bmatrix} 4000 \\ 700 \\ 800 \end{bmatrix}$ **21b.** $\begin{bmatrix} 29,000 \\ 15,700 \\ 8300 \end{bmatrix}$ **21c.** $\begin{bmatrix} 6000 \\ 4300 \\ 4200 \end{bmatrix}$

21d. $\begin{bmatrix} 25,000 \\ 15,000 \\ 7500 \end{bmatrix}$; Sample answer: The sum represents the combined size of the two libraries.

23. $\begin{bmatrix} 5 & 24 \\ \frac{167}{12} & -10 \end{bmatrix}$

25. $\begin{bmatrix} -8a & 32b & 8c - 8b \\ -104 & 80 & -40c \end{bmatrix}$

R36 | Selected Answers

Selected Answers and Solutions

Selected Answers and Solutions

35. a. The bonuses can be found by multiplying the number of cars sold by the amount of bonus given for each new and used car sold.

	Cars		Bonus
	27	49	
$C =$	35	36	$B = \begin{bmatrix} 1000 \\ 500 \end{bmatrix}$
	9	56	
	15	62	

$$CB = \begin{bmatrix} 27 & 49 \\ 35 & 36 \\ 9 & 56 \\ 15 & 62 \end{bmatrix} \cdot \begin{bmatrix} 1000 \\ 500 \end{bmatrix}$$ Write an equation.

$$= \begin{bmatrix} 27(1000) + 49(500) \\ 35(1000) + 36(500) \\ 9(1000) + 56(500) \\ 15(1000) + 62(500) \end{bmatrix}$$ Multiply columns by rows.

$$= \begin{bmatrix} 51,500 \\ 53,000 \\ 37,000 \\ 46,000 \end{bmatrix}$$ Simplify.

Westin earned the most, $53,000.

b. total amount on bonuses = 51,500 + 53,000 + 37,000 + 46,000 or $187,500

47b. $C(A + B) = \begin{bmatrix} a & b \\ c & d \end{bmatrix} \left(\begin{bmatrix} e & f \\ g & h \end{bmatrix} + \begin{bmatrix} j & k \\ m & n \end{bmatrix} \right)$

$= \begin{bmatrix} a & b \\ c & d \end{bmatrix} \begin{bmatrix} e+j & f+k \\ g+m & h+n \end{bmatrix}$ Substitution

$= \begin{bmatrix} a(e+j) + b(g+m) & a(f+k) + b(h+n) \\ c(e+j) + d(g+m) & c(f+k) + d(h+n) \end{bmatrix}$ Definition of matrix multiplication

$= \begin{bmatrix} ea + ja + gb + mb & fa + ka + hb + nb \\ ec + jc + gd + md & fc + kc + hd + nd \end{bmatrix}$ Distributive Property

$= \begin{bmatrix} ea + gb + ja + mb & fa + hb + ka + nb \\ ec + gd + jc + md & fc + hd + kc + nd \end{bmatrix}$ Commutative Property of Addition

$= \begin{bmatrix} ea + gb & fa + hb \\ ec + gd & fc + hd \end{bmatrix} + \begin{bmatrix} ja + mb & ka + nb \\ jc + md & kc + nd \end{bmatrix}$ Definition of matrix addition

$= CA + CB$ Definition of matrix multiplication

47b (continued). $(A + B)C = \left(\begin{bmatrix} a_{11} & a_{12} \\ a_{21} & a_{22} \end{bmatrix} + \begin{bmatrix} b_{11} & b_{12} \\ b_{21} & b_{22} \end{bmatrix} \right) \begin{bmatrix} c_{11} & c_{12} \\ c_{21} & c_{22} \end{bmatrix}$

$= \begin{bmatrix} a_{11} + b_{11} & a_{12} + b_{12} \\ a_{21} + b_{21} & a_{22} + b_{22} \end{bmatrix} \begin{bmatrix} c_{11} & c_{12} \\ c_{21} & c_{22} \end{bmatrix}$ Substitution

$= \begin{bmatrix} (a_{11} + b_{11})c_{11} + (a_{12} + b_{12})c_{21} & (a_{11} + b_{11})c_{12} + (a_{12} + b_{12})c_{22} \\ (a_{21} + b_{21})c_{11} + (a_{22} + b_{22})c_{21} & (a_{21} + b_{21})c_{12} + (a_{22} + b_{22})c_{22} \end{bmatrix}$ Definition of matrix multiplication

$= \begin{bmatrix} a_{11}c_{11} + b_{11}c_{11} + a_{12}c_{21} + b_{12}c_{21} & a_{11}c_{12} + b_{11}c_{12} + a_{12}c_{22} + b_{12}c_{22} \\ a_{21}c_{11} + b_{21}c_{11} + a_{22}c_{21} + b_{22}c_{21} & a_{21}c_{12} + b_{21}c_{12} + a_{22}c_{22} + b_{22}c_{22} \end{bmatrix}$ Distributive Property

$= \begin{bmatrix} a_{11}c_{11} + a_{12}c_{21} + b_{11}c_{11} + b_{12}c_{21} & a_{11}c_{12} + a_{12}c_{22} + b_{11}c_{12} + b_{12}c_{22} \\ a_{21}c_{11} + a_{22}c_{21} + b_{21}c_{11} + b_{22}c_{21} & a_{21}c_{12} + a_{22}c_{22} + b_{21}c_{12} + b_{22}c_{22} \end{bmatrix}$ Commutative Property of Addition

$= \begin{bmatrix} a_{11}c_{11} + a_{12}c_{21} & a_{11}c_{12} + a_{12}c_{22} \\ a_{21}c_{11} + a_{22}c_{21} & a_{21}c_{12} + a_{22}c_{22} \end{bmatrix} + \begin{bmatrix} b_{11}c_{11} + b_{12}c_{21} & b_{11}c_{12} + b_{12}c_{22} \\ b_{21}c_{11} + b_{22}c_{21} & b_{21}c_{12} + b_{22}c_{22} \end{bmatrix}$ Definition of matrix addition

$= AC + BC$ Definition of matrix multiplication

R38 | Selected Answers

---

37. $\begin{bmatrix} -10 - 4.5y & 36.75 \\ 2x + 4 + 3y^2 & -6x - 4.5y - 12 \\ 3.6y + 26 & -83.4 \end{bmatrix}$

39. $\begin{bmatrix} -1.5x - 1.5y + 15 \\ y^2 + xy - 3x - 6 \\ 1.2x + 1.2y - 39 \end{bmatrix}$

41. undefined

43. $\begin{bmatrix} 15x + 69y - 12 \\ -18y^2 + 42.75y - 18xy + 20.25x \end{bmatrix}$

45a. A: $421; B: $274; C: $150; D: $68

45b. A: $357.85; B: $232.90; C: $127.50; D: $57.80

47a. $c(A + B)$

$= c \left(\begin{bmatrix} a & b \\ d & e \end{bmatrix} + \begin{bmatrix} w & x \\ y & z \end{bmatrix} \right)$ Substitution

$= c \begin{bmatrix} a+w & b+x \\ d+y & e+z \end{bmatrix}$ Definition of matrix addition

$= \begin{bmatrix} ca + cw & cb + cx \\ cd + cy & ce + cz \end{bmatrix}$ Definition of scalar multiplication

$= \begin{bmatrix} ca & cb \\ cd & ce \end{bmatrix} + \begin{bmatrix} cw & cx \\ cy & cz \end{bmatrix}$ Definition of matrix addition

$= cA + cB$ Substitution

47c. $(AB)C = \left(\begin{bmatrix} a_{11} & a_{12} \\ a_{21} & a_{22} \end{bmatrix} \begin{bmatrix} b_{11} & b_{12} \\ b_{21} & b_{22} \end{bmatrix} \right) \begin{bmatrix} c_{11} & c_{12} \\ c_{21} & c_{22} \end{bmatrix}$ Substitution

$= \begin{bmatrix} a_{11}b_{11} + a_{12}b_{21} & a_{11}b_{12} + a_{12}b_{22} \\ a_{21}b_{11} + a_{22}b_{21} & a_{21}b_{12} + a_{22}b_{22} \end{bmatrix} \begin{bmatrix} c_{11} & c_{12} \\ c_{21} & c_{22} \end{bmatrix}$ Definition of matrix multiplication

$= \begin{bmatrix} (a_{11}b_{11} + a_{12}b_{21})c_{11} + (a_{11}b_{12} + a_{12}b_{22})c_{21} & (a_{11}b_{11} + a_{12}b_{21})c_{12} + (a_{11}b_{12} + a_{12}b_{22})c_{22} \\ (a_{21}b_{11} + a_{22}b_{21})c_{11} + (a_{21}b_{12} + a_{22}b_{22})c_{21} & (a_{21}b_{11} + a_{22}b_{21})c_{12} + (a_{21}b_{12} + a_{22}b_{22})c_{22} \end{bmatrix}$ Definition of matrix multiplication

$= \begin{bmatrix} a_{11}b_{11}c_{11} + a_{12}b_{21}c_{11} + a_{11}b_{12}c_{21} + a_{12}b_{22}c_{21} & a_{11}b_{11}c_{12} + a_{12}b_{21}c_{12} + a_{11}b_{12}c_{22} + a_{12}b_{22}c_{22} \\ a_{21}b_{11}c_{11} + a_{22}b_{21}c_{11} + a_{21}b_{12}c_{21} + a_{22}b_{22}c_{21} & a_{21}b_{11}c_{12} + a_{22}b_{21}c_{12} + a_{21}b_{12}c_{22} + a_{22}b_{22}c_{22} \end{bmatrix}$ Distributive Property

$= \begin{bmatrix} a_{11}b_{11}c_{11} + a_{11}b_{12}c_{21} + a_{12}b_{21}c_{11} + a_{12}b_{22}c_{21} & a_{11}b_{11}c_{12} + a_{11}b_{12}c_{22} + a_{12}b_{21}c_{12} + a_{12}b_{22}c_{22} \\ a_{21}b_{11}c_{11} + a_{21}b_{12}c_{21} + a_{22}b_{21}c_{11} + a_{22}b_{22}c_{21} & a_{21}b_{11}c_{12} + a_{21}b_{12}c_{22} + a_{22}b_{21}c_{12} + a_{22}b_{22}c_{22} \end{bmatrix}$ Commutative Property of Addition

$= \begin{bmatrix} a_{11}(b_{11}c_{11} + b_{12}c_{21}) + a_{12}(b_{21}c_{11} + b_{22}c_{21}) & a_{11}(b_{11}c_{12} + b_{12}c_{22}) + a_{12}(b_{21}c_{12} + b_{22}c_{22}) \\ a_{21}(b_{11}c_{11} + b_{12}c_{21}) + a_{22}(b_{21}c_{11} + b_{22}c_{21}) & a_{21}(b_{11}c_{12} + b_{12}c_{22}) + a_{22}(b_{21}c_{12} + b_{22}c_{22}) \end{bmatrix}$ Distributive Property

$= \begin{bmatrix} a_{11} & a_{12} \\ a_{21} & a_{22} \end{bmatrix} \begin{bmatrix} b_{11}c_{11} + b_{12}c_{21} & b_{11}c_{12} + b_{12}c_{22} \\ b_{21}c_{11} + b_{22}c_{21} & b_{21}c_{12} + b_{22}c_{22} \end{bmatrix}$ Definition of matrix multiplication

$= \begin{bmatrix} a_{11} & a_{12} \\ a_{21} & a_{22} \end{bmatrix} \left(\begin{bmatrix} b_{11} & b_{12} \\ b_{21} & b_{22} \end{bmatrix} \begin{bmatrix} c_{11} & c_{12} \\ c_{21} & c_{22} \end{bmatrix} \right)$ Definition of matrix multiplication

$= A(BC)$ Substitution

47d. $c(AB) = c \left(\begin{bmatrix} a_{11} & a_{12} \\ a_{21} & a_{22} \end{bmatrix} \begin{bmatrix} b_{11} & b_{12} \\ b_{21} & b_{22} \end{bmatrix} \right)$ Substitution

$= c \begin{bmatrix} a_{11}b_{11} + a_{12}b_{21} & a_{11}b_{12} + a_{12}b_{22} \\ a_{21}b_{11} + a_{22}b_{21} & a_{21}b_{12} + a_{22}b_{22} \end{bmatrix}$ Definition of matrix multiplication

$= \begin{bmatrix} c(a_{11}b_{11} + a_{12}b_{21}) & c(a_{11}b_{12} + a_{12}b_{22}) \\ c(a_{21}b_{11} + a_{22}b_{21}) & c(a_{21}b_{12} + a_{22}b_{22}) \end{bmatrix}$ Definition of scalar multiplication

$= \begin{bmatrix} ca_{11}b_{11} + ca_{12}b_{21} & ca_{11}b_{12} + ca_{12}b_{22} \\ ca_{21}b_{11} + ca_{22}b_{21} & ca_{21}b_{12} + ca_{22}b_{22} \end{bmatrix}$ Distributive Property

$= \begin{bmatrix} ca_{11} & ca_{12} \\ ca_{21} & ca_{22} \end{bmatrix} \begin{bmatrix} b_{11} & b_{12} \\ b_{21} & b_{22} \end{bmatrix}$ Definition of matrix multiplication

$= c \begin{bmatrix} a_{11} & a_{12} \\ a_{21} & a_{22} \end{bmatrix} \begin{bmatrix} b_{11} & b_{12} \\ b_{21} & b_{22} \end{bmatrix}$ Substitution

$= (cA)B$ Substitution

$c(AB) = c \left(\begin{bmatrix} a_{11} & a_{12} \\ a_{21} & a_{22} \end{bmatrix} \begin{bmatrix} b_{11} & b_{12} \\ b_{21} & b_{22} \end{bmatrix} \right)$ Substitution

$= c \begin{bmatrix} a_{11}b_{11} + a_{12}b_{21} & a_{11}b_{12} + a_{12}b_{22} \\ a_{21}b_{11} + a_{22}b_{21} & a_{21}b_{12} + a_{22}b_{22} \end{bmatrix}$ Definition of matrix multiplication

$= \begin{bmatrix} c(a_{11}b_{11} + a_{12}b_{21}) & c(a_{11}b_{12} + a_{12}b_{22}) \\ c(a_{21}b_{11} + a_{22}b_{21}) & c(a_{21}b_{12} + a_{22}b_{22}) \end{bmatrix}$ Definition of scalar multiplication

$= \begin{bmatrix} ca_{11}b_{11} + ca_{12}b_{21} & ca_{11}b_{12} + ca_{12}b_{22} \\ ca_{21}b_{11} + ca_{22}b_{21} & ca_{21}b_{12} + ca_{22}b_{22} \end{bmatrix}$ Distributive Property

$= \begin{bmatrix} a_{11}cb_{11} + a_{12}cb_{21} & a_{11}cb_{12} + a_{12}cb_{22} \\ a_{21}cb_{11} + a_{22}cb_{21} & a_{21}cb_{12} + a_{22}cb_{22} \end{bmatrix}$ Commutative Property

$= \begin{bmatrix} a_{11} & a_{12} \\ a_{21} & a_{22} \end{bmatrix} \begin{bmatrix} cb_{11} & cb_{12} \\ cb_{21} & cb_{22} \end{bmatrix}$ Definition of matrix multiplication

$= A(cB)$ Substitution

49. $a = 2$, $b = 1$, $c = 3$, $d = 4$ **51.** 12   **53.** H

55. $\begin{bmatrix} 42 & -24 \\ -42 & -31 \end{bmatrix}$ **57.** $\begin{bmatrix} -80 & 44 \\ 68 & -4 \end{bmatrix}$ **59.** (8, 3, 1)

61a. Sample answer: $y = 116.25x - 231,176.97$ **61c.** The value predicted by the equation is significantly lower than the one given in the graph.
61b. Sample answer: $2950.53

🖱 connectED.mcgraw-hill.com

connectED.mcgraw-hill.com

b. Let x = the number of people who will own CD players next year and let y = the number of people who will own digital audio players next year. Write a matrix equation.

$$\begin{bmatrix} 0.35 & 0.12 \\ 0.65 & 0.88 \end{bmatrix} \cdot \begin{bmatrix} 7748 \\ 17{,}252 \end{bmatrix} = \begin{bmatrix} x \\ y \end{bmatrix}$$

$$\begin{bmatrix} 4782 \\ 20{,}218 \end{bmatrix} = \begin{bmatrix} x \\ y \end{bmatrix}$$

So, about 20,218 people will own digital audio players next year.

c. Let x = the number of people who owned CD players last year and let y = the number of people who owned digital audio players last year. Write a matrix equation.

$$\begin{bmatrix} 0.35 & 0.12 \\ 0.65 & 0.88 \end{bmatrix} \cdot \begin{bmatrix} x \\ y \end{bmatrix} = \begin{bmatrix} 7748 \\ 17{,}252 \end{bmatrix}$$

Find the inverse of the coefficient matrix.

$$A^{-1} = \frac{1}{0.35(0.88) - 0.12(0.65)} \begin{bmatrix} 0.88 & -0.12 \\ -0.65 & 0.35 \end{bmatrix}$$

$a = 0.35,$
$b = 0.12,$
$c = 0.65,$
$d = 0.88$

$$= \frac{1}{0.23} \begin{bmatrix} 0.88 & -0.12 \\ -0.65 & 0.35 \end{bmatrix} \quad \text{Simplify.}$$

Multiply each side of the matrix equation by the inverse matrix.

$$\frac{1}{0.23} \begin{bmatrix} 0.88 & -0.12 \\ -0.65 & 0.35 \end{bmatrix} \cdot \begin{bmatrix} 0.35 & 0.12 \\ 0.65 & 0.88 \end{bmatrix} \cdot \begin{bmatrix} x \\ y \end{bmatrix} = \frac{1}{0.23} \begin{bmatrix} 0.88 & -0.12 \\ -0.65 & 0.35 \end{bmatrix} \cdot \begin{bmatrix} 4748 \\ 1002 \end{bmatrix}$$

$$\begin{bmatrix} 1 & 0 \\ 0 & 1 \end{bmatrix} \cdot \begin{bmatrix} x \\ y \end{bmatrix} = \frac{1}{0.23} \begin{bmatrix} 4748 \\ 1002 \end{bmatrix}$$

$$\begin{bmatrix} x \\ y \end{bmatrix} = \frac{1}{0.23} \begin{bmatrix} 20{,}643 \\ 4357 \end{bmatrix}$$

So, about 4357 people owned digital audio players last year.

39. The system would have to consist of two equations that are the same or one equation that is a multiple of the other.

41. Sample answer: $\begin{bmatrix} 2 & 3 \\ 4 & 6 \end{bmatrix} \cdot \begin{bmatrix} x \\ y \end{bmatrix} = \begin{bmatrix} 9 \\ 10 \end{bmatrix}$; any matrix that has a determinant equal to 0, such as $\begin{bmatrix} 1 & 0 \\ 0 & 0 \end{bmatrix}$ **43.** C **45.** $(\frac{1}{2}, \frac{1}{4})$ **47.** −54 **49.** 551

51. $\begin{bmatrix} -22 \\ -7 \end{bmatrix}$ **53.** 179 gal of skim and 21 gal of whole milk **55.** absolute value

Chapter 3 Study Guide and Review
1. unbounded **3.** constant matrix **5.** dimensions **7.** identity **9.** inconsistent **11.** (0, 2) **13.** (−3, 4) **15.** 4 h **17.** (−2, −7) **19.** (3, 5)

69b.
$$\begin{bmatrix} \$3.60 & \$5.40 & \$7.60 \\ \$3.60 & \$5.40 & \$7.60 \\ \$3.20 & \$6.00 & \$7.00 \\ \$3.80 & \$6.00 & \$8.20 \end{bmatrix}$$

69c.
$$\begin{bmatrix} \$0.90 & \$1.35 & \$1.90 \\ \$0.90 & \$1.35 & \$1.90 \\ \$0.80 & \$1.25 & \$1.75 \\ \$0.95 & \$1.50 & \$2.05 \end{bmatrix}$$

71.
$f(x) = -3|2x| + 4$

73. (−8, 2) **75.** (−3, −4)

Lesson 3-8
1. no **3.** yes **5.** $\begin{bmatrix} 0 & -1 \\ -\frac{1}{3} & -2 \end{bmatrix}$ **7.** $\begin{bmatrix} -\frac{3}{5} & 0 \\ \frac{5}{6} & \frac{1}{2} \end{bmatrix}$

9. (−2, 5) **11.** (1, −2) **13.** no **15.** no **17.** $\begin{bmatrix} \frac{1}{3} & 0 \\ 0 & \frac{1}{2} \end{bmatrix}$

19. $\begin{bmatrix} -\frac{5}{3} & 1 \\ -\frac{5}{3} & 1 \end{bmatrix}$

21. $\begin{vmatrix} -5 & -4 \\ 4 & 2 \end{vmatrix} = -5(2) - 4(-4)$ or 6 Find the determinant.

Since the determinant $\neq 0$, the inverse exists.

$A^{-1} = \frac{1}{ad - bc} \begin{bmatrix} d & -b \\ -c & a \end{bmatrix}$ Definition of inverse

$a = -5, b = -4,$
$c = 4, d = 2$

$$= \frac{1}{-5(2) - (-4)(4)} \begin{bmatrix} 2 & 4 \\ -4 & -5 \end{bmatrix}$$

$$= \frac{1}{6} \begin{bmatrix} 2 & 4 \\ -4 & -5 \end{bmatrix} \text{ or } \begin{bmatrix} \frac{1}{3} & \frac{2}{3} \\ -\frac{2}{3} & -\frac{5}{6} \end{bmatrix} \quad \text{Simplify.}$$

23. $\begin{bmatrix} \frac{9}{37} & \frac{5}{74} \\ \frac{2}{37} & \frac{7}{37} \end{bmatrix}$ **25.** $\begin{bmatrix} \frac{7}{22} & \frac{4}{11} \\ \frac{2}{11} & \frac{1}{11} \end{bmatrix}$ **27.** no solution

29. (−1, 5) **31.** no solution **33.** (−5, 0)

35. $(\frac{3}{4}, 3)$

37. a. Let x = the number of people who own CD players and let y = the number of people who own digital audio players. The following expressions represent the change in player ownership. Keeping or switching to CD players: $0.35x + 0.12y$. Keeping or switching to digital audio players: $0.65x + 0.88y$:

	From	
	CD	DAP
To CD	0.35	0.12
DAP	0.65	0.88

---

63. translation 4 units right and 3 units up
$f(x) = |x - 4| + 3$

65. translation 2 units left and 6 units down
$f(x) = (x + 2)^2 - 6$

Lesson 3-7
1. 26 **3.** −128
5. Rewrite the first two columns to the right of the determinant. Then find the products of the elements of the diagonals.

$3(2)(4) = 24$
$-2(-5)(-3) = -30$
$2(-4)(1) = -8$
$-3(2)(2) = -12$
$1(-5)(3) = -15$
$4(-4)(-2) = 32$

The sum of the first group is $24 + (-30) + (-8)$ or −14. The sum of the second group is $-12 + (-15) + 32$ or 5. The first sum minus the second sum is $-14 - 5$ or −19.

7. −284 **9.** 72 **11.** 182 **13.** (6, −3) **15.** (4, −1)
17a. 15.75 units² **17b.** 482,343.75 mi²
19. $(\frac{66}{7}, \frac{116}{7}, \frac{41}{7})$ **21.** (−4, −2, 8)
23. (−1, −3, 7) **25.** (4, 0, 8) **27.** 3 **29.** −135
31. −459 **33.** 0 **35.** 728 **37.** −952 **39.** (8, −5)

41.
$x = \frac{\begin{vmatrix} m & b \\ n & g \end{vmatrix}}{|C|} = \frac{\begin{vmatrix} -39 & -5 \\ 54 & 8 \end{vmatrix}}{\begin{vmatrix} -4 & -5 \\ 5 & 8 \end{vmatrix}}$
$= \frac{-39(8) - 54(-5)}{-4(8) - 5(-5)}$
$= \frac{-312 + 270}{-32 + 25} = \frac{-42}{-7}$ or 6

$y = \frac{\begin{vmatrix} a & m \\ f & n \end{vmatrix}}{|C|} = \frac{\begin{vmatrix} -4 & -39 \\ 5 & 54 \end{vmatrix}}{\begin{vmatrix} -4 & -5 \\ 5 & 8 \end{vmatrix}}$
$= \frac{-4(54) - 5(-39)}{-4(8) - 5(-5)}$
$= \frac{-216 + 195}{-32 + 25} = \frac{-21}{-7}$ or 3

The solution of the system is (6, 3).

43. (−3, −7) **45.** (4, −2, 5) **47.** 6 **49.** 2 m²
51. (4, 8, −5) **53.** $(\frac{6187}{701}, \frac{2904}{701}, \frac{4212}{701})$

55. a. Let x = the number of medium drinks. Then $2x$ = the number of small drinks. Let y = the number of large drinks.
number of drinks: $x + 2x + y = 1385 \rightarrow 3x + y = 1385$
total sales: $1.75x + 1.15(2x) + 2.25y = 2238.75$
$\rightarrow 4.05x + 2.25y = 2238.75$

$x = \frac{\begin{vmatrix} m & b \\ n & g \end{vmatrix}}{|C|} = \frac{\begin{vmatrix} 1385 & 1 \\ 2238.75 & 2.25 \end{vmatrix}}{\begin{vmatrix} 3 & 1 \\ 4.05 & 2.25 \end{vmatrix}}$
$= \frac{1385(2.25) - 2238.75(1)}{3(2.25) - 4.05(1)}$
$= \frac{3116.25 - 2238.75}{6.75 - 4.05}$
$= \frac{877.5}{2.7}$ or 325

$y = \frac{\begin{vmatrix} a & m \\ f & n \end{vmatrix}}{|C|} = \frac{\begin{vmatrix} 3 & 1385 \\ 4.05 & 2238.75 \end{vmatrix}}{\begin{vmatrix} 3 & 1 \\ 4.05 & 2.25 \end{vmatrix}}$
$= \frac{3(2238.75) - 4.05(1385)}{3(2.25) - 4.05(1)}$
$= \frac{6716.25 - 5609.25}{6.75 - 4.05}$
$= \frac{1107}{2.7}$ or 410

He sold 325 medium drinks, 2(325) or 650 small drinks, and 410 large drinks.

b. total sales = sales of small + sales of medium + sales of large
$= 1.25(650 - 140) + 1.75(325 + 125) + 2.25(410 + 35)$
$= 637.50 + 787.50 + 1001.25$
$= \$2426.25$

c. It seems like it was a good move for the vendor. Although he sold 140 fewer small drinks, he sold 125 more medium drinks and 35 more large drinks. On the whole, he made $2,426.25 − $2,238.75 or $187.50 more this week than in the previous week.

57. Sample answer: There is no unique solution of the system. There are either infinite or no solutions.

59. 0 **61.** Sample answer: Given a 2 × 2 system of linear equations, if the determinant of the matrix of coefficients is 0, then the system does not have a unique solution. The system may have no solution and the graphical representation shows two parallel lines. The system may have infinitely many solutions in which the graphical representation will be the same line. **63.** H **65.** D

67. no **69a.**
$$\begin{bmatrix} \$4.50 & \$6.75 & \$9.50 \\ \$4.50 & \$6.75 & \$9.50 \\ \$4.00 & \$6.25 & \$8.75 \\ \$4.75 & \$7.50 & \$10.25 \end{bmatrix}$$

R43

35a. y-int = 0; axis of symmetry: $x = \frac{5}{8}$; x-coordinate of vertex $= \frac{5}{8}$

35b.

x	$f(x)$
$-\frac{3}{4}$	-6
$\frac{1}{4}$	1
$\frac{5}{8}$	1.5625
1	1
2	-6

35c.

37a. y-int = 4; axis of symmetry: $x = -6$; x-coordinate of vertex $= -6$

37b.

x	$f(x)$
-10	-1
-8	-4
-6	-5
-4	-4
-2	-1

37c.

39a. y-int = -2.5; axis of symmetry: $x = -\frac{4}{3}$; x-coordinate of vertex $= -\frac{4}{3}$

39b.

x	$f(x)$
$-\frac{11}{3}$	3
$-\frac{8}{3}$	-2.5
$-\frac{4}{3}$	$-5\frac{1}{6}$
0	-2.5
1	3

39c.

41a. $f(x) = -x^2 + 6x + 475$ **41b.** D = $\{x \mid 0 \le x \le 25\}$, R = $\{y \mid 0 \le y \le 484\}$ **41c.** \$11; Because the function has a maximum at $x = 3$, it is in the domain. Therefore, three \$0.50 increases is reasonable. **41d.** \$484 **43.** max = 23 **45.** max = -0.10 **47.** max = -4.11

49 $a = -5$, so the graph opens down and has a maximum value. The maximum value is the y-coordinate of the vertex.

$x = -\frac{b}{2a}$ Equation of the axis of symmetry

$= -\frac{4}{2(-5)}$ $a = -5$ and $b = 4$

The x-coordinate of the vertex is 0.4. Find the y-coordinate of the vertex by evaluating the function for $x = 0.4$.

$f(x) = -5x^2 + 4x - 8$ Original function

$= -5(0.4)^2 + 4(0.4) - 8$ $x = 0.4$

$= -7.2$

The domain is all real numbers. The range is all real numbers less than or equal to the maximum value, or $\{f(x) \mid f(x) \le -7.2\}$.

The maximum value of the function is -7.2.

51. min = -9.375; D = [all real numbers], R = $\{f(x) \mid f(x) \ge -9.375\}$ **53.** min = -23.5; D = [all real numbers], R = $\{f(x) \mid f(x) \ge -23.5\}$

connectED.mcgraw-hill.com

---

19 **a.** $f(x) = ax^2 + bx + c$

$f(x) = 1x^2 - 3x - 10$ $a = 1, b = -3, c = -10$

The y-intercept is $c = -10$.

$x = -\frac{b}{2a}$ Equation of the axis of symmetry

$= -\frac{(-3)}{2(1)}$

$= \frac{3}{2}$ or 1.5 Simplify.

The equation of the axis of symmetry is $x = 1.5$. So, the x-coordinate of the vertex is 1.5.

b. Select five points, with the vertex in the middle and two points on either side of the vertex, including the y-intercept and its reflection.

x	$f(x)$	
0	-10	← y-intercept
1	-12	
1.5	-12.25	← vertex
2	-12	
3	-10	← reflection of y-intercept

c. Graph the five points from the table, connecting them with a smooth curve.

21a. y-int = 9; axis of symmetry: $x = 0.75$; x-coordinate of vertex = 0.75

21b.

x	$f(x)$
-1	-10
0	9
0.75	10.125
1.5	9
2.5	4

21c.

23. max = -12; D = [all real numbers], R = $\{f(x) \mid f(x) \le -12\}$ **25.** max = 13.25; D = [all real numbers], R = $\{f(x) \mid f(x) \le 13.25\}$ **27.** max = 7; D = [all real numbers], R = $\{f(x) \mid f(x) \le 7\}$

29. min = -9; D = [all real numbers], R = $\{f(x) \mid f(x) \ge -9\}$ **31.** min = -74; D = [all real numbers], R = $\{f(x) \mid f(x) \ge -74\}$ **33a.** y-int = -9; axis of symmetry: $x = 1.5$; x-coordinate of vertex = 1.5

33b.

x	$f(x)$
0	-9
1	-13
1.5	-13.5
2	-13
3	-9

33c.

3b.

x	$f(x)$
0	0
1	-3
2	-4
3	-3
4	0

3c.

5a. y-int = -3; axis of symmetry: $x = 0.75$; x-coordinate = 0.75

5b.

x	$f(x)$
-1	7
0	-3
0.75	-5.25
1.5	-3
2.5	7

5c. graph

7. max = 8; D = [all real numbers], R = $\{f(x) \mid f(x) \le 8\}$ **9.** min = $-\frac{1}{3}$; D = [all real numbers], R = $\{f(x) \mid f(x) \ge -\frac{1}{3}\}$ **11.** \$2.88 **13a.** y-int = 0; axis of symmetry: $x = 0$; x-coordinate = 0

13b.

x	$f(x)$
-2	-8
-1	-2
0	0
1	-2
2	-8

13c. graph

15a. y-int = 3; axis of symmetry: $x = 0$; x-coordinate = 0

15b.

x	$f(x)$
-2	7
-1	4
0	3
1	4
2	7

15c. graph

17a. y-int = 5; axis of symmetry: $x = 0$; x-coordinate = 0

17b.

x	$f(x)$
-2	-7
-1	0
0	5
1	2
2	-7

17c. graph

---

21. graph

23. graph

25. graph

27. \$480; 12 outdoor, 16 indoor **29.** (5, -5, -4)

31. $\begin{bmatrix} -3 & 27 \\ 9 & 12 \end{bmatrix}$ **33a.** buying price: $\begin{bmatrix} 15 \\ 25 \\ 30 \end{bmatrix}$; selling price: $\begin{bmatrix} 35 \\ 55 \\ 85 \end{bmatrix}$ **33b.** $\begin{bmatrix} 35 \\ 55 \\ 85 \end{bmatrix} - \begin{bmatrix} 15 \\ 25 \\ 30 \end{bmatrix} = \begin{bmatrix} 20 \\ 30 \\ 55 \end{bmatrix}$ **33c.**

Bracelets graph

35. $\begin{bmatrix} -12 & 17 \\ 82 & 26 \end{bmatrix}$ **37.** \$28.56 **39.** -44 **41.** (2, -3, 6)

43. $\frac{1}{2}\begin{bmatrix} 2 & -4 \\ -3 & 7 \end{bmatrix}$ **45.** does not exist **47.** (2, 1)

Chapter 4 Get Ready

1. 6 **3.** 4 **5.** 3 **7a.** $f(x) = 9x$ **7b.** 157,680 mi **9.** $(x + 8)(x + 5)$ **11.** $(2x - 1)(x + 4)$ **13.** prime **15.** $(x + 8)$ feet

Lesson 4-1

1a. y-int = 0; axis of symmetry: $x = 0$; x-coordinate = 0

1b.

x	$f(x)$
-2	12
-1	3
0	0
1	3
2	12

1c.

3a. y-int = 0; axis of symmetry: $x = 2$; x-coordinate = 2

R42 | Selected Answers

55. $f(x) = x^2 - 4x - 5$ **57.** $f(x) = x^2 - 6x + 8$

59. a. Let x = the number of 5-cent increases. Then $65 + 5x$ = price per can in cents and $600 - 100x$ = number of cans. Let $f(x)$ = income as a function of x.

Income = number of cans times the price per can

$f(x) = (600 - 100x) \cdot (65 + 5x)$

Solve for x in the equation.

$f(x) = (600 - 100x) \cdot (65 + 5x)$
$= 600(65) + 600(5x) + (-100x)(65) + (-100x)(5x)$ Distribute.
$= 39,000 + 3000x - 6500x - 500x^2$ Multiply.
$= 39,000 - 3500x - 500x^2$ Simplify.

So, the equation is $f(x) = 39,000 - 3500x + 39,000$ or $f(x) = -500x^2 - 3500x + 39,000$. Equation of the axis of symmetry

$x = -\dfrac{b}{2a}$

$x = \dfrac{-(-3500)}{2(-500)}$ or -3.5 $a = -500$ and $b = -3500$

$f(x) = -500(-3.5)^2 - 3500(-3.5) + 39,000$ $x = -3.5$
$= 45,125$ Simplify.

The coordinates of the vertex are $(-3.5, 45,125)$. The y-intercept is 39,000.

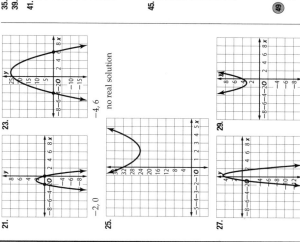

b. Let x = the number of 5-cent decreases. Then $65 - 5x$ = price per can and $600 + 100x$ = number of cans. Let $f(x)$ = income as a function of x.

Income = number of cans times the price per can

$f(x) = (600 + 100x) \cdot (65 - 5x)$

Solve for x in the equation.

$f(x) = (600 + 100x) \cdot (65 - 5x) + (100x)(65) + (100x)(-5x)$ Distribute.
$= 39,000 - 3000x + 6500x - 500x^2$ Multiply.
$= 39,000 + 3500x - 500x^2$ Simplify.

So, the equation is $f(x) = 39,000 + 3500x + 39,000$ or $f(x) = -500x^2 + 3500x + 39,000$. Equation of the axis of symmetry

$x = -\dfrac{b}{2a}$

$x = \dfrac{(3500)}{2(-500)}$ or 3.5 $a = -500$ and $b = 3500$

$f(x) = -500(3.5)^2 + 3500(3.5) + 39,000$ Original function
$= -500(3.5)^2 + 3500(3.5) + 39,000$ Replace x with 3.
$= 45,125$ Simplify.

The coordinates of the vertex are $(3.5, 45,125)$. So, Omar should have 3 price decreases and charge $65 - 3(5)$ or 50 cents. Or, he should have 4 price decreases and charge $65 - 4(5)$ or 45 cents.

c. $f(x) = -500x^2 + 3500x + 39,000$ Income function
$= -500(3)^2 + 3500(3) + 39,000$ Replace x with 3.
$= 45,000$ Simplify.

So, his income would be 45,000 cents or $450 per week.

61. Madison; sample answer: $f(x)$ has a maximum of -2. $g(x)$ has a maximum of 1. **63a.** $a = 22$; $b = 26$; $c = -6$; $d = 2$ **63b.** 0 **63c.** maximum **65.** Sample answer: A function is quadratic if it has no other terms than a quadratic term, linear term, and constant term. The function has a maximum if the coefficient of the quadratic term is negative and has a minimum if the coefficient of the quadratic term is positive. **67.** K **69.** C

71. $\begin{bmatrix} -\frac{1}{4} & \frac{1}{24} & \frac{1}{6} \\ 0 & & \end{bmatrix}$ **73.** 45 **75.** 0 **77.** No; it cannot be written as $y = mx + b$. **79.** Yes; it is written in $y = mx + b$ form, $m = 0$. **81.** -13

Lesson 4-2

1. no real solution

3. -4

5. Graph the related function $f(x) = x^2 - 3x - 18$. The equation of the axis of symmetry is $x = -\dfrac{(-3)}{2(1)}$ or 1.5. Make a table using x-values around 1.5. Then graph each point.

x	-3	0	1.5	3	6
f(x)	0	-18	-20.25	-18	0

The zeros of the function are -3 and 6. So, the solutions of the equation are -3 and 6.

7.

between -5 and -4, between 5 and 6

9.
no real solution

11.

$3, -\dfrac{4}{3}$

13. 5 seconds **15.** no real solution **17.** -2

19. $-3, 4$

21.
23.
$-4, 6$
no real solution

25.
$-2, 0$

27.
between -1 and 0, between 1 and 2

29.
no real solution

31. between 0 and 1; between 2 and 3

33. Let x = one of the numbers. Then $-15 - x$ = the other number.

$x(-x - 15) = -54$ The product is -54.
$-x^2 - 15x = -54$ Distributive Property
$-x^2 - 15x + 54 = 0$ Add 54 to each side.

Graph the related function $f(x) = -x^2 - 15x + 54$. The equation of the axis of symmetry is $x = \dfrac{-(-15)}{2(-1)}$ or -7.5. Make a table using x-values around -7.5. Then graph each point.

x	-20	-15	-10	-7.5	-5	0	5
f(x)	-46	54	104	110.25	104	54	-46

The zeros of the function are -18 and 3. So, the numbers are -18 and 3.

35. about -5.0 and 17.0 **37.** 11 and -19

39. about 3.4375 seconds

41.
-3, between 2 and 3

43.

between -3 and -2, between 1 and 2

45.
between -1 and 0, between 4 and 5

47.
between 3 and 4, between 8 and 9

49. Find t when $h_0 = 60$ and $h(t) = 0$.

$h(t) = -16t^2 + h_0$ Original equation
$0 = -16t^2 + 60$ $h(t) = 0$ and $h_0 = 60$

Graph the related function $f(t) = -16t^2 + 60$.

[0, 5] scl: 0.5 by [0, 100] scl: 10

Use the **Zero** feature in the **CALC** menu to find the positive zero of the function, since time cannot be negative. $x \approx 1.94$, so it would take the balloon about 1.94 seconds to hit the ground. Tony's brother should be 4.4 ft/s • 1.94 s or about 8.5 feet from the target when Tony lets go of the balloon.

51. 25 seconds **53.** $k = 8$ **55.** $f(x) = -5x^2 + 30x + 80$

57. $\dfrac{1}{20}$ **59.** H **61.** maximum, -12; D = [all real numbers], R = $\{f(x) \mid f(x) \le -12\}$

63. maximum, 15; D = [all real numbers], R = $\{f(x) \mid f(x) \le 15\}$ **65.** no **67a.** $\begin{bmatrix} 10 & 10 & 15 \\ 25 & 35 & 45 \end{bmatrix}$ **67b.** 2; $\begin{bmatrix} 20 & 20 & 30 \\ 50 & 70 & 90 \end{bmatrix}$ **67c.** $\begin{bmatrix} 10 & 10 & 15 \\ 25 & 35 & 45 \end{bmatrix}$; the number of additional shirts that he needs to stock

69. $(-3, 4)$ **71.** $x \le -\dfrac{8}{3}$ **73.** $x \le \dfrac{19}{9}$ **75.** 1

Selected Answers and Solutions

Lesson 4-3

1. $x^2 + 3x - 40 = 0$ **3.** $6x^2 - 11x - 10 = 0$
5. $(6x - 1)(3x + 4)$ **7.** $(x - 7)(x + 3)$ **9.** $(4x - 3)(4x - 1)$
11. $0, \frac{3}{2}$ **13.** $0, 9$ **15.** 6 **17.** $x^2 - 14x + 49 = 0$
19. $5x^2 - 31x + 6 = 0$ **21.** $17c(3c^2 - 2)$
23. $3(x + 2)(x - 2)$

25. $48cg + 36cf - 4dg - 3df$ Original expression
$= (48cg + 36cf) + (-4dg - 3df)$ Group terms with common factors.
$= 12c(4g + 3f) + (-d)(4g + 3f)$ Factor the GCF from each group.
$= (12c - d)(4g + 3f)$ Distributive Property

27. $(x - 11)(x + 2)$ **29.** $(5x - 1)(3x + 2)$
31. $3(2x - 1)(3x + 4)$ **33.** $(3x + 5)(3x - 5)$
35. $-\frac{5}{2}, 6$ **37.** $0, 9$ **39.** $8, -3$
41. $-2, -2$ **43.** $5, \frac{3}{4}$ **45.** 24 and 26 or -24 and -26
47. $x = 20$; 24 in. by 18 in. **49.** $-\frac{1}{5}, \frac{5}{2}$ **51.** $-\frac{3}{2}$
53. $6, -6$

55. To find the number of movie screens that produces a profit, first find the number of movie screens in which the profit is zero. Solve $-x^2 + 48x - 512 = 0$.
$ac = -1(-512)$ or 512
$m = 16$; $p = 32$ $mp = 512$ and $ac = 512$; $m + p = 48$ and $b = 48$
$-x^2 + 16x + 32x - 512 = 0$ Write the pattern.
$(-x^2 + 16x) + (32x - 512) = 0$ Group terms with common factors.
$-x(x - 16) + 32(x - 16) = 0$ Factor the GCF from each group.
$(-x + 32)(x - 16) = 0$ Distributive Property
$-x + 32 = 0$ or $x - 16 = 0$ Zero Product Property
$x = 32$ $x = 16$ Solve each equation.
The solutions are 16 and 32. When there are 16 or 32 movie screens, the profit is zero. Since a is negative, the graph of the function opens down and has a maximum value. So, $P(x)$ is nonnegative for $16 \le x \le 32$. When there are 16 to 32 movie screens, the company will not lose money.

57. $25x^2 - 100x + 51 = 0$ **59.** $-3, \frac{3}{2}$ **61.** $1, -\frac{5}{4}$
63. $-\frac{3}{2}, \frac{5}{6}$ **65.** $x^2 - 6^2$; $(x + 6)(x - 6)$ **67.** 20 in. by
15 in. **69.** 13 cm **71.** $2(3 - 4y)(3a + 8b)$
73. $6a^2b^2 - 12ab^2 - 18b^3$ Original expression
$= 6b^2(a^2 - 2a - 3b)$ Factor the GCF, $6b^2$.
75. $2(2x - 3y)(8a + 3b)$ **77.** $(x + y)(x - y)(5a + 2b)$
81. $5x^2(2x - 3y)(4x^2 + 6xy + 9y^2)$
79. Sample answer: Neither is correct; Gwen didn't have like terms in the parentheses in the third line; Morgan made a sign error in the fourth line.
83. Sample answer:
$(x - p)(x - q) = 0$ Original equation
$x^2 - px - qx + pq = 0$ Multiply.
$x^2 - (p + q)x + pq = 0$ Simplify.
$x = -\frac{b}{2a}$ Formula for axis of symmetry

$x = -\frac{-(p + q)}{2(1)}$ $a = 1$ and $b = -(p + q)$
$x = \frac{p + q}{2}$ Simplify.

x is midway between p and q. Definition of midpoint
85. Sample answer: Always; in order to factor using perfect square trinomials, the coefficient of the linear term, bx, must be a multiple of 2, or even. **87.** 192
square units **89.** H **91.** $-2, 4$ **93.** -2

95.

97a. $\begin{bmatrix} 72 & 49 \\ 68 & 63 \\ 90 & 56 \\ 86 & 62 \end{bmatrix}$ $\begin{bmatrix} 1.00 \\ 0.50 \end{bmatrix}$
97b. $\begin{bmatrix} 96.50 \\ 99.50 \\ 118 \\ 117 \end{bmatrix}$
97c. juniors **97d.** $\$431$ **99.** 16

Lesson 4-4

1. $9i$ **3.** 12 **5.** 1 **7.** $\pm 2i\sqrt{2}$ **9.** $3, -2$ **11.** $-3 + 2i$
13. $70 - 60i$ **15.** $\frac{1}{2} - \frac{1}{2}i$ **17.** $12 + 6j$ amps **19.** $13i$
21. $9i$ **23.** $-144i$ **25.** i **27.** -7 **29.** 9 **31.** $30 + 16i$
33. $1 + i$ **35.** $\frac{1}{3} - \frac{5}{3}i$

37. $3x^2 + 48 = 0$ Original equation
$3x^2 = -48$ Subtract 48 from each side.
$x^2 = -16$ Divide each side by 3.
$x = \pm\sqrt{-16}$ Square Root Property
$x = \pm 4i$ $\sqrt{-16} = \sqrt{16} \cdot \sqrt{-1}$ or $4i$

39. $\pm i\sqrt{5}$ **41.** $\pm 4i$ **43.** $2, -3$ **45.** $\frac{4}{3}, 4$ **47.** $25, -2$
49. $4i$ **51.** 8 **53.** $-21 + 15i$ **55.** $\frac{15}{13} + \frac{16}{13}i$
57. $11 + 23i$ **59.** $\frac{7}{7} - \frac{4\sqrt{3}}{7}$

61. $V = C \cdot I$ Electricity formula
$= (3 + 6j) \cdot (5 - j)$ $C = 3 + 6j$ and $I = 5 - j$
$= 3(5) + 3(-j) + 6j(5) + 6j(-j)$ FOIL Method
$= 15 - 3j + 30j - 6j^2$ Multiply.
$= 15 - 3j + 30j - 6(-1)$ $j^2 = -1$
$= 21 + 27j$ Simplify.
The voltage is $21 + 27j$ volts.

63. $(3 + j)x^2 + (-2 + i)x - 8i + 7$
65a. Sample answer: $x^2 + 9 = 0$
65b.

65c. Sample answer: $x^2 - 4x + 9 = 0$

65d.

65e. Sample answer: A quadratic equation will have only complex solutions when the graph of the related function has no x-intercepts.
67. $-11 - 2i$ **69.** Sample answer: $(4 + 2i)(4 - 2i)$
71a. $\triangle CBE \cong \triangle ADE$ **71b.** $\angle AED \cong \angle CEB$ (Vertical angles) $\overline{DE} \cong \overline{BE}$ (Both have length x.) $\angle ADE \cong \angle CBE$ (Given) Consecutive angles and the included side are all congruent, so the triangles are congruent by the ASA Property. **71c.** $EC \cong EA$ by CPCTC (corresponding parts of congruent triangles are congruent.) $EA = 7$, so $EC = 7$. **73.** J **75.** $-5, \frac{3}{2}$
77. $-\frac{1}{2}, \frac{4}{3}$ **79.** $3, 16$ **81.** $-9, -12$
83.

85. yes
87. no
89. yes

Lesson 4-5

1. $(-8.45, -3.55)$ **3.** $(-12.87, -5.13)$ **5.** 25 ft
7. 6.25; $(x - 2.5)^2$ **9.** $(2 - i\sqrt{5}, 2 + i\sqrt{5})$
11. $(-4.37, 1.37)$ **13.** $(-6.45, -1.55)$ **15.** $(-1.47, 7.47)$
17. $(-7.65, -2.35)$ **19.** $(-1, 3)$ **21.** $(4.67, 10.33)$
23. $(-0.95, 3.95)$ **25.** $(4, 5)$ **27.** 64; $(x + 8)^2$
29. 20.25; $(x + 4.5)^2$ **31.** $(-4.61, 2.61)$ **33.** $(1, 3)$
35. $\left\{\frac{3 - i\sqrt{31}}{4}, \frac{3 + i\sqrt{31}}{4}\right\}$

37. $3x^2 - 6x - 9 = 0$ Original equation
$x^2 - 2x - 3 = 0$ Divide by the coefficient of the quadratic term, 3.
$x^2 - 2x = 3$ Add 3 to each side.
$x^2 - 2x + 1 = 3 + 1$ Since $\left(\frac{-2}{2}\right)^2 = 1$, add 1 to each side.
$(x - 1)^2 = 4$ Write the left side as a perfect square.
$x - 1 = \pm 2$ Square Root Property
$x = \pm 2 + 1$ Add 1 to each side.
$x = 2 + 1$ or $x = -2 + 1$ Write as two equations.
$= 3$ $= -1$ Simplify.
The solution set is $\{-1, 3\}$ or $\{x \mid x = -1, 3\}$.

39. $\left\{\frac{-2 - i\sqrt{7}}{4}, \frac{-2 + i\sqrt{7}}{4}\right\}$ **41.** $\{5 - 2i, 5 + 2i\}$
43. $\left\{\frac{7 - i\sqrt{47}}{4}, \frac{7 + i\sqrt{47}}{4}\right\}$ **45.** $\{2.65 - i\sqrt{1.5775},$
$2.65 + i\sqrt{1.5775}\}$ **47.** $\{-0.89, 5.39\}$ **49.** $\{2.38, 4.62\}$

Lesson 4-6

1. $x = \frac{-b \pm \sqrt{b^2 - 4ac}}{2a}$ Quadratic Formula
$= \frac{-12 \pm \sqrt{12^2 - 4(1)(-9)}}{2(1)}$ $a = 1$, $b = 12$, and $c = -9$
$= \frac{-12 \pm \sqrt{144 + 36}}{2}$ Multiply.
$= \frac{-12 \pm \sqrt{180}}{2}$ Simplify.
$= \frac{-12 \pm 6\sqrt{5}}{2}$ $\sqrt{180} = 6\sqrt{5}$

51. $(-1.26, 0.26)$
53. a. The time in which the firework explodes is the t-coordinate of the vertex.
$x = -\frac{b}{2a}$ Equation of the axis of symmetry
$x = -\frac{(25)}{2(-1.5)}$ or $8\frac{1}{3}$ $a = -1.5$ and $b = 25$
So, the firework explodes after $8\frac{1}{3}$ seconds.
b. The time in which the firework explodes is the d-coordinate of the vertex.
$d = -1.5t^2 + 25t$ Original function
$d = -1.5\left(8\frac{1}{3}\right)^2 + 25\left(8\frac{1}{3}\right)$ $t = 8\frac{1}{3}$
$d \approx 104.2$ Simplify.
So, the firework explodes at a height of about 104.2 feet.
55. 2.56; $(x - 1.6)^2$
57a. A **57b.** $x = \frac{1 + \sqrt{5}}{2}$

57c.

CQ	x
2	$1 + \sqrt{5}$
3	$\frac{3 + 3\sqrt{5}}{2}$
4	$2 + 2\sqrt{5}$

57d. Sample answer: the x-values are multiples of $\frac{1 + \sqrt{5}}{2}$; $x = \frac{n(1 + \sqrt{5})}{2}$ **59.** $x = -\frac{b}{2a} \pm \frac{1}{9}; \frac{1}{4}; \left\{\frac{5}{6}, -\frac{1}{6}\right\}$
61. Sample answer: $x^2 - \frac{2}{3}x + \frac{1}{9}$ **69.** $\frac{4}{39} - \frac{19}{39}i$
63. E **65.** 125 **67.** $39 + 80i$ **73.** 1 $\$100$, 3 $\$50$, and
6 $\$20$ checks
71. $5x^2 - 28x - 12 = 0$ **73.1** $\$100$, 3 $\$50$, and
6 $\$20$ checks
75.

77.

79. $f(x) = \begin{cases} -8 \text{ if } x \le -2 \\ 2x - 6 \text{ if } -2 < x \le 5 \\ 2x - 18 \text{ if } x > 5 \end{cases}$
81. -4 **83.** -136

connectED.mcgraw-hill.com

R48 | Selected Answers

$= \dfrac{-12 + 6\sqrt{5}}{2}$ or $x = \dfrac{-12 - 6\sqrt{5}}{2}$ Write as two equations.

$= -6 + 3\sqrt{5}$ Simplify.

The solutions are $-6 + 3\sqrt{5}$ and $-6 - 3\sqrt{5}$.

3. $\left(\dfrac{5+\sqrt{57}}{8}, \dfrac{5-\sqrt{57}}{8}\right)$ **5.** $(1.5, -0.2)$

7. $\left(\dfrac{2+2\sqrt{7}}{3}, \dfrac{2-2\sqrt{7}}{3}\right)$ **9.** about 0.78 second

11a. -36 **11b.** 2 complex roots   **13a.** -76

13b. 2 complex roots   **15.** $\dfrac{-3 \pm \sqrt{15}}{2}$ **17.** $\dfrac{-7 \pm \sqrt{129}}{8}$

19. $\dfrac{-3 \pm i\sqrt{71}}{8}$

21 a. $b^2 - 4ac = 3^2 - 4(2)(-3)$ $a = 2, b = 3, c = -3$
$= 9 + 24$ or 33   Simplify.

b. The discriminant is positive and not a perfect square. So, there are 2 irrational roots.

c. $x = \dfrac{-b \pm \sqrt{b^2 - 4ac}}{2a}$ Quadratic Formula
$= \dfrac{-3 \pm \sqrt{3^2 - 4(2)(-3)}}{2(2)}$ $a = 2, b = 3,$ and $c = -3$
$= \dfrac{-3 \pm \sqrt{9 + 24}}{4}$ Multiply.
$= \dfrac{-3 \pm \sqrt{33}}{4}$ Simplify.

23a. 49   **23b.** 2 rational   **23c.** $\dfrac{3 \pm i\sqrt{87}}{6}$ **25a.** -87

25b. 2 complex   **25c.** $\dfrac{1}{6}, -1$ **27a.** 36

27b. 2 rational   **27c.** 1, $\dfrac{1}{5}$ **29a.** 1   **29b.** 2 rational

29c. $-1, -\dfrac{4}{3}$ **31a.** -16 **31b.** 2 complex   **31c.** $-1 \pm 2i$

33a. 0   **33b.** about 2.3 seconds   **35a.** 64

35b. 2 rational   **35c.** 0, $-\dfrac{8}{5}$ **37a.** 160   **37b.** irrational

37c. $\dfrac{-1 \pm \sqrt{10}}{6}$ **39a.** 13.48   **39b.** 2 irrational

39c. $\dfrac{-0.7 \pm \sqrt{3.37}}{0.6}$

41 a. $y = -0.26x^2 - 0.55x + 91.81$ Original equation
$= -0.26(10)^2 - 0.55(10) + 91.81$ Replace x with 10.
$= 60.31$ Simplify.

For 2010, the number of deaths per 100,000 is 60.31. For 2015, the number is 25.06.

b. $y = -0.26x^2 - 0.55x + 91.81$ Original equation
$50 = -0.26(15)^2 - 0.55(15) + 91.81$ Replace y with 50.
$= 25.06$ Simplify.

$0 = -0.26x^2 - 0.55x + 41.81$ Subtract 50 from each side.

$x = \dfrac{-b \pm \sqrt{b^2 - 4ac}}{2a}$ Quadratic Formula $a = -0.26,$
$= \dfrac{-(-0.55) \pm \sqrt{(-0.55)^2 - 4(-0.26)(41.81)}}{2(-0.26)}$ $b = -0.55,$ and $c = 41.81$

$= \dfrac{0.55 \pm \sqrt{43.7894}}{-0.52}$ Simplify.

$x \approx -13.8$ or $x \approx 11.7$

Since the number of years after 2000 cannot be negative, the solution is 11.58. So, 11.7 years after 2000, or in 2011, the death rate will be 50 per 100,000.

c. $y = -0.26x^2 - 0.55x + 91.81$ Original equation
$0 = -0.26x^2 - 0.55x + 91.81$ Replace y with 0.

$x = \dfrac{-b \pm \sqrt{b^2 - 4ac}}{2a}$ Quadratic Formula $a = -0.26,$
$= \dfrac{-(-0.55) \pm \sqrt{(-0.55)^2 - 4(-0.26)(91.81)}}{2(-0.26)}$ $b = -0.55,$ and $c = 91.81$

$= \dfrac{0.55 \pm \sqrt{95.7849}}{-0.52}$ Simplify.

$x \approx -19.88$ or $x \approx 17.76$

Since the number of years after 2000 cannot be negative, the solution is 17.76. So, 17.76 years after 2000, or in 2017, the death rate will be 0 per 100,000. Sample answer: This prediction is not reasonable because the death rate from cancer will never be 0 unless a cure is found. If and when a cure is found cannot be predicted.

43. Jonathan is correct; you must first write the equation in the form $ax^2 + bx + c = 0$ to determine the values of a, b, and c. Therefore, the value of c is -7, not 7. **45a.** Sample answer: Always; when a and c are opposite signs, then ac will always be negative and $-4ac$ will always be positive. Since b^2 will also always be positive, then $b^2 - 4ac$ represents the addition of two positive values, which will never be negative. Hence, the discriminant can never be negative and the solutions can never be imaginary. **45b.** Sample answer: Sometimes; the roots will only be irrational if $b^2 - 4ac$ is not a perfect square.
47. -0.75 **49.** B   **51.** 112.5 in^2 **53.** 42.25; $(x + 6.5)^2$

55. $\dfrac{4}{25}; \left(x + \dfrac{2}{5}\right)^2$ **57.** 4i   **59.** 27 hours of flight instruction and 23 hours in the simulator

61. $y = x^2 + 1$ **63.** $y = |x + 3|$

Lesson 4-7

1. $y = (x + 3)^2 - 7$ **3.** $y = 4(x + 3)^2 - 12$

5.

7.

9. $y = (x - 3)^2 - 6$

11 $y = x^2 + 2x + 7$ Original equation.
$y = (x^2 + 2x + 1) + 7 - 1$ Complete the square by adding $\left(\dfrac{2}{2}\right)^2$ or 1. Balance the equation by subtracting 1.

$= \dfrac{0.55 \pm \sqrt{43.7894}}{-0.52}$ Simplify.

$x \approx -13.8$ or $x \approx 11.7$

$y = (x + 1)^2 + 6$ Write $x^2 + 2x + 1$ as a perfect square.

13. $y = (x + 4)^2$ **15.** $y = 3\left(x + \dfrac{5}{3}\right)^2 - \dfrac{25}{3}$

17. $y = -4(x + 3)^2 + 21$ **19.** $y = -(x + 2)^2 + 3$

21. $y = -15(x - 8.5)^2 + 4083.75$

23.

25.

27.

29.

31.

33.

35. $y = 9(x - 6)^2 + 1$ **37.** $y = -\dfrac{2}{3}(x - 3)^2$

39. $y = \dfrac{1}{3}x^2 + 5$ **41.** $y = 3\left(x - \dfrac{2}{3}\right)^2 - \dfrac{10}{3}; \left(\dfrac{2}{3}, -\dfrac{10}{3}\right)$, opens up

$x = \dfrac{2}{3}$, opens up   **43.** $y = -(x + 2.35)^2 + 8.3225;$

$(-2.35, 8.3225), x = -2.35$, opens down

45. $y = \left(x - \dfrac{1}{3}\right)^2 - 3; \left(\dfrac{1}{3}, -3\right), x = \dfrac{1}{3}$, opens up

47 a. $S(t) = \dfrac{1}{2}at^2 + v_0 t$ Original equation
$S(t) = \dfrac{1}{2}(0.002)t^2 + 0.0097t$ $a = 0.002$ and $v_0 = 0.0097$ m/s
$S(t) = 0.001t^2 + 0.0097t$ Simplify.

$S(t) = 0.001(t^2 + 9.7t)$ Group $ax^2 + bx$ and factor, dividing by a.

$S(t) = 0.001(t^2 + 9.7t + 4.86^2) - 0.024$ Complete the square by adding 4.86^2 inside the parentheses. This is an overall addition of 0.024. Balance the equation by subtracting 0.024.

$S(t) = 0.001(t + 4.86)^2 - 0.024$ Write as a perfect square.

b. $a \cdot t = v$ acceleration · time = velocity

$0.002 \text{ mi/s}^2 \cdot t = (68 - 35) \text{ mi/h}$ Substitution

$0.002 \text{ mi/s}^2 \cdot t = 33 \text{ mi/h}$ Simplify.

$0.002 \text{ mi/s}^2 \cdot t = \dfrac{33 \text{ mi}}{1 \text{ h}} \cdot \dfrac{1 \text{ h}}{3600 \text{ s}}$ Convert hours to seconds.

$0.002 \text{ mi/s}^2 \cdot t = 0.009 \text{ mi/s}$ Simplify.

$t = \dfrac{0.009 \text{ mi}}{s} \div \dfrac{0.002 \text{ mi}}{s^2}$ Divide each side by 0.002 mi/s^2.

$t = \dfrac{0.009 \text{ mi}}{s} \cdot \dfrac{s^2}{0.002 \text{ mi}}$ Multiply by the reciprocal.

$t = 4.58$ s   Simplify.

It will take about 4.58 seconds for Valerie to accelerate from 35 mi/h to 68 mi/h.

c. Yes; if we substitute $\dfrac{1}{8}$ for $S(t)$ and solve for t, we get 7.346 seconds. This is how long Valerie will be on the ramp. Since it will take her 4.58 seconds to accelerate to 68 mph, she will be on the ramp long enough to accelerate at the average expressway speed.

49. The equation of a parabola can be written in the form $y = ax^2 + bx + c$ with $a \neq 0$. For each of the three points, substitute the value of the x-coordinate for x in the equation and substitute the value of the y-coordinate for y in the equation. This will produce three equations in three variables a, b, and c. Solve the system of equations to find the values of a, b, and c. These values determine the quadratic equation.
51. Sample answer: The variable a represents different values for these functions, so making $a = 0$ will have a different effect on each function. For $f(x)$, when $a = 0$, the graph will be a horizontal line, $f(x) = k$. For $g(x)$, when $a = 0$, the graph will be linear, but not necessarily horizontal, $g(x) = bx + c$.

53. B   **55.** D   **57.** $\dfrac{-15 \pm \sqrt{561}}{8}$ **59.** $\dfrac{3 \pm \sqrt{39}}{5}$

61. 0.0025   **63.** minimum, $9\dfrac{1}{3}$ **65.** minimum, -12

67.

Boxes of Brushes

Boxes of Paint

connectED.mcgraw-hill.com

Selected Answers and Solutions

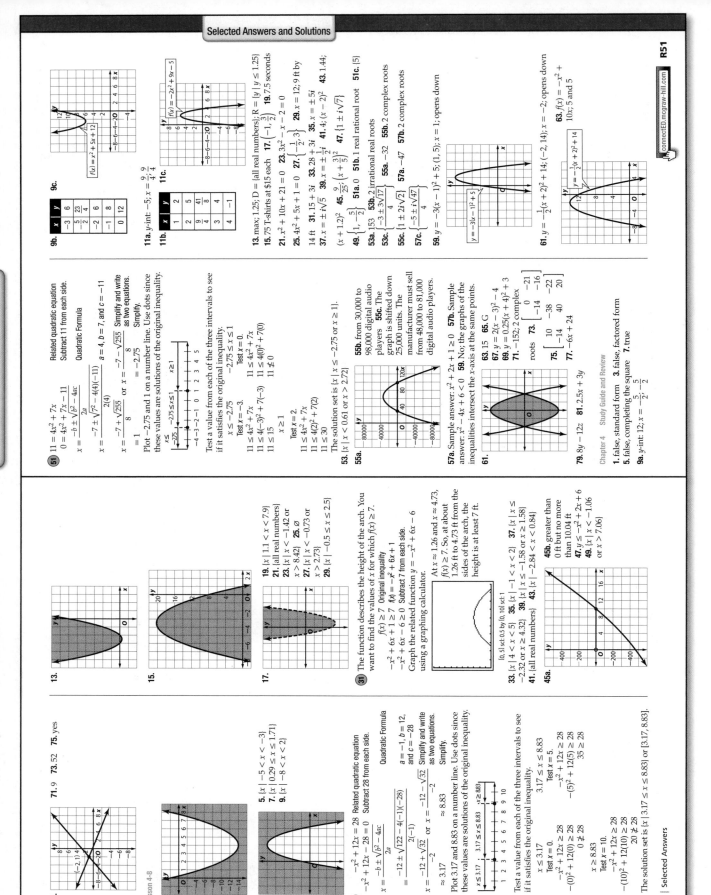

connectED.mcgraw-hill.com

9b.

x	y
−3	6
−5/2	23/4
−2	6
−1	8
0	12

11a. y-int: −5; x = 9/4, 9/4

11b.

x	y
1	2
2	5
9/4	41/8
3	4
4	−1

13. max; 1.25; D = {all real numbers}; R = {y | y ≤ 1.25} **15.** 75 T-shirts at $15 each **17.** (−1, 3/2) **19.** 7.5 seconds
21. x² + 10x + 21 = 0 **23.** 3x² − x − 2 = 0
25. 4x² + 5x + 1 = 0 **27.** {−1/2, 3} **29.** x = 12; 9 ft by
14 ft **31.** 15 + 3i **33.** 28 + 3i **35.** x = ±5i
37. x = ±i√5 **39.** x = ± 1/2 i **41.** 4; (x − 2)² **43.** 1.44;
(x + 1.2)² **45.** {−1, −5/2} **51a.** 0 **51b.** 1 irrational real root **51c.** {5}
53a. 153 **53b.** 2 irrational real roots
53c. {(−3 ± 3√17)/4} **55a.** −32 **55b.** 2 complex roots
55c. {1 ± 2i√2} **57a.** −47 **57b.** 2 complex roots
57c. {(−5 ± i√47)/4}
59. y = −3(x − 1)² + 5; (1, 5); x = 1; opens down
61. y = −1/2(x + 2)² + 14; (−2, 14); x = −2; opens down **63.** f(x) = −x² + 10x; 5 and 5

Lesson 4-8

69.

71. 9 **73.** 52 **75.** yes

13.

15.

17.

1.

3.

5. {x | −5 < x < −3}
7. {x | 0.29 ≤ x ≤ 1.71}
9. {x | −8 < x < 2}

19. {x | 1.1 < x < 7.9}
21. {all real numbers}
23. {x | x < −1.42 or x > 8.42} **25.** ∅
27. {x | x < −0.73 or x > 2.73}
29. {x | −0.5 ≤ x ≤ 2.5}

(11) −x² + 12x = 28 Related quadratic equation
−x² + 12x − 28 = 0 Subtract 28 from each side.
x = (−b ± √(b² − 4ac))/2a Quadratic Formula

a = −1, b = 12, and c = −28
x = (−12 ± √(12² − 4(−1)(−28)))/(2(−1))
x = (−12 + √32)/−2 or x = (−12 − √32)/−2 Simplify and write as two equations.
≈ 3.17 ≈ 8.83 Simplify.

Plot 3.17 and 8.83 on a number line. Use dots since these values are solutions of the original inequality.

Test a value from each of the three intervals to see if it satisfies the original inequality.

x ≤ 3.17   3.17 ≤ x ≤ 8.83   x ≥ 8.83
Test x = 0.   Test x = 5.   Test x = 10.
−x² + 12x ≥ 28   −x² + 12x ≥ 28   −x² + 12x ≥ 28
−(0)² + 12(0) ≥ 28 −(5)² + 12(5) ≥ 28 −(10)² + 12(10) ≥ 28
0 ≱ 28   35 ≥ 28   20 ≱ 28

The solution set is {x | 3.17 ≤ x ≤ 8.83} or [3.17, 8.83].

(31) The function describes the height of the arch. You want to find the values of x for which f(x) ≥ 7.

f(x) ≥ 7 Original inequality
−x² + 6x + 1 ≥ 7 f(x) = −x² + 6x + 1
−x² + 6x − 6 ≥ 0 Subtract 7 from each side.

Graph the related function y = −x² + 6x − 6 using a graphing calculator.

[0, 5] scl: 0.5 by [0, 10] scl: 1

At x ≈ 1.26 and x ≈ 4.73, f(x) ≥ 7. So, at about 1.26 ft to 4.73 ft from the sides of the arch, the height is at least 7 ft.

33. {x | 4 < x < 5} **35.** {x | −1 < x < 2} **37.** {x | x ≤ −2.32 or x ≥ 4.32} **39.** {x | x ≤ −1.58 or x ≥ 1.58}
41. {all real numbers} **43.** {x | −2.84 < x < 0.84}
45b. greater than 0 ft but no more than 10.04 ft
47. y ≤ −x² + 2x + 6 **49.** {x | x < −1.06 or x > 7.06}

45a.

(51) 11 = 4x² + 7x Related quadratic equation
0 = 4x² + 7x − 11 Subtract 11 from each side.
x = (−b ± √(b² − 4ac))/2a Quadratic Formula

a = 4, b = 7, and c = −11
x = (−7 ± √(7² − 4(4)(−11)))/(2(4)) Simplify and write as two equations.
x = (−7 + √255)/8 or x = (−7 − √255)/8 Simplify.
= 1 = −2.75

Plot −2.75 and 1 on a number line. Use dots since these values are solutions of the original inequality.

x ≤ −2.75   −2.75 ≤ x ≤ 1   x ≥ 1

Test a value from each of the three intervals to see if it satisfies the original inequality.

x ≤ −2.75   −2.75 ≤ x ≤ 1   x ≥ 1
Test x = −3.   Test x = 0.   Test x = 2.
11 ≤ 4x² + 7x   11 ≤ 4x² + 7x   11 ≤ 4x² + 7x
11 ≤ 4(−3)² + 7(−3) 11 ≤ 4(0)² + 7(0) 11 ≤ 4(2)² + 7(2)
11 ≤ 15   11 ≰ 0   11 ≤ 30

The solution set is {x | x ≤ −2.75 or x ≥ 1}.
53. {x | x < 0.61 or x > 2.72}

55a.

55b. from 30,000 to 98,000 digital audio players **55c.** The graph is shifted down 25,000 units. The manufacturer must sell from 48,000 to 81,000 digital audio players. **57b.** Sample answer: x² + 2x + 1 ≥ 0 **59.** No; the graphs of the inequalities intersect the x-axis at the same points.
57a. Sample answer: x² − 4x + 6 < 0
61.

Chapter 4 Study Guide and Review

1. false, standard form **3.** false, factored form
5. false, completing the square **7.** true
9a. y-int: 12; x = −5/2, −5/2

63. 15 **65.** G
67. y = 2(x − 3)² − 4
69. y = 0.25(x + 4)² + 3
71. −152; 2 complex roots **73.** [0 −21 / −14 −16]
75. [10 −38 −22 / −14 40 20]
77. −6x + 24
79. 8y − 12z **81.** 2.5x + 3y

41 $KE(v) = 0.5mv^2$ Original function
$= 0.5(171)(11)^2$ Replace m with 171 and v with 11.
$= 10{,}345.5$ Simplify.
The kinetic energy is 10,345.5 kg·m/s or 10,345 joules.

43. $p(-2) = -16$; $p(8) = 1024$ **45.** $p(-2) = -0.5$;
$p(8) = 3112$ **47.** D   **49.** A   **51.** $3a^3 - 24a^2 + 240a + 66$
53. $5a^6 - 298a^2 + 1008a - 928$

55a.

x	$p(x)$
-7	-585
-6	0
-4	240
-3	135
-2	0
-1	-105
1	-105
2	0
4	240
6	0
7	-585

55b. $-6, -2, 2, 6$ **55c.** Sample answer: 2000 and 6000 items   **55d.** Sample answer: The negative values should not be considered because the company will not produce negative items.

57 The degree, 4, is even and the leading coefficient, -5, is negative. So, $f(x) \to -\infty$ as $x \to +\infty$ and $f(x) \to -\infty$ as $x \to +\infty$.

59. $h(x) \to +\infty$ as $x \to -\infty$; $h(x) \to -\infty$ as $x \to +\infty$
61. $g(x) \to +\infty$ as $x \to -\infty$; $g(x) \to +\infty$ as $x \to +\infty$
63. Sample answer: Virginia is correct; an even function will have an even number of zeros and the double root represents 2 zeros.   **65.** Sample answer: $f(x) \to +\infty$ as $x \to -\infty$; $f(x) \to +\infty$ as $x \to +\infty$ and $f(x) \to +\infty$ as $x \to -\infty$; $f(x) \to +\infty$ as $x \to +\infty$ will become a 2-degree function with a positive leading coefficient.   **67.** Sometimes; a polynomial function with four real roots may be a sixth-degree polynomial function with two imaginary roots. A polynomial function that has four real roots is at least a fourth-degree polynomial.   **69.** Student A

71a.

71b. $t = 0.5c$

---

53. $2n^4 - 3n^3p + 6n^4p^4$ **55.** $b^3 + \dfrac{b}{a} + \dfrac{1}{a^2}$
57. $2n^5 - 14n^3 + 4n^2 - 28$ **59.** $64n^3 - 240n^2 + 300n - 125$ **61a.** $0.155x^2 + 8.818x + 835.8$
61b. $0.061x^2 - 10.57x + 112.4$ **63.** 9   **65.** $\dfrac{a^n}{a^n} = a^0 = a^{n-n} = a^n$ **67.** Sample answer: We would have a 0 in the denominator, which makes the expression undefined.   **69.** Sample answer: Astronomy deals with very large numbers that are sometimes difficult to work with because they contain so many digits. Properties of exponents make very large or very small numbers more manageable. As long as you know how far away a planet is from a light source, you can divide that distance by the speed of light to obtain how long it will take light to reach that planet.   **71.** D   **73.** D   **75.** $x > 5$ or $x < -8$

77.

79.

81. 42   **83.** 28   **85.** $\dfrac{7}{8}$ **87.** $\dfrac{1}{2}$ **89.** $\dfrac{5}{9}$
91. $4x(3ax^2 + 5bx + 8c)$ **93.** $(3y + 2)(4y + 3)$
95. $(2x - 3)(4a - 3)$

Lesson 5-2

1. $4y + 2x - 2$ **3.** $x - 8 - \dfrac{4}{x+2}$ **5.** $3z^3 - 15z^2 + 36z - 105 + \dfrac{309}{z+3}$ **7.** A   **9.** $6a + 6 + \dfrac{21}{3a-2}$
11. $3y + 5$ **13.** $x + 3y - 2$ **15.** $2a^2 + b - 3$
17. $3m - 6 + 7p$ **19.** $-w + 16 + \dfrac{1000}{w}$

21
$$
\begin{array}{r}
b^2 - 5b + 6 \\
b+1 \overline{\smash{)}b^3 - 4b^2 + b - 2} \\
(-)\,b^3 + b^2 \\
\hline
-5b^2 + b \\
(-)\,-5b^2 - 5b \\
\hline
6b - 2 \\
(-)\,6b + 6 \\
\hline
-8
\end{array}
$$
The quotient is $b^2 - 5b + 6$, and the remainder is -8. So, the expression equals $b^2 - 5b + 6 - \dfrac{8}{b+1}$.

23. $x^4 + 4x^3 + 12x^2 + 52x + 208 + \dfrac{832}{x-4}$
25. $g^3 + 2g^2 + g + 2 - \dfrac{14}{g-2}$
27. $2x^4 + x^3 - x + \dfrac{3}{2} - \dfrac{2}{9x+3}$ **29.** $b^2 - 4b + 8 - \dfrac{8}{b+1}$
31. $2y^5 - y^4 + y^3 + y^2 - y - 3$ **33.** $V(t) = t^2 + 5t + 6$

35 a.
$$
\begin{array}{r}
3500 \\
a^2 + 100 \overline{\smash{)}3500a^2} \\
(-)\,3500a^2 + 350{,}000 \\
\hline
-350{,}000
\end{array}
$$
The quotient is 3500, and the remainder is $-350{,}000$. So, the expression equals $3500 - \dfrac{350{,}000}{a^2+100}$.

65.

67.

Chapter 5   Get Ready

1. $-5 + (-13)$ **3.** $5mr + (-7mp)$ **5.** $20 + (-2x)$
7. $-3\beta^2 - 2b + 1$ **9.** $-\dfrac{9}{4}, \dfrac{9}{4}$ **11.** $-4, 2$ **13.** $-\dfrac{4}{3}, \dfrac{1}{2}$
15. about 1.77 seconds

Lesson 5-1

1. $-8a^5b^2$ **3.** $\dfrac{8a^6}{27b^3}$ **5.** yes, 1   **7.** no   **9.** $-2x^2 - 6x + 3$
11. $8ab + 10a$ **13.** $n^2 - 2n - 63$ **15.** $750 - 2.5x$
17. $-8b^5c^3$ **19.** $-yz^2$ **21.** $\dfrac{a^2c^2}{21}$ **23.** z^{18} **25.** yes; 3
27. no   **29.** $3\beta^2 + 6b - 5$ **31.** $8x^3 + 4xy$

33 $(a+b)(a^3 - 3ab - b^2)$
$= a(a^3 - 3ab - b^2) + b(a^3 - 3ab - b^2)$ Distributive Property
$= a(a^3) - a(3ab) - a(b^2) + b(a^3) - b(3ab) - b(b^2)$ Distributive Property
$= a^4 - 3a^2b - ab^2 + a^3b - 3ab^2 - b^3$ Multiply.
$= a^4 + a^3b - 3a^2b - 4ab^2 - b^3$ Simplify.

35. $10.3^{-3} - c^2 + 4c$ **37.** $12n^2b + 8a^2b^2 - 15ab^2 + 4b^2$
39. $4a^2x - 2a^2y + 10abx - 5aby + 6b^2x - 3b^2y$
41. $\dfrac{y^4}{81x^4}$ **43.** $\dfrac{x^6}{16y^{14}}$ **45.** b **47.** $\dfrac{7}{2}cd^4$ **49.** $\dfrac{1}{2}x^6y^3$

51 a. $d = rt$ distance = rate · time
    Solve the formula for time.
$t = \dfrac{d}{r}$
    Distance from Andromeda to Earth
$= \dfrac{2.367 \times 10^{21} \text{ m}}{3 \times 10^8 \text{ m/s}}$ Speed of light
$= \dfrac{2.367}{3} \cdot \dfrac{10^{21}}{10^8} \dfrac{\text{m}}{\text{m/s}}$ Separate to get powers of the same base.
$\approx 0.789 \cdot 10^{21-8}$ s   Subtract exponents.
$\approx 0.789 \times 10^{13}$ s   Simplify.
It takes about 7.89×10^{12} seconds or about 250,190.26 years.

b. $t = \dfrac{d}{r}$ Write the formula.
    Distance from the Sun to Mars
$= \dfrac{2.28 \times 10^{11} \text{ m}}{3 \times 10^8 \text{ m/s}}$ Speed of light
$= \dfrac{2.28}{3} \cdot \dfrac{10^{11}}{10^8} \dfrac{\text{m}}{\text{m/s}}$ Separate to get powers of the same base.
$= 0.76 \times 10^{11-8}$ s   Subtract exponents.
$= 0.76 \times 10^3$ s   Simplify.
$= 760$ s   Simplify.
It takes about 760 seconds or about 12.67 minutes.

b. $n = 3500 - \dfrac{350{,}000}{a^2 + 100}$ Write the equation.
$= 3500 - \dfrac{350{,}000}{15^2 + 100}$
$= 3500 - \dfrac{350{,}000}{325}$ Simplify.
≈ 2423 subscriptions
$a = 15$

37. $\dfrac{4c^2d - 3d}{2}$ **39.** $\dfrac{n^2 - n - 1}{2}$
41. $3z^4 - z^3 + 2z^2 - 4z + 9 - \dfrac{13}{z+2}$ **43.** Sample answer: Sharon; Jamal actually divided by $x + 3$.
45. Sample answer: The degree of the quotient plus the degree of the divisor equals the degree of the dividend.   **47.** $\dfrac{5}{x^2}$ does not belong with the other three. The other three expressions are polynomials. Since the denominator $\dfrac{5}{x^2}$ contains a variable, it is not a polynomial.   **49.** A   **51.** 360
53. $3x^3 + 2x^2 + x + 4$ **55.** $23a^2 - 24a$ **57.** $8x^5y^8z^3$
59. 10 to 10 ft or 24 to 34 ft   **61.** $4 \pm \sqrt{19}$ **63.** between -6 and -5; between -3 and -2 **65.** between -1 and 0; between 2 and 3   **67.** -21 **69.** -20 **71.** $-9d^2$

Lesson 5-3

1. degree = 6, leading coefficient = 11   **3.** not in one variable because there are two variables, x and y
5. $w(5) = -247$; $w(-4) = 104$ **7.** $4y^9 - 5y^6 + 2$
9. $1536a^3 - 426a^2 - 144a + 82$ **11a.** $f(x) \to -\infty$ as $x \to -\infty$; $f(x) \to +\infty$ as $x \to +\infty$ **11b.** Since the end behavior is in opposite directions, it is an odd-degree function.   **11c.** The graph intersects the x-axis at three points, so there are three real zeros.   **13.** not in one variable because there are two variables, x and y
15. degree = 6, leading coefficient = -12
17. degree = 4, leading coefficient = -5
19. degree = 2, leading coefficient = 3   **21.** degree = 9, leading coefficient = 2   **23.** $p(-6) = 1227$; $p(3) = 66$
25. $p(-6) = -156$; $p(3) = 78$

27 $p(x) = -x^3 + 3x^2 - 5$ Original function
$p(-6) = -(-6)^3 + 3(-6)^2 - 5$ Replace x with -6.
$= 216 + 108 - 5$ Simplify.
$= 319$ Simplify.

$p(x) = -x^3 + 3x^2 - 5$ Original function
$p(-6) = -(3)^3 + 3(3)^2 - 5$ Replace x with 3.
$= -27 + 27 - 5$ Simplify.
$= -5$ Simplify.

29. $18x^2 - 12a + 3$ **31.** $2b^4 - 4b^2 + 3$ **33.** $-64y^3 + 144y^2 - 104y + 25$ **35a.** $f(x) \to +\infty$ as $x \to -\infty$; $f(x) \to +\infty$ as $x \to +\infty$ **35b.** Since the end behavior is in the same direction, it is an even-degree function.   **35c.** The graph intersects the x-axis at four points, so there are four real zeros.   **37a.** $f(x) \to +\infty$ as $x \to -\infty$; $f(x) \to +\infty$ as $x \to +\infty$ **37b.** Since the end behavior is in opposite directions, it is an odd-degree function.   **37c.** The graph intersects the x-axis at one point, so there is one real zero.   **39a.** $f(x) \to +\infty$ as $x \to -\infty$; $f(x) \to +\infty$ as $x \to +\infty$.   **39b.** Since the end behavior is in the same direction, it is an even-degree function.   **39c.** The graph intersects the x-axis at two points, so there are two real zeros.

c. $d(x) \to +\infty$ as $x \to +\infty$; as x increases, $d(x)$ increases.

d. Sample answer: Since the diving board is only 10 feet long, x cannot be greater than 10. So, this trend cannot continue indefinitely.

35a. −2.5 (min), −0.5 (max), 1.5 (min) **35b.** −3.5, −1, 0, 3 **35c.** 4 **35d.** D = [all real numbers]; R = {$y \mid y \geq -3.1$} **37a.** −3.5 (min), −2.5(max), −2 (min), −1 (max), 1 (min) **37b.** −3.75, −3.25, −2, −1.75, −0.25, 2.9 **37c.** 6 **37d.** D = [all real numbers]; R = {$y \mid y \geq -5$} **39a.** −2 (max), 1 (min) **39b.** −3, −0.5, 2 **39c.** 3 **39d.** D = [all real numbers] R = [all real numbers]

41.

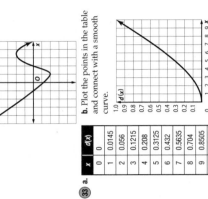

$1.25

43 **a.** Make a table of values and graph the function.

The value of $f(x)$ changes signs between $x = 1.5$ and $x = 2$. So, there is a zero between these values, at approximately 1.75. The x-intercept ≈ 1.75. The y-intercept is at −4. The value of $f(x)$ at $x \approx -1.25$ and at $x \approx 0.5$ is greater than the surrounding points, so $x \approx -1.25$ and $x \approx 0.5$ are turning points. The value of $f(x)$ at $x \approx -0.5$ and at $x \approx 1.25$ is less than the surrounding points, so $x \approx -0.5$ and $x \approx 1.25$ are turning points.

x	$f(x)$
−2	−16
−1.5	−4.5
−1	−4
−0.5	−4.6
0	−4
0.5	−3.3
1	−4
1.5	−3.5
2	8	← change in sign

b. The function does not have an axis of symmetry because the graph is not a parabola.

c. The function is increasing in the intervals $x \leq -1.25$, $-0.5 \leq x \leq 0.5$, and $x \geq 1.25$. The function is decreasing in the intervals $-1.25 \leq x \leq -0.5$ and $0.5 \leq x \leq 1.25$.

45a. no zeros, no x-intercepts, y-intercept: 5; no turning points **45b.** no axis of symmetry **45c.** decreasing: $x \leq -4$; constant: $-4 < x < 0$; increasing: $x > 0$ **47.** As the x-values approach large positive or negative numbers, the term with the largest degree becomes more and more dominant in determining the value of $f(x)$.

19a.

x	$f(x)$
−4	−176
−3	−77
−2	−22
−1	1
0	4
1	−1
2	−2
3	13
4	56

19b. between $x = -2$ and $x = -1$, between $x = 0$ and $x = 1$, and between $x = 2$ and $x = 3$

19c. rel. max: near $x = -0.3$; rel. min: near $x = 1.6$

21a.

x	$f(x)$
−4	372
−3	141
−2	36
−1	−3
0	−12
1	−3
2	36
3	141
4	372

21b. between $x = -2$ and $x = -1$ and between $x = 1$ and $x = 2$ **21c.** min: near $x = 0$ **23.** rel. max: $x = -2.73$; rel. min: $x = 1.34$, no rel. min

25. rel. min: $x = 0.73$

27. Sample answer:

29. Sample answer:

31. Sample answer:

33 **a.**

x	$d(x)$
0	0
1	0.0145
2	0.056
3	0.1215
4	0.208
5	0.3125
6	0.432
7	0.5635
8	0.704
9	0.8505
10	1

b. Plot the points in the table and connect with a smooth curve.

71c. 16 tables and 32 chairs **71d.** Sample answer: This can be determined by the intersection of the graphs. This point of intersection is the optimal amount of tables and chairs manufactured. **73.** $2x^4y^2 + 4x^4y^2z^2$ **75.** $6c^3 - 1 + 4a^5cd^2$ **77.** yes; 6 **79a.** $h(d) = -2d^2 + 4d + 6$; The graph opens downward and is narrower than the parent graph, and the vertex is at (1, 8). **79b.** $h(d) = -2(d - 1.25)^2 + 12.5$; it shifted the graph up 4.5 ft and to the right 3 in. **81.** $x \leq -\frac{2}{3}$ or $x \geq 2$ **83.** minimum; $-\frac{4}{3}$ **85.** maximum; 11

Lesson 5-4

1.

3.

5. between −2 and −1

7. between 0 and 1 and between 2 and 3

9. rel. max at $x \approx -1.8$; rel. min at $x \approx 1.1$; D = [all real numbers], R = [all real numbers]

11. rel. max at $x \approx 2.4$; rel. min at $x \approx 0.3$; D = [all real numbers], R = [all real numbers]

13a.

13b. Sample answer: Relative maximum at $x \approx 5$ and relative minimum at $x \approx 9.5$. $f(x) \to \infty$ as $x \to -\infty$ and $f(x) \to \infty$ as $x \to \infty$. The graph increases when $x < 5$ and $x > 9.5$ and decreases when $5 < x < 9.5$.

13c.

Sample answer: This suggests a dramatic increase in sales.

13d. Sample answer: No; with so many other forms of media on the market today, CD sales will not increase dramatically. In fact, the sales will probably decrease. The function appears to be accurate only until about 2005.

15 **a.** Since $f(x)$ is a third-degree polynomial function, it will have either 3 or 1 real zeros. Look at the values of $f(x)$ to locate the zeros. Then use the points to sketch the graph.

x	$f(x)$
−4	92
−3	41
−2	12
−1	−1
0	−4
1	−3
2	−4
3	−13
4	−36

← change in sign

b. The value of $f(x)$ changes signs between $x = -2$ and $x = -1$. So, there is a zero between −2 and −1.

c. The value of $f(x)$ at $x = 0$ is less than the surrounding points, so there must be a relative minimum near $x = 0$. The value of $f(x)$ near $x = 1$ is greater than the surrounding points, so there must be a relative maximum near $x = 1$.

17a.

x	$f(x)$
−4	−155
−3	−80
−2	−33
−1	−8
0	1
1	0
2	−5
3	−8
4	−3
5	16

17b. at $x = 1$, between −1 and 0, and between $x = 4$ and $x = 5$ **17c.** rel. max: $x \approx \frac{1}{3}$; rel. min: $x = 3$

49. Sample answer:

51. Sample answer: No; $f(x) = x^2 + x$ is an even degree, but $f(1) \neq f(-1)$. **53.** Sample answer: The degree will help determine whether the graph is even or odd and the maximum number of zeros and turning points for the graph. The leading coefficient determines the end behavior of the graph, and, along with the degree, builds the shape of the graph. The zeros and turning points allow for the plotting of specific points in the center of the graph. All of these things combine for an accurate sketch of the graph of a polynomial function. **55.** 95 **57.** C **59.** $f(x) \to -\infty$ as $x \to -\infty$, $f(x) \to +\infty$ as $x \to +\infty$. Since the end behavior is in the same direction, it is an even-degree function. The graph intersects the x-axis at six points, so there are six real zeros. **61.** $(x - 2)(x + 3)$ **63.** $2a^2 + a - 3 - \dfrac{3}{a - 1}$ **65.** $(x + 6)(x + 3)$ **67.** $(a + 8)(a - 2)$ **69.** $(3x - 4)(2x + 1)$

Lesson 5-5

1. $(a + b)(3x + 2y - z)$ **3.** prime **5.** $12q(w - q)(w^2 + qw + q^2)$ **7.** $x^2(a - b)(a^2 + ab + b^2)(a + b)(a^2 - ab + b^2)$ **9.** $(2c - 5d)(4c^2 + 12cd + 25d^2)$ **11.** $-4, -4, \pm\sqrt{3}$ **13.** $-3, \dfrac{3 \pm 3i\sqrt{3}}{2}$ **15.** 5 ft **17.** not possible **19.** $\sqrt{6}, -\sqrt{6}, 2\sqrt{3}, -2\sqrt{3}$ **21.** $x(4x + y)(16x^2 - 4xy + y^2)$ **23.** $y^2(x^2 + y^2)(x^4 - x^2y^2 + y^2)$ **25.** prime **27.** $(6x^2 - 5y^2)(2a - 3b + 4c)$ **29.** $8x^5 - 80x^4 - x^2y^3 + 200x^3 + (-25y^3 - x^2y^3 - 10xy^3)$ Original expression
$= (8x^5 + 80x^4 + 200x^3) + (-25y^3 - x^2y^3 - 10xy^3)$ Group to find a GCF.
$= 8x^3(x^2 + 10x + 25) - y^3(25 + x^2 + 10x)$ Factor the GCF.
$= 8x^3(x + 5)^2 - y^3(x + 5)^2$ Perfect squares
$= (8x^3 - y^3)(x + 5)^2$ Distributive Property
$= (2x - y)(4x^2 + 2xy + y^2)(x + 5)^2$ Factor the GCF.
31. $6, -6, \pm 2i\sqrt{5}$ **33.** $\pm\sqrt{7}, \pm i\sqrt{13}$ **35.** $-\dfrac{1}{4}, \dfrac{1 \pm i\sqrt{3}}{8}$ **37.** $-15(x^2)^2 + 18(x^2) - 4$ **39.** not possible **41.** $4(2x^5)^2 + 1(2x^5) + 6$ **43.** $4\sqrt{5}, \pm i\sqrt{2}$ **45.** $\pm\dfrac{2\sqrt{3}}{5}, \pm\dfrac{\sqrt{15}}{5}$ **47.** $\pm\dfrac{\sqrt{6}}{6}, \pm i\dfrac{\sqrt{3}}{3}$ **51.** $x(x + 2)(x - 2)(x^2 + 4)$ **53.** $(5x + 4y + 5z)(3a - 2b + c)$ **55.** $x(x + 3)(x - 3)(3x + 2)(2x - 5)$ **57.** $x = 8; 5, 8, 11$ **59.** $\pm\dfrac{2\sqrt{3}}{3}, \pm i\dfrac{\sqrt{2}}{3}$ **61.** $\dfrac{1}{3}, \pm i\dfrac{\sqrt{10}}{10}$ **63.** $3, -3, \pm i\dfrac{\sqrt{15}}{3}$
65. $x^6 - 26x^3 - 27 = 0$ Original equation
$(x^3)^2 - 26(x^3) - 27 = 0$ Let $u = x^3$.
$u^2 - 26u - 27 = 0$ Replace u with x^3.
$(u - 27)(u + 1) = 0$ Factor.
$u - 27 = 0$ or $u + 1 = 0$ Zero Product Property
$x^3 - 27 = 0$ or $x^3 + 1 = 0$ Replace u with x^3.
$x^3 = 27$
$(x - 3)(x^2 + 3x + 9) = 0$ Difference of Two Cubes

R56 | Selected Answers

$x - 3 = 0$ or $x^2 + 3x + 9 = 0$ Zero Product Property
$x = 3$ $x = \dfrac{-3 \pm \sqrt{3^2 - 4(1)(9)}}{2(1)}$
$= \dfrac{-3 \pm 3i\sqrt{3}}{2}$
$x^3 + 1 = 0$
$(x + 1)(x^2 - x + 1) = 0$ Sum of Two Cubes
$x + 1 = 0$ or $x^2 - x + 1 = 0$ Zero Product Property
$x = -1$ $x = \dfrac{-(-1) \pm \sqrt{(-1)^2 - 4(1)(1)}}{2(1)}$
$= \dfrac{1 \pm i\sqrt{3}}{2}$
The solutions are $-1, 3, \dfrac{1}{2}, \dfrac{-3 \pm 3i\sqrt{3}}{2}$, and $\dfrac{1 \pm i\sqrt{3}}{2}$.
67. $-1, 1, \pm\dfrac{1}{2}$ **69.** $\pm i\sqrt{5}, \pm i\sqrt{3}$ **71a.** 2 ft
71b. 176 ft² **71c.** 428 ft²
73 a. $f(x) = (x + 6)[x + x + (x + 2)] + (x + 2)] + x[x + (x + 2) + x(x + 2)] + x(x + 2)$
$= (x + 6)(4x + 4) + x(3x + 4) + x(x + 2)$
$= 4x^2 + 24x + 4x + 24 + 3x^2 + 4x + x^2 + 2x$
$= 8x^2 + 34x + 24$ Original function
b. $f(x) = 8x^2 + 34x + 24$
$1366 = 8x^2 + 34x + 24$ Replace $f(x)$ with 1366.
$0 = 8x^2 + 34x - 1342$ Subtract 1366 from each side.
$0 = 2(4x^2 + 17x - 671)$ Factor.
$0 = 2(4x + 61)(x - 11)$ Perfect squares
$4x + 61 = 0$ or $x - 11 = 0$ Zero Product Property
$x = -15.25$ $x = 11$ Simplify.
Since distance cannot be negative, $x = 11$ ft.
75. $(x + 2)^3(x - 2)^3$ **77.** $(x + y)^3(x - y)^3$ **79.** $(6x^{2n} + 1)^2$ **81.** Sample answer: $a = 1, b = -1$ **83.** Sample answer: The factors can be determined by the x-intercepts of the graph. An x-intercept of 5 represents a factor of $(x - 5)$. **85.** D **87.** D
89. rel. max. at $x \approx 1.5$, rel. min. at $x \approx 0.1$;

91. degree = 4; leading coefficient = 5 **93.** degree = 7; leading coefficient = -1 **95.** 18 skis and 10 snowboards **97.** $x + 2 - \dfrac{10}{x + 4}$ **99.** $8x^2 - 12x + 24 - \dfrac{42}{x + 2}$

Lesson 5-6

1. 58; -20 **3.** 12,526 **5.** $x + 4, x - 4$ **7.** $x - 5, 2x - 1$ **9.** 71; -6 **11.** $-435; -15$ **13.** $-4150; 85$ **15.** 647; -4 **17.** $(x - 1)^2$ **19.** $x - 4, x + 1$ **21.** $x + 6, 2x + 7$ **23.** $x + 1, x^2 + 2x + 3$ **25.** $x - 4, 3x - 2$

27 a.

	-0.04	0.8	0.5	-1	0
1		-0.04	0.76	1.26	0.26
	-0.04	0.76	1.26	0.26	0.26

$f(1) = 0.26$ ft/s; At 1 second, the speed of the boat is 0.26 ft/s.

	-0.04	0.8	0.5	-1	0
2		-0.04	-0.08	1.44	3.88	5.76
	-0.04	0.72	1.94	2.88	5.76

$f(2) = 5.76$ ft/s; At 2 seconds, the speed of the boat is 5.76 ft/s.

	-0.04	0.8	0.5	-1	0
3		-0.04	-0.12	2.04	7.62	19.86
	-0.04	0.68	2.54	6.62	19.86

$f(3) = 19.86$ ft/s; At 3 seconds, the speed of the boat is 19.86 ft/s.

b.

	-0.04	0.8	0.5	-1	0
6		-0.04	-0.24	3.36	23.16	132.96
	-0.04	0.56	3.86	22.16	132.96

$f(6) = 132.96$ ft/s; This means the boat is traveling at 132.96 ft/s when it passes the second buoy.
29. $x^2 + 2, x - 3, x^2 - x + 4$ **31a.** $g(x) = 9x^4 + 50x^3 + 51x^2 - 150x - 72$

31b.

x	$g(x)$
-5	-9922
-4	-4160
-3	-1242
-2	-112
-1	70
0	-72
1	-130
2	88
3	558
4	1040
5	1078
6	0

31c. There is a zero between $x = -2$ and $x = -1$ because $g(x)$ changes sign between the two values. There are also zeros between -1 and 0 and between $x = 1$ and $x = 2$ because $g(x)$ changes sign between the two values. There is also a zero at $x = 6$.

31d.

33.

2	1	k	$2k + 4$
		$k + 2$	
	1	$k + 2$	-17

$-17 + 2k + 4 = 3$ Write an equation.
$-13 + 2k = 3$ Simplify.
$2k = 16$ Add 13 to each side.
$k = 8$ Divide each side by 2.
35. -3 **37.** $\pm\sqrt{6}, \pm\sqrt{3}$ **39a.** $x - c$ is a factor of $f(x)$. **39b.** $x - c$ is not a factor of $f(x)$. **39c.** $f(x) = x - c$ **41.** Sample answer: $f(x) = -x^3 + x^2 + x + 10$ **43.** Sample answer: A zero can be located using the Remainder Theorem and a table of values by determining when the output, or remainder, is equal to zero. For instance, if $f(6)$ leaves a remainder if 2 and $f(7)$ leaves a remainder of -1, then you know that there is a zero between $x = 6$ and $x = 7$. **45.** 4 **47.** B
49. $\pm 3, \pm i\sqrt{3}$ **51a.** -1.5 (max.), 0.5 (min.), 2.5 (max.) **51b.** $-3.5, 3.75$ **51c.** 4 **51d.** D = {all real numbers}; $R = \{y \mid y \leq 4.5\}$ **53a.** -3 (min.), -1 (max.), 1 (min.) **53b.** $-0.25, 3$ **53c.** 4 **53d.** D = {all real numbers};

55.

57.

$R = \{y \mid y \geq -4.5\}$

59. $4a^2 - 8a + 16$ **61.** $79q^2 - 58a - 22$ **63.** $-4a^4 - 24a^2 - 48a - 22$

Lesson 5-7

1. $-2, 5; 2$ real **3.** $-\dfrac{3}{2}, \dfrac{3}{2}, -\dfrac{3}{2}i, \dfrac{3}{2}i; 2$ real, 2 imaginary **5.** 3 or 1; 0, 0 or 2 **7.** 1 or 3; 0 or 2, 0, 2, or 4 **9.** $-8, -2, 1$ **11.** $-4, 6, -4i, 4i$ **13.** $f(x) = x^3 - 9x^2 + 14x + 24$ **15.** $f(x) = x^4 - 3x^3 - x^2 - 27x - 90$ **17.** $-2, \dfrac{3}{2}; 2$ real **19.** $-1, \dfrac{1 \pm i\sqrt{3}}{2}; 1$ real, 2 imaginary **21.** $\dfrac{8}{3}, \dfrac{1}{2}; 2$ real **23.** $-\dfrac{5}{2}, \dfrac{5}{2}, -\dfrac{5}{2}i, \dfrac{5}{2}i; 2$ real, 2 imaginary **25.** $-2, -2, 0, 2; 5$ real
27 Find the number of sign changes for $f(x)$ and $f(-x)$.
$f(x) = x^4 - 5x^3 + 2x^2 + 5x + 7$
yes yes no no
Since there are 2 sign changes, the function has 0 or 2 positive real zeros.
$f(-x) = x^4 + 5x^3 + 2x^2 - 5x + 7$
no no yes yes
Since there are 2 sign changes, the function has 0 or 2 negative real zeros.
Since $f(x)$ has degree 4, the function has 4 zeros. So, the function could have the following.

2 positive real zeros, 2 negative real zeros, 0 imaginary zeros
2 positive real zeros, 0 negative real zeros, 2 imaginary zeros
0 positive real zeros, 2 negative real zeros, 2 imaginary zeros
0 positive real zeros, 0 negative real zeros, 4 imaginary zeros
29. 0 or 2; 1, 2 or 4 **31.** 0 or 2; 0 or 2; 2, 4, or 6 **33.** $-6, -2, 1$ **35.** $-4, -7, -5i, 5i$ **37.** 4, 4, $-2i, 2i$ **39.** $f(x) = x^3 - 2x^2 - 13x - 10$ **41.** $f(x) = x^4 + 2x^3 + 5x^2 + 8x + 4$ **43.** $f(x) = x^4 - x^3 - 20x^2 + 50x$ **45 a.** Find the number of sign changes for $P(x)$ and $P(-x)$.
$P(x) = -0.006x^4 + 0.15x^3 - 0.05x^2 - 1.8x$
yes yes no
Since there are 2 sign changes, the function has 0 or 2 positive real zeros.
$P(-x) = -0.006x^4 - 0.15x^3 - 0.05x^2 + 1.8x$
no no yes
Since there is 1 sign change, the function has 1 negative real zero.
Since $f(x)$ has degree 4, the function has 4 zeros.

Selected Answers and Solutions

27. $p(-2) = -25$; $p(x + h) = 3 - 5x^2 - 10xh - 5h^2 + x^3 + 3h^3 + 3h^2x + h^3$

29a.

29b. between -3 and -2, between -1 and 0, between 0 and 1, between 2 and 3

29c. rel. max: $x \approx 0$; rel. min: $x \approx 1.62$ and $x \approx -1.62$

31a.

31b. between -1 and 0, between 0 and 1, and between 1 and 2

31c. rel. max: $x \approx 0$; rel. min: $x \approx 1$ **33.** 2 relative maxima and 1 relative minima   **35.** prime

37. $(2y + z)(3q + 2b - c)$ **39.** $\pm\frac{\sqrt{3}}{2}, \pm\frac{\sqrt{2}}{2}$

41. $f(-2) = 1$; $f(4) = 13$ **43.** $f(-2) = 16$; $f(4) = 118$

45. $x + 2$ and $3x - 1$ **47.** $x + 3, x + 4$

49. positive real zeros: 0
negative real zeros: 4, 2, or 0
imaginary zeros: 4, 2, or 0
51. positive real zeros: 2 or 0
negative real zeros: 1
imaginary zeros: 4 or 2
53. $-2, -1 \pm \sqrt{2}$ **55.** $-2, \pm 2i$

CHAPTER 6

Inverses and Radical Functions and Relations

Get Ready

1. between 0 and 1, and between 3 and 4
3. between 1 and 2 seconds   **5.** $3x + 2 - \frac{20}{x + 4}$
7. $3x^3 - 4x^2 + 5x - 3 + \frac{6}{x - 3}$

Lesson 6-1

1. $(f + g)(x) = 4x + 1$; $(f - g)(x) = -2x + 3$;
$(f \cdot g)(x) = 3x^2 + 5x - 2$; $\left(\frac{f}{g}\right)(x) = \frac{x + 2}{3x - 1}, x \neq \frac{1}{3}$
3. $f \circ g$ is undefined, D = ∅, R = ∅; $g \circ f = \{(2, 8), (6, 13), (12, 11), (7, 15)\}$, D = \{2, 6, 7, 12\}, R = \{8, 11, 13, 15\}.   **5.** $[f \circ g](x) = -15x + 18$, R = [all multiples of 3];
$[g \circ f](x) = -15x - 6$, R = \{y | y = 5x - 6 for $-\infty \le x \le \infty$\].   **7.** Either way, she will have $228.95 taken from her paycheck. If she takes the college savings plan deduction before taxes, $76 will go to her college plan and $152.95 will go to taxes. If she takes the college savings plan deduction after taxes, only $62.70 will go to her college plan and $166.25 will go to taxes.
9. $(f + g)(x) = 6x - 3$; $(f - g)(x) = -4x + 1$;
$(f \cdot g)(x) = 5x^2 - 7x + 2$; $\left(\frac{f}{g}\right)(x) = \frac{x - 1}{5x - 2}, x \neq \frac{2}{5}$

Simplify.

b. $V = \pi r^3 + 6\pi r^2$ Volume of a cylinder
$160\pi = \pi r^3 + 6\pi r^2$ Substitute.
$0 = \pi r^3 + 6\pi r^2 - 160\pi$ Subtract 160π from each side.
$0 = r^3 + 6r^2 - 160$ Divide each side by π.

p: $\pm 1, \pm 2, \pm 4, \pm 5, \pm 8, \pm 20, \pm 32, \pm 40, \pm 80, \pm 160$
q: ± 1

possible rational zeros: $\frac{p}{q} = \pm 1, \pm 2, \pm 4, \pm 5, \pm 8, \pm 20, \pm 32, \pm 40, \pm 80, \pm 160$

Make a table and test some possible rational zeros.

r	1	6	0	-160
1	1	7	7	-153
2	1	8	16	-128
3	1	9	27	-79
4	1	10	40	0

$V(r) = 0$, so there is a zero at $x = 4$. Use the Quadratic Formula to find zeros of the depressed polynomial $r^2 + 10r + 40$.

$x = \frac{-b \pm \sqrt{b^2 - 4ac}}{2a}$ Quadratic Formula
$= \frac{-10 \pm \sqrt{10^2 - 4(1)(40)}}{2(1)}$ $a = 1, b = 10,$ and $c = 40$
$= \frac{-10 \pm \sqrt{100 - 160}}{2}$ Multiply.
$= \frac{-10 \pm \sqrt{-60}}{2}$ Simplify.
$= \frac{-10 \pm 2i\sqrt{15}}{2}$ $\sqrt{-60} = 2i\sqrt{15}$
$= -5 \pm i\sqrt{15}$ Simplify.

Since the radius cannot be an imaginary number, the only reasonable value is $r = 4$ in.

c. Since $r = 4$ in., $h = r + 6$ or 10 in.

43a. 30x^3 − 478x^2 + 1758x + 1092 = 0   **43b.** 1, 2, 3, 4, 6, 7, 12, 13, 14, 21, 26, 28, 39, 42, 52, 78, 84, 91, 156, 182, 273, 364, 546, 1092   **43c.** 2013   **43d.** No; Sample answer: Music sales fluctuate from 2005 to 2015, then increase indefinitely. It is not reasonable to expect sales to increase forever.   **45.** 2, 3, 3, −3, −4
47. Sample answer: $f(x) = x^4 - 12x^3 + 47x^2 - 38x - 58$
49. Sample answer: $f(x) = 4x^5 + 3x^3 + 8x + 18$
51. Sample answer: For any polynomial function, the constant term represents p and the leading coefficient represents q. The possible zeros of the function can be found with $\pm\frac{p}{q}$ where the fraction is every combination of factors of p and q. For example, if p is 4 and q is 3, then $\pm 4, \pm 2, \pm 1, \pm\frac{4}{3}, \pm\frac{2}{3}$, and $\pm\frac{1}{3}$ are all possible zeros.
53. J   **55.** 6   **57.** $f(x) = x^4 - 4x^3 + 11x^2 - 64x - 80$
59. $(x - 1)(x + 2)(x + 1)$ **61.** $(x - 3)(x + 4)(x - i)$
63a. about 3.5 s   **63b.** 430 ft   **65.** $3x^3 + 12x$ **67.** 32
69. $18x + 2$

Chapter 5   Study Guide and Review

1. true   **3.** false; depressed polynomial   **5.** true
7. true   **9.** true   **11.** $\frac{7x}{y^4}$ **13.** $r^2 + 8r - 5$
15. $m^3 - m^2p - mp^2 + p^3$ **17.** $3x^3 + 2x^2y^2 - 4xy$
19. $a^3 + 3a^2 - 4a + 2$ **21.** $x^2 + 3x - 40$ units^2
23. This is not a polynomial in one variable. It has two variables, x and y.   **25.** $p(-2) = -3$;
$p(x + h) = x^2 + 2xh + h^2 + 2x + 2h - 3$

So, the function could have the following.
2 positive real zeros, 1 negative real zero,
1 imaginary zero
0 positive real zeros, 1 negative real zero,
3 imaginary zeros
b. Negative zeros do not make sense in the problem because the number of computers produced cannot be negative. Since the zeros are values of x for which $P(x) = 0$, nonnegative zeros represent numbers of computers produced per hour which lead to no profit for the manufacturer.

47 Because $f(x)$ is of degree 3, there are at most 3 zeros. For $f(x)$, there is one sign change, so there are 1 or 0 positive real zeros. For $f(-x)$, there are two sign changes, so there are 2 or 0 negative real zeros. You can find the zeros by factoring, dividing, or using synthetic division.

$f(x) = 4x^3 + 2x^2 - 4x - 2$
$= 2x^2(2x + 1) - 2(2x + 1)$
$= (2x^2 - 2)(2x + 1)$
$= 2(x^2 - 1)(2x + 1)$

The zeros are $-1, -\frac{1}{2}$, and 1. Plot all three zeros on a coordinate plane. You need to determine what happens to the graph between the zeros and at its extremes. Use a table to analyze values close to the zeros. Use these values to complete the graph.

x	−2	−$\frac{3}{4}$	0	3
y	−18	0.4375	−2	112

49.

51. b   **53a.** 3 or 1, 0, 2 or 0
53b.

53c. 23.8; Sample answer: According to the model, the music hall will not earn any money after 2026.
55. 1 positive, 2 negative, 2 imaginary; Sample answer: The graph crosses the positive x-axis once, and crosses the negative x-axis twice. Because the degree of the polynomial is 5, there are 5 − 3 or 2 imaginary zeros.   **57.** Sample answer:
$f(x) = (x + 2i)(x - 2i)(3x + 5)(x + \sqrt{5})(x - \sqrt{5})$. Use conjugates for the imaginary and irrational values.
59a. Sample answer: $f(x) = x^4 + 4x^3 + 4$ **59b.** Sample answer: $f(x) = x^3 + 6x^2 + 9x$ **61.** C   **63.** H   **65.** $f(-8) =$

-1638; $f(4) = 342$ **67.** $f(-8) = -63,940$; $f(4) = 1868$
69. $(a^2 + b^2)(a^4 - a^2b^2 + b^4)$ **71.** $(a - 4)(a - 2)(5a + 2b)$
73. 0.25 s   **75.** $\pm 1, \pm 2, \pm 4, \pm 8, \pm\frac{1}{5}, \pm\frac{2}{5}, \pm\frac{4}{5}, \pm\frac{8}{5}$

3. 5 in. × 9 in. × 28 in.

Lesson 5-8

1. $\pm 1, \pm 2, \pm 3, \pm 4, \pm 6, \pm 8, \pm 12, \pm 24$

5 If $\frac{p}{q}$ is a rational zero, then p is a factor of 12 and q is a factor of 2.
p: $\pm 1, \pm 2, \pm 3, \pm 4, \pm 6, \pm 12$ q: $\pm 1, \pm 2$
possible rational zeros: $\frac{p}{q} = \pm 1, \pm 2, \pm 3, \pm 4, \pm 6, \pm 12, \pm\frac{1}{2}, \pm\frac{3}{2}$
There are no changes of signs for $f(x)$, so there are no positive real zeros. There are 4 changes of signs for $f(-x)$, so there are 0, 2, or 4 negative real zeros. Make a table for synthetic division and test the possible negative values.

$\frac{p}{q}$	2	11	26	29	12
−1	2	9	17	12	0
−2	2	7	12	5	2
−3	2	5	11	−4	24
−4	2	3	14	−27	120
−6	2	−1	32	−163	990
−12	2	−13	182	−2155	25,872
−$\frac{1}{2}$	2	10	21	18.5	2.75
−$\frac{3}{2}$	2	8	14	8	0

Because $f(-1) = 0$ and $f\left(-\frac{3}{2}\right) = 0$, there are zeros at $-\frac{3}{2}$ and -1.   **7.** $\frac{1}{2}, \frac{-5 \pm i\sqrt{23}}{8}$ **9.** $-\frac{3}{2}, \frac{1}{2}, 1 + 2i, 1 - 2i$
11. $\pm 1, \pm 2, \pm 4, \pm 7, \pm 8, \pm 14, \pm 28, \pm 56$

13 If $\frac{p}{q}$ is a rational zero, then p is a factor of 35 and q is a factor of 3.
p: $\pm 1, \pm 5, \pm 7, \pm 35$ q: $\pm 1, \pm 3$
possible rational zeros: $\frac{p}{q} = \pm 1, \pm 5, \pm 7, \pm 35, \pm\frac{5}{3}, \pm\frac{7}{3}, \pm\frac{35}{3}$
15. $\pm 1, \pm 2, \pm 3, \pm 6, \pm 7, \pm 14, \pm 21, \pm 42, \pm\frac{1}{2}, \pm\frac{3}{2}, \pm\frac{7}{2}, \pm\frac{21}{2}, \pm\frac{1}{4}, \pm\frac{3}{4}, \pm\frac{7}{4}, \pm\frac{21}{4}, \pm\frac{1}{8}, \pm\frac{3}{8}, \pm\frac{7}{8}, \pm\frac{21}{8}$
17. $\pm 1, \pm 2, \pm 4, \pm 8, \pm 16, \pm 32, \pm 64, \pm 128, \pm\frac{1}{2}, \pm\frac{1}{4}, \pm\frac{1}{16}$ **19.** $-5, -3, -2$ **21.** $-5, \frac{3}{4}, 5$ **23.** $-1, 2$
25. $-\frac{1}{4}$ **27.** $-7, 1, 3$ **29.** $2, -1, i, -i$ **31.** $0, 3, -i, i$
33. $-2, \frac{4}{3}, -\frac{3}{2}$ **35.** $3, \frac{1}{3}, -\frac{2}{3}, \frac{-3 \pm \sqrt{13}}{2}$
37. $-\frac{1}{2}, \frac{1}{3}, \frac{1}{3}$ **39a.** $V(x) = 324x^3 + 54x^2 - 19x - 2$
39b. 3, 1.05i, −4.22i; 3 is the only reasonable value for x. The other two values are imaginary.

41 **a.** $V = \pi r^2 h$ Volume of a cylinder
$V = \pi r^2(r + 6)$ $h = r + 6$
$V = \pi r^2(r) + \pi r^2(6)$ Distributive Property

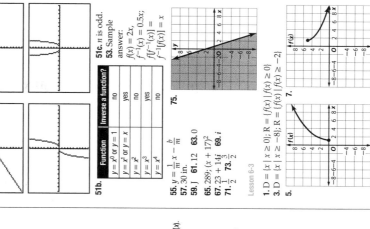

$= \sqrt{\dfrac{36}{\pi}}$ $A = 36$

≈ 3.39 cm Simplify.

43. yes **45.** no **47.** no **49a.** $F^{-1}(x) = \frac{5}{9}(x - 32)$;

$F[F^{-1}(x)] = \frac{9}{5}\left[\frac{5}{9}(x - 32)\right] + 32 = x - 32 + 32 = x$;

$F^{-1}[F(x)] = \frac{5}{9}\left[\frac{9}{5}x + 32 - 32\right] = \frac{5}{9}\left(\frac{9}{5}x + 0\right) = x$.

49b. It can be used to convert Fahrenheit to Celsius.

51a.

51b.

Function	Inverse a function?
$y = x^0$ or $y = 1$	no
$y = x^1$ or $y = x$	yes
$y = x^2$	no
$y = x^3$	yes
$y = x^4$	no

51c. n is odd. **53.** Sample answer:

$f(x) = 2x$;

$f^{-1}(x) = 0.5x$;

$f[f^{-1}(x)] = f^{-1}[f(x)] = x$

75.

55. $y = \frac{1}{m}x - \frac{b}{m}$ **57.** 30 in. **59.** J **61.** 12 **63.** 0

65. 289; $(x + 17)^2$ **69.** i **67.** $23 + 14i$ **71.** $\frac{1}{2}$ **73.** $\frac{3}{2}$

Lesson 6-3

1. $D = \{x \mid x \geq 0\}$; $R = \{f(x) \mid f(x) \geq 0\}$
3. $D = \{x \mid x \geq -8\}$; $R = \{f(x) \mid f(x) \geq -2\}$

5.

$D = \{x \mid x \geq 1\}$;
$R = \{f(x) \mid f(x) \geq 0\}$

7.

$D = \{x \mid x \leq \frac{5}{3}\}$;
$R = \{f(x) \mid f(x) \leq 5\}$

---

15. $f^{-1}(x) = x - 2$

17. $f^{-1}(x) = -\frac{x}{2} + \frac{1}{2}$

19. $f^{-1}(x) = -\frac{3}{5}(x + 8)$

21. $f^{-1}(x) = \frac{1}{4}x$

23. $f^{-1}(x) = \pm\sqrt{\frac{1}{5}x}$

25. $f^{-1}(x) = \pm\sqrt{2x + 2}$

27. no **29.** yes **31.** yes **33.** yes

35. $[f \circ g](x) = f[g(x)]$ Composition of functions
$= f(x + 3)$ Replace $g(x)$ with $x + 3$.
$= (x + 3)^2 - 9$ Substitute $x + 3$ for x in $f(x)$.
$= x^2 + 6x + 9 - 9$ FOIL
$= x^2 + 6x$ Simplify.

$[g \circ f](x) = g[f(x)]$ Composition of functions
$= g(x^2 - 9)$ Replace $f(x)$ with $x^2 - 9$.
$= (x^2 - 9) + 3$ Substitute $x^2 - 9$ for x in $g(x)$.
$= x^2 - 6$ Simplify.

The functions are not inverses because
$[f \circ g](x) \neq [g \circ f](x)$.

37. yes **39a.** $c(g) = 2.95g$ **39b.** $c(m) \approx 0.105m$

41. **a.** $A = \pi r^2$ Write the formula
$y = \pi x^2$ Write the formula using x and y.
$x = \pi y^2$ Exchange x and y in the equation.
$\frac{x}{\pi} = y^2$ Divide each side by π.
$\sqrt{\frac{x}{\pi}} = y$ Take the positive square root of each side.

So, replacing y with r and x with A, the inverse
is $r = \sqrt{\frac{A}{\pi}}$. Write the inverse of the function.

b. $r = \sqrt{\frac{A}{\pi}}$

---

numbers], $R = \{y \mid y = 128x^4 + 96x^3 + 18x^2 \text{ for } -\infty \leq x \leq \infty\}$; $[g \circ f](x) = 32x^4 + 6x^2$, $D = \{\text{all real numbers}\}$, $R = \{y \mid y = 32x^4 + 6x^2 \text{ for } -\infty \leq x \leq \infty\}$ **37a.** $p(x) = 0.65x$; $t(x) = 1.0625x$ **37b.** Since $[p \circ t](x) = [t \circ p](x)$, either function represents the price. **37c.** \$1587.75 **39.** $2[g \cdot f](x) = 2x^3 - 4x^2 - 30x + 72$; $D = \{\text{all real numbers}\}$ **41.** 25 **43.** 483 **45.** -5 **47.** $-30a + 5$ **49.** $-10a^2 + 10a + 1$

51. **a.** Let $w(x)$ represent the function for women and $m(x)$ represent the function for men. Addition of functions
$(w + m)(x) = w(x) + m(x)$
$= (1086.4x + 56,610) + (999.2x + 66,450)$ Substitution
$= 2085.6x + 123,060$ Simplify.

The equation $y = 2085.6x + 123,060$ models the total number.

b. $(f - g)(x) = $ the number of men employed in the U.S. $-$ the number of women employed in the U.S. So, the function represents the difference in the number of men and women employed in the U.S.

53. 0 **55.** 1 **57.** 256 **59.** Sample answer: $f(x) = x - 9$, $g(x) = x + 5$ **61a.** $D = \{\text{all real numbers}\}$
61b. $D = \{x \mid x \geq 0\}$ **63.** Sample answer: Many situations in the real world involve complex calculations in which multiple functions are used. In order to solve some problems, a composition of those functions may need to be used. For example, the product of a manufacturing plant may have to go through several processes in a particular order, in which each process is described by a function. By finding the composition, only one calculation must be made to find the solution to the problem.

65. G **67.** D **69.** $-3, 2, 4$ **71.** $-3, 5, \frac{1}{2}$ **73.** $1; 1; 2$
75. 2 or 0; 2 or 0; 4, 2, or 0 **77.** (1, 2, 3) **79.** (3, -1, 5)
81. $x = \frac{12 + 7y}{5}$ **83.** $x = \frac{15 - 8yz}{4}$ **85.** $k = \pm\sqrt{A - b}$

Lesson 6-2

1. {(10, -9), (-3, 1), (-5, 8)}

3. $f^{-1}(x) = -\frac{1}{3}x$

5. $h^{-1}(x) = \pm\sqrt{x + 3}$

7. no **9.** {(6, -8), (-2, 6), (-3, 7)}
11. {(-1, 8), (-1, -8), (-8, -2), (8, 2)}
13. {(-5, 1), (6, 2), (-7, 3), (8, 4), (-9, 5)}

---

11. $(f + g)(x) = f(x) + g(x)$ Addition of functions
$= (3x) + (-2x + 6)$ $f(x) = 3x$ and $g(x) = -2x + 6$
$= x + 6$ Simplify.

$(f - g)(x) = f(x) - g(x)$ Subtraction of functions
$= (3x) - (-2x + 6)$ Substitution
$= 5x - 6$ Simplify.

$(f \cdot g)(x) = f(x) \cdot g(x)$ Multiplication of functions
$= (3x)(-2x + 6)$ Substitution
$= -6x^2 + 18x$ Simplify.

$\left(\dfrac{f}{g}\right)(x) = \dfrac{f(x)}{g(x)}$ Division of functions
$= \dfrac{3x}{-2x + 6}$, $x \neq 3$ Substitution

13. $(f + g)(x) = x^2 + x - 5$; $(f - g)(x) = x^2 - x + 5$; $(f \cdot g)(x) = x^3 - 5x^2$; $\left(\dfrac{f}{g}\right) = \dfrac{x^2}{x - 5}$, $x \neq 5$
15. $(f + g)(x) = 4x^2 - 8x$; $(f - g)(x) = 2x^2 + 8x - 8$; $(f \cdot g)(x) = 3x^4 - 24x^3 + 8x^2 + 32x - 16$; $\left(\dfrac{f}{g}\right)(x) = \dfrac{3x^2 - 4}{x^2 - 8x + 4}$, $x \neq 4 \pm 2\sqrt{3}$ **17.** $f \circ g = \{(-4, 4)\}$, $D = \{-4\}$, $R = \{4\}$; $g \circ f = \{(-8, 0), (0, -4), (2, -5), (-6, -1)\}$, $D = \{-6, 0, 2\}$, $R = \{-5, -4, -1, 0\}$ **19.** $f \circ g$ is undefined, $D = \emptyset$, $R = \emptyset$; $g \circ f$ is undefined, $D = \emptyset$, $R = \emptyset$. **21.** $f \circ g$ is undefined, $D = \emptyset$, $R = \emptyset$; $g \circ f = \{(-4, 1), (1, 2)\}$, $D = \{-4, 1\}$, $R = \{0, 2\}$. **23.** $f \circ g = \{(4, 6), (3, -8)\}$, $D = \{3, 4\}$, $R = \{-8, 6\}$; $g \circ f$ is undefined, $D = \emptyset$, $R = \emptyset$. **25.** $f \circ g = \{(3, -1), (6, 11)\}$, $D = \{3, 6\}$, $R = \{-1, 11\}$; $g \circ f = \{(-4, 5), (-2, 4), (-1, 8)\}$, $D = \{-4, -2, -1\}$, $R = \{4, 5, 8\}$.

27. $[f \circ g](x) = f[g(x)]$ Composition of functions
$= f(x + 5)$ Replace $g(x)$ with $x + 5$.
$= 2(x + 5)$ Substitute $x + 5$ for x in $f(x)$.
$= 2x + 10$ Distributive Property

For $[f \circ g](x)$, $D = \{\text{all real numbers}\}$ and $R = \{\text{all even numbers}\}$.

$[g \circ f](x) = g[f(x)]$ Composition of functions
$= g(2x)$ Replace $f(x)$ with $2x$.
$= 2x + 5$ Substitute $2x$ for x in $g(x)$.

For $[g \circ f](x)$, $D = \{\text{all real numbers}\}$ and $R = \{\text{all odd numbers}\}$.

29. $[f \circ g](x) = 3x - 2$, $D = \{\text{all real numbers}\}$, $R = \{y \mid y = 3x - 2 \text{ for } -\infty \leq x \leq \infty\}$; $[g \circ f](x) = 3x + 8$, $D = \{\text{all real numbers}\}$, $R = \{y \mid y = 3x + 8 \text{ for } -\infty \leq x \leq \infty\}$ **31.** $[f \circ g](x) = x^2 - 6x - 2$, $D = \{\text{all real numbers}\}$, $R = \{y \mid y = x^2 - 6x - 2 \text{ for } -\infty \leq x \leq \infty\}$; $[g \circ f](x) = x^2 + 6x - 8$, $D = \{\text{all real numbers}\}$, $R = \{y \mid y = x^2 + 6x - 8 \text{ for } -\infty \leq x \leq \infty\}$ **33.** $[f \circ g](x) = 4x^3 + 7 \text{ for } -\infty \leq x \leq \infty$, $D = \{\text{all real numbers}\}$, $R = \{y \mid y = 4x^3 + 7, D = \{\text{all real numbers}\}, R = \{y \mid y = 64x^3 - 48x^2 + 12x + 1, D = \{\text{all real numbers}\}$; $[g \circ f](x) = 64x^3 - 48x^2 + 12x + 1 \text{ for } -\infty \leq x \leq \infty\}$ **35.** $[f \circ g](x) = 128x^4 + 96x^3 + 18x^2, D = \{\text{all real}$

connectED.mcgraw-hill.com

Selected Answers and Solutions

Selected Answers and Solutions

Lesson 6-4

1. $\pm 10q^4$

3. $\sqrt{(y-6)^8} = \sqrt{[(y-6)^4]^2}$
$= (y-6)^4$

5. $\pm 4iy^2$ **7.** 7.616 **9.** -2.122

11. about 4.088×10^8 m

13. $\pm\sqrt{225a^{16}b^{36}} = \pm\sqrt{(15a^8b^{18})^2}$
$= \pm 15a^8b^{18}$

15. $-4c_1^2|d|$ **17.** $-20x^{16}y^{20}$ **19.** $(x^2+6)^8$ **21.** $2a^2b^{p4}$

23. $3b^6c^4$ **25.** $\pm i(x+2)^4$ **27.** $|x^3|$ **29.** a^4

31. $(4x-7)^8$ **33.** $4|(5x-2)^3|$ **35.** $2a^3b^2$ **37.8** cm

39. -12.247 **41.** 0.787 **43.** -5.350 **45.** 29.573

47. $14|c^3|d^2$ **49.** $-3a^5b^3$ **51.** $20x^8|y^3|$ **53.** $4(x+y)^2$

55. about 141 million mi

57. bald eagle:
$P = 73.3\sqrt[3]{m^3}$
$= 73.3\sqrt[3]{4.5^3}$
≈ 226.5 Cal/d

komodo dragon:
$P = 73.3\sqrt[3]{m^3}$
$= 73.3\sqrt[3]{72^3}$
$= 1811.8$ Cal/d

Asian elephant:
$P = 73.3\sqrt[3]{m^3}$
$= 73.3\sqrt[3]{2300^3}$
$\approx 24{,}344.4$ Cal/d

golden retriever:
$P = 73.3\sqrt[3]{m^3}$
$= 73.3\sqrt[3]{30^3}$
≈ 939.6 Cal/d

bottlenose dolphin:
$P = 73.3\sqrt[3]{m^3}$
$= 73.3\sqrt[3]{156^3}$
≈ 3235.5 Cal/d

59. Kimi is correct; Ashley's error was keeping the y^2 inside the absolute value symbol.   **61.** Sample answer: Sometimes; when $x = -3$, $\sqrt[4]{(-x)^4} = |3|$ or 3. When $x = 3$, $\sqrt[4]{(-x)^4} = |3|$ or 3.   **63.** Sample answers: 1, 64   **65.** $2\sqrt{2xy}$ **67.** -0.1 **69.** $\frac{1}{625}$ **71.** G   **73.** B

75.

77. 3.41 kg and 49.53 cm
79. $4x^2 + 22x - 34$
81. $4a^4 + 24a^2 + 36$

83. $(-2, 3)$; $x = -2$; down

85. $(2, -2)$; $x = 2$; up

87. $x^2 + 9x + 20$ **89.** $a^2 - 7a - 18$ **91.** $x^2 + xy - 2y^2$

Lesson 6-5

1. $60^2c^2 - \sqrt{ac}$ **3.** $\frac{c^2\sqrt{cd}}{d^5}$ **5.** $60x$ **7.** $36xy$

9. $20\sqrt{2} + 13\sqrt{3}$ **11.** $12\sqrt{3} + 16\sqrt{5} + 40 + 6\sqrt{15}$

13. $\frac{15 - 5\sqrt{2}}{7}$ **15.** $-2 - \sqrt{2}$ **17.** $32 - 2\sqrt{3}$ cm

19. $3a^2b\sqrt{ab}$ **21.** $3|a^3|bc^2\sqrt{2bc}$ **23.** $\frac{\sqrt{70xy}}{10y^2}$

25. $\frac{\sqrt{28b^2c^3}}{2|b|}$ **27.** $32a^5b^3\sqrt{b}$ **29.** $25x^6y^4\sqrt[3]{2xy}$

31. $18\sqrt{3} + 14\sqrt{2}$ **33.** $9\sqrt{6} + 72\sqrt{2} - 7\sqrt{3}$

35. $A = \ell w$ Area of a rectangle
$= (8 + \sqrt{3})(\sqrt{6})$ $\ell = 8 + \sqrt{3}$ and $w = \sqrt{6}$
$= 8\cdot\sqrt{6} + \sqrt{3}\cdot\sqrt{6}$ Distributive Property
$= 8\sqrt{6} + \sqrt{18}$ Product Property
$= 8\sqrt{6} + 3\sqrt{2}$ ft^2 Simplify.

37. $56\sqrt{3} + 42\sqrt{6} - 36\sqrt{2} - 54$ **39.** 1260

41. $6\sqrt{3} + 6\sqrt{2}$ **43.** $\frac{20 - 7\sqrt{3}}{11}$ **45.** $2y^2\sqrt[4]{2y}$

47. $3|a|b^3\sqrt[4]{2a^2bc}$ **49.** $\frac{\sqrt{1500a^2b^3c^3y^2}}{5|a|b}$

51. $\frac{x+1}{\sqrt{x}-1} = \frac{x+1}{\sqrt{x}-1}\cdot\frac{\sqrt{x}+1}{\sqrt{x}+1}$ $\sqrt{x}+1$ is the conjugate of $\sqrt{x}-1$.
$= \frac{x(\sqrt{x}) + x(1) + 1(\sqrt{x}) + (-1)(1)}{\sqrt{x}(\sqrt{x}) + 1(\sqrt{x}) + -1(\sqrt{x}) + (-1)(1)}$ Multiply.
$= \frac{x\sqrt{x} + x + \sqrt{x} + 1}{x + \sqrt{x} - \sqrt{x} - 1}$ Simplify.
$= \frac{x(\sqrt{x}+1) + (\sqrt{x}+1)}{x-1}$ Simplify.
$= \frac{(x+1)(\sqrt{x}+1)}{x-1}$ Simplify.

53. $\frac{\sqrt{x^3}-x}{\sqrt{x}-1} = \frac{x+1}{\sqrt{x}-1}$, $\frac{\sqrt{x}+1}{\sqrt{x}+1}$ **55.** $|a|$ **57.** a^2

59a. $a^2 + b^2 = c^2$
$1^2 + 1^2 = c^2$
$2 = c^2$
$c = \sqrt{2}$ **59b.**

59c. $\sqrt{2} + \sqrt{2}$ units is the length of the hypotenuse of an isosceles right triangle with legs of length 2 units. Therefore, $\sqrt{2} + \sqrt{2} > 2$.

59d.

59e. The square creates 4 triangles with a base of 1 and a height of 1. Therefore the area of each triangle is $\frac{1}{2}bh = \frac{1}{2}(1)(1)$ or $\frac{1}{2}$; $4(\frac{1}{2}) = 2$. The area of the square is 2, so $\sqrt{2}\cdot\sqrt{2} = 2$.

61. $\left(\frac{-1-i\sqrt{3}}{2}\right)^3 = \left(\frac{-1-i\sqrt{3}}{2}\right)\cdot\left(\frac{-1-i\sqrt{3}}{2}\right)\cdot\left(\frac{-1-i\sqrt{3}}{2}\right)$.
$= \frac{(-1-i\sqrt{3})(-1-i\sqrt{3})(-1-i\sqrt{3})}{8}$

connectED.mcgraw-hill.com

9.

11.

13. $D = \{x \mid x \geq 0\}$; $R = \{f(x) \mid f(x) \leq 2\}$

15. The domain only includes values for which the radicand is nonnegative.
$x - 2 \geq 0$ Write an inequality.
$x \geq 2$ Add 2 to each side.
The domain is $\{x \mid x \geq 2\}$. Find $f(2)$ to find the lower limit of the range.
$f(2) = 4\sqrt{2-2} - 8$
$= 0 - 8$ or -8
The range is $\{f(x) \mid f(x) \geq -8\}$.

17. $D = \{x \mid x \geq 4\}$; $R = \{f(x) \mid f(x) \geq -6\}$

19.
$D = \{x \mid x \geq 0\}$;
$R = \{f(x) \mid f(x) \geq 0\}$

21.
$D = \{x \mid x \geq 8\}$;
$R = \{f(x) \mid f(x) \geq 0\}$

23.
$D = \{x \mid x \geq -3\}$;
$R = \{f(x) \mid f(x) \geq 2\}$

25.
$D = \{x \mid x \geq 5\}$;
$R = \{f(x) \mid f(x) \geq -6\}$

27.
$D = \{x \mid x \geq 1\}$;
$R = \{f(x) \mid f(x) \leq -4\}$

29. 1936 ft

31.

---

33. Graph the boundary $y = -4\sqrt{x+3}$. The domain is $\{x \mid x \geq -3\}$. Because y is greater than, the shaded region should be above the boundary and within the domain.

35.

37.

39a. $v = \sqrt{\frac{2E}{m}}$ **39b.** about 36.5 m/s   **39c.** about 85,135 m/s   **41.** $f(x) = \sqrt{x+6} - 6$

43. $f(x) = -\sqrt{x+6} - 6$

45a. $T = 2\pi\sqrt{\frac{L}{g}}$ Original function
$T = 2\pi\sqrt{\frac{L}{32}}$ Replace g with 32.

Make a table of values for $0 \leq L \leq 10$. Graph the points and connect with a smooth curve.

L	T
0	0
1	1.11
2	1.57
3	1.92
4	2.22
5	2.48
6	2.72
7	2.94
8	3.14
9	3.33
10	3.51

b. Use the table that you made in part **a**. (2, 1.57) means the period for a pendulum 2 feet long is about 1.57 seconds. (5, 2.48) means the period for a pendulum 5 feet long is about 2.48 seconds. (8, 3.14) means the period for a pendulum 8 feet long is about 3.14 seconds.

47. Sample answer: $y = -\sqrt{x+4} + 6$
49. Sample answer: $y = -\sqrt{x-8} + 14$
51. Molly: $y = \sqrt{5x} + 10$ has an x-intercept of -2 and would be at the right of the given graph.
53a. Sample answer: The original is $y = x^2 + 2$ and inverse is $y = \pm\sqrt{x} - 2$.   **53b.** Sample answer: The original is $y = \pm\sqrt{x} + 4$ and inverse is $y = (x-4)^2$.   **55.** G   **57.** E   **59.** no   **61.** $[d \circ h](m) = \frac{m}{1440}$ **63.** rational   **65.** rational

21. $f^{-1}(x) = \pm\sqrt{x}$

23. $1200   **25.** yes
27. no   **29.** no

31.

33.

35. $D = \{x \mid x \geq 0\}$;
$R = \{f(x) \mid f(x) \geq 0\}$

37. about 9.8 in.

$D = \{x \mid x \geq 7\}$;
$R = \{f(x) \mid f(x) \geq 0\}$

$D = \{x \mid x \geq 1\}$;
$R = \{f(x) \mid f(x) \geq 5\}$

39.

41. ± 11 **43.** 6   **45.** $(x^2 + 2)^3$ **47.** $a^2 \mid b^3 \mid$ **49.** 10 m/s
51. $12ab^2\sqrt{ab}$ **53.** $80\sqrt{2}$ **55.** $\dfrac{m^2\sqrt{6mp}}{p^6}$
57. $-\sqrt{15} - 3\sqrt{2}$ **59.** $x^{\frac{7}{6}}$
61. $\dfrac{d^{\frac{5}{12}}}{d}$ **63.** 3   **65.** $4a^{\frac{4}{3}}b^5c^2\pi$ units^2 **67.** $\dfrac{100}{9}$
69. 2   **71.** no solution   **73.** 3   **75.** $\dfrac{1}{3} \leq x < \dfrac{9}{10}$
77. $x \geq \dfrac{4}{3}$ **79.** no solution   **81.** $x > \dfrac{5}{2}$

connectED.mcgraw-hill.com

CHAPTER 7
Exponential and Logarithmic Functions and Relations

Chapter 7   Get Ready
1. a^{12} **3.** $\dfrac{-3x^6}{2y^3z^5}$ **5.** 5 g/cm^3

---

31. 9   **33.** $\dfrac{81}{16}$ **35.** 1 m   **37.** 3   **39.** 83   **41.** 61   **43.** 3
45. 18   **47.** 2   **49.** F   **51.** $x \geq 43$ **53.** no real solution
55. $\sqrt{d+3} + \sqrt{d+7} > 4$ Original inequality
$\sqrt{d+7} > 4 - \sqrt{d+3}$ Subtract $\sqrt{d+3}$ from each side.
$d + 7 > 16 - 8\sqrt{d+3} + d + 7$ Square each side.
$\dfrac{5}{2} < \sqrt{d+7}$ Simplify.
$\dfrac{25}{4} < d + 7$ Square each side.
$-\dfrac{3}{4} < d$ Subtract 7 from each side.
$d > -\dfrac{3}{4}$ Rewrite inequality.

57. $-\dfrac{5}{2} \leq y \leq 2$ **59.** $a > 8$ **61.** $0 \leq c < 3$

63. $M = \left(\dfrac{L}{0.46}\right)^3$

65. $A = \pi r^2$ Original formula
$250{,}000 = \pi r^2$ Replace A with 250,000.
$79{,}577.5 \approx r^2$ Divide each side by π.
$\sqrt{79{,}577.5} \approx r$ Take the square root of each side.
$282.1 \approx r$ Use a calculator.
The radius is about 282 ft.

67. $\sqrt{x+2} - 7 = -10$
69. never
$\sqrt{(x^2)^2} = x$
$\sqrt{x^2} = x$
$-x^2$
$x^2 = (x)(-x)$
$x^2 \neq -x^2$

71. They are the same number.   **73.** 3
75. Sometimes; Sample answer: When the radicand is negative, then there will be extraneous roots.   **77.** G
79. A   **81.** 81   **83.** $4x^2\sqrt{5}$ **85.** $y^{-1} = \dfrac{-x - 3}{2}$
87. $y^{-1} = \pm\frac{1}{2}\sqrt{x - \frac{3}{2}}$ **89a.** $f(x) \to +\infty$ as $x \to +\infty$,
$f(x) \to +\infty$ as $x \to -\infty$ **89b.** even   **89c.** 0   **91.** $\dfrac{1}{6}$
93. $\dfrac{5}{8}$ **95.** 16   **97.** $2\dfrac{1}{2}$

Chapter 6   Study Guide and Review
1. identity function   **3.** composition of functions
5. rationalizing the denominator   **7.** inverse relations
9. radical function
11. $[f \circ g](x) = x^2 - 14x + 50$; $[g \circ f](x) = x^2 - 6$
13. $[f \circ g](x) = 20x - 4$; $[g \circ f](x) = 20x - 1$
15. $[f \circ g](x) = x^2 + 4x$; $[g \circ f](x) = x^2 + 2x - 2$
17. $f^{-1}(x) = \dfrac{x+6}{5}$ **19.** $f^{-1}(x) = 2x - 6$

$= \dfrac{(1 + i\sqrt{3} + i\sqrt{3} + 3^2)(-1 - i\sqrt{3})}{8}$
$= \dfrac{(2i\sqrt{3} - 2)(-1 - i\sqrt{3})}{8}$
$= \dfrac{-2i\sqrt{3} - 6i^2 + 2 + 2i\sqrt{3}}{8}$
$= \dfrac{-6i^2 + 2}{8} = \dfrac{8}{8}$ or 1

63. $a = 1, b = 256; a = 2, b = 16; a = 4, b = 4; a = 8, b = 2$
65. Sample answer: It is only necessary to use absolute values when it is possible that n could be odd or even and still be defined.
It is when the radicand must be nonnegative in order for the root to be defined that the absolute values are not necessary.   **67.** G   **69.** C   **71.** $9ab^3$

73.

75. $-4, -4, -i, i$ **77.** $-4, 2 + 2i\sqrt{3}, 2 - 2i\sqrt{3}$
79. $\dfrac{3}{2}, -3 + 3i\sqrt{3}, -3 - 3i\sqrt{3}$ **81.** 9 small,
4 large   **83.** $\dfrac{1}{3}$ **85.** $\dfrac{7}{12}$ **87.** $\dfrac{5}{12}$

Lesson 6-6
1. $\sqrt{10}$ **3.** $15^{\frac{1}{3}}$ **5.** 7   **7.** 25
9. $\ell = A^{\frac{1}{2}}$ Write the formula.
$= 169^{\frac{1}{2}}$ $A = 169$
$= \sqrt{169}$ Write in radical form.
$= 13$ ft   Simplify.
11. $x^{\frac{3}{5}}$ **13.** $\sqrt{3g}$ **15.** $\dfrac{x - 2x^{\frac{1}{2}} + 1}{g - 1}$ **17.** $\sqrt[7]{16}$ **19.** $\sqrt{x^9}$
21. $63^{\frac{1}{4}}$ **23.** $5x^2$ **25.** 4   **27.** $\dfrac{1}{3}$ **29.** about 2.64 cm
31. $a^{\frac{25}{36}}$ **33.** $\dfrac{y^{\frac{5}{y}}}{y}$

35. $\dfrac{\sqrt[4]{27}}{\sqrt{3}} = \dfrac{27^{\frac{1}{4}}}{3^{\frac{1}{2}}}$
$= \dfrac{(3^3)^{\frac{1}{4}}}{3^{\frac{1}{2}}}$
$= \dfrac{3^{\frac{3}{4}}}{3^{\frac{1}{2}}}$ Rational exponents
$= 3^{\frac{3}{4} - \frac{1}{2}}$ Quotient of powers
$= 3^{\frac{1}{4}}$ Simplify.

37. $\sqrt[3]{9} \cdot \sqrt{g}$
$= 3^{\frac{1}{2}}$ Simplify.
$= \sqrt{3}$ Write in radical form.
39. $\dfrac{x + 4x^{\frac{3}{4}}}{x - 16}$ **41.** $28.27x^3y^5z^4$ units^2 **43.** $6 - 2 \cdot 4^{\frac{3}{4}}$ **45.** $x^{\frac{10}{3}}$ **47.** $y^{\frac{3}{20}}$
49. $\sqrt{6}$ **51.** $\dfrac{w^{\frac{1}{8}}}{w}$

53. $\dfrac{f^{-\frac{1}{4}}}{4f^{\frac{1}{2}} \cdot f^{-\frac{3}{4}}} = \dfrac{f^{-\frac{1}{4}}}{4f^{\frac{1}{2} - \frac{3}{4}}}$ Defn. of negative exponents
$= \dfrac{f^{-\frac{1}{4}} \cdot f^{\frac{3}{4}}}{4f^{\frac{1}{2}}}$ Add powers.
$= \dfrac{f^{\frac{1}{2}}}{4f^{\frac{1}{2}}}$ Multiply to rationalize the denominator.
$= \dfrac{f^{\frac{7}{12}}}{4f}$

55. $c^{\frac{1}{2}}$ **57.** $23\sqrt[3]{23}$ **59.** 3   **61.** $\dfrac{ab\sqrt{c}}{c}$ **63.** $2\sqrt{6} - 5$

65a.

x	$f(x)$	$g(x)$
-2	-8	-1.26
-1	-1	-1
0	0	0
1	1	1
2	8	1.26

65b.

65c. It is a reflection in the line $y = x$.   **67a.** Sample answer: $\sqrt[4]{(-16)^3} = \sqrt[4]{-4096}$; there is no real number that when raised to the fourth power results in a negative number.   **67b.** Sample answer: $\sqrt[4]{-1}$
69. Sample answer: It may be easier to simplify an expression when it has rational exponents because all the properties of exponents apply. We do not have as many properties dealing directly with radicals. However, we can convert all radicals to rational exponents, and then use the properties of exponents to simplify.   **71.** B   **73.** C   **75.** $9\sqrt{3}$ **77.** $6y^2z^3\sqrt{7}$
79. $-6; x + h - 2$ **81.** $-21; 6x + 6i + 3$ **83.** $\left\{-\frac{3}{4}, 4\right\}$
83. $20; x^2 + 2xh + h^2 - x - h$ **85.** $[0, 11]$ **87.** $\left\{-\frac{3}{4}, 4\right\}$
89. [3]   **91.** 5 in. by 4 in.   **93.** $3x - 4$ **95.** $x - 8\sqrt{x} + 16$
97. $9x + 6\sqrt{x} + 1$

Lesson 6-7
1. 20   **3.** 13

5. $\sqrt[3]{x - 2} = 3$ Original equation
$(\sqrt[3]{x - 2})^3 = 3^3$ Raise each side to the third power.
$x - 2 = 27$ Evaluate each side.
$x = 29$ Add 2 to each side.
7. 2   **9.** 49   **11.** $\dfrac{27}{2}$ **13a.** about 9.5 s   **13b.** about
324 feet   **15.** $-\dfrac{4}{3} \leq x \leq \dfrac{77}{3}$ **17.** $1 \leq y \leq 5$ **19.** $x > 1$
21. $x \leq -11$ **23.** 22   **25.** 3   **27.** no real solution   **29.** $\dfrac{1}{4}$

Selected Answers and Solutions

b. Sample answer: The graph of $f(x)$ appears to be the graph of $f(x) = b^x$ reflected across the x-axis. As the values of x increase, the output values decrease.
c. Sample answer: The graphs of $g(x)$ and $h(x)$ appear to be translated to the left.
d. Sample answer: $f(x)$ and $g(x)$ are growth and $h(x)$ is decay; the absolute value of the output is increasing for the growth functions and decreasing for the decay function.
35. Vince; the graphs of the function would be the same. **37.** Sample answer: 10 **39.** 12 **41.** H
43. 8 **45.** $x \geq 21$ **47.** $d > -\frac{3}{4}$ **49.** $\frac{y^{\frac{3}{5}}}{y}$ **51.** $3x^{\frac{5}{3}} + 4x^3$
53. $\sqrt{3}$ **55.** about 204.88 ft **57.** $\frac{1}{t^3}$ **59.** $8xy^4$

Lesson 7-2

1. 12 **3.** −10 **5a.** $c = 2^{\frac{1}{15}}$ **5b.** 16 cells **7.** $x \geq 4.5$ **9.** 0

11.
$81^{x+2} = 3^{3a+1}$ Original equation
$(3^4)^{x+2} = 3^{3a+1}$ Rewrite 81 as 3^4.
$3^{4(x+2)} = 3^{3a+1}$ Power of a Power
$3^{4x+8} = 3^{3a+1}$ Distributive Property
$4a + 8 = 3a + 1$ Property of Equality for Exponential Functions
$a + 8 = 1$ Subtract $3a$ from each side.
$a = -7$ Subtract 8 from each side.

13. $\frac{5}{3}$ **15a.** $y = 10,000(1.045)^x$ **15b.** about \$26,336.52
17. $y = 256(0.75)^x$ **19.** $y = 144(3.5)^x$ **21.** \$16,755.63
23. \$97,362.61 **25.** $b > \frac{1}{5}$ **27.** $d \geq -1$ **29.** $w < \frac{2}{5}$

31a.
$\frac{a}{w^{1.31}} = \frac{170}{45^{1.31}}$ Write a proportion
$a \cdot 45^{1.31} = w^{1.31} \cdot 170$ Cross Products Property
$a = \frac{45^{1.31} \cdot 170}{45^{1.31}}$ Divide each side by $45^{1.31}$.
$a = 1.16w^{1.31}$ Use a calculator.
b. $a = 1.16w^{1.31}$ Write the equation.
$= 1.16(430)^{1.31}$ $w = 430$
≈ 3268 Use a calculator.

33. $\frac{1}{7}$ **35.** $-\frac{13}{4}$ **37.** 1 **39a.** $d = 1.30h^{\frac{3}{2}}$ **39b.** about 1001 cm **41a.** 2, 4, 8, 16

41b.

Cuts	Pieces
1	2
2	4
3	8
4	16

41c. $y = 2^x$ **41d.** $y = 0.003(2)^x$
41e. about 3,221,225.47 in.

43. Sample answer: Beth; Liz added the exponents instead of multiplying them when taking the power of a power. **45.** Reducing the term will be more beneficial. The multiplier is 1.3756 for the 4-year and 1.3828 for the 6.5%. **47.** Sample answer: $4^x \leq 4^2$
49. Sample answer: Divide the final amount by the initial amount. If n is the number of time intervals that pass, take the nth root of the answer. **51.** F **53.** E

connectED.mcgraw-hill.com

21.

23.

25. D = {all real numbers}; R = {$f(x) | f(x) < 2$}

D = {all real numbers}; R = {$f(x) | f(x) > -6$}

D = {all real numbers}; R = {$f(x) | f(x) > -2$}

27a. decay; 0.9

27b. The y-intercept represents the number of pay phones in 1999. The asymptote is the x-axis. The number of pay phones can approach 0, but will never equal 0. This makes sense as there will probably always be a need for some pay phones. **29a.** $f(x) = 18(1.25)^{x-1}$
29b. growth; 1.25 **29c.** 134
31. $g(x) = 4(2)^{x-3}$ or $g(x) = \frac{1}{2}(2^x)$

33. a.

Pay Phones (millions) — Years Since 1999

---

7. $f^{-1}(x) = x + 3$

9. $f^{-1}(x) = 4x + 12$

11. $f^{-1}(x) = 3x - 12$
13. no

Lesson 7-1

1.

x	$f(x) = 3^{x-2} + 4$
−2	$3^{-2-2} + 4 = 4\frac{1}{81}$
−1	$3^{-1-2} + 4 = 4\frac{1}{27}$
0	$3^{0-2} + 4 = 4\frac{1}{9}$
1	$3^{1-2} + 4 = 4\frac{1}{3}$
2	$3^{2-2} + 4 = 5$
3	$3^{3-2} + 4 = 7$
4	$3^{4-2} + 4 = 13$

The domain is all real numbers and the range is all real numbers greater than 4.
D = {all real numbers}; R = {$f(x) | f(x) > 4$}

3. Make a table of values. Then plot the points, and sketch the graph.

D = {all real numbers}; R = {$f(x) | f(x) > 0$}

5.

D = {all real numbers}; R = {$f(x) | f(x) > -6$}

7. Number of Computers Infected / Minutes

9.

D = {all real numbers}; R = {$f(x) | f(x) < 5$}

11.

13.

D = {all real numbers}; R = {$f(x) | f(x) > 0$}

15.

D = {all real numbers}; R = {$f(x) | f(x) > 7$}

17.

D = {all real numbers}; R = {$f(x) | f(x) < 4$}

D = {all real numbers}; R = {$f(x) | f(x) > -5$}

19.
$y = a(1 + r)^t$ Equation for exponential growth
$y = 65(1 + 0.3)^t$ $a = 65$ and $r = 0.3$
$y = 65(1.3)^t$ Simplify.

Make a table of values. Then plot the points, and sketch the graph.

t	$y = 65(0.3)^t$
0	$y = 65(1.3)^0 = 65$
2	$y = 65(1.3)^2 \approx 110$
4	$y = 65(1.3)^4 \approx 186$
6	$y = 65(1.3)^6 \approx 314$
8	$y = 65(1.3)^8 \approx 530$
10	$y = 65(1.3)^{10} \approx 896$

Population / Number of Weeks

79. $\frac{27\sqrt{15}}{4}$ ft² **81.** batteries, $74; spark plugs, $58; wiper blades, $48 **83.** −1 **85.** $x \le -\sqrt{6}$ or $x \ge \sqrt{6}$

Lesson 7-4

1. 16 **3.** C **5.** $\{x \mid 0 < x \le \frac{1}{64}\}$ **7.** $\{x \mid 2 > x > \frac{4}{3}\}$ **9.** 3125 **11.** −2 **13.** 9

15. $\log_{12}(x^2 - 7) = \log_{12}(x + 5)$ — Original equation
$x^2 - 7 = x + 5$ — Property of Equality for Exponential Functions
$x^2 - x - 7 = 5$ — Subtract x from each side.
$x^2 - x - 12 = 0$ — Subtract 5 from each side.
$(x - 4)(x + 3) = 0$ — Factor.
$x - 4 = 0$ or $x + 3 = 0$ — Zero Product Property
$x = 4$ $x = -3$ — Solve each equation.

17. 5 **19.** −3 **21.** 318 mph **23.** $\{x \mid x \ge 256\}$

25. $\log_2 x \le -2$ — Original inequality
$0 < x \le 2^{-2}$ — Property of Inequality for Exponential Functions
$0 < x \le \frac{1}{4}$ — Simplify.
The solution is $\{x \mid 0 < x \le \frac{1}{4}\}$ or $(0, \frac{1}{4}]$.

27. $\{x \mid 0 < x < \frac{1}{7}\}$ **29.** $\{x \mid \frac{1}{2} < x \le 1\}$
31. $\{x \mid -\frac{5}{12} < x \le 1\}$ **33.** $\{x \mid x \ge 8\}$
35a. 37 **35b.** 61

37. a. $\beta = 10 \log_{10}\left(\frac{I}{10^{-12}}\right)$ — Original equation
$= 10 \log_{10}\left(\frac{10^{-12}}{10^{-12}}\right)$ — $I = 10$
$= 10 \log_{10} 10^{12}$ — Write $\frac{1}{10^{-12}}$ as 10^{12}.
$= 10(12)$ or 120 — Definition of logarithm
b. $\beta = 10 \log_{10}\left(\frac{I}{10^{-12}}\right)$ — Original equation
$= 10 \log_{10}\left(\frac{10^{-2}}{10^{-12}}\right)$ — $I = 10^{-2}$
$= 10 \log_{10} 10^{10}$
$= 10(10)$ or 100 — Quotient of Powers Property
— Definition of logarithm
c. Sample answer: The power of the logarithm only changes by 2. The power is the answer to the logarithm. That 2 is multiplied by the 10 before the logarithm. So we expect the decibels to change by 20.

39. $6\frac{17}{20}$ **41.** The logarithmic function of the form $y = \log_b x$ is the inverse of the exponential function of the form $y = b^x$. The domain of one of the two inverse functions is the range of the other. The range of one of the two inverse functions is the domain of the other. **43a.** less than **43b.** less than **43c.** no **43d.** infinitely many **45.** C **47.** B **49.** 4 **51.** 3 **53.** −3 **55.** $x \le 0$ **57.** $a \le -3$ **59.** −8 **61.** 5 ft **63.** x^8 **65.** $8p^6u^3$ **67.** $x^3 y^4$

Lesson 7-5

1. 2.085 **3.** 0.3685 **5.** Mt. Everest: 26,855.44 pascals; Mt. Trisuli: 34,963.34 pascals; Mt. Bonete: 36,028.42 pascals; Mt. McKinley: 39,846.22 pascals;

---

59. a. $\log\left(1 + \frac{0.24}{12}\right)\frac{A}{2000} = 12t$ — Original formula
$\log_{1.02}\frac{A}{2000} = 12t$ — Simplify.
$\frac{A}{2000} = 1.02^{12t}$ — Definition of logarithm
$A = 2000 \cdot 1.02^{12t}$ — Multiply each side by 2000.

Make a table of values. Then plot the points, and sketch the graph.

t	$A = 2000 \cdot 1.02^{12t}$
0	$A = 2000 \cdot 1.02^{12(0)} = 2000$
2	$A = 2000 \cdot 1.02^{12(2)} \approx 3217$
4	$A = 2000 \cdot 1.02^{12(4)} \approx 5174$
6	$A = 2000 \cdot 1.02^{12(6)} \approx 8322$
8	$A = 2000 \cdot 1.02^{12(8)} \approx 13,386$
10	$A = 2000 \cdot 1.02^{12(10)} \approx 21,530$

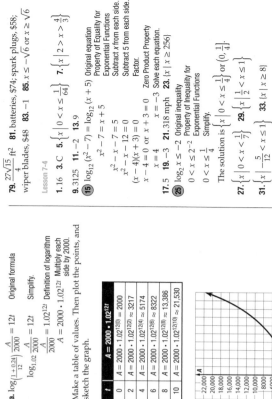

b. From the graph, $A = 4000$ at about $t = 3$. So, it will take approximately 3 years for the debt to double.
c. From the graph, $A = 6000$ at about $t = 4.5$. So, it will take approximately 4.5 years for the debt to triple.
61. Never; if zero were in the domain, the equation would be $y = \log_b 0$.
Then $b^y = 0$.
However, for any real number b, there is no real power that would let $b^y = 0$.
63. $\log_7 51$; Sample answer: $\log_7 51$ equals a little more than 2. $\log_6 61$ equals a little less than 2. $\log_8 71$ equals a little less than 2. Therefore, $\log_7 51$ is the greatest.
65. No; Elisa was closer. She should have $-y = 2$ or $y = -2$ instead of $y = 2$. Matthew used the definition of logarithms incorrectly. **67.** D
69. 80 **71.** $n > 5$ **73.** $n < 3$
75.
77.

47.

49a. 2 **49b.**
49c. less light; $\frac{1}{8}$

51. This represents a transformation of the graph of $f(x) = \log_2 x$.
$|a| = 4$: The graph expands vertically.
$h = 4$: The graph is translated 4 units to the right.
$k = 6$: The graph is translated 6 units up.

53.

55.

57a. $S(3) \approx 30$, $S(15) = 50$, $S(63) = 70$
57b. If $3000 is spent on advertising, $30,000 is returned in sales. If $15,000 is spent on advertising, $50,000 is returned in sales. If $63,000 is spent on advertising, $70,000 is returned in sales.

57c.

57d. Sample answer: Because eventually the graph plateaus, and no matter how much money you spend you are still returning about the same in sales.

Sales versus Money Spent on Advertising

$S(a) = 10 + 20 \log_4 (a + 1)$

55.

57. 4 **59.** 8.5 **61.** 5 **63.** 5 **65.** −1 **67.** booth 1, 190 lb; booth 2, 150 lb; booth 3, 100 lb **69.** $x^2 - 1$; $x^2 - 6x + 11$ **71.** $-15x - 5$; $-15x + 25$ **73.** $|x + 4|$; $|x| + 4$

Lesson 7-3

1. $8^3 = 512$ **3.** $\log_{11} 1331 = 3$ **5.** 2 **7.** 0
9.

11.

13. $2^4 = 16$ **15.** $9^{-2} = \frac{1}{81}$ **17.** $12^2 = 144$
19. $\log_9 \frac{1}{9} = -1$ **21.** $\log_2 256 = 8$ **23.** $\log_{27} 9 = \frac{2}{3}$
25. −2 **27.** 3 **29.** $\frac{1}{3}$ **31.** $\frac{1}{2}$

33. $\log_{\frac{1}{5}} 3125 = y$ — Let the logarithm equal y.
$3125 = \frac{1}{5}^y$ — Definition of logarithm
$5^5 = 5^{-1y}$
$5 = -1y$ — Property of Equality for Exponential Functions
$-5 = y$ — Divide each side by −1.
So, $\log_{\frac{1}{5}} 3125 = -5$.

35. 4
37.
39.
41.
43.
45.

connectED.mcgraw-hill.com

Selected Answers and Solutions

$10,000 = Pe^{(0.0475)(12)}$ $A = 10,000, r = 0.0475, t = 12$
$10,000 = Pe^{0.57}$ Simplify.
$\dfrac{10,000}{e^{0.57}} = P$ Divide each side by $e^{0.57}$.
$5655.25 \approx P$ Use a calculator.
You would need to deposit about $5655.25.

49. $4\ln 2 - 3\ln 5$ **51.** $\ln x + 4\ln y - 3\ln z$
53. -0.8340 **55.** 1.1301

57b. y-axis; $a(x) = -e^x$
57c. $\ln(-x)$ is a reflection across the y-axis. $-\ln x$ is a reflection across the x-axis.

57a.

57d. Sample answer: No; these functions are reflections along $y = -x$, which indicates that they are not inverses.

59. Let $p = \ln a$ and $q = \ln b$. That means that $e^p = a$ and $e^q = b$.
$ab = e^p \times e^q$
$ab = e^{p+q}$
$\ln(ab) = (p+q)$
$\ln(ab) = \ln a + \ln b$
61. Sample answer: $e^{\ln 3}$ **63.** B   **65.** G   **67.** 5.7279
69. $x < 7.3059$ **71.** $x \geq 5.8983$ **73.** 10 decibels
75. $x^2 + 2x + 3$ **77.** $\dfrac{2}{3}$ **79.** $-\dfrac{8}{3}$ **81.** $\dfrac{5}{3}$

Lesson 7-8
1a. 5.545×10^{-10} **1b.** $1,578,843,530$ yr   **1c.** about
1d. $3,750,120,003$ yr
30.48 mg

3b. $P(t) = 16,500$
3c. 16,500
3d. about 102 years

3a.

$3^2 = x$ Property of Equality for Logarithmic Functions
$9 = x$

71. $\log_3 27 = 3$ and $\log_{27} 3 = \frac{1}{3}$; Conjecture: $\log_a b = \dfrac{1}{\log_b a}$

Proof: $\log_a b = \dfrac{\log_b b}{\log_b a}$ Change of Base Formula
$\dfrac{\log_b b}{\log_b a} = \dfrac{1}{\log_b a}$ Inverse Property of Exponents and Logarithms
$\dfrac{1}{\log_b a} = \log_a b$ Original statement

73. B   **75.** G   **77.** 14   **79.** 15   **81.** 2   **83.** $-4, 3$
85. $32^{-3} + 8x^2 - 24x + 16$ **87.** $2^x = 5$ **89.** $5^2 = 25$
91. $6^4 = x$

Lesson 7-7
1. $\ln 30 = x$ **3.** $\ln x = 3$ **5.** $5.7\ln 2$ **7.** $\ln 17496$
9. 2.0794 **11.** 0.1352 **13.** 993.6527 **15.** $\{x \mid -25.0855$
$< x < 15.0855, x \neq -5\}$ **17.** $\{x \mid x > 3.3673\}$
19. about 58 min   **21.** $\ln 0.1 = -5x$ **23.** $5.4 = e^x$
25. $e^{36} = x + 4$ **27.** $e^7 = e^x$ **29.** $7\ln 10$
31. $7\ln\frac{1}{2} + 5\ln 2 = 7\ln 2^{-1} + 5\ln 2$ Rewrite $\frac{1}{2}$ as 2^{-1}.
$= \ln 2^{-7} + \ln 2^5$ Power Property of Logarithms
$= \ln(2^{-7})(2^5)$ Product Property of Logarithms
$= \ln 2^{-2}$ Simplify.
$= -2\ln 2$ Power Property of Logarithms

33. $\ln 81x^6$ **35.** 3.7955 **37.** 0.6931 **39.** -0.5596
41. $\{x \mid x \leq 2.1633\}$ **43.** $\{x \mid x > 8.0105\}$
45. $\{x \mid x < -239.8802$ or $x > 239.9802\}$
47.
a. $A = Pe^{rt}$ Continuous Compounding Formula
$= 800e^{(0.045)(5)}$ $P = 800, r = 0.045, t = 5$
$= 800e^{0.225}$ Simplify.
≈ 1001.86 Use a calculator.
About $1001.86 will be in the account.
b. $A = Pe^{rt}$ Continuous Compounding Formula
$1600 = 800e^{(0.045)t}$ $A = 2 \cdot 800$ or $1600, P = 800, r = 0.045$
$2 = e^{0.045t}$ Divide each side by 800.
$\ln 2 = \ln e^{0.045t}$ Property of Equality of Logarithms
$\ln 2 = 0.045t$ $\ln e^x = x$
$\dfrac{\ln 2}{0.045} = t$ Divide each side by 0.045.
$15.4 \approx t$ Use a calculator.
It would take about 15.4 years to double your money.
c. $A = Pe^{rt}$ Continuous Compounding Formula
$1600 = 800e^{r(9)}$ $A = 1600, P = 800, t = 9$
$2 = e^{9r}$ Divide each side by 800.
$\ln 2 = \ln e^{9r}$ Property of Equality of Logarithms
$\ln 2 = 9r$ $\ln e^x = x$
$\dfrac{\ln 2}{9} = r$ Divide each side by 9.
$0.077 \approx r$ Use a calculator.
You would need a rate of about 7.7%.
d. $A = Pe^{rt}$ Continuous Compounding Formula

$-3\log 11 = b\log 5 - b\log 11$ Subtract $b\log 11$ from each side.
$-3\log 11 = b(\log 5 - \log 11)$ Distributive Property
$\dfrac{-3\log 11}{\log 5 - \log 11} = b$ Divide each side by $\log 5 - \log 11$.
$9.1237 \approx b$ Use a calculator.

11. $\{p \mid p \leq 4.4190\}$ **13.** $\dfrac{\log 23}{\log 4} \approx 2.2618$
15. $\dfrac{\log 5}{\log 2} \approx 2.3219$ **17.** 1.0414 **19.** 0.9138 **21.** -1.3979
23. 1.7740 **25.** 5.9647 **27.** ± 1.1691 **29.** $\{n \mid n > 0.6667\}$
31. $\{y \mid y \geq -3.8188\}$ **33.** $\dfrac{\log 18}{\log 7} \approx 1.4854$
35. $\dfrac{\log 16}{\log 2} = 4$ **37.** $\dfrac{\log 11}{\log 3} \approx 2.1827$
$\dfrac{\log 16}{\log 2} = \dfrac{\log_{10} 12}{\log_{10} 4}$ Change of Base Formula
$= 62.737$ Use a calculator.
In 2010, there are about 62.737 thousand, or 62,737 pet owners.
39. a. $n = 35[\log_4(t+2)]$ Original equation
$= 35[\log_4(10+2)]$ $t = 10$
$= 35[\log_4(12)]$ Simplify.
$= 35 \cdot \dfrac{\log_{10} 12}{\log_{10} 4}$ Change of Base Formula
≈ 62.737
b. $n = 35[\log_4(t+2)]$ Original equation
$80 = 35[\log_4(t+2)]$ $n = 80$
$2.2857 = \log_4(t+2)$ Divide each side by 35.
$4^{2.2857} = t+2$ Definition of logarithm
$22 \approx t$
In 22 years after 2000, or in 2022, there will be 80,000 pet owners.

41. 3.3578 **43.** -0.0710 **45.** 4.7393 **47.** $\{x \mid x \geq 2.3223\}$
49. $\{x \mid x \leq 0.9732\}$ **51.** $\{p \mid p \geq 2.9437\}$
53. $\dfrac{\log 12}{\log 4} \approx 1.7925$ **55.** $\dfrac{\log 2}{\log 8} \approx 0.3333$
57. $\dfrac{\log 7.29}{\log 5} \approx 1.2343$ **59a.** 113.03 cents
59b. about 218 Hz
61. $4^{x^2-3} = 16$ Original equation
$4^{x^2-3} = 4^2$ Rewrite 16 as 4^2.
$x^2 - 3 = 2$ Property of Equality for Exponential Functions
$x^2 = 5$ Add 3 to each side.
$x = \pm\sqrt{5}$ Take the square root of each side.
$x = \pm 2.2361$ Use a calculator.
63. 3.5   **65.** -3.8188 **67a.** Yes; all methods produce the same result because you are starting with the same equation. If they do not, then an error was made.
67b. $(1.85, 13)$ **67c.** The solution is between 1.8 and 1.9. The solution is between 1.85.
69. $\log_{\sqrt{a}} 3 = \log_a x$ Original equation
$\dfrac{\log_a 3}{\log_a \sqrt{a}} = \log_a x$ Change of Base Formula
$\dfrac{\log_a 3}{\frac{1}{2}} = \log_a x$ $\sqrt{a} = a^{\frac{1}{2}}$
$2\log_a 3 = \log_a x$ Multiply numerator and denominator by 2.
$\log_a 3^2 = \log_a x$ Power Property of Logarithms

Mt. Logan: $41,261.82$ pascals   **7.** 2.4182 **9.** 2
11. 13.4403 **13.** 2.1610
15. $\log_4 \dfrac{4}{3} = \log_4 4 - \log_4 3$ Quotient Property
$= 1 - \log_4 3$ Inverse Property of Exponents and Logarithms
$\approx 1 - 0.7925$ Replace $\log_4 3$ with 0.7925.
≈ 0.2075 Simplify.
17. 1.5   **19.** 2.1606 **21.** 3.4818 **23.** 8   **25.** 2
27. a. $P = \log_{10}\left(1 + \frac{1}{d}\right)$ Original equation
$10^P = 1 + \dfrac{1}{d}$ Definition of logarithm
$10^P - 1 = \dfrac{1}{d}$ Subtract 1 from each side.
$d(10^P - 1) = 1$ Multiply each side by d.
$d = \dfrac{1}{10^P - 1}$ Divide each side by $10^P - 1$.
b. $d = \dfrac{1}{10^P - 1}$ Write the formula.
$= \dfrac{1}{10^{0.097} - 1}$
≈ 4 Use a calculator.
c. $P = \log_{10}\left(1 + \frac{1}{d}\right)$ Original equation
$= \log_{10}\left(1 + \frac{1}{1}\right)$ $d = 1$
$= \log_{10} 2$ Simplify.
≈ 0.30103 Replace $\log_{10} 2$ with 0.30103.
The probability is about 30.1%.
29. 2.1133 **31.** 0.1788 **33.** 1.7228 **35.** 2.0478 **37.** 3
39. 5   **41.** $85\frac{1}{3}$ **43.** $\left(x - \frac{2}{256}\right)$ **45.** $\sqrt{6}, -\sqrt{6}$ **47.** 5
49. false   **51.** false   **53.** false   **55.** true   **57.** false   **59a.**
59a. 10^{12} **59b.** 10^4 or about 10,000 times   **61a.** Sample
61b. Sample answer: $\log_b \dfrac{xz}{5} = \log_b x + \log_b z - \log_b 5$
answer: $\log_b m^4 p^6 = 4\log_b m + 6\log_b p$
61c. Sample answer: $\log_b \dfrac{p^8 k}{h^5} = 8\log_b j + \log_b k - 5\log_b h$
63a. $\log_b 1 = 0$, because $b^0 = 1$.   **63b.** $\log_b b = 1$, because $b^1 = b$.   **63c.** $\log_b b^x = x$, because $b^x = b^x$.
65. $\log_{b_7} 24 \neq \log_b 8 + \log_b 4$; all other choices are equal to $\log_b 24$.
67. $x^3 \log_2 x - \log_4 5 = x^{\log_2 5} - \log_4 5$
$= x^{\log_2 8 - \log_4 5}$
$= x^{\log_b \frac{8}{5}}$
$= \dfrac{8}{5}$
69. D   **71.** growing exponentially   **73.** $\dfrac{1}{2}$, 1
75. no solution   **77.** $2x$ **79.** 6.3 amps   **81.** no   **83.** $\dfrac{3}{5}$
85. 10   **87.** $x > 26$

Lesson 7-6
1. 0.6990 **3.** -0.3979 **5.** 3.55×10^{24} ergs   **7.** 0.8442
9. $11^{b-3} = 5^b$ Original equation
$\log 11^{b-3} = \log 5^b$ Property of Equality for Logarithmic Functions
$(b-3)\log 11 = b\log 5$ Power Property of Logarithms
$b\log 11 - 3\log 11 = b\log 5$ Distributive Property

---

5. a.

$y = 80e^{kt}$	Original formula
$675 = 80e^{k(30)}$	$y = 675, t = 30$
$8.4375 = e^{30k}$	Divide each side by 80.
$\ln 8.4375 = \ln e^{30k}$	Property of Equality for Logarithmic Functions
$\ln 8.4375 = 30k$	$\ln e^x = x$
$\dfrac{\ln 8.4375}{30} = k$	Divide each side by 30.
$0.071 \approx k$	Use a calculator.

b.

$y = 80e^{kt}$	Original formula
$6000 = 80e^{0.071t}$	$y = 6000, k \approx 0.071$
$75 = e^{0.071t}$	Divide each side by 80.
$\ln 75 = \ln e^{0.071t}$	Property of Equality for Logarithmic Functions
$\ln 75 = 0.071t$	$\ln e^x = x$
$\dfrac{\ln 75}{0.071} = t$	Divide each side by 0.071.
$60.8 \approx t$	Use a calculator.

The bacteria will reach a population of 6000 cells in about 60.8 minutes.

c.

$35e^{0.09781} > 80e^{0.071t}$	Formula for exponential growth
$\ln 35e^{0.09781} > \ln 80e^{0.071t}$	Property of Inequality for Logarithms
$\ln 35 + \ln e^{0.09781} > \ln 80 + \ln e^{0.071t}$	Product Property of Logarithms
$\ln 35 + 0.09781 > \ln 80 + 0.071t$	$\ln e^x = x$
$0.0268t > \ln 80 - \ln 35$	Subtract $0.071t +$ $\ln 35$ from each side.
$t > \dfrac{\ln 80 - \ln 35}{0.0268}$	Divide each side by 0.0268.
$t > 30.85$	Use a calculator.

The number of cells of this bacteria exceed the number of cells in the other bacteria in about 30.85 minutes.

7.

$y = ae^{-0.00012t}$	Equation for the decay of Carbon-14
$0.85a = ae^{-0.00012t}$	$y = 0.85a$
$0.85 = e^{-0.00012t}$	Divide each side by a.
$\ln 0.85 = \ln e^{-0.00012t}$	Property of Equality for Logarithmic Functions
$\ln 0.85 = -0.00012t$	$\ln e^x = x$
$\dfrac{\ln 0.85}{-0.00012} = t$	Divide each side by -0.00012.
$1354 \approx t$	Use a calculator.

The bone is about 1354 years old.

9. about 14.85 billion yr   **11.** about 20.1 yr

13a.

13b. The graphs intersect at $t = 20.79$. Sample answer: This intersection point indicates the point at which both functions determine the same population at the same time.

13c. Sample answer: The logistic function $g(t)$ is a more accurate estimate of the country's population since $f(t)$ will continue to grow exponentially and $g(t)$ considers limitations on population growth such

as food supply. **15.** $t \approx 113.45$ **17.** Sample answer: The spread of the flu throughout a small town. The growth of this is limited to the population of the town itself. **19.** C **21.** D **23.** $\ln y = 7$ **25.** $5.5^4 = e^9$
27. $\dfrac{1}{6}$ **29.** $\dfrac{5}{8}$ **31.** 16 **33.** $2\dfrac{1}{2}$

Chapter 7 Study Guide and Review

1. exponential growth **3.** common logarithms
5. change of base formula **7.** logarithmic function
9. natural logarithm

11.
$D = $ [all real numbers]
$R = \{f(x) \mid f(x) > -6\}$

13.
$D = $ [all real numbers]
$R = \{f(x) \mid f(x) > -1\}$

15.
$D = $ [all real numbers]
$R = \{f(x) \mid f(x) > 0\}$

17a. $f(x) = 120,000(0.97)^x$
17b. about 88,491
19. -7 **21.** $\dfrac{9}{41}$
23. $x \le -\dfrac{2}{5}$ **25.** $2^{-4} = \dfrac{1}{16}$
27. 4
29.
$33.\ 0 < x < 64$
31. $x = 64$
35. no solution
37. no solution
39. 1.2920 **41.** 0.8614
43. -0.4307 **45.** $a = 3$
47. $n = 5$ **49.** $x \approx 2.4650$
51. $m \approx 0.6356$
53. $n \approx 0.5786$
55a. about 8.2 years
55b. about 13.9 years **57.** $x \approx -1.9459$ **59.** $x > 1.9459$
61. $x < -2.8904$ **63.** 51054.69 **65.** about 4.1%

CHAPTER 8
Rational Functions and Relations

Chapter 8 Get Ready

1. $x = \dfrac{15}{14}$ **3.** $k = \dfrac{32}{5}$ **5.** 5.27 gallons **7.** $\dfrac{1}{18}$ **9.** $\dfrac{43}{6}$

11. $p = 27$ **13.** $k = 17.5$

Lesson 8-1

1. $\dfrac{x+3}{x+8}$ **3.** D

5. $\dfrac{a^2x - b^2x}{by - ay} = \dfrac{x(a^2 - b^2)}{y(b-a)}$ Factor.

$= \dfrac{x(a-b)(a+b)}{y(b-a)}$ Factor.

$= \dfrac{-x(b-a)(a+b)}{y(b-a)}$

$= \dfrac{-x(a+b)}{y}$ $a - b = -1(b - a)$

$= \dfrac{-x(a+b)}{y}$ Simplify.

7. $\dfrac{2x^2}{3ahy^2}$ **9.** $\dfrac{(a-b)(a+1)}{12(a-2)}$ **11.** 4 **13.** $\dfrac{x(x+6)}{x+4}$
15. $\dfrac{x}{(x+3)(x-3)}$ **17.** $\dfrac{x(x+2)}{6(x+5)}$ **19.** J **21.** $\dfrac{x^2}{x+6}$
23. $\dfrac{c+4}{c+5}$ **25.** $\dfrac{c}{4ab^2f^2}$ **27.** $\dfrac{32b}{3ac^3f^2}$ **29.** $\dfrac{5r^4c}{3b}$

31. $\dfrac{y^2 + 8y + 15}{y - 6} \cdot \dfrac{y^2 - y - 9}{(y-3)(y-6)}$

$= \dfrac{(y+3)(y+5)}{y-6} \cdot \dfrac{(y-3)(y-6)}{(y+3)(y+5)}$ Factor.

$= \dfrac{\overset{1}{(y+3)}(y+5)}{y-6} \cdot \dfrac{(y-3)\overset{1}{(y-6)}}{\underset{1}{(y+3)}(y+5)}$

$= y + 5$

33. $\dfrac{(x+4)(x+2)}{33}$ **35.** $\dfrac{(x-3)(x+1)}{6(x+7)}$ **37.** $-\dfrac{a^2(a+b)}{b^4}$
39a. $\dfrac{2(x-5)}{121}$ **39b.** $\dfrac{33+m}{121+a}$ **41a.** $T(x) = \dfrac{0.4}{x+3}$
43. $\dfrac{1}{4}$ **45.** $\dfrac{x(x+2)(x-1)}{(x+3)(x-7)}$

41b. about 3.9 mm thick

47. $\dfrac{20x^2 b^2 z^{-2}}{3a^3 c^2} \div \left(\dfrac{16x^3 y^3}{9acz}\right)^{-1}$

$= \dfrac{20x^2 b^2 z^{-2}}{3a^3 c^2} \cdot \left(\dfrac{16x^3 y^3}{9acz}\right)$

$= \dfrac{20x^2 b^2}{3a^3 c^2 z^2} \cdot \dfrac{9acz}{16x^3 y^3}$ Simplify.

$z^{-2} = \dfrac{1}{z^2}$

$= \dfrac{15y^3}{4u^2 cxz}$

49. $\dfrac{2(4x+1)(2x+1)}{15y^3}$ **51.** $\dfrac{2x+1}{x+1}$
53. $\dfrac{5(2x-1)(x+2)}{2(2x-1)(3x+1)}$ **55.** $\dfrac{-9x(x+2)}{(2x-8)(x+4)(x-2)(x-6)}$

57. a. 5 tracks $\cdot$ $\dfrac{2 \text{ miles}}{1 \text{ track}} \cdot \dfrac{5280 \text{ feet}}{1 \text{ mile}} \cdot \dfrac{1 \text{ car}}{75 \text{ feet}}$

$= 5 \overset{1}{\text{tracks}} \cdot \dfrac{2 \text{ miles}}{1 \overset{}{\text{track}}} \cdot \dfrac{5280 \text{ feet}}{1 \text{ mile}} \cdot \dfrac{1 \text{ car}}{75 \text{ feet}}$

$= \dfrac{1 \cdot 2 \cdot 352 \cdot 1 \text{ car}}{1 \cdot 1 \cdot 1}$

$= 704 \text{ cars}$

b. 5 tracks $\cdot \dfrac{2 \text{ miles}}{1 \text{ track}} \cdot \dfrac{5280 \text{ feet}}{1 \text{ mile}} \cdot \dfrac{1 \text{ car}}{75 \text{ feet}}$

$= 5 \overset{1}{\text{tracks}} \cdot \dfrac{2 \text{ miles}}{1 \overset{}{\text{track}}} \cdot \dfrac{15 \cdot 352 \text{ feet}}{1 \text{ mile}} \cdot \dfrac{1 \text{ car}}{5 \cdot 15 \text{ feet}}$

c. 704 cars $\cdot \dfrac{8 \text{ attendants}}{1 \text{ car}} \cdot \dfrac{45 \text{ s}}{1 \text{ attendant}} \cdot \dfrac{1 \text{ min}}{60 \text{ s}} \cdot \dfrac{1 \text{ h}}{60 \text{ min}}$

$= 704 \text{ cars} \cdot \dfrac{8 \text{ attendants}}{1 \text{ car}} \cdot \dfrac{45 \text{ s}}{1 \text{ attendant}} \cdot \dfrac{1 \text{ min}}{60 \text{ s}} \cdot \dfrac{1 \text{ h}}{60 \text{ min}}$

59. Sample answer: The two expressions are equivalent except that the rational expression is undefined at $x = 3$. **61.** $x^2 + x - 6$ **63.** Sample answer: Sometimes; with a denominator like $x^2 + 2$, in which the denominator cannot equal 0, the rational expression can be defined for all values of x. **65.** Sample answer: When the original expression was simplified, a factor of x was taken out of the denominator. If x were to equal 0, then this expression would be undefined. So, the simplified expression is also undefined for x.
67. J **69.** 4π **71.** 0.2877 **73.** 0.2747 **75.** $10^{1.7}$ or about 50 times **77.** $2\sqrt[3]{2}$ **79.** $2ab^2\sqrt{10a}$
81. $10a - 2b$ **83.** $-3y - 3y^2$ **85.** $x^2 + 9x + 18$

Lesson 8-2

1. $80x^3 y^3$ **3.** $3y(y-3)(y-5)$ **5.** $\dfrac{48y^4 + 25x^2}{20xy^3}$
7. $21b^4 - 2$ **9.** $\dfrac{9x+5}{(x+3)(x+6)}$ **11.** $\dfrac{3(x+2)(x-2)}{x-1}$
13. $\dfrac{14x-10}{36ab^3}$ **15.** $\dfrac{3y+2}{y+3}$ **17.** $\dfrac{2a+5b}{3b-8a}$
19. $180x^2 y^4 z^2$ **21.** $6(x+4)(2x-1)(2x+3)$
23. $\dfrac{28by^2z - 9bx}{15x^3 y}$ **25.** $\dfrac{20x^2 y + 120y + 6x^2}{240ab^3}$
27. $\dfrac{105x^3 y^4 z}{15b^3 + 100ab^2 - 216a}$

29. $\dfrac{6}{y^2 - 2y - 35} + \dfrac{4}{y^2 + 9y + 20}$

$= \dfrac{6}{(y-7)(y+5)} + \dfrac{4}{(y+4)(y+5)}$ Factor denominators.

$= \dfrac{6(y+4)}{(y-7)(y+5)(y+4)} + \dfrac{4(y-7)}{(y+4)(y+5)(y-7)}$ Multiply by missing factors.

$= \dfrac{6y+24+4y-28}{(y-7)(y+5)(y+4)}$ Add the numerators.

$= \dfrac{10y-4}{(y-7)(y+5)(y+4)}$ Simplify.

31. $\dfrac{-10x-10}{(2x-1)(x+4)(x-3)}$ **33.** $\dfrac{2x^2 + 32x}{3(x-2)(x+3)(2x+5)}$
35. $\dfrac{1000x + 800y}{x(x+2y)}$

37. $\dfrac{\frac{4}{x+5} + \frac{9}{x-6}}{\frac{5}{x-6} - \frac{8}{x+5}}$

$= \dfrac{\frac{4(x-6)}{(x+5)(x-6)} + \frac{9(x+5)}{(x+5)(x-6)}}{\frac{5(x+5)}{(x+5)(x-6)} - \frac{8(x-6)}{(x+5)(x-6)}}$

$= \dfrac{\frac{4x-24+9x+45}{(x+5)(x-6)}}{\frac{5x+25-8x+48}{(x+5)(x-6)}}$ Simplify the numerator and denominator.

$= \dfrac{\frac{13x+21}{(x+5)(x-6)}}{\frac{-3x+73}{(x+5)(x-6)}}$ Combine like terms.

$= \dfrac{13x+21}{(x+5)(x-6)} \div \dfrac{-3x+73}{(x+5)(x-6)}$ Write as a division expression.

59. $\dfrac{704 \cdot 8 \cdot 45 \cdot 1 \cdot 1 \, h}{1 \cdot 1 \cdot 60 \cdot 60}$
$= 70.4 \text{ hours}$

m.connectED.mcgraw-hill.com

R75

25.
$$f(x) = \frac{3}{2x - 4}$$
$D = \{x \mid x \neq 2\}$;
$R = \{f(x) \mid f(x) \neq 0\}$

27.
$$f(x) = \frac{2}{4x + 1}$$
$D = \left\{x \mid x \neq \frac{1}{4}\right\}$;
$R = \{f(x) \mid f(x) \neq 0\}$

29. **a.** $rt = d$ rate • time = distance
$rt = 60.5$ $d = 60.5$
$r = \frac{60.5}{t}$ Divide each side by t.

b. This represents a transformation of the graph of $f(x) = \frac{1}{x}$. There are asymptotes at $t = 0$ and $r = 0$. Since $a = 60.5$, the graph is stretched vertically.

c. $r = \frac{60.5}{t}$ Write the equation.
$0.48 = \frac{60.5}{t}$
$t = \frac{60.5}{0.48}$ $t = 0.48$
≈ 126 ft/s   Use a calculator.

31. $D = \{x \mid x \neq -2\}$;
$R = \{f(x) \mid f(x) \neq -5\}$;
$x = -2, f(x) = -5$

33. $D = \{x \mid x \neq 4\}$;
$R = \{f(x) \mid f(x) \neq 3\}$;
$x = 4, f(x) = 3$
$$f(x) = \frac{2}{x - 4} + 3$$

13.
$$f(x) = \frac{2}{x - 6}$$
$D = \{x \mid x \neq 6\}$;
$R = \{f(x) \mid f(x) \neq 0\}$

15. This represents a transformation of the graph of $f(x) = \frac{1}{x}$.
$a = 2$: The graph is stretched vertically.
$k = 3$: The graph is translated 3 units up. There is a horizontal asymptote at $f(x) = 3$. Domain: $D = \{x \mid x \neq 0\}$. Range: $R = \{f(x) \mid f(x) \neq 3\}$.

17.
$$f(x) = \frac{2}{x}$$
$D = \{x \mid x \neq 5\}$;
$R = \{f(x) \mid f(x) \neq 0\}$

19.
$$f(x) = \frac{9}{x + 3} + 6$$
$D = \{x \mid x \neq -3\}$;
$R = \{f(x) \mid f(x) \neq 6\}$

21.
$$f(x) = \frac{-6}{x + 4} + 2$$
$D = \{x \mid x \neq -4\}$;
$R = \{f(x) \mid f(x) \neq -2\}$

23a. $m = \frac{5000}{d}$

23b.
$$m = \frac{5000}{d}$$

23c. 13.7 mi

85.
$$y = 4(x + 3)^2 + 1$$

87.
$$y = \frac{1}{4}(x - 2)^2 + 4$$

89.
$$y = x^2 + 6x + 2$$

Lesson 8-3

1 $x - 1 = 0$
$x = 1$
$f(x)$ is not defined when $x = 1$. So, there is a vertical asymptote at $x = 1$.
From $x = 1$, as x-values decrease, $f(x)$ values approach 0, and as x-values increase, $f(x)$ values approach 0. So there is a horizontal asymptote at $f(x) = 0$. The domain is all real numbers not equal to 1 or $D = \{x \mid x \neq 1\}$. The range is all real numbers not equal to 0 or $R = \{f(x) \mid f(x) \neq 0\}$.

3.
$$f(x) = \frac{5}{x} - 4$$
$D = \{x \mid x \neq 0\}$;
$R = \{f(x) \mid f(x) \neq 0\}$

5.
$$f(x) = \frac{-1}{x - 2} + 4$$
$D = \{x \mid x \neq 2\}$;
$R = \{f(x) \mid f(x) \neq 4\}$

7. $x = -4, f(x) = 0$; $D = \{x \mid x \neq -4\}$;
$R = \{f(x) \mid f(x) \neq 0\}$ **9.** $x = -6, f(x) = -2$;
$D = \{x \mid x \neq -6\}$; $R = \{f(x) \mid f(x) \neq -2\}$

11.
$$f(x) = \frac{3}{x}$$
$D = \{x \mid x \neq 0\}$;
$R = \{f(x) \mid f(x) \neq 0\}$

$$= \frac{13x + 21}{(x + 5)(x - 6)} \cdot \frac{(x + 5)(x - 6)}{-3x + 73}$$ Multiply by the reciprocal of the divisor.
$$= \frac{13x + 21}{-3x + 73}$$ Simplify.

39. $-x^2 + 33x + 16$ **41.** $420x^5 y^4 z^3$
$12x^2 + 11x - 27$

43. $(x + 4)(x - 4)(2x + 1)(x - 7)$ **45.** $\frac{360u^2 + 5u - 36}{60u^2}$

47. $\frac{42x + 41}{6(3x - 1)(x + 8)(2x + 3)}$ **49.** 0   **51.** $\frac{5u - 11}{6}$

53. $(x - 3)(x + 2)$ **55.** $-\frac{3}{2}$ **57.** -1

59a. $y = \frac{70x}{x - 70}$ **59b.** Sample answer: When the object is 70 mm away, y needs to be 0, which is impossible.

61 **a.** $P_0\left(\dfrac{s_0}{s_0 - x}\right) - P_0\left(\dfrac{s_0}{s_0 - y}\right) = \dfrac{P_0 s_0}{s_0 - x} - \dfrac{P_0 s_0}{s_0 - y}$

$= \dfrac{P_0 s_0(s_0 - y)}{(s_0 - x)(s_0 - y)} - \dfrac{P_0 s_0(s_0 - x)}{(s_0 - x)(s_0 - y)}$

$= \dfrac{P_0 s_0(s_0 - y) - P_0 s_0(s_0 - x)}{(s_0 - x)(s_0 - y)}$

$= \dfrac{P_0 s_0 x - P_0 s_0 y}{(s_0 - x)(s_0 - y)}$

b. $\dfrac{P_0 s_0 x - P_0 s_0 y}{(s_0 - x)(s_0 - y)}$

$P_0 = 500, s_0 = 332, x = 70, y = 45$

$= \dfrac{(500)(332)(70) - (500)(332)(45)}{(332 - 70)(332 - 45)}$

$= \dfrac{4,150,000}{75,194}$ Simplify.

≈ 55.2 Hz   Simplify.

63. $\dfrac{-3x^3 - 2x^2 + 16x - 5}{4x^3 + 18x^2 - 6x}$ **65.** Sample answer: $20u^4 b^2 c$,
$15a b^6, 9abc$ **67.** D   **69.** F   **71.** $\dfrac{4bc}{33a}$ **73.** $(n + 3)(n - 6)$

75. $D = \{x \mid x \geq -0.5\}$,
$R = \{y \mid y \leq 0\}$
$$y = -\sqrt{2x + 1}$$

77. $D = \{x \mid x \geq -6\}$,
$R = \{y \mid y \geq -3\}$
$$y = \sqrt{x + 6} - 3$$

79. $D = \{x \mid x \geq 2\}$,
$R = \{y \mid y \geq 4\}$
$$y = \sqrt{3x - 6} + 4$$

81. $-\dfrac{8}{3}$; 1 real   **83.** 0, 3i,
$-3i$; 1 real, 2 imaginary

connectED.mcgraw-hill.com

7.

$$f(x) = \frac{x^2 - 12}{x + 4}$$

9.
$$f(x) = \frac{x^3}{8x - 4}$$

11.
$$f(x) = \frac{x^3 + 64}{16x - 24}$$

13.
$$f(x) = \frac{x}{x + 2}$$

15 Since $a(x) = 4$, there are no zeros. The function is undefined for $x = 2$, so there is a vertical asymptote at $x = 2$. Since the degree of the numerator is less than the degree of the denominator, there is a horizontal asymptote at $f(x) = 0$. The difference between the degree of the numerator and the degree of the denominator is 2, so there is no oblique asymptote.

17.
$$f(x) = \frac{1}{(x + 4)^2}$$
$$f(x) = \frac{4}{(x - 2)^2}$$

19.

$$f(x) = \frac{(x - 4)^2}{x + 2}$$

Lesson 8-4

1.
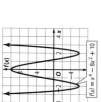
$$f(x) = \frac{x^2 - 2}{x^2 - 1}$$

3a.

$$P(x) = \frac{7 + x}{11 + x}$$

3b. The part in the first quadrant. **3c.** It represents his original field goal percentage of 63.6%. **3d.** $y = 1$; this represents 100% which he cannot achieve because he has already missed 4 field goals.

5 $x^2 + 8x + 20 = 0$ Set $a(x) = 0$.

Since $b^2 - 4ac = 8^2 - 4(1)(20)$ or -16, there are no real roots. So, there are no zeros.

$x + 2 = 0$ Set $b(x) = 0$.

$\quad x = -2$ Subtract 2 from each side.

There is a vertical asymptote at $x = -2$. The degree of the numerator is greater than the degree of the denominator, so there is no horizontal asymptote. The difference between the degree of the numerator and the degree of the denominator is 1, so there is an oblique asymptote.

$$\begin{array}{r} x + 6 \\ x + 2 \overline{)\, x^2 + 8x + 20} \\ \underline{(-)\, x^2 + 2x} \\ 6x + 20 \\ \underline{(-)\, 6x + 12} \\ -2 \end{array}$$

The oblique asymptote is $y = x + 6$.

$$f(x) = \frac{x^2 + 8x + 20}{x + 2}$$

61. rel. max at $x = 0$, rel. min at $x = -2$ and at $x = 2$; D = {all real numbers}, R = {$f(x) \mid f(x) \geq -6$}

$$f(x) = x^4 - 8x^2 + 10$$

---

35. D = {$x \mid x \neq 7$}; R = {$f(x) \mid f(x) \neq -8$}; $x = 7$, $f(x) = -8$

$$f(x) = \frac{-6}{x - 7} - 8$$

37a.

x	$a(x) = x^2$	$b(x) = x^{-2}$	$c(x) = x^3$	$d(x) = x^{-3}$
-4	16	$\frac{1}{16}$	-64	$-\frac{1}{64}$
-3	9	$\frac{1}{9}$	-27	$-\frac{1}{27}$
-2	4	$\frac{1}{4}$	-8	$-\frac{1}{8}$
-1	1	1	-1	-1
0	0	undefined	0	undefined
1	1	1	1	1
2	4	$\frac{1}{4}$	8	$\frac{1}{8}$
3	9	$\frac{1}{9}$	27	$\frac{1}{27}$
4	16	$\frac{1}{16}$	64	$\frac{1}{64}$

37b.
$$a(x) = x^2$$
$$b(x) = x^{-2}$$

37c. $a(x)$: D = {all real numbers}, R = {$a(x) \mid a(x) \geq 0$}; $x \to -\infty$, $a(x) \to \infty$, $x \to \infty$, $a(x) \to \infty$; At $x = 0$, $a(x) = 0$, so there is a zero at $x = 0$. $b(x)$: D = {$x \mid x \neq 0$}, R = {$b(x) \mid b(x) > 0$}; $x \to -\infty$, $b(x) \to 0$, $x \to \infty$, $a(x) \to 0$; At $x = 0$, $b(x)$ is undefined, so there is an asymptote at $x = 0$.

37d.
$$c(x) = x^3$$
$$d(x) = x^{-3}$$

37e. $c(x)$: D = {all real numbers}, R = {all real numbers}; $x \to -\infty$, $a(x) \to -\infty$, $x \to \infty$, $a(x) \to \infty$; At $x = 0$, $a(x) = 0$, so there is a zero at $x = 0$. $d(x)$: D = {$x \mid x \neq 0$}, R = {$b(x) \mid b(x) \neq 0$}; $x \to -\infty$, $a(x) \to 0$, $x \to \infty$, $a(x) \to 0$; At $x = 0$, $b(x)$ is undefined, so there is an asymptote at $x = 0$.

37f. For two power functions $f(x) = ax^n$ and $g(x) = ax^{-n}$, for every x, $f(x)$ and $g(x)$ are reciprocals. The domains are similar except that for $g(x)$, $x \neq 0$. Additionally, wherever $f(x)$ has a zero, $g(x)$ is undefined.

39a. The first graph has a vertical asymptote at $x = 0$ and a horizontal asymptote at $y = 0$. The second graph is translated 7 units up and has a vertical asymptote at $x = 0$ and a horizontal asymptote at $y = 7$. **39b.** Both graphs have a vertical asymptote at $x = 0$ and a horizontal asymptote at $y = 0$. The second graph is stretched by a factor of 4. **39c.** The first graph has a vertical asymptote at $x = 0$ and a horizontal asymptote at $y = 0$. The second graph is translated 5 units to the left and has a vertical asymptote at $x = -5$ and a horizontal asymptote at $y = 0$.

39d.

$$y - 7 = 4\left(\frac{1}{x} + 5\right)$$

41. Sample answer: $f(x) = \dfrac{2}{x - 3} + 4$ and $g(x) = \dfrac{5}{x - 3} + 4$

$$g(x) = \frac{5}{x - 3} + 4$$
$$f(x) = \frac{2}{\left(x - 3\right)} + 4$$

43. 4   **45.** B   **47.** B   **49.** $-2p$ **51.** $\dfrac{2x + y}{2x - y}$

53.
$$y = 5(2)^x$$

55.
$$y = 4\left(\frac{1}{3}\right)^x$$

D = {all real numbers}, R = {$y \mid y > 0$}

D = {all real numbers}, R = {$y \mid y > 0$}

57. $(f + g)(x) = 6x + 6$; $(f - g)(x) = -2x - 12$; $(f \cdot g)(x) = 8x^2 + 6x - 27$, $\left(\frac{f}{g}\right)(x) = \dfrac{2x - 3}{4x + 9}$, $x \neq -\dfrac{9}{4}$

59. $w = 4$ cm, $\ell = 8$ cm, $h = 2$ cm

connectED.mcgraw-hill.com

47. a. $F = G\frac{m_1 m_2}{d^2}$ Law of Universal Gravitation
$= (6.67 \times 10^{-11}) \frac{(7.36 \times 10^{22})(5.97 \times 10^{24})}{(3.84 \times 10^8)^2}$
$\approx 2 \times 10^{20}$ newtons

b. $F = G\frac{m_1 m_2}{d^2}$ Law of Universal Gravitation
$= (6.67 \times 10^{-11}) \frac{(1.99 \times 10^{30})(5.97 \times 10^{24})}{(1.5 \times 10^{11})^2}$
$\approx 3.5 \times 10^{22}$ newtons

c. $F = G\frac{m_1 m_2}{d^2}$ Law of Universal Gravitation
$= (6.67 \times 10^{-11}) \frac{(1000)(1000)}{(0.1)^2}$
$= 6.67 \times 10^{-3}$ newtons

49. a and c are directly related. **51.** Sample answer: The force of an object varies jointly as its mass and acceleration. **53.** D

55a.

Year	Length
1	10
2	13
3	16
4	19

55b. $f(x) = 3x + 7$ **56c.** 34 in.
57. asymptotes: $x = -2$, $x = -3$
59. hole: $x = -3$ **61.** 6 **63.** 3
65. $x + 3, x - \frac{1}{2}$ or $2x - 1$ **69.** $24x$
67. $2a(a + 1)$ **71.** $60ab^2$

1.11 **3.7**

5 The LCD for the terms is $(x - 5)(x - 4)$.

$\frac{8}{x - 5} - \frac{9}{x - 4} = \frac{5}{x^2 - 9x + 20}$ Original equation

$(x - 5)(x - 4)\frac{8}{x - 5} - (x - 5)(x - 4)\frac{9}{x - 4} = (x - 5)(x - 4)\frac{5}{x^2 - 9x + 20}$ Multiply by the LCD.

$\frac{1}{(x - 5)}(x - 4)\frac{8}{(x - 5)} - (x - 5)(x - 4)\frac{9}{(x - 4)} = \frac{5(x - 4)(x - 5)}{x^2 - 9x + 20}$

$(x - 4)(8) - (x - 5)(9) = 5$

$8x - 32 - 9x + 45 = 5$ Simplify.
$-x + 13 = 5$ Simplify.
$-x = -8$ Subtract 13 from each side.
$x = 8$ Divide each side by -1.

7.14

9a.

	pounds	price per pound	total price
dried fruit	10	$6.25	6.25(10)
mixed nuts	m	$4.50	4.5m
trail mix	$10 + m$	$5.00	5(10 + m)

9b. $62.5 + 4.5m = 50 + 5m$ **9c.** 25 **11a.** $\frac{1}{60}$ **11b.** $\frac{x}{60}$
11c. $\frac{1}{80}$ **11d.** $\frac{x}{80}$ **11e.** $\frac{x}{60} + \frac{x}{80} = 1$ **11f.** about 34.3 min
13. $c < 0$, or $\frac{13}{18} < c$ **15.** $b < 0$, or $\frac{35}{12} < b$ **17.** 2 **19.** 1
21. $\varnothing$
23 cost of 3 pounds of bananas for $0.90/pound = 0.9(3); cost of x pounds of apples for $1.25/pound = 1.25x; total weight = $3 + x$

---

43. Similarities: Both have vertical asymptotes at $x = 0$. Both approach 0 as x approaches $-\infty$ and approach 0 as x approaches ∞. Differences: $f(x)$ has holes at $x = 1$ and $x = -1$, while $g(x)$ has vertical asymptotes at $x = \sqrt{2}$ and $x = -\sqrt{2}$. $f(x)$ has no zeros, but $g(x)$ has zeros at at $x = 1$ and $x = -1$.
45. $f(x) = \frac{x}{a - b} + \frac{c(a - b)}{a - b}$ **47.** C **49.** 4
$= \frac{x + ca - cb}{a - b}$

51.

$f(x) = \frac{-5}{x + 2}$

53.

$f(x) = \frac{1}{x + 6} + 1$

55. $\frac{y(y - 9)}{(y + 3)(y - 3)}$ **57.** $\frac{-8d + 20}{(d - 4)(d + 4)(d - 2)}$ **59.** x^3 **61.** $a^{\frac{1}{9}}$

$D = \{x \mid x \neq -2\}$, $R = \{f(x) \mid f(x) \neq 0\}$

$D = \{x \mid x \neq -6\}$, $R = \{f(x) \mid f(x) \neq 1\}$

---

1. 21 **3.** −32 **5.** −48 **7.** 1.5 **9.** −56 **11.** $m = \frac{1}{6}w$

13 Joint variation

$\frac{a_1}{b_1 c_1} = \frac{a_2}{b_2 c_2}$

$\frac{-60}{-5(4)} = \frac{a_2}{4(-3)}$ $a_1 = -60$, $b_1 = -5$, $c_1 = 4$, $b_2 = 4$, $c_2 = -3$

$-60(4)(-3) = -5(4)(a_2)$ Cross multiply.
$720 = -20a_2$ Simplify.
$-36 = a_2$ Divide each side by -20.

15. −3 **17.** −22.5 **19.** 38 **21a.** $s = \frac{48}{t}$
21b. 3.2 hours **23.** −10 **25.** direct **27.** neither

29 Inverse variation

$\frac{x_1}{y_2} = \frac{x_2}{y_1}$ $x_1 = 16$, $y_1 = 5$, $y_2 = 20$

$\frac{16}{20} = \frac{5}{x_2}$ Cross multiply.
$16(5) = 20(x_2)$ Simplify.
$80 = 20x_2$ Divide each side by 20.
$4 = x_2$

31. −12 **33.** inverse; −2 **35.** combined; 10
37. direct; 4 **39.** direct; −2 **41.** inverse; 7 **43.** joint; 20

45a. $800 = rt$

45b. $44.\overline{4}$ mph

35.

$f(x) = \frac{2x^4 + 10x^3 + 12x^2}{x^2 + 5x + 6}$

37. a. total cost = phone cost + monthly usage charge
$= 150 + 40x$
average monthly cost $= \frac{\text{total cost}}{\text{number of months}}$
$f(x) = \frac{150 + 40x}{x}$

b. The vertical asymptote is $x = 0$. Since the degree of the numerator equals the degree of the denominator, the horizontal asymptote is at $f(x) = \frac{40}{1}$ or $f(x) = 40$.

c. Sample answer: The number of months and the average cost cannot have negative values.

d. $f(x) = \frac{150 + 40x}{x}$ Write the equation.
$45 = \frac{150 + 40x}{x}$ $f(x) = 45$
$45x = 150 + 40x$ Multiply each side by x.
$5x = 150$ Subtract 40x from each side.
$x = 30$ Divide each side by 5.

After 30 months, the average monthly charge will be $45.

39.

$f(x) = \frac{x + 1}{x^2 + 6x + 5}$

41.

$f(x) = \frac{6x^2 + 4x + 2}{x^2 + 2}$

21.

$f(x) = \frac{x + 1}{x^2 - 4}$

23.

$f(x) = \frac{3x^2 + 8}{2x - 1}$

25.

$f(x) = \frac{x^4 - 2x^2 + 1}{x^3 + 2}$

27a.

27b. $R_1 = -100$; no R_1-intercept; 1.2
27c. 0.5 amperes
27d. $R_1 \geq 0$ and $0 < I \leq 1.2$

29.

$f(x) = \frac{x^2 + 4x - 12}{x - 2}$

31.

$f(x) = \frac{x^2 - 64}{x - 8}$

33.

$f(x) = \frac{(x + 5)(x^2 + 2x - 3)}{x^2 + 8x + 15}$

(R81 — top section)

41b. midpoint of $\overline{XY}$ = (6, 0); midpoint of $\overline{YZ}$ = (1, −2); midpoint of $\overline{XZ}$ = (−1, 7) **41c.** The perimeter of △XYZ is $2\sqrt{29} + 14\sqrt{2} + 2\sqrt{85}$ units, perimeter = $\sqrt{29} + 7\sqrt{2} + \sqrt{85}$

41a.

41d. The perimeter of △XYZ is twice the perimeter of the smaller triangle. **43.** a circle and its interior with center at (5, 6) and radius 3 units **45.** The distance from A to B equals the distance from B to A. Using the Distance Formula, the solution is the same no matter which ordered pair is used first. **47.** 8.91

49. G **51.** −6, −2 **53.** $\frac{3}{2}$ **55.** 4.8362 **57.** 8.0086
59. $\{p \mid p \le 1.9803\}$ **61.** −20 **63.** $\frac{1}{3}$
65. $y = (x - 3)^2 - 8$; (3, −8); $x = 3$; up

Lesson 9-2
1. $y = 2(x - 6)^2 - 32$; vertex (6, −32); axis of symmetry: $x = 6$; opens upward **3.** $x = (y - 4)^2 - 27$; vertex (−27, 4); axis of symmetry: $y = 4$; opens right

5.

$y = (x - 4)^2 - 6$

7.

$y = -3x^2 - 4x - 8$

9. $y = \frac{1}{8}x^2 + 2$

$y = \frac{1}{8}x^2 + 2$

CHAPTER 9
Conic Sections

Chapter 9 Get Ready
1. (−7, −1) **3.** (3, 5) **5.** $\left\{-5, \frac{3}{2}\right\}$ **7.** $\left\{\frac{3}{4} \pm \sqrt{2}\right\}$
9. $A'(-2, -5), B'(-1, -1), C'(5, -1), D'(4, -5)$
11. $J'(-3, 6), K'(-2, -3), L'(-10, -4)$

13. (115, 60), (125, 60), (125, 50), and (115, 50)

Lesson 9-1
1. $\left(-\frac{1}{2}, 8\right)$ **3.** (14.5, 9.75) **5.** 11.662 units

7. $d = \sqrt{(x_2 - x_1)^2 + (y_2 - y_1)^2}$ Distance Formula
$= \sqrt{(3.5 - 0.25)^2 + (2.5 - 1.75)^2}$ $(x_1, y_1) = (0.25, 1.75)$ and $(x_2, y_2) = (3.5, 2.5)$
$= \sqrt{3.25^2 + 0.75^2}$ Simplify.
$= \sqrt{11.125}$ or about 3.335 units Simplify.

9. A **11.** (−4, −1) **13.** (7.3, 1) **15.** (−7.75, −4.5)
17. 16.279 units **19.** 16.125 units **21.** 21.024 units
23. 55.218 units **25.** (−1.5, 0); 185.443 units **29.** (8, 15);
136.953 units **31.** (−0.43, −2.25); 9.624 units

33. $\left(\dfrac{x_1 + x_2}{2}, \dfrac{y_1 + y_2}{2}\right)$ Midpoint Formula
$= \left(\dfrac{-\frac{5}{12} + \left(-\frac{17}{2}\right)}{2}, \dfrac{-\frac{1}{3} + \left(-\frac{5}{3}\right)}{2}\right)$ $(x_1, y_1) = \left(-\frac{5}{12}, -\frac{1}{3}\right)$ and $(x_2, y_2) = \left(-\frac{17}{2}, -\frac{5}{3}\right)$
$= \left(\dfrac{-\frac{107}{12}}{2}, \dfrac{-2}{2}\right)$ Simplify.
$\approx (-4.458, -1)$ Simplify.

35. (−4.719, 0.028); 17.97 units **37.** 14.53 km Use a calculator.

39. $d = \sqrt{(x_2 - x_1)^2 + (y_2 - y_1)^2}$ Distance Formula
$= \sqrt{(254 - 132)^2 + (105 - 428)^2}$ $(x_1, y_1) = (132, 428)$ and $(x_2, y_2) = (254, 105)$
$= \sqrt{122^2 + (-323)^2}$ Simplify.
$= \sqrt{119{,}213}$ or about 345 units Simplify.
345 units • 0.316 mi/unit ≈ 109 mi

(R80 — bottom section)

$\dfrac{\text{total cost}}{\text{total weight}} = 1$ Write an equation.
$\dfrac{0.9(3) + 1.25x}{3 + x} = 1$ Substitute.
$\dfrac{2.7 + 1.25x}{3 + x} = 1$ Simplify the numerator.
$\dfrac{(3 + x)(2.7 + 1.25x)}{3 + x} = (3 + x)(1)$ LCD is $(3 + x)$. Multiply by the LCD.
$\dfrac{(3 + x)(2.7 + 1.25x)}{3 + x} \cdot 1 = (3 + x)(1)$ Divide out common factors.
$2.7 + 1.25x = 3 + x$ Simplify.
$2.7 + 0.25x = 3$ Subtract x from each side.
$0.25x = 0.3$ Subtract 2.7 from each side.
$x = 1.2$ Divide each side by 0.25.
She must purchase 1.2 pounds of apples.

25. $x < 0$ or $x > 1.75$ **27.** $x < -2$, or $2 < x < 14$
29. $x < -5$ or $4 < x < \frac{17}{3}$ **31.** 55.56 mph **33a.** 1; yes; 3
33b.

33c. 1; no

33d. Graph both sides of the equation. Where the graphs intersect, there is a solution. If they do not, then the possible solution is extraneous.

35. ∅ **37.** all real numbers except 5, −5, 0
39. Sample answer: Multiplying both sides of a rational inequality can produce extraneous solutions.
41. J **43.** all of the points **45.** direct

47.

49.
51a. $s \cdot 4^x$ **51b.** 0.5 three-yr periods or 1.5 yr **53.** yes

Chapter 8 Study Guide and Review
1. complex fraction **3.** oblique **5.** rational equations
7. Joint variation **9.** Point discontinuity **11.** $-\dfrac{10yz^2}{9x}$
13. $\dfrac{x - 1}{x - 2}$ **15.** $\dfrac{x - 3}{x + 1}$ **17.** $\dfrac{27b + 10a^2}{12ab^2}$ **19.** $\dfrac{3ry^2 + 8y^3 - 5x}{6x^2y^2}$
21. $\dfrac{12x^2 - 10x + 6}{2(x + 2)(3x - 4)(x + 1)}$ **23.** $\dfrac{10x + 20}{(x + 6)(x + 1)}$

R80 | Selected Answers

(R80 — right column)

25. $D = \{x \mid x \ne 0\}$, $R = \{f(x) \mid f(x) \ne 2\}$

27. $D = \{x \mid x \ne 9\}$, $R = \{f(x) \mid f(x) \ne 0\}$

29. $D = \{x \mid x \ne -4\}$, $R = \{f(x) \mid f(x) \ne -8\}$

31. $x = -4$, $x = 0$ **33.** $x = 8$; hole: $x = -3$

35.

37.

39. $a = 15$ **41.** $y = -\frac{1}{3}$ **43.** $y = \frac{48}{5}$ **45.** $x = \frac{46}{17}$
47. $x = -7$ **49.** $x = 8$ **51.** $\frac{9}{10} < x < 0$

29. a. $(h, k) = \left(\dfrac{x_1 + x_2}{2}, \dfrac{y_1 + y_2}{2}\right)$ Midpoint Formula

$= \left(\dfrac{-12 + 12}{2}, \dfrac{16 + (-16)}{2}\right)$ $(x_1, y_1) = (-12, 16)$ and $(x_2, y_2) = (12, -16)$

$= (0, 0)$ Simplify.

$r = \sqrt{(x_2 - x_1)^2 + (y_2 - y_1)^2}$ Distance Formula

$= \sqrt{[0 - (-12)]^2 + (0 - 16)^2}$ $(x_1, y_1) = (-12, 16)$ and $(x_2, y_2) = (0, 0)$

$= \sqrt{12^2 + 16^2}$ Subtract.

$= \sqrt{400}$ Simplify.

$(x - h)^2 + (y - k)^2 = r^2$ Equation of a circle

$(x - 0)^2 + (y - 0)^2 = (\sqrt{400})^2$ $(h, k) = (0, 0)$ and $r = \sqrt{400}$

$x^2 + y^2 = 400$ Simplify.

b. $A = \pi r^2$ Area of a circle

$= \pi(\sqrt{400})^2$ $r = \sqrt{400}$

$= 400\pi$ Simplify.

≈ 1256.64 units^2 Use a calculator.

31. center: $(0, 0)$; radius: $5\sqrt{3}$

33. center: $(1, 4)$; radius: $\sqrt{34}$

35. center: $(5, -2)$; radius: 4

37. center: $(4, 0)$; radius: $\dfrac{\sqrt{8}}{3}$

---

35. The high beams should be placed at the focus. The y-coordinate of the focus is $k + \dfrac{1}{4a}$.

$k + \dfrac{1}{4a} = 0 + \dfrac{1}{4\left(\frac{1}{12}\right)}$ $k = 0, a = \dfrac{1}{2}$

$= \dfrac{12}{4}$ or 3   Simplify.

The filament for the high beams should be placed 3 units above the vertex.

37. Rewrite it as $y = (x - h)^2$, where $h > 0$. **39.** Russell; the parabola should open to the left rather than to the right. **41.** C **43.** D **45.** $5\sqrt{2} + 3\sqrt{10}$ units **47.** 1.7183 **49.** $x > 0.4700$ **51.** 0.5 **53.** 2^2 **55.** $\pm 1, \pm 2, \pm 3, \pm 6$ **57.** $\pm 1, \pm \frac{1}{3}, \pm \frac{1}{9}, \pm 3, \pm 9, \pm 27$ **59.** $3\sqrt{5}$ **61.** $16\sqrt{2}$

Lesson 9-3

1. $(x - 72)^2 + (y - 39)^2 = 10{,}000$

3. $(x - 1)^2 + (y + 5)^2 = 9$ **5.** $(x + 5)^2 + (y + 3)^2 = 90$

7. $x^2 + (y + 4)^2 = 20$

9. center: $(0, 7)$; radius: 3

11. center: $(2, -4)$; radius: 5

13. $(x + 3)^2 + (y - 1)^2 = 16$ **15.** $(x + 2)^2 + (y + 1)^2 = 81$

17. $(x - h)^2 + (y - k)^2 = r^2$ Equation of a circle

$(x - 0)^2 + [y - (-6)]^2 = (\sqrt{35})^2$ $(h, k) = (0, -6)$ and $r = \sqrt{35}$

$x^2 + (y + 6)^2 = 35$ Simplify.

19. $(x - 1)^2 + (y - 1)^2 = 4$ **21.** $x^2 + (y + 6) = 53$

23. $(x - 2)^2 + \left(y + \dfrac{3}{2}\right)^2 = \dfrac{25}{4}$ **25.** $\left(x - \dfrac{3}{2}\right)^2 + (y + 8)^2 = \dfrac{53}{4}$ **27.** $(x - 4)^2 + (y + 1)^2 = 20$

---

27. $y = \dfrac{1}{20}(x - 1)^2 + 8$

29. The directrix is a vertical line, so the equation of the parabola is of the form $x = a(y - k)^2 + h$. Since the vertex is equidistant from the focus and the directrix, the vertex is at $(6, 4)$.

$x = h - \dfrac{1}{4a}$ Equation of directrix

$10 = 6 - \dfrac{1}{4a}$ $x = 10$ and $h = 6$

$4 = -\dfrac{1}{4a}$ Subtract 6 from each side.

$16a = -1$ Multiply each side by $4a$.

$a = -\dfrac{1}{16}$ Divide each side by 16.

$x = a(y - k)^2 + h$ Equation of a parabola

$x = -\dfrac{1}{16}(y - 4)^2 + 6$ $a = -\dfrac{1}{16}$, $(h, k) = (6, 4)$

31. $y = \dfrac{1}{4}(x - 9)^2 + 6$

33a.

33b. $y = \dfrac{x^2}{192}$ and $x = \dfrac{y^2}{-192}$ **33c.** Sample answer: No; except for the direction in which they open, the graphs are identical.

---

11. $y = -\dfrac{1}{12}(x - 3)^2 + 5$

13a. $y = \dfrac{1}{24}x^2 - 6$

13b.

15. $y = 3(x + 7)^2 + 2$; vertex: $(-7, 2)$; axis of symmetry: $x = -7$; opens upward **17.** $y = -3\left(x + \dfrac{3}{2}\right)^2 + \dfrac{3}{4}$; vertex: $\left(-\dfrac{3}{2}, \dfrac{3}{4}\right)$; axis of symmetry: $x = -\dfrac{3}{2}$; opens downward **19.** $x = \dfrac{2}{3}(y - 3)^2 + 6$; vertex: $(6, 3)$; axis of symmetry: $y = 3$; opens right

21.

23.

25.

39. center: $(-2, 0)$; radius: $\sqrt{13}$

$x^2 + y^2 + 4x = 9$

41. center: $(-1, -2)$; radius: $\sqrt{14}$

$x^2 + y^2 + 2x + 4y = 9$

43. center: $(-7, -3)$; radius: $2\sqrt{2}$ units

$x^2 + y^2 + 6y = -50 - 14x$

45. center: $(1, -2)$; radius: $\sqrt{21}$

47a. $x^2 + y^2 = 841,000,000$

47b.

Satellite · 25,000 miles · Earth · 29,000 miles · Radius of Earth 4000 miles

49a. $(x + 1)^2 + (y - 4)^2 = 36 + 16\sqrt{5}$
49b. $(x + 1)^2 + (y - 4)^2 = 24 - 8\sqrt{5}$

49c.

51. $r = \sqrt{(x_2 - x_1)^2 + (y_2 - y_1)^2}$ Distance Formula
$= \sqrt{(19 - 9)^2 + [22 - (-8)]^2}$ $(x_1, y_1) = (9, -8)$ and $(x_2, y_2) = (19, 22)$
$= \sqrt{10^2 + 30^2}$ Subtract.
$= \sqrt{1000}$ Simplify.

$(x - h)^2 + (y - k)^2 = r^2$ Equation of a circle
$(x - 9)^2 + [y - (-8)]^2 = (\sqrt{1000})^2$ $(h, k) = (9, -8)$ and $r = \sqrt{1000}$
$(x - 9)^2 + (y + 8)^2 = 1000$ Simplify.

53. $(x - 8)^2 + (y + 9)^2 = 64$ **55.** $(x - 2.5)^2 + (y - 2.5)^2 = 6.25$ **57a.** circle **57b.** $x^2 + y^2 = 9$

57c. Solve the equation for y: $y = \pm\sqrt{49 - x^2}$. Then graph the positive and negative answers.

$y = \sqrt{49 - x^2}$
$y = -\sqrt{49 - x^2}$

57d. $y = \pm\sqrt{4 - (x - 2)^2} - 1$; because when you solve for y you must take the square root resulting in both a positive and negative answer, so you have to enter the positive equation as $Y1$ and the negative equation as $Y2$.

59. center: $\left(0, -\dfrac{9}{2}\right)$; radius: $\sqrt{19}$

$4x^2 + 4y^2 + 36y + 5 = 0$

61. center: $(-\sqrt{7}, \sqrt{11})$; radius: $\sqrt{11}$

$x^2 + 2\sqrt{7}x + 7 + (y - \sqrt{11})^2 = 11$

63. Circles with a radius of 8 and centers on the graph of $x = 3$.

65. Sample answer: $(x - 2)^2 + (y - 3)^2 = 25$ and $(x - 2)^2 + (y - 3)^2 = 36$

67. Quadrant I

$a > 0, b > 0, a = b, r > 0$

Quadrant II

$a < 0, b > 0, a = -b, r > 0$

Quadrant III

$a < 0, b < 0, a = b, r > 0$

Quadrant IV

$a > 0, b < 0, a = -b, r > 0$

69. A **71.** C

Sample answer: The circle is rotated 90° about the origin from one quadrant to the next.

73.

$4(x - 2) = (y + 3)^2$

75. $\left(1, \dfrac{7}{22}\right)$; $\sqrt{65}$ units **77.** $(0, 3)$; $\sqrt{29}$ units **79.** 64

81. $\dfrac{5}{2}$ **83.** -4 **85a.** The square root of a difference is not the difference of the square roots. **85b.** 34.1 ft/s

87. $\left\{\dfrac{3 \pm \sqrt{33}}{4}\right\}$

Lesson 9-4

1. $\dfrac{y^2}{25} + \dfrac{x^2}{9} = 1$ **3.** $\dfrac{(y + 1)^2}{25} + \dfrac{(x + 2)^2}{9} = 1$

5a. $a = 240, b = 160$ **5b.** $\dfrac{x^2}{57,600} + \dfrac{y^2}{25,600} = 1$

5c. about $(179, 0)$ and $(-179, 0)$

7. center $(5, -1)$; foci $(5, 5)$ and $(5, -7)$; major axis: 16; minor axis: ≈ 10.58

9

$\dfrac{(y + 1)^2}{64} + \dfrac{(x - 5)^2}{28} = 1$

$4x^2 + y^2 - 32x - 4y + 52 = 0$ Original equation
$4x^2 - 32x + y^2 - 4y = -52$ Commutative Property
$4(x^2 - 8x) + (y^2 - 4y) = -52$ Distributive Property
$4(x^2 - 8x + \blacksquare) + (y^2 - 4y + \blacksquare) = -52 + 4(\blacksquare) + (\blacksquare)$ Complete the squares.
$4(x^2 - 8x + 16) + (y^2 - 4y + 4) = -52 + 4(16) + (4)$
$(-4)^2 = 16$ and $(-2)^2 = 4$ Write as perfect squares.
$4(x - 4)^2 + (y - 2)^2 = 16$ Write as perfect squares.
$\dfrac{(x - 4)^2}{4} + \dfrac{(y - 2)^2}{16} = 1$ Divide each side by 16.

$h = 4$ and $k = 2$, so the center is at $(4, 2)$.
The ellipse is vertical. $a^2 = 16$, so $a = 4$, and $b^2 = 4$, so $b = 2$.
$c^2 = 16 - 4$ or 12, so $c \approx 3.46$.

R87 (right page)

19.
21.
23.

25. hyperbola   27. ellipse   29. ellipse

31. Because the center is at the origin, use the equation $\frac{x^2}{a^2} - \frac{y^2}{b^2} = 1$. The hyperbola intersects the x-axis at (10, 0), and one of the vertices is (10, 0). So, a = 10. The hyperbola also passes through (30, 100). Use a = 10, x = 30, and y = 100 to solve for b^2.

$\frac{x^2}{a^2} - \frac{y^2}{b^2} = 1$	Equation of an ellipse
$\frac{30^2}{10^2} - \frac{100^2}{b^2} = 1$	a = 10, x = 30, y = 100
$\frac{900}{100} - \frac{10,000}{b^2} = 1$	Evaluate exponents.
$9 - \frac{10,000}{b^2} = 1$	Simplify.
$-\frac{10,000}{b^2} = -8$	Subtract 9 from each side.
$-10,000 = -8b^2$	Multiply each side by b^2.
$\frac{-10,000}{-8} = b^2$	Divide each side by −8.
$1250 = b^2$	Simplify.

Substitute 1250 for b^2 in the equation $\frac{x^2}{a^2} - \frac{y^2}{b^2} = 1$. So, the equation of the path of the comet is $\frac{x^2}{100} - \frac{y^2}{1250} = 1$.

63. $4x^2 - 3xy - 6y^2$ 65. $y = -\frac{4}{5}x + \frac{17}{5}$
67. $y = -\frac{3}{5}x + \frac{16}{5}$

Lesson 9-5

1. $\frac{y^2}{36} - \frac{x^2}{28} = 1$ 3. $\frac{x^2}{64} - \frac{y^2}{25} = 1$
5. (graph)
7. (graph)

9. $\frac{x^2}{900} - \frac{y^2}{5500} = 1$

11. Since the vertices are equidistant from the center, the center is at (−8, 4). The value of a is the distance between a vertex and the center, or 4 units. The value of c is the distance between a focus and the center, or 8 units.

$c^2 = a^2 + b^2$	Equation relating a, b, and c for a hyperbola
$8^2 = 4^2 + b^2$	c = 8 and a = 4
$48 = b^2$	Subtract 4^2 from each side.
$\frac{(y-k)^2}{a^2} - \frac{(x-h)^2}{b^2} = 1$	Equation of a vertical hyperbola
$\frac{(y-4)^2}{4^2} - \frac{[x-(-8)]^2}{48} = 1$	(h, k) = (−8, 4), a = 4, b^2 = 48
$\frac{(y-4)^2}{16} - \frac{(x+8)^2}{48} = 1$	Simplify.

13. $\frac{(x+1)^2}{9} - \frac{(y-6)^2}{49} = 1$
15. (graph)
17. (graph)

R86 (left page)

foci: (4, 2 + 3.46) or (4, 5.46); (4, 2 − 3.46) or (4, −1.46)
major axis: 2 · 4 or 8
minor axis: 2 · 2 or 4

31. center (−2, 5); foci (−2, 7.83) and (−2, 2.17); major axis: ≈9.80; minor axis: 8
 $3x^2 + 2y^2 + 12x - 20y + 14 = 0$

33. $\frac{(y+2)^2}{25} + \frac{(x+5)^2}{9} = 1$
35. $\frac{(x-2)^2}{20} + \frac{(y-8)^2}{4} = 1$

37. length of major axis = 2a
10.9 = 2a
5.45 = a
length of minor axis = 2b
8.8 = 2b
4.4 = b

$\frac{x^2}{a^2} + \frac{y^2}{b^2} = 1$	Equation of an ellipse
$\frac{x^2}{5.45^2} + \frac{y^2}{4.4^2} = 1$	Substitute.
$\frac{x^2}{29.7025} + \frac{y^2}{19.36} = 1$	Simplify.

39a. (graph)

39b. Sample answer: The first graph is more circular than the second graph.
39c. first graph: 0.745; second graph: 0.943
39d. Sample answer: The closer the eccentricity is to 0, the more circular the ellipse.

41. Sample answer: $\frac{(x+4)^2}{40} + \frac{y^2}{24} = 1$
43. $\frac{y^2}{9} + \frac{(x-2)^2}{3} = 1$
45. For any point on an ellipse, the sum of the distances from that point to the foci is constant by the definition of an ellipse. So, if (2, 14) is on the ellipse, then the sum of the distances from it to the foci will be a certain value consistent with every other point on the ellipse. The distance between (−7, 2) and (2, 14) is $\sqrt{(-7-2)^2 + (2-14)^2}$ or 15. The distance between (18, 2) and (2, 14) is $\sqrt{(18-2)^2 + (2-14)^2}$ or 20. The sum of these two distances is 35.
The distance between (−7, 2) and (2, −10) is $\sqrt{(-7-2)^2 + [2-(-10)]^2}$ or 15. The distance between (18, 2) and (2, −10) is $\sqrt{(18-2)^2 + [2-(-10)]^2}$ or 15. The sum of these two distances is also 35. Thus, (2, −10) also lies on the ellipse.

47. B   49. 7   51. $(x-8)^2 + (y+9)^2 = 1130$
53. $(x+5)^2 + (y-4)^2 = 25$ 55. $\frac{5d+16}{(d+2)^2}$
57. $\frac{x^2-5x+3}{(x-5)(x+1)}$ 59. 4   61. $15a^2b^3 - 30a^4b^3 + 15a^5b^6$

11. $\frac{y^2}{100} + \frac{x^2}{36} = 1$ 13. $\frac{(x+5)^2}{49} + \frac{(y+4)^2}{25} = 1$
15. $\frac{(y-1)^2}{64} + \frac{(x+5)^2}{16} = 1$ 17. $\frac{(x-3)^2}{81} + \frac{(y-4)^2}{64} = 1$

19. The x-coordinate is the same for both vertices, so the ellipse is vertical.
length of major axis: 16 − 6 or 10 units, so a = 10
length of minor axis: 1 − (−2) or 3 units, so b = 3

$\frac{(y-k)^2}{a^2} + \frac{(x-h)^2}{b^2} = 1$	Equation of a vertical ellipse
$\frac{(y-6)^2}{10^2} + \frac{[x-(-2)]^2}{3^2} = 1$	(h, k) = (−2, 6), a = 10, b = 3
$\frac{(y-6)^2}{100} + \frac{(x+2)^2}{9} = 1$	Simplify.

21. $\frac{(y-4)^2}{100} + \frac{(x-4)^2}{64} = 1$ 23. $\frac{y^2}{73.96} + \frac{x^2}{53.29} = 1$

25. center (−6, 3); foci (−6, 7.69) and (−6, −1.69); major axis: ≈16.97; minor axis: ≈14.14
(graph)

27. center (−4, 0); foci (−4, 7.68) and (−4, −7.68); major axis: ≈17.32; minor axis: 8
(graph) $\frac{(x+4)^2}{16} + \frac{y^2}{75} = 1$

29. center (3, −3); foci (5.24, −3) and (0.76, −3); major axis: ≈8.94; minor axis: ≈7.75
 $3x^2 + 4y^2 - 18x + 24y + 3 = 0$

47. Sample answer: An ellipse is a flattened circle. Both circles and ellipses are enclosed regions, while hyperbolas and parabolas are not. A parabola has one branch, which is a smooth curve that never ends, and a hyperbola has two such branches that are reflections of each other. In standard form and when there is no xy-term: an equation for a parabola consists of only one squared term, an equation for a circle has values for A and C that are equal, an equation for an ellipse has values for A and C that are the same sign but not equal, and an equation for a hyperbola has values of A and C that have opposite signs. **49.** J **51.** B

$(0, 0)$; $(0, \pm 3)$; $6\sqrt{2}$; 6

$(4, -2)$;
$(4 \pm 2\sqrt{6}, -2)$;
10; 2

53. $\dfrac{y^2}{18} + \dfrac{x^2}{9} = 1$

55. $x^2 + 25y^2 - 8x + 100y + 91 = 0$

57. $f(x) = \dfrac{-2}{x+5}$

59a. Decay; the exponent is negative. **59b.** about 33.5 watts **59c.** about 402 days **61.** $\left(-\dfrac{1}{2}, \dfrac{3}{2}\right)$

Lesson 9-7

1. $(4, -5), (-4, 5)$ **3.** $(3, 6), (6, 42)$ **5.** no solution

7. Solve the second equation for x^2.
$y^2 - x^2 = 8 \rightarrow y^2 - 8 = x^2$

$x^2 + 2y = 7$ Write the first equation.
$y^2 - 8 + 2y = 7$ Substitute $x^2 - 8$ for x^2
$y^2 + 2y - 15 = 0$ Subtract 7 from each side.
$(y - 3)(y + 5) = 0$ Factor.
$y - 3 = 0$ or $y + 5 = 0$ Zero Product Property
$y = 3$ $y = -5$ Solve each equation.
Substitute 3 and -5 into one of the original equations and solve for x.

$x^2 + 2y = 7$ $x^2 + 2y = 7$
$x^2 + 2(3) = 7$ $x^2 + 2(-5) = 7$
$x^2 + 6 = 7$ $x^2 - 10 = 7$
$x^2 = 1$ $x^2 = 17$
$x = \pm 1$ $x = \pm\sqrt{17}$

connectED.mcgraw-hill.com

13a. parabola; $y = -0.024(x - 660)^2 + 10{,}500$
13b. about 1320 ft **13c.** 10,500 ft
15. $\dfrac{(x+4)^2}{32} + \dfrac{(y-5)^2}{24} = 1$; ellipse
17. $y = 8(x + 2)^2 - 4$; parabola
19. $(x - 3)^2 + (y + 4)^2 = 36$; circle
21. $\dfrac{(x+4)^2}{21} - \dfrac{(y-8)^2}{24} = 1$; hyperbola
23. $\dfrac{(x+4)^2}{64} - \dfrac{(y-3)^2}{25} = 1$; hyperbola

25. circle
27. $18x^2 - 16y = 12x - 4y^2 + 19$ Original equation
$18x^2 + 4y^2 - 12x - 16y - 19 = 0$ Standard form
$A = 18$, $B = 0$, and $C = 4$
$B^2 - 4AC = 0^2 - 4(18)(4)$ or -288
Since the discriminant is less than 0 and $A \ne C$, the conic is an ellipse.
29. hyperbola **31.** hyperbola **33.** hyperbola **35.** c
37. b **39.** b
41. Equation c can be written as $y = -16x^2 + 90x + 0.25$. Since this is an equation of a parabola that opens downward, it could be used to represent the height of a football above the ground after being kicked.
43a. $\dfrac{(x - 3)^2}{16} + \dfrac{(y + 2)^2}{4} = 1$
43b. $x^2 + 4y^2 - 6x + 16y + 9 = 0$
43c.
43d. $N(5, -2)$; 90° counterclockwise
45. Sample answer: Always; when a conic is vertical, $B = 0$. When this is true and $A = C$, the conic is a circle.

---

and lower limit to the values of the coordinates of an ellipse, while maximum x- and y-values for the coordinates of hyperbolas are infinite. When both of the x^2 and y^2 terms are on the same side of the equation, the equation is for an ellipse if the signs of their coefficients are the same. Otherwise it is a hyperbola. (This is true only for conics that are not rotated.)
51. J **53.** E **55.** $\dfrac{y^2}{100} + \dfrac{x^2}{36} = 1$
57. $(0, 3)$, 5 units $x^2 + y^2 - 6y - 16 = 0$

59a. $P(x) = \dfrac{6+x}{10+x}$

59b. the part in the first quadrant **59c.** It represents her original free-throw percentage of 60%. **59d.** $P(x) = 1$; this represents 100%, which she cannot achieve because she has already missed 4 free throws. **61.** $\dfrac{5}{3}$

63. $y = \sqrt{5x - 8}$

65. $y = \sqrt{6x + 2} + 1$

67. $y = \dfrac{4}{3}(x + 3)^2 - 4$

Lesson 9-6

1. $\dfrac{(x-3)^2}{36} - \dfrac{(y+2)^2}{9} = 1$; ellipse
3. $\dfrac{(y-1)^2}{16} - \dfrac{(x+2)^2}{9} = 1$; hyperbola
5. ellipse **7.** circle **9.** hyperbola **11.** ellipse

33a.

33b.

x	y
-12	-1.33
-10	-1.6
-8	-2
-6	-2.67
-4	-4
-2	-8
0	undef
2	8
4	4
6	2.67
8	2
10	1.6
12	1.33

33c. The asymptotes are $y = 0$ and $x = 0$.
33d. They are perpendicular. **33e.** For $xy = 25$, the vertices will be at $(5, 5)$ and $(-5, -5)$, and for $xy = 36$, they will be at $(-6, -6)$ and $(6, 6)$.
35. $\dfrac{x^2}{2{,}722{,}500} - \dfrac{y^2}{1{,}277{,}500} = 1$ **37.** $\dfrac{x^2}{64} - \dfrac{y^2}{100} = 1$
39. The vertices are equidistant from the center. The center is at $(2, -2)$. The value of a is the distance between a vertex and the center, or 4 units. The value of c is the distance between a focus and the center, or 8 units.
$c^2 = a^2 + b^2$ Equation relating a, b, and c for a hyperbola
$8^2 = 4^2 + b^2$ $c = 8$ and $a = 4$
$48 = b^2$ Subtract 4^2 from each side.
$\dfrac{(x - h)^2}{a^2} - \dfrac{(y - k)^2}{b^2} = 1$ Equation of a horizontal hyperbola
$\dfrac{(x - 2)^2}{4^2} - \dfrac{[y - (-2)]^2}{48} = 1$ $(h, k) = (2, -2)$, $a = 4$, $b^2 = 48$
$\dfrac{(x - 2)^2}{16} - \dfrac{(y + 2)^2}{48} = 1$ Simplify.
41. $\dfrac{x^2}{25} - \dfrac{y^2}{4} = 1$ **43.** $(2308, 826)$ **45.** $\dfrac{(x - 3)^2}{5} - \dfrac{(y + 2)^2}{5} = 1$
47. Sample answer: When 36 changes to 9, the vertical hyperbola widens (splits out from the y-axis faster). This is due to a smaller value of y being needed to produce the same value of x. The vertices are moved closer together due to the value of a decreasing from 6 to 3. The foci move farther from the vertices because the difference between c and a increased.
49. Sample answer: The graphs of ellipses are closed in, while the branches of the hyperbolas extend without bound. There is always an upper

R91

connectED.mcgraw-hill.com

39. $(0, 4)$; $(\pm 4\sqrt{2}, 4)$; 12; 4

41. $(-1, 2)$; $(2, 2)$; 10; 8

43. $(1, -1)$; $(-3, -1)$, $(5, -1)$; 10; 6

45. $\frac{x^2}{64} + \frac{y^2}{25} = 1$

47. $(2, -2)$, $(4, -2)$; $(3 \pm \sqrt{5}, -2)$; $y + 2 = \pm 2(x - 3)$

49. $(\pm 3, 0)$; $(\pm \sqrt{13}, 0)$; $y = \pm \frac{2}{3}x$

51. $\left(\frac{40 - 24\sqrt{5}}{5}, \frac{45 - 12\sqrt{5}}{5} \right)$

53. $\frac{x^2}{16} + \frac{y^2}{9} = 1$; ellipse

Chapter 9   Study Guide and Review

1. false, center   **3.** false, vertices   **5.** true   **7.** false, circle   **9.** true   **11.** $\left(-\frac{5}{2}, 5 \right)$ **13.** $\left(\frac{5}{24}, \frac{5}{24} \right)$ **15.** $\sqrt{85}$

17. $\frac{\sqrt{34}}{4}$ **19a.** $\sqrt{89} \approx 9.4$ miles   **19b.** $\left(4, -\frac{5}{2} \right)$

21.

23.

25. $y = 4(x - 2)^2 - 7$; vertex: $(2, -7)$; axis of symmetry: $x = 2$; opens upward   **27.** $x = (y + 7)^2 - 29$; vertex: $(-29, -7)$; axis of symmetry: $y = -7$; opens to the right   **29.** $(x + 1) + (y - 6)^2 = 9$
31. $(x - 1)^2 + (y + 4)^2 = 13$
33. $(3, -1)$; $r = 5$

35. $(-2, 1)$; $r = 4$

37. $(0, 0)$; $(0, \pm 3\sqrt{3})$; 12; 6

The solutions are $(-1, 3)$, $(1, 3)$, $(-\sqrt{17}, -5)$, $(\sqrt{17}, -5)$.
9. $(40, 30)$ **11.**

13.

15. no solution
17. $(-3, -6)$, $(3, 6)$
19. $(-1, -3)$, $(8, 5)$
21. $(1, -\sqrt{15})$, $(1, \sqrt{15})$
23. $(-\sqrt{2}, 4)$, $(\sqrt{2}, 4)$
25. $(5, -6)$, $(5, 6)$, $(-3, -2)$, $(-3, 2)$

27.

29.

31.

33.

35.

37. no solution
39a. $y = \pm 900\sqrt{1 - \dfrac{x^2}{(300)^2}}$
$y = \pm 690\sqrt{1 - \dfrac{x^2}{(600)^2}}$
39b. Sample answer: $(209, 647)$, $(-209, 647)$, $(-209, -647)$, $(209, -647)$

39c. Sample answer: The orbit of the satellite modeled by the second equation is closer to a circle than the other orbit. The distance on the x-axis is twice as great for one satellite as for the other.

41. $y = -0.0037x^2 + 1.77x - 1.72$
$\frac{3}{7}x - 128.6 = -0.0037x^2 + 1.77x - 1.72$
$-128.6 \approx -0.0037x^2 + 1.34x - 1.72$

Path of baseball equation
Substitute $\frac{3}{7}x$ for y.
-128.6 for y.
Substitute $\frac{3}{7}x$ from each side.

$0 \approx -0.0037x^2 + 1.34x + 126.88$ Add 128.6 to each side.

$x = \dfrac{-b \pm \sqrt{b^2 - 4ac}}{2a}$ Quadratic Formula

$x = \dfrac{-1.34 \pm \sqrt{1.34^2 - 4(-0.0037)(126.88)}}{2(-0.0037)}$ $a = -0.0037$, $b = 1.34$, $c = 126.88$

$x \approx \dfrac{-1.34 \pm \sqrt{3.67}}{-0.0074}$ Simplify.

$x \approx -78$ or $x \approx 440$ Use a calculator.

Since the vertical distance cannot be negative, the ball landed about 440 ft from home plate.

$y = \frac{3}{7}x - 128.6$ Original equation
$y = \frac{3}{7}(440) - 128.6$ Replace x with 440.
$y \approx 60$ Use a calculator.

The ball was 60 ft above the playing surface.

43. Sample answer: A circle with the equation $(x + 10)^2 + y^2 = 36$ has its center at $(-10, 0)$ and contains the point $(-4, 0)$. An ellipse with the equation $\frac{x^2}{16} + \frac{y^2}{36} = 1$ has its center at $(0, 0)$ and contains the point $(-4, 0)$. The circle and ellipse intersect only $(-4, 0)$.

45. Sample answer: $x^2 + y^2 = 1$ and $\frac{x^2}{16} - \frac{y^2}{36} = 1$
47. Sample answer: $\frac{x^2}{64} + \frac{y^2}{100} = 1$ and $x^2 - y^2 = 1$
49. Sample answer: No; if one player is in one of the shaded areas and the other player is in the other shaded area, they will not be able to hear each other.

51. $k = a$ or $k = b$ **53.** Sample answer: $\frac{y^2}{128} + \frac{x^2}{32} = 1$ and $\frac{x^2}{8} - \frac{y^2}{64} = 1$ **55.** $(-3, -4)$, $(3, 4)$ **57.** F   **59.** b   **61.** c
63. $(2, -2)$, $(2, 8)$;
$(2, 3 \pm \sqrt{41})$;
$y - 3 = \pm \frac{5}{4}(x - 2)$

65. $\frac{p + 5}{p + 1}$ **67.** $\frac{3(r + 4)}{r + 3}$

69.

$D = \{$all real numbers$\}$,
$R = \{y \mid y < 0\}$
71. $\frac{d}{t} = r$ **73.** $\frac{3V}{\pi r^2} = h$

75. $y = 0.5x + 1$
77. $y = 3x - 6$
79. $y = -\frac{1}{2}x + \frac{7}{2}$

73.

$f(x) = \dfrac{x^2 - 36}{x + 6}$

Lesson 10-2

1. 104 **3.** $a_n = 6n + 7$ **5.** 15, 24, 33 **7.** 1275 **9.** 4500
11. 8, 12, 16 **13.** C **15.** 248 **17.** −103 **19.** 14
21. $a_n = -14n + 45$

23.
$a_n = a_1 + (n - 1)d$ *n*th term of an arithmetic sequence
$21 = a_1 + (7 - 1)5$ $a_1 = 21, n = 7,$ and $d = 5$
$21 = a_1 + 30$ Multiply.
$-9 = a_1$ Subtract 30 from each side.

$a_n = a_1 + (n - 1)d$ *n*th term of an arithmetic sequence
$a_n = -9 + (n - 1)5$ $a_1 = -9$ and $d = 5$
$a_n = -9 + (5n - 5)$ Distributive Property
$a_n = 5n - 14$ Simplify.

25. $a_n = 4.5n - 21$ **27.** $a_n = 9n - 32$ **29.** $a_n = \frac{2}{3}n - 3$
31. $a_n = \frac{1}{2}n - \frac{23}{10}$ **33.** 19, 14, 9, 4 **35.** −21, −14, −7, 0
37. −21, −30, −39, −48, −57 **39.** 10,100 **41.** 10,000
43. 696 **45.** 1272 **47.** $4400

51.
$S_n = n\left(\dfrac{a_1 + a_n}{2}\right)$ Sum Formula
$2982 = 28\left(\dfrac{a_1 + 228}{2}\right)$ $S_n = 2982, n = 28, a_n = 228$
$2982 = 14a_1 + 3192$ Simplify.
$-210 = 14a_1$ Subtract 3192 from each side.
$-15 = a_1$ Divide each side by 14.

$a_n = a_1 + (n - 1)d$ *n*th term of an arithmetic sequence
$228 = -15 + (28 - 1)d$ $a_n = 228, a_1 = -15, n = 28$
$243 = 27d$ Add 15 to each side.
$9 = d$ Divide each side by 27.

$a_1 = -15, a_2 = -15 + 9$ or $-6, a_3 = -6 + 9$ or 3
The first three terms are −15, −6, 3.
53. −44, −30, −16 **55.** −33, −21, −9 **57.** 512 **59.** 324
61. $2250 **63.** $a_n = 13n - 1055$ **65.** $a_n = 9n - 88$
67a. 14, 18, 22

67b. $p_n = 4n + 2$ **67c.** No; there is no whole number *n* for which $4n + 2 = 100$.

69.
$a_n = a_1 + (n - 1)d$ *n*th term of an arithmetic sequence
$100,000 = 28,000 + (n - 1)4000$ $a_n = 100,000, a_1 = 28,000, d = 4000$
$72,000 = (n - 1)4000$ Subtract 28,000 from each side.
$18 = n - 1$ Divide each side by 4000.
$19 = n$ Add 1 to each side.
He will have a salary of $100,000 in the 19th year.

33. no
35. $\dfrac{18}{-27} = \dfrac{2}{3}$ and $\dfrac{-12}{18} = \dfrac{2}{3}$
Since the ratios of the consecutive terms are the same, the sequence is geometric.
37. no **39.** −8, 32, −128
41. 27, $\frac{81}{4}$, $\frac{243}{16}$
43. 27, 81, 243

45. Neither; there is no common difference or ratio.
47. Geometric; the common ratio is 3. **49.** Arithmetic; the common difference is $\frac{1}{2}$. **51.** 86 pg/day **53.** about 13,744 km **55.** Sample answer: A babysitter earns $20 for cleaning the house and $8 extra for every hour she watches the children. **57.** Sample answer: Neither; the sequence is both arithmetic and geometric.
59. Sample answer: If a geometric sequence has a ratio r such that $|r| < 1$, as n increases, the absolute value of the terms will decrease and approach zero because they are continuously being multiplied by a fraction. When $|r| \geq 1$, the absolute value of the terms will increase and approach infinity because they are continuously being multiplied by a value greater than 1. **61.** $421.85
63. H **65.** $(\pm 4, 5)$ **67.** no solution
69. hyperbola; $\dfrac{x^2}{4} - \dfrac{y^2}{1} = 1$

71.

$f(x) = \dfrac{6}{(x - 2)(x + 3)}$

Lesson 10-1

1. yes **3.** no
5. 42, 54, 66, 78 **7.** 5, 13, 21, 29

9a. $850 **9b.** 24 wk **11.** yes **13.** no
15. 128, 256, 512 **17.** $\frac{1}{9}, \frac{1}{27}, \frac{1}{81}$

19. Geometric; the common ratio is $-\frac{1}{2}$. **21.** no **23.** no
25. 8, 11, 14, 17 **27.** −29, −35, −41, −47
29. 2, $\frac{13}{5}$, $\frac{16}{5}$, $\frac{19}{5}$

31. The difference between any two consecutive rows is 2, so the common difference is 2. Use point-slope form to write an equation for the sequence.
Let $m = 2$ and $(x_1, y_1) = (1, 28)$. Solve for $x = 24$.
$(y - y_1) = m(x - x_1)$ Point-slope form
$(y - 28) = m(x - x_1)$ $m = 2, (x_1, y_1) = (1, 28)$
$y - 28 = 2(x - 1)$ Multiply.
$y - 28 = 2x - 2$ Multiply.
$y = 2x + 26$ Add 28 to each side.
$y = 2(24) + 26$ Replace x with 24.
$y = 48 + 26$ or 74   Simplify.
There are 74 seats in the last row of the theater.

---

55. $\dfrac{y^2}{9} - \dfrac{(x - 2)^2}{1} = 1$; hyperbola
57. hyperbola **59.** circle **61.** $(2, -2), (-2, 2)$
63. $(4, 0)$ **65.** $(1, \pm 5), (-1, \pm 5)$
67a. 1.5 seconds
67b. 109 feet

69.

$-9x^2 + y^2 + 36x - 45 = 0$

71.

73.

CHAPTER 10
Sequences and Series

Chapter 10 **Get Ready**

1. $x = -12$ **3.** $x = 3$ **5.** 9 rows
7.

9.
11. −30 **13.** $\dfrac{2}{729}$

Right column (R95)

$$\frac{\frac{21}{1000} + \frac{3}{10}}{1 - \frac{1}{10}}$$ Simplify. $a_1 = \frac{21}{1000}$ and $r = \frac{1}{100}$

$$= \frac{\frac{2100}{99,000} + \frac{3}{10}}{1 - \frac{3}{10}}$$ Simplify.

37. $\frac{24}{11}$ **39.** $\frac{601}{4950}$ $= \frac{\frac{107}{330} + \frac{3}{10}}{\frac{53}{10}}$ or $\frac{53}{165}$ Simplify.

41. $\frac{40}{3}$ **43.** 8000 hrs **45.** $\frac{45}{4}$

47. No sum exists. **49.** $-\frac{54}{35}$ **51.** 200 ft **53.** 1170 ft

55. $S = \frac{a_1}{1-r}$ Sum formula

$= \frac{500}{1 - 0.8}$ $a_1 = 500$ and $r = 0.8$

$= \frac{500}{0.2}$ or 2500 Simplify.

57. b **59.** a **61.** Sample answer: The sum of a geometric series is $S_n = \frac{a_1 - a_1 r^n}{1 - r}$. For an infinite series with $|r| < 1$, $r^n \to 0$ as $n \to \infty$. Thus, $S = \frac{a_1 - a_1(0)}{1 - r}$ or $\frac{a_1}{1 - r}$. **63.** Sample answer: An infinite geometric series has a sum when the common ratio has an absolute value less than 1. When this occurs, the terms will approach 0 as n approaches infinity. With the future terms almost 0, the sum of the series will approach a limit. When the common ratio is 1 or greater, the terms will keep increasing and approach infinity as n approaches infinity and the sum of the series will have no limit. **65.** Sample answer: $3 + 2 + \frac{4}{3} + \dots$. **67.** An arithmetic series has a common difference, so each term will eventually become more positive or more negative, but never approach 0. With the terms not approaching 0, the sum will never reach a limit and the series cannot converge. **69.** H **71.** C **73a.** 9, 18, 27, 36, 45, 54, 63, 72 **73b.** 42 meetings **75.** 12

Lesson 10-5

1. 16, 20, 24, 28, 32 **3.** 5, 17, 53, 161, 485
5. $a_{n+1} = 2a_n + 2$; $a_1 = 3$ **7a.** $a_n = 1.01a_{n-1} - 100$, $a_1 = 1500$ **7b.** $1415, $1329.15, $1242.44, $1154.87
7c. $77.08 **9.** -18, 74, -294 **11.** -52, -420, -3364
13. -9, -10, -12, -16, -24 **15.** -4, -7, -12, -21, -38

17. $a_{n+1} = 5a_n + 2n$ Recursive formula

$a_{n+1} = 5a_1 + 2(1)$ $n = 1$

$a_2 = 5(-2) + 2(1)$ or -8 $a_1 = -2$
$a_3 = 5(-8) + 2(2)$ or -36 $a_2 = -8$, $n = 2$
$a_4 = 5(-36) + 2(3)$ or -174 $a_3 = -36$, $n = 3$
$a_5 = 5(-174) + 2(4)$ or -862 $a_4 = -36$, $n = 4$

The first five terms are -2, -8, -36, -174, and -862.

19. 4, 5, 6, 8, 8 **21.** 3, 2x, 8x $-$ 9, 26x $-$ 36, 80x $-$ 117
23. 1, x, 3x + 6, 15x + 18, 63x + 90 **25.** $a_{n+1} = 1$
27. $a_{n+1} = 0.25a_n + 8$; $a_1 = 480$ **31.** $a_{n+1} = 2a_n - 32$;
$a_1 = 84$ **33.** 56, 680, 8168 **35.** -45, 273, -1635
37. -3, -18, -963 **39.** 43, 3484, 24,259,093 **41.** 4.25, 29.5625, 936.0664

connectED.mcgraw-hill.com

$$\frac{a_n(1 - r^n)}{r^{n-1}}$$

$$= \frac{\frac{1 - r^n}{r^{n-1}(1 - r)}}{= \frac{a_n(1 - r)}{r^{n-1} - r^n}}$$ Simplify.

Divide by $(1 - r)$.

Simplify.

67. Sample answer: $n - 1$ needs to change to n, and the 10 needs to change to a 9. When this happens, the terms for both series will be identical (a_1 in the first series will equal a_0 in the second series, and so on), and the series will be equal to each other. **69.** 234
71. Sample answer: $4 + 8 + 16 + 32 + 64 + 128$ **73.** B
75. $32,000 **77.** $1550 **79.** Arithmetic; the common difference is $\frac{1}{50}$.
81. (3, 1), 5 units

83. $(3, -7)$, $5\sqrt{2}$ units

85. 731 customers **87.** $-\frac{5}{7}$ **89.** $\frac{27}{2}$

Lesson 10-4

1. convergent **3.** divergent **5.** 880 **7.** No sum exists.
9. 12.5% **11.** -4 **13.** 2 **15.** $\frac{214}{333}$ **17.** convergent
19. divergent **21.** divergent **23.** No sum exists.

25. $r = \frac{6}{5} \div \frac{12}{5}$ or $\frac{1}{2}$ Divide consecutive terms.

Since $\left|\frac{1}{2}\right| < 1$, the sum exists.

$S = \frac{a_1}{1 - r}$ Sum formula

$a_1 = \frac{12}{5}$ and $r = \frac{1}{2}$

$= \frac{\frac{12}{5}}{1 - \frac{1}{2}}$

$= \frac{12}{5} \div \frac{1}{2}$ or $\frac{24}{5}$ Simplify.

27. No sum exists. **29.** No sum exists.

35. $0.3\overline{21}$ $0.3 + 0.021 + 0.00021 + \dots$

$= \frac{3}{10} + \frac{21}{1000} + \frac{21}{100,000} + \dots$

$S = \frac{a_1}{1 - r}$ Sum formula

Left column (R94)

71a.

n	S_n
1	4
2	10
3	18
4	28
5	40
6	54
7	70
8	88
9	108
10	130

71b.

71c.

71d. Sample answer: The graphs cover the same range. The domain of the series is the natural numbers, while the domain of the quadratic function is all real numbers, $0 \le x \le 10$.

71e. Sample answer: For every partial sum of an arithmetic series, there is a corresponding quadratic function that shares the same range. **71f.** $\sum_{k=1}^{x} 2k + 7$

73. 16 **75.** $4b - 3a$ **77.** $S_n = nx + y\left(\frac{n^2 + n}{2}\right)$

79. Sample answer: An arithmetic sequence is a list of terms such that any pair of successive terms has a common difference. An arithmetic series is the sum of the terms of an arithmetic sequence.

81. $S_n = (a_1 + a_n) \cdot \left(\frac{n}{2}\right)$ General sum formula

$a_n = a_1 + (n-1)d$ Formula for nth term

$a_n - (n-1)d = a_1$ Subtract $(n-1)d$ from both sides.

$S_n = [a_n - (n-1)d + a_n] \cdot \left(\frac{n}{2}\right)$ Substitution

$S_n = [2a_n - (n-1)d] \cdot \left(\frac{n}{2}\right)$ Simplify.

83. C **85.** A **87.** yes **89.** no

91.

93a. 4.8 cm/g **93b.** 24 cm; The answer is reasonable. The object would stretch the first spring 60 cm and would stretch the second spring 40 cm. The object would have to stretch the combined springs less than it would stretch either of the springs individually.

95.

$D = \{$all real numbers$\}$,
$R = \{f(x) \mid f(x) > 3\}$
97. 2.4550
99. 0.4341

Lesson 10-3

1. 1.2046 **3.** $a_n = 18 \cdot \left(\frac{1}{3}\right)^{n-1}$

5. $a_n = a_1 r^{n-1}$ nth term of a geometric sequence

$4 = a_1(3^{2-1})$ $a_n = 4$, $r = 3$, and $n = 2$

$4 = a_1(3)$ Evaluate the power.

$\frac{4}{3} = a_1$ Divide each side by 3.

$a_n = \frac{4}{3}(3)^{n-1}$ nth term of a geometric sequence

7. $a_n = 12(-8)^{n-1}$ **9.** 1, 5, 25 or -1, 5, -25
11. 4095 **13.** 512 **15.** 512 **17.** 93 in. **19.** 25
21. 512 **23.** $a_n = (-3)(-2)^{n-1}$ **25.** $a_n = (-1)(-1)^{n-1}$
27. $a_n = 8 \cdot \left(\frac{1}{4}\right)^{n-1}$ **29.** $a_n = 7(2)^{n-1}$ **33.** $a_n = 648\left(\frac{1}{3}\right)^{n-1}$
31. $a_n = \frac{1}{15,552}(6)^{n-1}$ **35.** 270, 90, 30 or -270, 90, -30
37. $\frac{7}{3}, \frac{14}{9}, \frac{28}{27}$ or $\frac{7}{3}, -\frac{14}{9}, \frac{28}{27}$ **39.** 15 and 75
41. 99.19% **43.** 31.9375 **45.** 9707.82 **47.** 2188
49. $-87,381$

51. $S_n = \frac{a_1 - a_1 r^n}{1 - r}$ Sum formula

$-2912 = \frac{a_1 - a_1(3^6)}{1 - 3}$ $S_n = -2912$, $r = 3$, and $n = 6$

$-2912 = \frac{a_1(1 - 3^6)}{1 - 3}$ Distributive Property

$-2912 = \frac{-728a_1}{-2}$ Subtract.

$-2912 = 364a_1$ Simplify.

$-8 = a_1$ Divide each side by 364.

53. 64 **55.** 0.25

57. $S_n = \frac{a_1 - a_1 r^n}{1 - r}$ Alternate sum formula

$= \frac{100 - 100(0.5)^5}{1 - (0.5)}$ Substitution

$= 193.75$ ft Use a calculator.

59. 524, 288 **61.** about 471 cm **63a.** $11.79, $30.58, $205.72 **63b.** $7052.15 **63c.** Each payment made is rounded to the nearest penny, so the sum of the payments will actually be more than the sum found in part b.

65. $S_n = \frac{a_1 - a_n r}{1 - r}$ Alternate sum formula

$a_n = a_1 \cdot r^{n-1}$ Formula for nth term

$\frac{a_n}{r^{n-1}} = a_1$ Divide both sides by r^{n-1}.

$S_n = \frac{\frac{a_n}{r^{n-1}} - a_n r}{1 - r}$ Substitution.

$S_n = \frac{\frac{a_n}{r^{n-1}} - \frac{a_n r}{r^{n-1}}}{\frac{r^{n-1}}{r^{n-1}} - \frac{1}{r^{n-1}}}$ Multiply by $\frac{r^{n-1}}{1}$.

connectED.mcgraw-hill.com

$$= \frac{k(k+1)(k+2)}{6} + \frac{(k+1)(k+1+1)}{2}$$

$$= \frac{k(k+1)(k+2)}{6} + \frac{3(k+1)(k+2)}{6}$$

$$= \frac{(k+1)(k+2)(k+3)}{6}$$

$$= \frac{(k+1)[(k+1)+1][(k+1)+2]}{6}$$

The last expression is the right side of the equation to be proved, where $n = k + 1$. Thus, the equation is true for $n = k + 1$. Therefore, $1 + 3 + 6 + \cdots + \frac{n(n+1)}{2} = \frac{n(n+1)(n+2)}{6}$ for all natural numbers n. **5.** Step 1: $4^1 - 1 = 3$, which is divisible by 3. The statement is true for $n = 1$.
Step 2: Assume that $4^k - 1$ is divisible by 3 for some whole number r.
Step 3: $4^k - 1 = 3r$

$$4^k = 3r + 1$$
$$4 \cdot 4^k = 4(3r + 1)$$
$$4^{k+1} = 12r + 4$$
$$4^{k+1} - 1 = 12r + 3$$
$$4^{k+1} - 1 = 3(4r + 1)$$

Since r is a whole number, $4r + 1$ is a whole number. Thus, $4^{k+1} - 1$ is divisible by 3, so the statement is true for $n = k + 1$. Therefore, $4^n - 1$ is divisible by 3 for all natural numbers n. **7.** $n = 1$ **9.** Step 1: When $n = 1$, the left side of the given equation is 2. The right side is $\frac{1[3(1) + 1]}{2}$ or 2, so the equation is true for $n = 1$.
Step 2: Assume that $2 + 5 + 8 + \cdots + (3k - 1) = \frac{k(3k+1)}{2}$ for some natural number k.
Step 3: $2 + 5 + 8 + \cdots + (3k - 1) + [3(k + 1) - 1]$

$$= \frac{k(3k+1)}{2} + [3(k+1) - 1]$$
$$= \frac{k(3k+1) + 2[3(k+1) - 1]}{2}$$
$$= \frac{3k^2 + k + 6k + 6 - 2}{2}$$
$$= \frac{3k^2 + 7k + 4}{2}$$
$$= \frac{(k+1)(3k+4)}{2}$$
$$= \frac{(k+1)[3(k+1)+1]}{2}$$

The last expression is the right side of the equation to be proved, where $n = k + 1$. Thus, the equation is true for $n = k + 1$. Therefore, $2 + 5 + 8 + \cdots + (3n - 1) = \frac{n(3n+1)}{2}$ for all natural numbers n. **11.** Step 1: When $n = 1$, the left side of the given equation is 1. The right side is $1[2(1) - 1]$ or 1, so the equation is true for $n = 1$.
Step 2: Assume that $1 + 5 + 9 + \cdots + (4k - 3) = k[2k - 1]$ for some natural number k.
Step 3: $1 + 5 + 9 + \cdots + (4k - 3) + [4(k + 1) - 3]$

$$= k[2k - 1] + [4(k + 1) - 3]$$
$$= 2k^2 - k + 4k + 4 - 3$$
$$= 2k^2 + 3k + 1$$
$$= (k + 1)(2k + 1)$$
$$= (k + 1)[2(k + 1) - 1]$$

13. Step 1: When $n = 1$, the left side of the equation is $4(1) - 1$ or 3. The right side is $2(1)^2 + 1$ or 3, so the equation is true for $n = 1$.
Step 2: Assume that $3 + 7 + 11 + \cdots + (4k - 1) = 2k^2 + k$ for some natural number k.
Step 3: $3 + 7 + 11 + \cdots + (4k - 1) + [4(k + 1) - 1]$

$$= 2k^2 + k + [4(k + 1) - 1]$$
$$= 2k^2 + k + 4k + 4 - 1$$
$$= 2k^2 + 5k + 3$$
$$= 2k^2 + 4k + 2 + k + 1$$
$$= [2(k+1)^2] + (k+1)$$

The last expression is the right side of the equation to be proved, where $n = k + 1$. Thus, the equation is true for $n = k + 1$. Therefore, $3 + 7 + 11 + \cdots + (4n - 3) = 2n^2 + n$ for all natural numbers n. **15.** Step 1: When $n = 1$, the left side of the given equation is 1^2 or 1. The right side is $\frac{1[2(1) - 1][2(1) + 1]}{3}$ or 1, so the equation is true for $n = 1$.
Step 2: Assume that $1^2 + 3^2 + 5^2 + \cdots + (2k - 1)^2 = \frac{k[2(k) - 1][2(k) + 1]}{3}$ for some natural number k.
Step 3: $1^2 + 3^2 + 5^2 + \cdots + (2k - 1)^2 + [2(k + 1) - 1]^2$

$$= \frac{k(2k-1)(2k+1)}{3} + [2(k+1) - 1]^2$$
$$= \frac{k(2k-1)(2k+1) + 3(2k+1)^2}{3}$$
$$= \frac{(2k+1)[k(2k-1) + 3(2k+1)]}{3}$$
$$= \frac{(2k+1)(2k^2 + 5k + 3)}{3}$$
$$= \frac{(2k+1)(k+1)(2k+3)}{3}$$
$$= \frac{(k+1)(2k+1)(2k^2 + 6k + 3)}{3}$$
$$= \frac{(k+1)[2(k+1) - 1][2(k+1) + 1]}{3}$$

17. Step 1: $5^1 + 3 = 8$, which is divisible by 4. The statement is true for $n = 1$.
Step 2: Assume $5^k + 3$ is divisible by 4 for some natural number k. This means that $5^k + 3 = 4r$ for some natural number r.
Step 3: $5^k + 3 = 4r$

$$5^k = 4r - 3$$
$$5^{k+1} = 20r - 15$$
$$5^{k+1} + 3 = 20r - 12$$
$$5^{k+1} + 3 = 4(5r - 3)$$

Since r is a natural number, $5r - 3$ is a natural number. Thus, $5^k + 3$ is divisible by 4, so the statement is true for $n = k + 1$. Therefore, $5^n + 3$ is divisible by 4 for all natural numbers n.

---

43. **a.** The number of blue triangles is 1, 3, and 9. The sequence is geometric because each term after the first can be found after multiplying by a common ratio, 3.

$a_n = r \cdot a_{n-1}$ Recursive Formula for geometric sequence
$a_n = 3a_{n-1}, a_1 = 1$ $r = 3$
b. $a_1 = 1, a_2 = 3, a_3 = 9$
$a_n = 3a_{n-1}$ Recursive formula
$a_4 = 3a_3$ $n = 4$
$\quad\;\; = 3(9)$ or 27 $a_3 = 9$
$a_5 = 3a_4$ $n = 5$
$\quad\;\; = 3(27)$ or 81 $a_4 = 27$
$a_6 = 3a_5$ $n = 6$
$\quad\;\; = 3(81)$ or 243 $a_5 = 27$

There will be 243 blue triangles in the sixth figure. **45.** No; the population of fish will reach 12,500. Each year, 20% of 12,500 or 2500 fish plus 10,000 additional fish yields 12,500 fish. **47a.** 11,000 **47b.** It converges to 7142.857. **47c.** Sample answer: They make it easier to analyze recursive sequences because they can produce the first 100 terms instantaneously; it would take a long time to calculate the terms by hand. **49.** Armando; Marcus included x_0 with the iterates and only showed the first 2 iterates. **51.** Sample answer: Sometimes; the recursive formula could involve the first three terms. For example, 2, 2, 2, 8, 20, … is recursive with $a_{n+3} = a_n + a_{n+1} + 2a_{n+2}$. **53.** Sample answer: In a recursive sequence, each term is determined by one or more of the previous terms. A recursive formula is used to produce the terms of the recursive sequence. **55a.** 160 m **55b.** 5.7 s **55c.** 11.4 s **57.** C **59.** $5\frac{14}{111}$ **61a.** 2, 3, 4.5, 6.75, 10.125 **61b.** the eighth session **61c.** during the ninth session **63.** dependent **65.** $x^2 + 4x - 12$ **67.** $4t^2 + 33t + 35$ **69.** $10g^2 + 19g - 56$

Lesson 10-6

1. $c^5 + 5c^4d + 10c^3d^2 + 10c^2d^3 + 5cd^4 + d^5$ **3.** $x^6 - 24x^5 + 240x^4 - 1280x^3 + 3840x^2 - 6144x + 4096$ **5.** $x^5 + 15x^4 + 90x^3 + 270x^2 + 405x + 243$ **7.** $\frac{3}{32}$ or 0.09375 **9.** $(x + 3y)^8 = \sum_{k=0}^{8} \frac{8!}{k!(8-k)!} x^{8-k} (3y)^k$

$$\frac{8!}{k!(8-k)!} x^{8-k} (3y)^k = \frac{8!}{4!(8-4)!} x^{8-4} (3y)^4$$ For the fifth term, $k = 4$.
$$= 70x^4(81y^4)$$ $(3y)^4 = 81y^4$
$$= 5670x^4y^4$$ Simplify.

11. $-108,864c^3d^5$ **13.** $243a^5$ **15.** $a^6 - 6a^5b + 15a^4b^2 - 20a^3b^3 + 15a^2b^4 - 6ab^5 + b^6$ **17.** $x^6 + 36x^5 + 540x^4 + 4320x^3 + 19,440x^2 + 46,656x + 46,656$ **19.** $16a^4 + 128a^3b + 384a^2b^2 + 512ab^3 + 256b^4$ **21.** Let w represent the number of women and m represent the number of men.

$$(w + m)^{10} = \sum_{k=0}^{10} \frac{10!}{k!(10-k)!} w^{10-k} m^k$$

To find the probability that 7 members are women, find the term in which the exponent of w is 7. Since $10 - 3 = 7$, find the term in which $k = 3$, the fourth term.

$$\frac{10!}{k!(10-k)!} w^{10-k} m^k = \frac{10!}{3!(10-3)!} w^{10-3} m^3$$ For the fourth term, $k = 3$.
$$= 120w^7m^3$$ $C(10,3) = 120$

The probability of choosing a woman is $\frac{1}{2}$ and the probability of choosing a man is $\frac{1}{2}$.

$$120w^7m^3 = 120\left(\frac{1}{2}\right)^7\left(\frac{1}{2}\right)^3$$ $w = \frac{1}{2}$ and $m = \frac{1}{2}$
$$= 120\left(\frac{1}{2}\right)^{10}$$ Product of Powers Property
$$= \frac{120}{1024}$$ $\left(\frac{1}{2}\right)^{10} = \frac{1}{1024}$
$$= \frac{15}{128}$$ Simplify.

The probability that 7 of the members will be women is $\frac{15}{128}$, or about 0.117. **23.** $84x^2z^2$ **25.** $5,7168a^2b^6$ **27.** $32,256x^5$ **29.** $x^5 + \frac{5}{2}x^4 + \frac{5}{2}x^3 + \frac{5}{4}x^2 + \frac{5}{16}x + \frac{1}{32}$ **31.** $32b^5 + 200b^4 + 50b^3 + \frac{5}{8}b^2 + \frac{5}{128}b + \frac{1}{1024}$ **33a.** 0.121 **33b.** 0.121 **33c.** 0.309 **35.** Sample answer: While they have the same terms, the signs for $(x + y)^n$ will all be positive, while the signs for $(x - y)^n$ will alternate. **37.** Sample answer: $\left(x + \frac{6}{5}y\right)^5$ **39.** A **41.** G **43.** −2, 3, 8, 13, 18 **45.** 4, 6, 12, 30, 84 **47.** $1\frac{1}{8}$ **49a.** $\frac{150}{x}$, $\frac{150}{x-10}$ **49b.** $\frac{150}{x} = \frac{150}{x-10}$ **51.** true; 3(1) + 5 = 8, which is even

Lesson 10-7

1. Step 1: When $n = 1$, the left side of the given equation is 1. The right side is 1^2 or 1, so the equation is true for $n = 1$.
Step 2: Assume that $1 + 3 + 5 + \cdots + (2k - 1) = k^2$ for some natural number k.
Step 3: $1 + 3 + 5 + \cdots + (2k - 1) + (2(k + 1) - 1)$

$$= k^2 + (2(k + 1) - 1)$$
$$= k^2 + (2k + 2 - 1)$$
$$= k^2 + 2k + 1$$
$$= (k + 1)^2$$

The last expression is the right side of the equation to be proved, where $n = k + 1$. Thus, the equation is true for $n = k + 1$. Therefore, $1 + 3 + 5 + \cdots + (2n - 1) = n^2$ for all natural numbers n. **3a.** 3, 6, 10, 15, 21 **3b.** $a_n = \frac{n(n + 1)}{2}$ **3c.** Step 1: When $n = 1$, the left side of the given equation is $\frac{1(1 + 1)}{2}$ or 1. The right side is $\frac{1(1 + 1)(1 + 2)}{6} = 1$, so the equation is true for $n = 1$.
Step 2: Assume that $1 + 3 + 6 + \cdots + \frac{k(k + 1)}{2} = \frac{k(k + 1)(k + 2)}{6}$ for some natural number k.
Step 3: $1 + 3 + 6 + \cdots + \frac{k(k + 1)}{2} + \frac{(k + 1)(k + 1 + 1)}{2}$

19. Step 1: $12^1 + 10 = 22$, which is divisible by 11. The statement is true for $n = 1$. Step 2: Assume that $12^k + 10$ is divisible by 11 for some natural number k. This means that $12^k + 10 = 11r$ for some natural number r. Step 3: $12^k + 10 = 11r$
$12^k = 11r - 10$
$12^{k+1} = 132r - 120$
$12^{k+1} + 10 = 132r - 110$
$12^{k+1} + 10 = 11(12r - 10)$
Since r is a natural number, $12r - 10$ is a natural number. Thus, $12^{k+1} + 10$ is divisible by 11, so the statement is true for $n = k + 1$. Therefore, $12^n + 10$ is divisible by 11 for all natural numbers n. **21.** $n = 2$ **23.** $n = 1$

25. In the sequence $1, 2, 3, 5, 8, ..., f_1 = 1, f_2 = 1,$
$f_3 = 2, f_4 = 3, f_5 = 5, f_6 = 8, ...$
$f_1 + f_2 + ... + f_n = f_{n+2} - 1$ Original equation
$f_1 = f_{1+2} - 1$ Let $n = 1$.
$f_1 = f_3 - 1$ Simplify.
Step 1: When $n = 1$, the left side of the given equation is f_1. The right side is $f_3 - 1$. Since $f_1 = 1$ and $f_3 = 2$, the equation becomes $1 = 2 - 1$ and is true for $n = 1$.
Step 2: Assume that $f_1 + f_2 + ... + f_k = f_{k+2} - 1$ for some natural number k.
Step 3: $f_1 + f_2 + ... + f_k + f_{k+1} = f_{k+2} - 1 + f_{k+1}$
$= f_{k+1} + f_{k+2} - 1$
$= f_{k+3} - 1$, since
Fibonacci numbers are produced by adding the two previous Fibonacci numbers.
The last expression is the right side of the equation to be proved, where $n = k + 1$. Thus, the equation is true for $n = k + 1$ for all natural numbers n.

27. Step 1: $18^1 - 1 = 17$, which is divisible by 17. The statement is true for $n = 1$.
Step 2: Assume that $18^k - 1$ is divisible by 17 for some natural number k. This means that $18^k - 1 = 17r$ for some natural number r.
Step 3: $18^k - 1 = 17r$
$18^k = 17r + 1$
$18^{k+1} = 18(17r + 1)$
$18^{k+1} - 1 = 306r + 18$
$18^{k+1} - 1 = 306r + 17$
$18^{k+1} - 1 = 17(18r + 1)$
Since r is a natural number, $18r + 1$ is a natural number. Thus, $18^{k+1} - 1$ is divisible by 17, so the statement is true for $n = k + 1$. Therefore, $18^n - 1$ is divisible by 17 for all natural numbers n. **29.** $n = 3$ **31.** Step 1: When $n = 1$, the left side of the given equation is $\frac{1}{1(1+2)}$ or $\frac{1}{6}$. The right side is $\frac{1(1+3)}{4(1+1)(1+2)}$ or $\frac{1}{6}$, so the equation is true for $n = 1$.
Step 2: Assume that $\frac{1}{1 \cdot 2 \cdot 3} + \frac{1}{2 \cdot 3 \cdot 4} + \frac{1}{3 \cdot 4 \cdot 5} + ... + \frac{1}{k(k+1)(k+2)} = \frac{k(k+3)}{4(k+1)(k+2)}$ for some natural number k.

Step 3: $\frac{1}{1 \cdot 2 \cdot 3} + \frac{1}{2 \cdot 3 \cdot 4} + ... + \frac{1}{k(k+1)(k+2)} + \frac{1}{(k+1)(k+2)(k+3)}$
$= \frac{k(k+3)}{4(k+1)(k+2)} + \frac{1}{(k+1)(k+2)(k+3)}$
$= \frac{k(k+3)^2}{4(k+1)(k+2)(k+3)} + \frac{4}{4(k+1)(k+2)(k+3)}$
$= \frac{k^3 + 6k^2 + 9k + 4}{4(k+1)(k+2)(k+3)}$
$= \frac{(k+1)(k+1)(k+4)}{4(k+1)(k+2)(k+3)}$
$= \frac{(k+1)(k^2 + 5k + 4)}{4(k+1)(k+2)(k+3)}$
$= \frac{(k+1)(k+4)}{4(k+2)(k+3)}$
$= \frac{(k+1)[(k+1)+3]}{4[(k+1)+1][(k+1)+2]}$
The last expression is the right side of the equation to be proved, where $n = k + 1$. Therefore, the equation is true for $n = k + 1$.

61. $193,536x^2y^5$

63. Step 1: When $n = 1$, the left side of the equation is equal to 2. The right side of the equation is also equal to 2. So the equation is true for $n = 1$.
Step 2: Assume that $2 + 6 + 12 + ... + k(k + 1) = \frac{k(k+1)(k+2)}{3}$ for some positive integer k.
Step 3: $1 \cdot 2 + 2 \cdot 3 + ... + k(k + 1) + (k + 1)(k + 2)$
$= \frac{k(k+1)(k+2)}{3} + (k+1)(k+2)$
$= \frac{k(k+1)(k+2)}{3} + \frac{3(k+1)(k+2)}{3}$
$= \frac{(k+1)(k+2) + 3(k+2)}{3}$
$= \frac{(k+1)(k+2)[k+3]}{3}$
$= \frac{(k+1)(k+2)(k+3)}{3} = \frac{(k+1)[(k+1)+1][(k+1)+2]}{3}$
The last expression is the right side of the equation to be proved, where $n = k + 1$. Thus, the equation is true for $n = k + 1$.
Therefore, $2 + 6 + 12 + ... + n(n + 1) = \frac{n(n+1)(n+2)}{3}$ for all positive integers n. **33.** $n(n + 1)$

65. Step 1: When $n = 1, 5^1 - 1 = 5$ or 4. Since 4 divided by 4 is 1, the statement is true for $n = 1$.
Step 2: Assume that $5^k - 1$ is divisible by 4 for some positive integer k. This means that $5^k - 1 = 4r$ for some whole number r.
Step 3: $5^k - 1 = 4r$
$5^k = 4r + 1$
$5^{k+1} = 20r + 5$
$5^{k+1} - 1 = 20r + 5 - 1$
$5^{k+1} - 1 = 20r + 4$
$5^{k+1} - 1 = 4(5r + 1)$
Since r is a whole number, $5r + 1$ is a whole number. Thus, $5^{k+1} - 1$ is divisible by 4, so the statement is true for $n = 1$.
Therefore, $5^n - 1$ is divisible by 4 for all positive integers n. **67.** $n = 2$ **69.** $n = 1$

Step 3: $\frac{1}{1 \cdot 2 \cdot 3} + \frac{1}{2 \cdot 3 \cdot 4} + ... + \frac{1}{k(k+1)(k+2)} + \frac{1}{(k+1)(k+2)(k+3)}$
$= \frac{k(k+3)}{4(k+1)(k+2)} + \frac{1}{(k+1)(k+2)(k+3)}$
$= \frac{4(k+1)(k+2)(k+3)}{k(k+3)}$ ... [partial]
$= \frac{4(k+1)(k+2)(k+3)}{k^3 + 6k^2 + 9k + 4}$
$= \frac{4(k+1)(k+2)(k+3)}{(k+1)^2(k+4)}$
$= \frac{4(k+1)(k+2)(k+3)}{(k+1)(k^2 + 5k + 4)}$
$= \frac{4(k+1)(k+3)}{(k+1)(k+4)}$
$= \frac{4(k+2)(k+3)}{(k+1)[(k+1)+2]}$
The last expression is the right side of the equation to be proved, where $n = k + 1$. Thus, the equation is true for $n = k + 1$. Therefore, $\frac{1}{1 \cdot 2 \cdot 3} + \frac{1}{2 \cdot 3 \cdot 4} + ... + \frac{1}{n(n+1)(n+2)} = \frac{n(n+1)(n+2)}{4(n+1)(n+2)}$ for all natural numbers n.

35. Sample answer: False; assume $k = 2$, just because a statement is true for $n = 1, 2$ and $n = 3$ does not mean that it is true for $n = 1$. **37.** $x = 3$ **39.** Sample answer: When dominoes are set up, after the first domino falls, the rest will fall as well. With induction, once it is proved that the statement is true for $n = 1$ (the first domino), $n = k$ (the second domino), and $n = k + 1$ (the next domino), it will be true for any integer value (any domino).

41. B **43.** 96 **45.** $160x^3y^3$ **47.** $-84x^6y^3$

49. $(6, -8), (12, -16)$ **51.** 56 **53.** 665,280 **55.** 70

57. 132 **59.** 28 **61.** 24

Chapter 10 | Study Guide and Review

1. true **3.** true **5.** true **7.** false, arithmetic sequence **9.** false, arithmetic means **11.** 48 **13.** -22 **15.** $-7,$ $-2, 3$ **17.** $8, 4, 0, -4$ **19.** 5480 **21.** 1040 **23.** -245 **25.** 629 **27.** -99 **29.** 99 **31.** $\frac{2187}{8}$ **33.** $\pm 24, 72, \pm 216$ **35.** \$1823.26 **37.** 12,285 **39.** 363 **41.** $\frac{6305}{45,927}$ **43.** 32 **45.** 6 **47.** $-3, 1, 5, 9, 13$ **49.** $1, 6, 11, 16, 21$ **51.** 7, 15, 31 **53.** 11, 65, 389 **55.** $a^3 + 3a^2b + 3ab^2 + b^3$ **57.** $-32a^5 + 240a^4 - 720a^3 + 1080a^2 - 810a + 243$ **59.** $x^5 - \frac{5}{4}x^4 + \frac{5}{8}x^3 - \frac{5}{32}x^2 + \frac{5}{256}x - \frac{1}{1024}$

---

how many people in the U.S. are interested in purchasing a hybrid vehicle; population: the people surveyed; sample survey questions: Do you currently own a hybrid vehicle? Are you planning on purchasing a hybrid vehicle? **9.** objective: to determine whether the protein shake helps athletes recover from exercise; population: all athletes; experiment group: athletes given the protein shake; control group: athletes given a placebo; sample procedure: The researchers could randomly divide the athletes into two groups: an experimental group given the protein shake and a control group given the placebo. Next, they could have the athletes exercise and then drink the protein shake or placebo. Later, the researchers could interview the athletes to see how they feel.

11. This is an observational study because the study group is going to observe the students' performance without directly affecting the students. The sample is the 80 physics students because they are the ones being observed. Both halves of the selected students are included because one half is going to be compared to the other half. The population is all college students that take a physics course.

13. survey; sample: people that receive the questionnaire; population: all viewers **15.** Survey; sample answer: The data will be obtained from opinions given by members of the sample population. **17.** Experiment; sample answer: Metal samples will need to be tested, which means that the members of the sample will be affected by the study. **19.** Biased; sample answer: The question only gives two options, and thus encourages a certain response. **21.** Biased; sample answer: The question encourages a certain response. The phrase "don't you agree" suggests that the students should agree. **23.** Sample answer: The flaw is that the experimental group consists of stores in the midwest, and the control group consists of stores on the west coast. On average, the temperature is higher on the west coast than in the midwest, and people use more sunscreen. Therefore, the sunscreen sales in stores located in those regions would likely be different and should not be compared in an experiment.

25. a. The sample group is the 8- to 18-year-olds who were actually surveyed. The population is represented by the sample, so the population is all 8- to 18-year-olds in the U.S.
b. average time
c. Interpret the bar graph for each group. The red bar represents talking and the blue bar represents texting. Sample answer: The 8- to 10-year-old group talked for about 10 minutes a day and did not text at all. The 11- to 14-year-old group talked for about 30 minutes a day and texted for about 70 minutes a day. The 15- to 18-year-old group talked for about...

CHAPTER 11
Statistics and Probability

Chapter 11 | Get Ready

1. 89 customers, 88 customers, no mode **3.** 7.7 touchdowns, 8 touchdowns, 10 touchdowns **5.** $\frac{5}{4}$ **7.** $a^4 - 8a^3 + 24a^2 - 32a + 16$ **9.** $16a^4 - 32b^3 + 240b^2 + x^4$ **11.** $243z^5 - 810x^4y + 1080x^3y^2 - 720x^2y^3 + 240xy^4 - 32y^5$ **13.** $\frac{a^5}{32} + \frac{5a^4}{8} + 5a^3 + 20a^2 + 40a + 32$

Lesson 11-1

1. survey; sample: the students in the study; population: the student body **3.** Observation study; sample answer: The scores of the participants are observed and compared without them being affected by the study. **5.** unbiased **7.** objective: to determine talked for about 40 minutes a day and texted for about 110 minutes a day.

d. Sample answer: A cell phone company might use a report like this to determine which age group to target in their ads.

27a. See students' work.
27b. Sample answer for Product A: ≈63.3%

Product A

Number	Frequency
0–6	卌 卌 卌 IIII
7–9	卌 I

Sample answer for Product B: ≈76.7%

Product B

Number	Frequency
0–7	卌 卌 卌 III
8–9	卌 II

27c. Sample answer: Yes; the probability that Product B is effective is 13.4% higher than that of Product A. **27d.** Sample answer: It depends on what the product is and how it is being used. For example, if the product is a pencil sharpener, then the lower price may be more important than the effectiveness, and therefore, might not justify the price difference. However, if the product is a life-saving medicine, the effectiveness may be more important than the price, and therefore, might justify the price difference. **29.** true

31 An invalid sampling method and type of sample can produce bias. For example, if a sample is not random, the person conducting the study can influence the results by selecting a specific sample of people. Also, if an experiment is used when an observational study is the more logical type of study to be used, the study can be unreliable. For example, if someone wants to analyze the speeds of vehicles on a specific stretch of highway and decides to place an empty police car on the side of the road, the data will be affected by the police car. The results of this study will show lower speeds than are normally driven on the highway. Biased survey questions and incorrect procedures can affect the reliability of a study as well. A survey question that is poorly written may result in a response that does not accurately reflect the opinion of the participant.

33. C **35.** G **37.** Step 1: $9^1 − 1 = 8$, which is divisible by 8. The statement is true for $n = 1$.
Step 2: Assume that $9^k − 1$ is divisible by 8 for some positive integer k. This means that $9^k − 1 = 8r$ for some whole number r.
Step 3: $9^k − 1 = 8r$
$9^k = 8r + 1$
$9^{k+1} − 1 = 72r + 9$
$9^{k+1} − 1 = 72r + 8$
$9^{k+1} − 1 = 8(9r + 1)$

Since r is a whole number, $9r + 1$ is a whole number. Thus, $9^{k+1} − 1$ is divisible by 8, so the statement is true for $n = k + 1$. Therefore, $9^n − 1$ is divisible by 8 for all positive integers n. **39.** $\left(\frac{3}{2}, \frac{9}{2}\right)$ **41.** $(−1, 2)$
41. no solution **43.** $(±8, 0)$ **45.** $3\sqrt{17}$ units
47. 25 units **49.** $\sqrt{70.25}$ units **51.** $−5c^5 b^3$ **53.** an
55. $−y^3 z^2$ **57.** $x^2 − 6x − 27 = 0$ **59.** $x^2 + x − 20 = 0$
61. 63

Lesson 11-2

1 a. First, press STAT ENTER and enter each data value. Then, press 2nd [STAT PLOT] ENTER ENTER and choose ▣. Finally, adjust the window to dimensions appropriate for the data.

[4, 32] scl: 4 by [0, 8] scl: 1

Since the majority of the data is on the left, and there is a tail on the right, the distribution is negatively skewed.
b. Sample answer: The distribution is skewed, so use the five-number summary. Press STAT ▲ ENTER ENTER and scroll down to display the statistics for the data set.

```
1-Var Stats
 n=24
 minX=7
 Q1=15.5
 Med=22.5
 Q3=26
 maxX=30
```

The range is 7 to 30 minutes. The median is 22.5 minutes, and half of the data are between 15.5 and 26 minutes.

3a.

Mr. Edmunds' Class
[5, 40] scl: 5 by [0, 8] scl: 1

Mrs. Johnson's Class
[5, 40] scl: 5 by [0, 8] scl: 1

[5, 40] scl: 5 by [0, 8] scl: 1

Mrs. Johnson's class, positively skewed; Mr. Edmunds' class, negatively skewed. **3b.** Sample answer: The distributions are skewed, so use the five-number summaries. The range for both classes is the same. However, the median for Mrs. Johnson's class is 17 and the median for Mr. Edmunds' class is 28. The lower quartile for Mr. Edmunds' class is 20. Since this is greater than the median for Mrs. Johnson's class,

this means that 75% of the data from Mr. Edmunds' class is greater than 50% of the data from Mrs. Johnson's class. Therefore, we can conclude that the students in Mr. Edmunds' class had slightly higher sales overall than the students in Mrs. Johnson's class.

5a.

[50, 200] scl: 25 by [0, 8] scl: 1
[50, 200] scl: 25 by [0, 5] scl: 1

negatively skewed
5b. Sample answer: The distribution is skewed, so use the five-number summary. The range is 53 to 179 points. The median is 138.5 points, and half of the data are between 106.5 and 157 points.

7a. Sophomore Year

[1200, 1900] scl: 100 by [0, 8] scl: 1

Junior Year

[1300, 2200] scl: 100 by [0, 8] scl: 1

both symmetric
7b. Sample answer: The distributions are symmetric, so use the means and standard deviations. The mean score for sophomore year is about 1552.9 with standard deviation of about 147.2. The mean score for junior year is about 1753.8 with standard deviation of about 159.1. We can conclude that the scores and the variation of the scores from the mean both increased from sophomore year to junior year.

9 a. Enter the tuitions for the public colleges as L1. Graph these data as Plot1 by pressing 2nd [STAT PLOT] ENTER ENTER and choosing ▣. Enter the tuitions for the private colleges as L2. Graph these data as Plot2 by pressing 2nd [STAT PLOT] ► ENTER ENTER and choosing ▣. For Xlist, enter L2. Adjust the window to dimensions appropriate for the data.

[3000, 18,000] scl: 1000 by [0, 5] scl: 1

For both sets of data, the whiskers are approximately equal, and the median is in the middle of the data. The distributions are symmetric.
b. Sample answer: The distributions are symmetric, so use the means and standard deviations. Press STAT ▲ ENTER ENTER to display the statistics for the public colleges.

```
1-Var Stats
 x̄=4037.5
 Σx=72675
 Σx²=300000777
 Sx=621.9260643
 σx=604.4034938
↓n=18
```

Press STAT ▲ ENTER 2nd [L1] ENTER to display the statistics for the private colleges

```
1-Var Stats
 x̄=12803.11111
 Σx=230456
 Σx²=3095025302
 Sx=2915.198879
 σx=2833.063985
↓n=18
```

The mean for the public colleges is $4037.50 with standard deviation of about $621.93. The mean for private colleges is about $12,803.11 with standard deviation of about $2915.20. We can conclude that not only is the average cost of private schools far greater than the average cost of public schools, but the variation of the costs from the mean is also much greater.

11a.

[0, 30] scl: 3 by [0, 5] scl: 1

Sample answer: The distribution is symmetric, so use the mean and standard deviation. The mean of the data is 18 with standard deviation of about 5.2 points.
11b.

[0, 30] scl: 3 by [0, 5] scl: 1

mean: 14.6;
median: 17

11c. Sample answer: Adding the scores from the first four games causes the shape of the distribution to go from being symmetric to being negatively skewed. Therefore, the center and spread should be described using the five-number summary.

13 a. Sample answer: Since the distribution is positively skewed, the median will be to the left of the mean closer to the majority of the data. An estimate for the median is 10. The mean will be more affected by the tail, and will be to the right of the majority of the data. An estimate for the mean is 14.

b. Sample answer: Since the distribution is negatively skewed, the median will be to the right of the mean closer to the majority of the data. An estimate for the median is 24. The mean will be more affected by the tail, and will be to the left of the majority of the data. An estimate for the mean is 20.

c. Sample answer: Since the distribution is symmetric, the mean and median will be approximately equal near the middle of the data. An estimate for the mean and median is 17.

15. Sample answer: The heights of the players on the Pittsburgh Steelers roster appear to represent a normal distribution.

Heights of the Players on the 2009 Pittsburgh Steelers Roster (inches)

75	74	71	70	74	75	77	72
71	72	70	70	75	78	71	75
77	71	69	70	77	75	74	72	75
77	71	73	76	71	74	72	73
75	70	70	74	77	76	79	77
71	69	70	70	77	80	75	77
67	74	69	76	76			

The mean of the data is about 73.61 in. or 6 ft 1.61 in. The standard deviation is about 2.97 in.

[66, 82] scl: 2 by [0, 15] scl: 3

The birth months of the players do not display central tendency.

Birth Months of the Players on the 2009 Pittsburgh Steelers Roster

1	12	10	3	11	1	10	5	
4	8	9	11	1	1	11	5	
8	6	11	4	3	4	3	5	
3	7	2	11	4	3	2	9	
1	1	6	1	6	8	11	9	
3	3	1	6	9	1	4	9	
6	5	10	11	11	12			

[0, 12] scl: 2 by [0, 15] scl: 3

17. D **19.** H **21.** unbiased **23.** Biased; sample answer: The question encourages a certain response. The phrase "Don't you hate" encourages you to agree that gas prices are too high. **25a.** (39.2, ±4.4) **25b.** No; the comet and Pluto may not be at either point of intersection at the same time. **25c.** $\left(-\frac{5}{3}, -\frac{7}{3}\right)$ (1, 3) **25d.** (3, ±4) (−3, ±4) **27.** combination; 28 **29.** permutation; 120

Lesson 11-3

1. The random variable X is the number of pages linked to a Web page. The pages are finite and countable, so X is discrete. **3.** The random variable X is the amount of precipitation in a city per month. Precipitation can be anywhere within a certain range. Therefore, X is continuous.

5a.

Sum	Frequency	Relative Frequency
4	1	$\frac{1}{64}$
6	2	$\frac{1}{32}$
7	2	$\frac{1}{32}$
8	3	$\frac{3}{64}$
9	4	$\frac{1}{16}$
10	5	$\frac{5}{64}$
11	4	$\frac{1}{16}$
12	7	$\frac{7}{64}$
13	4	$\frac{1}{16}$
14	7	$\frac{7}{64}$
15	4	$\frac{1}{16}$
16	5	$\frac{5}{64}$
17	4	$\frac{1}{16}$
18	4	$\frac{1}{16}$
19	2	$\frac{1}{32}$
20	3	$\frac{3}{64}$
22	2	$\frac{1}{32}$
24	1	$\frac{1}{64}$

5b.

Sum of Two Spins

Probability vs Sum (4 6 7 8 9 10 11 12 13 14 15 16 17 18 19 20 22 24)

5c.

Sum	Frequency	Relative Frequency
4	1	$\frac{1}{64}$
6	1	$\frac{1}{64}$
7	3	$\frac{3}{64}$
8	3	$\frac{3}{64}$
9	2	$\frac{1}{32}$
10	6	$\frac{3}{32}$
11	5	$\frac{5}{64}$
12	7	$\frac{7}{64}$
13	5	$\frac{5}{64}$
14	8	$\frac{1}{8}$
15	3	$\frac{3}{64}$
16	3	$\frac{3}{64}$
17	4	$\frac{1}{16}$
18	5	$\frac{5}{64}$
19	2	$\frac{1}{32}$
20	3	$\frac{3}{64}$
22	2	$\frac{1}{32}$
24	1	$\frac{1}{64}$

5d.

Sum of Two Spins

Probability vs Sum (4 6 7 8 9 10 11 12 13 14 15 16 17 18 19 20 22 24)

5e. 13.5 **5f.** 4.29 **7.** The random variable X is the number of diggs for a web page. The diggs are finite and countable, so X is discrete. **9.** The random variable X is the number of files infected by a computer virus. The files are finite and countable, so X is discrete.

11 Find the sum of the weighted values of each variable.

$0(0.1) = 0.00$
$1(0.1) = 0.10$
$2(0.15) = 0.30$
$3(0.15) = 0.45$
$4(0.25) = 1.00$
$5(0.1) = 0.50$
$6(0.08) = 0.48$
$7(0.05) = 0.35$
$+ 8(0.02) = 0.16$
$= 3.34$

$E(X) = \Sigma[X \cdot P(X)]$
$= 3.34$

13a.

Probability to Win Each Prize

P(X) vs Prize, X (no prize, $20, $10, $5)

13b. $1.20 **13c.** Sample answer: The expected value is positive, so a person buying a ticket can expect to win $0.20 even after the cost of the ticket is considered. Thus, a person would want to participate in this raffle. On the other hand, this raffle is guaranteed to lose money for the organizers and they should change the distribution of prizes or not do the raffle. **15a.** 4.34; Sample answer: The expected number is 4.34, so we can expect there to be 4 upsets. We cannot have 0.34 upsets, so we round to the nearest whole number. **15b.** 1.90

15c.

Number of Upsets, X	Frequency	Relative Frequency
0	1	0.02
1	2	0.04
2	7	0.14
3	4	0.08
4	9	0.18

Number of Upsets, X	Frequency	Relative Frequency
5	17	0.34
6	4	0.08
7	6	0.12
8	0	0

15d.

Number of Upsets

17. Find the expected value of each investment. Multiply the possible value of each fund by the associated probability. A profit is a positive value while a loss is a negative value.

Fund A:
$$0.30(1900) = 570$$
$$0.30(600) = 180$$
$$0.15(-200) = (-30)$$
$$+\ 0.25(-500) = (-125)$$
$$= 595$$

Fund B:
$$0.40(1600) = 640$$
$$0.10(900) = 90$$
$$0.10(-300) = (-30)$$
$$+\ 0.40(-400) = (-160)$$
$$= 540$$

The expected values of Funds A and B are $595 and $540, respectively. Calculate each standard deviation.

Fund A:
$$(1900 - 595)^2 \cdot 0.30 = 510{,}907.50$$
$$(600 - 595)^2 \cdot 0.30 = 7.50$$
$$(-200 - 595)^2 \cdot 0.15 = 94{,}803.75$$
$$+(-500 - 595)^2 \cdot 0.25 = 299{,}756.25$$
$$\Sigma[[X - E(X)]^2 \cdot P(X)] = 905{,}475.00$$
$$\sqrt{905{,}475} \approx 951.6$$

Fund B:
$$(1600 - 540)^2 \cdot 0.40 = 449{,}440.00$$
$$(900 - 540)^2 \cdot 0.10 = 12{,}960.00$$
$$(-300 - 540)^2 \cdot 0.10 = 70{,}560.00$$
$$+(-400 - 540)^2 \cdot 0.40 = 353{,}440.00$$
$$\Sigma[[X - E(X)]^2 \cdot P(X)] = 886{,}400.00$$
$$\sqrt{886{,}400} \approx 941.5$$

The expected value of Funds A and B is $595 and $540, respectively. The standard deviation for Fund A is about 951.6, while the standard deviation for Fund B is about 941.5. Since the standard deviations are about the same, the funds have about the same amount of risk. Therefore, with a higher expected value, Fund A is the better investment.

19. Sample answer: Liana; Shannon didn't consider every scenario in determining the total probability. For example, in calculating the probability of a sum of 5, she considered spinning a 3 then a 2, but not a 2 then a 3.

21. Ensure that each outcome is independent of the others and that the sum of the probabilities of the outcomes is one. Sample answer: A spinner with 5 equal-sided areas shaded red, blue, yellow, green, and brown.

Color	red	blue	yellow	green	brown
Probability	0.2	0.2	0.2	0.2	0.2

Color	red	blue	yellow	green	brown
Probability	0.2	0.2	0.2	0.2	0.2

23. Sample answer: A discrete probability distribution can be the uniform distribution of the roll of a die. In this type of distribution, there are only a finite number of possibilities. A continuous probability distribution can be the distribution of the lives of 400 batteries. In this distribution, there are an infinite number of possibilities. **25.** 2.4 **27.** H

29.

Peter's Articles

Paul's Articles

[0, 70] scl: 5 by [0, 10] scl: 1

[0, 80] scl: 5 by [0, 10] scl: 1

Peter's articles, positively skewed; Paul's articles, symmetric

29b. Sample answer: One of the distributions is symmetric and the other is skewed, so use the five-number summaries. The range for Peter's articles is 64, and the range for Paul's articles is 53. However, the upper quartile for Peter's is 33, while the lower quartile for Paul's is 34. This means that 75% of Paul's articles have more likes (and are more popular) than 75% of Peter's articles. Therefore, we can conclude that Paul's articles are more popular overall. **31.** Sample answer: This situation calls for a survey because the data will be collected from responses from members of a sample of the population. **33.** 0.5, 1.25, 3.125, 7.8125, 19.53125 **35.** 12, 4, $\frac{4}{3}$, $\frac{4}{9}$, $\frac{4}{27}$ **37.** 80, 100, 125, $\frac{625}{4}$, $\frac{3125}{16}$ **39.** 27 **41.** 3

43. $m^4 + 4m^3n + 6m^2n^2 + 4mn^3 + n^4$

Lesson 11-4

1. This experiment cannot be reduced to a binomial experiment because there are more than two possible outcomes. **3.** This experiment can be reduced to a binomial experiment. Success is yes, failure is no, a trial is asking a student, and the random variable is the number of yeses; $n = 30$, $p = 0.72$, $q = 0.28$.

5. Find the probability that the spinner lands on WIN three times out of five spins. Since the spinner is divided into eight equal spaces and WIN is on two of them, $p = \frac{2}{8}$ or 0.25. So, $q = 1 - 0.25$ or 0.75 and $n = 5$.
$$P(X) = {}_nC_x p^x q^{n-x}$$
$$P(3) = {}_5C_3(0.25)^3(0.75)^{5-3}$$
$$\approx 0.0879$$
The probability of Aiden receiving three prizes is about 0.088 or 8.8%. So, the correct answer is D.

7. This experiment can be reduced to a binomial experiment. Success is a day that it rains, failure is a day that it does not rain, a trial is a day, and the random variable is the number of days it rains; n = the number of days in the month, $p = 0.35$, $q = 0.65$.

9. This experiment cannot be reduced to a binomial experiment because the events are not independent. The probability of choosing the hat that covers the ball changes after each selection.

11. Sample answer:
Step 1 A trial is pulling out a marble. The simulation will consist of 20 trials.
Step 2 A success is pulling out a red marble. The probability of success is $\frac{5}{12}$ and the probability of failure is $\frac{7}{12}$.
Step 3 The random variable X represents the number of red marbles pulled out in 20 trials.
Step 4 Use a random number generator. Let 0–4 represent pulling out a red marble. Make a frequency table and record the results as you run the generator.

Outcome	Tally	Frequency
Red Marble	||| ||||	10
Other Outcomes	||| ||||	10

The experimental probability is $\frac{10}{20}$ or 50%. This is greater than the theoretical probability of $\frac{5}{12}$ or about 41.7%.

13. Sample answer:
Step 1 A trial is drawing a card from a deck. The simulation will consist of 20 trials.
Step 2 A success is drawing a face card. The probability of success is $\frac{3}{13}$ and the probability of failure is $\frac{10}{13}$.
Step 3 The random variable X represents the number of face cards drawn in 20 trials.
Step 4 Use a random number generator. Let 0–2 represent drawing a face card. Let 3–12 represent all other outcomes. Make a frequency table and record the results as you run the generator.

Outcome	Tally	Frequency
Face Card	II	2
Other Cards	|||| |||| |||| |||	18

The experimental probability is $\frac{2}{20}$ or 10%. This is less than the theoretical probability of $\frac{3}{13}$ or about 23.1%.

15. 0.183 or 18.3% **17.** 0.096 or 9.6%

19. Find the probability that the kicker makes 7 of his next 10 kicks from within 35 yards. A success is making a field goal, so $p = 0.75$, $q = 1 - 0.75$ or 0.25, and $n = 10$.
$$P(X) = {}_nC_x p^x q^{n-x}$$
$$P(7) = {}_{10}C_7(0.75)^7(0.25)^{10-7}$$
$$\approx 0.25$$
The probability of the kicker making 7 of his next 10 kicks from within 35 yards is about 0.25 or 25%.

21a. 0 own a laptop, 0.0006 or 0.06%; 1 owns a laptop, 0.007 or 0.7%; 2 own a laptop, 0.0343 or 3.43%; 3 own a laptop, 0.0991 or 9.91%; 4 own a laptop, 0.1878 or 18.78%; 5 own a laptop, 0.2441 or 24.41%; 6 own a laptop, 0.2204 or 22.04%; 7 own a laptop, 0.1364 or 13.64%; 8 own a laptop, 0.0554 or 5.54% 9 own a laptop, 0.0133 or 1.33%; 10 own a laptop, 0.0014 or 0.14% **21b.** 0.0701 or 7.01% **21c.** 5 **23a.** 0 own vinyl records, 0.001 or 0.1%; 1 owns vinyl records, 0.012 or 1.2%; 2 own vinyl records, 0.058 or 5.8%; 3 own vinyl records, 0.152 or 15.2%; 4 own vinyl records, 0.253 or 25.3%; 5 own vinyl records, 0.268 or 26.8%; 6 own vinyl records, 0.178 or 17.8%; 7 own vinyl records, 0.067 or 6.7%; 8 own vinyl records, 0.011 or 1.1% **23b.** 0.256 or 25.6% **23c.** 5 **25.** 0.015 or 1.5% **27a.** about 7.8% **27b.** Sample answer: They can roll a six-sided die.

29. Find the mean when $n = 8$ and $p = 0.6$.
$$\mu = np$$
$$= 8(0.6) \text{ or } 4.8$$
Since there cannot be a fraction of a success, the most likely number of successes is 5.

31. 5 **33.** 7 **35.** 0.603 or 60.3% **37.** 0.322 or 32.2% **39.** 0.99 or 99% **41.** 0.889 or 88.9% **43.** Sample answer: You should consider the type of situation for which the binomial distribution is being used. For example, if a binomial distribution is being used to predict outcomes regarding an athletic event, the probabilities of success and failure could change due to other variables such as weather conditions or player health. So, binomial distributions should be used cautiously when making decisions involving events that are not completely random. **45.** Sample answer: A full binomial distribution can be determined by expanding the binomial, which itself utilizes Pascal's triangle. **47a.** 0.003 or 0.3% **47b.** 0.056 or 5.6% **48c.** 0.056 or 5.6% **51.** The random variable X

48a. 0.25 or 25% **48b.** 0.00003 or 0.003% **48d.** 0.003 or 0.3% **49.** H is the number of customers at an amusement park. The customers are finite and countable, so X is discrete. **53.** The random variable X is the number of hot dogs sold at a sporting event. The hot dogs are finite and countable, so X is discrete.

55a. symmetric

[5, 25] scl 2 by [0, 5] scl 1

55b. Sample answer: The distribution is symmetric, so use the mean and standard deviation. The mean is about $16.02 with standard deviation of about $4.52. **57.** 6 **59.** −57 **61.** 20 **71a.** 0.36

65. $e^1 = e$ **67.** $x + 1 = \ln 9$ **69.** $e^{2x} = \ln 5$ **71a.** 0.36

71b. 0.42 **71c.** 0.05

Lesson 11-5

1. 251 < X < 581 **3a.** about 386 teens **3b.** 84%
5. 2.05 **7.** 3.08 **9.** X < 15.9 or X > 42.7 **11a.** 0.509
11b. 0.15% **13.** −1.33 **15.** 177.7

17a. Find the z-values associated with 90,000 and 110,000. The mean μ is 113,627 and the standard deviation σ is 14,266.

$$z = \frac{X - \mu}{\sigma}$$
$$= \frac{90,000 - 113,627}{14,266}$$
$$\approx -1.656$$

$$z = \frac{X - \mu}{\sigma}$$
$$= \frac{110,000 - 113,627}{14,266}$$
$$\approx -0.254$$

Use a graphing calculator to find the area between the z-values.

```
normalcdf(-1.656
,-.254)
       .3508869415
```

The value 0.35 is the percentage of batteries that will last between 90,000 and 110,000 miles. The total number of batteries in this group is 0.35 × 20,000 or 7000.

b. Find the z-value associated with 125,000.

$$z = \frac{X - \mu}{\sigma}$$
$$= \frac{125,000 - 113,627}{14,266}$$
$$\approx 0.797$$

We are looking for values greater than 125,000, so we can use a graphing calculator to find the area between $z = 0.797$ and $z = 4$.

```
normalcdf(.797,4
)
       .2126937641
```

The value 0.21 is the percentage of batteries that will last more than 125,000 miles. The total number of batteries in this group is 0.21 × 20,000 or 4200.

z. Find the z-value associated with 100,000.

$$z = \frac{X - \mu}{\sigma}$$
$$= \frac{100,000 - 113,627}{14,266}$$
$$\approx -0.955$$

We are looking for values less than 100,000, so we can use a graphing calculator to find the area between $z = -4$ and $z = -0.955$.

```
normalcdf(-4,-.9
55)
       .1697571582
```

19a. The middle 68% represents all data values within one standard deviation of the mean. Add ±$115 to $829. The range of rates is $714 to $944.

b. Find the z-value associated with 1000.

$$z = \frac{X - \mu}{\sigma}$$
$$= \frac{1000 - 829}{115}$$
$$\approx 1.487$$

We are looking for values more than 1000, so we can use a graphing calculator to find the area between $z = 1.487$ and $z = 4$.

```
normalcdf(1.487,
4)
       .0684757585
```

The probability that a customer selected at random will pay more than $1000 is about 6.8%. Out of 900 people, about 0.068 · 900 or 62 people will pay more than $1000.

c. Sample answer: I would expect people with several traffic citations to lie to the far right of the distribution where insurance costs are highest, because I think insurance companies would charge them more.

d. Sample answer: I think auto insurance companies would charge younger people more than older people because they have not been driving as long. I think they would charge more for expensive cars and sports cars and less for cars that have good safety ratings. I think they would charge a person less if they have a good driving record and more if they have had tickets and accidents.

21. Sample answer: Hiroko; Monica's solution would work with a uniform distribution. **23.** Sample answer: True; according to the Empirical Rule, 99% of the data lie within 3 standard deviations of the mean. Therefore, only 1% will fall outside of three sigma. An infinitely small amount will fall outside of six-sigma. **25.** Sample answer: The scores per team in each game of the first round of the 2010 NBA playoffs. The mean is 96.56 and the standard deviation is 11.06. The middle 68% of the distribution is 85.50 < X < 107.62. The middle 95% is 74.44 < X < 118.68. The middle 99.7% is 63.38 < X < 129.74. **27.** D **29.** 32.5 **31.** 17.3% **33.** The random variable X is the amount of precipitation in a city per month. Precipitation can be anywhere within a certain range. Therefore, X is continuous. **35.** greatest integer **37.** constant

Lesson 11-6

1. 0.096 **3.** $H_0: \mu \geq 2$; $H_a: \mu < 2$ (claim)
5. $H_0: \mu \geq 20$ (claim); $H_a: \mu < 20$ **7.** $H_0: \mu \geq 84$ (claim), $H_a: \mu < 84$; Do not reject H_0. The manufacturer's claim that the discs can hold at least 84 minutes cannot be rejected.

9. The sample size n is 76 and the standard deviation s is 5.2. The mean is not needed to calculate the maximum error of estimate. The z-values which correspond to 95% significance are ±1.96.

$$E = z \cdot \frac{s}{\sqrt{n}}$$
$$= 1.96 \cdot \frac{5.2}{\sqrt{76}}$$
$$\approx 1.17$$

The maximum error of estimate is about 1.17.

11. $H_0: \mu \geq 6$ (claim); $H_a: \mu < 6$ **13.** $H_0: \mu \leq 2$ (claim); $H_a: \mu > 2$ **15.** $H_0: \mu \geq 30$; $H_a: \mu < 30$ (claim); Do not reject H_0; There is not enough evidence to support the pizza chain's claim of a delivery time of less than 30 minutes. **17.** $H_0: \mu = 12$; $H_a: \mu \neq 12$; The mean of the sample data is 12.9 with a standard deviation of about 1.08. The z-statistic is about 5.27, which falls in the critical region at 1% significance. Therefore, the null hypothesis is rejected and the company should not make the claim on the label.

19a.

$n = 250$ ↔ 24.69 — 25.31
$n = 100$ ↔ 24.51 — 25.49
$n = 50$ ↔ 24.30 — 25.70
24 — 25 — 26

19b. Sample answer: With everything else held constant, increasing the same size will decrease the size of the confidence interval.

19c.

90% ↔ 24.60 — 25.40
95% ↔ 24.52 — 25.48
99% ↔ 24.37 — 25.63
24 — 25 — 26

19d. Sample answer: With everything else held constant, increasing the confidence level will decrease the size of the confidence interval.
19e. Sample answer: Expanding the confidence interval reduces the accuracy of the estimate.

21. The range of values for the mean weight is 19.932 to 20.008. Half of this range is the maximum error of estimate.

$$\frac{20.008 - 19.932}{2} = 0.038.$$

Use the formula for calculating the maximum error of estimate to determine the sample size. The z-values which correspond to 95% significance are ±1.96.

$$E = z \cdot \frac{s}{\sqrt{n}}$$
$$0.038 = 1.96 \cdot \frac{0.128}{\sqrt{n}}$$
$$\frac{0.038}{1.96} = \frac{0.128}{\sqrt{n}}$$
$$\frac{0.038}{1.96} \sqrt{n} = 0.128$$
$$\sqrt{n} = 0.128 \div \frac{0.038}{1.96}$$
$$(\sqrt{n})^2 = \left(0.128 \div \frac{0.038}{1.96}\right)^2$$
$$n = 44$$

23. Sample answer: You can use a statistical test to help you determine the strength of your decision.
25. C **27.** B **29.** 27 students **31.** 729 **33.** 1

35. parabola

37. $y = 0.8x$ **39.** $y = -4$ **41.** 25

connectED.mcgraw-hill.com

Left column (R108)

Chapter 11 Study Guide and Review

1. probability distribution **3.** parameter
5. observational study **7.** survey; sample: every tenth shopper; population: all potential shoppers
9. survey; sample: every fifth person; population: student body

11a.

positively skewed

[300, 330] scl: 1 by [0, 5] scl: 1

11b. Sample answer: The distribution is skewed, so use the five-number summary. Kelly's times range from 301 to 329 seconds. The median is 311, and half of the data are between 307 and 316 seconds.
13. The random variable X is the time it takes to run the race. Time can be anywhere within a certain range. Therefore, X is continuous. **15.** 1.88 snow days **17.** This experiment cannot be reduced to a binomial experiment because there are more than two possible outcomes. **19.** 34.3% **21a.** 68.5%
21b. about 2 games **23.** 63.5 ≤ μ ≤ 65.1 **25.** $H_0: \mu \geq$ 45 (claim); $H_a: \mu <$ 45

CHAPTER 12
Trigonometric Functions

Chapter 12 Get Ready

1. 11.7 **3.** 20.5 **5.** $x = 9, y = 9\sqrt{2}$
7. $x = 12, y = 12\sqrt{3}$

Lesson 12-1

1. $\sin B = \frac{4}{5}, \cos B = \frac{3}{5}, \tan B = \frac{4}{3}, \csc B = \frac{5}{4}, \sec B = \frac{5}{3},$
$\cot B = \frac{3}{4}$ **3.** $\sin A = \frac{\sqrt{33}}{7}, \csc A = \frac{7\sqrt{33}}{33},$
$\cos A = \frac{4\sqrt{33}}{33}, \sec A = \frac{7}{4}, \tan A = \frac{\sqrt{33}}{4}$

5. The side opposite the 60° angle is given. The missing measure is the hypotenuse. Use the sine function to find x.

$\sin 60° = \frac{\text{opp}}{\text{hyp}}$	Sine function
$\sin 60° = \frac{22}{x}$	$\theta = 60°$ and opp = 22
$\frac{\sqrt{3}}{2} = \frac{22}{x}$	$\sin 60° = \frac{\sqrt{3}}{2}$
$\sqrt{3}x = 44$	Cross multiply.
$x = \frac{44}{\sqrt{3}}$	Divide each side by $\sqrt{3}$.
$x \approx 25.4$	Use a calculator.

7. 8.3 **9.** 25.4 **11.** about 274.7 ft **13.** $\sin \theta = \frac{12}{13},$
$\cos \theta = \frac{5}{13}, \tan \theta = \frac{12}{5}, \csc \theta = \frac{13}{12}, \sec \theta = \frac{13}{5},$
$\cot \theta = \frac{5}{12}$ **15.** $\sin \theta = \frac{\sqrt{51}}{10}, \cos \theta = \frac{7}{10}, \tan \theta = \frac{\sqrt{51}}{7},$
$\csc \theta = \frac{10\sqrt{51}}{51}, \sec \theta = \frac{10}{7}, \cot \theta = \frac{7\sqrt{51}}{51}$

Middle column (R108)

17. $\sin A = \frac{8}{17}, \cos A = \frac{15}{17}, \csc A = \frac{17}{8}, \sec A = \frac{17}{15},$
$\cot A = \frac{15}{8}$ **19.** $\sin B = \frac{3\sqrt{10}}{10}, \cos B = \frac{\sqrt{10}}{10},$
$\csc B = \frac{\sqrt{10}}{3}, \sec B = \sqrt{10}, \cot B = \frac{1}{3}$ **21.** 12.7

23.

$\tan \theta = \frac{\text{opp}}{\text{adj}}$	Tangent function
$\tan 30° = \frac{x}{18}$	Replace θ with 30°, opp with x, and adj with 18.
$\tan 30° = \frac{\sqrt{3}}{3}$	
$\frac{\sqrt{3}}{3} = \frac{x}{18}$	Multiply each side by 18.
$18\frac{\sqrt{3}}{3} = x$	Use a calculator.
$10.4 \approx x$	

25. 8.7 **27.** 132.5 ft **29.** 30 **31.** 36.9 **33.** 32.5
35. 25.3 ft higher **37.** $x = 21.9, y = 20.8$ **39.** $x = 19.3,$
$y = 70.7$ **41.** 54.9 **43.** 20.5 **45.** 11.5

47.

$\tan \theta = \frac{\text{opp}}{\text{adj}}$	Tangent function
$\tan 74.5° = \frac{x}{12}$	Replace θ with 74.5°, opp with x, and adj with 12.
$12 \cdot \tan 74.5° = x$	Multiply each side by 12.
$43 \approx x$	Use a calculator.

So, the height of the bird's nest is $43 + 5$ or 48 feet.
49a. about 647.2 ft **49b.** about 239.4 ft
51. $m\angle A = 59°, a = 31.6, c = 36.9$ **53.** $m\angle A = 38.7°,$
$m\angle B = 51.3°, b = 7.5, c = 9.6$ **55.** True; $\sin \theta = \frac{\text{opp}}{\text{hyp}}$
and the values of the opposite side and the hypotenuse of an acute triangle are positive, so the value of the sine function is positive. **57.** Sample answer: The slope describes the ratio of the vertical rise to the horizontal run of the roof. The vertical rise is opposite the horizontal run of the roof. The vertical run is the adjacent side. So, the tangent of the angle that the roof makes with the horizontal. The horizontal run is the adjacent side. So, the tangent of the angle of elevation equals the ratio of the rise to the run, or the slope of the roof;
$\theta = 33.7°$. **59.** 24 **61.** G **63.** $H_0: \mu = 3$ (claim);
$H_a: \mu \neq 3$ **65a.** 99.85% **65b.** 2.5% **65c.** 64%
67. 22,704 feet **69.** 35 dollars **71.** $216\frac{2}{3}$ centimeters

Lesson 12-2

1.

140°

3.

390°

Right column area — Lesson 12-2 continued

5. Sample answer:
positive angle: 175° + 360° = 535°
negative angle: 175° − 360° = −185°
7. 45° **9.** $-\frac{2\pi}{9}$

11.

75°

13.

−90°

15.

295°

17.

240°

19. Sample answer: 410°, −310° **21.** Sample answer:
565°, −155° **23.** Sample answer: 280°, −440°

25. $330° = 330 \cdot \frac{\pi \text{ radians}}{180°}$
$= \frac{330\pi}{180}$ or $\frac{11\pi}{6}$ radians
27. −60° **29.** $\frac{19\pi}{18}$ **31.** about 12.6 ft **33.** 6.7 cm
35. 1 h 15 min **37.** Sample answer: 260°, −100°
39. Sample answer: $\frac{5\pi}{4}, -\frac{11\pi}{4}$

41. a.

165°

b. $165° = 165 \cdot \frac{\pi \text{ radians}}{180°}$
$= \frac{165\pi}{180}$ or $\frac{11\pi}{12}$ radians
c. $s = r\theta$ Formula for arc length
$= 6.5 \cdot \frac{11\pi}{12}$ $r = 6.5$ and $\theta = \frac{11\pi}{12}$
≈ 18.7 ft   Use a calculator.
d. The arc length would double. Since $s = r\theta$, if r is doubled and θ remains unchanged, then the value of s is also doubled.
43. 472.5° **45.** $\frac{10\pi}{9}$ **47a.** $\frac{\pi}{6}$ **47b.** 2.1 ft **49.** $x = 2$
51. Sample answer:
440° and −280°

53. A

$\frac{\theta}{2\pi} = \frac{s}{2\pi r}$	Substitute.
$2\pi r\theta = 2\pi s$	Find the cross products.
$r\theta = s$	Divide each side by 2π.

One degree represents an angle measure that equals $\frac{1}{360}$ rotation around a circle. One radian represents the measure of an angle in standard position that intercepts an arc of length r. To change from degrees to radians, multiply the number of degrees by $\frac{\pi \text{ radians}}{180°}$. To change from radians to degrees, multiply the number of radians by $\frac{180°}{\pi \text{ radians}}$.

55. A **57.** B **59.** $\sin \theta = \frac{\sqrt{259}}{22}, \cos \theta = \frac{\sqrt{259}}{22},$
$\tan \theta = \frac{\sqrt{259}}{15}, \csc \theta = \frac{22}{\sqrt{259}}$ or $\frac{22\sqrt{259}}{259}, \sec \theta = \frac{22}{15},$
$\cot \theta = \frac{15}{\sqrt{259}}$ or $\frac{15\sqrt{259}}{259}$ **61.** $H_0: \mu \geq 8$ (claim);
$H_a: \mu < 8$ **63a.** 50% **63b.** 815 **63c.** 25
65. $3\sqrt{41}$ **67.** $\sqrt{317}$

Lesson 12-3

1. $\sin \theta = \frac{2\sqrt{5}}{5}, \cos \theta = \frac{\sqrt{5}}{5}, \tan \theta = 2, \csc \theta = \frac{\sqrt{5}}{2},$
$\sec \theta = \sqrt{5}, \cot \theta = \frac{1}{2}$ **3.** $\sin \theta = -1, \cos \theta = 0,$
$\tan \theta = \text{undefined}, \csc \theta = -1, \sec \theta = \text{undefined},$
$\cot \theta = 0$
5. 65° **7.** $\frac{\sqrt{2}}{2}$ **9.** −2

11a.

115°

11b. 55°; cos 55° **11c.** 3.2 in.

11a.

125°

13.

(−6, 8)

$r = \sqrt{x^2 + y^2}$
$= \sqrt{(-6)^2 + 8^2}$
$= \sqrt{100}$ or 10
Use $x = -6, y = 8$, and $r = 10$.
$\sin \theta = \frac{y}{r}$ $\cos \theta = \frac{x}{r}$

29a. Sample answer:

29b. Sample answer: Use the Law of Cosines to find the measure of ∠A. Then use the formula Area = $\frac{1}{2}bc \sin A$. **29c.** 54.6 yd²

31.
$$\frac{\sin A}{a} = \frac{\sin B}{b} \quad \text{Law of Sines}$$
$$\frac{\sin 104°}{12.4} = \frac{\sin B}{8.1} \quad \text{Substitution}$$
$$\frac{8.1 \sin 104°}{12.4} = \sin B \quad \text{Multiply each side by 8.1.}$$
$$39° \approx B \quad \text{Use the } \sin^{-1} \text{ function.}$$

$$m∠C \approx 180 - (39° + 104°) \text{ or } 37°$$
$$\frac{\sin A}{a} = \frac{\sin C}{c} \quad \text{Law of Sines}$$
$$\frac{\sin 104°}{12.4} = \frac{\sin 37°}{c} \quad \text{Substitution}$$
$$c = \frac{12.4 \sin 37°}{\sin 104°} \quad \text{Solve for } c.$$
$$c \approx 7.7 \quad \text{Use a calculator.}$$

33. F ≈ 42°, G ≈ 72°, H ≈ 66° **35.** The longest side is 14.5 centimeters. Use the Law of Cosines to find the measure of the angle opposite the longest side; 102°. **37.** When two angles and a side are given or when two sides and an angle opposite one of the sides are given, you can use the Law of Sines to solve a triangle. When two sides and an included angle are given or when three sides are given, you can use the Law of Cosines to solve a triangle. **39.** G **41.** 4, $\frac{23}{15}$

43. 7.5 yd² **45.** $\sin θ = \frac{5\sqrt{89}}{89}$, $\cos θ = \frac{8\sqrt{89}}{89}$, $\tan θ = \frac{5}{8}$, $\csc θ = \frac{\sqrt{89}}{5}$, $\sec θ = \frac{\sqrt{89}}{8}$, $\cot θ = \frac{8}{5}$

47. $\sin θ = -\frac{3\sqrt{13}}{13}$, $\cos θ = \frac{2\sqrt{13}}{13}$, $\tan θ = -1.5$, $\csc θ = -\frac{\sqrt{13}}{3}$, $\sec θ = \frac{\sqrt{13}}{2}$, $\cot θ = -\frac{2}{3}$ **49.** $60, $50

51. hyperbola **53.**

55.

Lesson 12-6

1. $\cos θ = \frac{15}{17}$, $\sin θ = \frac{8}{17}$ **3.** 2 **5a.** 4 seconds

5b. Sample answer:

connectED.mcgraw-hill.com **R111**

41a. Sample answer:

41b. Sample answer: $\frac{\sin 66°}{4} = \frac{\sin 64°}{b}$ **43.** Cameron; R is acute and r > t, so there is one solution.

41c. about 11.5 mi

45. Sample answer:

Sin A = $\frac{opposite}{hypotenuse}$ Definition of sine

Sin A = $\frac{h}{c}$

c sin A = h   h = opposite side, c = hypotenuse

Area = $\frac{1}{2} \cdot$ base $\cdot$ height   Area of a triangle

Area = $\frac{1}{2}bh$ b = base, h = height

Area = $\frac{1}{2}bc \sin A$ Substitution

47. Sample answer: In the triangle, B = 115°. Using the Law of Sines, $\frac{\sin 50°}{a} = \frac{\sin 115°}{b}$. This equation cannot be solved because there are two unknowns. To solve a triangle using the Law of Sines, two sides and an angle must be given or two angles and a side opposite one of the angles must be given.

49. 2 **51.** G **53.** $-\frac{1}{2}$ **55.** $\frac{\sqrt{3}}{3}$ **57.** 328°, −392° **59.** 96 cm **61.** No sum exists. **63.** $\frac{8.714 \times 10^{15} + x^2}{y^2}$

$\frac{8.710 \times 10^{15}}{8.710 \times 10^{15}} = 1$ **65.** $(y + 2)^2$ **67.** 56.25 **69.** 26

Lesson 12-5

1. A ≈ 36°, C ≈ 52°, b ≈ 5.1 **3.** A ≈ 18°, B ≈ 29°, C ≈ 133° **5.** Sines; B ≈ 40°, C ≈ 33°, c ≈ 6.8

7. Since the lengths of two sides and the measure of the included angle are known, first use the Law of Cosines to find the missing side length.
$$r^2 = s^2 + t^2 - 2st \cos R \quad \text{Law of Cosines}$$
$$r^2 = 16^2 + 9^2 - 2(16)(9) \cos 35° \quad s = 16, t = 9, R = 35°$$
$$r^2 = 101.1 \quad \text{Use a calculator.}$$
$$r \approx 10.1 \quad \text{Take the positive square root of each side.}$$

$$\frac{\sin R}{r} = \frac{\sin T}{t} \quad \text{Law of Sines}$$
$$\frac{\sin 35°}{10.1} = \frac{\sin T}{9} \quad \text{Substitution}$$
$$\frac{9 \sin 35°}{10.1} = \sin T \quad \text{Multiply each side by 9.}$$
$$31° \approx T \quad \text{Use the } \sin^{-1} \text{ function.}$$

$m∠S = 180 - (35° + 31°) \text{ or } 114°$

9. A ≈ 70°, B ≈ 40°, c ≈ 3.0 **11.** A ≈ 31°, B ≈ 108°, C ≈ 41° **13.** a ≈ 6.9, B ≈ 41°, C ≈ 23° **15.** F ≈ 65°, G ≈ 94°, H ≈ 21° **17.** Sines; C ≈ 45°, A ≈ 85°, a ≈ 18.2 **19.** Cosines; A ≈ 27°, B ≈ 115°, C ≈ 38° **21.** Sines; A ≈ 27°, B ≈ 79°, b ≈ 6.9

23. $d^2 = 338^2 + 520^2 - 2(338)(520) \cos 70°$
$d^2 \approx 264,417.1$
$d \approx 514.2$ m **27.** about 13,148 yd²

25. 81°, 36°, 63°

---

angle and determine in which quadrant it is located. Then use the appropriate rule for finding its reference angle θ′. A reference angle is the acute angle formed by the terminal side of θ and the x-axis. Next, find the value of the trigonometric function for θ′. Finally, use the quadrant location to determine the sign of the trigonometric function value of θ. **53.** C **55.** E **57.** 330° **59.** 40.1° **61.** 66.0° **63.** $10.737, $18.23 **65.** $(x + 4)^2 + (y + 2)^2 = 73$ **67.** $\frac{x^2 + 7x - 35}{(x + 2)(x + 4)(x - 7)}$ **69.** $\frac{2(3x^2 + 2x - 12)}{3x(x + 4)(x - 6)}$ **71.** 2.5841 **73.** $\frac{1}{2}$ **75.** $\frac{1}{3125}$ **77.** 9

Lesson 12-4

1. 27.9 mm²

3. Area = $\frac{1}{2}bc \sin A$ Area Formula
= $\frac{1}{2}(11)(6) \sin 40°$ Substitution
≈ 21.22 cm²   Simplify.

5. E = 107°, d ≈ 7.9, f ≈ 7.0 **7.** F = 60°, f ≈ 12.3, h ≈ 9.1 **9.** no solution **11.** one; B = 90°, C = 60°, c ≈ 5.2 **13.** 10.6 km² **15.** 36.8 m² **17.** 5.9 ft² **19.** 65.2 m² **21.** C = 30°, b ≈ 11.1, c ≈ 5.8 **23.** L = 74°, m ≈ 4.9, n ≈ 3.1

25. $m∠K = 180 - (53 + 20) \text{ or } 107°$
$$\frac{\sin H}{h} = \frac{\sin J}{j} \quad \text{Law of Sines}$$
$$\frac{\sin 53°}{31} = \frac{\sin 20°}{j} \quad \text{Substitution}$$
$$j = \frac{31 \sin 20°}{\sin 53°} \quad \text{Solve for } j.$$
$$j \approx 13.3 \quad \text{Use a calculator.}$$
$$\frac{\sin H}{h} = \frac{\sin K}{k} \quad \text{Law of Sines}$$
$$\frac{\sin 53°}{31} = \frac{\sin 107°}{k} \quad \text{Substitution}$$
$$k = \frac{31 \sin 107°}{\sin 53°} \quad \text{Solve for } k.$$
$$k \approx 37.1 \quad \text{Use a calculator.}$$

27. B = 63°, b ≈ 2.9, c ≈ 3.0 **29.** one; B ≈ 25°, C ≈ 55°, c ≈ 5.8 **31.** one; B ≈ 32°, C ≈ 110°, c ≈ 32.1 **33.** two; B ≈ 53°, C ≈ 85°, c ≈ 7.4; B ≈ 127°, C ≈ 11°, c ≈ 1.4 **35.** no solution **37.** about 28°

39.

$m∠C = 180 - (40 + 112) \text{ or } 28°$
$$\frac{\sin A}{a} = \frac{\sin C}{c} \quad \text{Law of Sines}$$
$$\frac{\sin 112°}{a} = \frac{\sin 28°}{8} \quad \text{Substitution}$$
$$a = \frac{8 \sin 112°}{\sin 28°} \quad \text{Solve for } a.$$
$$a \approx 15.8 \quad \text{Use a calculator.}$$

Sirens B and C are about 15.8 miles apart.

$$= \frac{8}{10} \text{ or } \frac{4}{5}$$
$$\tan θ = \frac{y}{x}$$
$$= \frac{-8}{-6} \text{ or } \frac{4}{3}$$
$$\sec θ = \frac{r}{x}$$
$$= \frac{-10}{-6} \text{ or } \frac{5}{3}$$

$$\csc θ = \frac{r}{y}$$
$$= \frac{-6}{10} \text{ or } -\frac{3}{5}$$

$$\cot θ = \frac{x}{y}$$
$$= \frac{-6}{-8} \text{ or } \frac{3}{4}$$

15. $\sin θ = -1$, $\cos θ = 0$, $\tan θ = $ undefined, $\csc θ = -1$, $\sec θ = $ undefined, $\cot θ = 0$ **17.** $\sin θ = \frac{\sqrt{10}}{10}$, $\cos θ = -\frac{3\sqrt{10}}{10}$, $\tan θ = -\frac{1}{3}$, $\csc θ = \sqrt{10}$, $\sec θ = -\frac{\sqrt{10}}{3}$, $\cot θ = 3$ **19.** 75° **21.** $\frac{π}{4}$

23. 40°

25. −1 **27.** −√2

29. $\frac{1}{2}$ **31.** $\frac{2\sqrt{3}}{3}$

33.

The water reaches about 8.2 feet to the left of the sprinkler.

35. about 10.1 m **37.** $\cos θ = -\frac{3}{5}$, $\tan θ = -\frac{4}{3}$ **39.** $\sin θ = -\frac{4}{5}$, $\csc θ = \frac{5}{4}$, $\sec θ = -\frac{5}{3}$, $\cot θ = -\frac{3}{4}$

41. 0 **43.** $-\frac{1}{2}$ **45.** $\frac{\sqrt{3}}{2}$ **47.** No; for $\sin θ = -1$, the reference angle is 45°. However, for $\sin θ$ to be positive and $\tan θ$ to be negative, the reference angle must be in the second quadrant. So, the value of θ must be 135° or an angle coterminal with 135°. **49.** Sample answer: We know that $\cot θ = \frac{x}{y}$, $\sin θ = \frac{y}{r}$, and $\cos θ = \frac{x}{r}$. Since $\sin 180 = 0$, it must be true that y = 0. Thus, $\cot θ = \frac{x}{0}$, which is undefined. **51.** Sample answer: First, sketch the

19. amplitude: 3; period: 180°

21a. $h = 4 \sin \frac{2}{3}\pi t$

21b.

23. period: 360°

25. period: 180°

27. period: 180°

29. a. Since the frequency is 0.5, the period is $\frac{1}{0.5}$ or 2.

$\text{period} = \frac{2\pi}{|b|}$ Write the relationship between the period and b.

$2 = \frac{2\pi}{|b|}$ Substitution

$b = \pi$ Solve for b.

$y = a \sin b\theta$ General equation for the sine function

$h = 1 \sin \pi t$ Replace y with h, a with 1, b with π, and θ with t.

$h = \sin \pi t$ Simplify.

---

7. period: 360°

9. amplitude: 2; period: 360°

11. amplitude: 1; period: 180°

13. amplitude: 1; period: 720°

15. amplitude: $\frac{3}{4}$; period: 360°

17. amplitude: $|a| = \left|\frac{1}{2}\right|$ or $\frac{1}{2}$

period: $\frac{360°}{|b|} = \frac{360°}{|2|}$ $= 180°$

29c. Sample answer: The slope corresponds to the tangent of the angle. For $\theta = 120°$, the x-coordinate of P is $-\frac{1}{2}$ and the y-coordinate is $\frac{\sqrt{3}}{2}$; slope = $\frac{\text{change in } y}{\text{change in } x}$. Since change in $x = -\frac{1}{2}$ and change in $y = \frac{\sqrt{3}}{2}$, slope = $\frac{\sqrt{3}}{2} \div \left(-\frac{1}{2}\right) = -\sqrt{3}$ or about -1.7.

31. $\cos 45° - \cos 30° = \frac{\sqrt{2}}{2} - \frac{\sqrt{3}}{2} = \frac{\sqrt{2} - \sqrt{3}}{2}$

33. $\frac{5\sqrt{3}}{2}$ **35.** 1   **37.** Benita; Francis incorrectly wrote $\cos \frac{\pi}{3} = -\cos \frac{\pi}{3}$ **39.** Sometimes; the period of a sine curve could be $\frac{\pi}{2}$, which is not a multiple of π.

41. The period of a periodic function is the horizontal distance of the part of the graph that is nonrepeating. Each nonrepeating part of the graph is one cycle.

43. C   **45.** A   **47.** $A \approx 34°$, $C \approx 64°$, $c \approx 12.7$

49. $B \approx 33°$, $C \approx 29°$, $c \approx 9.9$ **51.** one solution; $B \approx 35°$, $C \approx 99°$, $c \approx 13.7$ **53.** 0.267   **55.** 7   **57.** \$46,794.34   **59.** $(5, 0)$, $(-4, \pm 6)$ **61.** 108

Lesson 12-7

1. amplitude: 4; period: 360°

3. amplitude: 1; period: 180°

5a. $\frac{1}{14}$ or about 0.07 second

5b. $y = \sin 28\pi t$

29b.

Angle	Slope
30	0.6
60	1.7
120	−1.7
150	−0.6
210	0.6
315	−1

---

7. $-\frac{\sqrt{3}}{2}$ **9.** $\cos \theta = \frac{3}{5}$, $\sin \theta = -\frac{4}{5}$

11. $P\left(\frac{\sqrt{3}}{2}, \frac{1}{2}\right) = P(\cos \theta, \sin \theta)$

$\cos \theta = \frac{\sqrt{3}}{2}$, $\sin \theta = \frac{1}{2}$

13. 3   **15.** 12   **17.** 180°

19a.

Average High Temperatures

19b. 12 mo or 1 yr   **21.** $\frac{1}{2}$ **23.** $\frac{\sqrt{2}}{2}$ **25.** $\frac{\sqrt{3}}{2}$

27. a. The period is the time it takes to complete one rotation. So, the period is 60 seconds ÷ 2.5 or 24 seconds.

b. Since the siren is 1 mile from Ms. Miller's house and the beam of sound has a radius of 1 mile, then the minimum distance of the sound beam from the house is 0 miles and the maximum distance is 2 miles. Draw a sine curve with the pattern repeating every 24 seconds. Sample answer:

29a.

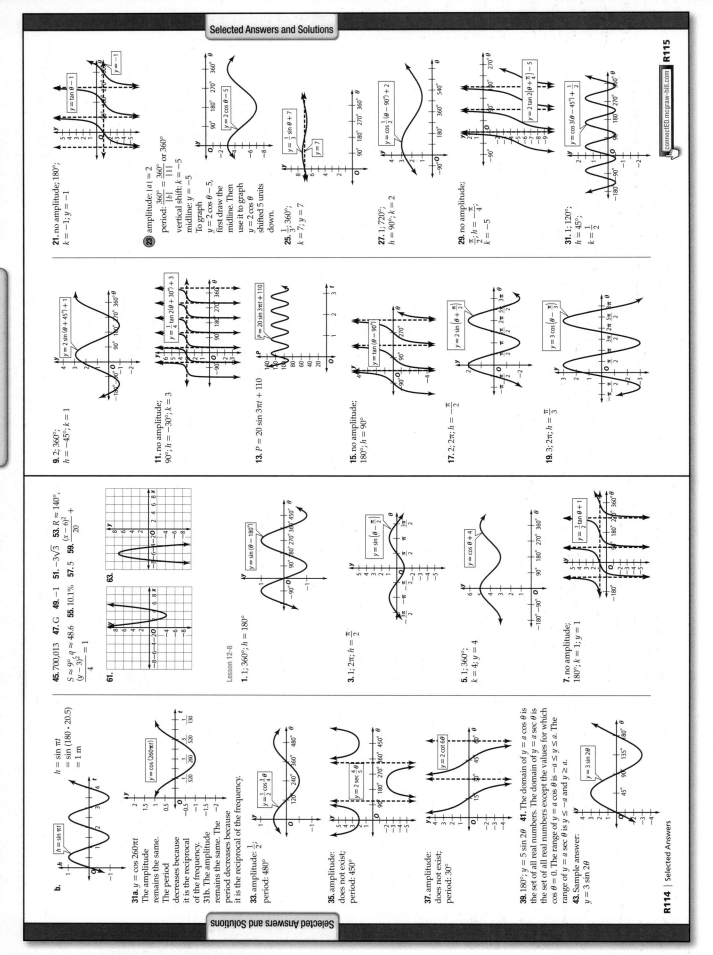

21. no amplitude; 180°; $k = -1; y = -1$

23. amplitude: $|a| = 2$
period: $\dfrac{360°}{|b|} = \dfrac{360°}{|1|}$ or 360°
vertical shift: $k = -5$
midline: $y = -5$
To graph $y = 2 \cos \theta - 5$, first draw the midline. Then use it to graph $y = 2 \cos \theta$ shifted 5 units down.

25. $\dfrac{1}{3}$; 360°; $k = 7; y = 7$

27. 1; 720°; $h = 90°; k = 2$

29. no amplitude; $\dfrac{\pi}{2}; h = -\dfrac{\pi}{4}; k = -5$

31. 1; 120°; $h = 45°; k = \dfrac{1}{2}$

9. 2; 360°; $h = -45°; k = 1$

11. no amplitude; 90°; $h = -30°; k = 3$

13. $P = 20 \sin 3\pi t + 110$

15. no amplitude; 180°; $h = 90°$

17. 2; 2π; $h = -\dfrac{\pi}{2}$

19. 3; 2π; $h = \dfrac{\pi}{3}$

45. 700,013 **47.** G **49.** −1 **51.** $-3\sqrt{3}$ **53.** $R \approx 140°$, $S \approx 9°; q \approx 48.6$ **55.** 10.1% **57.** 5 **59.** $\dfrac{(x-6)^2}{20} + \dfrac{(y-3)^2}{4} = 1$

61.

63.

Lesson 12-8

1. 1; 360°; $h = 180°$

3. 1; 2π; $h = \dfrac{\pi}{2}$

5. 1; 360°; $k = 4; y = 4$

7. no amplitude; 180°; $k = 1; y = 1$

b. $h = \sin \pi t$
$h = \sin (180 \cdot 20.5)$
$= 1$ m

31a. $y = \cos 260\pi t$
The amplitude remains the same. The period decreases because it is the reciprocal of the frequency.
31b. The amplitude remains the same. The period decreases because it is the reciprocal of the frequency.

33. amplitude: $\dfrac{1}{2}$; period: 480°

35. amplitude: does not exist; period: 450°

37. amplitude: does not exist; period: 30°

39. 180°; $y = 5 \sin 2\theta$ **41.** The domain of $y = a \cos \theta$ is the set of all real numbers. The domain of $y = a \sec \theta$ is the set of all real numbers except the values for which $\cos \theta = 0$. The range of $y = a \cos \theta$ is $-a \leq y \leq a$. The range of $y = a \sec \theta$ is $y \leq -a$ and $y \geq a$.
43. Sample answer: $y = 3 \sin 2\theta$

33. 3; 6π; $h = \frac{\pi}{2}$; $k = -2$

$$y = -2 + 3\sin\frac{1}{3}\left(\theta - \frac{\pi}{2}\right)$$

35.

$$d = 1.8\sin\frac{3\pi}{4}t + 12$$

min: 10.2 ft;
max: 13.8 ft

37. $y = \sin(x - 4) + 3$

39. $y = \tan(x - \pi) + 2.5$

41. a. The midline lies halfway between the maximum and minimum values. So, $y = \frac{55 + 37}{2}$ or 46. Since the midline is $y = 46$, the vertical shift is $k = 46$.

The amplitude is the difference between the midline value and the maximum value. So, $|a| = |55 - 46|$ or 9.

Since the carousel rotates once every 21 seconds and a horse on the carousel goes up and down 3 times per rotation, it goes up and down every $21 \div 3$ or 7 seconds. So, the period is 7 seconds.

$period = \frac{2\pi}{|b|}$ Write the relationship between the period and b.

$7 = \frac{2\pi}{|b|}$ Substitution

$b = \frac{2\pi}{7}$ Solve for b.

$y = a\sin b\theta + k$ General equation for the sine function

$h = 9\sin\frac{2\pi}{7}t + 46$ Replace y with h, a with 9, b with $\frac{2\pi}{7}$, θ with t, and k with 46.

b.

$$h = 9\sin\frac{2\pi}{7}t + 46$$

c. Sample answer: On the graph, when $t = 8$, $d \approx 53$. So after 8 seconds, the height is about 53 inches.

$h = 9\sin\frac{2\pi}{7}t + 46$

$= 9\sin\frac{2\pi}{7}(8) + 46$

≈ 53.0 in.

43. $\left(\frac{3\pi}{2}, 2\right)$ **45.** no maximum values **47.** The graphs are reflections of each other over the x-axis. **49.** The graphs are identical. **51.** 360°; Sample answer: $y = 2\cos(\theta + 90°)$

53. The midline lies halfway between the maximum and minimum values. So $y = \frac{4 + 2}{2}$ or 3. Since the midline is $y = 3$, the vertical shift is $k = 3$.

The amplitude is the difference between the midline value and the maximum value. So $|a| = |4 - 3|$ or 1.

Since the cycle repeats every 180°, the period is 180°.

$period = \frac{360°}{|b|}$ Write the relationship between the period and b.

$180° = \frac{360°}{|b|}$ Substitution

$b = 2$ Solve for b.

The graph is the sine curve shifted 45° to the right. So the phase shift is $h = 45°$.

$y = a\sin b(\theta - h) + k$ General equation for the sine function

$y = 1\sin 2(\theta - 45°) + 3$ Replace a with 1, b with 2, h with 45°, and k with 3.

$y = \sin 2(\theta - 45°) + 3$ Simplify.

55. 180°; no phase shift; $k = 6$

$$y = \cot\theta + 6$$

57. 120°; $h = 45°$; $k = 1$

$$y = \frac{1}{2}\csc 3(\theta - 45°) + 1$$

59. π; $h = -\frac{\pi}{2}$; $k = -3$

$$y = 4\sec 2\left(\theta + \frac{\pi}{2}\right) - 3$$

61. The graph of $y = 3\sin 2\theta + 1$ has an amplitude of 3 rather than an amplitude of 1. It is shifted up 1 unit from the parent graph and is compressed so that it has a period of 180°.

answer: $y = \tan^{-1}x$ is a relation that has a domain of all real numbers and a range of all real numbers except odd multiples of $\frac{\pi}{2}$. The relation is not a function. $y = \text{Tan}^{-1}x$ is a function that has a domain of all real numbers and a range of $-\frac{\pi}{2} \le y \le \frac{\pi}{2}$.

47. A **49.** G **51a.** 164; 164; 360°; 90°

51b. $y = 100[\sin(x - 90°)] + 100$ **53.** 2 **55.** 4 **57.** −1

59. $-\frac{\sqrt{3}}{2}$

Chapter 12 Study Guide and Review

1. false, Law of Sines **3.** true **5.** false, Arcsine function **7.** $a = 10.9$; $A = 65°$; $B = 25°$ **9.** $A = 15°$; $a = 4.0$; $c = 15.5$ **11.** $B = 55°$; $a = 12.6$; $b = 18.0$ **13.** about 8.8 feet **15.** 450° **17.** $-\frac{7\pi}{4}$ **19.** 295°, −425°

21. $\frac{4\pi}{15}$ **23.** $-\frac{\sqrt{3}}{3}$ **25.** 0 **27.** $\sin\theta = \frac{12}{13}$; $\cos\theta = \frac{5}{13}$;

$\tan\theta = \frac{12}{5}$, $\csc\theta = \frac{13}{12}$, $\sec\theta = \frac{13}{5}$, $\cot\theta = \frac{5}{12}$

29. about 17.1 meters **31.** two solutions; First solution: $C = 30°$, $B = 125°$; $b = 29.1$; second solution: $C = 150°$, $B = 5°$, $b = 3.1$ **33.** 98.9 ft **35.** Law of Sines; $B \approx 52°$, $C \approx 48°$, $c \approx 11.3$ **37.** Law of Sines; $B \approx 75°$, $C \approx 63°$, $c \approx 12.0$ or $B \approx 105°$, $C \approx 33°$, $c \approx 7.3$

39. 750.5 ft **41.** $-\frac{\sqrt{6}}{4}$ **43.** 0 **45.** 15 seconds

47. amplitude: 1; period: 720°

$$y = \cos\frac{1}{2}\theta$$

49. amplitude: not defined, period: 360°

51. amplitude: not defined, period: 720°

---

63. Sample answer: $y = 2\sin\theta - 3$

$$y = 2\sin\theta - 3$$

65. 1.25 **67.** F

69. amplitude: 2; period: 360°

$$y = 2\cos\theta$$

71. amplitude: 1; period: 180°

$$y = \sin 2\theta$$

75. experiment; sample: people that exercise for an hour a day; population: all adults **77.** observational study; sample: 100 students selected; population: all students that have part-time jobs **79.** 8 days **81.** 42° **83.** 37° **85.** 16°

Lesson 12-9

1. 30°; $\frac{\pi}{6}$ **3.** 180°; π **5.** 0 **7.** A **9.** −27.4°

11. Arctan $\frac{6.2}{18}$; 19°

13. KEYSTROKES: [2nd] [ENTER] [COS⁻¹] [2nd] [√ 13
[ )] [÷] [2] [ )] [ENTER] 30

So, Arccos $\left(\frac{\sqrt{3}}{2}\right) = 30° = \frac{\pi}{6}$

15. 60°; $\frac{\pi}{3}$ **17.** −30°; $-\frac{\pi}{6}$ **19.** −0.58 **21.** 0.87

23. 0.71 **25.** 64.2° **27.** 104.5° **29.** −11.3°

31. Arcsin $\frac{2.5}{24}$; 6°

33. $15 \text{ m/s} (\sin x) = 1 \text{ s}$

$\frac{9.8}{15}$

9.8 m/s^2 $\frac{9.8}{15}$

$\sin x = \frac{9.8}{15}$ Maximum height equals 1.

$x = \text{Sin}^{-1}\frac{9.8}{15}$ Solve for x.

$x \approx 40.8°$ Inverse sine function

35. π **37.** no solution **39.** $\frac{5\pi}{3}$, $\frac{\pi}{3}$ **41.** false; $x = 2\pi$

43. The domain of $y = \text{Sin}^{-1}x$ is $-1 \le x \le 1$. This is the same as the range of $y = \sin x$. Use a calculator. **45.** Sample

connectED.mcgraw-hill.com

Selected Answers and Solutions

53. vertical shift: up 1; amplitude: 3; period 180°; phase shift: 90° right

55. vertical shift: up 2 amplitude: not defined period: $\frac{2\pi}{3}$ phase shift: $\frac{\pi}{2}$ right

57. vertical shift: up 2 amplitude: $\frac{1}{3}$ period: 1080° phase shift: 90° right

59. 90°, $\frac{\pi}{2}$ **61.** 60°, $\frac{\pi}{3}$ **63.** 45°, $\frac{\pi}{4}$ **65.** $\sin^{-1}\frac{5}{10} = \theta$; 30° **67.** −0.71 **69.** −55.0° **71.** 65.8°

CHAPTER 13
Trigonometric Identities and Equations

Chapter 13 Get Ready

1. $-4a(a-1)$ **3.** prime **5.** $(x+2)$ in. **7.** $\{-7, 5\}$

9. $\{3, 4\}$ **11.** $\frac{\sqrt{2}}{2}$ **13.** $\frac{\sqrt{3}}{3}$ **15.** 45 ft

Lesson 13-1

1. $\frac{1}{2}$ **3.** $\sqrt{5}$ **5.** $\sin\theta\cos\theta$ **7.** $\cot^2\theta$ **9.** $\frac{5}{4}$ **11.** $\frac{4}{5}$

13. $-\frac{5}{4}$

(15) $\cot^2\theta + 1 = \csc^2\theta$ Pythagorean Identity
$\left(\frac{1}{4}\right)^2 + 1 = \csc^2\theta$ Substitute $\frac{1}{4}$ for $\cot\theta$.
$\frac{1}{16} + 1 = \csc^2\theta$ Square $\frac{1}{4}$.
$\frac{17}{16} = \csc^2\theta$ Add.
$\pm\frac{\sqrt{17}}{4} = \csc\theta$ Take the square root of each side.
Since θ is in the third quadrant, $\csc\theta$ is negative.
So, $\csc\theta = -\frac{\sqrt{17}}{4}$.

17. $-\frac{12}{13}$ **19.** $\frac{3}{5}$ **21.** $\sec^3\theta$ **23.** $\csc\theta$ **25.** 1

(27) $B = \frac{F\csc\theta}{l\ell}$ Original equation
$l\ell B = \frac{F\csc\theta}{l\ell} \cdot l\ell$ Multiply each side by $l\ell$.
$l\ell B = F\csc\theta$ Multiply each side by $\sin\theta$.
$l\ell B \cdot \sin\theta = F \cdot \frac{1}{\sin\theta} \cdot \sin\theta$ Multiply each side by $\sin\theta$.
$l\ell B \sin\theta = F$
The equation can be written as $F = l\ell B\sin\theta$.

29. $\sec\theta$ **31.** 2 **33.** $2\cos^2\theta$ **35a.** $\frac{\sqrt{65}}{9}$ **35b.** $\frac{4\sqrt{65}}{65}$

35c. $\frac{4}{9}$, $\frac{\sqrt{65}}{9}$, $\frac{4\sqrt{65}}{65}$ **37.** $\mu_k = \tan\theta$

(39) $\frac{\cos\left(\frac{\pi}{2}-\theta\right)-1}{1+\sin(-\theta)}$

$= \frac{\sin\theta-1}{1+\sin(-\theta)}$ $\cos\left(\frac{\pi}{2}-\theta\right) = \sin\theta$

$= \frac{\sin\theta-1}{1-\sin\theta}$ $\sin(-\theta) = -\sin\theta$

$= \frac{-1(\sin\theta-1)}{-1(\sin\theta-1)}$ Simplify.

$= \frac{1}{-1}$ or -1

41. $-\cot^2\theta$ **43.** Sample answer: $x = 45°$ **45.** The functions $\cos\theta$ and $\sin\theta$ can be thought of as the lengths of the legs of a right triangle, and the number 1 can be thought of as the measure of the corresponding hypotenuse. **47.** Sample answer: $\frac{\sin\theta}{\cos\theta} \cdot \sin\theta$ and $\frac{\sin^2\theta}{\cos\theta}$ **49.** $-\frac{4}{3}$ **51.** A **53.** D

55. 2.09 **57.** 0.52 **59.** 0.5 **61.** $d = 4 - \cos\frac{\pi}{4}t$ or

$d = 4 - \cos 90°t$ **63.** $\frac{1093}{9}$ **65.** −3, 2 **67.** 2

Lesson 13-2

1. $\cot\theta + \tan\theta \overset{?}{=} \frac{\sec^2\theta}{\tan\theta}$

$\cot\theta + \tan\theta \overset{?}{=} \frac{\tan^2\theta + 1}{\tan\theta}$

$\cot\theta + \tan\theta \overset{?}{=} \frac{\tan^2\theta}{\tan\theta} + \frac{1}{\tan\theta}$

$\cot\theta + \tan\theta = \tan\theta + \cot\theta$ ✓

3. $\sin\theta \overset{?}{=} \frac{\sec\theta}{\tan\theta + \cot\theta}$

$\sin\theta \overset{?}{=} \frac{\frac{1}{\cos\theta}}{\frac{\sin\theta}{\cos\theta} + \frac{\cos\theta}{\sin\theta}}$

$\sin\theta \overset{?}{=} \frac{\frac{1}{\cos\theta}}{\frac{\sin^2\theta + \cos^2\theta}{\cos\theta\sin\theta}}$

$\sin\theta \overset{?}{=} \frac{1}{\cos\theta} \cdot \frac{\cos\theta\sin\theta}{1}$

$\sin\theta = \sin\theta$ ✓

5. $\tan^2\theta\csc^2\theta \overset{?}{=} 1 + \tan^2\theta$

$\frac{\sin^2\theta}{\cos^2\theta} \cdot \frac{1}{\sin^2\theta} \overset{?}{=} \sec^2\theta$

$\frac{1}{\cos^2\theta} \overset{?}{=} \sec^2\theta$

$\sec^2\theta = \sec^2\theta$ ✓

(7) $\tan^2\theta + 1 \overset{?}{=} \frac{\sec^2\theta}{\tan^2\theta}$

$\tan^2\theta + 1 \overset{?}{=} \frac{\frac{1}{\cos^2\theta}}{\frac{\sin^2\theta}{\cos^2\theta}}$

$\sec^2\theta \overset{?}{=} \frac{1}{\cos^2\theta} \cdot \frac{\cos^2\theta}{\sin^2\theta}$ Invert the denominator and multiply.

$\sec^2\theta \overset{?}{=} \frac{1}{\sin^2\theta}$ Simplify.

$\sec^2\theta = \frac{1}{\sin^2\theta}$

The answer is D.

9. $\cot\theta(\cot\theta + \tan\theta) \overset{?}{=} \csc^2\theta$

$\cot^2\theta + \cot\theta \cdot \tan\theta \overset{?}{=} \csc^2\theta$

$\cot^2\theta + \frac{\cos\theta}{\sin\theta} \cdot \frac{\sin\theta}{\cos\theta} \overset{?}{=} \csc^2\theta$

$\cot^2\theta + 1 \overset{?}{=} \csc^2\theta$

$\csc^2\theta = \csc^2\theta$ ✓

11. $\sin\theta\sec\theta\cot\theta \overset{?}{=} 1$

$\sin\theta \cdot \frac{1}{\cos\theta} \cdot \frac{\cos\theta}{\sin\theta} \overset{?}{=} 1$

$1 = 1$ ✓

13. $\frac{1-2\cos^2\theta}{\sin\theta\cos\theta} \overset{?}{=} \tan\theta - \cot\theta$

$\frac{(1-\cos^2\theta)-\cos^2\theta}{\sin\theta\cos\theta} \overset{?}{=} \tan\theta - \cot\theta$

$\frac{\sin^2\theta}{\sin\theta\cos\theta} - \frac{\cos^2\theta}{\sin\theta\cos\theta} \overset{?}{=} \tan\theta - \cot\theta$

$\frac{\sin\theta}{\cos\theta} - \frac{\cos\theta}{\sin\theta} \overset{?}{=} \tan\theta - \cot\theta$

$\tan\theta - \cot\theta = \tan\theta - \cot\theta$ ✓

15. $\cos\theta \overset{?}{=} \sin\theta\cot\theta$

$\cos\theta \overset{?}{=} \sin\theta \cdot \frac{\cos\theta}{\sin\theta}$

$\cos\theta = \cos\theta$ ✓

17. $\cos\theta\cos(-\theta) - \sin\theta\sin(-\theta) \overset{?}{=} 1$

$\cos\theta\cos\theta - \sin\theta(-\sin\theta) \overset{?}{=} 1$

$\cos^2\theta + \sin^2\theta \overset{?}{=} 1$

$1 = 1$ ✓

19. $\sec\theta - \tan\theta \overset{?}{=} \frac{1-\sin\theta}{\cos\theta}$

$\frac{1}{\cos\theta} - \frac{\sin\theta}{\cos\theta} \overset{?}{=} \frac{1-\sin\theta}{\cos\theta}$

$\frac{1-\sin\theta}{\cos\theta} = \frac{1-\sin\theta}{\cos\theta}$ ✓

21. $\sec\theta\csc\theta \overset{?}{=} \tan\theta + \cot\theta$

$\frac{1}{\cos\theta} \cdot \frac{1}{\sin\theta} \overset{?}{=} \frac{\sin\theta}{\cos\theta} + \frac{\cos\theta}{\sin\theta}$

$\frac{1}{\cos\theta\sin\theta} \overset{?}{=} \frac{\sin^2\theta}{\cos\theta\sin\theta} + \frac{\cos^2\theta}{\cos\theta\sin\theta}$

$\frac{1}{\cos\theta\sin\theta} \overset{?}{=} \frac{\sin^2\theta + \cos^2\theta}{\cos\theta\sin\theta}$

$\frac{1}{\cos\theta\sin\theta} = \frac{1}{\cos\theta\sin\theta}$ ✓

23. $(\sin\theta + \cos\theta)^2 \overset{?}{=} 2 + \sec\theta\csc\theta$

$(\sin\theta + \cos\theta)^2 \overset{?}{=} \left(2 + \frac{1}{\cos\theta\sin\theta}\right) \cdot \frac{\cos\theta\sin\theta}{1}$

$(\sin\theta + \cos\theta)^2 \overset{?}{=} 2\cos\theta\sin\theta + 1$

$(\sin\theta + \cos\theta)^2 \overset{?}{=} \sin^2\theta + \cos^2\theta + 2\cos\theta\sin\theta + \sin^2\theta$

$(\sin\theta + \cos\theta)^2 = (\sin\theta + \cos\theta)^2$ ✓

25. $\csc\theta - 1 \overset{?}{=} \frac{\cot^2\theta}{\csc\theta + 1}$

$\csc\theta - 1 \overset{?}{=} \frac{\csc^2\theta - 1}{\csc\theta + 1}$

$\csc\theta - 1 \overset{?}{=} \frac{(\csc\theta - 1)(\csc\theta + 1)}{\csc\theta + 1}$

$\csc\theta - 1 = \csc\theta - 1$ ✓

27. $\sin\theta\cos\theta\tan\theta + \cos^2\theta \overset{?}{=} 1$

$\sin\theta\cos\theta \cdot \frac{\sin\theta}{\cos\theta} + \cos^2\theta \overset{?}{=} 1$

$\sin^2\theta + \cos^2\theta \overset{?}{=} 1$

$1 = 1$ ✓

29. $\csc^2\theta \overset{?}{=} \cot^2\theta + \sin\theta\csc\theta$

$\csc^2\theta \overset{?}{=} \cot^2\theta + \sin\theta \cdot \frac{1}{\sin\theta}$

$\csc^2\theta \overset{?}{=} \cot^2\theta + 1$

$\csc^2\theta = \csc^2\theta$ ✓

31. $\sin^2\theta + \cos^2\theta \overset{?}{=} \sec^2\theta - \tan^2\theta$

$1 \overset{?}{=} \tan^2\theta + 1 - \tan^2\theta$

$1 = 1$ ✓

33. yes

(35) $\cot(-\theta)\tan(-\theta) \overset{?}{=} \frac{1}{\tan(-\theta)} \cdot \tan(-\theta)$ $\frac{\tan(-\theta)}{\cot(-\theta)} = \frac{1}{\cot(-\theta)}$

$1 = 1$ Simplify.

37. 1 **39.** 1 **41.** $\cos\theta$ **43.** 2 **45.** $\sin\theta$ **47.** 1

49. $y = -\frac{gx^2}{2v_0^2}(1 + \tan^2\theta) + x\tan\theta$

(51) a. $h = \frac{v_0^2\sin^2\theta}{2g}$ Replace v_0 with 47 and g with 9.8.

$= \frac{47^2\sin^2\theta}{2(9.8)}$

$= \frac{2209\sin^2\theta}{19.6}$ Simplify.

$\frac{2209\sin^2 30°}{19.6} \approx 28.2$ m $\theta = 30°$

$\frac{2209\sin^2 45°}{19.6} \approx 56.4$ m $\theta = 45°$

$\frac{2209\sin^2 60°}{19.6} \approx 84.5$ m $\theta = 60°$

$\frac{2209\sin^2 90°}{19.6} \approx 112.7$ m $\theta = 90°$

b. Sample answer: Enter the equation $y = \frac{2209(\sin\theta)^2}{19.6}$.
Use the window
Xmin = 0, Xmax = 180,
Xscl = 10, Ymin = 0,
Ymax = 150, Yscl = 10,
Xres = 1.

[0, 180] scl: 10 by [0, 150] scl: 10

c. $\frac{v_0^2\tan^2\theta}{2g\sec^2\theta} \overset{?}{=} \frac{v_0^2\sin^2\theta}{2g}$

$\frac{v_0^2\left(\frac{\sin^2\theta}{\cos^2\theta}\right)}{2g\left(\frac{1}{\cos^2\theta}\right)} \overset{?}{=} \frac{v_0^2\sin^2\theta}{2g}$

$\frac{v_0^2\sin^2\theta}{2g} = \frac{v_0^2\sin^2\theta}{2g}$ ✓ $\tan^2\theta = \frac{\sin^2\theta}{1}$

$\sec^2\theta = \frac{1}{\cos^2\theta}$ Simplify.

53. $\tan^2\theta = \frac{\sin^2\theta}{\cos^2\theta} = \frac{1-\cos^2\theta}{\cos^2\theta} = \frac{1}{\cos^2\theta} - \frac{\cos^2\theta}{\cos^2\theta}$

$= \sec^2\theta - 1$

55. Sample answer: counterexample 45°, 30°

57. Sample answer: Sine and cosine are the trigonometric functions with which most people are familiar, and all trigonometric expressions can be written in terms of sine and cosine. Also, by rewriting complex trigonometric expressions in terms of sine and cosine it may be easier to perform operations and to apply trigonometric properties.

59. Using the unit circle and the Pythagorean Theorem, we can justify $\cos^2\theta + \sin^2\theta = 1$.

connectED.mcgraw-hill.com

Selected Answers and Solutions

$$= 1 - 2\cos(A - B) + 1$$
$$= 2 - 2\cos(A - B)$$

43. 9 **45.** H

47.

$$\frac{\sin\theta}{\tan\theta} + \frac{\cos\theta}{\cot\theta} \overset{?}{=} \cos\theta + \sin\theta$$
$$\frac{\sin\theta}{\frac{\sin\theta}{\cos\theta}} + \frac{\cos\theta}{\frac{\cos\theta}{\sin\theta}} \overset{?}{=} \cos\theta + \sin\theta$$
$$\sin\theta \cdot \frac{\cos\theta}{\sin\theta} + \cos\theta \cdot \frac{\sin\theta}{\cos\theta} \overset{?}{=} \cos\theta + \sin\theta$$
$$\cos\theta + \sin\theta = \cos\theta + \sin\theta \checkmark$$

49. $\sin^2\theta$ **51.** $\sec\theta$

53. Step 1: $4^1 - 1 = 3$, which is divisible by 3. The statement is true for $n = 1$.
Step 2: Assume that $4^k - 1$ is divisible by 3 for some positive integer k. This means that $4^k - 1 = 3r$ for some whole number r.
Step 3: $4^k - 1 = 3r$
$$4^k = 3r + 1$$
$$4^{k+1} - 1 = 12r + 4$$
$$4^{k+1} - 1 = 12r + 3$$
$$4^{k+1} - 1 = 3(4r + 1)$$
Since r is a whole number, $4r + 1$ is a whole number. Thus, $4^{k+1} - 1$ is divisible by 3, so the statement is true for $n = k + 1$. Therefore, $4^n - 1$ is divisible by 3 for all positive integers n. **55.** -1 **57.** no solution

Lesson 13-4

1. $\frac{\sqrt{15}}{8}, \frac{7}{8}, \sqrt{8 - 2\sqrt{15}}$ **3.** $\frac{120}{169}, \frac{119}{169},$
$\frac{3\sqrt{13}}{13}, \frac{2\sqrt{13}}{13}$ **5.** $\frac{240}{289}, \frac{161}{289}, \frac{4\sqrt{17}}{17}, \frac{\sqrt{17}}{17}$ **7.** $\frac{\sqrt{2 - \sqrt{2}}}{2}$

9a. $d = \frac{v^2\sin 2\theta}{8}$ **9b.** ≈ 81 ft

11. $(\sin\theta + \cos\theta)^2 \overset{?}{=} 1 + 2\sin\theta\cos\theta$
$\sin\theta + \cos\theta)(\sin\theta + \cos\theta) \overset{?}{=} 1 + 2\sin\theta\cos\theta$
$\sin^2\theta + 2\sin\theta\cos\theta + \cos^2\theta \overset{?}{=} 1 + 2\sin\theta\cos\theta$
$1 + 2\sin\theta\cos\theta = 1 + 2\sin\theta\cos\theta \checkmark$

13. $\frac{240}{289}, \frac{161}{289}, \frac{5\sqrt{34}}{34}, \frac{3\sqrt{34}}{34}$

15. Since θ is in the fourth quadrant, sine is negative. So,

$\sin\theta + \cos\theta$	Double-angle identity
$\sin^2\theta + \cos^2\theta = 1$	
$\cos\theta = \frac{1}{5}$	Subtract.

$\sin\theta = -\frac{2\sqrt{6}}{5}$ Take the square root of each side.

$\sin 2\theta = 2\sin\theta\cos\theta$ Double-angle identity

$\sin 2\theta = 2\left(-\frac{2\sqrt{6}}{5}\right)\left(\frac{1}{5}\right)$ $\sin\theta = -\frac{2\sqrt{6}}{5}$ and $\cos\theta = \frac{1}{5}$

$= -\frac{4\sqrt{6}}{25}$ Simplify.

$\cos 2\theta = 1 - 2\sin^2\theta$ Double-angle identity

$= 1 - 2\left(\frac{24}{25}\right)$ $\sin^2\theta = \frac{24}{25}$

$= -\frac{23}{25}$ Simplify.

$\sin\frac{\theta}{2} = \pm\sqrt{\frac{1 - \cos\theta}{2}}$ Half-angle identity

$= \pm\sqrt{\frac{1 - \frac{1}{5}}{2}}$ $\cos\theta = \frac{1}{5}$

---

35. $\cos(A + B) \overset{?}{=} \frac{1 - \tan A\tan B}{\sec A\sec B}$
$\cos(A + B) \overset{?}{=} \frac{1 - \frac{\sin A}{\cos A}\cdot\frac{\sin B}{\cos B}}{\frac{1}{\cos A}\cdot\frac{1}{\cos B}}$
$\cos(A + B) \overset{?}{=} \frac{1 - \frac{\sin A}{\cos A}\cdot\frac{\sin B}{\cos B}}{\frac{1}{\cos A}\cdot\frac{1}{\cos B}}\cdot\frac{\cos A\cos B}{\cos A\cos B}$
$\cos(A + B) \overset{?}{=} \left(1 - \frac{\sin A}{\cos A}\cdot\frac{\sin B}{\cos B}\right)\left(\frac{\cos A\cos B}{\cos A\cos B}\right)$
$\cos(A + B) \overset{?}{=} \cos A\cos B - \sin A\sin B$ Simplify.
$\cos(A + B) = \cos(A + B) \checkmark$ Difference identity

37. $\sin(A + B)\sin(A - B) \overset{?}{=} \sin^2 A - \sin^2 B$
$\sin^2 A - \sin^2 B$
$(\sin A\cos B + \cos A\sin B)(\sin A\cos B - \cos A\sin B) \overset{?}{=} \sin^2 A - \sin^2 B$
$(\sin A\cos B)^2 - (\cos A\sin B)^2 \overset{?}{=} \sin^2 A - \sin^2 B$
$\sin^2 B\cos^2 B - \cos^2 A\sin^2 B \overset{?}{=} \sin^2 A - \sin^2 B$
$\cos^2 A\sin^2 B \overset{?}{=} \sin^2 A\sin^2 B - \sin^2 A\sin^2 B -$
$\sin^2 A(\cos^2 B + \sin^2 B) - \sin^2 B(\sin^2 A + \cos^2 A) \overset{?}{=}$
$\sin^2 A - \sin^2 B$
$(\sin^2 A)(1) - (\sin^2 B)(1) \overset{?}{=} \sin^2 A - \sin^2 B$
$\sin^2 A - \sin^2 B = \sin^2 A - \sin^2 B \checkmark$

39. Sample answer: To determine wireless Internet interference, you need to determine the sine or cosine of the sum or difference of two angles. Interference occurs when waves pass through the same space at the same time. When the combined waves have a greater amplitude, constructive interference results. When the combined waves have a smaller amplitude, destructive interference results.

41. $d = \sqrt{(\cos A - \cos B)^2 + (\sin A - \sin B)^2}$
$d^2 = (\cos A - \cos B)^2 + (\sin A - \sin B)^2$
$d^2 = (\cos^2 A - 2\cos A\cos B + \cos^2 B) + (\sin^2 A - 2\sin A\sin B + \sin^2 B)$
$d^2 = \cos^2 A + \sin^2 A + \cos^2 B + \sin^2 B - 2\cos A\cos B - 2\sin A\sin B$
$d^2 = 1 + 1 - 2\cos A\cos B - 2\sin A\sin B$ $\sin^2 A + \cos^2 A = 1$ and $\sin^2 B$
$d^2 = 2 - 2\cos A\cos B - 2\sin A\sin B + \cos^2 B = 1$

Now find the value of d^2 when the angle having measure $A - B$ is in standard position on the unit circle, as shown in the figure below.

$[\cos(A - B), \sin(A - B)]$
$(1, 0)$

$d^2 = \sqrt{[\cos(A - B) - 1]^2 + [\sin(A - B) - 0]^2}$
$d^2 = [\cos(A - B) - 1]^2 + [\sin(A - B) - 0]^2$
$= [\cos^2(A - B) - 2\cos(A - B) + 1] + \sin^2(A - B)$
$= \cos^2(A - B) + \sin^2(A - B) - 2\cos(A - B) + 1$

23a. $y = 30.9\sin\left(\frac{\pi}{6}x - 2.09\right) + 42.65$ **23b.** The new function represents the average of the high and low temperatures for each month. **25.** $\sqrt{2} - \sqrt{6}$
27. $-2 + \sqrt{3}$ **29.** $2 - \sqrt{3}$

31a. Let X be the endpoint of the segment that is 8 inches long.
$\sin(m\angle BAC)$
$= \sin(m\angle BAX + m\angle XAC)$
$= \sin(m\angle BAX)\cos(m\angle XAC) + \cos(m\angle BAX)$
$\sin(m\angle XAC)$
$= \frac{8\sqrt{3}}{16}\cdot\frac{8}{10} + \frac{8}{16}\cdot\frac{6}{10}$ $\sin = \frac{\text{opp}}{\text{hyp}}$ and $\cos = \frac{\text{adj}}{\text{hyp}}$
$= \frac{4\sqrt{3}}{10} + \frac{3}{10}$ Multiply.
$= \frac{3 + 4\sqrt{3}}{10}$ Add.

b. Let X be the endpoint of the segment that is 8 inches long.
$\cos(m\angle BAC)$
$= \cos(m\angle BAX + m\angle XAC)$
$= \cos(m\angle BAX)\cos(m\angle XAC) - \sin(m\angle BAX)$
$\sin(m\angle XAC)$
$= \frac{8}{16}\cdot\frac{8}{10} - \frac{8\sqrt{3}}{16}\cdot\frac{6}{10}$ $\sin = \frac{\text{opp}}{\text{hyp}}$ and $\cos = \frac{\text{adj}}{\text{hyp}}$
$= \frac{4}{10} - \frac{3\sqrt{3}}{10}$ Multiply.
$= \frac{4 - 3\sqrt{3}}{10}$ Add.

c. $\cos(m\angle BAC) = \frac{4 - 3\sqrt{3}}{10}$
$m\angle BAC = \cos^{-1}\left(\frac{3 + 4\sqrt{3}}{10}\right)$
$\approx 96.9°$

d. Since $m\angle BAC \neq 90$, the triangle formed is not a right triangle.

33a.

A	B	$\sin A$	$\sin B$	$\sin(A + B)$	$\sin A + \sin B$
30°	90°	$\frac{1}{2}$	1	$\frac{\sqrt{3}}{2}$	$\frac{3}{2}$
45°	60°	$\frac{\sqrt{2}}{2}$	$\frac{\sqrt{3}}{2}$	$\frac{\sqrt{2} + \sqrt{6}}{4}$	$\frac{\sqrt{2} + \sqrt{3}}{2}$
60°	45°	$\frac{\sqrt{3}}{2}$	$\frac{\sqrt{2}}{2}$	$\frac{\sqrt{2} + \sqrt{6}}{4}$	$\frac{\sqrt{2} + \sqrt{3}}{2}$
90°	30°	1	$\frac{1}{2}$	$\frac{\sqrt{3}}{2}$	$\frac{3}{2}$

33b.

33c. No; a counterexample is: $\cos(30° + 45°) = \cos 30° + \cos 45°$, which equals $\frac{\sqrt{2}}{2} + \frac{\sqrt{2}}{2}$ or about 1.5731. Since a cosine value cannot be greater than 1, this statement must be false.

If we divide each term of the identity $\cos^2\theta + \sin^2\theta = 1$ by $\cos^2\theta$, we can justify $1 + \tan^2\theta = \sec^2\theta$.
$$\frac{\cos^2\theta}{\cos^2\theta} + \frac{\sin^2\theta}{\cos^2\theta} = \frac{1}{\cos^2\theta}$$
$$1 + \tan^2\theta = \sec^2\theta$$

If we divide each term of the identity $\cos^2\theta + \sin^2\theta = 1$ by $\sin^2\theta$, we can justify $\cot^2\theta + 1 = \csc^2\theta$.
$$\frac{\cos^2\theta}{\sin^2\theta} + \frac{\sin^2\theta}{\sin^2\theta} = \frac{1}{\sin^2\theta}$$
$$\cot^2\theta + 1 = \csc^2\theta$$

61. H **63.** G **65.** $\frac{\sqrt{5}}{3}$ **67.** $\frac{3}{5}$ **69.** $4

71. $(1, -6 \pm 2\sqrt{5})$;
$(1, -6 \pm 3\sqrt{5})$;

$y + 6 = \pm\frac{2\sqrt{5}}{5}(x - 1)$

73. $\frac{12 + 7\sqrt{2}}{23}$ **75.** $\sqrt{x} + 1$

Lesson 13-3

1. $\cos 165° = \cos(120° + 45°)$

3. $\frac{\sqrt{6} - \sqrt{2}}{4}$ **5.** $\frac{\sqrt{2}}{2}$ **7a.** 0 **7b.** The interference is destructive. The signals cancel each other completely.

9. $\cos\left(\frac{3\pi}{2} - \theta\right) \overset{?}{=} -\sin\theta$
$\cos\frac{3\pi}{2}\cos\theta + \sin\frac{3\pi}{2}\sin\theta \overset{?}{=} -\sin\theta$
$(0)\cos\theta + (-1)\sin\theta \overset{?}{=} -\sin\theta$
$-\sin\theta = -\sin\theta \checkmark$

11. $\sin(\theta + \pi)$
$\sin\theta\cos\pi + \cos\theta\sin\pi \overset{?}{=} -\sin\theta$
$(\sin\theta)(-1) + (\cos\theta)(0) \overset{?}{=} -\sin\theta$
$-\sin\theta = -\sin\theta \checkmark$

13. $\frac{\sqrt{2}}{2}$ **15.** $\frac{\sqrt{6} - \sqrt{2}}{4}$ **17.** $\frac{\sqrt{2} + \sqrt{6}}{4}$

19. $\cos\left(\frac{\pi}{2} + \theta\right) \overset{?}{=} -\sin\theta$
$\cos\frac{\pi}{2}\cos\theta - \sin\frac{\pi}{2}\sin\theta \overset{?}{=} -\sin\theta$
$(0)(\cos\theta) - (1)(\sin\theta) \overset{?}{=} -\sin\theta$
$-\sin\theta = -\sin\theta \checkmark$

21. $\cos(180° + \theta) \overset{?}{=} -\cos\theta$
$\cos 180°\cos\theta - \sin 180°\sin\theta \overset{?}{=} -\cos\theta$
$-1\cdot\cos\theta - 0\cdot\sin\theta \overset{?}{=} -\cos\theta$
$-\cos\theta = -\cos\theta \checkmark$

77. 17, 26, 35
79. −12, −9, −6
81.

$f(x) = \dfrac{1}{(x+3)^2}$

83.

$f(x) = \dfrac{x+2}{x^2-x-6}$

day of the year occurs around June 22, the days between February 19 and October 20 must increase in length until June 22 and then decrease in length until October 20. **23.** $\frac{3\pi}{4} + \pi k$ **25.** $\frac{\pi}{2} + \pi k$, $\frac{\pi}{6} + 2\pi k$, $\frac{5\pi}{6} + 2\pi k$ **27.** 0° + k · 45° or 0° + k · $\frac{\pi}{4}$ **29.** 90° + k · 180°, 60° + k · 360°, 300° + k · 360° **31.** 135°, 225° **33.** $\frac{\pi}{6}$ **35.** 210°, 330°

37.
$2\sin^2\theta = \cos\theta + 1$ Original equation
$2(1-\cos^2\theta) = \cos\theta + 1$
$2 - 2\cos^2\theta = \cos\theta + 1$ Simplify.
$-2\cos^2\theta - \cos\theta + 1 = 0$ Subtract cos θ + 1 from each side and simplify.
$(-2\cos\theta + 1)(\cos\theta + 1) = 0$ Factor.
$-2\cos\theta + 1 = 0$ or $\cos\theta + 1 = 0$ Zero Product Property
$\cos\theta = \frac{1}{2}$ $\cos\theta = -1$

The solutions of $\cos\theta = \frac{1}{2}$ are $\frac{\pi}{3} + 2\pi k$ and $\frac{5\pi}{3} + 2\pi k$.
The solution of $\cos\theta = -1$ is $\pi + 2\pi k$.

39. 0 + 2πk **41.** 0° + k · 180° **43.** 30° + k · 360°, 150° + k · 360°, 330° + k · 360° **45.** $\frac{7\pi}{6} + 2\pi k$, $\frac{11\pi}{6} + 2\pi k$ or **47.** 0 + 2πk, $\frac{\pi}{2} + k\pi$ **49.** 11 m **49a.** 7:00 A.M. and 7:00 P.M. **51.** $\frac{\pi}{6} + 2\pi k$, $\frac{5\pi}{6} + 2\pi k$, $\frac{5\pi}{4} + 2\pi k$, **53.** 120° + 360°k, 240° + 360°k **55.** $\frac{\pi}{6} + 2\pi k$, $\frac{5\pi}{6} + 2\pi k$

57.
$D = 0.5\sin(6.5x)\sin(2500t)$ Original equation
$0.01 = 0.5\sin(6.5 \cdot 500)\sin(2500t)$
$D = 0.01$ mm and $x = 0.5 \cdot 1000$ or 500 mm
$0.01 = 0.5\sin(3250)\sin(2500t)$ Simplify.
$0.1152 \approx \sin(2500t)$ Divide each side by 0.5 sin (3250).
$Sin^{-1}(0.1152) \approx 2500t$ Use the $\sin^{-1}$ function.
$6.6152 \approx 2500t$ Divide each side by 2500.
$0.0026 \approx t$
The time is about 0.0026 second.

59. $\frac{\pi}{3} < x < \pi$ or $\frac{5\pi}{3} < x < 2\pi$ **61.** Sample answer: All trigonometric functions are periodic. Therefore, once one or more solutions are found for a certain interval, there will be additional solutions that can be found by adding integral multiples of the period of the function to those solutions. **63.** 0, b, 2b **65.** A **67.** E

69. $\frac{\sqrt{2}-\sqrt{2}}{2}$ **71.** $\frac{\sqrt{2}-\sqrt{3}}{2}$

73.
$\cos(90° + \theta) \stackrel{?}{=} -\sin\theta$
$\cos 90°\cos\theta - \sin 90°\sin\theta \stackrel{?}{=} -\sin\theta$
$0 - 1\sin\theta \stackrel{?}{=} -\sin\theta$
$-\sin\theta = -\sin\theta$ ✓

75.
$\sin(90° - \theta) \stackrel{?}{=} \cos\theta$
$\sin 90°\cos\theta - \cos 90°\sin\theta \stackrel{?}{=} \cos\theta$
$1 \cdot \cos\theta - 0 \cdot \sin\theta \stackrel{?}{=} \cos\theta$
$\cos\theta = \cos\theta$ ✓

Chapter 13   Study Guide and Review

1. difference of angles identity **3.** trigonometric identity **5.** trigonometric equation **7.** reciprocal identities **9.** Pythagorean identity **11.** $-\sqrt{3}$ **13.** $-\frac{4}{5}$ **15.** First find the length of the diagonal: $75^2 + 110^2 = c^2$; $5625 + 12{,}100 = c^2$; $17{,}725 = c^2$; $c = 5\sqrt{709}$; $\sin\theta = \frac{75}{5\sqrt{709}} = \frac{15\sqrt{709}}{709}$ **17.** sec θ **19.** sec θ

21.
$\cos\theta + \dfrac{\sin\theta}{\tan\theta} \stackrel{?}{=} \sin\theta + \cos\theta$
$\cos\theta + \sin\theta \div \dfrac{\sin\theta}{\cos\theta} \stackrel{?}{=} \sin\theta + \cos\theta$
$\cos\theta + \sin\theta \cdot \dfrac{\cos\theta}{\sin\theta} \stackrel{?}{=} \sin\theta + \cos\theta$
$\cos\theta + \cos\theta \stackrel{?}{=} \sin\theta + \cos\theta$ ✓

23. $\tan^2\theta + 1 = \left(\sqrt{\frac{7}{3}}\right)^2 + 1 = \frac{7}{3} + 1 = \frac{7}{9} + \frac{9}{9} = \frac{16}{9}$; $\sec^2\left(\frac{4}{3}\right)^2 = \frac{16}{9}$

25. $\frac{\sqrt{6}+\sqrt{2}}{4}$ **27.** $\frac{\sqrt{6}+\sqrt{2}}{4}$ **29.** $\frac{-\sqrt{6}+\sqrt{2}}{4}$

31. $\sin\left(\frac{3\pi}{2} - \theta\right) \stackrel{?}{=} -\cos\theta$
$\sin\frac{3\pi}{2}\cos\theta - \cos\frac{3\pi}{2}\sin\theta \stackrel{?}{=} -\cos\theta$
$(-1)\cos\theta - (0)\sin\theta \stackrel{?}{=} -\cos\theta$
$-\cos\theta = -\cos\theta$ ✓

33. $\sin 2\theta = \frac{24}{25}$, $\cos 2\theta = \frac{7}{25}$, $\sin\frac{\theta}{2} = \frac{\sqrt{10}}{10}$, and $\cos\frac{\theta}{2} = \frac{3\sqrt{10}}{10}$ **35.** $\sin 2\theta = -\frac{4\sqrt{5}}{9}$, $\cos 2\theta = -\frac{1}{9}$, **37.** 60°, 300° $\sin\frac{\theta}{2} = \frac{3\sqrt{30}}{6}$, and $\cos\frac{\theta}{2} = \frac{\sqrt{6}}{6}$ **41.** $\frac{\pi}{3}$, $\frac{5\pi}{3}$ **39.** 90°, 210°, 270°, 330°

---

31. $\frac{24}{25}$, $\frac{7}{25}$, $\frac{24}{7}$ **33.** $-\frac{3}{5}$, $-\frac{4}{5}$, $\frac{3}{4}$ **35.** $-\frac{4\sqrt{21}}{25}$, $\frac{17}{25}$, $-\frac{4\sqrt{21}}{17}$

37. No; Teresa incorrectly added the square roots. He used the half-angle identity incorrectly. Nathan used sin 30° in the formula instead of first finding the cosine. **39.** If you are only given the value of $\cos\theta$, then $\cos 2\theta = 2\cos^2\theta - 1$ is the best identity to use. If you are only given the value of $\sin\theta$, then $\cos 2\theta = 1 - 2\sin^2\theta$ is the best identity to use. If you are given the values of both $\cos\theta$ and $\sin\theta$, then $\cos 2\theta = \cos^2\theta - \sin^2\theta$ works just as well as the other two.

41.
$1 - 2\sin^2\theta = \cos 2\theta$ Double-angle identity
$1 - 2\sin^2\frac{A}{2} = \cos A$ Substitute $\frac{A}{2}$ for θ and A for 2θ.
$\sin^2\frac{A}{2} = \frac{1-\cos A}{2}$ Solve for $\sin^2\frac{A}{2}$.
$\sin\frac{A}{2} = \pm\sqrt{\frac{1-\cos A}{2}}$ Take the square root of each side.

Find $\cos\frac{\theta}{2}$.
$2\cos^2\theta - 1 = \cos 2\theta$ Double-angle identity
$2\cos^2\frac{A}{2} - 1 = \cos A$ Substitute $\frac{A}{2}$ for θ and A for 2θ.
$\cos^2\frac{A}{2} = \frac{1+\cos A}{2}$ Solve for $\cos^2\frac{A}{2}$.
$\cos\frac{A}{2} = \pm\sqrt{\frac{1+\cos A}{2}}$ Take the square root of each side.

43. 22.5 **45.** G **47.** $\frac{\sqrt{2}}{2}$ **49.** $\frac{-\sqrt{6}-\sqrt{2}}{4}$ **51.** $\frac{\sqrt{3}}{2}$

53.
$\cot\theta + \sec\theta \stackrel{?}{=} \frac{\cos\theta + \sin\theta}{\sin\theta}$
$\cot\theta + \sec\theta \stackrel{?}{=} \frac{\sin\theta\cos\theta + \sin\theta}{\sin\theta\cos\theta}$
$\cot\theta + \sec\theta \stackrel{?}{=} \frac{\cos\theta}{\sin\theta} + \frac{1}{\cos\theta}$
$\cot\theta + \sec\theta \stackrel{?}{=} \cot\theta + \sec\theta$ ✓

55. sines; $B \approx 102°$, $C \approx 44°$, $b \approx 21.0$ or $B \approx 10°$, $C \approx 136°$, $b \approx 3.7$ **57.** sines; $A = 80°$, $a \approx 10.9$, $c \approx 5.4$ **59.** {−4, 7}

Lesson 13-5

1. 210°, 330° **3.** 60°, 180°, or 300° **5.** 150°, 210°

7.
$\cos 2\theta + 15\sin\theta - 8 = 0$ Original equation
$\cos 2\theta = 8 - 15\sin\theta$
$1 - 2\sin^2\theta + 15\sin\theta - 8 = 0$ Add 15 sin θ − 8 to each side.
$-2\sin^2\theta + 15\sin\theta - 7 = 0$ Double-angle identity
$2\sin^2\theta - 15\sin\theta + 7 = 0$ Simplify.
$(-2\sin\theta + 1)(\sin\theta - 7) = 0$ Factor.
$-2\sin\theta + 1 = 0$ or $\sin\theta - 7 = 0$ Zero Product Property
$\sin\theta = \frac{1}{2}$ $\sin\theta = 7$
$\theta = 30°$ or $150°$ no solution since $0 \le \sin\theta \le 1$

9. $\frac{\pi}{6} + 2\pi k$ or $\frac{5\pi}{6} + 2\pi k$ **11.** $\frac{3\pi}{2} + 2\pi k$ **13.** $\pi + 2\pi k$ **15.** 90° + k · 180° **17.** 45° + k · 90° **19.** 270° + k · 360°

21a. There will be $10\frac{1}{2}$ hours of daylight 213 and 335 days after March 21; that is, on October 20 and February 19. Every day from February 19 to October 20; sample explanation: Since the longest

---

$= \pm\sqrt{\frac{2}{5}}$ Simplify.
$= \pm\frac{\sqrt{2}}{\sqrt{5}} \cdot \frac{\sqrt{5}}{\sqrt{5}}$ or $\pm\frac{\sqrt{10}}{5}$ Rationalize the denominator.
If θ is between 270° and 360° $\frac{\theta}{2}$ is between 135° and 180°. So, $\sin\frac{\theta}{2}$ is $-\frac{\sqrt{10}}{5}$.

$\cos\frac{\theta}{2} = \pm\sqrt{\frac{1+\cos\theta}{2}}$ Half-angle identity
$= \pm\sqrt{\frac{1+\frac{1}{5}}{2}}$ $\cos\theta = \frac{1}{5}$
$= \pm\sqrt{\frac{3}{5}}$ Simplify.
$= \pm\frac{\sqrt{3}}{\sqrt{5}} \cdot \frac{\sqrt{5}}{\sqrt{5}}$ or $\pm\frac{\sqrt{15}}{5}$ Rationalize the denominator.
If θ is between 270° and 360° $\frac{\theta}{2}$ is between 135° and 180°. So, $\cos\frac{\theta}{2}$ is $-\frac{\sqrt{15}}{5}$.

17. $-\frac{4}{5}$, $-\frac{3}{5}$ **19.** $\frac{\sqrt{2+\sqrt{2}}}{2}$ **21.** $\sqrt{3} - 2$ **23.** $\sqrt{2} - 1$

25. $P = I_0^2 R\sin^2\theta t$
$P = I_0^2 R(\cos^2\theta t - \cos 2\theta t)$
$P = I_0^2 R\left(\frac{1}{2}\cos 2\theta t + \frac{1}{2} - \cos 2\theta t\right)$
$P = I_0^2 R\left(\frac{1}{2} - \frac{1}{2}\cos 2\theta t\right)$
$P = \frac{1}{2}I_0^2 R - \frac{1}{2}I_0^2 R\cos 2\theta t$

27. $1 + \frac{1}{2}\sin 2\theta \stackrel{?}{=} \frac{\sec\theta + \sin\theta}{\sec\theta}$
$\stackrel{?}{=} \frac{\frac{1}{\cos\theta} + \sin\theta}{\frac{1}{\cos\theta}}$
$\stackrel{?}{=} \frac{\frac{1}{\cos\theta} + \sin\theta}{\frac{1}{\cos\theta}} \cdot \frac{\cos\theta}{\cos\theta}$
$\stackrel{?}{=} 1 + \frac{1}{2} \cdot 2\sin\theta\cos\theta$
$\stackrel{?}{=} 1 + \frac{1}{2}\sin 2\theta$ ✓

29. $\tan\frac{\theta}{2} = \frac{\sin\theta}{1+\cos\theta}$
$\tan\frac{\theta}{2} = \frac{\sin 2\left(\frac{\theta}{2}\right)}{1 + \cos 2\left(\frac{\theta}{2}\right)}$
$\tan\frac{\theta}{2} = \frac{2\sin\frac{\theta}{2}\cos\frac{\theta}{2}}{1 + 2\cos^2\frac{\theta}{2} - 1}$
$\tan\frac{\theta}{2} = \frac{2\sin\frac{\theta}{2}\cos\frac{\theta}{2}}{2\cos^2\frac{\theta}{2}}$
$\tan\frac{\theta}{2} = \frac{\sin\frac{\theta}{2}}{\cos\frac{\theta}{2}}$
$\tan\frac{\theta}{2} = \tan\frac{\theta}{2}$ ✓

Glossary/Glosario

MultilingualeGlossary

Go to connectED.mcgraw-hill.com for a glossary of terms in these additional languages:

Arabic	Chinese	Hmong	Vietnamese
Bengali	English	Korean	Spanish	Tagalog
Brazilian Portugese	Haitian Creole	Russian	Urdu

English

A

absolute value (p. 27) A number's distance from zero on the number line, represented by |x|.

absolute value function (p. 103) A function written as $f(x) = |x|$, where $f(x) = \begin{cases} x \text{ if } x > 0 \\ 0 \text{ if } x = 0 \\ -x \text{ if } x < 0 \end{cases}$, for all values of x

algebraic expression (p. 5) An expression that contains at least one variable.

alternative hypothesis (p. 771) Mutually exclusive to the null hypothesis. It is stated as an inequality using $\neq$, $<$, $\leq$, $>$, or $\geq$.

amplitude (p. 837) For functions in the form $y = a \sin b\theta$ or $y = a \cos b\theta$, the amplitude is $|a|$.

angle of depression (p. 794) The angle between a horizontal line and the line of sight from the observer to an object at a lower level.

angle of elevation (p. 794) The angle between a horizontal line and the line of sight from the observer to an object at a higher level.

Arccosine (p. 853) The inverse of $y = \cos x$, written as $x = \text{Arccos } y$.

Arcsine (p. 853) The inverse of $y = \sin x$, written as $x = \text{Arcsin } y$.

Arctangent (p. 853) The inverse of $y = \tan x$ written as $x = \text{Arctan } y$.

arithmetic mean (p. 667) The terms between any two nonconsecutive terms of an arithmetic sequence.

arithmetic sequence (p. 659) A sequence in which each term after the first is found by adding a constant, the common difference d, to the previous term.

arithmetic series (p. 668) The indicated sum of the terms of an arithmetic sequence.

asymptote (p. 451) A line that a graph approaches.

axis of symmetry (p. 220) A line about which a figure is symmetric.

axis of symmetry

B

$\frac{1}{b^n}$ **(p. 422)** For any real number b and for any positive integer n, $b^{\frac{1}{n}} = \sqrt[n]{b}$, except when $b < 0$ and n is even.

bias (p. 723) An error that results in a misrepresentation of members of a population.

binomial distribution (p. 754) A distribution that shows the probabilities of the outcomes of a binomial experiment.

binomial experiment (p. 752) An experiment in which there are exactly two possible outcomes for each trial, a fixed number of independent trials, and the probabilities for each trial are the same.

Binomial Theorem (p. 699) If n is a nonnegative integer, then $(a + b)^n = 1a^n b^0 + \frac{n}{1} a^{n-1} b^1 + \frac{n(n+1)}{1 \cdot 2} a^{n-2} b^2 + \cdots + 1a^0 b^n$.

bivariate data (p. 92) Data consisting of pairs of values.

boundary (p. 117) A line or curve that separates the coordinate plane into two regions.

bounded (p. 154) A region is bounded when the graph of a system of constraints is a polygonal region.

box-and-whisker plot (p. 735) A diagram that divides a set of data into four parts using the median and quartiles. A box is drawn around the quartile values and whiskers extend from each quartile to the extreme data points.

Español

A

valor absoluto Distancia entre un número y cero en una recta numérica; se denota con |x|.

función del valor absoluto Una función que se escribe $f(x) = |x|$, donde $f(x) = \begin{cases} x \text{ si } x > 0 \\ 0 \text{ si } x = 0 \\ -x \text{ si } x < 0 \end{cases}$, para todos los valores de x.

expresión algebraica Expresión que contiene al menos una variable.

hipótesis alternativa Mutuamente exclusiva a la hipótesis nula. Se indica como usar de la desigualdad $\neq$, $<$, $\leq$, $>$, $0 \geq$.

amplitud Para funciones de la forma $y = a \sin b\theta$ o $y = a \cos b\theta$, la amplitud es $|a|$.

ángulo de depresión Ángulo entre una recta horizontal y la línea visual de un observador a una figura en un nivel inferior.

ángulo de elevación Ángulo entre una recta horizontal y la línea visual de un observador a una figura en un nivel superior.

arcocoseno La inversa de $y = \cos x$, que se escribe como $x = \text{arccos } y$.

arcoseno La inversa de $y = \sin x$, que se escribe como $x = \text{arcsen } y$.

arcotangente La inversa de $y = \tan x$ que se escribe como $x = \text{arctan } y$.

media aritmética Cualquier término entre dos términos no consecutivos de una sucesión aritmética.

sucesión aritmética Sucesión en que cualquier término después del primero puede hallarse sumando una constante, la diferencia común d, al término anterior.

serie aritmética Suma específica de los términos de una sucesión aritmética.

asíntota Recta a la que se aproxima una gráfica.

eje de simetría Recta respecto a la cual una figura es simétrica.

eje de simetría

B

$\frac{1}{b^n}$ Para cualquier número real b y para cualquier entero positivo n, $b^{\frac{1}{n}} = \sqrt[n]{b}$, excepto cuando $b < 0$ y n es par.

sesgo Error que resulta en la representación errónea de los miembros de una población.

distribución binómica Distribución que muestra las probabilidades de los resultados de un experimento binómico.

experimento binomial Experimento con exactamente dos resultados posibles para cada prueba, un número fijo de pruebas independientes y en el cual cada prueba tiene igual probabilidad.

teorema del binomio Si n es un entero no negativo, entonces $(a + b)^n = 1a^n b^0 + \frac{n}{1} a^{n-1} b^1 + \frac{n(n+1)}{1 \cdot 2} a^{n-2} b^2 + \cdots + 1a^0 b^n$.

datos bivariados Datos que consiste en pares de valores.

frontera Recta o curva que divide un plano de coordenadas en dos regiones.

acotada Una región está acotada cuando la gráfica de un sistema de restricciones es una región poligonal.

diagrama de caja y bigotes Diagrama que divide un conjunto de datos en cuatro partes usando la mediana y los cuartiles. Se dibuja una caja alrededor de los cuartiles y se extienden patillas de cada uno de ellos a los valores extremos.

connectED.mcgraw-hill.com **R125**

Glossary/Glosario

Glossary/Glosario

English (R126–R127)

break-even point (p. 136) The point at which the income equals the cost.

Cartesian coordinate plane (p. P4) A plane divided into four quadrants by the intersection of the x-axis and the y-axis at the origin.

census (p. 724) A survey in which every member of the population is polled.

center of a circle (p. 607) The point from which all points on a circle are equidistant.

center of a hyperbola (p. 624) The midpoint of the segment whose endpoints are the foci.

center of an ellipse (p. 615) The point at which the major axis and minor axis of an ellipse intersect.

central angle (p. 802) An angle with a vertex at the center of the circle and sides that are radii.

Change of Base Formula (p. 494) For all positive numbers a, b, and n, where $a \neq 1$ and $b \neq 1$, $\log_b n = \dfrac{\log_a n}{\log_a b}$.

circle (p. 607) The set of all points in a plane that are equidistant from a given point in the plane, called the center.

circular function (p. 830) A function defined using a unit circle.

coefficient matrix (p. 192) A matrix that contains only the coefficients of a system of equations.

combination (p. P11) An arrangement of objects in which order is not important.

combined variation (p. 565) When one quantity varies directly and/or inversely as two or more other quantities.

common difference (p. 659) The difference between the successive terms of an arithmetic sequence.

common logarithms (p. 492) Logarithms that use 10 as the base.

common ratio (p. 661) The ratio of successive terms of a geometric sequence.

completing the square (p. 257) A process used to make a quadratic expression into a perfect square trinomial.

complex conjugates (p. 249) Two complex numbers of the form $a + bi$ and $a - bi$.

complex fraction (p. 532) A rational expression whose numerator and/or denominator contains a rational expression.

complex number (p. 247) Any number that can be written in the form $a + bi$, where a and b are real numbers and i is the imaginary unit.

composition of functions (p. 387) A function is performed, and then a second function is performed on the result of the first function. The composition of f and g is denoted by $f \circ g$, and $[f \circ g](x) = f[g(x)]$.

compound event (p. P14) Two or more simple events.

compound inequality (p. 41) Two inequalities joined by the word *and* or *or*.

compound interest (p. 462) Interest paid on the principal of an investment and any previously earned interest.

conditional probability (p. P16) The probability of an event occurring given that another event has already occurred.

confidence interval (p. 769) An estimate of a population parameter stated as a range with a specific degree of certainty.

conic section (p. 632) Any figure that can be obtained by slicing a double cone.

conjugate axis (p. 624) The segment of length $2b$ units that is perpendicular to the transverse axis at the center.

conjugates (p. 418) Binomials of the form $a\sqrt{b} + c\sqrt{d}$ and $a\sqrt{b} - c\sqrt{d}$, where a, b, c, and d are rational numbers.

consistent (p. 137) A system of equations that has at least one ordered pair that satisfies both equations.

constant difference (p. 627) The absolute value of the difference between the distances from any point on the hyperbola to the foci of the hyperbola.

Español

el punto de equilibrio Cuando el punto la renta iguala la casualidad del coste.

C

plano de coordenadas cartesiano Plano dividido en cuatro cuadrantes mediante la intersección en el origen de los ejes x y y.

censo Examen en el cual voten a cada miembro de la población.

centro de un círculo El punto desde el cual todos los puntos de un círculo están equidistantes.

centro de una hipérbola Punto medio del segmento cuyos extremos son los focos.

centro de una elipse Punto de intersección de los ejes mayor y menor de una elipse.

ángulo central Ángulo cuyo vértice es el centro del círculo y cuyos lados son radios.

fórmula del cambio de base Para todo número positivo a, b y n, donde $a \neq 1$ y $b \neq 1$, $\log_d n = \dfrac{\log_a n}{\log_a d}$.

círculo Conjunto de todos los puntos en un plano que equidistan de un punto dado del plano llamado centro.

funciones circulares Funciones definidas en un círculo unitario.

matriz coeficiente Una matriz que contiene solamente los coeficientes de un sistema de ecuaciones.

combinación Arreglo de elementos en que el orden no es importante.

variación combinada Cuando una cantidad varía directamente e inverso como dos o más otras cantidades.

diferencia común Diferencia entre términos consecutivos de una sucesión aritmética.

logaritmos comunes El logaritmo de base 10.

razón común Razón entre términos consecutivos de una sucesión geométrica.

completar el cuadrado Proceso mediante el cual una expresión cuadrática se transforma en un trinomio cuadrado perfecto.

conjugados complejos Dos números complejos de la forma $a + bi$ y $a - bi$.

fracción compleja Expresión racional cuyo numerador o denominador contiene una expresión racional.

número complejo Cualquier número que puede escribirse de la forma $a + bi$, donde a y b son números reales e i es la unidad imaginaria.

composición de funciones Se evalúa una función y luego se evalúa una segunda función en el resultado de la primera función. La composición de f y g se define con $f \circ g$, y $[f \circ g](x) = f[g(x)]$.

evento compuesto Dos o más eventos simples.

desigualdad compuesta Dos desigualdades unidas por las palabras *y* u *o*.

interés compuesto Interés obtenido tanto sobre la inversión inicial como sobre el interés conseguido.

probabilidad condicional Probabilidad de un evento dado que otro evento ya ha ocurrido.

intervalo de la confianza Una estimación de un parámetro de la población indicado como gama con un grado específico de la certeza.

sección cónica Cualquier figura obtenida mediante el corte de un cono doble.

eje conjugado El segmento de $2b$ unidades de longitud que es perpendicular al eje transversal en el centro.

conjugados Binomios de la forma $a\sqrt{b} + c\sqrt{d}$ y $a\sqrt{b} - c\sqrt{d}$, donde a, b, c, y d son números racionales.

consistente Sistema de ecuaciones para el cual existe al menos un par ordenado que satisface ambas ecuaciones.

diferencia constante El valor absoluto de la diferencia entre las distancias de cualquier punto en la hipérbola a los focos de la hipérbola.

connectED.mcgraw-hill.com

constant function (p. 109) A linear function of the form $f(x) = b$.

constant matrix (p. 200) A matrix that contains the constants of a system.

constant of variation (pp. 90, 562) The constant k used with direct or inverse variation.

constant sum (p. 616) The sum of the distances from the foci to any point on the ellipse.

constant term (p. 219) In $f(x) = ax^2 + bx + c$, c is the constant term.

constraints (p. 29) Conditions given to variables, often expressed as linear inequalities.

continuous random variable (p. 742) The numerical outcome of a random event that can take on any value.

continuous relation (p. 62) A relation that can be graphed with a line or smooth curve.

convergent series (p. 683) An infinite series with a sum.

correlation coefficient (p. 94) A measure that shows how well data are modeled by a linear equation.

cosecant (p. 790) For any angle, with measure α, a point $P(x, y)$ on its terminal side, $r = \sqrt{x^2 + y^2}$, $\csc \alpha = \dfrac{r}{y}$.

cosine (p. 790) For any angle, with measure α, a point $P(x, y)$ on its terminal side, $r = \sqrt{x^2 + y^2}$, $\cos \alpha = \dfrac{x}{r}$.

cotangent (p. 790) For any angle, with measure α, a point $P(x, y)$ on its terminal side, $r = \sqrt{x^2 + y^2}$, $\cot \alpha = \dfrac{x}{y}$.

coterminal angles (p. 800) Two angles in standard position that have the same terminal side.

co-vertices (pp. 615, 624)
ellipse—The endpoints of the minor axis.
hyperbola—The endpoints of the conjugate axis.

Cramer's Rule (p. 192) A method that uses determinants to solve a system of linear equations.

critical region (p. 771) The range of values that suggests a significant enough difference to reject the null hypothesis.

cycle (p. 831) One complete pattern of a periodic function.

función constante Función lineal de la forma $f(x) = b$.

matriz constante Una matriz que contiene las constantes de un sistema.

constante de variación La constante k que se usa en variación directa o inversa.

suma constante La suma de las distancias de los focos a cualquier punto en la elipse.

término constante En $f(x) = ax^2 + bx + c$, c es el término constante.

restricciones Condiciones a que están sujetas las variables, a menudo escritas como desigualdades lineales.

variable aleatoria continua Resultado numérico de un evento aleatorio que puede tomar cualquier valor.

relación continua Relación cuya gráfica puede ser una recta o una curva suave.

serie convergente Serie infinita con una suma.

coeficiente de correlación Una medida que demuestra cómo los datos bien son modelados por una ecuación linear.

cosecante Para cualquier ángulo de medida α, un punto $P(x, y)$ en su lado terminal, $r = \sqrt{x^2 + y^2}$, $\csc \alpha = \dfrac{r}{y}$.

coseno Para cualquier ángulo de medida α, un punto $P(x, y)$ en su lado terminal, $r = \sqrt{x^2 + y^2}$, $\cos \alpha = \dfrac{x}{r}$.

cotangente Para cualquier ángulo de medida α, un punto $P(x, y)$ en su lado terminal, $r = \sqrt{x^2 + y^2}$, $\cot \alpha = \dfrac{x}{y}$.

ángulos coterminales Dos ángulos en posición estándar que tienen el mismo lado terminal.

co-cimas
(elipse)—Puntos finales del eje de menor importancia.
(hipérbola)—Puntos finales del eje conjugal.

regla de Cramer Método que usa determinantes para resolver un sistema de ecuaciones lineales.

región crítica Rango de valores que sugiere una diferencia suficientemente significativa para rechazar la hipótesis nula.

ciclo Un patrón completo de una función periódica.

D

decay factor (p. 454) In exponential decay, the base of the exponential expression, $1 - r$.

degree of a polynomial (p. 305) The greatest degree of any term in the polynomial.

dependent (p. 137) A system of equations that has an infinite number of solutions.

dependent events (p. P16) Events in which the outcome of one event affects the outcome of another event.

dependent variable (p. 64) The other variable in a function, usually y, whose values depend on x.

depressed polynomial (p. 354) The quotient when a polynomial is divided by one of its binomial factors.

descriptive statistics (p. P24) The branch of statistics that focuses on collecting, summarizing, and displaying data.

determinant (p. 189) A square array of numbers or variables enclosed between two parallel lines.

diagonal rule (p. 190) A method for finding the determinant of a third-order matrix.

dilation (p. 111) A transformation in which a geometric figure is enlarged or reduced.

dimensional analysis (p. 310) Performing operations with units.

dimensions of a matrix (p. 169) The number of rows, m, and the number of columns, n, of the matrix written as $m \times n$.

directrix (p. 599) See parabola.

direct variation (pp. 90, 562) y varies directly as x if there is some nonzero constant k such that $y = kx$. k is called the constant of variation.

discrete random variable (p. 742) The numerical outcome of a random event that takes on countable values.

discrete relation (p. 62) A relation in which the domain is a set of individual points.

discriminant (p. 267) In the Quadratic Formula, the expression $b^2 - 4ac$.

Distance Formula (p. 594) The distance between two points with coordinates (x_1, y_1) and (x_2, y_2) is given by
$$d = \sqrt{(x_2 - x_1)^2 + (y_2 - y_1)^2}.$$

divergent series (p. 683) An infinite geometric series that does not have a sum.

factor de decaimiento En decaimiento exponencial, la base de la expresión exponencial, $1 - r$.

grado de un polinomio Grado máximo de cualquier término del polinomio.

dependiente Sistema de ecuaciones que posee un número infinito de soluciones.

evento dependiente Eventos en que el resultado de uno de los eventos afecta el resultado de otro de los eventos.

variable dependiente La otra variable de una función, por lo general y, cuyo valor depende de x.

polinomio reducido El cociente cuando se divide un polinomio entre uno de sus factores binomiales.

estadística descriptiva Rama de la estadística cuyo enfoque es la recopilación, resumen y demostración de los datos.

determinante Arreglo cuadrado de números o variables encerrados entre dos rectas paralelas.

regla diagonal Método para encontrar el determinante de una matriz third-order.

homotecia Transformación en que se amplia o se reduce un figura geométrica.

análisis dimensional Realizar operaciones con unidades.

tamaño de una matriz El número de filas, m, y columnas, n, de una matriz, lo que se escribe $m \times n$.

directriz Véase parábola.

variación directa y varía directamente con x si hay una constante no nula k tal que $y = kx$. k se llama la constante de variación.

variable aleatoria discreta Resultado numérico de un evento aleatorio que toma valores contables.

relación discreta Relación en la cual el dominio es un conjunto de puntos individuales.

discriminante En la fórmula cuadrática, la expresión $b^2 - 4ac$.

fórmula de la distancia La distancia entre dos puntos (x_1, y_1) and (x_2, y_2) viene dada por
$$d = \sqrt{(x_2 - x_1)^2 + (y_2 - y_1)^2}.$$

serie divergente Serie geométrica infinita que no tiene suma.

E

domain (p. P4) The set of all x-coordinates of the ordered pairs of a relation.

dominio El conjunto de todas las coordenadas x de los pares ordenados de una relación.

dot plot (p. 92) Two sets of data plotted as ordered pairs in a coordinate plane.

diagrama del punto Dos conjuntos de datos graficados como pares ordenados en un plano de coordenadas.

e (p. 501) The irrational number 2.71828.... e is the base of the natural logarithms.

e El número irracional 2.71828.... e es la base de los logaritmos naturales.

element (p. 169) Each value in a matrix.

elemento Cada valor de una matriz.

elimination method (p. 139) Eliminate one of the variables in a system of equations by adding or subtracting the equations.

método de eliminación Eliminar una de las variables de un sistema de ecuaciones sumando o restando las ecuaciones.

ellipse (p. 615) The set of all points in a plane such that the sum of the distances from two given points in the plane, called foci, is constant.

elipse Conjunto de todos los puntos de un plano en los que la suma de sus distancias a dos puntos dados del plano, llamados focos, es constante.

empty set (p. 28) The solution set for an equation that has no solution, symbolized by { } or ∅.

conjunto vacío Conjunto solución de una ecuación que no tiene solución, denotado por { } o ∅.

end behavior (p. 324) The behavior of the graph as x approaches positive infinity (+∞) or negative infinity (−∞).

comportamiento final El comportamiento de una gráfica a medida que x tiende a más infinito (+∞) o menos infinito (−∞).

equation (p. 18) A mathematical sentence stating that two mathematical expressions are equal.

ecuación Enunciado matemático que afirma la igualdad de dos expresiones matemáticas.

event (p. P9) One or more outcomes of a trial.

evento Uno o más resultados de una prueba.

expected value (p. 745) The expected value of a discrete random variable is the weighted average of the values of the variable.

valor esperado El valor esperado de una variable aleatoria discreta es el promedio ponderado de los valores de la variable.

experiment (p. 723) Something that is intentionally done to people, animals, or objects, and then the response is observed.

experimento Algo se hace intencionalmente poblar, los animales, o los objetos, y entonces la respuesta se observa.

experimental probability (p. P13) What is estimated from observed simulations or experiments.

probabilidad experimental Qué se estima de simulaciones o de experimentos observados.

experimental probability distribution (p. 744) A distribution of probabilities estimated from experiments.

distribución de probabilidad experimental Distribución de probabilidades estimadas a partir de experimentos.

explicit formula (p. 692) Gives a_n as a function of n, such as $a_n = 3n + 1$.

un fórmula explícito Da en función de n, tal como $a_n = 3n + 1$.

exponential decay (p. 453) Exponential decay occurs when a quantity decreases exponentially over time.

desintegración exponencial Ocurre cuando una cantidad disminuye exponencialmente con el tiempo.

exponential equation (p. 461) An equation in which the variables occur as exponents.

ecuación exponencial Ecuación en que las variables aparecen en los exponentes.

exponential function (p. 451) A function of the form $y = ab^x$, where $a \neq 0$, $b > 0$, and $b \neq 1$.

función exponencial Una función de la forma $y = ab^x$, donde $a \neq 0$, $b > 0$, y $b \neq 1$.

exponential growth (p. 451) Exponential growth occurs when a quantity increases exponentially over time.

crecimiento exponencial El que ocurre cuando una cantidad aumenta exponencialmente con el tiempo.

exponential inequality (p. 463) An inequality involving exponential functions.

desigualdad exponencial Desigualdad que contiene funciones exponenciales.

extraneous solution (pp. 29, 429) A number that does not satisfy the original equation.

solución extraña Número que no satisface la ecuación original.

extrema (p. 331) The maximum and minimum values of a function.

extrema Son los valores máximos y mínimos de una función.

F

factored form (p. 238) The form of a polynomial showing all of its factors. $y = a(x - p)(x - q)$ is the factored form of a quadratic equation.

forma reducida La forma de un polinomio que demuestra todos sus factores. $y = a(x - p)(x - q)$ es la forma descompuesta en factores de una ecuación cuadrática.

Factor Theorem (p. 354) The binomial $x - r$ is a factor of the polynomial $P(x)$ if and only if $P(r) = 0$.

teorema factor El binomio $x - r$ es un factor del polinomio $P(x)$ si $P(r) = 0$.

family of graphs (p. 109) A group of graphs that displays one or more similar characteristics.

familia de gráficas Grupo de gráficas que presentan una o más características similares.

feasible region (p. 154) The intersection of the graphs in a system of constraints.

región viable Intersección de las gráficas de un sistema de restricciones.

Fibonacci sequence (p. 692) A sequence in which the first two terms are 1 and each of the additional terms is the sum of the two previous terms.

sucesión de Fibonacci Sucesión en que los dos primeros términos son iguales a 1 y cada término que sigue es igual a la suma de los dos anteriores.

H

histograma Un histograma usa barras para exhibir datos numéricos que han sido organizados en intervalos iguales.

asíntota horizontal Una línea horizontal a que un gráfico se acerca.

hipérbola Conjunto de todos los puntos de un plano en los que el valor absoluto de la diferencia de sus distancias a dos puntos dados del plano, llamados focos, es constante.

prueba de hipótesis Prueba que se usa para evaluar una conjetura específica acerca de la media.

I

identidad Ecuación en la cual su lado izquierdo es igual a su lado derecho para todos los valores de la variable para los cuales ambos lados están definidos.

unción identidad La función $f(x) = x$.

matriz identidad Matriz cuadrada que al multiplicarse por otra matriz, es igual a la misma matriz. Si A es una matriz de $n \times n$ e I es la matriz identidad de $n \times n$, entonces $A \cdot I = A$ y $I \cdot A = A$.

unidad imaginaria i, o la raíz cuadrada principal de -1.

inconsistente Un sistema de ecuaciones para el cual no existe par ordenado alguno que satisfaga ambas ecuaciones.

independiente Un sistema de ecuaciones que posee una única solución.

eventos independientes Eventos en que el resultado de uno de los eventos no afecta el resultado de otro evento.

variable independiente En una función, la variable, por lo general x, cuyos valores forman el dominio.

índice (de un radical) En las nth raíces, el valor de n en el símbolo $\sqrt{\ }$. Indica qué raíz se está llevando el valor bajo el radicando.

hipótesis inductiva El suponer que un enunciado es verdadero para algún entero positivo k, donde $k \geq n$.

---

histogram (p. 733) A histogram uses bars to display numerical data that have been organized into equal intervals.

horizontal asymptote (p. 553) A horizontal line which a graph approaches.

hyperbola (pp. 545, 624) The set of all points in the plane such that the absolute value of the difference of the distances from two given points in the plane, called foci, is constant.

hypothesis test (p. 771) A test used to assess a specific claim about the mean.

identity (p. 350) An equation in which the left side is equal to the right side for all values.

identity function (p. 109) The function $f(x) = x$.

identity matrix (p. 198) A square matrix that, when multiplied by another matrix, equals that same matrix. If A is any $n \times n$ matrix and I is the $n \times n$ identity matrix, then $A \cdot I = A$ and $I \cdot A = A$.

imaginary unit (p. 246) i, or the principal square root of -1.

inconsistent (p. 335) A system of equations with no ordered pair that satisfy both equations.

independent (p. 335) A system of equations with exactly one solution.

independent events (p. P16) Events in which the outcome of one event does not affect the outcome of another event.

independent variable (p. 64) In a function, the variable, usually x, whose values make up the domain.

index (p. 407) In nth roots, the value of n in the symbol $\sqrt{\ }$. Indicates to what root the value under the radicand is being taken.

induction hypothesis (p. 705) The assumption that a statement is true for some positive integer k, where $k \geq n$.

---

secuencia finitas Una secuencia que contiene un número limitado de términos.

resumen de cinco números Los tres cuartiles y los valores máximo y mínimo de un conjunto de datos.

focos
(elipse) Los dos puntos fijos de los cuales la suma de las distancias de un sistema de todos los puntos en un plano es constante.

(hipérbola) Los dos puntos fijos de los cuales la diferencia de las distancias de un sistema de todos los puntos en un plano es constante.

foco Véase parábola, elipse, hipérbola.

método FOIL El producto de dos binomios es la suma de los productos de los primeros (First) términos, los términos exteriores (Outer), los términos interiores (Inner) y los últimos (Last) términos.

fórmula Enunciado matemático que describe la relación entre ciertas cantidades.

frecuencia El número de ciclos en una unidad del tiempo dada.

función Relación en que a cada elemento del dominio le corresponde un solo elemento del rango.

notación funcional Una ecuación de y en términos de x puede escribirse en la forma $y = f(x)$. Por ejemplo, $y = 2x + 1$ puede escribirse como $f(x) = 2x + 1$.

G

forma general Una ecuación de una parábola en la forma $y = ax^2 + bx + c$.

media geométrica Cualquier término entre dos términos no consecutivos de una sucesión geométrica.

sucesión geométrica Sucesión en que cualquier término después del primero puede hallarse multiplicando el término anterior por una constante r, llamada razón común.

serie geométrica La suma de los términos de una sucesión geométrica.

función del máximo entero Una función etapa que se escribe $f(x) = [x]$, donde $f(x)$ es el máximo entero que es menor que o igual a x.

factor del crecimiento En el crecimiento exponencial, la base de la expresión exponencial, $1 + r$.

---

finite sequence (p. 659) A sequence containing a limited number of terms.

five-number summary (p. P26) The three quartiles and the maximum and minimum values in a set of data.

foci (pp. 615, 624)
ellipse—The two fixed points from which the sum of the distances from a set of all points in a plane is constant.

hyperbola—The two fixed points from which the difference of the distances from a set of all points in a plane is constant.

focus (p. 599) See parabola, ellipse, hyperbola.

FOIL method (pp. P6, 238) The product of two binomials is the sum of the products of F the first terms, O the outer terms, I the inner terms, and L the last terms.

formula (p. 6) A mathematical sentence that expresses the relationship between certain quantities.

frequency (p. 838) The number of cycles in a given unit of time.

function (p. P4) A relation in which each element of the domain is paired with exactly one element in the range.

function notation (p. 64) An equation of y in terms of x can be rewritten so that $y = f(x)$. For example, $y = 2x + 1$ can be written as $f(x) = 2x + 1$.

general form (p. 599) An equation of a parabola in the form $y = ax^2 + bx + c$.

geometric mean (p. 675) The terms between any two nonsuccessive terms of a geometric sequence.

geometric sequence (p. 661) A sequence in which each term after the first is found by multiplying the previous term by a constant r, called the common ratio.

geometric series (p. 676) The sum of the terms of a geometric sequence.

greatest integer function (p. 102) A step function, written as $f(x) = [x]$, where $f(x)$ is the greatest integer less than or equal to x.

growth factor (p. 453) In exponential growth, the base of the exponential expression, $1 + r$.

R134 English

inferential statistics (p. 769) Statistics like predictions and hypothesis testing are used to draw conclusions about a population by using a sample.

infinite geometric series (p. 683) A geometric series with an infinite number of terms.

infinite sequence (p. 659) A sequence that continues without end.

infinity (pp. 40, 685) Without bound, or continues without end.

initial side (p. 799) The fixed ray of an angle.

integer (p. 11) $\{\ldots, -3, -2, -1, 0, 1, 2, 3, \ldots\}$

interquartile range (IQR) (p. P27) The range of the middle half of a set of data. It is the difference between the upper quartile and the lower quartile.

intersection (p. 41) The graph of a compound inequality containing *and*.

interval notation (p. 40) A way to describe the solution set of an inequality.

inverse function (p. 393) Two functions f and g are inverse functions if and only if both of their compositions are the identity function.

inverse matrices (p. 198) Two $n \times n$ matrices are inverses of each other if their product is the identity matrix.

inverse of a trigonometric function (p. 853) The arccosine, arcsine, and arctangent relations.

inverse relation (p. 393) Two relations are inverse relations if and only if whenever one relation contains the element (a, b) the other relation contains the element (b, a).

inverse variation (p. 564) y varies inversely as x if there is some nonzero constant k such that $xy = k$ or $y = \frac{k}{x}$ where $x \neq 0$ and $y \neq 0$.

irrational number (p. 11) A real number that is not rational. The decimal form neither terminates nor repeats.

R134 Español

estadística deductiva La estadística como predicciones y la prueba de la hipótesis es utilizada para dibujar conclusiones sobre una población usando una muestra.

serie geométrica infinita Serie geométrica con un número infinito de términos.

secuencia infinita Una secuencia que continúa sin extremo.

infinito Sin límite, o continúa sin extremo.

lado inicial de un ángulo El rayo fijo de un ángulo.

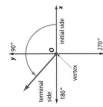

número entero $\{\ldots, -3, -2, -1, 0, 1, 2, 3, \ldots\}$

amplitud intercuartílica Amplitud de la mitad central de un conjunto de datos. Es la diferencia entre el cuartil superior y el inferior.

intersección Gráfica de una desigualdad compuesta que contiene la palabra *y*.

notación del intervalo Una manera de describir el sistema de la solución de una desigualdad.

función inversa Dos funciones f y g son inversas mutuas si y sólo si las composiciones de ambas son la función identidad.

matrices inversas Dos matrices de $n \times n$ son inversas mutuas si su producto es la matriz identidad.

inversa de una función trigonométrica Las relaciones arccoseno, arcoseno y arcotangente.

relaciones inversas Dos relaciones son relaciones inversas mutuas si y sólo si cada vez que una de las relaciones contiene el elemento (a, b), la otra contiene el elemento (b, a).

variación inversa y varía inversamente con x si hay una constante no nula k tal que $xy = k$ o $y = \frac{k}{x}$, donde $x \neq 0$ y $y \neq 0$.

número irracional Número que no es racional. Su expansión decimal no es ni terminal ni periódica.

R135 English

iteration (p. 694) The process of composing a function with itself repeatedly.

J

joint variation (p. 563) y varies jointly as x and z if there is some nonzero constant k such that $y = kxz$.

L

latus rectum (p. 599) The line segment through the focus of a parabola and perpendicular to the axis of symmetry.

Law of Cosines (p. 823) Let $\triangle ABC$ be any triangle with a, b, and c representing the measures of sides, and opposite angles with measures A, B, and C, respectively. Then the following equations are true.

$$a^2 = b^2 + c^2 - 2bc \cos A$$
$$b^2 = a^2 + c^2 - 2ac \cos B$$
$$c^2 = a^2 + b^2 - 2ab \cos C$$

Law of Large Numbers (p. 744) The variation in a data set decreases as the sample size increases.

Law of Sines (p. 815) Let $\triangle ABC$ be any triangle with a, b, and c representing the measures of sides opposite angles with measurements A, B, and C, respectively. Then

$$\frac{\sin A}{a} = \frac{\sin B}{b} = \frac{\sin C}{c}.$$

leading coefficient (p. 322) The coefficient of the term with the highest degree.

like radical expressions (p. 417) Two radical expressions in which both the radicands and indices are alike.

limit (p. 690) The value that the terms of a sequence approach.

linear correlation coefficient (p. 94) A value that shows how close data points are to a line.

linear equation (p. 69) An equation that has no operations other than addition, subtraction, and multiplication of a variable by a constant.

linear function (p. 69) A function whose ordered pairs satisfy a linear equation.

linear inequality (p. 117) An inequality that describes a half-plane with a boundary that is a straight line.

linear programming (p. 154) The process of finding the maximum or minimum values of a function for a region defined by inequalities.

linear relation (p. 69) A relation that has straight line graphs.

R135 Español

iteración Proceso de componer una función consigo misma repetidamente.

J

variación conjunta y varía conjuntamente con x y z si hay una constante no nula k tal que $y = kxz$.

L

latus rectum El segmento de recta que pasa por el foco de una parábola y que es perpendicular a su eje de simetría.

Ley de los cosenos Sea $\triangle ABC$ un triángulo cualquiera, con a, b, y c las longitudes de los lados y con ángulos opuestos de medidas A, B y C, respectivamente. Entonces se cumplen las siguientes ecuaciones.

$$a^2 = b^2 + c^2 - 2bc \cos A$$
$$b^2 = a^2 + c^2 - 2ac \cos B$$
$$c^2 = a^2 + b^2 - 2ab \cos C$$

ley de los grandes números La variación en un conjunto de datos disminuye a medida que aumenta el tamaño de la muestra.

Ley de los senos Sea $\triangle ABC$ cualquier triángulo con a, b y c las longitudes de los lados y con ángulos opuestos de medidas A, B y C, respectivamente. Entonces

$$\frac{\sin A}{a} = \frac{\sin B}{b} = \frac{\sin C}{c}$$

coeficiente líder Coeficiente del término de mayor grado.

expresiones radicales semejantes Dos expresiones radicales en que tanto los radicandos como los índices son semejantes.

límite El valor al que tienden los términos de una sucesión.

coeficiente de correlación lineal Valor que muestra la cercanía de los datos a una recta.

ecuación lineal Ecuación sin otras operaciones que las de adición, sustracción y multiplicación de una variable por una constante.

función lineal Función cuyos pares ordenados satisfacen una ecuación lineal.

desigualdad lineal Desigualdad lineal es una desigualdad que describe un semiplano con un límite que es una línea recta.

programación lineal Proceso de hallar los valores máximo o mínimo de una función lineal en una región definida por las desigualdades.

notación del intervalo Una manera de describir el sistema de la solución de una desigualdad.

connectED.mcgraw-hill.com

linear term (p. 219) In the equation $f(x) = ax^2 + bx + c$, bx is the linear term.

línea de la reflexión La línea excedente que una reflexión mueve de un tirón una figura.

término lineal En la ecuación $f(x) = ax^2 + bx + c$, el término lineal es bx.

line of fit (p. 92) A line that closely approximates a set of data.

recta de ajuste Recta que se aproxima estrechamente a un conjunto de datos.

line of reflection (p. 111) The line over which a reflection flips a figure.

Location Principle (p. 330) Suppose $y = f(x)$ represents a polynomial function and a and b are two numbers such that $f(a) < 0$ and $f(b) > 0$. Then the function has at least one real zero between a and b.

principio de ubicación Sea $y = f(x)$ una función polinómica con a y b dos números tales que $f(a) < 0$ y $f(b) > 0$. Entonces la función tiene por lo menos un resultado real entre a y b.

logarithm (p. 468) In the function $x = b^y$, y is called the logarithm, base b, of x. Usually written as $y = \log_b x$ and is read "y equals log base b of x."

logaritmo En la función $x = b^y$, y es el logaritmo en base b, de x. Generalmente escrito como $y = \log_b x$ y se lee "y es igual al logaritmo en base b de x."

logarithmic equation (p. 478) An equation that contains one or more logarithms.

ecuación logarítmica Ecuación que contiene uno o más logaritmos.

logarithmic function (p. 469) The function $y = \log_b x$, where $b > 0$ and $b \neq 1$, which is the inverse of the exponential function $y = b^x$.

función logarítmica La función $y = \log_b x$, donde $b > 0$ y $b \neq 1$, inversa de la función exponencial $y = b^x$.

logarithmic inequality (p. 479) An inequality that contains one or more logarithms.

desigualdad logarítmica Desigualdad que contiene uno o más logaritmos.

logistic growth model (p. 512) A growth model that represents growth that has a limiting factor. Logistic models are the most accurate models for representing population growth.

modelo logístico del crecimiento Un modelo del crecimiento que representa el crecimiento que tiene un factor limitador. Los modelos logísticos son los modelos más exactos para representar crecimiento de la población.

lower quartile (p. P26) The median of the lower half of a set of data, indicated by LQ.

cuartil inferior Mediana de la mitad inferior de un conjunto de datos, se denota con LQ.

M

major axis (p. 615) The longer of the two line segments that form the axes of symmetry of an ellipse.

eje mayor El más largo de dos segmentos de recta que forman los ejes de simetría de una elipse.

mapping (p. P4) How each member of the domain is paired with each member of the range.

transformaciones La correspondencia entre cada miembro del dominio con cada miembro del rango.

margin of error (p. 731) The limit on the difference between how a sample responds and how the total population would respond.

margen de error muestral Límite en la diferencia entre las respuestas obtenidas con una muestra y cómo pudiera responder la población entera.

mathematical induction (p. 705) A method of proof used to prove statements about positive integers.

inducción matemática Método de demostrar enunciados sobre los enteros positivos.

matrix (p. 169) Any rectangular array of variables or constants in horizontal rows and vertical columns.

matriz Arreglo rectangular de variables o constantes en filas horizontales y columnas verticales.

matrix equation (p. 200) A matrix form used to represent a system of equations.

ecuación matriz Forma de matriz que se usa para representar un sistema de ecuaciones.

maximum error of estimate (p. 769) The maximum difference between the estimate of the population mean and its actual value.

error máximo de estimación Diferencia máxima entre la estimación de la media de una población y su valor real.

maximum value (p. 222) The y-coordinate of the vertex of the quadratic function $f(x) = ax^2 + bx + c$, where $a < 0$.

valor máximo La coordenada y del vértice de la función cuadrática $f(x) = ax^2 + bx + c$, where $a < 0$.

mean (p. P24) The sum of the values in a set of data divided by the total number of values in the set.

media Suma de los valores en un conjunto de datos dividida entre el número total de valores en el conjunto.

measure of central tendency (p. P24) A number that represents the center or middle of a set of data.

medida de tendencia central Número que representa el centro o medio de un conjunto de datos.

measure of variation (p. P25) A representation of how spread out or scattered a set of data is.

medida de variación Número que representa la dispersión de un conjunto de datos.

median (p. P24) The middle value or the mean of the middle values in a set of data when the data are arranged in numerical order.

mediana Valor del medio o la media de los valores del medio en un conjunto de datos, cuando los datos están ordenados numéricamente.

midline (p. 846) A horizontal axis used as the reference line about which the graph of a periodic function oscillates.

línea media Eje horizontal que se usa como recta de referencia alrededor de la cual oscila la gráfica de una función periódica.

minimum value (p. 222) The y-coordinate of the vertex of the quadratic function $f(x) = ax^2 + bx + c$, where $a > 0$.

valor mínimo La coordenada y del vértice de la función cuadrática $f(x) = ax^2 + bx + c$, donde $a > 0$.

minor axis (p. 615) The shorter of the two line segments that form the axes of symmetry of an ellipse.

eje menor El más corto de los dos segmentos de recta de los ejes de simetría de una elipse.

mode (p. P24) The value or values that appear most often in a set of data.

moda Valor o valores que aparecen con más frecuencia en un conjunto de datos.

mutually exclusive (p. P14) Two events that cannot occur at the same time.

mutuamente exclusivos Dos eventos que no pueden ocurrir simultáneamente.

N

nth root (p. 407) For any real numbers a and b, and any positive integer n, if $a^n = b$, then a is an nth root of b.

raíz enésima Para cualquier número real a y b y cualquier entero positivo n, si $a^n = b$, entonces a se llama una raíz enésima de b.

natural base, e (p. 501) An irrational number approximately equal to 2.71828….

base natural, e Número irracional aproximadamente igual a 2.71828….

natural base exponential function (p. 501) An exponential function with base e, $y = e^x$.

función exponencial natural La función exponencial de base e, $y = e^x$.

natural logarithm (p. 501) Logarithms with base e, written ln x.

logaritmo natural Logaritmo de base e, el que se escribe ln x.

natural number (p. 11) {1, 2, 3, 4, 5, …}

número natural {1, 2, 3, 4, 5, …}

negative correlation (p. 92) When the values in a scatter plot are closely linked in a negative manner.

correlación negativa Cuando los valores en un diagrama de dispersión se ligan de cerca de una manera negativa.

negative exponent (p. 303) For any real number $a \neq 0$ and any integer n, $a^{-n} = \frac{1}{a^n}$ and $\frac{1}{a^{-n}} = a^n$.

exponente negativo Para cualquier número real $a \neq 0$ y cualquier entero positivo n, $a^{-n} = \frac{1}{a^n}$ y $\frac{1}{a^{-n}} = a^n$.

Glossary/Glosario

normal distribution (p. 760) A continuous, symmetric, bell-shaped distribution of a random variable.

Normal Distribution

null hypothesis (p. 771) A specific hypothesis to be tested. It is expressed as an equality and is considered true until evidence indicates otherwise.

O

oblique asymptote (p. 555) An asymptote that is neither horizontal nor vertical and is sometimes called a *slant asymptote*.

observational study (p. 723) Individuals are observed and no attempt is made to influence the results.

odds (p. P15) A ratio that compares the number of ways an event can occur to the number of ways that it cannot occur.

one-to-one function (p. 61) 1. A function where each element of the range is paired with exactly one element of the domain 2. A function whose inverse is a function.

onto function (p. 61) Each element of the range corresponds to an element of the domain.

open sentence (p. 18) A mathematical sentence containing one or more variables.

optimize (p. 156) To seek the optimal price or amount that is desired to minimize costs or maximize profits.

ordered triple (p. 161) 1. The coordinates of a point in space 2. The solution of a system of equations in three variables *x*, *y*, and *z*.

Order of Operations (p. 5)
Step 1 Evaluate expressions inside grouping symbols.
Step 2 Evaluate all powers.
Step 3 Do all multiplications and/or divisions from left to right.
Step 4 Do all additions and subtractions from left to right.

outcome (p. P9) The results of a probability experiment or an event.

R138 | Glossary/Glosario

---

distribución normal Distribución con forma de campana, simétrica y continua de una variable aleatoria.

Distribución normal

hipótesis nula Es una hipótesis específica que se probará. Se expresa como igualdad y se considera verdad hasta que la evidencia indica de otra manera.

O

asíntota oblicuo Una asíntota que es ni horizontal ni la vertical y a veces se llama una *asíntota inclinada*.

estudio de observación Observan a los individuos y no se hace ninguna tentativa de influenciar los resultados.

posibilidad Razón que compara el número de maneras en que puede ocurrir un evento al número de maneras en que no puede ocurrir dicho evento.

función biunívoca 1. Función en la que a cada elemento del rango le corresponde sólo un elemento del dominio. 2. Función cuya inversa es una función.

sobre la función Cada elemento de la gama corresponde a un elemento del dominio. Centro y un punto en el círculo. Indice del crecimiento continuo—la tarifa en la cual algo crece continuamente. El valor de *k* en la función exponencial del crecimiento, *f(x)* = *ae*.

enunciado abierto Enunciado matemático que contiene una o más variables.

optimice Buscar el precio óptimo o ascender que se desea para reducir al mínimo costes o para maximizar los beneficios.

triple ordenado 1. Las coordenadas de un punto en el espacio 2. Solución de un sistema de ecuaciones en tres variables *x*, *y* y *z*.

orden de las operaciones
Paso 1 Evalúa las expresiones dentro de símbolos de agrupamiento.
Paso 2 Evalúa todas las potencias.
Paso 3 Ejecuta todas las multiplicaciones y divisiones de izquierda a derecha.
Paso 4 Ejecuta todas las adiciones y sustracciones de izquierda a derecha.

resultados Lo que produce un experimento o evento probabilístico.

---

outlier (pp. P27, 93) A data point that does not appear to belong to the rest of the set.

P

parabola (pp. 219, 599) The graph of a quadratic function. The set of all points in a plane that are the same distance from a given point, called the focus, and a given line, called the directrix.

parallel lines (p. 85) Nonvertical coplanar lines with the same slope.

parameter (p. 723) A measure that describes a characteristic of a population.

parent function (p. 109) The simplest, most general function in a family of functions.

parent graph (p. 109) The simplest of graphs in a family.

partial sum (p. 668) The sum of the first *n* terms of a series.

Pascal's triangle (p. 699) A triangular array of numbers such that the $(n + 1)$th row is the coefficient of the terms of the expansion $(x + y)^n$ for $n = 0, 1, 2 \ldots$

period (p. 831) The least possible value of *a* for which $f(x) = f(x + a)$.

periodic function (p. 831) 1. A function with *y*-values that repeat at regular intervals. 2. A function is called periodic if there is a number *a* such that $f(x) = f(x + a)$ for all *x* in the domain of the function.

permutation (p. P9) An arrangement of objects in which order is important.

perpendicular lines (p. 85) In a plane, any two oblique lines, the product of whose slopes is −1.

phase shift (p. 845) A horizontal translation of a trigonometric function.

piecewise-defined function (p. 101) A function that is written using two or more expressions.

piecewise-linear function (p. 102) A function in which the equation for each interval is linear.

---

valor atípico Dato que no parece pertenecer al resto del conjunto.

P

parábola La gráfica de una función cuadrática. Conjunto de todos los puntos de un plano que están a la misma distancia de un punto dado, llamado foco, y de una recta dada, llamada directriz.

rectas paralelas Rectas coplanares no verticales con la misma pendiente.

parámetro Una medida que describe una característica de una población.

función del padre El más simple, la mayoría de la función general en una familia de funciones.

gráfica madre La gráfica más sencilla en una familia de gráficas.

suma parcial La suma de los primeros *n* términos de una serie.

triángulo de Pascal Arreglo triangular de números tal que la fila de $(n + 1)$ está compuesta de proporciona los coeficientes de los términos de la expansión de $(x + y)^n$ para $n = 0, 1, 2 \ldots$

periodo El menor valor positivo posible para *a*, para el cual $f(x) = f(x + a)$.

función periódica 1. Una función con *y*-valores aquella repetición con regularidad. 2. Función para la cual hay un número *a* tal que $f(x) = f(x + a)$ para todo *x* en el dominio de la función.

permutación Arreglo de elementos en que el orden es importante.

rectas perpendiculares En un plano, dos rectas oblicuas cualesquiera cuyas pendientes tienen un producto igual a −1.

desplazamiento de fase Traslación horizontal de una función trigonométrica.

función por trozos-definida Una función que se escribe usando dos o más expresiones.

función lineal por partes Una función en que la ecuación para cada intervalo es lineal.

point discontinuity (p. 556) If the original function is undefined for $x = a$ but the related rational expression of the function in simplest form is defined for $x = a$, then there is a hole in the graph at $x = a$.

point-slope form (p. 84) An equation in the form $y - y_1 = m(x - x_1)$ where (x_1, y_1) are the coordinates of a point on the line and m is the slope of the line.

polynomial function (p. 323) A function that is represented by a polynomial equation.

polynomial identity (p. 350) A polynomial equation that is true for any values that are substituted for the variables.

polynomial in one variable (p. 322)
$a_n x^n + a_{n-1} x^{n-1} + \cdots + a_2 x^2 + a_1 x + a_0$, where the coefficients $a_n, a_{n-1}, \ldots, a_1$, represent real numbers, and a_n is not zero and n is a nonnegative integer.

population (p. P24) An entire group of living things or objects.

positive correlation (p. 92) When the values in a scatter plot are closely linked in a positive manner.

power function (p. 320) An equation in the form $f(x) = ax^b$, where a and b are nonzero real numbers.

prediction equation (p. 92) An equation suggested by the points of a scatter plot that is used to predict other points.

prime polynomial (p. 342) A polynomial that cannot be factored.

principal root (p. 407) The nonnegative root.

principal values (p. 853) The values in the restricted domains of trigonometric functions.

probability (p. P13) A measure of the chance that a given event will occur.

probability distribution (p. 743) A function that maps the sample space to the probabilities of the outcomes in the sample space for a particular random variable.

probability model (p. P13) A mathematical model used to represent the outcomes of an experiment.

pure imaginary number (p. 246) The square roots of negative real numbers. For any positive real number b,
$\sqrt{-b^2} = \sqrt{b^2} \cdot \sqrt{-1}$, or bi.

---

discontinuidad evitable Si la función original no está definida en $x = a$ pero la expresión racional reducida correspondiente de la función está definida en $x = a$, entonces la gráfica tiene una ruptura o corte en $x = a$.

forma punto–pendiente Ecuación de la forma $y - y_1 = m(x - x_1)$ donde (x_1, y_1) es un punto en la recta y m es la pendiente de la recta.

función polinomial Función representada por una ecuación polinomial.

identidad polinómica Ecuación polinómica que es verdadera para cualquier valor por el que se reemplazan las variables.

polinomio de una variable
$a_n x^n + a_{n-1} x^{n-1} + \cdots + a_2 x^2 + a_1 x + a_0$, donde los coeficientes $a_n, a_{n-1}, \ldots, a_1$, representan números reales, a_n no es nulo y n es un entero no negativo.

población Un grupo entero de cosas o de objetos vivos.

correlación positiva Cuando los valores en un diagrama de la dispersión se ligan de cerca de una manera positiva.

función potencia Ecuación de la forma $f(x) = ax^b$, donde a y b son números reales.

ecuación de predicción Ecuación sugerida por los puntos de una gráfica de dispersión y que se usa para predecir otros puntos.

polinomio primero Un polinomio que no puede ser descompuesto en factores.

raíz principal La raíz no negativa.

valores principales Valores en los dominios restringidos de las funciones trigonométricas.

probabilidad Medida de la posibilidad de que ocurra un evento dado.

distribución de probabilidad Función que aplica el espacio muestral a las probabilidades de los resultados en el espacio muestral obtenidos para una variable aleatoria particular.

modelo de probabilidad Modelo matemático que se usa para representar los resultados de un experimento.

número imaginario puro Raíz cuadrada de un número real negativo. Para cualquier número real positivo b,
$\sqrt{-b^2} = \sqrt{b^2} \cdot \sqrt{-1}$, ó bi.

Q

quadrantal angle (p. 808) An angle in standard position whose terminal side coincides with one of the axes.

quadrants (p. P4) The four areas of a Cartesian coordinate plane.

quadratic equation (p. 229) A quadratic function set equal to a value, in the form $ax^2 + bx + c = 0$, where $a \neq 0$.

quadratic form (p. 345) For any numbers a, b, and c, except for $a = 0$, an equation that can be written in the form $u^2 + u + c = 0$, where u is some expression in x.

Quadratic Formula (p. 264) The solutions of a quadratic equation of the form $ax^2 + bx + c = 0$, where $a \neq 0$, are given by the Quadratic Formula, which is
$$x = \frac{-b \pm \sqrt{b^2 - 4ac}}{2a}$$

quadratic function (pp. 109, 219) A function described by the equation $f(x) = ax^2 + bx + c$, where $a \neq 0$.

quadratic inequality (p. 282) An inequality of the form
$y > ax^2 + bx + c$, $y \geq ax^2 + bx + c$,
$y < ax^2 + bx + c$, or $y \leq ax^2 + bx + c$.

quadratic term (p. 219) In the equation
$f(x) = ax^2 + bx + c$, ax^2 is the quadratic term.

quartic function (p. 324) A fourth-degree function.

quartiles (p. P26) The values that divide a set of data into four equal parts.

quintic function (p. 322) A fifth-degree function.

R

radian (p. 801) The measure of an angle θ in standard position whose rays intercept an arc of length 1 unit on the unit circle.

radical equation (p. 429) An equation with radicals that have variables in the radicands.

radical function (p. 400) A function that contains the root of a variable.

radical inequality (p. 431) An inequality that has a variable in the radicand.

radical sign (p. 407) In nth roots, the symbol $\sqrt[n]{}$.

radicand (p. 407) In nth roots, the value inside the symbol $\sqrt[n]{}$. Indicates the value that is being taken to the nth root.

Q

ángulo de cuadrante Ángulo en posición estándar cuyo lado terminal coincide con uno de los ejes.

cuadrantes Las cuatro regiones de un plano de coordenadas Cartesiano.

ecuación cuadrática Función cuadrática igual a un valor, de la forma $ax^2 + bx + c = 0$, donde $a \neq 0$.

forma de ecuación cuadrática Para cualquier número a, b, y c, excepto $a = 0$, una ecuación que puede escribirse de la forma $u^2 + u + c = 0$, donde u es una expresión en x.

fórmula cuadrática Las soluciones de una ecuación cuadrática de la forma $ax^2 + bx + c = 0$, donde $a \neq 0$, se dan por la fórmula cuadrática, que es
$$x = \frac{-b \pm \sqrt{b^2 - 4ac}}{2a}$$

función cuadrática Función descrita por la ecuación
$f(x) = ax^2 + bx + c$, donde $a \neq 0$.

desigualdad cuadrática Desigualdad cuadrática de la forma
$y > ax^2 + bx + c$, $y \geq ax^2 + bx + c$,
$y < ax^2 + bx + c$, o $y \leq ax^2 + bx + c$.

término cuadrático En la ecuación
$f(x) = ax^2 + bx + c$, ax^2 el término cuadrático es ax^2.

función quartic Una función del cuarto-grado.

cuartiles Valores que dividen un conjunto de datos en cuatro partes iguales.

función quintic Una función del quinto-grado.

R

radián Medida de un ángulo θ en posición normal cuyos rayos intersecan un arco de 1 unidad de longitud en el círculo unitario.

ecuación radical Ecuación con radicales que tienen variables en el radicando.

función radical Una función que contiene la raíz de una variable.

desigualdad radical Desigualdad que tiene una variable en el radicando.

signo radical El símbolo $\sqrt[n]{}$, que se usa par indicar la raíz cuadrada no negativo o el símbolo por raíz enésima.

radicando El número o la expressión que aparece debajo del signo radical.

English (R142)

radius (p. 607) Any segment whose endpoints are the center and a point on the circle.

random sample (p. 723) A sample in which every member of the population has an equal chance of being selected.

random variable (p. 742) The outcome of a random process that has a numerical value.

range (pp. P4, P25) 1. The set of all y-coordinates of a relation. 2. The difference between the greatest and least values in a set of data.

rate of change (p. 76) How much a quantity changes on average, relative to the change in another quantity, over time.

rate of continuous decay (p. 509) The rate at which something decays continuously. Represented by a constant k in the exponential decay function $f(x) = ae^{-kt}$, where a is the initial value, and t its time in years.

rate of continuous growth (p. 509) The rate at which something grows continuously. The value of k in the exponential growth function, $f(x) = ae^{kt}$.

rational equation (p. 570) Any equation that contains one or more rational expressions.

rational exponent (p. 423) For any nonzero real number b, and any integers m and n, with n > 1, $b^{\frac{m}{n}} = \sqrt[n]{b^m} = (\sqrt[n]{b})^m$, except when b < 0 and n is even.

rational expression (p. 529) A ratio of two polynomial expressions.

rational function (p. 553) An equation of the form $f(x) = \frac{p(x)}{q(x)}$, where p(x) and q(x) are polynomial functions, and q(x) ≠ 0.

rational inequality (p. 575) Any inequality that contains one or more rational expressions.

rationalizing the denominator (p. 416) To eliminate radicals from a denominator or fractions from a radicand.

rational number (p. 11) Any number $\frac{m}{n}$, where m and n are integers and n is not zero. The decimal form is either a terminating or repeating decimal.

Rational Zero Theorem (p. 367) Helps you choose some possible zeros of a polynomial function to test.

real numbers (p. 11) All numbers used in everyday life; the set of all rational and irrational numbers.

Español (R142)

radio (p. 607) Un segmento cuyos extremos son el centro y un punto del círculo.

muestra aleatoria (p. 723) Muestra en la cual cada miembro de la población tiene igual posibilidad de ser elegido.

variable aleatoria (p. 742) El resultado de un proceso aleatorio que tiene un valor numérico.

rango (pp. P4, P25) 1. Conjunto de todas las coordenadas y de una relación. 2. Diferencia entre el valor mayor y el menor en un conjunto de datos.

tasa de cambio (p. 76) Lo que cambia una cantidad en promedio, respecto al cambio en otra cantidad, por lo general en el tiempo.

índice de desintegración continua (p. 509) Ritmo al cual algo se desintegra continuamente. Representado por la constante k en la función de desintegración exponencial $f(x) = ae^{-kt}$, donde a es el valor inicial y t es el tiempo en años.

el índice del crecimiento continuo (p. 509) Es la tasa en la cual algo crece continuamente. El valor de k en la función exponencial del crecimiento, $f(x) = ae^{kt}$. La tasa en la cual algo crece continuamente.

ecuación racional (p. 570) Cualquier ecuación que contiene una o más expresiones racionales.

exponent racional (p. 423) Para cualquier número real no nulo b y cualquier entero m y n, con n > 1, $b^{\frac{m}{n}} = \sqrt[n]{b^m} = (\sqrt[n]{b})^m$, excepto cuando b < 0 y n es par.

expresión racional (p. 529) Razón de dos expresiones polinomiales.

función racional (p. 553) Ecuación de la forma $f(x) = \frac{p(x)}{q(x)}$, donde p(x) y q(x) son funciones polinomiales y q(x) ≠ 0.

desigualdad racional (p. 575) Cualquier desigualdad que contiene una o más expresiones racionales.

racionalizar el denominador (p. 416) La eliminación de radicales de un denominador o de fracciones de un radicando.

número racional (p. 11) Cualquier número $\frac{m}{n}$, donde m y n son enteros y n no es cero. Su expansión decimal es o terminal o periódica.

El teorema cero racional (p. 367) Ayudas usted elige algunos ceros posibles de una función polinómica para probar.

números reales (p. 11) Todos los números que se usan en la vida cotidiana; el conjunto de los todos los números racionales e irracionales.

Español (R143)

funciones recíprocas 1. Una función de la forma $f(x) = \frac{1}{a(x)}$, donde a(x) es una función linear y a(x) ≠ 0. 2. Funciones trigonométricas de la función que son recíprocals de uno a.

fórmula recursiva (p. 692) Cada término proviene de uno o más términos anteriores.

sucesión recursiva (p. 692) Una secuencia en la cual cada término es determinado por uno o más de los términos anteriores.

ángulo de referencia (p. 808) El ángulo agudo formado por el lado terminal de un ángulo en posición estándar y el eje x.

reflexión (p. 111) Transformación en que cada punto de una figura se aplica a través de una recta de simetría a su imagen correspondiente.

reca de regresión (p. 94) Una recta de óptimo ajuste.

relación (p. P4) Conjunto de pares ordenados.

frecuencia relativa (pp. P18, 743) Razón del número de observaciones en una categoría al número total de observaciones.

máximo relativo (p. 331) Punto en la gráfica de una función en donde ningún otro punto cercano tiene una coordenada y mayor.

mínimo relativo (p. 331) Punto en la gráfica de una función en donde ningún otro punto cercano tiene una coordenada y menor.

raíz (p. 229) Las soluciones de una ecuación cuadrática.

matriz fila (p. 169) Matriz que sólo tiene una fila.

S

muestra (pp. P24, 723) Parte de una población.

espacio muestral (p. P9) Conjunto de todos los resultados posibles de un experimento probabilístico.

escalar (p. 173) Una constante.

multiplicación por escalares (p. 173) Multiplicación de una matriz por una constante llamada escalar; producto de un escalar k y una matriz de m × n.

English (R143)

reciprocal function (pp. 545, 791) 1. A function of the form $f(x) = \frac{1}{a(x)}$, where a(x) is a linear function and a(x) ≠ 0. 2. Trigonometric functions that are reciprocals of each other.

recursive formula (p. 692) Each term is formulated from one or more previous terms.

recursive sequence (p. 692) A sequence in which each term is determined by one or more of the previous terms.

reference angle (p. 808) The acute angle formed by the terminal side of an angle in standard position and the x-axis.

reflection (p. 111) A transformation in which every point of a figure is mapped to a corresponding image across a line of symmetry.

regression line (p. 94) A line of best fit.

relation (p. P4) A set of ordered pairs.

relative frequency (pp. P18, 743) The ratio of the number of observations in a category to the total number of observations.

relative maximum (p. 331) A point on the graph of a function where no other nearby points have a greater y-coordinate.

relative minimum (p. 331) A point on the graph of a function where no other nearby points have a lesser y-coordinate.

root (p. 229) The solutions of a quadratic equation.

row matrix (p. 169) A matrix that has only one row.

sample (pp. P24, 723) A part of a population.

sample space (p. P9) The set of all possible outcomes of an experiment.

scalar (p. 173) A constant.

scalar multiplication (p. 173) Multiplying any matrix by a constant called a scalar; the product of a scalar k and an m × n matrix.

square matrix (p. 198) A matrix with the same number of rows and columns.

square root function (p. 400) A function that contains a square root of a variable.

square root inequality (p. 402) An inequality involving the square root of a variable expression.

Square Root Property (p. 247) For any real number n, if $x^2 = n$, then $x = \pm\sqrt{n}$.

standard deviation (p. P25) The square root of the variance.

standard form (pp. 70, 229, 599) **1.** A linear equation written in the form $Ax + By = C$, where A, B, and C are integers whose greatest common factor is 1, $A \geq 0$, and A and B are not both zero. **2.** A quadratic equation written in the form $ax^2 + bx + c = 0$, where a, b, and c are integers, and $a \neq 0$.

standard normal distribution (p. 762) A normal distribution with a mean of 0 and a standard deviation of 1.

standard position (p. 799) An angle positioned so that its vertex is at the origin and its initial side is along the positive x-axis.

statistic (p. 723) A measure that describes a characteristic of a sample.

statistical inference (p. 769) Use information from a sample to draw conclusions about a population.

step function (p. 102) A function whose graph is a series of line segments.

substitution method (p. 138) A method of solving a system of equations in which one equation is solved for one variable in terms of the other.

survey (p. 723) Used to collect information about a population.

synthetic division (p. 312) A method used to divide a polynomial by a binomial.

synthetic substitution (p. 352) The use of synthetic division to evaluate a function.

system of equations (p. 136) A set of equations with the same variables.

system of inequalities (p. 146) A set of inequalities with the same variables.

---

matriz cuadrada Matriz con el mismo número de filas y columnas.

función radical Función que contiene la raíz cuadrada de una variable.

desigualdad radical Desigualdad que presenta raíces cuadradas de un expresión con variables.

Propiedad de la raíz cuadrada Para cualquier número real n, si $x^2 = n$, entonces $x = \pm\sqrt{n}$.

desviación estándar La raíz cuadrada de la varianza.

forma estándar **1.** Ecuación lineal escrita de la forma $Ax + By = C$, donde A, B y C son enteros cuyo máximo común divisores 1, $A \geq 0$, y A y B no son cero simultáneamente. **2.** Una ecuación cuadrática escrita en la forma $ax^2 + bx + c = 0$, donde a, b, y c son números enteros, y $a \neq 0$.

distribución normal estándar Distribución normal con una media de 0 y una desviación estándar de 1.

posición estándar Ángulo en posición tal que su vértice está en el origen y su lado inicial está a lo largo del eje x positivo.

estadística Una medida que describe una característica de una muestra.

inferencia estadística Utiliza la estadística deductiva como las predicciones y la hipótesis que prueban información de uso de una muestra para dibujar las conclusiones acerca de una población.

función etapa Función cuya gráfica es una serie de segmentos de recta.

método de sustitución Método para resolver un sistema de ecuaciones en que una de las ecuaciones se resuelve en una de las variables en términos de la otra.

encuesta Reunía información acerca de una población.

división sintética Método que se usa para dividir un polinomio entre un binomio.

sustitución sintética Uso de la división sintética para evaluar una función polinomial.

sistema de ecuaciones Conjunto de ecuaciones con las mismas variables.

sistema de desigualdades Conjunto de desigualdades con las mismas variables.

---

scatter plot (p. 92) A set of data graphed as ordered pairs in a coordinate plane.

secant (p. 790) For any angle, with measure α, a point $P(x, y)$ on its terminal side, $r = \sqrt{x^2 + y^2}$, $\sec \alpha = \dfrac{r}{x}$.

second-order determinant (p. 189) The determinant of a 2×2 matrix.

sequence (p. 659) A list of numbers in a particular order.

series (p. 668) The sum of the terms of a sequence.

set-builder notation (p. 35) The expression of the solution set of an inequality, for example $\{x \mid x > 9\}$.

sigma notation (p. 669) For any sequence $a_1, a_2, a_3, \ldots$, the sum of the first k terms may be written $\sum_{n=1}^{k} a_n$ which is read "the summation from $n = 1$ to k of a_n." Thus, $\sum_{n=1}^{k} a_n = a_1 + a_2 + a_3 + \cdots + a_k$, where k is an integer value.

simple event (p. P14) One event.

simplify (p. 303) To rewrite an expression without parentheses or negative exponents.

simulation (p. 731) The use of a probability experiment to mimic a real-life situation.

sine (p. 790) For any angle, with measure α, a point $P(x, y)$ on its terminal side, $r = \sqrt{x^2 + y^2}$, $\sin \alpha = \dfrac{y}{r}$.

slope (p. 77) The ratio of the change in y-coordinates to the change in x-coordinates.

slope-intercept form (p. 83) The equation of a line in the form $y = mx + b$, where m is the slope and b is the y-intercept.

solution (p. 18) A replacement for the variable in an open sentence that results in a true sentence.

solving a right triangle (p. 815) The process of finding the measures of all of the sides and angles of a right triangle.

solving a triangle (p. 815) Using given measures to find all unknown side lengths and angle measures of a triangle.

---

gráfica de dispersión Conjuntos de datos graficados como pares ordenados en un plano de coordenadas.

secante Para cualquier ángulo de medida α, un punto $P(x, y)$ en su lado terminal, $r = \sqrt{x^2 + y^2}$, $\sec \alpha = \dfrac{r}{x}$.

determinante de segundo orden El determinante de una matriz de 2×2.

sucesión Lista de números en un orden particular.

serie Suma específica de los términos de una sucesión.

notación de construcción de conjuntos Escritura del conjunto solución de una desigualdad, por ejemplo, $\{x \mid x > 9\}$.

notación de suma Para cualquier sucesión $a_1, a_2, a_3, \ldots$, la suma de los k primeros términos puede escribirse $\sum_{n=1}^{k} a_n$, lo que se lee "la suma de $n = 1$ a k de los a_n" Así, $\sum_{n=1}^{k} a_n = a_1 + a_2 + a_3 + \cdots + a_k$, donde k es un valor entero.

evento simple Un solo evento.

reducir Escribir una expresión sin paréntesis o exponentes negativos.

simulación Uso de un experimento probabilístico para imitar una situación de la vida real.

seno Para cualquier ángulo de medida α, un punto $P(x, y)$ en su lado terminal, $r = \sqrt{x^2 + y^2}$, $\sin \alpha = \dfrac{y}{r}$.

pendiente La razón del cambio en coordenadas y al cambio en coordenadas x.

forma pendiente-intersección Ecuación de una recta de la forma $y = mx + b$, donde m es la pendiente y b la intersección.

solución Sustitución de la variable de un enunciado abierto que resulta en un enunciado verdadero.

resolver un triángulo rectángulo Proceso de hallar las medidas de todos los lados y ángulos de un triángulo rectángulo.

resolver un triángulo Usar medidas dadas de hallar todas las medidas de las longitudes y laterales desconocidas del ángulo de un triángulo.

T

tangent (p. 790) 1. A line that intersects a circle at exactly one point. 2. For any angle, with measure α, a point $P(x, y)$ on its terminal side, $r = \sqrt{x^2 + y^2}$, $\tan \alpha = \dfrac{y}{x}$.

term (p. 659) 1. The monomials that make up a polynomial. 2. Each number in a sequence or series.

terminal side (p. 799) A ray of an angle that rotates about the center.

theoretical probability (p. P13) What should occur in a probability experiment.

theoretical probability distribution (p. 743) A distribution of probabilities based on what is expected to happen.

third-order determinant (p. 190) Determinant of a 3×3 matrix.

translation (p. 110) A figure is moved from one location to another on the coordinate plane without changing its size, shape, or orientation.

transverse axis (p. 624) The segment of length $2a$ whose endpoints are the vertices of a hyperbola.

tree diagram (p. P9) A diagram that shows all possible outcomes of an event.

trigonometric equation (p. 901) An equation containing at least one trigonometric function that is true for some but not all values of the variable.

trigonometric functions (p. 790) For any angle, with measure α, a point $P(x, y)$ on its terminal side, $r = \sqrt{x^2 + y^2}$, the trigonometric functions of α are as follows.

$$\sin \alpha = \frac{y}{r} \quad \cos \alpha = \frac{x}{r} \quad \tan \alpha = \frac{y}{x}$$
$$\csc \alpha = \frac{r}{y} \quad \sec \alpha = \frac{r}{x} \quad \cot \alpha = \frac{x}{y}$$

trigonometric identity (p. 873) An equation involving a trigonometric function that is true for all values of the variable for which the function is defined.

trigonometric ratio (p. 790) Compares the side lengths of a right triangle.

trigonometry (p. 790) The study of the relationships between the angles and sides of a right triangle.

turning point (p. 331) Point at which a graph turns. The location of relative maxima or minima.

U

unbounded (p. 154) A system of inequalities that forms a region that is open.

uniform probability model (p. P13) An experiment for which all outcomes are equally likely.

union (p. 42) The graph of a compound inequality containing *or*.

unit analysis (p. 310) The process of including unit measurement when computing.

unit circle (p. 830) A circle of radius 1 unit whose center is at the origin of a coordinate system.

univariate data (p. P24) Data with one variable.

V

variable (pp. P24, 5) 1. A characteristic of a population that can assume different values called *data*. 2. A symbol, usually a letter, used to represent an unknown quantity.

variable matrix (p. 200) A matrix that only contains the variables of a system of equations.

variance (p. P25) The mean of the squares of the deviations from the arithmetic mean.

vertex (pp. 154, 220) 1. Any of the points of intersection of the graphs of the constraints that determine a feasible region. 2. The point at which the axis of symmetry intersects a parabola. 3. The point on each branch nearest the center of a hyperbola.

vertex form (p. 275) A quadratic function in the form $y = a(x - h)^2 + k$, where (h, k) is the vertex of the parabola and $x = h$ its axis of symmetry.

T

tangente 1. Recta que interseca un círculo en un solo punto. 2. Para cualquier ángulo, de medida α, un punto $P(x, y)$ en su lado terminal, $r = \sqrt{x^2 + y^2}$, $\tan \alpha = \dfrac{y}{x}$.

término 1. Los monomios que constituyen un polinomio. 2. Cada número de una sucesión o serie.

lado terminal Rayo de un ángulo que gira alrededor de un centro.

probabilidad teórica Lo que debería ocurrir en un experimento probabilístico.

distribución de probabilidad teórica Distribución de probabilidades basada en lo que se espera que ocurra.

determinante de tercer orden Determinante de una matriz de 3×3.

traslación Se mueve una figura de un lugar a otro en un plano de coordenadas sin cambiar su tamaño, forma u orientación.

eje transversal El segmento de longitud $2a$ cuyos extremos son los vértices de una hipérbola.

diagrama de árbol Diagrama que muestra todos los posibles resultados de un evento.

ecuación trigonométrica Ecuación que contiene por lo menos una función trigonométrica y que sólo se cumple para algunos valores de la variable.

funciones trigonométricas Para cualquier ángulo, de medida α, un punto $P(x, y)$ en su lado terminal, $r = \sqrt{x^2 + y^2}$, las funciones trigonométricas de α son las siguientes.

$$\sin \alpha = \frac{y}{r} \quad \cos \alpha = \frac{x}{r} \quad \tan \alpha = \frac{y}{x}$$
$$\csc \alpha = \frac{r}{y} \quad \sec \alpha = \frac{r}{x} \quad \cot \alpha = \frac{x}{y}$$

identidad trigonométrica Ecuación que involucra una o más funciones trigonométricas y que se cumple para todos los valores de la variable en que el función es definido.

razón trigonométrica Compara las longitudes laterales de un triángulo derecho.

trigonometría Estudio de las relaciones entre los lados y ángulos de un triángulo rectángulo.

momento crucial Un punto en el cual un gráfico da vuelta. La localización de máximos o de mínimos relativos.

U

no acotado Sistema de desigualdades que forma una región abierta.

modelo de probabilidad uniforme Experimento cuyos resultados son todos equiprobables.

unión Gráfica de una desigualdad compuesta que contiene la palabra *o*.

análisis de la unidad Proceso de incluir unidades de medida al computar.

círculo unitario Círculo de radio 1 cuyo centro es el origen de un sistema de coordenadas.

datos univariados Datos con una variable.

V

variable 1. Característica de una población que puede tomar diferentes valores llamados *datos*. 2. Símbolo, generalmente una letra, que se usa para representar una cantidad desconocida.

matriz variable Una matriz que contiene solamente las variables de un sistema de ecuaciones.

varianza Media de los cuadrados de las desviaciones de la media aritmética.

vértice 1. Cualqeiera de los puntos de intersección de las gráficas que los contienen y que determinan una región viable. 2. Punto en el que el eje de simetría interseca una parábola. 3. El punto en cada rama más cercano al centro de una parábola.

forma de vértice Función cuadrática de la forma $y = a(x - h)^2 + k$, donde (h, k) es el vértice de la parábola y $x = h$ es su eje de simetría.

connectED.mcgraw-hill.com R147

vertical asymptote (p. 553) If the related rational expression of a function is written in simplest form and is undefined for $x = a$, then $x = a$ is a vertical asymptote.

vertical line test (p. 62) If no vertical line intersects a graph in more than one point, then the graph represents a function.

vertical shift (p. 846) When graphs of trigonometric functions are translated vertically.

vertices (pp. 615, 624) ellipse—The endpoints of the major axis. **hyperbola**—The endpoints of the transverse axis.

W

weighted average (p. 572) A method for finding the mean of a set of numbers in which some elements of the set carry more importance, or weight, than others.

whole numbers (p. 11) $\{0, 1, 2, 3, 4, \ldots\}$

X

x-intercept (p. 71) The x-coordinate of the point at which a graph crosses the x-axis.

Y

y-intercept (p. 71) The y-coordinate of the point at which a graph crosses the y-axis.

Z

zero matrix (p. 169) A matrix in which every element is zero.

zeros (p. 229) The x-intercepts of the graph of a function; the points for which $f(x) = 0$.

z-value (p. 762) The number of standard deviations that a given data value is from the mean.

---

asíntota vertical Si la expresión racional que corresponde a una función racional se reduce y está no definida en $x = a$, entonces $x = a$ es una asíntota vertical.

prudba de la recta vertical Si ninguna recta vertical interseca una gráfica en más de un punto, entonces la gráfica representa una función.

cambio vertical Cuando los gráficos de funciones trigonométricas son translado verticales.

vértices **(elipse)** Son las puntos finales del eje principal. **(hipérbola)** Son las puntos finales del eje transversal.

W

promedio ponderado Un método para encontrar el medio de un sistema de los números en los cuales algunos elementos del sistema llevan más importancia, o peso, que otros.

números naturales $\{0, 1, 2, 3, 4, \ldots\}$

X

intersección x La coordenada x del punto o puntos en que una gráfica interseca o cruza el eje x.

Y

intersección y La coordenada y del punto o puntos en que una gráfica interseca o cruza el eje y.

Z

matriz nula matriz cuyos elementos son todos igual a cero.

ceros Las intersecciones x de la gráfica de una función; los puntos x para tos que $f(x) = 0$.

valor Z Número de variaciones estándar que separa un valor dado de la media.

Index

21st Century Skills, T27, 2, 58, 132, 216, 300, 382, 448, 526, 590, 656, 720, 786, 870

A

Absolute value, 27–31, 277, 642, 684
complex number, 254
equations, 27–31, 51
functions, 103–104, 106, 109–110
inequalities, 43–47, 52, 118–121, 122, 126
simplify using, 408

Absolute value equations, 27–31, 51

Absolute value functions, 103–104, 106, 109–110
graphing, 110

Absolute value inequalities, 43–47, 52, 118–121, 122, 126
graphing, 118, 126

Activities. *See* Algebra Labs; Graphing Technology Labs; Spreadsheet Labs; TI-Nspire Technology Labs

Acute angles, 791, 816

Addition
additive identity, 12
additive inverse, 12
Associative Property of, 12, 173
Closure Property of, 12
Commutative Property of, 12–13, 173
of complex numbers, 248
elimination using, 139–140
of fractions, 539
of functions, 385–387
of matrices, 172
of radical expressions, 417
of rational expressions, 539
solving inequalities, 34

Addition Property
of equality, 19
of inequality, 33

Additive identity, 12

Additive inverse, 12, 188

Algebraic expressions, 50
absolute value, 27, 103
containing powers, 303
equivalent, 501–502
evaluating, 5–10

rational, 529–530, 531, 539
simplifying, 13–16
to verbal sentence, 18, 22

Algebra Labs
Area Under a Curve, 681–682
Combinations and Pascal's Triangle, 704
The Complex Plane, 253
Correlation and Causation, 99–100
Dimensional Analysis, 310
Discrete and Continuous Functions in the Real World, 68
Interval Notation, 40
Investigating Ellipses, 614
Median Fit Lines, 99–100
Quadratics and Rate of Change, 281
Roots of Equations and Zeros of Functions, 75

Algorithms, 311

Alternative hypothesis, 773

Amortizing loans, 698

Amplitude, 837–842

Angles
acute, 791, 808
central, 802
congruent, P20–P21
coterminal, 800, 809
degrees of, 801
of depression, 794
difference of, 886–891, 908, 910
double-angle identities, 893–894, 896, 908, 910
of elevation, 794
exterior, 46
general, 807–813, 861
graphing, 809
half-angle identities, 894–895, 908, 910
Interior Angle Sum Formula, 707
measures of, 799–800, 854, 859, 860
negative, 873
quadrantal, 808
radians of, 801–804, 808, 832
reference, 808–812, 817
of rotation, 800
in standard position, 799–804, 808
sum of, 886–891, 908, 910
terminal sides, 808–809, 830
vertex of, 349, 365

Animations, 2, 12, 42, 58, 84, 94, 103, 132, 146, 156, 162, 180, 216, 247, 258, 277, 283, 300, 306, 324, 331, 360, 382, 387, 417, 448, 453, 469, 494, 526, 532, 557, 590, 641, 656, 659, 669, 720, 786, 799, 801, 808, 837, 870

Aphelion, 618

Apothem, 822

Applications. *See also* Real-World Careers; Real-World Examples; Real-World Links, P3, P6, P9, P10, P11, P12, P15, P17, P18, P19, P21, P23, P25, P26, P27, P28, P29, 2, 3, 6, 7, 8, 9, 10, 13, 14, 15, 16, 17, 21, 22, 23, 24, 25, 26, 28, 30, 31, 32, 36, 37, 38, 39, 44, 45, 46, 47, 48, 50, 51, 52, 53, 58, 59, 63, 64, 65, 66, 67, 71, 72, 73, 74, 75, 76, 77, 79, 80, 81, 82, 87, 88, 89, 90, 91, 93, 95, 96, 97, 98, 103, 104, 105, 106, 107, 112, 113, 114, 115, 116, 119, 120, 121, 123, 124, 125, 126, 127, 132, 133, 136, 141, 147, 149, 150, 151, 152, 156, 157, 158, 159, 160, 164, 165, 166, 167, 168, 169, 171, 174, 181, 183, 184, 185, 186, 191, 194, 195, 196, 201, 202, 203, 204, 216, 217, 223, 224, 225, 226, 227, 228, 233, 234, 235, 236, 242, 243, 244, 245, 250, 251, 252, 255, 260, 261, 262, 269, 270, 271, 272, 278, 279, 280, 284, 285, 286, 287, 288, 291, 292, 293, 295, 300, 301, 306, 307, 308, 309, 315, 316, 317, 323, 327, 328, 329, 333, 334, 335, 336, 337, 340, 346, 347, 348, 349, 353, 355, 357, 364, 365, 368, 369, 370, 371, 372, 374, 375, 376, 377, 382, 383, 388, 389, 390, 391, 392, 393, 396, 397, 398, 402, 403, 404, 405, 406, 409, 410, 411, 412, 414, 418, 419, 420, 421, 423, 425, 426, 427, 428, 432, 433, 434, 435, 439, 440, 441, 442, 443, 448, 449, 453, 455, 456, 458, 462, 464, 465, 467, 472, 473, 474, 475, 481, 482, 483, 484, 486, 488, 489, 490, 491, 493, 495, 496, 497, 498, 505, 506, 507, 511, 513, 514, 515, 518, 519, 520, 526, 527, 533, 534, 535, 536, 537, 541, 542, 543, 544, 548, 549, 550,

551, 552, 555, 557, 558, 559, 560, 566, 567, 568, 569, 572, 573, 574, 576, 577, 578, 583, 584, 586, 587, 590, 591, 596, 597, 598, 602, 603, 604, 605, 607, 610, 611, 613, 618, 619, 620, 621, 622, 623, 627, 628, 629, 630, 631, 634, 635, 636, 642, 643, 644, 645, 647, 648, 649, 650, 651, 654, 655, 656, 657, 660, 663, 670, 671, 672, 673, 674, 677, 678, 679, 680, 686, 687, 688, 689, 691, 695, 696, 697, 698, 701, 702, 703, 707, 708, 709, 711, 712, 713, 714, 715, 720, 727, 728, 729, 730, 731, 732, 736, 737, 739, 740, 741, 742, 747, 748, 749, 750, 752, 754, 755, 756, 757, 758, 759, 760, 761, 763, 765, 766, 767, 768, 772, 774, 775, 776, 778, 781, 786, 792, 794, 795, 796, 797, 798, 800, 802, 803, 804, 805, 810, 811, 812, 813, 818, 820, 821, 825, 826, 827, 828, 829, 831, 833, 834, 835, 836, 839, 841, 842, 843, 848, 849, 850, 852, 855, 856, 857, 858, 861, 862, 865, 871, 876, 877, 878, 879, 882, 883, 884, 885, 888, 889, 890, 891, 892, 896, 897, 898, 899, 902, 904, 905, 906, 907, 909, 910, 911

Archytas, 675

Arc length, 802

Area diagram, 681–682

Argand plane, 253

Arithmetic means, 667, 670

Arithmetic sequences, 659–660, 662–664, 666–667, 711
  common differences, 659–660, 666, 668
  recursive formula for, 692

Arithmetic series, 668–672, 711

Aryabhatta, 875

Assessment. T20–T21, 2C, 58E, 132E, 216E, 300E, 382E, 448E, 526C, 590E, 656E, 720C, 786E, 870C *See also* Chapter Test; Guided Practice; Mid-Chapter Quiz; Prerequisite Skills; Quick Checks; Spiral Review; Standardized Test Practice
  Crystal Ball, P6, P12, P21, 32, 67, 107, 152, 236, 280, 309, 365, 406, 475, 491, 537, 551, 636, 680, 750, 813, 879
  Diagnostic Assessment, 3, 59, 133, 217, 301, 383, 449, 527, 591, 657, 721, 787, 871

Extended Response, 17, 57, 121, 131, 167, 215, 252, 299, 329, 381, 412, 445, 447, 498, 525, 569, 589, 636, 655, 697, 719, 759, 785, 798, 869, 907, 915

Formative Assessment, P5, P6, P8, P12, P15, P19, P21, P23, P28, 5, 6, 7, 9, 10, 11, 12, 13, 14, 16, 17, 18, 19, 20, 21, 22, 24, 25, 26, 27, 28, 29, 30, 31, 32, 34, 35, 36, 38, 39, 40, 42, 43, 44, 45, 47, 48, 49, 56–57, 62, 63, 64, 65, 66, 67, 68, 69, 70, 71, 73, 74, 75, 76, 77, 78, 79, 81, 82, 84, 85, 86, 88, 89, 90, 91, 93, 95, 97, 98, 99, 101, 102, 103, 104, 106, 107, 108, 110, 111, 112, 113, 115, 116, 117, 118, 119, 120, 121, 122, 130–131, 135, 136, 137, 139, 140, 141, 144, 146, 147, 148, 149, 151, 152, 153, 155, 157, 156, 159, 160, 162, 163, 164, 165, 166, 167, 168, 170, 171, 172, 173, 174, 175, 177, 178, 179, 180, 181, 182, 183, 185, 186–187, 188, 189, 190, 191, 192, 193, 194, 196, 197, 199, 200, 201, 202, 203, 204, 205, 206, 214–215, 219, 221, 222, 223, 224, 226, 227, 228, 230, 231, 232, 233, 235, 236, 237, 238, 239, 241, 242, 244, 245, 246, 247, 248, 249, 250, 251, 252, 254, 255, 256, 257, 258, 259, 260, 261, 262, 263, 265, 266, 267, 268, 269, 271, 272, 274, 275, 276, 277, 278, 279, 280, 281, 282, 283, 284, 285, 287, 288, 289, 290, 298–299, 304, 305, 306, 307, 308, 309, 310, 311, 312, 313, 314, 315, 316, 317, 319, 321, 322, 323, 325, 326, 328, 329, 331, 332, 333, 334, 336, 337, 339, 340, 341, 342, 343, 344, 345, 346, 348, 349, 351, 353, 354, 355, 356, 357, 359, 360, 361, 362, 363, 364, 365, 366, 367, 368, 369, 371, 372, 373, 380–381, 385, 386, 387, 388, 389, 391, 392, 393, 395, 396, 397, 398, 399, 400, 401, 402, 403, 405, 406, 408, 409, 410, 411, 412, 413, 414, 415, 416, 417, 418, 419, 420, 421, 422, 423, 424, 425, 427, 428, 430, 431, 432, 434, 435, 437, 438, 446–447, 451, 452, 453, 454, 455, 457, 458, 460, 461, 462, 463, 464, 466, 467, 468, 469, 470, 471, 472, 474, 475, 477, 478, 479, 480, 482, 483, 484, 485, 486, 487, 488, 490, 491, 492, 493, 494, 495, 497, 498, 500, 501, 502, 503, 504, 505, 506, 507, 508, 509, 510, 511, 512, 503, 514, 505, 516,

517, 524–525, 529, 530, 531, 532, 533, 536, 537, 538, 539, 540, 541, 543, 544, 545, 546, 547, 548, 550, 551, 552, 554, 555, 556, 557, 559, 560, 561, 563, 564, 565, 566, 568, 569, 570, 571, 572, 573, 574, 575, 576, 577, 578, 580, 581, 588–589, 593, 594, 595, 597, 598, 600, 601, 602, 603, 604, 605, 606, 607, 608, 609, 610, 612, 613, 614, 616, 617, 618, 619, 621, 622, 623, 625, 626, 627, 628, 630, 631, 632, 633, 634, 635, 636, 638, 639, 640, 641, 642, 644, 645, 646, 654–655, 659, 660, 661, 662, 663, 664, 665, 666, 667, 668, 669, 670, 672, 673, 674, 675, 676, 677, 679, 680, 682, 684, 685, 686, 688, 689, 690, 691, 692, 693, 694, 695, 696, 697, 698, 699, 700, 701, 702, 703, 704, 705, 706, 707, 708, 709, 710, 718–719, 724, 725, 726, 727, 729, 730, 732, 734, 735, 736, 737, 738, 740, 741, 742, 743, 744, 745, 746, 747, 749, 750, 751, 752, 753, 754, 755, 756, 758, 759, 761, 762, 763, 764, 765, 766, 768, 770, 771, 772, 773, 774, 775, 776, 777, 784–785, 789, 790, 791, 792, 793, 794, 795, 797, 798, 799, 800, 801, 802, 804, 805, 806, 807, 808, 809, 810, 811, 812, 813, 814, 815, 817, 818, 820, 821, 822, 823, 824, 825, 827, 828, 829, 830, 831, 832, 833, 835, 836, 837, 838, 839, 840, 841, 842, 843, 844, 845, 846, 847, 847, 848, 849, 851, 852, 854, 855, 856, 857, 858, 859, 868–869, 874, 875, 876, 878, 879, 880, 881, 882, 884, 885, 886, 887, 888, 890, 891, 892, 893, 894, 895, 896, 897, 898, 899, 900, 901, 902, 903, 904, 906, 907, 908, 914–915

Gridded Response, 32, 48, 57, 82, 116, 131, 186, 212, 213, 215, 288, 299, 317, 381, 392, 447, 458, 467, 507, 525, 551, 560, 589, 613, 622, 655, 689, 703, 709, 750, 785, 813, 836, 852, 869, 891, 915

Multiple Choice, 56, 130, 298, 380, 446, 524, 588, 654, 718, 784, 868, 914

Name the Math, P23, 10, 82, 121, 145, 167, 197, 227, 245, 262, 317, 349, 372, 392, 421, 467, 483, 507, 560, 578, 598, 631, 665, 697, 709, 759, 766, 821, 836, 852, 899

SAT/ACT, 10, 17, 25, 32, 39, 48, 67, 74, 82, 89, 98, 107, 116, 121, 152, 160, 167, 186, 197, 204, 227, 236,

Index

245, 252, 262, 272, 280, 288, 309, 317, 329, 337, 349, 357, 365, 372, 392, 398, 406, 412, 421, 428, 435, 458, 467, 475, 483, 491, 498, 515, 537, 544, 551, 569, 598, 605, 622, 631, 636, 645, 665, 673, 680, 689, 730, 741, 750, 766, 776, 798, 805, 813, 821, 828, 836, 843, 879, 885, 891, 899, 907

Short Response, 10, 25, 39, 57, 67, 74, 89, 98, 107, 121, 131, 152, 160, 197, 204, 227, 236, 245, 262, 272, 280, 288, 299, 309, 329, 337, 349, 381, 398, 406, 421, 428, 435, 447, 475, 491, 525, 537, 544, 551, 578, 589, 598, 605, 631, 645, 655, 673, 680, 719, 730, 741, 766, 776, 785, 798, 805, 821, 828, 836, 843, 858, 869, 899, 912, 913, 915

Summative Assessment, 26, 49–52, 53, 91, 122 –126, 127, 168, 206–210, 211, 254, 290–294, 295, 340, 373–376, 377, 414, 438–442, 443, 484, 517–520, 521, 552, 581–584, 585, 623, 751, 646–650, 651, 691, 710–714, 715, 777–780, 781, 829, 859–864, 865, 892, 908–910, 911

Ticket Out the Door, P5, P15, P19, P28, 17, 39, 48, 74, 98, 116, 186, 204, 252, 288, 329, 357, 412, 435, 458, 515, 569, 613, 689, 703, 730, 741, 798, 843, 858, 891, 907

Yesterday's News, P8, 25, 89, 160, 178, 272, 337, 398, 428, 498, 544, 605, 622, 645, 673, 776, 805, 826, 885

Associative Property
of addition, 12, 173
of multiplication, 12, 183, 246, 248

Asymptotes, 545–549, 553–556
of exponential functions, 451
graphing with, 561
of hyperbolas, 624–626
of rational functions, 546

Audio Recording. *See* Teach with Tech

Auditory/Musical Learners. *See* Learning Styles

Augmented matrices, 205

Axes. *See also* Axis of symmetry
conjugate, 624
major, 615–620
minor, 615–620
transverse, 624–625

Axis of symmetry, 220–221, 223–226, 599–601, 609, 625

B

Bari, Nina Karlovna, 140

Bar notation, 685

Base
change of base formula, 494–495
of logarithms, 468, 469, 504, 520
natural, 501–507
of powers, 501–507

Bell curve, 762–770

Biased sample, 723

Big Ideas. *See* Essential Questions

Binomial distributions, 752–768

Binomial expansions, 700, 701

Binomial experiment, 754

Binomial probability, 754, 759

Binomials
difference of cubes, 342–343
difference of squares, P8, 239, 343
expansion of, 700, 701
multiplying, P6, 238
sum of cubes, 342–344

Binomial Theorem, 699–703, 710, 714
expanding binomial expressions, 699–701
in sigma notation, 701

Bivariate data, 92

Blog. *See* Teach with Tech

Boundaries, 117–118, 148

Bounded regions, 154–155

Box-and-whisker plots, 735, 737

Brahmagupta, 266

Break-even point, 136

C

Calculator, 866. *See also* Graphing Technology Labs

Carbon dating, 510

Careers. *See* Real-World Careers

Causation, 99–100

Celsius, 7, 397

Census, 724

Centers
of circles, 607–612
of ellipses, 615–621
of hyperbolas, 624–626

Central angle, 802

Central tendency. *See* Measures of central tendency

Challenge. *See* Higher-Order Thinking Problems

Change of Base Formula, 494–495

Chapter 0, P2–P29
Adding Probabilities, P13– P15
Congruent and Similar Figures, P20–P21
Counting Techniques, P9–P12
Factoring Polynomials, P7–P8
FOIL Method, P6
Measures of Center, Spread, and Position, P24–P28
Multiplying Probabilities, P16–P19
Pythagorean Theorem. P22–P23
Representing Functions, P4–P5

Chapter Project. *See* 21st Century Skills

Chapter Test, 53, 127, 211, 295, 377, 443, 521, 585, 651, 715, 781, 865, 911

Check for Reasonableness. *See* Study Tips

Circles, 607–613, 632–633, 646, 648
area, 397, 440, 441, 750
circumference, 801
concentric, 611
equations of, 606, 607–613
graphing, 606, 609
radius of, 607–608, 610–612, 801
unit, 830–832

Circular functions, 830–836, 859, 862
cosine, 830, 832–833
sine, 830, 832–833

Circumference, 801

Closed, 305, 306

Coefficient matrix, 192, 200–201, 205

Coefficients
in binomial expansions, 700
correlation, 94, 97
leading, 314, 322–324

Cofunction identities, 873

College Readiness, T22–T23

Column matrices, 169

Combinations, P11, 704

Combined variation 565, 567, 568

Common denominators, least, 538, 540, 570–575

Common difference, 666, 668–669

Common logarithms, 492–498

Common ratio, 661–662

Commutative Property
of addition 12, 173
of multiplication, 12, 246, 248
test of, 182

Completing the square, 256–262, 293

Complex conjugates, 249, 362

Complex Conjugates Theorem, 362

Complex fractions, 532–533, 540

Complex numbers, 246–252, 267, 290, 292
adding, 248
absolute value, 254
conjugates, 249, 362
dividing, 249
graphing, 251
multiplying, 249
subtracting, 248

Complex plane, 253

Composition of functions, 387–388, 694
iterations, 694–695, 710

Compound event, P14

Compound inequalities, 41–43, 52
conjunction (and), 41
disjunction (or), 42

Compound interest, 462–463, 508
continuous, 504

Computer Projector. *See* Teach with Tech

Concept Summary
binomial expansion, 701
degrees and radians, 801
expressions with rational exponents, 425
factoring techniques, 343
graph trigonometric functions, 847
properties of exponents, 303
real number properties, 12
simplifying radical expressions, 417
solving oblique triangles, 824
solving quadratic equations, 269
solving systems of equations, 140
standard forms of conic sections, 632
tests of significance, 771
transformations of functions, 112
transformations of quadratic functions, 277
zeros, factors, roots, and intercepts, 358

Conditional probability, P16

Cones
lateral area, 197
volume 6, 8

Confidence intervals, 769–776

Congruence statements, P20

Congruent angles, P20

Congruent figures, P20–P21

Congruent sides, P20

Conic sections, 632–636, 649
circles, 607–613, 632–633
classifying with the discriminant, 633
ellipses, 615–622, 632–633
equations of, 632–633
general form of, 632
graphing, 639
hyperbolas, 624–631, 632–633
parabolas, 599–605, 632–633
standard form of, 632

Conjectures, making, 281, 399, 516, 606, 614, 822

Conjugate axis, 624

Conjugates, 249, 362, 418

Conjunctions, 41

ConnectED
Animations, 2, 12, 42, 58, 84, 94, 103, 132, 146, 156, 162, 180, 216, 247, 258, 277, 283, 300, 306, 324, 331, 360, 382, 387, 417, 448, 453, 469, 494, 526, 532, 557, 590, 641, 656, 659, 669, 720, 786, 799, 801, 808, 837, 870
Chapter Readiness Quiz, 3, 59, 133, 217, 301, 383, 449, 527, 591, 657, 721, 787, 871
Graphing Calculator 90, 108, 135, 153, 187, 205, 228, 237, 263, 273, 289, 318, 320, 338, 341, 350, 368, 399, 413, 436, 459, 476, 499, 516, 561, 579, 606, 637, 639, 690, 731, 844, 900
Personal Tutor, P4, P6, P7, P9, P10, P11, P13, P14, P16, P17, P18, P24, P25, P26, P27, 5, 6, 11, 12, 13, 18, 19, 20, 21, 27, 28, 29, 34, 35, 36, 41, 42, 43, 44, 61, 63, 64, 69, 70, 71, 76, 77, 78, 84, 85, 86, 90, 93, 94, 101, 102, 103, 104, 108, 110, 111, 112, 117, 118, 135, 136, 137, 138, 139, 140, 146, 147, 148, 153, 155, 156, 162, 163, 164, 169, 170, 171, 172, 173, 174, 179, 180, 181,
182, 189, 190, 191, 192, 193, 199, 200, 201, 205, 219, 221, 222, 223, 228, 230, 231, 232, 237, 238, 239, 240, 241, 246, 247, 248, 249, 256, 257, 258, 259, 263, 265, 266, 267, 268, 273, 275, 276, 277, 282, 283, 284, 285, 304, 305, 306, 311, 312, 313, 314, 318, 320, 322, 323, 325, 330, 331, 332, 333, 338, 341, 342, 343, 344, 345, 350, 353, 354, 359, 360, 361, 362, 367, 368, 369, 382, 385, 386, 387, 388, 393, 394, 395, 399, 400, 401, 402, 408, 409, 413, 415, 416, 417, 418, 422, 423, 424, 429, 430, 431, 436, 451, 452, 453, 454, 455, 461, 462, 463, 468, 469, 470, 471, 476, 478, 479, 480, 485, 486, 487, 492, 493, 494, 495, 499, 501, 502, 503, 504, 508, 509, 510, 511, 512, 529, 530, 531, 532, 533, 538, 539, 540, 545, 546, 547, 548, 554, 555, 556, 557, 561, 563, 564, 565, 570, 571, 572, 573, 574, 575, 579, 593, 594, 595, 600, 601, 602, 606, 607, 608, 609, 616, 617, 618, 619, 625, 626, 627, 632, 633, 639, 640, 641, 642, 659, 660, 661, 662, 666, 667, 668, 669, 674, 675, 676, 677, 683, 684, 685, 690, 692, 693, 694, 698, 699, 700, 701, 705, 706, 724, 725, 726, 731, 732, 734, 735, 736, 737, 742, 743, 745, 746, 752, 753, 754, 755, 761, 762, 763, 767, 770, 771, 772, 773, 789, 790, 791, 792, 793, 794, 799, 800, 801, 802, 807, 808, 809, 810, 814, 815, 816, 817, 818, 823, 824, 825, 830, 831, 832, 837, 838, 839, 840, 844, 845, 846, 847, 848, 854, 855, 874, 875, 880, 881, 882, 886, 887, 888, 893, 894, 895, 896, 900, 901, 902, 903, 904
Self-Check Practice, 3, 7, 14, 22, 30, 36, 45, 53, 57, 59, 64, 71, 79, 86, 95, 104, 113, 119, 127, 131, 133, 141, 149, 157, 165, 175, 183, 194, 202, 211, 215, 217, 224, 233, 242, 250, 260, 269, 278, 285, 295, 299, 301, 307, 315, 326, 334, 346, 355, 363, 369, 377, 381, 383, 389, 396, 403, 409, 419, 425, 432, 443, 447, 449, 455, 464, 472, 480, 488, 495, 505, 513, 521, 525, 527, 533, 541, 548, 557, 566, 576, 585, 589, 591, 595, 603, 610, 619, 628, 634, 642, 651, 655, 657, 663, 670, 677, 686, 695, 701, 707, 715, 719, 721, 727, 738, 747, 756, 764, 774, 781, 785,

Index

787, 795, 802, 811, 818, 825, 833, 841, 849, 856, 865, 869, 871, 876, 882, 888, 897, 904, 911, 915

Connections. *See* Applications

Constant difference, 627

Constant functions, 109–110, 324
graphing, 110
graphs of, 110

Constant matrix, 200–201

Constant of variation, 90, 562

Constants, 78

Constant sum, 616

Constant terms, 219

Constraints, 29, 44, 138, 154, 156

Continuous functions, 68
absolute value, 103–104, 106, 109–110
cosine 840, 842–843
exponential, 451–458, 461, 462, 463, 471, 501–507, 509–511, 517, 518, 520, 662, 866
logarithmic, 468–475, 478, 479, 480, 501–507, 517, 518, 520
polynomial, 322–329, 330–337, 338–339, 344, 358–362, 367–369, 373, 374, 375, 376, 553, 554
quadratic, 109–110, 219–227, 275–279, 281, 291, 294, 324
sine, 830, 832–833
square root, 400–406, 407–412, 440

Continuously compounded interest, 504

Continuous probability distribution, 744
bell curve, 760–768
normal distributions, 762–768
skewed distribution, 733

Continuous relation, 62–65

Contraction. *See* Dilations

Control group, 723, 726

Convergent series, 683, 686–687

Converse of the Pythagorean Theorem, P23

Cooperative Groups, 40, 68, 75, 90, 99, 108, 135, 153, 169, 187, 205, 228, 237, 253, 263, 273, 281, 289, 310, 318, 320, 338, 341, 350, 366, 399, 413, 437, 459, 476, 499, 508, 516, 561, 579, 606, 614, 637, 639, 681, 690, 698, 704, 731, 768, 789, 806, 822, 844, 900

Coordinate graph. *See* Coordinate plane

Coordinate grid. *See* Coordinate plane

Coordinate plane, P4

Corollary to the Fundamental Theorem of Algebra, 359, 369

Corollary to the Rational Zero Theorem, 367

Correlation
negative, 92–97
positive, 92–97

Correlation coefficient, 94, 97

Corresponding elements, 169, 174

Cosecant, 790–791, 809, 840

Cosine, 790–791, 809, 830, 832–833, 837–838
law of, 823–828, 862

Cotangent, 790–791, 809, 840

Coterminal angles, 800, 809

Counterexamples, 706, 873

Cramer's Rule, 192–196

Critical region, 773

Critical Thinking. *See* Higher-Order Thinking Problems

Cross-Curriculum. *See* Applications

Cube root equations, 430

Cubes
difference of, 342–344
sides of, 410
sum of, 342–344

Cubic functions, 324

Cycles, 831, 837

Cylinders
surface area, 39, 703
volume, 9, 23, 50, 51, 53

Data *See also* Graphs; Statistics; Tables
bivariate, 92
interquartile range, P27
means of, P24, 735, 739, 762
medians of P24, 735, 762
modeling using polynomial functions, 338–339
modes of, P24, 726, 762
organizing with matrices, 169–171
organizing with spreadsheets, 170
outliers, P27, 93
range of, P4–P5
scatter plots of, 92–97, 122, 125

standard deviation of, P25, 733–737, 745–749, 762
univariate, P24
variance of, P25, 745

Data collection device, 289

Decay
exponential, 453–455, 509–510, 514
factor, 453–457

Decimals
repeating, 11, 685
terminating, 11
write as fractions, 685

Decision Making, 746, 748, 758, 775

Degrees
of angles, 801
convert to radians, 801
of a polynomial, 305, 307, 322–328, 553

Denominators, 532, 539
least common, 538–540, 570–571
monomials as, 539
polynomials as 532, 539
rationalizing, 416, 441

Dependent events, P16, 764

Dependent variables, 64, 660

Depressed polynomials, 354

Depression, angles of, 794

Descartes' Rule of Signs, 360

Descriptive statistics, P24–P28

Determinants, 189–196, 199–200
Cramer's Rule, 192–196
second-order, 189
third-order, 190

Diagnose readiness. *See* Prerequisite Skills

Diagnostic Assessment, 3, 59, 133, 217, 301, 383, 449, 527, 591, 657, 721, 787, 871

Diagonal rule, 190, 194

Diagonals, 826

Diagrams
area, 681–682
drawing, 617
tree, P9
Venn, P14, 11, 248

Diameter, 617

Difference of angles identity, 886–891, 908, 910

Difference of cubes, 342–343

Difference of squares, 239, 343

Differences, common, 659–660, 666, 668

Differentiated Homework Options, 7, 14, 22, 30, 37, 45, 65, 72, 79, 87, 96, 105, 113, 119, 141, 149, 158, 165, 175, 184, 195, 202, 224, 233, 242, 250, 260, 270, 278, 286, 307, 326, 334, 346, 355, 363, 370, 389, 396, 403, 410, 419, 426, 433, 456, 464, 472, 480, 488, 495, 505, 513, 534, 541, 549, 558, 566, 576, 596, 603, 611, 620, 628, 634, 643, 663, 670, 678, 686, 695, 701, 707, 727, 738, 747, 756, 764, 774, 795, 803, 811, 819, 826, 833, 841, 850, 856, 876, 882, 889, 897, 904

Differentiated Instruction, T12–T13, T22–T23, 2D, 4, 6, 8, 10, 12, 13, 17, 20, 21, 29, 32, 35, 39, 42, 44, 58F, 60, 64, 67, 71, 74, 78, 82, 84, 89, 94, 98, 103, 104, 111, 116, 121, 132F, 134, 137, 140, 145, 148, 150, 152, 156, 160, 164, 167, 174, 178, 180, 181, 190, 193, 197, 200, 204, 216F, 218, 223, 227, 232, 236, 239, 245, 247, 252, 259, 262, 265, 268, 277, 280, 285, 288, 300F, 302, 304, 309, 314, 317, 324, 329, 333, 337, 338, 344, 349, 354, 357, 361, 365, 369, 372, 382F, 384, 387, 388, 391, 395, 398, 404, 406, 409, 412, 418, 421, 423, 424, 431, 435, 448F, 450, 453, 458, 463, 465, 467, 469, 475, 479, 483, 487, 491, 494, 498, 503, 507, 510, 511, 515, 526D, 528, 532, 537, 540, 542, 546, 548, 555, 556, 560, 567, 569, 574, 578, 590F, 592, 594, 598, 600, 605, 608, 610, 613, 617, 618, 622, 625, 631, 633, 636, 641, 645, 656F, 658, 662, 665, 668, 671, 676, 680, 685, 689, 694, 697, 700, 703, 706, 709, 720D, 722, 725, 730, 735, 741, 745, 750, 755, 759, 762, 766, 771, 776, 786F, 788, 793, 798, 801, 805, 810, 813, 817, 821, 825, 828, 832, 836, 838, 843, 846, 852, 855, 858, 870D, 872, 874, 875, 879, 881, 885, 887, 891, 896, 899, 903, 907. *See also* Learning Styles

Dilations, 111–114
graphing, 111

Dimensional analysis, 310

Dimensions, 169–171, 179

Diophantus of Alexandria, 19

Directrix, 599

Direct variation, 90, 562–563, 565–568

Discontinuity, 556–557, 561

Discrete functions, 68

Discrete mathematics. *See* Combinations; Permutations; Probability; Sequences; Statistics

Discrete probability distributions, 743, 744, 745, 746, 754

Discrete relations, 62–65

Discriminants, 267–270, 293, 633

Disjunctions, 42

Distance Formula, 76, 573, 593–597, 608, 646, 647, 653

Distributions, 742–749
binomial, 752–758
continuous probability, 743
discrete probability, 743
full probability, 755
normal, 760–768
probability, 742–749
skewed, 733
uniform, P13

Distributive Property, P7, 12–13, 14, 173, 182, 183, 248

Divergent series, 683–684, 686–687

Division
of complex numbers, 249
of fractions, 531
of functions, 385–386
long, 311–312
of monomials, 303
of a polynomial by a monomial, 311–314
of polynomials, 311–316, 318–319, 374
of radicals, 416, 417
of rational expressions, 531
remainders, 312
solving inequalities, 35
synthetic, 312–314, 352

Division algorithm, 311

Division Property
of equality 19, 20
of inequalities, 34

Document Camera. *See* Teach with Tech

Domains, P4–P5
of composite functions, 387–389
of exponential functions, 461–463, 545
of linear functions, 108
of rational functions, 561
of relations, 61–62

of square root functions, 400–405
of trigonometric functions, 844

Dot plot. *See* Scatter plots

Double-angle identities, 893–894, 896, 908, 910

Double roots, 325

Draw a picture, 378

E

e, 501–507

Elements, 169–171, 174

Elevation, angles of, 794

Elimination method, 139–140

Ellipses, 614, 615–622, 632–633, 638, 646, 648
center of, 615–620
co-vertices of, 615–617, 619–620
eccentricity of, 621
equations of, 615–620
foci of, 615–616, 618–620
graphing, 618–619
major axis of, 615–620
minor axis of, 615–620
standard form of, 618
vertices of, 615–617, 619–620

Empirical Rule, 760

Empty set, 28–31, 147

End behavior, 324, 452, 471
of polynomial functions, 324

English Language Learners, 2D, 5, 6, 11, 13, 18, 20, 27, 29, 33, 35, 41, 42, 58F, 61, 69, 71, 76, 78, 83, 84, 92, 101, 103, 109, 111, 117, 121, 132F, 136, 146, 154, 161, 164, 171, 179, 189, 198, 216F, 219, 223, 229, 232, 238, 246, 247, 256, 259, 264, 265, 275, 277, 282, 300F, 303, 304, 311, 322, 330, 333, 342, 244, 352, 354, 358, 361, 367, 382F, 385, 388, 391, 393, 400, 404, 407, 415, 422, 423, 429, 431, 448F, 451, 453, 461, 468, 478, 485, 487, 492, 501, 509, 510, 526D, 529, 538, 540, 545, 548, 553, 556, 562, 570, 574, 590F, 593, 599, 607, 615, 618, 624, 632, 640, 656F, 659, 662, 666, 668, 674, 676, 683, 692, 699, 700, 705, 720D, 723, 725, 733, 735, 742, 745, 752, 755, 760, 769, 771, 786F, 790, 799, 807, 814, 817, 823, 825, 830, 837, 845, 853, 870D, 873, 880, 881, 886, 887, 893, 901

Equal matrices, 169

Index

Equations, 18, 135, 136–137, 138–140, 161–167, 200–203, 205
  absolute value, 27–31, 51
  of circles, 606, 607–613
  cube root, 430
  direct variation, 90, 562–563, 565–568
  ellipses, 615–620
  exponential, 459–460, 461–463, 493, 518
  of hyperbolas, 624–631
  inverse variation, 564–568
  with irrational roots, 257
  joint variation, 563, 565–567
  linear, 69–73, 83–89, 90, 124
  logarithmic, 478–483, 493, 499–500, 519
  matrix, 189–202
  multi-step, 20
  natural log, 501–507, 520
  of parabolas, 599–605
  point-slope form, 84
  polynomial, 342–349, 375
  prediction, 92–93, 96
  quadratic, 228, 229–236, 237, 238–245, 247, 256–269, 291, 292, 644
  quadratic form, 221, 345–347
  radical, 429–433, 436–437, 442
  rational, 570–574, 576–577, 579–580, 584
  with rational roots, 256
  regression, 228
  of relations, 61–67, 123
  roots of, 75, 238
  slope-intercept form, 83–84, 86–88
  solving, 18–24, 49, 51, 135, 136, 137, 138, 192, 193–196, 200–202, 205, 229–236, 237, 238–245, 246–259, 263–271, 290, 291, 292, 342–349, 373, 375, 429, 433, 436–437, 459–460, 461–463, 478–483, 493, 499–500, 518, 519, 570–571, 576–577, 579–580, 855, 900, 901–907, 910
  standard form of, 70, 72–73, 229, 242–244, 264, 607, 609
  systems of, 136, 640–644
  trigonometric, 900, 901–907, 910

Equilateral triangles, 679, 822, 852

Equivalent expressions, 501–502, 881

Error Analysis. *See* Higher-Order Thinking Problems

eSolutions Manual Plus, 2A–2B, 58A–58D, 132A–132C, 216A–216D, 300A–300D, 382A–382C, 448A–448C, 526A–526B, 590A–590C, 656A–656C, 720A–720B, 786A–786C, 870A–870B

Essential Questions, 31, 40, 81, 88, 144, 203, 226, 287, 336, 371, 391, 474, 514, 559, 577, 604, 679, 696, 758, 775, 827, 842, 884, 906

Essential Questions, T14–T15, 3, 8, 48, 59, 74, 98, 133, 142, 186, 217, 235, 272, 301, 337, 370, 383, 394, 402, 449, 455, 512, 527, 557, 566, 591, 627, 634, 657, 661, 697, 721, 749, 766, 787, 848, 858, 871, 883, 896

Even-degree functions, 324, 325

Events *See also* Probability
  compound, P14
  dependent, P16, 764
  independent, P16
  mutually exclusive, P14
  simple, P14

Expansion of binomials, 700, 701

Expected value, 745

Experiment, P9

Experimental probability, P9, P13, 745–746, 755

Experiments, 725

Explicit formula, 692

Exponential decay, 453–455, 509–510, 514

Exponential equations, 459, 460, 461–463, 493, 518

Exponential form, 422
  to logarithmic form, 469

Exponential functions, 517, 520, 662. *See also* Functions
  graphing, 451–458, 518
  inverse of, 471
  natural base, 501–507
  Property of Equality, 461
  Property of Inequality, 463
  transformations of, 452
  writing, 462

Exponential growth, 451–453, 509–511
  compound interest, 462–463, 504, 508

Exponential inequalities, 459–460, 463, 494, 518

Exponents, 6
  fractional, 422–428
  negative, 303
  properties of, 303
  rational, 422–428, 441
  zero, 470

Expressions, 417
  algebraic, 5–10, 13–16, 18, 22, 27, 50, 103, 303, 501–502, 529–530, 531, 539, 881
  containing powers, 303
  equivalent, 501–502
  logarithmic, 468–469
  radical, 415–418, 424, 425, 441
  rational, 424, 425, 529–530, 531, 539
  simplifying 303, 874–875, 881, 912
  trigonometric, 874–875
  verbal, 18, 22

Extended Response, 17, 57, 121, 131, 167, 252, 299, 329, 381, 412, 445, 447, 498, 525, 569, 589, 636, 655, 697, 719, 759, 785, 798, 869, 907, 915

Exterior Angle Inequality Theorem, 46

Extraneous solutions, 29, 43, 430, 571, 903

Extrapolation, 95

Extra Practice, R1–R13

Extrema, 331

F

Factored form, 238

Factorials, P10

Factoring, 354
  greatest common factor, 239, 343, 529
  polynomials, P7–P8, 239–244, 342, 343, 358, 532
  quadratic equations, 238–244, 269, 292
  techniques, 343
  trinomials, 239–240

Factors, greatest common, 239, 343, 529

Factor Theorem, 354, 373, 376
  proof of, 356

Fahrenheit, 7, 397

Failure, 754

Families of graphs, 109
  of linear functions, 108
  of parabolas, 273–274
  of quadratic functions, 275

Feasible regions, 154–155, 157–158

Index

Fibonacci sequence, 692

Financial Literacy, 7, 15, 23, 72, 87, 95, 97, 151, 157, 166, 196, 225, 243, 278, 309, 334, 355, 389, 423, 439, 455, 465, 474, 490, 504, 505, 506, 550, 559, 577, 644, 663, 671, 679, 687, 694, 695, 696, 748, 765, 805
  credit cards, 474, 490, 694
  interest, 462–463, 508
  loans, 696, 698
  money, 2, 13, 14, 30, 32, 37, 38, 44, 45, 50, 51, 52, 53, 202, 301, 464, 660

Finite sequence, 659

Five-number summary, P26, 736

Foci
  of ellipses, 615–616, 618–620
  of hyperbolas, 624–630
  of parabolas, 599

Focus of parabolas, 599

Focus on Mathematical Content, P9, P12, P13, P18, P24, P26, P27, 2E–2F, 20, 35, 43, 58G–58H, 63, 70, 77, 85, 93, 102, 132G–132H, 140, 156, 164, 172, 173, 183, 191, 193, 201, 216G–216H, 222, 231, 240, 249, 258, 265, 267, 284, 300G–300H, 304, 313, 344, 354, 359, 382G–382H, 386, 408, 416, 424, 430, 448G–448H, 453, 462, 470, 486, 494, 503, 511, 526E–526F, 531, 540, 547, 563, 571, 590G–590H, 596, 616, 626, 656G–656H, 662, 667, 669, 675, 677, 684, 701, 720E–720F, 724, 745, 755, 761, 772, 786G–786H, 792, 810, 817, 824, 839, 854, 870E–870F, 875, 883, 895, 903

FOIL method, 238–249

Foldables™ Study Organizer
  conic sections, 592, 646
  equations and inequalities, 4, 49
  exponential and logarithmic functions, 450, 517
  inverses and radical functions, 384, 438
  linear functions, 60, 122
  polynomials, 302, 373
  quadratic functions, 218, 290
  rational functions and relations, 528, 581
  sequences and series, 658, 710
  statistics and probability, 722, 777

  systems of equations and inequalities, 134, 206
  trigonometric functions, 788, 859
  trigonometric identities and equations, 872, 908

Formative Assessment, P5, P6, P8, P12, P15, P19, P21, P23, P28, 5, 6, 7, 9, 10, 11, 12, 13, 14, 16, 17, 18, 19, 20, 21, 22, 24, 25, 26, 27, 28, 29, 30, 31, 32, 34, 35, 36, 38, 39, 40, 42, 43, 44, 45, 47, 48,49, 56–57, 62, 63, 64, 65, 66, 67, 68, 69, 70, 71, 73, 74, 75, 76, 77, 78, 79, 81, 82, 84, 85, 86, 88, 89, 90, 91, 93, 95, 97, 98, 99, 101, 102, 103, 104, 106, 107, 108, 110, 111, 112, 113, 115, 116, 117, 118, 119, 120, 121, 122, 130–131, 135, 136, 137, 139, 140, 141, 144, 146, 147, 148, 149, 151, 152, 153, 155, 157, 156, 159, 160, 162, 163, 164, 165, 166, 167, 168, 170, 171, 172, 173, 174, 175, 177, 178, 179, 180, 181, 182, 183, 185, 186–187, 188, 189, 190, 191, 192, 193, 194, 196, 197, 199, 200, 201, 202, 203, 204, 205, 206, 214–215, 219, 221, 222, 223, 224, 226, 227, 228, 230, 231, 232, 233, 235, 236, 237, 238, 239, 241, 242, 244, 245, 246, 247, 248, 249, 250, 251, 252, 254, 255, 256, 257, 258, 259, 260, 261, 262, 263, 265, 266, 267, 268, 269, 271, 272, 274, 275, 276, 277, 278, 279, 280, 281, 282, 283, 284, 285, 287, 288, 289, 290, 298–299, 304, 305, 306, 307, 308, 309, 310, 311, 312, 313, 314, 315, 316, 317, 319, 321, 322, 323, 325, 326, 328, 329, 331, 332, 333, 334, 336, 337, 339, 340, 341, 342, 343, 344, 345, 346, 348, 349, 351, 353, 354, 355, 356, 357, 359, 360, 361, 362, 363, 364, 365, 366, 367, 368, 369, 371, 372, 373, 380–381, 385, 386, 387, 388, 389, 391, 392, 393, 395, 396, 397, 398, 399, 400, 401, 402, 403, 405, 406, 408, 409, 410, 411, 412, 413, 414, 415, 416, 417, 418, 419, 420, 421, 422, 423, 424, 425, 427, 428, 430, 431, 432, 434, 435, 437, 438, 446–447, 451, 452, 453, 454, 455, 457, 458, 460, 461, 462, 463, 464, 466, 467, 468, 469, 470, 471, 472, 474, 475, 477, 478, 479, 480, 482, 483, 484, 485, 486, 487, 488, 490, 491, 492, 493, 494, 495, 497, 498, 500, 501, 502, 503, 504, 505, 506, 507, 508, 509,

510, 511, 512, 503, 514, 505, 516, 517, 524–525, 529, 530, 531, 532, 533, 536, 537, 538, 539, 540, 541, 543, 544, 545, 546, 547, 548, 550, 551, 552, 554, 555, 556, 557, 559, 560, 561, 563, 564, 565, 566, 568, 569, 570, 571, 572, 573, 574, 575, 576, 577, 578, 580, 581, 588–589, 593, 594, 595, 597, 598, 600, 601, 602, 603, 604, 605, 606, 607, 608, 609, 610, 612, 613, 614, 616, 617, 618, 619, 621, 622, 623, 625, 626, 627, 628, 630, 631, 632, 633, 634, 635, 636, 638, 639, 640, 641, 642, 644, 645, 646, 654–655, 659, 660, 661, 662, 663, 664, 665, 666, 667, 668, 669, 670, 672, 673, 674, 675, 676, 677, 679, 680, 682, 684, 685, 686, 688, 689, 690, 691, 692, 693, 694, 695, 696, 697, 698, 699, 700, 701, 702, 703, 704, 705, 706, 707, 708, 709, 710, 718–719, 724, 725, 726, 727, 729, 730, 732, 734, 735, 736, 737, 738, 740, 741, 742, 743, 744, 745, 746, 747, 749, 750, 751, 752, 753, 754, 755, 756, 758, 759, 761, 762, 763, 764, 765, 766, 768, 770, 771, 772, 773, 774, 775, 776, 777, 784–785, 789, 790, 791, 792, 793, 794, 795, 797, 798, 799, 800, 801, 802, 804, 805, 806, 807, 808, 809, 810, 811, 812, 813, 814, 815, 817, 818, 820, 821, 822, 823, 824, 825, 827, 828, 829, 830, 831, 832, 833, 835, 836, 837, 838, 839, 840, 841, 842, 843, 844, 845, 846, 847, 847, 848, 849, 851, 852, 854, 855, 856, 857, 858, 859, 868–869, 874, 875, 876, 878, 879, 880, 881, 882, 884, 885, 886, 887, 888, 890, 891, 892, 893, 894, 895, 896, 897, 898, 899, 900, 901, 902, 903, 904, 906, 907, 908, 914–915

Formulas, 6, 50. *See also* Interest; Measurements; Rates
  95% confidence interval, 769
  area
    of circles, 17, 397, 440, 441
    of trapezoids, 21
    of triangles, 7, 9, 443, 475, 814
  binomial probability, 756
  change of base, 494–495
  circumference, 801
  confidence interval, 770
  for converting Fahrenheit to Celsius, 397
  for converting yards to feet, 439

distance, 76, 573, 593–597, 608, 646, 647, 653

Heron's, 9

interior angle sum, 707

lateral area
  of cones, 197

maximum error of estimate, 769

midpoint, 593, 595–596, 608, 646, 647

nth term of an arithmetic sequence, 666–667

quadratic, 264–267, 293

radius
  of spheres, 409, 426

recursive, 692–696, 710, 713

slope, 77

standard deviation, P25

standardized test practice, 652

surface area
  of cylinders, 39
  of rectangular prisms, 26, 32
  of spheres, 26

variance, P25

volume,
  of cones, 6, 8
  of cylinders, 9, 23, 50, 51, 53
  of pyramids, 8
  of rectangular prisms, 6
  of spheres, 26

z-value, 762

Fractals, 696

Fractional exponents, 422–428

Fraction bars, 6

Fractions, 221. *See also* Rational numbers
  adding, 539
  bar, 6
  complex, 532–533, 540
  dividing, 531
  write repeating decimals as, 685

Frequency, 838–839
  table, 743

Full probability distribution, 755

Function notation, 64

Functions
  absolute value, 103–104, 106, 109–110
  adding, 385–387
  arccosine, 853
  arcsine, 853
  arctangent, 853
  binomial distributions, 752–759
  circular, 830–836, 859, 862

composite, 387

composition of, 387–388, 694

constant, 109–110, 324

continuous, 68

cosecant, 790, 840

cosine, 790, 830, 832–833, 837

cotangent, 790, 840

cubic, 324

discrete, 68

domains of, 400, 545

evaluating, 64

even-degree, 325

exponential, 451–458, 461, 462, 463, 471, 501–507, 517, 518, 520, 662, 866

greatest integer, 102

identity, 109, 395

inverse, 393–397, 399, 438, 439, 864

linear, 69–73, 108, 110, 123, 324, 660

logarithmic, 468–475, 478, 479, 480, 501–507, 517, 518, 520, 866

logistic growth, 512

multiplying, 385–386

natural base, 501–507

notation, 61, 155

nth roots, 413

odd-degree, 325, 331

one-to-one, 61, 63–66

onto, 61, 63–66

operations on, 385–386, 389–392

periodic, 831–835, 838

piecewise-defined, 101–102, 104–105

piecewise-linear, 102–103

polynomial, 305, 307, 322–329, 330–337, 338–339, 344, 358–362, 367–369, 373, 374, 375, 376

power, 320–321, 323

quadratic, 109–110, 219–227, 229, 275–279, 281, 290–291, 294, 324

quartic, 324

quintic, 324

radical, 400

rational, 546, 553–560, 561, 563, 583

reciprocal, 545–551, 583, 791, 840

relations as, P4–P5, 61–67, 122, 123

representing, P4–P5

secant, 790, 840

sequences as, 659–660, 711

sine, 790, 830, 832–833, 837

square root, 400, 401–402, 407–412, 438, 440

step, 102–103

subtracting, 385–386

tangent, 790, 840

transformations of, 275–279, 290, 294, 401, 452, 470, 547

translations of, 845–852, 863

trigonometric, 790–794, 807–813, 831–832, 837–843, 844, 863, 866, 874

variation, 562–569, 584

zeros of, 229, 331, 358–362, 367–369, 376

Fundamental Counting Principle, P9

Fundamental Theorem of Algebra, 358–359, 369

G

GCF. *See* Greatest common factor (GCF)

General angles, 807–813, 861

General form
  of conic sections, 632
  of parabolas, 599–602, 603

General trinomials, 239, 343

Geometric means, 675, 677–678

Geometric sequences, 661–662, 674–680, 710, 712
  limits of, 690
  recursive formula for, 692

Geometric series, 676–679, 710, 712
  convergent, 683
  infinite, 683–689, 713
  summation formula, 684

Geometry, 393
  angles, 10, 46
  circles, 17, 397, 611, 750
  cones, 6, 8
  cubes, 48, 410, 498
  cylinders, 23, 39, 50, 51, 53, 534, 703
  diagonals, 186
  parallelograms, 17
  pentagons, 23
  prisms, 26
  rectangles, 37, 242, 243, 255, 418, 419, 441, 541, 871
  rectangular prisms, 6, 26, 32, 368, 374, 375, 377, 535, 551
  spheres, 410, 426, 689
  squares, 243, 605
  trapezoids, 21, 26, 398
  triangles, 7, 9, 25, 242, 243, 255, 398, 419, 428, 435, 475, 605, 679, 696, 730, 776, 885, 909
  triangular pyramids, 369

Index

Geometry Labs
  Areas of Parallelograms, 806
  Regular Polygons, 822

Germain, Sophie, 343

Get Ready for the Chapter. *See* Prerequisite Skills

Get Ready for the Lesson. *See* Prerequisite Skills

Graphing calculator, 255

Graphing Calculator. *See* Teach with Tech

Graphing Technology Labs
  Analyzing Polynomial Functions, 368
  Analyzing Quadratic Relations, 637–638
  Augmented Matrices, 205
  Choosing the Best Model, 476–477
  Cooling, 516
  Direct Variation, 90
  Dividing Polynomials, 318–319
  Equations of Circles, 606
  Families of Lines, 108
  Families of Parabolas, 273–274
  Graphing *n*th Root Functions, 413
  Graphing Rational Functions, 561
  Intersections of Graphs, 135
  Inverse Functions and Relations, 399
  Limits, 690
  Linear-Nonlinear Systems, 639
  Modeling Data Using Polynomial Functions, 338–339
  Modeling Motion, 289
  Modeling Real-World Data, 228
  Operations with Matrices, 187–188
  Polynomial Identities, 350–351
  Power Functions, 320–321
  Simulations and Margin of Error, 731–732
  Solving Exponential Equations and Inequalities, 459–460
  Solving Logarithmic Equations and Inequalities, 499–500
  Solving Polynomial Equations by Graphing, 341
  Solving Quadratic Equations, 263
  Solving Quadratic Equations by Graphing, 237
  Solving Radical Equations and Inequalities, 436–437
  Solving Rational Equations and Inequalities, 579–580
  Solving Trigonometric Equations, 900
  Systems of Linear Inequalities, 153
  Trigonometric Graphs, 844

Graphs *See also* Data; Statistics
  absolute value functions, 110
  absolute value inequalities, 118, 122, 126
  amplitude, 837–842
  angles, 808–809
  arithmetic sequence, 660
  boundaries, 117–118, 148
  box-and-whisker plots, 735, 737
  circles, 606, 608
  conic sections, 600–602, 608–609, 616, 618, 625–626
  constant functions, 109, 110
  dilations, 111
  ellipses, 618, 620
  end behavior of, 324, 471
  exponential functions, 451–458, 518
  families of, 108, 273–274, 275
  geometric sequences, 661–662
  hyperbolas, 626, 628–629
  inequalities, 34–37, 40, 41–47
  intersections, 135, 153
  inverse functions 394, 399
  linear functions, 110
  linear inequalities, 122, 126
  logarithmic equations, 499–500
  logarithmic functions, 469–470, 471
  logarithmic inequalities, 499–500
  parabolas, 600–603
  parent, 109–112, 115, 275
  piecewise-defined functions, 101
  point discontinuity, 556, 557, 561
  polynomial functions, 324–325, 330–337, 373, 375
  preparing for standardized tests, 296
  probability distribution, 744
  quadratic equations, 229–236, 237, 266–268, 291
  quadratic functions, 109, 219–227, 290, 291
  quadratic inequalities, 282, 283–287
  rational functions, 545–550, 553–559, 583
  reciprocal functions, 545–551, 583
  reflections, 111
  relations, 63
  scatter plots, 92–97, 122, 125
  square root functions, 401–402
  square root inequalities, 402
  systems of equations, 135, 137, 140
  systems of inequalities, 146–152, 155
  transformations, 547

  translations, 110, 845–852
  trigonometric functions, 837–843, 844, 847, 859, 863
  vertex form, 277

Greatest common factor (GCF), 529
  factoring, 239, 343

Greatest integer function, 102

Gridded Response, 32, 48, 57, 82, 116, 131, 186, 212, 213, 288, 299, 317, 381, 392, 447, 458, 467, 507, 525, 551, 560, 589, 613, 622, 655, 698, 703, 709, 750, 785, 836, 852, 869, 891, 915

Grouping, 343

Growth
  exponential, 451–453, 462–463, 504, 508, 509–511
  factor, 453
  rate of, 70

Guess and check, 586–587

Guided Practice, 5, 6, 11, 12, 13, 18, 19, 20, 21, 27, 28, 29, 34, 35, 36, 42, 43, 44, 62, 63, 64, 69, 70, 71, 76, 77, 78, 84, 85, 86, 93, 95, 101, 102, 103, 104, 110, 111, 112, 117, 118, 136, 137, 139, 140, 146, 147, 148, 155, 156, 162, 163, 164, 169, 171, 172, 173, 174, 179, 180, 181, 182, 189, 190, 191, 192, 193, 199, 200, 201, 219, 221, 222, 223, 230, 231, 232, 238, 239, 240, 241, 246, 247, 248, 249, 256, 257, 258, 259, 265, 266, 267, 268, 275, 276, 277, 282, 283, 284, 285, 304, 305, 306, 311, 312, 313, 314, 322, 323, 325, 330, 331, 332, 333, 342, 343, 345, 353, 354, 359, 360, 361, 362, 367, 368, 369, 385, 386, 387, 388, 393, 395, 400, 401, 402, 408, 409, 415, 416, 417, 418, 422, 423, 424, 425, 430, 431, 432, 451, 452, 453, 454, 455, 461, 462, 463, 468, 469, 470, 471, 478, 479, 480, 485, 486, 487, 492, 493, 494, 495, 501, 502, 503, 504, 509, 510, 511, 512, 529, 530, 531, 532, 533, 538, 539, 540, 545, 546, 547, 548, 554, 555, 556, 557, 563, 564, 565, 570, 571, 572, 573, 574, 575, 593, 594, 595, 600, 601, 602, 607, 608, 609, 616, 617, 618, 619, 625, 626, 627, 632, 633, 640, 641, 642, 659, 660, 661, 662, 666, 667, 668, 669, 674, 675, 676, 677, 684, 685, 692, 693, 694, 699, 700, 701, 705, 706, 724, 725, 726, 734,

Index

735, 736, 737, 742, 743, 744, 745, 746, 790, 791, 792, 793, 794, 799, 800, 801, 802, 807, 808, 809, 810, 814, 815, 817, 818, 823, 824, 825, 830, 831, 832, 837, 838, 839, 840, 845, 846, 847, 848, 854, 855, 874, 875, 880, 881, 882, 886, 887, 888, 893, 894, 895, 896, 901, 902, 903, 904

H

Half-angle identities, 894–895, 908, 910

Half-life, 509

Hands-On. *See also* Algebra Labs; Algebra Tiles; Manipulatives

Heron's Formula, 9

Higher-Order Thinking Problems

Challenge, 9, 16, 24, 31, 38, 47, 66, 73, 81, 88, 97, 106, 115, 120, 151, 189, 166, 185, 196, 203, 226, 235, 244, 251, 261, 271, 279, 287, 308, 316, 328, 336, 348, 356, 364, 371, 391, 397, 405, 411, 420, 427, 434, 457, 466, 474, 482, 490, 497, 506, 514, 536, 543, 550, 559, 568, 577, 597, 604, 612, 621, 630, 635, 644, 664, 672, 679, 688, 696, 702, 708, 740, 758, 765, 775, 797, 804, 812, 820, 827, 835, 842, 851, 857, 878, 884, 890, 898, 906

Error Analysis, 9, 16, 24, 31, 38, 47, 66, 81, 115, 120, 203, 226, 235, 244, 251, 261, 271, 287, 316, 328, 371, 391, 405, 411, 420, 427, 457, 466, 474, 482, 497, 536, 568, 604, 612, 621, 630, 664, 672, 688, 696, 749, 765, 775, 804, 820, 835, 857, 878, 898

Open Ended, 9, 16, 24, 31, 38, 47, 66, 73, 81, 88, 97, 106, 115, 120, 151, 159, 166, 185, 196, 203, 226, 234, 244, 251, 261, 271, 279, 287, 308, 316, 328, 336, 348, 356, 364, 371, 391, 397, 405, 411, 420, 427, 434, 457, 466, 474, 482, 490, 506, 514, 536, 543, 550, 559, 568, 577, 597, 604, 612, 621, 630, 635, 644, 664, 672, 679, 688, 696, 702, 708, 729, 740, 749, 758, 765, 775, 797, 804, 812, 820, 827, 835, 842, 851, 857, 878, 890, 898, 906

Proof, 185, 308, 397, 466, 490, 497, 506, 559, 672, 679, 688, 878, 890, 898

Reasoning, 9, 16, 24, 31, 38, 47, 66, 73, 81, 88, 97, 106, 115, 120, 151, 159, 166, 185, 196, 203, 226, 235, 244, 251, 261, 271, 279, 287, 308, 316, 328, 336, 348, 356, 364, 371, 391, 397, 405, 411, 420, 427, 434, 457, 466, 474, 482, 490, 497, 506, 514, 536, 543, 550, 559, 568, 577, 597, 604, 621, 630, 635, 644, 664, 672, 679, 688, 696, 702, 708, 729, 740, 749, 765, 775, 797, 804, 812, 820, 827, 835, 842, 851, 857, 878, 884, 890, 898, 906

Which One Doesn't Belong?, 16, 47, 73, 159, 316, 364, 434, 474, 490, 536, 550, 884

Writing in Math, 9, 16, 24, 31, 38, 47, 66, 68, 73, 81, 88, 97, 106, 115, 120, 151, 159, 166, 185, 196, 203, 226, 235, 244, 251, 261, 271, 279, 287, 308, 310, 316, 328, 336, 348, 356, 364, 371, 391, 397, 405, 411, 420, 427, 434, 437, 457, 460, 466, 474, 482, 490, 497, 506, 514, 536, 543, 550, 559, 568, 577, 597, 604, 612, 621, 630, 635, 664, 654, 672, 679, 688, 696, 702, 708, 729, 740, 749, 758, 765, 775, 797, 804, 812, 820, 827, 835, 842, 851, 857, 878, 884, 890, 898, 906

Histograms, P13, 735–736

History, Math, 19, 103, 140, 200, 266, 343, 418, 479, 494, 571, 625, 675, 745, 815, 875

Horizontal asymptotes, 553–554

Horizontal lines, 71

Horizontal translations, 401, 452, 470, 845

H.O.T. Problems. *See* Higher-Order Thinking Problems

Huygens, Christian, 745

Hypatia, 625

Hyperbola equations, 624–631
writing, 627–628
given asymptotes, 625
given vertices and foci, 624

Hyperbolas, 545–549, 624–631, 649
asymptotes, 624–626
axes of symmetry, 624
conjugate axis, 624
constant difference, 627
co-vertices of, 624, 626
equations of, 624–631
foci, 624–626

graphing, 626, 628–629
orientation, 624, 626
standard form of, 624, 626
transverse axis, 624–625
vertices of, 624–626

Hypotenuse, P22

Hypothesis testing, 771–773

I

Identities
additive, 12
cofunction, 873
difference of angles, 886–891, 908, 910
double-angle, 893–894, 896, 908, 910
half-angle, 894–895, 908, 910
negative angle, 873
odd-even, 873
polynomial, 350–351
Pythagorean, 873
quotient, 873
reciprocal, 873
sum and difference of angle, 886–891, 908, 910
trigonometric, 872, 873–879, 880–885, 888, 896, 908
verifying, 350–351, 880–884, 888, 896

Identity functions, 109, 115, 395
graphing, 109

Identity matrices, 198

Identity Property, 12

Imaginary axis, 253

Imaginary numbers
multiplying, 247

Imaginary unit, 246–248

Independent events, P16

Independent variables, 64, 660

Index, 407

Induction hypothesis, 705–706

Inequalities
absolute value, 43–47, 52, 118–121, 122, 126
compound, 41–43, 52
exponential, 459–460, 463, 494, 518
graphing, 34–37, 40, 41–47
interval notation, 40
linear, 117–120, 126
logarithmic, 479–483, 499–500, 503, 519
multi-step, 35

Index

one-step, 33

polynomial, 341

property of, 463

quadratic, 282–287, 294, 641

radical, 431–433, 436–437, 442

rational, 575–577, 579–580

solving, 33–38, 41–47, 49, 52, 146–152, 153, 155, 431, 436–437, 459–460, 499–500, 575–577, 579–580

square root, 402, 440

systems of, 146–152, 153, 155

writing, 36

Inferential statistics, 769

Infinite geometric series, 683–689, 713

Infinite sequence, 659

Infinity symbol, 40, 685

Initial side, 799

Instant Messaging. *See* Teach with Tech

Integers, 11

greatest, 102

Interactive Classroom, 2A–2B, 58A–58D, 132A–132C, 216A–216D, 300A–300D, 382A–382C, 448A–448C, 526A–526B, 590A–590C, 656A–656C, 720A–720B, 786A–786C, 870A–870B

Interactive Whiteboard, 2A–2B, 58A–58D, 132A–132C, 216A–216D, 300A–300D, 382A–382C, 448A–448C, 526A–526B, 590A–590C, 656A–656C, 720A–720B, 786A–786C, 870A–870B. *See also* Teach with Tech

Intercepts. *See* x-intercepts; y-intercepts

Interdisciplinary connections. *See also* Applications

anatomy, 45

anthropology, 514, 537

archaeology, 195, 280

architecture, 260, 286, 619, 630, 650, 813, 885

agriculture, 521

astronomy, 39, 308, 603, 636, 741, 759, 821

biology, 348, 411, 511, 513, 520, 541, 544, 678, 709

business, 96, 103, 114, 132, 133, 138, 156, 158, 159, 160, 166, 174, 185, 224, 272, 287, 291, 316, 317, 327, 328, 363, 365, 375, 551, 559

chemistry, 76, 337, 410, 549, 567, 572, 577, 679, 728

conservation and nature, 583, 696

environmental science, 489, 497

forestry, 513, 651

geography, 194, 795, 820, 897

music, 77, 95, 150, 203, 402, 481, 496, 565, 568, 674, 676, 759, 774, 906

paleontology, 513, 514

physical science, 255, 293, 295, 650, 713

physics, 233, 295, 301, 315, 327, 383, 404, 405, 410, 428, 432, 440, 442, 673, 687, 812, 842, 883

science, 112, 448, 455, 456, 462, 464, 465, 472, 481, 484, 486, 488, 495, 505, 506, 509, 510, 513

statistics, 39, 309, 349, 515, 821, 836, 843

weather, 8, 578, 610, 629, 779, 834, 889

zoology, 191, 347, 513

Interest,

compound, 462–463, 504, 508

Interior Angle Sum Formula, 707

Internet Connections. *See* ConnectED

Interpersonal Learners. *See* Learning Styles

Interpolation, 95

Interquartile range (IQR), P27

Intersections, 41

graphing, 135, 153

Interval notation, 40

Intervention, T22–T23, 2C, 3, 5, 11, 18, 26, 27, 33, 41, 53, 58E, 59, 61, 69, 76, 83, 91, 92, 101, 109, 117, 127, 132E, 133, 136, 146, 154, 161, 168, 169, 171, 172, 179, 189, 198, 211, 216E, 217, 219, 229, 238, 246, 255, 256, 264, 275, 282, 295, 300E, 301, 303, 311, 322, 330, 340, 342, 352, 358, 367, 377, 382E, 383, 385, 393, 400, 407, 414, 415, 422, 429, 443, 448E, 449, 451, 461, 468, 478, 484, 485, 492, 501, 509, 521, 526C, 527, 529, 538, 545, 552, 553, 562, 570, 585, 590E, 591, 593, 599, 607, 615, 623, 624, 632, 640, 651, 656E, 657, 659, 666, 674, 683, 691, 692, 699, 705, 715, 720C, 721, 723, 733, 742, 751, 752, 760, 769, 781, 786E, 787, 790, 799, 807, 814, 823, 829, 830, 837, 845, 853, 865, 870C, 871, 873, 880, 886, 892, 893, 901, 911

Intrapersonal Learners. *See* Learning Styles

Inverse cosine, 793

Inverse functions, 393–397, 399, 438, 439, 853–857, 864

graphing, 394, 399

Inverse matrices, 198–202

Inverse Property, 12, 394

Inverse relations, 393, 396, 399, 439

Inverse sine, 793

Inverse tangent, 793

Inverse variation, 564–568

Investigations. *See* Algebra Labs; Graphing Technology Labs; Spreadsheet Labs; TI-Nspire Technology Labs

IQR. *See* Interquartile range (IQR)

Irrational numbers, 11

Approximation, 257, 409

Iterations, 694–695, 710, 713

J

Joint variation, 563, 566–568

K

Key Concepts,

absolute value, 27

absolute value inequalities, 43

absolute value of a complex number, 254

adding and subtracting matrices, 172

adding rational expressions, 539

addition and subtraction properties of equality, 19

addition property of inequality, 33

addition rules for probability, P14

"and" compound inequalities, 41

angle measures, 799

arc length, 802

area of a triangle, 191, 814

arithmetic sequences and series, 710

basic trigonometric identities, 873

binomial experiments, 752

binomial probability formula, 754

binomial theorem, 700, 710

box-and-whiskers plots as distributions, 735

change of base formula, 494

characteristics of the standard normal distribution, 762

circles, 646

Index

circular and inverse trigonometric functions, 859

classify conics with the discriminant, 633

combinations, P11

completing the square, 258

complex conjugates theorem, 362

complex numbers, 247, 290

composition of functions, 387

compound interest, 462

conditional probability, P17

conducting binomial experiments, 753

confidence interval for the population mean, 770

continuously compounded interest, 504

convergent and divergent series, 683

convert between degrees and radians, 801

corollary to the Fundamental Theorem of Algebra, 359

cosecant, secant, and cotangent functions, 840

Cramer's Rule, 192

Cramer's Rule for a system of three equations, 193

definition of nth root, 407

Descartes' Rule of signs, 360

diagonal rule, 190

direct variation, 90, 562

discriminant, 268

distance formula, 594

dividing rational expressions, 531

division property of inequality, 34

double-angle identities, 893, 908

elimination method, 139

ellipses, 646

Empirical Rule, 760

end behavior of a polynomial function, 324

equations of circles, 607

equations of ellipses centered at (h, k), 616

equations of ellipses centered at the origin, 615

equations of hyperbolas centered at (h, k), 626

equations of hyperbolas centered at the origin, 624

equations of parabolas, 599

evaluate trigonometric functions, 809

expected value of a discrete random variable, 745

exponential functions, 517

exponential growth and decay, 509

expressions and formulas, 49

factor theorem, 354

factorial, P10

feasible regions, 154

FOIL method for multiplying binomials, 238

formula for z-values, 762

functions, 61

functions on a unit circle, 830

Fundamental Counting Principle, P9

Fundamental Theorem of Algebra, 358

geometric sequences and series, 710

graphing linear and absolute value inequalities, 122

graphing quadratic functions, 220, 290

graphing trigonometric functions, 847

half-angle identities, 894, 908

hyperbolas, 646

identity matrix for multiplication, 198

inverse and square root functions, 438

inverse functions, 395

inverse of a 2×2 matrix, 199

inverse relations, 393

inverse trigonometric functions, 853

inverse trigonometric ratios, 793

inverse variation, 564

joint variation, 563

law of cosines, 823

law of sines, 815

law of sines and law of cosines, 859

location principle, 331

logarithms and logarithmic functions, 517

logarithm with base b, 468

logistic growth function, 512

margin of error formula, 732

mathematical induction, 705, 710

maximum and minimum value, 222

maximum error of estimate, 769

mean of a binomial distribution, 754

measures of center, P24

measures of spread, P25

midpoint and distance formulas, 646

midpoint formula, 593

multiplication and division properties of equality, 19

multiplication property of inequality, 34

multiplying by a scalar, 173

multiplying matrices, 180

multiplying rational expressions, 531

natural base functions, 501

natural logarithms, 517

normal approximation of a binomial distribution, 767

normal distribution, 760

nth term of a geometric sequence, 674

nth term of an arithmetic sequence, 666

oblique asymptotes, 555

operations on functions, 385, 438

operations with polynomials, 373

optimization with linear programming, 156

"or" compound inequalities, 42

order of operations, 5

parabolas, 646

parallel and perpendicular lines, 85

parent function of absolute value functions, 103

parent function of exponential decay functions, 453

parent function of exponential growth functions, 451

parent function of logarithmic functions, 469

parent function of reciprocal functions, 545

parent function of square root functions, 400

parent functions, 109

partial sum of a geometric series, 676

partial sum of an arithmetic series, 668

permutations, P10

phase shift, 845

point discontinuity, 557

point-slope form, 84

polynomial functions and graphs, 373

possible triangles in SSA case, 816

power property of logarithms, 487

probability distributions, 743

probability of dependent events, P16

probability of independent events, P16

product property of logarithms, 485

product property of radicals, 415

properties of equality, 19

properties of matrix multiplication, 183

properties of matrix operations, 173

properties of real numbers, 49

property of equality for exponential functions, 461

property of equality for logarithmic functions, 478

property of inequality for exponential functions, 463

Index

property of inequality for logarithmic functions, 479, 480

property of inverses, 394

quadrantal angles, 808

quadratic form, 345

quadratic formula, 265

quotient property of logarithms, 486

quotient property of radicals, 416

rational exponents, 422, 423

rational zero theorem, 367

real nth roots, 408

real numbers (R), 11

recursion and iteration, 710

recursive formulas for sequences, 692

reference angles, 808

relations and functions, 122

remainder and factor theorems, 373

reminder theorem, 352

roots of real numbers, 438

roots, zeros, and the rational zero theorem, 373

scatter plots, 92

scatter plots and lines of regression, 122

second-order determinant, 189

sequences as functions, 659

sigma notation, 669

simplifying monomials, 304

sine and cosine functions, 837

slope-intercept form, 83

slope of a line, 78

solutions of a quadratic equation, 230

solving equations, 49

solving inequalities, 49

solving polynomial equations, 373

solving quadratic equations, 290

solving quadratic systems, 646

solving radical equations, 429

solving radical inequalities, 431

solving rational inequalities, 575

solving systems of inequalities, 146

special functions, 122

standard deviation of a probability distribution, 745

standard form of a linear equation, 70

steps for hypotheses testing, 772

study types, 723

substitution method, 138

subtracting rational expressions, 539

subtraction property of inequality, 33

suggestions for verifying identities, 881

sum and difference of angle identities, 908

sum and difference of cubes, 342

sum and difference identities, 886

sum of an infinite geometric series, 684

symmetric and skewed distribution, 733

synthetic division, 313

tangent functions, 839

theoretical and experimental probability, P13

transformations of exponential functions, 452

transformations of logarithmic functions, 470

transformations of reciprocal functions, 547

transformations of square root, functions, 401

transformations of quadratic graphs, 290

trigonometric functions in right triangles, 790

trigonometric functions of general angles, 807

trigonometric identities, 908

trigonometric values for special angles, 791

using exponential and logarithmic functions, 517

verifying identities by transforming one side, 350, 880

vertical and horizontal asymptotes, 553

vertical line test, 62

vertical shift, 846

zero product property, 239

zeros of even- and odd-degree functions, 325

Kinesthetic Learners. *See* Learning Styles

Kleiber's Law, 411

Labs. *See* Algebra Labs; Geometry Labs; Graphing Technology Labs; Spreadsheet Labs; TI-Nspire Technology Labs

Lateral area, 197

Latus rectum, 599, 601

Law of Cosines, 823–828, 862

Law of Large Numbers, 744

Law of Sines, 814–821, 861
ambiguous case, 816–817

LCD. *See* Least common denominators (LCD)

LCM. *See* Least common multiples (LCM)

Leading coefficients 314, 322–324

Learning Styles
Auditory/Musical Learners, 382F, 810, 896
Interpersonal Learners, 2D, 58F, 132F, 300F, 431, 487, 526D, 656F, 662, 668, 785, 881, 903
Intrapersonal Learners, 150, 277, 388, 786F, 817
Kinesthetic Learners, 2D, 58F, 132F, 216F, 395, 590F, 694, 706, 720D, 755, 786F, 870D
Logical Learners, 132F, 140, 200, 300F, 324, 369, 409, 448F, 469, 494, 510, 526D, 574, 656F, 685, 870D
Naturalist Learners, 590F, 832
Social Learners, 35, 641, 745
Verbal/Linguistic Learners, 71, 94, 103, 174, 259, 453, 556, 618, 725, 771, 887
Visual/Spatial Learners, 94, 216F, 232, 382F, 448F, 548, 594, 720D, 793, 838, 855

Least common denominators (LCD), 538, 540, 570–571

Least common multiples (LCM), 538, 541

Left Distributive Property, 183

Left Scalar Distributive Property, 173

Legs, P22

Lesson Objectives, 2A–2B, 58A–58D, 132A–132C, 216A–216D, 300A–300D, 382A–382C, 448A–448C, 526A–526B, 590A–590C, 656A–656C, 720A–720B, 786A–786C, 870A–870B

Like radical expressions, 417

Limits, 690

Linear correlation coefficient, 94, 97

Linear equations, 69–73. *See also* Equations
direct variation 90, 562–563, 566–568
point-slope form, 84
slope-intercept form, 83–84, 86–88
standard form of, 70, 71–73
writing, 83–89, 124

Linear functions, 69–73, 108, 123, 324. *See also* Functions
arithmetic sequence, 659–660

evaluating, 70
graphing, 110
identifying, 69

Linear inequalities, 117–120, 126

Linear-nonlinear systems, 640–645
linear-quadratic system, 640
quadratic inequalities, 641
quadratic-quadratic system, 641
quadratics with absolute value, 642

Linear programming, 154–160

Linear regression equation, 228

Linear relations, 69, 123

Linear terms, 219

Line of fit, 92–97

Line of reflection, 111

Lines. *See also* Functions; Slope
directrix, 599
of fit, 92–97, 99–100
horizontal, 71
number, 28, 34–37, 40, 41–47
parallel 85–86, 88
perpendicular, 85–86, 88
regression, 94, 476–477
slope of, 77–82, 85–86, 124
vertical, 62–63, 71

Loans
amortization, 698
principal, 698

Location Principle, 330–331

Logarithmic equations
solve
algebraically, 478–483, 493, 519
by graphing, 499–500
using a table, 499–500

Logarithmic expressions, 468–469

Logarithmic form
to exponential form, 468

Logarithmic functions, 468–475,
517, 518, 520, 866
graphing, 469–470, 471
inverse, 471
natural, 501–507, 520
Property of Equality, 478
Property of Inequality 479, 480
transformations of, 470

Logarithmic inequalities
with logarithms on each side, 480
solve
algebraically, 479–483, 519
by graphing, 499–500
using a table, 499–500

Logarithms, 468–475, 518
with base *b*, 468

solving exponential equations, 493
solving exponential inequalities, 494

Logical Learners. *See* Learning Styles

Logistic growth function, 512

Logistic growth model, 512

Long division, 311–312

Lower quartile, P26

Lurking variables, 99

M

Magnifications. *See* Dilations

Mapping, P4

Margin of error formula, 732

Materials
board, 289
books, 289
cardboard, 614
compass, 822
data collection device, 289
string, 614
thumbtacks, 614
toy car, 291

Materials, 2A–2B, 58A–58D, 90, 99,
108, 132A–132D, 135, 153, 169,
187, 205, 216A–216D, 228, 237,
263, 273, 289, 300A–300D, 318,
320, 338, 341, 350, 366,
382A–448C, 399, 413, 436,
448A–448C, 459, 476, 499, 508,
516, 526A–526B, 561, 579,
590A–590C, 606, 614, 637, 639,
656A–656C, 681, 690, 698,
720A–720B, 731, 767, 786A–786C,
789, 822, 844, 870A–870B, 900

Mathematical Background. *See*
Focus on Mathematical Content

Mathematical induction, 705–709,
714

Math History Link
Archytas, 675
Aryabhatta, 875
Bari, Nina Karlovna, 140
Brahmagupta, 266
Diophantus of Alexandria, 19
Germain, Sophie, 343
Huygens, Christian, 745
Hypatia, 625
Napier, John, 494
Seki Kowa, 200
Sperry, Pauline, 815
Taylor, Brook, 571
Theano, 418
Weierstrass, Karl, 103
Zhang Heng, 479

Math Online. *See* ConnectED

Matrices, 169
adding, 172, 187–188
augmented, 205
coefficient, 192, 200–201, 205
column, 169
constant, 200–201
Cramer's Rule, 192–196
determinants of, 189–196, 199–200
dimensions of, 169, 179
elements of, 169–171, 174
equations involving, 200–202
identity, 198
inverse, 198–202
multiplying, 173, 179–186
by a scalar, 173, 187–188
multi-step operations, 174
properties of 173, 183
rectangular, 169
row, 169
solving systems of equations,
198–205
subtracting, 172, 187–188
variable, 200

Matrix equations, 200–202

Maximum values, 154–155,
157–159, 222–226

Means
arithmetic, 667, 670
of data, P24, 723, 735, 745, 754,
760, 762
geometric, 675, 677–678

Measurement
of angles, 799–804, 854, 859, 860
area, 7, 9, 17, 21, 397, 440, 443, 475,
814
circumference, 801
degrees, 801
dimensional analysis, 310
lateral area, 197
radians, 801–804, 808, 832
surface area, 26, 32, 39
temperature, 7, 397, 850
volume, 6, 8, 9, 23, 26, 50, 51, 53

Measures of central tendency,
P24–P28

Measures of variation, P24–P28
range, P4–P5
standard deviation, P25, 733–737,
745–746, 760–763
variance, P25, 745

Median, P24, 760

mental math, 90, 228, 338, 339, 522

Mid-Chapter Quiz, 26, 91, 168, 255,
340, 414, 484, 552, 623, 691, 751,
829, 892

Midline, 846–850

Midpoint Formula, 593, 595–596, 608, 646, 647

Minimum values, 154–155, 157–159, 222–226

Mixture problem, 572

Mode, P24, 760

Monomial function, 320

Monomials
as denominators, 539
dividing, 303
polynomials by, 311–314
least common multiple of, 538
multiplying, 303
simplifying, 304

Multiple Choice, 10, 17, 21, 22, 25, 26, 32, 39, 48, 53, 55, 56, 67, 74, 82, 85, 86, 89, 91, 98, 107, 116, 121, 130, 140, 152, 160, 167, 168, 186, 197, 204, 214, 227, 236, 245, 252, 255, 262, 272, 276, 278, 280, 288, 295, 297, 298, 309, 312, 315, 317, 329, 337, 340, 349, 357, 365, 372, 377, 380, 392, 398, 406, 412, 414, 421, 428, 431, 432, 433, 435, 443, 445, 446, 458, 467, 475, 479, 480, 483, 484, 491, 498, 507, 515, 521, 523, 524, 530, 533, 534, 537, 544, 551, 552, 560, 569, 578, 588, 595, 598, 605, 613, 622, 631, 636, 645, 653, 654, 670, 673, 680, 689, 697, 703, 709, 717, 718, 730, 741, 750, 751, 754, 756, 759, 766, 776, 781, 783, 784, 798, 805, 813, 821, 828, 829, 836, 843, 852, 855, 856, 858, 865, 867, 868, 879, 881, 882, 885, 891, 892, 899, 907, 911, 913, 914

Multiple representations, 9, 16, 24, 31, 38, 66, 73, 81, 106, 226, 243, 251, 261, 308, 316, 328, 336, 356, 390, 391, 397, 404, 411, 420, 427, 457, 466, 481, 497, 506, 514, 536, 550, 577, 597, 604, 612, 621, 629, 635, 672, 687, 729, 749, 775, 804, 835, 857, 877, 890

Multiples, least common, 139, 538, 541

Multiplication
Associative Property of, 12, 183, 246, 248
of binomials, P6, 238
closure property of, 12
Commutative Property of, 12, 246, 248

of complex numbers, 249
elimination using, 139–140
FOIL method, P6, 238, 249
of functions, 385–386
identity matrix for, 198
of imaginary numbers, 247
of matrices, 173, 179–186
of monomials, 303
of polynomials, 306, 374
of probabilities, P16–P19
of radical expressions, 417–418
of rational expressions, 531
scalar, 173, 183, 187–188

Multiplication Property
of equality, 19, 20
of inequality, 34

Multiplicative identity, 12

Multiplicative inverse, 12

Multi-step operations, 174

Multi-step problems, 782

Mutually exclusive events, P14

N

Napier, John, 494

Natural base *e,* 501–507

Natural base exponential function, 501–507

Natural logarithm equations, 501–507, 520

Natural logarithm inequalities, 503

Natural numbers, 11, 659, 666

Naturalist Learners. *See* Learning Styles

Negative angle identities, 873

Negative correlation, 92–97

Negative Exponent Property, 303

Negatively skewed distribution, 733

Negative slopes, 92–97

Nonlinear relation, 69

Normal approximation, 760–768

Normal distribution, 760–768

Notation
bar, 685
function, 64
null hypothesis, 771
partial sum, 668
set builder, 35

Note taking. *See* Foldables™ Study Organizer

*n*th **roots,** 407–412, 413, 440, 866

*n*th **term**
of arithmetic sequence, 666, 667, 668–669
of geometric sequence, 674–675

Null hypothesis, 771

Number lines
absolute value equations, 28
graphs of inequalities, 34–37, 40, 41–47

Numbers
complex, 246–252, 267, 290, 292
imaginary, 246–247, 408
integers, 11
irrational, 11, 255, 409
natural, 11, 659, 666
rational, 11
real, 11–16, 43, 49, 50, 438, 659
whole, 11

Number Theory, P3, 74, 107, 233, 234, 242, 252, 271, 456, 483, 707

Numerators, 6, 532

O

Oblique asymptotes, 555–556

Oblique triangles, 824

Observational study, 723

Odd-degree functions 324, 325, 331

Odd-Even Identities, 873

Odds, P15

One-to-one function, 61, 63–66

Onto function, 61, 63–66

Open Ended. *See* Higher-Order Thinking Problems

Open sentence, 18

Opposite reciprocals, 86

Optimization, 156–160

Ordered triple, 161

Order of operations, 5

Orientation, 401

Origin, 360, 615, 624

Outcome, P9. *See also* Probability

Outliers, P27, 93

P

Pacing, 2A–2B, 58A–58D, 132A–132C, 216A–216D, 300A–300D, 382A–382C,

448A–448C, 526A–526B, 590A–590C, 656A–656C, 720A–720B, 786A–786C, 870A–870B

Parabolas, 219–220, 599–605, 646, 647
  axis of symmetry, 600–601
  constant sum, 616
  directrix of, 599
  equations of, 599–602
  families of, 273–274
  focus of, 599
  general form of, 599, 601
  graphing, 601
  latus rectum of, 599
  maximum values of, 154–155, 157–159, 222–226
  minimum values of, 154–155, 157–159, 222–226
  parent functions, 600–601
  standard form of, 599–600, 603
  vertices of, 220–225, 599–603
  writing equations for, 639–640

Parallel lines, 85–88
  slope of, 85–86

Parallelogram
  area, 806

Parameter, 723

Parent function, 109–111, 126
  of absolute value functions, 103
  of exponential decay functions, 453
  of exponential growth functions, 451–453
  of family of linear functions, 108
  of logarithmic functions, 469
  of parabolas, 600–601
  of reciprocal functions, 545
  of square root functions, 400
  of trigonometric functions, 838–839

Parent graphs, 109–112, 115, 275

Part. *See* Percents

Partial sums
  of arithmetic series, 668, 672
  of geometric series, 676

Pascal's triangle, 699–700, 704

Patterns
  arithmetic sequences, 659–660, 662–664, 666–667, 711
  Fibonacci sequence, 692
  geometric sequences, 661–664, 674–680, 710, 712
  looking for, 716, 903
  Pascal's triangle, 699–700, 704
  sequences, 659–665, 711

Pentagons
  perimeter, 23

Perfect square trinomials, P8, 239, 244, 343

Perihelion, 618

Periodic functions, 831–835, 848
  amplitude of, 837–842
  cosecant, 790–791, 809, 840
  cosine, 790–791, 809, 830, 832–833, 837–838
  cotangent, 790–791, 809, 840
  phase shift of, 845–847, 849–851
  secant, 790–791, 809, 840
  sine, 790–791, 809, 830, 832–833, 837–838, 859
  tangent, 790–791, 809, 839–840
  translations of, 845–852, 863

Periods, 831–835, 837–839

Permutations, P9

Perpendicular lines, 85–86
  slope of, 85–86

Personal Tutor. *See* ConnectED

Phase shift, 845–847, 849–851

Piecewise-defined functions, 101–102, 104–106

Piecewise-linear functions, 102–103

Planning Guide. *See* Pacing

Point discontinuity, 556–557, 561

Points
  break-even, 136
  foci, 615–616, 624–626
  midpoint, 593, 595–596, 608, 646, 647
  turning, 331
  vertices, 220–225, 599–600, 615–617, 619–620, 624–626

Point-slope form, 84

Polygons
  quadrilaterals, 88
  regular, 822
  triangles, P22–P23, 679, 789, 790–791, 822, 852, 860, 885, 909

Polynomial equations, 342–349, 375
  solving, 341

Polynomial functions, 322–329, 338, 339, 344, 374
  analyzing, 366
  cubic, 324
  degree of, 305, 307, 322–328, 553
  end behavior of, 324
  factors of, 358
  graphing, 324–325, 330–337, 373, 375, 554

quadratic, 109–110, 219–227, 275–279, 281, 290, 291, 294, 324
  roots of, 358–359, 376
  zeros of, 331, 358–362, 367–369, 376

Polynomial identities, 350–351

Polynomials
  binomials, P6–P8, 238, 239, 342–344, 700, 701
  cubic, 324
  degrees of, 305, 307, 322–328, 553
  as the denominator, 532, 539
  depressed, 354
  difference of cubes, 342–343
  difference of squares, P8, 239, 343
  dividing, 311–317, 318–319, 374
  end behavior of, 324
  even-degree, 325
  factoring, P7–P8, 239, 240, 242–244, 342, 343, 532
  least common multiples of, 538
  multiplying, 306, 374
  as the numerator, 532
  odd-degree, 325
  in one variable, 322
  operations with, 303–308, 373
  perfect square trinomials, P8, 239, 244, 343
  prime, 342–344, 346–347
  quadratic, 221, 345–346
  simplifying, 305
  standard form of, 322
  sum of cubes, 342–344
  trinomials, P7–P8, 239–240, 244, 343
  zeros of, 331, 358–362, 367–369, 376

Population, P24

Population mean, 769–771

Portable Media Player. *See* Teach with Tech

Positive correlation, 92–97

Positive skewed distributions, 733

Positive slope, 92–97

Power function, 320–321, 323

Power of a Power Property, 303

Power of a Quotient Property, 303

Power of Product Property, 303

Power Property of Logarithms, 487

Powers
  base of, 501–507
  expressions containing, 303
  power of, 303

product of, 303
of products, 303
properties of, 303
of quotients, 303
quotients of, 303
simplifying, 303

Pre-AP. *See* Standardized Test Practice

Prediction equations, 92–93, 96

Preparing for Advanced Algebra. *See* Chapter 0

Preparing for Standardized Tests. *See* Standardized Test Practice

Prerequisite skills
get ready for chapter, 3, 59, 133, 217, 301, 383, 449, 527, 591, 657, 721, 787, 871
Skills Review, 10, 17, 25, 32, 39, 48, 67, 74, 82, 89, 98, 107, 116, 121, 145, 152, 160, 167, 178, 186, 197, 204, 227, 236, 245, 252, 262, 272, 280, 288, 309, 317, 329, 337, 349, 357, 365, 372, 392, 398, 406, 412, 421, 428, 435, 458, 467, 475, 483, 491, 498, 507, 515, 537, 544, 551, 560, 569, 578, 598, 605, 613, 622, 631, 636, 645, 665, 673, 680, 689, 697, 703, 709, 730, 741, 750, 759, 766, 776, 798, 805, 813, 821, 828, 836, 843, 852, 858, 879, 885, 891, 899, 907

Prime polynomials, 342–344, 346–347

Principal roots, 407

Principal values, 853

Prisms, 6, 26, 32, 368, 374, 375, 377, 551

Probability, P13–P15, 745–751
adding, P13–P15
binomial, 754–758
combinations, P11–P12
compound events, P14
conditional, P14
continuous probability distributions, 743
dependent events, P16, 754
discrete probability distributions, 744
distributions, 743
expected values, 747
experimental, P9, P13, 744
failure, 752
full probability distribution, 755
Fundamental Counting Principle, P9
independent events, P16
multiplying, P16–P19

mutually exclusive events, P13
odds, P15
outcomes, P9
permutations, P9
sample spaces, P9
simple events, P14
simple probability, P14
simulations, 731–732
success, 754
theoretical, P13, 743
tree diagrams, P9
uniform distribution, P13
uniform model, P13

Probability distribution, 733–741, 752–759, 760–768

Problem-solving. *See also* Problem-Solving Tip
dimensional analysis, 310
four-step plan, 28, 36, 103, 138, 164, 181, 231, 362, 388, 409, 463, 486, 510, 572, 573, 574, 617–618, 627, 660, 674, 825, 887
work backward, 20

Problem-solving strategies. *See also* Problem-solving

Problem-Solving Tip. *See also* Problem-solving; Problem-solving strategies
check your answer, 304
make a model, 887
use a formula, 511
write an equation, 28

Product of Powers Property, 303

Product Property
of logarithms, 485
of radicals, 415–417

Products. *See* Multiplication

Professional Development, 2F, 58H, 132H, 216H, 300H, 382H, 448H, 526F, 590H, 656H, 720F, 786H, 870F

Proofs
counterexamples, 706, 873
divisibility, 706
mathematical induction, 705–709
summation, 705

Properties
addition of equality, 19
addition of inequality, 33
associative
of addition, 12, 173
of multiplication, 12, 183, 246, 248
closure
of addition, 12
of multiplication, 12

commutative
of addition, 12, 173
of multiplication, 12, 246, 248
distributive, P7, 12–13, 14, 173, 182, 183
division of equality, 19, 20
division of inequality, 34
of equality, 19
of exponents, 303
identity, 12
inverse, 12
left distributive, 183
left scalar distributive, 173
of logarithms, 478, 479, 480, 485–487, 519
of matrices, 173, 183
multiplication of equality, 19, 20
multiplication of inequality, 34
power
of a logarithm, 487
of powers, 303
of a product, 303
of a quotient, 303
product
of logarithms, 485
of powers, 303
of radicals, 416
quotient
of logarithms, 486
of powers, 303
of radicals, 416
of real numbers, 11–16, 50
of reciprocal functions, 546
reflexive, 19
right distributive, 183
right scalar distributive, 173
square root, 247, 256
substitution, 19
subtraction of equality, 19
subtraction of inequality, 33
symmetric, 19
transitive, 19
zero power, 303
zero product, 239, 241

Property of Equality
for exponential functions, 461
for logarithmic functions, 478

Property of Inequality
for exponential functions, 463
for logarithmic functions, 479–480

Property of Inverse Functions, 394

Pure imaginary numbers, 246–247, 408

Pyramids
triangular, 369
volume, 8

Pythagorean identities, 873

Pythagorean Theorem, P22, P23, 435

<hr>

Q

Quadrantal angle, 808

Quadrants, P4

Quadratic equations, 644
roots of, 229, 242–243
solve
by completing the square, 258, 260, 269, 293
by factoring, 241–243, 269, 292
by graphing, 229–236, 237, 269, 291
by inspection, 239
using a calculator, 232, 237, 263
using a table, 232
using Quadratic Formula, 263–271
using Square Root Property, 247, 260, 269
standard form of, 229, 242–244

Quadratic form, 221, 345–346

Quadratic Formula, 263–267, 269–271, 293
derivation of, 264

Quadratic functions, 109–110, 324.
See also Functions
families of, 273–274
graphing, 219–227, 290, 291
rate of change for, 281
transformations of, 275–279, 290, 294
zeros of, 229

Quadratic inequalities, 282–288, 294, 641
graphing, 282
solve
algebraically, 285–287
by graphing, 283–287
using a table, 287

Quadratic regression equation, 228

Quadratic relations, 637–638

Quadratic terms, 219

Quadrilaterals, 88

Quantitative reasoning. *See* Number Sense

Quartic function, 324

Quartiles, P26

Quick Check, 3, 59, 133, 217, 301, 383, 449, 527, 591, 657, 721, 787, 871

Quintic function, 324

Quotient identities, 873

Quotient of Powers Property, 303

Quotient Property
of logarithms, 486
of radicals, 416–417

Quotients. *See* Division

<hr>

R

Radians, 801–804, 808, 832

Radical equations, 429–433, 436–437, 442

Radical expressions, 417, 425, 441
adding, 417
dividing, 416–417
like expressions, 417
multiplying, 417–418
simplifying, 415–417, 424–425
subtracting, 417

Radical form, 422

Radical functions, 400

Radical inequalities, 431–433, 436–437, 442

Radical sign, 407

Radicands, 407

Radii, 607–612, 801

Random samples, 723

Random variable, 742

Range, P25
of composite functions, 387–391
of exponential function, 451
interquartile, P27
of linear functions, 108
of rational functions, 553–557
of relations, P4–P5, 61
of square root functions, 400
of trigonometric functions, 846

Rate of change, 76–77, 79–81, 124.
See also Rates
average, 77
constant, 76–77
for quadratic functions, 281

Rates
constant, 76–77, 90

of change, 77
of decay, 509–512
of growth, 509–512

Rational equations, 570–571, 576, 577, 579–580, 584

Rational exponents, 422–428, 441

Rational expressions
adding, 539
complex fractions, 532–533, 540
dividing, 531
exponents, 425
multiplying, 531
simplifying, 529–530
subtracting, 539

Rational functions, 561
graphing, 553–560, 583

Rational inequalities, 575–577, 579–580, 584

Rationalizing the denominator, 416, 418

Rational numbers, 11. *See also* Fractions; Percents

Rational zeros, 367–371

Rational Zero Theorem, 367, 373, 376

Ratios
rate of change, 76
rational expressions, 529
similar triangles, 789
slope, 77–78

Reading Math
angle of rotation, 800
arithmetic mean, 667
composition of functions, 387
determinants, 192
empty set, 147
exponents, 6
extrapolation, 95
function notation, 64, 155
geometric means, 675
interpolation, 95
labeling triangles, 791
latus rectum, 601
permutations, P10
plus or minus, 895
power functions, 423
predictions, 95
radian measures, 801
repeated roots, 359
scalar, 173
set-builder notation, 35

sigma notation, 669
significance level, 771
standard form, 625
theta prime, 808
translation, 110
within and *between,* 43

Real axis, 253

Real numbers, 11–12, 49, 50, 438, 659
absolute value inequalities, 43
irrational numbers, 11, 257, 409
natural numbers, 11, 659, 666
properties of, 11–16, 50
rational numbers, 11
whole numbers, 11

Real-World Careers
acoustical engineer, 493
aerospace technician, 617
economist, 423
electrical engineer, 247
electrician, 894
financial planner, 174
music management, 164
mutual fund manager, 746
operations manager, 156
politician, 772
recreation workers, 118
retail store manager, 13
sports science administrator, 855
travel agent, 548
truck driver, 306
U.S. Coast Guard, 555

Real-World Examples, P18, P24, P25, P26, 6, 13, 63, 70, 76, 77, 93, 94, 103, 112, 118, 138, 147, 156, 164, 171, 181, 191, 201, 223, 232, 241, 249, 284, 306, 323, 333, 344, 353, 368, 388, 402, 409, 418, 423, 453, 455, 462, 471, 486, 493, 504, 509, 510, 511, 548, 555, 565, 572, 573, 574, 594, 602, 607, 617, 627, 660, 674, 676, 694, 699, 725, 726, 734–737, 745–746, 755, 761, 763, 770, 800, 802, 810, 818, 825, 831, 839, 848, 855, 875, 887, 896, 902

Real-World Links, 28, 36, 44, 63, 70, 77, 93, 112, 147, 181, 191, 201, 223, 232, 241, 249, 284, 323, 333, 353, 388, 402, 409, 453, 455, 462, 471, 486, 504, 510, 512, 565, 574, 594, 602, 618, 627, 660, 694, 725, 734, 755, 761, 763, 770, 792, 794, 800, 810, 818, 825, 831, 839, 848, 896

Reasonableness, check for. *See* Study Tips

Reasoning. *See* Higher-Order Thinking Problems

Reciprocal functions, 791
graphing, 545–551, 583
properties of, 546

Reciprocal identities, 873

Rectangular prisms
surface area, 26, 32, 368, 374, 375
volume, 6

Recursive formula, 692–696, 713

Recursive sequence, 692

Reference angles, 808–812, 817

Reflections, 111–114
graphing, 111

Reflexive property, 19

Regions
bounded, 154–158
feasible, 154–158
intersecting, 146–147
separate, 147
unbounded, 154–155, 159
vertices of, 148, 154–158

Regression equations, 97, 228

Regression line, 94–97, 125, 476–477

Regular polygons, 822

Relations
continuous, 62–65
discrete, 62–65
domain of, P4–P5, 61
functions, P4–P5, 61–67, 122, 123
graphing, 63
inverse, 393, 396, 399, 439
mappings of, P4
quadratic, 637–638, 639
range of, P4–P5, 61

Relative frequency histogram, 743

Relative maximum, 331–335, 366

Relative minimum, 331–335, 366

Remainder, 312

Remainder Theorem, 352–354, 373, 376

Repeating decimals, 11
write as a fraction, 685

Review. *See* Guided Practice; ConnectED; Prerequisite Skills; Review Vocabulary; Spiral Review; Standardized Test Practice; Study Guide and Review

Review Vocabulary
completing the square, 600
complex conjugates, 362
composition of functions, 694
counterexample, 706
discriminant, 633
exponential function, 662
extraneous solutions, 571
inequality symbols, 34
infinity, 324
inverse function, 854
least common multiple, 139
oblique, 824
outlier, 93
perfect square, 239
pure imaginary numbers, 408
radicand, 265
relative frequency, P18
simulation and random number generator, 744
symmetry, 220
theoretical and experimental probability, 743
zero, 325

Right Distributive Property, 183

Right Scalar Distributive Property, 173

Right triangles, P22–P23, 789, 790–791, 792–793, 860, 885, 909

Rise, 78

Roots
complex, 263, 267
determining, 75
double, 325
estimating, 231
exact, 416
irrational, 257, 266
nth, 407–412, 413, 440, 866
one rational, 266
of polynomial functions, 358–359, 376
of quadratic equations, 229, 242–243
rational, 256
repeated, 359
square, 246, 866
of square root functions, 407–412
two rational, 265

Row matrices, 169

Run, 78

Index

S

Sample mean, 769–771

Samples, P24, 734
biased, 723
normally distributed, 760–768
unbiased, 723

Sample space, P9

Scaffolding Questions, 5, 11, 18, 27, 33, 41, 54, 61, 69, 76, 83, 92, 101, 109, 117, 128, 136, 146, 154, 161, 169, 179, 189, 198, 212, 219, 229, 238, 246, 256, 264, 275, 282, 296, 303, 311, 322, 330, 342, 352, 358, 367, 378, 385, 393, 400, 407, 415, 422, 429, 444, 451, 461, 468, 478, 485, 492, 501, 509, 522, 529, 538, 545, 553, 562, 570, 586, 593, 599, 607, 615, 624, 632, 640, 652, 659, 666, 674, 683, 692, 699, 705, 716, 723, 733, 742, 752, 760, 769, 782, 790, 799, 807, 814, 823, 830, 837, 845, 853, 866, 873, 880, 886, 893, 901, 912

Scalar multiplication, 171, 173, 183

Scalars, 171, 173

Scatter plots, 92–97, 122, 125
outliers, P27, 93
prediction equations, 92–93, 96

Scientific calculator, 866, 914

Scientific notation, 866

Secant, 790–791, 809, 840

Second-order determinant, 189

Seki Kowa, 200

Sense-Making. *See* Check for Reasonableness; Reasoning

Sequences
arithmetic, 659–660, 666–667, 711
Fibonacci, 692
as functions, 659–665, 711
geometric, 661–662, 674–680, 710, 712
terms of, 659–660

Series, 668
arithmetic, 668–672, 711
convergent 683, 686–687
divergent, 683
geometric, 676–679, 683–685, 710, 712
infinite, 683–689, 713

Set-builder notation, 35

Sets
empty, 28–31, 147
intersections of, 41
unions of, 42

Short Response, 10, 25, 39, 57, 67, 74, 89, 98, 107, 121, 131, 152, 160, 197, 204, 227, 236, 245, 262, 272, 280, 288, 299, 309, 329, 337, 349, 381, 398, 406, 421, 428, 435, 447, 475, 491, 525, 537, 544, 551, 578, 589, 598, 605, 631, 645, 652, 653, 655, 673, 680, 719, 730, 741, 766, 776, 785, 798, 805, 821, 828, 836, 843, 858, 869, 899, 912, 913, 915

Sides
congruent, P20
of cubes, 410
initial, 799
terminal, 799–801, 808–809, 830

Sigma notation, 669, 676, 685

Similar figures, P20–P21, 428, 789

Simple event, P14

Simple probability model, P14

Simplifying
expressions, 303, 912
expressions containing powers, 303
monomials, 304
polynomials, 305–306
using Distributive Property, 306

Sine, 790–791, 809, 830, 832–833, 837–838, 859

Skewed distribution, 735

Skills Review, 10, 17, 25, 32, 39, 48, 67, 74, 82, 89, 98, 107, 116, 121, 145, 152, 160, 167, 178, 186, 197, 204, 227, 236, 245, 252, 262, 272, 280, 288, 309, 317, 329, 337, 349, 357, 365, 372, 392, 398, 406, 412, 421, 428, 435, 458, 467, 475, 483, 491, 498, 507, 515, 537, 544, 551, 560, 569, 578, 598, 605, 613, 622, 631, 636, 645, 665, 673, 680, 689, 697, 703, 709, 730, 741, 750, 759, 766, 778, 798, 805, 813, 821, 828, 836, 843, 852, 858, 879, 885, 891, 899, 907

Slant asymptote, 555

Slope-intercept form, 83–84, 86–88

Slopes, 77–82, 124
common difference, 660
negative, 92–97
of parallel lines, 85–86
of perpendicular lines, 85–86
positive, 92–97

Social Learners. *See* Learning Styles

Solids. *See also* Three-dimensional figures
cones, 6, 8, 197
cylinders, 9, 23, 39, 50, 51, 53, 414, 703
prisms, 6, 26, 32, 368, 374, 375, 377, 551
pyramids, 8, 369
spheres, 26, 409, 426

Solutions, 18, 75
extraneous, 29, 429–430, 571, 903
imaginary, 259
infinite, 140, 161, 163, 192, 902
none, 29, 140, 161, 163, 192
no real, 230–231, 900
one, 29, 161–162
one real, 230
pure imaginary, 247
two real, 230

Solving triangles
using Law of Cosines, 823–828
using Law of Sines, 815–820

Sperry, Pauline, 815

Spheres
radius, 409, 426
surface area, 26
volume, 26

Spiral Review, 10, 17, 25, 32, 39, 48, 67, 74, 82, 89, 98, 107, 116, 121, 145, 152, 160, 167, 178, 186, 197, 204, 227, 236, 245, 252, 262, 272, 280, 288, 309, 317, 329, 337, 349, 357, 365, 372, 392, 398, 406, 412, 421, 428, 435, 458, 467, 475, 483, 491, 498, 507, 515, 537, 544, 551, 560, 569, 578, 598, 605, 613, 622, 631, 636, 645, 665, 673, 680, 689, 697, 703, 709, 730, 741, 750, 759, 766, 778, 798, 805, 813, 821, 828, 836, 843, 852, 858, 879, 885, 891, 899, 907

Spreadsheet Labs
Amortizing Loans, 698
Compound Interest, 508
Investigating Special Right Triangles, 789
Normal Approximation of Binomial Distributions, 767, 768
Organizing Data with Matrices, 169

Spreadsheets, 170

Square matrices 180

Square root functions, 400–406, 407–412, 413, 438, 440

Square root inequalities, 402, 440

Square Root Property, 247, 256

Square roots, 239, 866
of negative numbers, 246

Squares, 88
differences of, P8, 239, 343
perfect, P8, 239, 343

Staff Development. *See* Professional Development

Standard deviation, P25, 733–738, 745–747, 760–762

Standard form
of conic sections, 632
of ellipses, 615, 616
of equations of circles, 607, 609
of hyperbolas, 624–626
of linear equations, 70, 72–73
of parabolas, 599–600, 603
of polynomials, 322
of quadratic equations, 229, 242–244

Standardized Test Practice. *See also* Extended Response; Gridded Response; Short Response; Worked-Out Examples 10, 17, 25, 32, 39, 48, 54–57, 67, 74, 82, 89, 98, 107, 116, 121, 128–131, 145, 152, 160, 167, 175, 186, 197, 204, 212–215, 227, 236, 245, 252, 262, 272, 280, 288, 296–299, 309, 317, 329, 337, 349, 357, 365, 372, 378–381, 392, 398, 406, 412, 421, 428, 435, 444–447, 458, 467, 475, 483, 491, 498, 507, 515, 522–525, 537, 544, 551, 560, 569, 578, 586–589, 598, 605, 613, 622, 631, 636, 645, 652–655, 665, 673, 680, 689, 697, 703, 709, 716–719, 730, 741, 750, 759, 766, 776, 782–785, 798, 805, 813, 821, 828, 836, 843, 852, 858, 866–869, 879, 885, 891, 899, 907, 912–915

Standard position, 799–804, 808

Statistical inference, 769

Statistics, P24–P28, 777. *See also* Data; Graphs; Measures of central tendency
best fit lines, 92–97
bias, 725
box-and-whisker plots, 737
continuous probability distributions, 743
discrete probability distributions, 743
five-number summary, P26, 734–735
histograms, P13, 733–736
interquartile range, P27
margin of sampling error, 731–732

maximum error of estimate, P24
mean, P24, 733, 745, 754, 760
measures of central tendency, P24
measures of variation, P24
median, P24, 760
mode, P24, 760
normal distributions, 760–768
outliers, P27, 93
prediction equations, 92–93, 96
quartiles, P26
random sampling, 723
relative frequency histogram, 743–744
scatter plots, 92–97, 122, 125
skewed distributions, 733
standard deviation, P25, 733–735, 745, 760
testing hypothesis, 771–777, 780
unbiased sample, 723
univariate data, P24
variance, P25, 745

Step functions, 102–103

Student Built Glossary, 4, 60, 134, 218, 302, 384, 450, 528, 592, 658, 722, 788, 872

Student Response System. *See* Teach with Tech

StudentWorks™, 2A–2B, 58A–58D, 132A–132C, 216A–216D, 300A–300D, 382A–382C, 448A–448C, 526A–526B, 590A–590C, 656A–656C, 720A–720B, 786A–786C, 870A–870B

Study Guide and Review, 49–52, 122–126, 206–210, 290–294, 373–376, 438–442, 517–520, 581–584, 646–650, 710–714, 777–780, 859–864, 908–910

Study Organizer. *See* Foldables™ Study Organizer

Study Tips
A is acute, 816
absolute value, 277, 684
absolute value inequalities, 43
adding and subtracting equations, 140
adding and subtracting radicals, 417
additive and multiplicative inverses, 12
alternative methods, 305, 824
amplitude, 838
amplitude and period, 839
angle measure, 854
angles of elevation and depression, 794
answer checks, 343

asymptotes, 547
axis of symmetry, 609
bias in experiments, 726
binomial probability formula, 754
boundaries, 148
calculating quartiles, P26
calculators, 503, 626
carbon dating, 510
census, 724
center-radius form, 608
check for accuracy, 193
checking answers, 20
checking solutions, P8, 137, 162, 667
check your results, 84
choose a function, 792
choosing the sign, 895
combinations, 700
combined variation, 565
comparing standard deviations, P26
complex numbers, 248, 267
composition, 387
conjugates, 418
convergence and divergence, 684
corresponding elements, 172, 174
cycles, 831
degree, 331
deriving formulas, 894
diagonal rule, 190
direct and inverse variation, 564
discrete vs. continuous, 743
distance, 594
distance problems, 573
domain and range, 222, 401
double roots, 325
eliminating common factors, 531
end behavior, 471
evaluating permutations and combinations, P11
exact locations, 627
exponential decay, 454
expressing solutions as multiples, 902
factoring, 354
fractions, 221
functions, 395
graphing, 601
graphing angles, 809
graphing calculator, 554, 642, 700
graphing inequalities, 34
graphs of piecewise-defined functions, 102
greatest integer function, 102
grouping 6 or more terms, 344
identifying conics, 633
independent events, P17
independent quantities, 77
inequality symbols, 35

infinite and no solutions, 163
interest, 453
interquartile range, P27
intersections, 512
inverse, 201
joint marginal frequencies, P18
joint variation, 563
law of large numbers, 746
locating zeros, 361
major axis, 616
make a list, 888
maximum and minimum, 332
mean and expected values, 757
memorize trigonometric ratios, 791
midpoints, 595
multiple data sets, 738
multiplication and division properties of equality, 20
mutually and not mutually exclusive, P14
normal distributions, 763
notation, 846
oblique asymptotes, 556
odd functions, 331
odd index, 408
optional graph, 230
outliers, 737
periods, 838
plotting reflections, 220
plus or minus, 257
power of 1, 305
proof and counterexamples, 182
quadrantal angles, 808
quadrants, 874
quadratic form, 221, 345
quadratic formula, 265
radians, 832
radical expressions, 425
radical inequalities, 431
random number generator, 753
random number generators and proportions, 744
range of z-values, 763
rational inequalities, 575
rational values, 332
reading math, 248
real numbers, 12
reasonableness, P21, 156
reciprocal functions, 840
reference angles, 817
rejecting the null hypothesis, 772
repeating decimals, 685
return on investment, 746
roots, 268
scalar multiplication, 173

sequences and recursive formulas, 693
set-builder notation, 230
simplifying, 502, 875
simplifying expressions, 424
simplifying rational expressions, 539
sine and cosine, 832
slope, 78
slope and classifying systems, 137
slope is constant, 78
solution check, 430
solving inequalities, 494
solving quadratic inequalities algebraically, 285
solving quadratic inequalities by graphing, 283
solving trigonometric equations, 904
square root, 247
squared differences, P25
standard normal distribution, 764
study notebook, 269
symmetry, 764
tangent, 840
technology, 249
testing for zeros, 359
two solutions, 816
unbiased estimator, P25
undefined terms, 540
using color, 846
verifying a graph, 847
vertical and horizontal lines, 71
with repetition, P10
why subtract? P14
zero at the origin, 360
zero exponent, 470
z-test, 775

Substitution
method, 138–139, 140
property, 19
synthetic, 352–353, 355, 361

Subtraction
complex numbers, 248
elimination using, 139–140
of functions, 385–386
of matrices, 172
of radicals, 417
of rational expressions, 539

Subtraction Property
of equality, 19
of inequality, 33

Success, 752–754

Sum of angles identity, 886–891, 908, 910

Sum of cubes, 342–344

Summative Assessment, 49–52, 53, 122–126, 127, 206–210, 211, 290–294, 294, 373–376, 277, 438–442, 443, 517–520, 521, 581–584, 585, 646–650, 651, 710–714, 715, 777–780, 781, 859–864, 865, 908–910, 911

Sums. *See also* Addition
of angles, 886–891, 908, 910
of arithmetic series, 668, 671
of cubes, 342–344
of geometric series, 676, 684
partial, 668, 676
sigma notation, 669, 676, 685

Surface area
of cylinders, 39
of rectangular prisms, 26, 32
of spheres, 26

Surveys, 723–725
design, 725

Symbols
absolute value, 27
combinations, P11
for congruent to, P20
empty set, 28–31, 147
expected value, 745
greatest integer function, 102
grouping fraction bar, 6
inequalities 34, 35
infinity, 40, 685
inverse functions, 394
permutations, P10
for plus or minus, 257, 895
probability, 745
random variables, 742, 752
sigma, 669, 676, 685
for similar to, P20
variance, P25
z-test, 762

Symmetric Property, 19

Symmetry, 220, 366
axes of, 220–221, 223–226, 599–601, 609, 624

Synthetic division, 312–314, 319, 352

Synthetic substitution, 352–353, 355, 356, 361

System of three equations, 162–165, 193–196

System of two equations, 136–140, 192, 194–196

Systems of equations, 136
conic sections, 632–636, 649

Index

linear-nonlinear systems, 640–645
solve
  by graphing, 135, 137, 140
  by using a table, 136
  using a table, 140
  using elimination, 139–140
  using matrices, 200–202, 205
  using substitution, 138–139
in three variables, 161–167
writing, 164

Systems of inequalities, 146–152, 155
linear programming, 154–160

Tables, 219, 572
frequency, P18, 753
solving equations, 136, 140, 499–500
solving inequalities, 287

Tangent, 790–791, 809, 839–840

Taylor, Brook, 571

Teach with Tech
Audio Recording, 394, 565, 832
Blog, 21, 78, 103, 199, 284, 354, 409, 430, 463, 502, 609, 772, 903
Computer Projector, 220
Document Camera, P4, 44, 70, 139, 242, 323, 539, 575, 616, 677, 825, 840, 888
Graphing Calculator, 737
Instant Messaging, 662, 685, 761, 801
Interactive Whiteboard, P6, P11, P15, P18, P26, 12, 29, 62, 85, 190, 249, 258, 276, 305, 360, 387, 417, 423, 469, 486, 494, 530, 547, 595, 668, 694, 744, 791, 816, 847
Portable Media Player, P10, 118, 232
RSS, P8
Student Response System, P16, P23, 6, 116, 138, 268, 332, 634, 808
Video Recording, 162, 177, 181, 313, 345, 402, 479, 554, 601, 701, 706, 854, 895
Web Page, 627, 875
Web Search, 510, 755
Wiki, P21, 150

Teacher to Teacher, 382D, 590D, 656D

TeacherWorks Plus, T18–T19, 2A–2B,

58A–58D, 132A–132C, 216A–216D, 300A–300D, 382A–382C, 448A–448C, 526A–526B, 590A–590C, 656A–656C, 720A–720B, 786A–786C, 870A–870B

Teaching the Mathematical Practices,
Arguments, 47, 107, 151, 184, 244, 261, 279, 317, 364, 397, 466, 474, 490, 514, 543, 550, 597, 621, 644, 688, 727, 740, 775, 797, 827, 883
Critique, 9, 17, 31, 67, 81, 144, 196, 203, 226, 235, 251, 391, 411, 482, 497, 536, 568, 604, 630, 672, 749, 820, 835, 857, 898
Modeling, 15, 30, 38, 88, 94, 121, 141, 175, 223, 224, 234, 260, 269, 278, 402, 464, 481, 506, 542, 566, 572, 596, 620, 671, 701, 739, 796, 877, 888
Perseverance, 86, 139, 194, 202, 259, 347, 356, 409, 488, 513, 535, 541, 642, 669, 678, 695, 708, 747, 770, 818, 876, 881, 907
Precision, 22, 29, 155, 158, 233, 314, 419, 425, 452, 472, 496, 567, 595, 612, 686, 754, 793, 825, 882, 897
Reasoning, 8, 14, 84, 150, 166, 315, 355, 363, 369, 404, 432, 456, 489, 504, 534, 557, 802, 810, 815, 819
Regularity, 20, 79, 112, 307, 417, 532, 662, 677, 706, 903, 904
Sense-Making, 7, 24, 105, 114, 165, 183, 242, 270, 285, 324, 348, 395, 410, 433, 495, 505, 533, 548, 611, 617, 618, 635, 662, 685, 686, 757, 834, 856, 883, 888, 905
Structure, 6, 62, 65, 199, 240, 250, 368, 547, 576, 600, 628, 670, 801, 833
Tools, 95, 225, 487, 577, 641, 643, 736, 764

Technology. *See also* Applications; Graphing Technology Labs; ConnectED; Spreadsheet Labs; TI-Nspire Technology
standardized test practice, 522, 524

Technology, 2A–2B, 58A–58D, 132A–132C, 216A–216D, 300A–300D, 382A–382C, 448A–448C, 526A–526B, 590A–590C, 656A–656C, 720A–720B, 786A–786C, 870A–870B

Temperature, 7, 397, 850

Terminal side, 799–801, 807–810, 830

Terminating decimals, 11

Terms
constant, 219
linear, 219
nth, 666, 667, 668–669, 674–675
quadratic, 219
of sequences, 659–660
undefined, 540

Test Practice. *See* Standardized Test Practice

Test-Taking Tip
answer checks, 595
checking answers, 881, 914
Cramer's Rule, 192
definitions, 85
eliminate possibilities, 855
eliminating choices, 130, 530
failures, 754
the meaning of a, 276
multiple choice, 312
read the question, 21
solve a simpler problem, 669
substitute values, 431, 588
substitution, 56, 380, 479, 654
use a scientific calculator, 784
use technology, 524
use the Law of Sines, 868
work backward, 446
write an equation, 298

Theano, 418

Theorems
binomial, 699–703, 710, 714
complex conjugates, 362
exterior angle inequality, 46
factor, 354, 373, 376
fundamental, 359, 369
Pythagorean, P22–P23, 435
rational zero, 367–373, 376
remainder, 352–354, 373, 376

Theoretical probability, P13

Theta θ, 790

Third-order determinants, 190

Three-dimensional figures
cones, 6, 8, 197
cylinders, 9, 23, 39, 50, 51, 53, 703
prisms, 6, 26, 32, 368, 374, 375, 377, 551
pyramids, 8, 369
spheres, 26, 409, 426, 447

TI-Nspire Technology Labs
Analyzing Quadratic Relations, 637–638

Index

Dividing Polynomials, 318–319
Equations of Circles, 606
Polynomial Identities, 350–351
Solving Quadratic Equations, 263

Tips for New Teachers, P7, P10, P14,
P17, P20, P25, 13, 20, 28, 32, 43,
67, 72, 82, 85, 86, 87, 89, 93, 95,
98, 140, 147, 158, 162, 165, 173,
174, 182, 194, 197, 221, 223, 269,
285, 312, 315, 324, 348, 353, 395,
416, 425, 433, 470, 486, 487, 494,
505, 531, 533, 564, 574, 595, 601,
676, 684, 697, 734, 742, 744, 762,
763, 772, 773, 791, 796, 800, 826,
839, 847, 854, 876, 894, 902, 904

Transformations, 109–115, 126
dilations, 111–114
each side of an equation, 881–882
of exponential functions, 452
graphing, 547
identifying, 112
of logarithmic functions, 470
one side of an equation, 880
with quadratic functions, 275–279,
294
of reciprocal functions, 547
reflections, 111–114
of square root functions, 401
translations, 110, 112–115, 401, 452,
470, 600, 845–852, 863

Transitive Property, 19

Translations, 110, 112–115, 600
graphing, 110, 845–852
horizontal, 401, 452, 470, 845
of periodic functions, 845–852, 863
of trigonometric functions, 845–852,
863
vertical, 401, 452, 470, 846

Trapezoids, 21

Tree diagrams, P9

Triangles
30°-60°-90°, 789
45°-45°-90°, 789
area, 7, 9, 191, 443, 475, 814
equilateral, 679, 822, 852
legs of, P22–P23
missing measures, P22–P23, 776,
790–798
oblique, 824
Pascal's, 699–700, 704
right, P22–P23, 787, 789–797, 860,
885, 909
similar, 428, 789

solving, 815–820, 823–828
vertices of, 393, 596

Trigonometric equations, 900,
901–907, 910

Trigonometric functions, 790–797,
831–832, 866
cosecant, 790–791, 809, 840
cosine, 790–793, 809, 830, 832–835,
837–838, 859
cotangent, 790–791, 809, 840
evaluating, 790, 807, 809, 832
finding exact values of, 874, 886
of general angles, 807–810, 861
graphing, 837–843, 844, 845–852,
863
inverse, 853–858, 864
secants, 790–791, 809, 840
sine, 790–791, 809, 830, 832–835,
837–838, 859
tangent, 790–791, 809, 839–840

Trigonometric identities, 872,
873–879, 908, 909
double-angle identities, 893–894,
896, 908, 910
half-angle identities, 894–895, 908,
910
Pythagorean, 873
quotient, 873
reciprocal, 873
to simplify expressions, 874–875
sum and difference formulas,
886–891, 908, 910
verify, 880–885, 888, 896, 909

Trigonometric inequalities, 906

Trigonometric ratio, 790–791

Trigonometry, 790–798
angle measurement, 799–800, 854,
859, 860
arccosine, 853
arcsine, 853
arctangent, 853
circular functions, 830–836, 859, 862
cosecant, 790–791, 809, 840
cosine, 790–791, 809, 830, 832–833
cotangent, 790–791, 809, 839–840
double-angle identities, 893–894,
896, 908, 910
equations, 900, 901–907, 910
graphs, 837–843, 844, 845–852, 863
half-angle identities, 894–895, 908,
910
identities, 872, 873–879, 880–885,
886–891, 893–895, 896, 908, 909,
910
inverse functions, 853–855
Law of Cosines, 823–828, 862

Law of Sines, 814–821, 861
periodic functions, 831–835, 848
quadrantal angles, 808
reference angles, 808–812, 817
regular polygons, 822
right triangles, P22–P23, 789,
790–791, 792–793, 860
secants, 790–791, 809, 840
sine, 790–791, 809, 830, 832–833,
837–838, 859
solving triangles, 815–820, 823–828
sum and difference formulas,
886–891, 908, 910
tangent, 790–791, 809, 839–840
unit circles, 830–832

Trinomials
factoring, 239–240
general, 239, 341
perfect square, P8, 239, 244, 343
quadratic, P7–P8

Two-way frequency table, P18

Turning points, 331

Unbiased sample, 723
Unbounded regions, 154–155, 159
Uniform probability model, P13
Unions, 42
Unit analysis. *See* Dimensional
Analysis
Unit circles, 830–832
Univariate data, P24
Upper quartile, P26

Values
absolute, 27–31, 277, 642, 684
expected, 745

Variable matrices, 200

Variables, P24, 5
dependent, 64, 660
function values of, 323
independent, 64, 660
lurking, 99
polynomials in one, 322
random, 742
systems of equations in three, 161–167

Variance, P25, 747

Index

Variation. *See also* Measures of variation
  combined 565, 567–568
  constant of, 90, 562
  direct, 90, 562–563, 565–568
  functions, 562–569, 584
  inverse, 564–568
  joint, 563, 567–568

Variation functions, 562–569, 584

Venn diagrams, P14, 11, 248

Verbal expressions
  to algebraic expressions, 18, 22

Verbal/Linguistic Learners. *See* Learning Styles

Vertex form, 275–279

Vertical Alignment, P4, P6, P7, P9, P13, P16, P20, P22, P24, 2E, 5, 11, 18, 27, 33, 41, 58G, 61, 69, 76, 83, 92, 101, 109, 117, 132G, 136, 146, 154, 161, 169, 179, 189, 198, 216G, 219, 229, 238, 246, 256, 264, 275, 282, 300G, 303, 311, 322, 330, 342, 352, 358, 367, 382G, 385, 393, 400, 407, 415, 422, 429, 448G, 451, 461, 468, 478, 485, 492, 501, 509, 526E, 529, 538, 545, 553, 562, 570, 590G, 593, 599, 607, 615, 624, 632, 640, 656G, 659, 666, 674, 683, 692, 699, 705, 720E, 723, 733, 742, 752, 760, 769, 786G, 790, 799, 807, 814, 823, 830, 837, 845, 853, 870E, 873, 880, 886, 893, 901

Vertical asymptotes, 553–556

vertical lines, 71

Vertical line test, 62–63

Vertical shift, 846–849, 851

Vertical translations, 401, 452, 470, 846

Vertices
  of ellipses, 615–617, 619–620
  of hyperbolas, 624–626, 628–630
  of parabolas, 220–225, 599–600
  of a region, 148, 154–158

Video Recording. *See* Teach with Tech

Visual/Spatial Learners. *See* Learning Styles

Vocabulary Check, 49, 122, 206, 290, 373, 438, 517, 581, 646, 710, 777, 859, 908

Volume
  of cones, 6, 8
  of cylinders, 9, 23, 50, 51, 53, 703
  of pyramids, 8
  of rectangular prisms, 6
  of spheres, 26

Watch Out!
  arc length, 802
  common difference, 668
  each side, 258
  expected value, 745
  functions, P5
  holes, 557
  inferential statistics, 773
  inverse functions, 395
  logarithmic base, 469
  maxima and minima, 222
  parent functions, 848
  percents, 463
  positive square root, P23
  ratios, 661
  saving your place, 180
  sigma notation, 676
  standard deviation, 735
  synthetic division, 313
  synthetic substitution, 354
  zeros, 231
  zeros vs. vertical asymptotes, 556

Watch Out!, P4, P6, P8, P10, P14, P21, P23, P24, P26, 10, 12, 15, 17, 24, 24, 25, 42, 43, 47, 67, 121, 138, 139, 156, 163, 173, 180, 184, 191, 193, 195, 221, 222, 223, 235, 241, 244, 247, 248, 257, 258, 259, 266, 267, 276, 288, 304, 305, 313, 323, 328, 343, 345, 354, 355, 360, 368, 370, 386, 391, 404, 408, 417, 421, 424, 430, 432, 453, 467, 474, 482, 497, 502, 530, 539, 547, 563, 569, 571, 573, 574, 575, 605, 609, 613, 616, 617, 627, 630, 643, 660, 665, 670, 673, 675, 676, 685, 688, 693, 697, 739, 749, 766, 805, 808, 809, 815, 821, 825, 879, 899

Web Page. *See* Teach with Tech

Web Search. *See* Teach with Tech

Web site. *See* ConnectED

Weierstrass, Karl, 103

Weighted average, 572

Which One Doesn't Belong?. *See* Higher-Order Thinking Problems

Whisker. *See* Box-and-whisker plot

Whole numbers, 11

Wiki. *See* Teach with Tech

Work backward, 20, 445, 881

Worked-Out Solutions. *See* Selected Answers and Solutions

Work problems, 574

Writing in Math. *See* Higher-Order Thinking Problems

x-coordinates, P4, 78

x-intercepts, 71–73, 358

y-coordinates, P4, 78

y-intercepts, 71–73, 366

Zero Power Property, 303

Zero Product Property, 241

Zeros, 325
  as exponents, 470
  of functions, 75, 229, 331, 358–362, 367–369, 376
  locating, 361
  location principle, 330–331
  rational, 367–371

Zhang Heng, 479

z-test, 773

z-value, 762

Index